Proceedings of the XXIII International Conference on High Energy Physics

Proceedings of the XXIII International Conference on High Energy Physics

16–23 July 1986
Berkeley, California

Volume I

Stewart C. Loken, Editor

World Scientific

Published by

World Scientific Publishing Co Pte Ltd.
P. O. Box 128, Farrer Road, Singapore 9128

Library of Congress Cataloging-in-Publication data is available.

PROCEEDINGS OF THE XXIII INTERNATIONAL CONFERENCE ON HIGH ENERGY PHYSICS

ISBN 9971-50-183-X

Printed in Singapore by Singapore National Printers Ltd

INTERNATIONAL ADVISORY COMMITTEE

M. Blazek	Inst. Phys. Bratislavia
K. Berkelman	Cornell University
A. Donnachie	University of Manchester
T. Fujii	Kobe University
R. Klapisch	CERN
P. Lehmann	IN2P3
I. Mannelli	Pisa
P.K. Malhotra	TIFR, Bombay
V.A. Matveev	INR, USSR
L. Okun	ITEP, Moscow
C. Quigg	Fermilab
B. Richter	SLAC
P. Söding	DESY
K. Strauch	Harvard University
Zhou Guangzhao	Academia Sinica, Beijing

LOCAL ORGANIZERS AND PROGRAM COMMITTEE

M. Barnett
W. Carithers
M. Chanowitz
A.R. Clark
M.K. Gaillard
S.C. Loken (chairman)
R. Madaras
M. Ronan
J. Siegrist
M. Strovink
T. Trippe
B. Zumino

Scientific Secretaries

J. Burton (LBL)
M. Dugan (LBL)
J.J. Eastman (LBL)
H.S. Kaye (LBL)
L.G. Mathis (LBL)
H. Sonoda (LBL)

Foreword

The XXIII International Conference on High Energy Physics was held at the University of California, Berkeley, from 16–23 July, 1986. The conference was sponsored and supported by the International Union of Pure and Applied Physics, the Lawrence Berkeley Laboratory, the United States Department of Energy, and the National Science Foundation.

The meeting was organized as an open conference under similar guidelines to those for the first open conference at Madison in 1980. The conference was attended by 1531 delegates with 45% of the delegates from outside the United States.

The conference consisted of three days of parallel sessions followed by three days of plenary sessions. The parallel session organizers had the responsibility for the format of the sessions and for the selection of speakers. During the three days, more than 280 papers were presented. The plenary sessions consisted of 16 rapporteur talks.

It was a great disappointment that two invited speakers, A.M. Polyakov of ITEP, Moscow and E. Shuryak of Novosibirsk, were unable to attend the conference. Their talks were given by O. Alvarez of Berkeley and L. McLerran of Fermilab, respectively. A paper by Dr. Shuryak, written for these proceedings, is included in the session on High Energy Nuclear Physics.

Volume I of these proceedings contains the plenary talks. The parallel sessions begin in Volume I and continue in Volume II. The list of delegates and the list of contributed papers are in Volume II.

On behalf of the Organizing Committee, I want to thank the LBL conference staff for their efforts. In particular, I wish to thank our conference coordinator, Louise Millard; our database expert, Arlene Spurlock; and our technical editor, Loretta Lizama.

Stewart C. Loken
Conference Chairman

December 1986

Foreword

TABLE OF CONTENTS

Volume I

Plenary Sessions

Parallel Sessions

Session 1: Phenomenological Aspects of Unified Theories

Session 2: Superstrings

Session 3: Supergravity and Supersymmetry

Session 4: Substructure

Session 5: Cosmology and Elementary Particles

Session 6: General Properties of Field Theory

Session 7: Nonperturbative Methods in Quantum Field Theory (Including Lattice Gauge Theory)

Session 8: Searches for Quarks, Higgs Particles, Axions, Monopoles, Supersymmetric and Technicolored Particles

Session 9: Spectroscopy and Decays of Heavy Bound-Quark States

Session 11: Lifetimes and Weak Interactions of Heavy Quarks and Leptons

Volume II

Session 12: CP Violation and Weak Interactions of Light Quarks and Leptons

Session 13: Neutrino Masses and Neutrino Oscillations

Session 14: Electroweak Interactions and QED Tests

Session 15: New Phenomena in Elementary Particle Physics

Session 16: QCD Tests

Session 17: High Momentum-Transfer Reactions (Including Structure Functions)

Session 18: Jets and Fragmentation

Session 19: Gamma-Gamma Interactions

Session 20: Physics at Future High Energy Colliders

Session 22: Soft Hadron-Hadron and Photon-Hadron Reactions (Including Spin Physics)

Session 23: High Energy Nuclear Interactions

Session 24: Experimental Techniques

Plenary Session

Phenomenological Aspects of Unified Theories

R. Peccei (DESY)

Searches for New Particles

M. Davier (Orsay)

Chairman

P. Söding (DESY)

Scientific Secretaries

T. Barklow (SLAC)

M. Herrero (LBL)

PHENOMENOLOGICAL ASPECTS OF UNIFIED THEORIES

R.D. Peccei

DESY, D-2000 Hamburg, Fed. Rep. Germany

After some preliminary observations concerning attempts to go beyond the standard model, I briefly discuss two new phenomena of recent interest: the 5th force and variant axions. The former, for its elucidation, will require further gravitational experiments, but I conclude that variant axions are now definitly ruled out experimentally. Various aspects of superstring phenomenology are then addressed, including some of the generic predictions of superstrings and some of its generic problems. In particular, I discuss some of the phenomenological consequences of having an extra Z^0 boson and the circumstances under which this excitation is a genuine prediction of superstrings. Since it is likely that a more reliable relic of superstrings will be provided by the presence of superpartners at low energy ($\lesssim$ TeV), I discuss some of the bounds for squarks and gluinos obtained at the Sp$\bar{p}$S collider and the expectations for their production at the Tevatron. As a final topic, I touch upon some of the consequences that would result from having the Fermi scale arise from an underlying theory. Some aspects of the composite Higgs model and of the strongly coupled standard model are briefly reviewed.

I. PRELIMINARY OBSERVATIONS

The standard SU(3)xSU(2)xU(1) model of the strong and electroweak interactions works exceedingly well phenomenologically. This has been amply demonstrated again at this conference in the review talks by Altarelli[1] and Scott[2]. Yet theorists remain unhappy, even in the face of success, because they do not really understand the deep reasons that lie behind certain structural aspects of the standard model. Putting it rather succintly, theorists would like to understand three main points:

i) Why are these the forces we see, and where does gravity fit into this picture?

ii) Why does the matter we see, the quarks and leptons, have the peculiar transformation properties it has in the standard model?

iii) What fixes the dynamics which generates all masses?

This last point is associated with a further problem, that of the hierarchy of scales. In the standard model all masses are proportional to the scale of the SU(2)xU(1) breakdown, the Fermi scale: $\Lambda_F = (\sqrt{2}G_F)^{-1/2} \simeq$ 250 GeV. However, the constants of proportionality for fermions and the Higgs boson are unknown. Only for gauge bosons are their masses predictable, since they are related to Λ_F via the gauge coupling constant. Schematically one has

$$m \sim \begin{pmatrix} \Gamma_f \\ e \\ \lambda^{1/2} \end{pmatrix} \Lambda_F \qquad \begin{matrix} \text{Fermions} \\ W^\pm \\ \text{Higgs} \end{matrix} \qquad (I.1)$$

The hierarchy problem is two-fold. There is a "small" hierarchy problem connected with what physics forces the Yukawa couplings Γ_f to be so varied, so as to give the rather spread out quark and lepton mass spectrum we observe .However, the real hierarchy problem is why the Fermi scale itself is so different from the scale associated with gravitational phenomena, the Planck scale: $M_P = (G_N)^{-1/2} \sim 10^{19}$ GeV, where G_N is the Newtonian coupling. The Fermi scale, in the standard model, is not a dynamical scale like that associated with the running of the strong coupling constant, Λ_{QCD}. Thus its only natural value, in a world where the high energy cutoff is provided by gravity, is the Planck scale. Why is then $\Lambda_F \ll M_P$?

The theoretical landscape, in which one attempts to answer the open deep questions of the standard model, is quite vast. As Fig I.1 shows, one can adopt a bottom up approach, to try to answer the questions of forces, matter and masses, by focusing on issues connected with the dynamics which generates the Fermi scale. On the other end of the scale, however, one can adopt a top down approach and start with physics at the Planck scale; connected with gravity and superstrings, and attack in this way the standard model fundamental problems

The experimental landscape, shown in Fig I.2, is necessarily much more restricted. The next generation of accelerators, SLC, LEP, Tevatron and HERA will probe the 100 GeV region deeply, but we will have to wait for the SSC or a possible TeV e^+e^- collider (2TLC) to get into the TeV region. Non accelerator experiments (NAC) can provide crucial information about very high energy scales, but in rather restricted regions. Thus, most of the territory will never get direct experimental probing.

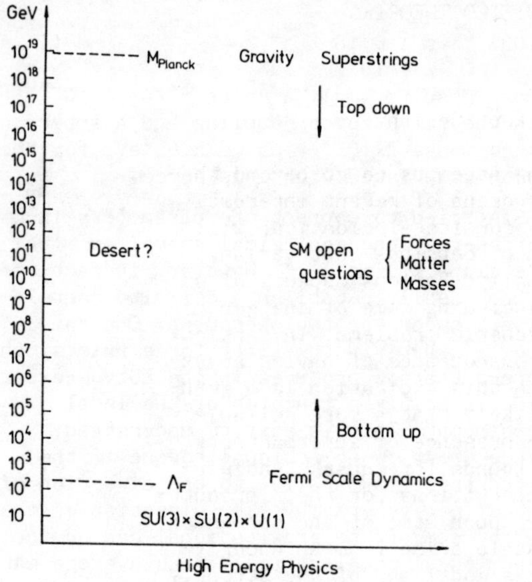

Fig I.1 Theoretical landscape to address the standard model open questions

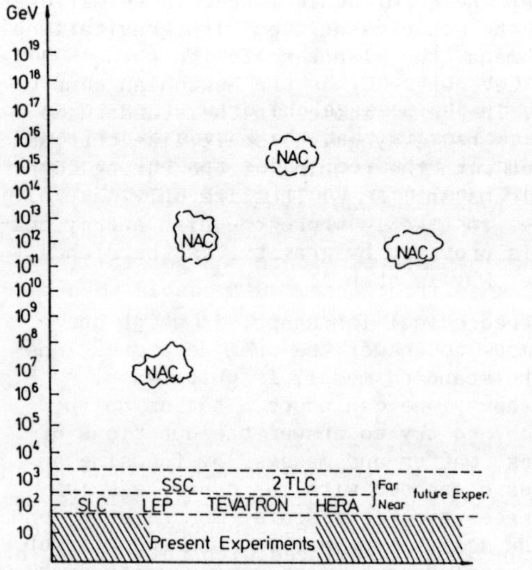

Fig I.2 Experimentally accessible landscape

Given this state of affairs, it is obvious that theories beyond the standard model, to make contact with reality, must predict some subTeV phenomena and/or some phenomena which will be accessible to non accelerator experiments. After all, physics is an experimental science! It is here that phenomenologists can play a useful role in trying to translate theoretical predictions of Planck scale physics, or coming out of Fermi scale dynamics, into possible signals of subTeV phenomena, which might be experimentally detectable. Before discussing some of the speculative novel effects studied by the Desert Trekkers and Moose Herders of Fig I.3 (which, if detected in future experiments, will point to physics beyond the standard model), let me first discuss two phenomena which already could signal new physics.

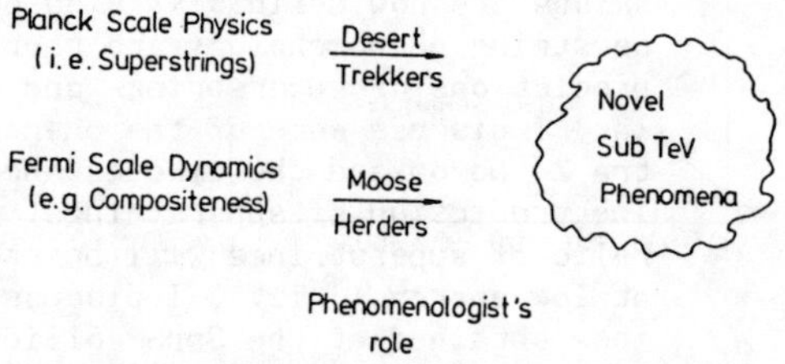

Fig I.3 Phenomenologist's role in modern day high energy physics

II. NEW PHENOMENA

a) The Fifth Force

Considerable excitement, and a certain amount of controversy and confusion, has surrounded a recent reanalysis by Fischbach, Sudarsky, Szafer, Talmadge and Aronson[3)] (FSSTA) of the classic Eötvos, Pekar and Fekete[4)] experiment on the equality of gravitational and inertial mass. What FSSTA found was a correlation in the data of the Eötvos experiment between the discrepancy in the torque measurements ΔK and the relative baryon number in the sample used $\Delta(B/\mu)$. This correlation is shown in Fig II.1, which is a slightly updated version[5)] of the figure presented in the original FSSTA paper. Bearing in mind that the data plotted was obtained almost three quarters of a century ago, and so is subject to difficult to quantify systematic uncertainties, the correlation found should be treated with some skepticism. However, optically, Fig II.1 looks quite impressive. FSSTA took the correlation they found seriously and suggested that it was an indication for a new, extremely weak, force in nature which coupled to baryon number or strong hypercharge Y - a fifth force. Since this force, if it existed, would give rise to a typical Yukawa potential between two bodies:

$$V_5 = f^2 \frac{B_1 B_2}{r} e^{-r/\lambda} \qquad \text{(II.1)}$$

with B_i here standing for either baryon number of hypercharge and λ being the range of the force, FSSTA tried to relate the Eötvos reanalysis to discrepancies found in geophy-

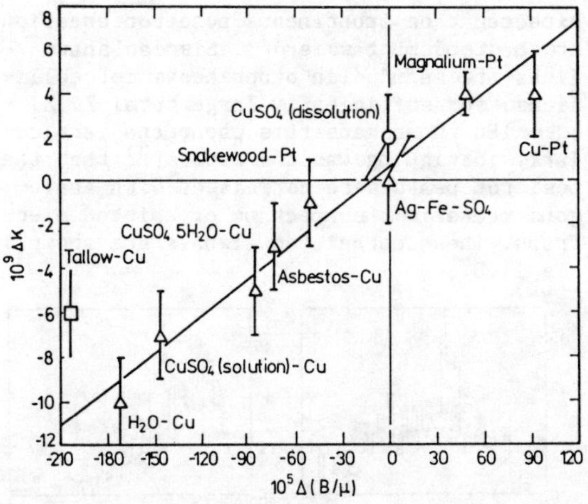

Fig II.1 Updated plot (from Ref 5) of the measured values of ΔK in the Eötvos experiment plotted against $\Delta(B/\mu) = B_1/\mu_1 - B_2/\mu_2$. Here μ_i is the mass of the sample expressed in terms of m(,H') and B_i is its baryon number.

sical determinations of the Newtonian constant G_N. It has been known for some time[6] that the value of G_N, inferred from measurements of the gravitational acceleration g in mines and boreholes, is about 1% larger than that determined from the classical Cavendish type laboratory measurement[7]: $G_N = 6.6720 \pm 0.0041 \times 10^{-11} m^3 kg^{-1} sec^{-2}$. For example, a particularly careful recent investigation in the Hilton mine in Queensland gives a value[8] (II.2)

$$(G_N)^{\text{Hilton Mine}} = (6.720 \pm 0.002 \pm 0.024) \times 10^{-11} m^3 kg^{-1} sec^{-2}$$

where the last error is an estimate of the possible systematic error, arising from a lack of precision in the density determination of the area surrounding the mine. This discrepancy (which, however, is only at the 2σ level) could be attributed to a non gravitational addition to the potential, so that

$$V = -G_N \frac{m_1 m_2}{r} \left[1 + \alpha e^{-r/\lambda} \right] \qquad \text{(II.3)}$$

where the last term above comes from the fifth force. The geophysical data is actually only sensitive to the combination of $\alpha\lambda$ and the discrepancy given in (II.2) gives the bounds[8] $0.004 \leqslant -\alpha\lambda \leqslant 10$m. However, satelite data[9] severely restrict large values of λ and one deduces[8]

$$0.035 \lesssim -\alpha \lesssim 0.15 \qquad \text{(II.4)}$$
$$1 \leqslant \lambda \leqslant 10^3 m$$

These results imply a value $f^2/4\pi \sim 10^{-40}$ for the fifth force coupling and a superlight mass, $2 \times 10^{-10} \leqslant m_5 \leqslant 2 \times 10^{-7}$eV, for the boson associated with this force.

FSSTA tried to connect the parameters inferred from the geological anomalies with the slope of Fig II.1. However, in fact one cannot really establish a definite correlation between the two phenomena. One can show[10] that, if the fifth force exists, the torque that it produces in the Eötvos experiment is mainly a function of the local topography. This is easy to understand, since there is no residual torque if the fifth force is parallel to the direction of the effective gravity (the direction of the vector sum of the gravitational and of the centrifugal acceleration). Thus average matter beneath the experiment is of little importance for the torque on the wire. But the results are crucially dependent on what nearby large buildings (and basements) existed at the time of the Eötvos experiment! Thus it seems pointless to compare the predictions of the fifth force - with the parameters fixed by the geological observations - with the slope of Fig II.1, although an attempt to do this is presented in Ref 5. Rather, the message one draws from these considerations is that the Eötvos experiment should be repeated, under careful controlled conditions, near the side of a mountain, so as to maximize the effect.

Although FSSTA put forth the possibility that the fifth force could couple to hypercharge, this suggestion can be ruled out by using the strong bound[11]

$$B(K^+ \to \pi^+ \text{Nothing}) < 3.8 \times 10^{-8} \qquad \text{(II.5)}$$

obtained some time ago at KEK. If the fifth force coupled to hypercharge a K^+ could decay into a π^+ emitting a hyperphoton - the vector boson associated with this force, of mass m_5. Since the coupling f is so small, one might think that the rate for this process would be negligible. However, as pointed out by Weinberg[12] more than 20 years ago, the fact that hypercharge is not conserved gives a contribution from longitudinally polarized hyperphotons proportional to f/m_5. Thus the process $K^+ \to \pi^+ \gamma_5$ is not negligible and one can bound this ratio. This has been studied recently by a number of authors[13] and a typical bound, taken from the paper of Suzuki[13] is

$$\frac{f^2}{4\pi m_5^2} \leq 7 \times 10^{-26}\ (\mathrm{eV})^{-2} \qquad \text{(II.6)}$$

This is the conflict with the values of f and m_5 given earlier. However, the bound is trivially avoided by supposing that the fifth force couples to baryon number only. In fact, this is much more reasonable, since strong hypercharge has only a meaning neglecting weak interactions.

My conclusions on the fifth force are twofold:

i) The whole subject of departures from gravity is very interesting and FSSTA should be given credit for having stimulated a variety of new experiments of the Eötvos and Galilei type, whose results should be soon forthcoming. However, until these results are in, very little can be settled. It remains an open question if there is a connection between the residual torque correlation, obtained by FSSTA, and the geological anomaly and indeed if either or both of these phenomena are real.

ii) Theoretical attempts to fit the fifth force in a grand picture[14], although useful and clever exercises, are probably premature.

b) Variant Axions

A second phenomena which elicited a great deal of attention this year was the sharp positron peaks[15] and the correlated e^+e^- signals seen in heavy ion collisions at GSI. As Fig II.2 shows, the spectrum of positrons in U-Cm collisions exhibits a narrow line at $T_{e^+} \sim 350$ KeV, above the continuum spectrum

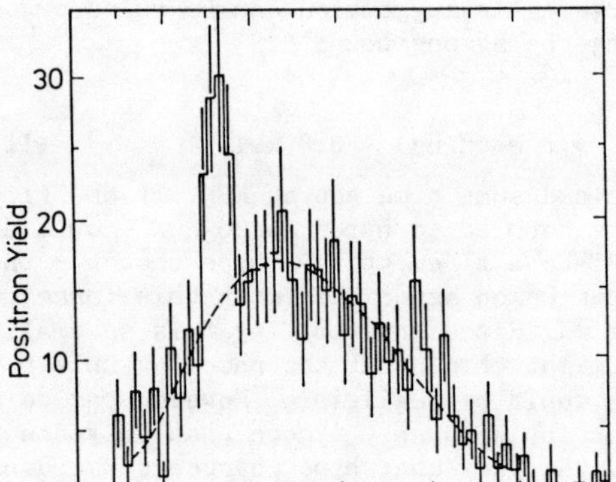

Fig II.2 Positron energy spectrum for U-Cm collisions at the Coulomb barrier. From Schweppe et al, Ref 15.

expected from spontaneous positron creation at the Coulomb barrier[17]. Similar sharp lines are seen[18] in other heavy ion collisions, for sufficiently large total Z: $Z_1 + Z_2 \geq 180$. What made this phenomena particularly intriguing was the report[16] that the positron peaks were correlated with analogous peaks in the spectrum of emitted electrons. These correlated signals are shown in Fig II.3.

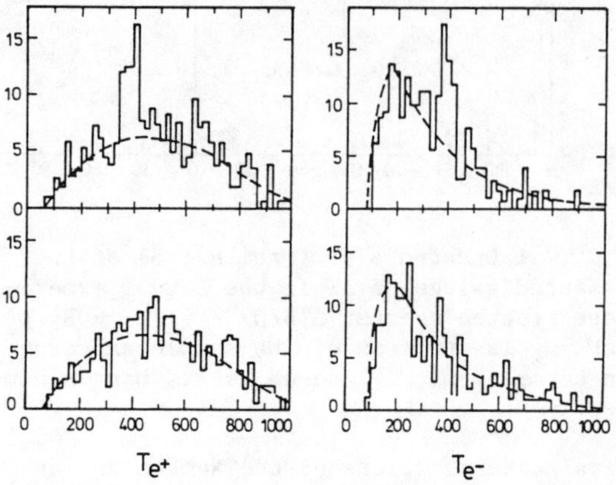

Fig II.3 Peaks in the positron (electron) spectum for electrons (positrons) with $340 \leq T_e \leq 420$ KeV. The bottom curves show these spectrums when the energy cuts correspond to the adjacent bins. From Ref 16.

These observations have a "trivial" kinematical explanation if in the heavy ion collision one produces, essentially at rest, a particle of mass $M_a \simeq 1.7$ MeV, which decays rapidly into e^+e^- pairs. Although no convincing mechanism has been invented to justify why such a particle should be produced at rest[19], the fact that M_a is so light naturally leads one to the speculation that this particle might be an axion. On the other hand, there is no way to associate a particle of mass as heavy as 1.7 MeV with the standard axion[20], since then it would have very enhanced couplings to either charm or bottom quarks, and so would be in violent conflict with the existing bounds on $\psi \to \gamma a$ or $\Upsilon \to \gamma a$[21]. It is, however, possible to construct variant axion models[22], where axions with masses above the e^+e^- threshold can exist, without being in direct contradiction with the quarkonia bounds.

Axions arise out of an attempt to solve the strong CP problem by imposing an additional global symmetry[20], $U(1)_{PQ}$, on the standard model Lagrangian. To achieve this it is necessary to have at least two Higgs doublets, Φ_1 and Φ_2, in the theory. Although the

$U(1)_{PQ}$ symmetry allows one to set to zero the overall coefficient of the CP violating $F_a^{\mu\nu} \tilde{F}_{a\mu\nu}$ term, where $F_a^{\mu\nu}$ is the gluon field strength, the fact that the fields Φ_i have non zero vacuum expectation value implies that this symmetry is spontaneously broken. The axion is the resultant Goldstone boson which, however, obtains a slight mass since the $U(1)_{PQ}$ symmetry is anomalous. For the standard axion model of Ref 20, one finds

$$M_a = \frac{m_\pi f_\pi}{\Lambda_F} \left(\frac{m_u m_d}{(m_u + m_d)^2}\right)^{1/2} N_F(x + \frac{1}{x})$$

$$\simeq 25\ N_F\ (x + \frac{1}{x})\ \text{KeV} \qquad (II.7)$$

where N_f is the number of families and $x = \langle\Phi_2\rangle/\langle\Phi_1\rangle$ is the ratio of the Higgs vacuum expectation values. Clearly for $M_a \simeq 1.7$ MeV x or x^{-1} must be very large. Since the coupling of the standard axion to charge 2/3 quarks (charge-1/3 quarks) is proportional to $x(x^{-1})$, one of the branching ratios: $B(\psi \to a\gamma) \sim x^2$ or $B(\Upsilon \to a\gamma) \sim x^{-2}$ is predicted to be very large, in contradiction with experiment[21].

Variant axion models[22] treat quarks asymmetrically under $U(1)_{PQ}$. Thus in the simplest model, for example, only the u quark and the electron have a coupling to the axion proportional to x, while all other fermion couplings are proportional to x^{-1}. If x is large, then <u>both</u> the ψ and Υ decays into γ axion will be suppressed. Furthermore, since the electron coupling is enhanced, the decay lifetime for the process $a \to e^+e^-$ will also be very short, making it possible for these kind of variant axions to escape some previous beam dump bounds, which assumed that axions were long lived.

Although variant axion models appeared for a while this year to be viable, and could be adduced as an "explanation" for the GSI positrons[22], their existence has proven ephemeral. Three distinct factors contributed to their demise:

i) Very recent experiments at GSI[24], although confirming the existence of the e^+e^- correlations, appear to see two distinct correlated e^+e^- signals whose origins, obviously, are difficult to reconcile with a single axion.

ii) New electron beam dump experiments, discussed by Davier[25] at this conference, are sensitive to the production and decay of variant axions. However, no signal for these excitations is seen in the data and all values for the ae^+e^- coupling, not in violation of g-2 bounds, are excluded[25].

iii) New experiments measuring axion deexcitations in hadronic transitions also rule out these models entirely[26].

Let me briefly discuss this last point here.

Variant axion models[22] are characterized by the number of quark doublets N_{PQ} which are active under the $U(1)_{PQ}$ transformation. This parameter replaces N_f in Eq(II.7) for the axion mass. Hence, for x large, and $M_a \simeq 1.7$ MeV, the combination $N_{PQ}x \simeq 70$ is fixed. In hadronic decays it is important to know to which extent the axion acts an an isovector or an isoscalar excitation. This is detailed in variant axion models by the axion's isovector and isoscalar mixing parameters, which depend again on x and N_{PQ}. One finds[27]

$$\lambda_3 \simeq \frac{x}{2} \left[1 - N_{PQ} \frac{(m_d - m_u)}{(m_d + m_u)}\right] \simeq \frac{x}{8} (4 - N_{PQ}) \qquad (II.8a)$$

$$\lambda_s \simeq \frac{x}{2} (1 - N_{PQ}) \qquad (II.8b)$$

Note that it is not possible for both of these parameters to be small since

$$\lambda_3 - \lambda_8 \simeq \frac{3}{8} (xN_{PQ}) \simeq 25 \qquad (II.9)$$

The strongest bound on λ_3 comes from a recent experiment at SIN on π^+ decay, where the rare process $\pi^+ \to e^+e^-e^+\nu_e$ was measured, thereby allowing a bound to be set on the process $\pi^+ \to ae^+\nu_e$[28]:

$$B(\pi^+ \to ae^+\nu_e) \leqslant (1\text{-}2)\times10^{-10} \qquad (II.10)$$

where the range given above depends on the precise value of the axion lifetime. Theoretically one computes[26]

$$B(\pi^+ \to ae^+\nu_e) \simeq 3\times10^{-9}(\lambda_3)^2 \qquad (II.11)$$

yielding $|\lambda_3| \leqslant 0.25$, in contradiction to Eq(II.8a) unless $N_{PQ} = 4$. However, such a value of N_{PQ} is not compatible with the recent result of an isoscalar, axion induced, nuclear deexcitation experiment in ^{10}B[29]. The axion to photon rate for the 3.59 MeV $2^+0 \to 3^+0$ rate is predicted to be[27)30]

$$\frac{\Gamma_a}{\Gamma_\gamma} \simeq 7.9 \times 10^{-4}(\lambda_s)^2 \qquad (II.12)$$

while experimentally one has the bound[29]

$$\frac{\Gamma_a}{\Gamma_\gamma} \leqslant 7.2 \times 10^{-3} \qquad (II.13)$$

so that $|\lambda_s| \lesssim 3$. For $N_{PQ} = 4$, however, one expects $\lambda_s \simeq 25$.

Given the above state of affairs, my conclusions are easily drawn:

i) The GSI phenomena, although very interesting by itself, has nothing to do with axions.

ii) Since neither variant or standard axions exist, if one insists in solving the strong CP problem by imposing a $U(1)_{PQ}$ symmetry, this symmetry must be broken at very high scales, leading to the invisible axion scenario[31].

III. SUPERSTRING INSPIRED PHENOMENOLOGY

Superstring theories were very much on the backburner at the time of the last International Conference of High Energy Physics in Leipzig. Indeed, I could find only one reference to them in the Conference proceedings, and that in the last paragraph of the summary talk of C. Callan[32]! However, after the publication of the anomaly cancellation paper of M. Green and J. Schwarz[33], superstrings have become a major theoretical industry. J. Schwarz[34], in this meeting, has thoroughly discussed the motivation and structure of these elegant theories. My job here is to summarize the status of the phenomenology which superstrings have inspired.

Superstrings are unfortunately not directly amenable to phenomenological study, since they are only consistent theories in a D = 10 dimensional space-time[34]. Therefore, if superstrings are to connect at all to reality, six of these ten dimensions must spontaneously compactify. Since these theories contain gravity, the scale associated with the compact dimensions is related to the Planck mass. Physics in four dimensional space-time, according to these theories, is set at the scale of compactification, $M_{comp} \sim M_P$, and depends on the geometry of the manifold K which compactified. At present there is no proof that this compactification actually takes place. Indeed, it is not even clear if one can expect this to happen for a unique space K, or if for a given superstring theory there is an infinity of such compact spaces.

These uncertainties notwithstanding, it has been argued by Candelas, Horowitz, Strominger and Witten[35] that particularly interesting manifolds for superstring compactification are provided by manifolds of SU(3) holonomy, which are known as Calabi Yau spaces[36]. Recall that the holonomy group is the group of all rotations generated when a vector, or spinor, is parallel transported around a closed loop in K. The fact that the holonomy group is not the full O(6) group, but only SU(3), means that there is at least one spinor which is not rotated under parallel transport. The existence of a covariantly constant spinor in the compact space assures that in the 4 dimensional theory an N = 1 supersymmetry is retained. Furthermore, one can show that the existence of the SU(3) holonomy allows a complex structure to exist in K, leading to a Kähler manifold whose metric is Ricci flat. This last circumstance is important because one can argue[37] that precisely such metrics will provide solutions to the string theory, since they preserve the conformal invariance of the two dimensional σ model on the string world sheet[38]. Since superstrings are endowed with a fixed gauge group[34], one must also specify the background gauge field in a consistent manner to guarantee these solutions. As Candelas et al show[35], the simplest consistent specification is to identify the background gauge fields, in an SU(3) subgroup of the superstring Yang Mills group, with the spin connection, ω, which transforms as an SU(3) matrix ($A = \omega$).

My discussion of superstring phenomenology will be restricted to the case when the compactification occurs in a Calabi-Yau space, in which the identification $A = \omega$ is made. Furthermore, I shall only consider the heterotic $E_8 \times E_8$ superstring[42], since the SO(32) superstring does not lead to realistic models[35]. There are other compactification possibilities which have been explored, including orbifolds[39] - which can be thought as limiting cases of Calabi-Yau spaces - and manifolds where the identification $A = \omega$ is not made[40]. Orbifolds are known to provide solutions to the string equations[39], but their phenomenology is largely unexplored. Manifolds with $A \neq \omega$, in general, can be shown not to provide solutions to the string equations, due to non perturbative effects[41]. These same effects, however, do not affect manifolds where $A = \omega$.

Let me begin by detailing the main features of the four dimensional theory which emerges from the $E_8 \times E_8$ superstring, after Calabi-Yau compactification[35]:

i) *The theory possesses an N = 1 supersymmetry.*
This is a very nice feature since supersymmetry allows naturally two different scales, like Λ_F and M_P, to coexist. So hierarchies

are not unnatural, except that for the moment the only scale of the theory is that of the compact dimensions: $M_{comp} \sim Mp$.

ii) *The full $\mathcal{E}_8 \times \mathcal{E}_8$ gauge group of the ten dimensional theory is reduced to $\mathcal{G}$ = $g \times \mathcal{E}_8$ where $g \subset \mathcal{E}_6$.*

One can understand this reduction by decomposing E_8 in terms of its maximal subgroup $E_6 \times SU(3)$. Since, in the compactification, the gauge fields associated with an SU(3) subgroup of E_8 were identified with the spin connection, clearly only an E_6 symmetry remains. In fact, if the manifold K is not simply connected, non trivial gauge configurations (Wilson loops) can be trapped in the manifold, even though the E_6 gauge field strength vanishes[43]. This can lead to a further breakdown of the E_6 group, with the flux trapping mechanism acting analogously to a breakdown induced by a Higgs field in the adjoint representation. It is important for phenomenology that at M_{comp} the remaining symmetry group g be not a GUT group, because otherwise one risks having a low energy group with unacceptable baryon number violations. Note that the second E_8 group is left untouched in the compactification and it provides a shadow matter world, which interacts only gravitationally with ordinary matter.

iii) *The matter representations are fixed by the properties of the manifold K.*

The four dimensional massless fermions which emerge correspond to chiral zero modes of the Dirac operator in K, and therefore have non trivial SU(3) properties in D = 10. Since the adjoint representation of E_8 decomposes under SU(3) x E_6 as

$$248 = (1,78) + (3,27) + (\bar{3},\overline{27}) + (8,1) \qquad \text{(III.1)}$$

one expects fermions in the 27 and $\overline{27}$ representations of E_6. Let me write these fermions as $n_f\, 27 + \delta(27 + \overline{27})$. The number n_f, which details the difference between 27's and $\overline{27}$'s, is entirely fixed by the topology of K[35] and it turns out to be one half of the Euler number χ of K:

$$n_f = \frac{1}{2}\,|\chi| \qquad \text{(III.2)}$$

Since, as I will detail below, the ordinary quarks and leptons fit in the 27 of E_6, we see that n_f is just the number of families and this number is topologically determined. If there is flux breaking, the number δ of $(27 + \overline{27})$ fermions need not remain in complete E_6 representations, but the n_f 27's must remain, since their number is fixed by (III.2). The 27 of E_6 can be decomposed with respect to its SO(10) and SU(5) subgroups, respectively, as

$$27 = 16+10+1 = (10+\bar{5}+1) + (5+\bar{5}) + 1 \qquad \text{(III.3)}$$

The ordinary quarks and leptons fit in the 16 of SO(10), including a right-handed neutrino. In addition there appears a new charge −1/3 quark and its antiquark plus a new electroweak doublet and its antidoublet, along with a total SU(5) and SO(10) singlet state. The new matter in the 27 is vector like, since ψ and ψ^c have conjugate transformation properties under the group. All the states in the 27 are displayed in Table III.1.

Table III.1 States in the 27 dimensional representation of E_6

SO(10)	SU(5)	states
16	10	$\begin{pmatrix} u \\ d \end{pmatrix}_L u^c_L\ e^c_L$
	$\bar{5}$	$d^c_L \begin{pmatrix} \nu \\ e \end{pmatrix}_L$
	1	ν^c_L
10	5	$g_L \begin{pmatrix} N^c \\ E^c \end{pmatrix}_L$
	$\bar{5}$	$g^c_L \begin{pmatrix} N \\ E \end{pmatrix}_L$
1	1	S_L

iv) *The Yukawa couplings in the four dimensional theory, at the scale of M_{comp}, are in principle computable.*

This point is simple to understand. A gauge fermion fermion coupling in D = 10 contains, when all fields are expanded in terms of four dimensional fields, also a scalar fermion fermion coupling, with the scalar fields corresponding to components of the gauge fields in the compact directions. This correspondence is shown schematically in Fig III.1.

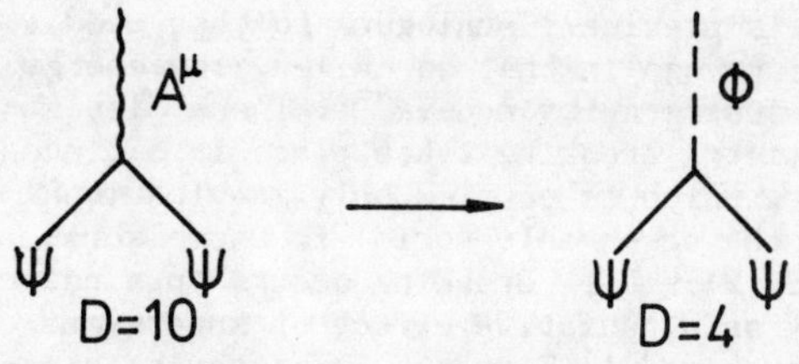

Fig III.1 Generation of Yukawa couplings on compactification.

It is impossible at the moment to really go ahead and compute all Yukawa couplings, although there have been some very interesting suggestions of how one might actually be able to achieve this by using topological considerations[44]. For some manifolds, however, even now one is able to infer, from the existence of certain discrete symmetries, that certain of the Yukawa couplings vanish[45].

In my opinion, properties i) - iv), along with the promise that superstrings may indeed provide one with a consistent quantum theory of gravity, are responsible for the extreme interest that these, otherwise rather remote, theories have stirred up in the high energy physics community. The above four results provide a reasonable starting point for answering the questions about hierarchy, forces, matter and mass dynamics, which I raised in my introductory remarks. However, one should remember that one must still transit from an energy scale of order $M_{comp} \sim M_P \sim 10^{19}$ GeV down to present accessible energies. What, if anything, survives of these beautiful patterns at 100 GeV?

The answer to this question depends a bit on how optimistic or pessimistic one is about the issue of supersymmetry breaking. Although the N=1 supersymmetry at the compactification scale makes hierarchies natural, it is clear that this supersymmetry must break down, allowing to split ordinary matter from its superpartners. If supersymmetry is to stabilize the low energy theory, so that a parameter like Λ_F is naturally 250 GeV and not 10^{19} GeV, the difference in mass between superpartners and ordinary matter must also be of this order of magnitude:

$$\tilde{m} - m \lesssim O(\Lambda_F) \sim \text{TeV} \qquad \text{(III.3)}$$

Superstring phenomenology is based on the idea that the shadow sector of the theory, that corresponding to the other E_8, triggers this breakdown and then transmits it to the visible sector. In this respect, this scenario is precisely analogous to that used a few years ago in the, so called, low energy N = 1 supergravity models[46]. There also supersymmetry breaking takes place in a hidden sector, which is coupled only gravitationally to the observable world. In these schemes, SU(2) x U(1) breaking occurs in a natural way as a radiative effect of supersymmetry breaking[47]. Thus the Fermi scale is intimately connected with the way one breaks supersymmetry.

There are, however, some important differences between the case of shadow sector supersymmetry breaking and that which occurs in the N = 1 supergravity models[48]. Even though supersymmetry is broken in both cases in a hidden sector, which is coupled to ordinary matter only gravitationally, for the superstring case the transmission of this breaking to the ordinary sector is not so straightforward[49]. For instance, it is usually assumed that the hidden sector breaking occurs through the formation of gluino condensates of the shadow E_8, $\langle\chi\chi\rangle$, and/or condensates involving the field strength $F_{\alpha\beta\gamma}$ of the second rank field $a_{\alpha\beta}$[50], crucial for the anomaly cancellation[33]. Although these condensates break the supersymmetry, leading to a gravitino mass $m_{3/2} \sim \langle\chi\chi\rangle/M_P^2$, their contributions cancel in the scalar potential. Thus, at tree level, even though supersymmetry is broken in the shadow sector, the observable world remains supersymmetric[49][50]. It turns out that also one loop radiative effects do not change this situation for the scalar fields[51], although gauginos can obtain a mass radiatively[52]. More generally, one can argue that quantum corrections involving heavy string modes[53] can be used to transmit the supersymmetry breaking from the hidden sector to the observable sector. However, one is then confronted with terms which destabilize the vacuum[54], unless one can find a mechanism to cancel the cosmological constant.

Clearly the present situation regarding supersymmetry breaking in superstring theories is unsatisfactory. In these circumstances a pessimist would argue that it is impossible to extrapolate down from the compactification scale, since there is no way to reliably generate any other scales in the theory, including the Fermi scale. An optimist, on the other hand, would argue that, in time, the matter of supersymmetry breaking will be resolved and that, for practical purposes, one can just assume that the supersymmetry breaking scenario will turn out to be just like that of the N = 1 supergravity theory[46]. Obviously, superstring inspired phenomenology is pursued by optimists!

The matter of supersymmetry breaking is not the only source of uncertainty in trying to connect superstrings to reality. Since one does not know precisely what the compact space K is, it is also necessary to make some assumptions on what the resulting four dimensional group g is. As I mentioned earlier, a necessary assumption of desert trekkers is that g must be smaller than SU(5), to avoid immediate problems with proton decay. Thus flux breaking must be allowed in the manifold K[43]. The pattern of sensible g's obtained after flux breaking has been studied by many people[55]. The results obtained depend crucially on whether one has or does not have an intermediate scale of symmetry breaking between M_{comp} and Λ_F. If there is no intermediate symmetry breaking then, due to flux breaking, E_6 breaks at M_{comp} to g and this is the surviving low energy group. It turns out, as I will demonstrate below, that g necessarily is <u>bigger</u> than the standard model SU(3) x SU(2) x U(1). If there is one, or more, intermediate scales of symmetry breaking, then the low energy group g obtained could be bigger than the standard model group, but it could also be precisely the standard model. Since flux tube breaking is equivalent to adjoint breaking and since matter and therefore also the Higgs fields are in 27's, one can characterize these two possibilities as:

i) Direct breaking (III.4)

$$E_6 \xrightarrow[M_{comp}]{\langle 78\rangle} g \qquad g > SU(3)xSU(2)xU(1)$$

ii) Intermediate scale breaking (III.5)

$$E_6 \xrightarrow[M_{comp}]{\langle 78\rangle} g' \xrightarrow[M_{int}]{\langle 27\rangle} g \qquad g \geqslant SU(3)xSU(2)xU(1)$$

Let me first discuss the case of direct breaking, Eq(III.4). If this happens, as I indicated above, g is necessarily bigger than the standard model group[55]:

$$g = SU(3) \times SU(2) \times U(1) \times \tilde{g}$$

$$\tilde{g} = \begin{cases} U(1) \\ U(1)^2 \\ SU(2) \times U(1) \end{cases} \qquad (III.6)$$

Thus one is lead to expect at least one extra neutral gauge boson, which survives at low energies, providing a characteristic signal for these patterns of compactification. The necessary existence of $\tilde{g}$ is rather easy to see by decomposing the 78 and 27 representations of E_6 in terms of the $SU(3)_c x SU(3)_L x SU(3)_R$ maximal subgroup[56]:

$$78 = (3,\bar{3},\bar{3})+(\bar{3},3,3)+(8,1,1)+(1,8,1)+(1,1,8) \qquad (III.7a)$$

$$27 = (3,3,1)+(1,\bar{3},3)+(\bar{3},1,\bar{3}) \qquad (III.7b)$$

If one does not want to break color, then only the last two terms in (III.7a) can contribute to the vacuum expectation of the 78. Now $\langle(1,8,1)\rangle \neq 0$ will break $SU(3)_L$ to $SU(2)_L \times U(1)_L$. However, it is not possible to identify $U(1)_L$ with the U(1) of the standard model since, according to (III.7b), the antiquarks which are singlets of $SU(3)_L$ would then have also no hypercharge. So the U(1) of the standard model gets contributions both from $U(1)_L$ and some $U(1) \subset SU(3)_R$. Thus it is not possible to break down $SU(3)_R$ completely and a non trivial $\tilde{g}$ ensues, which could be as large as SU(2) x U(1).

Besides having an extra Z^0, a further general property of direct breaking is that n_f 27's survive at low energy. This number, since it is a topological property of the compact space K, is not affected by flux breaking. Hence, in the case of direct breaking, all the exotic fermions of Table III.1 survive at low energy. Although the discovery of these states would provide evidence for superstring ideas, their presence at low energy, as I will discuss below, is far from being an unmitigated blessing.

The phenomenology of models with extra Z^0's has been studied by a great many authors in the last year and several papers on this topic have been submitted to this conference[57]. To discuss these models, it is particularly convenient to characterize the two extra U(1)'s in E_6, orthogonal to the usual hypercharge, via the decomposition[58]:

$$E_6 \to SO(10)xU(1)_\psi \to SU(5)xU(1)_\psi xU(1)_\chi$$

with the standard model group being embedded in the SO(10) group. The charges corresponding to $U(1)_\psi$ and $U(1)_\chi$ for the 27 dimensional E_6 representation can be read off from the tables in the review of Slansky[56] and are displayed in Table III.2. If E_6 is broken down to [SU(3)xSU(2)xU(1)]xU(1), the additional U(1) charge, in general, will be a linear superposition of these charges.

$$Q' = Q_\psi \sin\theta + Q_\chi \cos\theta \qquad (III.9)$$

However, if this breakdown is induced by flux breaking then the angle θ is fixed to be[59]: $\theta = \cos^{-1}\sqrt{3/8}$. This can be seen as follows. Instead of Q' and Y, the two U(1)'s can be described in terms of $U(1)_L$ and $U(1)_R$ in the $SU(3)_c x SU(3)_L x SU(3)_R$ decomposition of E_6. Their properly normalized charges are

$$Y_L = \sqrt{3/5}\begin{pmatrix} 1/6 & & \\ & 1/6 & \\ & & -1/3 \end{pmatrix}_L ;$$

$$Y_R = \sqrt{3/5}\begin{pmatrix} 1/3 & & \\ & -1/6 & \\ & & -1/6 \end{pmatrix}_R \qquad (III.10)$$

acting on the appropriate SU(3) group. The hypercharge Y is simply expressible in terms of Y_L and Y_R

$$Y = Y_L + 2Y_R \qquad (III.11)$$

and Q' is the orthogonal combination

$$Q' = -2Y_L + Y_R \qquad \text{(III.12)}$$

Comparing (III.12), with (III.9) for an u quark, for example, establishes $\theta = \cos^{-1}\sqrt{3/8}$. The value of Q' is also given in Table III.2.

Table III.2 Charge values for the 27 of E_6.

States	Q_ψ	Q_χ	Q'
10: $\begin{pmatrix} u \\ d \end{pmatrix}_L u^c_L e^c_L$	$-\frac{1}{2\sqrt{6}}$	$-\frac{1}{2\sqrt{10}}$	$-\frac{1}{\sqrt{15}}$
$\bar{5}$: $d^c_L \begin{pmatrix} \nu \\ e \end{pmatrix}_L$	$-\frac{1}{2\sqrt{6}}$	$\frac{3}{2\sqrt{10}}$	$\frac{1}{2\sqrt{15}}$
1: ν^c_L	$-\frac{1}{2\sqrt{6}}$	$-\frac{5}{2\sqrt{10}}$	$-\frac{5}{2\sqrt{15}}$
5: $\begin{pmatrix} N^c \\ E^c \end{pmatrix}_L g_L$	$\frac{1}{\sqrt{6}}$	$\frac{1}{\sqrt{10}}$	$\frac{2}{\sqrt{15}}$
$\bar{5}$: $\begin{pmatrix} N \\ E \end{pmatrix}_L g^c_L$	$\frac{1}{\sqrt{6}}$	$-\frac{1}{\sqrt{10}}$	$\frac{1}{2\sqrt{15}}$
1: S_L	$-\frac{2}{\sqrt{6}}$	0	$-\frac{5}{2\sqrt{15}}$

An extra Z^0, if sufficiently light, would give rise to departures of neutral current experiments from the predictions of the standard model. Thus one can use neutral current data to put bounds on the mass of this particle[60]. In particular for the $Z^{0\prime}$, which couples to Q', since its coupling to quarks and leptons is not as strong as that of the ordinary Z^0, the bounds obtained are rather weak, hovering around $M_{Z^{0\prime}} \gtrsim 100$ GeV. These bounds have been reviewed by Deshpande[61] in this conference. An illustration is provided by Fig III.2, where the limits on $M_{Z^{0\prime}}$, from its non observation at the CERN collider, are detailed. Clearly if the $Z^{0\prime}$ cannot decay into exotic matter, these bounds are stronger. Note that even for a $Z^{0\prime}$ with the same mass as the Z^0, $\sigma B(Z^{0\prime} \to e^+e^-)$ is much smaller than for the Z^0. Analyses for other possible $Z^{0\prime}$'s, like $Z(\theta)$, which couples to $Q_\psi \sin\theta + Q_\chi \cos\theta$, give similar bounds, although for particular θ-values these bounds can be rather weak[60].

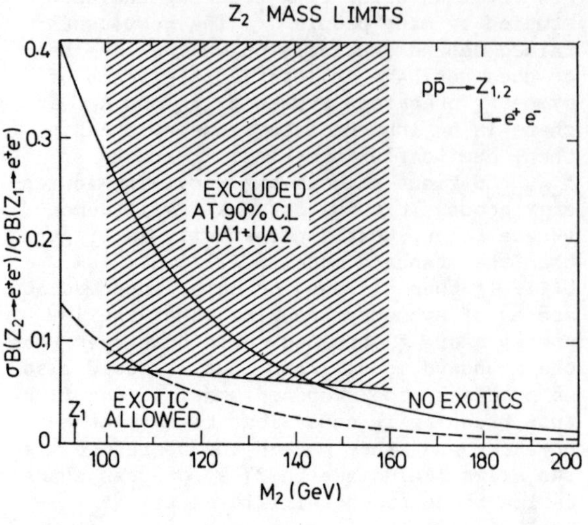

Fig III.2 Mass limits on $M_{Z^{0\prime}}$ from the CERN collider, taken from Barger et al, Ref 60.

Even though the $Z^{0\prime}$ is harder to produce than the ordinary Z^0, the Tevatron can push the mass limits for this excitation to near 300 GeV. For instance, London and Rosner[57] estimate a cross section times branching ratio into e^+e^-, for a $Z^{0\prime}$ of 200 GeV, of 1 pb at $\sqrt{s} = 1.8$ TeV. Such a signal should be detectable in a high luminosity run. Perhaps more favorable is the situation regarding the $Z^{0\prime}$ in e^+e^- collisions, since the presence of the $Z^{0\prime}$ can give rise to dramatic effects in various asymmetries, which will be measured at LEP and the SLC[57,62,63]. I illustrate this in Fig III.3, taken from Belanger and Godfrey[57], which shows the shift in the forward backward asymmetry and the left-right asymmetry expected at $\sqrt{s}=M_Z$ for various values of $M_{Z(\theta)}$. One sees that for the superstring $Z^{0\prime}$, corresponding to $\theta = \cos^{-1}\sqrt{3/8}$, the shifts in A_{FB} and A_{LR} are not as large as those for other values of θ. Nevertheless, these shifts are of a comparable order of magnitude to the expected effect in the standard model and much above the hoped for accuracy in these measurements[54]. If the $Z^{0\prime}$ is really as low as 150 - 200 GeV it will be visible directly in experiments at LEP 200, leading to large departures in the forward backward asymmetry for energies well below the resonance behaviour, as illustrated in Fig III.4. Indeed one should be sensitive to $Z^{0\prime}$ effects up to $M_{Z^{0\prime}} \lesssim 1$ TeV[63].

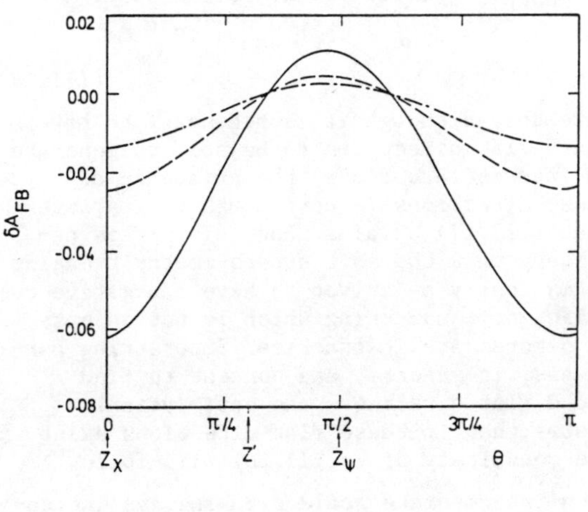

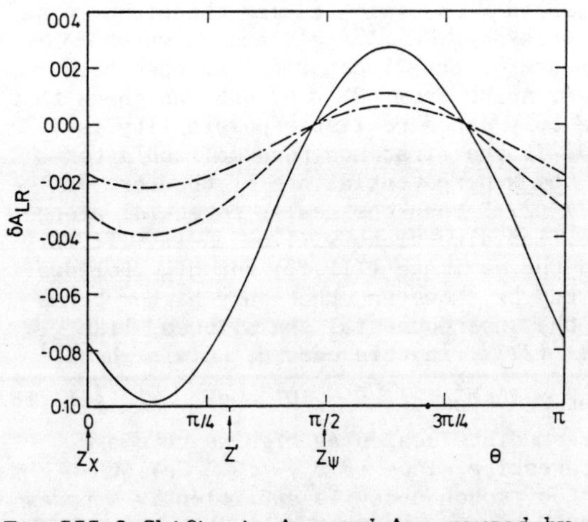

Fig III.3 Shifts in A_{FB} and A_{LR} caused by a new U(1) gauge boson $Z(\theta) = Z_\psi \sin\theta + Z_\chi \cos\theta$. The superstring gauge boson has $\theta = \cos^{-1}\sqrt{3/8}$. From Belanger and Godfrey, Ref 57.

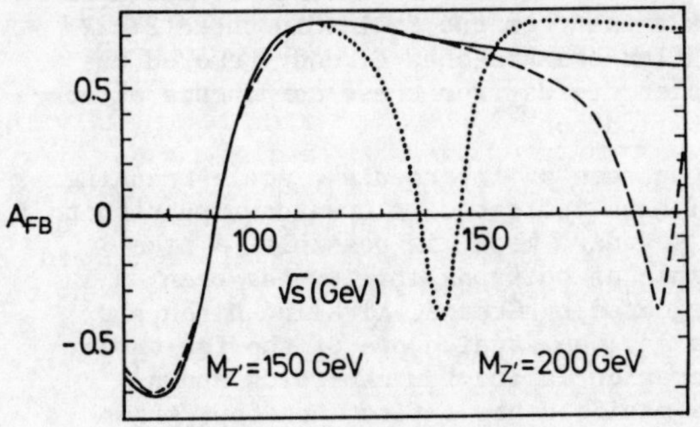

Fig III.4 Departures of A_{FB} from standard model expectations in the LEP 200 range due to a $Z^{0'}$. From Matsuoka et al, Ref 57.

Since there is no plausible motivation, except superstrings, for having an extra Z^0 at low energy, the detection of such a signal would provide strong corroboration for the superstring ideas. However, recall that if there is direct breaking, one also requires the presence at low energy of n_f full multiplets of 27's. The extra fermions in the 27, although providing further evidence for superstrings, are in themselves quite problematic[49]. Let me briefly discuss some of these potential difficulties:

i) Having both supersymmetry and 27, rather than 15 (or 16), fermions per family makes the SU(3) and SU(2) coupling constants run very differently from the case where one has only the usual quarks and leptons. The relevant β-functions, for scales above possible low energy thresholds read, in this case:

$$\beta_3 = -\frac{3g_3^3}{8\pi^2}(3 - n_f) \tag{III.13}$$

$$\beta_2 = -\frac{3g_2^3}{8\pi^2}(2 - n_f)$$

Hence for $n_f > 3$ both α_3 and α_2 grow at high energy. Indeed, one can convince oneself from (III.13) that if $n_f = 4$ $\alpha_3 > 1$ already at $q^2 \simeq 10^8$ GeV! Since α_3, α_2 and α_1 must unify at M_{comp}, in these schemes, and $M_{comp} \sim M_P$, it is clear that *direct breaking is only possible if $n_f = 3$*. For $n_f = 3$, α_3 does not run at high energy. Nevertheless it is possible to have SU(3)xSU(2)xU(1)xU(1) unified at scales of order 4×10^{17} GeV[65], with a resulting $\sin^2\theta_W \simeq 0.21$, which is adequate, if not spectacular.

ii) If the g quarks survive at low energy, there can be universality violations, and flavor changing neutral currents, induced by mixings of g with the other charge −1/3 quarks. Similarly the charged lepton E can mix with eμτ. Both of these circumstances are catastrophic, unless one can effectively suppress, to an arbitrary low level, all possible mixings. Although one can argue for the absence of these mixing effects[66], by allowing only certain Higgses to have vacuum expectation values, these arguments are not particularly compelling to me.

iii) The presence of the exotic fields in the 27 can lead to very rapid proton decay, through the presence of dimension four terms in the superpotential. On general grounds, one knows that the superpotential itself contains terms involving the $(27)^3$ of E_6. However, the coefficients of these terms are not fixed by E_6[43], since they get altered by flux breaking. The dangerous terms for proton decay are the ones connecting g to quarks and leptons:

$$g^cLQ\ ;\ ge^cu^c;\ gd^c\nu^c;\ gQQ;\ g^cu^cd^c \qquad (III.14)$$

Here $Q=\binom{u}{d}_L$ and $L=\binom{\nu}{e}_L$. One may argue[66] that only a subset of these couplings (the first three, or the last two) are nonvanishing because of topological reasons. In this case one can define a baryon number and there is no problem with proton decay. However, to my knowledge, no specific example has been found of a Calabi Yau manifold where the above can be demonstrated.

iv) Similar problems exist with neutrino masses. In principle, a coupling in the superpotential involving ν^c, L and the extra doublet $L^c = \binom{N^c}{E^c}$ exists. Since the scalar partners of L^c play the role of Higgs fields in the theory one is lead naturally to a Dirac mass for neutrinos, unless this coupling itself is absent.

In view of the above difficulties it is perhaps easier envisaging models with an intermediate scale breaking, since in these models one can, in principle, push some of the unwanted fermions in the 27 to high scales. The possibility of intermediate scale breaking requires more dynamical assumptions and obtains only if one has a certain number, δ, of chirally paired states $(\delta(27+\overline{27}))$. As pointed out by Dine et al[65] and Witten[40)43)], it is only consistent to have some of the scalar fields in the 27 acquire large vacuum expectation values, if there are directions in the scalar potential which are not affected by these expectation values (flat directions). In general, the scalar potential can be written as

$$V = |F|^2 + D^2 + \text{soft Susy breaking terms} + \text{non renorm. terms} \qquad (III.15)$$

Ignoring for the moment the last two terms, it is clear that intermediate scale breaking only will obtain if V is both D and F flat. D flat directions occur if the vacuum expectation value of a 27 component can be cancelled by that of a $\overline{27}$. Hence δ must be non vanishing. Furthermore, for example, since in the $(27)^3$ terms in the superpotential, no factors containing S^2 or S^3 appear one sees that $\langle S\rangle = \langle\bar{S}\rangle \neq 0$ is also an F-flat direction[67]. Hence the value of $\langle S\rangle = \langle\bar{S}\rangle$ is totally determined by the supersymmetry breaking and non renormalizable terms in V. A non zero and large value of $\langle S\rangle$ obtains if the soft supersymmetric breaking terms for S are, in fact, negative and act against the non renormalizable pieces in V. That is

$$V \simeq -\Lambda^2|S|^2 + \frac{|S|^6}{M^2_{comp}} \qquad (III.16)$$

where, presumably $\Lambda \simeq \Lambda_F$, and the scale which typifies the non renormalizable terms is M_{comp}. From (III.6) it follows that

$$\langle S\rangle = \langle\bar{S}\rangle = M_{int} \sim \sqrt{\Lambda_F M_{comp}} \sim 10^{10}\ \text{GeV} \qquad (III.17)$$

Note that although it is necessary to have D and F flat directions to be able to generate an intermediate scale, the existence of these directions is not enough to guarantee that (III.17) obtains. For this, it is necessary that the soft supersymmetry breaking terms really be driven to have a negative coefficient - something which is not so easy to demonstrate in practice. Superstring partisans, in general, are content to find F and D flat directions and optimistically assume that if these flat directions exist, the conspiracy of Eq(III.16) will follow[65].

With intermediate scale breaking, as indicated in Eq(III.5), the final group g can be bigger or equal to the standard model group. What pattern ensues depends both on what the manifold K is, which causes the first stage of breakdown: $E_6 \xrightarrow{\langle 78\rangle} g'$, and on which components of the 27 cause the further breakdown. Apart from $\langle S\rangle \neq 0$, one can check that the only other realistic possibility is $\langle\nu^c\rangle \neq 0$. If the first non renormalizable terms in the superpotential are of the form of $(27)^2(\overline{27})^2$ then the scalar potential along the flat directions will be as in (III.16) and the estimate (III.17) for M_{int} follows. It may be, however, that only higher terms in the superpotential are allowed, like $(27)^3(\overline{27})^3$. In this case M_{int} is higher[68)]

$$M_{int} \sim (\Lambda_F M^3_{comp})^{1/4} \sim 10^{14}\ \text{GeV} \qquad (III.18)$$

Intermediate scales as high as this are interesting since if $m_g \sim 10^{14}$ GeV, then g mediated proton decay is sufficiently suppressed. However, one must make sure that there remain in the theory light doublets to allow for a low energy breakdown of the standard model. The splitting of the triplets from the doublets is rather natural if the doublets arise from the fields in the $\delta(27+\overline{27})$ and flux breaking has already removed the triplet fields from these components at compactification[69].

In the case of intermediate scale breaking, as I have indicated, a breakdown pattern to the standard model is possible. A nice example of this possibility has been considered by Greene, Kirklin, Miron and Ross[70], who studied one of the few three generation Calabi Yau manifolds known[71]. This manifold has a first homotopy group $\pi_1(K)=Z_3$, so it admits flux breaking. Before flux breaking, the model has 3(27) and 6(27 + $\overline{27}$) multiplets. The two non trivial embeddings of Z_3 in E_6 give SU(3) x SU(3) x SU(3) and SU(6) x U(1), respectively, as the resulting group after flux breaking. Greene et al[70]) concentrate on the first possibility, since it can lead to a realistic theory. After flux breaking all the existing $SU(3)_c$ singlet fields in the chirally paired (27 + $\overline{27}$) representations survive, as well as 4 out of 6 of the $SU(3)_c$ triplet and antitriplet fields in these representations. This very unpleasant phenomenologial situation is remedied by a sequence of two intermediate scale breakings, triggered by ν^c and S vacuum expectation values:

$$SU(3)^3 \xrightarrow{\langle\nu^c\rangle} SU(3) \times SU(2) \times SU(2) \times U(1) \xrightarrow{\langle S\rangle} SU(3) \times SU(2) \times U(1) \quad \text{(III.19)}$$

Greene et al[70]), by studying the manifold's discrete symmetries show that the superpotential is F-flat to $O(27^3\overline{27}^3)$ for the first breaking. Thus $\langle\nu^c\rangle \sim 10^{14}$ GeV, while $\langle S\rangle \sim 10^{10}$ GeV, provided, of course, that the appropriate supersymmetry breakdown to trigger this sequence exists. This multiple breakdown gives high mass to almost all the remaining vector-like states in the theory. Remarkably, however, even though all g quarks are heavy, two doublet superfields stay light. These are precisely the Higgs multiplets necessary for a supersymmetric extension of the standard model[46]). In addition, the coupling of these multiplets to the quarks, possesses certain discrete symmetries which yield a reasonable structure for the Kobayashi-Maskawa matrix[70]).

The results of Greene et al[70]) are both encouraging and discouraging. I find it encouraging that there exist Calabi Yau manifolds with topological properties which can lead one to a model at low energy with many of the characteristics of the standard model. It is discouraging, however, that there is so little to show from the superstring superstructure, except certain interrelations among Yukawa couplings. Furthermore to get from the superstrings to the standard model one has had to make many assumptions, each of them hard to justify. So one is left in the ambivalent position of not being able to decide whether defects in the resulting theory are due to poor intermediate assumptions or are really signals of some profound sickness in the scheme. A case in point is provided by SU(3) x SU(2) x U(1) unification in the manifold studied by Greene et al[70]). Although one can get unification into $(SU(3))^3$ at 10^{14} GeV with a reasonable $\sin^2\theta_W$, the presence of so many matter fields beyond this scale drives the gauge couplings above unity much before M_{comp}. Is this a deadly defect, or can it be conveniently ignored? Questions of this ilk, unfortunately, abound in trying to bring superstring ideas down to laboratory energies.

My conclusions on superstring inspired phenomenology are two fold:

i) Most of the "predictions" of superstrings at low energy are strongly dependent on implicit assumptions made at the compactification scale and at possible intermediate scales. Although certain specific predictions are phenomenologically appealing, like the presence of an extra Z^0, none of these predictions are sure things.

ii) To motivate the existence of superstrings it is important to find evidence for the "super" aspect of these theories. The existence of low mass ($\tilde{m} \lesssim$ TeV) superpartners remains the best "smoking gun" for superstrings.

IV LOOKING FOR SUPERSYMMETRY

Long before superstrings became popular, there was considerable theoretical activity in low energy supersymmetry[72]). The physical motivation for considering supersymmetric extensions of the standard model was related to the hierarchy problem. Although supersymmetry cannot explain the existence of mass hierarchies, it allows hierarchies naturally to exist. Mass shifts in the scalar sector are no longer quadratically divergent since there is a cancellation between bosonic and fermionic contribution, so radiative corrections do not destabilize the theory. Although supersymmetry is not the only way to make the Fermi scale a natural parameter, it is obviously the solution chosen by superstrings. So, for these theories also, it is sensible to expect to have superpartners of the known excitations in a mass range below, say, 1 TeV. Unfortunately, since one does not know precisely how supersymmetry is broken, one cannot really pin down the masses of the superpartners. All that is necessary is that these masses be low enough to provide a credible mechanism for having Λ_F of O(250 GeV).

The usual assumption pursued is that supersymmetry is broken spontaneously. The favored scenario is that discussed earlier, which is based on an N=1 supergravity theory in which the supersymmetry breaking occurs in a hidden sector[46)48)]. In the low energy theory the manifestation of this breakdown is the appearance of soft breaking terms which give masses to the scalars and the gauginos and provide corrections to scalar vertices. The schematic structure of these terms is shown in Fig IV.1.

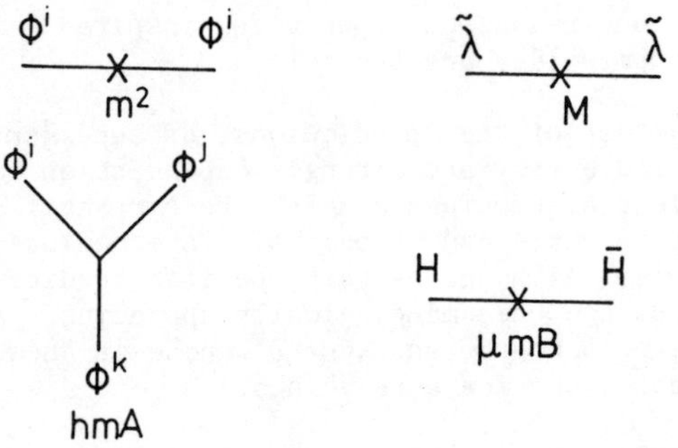

Fig IV.1 Soft supersymmetry breaking terms generated through hidden sector breaking. ϕ_i are scalars, $\tilde{\lambda}$ are gauginos and H,$\bar{H}$ are the Higgs doublets.

The resulting superpartner spectrum depends on detailed assumptions one makes on M, m, A and B at some high scale (M_{comp} or M_p), plus the renormalization group evolution of these parameters down from this scale. In general one takes the gaugino and scalar masses, M and m, to be universal at the high scale and one gets SU(2) x U(1) to break down when the soft Higgs mass squared is driven negative during the renormalization group evolution[47].

In the absence of precise knowledge of the superpartner spectrum, the most important quantity to know for phenomenology is which is the lightest supersymmetric particle (LSP). Supersymmetric theories are invariant under an additional symmetry, R parity[73]. R parity is a multiplicatively conserved quantity, given by $R = (-1)^{2J+3B+L}$, so that R = +1 for all the known particles and R = -1 for their superpartners. The existence of R parity implies that:

i) Sparticles are produced in pairs.

ii) The lightest sparticle, LSP, is absolutely stable.

Because of ii) all sparticle decay chains end in an LSP and so for supersymmetric searches one needs to know which is this particle. There are good astrophysical arguments[74] that an LSP cannot be charged or have strong interactions, if not it would have condensed in galaxies and planets and would have been detected in searches of matter with anomalous values of e/m.

The usual assumption made is that the LSP is the photino, $\tilde{\gamma}$. One can adduce a variety of astrophysical, cosmological and particle physics arguments in favor of this hypothesis [74][75]. For instance, if the gaugino soft masses M are universal, then the different renormalization behaviour for gluinos and photinos implies $m_{\tilde{g}} \simeq 7m_{\tilde{\gamma}}$. In certain models[76] it is possible that the sneutrino is the LSP. However, I shall not consider this possibility here. I shall also not discuss in detail the present status of bounds on supersymmetric particles, since Davier[25] has discussed this topic in his rapporteur talk here. Rather, I'll consider only one example of a supersymmetric particle search: that of squarks and gluinos at the CERN $Sp\bar{p}S$ collider, and how this will be extended at the Tevatron collider. If the photino is the LSP, the produced squarks and gluinos will decay to photinos through the chains

$$m_{\tilde{g}} > m_{\tilde{q}} \qquad \tilde{g} \to q\bar{q}\tilde{\gamma} \; ; \; \tilde{q} \to q\tilde{\gamma}$$

$$m_{\tilde{g}} < m_{\tilde{q}} \qquad g \to q\bar{q}\tilde{\gamma} \; ; \; \tilde{q} \to q\tilde{g} \to qq\bar{q}\tilde{\gamma} \qquad \text{(IV.1)}$$

Since the produced photino interacts weakly, it provides a missing energy signal. Hence the well known experimental signature to expect, in the case of gluino or squark production, is missing energy plus (multi) jets. How many jets, however, is a sensitive issue that depends crucially on experimental cuts.

The famous (infamous?) UA_1 monojet signal and its relation to supersymmetry were the hot subject in 1984-1985[77]. At this conference the UA_1 collaboration has presented[78] a very complete analysis of their missing energy signal, which consists of 53 monojets and 3 dijets. Already the preponderance of monojets, even with the UA_1 cuts, is a bad signal for squark or gluino production, since one would expect from these decays a sizeable fraction of multijet events[79]. As Honma reported[78], the missing energy events are essentially accounted for by standard model backgrounds. This allows the UA_1 collaboration to set rather strong bounds on the masses of squarks and gluinos. These bounds are shown in Fig IV.2. One sees from this figure that $m_{\tilde{q}} \lesssim 80$ GeV and $m_{\tilde{g}} \lesssim 60$ GeV are excluded. Actually, the UA_1 collaboration, cannot also exclude a light gluino window. However, this light gluino scenario has fallen in theoretical disrepute[80], and I have taken the liberty of removing this window from Fig IV.2.

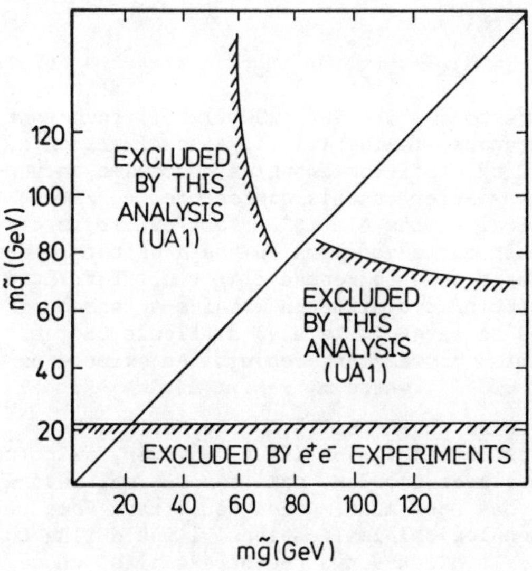

Fig IV.2 Limits on gluino and squark masses obtained by the UA_1 collaboration[78].

It would have been nice if the UA_1 collaboration could have given a bound also on the Wino. Although the mass matrix for the charged gauginos is model dependent, for models with a small supersymmetry breaking gaugino mass, one of the eigenstates, $\tilde{W}$, is lighter than the W[81]. So sequential decays like $W \to \tilde{W}\tilde{\gamma}$; $\tilde{W} \to q\bar{q}\tilde{\gamma}$ are possible. These processes give rise mostly to monojets, when one takes into account of the UA_1 cuts. So the preponderance of monojets in the missing energy signal is at least not unfavorable to Winos. In this conference, Arnowitt[32] estimated that a $\tilde{W}$ mass $M_{\tilde{W}} \gtrsim 40$ GeV is still compatible with the UA_1 signal. However, it is clear that a reliable analysis can only be done by the UA_1 collaboration itself.

Although the bounds on gluinos and squarks obtained at the CERN collider are impressive, it is important to emphasize that if these excitations exist near these bounds then they will be rather easily seen at the Tevatron. This point has been forcefully made by Baer and Berger[83] and Reya and Roy [84] and has been discussed at this conference by Reya[85]. For given sparticle masses the production cross section grows rapidly with energy in the range from $\sqrt{s}$ = 620 GeV to $\sqrt{s}$ = 2 TeV and the signal relative to standard model background also increases. As an example, I show in Fig IV.3, the event rate for producing gluinos at $\sqrt{s}$ = 1700 GeV for various cuts on the missing energy, for the case $m_{\tilde{q}} = 2m_{\tilde{g}}$. One sees that even for the very safe cut of $p_T^{miss} > 60$ GeV one gets nearly 5 events per 100 nb^{-1}, for gluinos as heavy as 100 GeV.

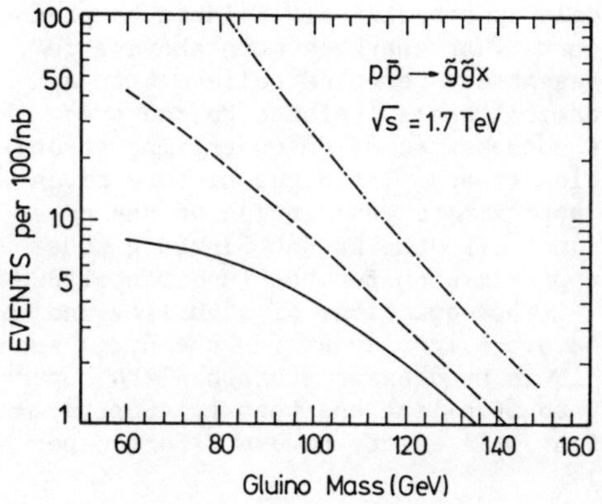

Fig IV.3 Gluino production at $\sqrt{s}$ = 1700 GeV, for $m_{\tilde{q}} = 2\ m_{\tilde{g}}$, as a function of various p_T^{miss} cuts (Dot dash: total; dashed > 40 GeV; solid > 60 GeV) from Ref 86. For similar curves see also Ref 83 and 84.

Secondly, at the Tevatron multijet events will be a dominant feature of squark and gluino production[83)84]. Not only dijets will dominate over monojets but also, for sufficiently large gluino masses, trijets become quite important, as shown in Fig IV.4.

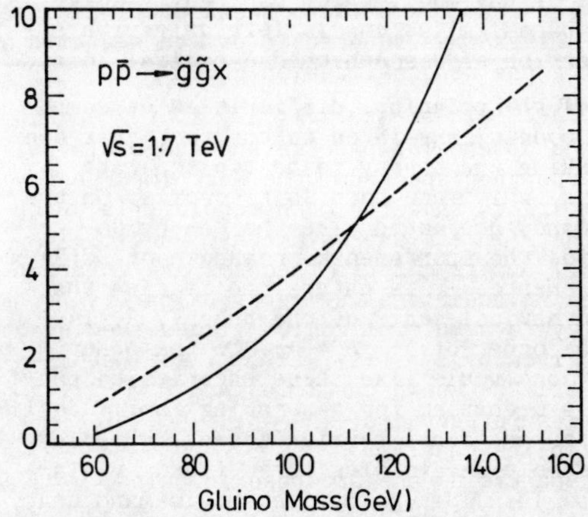

Fig IV.4. Fraction of dijets (dashes) and trijets (solid) to monojets for $\sqrt{s}$ = 1.7 TeV. From Ref 86.

Operating the Tevatron at $\sqrt{s}$ = 1.6 TeV one should be able to discover[83,85] squarks and gluinos if $m_{\tilde{g}}$, $m_{\tilde{q}} \lesssim 150$ GeV. This discovery range can be pushed to near 200 GeV when $\sqrt{s}$ is raised to 2 TeV[85].

V FERMI SCALE PHYSICS - CHALLENGES AND HOPES

There is a minority of theorists, to which I belong, who contend that the origin of the Fermi scale is not directly related to phenomena occurring at energies much above a TeV. These present day heretics believe that Λ_F is a dynamical scale, related to the presence of condensates of an underlying strong interaction theory. The Higgs picture gives only an approximate description of the true theory, just like the Landau Ginzburg model was an approximation for the fundamental BCS theory[87]. Thus questions of stability and naturalness are irrelevant for the Higgs sector and it is unnecessary to appeal to supersymmetry to stabilize the theory. (Supersymmetry might well exist, however, for deeper reasons).

One can imagine that the role of the underlying theory is just to provide a mechanism for generating the Fermi scale, as was the case for technicolor[88]. However, it is perhaps more reasonable to suppose that this strongly interacting theory is also responsible for producing quarks and leptons as composite bound states of more fundamental objects - preons[89]. The status of these theories has been summarized by Harari at this conference[90] in a very kind way, by pointing out that although the motivation for compositeness remains as good as ever, the major problems are largely unchanged! Particularly troublesome to me is the absence of any model which can really serve as a paradigm, so one is left only with a collection of disconnected dynamical ideas.

One of the principal difficulties of composite models is related to the fact that one is asking the theory to do two separate things, which are hard to reconcile. On the one hand, one would like the theory to provide the spontaneous breakdown of SU(2) x U(1). Hence, it is natural to imagine that the dynamical scale of the theory, Λ, is of the order of Λ_F: $\Lambda \simeq \Lambda_F$. On the other hand, one would like these theories to provide a mechanism for generating family replications and small fermion masses, while at the same time avoiding large flavor violations. This last point seems to demand that $\Lambda >> \Lambda_F$.

The physics of the underlying theory will generate, in general, effective non renormalizable terms which violate flavor and which must be added to the standard model Lagrangian[91]:

$$\mathcal{L}_{eff} = \mathcal{L}_{SM} + \sum_i \frac{\lambda_i}{\Lambda^2} O_i \qquad (V.1)$$

where the O_i are SU(3)xSU(2)xU(1) invariant operators. Taking $\lambda_i = 1$, an analysis of a variety of flavor changing processes, reported by Wyler at this conference[92], gives typical bounds $\Lambda \gtrsim 10^2 - 10^3$ TeV. It is difficult for a model which has Λ of this order of magnitude to generate $\Lambda_F \sim 0.1$ TeV. Even admitting two different scales Λ_F and Λ, if Λ is so large it is also difficult to get fermion masses large enough. An example being ETC[93], where $m_f \sim \Lambda_F^3 / \Lambda^2$.

There has been really no theoretical progress on the flavor issue. However, skirting the flavor problem, some theoretical advances has been made, which could have some phenomenological implications. I would like to briefly discuss two recent examples, whose lessons perhaps can be useful, even though they do not illuminate the role of flavor. One should keep in mind, in this respect, that it is perfectly possible for the compositeness scale of electrons and muons individually to be of O(TeV), and yet only to be able to probe the e-μ difference at distances of order $(10^3\text{TeV})^{-1}$ [90].

The first example which I will discuss is the, so called, composite Higgs model[94]. This model separates the scales Λ and Λ_F by making the Higgs bosons essentially Goldstone bosons. One imagines that, in the limit in which the electroweak interactions are turned off, the underlying theory posesses a global symmetry G which is broken down to another group H, producing certain bound state Goldstone bosons Φ. Turning on the SU(2)xU(1) couplings causes a realignement of the vacuum and Φ acquires a non zero vacuum expectation value

$$\langle\Phi\rangle = \Lambda_F = f(\alpha)\Lambda \qquad (V.2)$$

The function $f(\alpha)$ is dynamically determined and it is possible that $\Lambda >> \Lambda_F$. For the vacuum realignement to actually take place, it is necessary that the low energy group be bigger than SU(2)xU(1)[95]. Although H > SU(2)xU(1), the unbroken group H should not contain the full gauge theory, if one wants the vacuum to reorient itself when the gauge couplings are turned on. Amusingly enough, therefore, the simplest realization of these theories contains also an extra Z^0, although with characteristics quite different from the superstring Z^0's.

Dugan, Georgi and Kaplan[96]) studied a toy model where the weak group is H_W = SU(2)xU(1)x$U(1)_A$ and G = SU(5) while H = O(5). In this model the effective Higgs potential depends on a function of the ratio of the SU(2)xU(1) and $U(1)_A$ couplings:

$$c_0 = \frac{3g^2 + g'^2}{g_A^2} \qquad (V.3)$$

and one can establish that Λ_F vanishes in the limit as $c_0 \to 1$, so that

$$f(\alpha) \sim \sqrt{1-c_0} \qquad (V.4)$$

The Higgs effective potential is calculable in the model, and therefore the Higgs mass and that of the $\tilde{Z}$ boson are given as functions of c_0. The dependence of these quantities on c_0 is displayed in Fig V.1. To be in agreement with neutral current data $M_{\tilde{Z}}$ must be high enough, which in this case implies $c_0 > 0.6$[96]). For values of c_0 in the allowed range, it is easy to see from Fig V.1 that the Higgs mass lies in a band near 200 GeV.

The model discusses in Ref 96 is not quite realistic since, for example, one needs to introduce spectator fermions to cancel some of the anomalies associated with the extra axial $U(1)_A$ gauge interactions. Trying to make the composite Higgs model more realistic has engendered a growing set of bizzare and baroque, but clever, models: the moose models[97]). Unfortunately, these models are not yet ripe for phenomenology.

The second example I want to discuss is the strongly coupled standard model, which was reported on, in this conference, by B. Schrempp[98]). This model was proposed originally by Abbott and Farhi[99]) and has been recently reexamined in some detail by Claudson, Farhi and Jaffe[100]). The Lagrangian for the model is precisely that of the standard model, except that all left handed fields $\hat{\psi}_L$ and the Higgs doublet Φ are considered as preons. Most importantly, the SU(2) gauge group is supposed to confine and <u>not</u> suffer spontaneous breakdown. In the limit of vanishing g', and neglecting all Yukawa couplings, the theory possesses a global $SU(4n_f)\times SU(2)_W$ symmetry, where the last sym-

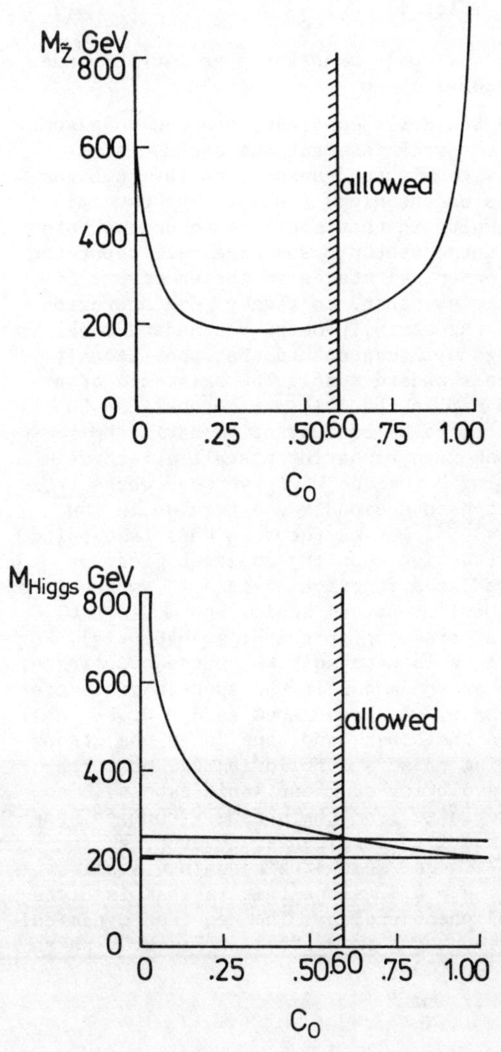

Fig V.1 Masses of the $\tilde{Z}$ and composite Higgs boson as a function of c_0, for the model of Ref 96. Phenomenologically $c_0 > 0.6$

metry group arises from the O(4) symmetry properties of the Higgs potential. This symmetry can be preserved in the binding if one can find a set of massless composite fermions to match the global symmetry anomalies [101]). This set is trivially provided by the left handed quarks and leptons constructed as

$$l_L, q_L \sim (\psi_L \ \Phi) \qquad (V.5)$$

which obviously transforms as $(4n_f, 2)$ under the global group.

Since the SU(2) confines, one has to assume that the weak interactions are mediated by composite $W^\pm$ and Z bosons, which are bound states of the Higgs field Φ. The crucial difficulty in this model is to demonstrate that these vector bosons are well separated from other J=1 states in the spectrum. If this is so, then, to a very good approximation, the strongly coupled standard model is analogous in content to the spontaneously broken standard model. The existence of a large gap in the J=1, T=1 channel can be seen[100] to be equivalent to having $M_W = M_Z \cos\theta_W$ and to having a small effective coupling $\bar{g}$ for the Wf_Lf_L vertex, where f_L is a left-handed bound state fermion in the theory[100]. Let me focus on this last point. The effective coupling constant $\bar{g}$ can be plotted as a function of $(\Lambda/\Lambda_F)^2$, where Λ is the SU(2) dynamical scale, and Λ_F is now just a parameter in the Higgs potential. For $\Lambda \ll \Lambda_F$ $\bar{g}$ is precisely the gauge coupling of the standard model in the spontaneously broken phase, which vanishes as $\Lambda \to 0$. For $\Lambda \gg \Lambda_F$, on the other hand, one is in the strong coupling phase and following the perturbative evolution of $\bar{g}$ one would expect it to become large. For the strongly coupled standard model to make sense, however, $\bar{g}$, for Λ^2/Λ_F^2 large, must attain again a small value, $\bar{g} \simeq 0.7$, since this is what is dictated by phenomenology. The required dynamical behaviour of $\bar{g}$ is plotted in Fig V.2, taken from Ref 100.

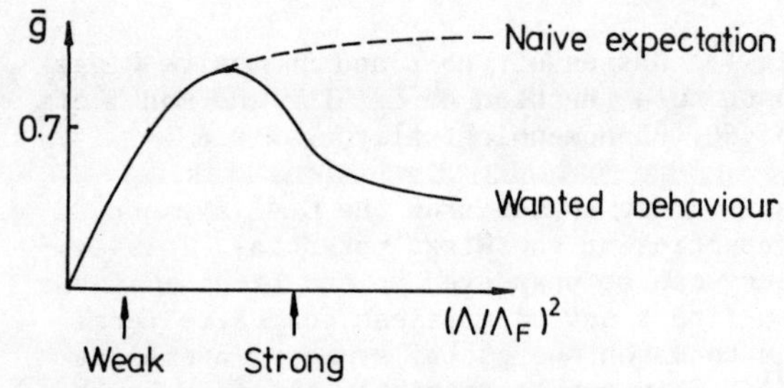

Fig V.2 Behaviour of the effective Wf_Lf_L coupling $\bar{g}$ needed for the strongly coupled standard model to be consistent.

It is an unsolved problem whether dynamically the strongly coupled standard model makes sense; i.e. if the behaviour shown in Fig V.2 for $\bar{g}$ really obtains. If this happens, however, one has a model which for most purposes is equivalent to the standard model but which, in addition, has a rich spectrum of composite bosons and fermions[100]. Some of the phenomenology of these states has been discussed in this conference by B. Schrempp[98]. Here I will only comment on a noteworthy class of candidates for the model: spin zero difermions. These states are bound states of two Ψ_L which are antisymmetric in the flavor indices a,b of the Ψ_L's, with a,b going from 1 to $4n_f$. That is

$$S^{[a,b]} \sim \Psi_L^a \Psi_L^b \qquad (V.6)$$

The $S^{[a,b]}$ contain both charge −1 dileptons, charge −1/3 leptoquarks and charge +1/3 antitriplet diquarks. Hence their coupling to the ordinary quark and lepton left handed doublets $D^a = \left\{ q_L \ , \ l_L \right\}$ is given by

$$\mathcal{L}_{eff} = \frac{\lambda}{2} S^+_{ab} D^{aT} C D^b + \text{h.c.} \qquad (V.7)$$

Note that since these scalar excitations are <u>not</u> Goldstone bosons, there is no reason to suppose that λ is proportional to the mass of the bound state fermions. Korpa and Ryzak[102] used data on neutrino nucleon scattering to put a bound on λ/M_S, where M_S is the mass of the difermions:

$$M_S > 275 \ \lambda \ \text{GeV} \qquad (V.8)$$

Hence if λ is of the same order of magnitude as $\bar{g}$, the leptoquarks in S would be in the discovery range of HERA. Indeed, in this case, as discussed by Wudka[103], the cross sections are very large, since the leptoquark production is a resonance process. Unfortunately the value of λ, like that of $\bar{g}$, is beyond our present computational capacity.

VI CONCLUDING REMARKS

I hope this report has demonstrated that theoretical ideas beyond the standard model are rich and varied, leading to potentially very interesting phenomenology. However, phenomenology needs phenomena! Thus we are all waiting with extreme interest for the results which shall be forthcoming in the coming years from the Tevatron, SLC, LEP and HERA, as well as from non accelerator experiments. Only then we shall know what kind of physics, if any, lies beyond the standard model. From this point of view, the tremendous theoretical investment in D>4 physics could be a bit premature. After all, Fermi scale physics may well turn out to be different!

REFERENCES

(1) G. Altarelli, these proceedings

(2) W. Scott, these proceedings

(3) E. Fischbach, D. Sudarsky, A. Szafer, C. Talmadge and S.H. Aronson, Phys. Rev.Lett. 56(1986)3; E. Fischbach, these proceedings

(4) R.V. Eötvos, D. Pekár and E. Fekete, Ann.Phys. (Leipzig)68(1922)11

(5) E. Fischbach, D. Sudarsky, A. Szafer, C. Talmadge and S.H. Aronson, Proceedings of the 2nd Conference on Intersections between Particle and Nuclear Physics, Lake Louise, May 1986

(6) F.D. Stacey and G.J. Tuck, Nature 292 (1981)230

(7) Review of Particle Properties, Phys. Lett. 170B(1986)1

(8) S.C. Holding, F.D. Stacey and G.J. Tuck, Phys.Rev. D33(1986)3487

(9) G.W. Gibbons and B.F. Whiting, Nature 291(1981)636

(10) M. Milgrom, IAS preprint; S.C. Holding, F.D. Stacey and G.J. Tuck, Ref 8; P. Thieberger, Phys.Rev.Lett. 56(1986)2347; A. Bizzeti, Florence University preprint; D.A. Neufeld, Phys.Rev.Lett. 56(1986) 2344; C. Talmadge, S.H. Aronson and E. Fischbach, in Proceedings of the XXI Rencontre de Moriond, Les Arcs, March 1986

(11) Y. Asano et al, Phys.Lett. 107B(1981) 159; 113B(1982)195

(12) S. Weinberg, Phys.Rev.Lett. 13(1964) 495

(13) M. Suzuki, Phys.Rev.Lett. 56(1986) 1339; C. Bouchiat and J. Iliopoulos, Phys.Lett.169B(1986)447; M. Lusignoli and A. Pugliese, Phys.Lett. 171B(1986) 468;S.H. Aronson, H.-Y. Cheng, E. Fischbach and W. Haxton, Phys.Rev.Lett. 56 (1986)1342

(14) I. Bars and M. Visser, Phys.Rev.Lett. 57(1986)25; I. Bars, these proceedings

(15) J. Schweppe et al, Phys.Rev.Lett. 51 (1983)2261; M. Clemente et al, Phys. Lett. 137B(1984)41;

(16) T. Cowan et al, Phys.Rev.Lett. 56 (1986)446

(17) J. Reinhardt, B. Müller and W. Greiner, Phys.Rev. A24(1981)173

(18) T. Cowan et al, Phys.Rev.Lett. 54 (1985)1761

(19) A. Schäfer et al, J.Phys.G: Nucl.Phys. 11(1985)L69; A. Chodos and L.C.R. Wijewardhana, Phys.Lett. 56(1986)302; J. Reinhardt et al, Phys.Rev. C33 (1986)194; K. Lane, Phys.Lett. 169B (1986)97

(20) R.D. Peccei and H.R. Quinn, Phys.Rev. Lett. 38(1977)1440; Phys.Rev. D16 (1977) 1791; S. Weinberg, Phys.Rev. Lett. 40(1978)223; F. Wilczek, Phys. Rev.Lett. 40(1978)279

(21) For a review of these bounds, see for example, S. Yamada in Proceedings of the International Symposium on Lepton Photon Interactions at High Energies, Cornell University, Ithaca, 1983

(22) R.D. Peccei, T.T. Wu and T. Yanagida, Phys.Lett. 172B(1986)435; L.M. Krauss and F. Wilczek, Phys.Lett. 173B(1986) 189

(23) W.A. Bardeen and S.H. Tye, Phys.Lett. 74B(1978)229

(24) J. Greenberg, Proceedings of the International Symposium on Weak and Electromagnetic Interactions in Nuclei, Heidelberg, July 1986

(25) M. Davier, these Proceedings

(26) W.A. Bardeen, R.D. Peccei and T. Yanagida, Nucl.Phys. B279(1987)401; L.M. Krauss and M. Wise, Phys.Lett. 176B(1986)483; M. Suzuki, Phys.Lett. 175B(1986)364

(27) W.A. Bardeen et al, Ref 26

(28) E. Eichler et al, Phys.Lett. 175B (1986)101

(29) F.W.N. de Boer et al, Groningen preprint, Phys.Lett. B to be published

(30) T.W. Donnelly et al, Phys.Rev. D18 (1978)1607

(31) For a review of this scenario and its astrophysical and cosmological constraints see for example, F. Wilczek in

How far are we from the Gauge Forces, Erice lectures 1983 (Plenum Press, New York 1985)

(32) C. Callan, Proceedings of the XXII International Conference of High Energy Physics, Leipzig 1984

(33) M. Green and J. Schwarz, Phys.Lett. 149B(1984)117

(34) J. Schwarz, these proceedings

(35) P. Candelas, G.T. Horowitz, A. Strominger and E. Witten, Nucl.Phys. B258 (1985)46; Proceedings of the Argonne Symposium on Anomalies, Geometry, and Topology, (World Scientific, Singapore)

(36) E. Calabi, Algebraic Geometry and Topology: a Symposium in Honour of S. Letschetz (Princeton University Press, Princeton, N.Y. 1957); S.T. Yau, Proc. Nat.Acad.Sci. 74(1977)1798

(37) P. Candelas et al, Ref 35; C.G. Callan, E. Martinec, D. Friedan and M. Perry, Nucl.Phys. B262(1983)593; A. Sen, Phys.Rev. D32(1985)2102; C.M. Hull and P.K. Townsend, Nucl.Phys. B274(1986)349

(38) Even though the β function for these σ models does not vanish at four loops (M.T. Grisaru, A. van de Ven and D. Zanon, Nucl.Phys. B277(1986)388, C.N. Pope, M.F. Sohnius and K. Stelle, Imperial College preprint) one can perperturbatively construct such conformally invariant σ models. See D. Nemeschansky and A. Sen, Phys.Lett. 178B (1986)365; D. Gross and E. Witten, Nucl.Phys. B277(1986)1

(39) L. Dixon, J.A. Harvey, C. Vafa and E. Witten, Nucl.Phys. B261(1985)678; B276 (1986)285; C.S. Hamidi and C. Vafa, Caltech preprint; L. Dixon, D. Friedan, E. Martinec and S. Shenker, Chicago preprint

(40) E. Witten, Nucl.Phys. B268(1986)79

(41) M. Dine, N. Seiberg, X.G. Wen and E. Witten, Princeton preprint; J. Ellis, C. Gomez, D.V. Nanopoulos and M. Quiros, Phys.Lett. 173B(1986)59

(42) D.J. Gross, J.A. Harvey, E. Martinec and R. Rohm, Phys.Rev.Lett. 54(1985) 502; Nucl.Phys. B256(1985) 253; B267; (1986)75

(43) E. Witten, Nucl.Phys. B258(1985)75; Y. Hosotani, Phys.Lett. 126B(1983)309; 129B(1983)193

(44) A. Strominger and E. Witten, Comm. Math.Phys. 101(1985)341; A. Strominger, Phys.Rev.Lett.55(1985)2547

(45) B.R. Greene, K.H. Kirklin, P.J. Miron and G.G. Ross, Oxford preprint

(46) For a review see H.P. Nilles, Phys.Rept. 110(1984)1

(47) L.E. Ibañez, Phys.Lett. 118B(1982)73; J. Ellis, D.V. Nanopoulos and K. Tamvakis, Phys.Lett.121B(1983)123; H.P. Nilles, Phys.Lett. 115B(1982)193; K. Inoue, A. Kakuto, H. Komatsu and S. Takeshita, Prog.Theo.Phys. 68(1982) 927; L. Alvaréz Gaume, J. Polchinski and M. Wise, Nucl.Phys. B221(1983)495

(48) E. Cremmer et al, Phys.Lett. 79B(1978) 231; Nucl.Phys. B147(1979)105; E. Cremmer et al, Phys.Lett. 116B(1982)231; Nucl.Phys. B212(1983)413; R. Barbieri, S. Ferrara and C.A. Savoy, Phys.Lett. 119B(1982)343; R. Arnowitt, A.H. Chamseddine and P. Nath, Phys.Rev.Lett. 49 (1982)970;50(1983)232

(49) L.E. Ibañez, these proceedings

(50) J.P. Derendinger, L.E. Ibañez and H.P. Nilles, Phys.Lett. 155B(1985)65; M. Dine, R. Rohm, N. Seiberg and E. Witten, Phys.Lett. 156B(1985)55

(51) J. Breit, B. Ovrut and G. Segré, Phys.Lett. 162B(1985)303; P. Binétruy and M.K. Gaillard, Phys.Lett. 168B (1986)347

(52) P. Binétruy, S. Dawson and I. Hinchliffe, LBL-21606; S. Dawson these proceedings

(53) J.P. Derendinger, L.E. Ibañez and H.P. Nilles, Nucl.Phys. B267(1986)365; L.E. Ibañez and H.P. Nilles, Phys.Lett. 169 B(1986)354

(54) M. Dine and N. Seiberg, Phys.Lett 162B (1985)299; and in Unified String Theories (World Scientific, Singapore 1986)

(55) E. Witten, Nucl.Phys. B258(1985)75, M. Dine, V. Kaplunovsky, M. Mangano, C. Nappi and N. Seiberg; Nucl.Phys. B259 (1985)549; G. Cecotti, J.P. Derendinger, S. Ferrara, L. Girardello and M.

Roncadelli, Phys.Lett. 156B(1985)318; J. Breit, B. Ovrut and G. Segre, Phys. Lett. 158B(1985)33; F. del Aguila, G. Blair, M. Daniel and G.G. Ross, Nucl. Phys. B272(1986)413; P. Binétruy, S. Dawson, I. Hinchliffe and M. Sher, Nucl.Phys. B273(1986)501; T. Matsuoka and D. Suematsu, Nagoya preprint; L.E. Ibañez and J. Mas, CERN preprint TH 4426/86; G.G Ross, these proceedings

(56) R. Slansky, Phys.Rept. 79(1981)1

(57) J.P. Ader, S. Narison and J. Wallet, Montpellier preprint (#7889); G. Belanger and S. Godfrey, Phys.Rev. D34(1986) 1309(#7730 and #7749); T. Matsuoka, H. Mino, D. Suematsu and S. Watanabe, Nagoya preprint (#7617); T.G. Rizzo, Iowa State preprint (#7773); D. London and J. Rosner, Phys.Rev. D34 (1986)1530 (#9288)

(58) S. Barr, Phys.Rev.Lett. 55(1985)2778

(59) E. Witten, Ref 43

(60) V. Barger, N.G. Deshpande and K. Wishnant, Phys.Rev.Lett. 56(1986)30; L.S. Durkin and P. Langacker, Phys. Lett. 166B(1986)436; J. Ellis, K. Enqvist, D.V. Nanopoulos and F. Zwirner, Nucl.Phys. B276(1986)14; F. del Aguila, G.A. Blair, M. Daniel and G.G. Ross, CERN TH4376; D. London and J. Rosner, Phys.Rev. D34(1986)1530

(61) N.G. Deshpande, these proceedings

(62) V.D. Angelopoulos, J. Ellis, D.V. Nanopoulos and N.D. Tracas, Phys.Lett. 176B(1986)203

(63) M. Cvetic and B.W. Lynn, SLAC PUB 3900; P.J. Franzini, SLAC PUB 3920; P.J. Franzini and F.J. Gilman, SLAC PUB 3932

(64) For an estimate of the accuracy expected for A_{FB} at LEP, see P. Baillon et al, Physics at LEP CERN 86-02

(65) M. Dine et al, Ref 55

(66) E. Cohen, J. Ellis, K. Enqvist and D.V. Nanopoulos, Phys.Lett. 165B (1985)76; J. Ellis, K. Enqvist, D.V. Nanopoulos and F. Zwirner, Ref 60 and Mod.Phys.Lett. A1(1986)57

(67) I am using here the same notation for scalars as that for the fermions in table III.1. I hope this is not confusing

(68) F. del Aguila et al, Ref 55

(69) E. Witten, Ref 43; J. Breit et al, Ref 55; A. Sen, Phys.Rev.Lett. 55(1985)33

(70) B.R. Greene, K.H. Kirklin, P.J. Miron and G.G. Ross, Ref 45; G.G. Ross, these proceedings

(71) S.T. Yau in Proceedings of the Argonne Symposium on Anomalies Geometry and Topology (World Scientific, Singapore 1985)

(72) See for example, D.V. Nanopoulos, Proceedings of the XXII International Conference of High Energy Physics, Leipzig 1984

(73) P. Fayet, Phys.Lett. 69B(1977)489; G. Farrar and P. Fayet, Phys.Lett. 76B (1978)575

(74) J. Ellis, J.S. Hagelin, D.V. Nanopoulos, K.A. Olive and M. Srednicki, Nucl. Phys. B238(1984)453

(75) L. Ibañez, Phys.Lett. 137B(1984)160; J.S. Hagelin, G.L. Kane and S. Raby, Nucl.Phys. B241(1984)638

(76) For a discussion, see for example, J. Ellis, K. Enqvist, D.V. Nanopoulos and F. Zwirner, Mod.Phys.Lett. A1(1986)57

(77) For a critical discussion see: J. Ellis in Proceedings of the 1985 International Symposium on Lepton and Photon Interactions at High Energies, Kyoto, Japan

(78) A. Homna, these proceedings

(79) R.H. Barnett, H.E. Haber and G.L. Kane, Nucl.Phys. B267(1986)625

(80) See for example, K. Hagiwara in The Quark Structure of Matter (World Scientific, Singapore 1986)

(81) S. Weinberg, Phys.Rev.Lett. 50(1983) 387; for a general discussion, see for example, H.E. Haber and G.L. Kane, Phys.Rept 117(1985)75

(82) R. Arnowitt, these proceedings

(83) H. Baer and E.L. Berger, Phys.Rev. D34 (1986)1361

(84) E. Reya and D.P. Roy, Phys.Lett. 166B

(1986)223 and Zeit.f.Phys.C to be published

(85) E. Reya, these proceedings

(86) R.M. Barnett and H.E. Haber, to be published

(87) V.L. Ginzburg and L. Landau, Zh.Eksp. Teor.Fiz. 20(1950)1064; J. Bardeen, L.N. Cooper and J.R. Schrieffer, Phys. Rev.108(1957)1175

(88) L. Susskind, Phys.Rev. D20(1979)2619; S. Weinberg, Phys.Rev. D19(1979)1277

(89) For a recent review see M.E. Peskin, Proceedings of the 1985 International Symposium on Lepton and Photon Interactions at High Energies, Kyoto, Japan

(90) H. Harari, these proceedings

(91) C. Burges and H. Schnitzer, Nucl.Phys. B228(1983)464; C.N. Leung, S. Love and S. Rao, Zeit.f.Phys. 31C(1986)433; W. Buchmüller and D. Wyler, Nucl.Phys. B268(1986)621

(92) D. Wyler, these proceedings

(93) S. Dimopoulos and L. Susskind, Nucl.Phys. B155(1979)237, E. Eichten and K. Lane, Phys.Lett. 90B(1980)125

(94) D.B. Kaplan, these proceedings; H. Georgi and D.B. Kaplan, Phys.Lett. B145(1984)216; H. Georgi, D.B. Kaplan and B. Galison, Phys.Lett. B143(1984) 152; M.J. Dugan, H. Georgi and D.B. Kaplan, Nucl.Phys. B254(1985)299; for and extensive discussion see H. Georgi, Proceedings of the 1985 Les Houches Summer School

(95) M.E. Peskin, Nucl.Phys. B175(1981)197; J. Preskill, Nucl.Phys. B177(1981)21

(96) M.J. Dugan, H. Georgi and D.B. Kaplan, Ref 94

(97) H. Georgi, Phys.Lett. B151(1985)57; H. Georgi and J. Preskill, Phys.Lett. B156(1985)369; see also H. Georgi, Ref 94

(98) B. Schrempp, these proceedings

(99) L. Abbott and E. Farhi, Phys.Lett. 101B(1981)69; Nucl.Phys. B189(1981)547

(100) M. Claudson, E. Farhi and R.L. Jaffe, Phys.Rev. D34(1986)873

(101) G. 't Hooft in Recent Development in Field Theory (Plenum Press, New York, 1980)

(102) C. Korpa and Z. Ryzak, MIT preprint

(103) J. Wudka, Phys.Lett. 167B(1986)337

SEARCHES FOR NEW PARTICLES

Michel DAVIER

Laboratoire de l'Accélérateur Linéaire
Université de Paris-Sud, Orsay, France

New experimental results on particle searches are reviewed. Two experimental techniques give powerful limits : single photon events in e^+e^- annihilation and missing transverse momentum in $\bar{p}p$ collisions. The standard model phenomenology still awaits the discovery of the top quark and of the Higgs boson; no fourth generation appears in sight and looks in fact unlikely. Axion searches are negative, despite a renewed interest due to anomalous positron production in heavy ion collisions which turns out not to be explainable within particle physics only. Supersymmetry remains an elegant theory without contact with experiment.

My duty is to review experimental searches for new particles, including possible new quarks and leptons, Higgs bosons, axions, supersymmetric particles, technicoloured particles, monopoles, etc.... This represents a wide-open field of experimentation employing a large spectrum of techniques. In fact, a mere 337 transparencies were shown on these topics in the course of two specialized parallel sessions. It would certainly be a challenging task to summarize all this information, if not for the fact that all contributions shared the common feature of obtaining a negative result. The talk might well stop here ; however, despite the lack of an exciting discovery, it remains a very interesting task to discuss the various limits achieved as some of them are quite constraining for the phenomenology at hand.

The review will be organized in increasing order of speculation. First, the missing pieces within the standard model : the top quark, possibly other generations of quarks and leptons, and the Higgs boson(s). Then, I discuss the recently renewed interest in the axion, which lies in the fringe of the standard phenomenology. Finally, I shall review searches for particles beyond the standard model, as predicted by supersymmetry or compositeness theories. The important subject of free quarks and monopoles was reviewed by H. Matis in the parallel sessions[1] : again no positive result was reported despite experimental techniques of ever increasing ingenuity.

1. TWO POWERFUL EXPERIMENTAL APPROACHES

Several interesting searches I am going to report on make use of two very useful techniques which I proceed to discuss first.

1.1. Single photons in e^+e^- colliders

The reaction

$$e^+e^- \to \gamma + \text{invisible state X} \qquad (1)$$

is a powerful tool to explore new physics. The invisible states can be ordinary neutrinos with a possibility to measure the number of their different types or they can be new kinds of particles, such as the photinos of supersymmetric theories. The method has been also used extensively on ψ and Υ resonances.

It is important for the detectors to be able to reject cosmic interactions and QED processes. The first task requires good timing and small granularity for the electromagnetic calorimeter with longitudinal sampling in order to insure that the shower points to the interaction volume. The second demands a good veto coverage of the calorimeter down to a small angle to the beams, to be sure of the momentum imbalance (Fig. 1).

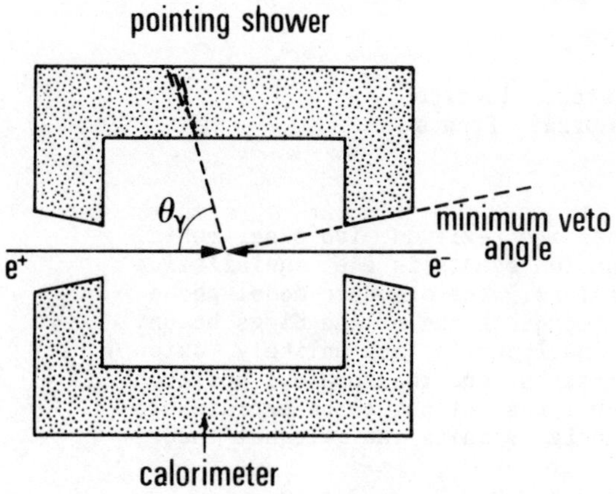

Fig. 1 - Schematic view of a detector for single photons in e^+e^- collisions

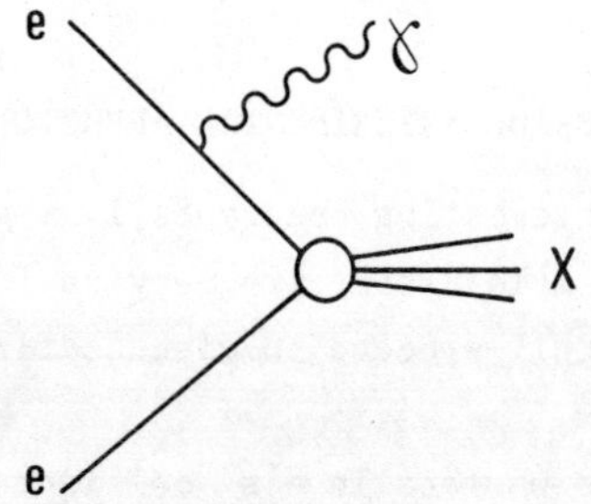

Fig. 2 - Radiative production of a state X. This process is particularly useful if X is a system of neutral, stable and weakly interacting particles which consequently escape detection

The angular and energy distributions of the photon emitted in reaction (1) has a simple form (Fig. 2) :

$$\frac{d\sigma_{\gamma X}(s)}{dE_\gamma\, d\cos\theta_\gamma} \sim \frac{2\alpha}{\pi}\,\frac{1}{E_\gamma}\,\frac{1}{\sin^2\theta_\gamma}\,\sigma_X(\hat{s}) \quad (2)$$

where the cross section σ_X has to be evaluated at the reduced energy squared $\hat{s}$

$$\hat{s} = s(1-x) = s\left(1 - \frac{E_\gamma}{E_{beam}}\right) \quad (3)$$

It is obvious from eq.(2) that one should try to decrease the p_T threshold of the detected photon in order to maximize the rate of any process of type (1). Correspondingly, this has to be matched with the minimum veto angle provided by the calorimeter, since the reaction

$$e^+e^- \to e^+e^-\gamma$$

is a most dangerous background when the electrons avoid detection in the calorimeter. The caracteristics of the contributing experiments[2] which have studied reaction (1) in the continuum are shown in Table 1. Experiments running on resonances do not need so stringent requirements in general.

detector	ASP	MAC	CELLO	Mark J
$\int\mathcal{L}dt$ (pb^{-1})	115	177	38	28
$\sqrt{s}$ (GeV)	29	29	42.6	42
P_T^{min} (GeV) for detected γ	1.	2.→4.5	1.2→2.1	6
θ^{min} (mr) for veto	20	66→175	50	87

Table 1 - Properties of detectors for single photons in the continuum

It must be emphasized that single photon experiments are background-free. Also, it is easy to monitor their efficiency with known QED processes giving additional tracks or showers. Finally, any sought-for process can be exactly calculated so that model-independent limits can be achieved.

In the continuum, all together the experiments have detected only 2 events. The ASP event is shown in figure 3. This small rate translates into very powerful limits I shall shortly discuss.

1.2. Missing transverse momentum in $p\bar{p}$ colliders

A comprehensive analysis of events with transverse momentum imbalance has been presented by the UA1 collaboration[3]. Events with such properties are indeed seen (besides the W sample) and a first analysis had been published[4] prompting a high level of activity among theorists[5] up to a point of claiming the discovery of supersymmetry. As such serious matters deserve detailed experimental scrutiny, a complete re-analysis was performed with the full data : 715 nb^{-1} at 630 GeV centre-of-mass energy.

Each event is characterized by its total transverse momentum measured in the calorimeter cells (Fig. 4)

$$E_T^M = |\sum_i E_i \sin\theta_i \vec{u}_i|$$

with an expected number of standard deviations (significance)

$$N_\sigma = \frac{E_T^M}{0.7\sqrt{E_T^M}}$$

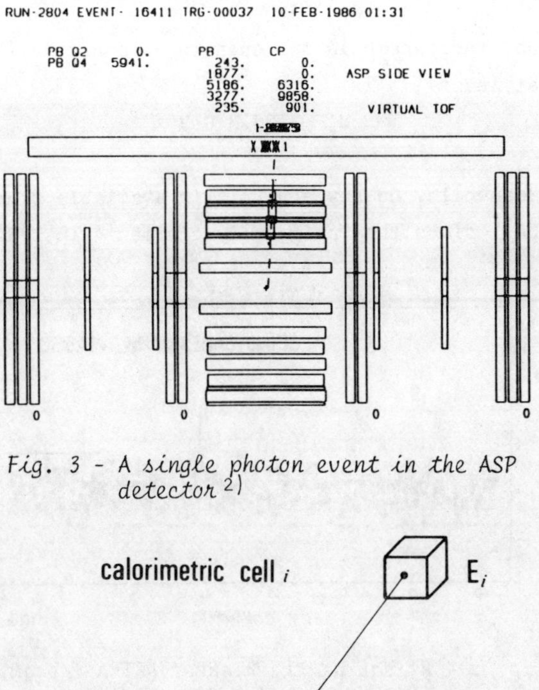

Fig. 3 - A single photon event in the ASP detector [2]

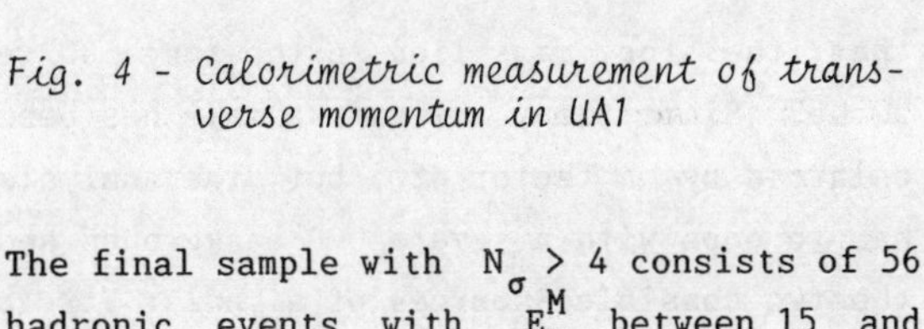

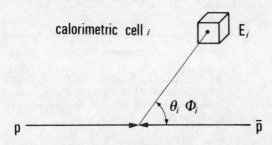

Fig. 4 - Calorimetric measurement of transverse momentum in UA1

The final sample with $N_\sigma > 4$ consists of 56 hadronic events with E_T^M between 15 and 70 GeV, including 53 monojets (jets are counted as such only if their transverse energy exceeds 12 GeV collimation cuts).

Figure 5 shows the N_σ distribution of monojets for $N_\sigma > 3$ and the simulated contributions from expected physics. It can be seen that for $N_\sigma > 4$ the main contributions are from $W \to \tau\nu$ decays and (jets + Z x $\to \nu\bar{\nu}$) events (also jet + $W \to e\nu$ with e-jet overlap). At that point, it is clear that any new physics must be at a small level at most.

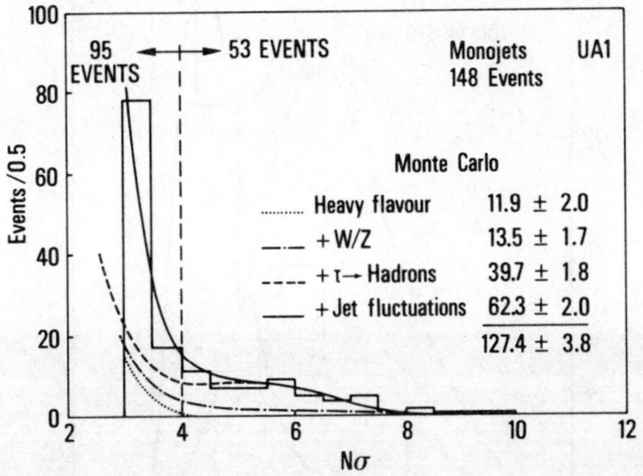

Fig. 5 - The significance distribution of monojets in UA1 and the simulated contributions from expected physics

The $W \to \tau\nu$ sample can be further identified by requiring a narrow jet with small multiplicity. A τ likelihood value L_τ can be ascribed to every event (Fig. 6) and a cut $L_\tau > 0$ isolates a rather pure τ sample. The τ events behave as expected ; in particular they allow a new test of lepton universality in the weak charged current (this was previously checked with the τ lifetime at smaller q^2)

$$\frac{BR(W \to \tau\nu)}{BR(W \to e\nu)} = 1.02 \pm .2 \pm .1$$

while providing an independent W mass determination

$$M_W = 89 \pm 3 \pm 6 \text{ GeV}$$

giving confidence in the analysis.

The sample of events with $L_\tau < 0$ (mostly non-τ events) is now open for investigation (Fig. 7).

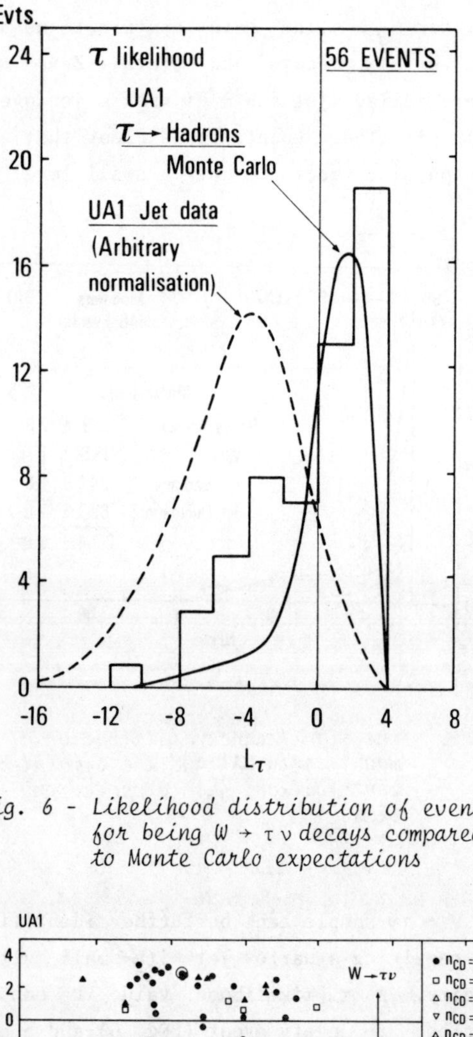

Fig. 6 - Likelihood distribution of events for being $W \to \tau\nu$ decays compared to Monte Carlo expectations

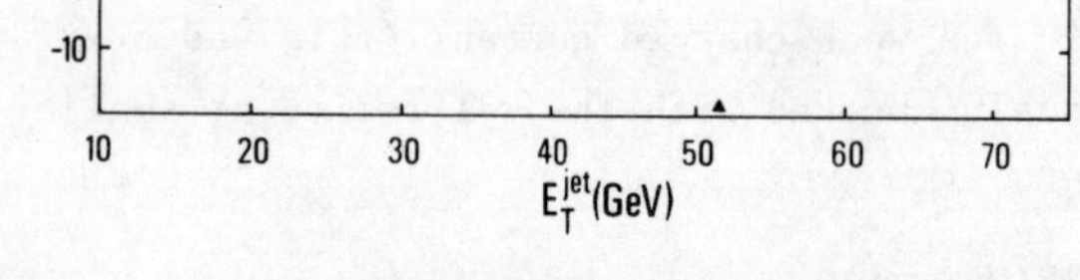

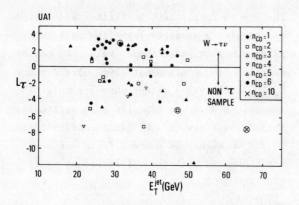

Fig. 7 - Scatter plot of L_τ versus E_T^{jet}. The 3 circled events have 2 jets; also indicated is the charged track multiplicity in the jets

2. SEARCHES FOR STANDARD PARTICLES

2.1. The top quark

The quest for the last member of the third generation of quarks and leptons was halted in 1984 at PETRA. At the maximum centre-of-mass energy of 46.8 GeV, no evidence had been seen in either the R scan for sharp resonances or the detailed event analysis (thrust, aplanarity, semi-leptonic events) for open top production. Figure 8 shows the compilation of all four PETRA experiments[6] for R in the last 7 GeV range allowed by the machine, with an indication of what would have been expected, should the top mass had been in that range. Quantitatively, no $t\bar{t}$ bound state is observed

$$M_{t\bar{t}} > 46.8 \text{ GeV}$$

and the threshold for open top production is at least

$$M_t > 23.3 \text{ GeV}$$

Meanwhile, no new analysis is available from UA1 since they originally claimed[7] in 1984

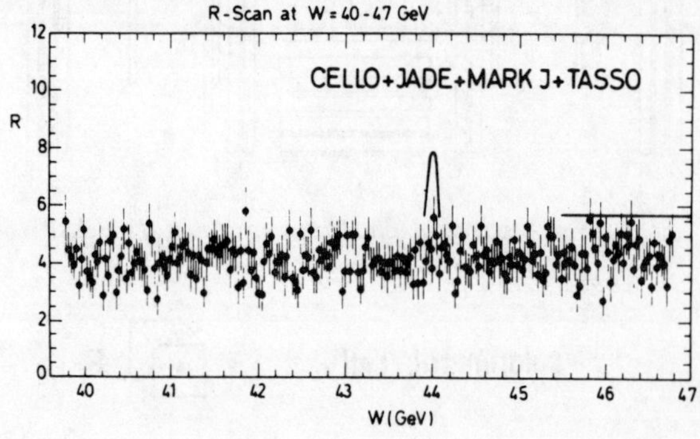

Fig. 8 - R scan at the highest PETRA energy. The effects of a $t\bar{t}$ bound state and the opening of the $t\bar{t}$ continuum are indicated

that the top mass lies in the range 30 to 50 GeV. Since then, the data sample has been enlarged by a factor of 3 but the analysis has to cope with a severe QCD background and the two possible sources of signal : $W \to t\bar{b}$ and $t\bar{t}$ production by gluon fusion. While the first source is more constrained, the second one is expected to dominate for top quark masses above 40 GeV. The results of the final analysis before ACOL data should be available soon[8].

2.2. New quarks of charge 1/3

At PETRA, the R scan is not quite sensitive enough to detect possible $b'\bar{b}'$ bound states, but from the event analysis a lower limit for the continuum $b'\bar{b}'$ production[6] can be obtained

$$M_{b'} > 22.7 \text{ GeV}$$

However, some excess of semi-leptonic events at the highest PETRA energies, $\sqrt{s} > 46.3$ GeV, have been reported by Mark J[9] in 1985 and JADE[10] in a re-analysis in 1986. The effect is seen in the reaction

$$e^+e^- \rightarrow \mu + \text{hadrons}$$

for small thrust ($T < .8$) and isolated muons ($\cos\delta < .7$) with the notation of figure 9,

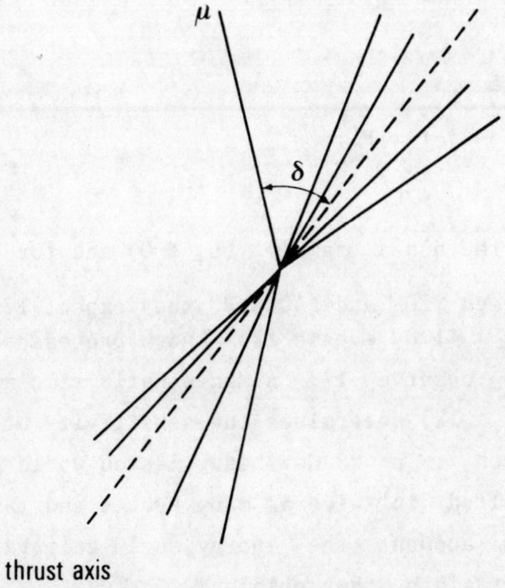

Fig. 9 - Semi-leptonic e^+e^- annihilation event with thrust axis and muon angle δ with respect to the axis

as compared with the expectation derived from Monte-Carlo simulation or, better, extrapolating from data at lower nearby energies. Figure 10 presents the Mark J data.

The results of all four PETRA experiments[9-12] are shown in Table 2. They show

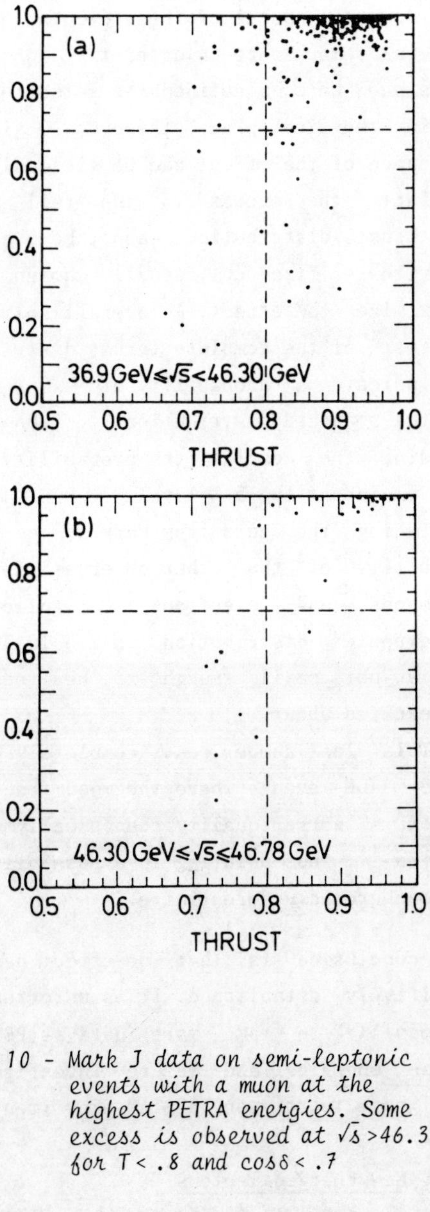

Fig. 10 - Mark J data on semi-leptonic events with a muon at the highest PETRA energies. Some excess is observed at $\sqrt{s} > 46.3$ GeV for $T < .8$ and $\cos\delta < .7$

Experiment	Mark J	JADE		CELLO		TASSO	
	μ	μ	e	μ	e	μ	e
expected (b,c,background)	.8	.6	.7	.7	.2	.3	.3
observed	7	5	0	1	0	1	0

Table 2 - Observed semi-leptonic events with small thrust and isolated lepton at PETRA ($T < .8$ and $\cos\delta < .7$)

some level of inconsistency between experiments for muon production and also between muon and electron production. It is hard to pin down the source of these events, however several remarks are in order :

1) since Mark J defined the selection cuts in the first place, the overall significance of the effect can be misleading. In fact, the excess in the tail of the thrust distribution cannot be evaluated with a fixed cut, usually chosen to maximize the effect. An overall Kolmogorov test of the complete thrust distribution indicate a probability of ~ 10 % to fit the standard distribution[13]. Even including the $\cos\delta$ cut the probability still remains at the 5 % level.
2) Taking the cuts from Mark J, the probability of the other observations (both muons and electrons) to follow the expected distribution is $2.5\ 10^{-2}$ which is not small enough to be too much excited about.
3) This last number rests completely on the 5 JADE events where the muon tracks are of a worse quality than usually observed[14], thus pointing to a possibility of background underestimate.

My conclusion is that no effect has been positively established. It is unfortunately impossible to take more data at PETRA at these energies and future investigations will have to rely on TRISTAN experiments.

2.3. New charged leptons

A new sequential charged lepton comes in a doublet (L, ν_L) with its corresponding neutrino and should decay into a pair of fermions with missing energy-momentum (Fig. 11).

The best limits in e^+e^- annihilation come from the non-observation of acoplanar jets in the process

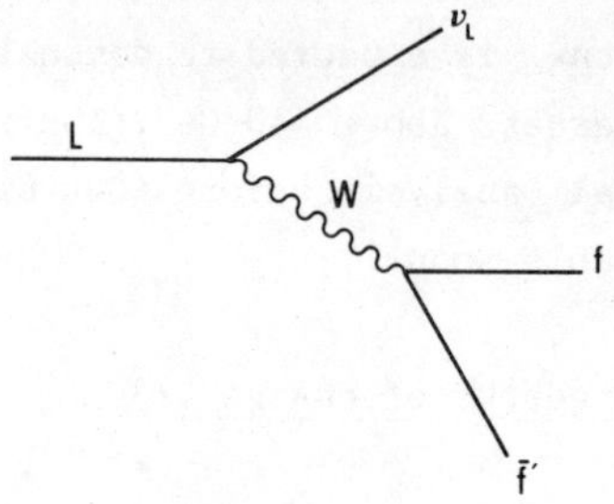

Fig. 11 - Decay of a sequential heavy lepton

$e^+e^- \to L^+L^-$
  ↳ jet + $\not{p}_T$
  ↳ jet + $\not{p}_T$

by CELLO[15] and JADE[16] and yield

$$M_L > 22.7 \text{ GeV} \quad (95\ \%\ \text{CL})$$

A qualitative improvement was presented at this conference by UA1[3] searching for the decay

$W \to L\ \bar{\nu}_L$
  ↳ jet + $\not{p}_T$

In the non-τ sample ($L_\tau < 0$) and for E_T^{jet} between 20 and 40 GeV, they expect $17.8 \pm 3.7 \pm 1.0$ events from known processes and they observe 17. A Monte-Carlo simulation (Fig. 12) determines the sensitivity of the search : a 23 GeV heavy lepton would have resulted in twice as many events and taking into account the energy scale calibration uncertainty, they obtain

$$M_L > 41 \text{ GeV} \quad (90\ \%\ \text{CL})$$

Note that a new lepton L and its neutrino would give two contributions to the monojet sample : first through $W \to L\bar{\nu}$ decay as just discussed and also through an additional neutrino channel in Z decays together with hard gluon radiation (see next section).

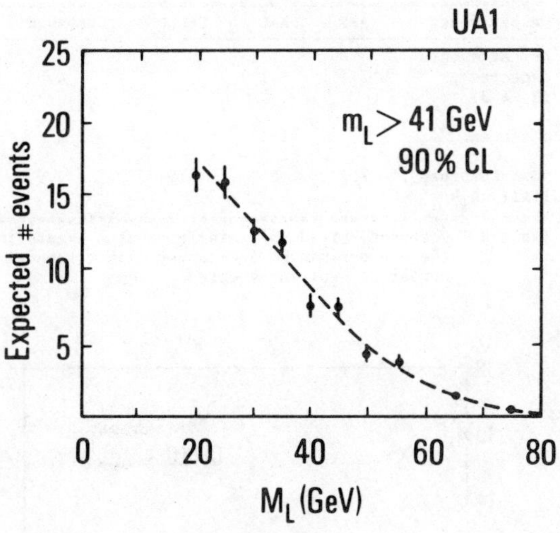

Fig. 12 - Expected yield of unbalanced jets in UA1 from $W \to L\bar{\nu}_L$ as a function of M_L

2.4. <u>Number of generations by neutrino counting</u>

The standard techniques to measure the number of light ($\lesssim$ 30 GeV) neutrino species rely on the direct measurement of Γ_Z, and on an indirect determination through the measurement of r

$$r = \frac{\sigma(p\bar{p} \to W + \ldots)\ BR(W \to \ell\bar{\nu})}{\sigma(p\bar{p} \to Z + \ldots)\ BR(Z \to \ell\bar{\ell})}$$

and

$$\Gamma_Z = \Gamma_W \frac{\Gamma(Z \to e^+e^-)}{\Gamma(W \to e\nu)} \frac{\sigma_Z}{\sigma_W}$$

The first method is limited by experimental resolution and the top mass uncertainty ; the second by uncertainties in estimating through QCD the ratio of W and Z cross sections and also by the top mass uncertainty since

$$\Gamma_Z = \Gamma_Z \text{ (3 generations)} + 0.17 \text{ GeV } (N_\nu - 3)$$

where the first term includes the top only it is sufficiently light. The second method is the most powerful and yields

$N_\nu = 2.8 \begin{array}{l} +\ 2.8 \\ -\ 2.2 \end{array}$ UA1[17)]

$N_\nu = 1.7 \begin{array}{l} +\ 2.3 \\ -\ 1.7 \end{array}$ UA2[18)]

with however an uncertainty of ± 2 neutrinos from the QCD calculation and the top mass uncertainty.

A novel method has been used by UA1 and presented at this conference[3)] : it relies on "gluon tagging" in order to measure invisible Z° decays (Fig. 13). It turns out that a significant part (~ 30 %) of the non-τ monojet events are QCD radiative Z events where the Z decays invisibly. In the sample ($L_\tau < 0$, $E_T^{jet} > 20$ GeV) one expects the following contributions

tail of $W \to \tau\nu$	8.0	
backgrounds	5.7	20.8
$Z \to \nu\bar{\nu}$ + jets	7.1	
($N_\nu = 3$)		

for a total of 24 observed events. This translates into a limit

$$N_\nu < 10 \qquad (90\ \%\ \text{CL})$$

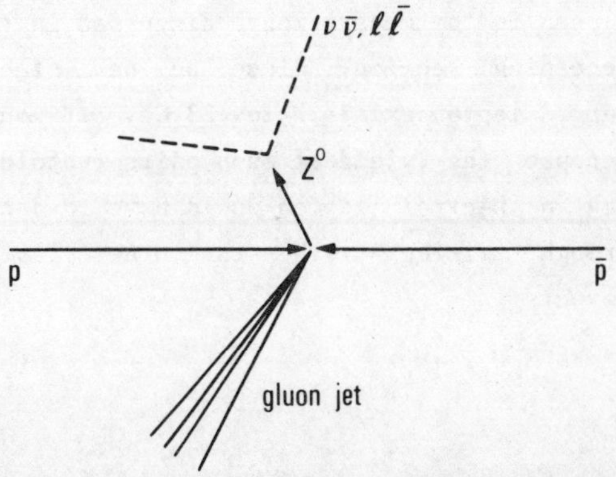

Fig. 13 - Gluon tagging in $p\bar{p}$ colliders to determine the branching fraction of Z to all neutrino species

Surprisingly enough, the single photon experiments at e^+e^- colliders are very competitive to measure N_ν even though they run at energies well below the Z mass. The measured process is

$$e^+e^- \to \gamma\ \nu\ \bar{\nu}$$

and it receives contributions from all neutrino species with a mass not exceeding

~ 10 GeV at PEP and PETRA energies. The relevant Feynman diagrams are shown in figure 14 : the $\nu_e \bar{\nu}_e$ final state is produced by both W and Z exchange in the t and s channel, respectively and their interference has to be taken into account. The calculation[19-20] is very reliable : only known couplings are involved and the result is insensitive to the Z width. The cross section is proportional to $N_\nu + 4$ in the local limit where $s \ll M_Z^2$. Results are given in Table 3: the combined limit from ASP[21], MAC[22] and CELLO[23] calculated by T. Lavine[24] reads

$$N_\nu < 4.9 \qquad (90~\%~\mathrm{CL})$$

Figure 15 compiles our knowledge on N_ν : clearly little ground is allowed beyond the 3 known neutrino types.

2.5. New unstable neutral leptons

Massive unstable neutral leptons would not be seen in the measurements discussed in the preceeding section. Since no new stable charged lepton exists below 23 GeV (it would increase the yield of muon pairs twofold), such a neutral lepton would have to decay through mixing with the other lepton flavours

$$|L^\circ\rangle = \sum_\ell U_{\ell L}\ |\nu_\ell\rangle \qquad (4)$$

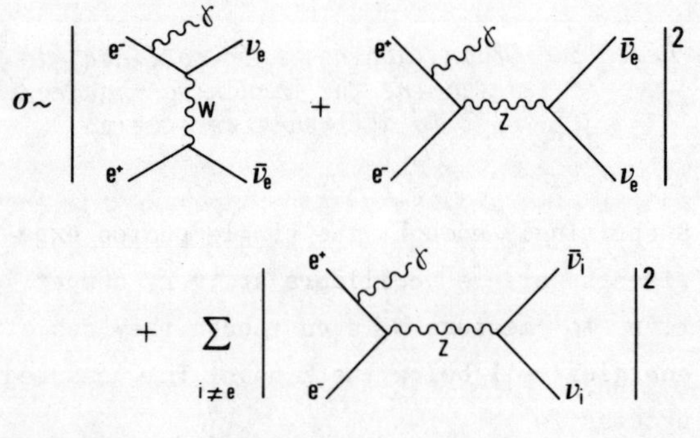

Fig. 14 - Cross section decomposition for the process $e^+e^- \to \gamma\, \nu\bar{\nu}$

experiment	ASP	MAC	CELLO	Combined[24]
nb. of events expected (N_ν = 3)	2.2	1.1	.7	4.0
observed	1	1	0	2
90 % CL upper limit on N_ν	7.5	16	15	4.9

Table 3 - Observed yield of single photon events in e^+e^- experiments and upper limits on the number of neutrino species

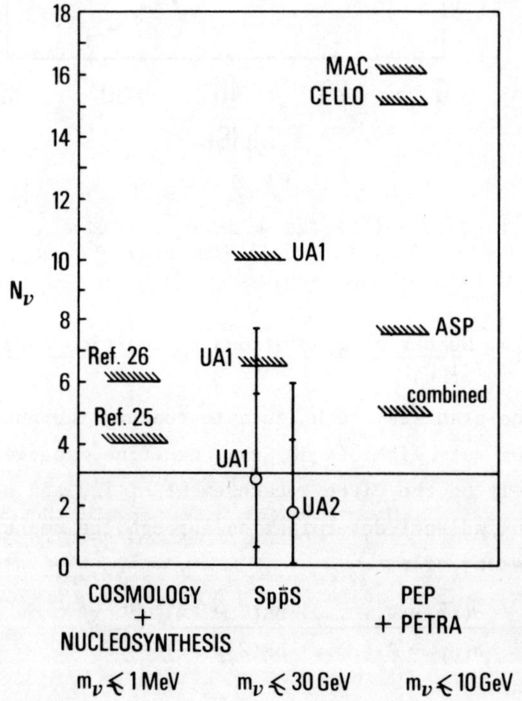

Fig. 15 - Compilation of N_ν values and upper limits

as shown, for example, on figure 16. They can be produced in π and K decays (seen by missing mass or by direct observation of

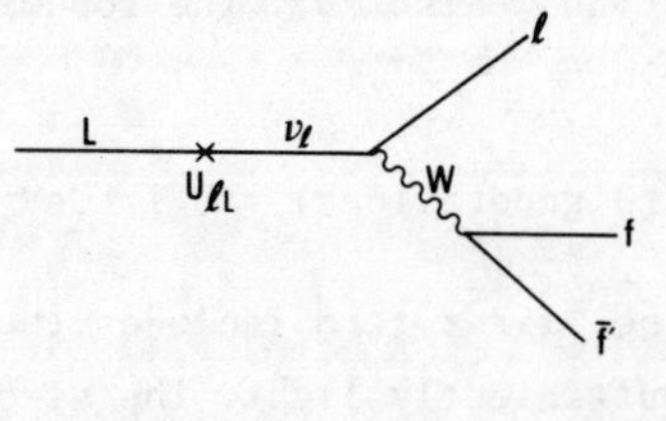

Fig. 16 - Decay of a heavy neutral lepton through mixing with lower mass lepton flavours

their decay in a neutrino beam) and many results have already been obtained[6,27].

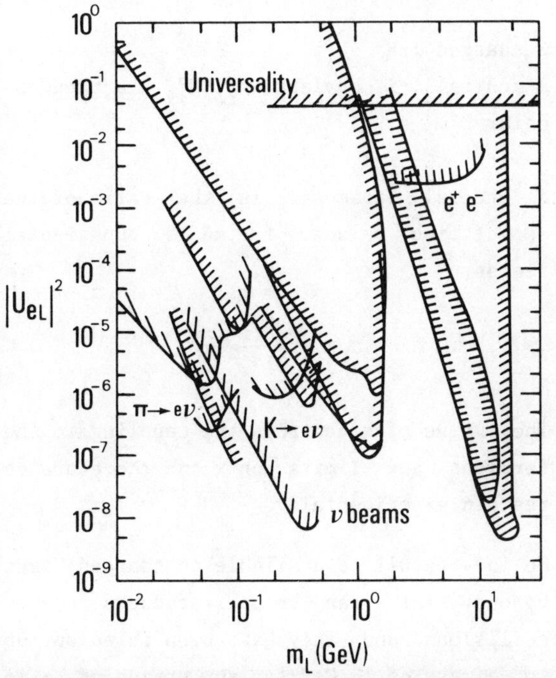

Fig. 17 - Excluded domains for heavy unstable neutral leptons decaying by mixing to the electron neutrino

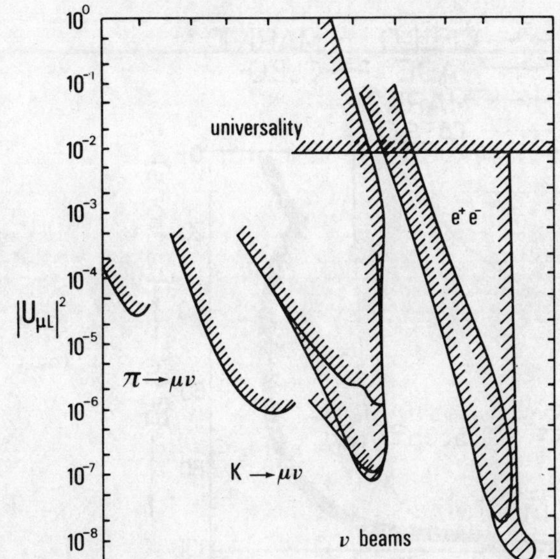

Fig. 18 - Excluded domains for heavy unstable neutral leptons decaying by mixing to the muon neutrino

It is also possible to produce these leptons in pairs through Z° exchange in e^+e^- annihilation with a known cross section. Small mixing (small $U_{\ell L}$ values) would lead to long lifetimes and one should be prepared to detect detached vertices. Such an analysis was performed by Mark II[28] and was recently extended by CELLO[29] up to masses ~ 20 GeV and mixing parameters $|U_{eL}|^2$ and $|U_{\mu L}|^2$ as small as 10^{-8} : no signal is seen and the corresponding limits can be inspected on fig. 17 and 18.

2.6. The Higgs boson sector

This is a poorly controlled aspect of the electroweak theory. Little information is available : we know, from the value of the ρ parameter (consistent with 1) that Higgs bosons are in weak isospin doublets. But, we do not know how many doublets are involved, and of paramount practical importance, the theory remains silent on the possible mass range up to a few TeV where a breakdown of the perturbation expansion is known to occur. The Higgs search is therefore wide open.

As far as the standard Higgs is concerned (the only remnant of the complex Higgs doublet needed to break SU(2) × U(1) symmetry), the only sensitive window we have is the decay of heavy quarkonium through the Wilczek mechanism, most efficiently applied to the ϒ system

$$\Upsilon \to \gamma H^\circ$$

with a rate calculated in 1st order QCD

$$\frac{BR(\Upsilon \to \gamma H^\circ)}{BR(\Upsilon \to \mu^+\mu^-)} = \frac{G_F}{\pi \alpha \sqrt{2}} m_b^2 \left(1 - \frac{M_H^2}{M_\Upsilon^2}\right) \qquad (5)$$

This yields a branching ratio of ~ 10^{-4} for Higgs masses up to a few GeV, a value within reach for high resolution photon detectors like CUSB or the Crystal Ball.

In fact, such a measurement was performed in 1985 by CUSB and a mass range up to ~ 5 GeV could be ruled out for the Higgs[30]. Results obtained by CLEO[31] and ARGUS[105] are not sensitive to a standard model Higgs. However, it was pointed out that QCD corrections have to be included, with the effect of reducing the expected branching ratio[32] by a factor of 2, thereby invalidating the previous limit. Furthermore, it was remarked at this conference[33] that large relativistic corrections have also to be taken into account, again reducing the expected rate by a factor of approximately 2. This is in contradiction with previous estimates[34] which however do not seem to extrapolate correctly to the non-relativistic case[35]. At this point, it is fair to say that the branching ratio for $\Upsilon \to \gamma H^{\circ}$ is beyond the sensitivity of the experiments at any Higgs mass value.

The situation could be eased a little if the basic Higgs structure required not one, but two doublets ϕ_1 and ϕ_2. This is the situation encountered in minimal supersymmetric models for example. After giving masses to the weak bosons there remains 5 states

2 charged : $H^{\pm}$
3 neutral : 2 scalars H°_1, H°_2 + a pseudoscalar H°_3.

A crucial parameter is the ratio of the expectation values of the 2 fundamental fields

$$x = \frac{\langle \phi_1 \rangle}{\langle \phi_2 \rangle}$$

The value of x controls the coupling to the fermions and limits on x can therefore be reached experimentally.

No new result is available on charged Higgs bosons. They can be pair-produced in e^+e^- collisions and they have been ruled out up to a mass of 18 GeV for any branching ratio values, except maybe for large hadronic branching fractions (Fig. 19). The possible

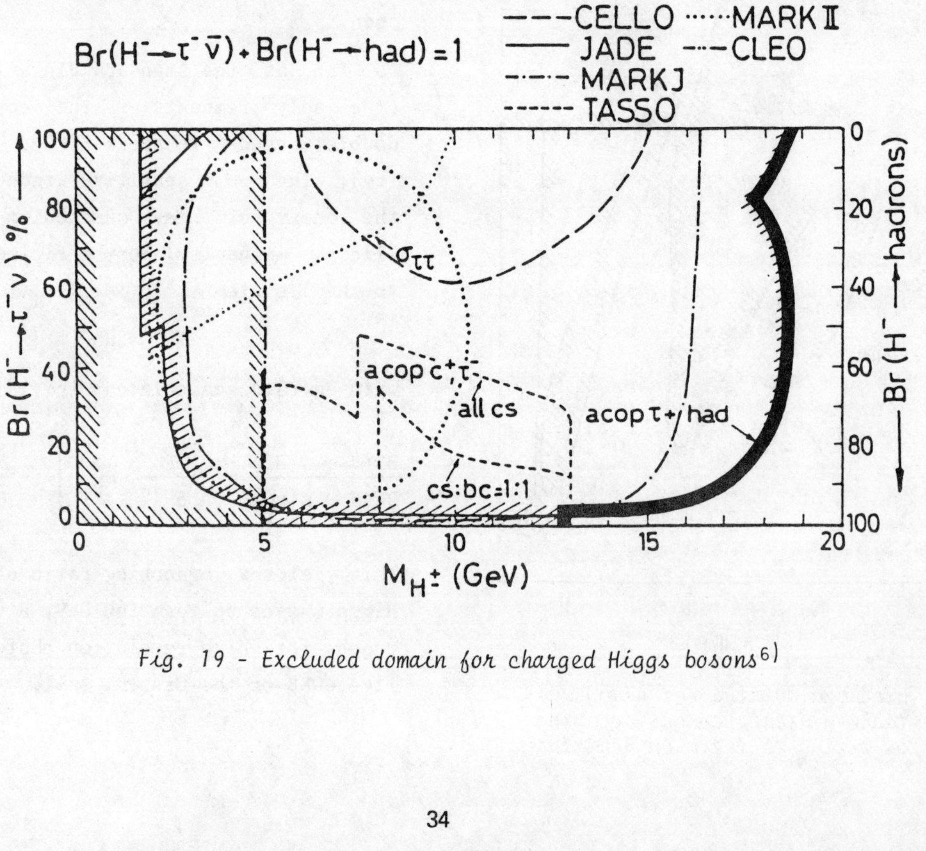

Fig. 19 - Excluded domain for charged Higgs bosons[6]

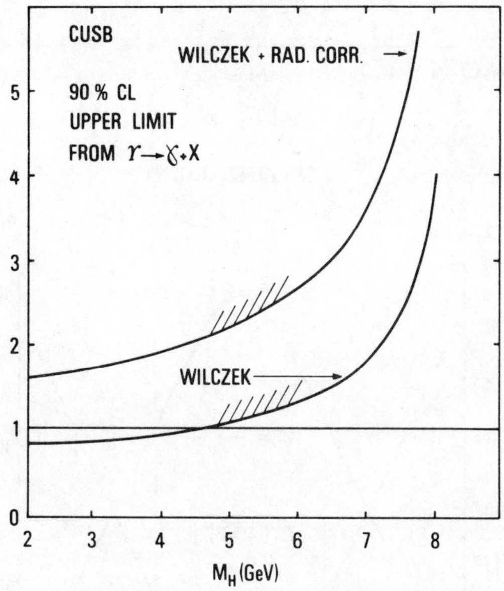

Fig. 20 - Excluded domain for light neutral Higgs masses versus the x parameter by CUSB[36]. The curves do not include possibly large relativistic effects[33]. The line at x = 1 corresponds to the minimal standard model

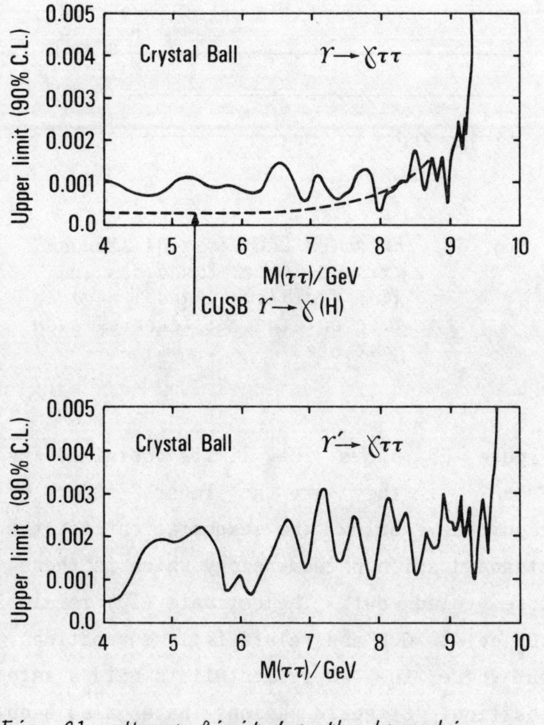

Fig. 21 - Upper limit for the branching ratio Υ or Υ' → γH_1^o with $H_1^o \to \tau^+\tau^-$ by the Crystal Ball[37]. The dashed curve shows the limit on Υ from CUSB[36] irrespective of the H_1^o decay mode

neutral states have been investigated in Υ decays

$$\Upsilon \to \gamma\, H_1^o$$

with inclusive photons in the case of CUSB[36] regardless of H decays (Fig. 20) and for a specific expected mode

$$H_1^o \to \tau^+\tau^- \to e\ \mu + \not{p}_T$$

in the case of the Crystal Ball[37] (Fig. 21). However, one should point out that the limits using $\tau^+\tau^-$ decays are not competitive with the global limit from CUSB, as shown in figure 21, as the selected τ modes comprise only 6 % of the total τ decays.

The experimental situation on Higgs searches in therefore not satisfactory : one badly awaits the exploration of Z° decays at SLC and LEP and possibly toponium decays at LEP.

3. THE AXION STORY

3.1. The standard axion phenomenology

In order to cure the strong CP and P violations occuring in QCD, an elegant solution has been found by Peccei and Quinn through a global U_{PQ} (1) symmetry imposed to the standard model[38]. To implement this symmetry, one needs to introduce 2 Higgs field doublets ϕ_1 coupled to up quarks and ϕ_2 coupled to down quarks. An important parameter is again the ratio of the vacuum expectation values

$$x = \frac{f_2}{f_1} \geqslant 0$$

$$\langle\phi_i\rangle = \frac{1}{\sqrt{2}}\, f_i$$

with $f = \sqrt{f_1^2 + f_2^2} = (\sqrt{2}\, G_F)^{-1/2} \sim 250$ GeV

The axion a is the Goldstone boson associated with the spontaneous breaking of this new symmetry.

The axion mass can be estimated :

$$m_a \simeq m_\pi \frac{f_\pi}{f} N\left(x + \frac{1}{x}\right) \qquad (6)$$

where N is the number of quark doublets, giving

$$m_a \simeq 75 \quad \left(x + \frac{1}{x}\right) \quad \text{keV}$$

The axion couplings to quarks and leptons are completely determined

$$\frac{m_q}{f} \; i \; \bar{q} \; \gamma_5 q \quad a \; . \begin{pmatrix} x \\ \frac{1}{x} \end{pmatrix} \quad \begin{matrix} \text{up quark} \\ \text{down quark} \end{matrix} \qquad (7)$$

$$\frac{m_e}{f} \; i \; \bar{e} \; \gamma_5 \; e \; a \; . \; \frac{1}{x} \qquad \text{electron}$$

resulting in a partial width

$$\Gamma_{a \to ee} = \frac{m_e^2 \left(m_a^2 - 4m_e^2\right)^{1/2}}{8 \; \pi \; x^2 \; f^2} \qquad (8)$$

For values of x near 1, the axion is expected to be long-lived decaying only in the $\gamma\gamma$ channel with a lifetime

$$\tau_{a \to \gamma\gamma} \simeq .7 \left(\frac{100 \text{ KeV}}{m_a}\right)^5 \qquad (9)$$

The expected behaviour of axions as a function of the free parameter x can be reconstructed from figure 22. For $x > 0.1$, the axion only decays to $\gamma\gamma$, while the dominant mode below is to e^+e^- with a fast rate.

Also indicated on figure 22 are the excluded domains of x from ψ and Υ decays into γ a. Since the ψ rate is proportionnal to x^2 while the Υ rate behaves as x^{-2}, it is easy to understand how the non-observation of these decays[39-41] can rule out any x value. Small values of x in fact have only been ruled out very recently[42-44] since one had to take into account the fast decay to e^+e^-, with however a very large expected branching ratio

$$\frac{BR(\Upsilon \to \gamma \; a)}{BR(\Upsilon \to \mu^+\mu^-)} = \frac{G_F \; m_b^2}{\pi \; \alpha \; \sqrt{2} \; x^2} \qquad (10)$$

$$\sim 7.5 \; 10^{-3} \; x^{-2}$$

STANDARD AXION

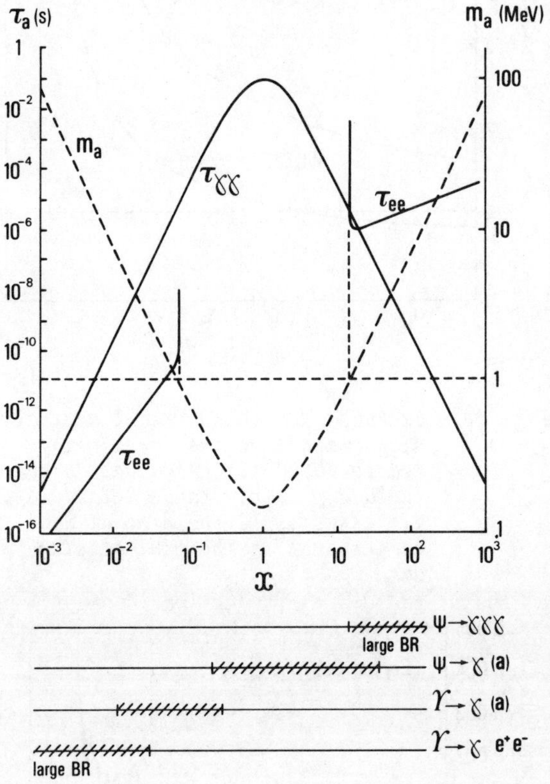

Fig. 22 - Expected behaviour of standard axions. The excluded domains from radiative ψ and Υ decays rule out the existence of such particles

Figure 23 shows the limits obtained by CUSB[42)] : they are at least 2 orders of magnitude below the expectation of the standard axion phenomenology which is therefore ruled out. The estimate (10) is also subject to QCD and relativistic corrections, but here the experimentalists hold a safe position. Standard axions have also been ruled out in some restricted x ranges by studies of nuclear transitions, beam dump experiments and bounds on $K \to \pi a$[45)].

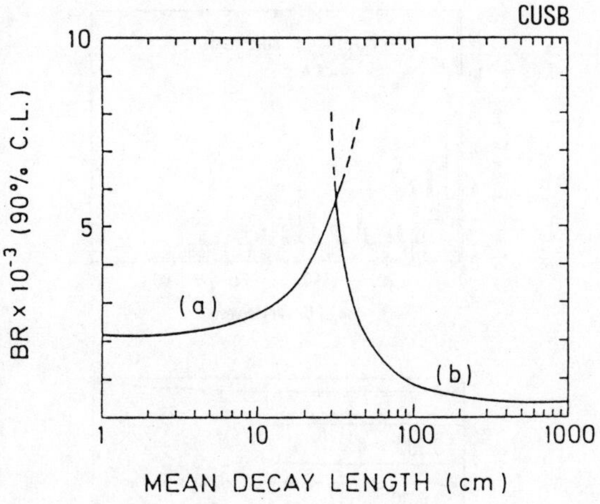

Fig. 23 - Branching ratio upper limits from CUSB[42] for ϒ decays into γ + axion. The expected ratio is more than two orders of magnitude larger

3.2. GSI peaks and non-standard axions

Although no experimental contribution was given at this conference, the observation of a sharp peak in e^+ production in heavy ion collisions[46-48] at GSI Darmstadt has stirred considerable interest. Positron production is expected in the strong Coulomb field generated when the two ions are just touching one another and this is in fact observed with an energy distribution about 1 MeV vide. The surprise came with the simultaneous observation of a peak

$$E_{e^+} \sim 350 \text{ KeV} \qquad \Gamma \sim 80 \text{ KeV}$$

which cannot be explained by the same mechanism with its characteristic interaction time typically 10 times shorter. Earlier this year, the observation of correlated e^+ and e^- peaks was reported[49] at the same energies

$$E_{e^\pm} \sim 350 - 400 \text{ KeV}$$

$$E_{e^+} + E_{e^-} = 760 \pm 20 \text{ KeV}$$

with a width of 80 MeV consistent with experimental resolution (Fig. 24).

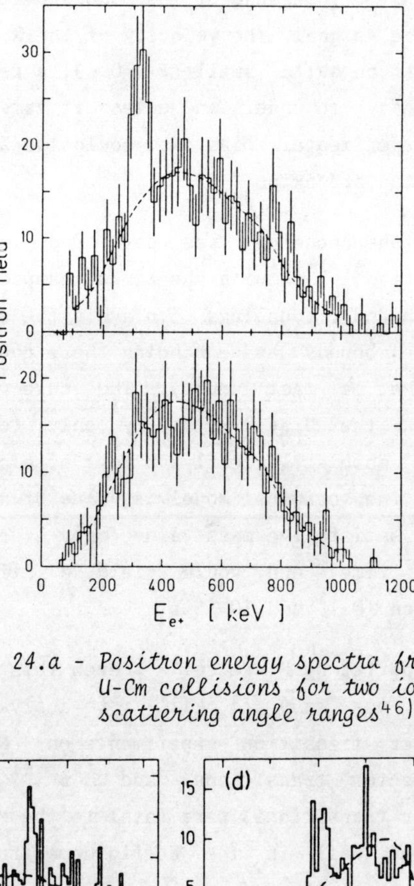

Fig. 24.a - Positron energy spectra from U-Cm collisions for two ion scattering angle ranges[46]

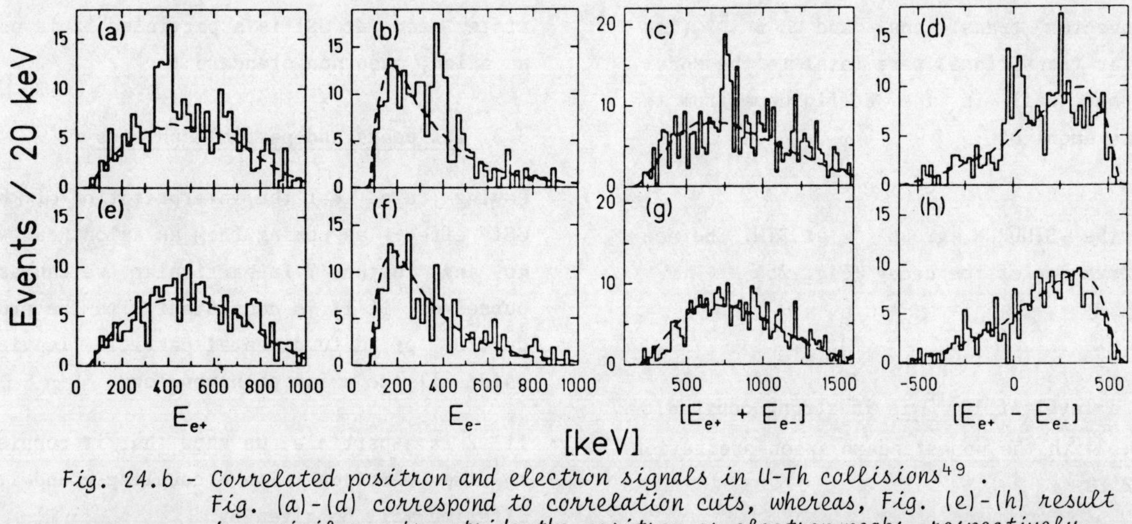

Fig. 24.b - Correlated positron and electron signals in U-Th collisions[49]. Fig. (a)-(d) correspond to correlation cuts, whereas, Fig. (e)-(h) result from similar cuts outside the positron or electron peaks, respectively

It is difficult to ascertain the experimental message from these apparently difficult experiments : the thin targets used are easily destroyed by the beam and the spectrometer design is very involved. More experiments will take data in coïncidence shortly so that we should have soon more informations.

It was tempting to interpret the GSI effect as originating in the production of a particle X of mass 1.8 MeV decaying into e^+e^- [50-51]. Because of the monochromaticity of the signal the velocity of the X would have to be quite small ($\beta < 0.6$), a feature not easy to understand unless it came from an experimental bias. Now could the X be a non-standard axion ?

Some phenomenology was developed in this direction[52-53] with the aim of suppressing the b and c couplings (to avoid the deadly ψ and Υ bounds) and enhancing the e coupling (to get a fast decay). The cure was to retain the $U_{PQ}(1)$ symmetry only for the (u,d) doublet, clearly a loss of elegance from the original model. A value of $x \sim 70$ would satisfy the mass value (only 1 doublet this time) and would yield a lifetime between 10^{-13} and 10^{-12} s.

As reported by R. Peccei[54], even this axion model was easy to kill : re-analyses of nuclear transition experiments on ^{14}N [55] (isovector transitions) and ^{10}B [56] (isoscalar transitions) were fatal to the model. Another nail in the coffin came from the measurement of

$$\pi^+ \to e^+ \nu e^+e^-$$

by the SINDRUM group[57] at SIN. The non-observation of the decay (Fig. 25)

$$\pi^+ \to e^+ \nu a, \quad a \to e^+e^-$$

at a level of 10^{-10} is in strong contradiction with the non-standard axion prediction of $2\ 10^{-6}$.

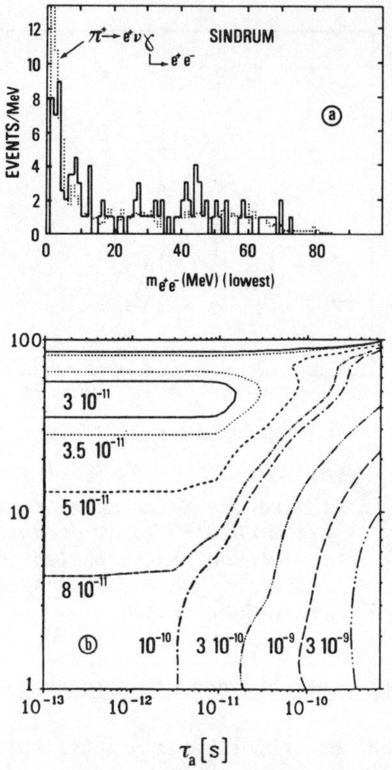

Fig. 25 - Search for the decay $\pi^+ \to e^+\nu\, a$ with $a \to e^+e^-$ (ref. 57)
(a) e^+e^- mass spectrum (lowest mass combination)
(b) branching ratio bounds in the mass-lifetime plane

In conclusion, killing the heavy quark couplings was not enough : the coupling to of an axion to light quarks is too large to be tolerated by experiments and should also be reduced ! This, of course, defeats the original purpose of introducing the axion to cure a problem related to quarks. If the X state seen at GSI is a particle, it is not an axion, even non-standard !

3.3. GSI peaks and particle physics

Having ruled out the interpretation of the GSI effect as coming from an axion, can we go any further ? In particular, we can ask ourselves if it is consistent with the production of a fundamental particle, leaving aside all the axion phenomenology.

If X is a particle, we know that it couples to the electron with a coupling g and its lifetime will be given by

$$\Gamma_X = \frac{1}{\tau_X} = \alpha_X \frac{m_X}{2} \left(1 - \frac{4\, m_e^2}{m_X^2}\right)^{1/2} \qquad (11)$$

where I assume for illustration that X is a 0^- state. Large values of α_X are ruled out by the agreement with QED of the measurements of g-2 for the electron (Fig. 26). In fact, a bound can be obtained[58-59]

$$\tau_X > 6\ 10^{-14}\ \text{s} \qquad (12)$$

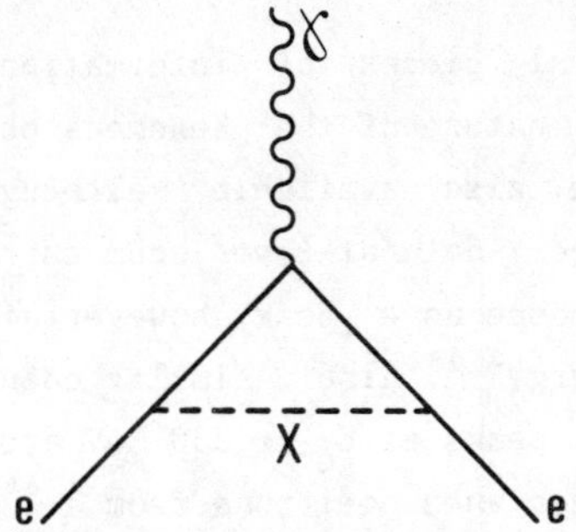

Fig. 26 - X contribution to the electron anomalous magnetic moment

in the absence of unforeseen cancellations between several contributions. Small values of α_X are excluded from previous beam dump experiments sensitive to long lifetimes. In particular, results from an electron beam dump[60] at 45 MeV with a flight path of at least 12m, yields

$$\tau_X < 2\ 10^{-11}\ \text{s} \qquad (13)$$

The remaining lifetime gap between 6 10^{-14} and 2 10^{-11} s has been most recently investigated by 4 experiments[63-66] all relying on X bremsstrahlung by electrons and observation of possible decays behind a beam dump. The cross section for X bremsstrahlung (Fig. 27) has been computed for massive X particles[61-62] and limits can therefore be expressed in the (m_X, g) plane or in a more direct way in the (m_X, τ_X) plane. This can be done in a model-independent way. To reach small lifetimes, short dumps are mandatory. Table 4 gives a list of the experiments with their basic parameters.

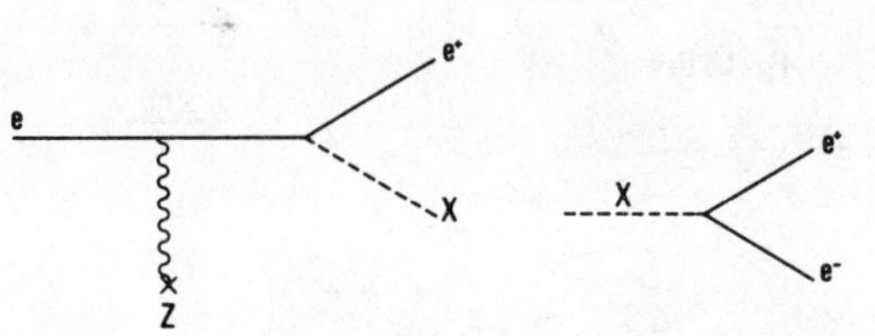

Fig. 27 - Bremsstrahlung of X particles in the Coulomb field of a nucleus followed by the decay of X into an e^+e^- pair

experiment	E_e(GeV)	minimum flight-path L(m)	τ_X range excluded (s)
KEK[63]	2.5	2.4	5.10^{-13}-1.10^{-7}
LAL Orsay[64]	1.5	0.1	6.10^{-14}-9.10^{-11}
SLAC[65]	9	0.1	1.10^{-14}-5.10^{-11}
FNAL[66]	800 (protons)	5.5	1.10^{-14}-1.10^{-11}

Table 4 - Electron beam dump experiments looking for low mass axion-like particles decaying in e^+e^-

These experiments are straightforward. Let me take the example of the LAL experiment[64]. The dump is made of two 5 cm-long tungsten pieces which can be separated in order to control separately the absorption length and the decay length (Fig. 28). Positrons produced in the dump or emitted by a short-lived particle in the 1 m-long decay volume behind are analysed in a magnetic spectrometer and identified in a lead-glass block. Photoproduced muons are also detected and serve as a useful monitor and for normalisation. The results with the 3 dump configurations shown in figure 29 can rule out the production of X particles in the complete lifetime range allowed by (12) and (13). Comparison of configurations I and II eliminate the possibility of short-lived X

$$1.5\ \text{cm} < \gamma\beta c\tau \lesssim 60\ \text{cm}$$

with a rate per incident electron

$$\frac{N}{N_0} < 3.2\ 10^{-14} \qquad (95\ \%\ \text{CL})$$

whereas configurations II and III rule out longer-lived X

$$60\ \text{cm} \lesssim \gamma\beta c\tau < 20\ \text{m}$$

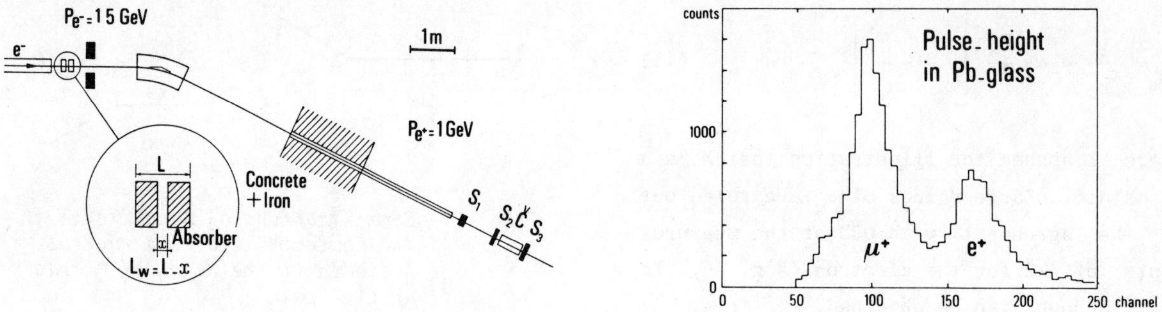

Fig. 28 - Setup of the LAL electron beam dump experiment to search for short-lived axion-like X particles[64]

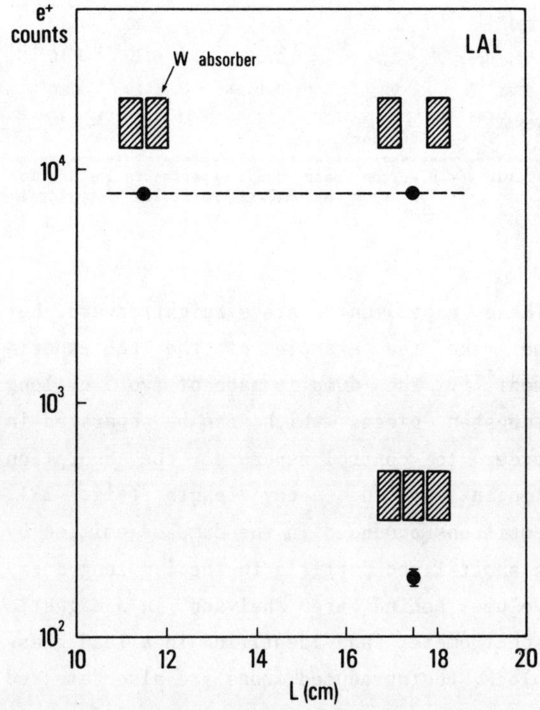

Fig. 29 - Positron rates in the LAL experiment for 3 configurations of the dump

with a rate

$$\frac{N}{N_o} < 2.4\ 10^{-14} \qquad (95\ \%\ CL)$$

The excluded domain in the mass-lifetime plane is given in figure 30 together with the results of the other experiments : at the mass of 1.8 MeV, all values of the coupling to electrons are excluded. The GSI effect does not seem to belong to particle physics.

Additional pieces of information on the complex nature of the phenomena observed at GSI are also available, although not yet published : data at lower beam energy at GSI also produce an e^+ peak, however at a different energy[67]. Also , similar coïncident e^+ and e^- peaks at $E_{e^\pm} \sim 330$ KeV are reported to be seen when positrons from a ^{68}Ga source

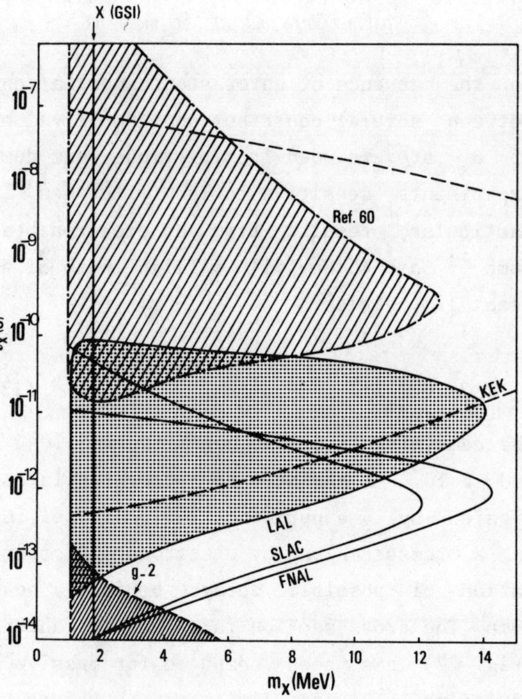

Fig. 30 - Excluded domain in the mass-lifetime plane for weakly interacting particles coupled to the electron and decaying into e^+e^- pairs

(1.9 MeV maximum energy) strike a Th foil ; surprisingly, no such effect is seen when the target is made of Ta [68].

4. SEARCHES FOR SUPERSYMMETRIC PARTICLES

4.1. Minimal particle content

Table 5 displays the minimal particle content of a theory obeying supersymmetry (SUSY)[69-71]. The dark boxes correspond to the particles we see everyday in our laboratories. The other boxes signal the new partners which have been hunted without mercy, but unfortunately without success, for the past 5 years. Note that the Higgs sector has the content corresponding to 2 Higgs field doublets.

SPIN	2	3/2	1	1/2	0
				ℓ	$\tilde{\ell}_L$ $\tilde{\ell}_R$
				q	$\tilde{q}_L$ $\tilde{q}_R$
			g	$\tilde{g}$	
			W	$\tilde{W}^{\pm}_{+}$ $\tilde{W}^{\pm}_{-}$	$w^{\pm}$
			γ	$\tilde{\gamma}$	
			$Z°$	$\tilde{Z}°_{+}$ $\tilde{Z}°_{-}$	z
				h	h h'
	G	$\tilde{G}$		(goldstino)	

Table 5 - Minimal particle content of a supersymmetric theory. States in the larger boxes can mix. In current supergravity models, the goldstino is absorbed into the gravitino $\tilde{G}$

If the states and their couplings are well defined in SUSY, the mass spectrum depends on details of the SUSY breaking mechanism and it cannot therefore be predicted on general grounds. Several models for symmetry breaking have been proposed, but it is up to experiment to decide. In general the mass splitting is expected to be of order M_W and some particles at least should be found with a mass below the W boson.

Due to the conservation of R-parity, SUSY particles should be produced in pairs and the lightest SUSY particle (LSP) must be stable. Cosmology suggests[72] that the LSP has no electric and colour charges, so that the candidates are

$$LSP \stackrel{?}{=} \tilde{\gamma},\ \tilde{\nu},\ \tilde{Z},\ \tilde{h},\ \tilde{G}$$

Since its interaction with matter involves a large mass (see Fig. 31) the LSP is weakly interacting and as such leaves the detectors un-noticed except for the corresponding missing energy and momentum (p).

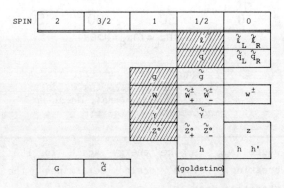

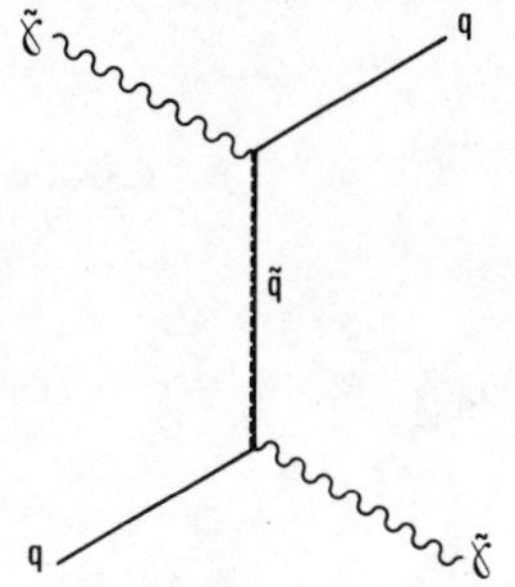

Fig. 31 - Interaction of a photino with matter

4.2. The case for the unstable photino : experimental constraints and scale of SUSY breaking

Since most analyses rely on the LSP being the photino, it is useful to have a check on that experimentally. Let us therefore consider the case where the photino is not the LSP and is consequently unstable. Table 6 considers the major decay modes which are available under the different assumptions for what particle is the LSP[71-73]. Except for the case where the LSP is one of the $\tilde{Z}$ states, the $\tilde{\gamma}$ decay involves

LSP	Dominant decay mode
$\tilde{\gamma}$	stable
$\tilde{\gamma}$	$\tilde{\gamma} \to \tilde{\nu}\,\bar{\tilde{\nu}}$
$\tilde{h}$	$\tilde{\gamma} \to \gamma\,\tilde{h}$
$\tilde{G}$	$\tilde{\gamma} \to \gamma\,\tilde{G}$
$\tilde{Z}_+$ or $\tilde{Z}_-$	$\tilde{\gamma} \to f\bar{f}\,\tilde{Z}$

Table 6 - Decay modes of an unstable photino

an unseen particle and a photon. The case where $\tilde{\gamma} \to \tilde{\nu}\nu$ is identical in practice to that of a stable $\tilde{\gamma}$.

The possibility of an unstable $\tilde{\gamma}$ is most directly adressed in studying the process (Fig. 32)

$$e^+e^- \to \tilde{\gamma}\,\tilde{\gamma} \to \gamma\gamma + \not{p}$$

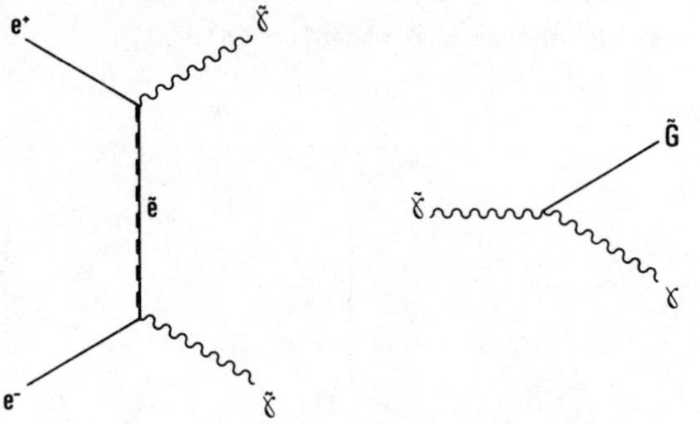

Fig. 32 - Pair production of photinos in e^+e^- annihilation, followed by their radiative decay

where new results are available from CELLO[74]. The strength of the cross section is controlled by the $\tilde{e}$ mass so that the absence of such events can be most usefully expressed as a bound in the $\tilde{e}$ mass-$\tilde{\gamma}$ mass plane (Fig. 33). For $\tilde{\gamma}$ masses larger than 300 MeV, their decay is expected bo be very fast (i.e. well inside the detector). For lower masses, only one or even no photon is seen, and in that case one has to consider radiative production of photinos, as will be discussed next, and the bound on the $\tilde{e}$ mass is reduced to 33 GeV. The message from figure 33 is that for searches confined to a domain $M_{\tilde{e}} < 100$ GeV and $M_{\tilde{\gamma}} < 15$ GeV, the photino can be considered as stable. This applies to all current searches.

In some SUSY breaking models, the LSP is the gravitino $\tilde{G}$ and the photino decays to $\tilde{\gamma} + \tilde{G}$ with a lifetime[75]

$$\tau_{\tilde{\gamma}} = \frac{8\pi\ \Lambda_{SB}^4}{M_{\tilde{\gamma}}^5} \tag{14}$$

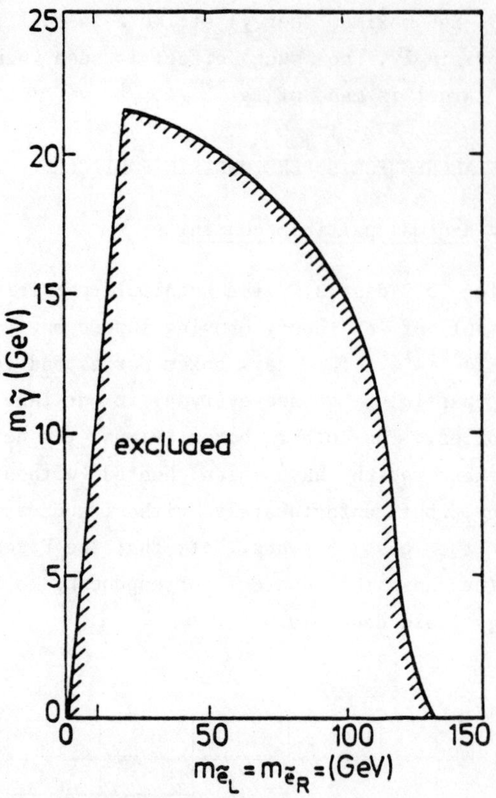

Fig. 33 - Excluded domain by CELLO[74] for the $\tilde{e}$ and $\tilde{\gamma}$ masses in the case of unstable photinos, decaying radiatively to the LSP

where Λ_{SB} is the energy scale for SUSY breaking. This scale determines the gravitino mass

$$M_{\tilde{G}} = \left(\frac{4\pi}{3} G_N\right)^{1/2} \Lambda_{SB}^2$$

$$= 1.7\ 10^{-6}\ \text{eV} \left(\frac{\Lambda_{SB}}{100\ \text{GeV}}\right)^2 \tag{15}$$

where G_N is the Newton gravitational constant. It is possible to use the data on single γ production in e^+e^- experiments to place a bound on the process (Fig. 34)

$$e^+e^- \to \tilde{\gamma}\,\tilde{G}$$
$$\hookrightarrow \gamma\,\tilde{G}$$

which would be the first process measured in particle physics to be proportional to G_N, only measurable if the $\tilde{G}$ mass is sufficiently small or Λ_{SB} not too large, as[76)]

$$\sigma(e^+e^- \to \tilde{\gamma}\,\tilde{G}) \sim \frac{G_N\,\alpha\,s}{M_{\tilde{G}}^2} = \frac{3\,\alpha\,s}{4\pi\,\Lambda_{SB}^4} \qquad (16)$$

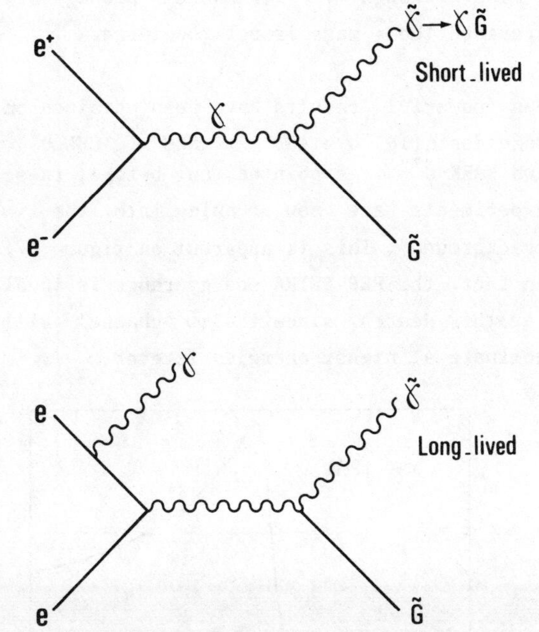

Fig. 34 - Photino-gravitino production in e^+e^- annihilation giving a single photon in the final state

Combining the results from ASP, MAC and CELLO on single γ searches, one gets the bounds on Λ_{SB} (or $M_{\tilde{G}}$) shown in figure 35 in the range of 200-400 GeV up to a photino mass of ~ 30 GeV. These limits are much more constraining than those obtained from quarkonium decays[77)]

$$\psi,\ \Upsilon \to \tilde{\gamma}\,\tilde{G}$$

detected through cascade decays i.e.

$\Upsilon' \to \pi\pi\ \Upsilon$
└→ unseen states or
γ + unseen states

A new result from ARGUS[44)] yields BR($\Upsilon \to$ unseen states) < 2.3 10^{-2} (90 % CL) and some data on γ + unseen states have been presented by the Crystal Ball[78)]. These limits on Λ_{SB} are independent on the $\tilde{e}$ mass. More stringent bounds can be obtained from

$$e^+e^- \to \tilde{\gamma}\,\tilde{\gamma}$$

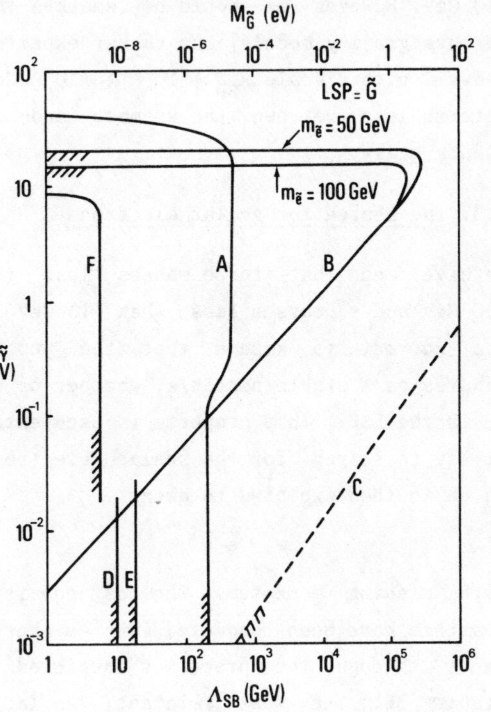

Fig. 35 - Lower bounds on the SUSY breaking scale Λ_{SB} as a function of the photino mass $M_{\tilde{\gamma}}$ in the case where the LSP is the gravitino $\tilde{G}$. The various bounds correspond to :

A : $e^+e^- \to \tilde{\gamma}\tilde{G}\gamma$ and $e^+e^- \to \tilde{\gamma}\tilde{G}$ followed by $\tilde{\gamma} \to \gamma\tilde{G}$, data from ASP[21)], MAC[22)] and CELLO[23)]

B : $e^+e^- \to \tilde{\gamma}\tilde{\gamma}$ followed by $\tilde{\gamma} \to \gamma\tilde{G}$, data from CELLO[74)]

C : Cosmological bound[75)]

D : $\psi \to \tilde{\gamma}\tilde{G}$ as analyzed in Ref. 77

E : $\Upsilon \to \tilde{\gamma}\tilde{G}$ using the new upper limit from ARGUS[44)]

F : $\Upsilon \to \tilde{\gamma}\tilde{G}$ followed by $\tilde{\gamma} \to \gamma\tilde{G}$, as deduced from data given by the Crystal Ball[78)]

for unstable $\tilde{\gamma}$, but they only hold if the $\tilde{e}$ mass is less than 120 GeV.

In conclusion, if the gravitino $\tilde{G}$ is the LSP and for $\tilde{\gamma}$ masses up to ~ 30 GeV, one can experimentally push the lower bound on the scale of SUSY breaking Λ_{SB} up to 400 GeV and possibly 10^5 GeV if the $\tilde{e}$ mass is less than

120 GeV. However, it should be remarked that in supergravity models, one rather expects a heavy $\tilde{G}$ with a scale $\Lambda_{SB} \sim 10^{10}$ to 10^{12} GeV, intermediate between the W mass and the Planck scale.

4.3. The search for scalar electrons

We have seen that for $\tilde{e}$ masses less than 100 GeV and $\tilde{\gamma}$ masses less than 15 GeV, it was correct to assume that the photino behaves as a stable particle, whether or not it is the LSP. This property is used extensively to search for the scalar electron $\tilde{e}$ which is then expected to decay

$$\tilde{e} \rightarrow e\,\tilde{\gamma}$$

with missing momentum. The most sensitive searches have been conducted in e^+e^- experiments through the processes depicted in figure 36. The most relevant, as far as current results are concerned are the pair-production

$$e^+e^- \rightarrow \tilde{e}\,\tilde{e} \quad \rightarrow \quad e^+e^-\,\tilde{\gamma}\,\tilde{\gamma} \qquad (17)$$

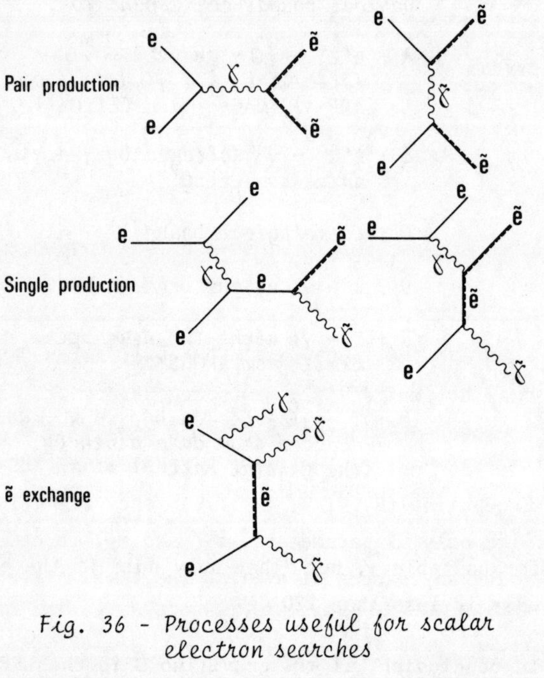

Fig. 36 - Processes useful for scalar electron searches

limited by the beam energy but sensitive to large $\tilde{\gamma}$ masses, and the radiative production of a pair of photinos

$$e^+e^- \rightarrow \gamma\,\tilde{\gamma}\,\tilde{\gamma} \qquad (18)$$

where the cross section is determined by the $\tilde{e}$ mass through the t-channel propagator, provided the $\tilde{\gamma}$ mass is not too large.

New powerful results have been obtained on reaction (18) by ASP[21], MAC[22], CELLO[23] and MARK J[79]. As pointed out before, these experiments are now running into the $\nu\bar{\nu}$ "background". This is apparent on figure 37: in fact, the PEP-PETRA energy range is ideal for this search, since the $\nu\bar{\nu}$ channel will dominate at higher energies forever.

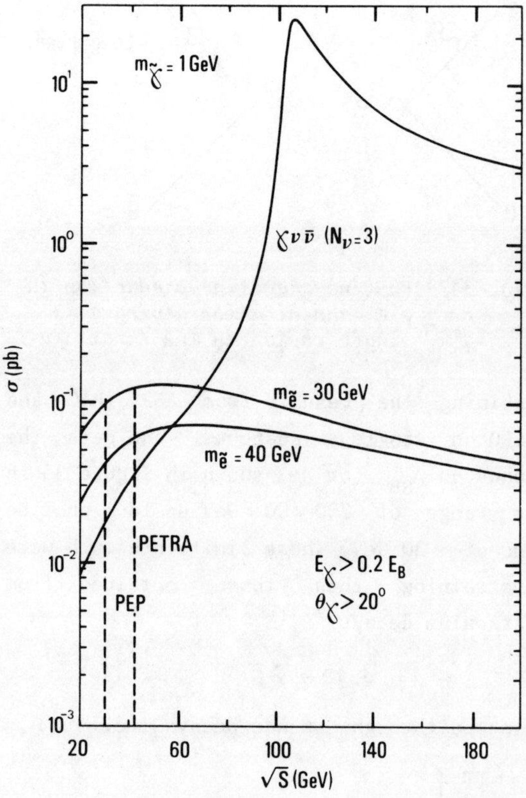

Fig. 37 - Cross sections for single γ production in e^+e^- annihilation for the processes $e^+e^- \rightarrow \gamma\tilde{\gamma}\tilde{\gamma}$ through $\tilde{e}$ exchange and $e^+e^- \rightarrow \gamma\,\nu\bar{\nu}$

The results of ASP and MAC, and a combined analysis[24] of ASP, MAC and CELLO data giving a $\tilde{e}$ lower mass limit of 84 GeV were presented at the conference. We now feel that the statistical analysis used is incorrect. The point is that the expected $\nu\bar{\nu}$ background was subtracted out, where actually the observed rate is smaller. The correct analysis[106] with the truncated a priori distribution of background events gives lower mass values. Table 7 shows the mass limits presented by the experiments at the conference, and the corresponding values obtained with the proper treatment of the background[107]. The combined $\tilde{e}$ lower mass limit from ASP, MAC and CELLO is 65 GeV at the 90 % CL.

experiment	ASP[21]	MAC[22]	CELLO[23]	Mark J[79]	ASP MAC CELLO combined
$m_{\tilde{e}}$ > (GeV)					
presented at the conference	66	50	38	33	84
reanalysis[107]	57	47	38		65

Table 7 - 90 % CL lower bounds on the scalar electron mass from single photon experiments for photino masses not exceeding a few GeV (degenerate $\tilde{e}_L$ and $\tilde{e}_R$ states)

The full picture, including both $\tilde{e}$ and $\tilde{\gamma}$ masses, is available on figure 38, with the same caveat as to the ASP and MAC analysis. Progress in the last 2 years on the $\tilde{e}$ search has been considerable and the limit is now pushed close to the W mass.

4.4. Gluinos and squarks

Many subprocesses capable of producing gluinos and/or squarks are available at $p\bar{p}$ colliders :

$gg \to \tilde{g}\,\tilde{g}$ $\qquad gg \to \tilde{q}\,\bar{\tilde{q}}$ $\qquad qg \to \tilde{q}\,\tilde{g}$

$q\bar{q} \to \tilde{g}\,\tilde{g}$ $\qquad q\bar{q} \to \tilde{q}\,\bar{\tilde{q}}$

$q\bar{q} \to \tilde{g}\,\tilde{\gamma}$ $\qquad qg \to \tilde{q}\,\tilde{\gamma}$

Despite their diversity, these processes possess a remarkable property : $\tilde{g}$ and $\tilde{q}$ are essentially decoupled, i.e. it is possible to produce gluinos even if the $\tilde{q}$ mass is large and inaccessible, and vice-versa. This property has its root in the existence of the non-abelian couplings of colour SU(3) as illustrated in figure 39.

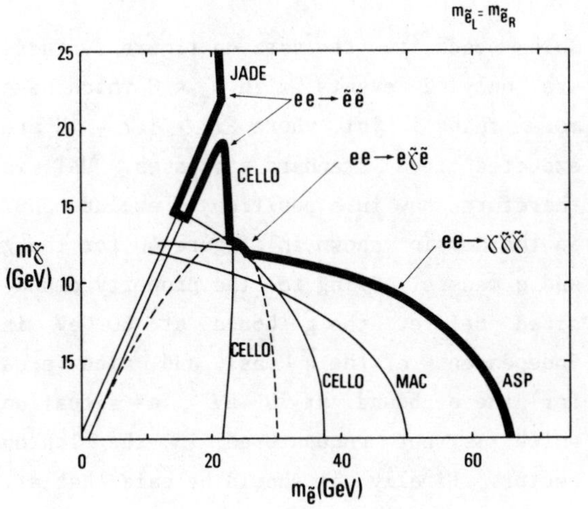

Fig. 38 - Bounds on $\tilde{e}$ and $\tilde{\gamma}$ masses from e^+e^- experiments

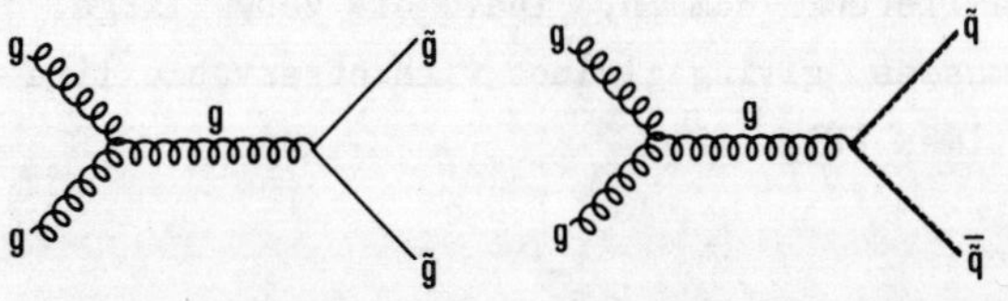

Fig. 39 - Two examples of subprocesses for $\tilde{g}$ and $\tilde{q}$ production at hadron colliders

Old and new data collected by UA1 have been analysed for this search. What to expect ? Assuming (1) the LSP is the photino (taken massless but probably a mass of a few GeV does not matter) (2) mass degeneracy of the 5 expected squarks ($m_{\tilde{u}} = m_{\tilde{d}} = m_{\tilde{s}} = m_{\tilde{c}} = m_{\tilde{b}}$) and (3) not taking into account other SUSY production ($\tilde{H}$, $\tilde{W}$, $\tilde{Z}$, $\tilde{\ell}$, $\tilde{t}$...), they are left with only 2 parameters, $m_{\tilde{q}}$ and $m_{\tilde{g}}$, to describe any excess of events. The final states depend on the mass hierarchy and the following decays are expected

if $m_{\tilde{q}} > m_{\tilde{g}}$ $\qquad \tilde{g} \to q\bar{q}\,\tilde{\gamma}$ $\qquad \tilde{q} \to q\tilde{g}$, $\;\tilde{g} \hookrightarrow q\bar{q}\tilde{\gamma}$

if $m_{\tilde{g}} > m_{\tilde{q}}$ $\qquad \tilde{g} \to \bar{q}\tilde{q}$, $\;\tilde{q} \hookrightarrow q\tilde{\gamma}$ $\qquad \tilde{q} \to q\tilde{\gamma}$

It is clear that multijet events should dominate ; in fact, a full simulation indicates that for $E_T^{jet} > 12$ GeV

$\langle n_{jet} \rangle \sim 3$

Going back to the data on figure 7, there are only 2 events with $L_\tau < 0$ which have more than 1 jet, where $2.8 \pm 1.7 \pm .3$ are expected from standard processes. UA1 is therefore now in a position to exclude SUSY in the domain shown in figure 40 for the $\tilde{g}$ and $\tilde{q}$ masses. Owing to the property emphasized before, the $\tilde{g}$ bound at 60 GeV is independent of the $\tilde{q}$ mass, and vice-versa for the $\tilde{q}$ bound at 70 GeV - a situation which is not encountered in the lepton sector. Finally it should be said that at, the moment, a light gluino (5 to 10 GeV) is claimed not to be ruled out for sure. Several results on light $\tilde{g}$ searches have been submitted[80-82] which are relevant to a different domain, that of very large $\tilde{q}$ masses giving gluinos with observable lifetimes (Fig. 41).

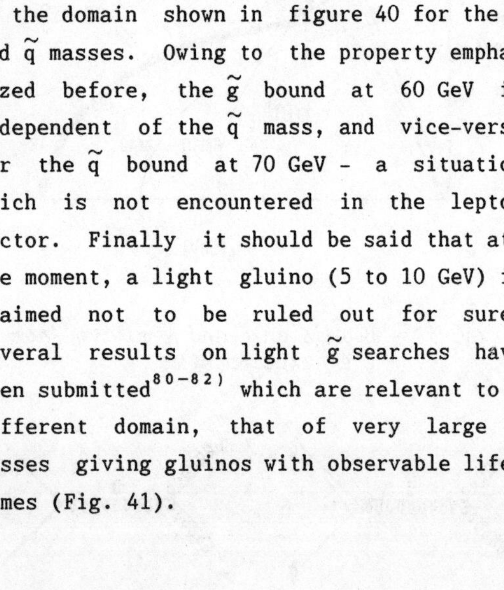

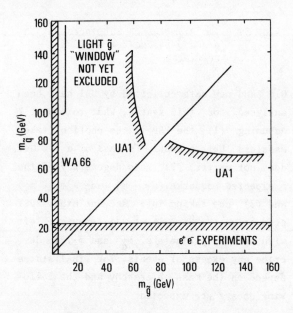

Fig. 40 - Excluded domain for gluinos and squarks with a massless $\tilde{\gamma}$ as the LSP

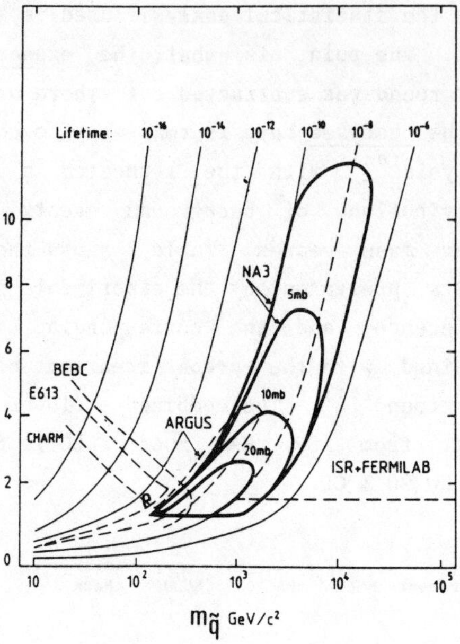

Fig. 41 - Excluded domains for light long-lived gluinos

4.5. Other gauginos

The phenomenology of the $\tilde{W}$ and $\tilde{Z}$ states depends significantly on their possible mixing. The mass eigenstates can be in general mixtures of the charged and neutral fields, respectively and therefore many possibilities exist reflecting into different experimental scenarios.

The lightest chargino state has to be heavier than 23 GeV regardless of what is the LSP, otherwise it would have been pair-produced at PETRA[83-85].

In the case where the LSP is the scalar electron neutrino $\tilde{\nu}$ and assuming the $\tilde{W}$ is a pure gaugino state a large domain can be ruled out in the $M_{\tilde{W}} - M_{\tilde{\nu}}$ plane (Fig. 42) from $\tilde{W}$ pair production and mostly single γ production[21-23,79] through the process

$$e^+e^- \rightarrow \gamma\, \tilde{\nu}\, \bar{\tilde{\nu}}$$

via $\tilde{W}$ exchange in the t-channel. On the other hand, if the LSP is the photino, the $\tilde{W}$

mass should certainly exceed 40 GeV from the non-observation of unbalanced monojets in UA1, where the analysis would be quite similar to the heavy lepton case :

$$W \to \tilde{\gamma}\,\tilde{W}$$
$$\hookrightarrow f\bar{f}'\tilde{\gamma}$$

Neutralinos have been searched for at PETRA[84,86,87] through the reaction

$$e^+e^- \to \tilde{\gamma}\,\tilde{Z}$$

via $\tilde{e}$ exchange in the t-channel. Various mass domains have been excluded in the $M_{\tilde{e}} - M_{\tilde{Z}}$ plane with typical masses up to 40 GeV for $\tilde{Z}$ and ~ 100 GeV for $\tilde{e}$. One should however remark that these bounds depend rather strongly on the $\tilde{\gamma}$ mass, and on the assumed mass hierarchy controlling the $\tilde{Z}$ decays and the importance of the various branching fractions.

4.6. Consistent SUSY analysis within N=1 supergravity models

Since a priori many possibilities exist for the SUSY mass spectrum, it is useful to try to take into account the various experimental bounds within a consistent framework. N=1 supergravity models are promising and they can be conveniently chosen for the purpose. Such an analysis was indeed presented at this conference[88].

In minimal N=1 supergravity models[89-92], the SUSY particles masses are expressed as a function of a minimal set of parameters : the gravitino mass $M_{\tilde{G}}$, a gaugino mass M, a higgsino mass μ, a parameter A related to the super-Higgs mechanism and the ratio x of the vacuum expectation values of the 2 Higgs fields. Renormalisation group equations are used to track the masses from the unification Planck mass down to the electroweak scale. If the top mass is not too heavy (~ 40 GeV), then x ~ 1 and the equations read

$$M_{\tilde{\gamma}} \simeq .5\,M$$
$$M_{\tilde{g}} \simeq 3\,M$$
$$M^2_{\tilde{f}} = M^2_{\tilde{G}} + C_f\,M^2$$

where the coefficients C_f are specified. Gaugino masses depend on different parameters of the theory (M, μ).

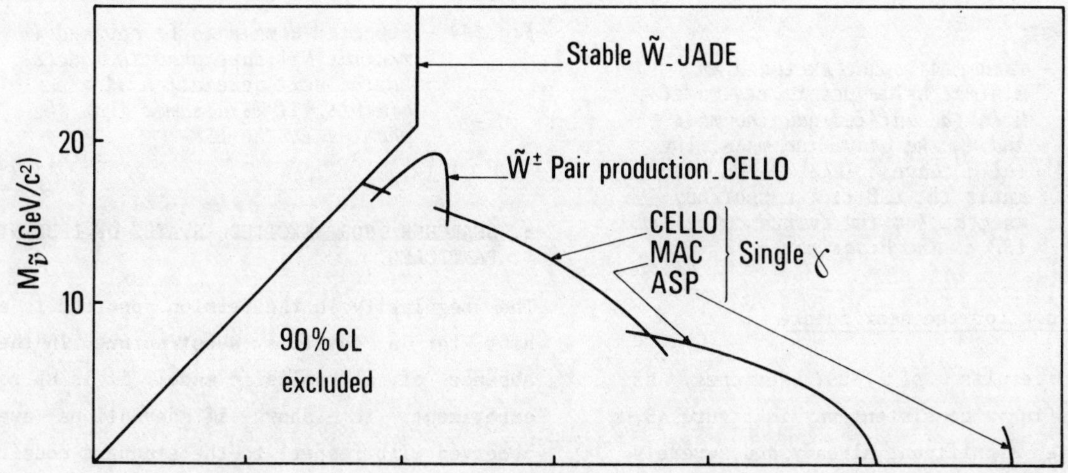

Fig. 42 - Excluded domain for $\tilde{W}$ and $\tilde{\nu}$ masses in the case where the LSP is the scalar electron neutrino $\tilde{\nu}$

Figure 43 summarizes the experimental constraints from $\tilde{e}$, $\tilde{\gamma}$, $\tilde{q}$ and $\tilde{g}$ searches. In the case where the LSP is the photino (the favoured case), one sees that the most constraining information is the $\tilde{g}$ bound by UA1 ruling out any $M_{\tilde{G}}$ for M less than 20 GeV. If, on the other hand, the LSP is the higgsino, then the most powerful constraint comes from the data on unstable $\tilde{\gamma}$ pairs from CELLO.

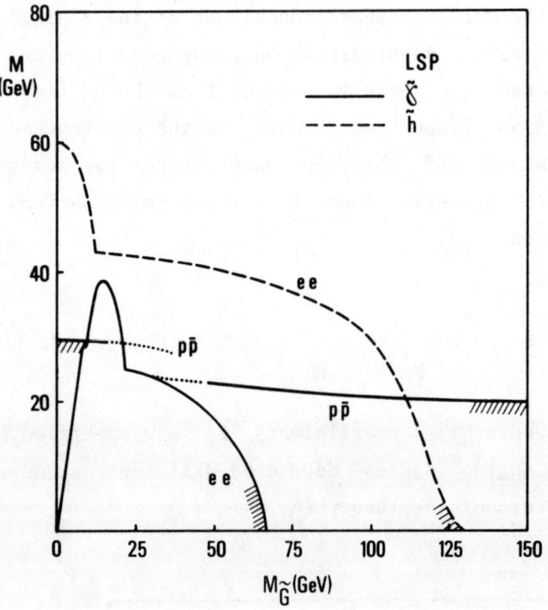

Fig. 43 - Parameter space excluded in minimal N=1 supergravity models. M is the unified gaugino mass and $M_{\tilde{G}}$ the gravitino mass. The solid curve refers to the case where the LSP is the photino, whereas for the dashed curve the LSP is the Higgsino

4.7. Outlook for the near future

Present results of SUSY searches, as analysed in a consistent way in figure 43, are quite significant already and severely limit the possibilities at future colliders. For example, the achieved bounds on the scalar electron mass make it impossible for the Z° to decay into $\tilde{e}$ pairs unless the $\tilde{\gamma}$ mass exceeds 10 GeV. Therefore progress will come first by increasing the bound on the gaugino mass M. This is also true of $\tilde{g}$ searches in $p\bar{p}$ colliders, where the Tevatron should push the $\tilde{g}$ mass bound to ~ 150 GeV[93-94]. The ground covered by the next generation of experiments is indicated on figure 44 : both M and $M_{\tilde{G}}$ will be explored till ~ 100 GeV or more. Will SUSY show up by then ?

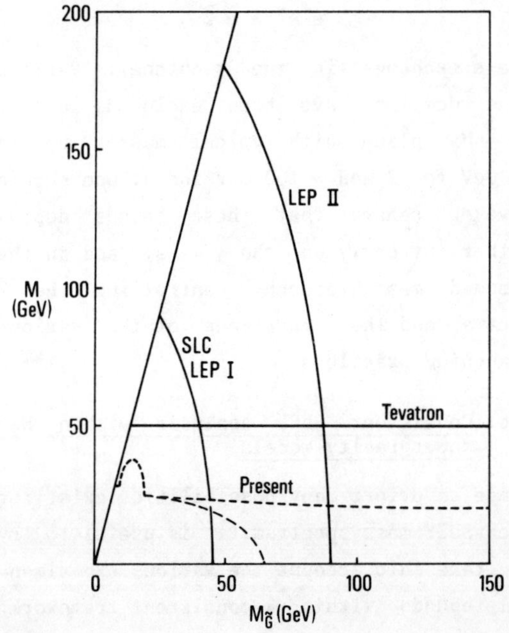

Fig. 44 - Expected domain to be covered in minimal N=1 supergravity models by the next generation of experiments. It is assumed that the photino is the LSP

5. SEARCHES FOR EXCITED STATES OF PRESENT PARTICLES

The regularity in the fermion spectrum is a hint for a possible substructure. In the absence of a realistic model, it is up to experiment to show if deviations are observed with respect to the standard model. The more general test, involving only one type of particle has been performed on Bhabha scattering[6] and the scale of substructure Λ has been found to be at least 1 TeV. Excited states f* of the known fermions f are generally expected on the same mass scale ; however it may well be possible that some states could lie at a smaller scale, well below Λ. Anyway it is the only one we can explore experimentally (this is known as the lamp-post technique).

The Lagrangian describing the $f f^* \gamma$ coupling is written as an effective interaction

$$\mathcal{L} = \lambda \frac{e}{2M^*} \bar{f}^* \sigma_{\mu\nu} f F^{\mu\nu} + \text{h.c.}$$

where λ parametrizes the strength of the coupling.

5.1. Excited leptons

New results have be presented on e* searches by CELLO[95-96] and a τ* search by JADE[97]. Pair production excludes these states up to ~ 23 GeV, independent of the value for λ. Larger masses are attainable for single production.

$$e^+e^- \to \ell\, \bar{\ell}^* + \bar{\ell}\, \ell^*$$

A particularly sensitive method for the excited electron is based on the virtual Compton configuration (see Fig. 45) where one of the electron (or the positron) scatters at a very small angle and stays undetected, corresponding to a γ e collision capable of forming the e* state[98-99]. No signal is seen and the best limits obtained are given in figure 46 and 47.

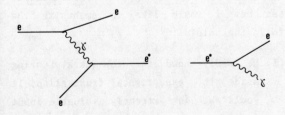

Fig. 45 - The virtual Compton configuration for producing an excited electron

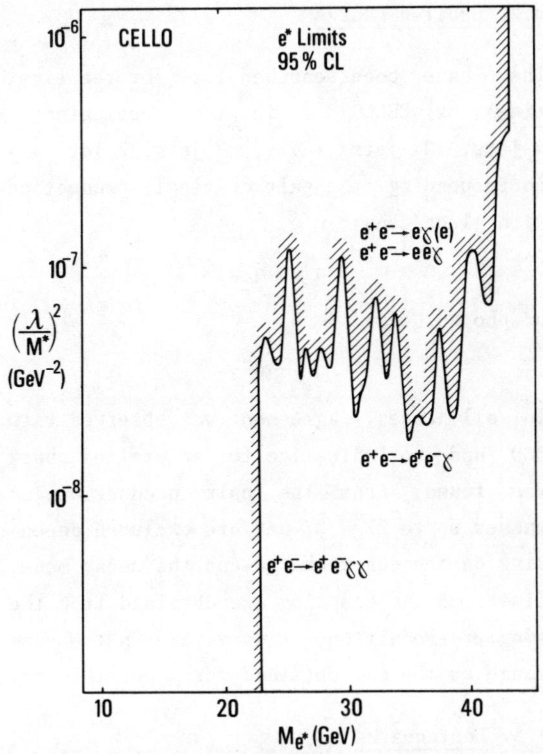

*Fig. 46 - Limit on the e*eγ coupling as a function of the e* mass*

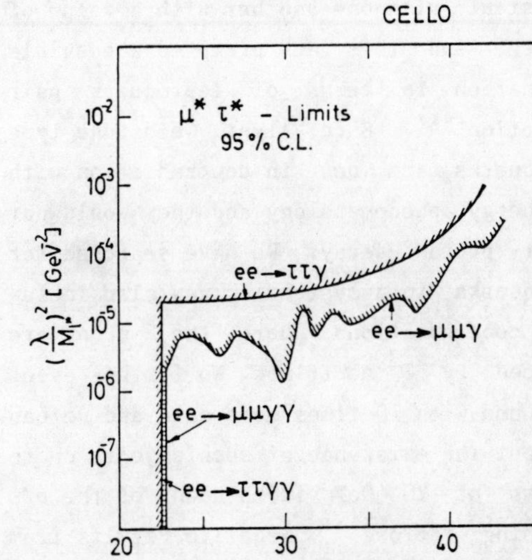

*Fig. 47 - Limits on the μ*μγ and τ*τγ couplings as a function of the μ* or τ* masses*

5.2. Excited quarks

They have been searched for (for the first time) by CELLO[100] in the topologies : 4 jets, 2 jets + 2γ, 3 jets, 2 jets + γ corresponding to pair or single production with gluon decay

$$q^* \to q\ g$$

or photon decay

$$q^* \to q\ \gamma$$

In all cases, agreement was observed with QCD and no indication for an excited quark was found. From the pair production, q* masses up to 21 - 23 GeV are excluded depending on the quark charge and the decay mode. Limits on the coupling are obtained from the single production : they are in the same range as the one obtained for a possible τ*.

5.3. Leptoquarks

A spectacular event with 2 muons opposite to 2 jets was observed by CELLO[101], each parton sharing about equally the total energy (Fig. 48). The μ-jet masses were consistent with one another with a value of 21 GeV, and this fact prompted a possible explanation in terms of leptoquark pair production[102]. Such light Goldstone-type leptoquarks are not in contradiction with low-energy phenomenology and they would not affect proton decay. We have searched for leptoquarks in a systematic way also including configurations where the 2 muons are replaced by 2 neutrinos. No further event was found with 10 times more data and we can rule out the existence of such objects up to a mass of 20 GeV independent of the μ/ν branching ratios[103]. Negative results from JADE are also available[104]. What is then the origin of the CELLO event ? The most economical (although unlikely) explanation is α^4 QED production (Fig. 49) : the probability to observe such an event with $M_{\mu\mu} > 15$ GeV and $M_{q\bar{q}} > 15$ GeV is 1.9 % with the complete statistics.

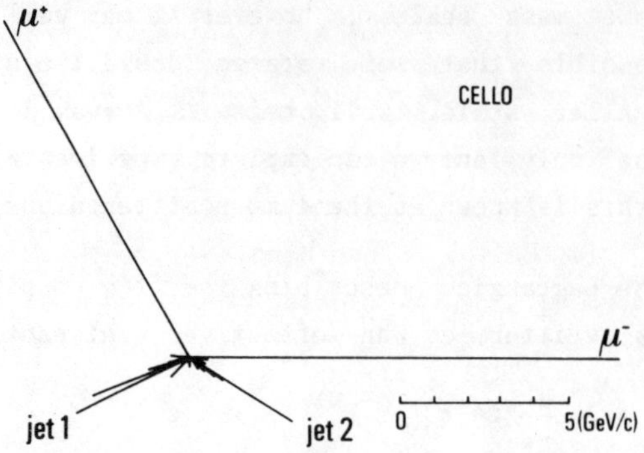

Fig. 48 - Di-muon di-jet CELLO event in momentum space

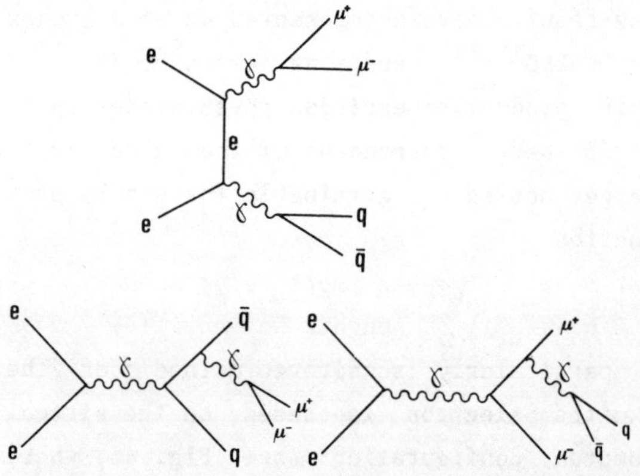

Fig. 49 - α^4 QED processes yielding 2 muons and 2 jets

6. CONCLUSIONS

Among many interesting, but unsuccessful, searches I would like to emphasize the following points :

(1) We badly need the top quark. Leaving aside the experimental frustration, it would be an extremely valuable input into our standard phenomenology (electroweak radiative corrections, understanding of Γ_Z and Γ_W, CP, etc...). First results from Z° decays from SLC should either discover the top or place a firm lower bound on its mass. Higher

masses will have to wait for LEP 200, and/or improved background rejection in the $p\bar{p}$ collider experiments.

(2) No fourth generation appears in sight. In fact, it looks as if it is becoming unlikely : on one hand, N_ν measurements seem to be closing down on 3, a new charged lepton is not allowed below 41 GeV and a new b′ quark has to be heavier than 23 GeV. On the other hand, we know that electroweak phenomenology would be perturbed if a new top quark mass exceeded ~ 300 GeV (the ρ parameter would deviate from 1 beyond experimental uncertainty). Putting all these facts together with the known patterns of the first 3 generations, I find unlikely that a 4th generation could sneak in.

(3) The experimental situation on the Higgs sector is frustrating. It has turned out that the Higgs boson of the minimal model is even more elusive than we originally thought. No sensitive enough experiment is known to exist today. The situation is expected to change after a million or so Z° are registered by experiments at SLC and LEP.

(4) Conspicuous $e^\pm$ peaks in heavy ion collisions at GSI have prompted a new wave of axion searches. They have all been negative. There is no axion with 1.8 MeV mass, and data from electron beam dump experiments and g-2 results rule it out as a fundamental particle. The GSI phenomenon begs for an explanation ; however, it is probably not within the scope of particle physics. Meanwhile, the "axion" scale f has to retreat from the electroweak range to some higher, yet not precisely defined, value : searches for light and weakly coupled axions are now in the domain of atomic physics !

(5) Searches for an experimental indication of supersymmetry have been unsuccessful despite a wide-open and intense activity. Mass limits are now in the W mass range : $m_{\tilde{\ell}} > 65$ GeV (for $m_{\tilde{\gamma}}$ less than a few GeV), $m_{\tilde{q}} > 70$ GeV, $m_{\tilde{g}} > 60$ GeV. The "monojets" of UA1 are now fully under control and are explained by known sources in the standard model. The next generation of collider experiments (Tevatron, SLC and LEP) should make spectacular steps past the electroweak scale. Exploration at the TeV scale will have to wait for the following generation of machines.

I wish to acknowledge valuable discussions with J.F. Grivaz, P. Lehmann, F. Pauss, R. Peccei, C. Rubbia and the parallel session organizers, R. Hollenbeek, S. Komamiya, W. Schlatter and S. Dawson. Many thanks to T. Barklow who acted as an efficient scientific secretary.

REFERENCES

(1) Matis, H., Searches for monopole and quarks, these proceedings

(2) Whitaker, S., these proceedings

(3) Honma, A., these proceedings

(4) Arnison, G. et al., UA1 collaboration, Phys. Lett. 139B, 115 (1984)

(5) At least 30 phenomenological papers can be located up to mid 1985 ; a complete list is given in R.M. Barnett, proceedings of the 1985 SLAC Summer Institute on Particle Physics

(6) Komamiya, S., Proceedings of the 1985 International Conference on Lepton and Photon Interactions at High Energies, Kyoto (1985)

(7) Arnison, G. et al., UA1 Collaboration, Phys. Lett. 172, n°3-4, 461 (1986)

(8) Rubbia, C., private communication

(9) Adeva, B. et al., Mark J Coll., MIT Technical Report 146 (1986)

(10) JADE Coll., Proceedings of the 21th Rencontre de Moriond (1986)

(11) Behrend, H.J. et al., CELLO Collaboration, contribution to this conference

(12) TASSO Collaboration, results quoted by J. Branson, this conference

(13) D'Agostini, G. and De Boer, W., private communication

(14) Felst, R., private communication

(15) Behrend, H.J. et al., CELLO Collaboration, contribution to this conference

(16) Bartel, W. et al., JADE Coll., Phys. Lett. 123B, 353 (1983)

(17) Value taken from a graph in Ref. 3

(18) Di Lella, L., private communication

(19) Ma, E. and Okada, J., Phys. Rev. Lett. 41, 287 (1978) ; Phys. Rev. D18, 4219 (1978)

(20) Gaemers, K.J.F., Gastmans, R. and Renard, F.M., Phys. Rev. D19, 1605 (1979)

(21) ASP Coll., results quoted in ref. 2

(22) Ford, W.T. et al., MAC Coll., SLAC-PUB 4003 (1986), contribution to this conference

(23) Behrend, H.J. et al., CELLO Coll., Phys. Lett. 176B, 274 (1986)

(24) Lavine, T., private communication and MAC note 748 (1986)

(25) Yang, J. et al., Astrophys. J. 281, 493 (1984)

(26) Ellis, J. et al., Phys. Lett. 167B, 457 (1986)

(27) Gilman, F., SLAC-PUB 3898 (1986), to appear in comments in Nuclear and Particle Physics

(28) Feldman, G., Proceedings of the 20th Rencontre de Moriond (1985)

(29) Behrend, H.J. et al., CELLO Collaboration, contribution to this conference

(30) Youssef, S. et al., CUSB Coll. 139B, 332 (1986), Lee-Franzini, J., Proceedings of the 20th Rencontre de Moriond (1985)

(31) CLEO Coll., Phys. Rev. D33, 300 (1986)

(32) Wysotsky, M.I., Phys. Lett. 97B, 159 (1980)

(33) Aznauryan, I.G., Grigorian, S.G. and Matinyan, S.G., Erevan preprint 882 (33)-86 (1986)

(34) Biswas, S.N., Goyal, A. and Pasupathy, J., Phys. Rev. 32D, 1844 (1985)

(35) Matinyan, S.G., private communication

(36) Tuts, P.M. et al., CUSB Coll., contribution to this conference

(37) Crystal Ball Coll., contribution to this conference

(38) For a nice review of the standard axion phenomenology, see Donnelly, T.W. et al., Phys. Rev. 18D, 1607 (1978)

(39) Edwards, C. et al., Crystal Ball Coll., Phys. Rev. Lett. 48, 903 (1982)

(40) Sivertz, M. et al., CUSB Coll., Phys. Rev. 26D, 717 (1982)

(41) Alam, M.S. et al., CLEO Coll., Phys. Rev. 27D, 1665 (1983)

(42) Mageras, G. et al., CUSB Coll., Phys. Rev. Lett. 56, 2672 (1986)

(43) Bowcock, T. et al., CLEO Coll., Phys. Rev. Lett. 56, 2676 (1986)

(44) Albrecht, H. et al., ARGUS Coll., contribution to this conference

(45) For a review, see Zehnder, A., Proceedings of the 1982, Gif-sur-Yvette Summer School

(46) Schweppe, J. et al., Phys. Rev. Lett. 51, 2261 (1983)

(47) Clemente, M. et al., Phys. Lett. 137B, 41 (1984)

(48) Cowan, T. et al., Phys. Rev. Lett. 54, 1761 (1985)

(49) Cowan, T. et al., Phys. Rev. Lett. 56, 444 (1986)

(50) Schaefer, A. et al., J. Phys. G11, L 69 (1985)

(51) Balantekin, A.B. et al., Phys. Rev. Lett. 55, 461 (1985)

(52) Peccei, R.D., Wu, T.T. and Yaganida, T., Phys. Lett. 172, n° 3-4 (1986)

(53) Krauss, L.M. and Wilczek, F., Phys. Lett. 173, n° 2, 189 (1986)

(54) Peccei, R.D., rapporteur's talk, these proceedings

(55) Savage, M.J. et al., Phys. Rev. Lett. 57, 178 (1986)

(56) Calaprice, F.P. et al., Internal Princeton Report (1986)

(57) Eichler, R. et al., contribution to this conference

(58) Reinhardt, J. et al., Phys. Rev. 33C, 194 (1986)

(59) Suzuki, M., Phys. Lett. 175B, 364 (1986)

(60) Bechis, D.J. et al., Phys. Rev. Lett. 42, 1511 (1979)

(61) Tsai, Y.S., SLAC-3926 (1986)

(62) Bilal, A. and Georges, A., private communication

(63) Konaka, A. et al., contribution to this conference

(64) Davier, M., Jeanjean, J. and Nguyen Ngoc, H., contribution to this conference, to appear in Phys. Lett.

(65) Arnold, R. et al., contribution to this conference

(66) Brown, C.N. et al., contribution to this conference

(67) Greenberg, J.S., private communication to Krauss, L.M.

(68) Erb, K., private communication

(69) Fayet, P., Phys. Rep. 105, 21 (1984)

(70) Haber, H.E. and Kane, G., Phys. Rep. 117, 75 (1985)

(71) Barnett, R.M., Proceedings of the 1985 SLAC Summer Institute on Particle Physics

(72) Ellis, J. et al., Nucl. Phys. B238, 453 (1984)

(73) Haber, H.E., Proceedings of the 1985 SLAC Summer Institute on Particle Physics

(74) Behrend, H.J. et al., CELLO Coll., contribution to this conference

(75) Cabibbo, N., Farrar, G.R. and Maiani, L., Phys. Lett. 105B, 155 (1981)

(76) Fayet, P., LPTENS 86.9 (1986)

(77) Fayet, P., Phys. Lett. 84B, 421 (1979); 86B, 272 (1979)

(78) Leffer, S. et al., Crystal Ball Coll., contribution to this conference

(79) Mark J results as quoted in Ref. 2

(80) Albrecht, M. et al., ARGUS Coll., Phys. Lett. 167B, 360 (1986)

(81) Badier, J. et al., NA3 Coll., Z. Phys. C31, 21 (1986)

(82) Tuts, P.M. et al., CUSB Coll., contribution to this conference

(83) Bartel, W. et al., JADE Coll., Z. Phys. C, 505 (1985)

(84) Adeva, B. et al., Mark J. Coll., Phys. Rev. Lett. 53, 1806 (1984)

(85) Behrend, H.J. et al., CELLO Coll., contribution to this conference

(86) Bartel, W. et al., JADE Coll., Phys. Lett. 146B, 126 (1984)

(87) Behrend, H.J. et al., CELLO Coll., contribution to this conference

(88) Grivaz, J.F., these proceedings

(89) Nilles, H.P. and Nusbaumer, M., Phys. Lett. 145B, 73 (1984)

(90) Nilles, H.P., Phys. Rep. 110C, 1 (1984)

(91) Arnowitt, R., Chamseddine, A.M. and Nath, P., Applied N=1 Supergravity, World Scientific, Singapore (1985)

(92) Ellis, J., rapporteur's talk at the 1985 International Symposium on Lepton and Photon Interactions at High Energies, Kyoto

(93) Barnett, R.M., Haber, H.E. and Kane, G.L., Nucl. Phys. B267, 625 (1986)

(94) Reya, E. and Roy, D.P., DO-TH 86/06 TIFR/TH 86-16, contribution to this conference

(95) Behrend, H.J. et al., CELLO Coll., Phys. Lett. 168B, 420 (1986)

(96) Behrend, H.J. et al., CELLO Coll., contribution to this conference

(97) Bartel, W. et al., JADE Coll., DESY 86-023 (1986)

(98) Courau, A., Kessler, P., Phys. Rev. D33, 2024 (1986)

(99) Hagiwara, K., Zeppenfeld, D., Komamiya, S., Phys. Lett. 158B, 270 (1985)

(100) Behrend, H.J. et al., CELLO Coll., contribution to this conference

(101) Behrend, H.J. et al., CELLO Coll., Phys. Lett. 141B, 145 (1984)

(102) Schrempp, B. and Schrempp, F., Phys. Lett. 153B, 101 (1985)

(103) Behrend, H.J. et al., CELLO Coll., to this conference, Phys. Lett. 178B, 452 (1986)

(104) Searches for similar topologies are reported in Ref. 6

(105) Albrecht, H. et al., ARGUS Coll., Phys. Lett. 154B, 452 (1985)

(106) Le Diberder, F., Grivaz, J.F., CELLO note 0-082 (1986)

(107) Grivaz, J.F., private communication

DISCUSSION

G. Morpurgo (INFN Genova)

Because Dr. Davier did not speak about the searches for free quarks and because Dr. Matis in the parallel session only showed a table for 1/10 of a second, I believe that it is important to mention the following result : the Rutherford group using our technique of ferromagnetic levitation and coating niobium spheres with iron explored a quantity of niobium five times larger than that explored by Fairbank, finding no fractionally charged quark.

S. Komamiya (SLAC)

I would like to give a short comment on the UA1 $\tilde{q}$ mass limit : UA1 assumed that scalar quarks with $m_{\tilde{q}_L} = m_{\tilde{q}_R}$ and 5-flavours are mass degenerate.

So, the assumed scalar quark cross section is a factor 10 larger than that for the single flavour $\tilde{q}$ without mass degeneracy ($m_{\tilde{q}_L} \lessgtr m_{\tilde{q}_R}$), for $m_{\tilde{q}} < m_{\tilde{g}}$. The $\tilde{q}$ mass, assuming single flavour and $m_{\tilde{q}_L} \lessgtr m_{\tilde{q}_R}$ (for example one of the $\tilde{t}$), is excluded up to about 21 GeV by JADE at PETRA.

M. Davier

This is true, but the PETRA limit would not improve significantly by including all flavours and assuming mass degeneracy, since it is essentially given by kinematic limitation.

V. Obraztov (IHEP, Serpukhov)

There had been a limit an standard model Higgs from $\eta' \to \eta\mu^+\mu^-$, namely $\lesssim$ 500 MeV. New calculations have shown this limit to be incorrect ; the experiments are a factor of $\sim$ 5 from seeing the standard Higgs at these masses.

Plenary Session

General Properties of Field Theory

O. Alvarez (Berkeley)

This paper is not available

Chairman

C. Quigg (Fermilab)

Scientific Secretary

M. Karliner (SLAC)

Plenary Session

Accelerator Technology
D. Edwards (Fermilab)

Spectroscopy of Light and Heavy Quarks
S. Cooper (SLAC)

Chairman
J. Sacton (Brussells)

Scientific Secretaries
D. Lambert (LBL)
P. Reutens (SLAC)

ACCELERATOR TECHNOLOGY

D. A. Edwards

Fermi National Accelerator Laboratory*
PO Box 500, Batavia, Illinois 60510

This article is a brief review of the state of accelerator technology and construction as it applies to High Energy Physics. It is divided into four parts. First, we look at those facilities that are being commissioned, namely, TRISTAN, the SLC, and Tevatron I. Then, we summarize the facilities that are being built, LEP, HERA, and UNK, followed by those that are in the proposal stage, the SSC, LHC, and RHIC. Finally, we comment on the future and the great interest in linear colliders as a subject of advanced accelerator R&D.

1. INTRODUCTION

The talk on which this paper is based was more a travelogue through specific projects both existing and virtual rather than a discussion of accelerator technology per se. I follow the same format here (though without the colored slides), in the hope that the state of the technology is best illustrated by its realization in our facilities rather than in the abstract.

The accelerators are divided into four categories: those in the process of being turned on, those in construction, the proposed facilities, and the possibilities for the future. Even with an arbitrary low energy cutoff at about 50 GeV in $\sqrt{s}$, there are nine projects in the first three headings.

This energy boundary was suggested by the subjects covered in the parallel session on accelerators at the conference, and is consistent with the orientation of reviews at earlier meetings in this series. It does have the consequence of excluding entire classes of accelerators from the discussion and some multi-hundred million dollar enterprises at that. But though I don't write about the next generation of synchrotron light sources, kaon factories, CEBAF, the very promising developments in medical applications of accelerators, or the new e^+e^- ring at Beijing, I do hope that the tehnologies relevant to HEP are touched on in this report. However, the focus here is supposed to be on high energy physics, with the emphasis on "high." So the the future receive the most attention.

*Operated by the Universities Research Association, Inc., under contract with the United States Department of Energy.

2. IN THE COMMISSIONING PHASE

Three projects are in this category; two synchrotron colliders, TRISTAN and Tevatron I, and the SLC. TRISTAN, at KEK, is an e^+e^- storage ring designed for the 25 to 35 GeV per beam energy range, the latter half of which necessitates the use of superconducting RF. Tevatron I is the $\bar{p}p$ variant of the Fermilab facility, and unites the superconducting magnet progress that culminated in the Tevatron with the $\bar{p}$ source technology pioneered at CERN. The SLAC Linear Collider is the first project of its kind, and, as such, represents a challenge to the twenty-five year hegemony of the synchrotron in high energy physics.

The design of TRISTAN[1] reflects the desire to obtain the highest energy in an e^+e^- collider within the boundaries of the KEK site. The main ring of 3 km circumference has about 1/3 of its perimeter reserved for the RF system. Conventional cavities will permit TRISTAN to achieve close to 30 GeV per beam. But clearly the likelihood of finding interesting physics improves with energy as the Z^o mass is approached from below, and the progressive installation of superconducting cavities is planned to raise the energy.

The e^+e^- interaction cross section passes through a minimum in the TRISTAN energy range. The interference between Z^o and photon exchange, leading to forward-backward asymmetry in fermion pair distributions, is a maximum in this range. High luminosity is a must. The maximum design luminosity is $8 \cdot 10^{31}$ cm^{-2}s^{-1}, nearly a factor of 3 higher than that achieved at CESR, PEP, or PETRA. A comparison of the TRISTAN parameters with other e^+e^- rings shows that this performance level is consistent with experience.[2] The maximum luminosity occurs at a collision energy of 54 GeV, and drops off by a factor of 4 by 60 GeV if RF power is held constant.

The $\bar{p}$ source at Fermilab[3] seeks to achieve a higher collection rate than its CERN counterpart by targetting at higher energy, 120 GeV, and by accepting a large $\bar{p}$ momentum bite through the use of a debuncher ring between the target and the accumulator. The source is intended to build a stack of $4.3 \cdot 10^{11}$ $\bar{p}$'s in 4 hours, from which $1.8 \cdot 10^{11}$ will be extracted for injection into the Tevatron.

With the $1.8 \cdot 10^{11}$ $\bar{p}$'s divided equally among three bunches and colliding with three proton bunches of like intensity, the luminosity is intended to be 10^{30} $cm^{-2}s^{-1}$ at 1.8 TeV in the center-of-mass. The initial shakedown run this Winter has the more modest goal of building up to 10^{29} $cm^{-2}s^{-1}$ over the three months scheduled for the exercise.

The smaller number of particles and the infrequent ramping, as compared to fixed target operation, make the superconducting magnets less susceptible to quenching, and the 1.8 TeV figure appears reasonable. The ring does quench without beam at slightly above 0.9 TeV per beam, and so the goal of 1 TeV operation must await either further component replacement or lower temperature capability.

Fermilab's hoped for edge in $\bar{p}$ rate is only temporary; with the addition of ACOL, a "collector" ring, upstream of the accumulator, CERN plans to come within a factor of two of the Fermilab design value. The Sp$\bar{p}$S has already operated with 6 proton and 6 antiproton bunches at the same time, with electrostatic separators preventing collisions throughout most of the ring. The combination of these improvements at CERN should raise the average luminosity to about the same level as the Fermilab design.

The linear collider seeks to avoid the synchrotron radiation problem for e^+e^- colliders by not bending the beams. If there is to be no penalty in luminosity, any resulting reduction in bunch-bunch collision frequency must be balanced by an increase in the bunch intensity and/or a reduction in the bunch cross section. The SLC[4] is the "proof of principle" project for this class of device, using a single linac and so tolerating a semicircular bend for each bunch.

The goal is to achieve the same collision energy as LEP Phase 1, 100 GeV, and 40% of the luminosity at a fraction of the cost. To do this, the SLAC group will collide intense low emittance bunches of $7 \cdot 10^{11}$ electrons focussed to a σ of 1.6 microns at a repetition rate of, eventually, 180 Hz. If you put this collection of parameters into the basic luminosity formula, the result is short of the $6 \cdot 10^{30}$ $cm^{-2}s^{-1}$ desired by about a factor of two. This final factor is to be obtained by turning the beam-beam interaction, hitherto a limitation, into a virtue, as the pinch effect reduces the bunch area and so enhances the luminosity.

The three facilities of this section are all scheduled to be operating for physics in 1987. They are being commissioned now, so anything written here may well be obsolete before the ink is dry.

3. UNDER CONSTRUCTION

Of the three projects here, two, LEP and HERA, are intended to be in operation before 1990. The construction schedule of the third, UNK, extends into the early 90's.

LEP is the mammoth e^+e^- collider being built at CERN.[5] Its initial phase calls for 100 GeV in the center of momentum and $1.6 \cdot 10^{31}$ $cm^{-2}s^{-1}$ at each of eight interaction points, though only four crossings will be equipped at the outset. The underground tunnel with a circumference of 27 km is a natural location for future developments beyond LEP.

The ring is scaled to permit an eventual upgrade to 100 GeV per beam. As in the case of TRISTAN, a conversion to superconducting RF is envisaged. The 180 MeV per turn energy loss at 50 GeV soars to nearly 3 GeV per turn at 100 GeV. Even the conventional accelerating system has a new twist; a bunch passes an RF cavity sufficiently rarely that it makes sense to store the energy in an auxiliary low loss storage resonator until it is needed.

The magnets, of course, head toward the other extreme. At 50 GeV, the field of a bending magnet with its radius of curvature of 3.1 km is only 0.06 Tesla. A standard iron yoke is no longer necessary. Rather, a solution that is more economical and completely adequate magnetically is provided by a stack of 1.5 mm steel laminations that are separated by 4 mm gaps filled with cement mortar.

One of the trends in accelerator technology that is illustrated by the LEP project is the requirement for ever more complex operational scenarios to be performed by the older components. The PS and SPS will turn into electron accelerators in addition to their other duties. For instance, interleaved operation of the SPS will call for a 450 GeV proton cycle for fixed target physics followed by four 20 GeV electron and positron cycles for LEP. Each of the four low energy cycles will be performed in a different magnetic environment as the remanent magnetization relaxes from that associated with the high energy ramp, a situation that would be difficult to confront only a few years ago.

After a long history as an unexercised "option," an ep collider is imminent. Of course, the problem with an ep facility is that you have to build one of just about everything. In the instance of HERA, DESY has undertaken to build an entire proton plant, complete with superconducting main ring, as well as the large electron storage ring.[6]

The HERA project will lead to the collision of 820 GeV protons with 30 GeV electrons and positrons, for $\sqrt{s}$ = 314 GeV. The luminosity goal is $2.5 \cdot 10^{31}$ at 4 crossing points. The circumference of the main enclosure is 6.3 km. PETRA becomes the final stage in both the proton and electron acceleration stages, delivering the former at 40 GeV and the latter at 14 GeV.

As one would expect, the collision optics is a bit more complex than that appropriate to a single ring. In order to avoid excitation of synchro-betatron oscillations, the collisions remain head-on, but now the geometry of two rings must be arranged to make this possible. Spin rotators are provided in the electron ring outboard of the collision point to turn the spin into both helicity states over the range from 28 to 35 GeV.

For their superconducting bending magnets, the HERA designers have selected a cold-iron collared coil type, which is being produced industrially. The nominal peak energy of 820 GeV corresponds to 4.73 T bend field, and prototypes have gone to well in excess of this figure. Clear distinctions from the Tevatron are a low injection field and a long 1000 second acceleration period. The similarities of this magnet to the SSC choice, in type, in acceleration ratio and rate, and in reliance on an industrial production base are worth noting; HERA will be the next major source of design input for larger superconducting storage rings.

Superconductivity will appear in the other ring as well. Conventional RF taken from PETRA will be adequate to achieve initial operation at some 26 GeV; a superconducting RF system will be installed to achieve the design figures of energy and luminosity.

Work on the IHEP Accelerating and Storage Complex, or UNK,[7] at Serpukhov in the Soviet Union continues, though not at the pace that this important project deserves. This large facility is to consist of a conventional proton synchrotron which can either operate in fixed target mode up to 600 GeV or as a storage ring at 400 GeV (Stage I), a 3 TeV superconducting proton ring in the same enclosure (Stage II), and possibly a second superconducting proton ring (Stage III). The Stage II synchrotron could perform as a fixed target source, or, in collision with protons of the Stage I accelerator, yield a center of momentum energy of 2.2 TeV. The existing 70 GeV Serpukhov synchrotron is to serve as the injector to Stage I.

Some 24 km of underground enclosure is needed, of which 5 km has been completed. The remainder is to be built at a rate of about 5 km per year until finished in 1990. Prototype magnets have been developed for Stage I and Stage II, and production of 100 of each is planned for 1986 and 1987. UNK has been in the works for quite a while; I recall that in 1978 the design of the Tevatron benefitted from studies carried out by the team working on the UNK.

3. PROPOSALS

The Superconducting SuperCollider (SSC) and the Relativistic Heavy Ion Collider (RHIC) are firm proposals under consideration by the U. S. Department of Energy. A potential CERN project, the Large Hadron Collider could become a firm proposition at any time, and because of the many common features that the LHC and the SSC share, the two are taken up here in parallel.

The SSC is designed to effect pp collisions at $\sqrt{s}$ = 40 GeV; the LHC as currently conceived has pp collisions at $\sqrt{s}$ = 17 GeV. The preference for pp as opposed to $\bar{p}p$ arises from the desire to achieve luminosity at the 10^{33} $cm^{-2}s^{-1}$ level for investigation of the TeV mass scale.

The combination of a luminosity at this level and a total effective cross section of 100 mb or so implies an event rate of about 100 Mhz. This last figure is ominous enough in itself; to make it at all tolerable means that the SSC and LHC be designed as "many bunch" colliders. The mean number of events per bunch-bunch collision is

$$\langle n \rangle = L\sigma\ (S_B/c) \simeq S_B(\text{in meters})/3 \qquad (1)$$

where L is the luminosity, σ the effective

total cross section, S_B is the bunch spacing, and c is the speed of light. The collision time of a pair of bunches is less than one nanosecond, so ⟨n⟩ should be limited to "a few." For instance, if the bunch spacing is 6 m, the average number of interactions per bunch-bunch hit would be ~2. The circumferences of the LHC and SSC rings are 27 km and 83 km respectively, so one is talking about thousands of bunches in each beam.

The number of particles in each bunch will be relatively small in comparison to present practice. The luminosity can be written in the form

$$L = \left(\frac{c}{S_B}\right) \frac{1}{\beta^*} \frac{\gamma}{4\pi\epsilon_N} \quad N_B^2 \qquad (2)$$

where N_B is the number of protons per bunch. The three groups of factors in front represent, in turn, the bunch-bunch collision frequency, the influence of the collider ring beam optics through the amplitude function at the interaction point, and the characteristics of the proton beam itself. In the last group, γ is the Lorentz factor of the the protons, and ϵ_N the normalized emittance. This emittance is a measure of the area in a one degree-of-freedom phase space for one of the transverse coordinates of the beam, and is the factor in the equation that conveys some sense of beam quality. If we take $\beta^* = 1$m, $\epsilon_N \equiv \gamma\sigma^2/\beta = 1$ mm mrad (where this σ characterizes the beam size), and the other quantities as in the preceding example, then $N_B \simeq 1.1 \cdot 10^{10}$.

This value of N_B is almost an order of magnitude less than the corresponding number for either existing lepton or hadron colliders. Without even bothering to calculate the head-on beam-beam tune shift, we therefore would expect to hear less about this famous effect for the SSC or LHC than for present colliders, even though the anticipated normalized emittance is a factor of four smaller than todays norm. But with the small bunch spacing, the beams must cross at an angle to avoid having a multiplicity of interaction points in the neighborhood of a detector, and so there will be a number of close passages of bunches on either side of the nominal collision point. As a result, we will hear more about the so-called long range beam-beam effect, which, though less intrinsically nonlinear than the head-on case, must nevertheless be taken into account.

The total number of particles in each beam is large, in the 10^{14} range. For the SSC at 20 TeV, the kinetic energy of each beam for the conceptual design parameters is 400 MJ. The main impact of these numbers is in the potential for energy deposition in the superconducting magnets, and, for the SSC, in the consequences of proton synchrotron radiation.

In the SSC, sychrotron radiation emerges as an important design consideration for a proton ring. The energy radiated per turn by a proton following an orbit with radius of curvature ρ is

$$W = 78 \left(\frac{E_{TeV}}{10}\right)^4 \frac{1}{\rho_{km}} \quad \frac{\text{kev}}{\text{turn}} \,. \qquad (3)$$

The bending magnets in the SSC design have a field of 6.6 T at 20 TeV. The radius of curvature is 10 km, and so the radiation per particle per turn is 125 keV. By electron synchrotron standards, this is trivial, as is the synchrotron radiation power of ~10 kW. But in the superconducting environment, this power must be removed at liquid helium temperature and represents about one-half of the 4 K load. Even if the refrigerators operate as high as 20% of ideal Carnot efficiency, a power input of 5 MW per ring is required.

The SSC designers consider their choice of bend field to conform to optimum cost; the LHC value of 10 T reflects a wish to obtain the highest reasonable energy in the LEP tunnel. Provided that NbTi remains the alloy of choice, the LHC field implies low temperature (i.e., ~2 K) operation. Other differences can be attributed to the circumstance that the housing for the LHC already exists and will still be occupied by the e^+e^- ring; for instance, CERN has opted to pursue the compact "two-in-one" magnet design, while the SSC employs the operationally more attractive alternative of separate rings in separate cryostats.

The biggest single material expense in storage rings such as these is the superconducting alloy, and the critical current specification and coil diameter are the primary parameters exerting leverage on this cost factor. When construction of Tevatron magnets got underway, the critical current was specified as 1800 A/mm^2 at 4.2 K and 5 T. Improvement in the NbTi alloy in just a few years makes it possible to use a figure of 2750 A/mm^2 at the same temperature and field today. Also, reductions have been made in the filament size that can be achieved in production quantities. The filament size typical of present SSC and LHC designs is 5 μm. Conductor has been made commercially at a filament diameter of 2.5

μm; use of filaments at this scale would greatly ameliorate the persistent current correction problem.

The choice of coil bore is a more visible (audible?) problem. Now, large synchrotrons do not work without a host of auxiliary magnets that make corrections and adjustments. A discussion of main magnet aperture and field quality requirements draws on a variety of inputs, including the scaling of field imperfections with magnet size and type, the design of the correction magnet system, a magnet measurement plan, installation procedures, and a performance specification. The analysis carried out by the SSC workers resulted in a 4 cm coil diameter; the provisional LHC parameters call for 5 cm. No doubt we will hear more on this issue before either project moves into a construction phase.

Apparently, RHIC[8] is formally a nuclear physics proposal. It arises from the happy conjunction of two circumstances. At an energy density of about 2 GeV/fm^3, nucleons are expected to undergo a phase change analogous to a Mott transition leading to a locally free constituent state, the so-called quark-gluon plasma. And at BNL, there is an unoccupied experimental facility suitable for a large superconducting collider.

The RHIC proposal calls for the collision of fully stripped heavy nuclei at an energy per beam up to, in the case of gold, 100 GeV per amu, and at a luminosity, again for gold, of $4.4 \cdot 10^{26}$ cm^{-2}s^{-1}. Two superconducting storage rings are required, but in contrast to the SSC/LHC situation, there is little need to strain the boundaries of technology, for the dependences of energy density and rapidity range are logarithmic with collision energy.

In the CBA enclosure, 100 GeV/amu for gold leads to a magnetic field of 3.5 T. This leads in turn to a relatively simple single shell coil magnet, prototypes of which have performed well. The two rings are horizontally displaced, and in separate cryostats.

Of course, the interest in RHIC is associated with its heavy ion collisions, but it would also have respectable performance as a pp collider. Protons would achieve 250 GeV per beam at a luminosity of 10^{31} cm^{-2}s^{-1}.

With the completion of RHIC, Brookhaven would be able to cover a wide and continuous range of c.m. energies, from 1.5 GeV/amu in the AGS to 100 GeV/amu in the new superconducting rings.

4. THE FUTURE

Suppose we take the optimistic view, and assume that one of the high energy pp colliders - the SSC or LHC - goes quickly into construction and that as a result the ~1 TeV mass scale is being explored by the mid-1990's. What next?

The next step may well be a linear e^+e^- collider, if the strong R&D program that is necessary can be mounted. The reason why synchrotron storage rings no longer make sense for electrons as one goes into the TeV range is doubtless known to everyone, but bear with me for a moment while I repeat the one line argument. In units appropriate to the scale, the energy loss per turn due to synchrotron radiation for an electron of energy E moving with radius of curvature ρ is

$$W = 88.5\, E_{TeV}^4 / \rho_{km} \quad \text{TeV/turn} \tag{4}$$

So an earth-girdling ring at 5 TeV would lose 10 TeV per turn, and one might as well think of pointing two linacs at each other.

But as a goal beyond the SSC/LHC, 10 TeV in $\sqrt{s}$ for e^+e^- collisions may be about right. To progress from the 1 TeV to the 5 TeV mass scale suggests a step in luminosity of 5^2, or $2.5 \cdot 10^{34}$ cm^{-2}s^{-1}. If we accept these two numbers and supplement them with a few other parameters, we can get some idea of the development needed to achieve them.[9]

First, write down four relationships. The basic form for the luminosity is

$$L = f \frac{N^2}{4\pi\sigma^2} , \tag{5}$$

where f is the frequency of collision of bunches containing N particles with a Gaussian distribution having standard deviation σ in each of the two transverse degrees of freedom. The standard deviation is related to the normalized emittance according to

$$\epsilon_N = \gamma\sigma^2/\beta^* , \tag{6}$$

as in the discussion of the SSC/LHC. The power in each beam is

$$P_b = fN\gamma mc^2 , \tag{7}$$

where mc^2 is the rest energy of the electron. The fourth relation gives the spread in energy caused by particles of one bunch emitting synchrotron radiation as they are deflected by the electromagnetic field of the other bunch. This process has been given the name "beamstrahlung." For the parameters to be considered here, it's necessary to use the quantum mechanical form for the spread, which I will approximate as[10]

$$\delta \simeq \left(\frac{\alpha^4 r_c N^2 \sigma_z}{4\pi\, \sigma^2 \gamma}\right)^{1/3} \qquad (8)$$

Here, α is the fine structure constant and r_c is the classical radius of the electron. Like L and γ, δ is a physics input.

Of the ten quantities in the four equations, three more can be specified. Let me choose ϵ_N, β^*, and σ_z as inputs from accelerator technology, all of which one would like to be as small as possible.

Then the solution for the transverse σ is immediate from the definition of normalized emittance. The number of particles per bunch is

$$N = \frac{5}{4} \cdot 10^{12} \left(\frac{\beta^* \epsilon_N}{\sigma_z}\right)^{1/2} \delta^{3/2} \qquad (9)$$

For the frequency, we have

$$f = 8 \cdot 10^{-24} (L\sigma_z/\gamma\delta^3) \quad , \qquad (10)$$

and for the total beam power

$$2P_b = 1.6 \cdot 10^{-24} L\ (\epsilon_N \beta^* \sigma_z)^{1/2} / \delta^{3/2} \quad , \qquad (11)$$

all in MKS units.

For the three accelerator technology inputs, which are supposed to be small, let us take $\epsilon_N = 10^{-8}$ m, $\beta^* = 10^{-3}$ m, and $\sigma_z = 10^{-6}$. Take $\delta = 0.3$ for the other physics input, a value that corresponds to a rms fractional deviation from the mean collision energy of about 10%. Also edge the energy up to 5.1 TeV, so γ is an even 10^7. Then we get $N = 6.5 \cdot 10^8$, $f = 7400$ Hz, and $2P_b = 7.7$ MW.

The number of particles per bunch looks all right; everything else implies a challenge. The normalized emittance is a factor of 3000 smaller than that which the SLC damping rings are to deliver. The β^* of 1 mm is to be compared with the 7.5 mm value for the SLC final focus with superconducting quadrupoles. Small bunch length is an advantage in the quantum beamstrahlung regime, whereas SLC will operate in the classical limit where long bunches are preferred. The production and preservation of σ_z at the μm level has yet to be tried.

The overall efficiency of converting plug power into beam power is as high as 1% in some linacs.[11] A factor of 5 or so improvement would be needed to reduce the power demand implied above to the same level as that projected for the SSC. The frequency f is the product of the linac repetition rate and the number of bunches per pulse - another subject for study.

Thus, quite aside from the question of how the physical scale of the collider can be reduced to reasonable proportions, there are a collection of R&D issues. The next section returns to the original problem.

Before going on, though, I ought to make a couple of comments. One pertains to the necessity to use the quantum mechanical form (8) for the energy spread. The scale parameter of the synchrotron radiation spectrum is a characteristic frequency ω_c, which if expressed as an equivalent photon energy E_c is

$$E_c = \frac{3}{2} \frac{\hbar c}{\rho} \gamma^3 \qquad (12)$$

In the classical limit, E_c is much less than the particle energy. The power spectrum rises with frequency to the neighborhood of ω_c and then drops off. But is E_c is larger than the particle energy, the classical description can no longer be valid. Rather, the spectrum follows the classical case at low frequencies, but falls abruptly as the beam energy is approached.

For a particle at a radius of 2σ opposite the center of a (trigaussian) bunch through which it is passing, the ratio of the beam energy, E_b, to E_c is

$$\frac{E_b}{E_c} = \frac{(2\pi)^{1/2}}{3} \frac{\alpha}{r_c^2} \frac{\sigma\sigma_z}{N\gamma} \quad , \qquad (13)$$

from which $E_b/E_c = 10^{-4}$, far outside of the classical regime.

The other comment has to do with the absence of any luminosity enhancement from the pinch effect, in contrast to the situation with the SLC. This is a consequence of being in the quantum beamstrahlung limit, with its preference for short bunches. The distortion of one bunch during the collision with another will vary with σ_z divided by the focal length of the lens presented by the target bunch. This ratio is the disruption parameter, $D = r_c N \sigma_z / (\epsilon_N \beta^*)$. For the numerical example worked out above, $D = 0.18$, which has a negligible effect on the luminosity.

5. APPROACHES TO HIGH GRADIENT

The SLAC linac is capable of 20 MV/m. A pair of linacs of this design would occupy at least 500 km, and would be prohibitively expensive. One would like an order of magnitude or two increase in accelerating gradient.

The SLAC linac is at a disadvantage in a cost comparison because it is a representative of an existing technology with known costs. It's hard to assign costs to technologies that are not yet developed; about all I can do is suggest the likely trends of some parameters.

The accelerating frequency will be higher than the familiar 3 GHz of the S-band linac. For anything resembling an RF driven linac, the stored energy, and, in the quantum beamstrahlung limit, the average power scales as G/ω^2 for a particular beam energy. Here G is the gradient and ω the angular frequency of the guide excitation. The average power per unit length goes like G^2/ω^2. The peak power per unit length varies as $G^2/\sqrt{\omega}$. The first of these surely must be reduced, and almost surely the second.

The gradient possible in a waveguide structure does increase with frequency. Some recent measurements[12] indicate the breakdown gradient at microwave frequencies varies as $\sqrt{\omega}$. At a wavelength of about 1 mm, surface melting becomes the limiting factor, and a different dependence on frequency appears, but ignore that for the moment. The S-Band cavity measurements also revealed that gradients up to 150 MV/m were possible, but with high field emission.

If we use the $\sqrt{\omega}$ scaling for gradient, both average power and average power per unit length decrease as frequency rises. Of course, it remains to be seen whether or not the technology converts those lower powers into lower costs as well. Further, given the encouraging high gradient measurements in the SLAC structure, it would seem reasonable to assume that gradients in the 200 MV/m to 500 MV/m will be possible at a frequency an order of magnitude higher.

All of this has inspired an inventive surge among the practitioners of advanced accelerator R&D. I don't claim that I can do justice to these ideas, but let me mention a few, separated into the near, intermediate and long term.

There are three microwave amplifier tubes that appear as candidates for power sources at the 3 cm wavelength, 100 MV/m gradient level. These are the klystron, its relative, the gyroklystron, and the lasertron.[13] These tubes would be used in conjunction with a high-efficiency pulse compression scheme.[14] The resulting linac would be relatively conventional, consisting of many independent sections, with the operational advantages that go with such a design.

In the next generation are a collection of devices that resemble linacs in that the particles undergoing acceleration find themselves travelling through holes in a succession of disks. The difference lies in the means by which the electric field sensed by the particles is generated.

In the Two Beam Accelerator,[15] microwave power is generated by a series of free electron lasers traversed by an electron beam running parallel to the beam that is being accelerated. The electron beam that is used for power generation would be periodically reaccelerated. In a series of experiments performed by an LBL-LLNL collaboration,[16] a 50 kW input signal at 34.6 GHz was amplified to a power level of 1 GW. An acceleration gradient of 180 MV/m was achieved in a 5-cell structure.

Wakefield accelerators use the fields left behind by one set of particles to increase the energy of particles of another set. For example, in an arrangement currently being studied at DESY,[17] a hollow cylindrical electron beam bunch flows past the outside edges of the disks comprising the accelerating structure. The wakefield excited by this bunch propagates inward toward the center of the disks, increasing in magnitude as the energy is compressed into a progressively smaller volume. Thus far the experiments at DESY have concentrated on the formation of the cylindrical driver beam. The goal of the present work is to produce a gradient in the 100 MV/m region.

The most adventurous propositions do away with the conventional accelerating structure. For instance, the plasma beat wave accelerator[18] replaces the structure entirely, substituting a plasma in its place. The accelerating gradient is to be achieved by directing two laser beams colinearly through the plasma, with their frequency difference at the plasma frequency. This results in the generation of a compressional wave, travelling in the direction of the laser beams, but with the electric field oriented in the longitudinal rather than transverse sense. So far, the space charge wave has been detected, with gradients at the 1 GV/m level.[19]

The discussion of this section is only supposed to suggest the fascinating variety of avenues to high gradient that are being explored; there are many more proposals than the few examples mentioned here.

6. CONCLUSIONS

There is at present a strong construction program that will be concluded by the end of this decade. Superconducting magnet and RF technology has matured, and is well represented in that program.

A major construction start is necessary now if the momentum of accelerator-oriented HEP is to be maintained. The technology is in place for a multi-TeV hadron collider, suitable for the exploration of the TeV mass scale.

A vigorous accelerator R&D program is needed, if a cost-effective collider technology suitable to the 5 TeV or higher mass scale is to be in hand by the turn of the century.

ACKNOWLEDGEMENTS

I have drawn freely on the presentations of the eight speakers of the parallel session: S. Ozaki, L. Lederman, H. Schopper, B. Richter, D. Lowenstein, M. Tigner, A. Sessler, and P. Soding. I would also like to thank G. Brianti, H. Edwards, E. Keil, K. Johnsen, P. Mantsch, R. Siemann, and V. Yarba for their help.

REFERENCES

1. Nishikawa, T., Proc. 12th International Conf. on High Energy Accelerators, Fermilab, 1983, p. 143.

2. Seeman, J. T., Proc. 12th International Conf. on High Energy Accelerators, Fermilab, 1983, p. 212.

3. Dugan, G., IEEE Trans. on Nucl. Sci., Vol. NS-32, 15482 (1985).

4. Ecklund, S. D., IEEE Trans. on Nucl. Sci., Vol. NS-32, 1592 (1985). This paper, though short, includes many references to all aspects of the SLC.

5. Schopper, H., IEEE Trans. on Nucl. Sci., Vol. NS-32, 1561 (1985).

6. Wiik, B. H., IEEE Trans. on Nucl. Sci., Vol. NS-32, 1587 (1985).

7. Balbekov, V. I., et al, Proc. 12th Int. Conf. on High-Energy, Accelerators, Fermilab, 1983, p. 40.

8. Lowenstein, D., these proceedings.

9. Rees, John, SLAC-PUB-3713, June 1985.

   Lawson, J. D., CERN 85-12, August 1985.

   Himel, T. and Siegrist, J., SLAC-PUB-3572, February 1985.

   Loew, G. A., SLAC-PUB-4038, July 1986, to be published in the Proceedings of the 13th Int. Conf. on High-Energy Accelerators.

10. This expression gets the scaling right but the numerical factors depend on the particle distribution. For a proper treatment, see, for example

    Noble, R. J., SLAC-PUB-3871, January 1986, to be published in Nucl. Inst. & Methods.

11. Siemann, R. et al, Report of the HEPAP Subpanel on Advanced Accelerator R&D and the SSC, DOE/ER-0255, December 1985.

12. Wang, J. W. et al, SLAC-PUB-3940, June, 1986.

    Tanabe, E. et al, "Voltage Breakdown at X-Band and C-Band Frequencies," presented at the 1986 Linear Accelerator Converence, SLAC, June 1986.

13. Garwin, E. L., et al, IEEE Trans. on Nucl. Sci., Vol. NS-32, 2906 (1985).

14. Farkas, Z. D., et al, SLAC-PUB-3694.

15. Sessler, A. M. and Hopkins, D. B., "The Two Beam Accelerator," 1986 Linear Accelerator Conference, SLAC, June 1986.

16. Orzechowski, T. J., et al, Phys. Rev. Lett. 57, 2172 (1986).

17. Weiland, T., IEEE Trans. on Nucl. Sci., Vol. NS-32, 3471 (1985).

18. Joshi, C., et al, Nature, 311, 525 (1984).

19. Joshi, C., et al, AIP Conference Proc. No. 130, p. 99 (1985).

Spectroscopy of Light and Heavy Quarks

S. Cooper *
SLAC

New results on various controversial light mesons are reviewed, including the glueball candidates $f_2(1720)$ and $\eta(1460)$, the $1^{++}-0^{-+}$ mass "coincidences" $f_1(1285)-\eta(1275)$ and $f_1(1420)-\eta(1420)$, as well as evidence for the $X(3100)\rightarrow\Lambda\bar{p}+n\pi$ and the $\rho(1480)\rightarrow\phi\pi$, which have quantum numbers not allowed for $q\bar{q}$. The $\gamma\gamma\rightarrow VV$ effects move out of the threshold region with data on $\gamma\gamma\rightarrow\omega\rho$. Statistically weak data on $\Gamma_{\gamma\gamma}(\eta_c)$ and the search for heavy quark $^1P_1$ states are presented. Γ_{ee}, $B_{\mu\mu}$, and Γ_{tot} for the $\Upsilon(1S)$, $\Upsilon(2S)$, and $\Upsilon(3S)$ are updated using new data and a consistent treatment of the radiative corrections for Γ_{ee}. New data on the mass splittings of the $\chi_b(2P)$ compare favorably with the scalar confinement model, which may however have new trouble.

This review was supposed to cover all the spectroscopy presented in 3 parallel sessions: $\gamma\gamma$ Interactions, Heavy Bound-Quark States, and a particularly full session on Hadron Spectroscopy. It was too much, even after the usual restriction to things I found interesting and could understand, and in desperation I decided to exclude baryons and $Q\bar{q}$ mesons. I sincerely hope the next conference offers those fields better treatment (and more time!)

The light mesons part of this paper may be more than the casual reader wants to digest. There are many details, and few firm conclusions. However to simplify would be to mislead. Heavy mesons are a lot easier to deal with, and the last pages are devoted with relief to them.

1 New Meson Names

Starting this year the Particle Data Group asks us to use their new naming scheme for hadrons [1]. New and old names are compared in Table 1. Much of this talk is devoted to the ^3P_J states, which have $J^{PC} = 0^{++}, 1^{++}$, and 2^{++}. The new scheme unifies them, with a subscript to label the spin. **f** stands for **I=0** and **a** for **I=1**. If you can remember that S^* and δ were 0^{++} mesons with I=0 and 1, you know that they are now f_0 and a_0. The first step is harder than the second, demonstrating the need for new names. However progress is always accompanied by some pain. f′ and Θ were a lot easier than $f_2(1525)$ and $f_2(1720)$.

2 $K\bar{K}$ and $\eta\eta$ Resonances

2.1 Glueball Production in Hadron-Hadron Scattering

The search for glueballs has occupied many people over the last several years. At first there was hope that a candidate would be found satisfying all the criteria naively expected of a pure glueball:

- no place in a $q\bar{q}$ nonet
- flavor symmetric decay (equal coupling to u, d, s quarks)
- copious production in glue-enhanced channels like radiative J/ψ decay
- not produced in $\gamma\gamma$ scattering
- not produced in hadron scattering

Of the various candidates[1] that have turned up: the $\eta(1460)\rightarrow K\bar{K}\pi$ and $f_2(1720)\rightarrow\eta\eta,K\bar{K}$ in radiative J/ψ decay, the $f_2(2050,2300,2350)\rightarrow\phi\phi$ [2] and the $f_0(1590)\rightarrow\eta\eta,\eta\eta'$ [3] in hadronic collisions, none clearly satisfies all of the above. The lack of an ideal glueball has not shaken our faith that QCD gives states of bound glue. Instead we have begun to question how valid the above criteria are. Inspired by a contribution to this conference [4], I would like to re-examine the last of the criteria: **Do we expect to see glueballs in hadron-hadron collisions?** The standard answer is **NO! Glueballs have no quark content, and therefore don't couple to ordinary hadrons.** Hadron scattering is thought of in terms of quark line diagrams, as in

*Supported by the Department of Energy, contract DE-AC03-76SF00515.

[1] Their original names were ι, Θ, g_T, and G.

Table 1: **New Meson Names**

J^{PC}	New Names (masses in MeV, if known)				Old Names			
0^{-+}	K (494)	π (140)	η (549)	η' (958)	K	π	η	η'
$^1S_0$ ↓↑	D_s (1971)	D (1860)	η_c (2981)		F	D	η_c	
	B_s	B (5271)	η_b		B_s	B	η_b	
1^{--}	K^* (892)	ρ (770)	ω (783)	ϕ (1020)	K^*	ρ	ω	ϕ
$^3S_1$ ↑↑	D_s^* (2109)	D^* (2010)	J/ψ (3097)		F^*	D^*	J/ψ	
	B_s^*	B^* (5325)	Υ (9460)		B_s^*	B^*	Υ	
1^{+-}	K_1 (1400)	b_1 (1235)	h_1 (1190)	h_1'	Q_2	B	H	
$^1P_1$ ↓↑			h_c					
			h_b					
0^{++}	K_0^* (1350)	a_0 (980)	f_0 (1300)	f_0 (975)	κ	δ	ϵ	S^*
$^3P_0$ ↑↑			χ_{c0} (3415)				χ	
			χ_{b0} (9860)				χ_b	
1^{++}	K_1 (1280)	a_1 (1270)	f_1 (1285)	f_1 (1420)	Q_1	A_1	D	E
$^3P_1$ ↑↑			χ_{c1} (3510)				χ	
			χ_{b1} (9892)				χ_b	
2^{++}	K_2^* (1425)	a_2 (1320)	f_2 (1270)	f_2' (1525)	K^*	A_2	f^0	f'
$^3P_2$ ↑↑			χ_{c2} (3556)				χ_c	
			χ_{b2} (9913)				χ_b	

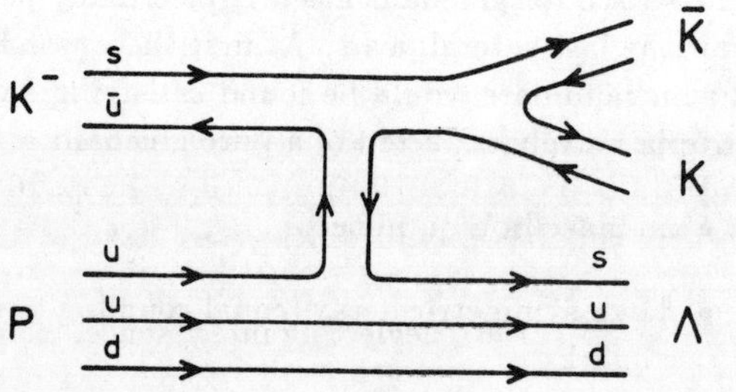

Figure 1: Quark diagram for $K^-p\rightarrow\bar{K}K\Lambda$.

Fig. 1. A glueball has no place in such a diagram, at least to the extent that α_s is small. The suppression is expected to be approximately the square root of the normal OZI suppression. It should lead to a narrow width for glueballs, since their only possible decay, to hadrons, is suppressed. However, a wide glueball has by definition a large coupling to hadrons. The strength of its decay to hadrons is related by time reversal to the strength of its formation in collisions of those hadrons.

In $K\bar{K}$ collisions we should see resonances which couple to $K\bar{K}$. The relativistic Breit-Wigner scattering formula is [5]

$$\sigma(W) = \frac{4\pi\ g}{p_i^2}\ \frac{M^2\ \Gamma_i\Gamma_f}{(W^2-M^2)^2+M^2{\Gamma_{tot}}^2} \qquad (1)$$

where p_i is the momentum of the initial particles in their center of mass frame and W is the total center of mass energy; M is the mass of the resonance, Γ_{tot} is its total width, and Γ_i and Γ_f its partial widths to the initial and final states. $g = (2J+1)/[(2s_1+1)(2s_2+1)]$ is the spin factor for a spin J resonance formed by spin s_1 and s_2 initial state particles.

To calculate the total production rate of a resonance in $K\bar{K}$ collisions we sum over the final states $\sum\Gamma_f = \Gamma_{tot}$, and integrate over W to obtain (for a narrow resonance)

$$\sigma_R = \int \sigma(W)\ dW = \frac{2g\pi^2}{p_K^2}\ \Gamma_{KK}. \qquad (2)$$

Thus the production rate of a resonance is determined by its partial width to the initial state, entirely independent of whether the resonance is $q\bar{q}$ or gg. Now we can correctly answer the question above with **Yes, we expect a *wide* glueball to be produced in hadronic collisions, at least when the input channel corresponds to one of its dominant decays.**

2.2 Properties of the $f_2(1720)$

Consider two resonances with the same quantum numbers which couple strongly to $K\bar{K}$: the $f_2(1525)$ and the $f_2(1720)$. You know these as the $f'(1525)$ $s\bar{s}$ resonance and the $\Theta(1720)$ glueball candidate seen in radiative J/ψ decays. The $f_2(1525)$ decays dominantly to $K\bar{K}$. A predicted [17] 13% branching ratio to $\eta\eta$ may have been observed [3]. Using Ref. [17] let us assume $\Gamma_{K\bar{K}}^{1525} = 0.8 \times \Gamma_{tot}^{1525} = 0.8 \times (70 \pm 10)$ MeV. The observed $f_2(1720)$ branching ratios are listed in Table 2. $K\bar{K}$ is ~70% of the total observed, giving $\Gamma_{K\bar{K}}^{1720} \sim 0.7 \times 147 = 100$ MeV, less if

Table 2: **Properties of the $f_2(1720)$ in Radiative J/ψ Decay.**

		Crystal Ball [6]	Mark II[7]	Mark III [8]	DM2 [12]	average
Number of Events		39±11	~50	192±25	410	
Mass in MeV		[1720]	1700±30	1720±10±10	1707±10	1711±8
Width in MeV		[130]	156±20	130±20	166±33	147±13
$10^4\times$ B $J/\psi\to\gamma f_2$, $f_2\to$	$\eta\eta$	2.6±0.8±0.7				2.6±1.1
	$K\bar{K}$		12±2±5	9.6±1.2±1.8	9.2±1.4±1.4	9.6±1.4
	$\pi\pi$	2.3±0.7±0.8	<3.2	2.4±0.6±0.5	1.8±0.3±0.3	2.0±0.4
	$\rho\rho$			<5.5 [9]		
	$\omega\omega$			<2.4 [10]		
	K^*K^*			<4.5 [11]		
	$\eta\eta'$			<2.1 [11]		
	$K\bar{K}\pi$			<2.8 [11]		

In the decay $J/\psi\to\gamma X$, I=0 is prefered for the state X, and 0^{++},0^{-+}, and 2^{++} are expected to dominate. For I=0 and X a meson anti-meson pair, $J^{PC}(X)$ = $(\text{even})^{++}$. The angular distributions of the 1720 prefer 2^{++} over 0^{++} [13,7,8,12], although sometimes not by as much as one would like [13,12,14]. This might be due to some production of the $f_0(1730)$ [4,15,1]. The branching ratios assume I=0, C=+, thus $K\bar{K} = \frac{1}{2}K^+K^- + \frac{1}{4}K_sK_s + \frac{1}{4}K_LK_L$ and $\pi\pi = \frac{2}{3}\pi^+\pi^- + \frac{1}{3}\pi^0\pi^0$. Upper limits are 90% confidence level. The average mass and width use K^+K^- data only. The K^+K^- results are all from fits to the $f_2(1525)$ + $f_2(1720)$, neglecting interference.

The $f_2(1720)$ was discovered [13] in $J/\psi\to\gamma\eta\eta$ with M = 1640±50 MeV, $\Gamma = 220^{+100}_{-70}$ MeV, and $B(J/\psi\to\gamma 1640, 1640\to\eta\eta) = (4.9\pm1.4)\times10^{-4}$. That is a lower mass and larger width than later seen in the higher statistics $\gamma K\bar{K}$ measurements. The $\gamma\eta\eta$ signal may be contaminated by the $f_2(1525)$, which is 2^{++} but with unknown rate to $\eta\eta$ [3], or by the $f_0(1590)$, which does decay to $\eta\eta$ but is 0^{++} [3]. Thus we should probably regard the original branching ratio as an upper limit for the $f_2(1720)\to\eta\eta$. Indeed a fit including the $f_2(1525)$ gave M = 1670±50, Γ = 160±80, and $B(J/\psi\to\gamma1670, 1670\to\eta\eta) = (3.8\pm1.6)\times10^{-4}$ [16]. The branching ratio in the table is from a third fit where the masses and widths were fixed at the Mark III values.

The $\pi\pi$ branching ratios are from fits to $f_2(1270)$ + $f_2(1720)$ + X(~2100), neglecting interference, which can reduce the $f_2(1720)$ branching ratio by at least a factor of 2 [11]. Only the $f_2(1270)$ spin has been determined in $J/\psi\to\gamma\pi\pi$, so the identification of the 1720 peak with the $f_2(1720)$ is tentative, and may be contradicted by $\pi\pi\to K\bar{K}$ data [4], as discussed in the text.

there are other decays not seen yet. Thus $K\bar{K}$ scattering should produce one of our favorite glueball candidates about as strongly as it does one of our better understood $q\bar{q}$ mesons. The equality of the quantum numbers and near equality of the masses reduces most kinematic effects. For example, the $1/p_K^2$ in the B-W gives a suppression of only 0.7 for $f_2(1720)/f_2(1525)$.

Although we expect to produce about the same number of $f_2(1525)$ and $f_2(1720)$ mesons, if we require the final state to be $K\bar{K}$, they will be seen in the ratio of their $K\bar{K}$ branching ratios. Furthermore if a fit has not been done to the spectrum, we have to rely on our eyes, which are more sensitive to peak height than to area. The peak height scales as $1/\Gamma_{tot}$ for equal area. Including the $1/p_K^2$ factors from the Breit-Wigner scattering formula, we find that the peak height of each f_2 is proportional to $B_{K\bar{K}}^2/p_K^2$. This gives a ratio $f_2(1720)/f_2(1525)$ ~1/2 (with a large error due to the poorly known branching ratios.)

Lacking a $K\bar{K}$ collider, we turn to the reaction of Fig. 1, $K^-p\to K\bar{K}\Lambda$, since the upper vertex looks like $K\bar{K}$ scattering. The K_sK_s final state can only be 0^{++}, 2^{++}, etc., so it is preferable to K^+K^- for the study of 2^{++} mesons. The new LASS data [18] is shown in Fig. 2. The $f_2(1525)$ is prominent in the spectrum, but the $f_2(1720)$ is in one of the lowest bins around. We would like to use this result to gain quantitative information on the f_2 branching ratios.

In the K exchange picture we have assumed, there is additional mass dependence of the effective virtual kaon flux which must be calculated correctly, and tested against data on known mesons. The Watson theorem [19] gives the total production of a resonance R in $K^-p\to R\Lambda$ as [20]

$$\sigma_R = \frac{2\pi M_R}{(p_K)^{2J+1}}\Gamma_{K\bar{K}} \,. \qquad (3)$$

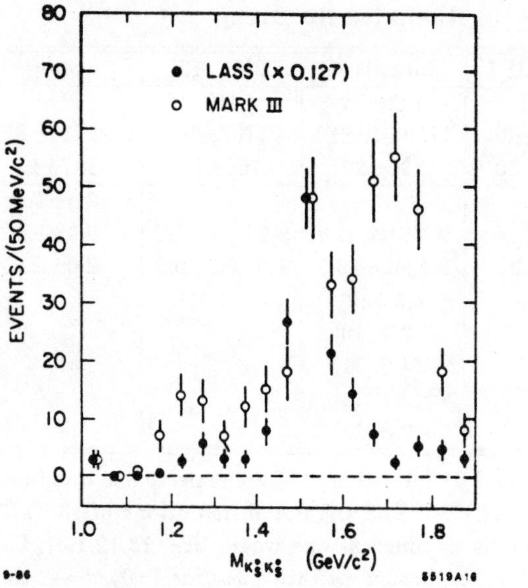

Figure 2: K_sK_s mass distributions [18] from (open circles) Mark III uncorrected $J/\psi \rightarrow \gamma K_s K_s$ data and (solid circles) LASS acceptance corrected $K^- p \rightarrow K_s K_s \Lambda$. The LASS data have been multiplied by 0.127 to match the Mark III data in the 1525 MeV bin. LASS used a 11 GeV/c K^- beam, and for this plot required $|t'| < 2\ GeV^2$, where $t' \equiv t - t_{min}$.

This leads to a suppression from the p_K factors of 0.4 rather than 0.7. I don't understand the origin of Eq. (3), but it gives the strongest suppression that has been suggested to me, so I quote it to show that even the extreme case is not very dramatic.

More critical is the validity of the K exchange picture itself. The analogous π exchange mechanism has been well tested, but K exchange has not. Data comparing production of the $u\bar{u}+d\bar{d}$ $f_2(1270)$ with the $s\bar{s}$ $f_2(1525)$ in $\pi^- p \rightarrow f_2 n$ and $K^- p \rightarrow f_2 \Lambda$ indicate a factor of two problem in the K exchange picture [20,21]. In contrast to the pion, the kaon may not be light enough to dominate over other strange meson exchanges. Fits to t-distributions, and polarised target data, may help to separate the various contributions [20].

If the strangeness exchange process can be understood, it could provide useful information on the absolute $f_2(1720)$ decay branching ratios. That, combined with the data in Table 2, would give the total rate for $J/\psi \rightarrow \gamma f_2(1720)$. The more the $f_2(1720)$ is produced in that glue-enriched channel, the more glueball character we are willing to attribute to it. In principle the total $J/\psi \rightarrow \gamma f_2(1720)$ rate could be measured from the inclusive γ spectrum in $J/\psi \rightarrow \gamma + X$ shown in Fig. 3. It would appear at $E_\gamma \sim 1070$ MeV with a full width of ~ 70

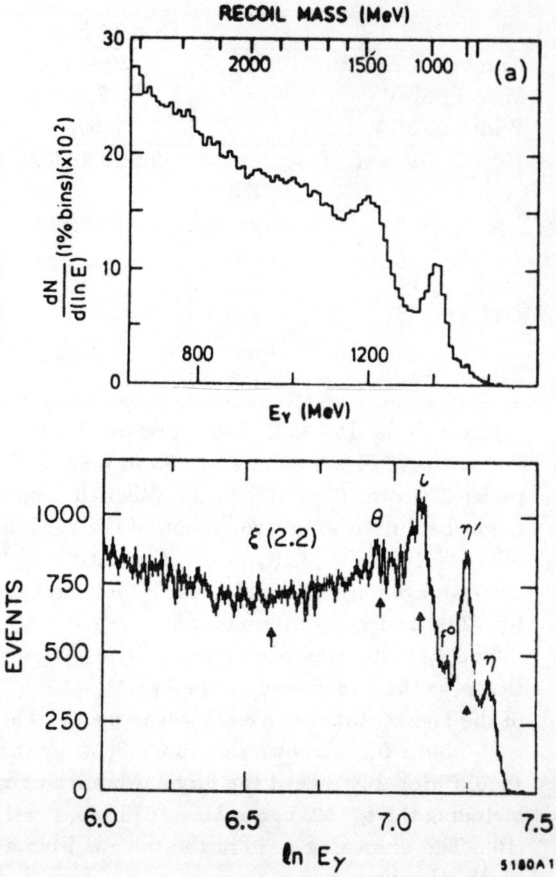

Figure 3: Crystal Ball inclusive γ spectra, without efficiency correction or background subtraction, from **(a)** $J/\psi \rightarrow \gamma$+hadrons [22], and **(b)** $J/\psi \rightarrow \gamma + X$ [23].

MeV. However here no spin-parity analysis is possible, so that it would be difficult to cleanly isolate the $f_2(1720)$ signal from all the other radiative J/ψ decays. Knowledge of the γ background from π^0 decays might help.

While one-kaon exchange may have its problems, one-pion exchange (OPE) is well established. One could therefore use the $J/\psi \rightarrow \gamma K\bar{K}$ data to determine the mass and Γ_{tot} of the $f_2(1720)$, and $\pi\pi \rightarrow \pi\pi$ to get $\Gamma^2_{\pi\pi}$, and thus $B(f_2(1720) \rightarrow \pi\pi)$. $\pi\pi \rightarrow K\bar{K}$ similarly yields $\Gamma_{\pi\pi}\Gamma_{K\bar{K}}$.

Longacre *et al.* [4] have done a partial wave analysis to extract the 2^{++} contribution from their new $\pi^- p \rightarrow K_s K_s n$ data (Fig. 4). Using OPE they extrapolate to the pion pole to get the $\pi\pi \rightarrow K\bar{K}$ 2^{++} intensity. They then make a simultaneous fit to that and to older 2^{++} $\pi\pi \rightarrow \pi\pi, \eta\eta$, and (assumed) $K\bar{K} \rightarrow K\bar{K}$ data, as well as to the total (not

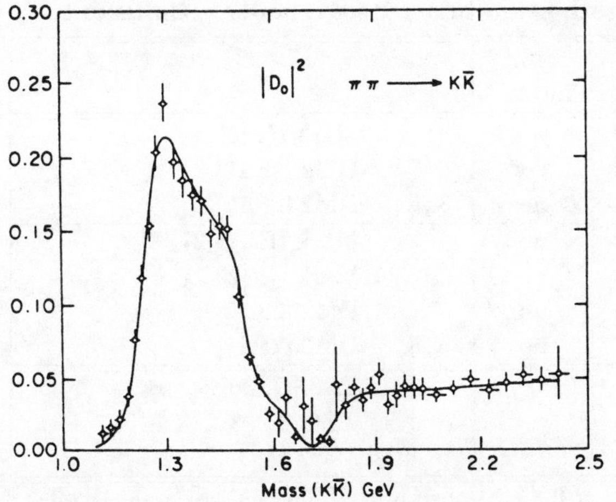

Figure 4: $\pi\pi \to K_sK_s$ 2^{++} intensity extrapolated using OPE from 22 GeV/c $\pi^- p \to K_sK_sn$ data with $|t'| < 0.1$ GeV^2 [4]. The curve is the fit described in the text. It has destructive interference between the $f_2(1525)$ and $f_2(1720)$ to give the maximum $f_2(1720) \to \pi\pi$ branching ratio.

spin-separated) $J/\psi \to \gamma K\bar{K}, \gamma\pi\pi, \gamma\eta\eta$. The hadronic scattering data have no need for the $f_2(1720)$, and an upper limit of 4% for the branching ratio $f_2(1720) \to \pi\pi$ is obtained.[2] This is in contradiction to the $\pi\pi$ fraction of the observed branching ratios in Table 2, which were however obtained without spin analysis and interference. Allowing interference can reduce $\pi\pi$ to $(8\pm3)\%$ of the observed decays. A glueball candidate which decays dominantly to $K\bar{K}$ and <4% to $\pi\pi$ isn't doing very well at passing the test of flavor symmetric decay. It would be interesting to repeat this fit with the newer $\pi\pi \to \eta\eta$ [3] and $K\bar{K} \to K\bar{K}$ [18] data, as well as spin-analysed radiative J/ψ decay data, should that become available. It may also be possible to use double-Pomeron scattering data [24], as done in Ref. [25] for the 0^{++} system.

DM2 [28,12] and Mark III [29,26,27] have compared tensor resonance production in $J/\psi \to \gamma X$ with $J/\psi \to \omega X$ and $J/\psi \to \phi X$. The first channel presumably goes via γgg and therefore enhances glueball production, while the presence of the ω and ϕ in the last two select $u\bar{u} \pm d\bar{d}$ and $s\bar{s}$ mesons respectively (see Fig. 5). They look for the $f_2(1720)$ in events where X is $K\bar{K}$ or $\pi\pi$. The Mark III spectra with $X = K^+K^-$ are shown in Fig. 6. The $f_2(1720)$ is seen clearly in the $J/\psi \to \gamma K^+K^-$ spectrum, where it has been determined to be 2^{++}. The ωK^+K^- spectrum shows a peak with the same mass and width, but no spin analysis has been done. Since

[2]The limit depends on the assumption of K exchange.

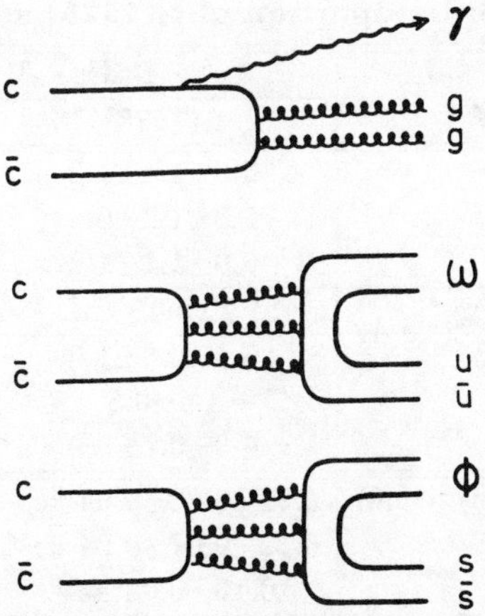

Figure 5: Quark diagrams for $J/\psi \to \gamma X$, $J/\psi \to \omega X$, and $J/\psi \to \phi X$.

there is evidence [1,4,15] for a 0^{++} $K\bar{K}$ resonance $f_0(1730)$, it is not safe to assume that the peak in $J/\psi \to \omega K^+K^-$ is the $f_2(1720)$. Let us call it X. Then $B(J/\psi \to \gamma f_2, f_2 \to K\bar{K}) \approx 2 \times B(J/\psi \to \omega X, X \to K\bar{K})$. The values are given in Table 3. It is rather peculiar that X is produced so strongly with ω, which should suppress $s\bar{s}$ resonances, yet decays to $K\bar{K}$. The $J/\psi \to \phi K^+K^-$ spectrum is dominated by the $f_2(1525)$ (expected here since it is $s\bar{s}$) with a shoulder on its high mass side. In looking for the $f_2(1720)$ here, not only do we have to worry about interference with the $f_2(1525)$, but also about the apparent (but low-statistics) disagreement between the K^+K^- and K_sK_s spectra[3] (see Fig. 7). If the same X is being seen in the DM2 ϕK_sK_s and the Mark III ωK^+K^- spectra, it is being produced with ϕ and ω at about the same rate. Using ϕK^+K^- gives considerably less. These data and those with $X = \pi\pi$, especially when spin-analysed, will provide a wealth of information.

2.3 f_2 States Near 2.2 GeV

There is quite a bit of action in the 2^{++} system above 2 GeV. There are three wide 2^{++} $\phi\phi$ resonances originally named[4] g_T:

g_T	$M = 2050^{+90}_{-50}$	$\Gamma = 200^{+160}_{-50}$	MeV
g_T	$M = 2300^{+20}_{-100}$	$\Gamma = 200^{+60}_{-50}$	MeV
g_T	$M = 2350^{+20}_{-30}$	$\Gamma = 270^{+270}_{-130}$	MeV

seen in $\pi^- p \to \phi\phi n$ [2]. The $g_T(2050)$ is the dom-

[3]In $J/\psi \to \phi X$, isospin conservation requires that X have I=0, and therefore $J^{PC} = 0^{++}, 2^{++}$, etc. for both K^+K^- and K_sK_s. However the K_sK_s spectrum is nearly background free, while K^+K^- is not.

[4]They should be called f_2. I use g_T here as a generic name for all three $\phi\phi$ resonances, and to avoid hopelessly confusing them with Mark III's $f_2(2230)$.

Table 3: **Comparison of $f_2(1525)$ and $f_2(1720)$ production in J/ψ decays and $\gamma\gamma$ Collisions.**

V	$10^4 \times B(J/\psi\to Vf_2, f_2\to K\bar{K})$		Spin		
	$f_2(1525)$	$f_2(1720)$	known?		
γ	6.0±1.4±1.2	9.6±1.2±1.8	yes		Mark III [8]
	5.0±1.2±0.8	9.2±1.4±1.4	yes		DM2 [28,12]
	6.0±1.2±1.2	12.4±2.8±2.4	no	K_sK_s	DM2 [28,12]
ω	<1.2 (90% C.L.)	4.5±1.2±1.0	no		Mark III [29,27]
ϕ	6.4±0.6±1.6	$\sim 1.4^\dagger$	no		Mark III [29,27]
	4.6±0.5		no		DM2 [28,12]
	4.3±0.7±0.9	3.6±0.7±0.7	no	K_sK_s	DM2 [28,12]
$\Gamma_{\gamma\gamma} \times B(f_2\to K\bar{K})$	0.11±0.02±0.04 keV	<0.3 keV			TASSO [30]
	0.12±0.07±0.04 keV	<0.2 keV			TPC/2γ [31]
	0.10±0.04 keV			(prel.)	Mark II [32]

Branching ratios have been converted to $K\bar{K}$ assuming $I = 0, C = + :\ K\bar{K} = \frac{1}{2}K^+K^- + \frac{1}{4}K_sK_s + \frac{1}{4}K_LK_L$. They were measured in K^+K^- unless indicated otherwise.

† Mark III and DM2 quote $B(J/\psi\to\phi f_2(1720))$ from fits to K^+K^- without $f_2(1720)$-$f_2(1525)$ interference. Their $f_2(1720)$ masses come out 1671 and 1643 MeV, respectively, which I find too far off to be meaningful. Interference moves the mass up some, but it stays below 1700. Mark III [27] say they can accomodate a "standard" $f_2(1720)$ with interference, for which B goes down a factor of ~2.5. From that I derive the ~1.4.

inant one, and its $\phi\phi$ decay is mostly L=0. The others are mostly L=2. The g_T are glueball candidates because production of $q\bar{q}$ resonances is suppressed by OZI in $\pi^-p\to\phi\phi n$. Three 2^{++} glueballs with similar masses can be described by theory, e.g. a strong coupling calculation [33]. However so far these states have not taught us much about glueballs, since they have only been seen in this production and this decay. For a new evaluation of the g_T's and the rest of the f_2 system see Ref. [34].

DM2 [12] and Mark III [29] have some $J/\psi\to\gamma\phi\phi$ events. Mark III [35] also have $J/\psi\to\gamma\phi\omega$. Further efforts at increasing the ϕ reconstruction efficiency may yield enough events to compare them to the g_T resonances and to Mark III's $f_2(2230)$.

If the matrix element for the g_T decay to $K\bar{K}$ were the same as that for $\phi\phi$, the p^{2L+1} phase space factor would give ~30 times more $K\bar{K}$ than $\phi\phi$ for the L=2 $g_T(2300)$, although this factor gets much smaller if damped by an interaction radius [20]. The measured [4] D_0 waves in $\pi^-p\to K\bar{K}n$ and $\pi^-p\to\phi\phi n$ are in the ratio ~20:1 and the $K\bar{K}$ is flat above 1.9 GeV. Very preliminary MIS ITEP data [15] on the 2^{++} K_sK_s in 40 GeV/c $\pi^-p\to K_sK_sn$ look different, going to zero at 2 GeV and rising again to a ~70 Mev wide peak at ~2230 MeV. However the region above 1.9 GeV is subject to ambiguities in the PWA, so that comparisons between experiments cannot be made before seeing all the waves and their phases [20,36].

GAMS have seen a narrow peak (Γ<80 MeV) at 2220±10 MeV in their $\eta\eta'$ spectrum from 38 and 100 GeV/c $\pi^-p\to\eta\eta'n$ data [37]. The anisotropy of the decay angular distribution suggests $J \geq 2$. If GAMS and MIS ITEP are seeing the same state, its branching ratio to $\eta\eta'$ is at least twice as large as to $K\bar{K}$ [15]. This would be evidence against its interpretation as an L=3 $s\bar{s}$ state, an explanation put forward [38] for Mark III's $f_2(2230)$.

The K_sK_s spectrum from the LASS $K^-p\to K\bar{K}\Lambda$ data shows evidence for a peak at ~2.2 GeV with spin ≥ 2, and the moments of their K^+K^- indicate a $J \geq 4$ state at 2193±25 GeV with a width of 83±101 GeV [18].

The above can be compared to the $f_2(2230)$ a narrow 2^{++} $K\bar{K}$ resonance[5] seen by Mark III [39] – but not by DM2 [28,12] – in $J/\psi\to\gamma K\bar{K}$. The discrepancy was not settled at this conference. Mark III [14] have shown that its spin is more likely ≥ 2 than 0. (Only J^{PC} = (even)$^{++}$ is allowed for $J/\psi\to\gamma K\bar{K}$.)

A search for the $f_2(2230)$ in $p\bar{p}$ annihilation gave a limit on $B(f_2\to p\bar{p}) \times B(f_2\to K\bar{K})$ ranging from $<2\times10^{-4}$ if its width is 35 MeV to twice that for a width of 7 MeV [40].

Both Mark III and DM2 see a wide (100-300 GeV) enhancement at 2100-2200 MeV in their K_sK_s data, which is nearly background-free. This wide resonance also seems to have $J \geq 2$ [28,14]. The $\pi\pi$ mass distributions in radiative J/ψ decay also have a wide enhancement at ~2100 MeV. This last is consistent in mass and width with the 4^{++} $f_4(2030)$, but such high spin resonances are not expected in radiative J/ψ decays.

[5] originally called ξ

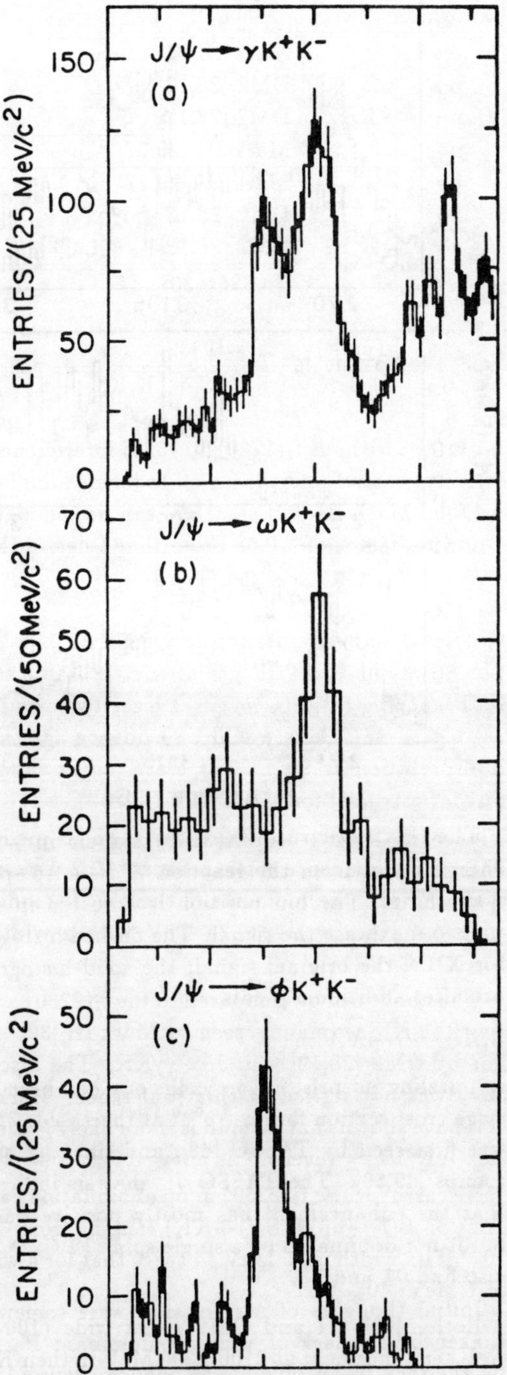

Figure 6: Mark III [27] K^+K^- mass spectra from (a) $J/\psi\to\gamma K^+K^-$, (b) $J/\psi\to\omega K^+K^-$, (c) $J/\psi\to\phi K^+K^-$.

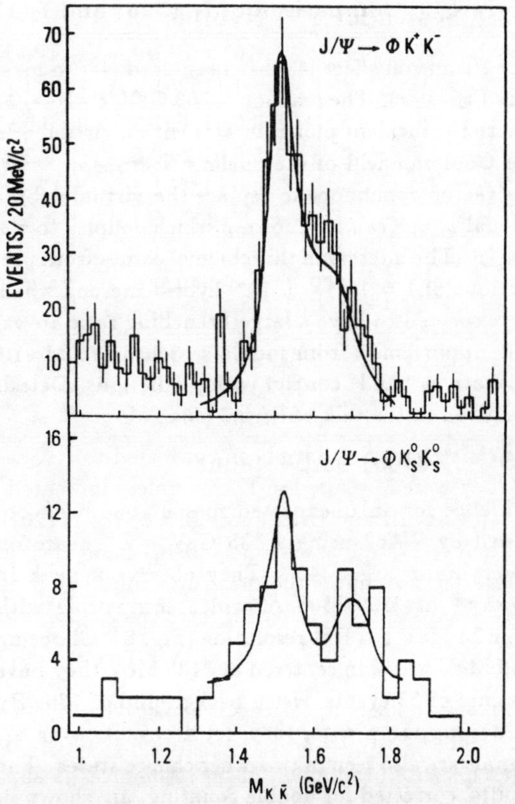

Figure 7: DM2 [28] $K\bar{K}$ mass spectra from (a) $J/\psi\to\phi K^+K^-$ and (b) $J/\psi\to\phi K_s K_s$.

2.4 $f_0(1590)$

A discussion of the GAMS $f_0(1590)$ glueball candidate[6] was given by Obraztsov [15,3,41]. Its decay rates are in the proportion $\eta\eta'$:$\eta\eta$:$K\bar{K}$:$\pi\pi$= ~3:1:<1:<1. Its t distribution in $\pi^-p\to\eta\eta n$ is in good agreement with dominant one-pion exchange, which can be used to show that the $f_0(1590)$ branching ratio to $\pi\pi$ must be between 3.6 and 6%.

3 Mystery Mesons

3.1 $\rho(1480)\to\phi\pi$

Lepton-F [15,42] have further studied their $\rho(1480)\to\phi\pi^0$ resonance[7]. It has M = 1480 ± 40 MeV, $\Gamma = 130 \pm 60$ MeV, $J^{PC}=1^{--}$, and has been seen in 32 GeV $\pi^-p\to\phi\pi^0 n$ with $|t'| < 0.2$ GeV^2. An I=1 $q\bar{q}$ resonance should decay much more readily to $\omega\pi^0$ than to $\phi\pi^0$, but comparing with GAMS data they find $B(1480\to\omega\pi^0) < 2\ B(1480\to\phi\pi^0)$. This resonance is a good candidate for a 4-quark or a $q\bar{q}g$ state.

[6] originally called G
[7] originally called C

3.2 Search for 1^{-+} Hybrids

The Primakoff effect [43] has been used [44] to measure $\Gamma(\rho \rightarrow \gamma\pi)$. The reaction is 200 GeV/c $\pi N \rightarrow \rho N$, where the incident pion interacts with a virtual γ in the Coulomb field of the nucleus N: $\pi\gamma \rightarrow \rho$. Using vector dominance to replace the virtual γ by a virtual ρ, one can look for a $\tilde{\rho}$ which couples to $\pi\rho$: $\pi\rho \rightarrow \tilde{\rho}$. The interest in this channel comes from predictions of $I = 1, J^{PC} = 1^{-+}$ hybrid mesons which are expected to have a large branching ratio to $\rho\pi$. The upper limits from such a search up to M$\sim$1.5 GeV are in "mild" conflict with predictions. Details are given in the talk of Ferbel [45].

3.3 X(3100)$\rightarrow \Lambda\bar{p}$+ pions

Evidence for an unexpected meson[8] has been obtained by WA62 using a 135 GeV/c Σ^- beam on a beryllium target [46]. They observe a peak in $\Lambda\bar{p}\pi^+\pi^+$ at 3100 MeV of width compatible with their 24 MeV FWHM resolution (Fig. 8). Choosing a 30 MeV wide bin centered at 3105 MeV they have a signal of 53 events over a background of 136. By tightening the $\bar{p}$ cuts, they get 46 events over 52. Signals are also seen in two other charge states. The results, corrected for double counting, are shown in Table 4, along with the Monte Carlo calculations of the probability that they are statistical fluctuations. These calculations take into account choosing the bin position to maximise the signal, but cannot correct for the effect of choosing the $\bar{p}$ cuts.

Now BIS-2 [47] also report seeing the X(3100), in their case centered at 3060$\pm$40 MeV. They use a 40 GeV neutral beam composed mainly of neutrons incident on H_2, C, Al, and Cu targets, and have so far analysed 1/4 of their data. They see their strongest signal in $X^0 \rightarrow \Lambda\bar{p}\pi^+$, where WA62 sees none, and nothing significant in $X^+ \rightarrow \Lambda\bar{p}\pi^+\pi^+$, WA62's best channel. This difference is puzzling, at best. The BIS-2 signals, after $\bar{p}$ and π^+ cuts, are shown in Table 4 and Fig. 9.

If the X^+ (seen only by WA62) decays strongly to $\Lambda\bar{p}\pi^+\pi^+$, it is a four quark state ($su\bar{d}\bar{d}$), and its narrowness is striking. If not, the absence of a strong decay must be explained.

When the WA62 result first became known, it was assumed that its strange dibaryon composition meant a hyperon beam was needed to produce it. Now that BIS-2 have seen it with a neutron beam, there are presumably many other experiments which can see this object (if it is real) and give more information about its nature.

[8] They named it the U(3100), but the PDG say unknown particles should be called X.

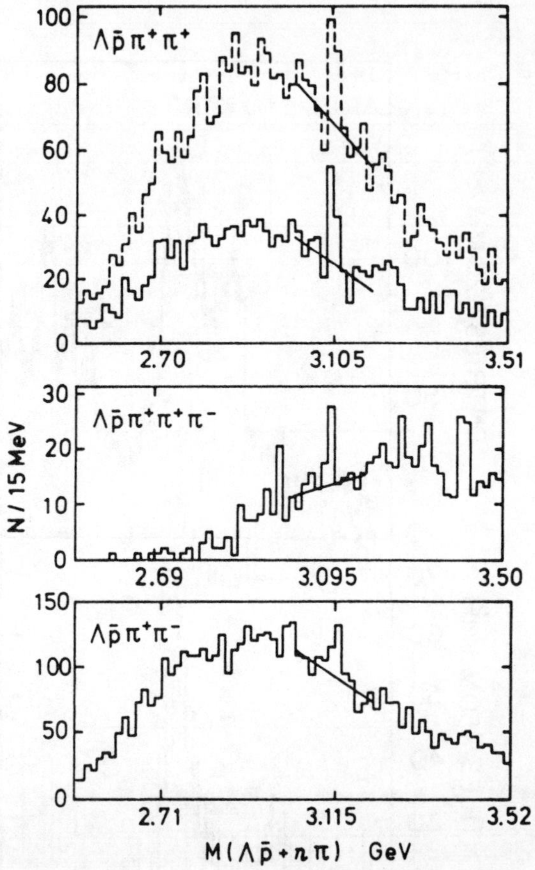

Figure 8: WA62 [46] X(3100) signals in +,0,− charge states from the reaction $\Sigma^- N \rightarrow \Lambda\bar{p}$+pions +anything. The bin position is adjusted in each plot to maximise the signal. The dashed histogram for X^+ is the original signal; the solid histograms are after additional $\bar{p}$ cuts.

3.4 $\gamma\gamma \rightarrow VV$

A puzzling surprise in $\gamma\gamma$ collisions has been the large cross section for $\gamma\gamma \rightarrow \rho^0\rho^0$ at threshold, which was first seen by TASSO [48], and since by other groups [49,50]. The TASSO J^P analysis indicated that the enhancement has mostly positive parity, but is not dominated by a single spin. TPC/2γ [50] also find 0^+ and 2^+.

Initial thoughts of a resonance were somewhat shaken by the lack of a clearly dominant J^P, and by the lack of an obvious resonance shape. (The continuation of the cross section below $\rho\rho$ threshold implies a strongly rising matrix element.) One naively wonders if this is not some kind of threshold effect, perhaps having something to do with the fact that ρ^0's are broad, and couple directly to γ's in the Vector Dominance Model.

The analyses are made difficult by the width of the ρ, the necessity to accomodate Bose-Einstein

Table 4: **X(3100) Signals**

Charge State	Decay Mode	WA62 [46]				BIS-2 [47]		
		Bin Center [MeV]	Signal Events	Background Events	Statistical Probability	Mass [MeV]	Signal Events	Background Events
X^+	$\Lambda\bar{p}\pi^+\pi^+$	3105	45	50	6×10^{-5}	~3060	10	16
X^0	$\Lambda\bar{p}\pi^+\pi^+\pi^-$	3095	19	28	0.2	~3060	10	7
	$\Lambda\bar{p}\pi^+$		<20	~115		~3050	73	168
X^-	$\Lambda\bar{p}\pi^+\pi^-$	3105	62	187	9×10^{-3}	~3070	32	49

BIS-2 say all their peaks are near 3060; the masses here are my own best guess from the histograms. They do not consider their X^+ signal significant. The WA62 uncertainty in the mass scale is ±20 MeV; that of BIS-2 is 40 MeV. No X^{++} or X^{--} has been seen.

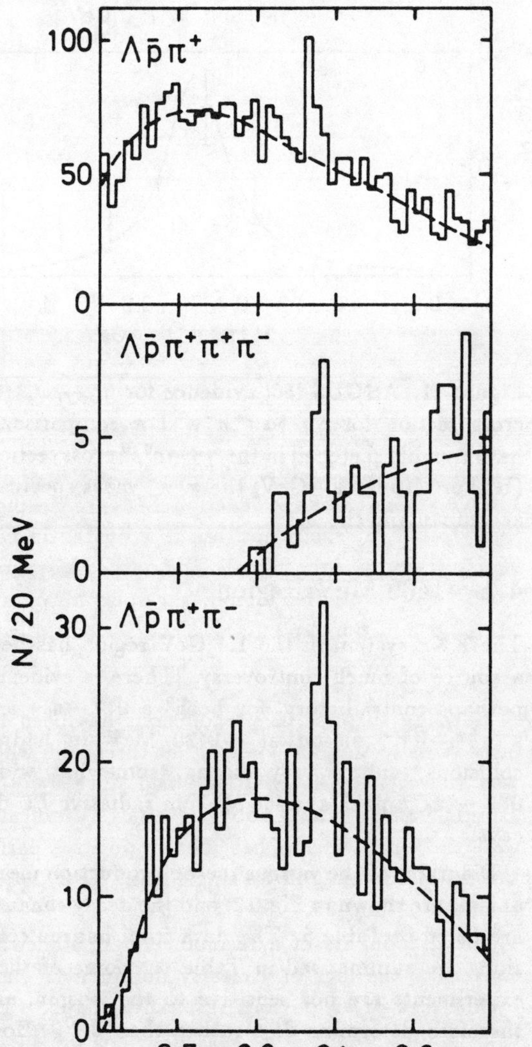

Figure 9: BIS-2 [47] X(3100) signals in 0,0,− charge states from $nN\to\Lambda\bar{p}$+pions+anything.

statistics among the 4 π's, and the presence of $\rho^0\pi\pi$ and 4π backgrounds to the $\rho^0\rho^0$. Interference between the various channels was ignored. It has been pointed out [51] that the $\rho^0\pi\pi$ channel must have the $\pi\pi$ in an L=odd state. This effect, which changes the shape of the $\rho^0\pi\pi$ background and also increases the chances that it should interfere with $\rho^0\rho^0$, was not included in the TASSO analysis [48]. Until its effect is studied, we must admit the possibility that a correct analysis might show a dominant J^P after all.

If the $\rho^0\rho^0$ enhancement is due to a resonance, some values of isospin can be ruled out. The lack of a similar enhancement in $\rho^+\rho^-$ [52] rules out an I=0 resonance as the explanation of the $\rho^0\rho^0$ signal. A neutral I=1 resonance is conveniently forbidden by its Clebsch-Gordan coefficient to decay to $\rho^0\rho^0$. I=2 and I=0 resonances interfering were predicted to cause a large $\rho^0\rho^0/\rho^+\rho^-$ ratio [51] and I=2 occurs naturally among $qq\bar{q}\bar{q}$ states. An alternate explanation [53] relating $\gamma\gamma\to VV$ to photon production of vector mesons and to nucleon-nucleon scattering, assuming t-channel exchange and factorisation, has been revised [54] to accomodate later $\gamma\gamma$ data.

For more information one looks to other $\gamma\gamma\to VV$ reactions. In $\gamma\gamma\to K^+K^-\pi^+\pi^-$, single K^*'s and ϕ's are seen [55]. ARGUS [56] have now reported a smooth $K^{*0}\bar{K}^{*0}$ cross section at the level of ~25% of the total $K^+K^-\pi^+\pi^-$ (Fig. 10).

More exciting is the new ARGUS [56] evidence for $\gamma\gamma\to\rho\omega$. The signal is among their $\gamma\gamma\to\pi^+\pi^-\pi^+\pi^-\pi^0$ events, for which they estimate a ~12% background from other processes. They extract the number of $\omega\to\pi^+\pi^-\pi^0$ decays as a function of total mass $M(5\pi)$, obtaining the cross section for $\gamma\gamma\to\omega\pi^+\pi^-$ shown in Fig. 11a. For events with $M(5\pi)>1.7$ GeV the mass distribution of the $\pi^+\pi^-$ not included in the ω is shown in Fig. 11b. A clear ρ is seen with no noticable background. The ω-sidebands show no such ρ signal. Thus above 1.7

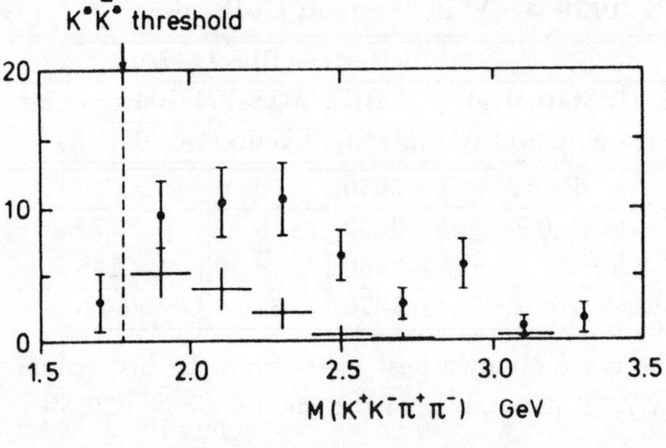

Figure 10: **Preliminary ARGUS [56] cross section for $\gamma\gamma\to K^+K^-\pi^+\pi^-$ with $\phi\pi^+\pi^-$ removed (•), and $\gamma\gamma\to K^{*0}\bar{K}^{*0}$ (+).**

GeV the $\omega\pi^+\pi^-$ cross section is dominantly $\omega\rho$.

What is intriguing here is that the $\rho\omega$ does not peak at threshold, but considerably above it. Its shape even resembles a resonance, unlike the $\rho^0\rho^0$.

Neither of the above models predicted this $\rho\omega$ behaviour; it will be interesting to see if they can accomodate it. A successful description should also deal with (or explain why not) other production mechanisms for VV, such as radiative J/ψ decay where there is new data on $\omega\phi$ [35] and $\rho^0\rho^0$ [28], central production in π^+p, where $\phi\phi$ and K^*K^* have been reported [57], and the $\phi\phi$ resonances seen in π^-p [2]. There is also evidence from $\bar{p}$ annihilation in deuterium for a $\rho\rho$ resonance at 1485 MeV [58], which has been compared to the $\gamma\gamma\to\rho^0\rho^0$ enhancement [59]. These VV reactions may not *all* be related, but it would be unaesthetic to propose a different explanation for *each*.

4 $K\bar{K}\pi$ and $\eta\pi\pi$ Resonances

Table 5: **Isobars for $K\bar{K}\pi$ and $\eta\pi\pi$.**

	G*	I	J^P				
			0^-	1^+	1^-	2^+	2^-
$K\bar{K}\pi$:							
$K^*_{ch}K + K^*_{neut}K$	+	0,1	P	S,D	P	D	P
$K^*_{ch}K - K^*_{neut}K$	−	0,1	P	S,D	P	D	P
$K\bar{K}\pi$ and $\eta\pi\pi$:							
$a_0(980)\pi$	+	0,1,2	S	P	-	-	D
$\eta\pi\pi$:							
$f_0\eta$	+	0	S	P	-	-	D
$\rho\eta$	+	1	P	S	P	D	P

* $C = G(-1)^I$

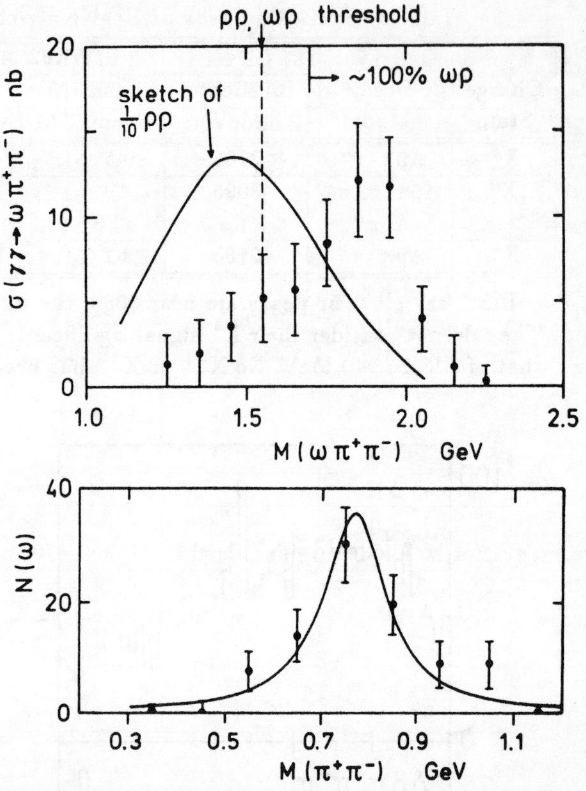

Figure 11: **ARGUS [56] evidence for $\gamma\gamma\to\rho\omega$. (a) cross section for $\gamma\gamma\to\omega\pi^+\pi^-$. For comparison I have roughly sketched in the $\gamma\gamma\to\rho^0\rho^0$ cross section. (b) For M(5π)>1.7 GeV, the $\pi^+\pi^-$ mass spectrum accompanying the ω.**

4.1 1400 MeV Region

The $K\bar{K}\pi$ system in the 1.4 GeV region has been a source of much controversy. There is evidence, perhaps contradictory, for both[9] a $0^{-+}\to a_0\pi$ and a $1^{++}\to\bar{K}K^*$ meson at ~1420 MeV in hadron collisions, and a very strong, somewhat wider $0^{-+}\to$"a_0"π peak at 1460 MeV in radiative J/ψ decays.

Diagrams of the various meson production mechanisms are shown in Fig. 12, and the decay channels are listed in Table 5. The data from hadron collisions are summarised in Table 6. Some of these experiments are not sensitive to the isospin, and therefore determine J^{PG} rather than J^{PC}. However all isospin measurements of the 1420 peak have given I=0, in which case C=G. The J^{PG} determinations here rely on the isobar model, i.e. the assumption that all 3 body decays go via two body intermediate states. Those allowed for the $K\bar{K}\pi$ final state are $a_0\pi$ and $\bar{K}K^*$. Since the $a_0(980)$ decays

[9]The old δ is now called $a_0(980)$.

Table 6: **Spin Analyses of $E \to K\bar{K}\pi$ peak at ~1420 MeV in Hadron Collisions**

Experiment	81 cm HBC CERN	2 m HBC CERN	Ω Spectr. CERN WA76	Multi-Particle Spectrometer BNL AGS-771	
Reference	Baillon [60]	Dionisi [61]	Armstrong [62]	Chung [63]	Reeves [64]
Process	$\bar{p}p \to E\pi\pi$	$\pi^-p \to E\,n$	$\pi^+p \to \pi^+ E\,p$ $p\,p \to p\,E\,p$	$\pi^-p \to E\,n$	$\bar{p}p \to E\,X$
$E \to$	$K\bar{K}\pi$	$K_sK^\pm\pi^\mp$	$K_sK^\pm\pi^\mp$	$K_sK^+\pi^-$	$K_sK^+\pi^-$
beam energy	at rest	3.95 GeV/c	85 GeV/c	8 GeV/c	6.6 GeV/c
Fit to all $K\bar{K}\pi$:					
E mass [MeV]	1425±7	1426±6	1425±2	1421±3	1424±3
E width [MeV]	80±10	40±15	62±5	70±8	60±10
no. events					
in E peak	~800	152±25	~1000	4240	620±50
background	~70	~200	~1000	~4000	~1700
Isospin	0	0			
$J^{PG}(E)$	0^{-+}	1^{++}	1^{++}(~10% 0^-)	0^{-+}	0^{-+}(<20% 1^{++})
for $M(K\bar{K}\pi)$	unbinned	1390-1470	40 MeV bins	20 MeV bins	40 MeV bins
J^{PG}'s tried	0^-,1^+,2^-	0^-,$1^\pm$, $2^\pm$	0^{-+}, $1^{+\pm}$	$0^{-\pm}$,$1^{+\pm}$,$1^{-\pm}$	0^{-+},1^{++},1^{+-}
$K\bar{K}\pi$ P.S.	no	yes	yes	yes	yes
Flatté a_0		yes	yes	yes	yes
$E \to a_0\pi/K\bar{K}\pi$	~0.5		0.02±0.02	dominant	> 75%
$E \to \bar{K}K^*/K\bar{K}\pi$	~0.5	0.86±0.12	0.98		

to both $K\bar{K}$ and $\eta\pi$, any signal in $a_0\pi$ leads us to look also in the $\eta\pi\pi$ channel, where the isobars[10] are $a_0\pi$, $\rho\eta$, and $f_0\eta$.

While discussing the indeterminateness of the spin, it is convenient to use the historic name E for the particle seen in hadronic collisions. A 1^{++} resonance would properly be called $f_1(1420)$, and a 0^{-+} would be $\eta(1420)$. Similarly the ι seen in radiative J/ψ decay is now called the $\eta(1460)$.

4.1.1 $\bar{p}p \to \eta(1420)\,\pi^+\pi^-$

The E(1420) was first seen in $\bar{p}p$ annihilation at rest, and determined to have $J^{PG}=0^{-+}$, I=0, with about equal decays to $a_0\pi$ and $\bar{K}K^*$[60]. More recently AGS-771 have observed the E in $p\bar{p}$ annihilation in flight with the Multiparticle Spectrometer (MPS) [64]. Their data also favor 0^{-+}, this time with the $a_0\pi$ decay dominant.

4.1.2 $\pi^-p \to \eta(1420)\,n$

Dionisi *et al.* [61] studied the E(1420) in a 4 GeV/c $\pi^-p \to (K_sK^\pm\pi^\mp)$ n bubble chamber experiment at CERN. They performed a Dalitz plot analysis in a 80 MeV wide $M(K\bar{K}\pi)$ bin centered on the E signal, and found it to be 1^{++} decaying into $\bar{K}K^*$.

In 1983 AGS-771 [65] accumulated an order of magnitude more statistics in nearly the same reaction: 8 GeV/c $\pi^-p \to (K_sK^+\pi^-)$ n. Last year they presented [66] a partial wave analysis (PWA) of this data, finding that the 1420 is dominantly 0^{-+} decaying into $a_0\pi$. Their analysis had a large advantage over that of Dionisi *et al.* because the higher statistics allowed the PWA to be done in 20 MeV bins of $M(K\bar{K}\pi)$. This showed a large step in the $1^{++}(\bar{K}K^*)$ at threshold, not resembling the 1420 peak in the total $K\bar{K}\pi$ distribution. (Could it however be a 1420 peak on a rising background?) There was a strong peak at ~1400 in $0^{-+}(a_0\pi)$, with some contribution from $0^{-+}(\bar{K}K^*)$, indicated primarily by its interference with the the $a_0\pi$. The 0^{-+} wave was well described by a Breit-Wigner with M = 1402 MeV and Γ = 47 MeV, both smaller than the usual fits to the full $K\bar{K}\pi$ spectrum listed in Table 6. There was also a substantial peak in $1^{+-}(\bar{K}K^*)$. The collaboration prefered the 0^{-+} as the resonance, since its relative phase showed resonance behaviour. However the $0-2\pi$ ambiguity allowed one to also interpret the phase behaviour as essentially flat. Part of the problem was the lack of a smooth background to compare the phase to.

At this conference Chung of AGS-771 [63] presented a new analysis including data taken in 1985, thus almost doubling the statistics. The results with the same cuts as used before ($|t'|<1$ GeV^2)

[10]The old $\epsilon \to \pi\pi$ resonance, at whatever mass it may be, is now called f_0.

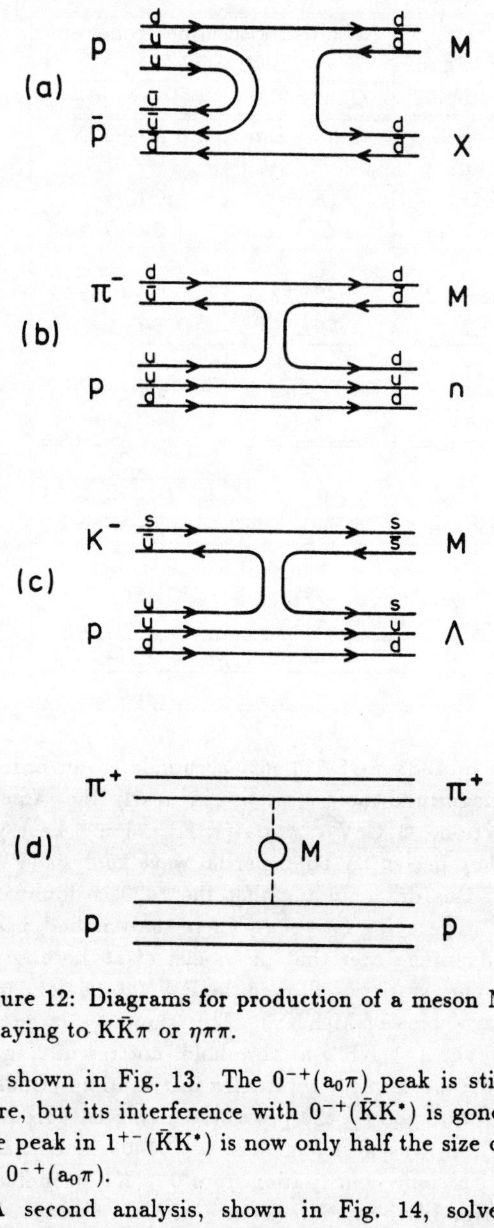

Figure 12: Diagrams for production of a meson M decaying to $K\bar{K}\pi$ or $\eta\pi\pi$.

are shown in Fig. 13. The $0^{-+}(a_0\pi)$ peak is still there, but its interference with $0^{-+}(\bar{K}K^*)$ is gone. The peak in $1^{+-}(\bar{K}K^*)$ is now only half the size of the $0^{-+}(a_0\pi)$.

A second analysis, shown in Fig. 14, solves the "too little background" problem by selecting $0.2<|t'|<1$ GeV^2. This gives a flatter $1^{++}(\bar{K}K^*)$ background, and the phase of the $0^{-+}(a_0\pi)$ against this shows unambiguous resonant behaviour. The $0^{-+}(a_0\pi)$ mass distribution and its phase are well described by a B-W with M = 1400 and Γ = 60 MeV. Chung suggests there might be a broader resonance at 1450 MeV in $0^{-+}(\bar{K}K^*)$. To my eye the 1^{++} looks equally interesting.

In this type of experiment the p and n spins are not measured. Because they are in principle measurable, there are two categories of waves (with and without nucleon spin flip), which do not interfere. Spin flip and non-flip correspond to different mesons being exchanged. For example, consider the case where a spin 0 meson is exchanged. Isospin conservation at the meson-nucleon vertex requires that the exchanged meson have I = 1. If the spin flips, the meson-nucleon orbital angular momentum must be 1, and the meson's parity must be −, so it is a π. If the nucleon spin is not flipped, L can be 0 and the meson parity even, so the $a_0(980)$ can be exchanged. a_0 exchange with a π beam is an appealing mechanism for producing an $a_0\pi$ resonance.

Originally AGS-771 required all waves of a given J^{PG} to be coherent, assuming they would come from a single resonance and a single production mechanism. When they release this requirement, they find that the fit prefers incoherent $0^{-+}(a_0\pi)$ and $0^{-+}(\bar{K}K^*)$. This corresponds to the $\eta(1400)\rightarrow a_0\pi$ and the $\eta(1450)\rightarrow\bar{K}K^*$ being produced by different meson exchanges. As usual, more data is needed to know for sure. They have more, taken in 1986, which is still to be analysed.

The decay mode $\eta\pi\pi$ has been investigated at KEK in 8 GeV/c $\pi^-p\rightarrow(\eta\pi^+\pi^-)$ n [67]. The results of their PWA analysis, which allowed incoherence between the spin flip and spin non-flip waves are shown in Fig. 15.

The $1^{++}(a_0\pi)$ contribution near 1420 is quite small. The $0^{-+}(a_0\pi)$ has a peak at 1420±5 MeV. The width, including the 25 MeV (FWHM) experimental resolution, is only 31±7 MeV. This peak is not very evident in the total $\eta\pi\pi$ mass spectrum, nor even in the total 0^{-+}. That is because the two isobars $a_0\pi$ and $f_0\eta$ interfere destructively in this region. Their relative phases are shown in Fig. 16.

GAMS [68] have data on 100 GeV/c $\pi^-p\rightarrow(\eta\pi^0\pi^0)$ n. They see a peak at ~1420 MeV, as well as a much stronger one at ~1285 MeV. Both have a large fraction of $a_0\pi$, and are produced preferentially at large t ($|t|>0.1$ GeV^2). However they have yet to do a spin analysis.

Thus the KEK and AGS-771 experiments both see the E in 8 GeV/c $\pi^-p\rightarrow$E n as a 0^{-+} particle decaying via $a_0\pi$ to $\eta\pi\pi$ and $K\bar{K}\pi$ respectively. The mass difference (1400 MeV for AGS-771 and 1420 for KEK) is a bit mysterious, as is the fact that AGS-771 sees the peak in their total $K\bar{K}\pi$ spectrum at 1420. However they do not quote an error on their 1400 MeV; the shift may not be significant. Neither see a significant signal for a 1^{++} resonance at 1420 MeV.

4.1.3 $K^-p\rightarrow(K\bar{K}\pi)\,\Lambda$

Comparison of production by K^- and π^- beams is a crucial test of a meson's quark content. There

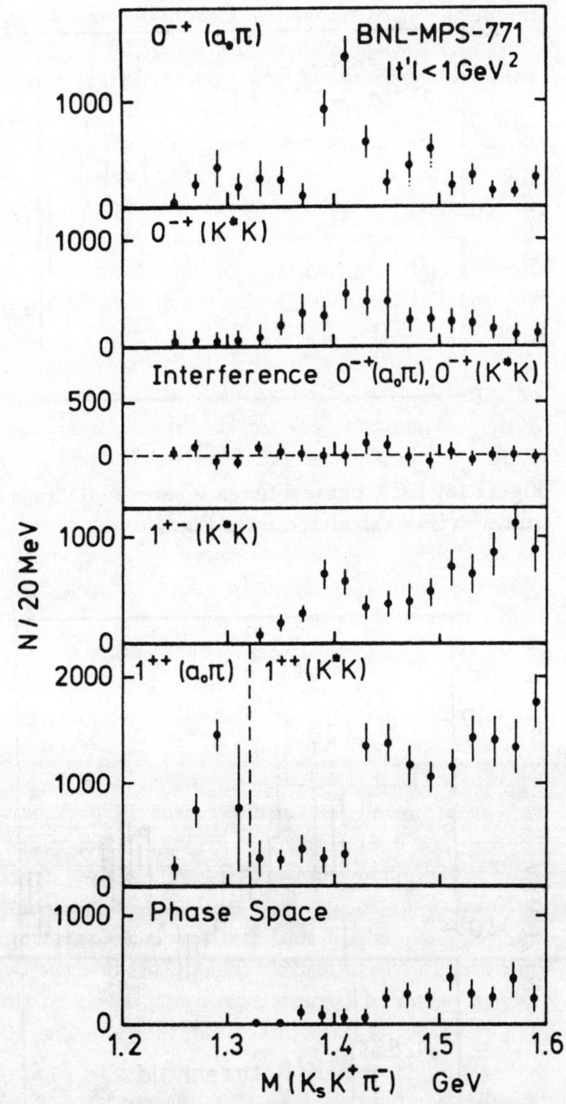

Figure 13: Partial wave analysis of AGS-771 1983+1985 data with $|t'| < 1$ GeV^2 [63].

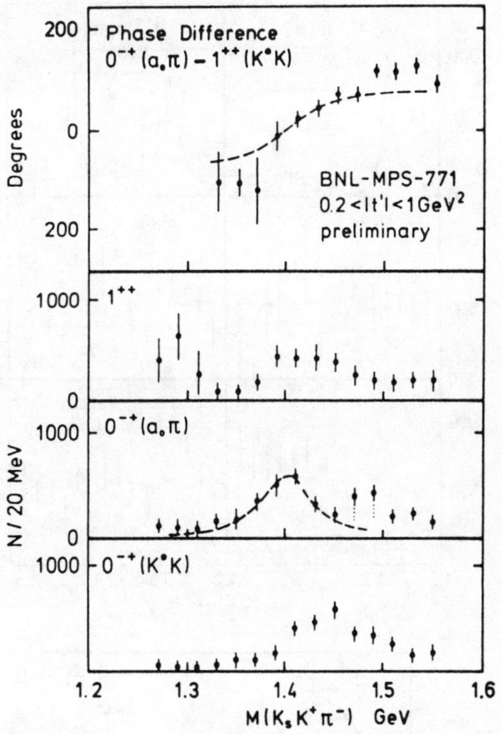

Figure 14: Preliminary PWA of AGS-771 1983+1985 data with 0.2<$|t'|$<1 GeV^2 [63]. The dashed curve corresponds to a 0^{-+} B-W with M = 1400 and Γ = 60 MeV.

should be two isosinglets in the ground state 1^{++} nonet. If they are ideally mixed, the lower mass one would be $u\bar{u}+d\bar{d}$ and the higher mass $s\bar{s}$. Both can decay to $K\bar{K}\pi$, but the former should be preferentially produced by a π^- beam, and the later by K^- (see Figs. 12b+c). The old D meson, now called $f_1(1285)$, is the undisputed candidate for the $u\bar{u}+d\bar{d}$ state. If there is a 1^{++} E, i.e. an $f_1(1420)$, it would presumably belong to this nonet and should be $s\bar{s}$. However it has a competitor for that place: the $f_1(1526)$ [69], seen with a K^- beam.

Lepton-F [70] have published data on 32.5 GeV/c $\pi^-p\to(K^+K^-\pi^0)$ n and $K^-p\to(K^+K^-\pi^0)$ Y. They perform no spin analysis, but see a peak at 1420 in the K^- beam data, and none with π^-. The ratio of the "E" production cross sections is $K^-/\pi^- > 10$. They see no sign of the $f_1(1526)$.

A different picture is presented by LASS's [18] new data on $K^-p\to(K\bar{K}\pi)\,\Lambda$, shown in Fig. 17. One is not overwhelmed by the E signal. It will be difficult to tell if it is there at all. The bins are 20 MeV wide, and the one to the left of 1420 is high. That is rather narrow for a resonance with Γ=60 MeV in an experiment with $\sigma_M \sim 12$ MeV. More convincing is the peak at 1526 MeV in 5 of the 20 MeV bins, corresponding well to the $f_1(1526)$ width of 107±15 MeV [69]. A preliminary partial wave analysis of the LASS data [71] indicates that the $K\bar{K}\pi$ spectrum up to ~1.8 GeV is dominated by 1^{++}. The statistics in this spectrum are rather disappointing, and clearly can't compete with the π^- data for determining the spin of the 1420 resonance(s). However it looks like this data will confirm the existence of the $f_1(1526)$, and it will certainly be interesting to see the final analysis. AGS-771 have also accumulated K^- data, and the analysis should be available soon.

4.1.4 $\pi^+p\to\pi^+ f_1(1420)$ p

WA76 used the Ω spectrometer at CERN to study the reactions (π^+ or p) p $\to$ (π^+ or p)$_f$ ($K_sK^\pm\pi^\mp$)

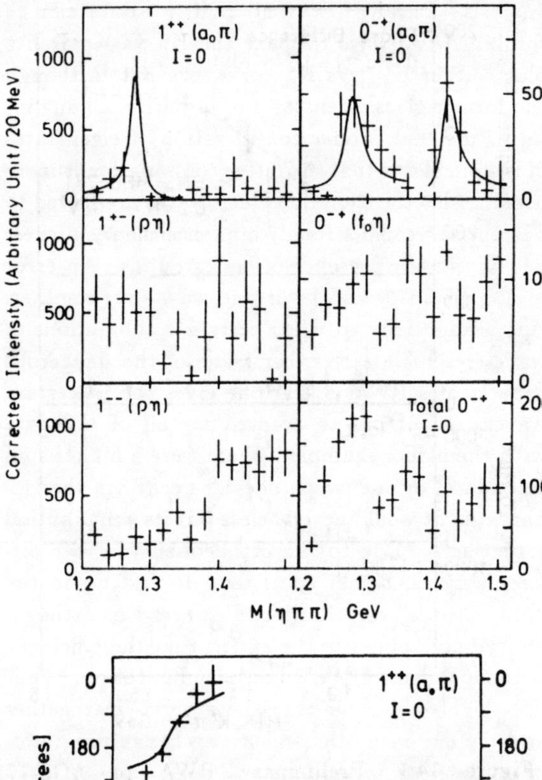

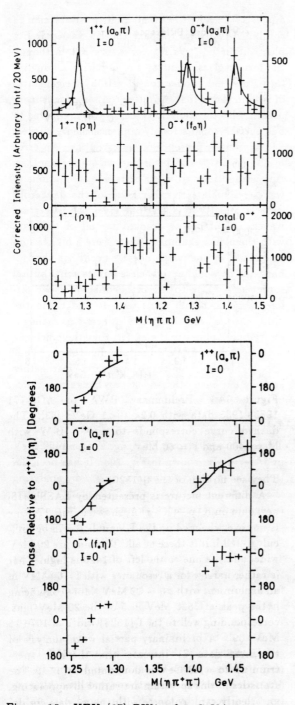

Figure 15: KEK [67] PWA of 8 GeV/c $\pi^- p \rightarrow (\eta\pi^+\pi^-)$ n.

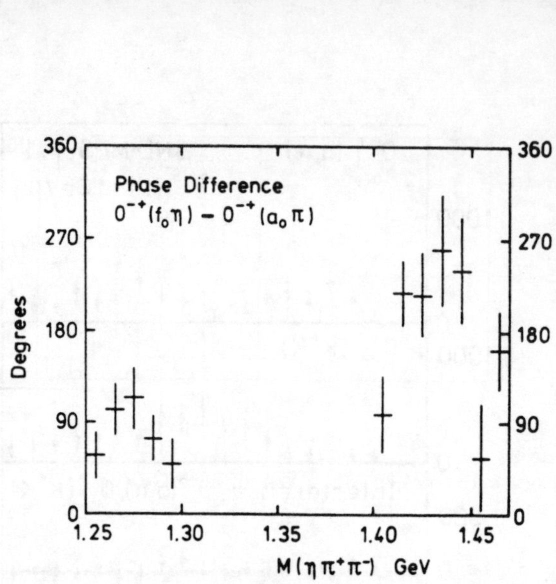

Figure 16: KEK phase difference between $0^{-+}(a_0\pi)$ and $0^{-+}(f_0\eta)$, calculated from Fig. 15.

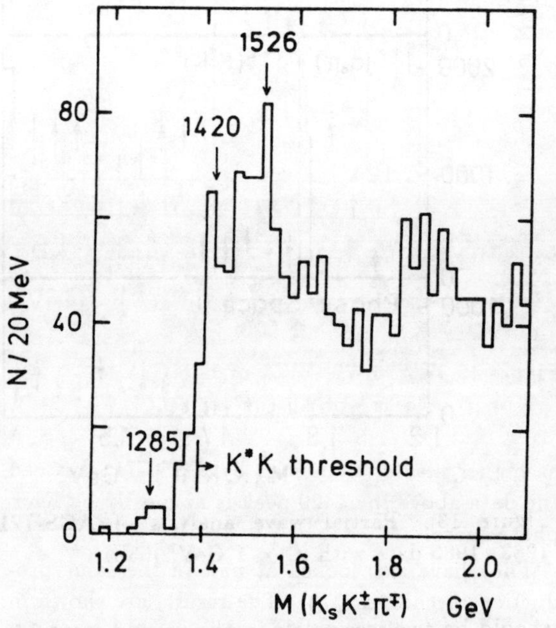

Figure 17: LASS [18] $K_s K^\pm \pi^\mp$ mass spectrum from 11 GeV/c $K^- p \rightarrow K_s K^\pm \pi^\mp \Lambda$.

p_s, where $_f$ and $_s$ denote the fastest and slowest particles in the lab frame. The $K\bar{K}\pi$ system is centrally produced (Feynman x~0), presumably by double exchange graphs as in Fig. 12d. A data selection with loose particle identification requirements is used to achieve an acceptance which is approximately flat over the Dalitz plot. Their published results [72] gave $1^{++}(\bar{K}K^*)$ for the 1420 region, for which they did a Dalitz plot analysis in the 1390 − 1470 $M(K\bar{K}\pi)$ bin. Since this wide bin could have masked a step in $1^{++}(\bar{K}K^*)$ as seen by AGS-771, they have now done the analysis in 40 MeV bins [63,62]. The results for the combined π^+ and p beam samples are shown in Fig. 18. (The analyses done separately for π^+ and p are similar.) The $1^{++}(\bar{K}K^*)$ peak looks very convincing. There may be an ~10% contribution from 0^{-+}, mostly as $\bar{K}K^*$. This reaction shows relatively less $1^{++}(\bar{K}K^*)$

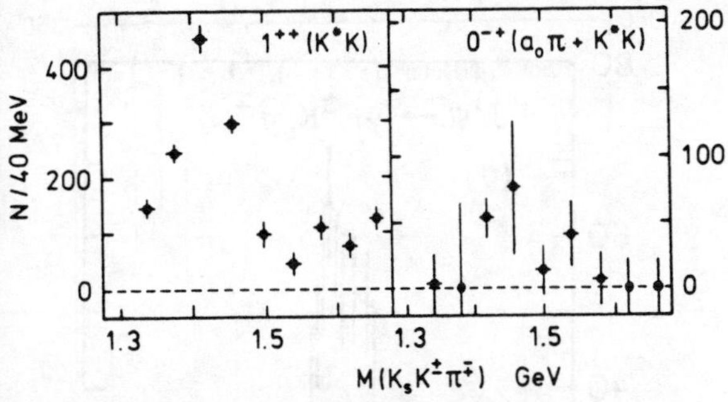

Figure 18: WA76 [62] PWA of 32.5 GeV/c (π^+ or p) p $\rightarrow$ (π^+ or p)$_f$ ($K_sK^{\pm}\pi^{\mp}$) p_s.

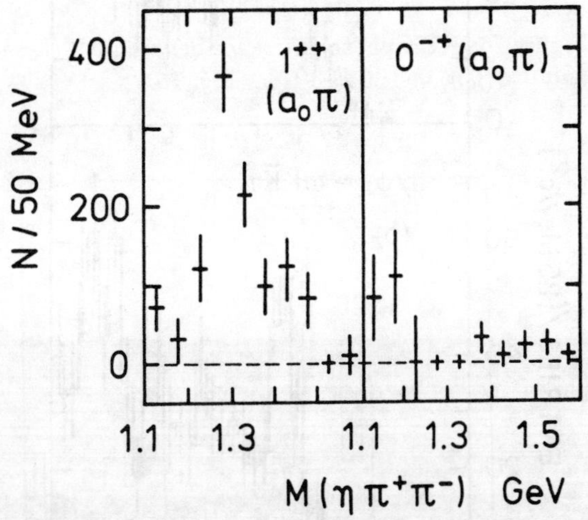

Figure 19: WA76 [73] PWA of 32.5 GeV/c (π^+ or p) p $\rightarrow$ (π^+ or p)$_f$ ($\eta\pi^+\pi^-$) p_s.

at higher masses than seen by AGS-771. Instead the data above the 1420 peak is fit mostly by $K\bar{K}\pi$ phase space.

They have also looked at $\eta\pi\pi$ in the same production mechanism [73]. The results are shown in Fig. 19. They see a slight enhancement at ~1420 MeV in the $1^{++}(a_0\pi)$ wave, from which they obtain (correcting for missing modes assuming I = 0):

$$\frac{f_1(1420)\rightarrow a_0\pi,\, a_0\rightarrow\eta\pi}{f_1(1420)\rightarrow K\bar{K}\pi} = 0.06 \pm 0.04 \ .$$

4.1.5 Remarks on E in Hadron Collisions

Thus it seems that two different particles are seen in two different production mechanisms:

$$\pi^- p \rightarrow \eta(1420)\ n$$
$$\pi^+ p \rightarrow \pi^+ f_1(1420)\ p\ .$$

Nevertheless there is still dispute between the collaborations on how best to handle details of the analysis. The $K_sK^+\pi$ and $K_sK^-\pi^+$ data can be combined two ways to make the $K\pi$ vs $K\pi$ Dalitz plot: either K^*_{neut} vs K^*_{ch}, or K^* vs $\bar{K}^*$. Although the former gives eigenstates of G-parity, Chung informs me that proper consideration of eigenstates of isospin shows that the latter combination is more useful. Also the choice of which waves are forced to be coherent could strongly influence the results, especially in the presence of backgrounds. An error in the efficiency can favor one wave over another; this is especially so when certain combinations of waves resemble each other. One of the dangerous aspects of a PWA is that one gives the fit certain waves, and it has to account for all of the data with them. For example if there were a higher spin resonance, or one which doesn't decay via the isobars, the fit would nevertheless put its contribution somewhere. Due to the orthogonality of the different waves, higher spins tend to end up in the "phase space" contribution. However this orthogonality only applies to the extent that the efficiency is independent of the 3-body configuration. Thus it is highly desirable that the groups get together and try out each others ideas, exchange programs, or analyse each others' data.

4.1.6 J/ψ Decays

There is a very strong $K\bar{K}\pi$ signal at ~1460 MeV in $J/\psi\rightarrow\gamma K\bar{K}\pi$, shown in Fig. 20a. It has been determined [74,12] to be dominantly 0^{-+} by analysing the orientation of the 3-body decay plane, which is independent of isobars. It contains a low-mass $K\bar{K}$ enhancement which corresponds approximately to the $a_0(980)$.[11]

Mark III [29] have noticed that the peak is not well fit by a single Breit-Wigner. Their Dalitz plots for the low and high mass sides look different, with the high side (1460 – 1580 MeV) showing clear K^* bands. The new ~1450 $0^{-+}(\bar{K}K^*)$ enhancement of AGS-771 [63] discussed in Section 4.1.2 is on the wrong side of 1460 to explain this. There may be two separate resonances appearing in the $J/\psi\rightarrow\gamma K\bar{K}\pi$, or one whose shape and branching ratios are distorted by the $\bar{K}K^*$ threshold, or a statistical fluctuation in the resonance shape.

There is no 1460 peak in $J/\psi\rightarrow\gamma a_0\pi\rightarrow\eta\pi\pi$ (see Fig. 21). Rather a peak at ~1390 MeV is seen by both DM2 [28] and Mark III [29,75] and the spectra drop sharply at ~1440 MeV. This might be the $\eta(1460)$ showing itself through interference, but no conclusion can be made until a spin-parity analysis is available.

---

[11] Have you remembered that this was the δ?

Using the technique of Fig. 5 to study the quark content of $K\bar{K}\pi$ and $\eta\pi\pi$ resonances, Mark III [29,27,76] and DM2 [28,12] have compared the systems X in $J/\psi\to\gamma X$, ωX, and ϕX, where $X = K\bar{K}\pi$ or $\eta\pi^+\pi^-$. The Mark III data are shown in Figs. 20 and 21. The $K\bar{K}\pi$ accompanying an ω has a clear peak with $M = 1440 \pm 7^{+10}_{-20}$ and $\Gamma = 40^{+17}_{-13} \pm 10$ MeV. The angular distributions do not look like 0^{-+}. Perhaps it is the $f_1(1420)$? The ratio of the $K_sK^\pm\pi^\mp$ to $K^+K^-\pi^0$ signals is consistent with the value 2 expected for I = 0. Assuming I = 0 yields $B(J/\psi\to\omega X, X\to K\bar{K}\pi) = (6.8^{+1.9}_{-1.6} \pm 1.7) \times 10^{-4}$.

A very similar peak is seen in $J/\psi\to\omega\eta\pi^+\pi^-$, correlated with an $a_0(980)\to\eta\pi$ subsystem, and with $M = 1421 \pm 8 \pm 10$, $\Gamma = 45^{+32}_{-23} \pm 15$ MeV, and $B(J/\psi\to\omega X, X\to a_0\pi\to\eta\pi\pi) = (9.2 \pm 2.4 \pm 2.8) \times 10^{-4}$. Here no spin information is available. If it is the same X as above, then it has a slightly larger branching ratio to $\eta\pi\pi$ than to $K\bar{K}\pi$. This is in contrast to WA76's ratio for the $f_1(1420)$ of 0.06 ± 0.04 [73]. Thus either the X, or one of the X's, is not the $f_1(1420)$.

The Mark III data for $K\bar{K}\pi$ accompanying a ϕ show no peak in the 1400 MeV region. The upper limit $B(J/\psi\to\phi X, X\to K\bar{K}\pi) < 1.1\times10^{-4}$, is obtained by fixing the X parameters to those of the peak seen accompanying an ω, or to the E(1420), and assuming isotropic angular distributions. The $\eta\pi\pi$ accompanying a ϕ has one high bin just below 1400 MeV. DM2 [28] see no signal in this region in $K\bar{K}\pi$ or $\eta\pi\pi$ accompanying ϕ's. We would have expected an $s\bar{s}$ $f_1(1420)$ to prefer to accompany ϕ's rather than ω's. There is also no sign of the $f_1(1526)$.

These J/ψ decays, even the purely hadronic ones, escape any attempt I make to explain them by resonances seen in hadron collisions. Perhaps I am not clever enough. Perhaps some of the results are wrong. Otherwise, there are a few more particles: X(1390), X(1440), and the split or whole $\eta(1460)$. The despair of the orderly $q\bar{q}$ modeller rises, along with the hopes of the gg, $gq\bar{q}$, and $q\bar{q}q\bar{q}$ fans.

4.1.7 $\gamma\gamma\to f_1(1420)$

Information on the 1400 MeV region is now coming from a new source: $\gamma\gamma$ collisions.

At an e^+e^- storage ring there are not only e^+e^- collisions, but also a great many "Bremsstrahlung" γ's emitted which can collide as shown in Fig. 22. As with normal Bremsstrahlung γ's, they are emitted preferentially at very small angles to the $e^\pm$ direction. In this case, the $e^\pm$ proceed down the beam pipe with $\theta\sim0$, and are not detectable in most experimental setups. The γ's are nearly massless. By

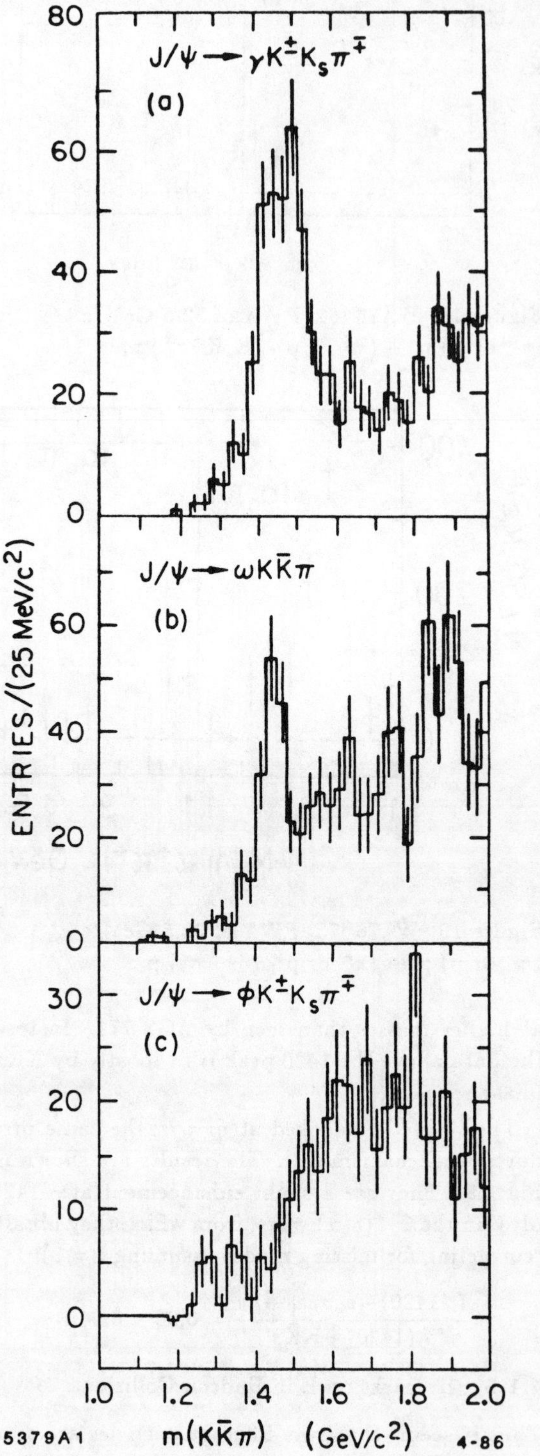

Figure 20: Mark III [27] $X \equiv K_sK^\pm\pi^\mp$ mass plots from **(a)** $J/\psi\to\gamma X$, **(b)** $J/\psi\to\omega X$ background subtracted, and **(c)** $J/\psi\to\phi X$.

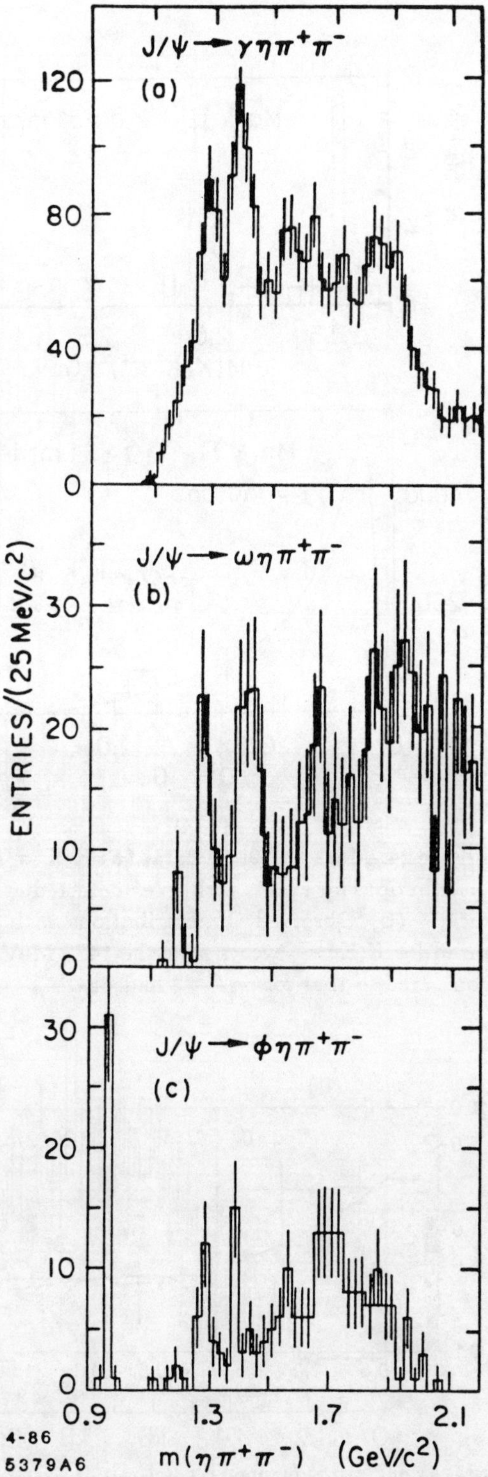

Figure 21: Mark III [27] $X \equiv \eta\pi^+\pi^-$ mass plots from **(a)** $J/\psi \to \gamma X$ for events with an $\eta\pi$ combination consistent with the $a_0(980)$, **(b)** $J/\psi \to \omega X$ consistent with $a_0(980)$ and background subtracted, and **(c)** $J/\psi \to \phi X$.

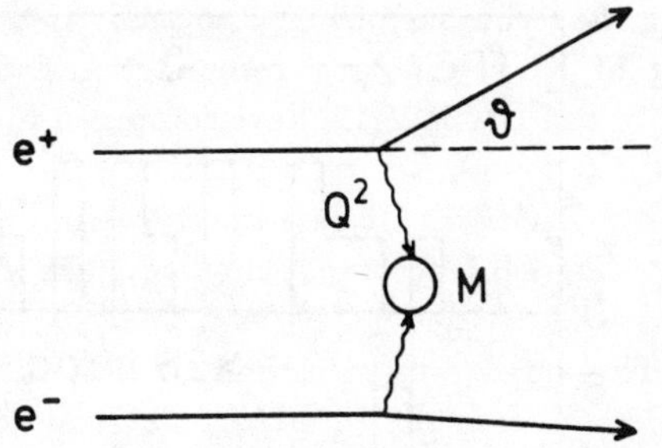

Figure 22: $\gamma\gamma$ collision at e^+e^- storage ring with one nearly massless and one massive γ.

Yang's theorem [77], two massless γ's cannot combine to form a J=1 state. Thus we don't expect to see spin one mesons formed in $\theta \sim 0$ $\gamma\gamma$ collisions. On the rare occasions when an $e^\pm$ is scattered at a large enough angle to be detected, the γ acquires a negative mass: $-M^2 \equiv Q^2 \approx 4EE'\sin^2(\theta/2)$, where E and E′ are the initial and final $e^\pm$ energies and θ is the scattering angle (see Fig. 22). Then Yang's theorem no longer applies, so that spin 1 mesons may be produced in the collision of a massless γ with a massive one. (The case where both γ acquire substantial mass is rare enough to ignore here.)

The TPC/2γ [78] collaboration have measured $\gamma\gamma \to K_s K^\pm \pi^\mp$ for both $\theta \sim 0$ and $\theta > 0$, shown in Fig. 23. The $K_s K^\pm \pi^\mp$ mass distribution from $\theta \sim 0$ data shows no structure, while that from $\theta > 25$ mrad shows a clear peak at ~ 1420 MeV. This effect has now been confirmed by Mark II [79]. In Fig. 24 the Mark II data with $\theta > 21$ mrad are shown. The observed Q^2 distribution for events with $M(K_s K^\pm \pi^\mp)$ between 1400 and 1500 MeV is clearly different from that for all $\gamma\gamma \to$ hadrons events. When quantified, this difference is evidence that the 1420 MeV meson being produced here has spin 1.

To make this concrete, TPC/2γ have compared the production of the 1420 peak as a function of the γ mass to that expected for spin 0 and spin 1 mesons. Massless γ's always have transverse polarisation (T). Massive ones can also have longitudinal polarisation (L). Neglecting the small probability of both γ's being massive, we have TT and LT scattering. A spin 0 meson cannot be formed by LT, so only TT contributes:

$$\sigma_{TT}(J=0) \sim \rho(Q^2) \tag{4}$$
$$\sigma_{LT}(J=0) = 0\,. \tag{5}$$

Here $\rho(Q^2)$ is the ρ-pole photon form factor from

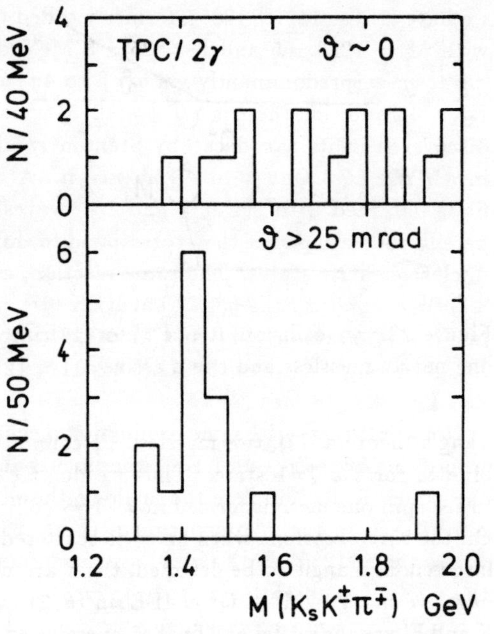

Figure 23: TPC/2γ [78] $K_sK^{\pm}\pi^{\mp}$ mass distributions from $e^+e^- \to e^+e^-\gamma\gamma$, $\gamma\gamma \to K_sK^{\pm}\pi^{\mp}$ for $e^{\pm}$ scattering angles **(a)** $\theta \sim 0$ (nearly massless γ's), and **(b)** $\theta > 25$ mrad (one massive γ).

VDM

$$\rho(Q^2) = \frac{1}{\left(1 - \frac{Q^2}{m_\rho^2}\right)^2},$$

which describes fairly well the Q^2-dependent production of other mesons (e.g. the $f_2(1270)$ [31]).

For spin 1 both TT and LT contribute:

$$\sigma_{TT}(J = 1) \sim \rho(Q^2)\,\frac{Q^4}{M^4} \qquad (6)$$

$$\sigma_{LT}(J = 1) \sim \rho(Q^2)\,\frac{Q^2}{M^2} \qquad (7)$$

where M is the meson mass. At small Q^2 LT will dominate. Fig. 25 shows a comparison of the $J = 0$ expectation and the $J = 1$ LT term with the TPC/2γ data on formation of the 1420 MeV peak as a function of Q^2. $J = 0$ disagrees strongly with the $Q^2 \sim 0$ point, while $J = 1$ gives a good description of the data.

A value for the strength of the $\gamma\gamma$ coupling of the spin 1 resonance can be obtained by extrapolating the fitted curve in Fig. 25b to $Q^2 = 0$. (This is better done after dividing out the Q^2/M^2 dependence). The result is

$$\lim_{Q^2 \to 0} \frac{M^2}{Q^2}\,\Gamma_{\gamma\gamma} \times B(K\bar{K}\pi) = 6 \pm 2 \pm 2 \text{ keV}.$$

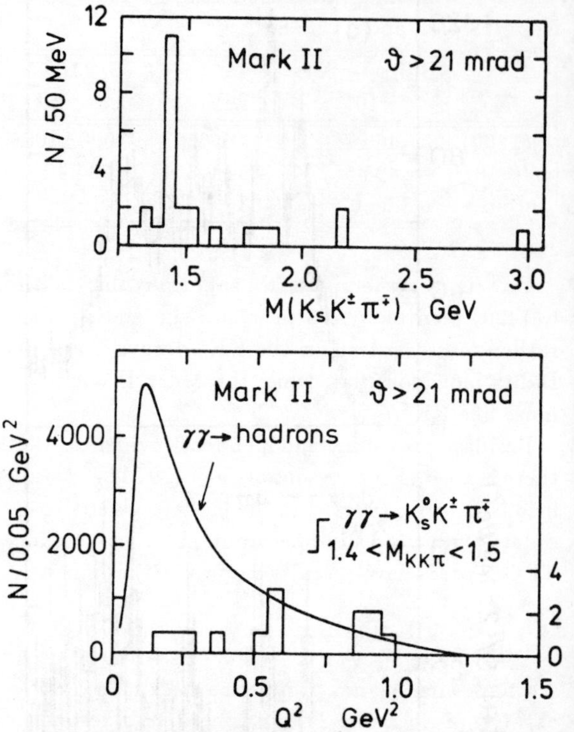

Figure 24: Mark II [79] $\gamma\gamma$ data. **(a)** $K_sK^{\pm}\pi^{\mp}$ mass distributions for events with an $e^{\pm}$ detected at $\theta > 21$ mrad. **(b)** Observed Q^2 distributions with $\theta > 21$ mrad for $K_sK^{\pm}\pi^{\mp}$ events near the 1420 MeV peak compared to that for $\gamma\gamma \to > 3$ hadrons.

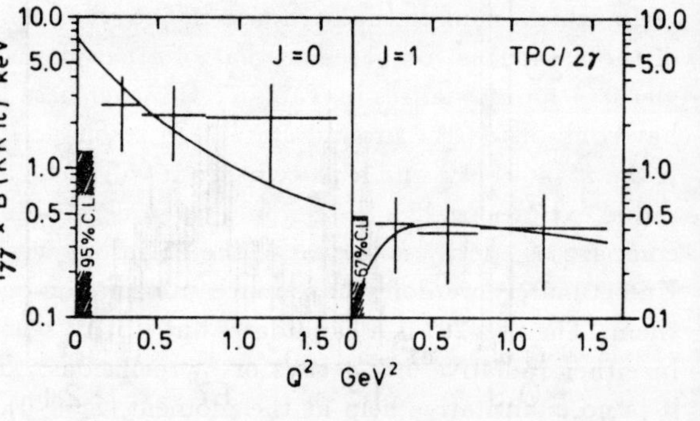

Figure 25: TPC/2γ [78] Q^2 dependence of $\gamma\gamma$ formation of the 1420 $K\bar{K}\pi$ peak. In **(a)** the data are corrected for the experimental acceptance using a $J = 0$ Monte Carlo, and in **(b)** using a $J = 1$ Monte Carlo. The curves are the best fits to the $Q^2 > 0$ points using Eqs. (4) and (7) respectively.

Table 7: **Stickiness of 0^{-+} mesons.**
The J/ψ branching ratios are from [1]. The $\Gamma_{\gamma\gamma}$ are from [78], [1], and Table 8.

X	$B(J/\psi\to\gamma X)$ $\times 10^4$	$\Gamma_{\gamma\gamma}$ [keV]	$S \times 10^4$
$\eta(1460)$	46 ± 7	< 1.6	$>4000\pm100$
$\eta'(958)$	42 ± 5	4.5 ± 0.5	23 ± 4
$\eta(549)$	8.6 ± 0.8	0.53 ± 0.03	6 ± 1

One is very tempted to say that this is the $f_1(1420)$ seen by WA76 in π^+p. The statistics are rather sparse to test for the $\bar{K}K^*$ decay. The TPC Dalitz plot looks flat, while the Mark II one looks more like $\bar{K}K^*$.

Besides providing independent evidence that there is a $1^{++}(K\bar{K}\pi)$ resonance at 1420 MeV, $\gamma\gamma$ collisions also give support to the $\eta(1460)$ glueball candidate by putting an upper limit on its $\gamma\gamma$ coupling. TPC/2γ [78] have recently improved the limit to

$$\Gamma_{\gamma\gamma}[\eta(1460)] \times B(K\bar{K}\pi) < 1.6 \text{ keV} \quad (95\% \text{ C.L.}).$$

Chanowitz [80] has proposed a stickiness variable as a test of glueballs. The stickiness S of a resonance is essentially the ratio of its production in radiative J/ψ decays to that in $\gamma\gamma$ collisions, with kinematic factors divided out:

$$S_X \equiv \frac{\Gamma(J/\psi\to\gamma X)}{\mathrm{LIPS}(J/\psi\to\gamma X)} \Big/ \frac{\Gamma(X\to\gamma\gamma)}{\mathrm{LIPS}(X\to\gamma\gamma)} \quad .$$

LIPS is the Lorentz Invariant Phase Space, which for $X=0^{-+}$ is p-wave for both channels. One advantage of this variable is that the decay branching ratio, which is unknown, divides out. The stickiness of the $\eta(1460)$ is compared to that of other isosinglet 0^{-+} $q\bar{q}$ resonances in Table 7. However here I have only listed the ground state (1S) resonances. If the $\eta(1460)$ can find a place as $q\bar{q}$ it will be as a radial excitation (2S). So it would be preferable to compare its stickiness to that of the 2S resonances. Unfortunately we don't have much information on them. The $\eta(1420)$ is a candidate, but it isn't seen in either radiative J/ψ decays or $\gamma\gamma$ collisions, so it is no quantitative help at the moment (S=0/0). The question of 2S 0^{-+} states is dealt with further in the following sections.

4.2 1280 MeV Region

4.2.1 Hadron Collisions

It may seem an unlikely coincidence to have a 0^{-+} and a 1^{++} $K\bar{K}\pi$ resonance at ~1420 MeV. However the same effect occurs at ~1280 MeV. Here we have [1] the 1^{++} $f_1(1285)$, formerly called the D, with $M = 1283 \pm 5$ and $\Gamma = 25 \pm 3$ MeV decaying to $\eta\pi\pi$ (predominantly via $a_0\pi$), to 4π, and to $K\bar{K}\pi$. In addition there is a 0^{-+} $\eta(1275)$ which was discovered in its $a_0\pi$ decay by Stanton *et al.* [81] in a PWA of 8 GeV $\pi^-n \to (\eta\pi^+\pi^-)$ n. A better fit is obtained with the 0^{-+} and 1^{++} waves incoherent, indicating that they correspond to different nucleon helicity states. The same reaction, also at 8 GeV/c, confirms the $\eta(1275)$ at KEK [67], shown in Fig. 15. Stanton *et al.* find $M = 1275 \pm 15$ and $\Gamma = 70 \pm 15$ MeV, while the KEK has $M = 1279 \pm 5$ and $\Gamma = 32 \pm 10$ MeV.

Note that this is the same production mechanism where the AGS-771 and KEK experiments also see the $\eta(1420)$. To complete the analogy, the $\eta(1275)$ is **not** seen in the π^+p experiment (Figs. 18&19), where the $f_1(1285)$ and $f_1(1420)$ are prominant and the $\eta(1420)$ is absent.

AGS-771 do not have evidence for the $\eta(1275)$ in their $K\bar{K}\pi$ data (Fig. 13), but can accomodate it by changing the $a_0(980)$ parameterisation [82].

4.2.2 J/ψ Decays

DM2 [28] and Mark III [29,27] see a peak at ~1285 MeV in the $\eta\pi\pi$ spectra accompanying a γ, ω, and ϕ in J/ψ decays, and in 4π accompanying a ϕ. DM2 also sees it in 4π accompanying a γ. In this channel they have analysed the 4π system and found the peak to be mostly $\rho^0\pi\pi$. Since this is one of the standard [1] decay modes of the $f_1(1285)$, it could be evidence for $J/\psi\to\gamma f_1(1285)$. Such a radiative decay of the J/ψ via $J/\psi\to\gamma gg$ with both gluons massless is forbidden by Yang's theorem. However massless vitual gluons is only an approximation. Also, the determination that the $f_1(1285)$ decays to $\rho^0\pi\pi$ was made before the $\eta(1275)$ was seen, and may not have allowed for the possibility that not all of the 1285 peak is due to the $f_1(1285)$. As usual, caution is advised before J^P measurements are available.

4.2.3 $\gamma\gamma\to\eta(1275)$

The $\eta(1275)$ is a candidate $\eta(2S)$, although $q\bar{q}$ models (e.g. [17]) would rather have it at a higher mass. Predictions for the $\gamma\gamma$ coupling of an $\eta(2S)$ are in the region of a few keV [17,83]. However Crystal Ball data [84] on $\gamma\gamma\to\eta\pi^0\pi^0$ (Fig. 26) show no sign of the $\eta(1275)$. The 90% C.L. upper limit, assuming $\Gamma_{tot} = 50$ MeV, is

$$\Gamma_{\gamma\gamma}[\eta(1275)] \times B(\eta\pi\pi) < 0.3 \text{ keV} \quad \text{(prel.)}$$

This was surprising [85], but perhaps can be understood on closer inspection of the possibilities for mixing among the 0^{-+} mesons.

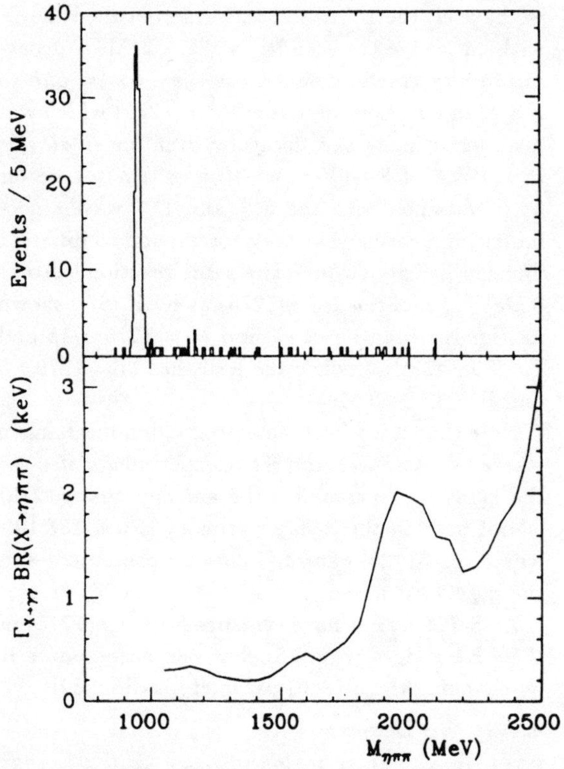

Figure 26: **(a)** Crystal Ball [84] $\eta\pi\pi$ mass distribution from $\gamma\gamma\to\eta\pi^0\pi^0$. The large peak is the $\eta'(958)$. **(b)** Preliminary upper limits for $\Gamma_{\gamma\gamma}(X)\times B(X\to\eta\pi\pi)$, for $\Gamma_{tot}(X)$=50 MeV. Although there are no events at high mass, the upper limit rises because the $\eta\pi^0\pi^0$ reconstruction efficiency and the $\gamma\gamma$ flux fall with increasing mass.

The $\gamma\gamma$ coupling of a meson is the coherent sum of the $\gamma\gamma$ couplings of its $q\bar{q}$ components, each of which is proportional to $e_q^2|\psi(0)|^2$, where e_q is the quark charge and $\psi(0)$ the $q\bar{q}$ spatial wave function at r=0. An isosinglet always has equal amounts of $u\bar{u}$ and $d\bar{d}$, so we can write the arbitrary isosinglet wave function as

$$\psi = v\,(u\bar{u} + d\bar{d}) + \sigma\, s\bar{s} \qquad |v|^2 + |\sigma|^2 = 1.$$

In the approximation that the spatial wave function is the same for all $q\bar{q}$ components we get

$$\Gamma_{\gamma\gamma} \sim \left|\frac{5v+\sigma}{9}\right|^2 .$$

Thus vanishing $\Gamma_{\gamma\gamma}$ is achieved for $\sigma = -5v$ [86]. However that means a large $s\bar{s}$ content for the $\eta(1275)$, which would make its production via a pion beam and its $a_0\pi$ decay improbable. Also, this mixing leaves a primarily $u\bar{u}+d\bar{d}$ partner, which should be seen in $\gamma\gamma\to\eta\pi\pi$. The data in Fig. 26

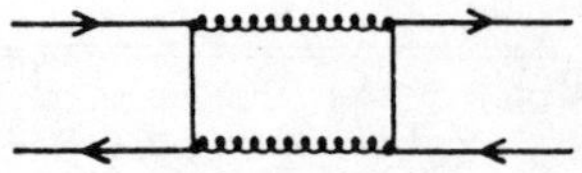

Figure 27: 1S-2S mixing for I=0 0^{-+} mesons.

don't leave much room for such a meson below ~1800 MeV.

The Stanton *et al.* [81] and KEK [67] data on the $\eta(1275)$ present an amusing cure for its non-appearance in $\gamma\gamma\to\eta\pi\pi$: They see destructive interference between the $a_0\pi$ and $f_0\eta$ isobars, with little sign of the $\eta(1275)$ in the total $\eta\pi\pi$ spectrum (see Figs. 15&16). So all the Crystal Ball have to do is perform a partial wave analysis of their 0 events to separate the interfering isobars!

Now let us be serious and look at what goes into the predictions of $\Gamma_{\gamma\gamma}$ for the 2S states. The wave function at the origin is smaller for 2S states than for 1S, but the p-wave phase space for the decay is larger due to their higher mass. In addition, the 2S and 1S states mix, via the diagram in Fig. 27, to form the physical states we observe. Expanding the notation used above, so that for a given isosinglet meson, v_{1S} is its $u\bar{u}+d\bar{d}$ 1S component, etc.,

$$\psi = (v_{1S} + v_{2S})\,(u\bar{u} + d\bar{d}) + (\sigma_{1S} + \sigma_{2S})\, s\bar{s}\ .$$

Using the individual $\gamma\gamma\to q\bar{q}$ couplings from the Godfrey&Isgur model [87] yields

$$\Gamma_{\gamma\gamma} = M^3\,|2.65v_{1S} - 1.12v_{2S} + 0.58\sigma_{1S} - 0.28\sigma_{2S}|^2\ ,$$

where inserting M in GeV gives $\Gamma_{\gamma\gamma}$ in keV.

Notice that the 2S states enter with a minus sign in this particular model, giving an opportunity to achieve $\Gamma_{\gamma\gamma}\sim 0$ without an $s\bar{s}$ component. Godfrey&Isgur's prediction for $\Gamma_{\gamma\gamma}[\eta(1440)]\approx 7$ keV was based on $v_{1S} = -0.26$, $v_{2S} = +0.79$, $\sigma_{1S} = -0.17$, $\sigma_{2S} = -0.44$. However the mixing among the 0^{-+} mesons is quite uncertain in their model due to the difficulty of calculating Fig. 27 at such low masses. The isosinglets are doubly difficult because they have octet-singlet mixing as well as 1S-2S. With the I=1 $\pi(2S)$ we should be on firmer ground. The candidate particle is the $\pi(1300)$ which has a width between 200 and 600 MeV, and decays to 3π, including $\rho\pi$. Godfrey&Isgur predict $\Gamma_{\gamma\gamma}[\pi(1300)]$=1 keV. Since the branching ratios are not known, a useful measurement of its $\Gamma_{\gamma\gamma}$ should include both the $\pi^0\pi^-\pi^-$ and the $3\pi^0$ channels. Next year?

Until we understand the $\Gamma_{\gamma\gamma}$ widths of the 2S mesons, we should be very cautious in using the non-observation of the glueball candidate $\eta(1460)$ in $\gamma\gamma$ collision to say that it can't be 2S $q\bar{q}$.

5 $\Gamma_{\gamma\gamma}$ of Other 0^{-+} Mesons

5.1 π^0 and η

At this conference new results were presented by Crystal Ball [92] using $\gamma\gamma\to\gamma\gamma$ scattering to measure $\Gamma_{\gamma\gamma}$ of the $\eta(548)$ and the π^0. The data are shown in Fig. 28, and the results compared to previous measurements in Table 8. Their $\Gamma_{\gamma\gamma}(\pi^0)$ agrees well with a very precise measurement using the decay length of high energy π^0's [91], and with the previous best measurement by Browman *et al.* [90] using the Primakoff effect. This agreement makes it all the more surprising that the new $\gamma\gamma$ measurement of the η disagrees with that of Browman *et al.*, although all the $\gamma\gamma$ measurements agree with each other (see Table 8).

In the Primakoff effect the η is formed by an incoming photon beam interacting with a virtual photon of the Coulomb field of a heavy nucleus:

$$\gamma\, N \to \gamma\gamma^*\, N \to \eta\, N\ .$$

The cross section for this process is

$$\sigma_{\text{Primakoff}} \propto \Gamma_{\gamma\gamma}\, |F_{em}(q)|^2\, E^4_{beam}\, \frac{\sin^2\theta}{q^4}, \qquad (8)$$

where θ is the angle of the η relative to the incoming γ direction. There is a minimum momentum transfer squared [97]

$$q^2 \equiv t_{min} \approx -\frac{M^4}{4E^2_{beam}}$$

for a photon of energy E_{beam} producing a meson of mass M. The t_{min}'s for the various experiments are listed in Table 8.

$F_{em}(q)$ is the electromagnetic form factor of the nucleus of charge Z. Its q-dependence must be known in order to extract $\Gamma_{\gamma\gamma}$ from Eq. 8. A physicist of the previous generation would be able to say something wise at this point about the size of the error this is likely to introduce. The best I can do is to point out that the correction is reduced as $|q^2|$ decreases for higher beam energies, and that the η needs a factor 16 higher beam energy than the π^0 to get down to the same $|q^2|$. Modern Primakoff experiments are performed at hundreds of GeV (e.g. Ref. [44]), but none measure the η.

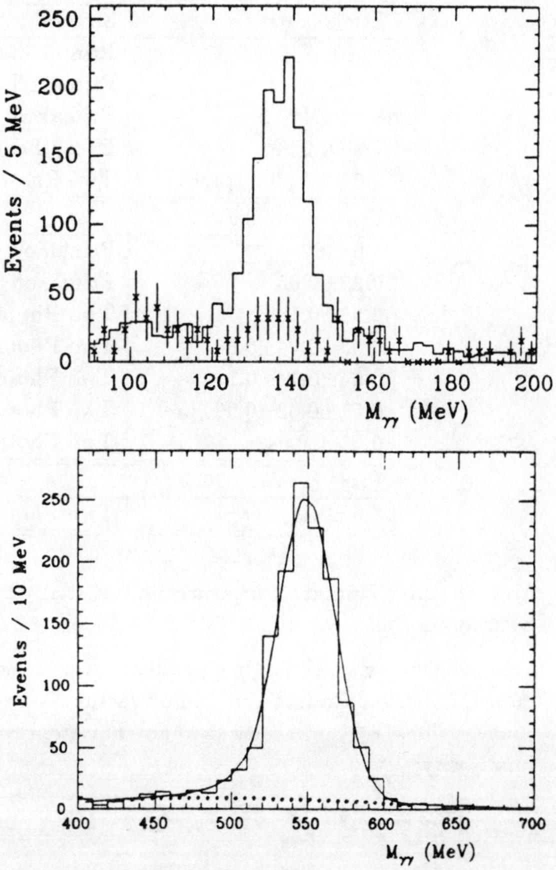

Figure 28: Crystal Ball [92] data on $\gamma\gamma\to\gamma\gamma$ vs. $M(\gamma\gamma)$ **(a)** near the π^0 and **(b)** near the η. The crosses in (a) are the normalised single beam data, which is too sparse in (b) to plot but is well represented by the dotted background curve. The $p_t(\gamma\gamma)$ requirement is <8.7 MeV in (a) and <45 MeV in (b).

Primakoff production of π^0's and η's has a background from π^0's and η's produced in the **hadronic** field of the nucleus, e.g.

$$\gamma\, N \to \gamma\rho^*\, N \to \eta\, N\ .$$

The Coulomb and hadronic contributions can be statistically separated by fitting the angular distributions, since that for production in the Coulomb field is peaked at smaller angles.

The most recent Primakoff experiment, that of Browman *et al.*, was able to take advantage of higher energies available by then to improve the signal to background (the Primakoff effect increases as E^4_{beam}), and to check the systematics of the separation by comparing results from a wide range of beam energies. Therefore their $\Gamma_{\gamma\gamma}(\eta)$ became the accepted value, although it differed substantially

Table 8: **Measurements of $\Gamma_{\gamma\gamma}$ of the π^0 and η.**

$\Gamma(\pi^0)$ (eV)	Method	t_{min} [MeV^2]	Reference
12±1	Primakoff	-83	Bellettini *et al.* [88]
7.3±0.6	Primakoff	-69	Kryshkin *et al.* [89]
8.0±0.4	Primakoff	-2	Browman *et al.* [90]
7.3±0.2±0.1	Decay Length		Cronin *et al.* [91]
7.8±0.4±0.9 (prel.)	Two Photon		Crystal Ball (DORIS) [92]
$\Gamma(\eta\to\gamma\gamma)$ [keV]			
1.0±0.2	Primakoff	-745	Bemporad *et al.* [93]
0.32±0.05	Primakoff	-172	Browman *et al.* [90]
0.56±0.16	Two Photon		Crystal Ball (SPEAR) [94]
0.53±0.04±0.04	Two Photon		JADE [95]
0.64±0.14±0.13	Two Photon		TPC/2γ [96]
0.51±0.02±0.06 (prel.)	Two Photon		Crystal Ball (DORIS) [92]
0.53 ± 0.04	Two Photon		average
$\Gamma(\eta')$ (keV)			
4.5±0.4	Two Photon		average [1]

from the first Primakoff measurement of the η by Bemporad *et al.*

In order to measure $\Gamma_{\gamma\gamma}$ in $\gamma\gamma$ collisions one relies on a QED calculation of the $\gamma\gamma$ flux factor $\mathcal{F}_{\gamma\gamma}$ to convert the e^+e^- luminosity $\mathcal{L}$ to an equivalent $\gamma\gamma$ luminosity:

$$N(e^+e^-\to e^+e^-M) = \mathcal{L}\ \mathcal{F}_{\gamma\gamma}\ \sigma(\gamma\gamma\to M).$$

Here $\sigma(\gamma\gamma\to M)$ is the Breit-Wigner resonance cross section as in Eq. (1), and is proportional to $\Gamma_{\gamma\gamma}(M)$. The efficiency for detecting this process is typically only a few %, because the events tend to have a large Lorentz boost, and is determined using a Monte Carlo simulation of the $e^+e^-\to e^+e^-M$ process. A possible background from beam-gas production of η's, i.e. $ep\to\eta+X$, must be checked. The four available measurements agree with each other quite well (Table 8). Two of them have errors approaching that of the latest Primakoff experiment. The $\gamma\gamma$ average is $\Gamma_{\gamma\gamma}(\eta)$=0.53±0.04, which is a factor of 1.7±0.3 larger than the Primakoff result. A graphical comparison of the $\Gamma_{\gamma\gamma}(\eta)$ results is presented in Fig. 29.

Both types of experiments have a small effect from the q^2 dependence of the $\gamma\gamma$ coupling to the η. $\Gamma_{\gamma\gamma}$ is defined for the decay $\eta\to\gamma\gamma$, where the γ's are real and massless. When we measure $\Gamma_{\gamma\gamma}$ in $\gamma\gamma\to\eta$, where the photons are not quite real, we expect a ρ-pole form factor for each γ, so that in the above equations $\Gamma_{\gamma\gamma}$ should be replaced by $\rho(q)\ \Gamma_{\gamma\gamma}$ for the Primakoff effect (where only one γ is virtual) and by $\rho(q_1)\ \rho(q_2)\ \Gamma_{\gamma\gamma}$ for $\gamma\gamma$ collisions. In the latter, the γ's have small q^2, especially after typical cuts on the net p_t of the observed η. The Crystal Ball [92] requires $p_t(\eta)<45$ MeV, for which the form factors reduce the observed cross section by less than 0.3%.

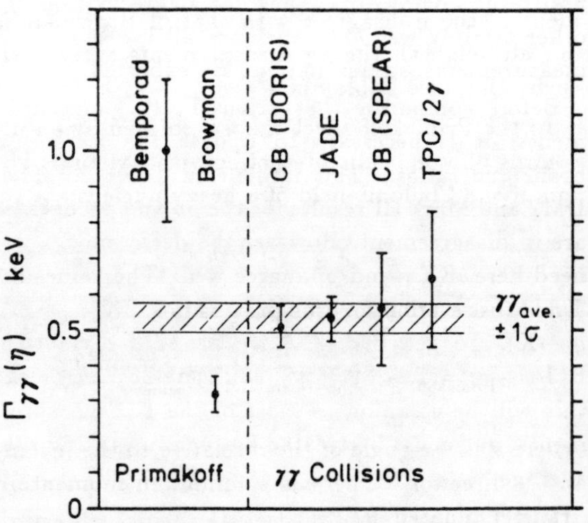

Figure 29: Comparison of Primakoff and $\gamma\gamma$ measurements of $\Gamma_{\gamma\gamma}(\eta)$. The points are not drawn in chronological order, but rather so as to put the most precise measurements next to each other.

We are left with two different types of precise measurements which don't agree with each other as well as we would like. It is tempting to take the $\gamma\gamma$ results as correct, saying that even the Browman *et al.* experiment was at too low a beam energy for a reliable η measurement. However it would be preferable to first obtain a realistic estimate of how much the uncertainty in the nuclear form factor affects the Primakoff results.

Table 9: η_c **Decay Branching Ratios.**
The $K\bar{K}\pi$ branching ratios assume I=0; thus $(K\bar{K}\pi)^0 = \frac{1}{3}K_sK^\pm\pi^\mp + \frac{1}{3}K_LK^\pm\pi^\mp + \frac{1}{6}K^+K^-\pi^0 + \frac{1}{6}K^0\bar{K}^0\pi^0$. The $2(\pi^+\pi^-)$ includes $\rho^0\rho^0$. Additional Mark III branching ratios are in Ref. [99].

X	$10^4 \times B(J/\psi\to\gamma\eta_c, \eta_c\to X)$		
	Mark III [99]	DM2 [100]	average
$K\bar{K}\pi$	6.4±1.4	7.5±1.6	6.9±1.1
$p\bar{p}$	0.14±0.07	0.13±0.04	0.13±0.03
$2(\pi^+\pi^-)$	1.6±0.6	1.3±0.3	1.4±0.3
$\rho^0\rho^0$	< 0.6	1.1±0.2	?
$\phi\phi$	1.02±0.29	0.41±0.12	?

5.2 $\Gamma_{\gamma\gamma}(\eta_c)$

The $\gamma\gamma$ flux falls with increasing mass, so that data on the heavier mesons is scarce. This year we have first results on $\gamma\gamma\to\eta_c$. They don't agree with each other terribly well, but they are all low statistics measurements, so fluctuations are expected.

Before comparing the various $\Gamma_{\gamma\gamma}(\eta_c)$ measurements, it is necessary to update the η_c decay branching ratios. This is done in Table 9. The DM2 and Mark III results for the $\rho\rho$ and $\phi\phi$ decays are in disagreement. However the decay modes we need here, $K\bar{K}\pi$ and $p\bar{p}$, agree well. The values in Table 9 are product branching ratios $B(J/\psi\to\gamma\eta_c, \eta_c\to X)$. To get $B(\eta_c\to X)$ we need to divide by $B(J/\psi\to\eta_c)=(1.27\pm0.36)\%$ [98]. The 28% error on this will be common to most of the $\Gamma_{\gamma\gamma}$ measurements, and must be treated as such in making the average.

PLUTO have published [101] their 7 observed $\gamma\gamma\to K_sK^\pm\pi^\mp$ events, shown in Fig. 30. TASSO are still working on their analysis, but reported a preliminary result this spring [102]. Now Mark II have a preliminary result [79] from 4 observed events in the η_c region (Fig. 31). All three $\Gamma_{\gamma\gamma}(\eta_c)$ values are listed in Table 10.

The MD-1 collaboration [104] at the VEPP-4 e^+e^- storage ring in Novosibirsk have looked for $e^+e^-\to e^+e^-\gamma\gamma\to e^+e^-\eta_c$. Instead of observing the η_c decay products, they measure the energies and angles of the scattered electrons and calculate the missing mass X in $e^+e^-\to e^+e^-X$. This is possible because, unlike the standard detector with solenoidal magnetic field parallel to the $e^\pm$ beams, MD-1 has a perpendicular field which bends off-energy electrons out of the beam. The analysis is done for events with the e^+ and e^- detected at scattering angles $\theta>0.5$ mrad, and two additional tracks (charged or neutral) at large angles to the beam.

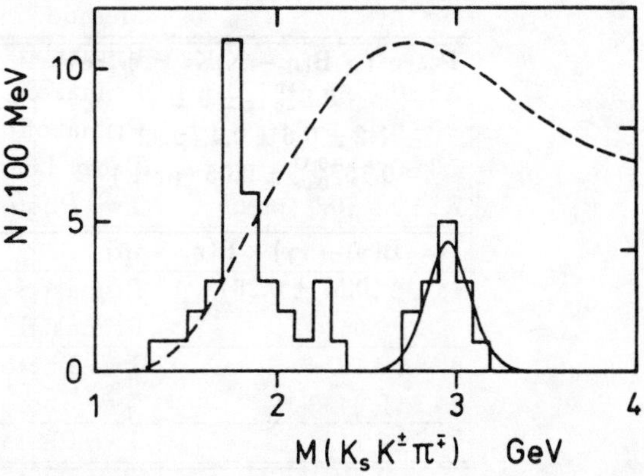

Figure 30: PLUTO evidence for $\gamma\gamma\to\eta_c$ [101]. K_s's are identified by their reconstructed $\pi^+\pi^-$ decay, requiring that the decay vertex be clearly separated from the interaction point. No particle identification is used for the charged particles, so each event appears twice, once for the $K_sK^+\pi^-$ hypothesis, once for $K_sK^-\pi^+$. The two hypotheses tend to lie close to each other in $M(K_sK^\pm\pi^\mp)$. Thus the 14 entries in the η_c peak are from only 7 events. Events consistent with K_sK_s have been rejected. The dashed curve shows the mass dependence of the efficiency times $\gamma\gamma$ flux.

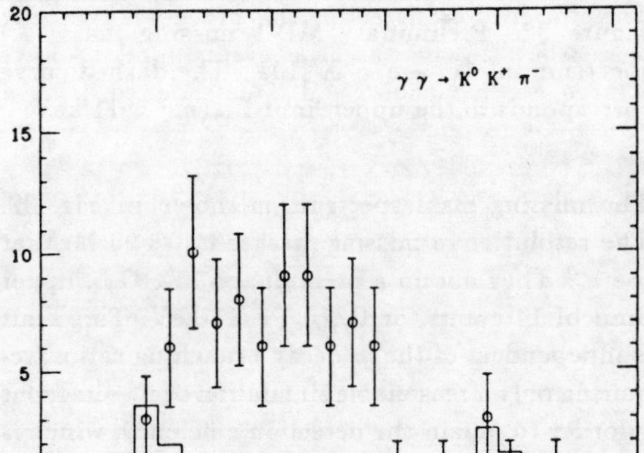

Figure 31: The points are the Mark II data on $\gamma\gamma\to\eta_c$ [79]. Only 6 out of the 84 events have more than one combination of tracks consistent with $K_sK^\pm\pi^\mp$. The solid histograms are Monte Carlo simulations of $\gamma\gamma\to\eta(1460)$ and $\gamma\gamma\to\eta_c$.

Table 10: **Measurements of the $\gamma\gamma$ width of the η_c**

	$\Gamma_{\gamma\gamma}(\eta_c)$ [keV]	Reference
$\Gamma_{\gamma\gamma}(\eta_c) \times B(\eta_c \to K_s K^\pm \pi^\mp)$ [keV]		
$0.5^{+0.2}_{-0.15} \pm 0.1$	$30 \pm 13^*$	PLUTO [101]
$1.2 \pm 0.6 \pm 0.4$ (prel.)	$71 \pm 43^*$	TASSO [102]
$0.15^{+0.11}_{-0.08} \pm 0.05$ (prel.)	$9 \pm 7^*$	Mark II [79]
	* neglecting $B(\eta_c \to K\bar{K}\pi)$ error	
$B(\eta_c \to \gamma\gamma) \times B(\eta_c \to p\bar{p})$		
$(0.57 \pm 0.26) \times 10^{-6}$	6 ± 4	R704 [103]
	neglecting $B(\eta_c \to p\bar{p})$ error	
	9 ± 4	average
	including all errors	
	< 11 (90% C.L., prel.)	MD-1 [104]

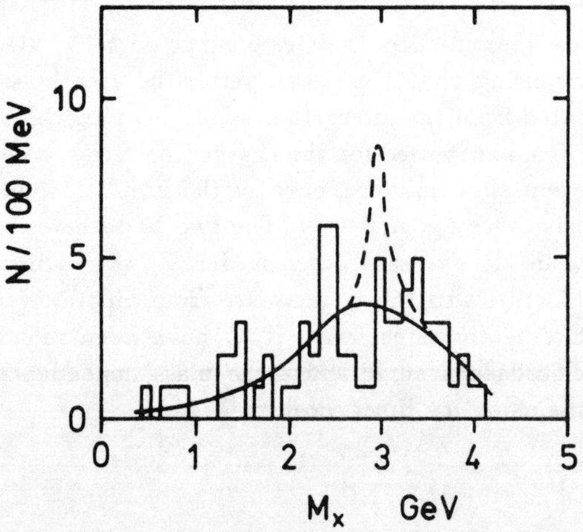

Figure 32: Preliminary MD-1 missing mass (X) spectrum in $e^+e^- \to e^+e^- X$ [104]. The dashed curve corresponds to the upper limit $\Gamma_{\gamma\gamma}(\eta_c) < 11$ keV.

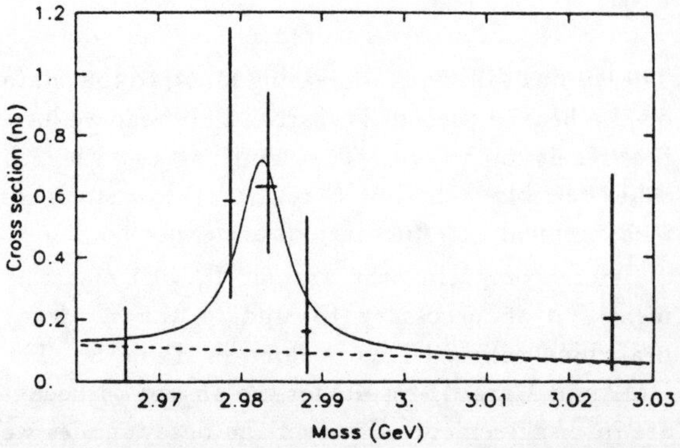

Figure 33: R704 [103] η_c data. Note that this is a scan so that where there are no points the cross section is not necessarily 0; rather there is no information there. The dashed line is the background level determined from a study of $p\bar{p} \to \pi^0\pi^0$ events.

The missing mass spectrum is shown in Fig. 32. The resolution in missing mass is $\sigma_m \approx 90$ MeV at the η_c. They obtain a preliminary 90% C.L. upper limit of 14 events, or $\Gamma_{\gamma\gamma}(\eta_c) < 11$ keV. This limit is independent of the η_c decay branching ratios, requiring only a reasonable simulation of $\eta_c \to$hadrons in order to obtain the detection efficiency, which is about 20%.

Yet another way of measuring $\Gamma_{\gamma\gamma}(\eta_c)$ has been developed by R704 at the ISR. They produce the η_c's directly in $p\bar{p}$ annihilation, scanning the beam energy over the η_c region. The ISR beam gives them a mass resolution of ~0.3 MeV. To measure $p\bar{p} \to \eta_c \to \gamma\gamma$, they must suppress the background from $p\bar{p} \to \pi^0\pi^0$ with two γ's outside the detector, or $\pi^0 \to \gamma\gamma$ with the γ's so close together that they are indistinguishable from a single γ. They were only able to collect small numbers of events before the ISR shut down. Slight changes in the cuts, which cause a few events more or less to be accepted, make noticable changes in $\Gamma_{\gamma\gamma}(\eta_c)$. Thus several values have been quoted as the analysis has progressed; however they are all compatible within the statistical error. Their η_c scan is shown in Fig. 33. The total width of the η_c cannot be very well determined from this data: a fit gives $\Gamma_{tot} = 7.0^{+7.5}_{-7.0}$ MeV; while the peak height is better determined: $B(\eta_c \to \gamma\gamma)B(\eta_c \to p\bar{p}) = 0.68^{+0.42}_{-0.31} \times 10^{-6}$. The value in Table 10 is from a fit with Γ_{tot} fixed at the Crystal Ball [98] value of 11.5±4.5 MeV.

The various $\Gamma_{\gamma\gamma}(\eta_c)$ measurements are compared in Fig. 34. They exhibit the usual effect that the first experiment that sees a significant signal is observing an upward fluctuation, and the value tends to decrease as more information comes in. The present average is $\Gamma_{\gamma\gamma}(\eta_c) = 9 \pm 4$ keV (χ^2/d.f. = 5/3). This may be a biased average, since experiments with even lower values may not bother to

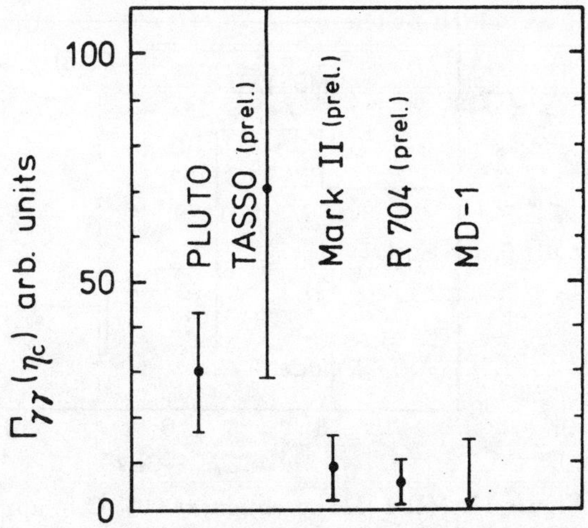

Figure 34: $\Gamma_{\gamma\gamma}(\eta_c)$ measurements. The comparison is complicated by common systematic errors. I have put no error for $B(\eta_c \to K\bar{K}\pi)$ on the 3 $K\bar{K}\pi$ points, and assigned the relative error between it and $B(\eta_c \to p\bar{p})$ to the single $p\bar{p}$ point. The MD-1 upper limit has been increased by the full error on $B(K\bar{K}\pi)$. This confuses the absolute y-scale, but is (I think) the fairest comparison between the measurements, which is the point here.

report them.

Clearly all the experiments need more statistics. The most promising opportunity is with experiment E760 [105] being built to repeat the $p\bar{p}$ annihilation measurement at the Fermilab $\bar{p}$ accumulator ring.

The expected $\Gamma_{\gamma\gamma}(\eta_c)$ can be related to $\Gamma_{ee}(J/\psi)$. Both the η_c and the J/ψ are bound $c\bar{c}$ states, the former is the $1\,^1S_0$ and the latter is the $1\,^3S_1$. The approximation where they have the same wave function at the origin gives [106]

$$\Gamma_{\gamma\gamma}(\eta_c) = 3\ e_q^2\ \Gamma_{ee}(J/\psi) \times \left(1 + 1.96\ \frac{\alpha_s}{\pi}\right),$$

where the last factor is the 1st order QCD correction. The measured [1] $\Gamma_{ee}(J/\psi) = 4.7\pm0.3$ keV leads to a predicted $\Gamma_{\gamma\gamma}(\eta_c)\sim 7$ keV for α_s=0.22. QCD sum rules [107] give ~4 keV.

6 $b\bar{b}$ Mesons

6.1 Limits on Other Narrow 1^{--} Resonances in the Υ Region

The Υ's are the $n\,^3S_1$ bound states of $b\bar{b}$ quarks. The ground state (n=1), called the $\Upsilon(1S)$, was discovered via its $\mu^+\mu^-$ decay in $p\,N \to \mu^+\mu^-\,X$ [108]. There indications were also seen for the first two excited states, the $\Upsilon(2S)$ and $\Upsilon(3S)$ [109]. Since then the Υ family has been largely the domain of the e^+e^- machines. However now E605 at Fermilab [110] have achieved ~0.3% mass resolution and clearly separated the first three Υ's in the reaction $p\,Cu \to \mu^+\mu^-\,X$, as shown in Fig. 35.

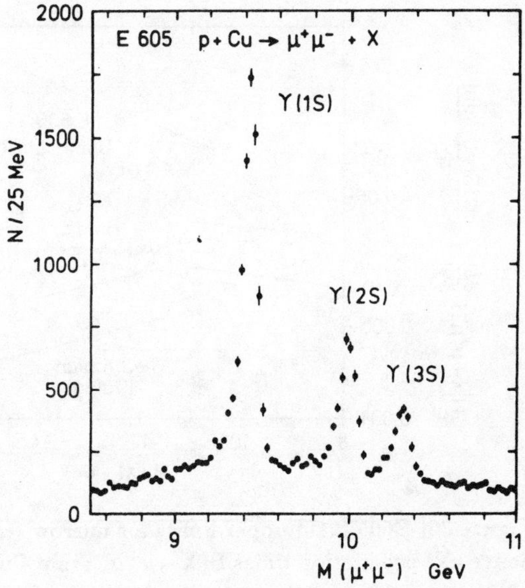

Figure 35: E605 [110] $M(\mu^+\mu^-)$ distribution from $p\,Cu \to \mu^+\mu^- X$.

These data have been used to extract upper limits for the production of other resonances in the Υ region (Fig. 36) relative to the Drell-Yan production of $\mu^+\mu^-$. Although they are not sensitive enough to rule out Higgs particle or Technipion production, the expected levels for which are also shown in Fig. 36, they are useful for testing other hypotheses that come up.

One such hypothesis was suggested by Tye and Rosenfeld [111] after the 1984 report by Crystal Ball [112] of a narrow resonance ζ seen at 8.3 GeV in radiative $\Upsilon(1S)$ decays.

Tye and Rosenfeld speculated that the ζ was the 1S bound state of a pair of scalar quarks, $\tilde{Q}\bar{\tilde{Q}}$, and that it was coming not from $\Upsilon(1S)$ decays, but from those of the $\tilde{Q}\bar{\tilde{Q}}(3P)\ 1^{--}$ state which would be close to the $\Upsilon(1S)$ in mass. This explained the ζ's non-appearance in $\Upsilon(2S)$ decays [112]; and also the fact that it wasn't seen [113] by CLEO or CUSB at the CESR e^+e^- machine, which has a ~4 MeV center of mass resolution compared to DORIS's ~8 MeV. The resolution difference was perhaps even greater, since the original Crystal Ball $\Upsilon(1S)$ data was accumulated in 3 short run periods, a situation not conducive to precise setting of the beam energy on the $\Upsilon(1S)$ mass.

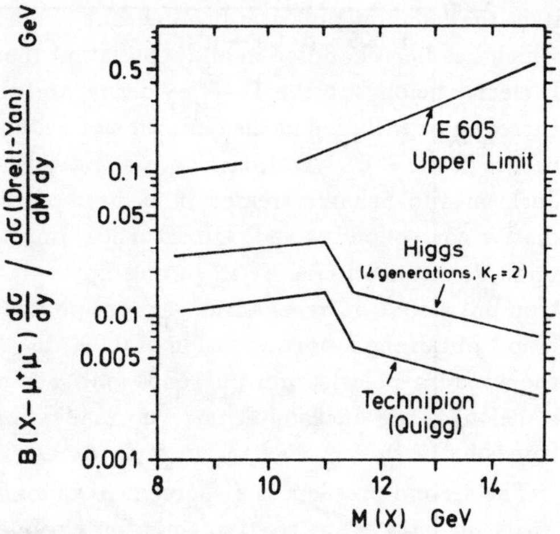

Figure 36: E605 [110] upper limits for narrow resonance (X) production times $B(X \to \mu^+\mu^-)$ in p Cu→ $\mu^+\mu^-$+ anything, relative to the Drell-Yan $\mu^+\mu^-$ cross section.

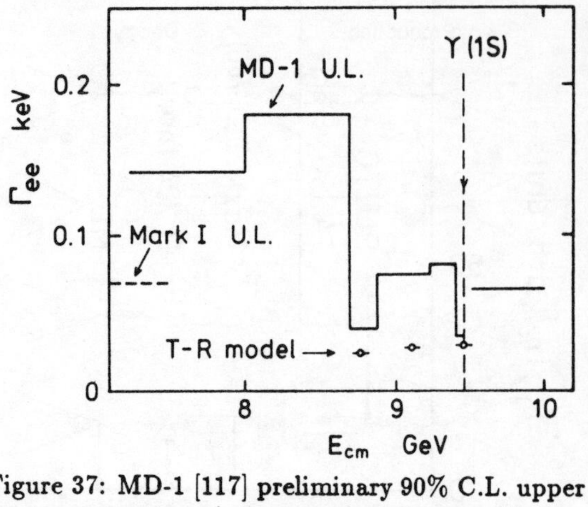

Figure 37: MD-1 [117] preliminary 90% C.L. upper limits on Γ_{ee} of a narrow resonance (Γ<4 MeV). The energy range is divided into 7 regions with equal integrated luminosity (excluding the $\Upsilon(1S)$); for each region the maximum Γ_{ee} is plotted. The Mark I [118] upper limit is also sketched. The LENA upper limit [119] is off scale. The points are the predicted masses and Γ_{ee}'s of the 1P, 2P and 3P scalar quark states in the Tye-Rosenfeld model [111].

Later in 1984 Crystal Ball took more data on the $\Upsilon(1S)$, taking great care to put the beam energy at the peak of the $\Upsilon(1S)$. No ζ was seen, and an upper limit set which contradicted the previous signal, assuming it came from the $\Upsilon(1S)$ [114,115]. Elimination of the $Q\bar{Q}$ explanation of the ζ signal was achieved in 1986 with data taken 12 MeV above and 12 MeV below the $\Upsilon(1S)$, where again no peak was seen [116]. Thus the original peak must have been a statistical fluctuation.

The idea of other bound states besides the Υ remains an interesting possibility however, and wherever one can look for new phenomena one should. MD-1 [117] have scanned the e^+e^- center-of-mass region from 7.2 to 10 GeV setting the limits shown in Fig. 37 on Γ_{ee} of a narrow resonance. Their data are a substantial improvement over what was previously available in this energy region.

6.2 Γ_{ee} and Radiative Corrections *Corrections*

A part of Crystal Ball's $\Upsilon(1S)$ running this year was a careful scan over the resonance to measure its leptonic width Γ_{ee}. That result is not ready yet, but I report on our progress on understanding how to extract Γ_{ee} from the scan data. In principle Γ_{ee} is proportional to the area of the e^+e^-→hadrons resonance curve, just as described previously for $K\bar{K}$ scattering (Eq. (2)). However in the case of the Υ, the resonance width is much smaller than the energy resolution Δ of the e^+e^- machine, and the Υ Breit-Wigner curve is replaced by the Gaussian of the machine resolution:

$$\begin{aligned} \bar{\sigma}(W) &= H_0\, e^{-(W-M)^2/2\Delta^2}, \text{ where} \\ H_0 &= \Gamma_{ee} \frac{\Gamma_{had}}{\Gamma_{tot}} \frac{6\pi^2}{M^2} \frac{1}{\Delta\sqrt{2\pi}} \end{aligned} \tag{9}$$

is the new peak height. The area of the resonance curve is not changed by this smearing.

Initial state radiation (Fig. 38d) adds to the Gaussian a high energy tail, as shown in Fig. 39, because at a nominal setting above the Υ, radiation of a photon before the e^+e^- annihilation can bring the effective energy down to the Υ. Until now, almost all Υ-experimenters relied on the formulation of Jackson and Scharre [120] for this radiative correction. Last year rumors reached us of a Russian paper by Kuraev and Fadin which claimed to find significant deviations from the Jackson-Scharre result. Even after translation into English [121] it remained a formidable obstacle. However now the VEPP-4 experimenters [122] have published a paper using an approximation to the Kuraev-Fadin form, and very considerately give their formula explicitly. It is this last which I shall use in the following.

As shown in Fig. 39, the new and old formulations agree quite well in the shape. However the normalisation is significantly affected. Evaluated for Δ=4 MeV, the peak height of the Jackson-Scharre curve is 0.645 H_0, while that of VEPP-4 is 0.584 H_0, a 10%

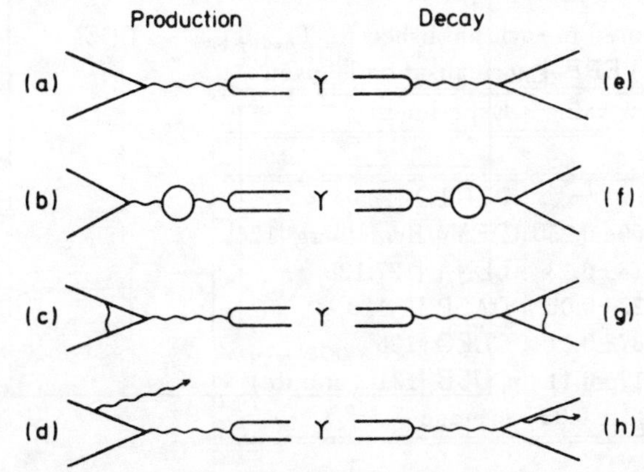

Figure 38: Diagrams for $e^+e^- \to \Upsilon$ and for $\Upsilon \to e^+e^-$.

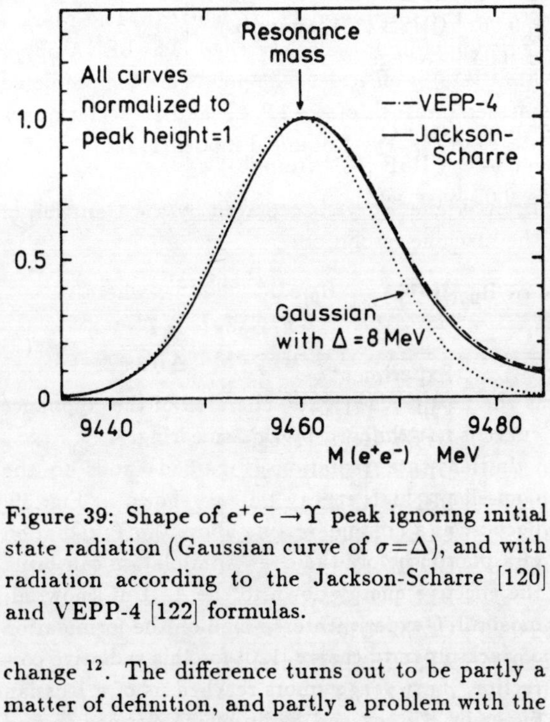

Figure 39: Shape of $e^+e^- \to \Upsilon$ peak ignoring initial state radiation (Gaussian curve of $\sigma = \Delta$), and with radiation according to the Jackson-Scharre [120] and VEPP-4 [122] formulas.

change [12]. The difference turns out to be partly a matter of definition, and partly a problem with the treatment of low energy photons.

The matter of definition was pointed out by Tsai [123] in 1983, but overlooked by the experimenters. Although we measure Γ_{ee} in $e^+e^- \to$hadrons, we use it to extract the Υ total width Γ_{tot} from $B(\Upsilon \to e^+e^-) \equiv \Gamma_{ee}/\Gamma_{tot}$. For this equivalence to hold, the definition of Γ_{ee} must include all the diagrams which contribute to the $\Upsilon \to e^+e^-$ decay, shown in Fig. 38. The $e^+e^-\gamma$ final state has been included in both the calculation and the experimental branching ratio, and is not the problem. Rather it is the vacuum polarisation diagram (Fig. 38b), which has been handled in an inconsistent manner. It clearly belongs to the $\Upsilon \to e^+e^-$ decay, and should therefore be included in the definition of Γ_{ee}. It also occurs in the $e^+e^- \to$hadrons process. However here Jackson and Scharre treated it as part of the radiative correction instead. CLEO [130] "improved" upon Jackson-Scharre by including not only electron but also μ, τ, and hadron vacuum polarisation loops, obtaining a normalisation of 0.693 H_0. With the vacuum polarisation put back into Γ_{ee} where it belongs, the Jackson-Scharre normalisation becomes 0.617 H_0.

The second problem is disagreement among theorists on how to do the "soft-photon exponentiation". Jackson-Scharre applied it only to diagram 38d, while current majority vote (e.g. [124,121]) favors applying it (or something very close to it [123]) to all terms. The proper resolution to the dispute will come from a calculation of the next order. For now I will take the majority opinion, with its peak normalisation of 0.584 H_0, and the caution that there remains an $\sim$5% theoretical uncertainty.

Since the shapes of all the curves agree so well, we can renormalise the published results to the new consistent definition of Γ_{ee}. As can be seen from Eq. (9), the observed peak height is proportional to $\Gamma_{ee}\ \Gamma_{had}/\Gamma_{tot}$ times the radiative corrections normalisation factor. Reducing that factor from its original 0.645 to 0.584 increases the extracted $\Gamma_{ee}\ \Gamma_{had}/\Gamma_{tot}$ values by 10%. The re-normalised values are given in Table 11.

In order to get Γ_{ee}, we need the $\Gamma_{had}/\Gamma_{tot}$ factor which accounts for the fact that in this measurement we use only decays $\Upsilon \to$hadrons. Assuming lepton universality and that the Υ decays only to charged lepton pairs or to hadrons gives $\Gamma_{tot} = 3\Gamma_{ee} + \Gamma_{had}$, or $\Gamma_{had}/\Gamma_{tot} = 1 - 3B_{\mu\mu}$.

New world averages for $B_{\mu\mu}$ are calculated in Table 12, including the new values presented at this conference: a new $B_{\mu\mu}(1S)$ measurement from ARGUS [137], and the first statistically significant $B_{\mu\mu}(3S)$, provided by CUSB [141]. In Table 13 the average $\Gamma_{ee}\ \Gamma_{had}/\Gamma_{tot}$ and $B_{\mu\mu}$ values are combined to get Γ_{ee} and Γ_{tot}. The results are significantly different from those in the latest PDG tables [1].

There are several applications of these measurements. Both Γ_{ee} and Γ_{tot} are proportional to the $Q\bar{Q}$ wave function at the origin; Γ_{tot} is also proportional to α_s^3. Thus Γ_{ee} is a good test of the $Q\bar{Q}$ potential, subject of course to the usual cautions when QCD gets in the picture. In $B_{\mu\mu}$, the ratio of the two, the effect of the wave function drops out leaving only the α_s dependence, but unfortunately

[12] Note that the lowest order QED corrections have reduced the peak height by $\sim$35% from the Gaussian's H_0, an astonishingly large effect for $\alpha = 1/137$.

Table 11: **Measurements of $\Gamma_{ee}\ \Gamma_{had}/\Gamma_{tot}$ (in keV)**

The type of radiative correction that was used in each published $\Gamma_{ee}\ \Gamma_{had}/\Gamma_{tot}$ value is listed, and the new value with VEPP-4 normalisation is given.

Published $\Gamma_{ee}\ \Gamma_{had}/\Gamma_{tot}$	Rad. corr.	new value	Experiment
$\Upsilon(1S)$			
?	Greco	–	PLUTO [125]
1.00±0.23	JS	1.09±0.25	DESY-Heidelberg [126]
1.10±0.07±0.11	Greco	1.13±0.13	LENA [127,128]
1.12±0.07±0.04	JS	1.23±0.09	DASP II [129]
1.17±0.05±0.08	JS, full δ_{vac}	1.37±0.11	CLEO [130]
1.04±0.05±0.09	JS	1.17±0.11	CUSB [131] (unpub.)
		1.22 ± 0.05	average
$\Upsilon(2S)$			
0.37±0.16	JS	0.41±0.18	DESY-Heidelberg [126]
$0.53 \pm 0.07^{+0.09}_{-0.05}$	Greco	0.54±0.12	LENA [132,128]
0.55±0.11±0.06	JS	0.60±0.14	DASP II [129]
0.49±0.03±0.04	JS, full δ_{vac}	0.58±0.06	CLEO [130]
0.53±0.03±0.05	JS	0.59±0.06	CUSB [131] (unpub.)
		0.57 ± 0.04	average
$\Upsilon(3S)$			
0.38±0.03±0.03	JS, full δ_{vac}	0.45±0.05	CLEO [130]
0.35±0.02±0.03	JS	0.39±0.04	CUSB [131] (unpub.)
		0.41 ± 0.03	average

Table 12: **Measurements of $B_{\mu\mu}$(in %)**

Reaction		Experiment
	$\Upsilon(1S)$	
$\Upsilon \to \mu^+\mu^-$	2.2±2.0	PLUTO [125]
$\Upsilon \to \mu^+\mu^-$	$1.4^{+3.4}_{-1.4}$	DESY-Heid. [126]
$\Upsilon \to \mu^+\mu^-$	3.2±1.3±0.3	DASP II [129]
$\Upsilon \to \mu^+\mu^-$	3.8±1.5±0.2	LENA [127]
$\Upsilon \to \mu^+\mu^-$	2.7±0.3±0.3	CLEO [133]
$\Upsilon \to \mu^+\mu^-$	2.7±0.3±0.3	CUSB [134]
$\Upsilon \to ee$	5.1±3.0	PLUTO [135]
$\Upsilon' \to \pi^+\pi^-\Upsilon,\ \Upsilon \to \mu^+\mu^-, e^+e^-$	2.84±0.18±0.20	CLEO [136]
$\Upsilon' \to \pi^+\pi^-\Upsilon,\ \Upsilon \to \mu^+\mu^-, e^+e^-$	2.39±0.12±0.14	ARGUS [137] (prel.)
$\Upsilon \to \tau\tau$	3.4±0.4±0.4	CLEO [138]
	2.63 ± 0.13	average
	$\Upsilon(2S)$	
$\Upsilon' \to \mu\mu$	1.8±0.8±0.5	CLEO [139]
$\Upsilon' \to \mu\mu$	1.9±0.3±0.5	CUSB [134]
$\Upsilon' \to \mu\mu$	1.0±0.6±0.5*	ARGUS [140]
$\Upsilon' \to \tau\tau$	1.7±1.5±0.6	CLEO [139]
	1.6 ± 0.4	average
	$\Upsilon(3S)$	
$\Upsilon'' \to \mu\mu$	3.3±1.3±0.7	CLEO [133]
$\Upsilon'' \to \mu\mu$	1.53±0.29±0.21	CUSB [141]
	1.6 ± 0.4	average

* The ARGUS 2S value is scaled to the average 1S value with
$B_{\mu\mu}(2S) = 1.57 \pm 0.59 \pm 0.53 + 2.1(B_{\mu\mu}(1S) - 2.9)$ (in %) [140].

Table 13: **Average values of $B_{\mu\mu}$, Γ_{ee} and Γ_{tot}.**

	New Values			PDG Values [1]		
Resonance	$B_{\mu\mu}$ [%]	Γ_{ee} [keV]	Γ_{tot} [keV]	$B_{\mu\mu}$ [%]	Γ_{ee} [keV]	Γ_{tot} [keV]
$\Upsilon(1S)$	2.63±0.13	1.33±0.06	51±3	2.8±0.2	1.22±0.05	43±3
$\Upsilon(2S)$	1.6±0.4	0.60±0.04	37±10	1.8±0.4	0.54±0.03	30±7
$\Upsilon(3S)$	1.6±0.4	0.43±0.03	27±6	3.3±1.5	0.40±0.03	12^{+10}_{-4}

also a large sensitivity to QCD problems. My not-yet-updated but still relevant thoughts on Γ_{ee} and $B_{\mu\mu}$ can be found in Ref. [142]. The latest Λ_{QCD} values were given by J. Lee-Franzini at this conference [141].

The interest in Γ_{tot} is more subtle, but pervasive. Experimenters measure Υ decay branching ratios, while theorists calculate partial widths. All comparisons between the two rely on Γ_{tot}!

6.3 $\chi_b(2P)$ States

The CLEO and CUSB experiments have been running on the $\Upsilon(3S)$. Their results presented at this conference represent about 1/3 of the total luminosity they hope to accumulate this year.

For this run the CUSB detector was upgraded to CUSB-II by the installation of a BGO inner detector. The energy resolution for electromagnetically showering particles is now $\sigma_E/E = 2.2\%/\sqrt[4]{E}$, for E up to ~1 GeV [141]. This resolution is about 3 MeV in the range $E_\gamma = 85 - 100$ MeV, sufficient to separate the three γ lines corresponding to $\Upsilon(3S)\to\gamma\ \chi_{bJ}(2P)$, $J = 2,1,0$ in the inclusive γ spectrum from $\Upsilon(3S)\to\gamma$+hadrons (Fig. 40). The γ energies obtained from a fit to that spectrum are:

$$\chi_{b2} \ : \ 86.5 \pm 0.7 \text{ MeV}$$
$$\chi_{b1} \ : \ 99.3 \pm 0.8 \text{ MeV}$$
$$\chi_{b0} \ : \ 124.2 \pm 2.3 \text{ MeV}$$

This spectrum, along with the data on the exclusive decays $\Upsilon(3S)\to\gamma\ \chi_b(2P)$, $\chi_b(2P)\to\gamma\Upsilon$, $\Upsilon\to\ell^+\ell^-$, provide several branching ratios which can be compared to those expected from potential model calculations, and also can be used to extract total widths of the $\chi_b(2P)$ states for comparison with QCD calculations. All of these tests will be significantly improved when he full data sample is available. The currect status was fully covered in the talk of J. Lee-Franzini at this conference. Here I will concentrate on the relative mass splitting of the $\chi_b(2P)$ states and what it means for the heavy quark potential.

Combining their inclusive and exclusive measurements of the $\chi_b(2P)$ states, CUSB obtain

$$\mathbf{r}(2P) \equiv \frac{M_2 - M_1}{M_1 - M_0} = 0.57 \pm 0.06,$$

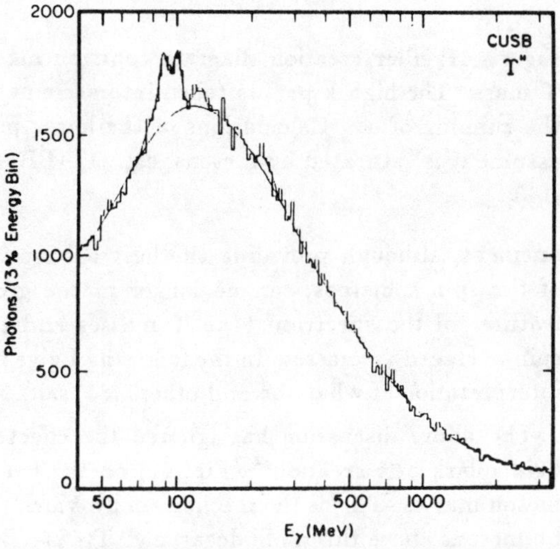

Figure 40: CUSB-II [141] inclusive γ spectrum from $\Upsilon(3S)\to\gamma$+hadrons.

where M_J is the mass of the spin J $\chi_b(2P)$ state. The parameter **r** is sensitive to the Lorentz form of the confining part[13] kr+C of the $Q\bar{Q}$ potential [143], which can be written approximately as

$$V(r) = -\frac{4}{3}\frac{\alpha_s}{r} + kr + C.$$

The α_s/r term is expected in perturbative QCD to come from the exchange of a gluon, which is a vector particle. However the confining part is non-perturbative and not that well understood, although most theorists expect it to behave as a scalar. Without any confining term in the potential, **r**=0.8. Adding scalar confinement reduces **r**, while vector confinement increases it. Although the earlier CUSB measurements [144] had large errors: $\mathbf{r}(2P) = 0.85\pm0.1\pm0.3$ and $\mathbf{r}(1P) = 0.93\pm0.1\pm0.2$, their tendency to favor vector confinement caused some consternation. The new more precise CUSB 2P value, last year's world average of $\mathbf{r}(1P) = 0.66\pm0.05$ [143], and the χ_c value of 0.48±0.01 are all well on the side of scalar confinement.

However now that the data is behaving, the theory is not. N. Byers [145] reported that scalar con-

[13] I use bold-face **r** for the ratio, and italic *r* for the distance in the potential *V(r)*.

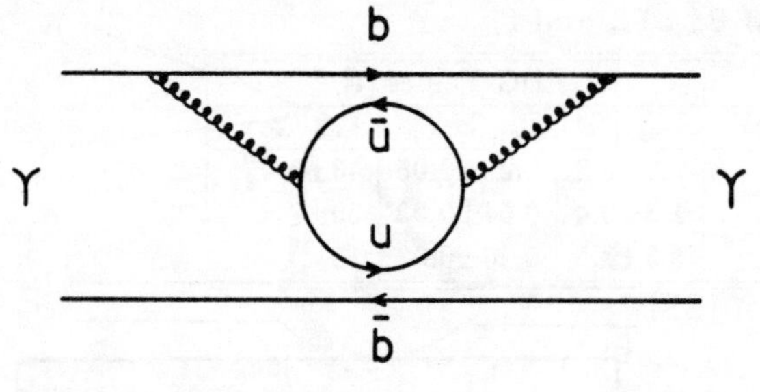

Figure 41: Pair creation diagram contributing to Υ mass. The high k part is taken into account by the running of α_s. Calculations of the low k part assume it is saturated by mesons, e.g. $\Upsilon \to B\bar{B} \to \Upsilon$ [145].

finement, although providing the best description of the spin splittings, can no longer fit the gross features of the spectrum: the Υ masses and the spin-averaged χ_b masses. In the following I give my interpretation of what she and others [85] said.

The above discussion has ignored the effect of light quark pair creation[14] (Fig. 41) on the heavy meson masses. This is the mechanism by which the $b\bar{b}$ mesons above threshold decay, e.g. $\Upsilon(4S) \to B\bar{B}$. The decay is forbidden for the lighter Υ's, but can exist as a virtual intermediate state, and thus affect the mass. Again it is clear how to treat the perturbative part. A gluon exchanged in the t-channel is what makes the α_s/r part of the potential. Exchanged in the s-channel as in Fig. 41, it creates a light-quark pair. The use of the same exchange particle in the t and s channels is called "crossing", and is uncontroversial in the perturbative case.

Less clear is how to calculate the pair creation by the confining part of the potential. Confinement is probably not due to single particle exchange, but to the collective action of many gluons. The term scalar confinement is to be understood as "effective", i.e. its Lorentz properties are as though a scalar particle were being exchanged. But this is not enough to tell us how to create pairs. Byers makes the simplest approximation: that a scalar particle really is being exchanged in the t channel for the potential, and in the s channel to make pairs. The scalar particle is described by the kr+C terms in the potential. C sets the zero of the mass scale, and is large and negative. Pair production by a scalar particle is proportional to the potential, and both the kr and the C terms contribute strongly, with opposite sign. The net shift in masses is large. No successful fit to the measured masses could be found with scalar confinement model when the mass shifts due to pair creation were included. (Previous fits ignored pair creation; then the type of confinement has no effect on the Υ and average χ_b masses.)

[14] Vacuum polarisation come back to haunt us again!

If confinement is via a vector particle exchange, the pair creation by that particle is proportional to the derivative of the potential, so the C term doesn't contribute. The mass shifts are still substantial: 126 MeV for the $\chi_b(1P)$ center of gravity. However they can be accomodated by changing the parameters of the potential. A good fit was found with the following potential (where I have put in the $\hbar c$'s, r is in fm, and V in GeV):

$$V(r) = \frac{-0.097}{r} + 1.57\, r - 0.97\ .$$

Thus Byers concludes that the spin-averaged masses require vector confinement, while the spin-splittings need scalar confinement, so that we are in trouble. This conclusion depends however on the "crossing" assumption. Models with different treatment of pair creation can fit the data. For example the $^3P_0$ model [85] finds only a ~20 MeV shift for the χ_b states using scalar confinement.

6.4 Search for the $^1P_1$

Still missing in heavy quark spectroscopy are the $^1P_1$ states of $c\bar{c}$ and $b\bar{b}$ [146]. They have $J^{PC}=1^{+-}$, and thus cannot be produced directly in e^+e^- annihilation or in radiative decays of the Υ's or ψ's.

The $^1P_1$ is an L=1 state like the ^3P_J χ_c and χ_b, but with the quark spins anti-parallel instead of parallel. Since the net quark spin S=0, there is no $\vec{L}\cdot\vec{S}$ force contributing to the $^1P_1$ mass. The $\vec{L}\cdot\vec{S}$ contribution to ^3P_J averages to 0 in taking the spin-weighted average (usually refered to as the center-of-gravity):

$$M(^3P_{cog}) \equiv \frac{5M(^3P_2) + 3M(^3P_1) + M(^3P_0)}{9}.$$

The only contribution to a mass difference between the $^3P_{cog}$ and the $^1P_1$ is the $s_1 \cdot s_2$ force. In lowest order it is proportional to the square of the wave function at the origin, which is 0 for P states. Thus we expect the $^1P_1$ to be very close to the $^3P_{cog}$.

R704 [147] have looked for the $c\bar{c}$ $^1P_1$ in $p\bar{p} \to {}^1P_1 \to J/\psi$+anything. They have 5 events at the right mass, corresponding to a 2.3σ effect. If this signal is real, their new experiment E760 at Fermilab should be able to confirm it with many more events, and provide a statistically reliable mass determination.

A possibility for observing the $b\bar{b}$ $^1P_1$ state is the reaction $\Upsilon(3S) \to \pi\pi\, ^1P_1$ [149]. The new CLEO [148] recoil mass spectrum against $\pi^+\pi^-$ in

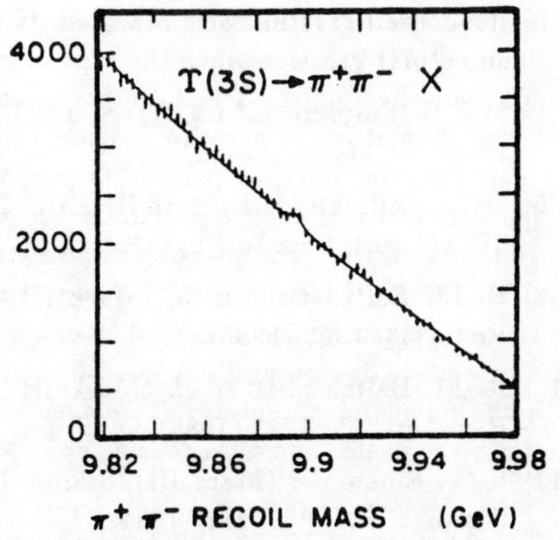

Figure 42: CLEO [148] missing mass spectrum against $\pi^+\pi^-$ from $\Upsilon(3S)\to\pi^+\pi^-$+hadrons.

$\Upsilon(3S)\to\pi^+\pi^-$+hadrons is shown for the $^1P_1$ region in Fig. 42. The peak near the expected $^1P_1$ mass contains ~330 events, but due to the large background is only a 2.5σ effect. Again, if it is real, more data will tell, and CLEO hopes to at least double the data sample by the end of this year. The present effect corresponds to a branching ratio $B(3S\to\pi^+\pi^-X)\sim0.4\%$ or an upper limit of 0.6%.

Theoretically the $\Upsilon(3S)\to\pi\pi\,^1P_1$ decay is treated as a two step process (Fig. 43): the $Q\bar{Q}$ system emits two gluons, which then turn into pions. Kuang and Yan [149] assume that the $gg\to\pi\pi$ transition occurs with probability 1. This assumption works well for the $\Upsilon(3S,2S)\to\pi\pi\Upsilon(1S)$ decays, but predicts too large a rate for $\Upsilon(3S)\to\pi\pi\Upsilon(2S)$, where the smaller energy available makes ignoring mass effects more risky. For $B(\Upsilon(3S)\to\pi\pi\,^1P_1)$ it gives ~1%.

Voloshin [150] has disputed the validity of this picture. The ITEP group have analysed the matrix elements $\langle\pi\pi|\theta|0\rangle$, where θ is the QCD stress tensor, containing products of color electric and magnetic fields, analogous to the stress tensor used in textbook electricity and magnetism. The $E1\cdot E1$ element, which is used in $\Upsilon(3S)\to\pi\pi\Upsilon(1S)$, is enhanced by the triangle anomaly. This enhancement does not apply to the $E1\cdot M1$ of $\Upsilon(3S)\to\pi\pi\,^1P_1$, so its rate should be relatively suppressed. Voloshin estimates $B(\Upsilon(3S)\to\pi\pi\,^1P_1)\leq 10^{-4}$ and $B(\Upsilon(3S)\to\pi^0\,^1P_1)\sim10^{-3}$.

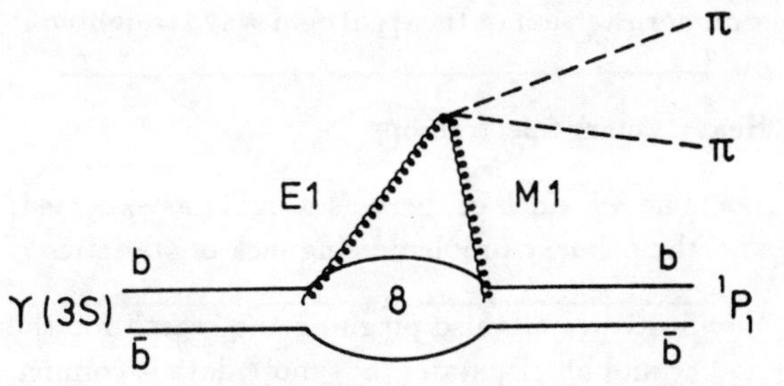

Figure 43: Diagram for $\Upsilon(3S)\to\pi\pi\,^1P_1$. Note that the gluons are emitted from the $Q\bar{Q}$ system as a whole. Gluons of such low energy don't have enough spatial resolution to see the individual quarks. Physical states must be color singlets, but since a gluon carries color, the $Q\bar{Q}$ after the first gluon emission is in a color octet state. Upon emitting the second gluon, it becomes a color singlet again.

Conclusions

Light mesons:

- Time reversal invariance tells us that mesons which decay to $K\bar{K}$ are also produced in $K\bar{K}$ collisions, even if they are glueballs.
- The 2.2 GeV region is populated by $\phi\phi$, $K\bar{K}$, and $\eta\eta'$ resonances, largely with $J^{PC}=2^{++}$. Along with the $f_2(1720)$ seen in radiative J/ψ decay, there are too many neutral 2^{++} mesons for a $q\bar{q}$ nonet.
- Systematic comparisons of the production of final states X in $J/\psi\to\gamma X$, ωX, and ϕX are very promising, but need spin-parity measurements.
- 2 high-statistics experiments have seen $\eta(1420)\to a_0\pi$; 1 has seen $f_1(1420)\to K^*K$. The production mechanisms are different, so it is quite possible that there are two particles at 1420 MeV. However cross-checks between the experiments would help us to be convinced, or not.
- Evidence from $Q^2>0$ $\gamma\gamma$ collisions supports the existence of an $f_1(1420)$. However in its hadronic production and decay it doesn't behave like the partner of the $f_1(1285)$. $f_1(1285)$, $f_1(1420)$, and $f_1(1525)$ are one too many I=0 1^{++} mesons for one $q\bar{q}$ nonet.
- $Q^2=0$ $\gamma\gamma$ collisions show no sign of any of the radially excited 0^{-+} candidates: $\eta(1275)$, $\eta(1420)$, $\eta(1460)$. Especially if the new $\eta(1450)$ is confirmed, we have too many for a $q\bar{q}$ nonet.

We have been frustrated in looking for a pure glueball candidate. But a complicated spectrum due to mixtures of $q\bar{q}$, gg, and $gq\bar{q}$ is a more reasonable expectation, and a better match to the chaos which is emerging (again) in light meson spectroscopy. Will a modern version of the eight-fold way straighten it out?

Heavy Quark Spectroscopy:

- The $\gamma\gamma$ width of the η_c is roughly as expected, the primary problem being lack of statistics.

- Statistics are also plaguing the search for the $c\bar{c}$ and $b\bar{b}$ $^1P_1$ states, but more data is coming. Calculations of $\Upsilon(3S)\rightarrow\pi\pi\,^1P_1$ should be done soon if they are to count as **pre**dictions.

- The leptonic branching ratio of the $\Upsilon(3S)$ has been measured, and the leptonic and total widths of the $\Upsilon(1S)$ - $\Upsilon(3S)$ updated with corrected use of radiative corrections.

- The $\chi_b(2P)$ mass splitting has now been well measured, and agrees with theorists' (previous) preference for scalar confinement. But now there are theoretical doubts in some quarters.

The heavy quark spectroscopy measured so far is a gratifying confirmation of the heavy quark potential model, and demonstrates that, at least in this domain, meson spectroscopy can be understood.

Acknowlegements

This talk is based on information presented by the speakers in the Two Photon, Heavy Quark, and Light Quark parallel sessions, and on discussions with many of them. For help in clarifying various issues after the conference, I thank C. Bemporad, F. Berends, N. Byers, R. Cester, S.-U. Chung, S. Godfrey, N. Isgur, K. Königsmann, L. Köpke, R. Longacre, D. Morgan, V. Obraztsov, B. Ratcliff, M. Reidenbach, H.-W. Siebert, Y. S. Tsai, and D. Williams. In particular I am grateful for many discussions with Helmut Marsiske, and for the drawings by Ursula Rehder.

References

[1] Review of Particle Properties, Particle Data Group, Phys. Lett. 170B (1986).

[2] A. Etkin *et al.*, Phys. Rev. Lett. 49, 1620 (1982) S. J. Lindenbaum, Proc. Int. Conf. on HEP, Bari, July 1985, p. 311.

[3] F. Binon (GAMS), talk 10FB; D. Alde *et al.*, Nucl. Phys. B269, 485 (1986).

[4] R. Longacre, talk 10RL; A. Etkin *et al.*, paper 1899, sub. to Phys. Lett. B.

[5] J. D. Jackson, Nuovo Cim. 34, 1644 (1964).

[6] R. A. Lee (Crystal Ball), Stanford Ph.D. thesis, SLAC-282 (1985).

[7] M.E.B. Franklin (Mark II), Stanford Ph.D. thesis, SLAC-254 (1982).

[8] D. M. Coffman *et al.* (Mark III), SLAC-PUB-3720 (1986), Sub. to Phys. Rev. D.

[9] R. M. Baltrusaitis *et al.* (Mark III), Phys. Rev. D33, 1222 (1986).

[10] R. M. Baltrusaitis *et al.* (Mark III), Phys. Rev. Lett. 55, 1723 (1985).

[11] K. F. Einsweiler (Mark III), Stanford Ph.D. Thesis, SLAC-272, May 1984.

[12] J. E. Augustin *et al.* (DM2), Orsay preprint LAL/85-27, July 1985.

[13] C. Edwards *et al.* (Crystal Ball), Phys. Rev. Lett. 48, 458 (1982).

[14] J. J. Becker *et al.* (Mark III), paper 3409.

[15] V. F. Obraztsov (GAMS), talk 10VO.

[16] E. D. Bloom and C. W. Peck (Crystal Ball), Ann. Rev. Nucl. Part. Sci. 33, 143 (1983).

[17] S. Godfrey and N. Isgur, Phys. Rev. D32, 189 (1985).

[18] D. Aston (LASS), talk 10DA.

[19] M. Gell-Mann and K. M. Watson, Ann. Rev. Nucl. Sci. 4, 219 (1954); K. M. Watson *et al.*, Phys. Rev. 101, 1159 (1956); Hamilton, *The Theory of Elementary Particles*, p. 358, Oxford Univ. Press (1959).

[20] R. Longacre, private communication.

[21] G. W. Brandenburg *et al.*, Nucl. Phys. B104, 413 (1976); M. Aguilar-Benitez *et al.*, Z. Phys. C8, 313 (1981).

[22] F. C. Porter (Crystal Ball), Proc. Summer Inst. on Particle Physics, Stanford, 1981, SLAC Report 245.

[23] E. D. Bloom (Crystal Ball), SLAC-PUB-3573, Proc. Aspen Winter Physics Conf., Jan. 1985.

[24] T. A. Armstrong *et al.* (WA76), Phys. Lett. 167B, 133 (1986). This data has peaks in $K\bar{K}$ at 1629 and 1742 MeV.

[25] K. L. Au, D. Morgan and M. R. Pennington, Rutherford preprint RAL-86-076 (1986).

[26] J. J. Becker *et al.* (Mark III), paper 3441.

[27] L. Köpke (Mark III), Proc. XXI Rencontre de Moriond, Les Arcs, 9-16 March 1986, Santa

Cruz preprint SCIPP 86/61; U. Mallik, Les Arcs, 16-22 March 1986, SLAC-PUB-3946.

[28] B. Jean-Marie (DM2), talk 10BJ.

[29] L. Köpke (Mark III), talk 10LK.

[30] M. Althoff *et al.* (TASSO), Phys. Lett. 121B, 216 (1983).

[31] H. Aihara *et al.* (TPC/2γ), Phys. Rev. Lett. 57, 404 (1986).

[32] D. Cords (Mark II), talk 19DC and contributed paper.

[33] T. D. Lee, Proc. 21st/ Int. School of Subnuclear Physics, Erice, 3-14 Aug. 1983. See also S. J. Lindenbaum, BNL 37412 and in Superstrings, Supergravity and Unified Theories, pub. by World Scientific, p. 570.

[34] R. Sinha, S. Okubo, S. F. Tuan, UH-511-592-86 (1986).

[35] J. J. Becker *et al.* (Mark III), paper 3468.

[36] A. Etkin *et al.*, Phys. Rev. D25, 1786 (1982).

[37] F. Binon (GAMS), talk 10FB; D. Alde *et al.*, paper 9822, CERN-EP/86-71, June 1986.

[38] S. Godfrey, R. Kokoski and N. Isgur, Phys. Lett. 141B, 439 (1984).

[39] R. M. Baltrusaitis *et al.* (Mark III), Phys. Rev. Lett. 56, 107 (1986).

[40] J. Sculli, talk 10JS; J. H. Christenson *et al.*, paper 5975.

[41] S. S. Gershtein, A. A. Likhoded and Yu. D. Prokoshkin, Z. Phys. C24, 305 (1984).

[42] S. I. Bityukov *et al.*, (Lepton-F), Serpukov preprint 86-110 (1986), sub. to Phys. Lett. B; JETP Lett. 42, 384 (1985); Sov. J. Nucl. Phys. 38, 727 (1983).

[43] H. Primakoff, Phys. Rev. 81, 899 (1951).

[44] J. Huston *et al.*, Phys. Rev. D33, 3199 (1986).

[45] T. Ferbel, talk 10TF.

[46] M. Bourquin *et al.* (WA62), Phys. Lett. 172B, 113 (1986); H.-W. Siebert, private communication.

[47] A. N. Aleev *et al.* (BIS-2), paper 11207.

[48] R. Brandelik *et al.* (TASSO), Phys. Lett. 97B, 448 (1980); M. Althoff *et al.*, Z. Phys. C16, 13 (1982).

[49] D. L. Burke *et al.* (Mark II), Phys. Lett. 103B, 153 (1981); H.J. Behrend *et al.* (CELLO), Z. Phys. C21, 205 (1984).

[50] W. G. J. Langeveld (TPC/2γ), talk 19WL.

[51] N. N. Achasov, S. A. Devyanin and G. N. Shestakov, Z. Phys. C16, 55 (1982) and Z. Phys. C27, 99 (1985); B. A. Li and K. F. Liu, Phys. Rev. D30, 613 (1984).

[52] J. E. Olsson (JADE), cont. paper to EPS Conf., Brighton, 1983. The data is shown in H. Kolanoski, Proc. Int. Symp. on Lepton Photon Interactions, Kyoto, 1985.

[53] G. Alexander, U. Maor and P. G. Williams, Phys. Rev. D26, 1198 (1982).

[54] G. Alexander, A. Levy and U. Maor, Z. Phys. C30, 65 (1986).

[55] H. Aihara *et al.* (TPC/2γ), Phys. Rev. Lett. 54, 2564 (1985); M. Althoff *et al.* (TASSO), Z. Phys. C32, 11 (1986).

[56] P. M. Patel (ARGUS), talk 19PP; H. Albrecht *et al.*, paper 7986.

[57] T. A. Armstrong *et al.* (WA76), K^*K^*: paper 7854; $\phi\phi$: Phys. Lett. 166B, 245 (1986).

[58] D. Bridges *et al.*, Phys. Rev. Lett. 56, 211 and 215 (1986); Phys. Lett. 180B, 313 (1986).

[59] K.-F. Liu and B.-A. Li, paper 2283, Stonybrook preprint ITP-SB-86-48 (1986).

[60] P. Baillon *et al.*, Nuovo Cim. 50A, 393 (1967); P. Baillon, CERN/EP 82-127, Proc. XXI Int. Conf. on High Energy Physics, Paris, 1982.

[61] C. Dionisi *et al.*, Nucl. Phys. B169, 1 (1980).

[62] T. A. Armstrong *et al.* (WA76), paper 7870.

[63] S. U. Chung (AGS-771), talk 10SC and private communication.

[64] D. F. Reeves *et al.* (AGS-771), Phys. Rev. D34, 1960 (1986).

[65] S. U. Chung *et al.* (AGS-771), Phys. Rev. Lett. 55, 779 (1985).

[66] D. Zieminska (AGS-771), Proc. Int. Conf. on Hadron Spectroscopy, College Park, Maryland, April 1985.

[67] A. Ando *et al.* (KEK), Phys. Rev. Lett. 57, 1296 (1986); K. Takamatsu, private communication.

[68] Th. Mouthuy (GAMS), Proc. Int. Conf. on HEP, Bari, July 1985, p. 320.

[69] Ph. Gavillet *et al.*, Z. Phys. C16, 119 (1982).

[70] S. I. Bityukov *et al.* (Lepton-F), Sov. J. Nucl. Phys. 39, 735 (1984).

[71] B. Ratcliff (LASS), talk at SLAC Summer Institute, 1986.

[72] T. A. Armstrong *et al.* (WA76). Phys. Lett. 146B, 273 (1984).

[73] O. Villalobos Baille (WA76), Proc. Int. Conf. on HEP, Bari, July 1985, p. 314.

[74] J. D. Richman (Mark III), CalTech Ph.D. Thesis, CALT-68-1231 (1985).

[75] J. J. Becker *et al.* (Mark III), paper 3433.

[76] J. J. Becker *et al.* (Mark III), paper 3476.

[77] L. D. Landau, Soviet Physics "Doklady" 60, 207 (1948). C. N. Yang, Phys. Rev. 77, 242 (1950).

[78] A. M. Eisner (TPC/2γ), talk 19AE; H. Aihara *et al.*, Phys. Rev. Lett. 57, 51 and 2500 (1986).

[79] G. Gidal (Mark II), talk 19GG and contributed paper.

[80] M. S. Chanowitz, Proc. VI Int. Workshop on Photon-Photon Collisions, Lake Tahoe, Sept. 1984.

[81] N. Stanton *et al.*, Phys. Rev. Lett. 42, 346 (1979).

[82] S. D. Protopopescu (AGS-771), Proc. XXI Recontre de Moriond, Les Arcs, 16-22 March 1986.

[83] M. Frank and P. J. O'Donnell, Phys. Rev. D32, 1739 (1985); S. B. Gerasimov and A. B. Govorkov, Z. Phys. C29, 61 (1985).

[84] R. Clare (Crystal Ball), Proc. XXI Rencontre de Moriond, Les Arcs, 16-22 March 1986.

[85] N. Isgur, private communication.

[86] S. Meshkov, W. F. Palmer and S. S. Pinsky, DOE/ER/01545-382 (1986), sub. to Phys. Rev. Lett.

[87] S. Godfrey and N. Isgur, private communication and Ref. [17].

[88] G. Bellettini *et al.*, Nuovo Cim. 40A, 1139 (1965).

[89] V. I. Kryshkin *et al.*, Soviet Physics JETP 30, 1037 (1970).

[90] A. Browman *et al.*, Phys. Rev. Lett. 32, 1067 (1974) and 33, 1400 (1974).

[91] H. W. Atherton *et al.*, Phys. Lett. 158B, 81 (1985).

[92] D. Williams (Crystal Ball), talk 19DG and private communication.

[93] C. Bemporad *et al.*, Phys. Lett. 25B, 380 (1967), corrected for $B(\eta\rightarrow\gamma\gamma)=(38.9\pm0.4)\%$ [1].

[94] A. Weinstein *et al.* (Crystal Ball), Phys. Rev. D28, 2869 (1983).

[95] W. Bartel *et al.* (JADE), Phys. Lett. 160B, 417 (1985).

[96] H. Aihara *et al.* (TPC/2γ), Phys. Rev. D33, 844 (1986).

[97] t_{min} occurs when the η is produced parallel to the incoming γ direction ($\theta=0$). Then $t = m_\eta^2 - 2E_\gamma(E_\eta - p_\eta)$. Approximating the nucleus as infinitely heavy, $E_\eta = E_\gamma$ in the lab frame. Expanding p_η according to $\sqrt{1-x} \approx 1 - x/2 - x^2/8$ yields $t_{min} \approx -M_\eta^4/4E_\gamma^2$.

[98] J. E. Gaiser *et al.* (Crystal Ball), Phys. Rev. D34, 711 (1986).

[99] R. M. Baltrusaitis *et al.* (Mark III), Phys. Rev. D33, 629 (1986).

[100] B. Jean-Marie (DM2), talk 9BJ

[101] Ch. Berger *et al.* (PLUTO), Phys. Lett. 167B, 120 (1986).

[102] U. Karshon (TASSO), Cont. to VIIth Int. Workshop on Photon-Photon Collisions, Paris, April 1986.

[103] C. Baglin *et al.* (R704), CERN/EP/0061P (1986).

[104] A. E. Blinov *et al.* (MD-1), Novosibirsk preprint 86-107.

[105] R. Cester, E760 spokesperson.

[106] R. Barbieri *et al.*, Phys. Lett. 95B, 93 (1980).

[107] L. J. Reinders, H. R. Rubinstein, and S. Yazaki, Phys. Lett. 113B, 411 (1982).

[108] S. W. Herb *et al.*, Phys. Rev. Lett. 39, 252 (1977).

[109] W. R. Innes *et al.*, Phys. Rev. Lett. 39, 1240 (1977); K. Ueno *et al.*, Phys. Rev. Lett. 42, 486 (1979).

[110] M. Alex (E605), talk 9MA.

[111] S.H.H. Tye and C. Rosenfeld, Phys. Rev. Lett. 53, 2215 (1984).

[112] C. Peck *et al.* (Crystal Ball), SLAC-PUB-3380 and DESY 84-064; H.-J. Trost, Proc. XXII Int. Conf. on HEP, Leipzig, July 1984.

[113] D. Besson *et al.* (CLEO), Phys. Rev. D33, 300 (1986). J. Lee-Franzini (CUSB), Proc. 5th Int. Conf. on Physics in Collision, Autun, France, July 1985.

[114] E. D. Bloom (Crystal Ball), SLAC-PUB-3686, Proc. 5th Topical Workshop on Proton Antiproton Collider Physics, St. Vincent, Italy, Feb. 1985.

[115] H. Albrecht *et al.* (ARGUS). Phys. Lett. 154B, 452 (1985); Z. Phys. C29, 167 (1985).

[116] U. Strohbusch (Crystal Ball), DESY seminar, 1986.

[117] A. E. Blinov *et al.* (MD-1), Novosibirsk preprint 85-99.

[118] J. Siegrist *et al.* (Mark I), Phys. Rev. D26, 969 (1982).

[119] B. Niczyporuk *et al.* (LENA), Z. Phys. C15, 299 (1982).

[120] J. D. Jackson and D. L. Scharre, Nucl. Inst. and Meth. 128, 13 (1975).

[121] É. A. Kuraev and V. S. Fadin, Sov. J. Nucl. Phys. 41, 466 (1985).

[122] S. E. Baru *et al.* (VEPP-4), Z. Phys. 30, 551 (1986).

[123] Y. S. Tsai, SLAC-PUB-3129 (1983).

[124] M. Greco, G. Pancheri-Srivastava, Y. Srivastava, Phys. Lett. 56B, 367 (1975).

[125] Ch. Berger *et al.* (PLUTO), Z. Phys. C1, 343 (1979). Only Γ_{ee} is quoted, not $\Gamma_{ee}\ \Gamma_{had}/\Gamma_{tot}$. The description of the radiative corrections used does not obviously correspond to any preseented here. Therefore the PLUTO value could not be corrected or included in the average.

[126] P. Bock *et al.* (DESY-Heid.), Z. Phys. C6, 125 (1980).

[127] B. Niczyporuk *et al.* (LENA), Phys. Rev. Lett. 46, 92 (1981). In Z. Phys. C15 299 (1982), this group revised their $\Gamma_{ee}(1S)$ value, normalising to their new measurement of the continuum R. However the R they originally used is closer to the current average R see Ref. [130]; therefore I use their original Γ_{ee}.

[128] B. Nicyporuk (LENA), private communication: The LENA Γ_{ee} measurements used a modified Jackson-Scharre radiative correction formual with full soft photon exponentiation.

[129] H. Albrecht *et al.* (DASP II), Phys. Lett. 116B, 383 (1982).

[130] R. Giles *et al.* (CLEO), Phys. Rev. D29 1285(1984). The values for $\Gamma_{ee}\ \Gamma_{had}/\Gamma_{tot}$ before correction for $B_{\mu\mu}$ are quoted in Ref. [133]. The value of $\delta^{\mathrm{hads}}_{\mathrm{vac}}$ used was 0.034, as quoted in R. K. Plunkett, Cornell Ph.D. thesis (1983).

[131] P. M. Tuts, Int. Symp. on Lepton and Photon Interactions at High Energy, Ithaca, N.Y. (1983), p. 284; and private communication.

[132] B. Niczyporuk *et al.* (LENA), Phys. Lett. 99B, 169 (1981).

[133] D. Andrews *et al.* (CLEO), Phys. Rev. Lett. 50, 807 (1983).

[134] J. E. Horstkotte *et al.*, CUSB-83-09, quoted in Ref. [131].

[135] Ch. Berger *et al.* (PLUTO), Phys. Lett. 93B, 497 (1980).

[136] D. Besson *et al.* (CLEO), Phys. Rev. D30, 1433 (1984).

[137] D. MacFarlane (ARGUS), talk at XXIII Int. Conf. on HEP, Berkeley, July 1986; and R. Waldi, private communication.

[138] R. Giles *et al.* (CLEO), Phys. Rev. Lett. 50, 877 (1983).

[139] P. Haas *et al.* (CLEO), Phys. Rev. D30, 1996 (1984).

[140] H. Albrecht *et al.* (ARGUS), Z. Phys. C28, 45 (1985).

[141] J. Lee-Franzini (CUSB), talk 9JL; and CUSB contributed papers: $B_{\mu\mu}$–T. Kaarsberg *et al.*, paper 7480; $\chi_b(2P)$ inclusive–C. Yanagisawa *et al.*, paper 7501; $\chi_b(2P)$ exclusive–T. Zhao *et al.*, paper 7498.

[142] S. Cooper, SLAC-PUB-3555, in Proc. 12th SLAC Summer Inst. on Particle Physics, Stanford, July 1984.

[143] I have given a more thorough discussion of this with references: S. Cooper, SLAC-PUB-3819, Proc. Int. Conf. on HEP, Bari, July 1985.

[144] C. Klopfenstein *et al.* (CUSB), Phys. Rev. Lett. 51, 160 (1983).

[145] N. Byers, talk 9NB; N. Byers and V. Zambetakis, UCLA/86/TEP/24.

[146] A recent review is R. H. Datitz and S. F. Tuan, Proc. 10th Hawaii Conf. on HEP, 1985.

[147] C. Baglin *et al.* (R704), Phys. Lett. 171B, 135 (1986).

[148] T. Bowcock *et al.* (CLEO), paper 6173, Cornell preprint CLNS-86/740.

[149] Y.-P. Kuang and T.-M. Yan, Phys. Rev. D24, 2874 (1981); G.-Z. Li and Y.-P. Kuang, Commun. Theor. Phys. 5, 79 (1986).

[150] M. B. Voloshin, Moscow preprint ITEP-166 (1985).

Questions and Comments

[The questions were posed in the old meson naming convention, I have modernised them here.]

R. Sinha (Univ. of Rochester) – The absence of the $f_2(1720)$ in $\pi^- p \to K_s K_s \Lambda$ could mean $B(f_2 \to K\bar{K})$ is small and $B(J/\psi \to \gamma f_2)$ is large. Profs. S. F. Tuan, S. Okubo and myself use GAMS results to derive a limit $B(f_2 \to \pi\pi)B(f_2 \to \eta\eta) < 1.3 \times 10^{-3}$ This is consistent with Crystal Ball and Mark III data only if $B(J/\psi \to \gamma f_2) > 6.7 \times 10^{-3}$.
Cooper – The J/ψ data on the $\pi\pi$ and $\eta\eta$ decays of the $f_2(1720)$ decays need to be treated with some caution, as I have tried to show in Table 2.

R. Longacre (BNL) – In your conclusions you said there is chaos in the 0^{++}, 1^{++} and radially excited 0^{-+} nonets. The 2^{++} is in chaos also.
Cooper – I meant to include 2^{++}.

R. L. Jaffe (MIT) – In one of the parallel sessions we heard a report by Siebert of the Heidelberg group of an intriguing narrow resonance at 3.1 GeV which decays into $\Lambda\bar{p}+n\pi$. Do you have any comment on this report?
Cooper – I have included the X(3100) in this written version of my talk (section 3.3).

K. Takamatsu (KEK) – We have measured the $\eta\pi\pi$ channel in the $\pi^- p$ charge exchange reaction. A partial wave analysis shows a clear peak at 1.42 GeV in the 0^{-+} wave. It is interesting that we see the $\eta(1420)$, because until now no experiment had seen its $\eta\pi\pi$ decay, although it is thought to decay through the $a_1(980)\pi$ intermediate state.

Cooper – Your result is now included in section 4.1.1.

G. Morpurgo (Univ. Genoa) – I am referring to the question of the Primkoff effect mentioned by the speaker in connection with the $\eta \to \gamma\gamma$ decay (section 5.1). If one takes the $\gamma\gamma$ determinations of the width as correct, that means that the Cornell experiment is incorrect and that the DESY one (Bellettini *et al.*) is 2 standard deviations off. But the Primakoff effect is O.K. if used properly, taking into account if necessary the nuclear form factor corrections.

T. Ferbel – I rise to the defense of the late Henry Primakoff. There are two possible problems that can occur using the Primakoff technique: One is production at too high momentum transfers which make the experimental result very sensitive to nuclear form factors. Two is if the experiment is performed at too low energy, in which case the result is too sensitive to other than photon exchanges. At high energy (e.g. Fermilab) and small momentum transfers there are no problems, as our group has clearly indicated over the past few years.
Cooper – Do you intend to measure the η?
T. Ferbel – No.

N. Isgur (Toronto) – I don't believe that there is any problem with scalar confinement. The problem with the decay channel couplings apparently arises if one assumes that the confinement can be described by an effective scalar exchange which is then crossed to create pairs. There is, however, not necessarily such a simple connection, and indeed if one uses a $^3P_0$ or string-breaking model (which is consistent with scalar confinement but not simply related by crossing), no such problem appears.

Plenary Session

Superstrings

J.H. Schwarz (Caltech)

Chairman

A. Salam (ICTP, Trieste)

Scientific Secretary

B. Ratra (SLAC)

Superstrings

John H. Schwarz
California Institute of Technology, Pasadena, CA 91125

String theories are promising candidates for perturbatively finite quantum theories of gravity. Mathematical consistency appears to require a supersymmetric theory in ten dimensions with gauge group $E_8 \times E_8$ or SO(32). The $E_8 \times E_8$ theory may be capable, after spontaneous compactification of six dimensions, of producing realistic low-energy physics. Superstring theory has been developing very rapidly in the last year, but it will probably take a long time to make a convincing case that this is the route that nature has chosen.

I. Introduction

Superstring theory is an approach to the unification of fundamental particles and forces that has attracted a great deal of attention in the last two years [1]. It offers the possibility of overcoming many of the shortcomings of the standard model and providing a unified theory free from much of the arbitrariness that is inherent in conventional point-particle theories.

The standard model of electroweak and strong forces has enjoyed an enormous amount of success. Indeed it appears to be consistent with all established particle physics experiments. This being the case, the first question one should ask is why one should even be looking for something better. Most criticisms of the standard model are based on the fact that it requires a number of arbitrary choices and fine-tuning adjustments of parameters. These features do not prove that it is wrong or even incomplete. However, given the history of successes in elementary particle physics, it is natural to seek a deeper underlying theory that can account for many of the arbitrary choices and parameters. These include the choice of gauge groups and representations, the number of families of quarks and leptons, the origins of the Higgs symmetry breaking mechanism, and the specific values of various parameters.

One bold attempt to simplify the picture and reduce this arbitrariness was provided by 'grand unification.' That program has had one striking success (the calculation of $\sin^2\theta_W$) in addition to providing some simplification of the group-theoretical structure of the model. Still, it represents only a slight improvement over the standard model and is also subject to the same list of criticisms. Even this modest success required a bold extrapolation to an extraordinarily high energy scale, about $10^{14} - 10^{16} GeV$, many orders of magnitude beyond what is likely to be experimentally accessible for a long, long time. Understandably, many physicists feel that it is preposterous to attempt to sort out the physics at such a scale, especially when all our experience suggests that there could be numerous surprises in store for us at intermediate energy scales. One couldn't reasonably expect to piece the puzzle together without an understanding of the phenomena that occur at all these scales. Superstring theory requires considering a somewhat higher energy scale and is therefore subject to the same criticism. Curiously, this concern has been voiced most strongly by some of the originators of grand unification.

The aesthetic shortcomings of the standard model or a grand-unified model do not prove that it is wrong or incomplete. The thing that does this most convincingly is the absence of gravity. There is a straightforward way to couple Einsteinian gravity to any relativistic field theory by following the dictates of general coordinate invariance, and at the classical level it poses no difficulties. However, the quantum theory, which is renormalizable before gravity is appended, inevitably acquires nonrenormalizable ultraviolet divergences. Many years of study strongly suggest that this is a generic feature of all quantum field theories based on point particles. Even Einsteinian gravity by itself has recently been shown to be singular at two loops [2]. Pure supergravity theories in four dimensions are finite at two loops, but are very likely to be singular at three loops. The addition of extra spatial dimensions (such as in eleven-dimensional supergravity) makes the divergences even more severe, and therefore represents a step in the wrong direction. This is where superstring theory comes to the rescue.

String theory was developed in the late 1960's and early 1970's in an attempt to describe the strong nuclear force [3]. The first string theory was plagued by a number of unrealistic features -- the absence of fermions, the presence of a tachyon and massless modes, and the necessity of 26-dimensional space-time. In an attempt to do better a second string theory was introduced in 1971 [4]. It incorporates a spectrum of fermions that was proposed by Pierre Ramond [5]. In the form first discussed this theory also contained a tachyon. It possesses two-dimensional superconformal symmetry on the string world sheet but not space-time supersymmetry. Five years later, Gliozzi, Scherk, and Olive proposed a truncation of the spectrum that eliminated the tachyon [6]. They noted that then there are an equal number of bosons and fermions at each mass level and conjectured that this version of the theory possesses space-time supersymmetry. This result was proved in 1980 by Michael Green and myself [7]. This theory is called type I superstring theory in the modern classification (see below). It was discovered in 1972 that superstring theories required ten-dimensional space-time for their consistency [8].

II. Strings for Unification

Neither the bosonic string theory nor superstring theory succeeded in giving a realistic description of hadron physics. The extra dimensions were an embarrassment as were the massless modes, which had no counterparts in the hadronic spectrum. In 1974 Joël Scherk and I demonstrated that the massless spin-2 mode that is present in both theories interacts in the standard Einsteinian way at low energies [9]. (This was also shown by Yoneya [10].) We therefore suggested that it be interpreted as a graviton rather than a hadron and that string theory be used as a basis for constructing a unified description of gravity and all the other forces rather than as a theory of hadrons. In other words, it was proposed that elementary particles are strings rather than points as in conventional quantum field theory.

The idea of using strings for unification offered a number of appealing features. First of all, the existence of gravity would be understood as necessary since it is a generic feature of all consistent string theories. String theories do not possess ultraviolet divergences of the conventional type, and therefore it appeared plausible that this type of theory could overcome the divergence problem discussed above. To state the case as forcefully as possible, string theory requires the existence of gravity, whereas the point-particle theories require that it does not exist. This conviction sustained my enthusiasm for the subject throughout the ten-year period prior to its widespread acceptance.

The use of string theory for unification rather than hadronic physics has a number of other advantages. The extra dimensions can acquire a sensible interpretation, because space-time geometry is dynamical in a gravity theory and there is a possibility that the unobserved dimensions could be spontaneously compactified. Also, string theories do not possess adjustable dimensionless parameters, or much freedom to choose groups and representations. So if a scheme of this type is successful, it should be much more predictive than the standard model. To put it differently, the consistent introduction of gravity in a unified theory is such a severe constraint that it could have dramatic predictive consequences for low-energy physics.

An obvious question that is often raised in discussions of string theory is why one should stop with one-dimensional objects. Why not consider elementary particles that are "membranes," or even higher dimensional. I do not know for sure that this is impossible, but string theory possesses many miracles that seem unlikely to generalize. For one thing, the two-dimensional world-sheet action

$$S \sim \int d^2\sigma \sqrt{-g}\, g^{\alpha\beta} \partial_\alpha X^\mu \partial_\beta X_\mu \, , \qquad (1)$$

is renormalizable, whereas a higher-dimensional analog would not be. Thus in considering membranes, one is simply replacing one nonrenormalizable theory (general relativity) with another one. Another virtue of strings is that the two-dimensional action has an infinite-dimensional conformal symmetry that helps to keep the mathematics tractable. This does not generalize to higher dimensions. To summarize the case, it is very difficult to construct consistent relativistic quantum theories for extended objects. I consider it very unlikely that they exist for objects of more than one dimension. Even in the case of strings there seem to be a very limited number of such theories, which is one reason why they are potentially very predictive.

String theory contains a fundamental length parameter, which is the characteristic size of a string. In order for the strength of gravity to emerge with the usual Newtonian value, it is necessary to associate this length with the Planck length -- a few times $10^{-33} cm$. Thus in the unification context the strings are taken to be some twenty orders of magnitude smaller than they were in the hadronic context.

III. Classification of String Theories

Strings have two possible topologies: open (with free ends) or closed (a loop without ends). The most interesting theories consist of oriented closed strings only. Type I superstring theory, which was the first one to be understood, consists of unoriented open and closed strings.

How many string theories are there? The answer to this question is not yet known. Ideally it will turn out that there is just one. This could happen if the ones that are presently known are either inconsistent or equivalent to one another. In order for a theory to be consistent it must be free from tachyons (states of negative mass squared) and ghosts (states of negative norm). It is also necessary that loop amplitudes be free from anomalies and possess modular invariance -- properties to which we will return. Additional criteria, perhaps of nonperturbative origin, may eventually be found to be necessary also.

Six string theories are known that satisfy the consistency properties mentioned above. Each of them requires that space-time has ten dimensions (nine space and one time) and that the two-dimensional world-sheet action has superconformal symmetry. These theories are listed in table 1. There is some evidence that the three

Name	D = 10 Supersymmetry	Yang–Mills Symmetry
Type I [11]	N = 1	SO(32)
Type IIA [12]	N = 2	–
Type IIB [12]	N = 2	–
Heterotic [13]	N = 1	SO(32)
Heterotic [13]	N = 1	$E_8 \times E_8$
Heterotic [14]	N = 0	SO(16) × SO(16)

Table 1. Six String Theories

heterotic theories are actually different phases of the same theory, which would reduce the number of different theories to be considered.

Superstring theories possess not only two-dimensional superconformal symmetry but local ten-dimensional super-Poincaré symmetry, a possibility first pointed out by Gliozzi, Scherk, and Olive [6]. There are three possibilities for D = 10 supersymmetry, each of which can be realized in a superstring theory. The minimal irreducible spinor in D = 10 satisfies simultaneous Majorana and Weyl properties and has sixteen independent real components. A theory with a single conserved Majorana-Weyl supercharge has N = 1 supersymmetry, a possibility realized by three of the six theories listed in the table. There are two distinct possibilities for theories with two conserved Majorana-Weyl supercharges. Either they have opposite chirality (type IIA) or they have the same chirality (type IIB). This is the maximum amount of supersymmetry that is possible in an interacting theory (corresponding to N = 8 supersymmetry in four dimensions). The type II theories are not promising for phenomenology because they do not accommodate elementary Yang–Mills fields. However, the type IIB theory is in some respects the most beautiful of all the theories, and it is not totally out of the question that it could form the basis of a realistic phenomenology.

An important fact about the standard model is the left–right asymmetry in the classification of quark and lepton multiplets (chirality). The most natural way to achieve this chiral asymmetry, starting from a theory in more than four dimensions, is for the higher-dimensional theory to be left-right asymmetric itself. This property is shared by all the theories in the table except for the type IIA.

Another important consideration in choosing among the theories is a desire to understand the origin of the hierarchy of mass scales. In particular, one wants to understand why radiative corrections to the mass of Higgs scalars do not destroy the enormous ratio between the electroweak scale and the unification scale. The most promising possibility seems to be "low-energy" supersymmetry spontaneously broken around the electroweak scale. This can be realized in superstring theories with space-time supersymmetry.

In addition to the string theories listed in the table, there are some others that are consistent in all respects except for the presence of a tachyon in the spectrum. The original D = 26 bosonic string is an example of such a theory, and a number of others have been discovered recently [15]. It is an open question whether any of these theories can be given a consistent interpretation by identifying a stable ground state that is free from tachyonic modes. Even if this is possible, one might have to pay an unacceptable price in terms of a large cosmological constant. But that is a threat to all theories once supersymmetry is broken.

IV. Feynman Diagrams

In conventional field theory, Feynman diagrams correspond to the distinct topological possibilities for connecting the world lines of elementary particles. The situation is analogous in string theory. Since strings are one-dimensional they sweep out a two-dimensional surface in space-time. Thus Feynman diagrams correspond to the topologically distinct world sheets representing the possible space-time histories of interacting strings. In thinking about these surfaces it is extremely convenient to imagine carrying out a Wick rotation and regarding the surfaces as being embedded in a space of Euclidean metric. This makes the corresponding path integrals well-defined. This rotation can probably be rigorously justified by arguing that it gives a unique unitary S matrix.

The classification of string Feynman diagrams is especially simple for the theories that consist of oriented closed strings only. In these theories there is a single fundamental string interaction, depicted in figure 1, that describes one string breaking into two or two strings joining to give a single one. The surface shown is topologically like a pair of pants. Since the surface is smooth the particular space-time point at which the interaction takes place is a frame-dependent question. If the time slices are taken using planes that are tilted relative to the ones shown a different interaction point would be identified.

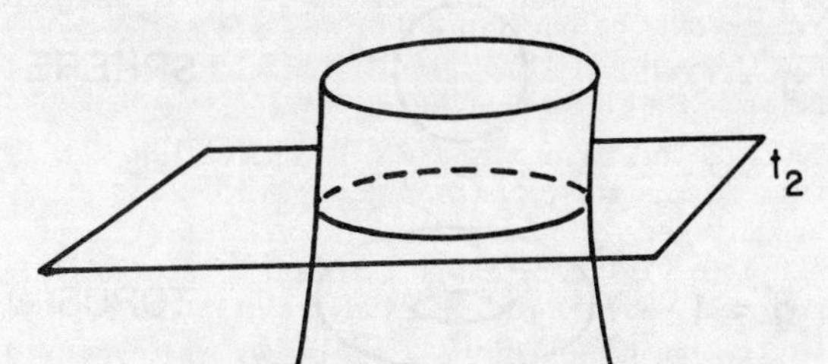

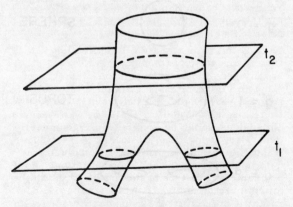

Figure 1. A piece of world sheet in the shape of pants. The slice at time t_1 shows two closed strings while the one at time t_2 shows one closed string.

Thus there is no preferred point on the surface and the existence of interaction is an inevitable feature of geometric origin, not something that needs to be appended to the theory in an ad hoc way.

In calculating S-matrix elements one should consider amplitudes for scattering closed strings that propagate to time $\pm\infty$, i.e., with tubes emerging from the surface that extend to infinity. However, the conformal symmetry of the underlying two-dimensional field theory [16, 17], implies that world sheets that differ by a conformal transformation are equivalent. In particular, one may map the asymptotic strings to finite coordinates. When this is done, they are represented by points on the world sheet. Altogether, the amplitude is then represented by a closed oriented surface with "punctures" representing the initial and final string states.

The topological classification of closed oriented two-dimensional surfaces is characterized by a single integer, the genus g, which can be thought of as the number of handles that is attached to a sphere. Thus, as shown in figure 2, the sphere has $g = 0$, the torus has $g = 1$, and so forth.

In the type I superstring theory the topological classification of diagrams is a good deal more complicated. Since the strings themselves can be open or closed and have no intrinsic orientation, the Feynman diagrams can have boundaries and need not be orientable. This makes the analysis of this theory more difficult. An important advantage of orientability is that an orientable surface can be regarded as a Riemann surface and powerful techniques of complex analysis can be utilized.

In the type II and heterotic string theories, the topology of a diagram is characterized by the genus g, which can be thought of as the number of loops in the diagram. Thus there is just a single diagram at each order of the perturbation expansion. This is a remarkable simplification

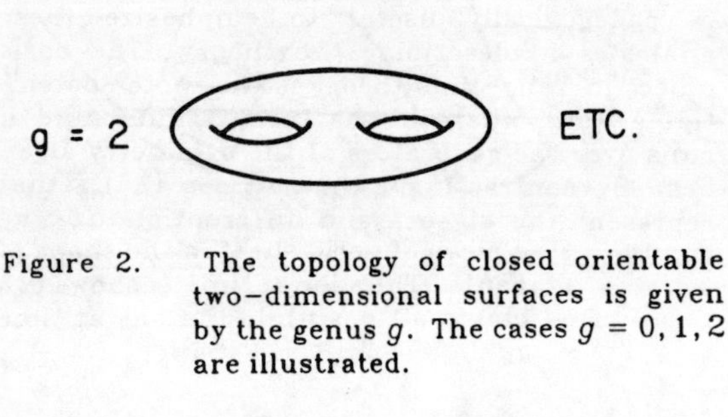

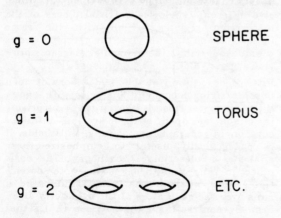

Figure 2. The topology of closed orientable two-dimensional surfaces is given by the genus g. The cases $g = 0, 1, 2$ are illustrated.

compared to conventional fields in which the number of diagrams at n loops is roughly of order $n!$. It has been shown by explicit computation that the one-loop amplitudes are finite [12, 13]. Arguments have been made that suggest that the loop amplitudes should be finite at every order, but this still requires a careful analysis to be definitively established. This is likely to be settled within a year or so. Assuming that the expected result emerges, we will then be in the happy position of having the first examples of perturbatively finite quantum theories of gravity.

The analysis of multiloop (genus $g > 1$) string amplitudes is a very rapidly developing subject in which much beautiful work has been done in the last year [18]. The details require state-of-the-art methods in the theory of Riemann surfaces and algebraic geometry. All I can realistically hope to do here is to sketch some of the basic ideas.

The loop amplitudes are given by integrals that represent a sum over all conformally inequivalent geometries of the given topologies. Fortunately, these are finite-dimensional integrals. $3g - 3 + n$ complex parameters describe the integration space M_g for a genus g amplitude with n external particles. (This result is due to Riemann.) The structure of an amplitude is then given by an expression of the form

$$\int_{M_g} d\mu \prod_{i<j} (F_{ij})^{k_i \cdot k_j} . \qquad (2)$$

In this expression the indices i and j refer to the external particles and the k_i are the momenta (assumed to be on-shell). The function $\log F_{ij}$ is proportional to the Green's function on the world sheet between the coordinates of particles i and j. Also, $d\mu$ represents a suitable integration measure on the "moduli space" M_g.

Not only the two-dimensional world sheet, but also the $2(3g - 3+n)$-dimensional moduli space has a complex structure. As a result it is possible to express the measure $d\mu$ and functions F_{ij} in terms of holomorphic functions on M_g. The space M_g is naturally expressed as a quotient of a space T_g, known as Teichmüller space, divided by an infinite discrete group of transformations. The expressions $d\mu$ and F_{ij} can be expressed as products of various terms that are well-defined on T_g but are multivalued ("line bundles") on M_g. In order for the loop amplitude to be well-defined it is necessary that when the relevant products are formed, expressions that are single-valued on M_g ("modular invariant") result. The fact that this works for each of the various string theories is highly nontrivial. In fact, it almost completely determines the various functions!

Moduli space M_g has a very rich and subtle topology. In fact, it even has boundary components that correspond to singular limits of the geometry in which the world sheet "degenerates." The relevant issue for finiteness is whether or not the measure $d\mu$ diverges on any of these boundary components so fast as to give a singular integral. There are two distinct ways in which the surface can degenerate. One, depicted in fig. 3a, involves the formation of a long thin tube that separates the diagram into two pieces. The second, shown in fig. 3b, involves the formation of a long thin tube that does not separate the surface.

Divergences associated with diagrams of the first type (fig. 3a) do occur in theories with tachyons. They can also occur if a massless scalar ("dilaton") develops a one-point function. (This happens in the SO(16) × SO(16) theory at one-loop.) In either case the divergence simply means that the vacuum has not been properly identified. By itself, it does not imply that the theory is sick. This should not happen in the supersymmetric string theories, if (as expected) supersymmetry is unbroken at all orders of the perturbation expansion.

Divergences are expected to occur in association with degenerate surfaces of the type shown in fig. 3b. However, it seems likely that they all have a physically sensible interpretation in terms of multiparticle thresholds.

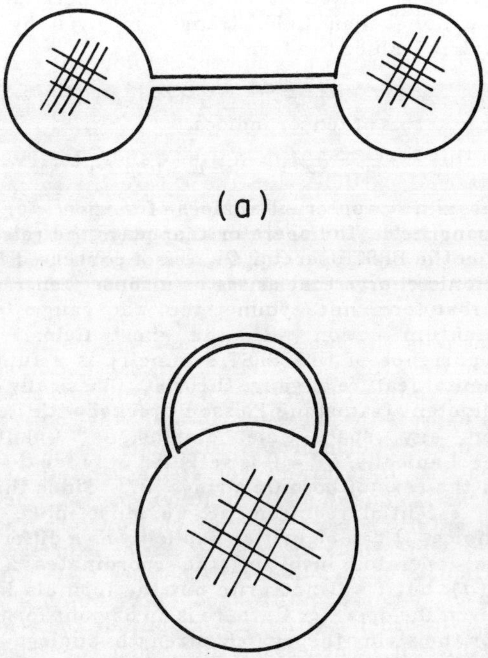

Figure 3. The boundary of moduli space corresponds to singular limits in which the surface "degenerates." This can occur by the formation of a long thin tube that splits the diagram as in (a) or that does not split it as in (b).

One might wonder whether there could be more complicated ways of generating singular behavior than those depicted in fig. 3. In particular, what is the string analog of the overlapping divergences of conventional field theory? The remarkable mathematical fact is that the holomorphic structure of moduli space ensures that there are no overlapping divergences, and only the cases depicted need to be considered.

V. String Field Theory

The history of string theory has many strange aspects. One of them is that the Feynman diagrams that I have been discussing were formulated before a field theory that gives rise to them. The development of string field theory was carried out in the light-cone gauge for bosonic strings in the 1970's [19] and for superstrings in the early 1980's [20]. However, what one really wants is a covariant gauge-invariant action that makes manifest all of the beautiful underlying symmetries of the theory in a way that is not tied to a particular choice of background fields. In short, one would like to know the string analog of the Hilbert-Einstein action of general relativity. This has not yet been achieved, but an enormous effort by many workers over the past year has brought us much closer to this goal. In particular, a beautiful field theory for open strings, correct at least at tree level, has been achieved. It is almost impossible to give adequate credit to all the workers who have made important contributions. Some of them are Siegel and Zwiebach [21], Banks and Peskin [22], Neveu and West [23], Hata, Itoh, Kugo, Kumimoto, and Ogawa [24]. I will give a brief description of the results in the form developed by Witten [25].

One important remark is that in field theory an individual Feynman diagram is given by an integral whose integration region is topologically trivial – some sort of hypercube. Thus, since moduli space M_g has a complicated topology, it cannot arise from a single diagram. What happens is that many different field theory diagrams give different topologically trivial pieces of the integration region M_g, all with the same integrand, and the sum gives the complete genus g amplitude. Thus the individual diagrams provide a "triangulation" of moduli space [26].

Witten's description of the field theory of open bosonic string has many analogies with Yang-Mills theory. This is not really surprising inasmuch as open strings are an infinite-component generalization of Yang-Mills fields. It is pedagogically useful to emphasize these analogies in describing the theory. The basic object in Yang-Mills theory is the vector potential $A_\mu^a(x^\rho)$, where μ is a Lorentz index and a runs over the generators of the symmetry algebra. By contracting with matrices $(\lambda^a)_{\alpha\beta}$ that represent the algebra and differentials dx^μ we can define

$$A_{\alpha\beta} = \sum_{a,\mu} A_\mu^a \, (\lambda^a)_{\alpha\beta} \, dx^\mu \, , \tag{3a}$$

a matrix of one forms. This is a natural quantity from a geometric point of view. The analogous object in open-string field theory is the string field

$$A\left[x^\rho(\sigma)\,,\,c(\sigma)\right]\,. \tag{3b}$$

This is a functional field that creates or destroys an entire string with coordinates $x^\rho(\sigma)\,,\,c(\sigma)$, where the parameter σ is taken to have the range $0 \leqq \sigma \leqq \pi$. The coordinates $c(\sigma)$ are anticommuting ghost degrees of freedom that arise in the first quantization of the action (1). They are essential so that when A is expanded in an infinite sequence of point fields, there is an appropriate set of auxiliary Stückelberg fields at each mass level. The details of such an expansion are quite complicated, but the mathematics of the complete string field is not so bad. The Yang-Mills field is one of the infinity of terms in such an expansion.

The string field A can be regarded as a matrix (in analogy to $A_{\alpha\beta}$) by regarding the coordinates with $0 \leqq \sigma \leqq \pi/2$ as providing the left matrix index and those with $\frac{\pi}{2} \leqq \sigma \leqq \pi$ as providing the right matrix index as shown in fig. 4a. One could also associate quark-like charges with the ends of the strings which would then be included in the matrix labels as well. This is a minor and inessential complication, which we will suppress. By not including such charges we are constructing the string generalization of U(1) gauge theory. U(1) gauge theory (without matter fields) is a free theory, but the string extension has nontrivial interactions, as we will see.

In the case of Yang-Mills theory, we can multiply two fields by the rule

$$\sum_\gamma A_{\alpha\gamma} \wedge B_{\gamma\beta} = C_{\alpha\beta}\,. \tag{4a}$$

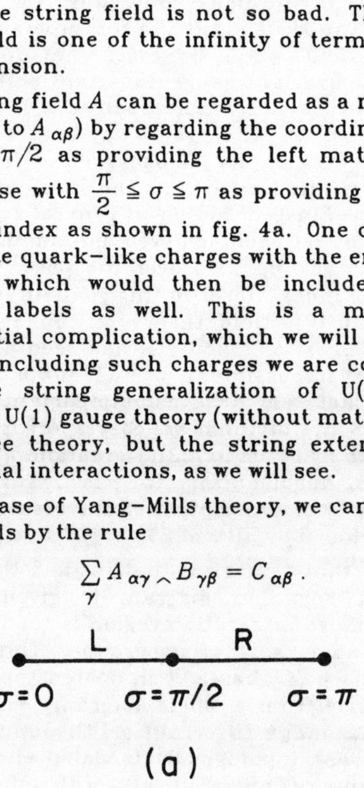

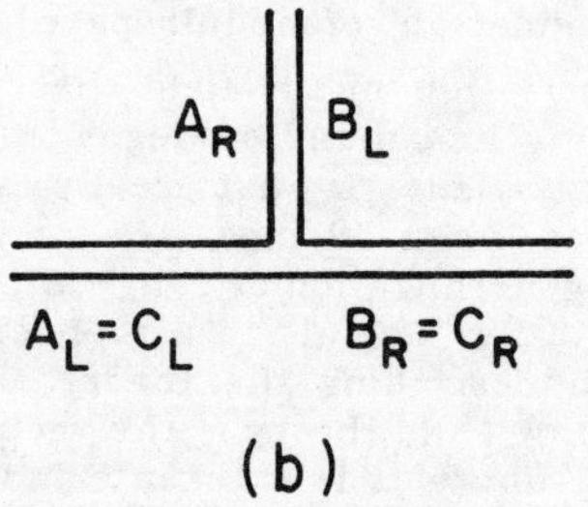

Figure 4. An open string has a left-hand ($\sigma < \pi/2$) and right-hand ($\sigma > \pi/2$) segment, depicted in (a), which ca be treated as matrix indices. The multiplication $A * B = C$ is depicted in (b).

This is a combination of matrix multiplication and antisymmetrization of the tensor indices (the wedge product of differential geometry). This multiplication is associative but noncommutative. A corresponding rule for string fields is given by a $*$ product,

$$A * B = C\,. \tag{4b}$$

This infinite-dimensional matrix multiplication is depicted in fig. 4b. One identifies the coordinates of the right half of string A with those of the left half of string B and functionally integrates over them. This leaves string C consisting of the left half of string A and the right half of string B. It is also necessary to include a suitable factor involving the ghost coordinates at the midpoint $\sigma = \pi/2$.

A fundamental operation in gauge theory is exterior differentiation $A \to dA$. In terms of components

$$dA = \frac{1}{2}(\partial_\mu A_\nu - \partial_\nu A_\mu)dx^\mu \wedge dx^\nu\,,$$

which contains the abelian field strengths as coefficients. Exterior differentiation is a nilpotent operation, $d^2 = 0$, since partial derivatives commute and vanish under antisymmetrization. The nonabelian field strength is given by the matrix-valued two form

$$F = dA + A \wedge A\,, \tag{5a}$$

or in terms of tensor indices,

$$F_{\mu\nu} = \partial_\mu A_\nu - \partial_\nu A_\mu + [A_\mu, A_\nu]\,.$$

Let us now construct analogs of d and F for the string field. The operator that plays the roles of d is the BRST operator Q. Q is a conserved fermionic charge that arises as a consequence of a global fermionic symmetry of the gauge-fixed quantum action with the ghost fields. The occurrence of this BRST symmetry is a fundamental feature of gauge theories. Classically Q is nilpotent (vanishing Poisson bracket with itself) for any space-time dimension. Quantum mechanically, $Q^2 = 0$ is satisfied only for $d = 26$ in the case of bosonic strings [27]. Since this is an essential requirement, we must make this choice. Q can be written explicitly as a differential operator involving the coordinates $X(\sigma)$, $c(\sigma)$, but I will not write out the formula here. Given the operator Q, there is an obvious formula for the string theory field strength, analogous to the Yang-Mills formula, namely

$$F = QA + A * A\,. \tag{5b}$$

An essential feature of Yang-Mills theory is gauge invariance. Infinitesimal gauge transformation can be described by a matrix of infinitesimal parameters $\Lambda(x^\rho)$, which are functions of the space-time coordinates x^ρ. The transformation rules for the potential and the field strength are then

$$\delta A = d\Lambda + [A\,,\Lambda] \tag{6a}$$

and

$$\delta F = [F, \Lambda] . \qquad (7a)$$

We can write down completely analogous formulas for the string theory, namely

$$\delta A = Q\Lambda + [A, \Lambda] \qquad (6b)$$

and

$$\delta F = [F, \Lambda] . \qquad (7b)$$

In this case $[A, \Lambda]$ means $A * \Lambda - \Lambda * A$, of course. Since the infinitesimal parameter $\Lambda[x^\rho(\sigma), c(\sigma)]$ is a functional, it can be expanded in terms of an infinite number of ordinary functions. Thus the gauge symmetry of string theory is infinitely richer than that of Yang–Mills theory, as required for the consistency of the infinite spectrum of high spin fields contained in the theory.

The next step is to formulate a gauge–invariant action. The key ingredient in doing this is to have a suitably defined integral. In the case of Yang–Mills theory we must integrate over space–time and take a trace over the matrix indices. Thus it is convenient to define $\int X$ as $\int d^4x \, Tr(X)$. In this notation the usual Yang–Mills action is

$$I \sim \int g^{\mu\rho} g^{\nu\lambda} F_{\mu\nu} F_{\rho\lambda} , \qquad (8)$$

which is easily seen to be gauge invariant. Requiring I to be stationary gives the classical equations of motion $D^\mu F_{\mu\nu} = 0$, where D^μ is a covariant derivative. Since F is quadratic in A, I describes cubic and quartic interactions.

The definition of integration appropriate to string theory is a "trace" that identifies the left and right segments of the string field Lagrangian. Thus in the case of string theory we define

$$\int X = \int \sum_{\sigma < \pi/2} \delta(x(\sigma) - x(\pi - \sigma)) \cdots e^{-\frac{3i}{2}\phi(\pi/2)} X . (9)$$

As indicated in fig. 5a, this identifies the left and right segments of X. A ghost factor has been inserted at the midpoint. (ϕ is a bosonized form of the ghost coordinates.) The ... signifies that analogous integrations should also be performed for the ghost coordinates. We now have the necessary ingredients to write a string action. If we try to emulate the Yang–Mills formula we run into a problem, namely no analog of the metric $g^{\mu\rho}$ has been defined. Rather than trying to find one, it proves more fruitful to look for a gauge–invariant action that does not require one. The simplest possibility is given by the Chern–Simons form

$$I \sim \int A * QA + \frac{2}{3} A * A * A . \qquad (10)$$

In the context of ordinary Yang–Mills theory the integrand is a three–form and therefore such a term can only be introduced in three dimensions, where it is interpreted as giving mass to the gauge field. In string theory the interpretation is different and the formula makes perfectly good sense. In fact, it gives rise to the deceptively simple field equation $F = 0$.

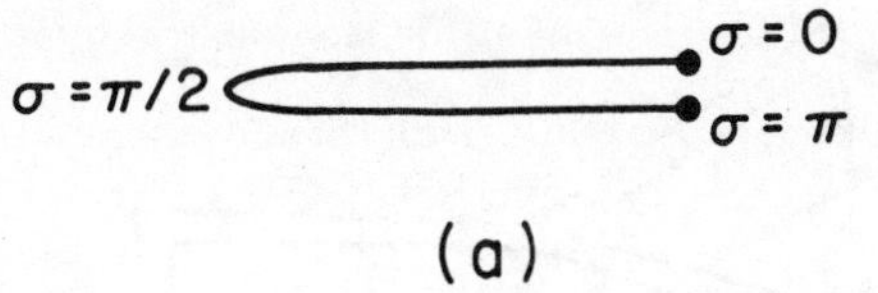

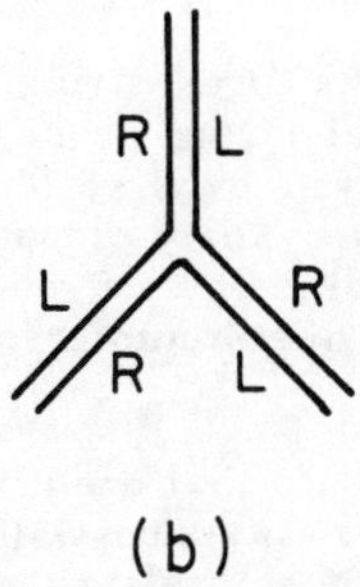

Figure 5. Integration of a string functional requires identifying the left– and right–hand halves as depicted in (a). The three–string vertex, shown in (b) is based on two multiplications and one integration and treats the three strings symmetrically.

The fact that the string equation of motion is $F = 0$ does not mean the theory is trivial. If we drop the interaction term the equation of motion for the free theory is $QA = 0$, which is invariant under the abelian gauge transformation $\delta A = Q\Lambda$ since $Q^2 = 0$. Once one imposes suitable restrictions on ghost number, one can show this precisely reproduces the known spectrum of the bosonic string.

The cubic string interaction is depicted in fig. 5b. Two of the segment identifications are consequences of the * products in $A * A * A$ and the third is a consequence of the integration resulting in a symmetric expression. As an alternative way of thinking about the interaction we can attach some flat world sheet corresponding to free propagation of each of the three strings in the interaction. This is depicted in fig. 6. The folds in the drawing do not imply intrinsic curvature of the surface and therefore have no physical significance. (They are introduced for ease of depiction only.) There is curvature at the interaction point, however. To see this imagine drawing a small circle of radius r around it. It has circumference $3\pi r$ (a contribution of πr coming from each side of the surface). Thus the surface is flat everywhere except at this one point where the curvature has a δ function singularity.

A standard theorem implies that a closed orientable two–dimensional surface of genus g has an integrated curvature proportional to $\chi = 2(1 - g)$. Thus, except for genus one, the surface does not admit a flat metric. In string theory one is only interested in equivalence

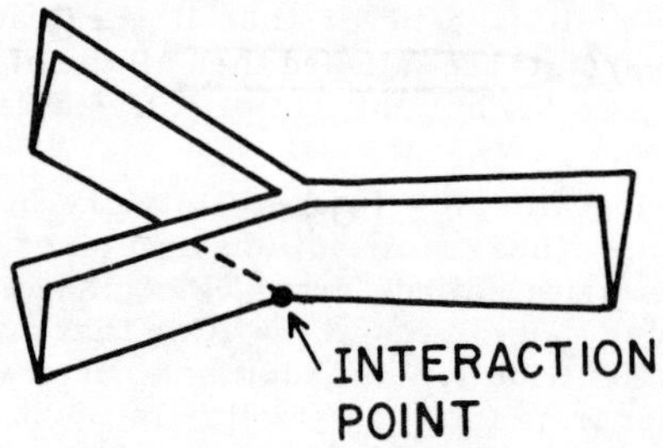

Figure 6. The three-string vertex with external string propagators attached is given by a world sheet that is flat everywhere except at the interaction point where a small circle of radius r has circumference $3\pi r$.

classes of metrics that are related by conformal mappings. It is always possible to find representatives of each equivalence class in which the metric is flat everywhere except at isolated points where the curvature is infinite. Such a metric describes a surface with conical singularities, which is not a manifold in the usual sense. In fact, it is an example of a class of surfaces called orbifolds. The string field theory construction of the amplitude automatically chooses a particular metric which, as we have indicated, is of this type. It is also possible to choose constant curvature metrics, but they do not arise from the string field theory.

The formulation of string field theory described above is certainly very beautiful. It is also deceptively simple, To really understand it in detail one must define the various functional integrations that appear very carefully. One way of doing this is to expand the fields in a Fock-space basis using an infinite number of harmonic oscillators corresponding to the normal modes of $x^\rho(\sigma)$ and $c(\sigma)$. The functional integrations can then be carried out explicitly leaving a Fock space expression for the interaction vertex that can be evaluated for any three states of the string spectrum. This calculation has been carried out in recent papers [28], demonstrating that the formula is in fact well-defined and possesses the properties that it should. Evidence has also been found that it reproduces the standard scattering amplitudes [29].

VI. Anomalies

The gauge invariance of classical gauge theories implies the existence of conserved gauge currents, whose form can be deduced by the standard Noether procedure. In Yang-Mills theories the associated conserved charges are the symmetry generators. In general relativity the conserved current is the energy-momentum tensor and the associated charges are the energy and momentum. The classical conservation of these currents can, under certain circumstances, be destroyed by quantum effects called anomalies. When this happens unphysical modes of the gauge fields become coupled in the S matrix leading to a breakdown of unitarity and causality. Thus, in general, all anomalies in gauge currents must cancel or else the theory must be rejected as inconsistent.

The Feynman diagrams that give rise to anomalies typically arise at one-loop order and involve a loop of chiral fermions. In the well-known case of a gauge theory in four dimensions the simplest diagram that can give rise to anomalies is a triangle diagram, as depicted in fig. 7a. At one vertex we have attached the current whose conservation we wish to study. Gauge fields are attached to the other two. The anomaly that occurs typically has the form

$$\partial^\mu J_\mu \sim \varepsilon^{\mu\nu\rho\lambda} F_{\mu\nu} F_{\rho\lambda}.$$

analogous phenomena can occur in higher dimensions, but then it is necessary th consider diagrams with more external lines. The reason is simply that the ε symbol has D indices in D dimensions. Thus in ten dimensions, for example, the simplest anomaly has the structure

$$\partial^\mu J_\mu \sim \varepsilon^{\mu_1\mu_2\cdots\mu_{10}} F_{\mu_1\mu_2} \cdots F_{\mu_9\mu_{10}}.$$

The corresponding Feynman diagram must have at least six legs (hexagon graph) as shown in fig. 7b.

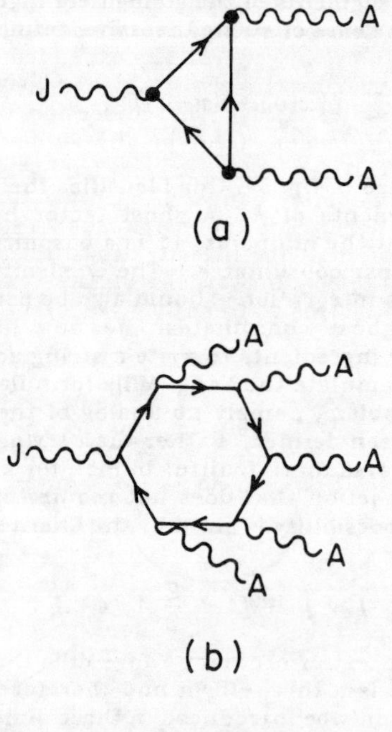

Figure 7. Gauge-current anomalies in $D = 4$ can occur in triangle graphs such as the one in (a). In $D = 10$ the simplest anomalous graph is the hexagon diagram shown in (b).

In the case of string theories the situation is similar, but one should consider a Feynman diagram appropriate to string theory, i.e. a two-dimensional surface. For example, in the case of a theory involving closed oriented strings only the relevant diagram is a torus diagram with external strings (one to represent the gauge current and five to represent gauge fields), as shown in fig. 8. In the case of type IIA superstrings all anomalies trivially cancel because of the left-right symmetry of the theory. In the case of the type IIB theory and the heterotic string theories the spectrum is not symmetric and the cancellation of anomalies is very non-trivial. In fact, the full-fledged string calculations have not yet been carried out. However, there is evidence (based on a low-energy expansion described below) that the cancellation does in fact take place for the theories listed in table I [11,14,30]. In the case of type I superstrings surfaces of various different topologies need to be studied. The two that are relevant to pure gauge anomalies were evaluated by Green and me in 1984 [31]. We found that they individually give anomalies but that the anomaly cancels from the sum for the group choice SO(32) only.

At low energies string theories can be approximated by point-particle theories. This is physically reasonable since when the wavelengths that occur are much longer than the strings, the spatial extension of the string becomes irrelevant. In practice one associates a quantum field with each massless mode of the string theory and writes an effective Lagrangian in terms of these fields only. The effects of massive string modes, and stringiness in general, are expanded out in a series of terms involving higher powers of derivatives and fields. The correction terms in such an expansion are typically suppressed by powers of E/M_{PL}, where M_{PL} is about $10^{19}\,GeV$. Since the anomaly cancellation must take place at each order in this expansion, such a formulation is sufficient for investigating the cancellations in the leading orders of the low-energy expansion. This already gives very severe constraints. It is believed that if the leading order constraints are satisfied then all the higher ones will also be, but this is not yet completely established.

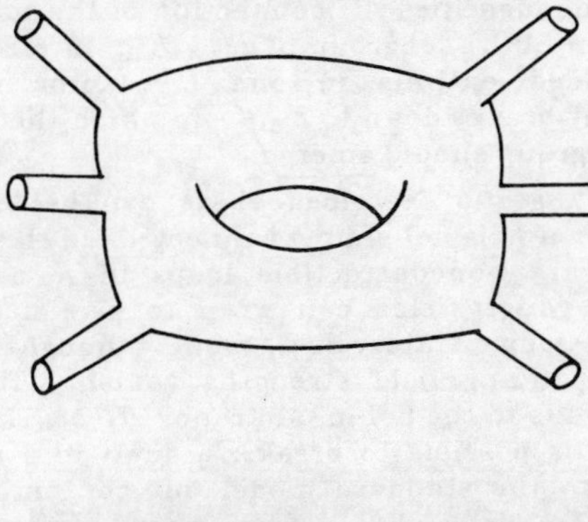

Figure 8. The simplest potentially anomalous diagram in a $D = 10$ theory of closed oriented strings is the torus with six external massless strings.

For theories with $N = 1$ supersymmetry in ten dimensions the contributions to hexagon anomalies arise entirely from chiral fermions. There is one Majorana-Weyl gravitino that makes a certain contribution, an additional Majorana-Weyl spinor from the supergravity multiplet, and n Majorana-Weyl spinors from the Yang-Mills supermultiplet for a gauge group with n generators. By considering the pure gravitational anomaly arising from hexagon diagrams with six external graviton lines one learns that a necessary condition for anomaly cancellation is that the gauge group have 496 generators. This requirement is satisfied, in particular, by SO(32) which has $\frac{1}{2} \cdot 32 \cdot 31 = 496$ generators and by $E_8 \times E_8$ which has $248 + 248 = 496$ generators [11]. Thus one prediction of superstring theory is that there are 484 new forces beyond the 12 that are already known. (This counting does not include gravity, which is not associated with a Yang-Mills generator.) Of course, if this is to be reconciled with nature, almost all these symmetries must be broken at a very high mass scale.

There are many groups that have 496 generators in addition to the two that have been mentioned. To obtain additional restrictions we consider anomaly diagrams involving Yang-Mills fields as well as gravitons. The cancellation of all such anomalies leads to the additional requirement that an arbitrary generator F of the algebra, expressed in the adjoint representation, should satisfy the equation [11]

$$Tr\,F^6 = \frac{1}{48}\,Tr\,F^2\,Tr\,F^4 - \frac{1}{14{,}400}(Tr\,F^2)^3.$$

Remarkably this equation, with exactly the right coefficients, is satisfied for both SO(32) and $E_8 \times E_8$. The only other possible solutions are $E_8 \times [U(1)]^{248}$ and $[U(1)]^{496}$, neither of which seems to correspond to a string theory. In the anomaly analysis an antisymmetric tensor field $B_{\mu\nu}$ that is part of the supergravity multiplet plays an important role. The details are given in the references.

The $SO(16) \times SO(16)$ heterotic string theory, listed in table 1, has only 240 generators. This is possible because it is not supersymmetric. It is chiral, however, and the cancellation of anomalies involves "miracles" analogous to those of the other theories [14].

The groups SO(32) and $E_8 \times E_8$, singled out by the anomaly cancellation analysis in superstring theories, had previously arisen in another context [32]. Specifically, mathematicians have investigated lattices, like those of solid-state physics, in higher dimensions. They were led to consider in particular lattices that are self-dual (i.e., are coincident with the dual lattice) for which the distance squared of each lattice site from the origin is an even integer. (A cubic lattice generated by orthogonal unit vectors is self-dual but not even.) It turns out that even self-dual lattices only occur in dimensions that are multiples of eight. In eight dimensions there is just one and it is generated by the root vectors of the Lie algebra of E_8. In sixteen dimensions there are two of them. One is obvious from the eight-dimensional construction, namely the root lattice generated by $E_8 \times E_8$. The second possibility is closely associated with SO(32). It is generated by root vectors of the algebra SO(32) and certain spinorial weights of its covering group spin(32). More precisely, it is the weight lattice of the group spin$(32)/Z_2$. This coincidence between self-dual lattices and the algebras singled out by the anomaly analysis was skillfully exploited in the construction of the heterotic string theories. The basic idea is that modes that travel clockwise (right-moving) on the string are described by the mathematics of a ten-dimensional superstring whereas those that travel counterclockwise (left-moving) are described by the mathematics of a 26-dimensional bosonic string. Ten of the 26 are paired with the right-moving coordinates to describe "ordinary" ten-dimensional space-time, whereas the remaining 16 are required to form a 16-dimensional torus that is conjugate to one of the two even self-dual lattices. These degrees of freedom, which are the origin of the Yang-Mills symmetry structure of the theory, can equivalently be described using fermionic coordinates instead [13].

VII. Compactification

It may be that our theoretical understanding of superstring theory is not yet sufficiently developed to be able to do correct phenomenology. If we nonetheless choose to plunge in with reckless abandon, then it is clear that a crucial question is what to do with the six extra spatial dimensions. The natural guess is that they curl up to form a six-dimensional space K that is sufficiently small to not have been observed. In fact, the only fundamental scale in the theory is the Planck length, and it is natural to suppose that the internal space K is roughly of this size. This means that it is about the same size that is characteristic of strings themselves.

Since string theory contains gravity it should determine space-time geometry dynamically, and so we must require that the "background geometry" $M_4 \times K$ corresponds to a solution of the classical equations of motion. That is the classical statement sometimes called "spontaneous compactification." Quantum mechanically, it should be determined as part of the characterization of the vacuum - the quantum state of lowest energy. It is generally assumed, since that is all we can do at this point, that a classical solution is a good approximation to a quantum ground state. This might not be true. Another potential pitfall arises from describing K in terms of classical differential geometry. In doing this it is implicit that geometry is determined entirely in terms of the gravitational field (metric tensor). However, this is just one of an infinity of modes of the string. Of course, it is singled out by being massless. The massive modes could play an equally important role, however, in characterizing the geometry and topology of K if K is not much larger than the characteristic string length scale. In this case a whole new type of geometry, let us call it "string geometry," may be required for a suitable description of K. Since this does not yet exist, one assumes that this is not necessary. I would not be surprised, however, if when these matters are better understood the prevailing opinion would change.

Whatever the correct language for describing it may turn out to be, once the vacuum of the theory is identified correctly we will be in a strong position for calculating many quantities of physical interest. The particle spectrum would be determined by a small oscillation analysis, i.e. by studying low-lying excitation modes. Also, as it turns out, many interesting quantities are controlled by the topology of K. Thus one can go a long way with qualitative topological information rather than quantitative geometric information. For example, in a large class of models that has been considered the number of generations of quarks and leptons is controlled by the Euler characteristic of K

$$N_{gen} = \frac{1}{2}\,|\chi(K)|\,.$$

A rather specific compactification scenario was proposed in the paper of Candelas, Horowitz, Strominger, and Witten [33]. They begin with the $E_8 \times E_8$ theory and argue that a classical solution is obtained if K is a Calabi-Yau space, a Kähler manifold of SU(3) holonomy (vanishing first Chern class). Such a manifold always admits a Ricci-flat metric. It has recently been realized that, beginning at fourth order in an expansion in the string length scale, this is not the metric that solves the string equations of motion [34]. However, it has been argued convincingly that there is always another "nearby" metric that does [35]. In formulating the solution one identifies the connection of the space K with an SU(3) subgroup of $E_8 \times E_8$. In fact it is embedded entirely in one E_8 factor which thereby breaks down to E_6, from which the usual gauge group should emerge.

The E_6 group described above can be broken further if K is not simply connected. In this case there are noncontractible loops in K around which gauge fields can wrap to give nonzero Wilson-loop integrals even though the corresponding field strengths vanish. (This is analogous to the Bohm-Aharanov effect.) In this way it is possible to break E_6 down to a group close to the standard model but containing at least an extra U(1). This would imply the existence of a second Z boson at the electroweak scale. There are several alternative scenarios, however.

One alternative "superstring-inspired model" has been described by G. Ross at this meeting [36]. It starts with a specific Calabi–Yau space K with $\chi = 6$ and fundamental group $\pi_1 = Z_3$. Wilson loops are used to break E_6 to $SU(3) \times SU(3) \times SU(3)$. Then by introducing some optimistic assumptions about a resulting potential, it is suggested that this could break to $SU(3) \times SU(2) \times U(1)$ at an intermediate scale by a standard Higgs mechanism. I am sure Ross and his collaborators would agree that this is unlikely to be the final correct model. Still, it is remarkable how close they can come to accounting for a large number of desiderata in this way. There are indications that the axions may not have the required properties, however [37].

VIII. Remaining Problems and Conclusions

Superstring theory is an enormously ambitious attempt to account for all properties of fundamental forces and particles. It starts by describing the physics at the Planck scale. From this all the physics at ordinary energies should be derivable. However, the gap is so large that ordinary physics is, in a sense, all hyperfine structure. Working out these connections is detail is surely going to be a very long struggle requiring the efforts of many clever people. Whether it will ultimately succeed is hard to foresee, even if we assume that the theory is correct. The experimental information that will be learned from the next generation of accelerators, especially as concerns supersymmetry and the Higgs sector, should provide very valuable clues. Without it we are unlikely to find our way.

There are a number of theoretical issues, less directly connected with phenomenology, that must also be addressed. We would like to know how many consistent superstring theories there are. Ideally, there is just one and it explains everything, but there is certainly no guarantee that is the case. Further work is required in formulating the theories. It seems likely that a deep and beautiful principle underlies string theory, but it remains to be elucidated. The work on string field theory is one approach that is being pursued. Some other may, however, be required. One alternative that looks interesting is based on the analytic geometric of conformal field theory [38]. Whatever the answer may be, it will serve to define the theory nonperturbatively and undoubtedly lead to new insights. It could happen that by the time the dust settles the subject will look radically different than it does today.

Superstring theory was developed primarily because it seemed promising for overcoming the divergence problems of quantum gravity. However, even classical general relativity has its problems. It has been proved that generic initial data lead to singularities. This is bad because the structure of space-time subsequent to the formation of the singularity is not determined. The theory is therefore incomplete. It is a plausible conjecture that classical superstring theory is not subject to the same problems [39]. They are short-distance effects, and it is at short distances that the theory is modified. The recent result [34] that a general solution of $R_{\mu\nu} = 0$ is not a solution of string theory is encouraging, since it implies that the Schwarzschild solution, with its singularity, is averted. The actual proof that string theory is a complete classical theory will undoubtedly require a lot of work. The discovery of the optimal formulation and underlying principles discussed above is probably a necessary preliminary.

Other questions that need to be studied are why a particular ground state of the form $M_4 \times K$ should be singled out, if that is the case. It would be very sad if there were thousands of theoretically acceptable vacua,. since then much predictivity would be lost despite the uniqueness of the underlying theory. We need to understand the origin of the mass hierarchy $m_W/m_{PL} \sim 10^{-17}$ and the origin of the supersymmetry breaking. These are tough problems, but the one that really worries us the most is to understand why the cosmological constant $|\Lambda| < 10^{-120}$ in natural units. Every little effect one can think of makes a much larger contribution to Λ than this. Why they should all precisely cancel is, at the moment, totally baffling.

In conclusion, the theoretical understanding of superstrings is progressing rapidly. The long-term prospects for the subject appear very bright. However, anyone who expects dramatic phenomenological successes in the short term is likely to become disillusioned. Developing this subject is an exhilarating intellectual experience, but it requires a major commitment of time and effort that is not appropriate for everyone. In any case, there is plenty of material to keep busy those who choose to make this commitment while we await the next round of experimental results.

Bibliography

1. J.H. Schwarz, "Superstrings. The first fifteen years of superstring theory," (World Scientific, 1985);
   M.B. Green, J.H. Schwarz, and E. Witten, "Superstring theory," (Cambridge Univ. Press) to be published.
2. M. Goroff and A. Sagnotti, Nucl. Phys. B266, (1986) 709.
3. M. Jacob, editor, "Dual theory," Physics Reports Reprint Volume I (North Holland, 1974);
   J. Scherk, "An introduction to the theory of dual models and strings," Rev. Mod. Phys. 47 (1975) 123.
4. A. Neveu and J.H. Schwarz, Nucl. Phys. B31 (1971) 86; Phys. Rev. D4 (1971) 1108;
   A. Neveu, J.H. Schwarz, and C.B. Thorn, Phys. Lett. 35B (1971) 529;
   C.B. Thorn, Phys. Rev. D4 (1971) 1112.
5. P. Ramond, Phys. Rev. D3 (1971) 2415.

6. F. Gliozzi, J. Scherk, and D. Olive, Phys. Lett. 65B (1976) 282; Nucl. Phys. B122 (1977) 253.
7. M.B. Green and J.H. Schwarz, Nucl. Phys. B181 (1981) 502.
8. P. Goddard and C.B. Thorn, Phys. Lett. 40B (1972) 235;
J.H. Schwarz, Nucl. Phys. B46 (1972) 61;
R.C. Brower and K.A. Friedman, Phys. Rev. D7 (1973) 535.
9. J. Scherk and J.H. Schwarz, Nucl. Phys. B81 (1974) 118.
10. T. Yoneya, Prog. Theor. Phys. 51 (1974) 1907.
11. M.B. Green and J.H. Schwarz, Phys. Lett. 149B (1984) 117.
12. M.B. Green and J.H. Schwarz, Phys. Lett. 109B (1982) 444.
13. D.J. Gross, J.A. Harvey, E. Martinec, and R. Rohm, Phys. Rev. Lett. 54 (1985) 502; Nucl. Phys. B256 (1985) 253; Nucl. Phys. B267 (1986) 75.
14. L. Dixon and J.A. Harvey, Nucl. Phys. B274 (1986) 93;
L. Alvarez-Gaumé, P. Ginsparg, G. Moore, and C. Vafa, Phys. Lett. 171B (1986) 155.
15. N. Seiberg and E. Witten, Princeton preprint.
16. L. Brink, P. Di Vecchia, and P. Howe, Phys. Lett. 65B (1976) 471;
S. Deser and B. Zumino, Phys. Lett. 65B (1976) 369.
17. A.M. Polyakov, Phys. Lett. 103B (1981) 207,211.
18. O. Alvarez, Nucl. Phys. B216 (1983) 125;
E. D'Hoker and D. Phong, Nucl. Phys. B269 (1986) 205;
A. Belavin and V. Knizhnik, Landau Institute preprint 32/7 (1986);
L. Alvarez-Gaumé, G. Moore, and C. Vafa, Harvard preprint HUTP-86/A017;
Yu. I. Manin, Pisma ZETP 43 (1986) 161; Phys. Lett. 172B (1986) 184;
J.B. Bost and P. Nelson, Harvard preprint HUTP-86-A014;
J.B. Bost and T. Jolicoeur, Phys. Lett. 174B (1986) 273;
R. Catenacci, M. Cornalba, M. Martellini, and C. Reina, Phys. Lett. 172B (1986) 328;
C. Gomez, Phys. Lett. 175B (1986) 32;
L. Alvarez-Gaumé, G. Moore, P. Nelson, C. Vafa, and J.B. Bost, Harvard preprint HUTP-86/A039;
G. Moore, J. Harris, P. Nelson, and I. Singer, Harvard preprint HUTP-86/A051;
A. Morozov, preprint ITEP 86-88;
A. Belavin, V. Knizhnik, A. Morozov, and A. Perelomov preprint ITEP 86-59;
A.A. Beilinson and Yu. I. Manin, Moscow preprints 1986;
A. Restuccia and J.G. Taylor, Phys. Lett. 174B (1986) 56.
19. E. Cremmer and J.L. Gervais, Nucl. Phys. B76 (1974) 209; Nucl. Phys. B90 (1975) 410;
M. Kaku and K. Kikkawa, Phys. Rev. D10 (1974) 1110, 1823.
20. M.B. Green and J.H. Schwarz, Nucl. Phys. B218 (1983) 43; Nucl. Phys. B243 (1984) 475;
M.B. Green, J.H. Schwarz, and L. Brink, Nucl. Phys. B219 (1983) 437.
21. W. Siegel, Phys. Lett. 151B (1985) 391, 396;
W. Siegel and B. Zwiebach, Nucl. Phys. B263 (1986) 105;
B. Zwiebach, MIT preprint (1985).
22. T. Banks and M. Peskin, Nucl. Phys. B264 (1986) 513;
M.E. Peskin and C.B. Thorn, Nucl. Phys. B269 (1986) 509.
23. A. Neveu and P.C. West, Phys. Lett. 165B (1985) 63 and CERN preprints.
24. H. Hata, K. Itoh, T. Kugo, H. Kunitomo, and K. Ogawa, Phys. Lett. 172B (1986) 186, 195, and Kyoto University preprints.
25. E. Witten, Princeton preprints (1985, 1986).
26. S.B. Giddings, E. Martinec, and E. Witten, Princeton preprint (1986).
27. M. Kato and K. Ogawa, Nucl. Phys. B212 (1983) 443;
S. Hwang, Phys. Rev. D28 (1983) 2614.
28. D. Gross and A. Jevicki, Princeton preprint (1986);
E. Cremmer, A. Schwimmer, and C.B. Thorn, Ecole Normale Supérieure preprint LPTENS-86-14.
29. S.B. Giddings, Princeton preprint (1986);
S.B. Giddings and E. Martinec, Princeton preprint (1986).
30. L. Alvarez-Gaumé and E. Witten, Nucl. Phys. B234 (1983) 269.
31. M.B. Green and J.H. Schwarz, Nucl. Phys. B255 (1985) 93.
32. See P. Goddard and D. Olive, p. 51 in "Vertex operators in mathematics and physics," eds J. Lepowsky et al. (Springer, 1985).
33. P. Candelas, G. Horowitz, A. Strominger, and E. Witten, Nucl. Phys. B258 (1985) 75.
34. M.T. Grisaru, A.E.M. Van de Ven, and D. Zanon, Phys. Lett. 173B (1986) 423;
M.D. Freeman and C.N. Pole, Phys. Lett. 174B (1986) 48;
D. Gross and E. Witten, Princeton preprint (1986).
35. E. Witten, Princeton preprint (1986);
D. Nemeschansky and A. Sen, SLAC preprint (1986).
36. B.R. Greene, K.H. Kirklin, P.J. Miron, and G.G. Ross, Oxford Univ. preprints (1986).
37. J.E. Kim, Seoul National University preprint (1986);
M. Dine and N. Seiberg, Nucl. Phys. B273 (1986) 109.
38. D. Friedan and S. Shenker, Enrico Fermi Institute preprints EFI 86-18A,B.
39. These remarks are mostly due to D. Gross.

Plenary Session

Status of the Electroweak Theory

G. Altarelli (Rome)

CP Violation and Weak Decays of Quarks and Leptons

M.G.D. Gilchriese (Cornell)

Chairman

H. Schopper (CERN)

Scientific Secretaries

P. Drell (LBL)

D.A. Herrup (LBL)

STATUS OF THE ELECTROWEAK THEORY

G.Altarelli

Dipartimento di Fisica, Università di Roma "La Sapienza"
INFN - Sezione di Roma
P.le Aldo Moro, 2 - 00185 Roma, Italy

In this talk I shall try to review the main developments over the last year in the physics of electro-weak interactions as they emerged from the parallel Sessions. The reader is referred to the excellent work of my predecessors, the rapporteurs L.Maiani at the Bari Conference (1) and P.Langacker at Kyoto (2), for summaries of the field which in most respects are still quite up to date. I have also profited of other recent review papers (3).

W/Z° PHYSICS

At present the total collected luminosity, useful for the analysis of the $W \to e\nu$ and $Z^\circ \to e^+e^-$ channels, has reached the values:(4)(5)

$$\int dt \mathcal{L} \simeq \left[729(UA1)+880(UA2)\right] nb^{-1} \qquad (1)$$

The number of observed events is as follows:

$$\begin{array}{llll} W \to e\nu & \sim 500 & (UA1+UA2) & \\ W \to \mu\nu & \sim \ \ 65 & (UA1) & (2) \\ W \to \tau\nu & \sim \ \ 30 & (UA1) & \end{array}$$

$$\begin{array}{lll} Z^\circ \to e^+e^- \sim 69 & (UA1+UA2) & \\ Z^\circ \to \mu^+\mu^- \sim 19 & (UA1) & (3) \end{array}$$

The updated mass values are given in Table 1. Note that the UA1 values do not yet include the results of the '85 run.

TABLE 1

	UA1 (old)	UA2
M_W(GeV)	$83.5^{+1.1}_{-1.0} \pm 2.7$	$80.1 \pm 0.8 \pm 1.3$
M_Z(GeV)	$93.0 \pm 1.4 \pm 3.0$	$92.1 \pm 1.1 \pm 1.5$
Γ_W(GeV)	< 6.5 (90%)	
Γ_Z(GeV)	< 8.3 (90%)	< 7.1 (90%)

The $W \to \tau\nu$ signal has been clearly extracted by the UA1 collaboration from the missing energy sample, as reported by S.Geer (4) and A.Honma (6). The value of M_W obtained from the 32 $W \to \tau\nu$ events, $M_W = 89 \pm 3 \pm 6$ GeV, is in good agreement with the results from $W \to e\nu$ and $W \to \mu\nu$. All observed properties are in agreement with lepton universality. For example:

$$\frac{\sigma B(\tau\nu)}{\sigma B(e\nu)} = 1.02 \pm 0.20 \pm 0.05 \qquad (4)$$

or, in terms of the weak couplings:

$$g_\tau / g_e \simeq 1.01 \pm 0.09 \pm 0.05 \qquad (5)$$

The UA2 collaboration has invested a large effort in trying to detect $W/Z^\circ \to$ jet-jet, as discussed by A.Roussarie (5). The status of the art can be inferred from fig.1, which was obtained after suppression of the formidable QCD background by suitable cuts.

With increasing statistics the production cross sections for W and Z° agree better and better with QCD expectations. This is displayed in fig.2 for $\sigma \cdot B$. The QCD predictions were obtained in ref.7. Independent but consistent calculations were presented at this conference by W.J.Stirling (8). The theoretical error is due to uncertainties on parton structure functions and on the scale to be inserted in the corrections of order α_s. It must also be noted that the branching ratios used in the calculation are those appropriate for $m_{top} \simeq 40$ GeV.

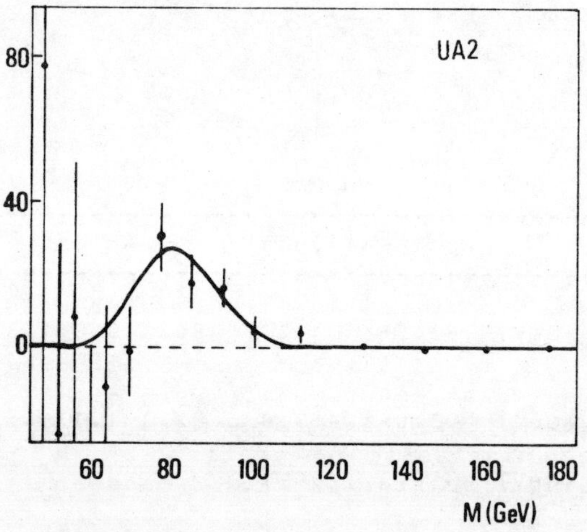

Fig.1

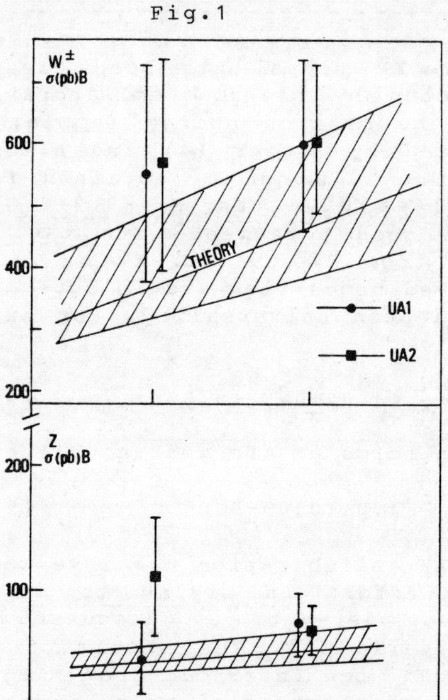

Fig.2

If m_{top} is larger $B(W \to e\nu)$ is increased by $\lesssim$ 20% and $B(Z \to e^+e^-)$ by $\lesssim$ 10%. The successful prediction of the total cross sections is a very important quantitative test of QCD in the domain of Drell-Yan type processes. Even more significant is the prediction of the P_T distribution (7) (also well confirmed by the data) which is determined by highly non trivial dynamical aspects of QCD beyond the parton model. The comparison of theory and experiment is shown in fig. 3. The agreement is extremely good in the region of P_T where the statistics is significant ($P_T \lesssim$ 25 GeV). From fig. 4, presented by E.Duchovni, (9) one might get the impression that the theory does not reproduce the data at larger values of P_T. Note that at small P_T where the theoretical predictions are quite well supported by experiment, the theory is complicated. On the other hand, at large P_T the much simpler perturbative treatment should be completely reliable. At a closer inspection one notes that there are only two events with $P_T >$ 50 GeV. Also, these two events (one $W \to e\nu$ and one $W \to \mu\nu$ event) show two well separated jets, with a di-jet mass of order M_W. This is apparently at odds with the expectation that single jet events should be dominant at large P_T. The total W-jet-jet mass is ~ 250-300 GeV. Two di-jet events from the missing energy sample interpreted as $Z(\to \nu\bar{\nu})$+jet+jet, with large di-jet masses, were also discussed, because of their (rough) similarity with the previous ones.

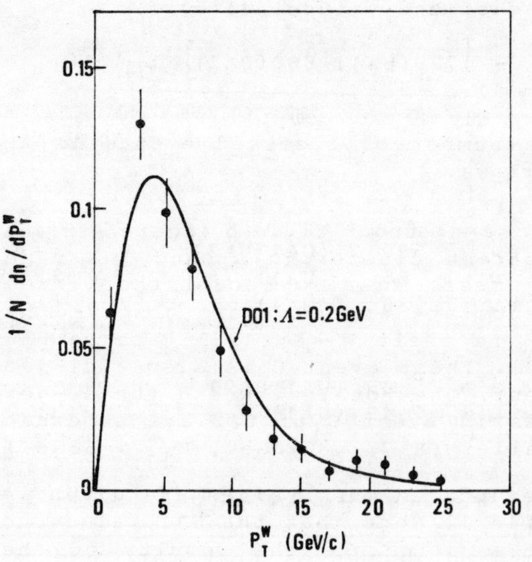

Fig.3

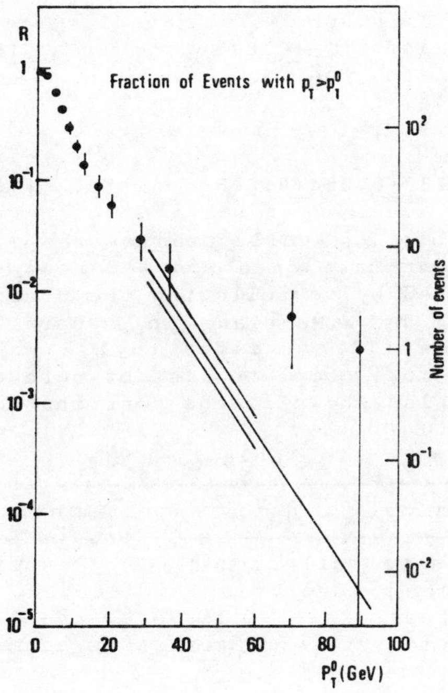

Fig.4

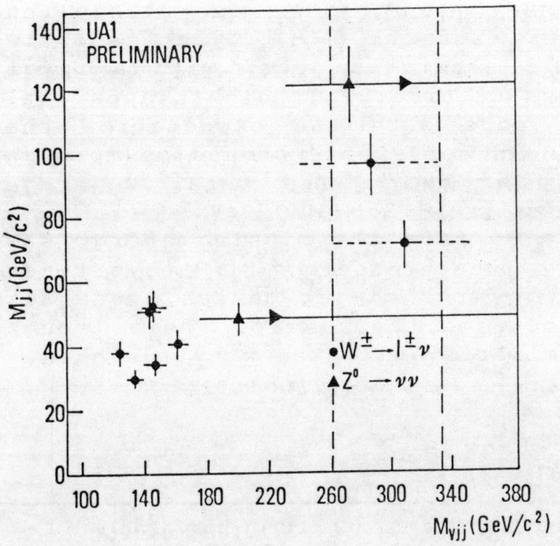

Fig.5

As seen from fig. 5 there is no apparent clustering around the same j-j mass. We also recall that three events at large P_T were observed by UA2 in their $W \to e\nu$ 1983 data sample (10). Their event C is also a di-jet event. The total mass and the di-jet mass in event C of UA2 are somewhat smaller ($M_{Wjj} \sim 160GeV$, $M_{jj} \sim 66GeV$) than for the two $W \to l\nu$ events of UA1. In conclusion, there are standard processes that can lead to these topologies, e.g. $p\bar{p} \to W+j_1+j_2+X$, $p\bar{p} \to W+W+X$. However the cross sections are quite small. W.J.Stirling (8) has reported on a calculation that shows that in the UA1 W sample one would expect something of the order of 0.1 events while 2 are observed. I think that this fact is not per se very impressive, because few events do not make a statistically significant sample. However, these events are obviously interesting. One should wait for more data (Tevatron) and see. Meanwhile one can look for possibly related signatures like anomalies in 4 jet distributions, photon plus jets etc.

RECENT RESULTS ON $\sin^2\theta_W$

In this section we always refer to the definition(11)

$$\sin^2\theta_W \equiv s^2 = 1 - \frac{M_W^2}{M_Z^2} \qquad (6)$$

All experimental values given in the following were radiatively corrected in order to lead to a determination of s^2 as defined in eq.6. New precise measurements of s^2 were discussed at this conference from a) $\overset{(-)}{\nu_\mu}$-N deep inelastic scattering b) $\overset{(-)}{\nu_\mu}$ p elastic scattering c) W/Z masses.

a) <u>ν_μ-N deep inelastic scattering.</u>

In these reactions s^2 is obtained from the measurement of the ratio R_ν of the neutral to charged current cross sections of neutrinos. New results were obtained recently at CERN by the CDHSW (12) and CHARM (13) collaborations and at FNAL by the CCFRR (14) and FMM (15) groups. These results are:

CDHSW:

$$s^2=0.225\pm0.005\pm0.003\pm\pm0.013\ (m_c-1.5(GeV)) \qquad (7)$$

CHARM:

$$s^2=0.236\pm0.005\pm0.003\pm\pm0.012\ (m_c-1.5(GeV)) \qquad (8)$$

CCFRR:

$$s^2=0.239\pm0.008\pm0.006\pm0.006 \qquad (9)$$

FMM:

$$s^2=0.244\pm0.012\pm0.013 \qquad (10)$$

In eqs.7-10 the last error is the theoretical error associated to our ignorance of the precise value of the charm quark mass. This error is stated explicitly in terms of m_c by CDHSW and CHARM, while it is computed for $m_c=(1.5\pm0.4)$ GeV by CCFRR and FMM.

b) $\overset{(-)}{\nu}_\mu$p elastic scattering. From BNL (16) we have the recent result:

BNL:

$$s^2=0.220\pm0.016\pm^{0.023}_{0.031} \qquad (11)$$

b) W/Z masses. As well known, there are two independent methods of deriving s^2 from the values of M_W and M_Z. The first method is to obtain s^2 from the ratio M_W/m_Z by directly using eq.6. The advantage is that the energy calibration error (which is responsible for most of the systematic errors shown in Table 1) drops away in the ratio and, of course, there are no radiative corrections to be applied. The drawback is that the statistical error is still very large, due to the present smallness of the Z° sample. The results obtained by this method are:

UA1 (old):

$$s^2=0.194\pm0.031 \qquad (12)$$

UA2:

$$s^2=0.242\pm0.023\pm0.009 \qquad (13)$$

The UA2 value given here also includes the analysis of the 1985 run.

The second method, which at the moment is the most precise, is to obtain s^2 from M_W (plus separately from M_Z) via the relation

$$s^2=\frac{\mu^2}{M_W^2}=\frac{\mu^2{}_{Born}}{1-\Delta r}\,\frac{1}{M_W^2} \qquad (14)$$

where

$$\mu^2_{Born}=\frac{\pi\alpha}{\sqrt{2}\,G_F}=(37.281\text{GeV})^2$$

and Δr is the effect of radiative corrections to be discussed later. By this method the results are:

UA1 (old):

$$s^2=0.214\pm^{0.005}_{0.006}\pm0.015 \qquad (15)$$

UA2:

$$s^2=0.232\pm0.004\pm0.008 \qquad (16)$$

In table 2 a collection of the most relevant data on s^2 is reported, as compiled by me following the presentation by W.Marciano in a parallel session (17). I refer to his paper for a more complete list of references. Also shown is the contribution

Table 2

Experiment	s^2	RAD.CORR.
ATOMIC P.V. (Cs) (Paris)(18)	0.230±0.030	+0.009
e-D ASYMMETRY (SLAC)(19)	0.218±0.020	-0.010
$\overset{(-)}{\nu}_\mu$ e (BNL+CHARM)(20)(21)	0.212±0.023	SMALL
$\overset{(-)}{\nu}_\mu$ P (ELASTIC) (BNL)	0.220±0.031	SMALL
ν_μ N DEEP IN. (CDHSW+CHARM+ +CCFRR+FMM)	0.233±0.003±0.006	-0.011
$M_{W,Z}$ (UA1+UA2)	0.227±0.003±0.008	+0.016
$1-\frac{M_W^2}{M_Z^2}$ (UA1+UA2)	0.218±0.022	-
COMBINED	0.227±0.003±0.006	

of radiative corrections to the result (computed for $m_{top}\sim$ 40GeV and $m_{Higgs}\sim M_Z$). For example, without radiative corrections, the central value of s^2 for ν_μ N scattering would be 0.244 instead of 0.233. For the most precise experiments, i.e. those which are dominant in computing the world average, the statistical and systematic (or theoretical) errors are kept separate. These errors were added linearly when fixing the weights for the average. The resulting combined value of s^2

from all the entries in table 2 is also reported in the last line of the same table. As the most precise experiments are affected by a comparable systematic error, I think that this same error is to be reported for the average value, as given in Table 2.

What is really tested by this impressive list of experiments all leading to compatible values of s^2? On one hand, very important constraints on the tree level structure of the theory are obtained. On the other hand, some quite valuable information is also derived on the one loop radiative corrections, which now start to be crucial in comparing different experiments. In fact, from table 2, even adopting the rather conservative procedure of adding all errors linearly, one obtains

$$s^2_{\nu_\mu N}-s^2_{M_W}=0.006\pm0.020 \text{ (rad.corr.)} \tag{17}$$

and

$$s^2_{\nu_\mu N}-s^2_{M_W}=0.033\pm0.020 \text{ (no rad.corr.)} \tag{18}$$

In the following I shall first briefly discuss the importance of the previous results within the context of the theory at tree level and then consider the radiative corrections.

In the minimal standard theory at the tree level three in principle completely different definitions of $\sin^2\theta_W$ are predicted to coincide. They are:

a) the quantity $\sin\theta_W=e/g$ defined as the ratio between the electric charge e and the $SU(2)_{Weak}$ gauge coupling g_2. This is obtained by measuring M_W from the relation:

$$\frac{G_F}{\sqrt{2}} = \frac{g_2^2}{8M_W^2} = \frac{e^2}{8M_W^2\sin^2\theta_W} \tag{19}$$

b) The quantity $\sin^2\theta_W$ derived from the gauge boson mass matrix, according to $\sin^2\theta_W=1-M_W^2/M_Z^2$, which is of course obtained from measuring the ratio M_W/M_Z.

c) The parameter $\sin^2\theta_W$ which appears in the neutral current couplings:

$$\mathcal{L}^{Z^0}_{eff} \sim 4\frac{G_F}{\sqrt{2}}\rho(J_3-\sin^2\theta_W J_{em})^2 \tag{20}$$

which is measured from ν-N, e-D, ... reactions.

The equality at the tree level of these three definitions of $\sin^2\theta_W$ is the signature of the minimal standard theory. Practically all conceivable departures from the minimal standard theory remove the degeneracy of the above three ways of defining $\sin^2\theta_W$. For example, in non gauge models (22), with a global SU(2) symmetry and vector dominance by three massive bosons W, W_3 the neutral current couplings are induced by $W_3-\gamma$ mixing. One has in general two independent parameters: e/g=s and the $W_3-\gamma$ mixing λ. These two parameters appear in the following way:

$$\frac{G_F}{\sqrt{2}} = \frac{e^2}{8M_W^2 s^2}$$

$$1-\frac{M_W^2}{M_Z^2} = \lambda^2 \tag{21}$$

$$\mathcal{L}^{Z^0}_{eff} \sim 4\frac{G_F}{\sqrt{2}}\rho(J_3-s\lambda J_{em})^2$$

Similarly the three ways of measuring $\sin^2\theta_W$ would also give different results as a consequence of non doublet Higgs bosons, of more W's and/or Z° (new gauge degrees of freedom, composite W/Z's (23)), of additional vertices in the lagrangian density (for example, residual non renormalizable interactions (24) from quark and lepton compositeness) and so on. Thus precise experiments comparing neutral current couplings with measurements of weak intermediate boson masses are crucial tests of the standard model and could well lead to the discovery of new physics.

The tree level relations connecting M_W or the neutral current measurements with $\sin^2\theta_W$ are also modified by radiative corrections (25). It was already mentioned that by now the experiments are sufficiently precise that these corrections cannot be ignored for a meaningful comparison. Most of the radiative corrections arise from known physics and are therefore not much interesting. For example, the corrective factor $(1-\Delta r)^{-1}$ which appears in eq.14 can be written in the form:

$$\frac{1}{1-\Delta r} \sim \frac{\alpha(M_W)}{\alpha}(1+\epsilon) \qquad (22)$$

The running of the electromagnetic coupling α accounts for a 7% effect (26)(27). This is completely determined by the photon vacuum polarization diagram, where the contributions of charged fermions with $m \ll M_W$ are dominant.

If $m_{top} \lesssim M_Z$ and $M_{Higgs} \lesssim$ few TeV, then ϵ is much smaller in the standard theory, its effect being of the order of a few per mille (27). But ϵ could become large for a variety of reasons: $m_{top} \gtrsim 100$ GeV, the existence of new families of quarks (28)(29), new physics (e.g. widely split multiplets of SUSY particles (30)), exotic W/Z self couplings (31) and other possibilities. For example, the dependence in the standard theory of Δr on m_{top} and m_H (the Higgs mass) is displayed in fig. 6 taken from the last of ref.27. For light m_{top} we have $\Delta r \sim 7\%$. Precisely for $m_{top} \sim$ 40GeV and $m_H \simeq M_Z$ one obtains (32):

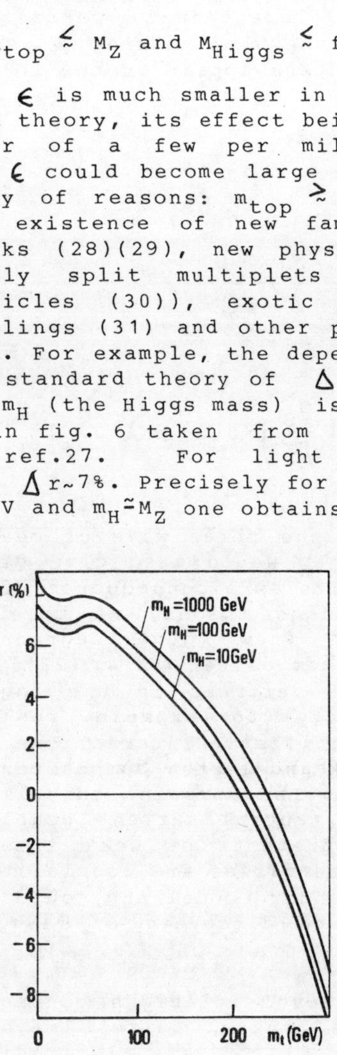

Fig.6

$$\Delta r = 0.0711 \pm 0.0018 \qquad (23)$$

While Δr is not much sensitive to m_H it rapidly decreases with m_{top} as soon as $m_{top} \gtrsim M_W$ and vanishes for $m_{top} \sim (210\text{-}240)$GeV. As the previous experimental indications for $m_{top} \simeq (30\div50)$GeV have not been confirmed, upper bounds on the top quark mass again become an interesting subject. It is well known that for large m_{top} the electro-weak radiative corrections are dominated by the quadratic divergences for $m_{top} \to \infty$ (28)(33). Thus the behaviour of Δr is not surprising. In fact an upper bound on m_{top} was already obtained from the measured value of the ρ parameter which appears in eqs.20 for $\mathcal{L}^Z_{eff}$. In the minimal standard model $\rho=1$ at tree level, but radiative corrections modify this prediction. From the ratio of neutral to charged currents in ν_μ-N scattering one obtains for $m_{top} \gg m_b \sim 0$ that $\rho_{\nu N} \simeq 1+\delta\rho$ with

$$\delta\rho = \frac{3\,G_F}{8\sqrt{2}\pi^2}\, m_{top}^2 \simeq 0.02\left(\frac{m_{top}}{250\text{GeV}}\right)^2 \qquad (24)$$

By taking into account the experimental value (2) of $\rho_{\nu N} \simeq 1.006 \pm 0.008$ and that ordinary radiative corrections predict $\rho_{\nu N} \simeq 0.99$, one derives a bound of about $m_{top} \lesssim$ 300GeV.

This is an old result. What is new is that now a comparable (if not better) limit on m_{top} can also be derived from the comparison of s^2 as derived from ν-N scattering and M_W. In fact we have seen that Δr decreases with m_{top}. Thus at fixed M_W the value of s^2 decreases with m_{top}.

On the other hand it has been shown (34) that the value of s^2 obtained from ν-N scattering is not much modified when m_{top} varies between M_W and 300GeV. This result is only true for s^2 defined by eq.6 and would not be true for $\sin^2\theta_W$ in the MS definition. Thus, when m_{top} is increased the agreement between the values of s^2 measured from M_W and from ν-N scattering is progressively spoiled. By this method from the results of ref.34 one obtains the bound

$$m_{top} \lesssim 250\text{GeV} \quad (25)$$

In conclusion relevant pieces of information are already obtained on the size of radiative corrections from the experiments on s^2. Similar constraints on one loop radiative corrections to charged current processes are also obtained. In fact these corrections if large would induce apparent violations of universality (35). Then from the present errors on the quark mixing matrix one can derive some interesting limits on possible forms of new physics as discussed by W.Marciano.

In the near future one expects progress in the domain of precision tests of the standard model from a) the measurement of s^2 from $\nu_\mu e$ scattering with a precision of ± 0.005. (36) With respect to ν-N scattering the advantage is the absence of all theoretical errors connected with hadronic physics. b) Better measurements of M_W and M_Z at the Tevatron and at ACOL. By the end of '88 one expects to have collected about ten times more statistics on $W \to e\nu$ and $Z \to e^+e^-$. One could obtain(37):

$$\delta\left(\frac{M_W}{M_Z}\right) = \pm 0.003 \pm 0.002 \quad (26)$$

$$\delta s^2 \simeq \pm 0.006 \pm 0.004 \text{ from } M_W/M_Z \quad (27)$$

$$\delta s^2 \simeq \pm 0.001 \pm 0.006 \text{ from } M_W \text{ and } M_Z \quad (28)$$

c) At LEP I (38) one can measure M_Z with precision $\delta M_Z \simeq \pm 50$MeV. If μ^2 in eq.14 was given, which to this level of precision is not the case, then this would correspond to $\delta s^2 \simeq \pm 0.0004$. One needs another experiment of comparable precision. At SLC and LEP-I the best possibilities are offered by the forward-backward asymmetry A_{FB}, the τ lepton helicity asymmetry A_{POL}, and possibly the left-right asymmetry A_{LR}, which of course needs longitudinally polarized electrons. It is estimated (38) that the precision that one can aim to is given by $\delta s^2 \simeq \pm 0.002$. d) At LEP 2 (39) one could measure the W mass with an error $\delta M_W \simeq \pm 200$MeV. Given M_Z, this corresponds to $s^2 \simeq \pm 0.004$.

Before closing this section I want to make some remarks on the electroweak tests from the measurement of the forward-backward asymmetry in $e^+e^- \to \mu^+\mu^-, \tau^+\tau^-$ etc. at PEP and PETRA (40). It is often mentioned that there are problems in that sector, so that it may be worthwhile to summarize the situation.

The measured asymmetry A_{FB} is given by:

$$A = -\frac{3}{8} a_e a_f \chi \quad (29)$$

where, for $\sqrt{s} \ll M_Z$

$$\chi = \frac{\rho G_F s}{2\sqrt{2}\pi\alpha}, \quad (30)$$

a_f is proportional to the neutral current axial coupling of the fermion f and ρ is defined in eq.20 ($a_f = +1$, $\rho = 1$ at the tree level). Thus what is directly measured is the product $\rho a_e a_f$. The experimental data for $e^+e^- \to \ell^+\ell^-$ are summarized in Table 3 (41). Within two standard deviations all entries in that table are in agreement with

Table 3

EXP	$\sqrt{s}$(GeV)	$\rho a_e a_\mu$	$\rho a_e a_\tau$
PEP	29	0.99 ± 0.12	0.88 ± 0.16
PETRA	34.5	1.27 ± 0.13	0.79 ± 0.22
PETRA	43.5	1.16 ± 0.13	0.85 ± 0.20
Combined		1.13 ± 0.07	0.85 ± 0.11
Combined+Universality Incl.Bhabha		1.05 ± 0.06 1.03 ± 0.04	

theoretical expectations. However, PETRA experiments show a systematic trend to larger values for muons and to smaller values for tauons than expected. When combined the average is in perfect agreement with the standard theory. Note that a_e is separately determined by $\overset{(-)}{\nu}_\mu$-e and $\overset{(-)}{\nu}_e$-e scattering to be $a_e = 0.99 \pm 0.05$. Thus the electron is OK, the muon shows a moderate over-fluctuation and the tauon a similar underfluctuation (only at PETRA and not at PEP). It can be added that (with less precision) nothing wrong is observed in the asymmetries of heavy quarks (42).

Sometimes the results are described in terms of $\sin^2\theta_W$ and M_Z^2 by writing χ, given in eq.30, in the form

$$\chi = \frac{\rho G_F s}{2\sqrt{2}\pi\alpha} = \frac{s}{4M_Z^2 \sin^2\vartheta_W \cos^2\theta_W} \tag{31}$$

Given M_Z one then obtains $\sin^2\theta_W$ by assuming $a_e a_f = 1$. The results are shown in fig.7 compared with the collider data and the standard relation (38) between M_Z and $\sin^2\vartheta_W$ (including radiative corrections and their error, for $m_{top} \sim 40$GeV). There we see the same situation in a different form: the muon data alone are within two σ's, the combined muon and tauon are within one σ. In conclusion I do not see much hope of new physics here.

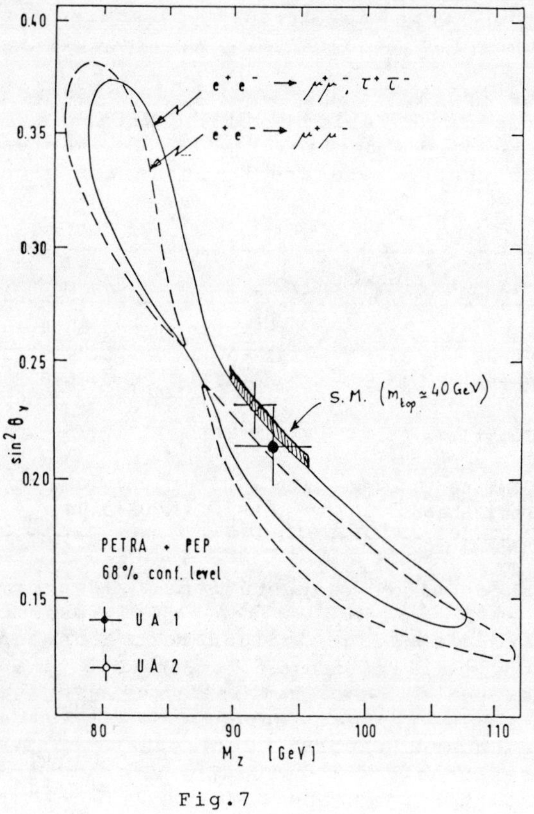

Fig.7

LIMITS ON NEW FAMILIES

An important lower bound on the mass of a new sequential charged lepton τ' was obtained by UA1 from the missing E_T sample (searches for heavy leptons were reviewed by M.Perl (43)). Assuming a normal branching ratio for $W \to \tau'\nu_{\tau'}$, with $\nu_{\tau'}$ of negligible mass, the following result was obtained (44):

$$m_{\tau'} \gtrsim 41\text{GeV} \tag{32}$$

This is a very important constraint on new sequential families. To better appreciate this fact the most effective way is to try to extrapolate the features of the known families. For example, consider the GUT inspired relation (45)

$$\frac{m_s}{m_\mu} \simeq \frac{m_b}{m_\tau} \tag{33}$$

which presumably should be better satisfied for heavier families and is empirically valid for the second and third ones. If this equality is extended to a possible fourth family, then from the bound eq.32 and $(m_{b'}/m_{\tau'}) \sim 3$ one obtains $m_{b'} \gtrsim 120$ GeV. The possibility $m_{t'} \gg m_{b'}$, in analogy with the other heavy families, is then already excluded by the bounds given in the previous section for m_{top} and obtained from $\rho_{\nu N}$ and $\sin^2\theta_W$.

A different set of constraints on new families is obtained by setting bounds on the number of neutrinos. In $p\bar{p}$ reactions one measures (4)(5)

$$R_{exp} = \frac{\sigma B(W \to e\nu)}{\sigma B(Z \to e^+e^-)} = \begin{cases} 8.9^{+1.6}_{-1.3} & \text{UA1} \\ 7.9^{+1.7}_{-1.3} & \text{UA2} \end{cases} \tag{34}$$

The Z° width is thus given by:

$$\Gamma_Z = R_{exp}\left(\frac{\sigma_Z}{\sigma_W}\right) \frac{\Gamma(Z \to e^+e^-)}{\Gamma(W \to e\nu)} \Gamma_W \tag{35}$$

σ_Z/σ_W is predicted by QCD (7) with

a smaller uncertainty than the individual cross sections (with a 7% error). The leptonic widths are quite safely computed. Also Γ_W is calculated theoretically. One must assume that no new charged lepton contributes to Γ_W. A new doublet $(\tau', \nu_{\tau'})$ with 41GeV $\lesssim m_{\tau'} \lesssim M_W$ and $\nu_{\tau'}$ light would evade the bound. Of course Γ_W and Γ_Z depend on m_{top}.

The bound on the number of additional neutrinos ΔN_ν is weakened by our ignorance of m_{top}. In fig.8 the dependence of the bound on N_ν from the value of m_{top} is shown. Without additional information on m_{top}, by this method one obtains the bound $\Delta N_\nu \lesssim 4$ (90%c.l.).

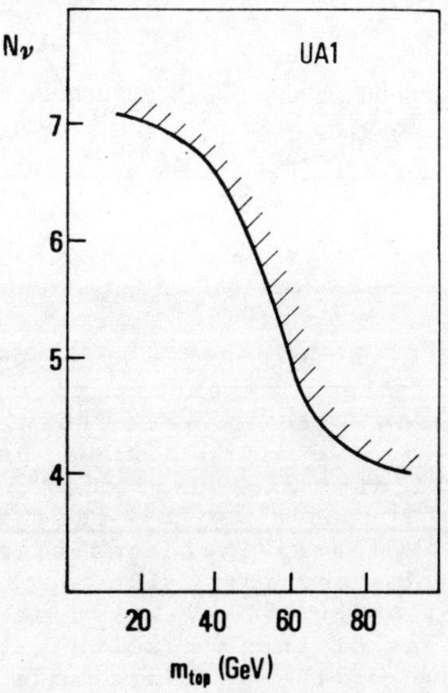

Fig.8

From the missing E_T sample one can put a limit on the production of Z°+jet followed by the decay $Z° \rightarrow \nu\bar{\nu}$. At 90% c.l. the result is $\Delta N_\nu \lesssim 7$ (44).

As discussed by Davier (46), very good limits on ΔN_ν are also obtained from $e^+e^- \rightarrow \gamma + X$. The ASP collaboration obtained $\Delta N_\nu \leq 4.5$ (for $m_\nu < 10$GeV). By combining the statistics collected by ASP, MAC and CELLO (47) the better limit $\Delta N_\nu \leq 2$ is derived. Finally we recall that recently the well known limit (48) on ΔN_ν from the observed amount of He and other light elements in the universe has been revised. From cosmology and nucleosynthesis, according to ref.49, one now obtains $\Delta N_\nu \leq 3$ (for $m_\nu < 1$MeV). From the group of ref.48 a better limit is defended (see ref.46).

CONSTRAINTS ON NEW GAUGE BOSONS

Some new results on limits for heavier weak gauge bosons were presented in the parallel sessions. First we consider the bounds obtained at the $p\bar{p}$ collider (4). The predicted number of events for $W' \rightarrow e\nu$ or $Z' \rightarrow e^+e^-$ is clearly determined by $\sigma \cdot B$. This product depends on the mass of W'/Z' and is proportional to the combinations v^2+a^2 of vector and axial vector couplings of the new gauge boson to light quarks and, separately, to leptons.

One can then indicatively state the lower bounds on $M_{W'}$ or $M_{Z'}$ which correspond to the same couplings to light quarks and leptons as the ordinary W or Z. The results are displayed in fig. 9 (50). By combining (51) the statistics of UA1 and UA2 one obtains roughly

$$M_{W'} \gtrsim 250\text{GeV} \qquad (36)$$
$$M_{Z'} \gtrsim 190\text{GeV}$$

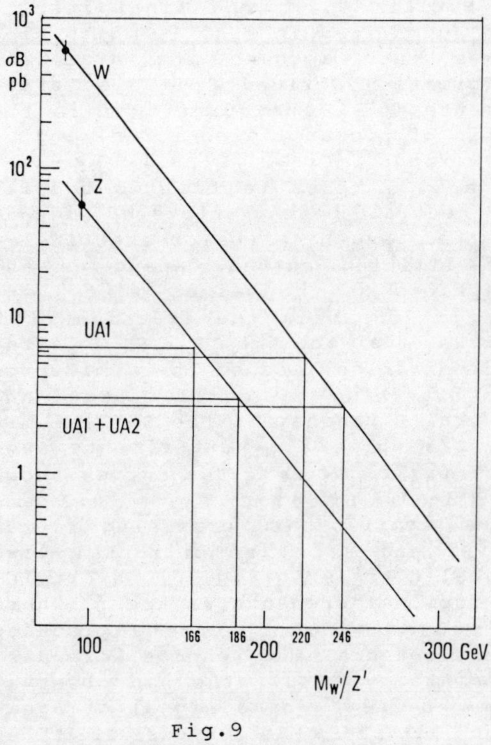

Fig.9

Comparable limits on $M_{Z'}$ are also obtained from the analysis of neutral current processes (52). Normally the interesting Z''s, i.e. those suggested by left-right theories or by some phenomenological speculations prompted by string theories and so on, are less coupled to light fermions than the ordinary Z°. As already mentioned, some other interesting limits on new gauge bosons were also derived from the study of the one loop radiative corrections to charged current processes (35). Finally updated results on the search for right handed currents in decay were presented by a group from TRIUMF (53). In the very restrictive assumption of a light right handed neutrino ($m_{\nu_R} \lesssim$ 6MeV) one obtains $M_{W_R} > 432$GeV.

In conclusion new W's and Z's could in principle be sufficiently nearby to be discovered for example by experiments at the Tevatron or even at ACOL.

B°$\bar{B}$° MIXING

UA1 has reported evidence in favour of a large amount of B°$\bar{B}$° mixing obtained from the measured ratio of equal to opposite sign dimuons. The experimental analysis was described by N.Ellis (54). The final results are summarized in fig. 10 which also takes into account the available information derived from e^+e^- experiments (55). The quantity r is the ratio of probabilities for mixing vs no mixing: $r=P(B\to\bar{B})/P(B\to B)$ $(0 \leq r \leq 1)$. r_d and r_s refer to the two species of neutral bottom flavored mesons $B_d \sim (b\bar{d})$ and $B_s \sim (b\bar{s})$ respectively. The limiting curves depend on assumptions on the values of f_{dd} and f_{ss} (f_{dd}(f_{ss}) is the fraction of b quarks that become B_d (B_s) after hadronization) and on the semileptonic branching ratios of charged and neutral B mesons. In the SU(3) limit f_{ss}=1/3 while e^+e^- experiments suggest $f_{dd}/f_{ss} \sim 2\div3$. The curves shown in fig.10 are for $f_{ss} \sim 0.2$ and equal semileptonic branching ratios of B° and B^+. The horizontal line (labeled Argus in fig.10) is obtained from measurements at the Υ' where only B_d and not B_s can be produced (less constraining results were also obtained by CLEO). The non observation at the Υ' of a signal of equal sign dileptons in excess of those expected from second generation decays: $b \to c \to \mu$ is translated into a limit on r_d: $r_d < 0.12$. The Mark II limit was instead obtained at PEP energies where both B_d and B_s can be produced.

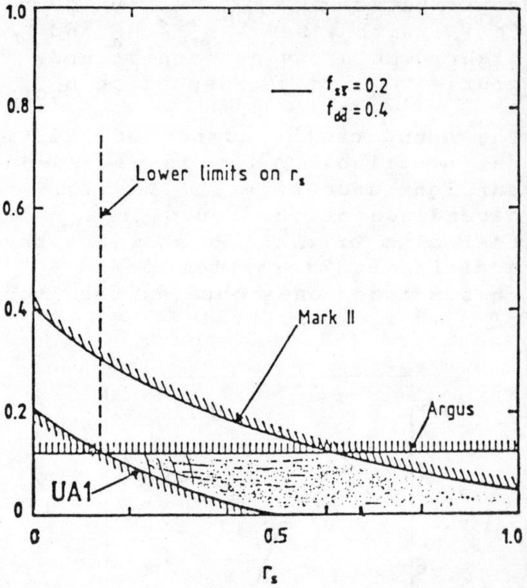

Fig.10

Contrary to the case of e^+e^- experiments, UA1 finds an excess of equal sign dimuons with respect to those expected in case of no mixing. Hence the UA1 limit excludes the point $r_d=r_s=0$. Together with the e^+e^- results it clearly indicates a large B°$\bar{B}$° mixing, probably with $r_s \gg r_d$. In the following I shall discuss the implications of these results within the context of the standard model.

Neglecting CP violation effects one has (56) $r=\bar{r}$, where

$$\bar{r} = \frac{P(\bar{B}\to B)}{P(\bar{B}\to\bar{B})}, \text{ and}$$

$$r = \bar{r} = \frac{(\frac{\Delta M}{\Gamma})^2 + (\frac{\Delta\Gamma}{2\Gamma})^2}{2 + (\frac{\Delta M}{\Gamma})^2 - (\frac{\Delta\Gamma}{2\Gamma})^2} \quad (37)$$

with $\Delta M=M_S-M_L$, $\Delta\Gamma = \Gamma_S - \Gamma_L$, $\Gamma = \frac{\Gamma_S + \Gamma_L}{2}$ where S(L) stands for "short" ("long"). While for kaons $\frac{\Delta M}{\Gamma} \sim \frac{\Delta\Gamma}{2\Gamma} \sim 0(1)$, for c and b quarks $\frac{\Delta\Gamma}{2\Gamma} \ll \frac{\Delta M}{\Gamma}$ because in presence of a moltitude of accessible final states the impact on the widths of CP selection rules is very small. Hence

$$r \simeq \frac{x^2}{2+x^2} \tag{38}$$

with $x = \frac{\Delta M}{\Gamma}$. The short distance contribution to ΔM, given by the box diagrams (57)(58) (fig.11), is expected to be the dominant term for sufficiently heavy quarks in the internal lines. Because of the GIM mechanism (59) the box contribution would vanish in the limit of degenerate internal quark masses. For B_s or B_d the box term increases approximately as

$$\Delta M \sim |V_{tb} V^*_{tq}|^2 \cdot \left(\frac{m_t}{M_W}\right)^2$$

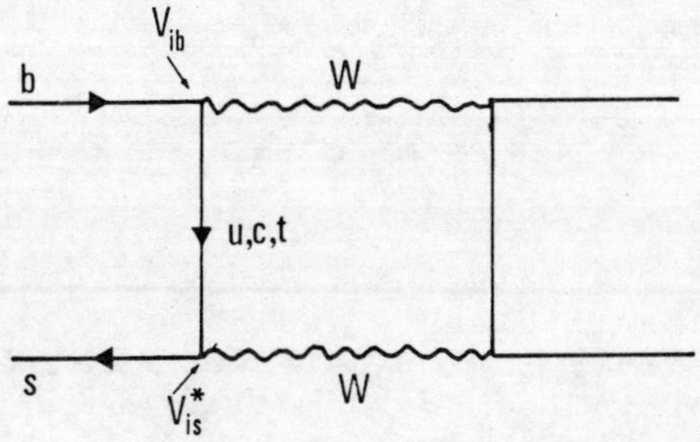

Fig.11

for a top quark t of sufficiently large mass (where q=d or s and V_{ij} are the K-M (60) matrix elements). For kaons $|V_{td} V^*_{ts}|^2$ is so small (61) in comparison with $|V_{cd} V^*_{cs}|^2$ that the charm term is dominant for all practical values of m_{top}. Thus for kaons non negligible long distance contributions to ΔM are quite possible (and indeed necessary (62) in view of $\Delta M_{Box}/\Delta M_{exp}$, if $B_k \lesssim 1$, where B_k is defined in analogy with eq.40). For the B system assuming 3 families one immediately obtains that the top quark is by far dominant (63), (at least in the B_s case), because $|V_{cb}V^*_{cs}| \sim |V_{tb}V^*_{ts}|$ (64) . As a consequence, the short distance approximation should be particularly reliable. Also r approaches unity very fast with increasing m_t. Finally one expects $r_d \ll r_s$ because $V_{td}/V_{ts} \lesssim 0(\theta_c)$ with θ_c the Cabibbo angle. For more families some of these results are no more necessarily true, but the situation is likely to remain qualitatively the same.

A precise calculation of ΔM_{box} for B_s leads to:

$$\frac{\Delta M}{\Gamma_B} = \frac{G_F^2 \, m_t^2}{6\pi^2 \, \Gamma_B} B_B \, f_B^2 \, m_B \cdot \tag{39}$$

$$\cdot |V^*_{ts} V_{tb}|^2 \frac{E(x_t)}{x_t} \eta_t$$

where B_B and f_B^2 are defined by:

$$\langle B^\circ | J \, J^+ | \bar{B}^\circ \rangle = B_B \langle B^\circ | J | 0 \rangle \langle 0 | J^+ | \bar{B}^\circ \rangle$$

$$= B_B \frac{4}{3} f_B^2 \, m_B \tag{40}$$

$x_t \sim \left(\frac{m_{top}}{M_W}\right)^2$, $\frac{E(x_t)}{X_t} \sim 0.9$ (58) and

$\eta_t \sim 0.85$ (62) is the result of QCD corrections. From the measured B lifetime (65) one obtains Γ_B. I have estimated

$$\frac{\Delta M}{\Gamma_B} \sim (1.2 \pm 0.5) \frac{B_B \, f_B^2}{(0.15\text{GeV})^2} \left(\frac{m_{top}}{40\text{GeV}}\right)^2 \tag{41}$$

From eq.38 one sees that $(\Delta M/\Gamma) \sim 1.2 \pm 0.5$ would correspond to $r_s \simeq 0.2 \div 0.6$.

However $B_B \, f_B^2$ is largely unknown and m_{top} could be much larger than 40 GeV.

In conclusion, in the standard model with 3 families $r_d \lesssim 10^{-3}$ and r_s is indeed expected to be in the range $0.1 \div 1$. r_s is dominated by m_{top} and is rapidly driven up to 1 by increasing m_{top}. Even if m_{top} was given, a precise prediction of r_s could not be derived because of the existing uncertainties on $B_B f_B^2$.

There are no new results on the CP violation parameters ϵ and ϵ'. The situation essentially remains the same as discussed by Langacker (2) at Kyoto last year. In particular it is to be noted that the theoretical prediction for ϵ also increases (66)(62) with m_{top}. Although the standard model is compatible with experiment even for $m_{top} \simeq 40GeV$ clearly there would be more room for the standard model if m_{top} is increased.

Finally there are no problems for the standard model as yet from the measured values of ϵ'/ϵ especially if penguin diagrams are not the explanation of the $\Delta I=1/2$ rule (67). An interesting result discussed by Donoghue at this Conference (68) is that isospin breaking and e.m. corrections tend to substancially decrease ϵ'/ϵ.

NEUTRINO MASSES AND OSCILLATIONS

Why should neutrinos be massless? If the lepton number L is conserved one obtains a natural mechanism for $m_\nu = 0$ by stipulating that ν_R does not exist. Then the Dirac mass terms $\bar{\nu}_L \nu_R$ or $\bar{\nu}_R \nu_L$ are obviously not possible and a Majorana mass term $\nu_L^T \nu_L$ is forbidden if L is conserved. But who really believes at present that L and B are exactly conserved? If L is broken even if ν_R does not exist and only doublet Higgs ϕ are present, still a (non renormalizable) coupling $(g/M)\nu_L^T \nu_L \phi\phi$ would be allowed by $SU(2) \otimes U(1)$ and would lead after symmetry breaking to a Majorana mass term.

In addition most GUTS predict the existence of ν_R (e.g. all GUTS such that $G_{GUT} \supset SO(10)$) together with lepton number non conservation. The so called see-saw mechanism (69) was originated in this context. The ν mass matrix is assumed to be given by a Dirac term $m_D(\bar{\nu}_L\nu_R+\bar{\nu}_R\nu_L)$ plus a Majorana term $M\nu_R^T\nu_R$ (a singlet under $SU(2)\otimes U(1)$). M is a large mass associated with the scale where L and possibly left-right symmetry, as is the case in SO(10), are broken.

Excluding or neglecting Majorana mass terms of the form $\nu_L^T\nu_L$ ($\nu_L^T\nu_L$ has weak isospin one, so that no invariant renormalizable Yukawa couplings with doublet Higgs are possible), then one obtains a mass matrix of the form (in ν_L, ν_R space):

$$M \simeq \begin{pmatrix} 0 & m_D \\ m_D & M \end{pmatrix} \qquad (42)$$

which leads to the eigenvalues

$$m_{\nu_{light}} \sim \frac{m_D^2}{M} \quad \text{and} \quad m_{\nu_{heavy}} \sim M.$$

Thus a natural reason for the smallness of $m_{\nu_{light}}$ would be that M is large, being the scale associated with L breaking. If M is flavour independent then one would expect

$$m_e : m_\mu : m_\tau = m_e^2 : m_\mu^2 : m_\tau^2 \qquad (43)$$

or $\qquad = m_u^2 : m_c^2 : m_t^2$

(Note that since the v.e.v of ϕ is proportional to m_D also $\frac{g}{M}\nu_L^T\nu_L\phi\phi$ would lead to $m \sim \frac{m_D^2}{M}$). It is amusing to observe that the present limits on m_{ν_e} ($\lesssim 30eV$, see later), m_{ν_μ} ($\lesssim 250KeV$) and m_{ν_τ} ($\lesssim 76MeV$)(70) are roughly equivalent if translated in terms of M via the relations (43).

Given this general context (see also the related discussions by Wolfenstein (71) and Weinberg (72)) let us now discuss some new interesting results on tritium β-decay experiments ($H^3 \rightarrow He^3 + e^- + \bar{\nu}$) aimed at directly measuring a possibly non vanishing m_{ν_e}. The experimental situation was exhaustively reviewed by Fackler (73), to which talk I refer the reader for more details and references. As well known, after the old limit by Bergkvist (74) $m_{\nu_e} < 60eV$, over the last few years the ITEP group (75) has claimed a positive signal for a non vanishing m_{ν_e}. The latest result obtained by also including new data (73) is given by $17eV < m_{\nu_e} < 40eV$.

In order to challenge this striking claim several new experiments are under way and some important results were already reported. The group from Zurich (76) has published a few months ago the bound $m_{\nu_e} \lesssim 18eV$ (95% c.l.) using a source of H^3 implanted on C. The group in Los Alamos (77) working with free gaseous molecular sources has reported the result of the first few runs: $m_{\nu_e} \lesssim 27.2eV$ (95%) and will give a better result in the near future. In Japan the INS group (73) finds $m_{\nu_e} \lesssim 33eV$ (95%). An experiment in Bejing was also mentioned (73), leading to the result $m_{\nu_e} \lesssim 30eV$. All the new experiments are still running. Thus more data will be available in the near future.

In conclusion for the time being we can only state that the ITEP result has not been confirmed and there is no clear evidence for $m_{\nu_e} \neq 0$.

Several new results from accelerator and reactor experiments on the search of ν oscillations were also reported at this conference (78) (79). A positive signal of ν oscillations would be a proof that not all ν masses are zero. However at the moment there is no evidence for the occurrence of ν oscillations. The Bugey reactor group (80) had reported a possible signal for oscillations. This has not been confirmed by other experiments (Gösgen (81), BEBC(82)) which are sensitive to the same domain of Δm^2 and mixing angles. An excess of electrons recorded by the experiment PS191 at CERN is now again confirmed by a new experiment (E816) of the same group at BNL (83). If this excess of electrons is interpreted as due to oscillations, then the corresponding domain of parameters would (entirely or in some cases almost so) fall in the region excluded by other experiments (BNL, BNL-BC, CHARM, BEBC)(78). Thus from ν oscillations in vacuum one does not yet obtain any positive clue on the existence of ν mass differences.

Some interesting developments have taken place on ν oscillations in matter which may offer an appealing solution to the long standing problem of solar ν's. As well known, the Davis experiment (84) which detects ν_e's by the reaction $\nu_e + {}^{37}Cl \rightarrow {}^{37}Ar + e^-$ (threshold energy 814KeV) has seen a signal of 2.1 ± 0.3 SNU while from the standard theory of the sun the expected number (85) is at present 5.9 ± 2.2 SNU. Some old and new possible ways out are listed in the following. (a) Modification of the theory of the sun. A small decrease of the temperature of the sun core would be sufficient. Note that the relatively high threshold of the clorine reaction is such that the Davis experiment is only sensitive to ν_e's from secondary branches (e.g. the ^8B reaction) of the sun energy generation chain. The expected rate of production of these ν_e's has a particularly steep dependence on T (T^4 for the primary pp reaction and $\sim T^{20}$ for the ^8B reaction (86)). This still remains a quite plausible explanation of the problem in my opinion. The proposed Ga experiments (Gallex (87), Baksan ν Obs. (88)) will be important in this respect because in this case the threshold is much lower (233 KeV) so that ν_e's from more conspicuous processes are involved. (b) ν oscillations in vacuum on the way between sun and earth(89). The probability for a ν_e to remain ν_e is given by

$$P_{ee} = 1 - \frac{1}{2}\sin^2 2\theta_\nu \left(1 - \cos 2\pi \frac{R}{L_\nu}\right) \quad (44)$$

where R is the distance sun-earth and $L_V = 4\pi E/\Delta m^2$ is the oscillation lenght in vacuum and θ_V is the mixing angle (for two ν's). The problem with this explanation is that it needs $\sin^2 2\theta_V \sim 1$ (and $\Delta m^2 \gtrsim (10^{-12} \div 10^{-10})eV^2$). This hurts against the theoretical prejudice, borrowed from the quark case, that mixing angles are small. (c) ν decay (90): $\nu_e \rightarrow \nu_x + \phi$ (d) ν magnetic moment (91) (the sun magnetic field would then sweep many ν's away). (e) ν oscillations in matter inside the sun (92). This is an elegant and interesting possibility which has attracted a lot of attention (93) over the last year. As all explanations (b) to (e) it needs massive ν's. It's merit is that it would also work for small mixing angles provided that some restrictions on Δm^2 are satisfied. A brief discussion of this possibility follows.

Let's for simplicity denote by A_{ee} the amplitude for ν_e to remain ν_e, $A_{ee}=A(\nu_e \to \nu_e)$ and similarly for $A_{\mu e}$, $A_{e\mu}$, $A_{\mu\mu}$ (again for simplicity we work with two ν's: μ can be either μ or τ). The formalism for ν oscillations can be phrased in terms of time evolution equations:

$$i\frac{d}{dt}\begin{pmatrix}A_{ee}\\ A_{\mu e}\end{pmatrix} = H\begin{pmatrix}A_{ee}\\ A_{\mu e}\end{pmatrix}, \tag{45}$$

$$H = \begin{pmatrix}H_{ee} & H_{e\mu}\\ H_{\mu e} & H_{\mu\mu}\end{pmatrix}$$

where the hamiltonian H can be replaced by

$$H = \begin{pmatrix}H_{ee}-H_{\mu\mu} & H_{e\mu}\\ H_{\mu e} & 0\end{pmatrix} \tag{46}$$

because an overall phase is irrelevant for probabilities. In vacuum $H=H_o$, while in matter the hamiltonian has an additional term $H=H_o+V$. V arises from charged current interactions that contribute to $V_{ee}-V_{\mu\mu}$ because in matter there are electrons and not muons or tauons (fig.12). Also $V_{e\mu}=V_{\mu e}=0$ because there are no flavor changing neutral currents. One then has

$$V_{ee}-V_{\mu\mu} = \frac{G_F}{\sqrt{2}}\bar{\nu}_e\gamma_\mu(1-\gamma_5)\nu_e\bar{e}\gamma^\mu(1-\gamma_5)e \tag{47}$$

For e^- at rest only the γ° part is important and $(1-\gamma_5)\nu=2\nu$, so that

$$V_{ee}-V_{\mu\mu} = \sqrt{2}\,G_F N_e \equiv \frac{2\pi}{L_o} \tag{48}$$

where N_e is the number of electrons per unit volume. Note that the sign is positive for ν and opposite for $\bar{\nu}$. The solution in matter is then given by:

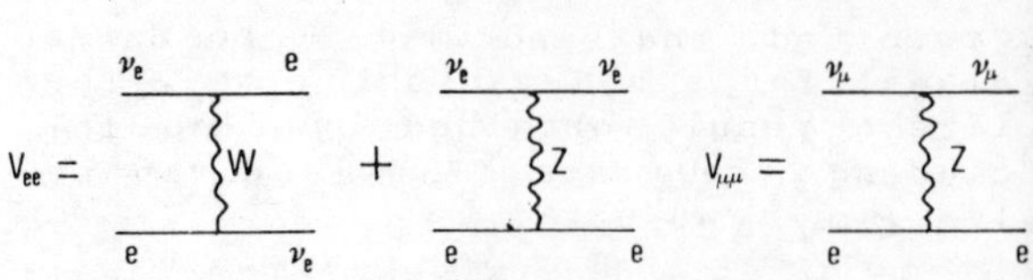

Fig.12

$$P_{ee}=1-\frac{1}{2}\sin^2 2\theta_m\left(1-\cos 2\pi\frac{R}{L_m}\right) \tag{49}$$

(compare with eq.44) with

$$\sin^2 2\theta_m = \frac{\sin^2 2\theta_V}{\left(\frac{L_V}{L_o}-\cos 2\theta_V\right)^2+\sin^2 2\theta_V} \tag{50}$$

$$L_m = \frac{L_V}{\left[\left(\frac{L_V}{L_o}-\cos 2\theta_V\right)^2+\sin^2 2\theta_V\right]^{1/2}} \tag{51}$$

(The minus signs in the denominators become plus signs for $\bar{\nu}$). A resonant structure is evident in the ν as opposite to $\bar{\nu}$ case. At resonance $\sin^2 2\theta_m \to 1$ irrespective of the value of θ_V. The resonance position and width are $L_V = L_o\cos 2\theta_V$ and $\sin^2 2\theta_V$ respectively. Near the center of the sun where the ν's are produced, the density is large and so is $V_{ee}-V_{\mu\mu}$. The ν_e level is higher than the ν_μ level. When the density decreases as the ν's move toward the surface, a level crossing takes place and the resonance occurs (it is needed that in vacuum ν_e is predominantly light, a possibility which is appealing to intuition, but not granted). Many detailed calculations (93) have been performed recently on the application of this idea to the explanation of Davis experiment and to the related implications for future experiments. For example, in figs.13a,b the results of a very recent work (94) on the Davis experiment (Clorine) and on the planned Gallium experiment are reported. The band within the dashed contour in figs. 13 corresponds to the region of Δm^2 and $\sin^2 2\Theta_V$ values which are selected if the Davis result is to be explained by this effect. As is seen, the mixing angle in vacuum can take a large range of values. Δm^2 is typically of order 10^{-4} eV^2, but more in general could take values in the range $(10^{-4} \div 10^{-7}) eV^2$. The corresponding predictions for the Ga experiment for each pair of values of Δm^2 and $\sin^2 2\Theta_V$ are shown in fig.13b. These results reiterate our interest for the Ga experiment (and other solar neutrino experiments like Icarus (95) and LVD (96)).

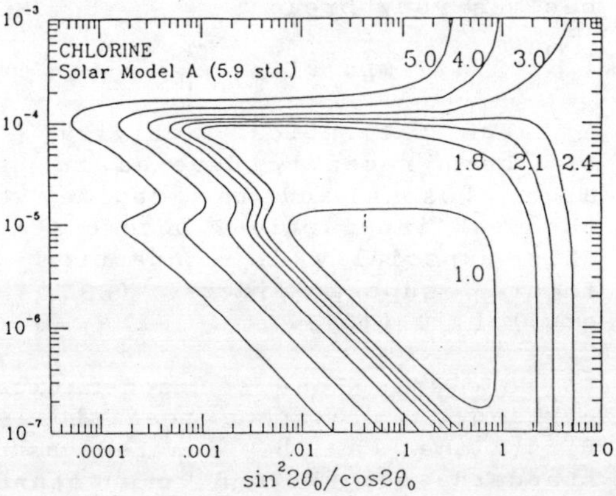

Fig.13a

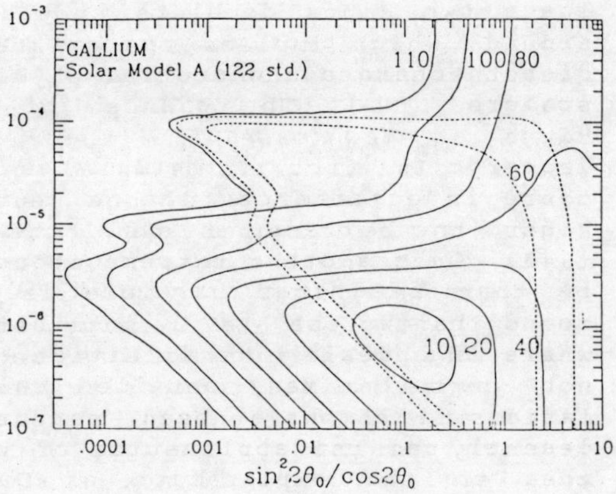

Fig.13b

The possible explanation of the solar ν problem in terms of ν oscillations in matter is interesting not only for its elegance but also because it could open a window into the physics at very large energy scales. In fact, if we go back to the idea that $m_\nu \sim (m_D^2/M)$ (see eqs. 42,43) then, from $\Delta m^2 \sim (10^{-4} \div 10^{-7})$ eV^2 one obtains $m_{\nu_x} \sim \sqrt{\Delta m^2} \sim (10^{-2} \div 10^{-3.5})$ eV. Here ν_x is either ν_μ or ν_τ and the mass of ν_e has been neglected. If m_D is identified with the mass of either the up quark or the charged lepton of the corresponding family, then for the scale M one is led to the values

$$\begin{aligned} M &\sim 19^9 \div 10^{12} \text{ GeV} \quad \text{for } \nu_e \to \nu_\mu \\ &\sim 10^{11} \div 10^{15} \text{ GeV} \quad \text{for } \nu_e \to \nu_\tau \end{aligned} \qquad (52)$$

Thus solar ν's can possibly provide us with a probe into energies of interest for GUTS. However it is only future experiment that can tell us whether oscillations in matter are really relevant for solar neutrinos.

THE HIGGS SECTOR

One cannot complete a talk on the status of the electro-weak theory without mentioning the well known and capital problem of the symmetry breaking sector of the theory, although over the last year little progress has to be reported. The Higgs mechanism of gauge symmetry breaking is without experimental support (the only indirect clue is the proven validity of the relation $M_W \sim M_Z \cos\Theta_W$ predicted by the minimal doublet-Higgs theory). Moreover assuming the validity of the standard theory up to the Planck mass leads to well known naturalness problems (e.g. the hierarchy problem). Thus new physics within the few TeV energy region is expected to complete (e.g. Susy) or to modify (e.g. compositeness) the standard picture of fundamental scalar Higgs particles. In any case the search for the standard Higgs particle(s) remains the best way to organize the crucial experimental exploration of the symmetry breaking sector of the theory.

It is to be remarked that the information that one has on the mass of the standard Higgs is really even less than commonly believed. The experimental lower bound (97) on the mass from atomic and nuclear physics (absence of long range forces):

$$m_H > 15 \text{ MeV} \tag{53}$$

is unchallenged. The limit $m_H \gtrsim 325$ MeV which was claimed in the past (98) to derive from $K^+ \to \pi^+ + H$ was later criticised (99). In a very recent reanalysis (100) of the problem the excluded domain was reported to be $50 < m_H < 140$ (MeV). However some model dependence is clearly unavoidable in this case (e.g. hadronic matrix elements). The limit $m_H > 4$ GeV would follow (3) from Υ decays if the branching ratio $\Upsilon \to H + \gamma$ was computed in lowest order (101). However, this limit is washed out by the existence of large QCD corrections (102) (which essentially make the exact branching ratio unpredictable). The lowest order QCD corrections have recently been independently rechecked and confirmed (103).

The theoretical lower bounds on the Higgs mass based on vacuum stability (104), $m_H \gtrsim$ few GeV, are only valid if one assumes no heavy fermions. For example, this limit is completely washed out for $m_{top} \sim M_W$ (105).

On the other hand, the theoretical upper bounds on m_H are all based, in one form or the other, on requiring perturbation theory to hold up to some large energy Λ. But this perhaps appealing requirement is in no way necessary. As well known the coupling of the quartic term $\lambda(\phi^\dagger\phi)^2$ in the Higgs potential increases with m_H^2, because $m_H^2 \sim \lambda/G_F$. Also for a given m_H, λ increases logarithmically with energy since the theory is not asymptotically free in the Higgs sector. Then, requiring perturbation theory to hold up to $= M_{Planck}$ leads to (106)

$$m_H \lesssim 200 \text{ GeV} \tag{54}$$

Similarly, from the requirement of avoiding problems due to the possible triviality of the $\lambda\phi^4$ theory, the limit was obtained (107)

$$m_H \lesssim 125 \text{ GeV} \tag{55}$$

However, if m_H is made to increase no physical contradiction is met. All what happens is that at $m_H \gtrsim$ 1TeV the Born amplitudes for longitudinal gauge boson scattering violate unitarity (108), manifesting the breakdown of perturbation theory. The helicity zero state of gauge bosons is obviously connected to symmetry breaking, because it does not exist for massless vector bosons. For $m_H \sim$ 1TeV the weak interactions become strong. The Higgs boson becomes very broad

($\Gamma_H \sim 1/2\, m_H^3$, with Γ_H, m_H in TeV).

A large theoretical activity (109) has been recently devoted to heavy Higgs bosons and the regime where the weak interactions become strong. This expecially in connection with future supercolliders (SSC, LHC etc.) (110)(111).

In conclusion one can schematically envisage two extreme scenarios. The first one can be denoted as the standard way beyond the standard model: the Higgs is elementary, it is rather light so that perturbation theory is valid up to M_{GUT} or M_{Planck} and naturality is restored by supersymmetry. For this, Susy partners should be discovered below the few TeV energy domain (112). At least two Higgs doublets should be around, which implies the existence of both charged and neutral physical scalars (113). The chances of one Higgs at low mass are then even larger. In fact in simple models there is a tendency for one neutral Higgs to be lighter than the Z° (113). On the other extreme it could be that the Higgs particle is not found below the TeV energy scale where the weak interactions become non perturbative (composite Higgs (114) and compositeness in general lean on this side). The gap between the Fermi and the Planck scale is then essentially not accessible to the present theory. In all cases input from experiment is badly needed. The unveiling of all misteries associated with the symmetry breaking sector of the electroweak gauge theory represents the main objective for the next decade of experimental research in particle physics.

It is a pleasure for me to thank the Local Organizing Committee for their kind invitation and the assistance in putting together my talk while in Berkeley. The hard and tough work of preparing this written version was made less painful by the extremely skilful and fast editing by Mrs. Angela Di Silvestro of INFN, Rome.

REFERENCES

(1) Maiani, L., Proceedings of the HEP-85 Conference, Bari, 1985, p.639

(2) Langacker, P., Proceedings of the 1985 Int.Symp. on Lepton and Photon Int. at High En., Kyoto, 1985, p.186

(3) Lee-Franzini, J., Proceedings of the XXI Rencontre de Moriond "Perspectives in Electro-Weak Interactions and Unified Theories". Barbiellini, G., Santoni, C., Rivista del N.Cim. 9(1986)1

(4) Geer, S., (UA1), these Proceedings

(5) Roussarie, A., (UA2), these Proceedings

(6) Honma, A., (UA1), these Proceedings

(7) Altarelli, G., Ellis, R.K., Greco, M., Martinelli, G., Nucl.Phys. B246 (1984) 12; Altarelli, G., Ellis, R.K., Martinelli, G., Zeit.f.Physik C27 (1985) 617; Phys.Lett. 151B (1985) 457, see also Altarelli, G., Martinelli, G., Rapuano, F., CERN-TH 4401 (1986)

(8) Stirling, W.J., these Proceedings; see also Kleiss, R., Stirling, W.J., CERN-TH 4490 (1986)

(9) Duchovni, E., these Proceedings

(10) UA2 Collaboration, Bagnaia, P., et al., Phys.Lett. 139B (1984) 105

(11) Sirlin, A., Phys.Rev. D22 (1980) 971; Marciano, W., Sirlin, A., Phys.Rev. D22 (1980) 2695

(12) Abramowicz, H., et al. (CDHS collaboration), Phys.Rev.Letters 57 (1986) 298, presented by Abramowicz

(13) Allaby, J.V., et al. (CHARM collaboration), to appear in Phys.Letters, presented by K.Winter

(14) Reutens, P.G., et al. (CCFRR collaboration), Phys.Lett. 152B (1985) 404 presented by F.Merritt

(15) Boyert, D., et al. (FMM collaboration) Phys.Rev.Lett. 55 (1985) 1964 presented by F.Merritt

(16) Abe, K., et al., Phys.Rev.Lett. 56 (1986) 1107

(17) Marciano, W., these Proceedings

(18) Bouchiat, M.A., et al., Phys. Lett. 134B (1984) 465; For experiments on Tl see also: Bucksbaum, P., et al., Phys.Rev.Lett. 46 (1981) 640; Phys.Rev. D24 (1981) 1134

(19) Prescott, C.Y., et al. Phys.Lett.B 77 (1978) 347; (1979) 524

(20) Ahrens, L.A., et al., Phys. Rev. Lett. 54 (1985) 18

(21) Bergsma, F., et al., Phys.Lett. 147B (1984) 481

(22) Bjorken, J.D., Phys.Rev. D19 (1979) 335; Hung, P.Q., Sakurai, J.J., Nucl.Phys. B143 (1978) 81

(23) Abbott, L.F., Fahri, E., Phys.Lett. 101B (1981) 69, Nucl.Phys. B189 (1981) 547; Fritzsch, H., Mandelbaum, G., Phys.Lett. 109B (1982) 224; Claudson, M., Fahri, E., Jaffe, R.L., MIT preprint, CTP n.1331 (1986); Kuroda, M., et al., Bielefeld preprint BI-TP 86/12 (1986). See also the contributions by Schrempp, B., and Harari, H., these Proceedings.

(24) See the contributions by Bars, I., and Wyler, D., these Proceedings

(25) For recent reviews see, for example: Marciano, W., Proceedings of the 1983 Int. Lepton/Photon Symposium, Cornell; Maiani, L., Proceedings of the Advanced Study Institute on Elementary Particle Physics, Ann Arbor, 1984; Hioki, Z., Tokushima 86-01, to appear in Acta Physica Polonica, (1986)

(26) Sirlin, A., Phys.Rev. D22 (1980) 971; Marciano, W., Sirlin, A., Phys.Rev. D22 (1980) 2695;
Sirlin, A., Phys.Rev. D29 (1984) 89;
Paschos, E.A., Nucl.Phys. B159 (1979) 285;
Cole, J., Penso, G., Verzegnassi, C., ISAS preprint, 19/85/EP, Trieste (1985);
Papadopoulos, N.A., et al., Nucl.Phys. B258 (1985) 1

(27) Consoli, M., Lo Presti, S., Maiani, L., Nucl.Phys. B223 (1983) 478; Lynn, D.W., Stuart, R.G., Nucl.Phys. B253 (1985) 216; Bardin, D.Yu., Riemann, S., Riemann, T., DUBNA E2-86-169 (1986) to appear in Zeit.Phys.C

(28) Veltman, M., Nucl.Phys. B123 (1977) 89

(29) Bertolini, S., Sirlin, A., Nucl.Phys. B248 (1984) 589

(30) Barbieri, R., Maiani, L., Nucl.Phys. B224 (1983) 32

(31) Suzuki, M., Phys.Lett. 153B (1985) 289

(32) This recent value is due to Jegerlehner, quoted in ref.17

(33) Chanowitz, M.S., Furman, M.A., Hinchliffe, I., Phys.Lett. 78B (1978) 285

(34) Stuart, R.G., CERN-TH4342 (1985)

(35) Marciano, W.J., these Proceedings; see also Marciano, W.J., Sirlin, A., Phys.Rev.Lett. 56 (1986) 22

(36) CHARM II Collaboration (in progress)

(37) See for example, Proceedings of the Workshop on Physics in the 90's at the SPS Collider, Zinal, 1985

(38) Altarelli, G., et al., in Physics at LEP, ed. by J.Ellis and R.Peccei, CERN 86-02

(39) Barbiellini, G., et al., ibidem

(40) Fesefeldt, H., these Proceedings. The difficulties of comparing theory and experiments are described in Cashmore, R.J., et al., Zeit.Phys. C30 (1986) 125

(41) Naroska, B., Proceedings of Physics in Collision V, Autun, France (1985), see also R.Marshall, RAL-85-078 (1985)

(42) Marayuma, T., these Proceedings

(43) Perl, M., these Proceedings

(44) Honma, A., these Proceedings

(45) Buras, A.J., Ellis, J., Gaillard, M.K., Nanopoulos, D.V., Nucl.Phys. B135 (1978) 66

(46) Davier, M., these Proceedings

(47) See the contributions by Whitaker, S. and Tuts, P.M., these Proceedings

(48) Yang, J., et al., Ap.J. 281 (1984) 493

(49) Ellis, J., Enqvist, K., Nanopoulos, D.V., Sarkar, S., CERN-TH 4303 (1985)

(50) The theoretical curves were obtained by G.Martinelli (unpublished) following the calculations of ref.7

(51) Actually I have simply divided by two the minimum cross-section indicated by UA1

(52) Durkin, L.S., Langacker, P., UPR-0287-T (1986)

(53) Carr, J., these Proceedings

(54) Ellis, N., these Proceedings

(55) The e^+e^- results on B mesons were discussed by M.G.D. Gilchriese, these Proceedings

(56) See for example, L.L.Chau, Phys.Rep. 95 (1983) 3

(57) Gaillard, M.K., Lee, B.W., Phys.Rev. D10 (1974) 897

(58) Inami, T., Lim, C.S., Progr.Theor.Phys. 65 (1981) 297

(59) Glashow, S., Iliopoulos, J., Maiani, L., Phys.Rev. D2 (1970) 1285

(60) Kobayashi, M., Maskawa, K., Progr.Theor.Phys. 49 (1973) 652

(61) For a recent discussion of the experimental status of the K-M matrix, see e.g. Kleinknecht, K., Renk, B., Phys.Lett. 130B (1983) 459 and Proceedings of the Heidelberg Conference on Heavy Flavour Decays, Heidelberg, 1986

(62) Chau, L.L., Keung, W.Y., Phys.Rev.D29 (1984) 592; Buras, A.J., Slominski, W., Steger, H., Nucl.Phys. B238 (1984) 529

(63) Bigi, I., Phys.Lett. 162B (1985) 383

(64) Wolfenstein, L., Phys.Rev.Lett. 52 (1983) 1945

(65) Gilchriese, M.G.D., these Proceedings

(66) Ginsparg, P.H., Glashow, S.L., Wise, M.B., Phys.Rev.Lett. 50 (1983) 1415

(67) Chau, L.L., Cheng, H.Y. and Keung, W.Y., Phys.Rev. D32 (1985) 1837; Gavela, M.B. et al., Phys.Lett. 148B (1984) 225; Alfaro, J., et al., Phys.Lett. 147B (1984) 357; Donoghue, J.F., Phys.Rev. D30 (1984) 1499; Pham, T.N., Phys.Lett. 145B (1984) 113; Dupont, Y. and Pham, T.N., Rev. D29 (1984) 1368; Buras, A.J. and Slominski, W., Nucl.Phys. B253 (1985) 231; Bilec, N., and Guberina, B., Phys.Lett. 150B (1985) 311 and Z.Phys. C27 (1985) 399

(68) Donoghue, J., these Proceedings

(69) Gell-Mann, M., et al., in Supergravity, eds. P. von Nieuwenhuizen and D.Z.Freedman (N.Holland, 1979) p.135; Yanagida, T., Prog.Th.Phys. 64 (1980) 1103; Witten, E., Phys.Lett. 91B (1980) 81

(70) HRS collaboration: Bylsma, B., these proceedings

(71) Wolfenstein, L., these Proceedings

(72) Weinberg, S., these Proceedings

(73) Fackler, O., these Proceedings

(74) Bergkvist, K.E., Nucl.Phys. B39 (1972) 317

(75) Lubimov, V.A. et al., Phys.Lett. 94B(1980)266 and Proceedings of the Leipzig Conference, vol. II, p.108; Boris, S., et al., Phys.Lett. 159B (1985) 217; see also Jain, A., these Proceedings.

(76) Fritschi, M. et al., submitted to Phys.Lett.

(77) Wilkerson, J.F., these Proceedings

(78) On oscillations from accelerator experiments (BEBC): Baldo Ceolin, M., Barbiellini, G. (CHARM), Gauthier, A. (E531), Vannucci, F. (E816), these Proceedings

(79) On oscillations from reactor experiments: Aleksan, R. (Bugey), Zaczek, V. (Gösgen), Greenwood, Z. (Irvine)

(80) Cavaignac, J.F. et al., Phys.Lett. 148B (1984) 387

(81) Gabathuler, R. et al., Phys.Lett. 138B (1984) 449

(82) Talk by Baldo-Ceolin, M., ref.78

(83) Talk by Vannucci, F., ref.78

(84) Rowley, R.K., Cleveland, B.T., Davis, R., AIP Conference Proc. 126 (1985) 1

(85) Bahcall, J.N. et al., Rev.Mod.Phys. 54 (1982) 767; AIP Conference Proc. 126 (1985) 60

(86) Schatzmann, E., Proceedings of the XXth Recontre de Moriond on massive neutrinos p;369, Moriond (1985);
Zatsepin, G.T., Proceedings of the 8th International workshop on weak interactions and neutrinos. Alicante (Spain). A. Morales, Ed.(1982)

(87) Bellotti, E., these Proceedings

(88) Contributions by A.E. Chudakov and G. Zatsepin, Proceedings of 1st Symposium on Underground Physics, St.Vincent (1985)

(89) Pontecorvo, B., Zh.Eksper.Fiz. 34(1958)247 and 53(1967)1725; see also, e.g.: Bahcall, J.N. and Frautschi, S., Phys.Lett. 29B(1969)623; Barger, V. et al., Phys.Rev. D24(1981)538. For a recent discussion, see Krauss, L. and Wilczek, F., Phys.Rev.Lett. 55(1985)122

(90) Bahcall, J.N., N.Cabibbo, Yahil, A., Phys.Rev.Lett. 28 (1972) 316, see also Pakvasa, S., Tennakone, K., Phys.Rev.Lett. 28 (1972) 1415

(91) Voloshin, M.B., Vysotsky, M.I., Preprint ITEP-1 (1986); Ligneros, A., Astrophys. and Space Science 10 (1971) 87; Kyuldjev, A., Nucl.Phys. B243 (1984) 387; Okun, L.B., ITEP-14 (1986), Okun, L.B. et al., ITEP-20 (1986)

(92) Mikheyev, S.P., Smirnov, A.Yu., Sov.Journ.Nucl.Phys. 42(6) (1985) 913, and submitted to N.Cimento; Wolfenstein, L., Phys.Rev. D20 (1979) 2634, Barger, V. et al., Phys.Rev. D22 (1980) 2718

(93) Bethe, H.A., Phys.Rev.Lett. 56 (1986) 1305; Langacker, P. et al., CERN-TH 4421 (1986); Barger, V. et al., MAD-PH-280 (1986); Rosen, S.P., Gelb, J.M., LA-UR-86-804 (1986); Kolb, E.W. et al., FNAL-PUB 86/69-A (1986); Kuo, T.K., Pantaleone, J., PURD-TH-86-11 (1986); Raghavan, R.S. et al., UH-511-590-86 (1986); Wolfenstein, L., paper 7200, submitted to this Conference

(94) Parke, S.J., FNAL-PUB-86/67-5 (1986); Parke, S.J., Walker, T.P., FNAL-PUB-86/107-JA (1986)

(95) Icarus: A proposal for the Gran Sasso Laboratory, INFN-AE-85-7

(96) Proposal for a Large Volume Detector (LVD) for the Gran Sasso Laboratory, Frascati (1984)

(97) Ellis, J., Gaillard, M.K., Nanopoulos, D., Nucl.Phys. B106 (1976) 292; Barbieri, R., Ericson, T.D., Phys.Lett. B57 (1975) 270

(98) Vainshtein, A.I. et al., Sov.Phys.Usp. 23 (1980) 429

(99) Pham, T.N., Sutherland, D.G., Phys.Lett. B151 (1985) 444; Willey, R.S., Yu, H.L., Phys.Rev. D26 (1982) 3287

(100) Willey, R.S., Phys.Lett. B173 (1986) 480

(101) Wilczek, F., Phys.Rev.Lett. 39 (1977) 1304

(102) Vysotsky, M.I., Phys.Lett. B97 (1980) 159; see also Ellis, J. et al., CERN-TH 4143/85 (1985)

(103) Nason, P., CU-TP-346 (1986)

(104) Linde, A.D., JETP Letters 23 (1976) 64; Weinberg, S., Phys.Rev.Lett. 36 (1976) 294

(105) Linde, A.D., Phys.Lett. B92 (1980) 119; see also Ansel'm, A.A. et al., Sov.Phys.Usp. 28 (1985) 113

(106) Maiani, L. et al., Nucl.Phys. B136 (1978) 115; Cabibbo, N. et al., Nucl.Phys. B158 (1979) 295

(107) Beg, M.A.B. et al., Phys.Rev. 52 (1984) 883

(108) Veltman, M., Acta Phys.Polonica B8 (1977) 475; Lee, B.W., Quigg, C., Thacker, H.B., Phys.Rev. D16 (1979) 1519

(109) Van Der Bij, J., Veltman, M., Nucl.Phys. B321 (1984) 205; Einhorn, M.B., Nucl.Phys. 246B (1984) 75; Hung, P.Q., Thacker, P.Q., Phys.Rev. D31 (1985) 2866; Casalbuoni, R. et al., Phys.Lett. 147B (1984) 419; 155B (1985) 95; UGVA-DPT 1986/01-492 (1986); Chanowitz, M.S., Gaillard, M.K., Nucl.Phys. B261 (1985) 379; Schrempp, B., Schrempp, F., DESY-85-096 (1985)

(110) Eichten, E. et al., Rev.Mod.Phys. 56 (1984) 579; Duncan, M.J., Kane, G.L., Repko, W.W., Phys.Rev.Lett. 55 (1985) 773, Nucl.Phys. B272 (1986) 517; Cahn, R.N., Dawson, S., Phys.Lett. 136B (1984) 196; Dawson, S., Nucl.Phys. B249 (1985) 42; Kane, G.L., Repko, W.W., Rolnick, W.B., Phys.Lett. 148B (1984) 367; Jayaramon, T. et al., MUTP-83/1 (1983); Lindfors, J., UCR-TH-84-3 (1984); Cahn, R.N., Nucl.Phys. B255 (1985) 341; Cahn, R.N., Chanowitz, M.S., LBL21037 (1986); Altarelli, G., Mele, B., Pitolli, F., submitted to Nucl.Phys.

(111) Chanowitz, M.S., these Proceedings

(112) Reya, E., these Proceedings

(113) For a recent discussion, see Gunion, J.F., Haber, H.E., Nucl.Phys. B272 (1986) 1 and UCD-86-12 (1986)

(114) Korplan, D., these Proceedings.

G. Barbiellini CERN-INFN

Assuming that the top quark mass is measured in the near future what bound can be put on the Higgs mass from high precision measurents of $\sin^2\theta_W$?

G. Altarelli

Unfortunately there is in general a small sensitivity of $\sin^2\theta_W$ on m_H. An example is the behavior of Δr vs. m_H displayed in fig.6. Typically varying m_H in the range 10 GeV$<m_H<$ 1 TeV changes $\sin^2\theta_W$ by about 1%.

CP VIOLATION AND WEAK DECAYS OF QUARKS AND LEPTONS

M. G. D. Gilchriese

Laboratory of Nuclear Studies, Cornell University
Ithaca, NY 14853

Recent experimental results on CP violation, τ decays and weak decays of light and heavy quarks are reviewed. Although there are a number of puzzles and discrepancies among experimental results, definitive disagreements with the predictions of the standard electroweak model do not exist.

INTRODUCTION

A large amount of new experimental data on weak decays of quarks and leptons has become available in the last year or so. In this article I will focus on recent measurements and points of disagreement rather than present a comprehensive review. Of the greatest importance in any experimental measurement in this area is to search for striking evidence that the Standard Model of electroweak interactions is incorrect. Although there are a number of puzzles in weak decays, no such obvious evidence now exists. Also of importance is to compare the parameters of the Standard Model (eg. KM angles) extracted from the various measurements and compare them for consistency. This is a somewhat more subtle and difficult test of the Standard Model but there is, so far, no convincing evidence of any inconsistency.

CP VIOLATION

In this section I will very briefly discuss the current experimental status of CP violation in the $K^0\overline{K}^0$ system, in particular the status of measurements of

$$\eta_{+-} = \frac{A(K_L \to \pi^+\pi^-)}{A(K_S \to \pi^+\pi^-)} \quad \text{and} \quad \eta_{00} = \frac{A(K_L \to \pi^0\pi^0)}{A(K_S \to \pi^0\pi^0)}$$

and the related quantities ϵ and ϵ' from $\eta_{+-} \approx \epsilon + \epsilon'$ and $\eta_{00} \approx \epsilon - 2\epsilon'$. As of this conference no change has occurred in the experimental status of measurements of ϵ'/ϵ. The two best measurements are still

	ϵ'/ϵ
Chicago - Saclay[1]	$-0.0046 \pm 0.0053 \pm 0.0024$
Yale - BNL[2]	$0.0017 \pm 0.0072 \pm 0.0043$

The follow-up to the Chicago-Saclay experiment was described by Y. Wah at this meeting[3] and expects to reach a statistical (systematic) precision for ϵ'/ϵ of 0.003(0.001) in data already obtained in a test run of E731 at Fermilab. In a subsequent run they hope to improve the precision to 0.001(0.0005). M. Holder[4] described an experiment (NA31) at CERN designed to attain comparable or better precision.

On the theoretical side, Donoghue et al. have studied the effects of electromagnetism and isospin breaking on the value of ϵ'/ϵ calculated in the context of the Kobayashi-Maskawa model.[5] They find that such effects reduce the predicted value of ϵ'/ϵ by a factor of two to three to $0.0007 \leqslant \epsilon'/\epsilon \leqslant 0.005$.

The NA31 experiment also reported on a new measurement of $K_L \to \gamma\gamma$ and a much improved limit on $K_S \to \gamma\gamma$. They found $B(K_L \to \gamma\gamma) = (5.9 \pm 0.25) \times 10^{-4}$ and $[\Gamma(K_S \to \gamma\gamma)/\Gamma(K_L \to \gamma\gamma)] < 12$ at 90% confidence level, an improvement by about a factor of 30 over the previous limit.

A preliminary measurement of η_{+-0} where

$$\eta_{+-0} = \frac{A(K_L \to \pi^+\pi^-\pi^0)}{A(K_S \to \pi^+\pi^-\pi^0)}$$

by E621 at Fermilab was also presented.[6] They find $|\eta_{+-0}| = 0.04 \pm 0.035$ based on a small fraction (3%) of their complete data set (for reference η_{+-} or η_{00} is about 2.3×10^{-3}).

LIGHT QUARK DECAYS

A few new results are available on the weak decays of light quarks. The SINDRUM Collaboration[7] has been the first to observe the radiative pion decay $\pi^+ \to e^+\nu_e e^+e^-$ and measure the associated form factors. New results have also been presented on radiative hyperon decays,[8,9] on a measurement of the Σ^- magnetic moment[10] and on form factors in $\Sigma^-\beta$ decay.[11] I will only briefly discuss one of these, radiative hyperon decays.

Recent measurements of the branching ratios for weak radiative hyperon decays have been made by

E619[8] at Fermilab and in the charged hyperon beam at CERN.[9] These recent results are summarized below

Group	Mode	Branching Ratio (in units of 10^{-3})
CERN hyperon[9]	$\Sigma^+ \rightarrow p\gamma$	$1.27^{+0.16}_{-0.18}$
CERN hyperon[9]	$\Lambda \rightarrow n\gamma$	1.02 ± 0.33
E619[8]	$\Xi^0 \rightarrow \Lambda\gamma$	1.1 ± 0.2
CERN hyperon[9]	$\Xi^- \Sigma^- \gamma$	in progress
CERN hyperon[9]	$\Omega^- \rightarrow \Xi^- \gamma$	<2.2

There have been a number of additional measurements of the $\Sigma^+ \rightarrow p\gamma$ branching fraction,[12] but the measurement by the CERN hyperon group is based on the largest data sample. Weak radiative decays probe the non-leptonic part of the weak interaction and are less plagued by strong interaction effects present in non-leptonic hyperon decays. There is a vast body of theoretical literature which describes the attempts to model and predict these branching ratios — see Biagi et al. in reference 9 and references therein. Models in which there is only a single quark transition are not able to explain the data; additional mechanisms are required, such as W exchange.

TAU LEPTON DECAYS

New results are available on decay modes of the tau lepton leading to one-charged-prong final states. Recent comparisons[13] between the sum of the exclusive one-prong branching ratios and the one-prong topological branching ratio (B_1) have suggested the existence of modes contributing to B_1 but which have not yet been identified in exclusive channels.[14] Since the 1986 Particle Data Group (PDG) compilation,[12] there have been a number of new measurements, including contributions to this conference, of exclusive branching ratios which contribute to the one-prong rate. These measurements along with the 1986 PDG averages[15] are given in Table 1. New measurements of the topological branching ratio B_1 are also given. The difference between B_1 and the *measured* sum of exclusive modes is now $(2.5 \pm 2.6)\%$, not significantly different from zero. Modes with multiple neutrals are difficult to measure and previous comparisons have tended to use predictions for such decays. The branching ratios for $\pi^\pm(\geqslant 2\pi^0)$ modes may be predicted, under certain assumptions, as outlined in the excellent paper by Gilman and Rhie.[14] The rate for $\pi^- 2\pi^0$ may be related to the $\tau^- \rightarrow \nu_\tau \pi^+ \pi^- \pi^-$ rate using isospin conservation

$$B(\tau^- \rightarrow \nu_\tau \pi^- \pi^0 \pi^0) \leqslant B(\tau^- \rightarrow \nu_\tau \pi^- \pi^- \pi^+)$$

New measurements have been made of $B(\tau^- \rightarrow \nu_\tau \pi^- \pi^- \pi^+)$ and are summarized in Table 2.

Table 1. A summary of τ branching ratios yielding one-charged-prong final states and of the one-prong topological branching ratio (B_1).

Mode			World Average
$\mu\nu\nu$	'86 PDG[12]	$17.53 \pm 0.69\%$	
	MARKII[16]	18.3 ± 1.2	
	TPC[17]	18.1 ± 1.3	
	JADE[18]	18.8 ± 1.1	17.99 ± 0.49
$e\nu\nu$	'86 PDG[12]	17.63 ± 0.58	
	MARKII[16]	19.1 ± 1.4	
	TPC[17]	$18.0^{+2.2}_{-2.0}$	
	JADE[18]	17.0 ± 1.1	17.70 ± 0.47
$\pi\nu$	'86 PDG[12]	10.9 ± 1.4	
	MARKII[16]	10.0 ± 1.8	
	MAC[19]	10.6 ± 0.9	
	JADE[18]	11.8 ± 1.3	10.9 ± 0.6
$K\nu$	'86 PDG[12]	0.7 ± 0.2	0.7 ± 0.2
$\rho\nu$	'86 PDG[12]	22.1 ± 2.5	
	MARKIII[20]	22.3 ± 2.1	22.2 ± 1.6
$\pi^\pm\pi^0$(non-res)	'86 PDG[12]	0.3 ± 0.3	0.3 ± 0.3
$K^*\nu$	'86 PDG[12]	1.7 ± 0.7	
	MARKII[21]	1.3 ± 0.4	
	TPC[17]	1.4 ± 0.9	1.4 ± 0.3
$\pi^\pm(\geqslant 2$"π^0")	'86 PDG[12]	9.0 ± 4.4	
"π^0" = neutrals	MARKII[16]	12.0 ± 2.9	
measured	TPC[22]	13.9 ± 2.8	12.3 ± 1.8
$\pi^\pm(\geqslant 2\pi^0)$ predicted			$(\leqslant 7.9 \pm 0.4)$
	SUM(predicted)		83.8 ± 2.6
	(measured)		$(\leqslant 79.4 \pm 2.0)$
one-prong-B_1	'86 PDG[12]	86.5 ± 0.3	
	TPC[17]	84.5 ± 1.3	
	MARKII[16]	86.7 ± 1.2	
	CELLO[23]	83.9 ± 1.4	86.3 ± 0.3
Difference (B_1 - SUM)(measured)			2.5 ± 2.6
(predicted)			$(\geqslant 6.9 \pm 2.0)$

Note that there is a substantial spread in these measurements; a precise measurement of this branching ratio would be very valuable.

Table 2. Measurements of $B(\tau^- \rightarrow \mu_\tau \pi^- \pi^- \pi^+)$.

	$B(\tau^- \rightarrow \nu_\tau \pi^- \pi^- \pi^+)$
CELLO[25]	$9.7 \pm 2.4\%$
MAC[26]	8.1 ± 0.8
DELCO[27]	5.0 ± 1.0
MARKII[16]	7.8 ± 0.9
ARGUS[28]	5.6 ± 0.7
Average	6.7 ± 0.4

The rate for $\tau^- \rightarrow \nu_\tau\pi^-(3\pi^0)$ may be related to the rate for $\tau^- \rightarrow \nu_\tau e^- \overline{\nu}_e$ by

$$B(\tau^- \rightarrow \nu_\tau\pi^-(3\pi^0)) = 0.055\ B(\tau^- \rightarrow \nu_\tau e^- \overline{\nu}_e)$$
$$= (0.97 \pm 0.026)\%$$

One may also predict the rate for modes with 4 or 5 π^0's from the measured topological branching ratio to five charged prongs (B_5)

$$B(\tau^- \rightarrow \nu_\tau\pi^- 4\pi^0) \leqslant 0.75\ B(\tau^- \rightarrow \nu_\tau + 5 \text{ charged prongs})$$
$$B(\tau^- \rightarrow \nu_\tau\pi^- 5\pi^0) \leqslant 1.29\ B(\tau^- \rightarrow \nu_\tau + 5 \text{ charged prongs})$$

Measurements of B_5 are discussed below and lead to $B(\tau^- \rightarrow \nu_\tau\pi^-(\geqslant 4\pi^0)) \leqslant (0.24 \pm 0.06)\%$. Thus the total rate for $\tau^- \rightarrow \nu_\tau\pi^-(\geqslant 2\pi^0)$ is predicted to be $\leqslant (7.9 \pm 0.5)\%$, clearly in conflict with the directly measured rate and leading to an apparent discrepancy between B_1 and the sum of the exclusive modes (see Table 1).

This could be explained by the presence of additional modes yielding neutrals, such as $\tau^- \rightarrow \nu_\tau\eta\pi^-\pi^0$, which are not included in the theoretical estimates. Since the η most often decays into an all neutral state, this would yield one-charged-prong final states. The HRS[29] and Crystal Ball[30] groups have presented preliminary evidence for inclusive η production in τ decays at this conference. In both cases the η's are reconstructed through the $\gamma\gamma$ decay mode. The results from the Crystal Ball are shown in Fig. 1b and from HRS in Fig. 1a.

There is clear evidence in the Crystal Ball data for an η signal but they have not yet computed efficiencies or backgrounds and therefore are unable to give an inclusive $\tau \rightarrow \eta + X$ branching ratio or in fact conclude that the signal is not all from non-τ backgrounds. There is less clear but still some evidence from the HRS group for η production. HRS quotes[29] a preliminary branching ratio $B(\tau^- \rightarrow \nu_\tau\pi^-\eta + X) = (5.0 \pm 1.8)\%$. This could yield a one-prong contribution as much as $(3.6 \pm 1.3)\%$, which would be added to the predicted one-prong rate and thereby remove much of the discrepancy. Although the evidence for η production in τ decay is not yet conclusive, it suggests a possible resolution of the discrepancy between the one-prong topological branching ratio and the sum of exclusive modes. Better measurements of the $\tau^- \rightarrow \nu_\tau\pi^-$ (+ many neutrals) and $\tau^- \rightarrow \nu_\tau\pi^-\pi^-\pi^+$ branching ratios will be needed to finally resolve the issue. At present there is no need to invoke the presence of new particles produced in τ yielding one-charged-prong final states.

Measurements of the τ 5-prong topological branching ratio are summarized below:

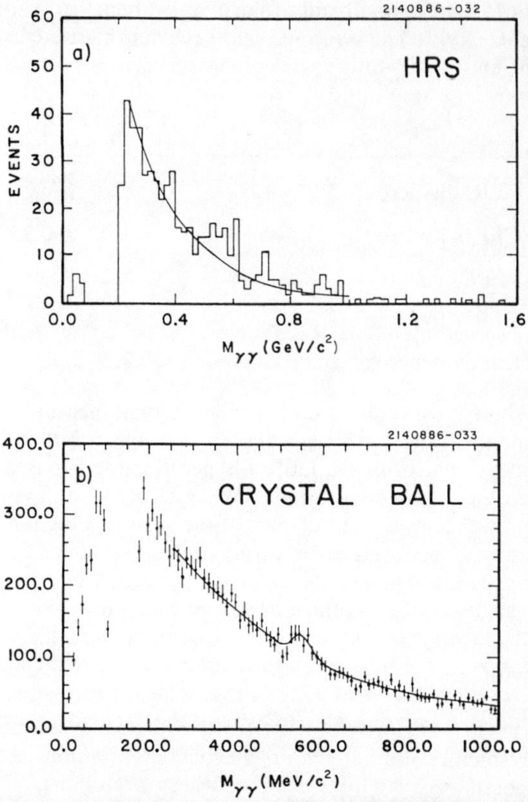

Figure 1. (a) The $\gamma\gamma$ invariant mass distribution observed by the HRS group in 1 vs 1 and 1 vs 3 τ decays. (b) The $\gamma\gamma$ mass seen by the Crystal Ball group in τ decays.

MARKII[31]	$0.16 \pm 0.09\%$
JADE[32]	0.30 ± 0.22
HRS[33]	0.10 ± 0.03
CLEO[34]	0.16 ± 0.06
Average	0.12 ± 0.03

In addition the HRS group[33] has measured the following branching ratios

$$B(\tau^- \rightarrow \nu_\tau 5\pi^\pm) = (5.1 \pm 2.0) \times 10^{-4}$$
$$B(\tau^- \rightarrow \nu_\tau 5\pi^\pm\pi^0) = (5.1 \pm 2.2) \times 10^{-4}$$
$$B(\tau^- \rightarrow \nu_\tau 7\pi^\pm + n\gamma) < 3.8 \times 10^{-4} \text{ at } 90\% \text{ CL}$$

Recent studies of the 3π invariant mass distribution in $\tau^- \rightarrow \nu_\tau\pi^-\pi^-\pi^+$ are reviewed by Burchat in her contribution to this conference.[16] The branching ratios are summarized in Table 2. In brief the 3π state is expected to be dominated by the $A_1(1270)$. Measurements of the 3π invariant mass in τ decays have shown this to be true but differ somewhat in the mass and width determined for the 3π resonance. Measurements of the A_1 mass (width) in πp interactions have found higher (narrower) values than studies in τ decay. As noted by Burchat, the difference in mass and width values obtained from tau decays may be a result of different parameterizations of the 3π invariant mass distribution by different experiments. Whether this also explains the discrepancy between hadronic experiments and τ decays is not clear.

The ARGUS[35] and HRS[29] groups have also presented evidence for ω^0 production in the decay $\tau^- \rightarrow \nu_\tau \pi^+\pi^-\pi^0\pi^-$. The $\pi^+\pi^-\pi^0$ mass distribution observed by ARGUS is shown in Fig. 2 and exhibits clear evidence for ω^0 production. ARGUS finds $B(\tau^- \rightarrow \nu_\tau \pi^+\pi^-\pi^0\pi^-) = (4.5 \pm 0.4 \pm 1.5)\%$ and $B(\tau^- \rightarrow \nu_\tau \omega^0 \pi^-) = (1.6 \pm 0.3)\%$. The interest in the decay $\tau^- \rightarrow \nu_\tau \omega^0 \pi^-$ is to look for second-class weak currents, which should be absent. For second class currents, J^{PG} of the $\omega\pi$ system would be 1^{++}(B(1235), for example) or 0^{-+}, whereas 1^{-+} can be produced via the allowed first class vector current. From an analysis of the angular distribution of the π^- (not from the ω decay) in the ω rest frame, the ARGUS group finds the J^P of the $\omega\pi$ system to be consistent with pure 1^- and therefore no evidence for second class currents. Although the $\omega\pi$ system is in a (mostly) vector state there is no evidence for narrow resonance structure in the $\omega\pi$ mass distribution. A possible explanation is a broad vector resonance such as the first radial excitation of the ρ.[36]

The Crystal Ball group[37] has improved the limits on two lepton number violating decays of the τ. They find

$$B(\tau^- \rightarrow e^-\gamma) < 3.2 \times 10^{-4} \text{ at 90\% CL}$$
$$B(\tau^- \rightarrow e^-\pi^0) < 4.4 \times 10^{-4}$$

The first of these decays may be interpreted to obtain a lower limit of 65 TeV on the compositeness mass scale of the τ.

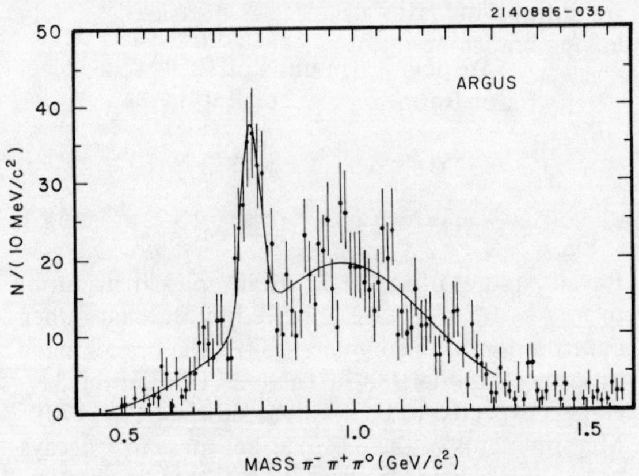

Figure 2. The $\pi^-\pi^+\pi^0$ mass distribution in $\tau^- \rightarrow \nu_\tau \pi^-\pi^+\pi^0\pi^-$ from the ARGUS collaboration.

Measurements of the τ lifetime are summarized in Fig. 3. There are now four measurements of about 10% precision from MARKII, MAC, HRS and CLEO. Substantial improvements in τ lifetime measurements will likely occur only upon the advent of the SLC. The average lifetime value is $\tau_\tau = (2.94 \pm 0.12) \times 10^{-13}$ sec. From this one may predict $B(\tau^- \rightarrow \nu_\tau e^- \overline{\nu}_e) = (18.5 \pm 0.8)\%$, just consistent with average of the measured values given in Table 1.

LIFETIMES OF CHARMED PARTICLES

There have been many measurements of charmed particle lifetimes within the last year. The most significant technical development is best exemplified by the results of the Tagged Photon Spectrometer group in Fermilab experiment E691.[39] The use of high precision vertex detectors, such as silicon microstrips in the case of E691, in conjunction with a good magnetic spectrometer with hadron identification now allow detailed studies of charm particle decays and lifetimes. With sufficient data very precise lifetime measurements are now possible and the ability to reconstruct charm particle secondary vertices will be a powerful tool in studies of rare D^0, D^+, $D_s(\equiv F)$ and Λ_c decay modes. An example of the precision that is now possible is shown in Fig. 4, the lifetime distribution of $D^0 \rightarrow K\pi$ candidates observed by the E691 group. The observed distribution is almost purely exponential reflecting the excellent secondary vertex resolution.

Recent measurements of the D^0, D^+, D_s and Λ_c lifetimes are summarized in Figs. 5–8, respectively. For the D^0 and D^+ I have taken as a starting point the

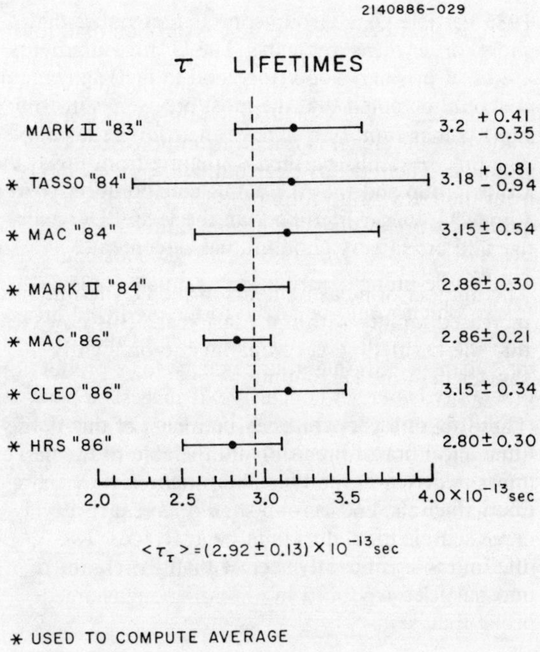

Figure 3. A summary of τ lifetime measurements.

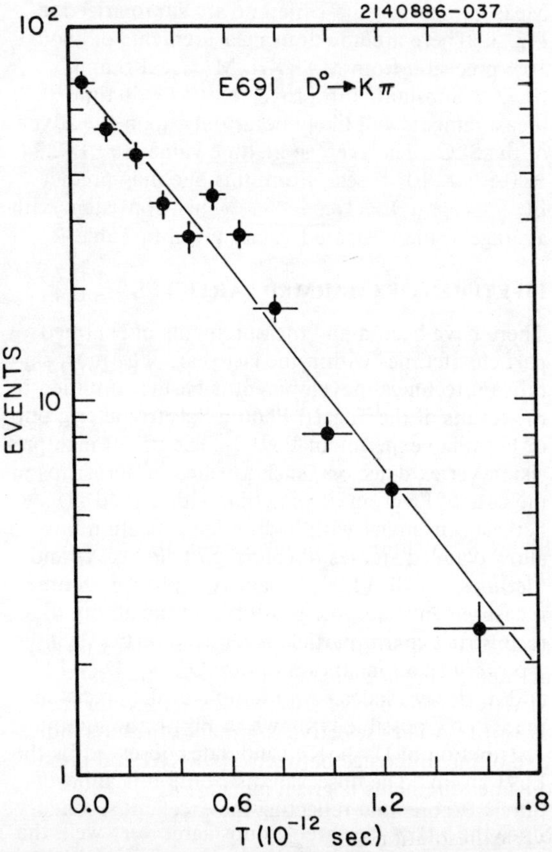

Figure 4. The lifetime distribution of $D^0 \to K^-\pi^+$ decays observed by the E691 group (Tagged Photon Spectrometer) at Fermilab.

1986 Particle Data Group compilation rather than show earlier measurements. The D^0 measurements are in surprisingly good (too good in fact) agreement. It should be noted that the most precise result from E691 comes from only 15% of their full data sample. The precise results on the D^+ lifetime from E691, the LEBC group and the ACCMOR collaboration yield a somewhat longer lifetime than the 1986 PDG average and are in very good mutual agreement.

The number of measurements of the D_s lifetime has increased rapidly within the last year. It is now clear that the D_s lifetime is comparable to or slightly shorter than the D^0 lifetime.

There are still very few measurements of the Λ_c lifetime and none of precision comparable to the best measurements of the D_s. This situation may change soon since the E691 group should have a relatively large sample of Λ_c decays to be analyzed. The Λ_c lifetime is significantly shorter than the charmed mesons lifetimes.

The charmed particle lifetime measurements and relevant ratios are summarized in Table 3.

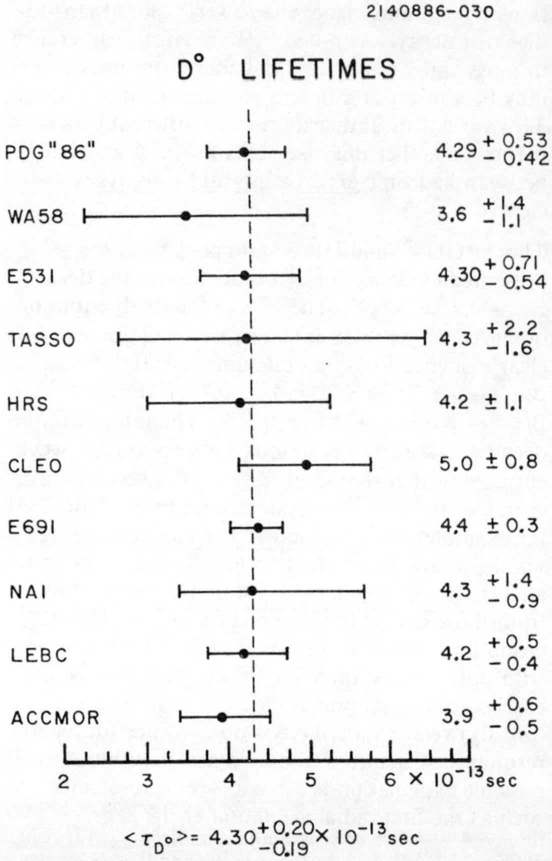

Figure 5. A summary of recent measurements of the D^0 lifetime.

CHARM DECAYS

Within the last year the MARKIII group has presented new measurements of D^0 and D^+ branching ratios from data accumulated at the $\Psi(3770)$.[44,45] They measured branching ratios by fitting to the

Table 3

Particle or Ratio	Lifetime (in units of 10^{-13} sec) or Ratio
D^0	$4.30^{+0.20}_{-0.19}$
D^+	$10.31^{+0.52}_{-0.44}$
D_s	$3.5^{+0.6}_{-0.5}$
Λ_c	$1.9^{+0.5}_{-0.3}$
D^+/D^0	2.40 ± 0.16
D_s/D^0	0.81 ± 0.13
Λ_c/D^0	0.44 ± 0.09

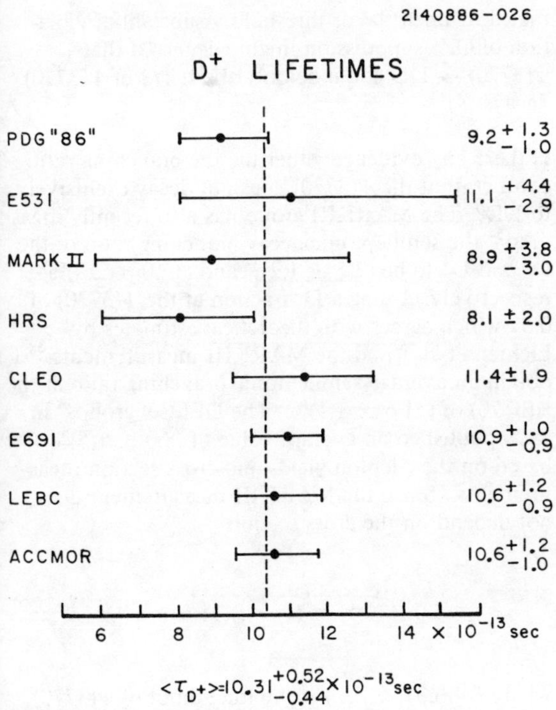

Figure 6. A summary of recent measurements of the D^+ lifetime.

observed number of single and double tagged (both D's reconstructed) events at the $\Psi(3770)$. In principle this should be more accurate than previous methods which relied on cross section measurements at the $\Psi(3770)$. The branching ratios measured by

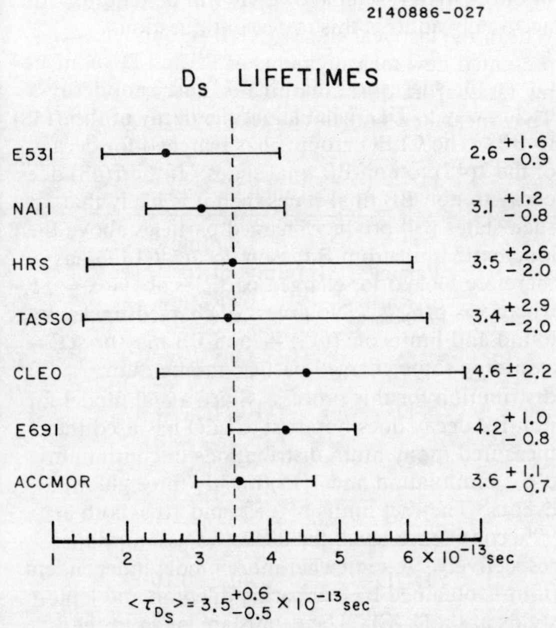

Figure 7. A summary of measurements of the D_s lifetime.

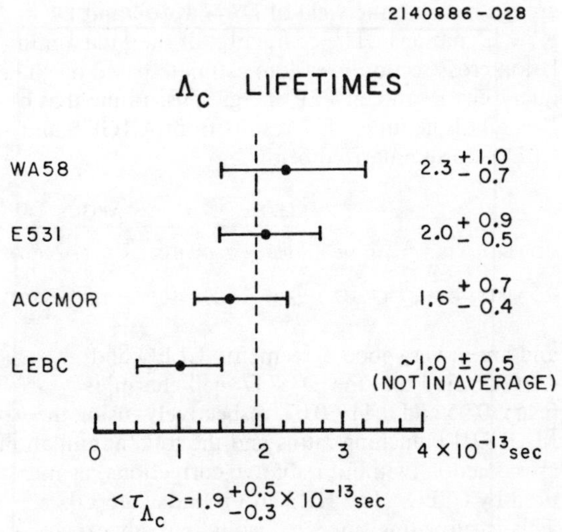

Figure 8. A summary of measurements of the Λ_c lifetime.

MARKIII are 50–100% larger than previous measurements. From their measured branching ratios, MARKIII may derive a total cross section for the $\Psi(3770)$. A very selective summary of results indicative of the difference between MARKIII and previous measurements is given in Table 4

Since the $\sigma{\cdot}B$ of the three groups agree very well, the problem lies either in the cross sections or in the branching ratios. There are three possibilities; the MARKIII branching ratios (and hence the inferred cross section) are wrong; the MARKIII branching ratios are correct but other measurements of the $\Psi(3770)$ cross section are wrong; or the $\Psi(3770)$ does not decay exclusively to $D\bar{D}$.

Is there any evidence, other than previous measurements of branching ratios, that the MARKIII branching ratios are too large? Preliminary measurements of D^0 and D^+ production in e^+e^- annihilation by the CLEO,[48] ARGUS[49] and HRS[50] groups *may* indicate that the branching ratios are too large. Both

Table 4. Measurements of D^0 and D^+ cross sections and branching ratios.

	Pb Glass Wall[46]	MARKII[47]	MARKIII[44]
σ_{D^0}(nb)	11.5 ± 2.5	8.0 ± 1.6	4.5 ± 0.5
$\sigma{\cdot}B(K^-\pi^+)$	0.25 ± 0.05	0.24 ± 0.02.	0.25 ± 0.02
$B(K^-\pi^+)$(%)	2.2 ± 0.6	3.0 ± 0.6	5.6 ± 0.5
σ_{D^+}(nb)	9.0 ± 2.0	6.0 ± 1.2	4.5 ± 0.5
$\sigma{\cdot}B(K^-\pi^+\pi^+)$	0.36 ± 0.06	0.38 ± 0.05	0.39 ± 0.03
$B(K^-\pi^+\pi^+)$(%)	3.9 ± 1.0	6.3 ± 1.5	11.6 ± 2.5

groups measure the yield of $D^0 \to K^-\pi^+$ and $D^+ \to K^-\pi^+\pi^+$ mesons. The $c\bar{c}$ fraction of the total annihilation cross section is easy to estimate based on the c quark charge and, at PEP energies, assuming that b $\to$ c all of the time. The results from ARGUS and CLEO are summarized below

	CLEO	ARGUS
$\sigma(D^0) \cdot B(D^0 \to K^-\pi^+)$	$0.046 \pm 0.002 \pm 0.004$	$0.042 \pm 0.006 \pm 0.005$
$\sigma(D^+) \cdot B(D^+ \to K^-\pi^+\pi^+)$	$0.047 \pm 0.006 \pm 0.004$	$0.050 \pm 0.010 \pm 0.005$

and are in very good agreement. CLEO and ARGUS find that the $[D^0 + D^+]/[\text{all charm}]$ is 0.45 ± 0.05 and 0.44 ± 0.07, respectively, using the MARKIII branching ratios and the total annihilation cross section (without radiative corrections) as measured by CLEO. The HRS group finds 0.46 ± 0.06. This is surprising since one would naively expect D^0 and D^+ to account for at least 3/4 of the charm rate. The only explanation is that charmed baryon + D_s production accounts for about 1/2 of the $c\bar{c}$ rate. Although charmed baryons and D_s have been observed in e^+e^- annihilation, the branching ratios for the observed modes are not known to sufficient accuracy to exclude copious baryon or D_s production. If one believes theoretical estimates of the $\Phi\pi$ branching ratio of the D_s, then charmed baryon production must be large, $\approx$ 40% of the $c\bar{c}$ rate. This hypothesis could be tested by comparing measurements of Λ and p yields in e^+e^- annihilation with expectations with and without an unusual amount of charm baryon production.

The $\Psi(3770)$ cross sections,[51] expressed in terms of the D^0 and D^+ cross sections, are tabulated in Table 5.

Even before the MARKIII measurements there was some lack of consistency among the cross section values. Motivated by this problem, attempts have been made to reexamine the possibility that $\Psi(3770)$ does not always go to $D\bar{D}$. Within the context of a potential model, Lane[52] claims that the E1 decay rates of the Ψ' are best fit by a value of 2S-1D mixing which implies a $\Psi(3770)$ cross section in agreement with the MARKIII cross section measurements. He also states that all but $\approx$ 5% of $\Psi(3770)$ decays must be to $D\bar{D}$. Lipkin,[53] however, claims

Table 5. D^0 and D^+ cross sections at the $\Psi(3770)$.

	σ_{D^0}(nb)	σ_{D^+}(nb)
Lead-glass wall	11.5 ± 2.5	9.0 ± 2.0
MARKII	$8.0 \pm 1.0 \pm 1.2$	$6.0 \pm 0.7 \pm 1.0$
Crystal Ball	6.8 ± 1.2	6.0 ± 1.1
MARKIII	$4.48^{+0.33}_{-0.29} \pm 0.37$	$3.35^{+0.44}_{-0.36} \pm 0.24$

that near open flavor threshold a substantial violation of OZI suppression might occur and that $\Psi(3770) \to D\bar{D}$ could be only about 1/2 of $\Psi(3770)$ decays.

Is there any evidence, other the the branching ratio conflict, that the $\Psi(3770)$ does not decay exclusively to $D\bar{D}$? The MARKIII group has also recently measured[54] the semileptonic decay branching ratio of the D^0 and D^+ to be $(7.5 \pm 1.2)\%$ and $(17.0 \pm 2.0)\%$, respectively. Using a D^0 fraction at the $\Psi(3770)$ of 0.57 which agrees with theoretical estimates by Eichten et al.[55] and the MARKIII measurements,[44] I obtain an average semileptonic branching ratio at the $\Psi(3770)$ of $(11.6 \pm 1.1)\%$. The DELCO group[56] in 1979 published an average value of $(8.0 \pm 1.5)\%$ based on their lepton yields and cross section measurements. Since the MARKIII measurement does not depend on the cross section

$$(8.0 \pm 1.5) = (1 - f)(11.6 \pm 1.1)$$

where $f = 0.31 \pm 0.14$ is the fraction of of $\Psi(3770)$ decays to non-$D\bar{D}$ final states. The CLEO group[57] has pointed out a similar conclusion comparing dilepton rates and single lepton rates as measured by DELCO with expectations from the well known D^0 and D^+ lifetimes (this is independent of MARKIII measurements). They find $f = 0.4 \pm 0.2$ in agreement with my simpler analysis. Neither of these observations represents statistically significant proof that $\Psi(3770)$ decays into non-$D\bar{D}$ final states. New measurements by the MARKIII group or reanalysis of old MARKII data at SPEAR will be required to accurately address this important question.

Given the suggestion that $\Psi(3770)$ may not decay exclusively to $D\bar{D}$, what about the decay of the $\Upsilon(4S)$ to $B\bar{B}$? The CLEO group[57] has searched for decays of the $\Upsilon(4S)$ to non-$B\bar{B}$ final states. If the $\Upsilon(4S)$ does decay to non-$B\bar{B}$ final states then it is likely that such states will produce charged particles above the kinematic limit from B meson decay. CLEO has therefore looked for charged particles above $x = 0.5$, where $x = p/E_{beam}$. No excess of charged tracks is found and limits on $\Upsilon(4S) \to$ non-$B\bar{B}$ may be set, assuming some charged particle momentum distribution for this process. Since a real model for non-$B\bar{B}$ decay does not exist, CLEO has used their measured momentum distributions in continuum e^+e^- annihilation and $\Upsilon(1S)$(mostly three gluon) events. They set limits of 3.8% and 10%, both at 90% confidence level, for these two assumptions, respectively. A somewhat more model independent limit is obtained by comparing dilepton and lepton yields at the $\Upsilon(4S)$. The details are given in the CLEO paper and yield a limit of 17% at 90% confidence level for the non-$B\bar{B}$ fraction of $\Upsilon(4S)$ decays.

$D \rightarrow K e \nu$ AND $D \rightarrow K \pi e \nu$

Direct measurements of the exclusive semileptonic decays of the D^0 and D^+ have been made by the MARKIII group.[58] They find

$$B(D^0 \rightarrow K^- e^+ \nu_e) = (4.1 \pm 0.6 \pm 0.6)\%$$
$$B(D^0 \rightarrow K^- \pi^0 e^+ \nu_e) = (1.4 \pm 0.6 \pm 0.2)$$
$$B(D^0 \rightarrow \overline{K}^0 \pi^- e^+ \nu_e) = (2.5 \pm 0.8 \pm 0.4)$$
$$B(D^+ \rightarrow \overline{K}^0 e^+ \nu_e) = (7.2 \pm 1.9 \pm 0.9)$$
$$B(D^+ \rightarrow K^- \pi^+ e^+ \nu_e) = (3.9 \pm 0.8 \pm 0.7)$$

A surprising aspect of the $(K\pi)$ modes is that there appears to be a large non-resonant (not K*(892)) component. For the sum of all $D \rightarrow K\pi e^+ \nu_e$ modes, the non-resonant s-wave fraction is found to be $0.45^{+0.14}_{-0.13}$ ie. about 55% percent only in K*(892).

THE DECAY $D^0 \rightarrow \Phi \overline{K}^0$

Evidence for this decay mode was first presented about a year ago by the ARGUS group.[59] At the time there was considerable uncertainty regarding the background shape in the K^+K^- mass distribution near threshold — sufficient uncertainty to cast doubt on the existence of the Φ. Within the last year additional data collected by ARGUS[60] and the results of the CLEO[61] and MARKIII[62] groups have shown that the $\Phi \overline{K}^0$ mode definitely is present. The branching ratios are summarized in Table 6.

D_s DECAY MODES

New results have been presented on the decay of the D_s^+ to $\overline{K}^{*0}K^+$ by ARGUS[63] and most convincingly by E691.[64] This decay cannot proceed through a simple spectator diagram and must occur via a color mixed or annihilation diagram. The data from E691 are presented in Fig. 9 and show convincing evidence for $D_s^+ \rightarrow \overline{K}^{*0}K^+$ and the Cabibbo suppressed decay of the D^+ to the same final state. The ratio $B(D_s^+ \rightarrow \overline{K}^{*0}K^+)/B(D_s^+ \rightarrow \Phi\pi^+)$ is measured to be $1.1^{+0.5}_{-0.4} \pm 0.15$ and 1.44 ± 0.37 by E691 and ARGUS, respectively. The results show that this decay mode is large, comparable to the color-favored decay into $\Phi\pi^+$.

Table 6

Group	$B(D^0 \rightarrow \Phi \overline{K}^0)$
ARGUS	$(1.29 \pm 0.27 \pm 0.19)\%$
CLEO	$1.18 \pm 0.40 \pm 0.17$
MARKIII	$1.1^{+0.7}_{-0.5}{}^{+0.3}_{-0.2}$
Average	1.17 ± 0.23

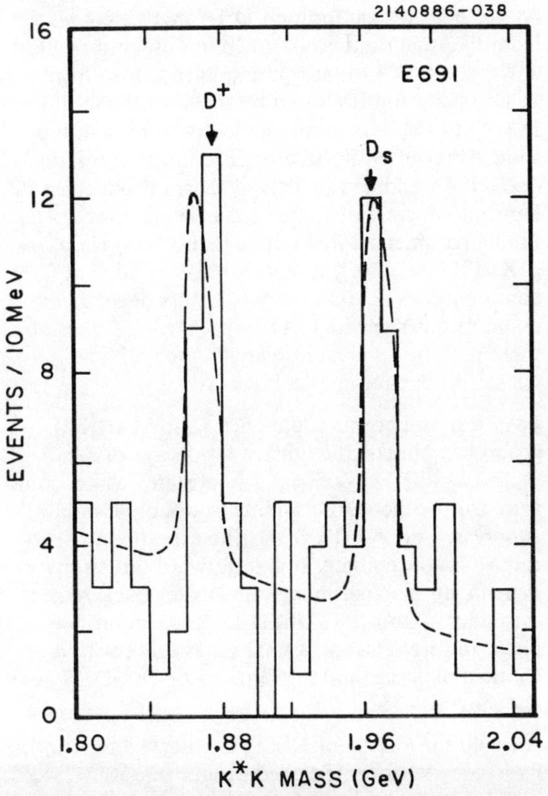

Figure 9. The $K^{*0}K^\pm$ mass distribution observed by E691 showing the Cabibbo suppressed decay of the D^+ and the decay of the D_s.

THE CHARM DECAY CONSTANT f_D

The MARKIII group[65] has obtained a limit on $B(D^+ \rightarrow \mu^+\nu)$ from which they infer a limit on the D decay constant f_D from the formula

$$\Gamma(D^+ \rightarrow \mu^+ \nu) = (1/8\pi) G_F^2 f_D^2 m_D m_\mu^2 |V_{cd}|^2 [1-(m_\mu/m_D)^2]^2$$

MARKIII finds $B(D^+ \rightarrow \mu^+\nu) < 8.4 \times 10^{-4}$ at 90% confidence level which implies $f_D < 340$ MeV at the same confidence level. Pseudoscalar decay constants have been predicted in the context of potential models,[66] bag models,[67] and QCD sum rules.[68] The upper end of predictions from potential models would seem to be excluded.

$D^0\overline{D}^0$ MIXING

Mixing of the neutral D mesons may occur through box diagrams[69] or via specific intermediate meson final states (dispersive or long range effects).[70] Predictions of the magnitude of $D^0\overline{D}^0$ mixing depend strongly on estimates of such dispersive effects, but still yield values ($\leqslant 10^{-3}$) which are not yet observable by any experiment. Alternatives to the Standard Model, however, might allow substantially greater mixing.[71]

At present the best limit on $D^0\overline{D}^0$ mixing comes from like-sign dimuon production observed in E615 at Fermilab.[72] Like-sign muon events arise from $D^0\overline{D}^0$ production followed by transformation of the $D^0(\overline{D}^0)$ to $\overline{D}^0(D^0)$ via mixing followed by semileptonic decay of both mesons. The limit on mixing derived depends on models of charm production in hadronic interactions. For a reasonable range of model parameters, E615 finds that mixing ($\Gamma(\overline{D}^0 \rightarrow \mu^+X)/\Gamma(\overline{D}^0 \rightarrow \mu^-X)$) is less than 4.7×10^{-3} at 90% confidence level. Because this results depends on assumptions regarding hadronic charm production, there remains interest in searching for mixing in charm production in e^+e^- reactions.

Last year and at this conference[73] the MARKIII group has presented evidence for decays of the $\Psi(3770) \rightarrow D^0\overline{D}^0 \rightarrow S = \pm 2$ final states which could arise from either $D^0\overline{D}^0$ mixing or doubly Cabibbo suppressed decays (DCSD). The experimental procedure is conceptually straightforward but strongly dependent on experimental details for background estimates. At the $\Psi(3770)$ $D^0\overline{D}^0$ production may result in three classes of final states each, with a characteristic signature for mixing or DCSD. These are shown in Table 7.

A signal for mixing or DCSD is observed only in the hadronic vs hadronic tagged events; one $K^+\pi^-$ vs $K^+\pi^-\pi^0$ event and two $K^+\pi^-\pi^0$ vs $K^+\pi^-\pi^0$ (and charge conjugate) events. No events consistent with mixing or DCSD are observed in the hadronic vs semileptonic or semileptonic vs semileptonic tags. These results and the estimated backgrounds are summarized in Table 8.

Given that the three events have the correct charge assignments, are they evidence for mixing or DCSD? The answer is not clear. Because of quantum statistics effects, as discussed by Bigi and Sanda,[74] if the two D's decay into identical final states then only mixing will yield $S = \pm 2$ final states. If the two final states are not identical then both mixing and DCSD can contribute. The problem is that two of the events could be either identical or non-identical final states. If one assumes that DCSD do not contribute, then mixing would be ≈ 0.01 in contradiction to the

Table 7. Signatures for $D^0\overline{D}^0$ mixing or doubly Cabibbo suppressed decays.

	Signature for mixing or DCSD
hadronic vs hadronic tags	
$D^0, \overline{D}^0 \rightarrow K\pi, K\pi\pi^0, K3\pi$	wrong (kaon) charge combination
hadron vs semileptonic	
$D^0 \rightarrow K\pi, K\pi\pi^0, K3\pi$ $\overline{D}^0 \rightarrow Ke\nu, K\mu\nu$	Q of hadronic kaon $\neq$ $-$Q of lepton
semileptonic vs semileptonic	
$D^0 \rightarrow Ke\nu, K\mu\nu,$ $\overline{D}^0 \rightarrow Ke\nu, K\mu\nu$	same charge dileptons (mixing only)

Table 8. Event candidates and backgrounds for mixing.

Tag type	# of normal events	# of mixing candidates	# of background events
hadronic vs hadronic	162	3	0.4
hadronic vs semileptonic	69	0	<0.1
semileptonic vs semileptonic	12	0	0

Standard Model. Attributing all three events to DCSD results in a rate $\approx 5\tan^4\Theta_c$ (where Θ_c is the Cabibbo angle), somewhat higher than expected,[74] but three events is not sufficient to make a definitive statement. The obvious conclusion is that more data are needed.

The ARGUS[75] and HRS[76] groups have also presented results at this conference on limits to $D^0\overline{D}^0$ mixing. The method uses $D^{*+} \rightarrow D^0\pi^+$, $D^0 \rightarrow K\pi$ or $K3\pi$. Mixing would be indicated by a kaon and a pion (from the D* decay) of the same sign. Identification of the D* decays is done in the usual way by computing the mass difference between the $D^0\pi$ system and the reconstructed D^0. The results from the ARGUS group are shown in Fig. 10 and yield a 90% confidence limit on mixing of 2.3%. Using the same method HRS finds a limit of 4%.

EXCLUSIVE DECAYS OF B MESONS

New results on exclusive decays and branching ratios of B^0 and B^+ mesons have been presented by the CLEO[77] and ARGUS[78] groups. B mesons are produced in pairs at the $\Upsilon(4S)$ resonance. A B meson then usually decays into a charm meson and other particles, mostly pions. CLEO reconstructs D^0, D^+ and D^{*+} mesons whereas ARGUS so far has only used D^{*+} mesons. The reconstructed D meson may be combined with 1, 2 or 3 pions (one of which may

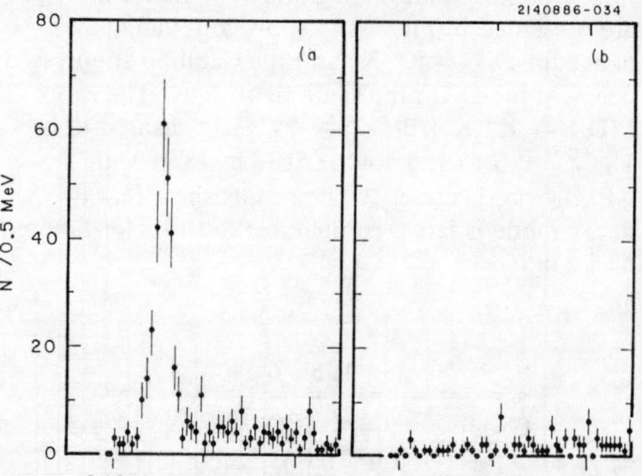

Figure 10. The $\Delta M = M(D^*) - M(D^0)$ for (a) right sign $K\pi$ charge combinations and (b) wrong sign $K\pi$ combinations. The data are from the ARGUS group.

be a π^0 in the case of ARGUS) to reconstruct a B meson. Both groups also reconstruct events containing a Ψ and a K or K*. Candidate events must satisfy a number of criteria, the most important being that the sum of the measured particle energies, ΣE_i, be close to one-half the mass of the $\Upsilon(4S)$ (the energy, E_{beam}, of the colliding e^+ or e^- beam). CLEO simply cuts on this energy difference, $\Delta E = \Sigma E_i - E_{beam}$, whereas ARGUS computes a χ^2 for each event and then cuts on χ^2. The methods yield very similar results.

For an event candidate which passes the criteria, the mass is computed using the beam energy, which is very well known, and the measured 3-momenta. The results from the CLEO group are shown in Fig. 11 for two different sets of ΔE criteria, "loose" (the dashed line) and "tight" (the numbered squares). In the case of the "loose" criteria, substantial feed-down from higher multiplicity modes to the observed mode may occur. Most of this feed-down comes from B decays to D* + pions, when a soft pion or photon is not observed from the decay of the D* to a D. Since the energy of the soft pion or photon is not large, the reconstructed mass will be very close to the actual B meson mass. Requiring "tight" cuts on ΔE eliminates most of this background.

The results from the ARGUS group are shown in Fig. 12, separating the B^0 and B^+ decays. Although

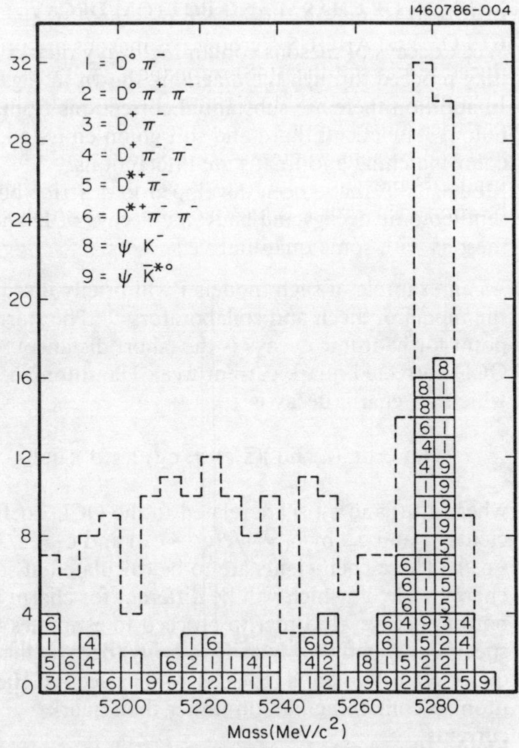

Figure 11. The mass distribution of B meson candidates observed by the CLEO collaboration. The dashed histogram is explained in the text.

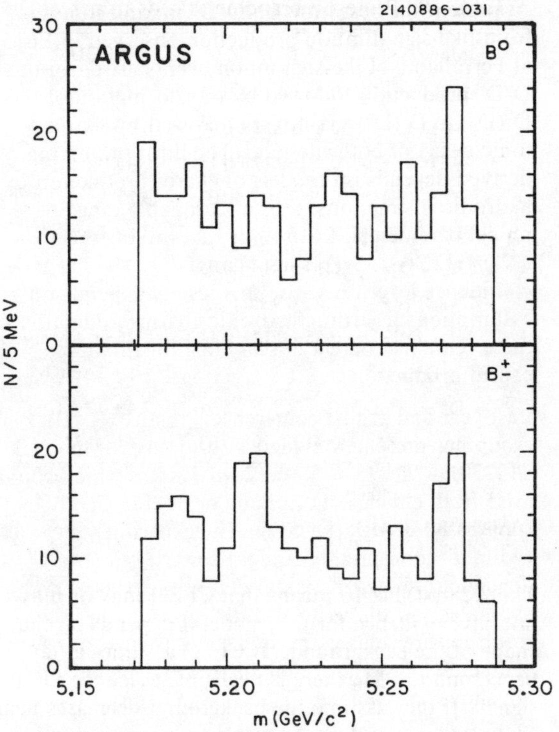

Figure 12. The mass distribution of B^0 and B^+ candidates decaying into $D^{*\pm}$ + (1–3 pions) observed by the ARGUS group.

there is good agreement between CLEO and ARGUS regarding the B^0 and B^+ masses, they do not agree on the shape of the background beneath the B peak. ARGUS finds the background to be diminishing at higher masses whereas CLEO finds the background to be flat. They may both be partially correct since most of the ARGUS signal is in high multiplicity modes whereas most of the CLEO signal is in lower multiplicity modes with smaller backgrounds.

Since the e^+ or e^- beam energy is used to compute the B mass, the final mass values depend on the beam energy calibration. In the summary below I have used the current CESR(CLEO) energy scale ($M(\Upsilon(4S)) = 10579.8$ MeV/c^2) and have modified the ARGUS results to conform to this scale. The B masses (in MeV/c^2) are summarized in Table 9.

The systematic error in the average mass is about ± 3 MeV/c^2 and ± 2 MeV/c^2 in the mass difference.

The branching ratios measured by CLEO and ARGUS are summarized in Table 10.

Table 9. Measurements of B meson masses in MeV/c^2.

	$M(B^0)$	$M(B^+)$	$M(B^0)$-$M(B^+)$
CLEO	$5281.0 \pm 0.9 \pm 3$	5277.9 ± 1.1	3.1 ± 1.4
ARGUS	$5279.6 \pm 1.0 \pm 3$	$5277.2 \pm 1.3 \pm 3$	2.4 ± 1.6
Average	5280.4 ± 0.7	5277.6 ± 0.8	2.8 ± 1.1

Table 10. B meson branching ratios in percent.

Mode	CLEO[a]	ARGUS[b]
$B^- \to D^0\pi^-$	$0.38 \pm 0.14 \pm 0.11$	
$B^- \to D^+\pi^-\pi^-$	$0.89 \pm 0.51 \pm 0.30$	
$B^- \to D^{*+}\pi^-\pi^-$	$0.31 \pm 0.17 \pm 0.11$	$0.4 \pm 0.2 \pm 0.2$
$B^- \to \Psi K^-$	$0.09 \pm 0.06 \pm 0.02$	<0.20
$B^- \to D^{*+}\pi^-\pi^-\pi^0$		$3.5 \pm 1.1 \pm 2.1$
$B^0 \to D^+\pi^-$	$0.14 \pm 0.19 \pm 0.05$	
$B^0 \to D^0\pi^+\pi^-$	$1.6 \pm 0.9 \pm 0.6$	
$B^0 \to D^{*+}\pi^-$	$0.35 \pm 0.14 \pm 0.11$	$0.25 \pm 0.15 \pm 0.15$
$B^0 \to D^{*+}\pi^-\pi^0$		$1.1 \pm 0.6 \pm 0.6$
$B^0 \to D^{*+}\pi^-\pi^-\pi^+$	???	$2.4 \pm 0.7 \pm 1.1$
$B^0 \to \Psi K^{*0}$	$0.41 \pm 0.19 \pm 0.03$	0.44 ± 0.27

[a]$B^0B^0{:}B^+B^-$ = 40:60

[b]$B^0B^0{:}B^+B^-$ = 45:55

The ??? in Table 10 means that CLEO may or may not have evidence for this mode; it depends on the shape of the background. If CLEO assumes a flat background in M_B there is negligible evidence for a signal. If they assume the background decreases near M_B, as does ARGUS, the resulting branching ratio would be consistent with the ARGUS value.

To compute branching ratios one must make an assumption about the relative fraction of $B^0\overline{B}^0$ to B^+B^- production at the $\Upsilon(4S)$. The uncertainty in the mass difference makes calculation of a precise value of this fraction difficult. CLEO and ARGUS assume slightly different values as shown in Table 10 and I have left the branching ratios as quoted rather than scale them to a common value of the neutral to charged fraction; the errors in individual measurements are too large to require this refinement. There is good agreement between ARGUS and CLEO where the measurements overlap. Both groups now find branching ratios which are considerably smaller than the original CLEO measurements in 1984.[79] Two reasons for this are known; the D branching ratios (from the MARKIII) now used are about a factor of two higher and the effects of backgrounds and feed-down in the original CLEO measurements were not well understood. The results from CLEO contain the data sample used in the 1984 analysis and more recent data.

The sum of the branching ratios in Table 10 is about 11%. Since purely hadronic decays of the B are expected to be about 75% of the total, only a small fraction of exclusive B decays has been observed. Since the mean charged multiplicity in B decay is about 6, this is not surprising.

The CLEO group[80] has also searched for a large number of rare or forbidden decays, within the Standard Model. The results are summarized in Table 11. Most of the decays searched for arise primarily or exclusively through penguin-type graphs. Calculations[81] of such graphs in B decay result in a suppression of 1–2 orders of magnitude over ordinary spectator decays (ie. those of Table 10). The possibility that such penguin induced decays are large, comparable to the spectator decays, is ruled out by the CLEO results.

$B \to \Psi + X$

The properties of this decay have been measured by CLEO[82] and ARGUS[83] and ARGUS now claims evidence for the decay $B \to \Psi' + X$. Part of the evidence for Ψ' production in shown in Fig. 13; there is also evidence of roughly equal statistical significance looking for $\Psi' \to \Psi\pi^+\pi^-$. The following branching ratios are found

$B(B \to \Psi + X) = 1.09 \pm 0.16 \pm 0.21\%$ CLEO

$B(B \to \Psi + X) = 1.15 \pm 0.24 \pm 0.21$ ARGUS

$B(B \to \Psi' + X) = 0.50 \pm 0.23$ ARGUS

The Ψ momentum distribution from the two experiments is shown in Fig. 14. In addition to contributions from the two-body ΨK and ΨK^* modes (see the section above on exclusive decays for branching ratios) there must be contributions from either Ψ' and/or ΨX_s, where X_s has mass greater than about 1 GeV/c^2. Roughly one-half of the Ψ events are two-body modes, ΨK or ΨK^*.

MODELS OF CHARM AND BOTTOM DECAY

Weak decays of mesons containing heavy quarks may proceed through the diagrams shown in Fig. 15. In addition there are substantial corrections from bound state effects, hard and soft gluon emission, color matching and final state interactions. Models[84,85,86] have been developed to describe both semileptonic decays and hadronic decays of D and B mesons with some quantitative success.

As an example of such models I will briefly discuss the model of Stech and collaborators.[84] The starting point for hadronic decays is the (short distance) QCD corrected quark current weak Hamiltonian which for charm decay is

$$H_W \propto \{c_1(\mu)(\overline{u}d')(\overline{s}'c) + c_2(\mu)(\overline{s}'d')(\overline{u}c)\}$$

where $c_1(\mu)$ and $c_2(\mu)$ are related to the QCD coefficients c_+ and c_- by $c_1 = 1/2(c_+ + c_-)$ and $c_2 = 1/2(c_+ - c_-)$. These coefficients are to be calculated at some energy scale, μ, which will be different for charm and bottom decays. In order to proceed to estimates for specific two-body (or quasi-two-body) decays, the quark Hamiltonian is replaced by an effective Hamiltonian containing hadron rather than quark currents

$$H_{eff} \propto \{a_1(\overline{u}d')_H(\overline{s}'c)_H + a_2(\overline{s}'d')_H(\overline{u}c)_H\}$$

Table 11.

Mode	Number of Events 90% CL limits	Number of B decays	Detection Efficiency	Branching Fraction 90% CL limits
$K^+\pi^-$	15.3	104 K	.46	3.2×10^{-4}
$K^0\pi^+$	5.3	156 K	.05	6.8×10^{-4}
$K^{0*}\pi^+$	6.8	156 K	.17	2.6×10^{-4}
$K^{+*}\pi^-$	2.3	104 K	.03	7.0×10^{-4}
ρK^+	10.1	156 K	.25	2.6×10^{-4}
ρK^0	3.4	104 K	.04	8.0×10^{-4}
ϕK^+	3.9	156 K	.12	2.1×10^{-4}
ϕK^0	3.9	104 K	.03	13.0×10^{-4}
ϕK^{0*}	3.9	104 K	.08	4.7×10^{-4}
ρK^{0*}	19.1	104 K	.16	11.5×10^{-4}
$K^{0*}\gamma$	22.0	104 K	.11	20.5×10^{-4}
$K^{+*}\gamma$	3.2	156 K	.014	18.4×10^{-4}
$K^+\mu^+\mu^-$	11.8	156 K	.24	3.2×10^{-4}
$K^+e^+e^-$	5.8	156 K	.18	2.1×10^{-4}
$K^0\mu^+\mu^-$	2.3	104 K	.05	4.5×10^{-4}
$K^0e^+e^-$	5.3	104 K	.03	6.5×10^{-4}
e^+e^-	2.3	104 K	.29	0.8×10^{-4}
$\mu^+\mu^-$	3.9	104 K	.40	0.9×10^{-4}
μe	2:3	104 K	.26	0.9×10^{-4}

Within this factorization hypothesis, the a coefficients are related to c_1 and c_2 by

$$a_1 = c_1 + \xi c_2\,, \quad a_2 = c_2 + \xi c_1$$

where ξ represents the color mismatch to form hadrons; $\xi = 1$ corresponds to no color suppression;

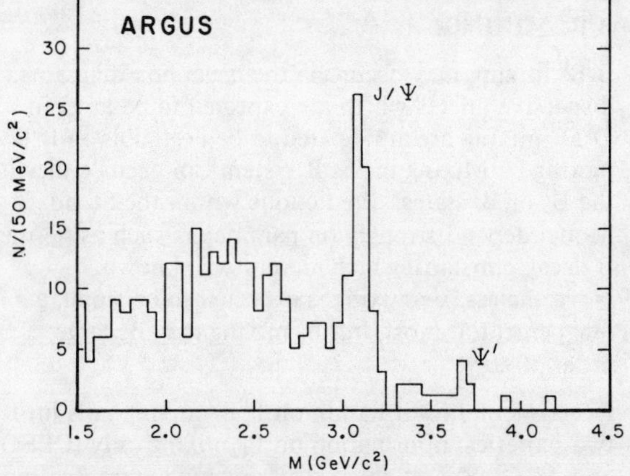

Figure 13. The dilepton mass distribution in B meson decay seen by ARGUS.

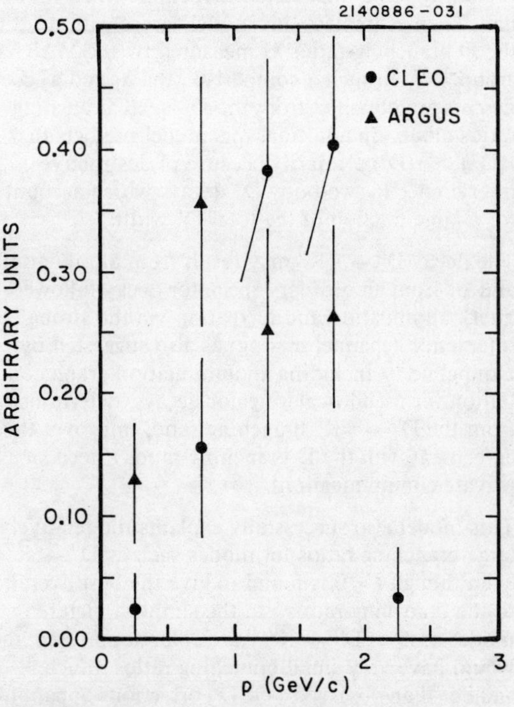

Figure 14. The Ψ momentum distribution from $B \to \Psi + X$.

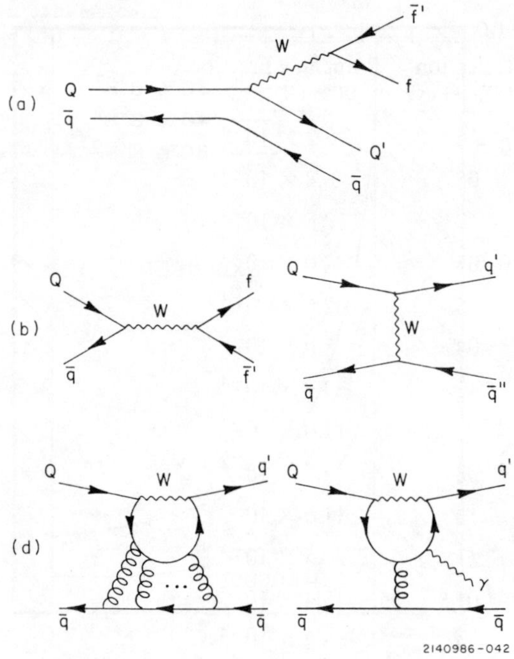

Figure 15. Decay modes of mesons containing heavy quarks (a) spectator; (b) annihilation; (c) exchange; and (d) penguin.

$\xi = 1/3$ to suppression expected from simple color counting; and $\xi = 0$ if only quarks from the same current combine to form hadrons, maximal color suppression.

Including some final state interactions when they have been estimated, this model agrees with most of the D branching ratios as measured by the MARKIII group.[44,45] In such a comparison the a_1 and a_2 coefficients are obtained from the $D \to \pi\overline{K}$ branching ratios alone. In addition, this model predicts that $\tau(D^+) > \tau(D^0)$ primarily because of destructive interference in two-body D^+ decays which account for a large fraction of the total D^+ width.

The decay $D^0 \to \Phi\overline{K}^0$ may result from annihilation and/or from an ordinary spectator decay followed by quark annihilation and recreation via the strong interaction (channel mixing) as also suggested by Donoghue.[87] Including an annihilation graph contribution for pseudoscalar-vector decays, determined from the $D^0 \to \Phi\overline{K}^0$ branching ratio, improves the agreement with the D branching ratios (Stech, private communication).

This model also successfully explains the relatively large branching ratios for modes such as $D^0 \to \overline{K}^0\pi^0$ even though $\xi = 0$ is found to give the best overall fit to the branching ratios. In the simplest picture, modes such as $D^0 \to \overline{K}^0\pi^0$ are color suppressed and would have very small branching ratios, much smaller than observed. QCD corrections apparently change this picture while retaining $\xi = 0$ in the context of the model; see ref. 86 for an explanation.

The relatively large branching ratio for $D_s \to K^*K$ cannot be explained by assuming factorization alone. Channel mixing (from $\Phi\rho^+$, for example) and/or annihilation must be present.

The short lifetime of the Λ_c relative to the D^0 results, for the most part, from the W exchange which can be substantial in baryon decays.

This type of model has also been applied to two-body B meson decays with some success, although the data are much poorer than in the charm case. Some predictions and data are compared in Table 12.

Table 12

	Data
$B(\overline{B}^0 \to D^+\pi^-) = 0.5a_1^2 \lvert V_{cb}/0.05 \rvert^2(\%) \approx 0.6\%$	$0.14 \pm 0.20\%$
$B(\overline{B}^0 \to D^{*+}\pi^-) = 0.4a_1^2 \lvert V_{cb}/0.05 \rvert^2 \approx 0.5$	0.31 ± 0.14
$B(\overline{B}^0 \to \overline{K}^{*0}\Psi) = 4.4a_2^2 \lvert V_{cb}/0.05 \rvert^2 \approx 0.3$	0.42 ± 0.16

where $a_1 \approx 1.1$ and $a_2 \approx -0.24$ for $\xi = 0$ and I have taken $V_{cb} = 0.05$ for convenience. Since the actual value of V_{cb} is uncertain to at least a factor of two, absolute agreement between model and data is not too meaningful. Nevertheless the model predicts branching ratios in the correct range. A better comparison awaits predictions for other decay modes which have been measured.

From a theoretical viewpoint these models are relatively simple, surprisingly successful and at least semi-quantitatively explain the characteristics of charm decays; branching ratios and lifetimes. The challenge to such models in the future is to improve their predictive power. This interface between QCD/strong interactions and weak interactions is of considerable importance in extracting KM matrix elements. In the B sector, as noted below, the uncertainty in the nature of the hadronic system in semileptonic decays is a limiting factor in the extraction of a limit on V_{ub} in the K-M matrix.

$B^0\overline{B}^0$ MIXING

$B^0\overline{B}^0$ mixing may occur via the usual box diagrams; dispersive effects which are expected to be large in $D^0\overline{D}^0$ mixing are anticipated to be negligible in $B^0\overline{B}^0$ mixing.[70] Mixing in the B system can occur either in the B_d or B_s states. Predictions within the Standard Model depend strongly on parameters such as f_B, the B decay constant, which are not well known. Nevertheless, B_d mixing is expected to be small, a few percent at most, but B_s mixing may be large, order 50%.[88]

Experimental information on $B^0\overline{B}^0$ mixing falls into two varieties; information on B_d mixing only (CLEO and ARGUS) and data on an unknown combination of B_d and B_s mixing (MARKII[89] at PEP and UA1). The measurable quantities are the same in all experiments; an excess of like-sign dileptons above background. Like-sign dilepton events may result from mixing and subsequent semileptonic decay of both B mesons. A large number of backgrounds may also result in like-sign dileptons. These backgrounds are enumerated in detail in contributions to this conference from CLEO,[90] ARGUS[91] and UA1.[91] An abbreviated summary of the results from these groups is given in Table 13.

Table 13. $B^0\overline{B}^0$ mixing results.

	CLEO	ARGUS	UA1
Like-sign observed	22	5	142
Like-sign background	17 ± 5	$2.5 \pm (2?)$	$(45 \pm 7)+(55 \pm 6)$
Opposite-sign observed	136	159	257
Opposite-sign background	19	15	45
Net signal for mixing	5.1 ± 5.9	2.5 ± 3	42 ± 15

Neither CLEO nor ARGUS have evidence for mixing but UA1 observes approximately a 3σ excess. The backgrounds in the case of CLEO and ARGUS include the (dominant) contribution from charm semileptonic decays; one B decays semileptonically, the other into charm which then undergoes semileptonic decay, yielding like-sign dileptons. The backgrounds in the UA1 experiment are discussed below.

The CLEO and ARGUS numbers may be converted to upper limits on B_d mixing with some assumptions. Because the data from both CLEO and ARGUS were obtained at the $\Upsilon(4S)$ one must know the fraction of $B^0\overline{B}^0$ events produced at the resonance. This has traditionally been taken as 40%, although it could easily be 30% or 50%. One must also know the ratio of the B^0 to $B^{\pm}$ semileptonic branching to set a limit. The CLEO group[90] using their dilepton data has set a limit on this ratio of $0.43 < B(B^0 \rightarrow l\nu X)/B(B^{\pm} \rightarrow l\nu X) < 2.05$ at 90% confidence level. Since this still leaves a wide range, the mixing limit is plotted against the ratio of branching ratios. The results are shown in Fig. 16. The limit from the ARGUS group is about a factor of two better than from CLEO. It has become customary to quote as a limit the value for the ratio of branching ratios equal to one and a 40% $B^0\overline{B}^0$ fraction; CLEO finds 24% and ARGUS 12%, both at 90% confidence level. These numbers *are not* true indicators of the actual limit. Predictions of $B(B^0 \rightarrow l\nu X)/B(B^{\pm} \rightarrow l\nu X)$ of 0.56 and above exist in the literature.[93] For this value, and a $B^0\overline{B}^0$ fraction of 30%, the ARGUS limit deteriorates to about 40%. This comment will become more relevant when discussing the UA1 results.

UA1 observes a large sample of dimuon events arising from Drell-Yan production, Υ decay and heavy flavor semileptonic decay. In order to study B semileptonic decays, they require the muons to be non-isolated, within jets, and require the muons to have a minimum P_T, to suppress contributions from charm semileptonic decays. The details are

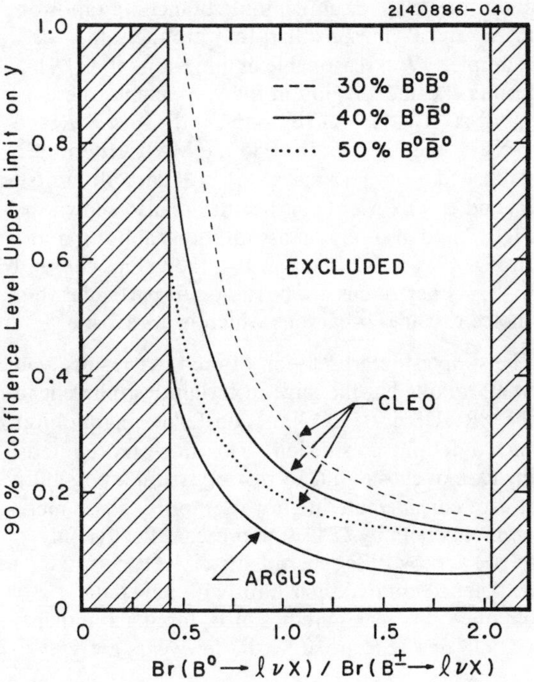

Figure 16. Limits on B_d mixing obtained by the ARGUS and CLEO groups vs the ratio of the B^0 to B^+ semileptonic branching ratio.

described in the contribution to this conference and in a forthcoming publication.[94] I have separated the "background" to the like-sign events into two categories. The first (45 events) represents the estimate of backgrounds from fake muons and sources other than secondary charm decays. Subtracting this background yields the ratio

$$R = N(\pm\pm)/N(+-) = 0.46 \pm 0.07 \text{ (statistical error only)} .$$

UA1 must resort to Monte Carlo simulation to estimate the value of R expected from secondary charm decays in the absence of mixing. Using a number of Monte Carlo generators they predict $R \approx 0.26 \pm 0.03$ in the absence of mixing and hence conclude that their measurements indicate the presence of $B^0\overline{B}^0$ mixing. A value of $R = 0.26 \pm 0.03$ yields a background from secondary charm decay of 55 ± 6 events. Adding this to the 45 events from other background sources one is left with $\approx 42 \pm 15$ like-sign events or about a 3σ effect. This level of statistical significance is in agreement with the more sophisticated analysis presented at the conference.[92]

This excess of like-sign events may be interpreted as a lower limit on the mixing from a combination of B_d and B_s decay. The fraction of B_s mesons compared to B_d mesons is, of course, unknown at the $Sp\overline{p}S$ collider so one must assume values for $f_s(f_d)$, the fraction of $B_s(B_d)$ mesons produced in b quark hadronization. One must also make an assumption about the relative semileptonic branching ratios of the B_s and B_d mesons; that they are the same, for example. With reasonable assumptions the UA1 results exclude a region in the y_d-y_s plane where $y_d(y_s)$ is $(B_d \to \mu^- X)/(B_d \to \mu^+ X)$ $[B_s \to \mu^- X)/(B_s \to \mu^+ X)]$ — see Fig. 17. In 1985 the MARKII group[89] published a similar analysis of PEP data on limits to B_d and B_s mixing. Their results are also shown in Fig. 17 and also rely on assumptions about f_s and f_d. In Fig. 17, $f_s = 0.2$ and $f_d = 0.4$; the values chosen by UA1. Other values are possible. In particular this choice excludes b-baryons which is unrealistic.

The simplest interpretation of the UA1 results would be B_s mixing but the large uncertainties inherent in the ARGUS or CLEO limits on B_d mixing prohibit such a definite conclusion. The much more interesting case of substantial B_d mixing is still a possibility. It is of considerable interest to improve these measurements both by UA1 (the present effect is only $\approx 3\sigma$) and by PEP experiments. ARGUS is likely to be able to improve their limit with new data obtained this year. Studies of B_s mixing at either DORIS or CESR are at least a few years away.

LIMITS ON V_{bu}

Measurements of the lepton momentum spectra in semileptonic B decay have been used to set limits on the $b \to u$ coupling. At the quark level, B semileptonic decay proceeds through either $b \to c l \nu$ or $b \to u l \nu$. Since the u quark mass is much smaller than the c quark mass, the $b \to u l \nu$ transition produces higher momentum leptons. In the free quark limit the lepton spectra from the two transitions are easily calculated and distinct. However, in B meson decay, bound state effects and hadronization may substantially modify the lepton spectrum. These effects have been taken into account by a variety of models for the $b \to c l \nu$ and $b \to u l \nu$ transitions. Published analyses by the CLEO[95] and CUSB[96] groups have used the (unmodified) model of Altarelli et al.[97] to obtain limits on $(b \to u)/(b \to c)$. Last year,[98] with more data, the CLEO group found that the unmodified model of Altarelli et al. did not adequately represent their data and pointed out the need for additional calculations of the expected lepton spectra. A number of different models now exist and have been compared with data as described below.

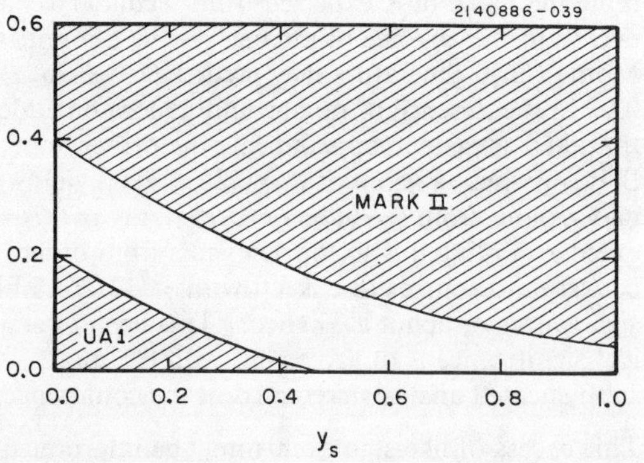

Figure 17. The lower (upper) limits on the combination of B_d and B_s mixing obtained by the UA1(MARKKII) group. The allowed region is not shaded. The regions shown are for $f_s = 0.2$ and $f_d = 0.4$ (see text).

New data are available from the CLEO,[99] ARGUS[100] and Crystal Ball[101] groups as shown in Table 14.

The integrated luminosities shown are at the $\Upsilon(4S)$. The number of leptons is after continuum subtraction. The ARGUS group has substantially better electron acceptance than CLEO. Since the data sample from the Crystal Ball group is much smaller than from CLEO and ARGUS, their limits on $b \to u$ are not as good as from CLEO or ARGUS and I will not discuss them further.

The CLEO group[99] has made an extensive comparison between their lepton spectra and model predictions. They use three different methods to extract limits on $b \to u$; (1) fits to model predictions over the entire accessible momentum range; (2) fits in the momentum range $2.2 < p < 2.6$ GeV/c where the $b \to c$ contribution is small; and (3) limits in the range $2.4 < p < 2.6$ GeV/c where the $b \to c$ contribution is essentially zero. Method (3) is completely insensitive to $b \to c$ models but very sensitive to the endpoint spectrum of the $b \to u$ model. Method (2) is slightly sensitive to the $b \to c$ model, but increasing the momentum range somewhat improves the statistical power of the limit. Method (1) is sensitive to both $b \to c$ and $b \to u$ models but uses the complete data sample and can be used to extract values for the B semileptonic branching ratio.

The models available for $b \to c$ are

Simple quark[102]

The parameters used were $m_b = 5.278$ GeV/c^2, $m_c = 2.05$ GeV/c^2 and $m_u = 1.0$ GeV/c^2

Altarelli et al.[97]

Includes bound state effects via a Fermi momentum (of 0.215 GeV/c in CLEO analysis — best fit to their data). CLEO assumes a c quark mass of 1.7 GeV/c^2 and a u quark mass of 0.15 GeV/c^2

Tye and Trahern[103]

Explicit decay into D* and D final states. CLEO takes $(B \to D^* l \nu)/(B \to D l \nu) = 0.16$

Table 14. Data sample for B semileptonic decay.

	Luminosity (pb^{-1})	Number of electrons (p>1.1 GeV)	Number of muons (p>1.2 GeV)
CLEO	79(e)	4675	
	120(μ)		6775
ARGUS	59	4950	not yet available
Crystal Ball	54	1575	----

Ali and Yang[104]
CLEO assumes $(B \rightarrow D^*l\nu)/(B \rightarrow Dl\nu) = 1.0$

Grinstein et al.[105]
Total rate is almost saturated by D* (71% of total) and D (19%) and is only slightly softer than the free quark spectrum. Parameters used by CLEO as in ref. 105.

Angelini et al.[106]
Parameters not modified in CLEO analysis

To fit the observed lepton spectra one must add to these models leptons from the decay of D mesons produced in B decay. The shape of the lepton spectrum from D decay is taken from the measured combination of $D \rightarrow Kl\nu$, $K^*l\nu$, $\pi l\nu$ and $\rho l\nu$. The number of leptons from this source, related to the average D semileptonic branching ratio, is a free parameter in the CLEO analysis. To determine the B semileptonic branching ratio, CLEO has fit their spectra using these six models, assuming $b \rightarrow u$ is negligible. The results are summarized in Table 15.

The value of the reduced $\chi^2(\chi_r^2)$ is very good for all except the Ali model. From this analysis, the CLEO group concludes that the B semileptonic branching ratio is $B(B \rightarrow l\nu X) = 0.11 \pm 0.01$, taking into account the model dependence (which dominates the error) and statistical errors. This conclusion is not changed if a small amount of $b \rightarrow u$ is allowed, consistent with the limits described below. This value for $B(B \rightarrow l\nu X)$ is consistent with previous measurements[12] and, in my opinion, represents the most accurate estimate rather than a weighted average of all measurements.

CLEO has chosen to use three models of $b \rightarrow u$ (simple quark, Altarelli and Grinstein) to extract limits. Using method (1), fitting over the accessible momentum range, yields the 90% confidence level limits given in Table 16. The limits range from 0.011 to 0.066. Some care must be used to interpret the results using the model of Grinstein et al. In their paper they predict a rate for $b \rightarrow u$ by summing over relatively low mass hadronic states which are important near the endpoint of the lepton spectrum. Their rate for $b \rightarrow u$ $(0.57 \times 10^{14} |V_{bu}|^2 \text{ sec}^{-1})$ is thus a fraction (0.48) of the total $b \rightarrow u$ rate expected in the free quark model $(1.18 \times 10^{14} |V_{bu}|^2 \text{ sec}^{-1})$. Since the fitting procedure used by CLEO (and other experiments) to determine $(b \rightarrow u)/(b \rightarrow c)$ is most sensitive to the spectrum near the endpoint, the value obtained for $(b \rightarrow u)/(b \rightarrow c)$ will likely be an underestimate of the true value, possibly by as much as $1/0.48 = 2.07$. In their analysis, the CLEO group obtained upper limits to $(b \rightarrow u)/(b \rightarrow c)$ using the spectral shape as given by Grinstein et al. If the rate calculations of Grinstein et al. are accurate, the remainder of the $b \rightarrow u$ process could produce higher mass hadronic states to which the fitting procedure would be insensitive. However, if one assumes that the predictions of Grinstein et al. accurately represent the shape of the *complete* $b \rightarrow u$ spectrum, then one obtains the limits given in Table 16.

Table 15. B semileptonic branching ratio for different models.

Model	$B(B \rightarrow l\nu X)$	$B(B \rightarrow D \rightarrow l\nu X)$	χ_r^2
Simple quark	0.107	0.099	0.58
Tye	0.112	0.089	0.53
Ali	0.126	0.059	2.50
Altarelli	0.106	0.101	0.74
Grinstein	0.108	0.097	0.74
Angelini	0.116	0.082	0.65

Table 16. Limits at 90% confidence level on $(b \rightarrow u)/(b \rightarrow c)$ for different $b \rightarrow c$ and $b \rightarrow u$ models. The entries in the last line ("None") are the limits independent of a $b \rightarrow c$ model. See the text for more explanation.

	$b \rightarrow u$ model		
$b \rightarrow c$ model	Simple Quark	Altarelli	Grinstein
Simple quark	0.029	0.028	0.038
Tye	0.015	0.015	0.020
Ali	0.049	0.047	0.066
Preparata	0.011	0.011	0.014
Altarelli	0.028	0.027	0.036
Grinstein	0.032	0.031	0.042
None ($2.4<p<2.6$)	0.038	0.035	0.054

Using method (2), restricting the momentum interval to 2.2 – 2.6 GeV/c, reduces the dependence on the $b \rightarrow c$ model somewhat as shown in Table 17. For this method the limits range from 0.023 to 0.060. The limit on $(b \rightarrow u)/(b \rightarrow c)$ may be calculated from

Table 17. Limits at 90% confidence level on $(b \rightarrow u)/(b \rightarrow c)$ for fits in the momentum range 2.2–2.6 GeV/c for different models.

		$b \rightarrow u$ model		
$b \rightarrow c$ model	$f_c(2.2–2.6)$	Simple Quark	Altarelli	Grinstein
Simple quark	7.9×10^{-3}	0.034	0.033	0.044
Tye	9.8×10^{-3}	0.024	0.023	0.031
Ali	5.8×10^{-3}	0.046	0.044	0.060
Preparata	10.8×10^{-3}	0.020	0.019	0.025
Altarelli	8.6×10^{-3}	0.030	0.029	0.039
Grinstein	7.8×10^{-3}	0.034	0.033	0.045

$$\frac{b \to u}{b \to c} = \frac{B(B \to l(2.2-2.6)\nu X) - f_c(2.2-2.6)B(B \to l\nu X)}{f_u(2.2-2.6)B(B \to l\nu X) - B(B \to l(2.2-2.6)\nu X)}$$

where $B(B \to l(2.2\text{-}2.6)\nu X)$ is the semileptonic branching ratio for lepton momenta between 2.2 and 2.6 GeV/c[$(12.16 \pm 2.12) \times 10^{-4}$], $f_c(2.2-2.6)$ is the fraction of the $b \to c$ spectrum with momenta between 2.2 and 2.6 GeV/c and $B(B \to l\nu X)$ is the total semileptonic branching ratio, 0.11. The values of $f_c(2.2-2.6)$ are given in Table 17. Again the results for the model of Grinstein et al. assume their calculated spectrum represents all of the $b \to u$ rate.

Restricting the momentum range to above 2.4 GeV removes the dependence on the model for $b \to c$, leaving only the dependence on the model for $b \to u$. CLEO determines the limit on the branching ratio for $B \to l\nu X$, for lepton momentum between 2.4 GeV/c and 2.6 GeV/c to be $B\ (B \to l(2.4-2.6)\nu X) \leqslant 2.2 \times 10^{-4}$ at 90% confidence level. From this number one may compute a limit on $(b \to u)/(b \to c)$ from

$$\frac{b \to u}{b \to c} = \frac{B(B \to l(2.4-2.6)\nu X)}{f_u(2.4-2.6)B(B \to l\nu X) - B(B \to l(2.4-2.6)\nu X)}$$

where $f_u(2.4-2.6)$ is the fraction of the $b \to u$ lepton spectrum between 2.4 and 2.6 GeV. For the quark model, Altarelli model and Grinstein et al., $f_u(2.4-2.6)$ is 0.055, 0.059 and 0.039, respectively. The $(b \to u)/(b \to c)$ limits are given in Table 16.

What values for $|V_{bu}/V_{bc}|$ may be obtained from these limits? For methods (1) and (2) the poorest limit comes from using the Ali model for $b \to c$ and the Grinstein et al. model for $b \to u$. The Ali model, however, is a significantly poorer fit to the data than all other models; I will therefore exclude it from consideration. The largest limit then comes from the model of Grinstein et al. Using this model we can extract limits on $|V_{bu}/V_{bc}|$ under at least three different assumptions about the absolute rate for $b \to u$; (1) as calculated by Grinstein et al.; (2) or as given by the free quark model but with the spectral shape of Grinstein et al.; or (3) as one-half the rate calculated by Grinstein et al., according to their estimate of a 50% uncertainty in their calculation. Under these three assumptions, limits on $|V_{bu}/V_{bc}|$ may be obtained using method (2), $2.2 < p < 2.6$ GeV/c, and method (3), $2.4 < p < 2.6$ GeV/c. The results are given in Table 18.

Table 18. Limits on $|V_{bu}/V_{bc}|$ for various assumptions about the $b \to u$ rate.

$b \to u$ rate from Grinstein et al.	$\lvert V_{bu}/V_{bc}\rvert$ 90% CL limits $2.2 < p < 2.6$	$2.4 < p < 2.6$ GeV/c
as free quark	0.148	0.162
as calculated	0.213	0.234
1/2 as calculated	0.303	0.332

For comparison the model of Altarelli et al. yields $|V_{bu}/V_{bc}| < 0.127$ at 90% confidence level. Given all these numbers what should be quoted as "the limit" on $|V_{bu}/V_{bc}|$? Given the large uncertainty in calculations of the lepton spectra, I would quote

$$|V_{bu}/V_{bc}| < 0.30 \text{ at 90\% confidence level}$$

with a warning that model dependence is the dominant factor in the reliability of the limit.

The ARGUS group has so far only analyzed the electron spectrum from B decay. Their spectrum is shown in Fig. 18. They have extracted upper limits by using a spectator model, with variable parameters, and by using Grinstein et al. They fit in the range $p > 1.6$ GeV/c to eliminate most of the contribution from secondary D decays (this is different than the CLEO analysis). Varying the parameters of the spectator model, keeping m_b-m_c, $m_{spectator}$ and $m_u \approx$ fixed, they compute a likelihood for each parameter set and value for $(b \to u)/(b \to c)$. Their results actually favor a finite value of $(b \to u)/(b \to c)$ of about 0.06 but the result is not statistically significant ($\approx 2\sigma$) and they set a 90% confidence limit of 0.12. Using the model of Grinstein et al. as given yields a limit of 0.06 at 90% confidence level. The preliminary ARGUS results therefore agree with the more extensive analysis by the CLEO group and have the promise of providing the best limits (or value) for $b \to u$ in the near future with the addition of muon spectra and data to be collected this year.

Another way to look for $b \to u$ is via exclusive decays to states without charm (or strangeness). Last year CLEO[107] presented upper limits on a number of decay modes and ARGUS[108] has presented preliminary data on some of the same modes. Some of the limits are summarized below. There is no evidence for exclusive decays of B mesons resulting from the $b \to u$ transition.

Mode	90% confidence level limits in % CLEO	ARGUS
$\overline{B}^0 \to \pi^+\pi^-$	0.02	0.03
$B^- \to \rho\pi^-$	0.02	0.05

In the model of Stech et al.[84] the branching ratio (in %) for $\overline{B}^0 \to \pi^+\pi^-$ is predicted to be $0.17a_1^2|V_{ub}/0.05|^2$ where $a_1(\approx 1.1)$ may be calculated in his model. Stech also predicts that the branching for $\overline{B}^0 \to D^{*+}\pi^-$ is $0.4a_1^2|V_{cb}/0.05|^2$. The weighted average of the CLEO and ARGUS values for the $\overline{B}^0 \to D^{*+}\pi^-$ branching ratio is (0.31 ± 0.14)%. Given the limit of 0.02% for the $\pi^+\pi^-$ mode, this leads to the limit $|V_{ub}/V_{cb}| \leqslant 0.4$–$0.5$. This limit is, of course, also model dependent.

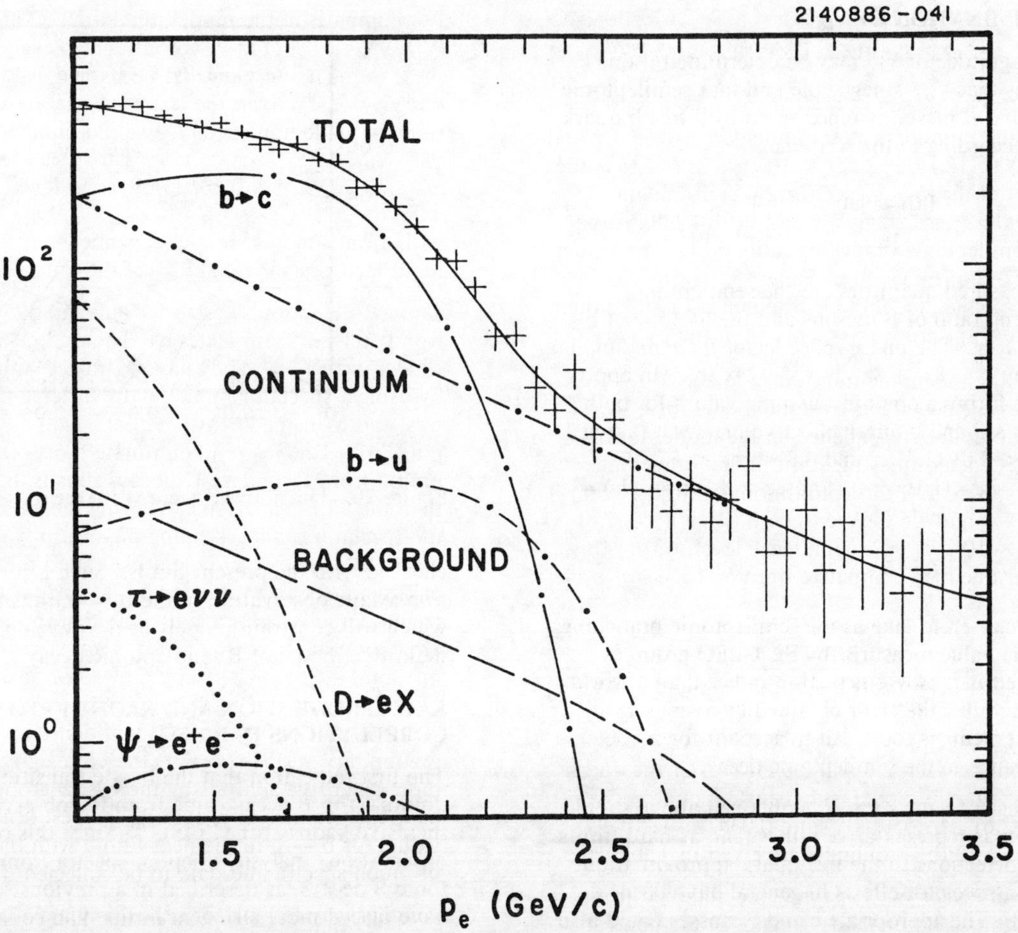

Figure 18. The electron spectrum at the ϒ(4S) observed by ARGUS. The contributions to the spectrum are as indicated.

B HADRON LIFETIME

Measurements of the b-hadron lifetime at PEP and PETRA are summarized in Fig. 19. Two new precise measurements from MAC and DELCO have been made this year. In computing an average value I have used the most recent values from each experiment as indicated by the * in Fig. 19. In some cases earlier measurements are not statistically independent of later results. Also the result from TASSO is an approximate average of the three methods they have used to determine the lifetime. The errors on the average value are now too good in the sense that differences of order 20% in lifetimes among B^0, B^+ mesons and b-baryons, which are certainly possible, become significant. In fact if charm baryon production is as large as suggested by the MARKIII branching ratios there would likely be a similarly large rate for b-baryons. If so then the present measurements of the b-hadron lifetime may have a substantial baryon component, which could be much shorter than the B meson lifetime as in the charm case. The very difficult experimental challenge in the future will be to measure separately the B^0, B^+ and b-baryon lifetimes; this is likely to take many years.

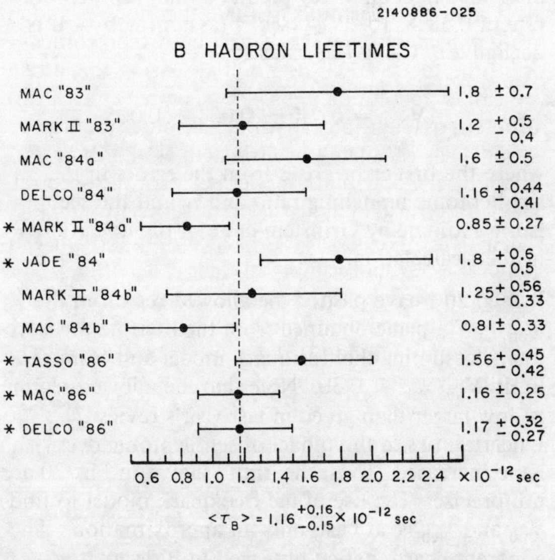

Figure 19. A summary of b — hadron lifetime measurements.

DETERMINATION OF V_{bc}

The magnitude of V_{bc} may be determined if one assumes that V_{bu} is negligible and that semileptonic decay of a B meson is represented by a free b quark decay according to the formula

$$\Gamma(b \rightarrow cl\nu) = \frac{B(b \rightarrow cl\nu)}{\tau_b} = \frac{G_F^2 m_b^5}{192\pi^3} f(m_c, m_b) |V_{bc}|^2$$

The measured quantities are the semileptonic branching ratio of B mesons and the lifetime of b hadrons, τ_b. The phase space factor $f(x = m_c/m_b)$ is given by $1 - 8x^2 + 8x^6 - x^8 - 24x^4 \ln x$. In applying this formula one must assume values for both the c quark and b quark effective masses. Taking $1.7 < m_c < 2.01$ GeV/c^2 and $4.8 < m_b < 5.279$ GeV/c^2, $\tau_b = (1.16 \pm 0.16)$ psec and $B(b \rightarrow cl\nu) = 0.11 \pm 0.01$ yields

$$0.035 < |V_{bc}| < 0.069$$

I have chosen to take as the semileptonic branching ratio the value measured by the CLEO group as described in a previous section rather than a world average value; the error obtained by averaging all measurements is too small to account for model dependence in the semileptonic decay.

The range of values for $|V_{bc}|$ obtained above should be taken as representative rather than as hard limits. QCD corrections to the free quark approximation and non-spectator effects in general have been ignored. The appropriate c and b masses could also be different than what I have assumed.[98]

Alternatively one may calculate $|V_{bc}|$ using the model of Grinstein et al. They predict a total semileptonic rate of $0.58 \times 10^{14} |V_{bc}|^2 \text{ sec}^{-1}$ (assuming $b \rightarrow u$ is negligible). Using this leads to

$$|V_{bc}| = 0.040 \pm 0.004 \pm 0.005$$

where the first error arises from the errors in the semileptonic branching ratio and τ_b and the second is the estimate by Grinstein et al. of the uncertainty in their calculation.

In Fig. 20 I have plotted the allowed region in the $V_{ub} - V_{cb}$ plane obtained from the lifetime measurements assuming the free quark model and from the limit $|V_{ub}/V_{cb}| < 0.30$. Note that the allowed region is now larger than given in last year's review.[98] Uncertainties in the model of semileptonic decay have increased. Even the limits shown in Fig. 20 are not precise. The use of the free quark model to find V_{ub} and V_{cb} is, at best, only an approximation. The apparent charm deficit observed in B decay (see below) suggests that all models of semileptonic B decay are incorrect. With all of this uncertainty, the consistency of the K-M parameterization to describe B decay and CP violation in K decay[98] cannot be checked with the present limits. Such a confrontation awaits observation of the $b \rightarrow u$ transition which will constrain models or the development of a definitive model of B semileptonic decay.

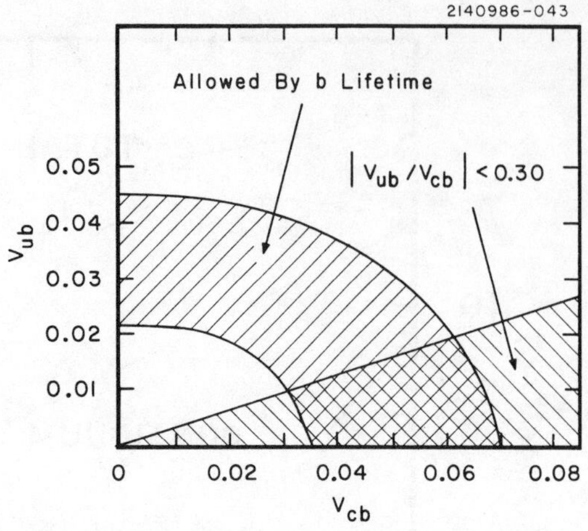

Figure 20. The allowed regions in the $V_{cb} - V_{ub}$ plane.

KAON PRODUCTION AND KAON-LEPTON CORRELATIONS IN B DECAY

The first indication that the $b \rightarrow c$ transition was dominant in B decay came from the observation of inclusive kaons at the $\Upsilon(4S)$.[110] Since this original observation, analysis of lepton spectra from semileptonic B decays, as described in a previous section, have placed more stringent limits than could be obtained via measurements of inclusive kaons. However, as also noted above, the model dependence of the $b \rightarrow u$ limit may be large. Also in the next section we will see that there may be evidence that $b \rightarrow c$ is not dominant, questioning the validity of the models for $b \rightarrow u$ semileptonic decay or measurements of D^0 and D^+ branching ratios. This puzzle has caused the CLEO group to reexamine kaon yields in B decay to extract a value for $(b \rightarrow c)/(b \rightarrow \text{all})$ which does not depend on D branching ratios or models of semileptonic decay.[111]

The analysis procedure relies on correlations in angle and in charge (for $K^{\pm}$) between leptons and kaons. For the $b \rightarrow c \rightarrow s$ transition chain, there will be a correlation in angle and charge between a lepton from the b semileptonic decay and the kaon produced in the $c \rightarrow s$ transition. For a $b \rightarrow u$ transition there will not be such a correlation. For the details of the analysis procedure refer to ref. 111. The results of the analysis are shown in Table 19 and compared with the Monte Carlo expectations (in parenthesis) assuming $(b \rightarrow c)/(b \rightarrow \text{all}) = 1$. There is very good agreement between the measurements and the predictions of the Monte Carlo for $(b \rightarrow c)/(b \rightarrow \text{all}) = 1$. From these numbers CLEO derives $(b \rightarrow c)/(b \rightarrow \text{all}) = 1.03 \pm 0.20$. The result does not depend on D exclusive branching ratios.

Table 19. Kaon and kaon-lepton yields.

$$\frac{\bar{B} \to l^- K^- X}{\bar{B} \to l^- X} = 0.59 \pm 0.12\ (0.50)$$

$$\frac{\bar{B} \to l^- K^+ X}{\bar{B} \to l^- X} = 0.09 \pm 0.06\ (0.03)$$

$$\frac{\bar{B} \to l^- K^0 X}{\bar{B} \to l^- X} = 0.41 \pm 0.07\ (0.51)$$

$$\frac{\bar{B} \to K^0 X}{\bar{B} \to \text{all}} = 0.65 \pm 0.10\ (0.71)$$

$$\frac{\bar{B} \to K^+ X}{\bar{B} \to \text{all}} = 0.19 \pm 0.06\ (0.13)$$

$$\frac{\bar{B} \to K^- X}{\bar{B} \to \text{all}} = 0.69 \pm 0.12\ (0.62)$$

INCLUSIVE B MESON DECAYS TO CHARM

The most controversial issue in B decays is the evidence presented by the CLEO group[112] of a deficit in charm production in B meson decay. As noted in the previous sections there is no evidence for the b → u transition and (model dependent) limits are about 10%. The dominant decay of B mesons therefore should be into charm particles. Ignoring any b → u component, the number of charm quarks, in mesons (including $c\bar{c}$ states) plus baryons, is expected to be about 1.15 per B meson decay. One charm quark is expected from the b → c transition and 0.15 from a virtual W decay to $c\bar{s}$. One would also expect most of the charm quarks to finally yield D^0 or D^+ mesons.

The experimental results may be divided into five pieces; limits on charmed baryon production; observation of D_s production in B decay; $c\bar{c}$ yields; D^0 yields; and D^+ yields. Charmed baryons have not yet been directly observed in B decay. Limits on the their production may be set by measurements of inclusive p and Λ production in B decay. Both CLEO[113] and ARGUS[114] have measured these inclusive rates. If one assumes that neutron + antineutron production is the same as proton + antiproton production, and attributes *all* of the baryon production to the decay of charmed baryons, then the branching ratio for B to charmed baryons could be about 10% as shown below

	CLEO	ARGUS
B → charmed baryons	≈ 0.09	0.09 ± 0.045

Since some baryon production is expected in the hadronization process in B decay, this is likely to be an overestimate. To be conservative, however, I fix the charmed baryon rate in B decay at 0.10 ± 0.05.

D_s production has now been observed by both CLEO[115] and ARGUS.[116] Both groups observe D_s through the $\Phi\pi$ mode. D_s production is expected to occur primarily through the decay chain $b \to W^- c$, $W^- \to \bar{c}s \to D_s$ or D_s^*. It is also possible to produce D_s through the production of an $s\bar{s}$ pair in the hadronization process of $b \to W^- c$, $W^- \to \bar{u}d$. Suzuki[117] for example predicts $B(B \to D_s + X) = (9 \pm 2)\%$ for the total of D_s production.

Since $B(D_s \to \Phi\pi)$ has not been directly measured, one must infer its value from other measurements. This has been done by both ARGUS[116] and CLEO[118] from measurements of D_s production in continuum e^+e^- annihilation events, assuming a value of 15% for the probability of popping an $s\bar{s}$ pair. Using this method ARGUS finds $B(D_s \to \Phi\pi) = (3.5 \pm 0.8 \pm 1.2)\%$ and CLEO is consistent with this number. Alternatively[115] one may use a theoretical value for the $D_s \to \Phi\pi$ width[119] $(11.24 \times 10^{10}\ \text{sec}^{-1})$ and the measured D_s lifetime $(3.5 \times 10^{-13}\ \text{sec}^{-13})$ to obtain $B(D_s \to \Phi\pi) \approx 3.9\%$.

For $B(B \to D_s + X)\, B(D_s \to \Phi\pi)$, CLEO finds 0.0038 ± 0.0010 and ARGUS 0.0024 ± 0.00092. *Assuming* $B(D_s \to \Phi\pi) = (3.5 \pm 1.5)\%$, this yields $B(B \to D_s + X)$ of 0.11 ± 0.06 and 0.07 ± 0.04, respectively.

A small fraction of the c (and $\bar{c}$) quarks from B decay will result in $c\bar{c}$ bound states. Following Kuhn et al.[120] I will assume that the relative contribution from $\eta_c{:}\Psi{:}\chi_1$ is 0.6:1.0:0.3 and that other $c\bar{c}$ states except Ψ', which is measured, are negligible. From the ARGUS measurement of Ψ' production in B decay, one may infer a value for direct Ψ production of $(0.84 \pm 0.24)\%$, ignoring the very small contribution from the decay of the χ_1. Hence the number of charm quarks from this source is $2(0.0050 + 0.0084 + 0.0025 + 0.0050) = 0.04 \pm 0.01$.

Most of the c quarks produced in B decay should finally yield D^0 or D^+ mesons. Both the CLEO and ARGUS groups have observed a large excess of $D^0 \to K^-\pi^+$ over continuum e^+e^- production at the $\Upsilon(4S)$. As an example the CLEO results are shown in Fig. 21. A similar excess of $D^+ \to K^-\pi^+\pi^+$ events is also seen by both groups. Subtracting the continuum contribution, the momentum spectrum of D mesons produced in B decay is obtained and is shown in Fig. 22. The CLEO group has also measured the spectrum of D^{*+} from B decay as shown in Fig. 22. The shape of the spectrum is somewhat softer than the previous measurements by CLEO[121] and is consistent with a spectator model wherein quarks from the virtual W in b→cW hadronize independently from the c and spectator quark (the solid line in Fig. 22). CLEO also finds that direct D production is small; most of the D's come from decay of D*'s. Making plausible assumptions, CLEO finds

$$\frac{D_{\text{direct}}}{D_{\text{direct}} + D^*} = 0.28 \pm 0.14^{+0.18}_{-0.11}$$

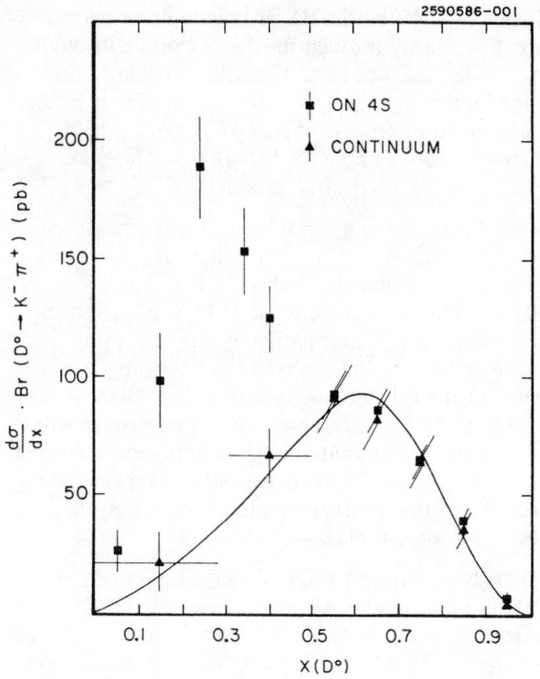

Figure 21. The x distribution ($x = p/E_{beam}$) of D^0 mesons observed at the $\Upsilon(4S)$ and nearby continuum by the CLEO collaboration. The large excess below x = 0.5 is from B meson decay.

in all B decays. The CLEO data sample is too small to derive the equivalent number for semileptonic B decays.

The number of D^0 and D^+ mesons produced in B decay may be extracted from the measured spectra using the MARKIII branching ratios for $D^0 \to K^-\pi^+$ and $D^+ \to K^-\pi^+\pi^+$. These numbers together with the other sources of charm described above are summarized in Table 20. One expects the number of c + $\bar{c}$ quarks produced in B decay to be about 1.15 if b → u is a negligible contribution. The CLEO result is clearly well below this number and the result from the ARGUS group is also low.

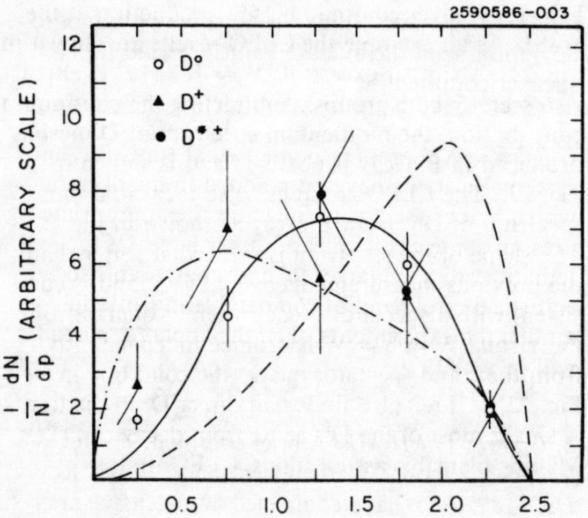

Figure 22. The D^0, D^+ and D^{*+} momenta spectra from B meson decay observed by CLEO.

There are many possible interpretations of this apparent charm deficit in B decay. If the b → u strength is at the limit, ≈ 0.12, obtained from analysis of semileptonic decay as described in a previous section, the charm measurement of ARGUS would be in agreement with the new expectations, ≈ 1.03, but the CLEO measurement would not be in agreement. Another explanation is that the MARKIII D branching ratios are too large. As noted in the section on charm decay above, measurements of D^0 and D^+ yields in continuum e^+e^- annihilation suggest but do not prove that the D^0 and D^+ branching ratios should be lower by about 50%. If this were true then the CLEO measurements would be in good agreement with 1.15 charm quarks per B decay and the ARGUS measurements somewhat higher.

Explanations which do not require experimental mismeasurement do require a substantial b → u coupling, larger than allowed by existing models of semileptonic decay, and/or $\Upsilon(4S)$ decays into non-$B\bar{B}$ final states (see the discussion in a previous section for the experimental limits to this possibility), and/or substantial contributions from annihilation graphs and/or penguin-type graphs. The last seems unlikely since a factor of ten enhancement, which would be required, is probably excluded by limits on exclusive B decays resulting from penguin graphs. The magnitude of an annihilation contribution also depends on the b → u coupling and other parameters such as f_B, the B decay constant. A combination of a relatively large b → u coupling and annihilation could account for the charm deficit. Decays of the $\Upsilon(4S)$ to non-$B\bar{B}$ final states by itself would not account for all of the deficit but could account for it in conjunction with one or more of the other possibilities.

In order to exclude the annihilation possibility or the possibility that $\Upsilon(4S)$ does not decay exclusively to

Table 20. Measurements of charmed particle production in B decay — see the the text for additional explanation.

	CLEO	ARGUS
D^0	$0.39 \pm 0.05 \pm 0.04$	$0.50 \pm 0.07 \pm 0.08$
D^+	$0.17 \pm 0.04 \pm 0.04$	$0.23 \pm 0.07 \pm 0.05$
D_s	0.11 ± 0.06	0.07 ± 0.04
$c\bar{c}$	0.04 ± 0.01	0.04 ± 0.01
Λ_c	0.1 ± 0.05	0.1 ± 0.05
Sum	0.81 ± 0.12	0.94 ± 0.15
Average	0.86 ± 0.09	
Expected	1.15	

$B\bar{B}$, CLEO has attempted to measure D production in semileptonic B decay. Prompt leptons from B decay arise from spectator diagrams and the annihilation contribution is negligible. The hard lepton positively tags a B decay. CLEO finds 0.35 ± 0.09 D^0 per B semileptonic decay. Unfortunately they do not have enough events to measure the D^+ rate in semileptonic decay. Since the D^0 rate is about the same as in all B decay, the simplest assumption is that the D^+ rate in semileptonic decay is the same as in all B decays.[122] Making this assumption yields a $D^0 + D^+$ rate of about 0.5 per B semileptonic decay. The contribution from D_s and charmed baryons should be small <0.1, so one expects 0.9–1.0 for the $D^0 + D^+$ rate. The absence of charm in semileptonic B decay rules out the annihilation graph or decays of the $\Upsilon(4S)$ to non-$B\bar{B}$ final states as the sole cause of the charm deficit in all B decays.

What are the possible conclusions from these measurements? CLEO is wrong and/or MARKIII is wrong and/or there is a large *and* anomalous $b \rightarrow u$ coupling, one not predicted by current models of semileptonic decay. This situation can only be clarified by additional measurements; the ARGUS group should have, soon, better measurements of D^0 and D^+ production in all and semileptonic B decays.

CONCLUSIONS

A considerable amount of new experimental data on weak decays has appeared within the last year. Although there are a number of interesting puzzles and discrepancies among experiments which I will briefly summarize below, there are no convincing conflicts with the expectations from the standard electroweak model.

Less than two years ago it appeared that measurements of Kobayashi-Maskawa angles might become so precise as to provide a serious test of the consistency of the Standard Model. This expectation has not been realized, in part because of the absence of new experimental results and also because of uncertainties in strong interaction effects that obscure precise measurements of weak parameters. In particular the present limit $|V_{ub}/V_{cb}| < 0.30$ is larger than previous limits because of these uncertainties. In the very likely absence of a definitive model of B meson semileptonic decay, substantial increases in experimental data on B decays will be necessary to improve limits on V_{ub}, or better yet find some evidence that $V_{ub} \neq 0$. It is also of crucial importance to improve the measurements of ϵ'/ϵ.

Although there is no obvious evidence that the Standard Model is incorrect in describing weak decays, there are a number of interesting puzzles to be solved. These are

(1) what are τ branching ratios to one-charged-prong + multi-neutral final states?

(2) are the MARKIII branching ratios correct and/or does the $\Psi(3770)$ decay into non-$D\bar{D}$ final states?

(3) if the MARKIII D branching ratios are correct, why is charmed baryon production in e^+e^- annihilation so large, $\approx 40\%$ of the total rate?

(4) is there really a charm deficit in B meson decay? If there is such a deficit, *all* models of B semileptonic decay are inadequate and the extraction of a limit on V_{ub} from lepton spectra in semileptonic decays must be completely revaluated.

(5) is the result from UA1 on like-sign dileptons the first indication of mixing in B mesons? If so, is it B_s mixing (as expected) or B_d mixing (much more interesting)?

Hopefully within the next year there will be answers to some or all of these questions.

QUESTIONS:

M. Ronan (Lawrence Berkeley Laboratory) You mentioned the need for measurements of $\tau^- \rightarrow \pi^- +$ several neutrals to help resolve the discrepancy in τ branching ratios. The TPC/Two Gamma collaboration has submitted to this conference a measurement of the inclusive branching ratio $\tau \rightarrow \pi^-\pi^0 + nh^0$, $n \geq 1$, where $h^0 = (\pi^0, \eta)$, of $(13.9 \pm 2.0^{+1.9}_{-2.1})\%$.

Gilchriese This is included in Table 1.

G. Kalmus (Rutherford Laboratory) Of the 7τ, 100^0, $8D^+$, $7D_s$ and $3\Lambda_c$ lifetime measurements presented by the speaker, only one was outside one standard deviation from the average value. Would the speaker comment.

Gilchriese A serious response is that statistical and systematic errors have been added in quadrature which makes the errors shown in the appropriate figures, in some cases at least, rather large. If the systematic errors are larger than or comparable to the statistical errors and of comparable magnitude among the experiments, then the apparent spread will be small.

S. Godfrey (TRIUMF) A brief comment on the $\tau \rightarrow \omega\pi\nu_\tau$ decay mode. Nathan Isgur and I (Phys. Rev. *D32*, 189(1985) have completed an extensive analysis of meson properties and we found that the width of the radial excitation of the ρ is fairly large, ≥ 300 MeV. Therefore, it is not likely that this state would appear as a resonance.

FOOTNOTES AND REFERENCES

1. R. Bernstein et al., Phys. Rev. Lett. *54*, 1631(1985).
2. J. K. Black et al., Phys. Rev. Lett. *54*, 1628(1985).
3. Y. Wah, presentation to this conference.
4. M. Holder, presentation to this conference; Also see D. Cundy et al., CERN/SPSC/81-110, SPSC/P174.
5. J. Donoghue et al., contribution to this conference and University of Massachusetts Preprint UMHEP-262(1986).
6. E621 Collaboration, contribution to this conference.
7. S. Egli et al., Phys. Lett. *B175*, 97(1986); Also Ch. Grab, contribution to this conference.
8. C. James et al. (E619), contribution to this conference.
9. S. F. Biagi et al., Z. Phys. *C28*, 495(1985); S. F. Biagi et al., Z. Phys. *C30*, 210(1985).
10. G. Zapalac et al., FERMILAB-Pub-86/96-E.
11. R. Winston, contribution to this conference.
12. Compilation of the Particle Data Group, Phys. Lett. *B170*, (1986).
13. K. K. Gan, in the Proceedings of the 1985 DPF Meeting, Eugene, Oregon.
14. F. Gilman and S. H. Rhie, Phys. Rev. *D31*, 1066(1985); F. Gilman, SLAC-PUB-3951(1986).
15. I have used the straight average values given in reference 12 rather than the fit values because the errors are somewhat larger. Even so it is likely that the errors are still too small and do not accurately reflect the experimental uncertainties.
16. P. Burchat (MARKII), contribution to these Proceedings and SLAC-Report-292(1986).
17. H. Aihara et al. (TPC), contribution to this conference.
18. W. Bartel et al. (JADE), contribution to this conference.
19. W. W. Ash et al. (MAC), contribution to this conference.
20. J. J. Becker et al. (MARKIII), contribution to this conference.
21. J. Dorfan et al., Phys. Rev. Lett. *46*, 215(1981).
22. H. Aihara et al. (TPC), contribution to this conference.
23. CELLO Collaboration, contribution to this conference.
24. H. J. Behrend et al., Z. Phys. *C23*, 103(1984).
25. E. Fernandez et al., Phys. Rev. Lett. *54*, 1624(1985).
26. W. Ruckstuhl et al., Phys. Rev. Lett. *56*, 2132(1986).
27. W. B. Schmidke et al., Phys. Rev. Lett. *57*, 527(1986).
28. H. Albrecht et al., (ARGUS Collaboration), "Measurement of Tau Decays into Three Charged Pions," contribution to this conference.
29. HRS Collaboration, "Observation of the Decays $\tau \rightarrow \eta X$, $\tau \rightarrow \omega X$," contribution to this conference.
30. Crystal Ball Collaboration, private communication.
31. P. R. Burchat et al., Phys. Rev. Lett. *54*, 2489(1985).
32. W. Bartel et al., Phys. Lett. *B161*, 188(1985).
33. HRS Collaboration, "Limit on τ Decay to 7-Charged Tracks," contribution to this conference.
34. CLEO Collaboration, private communication.
35. H. Albrecht et al., (ARGUS Collaboration), contribution to this conference.
36. See the comment in the question section at the end of this article.
37. Crystal Ball Collaboration, "Search for Exotic τ Decays," contribution to this conference.
38. τ lifetimes — statistical and systematic errors have been added in quadrature:
MARKII '83: J. Jaros et al., Phys. Rev. Lett. *51*, 955(1983)
TASSO '84: M. Althof et al., Phys. Lett. *B141*, 264(1984)
MAC '84: E. Fernandez et al., Phys. Rev. Lett. *54*, 1624(1985)
MARKII '84; J. Jaros, SLAC-PUB-3569(1984)
MAC '86: contribution to this conference
CLEO '86: "Measurements of Heavy Quark and Lepton Lifetimes at CLEO," contribution to this conference
HRS '86: contribution to this conference.
39. E691 Collaboration, contribution to this conference.
40. D^0 lifetimes — statistical and systematic errors have been added in quadrature:
PDG '86: reference 12
WA58: contribution to this conference
E531: N. Ushida et al., Phys. Rev. Lett. *56*, 1771(1986)

TASSO: M. Althoff et al., DESY Preprint 86-027
HRS: contribution to this conference
CLEO: see ref. 38
E691: reference 39
NA1: contribution to this conference
LEBC: contribution to this conference
ACCMOR: contribution to this conference.

41. D^+ lifetimes — statistical and systematic errors have been added in quadrature:
PDG '86: reference 12
E531: N. Ushida et al., Phys. Rev. Lett. *56*, 1767(1986)
MARKII: L. Gladney et al., SLAC-PUB-3947(1986)
HRS: contribution to this conference
CLEO: see ref. 38
E691: see ref. 39
LEBC: see ref. 40
ACCMOR: see ref. 40.

42. D_s lifetimes — statistical and systematic errors have been added in quadrature:
E531: see ref. 41
NA11: contribution to this conference
HRS: C. Jung et al., Phys. Rev. Lett. *56*, 1775(1986)
TASSO: contribution to this conference
CLEO: see ref. 40
E691: see ref.39
ACCMOR: see ref. 40.

43. Λ_c lifetimes — statistical and systematic errors have been added in quadrature:
WA58: contribution to this conference.
E531: see ref. 41
ACCMOR: see ref. 40
LEBC: see ref. 40.

44. R. M. Baltrusaitis et al., Phys. Rev. Lett. *56*, 2140(1986).

45. R. M. Baltrusaitis et al., Phys. Rev. Lett. *55*, 150(1985); R. H. Schindler et al., SLAC-PUB-3799.

46. I. Peruzzi et al., Phys. Rev. Lett. *39*, 1301(1977); D. L. Scharre et al., Phys. Rev. Lett. *40*, 74(1978).

47. R. H. Schindler, Phys. Rev. *D24*, 78(1981).

48. D. Bortoletto et al. (CLEO Collaboration), "Inclusive B Meson Decays into Charm," contribution to this conference.

49. Private communication from K. Schubert.

50. (HRS collaboration), "Charm D Production in e^+e^- Annihilation at 29 GeV," ANL-HEP-CP-86-75 and contribution to this conference.

51. Lead-glass wall: ref. 46 (I. Peruzzi et al.)
MARKII: R. H. Schindler et al., Phys. Rev. *D21*, 2716(1980)
Crystal Ball: H. Sadrozinski, in High Energy Physics — 1980, edited by Loyal Durand and Lee G. Pondrom, AIP Conference Proceedings No. 68, p. 681. The values I have quoted are taken from ref. 44 and were derived from the references above.

52. K. Lane, Harvard Preprint HUTP-86/A045.

53. H. J. Lipkin, Argonnne Preprint ANL-HEP-PR-86-43 and contribution to this conference.

54. R. M. Baltrusaitis et al., Phys. Rev. Lett. *43*, 1073(1985).

55. E. Eichten et al., Phys. Rev. *D21*, 203(1980).

56. W. Bacino et al., Phys. Rev. Lett. *45*, 329(1980).

57. S. Behrends et al. (CLEO Collaboration), contribution to this conference.

58. J. J. Becker et al. (MARKIII Collaboration), "Direct Measurement of the Branching Ratios $D \to Ke\nu_e$ and $D \to K\pi e\nu_e$," contribution to this conference.

59. H. Albrecht et al., Phys. Lett. *B158*, 525(1985).

60. H. Albrecht et al. (ARGUS Collaboration), "The Decay $D^0 \to \bar{K}^0\Phi$," contribution to this conference; also the talk of N. Kwak at this conference. I have used the value given by Kwak rather than the value given in the contribution of Albrecht et al. The difference is the branching ratio for $D^0 \to K^0\pi^+\pi^-$ used in the two cases; Kwak uses 8.3% and Albrecht et al. 7.6%. CLEO (the next reference) also uses 8.3%. Also see footnote 15 of reference 62.

61. C. Bebek et al., Phys. Rev. Lett. *56*, 1893(1986).

62. R. M. Baltrusaitis et al., Phys. Rev. Lett. *56*, 2136(1986).

63. H. Albrecht et al. (ARGUS Collaboration), "Observation of F Decays into $\bar{K}^*K$," contribution to this conference.

64. M. Witherell (E691), private communication.

65. J. J. Becker et al. (MARKIII Collaboration), "Search for $D^+ \to \mu^+\nu_\mu$ and the Pseudoscalar Decay Constant f_D," contribution to this conference.

66. H. Krasemann, Phys. Lett. *B96*, 397(1980); L. Maiani, Proceedings of the XXI International Conference on High Energy Phys, (editions de Physique, Le Ulis, France, 1982), pp.631– 657; P. J. O'Donnel, CERN-TH.4419/86; S. N. Sinha, Alberta Thy-3-86.

67. E. Golowich, Phys. Lett. *B91*, 271(1980); M. Claudson, HUTP-81/A1016(1982).

68. V. A. Novikov et al., Phys. Rev. Lett. *38*, 626(1977); V. S. Mathur and M. T. Yamawaki, Phys. Rev. *D29*, 2057(1984).

69. L. L. Chau, Phys. Rep. *95*, 1(1983); A. Datta and P. Kumbhakar, Z. Phys. *C27*, 515(1985); E. A. Paschos, Phys. Lett. *B128*, 240(1983).

70. J. F. Donoghue et al., Phys. Rev. *D33*, 179(1986); L. Wolfenstein, Phys. Lett. *B164*, 170(1985).

71. For example, A. Datta, Phys. Lett. *B159*, 287(1985).

72. W. C. Louis et al., Phys. Rev. Lett. *56*, 1027(1986).

73. R. H. Schindler, contribution to these Proceedings.

74. I. Bigi and A. I. Sanda, Phys. Lett. *B171*, 320(1986).

75. H. Albrecht et al., (ARGUS Collaboration), "An Upper Limit on D^0-$\overline{D}^0$ Mixing," contribution to this conference.

76. S. Abachi et al., (HRS Collaboration), "Search for Wrong Sign D^0 Decays with the HRS Detector," contribution to this conference.

77. C. Bebek et al., (CLEO Collaboration), "Exclusive Decays of B Mesons," contribution to this conference.

78. H. Albrecht et al., (ARGUS Collaboration), "Reconstruction of B Mesons" and "Determination of the Branching Ratio for the Decay $B^0 \rightarrow D^{*-}\pi^+$," contributions to this conference.

79. R. Giles et al., Phys. Rev. *D30*, 2279(1984).

80. P. Avery et al., (CLEO Collaboration), "Limits on Rare Exclusive Decays of B Mesons," contribution to this conference.

81. M. B. Gavela et al., Phys. Lett. *B154*, 425(1985)

82. H. Albrecht et al., IHEP-HD/86-3, contribution to this conference.

83. M. S. Alam et al., Cornell Preprint, CLNS 86/739, contribution to this conference.

84. M. Bauer and B. Stech, Phys. Lett. *B152*, 380(1985); M. Wirbel, B, Stech and M. Bauer, Z. Phys. *C29*, 637(1985); B. Stech, Proc. 5th Moriond Workshop, "Flavour Mixing and CP-Violation," 151(1985); B. Stech, HD-THEP-86-7,XXI Moriond meeting, "Perspectives in Electroweak Interactions and Unified Theories," 1986; B. Stech, private communication; R. Ruckl, contribution to this conference.

85. L. L. Chau and H. Y. Cheng, Phys. Rev. Lett. *56*, 1655(1986); H. Y. Cheng, in Proceedings of the Oregon Meeting of the DPF, edited by R. C. Hwa (World Scientific, Singapore, 1986).

86. A. J. Buras, J.-M. Gerard and R. Ruckl, Nucl. Phys. *B268*, 16(1986).

87. J. F. Donoghue, Phys. Rev. *D33*, 1516(1986).

88. A. Ali, DESY Preprint 86-107.

89. T. Schaad et al., Phys. Lett. *B160*, 188(1985).

90. A. Bean et al., Cornell Preprint CLNS 86/741.

91. H. Albrecht et al., (ARGUS Collaboration), "Search for $B_d\overline{B}_d$ Mixing in e^+e^- Annihilation at 10.6 GeV," contribution to this conference.

92. UA1 Collaboration, contribution to this conference.

93. See, for example, A. Soni, Phys. Rev. Lett. *53*, 1407(1984).

94. D. Cline, private communication.

95. A. Chen et al., Phys. Rev. Lett. *52*, 1084(1984).

96. C. Klopfenstein et al., Phys. Lett. *B130*, 444(1983).

97. G. Altarelli et al., Nucl. Phys. *B208*, 365(1982).

98. E. Thorndike, in Proceedings of the International Symposium on Lepton and Photon Interactions, Kyoto, Japan, 1985.

99. S. Behrends et al., (CLEO Collaboration), "Limits on $\Gamma(b\rightarrow u)/\Gamma(b\rightarrow c)$ from Semileptonic Decay of B Mesons," contribution to this conference.

100. K. Schubert, contribution to this conference.

101. Crystal Ball Group, contribution to this conference.

102. A. Ali, Z. Phys. *C1*, 25(1979).

103. S-H. H. Tye and G. Trahern, CLEO Collaboration Internal Note, CBX-85-45.

104. A. Ali and T. C. Yang, Phys. Lett. *B65*, 275(1976).

105. B. Grinstein, M. Wise and N. Isgur, Phys. Rev. Lett. *56*, 298(1986).

106. L. Angelini et al., Phys. Lett. *B172*, 447(1986).

107. CLEO Collaboration, private communication.

108. ARGUS Collaboration, private communication.

109. Measurements of the lifetime of b-hadrons. Statistical and systematic errors have been added in quadrature:
MAC '83: E. Fernandez et al., Phys. Rev. Lett. *51*, 1022(1983)
MARKII '83: N. Lockyer et al., Phys. Rev. Lett. *51*, 1316(1983)
MAC '84a: W. T. Ford, COLO-HEP-87-mc, presented at the Aspen Winter Physics Conf., Aspen, CO, Jan. 7–18, 1985
DELCO '84: Klem et al., Phys. Rev. Lett. *53*, 1873(1984)
MARKII '84a: taken from Ford above

JADE '84: W. Bartel et al., Z. Phys. *C31*, 349(1986)
MARKII '84b: taken from Ford above
MAC '84b: taken from Ford above
TASSO '86: contribution to this conference
MAC '86: Presentation by D. Ritson at this conference
DELCO '86: Presentation by D. Ritson at this conference.

110. A. Brody et al., Phys. Rev. Lett. *48*, 1070(1982).

111. M. S. Alam et al., (CLEO Collaboration), "Branching Ratios of B Mesons to K^+, K^-, and $K^0/\overline{K}^0$," contribution to this conference.

112. D. Bortoletto et al., Cornell Preprint CLNS 86/736, contribution to this conference.

113. M. S. Alam et al., Phys. Rev. Lett. *51*, 1143(1983).

114. Presentation by K. Schubert and contribution to this conference.

115. P. Haas et al., Phys. Rev. Lett. *56*, 2781(1986).

116. H. Albrecht et al., (ARGUS Collaboration), "Observation of Inclusive F Production in B Meson Decay," contribution to this conference.

117. M. Suzuki, Phys. Lett. *B142*, 207(1984); Phys. Rev. *D31*, 1158(1985).

118. A. Chen et al., Phys. Rev. Lett. *51*, 634(1985).

119. D. Farikov and B. Stech, Nucl. Phys. *B133*, 315(1978).

120. J. H. Kuhn et al., Z. Phys. *C5*, 117(1980).

121. J. Green et al., Phys. Rev. Lett. *51*, 347(1983); S. E. Csorna et al., Phys. Rev. Lett. *54*, 1894(1985).

122. Although the CLEO group does not quote a number for the D^+ yield in semileptonic B decay, their range of values is consistent with the D^+ yield in all B decay, private communication. Also the observation of D^{*+} in semileptonic B decay suggests that the D^0 yield should be larger than the D^+ yield in semileptonic decay.

Plenary Session

Nonperturbative Methods in Quantum Field Theory

P. Hasenfratz (Bern)

Chairman

Y. Yamaguchi (Tokyo)

Scientific Secretary

C. Bachas (SLAC)

NONPERTURBATIVE METHODS IN QUANTUM FIELD THEORY

P. Hasenfratz

Institute for Theoretical Physics
University of Bern
Sidlerstrasse 5, CH-3012 Bern, Switzerland

Session 7 of this Conference is dedicated to nonperturbative methods in quantum field theories. All, but one of the fifteen contributions at the parallel session dealt with lattice regularized field theories, their numerical studies, or closely related subjects. This might reflect a bias on the part of the session organizers. It is true, however that an overwhelming part of the contributed papers belonged to this cathegory (1).

This subject is not without its problems. The progress is slower than what was expected a few years ago. Nevertheless, it seems to be the only available method, which has a reasonable chance to recover nonperturbative aspects of quantum field theories in a controlled way.

Actually, this is not quite true. There exist limits in different theories, where, for certain questions, other methods become effective, like chiral perturbation theory in the chiral limit of QCD, or perturbation theory on the ultraviolet problems of asymptotically free theories, etc. There exist several phenomenological approaches also: different bag models, Skyrmions, QCD sum rules, and so on.

On certain low energy questions chiral perturbation theory (2) gives predictions, which could hardly be matched by numerical (or other) approaches in the near future (3). Its scope is limited, however. QCD sum rules (4) produced a lot of valuable results in the past few years (5). As it is usual with phenomenological approaches, the method expands horizontally: apart from questions in standard spectroscopy (heavy and light mesons, baryons, masses, widths and couplings) several recent papers (6) dealt with excited states, hybrids, exotic mesons, glueballs, weak matrix elements, etc. The method works well in many cases, in some cases it does not. It is difficult to improve systematically on the predictions (i.e. digging in vertically), but this is presumably not, what we should expect from this approach.

I will mainly talk about lattice regularized QFT's and results, where computation and numerical techniques played an important role. I want to return, however, to chiral perturbation theory and QCD sum rules, where important common questions are raised.

I will address the following points:

1. Regularization, scaling
2. SU(3) gauge theory
3. QCD (quenched)
4. QCD with dynamical light quarks
5. Higgs phenomenon

Before starting, let me make two remarks.

In discussing QCD results, usually we quote numbers and errors, wonder about consistencies, etc. Sometimes we forget that these calculations produced some very important qualitative results.

Numerical techniques can never prove anything. They can demonstrate phenomena with increasing precision. In this sense these calculations demonstrate quark confinement, the spontaneous breakdown of chiral symmetry and the existence of a transition to a new kind of matter, the quark-gluon plasma. These are important results. A few years ago there was a flood of papers dealing with these questions. When this flood dried out abruptly, we did not understand things much better than at the beginning. People just got bored, gave it up and turned to simpler problems, like spontaneous compactification, to name one.

The second remark concerns Section 5. The old Fermi theory of weak interactions works very well for many low energy phenomena. It is an effective theory only, and as such, it could predict within its own framework the scale, before which new physics should occur. Now, there are strong indications that the Weinberg-Salam model is an effective theory also. Again, it is expected that it can predict the scale before new physics should come. This is what makes the problem of Higgs phenomenon interesting.

1. REGULARIZATION, SCALING

Consider a regularized QFT. The action is expressed in terms of fields, and contains parameters: the bare couplings and masses.

It depends on the cut-off Λ^{cut} also.

An important property of QFTs is locality: there is interaction only between neighbouring variables. The elementary interactions contained in the action extend over distances of the order of the inverse cut-off $(\Lambda^{cut})^{-1}$. This is a very small distance, when the cut-off is large ($\Lambda^{cut} \to \infty$ at the end), and it is not so easy to arrange signals to propagate over finite distances (i.e. to get excitations with finite masses), when the elementary interactions extend over infinitesimal distances. In general, achieving that requires a careful tuning of the parameters.

In a pure Yang-Mills theory the parameter we tune is the bare coupling g^2

$$\left\{ \begin{matrix} g^2 \to 0 \\ \Lambda^{cut} \to \infty \end{matrix} \right\}$$

Although the bare coupling goes to zero, the renormalized coupling describing the interaction at some finite energy scale, can be kept finite. Its value depends on the precise way the limit $\left\{ \begin{matrix} g^2 \to 0 \\ \Lambda^{cut} \to \infty \end{matrix} \right\}$ is performed. This is in our hand, therefore, the renormalized coupling (or, equivalently Λ_{YM} is arbitrary, it is a free parameter.

Consider now QCD with one quark flavour. Here we have to tune two parameters: the bare coupling g^2 and the bare mass m_q.

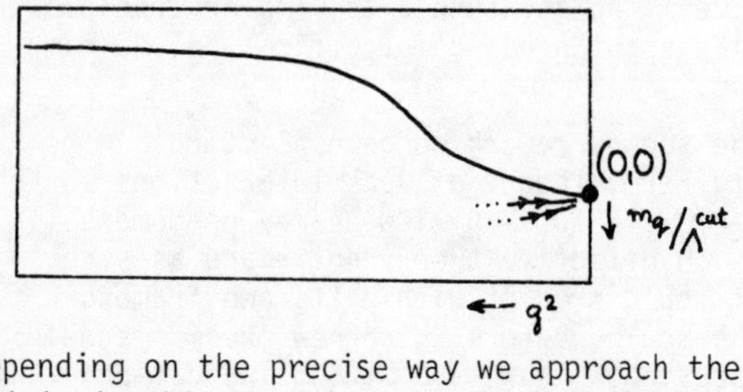

Depending on the precise way we approach the origin in this two-dimensional plane (when $\Lambda^{cut} \to \infty$), we obtain theories with different Λ_{QCD} and m_π/m_g . The theory contains two free parameters, which cannot be predicted.

In order to get a cut-off independent ("renormalized") theory at the end, the parameters should be tuned in a well-defined way.

In the first example, for instance, $g^2 = g^2(\Lambda^{cut})$. If everything is consistent, there must exist a function $g^2 = g^2(\Lambda^{cut})$, which assures that all the physical predictions are independent of the cut-off, when the cut-off is large. This is the way the cut-off disappears from the theory. The scaling region is given by those cut-off values (or, equivalently, coupling values), where this cut-off independence already appears. For very large cut-offs the functional dependence $g^2 = g^2(\Lambda^{cut})$ is known explicitly - it is determined by the leading terms of the (perturbatively calculable) β -function:

$$g^2(\Lambda^{cut}) = \frac{1}{2\beta_0 \ln \Lambda^{cut}} + \dots \qquad (1)$$

This behaviour is called asymptotic scaling (7).

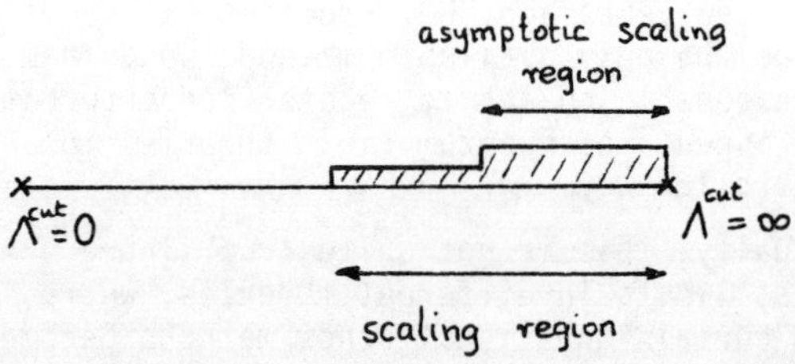

In the case of lattice regularization the procedure of removing the cut-off is called "the continuum limit". Cut-off independence ("scaling") is the # 1 consistency check in any calculation.

2. SU(3) GAUGE THEORY

This is a severely truncated version of QCD, where quarks enter as static sources only. Nevertheless, several interesting problems can be addressed already in this approximation: confinement, shape of the q-$\bar{q}$ potential, tension, spin-dependent potentials, glueballs, critical temperature, condensates, etc.

Most of these properties are contained in the asymptotic behaviour of appropriate Green's functions

$$\langle \mathcal{O}(0)\, \mathcal{O}(t) \rangle_c \underset{t \to \infty}{\sim} e^{-E_0 t} \qquad (2)$$

in Euclidean space, where E_0 is the energy of the lowest lying state in the channel characterized by the quantum numbers of $\mathcal{O}$. Using lattice regularization and finite volume, these expectation values are represented by a well defined, multidimensional integral over the field variables. The action is constructed from the gauge variables $U_{n\mu}$ and from the quark fields Ψ_n : $S = S(U, \Psi)$.

No details will be needed in the following. S is local, Ψ is associated with the sites, U with the links of the hypercubic lattice. In the static limit the mass of the quarks is sent to infinity.

2.1 Potential, tension, spin dependence, force between quark clusters

The potential at a distance r is the lowest energy state of a $q\bar{q}$ pair fixed at a distance r. Consider therefore the operator

, where the straight line connecting the quarks represents the flux line assuring gauge symmetry. We have

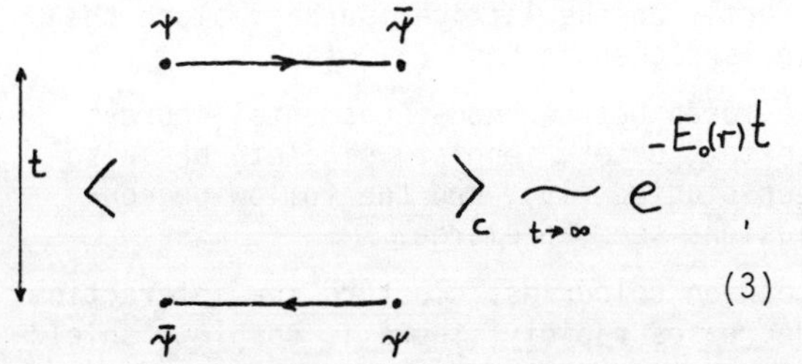

(3)

where $E_o = E_o(r)$ is the energy of the lowest energy state. Using the basic rules of Grassmann integration and the fact that the quarks are very heavy (sources) the left hand side of Eq. (3) gives

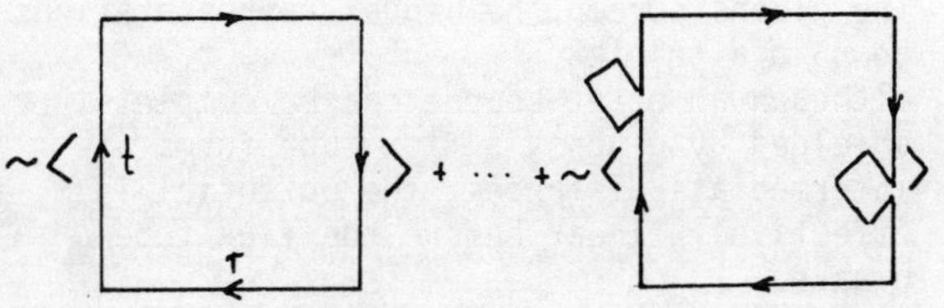

(4)

where the expectation values are evaluated in a pure gauge theory. For infinitely heavy quarks only the first, leading term survives, giving the well known result: the (spin independent) potential V(r) is directly related to the expectation value of gauge loops. The subleading terms, like the decorated loop in Eq. (4), are related to the spin dependent potentials (8). The practical problem is therefore to determine the expectation value of large Wilson loops in an SU(3) gauge theory.

Figure 1 collects the results on the string tension σ obtained from different MC calculations (9). The tension, as measured in units of the Λ parameter of the lattice formulation $\left(\Lambda^{latt} = \frac{1}{83.5}\Lambda^{MOM}\right)$,

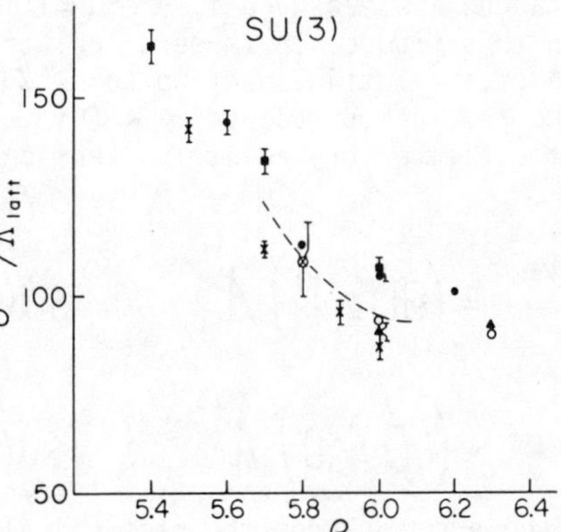

Figure 1: Monte Carlo results for the string tension.

should be independent of the cut-off (or, equivalently g^2) in the continuum limit. For this figure the tension was extracted from the bare Monte Carlo (MC) data under the assumption that at the coupling constant values considered asymptotic scaling (Eq. (1) already holds. According to Fig. 1 asymptotic scaling can hold (at best) only for $\beta \gtrsim 6.0$ $(\beta \equiv 6/g^2)$. As we discussed before, scaling might start somewhat earlier. This possibility is supported by Fig. 2,

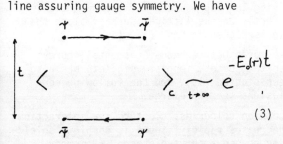

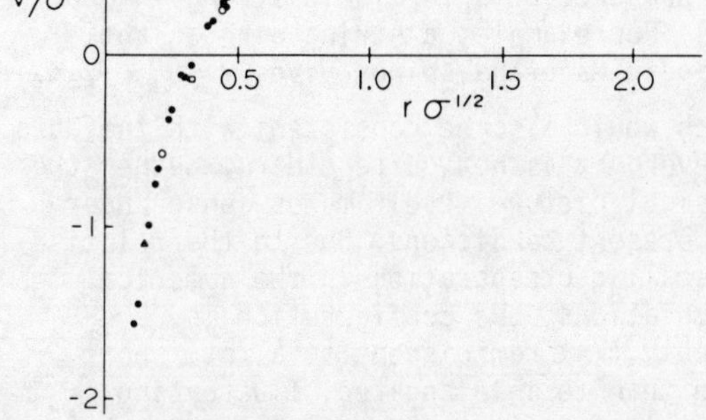

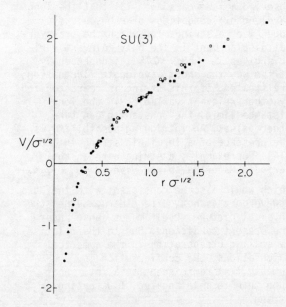

Figure 2: The q-$\bar{q}$ potential as obtained from different numerical simulations.

where predictions on the potential are collected from a large number of different MC studies. These calculations were performed at different β (or, Λ^{cut}) values ($\beta \in (5.7, 7.2)$) and the common, smooth curve formed by these points is a measure of scaling. The potential and distances are measured in $\sigma^{1/2}$ (≈ 0.42GeV) units in this figure. In these calculations the size of the lattice went up to $O(2\,fm)$, while the resolution reached $a \sim 0.1\,fm$. The best estimate for the string tension reads

$$\sigma^{1/2} = \left(90 \pm {}^{10}_{20}\right)\Lambda^{latt}, \quad SU(3)$$
$$= (40 \pm 10)\Lambda^{latt}, \quad SU(2)^{(10)}, \tag{5}$$

where the error includes the statistical error and a (subjective) estimate of the systematical errors involved. In a paper contributed to this conference Itoh et al. (11) obtained $\sim$ 30% higher value using a modified (imporved) action. As no other group used this action so far, it is difficult to guess the reason. One might mention that string tension estimates have a tendency to decrease as the statistical and systematical control is improved.

All the measurements show (as can be seen in Fig. 2 also) that in the long distance part of the potential σr cannot give the full story. An additional term $-c/r$ fits well the datapoints, where c $\sim$ 0.3 (12). Lüscher observed a few years ago (13) that the long wavelength fluctuations of a bosonic flux tube (i.e. a string, for which the transversal displacement is the only relevant variable gives $c_B = \pi/12 \sim 0.26$, independently of the way the short wavelength fluctuations are treated. If this string is characterized by other relevant variables in the long distance limit, the constant c can take other values. An intriguing possibility is the presence of a fermionic string in QCD (14). For example, a string with an anti-periodic Majorana spinor gives $c_{N-Sch} = 3/2\, c_B$, which would also be consistent with the data. However, it is not quite clear, whether the physical picture itself makes sense under the present conditions. Due to the relatively small q-$\bar{q}$ separation in the numerical calculations, the configuration is, presumably, more reminiscent to a fat, short tube than to an elongated, fluctuating string.

Keeping the $1/m_q$ corrections in Eq. (4), the decorated Wilson loops can be related to the spin dependent potentials. Four new, r dependent functions enter: spin-spin (V_{SS}), tensor (V_T) and two spin-orbit potentials (V_1 and V_2). The results are rather preliminary. The shape of the potentials V_{SS}, V_T and V_2 seems to be consistent with a perturbative, one-gluon exchange form (15-17), while $V_1(r)$ contains a significant non-perturbative contribution (17). This behaviour is in accord with theoretical expectations (18), and is welcome in heavy quark spectroscopy. At this moment, however, the lattice calculations of the spin dependent potentials have a problem with the overall normalization. This problem is related to the naive procedure usually followed: the formal continuum expressions relating the potentials to decorated Wilson loops (8) are rewritten in terms of lattice variables in some way. This is a non-unique procedure. The overall normalization (but not the shape) depends on the details of this step. I believe, the resolution of this problem can be found by deriving the expressions for the potentials directly on the lattice, perhaps along the line sketched in Eqs. (3), (4).

The force between non-fundamental sources (sextet, octet, tenplet, singlet) has also been studied (19), and the following conclusions seem to emerge:

- between colourless clusters the interaction decreases rapidly, there is complete shielding beyond a short distance. (Remember, there is no pion exchange in this truncated version of QCD.
- between two coloured sources with non-zero triality the potential increases linearly with a universal strength (which is equal to the fundamental string tension). The gluons screen the higher representations down to a triplet.
- although an octet source can be completely shielded by gluons, a flux tube forms at intermediate distances, the potential is effectively linear before the flux tube breaks.

These conclusions are consistent with the finite temperature study of Faber et al. (20).

In a contributed paper Irving et al. (21) reported results on the string tension obtained from an extensive Hamiltonian strong coupling expansion (up to g^{-28}):

$$\sigma^{1/2} = (140 \pm 30)\Lambda^{latt}, \quad SU(3)$$
$$= \left(35 \pm {}^{20}_{6}\right)\Lambda^{latt}, \quad SU(2). \tag{6}$$

Taking into account the difficulties of this approach, the results in Eqs. (5) and (6) can be considered as consistent with universality.

2.2 Deconfining temperature, latent heat

General arguments suggest that under extreme conditions (high temperature and/or baryon density) a new kind of matter, quark-gluon plasma is formed. There have been several talks on the possible experimental significance and phenomenology of this transition during this conference (22).

In a pure SU(3) gauge theory this phenomenon occurs as a spectacular first order phase transition. Tremendous effort was invested into this problem during the last two years (23, 24, 25). Large lattices (up to $14 \times (21)^3$) were treated and high statistical accuracy was achieved in simulations, which tested the limit of available supercomputers. The results on the critical temperature are summarized in Fig. 3.

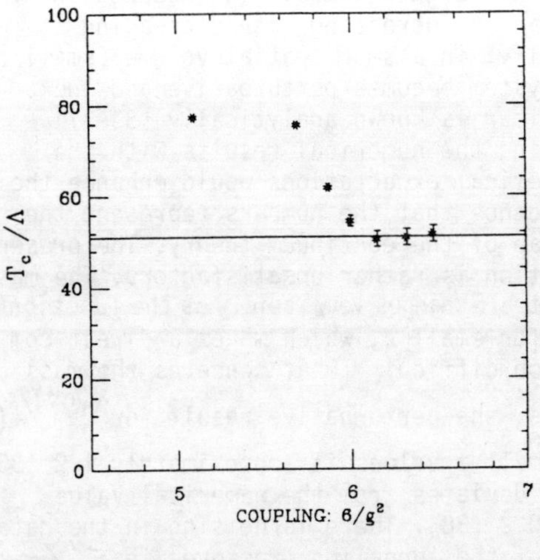

Figure 3: Monte Carlo results for the deconfining temperature in SU(3) gauge theory.

Asymptotic scaling seems to set in around $\beta \gtrsim 6.1$, giving

$$T_c = (50 \pm 1)\Lambda^{latt} \xrightarrow[\text{and Eq.(5)}]{\sigma^{1/2} = 0.42\,\text{GeV}} \qquad (7)$$

$$T_c = (230 {}^{+20}_{-70})\,\text{MeV}.$$

Results obtained on the Columbia dedicated machine (24, 26) are in good agreement with those given above, except one single case. (The results obtained on a 10×16^3 lattice differ beyond the statistical errors. The reason is not clarified yet.)

The latent heat (i.e. the energy density which has to be pumped into the system to make the first order jump) has also been estimated (27, 25)

$$\text{latent heat} \approx 1\ \text{GeV}/\text{fm}^3. \qquad (8)$$

An interesting connection is expected (28) to exist between the deconfining phase transition in d=4 gauge theories and the critical properties of simple d=3 spin models. The existence of this connection is supported by numerical studies (29), which enhances our confidence further that the finite temperature studies of the gauge sector are under control.

By comparing the graphs for T_c/Λ^{latt} and $\sigma^{1/2}/\Lambda^{latt}$ we see a rather similar behaviour, which suggests that ratios like $\sigma^{1/2}/T_c$ might be constant earlier than the point where asymptotic scaling sets in (scaling). One might try to check directly whether a function $g = g(\Lambda^{cut})$ exists, which makes the physical predictions cut-off independent.

2.3 β-function, Monte Carlo renormalization group methods

The question is the existence and properties of the function $g = g(\Lambda^{cut})$, or, equivalently, the β-function

$$\Lambda^{cut} \frac{d}{d\Lambda^{cut}} g(\Lambda^{cut}) = B(g) \qquad (9)$$

Deep in the continuum limit

$$B(g) \underset{g \to 0}{=} -b_0 g^3 - b_1 g^5 + \cdots \qquad (10)$$

given by two-loop perturbation theory. There exist numerical methods based on non-perturbative RG considerations, which determine the β-function of the theory. For some irrelevant technical reasons these calculations determine not the β-function itself, but some related function denoted here by $\Delta(\beta)$, which reflects the behaviour of the β-function between β and $\beta - \Delta(\beta)$. There is a one-to-one relation between $\Delta(\beta)$ and the β -function. The perturbative behaviour Eq. (10) corresponds to

$$\Delta(\beta) \underset{\beta\to\infty}{=} 0.579 + \frac{0.204}{\beta} + \mathcal{O}(\beta^{-2}), \quad SU(3). \tag{11}$$

Part of the MCRG results on $\Delta(\beta)$ are collected in Fig. 4 (30). The dotted line is the asymptotic scaling behaviour, Eq. (11). There is a definite violation of asymptotic scaling in the region, where most of the early calculations were performed. On the other hand, the consistency between the points obtained from different procedures suggests that there is a region, where the function $\Delta(\beta)$ exists (scaling), although its behaviour is not asymptotic yet. This conclusion is consistent with Fig. 5, where $T_c/\sigma^{1/2}$ is plotted against β. This ratio seems to be constant already for $\beta \gtrsim 5.7$.

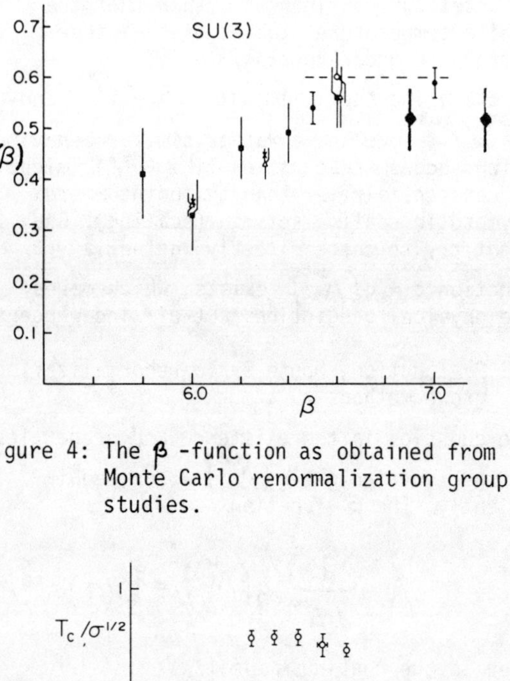

Figure 4: The β -function as obtained from Monte Carlo renormalization group studies.

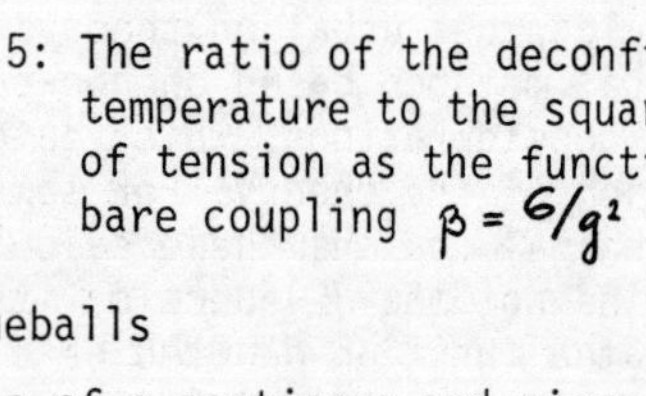

Figure 5: The ratio of the deconfining temperature to the square root of tension as the function of the bare coupling $\beta = 6/g^2$.

2.4 Glueballs

In spite of a continous and vigorous theoretical and numerical effort in this field, the situation remains rather unclear. Due to the fact that the available numerical techniques for glueball mass determination become increasingly poor as the resolution and/or the lattice volume is increased, the glueball is the "sick horse" of the lattice approach: all the possible systematic errors show up strongly and mixed in the numerical results.

In order to study the finite volume effects it is convenient to use the variable (31) $z = m_g L$, where L is the spatial size of the system, while m_g is the massgap of the corresponding finite volume continuum field theory. As shown by Lüscher (32), for large z (large volume) the massgap approaches its infinite volume value from below exponentially

$$m_g(z) \underset{z\to\infty}{=} m_g(\infty)\left[1 - G\,\frac{e^{-\frac{\sqrt{3}}{2}z}}{z}\right] \tag{12}$$

For small z the glueball is squeezed and the massgap is increasing with decreasing z. Actually, in a small spatial volume (small z) the system becomes perturbative and the z behaviour is known analytically (33-37). Matching the numerical results with the theoretical expectations would enhance the confidence that the numbers represent the massgap of the continuum theory. The present situation is rather unsatisfactory. The massgap itself changes very rapidly as the function of z for small z, which makes a direct comparison difficult. What concerns the mass ratios, the perturbative result for $m(0^{++})/m(2^{++})$ for small z values is approximately 1.2 (35), which deviates from the numerical value 1.6 + 0.2 (38). There is no sign in the data (38) of the tunneling crossover in $\sigma^{1/2}/m_g$ expected to occur around z = 1.6 (36). It is uncertain also, whether the volume dependence for larger z values can already be described by the asymptotic form Eq. (12) (39, 40). All this said, the infinite volume estimate (39, 41)

$$m_g(0^{++})/\sigma^{1/2} \approx 3 \quad (\to m_g(0^{++}) \sim 1300\,\mathrm{MeV}) \tag{13}$$

should be taken with reservations. The important question, whether the lightest glueball is 0^{++} or 2^{++} remains, at present, also unresolved (38, 42, 43).

Let us add here two recent analytic expansion results obtained in the Hamiltonian formulation:

$$m_g(0^{++})/\sigma^{1/2} \approx 3 \quad , \tag{14}$$

using the so-called "t-expansion" (44), and

$$m_g(0^{++})/\sigma^{1/2} = 2 \pm 1 \quad , \tag{15}$$

obtained in strong coupling ("linked cluster") expansion (21).

The quantities we discussed until now were related to long-distance correlations, which explains the technical difficulties. At the same time, certain quantities are measured with amazing precision. For example, the expectation value of the smallest Wilson loop (plaquette, p) is known to four digit accuracy at several β values. Unfortunately, there is no physics in these numbers. Or, is there ?

2.5 Condensates

By expanding the plaquette expectation value $\langle p\rangle$ for large cut-off values we get

$$1 - \langle p\rangle \sim \langle g^2 G^a_{\mu\nu} G^a_{\mu\nu} \rangle \,, \tag{16}$$

where the right hand side, the "gluon condensate", is one of the basic parameters in QCD sum rules !

Unfortunately this equation does not imply that we have a four digit determination of this fully non-perturbative number. Since the question of condensate values is the place, where the two approaches might meet, let me spend a few words on that.

Consider the correlation function of a conserved current with some quantum numbers (for example, the electromagnetic current). At large momenta one can write down an operator product expansion (OPE) of the form

$$i \int d^4x \, e^{iqx} \, T\big(j(x)\, j(0)\big) = \tag{17}$$

$$C_I(q)\, I + \sum_n C_n(q)\, O_n$$

where I is the unit operator, while the local operators O_n are arranged in order of increasing (engineering) dimension. The Wilson coefficients C_I, C_n contain increasing powers of $1/q^2$ for increasing dimension of O_n, therefore, by taking the vacuum matrix element of Eq. (17), a few vacuum expectation values dominate the right hand side ($\langle GG\rangle$, $\langle m\bar{q}q\rangle$, $\langle GGG\rangle$, $\langle \bar{q}\Gamma q\, \bar{q}\Gamma q\rangle$, ...) . The left hand side of this equation is related to a cross section via dispersion relations. At this side enter masses, couplings and widths of resonances. The idea is to calculate the Wilson coefficients in perturbation theory and, by a meticulous choice of q, predict the spectroscopical side in terms of a few condensate values (and usual QCD parameters) (4, 5).

The question of condensate values always attracted much attention in this approach. The validity of vacuum saturation, the actual value of the gluon condensate etc. are discussed in some of the contributed papers also (45).

At first sight, quoting a number like $\langle \alpha/\pi\, GG\rangle \sim (300\text{-}400\,\text{MeV})^4$ seems to be strange. Considering this expectation value in some generic scheme (PV, lattice,...) it will certainly contain pieces $\sim\mu^4$, where μ is the arbitrary renormalization point, spoiling the direct physical meaing of the condensate value. One might take the point of view that dimensional regularization (where this type of contributions are absent) is the regularization to be used, but a few people remained sceptic over the way OPE is used here (46). In several polemic papers Novikov et al. (47) dissipated partly the doubts by studying exactly solvable models or perturbative cases. In the studied cases the following conclusions seem to emerge. In the expectation value there is a clear cut separation between the physical ($\sim\Lambda^n_{QCD}$) piece and the non-physical contributions. The generic expected form of a condensate is

$$\langle \text{local operator, dim. } n\rangle = c\,\Lambda^n_{QCD} + \Lambda^{n-2}_{QCD}\mu^2\,(\text{power series in } g^2) + \dots + \mu^n\,(\text{power series in } g^2)\,, \tag{18}$$

where only the essential power dependence is indicated, the logarithms (due to the anomalous dimension of the operator in question) have been suppressed.

Therefore, there are in general (and specifically on the lattice) non-perturbative, non-physical pieces, which should also be subtracted. These contributions are coming from mixing with lower dimensional nontrivial operators. As a consequence, on the lattice the only operator, where subtracting the perturbative piece might be enough is

αGG (since there is no lower dimensional non-trivial operator). This procedure has been attempted a few years ago using the relatively poor data of that time (48). It might be useful to repeat the analysis with the uncomparably better data available now. What concerns, however, the higher condensates, it is not clear, how the lattice calculations and QCD sum rules can meet here.

3. QCD (QUENCHED)

3.1. On quenching

In the quenched approximation the contribution of virtual quark loops is artificially suppressed. For example in a $\langle \bar{q}q\,\bar{q}q \rangle$ amplitude

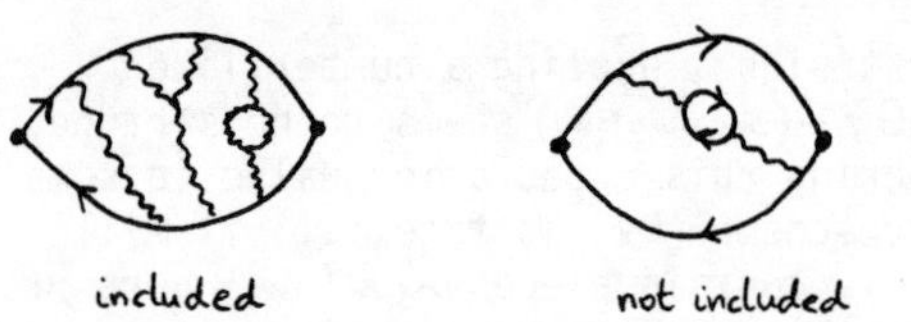

We all know the usual justifying arguments: Zweig rule, large N_c limit, success of quark models, etc. How good these handwaving arguments are - one can judge only after a quantitative comparison will be available between the quenched and full QCD results. Let us enumerate a few points here as a warning:

- The quenched approximation cannot account for the decay (therefore the width) of hadronic resonances. Additionally, it does certainly injustice to the important effect of pion cloud around the hadrons and to the specific infrared quark mass singularities due to virtual pions (3, 49).
- Apart from these direct physical effects there might exist technical problems associated with the quenched approximation.

  There are indications that it fails completely (i.e. produces physically unacceptable results) in the presence of a finite chemical potential (finite baryon density) (50).

  The quenched approximation puts the quark determinant equal to 1. This procedure gives undue weight to gauge configurations over which the quark propagator matrix has a small eigenvalue (these configurations would be suppressed by the fermion determinant). The possibility was raised recently (51) that, due to this problem, the quenched approximation is not even a well defined computational scheme.

Presumably we have to consider the framework only as a preparation for the full problem.

3.2 Chiral symmetry breaking, spectroscopy and weak matrix elements

The numerical calculations performed lately used significantly enlarged lattices and collected much better statistics than the early spectrum calculations. With the help of supercomputers lattices up to $16^3 \times 32$, $16^3 \times 48$ could be treated (with a typical resolution of $a \sim 0.1$ that implies $\sim (1.6\,fm)^3 \times 3.2\,fm$ lattice volume).

Chiral symmetry breaking

As mentioned already in the introduction, the calculations demonstrate clearly the spontaneous breakdown of chiral symmetry together with the accompanying Goldstone pions (52-54). Further information comes from finite temperature studies, where the restoration of chiral symmetry at some finite temperature is investigated (55). As an example, Fig. 6 shows $\langle \bar{\Psi}\Psi \rangle$ as the function of the quark mass as obtained in Ref. (53) at $\beta = 6.0$ on a $16^3 \times 32$ lattice

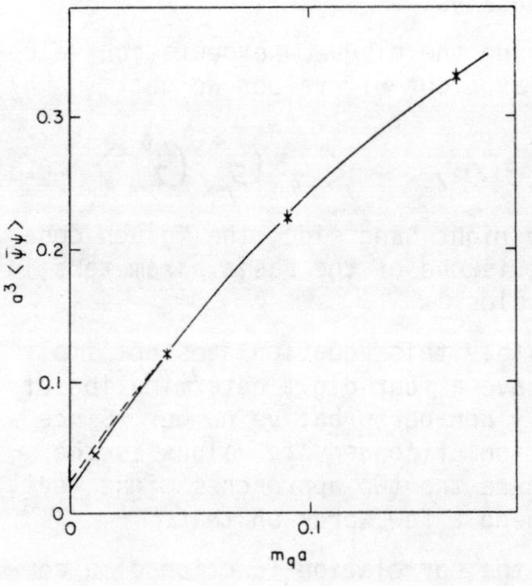

Figure 6: The chiral condensate as the function of the quark mass.

This calculation used staggered fermions, where part of the chiral symmetry is preserved. The non-zero extrapolated value of $\langle \bar{\Psi}\Psi \rangle$ at $m_q = 0$ is a sign of spontaneous chiral symmetry breaking.

Spectroscopy of low-lying states

The results of two recent large scale calculations (53, 56) are summarized in Table 1.

These results reflect a reasonable consistency, not only when we compare the two columns of Table 1, but also when compared with the results of the gauge sector. The nucleon tends to be heavy, but this is a, somewhat,

reference	(53)	(56)
type of fermions	Kogut-Susskind	Wilson
lattice size	$16^3 \times 32$	$16^3 \times 48$
lattice unit from m_ρ	$a^{-1} = 1.66(13)$ GeV	$a^{-1} \sim 1.6 - 1.8$ GeV
lattice from $\sqrt{\sigma}$	$a^{-1} = 1.9(2)$ GeV	$a^{-1} \sim 1.5$ GeV
masses (in MeV)	$m_N(\frac{1}{2}^+) = 1073(91)$, $m_N(\frac{1}{2}^-) = 1300(130)$ $m_S(0^{++}) = 1063(90)$, $m\ (1^{++}) = 1497(162)$ $f_\pi = 90 - 110$	$m_N = 1100(90)$ $m_\Delta = 1340(120)$
remark	different operators for the same channel give consistent masses	

Table 1

secondary point. (We do not know how heavy the proton should be in the quenched approximation. We expect it to be heavier than m_p^{exp} (by ~ 150 MeV ? (49)).

In a contributed paper Mütter et al. (51) analyzed the hadron spectrum on a $24^3 \times 48$ lattice. This is the largest lattice ever considered. The spectrum is calculated after two approximate blocking steps (i.e. on $6^3 \times 12$). Unfortunately, it is not known precisely, what kind of systematic errors are introduced with this blocking procedure. The authors observed a few anomalous configurations with very small eigenvalues in the fermion matrix. Their presence might be related to the quenching prescription, see 3.1.

The staggered fermion formulation has some unpleasant properties. Flavour symmetry is broken for finite "a", although - as discussed in a contributed paper (57) - this symmetry is restored to an increasing extent as the continuum limit is approached. There are tricky problems also with quantum number identification (58). Wilson's formulation is cleaner, but it has problems with chiral symmetry. For this reason it is of great importance that in a recent paper (59) a non-perturbative prescription has been suggested to identify the good axial currents and quark masses. Using this theoretical development Maiani and Martinelli (60) performed a numerical analysis with Wilson fermions and they could address questions, which were not possible (in a theoretically sound way) before: $\langle\bar{\psi}\psi\rangle$, the value of F_π and vector meson coupling constants

$$F_\pi = 110(20)\ \text{MeV} \quad (\text{expt. } 93)$$

$$1/f_\rho = 0.38(7)\ (\text{expt. } 0.29), \quad 1/f_\varphi = 0.30(4)\ (\text{expt. } 0.23) \qquad (19)$$

These results are consistent with those found earlier with staggered fermions (and, actually, with experiments), which supports the correctness of the theoretical analysis (59).

Understanding the chiral properties of Wilson fermions is relevant for the calculation of weak matrix elements also. There are several important questions in weak interaction physics ($\Delta I = 1/2$ rule, $K_0 - \bar{K}_0$ mixing, ϵ'/ϵ in CP non-conserving decays), where the answer crucially depends on the value of mesonic matrix elements of four-fermion operators. This is a non-perturbative problem, where very little a priori quantitative knowledge exists. Preliminary analyses (61-63) on the $\Delta I = 1/2 / \Delta I = 3/2$ ratio in $K \to \pi\pi$, and on the $K_0 - \bar{K}_0$, $\Delta S = 2$ matrix element indicate that a numerical precision could be achieved which makes this calculations relevant.

At the parallel session Martinelli (64) presented results obtained by measuring the $\Delta S = 2$, $K_0 - \bar{K}_0$ matrix element

$$\langle K_0 | (\bar{s}\gamma_\mu(1-\gamma_5)d)(\bar{s}\gamma_\mu(1-\gamma_5)d) | \bar{K}_0 \rangle$$

on a $10^3 \times 20$ lattice at $\beta=6.0$. The results look very reasonable. The matrix element is consistent with a $\sim m_K^2$ behaviour for small quark masses, as expected from chiral perturbation theory. The value of the B parameter is predicted to be

$$B = \frac{\langle K_0 | \Delta S=2 \text{ operator} | \bar{K}_0 \rangle}{(\text{vac. sat. value})} = 0.7 \pm 0.3 \quad (20)$$

consistent both with the vacuum saturation value (B=1) (65) and the value advocated by Donoghue et al. (66). This K_0-$\bar{K}_0$ matrix element can be connected with the $\Delta I = 3/2$ $K \to \pi\pi$ amplitude giving

$$\frac{\langle \pi^+\pi^0 | H_W | K^+ \rangle}{m_K} = (8 \pm 3) \cdot 10^{-8} \quad (\text{expt. } 3.7 \cdot 10^{-8}) \quad (21)$$

The results presented by A. Soni (67) on the same matrix element look much less satisfactory, especially for small quark masses. The reason might be related to the low β value, where this calculation is performed ($\beta=5.7$). At finite g^2 the measured operator contains a piece, which violates the correct chiral behaviour (goes to constant, rather than zero as $m_K^2 \to 0$). According to perturbation theory this contribution is small in this coupling constant region, but low order perturbation theory does not work very well even at $\beta=6.0$, therefore might be misleading at $\beta=5.7$ *).

4. QCD WITH DYNAMICAL LIGHT QUARKS

The technical difficulty, which makes this problem so difficult, is related to the presence of the fermionic determinant in the Boltzmann factor, which is a highly nonlocal expression.

4.1 Numerical methods for the computation of the fermion determinant

I will be very brief on this point. Doing so, I do complete injustice to this question, since this problem is very important, probably the most important single issue concerning the future of full QCD calculations. However, at the present stage, there is a pletora of different methods and tricks, without being clear, which method is winning, if any. For non-experts - like me - it is very difficult to follow the details. The different methods investiaged include the pseudo fermion method (68), Langevin equation with or without Fourier acceleration (69), microcanonical method (70), hybrid method (71), MC with unbiased noise (72) the Lanczos method (73) and their versions. There is a steady, continuous progress. without a phase transition.

*) I am indebted to G. Martinelli for a discussion on this point.

4.2 Commercial versus dedicated computers

Several groups decided to build dedicated computers to satisfy the increasing demand for computer time and memory (74, 75). This is an extremely demanding work. I am not in a position to decide, whether this is the right way to go. To get a feeling, let me just compare a few parameters which characterize a modern supercomputer and, say, the APE project (75).

CRAY 2

memory $\gtrsim$ 250 Mword ($\sim$ 2 Gbytes)
CPU: 4 independent unit, each has a 4.1 nsec clock period
peak performance: $\sim$ 1.9 Gflops
price $>$ 10^7 \$
significant programming effort is needed to get peak performance

APE (Array Processor with Emulator)
memory: 1 Gbytes
nominal speed: 1 Gflops
price: $\sim$ 0.5 Lire/1 flop (or 1 byte) (+ manpower)
16 fold parallelism

One quarter of the machine is already running. Assuming peak performance (?) it would update a 48×41^3 SU(3) gauge configuration 600 times a day.

4.3 Results

There exist a few preliminary works on spectroscopy (76) and on the effect of virtual quarks on the q-$\bar{q}$ potential (77). On the bare data the effect of virtual quark loops is quite significant. However, most of these effects just renormalize unphysical couplings, and when the dust settles, no effect is seen on the mass predictions - especially since the errors are large. One has to remark, however, that the pion is artificially heavy in these calculations and, as a consequence, the ρ is stable, so most of the effects are killed from the start.

Scaling

We discussed in detail the importance of this question. If the fermions are correctly simulated, the β-function is changed and the scaling behaviour should change accordingly. This effect is clearly seen in a recent study by Kogut (78) who investigated the scaling behaviour of the $q\bar{q}$ condensate (Fig. 7).

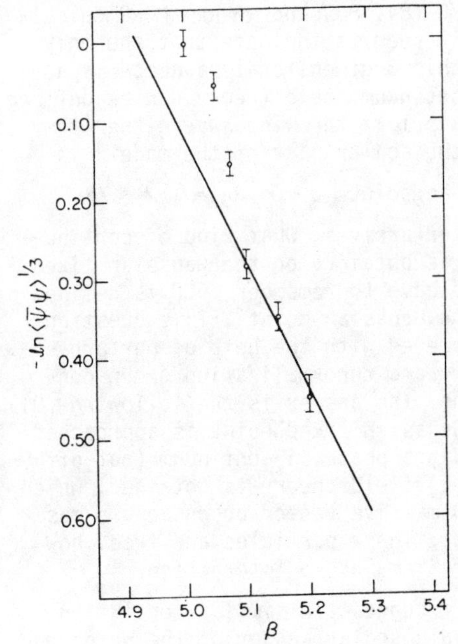

Figure 7: The scaling behaviour of the chiral condensate in QCD with four light flavours. The straight line indicates asymptotic scaling.

Beyond a narrow region, where $\langle\bar\psi\psi\rangle$ drops more rapidly than asymptotic scaling would predict *), the points follow the asymptotic straight line, which includes a 30% effect from the fermions ($n_f=4$).

Finite temperature studies

In the pure gauge sector, as we discussed, there is a strong, first order deconfining phase transition. When the effect of virtual quarks is included (Fig. 8), this transition becomes weaker. Whether the transition as a mathematical singularity disappears completely, is not settled yet (80, 71) (although most investigations seem to find that). What is clearly observed, however, is the chiral symmetry restoring phase transition for $m_q=0$ (78, 81). There is increasing theoretical (82) numerical (81) evidence for this transition being 1st order. The fate of this transition as the quark mass gets non-zero values is again not clear. It might be that for $n_f=3$ and with physical quark masses, strictly speaking, there is no phase transition between the hadronic phase and the chiral symmetry restored, deconfined quark-gluon plasma. There might be (and presumably will be) a rapid, but smooth transition. A strong first order phase transition with a significant latent heat, supercooling effects and alike - which would be welcome for experimental identification - is rather improbable.

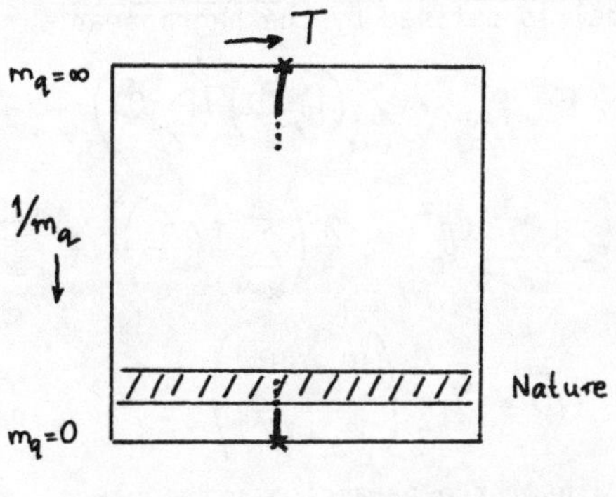

Figure 8: The schematic phase structure of QCD in the temperature - quark mass plane.

The qualitative results look quite promising. On a $10^3\times6$ lattice Kogut obtained (78)

$$T_c/\Lambda_{\overline{MS}} = 2.14 \pm 0.10\,, \quad n_f=4$$
$$\phantom{T_c/\Lambda_{\overline{MS}} =} 2.12\,, \quad \text{pure gauge}$$

$$T_c/\langle\bar\psi\psi\rangle^{1/3} = (0.85\pm0.05)\left(\frac{4\pi^2\beta}{33}\right)^{-4/33}, \quad n_f=4 \quad (\beta\approx5.2)$$

$$= (0.56\pm10)\left(\frac{4\pi^2\beta}{25}\right)^{-4/25}, \quad \text{quenched} \quad (\beta\approx6.) \qquad (22)$$

which indicates a rather small effect from firtual quarks in these ratios.

A puzzle related to QCD at finite baryon density

Totally unexpected, puzzling results were found (83) at finite baryon density in the quenched approximation. As discussed by Dagotto at the parallel session (50), the problem is presumably due to the quenching prescription. If this is so, this is the first complete failour of the quenched approximation.

5. HIGGS PHENOMENON

The importance of this issue has already been discussed in the Introduction. There was a large theoretical and numerical activity in this field during the last two

*) This is like the behaviour observed in the pure gauge sector (30) and seen also by Gavai and Karsch (79) in an MCRG study

years (84-91). I will present the problem and results on the model of an SU(2) gauge field with a Higgs doublet. This model forms a part of the standard model of electroweak interactions and was investigated by many groups in detail (84 - 87, 91).

SU(2) gauge theory with a doublet Higgs

The model is defined by the Lagrangean

$$\mathcal{L} = -\tfrac{1}{4} F^a_{\mu\nu} F^a_{\mu\nu} + \sum_{i=1}^{2} (D_\mu \Phi)^*_i (D_\mu \Phi)_i - -\tfrac{1}{2} r \sum_{\alpha=1}^{4} \varphi_\alpha^2 - \frac{\lambda}{4}\Big(\sum_{\alpha=1}^{4} \varphi_\alpha^2\Big)^2 \qquad (23)$$

where

$$\Phi = 1/\sqrt{2} \begin{pmatrix} \varphi_1 + i\varphi_2 \\ \varphi_3 + i\varphi_4 \end{pmatrix} \qquad (24)$$

It contains three parameters: the gauge coupling g^2, a mass parameter r and the selfcoupling λ . The phase structure of the model has been studied by several groups (84-87) on the lattice, where somewhat redefined parameters: κ (hopping parameter) and λ_{latt} are used to describe the scalar part of the Lagrangean.

The spontaneously broken (Higgs) phase is separated from the symmetric phase by a singular surface, whose position is quite well known from MC studies (84, 85).

Detailed MC (84, 85) and crude MCRG calculations (86) seem to indicate that the only point in this 3 dimensional parameter space, where a continuum field theory can be defined (without a priori throwing away either the gauge or the scalar part of the model) is the Gaussian point $g^2 = 0$, $\lambda_{latt} = 0$, $\kappa = 1/8$.

The question arrives, what kind of continuum theory is obtained on the Gaussian fixed point. (We have to remember, QCD is defined also on the Gaussian point). This question can be answered with the help of perturbation theory and renormalization group considerations. The answer is the following (91). When the Gaussian fixed point is approached from the Higgs phase, a continuum (cut-off independent) field theory is obtained, which contains 3 massive vector bosons and a massive scalar. These particles are free, however. There remains no interaction.

This result suggests a similar conclusion for the gauge scalar sector of the Weinberg-Salam model. How is this theory defined then? If the negative conclusion concerning the existence of a non-Gaussian fixed point remains unchanged, we are forced to consider this model as an effective theory: a theory

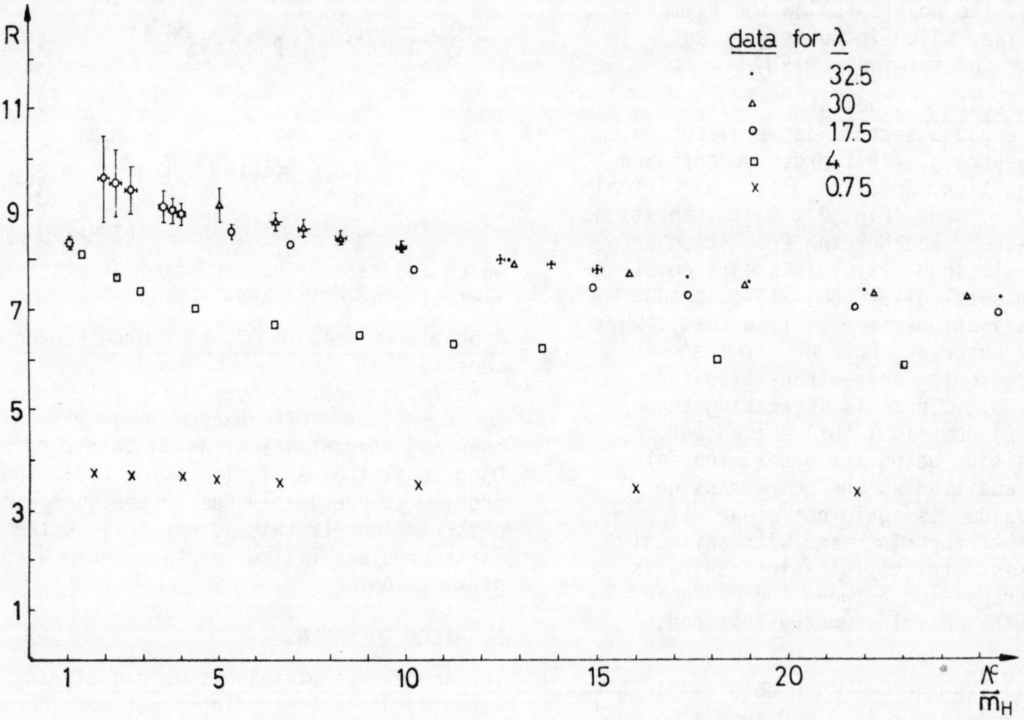

Figure 9: Upper bound on the Higgs mass as the function of the cut-off.

with a finite cut-off, which cannot be removed. This is not a catastrophe, since at scales much below the cut-off, the observable effects due to this cut-off are negligible.

This interpretation has an important consequence, however. One should be able to find the value of the cut-off (i.e. the scale, where the model breaks down, where new physics should enter) within the context of the model itself. The value of the cut-off depends on the parameters of the theory, most notably on the Higgs meson mass. The problem is to find the relation $\Lambda^{cut} = \Lambda^{cut}(m_H)$. This is a full fledged non-perturbative problem.

As it was discussed by Dashen and Neuberger (89) a few years ago, to a good precision this relation can be obtained within a pure, four component scalar field theory. Even that is a difficult problem. No direct MC attempt exists yet. There exist results, however, which were obtained (92) with the help of a non-perturbative approximate renormalization group relation (93), which performed very well on problems, where the answer is known from other sources. The result is shown in Fig. 9. If m_H/m_W is small, the cut-off can be very large and no useful prediction exists on the scale, where new physics should enter. For $m_H/m_W \gtrsim 6$, the maximum value of the cut-off approaches rapidly the value of the Higgs mass itself and for $m_H/m_W \gtrsim 10$, the model does not make sense anymore. This can be considered as an absolute upper bound.

These results are based on the assumption that the approximate RG recursion relation performs well. It illustrates how these bounds can be obtained by non-perturbative methods. I am sure, there will be direct MC analyses performed on this problem in the near future.

Acknowledgements

The author is indebted to B. Berg, J. Gasser, J. Govaerts, A. Hasenfratz, J. Kuti, H. Leutwyler, M. Locher, M. Lüscher, I. Montvay and G. Martinelli for discussions.

REFERENCES

(1) Summary papers on lattice gauge theories include
J.M. Drouffe, J.B. Zuber, Phys.Rep. 102 (1983) 1
M. Creutz, L. Jacobs, C. Rebbi, Phys. Rep. 95 (1983) 201
A. Hasenfratz, P. Hasenfratz, Ann.Rev. Nucl.Part.Sci. 35 (1985) 559

(2) R. Dashen, Phys.Rev. 183 (1969) 1245
R. Dashen, M. Weinstein, Phys.Rev. 183 (1969) 1291
H. Pagels, Phys.Rep. 16 (1975) 219

(3) For a systematic, extensive study in the meson sector, see
J. Gasser, H. Leutwyler, Ann.Phys. 158 (1984) 142; Nucl.Phys. B250 (1985) 465, 539, 517

(4) M.A. Shifman, A.I. Vainstein, V.I. Zakharov, Nucl.Phys. B147 (1979) 385, 448

(5) A recent summary paper:
L.J. Reinders, H.R. Rubinstein, S. Yazaki, Phys.Rep. 127 (1985) 1

(6) J.I. Latorre, S. Narison, D. Pascual, R. Tarrach, Phys.Lett. 147B (1984) 169
J. Govaerts, L.J. Reinders, H. Rubinstein, J. Weyers, Nucl.Phys. B258 (1985) 215
J. Govaerts, L.J. Reinders, J. Weyers, Nucl.Phys. B262 (1985) 575
C.A. Dominguez, N. Paver, Trieste prepr. IC/86/20 (1986)
C.B. Chiu, J. Pasupathy, S.L. Wilson, papers contributed to to this conference, #4715, 4707

(7) For further discussion on this point see
P. Hasenfratz, in Proc. of the NATO Adv. Summer Inst., Munich (1983), CERN prepr. TH-3737 (1983)

(8) E. Eichten, F.L. Feinberg, Phys.Rev. D23 (1981) 2724
M.E. Peshkin, SLAC prepr. SLAC-PUB-3273 (1983)
D. Gromes, Z.Phys. C22 (1984) 265; 26 (1984) 401
J. Pantaleone, S.-H. Henry Tye, Y.J. Ng, Phys.Rev. D33 (1985) 777

(9) A. Hasenfratz, P. Hasenfratz, U. Heller, F. Karsch, Z.Phys. C25 (1984) 191
D. Barkai, K.J.M. Moriarty, C. Rebbi, Phys.Rev. D30 (1984) 1283
J. Stack, Phys.Rev. D29 (1984) 1213
J.D. Stack, S. Otto, Phys.Rev.Lett. 52 (1984) 2328, E 53 (1984) 1028
A. Sommer, K. Schilling, Z.Phys. C29 (1985) 95
Ph. de Forcrand, G. Schierholz, H. Schneider, M. Teper, Phys.Lett. 160B (1985) 137
Ph. de Forcrand, C. Roisnel, Phys.Lett. 137B (1984) 213
K.C. Bowler, F. Gutbrod, P. Hasenfratz, U. Heller, F. Karsch, R.D. Kenway, I. Montvay, G.S. Pawley, J. Smit, D.J. Wallace, Phys.Lett. 163B (1985) 367
Ph. de Forcrand, unpublished, analyzed in M. Flensburg, C. Peterson, Phys.Lett.

153B (1985) 412

(10) F. Gutbrod, DESY prepr. 85-092 (1985)

(11) S. Itoh, Y. Iwasaki, T. Yoshié, Tokyo prepr. UTHEP-154, papers contributed to this conference # 7277, 1945

(12) See, for instance, the last reference in (9)

(13) M. Lüscher, Nucl.Phys. B180 (1981) 317

(14) P. Olesen, Phys.Lett. 160B (1985) 144
M. Caselle, R. Fiore, F. Gliozzi, R. Alzetta, Torino prepr. DFTT-7/86 (1986)

(15) C. Michael, P.E.L. Rakow, Nucl.Phys. B256 (1985) 640

(16) Ph. de Forcrand, J.D. Stack, Phys.Rev. Lett. 55 (1985) 1254

(17) C. Michael, Urbana prepr. P/85/12/196 (1985)
M. Campostrini, K.J.M. Moriarty, C. Rebbi, Brookhaven prepr. BNL-37980 (1986)

(18) See the last two references in (8)

(19) L.A. Griffith, C. Michael, P.E.L. Rakow, Phys.Lett. 150B (1985) 196
S. Ohta, M. Fukugita, A. Ukawa, Kyoto prepr. KEK-TH 118 (1986)

(20) M. Faber, H. Markum, M. Meinhart, paper contributed to this conference #1546

(21) A.C. Irving, T.E. Preece, C.J. Hamer, papers contributed to this conference #2852, 2860

(22) Talks at this conference by H. Satz and L.D. McLerrain
For recent summary papers see
B. Svetitsky, Phys.Rep. 132 (1986) 1
H. Satz, Ann.Rev.Nucl.Part.Sci. 35 (1985) 245
J. Cleymans, R.V. Gavai, E. Suhonen, Phys.Rep. 130 (1986) 219

(23) A.D. Kennedy, J. Kuti, S. Meyer, B.J. Pendleton, Phys.Rev.Lett. 54 (1985) 87
S.A. Gottlieb, A.D. Kennedy, J. Kuti, S. Meyer, B.J. Pendleton, D. Toussaint, R.L. Sugar, Phys.Rev.Lett. 55 (1985) 1958

(24) N.H. Christ, A.E. Terrano, Phys.Rev. Lett. 56 (1986) 111

(25) J. Kuti, talk at the parallel session

(26) S. Brown, talk at the parallel session

(27) F. Fucito, B. Svetitsky, Phys.Lett. 131B (1983) 165

(28) J. Polonyi, K. Szlachanyi, Phys.lett. 110B (1982) 395
B. Svetitsky, L. Yaffe, Nucl.Phys. B210 [FS6] (1982) 423.

(29) J. Kogut, M. Stone, H. Wyld, W. Gibbs, J. Shigemitsu, S. Shenker, D. Sinclair, Phys.Rev.Lett. 50 (1983) 393
R. Gavai, H. Satz, Phys.Lett. 145B (1984) 248
G. Curci, R. Trippicione, Phys.Lett. 151B (195) 145
J. Kiskis, paper contributed to this conference, 396
P. Suranyi, P. Harten, paper contributed to this conference, #1090

(30) A. Hasenfratz, P. Hasenfratz, U. Heller, F. Karsch, Phys.Lett. 143B (1984) 193
K.C. Bowler, A. Hasenfratz, P. Hasenfratz, U. Heller, F. Karsch, R.D. Kenway, H. Meyer-Ortmanns, I. Montvay, G.S. Pawley, D.J. Wallace, Nucl.Phys. B257 [FS14] (1985) 155
K.C. Bowler, F. Gutbrod, P. Hasenfratz, U. Heller, F. Karsch, R.D. Kenway, I. Montvay, G.S. Pawley, J. Smit, D.J. Wallace, Phys.Lett. B163 (1985) 367
K.C. Bowler, A. Hasenfratz, P. Hasenfratz, U. Heller, F. Karsch, R.D. Kenway, G.S. Pawley, D.J. Wall, Urbana prepr. ILL-(TH)-86-#44
R. Gupta, G. Guralnik, A. Patel, T. Warnock, C. Zemach, Phys.Rev.Lett. 53 (1984) 1721; Phys.Lett. 161B (1985) 352
A.D. Kennedy, J. Kuti, S. Meyer, B.J. Pendleton, Phys.Lett. 155B (1985) 414

(31) M. Lüscher, Phys.Lett. 118B (1982) 391

(32) M. Lüscher, Comm.Math.Phys. 104 (1986) 177

(33) M. Lüscher, Nucl.Phys. B219 (1983) 233

(34) M. Lüscher, G. Münster, Nucl.Phys. B232 (1984) 445

(35) P. Weisz, V. Ziemann, to be published

(36) J. Koller, P. van Baal, Nucl.Phys. B273 (1986) 387
P. van Baal, Nucl.Phys. B264 (1986) 548
P. van Baal, J. Koller, Stony Brook prepr. ITP-SB-86-31 and 76 (1986)

(37) A. Coste, A. Gonzales-Arroyo, C.P. Korthals-Altes, B. Söderberg, A. Tarancon, Marseille prepr. CPT-86/p.1876 (1986)

(38) B. Berg, A. Billoire, C. Vohwinkel, Phys.Rev.Lett. 57 (1986) 400

(39) Ph. de Forcrand, G. Schierholz, H. Schneider, M. Teper, in ref. (9)

(40) T. de Grand. C. Peterson, Colorado prepr. COLO-HEP-117 (1986)

(41) A. Patel, R. Gupta, G. Guralnik, G. Kilcup, S. Sharpe, San Diego prepr. UCSD-10P10-260 (1986)

(42) J. Smit, Nucl.Phys. B206 (1982) 309

(43) See, for instance, the discussion in the last ref. of (36)

(44) C.P. van-den Doel, D. Horn, paper contributed to this conference, #5827

(45) R. Bertlmann, C.A. Dominguez, M. Loewe, M. Perottet, E. de Rafael, paper contributed to this conference, #6416
J. Govaerts, L.J. Feinders, F. de Viron, J. Weyers, paper contributed to this conference, #7552

(46) H.R. Quinn, S. Gupta, Phys.Rev. D26 (1982) 499, ibid D27 (1983) 980
C. Taylor, B. McClain, Phys.Rev. D28 (1983) 1364
F. David, Nucl.Phys. B209 (1982) 433, ibid B234 (1984) 237

(47) V.A. Novikov, M.A. Shifman, A.I. Vainstein, V.I. Zakharov, Nucl.Phys. B249 (1985) 445
V.A. Novikov, M.A. Shifman, A.I. Vainstein, V.B. Voloshin, V.I. Zakharov, Phys.Rev. 116 (1984) 104

(48) A. Di Giacomo, G.C. Rossi, Phys.Lett. 100B (1981) 481
T. Banks, R. Horsley, H.R. Rubinstein, U. Wolff, Nucl.Phys. B190 [FS3] (1981) 692

(49) J. Gasser, H. Leutwyler, Phys.Rep. 87 (1982) 77

(50) E. Dagotto, A. Moreo, U. Wolff, Urbana prepr. ILL-(337)-86-#12 (1986)

(51) K.H. Mütter, K. Schilling, R. Sommer, paper contributed to this conference, #2518

(52) I.M. Barbour, P. Gibbs, J.P. Gilchrist, H. Schneider, G. Schierholz, M. Teper, Phys.Lett. 136B (1985) 385

(53) D. Barkai, K.J.M. Moriarty, C. Rebbi, Phys.Lett. 156B (1985) 385

(54) A. Billoire, R. Lacaze, E. Marinari, A. Morel, Nucl.Phys. B251 [FS13](1985) 581

(55) First ref. in (29)

(56) S. Itoh, Y. Iwasaki, T. Yoshié, prepr. UTHEP-155 (1986)

(57) R. Lacaze, contributed paper to this conference, #4499

(58) A. Morel, J.P. Rodrigues, Nucl.Phys. B247 (1984) 44
For a summary on baryon operators, see J. Smit, in Advances in Lattice Gauge Theory, p.1, eds. D.W. Duke, J.F. Owens, World Sci. (1985)

(59) M. Bochicchio, L. Maiani, G. Martinelli, G. Rossi, M. Testa, Nucl.Phys. B262 (1985) 331

(60) L. Maiani, G. Martinelli, CERN prepr. CERN-TH.4467/86 (1986)

(61) N. Cabibbo, G. Martinelli, R. Petronzio, Nucl.Phys. B244 (1984) 381

(62) R.C. Brower, G. Maturana, M.B. Gavela, R. Gupta, Phys.Rev.Lett. 53 (1984) 1318

(63) C. Bernard, T. Draper, G. Hockney, A.M. Rushton, A. Soni, Phys.Rev.Lett. 55 (1985) 2770

(64) G. Martinelli, talk at the parallel session

(65) M.K. Gaillard, B.W. Lee, Phys.Rev.Lett. 33 (1974) 108

(66) J.F. Donoghue, E. Golowich, B.R. Holstein, Phys.Lett. 119B (1982) 412

(67) A. Soni, talk at the parallel session

(68) F. Fucito, E. Marinari, G. Parisi, C. Rebbi, Nucl.Phys. B180 (1981) 369
F. Fucito, S. Solomon, in Advances in Lattice Gauge Theory, p. 64, Eds. D.W. Duke, J.F. Owens, World Sci. (1985)
R.V. Gavai, A. Gocksch, Phys.Rev.Lett. 56 (1986) 2659
R.V. Gavai, A. Gocksch, U. Heller, Brookhaven prepr. BNL 38449 (1986)

(69) G.G. Batrouni, G.R. Katz, A.S. Kronfeld, G.P. Lepage, B. Svetitsky, K.G. Wilson, Phys.Rev. D32 (1985) 2736
A. Ukawa, M. Fukugita, Phys.Rev.Lett. 55 (1985) 1854

(70) J. Polonyi, H.W. Wyld, Phys.Rev.Lett. 51 (1983) 2257
J. Polonyi, H.W. Wyld, J. Kogut, J. Shigemitsu, D.K. Sinclair, Phys. Rev.Lett. 53 (1984) 644
J. Kogut, in Advances in Lattice Gauge Theory, p. 19, eds. D.W. Duke, J.F. Owens, World Sci. (1985)

(71) S. Duane, Nucl.Phys. B257 FS14 (1986) 652
S. Duane, J. Kogut, Phys.Rev.Lett. 55 (1985) 2774
J. Kogut, Urbana prepr. ILL-(TH)-85-#79 (1985)

(72) A. Kennedy, J. Kuti, Phys.Rev.Lett. 54 (1985) 2473

(73) I.M. Barbour, N.-E. Behilil, P.E. Gibbs, G. Schierholz, M. Teper, DESY prepr. 84-087 (1984)
I.M. Barbour, in Advances in Lattice Gauge Theory, p. 55, eds. D.W. Duke, J.F. Owens, World Sci. (1985)

(74) N.H. Christ, A.E. Terrano, IEEE Transactions on Computers, c-33 (1984) 344

(75) P. Bacilieri et al. Amsterdam Conference on "Computing in High Energy Physics" (1985)
P. Bacilieri et al. prepr. Rom2F/85/6 (1985)
E. Marinari, prepr. ROM2F/86/005 (1986)

(76) F. Fucito, K.J.M. Moriarty, C. Rebbi, S. Solomon, Phys.Lett. 172B (1986) 235
E. Laermann, F. Langhammer, I. Schmitt, P.M. Zerwas, CERN prepr. TH.4394/86 (1986)
M. Fukugita, Y. Oyanagi, A. Ukawa, prepr. UTHEP-152 (1986)

(77) H. Joos, I. Montvay, Nucl.Phys. B225 (1983) 565
E. Laermann, F. Langhammer, I. Schmitt, P.M. Zerwas, CERN prepr. Th.4393/86 (1986)

(78) J. Kogut, Phys.Rev.Lett. 56 (1986) 2557, and Urbana prepr. ILL-(TH)-86- #35 (1986)

(79) R.V. Gavai, F. Karsch, Phys.Rev.Lett. 57 (1986) 40

(80) T. Banks, A. Ukawa, Nucl.Phys. B225 [FS9] (1983) 145
P. Hasenfratz, F. Karsch, I. Stamatescu, Phys.Lett. 133B (1983) 221
T. DeGrand, C. DeTar, Nucl.Phys. B225 [FS9] (1983) 590
J. Bartholomew, D. Hochberg, P. Damgaard, M. Gross, Phys..Lett. 133B (1983) 218
M. Ogilvie, Phys.Rev.Lett. 52 (1984) 1369
H. Matsuoka, Phys.Lett. 140B (1984) 233
F. Green, F. Karsch, Nucl.Phys. B238 (1984) 297, Phys.Rev. D29 (1984) 2986
T. Celik, J. Engels, H. Satz, Phys.Lett. 133B (1983) 427, Nucl.Phys. B256 (1985) 670
R.V. Gavai, F. Karsch, Urbana prepr. ILL-TH-85- #19 (1985)
Ph. de Forcrand, I.O. Stamatescu, paper contributed to this conference,
J. Kogut, D.K. Sinclair, Urbana prepr. ILL-(TH)-86- #46 (1986)
see also the second ref. in (68)

(81) R. Gupta, G. Guralnik, G.W. Kilcup, A. Patel, S.R. Sharpe, Los Alamos prepr LA-UR-86-3054 (1986)
D.K. Sinclair, talk at the Brookhaven conference (1986)

(82) R. Pisarski, F. Wilczek, Phys.Rev. D29 (1984) 338
H. Goldberg, Phys.Lett. 131B (1983) 133
A.J. McKane, M. Stone, Nucl.Phys. B163 (1980) 169
A. Margaritis, A. Patkos, Bonn prepr. HE-86-12 (1986)

(83) I. Babour, N.E. Behilil, E. Dagotto, F. Karsch, A. Moreo, M. Stone, H.W. Wyld, Urbana prepr. ILL-(TH)-86 #23 (1986)

(84) For a resent summary, see
J. Jersak, Aachen prepr. PITHA 85/25 (1985)

(85) J. Jersak, C.B. Lang, T. Neuhaus, G. Vones, Phys.Rev. D32 (1985) 2761
V.P. Gerdt, A.S. Ilchev, V.K. Mitrjushkin, I.K. Sobolev, A.M. Zadorozhny, Nucl.Phys. B265 [FS15] (1986) 145
V.P. Gerdt, A.S. Ilchev, V.K. Mitrjushkin, A.M. Zadorozhny, Z.Phys. C29 (1985) 363
H.G. Evertz, J. Jersak, D.P. Landau, T. Neuhaus, PITHA 85/23
K. Decker, I. Montvay, P. Weisz, Nucl. Phys. B268 (1986) 362
J. Jersak, in Advances in Lattice Gauge Theory, p. 241, eds. D.W. Duke, J.F. Owens, World Sci. (1985)

(86) D.J. Callaway, R. Petronzio, Nucl.Phys. B267 [FS12] (1986) 253

(87) I. Montvay, DESY 85-005 (1985)
W. Langguth, I. Montvay, Phys.Lett. 165B (1985) 135
I. Montvay, in Advances in Lattice Gauge Theory p. 266, eds. D.W. Duke, J.F. Owens, World Sci. (1985)
I. Montvay, W. Langguth, P. Weisz, DESY 85-138 (1985)

(88) I.-H. Lee, J. Shigemitsu, Phys.Lett. 169B (1986) 392, papers contributed to this conference, #8257, #8249
R.E. Schrock, Phys.Lett. 162B (1985) 165, Nucl.Phys. B267 (1986) 301, prepr. ITP-SB-86-41 (1986)
V.G. Gerdt, A.S. Ilchev, V.K. Mitrjushkin, paper contributed to this conference, #9407
V.P. Gerdt, V.K. Mitrjushkin, A.M. Zadorozhny, paper contributed to this conference, #9415

(89) R. Dashen, H. Neuberger, Phys.Rev.Lett. 50 (1983) 1897

(90) D.J.E. Callaway, Nucl.Phys. B223 (1984) 189
M.A. Beg, C. Panagiotakopoulos, A. Sirlin, Phys.Rev.Lett. 52 (1984) 883

A. Bovier, D. Wyler, Phys.Lett. 154B (1985) 43

(91) A. Hasenfratz, P. Hasenfratz, Tallahassee prepr. FSU-SCRI-86-30 (1986)

(92) P. Hasenfratz, J. Nager, Bern prepr. BUTP-86/20 (1986)

(93) A. Hasenfratz, P. Hasenfratz, Nucl. Phys. B270 (1986) 687

Plenary Session

Cosmology
M. Yoshimura (KEK)

High Energy Nuclear Interactions
L. McLerran (Fermilab)

Chairman
Huang Tao (Beijing)

Scientific Secretaries
P. Rankin (SLAC)
H. Yamamoto (LBL)

COSMOLOGY

M. Yoshimura

National Laboratory for High Energy Physics (KEK),
Oho-machi, Tsukuba, Ibaraki 305 Japan

A close interplay between particle physics and cosmology is reviewed. Recent development of particle physics serves to solve some outstanding problems of the standard big bang cosmology, while cosmology and astrophysics guide particle physics in a region of parameter space in which laboratory data are futile to yield useful information.

1. INTRODUCTION

Study of our universe, especially its evolution, has gained popularity in the scientific community only recently, dating back to the discovery of the microwave background radiation in 1965. Twenty years later at this time, the field of cosmology is rapidly expanding, at a rate totally unexpected in previous years. Contact with particle physics is still widening. A minimal knowledge of cosmology has become a must of particle physicists.

This report describes the present status of cosmology from a particle physics standpoint. Instead of being comprehensive in every aspect of an extremely rich and diverse set of subjects, I shall select topics that are most useful to particle physics, and present a variety of arguments that best illustrate the power of cosmology and astrophysics. Before starting, however, it is perhaps useful to recall successes and problems of the standard big bang cosmology.

The big bang cosmology[1] is defined by a few simple principles. The first of these is called cosmological principle and asserts that the universe is homogeneous and isotropic as a whole. This can be quantified by the Robertson-Walker metric which contains two parameters. The scale factor a(t) describes how the universe evolves as time goes on, and measures the proper distance, say, between two galaxies. The other parameter k tells a global topology of the universe, and distinguishes whether the universe is open (k = -1), flat (k = 0), or closed (k = 1).

The second input is the law of gravity which is taken to be Einstein's general relativity with the vanishing cosmological constant. It gives, combined with the cosmological principle, a set of differential equations for the scale factor, where the spacetime geometry is determined by the energy-momentum tensor;

$$\left(\frac{\dot{a}}{a}\right)^2 + \frac{k}{a^2} = \frac{8\pi G}{3}\rho , \qquad (1)$$

$$(\rho a^3)^{\bullet} = -3pa^2\dot{a} . \qquad (2)$$

When supplemented by an equation of state, a relation between the energy density ρ and the pressure p, this yields the scale factor as a function of time: $a \propto t^{1/2}$ for radiation dominated (RD) epoch with $\rho = 3p \propto a^{-4}$, and $a \propto t^{2/3}$ for the matter dominated (MD) epoch with $p \ll \rho \propto a^{-3}$ (effect of the curvature term ∝ k is ignored for simplicity). A useful mnemornic to keep in mind is the temperature-time relation valid in RD,

$$t \sim 10^{-6}\ \mathrm{sec}/\left(\sqrt{N_F}\ (T/\mathrm{GeV})^2\right) , \qquad (3)$$

where N_F is an effective number of degrees of freedom and is of the order, 10 - 100.

The big bang cosmology has been successful in explaining three fundamental facts of our universe; the Hubble expansion, presence of the black-body radiation at 3K, and a large abundance of light elements (^{4}He, D, ^{3}He, ^{7}Li) which is difficult to process in stellar environment. This success relies on the standard microphysics of strong, electromagnetic, and weak interactions, which is well established by laboratory experiments.

It is worthwhile to note that although these vintage successes are an integral part of our scientific achievements, some of the basic parameters in the big bang model are not accurately measured. The Hubble constant (actually is a function of time, but usually is denoted by its present value) H_o is $\sim$100 km $s^{-1}Mpc^{-1}$(1Mpc = 3 × 10^{24} cm), within an uncertainty factor of $\sim$2.[2] We shall denote it as H_o = 50 $h_{1/2}$ km s^{-1} Mpc^{-1}, where $1 \lesssim h_{1/2} \lesssim 2$. The age t_o of our universe is determined by two methods: (1) $t_o = (18 \pm 3) \times 10^9$ years from evolution of globular clusters[3] which are oldest objects in galaxies. (2) t_o = 11 $\sim$ 30 Gy from nucleochronology[4]. This value roughly satisfies the relation of matter dominated universe, $H_o t_o \simeq 2/3$. With a better measurement one can hope to check a consistency with the Ω parameter (defined shortly), $H_o t_o = f(\Omega)$.

The matter content of our universe is specified by Ω, defined as $\Omega \equiv \rho_o/\rho_c = 8\pi G\rho_o/3H_o^2$ where ρ_c is called closure density. The significance of this parameter is that the value of $\Omega = 1$ divides the closed ($\Omega > 1$) and the open ($\Omega < 1$) universe. Two conventional methods of estimation yield, (1) $\Omega h_{1/2}^2 \sim 0.14$ from the luminosity count[2] corrected by the mass to light (M/L) ratio, (2) 0.3 $\sim$ 0.5 from the infall of our galaxy towards the Virgo cluster (closest, local supercluster)[5]. More recent estimates, however, gives considerably larger values consistent with the flat value ($\Omega = 1$); (3) 0.85 ± 0.16 from the dipole anisotropy measured by IRAS galaxies[6], (4) 0.6 $\sim$ 2.0 from the number count of distant galaxies, using a photometric technique[6]. It seems that there is some dispute over these recent estimates, and one should perhaps wait for a few more years to settle the Ω value.

On the other hand, the microwave temperature is fairly accurately known[8]: $T_o = (2.75 \pm 0.05)$K. The measured fluctuation[9] is in the form of dipole with an amplitude of $\delta T/T \sim 1 \times 10^{-3}$. This is usually interpreted as a local motion of our galaxy.

The helium fraction Y by mass is measured to be $\sim$0.24 ± 0.02[10]. This quantity is sensitive to the expansion rate when the universe is a few minutes old (T $\sim$ 0.1 MeV). The more rapidly the expansion proceeds, the less neutron is transformed into proton, hence the more ^{4}He is synthesized. The number of neutrino species of mass less than 1 MeV contributes to the expansion rate, increasing Y by 0.013 per one species[11]. An upper bound of He abundance then gives the upper bound of light neutrino species: $N_\nu \lesssim 4$[11]. The CERN group[12] argues that one should allow more uncertainty factors in the input parameters, especially in the combined abundance of ^{3}He plus D (which is used to elliminate dependence on the baryon to photon ratio of $\sim(3\text{-}10) \times 10^{-10}$) to obtain a less restrictive bound of $\sim$5.5. This argument is however not without flaw[13]. In any case direct laboratory bounds, derived from the Z width and the process of $e^+e^- \to \gamma\nu\bar{\nu}$, may soon be improved[14] to a comparable level of cosmological argument.

The standard big bang cosmology thus gives a coherent picture of our universe with a parameter range of roughly, $t_o \simeq H_o^{-1}$ and $\Omega_o \simeq 1$, an impressive achievement. It has however left many fundamental problems unsolved, which is now discussed.

(1) Baryon Asymmetry. The universe is known to be asymmetric in baryons and antibaryons[15]. This asymmetry is quantified by the present number ratio of baryon to photon, roughly of $\sim 10^{-10}$. Whether this number is an initial condition imposed on our universe remains to be answered.

(2) Horizon Problem[16]. The homogeneity was assumed in the standard model, and this is indeed well satisfied by absence of the microwave background fluctuation. However, microwaves emitted from two regions in the sky, say, a few degrees apart have never been in causal contact in the standard model. The fact that regions well beyond particle horizon of $\sim$ct exhibit exactly the same temperature is very mysterious.

(3) Flatness or Oldness Problem[16]. The present mass density is very roughly equal to the closure density: $\rho_o = O(10^{-2} - 10)\rho_c$. This near equality becomes even more precise when one goes back to early times because $(\rho/\rho_c) - 1 \propto t$ (RD) or $t^{2/3}$ (MD). This is equivalent to saying that the curvature term, k/a^2, in the Einstein equation has never been important in all the history of our universe. The fine tuning to the critical density requires an accuracy of $\sim 10^{-60}$ at the Planck epoch (T $\sim$ 10^{19} GeV).

(4) Cosmological Constant. There is no principle that forbids presence of the cosmological constant, but observations tell that $|\Lambda|^{1/2} \lesssim O(10^{-32}\text{eV})$. This mass scale is extremely small compared to any masses that appear in particle physics.

(5) Formation of Structures. Galaxies and clusters of galaxies are believed to have a cosmological origin, yet the standard model has not been successful in accounting for this problem.

(6) Initial Singularity. The classical cosmology starts from a curvature singularity at the beginning where description in terms of the classical general relativity breaks down.

Recent developments of particle physics cosmology have vindicated that these outstanding puzzles of the big bang model are not something embarassing, but rather welcome hints to go beyond the standard gauge theory. This is particularly fortunate because laboratory data of particle physics are completely consistent with the standard theory and do not show any clue beyond what we already established. It is now time to review these new achievements of particle physics cosmology.

2. BARYON ASYMMETRY AND INFLATION

Let us now go beyond the standard SU(3) × SU(2) × U(1) theory and assume some kind of grand unified theories (GUT)[17]. The problem of baryon asymmetry is how to explain one excess of baryon over 10^{10} pairs of baryon and antibaryon. Suppose that the universe was initially symmetric between baryons and antibaryons, which is assumed here to be prepared either by inflation or by some sort of baryon nonconserving wash-out process.

It is well known that three conditions are necessary to generate the baryon asymmetry[18]: (1) Baryon nonconservation, (2) C-, CP-violation, (3) Departure from thermal equilibrium. The first two conditions are obeyed by any decent GUT models, while the third is provided by the expansion of our universe. To explain how nontrivial constraints arise from these conditions, take the decay process of X bosons[19] that mediate baryon nonconservation; $X \to \bar{q}\bar{q}$ and $q\ell$, $\bar{X} \to qq$ and $\bar{q}\bar{\ell}$. If this decay occurs simultaneously with the inverse process so that equilibrium abundance is diminished by the Boltzmann factor, the asymmetry resulting from the decay will be canceled by the inverse decay. For this not to happen, it is necessary that the decay starts only when the temperature becomes low enough to kenematically suppress the inverse decay. This yields a relation, $\alpha_x m_x \sim \sqrt{N_F} \cdot T^2/m_{p\ell}$ with $T < m_x$. Namely[19],

$$m_x > \alpha_x (N_F)^{-1/2} m_{p\ell} \ . \qquad (4)$$

A more careful analysis[20] gives the lower mass bound of $\sim 5 \times 10^{15}$ GeV for the gauge boson of the minimal SU(5) model. The corresponding bound for the Higgs boson H_x that mediates baryon nonconservation can be made much smaller, because its coupling to quarks and leptons tends to be smaller. A more quantitative approach[21] involves integration of time evolution of the baryon number, using a simplified set of Boltzmann equations.

The magnitude of the baryon asymmetry is computed by summing baryon numbers of each decay channel. The asymmetry ε per a pair of decays (X and $\bar{X}$) is then related to the present ratio of baryon to photon number by a calculable factor, $(n_B/n_\gamma)_o \sim 10^{-3} \cdot \varepsilon$. The dilution factor of $\sim 10^{-3}$ here is due to that at later epochs heavy particles (quarks, leptons, gauge bosons, ...) decay or annihilate each other to create more photons. For the simplest, two channel decay of X boson the asymmetry is given by

$$\varepsilon = (\tfrac{1}{3}\gamma_\ell - \tfrac{2}{3}\gamma_q - \tfrac{1}{3}\bar{\gamma}_\ell + \tfrac{2}{3}\bar{\gamma}_q)/(\gamma_\ell + \gamma_q)$$
$$= (\bar{\gamma}_q - \gamma_q)/(\gamma_\ell + \gamma_q) \ , \qquad (5)$$

where by CPT theorem $\gamma_\ell + \gamma_q = \bar{\gamma}_\ell + \bar{\gamma}_q$. In perturbation theory amplitudes of indivisual decay processes are expanded in powers of (several) coupling constants which are symbolically denoted by g_i for i - th power. The asymmetry is then roughly expressed as[22]

$$\varepsilon \sim 4 \ \mathrm{Im}(g_1 g_2^*) \cdot \mathrm{Im}\ (F_1 F_2^*)/(|g_1|^2 \cdot |F_1|^2). \qquad (6)$$

Besides complex coupling factors of $g_1 g_2^*$ reflecting a source of CP violation, a relative phase between the Born amplitude F_1 and a higher loop amplitude F_2 is needed to yield a nonvanishing asymmetry. This second phase is a generalization of the rescattering phase that arises between final decay products. What is important here is coexistence of two phase factors; a phase that does change its sign when a particle is transformed into its antiparticle, and the other phase that does not change. Appearance of an extra coupling ensures that the baryon to photon ratio is at most of the order, $10^{-3}\ \alpha$ with $\alpha \lesssim 10^{-2}$.

There have appeared in the literature many calculations[23] of the baryon asymmetry in specific GUT models. Only a few of them are mentioned here. The minimal SU(5) model with three generations yields too small a value[24], being in the third order of the Yukawa coupling $\alpha_H = f^2/4\pi$; $\varepsilon = O(\alpha_H^3) < 5 \times 10^{-17} \cdot (m_t/200\ \mathrm{GeV})$. This is due to that there is only one source of CP violation in this model, the Yukawa coupling matrix of one (5) Higgs multiplet. It is easy to cure this difficulty. One simply has to add another 5, which can result in the asymmetry of the first order α_H.

In models such as some of SO(10) GUT there is an intricate relation between CP violation and breaking of the left-right symmetry. By employing the seesaw mechanism[25)] of generating the neutrino mass, one can relate[26)] the baryon asymmetry to the neutrino mass. Roughly speaking, the more disparity between the right-handed (N_R) and the left-handed (ν_L) neutrino is generated, the more baryon asymmetry is produced. In the simplest model of this kind one obtains[27)] the largest mass of neutrino to be less than ~ 2 eV $(m_t/50 \text{ GeV})^{1.8}$.

In my opinion there is unfortunately no believable model that directly connects the baryon asymmetry and the CP violation in the low energy K meson system, despite many interesting attempts[28)]. The baryon asymmetry certainly gives useful constraints on model building, but there is no simple, straightforward relation with the proton lifetime. This makes some of us believe the general idea of GUT despite lack of the proton instability.

The inflationary universe scenario[29),30)] is an attempt to solve the horizon and the flatness problems. The idea of inflation is indeed very simple to make us believe that it has an element of truth.

The phase transition of gauge symmetries may undergo a stage of supercooling, namely the first order phase transition instead of a smooth second order transition. This happens when an effective potential for the order parameter (the Higgs vacuum expectation value in this case) has a symmetry preserving local minimum, separated from the symmetry breaking true vacuum by a barrier. The universe may then be trapped in the metastable state before making the phase transition. If this occurs, one has to take into account the Higgs contribution in the Einstein equation besides the usual thermal energy term,

$$\rho \text{ in } (1) = \rho_{th} + V_o ,$$

together with the field equation for the Higgs field,

$$\ddot{\phi} + 3(\dot{a}/a)\dot{\phi} = -\partial V/\partial\phi . \qquad (7)$$

Suppose that after a period of supercooling the Higgs potential is very flat near the origin of the Higgs field, a value favored by high temperature phase. Then an almost constant density V_o may dominate over the rapidly decreasing thermal term $\rho_{th}(\propto T^4 \propto a^{-4})$, and the scale factor may increase exponentially,

$$a(t) \sim \exp\left(\sqrt{8\pi G V_o/3}\cdot t\right) . \qquad (8)$$

The region within causal relation (particle horizon) is also exponentially stretched out. The horizon and the flatness problems are thus solved if the e-folding factor of (8) exceeds a number of ~ 65.

The exponential expansion is terminated by rapid oscillation of the Higgs field around the valley located at the true vacuum. This rapid oscillation induces particle creation[31)]. Produced particles get thermalized if interaction among them is strong enough, compared to the expansion rate.

In this picture our universe within the present horizon of $\sim ct$ is considered to have evolved from a tiny bubble of thermal contact prior to the phase transition. Much of information on the initial Friedmann phase has been lost during the inflationary period. For instance, magnetic monopoles that are usually created[32)] by the second order phase transition are diluted away, essentially to none. The baryon asymmetry must then be created after reheating. We shall not be able to know, for a long time to come, existence of many other universes that exist outside our own bubble universe. The inflation thus predicts that in the patch of our universe the present Ω value is effectively unity to a a very good accuracy.

More detailed investigation of the inflation has revealed some problems, which I now mention. First of all, the semiclassical description of the Higgs field ϕ in terms of a classical equation has been questioned[33)]. It was argued that near the critical temperature T_c a large local fluctuation of fields to magnitudes of $\sim \pm\phi_c$ (the true vacuum value) may occur and dynamics of the phase transition may then be governed by growth and coalescence of domain walls, invalidating the effective potential approach which only describes an average behavior of the field ϕ. This is a legitimate criticism if the fluctuation is not damped appreciably. However, in the inflationary universe the expansion gives a damping force against local fluctuations, and the real issue is which of the local fluctuation or the damping wins. Recently, it was shown[34)] that for a sufficiently weak coupling the damping is dominant and inflation is indeed realized.

Although inflation pushes essentially all inhomogeneities out of our horizon, it inevitably creates[35)] new quantum fluctuation which arises[36] in the exponentially expanding universe (de Sitter universe). The density fluctuation thus generated is scale invariant (independent of the wavelength of fluctuation) when it reenters the Hubble radius (H^{-1}). This spectrum, called Harrison-Zeldovich spectrum[37], has nice features when applied to galaxy formation. The problem, however, is that the magnitude of fluctuation is too large[35] (of ~10) to be compatible with observation, in the favored SU(5) model based on the Coleman-Weinberg (CW) mechanism of symmetry breaking[38].

Another problem with the CW SU(5) model is presence of competing phases[39]. Although the true vacuum of SU(3) × SU(2) × U(1) is at the absolute minimum of the effective potential, there is a steepest descent path towards the local minimum of SU(4) × U(1). It is likely that the universes is first trapped in the wrong metastable phase of SU(4) × U(1), which will then tunnel into the true vacuum, as indeed verified by numerical computations. This causes troubles such as gross inhomogeneity of the universe.

Problems with the large density fluctuation and the competing phase are not difficult to overcome. One should simply observe that the inflaton field which gives rise to inflation need not be the order parameter of a phase transition. For instance, the inflaton ϕ can be a gauge singlet[40] such that it is arranged to have a nice flat part of the potential, yielding a small density fluctuation. In this regard, supersymmetric models are of great help[41]. Cancellation of vacuum energy between bosons and fermions tends to give smaller couplings, hence smaller density fluctuation.

The inflation may even have nothing to do with a phase transition. A long time ago, even before Guth's proposal, Starobinsky[42] considered an inflationary model that uses a combination of $R^2(R^2_{\alpha\beta\gamma\delta}$, $R^2_{\alpha\beta}$, $R^2)$ terms. A completely different approach is Linde's chaotic inflation[43], in which the inflaton field is randomly distributed in different space regions in the beginning of universe. The inflaton field located away from the minimum then slowly rolls down towards the stable point. During this slow rollover a sufficient inflation may occur. A yet another possibility is that the inflaton field may be associated[44] with the presence of extra dimensions. In higher dimensional theories such as Kaluza-Klein and superstring theories the size b of the extra compact space is regarded as a scalar field ϕ in an effective four dimensional field theory, more precisely $\phi \sim \ln b$. During the process of compactification, $b(t) \to b_o$, the ordinary 3-space may be inflated. Although all these ideas are interesting in their own right, at this point of development it is difficult to pinpoint what the inflaton field is. The idea of inflation is, however, presumably general enough to have a lasting value. Essence of inflation lies in that a scalar field rolls down very slowly in an almost flat potential.

An important unsolved problem related to all inflationary models is how to exit to the hot Friedmann universe. Reheating requires some process of dissipation, or in abstract terms, diffusion of the vacuum energy into a larger phase space of a great many particles. A phenomenological approach[31] that uses a particular form of dissipative term, $\dot{\phi}\phi$, in the equation of motion for ϕ can give a parameter range which successfully describes a graceful exit to the Friedmann universe. Attempts[45] to compute the friction term from the first principle are however very much limited, and not realistic. One should note that unless a smooth transition to the hot universe is realized, inflation does not serve to resolve its original objective.

In many SUSY related models the reheating temperature, assuming that the reheating is ever possible, comes out to be rather low, 10^{10} GeV or even much less. It thus becomes important to have mechanisms of generating the baryon asymmetry at lower temperatures.

The most direct mechanism is baryon production via inflaton decay[46]. During the reheating process the inflaton may nonthermally produce X bosons even if the reheating temperature is not high enough to produce X and $\bar{X}$ in thermal abundance. When they decay, X boson can generate the baryon asymmetry. Assuming that the inflaton mainly decays into X-$\bar{X}$ pair, one finds a baryon asymmetry generated this way to be a factor of T_R/m_x (T_R = reheating temperature) less than the standard estimate. The reheating temperature can thus be low, of $\sim 10^{-4}\, m_x$ to give $n_B/n_\gamma \sim 10^{-10}$, if the asymmetry ε in (5) is of $\sim 10^{-3}$.

The severe constraint on the temperature scale of baryon generation is based on the out of equilibrium condition (4), which relies on the decay rate of X boson. If the decay interaction is of nonrenormalizable type, such as the Majorana lepton decay[47], $F \to qqq$ and $\overline{qqq}$, then the out of equilibrium condition gives an upper limit of the lepton mass, $m_F < (\tilde{G}^{-2} m^{-1}_{p\ell})^{1/3}$, instead of the lower bound.

This is not a severe constraint. A recent model[48] of this kind does not predict proton decay, but is rather contrived.

An entirely different approach was taken by Affleck and Dine[49]. They observe that SUSY models have many degenerate vacua prior to SUSY breaking and that scalar quark and scalar lepton condensates may not readily disappear if curvature of the potential is much less than the Hubble term. The scalar quark condensate carrys a baryon number, and when it eventually decays after SUSY breaking, this baryon number is transformed into a real baryon number of quarks. The universe in this picture remains cold until the condensate decays, then thermalizing to a rather low temperature of $\sim 10^4$ GeV. The baryon to photon ratio in this scenario can be considerably larger than that given by the usual estimate, depending on magnitude of the initial condensate. Further discussion of this mechanism and its application is found in Ref. (50).

Finally Kuzmin et al.[51] pointed out a possible mechanism of strong baryon nonconservation via electroweak interaction at finite temperatures. It has long been known since the work of 't Hooft[52], that instanton effect of electroweak origin gives rise to baryon number nonconservation, but at a completely negligible rate at zero temperature. At finite temperatures there may occur a second effect of barrier crossing over an unstable soliton, with a rate of $\exp(-2F/T)$. Kuzmin et al. claim that the rate of this process is obtained by modifying the soliton mass[53] at zero temperature with a minimal change, namely by allowing temperature dependence of parameters in the mass F. If this indeed is correct, baryon nonconserving processes may get thermalized at temperatures above 200 GeV, and the baryon asymmetry produced by a GUT process will be destroyed. Fukugita and Yanagida[54] however found an interesting possibility that even in this case a relic baryon number get transformed from a lepton number if the lepton number is created at higher temperatures by some non GUT process. This is possible because the electroweak baryon nonconservation preserves B-L conservation. Unfortunately the naive estimate of baryon non-conserving rate by Kuzmin et al. does not include the entropy factor, the determinant factor at finite temperatures, thus one is not sure of how probable this saddle point path of barrier crossing is. The problem raised here, however, calls for a further serious consideration.

3. CONSTRAINTS ON PARTICLE PHYSICS

We have shown how particle physics has contributed to solve some of the fundamental problems in cosmology. I will now discuss how cosmology guides particle physics in a parameter space region often inaccessible to laboratory experiments. The choice of topics included here is not extensive and only intended to illustrate a variety of different arguments.

Massive Neutrino

A massive stable particle can contribute to the energy density of the universe if their abundance is large. One can roughly limit the allowed mass density by the closure density ρ_c of $\sim 4.7 \times 10^{-30}$ gr $cm^{-3} h^2{}_{1/2}$. The inequality, $m_x n_x < \rho_c$, thus gives a bound on the mass m_x if one knows the number density n_x.

In the case of massive neutrino estimate of the number density differs in different mass region. If the neutrino is light and $m_\nu < 1$ MeV, then the neutrino decouples out of weak interaction at a temperature of ~ 1 MeV while it is still relativistic. Barring the lepton number degeneracy, the neutrino number density is roughly comparable to the photon density. More precisely, $n_\nu \sim 100$ cm^{-3}, which yields the well known mass bound[55]; $m_\nu < 30$ eV $h^2{}_{1/2}$.

If the neutrino mass is higher and $m_\nu > 1$ MeV, the number density at decoupling is suppressed by the Boltzmann factor $e^{-m/T}$. The relic abundance can be estimated[56] by comparing the interaction rate ($\nu\bar{\nu} \to e^+e^-$ is one of the main processes) with the expansion rate and determining when two rates become comparable. The relic abundance is thus given by

$$n_\nu/n_\gamma \approx \left(m_\nu m_{p\ell} \langle\sigma v\rangle_A\right)^{-1}, \qquad (9)$$

with $\langle\sigma v\rangle_A$ the average annihilation rate. In the mass region of 1 MeV $\ll m_\nu \lesssim G_F^{-1/2}$, $\langle\sigma v\rangle_A \sim G_F^2 m_\nu^2/2\pi$, whereas for $m_\nu \gg G_F^{-1/2}$, $\langle\sigma v\rangle_A \propto m_\nu^{-2}$. Thus, one finds an allowed mass range[57], 3 GeV $h_{1/2}^{-1} < m_\nu < 3$ TeV $h_{1/2}^{-1}$. As might sound paradoxical at first sight, the formula (9) shows that the weaker the interaction or the lighter the particle is, the more relic is left. This is one of the many reasons why cosmology gives information complementary to laboratory probe.

A similar argument can be extended[58] to the stable photino, the supersymmetric partner of the photon. Difference in this case is that the photino is a Majorana particle which has only axial-vector couplings to matter in the Fermi type of interaction, $\tilde{\gamma}\tilde{\gamma}\cdot\bar{f}f$. Thus, the amplitude of annihilation consists of a P-wave plus a new S-wave contribution proportional to the mass of fermion, m_f, unlike the mass independent S-wave term for the Dirac neutrino. This makes the annihilation rate depend on the photino mass in a quite different way from the Dirac case. Detail of the mass constraint is found in Ref. (58).

Unstable neutrinos can give rise to a variety of phenomena if their decay products contain a radiative component, γ or $e^{\pm}$, and their lifetime is short enough. For a mass range of $10 \sim 100$ eV the UV background gives a forbidden lifetime region[59] of $10^{13} \sim 10^{22}$ sec. In the 1 keV region absence of distortion in the microwave radiation spectrum limits[60] the lifetime to $\tau < 2 \times 10^6$ sec $\left(1 + (1\ \mathrm{keV}/m)^2\right)^{-1/2}$ for $m < 100$ keV. For a still larger mass range of $1 \sim 100$ MeV, one demands that photofission[61] of the primordial D and ^{4}He should not take place, yielding a limit, $\tau \lesssim O(10^3$ sec). In a small mass region of $1 \sim 10$ MeV supernova[62] give a very good bound, $\tau \lesssim O(10^{-3}$ sec). This comes about because otherwise the neutrino decays in supernova envelope, converting too much energy into the visible part.

These cosmology-astrophysics bounds become truly powerful when they are combined with complementary laboratory bounds. This is particularly the case for the tau neutrino. The ARGUS mass bound[63] (<70 MeV), the unsuccessful search for a massive neutrino in $\pi - e$ decay[64] and the beam damp search[65] of $\nu_\tau \to \nu + e^+ + e^-$, augmented by the photofission argument of light elements here, leave a small, or even no[66] parameter space of (m, τ).

Light Bosons

A variety of new light bosons have been introduced in particle physics, most of which are associated with a spontaneous breaking of some global symmetry. A popular example of these is the invisible axion[67] associated with the Peccei-Quinn (PQ) symmetry[68] breaking to solve the strong CP problem. Other examples include the majoron[69] associated with the lepton number nonconservation, and the familon[70] associated with the generation symmetry breaking.

A common feature of these bosons is that they have a weak pseudoscalar (PS) coupling to matter besides having a tiny (or vanishing) mass. In the case of the invisible axion the PS coupling is $\sim m_i/f_A$ ($\sim 10^{-13}$ for i = electron and the PQ symmetry breaking scale $f_A = 10^{10}$ GeV) and its mass is $\sim 10^{-3}$ eV$\cdot(10^{10}$ GeV$/f_A)$. One of the best laboratory bound[71] comes from the (g-2) factor of the electron; $|g_e| < 0.7 \times 10^{-5}$.

If the mass of these bosons is less than ~ 1 keV, a typical thermal energy available in stars, then astrophysics yields a far better bound on their coupling[72],[73]. The basic reason for this is that unless their coupling to matter is limited, these light bosons are emitted from the stellar core and may deplete stellar energy, thus destroying our present understanding of stellar evolution. The most reliable bound of this kind is set by demanding that the energy loss due to boson emission should not exceed the sun's luminosity. It was once thought[73] that the Primakoff process is dominant in the sun for the invisible axion, but it has recently been shown[74] that a Debye screening renders this contribution negligible and the bremsstrahlung process[75] off electron gives the most important effect. The electron PS coupling is then bounded from above; $|g_e| < 0.5 \times 10^{-10}$, which implies that the PQ symmetry breaking scale $f_A > 1 \times 10^7$ GeV. This bound is five orders of magnitude better than the laboratory bound, mainly due to the slow energy generation rate of $\sim$ several MeV/10^{10} years $\cdot$ nucleon.

Much better bounds are obtained by using a variety of arguments with white dwarfs, red giants and neutron stars. Low luminosity ($1 \sim 10^{-3} L_\odot$) white dwarfs become too short-lived in contradiction to observation unless $f_A > 1 \times 10^9$ GeV[76] for the invisible axion. Helium ignition may never occur in red giants unless $f_A > 4 \times 10^9$ GeV[77]. Cooling via axion emission overtakes that of the standard neutrino emission of neutron stars at $f_A = (0.6 \sim 3) \times 10^9$ GeV[78]. These bounds are based on reasonable arguments, but require a fair amount of astrophysical input besides raw data. An obvious conclusion of these model dependent arguments is that if the invisible axion with f_A below 10^9 GeV exists, axion processes hitherto neglected become very important and in certain cases dominant in stellar evolution.

The invisible axion is an example of many fields in particle physics zoo that almost decouple from matter. Another important example is the Polonyi field[79] that was invented to break the local supersymmetry in a hidden sector. Cosmological fate of these decoupled fields is very weird, and they sometimes cause too much troubles.

At early epochs of cosmological

evolution, there inevitably exists coherent mode of these fields with zero momentum. This particular mode usually disappears via interaction, but a unique feature of the decoupled field is that the coherent mode is not easily dissipated away precisely because of its decoupling. It may either yield too much mass density, or too much entropy if it decays.

A basic equation that governs evolution of a coherent field is given by

$$\ddot{\phi} + (3H + \Gamma_\phi)\,\dot{\phi} + m_\phi^2\,\phi = 0. \qquad (10)$$

This is an equation for damped oscillation, with damping given either by the Hubble expansion (H) or by a field decay (Γ_ϕ). Let us first discuss the case of invisible axion[80]. Decay lifetime of the invisible axion Γ_ϕ^{-1} is much larger than the age of universe of $\sim H^{-1}$ for an interesting range of axion parameters. When the temperature T >> the electroweak scale, an initial axion field, most presumably of order f_A, remains constant (time independent) because no mass term (m_A) develops. Below the electroweak scale the axion mass gradually turns on and damped oscillation fully sets in when $H \lesssim m_A$. This occurs when the temperature $T < T_D$ with $T_D \sim 0.7$ GeV $(f_A/10^{12}\ \mathrm{GeV})^{-1/6}$. Since the coherent mode has zero momentum, energy stored in this mode behaves like a nonrelativistic matter, showing a characteristic temperature dependence,

$$\rho_A \sim O(10)\cdot(f_A^2\, m_A/m_{pl}\; T_D)\; T^3. \qquad (11)$$

The usual type of constraint, $\rho_A <$ critical density at present, yields $f_A < 2 \times 10^{12}$ GeV[80] with a dependence on the Hubble parameter, $h_{1/2}^{12/7}$. The basic reason of getting an upper bound is that the weaker the coupling is, the less time is available for the field to relax, hence yielding a greater present mass density. This cosmological upper bound, taken together with the previous astrophysical lower bound, gives an intermediate scale of new physics in the range of $10^9 \sim 10^{12}$ GeV.

Fate of the Polonyi field somewhat differs from this case, since the field eventually decays[81]. The scale of local SUSY breaking which we denote by μ lies in an intermediate range, the most favorite choice being of $\sim\sqrt{m_w m_{pl}} \sim 10^{10}$ GeV. Mass and lifetime of the Polonyi field are: $m_\phi \sim \mu^2/m_{pl}$ (10 GeV), and $\Gamma_\phi \sim m_\phi^3/m_{pl}^2$ (10^{-10} sec^{-1}), where numbers in parentheses correspond to the choice of $\mu = \sqrt{m_w m_{pl}}$. The mass density stored by the coherent Polonyi field then dominates when the temperature T is in the range of $T_D < T < \mu$. The decay starts at T_D, which is determined by equating the decay rate to the expansion rate: $T_D \sim (\mu/m_{pl})^{8/3}\,\mu$ (10^{-12} GeV). Field decay at this late epoch creates too much entropy, roughly estimated to be $\sim T_R^3/T_D^3 \sim \rho_\phi^{3/4}/T_D^3 \sim (m_{pl}/\mu)^2$ (10^{17}) per one degree of freedom. The baryon asymmetry is diluted by this same factor, causing a serious problem. Many simple supergravity models are thus eliminated[81].

Gravitino

We shall now discuss the gravitino problem. It was once thought that the old mass limit of $m_{3/2} < 1$ keV (from mass density)[82] or $m_{3/2} > 10^4$ GeV (nucleosynthesis)[83] is removed by invoking inflation. But it was soon realized[84] that regeneration of gravitinos takes place after inflation, via two-body processes, $f\bar{f} \to f\tilde{G}$. The cross section of these processes is $\sim\alpha/m_{pl}^2$, and an estimate of relic abundance gives

$$Y_{3/2} \equiv n_{3/2}/n_\gamma \sim 0.1\; T_R/m_{pl}, \qquad (12)$$

with T_R the reheating temperature after inflation. Thus a constraint on abundance restricts the maximum reheating temperature, an interesting bound on inflation.

Note for reference that the lifetime of gravitino falls in a region of $10^2 \sim 10^{20}$ sec for the interesting mass range of 10^4 GeV $\sim$ 10 MeV, with variation of $m_{3/2}^3$. If the lifetime is larger than the age of the universe ($\sim 10^{17}$ sec), then the closure mass density yields $Y_{3/2} < 10^{-10} m_{100}^{-1} h_{1/2}^2$, with $m_{100} \equiv m_{3/2}/100$ GeV. With a shorter lifetime of $10^{13} \sim 10^{17}$ sec, the decay product γ contributes to a diffuse background of X-ray or γ, giving a bound of $Y_{3/2} < 10^{-11} m_{100}^3$ [85]. For the lifetime range of $10^5 \sim 10^{13}$ sec, distortion of the 3 K background radiation limits $Y_{3/2} < 10^{-9} m_{100}^{1/2}$ [85]. The best bound of this sort is derived from a possible destruction of primordially processed light elements (^{4}He, D, ^{3}He) by gravitino decay products, γ, $\bar{p}$, and $e^\pm$ if the lifetime < 10^{13} sec and the mass > 1 GeV: $Y_{3/2} < 3 \times 10^{-12} m_{100}^{-1}$ [85]. Implication for the reheating temperature is then

$$T_R < 10^8\ \mathrm{GeV}\cdot m_{100}^{-1} \quad \text{for } 1\ \mathrm{GeV} < m_{3/2} < 10^4\ \mathrm{GeV}.$$

As already discussed, this provides a severe constraint on the baryon generation.

Neutrino Oscillation in the Sun

Let us finally discuss a possibility of exciting positive result. It has been known for quite some time that the solar neutrino flux measured by Davis et al.[86] is as small as 1/3 of what the standard solar model predicts[87]. One possible explanation is the neutrino oscillation in vacuum between the sun and the earth. In this case neutrino masses must be at least larger than $\sim 10^{-5}$ eV, with a large mixing, for instance mixing of $\sim 1/\sqrt{3}$ among three neutrino species. A new solution, suggested last year by Mikheyev and Smirnov[88], uses neutrino oscillation within the solar medium.

The neutrino oscillation in matter can be quite different from vacuum oscillation, as first pointed out by Wolfenstein[89]. A new effect in matter is caused by a mass shift to ν_e via charged current weak interaction. Mass shifts due to neutral current interaction are common to all three neutrino species, without affecting the oscillation, but the electron neutrino receives an additional potential of the form, $\sqrt{2}\, G_F n_e$. Since the electron number density n_e varies as the neutrino propagates in the sun, this term appears as a time dependent contribution in the evolution equation for mixing amplitudes,

$$i\frac{d}{dt}\begin{pmatrix}\nu_e\\ \nu_\mu\end{pmatrix} = \frac{1}{2k}\begin{pmatrix} m_o^2 - \frac{\Delta}{2}\cos 2\theta + 2k\, V_e, & \frac{\Delta}{2}\sin 2\theta \\ \frac{\Delta}{2}\sin 2\theta, & m_o^2 + \frac{\Delta}{2}\cos 2\theta \end{pmatrix}\begin{pmatrix}\nu_e\\ \nu_\mu\end{pmatrix} \qquad (13)$$

When variation of the number density is slow, one can invoke an adiabatic approximation. The electron neutrino emitted from the core may slowly rotate, going through a resonance with maximal mixing, and then escape the sun as another kind of neutrino, say ν_μ.

The resonant amplification is an effect of level crossing, which occurs at a number density of

$$n_o = \Delta \cos 2\theta/(2\sqrt{2}\, G_F\, k)$$
$$\sim 800\, N_A \mathrm{cm}^{-3}(\Delta/10^{-4}\mathrm{eV}^2)\cdot(k/\mathrm{MeV})^{-1}, \qquad (14)$$

with N_A the Avogadro number. The effective mixing angle in matter is given by a resonance formula[88],[90],

$$\sin^2\theta_m = \sin^2 2\theta\left(\sin^2 2\theta + (\frac{L}{L_o} - \cos 2\theta)^2\right)^{-1}, \qquad (15)$$

with $L = 4\pi k/\Delta$ the vacuum oscillation length and $L_o = 2\pi/(\sqrt{2}\, G_F\, n_e) \sim 2\times 10^9$ cm $(n_e/N_A)^{-1}$. A sharper resonance results for a smaller mixing angle θ.

The parameter region that explains the Davis experiment is: (a)[91] $\Delta \sim 10^{-4}$ eV^2 and $\sin^2 2\theta \sim (10^{-3} - 10^{-1})$, or (b)[92] $\Delta\cdot\sin^2 2\theta \sim 10^{-7.5}$ eV^2. The Cl experiment is mainly sensitive to the higher energy neutrino from the ^{8}B decay. Forthcoming Ga experiment[93] explores the low energy neutrino from the pp chain. Two solutions, (a) and (b), predict[92] different yields for the Ga experiment: the solution (a) gives essentially the same prediction for the low energy neutrino flux as without neutrino oscillation, while the solution (b) gives a somewhat smaller flux.

It should be pointed out that these parameter ranges fit well[94] with GUT models once one goes beyond the simplest SU(5) model.

4. DARK MATTER AND FORMATION OF STRUCTURES

It is now well established that the bulk of matter in our universe is not luminous and only gravitationally detected. This is supported by numerous flat rotation curves[95]; the velocity of stars or gas $(\sim\sqrt{GM/r})$ = const. beyond the luminous part of galaxies, which indicates that a mass exists well beyond the luminous part. The mass to light ratio is also large for groups and clusters of galaxies: for instance, the infall of our Galaxy toward the Virgo supercluster suggests that M/L $\sim 100(M_\odot/L_\odot)\, h_{1/2}$, a considerably larger value[5] than in the solar neighborhood. The invisible mass is probably present also in dwarf galaxies[95] of much smaller mass $(\sim 10^6 M_\odot)$ and in the disk part[96] of our galaxy, although the evidence here is weaker. The total mass in the universe is of the order ($0.3 \sim 0.5$) or more of the closure density, as already mentioned.

The crucial question is whether the dark matter is baryonic or not. Nucleosynthesis[11] limits the baryonic component to ~20% of the closure density at most. Thus it is not clear from observations alone whether the dark matter is nonbaryonic or not. Note in this regard that the quark nugget[98] is not classified as a baryon, because it behaves independently of ordinary matter and radiation since the epoch of QCD phase transition.

There are however reasonable theoretical arguments that favor the nonbaryonic dark matter. In a baryon dominated universe growth rate of density fluctuation is slow. In the linear regime the density contrast $\delta\rho/\rho$ grows with the scale factor a[99], from the recombination epoch to $z = \Omega^{-1}$. Hence the growth rate is at most $\sim 10^3$. For adiabatic fluctuation[99] in which n_B/n_γ = const., the baryon fluctuation is related to the temperature fluctuation at recombination by $(\delta T/T)_r = \frac{1}{3}(\delta\rho/\rho)_r$. For a nonlinear structure such as a galaxy to develop until the present epoch, the initial temperature fluctuation must then be larger then $\sim 3 \times 10^{-4}$. This amount of anisotropy violates the present observational limit[100] of 3 K radiation at small angular scales: $(\delta T/T)_{3K} < 2 \times 10^{-5}$ at 4.5'. More detailed analysis[101], assuming a simple power law spectrum of fluctuation normalized to the galaxy correlation function, excludes the baryon dominated universe with $\Omega \leq 1$, irrespective of whether the fluctuation is adiabatic or not.

There is also the favored theoretical dogma of inflation, which asserts that $\Omega = 1$ to a very good accuracy, necessitating the nonbaryonic dark matter. One can summarize this situation by saying that the nonbaryonic dark matter is not of absolute necessity, but is certainly a very attractive idea to entertain such a possibility.

Let us now assume existence of some form of nonbaryonic dark matter. Nature of the nonbaryonic dark matter is divided into two broad classes[102]. The first class is called hot dark matter, and is characterized by having a relativistic velocity until a recent epoch close to recombination. The most famous example of this class is a stable massive neutrino in the mass range of 10 eV, which has been created in thermal equilibrium at early epochs and decoupled at a temperature of $\sim$1 MeV. The second class of dark matter, called cold dark matter, has a nonrelativistic velocity since well prior to recombination. Examples include the photino that decouples much earlier than the epoch of QCD phase transition, and the invisible axion that has never been in thermal equilibrium. There can be another, third class of dark matter, the warm dark matter that lies in between the hot and cold dark matter.

An important constraint follows in order to trap a fermionic hot dark matter into the galactic halo. When a system of noninteracting particles becomes gravitationally bound, it undergoes a process of violent relaxation[103] and reaches a state of Maxwellian distribution in velocity (not in energy). In this process the phase space density never increases, and by comparing maximal phase space densities of red shifted thermal Fermi distribution and Maxwellian velocity distribution, one obtains a mass bound[104] for the fermion,

$$m > 24 \text{ eV } (250\text{km s}^{-1}/\sqrt{\langle v^2\rangle})^{1/4} \cdot (20 \text{ kpc}/r)^{1/2} \quad , \qquad (16)$$

where $\sqrt{\langle v^2\rangle}$ is the velocity dispersion of the relaxed system and r the radius of galactic halo. The cold, and the hot bosonic dark matter do not suffer from this type of constraint.

Formation of structures is much affected by presence of the nonbaryonic dark matter. In a universe dominated by the hot dark matter, structure formation proceeds from a large to small scales[105]; top-down scenario. Relativistic noninteracting particles damp density perturbation within the particle horizon, due to free streaming. Subhorizon perturbations are thus erased, and the first structure formed exceeds the size[105] of $\sim 4 \times 10^{15}\, M_\odot\, (30 \text{ eV}/m)^2 \approx m_{pl}^3/m^2$. In the case of massive neutrino of our interest, this scale corresponds to the size of rich clusters of galaxies. The first object formed after recombination in the neutrino dominated universe is thus a cluster of galaxies. Baryons are later trapped into the gravitational potential well formed by the clustered neutrino. Gravitational collapse proceeds asymmetrically and the neutrino cluster collapses to a pancake-like structure[106]. The neutrino dominated universe thus has a nice feature in explaining large scale structures such as voids, sheets and filaments that seem to exist fairly commonly in the most recent red shift survey[107].

In a universe dominated by the cold dark matter[108], there is essentially no cutoff of the size of structures formed after recombination. This makes easier smaller structures to form first and a heirarchical clustering is a likely possibility. Small scale correlations such as galaxy-galaxy correlation[109] are easily explained in this picture. There are possibilities of generating isothermal perturbation (perturbation not coupled to fluctuation of radiation temperature)[110] in this scenario.

Detailed numerical simulation[111] of N-body problem under mutual gravitational force has been performed both for the cold and the hot dark matter. The result favors the cold dark matter, but agreement with observation is not complete. A disturbing problem in comparing numerical simulation with observation is the question of whether galaxies are a good tracer of the dynamical mass. It might happen that only high density peaks lead to bright galaxies easily detectable, in which case lower density peaks should be discarded in the comparison[112]. This gives the biased clustering scenario. With this sort of biasing galaxy formation based on the cold dark matter yields a better description[111] of our universe.

Besides the cold and the hot dark matter and their variants, there exist a few other proposals for galaxy formation; the nonradiative decay scenario[113], the cosmic string[114], and a series of explosive processes[115] similar to supernova explosion. Although there is no compelling theoretical motivation to introduce the string, galaxy formation based on the cosmic string seems very promissing.

Since there are many candidates of nonbaryonic dark matter, it is important to think of detecting them in terrestrial experiments. One can hope to detect the dark matter only when it is trapped in the galactic halo, or is subsequently accumulated in the sun or the earth, otherwise is directly emitted from the sun or from nearby energetic sources. The dark matter trapped in our galactic halo, the halo dark matter, should have an average mass density of $\rho_0 \sim 10^{-24}$ g$\cdot$cm^{-3} $\simeq$ 1 GeV$\cdot$cm^{-3}, and a velocity dispersion of $\sqrt{\langle v^2 \rangle} \sim 300$ km s^{-1} $\simeq 10^{-3}$ c. For a generic type of dark matter the mass of its elementary particle (weakly interacting massive particle, WIMP) and its interaction with ordinary matter are unknown parameters.

WIMP in the galactic halo may occasionally annihilate against each other and emit interesting high energy objects such as γ, e^+ and $\bar{p}$ among their decay products. This happens when the halo WIMP contains both particle and its antiparticle or when WIMP itself is a self-conjugate particle. Let us take a specific example of the photino annihilation: $\tilde{\gamma} + \tilde{\gamma} \rightarrow f + \bar{f}$ in which the fermion f ($\bar{f}$) decays into e^- (e^+), γ and ordinary quarks (u and d) and antiquarks. The annihilation rate depends on the exchanged SUSY particle $\tilde{f}$, whose mass is taken to be 50 GeV for illustration. An yet unknown parameter, $m(\tilde{\gamma})$, can be fixed by demanding the cosmological closure density, $\Omega(\tilde{\gamma}) = 1$, which yields[116] $m(\tilde{\gamma}) \simeq 3$ GeV and the halo number density $n_{\tilde{\gamma}} \simeq 0.1$ cm^{-3}. One can then estimate antiproton flux on earth. This flux is consistent[116] with the low energy flux measured by Buffington et al.[117] if one assumes a confinement time scale of charged particles in the glaxy to be $\sim 10^8$ years. Needless to say, it is crucial to measure energy spectrum of $\bar{p}$ in the relevant energy region.

Another potentially important process, $\tilde{\gamma} + \tilde{\gamma} \rightarrow \psi(3.1) + \gamma$[118], produces a unique monochromatic γ. The original estimate of its cross section, however, appears too optimistic, and a more realistic calculation[119] yields a rate, at best marginal to detection.

In the case of halo (invisible) axion a strong magnetic field may convert the axion mass into a microwave photon[120], whose detection however demands a very fine tuning of cavity frequency[121].

The halo dark matter occasionally passes through the sun, and it may be trapped[122] in the sun if it loses energy via elastic scattering off nuclei in the sun. When the captured dark matter accumulates within the sun, it may eventually start to annihilate each other in the sun. Abundance equilibrium of capture against annihilation may then be reached. For this to happen, evaporation at surface should not be large, and this requirement gives a mass bound, for example $m(\tilde{\gamma}) > 6$ GeV[123] in the case of photino. Most of the annihilation product will be mixed up with solar matter, but the neutrino immediately escapes the sun, hence is subject to a unique detection[124] on earth. Again by fixing the cross section with the closure density, one estimates the high energy neutrino flux from the photino annihilation to be roughly comparable with the atmospheric neutrino. This is a welcome feature for observation in proton decay detectors because the flux has a directionality towards the sun and an excess over the atmospheric neutrino is likely to appear in the high energy side of $\sim$ several GeV[124].

A similar consideration[125] applied to annihilation inside the earth actually rejects a mass range of certain WIMP. The earth mostly contains spinless nuclei and does not have a sizable rate of scattering with a particle such as the photino which couples with spin. On the other hand, a massive neutrino and its SUSY partner, scalar neutrino, has a vector coupling and their interaction rate is determined by the Fermi constant. A mass range of 9 $\sim$ 20 GeV for the Dirac ν and 12 - 20 GeV for the scalar ν would produce too much high energy neutrino above the atmospheric component, thus this range of the mass region is already ruled out[125]. No useful bound on the photino results from consideration of the earth.

Direct search for the halo dark matter has also been contemplated[126]. The principle of detection is very simple, but in practice it might have many background problems. The halo WIMP X deposits energy via elastic scattering in a detector up to $\epsilon_{max} \sim 2\ m_x^2 v^2/M \sim 0.7$ keV $(m_x/\mathrm{GeV})^2$ $(M/30\ \mathrm{GeV})^{-1}$ $(v/300\ \mathrm{km\ s^{-1}})^2$, where M is the mass of a target nucleus ($M >> m_x$ assumed). If the interaction rate is $\sim 10^{-38}$ cm^2 (of order the usual weak process), the event rate may become $\sim$ several events/kg·day. The first result[127)] of this type of search was reported recently. It employed a low background Ge detector of 135 cm^3, which gave a background rate of $\sim$ 100 counts/kg·day above an equivalent electron energy threshold of 4 keV. This detector is not sensitive to spin dependent coupling, but a mass range of Dirac and scalar neutrino, 20 GeV $\sim$ 5 TeV, is already excluded by assuming no significant signal of WIMP. This bound may be combined with the cosmological bound (>3 GeV) and with the annihilation argument in the earth, to leave a small window of 3 $\sim$ 9 GeV for a massive stable neutrino unless it is less than $\sim$30 eV.

Detection of the solar axion may also be feasible if the PQ symmetry breaking scale f_A is close to the solar luminosity bound of 1×10^7 GeV. It was recently pointed out[128)] that axion detection rate may be enhanced due to atomic binding effect for the keV region axion, much similar to the photoelectric effect. The axion flux from the sun[73),74)] is estimated $\sim 2 \times 10^{14}\ \mathrm{cm}^{-2}\mathrm{sec}^{-1}(f_A/10^7\ \mathrm{GeV})^{-2}$. The maximum flux is at the peak of the black-body distribution of 1 keV. Thus, the Homestake Ge detector[127)] can only see the high energy tail, and gives a direct bound of $\sim 0.5 \times 10^7$ GeV[129)].

More ambitious proposals of dark matter search are superconducting colloid detector[130)] and bolometric detection[131)] of the tiny energy deposit to be operated in mK region. If these become feasible, one may hope to drastically reduce the threshold mass of WIMP and probe much weaker interaction rate relevant to, for instance, the photino.

Finally, an interesting possibility of employing the accumulated WIMP to resolve the solar neutrino puzzle was pointed out[132)]. By its very nature of weak coupling WIMP (sometimes also called cosmion) rarely interacts, thus transports the core energy far away from the core region. This has an effect of reducing the temperature gradient near the core and hence of decreasing the high energy solar neutrino from ^{8}B detectable in the Davis experiment. Unfortunately, explanation of the observed ^{8}B neutrino requires a parameter range of mass and cross section which seems difficult to be satisfied by WIMP candidates so far considered[133)]. The slow decrease of the core temperature may however be directly tested by a careful analysis of the solar oscillation. A certain type of oscillation is known to be sensitive to the deep internal structure of the sun[134)]. One may thus hope that the helioseismology gives a means to test a certain type of WIMP, or at worst to check the standard solar model.

5. TOWARDS THE BEGINNING

Conventional study of cosmology relies on the classical general relativity and a state of matter, including quantum effects under a specified background metric. Understanding the beginning of our universe however requires quantum gravity at the Planck epoch. There are two recent developments in this subject. One is quantum cosmology which uses quantum theory of the ordinary Einstein gravity and mainly addresses the problem of the initial condition. The other approach is related to a new theory of gravity; the superstring theory[135)].

In quantum cosmology the universe is described in terms of a single wave functional, $\Psi(h_{ij}, \phi)$, where the argument is an infinite dimensional three-geometry $h_{ij}(x)$, the metric after subtracting gauge degrees of freedom, and an infinite dimensional matter field $\phi(x)$. The canonical quantization or the path integral formalism leads to a Schrödinger functional equation, called Wheeler-De Witt (WD) equation[136)]. This equation does not contain the time variable, which is merely a label of space-time point and does not reflect a physical degree of freedom. The WD equation is second order in derivatives and of a hyperbolic type, which expresses a bizarre feature of gravity but also causes a serious problem in probabilistic interpretation. In practice, the WD equation can be analyzed by reducing the infinite system to a finite tractable number of degrees of freedom, whose Hilbert space is called minisuperspace (having nothing to do with SUSY).

Quantum cosmology is expected to yield a useful information on the initial condition to the classical cosmology. Since the WD equation can not fix the boundary condition, it has to be introduced by physical consideration. The most natural choice of the boundary condition was given by Hartle and Hawking[137)], who defined the wave functional in terms of a Euclidean path integral over compact 4-manifolds bounded by the 3-geometry h_{ij}. In physical terms this definition has an effect that the wave functional is real and contain both of "outgoing" and "incoming" waves in the form of a standing wave. In this terminology the outgoing wave corresponds to an expanding universe, and the incoming wave to a contracting universe. Another choice of the boundary condition was advocated by Vilenkin[138], who proposes taking the outgoing wave alone, leading to a picture of creation of the universe from nothing.

Although the probabilistic interpretation is in doubt, what we need may only be a correlation of matter field and the metric. In a minisuperspace model in which one introduces the scale factor a and a homogenous scalar field ϕ, a semi-classical approximation[137] yields a rapidly oscillating wave function in a region of (a, ϕ) where these parameters are related by the de Sitter solution to the Hamilton-Jacobi equation for the semiclassical exponent S of $\Psi \sim \exp(iS)$. The inflation may thus have a quantum origin.

In my view the most important achievement of quantum cosmology is a new derivation[139] of the formula for the density perturbation in the inflationary universe. This method is conceptually superior to previous derivation[35] albeit its technical complexity. Study of quantum cosmology in higher dimensions has also been initiated[140].

Quantum gravity in the present form is however plagued by serious problems; nonrenormalizability and, even worse, presence of gravitational anomaly[141] which spoils at the quantum level the general covariance of classical gravity. A new theory of gravity, the superstring theory, seems to solve[142] both of these problems: the theory is anomaly free and presumably even a finite theory free of ultraviolet divergences. The superstring is based on one dimensional object, the string, unlike the point particle in the usual quantum field theory. Its basic formulation still lacks a clear-cut first principle to start with, although a set of computational rules[135] are well known at least at one loop level. At the present level of understanding the theory is consistent only when the spacetime dimensionality is ten and a particular gauge group such as $E_8 \times E_8$ and SO(32) is chosen[142].

Cosmology along theoretical ideas of the superstring theory has only recently been undertaken. There seem two important aspects in the superstring cosmology; existence of extra six dimensions and more degrees of freedom (infinite number of string excitations). Full implications of these aspects are of course not known at present, and I shall only sketch some recent progress.

Cosmology in higher dimensions[143] has been studied for quite some time even before the recent uprising of string theories. A large body of these works are associated with Kaluze-Klein type of theories which also use higher dimensions. The most important problem in higher dimensional cosmology is how to understand dynamically compactification of extra dimensions that are not observable at present. In one scenario[144] this proceeds as follows. Suppose that the universe was initially dominated by massless 9-dimensional ideal gas, and take for definiteness a product space of two maximally symmetric spaces of three (having any of $k = 0, \pm 1$) and six dimensions, which are characterized by two time dependent scale factors, $a_3(t)$ and $a_6(t)$. Two scales first expand simultaneously with a law of $a_i \propto t^{1/5}$ as time t increases. Difference in dimensionalities then starts to separate two scales, and a_6 with a higher dimensionality eventually collapses. Behavior of this collapse is described by a Kasner solution with $a_i(t) \propto (t_1-t)^{\gamma_i}$, in which $3\gamma_3 + 6\gamma_6 = 3\gamma_3^2 + 6\gamma_6^2 = 1$, hence $\gamma_6 > 0 > \gamma_3$. Thus the 3-space "inflates" according to a power law while the extra space shrinks. At a certain stage of the collapse the temperature T becomes of $\sim a_6^{-1}$ and the extra space is deexcited because the free energy of the system changes from the order of $T^{10}\, a_3^3\, a_6^6$ in the initial phase to the order of $T^4\, a_3^3$ for $T \ll a_6^{-1}$.

The collapse, eventually leading to a singularity, can be halted only by some kind of pressure forces. Popular candidates of the pressure have been field condensates[145] such as the gauge and antisymmetric tensor field, and quantum Casimir pressure[146]. Subsequent compactification follows if the theory allows a static solution of the form, $\text{Minkowski}_4 \times$ compact 6-space, although in some models the total (3+6)-space may enter the phase of exponential expansion[147], depending on the initial condition. The process of compactification entails a damped oscillation[148] around the static solution, which may provide a new mechanism of baryon production.

It has been argued[149] that a large entropy may be pumped from the extra space by adiabatic processes. The horizon problem is however unlikely to be solved by this kind of power law expansion halted by compactification.

The field theory limit of the superstring theory is a special class of supergravity theory in ten dimensions, modified by higher derivative terms such as a particular combination of curvature squared terms[150]. Many recent works[151] investigate cosmology at low temperatures, with new fields and new terms included, without leading to a compelling coherent picture.

It is perhaps more important to study effects of an infinite number of string excitations. Since the old days of the dual resonance model, the spectrum of massive string excitation is known to grow exponentially with mass[152], $\rho(m) \sim cm^{-a} e^{bm}$, with a, b, and c being constants fixed by models. This implies that there exists a limiting or a critical temperature[153] at $\sim b^{-1} \gtrsim m_{pl}$ since energy put in from outside goes to production of new states instead of raising kinetic energy of individual modes. Actually, what happens is emergence of a state of negative specific heat[153]. Because of a large fluctuation near the critical temperature the system exhibits an instability. The situation is much the same as in the black hole, which also behaves in similar fashions. This instability might have an important implication at the Plank epoch.

Finally there remains an important question of how to detect the extra dimensions in the present cold universe. In this respect it is convenient to regard the ten dimensional field theory limit of the superstring theory as an effective four dimensional theory. In general one may expand the metric and various fields in terms of harmonics[154] in extra coordinates. The zero mode sector of this expansion corresponds to the massless (or more precisely with mass $<< m_{pl}$) sector of the ordinary four dimensional theory. In this picture the ten dimensional metric already contains the four dimensional metric, a Brans-Dicke type of scalar field and a number of gauge fields depending on isometry of the extra compact space (this classification is exactly the same as in Kaluza-Klein theories[154]). This has an important consequence. It has long been known[155] in connection with the no-hair theorem of the black hole that presence of a nontrivial scalar field deprives the event horizon, leaving naked singularities or wormholes[156]. An exact and a complete set of vacuum solutions[157] with the three dimensional spherical symmetry has been worked out for the ten dimensional Einstein equation by assuming Ricci flat or toroidal compactification of the extra space, to confirm this general statement. Black holes may thus be not black. Dynamics of compactification is least understood, and it would take quite some time to write down an explicit solution in a realistic model. But it seems generally true that the black hole is a singular case (no excitation of the scalar mode or the extra size b = const. everywhere) of more general class of solutions. I personally do not believe in existence of singularities, which will presumably be elliminated by quantum effects or stringy consideration.

If the Ricci flatness persists at low energy scales, the Newton constant will change with time, at a rate almost as big[158] as in Dirac's large number theory. SUSY breaking and other considerations will however destroy the Ricci flat Calabi-Yau space[159], and the chance of this large variation is slim.

6. SUMMARY

Looking backward at the last several years of activities, one can not escape a sense of strong feeling that a substantial part of cosmology has been integrated into the field of particle physics phenomenology. This unification has two aspects; (1) contribution of particle physics to cosmology exemplified by the new scenario of generating the baryon asymmetry and inflation, and (2) useful constraints on particle physics from cosmology and astrophysics, which often gives information inaccessible to laboratory experiments. This mutual interaction has been both fruitful and cozy.

Looking forward to the future, I believe that this relation between particle physics and cosmology will become further closer, even calling for serious efforts of experimental physicists. A new result of the solar neutrino measurement will soon become available from Kamiokande[160], and the second generation of experiments using Ga has been funded[93]. Large scale experiments searching for the dark matter are also being discussed extensively. It should be kept in mind that we still lack a definite evidence of early relics and unification ideas beyond the standard theory.

I close this talk with following summary:

• GUT cosmology is a better description of our universe than the standard model, as has been illustrated by the baryon asymmetry and the inflation.
• Cosmology-Astrophysics yields useful constraints on weakly interacting and/or stable particles.
• A new solution of the solar neutrino puzzle was given.
• The cold dark matter is welcome both from observation and from galaxy formation.

• Useful laboratory bounds on the dark matter may soon be forthcoming.
• Elucidation of the beginning of our universe awaits a better understanding of the ultimate theory such as the superstring.

ACKNOWLEDGEMENTS

In preparing this report I have been aided by many people on various aspects of this subject. I would like to record my special thanks to those who helped me to clarify some technical points: M. Fukugita, H. Kodama, K. Olive, R. Peccei, S. Rudaz, M. Sasaki, H. Sato, K. Sato, K. Shizuya, J. Silk and G. Steigman. H. Yamamoto has been very helpful as my scientific secretary. Finally I would like to thank for a warm hospitality S. Rudaz and H. Suura at Minnesota where a part of this manuscript was written.

REFERENCES

1 For reviews, S. Weinberg, "Gravitation and Cosmology" (J. Wiley, N.Y. 1972); Ya.B. Zel'dovich and I.D. Novikov, "Relativistic Astrophysics II" (Univ. of Chicago Press, Chicago 1983).

2 For reviews, A. Sandage and G.A. Tammann; J.P. Huchra, in "Inner Space/Outer Space", ed. by E.W. Kolb et al. (Univ. Chicago Press, Chicago 1986).

3 I. Iben and A. Renzini, Phys. Rep. 105 (1984) 329.

4 B.S. Meyer and D.N. Schramm, FERMILAB-Pub-86/71-A (1986).

5 M. Davis, J. Tonry, J. Huchra and D.W. Latham, Ap. J. 238 (1980) L113.

6 A. Yahil, D. Walker and M. Rowan-Robinson, Preprint (1986).

7 E.D. Loh and E.J. Spillar, Ap. J. 307 (1986) L1.

8 J.B. Peterson, P.L. Richards and T. Timusk, Phys. Rev. Lett. 55 (1985) 332; G.F. Smoot et al., Ap. J. 291 (1985) L23; D. Meyer and M. Jura, Ap. J. 276 (1984) L1.

9 P.M. Lubin, G.L. Epstein and G.F. Smoot, Phys. Rev. Lett. 50 (1983) 616; D.J. Fixsen, E.S. Cheng and D.T. Wilkinson, Phys. Rev. Lett. 50 (1983) 620.

10 For a review, A.M. Boesgaard and G. Steigman, Ann. Rev. Astr. Astrophys. 23 (1985) 319.

11 J. Yang, M.S. Turner, G. Steigman, D.N. Schramm and K.A. Olive, Ap. J. 281 (1984) 493.

12 J. Ellis, K. Enqvist, D.V. Nanopoulos and S. Sarkar, Phys. Lett. 167B (1986) 457.

13 G. Steigman, K.A. Olive, D.N. Schramm and M.S. Turner, UMN-TH-562/86 (1986).

14 G. Altarelli and M. Davier, this Proceeding.

15 G. Steigman, Ann. Rev. Astr. Astrophys. 14 (1976) 339.

16 For example, R.H. Dicke and P.J.E. Peebles, in "General Relativity: An Einstein Centenary Survey" ed. by S. Hawking and W. Israel (Cambridge Univ. Press, Cambridge 1979).

17 For a review, P. Langacker, Phys. Rep. 72 (1981) 185.

18 A.D. Sakharov, Sov. Phys. J.E.T.P. Lett. 5 (1967) 24; M. Yoshimura, Phys. Rev. Lett. 41 (1978) 281, 42 (1979) 740 (E); A.Y. Ignatiev, N.V. Krasnikov, V.A. Kuzmin and A.N. Tavkhelidze, Phys. Lett. 76B (1979) 594. For reviews, M. Yoshimura, in "Grand Unified Theories and Related Topics" (World Scientific, Singapore 1981); E.W. Kolb and M.S. Turner, Ann. Rev. Nucl. Part. Sci. 33 (1983) 645.

19 D. Toussaint, S.B. Treiman, F. Wilczek and A. Zee, Phys. Rev. D19 (1979) 1036; S. Weinberg, Phys. Rev. Lett. 42 (1979) 850; M. Yoshimura, Phys. Lett. 88B (1979) 294.

20 M. Yoshimura, Ref. (19).

21 E.W. Kolb and S. Wolfram, Nucl. Phys. B172 (1980) 224; J.N. Fry, K. Olive and M.S. Turner, Phys. Rev. D22 (1980) 2953, 2977.

22 M. Yoshimura, 1st reference of (18).

23 J. Ellis, M.K. Gaillard and D.V. Nanopoulos, Phys. Lett. 80B (1979) 360, 82B (1979) 464 (E); D.V. Nanopoulos and S. Weinberg, Phys. Rev. D20 (1979) 2484; S. Barr, G. Segrè and H.A. Weldon, Phys. Rev. D20 (1979) 2494; T. Yanagida and M. Yoshimura, Nucl. Phys. B168 (1980) 534.

24 G. Segrè and M.S. Turner, Phys. Lett. 99B (1981) 399.

25 M. Gell-Mann, P. Ramond and R. Slansky, in "Supergravity" ed. by D.Z. Freedman and P. van Nieuwenhuizen (North Holland, Amsterdam 1979); T. Yanagida, in "Workshop on the Unified Theory and the Baryon Number in the Universe" ed. by O. Sawada and A. Sugamoto, KEK-79-18 (1979).

26 V.A. Kuzmin and M.E. Shaposhnikov, Phys. Lett. 92B (1980) 115; T. Yanagida and M. Yoshimura, Phys. Rev. D23 (1981) 2048.

27 M. Fukugita, T. Yanagida and M. Yoshimura, Phys. Lett. 106B (1981) 183.

28 A. Masiero, R.N. Mohapatra and R.D. Peccei, Phys. Lett. 108B (1982) 111; J.F. Nieves, Phys. Rev. D25 (1982) 1417; D. Chang, R.N. Mohapatra and G. Senjanović, Phys. Rev. Lett. 53 (1984) 1419; G. Branco and A.I. Sanda, Phys. Lett. 135B (1984) 383; X.G. He and S. Pakvasa, Phys. Lett. 173B (1986) 159; H.Y. Cheng, IUHET 122 (1986).

29 A.H. Guth, Phys. Rev. D23 (1981) 347; A.D. Linde, Phys. Lett. 108B (1982) 389; A. Albrecht and P.J. Steinhardt, Phys. Rev. Lett. 48 (1982) 1220.

30 For similar approaches, D. Kazanas, Ap. J. 241 (1980) L59; K. Sato, Month. Not. Roy. Astr. Soc. 195 (1981) 467.

31 A. Albrecht, P.J. Steinhardt, M.S. Turner and F. Wilczek, Phys. Rev. Lett. 48 (1982) 1437.

32 J.P. Preskill, Phys. Rev. Lett. 43 (1979) 1365; Ya.B. Zel'dovich and M.Yu. Khlopov, Phys. Lett. 79B (1978) 239.

33 G.F. Mazenko, W.G. Unruh and R.M. Wald, Phys. Rev. D31 (1985) 273.

34 A. Albrecht and R.H. Brandenberger, Phys. Rev. D31 (1985) 1225; A. Albrecht, R.H. Brandenberger and R.A. Matzner, Phys. Rev. D32 (1985) 1280; A.H. Guth and S.Y. Pi, Phys. Rev. D32 (1985) 1899.

35 S.W. Hawking, Phys. Lett. 115B (1982) 295; A.A. Starobinsky, Phys. Lett. 117B (1982) 175; A.H. Guth and S.Y. Pi, Phys. Rev. Lett. 49 (1982) 1110; J. Bardeen, P.J. Steinhardt and M.S. Turner, Phys. Rev. D28 (1983) 679; R. Brandenberger and R. Kahn, Phys. Rev. D29 (1984) 2172.

36 A. Vilenkin and L.H. Ford, Phys. Rev. D26 (1982) 1231; A.D. Linde, Phys. Lett. 116B (1982) 335.

37 E. Harrison, Phys. Rev. D1 (1970) 2726; Ya.B. Zel'dovich, Mon. Not. Roy. Astr. Soc. 160 (1972) 1.

38 S. Coleman and E. Weinberg, Phys. Rev. D7 (1973) 788.

39 J. Breit, S. Gupta and A. Zaks, Phys. Rev. Lett. 51 (1983) 1007; I.G. Moss, Phys. Lett. 128B (1983) 385; J. Kodaira and J. Okada, Phys. Lett. 133B (1983) 291; K. Sato and H. Kodama, Phys. Lett. 142B (1984) 18; M. Sakagami and A. Hosoya, Phys. Lett. 150B (1985) 342.

40 Q. Shafi and A. Vilenkin, Phys. Rev. Lett. 52 (1984) 691; S.Y. Pi, Phys. Rev. Lett. 52 (1984) 1725.

41 For a review, P. Binétruy, in Proc. of the 6th Workshop on Grand Unification, ed. by S. Rudaz and T. Walsh (World Scientific, Singapore 1985).

42 A.A. Starobinsky, Phys. Lett. 91B (1980) 99; A. Vilenkin, Phys. Rev. D32 (1985) 2511; L. Kofman, A. Linde and A.A. Starobinsky, Phys. Lett. 157B (1985) 361.

43 A.D. Linde, Phys. Lett. 129B (1983) 177.

44 Q. Shafi and C. Wetterich, Phys. Lett. 129B (1983) 387; Y. Okada, Phys. Lett. 150B (1985) 103.

45 A. Hosoya and M. Sakagami, Phys. Rev. D29 (1984) 2228; M. Morikawa and M. Sasaki, Progr. Theor. Phys. 72 (1984) 782.

46 A.D. Dolgov and A.D. Linde, Phys. Lett. 116B (1982) 327; L.F. Abbott, E. Farhi and M.B. Wise, Phys. Lett. 117B (1982) 29.

47 T. Yanagida and M. Yoshimura, Phys. Rev. Lett. 45 (1980) 71; J.A. Harvey, E.W. Kolb, D.B. Reiss and S. Wolfram, Nucl. Phys. B177 (1982) 456; R. Barbieri, D.V. Nanopoulos and A. Masiero, Phys. Lett. 98B (1981) 191.

48 M. Claudson, L.J. Hall and I. Hinchliffe, Nucl. Phys. 241B (1984) 309.

49 I. Affleck and M. Dine, Nucl. Phys. B249 (1985) 361.

50 A.D. Linde, Phys. Lett. 160B (1985)

243; B.A. Campbell, J. Ellis, D.V. Nanopoulos and K.A. Olive, CERN-TH, 4484/86 (1986).

51 V.A. Kuzmin, V.A. Rubakov and M.E. Shaposhnikov, Phys. Lett. 155B (1985) 36; K. Aoki, Phys. Lett. 174 (1986) 371.

52 't Hooft, Phys. Rev. Lett. 37 (1976) 8.

53 F.R. Klinkhamer and N.S. Manton, Phys. Rev. D30 (1984) 2212.

54 M. Fukugita and T. Yanagida, Phys. Lett. 174B (1986) 45.

55 S.S. Gershtein and Y.B. Zel'dovich, JETP Lett. 4 (1966) 174; R. Cowsik and J. McClelland, Phys. Rev. Lett. 29 (1972) 669; G. Marx and A. S. Szalay, in Proc. of Neutrino '72 (Technoinform, Budapest 1972).

56 Ya.B. Zel'dovich, Adv. Astr. Astrophys. 3 (1965) 241; H.Y. Chiu, Phys. Rev. Lett. 17 (1966) 712.

57 B.W. Lee and S. Weinberg, Phys. Rev. Lett. 39 (1977) 165.

58 H. Goldberg, Phys. Rev. Lett. 50 (1983) 1419; L.M. Krauss, Nucl. Phys. 227B (1983) 556; J. Ellis, J.S. Hagelin, D.V. Nanopoulos, K. Olive and M. Srednicki, Nucl. Phys. 238B (1984) 453.

59 R. Kimble, S. Bowyer and P. Jakobsen, Phys. Rev. Lett. 46 (1981) 80; A.L. Melott and D.W. Sciama, Phys. Rev. Lett. 46 (1981) 1369; F.W. Stecker, Phys. Rev. Lett. 45 (1980) 1460; A. De Rújula and S.L. Glashow, Phys. Rev. Lett. 45 (1980) 942; J. Maalampi, K. Mursula and M. Roos, Phys. Rev. Lett. 56 (1986) 1031.

60 M. Kawasaki and K. Sato, Phys. Lett. 169B (1986) 280; J. Silk and A. Stebbins, Ap. J. 269 (1983) 1; K. Sato and M. Kobayashi, Progr. Theor. Phys. 58 (1977) 1775.

61 D. Lindley, Ap. J. 294 (1985) 1.

62 S. Falk and D.N. Schramm, Phys. Lett. 79B (1978) 511; R. Cowsik, Phys. Rev. Lett. 39 (1977) 784.

63 H. Albrecht et al., Phys. Lett. 163B (1985) 404.

64 D.A. Bryman et al., Phys. Rev. Lett. 50 (1983) 1546; N. De Leener-Rosier et al., Phys. Lett. 177B (1986) 228.

65 F. Bergsma et al., Phys. Lett. 128B (1983) 361.

66 S. Sarkar, in Proc. of International Europhysics Conference on High Energy Physics, Bari (1985); L.M. Krauss, in Proc. of the Neutrino Mass Miniconference, Telemark, ed. by V. Barger and D. Cline (University of Wisconsin, Madison 1981).

67 M. Dine, W. Fischler and M. Srednicki, Phys. Lett. 104B (1981) 199; J.E. Kim, Phys. Rev. Lett. 43 (1979) 103; M.A. Shifman, A.I. Vainshtein and V.I. Zakharov, Nucl. Phys. 166B (1980) 493.

68 R.D. Peccei and H.R. Quinn, Phys. Rev. Lett. 38 (1977) 1440; S. Weinberg, Phys. Rev. Lett. 40 (1978) 223; F. Wilczek, Phys. Rev. Lett. 40 (1978) 279.

69 Y. Chikashige, R.N. Mohapatra and R.D. Peccei, Phys. Lett. 98B (1981) 265; G.B. Gelmini and M. Roncadelli, Phys. Lett. 99B (1981) 411.

70 F. Wilczek, Phys. Rev. Lett. 49 (1982) 1549; G.B. Gelmini, S. Nussinov and T. Yanagida, Nucl. Phys. B219 (1983) 31.

71 For a review, M. Yoshimura, in Proc. "Grand Unified Theories and Early Universe" KEK-TH 64 (KEK, Tsukuba 1983).

72 K. Sato and H. Sato, Progr. Theor. Phys. 54 (1975) 1564; D.A. Dicus, E.W. Kolb, V.L. Teplitz and R.V. Wagoner, Phys. Rev. D18 (1978) 1829, D22 (1980) 839; M.I. Vysotsskii, Ya.B. Zel'dovich, M.Yu. Khlopov and V.M. Chechetkin, JETP Lett. 27 (1978) 502.

73 M. Fukugita, S. Watamura and M. Yoshimura, Phys. Rev. Lett. 48 (1982) 1522; Phys. Rev. D26 (1982) 1840.

74 G.G. Raffelt, Phys. Rev. D33 (1986) 897.

75 L.M. Krauss, J.E. Moody and F. Wilczek, Phys. Lett. 144B (1984) 391.

76 G.G. Raffelt, Phys. Lett. 166B (1986) 402.

77 D.S.P. Dearborn, D.N. Schramm and G. Steigman, Phys. Rev. Lett. 56 (1986) 26.

78 N. Iwamoto, Phys. Rev. Lett. 53 (1984) 1198; A. Pantziris and K. Kang, Phys.

Rev. D33 (1986) 3509.

79 J. Polonyi, KFKI-1977-93 (1977); For a review, H.P. Nilles, Phys. Rep. 110 (1984) 2.

80 J. Preskill, M.B. Wise and F. Wilczek, Phys. Lett. 120B (1983) 127; L.F. Abbott and P. Sikivie, Phys. Lett. 120B (1983) 133; M. Dine and W. Fischler, Phys. Lett. 120B (1983) 137; M.S. Turner, F. Wilczek and A. Zee, Phys. Lett. 125B (1983) 35; J. Ipser and P. Sikivie, Phys. Rev. Lett. 50 (1983) 925; F. Stecker and Q. Shafi, Phys. Rev. Lett. 50 (1983) 928; M. Axemides, R. Brandenberger and M.S. Turner, Phys. Lett. 126B (1983) 178; M. Fukugita and M. Yoshimura, Phys. Lett. 127B (1983) 181.

81 G.D. Coughlan, W. Fischler, E.W. Kolb, S. Raby and G.G. Ross, Phys. Lett. 131B (1983) 59; A.S. Goncharov, A.D. Linde and M.I. Vysotsky, Phys. Lett. 147B (1984) 279; G. Germán and G.G. Ross, Phys. Lett. 172B (1986) 305.

82 H. Pagels and J.R. Primack, Phys. Rev. Lett. 48 (1982) 223.

83 S. Weinberg, Phys. Rev. Lett. 48 (1982) 1303.

84 J. Ellis, J.E. Kim and D.V. Nanopoulos, Phys. Lett. 145B (1984) 181.

85 J. Ellis, D.V. Nanopoulos and S. Sarkar, Nucl. Phys. B259 (1985) 175; M. Yu. Khlopov and A.D. Linde, Phys. Lett. 138B (1984) 265.

86 For a summary, R. Davis, in Proc. of 7th Workshop on Grand Unification, Toyama (1986).

87 J.N. Bahcall et al., Rev. Mod. Phys. 54 (1982) 767.

88 S.P. Mikheyev and A. Yu. Smirnov, INR Preprint, Moscow (1985).

89 L. Wolfenstein, Phys. Rev. D17 (1978) 2369, D20 (1979) 2634.

90 V. Barger, K. Whisnant, S. Pakvasa and R.J.N. Phillips, Phys. Rev. D22 (1980) 2718.

91 H.A. Bethe, Phys. Rev. Lett. 56 (1986) 1305.

92 S.P. Rosen and J.M. Gelb, Phys. Rev. D34 (1986) 969; W.C. Haxton, Phys. Rev. Lett. 57 (1986) 1271; S.J. Parke, Phys. Rev. Lett. 57 (1986) 1275; E.W. Kolb, M.S. Turner and T.P. Walker, Phys. Lett. 175B (1986) 478; V. Barger, R.J.N. Phillips and K. Whisnant; Phys. Rev. D34 (1986) 980; J. Bouchez et al., in Proc. of Neutrino '86, Sendai (1986).

93 T. Kirsten in Proc. of Neutrino '86, Sendai (1986).

94 M. Fukugita, T. Yanagida and M. Yoshimura, Ref. (27); P. Langacker, S.T. Petcov, G. Steigman and S. Toshev, CERN-TH. 4421 (1986); R. Johnson, S. Ranfone and J. Schechter, SU-4228-341 (1986).

95 S.M. Faber and J.S. Gallagher, Ann. Rev. Astr. Astrophys. 17 (1979) 135.

96 S.M. Faber and D.N.C. Lin, Ap. J. 266 (1983) L17, L20; M. Aaronson, Ap. J. 266 (1983) L11.

97 J.N. Bahcall, Ap. J. 276 (1984) 169.

98 E. Witten, Phys. Rev. D30 (1984) 272.

99 For reviews, S. Weinberg, Ref. (1); P.J.E. Peebles, "The Large-scale Structure of the Universe" (Princeton University Press, Princeton 1980).

100 J.M. Uson and D.T. Wilkinson, Nature 312 (1984) 427. For a review, D.T. Wilkinson in "Inner Space/Outer Space" ed. by E.W. Kolb et al. (Univ. Chicago Press, Chicago 1986).

101 M.L. Wilson and J. Silk, Ap. J. 243 (1981) 14.

102 For a review, J.R. Primack, in International School of Physics "Enrico Fermi", Varenna (1984).

103 D. Lynden-Bell, Mon. Not. Roy. Astr. Soc. 136 (1967) 101; F.H. Shu, Ap. J. 225 (1978) 83.

104 S. Tremaine and J. Gunn, Phys. Rev. Lett. 42 (1979) 407.

105 J.R. Bond, G. Efstathiou and J. Silk, Phys. Rev. Lett. 45 (1980) 1980; H. Sato and F. Takahara, Progr. Theor. Phys. 66 (1981) 508; F.R. Klinkhamer and C.A. Norman, Ap. J. 243 (1981) L1; I. Wasserman, Ap. J. 248 (1981) 1; A.G. Doroshkevich et al., Ann. New York Acad. 375 (1981) 32.

106 Ya. B. Zel'dovich, Astr. and Astrophys. 5 (1970) 84.

107 V. de Lapparent, M.J. Geller and J.P. Huchra, Ap. J. 302 (1986) L1.

108 Ref. (80); G.R. Blumenthal, H. Pagels and J.R. Primack, Nature 299 (1982) 37; P.J.E. Peebles, Ap. J. 277 (1983) 470.

109 M. Davis and P.J.E. Peebles, Ap. J. 267 (1983) 465.

110 M. Yoshimura, Phys. Rev. Lett. 51 (1983) 439; A.D. Linde, Phys. Lett. 158B (1985) 375.

111 M. Davis, G. Efstathiou, C.S. Frenk and S.D.M. White, Ap. J. 292 (1985) 371; S.D.M. White, in Proc. of 7th Workshop on Grand Unification, Toyama (1986).

112 N. Kaiser, Ap. J. 284 (1984) L9.

113 D. Dicus, E.W. Kolb and V. Teplitz, Ap. J. 223 (1978) 327; M. Davis, M. Lecar, C. Pryor and E. Witten, Ap. J. 250 (1981) 423; M.S. Turner, G. Steigman and L.M. Krauss, Phys. Rev. Lett. 52 (1984) 2090; M. Fukugita and T. Yanagida, Phys. Lett. 144B (1984) 386; G. Gelmini, D.N. Schramm and J.W. F. Valle, Phys. Lett. 146B (1984) 386; K. Olive, D. Seckel and E. Vishniac, Ap. J. 292 (1985) 1; Y. Suto, H. Kodama and K. Sato, Mon. Not. Roy. Astr. Soc. 218 (1986) 637.

114 Ya. B. Zel'dovich, Mon. Not. Roy. Astr. Soc. 192 (1980) 663; A. Vilenkin, Phys. Rev. Lett. 46 (1981) 1169, 1496 (E). For reviews, A. Vilenkin, Phys. Rep. 121 (1985) 263; N. Turok, this Proceeding.

115 J.P. Ostriker and L.L. Cowie, Ap. J. 243 (1981) L127; S. Ikeuchi, Publ. Astr. Soc. Japan 33 (1981) 211.

116 J. Silk and M. Srednicki, Phys. Rev. Lett. 53 (1984) 624; F.W. Stecker, S. Rudaz and T.F. Walsh. Phys. Rev. Lett. 55 (1985) 2622; J.S. Hagelin and G.L. Kane, Nucl. Phys. B263 (1986) 399.

117 A. Buffington, S. Schindler and C. Pennypacker, Ap. J. 248 (1981) 1179.

118 M. Srednicki, S. Theisen and J. Silk, Phys. Rev. Lett. 56 (1986) 263.

119 S. Rudaz, Phys. Rev. Lett. 56 (1986) 2128.

120 P. Sikivie, Phys. Rev. Lett. 51 (1983) 1415.

121 L.M. Krauss, J. Moody, F. Wilczek and D.E. Morris, Phys. Rev. Lett. 55 (1985) 1797.

122 W.H. Press and D.N. Spergel, Ap. J. 296 (1985) 679; L.M. Krauss, K. Freese, D.N. Spergel and W.H. Press, Ap. J. 299 (1985) 1001.

123 L.M. Krauss, M. Srednicki and F. Wilczek, Phys. Rev. D33 (1986) 2079.

124 J. Silk, K. Olive and M. Srednicki, Phys. Rev. Lett. 55 (1985) 257; M. Srednicki, K. Olive and J. Silk, UMN-TH 553/86 (1986); J.S. Hagelin, K.W. Ng and K.A. Olive, UMN-TH 566/86 (1986); T. Gaisser, G. Steigman and S. Tilav, BA-86-42 (1986).

125 L.M. Krauss, et al., Ref. (123); K. Freese, Phys. Lett. 167B (1986) 295.

126 M.W. Goodman and E. Witten, Phys. Rev. D31 (1985) 3059; I. Wasserman, Phys. Rev. D33 (1986) 2071.

127 S.P. Ahlen et al., CFA 2292 (1986).

128 S. Dimopoulos, G.D. Starkman and B.W. Lynn, Phys. Lett. 168B (1986) 145.

129 F.T. Avignone III, et al., SLAC-PUB-3872 (1986).

130 A. Drukier and L. Stodolsky, Phys. Rev. D30 (1984) 2295.

131 B. Cabrera, L.M. Krauss and F. Wilczek, Phys. Rev. Lett. 55 (1985) 25.

132 J. Faulkner and R.L. Gilliland, Ap. J. 299 (1985) 994; D.N. Spergel and W.H. Press, Ap. J. 294 (1985) 663; G. Steigman, C.L. Sarazin, H. Quintana and J. Faulkner, Ap. J. 83 (1978) 1050.

133 L.M. Krauss et al., Ref. (122); G.B. Gelmini, L. J. Hall and M.J. Lin, HUTP-86/A042 (1986).

134 J. Faulkner, D.O. Gough and M.N. Vahia, Nature 321 (1986) 226; W. Däppen, R.L. Gilliland and J. Christensen-Dalsgaard, Nature 321 (1986) 229.

135 For original references, J.H. Schwarz, Phys. Rep. 89 (1982) 223; "Superstrings" (2 Vols.) (World Scientific, Singapore 1985).

136 B.S. DeWitt, Phys. Rev. D160 (1967) 1113; J.A. Wheeler, in "Battelle

Rencontres", ed. by C. DeWitt and J.A. Wheeler (Benjamin, New York 1968).

137 J.B. Hartle and S.W. Hawking, Phys. Rev. D28 (1983) 2960; S. W. Hawking, Nucl. Phys. B239 (1984) 257.

138 A. Vilenkin, Phys. Rev. D27 (1983) 2848, D30 (1984) 509; A.D. Linde, Lett. Nuovo Cim. 39 (1984) 401.

139 J.J. Halliwell and S.W. Hawking, Phys. Rev. D31 (1985) 1777; W. Fischler, B. Ratra and L. Susskind, Nucl. Phys. B259 (1985) 730.

140 Y. Okada and M. Yoshimura, Phys. Rev. D33 (1986) 2164; Z.C. Wu, Phys. Lett. 146B (1984) 307; J.J. Halliwell, Nucl. Phys. B266 (1986) 228; T. Inami and S. Watamura, unpublished.

141 L. Alvarez-Gaumé and E. Witten, Nucl. Phys. B234 (1983) 269.

142 M.B. Green and J.H. Schwarz, Phys. Lett. 149B (1984) 117; 151B (1985) 21; E. Witten, 149B (1984) 351; D.J. Gross, J.A. Harvey, E. Martinec and R. Rohm, Phys. Rev. Lett. 54 (1985) 502.

143 For a review, M. Yoshimura, in Proc. of Takayama Workshop Toward Unification and its Verification, KEK-TH-107 (1984).

144 D. Sahdev, Phys. Lett. 137B (1984) 155; R.B. Abbott, S.M. Barr and S.D. Ellis, Phys. Rev. D30 (1984) 720; E.W. Kolb, D. Lindley and D. Seckel, Phys. Rev. D30 (1984) 1205.

145 P.G.O. Freund and M.A. Rubin, Phys. Lett. 97B (1980) 233; S. Randjbar-Daemi, A. Salam and J. Strathdee, Nucl. Phys. B214 (1983) 491.

146 T. Appelquist and A. Chodos, Phys. Rev. Lett. 50 (1983) 141; P. Candelas and S. Weinberg, Nucl. Phys. B237 (1984) 397; T. Koikawa and M. Yoshimura, Phys. Lett. 155B (1985) 137.

147 Y. Okada, Nucl. Phys. B264 (1986) 197; T. Koikawa and M. Yoshimura, Ref. (146) and Ref. (143).

148 D. Bailin, A. Love and C.E. Vayonakis, Phys. Lett. 142B (1984) 344; T. Koikawa and M. Yoshimura, Phys. Lett. 150B (1985) 107; KEK-TH 91 (Revised) (1984).

149 M. Yoshimura, Phys. Rev. D30 (1984) 344; Ref. (144); E. Alvarez and M.B. Gavela, Phys. Rev. Lett. 51 (1983) 931.

150 B. Zwiebach, Phys. Lett. 156B (1985)315; D.J. Gross, J. Harvey, E. Martinec and R. Rohm, Nucl. Phys. B267 (1986) 75; C.G. Callan, D. Friedan, E.J. Martinec and M.J. Perry, Nucl. Phys. B262 (1985) 593.

151 K. Maeda, Phys. Lett. 166B (1986) 59; M. Yoshimura, Progr. Theor. Phys. Suppl. 86 (1986) 208; K. Maeda, M.D. Pollock and C.E. Vayonakis, IC/86/5 (1986); P. Binétruy and M.K. Gaillard, UCB-PTH-86/15 (1986); J. Ellis, K. Enqvist, D.V. Nanopoulos and M. Quirós, CERN TH 4325 (1985); P. Oh, Phys. Lett. 166B (1986) 292; F.S. Accetta, M. Gleiser, R. Holman and E.W. Kolb, FERMILAB- Pub-86/38-A (1986).

152 K. Huang and S. Weinberg, Phys. Rev. Lett. 25 (1970) 895.

153 M.J. Bowick and L.C.R. Wijewardhana, Phys. Rev. Lett. 54 (1985) 2485; B. Sundborg, Nucl. Phys. B254 (1985) 583; S.H.H. Tye, Phys. Lett. 158B (1985) 388; E. Alvarez, Phys. Rev. D31 (1985) 418; P. Salomonson and B.S. Skagerstam, Göteborg 85-32 (1985).

154 A. Salam and J. Strathdee, Ann. Phys. (N.Y.) 141 (1982) 316.

155 J. Bekenstein, Phys. Rev. D5 (1972) 1239; C. Teitelboim, Phys. Rev. D5 (1972) 2941.

156 H. A. Buchdahl, Phys. Rev. 115 (1959) 1325; A.I. Janis, E.T. Newman and J. Winicour, Phys. Rev. Lett. 20 (1968) 878; A.G. Agnese and M. La Camera, Phys. Rev. D31 (1985) 1280.

157 M. Yoshimura, Phys. Rev. D34 (1986) 1021; A. Chodos and S. Detweiler, Gen. Rel. Grav. 14 (1982) 879.

158 Y.S. Wu and Z. Wang, Utah Preprint (1986).

159 P. Candelas, G. Horowitz, A. Strominger and E. Witten, Nucl. Phys. B258 (1985) 46.

160 A. Suzuki, in Proc. of Neutrino '86, Sendai (1986).

Report on the Parallel Session on High Energy Nuclear Interactions

Larry McLerran

Fermi National Laboratory

P.O. Box 500, Batavia, Illinois, 60510

The possibility that high energy nuclear collisions may give some new insight into the dynamics of QCD is discussed. Recent experimental data on the EMC is discussed in the perspective of various theoretical models for the effect. The prospects of making and observing a quark-gluon plasma in ultra-relativistic nuclear collisions is reviewed. The status of experimental programs at CERN and at BNL is assessed.

1 Introduction

Perhaps the most interesting aspect of high energy nuclear interactions is that it may allow for tests of unique features of QCD. These features reflect non-perturbative phenomenon such as confinement and chiral symmetry breaking. In this talk I shall first give an overview of current theoretical understanding of these non-perturbative phenomenon. The EMC effect gives a measure of these effects when particle number densities are of the order of those in nuclei, $\rho \sim 0.15$ Fm^{-3}. This density is less then the typical density scale set for QCD, $\rho \sim \Lambda^3_{QCD} \sim 1$ Fm^{-3}.

To study matter at densities of the order of and larger than those typical of QCD, we must study either the collisions of ultra-relativistic nuclei, or very high multiplicity fluctuations in hadron-hadron collisions. We shall see that simple arguments suggest that densities far in excess of those typical of ordinary nuclei may be achieved under such extreme conditions. I will briefly discuss a few suggested experimental probes of high density matter as it might be produced in such collisions.

There is now a major experimental effort under way at CERN and BNL to make and study ultra-relativistic nuclear collisions, as well as an effort at FNAL to study the extreme environment provided in high multiplicity fluctuations in $\overline{p}p$ collisions at the Tevatron. I shall briefly outline these programs, describing who is involved in these experiments, what they will attempt to measure, and when various experiments will be running.

An exciting subject which I shall not review in this talk is small-x physics. This problem has been studied in detail by the Leningrad groups of Gribov-Levin-Ryskhin and by Frankfurt and Strickman.[1,2] There were no representatives of these groups to present results at this meeting, and with my incomplete knowledge of the field, I do not feel competent to review it.

2 The Properties of High Energy Density Hadronic Matter

In this section I shall discuss the properties of hadronic matter at high energy density. The word high implies a scale for the measurement of the energy density. Such a scale may be provided by a variety of estimates, all of which agree on the order of magnitude of a typical density scale for hadronic matter. The first is the energy density of nuclear matter. With m the proton mass, R_A the nuclear radius, and A the nuclear baryon number, the density of nuclear matter is

$$\rho_A \sim \frac{Am}{\frac{4}{3}\pi R_A^3} \sim .14\ Gev/Fm^3 \qquad (1)$$

We can also use Eq. 1 to estimate the energy density inside a proton. If we use a proton radius of .8 Fm, Eq. 1 gives

$$\rho_p \sim .5\ Gev/Fm^3 \tag{2}$$

There is a good deal of uncertainty in this estimate of ρ_p. We might have instead used the MIT bag radius, or a proton hard core radius, corresponding to an order of magnitude uncertainty in Eq. 2. Finally, another estimate comes from dimensional grounds using the value of the QCD Λ parameter, suitably defined as Λ_{ms} or Λ_{mom}, as the dimensional scale factor. Using the Λ parameter, we find

$$\rho_{QCD} \sim \Lambda^4 \sim .2\ Gev/Fm^3 \tag{3}$$

Again there is an order of magnitude uncertainty both due to the lack of precise experimental knowledge of Λ, and differences induced by using alternative sensible definitions of Λ.

In all of the above energy density estimates, the typical scale was in the range of several hundreds of Mev/Fm^3 to several Gev/Fm^3. At energy densities low compared to this scale, we presumable have a low density gas of the ordinary constituents of hadronic matter, that is, mesons and nucleons. At densities very high compared to this scale, we expect an asymptotically free gas of quarks and gluons.[3] At intermediate energy densities, we expect that the properties of matter will interpolate between these dramatically different phases of matter. There may or may not be true phase changes at some intermediate densities.

To understand how such a transition might come about, consider the example of QCD in the limit of a large number of colors, N_C.[4] Recall that extensive quantities such as the energy density, ϵ, or entropy density, σ, measure the number of degrees of freedom of a system. The dimensionless quantities ϵ/T^4 or σ/T^3 should be of the order of the number of degrees of freedom. For hadronic matter, the number of degrees of freedom relevant at low density are the number of low mass hadrons. Since matter is confined at low density, the number of such degrees of freedom is $N_{dof} \sim 1$ in terms of the number of colors. At high energy density, the relevant number of degrees of freedom are those of unconfined quarks and gluons. The gluons dominate and give $N_{dof} \sim N_C^2$. Therefore in the large N limit, the number of degrees of freedom change by an infinite amount.

Assuming that the transition occurs at finite temperature in the large N_C limit, as is verified

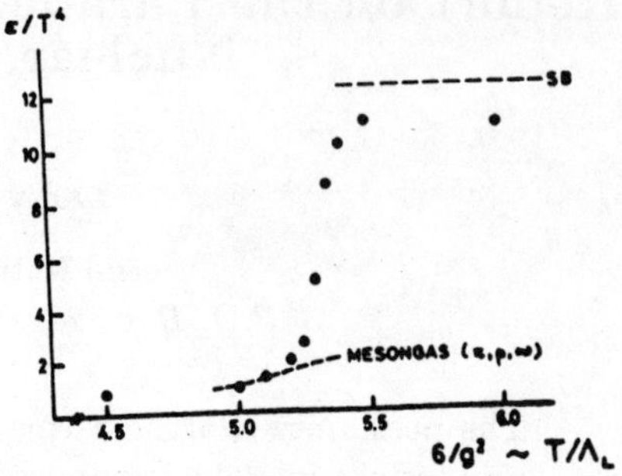

Figure 1: Energy density scaled by T^4 as a function of T

by Monte-Carlo simulation, this result can be interpreted in two ways.[5] From the vantage point of a high density world of gluons, the asymptotic energy density is finite, but at low energy density at some finite temperature the energy density goes to zero. The energy density itself is therefore an order parameter for a phase transition, and there is a limiting lowest temperature. Viewed from the low density hadronic world, there is some limiting temperature where the energy density and entropy density become infinite. Here there is a Hagedorn limiting temperature.[6]

For $N_C = 3$, the above statements are only approximate. The number of degrees of freedom of low mass mesons is

$$N_{dof} \sim N_F^2 \sim 4 \tag{4}$$

where we have taken the number of low mass quarks to be $N_F \sim 2$ for the up and down quarks. The number of degrees of freedom of a quark-gluon plasma is on the other hand

$$N_{dof} \sim 40 \tag{5}$$

The number of degrees of freedom might change in a narrow temperature range, or there might be a true phase transition where the degrees of freedom change by an order of magnitude, if our speculations concerning the large N_C limit are applicable.

Results of a Monte-Carlo simulation of the energy density are shown in Fig. 1.[7–12] These results are typical of the qualitative results arising from lattice Monte-Carlo simulation. The precise values of the energy density are difficult to estimate as is the scale for the temperature. The figure does make clear the essential point, on which all Monte-Carlo simulations agree, that the number of degrees of freedom of hadronic matter changes by an order of magnitude in a narrowly defined range of temperature. There is apparently a first order phase transition for SU(3) Yang-Mills theory in the absence of fermions, and a rapid transition which may or may not be a first order transition for SU(3) Yang-Mills theory with two or three flavors of massless quarks.

For Yang-Mills theory in the absence of dynamical quarks, there is a local order parameter which probes the confinement or deconfinement of a system. This order parameter measures the exponential of the free energy difference between the thermal system with and without the presence of a single static test quark inserted as a probe,

$$< L > = e^{-\beta F_q} \tag{6}$$

As originally proposed by Polyakov[13] and Susskind,[14] and developed in Monte-Carlo studies,[8,9] the Polyakov loop is a Wilson loop at the position of the quark which evolves only in time and is closed by virtue of the thermal boundary conditions which make the system have a finite extent in Euclidian time. The two phases of the theory are the confined and unconfined phases where

$$e^{-\beta F_q} \sim \text{ finite if confined, or } 0 \text{ if deconfined} \tag{7}$$

This quantity is an order parameter for a confinement-deconfinement in theories without fermions or in the large N_C limit in theories with fermions (in the fundamental representation of the gauge group). If there are fermions in the fundamental representation, in the 'confined phase' dynamical fermions may form a bound state with a heavy test quark, so the free energy is finite in what would be the confined phase.[15] Since it is already finite in the deconfined phase, the free energy of a static test quark does not provide an order parameter.

Although $< L >$ is not an order parameter, Monte-Carlo simulations with dynamical fermions show that $< L >$ changes very rapidly in a narrow range of temperatures. This is illustrated in Fig 2,[7] which is typical of lattice computations. For SU(3) lattice gauge theory without dynamical quarks, when $< L >$ is a true order parameter, there is a noticeable discontinuous change. It is

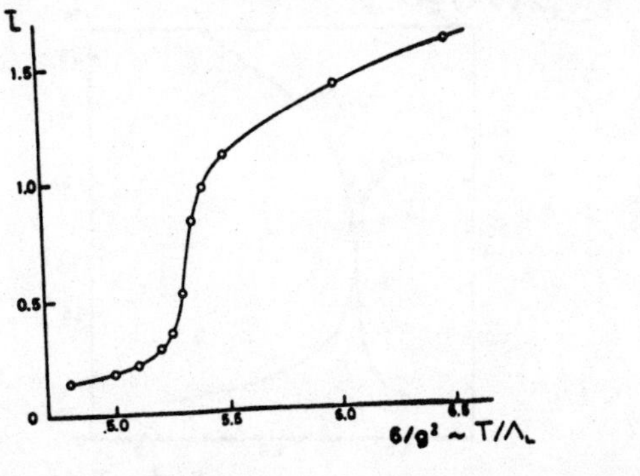

Figure 2: Exponential of free energy of isolated static quark as a function of T

not entirely clear whether there is a discontinuous change corresponding to a true phase change for the theory with fermions.

In the limit of large dynamical quark mass the quarks are no longer important at any finite temperature and decouple. In this limit the confinement-deconfinement phase transitions is a well defined concept with an order parameter which measures a phase change. At zero quark masses there is another phase transition which may be carefully defined, that is, the chiral symmetry restoration phase transition. Chiral symmetry is a continuous global symmetry of the QCD lagrangian in the limit of zero quark mass. Its realization would require that all non-zero mass baryons have partners of degenerate mass and opposite parity. Since this is far from true for the spectrum of baryons observed in nature, chiral symmetry must be broken. Breaking the continuous global symmetry generates a massless Goldstone boson, which we identify with the light mass pion. As a consequence of the breaking of chiral symmetry, the quarks acquire dynamical masses, which may be seen by computing $< \overline{\Psi}\Psi >$. For the chiral symmetric phase, $< \overline{\Psi}\Psi > = 0$, and is non-zero in the broken phase.

For not unreasonable values of the quark masses, $< \overline{\Psi}\Psi >$ is plotted in Fig. 3. There appears to be a rapid change in $< \overline{\Psi}\Psi >$ at about the same place where the order parameter $< L >$ changes rapidly. We conclude therefore that chiral symmetry is approximately restored at the same temperature where quarks stop being approximately confined. The word approximately is important here since absolute confinement or abso-

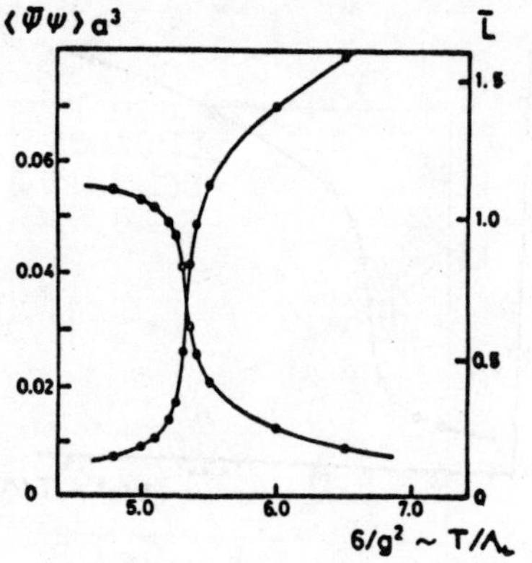

Figure 3: $\overline{\Psi}\Psi$ and free energy of isolated quark as function of T

lute chiral symmetry is impossible for finite mass dynamical quarks.

We can now conjecture on the phase diagram in the temperature mass plane. It is important to realize that we may physically vary the temperature, but not the masses of quarks. Theoretically in a Monte-Carlo simulation, these masses may be changed, but they cannot be changed in nature. It is also important to realize that the mass-temperature diagram represents an oversimplification to the case of equal mass quarks. With different mass quarks, the diagram has more variables and is more complicated.

To plot this diagram, we first discuss the limiting case $m = \infty$. Here there should be a first order confinement-deconfinement phase transition along the T axis. Since a discontinuous change will not be removed by a large but finite quark mass, this first order phase change must be a line of transitions in the m-T plane as shown in Fig. 4. Along the $m = 0$ axis there is a chiral symmetry restoration transition. By the arguments of Pisarski and Wilczek,[16] this transition is first order, and therefore must generate a line of transitions which extends into the m-T plane.

Of course, we do not know what happens with these two lines of transitions, whether they join or never meet, or pass through one another etc. There may be no true phase transition at the values of masses which are physically relevant, or there may be one or two which are the continuation of the chiral transition from zero mass and

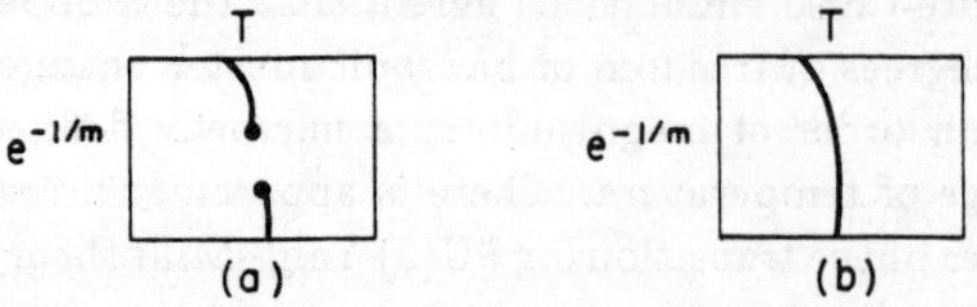

Figure 4: Phase diagrams in the T-m plane for a world where chiral and confinement phase transitions are (a) separate and (b) identified

the confinement-deconfinement transition from infinite mass. The weight of the evidence from Monte-Carlo numerical simulation suggests a very large transition in the properties of matter in a very narrow temperature range, and not much more than that can be said at present. There are a variety of conflicting claims as to whether or not there is a true first order transition at physically relevant masses.[17–22]

There have been serious attempts to obtain reliable quantitative measures of the properties of matter from Monte-Carlo simulation.[23,24] The only truly reliable numbers have been extracted for the unphysical case of $N_F = 0$, that is, no dynamical fermions. It has been shown that the critical temperature of the confinement-deconfinement transition is

$$T_C = 220 \pm 50\ Mev \tag{8}$$

by fitting the potential computed in these theories and comparing it with the potential which fits charmonium. This corresponds to an energy density of $1 - 2\ Gev/Fm^3$ required to make a quark-gluon plasma. These results now appear to be valid for the continuum limit, and seem to be fairly good.

The numerical situation for QCD with $N_F = 2 - 3$ is not nearly so good. The qualitative results have been summarized above, but it is premature to draw any firm conclusions about numbers.

3 The EMC Effect or Physics for $\rho \leq \Lambda^4_{QCD}$

The density of nuclear matter is $\rho \sim 150\ Mev/Fm^3$, a density which is not so small compared to the scale of densities appropriate for QCD. An outstanding question is whether the high energy density environment provided by nuclei can in anyway allow for an understanding of novel features of QCD. For example, is it possible to measure the effects of quarks in nuclear matter, as different from their presence in ordinary nucleons.

Certainly at some level it must be true that the structure functions of quarks in nucleons, $q_N(x)$ and those of nuclei, $q_A(x)$ must be different. The presence of a nuclear environment certainly allows the quarks to propagate over a larger distance scale than is true for nucleons. This may occur through multi-nucleon interactions where in some sense the quark degrees of freedom propagate through the medium of multi-nucleon forces, or it may happen because quark degrees of freedom may propagate more freely because the nucleon bag swells due to the presence of the nuclear medium. In any case, the shift in the spatial correlation length is correlated with fluctuations in the momentum space distributions because of the uncertainty principle. On quite general grounds, it is possible to show that such an increased freedom for quarks results in a degradation of the momentum of the quarks, that is, the structure functions in nuclei are softer than is the case for nucleons. (For very fast quarks, however, Fermi motion will promote some quarks to the region of $x > 1$. There are not many quarks with such large momentum, and this effect is fairly small, and difficult to measure. We are measuring the momentum distributions of quarks in terms of the energy per nucleon.)

The difficulty of using ordinary nuclei to measure the properties of high density matter are of course that the matter density is not so large, and more important, the variation in density between nuclei is quite small. Because of this small range of density variation in a range of densities probably significantly lower than is needed to study novel features of the phase diagram of QCD, it is difficult to assess the possibilities of producing such new phases of matter in high energy nuclear collisions from data on the EMC effect.

There has been much discussion of the EMC effect at this meeting [25–30]. Some claims have been made in lunchtime conversations that the effect has disappeared. In Fig. 5 the data prior to May 1986 for the ratio of structure functions of iron to deuterium is plotted. In Fig. 6, the corresponding new data from this year are plotted, taken from Ref. 31. The data in this figure are from Refs. 25-27 as given in Ref. 31. The EMC data has been

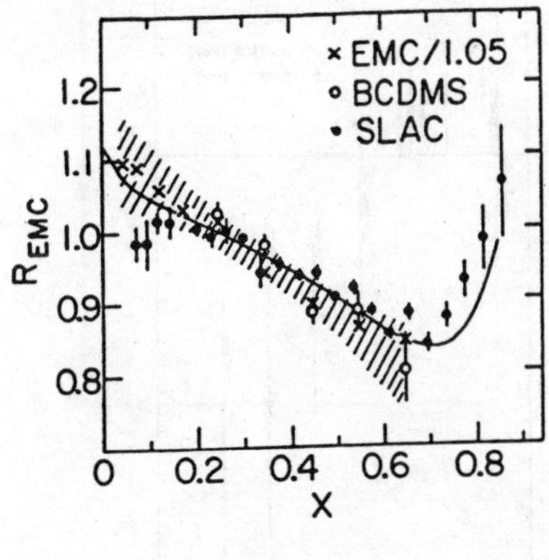

Figure 5: A compilation of data on the EMC effect prior to May 1986.

rescaled by 1.05, and the shaded band represents the EMC group's estimate of the systematic uncertainty. It is clear that the old data and the new data are consistent with one another, if the rescaling by a factor of 1.05 is done to the old EMC data[25]. This rescaling is consistent with quoted systematic uncertainties. Not shown on Fig. 6, is new data, from Ref. 32 where the ratio of $d\sigma/dx$ for Fe and deuterium have been measured with neutrinos, which is also of about the same size as would be inferred from muon measurements.

The controversy over the EMC effect arises not so much from new experimental data, which confirm the effect, but more from theorists enthusiasm based on estimates using the old EMC measurements not rescaled by 1.05. Theoretical speculations have included just about all possibilities from the most radical, that quarks are fully deconfined in nuclei, to the most conservative, where the effect is explained by conventional nuclear physics interactions within the nuclear matter. The size of the predicted effect is of course directly correlated with the radicalness of the theoretical description. The two most popular descriptions of the EMC effect have been explanations based on conventional nuclear physics,[33–35] and the description based on a variable scale parameter for QCD whose magnitude depends upon the density of the media in which it is measured.[36]

Very general arguments suggest that the conventional nuclear physics description, based on models of incoherent nucleons and mesons, must

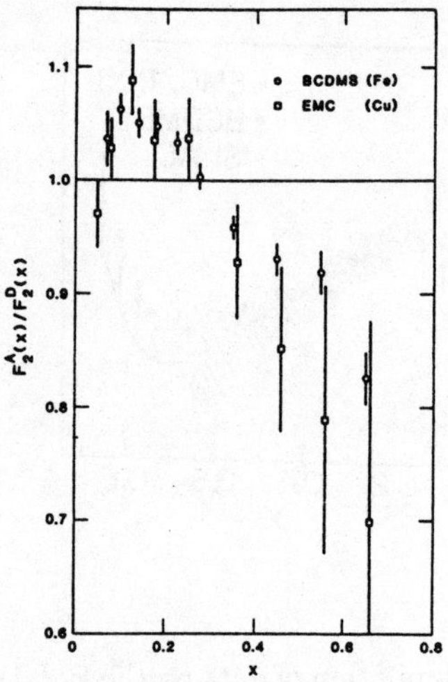

Figure 6: A compilation of new data on the EMC effect.

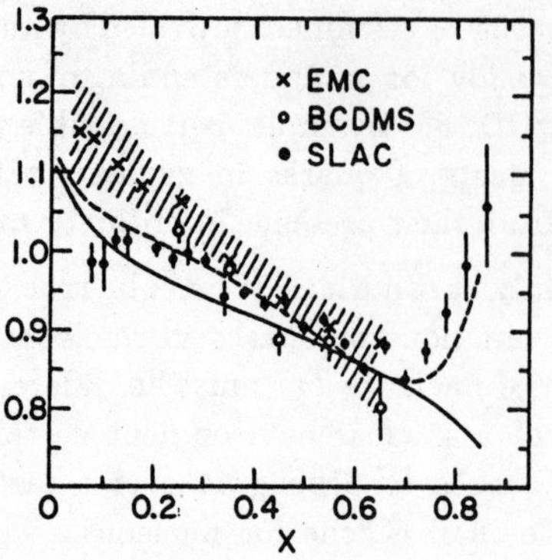

Figure 7: Comparison of nuclear physics model (upper dashed line) with Q^2 rescaling model (lower solid line).

break down at some density scale. Also, at any density, the scale of QCD must change due to the presence of nuclear matter. The real issue is which description adequately and economically describes the data at nuclear matter energy densities. At this range of densities, there is nothing which a priori forces one description to be right or another wrong, although the most exciting possibility would be that one could not describe the effect using conventional nuclear physics.

A comparison of between the Q^2 rescaling model[36] and a nuclear physics computation[34] are shown in Fig. 7. In this figure, the Q^2 rescaling model is evaluated for Q^2 values typical of the EMC experiment.The EMC data is the old data not rescaled down by a factor of 1.05

This comparison shows that the nuclear physics model may do a little better fitting the EMC data than does the Q^2 rescaling model. However, when Q^2 values appropriate for the SLAC data are used in this model, it seems to fit the data fairly well. I think the conservative conclusion based on this type of comparison is that a suitably tailored Q^2 rescaling model or nuclear physics model may be designed to fit the data. The data therefore do not yet warrant radically new phenomenon for their explanation, and a conventional nuclear physics model seems adequate.

Perhaps a better a good way of resolving the difference between the Q^2 rescaling model, and nuclear physics models may be provided by measuring anti-quark distributions in nuclei. The nuclear physics models typically have larger contributions from anti-quarks, arising from an enhanced meson contribution, relative to the rescaling model, where anti-quark distributions are not much enhanced by simply rescaling the distributions. In Fig. 8, a comparison of these two models is shown. The upper curve is a nuclear physics calculation and the lower comes from a rescaling model. The data points with the large error bars are from Ref. 28. The small error bars represent what might be gotten from the experiment FNAL-772.[37] With the results from FNAL-772, these two different models might be clearly resolved.

In conclusion, it seems there is indeed an EMC effect, although no compelling case has been made that it might not be explained as a conventional nuclear physics effect. Perhaps measurements of anti-quark distributions may improve the situation.

4 How to Make a Plasma

The collisions of ultra-relativistic nuclei and fluctuations in $\bar{p}p$ collisions provide the possibility of producing a quark-gluon plasma in a controlled experimental environment.[38,39] Such a collision is shown in Fig. 9 where two nuclei of transverse ra-

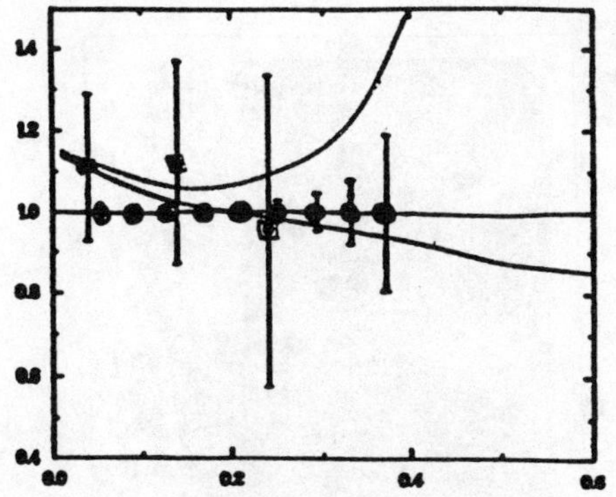

Figure 8: Anti-quark distributions in nuclei. Plotted is the ratio of anti-quark distributions in Fe to D as a function of Feynman x

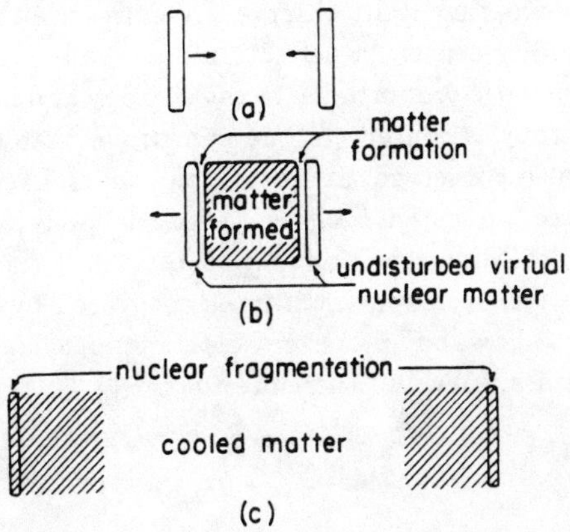

Figure 9: AA collision in the center of mass frame

dius R collide in the center of mass frame. The longitudinal size of the nuclei is Lorentz contracted.

There is a scale implicit in the Lorentz contraction. Once the nuclei have a large enough Lorentz gamma factor so that they would be contracted to a size less than some typical hadronic length scale, possibly a fermi, the Lorentz contraction of virtual quanta with energy corresponding to this length scale stops. Below the beam energy appropriate for this gamma factor, the nuclei Lorentz contract. This energy is

$$E^0_{CM} = m\gamma = \frac{mR}{l_0} = 7 - 70\ Gev \qquad (9)$$

for uranium nuclei and the hadronic distance scale $l_0 \sim .1 - 1\ Fm$. Here and in the rest of this paper, we shall quote the center of mass energy in Gev per nucleon in each nucleus.

We expect qualitative differences in the scattering above E^0_{CM}. Another equivalent estimate of E^0_{CM} is given by estimating the energy at which the fragmentation regions of the two nuclei separate. At energies greater than E^0_{CM} there will be a central region between the two colliding nuclei, which will have small net baryon number density.

An important fact to remember about the matter formed in the collision of two ultra-relativistic nuclei is that it is born expanding in the longitudinal direction. This is because particles are formed with a more or less uniform density in rapidity. Since these particles follow a trajectory which has its origin approximately at $x = t = 0$, and there is a large dispersion in particle velocities, there will be a large longitudinal velocity gradient built into the initial matter distribution. There should be no transverse expansion in the initial condition since we expect a random orientation in the transverse momentum of produced particles. It can be shown that if the distribution of produced particles is uniform in rapidity, the expansion is initially a 1+1 dimensional similarity expansion, and the density of particles decreases like 1/t.

The initial energy density may be estimated on dimensional grounds. The initial energy density should be proportional to the initial rapidity density per unit transverse area. The energy per particle should be of the order of the typical transverse momentum per particle. The longitudinal distance scale and p_T are correlated at early time by the uncertainty principle, since initially the matter appears in a quantum mechanical state, $p_T \sim 1/l_o$. We therefore have

$$\epsilon_i \sim \frac{dN}{dy} \frac{1}{\pi R^2} p_T^2|_{t=t_i} \qquad (10)$$

The initial time t_i will be chosen as the earliest time we believe that the matter may be described as approximately expanding as a perfect fluid.

If the matter expands approximately as a perfect fluid, then ϵ_i may be bounded by parameters which are experimentally measured at late times after the matter decouples, that is, after the pions present in the late state of evolution of the matter have stopped scattering from one another, and are experimentally observed. We first use that the rapidity density in perfect fluid hydrodynamic expansion is proportional to the entropy and because entropy is conserved, one can prove that dN/dy is also conserved, at least in the central region.[40] Since the system cools as it expands, p_T is a monotonically decreasing function of time. (Some of the transverse momentum is recovered by transverse flow, but p_T nevertheless monotonically decreases.) We find therefore that

$$\epsilon_i > p_t^2 \frac{1}{\pi R^2} \frac{dN}{dy} \tag{11}$$

In this equation, all quantities are experimentally observable.

Eq. 11 may be used in combination with experimental data from the JACEE collaboration cosmic ray experiment to estimate ϵ_i.[41] For average $\overline{p}p$ collisions at $E_{CM} \sim 100\ Gev$, $\epsilon_i \sim .6\ Gev/Fm^3$. If we take the average multiplicity for head-on collisions to be $2A^{1/3}$ as is consistent with the JACEE results and conservatively estimate p_T as the value appropriate for $\overline{p}p$ collisions, we find $\epsilon_i \sim 10\ Gev/Fm^3$.

To emphasize how difficult it is to make these cosmic ray measurements, I have shown in Fig. 10, a photo of the emulsion in a C Pb interaction at total energy of about 10^3 Gev. This is the most energetic interaction found in the JACEE experiment to date. If the reproduction appears to you to be only a black splot, you are not mistaken. The analysis of events with so many tracks in such a small emulsion chamber as can be put on a balloon is difficult.

The initial energy density might be much larger than this for a variety of reasons. In fluctuations in $\overline{p}p$ collisions, the multiplicity may be much larger. In nuclear collisions, the initial p_T may be much larger than is typical of the final state. This initial p_T may be determined by kinetic theory arguments, and might be in the range of $.4 - 2\ Gev$,[42,43] corresponding to uncertainty in the energy density of at least an order of magnitude. The initial transverse momentum, and correspondingly, the initial time, may even depend upon the nuclear baryon number A.[44-46] I think the best estimates of the achievable energy densities in central collisions of large nuclei is $2 - 200\ Gev/Fm^3$. This corresponds to an initial temperature in the range of $T_i \sim 200 - 700\ Mev$.

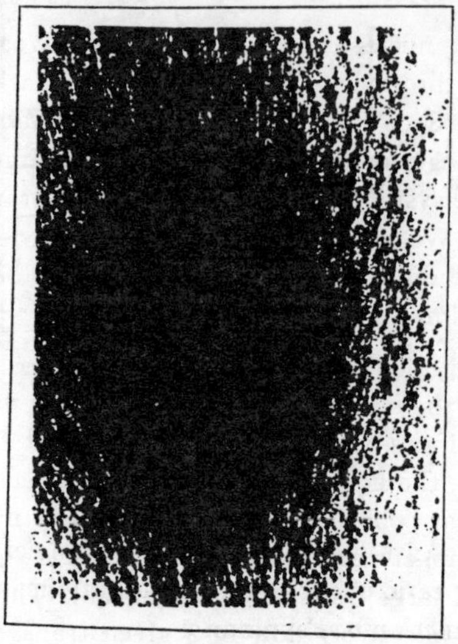

Figure 10: Emulsion photo of the most energetic cosmic ray interaction yet recorded in the JACEE experiment

Such a large uncertainty in the parameters which describe matter formed in ultra-relativistic nuclear collisions is unfortunately typical of the field, a field where there has been little experimental data. While the range of achieved energy densities seems sufficient to form a quark-gluon plasma, there is much reason for caution.

To make a convincing case that there is sufficient time for the formation and evolution of a quark-gluon plasma as an approximate perfect fluid, the expansion rate of the system should be compared to a typical particle collision time. When the collision time is much less than the expansion time, the system should expand approximately adiabatically as a perfect fluid. Since entropy is conserved, the initial and final times for expansion in d dimensions are related by

$$\left(\frac{t_f}{t_i}\right)^d = \frac{N^i_{dof}}{N^f_{dof}} \frac{T_i^3}{T_f^3} \sim 10 - 10^4 \tag{12}$$

where σ is the entropy density and N_{dof} are the number of particle degrees of freedom. At early time, the expansion is 1 dimensional, and later times becomes three dimensional. We estimate therefore that $t_f/t_i \sim 10 - 10^3$. Detailed hydrodynamic computations show that the final decoupling time is probably somewhere in the range of

$t_f \sim 20 - 50\ Fm/c$.[47,48]

Large nuclei are clearly the more favored system for producing and studying a quark-gluon plasma. This follows simply from the facts that the average energy density achieved is larger, and that the system is physically larger in transverse extent. We require $\lambda_{scat} << R_{nuc}$ in order for a perfect fluid hydrodynamic treatment to be sensible. Estimates of λ_{scat} give $.1 - 1\ Fm$.[42,43]

Experimental data exists which throws some light on the size of systems necessary for fluid dynamic effects to become important. At Bevalac energies, the flow of hadronic matter was studied in nuclear collisions.[49,50] In collisions of nuclei of small impact parameter, single particle collisions occur at large transverse momentum. The nuclei do not collectively flow in a given transverse direction unless there are subsequent rescatterings among the constituents of the nuclei. If these subsequent rescatterings do not occur, the transverse momentum of each particle is randomly oriented. To get collective flow, one needs rescattering, and this should be enhanced in collisions at small impact parameter, and collisions of large A nuclei.

In Fig. 11, the flow angle is plotted for various measures of the impact parameter (large impact parameters at the top and small at the bottom of the figure) for various nuclei (small on the left and large on the right). Little evidence of flow is shown for nuclei as large as calcium, and collective effects begin to become important for nuclei of the size of niobium.

5 Probes of the Quark-Gluon Plasma

In Table 1, various experimental probes of the quark-gluon plasma are presented.

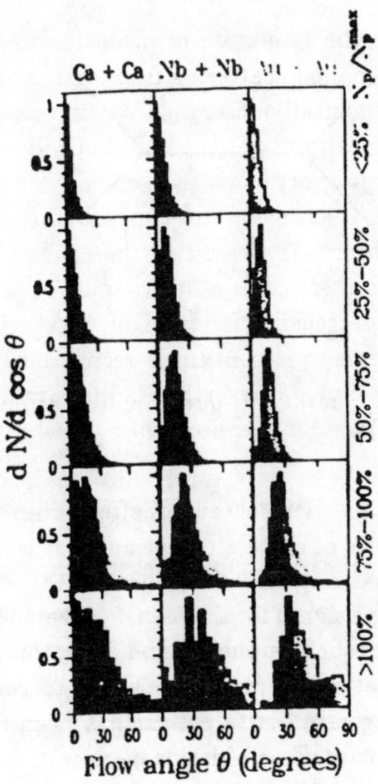

Figure 11: Flow distributions as measured by Gustafsson et. al.

Probes of the Quark-Gluon Plasma

Probe	Physics
Photons and Dileptons	T_i, T_{PT}, Plasma expansion, impact parameter meter, resonance melting
p_t distributions	Equation of state, Evidence of fluid flow
Strangeness	Dynamics of Expansion
Pion Correlations	Size and Lifetime of plasma
Jets	Scattering cross section of quarks or gluons with plasma and hadronic matter

We shall discuss in detail these probes in this section. The bottom line on all of these probes is that they all will involve correlations between several variables. For example, just the requirement of head-on, small impact parameter collisions requires a cut either on total multiplicity or nuclear

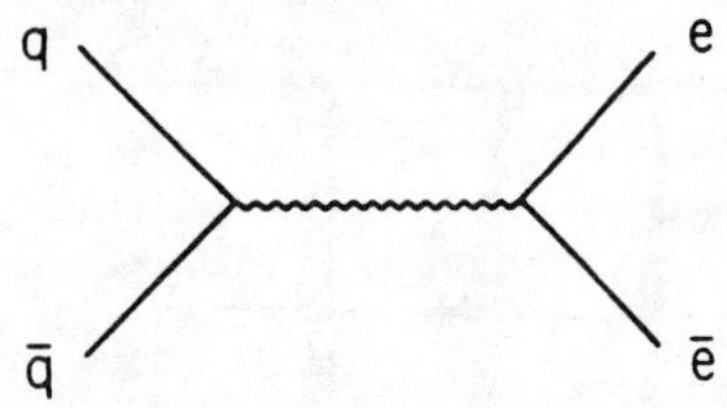

Figure 12: Quark anti-quark annihilation into a di-lepton pair

fragmentation. Because of this often times complicated analysis of correlated variables, it is difficult to argue that any one of the probes will yield an unambiguous signal for a plasma. Nevertheless, in several cases such as photon and di-lepton probes, with a little luck it may be possible to construct a convincing case that a plasma has been formed, and to measure some of its properties.

5.1 Photons and Dileptons

In Fig. 12, quark-antiquark annihilation to produce di-lepton pairs is shown. If we sum over all possible quark-gluon interactions in the initial and final state, then the overall rate for production of di-leptons and photons per unit time and volume is proportional to[51]

$$\frac{dN}{dt d^3x d^4q} \sim Im \int d^4x < J^\mu(x) J^\nu(0) > e^{iqx} \tag{13}$$

This assumes emission from a plasma at a fixed temperature T. The brackets $<>$ denote a thermal expectation value. The current $J^\mu(x)$ has a real, Minkowski time argument.

There are of course a variety of non-thermal sources for di-leptons and photons. There are backgrounds for photons from π^o decays, which in the low q region obscure the signal. There may also be backgrounds for the di-leptons arising from decays of charmed particles. For large q, hard scattering processes from the initially unthermalized beams of quarks and gluons presumably dominate. As the momentum is softened, the contributions arise from an ever more thermalized system which eventually may come from a plasma, provided backgrounds from soft hadronic decays do not become too large of a background. In this intermediate range of q, there are several thermal regions which contribute. At the higher q values, there is presumably a contribution from a quark-gluon plasma, at lower q a mixed phase of plasma and hadronic gas, and at the lowest q values larger than that for which background becomes important, there is a contribution from a hadronic gas.

To compute these distributions of photons and di-leptons, a knowledge of the space-time history of the evolution of the quark-gluon plasma is required.[52–55] Detailed estimates of the space-time evolution of matter produced in head-on collisions of nuclei at large A have now been carried out,[56–60] and the di-lepton distributions have been computed in detail. There has as yet been no attempt to treat non-zero impact parameter collisions. Techniques have also been developed to study the fragmentation region.[61–63] No attempt has been made to treat the pre-equilibrium region, although the cascade computation of Boal may be useful for this.[64] A treatment of the late stages in the evolution of the matter are best treated by cascade simulation of pion interactions, and again could easily be used to compute di-lepton and photon distributions.[65]

The general results of these analysis are the following:

1) For photons and di-leptons emitted from the plasma, the rapidity density of the electromagnetically produced particles is correlated with the rapidity density squared of hadrons. This has been shown to be a general feature of models where the electromagnetically produced particles are produced by final state interactions of hadrons.[66] A plot of this correlation computed in a 1+1 dimensional hydrodynamic model is shown in Fig. 13.[59]

2) Pion rapidity fluctuations are correlated with fluctuations in the di-lepton and photon production rate, at the same rapidity, for thermal emission. This correlation is much different from the case for Drell-Yan pair production where there is no such correlation.

3) The rate of thermal production may be as high as 10^2 times background for not unreasonable values of the temperature. The plasma contribution is most sensitive to the values of the initial temperature when the system becomes thermalized. In Figs. 14a-14b, these thermal distributions are compared to backgrounds from Drell-Yan, and

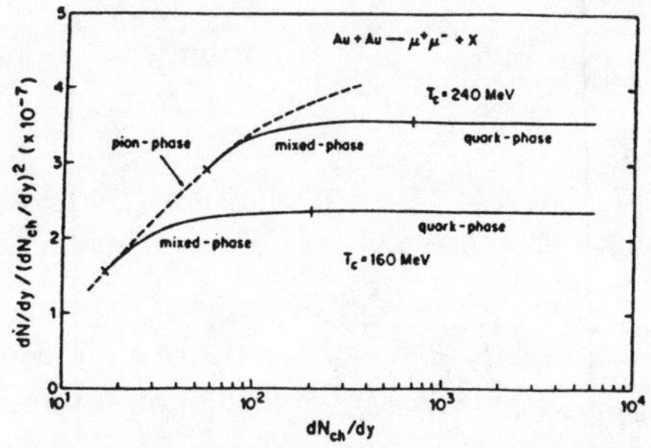

Figure 13: dN/dy of hadrons scaled by $(dN/dy)^2$ of hadrons for head on AA collisions as a function of dN/dy of hadrons

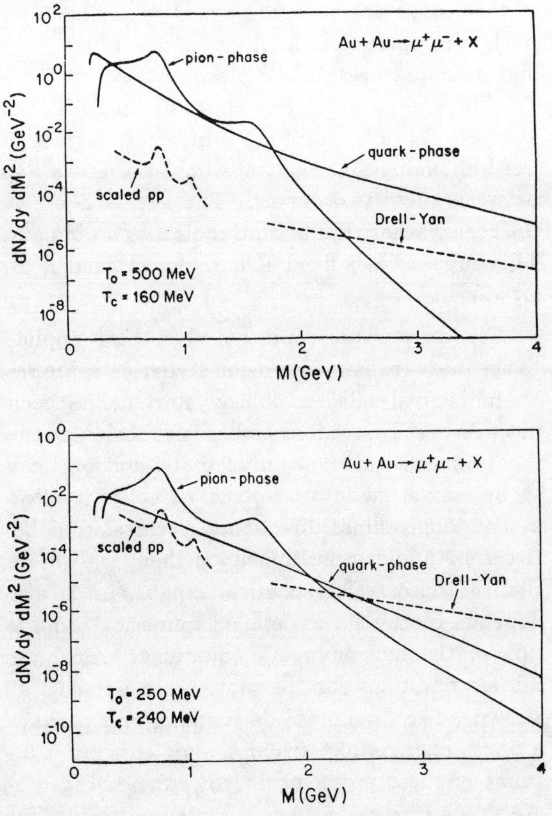

Figure 14: Di-leptons in ultra-relativistic nuclear collisions as a function of mass of di-lepton pair, (a) for an initial temperature of 500 Mev and (b) for 250 Mev.

a generous estimate of backgrounds from resonances and other low p_T phenomenon. For an initial temperature of 500 Mev, the thermal signal is always 10^2 times background for masses of 2-4 Gev, as shown in Fig. 14a. For initial temperature of 240 Mev, the di-lepton spectrum is shown in Fig. 14b. Here the plasma contribution is of the same order as the Drell-Yan contribution for masses of 2-4 Gev.

4) The shape of the thermal di-lepton distribution is fairly sensitive to T_i, the largest value of the temperature for which there is a thermal distribution. The effects of a pre-equilibrium distribution of quarks and gluons has not yet been included so this conclusion is a little soft.

5) For a quark-gluon plasma at high temperature, the distribution of di-leptons is a function only of the transverse mass, $M_t = \{M^2 + p_T^2\}^{1/2}$ There should be a strong correlation between M and p_T, a correlation not present in the Drell-Yan distribution for intermediate mass pairs.

6) The distribution of di-leptons in no simple way reflects the transition temperature. This is a consequence of doing a proper 3+1 dimensional hydrodynamic computation. In 1+1 dimensional computations, the transition temperature controls the distribution in the region of $M \sim 1 - 2$ *Gev*. The shape does of course weakly reflect the transition temperature, but there seems no obvious or convincing way to extract it.

7) The proposed melting of low mass resonances such as the ρ and ω, characteristic of 1+1 dimensional hydrodynamic simulations, [67–69] is not verified in 3+1 dimensional computations. In 1+1 dimensions, the ρ and ω disappear as a resonance in the mass spectrum at large p_T since di-leptons at large p_T are emitted from a high temperature plasma. A high temperature plasma has no ρ or ω resonance. This effect disappears in the 3+1 dimensional computations because transverse expansion makes a large amount of rapidly expanding hadron gas. This transversely expanding hadron gas dominates the spectrum for masses of $M \sim 1$ *Gev* and large p_T. The melting phenomenon is presumably still effective for large mass resonances such as the J/ψ.[70]

Some evidence for what may be thermal di-leptons has been proposed in the ISR experiment R807/808[71]. They have attempted to correlate the ratio of single electrons to pions with total charged

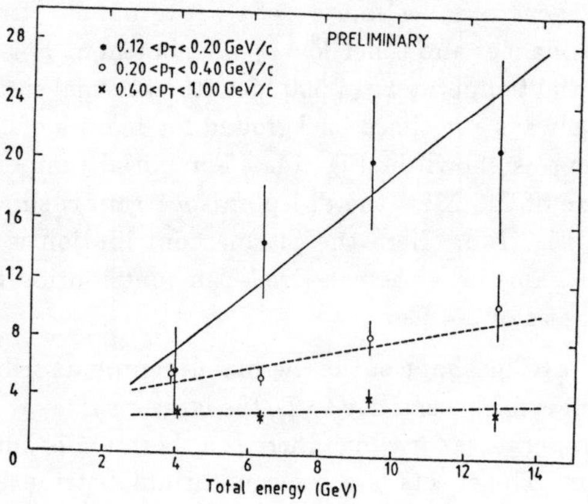

Figure 15: The correlation between e/π in experiment R807/808.

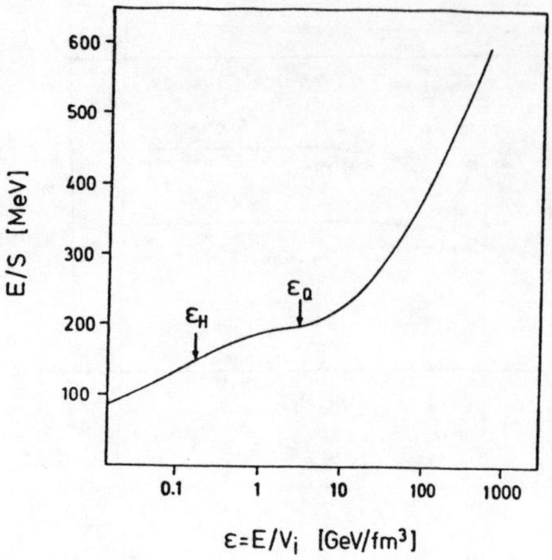

Figure 16: E/S vs ϵ in the MIT bag model

particle multiplicity. In Fig. 15, this correlation is plotted. This ratio should go like dn/dy of charged particles for a thermal source. The data seem to point in this direction, but to have a reliable indicator, it would be useful to have many of the other variables mentioned above measured.

5.2 The Correlation of p_T and $\frac{dN}{dy}$

The correlation between p_T and dN/dy reflects properties of the equation of state of matter.[72,73] This is easily seen from the example of a spherically expanding gas. We assume that at some initial time, there is a spherically symmetric drop of hadronic matter of uniform density matter at rest. We then allow the system to hydrodynamically expand. We assume we know the volume of the initial system, V_o. We measure the total energy of all particles and the total multiplicity of particles in the final state. Since the system is slowly expanding at late times, the entropy of particles in the final state is known assuming the particles were produced thermally from a weakly interacting gas. Since energy and entropy are conserved in the expansion of a perfect fluid, the energy and entropy of the final state is that of the initial state. We can therefore experimentally measure the correlation between say p_T, which is proportional to E/S, and the energy density.[74,75] We can compare this to a theoretically predicted correlation determined by knowing the equation of state.

A plot of E/S verse ϵ is shown in Fig. 16 for a bag model equation of state. The generic features of this curve are straightforward to understand. At low temperature, in the pion gas phase, and high temperatures, in the plasma phase, $E/S \sim T$. The energy density in these two phases goes as $\epsilon \sim N_{dof}T^4$. Since the number of degrees of freedom changes at the transition, there is a gap between these two curves. The gap is filled by the region where the plasma cools into a pion gas. This happens at a fixed T, and almost fixed E/S, for varying ϵ.

There are several problems when this is applied to the more realistic expansion scenarios appropriate for central collisions of heavy nuclei. First p_T is not conserved since longitudinal expansion causes the transverse momentum of individual particles to be converted into un-observed collective flow in the longitudinal direction. A correlation between p_T and say multiplicity is therefore weaker than is the case for spherical expansion. It also depends more on the detailed numerical simulation of the hydrodynamic equations. Also, the initial conditions for the matter are not so well known. The final state de-coupling and perhaps a phase change may produce some entropy. Fortunately these problems do not appear to generate much dispersion in the numerical results for such a correlation.[55] Finally, a severe limitation of present hydrodynamic simulations is that they are limited to the central region of impact parameter zero collisions. If we only have a multiplicity trigger to measure the degree to which collisions occurred at zero impact parameter, then the low

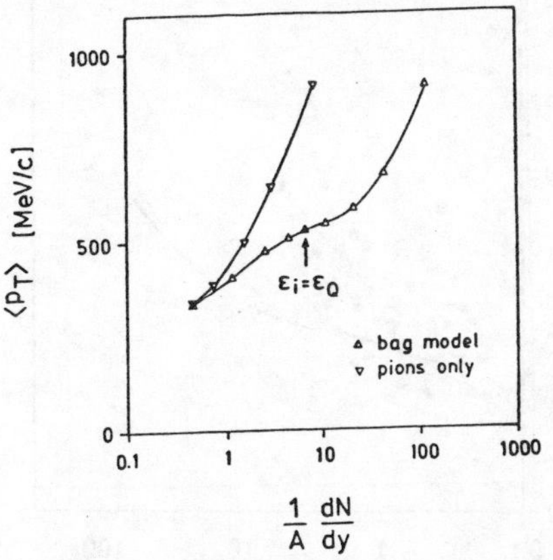

Figure 17: p_T vs multiplicity in head on heavy ion collisions for an ideal gas equation of state (upper curve) and a bag model (lower curve)

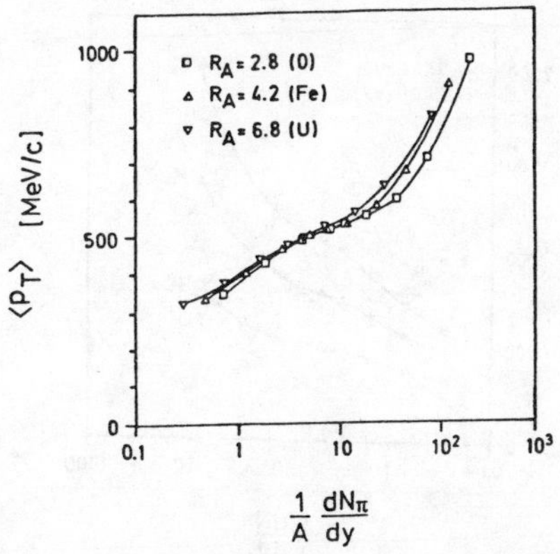

Figure 18: p_T vs dN/dy scaled by 1/A for a variety of A

multiplicity events will always be dominated by large impact parameter, and their contributions have not been computed. The present computations may therefore only provide information on head-on collisions and their fluctuations. Since the number of particles is already large, the fractional fluctuations in the multiplicity for such head-on collisions is small.

There is also the potential problem of backgrounds from conventional processes such as mini-jets obscuring the p_T enhancement from a quark-gluon plasma.[76] At energies typical of the SPS collider, production of mini-jets is presumably responsible for the high multiplicity events. In nuclear collisions at energies less than or equal to those proposed at RHIC, mini-jets are not expected to be a large background since the beam energy is low. Moreover, mini-jets should thermalize in the high multiplicity environment typical of central collisions of large nuclei, thus changing the initial conditions by making the matter initially a little hotter, but yielding a correlation between p_T and dN/dy which may be computed by hydrodynamics.

In Fig. 17, the results of a hydrodynamic computation of p_T vs dN/dy is shown for an equation of state typical of the bag model and a pion gas equation of state. The difference between these curves is large suggesting that an experimental probe of this correlation can resolve various equations of state. A general feature is that the softer is the equation of state, the softer is the p_T. A quark-gluon plasma produces lower p_T particles at fixed multiplicity than does a pion gas.

In Fig. 18, the same correlation is shown for head-on collisions of various nuclei. The curves approximately scale as a function of $1/A \;\; dN/dy$. The factor of $1/A^{2/3}\, dN/dy$ arises because the result must be proportional to the multiplicity per unit area. An additional suppression by a factor of $A^{1/3}$ arises due to the softening effects of longitudinal expansion.

As has been argued by Shuryak,[72] heavy particles should show the effect of collective transverse expansion more strongly than do light particles. This is shown in Fig. 19 where p_T is computed for pions, kaons and nucleons as a function of multiplicity. The physical origin of this effect is that in fluid expansion, there is a collective fluid velocity. Heavier particles have larger masses and therefore $p = mv\gamma$ is correspondingly larger.

In Fig. 20, the p_T distributions of pions, kaons and nucleons are shown. The distribution of nucleons clearly shows the effects of collective flow with the local maximum in dN/d^2p_T at $p_T \sim 1$ *Gev*.

In Fig. 21, an attempt is made to fit the experimentally observed correlation between p_T and transverse energy per unit rapidity as seen in the JACEE collaboration.[41] The JACEE data rises too rapidly to be explained by a quark-gluon plasma. The data does seem to be fit by a pion gas model (dashed line), but the temperatures where the system would be required to be in an ideal pion

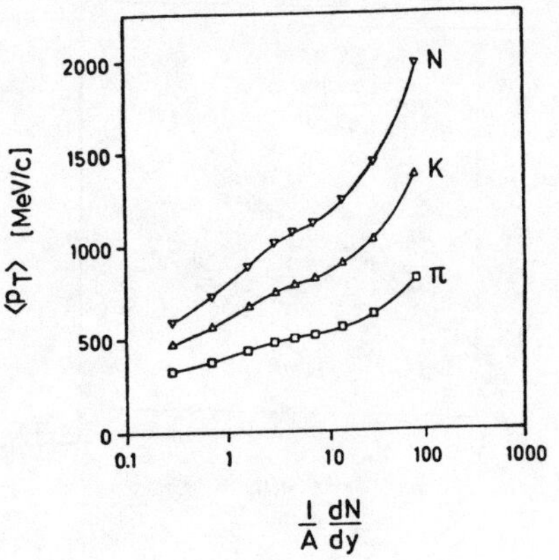

Figure 19: Average p_T vs dN/dy for a variety of particles

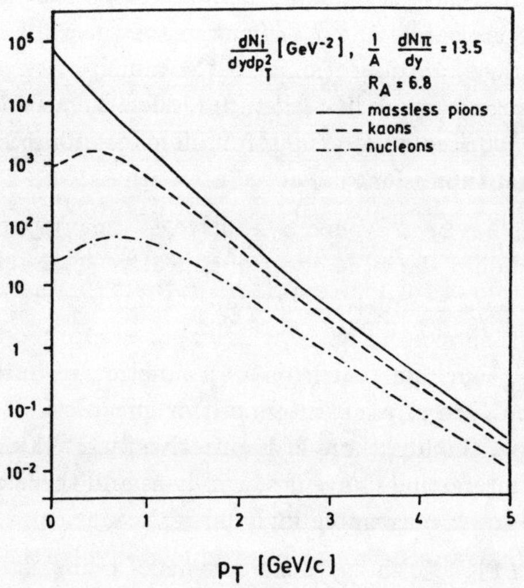

Figure 20: p_T distributions for a variety of particles

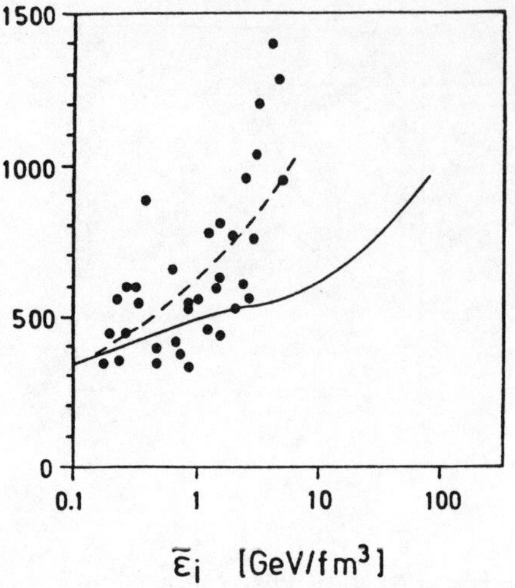

Figure 21: An attempt to fit the JACEE cosmic ray data with a bag model and ideal gas equation of state. The upper curve is the ideal gas

gas are quite large, and we consider this explanation unlikely. Either there is some non-thermal source of high p_T particles in the JACEE data, something is wrong with the space-time picture of the collisions,[74] or something is wrong with the data analysis.

There has been some recent data from p-Pb collisions which indicate that there may be an enhancement in the E_T distribution. [77] In Fig. 22, the E_T distribution of p-Pb collisions is compared to that of pp collisions, and a Glauber theory based multiple scattering model, the Hi-Jet Monte-Carlo.[78] and by Ranft et. al.[79] The theoretical computations do not give nearly the spread in E_T which is experimentally observed. The implications of this E_T enhancement for nucleus-nucleus collisions is not yet known, except that if this enhancement appears in nuclear collisions, the energy densities achieved may be higher than would be expected from conservative estimates.

5.3 Strange Particle Production

Strangeness has been widely suggested as a possible signal for the production of a quark-gluon plasma.[80,81] The argument for large strangeness in its most naive form follows from the observation that there are equal numbers of up, down and strange quarks in the plasma. One might naively

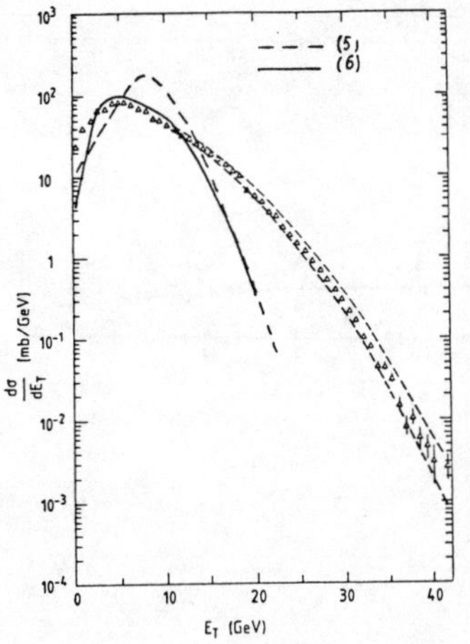

Figure 22: The E_T distribution for p-Pb collisions measured in the HELIOS experiment compared to the predictions of Hi-Jet and Ranft et. al.

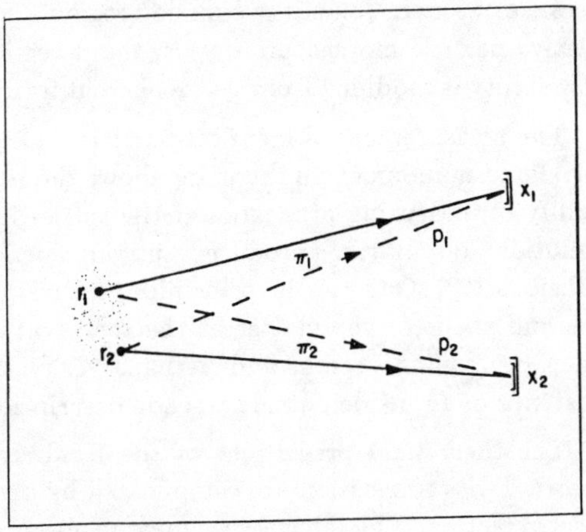

Figure 23: The paths which two particles may take to coincidence detectors

expect that there would be roughly equal numbers of kaons and pions produced, and that the ratio of strange to non-strange baryons would be proportional to their statistical weight, $N_S/N_{NS} \sim 2/3$.

For the case of mesons, the above argument may be easily seen to be false.[82,83] In the expansion of the quark-gluon plasma, and later the hadron gas, entropy is conserved, and the pions are a result of this entropy. A better measure of the strangeness of a plasma is therefore the K/S ratio, where S is the entropy. This may be computed and shown to be smaller in a plasma than in a hadron gas for all temperatures larger than 100 Mev. The K/π ratio is therefore not a direct signal for a plasma. Further, the K/π ratio may be computed in a variety of hydrodynamic scenarios.[83–86] The result is typically $K/\pi \sim .3$. This number is a little larger than is typical of $\overline{p}p$ interactions. As has been suggested by Rafelski and Muller, perhaps only if a plasma is formed will the dynamics allow for such a large K/π ratio, and therefore is a signal of interesting dynamics, or perhaps even the production of a plasma.[87]

Strange baryons and anti-baryons may also provide a signal. Direct computations of the ratio of the ratios of strange to non-strange baryons in a plasma to that in a hadronic gas shows however that a hadronic gas is (if at all) only a little less strange than a plasma.[82,88] These estimates are done for net baryon number zero plasma, and an enhancement may exist for the plasma in the baryon number rich region. At RHIC and SPS energies, the baryon number density is effectively small at all rapidities, and this should be a good approximation. Again, although this ratio of ratios indicates a lack of a signal for equilibrium quark-gluon plasmas, the ratio of non-strange to strange baryons is large, .3-2, in either scenario for $100Mev < T < 300Mev$. This number is far larger than is typical of $\overline{p}p$ interactions, and again by the arguments of Rafelski and Muller, perhaps the only way to dynamically achieve this is by production of the plasma.[87] This ratio is therefore interesting for dynamical reasons.

I conclude therefore that a large strangeness signal is not a direct signal for production of a quark-gluon plasma. It is almost certainly a signal for interesting dynamics, and it may be true that the only reasonable dynamical scenarios where large strangeness may be produced involve the formation of a quark-gluon plasma.

5.4 Hanberry-Brown-Twiss

The Hanberry-Brown-Twiss effect arises from the interference of the matter waves of identical particles as they are measured in coincidence experiments. In Fig. 23, the two possible paths of particles from emission to two coincidence detectors are shown. If the amplitudes for this process are summed and squared, even for incoherent emission amplitudes, the result depends on the distance of separation of the emission regions. For relative particle momentum $k \leq R$, the detection probability is modified from its incoherent form.

The measurement of identical particles closely correlated in momentum therefore allows the possibility of measuring properties of the space-time evolution of matter produced in heavy ion collisions.[89-93] One can in principle measure the size and shape of the matter at the temperature when decoupling occurs, and perhaps verify the existence of an inside-outside cascade description.

The theoretical predictions of the Hanberry-Brown-Twiss correlation are complicated by a variety of factors. The interference may be obscured by final state hadronic interactions which are difficult to compute. The space-time profile of decoupling is not yet so well known, and depends on details of the hydrodynamic simulations as well as the details of decoupling. Assuming that decoupling occurs at late times and large transverse sizes, t, $r_T \leq R$, the correlation occurs only for very small relative momentum, and is very difficult to measure.

5.5 Jets

The rescattering of jets after production in a quark-gluon plasma in principle provides a probe of the plasma and hadronic matter as the jet plows through the evolving system.[94-96] The jets will scatter from the constituents of the plasma as well as the constituents of hadronic matter which forms later. The degree of scattering is a measure of the quark-matter or gluon-matter cross section.

This scattering can dramatically change quantities such as the jet acoplanarity, and can produce phenomenon such as single jets. Theoretical predictions of jet acoplanarity for a variety of jet p_T for an $A = 100$ nucleus are shown in Fig. 24. The dashed curve represents the theoretical prediction in the absence of a hadronic matter distribution. The solid line includes rescattering. For jets of mass 10 Gev, the difference is striking, and the rescattering removes the planar nature of the jets. Even at jet mass of 20 Gev, the difference is still significant, and the jets are remarkably planar. In fact at these masses, the jets are probably largely extinguished.

The experimental measurement of this acoplanarity is very difficult. Particles with low rapidi-

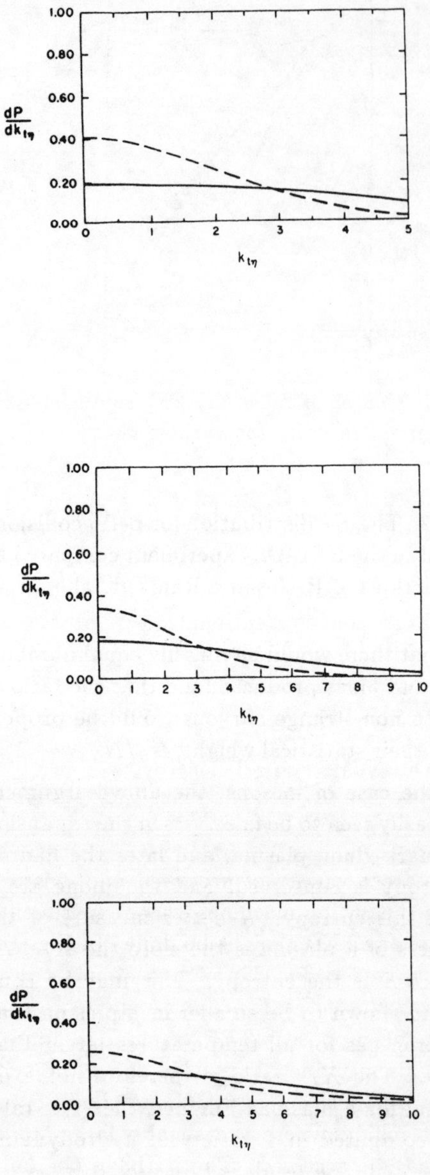

Figure 24: Acoplanarity distributions for jets in head on A=100 collisions (a) Q=10 Gev, (b) Q=20 Gev (c) Q=40 Gev

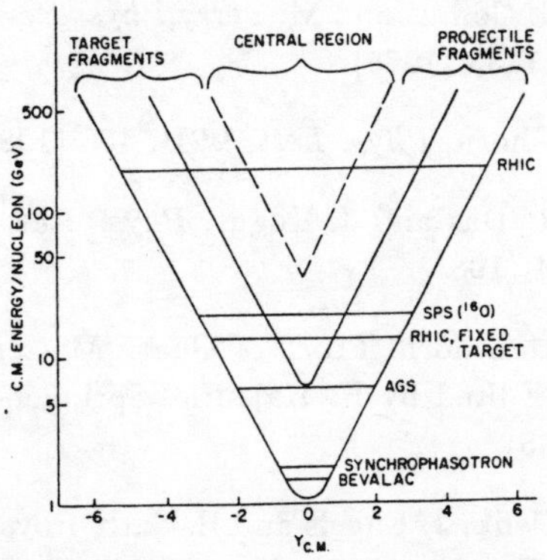

Figure 25: Center of mass energy per nucleon vs center of mass rapidity for various heavy ion accelerators

ties along the jet axis, $y < 2$, must be somehow removed from the sample of particles contributing to the acoplanarity distribution. These low p_T particles arise from conventional low p_T processes, and have little in common with the high p_T particles associated with the jet.

6 Who, What and When

There are a variety of proposed and existing relativistic heavy ion machines where experiments of one sort or another might be done. In Fig. 18, the rapidity gap produced in such machines is plotted against allowed center of mass energy.[97] (On this plot, the proposed ITEP machine is not included. This machine falls a little above the synchophasetron.) The AGS, RHIC and the SPS are the only machines where a reasonably large rapidity gap may be accessed. The RHIC is the only machine which may achieve truly asymptotic energies where a central region opens up.

In addition to beam energy, an important factor for these machines is the A of nuclei which will be accelerated. The AGS in the near future, and the SPS for the foreseeable future will accelerate light ions. In view of the Bevalac data on flow angles, this may be a dangerous thing to do. The collisions at the SPS and the AGS can involve light nuclei on heavy targets, but this considerably complicates any theoretical analysis. Perhaps some hint of the formation of a quark-gluon plasma may be extracted from such collisions, or if there is much good luck, a compelling case. A more important concern is however to see what can and cannot be measured in the dirty experimental environment provided by ultra-relativistic nuclear collisions.

In Table 2, the number of experiments and number of

People and Experiments

	AGS	SPS	TOTAL
Experiments	12	5	17
Physicists	159	208	367
University	93	115	208
Lab	66	93	159
High Energy	23	109	132
Nuclear	136	99	235
US	99	71	170
Non-US	60	140	200

experimentalists involved is shown for the experimental programs at the SPS and the AGS.[98] There are 5 major experiments which will analyze heavy ion collisions at the SPS and 12 experiments at the AGS. About 159 physicists are involved in the AGS program, and 208 at the SPS. The nuclear experimentalists outnumber the high energy by 235 to 132, but there is nevertheless a large commitment from both communities.

Not shown in Fig 25, or listed in Table 2 is the experimental work done at FNAL. The experiment C0 is a dedicated quark-gluon plasma experiment at the Tevatron, involving 27 people.[98] There will also be a small effort with CDF and perhaps D0 to look at high multiplicity, soft processes. These experiments are to be done at very high energy, and of course only with $\bar{p}p$ collisions. The emphasis will be on high multiplicity fluctuations in these collisions, where almost nothing is known about collective effects, or the degree of applicability of a hydrodynamical description.

Ultra-relativistic nuclear physics begins at the AGS and SPS with light ions in the fall of 1986. By 1989, the AGS with a booster should be able to accelerate heavy ions, such as gold. The RHIC project at BNL has R and D money as of 1986.

The largest experiments at the AGS are E802, E810 and E814. [99] E802 will measure inclusive cross sections with full particle identification over a complete kinematic range, with global event trigger. E810 will measure global properties of events. E814 will measure fragmentation with global event triggers.

At the SPS, the major experiments are NA38, NA35, NA36, WA80 and NA34.[100] NA38 is a muon pair experiment. NA35 has a 4π calorimeter and a 2π streamer chamber. NA36 involves a TPC and 2π calorimeter. NA34 has a 4π calorimeter, an external spectrometer, and will measure photons and muon pairs.

At FNAL, C0 will measure multiplicity in the central region, inclusive cross sections and has particle identification over a wide kinematic range.

7 Acknowledments

I gratefully acknowledge the rapid tutorials I received on the EMC effect, a subject the subtleties of which I am largely ignorant, before and after this meeting by E. Berger, R. Jaffe, C. Llewellyn Smith and D. Roberts.

8 Questions

Morris Pripstein (LBL): In your discussion of the EMC effect, you showed that the effect still exists, but then you posed the question 'Is the effect of interest?' Could you be more explicit in your answer?

McLerran: Interest is of course a personal issue. I would be interested if the EMC effect showed an effect which is beyond the ability of conventional nuclear physics to describe. It appears that the data is describable by conventional nuclear physics. This might not be the case after a careful measurement of anti-quark distributions as a function of A, and then I would be interested.

References

[1] L. Gribov, E. Levin and M. Ryskin, *Phys. Rep.* **100**, 1 (1983); E. Levin and M. Ryskin, *Leningrad Preprint 1147* (1985); E. Levin and M. Ryskin, *Yad. Phys.* bf 41, 472 and 1622 (1985).

[2] L. Frankfurt and M. Strikman, *Sov. Phys. Usp.* **28**, 281 (1985); *Nuc. Phys.* **B250**, 143 (1985).

[3] J. C. Collins and M. Perry,*Phys. Rev. Lett.* **34**, 1353 (1975).

[4] C. Thorn, *Phys. Lett.* **99B**, 458 (1981).

[5] S. R. Das and J. Kogut, *Phys. Rev.* **D31**, 2704 (1985).

[6] R. Hagedorn, *Proc. of Quark Matter 84*, p 53, Edited by K. Kajantie, Springer-Verlag (1985).

[7] T. Celik, J. Engels and H. Satz, *Phys. Lett.* **133B**, 427 (1983); *Nuc. Phys.* **B256**, 670 (1985).

[8] L. McLerran and B. Svetitsky,*Phys. Lett.* **98B**, 195 (1981), *Phys. Rev.* **D24**, 450 (1981).

[9] J. Kuti, J. Polonyi, and K. Szlachanyi, *Phys. Lett.* **98B**, 199 (1981).

[10] K. Kajantie, C. Montonen and E. Pietarinen, *Zeit. Phys.* **C9**, 253 (1981).

[11] J. Engels, F. Karsch, I. Montvay and H. Satz,*Phys. Lett.* **101B**, 89 (1981); *Nuc. Phys.* **B205**, 545 (1982).

[12] T. Celik, J. Engels, and H. Satz, *Phys. Lett.* **125B**, 411 (1983); *Phys. Lett.* **129B**, 323 (1983).

[13] A. M. Polyakov, *Phys. Lett.* **72B**, 427 (1978).

[14] L. Susskind, *Phys. Rev.* **D20**, 2610 (1979).

[15] E. Fradkin and S. Shankar, *Phys. Rev.* **D19**, 3682 (1979).

[16] R. D. Pisarski and F. Wilczek,*Phys. Rev.* **D29**, 338 (1984).

[17] F. Fucito, and S. Solomon, *Phys. Lett.* **140B**, 381 (1984); *Phys. Rev. Lett.* **55**, 2641 (1985).

[18] F. Fucito, C. Rebbi, and S. Solomon, *Nuc. Phys.* **B248**, 615 (1984); *Phys. Rev.* **D31**, 1460 (1985).

[19] R. Gavai, M. Lev and B. Petersson, *Phys. Lett.* **140B**, 397 (1984); *Phys. Lett.* **149B**, 492 (1984).

[20] M. Fukugita, S. Ohta, and A. Ukawa, *Phys. Rev. Lett.* **57**, 503 (1986); R. Gupta, G. Guralnik, G. Kilcup and A. Patel, *Los Alamos Preprint LA-UR-86-3054* (1986).

[21] J. Kogut, J. Polonyi, H. Wyld and D. Sinclair, *Nuc. Phys.* **B265**, 293 (1986); *Phys. Rev. Lett.* **54**, 1475 (1985).

[22] R. Gavai and F. Karsch, *Nuc. Phys.* **B261**, 273 (1985).

[23] A. Kennedy, B. J. Pendleton, J. Kuti and K. S. Meyer, *Phys. Lett.* **155B**, 414 (1985).

[24] N. Christ and A. Terrano, *Phys. Rev. Lett.* **56**, 111 (1986).

[25] J. J. Aubert at. al *Phys. Lett.* **123B**, 275 (1983).

[26] A. Bodek at. al. *Phys. Rev. Lett.* **50**, 1431 (1983) and **51**, 534 (1983); R. G. Arnold et. al. *Phys. Rev. Lett.* **52**, 727 (1984).

[27] G. Bari et. al. *Phys. Lett.* **163B**, 282 (1985).

[28] H. Abramowicz et. al *Z. Phys.* **C25**, 29 (1984); M. Parker et. al. *Nuc. Phys.* **B232**, 1 (1984); A. M. Cooper *Phys. Lett.* **141B**, 133 (1984); J. Hanlon et. al. *Phys. Rev.* **D32**, 2441 (1985); V. V. Ammosov et. al. *JETP Lett.* **39**, 393 (1984); A. E. Asratian et. al., *Sov. J. Nucl. Phys.* **43**, 380 (1986).

[29] EMC collaboration, K. Rith, *International Symposium on Weak and Electromagnetic Interactions in Nuclei*, Heidelberg, July (1986); P. Norton, *XXIII International Conference on High Energy Physics*, Berkeley, July (1986).

[30] BCDMS collaboration, A. Milsztajn, *International Symposium on Weak and Electromagnetic Interactions in Nuclei*, July (1986); R. Voss, *XXIII International Conference on High Energy Physics*, Berkeley, July (1986).

[31] E. Berger, *International Conference on High Energy Physics*, Berkeley, July (1986).

[32] E 632 collaboration, *Neutrino 86, 12'th Conference on Neutrino Physics and Astrophysics*, Sandai, Japan, July (1986).

[33] J. D. Sullivan, *Phys. Rev.* **D5**, 1732 (1972); C. H. Llewellyn Smith, **Phys. Lett. 128B**, 107 (1983); M. Ericson and A. W. Thomas, *Phys. Rev.* **126B**, 97 (1983); J. Szwed, *Phys. Lett.* **128B**, 245 (1983).

[34] E. L. Berger and F. Coester, *Phys. Rev.* **D32**, 1071 (1985); E. L. Berger, F. Coester and R. Wiringa, *Phys. Rev.* **D29**, 398 (1984).

[35] S. V. Akulinichev et. al. *Phys. Rev.* **158B**, 485 (1985); S. V. Akulinichev et. al. *Phys. Rev. Lett.* **55**, 2239 (1985); B. L. Birbair et. al. *Phys. Lett.* **166B**, 119 (1986).

[36] F. E. Close, R. G. Roberts and G. C. Ross, *Phys. Lett.* **168B**, 400 (1986); F. E. Close et. al. *Phys. Lett.* **129B**, 346 (1983); R. L. Jaffe et. al. *Phys. Lett.* **134B**, 449 (1984); F. E. Close et. al. *Phys. Rev.* **D31**, 1004 (1985).

[37] Fermilab Experiment 772, Private Communication.

[38] J. Bjorken, *Lectures at the DESY Summer Institute*, (1975). Proceedings edited by J. G. Korner, G. Kramer, and D. Schildnecht (Springer, Berlin, 1976).

[39] R. Anishetty, P. Koehler and L. McLerran, *Phys. Rev.* **D22**, 2793 (1980); L. McLerran, *Proc. of 5th High Energy Heavy Ion Study*, Berkeley, Ca. (1981).

[40] J. Bjorken, *Phys. Rev.***D27**, 140 (1983).

[41] T. Burnett at. al., *Phys. Rev. Lett.* **50**, 2062 (1983).

[42] A. Hosoya and K. Kajantie, *Nuc. Phys.* **B250**, 666 (1985).

[43] P. Danielowicz and M. Gyulassy, *Phys. Rev.* **D31**, 53 (1985).

[44] H. von Gersdorff, L. McLerran, M. Kataja, and P. V. Ruuskanen, *Phys. Rev.* **D34**, 794 (1986).

[45] A. Kerman, T. Matsui, and B. Svetitsky, *Phys. Rev. Lett.* **56**, 219 (1986).

[46] M. Gyulassy and A. Iwazaki, *LBL Preprint LBL-20318* (1985).

[47] G. Baym, B. Friman, J.-P. Blaizot, M. Soyeur and W. Czyz, *Nucl. Phys.* **A407**, 541 (1983)

[48] A. Bialas and W. Czyz, *Acta. Phys. Pol.* **B15**, 229 (1984).

[49] G. Buchwald, G. Graebner, J. Theis, S. Maruhn, W. Greiner and H. Stocker, *Phys. Rev. Lett.* **52**, 1594 (1984); *Nucl. Phys.* **A418**, 625 (1984).

[50] H. A. Gustafsson et. al. *Phys. Rev. Lett.* **52**, 1590 (1984).

[51] E. L. Feinberg, *Nuovo. Cim.* **34A**, 391 (1976).

[52] E. V. Shuryak and O. Zhirov, *Yadern. Fiz.* **24**, 195 (1976); E. V. Shuryak, *Phys. Lett.* **78B**, 150 (1978); E. V. Shuryak; *Sov. J. Nuc. Phys.* **28**, 408 (1978).

[53] L. McLerran and T. Toimela, *Phys. Rev.* **D31**, 545 (1985).

[54] R. Hwa and K. Kajantie, *Phys. Rev.* **D32**, 1109 (1985).

[55] H. von Gersdorff, L. McLerran, M. Kataja, and P. V. Ruuskanen, *FNAL Preprint Fermilab-Pub-86/73T* (1986).

[56] B. Friman, K. Kajantie and P. V. Ruuskanen, *Nucl. Phys.* **B266**, 468 (1986).

[57] J. P. Blaizot and J. Y. Ollitrault, *Saclay Preprint* (1986).

[58] O. D. Chernavskaya and D. C. Chernavskaya, *Kiev Preprint ITP-86-66* (1986).

[59] K. Kajantie, J. Kapusta, L. McLerran, and A. Mekjian, *U. of Minn. Preprint Print-86-0414* (1986).

[60] K. Kajantie, M. Kataja, L. McLerran, and P. V. Ruuskanen, *Phys. Rev.* **D34**, 811 (1986).

[61] K. Kajantie and L. McLerran, *Phys. Lett.* **119B**, 203 (1982); *Nucl. Phys.* **B214**, 261 (1983).

[62] K. Kajantie, R. Raitio and P. V. Ruuskanen, *Nucl. Phys.* **B222**, 152 (1983).

[63] L. Csernai and M. Gyulassy *LBL Preprint* (1986).

[64] D. Boal, *Proceedings of BNL RHIC Workshop*, p. 349 (1985).

[65] G. Bertsch, L. McLerran, P. V. Ruuskanen, and E. Saarkinen (in preparation).

[66] J. Pisut, *Proceedings of 25'th Crakow School of Theoretical Physics*, Zakopane, Poland (1985); V. Csernai, P. Lichard, and J. Pisut, *Z. Phys.* **C31**, 163 (1986).

[67] R. Pisarski, *Phys. Lett.* **110B**, 1551 (1982).

[68] A. I. Bochkarov and M. E. Shaposhnikov, *Nuc. Phys.* **B268**, 220 (1986).

[69] P. Siemans and S. A. Chiu, *Phys. Rev. Lett.* **55**, 1266 (1986).

[70] T. Matsui and H. Satz, *BNL Preprint BNL-38344* (1986).

[71] T. Akesson, M. Albrow et. al. *XXIII International Conference on High Energy Physics*, Berkeley, July (1986)

[72] E. V. Shuryak and O. Zhirov, *Phys. Lett.* **89B**, 253 (1979); *Yadern. Fiz.* **21**, 861 (1975).

[73] L. van Hove, *Phys. Lett.* **118B**, 138 (1982).

[74] H. von Gersdorff, J. Kapusta, L. McLerran and S. Pratt, *Phys. Lett.* **163B**, 253 (1985).

[75] K. Redlich and H. Satz, *Phys. Rev.* **D33**, 3747 (1986).

[76] J. C. Collins, *SSC Workshop*, Los Angeles (1986).

[77] T. Akesson, Y. Choi et. al. *XXIII International High Energy Physics Conference*, Berkeley, July (1986).

[78] T. Ludlam et. al. *Procedings of RHIC Workshop*, BNL51921, p 373 (1985)

[79] J. Ranft et. al. *Z. Phys.* **C20**, 347 (1983).

[80] T. Biro, H. Barz, B. Lukacs and J. Zimanyi, *Nuc. Phys.* **A386**, 617, (1982).

[81] B. Muller and J. Rafelski, *Phys. Rev. Lett.* **48**, 1066 (1982).

[82] K. Redlich, *Z. Phys.* **C27**, 633 (1985).

[83] N. Glendenning and J. Rafelski, *Phys. Rev.* **C31**, 823 (1985).

[84] J. Kapusta and A. Mekjian, *Phys. Rev.* **D33**, 1304 (1986).

[85] T. Matsui, L. McLerran, and B. Svetitsky, *Phys. Rev.* **D34**, 783 (1986); *MIT preprint MIT-CTP-1344* (1986).

[86] K. Kajantie, M. Kataja and P. V. Ruuskanen, *Jyvaskyla preprint, JYFL-9/86.*

[87] J. Rafelski and B. Muller, *GSI Preprint GSI-86-7* (1986).

[88] L. McLerran, *Proceedings of Quark-Matter 86*, Asilomar, Ca. (1986).

[89] G. Goldhaber, S. Goldhaber, W. Lee and A. Pais, *Phys. Rev.* **120**, 300 (1960).

[90] G. Kopylov and M. Podgoretsky *Sov. J. of Nuc. Phys.* **15** 3, 18, 219 (1972); **2**, 336 (1974).

[91] M. Gyulassy, S. Kauffmann, and L. W. Wilson, *Phys. Rev.* **C20**, 2267 (1979).

[92] S. Pratt, *Phys. Rev. Lett.* **53**, 1219 (1984).

[93] W. A. Zajc, *Proceedings of RHIC Workshop, BNL-51921.*

[94] J. D. Bjorken, *FNAL Preprint, Fermilab-Pub-82159-T* (1982).

[95] D. Appel, *Phys. Rev.* **D33**, 717 (1986).

[96] J. P. Blaizot and L. McLerran, *FNAL preprint Fermilab-Pub-86/56-T* (1986).

[97] T. Ludlam, *Proceedings of XXIII Int. Conf. on High Energy Physics*, Berkeley Ca., USA, (1986).

[98] C. Hojvat *Proceedings of XXIII Int. Conf. on High Energy Physics*, Berkeley Ca., USA, (1986).

[99] R. Ledeux, *Proceedings of XXIII Int. Conf. on High Energy Physics*, Berkeley Ca., USA, (1986).

[100] D. Lissauer, *Proceedings of XXIII Int. Conf. on High Energy Physics*, Berkeley Ca., USA, (1986).

Plenary Session

Experimental Techniques
F. Sauli (CERN)

Non-Accelerator Experiments
M. Goldhaber (BNL)

Chairman
K. Strauch (Harvard)

Scientific Secretaries
D. Coupal (SLAC)
J. Haggerty (LBL)

EXPERIMENTAL TECHNIQUES

Fabio Sauli

CERN, Geneva, Switzerland

This is a review of the contributions submitted to the session on experimental techniques, enlarged and completed by recent developments in the field of detectors for particle physics. Three main topics are discussed: particle tracking (mostly with gas detectors), particle identification with calorimetry and Cherenkov ring imaging, and some visualization of complex events with optical fibres and imaging chambers.

1. INTRODUCTION

The two dozen papers submitted to this conference for the session on Experimental Techniques (about a half of which have been accepted for oral presentation) cover a wide range of arguments giving a rather suggestive although fragmentary view of the recent developments in the field. To make this review more comprehensive to the occasional reader, I have added some related works and various 'missing links' to help in understanding the evolution and progress of instrumentation used in experimental particle physics. Also, taking advantage of the traditional rapporteur's privilege, I have included some (but not too much) of my own work in the field.

The various contributions have been grouped in three broad sections: particle tracking, particle identification, and optical imaging. This is a rather arbitrary classification, since some of the papers submitted cover more than one topic and others do not fit into any of the above-mentioned categories. A modern detector used for particle physics consists in general of a complex interleaving of the various techniques, designed to suit the often conflicting experimental needs in the best way.

2. TRACKING WITH GAS DETECTORS

Most of the devices described in this section derive from the original Multiwire Proportional Chamber (MWPC) a fast, position sensitive gaseous detector developed by Charpak and collaborators (1). Since then, and triggered by the experimental needs, many different devices have been designed and operated, exploiting the drift and multiplication properties of ionization electrons in gases. One of the popular tracking detectors for colliders is the Jet drift chamber [so-called because of the nature of the events analysed in the first operational instrument of this kind (2)]. A Jet chamber being built for the new Mark II detector at SLAC has been described by Bartelt (3). It consists (see Fig. 1) of many layers of drift wires in a common gas enclosure, with a pattern repeating itself with cylindrical symmetry. Rows of alternating sense and field wires, mounted radially between cathode wires, are used to collect, multiply and

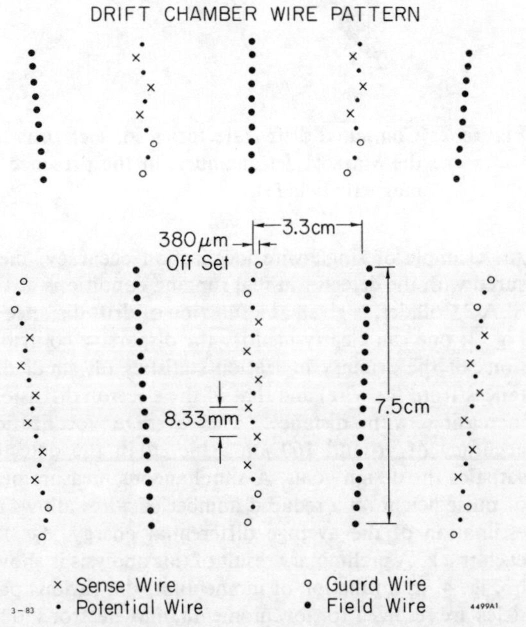

Figure 1: Cross-section of the Jet chamber for Mark II at SLAC (3).

detect the ionization produced in the gas by charged particles; the time of collection (or drift) from the instant of interaction is proportional to the space coordinate of each segment of track. As shown in the figure, sense wires are staggered (by 380 μm) to resolve the right–left ambiguity intrinsic in a symmetric drift cell; moreover, axial modules alternate with modules at a small stereo angle to permit reconstruction of tracks in space: the complete detector has 12 independent pairs of layers. Because of the presence of an axial magnetic field, the drift trajectories of electrons do not correspond to the electric field lines; they can be computed, as shown in Fig. 2, and have to be taken into account when associating a measured drift time with a space coordinate.

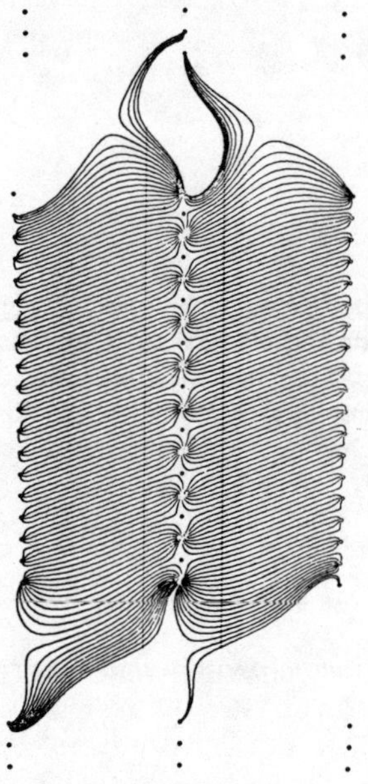

Figure 2: Computed drift trajectories for electrons for the Mark II Jet chamber, in the presence of magnetic field (3).

An example of single-wire localization accuracy, measured with the detector in real running conditions at the SLAC Collider, is given as a function of drift distance in Fig. 3; one can clearly identify the dispersive contributions of the primary ionization statistics (at small distances from the wire) and that of the electron diffusion, increasing with distance. The average localization accuracy of around 160 μm achieved in the detector satisfies the design goals. A simultaneous measurement of pulse height on a reduced number of wires allows an estimation of the average differential energy loss for each track. A preliminary result of this analysis is shown in Fig. 4, as a function of momentum; the various particles are resolved for low momenta, and electrons identified in the whole range.

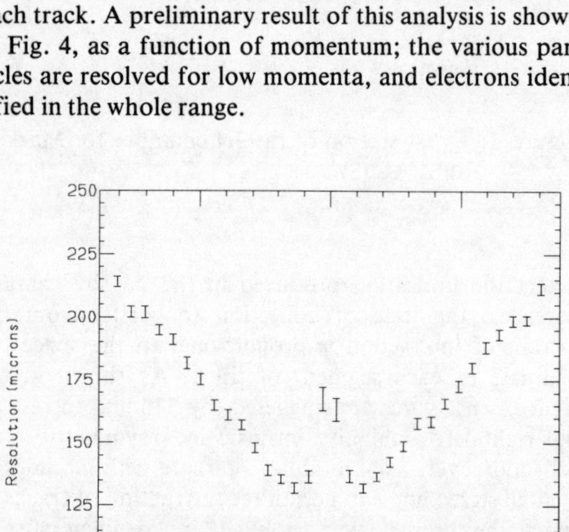

Figure 3: Single-wire localization accuracy as a function of drift distance measured, in running conditions, for the Mark II Jet chamber (3).

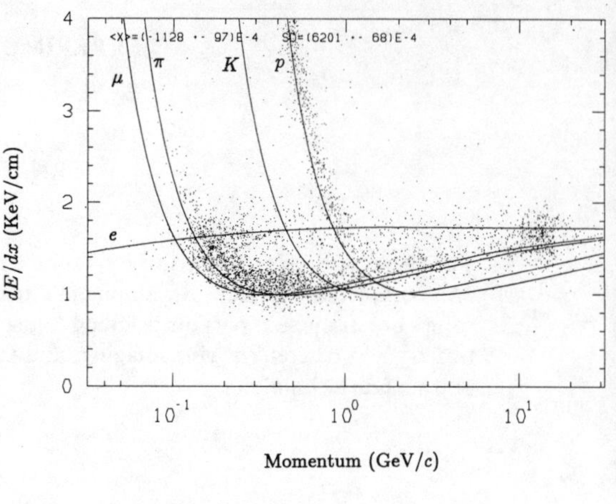

Figure 4: Differential energy loss measured, as a function of momentum, in the Mark II Jet chamber (3).

A recent work of the SLD group at SLAC has shown that the standard Jet chamber design can be modified to get better localization accuracy, making use of focusing wires that help in reducing the dispersions in the arrival time of electrons at the anodes; see Fig. 5 (4). The localization accuracy can be further improved using for the drift time measurement, instead of the conventional discriminators and time-to-digital converters (TDCs), a set of signal waveform encoders or flash analog-to-digital converters (flash ADCs); as shown in Fig. 6, an average localization accuracy of around 50 μm has been measured (5) with a prototype chamber in somewhat ideal conditions (a beam parallel to the wire plane); this result may prove difficult to preserve in a complete detector, but is certainly very encouraging.

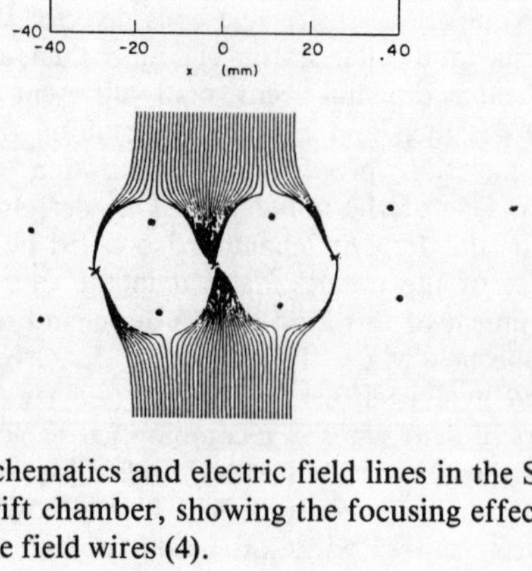

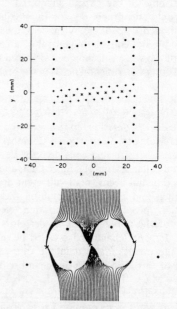

Figure 5: Schematics and electric field lines in the SLD drift chamber, showing the focusing effect of the field wires (4).

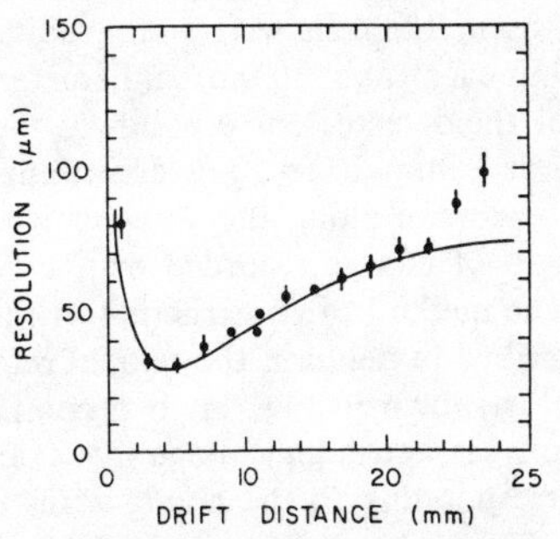

Figure 6: Single-wire localization accuracy as a function of drift distance, measured in the prototype SLD drift chamber (4).

Together with the single-track localization accuracy, a very important design goal in gaseous detectors is the two-track resolution. In drift chambers, the resolution is mainly limited by the width of the detected charge signal; for favourable geometries (tracks perpendicular to the anode wire, almost isochronous at detection), this width cannot be reduced below 50–60 ns, owing to the avalanche and pulse-formation mechanisms, and for normal drift conditions this correponds to around 3 mm resolution. A detailed analysis of the detected signal, possible when using waveform digitizers, allows this limit to be brought down to about 1 mm (6), but this is of course a rather expensive approach both in terms of hardware and of software. A simpler way of achieving similar resolutions has been discussed by Stanton (7). It consists in digitizing, with conventional TDCs, not only the time of the leading edge of a pulse, but also that of the trailing edge (see Fig. 7). A distribution of the time interval between the two measurements, for single tracks and for interacting-events tracks, is shown in Fig. 8; multiple tracks generate a long tail in the width as expected. One can therefore identify each measurement having a width exceeding the one given by single particles as a double track, at a distance that depends in a simple way on the measured width. Figure 9 shows an example of tracking done with such an algorithm; tracks are correctly resolved down to separations of around 0.6 mm.

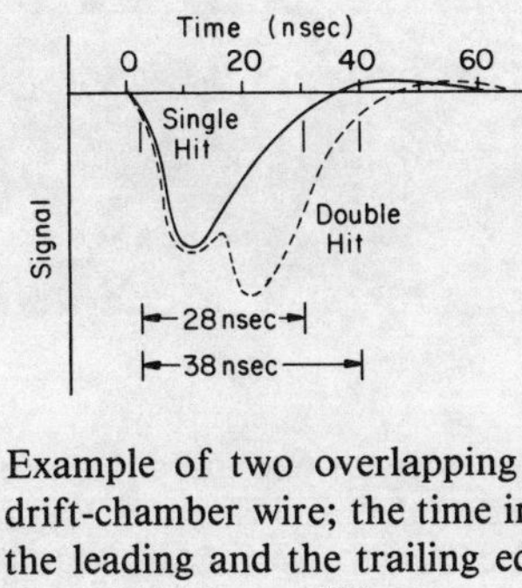

Figure 7: Example of two overlapping pulses from a drift-chamber wire; the time interval between the leading and the trailing edge depends on the drift-time difference between close tracks (7).

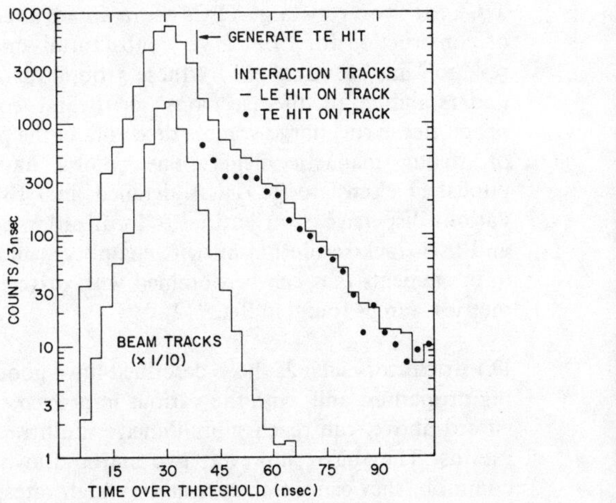

Figure 8: Experimental distribution of the pulse width for single and interacting particles; double tracks can be identified from the measurement of the width down to ~ 0.6 mm (7).

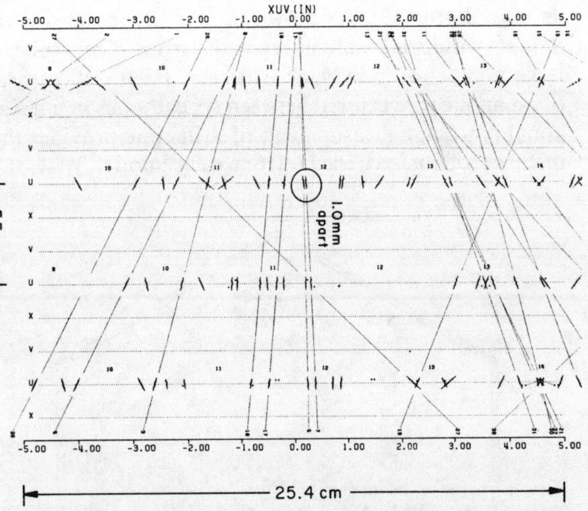

Figure 9: Enlarged view of a reconstructed event in the multisampling drift chambers; the short segment along horizontal lines represents the drift-time measurements in 4 parallel chambers (without removal of the right–left ambiguity). Pairs of close coordinates (one of which is circled in the figure) have been reconstructed with the algorithm described in the previous figure (7).

There was no contribution presented at this conference on the other popular model of the large-volume tracking detector, the Time Projection Chamber (TPC), often considered as the direct competitor to the Jet chamber design, although with different rate and resolution properties. Let me recall that, after the first successfully operational detectors of this kind at SLAC and at

TRIUMF, two very large TPCs are in an advanced stage of construction for LEP; a very substantial amount of research has been realized by these groups for a better understanding of the electrons' drift and collection properties in such large-volume detectors in the presence of strong magnetic fields. The results have been published elsewhere (8–11). A detailed analysis of the various dispersive contributions to localization accuracy and two-track resolution in drift chambers, and of the improvements that can be obtained with different geometries, can be found in Ref. (12).

Drift detectors such as those described have good tracking properties, and, with the various improvements discussed above, can reach submillimetre multitrack resolutions. They have, however, two major drawbacks in common: they cannot operate at very high rates, owing to the memory time intrinsic in the drift process, and they tend to be very sensitive to local malfunctioning (a broken wire in the Jet or TPC chambers in general impairs the whole detector). I have described in the parallel session the recent development in our group at CERN of a multidrift detector that, with around 30 ns resolution time, has very good tracking capability and is intrinsically more reliable than the previous devices because of its modularity (13). As shown in Fig. 10, the idea is to mount within a small hexagonal carbon-fibre tube a large number of individual drift cells. Each cell consists of an anode wire centred between six cathodes in a hexagonal pattern; a measurement of drift-time provides the distance of ionized tracks from each anode. With the geometry chosen, 128 cells with 1 mm radius have been stacked within each tube (30 mm in diameter). The time resolution of the detector corresponds to the maximum drift-time over 1 mm, 20 to 30 ns depending on the gas used; the coordinate along the wires can be measured from the ratio of charge recorded on the two sides of each wire. The method of construction of the multidrift tubes is described in detail in the quoted reference. The final result, as shown in Fig. 11, is a compact module with common gas and high voltage inlets; the array of contacts corresponding to the anode wires is visible in the picture. Figure 12 shows an example of two measured close tracks (1 mm apart on one edge) and still well resolved; indeed, even though each wire can only record one track (the earliest) within its resolution time, the staggering of wires guarantees that close tracks are detected by alternate anodes. Being essentially an array of individual proportional counters, the tube can be operated with virtually any gas filling. In the early prototypes, which had some external breakdown problems, the authors preferred to use a low quencher concentration (isobutane, methylal) in argon to keep the operating voltage low; the localization accuracy appeared hovever to be spoiled by the statistics of energy loss by fast particles: indeed, in argon at atmospheric pressure the average distance between primary interactions is around 300 μm. Since ionization trails are produced very close to the anodes (1 mm away at most) this is the dominant source of dispersions: a resolution of around 100 μm r.m.s. was measured on each wire (13). Recently, the group has improved the localization accuracy using as filling gas pure dimethylether (DME). Originally used in drift chambers because of its very low diffusion coefficient (14), this gas has about the same density as argon, but lower average atomic number: for the same ionization energy loss, it is expected to have about twice the density of primary interactions (15). The gas also gives one of the longest MWPC lifetime under strong irradiation (16). The first results of mea-

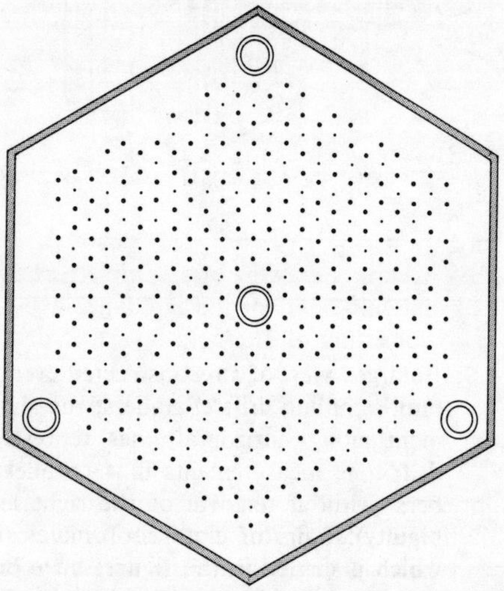

Figure 10: Schematic cross-section of the multidrift detector. Individual hexagonal drift cells, 1 mm in diameter, consisting of one anode and six cathode wires, are wired within a thin carbon fibre tube (30 mm in diameter); recording of the drift-time on all anodes provides on an average ten measured points for each ionizing track. The coordinate along the wires is obtained through current division (13).

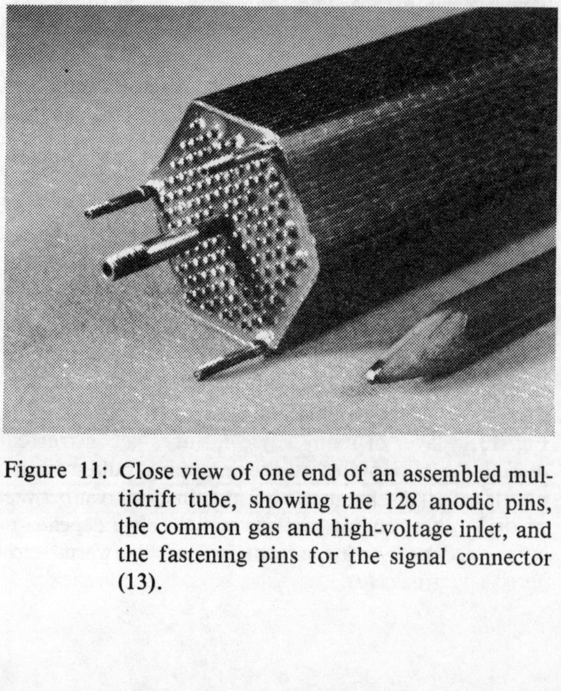

Figure 11: Close view of one end of an assembled multidrift tube, showing the 128 anodic pins, the common gas and high-voltage inlet, and the fastening pins for the signal connector (13).

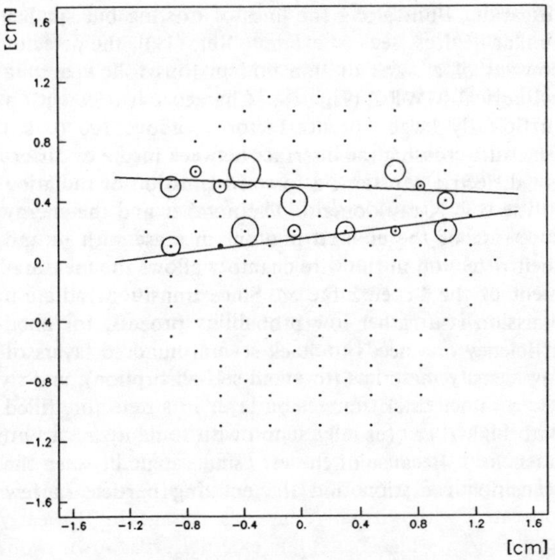

Figure 12: A double-track event recorded with the multidrift detector; the distance between the tracks is 1 mm at the right edge of the tube (10).

surements seem to confirm the expectations, with a measured localization accuracy of around 70 μm r.m.s. With this point accuracy and a two-track resolution better than a millimetre, the multidrift tube seems well suited for use as vertex detector in a high-rate environment; several modules can be assembled in a matrix around an intersect, such that each element can be easily removed and replaced in case of failure. A better localization accuracy can be obtained operating the modules at pressures higher than the atmospheric.

The gaseous detectors discussed so far have been designed for maximum performance, essential for handling the high event rates and particle multiplicities usual at modern accelerators and colliders. In the study of rare decays or cosmic-ray interactions, where what counts is active volume, the main goals are instead simplicity of operation and reduced costs. In a paper submitted to this conference (17), the authors describe the design and performance of a large system of drift tubes, embedded in a matrix of steel sheets, to perform calorimetry and tracking in the upgrade of the Soudan nucleon-decay experiment. The active elements of the system are individual drift tubes, 14 mm in diameter, realized using a resistive foil with coarse external field-shaping electrodes and capable of drifting electrons over a metre towards common anode wires. As shown in Fig. 13, a layer of many adjacent tubes is constructed, assembling mylar sheets, copper strips, and resistive plastic tubes. A long anode wire, at one end of the stack of drift tubes, collects and detects the drifting charge; corrugated steel sheets, separating successive and staggered layers of tubes, provide the conversion matter in

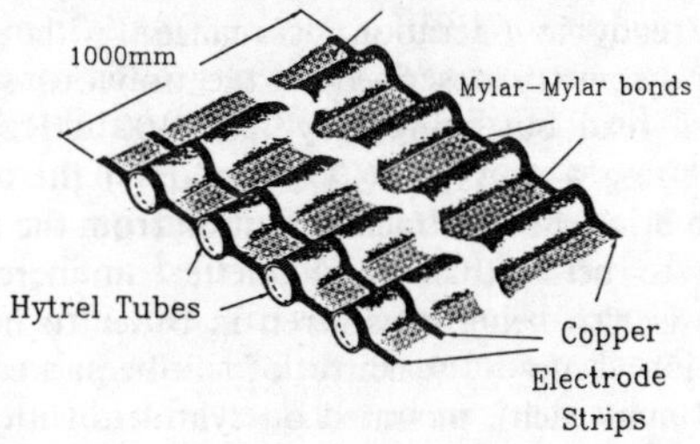

Figure 13: Sketch of the assembly of drift tubes for the Soudan 2 experiment, consisting of mylar sheets, copper drift electrodes, and resistive tubes (17).

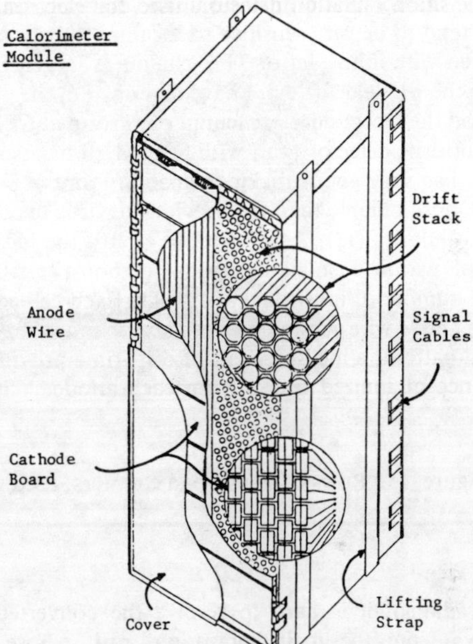

Figure 14: Cutaway drawing of a complete calorimeter module built for the Soudan 2 experiment, consisting of alternate layers of drift tubes and corrugated steel sheets (17).

the calorimeter (see Fig. 14). Readout is accomplished with flash ADCs recording the pulse waveforms on wires and cathode strips facing the wires. Preliminary measurements on prototypes show that electrons can be drifted along the tubes with an attenuation length of around 130 cm (mostly accounted for by diffusion losses); the position resolution, along the drift direction, is about 3 mm r.m.s.

3. PARTICLE IDENTIFICATION

Weidberg [18] described the upgrade program of the UA2 detector at the CERN $p\bar{p}$ Collider that includes several novel technologies both for tracking and for particle identification. The vertex detector of the experiment, already in operation, is realized with the Jet chamber geometry described in the previous section; wires are read out using very fast (100 MHz) flash ADCs. Using a slow gas (CO_2–isoC_4H_{10}) the authors expect to bring the two-track resolution from the present 2.5 mm to better than a millimetre; an increase of pressure is also being considered in order to improve localization accuracy. A matrix of silicon pad counters (9×40 mm^2 each), mounted on cylinders with 14 cm radius, surrounds the vertex chamber; it is used to help pattern recognition in the vertex chamber and through a measurement of the differential energy loss, to resolve electrons from γ conversions (see Fig. 15).

Perhaps the most ambitious part of the upgrade, and the reason to describe it in this section, is a large assembly of transition radiation detectors used for electron identification. Built along the lines of existing but smaller similar devices [see for example Ref. (19)], the detector consists of a stack of thin lithium foils followed by a cylindrical MWPC (Fig. 16). Charged particles with a sufficiently large Lorentz factor γ (above 100 or so) generate, crossing the interface between media of different dielectric constant, a forward emission of radiation in the soft X-ray domain. The number and the energy spectrum of the emitted photons increase with γ, and their detection in the wire chamber allows the measurement of the Lorentz factor. Since transition radiation emission is a rather low probability process, for good efficiency one needs to stack several hundred layers of low-density materials (to avoid self-absorption), and to use a rather thick conversion layer in a detector, filled with high-Z gas (usually xenon with some hydrocarbon quencher). Because of the very small angle between the transition radiation and the emitting particle (a few

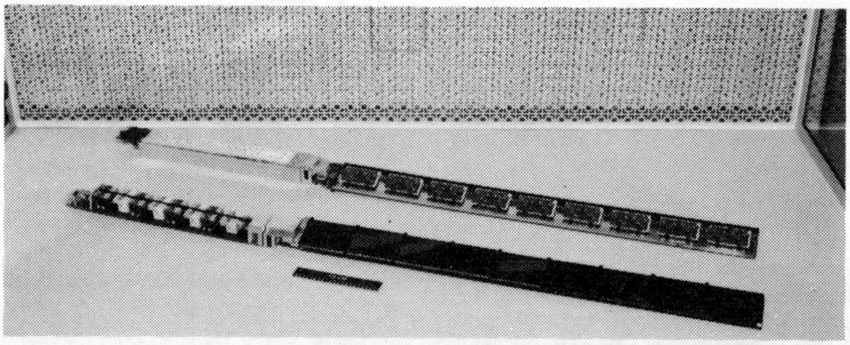

Figure 15: Rows of silicon pad counters, used for tracking and rejection of conversion γ's in the UA2 upgrade (18).

milliradians), it is hard to resolve the converted soft X-rays from the main ionization trail, whose total charge may exceed that generated by the photons (the most probable energy loss of fast particles in 1 cm of xenon is 4 keV, comparable with the lowest energy of X-rays that can be detected). The solution adopted by the group is to record the full charge profile versus time on each wire, using a set of flash ADCs, and to perform a detailed signal analysis to identify the localized clusters corresponding to X-ray conversions (see Fig. 17). Preliminary measurements using the described set-up indicate an e/π rejection power of 30 at 40 GeV/c, at 80% efficiency for electrons.

I suggest that a better way of instrumenting transition radiation detectors might be to use a row of the multidrift tubes described in the previous section. With a xenon filling, because of the small wire spacing, the energy loss of charged particles has a most probable value of 0.3 keV in each sensitive cell, compared to the minimum X-ray energy of several keV; a simple discriminator with high threshold setting on each anode wire could be used to count the photons, with a very good rejection against the direct energy loss. Moreover, the multidrift tube has a very good time resolution (30 ns or so) permitting its use in a high-rate environment.

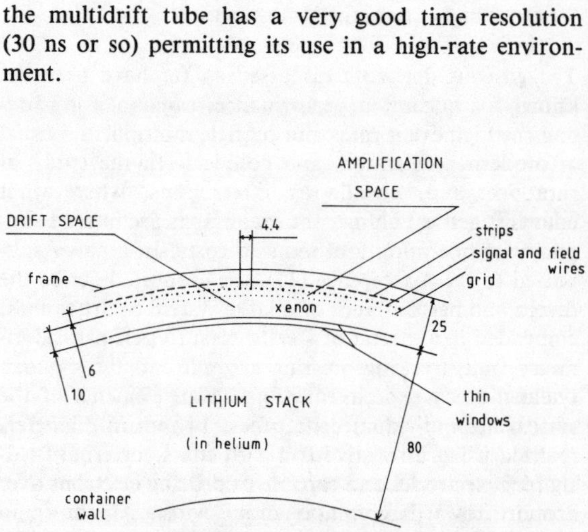

Figure 16: Schematics of the transition radiation detector adopted for electron identification in the UA2 upgrade. A radiator made with a stack of thin lithium foils in helium is followed by a MWPC with xenon filling (18).

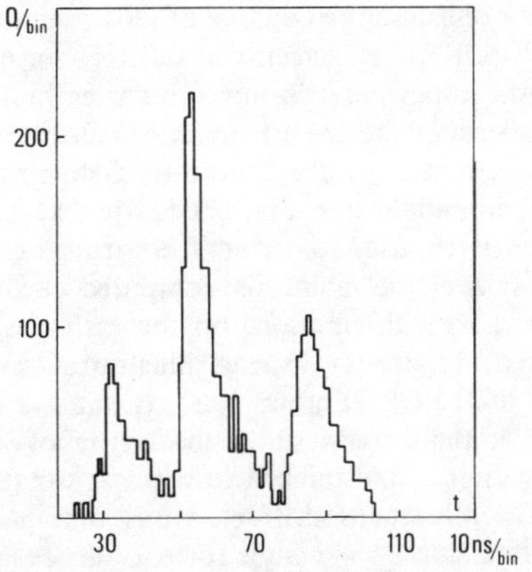

Figure 17: Example of the signal recorded with a flash ADC in the transition radiation detector; the charge delivered by two conversion X-rays is shown. The ionization losses of the particle contribute to the peak at short times (owing to the MWPC geometry) and to uniform backround, with fluctuations, in the tail.

Considerable progress has been made on a method for charged-particle identification based on detection and localization of photons, emitted by Cherenkov effect, in multiwire chambers; named RICH or CRID (acronyms for Cherenkov Ring Imaging Detector), the technique has been discussed in two contributions (20, 21). The essentials of the method are as follows. When a fast charged particle of velocity β traverses a transparent medium of index of refraction n, photons are coherently emitted by Cherenkov effect in a cone of opening given by $\cos\theta = 1/n\beta$ (for $\beta > 1/n$). A spherical mirror at one end of a radiating volume reflects the photons in a circular pattern on the image plane; a measurement of the ring's radius provides a direct estimate of β.

Known since many decades, this method of identification could not be applied for large surfaces of detection before the development, about ten years ago, of the first photosensitive MWPCs. Two problems had to be solved, however, before obtaining a practical device: to find an input window-photosensitive vapour combination giving a good enough quantum efficiency, and to develop a chamber geometry allowing large multiplication factors to be reached for detecting single photoelectrons; indeed at high gains photons emitted in the avalanche process may reconvert either on the photosensitive gas or on the cathodes thus inducing various secondary processes, increasing the noise counts or leading to discharge. One solution, based on the Multi-Step Proportional Chamber (MSPC) (22) was adopted in the first operational ring imager (23). In the MSPC, the electrodes and the field structure are such as to allow preamplification, by a factor of 100 or so, of the charge detected, before entering a regular MWPC for further multiplication. Large gains, in excess of 10^6, can then be realized without secondary feedback effects; localization is therefore easily accomplished measuring the direct and induced signals on anodes and wire cathodes conveniently angled. With a CaF_2 window, and using a gas mixture containing triethylamine (TEA) as photosensitive vapour, the device is capable of detecting and localizing with submillimetre accuracy single photoelectrons in the wavelength range between 120 and 160 nm; appropriate radiators in this region are the noble gases. As an example, Fig. 18 shows a ring consisting of five photons and reconstructed from the measurement of induced charge on two perpendicular sets of cathode wires and on the anodes; the central point corresponds to the position of crossing of the charged particle itself (23). The figure also illustrates the basic limitation of the method: for a larger number of photons, or for many particles, signals in projections overlap and cannot be resolved. The very good time resolution of the MSPC (40–50 ns), together with the quoted limit in the number of detected photons and in the choice of radiators, makes this instrument suitable for identification of high-rate, very-high-energy but low-multiplicity events.

For lower momenta, one has to use radiators having a higher index of refraction, such as gaseous or liquid isobutane, fluorocarbons, and others, with a transparency cut-off at longer wavelengths. This requires the use of a photosensitive vapour with lower ionization potential: tetrakis(dimethylamine)ethylene (TMAE) has been found to be appropriate, with a threshold at 220 nm. A

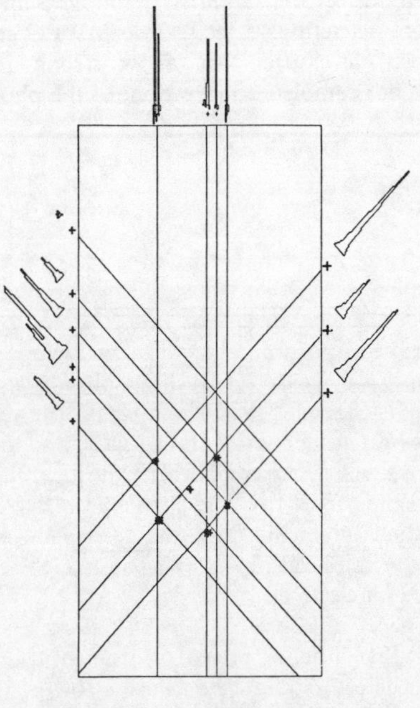

Figure 18: Example of a 5-photon Cherenkov ring generated by a fast particle, and measured with the MSPC at Fermilab. The coordinates of each point are reconstructed from the measurement of the charge profiles on anode and cathode wires; the central point corresponds to the crossing of the particle itself (23).

fused silica window (cut-off at 160 nm) can be used, being cheaper and more rugged than the fluorides. Because of the very low vapour pressure of TMAE (0.4 Torr at room temperature), which implies the use of gas layers several centimetres thick for efficient conversion, such a choice leads to modest time resolution (a few microseconds); moreover, the charges released in the gas by ionizing particles (hundreds of ion pairs) are a severe source of problems, as discussed later.

Instead of being directly detected after conversion, as in the previous detector, photoelectrons can be drifted in a long, TPC-like conversion space towards a small size MWPC [see Fig. 19 (24)]. The drift-time measurement then provides one coordinate of the conversion point, the wire number, whilst the centre of gravity of induced signals on cathode strips provides the other two. This method has the considerable advantage of requiring a small amount of electronics, and has moreover a better multi-event capability than the previous one; on the other hand, the time resolution is several tens of microseconds, which is appropriate for low-rate colliders but not in other applications.

As mentioned, the low vapour pressure of TMAE has two major consequences. The photons emitted in the avalanche during multiplication, copiously at high gains, may reconvert centimetres away from the anodes and feed back faking a good signal; multiplication of the several hundred electrons generated by the charged particle by direct ionization has, for the same reason, even more deleterious effects. Two MWPC geometries have been developed that allow strong reduction, although not complete suppression, of the photon feedback. The one described by Ekelöf (20), and adopted for the DELPHI RICH detector at CERN, consists in enclosing the anode wires in thin tubular cathodes, with a narrow entrance slit; the efficiency of collection for electrons is increased by the use of field-shaping wires at suitable potentials [see Fig. 20 (24)]. The coordinate along the wires, used to correct the parallax error in the photon conversion point, is computed recording the induced charge distributions on the cathodes, suitably segmented. Figure 21 instead illustrates the solution adopted by the CRID group at SLAC, and discussed by Toge (21); the anodes sit in the centre of a scallop-shaped cathode, and thick field wires at suitable potentials efficiently focus drift electrons onto the anodes while reducing the solid angle for photon feedback. The longitudinal coordinate is in this case computed recording the ratio of currents flowing from the two sides of each wire; for better sensitivity, the group has used as

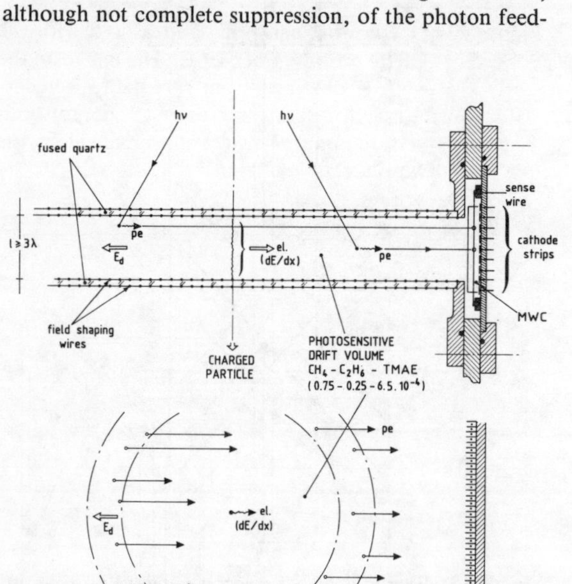

Figure 19: Principle of construction of the DELPHI RICH detector. Electrons produced by photoelectric effect in the gas drift along a thin drift chamber layer towards the MWPC, where they are detected and localized (24).

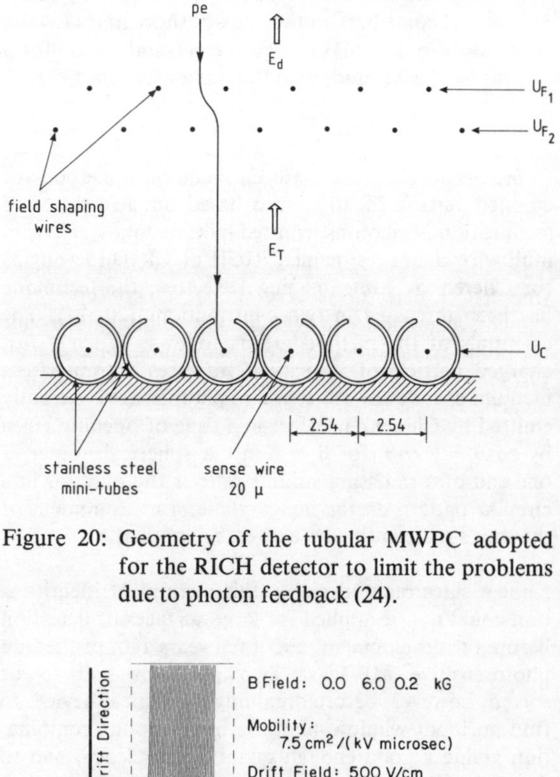

Figure 20: Geometry of the tubular MWPC adopted for the RICH detector to limit the problems due to photon feedback (24).

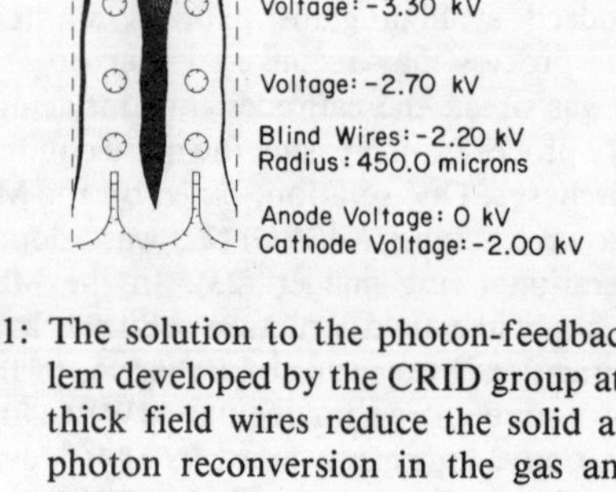

Figure 21: The solution to the photon-feedback problem developed by the CRID group at SLAC; thick field wires reduce the solid angle for photon reconversion in the gas and focus drifting electrons to the anodes (21).

anodes very thin (7 μm diameter) carbon filaments with a resistivity of 1.6 kΩ/cm. This choice also allows large proportional gains (2×10^5) to be reached at low operating voltages.

The same detector can be used to image rings produced by two radiators of different indexes of refraction, one on each side of the drift region; this allows coverage of a wider region of velocities. A combination of a gaseous radiator and of a thin liquid radiator, adopted by both groups, is shown in Fig. 22 (24); the image from the liquid forms directly, without the need for reflecting mirrors. Figure 23, from the same reference, shows the electronic image obtained with a prototype RICH in a monoenergetic and collinear beam; many events have been added in the display, and the gas radiator ring (the small circle), the liquid radiator ring and the beam-produced activity are clearly recognized. Figure 24 (21) shows the improvement in resolution obtained making use of a good measurement of the coordinate along the anode wire (which eliminates the parallax error due to the thickness of the conversion layer, see also Fig. 22); a gas radiator was used for this measurement, and moreover the beam did not cross the detector. The fitted radius of the ring, for single particles, is 27.3 mm with an error of 2 mm r.m.s.; the average number of detected photons is 8.6 per event, consistent with the estimates. Similar results have been reported by the CERN group.

The construction of a large acceptance CRID imager (the RICH detector is very similar in conception) is shown schematically in Fig. 25. The solid angle around the collision region is covered, combining a large-angle system with cylindrical symmetry with two circular detectors in the forward directions.

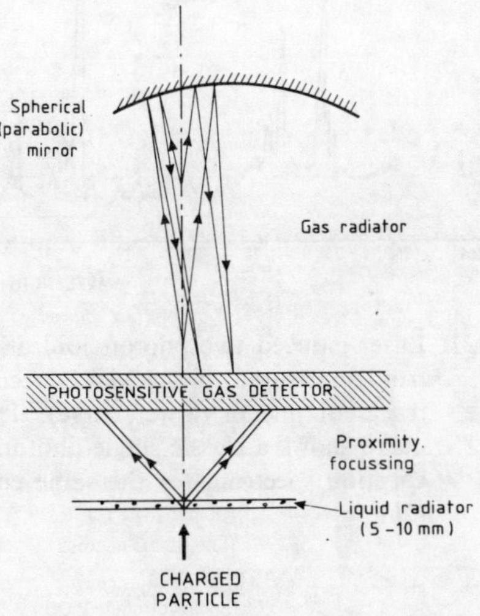

Figure 22: Schematics of the two-radiator ring imager. A gas radiator and a mirror focus the photons on a circular pattern on the detector; the image generated by the liquid radiator instead is directly projected (24).

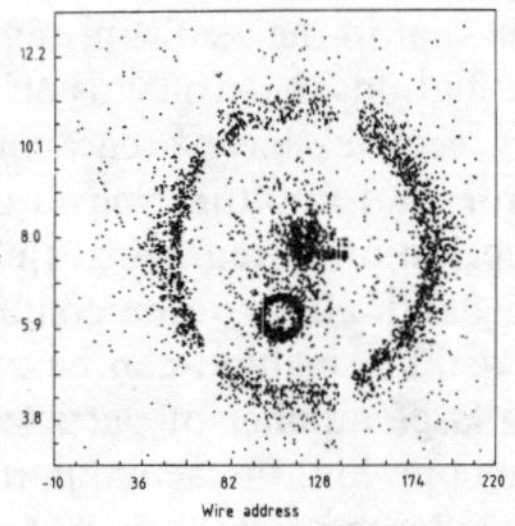

Figure 23: Overlap, on a computer display, of several hundred images obtained with a prototype RICH detector at CERN for collinear particles. The small, off-centre ring is generated by the gas radiator, and the large one by the liquid radiator; the counts in the central region are due to the beam crossing and associated feedback photons (24).

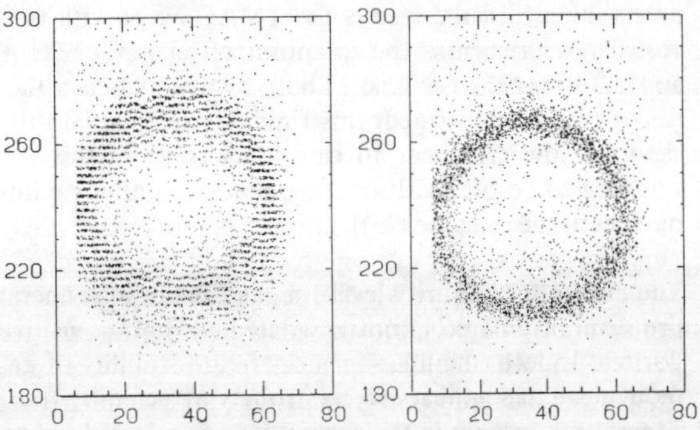

Figure 24: Overlap of many rings, measured with the CRID prototype at SLAC. The image on the right is obtained from raw data after correcting for parallax error on the photon conversion point, using the coordinate along the wires (21).

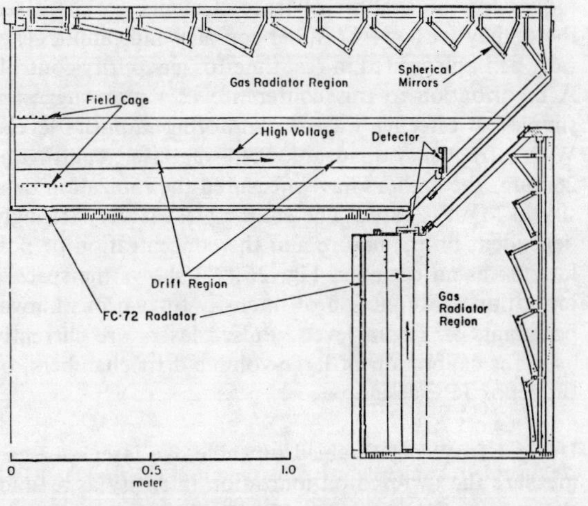

Figure 25: Schematics of a full Cherenkov ring particle identifier, covering most of the solid angle around an intersect (21).

Let me point out that in the results presented in both of the quoted contributions the problems originated by any residual photon feedback have been avoided, either by tilting the mirror to form the rings well outside the beam-crossing region or by removing, through software analysis of the data, all hits not contained within a fiducial region; neither method can be used in the real detector, with a large number of particles and overlapping rings. In my opinion, the actual particle identification power of a large-acceptance instrument in real experimental conditions remains to be demonstrated.

For a good estimate of the identification properties of an imager and for a stable operation, it is essential to know the quantum efficiency of the photo-ionizing vapour and its sensitivity to impurities. In a paper submitted to this conference, Holroyd et al. (25) describe various purification methods for TEA and TMAE, and present their measurements of absorption length and quantum efficiency for both vapours. It is interesting to note that, whilst the results for TMAE agree with previous measurements, the quantum efficiency of TEA appears lower than published data by as much as a factor of two; if confirmed, this finding could explain the reported low efficiency in ring imagers using this gas [thought to be due to absorption losses on unknown impurities in the radiator (23)].

The control of impurity levels in gas fillings is a general problem in MWPCs, not restricted to photosensitive devices. Indeed, the long-term operation stability of gaseous detectors appears to be strongly affected by trace compounds present in the gas mixture or released by the assembly materials; under irradiation, these compounds aggregate in a somewhat unpredictable way, building up solid or liquid deposits on electrodes (polymers) that quickly deteriorate the behaviour of the detector. The problem of MWPC ageing, or radiation damage, was not mentioned at this conference, but is of course of primeval importance. For an extensive collection of contributions on the subject, see Ref. (16). Various techniques for the analysis of trace elements are discussed there; they are however rather complex and cannot generally be implemented in real time for gas-purity control. A contribution to this conference (26) may suggest a simple but effective way of monitoring impurity levels. With a frequency quadrupled Nd–yag laser, emitting at 266 nm, the authors have measured the ionization density in MWPC gases. The charge yield is very strongly dependent on the nature and the concentration of pollutants; as an example, Fig. 26 (27) shows the specific ionization as a function of intensity for various known pollutants at 1 ppm levels. Pulsed lasers are currently used for calibration of large-volume drift chambers, of the Jet or TPC design.

Using a frequency-doubled tuneable dye laser, one can measure the two-photon ionization intensity as a function of wavelength; Fig. 27 shows (lower curve) an example obtained by the authors of Ref. (26) in a proportional counter with a typical gas filling, containing traces of phenol (a common plasticizer). The upper curve in the figure is the result of a classic UV absorption measurement: the similarity of the results proves the sensitivity of the ionization method to the chemical structure of the pollutant. Further work in this direction

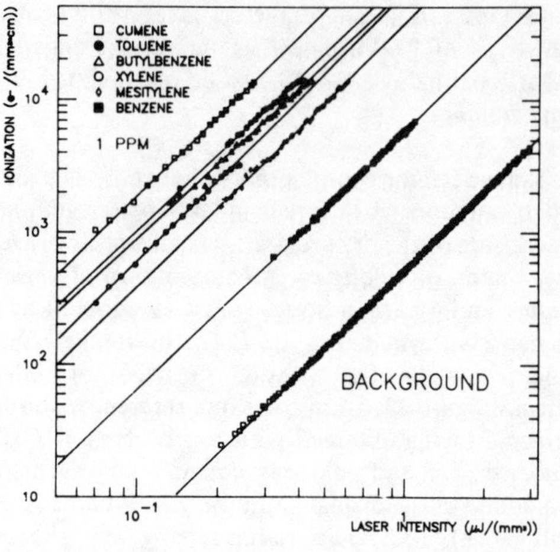

Figure 26: Ionization density measured as a function of laser intensity in a MWPC gas with various additives, all at 1 ppm concentrations. The laser emits at 266 nm, and ionization is obtained by double excitation (27).

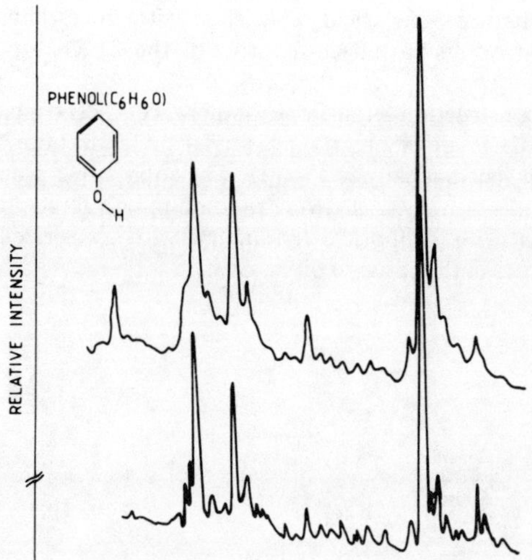

Figure 27: Laser-induced two-photon ionization spectrum measured in a MWPC containing traces of phenol (lower curve). The upper curve shows a classic single-photon UV absorption spectrum for the same compound (26).

is certainly encouraged, and may lead to a simple method of on-line monitoring of trace impurities in normal MWPC systems, or of quantum efficiency in RICH counters.

Calorimetry is the third and last method for particle identification to be briefly reviewed here. A sampling calorimeter consists of a stack of high-density converter layers alternated with energy-measuring layers, for example liquid-argon ionization chambers or gas MWPCs. Because of the nature of the interactions involved in the shower development, the energy resolution of electromagnetic calorimeters is far better than that of hadronic calorimeters, mostly because of the energy loss by strong processes that remains hidden (e.g. neutron production). The use of uranium as a converter partly restitutes, by various nuclear interactions, this hidden energy into detectable charge, thus producing a compensation effect and an improvement in energy resolution for hadrons (28). Tuts (29) described at this conference the expected performance of a large-size uranium–liquid argon calorimeter designed for the D0 detector at Fermilab; in a prototype of this instrument, the ratio of fractional energy losses between electron and pion showers has indeed been measured to be close to unity (1.1 ± 0.1). The group expects an e/π rejection of around 700/1 for the full instrument.

The use of liquid argon as ionization medium is difficult because of the cryogenic technology involved; a large amount of effort has been invested in recent years to find a room-temperature liquid that satisfies the ionization-chamber requirements, i.e. a large free electron lifetime and a high electron mobility to avoid recombination and allow fast collection. Figure 28, from the contribution of Sass to this conference (30), collects the

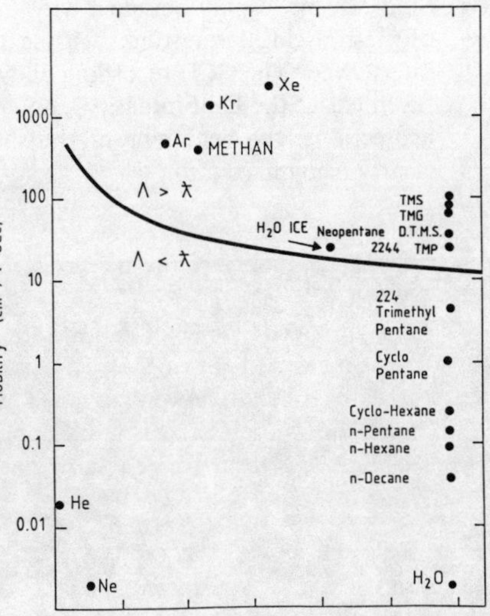

Figure 28: Electron mobility in various liquids. Electrons remain free only for liquids shown above the full line in the figure (30).

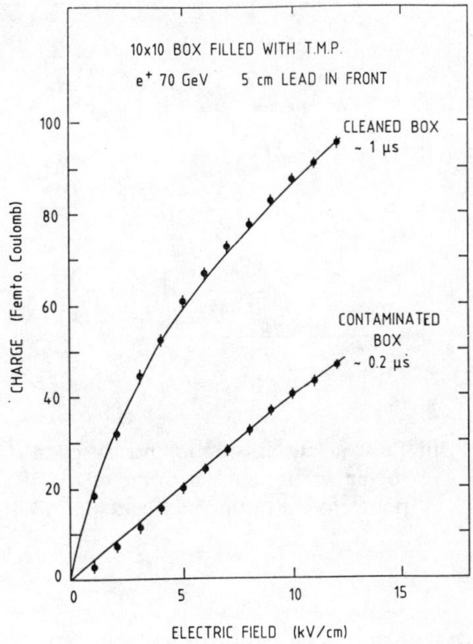

Figure 29: Charge collected in a prototype calorimeter filled with TMP, as a function of electric field. The two curves correspond to different impurity contents (30).

values of mobility as a function of temperature for various liquids. The full line in the figure indicates the boundary above which electrons can exist free (mean free path Λ larger than the de Broglie wavelength). It can be seen that, apart from the liquefied rare gases and methane, some compounds such as tetramethylsilane (TMS) and tetramethylpentane (TMP) satisfy the condition of having free electrons with large enough mobility; the whole problem is that of obtaining these liquids pure enough to avoid electron capture or recombination. For example, 200 ppb of oxygen in TMP results in an electron lifetime of 1 μs, just about at the limit for acceptable operation; other pollutants have even worse effects. The authors describe the results of measurements realized with a prototype calorimeter consisting of a stack of stainless-steel converter plates with 12 gaps, 12 mm thick each, filled with TMP in a containment vessel. Figure 29 shows the dependence of detected charge on the collection field, measured with the prototype for 70 GeV positrons, in two conditions of cleaning of the container; the best results correspond to a lifetime of drifting electrons of 1 μs. Note that even at the higher fields attained a full collection efficiency is not yet reached. The actual pulse-height distribution measured in these conditions is shown in Fig. 30: these results are very encouraging, and the group pursues its plan to instrument the UA1 detector upgrade with a warm liquid calorimeter in the near future.

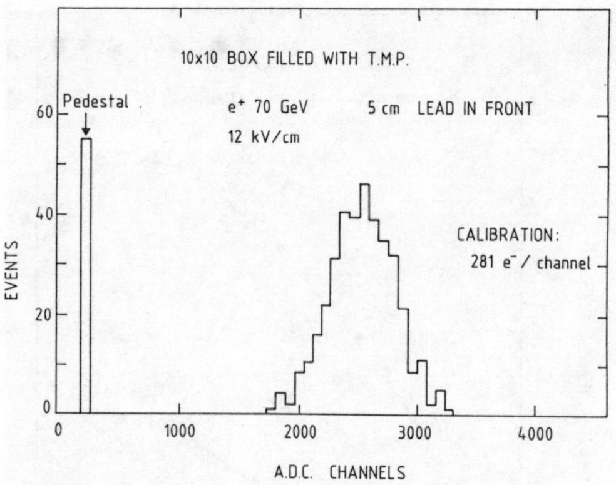

Figure 30: Pulse-height distribution measured in a prototype warm liquid calorimeter for 70 GcV positrons converting on a lead plate (30).

4. VISUALIZATION OF EVENTS

The previously described electronic tracking detectors have typical multitrack resolutions of several millimetres, and cannot in general cope with the pattern recognition problems arising in the case of large multiplicities. A definite need exists for an instrument that could visualize, without ambiguities and with very good resolution, complex patterns of interactions, much in the same way as emulsions or bubble chambers used to do but at higher event rates and with direct electronic recording. I will describe here two developments in this direction, aiming at rather different applications: the scintillating fibres and the imaging chambers. Both have the property of providing a two-dimensional projection of complex patterns, which can be observed or recorded using a television system or a digital image encoder for further analysis.

The first approach is based on the use of bundles of scintillating fibres, coupled to an image intensifier system, so as to provide a longitudinal or a transverse projection of an interaction. Each fibre consists of a core of scintillating material with a thin cladding having a lower refraction index, acting as a light-guide for photons all the way down to the detector; an absorbing layer is often added on the outer surface to avoid cross talk between closely packed fibres. The fibres used for tracking must have sufficient photon yield and transmission efficiency to provide at least one photoelectron at the image intensifier photocathode for minimum ionizing particles.

Two basic designs have been presented at this conference, using plastic scintillating fibres about 1 mm in diameter (18, 31), or 20–30 μm cerium glass fibres (32, 33). The last solution has obviously a better intrinsic accuracy, but is probably more expensive and moreover, at least with the present state of technology, does not allow the realization of long detectors because of the short attenuation length for photons in the scintillating glass (several centimetres in the best case). Figure 31 shows a single event recorded in a stack of plastic scintillating fibres, with two lead plates inserted to initiate an electromagnetic shower (18). Measuring the light intensity with a digital charge-coupled device (CCD) camera, one can also estimate the energy loss within the shower, though with a limited dynamic range, as represented in the histograms; the increase of photon yield at the position of the lead plates is clear in the figure. Figure 32 shows another example of image obtained for a particle interacting in the bundle, by the authors of the contributed paper quoted in Ref. 31. Figure 33 represents schematically the proposed installation of the scintillating fibre detector for the UA2 upgrade; bundles of fibres are mounted along a cylindrical surface, either in an axial direction or at a small angle to it for stereoscopy, and are interleaved with thin lead plates. The

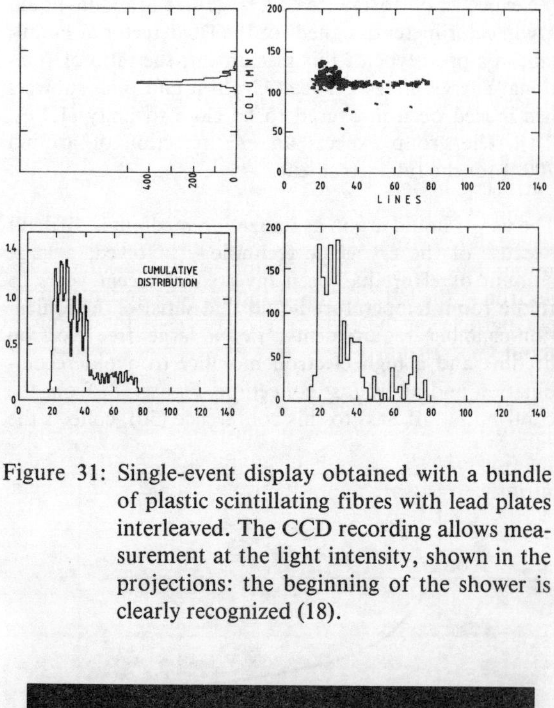

Figure 31: Single-event display obtained with a bundle of plastic scintillating fibres with lead plates interleaved. The CCD recording allows measurement at the light intensity, shown in the projections: the beginning of the shower is clearly recognized (18).

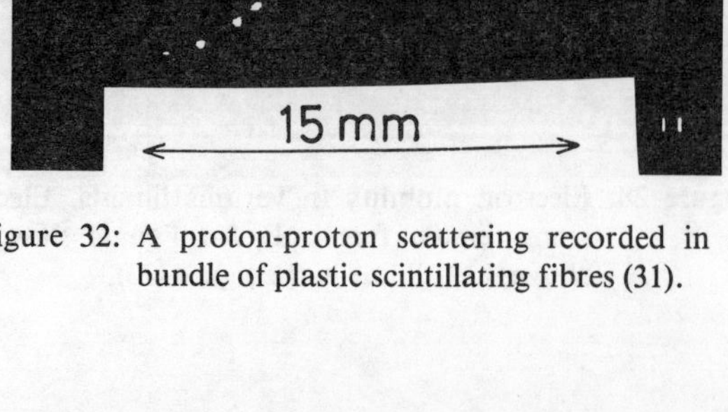

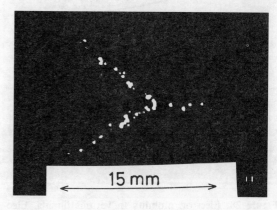

Figure 32: A proton-proton scattering recorded in a bundle of plastic scintillating fibres (31).

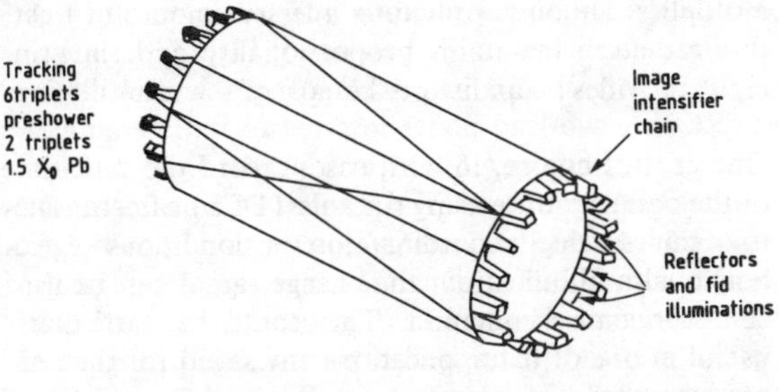

Figure 33: Assembly of the scintillating fibre detector, part of the UA2 upgrade. Bundles are mounted at angles to provide a stereo view, and are interleaved with thin lead sheets (18).

detector, 2 m long and 80 cm in diameter, will serve the double purpose of improving tracking and as pre-shower counter.

Fisher (32) described the results obtained with a bundle of very thin, Ce_2O_3-doped scintillating glass fibres; as shown in the microphotograph of Fig. 34 the fibres are bound together by a thermal process into a square matrix, each fibre having about a 30 × 30 μm^2 cross-section. Coupled to an image intensifier system, the device allows visualization of tracks with very good resolution; Fig. 35 gives an example with one interacting electron at the left, starting a shower, accompanied by a fast particle on the right; the real size of the frame is about 15 × 20 mm^2.

The resolution, efficiency, and multitrack capability of scintillating fibres are well illustrated by the results mentioned here. There is no doubt that this technique, intrinsically simple and reliable, will take a prominent

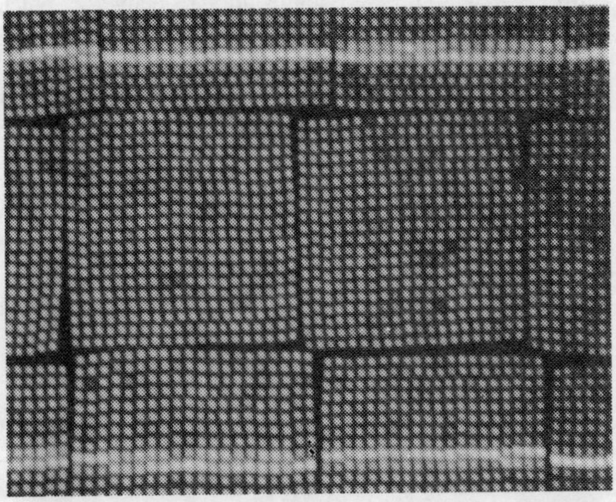

Figure 34: Microphotograph of a scintillating glass-fibre bundle; each fibre is about 30 × 30 μm^2 (32).

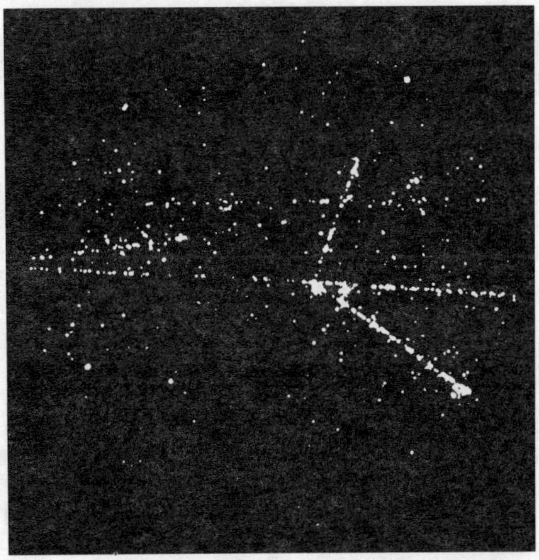

Figure 35: Image obtained with the glass scintillating fibres, for several traversing particles, one of which interacts. The size of the picture is about 15 × 20 mm^2 (32).

place in detectors which have to cope with local very high multiplicities; their use as an active target is also to be foreseen.

The last subject I would like to discuss here is the so-called imaging chamber, developed by our group at CERN (34). The detector is rather similar to a TPC, having a large volume of gas with uniform electric fields, constituting the sensitive drift region, followed by a MWPC or a parallel-plate chamber used for charge multiplication; however, instead of exploiting the electronic signal for localization, one tries to obtain enough photon emission in the avalanches in order to visualize the tracks, in projection, using an image intensifier system (see Fig. 36). While it is well known that photons are copiously produced in electron avalanches, they are mostly emitted at very short wavelengths and therefore strongly reabsorbed by the gas mixture and difficult to image (common image intensifiers only operate in the close ultraviolet and in the visible range). A systematic research by the group has identified several gas mixtures in which enough photons are emitted close to the visible, and can be detected; one of them is argon–methane–TEA, in which a strong peak of emission is observed around 300 nm in avalanching conditions (35). Figure 37 gives two examples of cosmic-ray events, interacting in a small gas volume followed by a high-gain light emitter; each picture covers about 5 × 10 cm^2. The upper one has been obtained using as multiplier a MWPC with thick (35 μm) anode wires, operated in the limited streamer mode; allowing very large gains, and therefore large light yields, this mode results however in a loss of proportionality and in a quantization of the avalanches in the wire position. In the second picture, instead, the MWPC has been replaced with a thin-gap parallel-plate

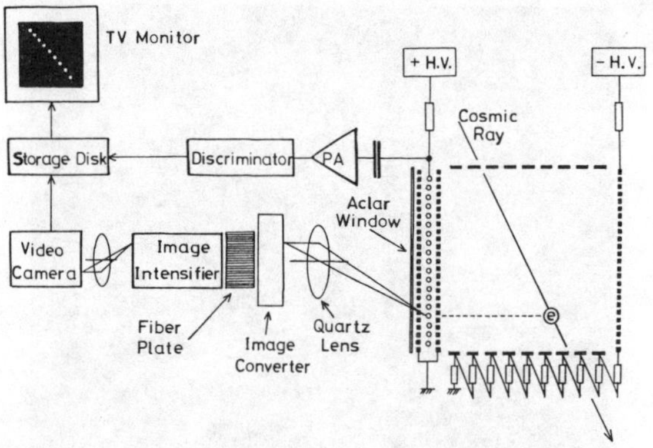

Figure 36: Schematics of the imaging chamber. Ionized trails formed in the conversion volume drift to the multiplying structure, a MWPC or a parallel-plate chamber. An image intensifier system amplifies and records the projected image (34).

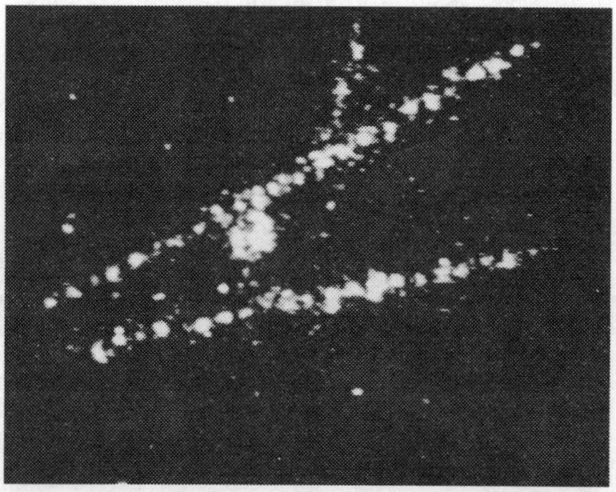

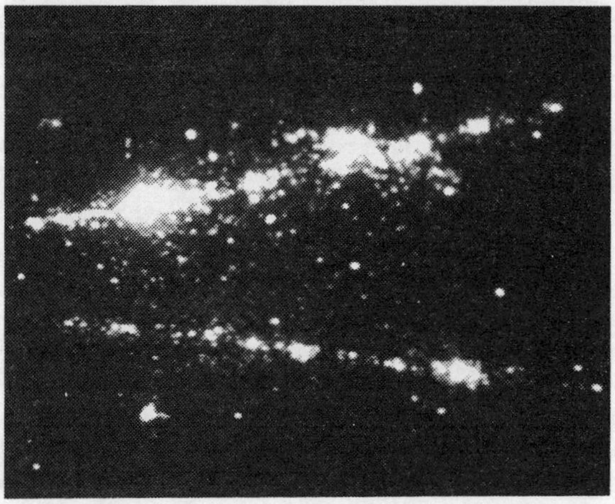

Figure 37: Examples of images obtained with cosmic rays, in a limited streamer mode of operation (upper picture) and in a proportional regime (lower picture) (34).

multiplier; although providing a lesser amount of light, this geometry maintains proportionality and, in principle, provides an undistorted image of the tracks.

One can, of course, in both cases record the drift-time or the charge profile using suitable TDCs or flash ADCs to obtain the depth coordinate; in the conditions of proportional multiplication, the charge signal can be used for energy discrimination. This could be particularly useful in one of the applications envisaged for the imager, the study of double beta decay in xenon. Other applications of the device include the analysis of low-rate, complex events and, coupled to a thick drift converter—such as the one described in Ref. (36)—for detection and imaging of electromagnetic and hadron showers. The group has also shown that one can reach full detection efficiency for single electrons, such as those pro-duced in the gas by Cherenkov radiation: the possibility of imaging of Cherenkov rings containing many hundreds of photons opens the way for identification of relativistic heavy ions.

Let me recall here that, as for the scintillating fibres, the imaging chamber is a continuously active device, and the rate limitations arise only from the use of image intensifiers with long persistence or of standard scan frequencies for the recorders.

REFERENCES

(1) Charpak, G., Annu. Rev. Nucl. Sci. **20,** 195 (1970).

(2) Drumm, H. et al., Nucl. Instrum. Methods **176,** 333 (1980).

(3) Bartelt, J., The new drift chamber for the Mark II detector, presented at this conference.

(4) Young, C.C. et al., IEEE Trans. Nucl. Sci. **NS-33,** 176 (1986).

(5) Hodges, C. et al., IEEE Trans. Nucl. Sci. **NS-33,** 167 (1986).

(6) Schaile, D., O. Schaile and J. Schwarz, Nucl. Instrum. Methods **A242,** 247 (1986).

(7) Krivatch, S.F., N.W. Reay, R.A. Sidwell and N.R. Stanton, Design and performance of large multisampling drift chambers for Fermilab Experiment 653, presented at this conference.

(8) Vilanova, D., Nucl. Instrum. Methods **A235,** 285 (1985).

(9) Amendolia, S.R. et al., Nucl. Instrum. Methods **A244,** 516 (1986).

(10) Brand, C. et al., Results of space measurement accuracy from tests of a half-scale DELPHI TPC prototype, Proc. Wire Chamber Conf., Vienna, 1986 (to be published in Nucl. Instrum. Methods).

(11) Blum, W., U. Stiegler, P. Gondolo and L. Rolandi, Measurement of avalanche broadening caused by the wire E × B effect, Proc. Wire Chamber Conf., Vienna, 1986 (to be published in Nucl. Instrum. Methods).

(12) Va'vra, I., Nucl. Instrum. Methods **A244,** 391 (1986).

(13) Bouclier, R., G. Charpak, W. Gao, P. Mine, A. Peisert, J.C. Santiard, F. Sauli and N. Solomey, A modular multidrift vertex detector, Proc. Wire Chamber Conf., Vienna, 1986 (to be published in Nucl. Instrum. Methods).
(14) Villa, F., Nucl. Instrum. Methods **217**, 273 (1983).
(15) Huth, J. and D. Nygren, Nucl. Instrum. Methods **241**, 375 (1985).
(16) Proc. Workshop on Radiation Damage to Wire Chambers, Berkeley, 1986 (ed. J. Kadyk), LBL-21170 (1986).
(17) Allison, W.W.M. et al., The Soudan 2 detector as a time projection calorimeter, presented at this conference.
(18) Weidberg, A., Upgrade for UA2, presented at this conference.
(19) Fabjan, C.W. et al., Nucl. Instrum. Methods **185**, 119 (1981).
(20) Leck, L.O. et al., Operation of a gaseous radiator RICH counter for low p_T electron identification at the CERN $p\bar{p}$ Collider, presented at this conference.
(21) Ashford, V. et al., Development of the Cherenkov Ring Imaging Detector for the SLD Experiment, presented at this conference.
(22) Charpak, G. and F. Sauli, Phys. Lett. **78B**, 523 (1978).
(23) Adams, M. et al., Nucl. Instrum. Methods **217**, 237 (1983).
(24) Arnold, R. et al., Photosensitive gas detectors for the RICH technique, Proc. Wire Chamber Conf., Vienna, 1986 (to be published in Nucl. Instrum. Methods).
(25) Holroyd, R., J. Press, C. Woody and R. Johnson, Measurement of the quantum efficiency of TMAE and TEA from threshold to 120 nm, presented at this conference.
(26) Cahill, J. et al., Laser induced ionization for wire chamber calibration, presented at this conference.
(27) Bamberger, A., R. Isele, J. Schlupmann and M. Stegle, Aromatic seeding agents for laser ionization in counting gases, Proc. Wire Chamber Conf., Vienna, 1986 (to be published in Nucl. Instrum. Methods).
(28) Fabjan, C.W. et al., Nucl. Instrum. Methods **141**, 61 (1977).
(29) Tuts, M., Uranium-liquid argon calorimeter performance, presented at this conference.
(30) Sass, J., Warm liquid calorimetry for UA1, presented at this conference.
(31) Konaka, A. et al., Plastic scintillating fibers for track detection, presented at this conference.
(32) Fisher, C., Development of a microvertex detector using scintillating glass fibres, presented at this conference.
(33) Ruchti, R. et al., A cerium glass active target for vertex detection and tracking applications, presented at this conference.
(34) Suzuki, M. et al., The optical readout of gas avalanche chambers and its applications, presented at the 3rd Pisa Meeting on Advanced Detectors, Castiglione, 1986.
(35) Suzuki, M., P. Strock, F. Sauli and G. Charpak, preprint CERN-EP/85-205 (1985), submitted to Nucl. Instrum. Methods.
(36) Jeavons, A., G. Charpak and R. Stubbs, Nucl. Instrum. Methods **124**, 491 (1975).

NON-ACCELERATOR EXPERIMENTS

Maurice Goldhaber
Brookhaven National Laboratory
Upton, N.Y. 11973

IBM Collaboration

R.M. Bionta,[12] G. Blewitt,[4] C.B. Bratton,[5] D. Casper,[14] A. Ciocio,[14] R. Claus,[2] M. Crouch,[9] S.T. Dye,[6] S. Errede,[10] G.W. Foster,[15] W. Gajewski,[12] K.S. Ganezer,[1] M. Goldhaber,[3] T.J. Haines,[1] T.W. Jones,[7] D. Kielczewska,[8] W.R. Kropp,[1] J.G. Learned,[6] J.M. LoSecco,[13] J. Matthews,[2] H.S. Park,[11] F. Reines,[1] J. Schultz,[1] S. Seidel,[2] E. Shumard,[16] D. Sinclair,[2] H.W. Sobel,[1] J.L. Stone,[14] L. Sulak,[14] R. Svoboda,[1] G. Thornton,[2] J.C. van der Velde,[2] and C. Wuest[12]

I. INTRODUCTION

The title of my talk, "Non-accelerator Experiments" sounds more like a budget item than a coherent field of physics. The aims pursued in these experiments overlap often with aims pursued in accelerator experiments. Clearly, we are dealing with a lot of good questions, sometimes disguised as theories.

Table I gives a partial list of non-accelerator experiments. Most of these experiments have been discussed at this Conference. Since some have been covered by other rapporteurs, I shall confine myself to a subset.

TABLE I

SOME NON-ACCELERATOR EXPERIMENTS

- Proton decay
- $n - \bar{n}$ oscillations
- Cygnus X-3
- ν masses
- $\beta\beta$ decay (with and without 2ν)
- ν oscillations
- Axions
- WIMPS
- Atmospheric ν's
- Solar ν's
- Extra solar ν's
  - a. Supernovae
  - b. H.E. ν's
- New particles (KGF)
- Dark matter
- Gravitational waves
- Neutral current effects in atoms
- Fifth force
- Electric dipole (n,e, atoms)
- Free quarks
- Magnetic monopole

Much physics done in the eighties, especially in the non-accelerator field, can be characterized as the "physics of pushing limits". Some of these limits (masses, mixing angles, abundances, etc.) may be too small to be detected with present day techniques. They may even be zero, a number that experiment can approach, but not reach; it can only be a theoretical prediction. Experimenters, therefore, must beware of "false positives"!

I belong to the old school that was brought up to talk of experiments in the past tense. However, as rapporteur for non-accelerator experiments, I cannot avoid talking of experiments in progress or planned, though I shall try to minimize this.

II. PROTON DECAY SEARCHES AND THE ATMOSPHERIC NEUTRINO BACKGROUND

Of the few definitive results in the non-accelerator field I shall concentrate on two in particular: (1) Proton decay searches, where the results contradict the minimal SU(5) theory, thus forcing theory into new directions; (2) the observation of atmospheric neutrinos resulting from interactions of primary cosmic-ray protons in the atmosphere, found in agreement with expectations from π, K, and μ decays.

---

1. The University of California, Irvine; 2. The University of Michigan; 3. Brookhaven National Laboratory; 4. California Institute of Technology; 5. Cleveland State University; 6. The University of Hawaii; 7. University College, London; 8. Warsaw University; 9. Case Western Reserve; 10. The University of Illinois; 11. Lawrence Berkeley Laboratory; 12. Lawrence Livermore National Laboratory; 13. Notre Dame University; 14. Boston University; 15. Fermi National Accelerator Laboratory; and 16. A.T.T. Bell Laboratory.

Fig. 1 shows the known particle states into which a proton could decay. This could happen in many ways. In fact, if you count all individual members of the singlets, doublets, triplets and quadruplets shown in Fig. 1 you find about 30 different potential two-body-decay modes. I always use "proton decay" as a kind of code word meant to include bound neutron decay, because the term "neutron decay" is ambiguous since it is already used for the beta-decay of a free neutron to a proton. The masses of the particles shown are all well known except for the masses of the neutrinos, for which we only know upper limits.

Table II gives the expected branching ratios for the minimal SU(5) theory, as compiled by W. Lucha from many theoretical estimates with a variety of models used for the nucleon.

To look for proton decay you might like to build a detector that is sensitive to all possible decay modes. But it is nearly impossible to build a detector that is equally sensitive to all modes, and so you have to make a decision which mode you would like to emphasize. The strongest branch in Table II is $e^+\pi^0$, and this is the mode which the IMB experiment was designed to emphasize. It is not widely appreciated, however, that our detector is also sensitive to most other potential modes, though usually less so.

When you build a proton decay detector you have to make a number of compromises: How deep are you going to go; how large is it going to be; what is it going to be made of; how good an energy resolution you should aim at; how well you can recontruct the vertex, and so on. Different researchers have made different compromises. There is a great variation in the depth chosen for the ongoing and proposed experiments. The advantage of going deeper lies in the fact that the flux of cosmic ray muons decreases very rapidly with depth. The IMB detector is located deep enough for us to be able to handle the number of muons, and to distinguish them from a potential proton decay, because of their large energy deposition.

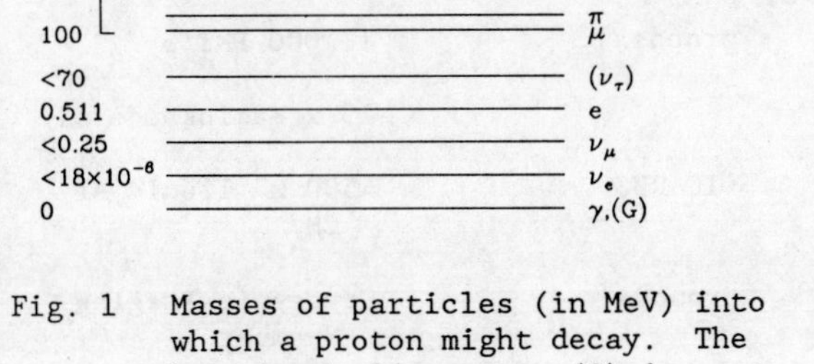

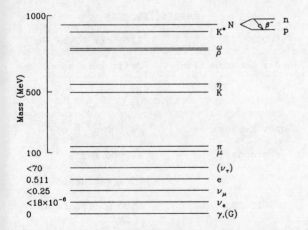

Fig. 1 Masses of particles (in MeV) into which a proton might decay. The hypothetical graviton (G) is included.

TABLE II

Theoretical Branching Ratios for Proton Decay

DECAY MODE	BRANCHING RATIO (%)
$p \rightarrow e^+\pi^0$	31 - 43
$p \rightarrow e^+\eta$	0 - 8
$p \rightarrow e^+\rho^0$	2 - 18
$p \rightarrow e^+\omega$	15 - 29
$p \rightarrow \bar{\nu}_e\pi^+$	11 - 17
$p \rightarrow \bar{\nu}_e\rho^+$	1 - 7
$p \rightarrow \mu^+K^0$	1 - 20
$p \rightarrow \bar{\nu}_\mu K^+$	0 - 1

At this conference T.J. Haines of our group reported on new a priori calculations of the background due to atmospheric neutrinos, the one background that cannot be reduced, no matter how deep you go. He has calculated the expected event distribution for neutrinos interacting with the water of our detector, yielding a visible energy E_c, which is what we measure. I shall not go into the details of his calculations, as they appear in these proceedings.

Starting from the predictions of Gaisser, Stanev, and others for the flux of atmospheric neutrinos, ν_e and ν_μ, as a function of energy he considers all important interactions due to either charged or neutral currents. For single pions he uses the Fogli and Nardulli model and for multiple pions the parton model. This gives a systematic error of about 30%. Since the agreement for energies beyond the proton mass is quite good, we can have confidence in the calculations also below that. Independent calculations on a different model by T.W. Jones and by other members of our group using calculations based on Gargamelle and other bubble chamber data also agree with Haines'calculations; he therefore has the courage to subtract neutrino backgrounds from the observed proton candidate events, thus obtaining somewhat better lifetime limits for most decay branches. The important branch $p \to e^+\pi^0$ is not improved, because it had no candidate.

G. Blewitt, and previously also the Kamiokande collaboration, as well as S. Seidel of our collaboration, studied invariant masses for two-body nucleon decay candidates and in this way some ambiguous candidates are removed. From minimal SU(5) theory W. Marciano predicts $4.5 \times 10^{29 \pm 1.7}$ years for the partial lifetime of the proton decay branch $e^+\pi^0$. Our experiment, however, gives a lower limit of 2.5×10^{32} years, and if you add the results of the Kamiokande detector you can get a combined lower limit of 3.3×10^{32} years. (See Fig. 2.) H. Meier (Heidelberg Conference, July 1986) adds also the Fréjus result and gets a lower limit of 4×10^{32} years. The experimental limits thus contradict the minimal SU(5) theory. If we believe in the general idea of grand unification, we would expect the proton decay to be governed by a grand unification mass not larger than the Planck mass. This would lead to a lifetime of around 10^{47} years, which would yield of the order of a few decays per hour for the whole earth!

A number of proton decay detectors have recently been improved and a number of new detectors are being proposed. (See Tables III and IV, bases on a compilation by H. Meier.)

As the resolution improves, proton decay candidates can be better resolved from neutrino background. Of the presently active detectors, Fréjus has the best resolution and reports the lowest number of candidates per ton-year of observation.

III. NEUTRINOS

R. Svoboda reported on the search for energetic neutrinos that come through the earth and produce muons that traverse our detector. He sees no indication of point sources in the sky; the upward going muons

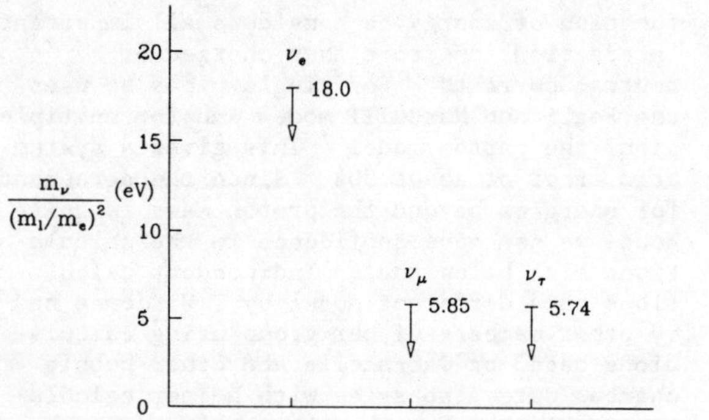

Fig. 2 Limits for partial lifetime $p \to e^+\pi^0$.

TABLE III

RECENTLY IMPROVED DETECTORS

Kolar II	10 mm → 6 mm
	130 ton → 260 ton
IMB II	All 5" PMT → 8"ϕ
	wavelength shifter plates on all PMT's
	factor 4 gain in light collection
Kamiokande II	Time digitizer on all PMT's
	4π veto shield (H_0O active)
	Factor 4-6 of vertex reconstruction

observed are compatible with the rates expected from atmospheric neutrinos.

Neutrinos were discussed in general by Prof. Altarelli. Let me just add a few remarks.

One can make some "reasonable" conjectures about neutrino masses:

1. Either all three masses are zero, or they are all finite.

2. If the masses are finite, no two are equal, i.e., all mass differences are finite.

TABLE IV

Future Projects

Soudan II	1200 ton, 2 mm Fe
under construction	Drift tube sampling
	$\frac{dE}{dx}$
Super Kamiokande	46,000 m^3 H_2O
proposal	11,000 PMT's
	15 x Kamiokande II!
ICARUS	4500 m^3 liquid Ar, CH_4
proposal	Cryogenic Imaging Chamber
	0.5T magnetic field

3. All mixing angles are either zero, or they are all finite.

GUTS which have right-handed heavy neutrinos, e.g., SO(10), lead to the Gell-Mann - Ramond - Slansky - Yanagida "see-saw" relation:

$$m_{\nu_e} : m_{\nu_\mu} : m_{\nu_\tau} \approx m_e^2 : m_\mu^2 ; m_\tau^2$$

If we make use of this relation, we can obtain from the experimentally established upper limits for the neutrino masses a "figure of merit" for each neutrino: $m_\nu/(m_\ell/m_e)^2$ (see Fig. 3). If this relation is correct, an improvement in the measurement of the mass of the electron-neutrino to ~ 5 eV, the best that could be expected from the ongoing tritium experiments, would make the three figures of merit approximately equal.

The solar neutrino problem is still with us. There are some new ideas: The Mikheev-Smirnov-Wolfenstein matter oscillations have been invoked to explain the signal obtained in the chlorine experiment by Davis, Rowley, and Cleveland, who find about one-third the amount of ^{37}A expected from the standard solar model. Since there are several possible solutions for the neutrino oscillation case, the proposed gallium experiment, which is now getting underway in the Gran Sasso Laboratory, may or may not be able to decide whether the amount of ^{8}B produced in the sun is compatible with the standard model. If neutrinos oscillate into "infertile" ones, they may oscillate back into fertile ones, as they traverse the earth, according to A. Baltz and J. Weneser (Phys. Rev. D15, to be published). A number of "on-line" solar neutrino searches have been proposed; the Kamiokande detector is ready for such a search, and it will be interesting to see whether a signal above background will be obtained.

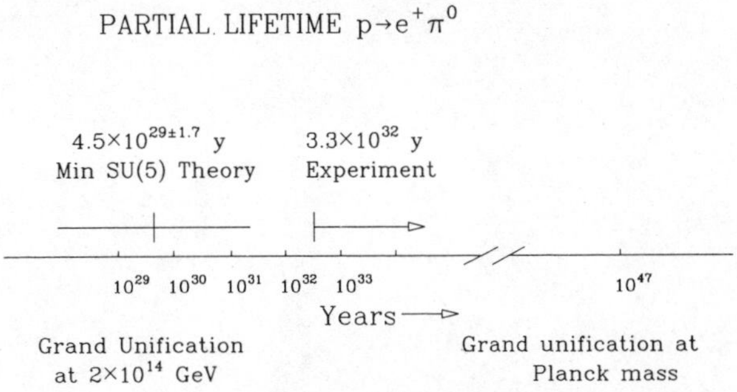

Fig. 3. "Figure of merit" for neutrinos, according to the see-saw relation.

IV. CYGNUS X-3

Cygnus X-3, a double-star with a period of 4.8 hr., that "accelerator in the sky" which just now is a "non-accelerator", has been studied with many detectors, with only Soudan I consistently claiming that muons from the direction of Cygnus X-3 can be detected underground at the "correct" phase. M. Marshak gave a detailed account of the present status of these observations and emphasized that the results of different observers cannot be reconciled. This may partly be due to the great variability of the source, and we must await further observations, when Cygnus X-3 "switches on" again.

V. DOUBLE-BETA-DECAY

In double-beta-decay experiments, reported by D.O. Caldwell, no evidence for no-neutrino double-beta-decay has yet been found. Should it be found, lepton conservation would have to be given up. From the limits obtained, one can deduce limits for the masses and interactions of some hypothetical intermediary particles which would make neutrinoless double-beta-decay possible. But Caldwell emphasized that considerable uncertainties remain, because those limits depend on uncertain estimates of the nuclear matrix elements involved.

Plenary Session

QCD
W. Scott (RAL)

This paper is not available

Physics at Future High Energy Colliders
C.H. Llewellyn-Smith (Oxford)

ICFA Report
V. Telegdi (Zurich)

This paper is not available

Chairman
D. Horn (Tel Aviv)

Scientific Secretaries
I. Juricic (LBL)
C. Klopfenstein (LBL)

PHYSICS AT FUTURE HIGH ENERGY COLLIDERS

C.H.Llewellyn Smith

Department of Theoretical Physics
1, Keble Road,
Oxford, OX1 3NP,
England.

A brief review is given of the potential capabilites of future $e\bar{e}$, pp and ep colliders as functions of energy and luminosity.

1. INTRODUCTION

With only 30 minutes available, I shall give a coarse-grained overview of the physics capability, or "reach," as a function of energy and luminosity of colliders beyond those under constuction. Detectors are already being built for the approved colliders (the Tevatron, Tristan, SLC, LEP, and Hera) which will soon produce real data. These data will presumably alter our perceptions. Speculation about the capabilities of subsequent colliders is therefore evidently a risky business but it is necessary as a guide for long term planning.

History suggests that if future colliders extend the mass/energy range that can be explored by a factor of ten, there is a good chance that they will uncover new phenomena. A factor of ten would take us to a TeV at the constituent level, at which energy it is highly likely that experiments will reveal the origin of electro-weak symmetry breaking. It is scandalous that in spite of a general belief that weak and electromagnetic interactions are unified, we do not know why the W and Z are massive while the photon is presumably massless (experimentally $M_W > 10^{26} M_\gamma$). This scandal has an intrinsic energy scale, M_W or the "Fermi scale" $2M_W/g \simeq 250$ GeV, and there are compelling reasons to believe that the secret of symmetry breaking is associated with new phenomena at this scale, say - at or below 1TeV.

2. e^+e^- PHYSICS AT TeV ENERGIES

No one knows whether TeV $\bar{e}e$ colliders can be built with adequate luminosity to do interesting physics.[1] I shall assume that a luminosity (L) of at least $1.3(E(\text{TeV}))^2 \times 10^{32} cm^{-2} sec^{-1}$ will be achieved, which would yield one event per day for processes with a point-like cross-section $\sigma_{pt} = 0.087/(E(\text{TeV}))^2$pb - the Born approximation cross-section for $\bar{e}e \to \bar{\mu}\mu$.

Various phenomena might yield cross-sections bigger than σ_{pt}:

1. If electrons have a size $1/\Lambda$, the cross-section would presumably flatten out for $E > \Lambda$ as sketched in fig. 1A, but σ would still be of order σ_{pt} at the threshold to see this effect.

2. A new t channel exchange

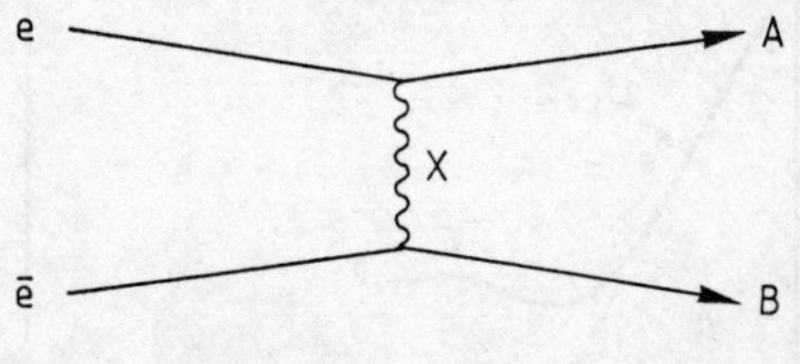

could generate an increased cross-section as shown in fig. 1B if $M_X \ll M_A + M_B$ (an example is provided by the case $X = \nu_e$, $A = W^-$, $B = W^+$); the behaviour depends on the spin of X, the most auspicious case being $J_X = 1$ which would lead to $\sigma \sim g^2/M_X^2$ well above threshold.

3. A new s channel exchange (Z') would lead to a peak as sketched in

fig. 1C. If the Z' branching ratios are similar to those of the Z then

$$R_{vis} \equiv \frac{\sigma(\bar{e}e \rightarrow \text{visible channels})}{\sigma_{pt}}$$

would rise to about 4000 on the peak. This large enhancement may be reduced, however, if the energy spread δE is much larger than $\Gamma_{Z'}$ e.g. due to effects such as beamstrahlung (the radiation generated by the beam - beam interaction at the collision point in linear colliders). For $\delta E >> \Gamma_{Z'}$, the peak value would be

$$\langle R_{vis} \rangle = \frac{9}{2\alpha^2} \frac{\sqrt{\pi}}{\sqrt{2}} \frac{\Gamma(Z' \rightarrow \bar{e}e)\Gamma(Z' \rightarrow vis)}{\Gamma_{Z'} \delta E}$$

which is about $60E/\delta E$ if the Z'couples to fermions in the same way as the Z, but is less by a factor of order $\sin^2\theta_W$ for "interesting" Z's[2], such as those which may or may not[3] be predicted by superstring inspired models.

There may be machine associated backgrounds from beamstrahlung and there will be "physics" backgrounds from $\bar{e}e \rightarrow WW$ and $\gamma\gamma$ collisions but it seems that they will not present a serious problem.[4] The physics "reach" for possible discoveries is therefore easily estimated[5], with the following results:

Charged Particles (quarks, leptons, W's, Higgs...) will be found for masses up to nearly E/2.

Supersymmetic Particles will be found up to nearly E/2 if they are charged; the processes $\bar{e}e \rightarrow \tilde{\bar{e}}\tilde{e}\tilde{\gamma}$ and $\bar{e}e \rightarrow \gamma\tilde{\gamma}\tilde{\gamma}$, mediated by scalar electron ($\tilde{e}$) exchange, will provide a reach for $\tilde{e}$ which will extend further if the photino ($\tilde{\gamma}$) is relatively light, the exact limit depending on the luminosity.

The tail of a new neutral vector boson Z' will be seen interfering with the γ and Z even if it cannot be produced. If, very optimistically[6], we assume that it has a vector coupling of order e, it would modify R by a factor $1+O(2E^2/M_Z{}^2)$, an effect which would be seen if $M_{Z'}<4E$. Note that a Z' could have a significant branching ratio into W^+W^- and a detailed analysis should consider this channel.

If electrons have substructure there would be an effective four fermion "contact interaction" between electrons and other fermions with which they share consituents. Characterising this interaction by $4\pi/\Lambda^2$, where Λ is usually called the "compositeness scale" and the strength corresponds to an effective coupling $g^2/4\pi=1$, the effects of interference with one photon exchange might be visible for $\Lambda \lesssim 45E$.

An analysis of the potential for discovering the standard Higgs boson (H) is less straightforward. Recall that if $M_H<2M_W$, H will decay to the heaviest quarks and leptons available $-t\bar{t}$ if $2M_t<M_H$, $b\bar{b}$ and $\tau\bar{\tau}$,

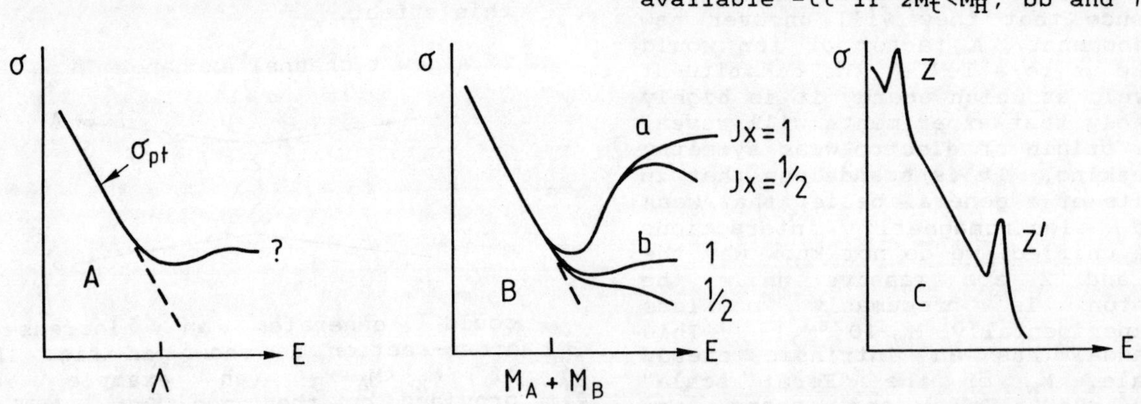

Fig. 1 Possible behaviour of the $\bar{e}e$ annihilation cross-section (σ), on a logarithmic scale, as a function of energy (E) in the cases of A) a finite electron size $1/\Lambda$, B) exchange of a new particle X with spin J_X that couples e and $\bar{e}$ to new particles A and B for a) $M_X << M_A + M_B$ and b) $M_X \approx M_A + M_B$, and C) a new vector boson Z'.

with a branching ratio of order $3M_b^2/M_\tau^2$, otherwise. If $M_H>2M_W$, it will decay to WW and ZZ with, approximately[7],

$$\Gamma(H\to WW)\simeq 2\Gamma(H\to ZZ)\simeq 40\left[\frac{M_H}{500\text{GeV}}\right]^3 \text{ GeV}.$$

Note that for $M_H>1.4$TeV, $\Gamma_H>M_H$ and the Higgs boson would hardly be a particle.

H may be found in Z and toponium ($t\bar{t}$) decay at SLC and LEP, the "reach" with the LEP luminosity being expected to be[8]
$M_H\lesssim 50$GeV for $Z\to H\mu\bar{\mu}$
$M_H\lesssim 0.8M_{t\bar{t}}$ for $(t\bar{t})\to H\gamma$.

At LEP2, it may be possible[8] to find H's produced in the reaction $\bar{e}e\to HZ$ for $M_H\lesssim 90$ GeV.

Going up in energy to, say, 500 GeV, we find[9] $\sigma(\bar{e}e\to ZH)/\sigma_{pt} >0.13$ for $M_H\lesssim 2M_W$ and it should be easy to see $H\to t\bar{t}$ above the $\bar{e}e\to Z\bar{q}q$ background[10], except perhaps if $M_H\simeq M_Z$ in which case it may be necessary to suppress the background by making a sample of enriched $Zt\bar{t}$ events (e.g. by looking for "fat" quark jets). If $M_H>2M_W$, the process $\bar{e}e\to HZ$ followed by $Z\to 2$ jets, $H\to WW/ZZ\to 4$ jets will generate six jet events for which the backgrounds, which have not been studied systematically, seem to be small (e.g. the orders of magnitude of various contributions to σ/σ_{pt} are $(\alpha_s/\pi)^4$ (6QCD jets), $\alpha_W\alpha_s^2/\pi^3$ (4QCD jets and one W), and α_W^2/π^2 ($q\bar{q}WW$)). It may, therefore, be possible to detect a very small signal and hence reach large masses e.g. if 1 event in 10 days could be detected, discovery would be possible for $M_H<350$GeV with $L=10^{32}\text{cm}^{-2}\text{sec}^{-1}$.

Going up again in energy to 1 TeV and above, the WW fusion mechanism[11]

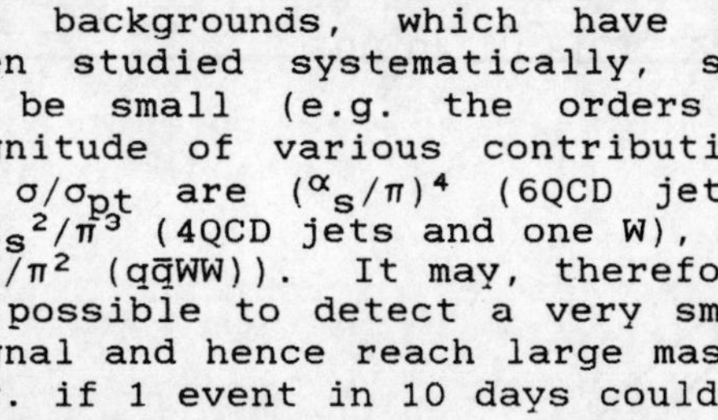

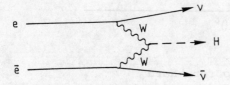

dominates. There is a background[12] for seeing $H\to WW$, and also for $H\to ZZ$ if the Z's decay hadronically, from the two photon process $e\bar{e}\to e\bar{e}$ WW in which the final electrons and positrons go down the beam pipe. The $\gamma\gamma$ process is forward/backward peaked and can probably be suppressed to an acceptable level by making a cut on $\cos\theta^*$. Requiring $\sigma/\sigma_{pt}>1/5$ after making a cut $|\cos\theta^*|<0.5$, a 1TeV machine would be sensitive to $M_H\leq 450$ GeV while a 1.7 TeV machine would be sensitive up to $M_H=1$ TeV.

A special feature of the W's which participate in the "WW fusion" process is that they have a high degree of longitudinal polarization, in contrast to the W's produced in the process $e\bar{e}\to WW$ which are dominantly transverse. This may make it possible to study the scattering of longitudinal W's and Z's at centre of mass energies of 1 TeV and above, e.g. in the process

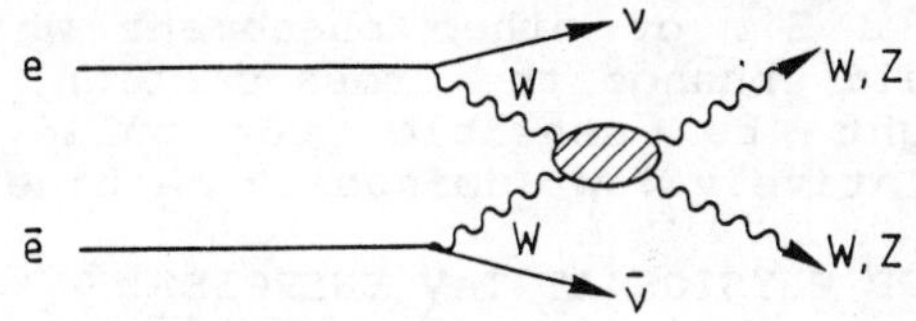

The existence of longitudinal polarization states distinguishes the W and Z from the photon, and their nature holds the key to the origin of electro-weak symmetry breaking. In the standard model these states are supplied by the Higgs field - they are the Higgs bosons that are "eaten." If $M_H>1$ TeV, the interaction between these states, which is controlled by the Higgs self-coupling $\lambda\ \alpha\ G_F\ M^2{}_H$, becomes strong at energies of 1 TeV and above. Strong interactions between longitudinal W's and Z's must also develop at these energies if there is no Higgs boson.[13] For example, in "technicolour" models[14] in which W_L and Z_L are $Q\bar{Q}$ bound states of "techniquarks" (Q) that develop dynamical masses of order 1 TeV, W_LW_L scattering would exhibit resonances ("technirhos") etc. In models[15] with totally composite W's and Z's, all polarization states would presumably develop strong interactions at around 1 TeV.

If a light Higgs boson is not found, experimental investigations of the interactions of longitudinal W's and Z's will obviously be very important. In order to eliminate the two photon background, it may be necessary to consider the channel $W_LW_L\to Z_LZ_L$, with one Z decaying leptonically. A conservative model[13] for the strong $W_L W_L \to Z_L Z_L$ amplitude gives[16]

$$\frac{1}{\sigma_{pt}} \int \frac{d\sigma}{dM_{WW}} (e\bar{e} \to \nu\bar{\nu} W_L W_L)$$

$$\simeq 0.59 \text{ for } E = 3\text{TeV}$$
$$\simeq 6.4 \text{ for } E = 5\text{TeV}$$
$$\simeq 75 \text{ for } E = 10\text{TeV}.$$

It may, therefore, be possible to study this process for $E \geqslant 5$TeV.

To summarize: $e\bar{e}$ colliders provide a clean way to study physics up to E/2, and beyond for some phenomena. Unfortunately, however, it is not at present clear whether TeV machines can be built with the luminosity needed, except perhaps for an astronomical price. Note that if there was a firm theoretical case for the existence of a Z', or other phenomena which would enhance the cross-section, it might be sensible to build a relatively low luminosity machine.

3 PP PHYSICS AT TeV ENERGIES.

It is certainly possible to build multi-TeV pp colliders but the quarks and gluons, whose interactions we wish to study, carry only a fraction of the beam energy. Consider a parton-parton induced process generated by a "hard" parton-parton interaction with a cross-section $\hat{\sigma} = {}^{C}/\hat{s}$, where $\hat{s} = \hat{E}^2$ is the parton-parton centre of mass energy squared. The differential cross-sections $\frac{d\sigma}{d\tau}$,

where $\tau = \hat{E}^2/E^2$, are shown in fig. 2 as functions of $\hat{s}$ for various cases, taking a value 10^{-2} for the constant c, which might be typical of a QCD sub-process (10^{-4} might be typical of an electroweak process). To get a feeling for the range of $\hat{E}$ which is accessible, note that one event/day in a $\Delta\tau = 0.01$ bin corresponds to $d\sigma/d\tau = 10^{-3}$nb. for a luminosity of 10^{33} cm^{-2} sec^{-1}.

From fig.2 we conclude that [17,18]
1. Gluon-gluon induced reactions will occur much more copiously than quark-quark induced reactions, except at the end of the accessible range.

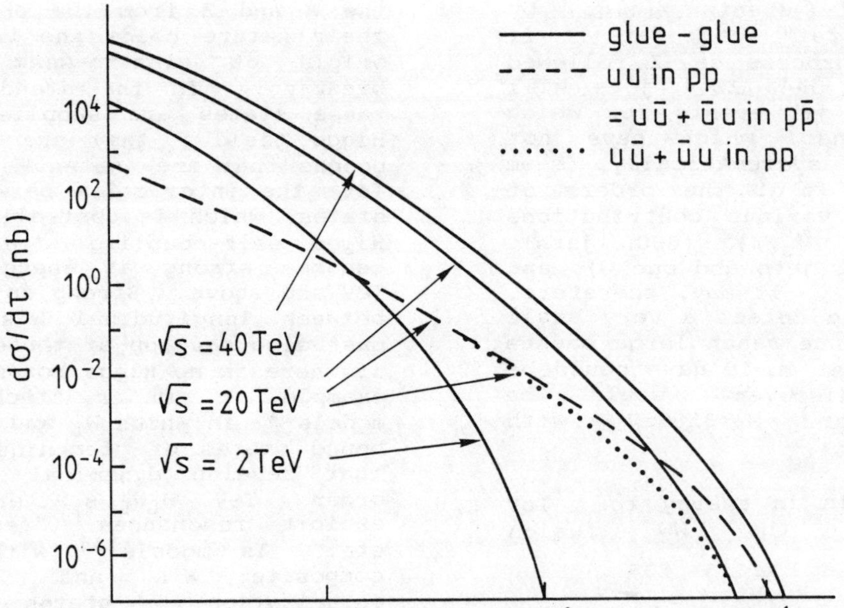

Fig. 2 Differential cross-sections $\frac{d\sigma}{d\tau}$ for "hard" sub-processes at sub-energy $\sqrt{\hat{s}}$ with typical strong interaction cross-sections $\hat{\sigma} = 10^{-2}/\hat{s}$, where $\tau = \hat{s}/s$. The solid curves are for glue-glue induced processes at $\sqrt{s}$ = 2, 20 and 40 TeV; the dashed curve for a uu induced process in pp collisions [$u\bar{u}$ in $p\bar{p}$] at $\sqrt{s}$ = 20 TeV and the dotted curve for a $u\bar{u}$ induced process [uu in $p\bar{p}$] at $\sqrt{s}$ = 20 TeV.

2. Quark-antiquark collisions will occur almost as frequently in pp as in $p\bar{p}$ collisions, except near the end of the accessible range where a $p\bar{p}$ luminosity of at least 10% of the pp luminosity is needed to exploit the advantage; there is therefore no clear physics case for building a $p\bar{p}$ collider, although there is an economic case which would be strengthened if detectors capable of exploiting the higher luminosity of a pp machine (10^{33} or more as opposed to perhaps 10^{32} for $p\bar{p}$) cannot be built.

3. The rate at a given $\hat{E}$ can be increased either by increasing the energy E or the luminosity L. Towards the end of the accessible range, a factor of 2 in E is equivalent to a factor of about 15 in L, assuming that larger L can be used.

While it should certainly be possible to build detectors capable of operating at $L = 10^{32}cm^{-2}sec^{-1}$, it is not clear that vertex detectors can be built which would work at 10^{33}. On the other hand, it should be possible[19] to dispense with a vertex detector, and perform experiments with only a calorimeter and muon detectors at luminosities as high as 10^{34}. I shall give "reaches" and rates for $L=10^{33}$ in a form that can be scaled to lower (or higher) L.

A crude average of the reaches for the various processes considered below gives a value

$$1.9 \left[\frac{E}{20}\right]^{0.65} \left[\frac{L}{10^{33}}\right]^{0.17} \text{TeV}$$

where E is in TeV and L in $cm^{-2}sec^{-1}$, the units used henceforth. This scales up to 3 TeV at the SSC energy[20] of 40 TeV for $L = 10^{33}$, and it scales down to 50GeV at the energy and luminosity of the CERN $p\bar{p}$ collider, whch is not an unreasonable number. If we neglect scaling violations, which is a good first approximation at high energy, the cross-section to produce or study a mass M in a hard process at energy E behaves as $\sigma = M^{-2}f(M/E)$. If the value of M at which σ reaches a minimal observable rate behaves as E^p over some range of E for fixed L, then generally the reach will behave as $E^pL^{(1-p)/2}$ (this connection between E and L scaling fails if either the signal or background involves an additional scale).

There will be severe backgrounds for many reactions from standard QCD processes and trigger rates will be high even at large P_T at TeV pp colliders. For example, with E=20, $L=10^{33}$ it is expected that there will be 10^5 jets per day with $P_T > 0.5$ TeV and 10 with P_T >2.3 TeV.

The physics reach for various processes can be inferred from the literature with the following results, which assume one hundred days continuous running at full luminosity (≃1 year real time?):

<u>Heavy quarks</u> might be seen up to

$$M \lesssim 1.1\ (E/20)^{0.6}\ (L/10^{33})^{0.15}\ \text{TeV}.$$

This limit is for the production of a colour triplet B quark by gluon fusion ($gg \to B\bar{B}$) followed by $B \to tW^-$, $\bar{B} \to \bar{t}W^+$ assuming that discovery would require fifty events in which both W's decay to $e\nu$ or $\mu\nu$. The limit must be treated with some caution since the strong coupling to the Higgs field that is needed to generate large fermion masses by the conventional mechanism leads to a failure of perturbation theory for quark [lepton] masses of order 500 GeV [1TeV] or more.[21] These strong interactions will alter the production amplitude, quite probably leading to a substantial suppression. The same comment applies to the production in $\bar{e}e$ annihilation of heavy quarks and leptons whose masses are generated by the Higgs mechanism.

<u>Heavy leptons</u> may, surprisingly, be copiously produced by gluon fusion.[22] It is reasonable to assume that a new lepton (L) would be accompanied by a new quark (Q) with $M_L \simeq M_Q$, in which case the amplitude corresponding to

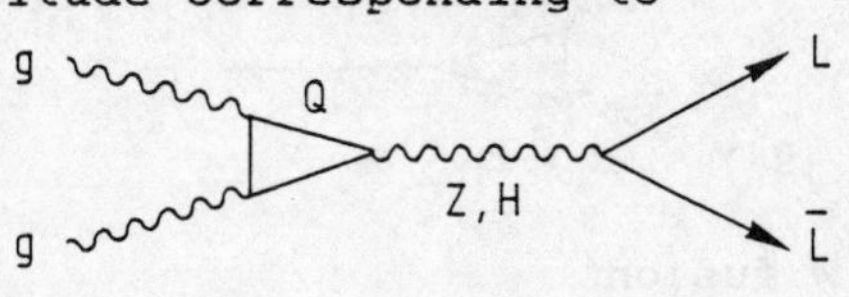

can become very large. For example, if $M_L=M_Q=1$ TeV, it gives a cross-section of 0.2pb at E=40 TeV, corresponding to 2000 events in the canonical 100 days for $L=10^{33}$, which should be easy to detect, the decay mode being $L \to \nu_L W$. The earlier comment that strong interactions with the Higgs field may suppress the cross-section is especially relevant in this case since they generate the strong coupling to longitudinal Z's that is responsible for the large cross-section.

Supersymmetric particles should be easy to detect through the missing P_T signal. If a rate of 100 gluino pairs in 100 days is detectable, which seems very plausible given the stringent limits from the CERN collider,[23)] the reach would be $M_{\tilde{g}}<2.7(E/20)^{0.5}\,(L/10^{33})^{0.25}$

New Vector bosons (W',Z') with a 10% branching ratio to easily visible modes, such as $e\nu$, $\mu\nu$ and $\bar{\mu}\mu$, would be found up to high masses, the reach for a W' corresponding to ten easily visible events being

$$M_{W'}<5.5(E/20)^{0.75}\,(L/10^{33})^{1/8}$$

Substructure would generate effective contact interactions which would interfere with the QCD amplitudes and produce deviations from the standard predictions for large P_T jet production. The reach for seeing a quark contact term, characterised by $4\pi/\Lambda^2$, would be

$$\Lambda<15\ (E/20)^{0.5}\ (L/10^{33})^{0.17},$$

requiring that the nominal ratio between the interference term and the pure QCD term $[P_T^2/(\alpha_s\Lambda^2)]$ is at least 0.5 at a value of P_T above which there are fifty events.

There is no way known to find a Higgs boson with $M_H<2M_W$ in TeV pp collisions,[24)10)25)] so I shall assume that $M_H>2M_W$ and that consequently $H\to WW/ZZ$. Gluon fusion

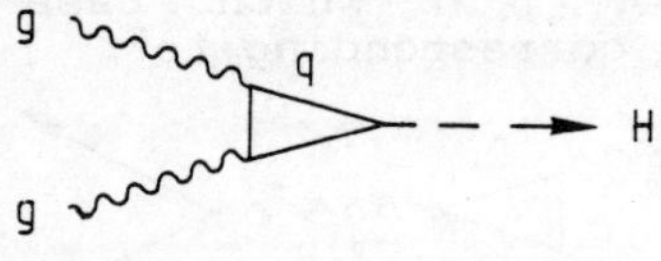

and W fusion

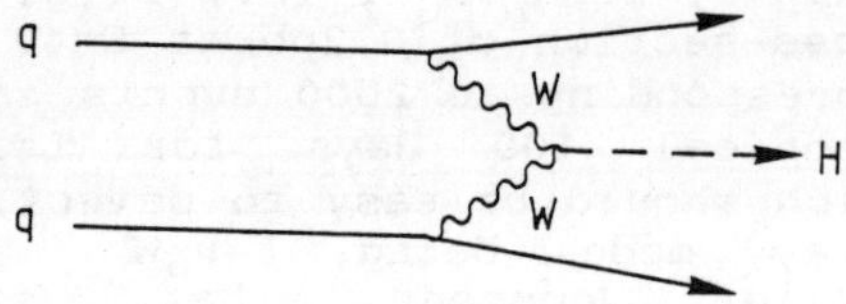

are likely to be the dominant production processes. If M_t=35 GeV and there are no heavier quarks, the former dominates up to $M_H\simeq300$ GeV and the latter for larger M_H; if M_t=150 GeV, however, gluon fusion dominates up to M_H=1 TeV.

The question of backgrounds for the detection of Higgs bosons in pp collisions is complicated, especially if the mass is of order 1TeV or more in which case the decay would generate a very broad enhancement in the WW or ZZ invariant mass distribution rather than a peak. Decays in which one or both of the W's or Z's decays to hadrons will be very hard to detect above the $pp\to W/Z+$ two jet background[26)], as is easy to appreciate when it is recalled how hard it is to see $p\bar{p}\to W(\to 2$ jets) above the $p\bar{p}\to$two jet background. Gunion and Soldate[27)] have found cuts which suppress the background without reducing the signal too dramatically and may make it possible to detect $H\to WW/ZZ\to$ jets+$(L\nu)/(LL)$, but more work is needed to discover whether the favourable signal/noise ratio will survive the effects of parton hadronization and simulation of a realistic detector.

The branching ratios W→hadrons: W→leptons $\simeq$3:1 and Z→hadrons: Z→neutrinos: Z→charged leptons $\simeq$8:2:1 show that the price to be paid for considering the much cleaner leptonic decay channels may not be prohibitive. The results of EHLQ[18)28)] suggest that a 5σ effect would be seen in the channel $H\to ZZ$ with both Z's decaying to $\bar{e}e$ or $\bar{\mu}\mu$ for

$$M_H\lesssim0.65\ (E/20)^{0.65}\ (L/10^{33})^{0.4}.$$

However, this conclusion is based on a calculation[29)] in which the signal was integrated over all M_{ZZ} while the background was only integrated over
$M_H - \Gamma_H/2\leqslant M_{ZZ}\leqslant M_H+\Gamma_H/2$.

If the same limits are placed on M_{ZZ} for the signal, as should be done, the significance drops from 5σ to less than 3σ at the upper limit for M_H quoted above.

For E $\simeq$ 40 TeV or greater, a more promising channel [30,31)] is $H\to ZZ$ followed by $Z\to(\bar{e}e+\mu\bar{\mu})+Z\to\nu\bar{\nu}$

for which the rate is six times larger than for $(\bar{e}e+\mu\bar{\mu})^2$. Since H is produced with small transverse momentum in WW fusion, this decay will generate a Jacobian peak at M_H in the transverse mass $M_T = (M_Z^2 + P_T^2)^{1/2}$ where P_T is the transverse momentum of the Z that can be fully reconstructed; this is exactly analogous to the peak in M_T which signals $W \to \mu\nu, e\nu$. Although the sum of the signal and background will not show a peak, the background should be understood well enough for it to be possible [32] to discover a Higgs boson with mass 1TeV or more at 40TeV with L = 10^{33}. At 20TeV, however, the power of this method is lost[32].

The potential for discovering a Higgs boson with a mass of order 1TeV or more is clearly uniquely sensitive to both the energy and the luminosity. The same is true of attempts to study possible strong interaction effects in the scattering of longitudinal vector bosons in pp collisions[31]. In the channel $W_LW_L \to Z_LZ_L \to (ee + \mu\mu) + (\nu\nu)$, an integrated luminosity of $10^{-4}pb^{-1}$ at 40 TeV would yield 20 events with $M_T > 0.9$TeV, $|y_{LL}|<1.5$, according[31] to the conservative model of ref 13, relative to a background of 7, from which some information could be obtained. At 20 TeV, however, the signal/background falls to 2/3.

<u>To summarize</u>: multi-TeV pp colliders can be built and the average reach, in so far as such an average makes sense, is

$$1.9 \left[\frac{E}{20}\right]^{0.65} \left[\frac{L}{10^{33}}\right]^{0.17} \text{TeV}.$$

It is not clear that vertex detectors can be built that can handle L = 10^{33}, but - with the exception of Higgs boson production - the loss in going to 10^{32} is not dramatic, being 30% for the "average" reach. It may, however, be possible to do "muon only" experiments at 10^{34}, which could detect the Higgs boson among other things. Note that if a new W' or Z' is in reach, study of its decays would open up a new window e.g. on heavy leptons, heavy neutrinos and supersymmetric leptons.

4. EP PHYSICS AT TEV ENERGIES

Possible ways to study ep physics beyond the reach of Hera, which has a design luminosity of 2 x 10^{31} cm^{-2} sec^{-1} at E = 0.31 TeV, include
a) colliding a possible 8.5 TeV proton beam in the LEP tunnel with LEP at 50 GeV, giving E = 1.3 TeV and a design luminosity of 2 x 10^{32}, or at 100 GeV, giving 1.8 TeV and 2 x 10^{31};
b) colliding an electron beam of, say, 20, 50 or 100 GeV with an SSC beam of 20 TeV giving E = 1.3, 2 or 2.8 TeV.
I shall therefore concentrate on energies of order one or two TeV and luminosities of order 10^{32}.

Experiments at ep colliders will be relatively clean[33], typical large Q^2 events consisting of a struck-quark jet and a lepton which will be well separated provided E_p/E_e is not too large:

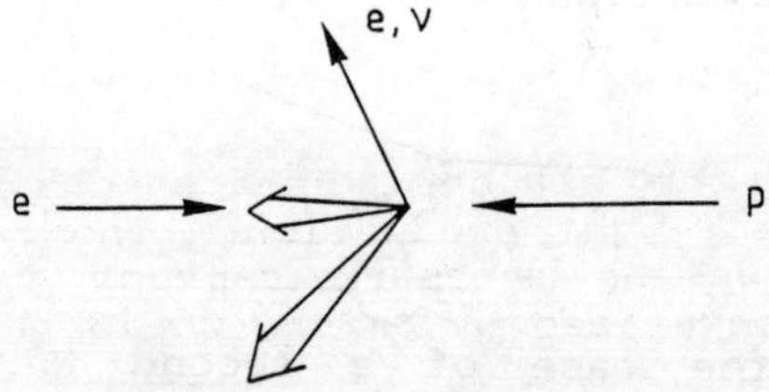

Both neutral and charged current events will be easy to pick up and measure.

The standard inclusive neutral current (ep→ex) and charged current (ep→νx) cross-sections are shown in fig. 3 integrated over $Q^2 \geq Q_o^2$; they "scale" to an excellent approximation for $Q^2 >> M^2_{W,Z}$ i.e. so effectively depends only on Q_o^2/s. This figure implies that in the canonical 100 days continuous running at full luminosity, used for the estimates of reaches below, there will be 1000 neutral current events for

$$\sqrt{Q^2} \gtrsim 0.27\, E^{0.5} \left[\frac{L}{10^{32}}\right]^{0.25} \text{TeV},$$

allowing precision measurements, and 50 events for

$$\sqrt{Q^2} \gtrsim 0.46\, E^{0.67} \left[\frac{L}{10^{32}}\right]^{0.17} \text{TeV},$$

allowing exploratory experiments. For charged current events the coefficients 0.27 and 0.46 are replaced by 0.30 and 0.50 respectively.

Electron-proton colliders are not particularly well suited for the production of new particles, except some exotic leptons and lepto-quarks. The reach is poor for standard quarks (0.085 $E^{0.6}[L/10^{32}]^{0.25}$ TeV on the basis of 1000 events) and even worse for standard leptons. It is more respectable for supersymmetric electrons and quarks which can be produced in the process $eq \to \tilde{e}\tilde{q}$ with a reach

$$M_{\tilde{e}} + M_{\tilde{q}} < 0.32\ E^{0.6} \left[\frac{L}{10^{32}}\right]^{0.25}$$

on the basis of 100 events.

The real strength of electron-proton colliders is their unique capability for looking for surprises in the charged current in the reaction

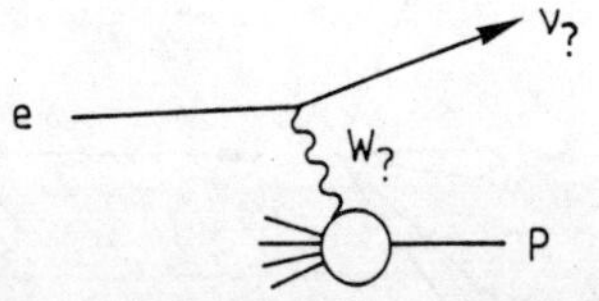

In the case of a second W that couples e to ν_e, interference with the standard W_1 gives a reach

$$M_{W_2} < 1.0\ E^{0.67} \left[\frac{L}{10^{32}}\right]^{0.17} \text{ TeV},$$

if W_2 couples with the same strength as W_1, on the basis of a 50% effect at a value of Q^2 above which 50 events are expected in a 100 days according to the model. The reach for a second Z which couples like the first is somewhat less. A more interesting possibility is that there exists a "right handed" W_R coupled to a right-handed ν_R, which would generate a non-zero charged current cross-section for incident right-handed electrons. Requiring a 3σ effect in the cross-section for $\sqrt{Q^2}>0.25\ \sqrt{s}$ in 100 days running divided equally between e^-_L and e^-_R (a choice of conditions which has not been optimised), a limit of

$$M_{W_R} < 0.70\ E^{0.75} \left[\frac{L}{10^{32}}\right]^{0.13} \text{ TeV}$$

could be set with 50% longitudinal polarization, assuming that W_R couples with the same strength as W_1 and that $M_{\nu_R} \ll M_{W_R}$. With 80% polarization the coefficient 0.70 increases to 0.85.

A really spectacular signature would be generated by a Majorana neutrino, which might be produced by the exchange of a new $\hat{W}$: it would decay equally to e^+X and e^-X. If five events in which a large p_T positron is produced by an incident electron could be seen in 100 days, the reach would be

$$M_{\hat{W}} < 1.5\sqrt{E} \left[\frac{L}{10^{32}}\right]^{1/4}$$

assuming standard couplings and that $M_{\nu_R} \ll M_{\hat{W}}$. With $M_{\nu_R} = M_{\hat{W}}$, the reach would be

$$M_{\nu} \leqslant 0.50\ E^{0.8} \left[\frac{L}{10^{32}}\right]^{0.13}.$$

Using the same criterion used for W_2, there would be a reach of

$$\Lambda < 12\ E^{0.67} \left[\frac{L}{10^{32}}\right]^{0.17}$$

for substructure if quarks and

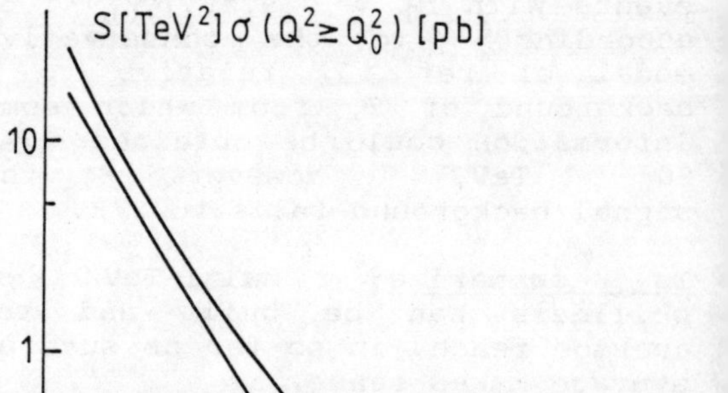

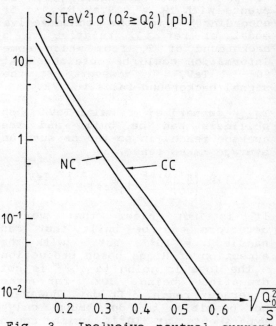

Fig. 3 Inclusive neutral current [NC] and charged current [CC] cross-sections (multiplied by s) integrated over $Q^2 \geqslant Q^2{}_0$ for $\sqrt{s} \gg M_{W,Z}$; these curves are for $\sqrt{s}$=3.16 TeV but the cross-sections scale to a good approximation and results calculated for $\sqrt{s}$=1.4 TeV proved to be essentially indistinguishable.

electrons share constituents that generate a contact interaction characterized by $\frac{4\pi}{\Lambda^2}$, which would produce an anomalous cross-section in deep inelastic scattering.

Experiments at ep colliders will not be particularly adept at discovering the Higgs boson. The most auspicious production mechanism is WW fusion[34] in the reaction $eq \to \nu q'H$. The neutrino tends to go down the beam pipe so the dominant $H \to WW/ZZ \to$ 4jet channel would generate four jet events without the large p_T electron or neutrino that is characteristic of deep inelastic scattering. The $\gamma\gamma \to WW$ background is significant but it appears that it can be reduced sufficiently by making a cut $|\cos\theta^*|<0.5$. A back of an envelope estimate suggests that the $\gamma p \to$ 4jet background may present no problem. If, after the $\cos\theta^*$ cut, 50 events are needed in 100 days running with $L = 10^{32}$, a 1 TeV machine would not see the Higgs boson, but if 5 events are sufficient the reach would be 0.24 TeV. The corresponding reaches for a 5 TeV machine are 0.44 for 50 events rising to a respectable 1.1 TeV for 5 events.

Finally, ep collisions are uniquely suited for searching for lepto-quarks (LQ) which could be produced in electron-quark fusion, $eq \to$ (LQ), or show up below threshold as an effective contact interaction in $eq \to (LQ) \to eq/\nu q'$. Lepto-quarks are present in extended technicolour theories, they occur in superstring theories (although superstring-leptoquarks must be banished to high masses or prevented from coupling to diquarks in order to avoid rapid proton deçay), and they must also exist, with masses expected to be of order 250 GeV, according to the very interesting "strongly coupled" version of the standard model due to Abbott and Farhi.[35] Lepto-quark production would appear as a new contribution to deep inelastic scattering with fixed $x = M_{LQ}^2/2s$ and a flat y distribution, which would be very easy to detect[36]. Unless the eu[d] coupling is much smaller than electro-weak (as happens, for example, in extended technicolour models in which LQ couples mainly to $\tau\bar{t}$, the $e\bar{t}$ coupling being suppressed and the $e\bar{u}[\bar{d}]$ coupling doubly suppressed), the production cross-section would be sufficient to ensure detection for $M_{LQ} \leqslant 0.95$ E. The effect of a virtual lepto-quark with a coupling g_W could be detected for[37]

$$M_{LQ} < 0.7 \sqrt{E} \left[\frac{L}{10^{32}}\right]^{1/4}.$$

To summarize: ep machines are sometimes regarded as poor sisters of e^+e^- and pp machines, but we should remember the story of Cinderella: poor sisters may strike rich and Hera or subsequent ep machines may be spectacularly successful if there are major surprises in the charged current or if lepto-quarks exist.

5. CONCLUDING REMARKS.

There are reservations which must qualify any discussion of future colliders. First, it is not known whether TeV $e\bar{e}$ machines can be built with adequate luminosity. Second, it is not clear what luminosity can be fully exploited at pp colliders. Third, as stressed in the introduction, guessing the physics questions that will remain interesting after the Tevatron, Tristan, SLC, LEP and Hera have operated is risky. In particular, experiments at these machines may solve the mystery of electroweak symmetry breaking, which would weaken the theoretical argument that further progress would be guaranteed by studying physics at 1 TeV. This could conceivably happen if

a) there is a light Higgs boson, which might be found at SLC for example, and

b) the supersymmetric particles that are presumably then needed to stabilize the mass of the W and Z are sufficiently light that the full spectrum of scalar leptons, winos, gluinos, scalar quarks etc. could be explored at LEP2 and the Tevatron (which will have a reach of about 150 GeV [200 GeV] for gluinos and scalar quarks when operating at 1.6 TeV [2 TeV][38]). Finally, the catalogue of possible discoveries reviewed above is clearly not complete, obvious omissions being charged Higgs bosons[25] and technicolour.[39]

But the sample of processes considered should give a reasonable feeling for the reaches of different machines.

It is tempting to try to compare the exploratory capabilites of different colliders. On the basis of present knowledge, it would be hard to argue that an ep collider should be chosen as the first machine to explore a new domain. The choice between e^+e^- and pp would depend on the projected energies and luminosities. If it is supposed that a 40 TeV machine has an average reach of 3 TeV, its capabilities might be comparable to those of a 4or 5 TeV e^+e^- collider, but it is potentially misleading to think that there is a unique, well defined, equivalent energy. In fact, $\bar{e}e$ and pp machines have different strengths and their relative capabilities depend on physics that will only be known after they have operated!

The scaling laws can be used to compare the "physics windows" that might be opened by different pp colliders. Possible windows are sketched in fig. 4 for the 40 TeV SSC and for a possible 17 TeV large hadron collider (LHC) in the LEP tunnel, for luminosities of 10^{32} and 10^{33} cm^{-2} sec^{-1}. Any of these windows would allow very valuable observations of physics at the TeV scale, but the SSC window is obviously bigger and would allow a clearer view of the region beyond 1 TeV. The significance of the extra reach of the SSC depends on what new phenomena await discovery in the 1 to 3 TeV region (and if we knew that we would not have to build the machines). If there are none, comparison of the SSC and LHC might appear to a future historian to be similar to a comparison of PETRA and PEP. If, however, there is a multitude of important phenomena that can only be studied at the SSC, the comparison might be more like that between SPEAR and ADONE.

The SSC proposal is now being considered by the US administration. I wish the project the best of luck and express the hope, which I am sure is shared by all of us here, that it is approved forthwith.

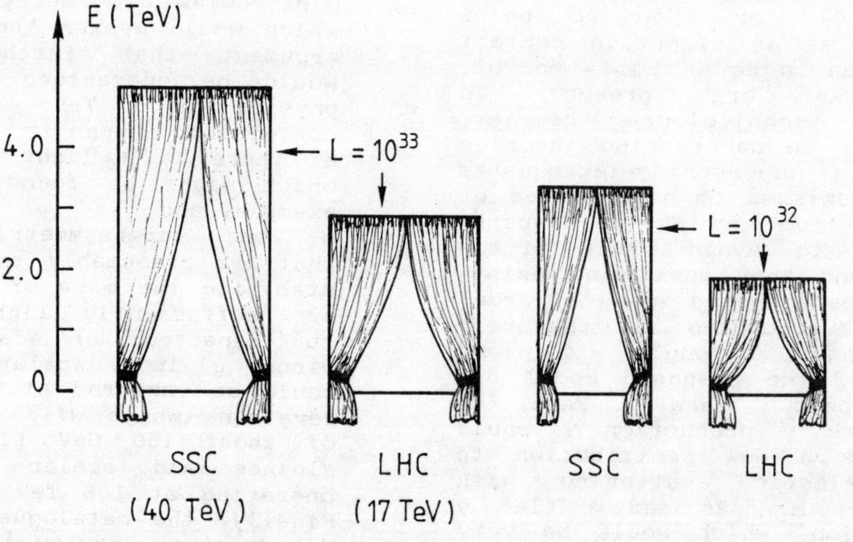

Fig. 4 Sketch representing the relative physics "windows" for E = 40 and 17 TeV, L = 10^{33} and 10^{32} cm^{-2} sec^{-1}. The range of phenomena that can be explored and the detail that can be seen decrease with increasing mass/energy.

FOOTNOTES AND REFERENCES

(1) For discussions of the prospects for constructing TeV $e\bar{e}$ colliders see D. Edwards and B. Richter, these proceedings.

(2) The adjective is Altarelli's (these proceedings). See also the contributions of R. Peccei and N. Deshpande and references therein.

(3) G. G. Ross, these proceedings and references therein.

(4) B. Richter, these proceedings; F. Bulos et al. Proc. 1982 DPF Summer Study, ed. R. Donaldson, R. Gustafson and F. Paige, p71.

(5) Apart from some brief discussions (e.g. F. Bulos et al. [loc.cit.];M. Peskin, "The State of High Energy Physics" ed. M. Month, P. F. Dahl and M. Dienes, AIP Conference Proceedings 134 (1985) 122), little has been published on $e\bar{e}$ physics at TeV energies, but the "reach" for various processes is easily estimated and can be compared with what has been achieved at existing colliders - see M. Davier, these proceedings. For the theoretical motivation for the possible discoveries considered in this talk see the talks by R. Peccei and G. Altarelli and contributions to the associated parallel sessions.

(6) In the case of the Z the square of the coupling to fermions $g_Z{}^2$ is of order $e^2/3$ (and in the total cross-section interference with one photon exchange is suppressed; it would vanish for unpolarized beams if $\sin^2\theta_W$ were 0.25) for "interesting" vector bosons[2]), $g_{Z'}{}^2$ is of order $\sin^2\theta_W e^2/3$. The estimated reaches for $M_{Z'}$ and for the compositeness scale Λ correspond to a 3σ effect for 500 annihilation events.

(7) The exact formulae (B.W.Lee et al. Phys. Rev. D 16 (1977), 1519) contain an extra factor $(1 - a + 0.75a^2)\ (1 - a)^{1/2}$, where $a = 4M_{W,Z}{}^2/M_H{}^2$, which is 0.76, 0.91 and 0.98 for M_H = 0.3, 0.5 and 1 TeV respectively.

(8) See F. Schrempp, these proceedings. For detailed discussions see "Physics at LEP", ed. J. Ellis and R. Peccei, CERN Yellow Report 86-02, pages 256-265 and 304-332 in Vol.1 and 42-46 in Vol.2.

(9) The complete formula (B.W.Lee et al., loc. cit.) is

$$\sigma/\sigma_{pt} \simeq p\ [3m + p^2]/(1 - m)^2$$

in the case $\sin^2\theta_W = 0.23$, where $m = M_Z{}^2/s$ and $p = [1 - 2m - 2\mu + (m - \mu)^2]^{1/2}\ /2$ with $\mu = M_H{}^2/s$.

(10) J. F. Gunion et al. Phys. Rev. D34 (1986) 101.

(11) D.R.T. Jones and S. T. Petcov, Phys. Lett. 84B (1979) 400; R.N.Cahn and S.Dawson. Phys. Lett. 136B (1984) 196 (E: 138B, 464); S. Dawson and J. L.Rosner. Phys. Lett. 148B (1984) 497. Using the effective WW approximation and making the narrow width approximation for the Higgs boson, $\sigma(e\bar{e} \rightarrow \nu\bar{\nu}H)/\sigma_{pt} = 1.41\ (E(\mathrm{TeV}))^2 K(\tau)$ where $\tau = M_H{}^2/s$ and $K = -(1 + \tau)\ln\tau + 2(\tau - 1)$.

(12) K. Hikasa. Phys. Lett. 164B (1985) 385. Dawson and Rosner suggest that the $e\bar{e} \rightarrow WW\gamma$ background, with the γ going forward, will be a problem but it is easily removed by making a cut against small missing mass and is in also greatly suppressed by the $\cos\theta^*$ cut necessary to remove the two photon background.

(13) See M. Chanowitz and M.K.Gaillard, Nucl. Phys. B261 (1985) 379, who have constructed models of the strong interactions and studied their effects in WW fusion in pp collisions. For a review of recent extensions of this work see M. Chanowitz, these proceedings.

(14) For references to technicolour, and also a discussion of models with composite Higgs bosons, see D.Kaplan, these proceedings.

(15) B. Schrempp, these proceedings.

(16) M.C.Bento and C.H.Llewellyn Smith, in preparation.

(17) C.H.Llewellyn Smith, "Large Hadron Collider in the LEP Tunnel" ed. M.Jacob, ECFA 84/85, CERN Yellow Report 84-10 (2 Vols.), p27. More detailed figures are given by I.Hinchcliffe, these proceedings, and in ref. 18, on which the results in this section are mainly based.

(18) E.Eichten, I.Hinchcliffe, K.Lane and C.Quigg. Rev. Mod. Phys. 56 (1984) 579 (and erratum to be published).

(19) R.Diebold and R.Wagner "Proc. 1984 Summer Study on the Design and Utilization of the SSC", eds. R. Donaldson and J.G.Morfin, p575; R.Diebold, to appear in Proc. 1985 FNAL Workshop or Triggering etc.

for High Luminosity Hadron Colliders H.Rykaczewski, contribution to this conference.

(20) I adjusted the coefficient in the reach formula to give this value, which seems to be generally accepted. It is obvious that the notion of an average reach is ill defined and must be treated with great caution. The scaling laws in this and the subsequent section were derived empirically from the results of detailed calculations. They should hold to a reasonable approximation (10%?) for variations in E and L such that the reach divided by the energy varies by less than or of order 50%.

(21) M.S.Chanowitz, M.A.Furman and I.Hinchcliffe, Nucl. Phys. B153 (1979) 402. These limits are for conventional doublets; the exact limit depends on the mass splitting.

(22) Scott S.D. Willenbrock, these proceedings and references therein; this paper also contains a discussion of the production of heavy quarks by W-gluon fusion.

(23) See A.Honma and M.Davier, these proceedings.

(24) B.Cox and F.J.Gilman "Design and Utilization of the SSC", ed. R.Donaldson and J.G.Morfin, p87.

(25) For a review of some aspects of Higgs boson production in pp collisions, see J.F.Gunion, these proceedings.

(26) S.D.Ellis, R.Kleiss and W.J.Stirling. Phys. Lett. 163B (1985) 261 J.F.Gunion, Z.Kunszt and M.Soldate. Phys. Lett. 163B (1985) 389 (E: 168B, 427)

(27) J.F.Gunion these proceedings; J.F.Gunion and M.Soldate. UCD-86-09-Fermilab-Pub-86/46-T.

(28) A similar conclusion was reached by J.F. Gunion and M.Soldate (UCD-85-13), to appear in Proc. FNAL SSC Trigger Workshop on the basis of a more detailed discussion (they suggest normalizing the background by studying WZ production); they used the EHLQ cross-sections, however, which are subject to the criticism below.

(29) This criticism applies to the signal in figs. 151-155 of EHLQ (I.Hinchcliffe, private communication).

(30) R.N.Cahn and M.Chanowitz, LBL preprint 21037 (1986).

(31) M.Chanowitz, these proceedings.

(32) Integrating over $M_T > 0.9$ TeV with $|y_{LL}| < 1.5$ gives a signal/background event ratio of 43/7 for $M_H = 1$ TeV. At 20 TeV results supplied by R.Cahn (private communication) imply ratios of 23/23 and 7/6 for $M_T > 0.5$ TeV and 0.7 TeV with $M_H = 0.6$ TeV, and 13/23 and 8.5/6 with $M_H = 0.8$ TeV.

(33) See R.Cashmore, these proceedings. The reaches in the remainder of this paper are mainly derived from results quoted in Cashmore's talk and from G.Altarelli, B.Mele and R.Ruckl in "Large Hadron Collider in the LEP Tunnel" ed. M. Jacob, ECFA 84/85, CERN Yellow Report 84-10 (2 Vols.), p549, and J.A.Bagger and M.Peskin, Phys. Rev. D31 (1985) 2211.

(34) D.A.Dicus and S.D. Willenbrock Phys. Rev. D32 (1985) 1642.

(35) L.F.Abbott and E.Farhi, Phys. Lett. 101B (1981) 69; Nucl. Phys. B189 (1981) 547. M.Claudson, E.Farhi and R.L.Jaffe, MIT preprint CTP 1331 (1986); B.Schrempp, these proceedings.

(36) J.Wudka Phys. Lett. 167B (1986) 337.

(37) This is on the basis of the earlier estimate of the reach for a contact interaction and Wudka's normalization, with which the s channel LQ exchange would generate a contact interaction with $4\pi/\Lambda^2 = g_W{}^2/8M_{LQ}{}^2$.

(38) E.Reya and D.P.Roy - see E.Reya, these proceedings. R.M.Barnett and H.E.Haber, contributed paper. H.Baer and E.L.Berger, Argonne preprint ANL-HEP-PR-86-20.

(39) E.Eichten et al. Fermilab Pub. 85/-145T-LBL-20304; P.Arnold and C.Wendt, SLAC-PUB-3827.

DISCUSSION

M.S. Chanowitz (LBL). Your formula for the average reach of a pp collider has a moderate 2/3 power dependence on the energy, so that the difference between 20 and 40 TeV seems rather small. As I believe you have already remarked, the differences can be much greater for phenomena crowding the edge of phase space. For the 1 TeV Higgs boson, which is a generic example of possible new TeV-scale physics associated wwith electroweak symmetry breaking, Cahn and I found for Higgs decays to electron or muon pair plus neutrino pair that the observable signal scales more like the cube of the energy while the background from quark-antiquark annihilation scales roughly linearly. With appropriate cuts and a 10,000 inverse picobarn run, we found a signal of 43 events over a background of 7 at 40 TeV compared to 6 over 3 at 20 TeV.

C.H. Llewellyn Smith. I agree that, as I hope I made clear, the reaches for discovering a Higgs boson with mass of order 1 TeV or surprises in WW scattering for $M_{WW} \gtrsim 1$ TeV are uniquely sensitive to energy and luminosity in the energy range that I have discussed.

Plenary Session

Summary and Outlook

S. Weinberg (Texas)

Chairman

I. Mannelli (Pisa)

Scientific Secretary

C. Kounnas (LBL)

SUMMARY AND OUTLOOK

Steven Weinberg
Theory Group
Physics Department
University of Texas
Austin, TX 78712

I'd like to offer my own thanks to the members of the organizing committee for bringing us together here. It is really very useful to get off Calabi-Yau space for a while, as I suppose it is good also to get away from the beam line or the salt mine, and find out what our friends are doing.

This conference also has a certain nostalgia value for those of us who were here in Berkeley in 1966 for the Thirteenth International Conference on High Energy Physics, the first time it was held in the United States after it left Rochester.

Current algebra was the big news that year, although its success wasn't well understood. There was little realization that what we were really doing was exploiting the fact that the strong interactions had a symmetry, chiral SU(2) x SU(2), that was hidden from us by spontaneous symmetry breaking. Spontaneous symmetry breaking was a fairly new and not terribly popular idea. A lot of attention was given, as it had been for the previous few years, to the phenomenology of hadron resonances and difraction scattering, and their connection through Regge pole theory. But we were terribly perplexed about what that had to do with the fundamentals of the strong interactions, in particular with the quark model. We had a good working phenomenological theory of the weak interactions, the famous V minus A Fermi interaction, but it was clear that it couldn't be applied beyond the lowest order of approximation. CP violation was a new mystery. Finally, I don't believe any of us talked about quantum gravity. Anyone who raised questions about quantum gravity would have been directed to another conference.

We have I think made great progress in the past twenty years. We now have a standard model, which provides a perfectly satisfactory description of all physics that is accessible to us at present accelerators. It's a renormalizable quantum field theory, very much like the quantum electrodynamics that was completed in the late 1940's. (That's an idea that would have seemed totally reactionary to the conference in 1966.) Instead of one photon we have twelve; three of them have acquired masses from a spontaneous symmetry breaking, and eight of them are trapped. Instead of one electron, we have a whole menu of quarks and leptons defined by their representations with respect to the electroweak and strong gauge group, and this menu is replicated three times: there are three generations.

As Gilchrist* said here, "This standard model is now not seriously questioned as a description of acceptable physics". However it has, we are all painfully aware, both holes and loose ends, much like a ball of string.

The least troublesome and most obvious hole is that we haven't yet found the top quark. As Davier said, its mass is above 23 GeV, but we really don't know what it

---

*I will follow the practice here for the most part of just quoting the rapporteurs who spoke in plenary sessions. References to the parallel sessions and the physics literature can be found in these rapporteur's reports.

should be. I think the general expectation, if you have to bet, is that it would be somewhere around the W, because why should it be anything different? In fact, I would say the real mystery is why all the other quarks and leptons are so much lighter than the W. So the fact that the top quark hasn't been pinned down is not in any way a worry yet.

Much more troublesome is the fact that we don't really have a clear view yet of the Higgs sector: the mechanism for the spontaneous breaking of the electroweak symmetry. This Higgs mass could be anywhere between a TeV and a few GeV. There is a prejudice, that many theorists now share, that there are probably just two doublets of Higgs. (The reason for this is that supersymmetry needs two; more than two raises problems with flavor changing neutral currents; and also more than two opens up new possibilities of CP violation, which might raise problems for the ε'/ε ratio in neutral K decay.) We don't know where the Higgs is, we're not even really sure that there is a Higgs as an elementary particle, but we are reassured that as Hasenfratz and also Llewellyn Smith told us this morning, something must show up by the time we get up to available energies of a TeV: either a Higgs, or new kinds of strong interactions, or entirely new physics.

I am not listing here as a hole in our understanding of the standard model the fact that we can't solve quantum chromodynamics and predict all phenomena having to do with hadron physics. An analogy may be useful between quantum chromodynamics and another science which I think it will increasingly come to resemble, that of hydrodynamics. In both cases, we think we know the underlying equations; the Navier-Stokes equations in hydrodynamics, the SU(3) Yang-Mills equations for quantum chromodynamics. In both cases there are fascinating important hard problems which have not yet been solved. In hydrodynamics there are problems involving flow at high Reynolds numbers - phenomena like turbulence and chaos. In quantum chromodynamics there is everything having to do with low energies and long distances: glueballs, confinement, phase transitions - problems discussed here experimentally by Cooper and theoretically by Hasenfratz. Turbulence is a fascinating subject, and will go on interesting physicists for many years, but there is one reason that we do not give for studying turbulence: we do not study turbulence in order to test the Navier-Stokes equation. In the same way I think that all the really interesting problems of quantum chromodynamics have nothing to do with testing quantum chromodynamics. Critical tests of quantum chromodynamics will come at high energy; my own opinion is that they will be found most beautifully in high energy annihilation of electron-positron pairs into jets, where we will be able to avoid all the "interesting" aspects of quantum chromodynamics.

The loose ends in the standard model are much more disturbing than the holes. Why SU(3) x SU(2) x U(1)? Why just the quark and lepton representations we know about? Why three generations? (Well, we don't really know there are just three generations, but it's looking increasingly as if that's the case. We heard evidence from Davier here that the number of neutrino species can't be much bigger than three, and the mass of any new charged lepton would have to be bigger than 41 GeV.) Finally, although SU(3) x SU(2) x U(1) does a very good job of pinning down the structure of the standard model in terms of a finite number of free parameters, that finite number is, embarrassingly, a large finite number. There are the gauge couplings, the mixing angles, the Yukawa coupling of the quarks and leptons to the Higgs sector, the Higgs self coupling, the mass of the Higgs, and so on. We don't have any good idea of why these parameters have the values that they have. There are undoubtedly wonderful lessons to be learned in these ratios of masses that we've been looking at throughout our lives, but we haven't yet read the lessons.

In order to try to tie up these loose ends a number of ideas have been proposed over the last decade, ideas of technicolor, of quark and lepton substructure, supersymmetry, grand unification, etc., etc. You know, just listening to myself reciting that list, I feel a sense of tremendous frustration. We've been working on these ideas for more than a decade, since the mid-1970's, and we have almost nothing to show for it in terms of hard agreement between predictions and experiments. Only perhaps the prediction of $\sin^2\theta$ stands up as a robust and verified prediction.

I'm not going to survey all this, partly because it's too painful, but also because I think it's - someone used the word this morning - presumptous. There are new facilities just about to come into operation. Edwards listed them in this order: Tristan and SLC (both electron-positron machines), and the Tevatron proton-antiproton collider. These I think are going to make many of our speculations about loose ends obsolete. In particular one point repeated in many discussions is that supersymmetry, if it exists at all as a low energy approximate symmetry, should begin to show up at the Tevatron collider. In fact it may be that the two dijets with large missing momentum transfer mentioned by Altarelli that have been seen at UA1 give us not only a lower bound on the mass of the squark, but actually are squarks, though of course it's too early to say.

I have now completed the "general survey" part of my talk. In the remaining time I can only cover a few topics in detail, so I will confine myself here to just two topics where I think exciting progress is being made right now: solar neutrinos and superstrings. Of course I can't be sure that this is a good choice of topics. It may well be that the future will look back at the Twenty-third Conference and recall that the most exciting thing reported here was the discovery of a weird hadronic resonance at 3.1 GeV (or whatever it is), or speculations about new long range feeble forces. But solar neutrinos and superstrings are two topics about which I wanted to express some opinions.

First, the solar neutrino problem. (This by the way was reviewed by Altarelli and Goldhaber, whose printed reports will doubtless be more detailed than my remarks.) The problem is that in experiments looking for the capture of electron neutrinos from the sun in the reaction $\nu_e + {}^{37}Cl \rightarrow e^- + {}^{37}A$, Davis has found a rate of 2.0 ± 0.3 solar neutrino units, whereas the theoretical prediction by Bahcall and his colleagues is 5.8 ± 2.2 solar neutrino units. For a long time it has been realized that one possible explanation for this is neutrino flavor oscillations: the neutrinos may start in the sun as electron type neutrinos, but turn into an incoherent mixture of electron, muon, and tau type by the time they reach the earth. The muon and tau type are not detected in this kind of experiment, and therefore one sees a reduced rate. Now this is a perfectly plausible explanation as far as the neutrino mass difference that is required. For a solar neutrino of energy 5 MeV, there will be many flavor oscillations between the earth and the sun provided the neutrino mass difference is much greater than $(5\ \mathrm{MeV}/1\ \mathrm{A.U.})^{1/2}$, or 10^{-6}eV. As I'll indicate a little later, this is a reasonable mass difference even if the masses arise from lepton nonconservation at a grand unification scale. The problem is that with N species of neutrinos, the maximum reduction you can possibly get from this kind of oscillation (without fine tuning the distance between the earth and the sun) is 1/N, and that means that the mixing angles have to just so arrange themselves to frustrate Davis. A more natural explanation, it had seemed to many of us, was that either the solar theory was wrong or the experiment was wrong, or both.

The plausibility of solar neutrino oscillations as an explanation for the reduced neutrino rate has recently been greatly increased as a result of a remarkable suggestion made in June 1985, at the conference at Savonlinna in Finland, by Mikheyev and Smirnov, based on earlier work by Wolfenstein. (Hence the name, MSW effect.) The MSW effect describes how neutrinos starting in the sun as electron neutrinos can, even if the mixing angle is quite small, and for quite a broad range of masses and mixing angles, turn into a different kind of neutrino, μ or τ, with nearly 100% efficiency, through a resonance induced by interaction with electrons in the sun.

To see how this works, all you need is a pair of weakly coupled mechanical oscillators, one whose frequency is variable, corresponding to ν_e, and the other with fixed frequency, corresponding, say, to ν_τ. (An apparatus of this sort was demonstrated in the lecture at Berkeley.) My reason for guessing that it is the ν_τ rather than the ν_μ that is relevant will be explained shortly; however, it would make no difference in this discussion if it were ν_μ rather than ν_τ.

First, start with the frequencies unequal, and with the ν_e mode excited. This corresponds to the situation at the center of the sun; the effective ν_e and ν_τ masses are different here both because we assume that the vacuum masses are unequal, and also because the electron neutrinos have a W-exchange interaction with electrons that is absent for τ (or μ) neutrinos.

Next, slowly change the frequency of the ν_e mode. This corresponds to what happens as ν_e rises through the sun -- its effective mass changes because it encounters lower electron densities.

Eventually, at some distance from the sun's center, the frequencies of the two modes may become equal. In this resonant case, the ν_τ mode becomes strongly excited. Continue slowly changing the frequency of the ν_e mode, and stop when the frequencies are again quite different, corresponding to the neutrino leaving the sun, where the effective masses reach their vacuum values. If the passage through the resonance has been sufficiently slow, most of the oscillation energy will be found in the ν_τ mode. If not, it stays in the ν_e mode. (This actually works.)

Now as you can see from this demonstration there are a number of conditions that have to be met to have a sizable reduction in the ν_e counting rate. First, the mass difference has to be small enough (and of the right sign: $m(\nu_\tau) > m(\nu_e)$) so that the resonance actually occurs in the sun. If the mass difference is too great the sun's density will not be large enough to compensate for it. So there's an upper bound on the neutrino mass difference, roughly 10^{-2} eV, for the resonance to occur in the sun. Also the transition has to be slow: the neutrino must go through the resonance sufficiently slowly, so as to follow one eigenmode continuously. This requires that the mixing angle, which determines the width of the resonance, can't be too small, because if it's too small the resonance is too narrow, and the neutrino zips right through it without feeling the effect of the resonance. Finally, the mixing angle can't be too big, because if it's too large then even if the MSW effect occurs with 100% efficiency the neutrinos wind up at the earth with still a strong mixture of the electron mode.

These three conditions define a roughly triangular-shaped region in the mass-difference/mixing angle plane within which there is a large (say, a factor 3) supression in the ν_e counting rate in the $^{37}Cl \rightarrow {}^{37}Ar$ reaction. The boundaries of this region depend upon the particular reaction used to detect the neutrinos, because different reactions are sensitive to neutrinos of different energy, and the efficiency of the MSW effect depends on the energy of the neutrino because there is a large relativistic time dilation. Thus for instance there is a large reduction in the counting rate in the gallium as well as the chlorine experiments for parameters on the low-mixing-angle side of the triangle for the chlorine experiments.

Now one would of course like to know whether the MSW effect is really responsible for the low ν_e rate in chlorine. Assuming that someone hasn't made a terrible error, there are two leading hypotheses that we would like to test against each other. One, hypothesis A, is the one that I've just described - that the electron neutrino has changed into some other neutrino species with high efficiency, perhaps 60% to 70% efficiency, through resonant oscillations in the sun. The other hypothesis, B, is that most of the electron neutrinos with enough energy to show up in the chlorine reaction are simply missing: they never were produced in the sun. The usual way of understanding this is that, since the electron neutrinos to which the Davis experiment is sensitive are primarilly from the beta decay of ^8B, all we have to explain is why the ^8B is not produced, which one can arrange by supposing that the temperature at the center of the sun is a little lower than we had thought.

Now, if one looks for solar neutrinos in reactions which arise only from neutral currents, (but which have the same sort of energy threshold as the ^{37}Cl reaction) then the Davis result leads us to expect rates that differ between these two hypotheses by about a factor of 3. Hypotheses B says that there are effectively only a third as many neutrinos of appropriate energy coming from the sun as we had thought, while hypotheses A says that there are just as many neutrinos coming from the sun as we had thought, they've just changed their nature, but any kind of neutrino is equally efficient as far as the neutral currents are concerned. (The ratio of rates under these two hypotheses can be somewhat greater or less than 3, for neutral current reactions that are relatively more or less sensitive to the higher energy neutrinos, respectively.)

So it has now become terribly important to do experiments looking for solar neutrinos, in which one looks not for charge current reactions as in chlorine and gallium, but for neutral current reactions. This has not been much discussed here as far as I know, so I've prepared a little list of experiments of this sort that have been proposed so far.

Table 1 - Proposed Neutral Current Solar Neutrino Experiments

Reaction	Signal	Authors*
$\nu N \to \nu N$	bolometric	Drukier & Stodolsky
$\nu N \to \nu N$	phonons	Cabrera, Mastoff, & Neuhauser
$\nu d \to \nu p n$	$n + d \to {}^3H + \gamma$	Chen (SUDBURY)
$\nu + {}^{11}B \to \nu + {}^{11}B^*$	${}^{11}B^* \to {}^{11}B + \gamma$	Ragavan, Pakvasa, & Brown
$\nu e^- \to \nu e^-$	Cerenkov	Bahcall, Baldo-Ceolin, Cline & Rubbia; Koshiba (ICARUS, KAMIOKA, SUDBURY)

*The list of authors is meant to be illustrative rather than exhaustive. Also indicated are some facilities where the proposed experiment might be sited.

Of these possible experiments, the first four have the disadvantage that there is no angular correlation between the signal and the direction of the incoming neutrino, which could be used to confirm that observed events are really caused by something from the sun. (For nuclear excitation reactions like that in ${}^{11}B$, this is a consequence of the V, A structure of weak interactions and the TP invariance of the electromagnetic interaction.) On the other hand, these are pure neutral current experiments, so there is a large difference expected in the counting rates for (A) neutrino oscillations, and (B) missing neutrinos. The last process has the advantage that the direction of the electron recoil is correlated with that of the initial neutrino, but some of the reaction rate comes from charged-current contributions, and so the difference in the counting rates for hypotheses (A) and (B) is somewhat reduced. It would be nice to design a pure neutral current experiment that can observe correlations with the direction to the sun. Eventually, one could also hope to verify the solar origin of neutrino events by observing seasonal variations due to the ellipticity of the earth's orbit, or diurnal variations due to resonant neutrino oscillations in the earth.

I am emphasizing the importance of an experimental check of the MSW idea because there is a great deal more at stake here than just the interesting question of how the sun works. Neutrino oscillation whether resonant or not requires a neutrino mass difference to begin with, and the confirmation of a neutrino mass would illuminate some of the deepest questions of particle physics.

Someone said earlier that we don't know any reason why the neutrino has zero mass, and this is true, but we do know a good reason why it should have a very small mass. The reason is that in the standard model, with just the usual quarks, leptons, and gauge bosons, there is no possible renormalizable interaction that can violate lepton conservation and give the neutrino a mass. However, even in the standard model, there can be non-renormalizable lepton - number - violating interactions. Non-renormalizable interactions necessarily are operators of dimensionality (counting powers of mass, with $\hbar = c = 1$) greater than four, and since the whole Lagrangian density must have dimensionality four, the non-renormalizable terms must be multiplied with coupling constants that are negative powers of some mass, presumably a very large mass, perhaps a grand unification mass like 10^{15} GeV. Clearly, the dominant effects at ordinary energies will be produced by the terms of minimum dimensionality.

In the standard model, the lepton - nonconserving interaction of minimum dimensionality is between a pair of lepton doublets and a pair of Higgs doublets:

$$\frac{g^2}{M}\,(\nu\phi^0 + e^-\phi^+)^2$$

This operator has dimensionality 5, so its coupling constant is written as the reciprocal of a large mass M characterizing the energy scale of lepton nonconservation, times some dimensionless constant g^2 that depends on the mechanism of lepton nonconservation. Now, this is not itself a neutrino mass, but it produces a Majorana neutrino mass when SU(2) x U(1) is broken:

$$m_\nu = \frac{g^2}{M} \langle\phi^0\rangle^2$$

For instance, if g is the SU(2) gauge coupling then $m_\nu = m_W^2/M$, while if g is the Yukawa coupling of ϕ to some quark q then $m_\nu = m_q^2/M$. Either way (since the top quark can't be much lighter than the W) it is plausible to expect the heaviest neutrino mass to be of order $(100\ \text{GeV})^2/10^{15}$ GeV, or 10^{-2} eV, just about what is needed to give an MSW resonant neutrino oscillation in the sun. It is natural to guess that the neutrino masses will roughly parallel the corresponding charged lepton masses, as is the case for quarks of charge 2/3 and -1/3. If this expectation is correct, then it is the τ neutrino that is heaviest, and has the best chance of being in the mass range required for a resonant oscillation with electron neutrinos in the sun.

The estimate $m_\nu \approx m_q^2/M$ is just that expected on the basis of the celebrated "see-saw" mechanism of Gell-Mann, Ramond, and Slansky, and Yanagida. The argument given here shows that a similar result would be expected for more general mechanisms than the see-saw mixing of a massless neutrino with a superheavy Majorana fermion. However, the argument given here also indicates that this result depends on the existence of an elementary Higgs scalar. Of course, even without elementary Higgs scalars we can write lepton-violating interactions by replacing ϕ's with bilinears $\bar{\psi}\psi$, where ψ is some fermion field with extra-strong as well as electroweak interactions. The trouble is that $\bar{\psi}\psi$ has dimensionality 3 rather than 1, so any such interactions would have high dimensionality and hence be strongly suppressed. For instance, replacing each ϕ with a single $\bar{\psi}\psi$ gives a lepton-violating interaction of dimensionality 9

$$\frac{g^2}{M^5} \left(\nu(\bar{\psi}\psi)^0 + e^-(\bar{\psi}\psi)^+\right)^2$$

For $\langle\bar{\psi}\psi\rangle$ of order $(300\ \text{GeV})^3$, g of order 1, and M of order 10^{15} GeV, this gives m_ν of order $(300\ \text{GeV})^6/(10^{15}\ \text{GeV})^5$, or 10^{-51} eV, which is too small to be any good to anyone. That is, if lepton nonconservation takes place at 10^{15} GeV, then without elementary scalars there is no way for this nonconservation to leak down to the ordinary 300 GeV scale. To put this another way, in order to get a neutrino mass as large as 10^{-4} eV, needed for the MSW mechanism, without elementary scalars the lepton nonconservation scale M would have to be no bigger than about $3 \cdot 10^5$ GeV.

There are other possible explanations of the reduced solar neutrino flux seen in the Davis experiment, but they're also very interesting. As I said earlier, this reduced flux could be understood if the temperature at the center of the sun is a little lower than had been thought. But why should the temperature not follow the standard solar model? Well, as Yoshimura told us, there may exist new weakly interacting massless particles, or "WIMPS". Since they're weakly interacting they would have long mean free paths, and long mean free paths means that they would be very efficient at transporting heat. That means that they would smooth out the temperature profile at the center of the sun, giving an essentially isothermal region near the very center of the sun. With a WIMP to baryon ratio as small as 10^{-11}, you can lower the temperature of the center of the sun by 10% and reduce Davis' counting rate from six to two SNU's. This idea can be tested by the methods of helioseismology, measuring the frequencies of little tremors in the sun. Coffee-break rumors at this meeting suggest that these frequencies are observed to be shifted in the manner expected for a slightly cooler solar core, but I am not able to assess the reliability of this conclusion.

These are evidently very important issues. On one hand, with the MSW proposal, we now have the possibility that we are seeing the first sign of something beyond the standard model, coming down to us from the 10^{15} GeV scale about which we've been dreaming all these years. We had been hoping that such a sign would come to us from observations of proton decay, but so far it hasn't. Maybe we are now going to learn something about physics at very high energy scales from neutrino masses that show up in solar neutrino oscillations. On the other hand there's a possibility that there are new sorts of weakly interacting massive particles in the center of the sun. Either way, it's pretty exciting. I would strongly stress that these experiments to look for neutral-current effects of solar neutrinos (and also charged-current experiments like that in gallium, which will fill out our picture of what's going on in the sun) are as important to the progress of particle physics as anything else right now in experimental physics, and should be fully supported.

Well, now for something completely different - I turn to superstring theory. I would like in a brief time to try to express here why I feel enthusiastic about this development. Schwarz told us that the basic idea is that the fundamental entities of the universe are not particles; they're 1-dimensional strings. At first sight that's not a completely compelling picture. There is another way of looking at this which I find more attractive, due to Polyakov and others before and after him, which is undoubtedly completely equivalent to the approach that Schwarz described, but which I'd like to tell you about, because it is this aspect of the theory that I find so exciting.

Many quantum field theories have been discussed over the years - both realistic ones, and toy models. For some time it has been known that there's a certain class of apparently unrealistic toy model quantum field theories with truly remarkable mathematical properties. These are the 2-dimensional conformal field theories, discussed here by Alvarez. (As Alvarez explained, these theories are actually realistic as applied to the study of the surfaces of solids near phase transitions.) Two dimensions means there are two coordinates, say σ^0 and σ^1, and a 2 X 2 metric tensor $g_{\alpha\beta}(\sigma)$, (with α and β running over the values 0,1) which just as in general relativity allows us to write everything in a generally covariant form. We also include some "matter" fields, which in the simplest case are 2-dimensional real scalars $x(\sigma)$. There may be several of these, so we attach a label μ, and call them $x^\mu(\sigma)$. So far, we could imagine an infinite variety of possible theories, but now we introduce two crucial symmetries. One is invariance under translations:

$$x^\mu(\sigma) \to x^\mu(\sigma) + a^\mu \qquad g_{\alpha\beta}(\sigma) \to g_{\alpha\beta}(\sigma)$$

The other is Weyl invariance, invariance under local re-scalings of lengths

$$g_{\alpha\beta}(\sigma) \to f(\sigma) g_{\alpha\beta}(\sigma) \qquad x^\mu(\sigma) \to x^\mu(\sigma)$$

with $f(\sigma)$ arbitrary. Now there is just essentially one possible theory, with action

$$I[g,x] = -\frac{1}{2}\int d^2\sigma \sqrt{\mathrm{Det}\ g(\sigma)}\ g^{\alpha\beta}(\sigma)\ \frac{\partial x^\mu(\sigma)}{\partial\sigma^\alpha}\ \frac{\partial x^\nu(\sigma)}{\partial\sigma^\beta}\ \eta_{\mu\nu}$$

This is Weyl invariant because, under a Weyl transformation, $g^{\alpha\beta}$ (the reciprocal of $g_{\alpha\beta}$) picks up a factor f^{-1}, while in 2 dimensions Det g picks up a factor f^2; however the inclusion of more derivatives or higher derivatives in the action would require the introduction of more factors like $g^{\alpha\beta}$, spoiling the cancellation of $f(\sigma)$. It is a remarkable and I believe unique feature of these theories that their dynamics is completely determined by their symmetries.

Up to this point, $\eta_{\mu\nu}$ is just some unknown constant symmetric real matrix. By discarding linear combinations of the x^μ that do not appear in the action, we can reduce $\eta_{\mu\nu}$ to a non-singular matrix, and then by taking suitable linear combinations of the remaining x^μ as our fields we can reduce $\eta_{\mu\nu}$ to a diagonal form, with just +1's and -1's on the main diagonal. There are consistency arguments (based on the need to exclude ghosts) that there can be at most one -1 on the diagonal. In any case, we can either appeal to causality or flip a coin, and arrive at the Minkowski form

$$\eta_{\mu\nu} = \begin{bmatrix} +1 & & & & \\ & +1 & & & \\ & & \cdot & & \\ & & & \cdot & \\ & & & & \cdot \\ & & & & & -1 \end{bmatrix}$$

So at least in this purely bosonic case, even Lorentz invariance is a consequence of the other symmetries of the theory.

I should explain at this point that in deriving this form for $\eta_{\mu\nu}$, we have had to choose appropriate units for $x^\mu(\sigma)$. If we want to measure $x^\mu(\sigma)$ in centimeters instead of these string units, then we must insert a dimensional constant T, the "string tension", as a factor in our formula for I. This is the only free parameter of the theory.

This is a quantum field theory, so it is to be used to evaluate quantum averages. The expectation value of an operator F[x,g] (any functional of $x^\mu(\sigma)$ and $g_{\alpha\beta}(\sigma)$) is to be calculated as a path integral over all functions $x^\mu(\sigma)$ and $g_{\alpha\beta}(\sigma)$:

$$\langle F\rangle \propto \int[dx]\int[dg]F[x,g]\exp(I[x,g])$$

The functional integrals over $x^\mu(\sigma)$ are always easy, because x^μ appears just quadratically in the action; that is, as far as the $x^{\mu}(\sigma)$ are concerned, this is a free field theory. At first sight, the functional integral over $g_{\alpha\beta}(\sigma)$ looks far more difficult, but here we are aided by some celebrated mathematical theorems about two dimensional (and only two dimensional) surfaces. The topology of such a surface is just that of a sphere to which are attached γ handles, and is completely specified by the "genus" γ. (This is for the specially important case of closed oriented surfaces; similar remarks apply in other cases.) Furthermore, the geometry of the surface is completely specified (up to coordinate and Weyl transformations) by a finite number of complex "Teichmuller" parameters; the number is 0 for $\gamma=0$; 1 for $\gamma=1$; and $3\gamma-3$ for $\gamma\geq 2$. The functional integral over metrics therefore amounts to a sum over γ, and for each γ, an ordinary integral over a finite number of Teichmuller parameters. This isn't easy, especially for $\gamma\geq 2$, but it is infinitely easier than the functional integrals in, say, four-dimensional general relativity.

But what does this have to do with reality? More specifically, what operators F are we supposed to average, and what physical interpretation do we give to the results?

I mentioned earlier that the action $I[x,g]$ is assumed to be invariant under translations $x^{\mu}(\sigma) \to x^{\mu}(\sigma) + a^{\mu}$, and it is then automatically also invariant under Lorentz transformations: $x^{\mu} \to \Lambda^{\mu}{}_{\nu} x^{\nu}$. We can construct vertex operators $V[x,g;p,j]$ to transform under the inhomogeneous Lorentz group like particles of momentum p and spin j; for instance, for j=0 the vertex operator is

$$V[x,g;p,0] \propto \int e^{ip\cdot x(\sigma)} \sqrt{\mathrm{Det}\, g(\sigma)}\, d^2\sigma$$

The quantum average of a product of such V's will then have just the right Lorentz transformation properties to serve as a candidate for an S-matrix element in the space with coordinates x^{μ}:

$$S_{12\cdots \to 34\cdots} \propto \langle V_1 V_2 \cdots V_3 V_4 \cdots \rangle$$

In order for such a quantum average to make sense, we must respect invariance under Weyl as well as coordinate transformations of the 2-metric $g_{\alpha\beta}(\sigma)$, and when we do, we find a major miracle: our candidate S-matrix elements, constructed to have the right Lorentz transformation properties, then turn out to have the right unitarity properties as well! (Strictly speaking, unitarity works only with an appropriate normalization for the vertex operators and in the quantum average, and serves to fix this normalization, up to one additional free parameter, a coupling constant g; the genus γ contribution to an S-matrix element involving n particles has a factor $g^{2\gamma+n-2}$). This then is the link between the underlying two-dimensional quantum field theory and physics in the higher-dimensional space described by x^{μ}: quantum averages in the former yield S-matrix elements in the latter.

As I have already mentioned, all this depends crucially on Weyl invariance. Otherwise in performing path integrals over $g_{\alpha\beta}(\sigma)$, we would be integrating over non-denumerably many parameters, and nothing would work. But one lesson we have learned again and again, since the discovery of the Adler-Bell-Jackiw anomaly in the 1960's, is that the symmetries of the action may be destroyed by quantum corrections: the so-called "anomalies". The original ABJ triangle anomaly, which governs the rate for $\pi^0 \to 2\gamma$, depends on the number of colors of quarks that circulate around the triangle graph, and as we all know, this allows us to infer that there are just three colors. There is also an anomaly that can destroy Weyl invariance, and just as for ABJ, it depends on the numbers of fields of each type in the theory. There is a contribution from each $x^{\mu}(\sigma)$, and a contribution from the metric that happens to have a value which is -26 times as large. Adding them up, we see that the anomaly cancels if there are 26 x's; so when we re-interpret the two-dimensional theory as a theory of the S-matrix, it is an S-matrix in 26 dimensions.

The vertex operators also introduce anomalies of their own. The factor $e^{ip\cdot x}$ which must be included to give the vertex operator the right translation properties actually depends on the metric, because it is a meaningless operator unless one introduces a regulator, and to do this in a coordinate invariant way requires that one uses the metric in the construction. This makes $e^{ip\cdot x}$ effectively proportional to $(\mathrm{Det}\, g)^{p^2/16\pi T}$. There is also a $(\mathrm{Det}\, g)^{1/2}$ from the integral, and with 2N derivatives of x's in the vertex function these are N factors of $g^{\alpha\beta}$, so Weyl invariance requires (in the $\gamma=0$, or "tree", approximation):

$$p^2/8\pi T = 1 - N$$

For N=0 the squared mass is $-8\pi T$, so this is a tachyon, a serious problem for this purely bosonic theory. For N=1 the mass is zero, and since the vertex operator contains two $\partial x/\partial\sigma$ derivatives this particle transforms as a tensor; it is a combination of a graviton, a massless scalar, and an antisymmetric tensor gauge particle. The interpretation of the massless spin 2 particle here as a graviton requires that the string scale $\sqrt{T}$ be comparable with the Planck scale of 10^{19} GeV. The higher states with $N \geq 2$ are then inaccessibly heavy.

I want to mention in passing a suggestion in Alvarez's talk that I find quite intriguing. The old anomalies like the ABJ anomaly can be given a purely geometric interpretation, so that they can be derived without regulators and all that. Perhaps the same can be done for these anomalies in Weyl transformations, that play such a crucial role in string theories.

The 2-dimensional theory that I have so far been describing can be interpreted as the theory of a closed string (a loop) moving through 26 space-time dimensions. (A set of such strings, each starting as a point and then growing until they merge with other strings, then breaking apart and finally shrinking until each disappears as a point again, sweeps out a 2-dimensional closed surface.) Of course, there are other two dimensional theories, corresponding to other sorts of strings. There are open strings, which again are afflicted with tachyons and require 26 spacetime dimensions, but in which the massless particles are vector bosons, the gauge bosons of U(n), O(n), or Sp(n) groups. Also, there are theories with 2-dimensional spinor fields, both left and right-handed. To avoid ghosts one needs a 2-dimensional supersymmetry for each spinor with a time index, which imposes a relation between the numbers of spinors and the spacetime dimensionality, the number of x's. Each spinor field makes its own contribution to the Weyl anomaly, and it turns out that here one needs 10 rather than 26 spacetime dimensions to cancel the anomalies. Here too one can choose open or closed strings, and there are other, more subtle, choices one can make regarding boundary conditions. As Schwarz described in his talk, there are now just six known consistent tachyon-free superstring theories.

I find this amazing. There may be a few more superstring theories left to be discovered (and it may be that some of these theories are equivalent) but the number of consistent theories is surely small. Furthermore, the number of free parameters is incredibly limited. There is the string tension T, which just sets the relation between c.g.s. units and string units of length or mass. And there is the coupling constant g, which as I mentioned for closed string theories appears in n-particle S-matrix elements in a factor $g^{n+2\gamma-2}$. In fact, this coupling is probably not an independent free parameter; it always appears multiplied with the expectation value of a scalar field (one of the massless modes of the closed string), and this product is determined dynamically. This has led to a certain amount of despair, because there apparently is no small dimensionless free parameter in which we can carry out perturbative expansions. However, the same is true of quantum chromodynamics and hydrodynamics, both theories with lots of predictive power. After all, if we are aiming at a really fundamental theory, we would not be happy to find in it an adjustable small parameter.

Much of the work that has been discussed at this meeting, and I would say most of the work that is being carried on now by the real string experts (among whom I can't count myself) is in the development of a second quantized string field theory. In such theories one tries to interpret the description I've given here in a more fundamental way, supposing that the basic quantum operators, instead of being fields as in ordinary quantum field theories, are functionals. Fields are functions of the position in space-time of a particle; they are to be replaced with functionals of the configuration in space-time of a string.

This is not how calculations are done. Almost all the calculations that have been done have been done using more or less the framework I've already described to you. The effort is made to develop a string field theory for two reasons, one of which I agree with, and the other of which I don't.

One of the reasons for work on string field theory is that we would like to explain the mysteries in string theories. These theories have things going on behind the blackboard that we don't understand; they are smarter so far than we are. They know enough so that although they are formulated in a way that makes no mention of 10-dimensional or 26-dimensional general covariance, nevertheless they turn out to have massless particles which behave just like gravitons, and, likewise, massless particles that behave just like Yang-Mills quanta. We don't know really where general covariance and non-Abelian gauge invariance come from in these theories, although progress has been made on this. Also, there's the miracle of unitarity. How in the world do these theories know they're supposed to produce an unitary S-matrix? And related to both is the absence of hexagon anomalies. These are not the Weyl anomalies discussed earlier; they are anomalies that would interfere with general covariance or gauge invariance in 10-dimensions. It was the discovery of the cancellation of these anomalies in O(32) open superstring theories by Green and Schwarz that started the current wave of excitement over string theories. (All of these matters are believed to be related to a symmetry, a kind of discrete version of coordinate and Weyl invariance, called modular invariance. You can describe a Riemann surface, like a donut, by cutting it up and flattening it down on a plane, but the physics shouldn't depend on how you slice it. No matter how you slice it, it's still a donut - or it was.) With this aim, of explaining the mysteries behind the blackboard in string theories, I'm completely sympathetic.

The other reason often given for developing a second quantized string field theory is that it will be needed in order to understand nonperturbative effects, which we know are physically important, nonperturbative effects like instantons and trapping that we've already encountered in quantum field theory, and possible stringy analogs of these effects. About this, I'm not so sure. It seems to me that the 2-dimensional field theory approach which I described earlier may be fully capable of describing all the nonperturbative effects that are physically relevant. In particular there is recent work on surfaces of infinite genus by Friedan, which I think may prove very helpful in this direction.

However we formulate them, the real reason so far for being interested in superstring theories is that they are mathematically consistent finite or at least renormalizable relativistic quantum theories, of which there are pretty few, and these are the only ones we know that contain gravity. Unfortunately the phenomenological success of superstring theory is not part of its justification so far.

Superstring phenomenology was reviewed here by Schwarz and Peccei, and is summarized in Table 2.

You first of all have to choose a theory. Well, that's not so hard. There are about six, and the only ones that look promising are the $E_8 \times E_8$ heterotic superstring, and the new one based on $O(16) \times O(16)$. Of course, these are all 10-dimensional theories, but we live in 4-dimensions, so six of the dimensions have to curl up. You have to choose the shape of this tiny 6-dimensional manifold, and also choose the pattern of the gauge field expectation values on the manifold, that allows the symmetries to break in a variety of different ways. Then you have to choose the way the flux loops thread through the handles in the manifold. That leads to a further symmetry breaking. The first symmetry-breaking step determines the contents of the theory at accessible energies, the menu of chiral quarks and leptons, which is of course a very exciting thing to be able to predict. But then you have to choose the flux loops, which leads to a further multiplicity of theories. There are several lines indicated in the figure which have been particularly popular; an older one initiated by Candelas, Horowitz, Strominger and Witten, and a new one which branches off from the older one, by Ross and his collaborators at Oxford, but there are many others.

I have mentioned several choices that must be made in deriving the physical consequences of these theories. You may well be bothered by that, because these are supposed to be specific parameter-free theories, so what am I doing choosing solutions? The theory should determine its own solutions. Well, up to this point it can't, because so far all of these choices of six dimensional space and flux loops and all that give models with vacuum energy zero; so you can't tell which is the right one by looking for the one with the smallest vacuum energy. In order to find the non-zero part of the vacuum energy, it may be necessary to consider nonperturbative effects which are also responsible for breaking supersymmetry, and this problem is just terribly murky. Thus at present it's not at all clear how we get down to the standard model from a fundamental superstring theory.

This superstring picture does have some qualitative successes. First, it has made it natural for there to be a "hidden sector". In the last five years or so physicists who had been thinking about the applications of supergravity to physics had already come to the conclusion that the neatest way of understanding the apparent absence of supersymmetry at low energies would be to assume

Table 2 -- Some Choices for Superstring Phenomenology

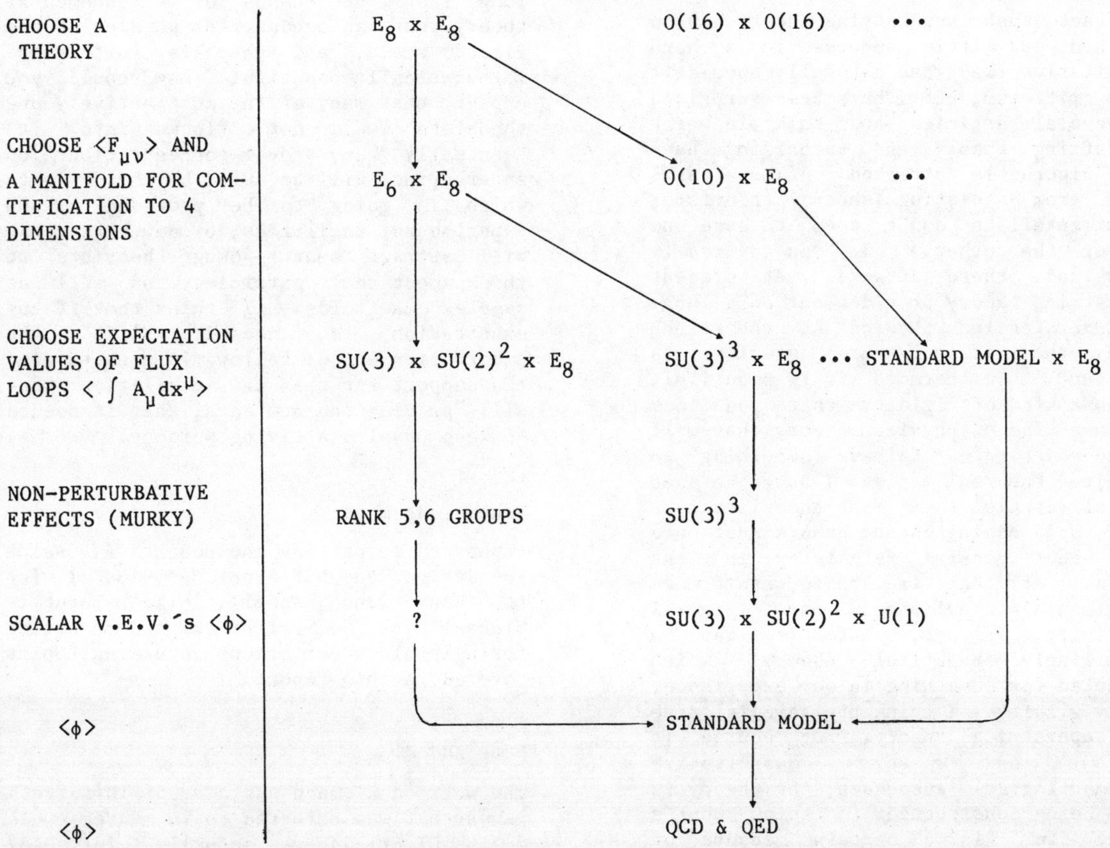

the supersymmetry is strongly broken in a hidden sector of fields, that don't interact with ordinary observable fields except gravitationally and supergravitationally. Here, it's automatic: there are two E_8's, and each fermion is either neutral under the first or the second E_8, so that there's automatically no communication between the two sectors except through gravitation. Another qualitative success, especially in the heterotic theories, is that it is natural for there to be chiral fermions, that is, fermions whose left-handed states form a complex representation of the gauge group. This has been a sticking point of the older Kaluza-Klein models, and it is gratifying to see here that superstring models avoid this difficulty. Also, there's a natural Pecci-Quinn symmetry here of the kind needed to solve the strong CP problem, although as Yoshimura said the axion that this theory gives you may not be the kind of axion you really want. However, the argument against this axion is cosmological, and I'm prepared to believe that there are cosmological surprises waiting beyond the horizon. Finally, various superstring theories continue to explain in a natural way why $\sin^2\theta=0.22$, in much the same way that a variety of grand unification theories did.

Set against these successes is a tremendous disappointment. These theories so far have given no clue as to why the cosmological constant is experimentally 122 orders of magnitude smaller than one would a priori expect. The value of the cosmological constant is one of the biggest mysteries in theoretical physics, and warns us that there's something fundamental we don't yet undertand.

There has been a lot of quantitative theoretical work, which trys to go beyond these qualitative remarks. A great deal of it takes the form of working out the predictions of supergravity in the special case of an E_6 grand unified theory. This is worth doing, but should definitely not be regarded as a crucial test of superstring theory. I would like to repeat and underline a remark made here by both Schwarz and Peccei, that there are no decisive tests of this theory in sight.

The fact that superstring theory has so far had so little success in a hard quantitative way has naturally opened it up to criticism, but I have been surprised at several articles and talks in which superstring theory and theorizing have been vigorously attacked. (The attacks are from distinguished theorists; experimentalists don't seem to care one way or the other.) I don't really understand these attacks. At present superstring theory provides our only hope of understanding physics at the Planck scale, the scale where gravity is important. Furthermore it is beautiful, it has a kind of rigidity that you look for in a kind of physical theory that will in the end turn out to have something to do with the real world. I have the same sort of reaction to it that theorists like Pauli and Eddington and others must have felt about general relativity in the 1920's. At that time, there wasn't much experimental evidence for general relativity either, but it was a compellingly beautiful theory which accounted for something in our experience, namely gravity. I think the same is true of superstring theory. Of course it hasn't had any quantitative phenomenological successes; the theory is still being constructed. I think what's called for is a certain measure of patience. We don't know all the details of what this theory is and why it works the way it does yet. And we're a long way from being able to solve the nonperturbative part of the theory, just as we're a long way from being able to understand turbulence.

One point however, that has been made by the critics of superstring theory, does I think have some validity. There is a high entrance fee to be paid to work on superstring theory. You have to learn mathematics of a sort which is not part of the ordinary arsenal of theoretical physicists; mathematics having to do with Riemann surfaces and Virasoro algebras and lattice theory and algebraic geometry and so on. It's beautiful mathematics and in my view, it's worth paying the price in learning it, but unfortunately it seems that some of our graduate students and postdocs are learning nothing else. Their education is getting extremely deep, but is also extremely narrow. It is as deep and narrow as a grave. In particular many of them have never had the delightful experience of talking to experimentalists and having something useful to be said on both sides.

I doubt that this real problem can be solved by exhortation. When we have at long last the chance of a fundamental theory that can account for physics at the Planck scale, and that is furthermore mathematically beautiful, how could you expect that many of the most active young theorists would not flock into it? Eventually any over-concentration on superstrings will be cured by the new data which is going to be provided by new experimental facilities, by new data that will attract smart young theorists to think about real particles, as well as complex manifolds. I think that if any exhortation is needed, it's the exhortation of our fellow citizens to give the support for the new facilities that will provide the new data, that is needed to keep physics a living science.

Acknowledgement

Supported in part by the Robert A. Welch Foundation and NSF Grant 8605978. I wish to thank John Bahcall, Mike Barnett, Michael Dine, Joe Polchinski and Ed Witten for helpful conversations regarding topics covered in this report.

Note Added

The work on Riemann surfaces of infinite genus that was referred to in my report as due to D. Friedan was actually jointly to D. Friedan and S. Shenker.

Parallel Session 1

Phenomenological Aspects of Unified Theories

Organizers:
L.E. Ibañez (CERN)
P. Roy (Tata)

Scientific Secretaries:
D. Klem (SLAC)
H. Yamamoto (SLAC)

SUPERSYMMETRY AND SUPERGRAVITY PHENOMENOLOGY

E. Reya

Institut für Physik, Universität Dortmund
D-4600 Dortmund 50, West-Germany

Supergravity models are briefly reviewed as well as their phenomenological consequences for gaugino (wino, zino, gluino) and squark production, most relevant for future e^+e^- (SLC, LEP) and $p\bar{p}$ (Tevatron) experiments. The existence of upper bounds for supersymmetric Higgs masses is reemphasized and the experimental observation of a rather light Higgs boson will be essential for the viability of minimal supergravity models.

1. INTRODUCTION

Effective N=1 supergravity (SUGRA) models are briefly reviewed with particular emphasis on their implications for supersymmetric (SUSY) particle masses. Then various phenomenological consequences and expectations for gaugino and squark production, most relevant for future e^+e^- (SLC, LEP) and $p\bar{p}$ (Tevatron) experiments, are critically discussed and compared with expectations at $\sqrt{s}$=630 GeV at CERN's $p\bar{p}$ collider. In particular, the distinctly different (observable?) signals of electroweak gauginos ($\tilde{W}^{\pm}, \tilde{Z}^{o}$) and strong gluinos ($\tilde{g}$) are discussed which allow for a clear experimental distinction between these particles. Finally, various upper bounds for supersymmetric Higgs boson masses are presented and the experimental observation of such a rather light Higgs boson will be essential for the viability of the minimal SUGRA model.

2. SUGRA MODELS

The minimal N=1 SUGRA extension of the standard model is characterized by having the smallest amount of additional new (s)particles required for constructing a theoretically consistent model and consists of the following chiral and vector supermultiplets:

$$\begin{pmatrix} \ell, q \\ \tilde{\ell}, \tilde{q} \end{pmatrix} \quad \begin{pmatrix} \gamma, W^{\pm}, Z^{o}, g \\ \tilde{\gamma}, \tilde{W}^{\pm}, \tilde{Z}^{o}, \tilde{g} \end{pmatrix} \quad \begin{pmatrix} H_{1,2} \\ \tilde{H}_{1,2} \end{pmatrix} \qquad (1)$$

where all particles carry the usual $SU(3)_c \times SU(2)_L \times U(1)$ quantum members. Note that here we need at least two Higgs doublets in order to guarantee the cancellation of the anomalies caused by their fermionic SUSY Higgsino partners $\tilde{H}_i$ and, furthermore, to give masses to all quarks and leptons. At an energy scale around the Planck mass $M_P \simeq 2.4\times10^{18}$ GeV, the Lagrangian reads (1)

$$\mathcal{L} = \text{kinetic terms} - \frac{1}{2}M(\tilde{g}_a\tilde{g}_a + \tilde{W}_i\tilde{W}_i + \tilde{B}\tilde{B}) - V \qquad (2)$$

where the tree level Majorana mass terms for the gauginos have already been included. The kinetic terms include the contributions from the gauge fields and the scalar fields. The former ones are proportional to $F^a_{\mu\nu}F^{b\mu\nu}$ which in general is multiplied by an arbitrary analytic function $f_{ab}(\phi)$ with ϕ referring to the various scalar fields contained in the chiral multiplets. A real gauge invariant function $G(\phi,\phi^*)$, the so called Kähler potential, defines the scalar kinetic terms

$$G^i_j\,(\partial_\mu\phi^j)\,(\partial^\mu\phi^*_i) \qquad (3)$$

where $G_i \equiv \partial G/\partial\phi^i$, $G^i \equiv \partial G/\partial\phi^*_i$. Clearly the functions $G(\phi,\phi^*)$ and $f_{ab}(\phi)$ largely determine the physics of the N=1 SUGRA models and the various arbitrary, but legitimate choices are mainly responsible for the vastly different predictions for SUSY masses of the various presently studied models. Indeed, the scalar potential V in Eq. (2) consists of two terms, $V = V_c + V_g$. The 'gauge' potential reads

$$V_g = \frac{1}{2}(\mathrm{Re}f^{-1}_{ab})D^aD^b\ , \qquad (4)$$

with the 'D-terms' given by

$$D^a = g_a G_j (T^a)_i^j \phi^i \quad , \tag{5}$$

and the 'chiral' potential is given by

$$V_c = e^G [G_i G^j (G_i^j)^{-1} - 3] \quad . \tag{6}$$

The minus sign in this latter expression is the crucial difference between local and global SUSY models and implies, unlike in the global case, a scalar potential which is not positive definite. This is very fortunate since one can now obtain spontaneous breakdown of local SUSY (SUGRA) with <V> = 0.

The simplest choice for the functions G and f_{ab} lead to the so called canonical SUGRA models where

$$G_j^i = \delta_i^j \quad , \quad f_{ab} = \delta_{ab} \tag{7}$$

corresponding to a flat Kähler manifold. Thus one usually writes (setting M_P = 1)

$$G(\phi,\phi^*) = \phi^i \phi_i^* + \ln|f(\phi)|^2 \tag{8}$$

where $f(\phi)$ denotes the gauge invariant superpotential and we have still the freedom to choose it at our will. [Note that ϕ_i denote generically the superheavy hidden fields as well as the 'light' observable fields of mass $O(M_W)$]. Inserting Eq. (8) into Eq.(6) yields

$$V_c = \exp(\sum_i |\phi^i|^2) \times$$
$$[\sum_i |\frac{\partial f(\phi)}{\partial \phi^i} + \phi_i^* f(\phi)|^2 - 3|f(\phi)|^2], \tag{9}$$

which explicitly shows that we can have a vanishing cosmological constant $\Lambda_{cosmo} = \langle V \rangle \overset{!}{=} 0$ and simultaneously break spontaneously local SUSY via $f_{min}(\langle\phi^{hidden}\rangle) \neq 0$. For this to hold, some fine-tuning of the parameters in G is clearly necessary.

This latter fine-tuning can be somehow avoided in the so called 'no-scale' models (1-3) which are non-canonical and result from imposing a vanishing chiral potential V_c, which in turn fixes the Kähler potential G: V_c in Eq. (6) can be rewritten as

$$V_c = 9e^{\frac{4}{3}G} \, G^{-1}_{\phi\phi^*} \, \frac{\partial^2}{\partial\phi\partial\phi^*} \, e^{-\frac{1}{3}G} \overset{!}{=} 0 \tag{10}$$

which vanishes for all ϕ if

$$G = -3 \ln(\phi + \phi^*) \, . \tag{11}$$

This specific form is clearly different from the canonical one in Eq.(8) and does not allow for scalar mass terms and Yukawa terms in the scalar potential at an energy scale M_P, i.e. $m_o = A = B = 0$ (these parameters will be discussed below in more detail). The only remaining nonvanishing tree-level mass is the gaugino mass M in Eq. (2) which becomes now the driving term for global SUSY breaking required for constructing an acceptable low energy theory (observable sector). Thus gaugino masses are here expected to be larger, $M \geq O(M_W)$, than in conventional canonical minimal models and, inspite of dealing with a flat potential at the tree level, finite scalar masses will be generated dynamically via higher order radiative corrections. This enables us to explain, at least in principle, the scale hierarchy dynamically. I will not go into further details (3) but will eventually refer to 'no-scale' expectations for SUSY masses in order to compare them with the ones of the simplest canonical minimal model.

Let me return to the canonical model characterized by Eq. (9). The minimal version of it will serve us as a guide for the expectations for the masses of SUSY particles which will be helpful for discussing the various (hopefully realistic) phenomenological consequences of SUGRA for present and future collider experiments. Strictly speaking, the potential in Eq. (9) applies at an energy scale $E \simeq M_P$. At lower energies $E = M_X < M_P$ one takes the so called 'flat' limit of $V = V_c + V_g$ where only the 'light' observable fields are kept explicitly which corresponds to expanding in powers of M_P^{-2} but keeping the gravitino mass $m_{3/2} \equiv \langle\exp(G/2)\rangle$ fixed (4)

$$V = \sum_i |\frac{\partial f}{\partial \phi_i}|^2 + \frac{1}{2}D^2 + m_o^2 \sum_i |\phi_i|^2 +$$
$$+ m_o [Af^{(3)} + Bf^{(2)} + h.c.] \tag{12}$$

with $m_o = m_{3/2}$ being the scale of the additional (soft) terms which break SUSY explicitly. The n-th order terms of $f(\phi)$ are denoted by $f^{(n)}$ which are characterized by just two unknown parameters A and B, regardless of how complicated the hidden sector is. [Note that in 'no-scale' models $m_o \simeq O(M)$ which is expected (3) to be larger ($\gtrsim M_W$) than in the present case]. The last remaining ambiguity is the choice of the superpotential $f(\phi)$: For the minimal particle content in Eq. (1) the most general allowed form reads

$$f = h_E \hat{L}\hat{E}\hat{H}_1 + h_D \hat{Q}\hat{D}\hat{H}_1 + h_U \hat{Q}\hat{U}\hat{H}_2 + \mu \hat{H}_1 \hat{H}_2 \qquad (13)$$

where the superfields, representing the supermultiplets in (1), are denoted by an additional ^ and carry the usual $SU(3)_c \times SU(2)_L \times U(1)$ quantum numbers. All generation indices have been suppressed, but in practice only the top quark coupling $h_{U=t}$ will dominate with $m_t = h_t \langle H_2^o \rangle$. Note that the bilinear mixing term $\hat{H}_1 \hat{H}_2$ of the two Higgs doublets is crucial for inducing $SU(2)_L \times U(1)$ breaking and to avoid the presence of an axion. In going from the scale $E = M_X$ down to present energies $E = M_W$, all mass parameters and couplings in Eqs. (2), (12) and (13) will get strongly and differently renormalized by standard renormalization group (RG) effects (5) which also provide us with an appropriate breaking (6,7) of the weak gauge symmetry. To correctly obtain such a radiative breaking of the electroweak symmetry, together with the requirement that the vacuum in the absolute $SU(2)_L$ breaking minimum conserves $SU(3)_c \times U(1)_{em}$, strongly constrains (7,8) the remaining five free input parameters at M_X:

$$m_o,\ M,\ \mu,\ A,\ B \qquad (14)$$

where $A, B = O(1)$ and in particular $\langle H_1^o \rangle \simeq \langle H_2^o \rangle$ if $m_t \lesssim 50$ GeV. In practice, however, most predictions for the light five flavor sparticle mass and gaugino mass spectra depend mainly on $m_o = m_{3/2}$, M and μ, except the ones for $\tilde{t}$.

The 'predictive' power of the minimal SUGRA model derives from the input assumption that at $M_X \simeq 3\times10^{16}$ GeV all smass parameters in Eqs. (2) and (12) are equal

$$M_1 = M_2 = M_3 \equiv M \ , \quad m^2_{\tilde{\ell}} = m^2_{\tilde{q}} = m^2_o \qquad (15)$$

where M_n refers to the $SU(3)_c \times SU(2)_L \times U(1)$ gauginos (note that this input does not necessarily hold in a SUGRA GUT). Extrapolating to $M_W \simeq 82$ GeV via the RG yields (5-7,9)

$$M_{\tilde{\gamma}} = \frac{8}{3}\frac{\alpha(M_W)}{\alpha_{GUT}} M \simeq 0.5\ M \ ,$$

$$M_{\tilde{g}} = \frac{\alpha_s(M_W)}{\alpha_{GUT}} M \simeq 3\ M \simeq 6 M_{\tilde{\gamma}}$$

$$m^2_{\tilde{e}_{R,L}} = m_o^2 + {0.15 \atop 0.44} M^2 \ , \quad m^2_{\tilde{\nu}_L} = m_o^2 + 0.53\ M^2$$

$$m^2_{\tilde{q}_L} = m_o^2 + 7.6\ M^2 \ , \quad \text{etc.} \qquad (16)$$

where $\alpha_{GUT} \simeq 1/24$. The gaugino mass input M is expected to be generated radiatively via the exchange of heavy fields of mass M_X of the GUT sector which gives rise to (1) $M \simeq C \alpha_{GUT} m_{3/2}$ with $C(\underline{5}) = 1/2$, $C(\underline{24}) = 5$, etc. Thus we expect gluino masses for example to be around or even below 100 GeV which are likely to be lighter than squarks, which are expected to have masses $\gtrsim m_o = m_{3/2}$. Remember that $m_o \lesssim O(M_W)$ in order to maintain huge mass hierarchies and to implement the intuitive requirement of 'naturalness', since our whole purpose in introducing SUSY was to prevent any contributions to the standard model Higgs mass larger than the weak scale [$\delta m_H^2 \lesssim m_H^2 = O(M_W^2)$]. On the other hand, Eq. (16) implies useful mass sum rules (10,11) which depend only on experimentally measurable quantities

$$m^2_{\tilde{q}} = m^2_{\tilde{e}_R} + 0.83\ M^2_{\tilde{g}} \qquad (17)$$

(for 4 generations the RG factor 0.83 turns into 1.6) which, when combined with experimental limits, imposes restrictive bounds on scalar and gluino masses. Furthermore, Eq. (16) suggests that the photino could be the lightest, and hence stable, SUSY particle (LSP) which I will assume for the time being.

The situation for the other gauginos is slightly more involved because of mixing. The masses of the charginos $(\tilde{W}^{\pm},\tilde{H}^{\pm})$ correspond to the eigenvalues of a 2×2 matrix

$$\begin{array}{c} \quad \tilde{W}^{+} \quad \tilde{H}_2^{+} \\ \begin{pmatrix} M_2 & M_W \\ M_W & \mu \end{pmatrix} \begin{array}{c} \tilde{W}^{-} \\ \tilde{H}_1^{-} \end{array} \end{array} \qquad (18)$$

where $\langle H_1^o\rangle \simeq \langle H_2^o\rangle = M_W/g_2$ has been used. In particular Weinberg (12) has emphasized that in a wide class of models (i.e. where M_2 and/or μ are small compared to M_W) Eq. (18) implies the existence of charged fermions (winos or Higgsinos) lighter than M_W with masses satisfying

$$M_{\tilde{W}^{\pm}} M_{\tilde{W}_h^{\pm}} = M_W^2 \qquad (19)$$

and, from a 4×4 mixing matrix, a similar mass relation can be obtained for the neutral gauginos 'zinos', $M_{\tilde{Z}^o} M_{\tilde{Z}_h^o} = M_Z^2$. This observation is rather important since it is one of the few relatively model independent statements which can be made about SUSY masses! Thus if $M_{\tilde{W}^{\pm}} < M_W$, winos should be seen in processes like (12,13) $W^{\pm} \to \tilde{W}^{\pm}\tilde{\gamma}$ to which I will turn in the next sections.

To summarize, the following mass spectrum can be anticipated

$$\tilde{\gamma}(\tilde{H}) < \tilde{W}^{\pm},\tilde{Z}^o \lesssim \tilde{\ell},\tilde{\nu} \overset{?}{\lesssim} \tilde{g} \lesssim \tilde{q} \qquad (20)$$

and, although the absolute mass scale is hard to guess, it is fair to say at least some SUSY particles should have masses $\lesssim$ 200-300 GeV $[\sim G_F^{-1/2}]$. For comparison, no-scale models indicate that (3)

$$\tilde{\gamma}(\tilde{H}) < \tilde{\ell},\tilde{q},\tilde{g} \gtrsim 200\text{-}400 \text{ GeV} . \qquad (21)$$

Finally, let me mention non-minimal SUGRA models where one assumes the existence of additional particles than stated in Eq. (1). For example, an additional $SU(3)_c \times SU(2)_L \times U(1)$ singlet field N is required in the minimal superpotential f in Eq. (13),

$$f = \text{minimal} + \lambda \hat{H}_1 \hat{H}_2 \hat{N} \qquad (22)$$

in order to achieve the appropriate symmetry breakings at the tree level (14,1) which are the so called 'tree breaking' models. Here, expectations for gaugino masses are similar to the ones in Eqs. (16), (18) and (19) whereas scalar masses are $m^2_{\tilde{\ell},\tilde{q}} = m^2_{3/2} \pm$ D-terms since now one can have $\langle H_1^o\rangle \neq \langle H_2^o\rangle$ even for $m_t \lesssim 50$ GeV. Because of the additional free parameter λ in the trilinear $\hat{H}_1\hat{H}_2\hat{N}$ term in Eq. (22), there is no upper bound for Higgs masses, in contrast to the minimal model, as we shall see later. Another example of non-minimal models is a scenario where SUSY is broken by pure 'F-type' mass splittings among the scalars (15, 1). Here the weak gauginos are, as in Eq. (19), expected to be lighter than $W^{\pm}$ and Z^o, but

$$\text{Higgs},\tilde{\gamma},\tilde{e},\tilde{q} \simeq 10\text{-}300 \text{ GeV}\ldots < M_{\tilde{g}} . \qquad (23)$$

From the preceding discussion one can say in conclusion that it would be hard to swallow for any presently known model if no sparticles exist with masses $\lesssim$ 200 GeV!

Based on these general expectations, I will now turn to a few further phenomenological consequences: Although there exists a vast variety of phenomenological implications (1,16,17), I will concentrate on those selected topics which appear to be most relevant for present and future collider experiments (CERN $Sp\bar{p}S$, Tevatron, SLC, LEP).

3. SPARTICLE PRODUCTION IN e^+e^-

As we have seen, the masses of the weak gauginos are expected to be well below M_W and thus the search for zinos and winos will be of utmost importance at SLC and LEP. The 'gold plated' test of SUGRA models is clearly provided by the process (18)

$$e^+e^- \to \tilde{\gamma}\tilde{Z}^o \qquad (24)$$
$$\quad\quad \hookrightarrow \ell^+\ell^-\tilde{\gamma}, q\bar{q}\tilde{\gamma}$$

since it gives rise to the cleanest SUSY signal of a $\ell^+\ell^-$ (or $q\bar{q}$) pair in one hemisphere and nothing in the other one, where 'nothing' refers to the missing transverse momentum, $\not{p}_T$, due to the two unobserved LSP photinos. In particular, this reaction is free of any standard model background! Since there is just one heavier neutralino in the final state of Eq. (24), its production should be kinematically accessible at SLC and

LEP I energies of $\sqrt{s} \simeq 100$ GeV. However, the reaction (24) is a pure t-channel process with selectron $\tilde{e}$ exchange which unfortunately suppresses the production rates if its mass becomes too large. If, on the other hand, $m_{\tilde{e}} \lesssim 50\text{-}70$ GeV clear signals are expected (19): $\sigma\times BR(\tilde{Z}^o \to \ell^+\ell^-\tilde{\gamma}) \simeq 0.1$ pb for $M_{\tilde{\gamma}} \lesssim 10$ GeV and $M_{\tilde{Z}} = 30\text{-}60$ GeV for $\sqrt{s} \gtrsim 90$ GeV where this rate is practically independent of $M_{\tilde{Z}}$.

If indeed $M_{\tilde{W}^\pm} \lesssim M_{Z^o}/2$ then also $\tilde{W}^\pm$ pair production (20,19) will be possible at SLC and LEP where obviously the s-channel Z^o-resonance gives the dominant contribution

$$e^+e^- \to Z^o \to \tilde{W}^+\tilde{W}^- \qquad (25)$$

where $\tilde{W}^\pm \to \ell\bar{\nu}\tilde{\gamma}, q\bar{q}'\tilde{\gamma}$. Again the predicted rates will be enormous (19) for $M_{\tilde{W}} \simeq 40$ GeV and $M_{\tilde{\gamma}} \lesssim 10$ GeV, $m_{\tilde{e}} \simeq 70$ GeV: $\sigma \simeq 10^4$ pb or $\sigma\times BR^2(\tilde{W}\to\ell\bar{\nu}\tilde{\gamma}) \simeq 10^2$ pb where for the latter more realistic prediction an acceptance and p_T cut has been incorporated.

The situation for $e^+e^- \to \tilde{e}^+\tilde{e}^-, \tilde{\mu}^+\tilde{\mu}^-$ is obvious and has been widely studied (19). However, since present experimental bounds on $m_{\tilde{e}}$ are already pretty high (17), depending on $M_{\tilde{\gamma}}$ of course, slepton pair production appears to be not very promising anymore for observing a definite SUSY signal at SLC and LEP. It should, however, be emphasized that photoproduction of single selectrons (21), $\gamma e \to \tilde{\gamma}\tilde{e}$, can probe much higher selectron masses $m_{\tilde{e}} \lesssim \sqrt{s} - M_{\tilde{\gamma}}$. Such an experiment appears to be feasible at SLC where the hard real photons ($E_\gamma \simeq 40$ GeV) can be obtained from backward Compton scattering of LASER light which is focused on the electron beam. In this way, selectron masses up to 80 GeV (for $M_{\tilde{\gamma}} \lesssim 10\text{-}15$ GeV) will be accessible at SLC!

4. SPARTICLE PRODUCTION IN $p\bar{p}$

The main advantage of hadron colliders is of course the much higher c.m. energies available as compared to e^+e^- machines, although their hadronic and standard model (SM) background is much larger and dirtier and thus SUSY signals are much harder to be identified. Let me start with weak gauge boson production and their SUSY decays. The decay mode (22) $W^\pm \to \tilde{e}^\pm\tilde{\nu}$ is likely to be phase space suppressed since according to Eq. (16) we expect $m_{\tilde{e}} \simeq m_{\tilde{\nu}}$ with the experimental bound (17) $m_{\tilde{e}} \gtrsim 40$ GeV for $M_{\tilde{\gamma}} \simeq 10$ GeV. Therefore the dominant observable channels appear to be again the gaugino modes (12,23)

$$p\bar{p} \to \begin{cases} W^\pm \to \tilde{W}^\pm\tilde{\gamma},\ \tilde{W}^\pm\tilde{Z}^o \\ Z^o \to \tilde{W}^\pm\tilde{W}^\mp \end{cases} \qquad (26)$$

where the decays $\tilde{W}^\pm \to \ell\bar{\nu}\tilde{\gamma}, q\bar{q}'\tilde{\gamma}$ and $\tilde{Z}^o \to \ell\bar{\ell}\tilde{\gamma}, q\bar{q}\tilde{\gamma}$ give rise to the following event structure in the final state

$$1j\ (\text{dominant}), 2j, \mu\mu, \mu\mu\mu, \ldots + \not{p}_T\,. \qquad (27)$$

In particular the multi(tri)lepton events would be essentially free from any SM background (24) and could serve as a clean signature of SUSY. The process and expected events in Eqs. (26) and (27) have recently been analyzed (25,26) with particular emphazis on $Sp\bar{p}S$ and Tevatron energies. Figure 1 shows the expected rates for the dominant hadronic 1j events for CERN and Tevatron energies where UA1 cuts have been incorporated (26). At $\sqrt{s} = 2$ TeV we expect only about 3 times as many 1j events than at 630 GeV which is due to the slowly increasing

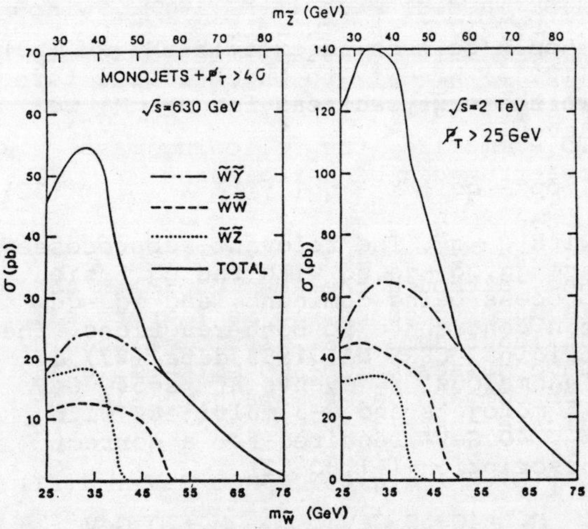

Figure 1: Contributions from $\tilde{W}\tilde{\gamma}$, $\tilde{W}\bar{\tilde{W}}$ and $\tilde{W}\tilde{Z}$ production to hadronic monojet cross sections for CERN $Sp\bar{p}S$ and Fermilab Tevatron energies (26). UA1 cuts and $M_{\tilde{\gamma}} = 8$ GeV have been used.

$q\bar{q}$ luminosity $\mathcal{L}_{q\bar{q}}$. This slow increase can be used to distinguish experimentally the weak gaugino events from purely hadronic ones ($\tilde{g}$ and/or $\tilde{q}$) which increase much faster due to strongly increasing gluon-gluon luminosity L_{gg} as we shall see below. Moreover, very few dijets are expected (25,26)

$$\# 2j \simeq \frac{1}{4} (\# 1j) \qquad (28)$$

which again is strikingly different to purely hadronic sparticle production ($\tilde{g},\tilde{q}$). For $M_{\tilde{W}} \gtrsim 50$ GeV the SUSY signal in Fig. 1 comes entirely from $W \to \tilde{W}\tilde{\gamma}$ in Eq. (26). In this region the SUSY rates become comparable to SM predictions for $W \to L\nu_L$ and it would be difficult to tell a $\tilde{W}$ from a new 4th generation heavy lepton L. Furthermore, as can be seen from Fig. 1, we expect for $M_{\tilde{W}} \simeq 40$-50 GeV a handful of monojet events (and practically no dijet events) at $\sqrt{s}$=630 GeV. It has been stressed in Ref. (25) that the UA1 $\not{p}_T$ data (27-29), taken at face value as an indication for new physics, are compatible with a wino mass of about 40 GeV and $M_{\tilde{\gamma}} \simeq 10$ GeV.

Alternatively one can search at $p\bar{p}$ colliders for purely hadronic SUSY particles such as gluinos and squarks. If $M_{\tilde{g}} < m_{\tilde{q}}$ the dominant process is (30)

$$p\bar{p} \to \tilde{g}\tilde{g} + X \to q\bar{q}q\bar{q}(\tilde{\gamma}\tilde{\gamma}) + X \qquad (29)$$

with $\tilde{g} \to q\bar{q}\tilde{\gamma}$, whereas for $m_{\tilde{q}} < M_{\tilde{g}}$ we have (31)

$$p\bar{p} \to \tilde{q}\bar{\tilde{q}} + X \to q\bar{q}(\tilde{\gamma}\tilde{\gamma}) + X \qquad (30)$$

with $\tilde{q} \to q\tilde{\gamma}$. The relevant subprocesses are $gg, q\bar{q} \to \tilde{g}\tilde{g}, \tilde{q}\bar{\tilde{q}}$ with the gg fusion process being dominant, and $gq \to \tilde{g}\tilde{q}$ can contribute to both reactions. The original CERN UA1 1983 data (27) of 'anomalous' $\not{p}_T$ events at $\sqrt{s}$=540 GeV (5 monojets and 2-3 multijets with $\not{p}_T \gtrsim 40$ GeV) required for a correct description (11,30-32)

$$M_{\tilde{g}} = 40\text{-}50 \text{ GeV}, \quad m_{\tilde{q}} \gtrsim M_{\tilde{g}}+20 \text{ GeV} \qquad (29')$$

or

$$m_{\tilde{q}} = 40\text{-}50 \text{ GeV}, \quad M_{\tilde{g}} \gtrsim 100 \text{ GeV} \qquad (30')$$

for the two alternatives Eq. (29) or (30), respectively. The 1984 run at $\sqrt{s} = 630$ GeV yielded (28) about 4 additional 1j events which would pass a $\not{p}_T > 40$ GeV cut and which do not disagree with the 5-15 monojet events predicted (33,34) from the smasses given in Eqs. (29') or (30'); however, no further dijet events have been observed whereas one expects about 10-20 of such events on account of Eqs. (29') and (30'). This is a very typical feature of purely hadronic SUSY events that for the smass ranges in Eqs. (29') and (30') one expects always <u>at least twice as many dijets</u> than monojets in contrast to the weak gaugino case in Eq. (28)! Thus the interpretation of the 'anomalous' $\not{p}_T$ UA1 events in terms of gluinos or squarks appears to be unlikely. Moreover, it has been demonstrated in the meantime that the SM background due to $W \to \tau\nu$, $Z^0(\to\nu\bar{\nu})$+g, etc. accounts for about (35) 4-7 events/100 nb^{-1} for $\not{p}_T < 40$ GeV and 1-2 events/100 nb^{-1} for $\not{p}_T > 40$ GeV which can fully explain the observed $\not{p}_T$ events. If we accept this latter attitude then one can derive only <u>lower bounds</u> on smasses (33,34), in contrast to Eqs. (29') and (30'), from the absence of any definite SUSY signal:

$$M_{\tilde{g}} > 70\text{-}80 \text{ GeV}, \quad m_{\tilde{q}} > 100 \text{ GeV} \qquad (29'')$$

or

$$m_{\tilde{q}} > 80\text{-}90 \text{ GeV}, \quad M_{\tilde{g}} > 120 \text{ GeV}. \qquad (30'')$$

Further few monojet events in the 1985 UA1 data (29), which are not inconsistent with the above SM expectations, seem to indicate the persistence of the $\not{p}_T$ signals.

Although the standard model, stretched to its extreme, can account for the presently observed missing p_T events, it is nevertheless conceivable that these events are indicative for the appearance of new physics: SUSY may be lurking at the edge of CERN's observability. Again, Fermilab's Tevatron will be of prime importance to clear up the situation. The main advantage of the Tevatron collider for SUSY signals is due to the fact that the gluon-gluon luminosity L_{gg} increases much faster with $\sqrt{s}$ than $L_{q\bar{q}}$:

$$\begin{aligned} L_{gg}(2 \text{ TeV})/L_{gg}(0.54 \text{ TeV}) &\simeq 100 \\ L_{q\bar{q}}(2 \text{ TeV})/L_{q\bar{q}}(0.54 \text{ TeV}) &\simeq 10 \end{aligned} \qquad (31)$$

and thus SUSY signals, entirely dominated by $gg \to \tilde{g}\tilde{g}$ (or $gg \to \tilde{q}\bar{\tilde{q}}$) become strongly enhanced for large energies over the SM background controlled by $q\bar{q}' \to W^{\pm}$ and $q\bar{q} \to Z^0$. Figure 2 shows the SUSY expectations (36,37) for the highest Tevatron energy $\sqrt{s} = 2$ TeV and and top luminosity: If $M_{\tilde{g}} \simeq 50$ GeV we expect a total of at least 2000 events with missing p_T, i.e. at least 50 times as many as expected at $\sqrt{s} = 630$ GeV at CERN's $Sp\bar{p}S$. A similar situation holds (at least 500 events) for the first Tevatron run at $\sqrt{s} = 1.6$ TeV (36, 37). It should be emphasized that the SUSY signal to SM background ratio is here about 10:1. Moreover the $\not{p}_T$ distribution of the SM background is much steeper than the ones for gluino and squark production (37) and is therefore strongly sensitive to $\not{p}_T$ cuts and thus less serious as compared to present CERN $p\bar{p}$ collider energies. This is demonstrated in Fig. 2 by varying the $\not{p}_T$ cut which affects the SM background strongly (reduction by an order of magnitude) but leaves the SUSY signal almost unchanged. This strong increase of the event rate at Tevatron energies could serve as a very distinctive test of purely hadronic sparticles (gluinos and/or squarks) versus SUSY signals resulting from weak gaugino production (wino and/or zinos) where the increase is much less pronounced as shown in Fig. 1.

On the other hand if the CERN data have nothing to do with 'new physics', then the Tevatron can significantly improve the lower bounds in Eqs. (29'') and (30'') and can probe (or observe) gluinos and/or squarks with masses up to 200 GeV at $\sqrt{s} = 2$ TeV as is evident from Fig. 2 (33,36,37).

So far it has been always assumed that the photino $\tilde{\gamma}$ is the LSP. Let me finally mention a few alternatives and consequences if this is not the case (38). If $\tilde{\nu}$= LSP, then $\tilde{\gamma} \to \nu\tilde{\nu}$ (at one loop); since both ν and $\tilde{\nu}$ escape, nothing in the above analysis changes and the same mass limits hold. If $\tilde{H}^0$ = LSP, two possibilities occur, depending on how $\tilde{\gamma}$ decays which is model dependent: If $\tilde{\gamma} \to \nu\tilde{\nu}$ with $\tilde{\nu} \to \nu\tilde{H}^0$, then the $\tilde{\gamma}$ or its decay products still escape, and everything discussed so far remains unchanged. If, however, $\tilde{\gamma} \to \gamma\tilde{H}^0$ then some of the $\tilde{\gamma}$ energy shows up as a (detected) γ and thus there is less missing p_T per event and fewer events pass the $\not{p}_T$ cuts discussed so far. Thus the above smass limits are weakened, i.e. reduced (34, 38). This situation could be easily checked experimentally, since $\tilde{\gamma} \to \gamma\tilde{H}^0$ gives rise to about as many events with an isolated hard γ as with large $\not{p}_T$. Furthermore a LSP $\tilde{H}^0$ can also contribute to the neutrino counting experiment: Since $\Gamma(Z^0 \to \tilde{H}^0\tilde{H}^0)/\Gamma(Z^0 \to \nu\bar{\nu}) = ((v_1^2 - v_2^2)/(v_1^2 + v_2^2))^2$, where $v_i \equiv \langle H_i^0 \rangle$, can assume values between 0 (radiatively broken minimal SUGRA model)

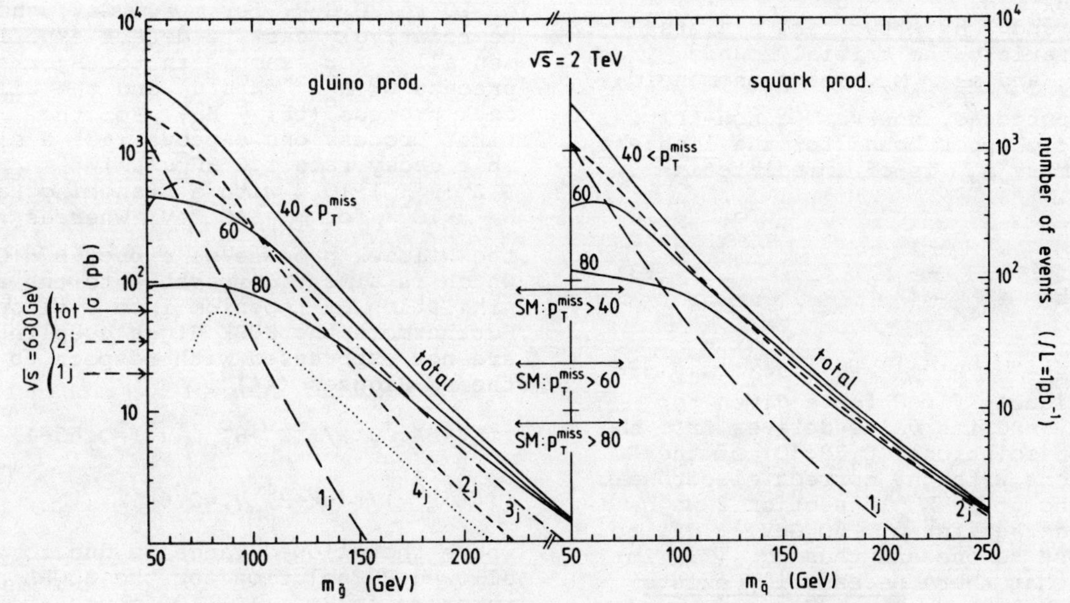

Figure 2: Predicted rates of large missing-p_T SUSY events at the Tevatron collider according to Ref. (37). The standard model (SM) background, dominated by 1j events, is given as well for the respective $\not{p}_T$ cuts. For comparison the expected rates at the CERN energy $\sqrt{s} = 630$ GeV is shown for $M_{\tilde{g}} \simeq 50$ GeV and $\not{p}_T > 40$ GeV. UA1 cuts have been incorporated.

and 1, $\tilde{H}^o$ will at best count as one extra neutrino.

5. SUSY HIGGS BOSONS

As we have seen in section 2, the minimal N=1 SUGRA extension of the standard model requires at least two Higgs isodoublets. The relevant Higgs potential is obtained from inserting Eq. (13) into (12),

$$V_{Higgs} = \mu_1^2|H_1|^2+\mu_2^2|H_2|^2+B\mu m_o(H_1H_2+h.c.) + \frac{g_1^2+g_2^2}{8}(|H_2|^2-|H_1|^2)^2 \qquad (32)$$

with $\mu_1^2=\mu_2^2=m_o^2+\mu^2$. This potential is the SUSY version of the celebrated SM Higgs potential ($\mu^2|\phi|^2+\lambda|\phi|^4$) but with the strengths of the quartic H^4 terms (D-terms) being now fixed by the gauge couplings g_i which allows us to derive rather stringent bounds (39,40) on the masses of the physical Higgs scalars. This is in contrast to the SM where λ is a priori arbitrary or to non-minimal SUGRA models with an additional Higgs singlet as, for example, in Eq. (22). After the Higgs mechanism has taken place at the scale $E=M_W$, the masses of the five remaining physical Higgs scalars (41) are constrained as follows: $m_{H^\pm} \geq M_W$ and $m_{H_b} \leq m_{H_c} \leq m_{H_a}$ where H_c^o is a pseudo-scalar and $H_{a,b}^o$ two real scalars which satisfy the trivial bounds $m_{H_b} < M_Z$ and $m_{H_a} > M_Z$. For phenomenological purposes, however, a non-trivial realistic upper bound for the lightest Higgs mass m_{H_b} is of immediate interest,

$$m_{H_b} \leq \frac{\omega_{max}^2 - 1}{\omega_{max}^2 + 1} M_Z \qquad (33)$$

with $\omega_{max}=\omega(h_t(M_W))_{max} \equiv (v_2/v_1)^2_{max}$ being uniquely fixed for a given top mass m_t and its value derives from the allowed solutions (8,39,40) of the RG equations with the correct electroweak breaking at $E=M_W$. In section 2 we have learned that for $m_t \lesssim 50$ GeV, ω has to be close to one and thus Eq. (33) implies that there necessarily exists one very light neutral Higgs. The resulting most general upper bound on m_{H_b} is shown by the solid curve of Fig. 3 as a function of m_t, whereas the dashed curve represents the upper bound for small gaugino and Higgs mixing mass parameters M and μ.

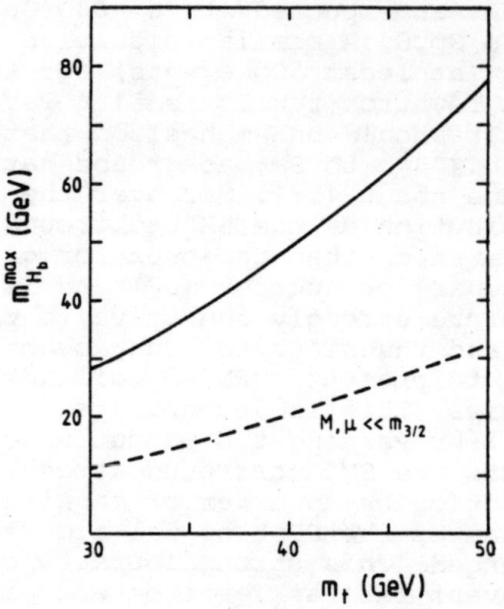

Figure 3: Upper bounds on the mass of the lightest neutral Higgs boson H_b^o in the radiatively broken minimal N=1 SUGRA model (Ref. (40)).

Even tighter bounds are obtained (42) in 'no-scale' models (m_o=A=B=0 at M_X) where $m_{H_b} \lesssim 10$-40 GeV for $40<m_t \lesssim 50$ GeV and furthermore $m_{H_c} \simeq M_{\tilde{g}}/4$.

These upper bounds are encouragingly below M_Z and a Higgs boson with a mass of 20 to 30 GeV, for $m_t \simeq 40$ GeV, should be relatively easy to detect experimentally, for example in the Bjorken process $Z^o \to H_b^o Z^{o*} \to H_b^o \mu^+\mu^-$ and the Wilczek process $(t\bar{t}) \to H_b^o \gamma$. For the first process one expects (43) a sizeable decay rate $\Gamma(Z^o \to H_b^o \mu^+\mu^-)/\Gamma(Z^o \to \mu^+\mu^-) \sim 10^{-3}$ with a branching ratio $B \sim 3\times10^{-5}$ for $m_{H_b} \lesssim 40$ GeV, whereas for the Wilczek process we expect $B \sim 10^{-2}$ which is almost comparable to the annihilation of toponium into 3 gluons. Furthermore the SUSY Higgs couplings are not suppressed with respect to the SM Higgses (44)

$$\Gamma(Z^o \to H_b^o \mu^+\mu^-)/\Gamma(Z^o \to H_{SM}^o \mu^+\mu^-) = 0.85\text{-}1$$
$$\Gamma(t\bar{t} \to H_b^o \gamma)/\Gamma(t\bar{t} \to H_{SM}^o \gamma) = 0.6\text{-}1 \qquad (34)$$

where the allowed range is due to the allowed RG solution for the SUGRA parameters in Eq. (14). Of course both ratios can be accidentally 1 but any possible deviation from 1 might suffice to distinguish experimentally between H_b^o and H_{SM}^o. Furthermore, any measurement of the decay rates in

Eq. (34) will in turn strongly constrain the parameters of the minimal SUGRA model. A search for H_b^o in the mass range indicated in Fig. 3 should be mandatory for the viability of the radiatively broken minimal SUGRA model. On the other hand, the absence of such a particle would rule out the minimal supersymmetric extension of the standard model.

This work has been supported in part by the Bundesministerium für Forschung und Technologie, Bonn.

REFERENCES

(1) For recent reviews see:
D.N.V. Nanopoulos, Proceedings of the XXII Int. Conf. on High Energy Physics, Leipzig 1984 (ed. by A. Meyer and E. Wieczorek), vol. II, p.36;
H.P. Nilles, Phys. Rep. 110 (1984) 1;
P. Nath, R. Arnowitt and A.H. Chamseddine, in 'Supersymmetry and Supergravity, Non-perturbative QCD', vol. 208 of Lecture Notes in Physics (Springer, N.Y. 1984), p.113;
L.J. Hall, ibid., p. 197;
J. Ellis, 28th Scottish Universities Summer School in Physics, Edinburg, 1985 (CERN-TH.4255/85);
E. Reya, in 'The Quark Structure of Matter', ed. by M. Jacob and K. Winter (World Scientific, Singapore, 1986), p.569.

(2) N.P. Chang, S. Ouvry and X. Wu, Phys. Rev. Lett. 51 (1983) 327;
E. Cremmer, S. Ferrara, C. Kounnas and D.V. Nanopoulos, Phys. Lett. 133B (1983) 61;
J. Ellis, A.B. Lahanas, D.V. Nanopoulos and K. Tamvakis, Phys. Lett. 134B (1984) 429;
J. Ellis, C. Kounnas and D.V. Nanopoulos, Nucl. Phys. B247 (1984) 373

(3) For a recent 'no-scale' review see:
A.B. Lahanas and D.V. Nanopoulos, CERN-TH.4400/86

(4) R. Barbieri, S. Ferrara and C.A. Savoy, Phys. Lett. 119B (1982) 343;
L. Hall, J. Lykken and S. Weinberg, Phys. Rev. D27 (1983) 2359

(5) K. Inoue, A. Kakuto, H. Komatsu and S. Takeshita, Prog. Theor. Phys. 68 (1982) 927; 71 (1984) 413

(6) L. Alvarez-Gaumé, J. Polchinski and M.B. Wise, Nucl. Phys. B221 (1983) 495;
J. Ellis, J.S. Hagelin, D.V. Nanopoulos and K. Tamvakis, Phys. Lett. 125B (1983) 275

(7) L.E. Ibáñez and C. López, Nucl. Phys. B233 (1984) 511

(8) L.E. Ibáñez, C. López and C. Muñoz, Nucl. Phys. B256 (1985) 218;
A. Bouquet, J. Kaplan and C.A. Savoy, Nucl. Phys. B262 (1985) 299

(9) C. Kounnas, A.B. Lahanas, D.V. Nanopoulos and M. Quiros, Nucl. Phys. B236 (1984) 438

(10) L.J. Hall and J. Polchinski, Phys. Lett. 152B (1985) 335;
S. Nandi, Phys. Rev. Lett. 54 (1985) 2493

(11) M. Glück, E. Reya and D.P. Roy, Phys. Lett. 155B (1985) 284

(12) S. Weinberg, Phys. Rev. Lett. 50 (1983) 387;
R. Arnowitt, A.H. Chamseddine and P. Nath, Phys. Rev. Lett. 50 (1983) 232; Phys. Lett. 129B (1983) 445

(13) D.A. Dicus, S. Nandi and X. Tata, Phys. Lett. 129B (1983) 451;
B. Grinstein, J. Polchinski and M. Wise, Phys. Lett. 130B (1983) 285

(14) A.H. Chamseddine, R. Arnowitt and P. Nath, Phys. Rev. Lett. 49 (1982) 970;
R. Barbieri, S. Ferrara and C.A. Savoy, Phys. Lett. 119B (1982) 343

(15) U. Ellwanger, N. Dragon and M.G. Schmidt, Z. Physik C29 (1985) 209

(16) For reviews see:
H.E. Haber and G.L. Kane, Phys. Rep. 117 (1985) 75;
R.M. Godbole, in 'Supersymmetry and Supergravity, Nonperturbative QCD', vol. 208 of Lecture Notes in Physics (Springer, N.Y. 1984), p.263;
S. Dawson, E. Eichten and C. Quigg, Phys. Rev. D31 (1985) 1581;
I. Hinchliffe, Les Houches lectures, 1985 (Berkeley LBL-20747);
J. Ellis, Lake Louise Winter Institute, Alberta, Canada, 1986 (CERN-TH.4391)

(17) J.F. Grivas, these Proceedings (parallel session 8)

(18) E. Reya, Phys. Lett. 133B (1983) 245;
D. Dicus, S. Nandi, W. Repko and X. Tata, Phys. Rev. D29 (1984) 1317; D30 (1984) 1112;
J. Ellis, J.-M. Frère, J.S.

Hagelin, G.L. Kane and S.T. Petcov, Phys. Lett. 132B (1983) 436

(19) 'Physics at LEP' study group, H. Baer et al., CERN 86-02, vol. 1, p. 297

(20) D. Dicus, S. Nandi, W. Repko and X. Tata, Phys. Rev. Lett. 51 (1983) 1030;
T. Schimert, C. Burgess and X. Tata, Phys. Rev. D32 (1985) 707;
V. Barger, W.Y. Keung, R.W. Robinett and R.J.N. Phillips, Phys. Lett. 131B (1983) 372;
A. Bartl, H. Fraas and W. Majerotto, Z. Phys. C30 (1986) 441;
X. Tata and D.A. Dicus, Univ. Wisconsin MAD/PH/281 (1986)

(21) M. Glück, Phys. Lett. 129B (1983) 255

(22) R. Barbieri, N. Cabibbo, L. Maiani and S. Petrarca, Phys. Lett. 127B (1983) 458;
R. Barnett, H.E. Haber and K. Lackner, Phys. Rev. Lett. 51 (1983) 176

(23) M. Mangano, CERN-TH.3717 (1983)

(24) H. Baer, J. Ellis, D.V. Nanopoulos and X. Tata, Phys. Lett. 153B (1985) 265;
H. Baer and X. Tata, Phys. Lett. 155B (1985) 278

(25) A.H. Chamseddine, P. Nath and R. Arnowitt, Phys. Lett. 174B (1986) 399; R. Arnowitt and P. Nath, Northeastern Univ. NUB#2699 (1986); R. Arnowitt, these Proceedings (parallel session 15)

(26) H. Baer, K. Hagiwara and X. Tata, Phys. Lett. 57 (1986) 294; Univ. Wisconsin MAD/PH/296 (1986)

(27) UA1 collab., G. Arnison et al., Phys. Lett. 139B (1984) 115

(28) UA1 collab., C. Rubbia, Proceedings of the Int. Symp. on Lepton Photon Interactions at High Energies, Kyoto, 1985, eds. M. Konuma and K. Takahashi (Kyoto Univ., 1986), p. 242; J. Colas, in 'The Quark Structure of Matter', eds. M. Jacob and K. Winter (World Scientific, Singapore, 1986) p.3

(29) UA1 collab., A. Honma, these Proceedings (parallel session 15)

(30) E. Reya and D.P. Roy, Phys. Lett. 141B (1984) 442; Phys. Rev. Lett. 53 (1984) 881; Phys. Rev. D32 (1985) 645;
J. Ellis and H. Kowalski, Phys. Lett. 142B (1984) 441

(31) J. Ellis and H. Kowalski, Nucl. Phys. B246 (1984) 189; B259 (1985) 109;
V. Barger, K. Hagiwara and W.Y. Keung, Phys. Lett. 145B (1984) 147;
A.R. Allan, E.W.N. Glover and D.A. Martin, Phys. Lett. 146B (1984) 247

(32) V. Barger, K. Hagiwara, W.Y. Keung and J. Woodside, Phys. Rev. D31 (1985) 528;
F. Delduc, H. Navelet, R. Peschanski and C.A. Savoy, Phys. Lett. 155B (1985) 173

(33) E. Reya and D.P. Roy, Phys. Lett. 166B (1986) 223

(34) R.M. Barnett, H.E. Haber and G.L. Kane, Nucl. Phys. B267 (1986) 625

(35) S.D. Ellis, R. Kleiss and W.J. Stirling, Phys. Lett. 158B (1985) 341; 167B (1986) 464

(36) H. Baer and E.L. Berger, Phys. Rev. D34 (1986) 1361

(37) E. Reya and D.P. Roy, Univ. Dortmund/Tata Inst. DO-TH 86/06, TIFR/TH/86-16, to appear in Z. Phys. C.

(38) H.E. Haber, SLAC Summer Institute 1985 (SLAC-PUB-3834), and references therein;
G.L. Kane, XXI Rencontre de Moriond, Les Arcs, 1986 (CERN-TH. 4433/86, and references therein

(39) P. Majumdar and P. Roy, Phys. Rev. D30 (1984) 2432; D33 (1986) 2674;
H.P. Nilles and M. Nusbaumer, Phys. Lett. 145B (1984) 73;
M. Drees, M. Glück and K. Grassie, Phys. Lett. 159B (1985) 118

(40) E. Reya, Phys. Rev. D33 (1983) 773

(41) K. Inoue, A. Kakuto, H. Komatsu and S. Takeshita, Prog. Theor. Phys. 67 (1982) 1889;
R.A. Flores and M. Sher, Ann. Phys. 148 (1983) 95

(42) P. Majumdar and P. Roy, Phys. Rev. D34 (1986) 911;
P. Roy, these Proceedings (parallel session 3)

(43) For a recent review see:
A.S. Schwarz, SLAC-PUB-3665 (1985)

(44) M. Drees and M. Glück, Univ. Dortmund DO-TH 86/08, to appear in Phys. Lett. B

PHENOMENOLOGY OF REAL GOLDSTONE PARTICLES IN UNIFIED GAUGE THEORIES†

Rabindra N. Mohapatra*
Department of Physics and Astronomy
University of Maryland, College Park, MD 20742

We discuss the phenomenology of various kinds of real Nambu-Goldstone particles that can arise in unified gauge theories of electroweak interactions such as the Majoron, Familon, etc. It is also pointed out that, in conjunction with strong CP-violating parameter θ that arises in QCD, these Goldstone bosons can lead to attractive long range forces coupled to baryon and lepton numbers with strength as big as $G_{Newton} \times 10^{-3}$ and a range of up to a thousand kilometers.

1. INTRODUCTION

Real Goldstone bosons were, for a long time, considered undesirable in any realistic model of elementary particle interactions for the fear that they might lead to observable long range forces of strength bigger than gravitational forces. It was, however, pointed out in 1980 by Chikashige, Peccei and this author[1] that, real Goldstone bosons coupling to the same species of fermions (or diagonal couplings) such as quarks and leptons are always of γ_5-type and therefore in the non-relativistic limit lead to spin-dependent $1/R^3$ type forces which are almost completely invisible between unpolarized macroscopic objects regardless of their strength. Thus models with spontaneously broken global symmetries are quite acceptable phenomenologically. Since the standard model of electro-weak interactions contains a large symmetry G in the gauge sector, where

$$G = U(1)_{B-L} \times SU(N_g)_L \times SU(N_g)_R^{(u)} \times SU(N_g)_R^{(d)} \times SU(N_g)_L^{\ell} \times SU(N_g)_R^{\ell}$$

a new area of investigation opened up in which attempts were made to break each of these global symmetries to resolve the various unresolved issues of the standard model such as possible non-vanishing neutrino mass, quark and lepton masses and mixings, etc. All these theories lead to a plethora of real Goldstone particles, whose phenomenological implications we wish to study in this paper. This review is organized as follows: in sec. 2 we discuss some general properties of the Goldstone bosons; in sec. 3 we discuss the properties of the Majoron,[1,2,3] which is the Goldstone boson associated with global lepton number symmetry; sec. 4 is devoted to discussion of astrophysical constraints on these models; in sec. 5 we briefly mention the models with spontaneously broken global family symmetry.[4] In sec. 6 we discuss a novel implication of models with real Goldstone bosons, which, when combined with the QCD θ-vacuum structure, lead to observable, attractive long range forces coupled approximately to the baryon number.

2. COUPLING OF GOLDSTONE BOSONS TO MATTER

Since the Goldstone bosons are associated with spontaneous breaking of a global symmetry, under the symmetry transformation by a parameter α, the goldstone boson field (to be denoted by χ henceforth) undergoes a shift:

$$\chi \to \chi + \alpha \tag{1}$$

Since the Lagrangian is invariant under the global symmetry transformations, the interactions of χ must respect this symmetry.[5] This has the following important implications:

a) The diagonal couplings of χ to fermions is always γ_5-type. In the non-relativistic limit, it leads to spin-dependent potential of type:

$$V(R) = \{\vec{\sigma}_1 \cdot \vec{\sigma}_2 - 3(\vec{\sigma}_1 \cdot \hat{R})(\vec{\sigma}_2 \cdot \hat{R})\}/M^2R^3 \tag{2}$$

where M is the scale of Global symmetry breaking. When we consider its contribution to the force between two unpolarized macroscopic objects, it becomes negligible, i.e. 10^{-46} (square of $N_{Avagadro}$ inverse) times the force between two polarized protons. Thus, the obvious concern about the existence of real Goldstone bosons is relieved. There are, however, atomic experiments due to Code and Ramsay,[6] which imply, $M \geq 100$ GeV.

b) A second implication of the above "shift" invariance is that, the Higgs coupling to χ must involve derivatives of χ-field and are therefore suppressed one power of mass. For instance, a typical coupling for Higgs decay to two Goldstone bosons is of the form: $H\partial_\mu\chi\partial_\mu\chi/M$. One can also prove using purely group theoretical arguments that the Goldstone bosons do not couple non-derivatively.[7]

3. PHENOMENOLOGY OF THE MAJORON MODELS

The first group of models[1,2,3] with Goldstone bosons were constructed with spontaneous breaking of lepton number. The associated Goldstone boson was called[1] Majoron. These models were constructed by simple extension of the standard model by adding new Higgs bosons with lepton number. These additional Higgs multiplets were chosen to have non-zero v.e.v.'s leading to the breakdown of global lepton number symmetry. Depending on whether the new Higgs multiplet transforms as a singlet, doublet or triplet under the $SU(2)_L$ group, the associated Majoron will be called the singlet,[1] doublet[3] or triplet[2] type. We introduce these models below and study their properties and possible experimental tests.

Let us remind the reader about the leptonic and Higgs sector of the standard model. It consists of the leptonic doublet $\ell \equiv (\nu, e^-)_L$ and right handed singlet e_R^- and the Higgs doublet $\varphi \equiv (\varphi^+, \varphi^0)$.

(i) The Singlet Majoron:[1] It consists of adding the right-handed neutrino, N_R, which has lepton number 1 and a singlet Higgs boson Δ^0 , with lepton number -2. If we include all possible couplings among the fields that respect gauge invariance, the model has exact B-L symmetry, which is spontaneously broken, when $\langle\Delta^0\rangle = M \neq 0$. The associated Majoron $\chi \equiv \mathrm{Im}\Delta^0$ couples in the tree approximation to right-handed neutrino with maximal strength (coupling parameter $h \sim 1$) and to left-handed neutrino with coupling of order (m_ν/M) . Coupling to charged leptons and quarks arise only in the one loop approximation and is given by:

$$f_{\psi\psi\chi} = \frac{G_F m_\nu m_\psi}{\pi} , \quad \psi = e,u,d \qquad (3)$$

$f_{\psi\psi\chi}$ is typically of order 10^{-15}-10^{-16} for ψ = e,u or d.

(ii) Triplet Majoron: Historically, the triplet Majoron model was constructed[2] soon after the singlet Majoron model. It consisted of adding only a triplet Higgs field to the standard model with Y = +2. It has a gauge invariant coupling to leptonic doublet of the form $h\ell^T\tau_2\vec{\tau}\cdot\vec{\Delta}\ell$. The global B-L symmetry is spontaneously by $\langle\Delta^0\rangle = V_T/\sqrt{2} \ll m_W$, leading to a Majorana neutrino mass $m_\nu = hv_T$. This model was investigated in great detail in ref. 8. For our purpose, we simply note that the Majoron $\chi \equiv (\mathrm{Im}\ \Delta^0 + \sqrt{2}\ v_T/\kappa\ \mathrm{Im}\varphi^0)$ $(\kappa = m_W/g)$ in this model, χ couples to left-handed neutrinos with maximal strength, whereas its coupling to charged leptons and quarks is given by $4G_F m_\psi v_T$, where $\psi = e^-$,u,d.

(iii) Doublet Majoron: It was pointed out[3] in 1982 that a novel possibility for spontaneous breaking of global lepton number arises in supersymmetric models using the scalar $SU(2)_L$ doublet which is the supersymmetric partner of the leptonic doublet. Denoting this doublet by $(\tilde{\nu},\tilde{e})$, it was noted that one can arrange to have $\langle\tilde{\nu}\rangle = v_\nu \neq 0$, leading to the doublet Majoron. It couples to all fermions with approximate strength of order $\approx 2\sqrt{2}\ G_F m_\psi v_\nu$, ψ = e,u,d, as in the case of the triplet Majoron.

Before proceeding to discuss the constraints on the v.e.v.'s v_T and v_ν , several phenomenological implications can be outlined. It is clear from the above discussion that matter couplings to Majoron are invisible except for the triplet case, which has maximal coupling to the left-hand neutrino and makes important contributions to neutrinoless double beta decay and corrections to universality in $\pi \to \ell\nu_\ell$ decay. In the case of $(\beta\beta)_{o\nu}$ decay, the diagram with Majoron emission has peak in its differential electron energy spectrum at a point where the $(\beta\beta)_{2\nu}$-energy spectrum has a very small value. This enables experimentalists[9] studying double β-decay to set precise limits on the parameter h, typically at the level of $h \leq 10^{-3}$. Similarly, corrections to universality in $\pi \to e\nu_e$ and $\pi \to \mu\nu_\mu$ decay enables[10] one to set limits on $h \leq 10^{-3}$.

An important test of the existence of doublet and the triplet Majoron will come from precise measurements of the width of the Z-boson at LEP. The reason is that Z can decay to triplet and doublet Majorons with widths, Γ(Z → Majoron) given by

$$\frac{\Gamma(Z \to \text{Majoron})}{\Gamma(Z \to \nu\bar{\nu})} = 2(\text{triplet}) \text{ or } \frac{1}{2}(\text{doublet}) \,. \quad (4)$$

Since LEP is expected to yield Z-width up to a precision of 50 Mev, the doublet and triplet Majorons (contributing to Z widths of order 135 Mev or 350 Mev respectively) can either be discovered or ruled out.

4. ASTROPHYSICAL CONSTRAINTS

The coupling of weakly interacting massless particles (WIMPS) such as these Goldstone bosons, are highly constrained by various astrophysical considerations as was first noted by Dicus, Kolb, Teplitz and Wagoner.[11] They pointed out that, once these particles are emitted by the electrons, protons and neutrons in the stars, because of their weak couplings, the star appears quite transparent to them and they escape, carrying away stellar energy. From the known rate of stellar energy loss, one can therefore bound the coupling of Majorons to electrons and to quarks. These bounds are roughly at the level of 10^{-12} for $f_{\phi\phi\chi}$ leading to bounds[12] on v_T and $v_\nu : v_{T,\nu} \lesssim$ 10 kev. Note that the singlet Majoron couplings being of order 10^{-15}-10^{-16} automatically satisfy these couplings. Similar constraints also arise from the study of hydrogen burning in the sun.[13] It is amusing to note that $m_{\nu_e} = h\, v_T \leq 10$ ev.

A second cosmological constraint that correlates the number of Goldstone particles with its coupling to electrons and neutrinos arise from nucleo-synthesis. The point is that the precise temperature T_D for decoupling of weak interactions processes involving neutrinos is a major determining factor in the ratio of helium to hydrogen in the universe. This temperature T_D, on the other hand, depends on the expansion rate of the universe, which is a function of the number of particle species in equilibrium at the epoch of nucleosynthesis. A particle out of equilibrium can also contribute to the expansion rate at a reduced level depending on when it went out of equilibrium. Since the triplet Majoron has maximal strength of interaction with neutrinos, it is in equilibrium at $T \simeq 1$ Mev and contributes as two independent Bose particles or almost as much as a neutrino. To the extent that present Helium abundance allows for only one extra species of neutrino, the triplet Majoron can barely be accomodated. The singlet and doublet Majorons on the other hand are so weakly coupled to matter that as many as 24 of them can be allowed without any conflicts with cosmology.[14] This question becomes particularly important for the case of spontaneous breaking of global family symmetry, which we touch on briefly now.

5. THE FAMILON

The same ideas have been extended in ref. 4 to discuss spontaneous breaking of global horizontal symmetry; the associated Goldstone particles are called familon. As yet, there exist no fully realistic model for the familon (in the sense of realistic fermion masses and mixings); nevertheless, the general qualitative idea leads to predictions for new decay modes such as μ → e + familon, K → π + familon. Recent LBL-Triumf experiments on muon decay looked for the first decay mode and put bound on the scale of family symmetry breaking to be bigger than[15] 6.5×10^9 Gev. The cosmological constraints on the number of Goldstone bosons can limit the nature of family symmetry.

6. LONG RANGE 1/R-TYPE FORCES INDUCED BY GOLDSTONE BOSONS

A rather interesting implication of Goldstone bosons is that, in the presence of strong CP-violating parameter θ, the diagonal γ_5-coupling of Goldstone particles gets converted into a scalar coupling, which can, then, lead to attractive spin independent long range forces, with finite range and strength about a thousand times weaker than gravity.[16] The strength of the scalar Goldstone boson coupling f_s to matter is of order

$$f_s \simeq f_{\psi\psi\chi} \cdot \theta \left(\frac{\mu}{m_\psi}\right) \qquad (5)$$

where $\psi = u,d$ and $\mu \approx m_u m_d/(m_u+m_d)$. Using the bound on the strong CP-parameter $\theta \leq 10^{-9}$, we find for triplet Majoron, $f_s \leq 10^{-21}$, leading the strength of the new force of about $\sim 10^{-42}$. The scalar interaction, induces a mass for the Goldstone boson of order $f_s \cdot m_q/4\pi \approx 10^{-22}$ Gev $\simeq 10^{-12}$ ev corresponding of about 1000 km. Thus, these forces are negligible on the scale of distances between the earth and the sun but are important for Eötvös type experiment that tests equivalence principle in the scale of distances of the order the radius of the earth.

Another interesting aspect of this new result is the prediction of T-violating spin-dependent forces of $\sigma \cdot \hat{R}/R^2$ type with strength of order 10^{-34} Gev^{-1}. It may be pointed out that present bounds[17] on forces of this type arise from the precision of frequency measurement of the 21 cm line of hydrogen and is at a level of 10^{-27} Gev^{-1}. Therefore, search for these new forces are also likely to throw light on the reality of these Goldstone bosons.

REFERENCES

1. Y. Chikasige, R.N. Mohapatra and R.D. Peccei, Phys. Lett. 98B, 265 (1981).
2. M. Gelmini and M. Roncadelli, Phys. Lett. 99B, 411 (1981).
3. C.S. Aulakh and R.N. Mohapatra, Phys. Lett. 119B, 136 (1982); 121B, 147 (1983).
4. D. Reiss, Phys. Lett. 115B, 217 (1982); F. Wilczek, Phys. Rev. Lett. 49, 1549 (1982); G. Gelmini, S. Nussinov and T. Yanagida, Nucl. Phys. B219, 31 (1983).
5. For a discussion of this see R.N. Mohapatra, "Unification and Supersymmetry," Springer-Verlag (1986), Ch. 2, p. 26-27.
6. R. Code and N.F. Ramsay, Phys. Rev. A4, 1945 (1971).
7. R. Barbieri, R.N. Mohapatra, D. Nanopoulos and D. Wyler, Phys. Lett. 107B, 80 (1981).
8. H. Georgi, S.L. Glashow and S. Nussinov, Nucl. Phys. B193, 297 (1981).
9. For a recent review of the situation, see D. Caldwell, Proceedings of ν' 86 Conference, held in Sendai, June (1986).
10. V. Barger, W.Y. Keung and S. Pakvassa, Phys. Rev. D25, 907 (1982); T. Goldman, E. Kolb and G. Stephenson, Jr., Phys. Rev. D26, 2503 (1982).
11. D.A. Dicus, E. Kolb, V. Teplitz and R. Wagoner, Phys. Rev. D18, 1829 (1978).
12. M. Fukugita, S. Watamura and M. Yoshimura, Phys. Rev. Lett. 48, 1522 (1982).
13. D. Dearborn, G. Steigman and D. Schramm, Phys. Rev. Lett. 56, 26 (1986).
14. G. Gelmini, et al., ref. 4; D. Chang, P. Pal and G. Senjanović, Phys. Lett. 153B, 407 (1985).
15. H. Steiner, Proceeding of Moriond Workshop (1985), on "Flavor Mixing and CP-Violating," ed. by J. Tran Than Van, Editions Frontieres, France, p. 395.
16. D. Chang, R.N. Mohapatra and S. Nussinov, Phys. Rev. Lett. 55, 2825 (1985).
17. J. Leitner and S. Okubo, Phys. Rev. 136, B1542 (1964).

*Work supported by a grant from the National Science Foundation.

LOW ENERGY TESTS OF SUPERSYMMETRIC MODELS

A. Masiero

Physics Department, New York University
New York, N.Y. 10003

We study the implications for spontaneously broken N=1 supergravity theories of three classes of rare processes for which new experiments are planned: $K^{\pm}\to\pi^{\pm}$+ "nothing" (extending the analysis to the heavier mesonic systems B and D), electron-and muon-lepton number violating reactions and the electric dipole moment of the electron. In particular we show that in supersymmetric models where neutrinos get a Majorana mass through the "see-saw" mechanism we expect experimentally interesting rates for $\mu\to e\gamma$ and μ-e conversion in nuclei.

There are several proposed or planned experiments to improve the bounds on low energy rare processes. While waiting for the forthcoming high energy machines, these low energy tests could provide some clue on the eagerly looked for new physics. We examine here the implications for low energy supersymmetric (SUSY) models from i) $K^{+}\to\pi^{+}$+ "nothing" (with extensions to the heavier B and D mesonic systems), ii) muon- and electron-lepton number violating processes, such as $\mu\to e\gamma$, $\mu\to ee\bar{e}$ and $\mu\to e$ conversion in nuclei and iii) the electric dipole moment of the electron (d_e). We perform our analysis in the context of minimal spontaneously broken N=1 supergravity models (1). The key ingredient in the analysis of the processes i) and ii) is that in these SUSY models there exist tree level flavour changes in the neutralino-sfermion-fermion vertices (2), where the neutralinos are the fermionic partners of the neutral vector or scalar bosons and the sfermions are the scalar partners of the fermions. This peculiarity of SUSY models is a consequence of the different renormalization effects on the fermion and sfermion mass materices: due to the presence of soft breaking terms of the residual N=1 global SUSY, after renormalization down to the Fermi scale, the sfermion mass matrices are generally not diagonal in the basis in which the corresponding fermion mass matrices are diagonal. The mixing angles which appear at these flavour changing vertices in the minimal models that we are considering are just the familiar angles of the Kobayashi-Mashawa matrix.

The BNL experiment (3) of this year plans to reach a sensitivity of 10^{-10} for the branching ratio $K^{+}\to\pi^{+}$+ "nothing". Since the Standard Model (SM) with three generations of fermions predicts the BR for $K^{+}\to\pi^{+}\nu\bar{\nu}$ to be less than 10^{-10}, a positive evidence in the BNL experiment would signal the presence of new physics. Low energy SUSY can play a twofold role in enhancing the $K^{+}\to\pi^{+}$+ "nothing" decay rate (4): a) contributions to $K^{+}\to\pi^{+}\nu\bar{\nu}$ through the exchange of virtual SUSY particles; b) direct emission of a pair of invisible SUSY particles, for instance $K^{+}\to\pi^{+}\tilde{\gamma}\tilde{\gamma}$ ($\tilde{\gamma}$≡photino). Making use of the abovementioned gluino($\tilde{g}$)-quark-squark flavour changing vertices, one can readily construct superpenguin diagrams which provide a leading a) contribution. In spite of the presence of the strong couplings at the gluino vertices, these contributions do not exceed the SM contribution with three generations. There are two reasons: the superGIM cancellation due to the exchange of squarks of different generation is quite efficient and, more important, the superpenguins must contain some seed of SU(2)xU(1) breaking in order to avoid the q^2/M_Z^2 suppression (with $q^2=m_k^2$) in the Z^o propagator. This entails the presence of $\tilde{q}_L-\tilde{q}_R$ mixing terms which are proportional to m_q. Other superpenguins with charged higgsino exchange and diagrams coming from the direct supersymmetrization of the SM contributions also fail to yield a BR larger than 10^{-10}.

On the other hand, SUSY can play a major role in enhancing the $K^{+}\to\pi^{+}$+ "nothing" rate if "nothing" denotes a pair of light photinos. The presence of photino-quark-squark flavour changing vertices, allows for the $K^+ \to \pi^+ \tilde{\gamma}\tilde{\gamma}$ with the exchange of a Q= - 1/3 squark. The amplitude is:

$$A(K^+ \to \pi^+ \tilde{\gamma}\tilde{\gamma}) = \frac{e^2}{g} \frac{c\ m_t^2}{m^4 + c\ m_t^2 m^2} U^*_{ts} U_{td}$$

$$(\bar{s}_L \gamma_\mu d_L)\ (\bar{\tilde{\gamma}}\gamma^\mu \gamma^5 \tilde{\gamma}), \qquad (1)$$

where m_t denotes the top mass, m is the scale of low energy supersymmetry breaking, U_{ts} and U_{td} are the corresponding entries in the Kobayashi-Maskawa matrix and, finally, c is a model dependent parameter that we take here to be 0.1. The presence of cm_t^2 in (1) comes from the SuperGIM suppression: the exchange of $\tilde{b}$ and $\tilde{s}$ yields a factor $m_{\tilde{b}}^2 - m_{\tilde{s}}^2 \simeq cm_t^2$. Obviously the major objection to $K^+ \to \pi^+ \tilde{\gamma}\tilde{\gamma}$ is that it is theoretically quite unlikely that $\tilde{\gamma}$ is so light as to allow kinematically for this process. However, experimentally, we cannot rule out this possibility so far. If the process exists, already the present experimental bound ($BR(K^+ \to \pi^+$ + "nothing" $< 10^{-7}$) implies down squark masses $\gtrsim$ 30 GeV and a negative result of the BNL experiment would put a lower bound of at least 70 GeV on these masses. This same kind of decays may be more interesting for heavier mesonic systems. For instance, taking $m_{\tilde{\gamma}}$=1GeV and m=40GeV, the decays of B^+ and B^0 into $\tilde{\gamma}\tilde{\gamma}$ + hadrons with s quark yield (5):

$$BR\ (B^+ \to X^{(s)} \tilde{\gamma}\tilde{\gamma}) \simeq 9.10^{-4} \qquad (2)$$

$$BR\ (B^0 \to X^{(s)} \tilde{\gamma}\tilde{\gamma}) \simeq 7.10^{-4}$$

These values might be of interest for the forthcoming machines. The analogous decays for the D system are more suppressed.

Coming now to the electron-and muon-lepton numbers (L_e and L_μ), it is immediate to see that, analogously to what happens in SM, in its usual supersymmetrization with conserved R-parity (SSM), L_e and L_μ are conserved to any order in perturbation theory. However, if we supersymmetrize extensions of SM obtained by embedding SM in a unified scheme or simply by adding new particles and couplings in general we expect violations of L_e and L_μ (6). They become significantly violated and, indeed, within the reach of the present or planned experiments in extensions of SSM (7) where light Majorana neutrinos are obtained involving the "see-saw" mechanism. This mechanism is implemented by adding to the usual SSM superpotential the two terms h LHN and MNN, where N is a superfield neutral under the SM gauge symmetry and L and H contain the leptonic isodoublet and the Higgs isodoublet, respectively. The h LHN term gives rise to the "Dirac entry" in the neutrino (ν) mass matrix, $h\bar{\nu}_L \nu_R \langle H^0 \rangle \equiv m_\nu^D \bar{\nu}_L \nu_R$, whereas MNN generates the NN Majorana entry of order M. The light mainly left-handed Majorana neutrino has a mass $m_\nu \simeq (m_\nu^D)^2/M$. Clearly m_ν^D is much larger than m_ν. For instance, to get m_ν=1eV, m_ν^D should be ~ 300 GeV is some grand unified scheme where M $\simeq 10^{14}$ GeV. The presence of h LHN induces a renormalization effect on the mass of the slepton $\tilde{\ell}_L$ proportional to hh^+, i.e. to $m_\nu^D m_\nu^{D+}$. The $\tilde{\ell}_L^+ \tilde{\ell}_L$ entry becomes:

$$\tilde{\ell}_L^+\ (m^2 + m_\ell m_\ell^+ + c\ m_\nu^D m_\nu^{D+})\ \tilde{\ell}_L, \qquad (3)$$

where m_ℓ is the lepton mass matrix and we take c = 0.5. From (3) it is apparent that when we work in the basis in which m_ℓ is diagonal, the matrix $\tilde{\ell}_L^+ \tilde{\ell}_L$ is not diagonalized and the off-diagonal amount responsible for the leptonic flavour change is $\Delta = c\ U\ m_\nu^D m_\nu^{D+} U^+$, where U is the unitary matrix which diagonalizes $m_\ell m_\ell^+$. Here comes the crucial point: the L_e and L_μ violations turn out to be proportional to m_ν^D (not m_ν, as in models with Dirac neutrinos). We obtain:

$$BR\ (\mu \to e\gamma) = \frac{\alpha^3}{G_F^2}\ 12\pi\ \frac{\Delta_{12}^2 \cdot F(x)^2}{m^8}, \qquad (4)$$

where

$$F(x) = \frac{1}{12} \frac{1}{(1-x)^5} \{17x^3 - 9x^2 - 9x + 1 - 6x^2(x+3)\ln x\},$$

$$x = \frac{m_{\tilde{\gamma}}^2}{m^2}$$

Taking, for definiteness, $\Delta_{12} = (U^+ m_\nu^D m_\nu^{D+} U)_{12} \simeq 0.2.\ m_\nu^{D2}$ and F(x) = 1/20, even for m as large as 150 GeV, we still get BR ($\mu \to e\gamma$) $\simeq$ $10^{-12} \div 10^{-13}$ for m_ν^D ~ 10÷20GeV. The present experimental bound, BR ($\mu \to e\gamma$) < $4.7.10^{-12}$, for the above choice of the parameters implies that m_ν^D<30GeV for m $\simeq$ 100 GeV. For $\mu \to eee$ and R_{en} ($R_{en} \equiv w(\mu N \to eN)/w(\mu N \to \nu N)$), we obtain:

$$R_{en}/BR(\mu \to e\gamma) \simeq 0.023,\quad R_{en}/BR(\mu \to eee) \simeq 26.42 \qquad (5)$$

The ratios in (5) are independent from Δ_{12}; more over the $x=m_{\tilde{\gamma}}^2/m^2$ dependence disappears in $R_{en}/BR(\mu\to ee\bar{e})$. Taking into account that the present bounds on R_{en} and $BR(\mu\to ee\bar{e})$ are $R_{en} < 4.5.10^{-12}$ and $BR(\mu\to ee\bar{e}) < 2.4.10^{-12}$, we see that in particular $\mu\to e\gamma$ and $\mu N\to eN$ are excellent tests for the class of SUSY models that we are considering. In SO(10) $m_\nu^D = m_u$ ($m_u \equiv$ up quark mass matrix) at the grand unification scale. Unfortunately, due to the smallness of the product $U_{23}\ U_{31}$, where U now is the Kobayaski-Maskawa matrix, no interesting upper bound on the top mass can be inferred from the present experimental bound on $\mu\to e\gamma$ (I thank L. Hall for pointing this out to me).

Fortson and his collaborators have recently proposed to improve the current bound (10^{-24}ecm) on d_e by up to four orders of magnitude. In SUSY models there can be one loop contributions to d_e with the exchange of a $\tilde{\gamma}$ and a $\tilde{e}$ in the loop (8). The helicity flip is realized by the photino mass insertion, $m_{\tilde{\gamma}}$. Consequently, in the $\tilde{e}$ propagator a mass insertion $m^2_{\tilde{\ell}_L\tilde{\ell}_R}$ must be present. Both $m_{\tilde{\gamma}}$ and $m^2_{\tilde{\ell}_L\tilde{\ell}_R}$ are in general complex parameters and thus they can be the necessary source of CP violation to give rise to a nonvanishing d_e. Taking $m_{\tilde{\ell}} \sim m_{\tilde{\gamma}} \sim$ 100GeV, we find $d_e = 2.10^{-24}$ecm. times the CP violating phases. If we assume that $m_{\tilde{g}} \simeq m_{\tilde{\gamma}}$ and the same CP violation is present for d_e and the electric dipole moment of the neutron, d_n, then from the experimental bound on d_n, we can infer $d_e < 4.10^{-27}$ ecm. If no extra CP violating phase is present in the SUSY version apart from the usual Kobayashi-Maskawa phase, then d_e is in general quite small, with the possible remarkable exception of the SUSY models with Majorana massive neutrinos that we have discussed in connection with the L_e and L_μ violation.

Acknowledgements

I thank my collaborators S. Barr, S. Bertolini, F. Borzumati, G.F. Giudice and A. Sanda for sharing with me their insights on the topics of this talk. Work supported by the NSF under Grant No. PHY 8116102.

References

1) E. Cremmer, S. Ferrara, L. Girardello and A. van Proeyen, Phys. Lett. 116B (1982) 231; Nucl. Phys. B212 413 (1983)

2) M.J. Duncan, Nucl. Phys. B221 285 (1983) J.F. Donoghue, H.P. Nilles and D. Wyler, Phys. Lett. 128B, 55 (1983)

3) Y. Asano et al., Phys. Lett. 107B, 159 (1981); L.S. Littenberg, Report No. BNL - 35086 (unpublished)

4) S. Bertolini and A. Masiero, Phys. Lett. 174B, 343 (1986)

5) G.F. Giudice, preprint NYU/TR6/86.

6) L.J. Hall, V.A. Kostelecky and S. Rabi, preprint HUTP - 85/A063

7) F. Borzumati and A. Masiero, Phys. Rev. Lett. 57, 961 (1986)

8) F. del Aguila, M.V. Gavela, J.A. Grifols and A. Mendez, Phys. Lett. 126B, 71 (1983);
S. Barr and A. Masiero, BNL preprint (1986).

SUPERSTRING INSPIRED MODELS AND PHENOMENOLOGY

G.G. Ross
Department of Theoretical Physics, University of Oxford
1 Keble Road, Oxford
and
Rutherford Appleton Laboratory, Chilton, Didcot
Oxon, UK

An investigation of the effective low-energy theory resulting from the superstring is given. The possible light gauge and chiral supermultiplet structure is considered and a specific model leading to a SU(3)xSU(2)xU(1) gauge group is presented. Phenomenological implications for such models are briefly discussed.

1. Introduction:

Superstring theories[1] offer the prospect of a consistent, finite, theory for the strong, electromagnetic, weak and gravitational interactions and may be the ultimate "Grand Unified Theory". However considerable obstacles lie between their elegant theoretical structure, relevant at the Planck scale, and their physics at a low scale, relevant for laboratory experiments and crucial if there is ever to be a test of the validity of such theories. The most obvious difficulty follows since superstrings are defined in a world with ten space-time dimensions and it is necessary to break this space-time symmetry in such a way as to leave just four dimensions at low scales. This process of compactification requires the introduction of a manifold describing how six dimensions are curled up on the compactification scale, M_c, (probably close to the Planck scale[2]). Once the manifold is specified the multiplet structure of the states light after compactification is determined as are all the couplings in the theory in terms of a single gauge coupling. In principle, at least, the structure of the low energy world is determined by physics at the Planck scale.

However, our understanding of compactification is not yet good enough to determine this manifold on energetic grounds so, at present, we must make a specific choice to make definite predictions about the low energy world; the uncertainty in this choice giving a corresponding uncertainty in the low energy predictions. Moreover there must be several stages of spontaneous breakdown of the internal symmetries of the model between the Planck scale and the laboratory scale in order to give a realistic theory and this also makes it more difficult to extract definite predictions from the superstring. In this talk, I will concentrate on an analysis of the breaking of space-time and internal symmetries to be expected in superstring models in order to discuss what may be expected in our low-energy world. I will concentrate on the question whether this can be close to the phenomenologically successful "standard model" and, if so, what predictions beyond the standard model result.

2. Compactification and Flux breaking

The most promising compactification schemes to date are based on Calabi-Yau manifolds[3]. These leave unbroken an N=1 supersymmetry in the four dimensional world, useful in maintaining an hierarchy between the electroweak breaking scale and the Planck scale. The first attempts[3] identified the spin connection of the six compactified dimensions with an SU(3) holonomy group embedded in the gauge group. As a result the $E_8 x E_8$ gauge symmetry of the superstring in ten dimensions is reduced to $E_6 x E_8$ in four dimensions. It is thought that the E_8 sector will be confining, with no low energy relics[4], leaving the visible world to come from the E_6 gauge group sector. The number of E_6 representations left light after compactification are determined in terms of Hodge numbers of the compactification manifold for they can (apart from some E_6 singlet fields) be identified with forms on the compactification manifold[2 5]. This is shown in Table 1, and may be seen to give an effective low energy theory with an E_6 gauge group, $N_g = \frac{1}{2}|\chi| = \frac{1}{2}|h^{1,2} - h^{1,1}|$ generations of chiral superfields transforming as the 27, and $h^{1,1}$ copies of$(27+\overline{27})$ representations together with further E_6 singlet fields. However, this is not a suitable starting point for an acceptable phenomenology since spontaneous breaking of E_6 by scalars transforming as

27 or 27 will leave at least an SU(5) group unbroken down to the electroweak scale giving unacceptable rapid proton decay etc. For non-simply-connected manifolds there is a way out, for there may be field configurations corresponding to non-trivial Wilson lines of flux round the holes in such manifolds which further break the E_6 gauge group at the compactification scale[6]. For example, the manifold $K=K_0/G$, where K_0 is simply connected and G is a freely-acting discrete group, can give rise to non-trivial flux loops transforming as a representation $\bar{G}$ of G in E_6. The residual light states are $G+\bar{G}$ singlets, leading to a reduction in the gauge group for non-trivial G. The minimal group left unbroken after flux breaking has rank five[7], if it is to include the standard model, and thus the low energy structure will include, at least, one new (neutral) gauge boson.

Much work[8] has gone into parameterizing the effect of this boson and analysing neutral current data but it should be emphasised that there is no evidence for any additional gauge structure beyond the standard model. In the next section we will discuss whether such additional structure is a necessary feature of low-energy models following from the superstring.

There have been several other schemes suggested for compactification which may lead to viable low energy models. Starting with Calabi-Yau spaces it is possible to envisage field configurations which break E_8 to SO(10) or SU(5), potentially more reasonable starting points[9,10]. Simple schemes of this type are unstable to non-perturbative corrections[11], but I am assured that this is not a general feature[12]. At present there are no known examples of such SO(10) or SU(5) models, so the predictive power following from choice of a definite manifold is lost. Another promising compactification scheme involves the use of orbifolds[13], formed by dividing a manifold by a non-freely acting group. However, as far as I know, no phenomenologically realistic models of this type have yet been constructed and I will not consider orbifold models further here.

3. Intermediate scale breaking

It is possible that the gauge group will be broken further by scalar components of the chiral super-multiplets, left light after compactification and flux breaking, acquiring vacuum expectation values (vevs)[14,15]. Such vevs break supersymmetry so they require supersymmetry breaking to develop and it is thought that supersymmetry breaking will be triggered by a gaugino condensate in the E_8 "hidden" sector[4]. This supersymmetry breaking will then be communicated to the E_6 "visible" sector by gravitational corrections[16]. This is the aspect of the theory most poorly understood and, in the absence of a realistic calculation of these effects, I will just assume that the light states in the theory may acquire supersymmetry breaking masses m_i; the magnitude of these masses are limited by the hierarchy problem to be $\lesssim O(1\ \text{TeV})$, otherwise electroweak breaking will occur at too large a scale. The resulting potential for light scalar components ϕ_i in the compactified four dimensional world has the form

$$V = \sum_i m_i^2 \phi_i^+ \phi_i + \sum_i \left|F_{\phi_i}\right|^2 + \left|D\right|^2 \tag{3.1}$$

If m_i^2 is negative, a vev for ϕ_i will develop. The magnitude of this vev depends on the stabilising D and F terms. If there are light 27 and $\overline{27}$ fields then there will be an energetically favourable, D-flat, direction with $\langle\phi_{27_i}\rangle = \langle\phi_{\overline{27}_i}\rangle$ giving

$$|D|^2 = \frac{1}{2} \langle \sum_\alpha \left| g_\alpha \sum_i (\phi^+_{27_i} \lambda^\alpha \phi_{27_i} - \bar{\phi}^+_{\overline{27}_i} \lambda^\alpha \bar{\phi}_{\overline{27}_i}) \right|^2 \rangle = 0 \tag{3.2}$$

Along this direction the stabilising term is the F_{term}, $\sum_i |F_{\phi_i}|^2 = |\frac{\partial P}{\partial \phi_i}|^2$,

where the superpotential, P, is an effective superpotential describing the theory below the compactification scale

$$P=\alpha_1 \phi_{27}^3 + \alpha_1' \phi_{\overline{27}}^3 + \frac{\alpha_2}{2M_c} \phi_{27}^2 \phi_{\overline{27}}^2 + \frac{\alpha_3}{3M_c^2} \phi_{27}^3 \phi_{\overline{27}}^3 + ..$$

The first two, renormalisable, terms have vanishing F terms along the direction $\langle[\phi_{27}]_N\rangle = \langle[\phi_{27}]_N\rangle$ where the subscript denotes the E_6 component, singlet under a SO(10) subgroup of E_6. Along this direction the third (non-renormalisable) term is the first stabilising term giving

$$V=m^2[\phi_{27}]_N^2 + m'^2[[\phi_{\overline{27}}]_N^2 + \frac{1}{2} g^2 |[\phi_{27}]_N^2 - [\phi_{\overline{27}}]_N^2|^2 \tag{3.4}$$

$$+\left|\frac{\alpha_2}{M_c} [\phi_{27}]_N [\phi_{\overline{27}}]_N^2\right|^2 + \left|\frac{\alpha_2}{M_c} [\phi_{27}]_N^2 [\phi_{\overline{27}}]_N\right|^2 + \ldots$$

For m^2 negative this has a minimum

$$\langle[\phi_{27}]_N\rangle = \langle[\phi_{\overline{27}}]_N\rangle = \left[\frac{m^2 M_c^2}{6\alpha_2^2} \right]^{1/4},$$

which for $\alpha=O(1)$, $m=O(1\ \text{TeV})$, $M_c=O(M_{\text{Planck}})$ gives $\langle[\phi_{27}]_N\rangle = O(10^{11}\ \text{GeV})$.
The point of all this is that the gauge group may be broken at an intermediate scale (I.S) much larger than M_W, provided after compactification there are left light $\phi_{\overline{27}}$ components allowing for D flat directions. This may then reduce the rank of the low energy gauge group. However,

when we consider flux breaking, we find that if flux breaking initially reduces the rank to five there are no $\phi_{\overline{27}}$ components left light if $h_{1,1}=1$, the minimum possible in Calabi Yau models[15]. In this case there is a unique low-energy gauge group SU(3)xSU(2)x U(1)xU(1)', and the U(1)' has definite couplings[7]. The phenomenology of this model has been extensively discussed[17]. If the flux breaking does not reduce the rank, then there are ϕ_{27} components left light and IS breaking can reduce the rank to five. If $h_{1,1}=1$, there are not further $\phi_{\overline{27}}$ left light, and no further IS breaking is possible.

If $h_{1,1}>1$, then IS breaking may reduce the rank of the group to four (or even less!) giving the standard model SU(3)xSU(2)x U(1)[15]. This means that new neutral currents are not an inevitable consequence of superstring models.

The scale of IS breaking plays an important role in the low energy phenomenology of proton decay and neutrino masses. From eq (3.4) we argued that it should by $O(10^{11}$ GeV), but much higher scales are possible in certain models. One of the features of the effective theory following from compactification of higher dimensions is that there are residual discrete symmetries determined by the manifold. These discrete symmetries may forbid the α_2 term in eq (3.4) for certain multiplets, and in this case the first stabilising term will occur at $O(\phi^3_{27}\phi^3_{\overline{27}})$ leading to an IS vev of $O(10^{14}$ GeV) for $\alpha_3=O(1)$[15].

Even higher scales are possible if our assumption $\alpha_1=O(1)$ is incorrect[11]. The higher dimension terms in P are generated by radiative corrections and also at tree level via heavy Kaluza-Klein mode exchange. The former are small, vanishing when supersymmetry is unbroken. The latter may vanish in certain directions to all orders in perturbation theory[9]. In this case the dominant contribution to α_2 will be via non-perturbative effects which give $\alpha_2=O(\exp \frac{1}{g}2)$[11]. These can be very small allowing for a much higher IS than estimated above. Another effect which leads to higher IS may arise if there is a cancellation of different contributions to the F term from different scalar fields. Such a cancellation usually requires a relation between various Yukawa couplings but, since these couplings are in turn determined by continuous parameters specifying the manifold, minimisation of the potential will automatically select just that manifold with the (necessary Yukawa for) cancellation and corresponding large IS vev. To summarise, it has been found that a combination of IS breaking and flux breaking may reduce the rank of the original E_6 gauge symmetry by one or more. This means the low energy effective theory following from the superstring may just be the standard model, the breaking from E_6 may occur at a scale which may be close to the compactification scale.

Our discussion in this section has assumed that only 27 and $\overline{27}$ fields are present in the low energy theory. However from Table 1, we see that E_6 singlet fields should also be light. Terms in the superpotential of the form $1 27 \overline{27}$ can spoil the F flatness needed for large IS breaking, but analysis of specific models shows that even with such terms large IS breaking is possible following as a result of the discrete symmetries and cancellations discussed above[18].

4. A superstring-inspired standard model

The form of the low-energy effective theory depends on the pattern of space-time and internal symmetry breaking. This, in turn, should be determined by the underlying theory as it corresponds to the vacuum state of lowest energy. Unfortunately we are not yet able to determine this state so, in exploring the structure to be expected at low energies, the best we can do at present is to assume a definite compactification manifold, and use its structure to determine the subsequent stages of symmetry breaking.

Although there are an infinite number of possible compactification manifolds, reasonable criteria rapidly reduce this to a small number of viable candidates. As discussed in the introduction, the requirement that there should be a low energy N=1 supersymmetry in four dimensions points towards Calabi-Yau manifolds[3]. To reproduce our low-energy world we should look for manifolds with three generations (up to four generations may be possible, but they have problems with $\sin^2\Theta_W$). Only a small number of three-generation Calabi-Yau manifolds exist[19]; three explicit examples were constructed by Yau[20]. Of these three only two can accommodate flux breaking and lead to viable low energy models. I wish to report on a detailed study of one of these models that we at Oxford have carried out[18,21].

The initial light multiplet structure before flux or intermediate scale breaking is given in Table 1 with $h^{1,2}=9$ and $h^{1,1}=6$. In addition to the E_6 symmetry, there are a number of discrete symmetries[23] which follow from the manifold and which we have determined[21]. These limit the possible couplings in the superpotential and have a bearing on the existence of F flat directions and hence the scale of I.S. breaking. In addition they place constraints on the light fermion (quark and lepton) mass matrices. The analysis of the model proceeds in three steps:-

We first identify a flux breaking pattern $E_6 \to [SU(3)]^3$, and establish that this still leaves light the leptonic components of the six $(27+\overline{27})$ chiral multiplets. This allows for D flat directions and, using the discrete symmetries, we show there exist F flat directions which will allow for further breaking of the gauge group at an I.S. $\gtrsim O(10^{14}$ GeV) reducing the rank by one. After this first stage of I.S. breaking we find there still remain light leptonic components of the $(27+\overline{27})$ chiral multiplets allowing for yet another stage of I.S. breaking. Thus the pattern of symmetry breaking at a high scale is

$$E_6 \xrightarrow[\text{Breaking}]{\text{Flux}} [SU(3)]^3 \xrightarrow[\langle N\rangle_{27}=\langle\bar{N}\rangle_{\overline{27}}]{} SU(3)xSU(2)xSU(2)xU(1) \xrightarrow[\langle\nu_R\rangle_{27}=\langle\bar{\nu}_R\rangle_{\overline{27}}]{} SU(3)xSU(2)xU(1) \qquad (4.1)$$

leading to an effective gauge theory below the IS which is just the standard model.

Having established the pattern of symmetry breaking, it is straightforward to work out which states can become massive as a result of the gauge and Yukawa couplings allowed by the symmetries of the theory. We find that the states left light are just those chiral multiplets needed to contain three families of quark and leptons together with the two SU(2) doublets of Higgs scalars needed for electroweak breaking. The former is what is expected, since we started with a manifold with three generations, but the lightness of the Higgs states results only as a result of a discrete symmetry following from the manifold. Of the vast number of other states in the original theory given in Table 2, most are superheavy, of the order of the IS ($\gtrsim 10^{14}$ GeV) or, from $27^2\overline{27}^2$ terms in eq (3,3), of order $(IS)^2/M_c$ ($\gtrsim 10^{11}$ GeV). The only new states, in addition to those needed for the minimal supersymmetric standard model, which are relatively light (of O(1 TeV)) are a charged (SU(2) singlet) lepton and two neutral lepton supermultiplets. As we will discuss in the next section, this means that the phenomenology of this model will be very similar to that of the minimal supersymmetric standard model.

The final stage of analysis is to construct the quark and lepton mass matrices. Since we know the composition of the light quarks and leptons and the Higgs fields we may write down the Yukawa couplings allowed by the discrete symmetries of the model and hence determine the structure of the mass matrices. Diagonalising these matrices then yields the quark and lepton masses and the Kobayashi Maskawa mixing matrix. This is given in Table 2. Remarkably, the pattern emerging is quite consistent with the experimental measurements for reasonable choices of parameters. There is even a prediction for a combination of the KM angles. The angle Θ_1, almost the Cabibbo angle, is given by

$$\Theta_1 = \sqrt{\frac{m_d}{m_s}} - \cos\delta\,\frac{\Theta_2}{\Theta_2}$$

Experimentally[22] Θ_2/Θ_3 is consistent with zero ($\frac{\Theta_2}{\Theta_3} < 0.14$ at 1σ), giving Θ_1 consistent with the experimental value.

This illustrates the predictive power of superstring models. Since there are, in principle, no arbitrary couplings[24], the light masses and mixing angles should be predicted. At present we have only been able to use the information coming from the discrete symmetries but work is proceeding to try to evaluate the Yukawas for this particular manifold.

5. Phenomenological implications of superstring-inspired models

In this section we concentrate on models with an underlying E_6 symmetry[25] before flux breaking. Models with SO(10) or SU(5) symmetry will have fewer new states, with a corresponding reduction in the new phenomena to be expected.

At low energies, as we have discussed, there may be new states associated with the original four-dimensional E_6 symmetry. The low energy gauge group will be a sub-group of E_6, the main phenomenological constraint being that the new light gauge bosons should not mediate proton decay. Detailed studies of the allowed structures have been carried out, and analyses of neutral current data[8,17] show that, while there is absolutely no need for an additional neutral current, experimental data is consistent with the existence of a light gauge boson within E_6, whose mass is only $\gtrsim O(100$ GeV)[8]. The reason such a low mass is possible is that, with a definite choice for the new neutral boson within E_6, the neutral current couplings to quarks are strongly suppressed. However, as we have emphasised above, after IS breaking the low energy group may just be the standard model SU(3)xSU(2)xU(1), so additional gauge boson structure is not a definite prediction of superstring inspired models.

There may be further low-energy structure corresponding to relics of the chiral super-multiplet transforming as 27 or $\overline{27}$ under E_6. Of course, the requirement that the low energy theory be supersymmetric means that there will be light superpartners of quarks, leptons and Higgs. At present, our poor understanding of the supersymmetry breaking mechanism[16] means the masses of the superpartners are not determined, but expected by $\lesssim O(1\ \mathrm{TeV})$. More interestingly, recent papers suggest that the Higgs sector is strongly constrained in superstring models with a light Higgs scalar of mass < 100 GeV.

In addition to the states of the minimal supersymmetric standard model, the relics of the 27 and $\overline{27}$ supermultiplets may leave light new charge $-\frac{1}{3}$ quarks, new charged and new neutral leptons. Much phenomenological work has been done studying their phenomenology. However, as our specific three generation model explicitly demonstrates, with large IS breaking the expectation is that most of such states will acquire large mass. The reason is that these new states may acquire large SU(3)xSU(2)xSU(1) invariant mass and will in general do so, unless there are additional gauge or discrete symmetries protecting these states. In our example of section four, there is one additional light charged lepton, because discrete symmetries forbid it from acquiring a mass until $O\left(\frac{27^3\overline{27}^3}{M_{Planck}^3}\right)$ in the superpotential.

Finally, we turn to the question of new phenomena in superstring models. The most important are baryon- and lepton-number violation and neutrino masses and E_6 based models have particular problems with these processes due to the new states in a 27 or $\overline{27}$ supermultiplet.

All supersymmetric models suffer from possible unsupressed contribution in B or L violation from dimension four operators[27]. In realistic models it is necessary to forbid these operators by a discrete symmetry such as matter parity. In superstring models the constraint is much more severe as the discrete symmetry should come from the structure of the manifold or from the gauge symmetry and not be input by hand. Remarkably, there are examples of models in which this actually happens. In models with an $SU(2)_R$ left unbroken to low scales there is automatically a matter parity for the discrete gauge group element $U_z = \exp\left[i\pi(T_{3_L} + T_{3_R})\right]$ plays the role of matter parity

$$q \equiv \begin{bmatrix} u \\ d \\ D \end{bmatrix}_L \rightarrow \begin{bmatrix} -u \\ -d \\ D \end{bmatrix}_L \; ; \; Q \equiv \begin{bmatrix} u^c \\ d^c \\ D^c \end{bmatrix}_L \rightarrow \begin{bmatrix} -u^c \\ -d^c \\ D^c \end{bmatrix}_L$$

$$L \equiv \begin{bmatrix} H_1^0 & H_2^+ & \ell^+ \\ H_1^- & H_2^0 & \ell^0 \\ \ell^c & \nu_R & N \end{bmatrix}_R \rightarrow \begin{bmatrix} H_1^0 & H_2^+ & -\ell^+ \\ H_1^- & H_2^0 & -\ell^0 \\ -\ell^c & -\nu_R & N \end{bmatrix}_R$$

In models with just the standard model after I.S. breaking SU(3)xSU(2)xU(1) we must lower the rank by two and it is necessary to give IS vevs to both ν_R and N components so U_Z is broken. It is possible, however, to combine U_Z with a discrete Z_2 group, D, of the manifold leaving the product DU_Z unbroken. This follows if the IS vevs occur only along $(\lambda_+)_N$ and $(\lambda_-)_{\nu_R}$ directions where ± refer to the D properties of the superfields[28]. In the three generation model of section 4 this is just what happens[18] with the discrete symmetry being identified with a Z_2 discrete symmetry following from the manifold[21]. The resultant exact symmetry DU_Z plays the role of matter parity and automatically forbids the dangerous dimension four operators. In other models matter parity may come from the discrete symmetries alone but, to date, no specific examples of manifolds with this property have been found.

Once dimension four contributions are eliminated, the dominant baryon and lepton number violation comes from dimension five operators. The limits on proton decay require that the heavy D-quark, mediating this contribution in E_6 models, should have mass $\gtrsim O(10^{14}\ \mathrm{GeV})$, so if the I.S. scale is this large acceptable rates for proton decay may results

We turn now to the expectation for neutrino masses[14,15]. Since E_6 based models have new right-handed neutrino components, ν_R, there is the expectation that there will be Dirac masses for neutrinos comparable to the charged lepton masses, m_ℓ. [It is possible some discrete symmetries may forbid such terms but no example is known]. Allowing for the possibility of SU(3)xSU(2)xU(1) invariant Majorana mass terms m_* for the ν_R states the usual diagonalisation of the neutrino mass matrix leaves light neutrinos with Majorana mass of order m_ℓ^2/m_*. In models with large IS breaking, Majorana masses occur at

$O\left(\frac{27^2\overline{27}^2}{M_{Planck}}\right)$ so $m_x = O\left(\frac{(IS)^2}{M_{Planck}}\right)$. For an I.S. greater than 10^{14} GeV this leads to the expectation that light neutrinos will have mass of order (10^{-7}ev to 10^{-1}ev). This indeed happens for the specific model presented in section 4[18].

Another low-energy prediction of superstring inspired models follows from the existence of an underlying grand-unified group relating the gauge couplings at the compactification scale. At low energies radiative corrections change these relations in a calculable manner, depending on the mass spectrum of the theory[14]. For example, the three generation model of section 4 has $\sin^2\theta_W = 0.23–0.25$ following from the $[SU(3)]^3$ structure at the I.S. Above the I.S. the large number of matter multiplets of the model means the strong coupling grows rapidly, close to the compactification scale. Such behaviour may be expected in any model with a large value for $h^{1,1}$.

6. Conclusions

Compactification of space-time symmetries and spontaneous breakdown of gauge symmetries can give phenomenologically realistic low-energy effective theories. Indeed specific examples have been constructed which are close to the supersymmetric standard model, showing that new phenomena such as new neutral currents or new quarks or leptons are not inevitable features of such models. Specific compactification schemes hold the promise of predictions beyond the standard model for the multiplet structure, couplings, masses and mixing angles. Although much work remains, particularly on the subject of supersymmetry breaking, I am encouraged by the remarkable realism of the simplest schemes investigated so far.

	E_6	$SU(3)_H$	No. Massless fields (in terms of Hodge numbers)
Vector	78	$(0,0)^1$	$h_1^{0,0} = 1$
super-	27	$(1,0)^3$	$h_3^{1,0} = 0$
fields	27	$(0,1)^3$	$h_3^{0,1} = 0$
	1	$(1,1)^8$	$h_8^{1,1} = h^2-1$
Chiral	78	$(1,0)^3$	$h_3^{1,0} = 0$
super-	27	$(2,1)^3+(1,2)^6$	$h_6^{1,2}$ $(h_3^{2,3} = 0)$
fields	27	$(1,1)^1+(1,1)^8$	$h_1^{1,1}+h_8^{1,1} = h^2$
	1	$(2,1)^3+(2,1)^6+15$	$h_6^{1,2}+ \#(15)$

Table 1 E_8 Yang-Mills supermultiplet decomposition. The notation $(p,q)^r$ refers to (p,q) forms transforming as a irreducible representation r of the holonomy group $SU(3)_H$.

$$U_{km}{}^u = e^{i\delta}\begin{pmatrix} 1 & \beta_1/\beta_2 & 0 \\ \beta_1/\beta_2 & 1 & 0 \\ 0 & 0 & 1 \end{pmatrix}$$

$$U_{km}{}^d = \begin{pmatrix} 1 & \sqrt{(m_d/m_s)} & \beta'_1 \\ \sqrt{(m_d/m_s)} & 1 & \beta'_2 \\ \beta'_1 & \beta'_2 & 1 \end{pmatrix}$$

$$U_{km} = U_{km}{}^{u+}U_{km}{}^d \equiv \begin{pmatrix} 1 & \Theta_1 & \Theta_2 \\ \Theta_1 & 1 & \Theta_3 \\ \Theta_2 & \Theta_3 & 1 \end{pmatrix}$$

Table 2 The KM matrix in terms of the contribution of up and down quarks. β_i, β'_i are combinations of Yukawa couplings and vevs with $\beta_1/\beta_2 = \beta'_1/\beta'_2$.

References

1. P.Ramond, Phys. Rev. D3, (1971) 2415
   A. Neveu and J. Schwarz, Nucl. Phys. B31, 86 (1971); Phys. Rev. D4, (1971) 1109
   M. Green and J. Schwarz, Phys. Lett. 149B, (1984) 117
   D. Gross, J. Harvey, E. Martinec, and R. Rohm, Phys. Rev. Lett. 55, (1985) 502; Nucl. Phys. B256, (1985) 253
2. V. Kaplunovsky, Phys. Rev. Lett.55, (1985) 1036
   M. Dine and N. Seiberg, Phys. Rev. Lett. 55, (1985) 366 and Phys. Lett. 162B, (1985) 299
3. P. Candelas, G. Horowitz, A. Strominger and E. Witten, Nucl. Phys. B258, (1985) 46
4. J-P. Derendinger, L.E. Ibanez and H.P. Nilles, Phys. Lett. 155B, (1985) 65
   M. Dine, R. Rohm, N. Seiberg and E. Witten Phys. Lett. 156B, (1985) 55
5. T. Hubsch, Maryland Univ. preprint 86-149 (1986)
6. Y. Hosotani, Phys. Lett. 126B, (1983) 309; 129B, (1983) 193
   E. Witten, Phys. Lett. 126B, (1984) 351
7. E. Witten, Nucl. Phys. B258, (1985) 75
   J.D. Breit, B. Ovrut and G. Segre, Phys. Lett. 158B, (1985) 33
8. F. del Aguila, G. Blair, M. Daniel and G.G. Ross, CERN preprint TH 4786 (1986) and Nucl. Phys. (to appear)
   S.M. Barr, Phys. Rev. Lett. 55, (1985) 2778
   V. Barger, N.G. Deshpande and K. Whisnant, Phys. Rev. Lett. 56, (1986) 30
   R.W. Robinett, Phys. Rev. D33, (1986) 1908; V. Barger, N.G. Deshpande, R.J.N. Phillips and K. Whisnant, Phys. Rev. D33, (1986) 1912.
   See also M. Deshpande, these proceedings
9. E. Witten, Nucl. Phys. B268, (1986) 79
10. B.R. Greene, K.H. Kirklin and P.J. Miron, Nucl. Phys. B274, (1986) 574
    D. Bailin, A. Love and S. Thomas Univ of Sussex preprint (1986)
11. M. Dine, N. Seiberg, X-G. Wen and E. Witten, Princeton preprint (1986)
12. J. Bagger, communication at this conference
13. L. Dixon, J.A. Harvey, C. Vafa and E. Witten, Nucl. Phys. B261, (1985) 678
    Nucl. Phys. B274, (1986) 285
    See also J. Bagger, these proceedings
14. M. Dine, V. Kaplunovsky, M. Mangano, C. Nappi and N. Seiberg, Nucl. Phys. B259, 519 (1985)
15. F. del Aguila, G. Blair, M. Daniel and G.G. Ross, Nucl. Phys. B272, (1986) 413
16. M. Quiros, Phys. Lett. 173B, (1986) 265
    Y.J. Ahm and J.D. Breit, Nucl. Phys. B273, (1986) 75
    J.D. Breit, B. Ovrut and G. Segré, Phys. Lett. 162B, (1985) 303
    P. Binetruy and M.K. Gaillard, Phys. Lett. 168B, (1986) 347
    S. Dawson - these proceedings
    G. Segre - these proceedings
17. E. Cohen, J. Ellis, K. Enqvist and D.V. Nanopoulos, Phys. Lett. 165B, (1985) 76; J. Ellis, K. Enqvist, D.V. Nanopoulos and F. Zwirner, Nucl. Phys. B276, (1986) 14 and Mod. Phys. Lett. A1 (1986) 57
18. B.R. Greene, K.H. Kirklin, P.J. Miron and G.G. Ross, University of Oxford preprint 39/86 (1986) to be published in Phys. Lett; and in preparation
19. P. Candelas et al. have constructed of the order of twenty such models - private communication
20. S-T. Yau in Proceedings of the Argonne Symposium on Anomalies, Geometry and Topology (World Scientific, 1985)
21. B.R. Greene, K.H. Kirklin, P.J. Miron and G.G. Ross, Nucl. Phys. B278, (1986) 667
22. E.H. Thorndike, Proceedings of 1985 International Symposium on Lepton and photon interactions at high energies, Kyoto 1985
23. M. Goodman and E. Witten, Nucl. Phys. B271 (1986) 21
24. A. Strominger and E. Witten, Comm. Math. Phys. (1986)
    A. Strominger, Phys. Rev. Lett. (1986)
25. F. Gursey, P. Ramond and P. Sikivie, Phys. Lett. 60B, (1976) 177; Y. Achiman and B. Stech, Phys. Lett. 77B, (1978) 389; Q. Shafi, Phys. Lett. 79B (1978) 301; H. Ruegg and T. Schuler, Nucl. Phys. B161, (1979) 388
    R. Barbieri and D.V. Nanopoulos, Phys. Lett. 91B, (1980) 369
26. P. Roy, these proceedings
    H. Haber and M. Sher, Santa Cruz preprint (1986)
27. S. Weinberg, Phys. Rev. D26, (1982) 287
    N. Sakai and T. Yanagida, Nucl. Phys. B197, (1982) 533
28. M. Bento, L. Hall and G.G. Ross, in preparation

LOW ENERGY FOUR DIMENSIONAL EFFECTIVE POTENTIAL FROM SUPERSTRINGS†

Gino Segrè

Department of Physics, University of Pennsylvania
Philadelphia, Pennsylvania 19104-6396

There has been recently a great deal of work on what I will call "superstring inspired" phenomenology (1). This means analyzing the spectrum of states and couplings of particles in four dimensions which are presumed to be constrained by the existence of an underlying superstring theory. The approach is very exciting, but clearly there are still a great many uncertainties.

The pioneering paper is that of Candelas et al. (2), who wrote down a modified Chapline-Manton (3) supergravity super Yang-Mills Lagrangian in d = 10 (ten dimensions. The gauge group was $E_8 \times E_8'$. Compactifying then on K, a so-called Calabi-Yau space (4) with discrete symmetries, they formulated a resulting four dimensional theory. For a general complex manifold in six dimensions, the holonomy group is U(3), but the Calabi-Yau spaces are Ricci flat and have SU(3) holonomy, so there exists on them a covariantly constant spinor. This in turn implies an unbroken N = 1 supersymmetry in four dimensions and hence we have an N = 1 local supergravity theory in d = 4.

The gauge group $E_8 \times E_8'$ is broken to $E_6 \times E_8'$ by background fields with non vanishing v.e.v.'s (2). It is then assumed that a further dynamical breaking occurs because of the discrete symmetries on K so $E_6 \times E_8'$ is broken to $\mathcal{G} \times E_8'$ (5). $\mathcal{G}$ is a group at least as big as the phenomenologically successful $SU(3)_c \times SU(2) \times U(1)$.

Several caveats need to be stated at this point. First of all the idea that we can meaningfully discuss an effective ten dimensional field theory derived from the superstring may be wrong. It is known that the characteristic scale of string theory is of the same order of magnitude as the Planck scale and the compactification scale.

$$\left(\frac{1}{\alpha'_{string}}\right)^{\frac{1}{2}} \sim M_{Planck} \sim M_{compact} \qquad (1)$$

so that in fact it may not make sense to speak of a d = 10 field theory (6). In many ways, it would in fact be desirable to proceed directly from string theory to d = 4 field theory since the problem of selecting the correct vacuum in a d = 10 effective field theory is tricky, to say the least.

A second caveat is that we do not at this point possess a full super Yang Mills supersymmetric field theory in d = 10. What we do have is a so-called on mass shell field theory expressed in terms of physical fields. It would be desirable to have the full off-shell theory in which the auxiliary fields have not yet been eliminated using equations of motion (7). One could then study the supersymmetry of the Lagrangian with the antisymmetric tensor modified à la Green and Schwartz by the additional Chern-Simons term.

We will assume however the existence of a d = 10 Lagrangian. The problems we wish to examine are those of low energy physics, namely how is supersymmetry breaking displayed in the low energy sector and how is the gauge group $\mathcal{G}$ broken down to $SU(3)_c \times U(1)_{e.m.}$, the unbroken gauge group of low energy physics.

The compactification on the Calabi-Yau manifold is too difficult to answer these questions, so we employ a model compactification suggested by Witten (8). The ansatz consists of assuming the light mass fields
i) do not depend explicitly on the internal coordinates x^m, m = 5, 6, 7, 8, 9, 10.
ii) They are invariant under $G + \bar{G}$ where G is the SU(3) holonomy group and $\bar{G}$ is the gauged SU(3) in the decomposition of E_8 into $SU(3) \times E_6$. As an example consider the d = 10 graviton fields g_{MN}: only $g_{\mu\nu}$, μ, ν = 1,2,3,4 and $\delta_{mn}\, e^{\sigma}$ survive. Fields such as $g_{\mu m}$ etc. are super-massive. The E_8 gauge super-multiplet is $A_M^{(\alpha)}$ where (α) denotes the decomposition of the adjoint of E_8 into $SU(3) \times E_6$

$$(\alpha) = (8,1) + (3,27) + (\bar{3},\overline{27}) + 1,78). \qquad (2)$$

Using the ansatz described above we find that the light fields which arise from the gauge multiplet are

$$A_\mu^{(1,78)},\, A_{\bar{a}}^{(3,27)} \rightarrow C^{(27)},\; A_a^{(\bar{3},\overline{27})} \rightarrow C^{(\overline{27})} \qquad (3)$$

where a = 1,2,3. The notation above means

---

†Invited talk at XXIII International Conference on High Energy Physics, Berkeley, California, July 1986.

that $A_{\bar{m}}$, the 6 of O(6) decomposes into a 3 and $\bar{3}$ under the SU(3) of holonomy and that $A_m^{(27)}$ then combine with the gauged SU(3) 3 and $\bar{3}$ to form singlets. The result is the E_6 d = 4 gauge supermultiplet $A_\mu^{(78)}$ and a complex 27 dimensional matter field C^x. In addition to these fields and their superpartners, we have two dilatons σ coming from $g_{mn} \to \delta_{mn} e^{\sigma}$, ϕ from the original supergravity multiplet and two pseudoscalars θ and η from the decomposition of the antisymmetric tensor present in the gravity super-multiplet. Witten (8) showed that by defining fields S and T

$$S = \phi^{-3/4} e^{3\sigma} + i\theta$$
$$T = \phi^{3/4} e^{\sigma} + i\eta + |C^x|^2 \qquad (4)$$

the Kahler potential could be written in the form ($\hat{S}$ and $\hat{Q}$ are defined by the equation)

$$K = -\ln(S + \bar{S}) - 3\ln(T + \bar{T} - 2|C|^2)$$
$$= -\ln \hat{S} - 3\ln \hat{Q} \ . \qquad (5)$$

The accompanying superpotential is

$$W = W(S) + W(C) = W(S) + d_{xyz} C^x C^y C^z \qquad (6)$$

where d_{xyz} is the E_6 coupling of three 27's to form a singlet. Obviously this model is a gross distortion of the true d = 4 Lagrangian, e.g. it has only one family, but it does allow us to go further in our calculations. It is of course exciting, but not entirely unexpected that the resulting Kahler potential is that of no-scale models (9). The similarities to the Kahler potential derived from so-called 16/16 supergravity models are also intriguing (10).

We can at this point introduce the standard machinery of supergravity (11). The scalar potential is given by

$$\frac{V}{e} = \frac{M^4}{\hat{S}\,\hat{Q}^3}\left[\frac{\hat{S}^2}{M^2}\left|D_S W\right|^2 + \frac{\hat{Q}}{6M}\left|\frac{\partial W}{\partial C}\right|^2\right] + 18\, g^2 \frac{M^3}{\hat{Q}^3}\,\mathrm{Re}\, S^{-1}\left\{C_x (T^a)^x_y\, C^y\right\}^2 \qquad (7)$$

where e is the determinant of the vierbein, g is the gauge coupling, M is the Planck mass and $D_S W = \frac{\partial W}{\partial S} - \frac{W}{\hat{S}}$.

So far we have only one scale, the Planck mass, and unbroken supersymmetry. To proceed further we must introduce supersymmetry breaking. Two ways have been introduced: the first involves supposing that a gluino condensate occurs in the E_8' sector (12) and the second that the antisymmetric tensor field has a non-vanishing v.e.v.

$$H_{mnp} = c\,\varepsilon_{mnp} \ .$$

The optimal scenario may be one in which both occur (13) in such a way that the cosmological constant is still kept equal to zero. In this case the minimum of the potential is at <V> = 0, at which point we have $\langle C^x \rangle = 0$ and $\langle D_S W \rangle = 0$, though $\langle W(s) \rangle$ itself is not equal to zero. The value of $\langle \hat{T} \rangle = \langle T + \bar{T} \rangle$ at the minimum is not determined, nor is the value of the gravitino mass (14)

$$m_{3/2}^{\ 2} = \frac{\langle W(S) \rangle^2}{\langle \hat{S} \rangle \langle \hat{T} \rangle^3} \qquad (8)$$

We (14) find no mass term for the C fields and hence no Higgs mechanism for the breaking of SU(2). To study this problem further, we (14) and several others (15,16,17) have used the effective d = 4 field theory to study the one loop potential, in the hope of fixing <T> and find masses for the C fields.

The field theory is divergent, but that's all right since it is only an effective field theory. We are assuming a third scale Λ at which we cut off the divertent integrals

$$\Lambda_{\text{compactif}} \sim M_{\text{Planck}} >> \Lambda_{\substack{\text{SUSY}\\\text{breaking}}} >> \Lambda. \qquad (9)$$

One might expect mass terms of the form

$$\Delta V \sim \frac{\Lambda^2 \left(m_{3/2}\right)^2}{M^2} |C|^2 \qquad (10)$$

In fact this doesn't happen as the dependence on C^2 in the one loop potential appears in the form

$$V_{1\ \text{loop}} = f(\hat{T} - 2|C|^2, S\ldots..)$$
$$m_c^{\ 2} = \left.\frac{\partial^2 V}{\partial C \partial \bar{C}}\right|_{\substack{\langle C \rangle = 0\\ T = \langle T \rangle}} = -2\left.\frac{\partial V}{\partial T}\right|_{T = \langle T \rangle} = 0 \qquad (11)$$

i.e. the condition for minimizing V to fix $\langle T \rangle, \frac{\partial V}{\partial T} = 0$, is the same expression as the mass term for C. By itself, this would not be disastrous, since one could imagine generating C mass terms at the two loop level, etc. The real problem is that of minimizing the one loop potential. What is found is that potential, zero at tree level, is at the one loop level a monotonically decreasing function of <T>. As $\langle T \rangle \to 0$, $V_{1,\text{loop}} \to -\infty$, so the minimum corresponds to an infinite cosmological constant. Of course, at the same time $\langle T \rangle \to 0 \quad (m_{3/2})^2 \to \infty$, i.e. our series of approximations breaks down. Keeping <S> fixed, not only do we have an infinite cosmological constant as <T> $\to$ 0, but referring back to (3), we see that $\sigma \to -\infty$ and $\phi \to 0$ at the minimum of the potential. This corresponds to an ultra-compactified (radius of compactification $\sim e^{-\sigma}$), strongly coupled theory, i.e. (9) is nonsense.

Our conclusion is that our naive model does not work. This is disappointing, but not surprising. What lesson should we learn from it? It appears that a successful superstring inspired phenomenology will require some intermediate scale Λ_{int}, to set a smaller scale of electroweak breaking at $\Lambda \sim \frac{(\Lambda_{int.})^2}{\Lambda_{comp}}$ and to break to $SU(3)_c \times SU(2) \times U(1)$. We have simply introduced this intermediate scale by breaking supersymmetry in a kind of hidden sector, but clearly there are other phenomena that one needs to consider. Contributions of higher mass Kaluza-Klein states and stringy states to the effective potential need to be evaluated. In this regard one has to remember the non-renormalization theorems (18). Nonperturbative effects are particularly likely to be interesting; so called "world sheet instantons" can modify the superpotential (19). The general idea is to look for flat directions in the potential for a field and then assume that there are some soft SUSY breaking terms of order M_W^2. The potential thus generated

$$V(\psi) \sim -M_W^2 |\psi|^2 + \frac{|\psi|^6}{M_p^2} + \ldots \quad . \qquad (12)$$

leads to

$$\langle\psi\rangle \sim M_I \sim \sqrt{M_W M_p} \qquad (13)$$

All this is questionable. Clearly a great deal of work remains to be done. We note however that assuming such a scenario exists can lead to encouraging phenomenology (20).

I would like to thank Burt Ovrut for teaching me much of what I know about this subject and Jik Ahn, John Breit and Burt Ovrut for working with me on these problems.

REFERENCES

(1) A few recent references are:
L. E. Ibañez and J. Mas, "Low Energy Supergravity and Superstring Inspired Models", CERN preprint TH 4426 (1986).
F. del Aguila, G. Blair, M. Daniel and G. G. Ross, "Superstring Inspired Models", CERN preprint Th 4336 (1986).
J. Ellis, K. Enqvist, D.V. Nanopoulos and F. Zwirner, "Observables in Low Energy Superstring Models" CERN preprint Th4350 (1986).
M. Dine, V. Kaplunovsky, M. Mangano, C. Nappi and N. Seiberg, Nucl. Phys. B259, 549 (1984).
S. Cecotti, J. P. Derendinger, S. Ferrara, L. Girardello and M. Roncadelli, Phys. Lett. 156B, 318 (1985).
J. P. Derendinger, L. E. Ibañez and H. P. Nilles, Nucl. Phys. B267, 365 (1986).

(2) P. Candelas, G. Horowitz, A. Strominger, and E. Witten, Nucl. Phys. B258, 46 (1985).

(3) G. Chapline and N. S. Manton, Phys. Letts. B120, 105 (1983).

(4) E. Calabi in "Algebraic Geometry and Topology", A Symposium in Honor of S. Lefshetz", Princeton Univ. Press 1957.
S. T. Yau, Proc. Nat. Acad. Sci. 74, 177 (1978).

(5) E. Witten, Nucl. Phys. B258, 75 (1985).
J. D.Breit, G. Segrè and B.A. Ovrut, Phys. Lett. 158B, 33 (1985).
A. Sen, Phys. Rev. Letts. 55, 33 (1985).

(6) M. Dine and N. Seiberg, Phys. Rev. Letts. 55, 366 (1985). V. Kaplunovsky, Phys. Rev. Letts. 55, 1036 (1985).

(7) For some progress in this direction see e.g., R.. Kallosh and B.E.Nilsson, Phys. Lett. 167B, 46 (1986); B. E. Nilsson and A. K. Tollsten, Phys. Lett. 169B, 369 (1986).

(8) E. Witten, Phys. Lett. 155B, 151 (1985).

(9) A. B. Lahanas and D. V. Nanopoulos, "The Road to No Scale Supergravity", to be published in Phys. Repts. CERN Th 4400/86 preprint for a review and references.

(10) W. Lang, J. Louis and B.A. Ovrut, "16/16 Supergravity Coupled to Matter: The Low Energy Limit of the Superstring" Unif. of Pennsylvania preprint UPR-0280 (1985).

(11) E. Cremmer, B. Julia, J. Scherk, S. Ferrara, L. Girardello and P. van Nieuwenhuizen, Nuc. Phys. B147, 105 (1979).

(12) J. P. Derendinger, L. E. Ibañez, and H. P. Nilles, Phys. Lett. 155B, 65 (1985).

(13) M. Dine, R. Rohm, N. Seiberg, and E. Witten, Phys. Lett. 156B, 55 (1985).

(14) We are following here the discussion in our printed paper: J. D. Breit, B. A. Ovrut and G. Segrè, Phys. Lett. 162B, 303 (1985).

(15) M. Mangano, Zeit fur Physik, C28, 613 (1985).

(16) P. Binetruy and M. K. Gaillard Lawrence Berkeley preprint (1985).

(17) Y. Ahn and J. D. Breit, Nucl. Phys. B273 75 (1986); M. Quiros, CERN preprint TH 4363 (1986).

(18) E. Witten, "New Issues in Manifolds of SU(3) Holonomy" Princeton preprint (1985) to be published in Nucl. Phys.

(19) M. Dine, N. Seiberg, X. G. Wen and E. Witten, "Non-Perturbative Effects of the String World Sheet" Princeton preprint (1986); J. Ellis, C. Gomez, D. V. Nanopoulos and M. Quiros, Phys. Letts. B173,

59 (1986).
20) For some recent promising results see, e.g., B. R. Greene, K. Kirklin, P. J. Miron and G. G. Ross, "A Superstring Inspired Standard Model" Oxford preprint (1986).

EXTRA GAUGE BOSONS AND E_6 FERMIONS FROM SUPERSTRINGS

N.G. Deshpande

Institute of Theoretical Science, University of Oregon
Eugene, OR 97403

We present mass limits, production rates and forward-backward asymmetries of extra gauge bosons that could occur in some string inspired E_6 models. We also discuss the properties of the new fermions in the 27 representation.

Recent work on superstring theories[1] indicates that the $E_8 \times E'_8$ superstring theory in 10 dimensions may yield, after compactification, a four-dimensional broken E_6 gauge group of the strong and electroweak interactions coupled to N=1 supergravity. The compactification on a manifold with SU(3) holonomy leads also to breaking of the E_6 by the Wilson loop mechanism to a low energy group which could be rank 5 or rank 6. If the gauge group is rank 5, the low energy roup has the structure

$$SU(3)_C \times SU(2)_L \times U_Y(1) \times U'(1) \qquad (1)$$

with the Z' associated with U'(1) having uniquely determined couplings to all fields.[2] It is also possible that the breaking occurs to rank 6 group. We shall consider the low energy group in this case to be

$$SU(3)_C \times SU(2)_L \times U_Y(1) \times U_\psi(1) \times U(1)_\chi \qquad (2)$$

where quantum numbers of the extra U(1)'s are specified by the breaking $E_6 \to SO(10) \times U(1)_\psi \to SU(5) \times U(1)_\chi \times U(1)_\psi$, where SU(5) contains the usual $SU(3)_C \times SU(2)_L \times U(1)_Y$. We shall consider the case of one light extra boson in the rank 6 scenario, with the other gauge boson being very massive. Both the rank 5 and the rank 6 cases can be considered simultaneously[3] if we define

$$|Z(\alpha)\rangle = \cos\alpha\,|Z_\psi\rangle + \sin\alpha\,|Z_\chi\rangle \qquad (3)$$

with α arbitrary for rank 6, while for rank 5 case $\cos\alpha = (5/8)^{1/2}$.

In an E_6 theory, each generation of fermions belongs to a 27 representation. The decomposition of 27 into SO(10), SU(5) and SU(3) multiplets and the fermion quantum numbers Q(charge), I_{3L}(weak isospin) and $\tilde{Q}$ (extra U(1) charge) are given in Table 1. Note that Q depends on the mixing angle α. In addition to the usual fermions u, $\bar{u}$, d, $\bar{d}$, $e^\pm$ and ν_e there is a charge $-1/3$ quark isosinglet h, charged leptons $E^\pm$ and neutral leptons ν_E, N_E, N_e and n.

We shall now consider the mass limits on the extra Z as a function of α from the low energy data as well as the UA(1)/UA(2) limits. The special case $\cos\alpha = (5/8)^{1/2}$ had been worked out earlier.[4] The neutral current Lagrangian for the E_6 models with one extra Z is

$$L_{NC} = eA_\mu J^\mu_{em} + g_Z Z_\mu J^\mu_Z + g' Z(\alpha)_\mu J^\mu_{Z(\alpha)} \qquad (4)$$

where J^μ_{em} and $J^\mu_Z (\equiv J^\mu_3 - x_W Q^\mu)$ are the usual electromagnetic and Z boson currents and

$$J^\mu_{Z(\alpha)} = \tfrac{1}{2} \sum_f \bar{f}\gamma^\mu (1-\gamma_5)\tilde{Q} f \qquad (5)$$

$SO(10)$	$SU(5)$	Left-handed state	$SU(3)$	Q	I_{3L}	$\tilde{Q}$
16	10	e^{-c}	1	1	0	
		d	3	$-\frac{1}{3}$	$-\frac{1}{2}$	$a_1 = \frac{1}{3}\left(\sqrt{\frac{5}{8}}\cos\alpha + \sqrt{\frac{3}{8}}\sin\alpha\right)$
		u	3	$\frac{2}{3}$	$\frac{1}{2}$	
		u^c	3^*	$-\frac{2}{3}$	0	
	5^*	d^c	3^*	$\frac{1}{3}$	0	
		e^-	1	-1	$-\frac{1}{2}$	$a_2 = \frac{2}{3}\left(\sqrt{\frac{5}{32}}\cos\alpha - \sqrt{\frac{27}{32}}\sin\alpha\right)$
		ν_e	1	0	$\frac{1}{2}$	
	1	N_e^c	1	0	0	$a_3 = \frac{\sqrt{10}}{3}\left(\frac{1}{4}\cos\alpha + \frac{\sqrt{15}}{4}\sin\alpha\right)$
10	5^*	h^c	3^*	$\frac{1}{3}$	0	
		E^-	1	-1	$-\frac{1}{2}$	$a_4 = \frac{2}{3}\left(-\sqrt{\frac{5}{8}}\cos\alpha + \sqrt{\frac{3}{8}}\sin\alpha\right)$
		ν_E	1	0	$\frac{1}{2}$	
	5	h	3	$-\frac{1}{3}$	0	
		E^{-c}	1	1	$\frac{1}{2}$	$a_5 = \frac{2}{3}\left(-\sqrt{\frac{5}{8}}\cos\alpha - \sqrt{\frac{3}{8}}\sin\alpha\right)$
		N_E^c	1	0	$-\frac{1}{2}$	
1	1	n	1	0	0	$a_6 = \frac{\sqrt{10}}{3}\cos\alpha$

Table 1. Decomposition of 27 and fermion quantum numbers. The $\tilde{Q}$ charge a_i are given as an amplitude times a factor which varies with α over the range -1 to +1.

where the summation is over left-handed states of Table 1. A coupling of left-handed charge conjugated state is equivalent to the negative of the corresponding right-handed state (eg. the u_R coupling to $Z(\alpha)$ is $g_R(u) = -g_L(u^c) = -a_2$). The coupling constant g', with our normalization of $\tilde{Q}$ charges, and using the renormalizaiton group to evolve the U(1) couplings, is

$$g' = g_Z(x_W)^{1/2} = \frac{e}{(1-x_W)^{1/2}} \tag{6}$$

Where $x_W = \sin^2\theta_W$. In this analysis we shall ignore $Z - Z(\alpha)$ mixing, although the mixing has been included in Ref. 4 and a forthcoming paper[5], the effect on the limits is not significant. The limits from UA(1)/UA(2) depend on the B.R.($Z(\alpha) \rightarrow e^+e^-$) which in turn depends on whether the extra fermions are light or heavy. The $Z(\alpha)$ total width is given by

$$\Gamma(Z(\alpha)) = \alpha_{em} M_{Z(\alpha)} [2(1-x_W)]^{-1}$$
$$\times\ [10a_1^2+5a_2^2+n_G(15-10a_1^2-5a_2^2)/3] \tag{7}$$

where n_G is the number of generations of exotic fermions that contribute in the decay. The individual fermion contribution to the partial width is

$$\Gamma(Z(\alpha) \rightarrow f\bar{f}) = \alpha_{em} M_{Z(\alpha)} [6(1-X_W)]^{-1}$$
$$\times\ [g_L^2+g_R^2]c_f \tag{8}$$

where $c_f = 3$ for quarks and $c_f = 1$ for leptons. The B.R. for different modes as a function of α, and the total width for $n_G = 0$ are given in figure 1. These ratios will be useful in distinguishing between different values of α.

The rate for $p\bar{p} \to Z(\alpha) \to e^+e^-$ is given by

$$\sigma(Z(\alpha)) = \frac{2\pi^2\alpha_{em}}{3s(1-x_W)} \sum_q [g_L(q)^2 + g_R(q)^2]H_q^+ \times \mathrm{B.R.}(Z(\alpha) \to e^+e^-)K \qquad (9)$$

where K is the K factor taken to be $(1 + 8\alpha_s(s)/\pi)$ and

$$H_q^{\pm}(m^2, \sqrt{s}) = \int dy\, G_q^{\pm}(y, m^2, \sqrt{s}) \qquad (10)$$

where

$$G_q^{\pm}(y, m^2, \sqrt{s}) = f_{q/A}(x_A) f_{\bar{q}/B}(x_B) \pm f_{\bar{q}/A}(x_A) f_{q/B}(x_B). \qquad (11)$$

We use set 1 of the Duke and Owens structure function[6] and the lower limit of the mass of $Z(\alpha)$ is derived from the combined UA1 - UA2 upper limit of $\sigma B < 3$pb at 90% CL.[7] We also use the low energy data (see data set discussed in V. Barger et al. in Ref. 4) to derive the lower limit of $Z(\alpha)$ mass again at 90% CL. These are superimposed on figure 2.

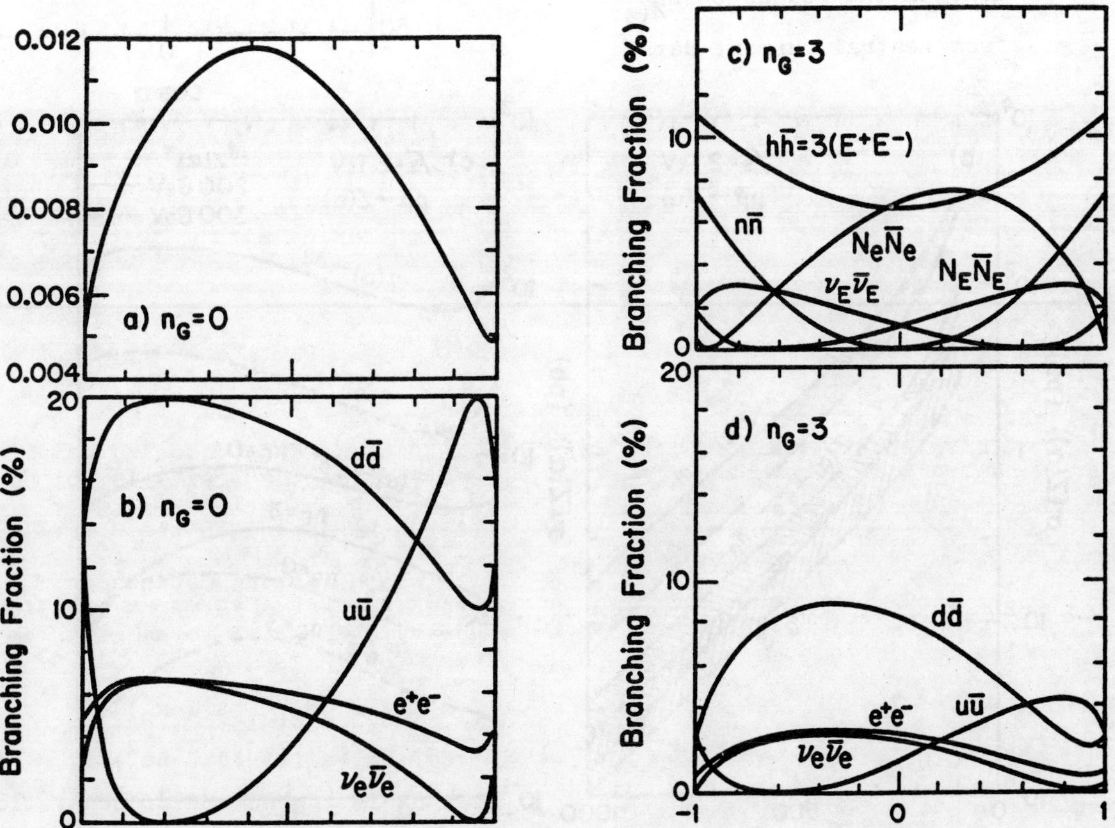

Fig. 1. Properties of $Z(\alpha)$ decays versus cos α: a) total rate and b) branching fractions of known fermions for $n_G = 0$; branching fractions of c) exotic fermions and d) known fermions for $n_G=3$.

We note that for cos $\alpha \approx 0$ the neutral current data is more restrictive, while for cos $\alpha \approx 1$ the $p\bar{p}$ data is more restrictive.

We present production cross-sections for $Z(\alpha)$ at Tevatron and SSC facility planned to operate at $\sqrt{s} = 40$ TeV and Luminosity of 10^4 pb per operating year in Figure 3 and 4 respectively.

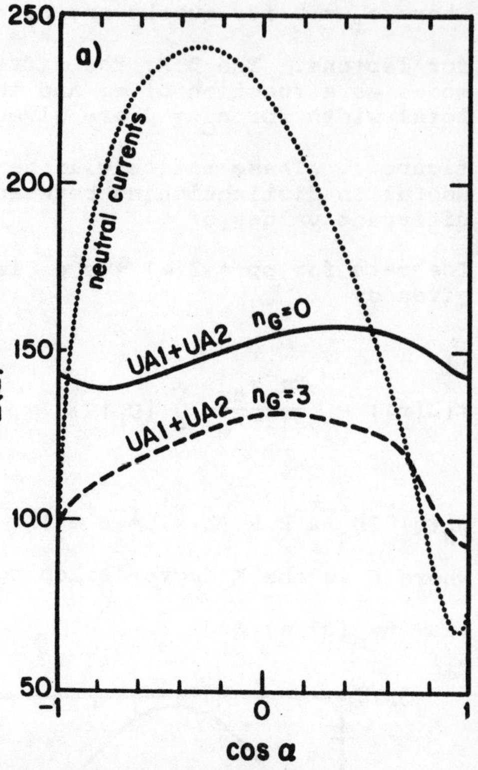

Fig. 2. Lower bounds on extra Z boson in E_6 versus cos α deduced from UA1/UA2 searches for $Z \to e^+e^-$ at $\sqrt{s} = 630$ GeV. The solid (dashed) curve assumes $n_G = 0(3)$. The dotted curve denotes the bounds on $M_{Z(\alpha)}$ from neutral current data.

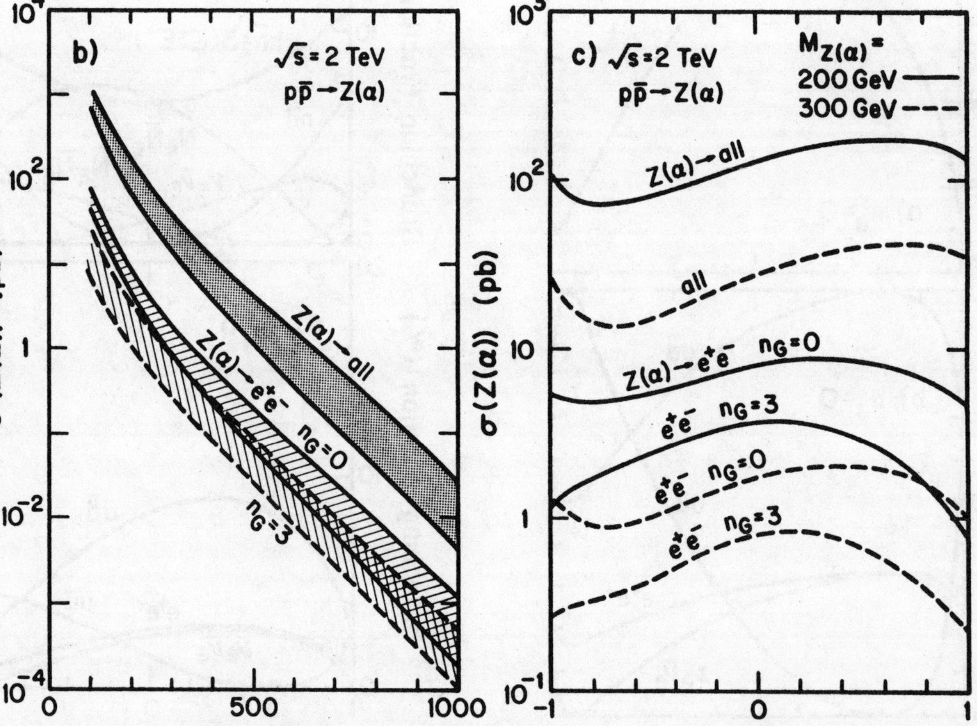

Fig. 3. Prediction for $Z(\alpha)$ production at 2 TeV Fermilab $p\bar{p}$ collider (a) σ of $Z(\alpha)$ and $\sigma B(Z(\alpha) \to e^+e^-)$ versus $M_{Z(\alpha)}$, for two extreme cases $n_G = 0$ and 3. The shaded regions correspond to the range allowed by varying α. (b) σ and σB for $M_{Z(\alpha)} = 200$ and 300 GeV shown versus cos α.

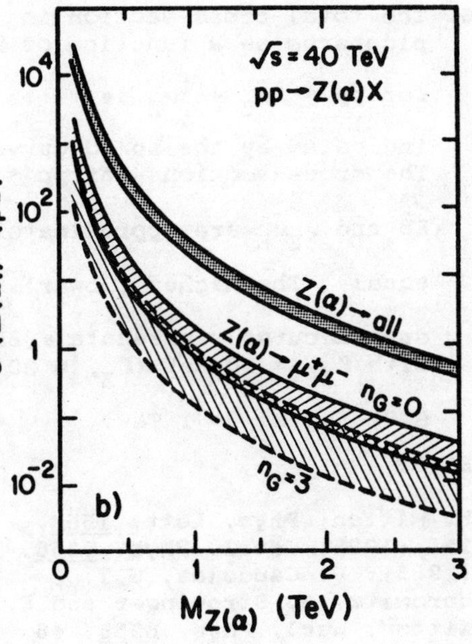

Fig. 4. Predictions for σ and $\sigma B(Z(\alpha) \to e^+e^-)$ for producing $Z(\alpha)$ versus $M_{Z(\alpha)}$; the shaded regions correspond to the range allowed by α

A way to determine which $Z(\alpha)$ is being produced is to study the forward-backward asymmetry as a function of the rapidity y.

In a pp machine this is given by

$$A^{FB}(y) = \frac{3}{4} \frac{\left[g_R^2(e) - g_L^2(e)\right]}{\left[g_R^2(e) + g_L^2(e)\right]} \tag{12}$$

$$\times \frac{\sum_q \left[g_R^2(q) - g_L^2(q)\right] G_q^-}{\sum_q \left[g_R^2(q) + g_L^2(q)\right] G_q^+}$$

We plot this asymmetry in Figure 5 for some sample values of α to show that it is a rather sensitive function of α. To study the asymmetry one will of course need a large sample of $Z(\alpha)$ events.

We now discuss briefly the properties of the exotic fermions. The Lagrangian and decay rates of these fermions may be found in Ref. 8.

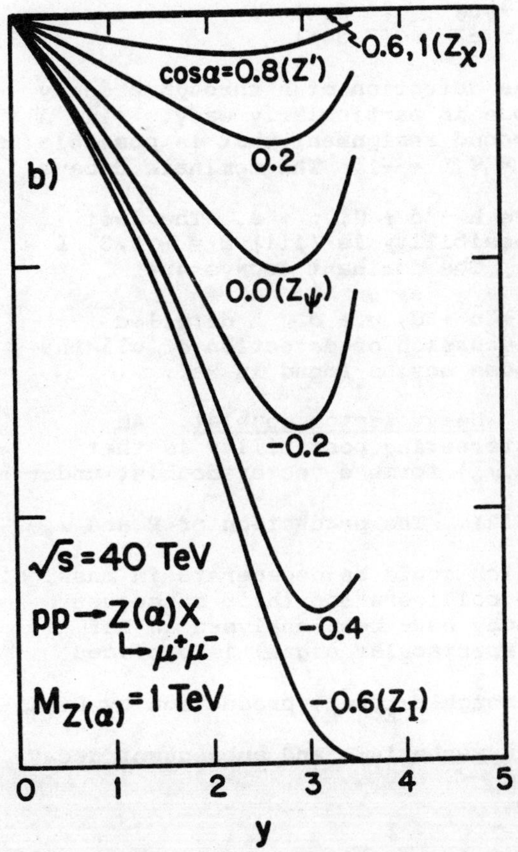

Fig. 5 Forward-backward asymmetries for the reaction $pp \to Z(\alpha)X$; the symmetry is plotted versus y for representative cos α for pp at $\sqrt{s}$=40TeV, $M_{Z(\alpha)}$ = 1 TeV.

Much of the above discussion is based on the work done by the author in collaboration with Barger, Rosner and Whisnant in a recent University of Wisconsin preprint MAD/PH/299. The work on fermions which will now be presented was carried out at a SSC Workshop in Snowmass in June 1986. See in particular Ref. 9 and Ref. 10 for more details.

A) <u>The h quark</u>. This quark is an SU(2) singlet with charge $Q = -\frac{1}{3}$. Its properties depend on the discrete symmetry obeyed by the interactions. Three assignments are possible. (i) Baryon number (B) = $\frac{1}{3}$, Lepton number (L) = 0. In this case h decays through h − d mixing. The dominant decays of the h quarks, assuming $m_h > m_Z$, are

$h \to uW^-$ (~ 67%)
$h \to dZ$ (~ 33%) (13)

The detection of h through Z decay mode is particularly easy. (ii) A second assignment that is possible is $B = \frac{1}{3}$ $L = -1$. The dominate decays are $h \to d + \bar{\nu}$, $u + \bar{e}$. The last possibility is (iii) $B = -2/3$, $L = 0$. The dominant decays are $h \to \bar{u} + \bar{\bar{d}}$, $\bar{\bar{u}} + d$. A detailed discussion of detection of all the modes may be found in Ref. 9.

B) Heavy lepton doublet. An interesting possibility is that (E, ν_E) forms a vector doublet under SU(2). The production of E and ν_E, which could be degenerate in mass, at pp colliders and their subsequent decay have been analysed in Ref. 10. A spectacular signal is provided through $\bar{E}\nu_E(E\bar{\nu}_E)$ production by Drell-Yan mechanism, and subsequent decay via

$$E^{\pm} \to e^{\pm} + Z$$
$$\nu_E \to e + W \quad (14)$$

These modes have almost 100% branching ratio, and provide excellent signature production cross-section is given in Figure 6.

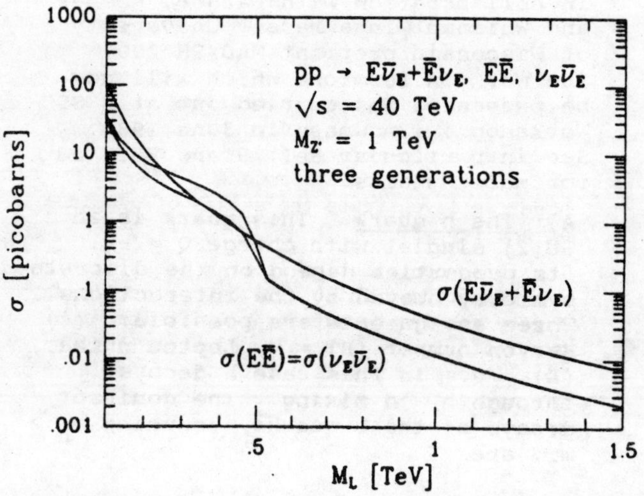

Fig. 6. The total cross-section in picobarns as a function of M_L for $pp \to E\bar{\nu}_E + \bar{E}\nu_E$ is indicated by the solid curve. The cross-sections for $pp \to E\bar{E}$ and $\nu_E\bar{\nu}_E$ are approximately equal. The higher (lower) dashed curve represents $\sigma(E\bar{E})$ with $\Gamma_{Z'} = 10$ GeV ($\Gamma_{Z'} = 20$ GeV) for $M_{Z'} = 1$ TeV.

REFERENCES

(1) E. Witten, Phys. Lett. 155B, 151 (1985); Nucl. Phys. B258, 75 (1985); P. Candelas, G.T. Horowitz, A. Strominger and E. Witten, Nucl. Phys. B258, 46 (1985); J.D. Breit, B.A. Ovrut and G. Segre, Phys. Lett. 158B, 33 (1985).

(2) J.L. Rosner, Comments on Nucl. and Part. Phys. 15, 195 (1986).

(3) D. London and J.L. Rosner, Phys. Rev. D34, 1530 (1986).

(4) V. Barger, N.G. Deshpande and K. Whisnant, Phys. Rev. Lett. 56, 30 (1986), L.S. Durkin and P. Langacker, Phys. Lett. 166B, 436 (1986); E. Cohen, J. Ellis, K. Enquist and D.V. Nanopoulos, CERN report TH. 4222/85.

(5) V. Barger, N.G. Deshpande and K. Whisnant, Univ. of Wisconsin preprint, MAD/PH/307 (86).

(6) D.W. Duke and J.F. Owens, Phys. Rev. D30, 49 (1984).

(7) S. Geer, report at XXIII International Conference on High Energy Physics, Berkeley, CA (1986).

(8) V. Barger, N.G. Deshpande, R.J.N. Phillips and K. Whisnant, Phys. Rev. D33, 1912 (1986).

(9) V. Barger et al, "Signatures for Exotic Quark Singlets from Superstrings", University of Wisconsin report MAD/PH/308.

(10) V. Barger et al, University Wisconsin preprint, MAD/PH/302 (1986).

GAUGINO MASSES IN SUPERSTRING INSPIRED MODELS*

S. Dawson

Lawrence Berkeley Laboratory
University of California
Berkeley, California 94720

Contributions to the masses of the gauginos in models arising as low energy limits of superstring theories are considered.

*This work was done in collaboration with P. Binétruy and I. Hinchliffe

Low-energy models based on superstring theories have received much attention in the last several years. This is due to the fact that if 10-dimensional superstring models based on the gauge group $E_8 \times E_8$ are compactified, it is possible to obtain a low energy theory with a spectrum of quark and lepton fields compatible with that observed.[1] After compactification, the theory is described by an unbroken N=1 supersymmetry and a gauge group of the form K x G. G describes the interactions of the observable world of quarks and leptons which are singlets under K. Fields in the adjoint of K are singlets under G and form a hidden sector. The sectors K and G are coupled only by gravitational interactions.

Our model is completely determined by the gauge fields kinetic term, the Kähler potential $\mathcal{G}$, and the effective superpotential W. With these three ingredients, we are left with the standard N=1 theory of supergravity coupled to ordinary matter.

One field of particular relevance for our results is the dilaton field common to all superstring theories.[2] As a result of compactification, the dilaton combines with some fields related to the Kähler structure to form the scalar component, S, of a chiral supermultiplet $\mathcal{S}$. This field couples in identical ways to the gauge multiplets of the two sectors,

$$\mathcal{L} = \tfrac{1}{4} \int d^2\theta \; S W^{\alpha} W_{\alpha} + \text{h.c.} \tag{1}$$

where we use a superfield notation in which W^{α} denotes both the K and G gauge supermultiplets.

A simple truncation of the string theory which is believed to reproduce the basic properties of compactification on a Calabi-Yau manifold gives for the Kähler potential,

$$\mathcal{G} = -\ln(S+S^*) - 3\ln(T+T^* - 2\sum_i |\phi_i|^2) + \ln|W|^2 \tag{2}$$

where T is another combination of the dilaton and fields related to the Kähler structure, and the ϕ_i are the scalar fields of the observable sector. The superpotential W is a function of degree three or higher of the ϕ_i fields only.

If the gauge group K becomes strong at some scale Λ, then a condensate of the gauginos of the group K can occur. A non-zero gravitino mass $m_{3/2}$ is generated, along with a non-zero cosmological constant. After integrating out the gaugino condensates in the hidden sector, we are left with the effective superpotential,[3]

$$W = c + h e^{-3S/2b_o} \tag{3}$$

where b_o is the coefficient of the one loop beta function of the hidden sector , h and c are constants, and we set $\phi_i=0$ in what follows.

The observable sector of this model is unaware of supersymmetry breaking at this order. The scalar masses in the observable sector remain zero, as do the gaugino masses. Even one loop corrections fail to generate non-zero scalar masses.[4] The low energy gauge symmetry in these models is broken only if some scalar mass-squared becomes negative. Since the renormalization group equations for scalar and gaugino masses are coupled, a non-zero gaugino mass can trigger the symmetry breaking process. In this paper we investigate the masses of gauginos at one loop.[5]

The first contribution is from loops involving gravitinos, which will be present in any theory coupled to supergravity.[6] The diagrams of Fig. 1 give an equal contribution to all gaugino masses. Both the logarithmic and quartic divergences cancel and the result is a finite contribution to gaugino masses,

$$m_1 = \frac{2m_{3/2}^3}{\pi M_p^2} . \quad (4)$$

We turn now to the contributions specific to the models which are believed to arise from superstring theories. The first contribution involves χ, the spin $\frac{1}{2}$ partner of S. The contributions to the gaugino masses are given, as in the gravitino case, by the tadpole and gauge loop diagrams of Fig. 1 with ψ_μ replaced by χ. Again the divergences cancel and we are left with the finite contribution,

$$m_2 = \frac{m_{3/2}^3 \, \omega^3}{\pi M_p^2} \quad (5)$$

where $\omega \equiv \frac{3}{b_o}$ Re <S> and b_o is the one-loop beta function of the hidden sector K.

Finally there are contributions to the gaugino masses involving the scalar field S. These are more difficult to evaluate since the S field kinetic term is not canonical and, as in the non-linear σ-model, cannot be put in canonical form by a redefinition of the fields,

$$\mathcal{L}_{K.E.} = \frac{1}{4(\mathrm{ReS})^2} \partial_\mu S \partial^\mu S^* . \quad (6)$$

Our result must not depend on the particular parametrization that we choose for our fields. The simplest way to achieve this is to keep the field redefinition invariance explicit at each order of the quantum expansion. The procedure for doing this is explained in Ref. (5).

We evaluate the Feynman diagrams of Fig. 2 to obtain the contribution of the S scalars to the gaugino masses. The quadratic divergences cancel between these contributions leaving a residual logarithmic divergence. This divergence is cut off at the scale of the gaugino condensate in the hidden sector (Λ). Above this scale supersymmetry is unbroken and the S scalar is massless. The result is,

$$m_3 = \frac{m_{3/2}^3 \, \omega(\omega-3)}{2\pi M_p^2} \left[4\omega \, \ln \Lambda/m_{3/2} + (1+\omega^2) \, \ln\left|\frac{1-\omega}{1+\omega}\right| - 2\omega \, \ln (1-\omega^2)\right]. \quad (7)$$

The total contribution to gaugino masses at one loop is therefore given by Eqs. (4, 5 and 7),

$$m_{1/2} = \frac{m_{3/2}^3}{\pi M_p^2} \left\{2+\omega^3+\omega(\omega-3)\left(2\omega \, \ln \frac{\Lambda}{m_{3/2}} + \tfrac{1}{2}(1+\omega^2)\ln\left|\frac{1-\omega}{1+\omega}\right| - \omega \ln(1-\omega^2)\right)\right\} . \quad (8)$$

It is non-zero and gives an equal contribution for all the gauginos in the observable sector.

All masses in the observable sector-gauginos and, through the coupled renormalization group equations, scalars - scale like $m_{3/2}^3 / M_p^2$. Radiative corrections from the Yukawa couplings of the quark, lepton, and Higgs fields induce a negative value of the mass-squared of the Higgs doublet. The requirement that the Higgs vacuum expectation value be stable with respect to higher order corrections fixes the value of $m_{1/2}$ to be of order M_W. We are then led to a value of $m_{3/2}$ of order 10^{13} GeV.

Acknowledgement

This work was supported by the Director, Office of Energy Research, Office of High Energy and Nuclear Physics, Division of High Energy Physics of the U.S. Department of Energy under Contract DE-AC0376SF00098.

REFERENCES

(1) M. Green and J. Schwarz, Nucl. Phys. B218 (1983) 43; Phys. Lett. B149 (1984) 117; D. Gross, J. Harvey, E. Martinec, and R. Rohm, Phys. Rev. Lett. 54 (1985) 502; Nucl. Phys. B256(1985) 253; Nucl. Phys. B267 (1986) 75; P. Candelas, G. Horowitz, A. Strominger, and E. Witten, Nucl. Phys. B258 (1985) 46.

(2) E. Witten, Phys. Lett. 155B (1985) 51.

(3) M. Dine, R. Rohm, N. Seiberg, and E. Witten, Phys. Lett. 156B (1985) 55.

(4) P. Binétruy and M.K. Gaillard, Phys. Lett. 168B (1986) 347; J. Breit, B. Ovrut, and G. Segré, Phys. Lett. 162B (1985) 303.

(5) P. Binétruy, S. Dawson, and I. Hinchliffe, LBL-21606/EFI-86-28, May 1986.

(6) R. Barbieri, L. Girardello, and A. Masiero, Phys. Lett. 127B (1983) 429.

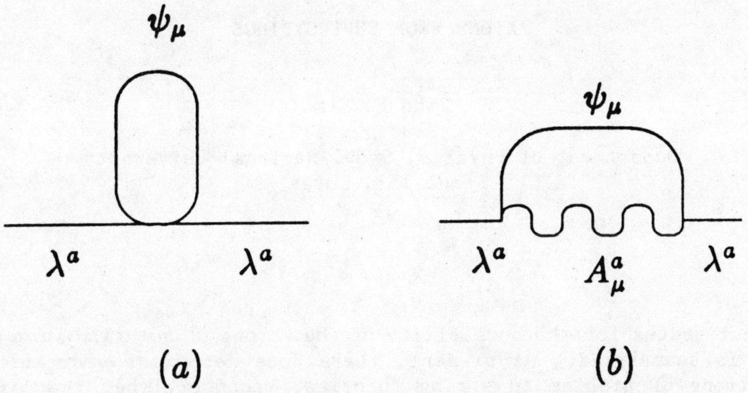

Figure 1. Contribution to gaugino masses from the spin-3/2 graviton, ψ_μ. A^a is the gauge boson associated with the gaugino, λ^a.

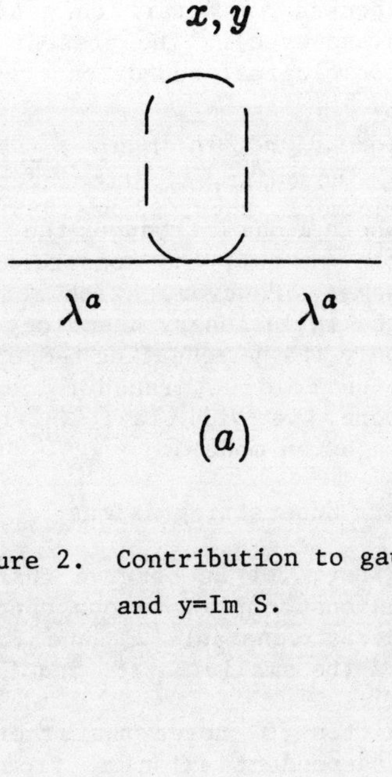

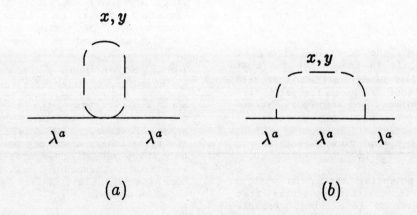

Figure 2. Contribution to gaugino masses from the S scalar. x=Re S and y=Im S.

AXIONS FROM SUPERSTRINGS

Jihn E. Kim

Department of Physics, Seoul National University
Seoul 151, Korea

The current status for the possibility of the strong CP solution in superstring theories is summarized. At present, there does not exist a decent solution of the strong CP problem in string theories, because either the light axion degrees are not enough or there exist cosmological problems.

1. Introduction

The known fact of quantum chromodynamics is that the $\bar{\theta}$ parameter is extremely small(1)

$$|\bar{\theta}| < 10^{-9} \tag{1}$$

which has led to the so-called strong CP problem. Also, it is known that the most attractive solution of this strong CP problem is the existence of axion. In an effective field theory language, the axion a is defined by

$$\mathcal{L} = \frac{1}{2}(\partial_\mu a)^2 + \frac{1}{32\pi^2}\frac{a}{F_a} F^i_{\mu\nu}\tilde{F}^{i\mu\nu} \tag{2}$$

without other potential terms, or with a negligible contribution to the potential from other sources even if it existed. Because the axion potential looks like Fig.1, the correct vacuum chooses $\bar{\theta}=0$, thus solving the strong CP problem dynamically.

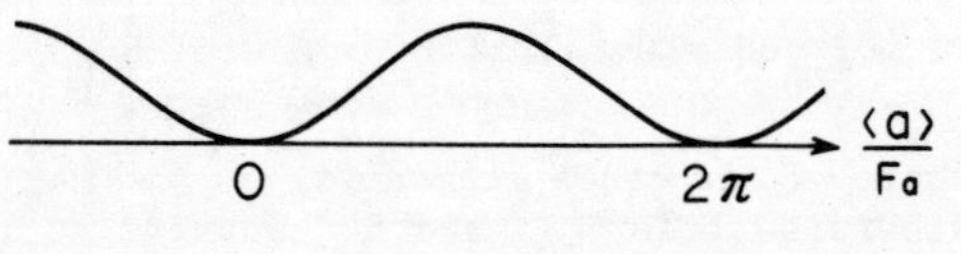

Fig.1

There are several axion models: the standard axion, the invisible axion, the composite axion, the supergravity axions, and axions in superstring models.

In all these axion models, there are two model independent parameters which are the axion decay constant F_a and the domain wall number N_{DW} (Other parameters such as the detailed axion-lepton or the axion-quark couplings are model dependent and they can be checked in detail only after the axion is discovered). The present astrophysical and cosmological bounds on these parameters are

$$10^8\,\mathrm{GeV} < F_a < 10^{12}\,\mathrm{GeV}, \quad N_{DW} = 1 \tag{3}$$

Whether one introduces the inflationary scenario or not, the constraint on F_a stays the same. However, axion models with $N_{DW} \neq 1$ in the inflationary cosmology may be acceptable only if the reheating temperature after inflation does not reach F_a. In our discussion here, we will take $N_{DW}=1$ as a constraint on axion models.

2. Superstring Axions

First, let us observe that for two would-be axions coupled to one nonabelian force with decay constants F_1 and F_2 the axion scale is the smaller of F_1 and F_2 (2).

Witten (3) noted that there arises a model-independent axion a_1 from the zero mode of B_{MN} when both M and N take the M_4 indices, e.g. μ, ν, ρ, etc. (Here, our notation is that M,N,P represent D=10 indices, μ, ν, ρ represent M_4 indices, and m,n,p,q represent the internal space (K) indices.). The a_1 coupling comes from the Chern-Simons term

$$dH = \mathrm{tr}\, R^2 - \frac{1}{30}\,\mathrm{Tr}\, F^2 \tag{4}$$

where H is the 3-form field strength constructed from antisymmetric tensor field B_{MN}. The equation of motion of a_1 is

$$\partial^2 a_1 = \frac{1}{32\pi^2}\frac{1}{F_1}\left(\mathrm{tr}R_{\mu\nu}\tilde{R}^{\mu\nu} - \frac{1}{30}\mathrm{Tr}F_{\mu\nu}\tilde{F}^{\mu\nu}\right) \quad (5)$$

where $F_1 = g^2 M_{P1}/192\pi^{5/2} \approx (10^{15} - 10^{16}\ \mathrm{GeV})$(2). The value F_1 is too large compared to the cosmological bound Eq.(3). It is desirable to have another would-be axion to bring the axion scale down along the scheme mentioned in the beginning of this section. The stituation for this scenario is the following. The O(32) model with internal space monopole compactification works too much, i.e. the axion scale comes down to 300 MeV, and it is difficult to give realistic masses to Q=2/3 quarks (4). With Calabi-Yau space compactification, the axion scale cannot be lowered (2), but in non-Calabi-Yau spaces it is possible to bring the axion scale to the desired range (5).

This brings us to consider the model-dependent axion a_2 (6). The number of the pseudo-scalar zero modes in B_{MN} is counted by the 0th Betti number ($B_{\mu\nu}$) and the 2nd Betti number (B_{mn}). The 0th Betti number is 1 for the connected manifold and the related axion has been discussed as the model-independent one. The 2nd Betti number is greater than zero for Calabi-Yau spaces, and we anticipate at least one more pseudo-scalar degree. For this additional degree, say a_2, to be interpreted as an axion, it must have $a_2 FF$ coupling as explained in Introduction and the coupling must be independent from that of a_1. In contrast to a_1, a_2 coupling arises from the Green-Schwarz term BX_8 (7) where

$$X_8 = \frac{1}{24}\mathrm{Tr}\,F^4 - \frac{1}{7200}(\mathrm{Tr}F^2)^2 - \frac{1}{240}\mathrm{Tr}F^2\mathrm{tr}R^2 + \frac{1}{8}\mathrm{tr}R^4 + \frac{1}{32}(\mathrm{tr}R^2)^2 \quad (6)$$

Because the second axion might be relevant only for $E_8 \times E_8'$ models because of a possibility of two surviving nonabelian groups, we restrict ourselves to $E_8 \times E_8'$ models for Calabi-Yau space compactification. For E_8, the quartic Casimir is not independent, but is related to the quadratic Casimir as $\mathrm{Tr}F^4 = (\mathrm{Tr}F^2)^2/100$, which simplifies the algebra. The D=4 operators obtained from Eq.(6) through compactification shift the vacuum angle, and hence are not of our interest. The D=5 and D=3 operators are potentially important, because they show the axion couplings and the Higgs phenomena. The D=5 terms are

$$\begin{aligned} BX_8 \propto\ & \varepsilon^{\mu\nu\alpha\beta}\varepsilon^{mnpqrs}B_{mn} \\ & \cdot\{2\,\mathrm{Tr}F_{\mu\nu}F_{\alpha\beta}\mathrm{Tr}\langle F_{pq}\rangle\langle F_{rs}\rangle \\ & + 2\,\mathrm{Tr}F'_{\mu\nu}F'_{\alpha\beta}\,\mathrm{Tr}\langle F'_{pq}\rangle\langle F'_{rs}\rangle \\ & - \mathrm{Tr}F_{\mu\nu}F_{\alpha\beta}\mathrm{Tr}\langle F'_{pq}\rangle\langle F'_{rs}\rangle \\ & - \mathrm{Tr}F'_{\mu\nu}F'_{\alpha\beta}\mathrm{Tr}\langle F_{pq}\rangle\langle F_{rs}\rangle \\ & - 15(\mathrm{Tr}F_{\mu\nu}F_{\alpha\beta} + \mathrm{Tr}F'_{\mu\nu}F'_{\alpha\beta})\mathrm{tr}\langle R_{pq}\rangle\langle R_{rs}\rangle\} \\ \propto\ & B_{mn}\,(F\tilde{F} - F'\tilde{F}') \end{aligned} \quad (7)$$

where we have used $\mathrm{Tr}F^2 = 30\mathrm{tr}R^2$. Therefore, a_2 or B_{mn} has the axion coupling which is independent from that of a_1 (7). We therefore interpret a_2 as a model-dependent axion. As we mentioned before, there should not be additional couplings to interpret both $B_{\mu\nu}$ and B_{mn} as axions. The other couplings of light degrees $B_{\mu\nu}$ and B_{mn} come from $(H_{MNP})^2$ and D=3 terms of BX_8. These are

$$\begin{aligned} H_{\mu mn}^{\ 2} = \{ & \sum_{A=1} \partial_\mu B^A C^A_{mn} - \frac{1}{30}\mathrm{Tr}(A_\mu\langle\partial_m A_n - \partial_n A_m\rangle \\ & + (A^i\partial_\mu A^j - A^j\partial_\mu A^i)C^i_m C^j_n\}^2 \end{aligned} \quad (8)$$

and

$$\begin{aligned} & \varepsilon^{\mu\nu\alpha\beta}\varepsilon^{mnpqrs}B_{\mu\nu} \\ & \cdot\{4\,\mathrm{Tr}\,F_{\alpha\beta}\langle F_{mn}\rangle\mathrm{Tr}\langle F_{pq}\rangle\langle F_{rs}\rangle \\ & + 4\,\mathrm{Tr}\,F'_{\alpha\beta}\langle F'_{mn}\rangle\mathrm{Tr}\langle F'_{pq}\rangle\langle F'_{rs}\rangle \\ & - 2\,\mathrm{Tr}\,F_{\alpha\beta}\langle F_{mn}\rangle\mathrm{Tr}\langle F'_{pq}\rangle\langle F'_{rs}\rangle \\ & - 2\,\mathrm{Tr}\,F'_{\alpha\beta}\langle F'_{mn}\rangle\mathrm{Tr}\langle F_{pq}\rangle\langle F_{rs}\rangle \\ & - 30(\mathrm{Tr}F_{\alpha\beta}\langle F_{mn}\rangle + \mathrm{Tr}F'_{\alpha\beta}\langle F'_{mn}\rangle). \\ & \mathrm{tr}\langle R_{pq}\rangle\langle R_{rs}\rangle\} \end{aligned} \quad (9)$$

It looks like that both $B_{\mu\nu}$ and B_{mn} couple to the gauge field A_μ. But the inspection of Eq.(8) and Eq.(9) shows that only one combination couples to A_μ if

$$[T, \langle F_{mn}\rangle] = 0, \quad \mathrm{Tr}\ T\langle F_{mn}\rangle \neq 0 \quad (10)$$

where T is the generator of U(1) group. Otherwise both $B_{\mu\nu}$ and B_{mn} survive as light degrees (axions). If Eq.(10) is satisfied, one combination becomes the longitudinal degree of the corresponding U(1) gauge boson, and there survives a global $U(1)_{PQ}$ symmetry below the compactification scale, which is reminiscent of the 't Hooft mechanism. This affair is summarized as Eq.(10); If Yang-Mills holonomy has (an) invariant abelian subgroup, there exists $U(1)_{PQ}$, and if Yang-Mills holonomy does not have an invariant abelian subgroup, the light degrees are present below the compactification scale. The former case arises in type I O(32) model compactified with monopole configuration (3), and the latter case arises in the Calabi-Yau compactification of $E_8 \times E_8'$ heterotic string.

Therefore, the $E_8 \times E_8'$ model with the Calabi-Yau compactification has at least two axions a_1 (from $B_{\mu\nu}$) and a_2(from B_{mn}). For the minimal case ($b_{2,0}=b_{0,2}=0$, $b_{1,1}=1$), we have already calculated the axion couplings of a_1 and a_2. This conclusion does not take into account the world sheet instanton effect (9).

3. Domain Walls

To determine N_{DW}, we need a full information on the periodicity of the axion field variable (10). However, for two axions we can say something like (integer)+(integer)=(integer) and (half integer)+(half integer)=(integer) even though we do not have the full information (11). We have calculated the axion interactions for $E_8 \times E_8'$ model with the Calabi-Yau compactification,

$$\mathcal{L}[a_1,a_2] \ni \frac{1}{2}(\partial_\mu a_1)^2 + \frac{1}{2}(\partial_\mu a_2)^2 - \frac{g^2}{32\pi^2}\left(\frac{a_1}{F_1} + \frac{a_2}{F_2}\right)F\tilde{F} - \frac{g'^2}{32\pi^2}\left(\frac{a_1}{F_1} - \frac{a_2}{F_2}\right)F'\tilde{F}' \tag{11}$$

Let us assume the periodicities of a_1 and a_2 are

$$a_1 = a_1 + 2\pi N_1F_1, \quad a_2 = a_2 + 2\pi N_2F_2 \tag{12}$$

Since the integrals are

$$\int\frac{1}{32\pi^2}F\tilde{F}\, d^4x = \text{integer}, \qquad \int\frac{1}{32\pi^2}F'\tilde{F}'\, d^4x = \text{integer} \tag{13}$$

we require

$$\frac{\langle a_1\rangle}{F_1} + \frac{\langle a_2\rangle}{F_2} = 2\pi\ (\text{integer}), \qquad \frac{\langle a_1\rangle}{F_1} - \frac{\langle a_2\rangle}{F_2} = 2\pi\ (\text{integer}) \tag{14}$$

The number of vacuum states is counted by

$$\left\{\frac{\langle a_1\rangle}{F_1}, \frac{\langle a_2\rangle}{F_2}\right\} \tag{15}$$

states consistent with Eq.(14) but within the periods of Eq.(12). These are

$$\{2n\pi,\ 2m\pi\}, \quad \{(2n+1)\pi,\ (2m+1)\pi\} \tag{16}$$

where $n=0,1,\ \ldots,\ N_1-1$ and $m=0,1,\ldots,N_2-1$. The case for $N_1=N_2=1$ is drawn in Fig.2. The black dots count the number of degenerate vacua.

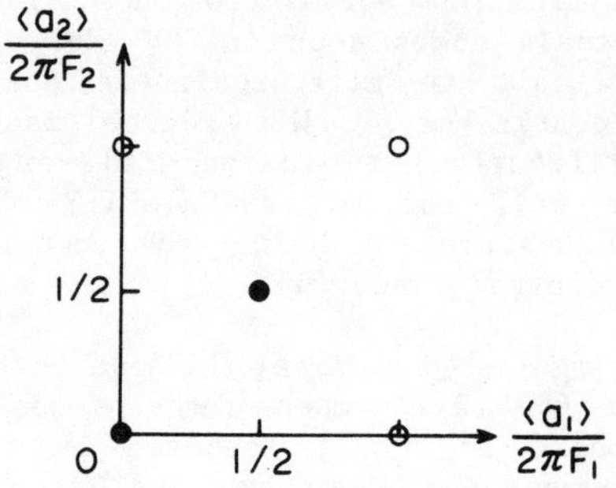

Fig.2

Counting the domain wall number according to Eq.(16), we obtain

$$N_{DW} = 2N_1N_2 \geq 2 \tag{17}$$

which does not satisfy the second condition in Eq.(3).

4. Conclusion

We have discussed the axion degrees arising in superstring models. For these to be interpreted as axions, these light degrees should not have contributions to their potential except those arising from $a_iF\tilde{F}$ (i=1,2) and $a_iF'\tilde{F}'$ (i=1,2) terms. However, it is found recently that the effects of world sheet instantons in some Calabi-Yau spaces contribute to the potential of the model-dependent axion a_2(9). Then a_2 is not an axion. Therefore, at present it looks like that we have not yet found a decent mechanism for the strong CP solution in superstring models due to the effect of the world sheet instantons. This statement applies to most Calabi-Yau spaces, but not to all. Even though the world sheet instanton effects are negligible, there are cosmological problems related to the superstring axions, which are discussed in (12). Therefore, it is of utmost interest to find a mechanism for the strong CP solution in string theories or find a working inflation. The axion problem in string theories is an excellent hurdle for the verification of string theory.

REFERENCES

(1) V. Baluni, Phys. Rev. D19(1979) 2227; R. Crewther et al, Phys. Lett. 89B(1979) 123.

(2) K. Choi and J.E. Kim, Phys. Lett. 154B (1985) 393.

(3) E. Witten, Phys. Lett. 149B(1985) 351.

(4) P.H. Frampton, H. Van Dam and K. Yamamoto, Phys. Rev. Lett. 54(1985) 1114.

(5) S.M. Barr, Phys. Lett. 158B(1985) 397.

(6) E. Witten, Phys. Lett. 155B(1985) 151.

(7) K. Choi and J.E. Kim, Phys. Lett. 165B (1985) 71; J.-P. Derindinger, L. Ibanez and H.P. Nilles, Nucl. Phys. B267(1986) 365.

(8) K. Choi, Ph.D. Thesis, Seoul National University, Dec.(1985).

(9) X.-G. Wen and E. Witten, Phys. Lett. 166B(1986) 397; M. Dine, N. Seiberg, X.-G. Wen and E. Witten, Princeton University preprint (1986).

(10) P. Sikivie, Phys. Rev. Lett. 48(1982) 1156.

(11) K. Choi and J.E. Kim, Phys. Rev. Lett. 55(1985) 2637.

(12) S.M. Barr, K. Choi and J.E. Kim, BNL preprint 37467 (1986).

Parallel Session 2

Superstrings

Organizers:
J. Bagger (IAS & SLAC)
P. West (King's College)

Scientific Secretaries:
J.J. Atick (SLAC)
P. Windey (LBL)
J. Yamron (LBL)

Superstring Field Theory and the Covariant Fermion Emission Vertex

H. Nicolai
Institut für Theoretische Physik
Universität Karlsruhe

1. Introduction

In superstring field theory (see e.g.[1-3]), there arises the need for an operator that converts Neveu Schwarz (NS) bosons into Ramond (R) fermions and vice-versa. Such an operator is necessary to exhibit the space-time supersymmetry of the G.O.S.-projected NSR model and for the computation of scattering amplitudes involving bosons and fermions (for a review containing many further references, see [4]).

To be of use in covariant superstring field theory this fermion emission vertex must furthermore be covariant. This means that it should also properly describe the conversion of the unphysical longitudinal and timelike bosonic and fermionic degrees of freedom. More precisely, it should be a physical operator in the sense that it (anti)commutes with the BRST-operator, i.e. we must require

$$[Q_{BRST}, V_{phys}] = 0 \qquad (1)$$

If (1) is satisfied, the operator V_{phys} converts a physical state (which is annihilated by Q_{BRST}) into another physical state. Physical operators are usually constructed as complex line integrals according to

$$V_{phys} = \int \frac{dz}{2\pi i z} V_{phys}(z) \qquad (2)$$

where $V_{phys}(z)$ obeys

$$[Q_{BRST}, V_{phys}(z)] = z\frac{d}{dz}W(z) \qquad (3)$$

with some other operator $W(z)$. In order for (3) to be valid, $V_{phys}(z)$ must be a conformal field [5] of dimension one, or, in other words, we must have

$$[L_m, V_{phys}(z)] = z^m\left(z\frac{d}{dz} + mJ\right)V_{phys}(z) \qquad (4)$$

with J=1 (L_m are the Virasoro generators). For instance, the tachyon emission vertex :exp ik.X(z): is physical only for $m^2=-k^2=-2$. In general, the operator $V_{phys}(z)$ may not only depend on the orbital oscillators but also on the ghost fields of the theory (by "orbital oscillators" we mean the oscillators associated with the matter fields $X^\mu(z)$ or $\lambda^\mu(z)$).

In this contribution, we review some recent work on the covariant fermion emission vertex and its application to superstring field theory. This vertex was first obtained in [5] in a version with bosonized ghosts. Here we will concentrate on an alternative and more direct construction of this operator which was given in [3] (for related work see also [6]).

2. Ghosts in Superstring Theory

In the framework of BRST quantization of string theories [7,8] one introduces ghosts and antighosts for all the symmetries of the theory such that commuting (anticommuting) symmetry generators are associated with anticommuting (commuting) ghosts. The symmetry underlying superstring theory is the superconformal extension of the conformal group in two dimensions which is generated by the operators L_m, F_m and L_m, G_r in the R-and NS-sectors, respectively. Thus, we have the following ghosts

$$\left.\begin{array}{l} L_m \leftrightarrow c_m, \overline{c}_m \\ F_m \leftrightarrow e_m, \overline{e}_m \\ G_r \leftrightarrow e_r, \overline{e}_r \end{array}\right\} \begin{array}{l} m \in Z \\ r \in Z + \frac{1}{2} \end{array} \qquad (5)$$

with (anti)commutation relations

$$\{c_m, \overline{c}_n\} = [e_m, \overline{e}_n] = \delta_{m+n,0}$$
$$[e_r, \overline{e}_s] = \delta_{r+s,0} \qquad (6)$$
all others = 0

(for a more complete discussion, see e.g.[8]). Note the appearance of a commuting zero-mode ghost oscillator in the R-sector which is the source of many complications.

The BRST operator is constructed according to the standard prescription[8]. For instance, for the bosonic string, we have

$$Q_{BRST} = \sum_m c_{-m} (L_m - \alpha_0 \delta_{m,0})$$
$$- \frac{1}{2} \sum_{m,n,p} f_{mn}{}^p : c_{-m}\, c_{-n}\, \overline{c}_p : \qquad (7)$$

where $f_{mn}{}^p = (m-n)\delta_{m+n,p}$ are the structure constants of the Virasoro algebra and the colons denote normal ordering. As is well konwn, $Q_{BRST}^2=0$ in the quantum theory only if D=26 and $\alpha_0=1$ [7]. Analogous results hold for the NS-and R-sectors of the spinning string [8,9].

To utilize this formalism in string field theory, we need to specify the vacuum state. The most natural choice is the one with respect to which Q_{BRST} is normal ordered, namely

$$c_m|0\rangle = \overline{c}_m|0\rangle = e_m|0_R\rangle = \overline{e}_m|0_R\rangle =$$
$$= e_r|0_{NS}\rangle = \overline{e}_r|0_{NS}\rangle = 0$$
$$\text{for } m,r > 0$$
$$\overline{c}_0|0\rangle = \overline{e}_0|0_R\rangle = 0 \qquad (8)$$

In addition, we need another vacuum $|\tilde{0}_R\rangle$ in the R sector which is defined by

$$\overline{e}_0|\tilde{0}_R\rangle = 0 \qquad (9)$$

and whose ghost number differs by one from the ghost number of $|0_R\rangle$. In fact, this is just the tip of the iceberg: there exists an infinity of "picture changed" vacua in both the NS- and the R-sectors [5]. It is not clear what their relevance is in superstring field theory.

3. The Covariant Fermion Emission Vertex

The orbital part of the fermion emission vertex has been known for a long time [10]. It can be written as

$$W_{orb}(z)=\sqrt{z}\,\exp(-zL^{R}_{-1})\,\tilde{W}(z) \qquad (10)$$

where

$$W(z)=\langle 0_{NS}|\exp \sum_{\substack{m\geq 1/2\\ r\geq 1/2}} \Gamma^{\mu}_{-m}B_{mr}(z)\gamma^{*}b_{r\mu}|0_{R}\rangle$$

$$\times \exp \tfrac{1}{2}\sum_{r,s\geq 1/2} b^{\mu}_{r}\,A_{rs}(z)\,b_{s\mu} \qquad (11)$$

The coefficients $B_{mr}(z)$ and $A_{rs}(z)$ are given by [10]

$$A_{rs}=$$

$$=\tfrac{1}{2}z^{-r-s}\,\frac{s-r}{r+s}(-1)^{r+s+1}\begin{pmatrix}-\frac{1}{2}\\ r-\frac{1}{2}\end{pmatrix}\begin{pmatrix}-\frac{1}{2}\\ s-\frac{1}{2}\end{pmatrix}$$

$$B_{mr}(z)=\frac{1}{\sqrt{2}}z^{m-r}\begin{pmatrix}m-\frac{1}{2}\\ r-\frac{1}{2}\end{pmatrix}(-1)^{m-r+1/2} \qquad (12)$$

and the NS- and R-oscillators appear in the NS- and R-fields as

$$H^{\mu}(z)=\sum_{r\in Z+\frac{1}{2}} b^{\mu}_{-r}z^{r}=H^{\mu}(z^{-1})^{+}$$

$$\Gamma^{\mu}(z)=\gamma^{\mu}+i\sqrt{2}\gamma^{*}\sum_{m\neq 0} d^{\mu}_{-m}\,z^{m}$$

$$\equiv\sum_{m=-\infty}^{+\infty}\Gamma^{\mu}_{-m}z^{m}=\gamma^{0}\Gamma^{\mu}(z^{-1})^{+}\gamma^{0} \qquad (13)$$

The operator $\tilde{W}(z)$ has been designed to convert the NS-field $H^{\mu}(z)$ into the R-field $\Gamma^{\mu}(z)$ and vice-versa

$$\tilde{W}(z)\gamma_{*}\frac{H^{\mu}(y)}{\sqrt{y}}=-\frac{i}{\sqrt{2}}\frac{\Gamma^{\mu}(y-z)}{\sqrt{y-z}}\,\tilde{W}(z) \qquad (14)$$

It is known that

$$[L_{m},W_{orb}(z)]=$$

$$=z^{m}\left(z\frac{d}{dz}+\frac{5m+1}{8}\right)W_{orb}(z) \qquad (15)$$

$W_{orb}(z)$ is not a physical operator as it has conformal dimension5/8, and we must therefore look for another operator of dimension 3/8. This operator comes from the ghost sector and is given by [3]

$$W_{gh}(z)=\langle 0_{NS}|\hat{W}_{gh}(z)|\tilde{0}_{R}\rangle \qquad (16)$$

with

$$W_{gh}(z)=$$

$$=\exp\Big(-\sum_{r,s\geq\frac{1}{2}} z^{-r-s}\,e_{r}\,\alpha_{rs}\,\bar{e}_{s}$$

$$+i\sum_{r\geq\frac{1}{2}} z^{-r+m}\,e_{r}\,\beta_{\ell m}\,\bar{e}_{-m}$$

$$+i\sum_{r\geq\frac{1}{2}} z^{m-s}\,e_{-m}\,\delta_{ms}\,\bar{e}_{s}$$

$$+\sum_{\substack{m\geq 1\\ n\geq 0}} z^{m+n}\,e_{-m}\,\gamma_{mn}\,\bar{e}_{-m}\Big) \qquad (17)$$

It is absolutely crucial that the $\hat{W}_{gh}(z)$ is multiplied by $|\tilde{0}_{R}\rangle$ and <u>not</u> $|0_{R}\rangle$ from the right; observe that the summation range in the exponent of (17) is natural for this choice of vacuum. The coefficients in (17) are determined from the generating functions

$$\alpha(x,y)\equiv\sum_{r,s\geq\frac{1}{2}} x^{r}\alpha_{rs}\,y^{s}$$

$$=\frac{x^{\frac{1}{2}}y^{\frac{1}{2}}}{x-y}\left[1-\left(\frac{1-y}{1-x}\right)^{1/2}\right]$$

$$\beta(x,y)\equiv\sum_{\substack{r\geq\frac{1}{2}\\ n\geq 0}} x^{r}\,\beta_{rn}\,y^{n}$$

$$=\frac{x^{\frac{1}{2}}}{1-xy}\left(\frac{1-y}{1-x}\right)^{1/2} \qquad (18)$$

and

$$\gamma^{mn}\equiv\alpha_{m-1/2,\,n+1/2}$$

$$\delta_{ms}\equiv\beta_{m-1/2,\,s-1/2} \qquad (19)$$

It is now a matter of lengthy calculations to work out the commutator of the full vertex

$$W(z)\equiv W_{orb}(z)W_{gh}(z) \qquad (20)$$

with the BRST operator. One first shows that indeed,

$$[L_m, W_{gh}(z)] = z^m\left(z\frac{d}{dz} + \frac{3m-1}{8}\right)W_{gh}(z) \qquad (21)$$

and, finally,

$$[Q_{BRST}, W(z)] = z\,\frac{d}{dz}\left(c(z)W(z)\right) \qquad (22)$$

as required.

4. Superstring Field Theory

The operator briefly described above may be utilized in superstring field theory. A string field is a functional of the string coordinates and the ghosts [11]; so, for example, in the NS-sector, we consider the expression

$$\Psi_{NS}\left(X^\mu(z), \lambda^\mu(z), c(z), \bar{c}(z), e(z), \bar{e}(z)\right)$$

$$\equiv \Big\{\Psi^{\circ}_{NS}\left(X^\mu(z), \lambda^\mu(z)\right) +$$

$$+ \sum_{r,s \geq \frac{1}{2}} e_{-r}\,\bar{e}_{-s}\,S^{rs}\left(\dot{X}^\mu(z), \lambda^\mu(z)\right)$$

$$+ \ldots\Big\}|0_{NS}\rangle \qquad (23)$$

which is furthermore constrained to have a fixed ghost number=$-1/2$. Note that the coefficients in this expansion are still functionals of the "orbital" variables $X^\mu(z)$ and $\lambda^\mu(z)$ and thus still contain infinitely many ordinary quantum fields. There is a similar expansion for the string functional in the R-sector but we will not write it out here, for lack of space, see [1-3]. On the basis of (22) one might now try to define the supercharge operator as a line integral of W(z) but inspection of the various expressions shows that W(z) has a square root branch cut. To make it well-defined one introduces the G.O.S. projector [12]

$$P_{NS} \equiv \frac{1}{2}\left\{1 - (-1)^{\sum_{r \geq \frac{1}{2}}\left(b^\mu_{-r}b_{r\mu} + \bar{e}_{-r}e_r + e_{-r}\bar{e}_r\right)}\right\} \qquad (24)$$

and a corresponding one in the R-sector. Thus, combining (2) and (22), we get the single-valued supercharge operator

$$W_{SUSY} \equiv \int\frac{dz}{2\pi iz}\,W(z)\,P_{NS} \qquad (25)$$

which may how be used to define space-time supersymmetry transformations. For instance, the variation of the NS field (23) is given by

$$\delta\Psi_{NS} = W^{+}_{SUSY}\,\Psi_R \qquad (26)$$

where W^{+}_{SUSY} is the adjoint of the operator (25) which takes an R-state into an NS-state. In the zero-slope limit, (26) reduces to

$$\delta A_\mu = \varepsilon\gamma_\mu\lambda \qquad (27)$$

i.e. the well-known transformation-law of D=10 super-Yang Mills theory [12]. A similar transformation rule for $\delta\Psi_R$ may be given but it is more complicated since it involves "picture changing", see however [2,3]. Using well-known technology [10] one can also compute the commutator of two supersymmetry transformations. For instance, in the NS-sector one has [2,13]

$$[\delta_1, \delta_2]\Psi_{NS} = \varepsilon_1\gamma^\mu\varepsilon_2\partial_\mu\Psi_{NS} +$$

$$+ \left\{Q_{BRST}, X\right\}\,\Psi_{NS} \qquad (28)$$

where $X \propto \sum_r \bar{\varepsilon}_1\,\gamma_\mu\,\varepsilon_2\,b^\mu_{-r}\,\bar{e}_r$.

Thus, the supersymmetry algebra closes only on-shell, i.e. up to a BRST-commutator.

Acknowledgment: The work described here has been done in collaboration with Y. Kazama, A. Neveu and P.C. West.

References

[1]Y. Kazama, A. Neveu, H. Nicolai and P. West, CERN preprint TH.4301/85 (1985), to be published in Nucl. Phys. B.
T. Banks, M. E. Peskin, C.R. Preitschopf, D. Friedan and E. Martinec, SLAC preprint 3853 (1985);
A. LeClair and J. Distler, Harvard preprint HUTP-86/A008(1985);
H. Terao and S. Uehara, Hiroshima preprint RRK 86-3 (1986);
C. D. Date', M.Günaydin, M. Pernici, K. Pilch and P. van Nieuwenhuizen, Stony Brook preprint ITP-SB-86-3 (1986);
S. P. de Alwis and N. Ohta, Texas preprint UTTG-06-86 (1986);
D. Pfeffer, P. Ramond and V. Rogers, Florida preprint UFTP-85-19 (1985)

[2]E. Witten, Princeton preprint (1986)

[3]Y. Kazama, A. Neveu, H. Nicolai and P. West, CERN preprint TH 4418/86

[4]J. H. Schwarz, Phys. Rep. 89C (1982) 223

[5]D. Friedan, E. Martinec and S. Shenker, Nucl. Phys. B271 (1986) 93

[6]H. Terao and S. Uehara, Hiroshima preprint RRK 86-14 (1986)

[7]M. Kato and K. Ogawa, Nucl. Phys. B212 (1983) 443;
S. Hwang, Phys. Rev. D28 (1983) 2614

[8]J. H. Schwarz, preprint CALT-68 -1304 (1985)

[9]N. Ohta, Phys. Rev. D33 (1986) 1681;
M. Itō, T. Morozumi, S. Nojiri and S. Uehara, Progr. Theor. Phys. 75 (1986) 934

[10]E. Corrigan and D. Olive, Nuovo Cim. 11A (1972) 749;
E. Corrigan and P. Goddard, Nuovo Cim. 18A (1973) 339

[11]W. Siegel, Phys. Lett. 151B (1985) 391

[12]F. Gliozzi, J. Scherk and D. Olive, Nucl. Phys. B122 (1977) 253

[13]Y. Kazama, A. Neveu and H. Nicolai, work in progress.

GAUGE COVARIANT STRING THEORY

P.C. West

C.E.R.N., CH1211, Geneva 23, Switzerland.[1]

An outline of gauge covariant string theory is given.

Since there are a number of talks on this subject at this conference I will first discuss some of the relavent history of string theory. Most of the important developments in modern physics have resulted from a deeper understanding of a symmetry principle. Given this symmetry principle, say general coordinate transformations for the case of general relativity, we can construct invariant actions which are then used to weight the Feynman path integral and derive the S-matrix.

In contrast, the first step in string theory was to guess the four point[1] and then the n point[2] S-matrix at the tree level. Of course, had their guess included the entire S-matrix, including loop and non-perturbative effects, there would be nothing left to do. The dual model[3] was the first step in the understanding of loop effects. Although this model consists of vertices and propagators the combinatoric rules which specified how these pieces should be assembled to compute the S-matrix are not those of field theory. The old dual model, however, did not incorporate Faddeev-Poppov ghosts so when calculating loops the effect of the ghosts had to be inserted by hand with projectors.

From the dual model it was realized that string theory processes could be viewed as a weighted sum over the world sheet of the string. The appropriate weight in this Feynman path integral being provided by the correctly gauge fixed Nambu action. Although one could calculate loop effects the weight of each diagram had to be seperately determined by unitarity.

Despite extensive progress in the above approaches, there was no attempt to find the deep symmetrics which underlie string theory and are responsible for the many miraculous results of string theory. This symmetry must include in the relevant strings general coordinate, Yang-Mills and supersymmetry transformations.

Gauge covariant string theory is the search for a formulation of string theory in which the string symmetry is manifest. At present, this formulation is a string field theory; that is the primary objects are functionals of the string co-ordinate $x^{\mu}(\sigma)$. This approach embodies the possibility of computing non-perturbative effects as does point particle field theory.

It is not entirely clear from the beginning whether field theory would lead to a good formulation of string theory, but as we will see the results found so far have a very elegant and simple appearance. The gauge covariant free theory has been understood for a number of months and we begin by giving a brief account of the free open bosonic string.

As a result of its two dimensional reparameterization invariance the open bosonic string is a system which classically has the constraints

----

[1]Permanent Address: King's College (KQC), London WC2.

$$L_n \equiv \frac{\pi\alpha'}{2}\int_{-\pi}^{\pi} d\sigma\, e^{-in\sigma}\, (\mathcal{P}^\mu)^2 = 0 \quad (1)$$

where

$$\mathcal{P}^\mu \equiv P^\mu - \frac{1}{2\pi\alpha'}\frac{dx^\mu}{\partial\sigma} \quad (2)$$

and P^μ is the momentum conjugate to x^μ and is easily computed from the Nambu action for the classical string. These constraints generate the two dimensional conformal group, namely

$$\{L_n, L_m\} = -i(n-m)L_{n+m} \quad (3)$$

The correct constraints[4] to impose on the quantum system, after the usual operator replacements, are

$$L_n|\psi\rangle = 0, n \geq 1; (L_0 - 1)|\psi\rangle = 0 \quad (4)$$

These constraints lead to a ghost free spectrum when the space-time dimension D is 26[5]. In equation (4) one may view $|\psi\rangle$ as a functional of $x^u(\sigma)$ or think of it as being in any other basis such as the oscillator basis. An alterntive way of finding the on-shell states of string theory is to solve the constraints of equation (1) by going to the light cone-gauge which has coordinates $x^i(\sigma)$, x^- and $x^+ = \tau$. The functionals $\psi(x^i(\sigma), x^-, \tau)$ i = 1,...,24, then only contains the physical degrees of freedom.

The open string contains Yang-Mills fields and equation (4) contains within it the constraints $\partial\mu A^\mu = 0$, $\partial^2 A^\mu = 0$.

The problem of free gauge covariant theory is to find a local action which leads to the correct on-shell spectrum discussed above. For the Yang-Mills fields the result is of course $\partial^\mu\partial_\mu A_\nu - \partial_\nu\partial^\mu A_\mu = 0$ which possess the gauge invariance $\delta A_\mu = \partial_\mu\Lambda$. The string however contains particles of ever increasing spin with increasing mass. Unlike spins 0, $\frac{1}{2}$ and 1, all higher spins do not possess Lorentz invariant actions which can be constructed <u>only</u> from the fields that occur in the on-shell conditions and which lead to the correct on-shell conditions. Above spin 1, one must include in the Lorentz covariant action new fields that disappear from the on-shell conditions.

For example massive spin 2 is described by the on-shell conditions $h_\mu{}^\mu = (\partial^2 - m^2)h_{\mu\nu} = \partial^\mu h_{\mu\nu} = 0$. To find an action which leads to this on-shell system we must introduce a scalar which can be identified as the trace of $h_\mu{}^\mu$. We call these additional fields, supplementary fields.

Since the string contains an infinite number of higher spin particles we must introduce an infinite number of supplementray fields which must be identified in terms of string functionals. The most convenient formulations of gauge covariant string theory in fact contain an infinite number of supplementary string functionals. The problem now arises as to how to handle these additional fields. The answer is to introduce two additional anti-commuting co-ordinates $c(\sigma)$ and $\bar{c}(\sigma)$ making with $x^u(\sigma)$ a 28 dimensional space. The encoding works much the same way as it does in the superspace formulation of supersymmetric theories, except in this case an algebraic constraint is used to eliminate all the anticommuting and even some of the commuting component string functions. We therefore consider functionals $\chi[x^u(\sigma), c(\sigma), \bar{c}(\sigma)]$. The free action is of the form

$$\langle\chi|Q|\chi\rangle = \int \mathcal{D}x^\mu(\sigma)\mathcal{D}c(\sigma)\mathcal{D}\bar{c}(\sigma)\, \chi\, Q\, \chi \quad (5)$$

where

$$Q \equiv Q\left[x^\mu(\sigma), c(\sigma), \bar{c}(\sigma), \frac{\delta}{\delta x^\mu(\sigma)}, \frac{\delta}{\delta c(\sigma)}, \frac{\delta}{\delta \bar{c}(\sigma)}\right]$$

The object Q has in fact a deep group theoretic interpretation. Since this object has been met before in other contexts and is likely to play an important part in future developments I will give a general definition. Consider a Lie group G whose generators are represented by L_n and which obey the commutation relations

$$[L_n, L_m] = f_{nm}{}^p L_p \quad (6)$$

If we are dealing with a constrained system, the constraints which generate G will satisfy in the classical theory a Poisson Bracket relation as they do for the string (see equation (3)).

We now introduce two anti-commuting oscillators $\beta_n, \bar{\beta}_n$ for each generator, L_n of G. These oscillators obey the relations $\{\beta_n, \bar{\beta}_m\}$ =

$\delta_{m+n,0}$ while all other anti-commutators vanish. We define

$$\hat{Q} = \sum_n \beta_n L_{-n} - \frac{1}{2}\sum_{n,m,p} \bar{\beta}_p f_{nm}{}^p \beta_{-n}\beta_{-m} \quad (7)$$

This object arises in the B.R.S.T. quantization of gauge theories. It is straightforward to demonstrate that $Q^2 = 0$. For the string we take G to be the D = 2 conformal group and due to normal ordering ambiguities in the quantum theory we actually use the object.

$$Q = :\hat{Q} - \beta_0: \quad (8)$$

In reference [6] it was shown that $Q^2 = 0$ in D = 26.

In fact, the Q in equation (8) is the Q given above in the expression for the free string action and the $\beta_n, \bar{\beta}_m$ are constructed from the coordinates as follows

$$\bar{\beta}(\sigma) = \frac{\delta}{\delta c(\sigma)} - \frac{1}{2\pi}\bar{c}(\sigma) = \frac{1}{\pi}\frac{1}{\sqrt{2}}\sum_{n=-\infty}^{\infty} \bar{\beta}_n e^{in\sigma}$$

$$\beta(\sigma) = -\frac{\delta}{\delta \bar{c}(\sigma)} + \frac{1}{2\pi} c(\sigma) = \frac{1}{\pi}\frac{1}{\sqrt{2}}\sum_{n=-\infty}^{\infty} \beta_n e^{in\sigma} \quad (9)$$

Clearly the above free action which was found in references [7], [8] and [9] is invariant under

$$\delta|\chi\rangle = Q|\chi\rangle \quad (10)$$

provided the dimension of space-time is 26. By using the oscillator formalism one can work out the action of equation (5) in terms of component fields. In particular, one can show that it contains the free Yang-Mills action, if in somewhat an unusual first order form.

In fact, it is not difficult to demonstrate[7] that the above action does lead to the correct on-shell spectrum of string theory. In this proof one finds that the light-cone count in 24 dimensions emerges as 24 = 26-1-1. This is not just a numerical coincidence, but the result of the bosonic partition function in 26 dimensions being multiplied by the fermionic partition functions of c and $\bar{c}$. Thus c and $\bar{c}$ act in the on-shell count, so as to reduce the 26 dimensions of $x^u(\sigma)$ by two. This suggests a type of Parsi-Sourlas type mechanism. We will return to this point latter.

Yet a further method of arriving at the addition of $c(\sigma)$ and $\bar{c}(\sigma)$ to $x^u(\sigma)$ is to consider the first quantized string. In order to gauge fix the string we must introduce two sets of ghosts corresponding to the two dimensional general coordinate group. The associated BRST conserved charge is none other than the Q described above. What is not so clear in this approach is that the <u>first quantized BRST theory</u> should provide all the necessary tools for the <u>gauge covariant</u> second quantized field theory.

The general procedure for the description of first quantized constrained relativistic systems in terms of Q was given in an important, but up till now, rather ignored paper, of reference [10]. From the example of the string we see that the power of this formalism generalizes to the descripton of the second quantized field theory. It would seem likely that this procedure will lead to a deeper understanding of second quantized field theory itself.

The other free gauge covariant string theories with the exception of the closed superstring, can be constructed along the same lines as for the open bosonic string given above. For more details of the latter and a review of the former the reader may consult reference [11] where further references may also be found.

We now consider the interacting open bosonic string as formulated in references [12] and [13] and independently in reference [14]. An alternative description can be found in reference [8]. Given two strings, the most obvious way to join them to form a third is to join them at their end points. Clearly, the length of the final string is the sum of the lengths of the original two strings.

In fact it is this simple picture that emerges from the one gauge in which the second quantized field theory of strings is known, namely the light cone gauge[15]. Of course this gauge is very special, but one may expect some of its features to be generic. We adopt the above picture illustrated in the figure below. We assign a length α_r to the rth string and demand, in the interaction, that the lengths be preserved. In the light cone gauge, the lengths are proportional to p^+ and so are automatically preserved.

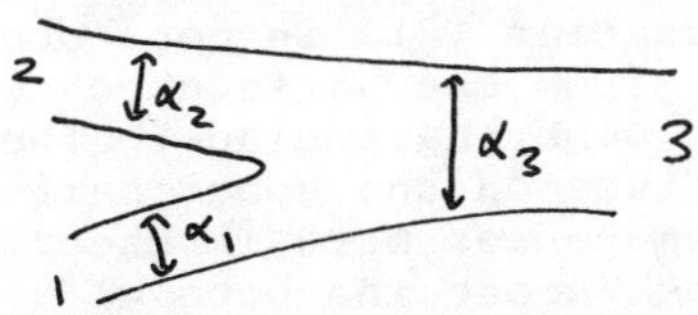

The above picture has a well defined meaning; at the interaction time we identify string 1 with the lower part of 3 and string 2 with the upper part of string 3. That the vertex is given by

$$V \sim N\,\delta\big(z^3(\eta_3) - \theta(\eta^{int}-\eta_3)\, z^2(\eta_2) - \theta(\eta_3-\eta^{int})\, z^1(\eta_1)\big) \tag{11}$$

where η_r parameterizes the rth string, η_{int} is the interaction point measured on string three and $z^r = (x, c^V, \bar{c}^r/\alpha^r)$. The factor N, as we will see, is necessary and is in fact proportional to the ghost coordinate.

The generic features of the interacting theory were given in reference [16]. The action is of the form

$$A = \tfrac{1}{2}\langle x|Q|x\rangle + g/_3\, V|x\rangle_1|x\rangle_2|x\rangle_3 \tag{12}$$

+ a possible quartic term.

The above is written in oscillator form and so V is of the form

$$V = {}_1\langle 0|\,{}_2\langle 0|\,{}_3\langle 0|\, f(\alpha^{\mu r}_n, \alpha^r) \tag{13}$$

It must contain among other terms the usual Yang-Mills term: $g(\partial_\mu \underline{A}_\nu - \partial_\nu \underline{A}_\mu)\cdot(\underline{A}^\mu \times \underline{A}^\nu)$. To find this term is just a matter of oscillator algebra once one takes $|x\rangle$ to have the generic form $\alpha^{\mu\dagger}_1 A_\mu|0\rangle$. In functional language the interaction term takes the form

$$\int \prod_{r=1}^{3} \mathcal{D}z^r\, V\, \chi(z^1(\eta_1))\,\chi(z^2(\eta_2))\,\chi(z^3(\eta_3)) \tag{14}$$

where V is of the generic form of equation (11).

The transformation law will contain the inhomogeneous given above as well as a term which is linear in the field $|x\rangle$ and parameter $|\Lambda\rangle$. It is given by

$$\delta\,{}_3\langle x| = \frac{1}{g}\,{}_3\langle\Lambda|Q^3 + V\Big(\frac{\alpha_2}{\alpha_3}|x\rangle_1|\Lambda\rangle_2 - \frac{\alpha_1}{\alpha_3}|\Lambda\rangle_1|x\rangle_2\Big). \tag{15}$$

This law will contain the usual Yang-Mills result; namely

$$\delta A_\mu = \frac{1}{g}\partial_\mu\Lambda + [\Lambda, A_\mu].$$

We can vary the action under the transformation law of equation (15) and keeping only terms of order g^0 we find that the condition for invariance is given by

$$V\sum_{r=1}^{3}\frac{Q^r}{\alpha^r} = 0 \tag{16}$$

One may also test the commutator of transformations, keeping the order g^{-1} terms we find that

$$[\delta_{\Lambda'},\delta_\Lambda]\,{}_3\langle x| = \frac{1}{g}V\Big[\frac{\alpha_2}{\alpha_3}Q^1|\Lambda'\rangle_1|\Lambda\rangle_2 - \frac{\alpha_1}{\alpha_3}Q^2|\Lambda\rangle_1|\Lambda'\rangle_2\Big] - (\Lambda\leftrightarrow\Lambda') \tag{17}$$

We must recognise this result as a transformation of the type we already have. It can only be of the form

$$[\delta_{\Lambda'},\delta_\Lambda]\,{}_3\langle x| = \frac{1}{g}\,{}_3\langle\Lambda^T|Q^3 \tag{18}$$

where

$${}_3\langle\Lambda^T| = -\frac{\alpha_1\alpha_2}{\alpha_3^2}V\big(|\Lambda'\rangle_1|\Lambda\rangle_2 - \Lambda\leftrightarrow\Lambda'\big) \tag{19}$$

It is easily verified that this will be the case provided equation (16) holds.

One could use equation (16) to determine V, but in reference (14), V was found by utilizing the overlap condition. The result is given by

$$V = {}_1\langle 0|\,{}_2\langle 0|\,{}_3\langle 0|\exp\Big\{\tfrac{1}{2}\alpha^{\mu r}_n N^{rs}_{nm}\alpha^{\mu s}_m + N^{rs}_m\alpha^{\mu r}_m\alpha^{\mu s}_0\Big\}$$

$${}_1\langle +|\,{}_2\langle +|\,{}_3\langle +|\exp\Big\{-\beta^r_n(nN^{rs}_{nm})\bar\beta^s_m + R^{rs}_n\,\beta^r_n\,\bar\beta^s_0\Big\}$$

$$\cdot\exp\Big[\tau_0\Big(\frac{L^r_0-1}{\alpha^r}\Big)\Big]$$

where N^{rs}_{nm}, N^{rs}_m and R^{rs}_m are complicated functions of the string lengths which can be found in references [13] and [14]. This form for V was then shown to satisfy equation (16).

Given the vertex in the oscillator basis we can, using the appropriate 'transition function', find it in any basis. In particular, in the coordinate basis we find that[14]

$$V|z\rangle = \Big[c_0^3 + \sum_{n=1}^{\infty} e_n c_n^3\Big]\,\delta\big[z^3(\eta_3) - \theta(\eta^{int}-\eta_3)\,z^2(\eta_2) - \theta(\eta_3-\eta^{int})\,z^1(\eta_1)\big] \tag{20}$$

where e_n is a complicated function of

the string lengths.

This demonstrates that the beautiful picture of the splitting and joining of strings at their end-points is contained within the above vertex. Presumably, the additional factor is related to the ghost co-ordinate c at the interaction point when made well defined by a suitable limiting process.

The on-shell scattering of three strings at the tree level has been known[17] for many years and we must check that the above vertex does recover this result on-shell. In fact, it can be shown[14] that V satisfies the equation

$$V = V^{csv} \exp\left(\sum_{r=1}^{3}\sum_{n=1}^{\infty} f_r(\alpha^r) L_n^r\right) \tag{21}$$

where V^{csv} is the known result. Clearly, for on-shell states (i.e. $L_n|x\rangle = 0;\ n \geq 1$)

$$V|x\rangle_1|x\rangle_2|x\rangle_3 = V^{csv}|x\rangle_1|x\rangle_2|x\rangle_3. \tag{22}$$

The calculation of the four point function goes much the same way as in the light-cone gauge. The extra anticommuting oscillators in the covariant theory just correct for the two additional bosonic integrations that are not present in the light cone gauge.

We also observe that all the string length, α^r dependence in V is contained in the conformal mapping term on the right hand side of equation (21) and so disappears for on-shell quantities. Given the complicated dependence of V on the string lengths α^r this result requires a considerable conspiracy. It leads one to strongly suspect that all scattering amplitudes are α^r independent. This lack of α^r dependence can be explained by an underlying local symmetry in an extended formalism [18]. In fact one must add additional bosonic and ghost dimensions. The α's are then the "p_+" light cone variables in this higher space. The extra dimensions disappear from physical amplitudes as a result of Parsi-Sourlas mechanism. These additional dimensions should be a consequence of a careful application of the principles of quantum mechanics to the classical actions. Even for the point particle, we find for example that the metric which enforces the constraint $p_\mu^2 + m^2$ in the action in such a coordinate [18].

The conformal mapping that occurs in equation (21) is none other than the map from the world sheet strip to the upper half plane. The V^{csv} which lives on the upper half plane, is the vertex used in the dual model. Since V contains both α_n^μ and β_n, $\bar{\beta}_m$ oscillators carrying out the conformal mapping to the upper half plane on the gauge covariant vertex V given above, we obtain not only V^{csv} but also a ghost oscillator piece. This total vertex forms the basis for a new dual model that incorporates ghosts[14].

The analysis described above can be extended to find the interacting closed bosonic string. In contrast to the open bosonic string, the closed bosonic string only possesses a cubic term. For an account of this theory and further details of the open bosonic string the reader may consult references [13] and [14].

We now briefly consider some future developments. At the beginning of this article we stressed the importance of finding the symmetry group which underlies gauge covariant string theory. At first sight one might wonder where such a large symmetry group comes from since local symmetries are usually associated with massless particles and these only occur at level one. In fact, in the covariant theory the massive particles gain their mass using Goldstone bosons by a type of Higgs mechanism. The string as an overall object is massless as indicated by a drop of two degrees of freedom in the on-shell count of states. This group for the open bosonic string is determined by equation (19). This equation in particular, contains the parameter composition law of the Yang-Mills gauge group (i.e. $\underline{\Lambda}^T = \underline{\Lambda} \times \underline{\Lambda}'$). It would, however, be useful to put a name to this gauge group. A similar relation to equation (19) holds for the closed bosonic string. There it may be ascertained that the group has rank 26 and has as many generators as there are component fields in $\Lambda[x^n, c, \bar{c}]$. On a torus one also finds within the closed bosonic vertex the vertex generator of the Lie group associated with the appropriate lattice which corresponds to the torus. On a certain torus one finds as many as three E_8's. Consequently, it is natural to suppose that for the closed bosonic string the gauge group is associated in some way with the self dual Lorentzian lattice in 26 dimensions. Work on this subject is in progress.

Recent work on the interacting gauge covariant open superstring can be found in references [19] and [20].

Finally, we give some speculative remarks. Despite the considerable elegance of the above formulation of gauge covariant string theory it has several drawbacks. On a practical level, it shares with point field theory, an explosion of graphs resulting from the combinatoric rules of the Feynman graph perturbative expansion. In the case of string theory many of these graphs are similar and may be combined into a single term. The individual graphs each supplying part of an integration range of the integrals in the final expression. In the dual model there are much fewer graphs, indeed one graph for each of the "combined" field theory graphs. Considerable progress was made in the old days in calculating multi-loop graphs in the dual model[3]. It would be interesting to investigate how in the gauge covariant field theory the graphs do combine to yield new dual model graphs and to re-do the old multi-loop calculations using the new dual model vertices given in reference [14]. From a symmetry point of view this is related to demonstrating explicitly the existence of duality ensuring symmetries in the gauge covariant field theory. It is also related to the existence of the symmetries that remove the dependence of on-shell quantities on the string lengths. It is perhaps worth noting that duality is a dynamical criterion which is ensured by a symmetry transformation on integration variables and not external quantities. In this sense it appears to be a new kind of symmetry principle.

Another point of concern with the present formulation of gauge covariant string theory is the fact that the primary objects $x^u(\sigma)$ are so utterly removed from any experimental measurements. This suggests that one should reformulate gauge covariant string field theory in terms of other objects. Presumably the oscillator vertices will play an important role in this. As first observed in reference [16] the vertices of string field theory not only have a dynamical role, but also determine the symmetry group of the string. In this way there is a very close interplay between dynamics and group theory. Indeed, there is a considerable amount to be learnt from vertices, only the two point vertex for the emission of the tachyon on the third leg has been so far discussed in any length. Although the three point vertex appears complicated, it was recently shown to be a very straight forward consequence of demanding a duality ensuring symmetries[21] and its group theoretic properties will be reported elsewhere. It is possible that given a proper understanding of string symmetries one can in a straightforward way deduce n point functions and indeed the S-matrix.

<u>References</u>

1. G. Veneziano, Nuovo Cimento <u>57A</u> 190 (1968).

2. K. Bardakci and H. Ruegg, Phys. Lett. <u>28</u>B, 342 (1968), Phys. Rev. <u>182</u> 1884 (1969).
   M. Virasoro, Phys. Rev. <u>177</u> (1969) 2309.
   H. Chan, Phys. Lett. <u>28</u>B (1968) 425.
   C. Goebel and B. Sakita, Phys. Rev. Lett. <u>22</u>, 259, (1969).

3. For a review of the dual model see V. Alessandini, D. Amati, M. Le Bellac and D.I. Olive, Phys. Reports <u>1C</u> 70 (1971).

4. M. Virasoro, Phys. Rev. <u>D1</u> 2933 (1970).

5. R.C. Brower, Phys. Rev. D6 1655 (1972).
   P. Goddard and C.B. Thorn, Phys. Lett. <u>40B</u> 235 (1972).

6. K. Kato and K. Ogawa, Nucl. Phys. <u>B212</u> 443 (1983).

7. A. Neveu, H. Nicolai, and P. West, Phys. Lett. <u>167B</u> 307 (1986).

8. E. Witten, Non Commutative Geometry and String Field Theory. Princeton preprint.

9. A. Restuccia and J.G. Taylor. Unpublished.

10. Fradkin, Vilkovisky, Phys. Lett. 55B 224 (1975).

11. P. West, Gauge-Covariant String Field Theory. CERN preprint-TH4460/86. Proceeding of 1986 Trieste Spring School to be published by World Scientific.

12. A. Neveu and P. West, Phys. Lett. 168B 192 (1985).

13. A. Neveu and P. West, Symmetries of the Interacting Bosonic String. CERN preprint TH4358/86 (1986) Nucl. Phys. B to appear.

14. H. Hata, K. Itoh, T. Kugo, H. Kunitomo and K. Ogawa, Kyoto preprints.

15. M. Kaku and Kikawa, Phys. Rev. D10 110 1823 (1974).

16. A. Neveu and P. West, Nucl. Phys. 268B 125 (1986).

17. S. Scuito, Lett. Nuovo Cimento 2, 411 (1969).
L. Caneschi, A. Schwimmer and G. Veneziano, Phys. Lett. 30B 351 (1969).

18. A. Neveu and P. West, to be published.

19 E. Witten, Princeton preprint.

20. A. Neveu and P. West, The Cyclic Symmetric Vertex for the Scattering of three Neveu-Schwarz String. CERN preprint.

21. A. Neveu and P. West, CERN preprint TH4507/86.

TWISTED STRINGS AND ORBIFOLDS*

JONATHAN A. BAGGER

The Institute for Advanced Study, Princeton, New Jersey 08540
and
Stanford Linear Accelerator Center, Stanford University, Stanford, California 94305

Orbifold compactifications provide a practical approach to string symmetry breaking. They have the potential to bridge the gap between string theory and the physics of the standard model.

As is by now well-known, string theories contain an enormous number of symmetries. For example, in their simplest form, heterotic strings describe ten-dimensional supergravity coupled to ten-dimensional super-Yang-Mills theory, with gauge group $E_8 \times E_8$ or Spin(32)/Z_2.

How can these symmetries be broken to SU(3) × SU(2) × U(1) in four dimensions? One powerful approach to string symmetry breaking was proposed in a beautiful paper by Candelas, Horowitz, Strominger and Witten [1]. This group advocated compactifying the heterotic string on $M_4 \times K$, where M_4 is four-dimensional Minkowski space, and K is a compact six-dimensional manifold of SU(3) holonomy, a so-called Calabi-Yau space. Topological methods were used to show that compactifications on $M_4 \times K$ give rise to chiral fermions in four dimensions.

The problem with Calabi-Yau spaces is that they are very complicated. They are usually described as algebraic varieties in complex projective space. Their metrics are hard to find, and it is very difficult to compute the masses and mixings of the physical spectrum [2].

An alternative approach to string symmetry breaking is provided by *orbifolds* [3, 4, 5]. Orbifolds can be used to describe:

- toroidal compactification of strings on $M_{10-d} \times T^d$,
- a singular limit of Calabi-Yau compactification, and
- gauge symmetry breaking by Wilson lines and their generalizations.

As we shall see, orbifolds are very practical spaces for string compactification. The cases we consider give *exact* solutions to the classical string equations of motion. This is in striking contrast to Calabi-Yau spaces, which are solutions only if their metrics are adjusted order-by-order in the string tension α' [6].

*Work supported by the Department of Energy, contract numbers DE-AC02-76ER02220 and DE-AC03-76SF00515.

In the rest of this talk I will give a simple introduction to orbifolds. What I have to say is well-known to string experts, but it is time to explain orbifolds to the community at large. I will try to do this by stepping through a series of four examples, of gradually increasing complexity. I hope to show that – despite their name – orbifolds are, in fact, very simple objects.

To begin, let us define an *orbifold* $\mathcal{O}$ to be the quotient space formed by dividing a manifold $\mathcal{M}$ by the action of a discrete group $\mathcal{G}$: $\mathcal{O} = \mathcal{M}/\mathcal{G}$. For our purposes, we will take $\mathcal{M}$ to be flat, either R^d or T^d. If $\mathcal{G}$ acts freely on $\mathcal{M}$, the resulting orbifold $\mathcal{O} = \mathcal{M}/\mathcal{G}$ is a smooth manifold. If the action of $\mathcal{G}$ has fixed points, $\mathcal{O}$ is an orbifold, with singularities located at the fixed point sets.

For our first example, I would like to consider the orbifold $\mathcal{O} = R^2/Z \times Z$. The group $Z \times Z$ is generated by the lattice translations

$$g_1 = e^{2\pi i P_1 R_1}, \qquad g_2 = e^{2\pi i P_2 R_2}, \tag{1}$$

where R_1 and R_2 are two vectors on the plane. The group action has no fixed points, so the orbifold $\mathcal{O}$ is a smooth manifold. In this case, it is obvious that the orbifold $\mathcal{O}$ is the torus T^2 (see Figure 1).

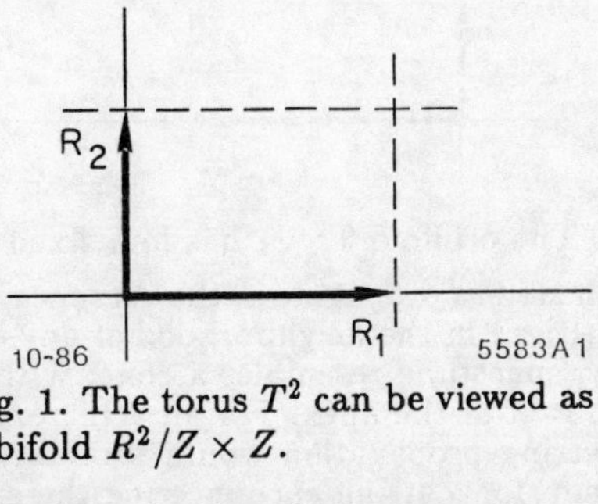

Fig. 1. The torus T^2 can be viewed as the orbifold $R^2/Z \times Z$.

Let us now consider the propagation of closed strings on this space. Clearly, closed strings can propagate consistently on the covering space R^2.

However, not all string configurations on R^2 are legal string configurations on T^2. The only legal configurations on the torus are the *translationally invariant* configurations on the plane. In the language of quantum mechanics, the physical states must be invariant under g_1 and g_2:

$$e^{2\pi i P_i R_i} \, |\, phys \rangle \; = \; |\, phys \rangle \; , \qquad (2)$$

for $i = 1$ or 2. The condition (2) forces the momenta to be quantized, with eigenvalues $P_i = M_i/R_i$, for $M_i \in Z$.

For point particles and open strings, that is the end of the story. The physical states on the torus are the translationally-invariant states on the plane. For closed strings, however, there is more to be done. Extra sectors must be added to the string Hilbert space. These sectors describe *shifted strings* – strings that are *open* on the plane but *closed* on the torus. The shifted strings obey the boundary conditions

$$X^i(\pi) \; = \; X^i(0) \; + \; N^i R_i \; , \qquad (3)$$

for $N^i \in Z$. The $N^i = 0$ sector contains to honest-to-God closed strings, on the plane and on the torus. The $N^i \neq 0$ states are open on the plane but closed on the torus. They are "soliton" states, and they are absolutely necessary for the modular invariance of the string. For $\mathcal{O} = R^2/Z \times Z$, there are an infinite number of soliton sectors, labelled by the winding numbers N^1 and N^2. In each sector of Hilbert space, the physical states must be invariant under g_1 and g_2.

Thus we have seen that string propagation on the torus can be identified with string propagation on the orbifold $R^2/Z \times Z$. For a less trivial example, let us now discuss the orbifold $\mathcal{O} = T^2/Z_2$, where T^2 is the torus generated by R_1 and R_2, and Z_2 acts on the torus by a π rotation about the origin. As shown in Figure 2, this rotation leaves four points invariant. At each fixed point, there is a conical singularity of deficit angle $\Delta = \pi$.

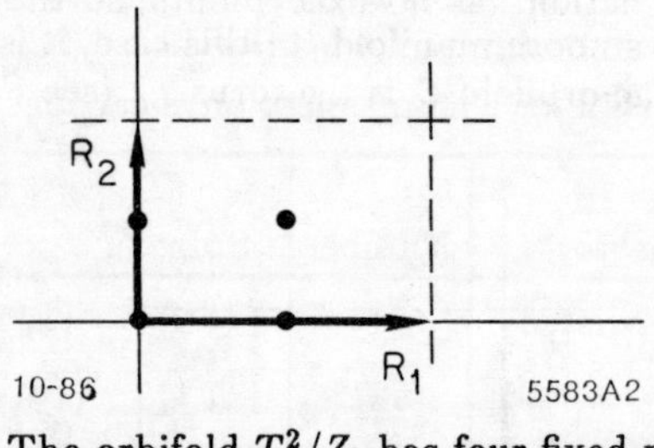

Fig. 2. The orbifold T^2/Z_2 has four fixed points.

How can strings propagate in the presence of these singularities? In the neighborhood of any one singularity, spacetime resembles a cone, with deficit angle $\Delta = \pi$ at the apex. For an arbitrary deficit angle, string propagation would probably be inconsistent, for a string encountering the singularity would develop a kink. However, for the special deficit angles $\Delta = 2\pi - 2\pi/N$, this is not so. For these special angles, N copies of the cone exactly cover the plane. Because of the symmetry restriction, the N-fold symmetric string configurations on the plane are legal string configurations on the cone.

To illustrate this, let us return to the case $N = 2$, or $\Delta = \pi$. Then two copies of the cone tile the plane, and rotationally-invariant string configurations of the plane are legal configurations on the cone. Because of the rotational symmetry, strings slip smoothly across the singularity, preserving the winding number about the singularity, modulo two. This is illustrated in Figure 3.

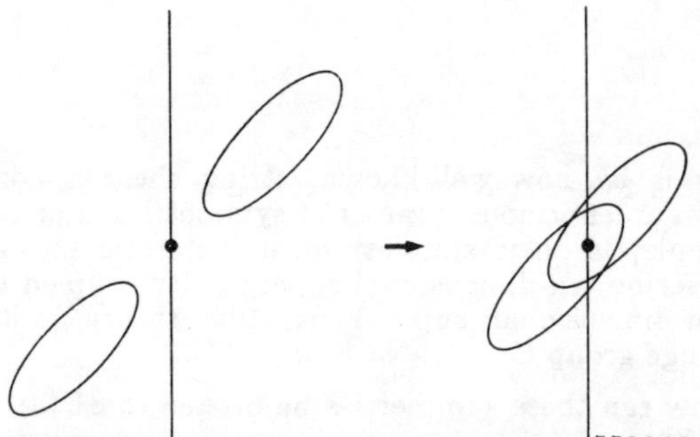

Fig. 3. Rotationally-invariant configurations on the plane are legal configurations on a cone of deficit angle $\Delta = \pi$. String propagation preserves winding number, modulo two.

As before, we must also consider *twisted sectors*. The twisted sectors on the cone are analogs of the soliton sectors on the torus. For the case at hand, the twisted sectors obey the boundary condition

$$X^i(\pi) \; = \; g \cdot X^i(0) \; , \qquad (4)$$

where g generates a rotation by $\Delta = \pi$. The boundary condition (4) fixes the center of mass of the string to lie at the apex of the cone. A typical twisted string is shown in Figure 4. Note that it has winding number one, modulo two. Twisted strings are open strings on the plane, but closed on the cone. For the orbifold $\mathcal{O} = T^2/Z_2$, there are twisted states located at each of the four fixed points of Figure 2.

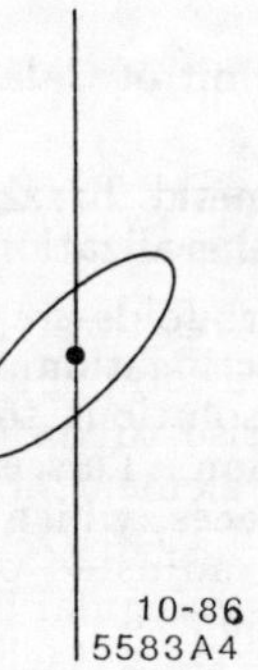

Fig. 4. Twisted strings wrap once around the singularity at the apex of the cone, modulo two.

Other orbifolds $\mathcal{O}$ are constructed by dividing a torus T^d by a group $\mathcal{G}$ of automorphisms of T^d. The group $\mathcal{G}$ is a *point group* of the torus, and its action typically leaves fixed points or even fixed tori. By appropriately choosing the torus T^d and the point group $\mathcal{G}$, many interesting compactifications can be studied. All one has to do is follow the general procedure, valid for all orbifolds $\mathcal{O} = \mathcal{M}/\mathcal{G}$:

(1) First, pass to the covering space $\mathcal{M}$.

(2) Then construct all strings that obey the boundary conditions $X^i(\pi) = g \cdot X^i(0)$, for each element $g \in \mathcal{G}$.

(3) Finally, project onto the $\mathcal{G}$-invariant subspace of states.

The twisted sectors are necessary for the modular invariance of the string.

Let us now move on to discuss our third example, the orbifold $\mathcal{O} = T^6/Z_3$. This space is known as the Z-orbifold [3]. When $M_4 \times \mathcal{O}$ is used as a background for the heterotic string, both gauge *and* spacetime symmetries are broken. The Z-orbifold produces a quasi-realistic spectrum, with $N = 1$ supersymmetry in four dimensions, and chiral fermions in 27-dimensional representations of E_6.

We shall begin by taking the six-torus T^6 to be the direct product of three identical two-tori. One of the two-tori is shown in Figure 5. We choose to describe T^6 by three complex coordinates, $(z_1,\ z_2,\ z_3)$. In terms of these coordinates, the Z_3 generator g takes the following form:

$$g \;=\; \text{diag}\,(e^{2\pi i/3},\ e^{2\pi i/3},\ e^{2\pi i/3})\ . \qquad (6)$$

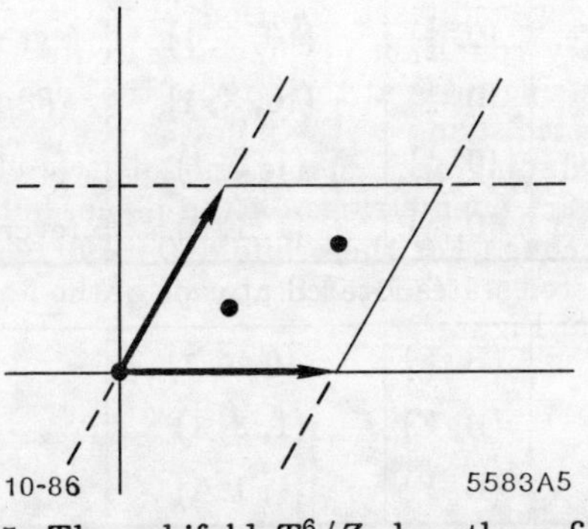

Fig. 5. The orbifold T^6/Z_3 has three fixed points in each plane.

The action of g leaves three fixed points in each plane, so there are a total of 27 fixed points. Each fixed point gives rise to its own twisted sector. Note that g is an element of SU(3), so the orbifold $\mathcal{O} = T^6/Z_3$ has discrete SU(3) holonomy. Therefore T^6/Z_3 is a singular limit of a Calabi-Yau space. It produces a tachyon-free spectrum, with unbroken $N = 1$ supersymmetry in four dimensions [3].

To describe gauge symmetry breaking, we associate an E_8 transformation $h \in E_8$ with each element $g \in \mathcal{G}$, and we project onto states invariant under $g' = gh$. For the case at hand, we choose h to lie in the center of the SU(3) subgroup defined by $E_8 \to E_6 \times SU(3)$. This breaks the gauge symmetry to $E_6 \times SU(3)$, and is the orbifold analog of symmetry breaking by Wilson lines.

The massless spectrum for the Z-orbifold is collected in Table 1. As expected, the states form $N = 1$ supersymmetry multiplets. The untwisted states contain the spin $(\frac{3}{2}, 2)$ gravitational multiplet and the spin $(\frac{1}{2}, 1)$ gauge field multiplets, with unbroken gauge group $E_6\times SU(3) \times E_8'$. There are also spin $(0, \frac{1}{2})$ matter multiplets, in various representations of the gauge group. The twisted states are localized at each of the 27 fixed points in the internal space. They also form $N = 1$ supersymmetry multiplets. As seen in Table 1, this simple example gives 36 generations of ordinary quarks and leptons – plus lots of extra particles. This spectrum is not ideal, but neither is it absurd. One might hope that more complicated orbifolds will give more realistic results.

For our final example, we investigate string propagation on the orbifold $\mathcal{O} = T^8/Z_6$. This is a particularly interesting example, because $M_2 \times \mathcal{O}$ describes a four-dimensional cosmic string embedded in a Z-orbifold background [7]. The question of strings propagating on a cosmic string background is of interest for its own sake, and also because it gives rise to various subtle issues relating to compactification on manifolds of SU(4) holonomy.

To describe the orbifold $\mathcal{O}$, we use complex coordinates (z_1, z_2, z_3, w). The z_i are as above, and w describes the xy-plane of four-dimensional spacetime. In terms of these coordinates, the Z_6 element g is taken to be

$$g \;=\; \text{diag}\,(e^{i\pi/3},\ e^{i\pi/3},\ e^{i\pi/3},\ -1)\ . \qquad (7)$$

Table 1

Z–Orbifold: The massless physical spectrum.

Sector	Number	Spin	$E_6\times SU(3) \times E_8'$
Untwisted	1	$(\frac{3}{2}, 2)$	(1, 1, 1)
	1	$(\frac{1}{2}, 1)$	(78, 1, 1)
	1	$(\frac{1}{2}, 1)$	(1, 8, 1)
	1	$(\frac{1}{2}, 1)$	(1, 1, 248)
	3	$(0, \frac{1}{2})$	(27, 3, 1)
	10	$(0, \frac{1}{2})$	(1, 1, 1)
Twisted $g,\ g^2$	27	$(0, \frac{1}{2})$	(27, 1, 1)
	81	$(0, \frac{1}{2})$	$(1, \bar{3}, 1)$

This group element gives rise to a conical singularity of deficit angle $\Delta = \pi$, located at the origin of the w-plane. If we take the tori in the z_i-directions to be tiny, and that in the w-direction to be huge, this background looks, for all intents and purposes, like the exterior spacetime surrounding an infinitesimally thin cosmic string source, of tension $\mu = 1/8G$. The cosmic string runs up and down the z-axis, and is located at the origin of the xy-plane in four-dimensional spacetime.

The group element g lies in SU(4), so this background is an eight-dimensional version of a Calabi-Yau space. As such, we expect it to be supersymmetric and tachyon-free. The complete string spectrum can be calculated as described above. The tachyon is indeed absent, so the cosmic orbifold is stable at tree level. Furthermore, the massive spectrum turns out to be supersymmetric, with unbroken gauge group O(10) × SU(3).

Table 2

Cosmic orbifold: the up–moving, massless physical spectrum.

Number	Spin	O(10) × SU(3) × E_8'	L_Z
Untwisted Sector			
1	$(\frac{3}{2}, 2)$	(1, 1, 1)	(odd, even)
1	$(\frac{1}{2}, 1)$	(45, 1, 1)	(even, odd)
1	$(\frac{1}{2}, 1)$	(1, 8, 1)	(even, odd)
1	$(\frac{1}{2}, 1)$	(1, 1, 248)	(even, odd)
1	$(\frac{1}{2}, 1)$	(16, 1, 1)	(odd, even)
1	$(\frac{1}{2}, 1)$	$(\overline{16}, 1, 1)$	(odd, even)
1	$(\frac{1}{2}, 1)$	(1, 1, 1)	(even, odd)
3	$(0, \frac{1}{2})$	(16, 3, 1)	(even, odd)
3	$(0, \frac{1}{2})$	(10, 3, 1)	(odd, even)
3	$(0, \frac{1}{2})$	(1, 3, 1)	(odd, even)
10	$(0, \frac{1}{2})$	(1, 1, 1)	(even, odd)
Twisted Sector g, g^5			
–	–	–	–
Twisted Sector g^2, g^4			
27	$(0, \frac{1}{2})$	(16, 1, 1)	(even, odd)
27	$(0, \frac{1}{2})$	(10, 1, 1)	(odd, even)
27	$(0, \frac{1}{2})$	(1, 1, 1)	(odd, even)
81	$(0, \frac{1}{2})$	$(1, \bar{3}, 1)$	(even, odd)
Twisted Sector g^3			
1	F	(10, 1, 1)	–

The computation of the massless spectrum is a little more subtle. This is because the massless spectrum is *different* for states moving up and down the cosmic string. The crucial point is that the cosmic string breaks four-dimensional Lorentz invariance. Massive states moving up the z-axis can be reversed by an unbroken Lorentz transformation, so the massive up- and down-moving spectra are identical. Massless states cannot be turned around, so the up- and down-moving spectra are free to differ – as indeed they do.

The massless physical spectrum for the cosmic orbifold is presented in Tables 2 and 3. The states are organized into representations of O(10) × SU(3), and their spins and multiplicities are indicated as well. Note that strings in sectors twisted an odd number of times have no coordinate zero modes. They are effectively bound to

Table 3

Cosmic string: the down-moving, massless physical spectrum.

Number	Spin	O(10) × SU(3) × E_8'	L_Z
Untwisted Sector			
1	$(\frac{3}{2}, 2)$	(1, 1, 1)	(even, even)
1	$(\frac{1}{2}, 1)$	(45, 1, 1)	(odd, odd)
1	$(\frac{1}{2}, 1)$	(1, 8, 1)	(odd, odd)
1	$(\frac{1}{2}, 1)$	(1, 1, 248)	(odd, odd)
1	$(\frac{1}{2}, 1)$	(16, 1, 1)	(even, even)
1	$(\frac{1}{2}, 1)$	$(\overline{16}, 1, 1)$	(even, even)
1	$(\frac{1}{2}, 1)$	(1, 1, 1)	(odd, odd)
3	$(0, \frac{1}{2})$	(16, 3, 1)	(even, even)
3	$(0, \frac{1}{2})$	(10, 3, 1)	(odd, odd)
3	$(0, \frac{1}{2})$	(1, 3, 1)	(odd, odd)
10	$(0, \frac{1}{2})$	(1, 1, 1)	(even, even)
Twisted Sector g, g^5			
1	(B, F)	(16, 1, 1)	–
3	(B, F)	(1, 3, 1)	–
12	(B, F)	(1, 1, 1)	–
Twisted Sector g^2, g^4			
27	$(0, \frac{1}{2})$	(16, 1, 1)	(even, even)
27	$(0, \frac{1}{2})$	(10, 1, 1)	(odd, odd)
27	$(0, \frac{1}{2})$	(1, 1, 1)	(odd, odd)
81	$(0, \frac{1}{2})$	$(1, \bar{3}, 1)$	(even, even)
Twisted Sector g^3			
–	–	–	–

the cosmic string, and behave like genuine two-dimensional objects. Therefore we do not indicate their spins, only whether they are bosons or fermions. On the other hand, strings in sectors twisted an even number of times *do* have coordinate zero modes in the xy plane. States in these sectors are ordinary four-dimensional massless particles. They are not bound to the string, so we are free to list their spins.

It is important to remember that the coordinate zero-mode wave functions transform under the holonomy group. This implies that there are different sets of states associated with even and odd angular momenta about the z-axis. In a compactification down to two dimensions, where the dimensions transverse to the string are "small," the states of non-zero angular momentum are viewed as having finite mass, and are not included in the massless spectrum. In a cosmic string interpretation, where two of the transverse dimensions are "large," all angular-momentum states are treated on the same footing.

In Tables 2 and 3, we have classified the states according to their spins and their $O(10) \times SU(3)$ representations. We see that the up-moving states are *not* supersymmetric, but that the down-moving states are. This is a generic feature of chiral strings compactified on manifolds of $SU(4)$ holonomy. As discussed earlier, there is no problem with this, since the cosmic string breaks four-dimensional Lorentz invariance.

Since we are describing a cosmic string embedded in the Z-orbifold background, we expect states far from the string to be identified with those of the Z-orbifold. This suggests that states in the even-twist sectors should fall into multiplets of $N = 1$ supersymmetry, with gauge group $E_6 \times SU(3)$. A glance at the tables shows that if we ignore the distinction between even and odd orbital angular momenta, as is appropriate for states far from the string, the even-twist states do fall into $E_6 \times SU(3)$ representations. The states are precisely those of the Z-orbifold.

This spectrum as an interesting, almost realistic example of a cosmic string that can be built in string theory. It is very different from the type of string expected in grand unification models, for there is no topology to guarantee the stability of the solution. The fact that supersymmetry is broken for the massless up-movers can be shown to induce a non-vanishing contribution to the vacuum energy, once string loops are taken into consideration. This contribution is properly interpreted as a correction to the tension μ of the cosmic string. This correction acts as a line source for the dilaton field, and results in dilaton emission.

What then is the final fate of the cosmic string solution when string loop corrections are included? There are at least two possibilities. One is that the configuration decays by dilaton emission to a configuration with no deficit angle. Another is that there might be a solution to the string equations of motion with a renormalized but non-zero deficit angle, and a spatially varying dilaton field.

It would be very interesting to find such a solution. It might help develop an understanding of how the cosmological constant and dilaton vacuum expectation value are determined once supersymmetry is broken. In cosmic string compactifications, supersymmetry is broken in the most innocuous possible way – only the massless modes are not supersymmetric. Analyzing radiative corrections and their effect on the dilaton field should be much simpler here than in a string theory where supersymmetry breaking affects all string modes. In addition to providing a useful laboratory for addressing these purely string-theoretic questions, it is possible that the renormalized values of the string energy density and deficit angle might be such that these strings are of cosmological interest.

In this talk I have given a simple introduction to orbifolds. The orbifolds presented here are consistent, exact solutions to the classical string equations of motion. I have shown how the singularities in orbifolds can be thought of as cosmic-string-like singularities in spacetime. Much work needs to be done to more fully explore orbifold compactifications of string theory. As far as I know, there is still no acceptable orbifold compactification with gauge group $SU(3) \times SU(2) \times U(1)$ in four dimensions. It would be wonderful to arrive at a standard-model orbifold, in order to make some connection between string theory and the world in which we live.

REFERENCES

1. P. Candelas, G. Horowitz, A. Strominger and E. Witten, *Nucl. Phys. B258* (1985) 46.
2. A. Strominger and E. Witten, *Comm. Math. Phys. 101* (1985) 341; A. Strominger, *Phys. Rev. Lett. 55* (1985) 2547.
3. L. Dixon, J. Harvey, C. Vafa and E. Witten, *Nucl. Phys. B261* (1985) 678; *Nucl. Phys. B274* (1986) 285.
4. C. Vafa, *Nucl. Phys. B273* (1986) 592.
5. For a review, see J. Harvey, in *Unified String Theories*, eds. M. Green and D. Gross (World Scientific, 1985).
6. See the contribution of D. Nemeschansky in this volume.
7. J. Bagger, C. Callan and J. Harvey, *Nucl. Phys. B278* (1986) 550.

STRINGS IN FOUR-DIMENSIONS AND MODULAR SUBGROUP INVARIANCE

L. Dolan

The Rockefeller University
New York, N.Y. 10021

The ten-dimensional Neveu-Schwarz string is compactified to four dimensions using a new construction of the affine Kac-Moody Lie algebra. A suggestion is made how this compactification might be incorporated to derive an interacting closed oriented superstring in four dimensions. The one-loop amplitude is invariant under the theta subgroup of modular transformations. The connection between modular invariance, chiral anomalies, Lorentz invariance and unitarity is discussed.

1. INTRODUCTION

This talk describes a new compactification technique[1] which reduces strings in ten dimensions to four dimensions in order to establish contact with the low energy spectrum of unified theories. It makes use of a new construction of the affine Kac-Moody Lie algebra given in terms of the half integrally moded operators of the Neveu-Schwarz model rather than the integrally moded operators of the Veneziano model. It gives rise to one-loop amplitudes of the interacting string model which are invariant under a subgroup of the modular group. This theta subgroup also divides the upper half plane into fundamental regions, so the one-loop amplitudes are still integrated over a finite domain. The unitarity, Lorentz invariance of the one-loop theory, and further consistency properties of this model are under current investigation.

The motivation for suggesting this mechanism is that other compactification techniques involve starting with a string model which already has a rather large gauge group (rank 16) in ten dimensions, and in some cases taking the field theory limit and then compactifying the point-like theory to four dimensions. These techniques 1) do not add gauge symmetry, but rather break it, and often 2) lose the string structure in four-dimensions, and 3) mix the two different reduction methods (algebraic 26→10 and Kaluza-Klein 10→4).

Affine Kac-Moody Lie algebras were originally introduced into particle field theory as infinite-dimensional symmetries of the two-dimensional sigma models.[2] This invariance may eventually prove to be an additional global symmetry of Yang-Mills theory and serve as a guide for a non-perturbative approximation of the strong interactions.[3]

Most recently it has become important that the physical particle creation operators of the dual string model on compactified space are generators of the affine Kac-Moody algebra; i.e. the one-particle states at fixed helicity lie in the infinite-dimensional irreducible basic representation of this infinite dimensional algebra.[4]

It turns out that this whole area of research of string vertex operators (i.e. the description of interacting strings) as functions of discrete momenta in compactified space is not only connected to the representation theory of affine algebras but also related to the theory of finite sporadic groups. In this way the large invariance group of the string may involve the Monster symmetry.

In Section II, the algebraic construction is derived in light-cone gauge for the free bosonic, open Neveu-Schwarz string. After compactification, in four dimensions, the SO(2) spin states carry infinite-dimensional irreducible highest representations of affine SO(18). Only the $SU(2)^6$ sub-algebra of SO(18) however, is generated by vertex operators of conformal spin one and commutes with the SO(3,1) Lorentz algebra.

In Section III, a closed orientable string is defined whose right moving and left-moving components are the operators of the Green-Schwarz superstring and the Neveu-Schwarz bosonic string. In ten dimensions the low energy limit of this model is N=1 supergravity which suggests a gravitational anomaly. In four dimensions, with the construction of Section II, the field content of the massless sector is N=4 supergravity coupled to N=4 supersymmetric Yang Mills theory with the gauge group $SU(2)^6$. In four-dimensions there are no gravitational anomalies, and the extended supergravity excludes chiral fermions. The implication of the submodular invariance of the one-loop amplitudes is discussed.

The connection between anomaly free string theories and modular invariance has been a puzzle. We remark that the four-point one-loop amplitude (external tachyons) for the closed bosonic Neveu-Schwarz model (no GSO projection) has the same theta-subgroup of modular invariance.[5] Since it has no fermions, it has no gravitational or chiral gauge anomalies. It thus provides a simpler example in which to study the effects of submodular invariance on unitarity. Certainly the identification of the full symmetry group of consistent interacting strings will reflect an understanding of the relationship among the requirements of conformal invariance, general covariance, non-abelian gauge invariance and the reparameterization symmetry of the n-genus surfaces of the n-loop calculations.

2. Free Open Neveu-Schwarz String

In light-cone gauge, the independent degrees of freedom of the open Neveu-Schwarz model in ten dimensions are given by the oscillators

$$A_n^i \text{ and } b_s^i \text{ where} \tag{1}$$

$$[A_n^i, A_m^j] = n\delta_{n,-m}\delta^{ij}$$

$$\{b_s^i, b_{s'}^j\} = \delta_{s,-s}\delta^{ij}$$

$$i = 1,\ldots 8,\ n\varepsilon Z,\ s\varepsilon Z + 1/2. \tag{2}$$

The momentum operator, whose internal components will take on discrete eigenvalues in the compactified model is

$$p^i = \frac{1}{\sqrt{2\alpha'}} A_o^i . \tag{3}$$

The dependent degrees of freedom are expressed in terms of Virasoro generators and will be useful in studying the commutation properties of the affine generators with the Lorentz algebra; for $n \neq 0$:

$$A^-_n = \frac{1}{\sqrt{2\alpha'}\, p^+} L_n \tag{4}$$

$$L_n = \frac{1}{2}\sum_m : A^i_{n-m} A^i_m : + \frac{1}{2}\sum_s (s - \frac{n}{2}) b^i_{n-s} b^i_s . \tag{5}$$

The mass operator is

$$\alpha' m^2 = N - \frac{1}{2} \tag{6}$$

$$N = \sum_{n=1}^{\infty} A^i_{-n} A^i_n + \sum_{s=\frac{1}{2}}^{\infty} s b^i_{-s} b^i_s = 0, \tfrac{1}{2}, 1, \ldots \tag{7}$$

The uncompactified theory has no non-abeliar gauge group in ten-dimensions. For the compactified model, we split transverse components as

$$i = \begin{cases} \hat{i} & = 1,2 \\ I & = 1,2,\ldots 6 . \end{cases} \tag{8}$$

The mass operator in four dimensions is (sum on I):

$$\alpha' m_4^2 = N - \frac{1}{2} + \alpha' (p^I)^2 \tag{9}$$

where the internal momentum components are given by discrete eigenvalue labelled by points on a six dimensional lattice Λ^I with basis vectors $\alpha_L{}^I$:

$$\sqrt{2}\alpha' p^I = \Lambda^I = \sum_{L=1}^{6} N^L \alpha_L{}^I . \tag{10}$$

Since the compactification is to insert a non-abelian group, at m=0 the charged bosons must come from the Kaluza-Klein tower. This requires:

$$\alpha' (p_1{}^I)^2 = \frac{1}{2} \tag{11}$$

where $p_1{}^I$ is a set of lattice points with equal length squared. Eq. (11) is satisfied if we choose

$\alpha_L{}^I = \delta_L{}^I$, which fixes $\Lambda^I = Z^6$,

the hypercubic lattice in six dimensions which is the root + vector conjugacy class weight lattice of SO(12). The theta function for this lattice is $\theta(q) = \theta_3(o|\tau)^6 = \sum_{n=-\infty}^{\infty} \left(q^{n^2}\right)^6$; $q \equiv e^{i\pi\tau}$. Thus the number of points $p_1{}^I$ with length squared one $(2\alpha'(p^I)^2 = 1)$ is 12, the number of points with length squared two $(2\alpha'(p_2^I)^2 = 2)$ is 60 etc.

To identify the non-abelian group G, we require the generators of G to commute with the SO(3,1) subgroup of the SO(9,1) Lorentz generators:

$$[G, SO(3,1)] = 0 \quad . \tag{12}$$

Since G is constructed from internal oscillators, the only SO(3,1) generator which must be checked explicitly is

$$M^{\hat{i}-} = \ell^{\hat{i}-} - i \sum_{n=1}^{\infty} A^{\hat{i}}_{-n} A^{-}_{n} - A^{-}_{-n} A^{\hat{i}}_{n}$$

$$\frac{-i}{\sqrt{2\alpha'}\, p^{+}} \sum_{s=\frac{1}{2}}^{\infty} (b^{\hat{i}}_{-s} G_{s} - G_{-s} b^{\hat{i}}_{s}) . \tag{13}$$

where $G_s = \sum_n A^{i}_{n} b^{i}_{s-n}$.

The key is to define G from an object with conformal spin one. We choose the vertex operator for the emission of the Neveu-Schwarz tachyon:

$$V(r,z) = r^{I} \sum_{s \in Z + \frac{1}{2}} b^{I}_{-s} z^{s} V_{0}(r,z) \equiv r^{I} V^{I}(r,z) \tag{14}$$

where $V_0(r,z) = :e^{\frac{ir^I}{\sqrt{2\alpha'}} X^I(z)}:$ has the form of the usual bosonic tachyon vertex operator, and the momenta $r^I = \sqrt{2\alpha'}\, k^I$ are discrete. When $(r^I)^2 = 1$, then

$$[L_n, V(r,z)] = z\frac{d}{dz}\,(z^n V(r,z)) \tag{15}$$

$$\{G, V(r,z)\} = z\frac{d}{dz}\,(z^{2s} V_0(r,z)) \tag{16}$$

Therefore,

$$[M^{\hat{i}-}, \oint \frac{dz}{z} V(r,z)] = 0 \; . \tag{17}$$

The generators of G are $A_0{}^I$ and $X_0(r) \in SU(2)^6$.

$$X_0(r) = \frac{c_r}{2\pi i} \oint \frac{dz}{z} V(r,z) \tag{18}$$

where c_r satisfies $c_r c_{r'} = (-1)^{r.r'+1} c_{r'} c_r$

$$c_r c_{-r} = 1 \; . \tag{19}$$

When $r^2=1$, the integrand in (18) is single-valued when $r.\sqrt{2\alpha'}p \in Z$. Therefore $r.r' = 0, \pm 1$, which now requires $\Lambda^I = Z^6$. There is also an SO(18) algebra whose generators are

$A_0{}^I$, $X^I_0(r)$, M^{IJ}, $X_0(r+r')$ for $r.r'=0$:

$$X_0{}^I(r) = \frac{c_r}{2\pi i} \oint \frac{dz}{z} V^I(r,z) \tag{20}$$

$$M^{IJ} = \sum_{s=\frac{1}{2}}^{\infty} (b_{-s}{}^{I} b_{s}{}^{J} - b_{-s}{}^{J} b_{s}{}^{I}) \quad \varepsilon\ SO(6) \tag{21}$$

$$X_0(r+r') = \frac{c_{r+r'}}{2\pi i} \oint \frac{dz}{z} V_0(r+r',z) \quad \text{for } r.r'=0 \tag{22}$$

$X_0(r+r')$ are the 60 step operators of SO(12)

The commutation relations are

$$[X_0{}^I(r), X_0{}^J(r')]$$

$$\begin{aligned} &= \delta^{IJ} \frac{c_{r+r'}}{2\pi i} \oint \frac{dz}{z} V_0(r+r',z) && r.r' = 0 \\ &= 0 && r.r' = 1 \\ &= M^{IJ} + \delta^{IJ} r.A_0 \; . && r.r' = -1 \end{aligned} \tag{23}$$

Eq. (23) is easily extended to affine SO(18):

$$M_n{}^{IJ} = \sum_{s=\frac{1}{2}}^{\infty} (b_{n-s}{}^{I} b_{s}{}^{J} - b_{n-s}{}^{J} b_{s}{}^{I}),\ A_n{}^I,$$

$$X_n{}^I(r) = \frac{c_r}{2\pi i} \oint \frac{dz}{z} z^n V^I(r,z) \quad \text{etc..}$$

The SO(2) spin states carry two infinite-dimensional irreducible representations of affine SO(18): the highest weight singlet and the highest weight vector: 1, 153, ... and 1, 18 + 816,...respectively.

Since $X_0{}^I(r)$ does not have conformal spin one, however, only the $SU(2)^6$ subgroup of SO(18) commutes with the four-dimensional Lorentz algebra, which means that for $m \neq 0$, the SO(3) spin multiplets break the SO(18) symmetry to $SU(2)^6$. From (23), the commutation relations for

$A_0{}^I$ and $X_0(r) = r^I X^I_0(r)$ are

$$[X_0(r), X_0(-r)] = r.A_0 \quad \varepsilon\ SU(2)^6 . \tag{24}$$

3. Closed Oriented Superstring x Bosonic Neveu-Schwarz Model

A suggestion to incorporate the compactification of Section 2 in the context of a closed orientable string is to combine a right-moving superstring and a left-moving bosonic Neveu-Schwarz model. The degrees of freedom are:

$$X^{\hat{i}}(\tau-\sigma) = x^{\hat{i}} + \alpha' p^{\hat{i}}(\tau-\sigma) + i\frac{\sqrt{2\alpha'}}{2} \sum_{n\neq 0} \frac{1}{n} A_n^{\hat{i}} e^{2in(\tau-\sigma)} \quad (25)$$

$$X^I(\tau-\sigma) = \bar{x}^I + 2\alpha' \bar{p}^I(\tau-\sigma) + i\frac{\sqrt{2\alpha'}}{2} \sum_{n\neq 0} \frac{1}{n} A_n^I e^{-2in(\tau-\sigma)} \quad (26)$$

$$S^a(\tau-\sigma) = \sum_n S_n^a e^{-2in(\tau-\sigma)} \quad (27)$$

and

$$\tilde{X}^{\hat{i}}(\tau+\sigma) = \frac{x^{\hat{i}}}{2} + \alpha' p^{\hat{i}}(\tau+\sigma) + \frac{i\sqrt{2\alpha'}}{2} \sum_{n\neq 0} \frac{1}{n} \tilde{A}_n^i e^{-2in(\tau+\sigma)} \quad (28)$$

$$\tilde{X}^I(\tau+\sigma) = \tilde{x}^I + 2\alpha' \tilde{p}^I(\tau+\sigma) + i\frac{\sqrt{2\alpha'}}{2} \sum_{n\neq 0} \frac{1}{n} \tilde{A}_n^I e^{-2in(\tau+\sigma)} \quad (29)$$

$$b^i(\tau+\sigma) = \sum_s b_s^i e^{-2is(\tau+\sigma)} . \quad (30)$$

The mass operator is

$$\alpha' m_4^2 = N + \tilde{N} - \frac{1}{2} + \alpha'(\bar{p}^I)^2 + \alpha'(\tilde{p}^I)^2 = 0,1,2,\ldots \quad (31)$$

The left-right light-cone gauge constraint is

$$N + \alpha'(\bar{p}^I)^2 = N + \alpha(\tilde{p}^I)^2 - \frac{1}{2} \quad (32)$$

where

$$N = \sum_{n=1}^{\infty} A_{-n}^i A_n^i + \sum_{n=1}^{\infty} \frac{n}{2} S_{-n}\gamma^- S_n$$

$$\tilde{N} = \sum_{n=1}^{\infty} \tilde{A}_{-n}^i \tilde{A}_n^i + \sum_s \frac{s}{2} b_{-s}^i b_s^i \quad (33)$$

and

$$\sqrt{2}\alpha' \bar{p}^I = \sum N^L \delta_L^I , \quad \sqrt{2}\alpha' \tilde{p}^I = \sum N^L \delta_L^I . \quad (34)$$

In four dimensions, the field content of the massless sector is the N=4 supergravity multiplet and the N=4 supersymmetric Yang-Mills multiplet in $SU(2)^6$ respectively.

The vertex operators and propagator and the tree amplitude for this model are given in reference 1. The four-point one-loop amplitude for external charged vector mesons is given by

$$A_{1\text{-loop}}(1234) = \left(\frac{\alpha' g}{2\pi}\right)^4 \cdot \frac{\kappa}{8\alpha'} \cdot \tilde{\varepsilon} \cdot (2\pi)^8$$

$$\int d^2\tau \int \prod_{M=1}^{3} d^2\nu_M (\mathrm{Im}\,\tau)^{-2} \prod_{1\leq I<J\leq 4} (\chi_{IJ})^{\alpha' k_I^{\hat{k}} \cdot k_J^{\hat{k}}}$$

$$e^{i\pi\bar{\tau}} (f(e^{-2i\pi\bar{\tau}}))^{-8} \prod_{I<J} (\psi_{IJ})^{2\alpha' \tilde{k}_I^k \tilde{k}_J^k}$$

$$\cdot \; (\phi(e^{-2\pi i\bar{\tau}}))^8 \, .$$

$$[\tilde{k}_1^i \tilde{k}_2^i \cdot \tilde{k}_3^j \tilde{k}_4^j \; \chi_{43}^+ \chi_{21}^+$$

$$- \tilde{k}_1^i \tilde{k}_3^i \cdot \tilde{k}_2^j \tilde{k}_4^j \; \chi_{42}^+ \chi_{31}^+$$

$$+ \tilde{k}_1^i \tilde{k}_4^i \cdot \tilde{k}_2^j \tilde{k}_3^j \; \chi_{41}^+ \chi_{32}^+] \quad (35)$$

where

$$\chi_{IJ} = 2\pi \left| \frac{\theta_1(\nu_{JI}|\tau)}{\theta_1'(o|\tau)} \right| e^{\frac{-\pi\, \mathrm{Im}\,\nu_{JI}}{\mathrm{Im}\,\tau}}$$

$$f(w) = \prod_{n=1}^{\infty} (1-w^n) \quad ; \quad w \equiv e^{2\pi i\tau}$$

$$\psi_{IJ} = 2\pi i \; e^{\frac{-i\pi}{\bar{\tau}}(\bar{\nu}_{JI})^2} \; \frac{\bar{\theta}_1(\bar{\nu}|\bar{\tau})}{\bar{\theta}_1'(o|\bar{\tau})}$$

$$\mathcal{L}' = \sum_{\sqrt{2}\alpha' \bar{p}^I \in Z^6} w^{\alpha'(\bar{p}^I)^2} = (\theta_3(o|\tau))^6$$

$$\mathcal{L} = \sum_{\sqrt{2}\alpha' \bar{p}^I \in Z^6} e^{\alpha' \ln \bar{w} (\tilde{p}^I - \sum_{M=1}^{4} \frac{\ln \bar{z}_M}{\ln \bar{w}} Q_M^I)^2}$$

$$Q_{M+1}^I = \sum_{N=1}^{M} k_N^I$$

$$\phi(w) = \prod_{s=\frac{1}{2}}^{\infty} (1+w^s) = \left(\frac{\theta_3(o|\tau)}{f(w)}\right)^{\frac{1}{2}}$$

$$\chi^+_{JI} = \sum_{s=\frac{1}{2}}^{\infty} \frac{\bar{\nu}_{JI}{}^s + \left(\frac{\bar{w}}{\bar{\nu}_{JI}}\right)^s}{1+\bar{w}^s}$$

$$= \frac{i}{2}\ \bar{\theta}_2(o|\bar{\tau})\ \bar{\theta}_4(o|\bar{\tau})\ \frac{\bar{\theta}_3(\bar{\nu}_{JI}|\bar{\tau})}{\bar{\theta}_1(\bar{\nu}_{JI}|\bar{\tau})} \qquad (36)$$

For further notation see reference 1.

The amplitude (35) is invariant under the theta subgroup of the modular group. The full modular transformations are

$$\tau \to \frac{\alpha\tau + b}{c\tau + d}$$

where $a,b,c,d \varepsilon Z$ and $ad-bc = 1$.

They represent global reparameterizations of the torus since a 2-torus is characterized by a 2-dimensional lattice $\Gamma^I = \sum_{L=1}^{2} N^L \alpha^I_L$, i.e. $\frac{R^2}{\Gamma}$. Define

$$w_1 = \alpha_1^{\ 1} + i\ \alpha_1^{\ 2} \qquad (37)$$
$$w_2 = \alpha_2^{\ 1} + i\ \alpha_2^{\ 2}.$$

Then $\Gamma = \Gamma^1 + i\Gamma^2 = N^1 w_1 + N^2 w_2$.

Under the transformations $w \to w'$

$$\begin{matrix} w_1{}' \\ w_2{}' \end{matrix} = A \begin{matrix} w_1 \\ w_2 \end{matrix} \qquad (38)$$

where $A = \begin{pmatrix} a & b \\ c & d \end{pmatrix}$,

$$\text{Then} \quad \Gamma = N^L w_L = N^{L'} w_L{}' . \qquad (39)$$

Since det A = 1, and $a,b,c,d\ \varepsilon\ Z$, then $N^{L'}\ \varepsilon\ Z$ and (38),(39) represent a change of basis vectors which leaves the points on the lattice invariant.

The set of all matrices A form a group, the modular group SL(2,Z). Any such matrix can be written as $A = U^{n_1}\ T\ U^{n_2}\ T \ldots$

for $n_i\ \varepsilon\ Z$,

$$T \equiv \begin{bmatrix} 0 & -1 \\ 1 & 0 \end{bmatrix},\quad U \equiv \begin{bmatrix} 1 & 1 \\ 0 & 1 \end{bmatrix} : T\tau = -\frac{1}{\tau},\ U\tau = \tau + 1.$$

T and U are the generators of the modular group, and invariance under these two particular transformations implies invariance under the full group. Fundamental regions on the upper half τ-plane are mapped into each other under (38). The basic region is $|\tau|>1$, $-1/2 \le \text{Re}\ \tau < 1/2$. The theta subgroup which leaves (35) invariant is generated by T and U^2, i.e. $\tau \to -1/\tau$, and $\tau \to \tau + 2$. This group is a subset of all matrices A which is congruent to $\Gamma°(2) = \{\begin{pmatrix} a & b \\ c & d \end{pmatrix}$; b even integer, $a,c,d \varepsilon Z$; $ad-bc = 1\}$: $\Gamma_\theta = U\Gamma°(2)U^{-1}$. The basic region in the τ-plane is now $|\tau| \geq 1, -1 \leq \text{Re}\,\tau \leq 1$. This periodicity together with $\nu_I \to \nu_I + 1$ and $\nu_I \to \nu_I + \tau$ suggests the integration region of (35) to be $-\frac{1}{2} \le \text{Re}\ \nu_I \le \frac{1}{2}$, $0 \le \text{Im}\ \nu_I \le \text{Im}\ \tau$, and $|\tau|>1$ with $-1 \le \text{Re}\ \tau \le 1$.

The perturbative unitarity of this amplitude remains to be checked.

References:

1. R. Bluhm and L. Dolan, Phys. Lett. B169, 347(1986).
2. L. Dolan, Phys. Rev. Lett. 47, 1371, (1981).
3. L. Dolan, Phys. Rep. 109, 1 (1984).
4. L. Dolan and R. Slansky, Phys. Rev. Lett. 54, 2075 (1985).
5. L. Dixon and J. Harvey, Princeton Preprint, February 1986.

OPERATOR FORMULATION OF WITTENS STRING FIELD THEORY*

Antal Jevicki+
Physics Department
Brown University
Providence, RI 02912

We summarize some recent developments on the construction of gauge invariant string field theory. The approach is based on Wittens differential geometric description of interactions and symmetries. We present the explicit construction of the interaction vertex and discuss its properties.

* A talk given at the XXIII International Conference on High Energy Physics, July 16-28, 1986, Berkeley, California.
+ Supported in part by the U.S. Department of Energy under contract No. DE-AC02-76ER03130 A009 Task A.

1. INTRODUCTION

A construction of gauge invariant string field theory represents a most interesting theoretical problem. It is expected to provide a much deeper insight into the symmetry structure of the theory and give a concrete nonperturbative framework for solving the outstanding theoretical questions. Recently, Witten has introduced a novel approach to interacting open string fields (1) based on loop space noncommutative differential geometry. The outlines of this are the following. In the BRST approach one has (2) the string field $\psi(x^{\mu}(\sigma),c(\sigma),b(\sigma))$ where $x^{\mu}(\sigma),\mu=1,...,26$ are the string coordinates and $c(\sigma),b(\sigma)$ are the ghost and anti-ghost coordinates respectively. These can be replaced through bosonization by a single bosonic coordinate $\phi(\sigma)$ (we shall often write the ghost as the 27th coordinate $x_{\mu=27}=\phi$). The BRST charge operator Q that plays a central role obeys $Q^2=0$ and as such has an interpretation of a derivative in string space (3). Witten has introduced an integration

$$\int A=\int \prod_{\mu=1}^{27} dx_{\mu}\, e^{-\frac{3}{2}i\phi(\pi/2)} \qquad \delta\left[x_{\mu}(\sigma)-x_{\mu}(\pi-\sigma)\right]A$$

which is designed to obey the required property $\int QA=0$. The phase factor gives the integration a ghost number of 3/2. The next basic operation is a multiplication of two string functionals A and B: A*B=C, where geometrically

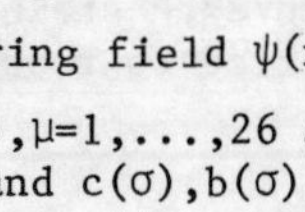

Based on these operations one can define a Lagrangian

$$S=\int(\psi*Q\psi+2/3\psi*\psi*\psi)$$

which is easily seen to be gauge invariant

$$\delta\psi=Q\Lambda-\Lambda*\psi+\psi*\Lambda$$

provided the basic axioms for the system Q, $\int$,* are obeyed. These are the BRST invariance of $\int$ as * and the associativity A*(B*C)=(A*B)*C. Originally, Witten has given arguments for these based on analogue first quantized Riemanian surface methods.

2. OPERATOR FORMULATION

In what follows we will present an explicit operator construction of the theory (4). It shows agreement with the dual models and provides a concrete computational framework in the string Hilbert space. As such it gives a basis for establishing the symmetry properties of the theory and of the axioms introduced by Witten. Its purpose however are concrete calculations.

We begin with the integration $\int$. In the string Hilbert space it is a vector $|I\rangle$ such that for a string state $|A\rangle$ the scalar product $\langle I|A\rangle$ would equal the integral $\int A$. The δ-function in the definition of the integral gives the overlap equations obeyed by $|I\rangle$:

$$(x_{\mu}(\sigma)-x_{\mu}(\pi-\sigma))\,|I\rangle=0$$

Using the creation-annihilation operator of the string Hilbert space

$$x(\sigma)=x_0+i\sum_{n=1}^{\infty}\frac{1}{\sqrt{n}}(a_n-a_n^+)\cos n\sigma$$

one looks for the integration as a quadratic form

$$|I\rangle=\exp\left[-\frac{1}{2}\sum_{n,m}a_n^+M_{nm}a_m^+\right]|0\rangle$$

The overlap equation translates into a matrix equation $(1-C)(1+M)=0$ with $C_{n,m}=(-)^n\delta_{n,m}$ so that the solution is M=C. This gives the explicit form for integration. We will in Section 3 give a description of the ghost contribution and the symmetries obeyed by this construction.

Much less trivial is the construction of the vertex operator which corresponds to the interaction term in the Lagrangian and also the multiplication *. The definition of the vertex and the corresponding overlap equations are as follows. One now has three strings with coordinates $x_\mu^r(\sigma)$, r=1,2,3; and the associated Hilbert spaces. The vertex is an operator in the direct product Hilbert space and the defining overlap equations are

$$x_\mu^r(\sigma)-x_\mu^{r+1}(\pi-\sigma)\sigma \quad |V\rangle =0$$

There are also three additional equations for the overlap of conjugate momentum variables $P_\mu^r(\sigma)$ (they come with the opposite sign). We exploit the high degree of symmetry characterizing the problem by introducing Z_3 linear combinations:

$$Q_k=1/\sqrt{3}\sum_{r=1}^{3} x^r e^{\frac{2\pi ikr}{3}}$$

They can be shown to obey the overlaps

$$Q_3(\sigma)=Q_3(\pi-\sigma)$$

$$Q(\sigma)=\begin{cases} e^{2\pi i/3}Q(\pi-\sigma) & 0\leq\sigma\leq\pi/2 \\ e^{-2\pi i/3}Q(\pi-\sigma) & \pi/2\leq\sigma\leq\pi \end{cases}$$

where $Q=Q_1=\bar{Q}_2$. The problem therefore separates with the Q_3 degree of freedom being identical to the integration case. The vertex is represented by a quadratic ansatz

$$|V_3\rangle =\exp(-\tfrac{1}{2}A_3^+\cdot C\cdot A_3^+-A^+\cdot U\cdot\bar{A}^+) \quad |0\rangle_{123}$$

The nontrivial overlap equations translate into

$$(1-Y)\sqrt{2/N}(1+U)=0$$

$$(1+Y)\sqrt{N/2}(1-U)=0$$

where the overlap matrix Y is

$$Y_{nm} = -\tfrac{1}{2}\delta_{n,m}+i\frac{\sqrt{3}}{2\pi}(-)^{\frac{n-m-1}{2}}\left(\frac{1}{n+m}+\frac{(-)^m}{n-m}\right)$$

A solution for the overlap equations of an inverse matrix form can be given (4).

An explicit solution for the vertex is achieved with use of the conformal mapping $\rho=\rho(z)$ that relates the problem to the dual model geometry (the upper half plane). It is relevant in its own right since it establishes the on-shell equivalence of the new formulation. This mapping relevant to the present rearrangement process of three equal length strings goes as follows. Using the basic Schwarz-Christoffel form

$$\rho=\sum\alpha i\ \ln(z-z_i)$$

and doubling the number of strings (to six) one distributes them symmetrically on the unit circle with alternating lengths $\alpha=\pm1$. This gives

$$\rho=\ln\left(\frac{z^3-i}{z^3+i}\right)-\frac{i\pi}{2}$$

One goes to three strings by identification z,-z. This essentially means that we are mapping not from circle but from the cone which is what it should be since in Witten's scattering process one has a curvature at $\pi/2$:

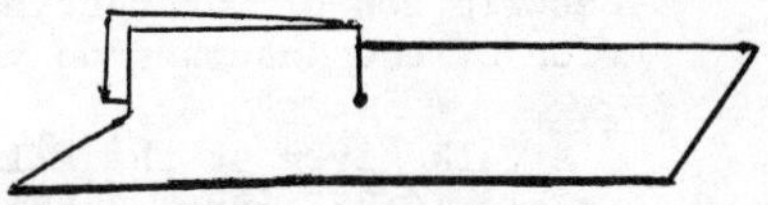

The string rearrangement induced by the above mapping is easily seen to give the required picture.

The vertex is essentially given by constructing the Neumann function $N(\rho,\rho')$ on the above scattering domain since

$$|V\rangle =e^{\frac{1}{2}\sum P_n^r N_{nm}^{rs} P_m^s}$$

with N_{nm}^{rs} being the Fourier coefficients on strings r,s respectively. Since the Neumann function $N(z,z')$ on the unit circle is known one uses the inverse mapping $z=z(\rho)$ to compute the nontrivial Neumann function and coefficients needed. On each string the conformal mapping is inverted to read

$$z=z_a\left(\frac{1+ie^\rho}{1-ie^\rho}\right)^{\frac{1}{3}}$$

The Fourier coefficients A_n that enter the vertex are then obtained

$$\left(\frac{1+x}{1-x}\right)^{\frac{1}{3}}=\sum_n a_n x^n$$

and similarly one has the coefficients b_n generated by the function $[(1+x)/(1-x)]^{\frac{2}{3}}$. They can all be found explicitly, they also obey simple recursion formulae of the type

$$\tfrac{2}{3}a_n=(n+1)a_{n+1}-(n-1)a_{n-1}$$

In terms of the six string Neumann coefficients $N_{nm}^{r,\pm s}$ the three string Neumann coefficients are simply

$$N_{nm}^{rs}=N_{nm}^{rs}(6)+N_{nm}^{r,-s}(6)$$

since we identify strings at z_a and z_{-a}. The structure of the Neumann coefficients found agrees with the general structure following from the δ-function overlap equations. The exact correspondence was also established in (4). The form of the vertex which we have constructed is then as follows.

$$|V> = \exp\left(\frac{1}{2}\alpha^r_{-m} N^{rs}_{nm}\alpha^s_{-m} + P^r_0 N^{rs}_{0m}\alpha^s_{-m} + \frac{1}{2}N_{00}\sum_{r=1}^{3} P_r^2\right)|0_{123}>$$

The form in the string index space (r,s=1,2, 3) is very simple and is given by the matrix representation

$$N = -\frac{1}{6}\left[(\widetilde{C}+U+\overline{U})\,1 + \left(\widetilde{C} - \frac{U+\overline{U}}{2}\right)\begin{pmatrix}0&1&1\\1&0&1\\1&1&0\end{pmatrix} + i\frac{\sqrt{3}}{2}(U-\overline{U})\begin{pmatrix}0&1&-1\\-1&0&1\\1&-1&0\end{pmatrix}\right]$$

$\widetilde{C}, U, \overline{U}$ are matrices in the n,m space and are given by

$$\widetilde{C}_{nm} = 2\frac{(-)^n}{n}\delta_{n,m}$$

$$(U+\overline{U})_{nm} = -2(-)^n\left(\frac{A_nB_m+B_nA_m}{n+m} + \frac{A_nB_m-B_nA_m}{n-m}\right)$$

$$(U-\overline{U})_{nm} = -2i\left(\frac{A_nB_m-B_nA_m}{n+m} + \frac{A_nB_m+B_nA_m}{n-m}\right)$$

These are valid when n=0 reducing to

$$U_{0m} = -\frac{1}{m}A_{m=2k} - iA_{m=2k+1}$$

Also

$$U_{00} = -\frac{1}{2}\ln\frac{3^3}{2^4}$$

The above list is valid even for n=m with $(U-\overline{U})_{nn}=0$ and a nontrivial limit involved in $(U+\overline{U})_{nn}$. This can also be found and the explicit result was given (4). This is then a complete description of the gauge invariant vertex.

In ref. (4) we have shown explicitly that the above vertex solves the overlap equations and is therefore a correct realization of the δ-function interaction. This proof is important since a formal argument which one could base on a naive limiting procedure cannot account for additional terms that could be in principle induced at the midpoint π/2. We have also shown agreement with the three particle dual model amplitudes and considered ghosts and the symmetry properties of the interaction.

In terms of the interaction vertex $|V_3>$ we can define the multiplication * in the following way. One considers two string states $|A>_1$, $|B>_2$ which are in two different Hilbert spaces (1 and 2 refer to strings 1 and 2). The multiplication $|A*B>$ is to give a new state $|C>_3$ which is in the Hilbert space of the third string. This operation is explicitly given by

$$|C>_3 = {}_1\!<0|{}_2\!<0|\hat{V}(a_1,a_2,-Ca_3^+)|0>_3|A>_1|B>_2$$

One can now directly check the properties obeyed by the multiplication *. For example the first property that we have explicitly checked is associativity. In general we have developed methods for the construction of general N-point overlap vertex operators $|V_N>$. The integration $|I>$ corresponds to N=1, $|V_2>$ would give the scalar product for two string states, N=3 is the interaction vertex etc. The four-string overlap $|V_4>$ is quite useful. One example is the proof of associativity, this is essentially the property that:

$$V_4(a_1^+a_2^+a_3^+a_4^+) = {}_0\!<0\; V_3(a_1^+, a_2^+, -Ca_0)\hat{V}_3(a_0^+, a_3^+, a_4^+)|0>_0$$

In turn one can also obtain $\hat{V}_3$ from $\hat{V}_4$ since

$$V_3(a_1^+a_2^+a_3^+) = {}_4\!<I|\hat{V}_4(a_1a_2a_3a_4)|0>_4$$

and in general $|V_{N-1}> = I|V_N>$.

3. GHOSTS AND SYMMETRIES

In considering the symmetry properties of the above construction we have shown in ref. 4 the appearance of certain anomalies at the midpoint π/2 and carefully established certain invariances that hold. The point π/2 where the whole interaction takes place is special; only a careful evaluation of ghost contributions which we have done gives a consistent theory.

One can formulate the effect of ghosts in either the fermionic or in the bosonized language. For the ghosts considered for instance in the bosonized form one only has the additional phase factors $e^{\mp\frac{3i}{2}\phi(\pi/2)}$. In the oscillator representation they give simple additional linear terms. For the integration

$$|I^\phi> = e^{-3\sum_{n=2k}\frac{(-)^{n/2}}{\sqrt{n}}a_n^+}\, e^{-\frac{1}{2}a^+\cdot C\cdot a^+}|0, P^\phi = \frac{3}{2}>$$

In the exponent the factor 2 comes from the annihilation operator contribution. The effect of the phase factor in the vertex is analogous since at the midpoint one has

$\phi(\pi/2)=1/3(\phi^1(\pi/2)+\phi^2(\pi/2)+\phi^3(\pi/2))$; this degree of freedom was seen to separate and have the form identical to $|I\rangle$. So in general the effect of the phase factor is straightforward to take into account.

Besides the Q symmetry mentioned the theory has to be invariant under the following reparametrization subgroup of the Virasoro algebra

$$K_n=L_n-(-)^n L_{-n}$$

This is a maximal subgroup with no central charge and naively it leaves the midpoint overlaps invariant. In the operator treatment the above is not true and one finds for example

$$K^x_{2N}\ |I^x\rangle = -\frac{D}{2}N(-)^N\ |I^x\rangle$$

These anomalies are of the form consistent with the algebra obeyed by the generators $K_n^{x,\phi}$:

$$K^x_n, K^x_m = (n-m)K^x_{n+m} - (-)^m (n+m) K^x_{n-m}$$

From here it follows that in general one must have

$$K_{2N+1}=0$$

$$K_{2N}=C(-)^N N$$

These anomalies are seen by direct evaluation of the operators $\hat{K}_n$ on the vertices, their origin is in operator ordering. Namely K^x_{2N} has a term

$$\frac{1}{2}N(\hat{X}_N\hat{P}_N+\hat{P}_N\hat{X}_N)$$

which induces the effect.

These anomalies were shown to cancel only when the ghost effect was taken into account. Here one must be careful in evaluating the effect of the phase factor, this even though it is at the point $\pi/2$ is not left invariant by the subgroup K_n. For the integration $|I^\phi\rangle$ this gives an additional effect of $9\ \frac{N}{2}(-)^N$. Also the ghost Virasoro generators themselves contain additional linear terms of the form $-\frac{3n}{2}\alpha_n^\phi$ which contributes $18\ \frac{N}{2}(-)^N$. Altogether one has that

$$K^\phi_{2n}|I^\phi\rangle = (-1+18+9)\frac{N}{2}(-)^N$$

and there follows $K^x+K^\phi=0$.

In ref. (4) we have also considered the symmetry properties of $|V_4\rangle$. This implies invariance of the interaction vertex $|V_3\rangle$ since $V_3=IV_4$ and I was just shown to be invariant. One can also consider the interaction vertex directly. For the ghost

$$|V_3{}^\phi\rangle = \delta(p^\phi+\frac{3}{2})\ e^{\frac{3}{\sqrt{3}}\sum_{n=2,4...}\frac{(-)}{\sqrt{n}}A^+_{3,n}} \cdot V(A_3^+,A^+,\bar{A}^+)$$

The coordinate part of the vertex again gives an anomaly. It is given by

$$k_n = \frac{1}{2}\sum_{m=1}^{n-1} m(n-m)\ N^{rr}_{m\,n-m}$$

with the value $\frac{5}{18}N(-)^N$. Again the above ghost vertex can be seen to give a canceling effect.

For establishing the Q invariance (which is closely related to the above established $\hat{K}_n$ invariance) we have worked out the ghost vertices in fermionic representation. They follow from the overlap equations for the ghost and anti-ghost coordinates $C(\sigma)$ and $b(\sigma)$. The ghost is to be treated as a momentum and b as a coordinate with the opposite for their conjugate variables: From the overlap equations we can deduce

$$|V^{gh}\rangle = \exp \sum b^r_n\ \tilde{N}^{rs}_{nm}\ mC^s_m$$

with the new Neumann coefficients $\tilde{N}_{nm}$ again given in terms of the coefficients A_n,B_n that enter the coordinate vertex. These can be used to demonstrate the BRST invariance. It is seen as a consequence of the overlap equations and K_n invariance.

The physical features of the interaction vertex are the following. When on shell it gives the same amplitudes as the old dual model amplitudes. This was checked by explicit evaluation in low orders; in general, there follows the fact that it is related through L_n transformations to the dual model three reggeon vertex. Namely, it was seen that the interaction is related through a conformal mapping to the dual model upper half plane; this implies that the operator vertices are explicitly related.

One of the most immediate problems is to derive the fixed gauge Feynman rules from the above field theory. In the analogue first quantized path integral there are the interesting results of ref. (5).

Related works that appeared are as follows: Cremmer, Schwimmer and Thorn (7) have also reached the Neumann function form of the vertex using a mapping that essentially corresponds to transformation $Z \to \sqrt{Z}$ in the above; S. Samuel has computed low level terms of the vertex (6). Ref. (8) attempted to solve the overlap equations.

REFERENCES

1. E. Witten, Nucl. Phys. B268, 253 (1986).

2. W. Siegel, Phys. Lett. 142B, 276 (1984); 149B, 157 (1984).

3. For review and also references see the talk by T. BANKS at this conference.

4. D.J. Gross and A. Jevicki "Operator Formulation of Interacting String Field Theory" Princeton University preprint (May 1986) and preprint in preparation.

5. S. Giddings, Princeton preprint (Feb. 1986) S. Giddings and E. Martinec, Princeton preprint (1986), S. Giddings, E. Martinec and E. Witten, Princeton preprint (1986).

6. S. Samuel, CERN preprint TH-4365/86 rev. (June 1986).

7. E. Cremmer, A. Schwimmer and C. Thorn, University of Florida preprint UFTP-86-8 (July 1986).

8. N. Ohta, Texas preprint (July 1986).

THE n-LOOP BOSE-STRING AMPLITUDE

Stanley Mandelstam
Department of Physics
University of California
Berkeley, California 94720, U.S.A.

An outline is given of the derivation, in the light-cone frame, of the formula for the n-loop amplitude of closed Bose strings. All factors in the formula are explicit, the final formula is manifestly Lorentz invariant, and the method used ensures that the S-matrix is unitary.

INTRODUCTION

In this talk I shall review the derivation of an explicit perturbation series for the Bose-string scattering amplitudes. I shall concentrate on the light-cone approach which I have used personally and which, in my opinion, leads most directly to explicit formulas. I shall also mention some of the work on the Polyakov approach and shall indicate points of comparison between the two methods.

The perturbation series for the Bosonic string is now fairly well understood. In fact, the formula had almost been obtained during the previous incarnation of string theories (1), but the models which were studied possessed ghosts. I gave the formula for the current Bosonic string models in my Santa Barbara lectures last year (2); one point in the proof remains to be tightened up, but I believe the result is correct. Though the light-cone approach was used, the final formula is manifestly covariant. D'Hoker and Phong (3), Belavin and Knizhnik (4) and Manin (5) have presented formulas based on the Polyakov approach. Their expressions are mathematically well defined, but an explicit formula for some of the factors in the integrands is not known at present.

No explicit formula for the n-loop superstring amplitude has been obtained to date. In the light-cone approach the difficulty was due to the fact that the functional integral contained operators at the joining points of the strings. A new formulation by Berkovits (6) shows that this difficulty can be avoided if one integrates over supersheets instead of ordinary sheets. It should now be fairly straightforward to obtain an explicit formula for the n-loop amplitude and, if the external particles are vector bosons, the result should be manifestly covariant.

Some of the above-mentioned results on the Polyakov Bose string have been extended to the superstring by D'Hoker and Phong (7), Friedan, Martinec and Shenker (8), and Nelson, Moore and Polchnski (9).

Approaches to String Perturbation Theory

Whether one uses the light-cone or Polyakov approach, one first performs the Gaussian functional integral for the general perturbation term; one thereby obtains the formula (for the Bose string)

$$A=-i(ig^2/4\pi)\int d^2Z_1\ldots d^2Z_N d^{3g-3}\nu \; M(Z_1,\ldots,Z_N,\nu)\; \mathrm{Exp}\{-\Sigma\; P_i\cdot P_j\; N(Z_i,Z_j)\}. \quad (1)$$

The points Z_i ($1\leq i\leq N$) represent the N external particles, and the integration is over the Riemann surface, parametrized by a complex variable z, which represents the world-sheet traced out by the string. (We shall write all our formulae for the case of closed strings.) The factor $P_i\cdot P_j$ in the exponential is the *d-vector product* (even in the light-cone approach) between the ith and j^{th} external momenta, and the function N is the Green's function of the Laplacian between the points Z_i and Z_j.

We still have to discuss the variables ν and the factor M; all recent work has been concerned with these factors. A Riemann surface of genus g, corresponding to a g-loop amplitude, is parametrized to within conformal equivalence by 3g-3 complex parameters if $g\geq 2$ (and by one complex parameter if g=1; all surfaces with g=0 are conformally equivalent). This fact was originally discovered by Riemann himself, but the parameters are known as Teichmüller parameters, since Teichmüller initiated the recent mathematical work on the subject. There is no universally accepted "best way" of specifying the Teichmüller parameters, and the formula (1) leaves this question open. The measure function M depends on the choice of the ν's. The only difference between the light-cone and Polyakov approaches in the calculation of M.

It must be emphasized that the Polyakov S-matrix, unlike that obtained from the light-cone approach, is not manifestly unitary. It is clearly necessary to prove unitarity before we know that we have an acceptable theory.

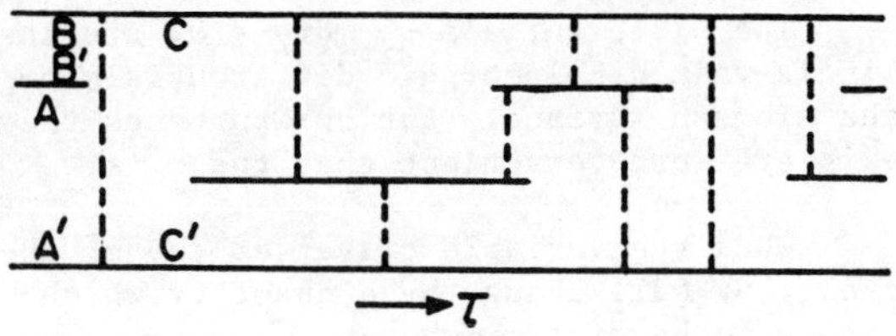

Figure 1. A string diagram.

In the light-cone interacting-string picture one treats the strings as an ordinary quantum-mechanical system. In Fig. 1, one cuts the plane along the horizontal lines and identifies points above one another on adjacent horizontal lines (e.g., AA',BB',CC'). The diagram then represents an interacting closed-string system; σ parametrizes the string itself, while τ is the light-cone time. We regard σ+iτ, with τ Wick-rotated, as a single complex variable. The precise process depicted in the diagram is a two-to-three scattering process with two loops. Along each dotted line one breaks the diagram, displaces the string on one side by an arbitrary twisting angle θ, and reidentifies the points.

In terms of the string-diagram variables, the measure is simply

$$\mathcal{N}\int \prod d\tilde{\tau}_i \prod d\tilde{\alpha} \prod d\theta_i \; |\Delta|^{-(d-2)/2}, \qquad (2)$$

where the $\tilde{\tau}_i$'s are the time co-ordinates of all joining points but one, the $\tilde{\alpha}_i$'s are the lengths of one of the strings in each loop, and the θ_i's are the twisting angles. The factor $|\Delta|^{-(d-2)/2}$, where $|\Delta|$ is the determinant of the Laplacian for the string diagram, results from the original Gaussian functional integral. $\mathcal{N}$ is the Feynman normalization factor. Throughout this talk we shall ignore external-line factors; they are exactly the same as for trees, and have been treated in ref. 2. It is easily checked that the number of variables of integration is 2N+6g-6, corresponding to the N+3g-3 complex variables of integration in Eq. (1).

As we shall see shortly, it is convenient to express the string-diagram variables in terms of a different set of variables. The measure will then contain the Jacobian J of the transformation as an additional factor. Since the "lengths" of the strings are proportional to the momentum in the + direction, the shape of the string diagram will depend on the Lorentz frame and neither of the factors $|\Delta|^{-(d-2)/2}$ or J will be Lorentz invariant. The product is Lorentz invariant if and only if d=26.

Convenient Variables for the n-Loop Problem

Now let us examine the interacting-string picture in a little more detail. We wish to replace the string-diagram variables by new variables Z_i, which have the property that the functions occurring in (1) can be calculated explicitly. A canonical set of such variables makes use of the theory of *automorphic functions* (10). On the string diagram one draws g *A-cycles* A_r and g *B-cycles* B_r which have the property that A_r intercects B_r once, but no other pairs of cycles intercept. One then cuts the diagram along the A-cycles and transforms it conformally

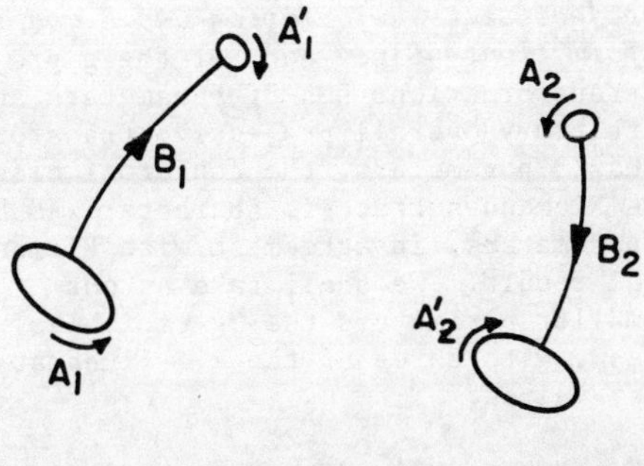

Figure 2. The z-plane of the conformally transformed string diagram.

onto the complex plane with holes corresponding to the cut cycles (Fig. 2). A B-cycle thus takes us from one hole to another. One identifies corresponding points z and z' on corresponding A-cycles by *projective transformations*, one for each A_r ($1 \leq r \leq g$)

$$T_r: \qquad z' = \frac{Az+B}{Cz+D}. \qquad (3)$$

The g transformations T_r thus correspond to the B-cycles. They and their reciprocals generate an infinite group of projective transformations; we denote a general member of the group by V_m ($1 \leq m < \infty$). One is only interested in those groups of transformations whose fundamental region is a region exterior to 2g holes as in Fig. 2. Such groups are called *Schottky* groups.

The transformation T_r depends only on the ratios of the constants A, B, C and D; one usually fixes them (to within a sign) by setting AD-BC=1. Each transformation thus depends on three complex parameters. The generic projective transformation can be written in the form:

$$\frac{z'-z_1}{z'-z_2} = w \frac{z-z_1}{z-z_2} . \qquad (4)$$

The parameter w is known as the *multiplier*, z_1 and z_2 are known as the *invariant points*. By interchanging z_1 and z_2 if necessary we can ensure that $|w| \leq 1$, and we shall always do so.

Given one set of generators $T_1, \ldots T_g$, one may change them in two ways without changing the string diagram. One may subject them all to a projective transformation, i.e., one may define $T_r' = A^{-1} T_r A$, where A is a fixed projective transformation. One may also take a new set of A- and B-cycles; the corresponding transformations of the T's are the *modular* transformations.

Poincaré and Klein made the fundamental conjecture, which was subsequently proved by Koebe (10), that *a conformal transformation from a Riemann surface (and, in particular, from a string diagram) into the complex plane in the manner discussed above is always possible and is unique up to an overall projective transformation and a modular transformation.* Since each of the g projective transformations has 3g parameters and the arbitrary overall projective transformation has 3 parameters, the conformal class of the Riemann surface is characterized by 3g-3 parameters, in agreement with Riemann's general result. We shall take as our Teichmüller parameters the 3g variables w_r, z_{1r}, z_{2r}, with three of the z's fixed at arbitrary values.

Functions defined on the string diagram must be unchanged when we traverse an A- or a B-cycle. In the z-plane, they must be unchanged when z is subjected to a projective transformation in the group. As in the theory of elliptic functions, it is of interest to consider multi-valued functions which have simple transformation properties when we traverse a cycle. In the z-plane, they must have simple transformation properties when z traverses an A-cycle or when it is subjected to a projective transformation in the group. One constructs such functions in the same way as one constructs the infinite series for the logarithm of the Jacobi θ-function (or, alternatively, the infinite product for the θ-function itself). One starts with a given function, subjects it to a transformation in the group, and sums over all group elements. We shall write down the series for the functions we require; the verification that they have the desired properties is not difficult.

It is known that there exist g linearly independent functions which change by a constant when the variable traverses an A- or a B-cycle. The canonical basis for such functions is formed by the functions $v_r(z)$, where v_r changes by $2i\pi\delta_{rs}$ when the variable traverses the sth A-cycle. The formula for v_r is:

$$v_r(z) = \sum_m^{(r)} \ell n \frac{z-V_m z_{1r}}{z-V_m z_{2r}}, \qquad (5)$$

the superscript (r) indicating that we *omit* those values for m for which V_m, when expressed as a product of the generators T_s, has a factor T_r or T_r^{-1} at its right-hand end.

The differentials $\frac{1}{2i\pi} dv_r = \omega_r$ are the single-valued holomorphic differentials on the Riemann surface. For our purposes the v_r's are more convenient that the ω_r's.

When the variable traverses the sth B-cycle, v_r will change by a quantity which we denote to be $2i\pi\tau_{rs}$, where

$$\tau_{rs} = \frac{1}{2i\pi} \{\sum_m^{(r,s)} \ell n \frac{(z_{1s}-V_m z_{1r})(z_{2s}-V_m z_{2r})}{(z_{1s}-V_m z_{2r})(z_{2s}-V_m z_{1r})} + \delta_{rs} \ell n\, w_r\}, \qquad (6)$$

the superscript (r,s) indicating that we omit those V_m's which have $T_r^{\pm 1}$ as their left-most member; we must also omit the identity transformation if r=s. The matrix of the τ's is known as the *period matrix*.

Finally the Green's function is given by the formula:

$$N(z,z') = \ell n|\phi(z,z')|, \qquad (7a)$$

$$\ell n\, \phi'(z,z') = \ell n \phi'(z,z) - (2\pi)^{-1} \sum_{r,s} \mathrm{Re}\{v_r(z) - v_r(z')\}\{(\mathrm{Im}\tau)^{-1}\}_{rs} \quad \mathrm{Re}\{v_s(z)-v_s(z')\}, (7b)$$

$$\ell n\, \phi(z,z') = \ell n(z-z') + \tfrac{1}{2} \sum_{m \neq 1} \ell n \frac{(z-V_m z')(z'-V_m z)}{(z-V_m z)(z'-V_m z')} . \qquad (7c)$$

It is not quite true that N remains unchanged when z is subjected to a projective transformation T_r' the change will be

$$\delta_{B_r},z = - \ell n|C_r z + D_r|, \qquad (8)$$

where C_r and D_r are the C- and D-parameters of T_r. The right-hand side of (8) does not depend on z', and as a consequence, it is not difficult to show that one obtains the correct result if one uses N, defined by Eq. (7), in Eq. (1). The argument depends on momentum conservation.

We notice that the right-hand side of (7b) is not analytic. The presence of a zero mode requires us to define N by the equation

$$\Delta_z N = - 2\pi\delta^2(z-z') + \not{\delta}(z). \quad (9)$$

There exists no function satisfying (9) with $\not{\delta}=0$.

The series (5), (6) and (7c) are known to converge absolutely in a sub-region of the (3g-3)-dimensional space of the Schottky region. The question whether they converge conditionally outside this sub-region has not yet been answered. If not, they must be defined outside the sub-region by analytic continuation. In fact, if one multiplies all the w_r's by a factor λ, the series will converge as long as λ is sufficiently small. For larger values of λ the functions can be defined by a Padé approximation in the single variable λ.

Measure Factor in the Interacting-String Picture.

Now let us outline the calculation of the factor M in (1) and, in particular, of $|\Delta|$. We use the formula

$$|\Delta| = \text{Exp}\,\{\text{Tr}\,\ell n\,\Delta\}. \quad (10)$$

The operator $\ell n\,\Delta$ is singular when $\rho=\rho'$, ρ being any local co-ordinate on a Riemann surface:

$$\ell n\,\Delta = \pi^{-1}|\rho-\rho'|^{-2} - (12\pi)^{-1}\,R\,\ell n|\rho-\rho'| + \text{non-singular terms}, \quad (11)$$

where R is the scalar curvature. To regularize we evaluate $\ell n\,\Delta$ at small values of $\rho-\rho'$, subtract the first two terms of (11), and take the limit $\rho-\rho'=0$. We also replace the joining point by a small region of large but finite curvature. The regularization adds to the energy of the string a term proportional to its "length" (i.e., to the + momentum), and also renormalizes the coupling constant. Neither of these changes is physically significant.

Another infinite contribution to $\ell n\,\Delta$ arises from the zero mode. The correct prescription is to replace the zero eigenvalue by $(1/2\pi\mathcal{A})$, where $\mathcal{A}$ is the area of the string diagram.

We evaluate $|\Delta|$ by examining its change under an infinitesimal change of the metric. The effect of *conformal* changes was considered several years ago by McKean and Singer (11); formulas based on their work are given in Refs. (12) and (13). To calculate the effect of a Teichmüller transformation, we shall consider a general diffeomorphism

$$\rho \to \rho + \delta\rho(\rho,\bar{\rho}). \quad (12)$$

We do *not* require the V's to be single-valued, but we allow them to change when the variables traverse a B-cycle. The change (12) is then sufficiently general to include the Teichmüller transformations.

We restrict ourselves to conformally flat metrics

$$g_{\mu\nu} = e^{2\phi}\,\delta_{\mu\nu}. \quad (13)$$

The Laplacian and Green's functions are defined as follows:

$$\Delta = - g^{-\frac{1}{4}}\,\partial\rho\partial\bar{\rho}\;g^{-\frac{1}{4}}, \quad (14a)$$

$$N_g = g^{\frac{1}{4}}\,N g^{\frac{1}{4}}, \quad (14b)$$

where N is the Green's function of Eq. (7). The measure on the Riemann surface will then be $d^2\rho$, without a factor $g^{\frac{1}{2}}$.

Under a transformation (12), together with a Weyl Conformal transformation $\phi\to\phi+\sigma\phi$, $\text{Tr}\,\ell n\,\Delta$ changes as follows:

$$\delta(\text{Tr}\,\ell n\,\Delta) = \text{Tr}\{\partial_\rho(\mu\partial_\rho N) + \partial_{\bar{\rho}}(\bar{\mu}\partial_{\bar{\rho}}N) + \nu\,\Delta\,N + \Delta\,\nu\,N\} + M, \quad (15a)$$

where

$$\mu = \partial_{\bar{\rho}}(\delta\rho), \quad (15b)$$

$$\nu = - \tfrac{1}{2}\,\partial_\rho(\delta\rho) - (\delta\rho)\partial_\rho\phi - \delta\phi, \quad (15c)$$

and M is an extra term, which we shall not specify, and which is necessary because of the limiting procedure used to define $\text{Tr}\,\ell n\,\Delta$. The quantity μ is known as the infintesimal *Beltrami differential.*

We shall take the uniformizing variables z of the previous section as our co-ordinates. The will of course vary discontinuously as we cross an A-cycle.

We now insert the expression (7) for N into Eq. (15). Let us first consider the summation on the right of (7c), excluding the term $\ell n(z-z')$. As all the terms except $\ell n(z-z')$ are non-singular when z=z', we can simply set z equal to z'. The trace can then be evaluated by a slight adaptation of a resummation procedure due to Selberg [14]. On integration we find that

$$|\Delta|_1 = \prod|1-w_m|^4 \quad (16)$$

where the product is over all *conjugacy classes* of elements V_m of the Schottky group of projective transformations, excluding the identity transformation. (All elements in the same conjugacy class have the same value of w.) Our remarks about the convergence of our previous summations apply equally to the logarithim of (16).

The contribution from the second term on the right of (7b) is also evaluated without difficulty. The result is:

$$|\Delta| = |\mathrm{Im}\tau|^{-1}, \tag{17}$$

where $|\mathrm{Im}\tau|$ is the determinant of the imaginary part of the period matrix.

The contribution from the first term on the right of (7c), though in principle the most straightforwad, requires the most care because of the limiting processes involved. One must separate ρ and ρ' (or z and z'), include the term M on the right of (15a), subtract the changes in the regularization terms on the right of (11), pass to the limit $\rho=\rho'$, and integrate over the Riemann surface. We find:

$$\ell n|\Delta|_3 = \frac{1}{12\pi}\{\int d^2z\phi\partial_z\partial_{\bar{z}}\phi - i\oint dz\ \ell n(C_r z+D_r)\partial_z\phi - i\oint\partial\bar{z}\ \ell n(\bar{C}_r\bar{z}+\bar{D}_r) - 2i\pi\phi(z_1)\}, \tag{18a}$$

where the first integration is taken over one fundamental region of the z-plane, while the last two are taken over the unprimed A-cycles (Fig. 2). The quantities C_r and D_r are the parameters in the appropriate projective transformations. The factor $\phi(z_1)$ in the last term is evaluated at the initial points of the contour integrals, so that the complete expression is independent of the choice of the initial point.

The integrand in the first term on the right of (18a) is zero if the Riemann surface is flat, and the only contributions in a string diagram are from the joining points. On evaluating these contributions as in ref. 2., we obtain the equation:

$$\int d^2z\phi\partial_z\partial_{\bar{z}}\phi = \pi\sum \ell n\left|\frac{d^2\rho}{dz^2}\right|, \tag{18b}$$

the summation being taken over all joining points. (We recall that $\partial\rho/dz=0$ at a joining point.) In Eq. (18b) we have dropped constant terms from the joining points, since they can be absorbed in the coupling constant.

The other factor in the measure M, Eq. (1), is the Jacobian of the transformation from the string-diagram variables σ and τ to the new variables z. As far as we can see, the Jacobian is too complicated to calculate directly, but we can obtain it by making use of its analytic properties. The 2N+6g-6 variables of integration in (2) cannot be replaced by N+3g-3 complex variables, so we proceed in two stages. We first transform to a "new string diagram", conformally equivalent to the old, where the variables α_i are regarded as fixed but the time intervals above and below the loops can differ. This means that the net twist θ on going around the loop is complex, and we can replace our variables by N+2g-3 complex joining points $\tilde{\rho}$ $(=\tilde{\tau}+i\tilde{\sigma})$ and g complex twists. The Jacobian from the old to the new string variables can be calculated explicitly.

The transformation from the new string variables to the variables Z_i, w_r, z_{1r}, z_{2r} (with three z_r's held fixed) is analytic except for isolated singularities. The Jacobian is thus the square of the modulus of an analytic function j, again except for isolated singularities. The right-hand side of Eq. (18a) is the real part of a function which is analytic except for isolated singularities and which we denote by $2\ \ell n\ \mathfrak{f}$. By examining all possible isolated singularities one finds that, apart from some simple factors, the singularities of j precisely cancel those of $\mathfrak{f}$, provided d-2=24. (The factors of 12 in the denominator of (18) are thus cancelled.) Furthermore, the product $j\mathfrak{f}$ is invariant under modular transformations; the proof of this fact is tricky and, at the moment, not rigorous. Since an analytic, singularity-free, modular-invariant function on the Teichmüller space is a constant, these properties determine the result:

$$J' = J|\Delta|_3^{-(d-2)/2} = 2^{2-g}\pi|\mathrm{Im}\tau|^{-1}|(z_a-z_b)(z_b-z_c)(z_c-z_a)|^2 \times \prod_r |w_r(z_{1r}-a_{2r})|^{-4}\ \prod\alpha_i|, \tag{19}$$

where z_a, z_b and z_c are the three z's which are kept fixed. The factor $|\mathrm{Im}\tau|^{-1}$, which is not the modulus of an analytic function, arises from the transformation from the old to the new string variables.

Our final formula for M, Eq. (1) is thus:

$$M = 2^{-g}\{|\Delta|_1|\Delta|_2\}^{-12}J', \tag{20}$$

the factors being given by Eqs. (16),(17) and (19). The extra factor 2^{-g} corrects for some double-counting we have performed (15), about which we shall not elaborate.

One must integrate the Teichmüller parameters over one fundamental region of the modular group, since different such regions correspond to the same string-diagram configurations with different choices of the A- and B-cycles. There is no simple formula for the change of our parameters w_r, z_{1r} and z_{2r} under a modular transformation. The period matrices change by a Siegel modular transformation:

$$\tau' = (A\tau + B)(C\tau + D)^{-1}, \tag{21}$$

where A,B,C and D are g×g matrices with integral entries such that AD-BC=DA-CB=1. The τ's must be calculated from our parameters using Eq. (6); one must then restrict the integration region to avoid two or more period matrices related by (21).

Extension to Superstrings

As we mentioned in the introduction, both approaches to the g-loop amplitude can be generalized to superstrings, though no explicit results have been obtained to date. All variables in (1) become replaced by supervariables, the Teichmüller space becomes replaced by a super-Teichmüller space, automorphic functions become replaced by superautomorphic functions, and so on. Obtaining a manifestly covariant amplitude with external fermions may possibly be more than a straightforward technical problem. The deepest problem is probably to obtain an amplitude with manifest Lorentz invariance *and* manifest space-time supersymmetry. Apart from that, we see no *fundamental* unsolved problems in the perturbation expansion of string amplitudes.

REFERENCES

(1) Alessandrini, V., Nuovo Cimento 2A, 321 (1971). Alessandrini, V., and Amati, D., Nuovo Cimento 4A, 793 (1971). Kaku, M. and Yu, L.P., Phys Lett. 33B, 166 (1970); Phys. Rev. D3, 2992,3007, 3020 (1971). Lovelace, C., Phys. Lett. 32B, 703 (1970).

(2) Mandelstam, S. *Unified String Theories*, Proceedings of the Santa Barbara Workshop, August 1985, pp. 46, 577. (World Publishing, Singapore, 1986.)

(3) D'Hoker, E. and Phong, D.H., Multiloop Amplitudes for the Bosonic Polyakov String. Columbia preprint.

(4) Belavin, A.A. and Knizhnik, V.G., Algebraic Geometry and the Geometry of Quantum Strings. Landau Institute preprint.

(5) Manin, Yu I., The Partition Function of the Polyakov String can be Expressed in Terms of Theta Functions. Preprint.

(6) Berkovits, N., Calculation of Scattering Amplitudes for the Neveu - Schwarz Model using Supersheet Functional Integration. Berkeley preprint.

(7) D'Hoker, E. and Phong, D.H., Loop Amplitudes for the Fermionic String. Columbia preprint.

(8) Friedan, D., Shenker, S. and Martinec, E., Covariant Quantization of Superstrings. Chicago preprint.

(9) Nelson, P., Moore, G. and Polchinsky, J. Strings and Supermoduli. Harvard preprint.

(10) Ford, L.R., *Automorphic Functions*. (Chelsea, New York, 1951).

(11) McKean, H. and Singer, I.M., J. Diff. Geom. 1, 43 (1973).

(12) Alvarez, O., Nucl. Phys. B216, 125 (1983).

(13) Durhuus, B. Nielsen, H.B., Olesen, P. and Petersen, J.L., Nucl. Phys. B196. 498 (1982). Durhuus, B., Olesen, P. and Petersen, J.L., Nucl. Phys. B198, 157; B201, 176 (1982).

(14) Selberg, A., Journ. Ind. Math Soc. 20, 47 (1956).

(15) Polchinski, J., Evaluation of the One-Loop String Path Integral. Texas preprint.

STRUCTURE OF MULTILOOP DIVERGENCES IN THE CLOSED BOSONIC STRING

E. Gava[a], R. Iengo[b,a,c], T. Jayaraman[d] and R. Ramachandran[e]

a) Istituto Nazionale di Fisica Nucleare (INFN), Sezione di Trieste, I-34127 Trieste, Italy.
b) International Centre for Theoretical Physics, I-34014 Trieste, Italy
c) International School for Advanced Studies (ISAS), I-34014 Trieste, Italy
d) Institut für Theoretische Physik, Universität Karlsruhe, D-7500 Karlsruhe, Fed. Rep. Germany
e) Department of Physics, Indian Institute of Technology, Kanpur 208016, India

The sources of divergences in the multiloop vacuum diagrams of the closed bosonic string are discussed, within the covariant Polyakov formalism. In particular, the infrared divergence due to the dilation tadpole is exploited by a direct computation.

The purpose of this work is to discuss the structure of divergences appearing in multiloop vacuum diagrams of the closed bosonic string. Some early attempts to understand this problem were made in the context of the old dual models (1). Here we will work in the framework of the Polyakov covariant formulation of string theory (2), where a h-loop vacuum amplitude is represented as sum over Riemann surfaces $\mathfrak{M}$ of genus h. We will choose critical spacetime dimensions D=26 and consider $h \geq 2$. The path integral is computed to give (3)(4):

$$Z_h = V \int_{\text{Moduli}} d\mu \; (\det{}' P_1^+ P_1)^{1/2}_{\hat g} \times$$

$$\times \left(\frac{2\pi \det{}' \Delta \hat g}{\int d\sigma \sqrt{\hat g}} \right)^{-D/2} \qquad (1)$$

In (1) V is the spacetime volume, $(\det{}' \Delta \hat g)^{-D/2}$ comes from the integration over the embeddings X^μ, $(\det{}' P_1^+ P_1)^{1/2}$ is the Faddeev-Popov determinant with P_1, P_1^+ the operators introduced by O. Alvarez , the prime denotes deletion of possible zero modes and $\hat g$ is some reference metric on $\mathfrak{M}$. For $h \geq 2$, one can represent $\mathfrak{M}$ by a fundamental domain H/Γ in the upper half plane H, Γ being a hyperbolic Fuchsian group of 2h generators. $\hat g$ will then be induced on $\mathfrak{M}$ by the Γ-invariant Poincaré metric on H, $ds^2 = dz d\bar z (\text{Im} z)^{-2}$, which has constant curvature -1.

The integration measure $d\mu$ is the Weil-Petersson measure on the (6h-6) dimensional Teichmüller space. Everything being invariant under the modular group representing global diffeomorphisms of $\mathfrak{M}$, one restricts the integration region in (1) to the moduli space, a fundamental domain of the modular group in Teichmüller space.

With the above setting, determinants can be expressed by means of Selberg's zeta function (4)(5)(6)(7):

$$Z(s) = \prod_{\{p\}} \prod_{n=0}^{\infty} \{ 1 - \exp[-(n+s) \ell_p] \} \qquad (2)$$

Here ℓ_p are the hyperbolic minimal lengths, corresponding to the set of inconjugate primitive elements of Γ, of closed curves belonging to definite homotopy classes. The number of ℓ_p's in infinite and they are functions of 6h-6 real parameters.

The final result for Z_h is:

$$Z_h = V \int d\mu \; (Z'(1))^{-13} Z(2) \qquad (3)$$

where $Z'(1) = (dZ(s)/ds)|_{s=1}$. From (2)(3) it is apparent that potential diverges in Z_h arise when $\ell\gamma \to 0$ for some geodesic γ since, in that case $Z(s) \to 0$. So we have to understand the degenerating process i.e. the process by which a smooth Riemann surface becomes singular. When γ shrinks and pinches the surface, it produces a node, either by splitting $\mathfrak{M}$ into two parts or by reducing the number of handles by one, depending on whether γ is homologically trivial or not. This process is constrained essentially by the Gauss-Bonnet theorem: the hyperbolic area of $\mathfrak{M}$ remains constant during the deformation. More precisely, take γ to lie along the imaginary axis in H and consider an annular piece of the surface containing it, called a collar (fig. 1):

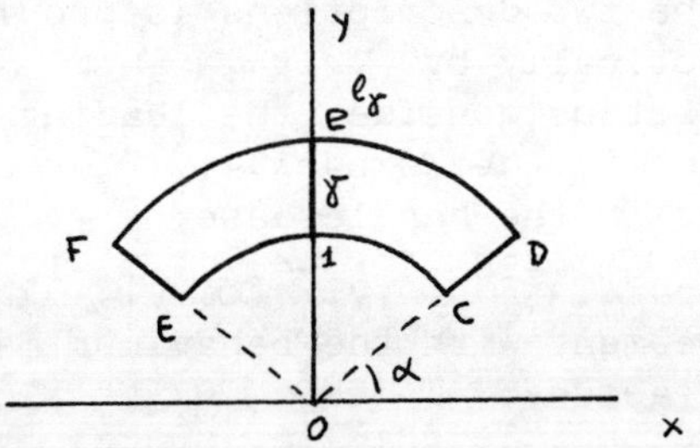

Fig. 1

Then it is a result that there always exists a collar with strictly positive area (8). Since the area of the collar (EFCD) is $2\ell\gamma \cot g\alpha$, $\alpha \sim \ell\gamma$ as $\ell\gamma \to 0$, implying that the geodesic intersecting γ must have length $|\ln \ell\gamma| \to \infty$ and that CD, EF must have finite lengths in the limit. Hence only non intersecting geodesics can be shrunk simultaneously and there are 3h-3 of them at most, partitioning $\mathfrak{M}$ into 2h-2 pieces called "pants". They in turn provide 6h-6 real, Fenchel-Nielsen, coordinates of Teichmüller space (9): 3h-3 are given by their lengths ℓ_i and the remaining 3h-3 by twist angles θ_i defined on them. Finally in a given partition there can be at most 2h-3 homologically trivial (dividing) geodesics.

Let us first consider a particular dividing geodesic γ and evaluate the behaviour of the integrand in (3) as $\ell\gamma \to 0$ (7). The use of Jacobi inversion formula gives:

$$Z(2) \xrightarrow[\ell\gamma \to 0]{} \text{const}\,\ell\gamma^{-3} \exp(-\pi^2/3\ell\gamma)$$

$$\prod_{n=1}^{\infty} [1-\exp(-\frac{4\pi^2 n}{\ell_\gamma})]^2 \tag{4}$$

Next, to evaluate $(Z'(1))^{-D/2}$, we go back to its original, path integral definition and divide $\mathfrak{M}$ into three parts, with a collar Ω around γ sandwiched bewtween $\mathfrak{M}_1, \mathfrak{M}_2$ whose boundaries are $\partial\mathfrak{M}_{1,2} = \sigma_{1,2}$ (fig. 2):

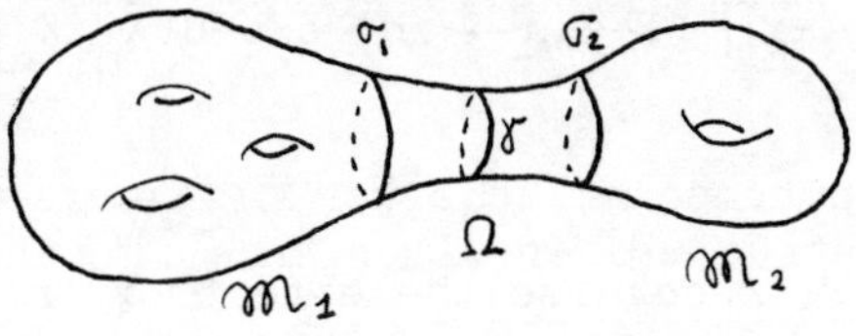

Fig. 2

By our previous discussion the area of Ω is bounded from below by a positive constant as $\ell\gamma \to 0$ and the lengths of $\sigma_{1,2}$ stay non zero in the limit; we can then represent $(Z'(1))^{-D/2}$ by:

$$V[Z'(1)|^{-D/2} = \int \mathcal{D}\bar{X}_1 \mathcal{D}\bar{X}_2 \int_{\mathfrak{M}_1} \mathcal{D} X \exp(X\Delta_{\hat{g}}X)$$

$$\times \int_{\Omega} \mathcal{D} X \exp(X\Delta_{\hat{g}}X) \int_{\mathfrak{M}_2} \mathcal{D} X \exp(X\Delta_{\hat{g}}X). \tag{5}$$

where $\bar{X}^{\mu}_{1,2}(\sigma)$ are boundary values of X^{μ} on $\sigma_{1,2}$.

Transforming Ω to a flat cylinder by conformal and Weyl transformations, after checking the irrelevance of the anomaly factors we get (10):

$$\int \mathcal{D} X \exp(X\Delta X) = \text{const.}(1/\lambda)^{D/2}$$

$$\times \exp(\lambda\pi D/6) \prod_{n=1}^{\infty} [1-\exp(-4\pi\lambda n)]^{-D}$$

$$\times \exp[-(\bar{X}_{01} - \bar{X}_{02})^2/\lambda] \exp[-S(\bar{X}_{n1}, \bar{X}_{n2}]], \tag{6}$$

where $\lambda = \pi/\ell\gamma$ and $\bar{X}_{n;1,2} = \int d\sigma e^{2\pi i n\sigma} \bar{X}_{1,2(\sigma)}$. It is important to observe that the integrals over $\mathfrak{M}_1, \mathfrak{M}_2$ do not depend on $\bar{X}_{0;1,2}$ by translational invariance.

Finally the Weil-Petersson measure in Fenchel-Nielsen coordinates is given by (11) $d\mu = \prod_{i=1}^{3g-3} \ell_i d\ell_i d\theta_i$; collecting the various terms we have, putting $\Delta X = X_{01} - X_{02}$:

$$Z_h \sim V \int d^{26}(\Delta X) \quad d\lambda \quad \lambda^{-13} \exp \frac{-(\Delta X)^2}{\lambda} \times$$
$$\times \exp 4\pi\lambda \prod (1-\exp -4\pi\lambda n)^{-24} \exp -S(\bar{X}_{n1}, X_{n2}). \tag{7}$$

Here we recognize the effect of "new zero modes" in $(Z'(1))^{-D/2}$ as coming from the ΔX integration, which produces a power λ^{13}. Consequently we get the following asymptotic behaviour of Z_h:

$$Z_h \sim \int d^{26}(\Delta X) \frac{d\lambda}{\lambda^{13}} \sum_{N=0}^{\infty} C_N \exp(4\pi\lambda(N-1) - \frac{(\Delta X)^2}{\lambda})$$

$$\sim V \left\{ \frac{C_0}{p^2 - 4\pi} + \frac{C_1}{p^2} + \frac{C_2}{p^2 + 4\pi} + \ldots \right\} \Big|_{p=0} \tag{8}$$

in which the spectrum of scalar states propagating on the cylinder has been exploited. In particular the second term in (8) is the expected infrared singularity due to the massless dilaton going into the vacuum (12).

A similar discussion in the handle case can be done and one can see that in such a case there is no additional λ^{13}, because there is a (ΔX) dependence also in the remaining part of the functional integral, hence there is no divergence for $\lambda \to \infty$, apart from the usual tachyonic one.

It is also possible to see the relation between our discussion, which employs a real description of moduli space, and the complex treatment of (13). There the partition function is written as:

$$Z_h = \int \prod_{i=1}^{3g-3} \frac{dy_i \wedge d\bar{y}_i}{(\mathrm{Det\ Im}\Pi)^{13}} \quad |F(y_i)|^2 \tag{9}$$

y_i being natural complex coordinates for moduli space, $\{\pi\}$ is the hxh period matrix, F is a holomorphic function of y_i, with $F \sim 1/y_i^2$ as $y_i \to 0$, the degeneration limit. As $y_i \to 0$, Det Imπ is bounded in the dividing case whereas goes like $\ell n|y_i|$ in the handle case. The relation between the two descriptions is provided asymptotically by $\lambda + i\theta \sim -\frac{1}{2\pi} \ell n y$, as $y \to 0$. Let us consider the leading term in (8): $d\lambda\, d\theta \exp 4\pi\lambda \sim |y|^{-4} dy\, d\bar{y}$, whereas in the handle case:

$d\lambda d\theta \exp 4\pi\lambda \quad \lambda^{-13} \sim |y|^{-4} (\ell n|y|)^{-13} dy\, d\bar{y}$, in agreement with the behaviour mentioned above.

In conclusion we have isolated the sources of divergences in the bosonic string: the tachyon and the dilaton; it is of course tempting to argue that superstring models (14) in which there is no tachyon and the dilaton tadpole is zero at one loop level, are completely finite to all orders. However, a proof of this, presumably, related to the understanding of how to generalise GSO projection (15) at higher loops, has yet to be found.

REFERENCES

(1) V. Alessandrini, Nuovo Cimento 2A (1971) 321; V. Alessandrini and D. Amati, Nuovo Cimento 4A(1971) 793, C. Lovelace, Phys. Lett. 34B (1971) 500; M. Kaku and J. Scherk, Phys. Rev. D3(1971)430,2000.

(2) A.M. Polyakov, Phys. Lett. 103B (1981)207, 211.

(3) O. Alvarez, Nucl.Phys. B216(1983) 125.

(4) E. D'Hoker and D.H. Phong, Columbia University preprint CU-TP-323 (1985); M.A. Baranov, A.S. Shvarts, JETP Letters 42(1985) 219.

(5) H.P. Mckean, Comm. Pure Applied Math. 25(1972)225; D.A. Hejhal, The Selberg trace formula for PSL(2,R), Vol. 1, Springer Lecture Notes in Mathematics, Vol. 548 (Springer, Berlin, 1976).

(6) S. Mandelstam, Lectures Workshop on Unified string theories (Santa Barbara), preprint UCB-PTH-85/47 (October 1985); A. Restuccia and J.G. Taylor, Phys. Lett. 162B (1985) 109; King's College preprint (October 1985); G. Gilbert University of Texas preprint UTTG-23-85 (1985); M.A. Namazie and S. Rajeev, CERN-TH. 4327/85.

(7) E. Gava, R. Iengo, T. Jayaraman, R. Ramachandran, Phys. Letters 168B(1986)207.

(8) L. Keen, Collars on Riemann surfaces, in: Discontinuous groups and Riemann surfaces, ed. L. Greenberg (Princeton, U.P., Princeton, 1974); S. Wolpert, Ann. Math. 109 (1979)323.

(9) W. Abikoff, The real analytic theory of Teichmüller space, Springer Lecture Notes in Mathematics, Vol. 820 (Springer, Berlin, 1980).

(10) S. Wolpert, Am.J. Math. 107(1985) 969.

(11) M. Ademollo, A. D'Adda, R. D'Auria, F. Gliozzi, E. Napolitano, S. Sciuto and P. Di Vecchia, Nucl. Phys. B94(1975)221; J. Shapiro, Phys. Rev. D11 (1975) 2937.

(12) A.A. Belavin, V.G. Knizhnik, Phys. Letters 168B(1986)201; R. Catenacci, M. Cornalba, M. Martellini, C. Reina, Phys. Lett. 172B(1986)328; J.B. Bost, T. Jolicoeur, Phys. Letters 174B(1986) 273.

(13) J.H. Schwarz, Phys. Rep. 89(1982) 223; M.B. Green, Surv. High Energy Phys. 3(1983)127; D.J. Gross, J.A. Harvey, E. Martinec and R. Rohm, Nucl. Phys. B256(1985) 253.

(14) F. Gliozzi, J. Scherk, D. Olive, Nucl. Phys. B122(1977)253.

CONFORMAL INVARIANCE ON CALABI-YAU SPACES*

DENNIS NEMESCHANSKY

Stanford Linear Accelerator Center
Stanford University, Stanford, California, 94305

The possibility of superstring compactification on Calabi-Yau manifolds is analyzed. Despite the apparent non-zero β function at four loop order, it is possible to construct a conformally invariant sigma model on a Calabi-Yau manifold. The background metric is not Ricci flat, but is related to the Ricci flat metric through a (non-local) field redefinition.

During the last few months it has become clear that $N = 2$ supersymmetric models on Calabi-Yau manifolds with a Ricci flat Kahler metric have a nonvanishing β-function at the four loop level (1,2). This destroys the expectation that such models have a vanishing β-function to all orders in perturbation theory. Despite the four loop contribution, I will show that it is always possible to choose a Kahler metric on a Calabi-Yau space such that the β-function vanishes to all orders in perturbation theory. This talk summarizes a recent paper with A. Sen (3).

These theories are of current interest since they provide us with possible solutions of the classical string equations (4). This is based on the equivalence between the equations of motion of the massless fields and conformal invariance of the two dimensional sigma-model (5). In this talk I will only consider the equation of motion for the graviton. I will set the antisymmetric tensor field to zero and I will assume that the dilaton field is constant. It is a trivial generalization to include other background fields. My talk is divided into two parts. First I will show how the contribution to the β-function arises at four loops. In the second part of the talk I will sketch how to choose the new Kahler potential.

The classical equation of motion for the string can be derived from the effective action for the massless fields (2). To find this effective Lagrangian one has to consider tree level string scattering amplitudes. The three graviton scattering is that of general relativity with no corrections coming from the string theory. The situation changes for four gravitons (2). In terms of an ordinary Feynman diagram the four graviton scattering can be represented in the form shown in fig. 1. The sum is over all string states. Since we are constructing the effective action for the massless fields we have to subtract off the massless poles. This then leaves us with intermediate states with masses of the order of the Planck scale. For energies much below the Planck scale, the propagator $(p^2 + m^2)^{-1}$ of the intermediate states in fig. 1 can be expanded in powers of the momenta p^2, giving rise to effective four graviton couplings. In order to reproduce the S-matrix elements of the string scattering amplitudes, one has to add to the effective Lagrangian new local four point vertices order by order in α'. This procedure can then be repeated for five and higher graviton couplings to yield, in principle, the effective Lagrangian to all orders.

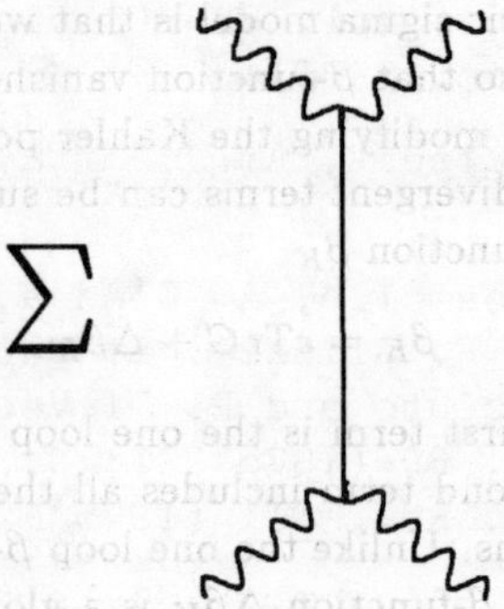

Fig. 1. Four graviton scattering.

The effective Lagrangian is not unique since any local field redefinition of the fields will not affect the S-matrix. However, the equations of motion derived from the effective Lagrangian do not change under this local field redefinition.

As mentioned earlier, there are no additional three graviton interactions coming from strings to Einstein's general relativity. Therefore we may

* Work supported by the Department of Energy, contract DE–AC03–76SF00515.

conclude that in the generic case of Ricci flat metric the β-function is zero at one, two and three loop order. However, since the four graviton scattering was modified by the exchange of heavy states the four loop β-function does not vanish (2,6) for the generic Ricci flat manifold. This result agrees with an explicit calculation of the four loop β-function for an $N = 2$ supersymmetric sigma model (1).

Having shown how the four loop β-function arises I will spend the rest of the talk proving that despite the apparent nonzero contribution to the four loop β-function the theory can be made to have a vanishing β-function to all orders in perturbation on a Calabi-Yau space (3).

Let me start the argument with the observation that the vanishing at the β-function takes different forms under the redefinition of the coupling constant of the two dimensional sigma model. Remember that the metric G_{ij} is nothing else than the coupling constant of the theory. Under the replacement $G_{ij} \to G_{ij} + T_{ij}$, the β-function is given by the Ricci tensor calculated from the new metric $G_{ij} + T_{ij}$. To linear order in T this is given by

$$R_{ij}(G) + \tfrac{1}{2}\left(T_{ij;m}{}^{m} + T^{m}{}_{m;ij} - T_{im;j}{}^{m} - T_{jm;}{}^{m}\right) \quad (1)$$

For a specific example $T_{ij} = R_{imnp}R_{j}{}^{mnp}$ the second term in eq. (1) gives a three loop contribution. Changing the metric in the sigma model corresponds to adding finite counter terms. Therefore the vanishing of the β-function depends on how we subtract the ultraviolet divergences in the theory.

What we now have to show for the $N = 2$ supersymmetric sigma model is that we can modify the metric so that β-function vanishes. This corresponds to modifying the Kahler potential. The ultraviolet divergent terms can be summarized in a single β-function β_K

$$\beta_K = c\,\mathrm{Tr}\, G + \Delta\beta_K \quad (2)$$

where the first term is the one loop contribution and the second term includes all the higher loop contributions. Unlike the one loop β-function the higher loop β-function $\Delta\beta_K$ is a globally defined scalar on the Calabi-Yau manifold (3).

It is very easy to show that for a Ricci flat metric $\widetilde{G}_{ij}$ there exists a Kahler potential K such that the β-function vanishes (3). To find K one has to solve the differential equation

$$c\mathrm{Tr}\,\ell n G + \Delta\beta_K = c\mathrm{Tr}\ell n\widetilde{G}\ . \quad (3)$$

The most convenient way to solve this equation is to iterate it powers of α'. To lowest order in α' eq. (3) takes the form $\Box(K - \tilde{K}) = \Delta\beta_{\tilde{K}}$. The solution of eq. (3) gives a metric in each coordinate patch. In order for the solution to be an admissible metric one must show that the metric is globally well defined. This means that when we calculate G_{ij} in two different coordinate patches G_{ij} must transform like a tensor. Since $\Delta\beta_K$ is a globally well defined scalar it can be shown that G_{ij} also has the correct transformation property to define a metric on the whole manifold (3).

Thus I have shown to you that given a Calabi-Yau manifold we can always construct to all orders in perturbation theory a conformally invariant sigma model. The metric is not Ricci flat but it is related to it by a nonlocal (in space-time) field redefinition. The new metric is a globally defined tensor on the manifold and hence a valid choice of the metric.

There is one more question that I would like to address. Does the procedure I have outlined converge? For a string theory we cannot stop at any order of perturbation theory because this would ruin the conformal invariance of the sigma model. Today I have shown that the theory is conformally invariant to all orders in perturbation theory. Therefore a violation of conformal invariance can at most be an exponential. But from the work of Dine, Seiberg, Wen and Witten (7) on the effective four dimensional theory we know that a (2,2) supersymmetric sigma model is conformally invariant even when the nonperturbative effects are included. Hence for the case I have studied this means that the procedure converges (8). They also considered the (2,0) supersymmetric sigma model and they showed that conformal invariance is violated by nonperturbative effects. Therefore, if one repeated the arguments I have presented here for the (2,0) case, the series could not be summed.

Let me conclude by discussing the implication of our results for the superstring theory. Our results tells us that given a Calabi-Yau manifold we can always find a background vacuum expectation value of th metric which satisfies the equation of motion of the string theory. The background metric is obtained by solving eq. (3).

For some purposes (e.g. the study of the four dimensional effective field theory obtained after compactification) we do not need to know what the metric that solves the equation of motion looks like in terms of the Ricci flat metric. We take the Calabi-Yau metric as the background metric and add finite local counterterms in each order of perturbation theory in order to have a vanishing β-function. The result is a two dimensional conformally invariant field theory. We may then calculate the particle spectrum and the interaction in the effective four dimensional theory by identifying operators of conformal dimension (1,1) as vertex operator and calculate their correlation functions in the two dimensional field theory obtained this way. It is in this scheme that Witten's (9) general argument showing the vanishing of the β-function on Ricci flat Kahler manifolds works. This argument is based on a study at the effective four dimensional field theory and does not specify the renormalization scheme in which his proof should work.

REFERENCES

(1) M. T. Grisaru, A. van de Ven and D. Zanon, preprints HUTP-86/A020, HUTP-86/A026, HUTP-86/A027 (1986).

(2) D. Gross and E. Witten, Princeton preprint (1986).

(3) D. Nemeschansky and A. Sen, preprint SLAC-PUB-3925.

(4) P. Candelas G. Horowitz, A. Strominger and E. Witten, Nucl. Phys. B258 (1985) 46.

(5) C. Callan, E. Martinec, D. Friedan and M. Perry, Nucl. Phys. B262 (1985) 593. A. Sen, Phys. Rev. D32 (1985) 2102, Phys. Rev. Lett. 55 (1985) 1846.

(6) M. Freeman and C. Pope, preprint Imperial/TP/85-86/17.

(7) M. Dine, N. Seiberg, X. Wen and E. Witten, Princeton preprint (1986).

(8) N. Seiberg, private communications.

(9) E. Witten, Princeton preprint (1985).

VERTEX OPERATORS, VIRASORO CONDITIONS AND STRING DYNAMICS IN CURVED SPACE

Spenta R. Wadia

Tata Institute of Fundamental Research*
Homi Bhabha Road, Bombay 400 005, India &

International Centre for Theoretical Physics
34100 - Trieste, Italy

String propagation in a background metric and dilaton field are considered in the context of conformal invariant field theory. A perturbatively renormalized tachyon vertex in the presence of these background fields is presented. This generalises the Berezinsky-Kosterlitz-Thouless construction. The equations of motion for the background fields and the wave equation for the vertex function emerge upon imposing the Virasoro gauge conditions on the vertex operator. This is equivalent to calculating the equation of motion $Q|\psi\rangle = 0$ in the BRST approach.

1. INTRODUCTION AND GENERAL FORMULATION

A basic issue in string theory is that we do not yet have a theory of strings in the sence of having invariance principles and a lagrangian which enables us to solve for the ground state and a spectrum of excitations. At present the formulation of the theory is closer to the S-matrix approach. There is a perturbative prescription for calculating the S-matrix in terms of properties of a 2-dim. conformal invariant field theory (CIFT).

Let us briefly describe this prescription [1], restricting ourselves to closed bosonic strings. Denote by x_a the co-ordinates of the target space in which the string propagates. Then the map $(z,\bar{z}) \to x_a(z,\bar{z})$, from a Riemann surface a genus n into the target space describes a string configuration at order n in string perturbation theory. A 2-dim. metric $g_{\mu\nu}$ characterizes the Riemann surface. The states of the string are in correspondence with vertex operators $\bar{V}_\alpha(x(z,\bar{z}), g(z,\bar{z}))$ which are scalars under reparametrizations of the Riemann surface. α specifies the quantum numbers of the associated particle.

The formula for the S-matrix of these particle states is given by

$$S(\alpha_1,\ldots,\alpha_N) = \frac{1}{Z}\sum_{n=0}^{\infty} g^n\Big(\int Dg_{\mu\nu} Dx_a\, e^{-A(x,g)} \int \prod_{i=1}^{N} \bar{V}(x(z_i),g(z_i))\sqrt{g(z_i)}\,d^2z_i\Big) \qquad (1)$$

$A(x,g)$ is a reparametrization invariant action and g is the string coupling constant. For asymptotically flat directions on the target space we admit the boundary condition $V_\alpha(x \to \infty) = h(k)_{a,b,\ldots} \exp(ik.x)$. k and $h(k)_{a,b,\ldots}$ are the momentum and polarization tensor of the associated particle.

Reparametrization invariance in (1) enables locally the choice of the conformal gauge $g_{\mu\nu} = e^\phi \delta_{\mu\nu}$. $\phi(z,\bar{z})$ is the Liouville field. The formula for the S-matrix becomes

$$S(\alpha) = \frac{1}{Z}\sum_{n=0}^{\infty} g^n\Big(\int d\mu(m) D\phi Dx_\phi e^{-A(x)} \int \prod_{i=1}^{N} \bar{V}(x(z_i),g(z_i)) e^{\phi(z_i)} d^2z_i \qquad (2)$$

$d\mu(m)$ denotes the measure over moduli space and Dx_ϕ includes the Faddeev-Popov factor. In the conformal gauge the residual reparametrizations are analytic/anti-analytic conformal transformations: $z \to w(z)$. When the central charge of the Virasoro or conformal algebra $C = 26$, then Dx_ϕ is in fact independent of ϕ. We call it Weyl invariant. We further require the vertex $V(x;\phi) = \bar{V}(x;\phi)e^\phi$ to be Weyl invariant. $V(x;\phi+\alpha) = V(x;\phi)$. Now performing a combined reparametrization $z \to w(z)$ and a Weyl transformation $\phi \to \phi + \ell n|w'|^2$ in $V(x;\phi)$ we conclude that $V(x)$ (we have dropped the ϕ dependence) transforms as a (1,1) conformal tensor: $V'(x'(w)) = |w'|^{-2} V(x(z))$. Hence $d^2zV(x(z))$ is a scalar. With the elimination of the Liouville mode from (2) the only vestige of of the 2-dim. metric in (2) is the integration over the moduli of the Riemann surface. We finally rewrite (2) manifestly independent of ϕ

$$S(\alpha) = \frac{1}{Z}\sum_{n=0}^{\infty} g^n\Big(\int d\mu(m) Dx\, e^{-A(x)} \int \prod_{i=1}^{N} V(x(z_i)) .d^2z_i\Big) \qquad (3)$$

The local statement on the Riemann surface that the vertex operator $V(x)$ transforms as a (1,1) conformal tensor is infinitisimally expressed as $\delta_\varepsilon V = \varepsilon' V + \varepsilon V'$, where $w(z) = z + \varepsilon(z)$. This infinitisimal conformal transformation is generated by the traceless part of the holomorphic stress tensor $T(z) = T_{zz} = T_{11} - T_{22} + 2iT_{12}$.

$$\delta_\varepsilon V = \oint \varepsilon(z)T(z)V(w) = \frac{d}{dw}(\varepsilon(w)V(w)) \qquad (4)$$

since the stress tensor is conserved this transformation law of the vertex is equivalent to the operator product expansion (OPE) [2]

$$T(z)V(w) = \frac{V(w)}{(z-w)^2} + \frac{V'(w)}{(z-w)} + \text{reg. terms} \qquad (5)$$

From (4) and the associativity of the OPE we may infer the closure of the OPE of T(z) upto a central extension

$$T(z)T(w) = \frac{2T(w)}{(z-w)^2} + \frac{T'(w)}{(z-w)} + \frac{C}{2(z-w)^4} + \ldots, \quad C = 26 \qquad (6)$$

It is important to note that the OPE (5) and (6) are valid on Riemann surfaces of arbitrary genus. They embody the local information of string propagation. The only other general requirement of (3) is that the functional integral in (3) is modular invariant.

2. VIRASORO GAUGE CONDITIONS AND CLASSICAL EQUATIONS OF MOTION OF STRING

Having explained the general formulation for our present purposes we restrict ourselves to the genus zero case: the sphere. There are no Teichmuller deformations and hence no integration over moduli. The expression for the tree level S-matrix is simply expressed in terms of the vertex operator correlation functions:

$$S(\alpha) = \int \prod_{i=1}^{N} \langle d^2 z_i V_\alpha(x(z_i))\rangle \qquad (7)$$

We regard the sphere as the complex plane with points at ∞ identified. Then performing the conformal transformation $z = e^{\tau+i\sigma}$, we identify $\tau = \ln|z|$ as the radial time for the evolution of string states. $\tau = -\infty$ corresponds to $z = 0$. In this picture we introduce the standard Virasoro generators $L_n = \oint z^{n+1} T(z)$ (the contour is evaluated around $z = 0$). The OPE (5) and (6) can be used to derive the commutation relations

$$[L_n, V(z)] = z^{n+1}\frac{d}{dz}V + (n+1)z^n V \qquad (8)$$

$$[L_n, L_m] = (n-m)L_{n+m} + \frac{C}{12}(n^3-n)\delta_{n+m,0}, \; C=26 \qquad (9)$$

(9) is the Virasoro algebra. At $\tau = -\infty$ or $z = 0$, (8) can be used to derive the Virasoro gauge conditions on a string state created by a vertex operator

$$L_o|\psi\rangle = |\psi\rangle, \; L_n|\psi\rangle = 0 \quad n > 0,$$
$$|\psi\rangle = V(x(o))|o\rangle \qquad (10).$$

Equation (5) or equivalently (10) are the basic classical equations of motion of the string. They say that string states are highest weight states of the Virasoro algebra with eigenvalue 1. The Virasoro algebra is characterized by C = 26. As is well known for strings propagating in flat spacetime these requirements lead to a unitary S-matrix which satisfies the properties of factorization and duality [3]. In what follows we will demonstrate that they also characterize the classical vacua of the string theory apart from determining the form of the vertex function V(x)[4].

3. STRING PROPAGATION IN CLASSICAL VACUA

The conformal invariant field theory relevant to the string model is the one which describes string propagation in the presence of non-trivial background fields corresponding to condensates of its massless modes namely the metric $G_{ab}(x)$, the dilaton $\Phi(x)$ and the anti-symmetric Kalb-Ramond field $A_{ab}(x)$[5]. The string action is given by

$$A = \int\frac{d^2z}{4\pi}\Big(\frac{1}{2\lambda^2}\sqrt{g}\, g^{\mu\nu}G^{ab}(x)\partial_\mu x_a \partial_\nu x_b + \frac{1}{2\lambda^2}A^{ab}(x)dx_a \Lambda dx_b - \sqrt{g}\,R^{(2)}\,\Phi(x)\Big) \qquad (11)$$

These couplings constitute the most general reparametrization invariant renormalizable lagrangian in 2-dim. $\lambda^2 \sim \alpha'$ the slope parameter. The stress tensor is given by the standard formula $T_{\mu\nu} = -\frac{4\pi}{\sqrt{g}}\frac{\delta A}{\delta g^{\mu\nu}}$. As mentioned earlier when the central charge C = 26 the Liouville mode of the 2-dim. metric decouples from the S-matrix. Hence we calculate $T_{\mu\nu}$ and then set $\phi = 0$. Evaluating the action and the traceless part of the stress tensor we get (Hence forth we set $A_{ab} = 0$ for simplicity)

$$A = \frac{1}{2\lambda^2}\int\frac{d^2z}{4\pi}\; G^{ab}(x)\partial_\mu x_a \partial_\mu x_b \qquad (12)$$

$$T_{zz} = -\frac{1}{2\lambda^2}\; G^{ab}\;\partial_z x_a \partial_z x_b + \partial_z^2\,\Phi \qquad (13)$$

4. VERTEX OPERATORS

As a prelude we study the case of free fields: $G_{ab} = \delta_{ab}$ and $\Phi = 0$. Here the central charge $C = D$, the number of free bose fields. Hence D = 26. An example is the tachyon vertex $V(x) = e^{ik.x}$. As an operator V(x) is ultra-violet (UV) divergent due to operator products at the same point z.

To find the renormalized operator with UV finite insertions in the greens functions $<\prod_i x(z_i)_{a_i}>$ we evaluate the expectation value of V in the presence of a source $j_a(z)$ coupled to $x_a(z)$. UV divergences are regularized by continuing above 2 dim. $d = 2+2\varepsilon$. Infra-red (IR) divergences are regularized by a mass term $m^2 x_a^2$ in the action. Then

$$<e^{ik.x(w)}> = e^{-\frac{k^2}{2\varepsilon}\Delta_\varepsilon(o)}\, e^{i\int d^2 z k^a j^a(z)\Delta(z-w)} \tag{14}$$

$$\Delta(z-w) = -\ln(V\mu^2|z-w|^2)(1+o(m^2), \; V=\frac{e^\gamma m^2}{4\pi\mu^2}$$

$$\Delta_\varepsilon(o) = -(\frac{1}{\varepsilon} + F + \ldots), \; F = \ln V$$

μ^2 is the scale that enters the dim. regularization procedure and γ is Euler's Constant. Now defining the renormalized vertex

$V_R = e^{-\frac{k^2}{2\varepsilon}} e^{ik.x}$ (this is equivalent to minimal subtraction)

$$<V_R>_j = (\frac{m^2}{\mu^2})^{\frac{k^2}{2}} e^{i\int k.j\,\Delta} \tag{15}$$

For $k^2 \neq 0$, $<V_R>_j = 0$ at $m = 0$ as expected for a operator that develops anomalous dimensions. Correlation functions of vertex operators have non-zero thermodynamic limits (i.e. are non-zero as $m \to 0$ provided certain conditions are satisfied:)

Let us demand that V_R is a conformal field of dim. (1,1). Since

$$<T(z)V_R(k,w)> = V^{\frac{k^2}{2}}(\frac{k^2}{2}\frac{1}{(z-w)^2} + m^2k^2\frac{(z-w)}{(z-w)^3}$$

$$.[1 + \ln v^2|z-w|^2] + o(m^4)\ldots)$$

$$<V_R(w,k)> = V^{\frac{k^2}{2}}$$

for small but non-zero m we conclude that the Virasoro condition

$$<T(z)V_R(w,k)> = \frac{k^2}{2}\;\frac{<V_R(k,w)>}{(z-w)^2} \tag{16}$$

implies the well known result $k^2 = 2$.

In order to construct perturbative highest weight representations of the Virasoro algebra we take guidance from free fields because perturbation theory in λ^2 implicitly assumes the curvature of the target space to be small. Candidate vertex operators generalizing the tachyon and the gravitational multiplet are $V(x)$ and $h_{ab}(x)\partial_\mu x^a \partial_\mu x^b$ where $V(x)$ is a scalar and $h_{ab}(x)$ is a tensor on the target space.

5. PERTURBATION THEORY IN THE LIMT OF ZERO SLOPE

The perturbation theory is constructed following Friedan [6] by parametrizing the string around a point X_a in target space: $x_a = X_a + \pi_a$. The fluctuation π is replaced by the Riemann normal variables (which transforms as a vector at point X) and A and T_{zz} are

$$A = \int\frac{d^2z}{4\pi}[\frac{1}{2}\partial_\mu\eta^A\partial_\mu\eta^A + \frac{\lambda^2}{6}(R_{ACDB}\,\eta^C\eta^D + \frac{1}{\varepsilon}R_{AB})\partial_\mu\eta^A\partial_\mu\eta^B + \frac{m^2}{2}(\delta_{AB} - \frac{1}{3\varepsilon}R_{AB}) .\eta^A\eta^B + o(\lambda^3)] \tag{17}$$

$$T_{zz} = -\frac{1}{2}\partial_z\eta^A\partial_z\eta^A - \frac{\lambda^2}{6}(R_{ACDB}\eta^C\eta^D + \frac{R_{AB}}{\varepsilon}) .\,\partial_z\eta^A\partial_z\eta^B + \lambda D_A\Phi\partial_z^2\eta^A + \lambda^2 D_A D_B\Phi(\partial_z\eta^A\partial_z\eta^B + \eta^A\partial_z^2\eta^B)\ldots \tag{18}$$

Here $\eta^A = e_a^A(X)\eta^a$ where $e_a^A(X)$ is the Vierbein at X and the Riemann tensor R_{ACDB}, the Ricci tensor $R_{AB} = R_{ACCB}$, $D_A\Phi$, $D_A D_B\Phi$ are all evaluated at X. The mass term and its counterterm in (17) regulate infrared divergences. Since the point X is independent of z the perturbation expansion is manifestly translation invariant in 2-dim.

6. RENORMALIZATION OF SCALAR VERTEX IN BACKGROUND METRIC AND DILATON

The target space scalar V(x) can also be expanded

$$V(x) = \sum\frac{1}{n!}\lambda^n\eta^{A_1}\ldots\eta^{A_n} D_{A_1}\ldots D_{A_n}V(X)$$

$$= \exp(\lambda\eta^A D_A)V(X) \tag{19}$$

With hindsight from free fields, V(x) describes the emission of a massive particle. Hence we expect the gradient of the wavefunction $D_a V(x) \sim \lambda^{-1}$. Introducing the notation $V_a = \lambda D_a V$ etc $V(X) = \sum\frac{1}{n!}\eta^{A_1}\ldots\eta^{A_n} V_{A_1\ldots A_n}$ and each term is of order one in the coupling constant.

This operator has divergent matrix elements. We renormalize it by adding counterterms in the minimal subtraction scheme to make all Greens functions $\langle V(x)\eta^{A_1}(z_1)\ldots\eta^{A_n}(z_n)\rangle$ finite to order λ^2. Basically there are two sources of divergences. The first source is internal contractions within V(x), without the interaction term. This is similar to the case of free fields but the procedure is vastly complicated by the fact that co-variant derivatives do not commute: $[D_A, D_B]V_C = R_{CDAB}V_D$. The second source is the interaction term in the action. Each successive n-point function within which the insertion is made leads to a new set of counterterms, resulting in a infinite series which ultimately exponentiates. This generalizes the Berzinsky-Kosterlitz-Thouless construction.

The procedure to construct the counterterm (CT) is inductive. First construct the zero point function $\langle V\rangle_o$ and form the CT in the minimal subtraction scheme. Then construct CT to $\langle V\eta\rangle_o$. Call it $W^{1,1}$ and so on for the 2 point function. Call its CT $W^{2,2}$. Now it so happens that $\langle W^{2,2}\rangle$ is divergent so we introduce a descendent CT $W^{2,0}$ and so on. We have constructed the table $W^{n,m}$ of CTs and descendent CTs upto the 6-point function. This is sufficient for the calculation of the Virasoro conditions. All these CTs have a dependence on the infrared cutoff and by themselves (i.e. individually) do not exponentiate. However their sum exponentiates and the m^2 dependence cancells. These calculations are long and difficult. We quote the final result for the renormalized operator

$$V_R = e^{\frac{\lambda^2}{2\varepsilon} D_A D_A + \lambda\eta^A D_A}\Big[V(X) - \frac{\lambda^2}{4\varepsilon^2} R_{AB}V^{AB} + \frac{\lambda^2}{6\varepsilon} R_{AC}V^{A}{}^{C} - \frac{\lambda^2}{2\varepsilon} R_{ACDB}V^{AB}\eta^C\eta^D\Big] \tag{20}$$

7. VIRASORO CONDITIONS ON THE RENORMALIZED VERTEX AND EQUATIONS OF MOTION

We present the results of the calculation of $\langle V_R(w)\rangle$ and $\langle T_{zz}V_R(w)\rangle$ to $o(\lambda^2)$

$$\langle V_R(w)\rangle = v^{-\frac{\lambda^2}{2}D_A D_A}\Big[V(X) - \lambda^2\Big(\frac{F}{6} + \frac{F^2}{4}\Big)R_{AB}V_{AB}\Big] \tag{21}$$

$$\langle T_{zz}V_R(w)\rangle = \frac{v^{-\frac{\lambda^2}{2}D_A D_A}}{(z-w)^2}\Big[-\frac{1}{2}V_{AA}$$

(equation continues in the next column..)

$$+ \lambda D_A\Phi V_A$$

$$+ \frac{\lambda^2}{2}\Big(\frac{F}{6} + \frac{F^2}{4}\Big)R_{AB}V_{CCAB}$$

$$+ (1 - \ell n\mu^2|z-w|^2)\Big(D_A D_B\Phi - \frac{1}{2}R_{AB}\Big)V_{AB}\Big] \tag{22}$$

Then the Virasoro condition

$$\langle T_{zz}V_R(w)\rangle = \frac{\langle V_R\rangle}{(z-w)^2} \tag{23}$$

is satisfied provided

$$R_{AB} = 2D_A D_B\Phi \tag{24}$$

$$\Big(-\frac{1}{2}D_A D_A V + D_A\Phi D_A V\Big) = \frac{1}{\lambda^2}V \tag{25}$$

This means that for vertex functions satisfying (25) the space $V_R(o)|o\rangle$ carries a representation of the Virasoro algebra provided the background fields satisfy (24), previously obtained by Callan et al[7].

It remains to compute the central charge C. This is done by explicitly evaluating $\langle T_{zz}T_{ww}\rangle$ using (18). We find

$$\langle T_{zz}T_{ww}\rangle = \frac{c}{2(c-w)^4}$$

with $C = D + 3\lambda^2[R + 4(D_A\Phi)^2 - 4D_A D_A\Phi]$ implying D = 26 and

$$R + 4(D_A\Phi)^2 - 4\,D_A D_A\Phi = 0 \tag{26}$$

It is well known [7] that (24) and (26) are equations of motion that follow from the classical action $S = \int d^{26}x\,\sqrt{G}\,\exp(-2\Phi)[R + 4(D\Phi)^2]$. The anomalous dimension eigenvalue equation (25) defines the wave function of the tachyon emitted by a string propagating in a metric and dilaton background.

The vertex function $h_{ab}(X)$ (which appears in the vertex $\partial_z x^a \partial_{\bar z} x^b\, h_{ab}(x)$) satisfies the linearized equation (24). This because the emission amplitudes are given by the replacement $G_{ab} \to G_{ab} + h_{ab}$, and the Virasoro conditions demand that the shifted metric $G_{ab} + h_{ab}$ also satisfy the equation (24). The traceless part of h_{ab} is the graviton wave function in the presence of curved geometry and a dilaton background.

8. BRST FORMULATION

We comment that our entire discussion can be recast into the BRST language since the Virasoro gauge conditions (10) are equivalent to the statement of BRST invariance $Q|\psi\rangle \otimes C_1|0\rangle = 0$. Q is the BRST charge and C_1 is the mode of the ghost field C^z which has dim. (-1,0). Once we have a representation of the Virasoro algebra and C = 26, the BRST charge is automatically nilpotent[8].

9. HIGHER LOOPS

In section 1 we had emphasized that the OPE version of the Virasoro gauge condition is valid on a surface of arbitrary genus. Since these are local equations on the string worldsheet we would expect the equations for the classical backgrounds to be the same on a surface of any genus. Clearly this is a paradoxical situation since we expect and know that higher loops modify the S-matrix. A possible resolution of this puzzle is that the σ-model perturbation theory beyond the leading order is renormalization scheme dependent. The freedom of adding counterterms to the σ-model action beyond the minimal subtraction scheme may be used to adjust the background field equations to match with the high loop S-matrix calculation. From this viewpoint one wonders whether the sum over Riemann surfaces in (1) makes sence when we treat the associated σ-model as a cutoff field theory. Work in this direction is in progress.

ACKNOWLEDGEMENT

I thank Sanjay Jain for discussions on the Virasoro conditions in the context of Polyakov's approach. I thank Sumit Das, Gautam Mandal and Curtis Callan for discussions on the higher loops puzzle.

* Permanent Address

REFERENCES

(1) See e.g. 'Unified String Theories', Proceedings of the Santa Barbara Workshop (1985) ed. Green, M. and D. Gross, World Scientific

(2) Belavin, A.A., A.M. Polyakov and A.B. Zamolodchikov, Nucl. Phys. B241 (1984) 333; Friedan, D., Z. Qiu and S. Shenker in Vertex Operators in mathematics and physics, ed. Lepowski, J.S. Mandelstam and I. Singer (Springer,1984).

(3) See e.g. Scherk, J., Rev. of Mod. Phys. 47, No.1 (1975).

(4) Jain, S., G. Mandal and S.R. Wadia, Preprint TIFR/TH/86-25. This talk is based on this work.

(5) Jain, S., R. Shankar and S.R. Wadia, Phys. Rev. D32 (1985) 2713; Lovelace, C., Phys. Lett. 135B (1984) 75; Fradkin, E.S. and A.A. Tseytlin, Phys. Lett. 158B (1985) 316; Candellas, P., G.T. Horowitz, A. Strominger and E. Witten, Nuc. Phys. B258 (1985) 46; Callan, C.G., D. Friedan, E.J. Martinec and M.J. Perry, Nuc. Phys. B262 (1985) 593; Sen, A., Phys. Rev. D32 (1985) 2102.

(6) Friedan, D., Phys. Rev. Lett. 45 (1980) 1057; Ann. Phys. 163 (1985) 318.

(7) Callan, C.G., et al in ref. 5.

(8) Nemeschansky, D., T. Banks and A. Sen, SLAC preprint (1986). These authors derived the classical equations for the background by imposing the nilpotency condition $Q^2 = 0$ for the BRST charge. See also Akhoury and Y. Okada, Univ. of Michigan preprint (1986) and Callan, C., and Z. Gan, Princeton Preprint (1986).

THE OSCILLATOR REPRESENTATION OF WITTEN'S THREE OPEN STRING VERTEX FUNCTION

Charles B. Thorn

Department of Physics, University of Florida, Gainesville, FL 32611

Using conformal mapping techniques, we show how to calculate the three open string vertex function in Witten's formulation of interacting string field theory. We obtain the vertex for the three strings in arbitrary off shell energy-momentum eigenstates, and we confirm BRS invariance for selected states.

There are two distinct proposals for an interacting string field theory [1,2]. Here we focus on Witten's version [1] and show how to evaluate the vertex function for three strings in arbitrary (off mass shell) energy momentum eigenstates in oscillator basis. This work was carried out by Cremmer, Schwimmer and me in Paris [3]. I shall try to highlight here the differences between our method and that of Gross and Jevicki [4], who gave an independent evaluation of the vertex.

We use the conformal mapping technique developed by Mandelstam [5] many years ago. We stress that we calculate the three string vertex <u>directly</u>, and we find no need for the device used in Ref. [4], of first calculating the six string vertex.

As is clear from Ref. [1,5] our calculation requires the inverse $z(\rho)$ of the map from the upper half z plane to the interacting string diagram obtained by replacing the string fields in the overlap integral by string propagators represented as functional integrals over space-time coordinates $x^\mu(\rho)$ and bosonized ghost field $\phi(\rho)$ defined on semi infinite strips. We draw this diagram on two Riemann sheets joined along a cut on the positive real axis (see Fig. 1).

In Fig. 1, dotted lines are on the second sheet and the line OB' on the second sheet is identified with the line OB, i.e. $x(A) = x(A')$, $p(A) = -p(A')$, where $p(A)$ is con jugate to $x(A)$. To achieve this identification, we notice that in the $\chi = \rho^{2/3}$ plane, the image of OB' concides with that of OB (see Fig. 2), so we need simply require that the <u>string coordinates and their conjugates be continuous in the variable χ</u>. The unique mapping to the upper half plane which maps string 1 at $\rho=-i\infty$ to $z=1$, string

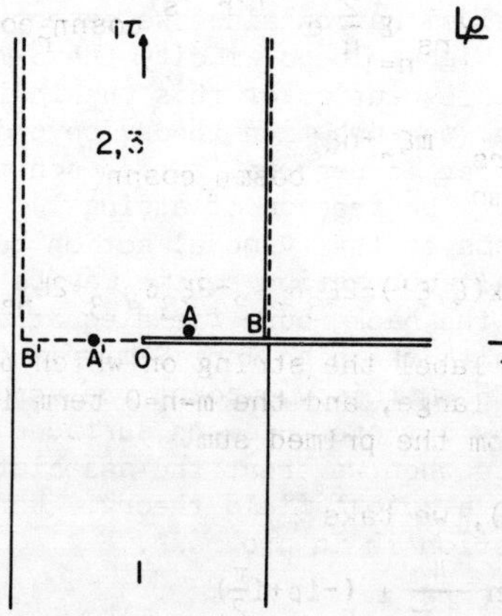

Figure 1: Interacting string diagram for the three open string vertex.

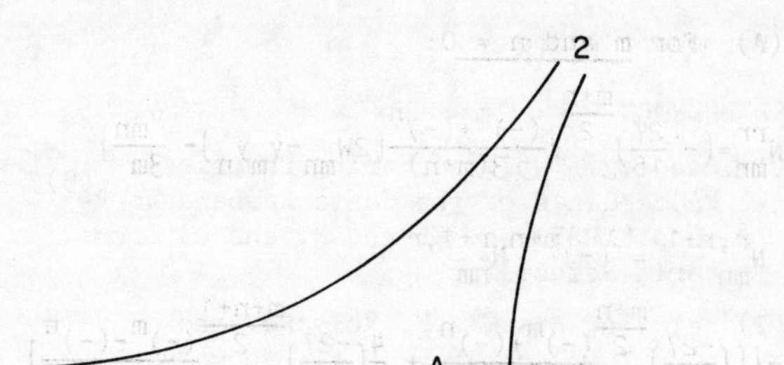

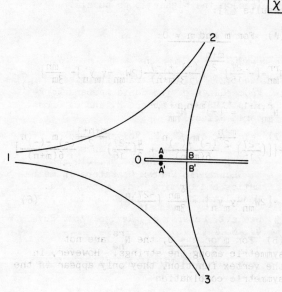

Figure 2: Image of Fig. 1 in the $\chi = \rho^{2/3}$ plane.

2 at $\rho=+i\infty_1$ to z=0, and string 3 at $\rho=+i\infty_2$ to z=∞ is [3]

$$z(\zeta)=\frac{1}{2}+\frac{i\sqrt{3}}{4}\left(\zeta+\frac{1}{\zeta}\right)+\frac{i\sqrt{3}}{4}\left(\zeta-\frac{1}{\zeta}\right)\left[Y_+(\zeta)e^{-\pi i/3}+Y_-(\zeta)e^{\pi i/3}\right] \qquad (1)$$

where $\zeta = \exp(-i\rho)$ and

$$Y_+(\zeta) = Y_-(-\zeta) = \left(\frac{1-\zeta}{1+\zeta}\right)^{1/3}. \qquad (2)$$

Mandelstam [5] has explained how to construct the oscillator representation for the coordinate part of a vertex in terms of the coefficients in the expansion of the Neumann function for large ρ, ρ':

$$N(\rho,\rho')\equiv \ln|z(\rho)-z(\rho')|+\ln|z(\rho)-z(\rho')^*|$$

$$= -\delta_{rs}\sum_{n=1}^{\infty}\frac{2}{n}e^{-n|\xi_r-\xi'_s|}\cos n\eta_r \cos n\eta'_s$$

$$+\sum_{m,n}{}' \, 2N^{rs}_{mn}\, e^{m\xi_r+n\xi'_s}\cos m\eta_r \cos n\eta'_s$$

$$+2\delta_{rs}\max(\xi,\xi')-2\xi_3\delta_{r,3}-2\xi_3\delta_{s,3}+2b_{rs} \qquad (3)$$

where r,s label the string on which ρ,ρ' are taken large, and the m=n=0 term is absent from the primed sum.

In Eq. (3), we take

$$\xi_r+i\eta_r=\ln\frac{4}{3\sqrt{3}} \pm \left(-i\rho+i\frac{\pi}{2}\right) \qquad (4)$$

with +(-) if $\tau\equiv \mathrm{Im}\rho\to-\infty(+\infty)$, then $b_{rs}=0$ as in Ref. [5]. The evaluation of the N^{rs}_{mn} is somewhat tedious, and we just quote the results [3]:

(A) For <u>m and n ≠ 0</u>:

$$N^{rr}_{mn}=\left(-\frac{27}{16}\right)^{\frac{m+n}{2}}\left\{\frac{(-)^m+(-)^n}{3(m+n)}\left(2W_{mn}-y_my_n\right)-\frac{\delta_{mn}}{3m}\right\} \qquad (5)$$

$$N^{r,r+1}_{mn} = (-)^{m+n}N^{r+1,r}_{nm} =$$

$$-\left\{\left[\left(\frac{-27}{16}\right)^{\frac{m+n}{2}}\frac{(-)^m+(-)^n}{6(m+n)}+\frac{4}{3}\left(\frac{-27}{16}\right)^{\frac{m+n+1}{2}}\frac{(-)^m-(-)^n}{6(m+n)}\right]\right.$$

$$\left.\cdot\left[2W_{mn}-y_my_n\right]+\frac{\delta_{mn}}{3m}\left(\frac{-27}{16}\right)^n\right\} \qquad (6)$$

(B) For <u>m or n = 0</u>, the N^{rs}_{n0} are not symmetric among the strings. However, in the vertex function, they only appear in the symmetric combination

$$\sum_s N^{rs}_{n0}P_s=\begin{cases}P_rN^{11}_{n0} & n \text{ even}\\ (P_{r+1}-P_{r+2})N^{11}_{n0} & n \text{ odd}\end{cases} \qquad (7)$$

where P_r is the energy momentum, and with

$$N^{11}_{n0}=\begin{cases}\dfrac{y_n}{n}\left(-\dfrac{27}{16}\right)^{n/2} & n \text{ even}\\[2ex] \dfrac{4y_n}{9n}\left(-\dfrac{27}{16}\right)^{(n+1)/2} & n \text{ odd}\end{cases} \qquad (8)$$

In the above formulae,

$$W_{mn}=\begin{cases}-\dfrac{3}{2}\,\dfrac{m(n+1)y_my_{n+1}-n(m+1)y_{m+1}y_n}{m-n}, & m\neq n\\[2ex] \displaystyle\sum_{k=0}^{n}(-)^{k+n}y_k^2, & m=n\end{cases} \qquad (9)$$

and

$$\left(\frac{1-x}{1+x}\right)^{1/3}=\sum_{n=0}^{\infty}y_nx^n. \qquad (10)$$

For the Fadeev-Popov ghost part of the vertex one also needs the coefficients in the expansion of

$$N(\rho,0)=\frac{2}{3}\sum_{\substack{n=2\\ n,\text{even}}}^{\infty}\left(\frac{-27}{16}\right)^{\frac{n}{2}}\frac{y_n-2}{n}e^{n\xi_r}\cos n\eta_r$$

$$+2\begin{cases}\displaystyle\sum_{n\text{ odd}}e^{n\xi_1}\cos n\eta_1\,N^{11}_{n0} & r=1\\ \displaystyle-\sum_{n\text{ odd}}e^{n\xi_2}\cos n\eta_2\,N^{11}_{n0} & r=2\\ \displaystyle\sum_{\substack{n\text{ even}\\ \neq 0}}e^{n\xi_3}\cos n\eta_3\,N^{11}_{n0}-\xi_3 & r=3,\end{cases} \qquad (11)$$

in order to evaluate the contribution of the ghost insertion needed to compensate for the violation of ghost number conservation [1]:

$$\sum_r G_r = -3/2. \qquad (12)$$

The asymmetries among the strings evident in Eq. (11) cancel in the vertex when use is made of Eq. (12).

We now quote the oscillator form of Witten's vertex. For the rth string, we use coordinate operators $a^{r\mu}_{-n}=a^{r\mu\dagger}_n$ and bosonized ghost operators

$$b^r_n=-b^{r\dagger}_{-n}=\sum_k :\gamma_{n-k}\gamma_k: -\frac{1}{2}\delta_{n0} \qquad (13)$$

where $\gamma_n(\overline{\gamma}_n)$ are the fermionic ghost (anti-ghost) operators. These operators satisfy

$$[a_m^{r\mu}, a_n^{s\nu}] = m\delta_{m+n}\delta_{rs}\, g^{\mu\nu} \tag{14}$$

$$[b_m^r, b_n^s] = m\delta_{m+n}\delta_{rs}. \tag{15}$$

Let the energy momentum and ghost number of the rth string be P_r^μ and G_r respectively. Then the vertex for the three strings in states $|\psi_1\rangle$, $|\psi_2\rangle$, $|\psi_3\rangle$ is given by

$$V = g\langle 0|\mathrm{Exp}\{\frac{1}{2}\sum_{m,n=1}^{\infty}[a_{mr}\cdot a_{ns} + b_{mr}b_{ns}]N_{mn}^{rs}$$

$$+ \sum_{\substack{n=1\\ \text{odd}}}^{\infty}[a_{nr}\cdot\sqrt{2}(P_{r+1}-P_{r+2}) + b_{nr}(G_{r+1}-G_{r+2})]N_{n0}^{11}$$

$$+ \sum_{\substack{n=2\\ \text{even}}}^{\infty}[a_{nr}\cdot\sqrt{2}P_r + b_{nr}(G_r N_{n0}^{11} + (\frac{-27}{16})^{\frac{n}{2}}\frac{y_n - 2}{2n})]\} \tag{16}$$

$$\{\frac{4}{3\sqrt{3}}\}^{\sum_r (L_0^r - 1)} \prod_r |\psi_r\rangle$$

where it is understood that $\sum_r P_r = \sum_r G_r + \frac{3}{2} = 0$, and where

$$L_0^r = \sum_{n=1}^{\infty}(a_{-n}^r\cdot a_n^r + b_{-n}^r b_n^r) + \frac{G_r^2}{2} + P_r^2$$

is the zeroeth Virasoro generator for the rth string.

We close with two comments. First, our results are in agreement with those of Ref. [4] though our method is somewhat different [6]. Secondly, although we have not checked BRS invariance for general states $\prod_r |\psi_r\rangle$, we have confirmed that the state

$$(Q_1+Q_2+Q_3)\overline{\gamma}_{-n}^{(1)}|0,-\tfrac{1}{2}\rangle \times |0,-\tfrac{1}{2}\rangle \times |0,-\tfrac{1}{2}\rangle, \tag{17}$$

with all string states off or on shell does indeed decouple for n=1 and 2. For the case n=2, we note that D=26 is essential for this decoupling. In Eq. (17) the ket $|0,G\rangle$ is a single string state annihilated by a_n^μ and b_n for n>0 and is an eigenstate of a_0^μ, b_0 with eigenvalues $\sqrt{2}\,P^\mu$, G. The momentum eigenvalue P^μ is suppressed.

REFERENCES

[1] Witten, E., Nuclear Physics B268 (1986) 253.

[2] Siegel, W., Phys. Lett. 149B (1984) 157, 162; 151B (1985) 391, 396. A. Neveu and P. West, Phys. Lett. 168B (1986) 192. H. Hata, K. Itoh, T. Kugo, H. Kumimoto, and K. Ozawa, Phys. Lett. 172B (1986) 186, 195.

[3] Cremmer, E., A. Schwimmer and C. Thorn, Univ. of Florida Preprint, UFTP-86-8/ LPTENS-86-14 (1986), to be published in Physics Letters B.

[4] Gross, D. and A. Jevicki, Princeton Univ. Preprint (1986) see also the talk by A. Jevicki at this conference.

[5] Mandelstam, S., Nuclear Physics B64 (1973) 205.

[6] See also N. Ohta, University of Texas Preprint UTTG-18-86.

Parallel Session 3

Supergravity and Supersymmetry

Organizers:

S.J. Gates (Maryland)

H. Nicolai (Karlsruhe)

Scientific Secretaries:

N. Berkovits (LBL)

A. Sagnotti (LBL)

LOW-ENERGY SUPERGRAVITY AND SUPERSTRINGS

L.E. Ibañez

Theory Division, CERN
1211 Geneva 23, Switzerland

Low-energy N = 1 supergravity provides us with elegant supersymmetric models with a stable hierarchy of scales. This general class of models may be obtained from the $E_8 \times E_8$ heterotic string after appropriate compactification of the six extra dimensions on a Kähler manifold of SU(3) holonomy. In some cases, the residual low-energy theory may contain some traces of its string origin through extra quarks, leptons or gauge bosons.

During the years 1980-1985 much effort was dedicated to the construction of phenomenologically consistent models with low-energy supersymmetry (1). The final aim of these models is to use the non-renormalization properties of supersymmetry in order to understand the smallness of the weak scale M_W compared to the Planck mass M_P. From this point of view the "low-energy supergravity models" are specially attractive. These are N = 1, d = 4 supergravity models coupled to quarks, leptons and gauge bosons. The minimal model is just an SU(3)×SU(2)×U(1) theory coupled to three usual families of chiral superfields plus Higgs particles. In these theories, supersymmetry breaking takes place in a "hidden sector" of the theory (2) and it is transmitted to the observable quark-lepton-Higgs world only through gravitational interactions. After SUSY-breaking, one is left at low energies with a theory with softly-broken supersymmetry having soft parameters (2),(3) m (universal scalar masses), M (universal gaugino masses) and other soft couplings proportional to the superpotential. A remarkable fact of this class of models is that the $SU(2)_L \times U(1)_Y$ symmetry is broken in a natural way as a radiative effect of supersymmetry breaking (4). Unfortunately, N = 1 supergravity theories are not very much constrained. The particle content is not fixed and the Lagrangian is specified by two arbitrary functions (5), the Kähler potential

$$G(z,z^*) = K(z,z^*) + \log|W(z)|^2 \qquad (1)$$

and the chiral gauge function $f_{ab}(z)$. W(z) is the superpotential and z denotes all the scalars in the theory. Furthermore, N = 1 is a non-renormalizable theory and hence makes no sense for energies >> M_P. Thus, in order to improve our understanding of the theory, one hopes that it is embedded into a renormalizable (possibly finite) theory which unifies all interactions including gravity.

In the last two years, it has become clear that superstrings (6) are serious candidates to be the required unified theories. Particularly, the $E_8 \times E_8$ heterotic string (7) (or some "twisted" version of it) (8) is specially interesting from the phenomenological point of view. If one compactifies the six extra dimensions (9) on a Kähler manifold of SU(3) holonomy (or some orbifold limit), one may obtain at low energies a GUT model [e.g., E_6, SO(10) or SU(5)] or even the usual 3×2×1 standard model, coupled to N = 1 supergravity. The compactified theory has a built-in "hidden sector" formed by some singlets plus some gauge fields coming from the other E_8 symmetry of the theory. Furthermore, the breaking of the residual N = 1 supersymmetry will, in general, occur through gaugino condensation (10) in the hidden sector. Thus, we see that the $E_8 \times E_8$ heterotic string seems to lead to a structure which is rather reminiscent of the "low-energy supergravity models" we were describing above.

In principle, once we knew the "correct" compact six-dimensional mani-(orbi)fold we could calculate all the relevant low-energy couplings of the residual "low-energy supergravity model", in particular, the Kähler potential $G(z,z^*)$ and the gauge kinetic function $f_{ab}(z)$ which determine it. In practice, Calabi-Yau are very complicated manifolds with unknown metrics and hence the above-mentioned computations are not possible at present. However, one can easily obtain information about what are the massless modes of the theory via topological arguments. Thus, for example, the Betti-Hodge numbers $b_{1,2}$ and $b_{1,1}$ tell us the number of light families and antifamilies which remain at low energies (9). Sometimes, one can also compute explicitly the Yukawa couplings of the theory using only topological considerations. Unfortunately, one cannot, in general, compute in this way the normalization matrix of the fields (D-terms) so that these Yukawa couplings are not the physically relevant ones. The situation is, in principle, much better for the case of orbifolds (8) since generally one can compute everything in the twisted torus. We still have to wait until realistic examples of orbifolds are found. Even without doing a proper string compactification, one can, however, extract some information about the general form of the low-energy Lagrangian by taking advantage of the classical scale invariances of the theory (13)-(16). We briefly discuss this in what follows.

It is well known that amongst the massless fields (at least at tree level) there are a couple of dilaton-axion complex scalar fields (12)

$$S = \phi^{-3/4} e^{3\sigma} + i\theta$$
$$T = e^{\sigma}\phi^{3/4} + i\eta \qquad (2)$$

where $\phi(x)$ comes from the D = 10 supergravity dilaton and e^{σ} measures the size of the compact manifold. θ and η are zero modes from the B_{MN} ten-dimensional antisymmetric tensor. The pseudoscalars θ and η only have derivative couplings since the field B_{MN} always appears through its field strength $H_{MNP} = \partial_{[M}B_{NP]} + \dots$. Thus there are two Peccei-Quinn invariances under $\theta \to \theta+c$, $\eta \to \eta+c'$. This implies that the Kähler potential G depends on S and T only through the combinations (S+S*) and (T+T*). Furthermore, the low-energy D = 4 theory has a classical scale invariance (12)-(14)

$$\begin{aligned} &S \to \lambda S \\ &T, C_x \to T, C_x \qquad \to e^{-1}\mathcal{L} \to \lambda^{-1}e^{-1}\mathcal{L} \qquad (3) \\ &g_{\mu\nu} \to \lambda g_{\mu\nu} \\ &\quad + \text{fermion transf.} \end{aligned}$$

which is directly hired from a similar dilation invariance of the ten-dimensional Lagrangian (12). In Eq. (3), C_x refers to the matter scalars (e.g., 27's of E_6) and $g_{\mu\nu}$ is the four-dimensional Minkowski metric. One can immediately check that this scale invariance dictates the dependence of the Kähler metric G and the function f_{ab} on the field S:

$$G_s = -\log(S+S^*) \qquad (4a)$$
$$f_{ab} = \delta_{ab}\, S \qquad (4b)$$

There is another classical scale invariance related to the fact that the overall scale of compactification is undetermined (13), (14)

$$\begin{aligned} &T \to (\lambda')^2 T \\ &C_x \to \lambda' C_x \qquad \mathcal{L} \to \mathcal{L} \qquad (5) \\ &S \to S \end{aligned}$$

Then if the Kähler potential G is to contain the standard piece $\log|W|^2$ with W trilinear, the dependence of G on T and C_x has to be of the form

$$G_{T,C_x} = -3\log(T+T^*+\alpha_i|C_i|^2) + \log|W|^2 \qquad (6)$$

where α_i are unknown constants which may be different for different SU(3)×SU(2)×U(1) representations. Thus as a whole, one obtains

$$\begin{aligned} G_0 = &-\log(S+S^*) - 3\log(T+T+\alpha_i|C_x|^2) \\ &+ \log|W|^2 + F((T+T^*)/|C_x|^2) \end{aligned} \qquad (7)$$

where we have added (14) an arbitrary function F of the scale-invariant combination $(T+T^*)/|C_x|^2$. The second term in Eq. (7) is very similar to the one appearing in the so-called "no-scale models", but there is an important difference [apart from the existence of the other terms in Eq. (7)]. The above result is just a tree level result and it is known that the radiative corrections and non-perturbative effects spoil its structure (since it has its origin in a classical invariance). This is to be contrasted with the very assumptions of the "no-scale" idea, in which the second term in Eq. (7) is assumed to represent the exact Kähler potential (including radiative gravitational corrections).

There are several sources of corrections to Eq. (7). There are corrections to the Lagrangian from higher derivative terms in the ten-dimensional theory (higher order terms in the string σ-model expansion). These ones should respect the classical invariances described above since they are tree-level terms (13),(18). There are also non-perturbative corrections (both from the space-time point of view and the σ-model point of view). These will violate those classical invariances. Furthermore, there will be perturbative corrections to the results of Eq. (7) coming from radiative corrections. Some of these corrections may involve the effect of heavy string modes. This is the case of the axion-like couplings of the field $\eta \equiv \text{Im}T$ which have their origin in the one-loop counterterms which are needed to cancel the ten-dimensional anomalies. These include, e.g., terms of the form B^{MN} $(F^{PQ}F^{RS})$ $(R^{TV}R^{VW})$ which after compactification induce the mentioned axion couplings (18)

$$B^{mn}(F^{\mu\nu}F^{\rho\sigma})\langle R^{pq}R^{rs}\rangle \rightarrow \sim \frac{1}{(2\pi)^5}\eta(F\tilde{F}) \qquad (8)$$

In the $E_8 \times E_8$ case, one can check that the coefficients of the observable E_6 and the hidden E_8 (or subgroups) axion couplings are equal and opposite (15),(20),(21). Since one assumes that there is an unbroken supersymmetry, the gauge kinetic functions get a one-loop correction (15),(20):

$$f^8_{ab} = \delta_{ab}(S+\varepsilon T); \quad f^6_{ab} = \delta_{ab}(S-\varepsilon T) \qquad (9)$$

where $\varepsilon \sim (2\pi)^{-5}$. Notice that the new one-loop terms generated in the Lagrangian will scale as:

$$\text{S-scale invariance: (one-loop)} \rightarrow \lambda^{-2}\,\text{(one-loop)}$$
$$\text{T-scale invariance: (one-loop)} \rightarrow \lambda'\,\text{(one-loop)} \qquad (10)$$

and hence explicitly break those invariances. In general, similar corrections (15) will appear also for the Kähler potential. In fact, the most general one-loop correction of this type to the Kähler potential allowed by the transformation properties of Eq. (10) is

$$G = G_0 + \delta G; \quad \delta G = -a\varepsilon\,\frac{(T+T^*)}{(S+S^*)} \qquad (11)$$

where a is a number O(1). Notice that Eqs. (9) and (11) substantially modify the results of Eqs. (4b) and (7). In particular, one can check that, at least at this level, the scalar potential is no longer positive-definite, as it was in the original truncation of Witten (15). This is a problem since then the scalar potential will in general be unbounded below and in any cases there will be a non-vanishing cosmological constant. Probably the meaning of all this is that it does not make much sense to use S and T as undetermined dynamical variables at low energies and ignore the effect of string heavy modes and full string dynamics. On the optimistic side, one may think of several aspects of the low-energy analysis which may survive. Gaugino condensation in the hidden sector seems a rather appealing mechanism for breaking the residual supersymmetry and obtaining a hierarchically small gravitino mass $m_{3/2} \sim \langle\chi\bar{\chi}\rangle/m_p^2$ which would presumably be related to the weak scale. It is well known that, even after this supersymmetry breaking, the observable sector remains supersymmetry at the tree level ($m = M = A = 0$). In particular, for the gaugino masses one has $M = f'_s F_s = 0$, since the S-auxiliary field F_s vanishes upon minimization of the scalar potential. This situation changes after including loop effects. Thus, e.g., it is obvious from Eq. (9) that even if F_s vanishes, there will be gaugino masses

$$M = f'_T F_T \sim \varepsilon m_{3/2}. \qquad (12)$$

It is obvious from this that the gravitino mass cannot exceed $\sim 10^7$ GeV since otherwise large masses would be generated radiatively for the Higgses and the gauge hierarchy would be spoiled. However, for the moment we cannot say much more about the form of the Kähler potential and low-energy physics Lagrangians since the conclusion of the above discussion is that the tree-level symmetries imply a Lagrangian which does not work. One has to include perturbative and non-perturbative corrections whose effect is at the moment unknown.

We have discussed above mostly the singlet "hidden" sector of the theory. Since the conclusions were not so positive, one may be led to the conclusion that nothing can be said about the low-energy limit of the $E_8 \times E_8$ string after compactification. This is not correct. There are some general features expected for the low-energy "observable" theory which are more or less independent of the details of the compactification. Thus, for example, if we compactify on a "Calabi-Yau" manifold embedding the gauge connection into the spin connection (9), we know that we obtain an E_6 (or some subgroup) model with several families of $\underline{27}$'s (or some subset of states). Also, some pairs of $(\underline{27}+\overline{\underline{27}})$ could be present to start with. One can further lower the rank of the group through the Wilson-loop mechanism (22) and/or by giving v.e.v.s to the $\underline{27}+\overline{\underline{27}}$ respectively (20),(23). The latest symmetry breaking may occur at an intermediate scale (e.g., 10^{10}-10^{13} GeV) by radiative corrections. There are other possible compactification schemes which may lead to a variety of low-energy models. If one goes beyond the usual recipe of identifying spin and gauge connections, one may find manifolds which break the original E_8 directly down to an SO(10) or SU(5) subgroup (9). These may be further broken through the Wilson-loop mechanism down to some rank 5 subgroup or the standard model. Low energy SO(10) or SU(5) subgroups may also be obtained compactifying on certain classes of "orbifolds". The orbifolds are obtained by modding out some six tori by a discrete subgroup of SU(3) (so that supersymmetry is preserved). If the original torus has some Wilson lines, one may break E_8 directly down to SU(5).

It seems clear that, from the low-energy limit point of view, there are a lot of possibilities (22)-(25). One may have just the standard model or one may have some extension of it including extra gauge bosons and/or quarks and leptons (e.g., like the ones included in a 27 of E_6). The simplest extensions of the standard model will include some possible extra Z^0 corresponding to the extra Cartan generators inside E_6. The possible experimental signatures of these Z^0''s are discussed in the contribution of Deshpande in these proceedings. Let me end up by making a comment concerning these "superstring-inspired" low-energy supergravity models. It seems that the extensions of the standard model have in general problems with fast proton decay, neutrino masses, flavour-changing neutral currents, etc., unless some Yukawa couplings are forbidden by some symmetry. The symmetries required to forbid all of the dangerous couplings are rather ad hoc. It would be much more satisfactory to have at low energies just the standard model. This may be achieved in specific Calabi-Yau models if there is an intermediate scale of symmetry breaking (see, e.g., the talk of G.G. Ross in this proceedings). I, however, think that the optimum solution would be to have compactification schemes in which the $SU(3)\times SU(2)\times U(1)$ symmetry is directly obtained at the Planck mass. From this point of view, orbifolds with pre-existing Wilson lines seem rather promising candidates. A more detailed discussion of the topics discussed in this talk may be found in Ref. (18).

REFERENCES

(1) For reviews, see, e.g.,
Nilles, H.P., Physics Reports 110, 1 (1984);
Ibañez, L.E., Proceedings of the 13th International Winter Meeting on Fundamental Physics, edited by J.E.N. Madrid, (1984);
Hall, L., Proceedings of the Winter School in Theoretical Physics, Mahabaleshwar, India (1984), Springer Verlag Lecture Notes in Physics 208, 197 (1984).

(2) Ibañez, L.E., Phys. Lett. 118B, 73 (1982);
Nilles, H.P., Phys. Lett. 115B, 193 (1982); Nucl. Phys. B217, 366 (1983);
Nath, P., Arnowitt, R., and Chamseddine, A.H., Phys. Rev. Lett. 49, 970 (1982);
Barbieri, R., Ferrara, S., and Savoy, C.A., Phys. Lett. 199B, 343 (1982).

(3) Ibañez, L.E., Nucl. Phys. B218, 54 (1983);
Ibañez, L.E. and Lopez, C., Phys. Lett. 126B, 54 (1983);
Alvarez-Gaumé, L., Polchinski, J. and Weise, M., Nucl. Phys. B221, 495 (1983);
Ellis, J., Hagelin, J., Nanopoulos, D.V. and Tamvakis, K., Phys. Lett. 125B, 275 (1983).

(4) Ibañez, L.E. and Ross, G.G., Phys. Lett. 110B, 215 (1982);
Alvarez-Gaumé, L., Claudson, M. and Wise, M., Nucl. Phys. B207, 16 (1982);
Ovrut, B. and Nappi, C., Phys. Lett. B113, 65 (1982).

(5) Cremmer, E., Ferrara, S., Girardello, L., and Van Proeyen, A., Nucl. Phys. B212, 413 (1983).

(6) Schwarz, J.H., Physics Reports 89, 223 (1982);
Green, M.B., Surveys in High Energy Phys. 3, 127 (1982).

(7) Gross, D., Harvey, J., Martinec, E., and Rohm, R., Phys. Rev. Lett. 55, 502 (1985); Nucl. Phys. B256, 253 (1985).

(8) Dixon, L., Harvey, J., Vafa, C. and Witten, E., Nucl. Phys. B261, 651 (1985) and Princeton preprint (1986).

(9) Candelas, P., Horowitz, G., Strominger, A. and Witten, E., Nucl. Phys. B258, 46 (1985);
Witten, E., New issues in manifolds of SU(3) holonomy Princeton preprint (1985).

(10) Derendinger, J.-P., Ibañez, L.E., and Nilles, H.P., Phys. Lett. 155B, 65 (1985);
Dine, M., Rohm, R., Seiberg, N. and Witten, E., Phys. Lett. 156B, 55 (1985).

(11) Strominger, A., "Yukawa couplings in superstring compactification", Santa Barbara preprint NSF-ITP-85-109 (1985).

(12) Witten, E., Phys. Lett. 155B, 151 (1985).

(13) Ibañez, L.E., Phenomenology from superstrings, CERN preprint TH.4308/85 (1985), to appear in the Proceedings of the 1st Torino Meeting on Superunification and Extra Dimensions, World Scientific, Singapore.

(14) Burgess, C., Font, A. and Quevedo, F., Texas preprint UTTG-31-85 (1985).

(15) Ibañez, L.E. and Nilles, H.P., Phys. Lett. 169B, 354 (1984).

(16) Nilles, H.P., CERN preprint TH.4522/86 (1986).

(17) Cremmer, E., Ferrara, S., Kounnas, C. and Nanopoulos, D.V., Phys. Lett. 133B, 61 (1983);
Ellis, J., Lahanas, A., Nanopoulos, D.V. and Tamvakis, K., Phys. Lett. 134B, 429 (1984);
Ellis, J., Kounnas, C. and Nanopoulos, D.V., Nucl. Phys. B241, 406 (1984); B274, 373 (1984).

(18) Ibañez, L.E., "Some topics in the low-energy physics from superstrings", CERN preprint TH.4459/86 (1986), to appear in the Proceedings of the Nato Workshop on "Super Field Theories", Vancouver, July 1986, to be published by Plenum Press.

(19) Witten, E., Phys. Lett. 153B, 243 (1985).

(20) Derendinger, J.-P., Ibañez, L.E. and Nilles, H.P., Nucl. Phys. B267, 365 (1986).

(21) Choi, K. and Kim, J., Phys. Lett. 165B, 71 (1985).

(22) Witten, E., Nucl. Phys. B258, 75 (1985).

(23) Dine, M., Kaplunovsky, V., Mangano, M., Nappi, C. and Seiberg, N., Nucl. Phys. B259, 549 (1985).

(24) Ibañez, L.E. and Mas, J., CERN preprint TH.4426/86 (1986).

(25) Ellis, J., Enqvist, K., Nanopoulos, D.V. and Zwirner, F., CERN preprint TH.4323/85 (1985).

NUCLEON DECAY IN SUPERGRAVITY UNIFIED THEORIES

Pran Nath and R. Arnowitt

Department of Physics, Northeastern University
Boston, Massachusetts 02115

Signatures of supersymmetry in nucleon decay within the framework of N=1 Supergravity unified models are analysed. It is shown that at least two branches for nucleon decay exist in these models. The first is the conventional one where the $\bar{\nu}K$ modes dominate. The second is the more recently discovered branch where the non-strange modes $\bar{\nu}\pi$, $\bar{\nu}\eta$, $\bar{\nu}\rho$, $\bar{\nu}\omega$ dominate or are comparable to the $\bar{\nu}K$ modes. The first branch appears to be already in difficulty with existing data. The domain of validity of the second branch is exhibited and is shown to be non-negligible. Effects of a possible fourth generation of matter and SUSY-matter on nucleon decay are analyzed.

1. INTRODUCTION

Current experimental data on proton decay gives a combined lower limit for the $p\to e^+\pi^o$ mode of (1)

$$\tau_{p\to e^+\pi^o} \geq 3.6 \times 10^{32} \text{ yr} \qquad (1)$$

The minimal SU(5) GUT model gives a theoretical upper limit of $\cong 2\times 10^{31}$ yr for this mode (2). The minimal GUT theory is thus ruled out experimentally. Essentially similar conclusions hold for other GUT models as well.

Supersymmetric (SUSY) GUT theories avert conflict with Eq. (1) since in SUSY GUTS the unification mass M_X is enlarged by a factor of ≈ 40 over the Standard Model GUT unification mass. This arises due to the contributions of the superpartners of the Standard particles to the beta functions in the renormalization group equations which determine the approach to the unification scale. Since the nucleon decay life-time mediated by the heavy gauge bosons scales by $\tau_N \sim O(M_X^4)$ one has $\tau_N \sim O(10^{37}$ yr). This lifetime is far beyond the realm of observation since the cosmic ray and the atmospheric neutrino interactions introduce an upper limit of $O(1)\times 10^{33}$ yr on the experimentally accessible nucleon lifetime. Interesting nucleon decay in SUSY GUTS arises, however, through Yukawa interactions mediated by heavy Higgs fields (3). These interactions generate baryon number violating dimension five operators which are scaled by the inverse power of heavy Higgs mass M. When dressed by gaugino exchange diagrams, the dimension five operators generate dimension six operators which lead to nucleon decay at experimentally accessible rate.

2. SU(5)×N=1 SUPERGRAVITY THEORY

For definiteness we shall discuss here an SU(5) theory governed by the superpotential g_Y of Yukawa interactions where

$$g_Y = -\frac{1}{8}\epsilon_{uvwxy} H^u M_i^{vw}(f_1^\dagger)_{ij} M_j^{xy} + H'_x M'_{yi}(f_2^\dagger)_{ij} M_j^{xy} \qquad (2)$$

In Eq. (2) H^x and H'_x are 5_L and $\bar{5}_L$ of Higgs fields, and M^{xy} and M'_y are 10_L and $\bar{5}_L$ of quark-lepton fields. f_1, f_2 are Yukawa-coupling-constant matrices and i,j=1,2,3 are the generation indices. We assume that the theory supports spontaneous breaking at the GUT scale which reduces SU(5) to SU(3)×SU(2)×U(1) and arranges the Higgs doublets to be light (either through fine tuning on through the missing partner mechanism). Elimination of the heavy Higgs triplet fields H_a and H'^a (a=1,2,3) with mass M generates the baryon number violating dimensions five operators and one gets

$$\mathcal{L}_5 = \mathcal{L}_5^L + \mathcal{L}_5^R \qquad (3)$$

$\mathcal{L}_5^L$ contains left-handed matter and SUSY-matter fields in the form LLLL and one has

$$\mathcal{L}_5^L = \frac{\epsilon_{abc}}{M}(Pf_1^uV)_{ij}(f_2^d)_{k\ell}\{\tilde{u}_{Lbi}\tilde{d}_{cj}\bar{e}^c_{Lk}(Vu_L)_{a\ell} + ..\} + \text{h.c.} \qquad (4)$$

Similarly $\mathcal{L}_5^R$ contains right-handed matter and SUSY-matter fields in the form RRRR and one has

$$\mathcal{L}_5^R = -\frac{\epsilon_{abc}}{M}(V^\dagger f^u)_{ij}(PVf^d)_{k\ell}(\bar{e}^c_{Ri}u_{Raj}\tilde{u}_{Rck}\tilde{d}_{Rb\ell} + ..) + \text{h.c.} \qquad (5)$$

In Eqs. (4) and (5), V is the K-M matrix and

P is a diagnonal phase matrix while $f^{u,d}$ are the diagonal matrices related to the up- and down-quark mass matrices m^u and m^d.

Elimination of the squark and slepton fields in Eqs. (4) and (5) leads to dimension six operators. Conventionally such eliminations have used Wino exchange and two generations of squarks and sleptons in the dressing loop diagrams. The dominant factors determining some typical decay modes are exhibited in Table I. Here one finds that the Kaon modes dominate over the pion modes by the inverse of an off-diagonal K-M matrix element which is <1. Similarly, the neutrino modes are seen to dominate over the charged lepton modes since m_c/m_u >>1. Thus overall one finds that the $\bar{\nu}K$ modes dominate the nucleon decay. This is the well-known conventional result.

Table I

Decay Mode	K-M and Quark Mass Factors
$p\to k^+\bar{\nu}_\mu$	$m_c m_s\ V_{21}^\dagger V_{21}$
$p\to\pi^+\bar{\nu}_\mu$	$m_c m_s\ V_{21}^\dagger (V_{21})^2$
$p\to K^o\mu^+$	$m_u m_s$
$p\to\pi^o\mu^\dagger$	$m_u m_s\ V_{21}^\dagger$

The conventional analysis (3) discussed above leaves out a number of important effects. These include (i) third generation effects, (ii) L-R mixing effects on squark and slepton masses, (iii) contributions of RRRR dimension five operators and (iv) gluino and Zino exchange terms. Recently a full analysis of these effects has been given (4). We exhibit here the dimension six operator for the process $N\to\bar{\nu}_i K(i=e,\mu,\tau)$:

$$\mathcal{L}_6(N\to\bar{\nu}_i K) = [(\alpha_2)^2(2MM_W^2\sin 2\alpha_H)^{-1}$$
$$P_2 m_c m_i^d V_{i1}^\dagger V_{21}V_{22}][F(\tilde{c};\tilde{d}_i;\tilde{W})$$
$$+ F(\tilde{c};\tilde{e}_i;\tilde{W})] \times \{[1+y_i^{tK}+(y_{\tilde{g}}+y_{\tilde{Z}})\delta_{i2}$$
$$+ \Delta_i^K]\alpha_i^L+[1+y_i^{tK}-(y_{\tilde{g}}-y_{\tilde{Z}})\delta_{i2}]\beta_i^L$$
$$+ (y_1^{(R)}\alpha_3^R+y_2^{(R)}\beta_3^R)\delta_{i3}\} \quad . \qquad (6)$$

In Eq. (6) α_H is the model dependent parameter of SU(2)×U(1) breaking (See Ref. (5) for a review of low energy Supergravity models obeying $-\pi<\alpha_H\leq\pi$ and $\tan\alpha_H=\langle H_5'\rangle/\langle H^5\rangle$, $\alpha_i^L=\epsilon_{abc}(d_{aL}\gamma^o u_{bL})\ (s_{cL}\gamma^o\nu_{iL})$, α_i^R is α_i^L with $(d_L,u_L)\to(d_R,u_R)$ and $\beta_i^{L,R}$ is $\alpha_i^{L,R}$ with $d\leftrightarrow s$. y_i^{tK} is the contribution from the third generation Wino dressings and is given by

$$y_i^{tK} = \frac{P_3}{P_2}\left[\frac{m_t V_{31}V_{32}}{m_c V_{21}V_{22}}\right]\left[\frac{F(\tilde{t};\tilde{d}_i;\tilde{W})+F(\tilde{t};\tilde{e}_i;\tilde{W})}{F(\tilde{c};\tilde{d}_i;\tilde{W})+F(\tilde{c};\tilde{e}_i;\tilde{W})}\right] \quad (7)$$

where $F(\tilde{t};\tilde{d}_i;\tilde{W})$ etc., in Eqs. (6) and (7) are the dressing loop integrals. $y_{\tilde{g}}$ and $y_{\tilde{Z}}$ in Eq. (6) are the gluino and the Zino exchange contributions while y_1^R and y_2^R in Eq. (6) are the contributions from the RRRR dimension five operators.

If one neglects all the correction terms, i.e., y_i^t, $y_{\tilde{g}}$, etc., in Eq. (6) one finds the conventional result, i.e., the dominance of the $\bar{\nu}K$ modes in nucleon decay. However, the correction terms can be significant. For a heavy top quark, i.e., m_t>40 GeV one finds (4) that there exist domains where $|y_i^{tK}|\sim O(1)$. Further, with P_3/P_2=+1 the sign of y_i^{tK} is negative leading to a destructive interference between the second generation and the third generation Wino dressing effects leading to a suppression of the $\bar{\nu}_i K$ modes. Simultaneously, for precisely the same domains where $\bar{\nu}_i K$ modes are suppressed one finds an enhancement of the $\bar{\nu}_i\pi$ modes due to a constructive interference between the second generation and the third generation Wino dressing contributions to these modes. A qualitative analysis of the domains where the $\bar{\nu}_i K$ modes are suppressed and the $\bar{\nu}_i\pi$ modes are enhanced was given in Ref. (4). A quantitative analysis of these domains was given in Ref. (6) where it was shown that these domains are non-negligible. Results of Ref. (6) are shown in Fig. (1) where the contour maps of the ratio $\Gamma(p\to\bar{\nu}\pi^+)/\Gamma p\to\bar{\nu}K^+)$ are given for the input parameters $P_1/P_2=-P_2/P_3=-1$, $\tilde{m}_\gamma$=3 GeV, μ=20

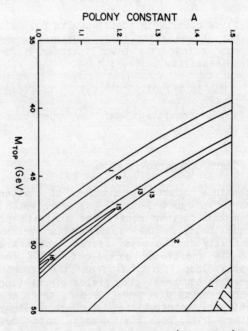

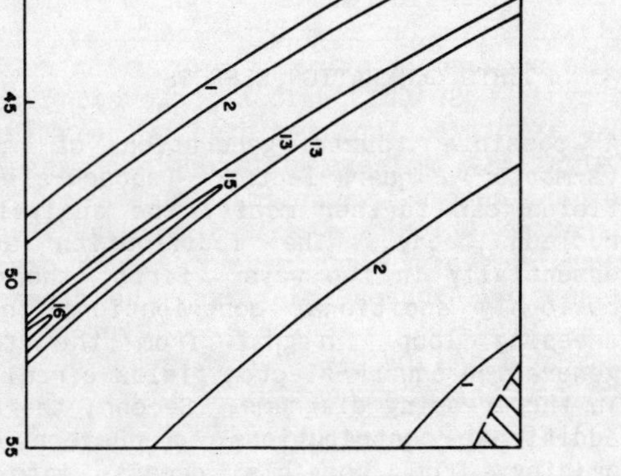

Fig. 1. The ratio $\Gamma(b\to\bar{\nu}\pi^+)/\Gamma(b\to\bar{\nu}K^+)$.

GeV and $m_{3/2}$=178 GeV. We note the significant domain of the parameter space in Fig. (1) where the $\bar{\nu}\pi$ modes are comparable to or exceed the $\bar{\nu}K$ modes. It was further shown in Ref. (6) that in the domains where the $\bar{\nu}\pi$ modes dominate the $\bar{\nu}K$ modes, one also has an enhancement of the $\bar{\nu}\rho$ and $\bar{\nu}\omega$ modes and a relative suppression of $\bar{\nu}K^*$ modes.

SUSY GUT theories do not predict the absolute decay lifetime of the nucleon due to the model dependence of the Higgs triplet mass. One may, however, use the experimental lower bound for one of the decay modes and predict the theoretical lower bounds for the remaining decay modes. This analysis was carried out in Refs. (4) and (6). It was seen there that the predicted theoretical lower bounds are close to and generally somewhat larger than the experimental lower bounds for these modes (7).

Recently it was pointed out by Enqvist, Masiero and Nanopoulos (8) that the Higgs triplet mass is not totally arbitrary. For both the five tuning and the missing partner scenario (which guarantee a light Higgs doublet) they found an upper bound from the heavy Higgs sector for the Higgs triplet mass of

$$M \leq M_X \tag{8}$$

where $M_X=(0.5\text{-}1.6)\times 10^{16}$ GeV. However, in models where $\bar{\nu}K$ modes dominate, the lower limit on M from the current experimental lower limit on the $\bar{\nu}K$ partial lifetime is

$$M \geq (7.0\text{-}7.0) \times 10^{16}\ \text{GeV} \tag{9}$$

Eq. (9) is clearly in conflict with Eq. (8). In contrast one finds that in models where $\bar{\nu}\pi$ modes dominate the lower limit on M for nucleon stability is (4,6)

$$M \geq (0.7\text{-}7.0) \times 10^{16}\ \text{GeV} \tag{10}$$

Eq. (10) is consistent with the upper bound on M of Eq. (8).

3. FOURTH GENERATION EFFECTS

A possible fourth generation of matter (s-matter) quark-lepton (squark-slepton) fields can further modify the analysis of nucleon decay. The modification arises essentially in two ways. First, there are obviously additional contributions to the dressing loop integral from the fourth generation squark-slepton fields circulating in the dressing diagrams. Second, there are additional contributions to nucleon decay arising from possible decays into new channels such as $\bar{\nu}_{\tau'}K, \bar{\nu}_{\tau'}\pi, \bar{\nu}_{\tau'}\eta$, etc. Dimension six operators analogous to Eq. (6) including the effects of the fourth generation can be obtained in a straightforward fashion (9). Here we exhibit only the dominant factors of quark mass and K-M matrix elements to estimate the size of these contributions. For example one has

$$\frac{\Gamma(N\to\bar{\nu}_{\tau'}K)}{\Gamma(N\to\bar{\nu}_{\tau}K)} \approx \left(\frac{m_{b'}V^{\dagger}_{41}}{m_b V^{\dagger}_{31}}\right)^2 \times \left(\begin{matrix}\text{Ratio of Loop}\\ \text{Integrals}\end{matrix}\right)^2 \tag{11}$$

Since the lower limit on M of Eq. (10) arising from the three generation analysis essentially saturates the bound of Eq. (8), there appears to be little room for a large fourth generation contribution. Thus we demand

$$(m_{b'}V^{\dagger}_{41})/(m_b V^{\dagger}_{31}) \leq O(1) \tag{12}$$

Further, since essentially in all models of a fourth generation $m_{b'}>>m_b$, one infers from Eq. (12) the existence of either small fourth generation K-M matrices or the existence of a sufficiently heavy fourth generation neutrino to forbid nucleon decay into the new channels.

REFERENCES

(1) Y. Totsuka, in Proc. of the 1985 International Symposium on lepton and Photon Interactions at High Energy, edited by M. Konuma and K. Takahashi.

(2) For a review see W. Lucha, Fortschr. Phys. 33, No. 10 (1985) and UWthPh-1985-24.

(3) S. Weinberg, Phys. Rev. D26, 287 (1982); N. Sakai and T. Yanagida, Nucl. Phys. B197, 533 (1982). See Ref. (2) for a more complete set of references.

(4) R. Arnowitt, A.H. Chamseddine and P. Nath, Phys. Lett. 156B, 215 (1985); P. Nath, A.H. Chamseddine and R. Arnowitt, Phys. Rev. 32D, 2348 (1985).

(5) P. Nath, R. Arnowitt and A.H. Chamseddine, "Applied N=1 Supergravity," (World Scientific Pub. Co., Singapore, 1984).

(6) T.C. Yuan, Phys. Rev. D33, 1894 (1986).

(7) M. Koshiba, in Proc. of the XXII International Conference on High Energy Physics, Leipzig, 1984.

(8) K. Enqvist, A. Masiero, and D.V. Nanopoulos, Phys. Lett. 156B, 209 (1985).

(9) A full analysis of the fourth generation effect on nucleon decay will be presented elsewhere.

TOP QUARK AND LIGHT HIGGS SCALAR MASS BOUNDS IN NO-SCALE SUPERGRAVITY

Probir Roy
Lawrence Berkeley Laboratory
University of California
Berkeley, CA 94720
and
Tata Institute of Fundamental Research
Bombay, India*

No-scale supergravity theories with the minimal low-energy particle content are shown to become untenable for a top quark mass m_T much less than 40 GeV. For $m_T < 55$ GeV, a stringent upper bound operates on the mass of the lowest-lying Higgs scalar. Further , the Higgs pseudoscalar is constrained to be nearly a quarter as massive as the gluino.

No-scale $N = 1$ supergravity theories[1)] hold a lot of interest these days. Such a theory is likely to be the effective low-energy limit[2)] of the heterotic superstring[3)]. The scalar potential in this type of a theory is flat. That is ensured by a non-compact $SU(n,1)/SU(n)\times U(1)$ symmetry in the Kähler sector. Consequently, the Kähler potential is characterized by a particular logarithmic form, specifically $-3\ln[f(z) + f^*(z) + g(\phi^{\dagger\bar{i}}, \phi^j)]$, where z is a generic gauge singlet scalar and ϕ^j's are $n-1$ gauge nonsinglet ones, f and g being analytic and real functions, respectively. The gravitino mass gets decoupled from the scale of global supersymmetry breaking at laboratory energies. The latter is seeded by a universal gaugino mass M at the grand unifying scale M_{GUT}. Since such theories have difficulties[4)] admitting a fourth generation, I shall consider only no-scale supergravity theories with the minimal low-energy particle content — namely, three fermionic generations, the 3-2-1 gauge bosons, two Higgs doublets and superpartners for all.

An important question concerns the requirements of stability and electroweak symmetry breadkown in the light Higgs sector. These tightly constrain the squared mass parameters μ_i^2 defined through the quadratic part of the Higgs potential in terms of the $Y = \pm 1$ doublets $\phi_\uparrow, \phi_\downarrow$ as

$$V_H^{(2)} = (\phi_\downarrow\ \phi_\uparrow^*)^* \begin{pmatrix} \mu_1^2 & -\mu_3^2 \\ -\mu_3^2 & \mu_2^2 \end{pmatrix} \begin{pmatrix} \phi_\downarrow \\ \phi_\uparrow^* \end{pmatrix}$$

Radiative effects make $\mu_i^2 \equiv \mu_i^2(t)$, t being $\ln(M_{GUT}^2 Q^{-2})$ with Q as the energy scale. These functions of t evolve[5)] to $t_W = \ln(M_{GUT}^2 M_W^{-2})$ from their boundary values at $t = 0$. Extrapolations, based on the observed values of the Weinberg angle and the rationalized fine structure constant, imply $M_{GUT} \sim 3.2 \times 10^{16}$ GeV and $t_W \simeq 66.95$. The boundary conditions at $t = 0$ are, however, determined from the flavor-independence and the universality of gravitational interactions in the underlying supergravity theory. When the latter is of the no-scale type, all global supersymmetry breaking constants — except M — vanish at $t = 0$. Consequently, very stringent experimentally verifiable restrictions[6)] emerge. In effect, these constraints provide laboratory tests of the type of theories considered here.

Recall that, in the simplest softly broken supersymmetric extension of the standard 3-2-1 theory, the matter fields of the latter are extended into chiral superfields. Thus

$$\begin{pmatrix} u \\ d \end{pmatrix}_{iL} = q_i \to \hat{Q}_i \qquad \begin{pmatrix} \nu \\ e \end{pmatrix}_{iL} = \ell_i \to \hat{L}_i$$

$$u_{iL}{}^C \to \hat{U}_i^C \qquad e_{iL}{}^C \to \hat{E}_i^C$$

$$d_{iL}{}^C \to \hat{D}_i^C \qquad \phi_\uparrow \to \hat{\Phi}_\uparrow$$

$$\phi_\downarrow \to \hat{\Phi}_\downarrow$$

The general form of the superpotential is

$$f = \lambda_U^{ij}\hat{Q}_i\hat{U}_j^C\hat{\Phi}_\uparrow + \lambda_D^{ij}\hat{Q}_i\hat{D}_j^C\hat{\Phi}_\downarrow + \lambda_E^{ij}\hat{L}_i\hat{E}_j^C\hat{\Phi}_\downarrow + \mu\hat{\Phi}_\uparrow\cdot\hat{\Phi}_\downarrow \qquad (1)$$

where λ's are Yukawa couplings and μ is a Higgs mass-mixing parameter. The soft supersymmetry breaking part of the Lagrangian is characterized by masses M_a of the gaugino fields λ_a, scalar masses m_{z_i} as well as mass parameters μ_3 and $\bar{A}^{ij}$

* Permanent address.

$-\frac{1}{2}\sum_a M_a \bar{\lambda}_a \lambda_a + \mu_3^2(\phi_\uparrow \cdot \phi_\downarrow + h.c.) - \sum_i m_{z_i}^2 \mid z_i \mid^2$
$-[(\lambda_U \bar{A}_U)^{ij} Q_i U_i^C \phi_\uparrow + (\lambda_D \bar{A}_D)^{ij} Q_i D_i^C \phi_\downarrow$
$+(\lambda_E \bar{A}_E)^{ij} L_i E_j^C \phi_\downarrow]$ (2)

Returning to the Higgs potential, let me recapitulate the conditions[7] imposed by minimization and stability, i.e.,

$$\cot^{-1}\frac{\langle\phi_\downarrow^0\rangle}{\langle\phi_\uparrow^0\rangle} \equiv \theta = \sin^{-1}\frac{2\mu_{3W}^2}{\mu_{1W}^2+\mu_{2W}^2} \quad real, \quad (3a)$$

$$\frac{1}{2}M_Z^2 = -\mu_{1W}^2 + (\mu_{1W}^2 - \mu_{2W}^2)\frac{\cos^2\theta}{\cos^2\theta - \sin^2\theta} \quad (3b)$$

and that required by electroweak symmetry breadkown, namely

$$\det \mu_{Wij}^2 \equiv \mu_{1W}^2\mu_{2W}^2 - \mu_{3W}^4 < 0 \qquad (4)$$

The four physical Higgs masses $m_{\pm,a,b,c}$ obey the constraints[7]

$$m_c^2 = \mu_{1W}^2 + \mu_{2W}^2 = m_a^2 + m_b^2 - m_Z^2 = m_\pm^2 - M_W^2$$

$$m_b \geq M_Z \mid \cos 2\theta \mid \geq M_Z \mid (m_{1W}^2 - \mu_{2W}^2)(\mu_{1W}^2 + \mu_{2W}^2)^{-1} \mid$$

$$m_a \leq \max(M_Z, m_c)$$

The boundary conditions at $t = 0$ are $m_{z_i}(0) = \overline{A}^{ij}(0) = \mu_{30}^2 = 0$, $\mu_{10}^2 = \mu_{20}^2 = \mu_0^2$, $M_a(0) = M$ and $\tilde{\alpha}_{2,3}(0) = 5/3\tilde{\alpha}_1(0) = \tilde{\alpha}(0) \simeq 1/96\pi$. Renormalization group evolution[8] makes μ_1^2 and μ_2^2 increase and decrease with t, respectively, the main driving term being $\lambda_T = \lambda_U^{33}$. Thus (3b) implies $1/2 < \cos\theta < 1$. Moreover, for $m_T < 55$ GeV, the Alvarez-Gaumé, Polchinski, Wise analysis[9] showing that $\langle\phi_\uparrow^0\rangle \sim \langle\phi_\downarrow^0\rangle$ and $1/2 \sim \cos^2\theta$ holds so that all Yukawa couplings except $Y_T = (\lambda_T/4\pi)^2$ can be safely neglected in the 1-loop evolution equations. The latter are shown in Table 1. These are now analytically integrable. The solutions can best be expressed in terms of certain evolution functions displayed in Table 2.

Recall first that $\tilde{\alpha}_{aW}\tilde{\alpha}_{a0}^{-1} = M_{aW}M^{-1} = (1 + \tilde{\alpha}_{a0}b_a t_W)^{-1}$ where the b_a's are given in Table 1. The other relevant analytic solutions to the 1-loop renormalization group equations at $t = t_W$ may be written in terms of the evolution functions of the table and the dimensionless ratio $3F_W(2\sqrt{2}\pi^2 E_W \cos^2\theta)^{-1}G_F m_T^2 = \beta(\theta)$.

$$\begin{array}{l}
\frac{d}{dt}(M_a\tilde{\alpha}_a^{-1}) = 0,\ \frac{d}{dt}\tilde{\alpha}_a^{-1} = b_a,\\
(b_1, b_2, b_3,) = (-3, 1, 11)\\
(\frac{d}{dt} + 6Y_T - \frac{16}{3}\tilde{\alpha}_3 - 3\tilde{\alpha}_2 - \frac{13}{9}\tilde{\alpha}_1)Y_T = 0\\
(\frac{d}{dt} + 3Y_T - 3\tilde{\alpha}_2 - \tilde{\alpha}_1)\mu^2 = 0\\
\frac{d}{dt}\mu_1^2 = \frac{d}{dt}\mu^2 + 3\tilde{\alpha}_2 M_2^2 + \tilde{\alpha}_1 M_1^2\\
(\frac{d}{dt} + 6Y_T)(\mu_1^2 - \mu_2^2) = 3Y_T\bar{A}_T^2 + Y_T\cdot\\
\cdot\left[\frac{16}{3}(M_3^2 - M^2) + 9(M^2 - M_2^2) + \frac{13}{33}(M^2 - M_1^2)\right]\\
(\frac{d}{dt} + \frac{3}{2}Y_T - \frac{3}{2}\tilde{\alpha}_2 - \frac{1}{2}\tilde{\alpha}_1)\mu_3^2\\
= \mu(3\bar{A}_T - 3\tilde{\alpha}_2 M_2 - \tilde{\alpha}_1 M_1)\\
(\frac{d}{dt} + 6Y_T)\bar{A}_T = \frac{16}{3}\tilde{\alpha}_3 M_3 + 3\tilde{\alpha}_2 M_2 + \frac{13}{9}\tilde{\alpha}_1 M_1
\end{array}$$

Table 1: 1-loop evolution equations with nontop Yukawa couplings neglected

$$\begin{array}{l}
E(t) = \left(1 - \frac{t}{32\pi}\right)^{-\frac{16}{9}}\left(1 - \frac{t}{96\pi}\right)^{3}\left(1 + 11\frac{t}{160\pi}\right)^{\frac{13}{99}}\\
F(t) = \int_0^t d\tau E(\tau),\\
H(t) = \frac{t}{E}\frac{dE}{dt},\ \bar{H}(t) = FH - tE + F\\
H'(t) = \frac{16}{3}\left[(1 - \frac{t}{32\pi})^{-2} - 1\right] + 6\left[1 - (1 + \frac{t}{96\pi})^{-2}\right]\\
-\frac{2}{9}\left[1 + 11\frac{t}{160\pi})^{-2}\right]\\
\bar{F}(t) = \frac{8}{9}\left[(1 - \frac{t}{32\pi})^2 - 1\right] + \frac{8}{99}\left[1 - (1 + 11\frac{t}{160\pi})^2\right]\\
g(t) = \frac{3}{2}\left[1 - (1 + \frac{t}{96\pi})^{-2}\right] + \frac{1}{22}\left[1 - (1 + 11\frac{t}{160\pi})^{-2}\right]\\
G(t) = \int_0^t d\tau E(\tau)[\bar{F}(\tau) - \frac{1}{6}H'(\tau)]\\
-\frac{1}{6F}[4FH^2 - 4H\bar{H} - 2\int_0^t d\tau E(\tau)H^2(\tau)]\\
K(t) = H(H - \frac{\bar{H}}{F}) + \frac{3}{2}G,\ L(t) = (H - \frac{\bar{H}}{F})^2,\\
S(t) = \frac{t}{32\pi}\left[1 + \frac{t}{96\pi})^{-1} + \frac{1}{5}(1 + 11\frac{t}{160\pi})^{-1}\right]
\end{array}$$

Table 2: Useful evolution functions

$$\begin{aligned}
\mu_W^2 &= \mu_0^2\left(1 + \frac{t_W}{96\pi}\right)^2\left(1 + 11\frac{t_W}{160\pi}\right)^{\frac{1}{11}}(1 - \beta(\theta))^{\frac{1}{2}}\\
\mu_{1W}^2 &= \mu_W^2 + g_W M^2\\
\mu_{2W}^2 &= \mu_{1W}^2 - \beta(\theta)(K_W - \frac{1}{2}L_W\beta(\theta))M^2\\
\mu_{3W}^2 &= \left\{\frac{1}{2}\beta(\theta)\left(\frac{t_W E_W}{F_W} - 1\right) - S_W\right\}M\mu_W
\end{aligned} \qquad (5)$$

The substitution of μ_{iW}^2 from (5) into inequality (3), together with the numerical evolution of the evolution constants up to the third decimal place, leads to

$$\eta^2 + (0.707 - 0.223\frac{w}{\cos^2\theta} + 0.002\frac{w^2}{\cos^4\theta})\eta$$

$$+0.284 - 0.149\frac{w}{\cos^2\theta} + 0.002\frac{w^2}{\cos^4\theta} < 0 \qquad (6)$$

In (6) $\eta \equiv M^{-2}\mu_W^2$ and $w = (m_T/40\ GeV)^2$. For $1/2 < \cos^2\theta$ and $m_T < 55\ GeV$, the real non-negativity at η turns out to necessarily imply the negativity of the η-independent part in (6). (Note that a vanishing μ_0, and hence μ_W, as suggested by the dimensional reduction[2] of superstring theories, yields this result directly.) The consequences are twofold:

(1) $0.5 < \cos^2\theta < 0.5085w$

i.e., $\frac{m_b}{M_Z} <|\cos 2\theta| < 1.017\left(\frac{m_T}{40GeV}\right)^2 - 1$ (7)

(2) $0.983 < w$, i.e., $39.6\ GeV \lesssim m_T$ (8)

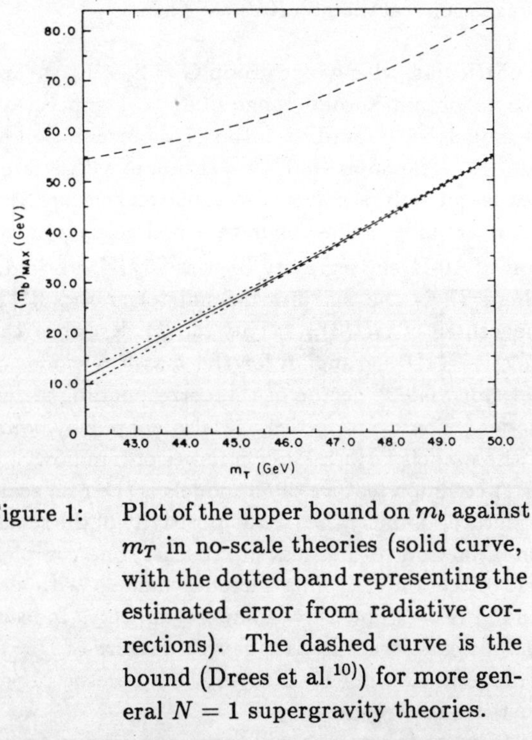

Figure 1: Plot of the upper bound on m_b against m_T in no-scale theories (solid curve, with the dotted band representing the estimated error from radiative corrections). The dashed curve is the bound (Drees et al.[10]) for more general $N = 1$ supergravity theories.

Figure 1 shows the upper bound (7) plotted against m_T in comparison with the earlier weaker result[10] for more general $N = 1$ supergravity theories. For $m_T < 42\ GeV$, the error due to radiative corrections (dotted band) becomes quite large. As m_T goes below 39.6 GeV, the bound becomes imaginary. On the other side, it saturates at unity as m_T exceeds[9] 55 GeV. The pseudoscalar mass is given by

$$m_c^2 = \mu_{1W}^2 + \mu_{2W}^2$$
$$= M^2(2\eta + 1.066 - 0.279w\cos^{-2}\theta$$
$$+ 0.004w^2\cos^{-4}\theta)$$

with η varying between 0 and $\eta_+(\cos\theta, w)$ which is the higher root of the quadratic equation corresponding to (6) being an equality. For $m_T < 50\ GeV$, the absolute upper bound on η is $\eta_+(0.15, 1.25) = 0.097$ so that one finds by use of $m^2 \simeq 0.112m_{\tilde{g}}^2$ that $0.21m_{\tilde{g}} < m_c < 0.28m_{\tilde{g}}$, where $m_{\tilde{g}}$ is the gluino mass.

It is my conclusion that no-scale $N = 1$ supergravity theories with the minimal low-energy particle content face rather critical experimental tests in the near future through (7) and (8).

This work was done in collaboration with P. Majumdar. I thank Mary K. Gaillard for the hospitality of the Theory Group at LBL where this talk was written up.

REFERENCES

[1] E. Cremmer et al., Phys. Lett. 133B (1983) 61; N.P. Chang et al., Phys. Rev. Lett. 51 (1983) 327; J. Ellis et al., Phys. Lett. 134B (1984) 429 and Nucl. Phys. B241 (1984) 406, B247 (1984) 373.

[2] E. Witten, Phys. Lett. 155B (1985) 151.

[3] D. Gross et al., Phys. Rev. Lett. 52 (1985) 502.

[4] K. Enqvist et al., Phys. Lett. 167B (1986) 73.

[5] For any function $f(t)$ we use f_0 and f_W to mean $f(0)$ and $f(t_W)$, respectively.

[6] P. Majumdar and P. Roy, to be published in Phys. Rev. D.

[7] K. Inoue et al., Prog. Theor. Phys. 67 (1982) 1889, 68 (1982) 927; R. Flores and M. Sher, Ann. Phys. (N.Y.) 148 (1983) 95.

[8] K. Inoue et al., Prog. Theor. Phys. 67 (1982) 1889; L. Ibañez et al., Nucl. Phys. B233 (1984) 511; B256 (1985) 218.

[9] L. Alvarez-Gaumé et al., Nucl. Phys. B221 (1983) 495.

[10] P. Majumdar and P. Roy, Phys. Rev. D30 (1984) 2432; 33 (1986) 2674.; P. Roy, in *Design and Utilization of the Superconducting Supercollider* (Snowmass 1984, eds R. Donaldson and J.G. Morfin) p.807; H.P. Nilles and M. Nusbaumer, Phys. Lett. 145B (1984) 73; M. Drees et al., *ibid.* 159B (1985) 118.; E. Reya, Phys. Rev. D33 (1986) 773.

D = 4 NO SCALE SUPERGRAVITIES FROM D = 10 SUPERSTRINGS.

Costas Kounnas
Physics Department and Lawrence Berkeley Laboratory
University of California, Berkeley, CA 94720

Dimensional reductions of supergravity theories are shown to yield to specific classes of four dimensional no scale models with N = 4, 2 or 1 residual supersymmetry. N = 1 "maximal" supergravity lagrangian, corresponding to the "untwisted" sector of orbifold compactification of superstrings, contains nine families and has a no scale structure based on the Kähler manifold $\frac{SU(3,3+3n)}{SU(3)\times SU(3+3n)} \times \frac{SU(1,1)}{U(1)}$. The quantum consistency of the resulting theories give informations on the non Kaluza–Klein (string) "twisted" sector.

In recent time ten-dimensional N = 1 chiral supergravity coupled to matter[1] received a lot of attention for different reasons: anomaly cancellation with $E_8 \times E_8$ or SO(32) gauge symmetry,[2] possibility of compactification with chiral fermions in four dimensions with realistic gauge group and vanishing cosmological constant, and its relation with the point field limit of superstring theories[3].

In this respect of particular interest are those compactifications preserving some supersymmetry, especially the N = 1 case, the only one which allows chiral families. Examples of six-dimensional compact manifolds with a residual N = 1 supersymmetry are the Calabi-Yau spaces[3] and some singular limits of them e.g. orbifolds.[4]

A guideline for approximating the effective D = 4 supergravity lagrangian describing the interactions of the massless modes of the theory has been proposed by Witten.[5] More specifically the strategy consists in a dimensional reduction of the 10-D supergravity Lagrangian and then performing a consistent truncation keeping only singlets with respect to some suitable group, isomorphic to a subgroup of the rotation group of the internal manifold and maintaining only one supersymmetry.

For instance the example worked out in ref. [5], is just one extreme case of this procedure, obtained by keeping only the singlets under the action of $SU(3)_D = SU(3)_1 \oplus SU(3)_2$ in which $SU(3)_1$ is a subgroup of the rotation group SO(6) of the internal coordinates X_I and $SU(3)_2 \times G' \subset G$, where G is one of the anomaly free gauge groups of the D = 10 theory.

In this talk I present a summary of the results of ref.[6], where we have generalized the strategy of ref.[5] in order to obtain more general $D = 4$ effective lagrangians preserving some supersymmetry. In particular, with gauge group $G = E_8 \times E_8'$ we are able to obtain a wide range of N = 1 supergravity models with families in the <u>27</u> representation of $E_6 \subset E_8$. The "maximal" N = 1 model with 9 families is not only a result of a consistent truncation but it actually corresponds to a real compactification of 10-D supergravity on the T^6/Z^3 orbifold,[4] where T^6 is the six dimensional torus and Z^3 is the center of $SU(3)_D$. This model describes the effective 4–D lagrangian for the massless modes of the "untwisted" sector of the corresponding string model.[4] These properties are also shared by some of the models with a lower number of families.

A common feature of all models is their no scale structure,[7] namely the (semi-)positivity of the scalar potential and its flatness in, at least, one complex field direction(s). This includes models with extended N = 2 and N = 4 supersymmetry. Indeed, the no scale structure is a general feature of D = 4 compactification of 10-D supergravity preserving at least one supersymmetry.

In what follows we will focus our discussion on the bosonic part of the Lagrangian. In fact in all supersymmetric theories the fermionic part is uniquely fixed by supersymmetry, once the bosonic sector is known.

In general we split the 10-D indices $\hat{\mu} = \mu, I(\mu = 1, \cdots 4, I = 1, \cdots, 6)$ so the D = 10 bosonic supergravity fields are

$$g_{\mu\nu}, g_{\mu I}, g_{IJ}, B_{\mu\nu}, B_{\mu I}, B_{IJ}, \phi \tag{1}$$

where $g_{\hat{\mu}\hat{\nu}}, B_{\hat{\mu}\hat{\nu}}$ and ϕ are respectively the metric tensor, the antisymmetric tensor and the dilaton of D = 10 supergravity.

The D = 10 gauge non singlet bosonic fields are

$$A^\alpha_\mu, A^\alpha_I \tag{2}$$

where $A^\alpha_{\bar\mu}$ are the D = 10 gauge vector fields.

The simplest dimensional reduction corresponds to a T^6 torus compactification. This demands to keep all modes given by eqs. (1, 2) (independent of the internal coordinates X_I) and results in an N = 4, D = 4 supergravity coupled to 502 N = 4 (vector) matter supermultiplets with gauge symmetry $U(1)^6 \times E_8 \times E_8'$. Due to the six graviphotons of the N = 4 supergravity multiplet ($g_{I\mu}$) the over all gauge symmetry is $U(1)^{12} \times E_8 \times E_8'$. The scalar manifold has dimension

$$D_S = 2 + 6(6+n), \quad n = \dim G = 496 \qquad (3)$$

The geometry of the scalar manifold is[6]

$$\frac{SO(6,6+n)}{SO(6)\times SO(6+n)} \times \frac{SU(1,1)}{U(1)} \qquad (4)$$

as expected from general results on N = 4 supergravity couplings.[8].

The gauging of the group $U(1)^{12} \times G$, i.e. the fact that there are six N = 4 gauge multiplets singlet under G, is essential to give rise to an N = 4 no scale positive semidefinite potential.

We will consider now consistent truncations (compactifications) of the D = 10 theory which reduce the number of supersymmetry. The gravitinos transform as $\underline{4}$ under $SU(4) \simeq SO(6)$. So, any subgroup of SU(4) under which some gravitinos are singlets correspond to a supersymmetric consistent truncation and, in some cases, to a real compactification. The number of singlet gravitinos correspond to the number of unbroken supersymmetries. For instance we may obtain an N = 2 supergravity[9] decomposing $SU(4) \to SU(2)' \times SU(2) \times U(1)$ and keeping only Z_2 singlets of the center of $SU(2)'$. To obtain a spectrum with charged matter fields we must retain singlets under the center of $SU(2)_D = SU(2)' \oplus SU(2)''$ where $SU(2)''$ is a subgroup of E_8. This model corresponds to an orbifold compactifiction on T_6/Z_2 with a maximal remaining gauge group $E_7 \times SU(2)'' \times E_8$. (See Ref. 6 for more details.)

To obtain the "maximal" N = 1 model we pick up the smallest group under which there is only one singlet gravitino. This is the center Z_3 of $SU(3)' \subset SU(4)$ considered in ref. [5]. The Z_3 singlet scalar fields coming from the 10-D supergravity sector are

$$g_{i\bar j}, B_{i\bar j}, \phi, D \qquad (5)$$

where we used complex notation for the internal coordinates corresponding to the decomposition $6 \to 3 + \bar 3$ of SU(4) into $SU(3)'$. The charged fields decomposition under $SU(3) \times E_6 \subset E_8$ is

$$\begin{aligned} A^\alpha_I \;&\to C_3^{(\bar 3,\bar{27})} + C_3^{(3,27)} + C_3^{(1,78)} \\ &+ C_3^{(8,1)} + (h.c.), \quad \alpha \epsilon E_8. \end{aligned} \qquad (6)$$

The diagonal Z_3 singlets (center of $SU(3)_D = SU(3)' \oplus SU(3)$) are the complex fields

$$\begin{aligned} C_i^{n,a} \quad &\text{with } a\epsilon \underline{27} \text{ of } E_6,\ n\epsilon\, \underline{3} \text{ of SU(3)}, \\ &i\epsilon\, \underline{3}\ of\ SU(3)'. \end{aligned} \qquad (7)$$

The surviving N = 1 gauge fields $A_\mu^{(1,78)}, A_\mu^{(8,1)}$ are in the adjoint of the SU(3) × E_6 gauge group.[4] The present model contains nine $\underline{27}$ of families which are triplets under the horizontal gauge SU(3) symmetry. The gravity sector corresponds to 10 chiral multiplets which are singlets under the surviving SU(3) $\times E_6 \times E_8'$ gauge group.

This model, containing chiral fermions in the $\underline{3}$ of the SU(3) gauge group, is anomalous. The simplest way to cancel this anomaly is to have 81 additional chiral multiplets in the $\overline{\underline{3}}$ of SU(3). However from the 10–D field theory the $\overline{\underline{3}}$ chiral fermions are also in the $\overline{\underline{27}}$ of E_6. It is therefore impossible to obtain an anomaly free model with a net number of chiral generations. A way out of this problem is to eliminate the SU(3) vector multiplets which is a totally arbitrary procedure. On the other hand, the additional chiral fermions needed to cancel the anomaly come naturally in the orbifold (T^6/Z^3) string compactifacation. They arise from the "twisted" sectors of the theory and they are in the $(\bar 3, 1)$ of $SU(3) \times E_6$.[4] There are precisely 27 copies of them. The remaining additional modes are singlets under SU(3) transforming as $\underline{27}$ under E_6. There are equally 27 copies of these extra families.[4]

Our analysis is based on 10–D field theory so we are only able to give a complete description of the lagrangian which would correspond, in the string theory framework, to the massless modes of the untwisted sector. Beside this we have pointed out the need for some additional fields, not obtainable from the 10–D field theory, and which are indispensable for the consistency of the model.

The N = 1 supergravity Lagrangian describing the effective interactions of the massless modes is entirely specified[11] in terms of the Kähler potential $J(Z, Z^*)$ of the chiral multiplet sector, the superpotential $g(Z)$ (which is an analytic function of the complex scalar fields Z_A) and the Yang-Mills metric $f_{AB}(Z)$ (analytic in the scalar fields and symmetric in the adjoint of $SU(3) \times E_6 \times E_8'$). Actually there are only two relevant functions; $G(Z, Z^*)$ and f_{ab} where[10] $G(Z, Z^*) = J(Z, Z^*) + \log|g(Z)|^2$. In order to bring the N = 1 supergravity Lagrangian in a standard supergravity form we define the complex fields $T_{i\bar{j}}$ and S

$$T_{i\bar{j}} = g_{i\bar{j}} e^{-\phi} + C_i^{ma} C_{\bar{j}}^{*\bar{m}\bar{a}} - i\sqrt{2} B_{i\bar{j}} \qquad (8a)$$

$$S = \det g_{i\bar{j}} \cdot e^{\phi} + 3i\sqrt{2}D. \qquad (8b)$$

Then, the N = 1 Kähler potential is, $Z = (S, T_{ij}, C_j^{ma})$

$$J(Z, Z^*) = -\log(S + S^*) - \log\det(T_{i\bar{j}} + T^*_{i\bar{j}} - 2C_i^{ma} C_{\bar{j}}^{*\bar{m}\bar{a}}), \qquad (9)$$

the superpotential is,

$$g(Z) = d_{abc} \epsilon_{mn\ell} \epsilon^{ijk} C_i^{ma} C_j^{nb} C_k^{\ell c} \qquad (10)$$

where $\epsilon_{mn\ell}$ is the antisymmetric tensor of SU(3) and d_{abc} the symmetric tensor of the $\underline{27}$ of E_6; the Yang-Mills $f_{AB}(Z)$ function is

$$f_{AB}(Z) = S \cdot \delta_{AB}, \quad A, B, \epsilon SU(3) \times E_6 \times E_8' \qquad (11)$$

The "maximal" nine family Kähler manifold defined from eq. (17) is

$$\frac{SU(1,1)}{U(1)} \times \frac{SU(3, 3+3n)}{SU(3) \times SU(3+3n) \times U(1)}, \quad n = 27. \qquad (12)$$

Obviously this manifold is a truncation of the N = 4 model given in eq. (4).

The complete Kähler manifold of the 4-D effective action, corresponding to the orbifold compactified string, must then be a Kähler space which contains that of eq. (19) as a proper submanifold. (see ref. 6 for more details).

N = 1 supersymmetric models with lower number of families and T singlets are obtained by demanding the massless modes to be singlets under larger symmetry than Z_3. Needless to say, they must all be thought as consistently embedded in string theories on orbifolds, and describing its untwisted sectors.

We can consider for instance singlets under Z_6 discrete group which is generated by the 3×3 matrix M = diag $(-e^{\frac{2i\pi}{3}}, -e^{\frac{2i\pi}{3}}, e^{\frac{2i\pi}{3}})$; notice that $M^6 = \mathbf{1}$. The Z_6 invariance keeps only singlets and triplets under $SU(2)_D \subset SU(3)_D$. The gauge group in this case becomes $U(1) \times SU(2) \times E_6 \times E_8'$. The model contains five families, five singlets and the universal S singlet. The five family Kähler potential is[6]

$$J^{(5)}(Z, Z^*) = -\log(S + S^*) - \log\det(w_{ij} + w^*_{ij} - 2C_i^{ma} C_{\bar{j}}^{*\bar{m}\bar{a}}) - \log(x + x^* - 2C_x^a C_{\bar{x}}^{\bar{a}}), \qquad (13)$$

where now w_{ij} is $2 \times \underline{2}$ hermitian matrix and $m\epsilon\underline{2}$ of $SU(2)_D$. x is a complex field, and $a\epsilon$ $\underline{27}$ of E_6. The corresponding superpotential is

$$g^{(5)}(Z) = d_{abc} \epsilon_{\ell m} \epsilon^{ij} C_i^{\ell a} C_j^{mb} C_x^c. \qquad (14)$$

where now $\epsilon_{\ell m}$ is the antisymmetric tensor of $SU(2)_D$. The $f_{AB}(Z)$ function is model independent, (see eq. 11).

The geometry of the scalar manifold is

$$\frac{SU(1,1)}{U(1)} \times \frac{SU(2, 2+2n)}{SU(2) \times SU(2+2n) \times U(1)} \times \frac{SU(1, 1+n)}{SU(1+n) \times U(1)}, \quad n = 27 \qquad (15)$$

Another example, leading to a three family model and $U(1) \times U(1) \times E_6 \times E_8'$ as gauge group is obtain by retaining singlets under Z_{12}. Z_{12} is the discrete group which is generated by the 3 × 3 matrix $M = diag(ie^{\frac{2i\pi}{3}}, -ie^{\frac{2i\pi}{3}}, e^{\frac{2i\pi}{3}})$ $(M^{12} = \mathbf{1})$. In this case the Kähler manifold is

$$J^{(3)}(Z, Z^*) = -\log(S + S^*) - \sum_{i=1}^{3} \log(T_i + T_i^* - 2C_i^a C_i^{*\bar{a}}) \qquad (16)$$

and the superpotential

$$g^{(3)}(Z) = d_{abc} C_1^a C_2^b C_3^c \qquad (17)$$

The unbroken gauge group is U(1) × U(1) $\times E_6 \times E_8'$ and the Kähler manifold is

$$\frac{SU(1,1)}{U(1)} \times \left[\frac{SU(1, 1+n)}{SU(1+n) \times U(1)}\right]^3 \qquad (18)$$

Note that under the two $U(1)$ the three families have charges (1, 1), (-1, 1), (0, -2) respectively.

The five and three family models are also related to orbifold compactifications, namely on T^6/Z_6 and T^6/Z_{12} respectively.

In this talk I presented some results of ref. (6) where we derived various models coming from dimensional reduction of 10-D supergravity. We show that those models which correspond to orbifold compactifications gives rise to the complete low energy lagrangian for the "untwisted" sector of the corresponding string theory. In this approach in particular, all the non polynomial σ–model interactions are taken into account. It is worth stressing that from our analysis we were able to find another motivation twin of the request of modular invariance in string theory on orbifolds[4,11] for the presence of "twisted" sectors: The anomaly cancelation in the four dimensional massless sector. Also, we get some restrictions on the possible form for the lagrangian of the "twisted" sectors.

References

1. F. Bergshoeff, M. de Roo, B. de Wit and P. van Nieuwenhuizen, Nucl. Phys. B195 (1982) 97. G. Chapline and N. Manton, Phys. Lett. 120B (1983) 105.

2. M. B. Green and J. H. Schwarz, Phys. Lett. 149B (1984)117.
D. J. Gross, J. A. Harvey, E. Martinec and R. Rohm, Phys. Rev. Lett. 54(1985) 502; Nucl. Phys. B260 (1985) 569.

3. P. Candelas, G. T. Hozowitz, A. Strominger and E. Witten, Nucl. Phys. B258 (1985) 46.

4. L. Dixon, J. A. Harvey, C. Vafa and E. Witten, Nucl. Phys. B261 (1985) 651; Princeton Preprint (1985) String on Orbifolds II.

5. E. Witten, Phys. Lett. 155B (1985) 151.

6. S. Ferrara, C. Kounnas and M. Porrati, UCB–PTH–86/5 preprint, to appear in Phys. Lett. B.

7. E. Cremmer, S. Ferrara, C. Kounnas and D. V.Nanopoulos Phys. Lett. 133B (1983) 61;
J. Ellis, A. B. Lahanas, D. V. Nanopoulos and K. Tamvakis, Phys. Lett. 134B (1984) 42;
J. Ellis, C. Kounnas and D. V. Nanopoulos, Nucl. Phys. B241 (1984) 406; Nucl. Phys. B247 (1985) 373.

8. J. -P. Derendinger and S. Ferrara, "In Supersymmetrry and Supergravity 84" ed. B. de Wit, P. Fayet and P. van Nieuwenhuizen. (World Scientific, Singapore, 1984), page 159. M. de Roo, Phys. Lett. 156B(1985)331; Nucl. Phys. B255 (1985)515; E. Bergshoeff, G. T. Koh and E. Sezgin, Nucl. Phys. B245 (1984) 89.

9. B. de Wit and A. van Proeyen, Nucl. Phys. B245 (1984) 89.

10. E. Cremmer, S. Ferrara, L. Girardello and A. van Proeyen, Phys. Lett. 116B (1982) 231; Nucl. Phys. 212 (1983) 413.

11. C. Vafa, Harvard Preprint HUTP–86/A011 (1986).

Geometrical-Integrability Constraints and Equations of Motion in Four Plus Extended Super Spaces

LING-LIE CHAU

Department of Physics
University of California
Davis, California 95616

and

Physics Department
Brookhaven National Laboratory
Upton, NY 11973

It is pointed out that many equations of motion in physics, including gravitational and Yang-Mills equations, have a common origin: i.e. they are the results of certain geometrical integrability conditions. These integrability conditions lead to linear systems and conservation laws that are important in integrating these equations of motion.

Recently it has been shown that integrability conditions along all light-like directions in extended curved superspace do lead to supergravity equations of motion[1] in four dimensional space with superspace dimension $n > 4$, i.e. n = 5, 6, 7, 8, and further these equations of motion have the powerful linear system and infinite conservation laws.[2] The results have two-fold significance: First, together with the previous results for supersymmetric Yang-Mills theories,[3] the extended superspace has the power of giving a universal framework of deriving equations of motion in physics; second, since the equations of motion are consequences of light-like integrabilities in extended superspace, such a formulation also provides means to integrate the equations of motion through the linear systems and conservation laws. It is the purpose of this paper to show that such formulated supergravity equations poses the important integrability properties shared by many other nonlinear systems[4] including the supersymmetric Yang-Mills equations.[5] Because of space limitations, I shall only list briefly the results.

Light-like integrability in extended superspace states that the anticommutation relations:[6]

$$\{D_{\underline{\alpha}}^{s}, D_{\underline{\beta}}^{t}\} = 0,\ \{\overline{D}_{\underline{\dot\alpha}s}, \overline{D}_{\underline{\dot\beta}t}\} = 0,\ \{D_{\underline{\alpha}}^{s}, \overline{D}_{\underline{\dot\beta}t}\} = -2i\delta_t^s \partial_{\underline{\alpha}\underline{\dot\beta}} \quad (1a)$$

hold true even in the presence of gravitational field:

$$\{\nabla_{\underline{\alpha}}^{s}, \nabla_{\underline{\beta}}^{t}\}_{\triangle}^{\square} = 0,\ \{\overline{\nabla}_{\underline{\dot\alpha}s}, \overline{\nabla}_{\underline{\dot\beta}t}\}_{\triangle}^{\square} = 0,$$

$$\{\nabla_{\underline{\alpha}}^{s}, \overline{\nabla}_{\underline{\dot\beta}t}\}_{\triangle}^{\square} = -2i\delta_t^s \nabla_{\underline{\alpha}\underline{\dot\beta}}{}_{\triangle}^{\square} \quad (1b)$$

where $D_{\underline{\alpha}} \equiv \lambda^\alpha D_\alpha$, $D_{\underline{\dot\alpha}} \equiv \overline{\lambda}^{\dot\alpha} D_{\dot\alpha}$; λ^α, $\overline{\lambda}^{\dot\alpha}$ are two-component C-number spinors (here we introduce ${}_{\triangle}^{\square}$ to explicitly denote the indefinite matrix elements) of the generators or group elements, though in general

$$\{\nabla_A, \nabla_B\}_{\triangle}^{\square} = R_{AB}{}_{\triangle}^{\square} + T_{AB}^{\ \ C}\nabla_C{}_{\triangle}^{\square},\ \text{where } A, B \equiv {}_{\underline{\alpha}}^{s}\ or\ \underline{\dot\beta}t, \quad (2)$$

and

$$R_{AB}{}_{\triangle}^{\square} = \partial_{(A}\Omega_{B)}{}_{\triangle}^{\square} + \Omega_{(A}{}_{\triangle}^{\times}\Omega_{B)}{}_{\times}^{\square} + C_{(AB)}^{\ \ \times}\Omega_{\times}{}_{\triangle}^{\square},\ \text{is the curvature},$$

$$T_{A\,B}^{\ \ C} = C_{(AB)}^{\ \ C'} + \Omega_{(AB)}^{\ \ C'},\ \text{is the torsion},$$

$\Omega_{AB}^{\ \ C}$ is the spin connection, and $C_{(AB)}^{\ \ \times}\partial_\times = \{\partial_A, \partial_B\}$, is the anholonomy term. (we use "×" to indicate summation over indices and round bracket on the indices to mean symmetrization.) Taking $\lambda^\alpha = \binom{1}{\lambda}$, $\overline{\lambda}^{\dot\alpha} = (1, \lambda')$, i.e. $\nabla_{\underline{\alpha}}^{s} = \nabla_{1}^{s} + \lambda\nabla_{2}^{s}$, $\nabla_{\underline{\dot\beta}t} = \nabla_{\dot 1t} + \lambda'\nabla_{\dot 2t}$, Eq. (1b) implies

$$R_{\substack{s\,t\\1\,1}} = 0,\quad T_{\substack{s\,t\\1\,1}}^{\ \ C} = 0;\quad R_{\substack{s\,t\\2\,2}} = 0,\quad T_{\substack{s\,t\\2\,2}}^{\ \ C} = 0; \quad (2a)$$

$$R_{\substack{s\,t\\1\,2}} + (s \leftrightarrow t) = 0,\quad T_{\substack{s\,t\\1\,2}}^{\ \ C} + (s \leftrightarrow t) = 0; \quad (2b)$$

same equations for $1 \to \dot 1$, $2 \to \dot 2$; (2c, d)

$$R_{\substack{s\\\alpha}\dot\beta t} = 0,\quad T_{\substack{s\\\alpha}\dot\beta t}^{\ \ C} = -2i\delta_t^s\delta_{\alpha\dot\beta}^{\ \ C}. \quad (2e)$$

Note that the independence of λ, λ' is crucial for Eqs. (2a-e), especially Eq. (2e).

It is shown in Ref. (1) after an extremely long and tedious calculation,[1] we first find all the physical fields from the constraints and Bianchi: the spin-$\frac{1}{2}$ fields $\overline{\lambda}^{[ijk]}{}_{\dot\alpha}$, the spin-one fields $\overline{F}^{[ij]}_{\{\dot\alpha\dot\beta\}}$, the spin-$\frac{3}{2}$ field $\overline{\Sigma}^{i}_{\{\dot\alpha\dot\beta\dot\gamma\}}$, and the spin-two field $\overline{V}_{\{\dot\delta\dot\gamma\dot\beta\dot\gamma\}}$, we then establish that they satisfy, in the linearized case, the following equations of motion:

$$(N-2)(N-3)(N-4)\square\partial_\alpha^{\dot\alpha}\,\overline{\lambda}_{\dot\alpha}^{[ijk]} = 0,$$

$$(N-2)(N-3)(N-4)\partial_\alpha^{\dot\alpha}\,\partial_\beta^{\dot\beta}\,\overline{F}^{[ij]}_{\{\dot\alpha\dot\beta\}} = 0,$$

$$(N-2)(N-3)(N-4)\partial^{\dot{\alpha}}_{\alpha}\,\partial^{\dot{\beta}}_{\beta}\,\overline{\Sigma}^{i}_{\{\dot{\alpha}\dot{\beta}\dot{\gamma}\}}=0,$$
$$(N-2)(N-3)(N-4)\partial^{\dot{\alpha}}_{\alpha}\,\partial^{\dot{\beta}}_{\beta}\,\overline{V}_{\{\delta\dot{\gamma}\dot{\beta}\dot{\alpha}\}}=0. \qquad (3)$$

Note that these equations are conformal-like and are equations in higher order in differentiation. However if we find solutions such that an intermediate field[1] $X^{[ijk]}_{i}$ becomes zero, we can show that all these equations of motion Eqs. (3) become first order differential equations. Just as the $N=4$ supersymmetric Yang-Mills equations contain solutions of the $N=0$, $N=1$, and $N=2$ Yang-Mills fields, the extended supergravity equations do contain solutions to gravitational fields of lower N.

These results put extended supergravity in four-dimensional space with $N>4$ on the same footing as supersymmetric Yang-Mills in four dimensions with $N>2$.

Now we shall derive the linear systems from Eq. (1b). The preservation of the form of Eq. (1a) in the presence of fields implies that there exist similarity transformations by some group element $\phi(\lambda,\ \lambda',\ x,\ \theta,\ \overline{\theta})$ of Poincare transformation such that

$$\nabla_A=\phi(\lambda,\ \lambda')\circ D_A\circ\phi^{-1}(\lambda,\ \lambda), \text{in operator form;} \quad (3a)$$

$$\nabla_{A\triangle}^{\ \square}=\phi^{\times}_{\triangle}(\lambda,\ \lambda')\circ\phi^{\otimes}_{A}(\lambda,\ \lambda')D_{\otimes}=\delta^{\times}_{A}D_{\times}=D_A \quad (3b)$$

in matrix-element operator form;

$$=[\phi(\lambda,\ \lambda')D]_A\,\delta^{\ \square}_{\triangle}+\phi^{\ \times}_{\triangle}(\lambda,\ \lambda')\left[[\phi(\lambda,\ \lambda')D]_A\,\phi^{-1}\,{}^{\square}_{\times}(\lambda,\ \lambda')\right],$$

in differential form (note that we use "o" between operators when in operator form); or more concisely,

$$\nabla_{A\triangle}^{\ \square}=\partial_A\circ\delta^{\square}_{\triangle}+\Omega_{A\triangle}^{\ \square},\ \text{with}\ \partial_A=[\phi(\lambda,\ \lambda')D]_A,$$
$$\Omega_{A}{}^{\square}_{\triangle}=\phi^{\times}_{\triangle}(\lambda,\ \lambda')[\partial_A\,\phi^{-1}\,{}^{\square}_{\times}(\lambda,\ \lambda')], \quad (3c)$$

where $A=\underset{\dot{\alpha}}{s}$, or $\underset{\sim}{\beta}t$, and we use ∂_A to denote curved differentiation, D_A the flat differentiation. Note that D_A gets transformed by ϕ from Eq. (3a) to Eq. (3b). This is the major difference of the gravity case from the external group gauge-fields case, like the Yang-Mills fields. One can explicitly check that the vielbein and spin connections given by Eq. (3c) do satisfies Eq. (2a-e). Of special interest is that the anhonolomy terms are related to the spin connections from the torsionless conditions in Eq. (2a-e)

$$C_{(\underset{\sim}{\alpha}\,\underset{\sim}{\beta})}^{\ s\ t\ \ \square}+(s\leftrightarrow t=-\Omega_{(\underset{\sim}{\alpha}\,\underset{\sim}{\beta})}^{\ s\ t\ \ \square}+(s\leftrightarrow t),$$

$$C_{(\dot{\alpha}s\underset{\sim}{\beta}t)}^{\ \ \square}+(s\leftrightarrow t)=-\Omega_{(\dot{\alpha}s\underset{\sim}{\beta}t)}^{\ \ \square}+(s\leftrightarrow t),$$

$$C_{(\underset{\sim}{\alpha}\dot{\beta}t)}^{\ s\ \ \square}+(s\leftrightarrow t)=-\Omega_{(\underset{\sim}{\alpha}\dot{\beta}t)}^{\ s\ \ \square}-2i\delta^{s}_{t}\delta_{\underset{\sim}{\alpha}\dot{\underset{\sim}{\beta}}}^{\ \square}+(s\leftrightarrow t). \quad (4a)$$

These relations are very important. Since $\Omega_{\underset{\sim}{\alpha}\,\underset{\sim}{\beta}}^{\ s\ t\ \square}$'s involve only $\square$ in the super directions, so do the anholonomy terms on the lefthand-side. Thus in the equations for the curvatures of Eqs. (2a-2e), only the spinor connections in the super directions are involved. These are then the nonlinear different equations they are restricted to satisfy. And thus from Eq. (2) and (3c)

$$\{\nabla_{\underset{\sim}{\alpha}}^{s},\ \nabla_{\underset{\sim}{\beta}t}\}_{\triangle}^{\ \square}=\partial_{(\underset{\sim}{\alpha}}^{\ s}\Omega_{\underset{\sim}{\beta}t)}{}_{\triangle}^{\ \square}+\Omega_{(\underset{\sim}{\alpha}}^{\ s}{}_{\triangle}^{\ \times}\Omega_{\underset{\sim}{\beta}t)}{}_{\times}^{\ \square}$$
$$+\Omega_{(\underset{\sim}{\alpha}\underset{\sim}{\beta}t)}^{\ s\ \ \times}\Omega_{\times}{}_{\triangle}^{\ \square}-2i\delta^{s}_{t}\partial_{\underset{\sim}{\alpha}\underset{\sim}{\beta}}\delta_{\triangle}^{\ \square},$$

$$\text{where }\partial_{\underset{\sim}{\alpha}\underset{\sim}{\beta}}\equiv\phi^{\times}_{\underset{\sim}{\alpha}}\phi^{\otimes}_{\underset{\sim}{\beta}}D_{\times\otimes} \quad (4b)$$

which is required to be, from Eq. (1b),

$$\left\{\nabla_{\underset{\sim}{\alpha}}^{s},\ \nabla_{\underset{\sim}{\beta}t}\right\}_{\triangle}^{\ \square}=0-2i\delta^{s}_{t}\left(\partial_{\underset{\sim}{\alpha}\underset{\sim}{\beta}}\circ\delta_{\triangle}^{\ \square}+\Omega_{\underset{\sim}{\alpha}\underset{\sim}{\beta}}{}_{\triangle}^{\ \square}\right), \quad (4c)$$

Thus this relation defines special direction of the spin connection in terms of the spin connections in the super directions:

$$-2i\delta^{s}_{t}\Omega_{\underset{\sim}{\alpha}\underset{\sim}{\beta}}{}_{\triangle}^{\ \square}\equiv\partial_{(\underset{\sim}{\alpha}}^{\ s}\Omega_{\underset{\sim}{\beta}t)}{}_{\triangle}^{\ \square}+\Omega_{(\underset{\sim}{\alpha}}^{\ s}{}_{\triangle}^{\ \times}\Omega_{\underset{\sim}{\beta}t)}{}_{\times}^{\ \square}+\Omega_{(\underset{\sim}{\alpha}\underset{\sim}{\beta}t)}^{\ s\ \ \times}\Omega_{\times}{}_{\triangle}^{\ \square}. \quad (4d)$$

Now we shall find out the vielbein and spin connections in the super direction more explicitly. From Eq. (3c) we obtain in the limit of $\lambda=0,\ \infty$, respectively

$$\partial_{\underset{1}{s}}=[\phi(0,\lambda')D]_{\underset{1}{s}},\ \Omega_{\underset{1}{s}}{}_{\triangle}^{\ \square}=\phi^{\times}_{\triangle}(0,\lambda')[\partial_{\underset{1}{s}}\phi^{-1}{}^{\square}_{\times}(0,\lambda')], \quad (5a)$$

$$\partial_{\underset{2}{t}}=[\phi(\infty,\lambda')D]_{\underset{2}{t}},\ \Omega_{\underset{2}{t}}{}_{\triangle}^{\ \square}=\phi^{\times}_{\triangle}(\infty,\lambda')[\partial_{\underset{2}{t}}\phi^{-1}{}^{\square}_{\times}(\infty,\lambda')]. \quad (5b)$$

Note that these combined fields in $\phi(0,\lambda')$ and $\phi(\infty,\lambda')$ should be independent of λ'. Similarly for $\lambda'=0,\ \infty$ respectively we obtain

$$\partial_{\dot{1}s}=[\phi(\lambda,0)D]_{\dot{1}s},\ \Omega_{\dot{1}s}{}_{\triangle}^{\ \square}=\phi^{\times}_{\triangle}(\lambda,0)\left[\partial_{\dot{1}s}\phi^{-1}{}^{\square}_{\times}(\lambda,0)\right], \quad (5c)$$

$$\partial_{\dot{2}t}=[\phi(\lambda,\infty)D]_{\dot{2}t},\ \Omega_{\dot{2}t}{}_{\triangle}^{\ \square}=\phi^{\times}_{\triangle}(\lambda,\infty)\left[\partial_{\dot{2}s}\phi^{-1}{}^{\square}_{\times}(\lambda,\infty)\right] \quad (5d)$$

Here these combined fields of $\phi(\lambda,0)$ or $\phi(\lambda,\infty)$ should have no λ dependence.

Since all the vielbein and spin connections on the left hand sides of Eqs. (5) are independent of λ, λ', without losing generality we can express them in terms of group elements g, and h which are independent of λ, λ':

$$\partial_{\underset{1}{s}}=[gD]_{\underset{1}{s}},\quad \Omega_{\underset{1}{s}}{}_{\triangle}^{\ \square}=g^{\ \times}_{\triangle}[\partial_{\underset{1}{s}}\,g^{-1}{}_{\triangle}^{\ \square}]; \quad (6a)$$

$$\partial_{\underset{2}{t}}=[hD]_{\underset{2}{t}},\quad \Omega_{\underset{2}{t}}{}_{\triangle}^{\ \square}=h^{\ \times}_{\triangle}[\partial_{\underset{2}{t}}\,h^{-1}{}_{\times}^{\ \square}]; \quad (6b)$$

$$\partial_{\dot{1}s}=[gD]_{\dot{1}s},\quad \Omega_{\dot{1}s}{}_{\triangle}^{\ \square}=g^{\ \times}_{\triangle}\left[\partial_{\dot{1}s}\,g^{\ \square}_{\times}\right]; \quad (6c)$$

$$\partial_{\dot{2}t}=[hD]_{\dot{2}t},\quad \Omega_{\dot{2}t}{}_{\triangle}^{\ \square}=h^{\ \times}_{\triangle}\left[\partial_{\dot{2}s}\,h^{-1}{}_{\times}^{\ \square}\right]; \quad (6d)$$

where the third integrability condition of Eq. (1b) requires that the g's in Eq. (6a) and Eq. (6c) are the same, so are the h's in Eq. (6b) and Eq (6d).

Using $\phi^{\times}_{\triangle}(\lambda,\lambda')\left[\partial_A\phi^{-1}{}^{\square}_{\times}(\lambda,\lambda')\right]=-[\partial_A\phi^{\times}_{\triangle}(\lambda,\lambda')]\phi^{-1}{}^{\square}_{\times}(\lambda,\lambda')$ for $\Omega_{A}{}^{\square}_{\triangle}$ and multiplying $\phi^{\ \square}_{\times}(\lambda,\lambda')$ from both sides of Eq.(3c) from the right, we obtain the linear systems,

$$\nabla_{A}{}^{\times}_{\triangle}\phi^{\square}_{\times}(\lambda,\ \lambda')=\left(\partial_A\circ\delta^{\times}_{\triangle}+\Omega_{A}{}^{\times}_{\triangle}\right)\phi^{\square}_{\times}(\lambda,\lambda')=0, \quad (7)$$

or more explicitly

$$\nabla_{\underset{\sim}{\alpha}}^{s}{}_{\triangle}^{\ \times}\phi^{\ \square}_{\times}(\lambda,\ \lambda')=[\partial_{\underset{1}{s}}\circ\delta^{\ \times}_{\triangle}+\Omega_{\underset{1}{s}}{}_{\triangle}^{\ \times}+\lambda(\partial_{\underset{2}{s}}\circ\delta^{\ \times}_{\triangle}+\Omega_{\underset{2}{s}}{}_{\triangle}^{\ \times})]$$
$$\times\,\phi^{\ \square}_{\times}(\lambda,\ \lambda')=0, \quad (7a)$$

$$\nabla'_{\underset{\sim}{\beta}t}{}^{\times}_{\triangle}\,\phi^{\square}_{\times}(\lambda,\ \lambda')=[\partial_{\dot{1}t}\circ\delta^{\times}_{\triangle}+\Omega_{\dot{1}t}{}^{\times}_{\triangle}+\lambda'(\partial_{\dot{2}t}\circ\delta^{\times}_{\triangle}+\Omega_{\dot{2}t}{}^{\times}_{\triangle})]$$
$$\phi^{\square}_{\times}(\lambda,\ \lambda')=0. \quad (7b)$$

Their integrability conditions are precisely theose constraint equations Eq. (2a-2e) (which are second order nonlinear differential equations in the super directions for g and h), i.e., the existence conditions of $\phi(\lambda,\ \lambda')$ are precisely the conditions Eqs (1b) and (2a-e): the intregrability among the equations of Eq. (7a) for different s gives Eq. (2a,b); the intregrability among the equation of Eq. (7b) for different t give Eq. (2c,d). The integrability among equations between Eq. (7a) and Eq. (7b) gives

$$\{\nabla_{\underset{\sim}{\alpha}}^{s},\ \nabla_{\dot{\underset{\sim}{\beta}}t}\}\phi(\lambda,\ \lambda')=0, \quad (8)$$

which leads to

$$\partial_{(\underset{\alpha}{s}}\partial_{\dot{\beta}t)}{}_{\triangle}^{\square} + \Omega_{(\underset{\alpha}{s}}{}_{\triangle}^{\times}\Omega_{\dot{\beta}t)}{}_{\times}^{\square} + \Omega_{(\underset{\alpha}{s}\dot{\beta}t)}{}^{\times}\Omega x{}_{\triangle}^{\square}$$

$$= 2i\delta_t^s[\partial_{\alpha\dot{\beta}}\phi_{\triangle}^{\times}(\lambda,\ \lambda')]\phi^{-1}{}_{\times}^{\square}(\lambda,\ \lambda')]. \qquad (8a)$$

This implies that we can consistently define, from the linear systems (7a), (7b), a spin connection in the spacial direction

$$-2i\delta_s^t\Omega_{\alpha\dot{\beta}}{}_{\triangle}^{\square} \equiv Eq.\ (8a) \qquad (8b)$$

which also is exactly Eq. (4d). Through Eq. (8b), with this definition for $\Omega_{\alpha\dot{\beta}}{}_{\triangle}^{\square}$, we obtain a third linear system

$$\nabla_{\underset{\alpha}{s}\dot{\beta}t}{}_{\triangle}^{\times}\phi_{\times}^{\square}(\lambda,\ \lambda') = [\partial_{1\dot{1}} + \Omega_{1\dot{1}} + \lambda'(\partial_{1\dot{2}} + \nabla_{1\dot{2}})$$

$$+ \lambda(\partial_{2\dot{1}} + \Omega_{2\dot{1}}) + \lambda\lambda'(\partial_{2\dot{2}} + \Omega_{2\dot{2}})_{\triangle}^{\times}\phi_{\times}^{\square}(\lambda,\ \lambda') = 0 \qquad (7c)$$

Reversely, the third linear system serves the function of defining the spin connection in the spaical direction. However, I would like to emphasize that Eq. (7c) follows from Eq. (7a,b), and is not an independent condition.

Note that $\Omega_{\underset{1}{s}}$, $\Omega_{\dot{1}s}$ are of the same pure gauge, we can gauge them away. We shall call such gauge the J gauge. Obviously in the J gauge, the equations of motion, and the linear system are much simplified, with

$$J \equiv h^{-1}g, \qquad (9)$$

$$\partial'_{\underset{1}{s}} = D_{\underset{1}{s}},\ \Omega'_{\underset{1}{s}}{}_{\triangle}^{\square} = 0;$$

$$\partial'_{\underset{2}{t}} = [JD]_{\underset{2}{t}},\ \Omega'_{\underset{2}{t}}{}_{\triangle}^{\square} = J^{-1}{}_{\triangle}^{\times}\left[\partial'_{\underset{2}{t}}J_{\times}^{\square}\right]; \qquad (9a)$$

$$\partial'_{\dot{1}s} = D_{\dot{1}s},\ \Omega'_{\dot{1}s}{}_{\triangle}^{\square} = 0;$$

$$\partial'_{\dot{2}t} = [JD]_{\dot{2}t},\ \Omega'_{\dot{2}t}{}_{\triangle}^{\square} = J^{-1}{}_{\triangle}^{\times}\left[\partial'_{\dot{2}t}J_{\times}^{\square}\right]; \qquad (9b)$$

The linear systems Eqs. (7a, b) become, with

$$\chi'(\lambda,\ \lambda') \equiv g^{-1}\phi(\lambda,\ \lambda'), \qquad (10)$$

$$\nabla'_{\underset{\alpha}{s}}{}_{\triangle}^{\times}\chi_{\times}^{\square}(\lambda,\lambda')$$

$$= \left[\partial'_{\underset{1}{s}}\delta_{\triangle}^{\times} + \lambda\left(\partial'_{\underset{2}{s}}\delta_{\triangle}^{\times} + \Omega'_{\underset{2}{s}}{}_{\triangle}^{\times}\right)\right]\chi_{\times}^{\square}(\lambda,\lambda') = 0, \qquad (11a)$$

$$\nabla'_{\dot{\beta}t}{}_{\triangle}^{\times}\chi_{\times}^{\square}(\lambda,\lambda')$$

$$= \left[\partial'_{\dot{1}t}\delta_{\triangle}^{\times} + \lambda'(\partial'_{\dot{2}t}\delta_{\triangle}^{\times} + \Omega'_{\dot{2}t}{}_{\triangle}^{\times})\right]\chi_{\times}^{\square}(\lambda,\lambda') = 0. \qquad (11b)$$

Because now $\Omega'_{\underset{1}{s}} = 0$, $\Omega'_{\dot{1}t} = 0$, the equations obtained from the integrability are much simplified:

$$\partial'_{\underset{1}{s}}\Omega'_{\underset{2}{t}}{}_{\triangle}^{\square} + \Omega'_{\underset{2}{t}}{}_{\underset{1}{s}}^{\overset{u}{2}}\Omega'_{\underset{2}{u}}{}_{\triangle}^{\square} + (s \leftrightarrow t) = 0, \qquad (12a)$$

from integrability among the equations of Eq. (11a) with different s; and

$$\partial'_{\dot{1}s}\Omega'_{\dot{2}t}{}_{\triangle}^{\square} + \Omega_{\dot{2}t}{}_{\dot{1}s}^{\dot{2}u}\Omega_{\dot{2}u}{}_{\triangle}^{\square} + (s \leftrightarrow t) = 0, \qquad (12b)$$

from integrability among the equations of Eq. (11b) with different t. We can see that the nonlinear equations, Eqs. (12a,b), in the J-gauge is much simplified in comparing to those given by Eqs. (2a,b) in the general gauge.[7]

The integrability conditions between Eqs. (11a) and (11b),

$$\{\nabla'_{\underset{\alpha}{s}}, \overline{\nabla}'_{\dot{\beta}t}\}_{\triangle}^{\times}\chi_{\times}^{\square}(\lambda,\ \lambda') = 0, \qquad (13)$$

gives

$$[\partial'_{(\underset{\alpha}{s}}\Omega'_{\dot{\beta}t)}{}_{\triangle}^{\times} + \Omega'_{(\underset{\alpha}{s}}{}_{\triangle}^{\otimes}\Omega'_{\dot{\beta}t)}{}_{\otimes}^{\times} + \Omega'_{(\underset{\alpha}{s}\dot{\beta}t)}{}^{\otimes}\Omega'_{\otimes}{}_{\triangle}^{\times}$$

$$- 2i\delta_t^s\partial'_{\alpha\dot{\beta}} \circ \delta_{\triangle}^{\times}]\chi_{\times}^{\square}(\lambda,\ \lambda') = 0, \qquad (13a)$$

where $\partial'_{\alpha\dot{\beta}} = \chi_{\alpha}^{\times}\chi_{\dot{\beta}}^{\otimes}\partial_{\times\otimes}$, or from Eq. (9a-b),

$$\partial_{1\dot{1}} = D_{1\dot{1}},\ \partial'_{1\dot{2}} = J_{\dot{2}}^{\times}D_{1\times},\ \partial_{2\dot{1}} = J_2^{\times}D_{\times\dot{1}},\ \partial_{2\dot{2}} = J_2^{\times}J_{\dot{2}}^{\otimes}D_{\times\otimes} \qquad (13b)$$

by rewriting the equation in the following form

$$\partial'_{(\underset{\alpha}{s}}\Omega'_{\dot{\beta}t)}{}_{\triangle}^{\square} + \Omega'_{(\alpha}{}_{\triangle}^{s\ \square}\Omega'_{\dot{\beta}t)}{}_{\times}^{\square} + \Omega'_{(\underset{\alpha}{s}\dot{\beta}t)}{}^{\times}\Omega'_{\times}{}_{\triangle}^{\square}$$

$$= 2i\delta_t^s\left[\partial'_{\alpha\dot{\beta}}\chi_{\triangle}^{\times}(\lambda,\ \lambda')\right]\chi^{-1}{}_{\times}^{\square}(\lambda,\ \lambda'). \qquad (13c)$$

we see that the linear systems guarantees the left hand side is proportional to δ_t^s and has the correct λ, λ' dependence. Therefore we can define it to be

$$-2i\delta_t^s\Omega'_{\alpha\dot{\beta}}{}_{\triangle}^{\square} = Eq.\ (13c), \qquad (14)$$

i.e. more explicitly,

$$\Omega'_{1\dot{1}}{}_{\triangle}^{\square} = 0 = -[\partial_{1\dot{1}}\chi_{\triangle}^{\times}(0,0)]\chi^{-1}{}_{\times}^{\square}(0,0) \qquad (14a)$$

$$\Omega'_{2\dot{2}}{}_{\triangle}^{\square} = \frac{-1}{2i\delta_t^s}[\partial'_{(\underset{2}{s}}\Omega'_{\dot{2}t)}{}_{\triangle}^{\square} + \Omega'_{(\underset{2}{s}}{}_{\triangle}^{\times}\Omega'_{\dot{2}t)}{}_{\times}^{\square} + \Omega'_{(\underset{2}{s}\dot{2}t)}{}^{\dot{2}u}\Omega'_{\dot{2}u}{}_{\triangle}^{\square}]$$

$$= -\left[\partial_{2\dot{2}}\chi_{\triangle}^{\times}(\infty,\ \infty)\right]\chi^{-1}{}_{\times}^{\square}(\infty,\ \infty) \qquad (14b)$$

$$\Omega'_{1\dot{2}}{}_{\triangle}^{\square} = \frac{-1}{2i\delta_t^s}\left[\partial'_{\underset{1}{s}}\Omega'_{\dot{2}t}{}_{\triangle}^{\square} + \Omega'_{\dot{2}t}{}_{\underset{1}{s}}^{\dot{2}u}\Omega'_{\dot{2}u}{}_{\triangle}^{\square}\right]$$

$$= -\left[\partial'_{1\dot{2}}\chi_{\triangle}^{\times}(\infty,\ \infty)\right]\chi^{-1}{}_{\times}^{\square}(\infty,\ \infty) \qquad (14c)$$

$$\Omega'_{2\dot{1}}{}_{\triangle}^{\square} = \frac{-1}{2i\delta_t^s}\left[\partial'_{\dot{1}t}\Omega'_{\underset{2}{s}}{}_{\triangle}^{\square} + \Omega'_{\underset{2}{s}}{}_{\dot{1}t}^{\dot{2}u}\Omega'_{\dot{2}u}{}_{\triangle}^{\square}\right]$$

$$= -\left[\partial'_{2\dot{1}}\chi_{\triangle}^{\times}(\infty,\ 0)\right]\chi^{-1}{}_{\times}^{\square}(\infty,\ 0) \qquad (14d)$$

Therefore the compatibility between the two sets of linear systems Eq. (11a) and Eq. (11b) lead to consistent definition of the spin connections in the spacial direction, as given by Eqs. (14). All these can be summarized into a "third" linear equation, which is a consequence of the linear systems (11a), (11b):

$$\nabla'_{\alpha\dot{\beta}}{}_{\triangle}^{\times}\chi_{\times}^{\square}(\lambda,\lambda')$$

$$= (\partial'_{1\dot{1}} \circ \delta + \lambda'\nabla'_{1\dot{2}} + \lambda\nabla'_{2\dot{1}} + \lambda'\lambda\nabla'_{2\dot{2}})_{\triangle}^{\times}\chi_{\times}^{\square}(\lambda,\lambda') = 0 \qquad (15c)$$

with $\Omega'_{\alpha\dot{\beta}}$ given by Eqs. (14a-d).

So the strategy of obtaining the spacial spinor connections $\Omega_{\alpha\dot{\beta}}$ and the spacial vielbeined derivatives $\partial_{\alpha\dot{\beta}}$ which satisfy the gravitational equation of motion derived in Ref. (1) is to solve for J, which satisfy the second order nonlinear differential equations for J, Eqs. (12a-b), via the linear system (11a), (11b), and then use Eq. (13b) and Eq. (14a-d).

From (11a) and (11b) we derive the following non-local conservation laws by expanding in the parameters λ or λ' near zero respectively,

$$\partial'_{\underset{1}{s}}\, \chi^{(n,\ )}(0,\lambda') = -\nabla'_{\underset{2}{s}}\chi^{(n-1,\ )}(0,\lambda'), \qquad (16a)$$

$$\partial'_{1t}\chi^{(\ ,n)}(\lambda,0) = -\nabla'_{\dot{2}t}\chi^{(\ ,n-1)}(\lambda,0), \qquad (16b)$$

and a third set of nonlocal conservation laws by expanding in both λ, λ' using Eq. (11c):

$$\partial'_{1\dot{1}}\, \chi^{(n,n)} + \nabla'_{2\dot{2}}\, \chi^{(n-1,n-1)} + \nabla'_{1\dot{2}}\, \chi^{(n,n-1)} + \nabla'_{2\dot{1}}\, \chi^{n-1,n} = 0, \qquad (16c)$$

where $\chi^{(m,n)} \equiv \chi^{(m,n)}(0,0)$ (for simplicity we have dropped the indices ◽, △, ×, etc). The conservation laws can also be derived for the previous linear systems (7a, b, c) in the general gauge by expanding in λ, λ', and both λ, λ' respectively. However they appear more complicated.

In conclusion, here I have shown that the gravitational equations from extended four plus extended superspace light-like integrability possess the powerful linear systems and infinite non-local conservation laws. This puts gravitational fields on a very similar footing to the other integrable systems. And the most surprising overall picture is that major equations in physics[6] possess such geometrical-integrable properties.

Acknowledgment: I would like to thank J. Gates for many enlightening and helpful discussions. My appreciation goes to Madelin Cameron for her artistry and expertise in typing this manuscript, and to the staff members of the Physics Department at U.C. Davis for their support, which make this work possible.

References

1. L.-L. Chau and C.-S. Lim, Phys. Rev. Lett. 56 294 (1986); and L.-L. Chau "Geometrical Integrability and Equations of Motion in Physics: An Unifying View." To appear in Proceedings of Seminaire de Mathematiques, Montreal, July 29 - Aug. 16, 1985.
2. L.-L. Chau, "Linear Systems and Conservation Laws of Gravitational Fields in Four Plus Extended Super Space", U.C. Davis preprint, 1986.
3. E. Witten, Phys. Lett. 77B 394 (1978); M. Sohnius, Nucl. Phys. B136, 461 (1978); L.-L. Chau, M.-L. Ge, C.-S. Lim, Phys. Rev. D33 1056 (1986).
4. For a review, see L.-L. Chau, in Nonlinear Phenomena, Lecture Notes in Physics, Vol. 189, e.d K.B. Wolf (Springer-Verlag, New York, 1983).
5. For super Yang-Mills fields, see I.V. Volovlich, Phys. Lett. 129B, 429 (1983); Teor. Mat. Fiz. 54, 39 (1983); C. Devchand, "An Infinite Number of Continuity Equations and Hidden Symmetries in Supersymmetric Gauge Theories", Nucl. Phys. B238, 333 (1984). L.-L. Chau, M.-L. Ge, anld Z. Popowicz, "Riemann-Hilbert Transforms and Bianchi-Băcklund Transformations for the Supersymmetric Yang-Mills Fields," Phys. Rev. Lett. 52, 1940 (1984). L.-L. Chau, "Supersymmetric Yang-Mills Fields as an Integrable System and Connections with Other Linear Systems", Proceedings of the Workshop on Vertex Operations in Mathematics and Physics, Berkeley, 1983, edited by J. Lepowsky, S. Mandelstam, and I.M. Singer (Springer, New York, 1984).
6. E. Witten in Ref. (2); J. Isenberg, P. Yasskin, P.J. Green, Phys. Lett. 78B 462 (1978); A. Ferber, Nucl. Phys. B132 55 (1978); and Crispim-Romão, A. Ferber, and P.G.O. Freund, Nucl. Phys. B182, 45 (1981).
7. This is equivalent to the zero gauge discussed in L.-L. Chau, B.-Y. Hou, Phys. Lett. 145B, 347 (1984).

NEW PROPERTIES OF UNIDEXTEROUS SUPERSYMMETRIC THEORIES

S. James Gates, Jr.
Department of Physics and Astronomy, University of Maryland
College Park, MD 20742

Unidexterous (one-handed) supersymmetry has been found to play an important role in the formulation of heterotic strings [1]. This type of supersymmetry was first discussed by Sakamoto [2] and was later studied [3] in the context of σ-model formulations of the heterotic string. In a previous work [4] the first complete superspace formulation of these theories was developed. (Similar results have also been found by other authors [5].) However, at this time little work [6] has been carried out on the superfield quantization of these theories. Within this note we want report on progress in this area.

II. QUANTIZATION OF SIMPLE UNIDEXTEROUS MATTER

There are two forms of simple unimanual matter [2]. These two supermultiplets correspond to scalar and spinor superfields. We denote these by the symbols $X(\zeta^{+},\tau,\sigma)$ and $\eta_{-}(\zeta^{+},\tau,\sigma)$ where ζ^{+} is a single Grassmannian coordinate, while τ and σ are timelike and spacelike coordinates, respectively. (We also use the notation $\sigma^{++} \equiv \tau + \sigma$ and $\sigma^{--} \equiv -\tau + \sigma$.) As usual in supersymmetrical theories, superfields can be expanded in terms of component fields

$$X(z) = x(\tau,\sigma) + \zeta^{+}\psi_{+}(\tau,\sigma) , \qquad (2.1)$$

$$\eta_{-}(z) = \lambda_{-}(\tau,\sigma) + i\zeta^{+}F(\tau,\sigma) . \qquad (2.2)$$

The actions for the free scalar and spinor superfields are given by

$$S(X) = \int d^2\sigma \, d\zeta^{-}[i \tfrac{1}{2}(D_{+}X)(\partial_{--}X)] , \qquad (2.3)$$

$$S(\eta_{-}) = \int d^2\sigma \, d\zeta^{-}[- \tfrac{1}{2} \eta_{-}D_{+}\eta_{-}] . \qquad (2.4)$$

These superfields may be quantized by use of a path integral formalism. Concentrating first on the scalar multiplet, a generating function for connected Green's functions is defined by

$$\exp[iW(J)] \equiv$$
$$\int[\mathcal{D} X]\exp[iS(X) + \int d^2\sigma d\zeta^{-}XJ_{-}] , \qquad (2.5)$$

where J_{-} is a spinorial source function. Following the standard procedure then leads to

$$W(J) = \tfrac{1}{2} \int d^2\sigma \, d\zeta^{-}(J_{-} \frac{D_{+}}{\Box} J_{-}) , \qquad (2.6)$$

with $\Box \equiv \partial_{++}\partial_{--}$. In order to obtain a propagator, functional differentiation in unidexterous superspace must be defined. This in turn depends on the existence of a superdelta function, $\delta_{-}^{5/2}(z-z')$, where

$$\delta_{-}^{5/2}(z-z') =$$
$$(\zeta-\zeta')^{+}\delta(\tau-\tau')\delta(\sigma-\sigma') . \qquad (2.7)$$

This definition implies the following properties

$$\int d^{5/2}z^{-}\delta_{-}^{5/2}(z-z')f(z) = f(z') ,$$

$$[\delta_{-}^{5/2}(z-z')]^{*} = \delta_{-}^{5/2}(z-z') , \qquad (2.8)$$

in addition to the fact that the unimanual superdelta function is spinorial. Due to the usual relation between spin and statistics, the superdelta function transforms as a spinor. This is a problem since a quantity like $\delta J_{-}(z)/\delta J_{-}(z')$ should be a Lorentz scalar. Since it is desirable to keep Lorentz covariance manifest at each step of quantization, we define

$$\frac{\delta J_{-}(z)}{\delta J_{-}(z')} \equiv \frac{D_{+}}{\Box} \delta_{-}^{5/2}(z-z') , \qquad (2.9)$$

and thus

$$\int d^{5/2}z^{-}f(z) \frac{\delta J_{-}(z)}{\delta J_{-}(z')} =$$
$$- \frac{D'_{+}}{\Box} f(z') . \qquad (2.10)$$

On calculating a functional derivative of

(2.5) we find

$$\frac{\delta W(J)}{\delta J_-(z')} \equiv ie^{-iW}\int\int[\mathcal{D}X][\frac{D_+X}{\Box}]\exp[iS(X) + \int d^{5/2}z^- XJ_-] \ . \qquad (2.11)$$

Next by taking the limit as J_- approaches zero, the rhs above is seen to be proportional to $\langle 0|T[\frac{D_+X}{\Box}]|0\rangle$. But the functional derivative should be proportional to $\langle 0|T[X]|0\rangle$. To achieve this we alter (2.5) to read

$$\exp[iW(J)] \equiv \int\int[\mathcal{D}X]\exp[iS(X) + \int d^{5/2}z^- X\partial_{--}D_+J] \ , \qquad (2.12)$$

and the standard evaluation procedure yields

$$W(J) = i\,\tfrac{1}{2}\int d^{5/2}z^-(J\partial_{--}D_+J) \ . \qquad (2.13)$$

The functional derivative of W(J) taken in the limit of vanishing source now corresponds to

$$\lim_{J\to 0}\frac{\delta W(J)}{i\delta J(z)} = \langle 0|T[X(z)]|0\rangle \ . \qquad (2.14)$$

The scalar superfield propagator, $\Delta_F(z-z')$, is then obtained as

$$\lim_{J\to 0}\frac{\delta^2 W(J)}{i\delta J(z')i\delta J(z)} = i\,\frac{D_+}{\Box}\,\delta_-^{5/2}(z-z') \ , \qquad (2.15)$$

$$\equiv i\Delta_F(z-z') \ . \qquad (2.16)$$

Having worked out the peculiar features (2.9, 2.12) of unidexterous superspace, the quantization of the spinor multiplet proceeds along similar lines. The generating functional is defined by

$$\exp[iW(J_+)] \equiv \int\int[\mathcal{D}\eta_-]\exp[iS(\eta_-) + \int d^{5/2}z^-\eta_-\partial_{--}D_+J_+] \ , \qquad (2.17)$$

and evaluation of the Gaussian integral yields

$$W(J_+) = \tfrac{1}{2}\int d^{5/2}z^- J_+(\partial_{--})^2 D_+J_+ \ . \qquad (2.18)$$

This leads to superpropagator $S_F(z-z')$ given by

$$S_F(z-z') = -i\,\frac{\partial_{--}D_+}{\Box}\,\delta_-^{5/2}(z-z') \ . \qquad (2.19)$$

The most general interacting matter theory can be written as $S_T = S(X) + S(\eta_-) + S_{Int}(X,\eta_-)$ where

$$S_{Int} = \int d^{5/2}z^- \mathcal{L}_{Int-}(X,\eta_-) \ . \qquad (2.20)$$

The generating functional for the interacting theory thus takes the following form

$$\exp[iW(J_-J_+)] = \exp[iS_{Int}(\frac{1}{i}\frac{\delta}{\delta J}, \frac{1}{i}\frac{\delta}{\delta J_+})]\exp[i(W(J) + W(J_+))] \ . \qquad (2.21)$$

where $W(J)$ and $W(J_+)$ are defined in (2.14) and (2.18). By use of a superfunctional Legendre transform, the effective action $\Gamma(X,\eta_-)$ is defined

$$\Gamma(X,\eta_-) = W(J,J_+) + i\int d^{5/2}z^-[X\partial_{--}D_+J + \eta_-\partial_{--}D_+J_+] \qquad (2.22)$$

in terms of $W(J,J_+)$.

III.) Quantization of Gauge Superfields

Simple unidexterous supersymmetric theories are similar to other supersymmetrical theories, there exist supermultiplet that extend Yang-Mills and d = 2 gravity theories. Of course, these multiplets possess no dynamical degrees of freedom. But the structure of the ghosts associated with the gauge-fixing of these theories is of some interest. The structure of the classical theory is specified by giving a set of constraints on a covariant derivative $\nabla_A \equiv (\nabla_+,\nabla_{++}\nabla_{--})$ and $\nabla_A = E_A{}^M D_M + \omega_A{}^M - ig\Gamma_A{}^{\hat{I}} t_{\hat{I}}$ where $E_A{}^M$, ω_A , and $\Gamma_A^{\hat{I}}$ are vielbein, spin-connection, and gauge-connection superfields, respectively.

$$[\nabla_+,\nabla_+\} = i2\nabla_{++} \ , \quad [\nabla_+,\nabla_{++}\} = 0 \ ,$$

$$[\nabla_+,\nabla_{--}\} = -i2\Sigma^+ M - igW_-{}^{\hat{I}}t_{\hat{I}} \ ,$$

$$[\nabla_{++},\nabla_{--}\} = -[\Sigma^+\nabla_+ + \mathcal{R}M - ig\mathcal{F}^{\hat{I}}t_{\hat{I}}] \ . \qquad (3.1)$$

The superfields $W_-{}^{\hat{I}}$ and Σ^+ are Yang-Mills and supergravity field strengths. The constraints are solved in terms of $\Gamma_+{}^{\hat{I}}$ and

$\Gamma_{--}{}^{\hat{I}}$, (Yang-Mills) and $H_{+}{}^{=}$, $H_{=}{}^{\neq}$, H^{+}, L, and Ψ (supergravity).

The quantization of the Yang-Mills multiplet incorporates the features of the matter quantization into the standard procedure. The generating function takes the form

$$\exp[iW(J_{-},J_{++})] = \iint[\Pi_{\phi} \mathcal{D}\phi]\exp[i\, S_{eff}] ,$$

$$S_{eff} = S_{YM} + S_{GF} + S_{FP} + S_{J} ,$$

$$S_{YM} = \int d^{5/2}z^{-}[-i\, \tfrac{1}{2}\, Tr(W_{-}\mathcal{F})] ,$$

$$S_{GF} = -\tfrac{1}{2}\,\alpha\int d^{5/2}z^{-}[(D_{+}\beta_{-}{}^{\hat{I}})(\beta_{-}{}^{\hat{I}} + \frac{1}{\lambda_1}(D_{+}\Gamma_{--}{}^{\hat{I}}) - \frac{1}{\lambda_2}(\partial_{--}\Gamma_{+}{}^{\hat{I}}))] ,$$

$$S_{FP} = \tfrac{1}{2}\,\alpha\int d^{5/2}z^{-} C'^{\hat{I}}[\frac{1}{\lambda_1}(D_{+}\nabla_{--}C^{\hat{I}}) - \frac{1}{\lambda_2}(\partial_{--}\nabla_{+}C^{\hat{I}})] ,$$

$$S_{J} = -\int d^{5/2}z^{-}[\Gamma_{+}{}^{\hat{I}}\nabla_{--}\nabla_{+}J_{-}{}^{\hat{I}} - \Gamma_{--}{}^{\hat{I}}\nabla_{--}\nabla_{+}J_{++}{}^{\hat{I}}] \qquad (3.2)$$

where $C^{\hat{I}}$, $C'^{\hat{I}}$ and $\beta_{-}{}^{\hat{I}}$ correspond to ghost, anti-ghost and BRST auxiliary superfields, respectively. Considering only the quadratic part of S_{eff} gives

$$S_{(2)} = \int d^{5/2}z^{-}\{i\, \tfrac{1}{2}(1+\frac{\alpha}{4\lambda_1^2})\, \Gamma_{--}{}^{\hat{I}}D_{+}\partial_{++}\Gamma_{--}{}^{\hat{I}} - i\,(1+\frac{\alpha}{4\lambda_1\lambda_2})\,\Gamma_{--}{}^{\hat{I}}\,\Gamma_{+}{}^{\hat{I}} + \tfrac{1}{2}\,(1+\frac{\alpha}{4\lambda_2^2})\,\Gamma_{+}{}^{\hat{I}}(\partial_{--})^2D_{+}\Gamma_{+}{}^{\hat{I}} + \tfrac{1}{2}\,\alpha\,(\frac{\lambda_2-\lambda_1}{\lambda_1\lambda_2})\, C'^{\hat{I}}\partial_{--}D_{+}C^{\hat{I}}\} . \qquad (3.3)$$

A variety of gauge choices are possible (including asymmetrical ones $\lambda_1 \neq \lambda_2$). The simplest choice results from $\alpha = \lambda_1 = -\lambda_2 = -\frac{1}{4}$ which lead to the following nontrivial propagators

$$\langle 0|T[\Gamma_{+}{}^{\hat{I}}(z)\Gamma_{--}{}^{\hat{J}}(z')]|0\rangle = -\tfrac{1}{2}\,\delta^{\hat{I}\hat{J}}\,\frac{1}{\Box}\,\delta_{-}^{5/2}(z-z') ,$$

$$\langle 0|T[C^{\hat{I}}(z)C'^{\hat{J}}(z')]|0\rangle = \delta^{\hat{I}\hat{J}}\,\frac{D_{+}}{\Box}\,\delta_{-}^{5/2}(z-z') . \qquad (3.4)$$

The quantized version of this multiplet exhibits an interesting behavior with respect to BRST and anti-BRST symmetries [7]. Denoting generators of BRST and anti-BRST symmetries by s and $\bar{s}$, these may be realized

$$s\Gamma_{A}{}^{\hat{I}} = -i(-)^{A}\nabla_{A}C^{\hat{I}},\quad \bar{s}\Gamma_{A}{}^{\hat{I}} = -i(-)^{A}\nabla_{A}C'^{\hat{I}},$$

$$sC^{\hat{I}} = i\,\tfrac{1}{2}\,gf^{\hat{I}}{}_{\hat{J}\hat{K}}\,C^{\hat{J}}C^{\hat{K}} ,$$

$$\bar{s}C^{\hat{I}} = i\nabla_{+}\beta_{-}{}^{\hat{I}} + igf^{\hat{I}}{}_{\hat{J}\hat{K}}\,C^{\hat{J}}C'^{\tilde{K}} ,$$

$$sC'^{\hat{I}} = iD_{+}\beta_{-}{}^{\hat{I}},\quad \bar{s}C'^{\hat{I}} = i\tfrac{1}{2}gf^{\hat{I}}{}_{\hat{J}\hat{K}}\,C'^{\hat{J}}C'^{\hat{K}},$$

$$s\beta_{-}{}^{\hat{I}} = 0 ,\quad \bar{s}\beta_{-}{}^{\hat{I}} = igf^{\hat{I}}{}_{\hat{J}\hat{K}}\,C'^{\hat{J}}\beta_{-}{}^{\hat{K}} . \qquad (3.5)$$

These satisfy the relations $\{s,s\} = \{\bar{s},\bar{s}\} = 0$ and imply

$$S_{GF} + S_{FP} = i\,\tfrac{1}{2}\,\alpha\int d^{5/2}z^{-}\; s[C'^{\hat{I}}(\beta_{-}{}^{\hat{I}} + \frac{1}{\lambda_1}(D_{+}\Gamma_{--}{}^{\hat{I}}) - \frac{1}{\lambda_2}(\partial_{--}\Gamma_{+}{}^{\hat{I}}))] . \qquad (3.6)$$

Thus, the effective action possesses BRST invariance. But it is <u>not</u> anti-BRST invariant. Furthermore, the anticommutator of s and $\bar{s}$ takes the form

$$[s,\bar{s}\}\,\Gamma_{A}{}^{\hat{I}} = -g\nabla_{A}(f^{\hat{I}}{}_{\hat{J}\hat{K}}\,\Gamma_{+}{}^{\hat{J}}\beta_{-}{}^{\hat{K}}) ,$$

$$[s,\bar{s}\}\,C^{\hat{I}} = -ig\nabla_{s}(f^{\hat{I}}{}_{\hat{J}\hat{K}}\,\Gamma_{+}{}^{\hat{J}}\beta_{-}{}^{\hat{K}}) ,$$

$$[s,\bar{s}\}\,C'^{\hat{I}} = -ig\nabla_{\bar{s}}(f^{\hat{I}}{}_{\hat{J}\hat{K}}\,\Gamma_{+}{}^{\hat{J}}\beta_{-}{}^{\hat{K}}) ,$$

$$[s,\bar{s}\}\,\beta_{-}{}^{\hat{I}} = -\tfrac{1}{2}\,D_{+}(f^{\hat{I}}{}_{\hat{J}\hat{K}}\,\beta_{-}{}^{\hat{J}}\beta_{-}{}^{\hat{K}}) , \qquad (3.7)$$

where $\nabla_{s} = s - gC^{\hat{I}}t_{\hat{I}}$ and $\nabla_{\bar{s}} = \bar{s} - gC'^{\hat{I}}t_{\hat{I}}$.

This is not the usual relationship for the realization of BRST and anti-BRST generators. This behavior clearly requires further study.

Coupling the global matter multiplets to supergravity is accomplished by the usual minimal coupling procedure,

$$D_A \to \nabla_A \ , \quad \int d^{5/2}z^- \to \int d^{5/2}z^- E^{-1} \ . \quad (3.8)$$

Therefore the local versions of (2.3) and (2.4) become

$$S_L = \int d^{5/2}z^- E^{-1}[i \tfrac{1}{2} (\nabla_+ X)(\nabla_{--} X) - \tfrac{1}{2} (\eta_- \nabla_+ \eta_-)] \ . \quad (3.9)$$

The supergravity field strength Σ^+ when integrated yields the number of holes minus the number of handles in the two dimensional τ-σ manifold,

$$\int d^{5/2}z^- E^- \Sigma^+ = - \tfrac{1}{2} \int d^2\sigma e^{-1} r(\omega(e,\psi)) \ . \quad (3.10)$$

The supergravity superfields H^+ and L may be "gauged away" algebraically. The action in (3.9) may be quantized leading to a partition function

$$Z = \int\int [\prod_\phi \mathcal{D} \phi] \exp[iS_L] \quad (3.11)$$

where now the product $\prod_\phi$ includes the supergravity superfields $H_+{}^=$, $H_={}^{\neq}$ and Ψ. Such a quantization requires gauge fixing of these superfields. The most convenient and familiar gauge choice is $H_+{}^= = H_={}^{\neq} = 0$. Following the usual Fadeev-Popov procedure this gauge choice leads to a dynamical ghost action that eliminate the redundancy in the measure [$H_+{}^=$][$H_={}^{\neq}$] . The explicit form this ghost action is

$$S_{Ghost} = -\int d^{5/2}z^- E^{-1}[B_+{}^= \nabla_= C_= + B_={}^{\neq} \nabla_+ C_{\neq}] \ , \quad (3.12)$$

where E^{-1}, $\nabla_=$ and ∇_+ may correspond to any choice of gauges for the supergravity fields and may be explicitly calculated by substituting $H_+{}^= = H_={}^{\neq} = 0$ into the results in ref. [4]. The action $S_L + S_{Ghost}$ represents the partition function for the heterotic string when the number of scalar and spinor multiplets is ten and sixteen respectively. (The resulting expression is free of local anomalies but not global ones, i.e. supermodili have not been introduced.)

References

[1.] D.J. Gross, J. Harvey, E. Martinec and R. Rohm, Nucl. Phys. B256 (1985) 253.

[2.] M. Sakamoto, Phys. Lett. 151B (1985) 115.

[3.] C.M. Hull and E. Witten, Phys. Lett. 160B (1985) 398.

[4.] R. Brooks, F. Muhammad, and S.J. Gates, Nucl. Phys. B268 (1986) 599.

[5.] G. Moore, P. Nelson, and J. Polchinski, Phys. Lett. 169B (1986) 53; M. Evans and B. Ovrut, Phys. Lett. 171B (1986) 177.

[6.] C.M. Hull and P.K. Townsend, DAMTP preprint, March, 1986.

[7.] C. Becchi, A. Rouet, and R. Stora, Comm. Math. Phys. 42, (1975) 127; ibid. Ann. of Phys. 98 (1976) 98; I.V. Tyukin, Int. Report FIAN 39 (1975); G. Curci and S. Ferrara, Phys. Lett. 63B (1976) 51; ibid. Nuovo Cim. 32A (1976) 151; ibid. 47A (1978) 555; I. Ojima, Prog. Theor. Phys. 64 (1980) 625.

NONCOVARIANT SUPERGAUGES

Wolfgang Kummer

Institut für Theoretische Physik, Technische Universität Wien
A-1040 Wien, Austria

The powerful methods of superfields and superspace are extended to general linear gauges in which supersymmetry is not manifest. Such a program is motivated by the fact that 1) the renormalization program for nonabelian super-Yang-Mills fields is well-known to be ill-defined without breaking of supersymmetry in the gauge; 2) it is desireable also to implement "physical" gauges like the Wess-Zumino gauge, into a superfield formalism.

1. INTRODUCTION

Somewhat surprisingly, more than 10 years after the discovery of supersymmetric gauge theories [1] the proper quantum field theoretical treatment of those theories in a manifestly supersymmetric way is still in the process of elaboration. This is related to the fact that the ordinary vector gauge field becomes a member of a dimensionless scalar superfield $V = V^+$. As an immediate consequence in a supersymmetrically covariant gauge the dimensionless scalar component of V exhibits a new type of infrared singularity in its propagator, yielding illdefined Green's functions. Even when this singularity is made to vanish in one-loop order in a certain gauge, it reappears in higher orders [2]. Thus in the general treatment of the renormalization problem one cannot evade the choice of a gauge which breaks supersymmetry, at least softly [3]. This suggests the study of general noncovariant supergauges which, on the other hand, allow the inclusion of celebrated gauges like the Wess-Zumino gauge in a supergraph formalism [4,5,6].

2. GAUGE FIXING

The supersymmetric Yang-Mills Lagrangian (our conventions for the covariant derivatives of supersymmetry follow from $\{D_\alpha, \bar{D}_{\dot\alpha}\} = 2i\sigma^n_{\alpha\dot\alpha}\partial_n = 2i\not\partial_{\alpha\dot\alpha}$)

$$L = \frac{\rho}{16g^2}\,\mathrm{Tr}\int d^6x_+\, W^\alpha W_\alpha \qquad W_\alpha = \bar{D}^2 e^{-gV} D_\alpha e^{gV} \tag{1}$$

has the chiral gauge-invariance

$$e^{gV'} = e^{-i\Lambda_+^+}\, e^{gV}\, e^{i\Lambda_+} \tag{2}$$

Inhomogeneous covariant gauges are characterized by a gauge breaking Lagrangian

$$L_{gb} = \frac{\rho}{16\alpha}\int d^8x\,(D^2V)(\bar{D}^2V) = \frac{\rho}{2\alpha}\int d^8x\, V(1-P_T) \tag{3}$$

$(P_+ = -(16\Box)^{-1}\bar{D}^2D^2, P_- = -(16\Box)^{-1}D^2\bar{D}^2,$
$P_+ + P_- + P_T = P_L + P_T = 1,\ P^2 = P)$

For $\alpha = 1$ this produces

$$L^{(2)} = \frac{\rho}{2}\int d^8x\, V\Box V \tag{4}$$

Unfortunately, the $1/k^4$-singularity reappears in higher loop orders in such a gauge too [2]. The fact that some sort of supersymmetry breaking cannot be evaded in the treatment of quantized superfields considerably complicates the general proofs of renormalization and gauge-independence of physical quantities of the superfield formalism so that only recently a solution with "minimal" breaking has been found [7]. Another practical argument for a study of general concovariant supergauges within the context of superfields is the wish to incorporate gauges like the Wess-Zumino gauge in the supergraph formalism as well. Therefore, a general gauge-breaking Lagrangian

$$L_{gb} = \mathrm{Tr}\int d^8x\,\{(BK + \bar{B}\bar{K})V - \alpha B\bar{B}\} \tag{5}$$

is considered, depending on a nonsupersymmetric operator K ($\partial_m K = 0$) and on an auxiliary chiral field B. The field equation for the B-field may be used to eliminate it from (5)

$$L_{gb} = \alpha^{-1}\int d^8x\,(K\_V)(\bar{K}\_V) \tag{6}$$

With the special covariant choice $K_- \propto \bar{D}^2$ the (inhomogeneous) gauge breaking (4) can be reproduced. In the following we consider the special case of homogeneous gauges 6 $\alpha = 0$ in (5). This represents still a large gauge-family parameterized by K and it contains also e.g. the Wess-Zumino gauge $c = \chi = M = = \ell^m v_m = 0$ in $V = c + (\Theta\sigma^m\bar{\Theta})v_m + \frac{1}{4}\Theta^2\bar{\Theta}^2 + + [(\Theta\chi) + \frac{\Theta^2}{2}M + \frac{\Theta^2}{2}(\bar{\Theta}\bar{\lambda}) + h.c.]$.

Alternative versions of (5) with $\alpha = 0$ are

$$L_{gb} = Tr\int d^6x_+ \, B\bar{B}^{-2}KV + h.c. = \tag{7}$$

$$= Tr\int d^6x \, (B + \bar{B})(K_- + \bar{K}_-)V$$

It shows that the longitudinal part, as projected by $P_L = P_+ + P_-$, is relevant for the determination whether K is admissible or not. (7) implies the gauge conditions

$$\bar{D}KV = D^2\bar{K}V = 0 \tag{8}$$

and hence by the standard argument the Faddeev-Popov Lagrangian (u' and u are chiral anticommuting superfields)

$$L_{F.P.} = Tr\int d^8x \, (u'K - \bar{u}'\bar{K})(Ru + \bar{R}\bar{u}) \tag{9}$$

The subsequent derivations are greatly facilitated by a supersymmetric generalization of the elegant compact notation of deWitt 8

$$\int d^8x \, a^i(x)a^i(x) \qquad a^i a^i$$

$$\int d^6x_+ \, a_+^\alpha a_+^\alpha \qquad a^{\alpha+} a^{\alpha+} \tag{10}$$

$$\int d^6x_- \, \bar{a}_+^{-\alpha} a_+^{-\alpha} \qquad a^{\bar{\alpha}+} a^{\bar{\alpha}+} \quad \text{etc.}$$

Summation with respect to i,j etc. involves $\int d^8x$ and the sum with respect to the index of the internal symmetry. A sum with a Greek letter $\alpha, \beta, \ldots$ from the beginning of the alphabet implies a positive chiral integral, whereas $\bar{\alpha}, \bar{\beta}, \ldots$ denote negative chiral integrals. It seems useful to write all fields (Yang-Mills-fields V^i, and matter fields φ_+ and φ_-) as components of one "vector"

$$\Phi^A = (V^i, \varphi_+^\alpha, \varphi_-^{\bar{\alpha}}) = (V^i, \varphi^\rho) \tag{11}$$

where a Greek index from the middle of the alphabet denotes a positive chiral plus a negative chiral integral. We note that in the gauge-parameter

$$(\Lambda^i, \Lambda_+^{i+}) \longrightarrow (\Lambda^\alpha, \Lambda^{\bar{\alpha}}) \longrightarrow \Lambda^\rho$$

α contains the index of the original adjoint representation of Λ. The infinitesimal form of the gauge transformation (2), supplemented by the matter-field transformations becomes

$$\delta\Phi^A = R^{A\alpha}\,\delta\Lambda^\alpha - \bar{R}^{A\bar{\alpha}}\delta\Lambda^{\bar{\alpha}} = R^{A\rho}\,\delta\Lambda^\rho . \tag{12}$$

With the definitions

$$\begin{array}{l} \bar{D}^2K \longrightarrow K^{\alpha i} \\ D^2\bar{K} \longrightarrow K^{\bar{\alpha} i} \end{array} \tag{13}$$

the total Lagrangian formally looks like the one in ordinary gauge theory

$$L = L_{inv} + B^\rho K^{\rho A}\,\Phi^A + u'K^{\rho A}\,R^{A\sigma}u^\sigma \tag{14}$$

the main difference being the nonpolynomial dependence R(V). This notation is also well suited for the evaluation of the V-propagator. The propagator requires the inversion of

$$\Gamma = \begin{pmatrix} \rho(\Box\, P_T)^{ij} & (K^T)^{i\rho} \\ K^{\rho i} & 0 \end{pmatrix} \tag{15}$$

in the space $(V^i, B^\rho_{B\bar{B}}) = (V^i, B^\alpha, B^{\bar{\alpha}})$. Simple algebra yields $\Delta^{BB} = 0$, $\Delta^{VB} = P_L K^T U$ and

$$\Delta^{VV} = (\rho\,\Box)^{-1}(P_T + P_T K^T UKP_L + P_L K^T UKP_T + + P_L K^T UKP_T K^T UKP_L) \tag{16}$$

where

$$U = U^T = -(KP_L K^T)^{-1} \tag{17}$$

The existence of U in the chiral space $(B, \bar{B})$ is the criterion for an admissible gauge K. Since the inversion (17) for U is relatively simple in practical cases, the full knowledge of the algebra enlarged by K is not required.

3. SPECIAL CHOICES

The simplest choice is to make K a scalar superfield depending on Θ and $\bar{\Theta}$ only. E.g. $K = 1 - \frac{\mu^2}{2}\Theta^2\bar{\Theta}^2$ has the property that the c-field conjugates with mass μ. Thus, any mass-term for c is an artefact of gauge-fixing, a result also obtained in [3] .
A "local" ansatz for K is not sufficient to produce a Wess-Zumino type gauge. We therefore discuss the most general "bilocal" $K = K(\Theta, \bar{\Theta}, \Theta', \bar{\Theta}')$. It is straightforward but lengthy to write down this expression which in fact is just an N = 2 scalar superfield. An alternative way to write K uses the derivative operator $\hat{K} = \hat{K}(\Theta, \bar{\Theta}, \partial_\alpha, \bar{\partial}_\alpha)$. It clearly has the same number of components and is, in fact, simply related to K by $K = \hat{K}\delta^4(\Theta, \Theta')$. Historically, the first noncovariant supergauge was the "N^2-gauge" [4] . Defining (n_m is a fixed Lorentz-vector)

$$N_\alpha = \partial_\alpha - i(\not{n}\bar{\Theta})_\alpha$$

$$\hat{N}_\alpha = \partial_\alpha + i(\not{n}\bar{\Theta})_\alpha$$

and $K = N^2$, the algebra of the covariant derivatives is enlarged in a very transparent manner. Here the c-field-component propagates with a mass n^2 and $n^m v_m = 0$ (axial gauge). The advantage of the "FD-gauge" [5]

$$K = F^\alpha D_\alpha$$

$$F_\alpha = m_\alpha - i(f\bar{\Theta})_\alpha$$

is its simple set of new projection operators

$$P = P^2 = \frac{i\,\bar{D}^2(FD)}{8(f\partial)} \quad \text{etc.}$$

The gauge condition (8) yields an IR-regular gauge, very similar to the Wess-Zumino gauge. There are infinitely many choices of K yielding the Wess-Zumino gauge, but differing in the sector of the auxiliary field B. A very simple choice for K is [6]

$$K = \bar{\Theta}^2[\,1 + i\partial^2\bar{\partial}^2(\Theta\,\not{f}\,\bar{\Theta})\,]$$

Recently, Johanson [9] has proposed a K of a general chiral type.

4. BRS-IDENTITIES AND RENORMALIZATION

The similarity of (15) to the Lagrangian of ordinary gauge theories immediately allows the introduction of a BRS-transformation [10] with a special gauge transformation $= iu^\rho\delta\lambda$ involving the Faddeev-Popov field and the anticommuting quantity $\partial\lambda$:

$$\begin{aligned} \delta\Phi^A &= iR^{A\rho}u^\rho\delta\lambda = s\,\phi^A\delta\lambda \\ \delta u^\rho &= -\frac{i}{2}f_{\rho\sigma\tau}\,u^\sigma u^\tau\delta\lambda = su^\rho\delta\lambda \\ \delta u'_\rho &= -\,iB^\rho\delta\lambda = su'_\rho\delta\lambda \\ \delta B &= 0 \end{aligned} \qquad (21)$$

so that $s^2 = 0$ and $f_{\alpha\beta\gamma} = f_{\bar{\alpha}\bar{\beta}\bar{\gamma}}$ are the structure constants in our supercompact notation. Following the usual methods one obtains the standard relations which provide the basic tools for the treatment of renormalization and for the proof of gauge-independence of physically observable quantities. For covariant gauges we have already mentioned the spurious infrared singularity of the scalar component of V which may be traced back to the fact that V is dimensionless. Another complication from this property of V is that during renormalization also a redefinition of $V \rightarrow F(V)$ must be permitted, such a redefinition being a gauge-effect [7]. For the general noncovariant supergauges K the approach is clear, because the way has been paved already in [7], where a special breaking of supersymmetry had to be assumed as well. The important point is that the breaking of supersymmetry must be controlled by identities from a BRS-invariance. The proof of K-independence for a suitably defined (on-shell) S-matrix-element between physical sources (neglecting the standard IR-problem of ordinary gauge-theories or considering a suitably spontaneously broken theory) is a rather straightforward transcription of the proof in ordinary gauge theories, although some modifications for the definition of physical sources of chiral (matter) fields are necessary.

REFERENCES:

[1] S. Ferrara and B. Zumino, Nucl.Phys. B79(1974)413;
A.Salam and J. Strathdee, Phys.Lett.51B (1974)353;
B. deWit and D. Freedman, Phys.Rev.D12 (1975)2286

[2] L.F. Abott, M.T. Grisaru and D. Zanon, Nucl.Phys.B244(1984)454

[3] O.Piguet and K. Sibold, Nucl.Phys. B248(1984)336; ibid. B249(1985)396

[4] W. Kummer and M.Schweda, Phys.Lett. 141B(1984)363

[5] T. Kreuzberger, W. Kummer, O.Piguet, A.Rebhan and M.Schweda, A simple implementation of Wess-Zumino-like gauges within the superfield technique, Phys. Lett. 167(1986)393

[6] T. Kreuzberger, W. Kummer, H.Mistelberger, P.Schaller, M.Schweda, Super-Yang-Mills-Fields in noncovariant supergauges, Nucl.Phys.B, to be publ.

[7] O.Piguet and K.Sibold, Nucl.Phys. B248 (1984)336; B249(1985)396

[8] B.S. de Witt, Phys.Rev. 162(1967)1195

[9] A.A. Johanson, Superfields in the noncovariant supergauges, Leningrad prep. 1985

[10] C. Becchi, A. Rouet and R.Stora, Phys. Lett. 52B (1974) 344.

PRESENT STATUS OF HARMONIC SUPERSPACE

A. Galperin[x)], E. Ivanov[xx)], V. Ogievetsky[xx)] and E. Sokatchev[xxx)]

x) Institute of Nuclear Physics, Tashkent, USSR.

xx) Laboratory of Theoretical Physics, Joint Institute for Nuclear Research, Dubna, USSR

xxx) Institute for Nuclear Research and Nuclear Energy, Sofia, Bulgaria

1. Preliminaries. 2. The most general self-coupling of N=2 matter and introducing hyper-Kähler potential. 3. A closed form of N=2 SYM action in terms of analytic harmonic connection. 4. Quantization in harmonic superspace. A proof of ultraviolet finiteness of d=2 hyper-Kähler supersymmetric σ models. 5. Conformal and Einstein N=2,3 supergravities in harmonic superspace. 6. New trends: applications to superparticle and superstring.

1. PRELIMINARIES

Harmonic superspace (SS) has been invented by us two years ago /1,2/ in searching for a manifestly invariant description of theories with extended SUSY. The first successes of this approach (unconstrained superfield (SF) formulations of matter hypermultiplets and N=2 super Yang-Mills (SYM) theory, finding analytic SS group and prepotentials of the Einstein N=2 supergravity (SG), the first off-shell formulation of N=3 (SYM) have been reported at the preceding conference /3/. During the next two years, an essential progress has been achieved both in understanding the basics of the method and in its further applications. In this talk we review these developments and outline the problems to be solved.

Recall the basic motivation which has led us to harmonic SS. We have realized in 1984 that an ordinary SS with finite sets of auxiliary fields is inadequate to extended SUSY's. There were several indications of this. First, one faced the familiar "no-go" theorems /4/. Second, there were serious difficulties in constructing unconstrained SF formulations of N=2 theories (analogous to formulations known in the case N=1).

The way out we suggested was as follows /1-3/. One has to extend a conventional SS $(x^{\alpha\dot\alpha}, \theta^\alpha_i, \bar\theta^{\dot\alpha i})$ by adding new purely internal coordinates $u^{(a)}_i$ which parametrize a coset G/H, G being the automorphism group of the relevant superalgebra, H its certain subgroup. These coordinate are lowest harmonics on G/H. The SF's defined on harmonic SS $\mathbb{HR}$ thus constructed contain infinite sets of conventional SF's appearing as coefficients of harmonic expansions in powers of $u^{(a)}_i$. Among these general harmonic SF's there can be found SF's of a lower Grassmann dimension, the analytic SF's. These live as unconstrained objects on an analytic subspace $\mathbb{AR}$ of $\mathbb{HR}$. Just the analytic SF's provide us with manifestly invariant unconstrained geometric formulations of SUSY theories with $N \geq 2$. Such formulations became possible due to the following radically new property. The number of auxiliary and/or gauge degrees of freedom in analytic SF's (appearing in their harmonic expansions) is infinite. This is just the point where the above no-go theorems fail.

Now we turn to listing the main results obtained within the harmonic SS approach for the last two years.

2. MOST GENERAL SELF-COUPLING OF N=2 MATTER

One of the important problems in SUSY is to identify the "ultimate" off-shell representation which yields the most general matter self-coupling. In the case N=1, it is chiral multiplet described by unconstrained SF on a complex N=1 SS $\mathbb{C}^{4|2} = (x_L^{\alpha\dot\alpha}, \theta^\alpha)$. The genuine N=2 analogue of N=1 chiral SF is a complex analytic N=2 SF $q^+(\zeta, u)$ living as an unconstrained function on analytic N=2 SS /1/ ($\pm$ denote U(I)-charge):

$$\mathbb{AR}^{4+2|4} = (\zeta, u) \equiv (x^{\alpha\dot\alpha}_A, \theta^{+\alpha}, \bar\theta^{+\dot\alpha}, u^\pm_i)$$
$$u^{+i}u^-_i = 1, \; u^\pm_i \in SU(2), \; i=1,2 \qquad (1)$$

The most general q^+ -action is /5,6/

$$S_q=\frac{1}{\varkappa^2}\int d\zeta^{(-4)}du\,\mathcal{L}^{(+4)}(q^+_A,\bar{q}^+_A,\dots u^+_i,u^-_i)$$

$$[\varkappa]=cm^2,\ [q^+]=cm^0 \qquad (2)$$

where integration goes over analytic SS(1). An analytic density $\mathcal{L}^{(+4)}$ arbitrarily depends on $q^+_A,\bar{q}^+_A$ (A=1,2,...), on its harmonic derivatives of any order (these are dimensionless and preserve analyticity) and may include explicitly harmonics u^+_i, u^-_i.

There were many attempts to describe N=2 matter by off-shell multiplets having a finite number of auxiliary fields (familiar tensor and relaxed multiplets and some newly proposed ones /6-9/). All these multiplets are most easily repre - sented by some analytic SF's /6/. In /10-12,6/ we have defined an N=2 duality transformation and have shown with the help of it that all the self-couplings of above multiplets are equivalent to certain subclasses of the general q^+-action (2)/12,6/ (e.g.,having special isometries, etc). Thus, the latter presumably describes the most general matter self-coupling in rigid N=2 SUSY [x]. This is apparently due to an infinite number of auxiliary fields in q^+. As has been shown recently by Howe, Stelle and West /8/, infiniteness of the number of auxiliary fields is inavoidable when extending off-shell a complex form of a hypermultiplet.

According to Alvarez - Gaumé and Freedman /14/ any N=2 matter action produces a hyper-Kähler σ model in the physical boson sector. Corres-ponsingly, we may call $\mathcal{L}^{(+4)}$ in (2) the "hyper-Kähler potential" /6/ (by analogy with the Kähler potential of N=1 case) /15/. An analytic SS formulation suggests a new way of explicit calculation of hyper-Kähler metrics. Given an arbitrary $\mathcal{L}^{(+4)}$, one may eliminate auxiliary fields by their equations of motion to obtain a metric on the manifold of physical bosons which is guaranteed to be hyper-Kähler. The simplest example is a familiar Taub-NUT manifold which is coded in the action /5/

$$S_{TN}=\frac{1}{\varkappa^2}\int d\zeta^{(-4)}du\left[\bar{q}^+D^{++}q^+ +\lambda(q^+)^2(\bar{q}^+)^2\right] \qquad (3)$$

---
[x] Extension to nonzero central charges /1/ or to d =6 /8,13/ is straightforward.

In ref./16/we have found analyticSF actions leading to other interesting metrics. These are the Eguchi - Hanson and multi-Eguchi - Hanson metrics (manifolds of dimension 4), Calabi and multi-Calabi ones (dimension 4n), metrics on the cotangent bundle of a 2nm-dimensional Grassmann manifold (dimension 4 nm, m $\geqslant$ 1).

Thus, we arrive at a suggestive idea of classifying hyper-Kähler metrics according to relevant hyper-Kähler potentials $\mathcal{L}^{(+4)}$. Intriguing questions are what is the precise mathematical meaning of $\mathcal{L}^{(+4)}$ and how the latter is connected with the primary principles of the hyper-Kähler geometry.

3. A CLOSED FORM OF THE N=2 SYM ACTION

An interesting development of our geometric formulation of N=2 SYM /1/ has been made by Zupnik/13/. He has found a closed compact form for the action in terms of the analytic harmonic connection $V^{++}(\zeta,u)$. It heavily uses harmonic distributions introduced by us in /17/ (analogs of the N=0 distributions $1/x^n$, $\delta(x)$) and is written as /13/ (in the central basis of $\mathbb{R}^{4+2|8}$):

$$S_{SYM_2}=-\frac{1}{g^2}\int d^{12}z\, \mathrm{Tr}\ln(1+K_V)(z),\ [g]=cm^0 \qquad (4)$$

where

$$K_V^{(-1,1)}(1,2)=\frac{V^{++}(z,u_2)}{u^{+i}_1\,u^+_{2i}} \qquad (5)$$

and Tr is taken both over discrete indices of the adjoint representation of gauge group and harmonic arguments $u_1,\dots u_n$ (the latter are regarded as continuous matrix indices with the harmonic integration over them instead of ordinary summation). The n-th term in the expansion of (4) in powers of V^{++} is as follows:

$$S^n_{SYM_2}=\frac{1}{g^2}\frac{(-i)^n}{n}\int d^{12}z\,du_1\dots du_n\frac{\mathrm{Tr}\,V^{++}(z,u_1)\dots V^{++}(z,u_n)}{(u^+_1u^+_2)\dots(u^+_nu^+_1)} \qquad (6)$$

In eqs. (5), (6), $1/u^+_1u^+_2$ is a particular example of harmonic distributions, viz. the harmonic Green function.

$$D^{++}_1\frac{1}{u^+_1u^+_2}=\delta^{(1,-1)}(u_1,u_2)\ ,\ D^{++}=u^{+i}\frac{\partial}{\partial u^{-i}}$$

where $\delta^{(1,-1)}(u_1,u_2)$ is one of the variety of harmonic δ-functions /17/. As in the N=1 case, the N=2 SYM action is non-polynomial. Essentially new features are the nonlocality in harmonics and absence of spinor derivatives in interaction vertices. Harmonic nonlocalities do not create problems when quantizing N=2 SYM and are not present in final answers for SF amplitudes /18/. Note that the Lagrangian density in (4) is gauge invariant only up to full harmonic derivatives and is thus of the Chern-Simons type (in contradistinction to the tensor density in the standard representation of N=2 SYM action via the constrained chiral strength).

4. QUANTIZATION IN HARMONIC N=2 SUPERSPACE

One of the main incentives to construct the harmonic SS formulations of theories with extended SUSY was the desire to have a manifestly supersymmetric quantization scheme in terms of unconstrained SF's /1-3/. Now this problem is completely solved for N=2 matter and SYM theories: in papers /17,18/ we have given an extensive exposition of relevant harmonic superspace Green functions and Feynman rules as well as the first examples of manifestly N=2 supersymmetric quantum calculations$^{x)}$. The main lesson is that these SF techniques are not more difficult than those in the case N=1. Let us quote, e.g., the SF propagators of q^+-hypermultiplet and of V^{++}(in the Feynman gauge and in the central basis of $\mathbb{HR}^{4+2|8}$):

$$\langle q^+_a(1)\, q^+_b(2)\rangle \sim \frac{i}{p^2}\frac{(D_1^+)^4(D_2^+)^4}{(u_1^+u_2^+)^3}\delta^8(\theta_1-\theta_2)\delta_{ab} \quad (7)$$

$$\langle V^{++}_a(1)\, V^{++}_b(2)\rangle \sim \frac{i}{k^2}(D_1^+)^4\delta^8(\theta_1-\theta_2)\delta^{(2,2)}(u_1,u_2)\delta_{ab}$$

where the operators $(D^+)^4$ ensuring analyticity and harmonic distributions $1/(u_1^+u_2^+)^3$ and $\delta^{(-2,2)}(u_1,u_2)$ appear. Expressions for vertices are also simple.

$^{x)}$Some of these studies have been performed in parallel with us by Kubota and Sawada and Ohta and Yamaguchi /19/ (in the latter paper for a nonzero central charge).

Now we shall sketch the most important features of harmonic supergraph techniques.

A. Harmonic coordinates (in contrast, e.g., to extra coordinates in standard Kaluza - Klein theories) produce no new divergences.

B. Harmonic nonlocalities disappear if external legs of a diagram are placed on-shell. All the harmonic integrals can be computed by simple algebraic manipulations.

C. Quantum corrections can always be written as integrals with the full Grassmann measure $d^8\theta$.

D. No ghosts-for-ghosts are needed when quantizing N=2 SYM.

The fact that the effective action is an integral of the type $d^8\theta$ is known to yield significant improvements in the ultraviolet behaviour. The most striking application /18/ is a convincing and simple proof of off-shell finiteness of hyper-Kähler supersymmetric σ models in two dimensions. Indeed, in d=2 $[x] = \text{cm}^0$ and $[q^+(p,\theta,u)] = \text{cm}^2$ so the n-particle contribution to the effective action has the generic form:

$$\Gamma_n = \int d^8\theta\, du\, (d^2p)^{n-1}[q(p,\theta,u)]^n\, I(p) \quad (8)$$

in accord with the property C. One easily observes that $[I(p)] = \text{cm}^2$, and hence $I(p)$ is convergent$^{x)}$.

Now we are concerned with quantization of N=3 SYM /2/ along similar lines. It seems especially urgent to work out a proper background field method.

5. N=2,3 SUPERGRAVITIES IN HARMONIC SUPERSPACE

In ref./1/ we have defined the harmonic SS group of Einstein N=2 SG (in its first version) and corresponding unconstrained analytic SG pre-prepotentials. Now we know these for conformal N=2,3 SG too /21 /. The gauge groups of the latter preserve the fundamental concepts of analytic subspace and U(I)-charges, as well as the unitarity and unimodularity conditions of harmonics. These are local extensions of rigid conformal supergroups SU(2,2/2) and SU(2,2/3). In the case N=2 /21/ :

$^{x)}$This proof is specific just for hyper-Kähler N=4,d=2 σ models. No similar reasoning can be given for general Ricci-flat N=2,d=2 σ models, in accord with the recent observation concerning the σ models on Calabi- Yau manifolds /20/ .

$$\delta\zeta^M = \lambda^M(\zeta,u),\ \delta u_i^+ = \lambda^{++}(\zeta,u)u_i^-,$$
$$\delta u_i^- = 0 \qquad (M = m, \mu^+, \dot{\mu}+) \qquad (9)$$

The fundamental unconstrained geometric quantities which represent the Weyl multiplet are ++ components of the analytic vielbein $H^{++M}(\zeta,u)$, $H^{++++}(\zeta,u)$ covariantizing the derivative D^{++}

$$\mathcal{D}^{++} = \partial^{++} + H^{++M}(\zeta,u)\partial_M + H^{++++}(\zeta,u)\partial_{++}$$
$$\partial^{++} = u^{+i}\frac{\partial}{\partial u^{-i}},\ \partial_{++} = u^{-i}\frac{\partial}{\partial u^{+i}}$$

Generalization to N=3 is straightforward; one has only to take into account the presence of two independent complex analytic directions in SU(3)/U(I) x U(I) /2/ instead of one (++) in SU(2)/U(I).

Following the standard compensation ideology we may, in principle, construct action for any version of the minimal Einstein N=2 SG as a sum of actions of compensating Maxwell $(V^{++5}(\zeta,u))$ and matter N=2 multiplets in the conformal SG background /22/ . At present we dispose of the analytic SS description of all the matter compensators known before /1,21,12/. So, to construct the complete SF actions for all the versions of minimal Einstein SG, it remains to find the action for $V^{++5}(\zeta,u)$, and this is in progress now. A new possibility is to use as a compensator the basic unconstrained analytic SF q^+ (or ω /21/). The corresponding version of Einstein SG will contain an infinite number of auxiliary fields. We expect it to be very promising.

An interesting problem ahead is to construct off-shell Einstein N=3 SG. A component consideration /23/ implies that it can be obtained by coupling three Maxwell N=3 multiplets (or one SO(3)- N=3 SYM multiplet) to conformal N=3 SG. Thus, what one needs is to extend the analytic SF action of N=3 SYM /2/ to local conformal SUSY.

6. NEW TRENDS

A most modern area of applications of the harmonic SS approach is the superparticle and superstring theories. Recently, the "light-cone harmonic SS" has been constructed /24/ which extends ordinary N=1 SS in D dimensions by adding harmonics on the coset SO(1,D-1),/SO(1,1) x x SO(1,D-2), SO(1,D-1) being the Lorentz group of M^D. In such a harmonic SS there is an invariant analytic subspace involving half the original spinor variables. The superparticle action can be reformulated in this light-cone analytic subspace so that no local fermionic invariance is needed and the Lorentz symmetry is preserved (due to the presence of new harmonic variables which carry Lorentz vector indices). Another application is the 10-dimensional SYM theory. Its on-shell constraints can be rewritten as the integrability conditions for the existence of the light-cone analytic SF's. There are, however, serious difficulties with the off-shell formulation (as distinct from the N=3, d=4 SYM /2/). We would like to point out that other harmonizations of d=10 N=1 SS are also possible, corresponding to different choices of the coset of SO(1,9). These conceal potentialities which may have utterly unexpected manifestations in superstrings and d=10, N=1 (d=4, N=4) SYM theories. An important point is that the invariant analytic subspaces arising with this type of harmonization contain reduced numbers both of Grassmann and ordinary bosonic (x) coordinates , as is illustrated already by the simple example of ref. /24/ .

Another line of extending the harmonic SS business has been proposed by Kallosh /25/ . She gave some reasonings /25a/ that the harmonic SS with the even part M^{10} x x $(E_8 \times E_8/U_1(1)\times\ldots\times U_{16}(1))$ presumably can be used for the off-shell formulation of d=10, $E_8\times E_8$ Yang-Mills - supergravity, associated with the heterotic string. Also, a complete tensor apparatus of N=3 SYM theory in the harmonic SS /2/ has been constructed /25b/.

For these two years, some mathematical developments of the harmonics SS approach have been made. In particular, Rosly and Schwarz in ref. /26/ treated it with the accent on its affinity with twistors (recall that our geometric formulation of N=2 SYM essentially incorporates the interpretation of N=2 SYM constraints given for the first time by Rosly /27/ .

We end this rather schematic survey by indicating the principal directions of further studies. The most urgent problem now is, in our opinion, to find a closed off-shell geometric formulation of the most intriguing of d=4 SYM theories, the N=4 SYM theory. It cannot be achieved by a simple prolongation of formulations found by us for N=2 and N=3 theories (see analysis in /28/) and thus seems to require an essentially new look. We believe that the solution of this task will naturally lead us to superstrings, keeping in mind a familiar correspondence between N=4, d=4 and N=1, d=10 SYM theories.

REFERENCES

/1/ Galperin, A., E.Ivanov, V.Ogievetsky and E.Sokatchev, JETP Lett., 40 (1984) 912.
Galperin, A., E.Ivanov, S.Kalitzin, V.Ogievetsky and E.Sokatchev. Class.Quantum Grav. 1 (1984) 469.

/2/ Galperin, A., E.Ivanov, S.Kalitzin, V.Ogievetsky and E.Sokatchev, Phys.Lett., 151B (1985) 215; Class.Quantum Grav. 2 (1985) 155.

/3/ Galperin, A., E.Ivanov, S.Kalitzin, V.Ogievetsky and E.Sokatchev, In: Proc. XXII Int.Conf. on High Energy Physics, Leipzig, July 19-25 1984, v.I, p.37, 1984.

/4/ Roček, M. and W.Siegel. Phys. Lett., 105 B (1981) 278; Rivelles, V. and J.Taylor. Phys. Lett., 121B (1983) 37.

/5/ Galperin, A., E.Ivanov, V.Ogievetsky and E.Sokatchev, Commun. Math.Phys. 103 (1986) 515

/6/ Galperin, A., E.Ivanov and V.Ogievetsky, Preprint JINR E2-86-277, Dubna, 1986

/7/ Yamron, J.P. and Siegel W. Nucl. Phys., B263 (1986) 70

/8/ Howe, P., K.Stelle and P.Weat, Class.Quantum Grav., 2 (1985) 815

/9/ Ketov, S.V., B.B.Lokhvitsky, K.E. Osetrin and I.V.Tyutin. Self-interacting N=2 matter in N=2 superspace. Preprint N 31, Tomsk, 1985 (in Russian)

/10/ Galperin, A., E.Ivanov, V.Ogievetsky and E.Sokatchev, Preprint JINR E2-85-128, Dubna, 1985

/11/ Ivanov, E. Lectures given at the XXI Winter School of Theoretical Physics in Karpacz, Poland 1985. In: "Spontaneous Symmetry Breakdown and Related Subjects", World Scientific, Singapore, 1985, p.413.

/12/ Galperin, A., E.Ivanov and V.Ogievetsky, Preprint JINR E2-85-897, Dubna, 1985.

/13/ Zupnik, B.M. Yadern.Fiz., 44 (1986) 1781

/14/ Alvarez-Gaumé, L. and D.Z.Freedman, Comm.Math.Phys. 80 (1981) 443

/15/ Zumino, B. Phys.Lett., 87B (1979) 203

/16/ Galperin, A., E.Ivanov, V.Ogievetsky and P.K.Townsend. Class. Quantum Grav., 3 (1986) 625

/17/ Galperin, A., E.Ivanov, V.Ogievetsky and E.Sokatchev, Class.Quantum Grav., 2 (1985) 601

/18/ Galperin, A., E.Ivanov, V.Ogievetsky and E.Sokatchev, Class.Quantum Grav., 2 (1985) 617

/19/ Kubota, T. and S.Sawada. Progr. Theor.Phys. 74 (1985) 1329
Ohta, N. and H.Yamaguchi, Phys. Rev. D32 (1985) 1954

/20/ Grisaru, M., A.E.M.Van de Ven and D.Zanon, Phys.Lett., 173B (1986) 423;
Pope, C.N., M.F.Sohnius and K.S.Stelle. Imperial College prepr. ITP/85-86/16;
Howe, P.S., G.Papadopoulos and K.S.Stelle. Phys.Lett., 174B (1986) 405

/21/ Galperin, A., E.Ivanov, V.Ogievetsky and E.Sokatchev, Prepr. JINR, E2-85-313, Dubna, 1985

/22/ de Wit B., J.W.van Holten and A.van Proyen, Nucl.Phys. B184 (1981) 77

/23/ Howe, P., K.Stelle and P.K.Townsend, Nucl.Phys., B192 (1981) 332

/24/ Sokatchev, E. Prepr. IC/85/305, Trieste, 1985; Phys.Lett. 169B (1986) 209

/25a/ Kallosh, R.E. Preprint CERN-TH. 4188/85

/25b/ Kallosh, R.E. JETP Lett., 41 (1985) 172

/26/ Rosly, A.A. and A.S.Schwarz, In: Proc.III Int.Seminar "Quantum Gravity", World Scientific, Singapore, 1985, p.308.

/27/ Rosly, A.A. In: Proc.Int.Seminar on Group Theoretical Methods in Physics, Nauka, Moscow, 1982, v.1, p.263

/28/ Ahmed, E., S.Bedding, C.T.Card, M.Dumbrell, M.Nouri-Moghadam and J.G.Taylor, J.Phys. A: Math.Gen. 18 (1985) 2095

HARMONIC SUPERSPACE FORMALISM AND THE CONSISTENT CHIRAL ANOMALY

Wenzhou Li*

Lawrence Berkeley Laboratory
University of California
Berkeley, California 94720 U.S.A.

The harmonic superspace formalism has been used to construct the consistent chiral anomaly in $N = 1$, $d = 6$ supersymmetric Yang-Mills theory. The expressions of the gauge anomaly Δ_s^ϕ and of the supersymmetric anomaly Δ_{SUSY}^ϕ are given together with the consistent condition.

The consistent chiral anomaly in $N = 1, d = 4$ supersymmetric Yang-Mills theory has been studied intensively. For higher dimension, due to the difficulties to get an unconstrained superspace formulation, it has been studied only in the Wess-Zumino gauge by Itoyama *et al* [1], who have found an $N = 1, d = 6$ nontrivial supersymmetric anomaly. The aim of this work is to get the $N = 1$, $d = 6$ supersymmetric consistent anomalies by using the harmonic superspace formulation[2),3)].

For six dimensions, the use of the $d = 6$ $SU(2)$ Majorana-Weyl spinors [4),5)] leads to a manifestly $SU(2)$ covariant formulation, which is naturally incorporated into the harmonic superspace formalism.

The $N = 1, d = 6$ harmonic superspace is parametrized by

$$\{z^M, u_i^\pm\} \tag{1}$$

in the central basis, where $z^M = (x^m, \theta_\alpha^i)$, $(m = 0, \cdots 5, i = 1, 2, \alpha = 1, \cdots 4)$ parametrize the ordinary superspace, $u_i^\pm$, the harmonic variables, parametrize the coset space.

The SUSY transformation is given by

$$\begin{aligned} \delta x^m &= \frac{i}{2}\epsilon_i^\alpha \Sigma_{\alpha\beta}^m \theta^{\beta i} \\ \delta\theta_{\alpha i} &= \epsilon_{\alpha i} \\ \delta u_i^\pm &= 0 \end{aligned} \tag{2}$$

where $\Sigma_{\alpha\beta}^m$ are 4×4 antisymmetric matrices.

*On leave from Department of Physics, Zhejiang University, Hangzhou, China.

We can pass from the central basis $\{z^M, u_i^\pm\}$ to the analytic basis, defined by

$$A.B. : \{z_A^M = (x_A^m, \theta_\alpha^+, \theta_\alpha^-), u_i^\pm\}, \tag{3}$$

where the new independent variables are related to the old ones as follows

$$\begin{aligned} x_A^m &= (x^m + \frac{i}{2}\theta_{(i}^\alpha \Sigma_{\alpha\beta}\theta_{j)}^\beta u^{+i}u^{-j}), \\ \theta^{\pm\alpha} &= u^{\pm i}\theta_i^\alpha \end{aligned} \tag{4}$$

In the analytic basis, the $N = 1, d = 6$ SUSY is realized as

$$\begin{aligned} \delta x_A^m &= \frac{i}{2}(\epsilon_i^\alpha \Sigma_{\alpha\beta}\theta^{+\beta})u^{-i} \quad \delta\theta_\alpha^+ = \epsilon_\alpha^i u_i^+ \\ \delta\theta_\alpha^- &= \epsilon_\alpha^i u_i^- \quad \delta u_i^\pm = 0. \end{aligned} \tag{5}$$

Since $x_A^m, \theta_\alpha^+, u_i^\pm$, form a subset closed under $N = 1, d = 6$ SUSY transformations,we get the analytic subspace,

$$(\varsigma_A^M = (x_A^m, \theta_\alpha^+), u_i^\pm). \tag{6}$$

In the analytic basis, the harmonic derivative D^{++} becomes

$$D^{++} = u^{i+}\frac{\partial}{\partial u^{i-}} = \frac{\partial}{\partial u^{i-}} + \frac{i}{2}\theta^{+\alpha}\Sigma_{\alpha\beta}^m \theta^{+\beta}\frac{\partial}{\partial x^m} + \theta_\alpha^+ \frac{\partial}{\partial\theta_\alpha^-} \tag{7}$$

and the spinor derivatives are decomposed into $D^\pm$ parts:

$$D_\alpha^+ = \frac{\partial}{\partial\theta^{-\alpha}}, \quad D_\alpha^- = \frac{\partial}{\partial\theta^{+\alpha}} + i\Sigma_{\alpha\beta}^m \theta^{-\beta}\frac{\partial}{\partial x_A^m} \tag{8}$$

The fact that D_α^+ are reduced to simple derivatives with respect to θ^- reflects the existence of the invariant analytic subspace. Therefore in the analytic basis, we can define superfields which do not depend on θ^- and are analytic automatically,

$$F^{(q)}(x_A, \theta^+, u^\pm), \quad D^+_\alpha F^{(q)} \equiv 0. \tag{9}$$

The unconstrained off-shell formulation just relies on the use of the real analytic superfields.

The exterior differential d in the harmonic superspace can be split as follows

$$d = \hat{d} + D + \hat{D}$$

where $\hat{d} = e^{\alpha\beta}\partial_{\alpha\beta}, D = e^{\alpha i}D_{\alpha i}, \hat{D} = e^a D_a$, with $D_a = (D++, D--)$. The rigid vielbeins e^A are 1-superforms, with the property

$$de^A = t^A_{BC} e^C e^B$$

$$t^{\gamma\gamma'}_{\alpha_i \beta_j} = -i\epsilon_{ij}\delta^{\gamma\gamma'}_{[\alpha\beta]}.$$

From the constraints $F^{++}_{\alpha\beta} = 0$, $F^{++,--} = 0$ we can solve for φ^+_α and $\varphi^{++}, \varphi^{--}$

$$\varphi^+_\alpha = e^{-U} D^+_\alpha e^U$$

$$\varphi^{++} = e^{-V} D^{++} e^V$$

$$\varphi^{--} = e^{-V} D^{--} e^V$$

From $F^{+\,++}_\alpha = 0$ we can get the important relation

$$D^+_\alpha V^{++} = 0 \tag{10}$$

$$V^{++} = e^U e^{-V} D^{++}(e^V e^{-U})$$

As φ_A have been obtained, the gauge transformations can be expressed as

$$e^U \to e^X e^U e^K, \quad D^+_\alpha X = 0, \tag{11}$$

$$e^V \to e^Y e^V e^K, \quad D^{++} Y = 0. \tag{12}$$

where K is the gauge transform parameter. By using a K gauge transformation the V can be set to zero, $V = 0$, and hence $\varphi^{++} = \varphi^{--} = 0$. Now (10) becomes $V^{++} = e^U D^{++} e^{-U}$ which can be reduced to

$$(D^{++} + V^{++})e^U = 0 \tag{13}$$

Equation (13) can be solved to express the prepotential U in terms of the unconstrained analyltic superfield V^{++}.

The gauge transformations can be expressed by the exterior differential operator in gauge group space[6),7)]

$$s = \sum_i dt_i \frac{\partial}{\partial t^i} \tag{14}$$

where t^i are all parameters the gauge group elements may depend on.

The following notations are introduced to express the gauge transformed quantities

$$e^v = e^X e^V e^K$$

$$e^u = e^X e^U e^K$$

$$\phi = e^{-K}\varphi e^K + e^{-K} d e^K$$

$$\mathcal{F} = d\phi + \phi\phi$$

We denote $c = e^{-K} s e^K$, $c' = e^X s e^{-X}$ where c, c' are $\mathcal{G}$-valued 1-forms in t space.

We can get

$$se^v = s(e^X e^V e^K) = -c'e^v + e^v c \tag{15}$$

$$se^u = s(e^X e^U e^K) = -c'e^u + e^u c \tag{16}$$

$$s\phi = [\phi, c] - dc$$

$$sc = -c^2 \tag{17}$$

which coincide with B.R.S. transformations.

The transgression formula [6)] is therefore still true, and we consider the following for $d = 6$,

$$Tr(\mathcal{F}\mathcal{F}\mathcal{F}\mathcal{F}) = (d+s)\Omega \tag{18}$$

$$\Omega = \int_0^1 dt$$

$$str((\phi{+}c), \hat{\mathcal{F}}_t, \hat{\mathcal{F}}_t, \hat{\mathcal{F}}_t) = t(d{+}s)(\phi{+}c) + t^2(\phi{+}c)(\phi{+}c) \tag{19}$$

The operator S_{SUSY} corresponding to rigid supersymmetry is defined by acting on superfields ψ integrated over space-time,

$$S_{SUSY} \int_x \psi = \int_x \epsilon^A D_A \psi = \int_x (\epsilon^{\alpha i} D_{\alpha i})\psi \tag{20}$$

where $\epsilon^A = (\epsilon^{\gamma\gamma'}, \epsilon^{\alpha i}, \epsilon^a) \equiv (0, \epsilon^{\alpha i}, 0)$ are parameters with the same commuting properties as dz^A, i.e., the spinor like $\epsilon^{\alpha i}$ are commuting among themselves. The reason we put $\epsilon^a = 0$ is that supersymmetry is realized with $\delta u^\pm_i = 0$; also we may set the space-time like $\epsilon^{\gamma\gamma'} = 0$ because they occur under space-time integration. Due to the commuting property of $\varepsilon^{\alpha i}$, S_{SUSY} is nilpotent when acting on space-time integrals of superfields,

$$S^2_{SUSY} = 0 \tag{21}$$

To construct consistent SUSY anomaly, we start from the transgression formula (19). The 7-superform

Ω can be split according to the powers of c, and by the number of factors of the vielbeins e^A with different dimensions, i.e., $q_1, q_2, (7-p)-q_1, -q_2$, which denote the number of the factors of the spinor-like vielbeins, $e^{\alpha i}$, the harmonic-like vielbeins e^a, and the space-time like vielbein $e^{\gamma\gamma'}$ respectively. Therefore, equation (18) contains different sets of identities, one for each sector. We depict the following,

$$p=0, q_1=2, q_2=0:$$

$$str(\mathcal{F}\mathcal{F}\mathcal{F}\mathcal{F})_{6,2,0} = d\omega^0_{5,2,0} + [D\omega^0_{6,1,0}]_{6,2,0} \quad (22)$$

$$p=1, q_1=0, q_2=0:$$

$$0 = s\omega^0_{6,1,0} + \hat{d}\omega^1_{5,1,0} + [D\omega^1_{6,0,0}]_{6,1,0} \quad (23)$$

$$p=2, q_1=0, q_2=0 : 0 = s\omega^1_{6,0,0} + \hat{d}\omega^2_{5,0,0} \quad (24)$$

After some calculation, we can get the final result,

$$s\Delta^\phi_S = 0 \quad (25)$$

$$S_{SUSY}\Delta^\phi_{SUSY} = 0 \quad (26)$$

$$s\Delta^\phi_{SUSY} + S_{SUSY}\Delta^\phi_S = 0 \quad (27)$$

where

$$\Delta^\phi_S = \int_x \omega^1_{6,0,0}$$

$$\Delta^\phi_{SUSY} = \int_x i_\epsilon \hat{\omega}_{6,1,0}$$

The notation $i_\epsilon \chi_m$ denotes the interior product of the m-superforms χ_m with respect to ϵ^M, where

$$\epsilon^M = \epsilon^A e^M_A \quad (28)$$

and e^M_A are the inverse vielbeins,

$$i_\epsilon \chi_{m=m} \epsilon^{A_1} e^{A_2} \cdots e^{Am} \chi_{Am} \cdots A_1$$

Equations (25), (26) and (27) are the consistency conditions and $\Delta^\phi_S, \Delta^\phi_{SUSY}$ are the gauge anomaly and supersymmetric anomaly respectively. Δ^ϕ_S, Δ^ϕ_{SUSY} can be expressed by

$$\Delta^\phi_S = 12 \int_x \int_0^1 dt(1-t)[str(cd\underline{\phi}\underline{\mathcal{F}}t\underline{\mathcal{F}}t)]_{\theta=0} \quad (29)$$

$$\Delta^\phi_{\mathrm{SUSY}} = -12 \int_x \int_0^1 dt$$

$$\mathrm{str}\,\{\epsilon^{\underline{\alpha}}[\ \underline{\phi}(\mathcal{F}_t)_{\underline{\alpha}_i}\underline{\mathcal{F}}_t\underline{\mathcal{F}}_t - \frac{1}{3}\phi_{\underline{\alpha}}\underline{\mathcal{F}}_t\underline{\mathcal{F}}_t\underline{\mathcal{F}}_t$$

$$-\frac{i}{64}\,(\mathcal{F}_{\underline{\alpha}_i}\underline{\mathcal{F}}\mathcal{F}_{\underline{\alpha}_2 i_2}\epsilon^{i_1 i_2}\underline{\delta}^{[\alpha_2,\alpha_1]}\mathcal{F}_{\underline{\alpha}_1 i_1})]\}_{\theta=0} \quad (30)$$

where

$$\underline{\phi} = e^{\gamma\gamma'}\phi_{\gamma\gamma'},$$

$$\mathcal{F}_{\underline{\alpha}_i} = e^{\gamma\gamma'}\mathcal{F}_{\underline{\alpha}_i\gamma\gamma'},$$

$$\underline{\mathcal{F}} = e^{(\gamma\gamma')_1}e^{(\gamma\gamma')_2}\mathcal{F}_{(\gamma\gamma')_2(\gamma\gamma')_1}$$

$$\underline{\delta}^{[\alpha_2,\alpha_1]} = e^{\gamma\gamma'}\delta^{[\alpha_2,\alpha_1]}_{\gamma\gamma'}. \quad (31)$$

and similar definitions for $\mathcal{F}_t$ and $(\mathcal{F}_t)_{\underline{\alpha}i}$.

I would like to thank Professor B. Zumino for reading the manuscript, encouragement and kind help, Professor Y.S. Wu and Professor Orlando Alvarez for useful discussions and Professor Mary K. Gaillard, LBL and U.C. Berkeley for warm hospitality.

REFERENCES

[1] H. Itoyama, V. P. Nair and Hai-cang Ren, Phys. Lett. 157B (1985) 179, Fermilab-PUB-85-139-T.

[2] A.Galperin, E. Ivanov, S. Katitzin V. Ogievetsky and E. Sokatchev, Class. Quantum Grav. 1 (1984) 469.

[3] P.S. Howe, K.S. Stelle and P.C. West, Class. Quantum Grav. 2 (1985) 815.

[4] T. Kugo and P.K. Townsend, Nucl. Phys. B221 (1983) 357.

[5] P.S. Howe, G. Sierra and P.K. Townsend, Nucl. Phys. B211 (1983) 331.

[6] B. Zumino, *Chiral Anomalies and Differential Geometry* in Les Houches Summer School, 1983, edited by B.S. DeWitt and R. Stora, North Holland, Amsterdam.

[7] Zi Wang, Yong-shi Wu, Phys. Lett. 164B (1985) 305.

FOUR-LOOP σ-MODEL β-FUNCTIONS AND IMPLICATIONS FOR SUPERSTRINGS

M. T. Grisaru
Brandeis University, Waltham, Massachusetts 02254, USA
A. M. van de Ven
University of Maryland, College Park, Maryland 20742, USA
D. Zanon
Harvard University, Cambridge, Massachusetts 02138, USA

We present results of four-loop calculations for supersymmetric non-linear σ-models in two dimensions. The four-loop gravitational β-function computed for both N=1 and N=2 models does not vanish on Ricci-flat Kahler manifolds. We derive the action whose variation reproduces the N=1 equations $\beta_{ij} = 0$, interpreted as equations of motions for the graviton. This action agrees with the low-energy effective action derived by a direct analysis of superstring scattering amplitudes.

According to fairly well accepted wisdom, the low-energy physics of strings can be extracted from a study of two-dimensional non-linear σ-models (1). In this approach one takes as target manifold 10 or 26 dimensional curved space and requires that the β-function corresponding to the renormalization of the σ-model metric (and other fields describing the massless modes of the string) vanish. The resulting equations are interpreted as equations of motion for these fields and can be used to determine both the vacuum configurations (e.g. possible compactifications to four dimensions) and string amplitudes.

Calculation of the metric (graviton) β-function at the one-loop level reveals that it is proportional to the Ricci tensor; requiring that it vanish leads to the Einstein equations, which is what one would expect. One may perhaps expect higher order modifications of this result, since in this approach one has integrated out the massive modes of the string. However, for the case of N=2 supersymmetry at least, it was thought that the corresponding corrections would still vanish for $R_{ij} = 0$ so that Ricci-flat Kahler (because of N=2 supersymmetry) manifolds would be vacuum solutions of the superstring.

We have computed the four-loop β-function for both N=1 and N=2 supersymmetric two-dimensional σ-models (2). For the N=2 case we obtained a result which does not vanish on Ricci-flat manifolds. For the N=1 case the equation $\beta_{ij} = 0$ leads to results in agreement with an analysis of superstring graviton scattering amplitudes through order $(\alpha')^3$ (3).

Since we are dealing with supersymmetric theories, the most efficient way to compute quantum corrections is through the use of superfields and supergraphs. The σ-model actions for the N=1 and N=2 theories are

$$S_1 = \frac{1}{4}\int d^2x d^2\theta \; g_{ij}(\phi) D_\alpha \phi^i D^\alpha \phi^j$$

$$S_2 = \int d^2x d^2\theta d^2\bar{\theta} \; K(\phi^\mu, \bar{\phi}^{\bar{\nu}}). \tag{1}$$

Here ϕ^i, $i = 1,\ldots,10$, are real N=1 unconstrained superfields and $g_{ij}(\phi)$ is the σ-model metric. For the N=2 case the natural description is in terms of chiral superfields ϕ^μ, satisfying the chirality constraint $\bar{D}_\alpha \phi^\mu = 0$. The σ-model geometry is described now by the Kahler potential K. In complex coordinates the corresponding metric is

$$g_{\mu\bar{\nu}} = \frac{\partial^2 K}{\partial\phi^\mu \partial\bar{\phi}^{\bar{\nu}}}. \tag{2}$$

We calculate radiative corrections and the divergent part of the effective action by using the background field method. In the N=1 case the quantum-background splitting is by means of the normal-coordinate expansion; the calculation and result are then manifestly covariant. For the N=2 case we cannot use the normal coordinate expansion because of the chirality constraint on ϕ^μ. We perform a linear quantum-background splitting. The intermediate stages of the calculation are not covariant; however, because of the manifest N=2 supersymmetry of our approach, we can show that the final result will be covariant.

For the N=1 case, the quantum-background splitting leads to

$$S = \int d^2x d^2\theta \Big\{ \frac{1}{4} g_{ij} \nabla_\alpha \xi^i \nabla^\alpha \xi^j + \frac{1}{4} R_{ikmj} D_\alpha \phi^i D^\alpha \phi^j \xi^k \xi^m \tag{3}$$

$$+ \frac{1}{12} R_{ikmj;n} D_\alpha \phi^i D^\alpha \phi^j \xi^k \xi^m \xi^n$$

$$+ \cdots \Big\}, \qquad (3)$$

where ξ^i are quantum fields, R_{ijmn} is the Riemann tensor, and the spinor derivative ∇_α is defined as

$$\nabla_\alpha \xi^i \equiv D_\alpha \xi^i + \Gamma^i_{jk} D_\alpha \phi^k \xi^j. \qquad (4)$$

For N=2 the corresponding expansion is

$$S = \int d^2x\, d^2\theta\, d^2\bar\theta \Big\{ K_{\mu\bar\nu} \phi^\mu \bar\phi^{\bar\nu} + $$

$$+ \frac{1}{2} K_{\mu\nu} \phi^\mu \phi^\nu + \frac{1}{2} K_{\bar\mu\bar\nu} \bar\phi^{\bar\mu} \bar\phi^{\bar\nu}$$

$$+ \frac{1}{2} K_{\mu\nu\bar\rho} \phi^\mu \phi^\nu \bar\phi^{\bar\rho} + \cdots \Big\}, \qquad (5)$$

where

$$K_{\mu\nu\cdots\bar\mu\bar\nu\cdots} = \frac{\partial K}{\partial\phi^\mu \partial\phi^\nu \cdots \partial\bar\phi^{\bar\mu} \partial\bar\phi^{\bar\nu} \cdots} \qquad (6)$$

In principle it is sufficient to calculate the β-function for the N=1 case and then specialize to a Kahler manifold; this gives the N=2 β-function up to terms corresponding to certain field redefinitions. However, because the calculation is much simpler, and because it provides a good check of our methods, we have done the N=2 calculation as well.

We use conventional supergraph and D-algebra techniques, and regularization by dimensional reduction. At the four-loop level we find a contribution represented by the diagram in Figure (1), which, after subtracting subdivergences, leads to a $1/\varepsilon$ divergence and a four-loop N=1 β-function given by

$$\beta_{ij} = \zeta(3)/3(4\pi)^4 \quad T_{(ij)}, \qquad (4)$$

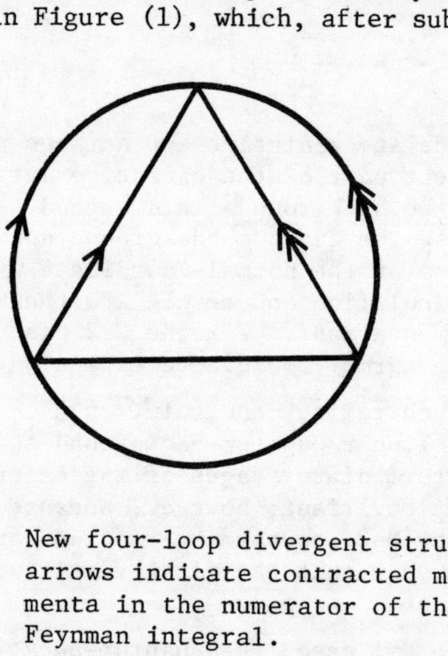

Figure 1: New four-loop divergent structure; arrows indicate contracted momenta in the numerator of the Feynman integral.

where

$$T_{ij} = 2R_{nijm;hk} (R^{msrk} R^n{}_{(sr)}{}^h$$

$$+ R^{msrn} R^k{}_{sr}{}^h)$$

$$+ 4R_{jkmn;[ih]} R^{m(rk)s} R^n{}_{sr}{}^h$$

$$+ 3(R_{ikht;r} R^{tsrq} R_j{}^{hk}{}_{q;s}$$

$$+ R_{irkt;h} R^t{}_s{}^{rq} R_j{}^{sh}{}_q{}^k{}_;$$

$$+ 2R_{ikht;r} R^t{}_s{}^{rq} R_j{}^{sk}{}_q{}^h{}_;)$$

$$+ (2R_{rqst;l} - R_{rsqt;l}) R^t{}_h{}^{rk} R^{hqs}{}_{k;j}$$

$$- 12R_{mhkl} R_{jrt}{}^m (R^k{}_{qs}{}^r R^{tqsh}$$

$$+ R^k{}_{qs}{}^t R^{hrsq}). \qquad (8)$$

Specializing to a Kahler manifold and making a field redefinition, or by direct computation in the N=2 case, we find

$$\beta_{\mu\bar\nu} = \frac{4}{3} \frac{\zeta(3)}{(4\pi)^4} \nabla_\mu \nabla_{\bar\nu} \Delta K, \qquad (9)$$

where

$$\Delta K = R_\lambda{}^\tau{}_\rho{}^\kappa R_\tau{}^\sigma{}_\kappa{}^\omega R_\sigma{}^\lambda{}_\omega{}^\rho$$

$$+ R_\lambda{}^\rho{}_\tau{}^\kappa R_\kappa{}^\tau{}_\sigma{}^\omega R_\omega{}^\sigma{}_\rho{}^\lambda \qquad (10)$$

We note that ΔK is proportional to the Euler density in 6 dimensions. As mentioned earlier, the β-function does not vanish for a Ricci-flat Kahler metric. It does vanish for a hyperkahler metric (N=4), but not for N=1 locally symmetric spaces. It would appear therefore that the only finite two-dimensional σ-models are the ones with N=4 supersymmetry (or with N=2 supersymmetry on Ricci-flat locally symmetric spaces).

Given the N=1 gravitational equations of motion $\beta_{ij} = 0$, one can construct the corresponding action through order $(\alpha')^3$ (4). It is given by

$$S = \int \sqrt{g}\, (-R + \frac{\alpha'^3 \zeta(3)}{8} [L_1 - 2L_2 + \frac{2}{9} L_3]) \qquad (11)$$

where

$$L_1 = R_{hmnk} R_p{}^{mn}{}_q R^{hrsp} R^q{}_{rs}{}^k$$

$$+ \frac{1}{2} R_{hkmn} R_{pq}{}^{mn} R^{hrsp} R^q{}_{rs}{}^k$$

$$L_2 = R^{hk}[\frac{1}{2} R_{htrk} R^{msqt} R_{msq}{}^{r}$$
$$+ \frac{1}{4} R_{htmn} R_{k}{}^{tqs} R_{qs}{}^{mn}$$
$$+ R_{hmnp} R_{kqs}{}^{p} R^{nqsm}]$$
$$L_3 = R[\frac{1}{4} R_{htmn} R^{htqs} R_{qs}{}^{mn}$$
$$+ R_{hmnp} R^{h}{}_{qs}{}^{p} R^{nqsm}]. \qquad (12)$$

The action in (11) can be compared with a direct superstring analysis of the four-particle dual-model S-matrix. For this comparison only L_1 is relevant (since L_2 and L_3 vanish "on-shell). One can show that the combination of Riemann tensors that appears in L_1 is the unique combination that vanishes on a Kahler manifold. The superstring analysis also leads to an effective Lagrangian that vanishes on a Kahler manifold (3,5), and this feature establishes the agreement between the σ-model approach and the superstring result.

The next step in the σ-model approach is a study of the β-functions associated with the other massless modes of the string, the dilaton and the antisymmetric tensor. The determination of the dilaton β-function is of particular interest because it is supposed to equal the effective Lagrangian describing these massless modes (6). However, its direct computation requires going one loop higher than the computation of the gravitational β-function. We have taken the first step in this direction, and obtained the five-loop gravitational contributions to the dilaton β-function on a Kahler manifold (N=2) (7). However, because of this restriction, it does not provide us with the complete information we would like to have. We hope to carry out the corresponding calculation for the N=1 case in the near future.

The implications of our results are as follows:

a) The σ-model approach to the superstring low-energy effective action appears to be in agreement with a direct analysis.

b) The σ-model metric β-function does not vanish on a Ricci-flat Kahler manifold and this seems to bring into question the possibility of compactification on a Calabi-Yau manifold. However, it appears that the physical metric, i.e. the solution of the equations $\beta_{ij} = 0$ is an acceptable metric for a Calabi-Yau manifold so that the corresponding compactification may still be viable (8).

REFERENCES

(1) C. Lovelace, Phys. Lett. 135B (1984) 75
C. G. Callan, D. Friedan, E. J. Martinec, and M. J. Perry, Nucl. Phys. B262 (1985) 593;
A. Sen, Phys. Rev. D32 (1985) 2102; Phys. Rev. Lett. 55 (1985) 1846;
E. S. Fradkin and A. A. Tseytlin, Phys. Lett. 158B (1985) 316; Nucl. Phys. B261 (1985) 1;
B. E. Fridling and A. Jevicki, Phys. Lett. 174B (1986) 75.

(2) M. T. Grisaru, A. van de Ven, and D. Zanon, Phys. Lett. 173B (1986) 423; Nucl. Phys. B277 (1986) 388; Nucl. Phys. B277 (1986) 409.

(3) D. Gross and E. Witten, Princeton preprint (1986);
M. D. Freeman and C. N. Pope, Phys. Lett. 174B (1986) 48.

(4) M. T. Grisaru and D. Zanon, Harvard preprint HUTP-86/A046 (1986).

(5) M. D. Freeman, C. N. Pope, M. F. Sohnius, and K. S. Stelle, Imperial/TP/85-86/27 preprint (1986).

(6) C. G. Callan, I. R. Klebanov, and M. J. Perry, Princeton preprint (1986).

(7) M. T. Grisaru and D. Zanon, to be published.

(8) D. Nemeschansky and A. Sen, SLAC-PUB-3925 preprint (1986).

PARTIAL SUPERSYMMETRY BREAKING IN N=4 SUPERGRAVITY

M. de Roo and P. Wagemans

Institute for Theoretical Physics, P.O. Box 800
9700 AV Groningen, the Netherlands

The structure of matter coupled N=4 supergravity is discussed. It is shown that gauged N=4 supergravity with matter can exhibit breaking to N=1,2 or 3 supersymmetries. In all examples the cosmological constant is negative.

The structure of N=1 supergravity with matter can be considered well-established at present. The potential phenomenological significance of models based on N=1 was discussed in this session by Ibáñez [1]. In our opinion it is interesting to study these matter coupled systems also beyond the context of N=1 supergravity, and in particular to investigate, in extended supergravity theories, the phenomenon of spontaneous breaking of local supersymmetry. Because of its non-chiral structure, extended supergravity has no immediate phenomenological applications. However, it does appear naturally in compactifications of higher-dimensional supergravity. For example, N=4 supergravity coupled to matter arises from certain compactifications of the ten-dimensional N=1 supergravity theory with matter [2], which is the low-energy limit of superstring theories. Also, the relation between Poincaré and conformal supergravity theories, which is important in matter coupling, deserves further attention.

In four dimensions matter multiplets (i.e. multiplets with physical fields of spin ≦ 1 only) exist only for N ≦ 4. For all these values of N the coupling to the corresponding supergravity theory has now been achieved. For N=2 this was done using the superconformal tensor calculus, which employs the known off-shell structure of the N=2 multiplets. For N=3 and 4 the auxiliary field structure is not known (and may not exist), so that one has to resort to other methods. Recently, matter coupling in N=3 supergravity was constructed using the group manifold approach [3].

The coupling of N=4 supergravity to N=4 vector multiplets (see Table 1) was obtained using superconformal methods [4]. For N=4 the off-shell conformal supergravity (Weyl) multiplet is known [5] and only for the matter multiplet is the off-shell extension absent. It is nevertheless possible to put the vector multiplet in a superconformal background. This requires the extension of the global N=4 supersymmetry to a set of local transformations which satisfy the superconformal algebra. The non-closure terms are then identified as equations of motion, and the corresponding Lagrangian density can be constructed. The resulting action is invariant under the full superconformal symmetry and can easily be extended to an arbitrary number of vector multiplets. The theory can be cast in the more conventional Poincaré supergravity form by breaking the superconformal dilations and S-supersymmetries with appropriate gauge choices and by eliminating auxiliary fields.

The fields of N=4 Poincaré supergravity coupled to n vector-multiplets are presented in Table 2. The metric η_{IJ} is given by

$$\eta_{IJ} = \mathrm{diag}(-,-,-,-,-,-,+,+,\ldots,+). \quad (1)$$

The signature of η is determined by the absence of ghosts. The theory has a global SO(6,n) symmetry of the action, under which all fields with indices I transform in the fundamental representation. There is an SU(1,1) symmetry of the equation of motion, of which SO(1,1) is in fact a symmetry of the action as well.

field			restrictions (i=1,...,4)
ϕ_{ij}	0	6	$\phi_{ij} = -\phi_{ji}$ $\phi^{ij} \equiv (\phi_{ij})^* = -\frac{1}{2}\varepsilon^{ijk\ell}\phi_{k\ell}$
ψ_i	$\frac{1}{2}$	$\bar{4}$	$\psi_i = -\gamma_5\psi_i$ $\psi^i \equiv (\psi_i)^* = \gamma_5\psi^i$
A_μ	1	1	gauge field; real

Table 1. The fields of the N=4 vector multiplet. The numbers indicate the spin, and SU(4) representation, resp.

field			restrictions
ϕ_α	0	1	$\alpha = 1,2$; $\phi^\alpha\phi_\alpha = 1$ $\phi^1 \equiv (\phi_1)^*$; $\phi^2 \equiv -(\phi_2)^*$
ϕ_{ij}^I	0	6	$I,J = 1,\ldots,6+n$ $\phi_{ij}^I\eta_{IJ}\phi^{k\ell J} = -\frac{1}{2}\delta^k_{[i}\delta^\ell_{j]}$
Λ_i	$\frac{1}{2}$	$\bar{4}$	$\Lambda_i = \gamma_5\Lambda_i$ $\Lambda^i \equiv (\Lambda_i)^* = -\gamma_5\Lambda^i$
ψ_i^I	$\frac{1}{2}$	$\bar{4}$	$\phi_{ij}^I\eta_{IJ}\psi^{kJ} = 0$
A_μ^I	1	1	gauge field; real
ψ_μ^i	$\frac{3}{2}$	4	$\psi_\mu^i = \gamma_5\psi_\mu^i$; gravitino
e_μ^a	2	1	vierbein

Table 2. The fields of N=4 Poincaré supergravity with n vector multiplets. The numbers indicate the spin and SU(4) representation, resp. The metric η_{IJ} is given in (1).

The scalar fields ϕ_α are an SU(1,1) doublet, and transform under a local U(1) symmetry in such a way that they parametrize the SU(1,1)/U(1) coset space. The theory also has a local SU(4) symmetry, of which the transformation character is indicated in Table 2. These local symmetries do not have independent gauge fields. The fields $\phi_{ij}{}^I$ and $\psi_i{}^I$ satisfy constraints which arise from the superconformal gauge choices and the elimination of auxiliary fields. The scalars $\phi_{ij}{}^I$ (inert under local U(1)) parametrize the manifold SO(6,n)/SO(6) × SO(n). For n=0 one recovers pure N=4 Poincaré supergravity. In this case $\phi_{ij}{}^I$ and $\psi_i{}^I$ do not represent physical degrees of freedom.

It is possible to extend the gauge invariance associated with the vector fields $A_\mu{}^I$ to a non-abelian group [6,7]. The vector multiplets transform under the adjoint representation of G, which must therefore be embedded in SO(6,n). The metric η_{IJ} must be invariant under G:

$$f_{KL}{}^I\eta_{IJ} + f_{KJ}{}^I\eta_{LI} = 0 \qquad (2)$$

where $f_{KL}{}^I$ are the structure constants of G. This imposes no constraint on the compact simple factors of G (for n large enough), but only a small number of non-compact simple factors can be gauged [7]. The gauging of internal symmetries requires modifications to preserve supersymmetry, both in the Lagrangian and in the supersymmetry transformation rules. These include a scalar potential and fermion mass terms.

Gauged N=4 supergravity without additional matter fields (n=0) has been known for some time. There are three inequivalent versions, all with gauge group G = SO(3) × SO(3) [8,9,10]. For vanishing coupling constants these theories are related by a duality transformation of three of the six vector fields, associated with one of the SO(3) factors in G. The duality transformation belongs to the SU(1,1) symmetry of the equations of motion. The gauging procedure explicitly breaks this symmetry, and the resulting theories are inequivalent.

In our construction the freedom of choosing different SU(1,1) orientations before gauging is extended to matter multiplets coupled to supergravity, and arises as follows. Each matter multiplet, labelled by I, is coupled to the N=4 Weyl multiplet, which contains the SU(1,1) doublet ϕ_α [5]. For each value of I one may introduce a parameter α_I, and replace ϕ by [7]:

$$\begin{pmatrix} \phi_{1(I)} \\ \phi_{2(I)} \end{pmatrix} \equiv \begin{pmatrix} e^{-i\alpha_I} & 0 \\ 0 & e^{+i\alpha_I} \end{pmatrix} \begin{pmatrix} \phi_1 \\ \phi_2 \end{pmatrix} . \quad (3)$$

Gauging a semi-simple group G then leads to inequivalent theories, depending on the α_I. Supersymmetry requires that the α_I must be the same for values of I associated with the same simple factor of G.

The scalar potential of the resulting theory reads

$$V = \frac{4}{9}|X_{ij}|^2 - \frac{1}{2}\eta_{IJ}\,|\Phi_{(I)}|^2\, W_i^{\,jI}\, W_j^{\,iJ} , \quad (4)$$

where

$$W_i^{\,jI} = f_{KL}{}^{I}\,\phi_{ik}{}^{K}\,\phi^{kjL} ;$$

$$\left(W_i^{\,jI}\right)^* = -W_j^{\,iI} , \quad (5)$$

$$X_{ij} = \eta_{IJ}\,\Phi^*_{(I)}\,\phi_{ik}{}^{I}\,W_j^{\,kJ} , \quad (6)$$

$$\Phi_{(I)} = e^{i\alpha_I}\phi^1 + e^{-i\alpha_I}\phi^2 \quad (7)$$

and η_{IJ} is given by (1). Note that the potential is not positive definite, in contradistinction to globally supersymmetric theories. The negative contribution arises from the six compensating multiplets, which are associated with the minus-signs in η_{IJ}. The potential is a sum of squares, some of which have negative coefficients. One expects a wealth of stationary points, which may have broken, unbroken or partially broken supersymmetry.

The value of the scalar potential in the stationary point corresponds to the cosmological constant. In the absence of ghosts, a positive cosmological constant necessarily implies completely broken supersymmetry [11] (also in global supersymmetry a positive value of the scalar potential breaks all supersymmetries). Therefore (partially) unbroken supersymmetry is possible only for a cosmological constant that is negative (anti-de Sitter space) or zero (Minkowski space).

It is not difficult to construct theories with completely unbroken supersymmetry in Minkowski space. If one couples N=4 Yang-Mills multiplets with a compact gauge group G to ungauged N=4 supergravity, i.e. $f_{KL}{}^{I} = 0$ for index values ≤ 6, the potential V is positive semi-definite, with a stationary point at V = 0. Note that V = 0 implies that $W_i^{\,jI}$ and X_{ij} must vanish. This can be realized for example by $\phi_{ij}{}^{I} = 0$ for $I \geq 7$, which is compatible with the constraint on $\phi_{ij}{}^{I}$ (see Table 2).

Completely broken supersymmetry in Minkowski space is also possible. There is one case where the potential (4) vanishes completely: $V \equiv 0$. This requires one matter multiplet (n=1) and gauge group G = SO(3) X SO(2,1), with coupling constants g_1 and g_2 satisfying $g_1^2 = g_2^2$. This is the only situation in which the potential (4) becomes flat, as can be shown by evaluation of the scalar potential for a few carefully chosen field configurations. Such flat potentials were previously found in N=1 [12] and N=2 [13] supergravity.

To establish the existence of N=4 theories with partial supersymmetry breaking [14] we have considered gauge groups $G = [SO(3)]^K$, $K > 2$. This choice simplifies many of the calculations. These gauge groups may however not be devoid of interest, as they occur in certain compactifications of string theories [15]. To further simplify the problem we have limited ourselves to ground state field configurations which have an isometry group H. With H = SO(3) it is possible to find an embedding in

SU(4) such that the four gravitinos transform as $3 \oplus 1$. Then there are always three equal gravitino masses, and breaking to N=1 and N=3 can be realized, including some degenerate situations with four surviving supersymmetries. Two equal gravitino masses can be obtained with H = SO(2), and we have constructed examples with unbroken N=2 (for $K \geq 4$ only). Explicit calculations can be found in [14]. In all cases the cosmological constant is negative. It remains an open problem, whether or not partial breaking in N=4 is possible in Minkowski space.

Let us finally mention that stationary points with partial supersymmetry breaking can be found without explicitly analyzing the scalar potential [16]. The super-Higgs effect occurs in a stationary point of the theory if the matter fields at this point are not invariant under a supersymmetry transformation. In a constant background without fermions condensation this means that

$$\langle \delta \chi_i \rangle = \langle A^j_{ij} \rangle \varepsilon^j \neq 0 \ , \qquad (8)$$

where χ_i is a generic spin-$\frac{1}{2}$ field, and A_{ij} a function of the scalar fields. Partial breaking occurs whenever $\langle A_{ij} \rangle$ has one or more zero eigenvalues. It was recently demonstrated [16] that a zero eigenvalue of A_{ij} for a particular field configuration implies that this configuration is a stationary point of the theory. This observation greatly facilitates the search for (partially) unbroken supersymmetry. It has also been employed succesfully in the analysis of matter coupled N=3 supergravity [17].

REFERENCES

1. L. Ibáñez, "Status Report on Low Energy Supergravity", these Proceedings.
2. C. Kounnas, "Reduction of D=10 Supergravity and Superstrings", these Proceedings;
   S. Ferrara, C. Kounnas and M. Porrati, preprint UCB-PTH-86/5; UCLA/86/TEP/8.
3. L. Castellani, A. Ceresole, S. Ferrara, R. D'Auria, P. Fré and E. Maina, Nucl. Phys. B268 (1986) 317.
4. M. de Roo, Nucl. Phys. B255 (1985) 515.
5. E. Bergshoeff, M. de Roo and B. de Wit, Nucl. Phys. B182 (1981) 173.
6. M. de Roo, Phys. Lett. 156B (1985) 331;
   E. Bergshoeff, I.G. Koh and E. Sezgin, Phys. Lett. 155B (1985) 71.
7. M. de Roo and P. Wagemans, Nucl. Phys. B262 (1985) 644.
8. A. Das, M. Fischler and M. Roček, Phys. Rev. D16 (1977) 3427.
9. S.J. Gates, Jr. and B. Zwiebach, Phys. Lett. 123B (1983) 200.
10. D.Z. Freedman and J.H. Schwarz, Nucl. Phys. B137 (1978) 333.
11. K. Pilch, P. van Nieuwenhuizen and M. Sohnius, Comm. Math. Phys. 98 (1985) 105;
    J. Lukierski and A. Nowicki, Phys. Lett. 151B (1985) 382.
12. E. Cremmer, S. Ferrara, C. Kounnas and D.V. Nanopoulos, Phys. Lett. 133B (1983) 61;
    N.-P. Chang, S. Ouvry and X. Wu, Phys. Rev. Lett. 51 (1983) 327.
13. E. Cremmer, C. Kounnas, A. van Proeyen, J.P. Derendinger, S. Ferrara, B. de Wit and L. Girardello, Nucl. Phys. B250 (1985) 385.
14. M. de Roo and P. Wagemans, Phys. Lett. B, to be published.
15. L. Dolan, "Modular Subgroup Invariance in a Four-Dimensional String", these Proceedings.
16. S. Cecotti, L. Girardello and M. Porrati, Nucl. Phys. B268 (1986) 295.
17. S. Ferrara, P. Fre and L. Girardello, Nucl. Phys. B274 (1986) 600.

Comment on Nonlinear Aspects of Kaluza Klein Theories

H. Nicolai
Institut für Theoretische Physik
Universität Karlsruhe

Compactification of extra dimensions has become an essential ingredient in all recent attempts at a unification of fundamental interactions. These extra dimensions parametrize a (usually compact) manifold whose size is sufficiently small to prevent its experimental detection at presently accessible energies. The groundstate has the topology of a product manifold

$$M_D \longrightarrow M_4 \times M_{D-4} \qquad (1)$$

where D is the dimension of the original theory and M_4 is the four dimensional space-time, which should be preferably Minkowski space. The coordinates are split accordingly

$$z^M \longrightarrow (x^\mu, y^m) \qquad (2)$$

and the four dimensional fields are related to the x-dependent coefficient functions in the expansion of the D-dimensional fields in terms of a suitable complete set of functions of y, i.e.

$$\phi(x,y) = \sum_n \phi^n(x) \; Y^n(y) \qquad (3)$$

It is most convenient to choose the functions $Y^n(y)$ to be eigenfunctions of a certain mass operator (which is determined through the higher dimensional theory) such that the fields ϕ^n are associated with certain eigenvalues m_n of this mass operator. When analyzing a Kaluza Klein theory from this point of view, one is usually interested in an effective low energy theory that describes the interactions of the zero mass fields. At energies which are small compared to the size of the internal manifold, one thus truncates the expansion (3) by setting

$$\phi(x,y) = \phi^o(x) \; Y^o(y) \qquad (4)$$

where ϕ^o schematically denotes all the zero mass fields (quarks, leptons, SU(3)xSU(2)xU(1) gauge bosons, etc.). Given this set of fields, one then tries to work out the low energy interactions (i.e., those involving no massive particles) and their couplings. Naively one might think that this can be accomplished by substituting (4) back into the interaction vertices of the higher-dimensional theory and performing the integration over the internal manifold. However, in general this procedure will <u>not</u> lead to the correct results. This is so because, in general, (4) will have to be replaced by the formula

$$f[\phi(x,y)]=\phi^{o}(x)\ Y^{o}(y) \qquad (5)$$

where f is some non-linear function such that

$$f(t) = t + O(t^2) \qquad (6)$$

The necessity of such modifications has been demonstrated recently in an analysis of the consistency of the S^7 truncation in D=11 supergravity (see [1] and references therein). To see what their effect is on higher order interactions, we solve (5) for $\phi(x,y)$ by inverting (6). The result can be re-expanded in terms of the linearized eigenmodes in (3) such that

$$\phi(x,y)=\sum_n f_n[\phi^{o}(x)]Y^{n}(y) \qquad (7)$$

with new(non-linear) functions f_n.
The fact that $\phi^{o}(x)$ is a zero mass field guarantees that f_0 starts off linearly

$$f_0[\phi(x)]=\phi^{o}(x) + O[\phi^{o}(x)^2] \qquad (8)$$

while the other f_n's have no linear term, i.e.,

$$f_n[\phi^{o}(x)] = a_n\phi^{o}(x)^2 + O[\phi^{o}(x)^3] \qquad (9)$$

Substituting the correct non-linear ansatz (7) back into the higher dimensional action, we see that additional contributions to the interactions of the zero mass fields arise from the higher modes in (7) after integration over the internal manifold: From (8), (9) it follows that these modifications exist for all k-point interactions $\phi^{o}(x)^k$ for $k\geq 4$. They are only absent for three-point vertices which are the only ones that can be safely calculated by the naive method. Since the effective low energy requires knowledge of all renormalizable couplings, the low energy theory is obviously sensitive to these non-linear effects. More specifically,this concerns the quartic interactions of the scalar fields which play an essential role in the final step of the symmetry breaking from $SU(2)_W \times U(1)_Y$ to $U(1)_{em}$. The relevant couplings are not suppressed by inverse powers of the Planck scale but on the contrary, sensitive to the non-linear ansätze described above. The problem becomes even more acute when one attemps to extract <u>non-polynomial</u> interactions from the higher dimensional theory as in currently popular N=1 supergravity models;
although, in this case, many coupling constants are interrelated by supersymmetry, the correct determination of the superpotential will still require the identification of all nonlinear modifications. Thus, conclusions about higher order interactions in the effective four dimensional theory which are based on linear truncations as in (4) are inherently unreliable.
It should be emphasized that the above problem arises even before one considers modifications of the four dimensional effective action through quantum effects due to massive particles circulating in loops. The latter can be computed in principle

by means of

$$\exp iS_{eff}(\phi^{o}) = \int \prod_{n\neq o} d\phi^{n} \exp iS(\phi^{o},\phi^{n}) \tag{10}$$

The nonlinear modifications induced by (5) are already present at the classical level and must therefore be distinguished from the ones induced by (10). One may anticipate that a correct computation of the full quantum action by use of (10) is only possible if the nonlinear effects at the classical level have been properly taken into account.

[1] B. de Wit and H. Nicolai,"The Consistency of the S^7-Truncation in D=11 Supergravity", preprint CERN—TH 4359(1986), to appear in Nucl. Phys. B.

N - EXTENDED d=2 SUPERCONFORMAL ALGEBRAS[+]

R. Gastmans[*], A. Sevrin[**], W. Troost[***] and A. Van Proeyen[***]
Instituut voor theoretische fysica, University of Leuven
B-3030 Leuven, Belgium

We construct new infinite dimensional superalgebras which contain the Virasoro algebra and N fermionic dimension 3/2 operators of the "Neveu-Schwarz type". A well defined set of assumptions leads to the known N=2 and N=4 algebras.

1. INTRODUCTION

Superconformal algebras are useful tools to obtain matter couplings in supergravity theories(1). In string theories the conformal symmetry which acts in the 2-dimensional world sheet is even more important, as it is required for the consistency of the theory. In the search for new models a clear knowledge of the possible superconformal algebras is therefore important. Superconformal algebras with N=1,2 and 4 supersymmetries are known (2) but it is neither well understood whether, nor why, these are the only algebras. A uniqueness proof has been given by Ramond and Schwarz (3), but we want to make the underlying assumptions more explicit. Our method simply consists in analyzing the Jacobi identities. Details will be given in a separate publication (4).

2. PRELIMINARIES

The conformal algebra in d dimensions is s o(d,2), but in 2 dimensions it consists of an infinite dimensional algebra with operators L_n and $\bar{L}_n$ (n=0, ±1, ± 2,...) of which $s\,o(2,2)$ is a subalgebra. The barred and unbarred sectors split completely so that we can further restrict ourselves to one sector, obeying the Virasoro algebra

$$[L_m, L_n] = (m-n)L_{m+n} + \frac{c}{12}(m^3-m)\delta_{m+n} \quad (1)$$

where c is a central extension. The superalgebra of Neveu and Schwarz(5) contains fermionic operators G_r (r=±1/2, ±3/2,...)

$$[L_n, G_r] = (\tfrac{1}{2}n-r)G_{n+r}$$

$$\{G_r, G_s\} = 2L_{r+s} + \frac{c}{3}(r^2-\tfrac{1}{4})\delta_{r+s} \quad (2)$$

The "supersymmetries" G_r are in a sense "square roots" of the space-time symmetries. In particular $G_{-1/2}$ is the usual supersymmetry which is the square root of L_{-1}, associated with translations. Another fermionic extension is the Ramond algebra (6) which is the same as eq.(2) but then r and s are integers. This implies that there is no square root of translations and we will not consider that case further. The N-extended superalgebras of Ademollo et al. (2), contain several operators G^i_r (i=1,...,N) which behave under the conformal transformations as supersymmetries

$$[L_n, G^i_r] = (\tfrac{1}{2}n-r)G^i_{n+r} \; . \quad (3)$$

They found an N=2 superalgebra in which they needed an extra bosonic operator V_n which forms a U(1) Kac-Moody algebra and an N=4 algebra which needs operators V^a_n (a=1,2,3) constituting an SU(2) Kac-Moody algebra.

Before describing our work we want to repeat the concept of conformal dimensions for operators X_m ($m=m_\circ$, $m_\circ \pm 1, m_\circ \pm 2$, ..). If an operator transforms under the conformal group as

$$[L_n, X_m] = (n(D-1)-m)X_{n+m} \; , \quad (4)$$

(where D is any number) we say that it is a primary operator of conformal dimension D. A secondary operator of dimensions D is an operator $X_m^{(2)}$ which transforms as

$$[L_n, X_m^{(2)}] = (n(D-1)-m)X_{m+n}^{(2)} + \text{terms with the primary operator } X_{n+m} \,. \tag{5}$$

This language stems from the theory of operator product expansions of conformal fields X(z) which is nicely explained in (7). However, we have to warn that for a general operator transforming under the conformal group, its commutator with L_n is not necessarily of the form (4) or (5).

3. GENERAL ALGEBRAS

Our work consists in looking for the general N extended superconformal algebras. This means that there are exactly N families of fermionic operators G_r^i ($r=\pm 1/2, \pm 3/2, \ldots; i=1,\ldots, N$) which are supersymmetries, i.e., they behave under the conformal group as in eq. (3). In other words, we have N dimension 3/2 (and no other) fermionic operators. In previous works fermionic operators of other conformal dimensions have been considered. Ademollo et al. (2) constructed algebras with 1 dimension 2, N dimension 3/2, $\binom{N}{2}$ dimension 1, ..., $\binom{N}{N}$ dimension 2-N/2. So for N>3 they contained dimension 1/2 and for larger N also negative dimension fermionic operators. Ramond and Schwarz (3) proved that under certain conditions there are no other algebras than N=2 and N=4 if one allows 1 dimension 2, N dimension 3/2, x dimension 1, y dimension 1/2 and no other operators. In a recent work of Bershadsky (8) fermionic operators of dimension 5/2 (and possibly higher) occur.

So far, we have already the commutation relations (1) and (3). The unknown quantity is the anticommutator of the fermionic operators

$$\{ G_r^i, G_s^j \} = X_{rs}^{ij} \,. \tag{6}$$

Its structure and commutation relations have to be determined by Jacobi identities. We obtain easily its behaviour under the conformal group

$$[L_n, X_{rs}^{ij}] = (\tfrac{1}{2}n-s)X_{r,s+n}^{ij} + (\tfrac{1}{2}n-r)X_{r+n,s}^{ij} \,. \tag{7}$$

So X_{rs} are operators where two indices run over an infinite range, and they do not have a conformal dimension. Further analysis of other Jacobi identities leads to the r,s,t, dependence of the $[X_{rs}, G_t]$ commutator. We obtain

$$[X_{rs}^{ij}, G_t^k] = ((s-r)O^{[ij]k}{}_{\ell} + (2t-r-s)O^{(ij)k}{}_{\ell}) G_{r+s+t}^{\ell} \tag{8}$$

where $O^{ijk}{}_{\ell}$ are analogous to "structure constants" which will determine the complete superalgebra. The remaining Jacobi identities give first a linear symmetry relation between these constants

$$O^{(ij)k}{}_{\ell} = O^{k(ij)}{}_{\ell} \,, \tag{9}$$

secondly several quadratic identities of the form

$$O^{ijk}{}_{h}\, O^{h\ell u}{}_{v} + \ldots = 0, \tag{10}$$

and thirdly relations on the X-operators

$$O^{ijk}{}_{h}\, X_{rs}^{h\ell} = Y_{r+s}^{ijk\ell} + (r-s)(V_{r+s}^{ijk\ell} + \tfrac{1}{2}(r+s)Z_{r+s}^{ijk\ell}) - (r^2+s^2-\tfrac{1}{2})Z_{r+s}^{ijk\ell} \,. \tag{11}$$

Here $Y_n^{ijk\ell}$, $V_n^{ijk\ell}$ and $Z_n^{ijk\ell}$ (the subscripts are integers) are new operators which still satisfy symmetry relations on their upper indices. If these 3 sets of relations are solved by tensors $O^{ijk}{}_{\ell}$ and operators X_{rs}^{ij} then all commutation relations are given and we have a consistent infinite dimensional superalgebra.

The specific r and s dependence of the r.h.s. in equation (11) leads in general to restrictions on the operators $X_{r,s}$. In fact, if $O^{ijk}{}_{\ell}$, seen as a $N^3 \times N$ matrix, has rank N then we can invert equation (11), which then implies

$$X^{ij}_{rs} \equiv \{G^i_r, G^j_s\} = Y^{ij}_{r+s} + (r-s)V^{ij}_{r+s} -$$

$$-(r^2+s^2-\frac{1}{2})Z^{ij}_{r+s} \qquad (12)$$

All commutation relations of the new operators follow from the previous results. In particular we find

$$[L_n, Y^{ij}_m] = (n-m)Y^{ij}_{n+m} - (n^3-n)Z^{ij}_{n+m}$$

$$[L_n, V^{ij}_m] = -m\, V^{ij}_{n+m}$$

$$[L_n, Z^{ij}_m] = -(n+m)Z^{ij}_{n+m} \qquad (13)$$

Comparing with equations (4) and (5) this implies that Y_m, V_m and Z_m have resp. conformal dimensions 2,1 and 0. Y_m is a secondary of the primary Z_m if the latter is not zero. We still have to obey the symmetry relations on $Y^{ijk\ell}_m$, $V^{ijk\ell}_m$ and $Z^{ijk\ell}_m$ which often (but not always) imply

$$Z^{ij}_m = z^{ij}\,\delta_m, \qquad (14)$$

where z^{ij} is then a central extension.

We obtained all the solutions of the equations for the case N=2. Apart from the known N=2 simple superalgebra from ref. (2) there also exist non simple algebras. Some of them contain an unrestricted operator X_{rs}, or a dimension zero operator. These algebras are all explicitly written down in ref. (4).

4. SPECIAL CASES

We now introduce more assumptions which will lead to the known N=2 and N=4 superalgebras. First we impose that there is no other dimension 2 operator apart from L_n. This leads to superalgebras where the anticommutator of the fermionic generators is

$$\{G^i_r, G^j_s\} = b^{ij}(L_{r+s} + \frac{c}{6}(r^2-\frac{1}{4})\delta_{r+s}) +$$

$$+(r-s)V^{ij}_{r+s}, \qquad (15)$$

where b^{ij} is a symmetric matrix which by redefinitions can be brought to a standard form

$$\begin{pmatrix} 1 & & & & 0 \\ & \ddots & & & \\ & & 1 & & \\ & & & 0 & \\ 0 & & & & \ddots \\ & & & & & 0 \end{pmatrix}. \qquad (16)$$

This still allows several possibilities for superalgebras with arbitrary N. They always have N-1 dimension 1 operators.

One example is

$$\{G^1_r, G^1_s\} = -2L_{r+s} - \frac{c}{3}(r^2-\frac{1}{4})\delta_{r+s},$$

$$\{G^1_r, G^a_s\} = (r-s)V^a_{r+s}, \quad a=2,\ldots,N$$

$$[V^a_n, G^1_r] = G^a_{n+r}, \qquad (17)$$

and, apart from the before mentioned commutators with L_n, the other (anti) commutators vanish.

Now we use one more input namely that each fermionic operator G is a "square root of L_n". By this we mean that in equation (15) the matrix b^{ij} is non singular and can thus be put equal to δ^{ij}. If these conditions are satisfied then we obtain the Ramond-Schwarz (3) result that only N=2 and N=4 extensions are possible.

5. CONCLUSIONS

We recapitulate the assumptions which lead to only N=2 and N=4
- there are only dimension 3/2 fermionic operators G^i_r
- only L_n exist as dimension 2 operators
- all G^i_r are "square roots" of L_n.

However we found that also other N extended superalgebras with only dimensions 3/2 fermionic operators exist. Some of them can contain bosonic operators X_{rs} where both indices run separately over the infinite range $r,s=\pm 1/2, \pm 3/2, \ldots$. Some contain also dimension zero operators. The open question is whether some of them are useful as underlying algebras for new string theories.

REFERENCES

+ Presented at the conference by A. Van Proeyen
* Onderzoeksleider NFWO, Belgium
** Onderzoeker IIKW, Belgium
***Bevoegdverklaard navorser NFWO, Belgium

(1) M.Kaku, P. Townsend and P. van Nieuwenhuizen,Phys. Rev. D17 (1978)3179
B. de Wit in "Supersymmetry and Supergravity 1982", eds. S. Ferrara, J.G. Taylor and P. van Nieuwenhuizen (World Scientific, 1983)
A. Van Proeyen, in "Supersymmetry and Supergravity 1983", ed. B. Milewski (World Scientific, 1983).

(2) M. Ademollo et al., Phys.Lett. B62 (1976)105; Nucl. Phys.B111 (1976)77;Nucl. Phys.B114(1976) 297

(3) P. Ramond and J.H. Schwarz, Phys.Lett.B64(1976)75

(4) R. Gastmans, A. Sevrin, W. Troost and A. Van Proeyen , Preprint Leuven, KUL-TF-86/6

(5) A. Neveu and J.H. Schwarz, Nucl. Phys. B31(1971)86

(6) P. Ramond, Phys. Rev.D3(1971) 2415

(7) A.A. Belavin, A.M. Polyakov and A.B. Zamolodchikov, Nucl.Phys. B241(1984) 333

(8) M. Bershadsky, Phys.Lett.174B (1986)285.

DUALITY TRANSFORMATIONS AND KÄHLER GEOMETRY IN SUPERSYMMETRIC THEORIES*

K.T. Mahanthappa
Dept. of Physics, University of Colorado
Boulder, CO 80309

G. M. Staebler
Dept. of Physics, VPI & SU
Blacksburg, VA 24061

The duality transformation between different off-shell representations of the scalar supermultiplet is shown to not always be invertible. The geometry of the supersymmetric non-linear σ-model in 3+1 dimensions is not necessarily Kähler, even on-shell, when the duality transformation fails.

1. INTRODUCTION

In 3+1 dimensions, the non-linear σ-models with global supersymmetry are known to have a Kähler geometry, if the chiral superfield representation of the scalar supermultiplet $(0^+, 0^-, \frac{1}{2})$ is used [1]. It is also known that the other off-shell representations of the scalar supermultiplet can be mapped into the chiral superfield one by a duality transformation [2], and hence are equivalent. This is the basis for the widely held belief that N=1 supersymmetry requires a Kähler geometry for the manifold of scalar fields in 3+1 dimensions. In this paper it will be shown that the duality transformation can fail to be invertible for special cases of the non-linear σ-model. This opens up the possibility of having a non-Kähler geometry, and new types of supersymmetric models, which are not equivalent to the ones based on chiral superfields, but which are found to be equally phenomenologically viable.

2. DUALITY AND GEOMETRY

The superspace action for the general non-linear σ-model for a set of chiral superfields [3] $\{S_a\}$ is given by [1]:

$$A = \int d^4x d^2\theta d^2\bar{\theta}\, K(S_a, \bar{S}_{\bar{a}}), \qquad (1)$$

where K is an arbitrary real function. Due to the chirality constraint on the superfields $\{S_a\}$ $(\bar{D}_{\dot\alpha} S_a = 0)$, this action is invariant under the Kähler transformation

$$\delta_g K(S_a, \bar{S}_{\bar{a}}) = g(S_a) + \bar{g}(\bar{S}_{\bar{a}}), \qquad (2)$$

where g is an arbitrary holomorphic function. Because of this invariance the component action can only depend on the Kähler metric

$$K_{a\bar{b}} \equiv \left.\frac{\partial^2 K}{\partial S_a \partial \bar{S}_{\bar{b}}}\right|_{\theta=\bar{\theta}=0}, \qquad (3)$$

and its derivatives, but not on the first or second derivatives; K_a, $K_{\bar{a}}$, K_{ab}, $K_{\bar{a}\bar{b}}$, as these are not invariant under (2). The metric $K_{a\bar{b}}$ is Kähler, since it is a Hermitian metric which can be expressed (locally) as the second derivative of some Kähler potential K.

Another off-shell representation of the $(0^+, 0^-, \frac{1}{2})$ multiplet is the complex linear superfield Σ_a, which satisfies the constraint $\bar{D}^2\Sigma_a = 0$. The non-linear σ-model is given by

$$A = \int d^4x d^2\theta d^2\bar{\theta}\, K(\Sigma_a, \bar{\Sigma}_{\bar{a}}), \qquad (4)$$

where K is again an arbitrary real function. This action is not invariant under Kähler transformations like (2), but only under the transformation

---

*talk given by G.M. Staebler

$$\delta_\lambda K(\Sigma_a, \bar{\Sigma}_{\bar{a}}) = \lambda^a \Sigma_a + \bar{\lambda}^{\bar{a}} \bar{\Sigma}_{\bar{a}} ,$$

$$\lambda_a = \text{constant}. \qquad (5)$$

This implies that the component action depends only on the metric

$$(K) \equiv \begin{pmatrix} K_{ab} & K_{a\bar{b}} \\ K_{\bar{a}b} & K_{\bar{a}\bar{b}} \end{pmatrix} , \qquad (6)$$

and its derivatives, but not on K_a or $K_{\bar{a}}$. The metric (6) appears non-Kähler, since it is not Hermitian. However, if we can perform a duality transformation to a chiral model, the metric will be simultaneously transformed into its Hermitian (Kähler) form.

The equations of motion for Σ_a are $\bar{D}_{\dot{\alpha}} K_a(\Sigma_a, \bar{\Sigma}_{\bar{a}}) = 0$, which are equivalent to the <u>duality transformation</u> equations

$$K_a(\Sigma_a, \bar{\Sigma}_{\bar{a}}) = S_a, \qquad (7)$$

where S_a are chiral superfields. If we can invert (7) to get $\Sigma_a = \Sigma_a(S_a, \bar{S}_{\bar{a}})$, then $K(\Sigma_a, \bar{\Sigma}_{\bar{a}}) = K(S_a, \bar{S}_{\bar{a}})$ implicitly, giving a chiral model. Under what conditions can the duality transformation (7) be inverted? Taking the differential of (7) and its conjugate gives

$$\begin{pmatrix} dS_a \\ d\bar{S}_{\bar{a}} \end{pmatrix} = \begin{pmatrix} K_{ab} & K_{a\bar{b}} \\ K_{\bar{a}b} & K_{\bar{a}\bar{b}} \end{pmatrix} \begin{pmatrix} d\Sigma_b \\ d\bar{\Sigma}_{\bar{b}} \end{pmatrix} . \qquad (8)$$

Provided that the Jacobian determinant is non-singular,

$$\det \begin{pmatrix} K_{ab} & K_{a\bar{b}} \\ K_{\bar{a}b} & K_{\bar{a}\bar{b}} \end{pmatrix} \neq 0, \qquad (9)$$

there exists a <u>unique</u> inverse matrix $(-\hat{K})$, such that $(K)\cdot(\hat{K}) = -1$, and

$$\begin{pmatrix} d\Sigma_a \\ d\bar{\Sigma}_{\bar{a}} \end{pmatrix} = - \begin{pmatrix} \hat{K}_{ab} & \hat{K}_{a\bar{b}} \\ \hat{K}_{\bar{a}b} & \hat{K}_{\bar{a}\bar{b}} \end{pmatrix} \begin{pmatrix} dS_b \\ d\bar{S}_{\bar{b}} \end{pmatrix} . \qquad (10)$$

Since (K) is a symmetric matrix so is its inverse $(-\hat{K})$, which implies

$$-\hat{K}_{ab} = \frac{\partial \Sigma_a}{dS_b} = \frac{\partial \Sigma_b}{\partial S_a} , \qquad (11a)$$

$$-\hat{K}_{a\bar{b}} = \frac{\partial \Sigma_a}{\partial \bar{S}_{\bar{b}}} = \frac{\partial \bar{\Sigma}_{\bar{b}}}{\partial S_a} , \qquad (11b)$$

and their Hermitian conjugates. These integrability conditions assure the existence of a potential functional $\hat{K}(S_a, \bar{S}_{\bar{a}})$ from which Σ_a can be locally derived by differentiation

$$\Sigma_a = - \frac{\partial \hat{K}}{\partial S^a} , \quad \bar{\Sigma}_{\bar{a}} = - \frac{\partial \hat{K}}{\partial \bar{S}^{\bar{a}}} . \qquad (12)$$

It is easy to verify that the Legendre transform functional

$$\hat{K}(S_a, \bar{S}_{\bar{a}}) = K(\Sigma_a, \bar{\Sigma}_{\bar{a}}) - \Sigma^a S_a - \bar{\Sigma}^{\bar{a}} \bar{S}_{\bar{a}} \qquad (13)$$

is the appropriate potential. It is unique up to Kähler transformations, since $(-\hat{K})$ is the unique inverse of (K). It can be shown [4], that all of the component equations of motion can be converted to Kähler form (chiral model) by the duality transformation, provided the Jacobian determinant satisfies (9) (non-singular). This is the only condition needed in order for the duality to go through. In order to obtain new models, which are not equivalent to chiral models, it is necessary to have the duality transformation fail. The necessary and sufficient condition for the duality transformation to fail is that the Jacobian determinant vanish

$$\det \begin{pmatrix} K_{ab} & K_{a\bar{b}} \\ K_{\bar{a}b} & K_{\bar{a}\bar{b}} \end{pmatrix} = 0 . \qquad (14)$$

Even though the duality transformation fails when (14) is satisifed, there are special cases [5] where the on-shell model is nevertheless equivalent to a chiral one, so (14) is only one of the conditions needed in order to obtain new types of couplings for scalar supermultiplets.

There are three classes of models for which the determinant vanishes so that they cannot be transformed by duality to Kähler form, but rather retain the non-Hermitian metric of the complex linear model. the first is the class where the determinant (14) vanishes <u>globally</u> due to the functional form of K. The second is when the determinant is forced to vanish globally by the equations of motion. For both cases where (14) holds globally a non-propogating superfield is required. For a non-trivial model, therefore, at least one additional propagating multiplet must be present. The third class is when only the classical vacuum expectation value of the determinant vanishes. In this case the complex linear model will have a vacuum which is not an allowed state of the dual chiral theory. Thus the two theories will be inequivalent. These three classes of models are discussed in greater detail elsewhere [5]. Similar conditions exist for the breakdown of duality transformations between other off-shell representations of the scalar supermultiplet. We have investigated the duality for complex linear superfields coupled to "old-minimal" supergravity, and found the most general coupling for combined complex linear and chiral superfields [5]. The scalar potential has been found, and the condition (14) still holds for the breakdown of duality. Examples of all three classes of models with failed duality are given in ref. 5, as well as the conditions for a vanishing cosmological constant. In the supergravity coupled nonlinear σ-models it is found that it is possible to have simultaneously: i) supersymmetry broken in vacuum, ii) a zero cosmological constant, iii) duality broken in vacuum. These models are thus phenomenologically viable, but due to the broken duality they result in non-Kähler terms in the low energy theory which would not be present in chiral models.

The explicit examples in ref. 5, which have a non-Kähler geometry also have broken supersymmetry. It has not been proven that it is possible to have unbroken supersymmetry and non-Kähler geometry globally in 3+1 dimensions. The determinant vanishing condition (14) is only necessary but not sufficient for this to occur. We have shown that phenomenological supergravity models based on chiral superfield couplings [6] are not the most general possible, and that other off-shell representations could lead to different physics if the duality transformation breaks down.

REFERENCES

[1] B. Zumino, Phys. Lett. 87B, 203 (1979).

[2] U. Lindström and M. Roček, Nucl. Phys. B222, 285 (1983).

[3] For conventions and notation see: J. Wess and S. Bagger, "Supersymmetry and Supergravity", Princeton Univ. Press, Princeton, New Jersey, U.S.A. (1983).

[4] G.M. Staebler, K.T. Mahanthappa, in preparation.

[5] G.M. Staebler, K.T. Mahanthappa, VPI preprints VPI-HEP-85/6; VPI-HEP- 85/7.

[6] E. Cremmer et al., Nucl. Phys. B147, 105 (1979); B212, 413 (1983).

Parallel Session 4

Substructure

Organizers:
W. Buchmuller (CERN)
H. Harari (Weizmann)

Scientific Secretaries:
O. Cheyette (LBL)
D. Karlen (SLAC)

TESTS FOR COMPOSITE QUARKS AND LEPTONS

Itzhak Bars

Department of Physics, University of Southern California
Los Angeles, CA 90089-0484, USA

Tests of composite quarks and leptons are reviewed. Limits on the compositeness scale are obtained from rare flavor changing processes, muon decay, anomalous magnetic moment, and electron- positron scattering. Tests are proposed at higher energies at the SLC, LEP-I, LEP-II and the SSC.

INTRODUCTION

In this talk I will concentrate on tests of compositeness that rely only on general properties of the theory of composite quarks and leptons. I will take the $W^{\pm}, Z$ as elementary, at least at the scale of preons. The properties that I will explore are based on the assumptions that 1)preons are confined by a new confining force that we call Precolor and that 2)at the confinement scale M ($\geq$ few TeV), the symmetries of the vacuum state permit unbroken chiral symmetries that include SU(3)xSU(2)xU(1).

At low energies a model independent approach is the method of effective Lagrangians. For composite models we expect

$$L_{eff} = L_{standard} + L_{non-renormalizable},$$

where the first part describes the standard model (without the Higgs, in this approximation), while the second part is suppressed at low energies by powers of 1/M, for dimensional reasons. Nevertheless if M is sufficiently low the second part can give rise to a number of effects that represent deviations from the standard model. These include rare decay modes and mixings, anomalous magnetic moments, μ dacay, and deviations in angular distributions in scatterings such as $e^+e^- \rightarrow \mu^+\mu^-$, $b\bar{b}, t\bar{t}$. I will argue that there is evidence for M$\geq$3 TeV, and that at SLC, LEPI and LEPII we can be sensitive to scales as high as 20-25 TeV.

At energies close to or above the scale M we expect a dramatic manifestation of the new strong interaction with cross sections much above the QCD background, and with the possible appearance of a whole new set of heavy states that can cause resonances in cross sections. I will argue that if $5 < M < 20$ TeV, the SSC is expected to easily see detailed evidence of compositeness. If M is larger, the SSC remains sensitive up to $M \sim 100$ TeV.

In this analysis it is useful to make analogies between the new confining force and QCD . On this basis we may expect confinement through precolor flux tubes, and thus get an idea of the expected spectrum of heavy states by assuming that they lie on approximately linear Regge trajectories [1]. We will choose M to correspond to the mass of the first heavy vector meson composed of preon-antipreon. This is the analog of the rho in QCD. Furthermore, in the energy regime $E \sim M$ we may model our strongly interacting amplitudes by analogies to low energy strong interactions involving the concepts of resonances, Regge exchanges, Pomeron (precolor glueballs), all of which are approximately representable by a version of the dual resonance model [1]. The low energy limit of these amplitudes are arranged to reproduce the effective non-renormalizable interactions that appear in the effective lagrangian above. Thus, the low and high energy analysis of composite models are corolated. At energies $E \sim M$, QCD and the electroweak interactions are negligible relative to the strong precolor interaction. Therefore they can be accounted for perturbatively in the same way that the electroweak interactions are a perturbation on QCD at energies of a few GeV. By contrast, at low energies the effects of the short range (confined) precolor interaction is negligible, and the standard model interactions dominate, as observed.

In both the low energy as well as the high energy analysis the assumed (nearly) unbroken chiral symmetry plays a major role. First, since it is an (nearly) unbroken symmetry, it helps classify the massless as well as the massive states. For the massive states, this implies the existence of nearly degenerate particles with different flavor-color-family quantum numbers, since the breaking scale in these quantum numbers for the usual quarks and leptons is much smaller than M. For an SSC scale analysis this has some obvious implications. It would suggest that leptoquarks, family changing vector bosons, colored vector mesons, etc. would be at approximately similar masses, and could give rise to interesting signals [2]. Second, at both low and high energies, the unbroken chiral symmetry provides a method for classifying the interactions. For example, in the effective low energy lagrangian, the 4-fermi interactions must obey chirality and also must be invariant under the surviving non-abelian chiral transformations. The same is true at high energies. The invariant amplitudes must be classified according to chirality and the non-abelian symmetries, as in ref1. The low energy limit of these amplitudes reproduce the 4-fermi interactions, thus insuring compatibility of the low and high energy analyses.

LIMITS AND TESTS AT LOW ENERGY

1. Rare Processes

We start with the 4-fermi interactions in the effective Lagrangian.

$$\frac{\lambda_{1234}^2}{2M^2}(\bar{\psi}_1\Gamma\psi_2)(\bar{\psi}_3\Gamma\psi_4)$$

which must be taken consistent with the symmetries. When these symmetries allow the 4-fermi terms listed below, then the corresponding limits on the table are obtained[4,5].(These are the most severe restrictions, for additional processes see the talk by D.Wyler, these proceedings.)

4-fermi	Process	Limit M≥
a) $\bar{e}d\bar{u}u^c$	proton decay	λx10^{13} TeV
b) $\bar{s}d\bar{s}d$	$K^0 - \bar{K}^0$ mixing	λx400 TeV
c) $\bar{c}u\bar{c}u$	$D^0 - \bar{D}^0$ mixing	λx50 TeV
d) $\bar{s}d\bar{e}\mu$	$K \to \pi\mu\bar{e}$, $K_L \to \mu\bar{e}$	λx30 TeV

where λ stands for λ_{1234} for the appropriate process. When the symmetries allow $\lambda \neq 0$ the underlying strong precolor interaction suggests the magnitude $\lambda^2/4\pi \approx 2$. It is clear that, for a model whose symmetries do not eliminate all of the processes above, M would be required to be above at least 100-150 TeV. Such a model would not be testable in the foreseeable future. Therefore it is crucial to hunt for models whose symmetries require $\lambda = 0$ for all the rare processes above. Although a), b), c) can be avoided by symmetries in classes of models, it is very hard to suppress d), as discussed in refs[4,5]. This problem is intimately connected to family replication. Whatever mechanism replicates the electron to the muon is likely to replicate the down quark to the strange quark. Then family, lepton and downess quantum numbers will be separately conserved by d), and therefore this term will be present with full strength. *This may turn out to be the case for most "realistic" models.* There may be ways of avoiding the problem either by dissociating the replication of quarks from the replication of leptons (this means usually more preons), or by appealing to lucky accidents in the mass matrix in the presence of extra families, as discussed in refs.[6,7]. I am not aware of a completely acceptable model that includes a realistic scheme of masses. Nevertheless it may exist, and therefore for the rest of this talk I shall concentrate on a model whose scale M is reasonably low so that it has measurable consequences at low energies or at least at the SSC.

2. Muon Decay

If M is sufficiently low it can modify the rate for μ-decay. The 4- fermi interaction

$$\frac{\lambda^2}{2M^2}(\bar{\mu}\gamma_\mu\nu_\mu)(\bar{\nu}_e\gamma^\mu e)$$

contributes to the decay rate. We find

$$\frac{\Delta\Gamma}{\Gamma} \approx \frac{\lambda^2}{4\pi}(1TeV^2/M^2)$$

The precision of the Berkeley-Triumph experiment [8] puts a limit M>2.9 TeV.

3. Anomalous Magnetic Moment of Muon

This has been discussed many times before [9]. The effective term responsible for this effect can arise only after the mass generating mechanism is turned on. Therefore, the term is proportional to the mass of the muon, and would vanish in the exact chiral symmetry limit. Dimensional considerations require the form

$$A^2 e\frac{m_\mu}{M^2}\bar{\mu}\sigma^{\mu\nu}\mu F_{\mu\nu}$$

where e, the electromagnetic charge is present since an interaction with the photon is involved, and A is a model dependent form factor. The contribution to the anomalous magnetic moment is proportional to $(m_\mu/M)^2$, and therefore is suppressed. The Hughes-Kinoshita analysis [10] leads to the limit $M > A$x0.72 TeV. We cannot estimate

A reliably, but it probably is of order 1. So this is not a very severe limit.

4. Electron-Positron Scattering

The scattering cross sections for $e^+e^- \to f\bar{f}$, where f is any fermion, would receive corrections from 4-fermi terms of the following type

$$[a_{LL}(\bar{e}_L\gamma_\mu e_L)(\bar{f}_L\gamma^\mu f_L) - c_{LR}(\bar{e}_L\gamma_\mu e_L)(\bar{f}_R\gamma^\mu f_R) + b_{LR}(\bar{e}_L\gamma_\mu f_L)(\bar{f}_R\gamma^\mu e_R) + d_{LR}(\bar{e}_L e_R)(\bar{f}_L f_R) + e_{LR}(\bar{e}_L f_R)(\bar{f}_L e_R) + (L \leftrightarrow R)] \times \frac{\lambda^2}{2M^2}$$

where the coefficients a,b,c,d,e are generally model dependent. They are probably of order 1 if allowed by the symmetry structure of the model. Note that the last five terms obtained by interchanging left with right have coefficients that are generally different than the ones appearing in the first five terms. Thus, generally parity and charge conjugation invariance is violated since the underlying left-right asymmetric preon model violates these symmetries through its precolor interactions.It is very interesting to hunt for such P or C violating effects that are expected to increase with energy and eventually overcome the P and C violation effects of the weak interactions. The coefficients b,d,e tend to violate chirality, therefore they would be identically zero in many models (i.e. they are of order m_f/M x m_e/M to some power). However there would also be many models for which chirality operations should simultaneously be applied to the electron and some other fermion (e.g muon) because they may share the same preons. For those cases, even though masses would be prevented, certain 4-fermi interactions, like the coefficients of b would be allowed. It is hard to imagine that d or e-type interactions would be allowed. Therefore, we will assume that d and e are negligible, but will include b in our analysis (keeping in mind that for certain models it will be zero). The coefficients a,c are always allowed since they cannot be prevented by any symmetry. It is possible to analyse models by drawing preon diagrams (analogous to duality diagrams) and obtain relative ratios between a/b/c that correspond to a simple counting of the number of diagrams contributing to each chirality projection above. This kind of analysis was introduced [1] at the level of the higher energy amplitudes but can equally apply at the low energy limit.

Eichten, Lane and Peskin [11] estimated the deviations in Bhabba scattering that would be caused by a model with nonzero a_{LL}. They found observable deviations at low energies (PEP, PETRA) if the scale M is low. For agreement with measurement it has been found that M>2-3 TeV. In collaboration with Gunion and Kwan [12], we extended this analysis to include effects of polarization and analysed the more general model with all allowed coefficients . We found that if sufficiently precise measurements of angular distributions are performed the sensitivity of the SLC and LEP-I to compositeness would extend to M<7 TeV. Such measurements, if they yield any deviations from the standard model, would be capable of also disentangling partially the coefficients a,b,c and measuring the possible left-right asymmetry in these coefficients. The analysis of ref.12 was not very sensitive to the amplitude b which, as remarked above may be present in some models. Together with Eilam and Gunion [13] we suggested polarized beam processes that are sensitive to this amplitude. Since there is no standard model background for this amplitude any effect at all would be interpreted as new physics. It appears, however, that after taking into account various restrictions from SU(2)xU(1), and estimating possible contributions from the amplitude b to the electron mass, it is not easy to get a measurable effect that is due solely to this amplitude, although it may be worth a try [see ref 13]. In our work in ref.[13] we also extended the analysis of refs.11,12 to the energies applicable to LEP-II. As illustrated in table 1, if M<25 TeV, the *unpolarized* cross section would show impressive deviations from the standard model at the higher center of mass energies. The quantity that is tabulated is Δ, as defined below, which is the percentage deviation of the cross section from the standard model in the reaction $\bar{e}e \to \bar{t}t$ (similarly for other heavy quark pairs),

$$\Delta(cos\theta) = \frac{(d\sigma/d\Omega)}{(d\sigma/d\Omega)_{standard}} - 1$$

The deviation Δ is given at center of mass energies ranging from 100 to 200 GeV and at certain backward angles ($\cos\theta$), where it is large. It is even larger at larger backward angles.

		$M(TeV)$				
$cos\theta$	E	5	10	15	20	25
−0.4	100	−0.003	−0.001	−0.001	"0"	"0"
−0.7	110	−0.017	−0.010	−0.005	−0.003	−0.002
−0.8	120	+0.077	−0.003	−0.003	−0.002	−0.002
−0.8	150	+1.35	+0.192	+0.074	+0.039	+0.024
−0.7	200	+3.55	+0.49	+0.185	+0.098	+0.061

It seen that, even for compositeness scales as high as 20 TeV, LEP-II can detect 9.8 % deviation from the standard model cross section at center of mass energies of 200 GeV and cosθ=-0.7, and larger deviation at larger backward angles. Therefore, *such measurements will set the mood for prospects of discovering compositeness at the SSC.*

TESTS AT THE SSC

Next, I would like to report on studies that relate to the SSC by concentrating on the reactions pp or $p\bar{p} \rightarrow$ jets or lepton pairs. In ref.[14] EHLQ did the analysis under the assumption that compositeness interactions are represented by 4-fermi interactions. This is true if the SSC (parton+parton) energies would be far below the compositeness scale Provided we assume this, their work produced plots of cross sections that decrease with energy, and in which the compositeness correction remains above the standard model and detectable up to a scale of M~20 TeV.

However, if M is as low as 20 TeV the 4-fermi description is no longer correct, and a model of energy and angle dependent amplitudes is necessary. For this purpose a model for the *scattering amplitudes* of the processes among the partons (quarks, gluons $\rightarrow$ quark pairs, gluon pairs, lepton pairs) as composed from preons was introduced [1]. The assumption was that at the SSC there would not be sufficient energies to get to the assymptotically free region of precolor interactions, so that precolor interactions should behave like the familiar strong interactions at low energies. So, we would expect resonances, regge behaviour (including a Pomeron= precolor glue balls), duality (reflecting string-like precolor confinement) and finally a low energy behaviour equivalent to the 4-fermi interaction. Furthermore the amplitudes must be consistent with the surviving chiral symmetries.

These properties are accomodated by building the following chirality projected amplitudes to represent the reaction $f_1\bar{f}_2 \rightarrow f_3\bar{f}_4$, where the f's are any set of fermions [1]

$$\begin{aligned}
&[A_{LL}(\bar{u}_{3L}\gamma_\mu u_{1L})(\bar{v}_{2L}\gamma^\mu v_{4L}) \\
&+ B_{LR}(\bar{u}_{3L}\gamma_\mu u_{1L})(\bar{v}_{2R}\gamma^\mu v_{4R}) \\
&- C_{LR}(\bar{u}_{3L}\gamma_\mu v_{4L})(\bar{v}_{2R}\gamma^\mu u_{1R}) \\
&+ D_{LR}(\bar{u}_{3L}u_{1R})(\bar{v}_{2L}v_{4R}) \\
&+ E_{LR}(\bar{u}_{3L}v_{4R})(\bar{v}_{2L}u_{1R}) \\
&+ (L \leftrightarrow R)]
\end{aligned}$$

Any 4-fermi amplitude can be put into this form after Fierz transformations. On the basis of chiral symmetry the D and E amplitudes were taken equal to zero, while the A, B, C amplitudes were taken to represent the sum of (chirality projected) preon-duality diagrams that contribute to the elementary parton process. A table of the relevant diagrams and their contributions is found in ref[1]. The amplitudes for fermion + fermion $\rightarrow$ fermion + fermion is obtained from the above expression by crossing symmetry, and similarly the gluonic parton amplitudes can be built [3]. Each individual preon duality diagram was represented by a Veneziano model type beta function . Similar diagrams (except for their quantum numbers) were represented by the same function. There were only a few independent such functions, but as a further first approximation all of them were taken to be identical, and equal to a beta function. For the channels allowing the vacuum quantum numbers a "Pomeron" contribution was added in such a way as to insure crossing symmetry (see ref 15 for a correction). With this construction the amplitude was guarantied to satisfy all the desired properties of resonances, Regge behaviour, duality , crossing symmetry and the correct normalization that reduces to the 4-fermi interaction at low energies.

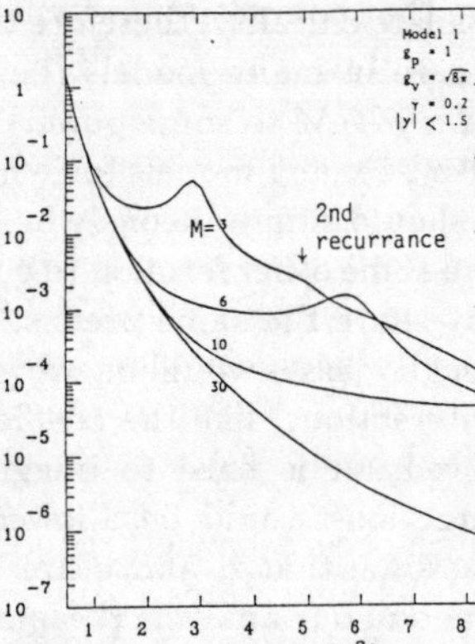

The elementary partonic cross sections built from these amplitudes were folded with the parton distributions of EHLQ of ref.[14]. This was done numerically in collaboration with Hinchliffe [15] in the case of jet final states, and in collaboration with Gunion and Kwan [16] in the case of leptonic final states. The result is a series of plots of cross sections that show a rich structure of bumps and shoulders and, most notably, *very large cross sections, far above the QCD background.* An example of jet final states plotted versus the jet invariant mass for various values of M=3,6,10,20,30,∞ TeV is shown in the figure ,taken from ref.[15]. The limit of M=∞ is equal to the pure QCD cross section.

This figure is produced under the assumption that the width of the jet-jet resonance is about 1/5 its mass, similar to the rho-meson in QCD. The width is model dependent and also depends on the quantum numbers of the resonance. Estimates for the widths in various models appear in ref.[16]. Resonances whose widths are larger than 1/2 its mass flatten out to a degree that they no longer show as a noticeable bump. If the width is smaller than 1/5 its mass, then higher regge recurences also begin to show, as can be noticed slightly in the figure. Ref.[15] found out that gluonic partons were negligible, mainly because a gluon's interaction introduces powers of the QCD coupling constant, which is small compared to the strong precolor interaction experienced by quark partons. Even though the gluonic partons are much more abundant at the SSC energies, the strength of the new strong interaction completely overwhelms this factor.

Similar features apply also to lepton final states, as discussed in ref.[16].

From these results it is seen that if M is not too much larger than 20-25 TeV, then the cross section will be so large that compositeness will be easy to see at the SSC ,and its features will be completely distinguishable from other "new physics" signatures at a comparable scale. In refs.[15,16] the larger values of M are also discussed. According to reasonable criteria for observing "new physics" at the SSC, it was concluded that the SSC would remain sensitive to compositeness up to scales of the order 100 to 300 TeV, and at the larger scales the lepton pairs would be a more sensitive probe.

A question which occurs is whether to expect multilepton or multijet final states to suddenly swamp the cross section if the compositeness threshold is crossed. We have argued that this could not be expected at the SSC, on the basis of confinement through precolor flux tubes [16]. The point is that there would not be enough energy available to the preons to be able to break the confining string more than one time. Furthermore, phase space provides a big suppression factor for multibody final states even if the final particles are (almost) massless. Therefore we expect that 2-body final states will dominate the total cross sections although, of course, there would be a small fraction of multibody final states whose compositeness cross section will be significantly above the QCD background for the ranges of M under discussion. Further discussion on multibody final states can be found in refs.[3,17,18].

REFERENCES

[1] I.Bars,*Study on the Design and Utilisation of the SSC, Snowmass 1984* , Eds.R.Donaldson and J.Morfin , page 38. [2] G.'tHooft, in *Recent Developments in Gauge Theories*, Ed. G. 'tHooft, Plenum Press, NY 1980.

[3] C.Albright and I.Bars, *Summer Study on the Design and Utilisation of the SSC, Snowmass 1984*, Eds.R.Donaldson and J.Morfin, page 34.

[4] I.Bars, Nucl.Phys B208, 77 (1982)

[5] I.Bars, Proc. Rencontres de Moriond 1982, *Quarks, Leptons and Supersymmetry*, Ed. Tran Thanhn Van, p.541.

[6] I.Bars, *Summer Study on the Design and Utilisation of the SSC, Snowmass 1984*, page 832.

[7] W.O.Greenberg, R.Mohapatra and N.Nussinov, Maryland preprint 1984.

[8] Berkeley-Triumph collaboration,

[9] see e.g. M.Peskin, Proc. 22nd International Conference on High Energy Physics, Tokyo, Jap , (1985).

[10] V.Hughes and Kinoshita

[11] E.Eichten, K.Lane and M.Peskin, Phys. Rev. Lett. 45, 255 (1983)

[12] I.Bars, J.Gunion and M.Kwan, Nucl. Phys. (1986)

[13] I.Bars, G.Eilam and J.Gunion , USC preprint 86/09

[14] E.Eichten, I.Hinchliffe, K.Lane and C.Quigg, Rev. Mod. Phys. (1985)

[15] I.Bars and I.Hinchliffe, Phys. Rev. D33, 704 (1986)

[16] I.Bars, J.Gunion and M.Kwan, Nucl. Phys. (1986)

[17] C.Albright, I.Bars, B.Blumenfeld, K.Braune, M.Dine, T.Ferbel, H.J.Lubatti, W.R.Molzon, J.F.Owens, M.Schmidt, G.Snow,*Summer Study for the Design And Utilization of the SSC*, Snow mass 1984, Eds. R.Donaldson and J.Morfin, p.27.

[18] G.Domokos and J.Domokos, Johns Hopkins preprint, contributed to this conference.

RARE PROCESSES AND NEW STRUCTURE

Daniel Wyler

Theoretische Physik, ETH-Hönggerberg, CH-8093 Zürich, Switzerland

The construction of effective operators due to new physics is outlined. They are then used to derive various bounds on the scales of new interactions and particle masses from experimental limits. I then discuss the implications for models, in particular those which include composite fields.

1. INTRODUCTION

The standard SU(3) x SU(2) x U(1) model [1] with the effective Higgs field breaking mechanism gives a convincing description of essentially all data. This includes even non-tree level considerations such as in the Kaon system [2]. However, the large number of arbitrary parameters may indicate that the model is a low energy remnant of a more fundamental theory, with new particles and interactions at higher energies. As a consequence, one may observe new particles and interactions at higher energies. If the masses of new particles are too high to allow their production, there will be only new interactions among the known (light) particles through exchanges of heavy states etc.

The quantitative success of the standard model implies that the new interactions give small effects and high precision is required to detect them. This favors low energy experiments as illustrated by the following example [3]. If there exists a lepton number violating four fermion interaction $(\bar{e}e)(\bar{\mu}e)$ with coupling constant G_x, then we have, roughly, the following count rates (N = rate of muons, s = energy, T = time):

$$\mu\text{-decay: } \#\text{ of events} = N\left(T\cdot\frac{G_x}{G_F}\right)^2$$

$$\begin{array}{l}e^+e^-\text{-annih.}\\ \text{at larger energies}\end{array} : \#\text{ of events} = s\int_0^T \text{Luminosity}\cdot G_x^2\,, \qquad (1)$$

and thus

$$\frac{\#\,(ee\to\mu e)}{\#\,(\mu\to eee)} \simeq \frac{S\cdot T\cdot \text{Lum}}{N\cdot T}\, G_F^2 \simeq 10^{-9} \qquad (2)$$

for typical values of the quantities involved. This estimate requires N to be not much larger than τ_μ^{-1} and that $G_x^{-1/2} \gtrsim s$.

Similar estimates, although usually less drastic, can be made for other interactions; they underline the importance of the low energy tests.

2. DESCRIPTION OF THE NEW INTERACTIONS

Depending on the "basic" theory (composite [4], technicolor [5], supersymmetry [6] etc) the new energy scales, denoted by Λ_i, may represent characteristic energies of new interactions, masses of new particles etc. In any case, in the spirit of low energy effective Lagrangians [7] and of the short distance expansion [8] and decoupling theorem [9], the Λ_i (and thus the new physics) appear only in coefficients of operators formed with the light particles, e.g. those of the standard model. Apart from renormalization effects containing $\log\Lambda_i$, only a dimensional analysis is required, which yields

$$L_{eff} = \sum \frac{c_i}{\Lambda_i} O_i^5 + \sum \frac{d_j}{\Lambda_j^2} O_j^6 + \ldots.. \qquad (3)$$

where the O_i^n are operators of light fields of dimension n, and the c_i, d_i are constants.

Since the typical logarithms of short distance coefficients change the naive results by at most a factor of order 1 and we are clearly not after a high (theoretical) accuracy, I feel that they can safely be neglected; in particular in view of the perturbative character of the standard model.

If we take $\Lambda_i >> G_F^{-1/2}$, $c_i, d_i \sim O(1)$, then the O_i^n are $SU(3) \times SU(2) \times U(1)$ invariant. This is, because otherwise fields with a mass of order Λ_i would break the $SU(2) \times U(1)$ symmetry, an unlikely situation. Then, given the light fields, one can construct all possible O^n. The c_i, d_i can only be computed if one has a model for the new physics, otherwise they are arbitrary. To give a consistent definition of the Λ_i we have used the following convention. Every measurable quantity MQ will take the form

$$MQ - MQ_o = \frac{\sum k_i c_i}{\Lambda_i} + \frac{\sum \ell_i d_i}{\Lambda_i^2} + \ldots \tag{4}$$

where MQ_o is the value in the standard model and k_i, ℓ_i are coefficients which are calculated (see ref [10]). As we will see, only the second term contributes and we set $\sum \ell_i d_i = 1$ for each MQ. If one operator contributes, this amounts to setting $d_i = 1$. (Many authors have used $d_i = 4\pi$ (for composite models); if necessary, we have rescaled their results).

With $\Lambda_i >> G_F^{-1/2}$, only the O_i^5, O_i^6 are important. Since operators giving rise to proton decay or Majorana ν masses must be suppressed by very large $\Lambda_i \gtrsim 10^{10}$ GeV, we ommit them here, that is we impose baryon and lepton number conservation on Eq. (3). Then, no $SU(3) \times SU(2) \times U(1)$ O_i^5 can be constructed and only the O_i^6 remain. Since we are interested in calculating only their matrix elements for on-shell particles, we can use the classical equations of motion to limit their number. The complete analysis gives, apart from various flavor possibilities, 80 operators. The systematic derivation is described in ref. [10]; see also the earlier work in refs. [11].

3. BOUNDS ON THE Λ_i

Some of the new operators will give rise to new effects not present in the standard model, in particular to processes involving neutral flavor violations. Experimental bounds (or small observed effects, compatible with the standard model including radiative corrections) translate into bounds on the various Λ_i. The table below summarizes the present situation [12].

A comment is in order. The scale Λ' is derived by replacing in operators of the form $\bar{\psi}_{1L} \Gamma \psi_{2R} B$, $\phi_{L,R} = \frac{1}{2}(1 \pm \gamma_5)$, where Γ is some matrix and B a bosonic field, the coupling $1/\Lambda^2$ by

$$\frac{m_{\psi_1} + m_{\psi_2}}{v \cdot \Lambda^2} \; ; \quad (v = \text{v.e.v. of Higgs field})$$

.In this way the usual mass suppression of helicity flip operators is obtained. As to the other operators it is not clear, whether factors of (m_ψ/v) should be built in. Clearly, if the operators arise through exchanges of a heavy Higgs-type fields, one might multiply their couplings with such a factor. In the case of a four Fermion operator, we can expect

$$\Lambda' = \frac{\Lambda}{v} \sqrt{(m_1+m_2)(m_3+m_4)} \tag{6}$$

where the exchange is between the fermions pairs (1,2) and (3,4). Comparing then, for instance, $K \to e\mu$ with $B \to \tau\mu$, or $\mu \to 3e$ with $\tau \to 3\mu$ one has

$$\frac{\Lambda'(K)}{\Lambda'(B)} \simeq \frac{BR(B)^{1/4}}{BR(K)^{1/4}} \sqrt{\frac{m_s m_\mu}{m_b m_\tau}} \simeq \frac{BR(B)^{1/4}}{BR(K)^{1/4}} \cdot 4 \cdot 10^{-1}$$

$$\frac{\Lambda'(\mu)}{\Lambda'(\tau)} \simeq \frac{BR(\tau)^{1/4}}{BR(\mu)^{1/4}} \sqrt{\frac{m_e}{m_\tau}} \simeq \frac{BR(\tau)^{1/4}}{BR(\mu)^{1/4}} \; 1 \cdot 4 \cdot 10^{-2} \tag{7}$$

which illustrates the modest gain when going to heavier particles. However, certain interactions may affect only heavy particles.

4. IMPLICATIONS FOR THEORY

Besides getting bounds on new particles etc. from the results in table 1, one might be more ambitious and try to find combinations of bounds which might rule out a model entirely. Unfortunately, so far no such strong statement can be made and only some general conclusions can be drawn.

i) We see immediately that there are very stringent bounds from flavor violating processes, Λ_{FV} and less

Observ.	Value	Proposed	$\Lambda(\Lambda')$ in TeV	Λ_{Prop} in TeV	Comments
$\mu \to e\gamma$	$<4.9.10^{-11}$ [13]	10^{-13} } LA 969(A)	10^4 (250)	10^5 (2500)	
$\mu \to e\gamma\gamma$	$<7.2.10^{-11}$ [14]	10^{-12} } LA 969(A)			
$\mu \to eee$	$<2.4.10^{-12}$ [15]		150		
$\tau \to \mu\gamma, e\gamma$	$<6.10^{-4}$ [16]		60		
$c \to uG$	<0.03 [10]		(6)		
$(g-2)$ e	$\delta < 2.10^{-10}$		40 (0.1)		Bhabha 2 scattering 0.4 TeV
$(g-2)$ μ	$\delta < 2.10^{-8}$ [17]		50 (1.5)		
$\mu^- N \to e^- N$ (Titanium)	$<4.10^{-12}$ [18]	10^{-13} SIN	290	600	
$\mu^+ e^- \to \mu^- e^+$	<0.04 [19]	10^{-5} LA 985			
$K_L \to \mu e$	$<10^{-8}$ [20]	10^{-10} AGS E780 (R) 5.10^{-13} AGS E791(A) 10^{-10}	35	700	diff. helicit. $\simeq$ 100 x sensit
$K^+ \to \pi^+ \mu e$	$<4.8.10^{-9}$ [21]	10^{-11} AGS E77 (R)	20	100	$\ell \leftrightarrow$ down
$K^+ \to \pi^+ \nu\nu$	$<1.4.10^{-7}$ [16]	2.10^{-10} AGS 781(A)	10	60	$\nu \leftrightarrow$ down LQ
Δm K	<exp.value [16]		1000		helicity $\to$ stronger Bounds ($\sim$ 3-5 x)
Δm D	<exp.value [22]		150		
Δm B	<standard [23]		200		
Scalar contr. μ-Dec	<0.076 [24]		0.65		
Cabibbo-Univer-sality	<0.003 [25]		5		
$\delta\rho$	<0.02 [26]		2	6	$\Theta<^{-9}$ leads to $\Lambda>10^5$TeV
d_n	$d_n < 10^{-25}$ ecm [16]		10^7		
$D^o \to \mu^+\mu^-$	$<1.0.10^{-5}$ [27]		~ 2		
$D^o \to \mu^{\pm} e^{\mp}$	$<1.5\ 10^{-3}$ [28]		1/2		
$B \to \mu^+\mu^-$	$<0.9.10^{-4}$				
$B \to \mu^+ e^-$	$<0.9.10^{-4}$ [29]		2		
$B \to e^+ e^-$	$<0.8.10^{-4}$				
$B \to \tau\mu$	$\lesssim 10^{-3}-10^{-4}$ [30]		2		
$B \to \tau e$	[30]				
$B \to K\mu^+\mu^-$	$<3.10^{-4}$ [29]		2		

Table 1. Limits on Λ from various experiments; as well as values Λ_{Prop} obtainable from proposed experiments. $\Lambda' = (\frac{m_1+m_2}{v})^{1/2}$ where m_1, m_2 are the masses of the two fermions involved and v = 175 GeV. Note that all decay branching rations scale with Λ^{-4}, while g-2, Δm, and the weak interaction parameters scale with Λ^{-2}. In the column on proposed experiments LA stands for Los Alamos, AGS for Brookhaven. The experiments are running (R) or accepted (A). Note that $\mu \to e\gamma\gamma$ cannot proceed through any of the O_i^6.

stringent ones, Λ_{FC} from processes testing the gauge structure. This suggests that different physics is involved in the two cases (a point often stressed by Harari for composite models), although in non-composite models, small couplings suppress FV. Bounds on Λ_{FC} are very interesting as the corresponding operators occur in all models beyond the standard model at tree level; in this case the radiative corrections of the standard model must be taken into account.

ii) Some composite models have stable particles with the characteristic scale Λ. Then, cosmological arguments lead to [31]

$$\Lambda < 370 \text{ TeV}. \tag{8}$$

If this value applies to models with flavor changes, BR for $K_L \to \mu e$ or $K^+ \to \pi^+ \nu\nu$ of the order of $\sim 10^{-12}$ - 10^{-13} would rule them out; as to $\mu \leftrightarrow e$ transitions ($\mu \to e\gamma$), the present bound is already in conflict with (8). However, it is more likely that the heavy states connected to FV are unstable and decay and only those connected to FC can be stable.

iii) It is important to improve the bounds on various processes. For example, $K \to \mu e$ $D^+ \to \pi^+ \nu\nu$ does not limit the existence of SU(5)-type leptoquarks (but strongly affects SU(4)-type leptoquarks [32,33]) whereas $K^+ \to \pi^+ \nu\nu$ or $D \to \mu e$ affects these models in just the opposite way.

A more general analysis [10] on FV shows that absence of neutral current FV requires that heavy leptoquarks or diquarks couple only to one pair of fermions. This result is intuitively clear; if a state would couple to $\bar{s}\mu$ and to $\bar{d}e$, it would give $K \to \mu e$. However, for the quarks in the SU(2) doubletts, this condition cannot be imposed for both quarks in the doublett, due to the non-vanishing Cabibbo-KM-angles.

Thus, simultaneous absence of $K \to \mu e$ and $D \to \mu e$ implies no leptoquark couplings to left-handed quarks, which is rather unlikely in composite models which try to explain the quark-lepton similarity [34].

A recent attempt to circumvent these problems [35] proposes a model in which the third and a (hypothetical) fourth generation have large FV. In this model, $K \to \mu e$ is suppressed by ε^4, where ε is $\sim 10^{-2}$. However, the processes $B_d \to \tau e$, $B_s \to \tau\mu$, ... go with ε^2; using the estimates of [36], I obtain 10^{-1} - 10^{-5} for the respective branching ratios - values easily accessible to experiments (see table 1). Furthermore, the ρ-parameter limits this model strongly.

iv) The bounds depend strongly on the spins of the exchanged particles because of helicity suppression / enhancements. For example, a scalar leptoquark giving rise to $M \to \ell_1 + \bar{\ell}_2$, M being a pseudoscalar meson with quarks q_1, $\bar{q}_2$, leads to a rate which is enhanced by $M_M^4 / (m_{q_1} + m_{q_2})^2 (m_{\ell_1}^2 + m_{\ell_2}^2)$ over the usual one. For $K \to \mu e$, this factor is ~ 220, leading to an increase in Λ by about a factor of 4.

v) We have seen that present and proposed experiments limit more or less strongly individual models. Here, I sketch a possible way to find a "critical" experimental value for many models, which may apply to composite models and others. It is based on the idea that the light fermions obtain their mass indirectly. In Fig. 1 the mass generation in SU(4)-type composite [35] and technicolor models is shown.

In all cases, there is a chirality breaking fermion bilinear of the inner fermions. If chiral symmetry is spontaneously broken, there exists a boson, H, which couples to the corresponding fermions, and thus, via Fig. 1, to the light fermions. If the chirality breaking term is just a mass term (no spontaneous breaking of a symmetry) it is a gauge singlet and breaking of SU(2) x U(1) by the light fermion mass terms requires a boson, H, coupled to the boson line, and thus to the light fermions.

Consider now the flavor structure in Fig. 1. In the basis where all internal lines are diagonal (e.g.

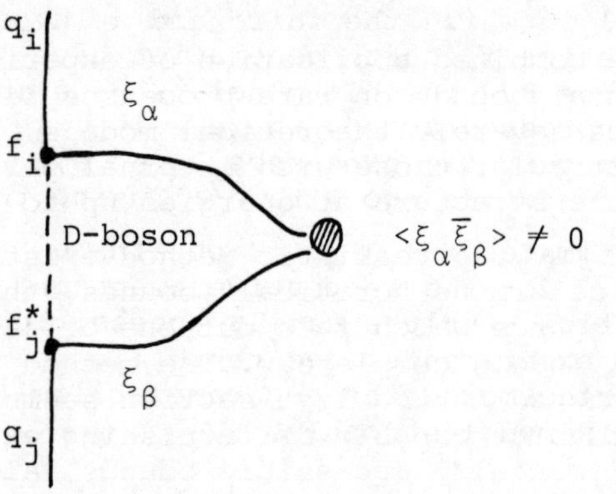

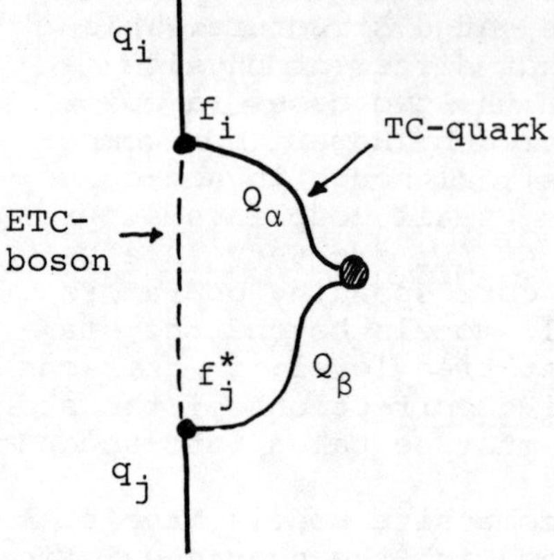

Fig. 1 mass generation in the PS and TC models. Similar, "radiative" sources for fermion masses are possible in ordinary gauge model. f_i, f_j^* are the couplings.

$\alpha = \beta$ in Fig. 1, and similarly for the bosons), the mass matrix for external fermions is

$$m_{ij} = \sum_{\alpha=1}^{A} K_\alpha f_i^\alpha f_j^{*\alpha} , \tag{9}$$

where α denotes the various internal boson-fermion pairs and f_i^α, $f_j^{*\alpha}$ their couplings to q_i and q_j. The couplings of the H-bosons above will be proportional to the terms in (9). Since, in general, every α will have a different scalar associated with it, we have several bosons, H'_α, coupled with a matrix proportional to $K_\alpha f_i^\alpha f_j^{*\alpha}$.

If we consider only one α, then m_{ij} has the eigenvalues $(0,\ldots 0,m)$ [37]; the coupling of H^α is proportional to m and thus flavor conserving. In order to have non-zero masses for all flavor we must have several α. Now, the couplings of the H^α are flavor conserving only, if the $f_i^\alpha\ f_j^{*\alpha}$ are simultaneously diagonalizable. But then, the eigenvalues of m are still $(0,\ldots 0,m)$. Thus, non-zero masses require flavor violating H_α couplings. (The argument fails if $f_i^\alpha\ f_j^{*\alpha}$ are simultaneously diagonalizable but with (ordered) eigenvalues $(0,\ldots 0,m_1)$; $(0,\ldots,0,m_2,0)$ etc. We exclude this here.)

In order to estimate the effect, we note that in the simple case of two flavors, $m_{heavy} \simeq K_1|f_1|^2$, $m_{light} \simeq K_2|f_2|^2$. Flavor changing couplings are proportional to $K_2|f|^2$ ($|f^\alpha|^2$ means typical values of $|f_i^\alpha\ f_j^{*\alpha}|$).

The exchange of an H^α with (mass)^2 m_α^2 will give rise to a flavor violating amplitude (see Fig. 2)

$$A_{FV} \sim \frac{(K_2|f^2|)\ m_j}{M_\alpha^2} \tag{10}$$

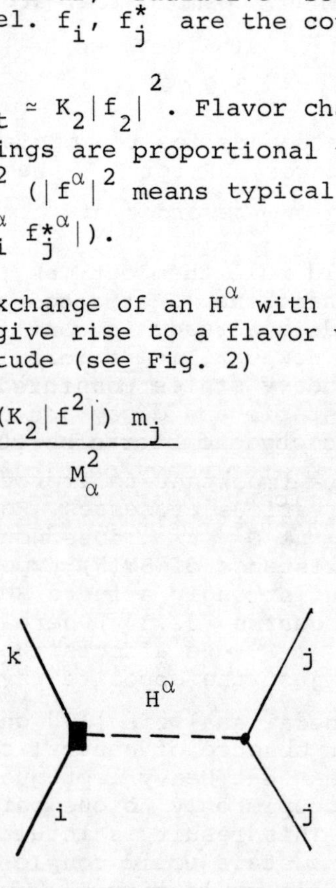

Fig. 2 FV process $i \to k + j + j$. The box stands for the flavor violating coupling $\sim K_2|f|^2$.

Taking $K_2|f^2| \approx m_{light}$, we get

$$A_{FV} \simeq \frac{m_{light}}{M_s^2} G_F\ m_j . \tag{11}$$

In obtaining (11) we have made the assumption that the chiral condensate in Fig. 1 (or equally the v.e.v. associated with H) is related to the breaking of SU(2) x U(1) and thus of the order of $G_F^{-1/2}$.

(11) then yield a typical branching ratio for a flavor violating process of

$$BR \simeq \left(\frac{m_{light} m_j}{m_\alpha^2} \right)^2 . \qquad (12)$$

In the spirit of the previous discussion, we now take $m_\alpha^2 \simeq 100$ GeV. With $m_{light} \simeq 10^{-2}$ GeV, we have

$$BR \simeq 10^{-16} \qquad (13)$$

for a typical process with mainly first generation particles. We should mention that the choice $K_2|f^2|$ is quite pessimistic. In the case of the d-s system,

$\theta_c \simeq \sqrt{\frac{m_\alpha}{m_s}}$, $K_2|f^2| \simeq \sqrt{m_\alpha m_s}$; this improves (13) by a factor of ~ 20. Furthermore, for heavy particles we have significant enhancements. For instance, above reasoning yields $BR \sim 10^{-10} - 10^{-9}$ for $B_d \to K + \tau^+ \tau^-$ etc. Unfortunately, these numbers are beyond experimental reach, a more detailed analysis seems required.

5. CONCLUSIONS

The combined application of experimental bounds on rare processes limits severely theoretical models, in particular those with internal particle structure at energies up to 10^3 TeV. No realistic method exists to determine "critical" bounds, that is bounds which rule out one or several models completely. The method sketched in the last section seems not to be capable of furnishing experimentally accessible bounds, although it raises the hope that a better procedure might be found. In that case, low energy precision tests would continue to be of greatest importance.

ACKNOWLEDGEMENTS

I am indebted to W. Buchmüller for enjoyable collaborations on the topics discussed here. I thank H. K. Walter, R. Eichler and R. Engfer for several discussions on the experimental aspects.

REFERENCES

[1] See e.g. T.P. Cheng and L.F. Li, Gauge theory of elementary particle Physics, Clarendon Press, Oxford, 1984.
[2] M.K. Gaillard and B.W. Lee, Phys. Rev. D10 (1974) 897.
[3] C.M. Hoffman, LAMPF preprint LA-UR-84-1327 (1984); in Proc. Int. School of Physics, Erice 1984.
[4] For a recent review, see W. Buchmüller, Schladming 1985, Acta Physica Aust., Suppl. XXVII, (1985) 517.
[5] See, M.A.B. Bég, in the Proc. of the Mexican School of particles and field, Amer. Inst. of Phys., New York 1986.
[6] H.P. Nilles, Phys.Rep. 117 (1985) 75.
[7] See, e.g. J. Gasser and H.Leutwyler, Ann.Phys.(NY), 158 (1984) 142.
[8] K. Wilson, Phys. Rev. 179 (1969) 1499.
[9] T. Appelquist and J. Carrazone Phys. Rev. D11 (1975) 2856.
[10] W. Buchmüller and D. Wyler, Nucl. Phys. B268 (1986) 621.
[11] C.J.C. Burges and H.J. Schnitzer, Nucl. Phys. B228 (1983) 464. C.Leung, S. Love and S. Rao, to be published.

[12] Recent reviews on rare decays are:
H.K. Walter, Nucl. Phys. A434 (1985) 409c;
H.K. Walter, in Summer School on Nucl.Structure, Dronten, 1985; SIN preprint;
L.S. Littenberg, BNL-Preprint No. 35086;
D. Bryman, TRIUMMF-Preprint (1986);
R. Engfer, Univ. of Zürich, Preprint, 1986.
[13] R.D. Bolton et. al. Draft paper, LAMPF, 1986.
[14] H. Blümer, Proc. Int. Conf. on High Energy Phys. Bari, p.429 (1985). See also C. Hoffman, this conference.
[15] W. Bertl et. al, Nucl.Phys. B260 (1985) 1.
[16] Rev. of Particle properties, Rev. Mod. Phys. 56 (1984).
[17] T. Kinoshita, B. Nizic and Y. Okamoto, Phys. Rev. Lett. 52 (1984) 717, the bound on Bhabba scattering is from E. Eichten, K. Lane and M. Peskin, Phys. Rev. Lett. 50 (1983) 811.
[18] D. Bryman, Workshop on Fundamental Muon Physics, LAMPF, 1986; also this conference.
[19] G.M. Marshall et. al. Phys. Rev. D25 (1982) 1174.
[20] See Ref. [16]; value multiplied by a factor of five; see first reference in [12].
[21] A. Diamant-Berger et.al Phys. Lett. B62 (1976) 485.
[22] A.C. Benvenuti et. al. Phys. Lett. B158 (1985) 531.
[23] The calculation uses the standard vacuum insertion and $f_B \simeq f_\pi$.
[24] K. Mursula and F. Scheck, Nucl. Phys. B253 (1985) 189.
[25] G. Barbiellini and G. Santoni, Rev. Nuov.Cim. 9(2) (1982); see also W. Marciano, this confer.
[26] See, e.g. W. Marciano, this conference. The proposed value ($\Lambda_{Prop} \sim 6$ TeV) refers to the precise measurements possible at LEP.
[27] W.C. Louis et.al. Phys.Rev.Lett. 56 (1986) 1027.
[28] ACCMOR preliminary results; H. Palka, private communication.
[29] R. Poling, CLEO, Collaboration, this conference, see A.G. Deshpande et.al.,Phys.Rev.Lett. 57 (1986) 1106.
[30] My guess from the presentations of R. Poling, A. Jawahery, this conference. I thank these authors for discussions.
[31] I. Bars, M. Bowick and K. Freese, Phys. Lett. B138 (1984) 159.
[32] J.C.Pati and A. Salam, Phys.Rev. D10 (1974) 275;
[33] W. Buchmüller and D. Wyler, Cern Preprint, CERN-TH 4461/86 (1986).
[34] The result is based on the assumption that each exchange conserves flavor individually ("Natural flavor conservation").
[35] J. Pati, Phys. Lett. 144B (1984) 375.
[36] J. Pati and H. Stremnitzer, Phys. Lett. 172B (1986) 441.
[37] U. Baur and H. Fritzsch, Phys. Lett B134 (198) 105.

UNIVERSAL W,Z SCATTERING THEOREMS AND NO–LOSE COROLLARY FOR THE SSC

Michael S. Chanowitz

Lawrence Berkeley Lab., University of California, Berkeley, CA 94720

The experimentally verified relationship $M_W = M_Z \cos\theta_W$ implies that the physics underlying electroweak symmetry breaking must become visible at or below 1 TeV in the following sense: either there are spin zero bosons much lighter than 1 TeV or there are enhanced cross sections for W, Z pair production. The latter enhancements would signal a new sector of strongly interacting particles at the 1 TeV scale or higher and could be observed at the SSC even if the associated new spectrum were so much heavier than 1 TeV that it could not be produced and studied directly.

I. Introduction

The purpose of this talk is to identify a general experimental signal that can tell us about the scale of the new physics responsible for breaking the electroweak gauge symmetry and giving the W and Z bosons their masses. This signal could be observed at a pp collider with the design parameters of the SSC, *i.e.*, $\sqrt{s} = 40$ TeV and $\mathcal{L} = 10^{33}$cm.$^{-2}$ sec.$^{-1}$. A pp collider with $\sqrt{s} = 20$ TeV and $\mathcal{L} = 10^{33}$cm.$^{-2}$ sec.$^{-1}$ would not suffice nor would a 40 TeV collider with $\mathcal{L} = 10^{32}$cm.$^{-2}$ sec.$^{-1}$.

The discussion is in two parts. The first part is a low energy theorem for the scattering of longitudinally polarized W and Z bosons (W_L and Z_L) that holds for $M_W^2 << s << M_{SB}^2$, where M_{SB} is the typical mass scale of the symmetry breaking sector. The theorem asserts that W_L and Z_L scattering amplitudes have a universal form (precisely the old Weinberg $\pi\pi$ low energy amplitudes with $F_\pi = 93$ MeV replaced by $v = 250$ GeV) provided A) that the theory correctly reproduces the experimentally verified relationship $\rho \equiv (M_W/M_Z \cos\theta_W)^2 = 1$ and B) that there are no $J = 0$ particles in the spectrum of the symmetry breaking sector that are much lighter than M_{SB}. This result extends previous work with M. Gaillard (1) and was obtained in collaboration with M. Golden and H. Georgi (2).

The second part of the talk is the "No-Lose Corollary" for the SSC. The point is that the universal W_L and Z_L scattering amplitudes imply enhanced WW, WZ, and ZZ yields in pp scattering, which are measurable at a collider with the energy and luminosity proposed for the SSC. If the enhanced signals are not observed we can conclude that there must be light (<< 1 TeV) $J = 0$ particles in the spectrum of the symmetry breaking sector — either Higgs scalars or pseudo–Goldstone bosons — which can be produced and studied directly. If on the other hand the enhanced gauge boson pair signals are observed we learn that the symmetry breaking sector is strongly interacting and has a mass scale $M_{SB} \gtrsim 1$ TeV. These signals will be observable at the SSC even if $M_{SB} >> 1$ TeV and the spectrum is too heavy to observe directly. These experimental implications extend earlier work with M. Gaillard (1) and a more recent study with R. Cahn (3).

2. Universal W_L, Z_L Scattering.

Suppose $\mathcal{L}_{SB}$ is the lagrangian of the symmetry breaking sector and neglect for the moment the electroweak gauge interactions. We require $\mathcal{L}_{SB}$ to have a global symmetry group G that breaks spontaneously, leaving an unbroken subgroup H. At this stage the spectrum of $\mathcal{L}_{SB}$ contains massless Goldstone bosons corresponding to the broken generators of G. We denote these bosons as $w^\pm$, z, and $\{\phi_i\}$. Introducing the $SU(2)_L \times U(1)$ gauge interactions, we select $w^\pm$ and z as the bosons which couple to the weak gauge currents and "become" the longitudinal gauge bosons, $W_L^\pm$ and Z_L, by virtue of the Higgs mechanism. We arrange things so that the $\{\phi_i\}$ (if any) acquire masses from the gauge interactions; they are then pseudo–Goldstone bosons with masses below the typical scale M_{SB}.

The identification of the Goldstone bosons

$w^{\pm}, z$ with the longitudinal gauge bosons $W_L^{\pm}, Z_L$ is made precise by the $w - W_L$ equivalence theorem, proved in ref. (1) to all orders in both the (possibly strong) coupling λ_{SB} of $\mathcal{L}_{SB}$ and the electroweak coupling g. The theorem asserts for W boson energies $E_i >> M_W$ that scattering amplitudes with external W_L, Z_L bosons can be computed in an R–gauge with external w, z Goldstone bosons substituted for the corresponding longitudinal gauge bosons. That is,

$$\mathcal{M}(W_L(p_1), W_L(p_2) \cdots)_U = \mathcal{M}(w(p_1), w(p_2) \cdots)_R + O(M_W/E_i) \qquad (1)$$

where U and R denote U–gauge and R–gauge.

With the equivalence theorem we immediately obtain the universal W_L, Z_L amplitudes under the assumption of ref. (1) that G contains $SU(2)_L \times SU(2)_R$ and H contains $SU(2)_{L+R}$ (sometimes referred to (4) as the "custodial" $SU(2)$). In this case we have an $SU(2)_L \times SU(2)_R$ current algebra with pseudoscalar Goldstone bosons $w^{\pm}$ and z related to the axial currents by PCAC. Consequently we simply take over Weinberg's $\pi\pi$ low energy theorems (5) with $F_\pi = 93$ MeV replaced by the vacuum expectation value v = 250 GeV. Provided there are no light ($<< M_{SB}$) particles in the spectrum of $\mathcal{L}_{SB}$ (other than w and z) we have for $s << M_{SB}^2$ the amplitudes

$$\mathcal{M}(w^+w^- \to zz) = s/v^2 \qquad (2a)$$

$$\mathcal{M}(w^+w^- \to w^+w^-) = -u/v^2 \qquad (2b)$$

$$\mathcal{M}(zz \to zz) = 0 \qquad (2c)$$

and other amplitudes related by crossing.

The relation $v = 2M_w/g$ is valid up to corrections of order g and M_W^2/M_{SB}^2. Therefore we may use the $w - W_L$ equivalence theorem to write the universal amplitudes for the low energy domain $M_W^2 << s << M_{SB}^2$:

$$\mathcal{M}(W_L^+W_L^- \to Z_LZ_L) = g^2 s/4M_W^2 \qquad (3a)$$

$$\mathcal{M}(W_L^+W_L^- \to W_L^+W_L^-) = -g^2 u/4M_W^2 \qquad (3b)$$

$$\mathcal{M}(Z_LZ_L \to Z_LZ_L) = 0 \qquad (3c)$$

The new derivation, a version of which is sketched below, removes the assumption about the content of G and H, replacing it with the experimental constraint that $\rho = 1$. This is not unexpected since the existence of a custodial $SU(2)$ is known to protect $\rho = 1$ against potentially large quantum corrections from $\mathcal{L}_{SB}(4)$. However the fact that the converse has not been proved provides the motivation for a derivation of the universal amplitudes eq. (3) which does not assume the existence of a global $SU(2)_L \times SU(2)_R$ that breaks to $SU(2)_{L+R}$.

We have verified the more general result in three ways (2): by a U–gauge argument presented in outline below, by a current algebra R–gauge argument analogous to Weinberg's (5) derivation, and by effective Lagranian methods (6).

In the U–gauge method we begin by decomposing the tree amplitude into a pure gauge–sector piece and a piece involving particle exchange from $\mathcal{L}_{SB}$, for instance,

$$\mathcal{M}(W_L^+W_L^- \to Z_LZ_L) = \mathcal{M}_{gauge} + \mathcal{M}_{SB}. \qquad (4)$$

$\mathcal{M}_{gauge}$ is given by u and t channel W exchanges and the four point contact term. For $s >> M_W^2$ it is

$$\mathcal{M}_{gauge} = g^2 s/4\rho M_W^2. \qquad (5)$$

It has "bad" high energy behavior that must be cancelled by $\mathcal{M}_{SB}$ as $s \to \infty$. For example in the standard Higgs model, we have $\rho = 1$ and $\mathcal{M}_{SB}$ is given by the lowest order s–channel Higgs exchange, $-(g^2 s/4M_W^2)\cdot(s/s - m_H^2)$, which cancels $\mathcal{M}_{gauge}$ at $s >> m_H^2$ *but is negligible for* $s << m_H^2$.

This decoupling of $\mathcal{M}_{SB}$ for $s << m_H^2$ is a general feature that holds to any finite order in the potentially strong coupling λ_{SB} of $\mathcal{L}_{SB}$. By power counting we verify that the only $O(\lambda_{SB})$ effects of $\mathcal{L}_{SB}$ that do not decouple at $s << M_{SB}^2$ are the strong renormalizations of M_W and M_Z or, equivalently, of M_W and ρ. In the absence of light $J = 0$ particles, all other effects of $\mathcal{L}_{SB}$ are nonleading in g^2 or in s/M_{SB}^2. Therefore to leading order in g and in the absence of light particles, the amplitudes at $M_W^2 << s << M_{SB}^2$ are given just by $\mathcal{M}_{gauge}$ where ρ and M_W have their physical values, including possible strong corrections from $\mathcal{L}_{SB}$. The results are

$$\mathcal{M}(W_L^+W_L^- \to Z_LZ_L) = g^2 s/4\rho M_W^2 \qquad (6a)$$

$$\mathcal{M}(W_L^+W_L^- \to W_L^+W_L^-) = -g^2 u(4\rho - 3)/4\rho M_W^2 \qquad (6b)$$

$$\mathcal{M}(Z_LZ_L \to Z_LZ_L) = 0. \qquad (6c)$$

With the experimental constraint $\rho = 1$, eq. (6) reduces to the universal scattering amplitudes of eq. (3).

3. Experimental Implications.

First, an important qualitative remark: WW fusion, shown in fig. (1), provides a significant enhancement over the $\bar{q}q \to WW$ background if and only if W_LW_L scattering (the shaded blob in fig. (1)) is strong, $O(\lambda_{SB}) >> O(g^2)$, *i.e.*, if $\mathcal{L}_{SB}$ has strong interactions. Quantitatively

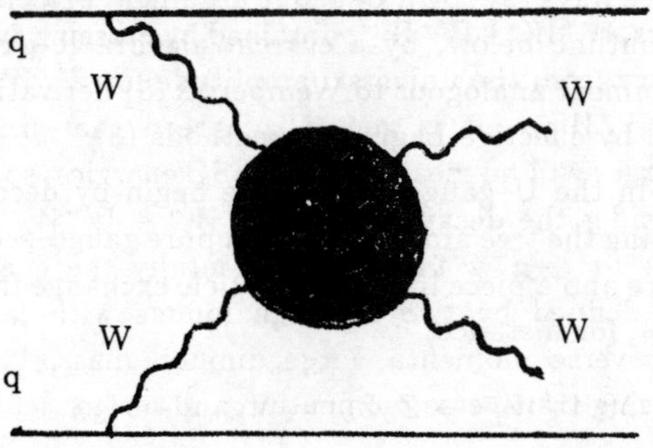

Figure 1: Production of W pairs by WW fusion.

we may use the universal scattering amplitudes of eq. (3) to estimate the magnitude of strong W_LW_L scattering. The prescription of ref. (1) is to extrapolate the universal amplitudes to the unitarity limit (which occurs from 1.8 to 4.3 TeV for the three relevant partial wave amplitudes) and to saturate the partial wave amplitudes at higher energies. This is a conservative ansatz for the following reasons:

1) It neglects possible (likely?) resonant enhancements, *e.g.*, it underestimates the yield from the 1 TeV standard Higgs by 50%.

2) It neglects all partial waves that do not contribute to the low energy theorems. Some of these would certainly make important contributions at energies for which the lower partial waves saturate unitarity.

3) It gives correct orders of magnitude for $\pi\pi$ scattering data and for ϕ^4 models solved in a large N approximation (1).

For instance for the isocalar amplitude $W_L^+W_L^- \to Z_LZ_L$ the prescription for the $J = 0$ partial wave amplitude is

$$a_{00} = \frac{s}{16\pi v^2}\theta\left(v^2 - \frac{s}{16\pi}\right) + \theta\left(\frac{s}{16\pi} - v^2\right). \quad (7)$$

The extrapolated form of the universal amplitude, eq. (3a), is then

$$\mathcal{M}(W_L^+W_L^- \to Z_LZ_L) = \frac{g^2 s}{4M_W^2}\theta\left(\frac{4M_W^2}{g^2} - \frac{s}{16\pi}\right) + 16\pi\theta\left(\frac{s}{16\pi} - \frac{4M_W^2}{g^2}\right). (8)$$

We use eq.(8) to estimate the yield of ZZ pairs in a pp collider with $\sqrt{s} = 40$ TeV and integrated luminosity of 10^{33}cm.$^{-2}$ sec.$^{-1} \cdot 10^7$ sec. = 10^4pb^{-1}. Requiring $M_{ZZ} \gtrsim 1$ TeV and $|y_Z| < 1.5$ to reduce the background from $\bar{q}q \to ZZ$ we find (1) a signal of 470 events over a background of 370. To the uninitiated this may sound like a big signal, but in the cleanest decay channel, $ZZ \to e^+e^-/\mu^+\mu^- + e^+e^-/\mu^+\mu^-$, it corresponds to a signal of only 1.7 events over a background of 1.3.

We therefore turn to another decay channel (3), $ZZ \to e^+e^-/\mu^+\mu^- + \bar{\nu}\nu$, with a (6 times) larger branching ratio and with further enhancement for a high p_T signal over a rapidly falling background. The signal is defined by A) one $Z \to e^+e^-/\mu^+\mu^-$ observed at large p_T in the central region, $|y_Z| < 1.5$, B) by large missing p_T, and C) by an absence of hot jet activity. For large p_T this is a very clean signal with little background, quite analogous to the $W \to e\bar{\nu}/\mu\bar{\nu}$ channel used for the discovery (and mass measurement) of the W at the SPS collider.

Defining the transverse mass in terms of the p_T of the observed Z, $m_T = 2\sqrt{p_T^2 + M_Z^2}$, a crude estimate using eq. (8) gives a signal of 20 events over a $\bar{q}q \to ZZ$ background of 7 for $M_T > 0.9$ TeV. A more careful calculation; like that already performed (3) for the 1 TeV Higgs boson, is in progress. The calculation for the 1 TeV Higgs boson gave a signal of 43 events for the same cuts over the background of 7. This background estimate for $\bar{q}q \to ZZ$ should be reliable to $\lesssim$25% and will be tested experimentally by "calibration" studies (7) at lower p_T. Given the absence of other significant backgrounds, observation of a signal of the expected magnitude would be an unmistakable sign of new physics and quite probably of strong interactions in the symmetry breaking sector.

Figure 2 shows the transverse mass distribution for standard Higgs bosons of mass 0.6 TeV, 0.8 TeV, and 1.0 TeV compared with the steeply falling background from $\bar{q}q \to ZZ$.

For a collider with $\sqrt{s} = 20$ TeV and the same integrated luminosity and with m_T and y cuts as above, the 1 TeV standard Higgs boson gives a signal of 6 events over a background of 3, compared to 43 events over 7 for the 40 TeV collider. Using the extrapolated universal amplitude, eq. (8), the crude estimate for the signal with a 20 TeV collider is only $\sim$ 2 events over the background of 3, compared to the crudely estimated signal of $\sim$ 20 events over 7 for the 40 TeV collider. It is clear that the full strength of the SSC design parameters is essential in order to see this physics.

Similar signals are present in other leptonic decay channels using the extrapolated universal amplitudes (1). For $W_L Z_L \to \ell\bar{\nu} + e^+e^-/\mu^+\mu^-$ my crude estimate is a signal of $\sim$ 15 events over a $\bar{q}q$ annihilation background of $\sim$9, for $|y_{W,Z}| < 1.5$ and $M_{WZ} \gtrsim 1$ TeV, with $\sqrt{s} = 40$ TeV and $10^4 pb^{-1}$. For $\sqrt{s} = 20$ TeV and the same cuts and luminosity there are only $\sim$ 2 events in the signal over a background of $\sim$4.

Another interesting signal is like–charged W pairs, $W^+W^+ + W^-W^-$, obtained by crossing symmetry from the universal amplitude for $W_L^+ W_L^- \to W_L^+ W_L^-$. It is generally agreed that muon charges will be measurable at SSC energies, so we consider the decay channel $W^+W^+ + W^-W^- \to \mu^+\mu^+\nu\nu + \mu^-\mu^-\bar{\nu}\bar{\nu}$. Experimentally the events are defined by two like–sign muons with large transverse momenta, large dimuon mass, large missing transverse momentum and no hot jet ac-

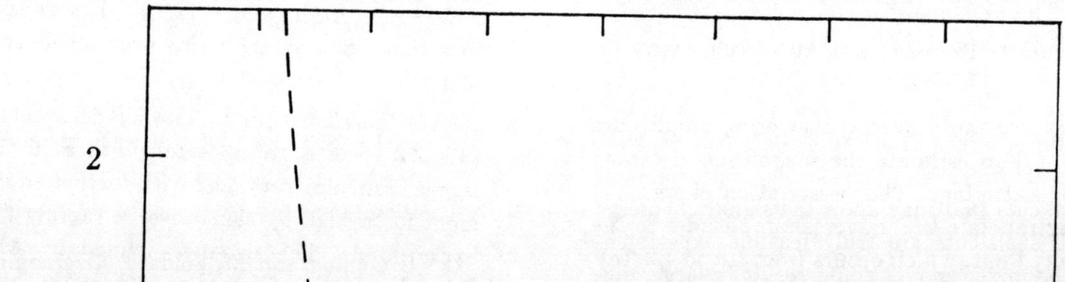

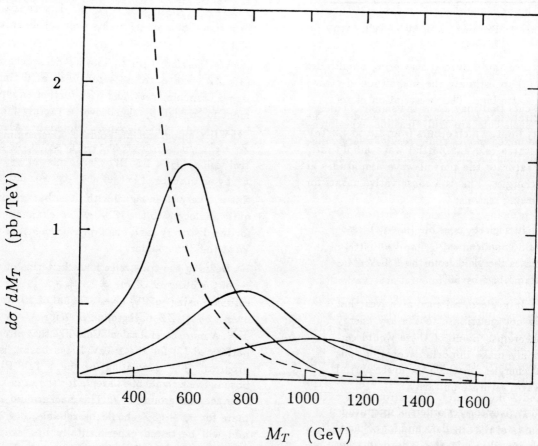

Figure 2: The transverse–mass distribution of the background and signal for $pp \to ZZ + ...$ for $\sqrt{s}$ =40 TeV. The transverse mass is defined in terms of the transverse momentum of the observed Z, $m_T = 2\sqrt{p_T^2 + M_Z^2}$. The signals shown correspond to M_H =600, 800, and 1000 GeV. The dashed curve corresponds to the background from $\bar{q}q \to ZZ$. The observed Z has a rapidity with magnitude less than 1.5.

tivity. Since there is no background from $\bar{q}q$ annihilation we can relax the rapidity cut, typically $|y_W| < 1.5$ above, to allow $|y_\mu| <5$.

A calculation of the yield for such an experimentally defined signal has not yet been done, but we can get a crude conservative estimate by considering W–pair invariant masses greater than 1 TeV with $|y_W| < 4$. The estimate (1) for this (non–implementable) signal is 9 events per 10^4pb^{-1} at $\sqrt{s} = 40$ TeV (and 1 event at $\sqrt{s} = 20$ TeV). I suspect that this yield is several times smaller than the true experimentally useable signal. The question is under study. (If electron charges are also measuable the yield increases four–fold.)

The prospects for detecting gauge boson pairs in "mixed" modes, where one boson decays leptonically and the other hadronically , is also under study. These channels face formidable QCD backgrounds from production of one gauge boson in association with strong production of two quark or gluon jets, which with dijet mass cuts alone are ~ 50 times larger than the signal. Nevertheless studies at the partonic level suggest possible additional cuts that may allow the signal to emerge (8). If such cuts are truly feasible experimentally the mixed modes could result in larger signal yields than the purely leptonic modes, though they would also have large QCD backgrounds.

4. Conclusion

The principal conclusion is that given a pp collider with energy and luminosity as proposed for the SSC, we cannot fail to find signs of the new physics responsible for electroweak symmetry breaking. The experimental observation that $\rho \equiv (M_W/M_Z \cos\theta_W)^2 = 1$ means that either there are new spin zero particles with masses much less than 1 TeV, that can be produced and studied directly, and/or there is a new sector of strongly interacting particles at or above 1 TeV, that will be signaled at the SSC by enhanced production of W and Z boson pairs. This strong interaction signal will be observable at the SSC even if the new quanta of the strongly interacting sector are much heavier than 1 TeV and too heavy to study directly.

The SSC as proposed fulfills the minimum requirements to observe this general signal of new strong interaction, TeV scale physics. Colliders with a factor two less energy or a factor ten less luminosity would not suffice.

References

1. M. Chanowitz and M. Gaillard, Nuc. Phys. B261 (1985) 379. See also Phys. Lett. 142B (1984) 85.
2. M. Chanowitz, H. Georgi, and M. Golden, in preparation.
3. R. Cahn and M. Chanowitz, Phys. Rev. Lett. 56 (1986) 1327.
4. P. Sikivie, L. Susskind, M. Voloshin, and V. Zakharov, Nuc. Phys. 173 (1980) 189.
5. S. Weinberg, Phys. Rev. Lett. 17 (1966) 616.
6. For a pedagogical presentation with references, see H. Georgi, Proc. 1985 Les Houches Summer School, to be published.
7. M. Chanowitz, Weak Interactions at the SSC, LBL–21290 and to be published in the Proceedings of the UCLA workshop on Standard Model Physics at the SSC, Jan. 15-24, 1986.
8. J. Gunion and M. Soldate, Overcoming a Critical Background to Higgs Detection, UCD-86-09.

THE COMPOSITE HIGGS MECHANISM

David Kaplan
Lyman Laboratory of Physics, Harvard University
Cambridge, Massachusetts 02138, USA

It is shown that one can construct theories in which the weak interactions are broken by a composite Higgs doublet. These theories differ from the standard model in that SU(2) xU(1) breaking is dynamical. They also differ from technicolor models in that the spectrum contains a true Higgs boson, and because the scale of the confining interactions may be much greater than the weak scale. Predictions for the Higgs mass follow, as well as predictions for new neutral current effects.

I. INTRODUCTION

The only thing we know about the breaking of SU(2) xU(1) is that three goldstone bosons are somehow produced and eaten by the W and Z, as well as the fact that the rho parameter is nearly equal to 1. How these goldstone bosons are produced is completely unknown. The standard scenario proposes that a scalar doublet gets a Vacuum Expectation Value (VEV). This implies the existence of a neutral scalar--the Higgs boson--whose mass is unknown. A second scenario--that of technicolor [1] proposes that some strongly interacting "techniquarks" condense at several hundred GeV in a weak doublet channel, breaking SU(2) x U(1). In order to predict the correct W and Z masses, the scale of this condensate is fixed so that the parameter f (corresponding to f_π in QCD) takes the usual value of 250 GeV. The particle spectrum does not contain a true Higgs boson.

A third scenario has been recently proposed: that of the composite Higgs mechanism [2-4]. In this scenario, strong interactions called "ultracolor" cause fermions to condense at some large (>>250 GeV) scale but do <u>not</u> break the weak interactions (distinguishing these theories from technicolor). They do produce some "ultrapions" however: pseudo-goldstone bosons of broken ultrafermion flavor symmetries. We craft the model so that these pseudo-goldstone bosons include something with the weak interactions properties of the Higgs boson. The interactions of these pseudos are calculable in terms of a chiral lagrangian. We show that by introducing a weak U(1) gauge force (which gets broken at the ultracolor scale Λ_{UC} >> 250 GeV), it is possible to force our Higgs-pseudo to develop a VEV, breaking the weak interactions. This is the composite Higgs mechanism. It is possible to calculate the Higgs boson mass in some models, and to restrict its range in others.

It is important to note that we will assume no bizarre dynamics for the ultracolor sector: we consider only vector-like theories which are quite well understood, thanks both to the example of QCD, as well as to recent work on the properties of such theories [5]. It is also important to note that we may calculate the Higgs potential even though the Higgs is a bound state of strongly interacting fermions--this is because it is a pseudo-goldstone boson, for which the techniques of current algebra work admirably.

II. THE SU(3) x SU(3) MODEL

We discuss now a model fashioned after QCD. For a more complete exposition the reader may refer to the literature [3]. Suppose there is an $SU(N)_{UC}$ ultracolor gauge force which gets strong at a large scale Λ_{UC}, and and that we have three flavors of massless Dirac fermions transforming as N's of $SU(N)_{UC}$. This model possesses an SU(3) x SU(3) global flavor symmetry, and looks like a high energy version of the u, d and s quarks of QCD. At Λ_{UC} a condensate forms, breaking SU(3) x SU(3) down to the diagonal subgroup, causing an octet of goldstone bosons to appear (the "ultrapions"). Nothing new so far. Now suppose we gauge the diagonal SU(2) x U(1) subgroup of SU(3)--namely isospin x hypercharge--and identify this with the SU(2) x U(1) of the weak interactions. I.e., the ultraquarks transform as:

$$\begin{pmatrix} U \\ D \end{pmatrix} = 2_{1/2}, \; S = 1_{-1} \tag{1}$$

under SU(2) x U(1). The ultrapions will now transform as an SU(2) triplet and be massive pseudo-goldstone bosons. The ultrakaons will transform as a weak charged doublet (just like the Higgs!) and also acquire a positive mass from the SU(2) x U(1) interactions. The ultraeta will remain massless, an exact goldstone boson of the broken axial hypercharge symmetry. To find the masses of these particles, one constructs the relevant chiral lagrangian (see ref. [6], for example). Define $\Sigma = \exp(2i\pi_a T_a/f)$ where π_a are the eight (pseudo) goldstone bosons, T_a are the SU(3) generators with $\mathrm{Tr}\, T_a^2 = 1/2$, and f corresponds to the f_π = 93 MeV of QCD. Under SU(3) x SU(3) rotations Σ transforms like $\Sigma \to L\Sigma R^+$. The chiral lagrangian then looks like:

$$\mathscr{L} = \frac{f^2}{4}\,\mathrm{Tr}\; D\Sigma D\Sigma^+ + kg^2f^4\,\mathrm{Tr}\,(Q_L^\alpha \Sigma Q_R^\alpha \Sigma^+) + \cdots \qquad (2)$$

where the D are SU(2) x U(1) gauge covariant derivatives, g_α are the gauge couplings and Q_L^α, Q_R^α are the gauge couplings to the right- and left-handed ultraquarks. Since the SU(2) x U(1) which we gauged was a vector symmetry, $Q_L^\alpha = Q_R^\alpha$. The parameter k is dimensionless and of order 1--we know it is positive [7] but cannot calculate it. We can measure the analogous parameter in QCD since the second term in Eq. (2) occurs in QCD and gives the $\pi^+ - \pi^0$ mass splitting--one finds k = .8 [3].

The second term of Eq. (2) is very important to us: it gives the potential for the pseudo-goldstone bosons. Expanding Σ to second order in π_a, we find that both the ultra-pions and -kaons get positive masses of order $g_2^2 f^2$, g_2 being the SU(2) coupling constant.

So far we haven't broken SU(2) x U(1). To do so we need to generate a negative mass for the ultrakaon. Eq. (2) tells us how to do this: gauge an <u>axial</u> subgroup of SU(3) x SU(3). Then $Q_L^\alpha = -Q_R^\alpha$ and a <u>negative</u> mass for the pseudos is generated. So we gauge an additional U(1): <u>axial</u> hypercharge $U(1)_A$ with coupling g_A. Eq. (1) becomes modified: under SU(2) x U(1) x U(1) the ultraquarks transform as:

$$\begin{pmatrix}U\\D\end{pmatrix} = 2_{\frac{1}{2},\frac{1}{2}}, \quad \begin{pmatrix}U\\D\end{pmatrix}_R = 2_{\frac{1}{2},-\frac{1}{2}} \qquad (3)$$

$$S_L = 1_{-1,-1}, \quad S_R = 1_{-1,+1}.$$

Indeed, the $U(1)_A$ as it stands is anomalous, but we can cancel the anomalies with spectators, ultracolor charged or otherwise. Note that the $U(1)_A$ symmetry is broken by the condensate at Λ_{UC} and the ultra-eta is eaten by the gauge boson. One can show that if normal quarks and leptons are to couple to the composite Higgs, then they will have to carry $U(1)_A$ charges as well.

The potential for Σ is now given by

$$V(\Sigma) = -kf^4\left\{g_2^2 \sum_{a=1}^{3} \mathrm{Tr}\; T^a\Sigma T^a\Sigma^+ + 3(g_1^2 - g_A^2)\; \mathrm{Tr}\; T_8\Sigma T_8\Sigma^+\right\}. \qquad (4)$$

Calculation shows that the ultrakaon gets a negative mass for large enough g_A. Solving for $\langle\Sigma\rangle$ we find that K^0 develops a VEV $\langle K^0\rangle = v$, so that

$$\langle\Sigma\rangle = \exp \frac{2i}{f}\begin{pmatrix}0 & 0 & 0\\ 0 & 0 & v\\ 0 & v & 0\end{pmatrix}/\sqrt{2} \qquad (5)$$

where the VEV v minimizes the potential

$$V(\Sigma) = kf^4\,\frac{g_2^2}{c_0}\,(\cos v/f - c_0)^2 \qquad (6)$$

where

$$c_0 = 2g_2^2/(3g_A^2 - 3g_1^2 - g_2^2). \qquad (7)$$

Thus for large enough g_A, $0 < c_0 < 1$ and our ultrakaon develops a VEV:

$$\left\langle \begin{pmatrix}K^+\\K^0\end{pmatrix}\right\rangle = \begin{pmatrix}0\\v\end{pmatrix}/\sqrt{2}.$$

We have succeeded in breaking SU(2) x U(1) in the proper way.

Further analysis of this model reveals that by fixing M_W, the only free parameter in the theory is c_0. Since this model posesses no custodial SU(2) symmetry, the ρ parameter is not exactly 1 at tree level, but rather is given by

$$\rho = 1 + \left(\frac{1-c_0}{1+c_0}\right)$$

To be compatible with experiment, this restricts one to the range $.98 < c_0 < 1$. Note that for $c_0 \simeq 1$, f >> v; i.e. the confining scale Λ_{UC} is >> M_W. By thus fixing

c_0 we have reduced the theory to a zero-parameter one, and so the Higgs mass is calculable! One finds the Higgs mass to be

$$m_H = \frac{3}{N} \cdot 1.7\, M_W \tag{8}$$

where N is the size of the $SU(N)_{UC}$ group. The details of how to calculate in this model are given much more fully in references [3].

III. THE SU(5)/SO(5) MODEL

It is interesting to ask whether one can construct a composite Higgs model with a custodial $SU(2)_C$ such that the parameter $c_0 = \cos v/f$ need not be so close to 1 (i.e. $f \gg v$). In such a case the scale of compositeness might be experimentally within reach, and not only would one see ultra-hadrons, but also lighter particles: the other pseudos besides the Higgs, and the massive $U(1)_A$ gauge boson--which one can show will couple to matter.

Such models are possible to build. One model proposed in references [4] is based on an SU(5)/SO(5) goldstone boson manifold, as opposed to the SU(3) x SU(3)/SU(3) manifold of the previous section. This is achieved by proposing an ultracolor group G_{UC}, under which we have five flavors of left-handed fermions transforming as a real representation of G_{UC}. This model, in the absence of weak gauge forces, possesses an SU(5) global symmetry. We gauge now an $SU(2) \times U(1) \times U(1)_A$ subgroup as in the previous model. The ultrafermions transform as in the previous model. The ultrafermions transform as:

$$\begin{pmatrix} \tilde{\psi} & 2_{-1/2,1} \\ \psi = & 2_{1/2,1} \\ S & 1_{0,-4} \end{pmatrix} . \tag{9}$$

The condensate will be symmetric in SU(5) flavor indices (since the fermions are real under G_{UC}) breaking SU(5) down to SO(5). Since the ψ can condense with $\tilde{\psi}$, and s with itself, the condensate need not break the SU(2) x U(1) interactions (but it will break the $U(1)_A$). Once again one finds a pseudo with quantum numbers of the Higgs ($\phi \sim \psi s, \tilde{\phi} \sim \tilde{\psi} s$) which has a negative mass due to $U(1)_A$ interactions, breaking SU(2) x U(1) at a much lower scale than $f \simeq \Lambda_{UC}$. This model contains a custodial $SU(2)_C$, however, and $\rho=1$ at tree level. This allows c_0 to be less than one, and hence the confinement scale may be reachable in future experiments. The phenomenology of this model is very rich, and is explored in detail in references [4]. The Higgs mass is bounded by $200\ \text{GeV} < M_H < 300\ \text{GeV}$ in this model.

IV. FERMION MASSES

As with all dynamical models of SU(2) x U(1) breaking, it is difficult to see how the quark and lepton mass matrices are generated in the composite higgs scenarios. Two possibilities for producing the needed Yukawa couplings come to mind: The first possibility is to adopt the extended technicolor scenario [1] and posit some "extended ultracolor" gauge group. The problems with this approach are as difficult as with technicolor--with the important exception that by making Λ_{UC} large, problematic flavor changing interactions can be made arbitrarily small. A more intriguing possibility for generating Yukawa couplings is to have the fermions be composite as well. Then Yukawa couplings can be generated much in the same way the pion-nucleon coupling is generated by QCD. Unfortunately, we know too little about the dynamics of composite fermions to construct realistic models. Progress can be made by ignoring dynamics and only considering what low energy symmetries have to tell us [8], or by making certain assumptions about the needed dynamics based on our understanding of QCD [9]. Generating Yukawa couplings between composite Higgs and composite fermions seems like a far more natural and promising approach than the extended ultracolor scenario.

V. CONCLUSIONS

Composite Higgs models offer a third viable alternative to the breaking of SU(2) x U(1), after the fundamental Higgs mechanism and technicolor. They offer a way to break SU(2) x U(1) dynamically by forces which get strong in the TeV range, rather than hundreds of GeV. The phenomenology of such models offer a Higgs boson of calculable mass, a host of heavier pseudos, and a heavy $U(1)_A$ gauge boson which will couple to ordinary matter and mediate neutral currents (possibly at levels in reach of experiments in the near future.) Fermion masses are probably most likely to be explained in these theories by having the fermions be composite as well as the Higgs.

REFERENCES

[1] E. Farhi and L. Susskind, Phys. Rep. 74 (1981) 277, and references therein.

[2] D. B. Kaplan and H. Georgi, Phys. Lett. 136B (1984) 183; D. B. Kaplan, H. Georgi, and S. Dimopoulos, Phys. Lett. 136B (1984) 187.

[3] T. Banks, Nucl. Phys. B243 (1984) 125; H. Georgi, P. Galison, and D. B. Kaplan, Phys. Lett. 143B (1984) 152.

[4] H. Georgi and D. B. Kaplan, Phys. Lett. 145B (1984) 216; M. Dugan, H. Georgi, and D. B. Kaplan, Nucl. Phys. B260 (1985) 215.

[5] D. Weingarten, Phys. Rev. Lett. 51 (1983) 1830; S. Nussinov, Phys. Rev. Lett. 51 (1983) 2081; C. Vafa and E. Witten, Nucl. Phys. B234 (1984) 173; D. Kosower, Phys. Lett. 144B (1984) 215.

[6] See, for example, H. Georgi, Weak Interactions and Modern Particle Physics, Benjamin/Cummings Publishing Co., Inc. (1984).

[7] E. Witten, Phys. Rev. Lett. 51 (1983) 2351.

[8] H. Georgi, Phys. Lett. 151B (1985) 57.

[9] H. Georgi, Nucl. Phys. B266 (1985) 266; H. Georgi, Les Houches Lectures, Summer (1985).

COMPOSITE VECTOR BOSONS

Barbara Schrempp

Institut für Theoretische Physik, Universität München
D-8000 München, Fed. Rep. of Germany

As main recent activities I review i) the efforts to establish the viability of the concept of composite W and Z bosons in the light of experiments becoming sensitive to one-loop radiative corrections, ii) lower mass bounds for further composite vector bosons and iii) the strongly coupled standard model as a respectable composite model, faking the low energy properties of the standard model and predicting 'colorful' deviations at high energies.

1. INTRODUCTION

The heretical suggestion (1-4) that not only quarks and leptons but also the W and Z bosons could be composite, was put forward in 1980, i.e. before the experimental confirmation of W and Z. Meanwhile we face fairly precise measurements of the W and Z masses combined with improved low energy charged and neutral current data. Within errors they are in good agreement with the standard model.

Clearly, the foremost question is whether the concept of composite W and Z is still viable in 1986. In particular, I shall consider three aspects: i) radiative corrections in nearby compositeness, ii) with $\rho = 1.006 \pm 0.008$, is there any "room" in the data for further composites? iii) implications of a high precision universality test. The global answer to the question of viability will be largely 'yes'. It implies somewhat sobering reflections on the extent to which the standard model is actually tested today.

A clearcut expectation in nearby compositeness is a spectrum of further composites, new composite ground states, mostly with exotic quantum numbers, excited states of the known q,l,W,Z etc.. Most of these states should have masses below or around 1 TeV. As a respectable reference model I shall discuss the strongly coupled standard model (SCSM) which predicts a large number of exotic composite isoscalar vector bosons.

2. CONSISTENCY AT LOW ENERGIES

A theory of weak interactions among composite q,l,W,Z is an <u>effective</u> theory. As in strong interactions among composite hadrons, the <u>fundamental</u> gauge theory will have to be looked for on the level of the constituents; the SCSM is a particularly convincing example.

Let us first concentrate on the effective theory at "low energies", i.e. at energies below the Fermi scale

$$\Lambda \sim (\sqrt{2}G_F)^{-1/2} \sim 250 \text{ GeV}. \quad (1)$$

It may be described by an effective Lagrangian $\mathcal{L}_{eff}$ in terms of q,l,W,Z fields involving effective couplings $g_{f\bar{f}W}, g_{WWZ}, \ldots (f = q,l)$. The point of this section is to establish a set of trustworthy conditions under which at low energies $\mathcal{L}_{eff}$ mimics the standard model to the extent to which it has been tested so far.

An obvious recipe is the revival of concepts which have been successful in low energy physics of composite hadrons, in particular of vector dominance (5,6) and of current algebra (7). With (8) as a role model this procedure culminates (6) in the following clearcut statement. Any composite model which provides

i) $q,l,W^{\pm},Z$ as lowest mass composites (implying a "natural" mechanism for $m_{q,l} \ll \Lambda$),

ii) <u>global</u> SU(2) symmetry of weak isospin (for e=0) and <u>local</u> $U(1)_{em}$ gauge invariance,

iii) vector dominance in the strong formulation as current-field identity

$$j_\mu = \frac{m_W^2}{g_W} W_\mu^0 \quad (2)$$

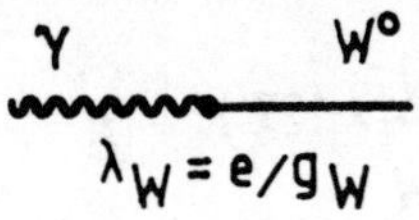

leads to the low energy Lagrangian

$$\mathcal{L}_{eff} \underset{E \lesssim \Lambda}{=} \quad (3)$$

$$\mathcal{L}_{SM}(\text{in the unitary gauge without physical Higgs}) \; + \; \mathcal{L}'_{dim>4}$$

which perfectly mimics the standard model Lagrangian $\mathcal{L}_{SM}$. More precisely, for e=0, $\mathcal{L}$ has <u>massive Yang-Mills form</u> (including the higher dimensional operators in $\mathcal{L}'$), i.e. <u>local</u> SU(2) gauge invariance in terms of the quasi-gauge coupling g_W, only broken by the W mass term; for e≠0, $\mathcal{L}$ results from the substitution $W^o_\mu \rightarrow W^o_\mu + \frac{e}{g} A\mu$. The result (3) was recently rederived (9) within the context of the SCSM in the intuitive language of one-pole approximation in a dispersion theoretical approach. Ref. (9) also contains a careful discussion of dimension 6 and 8 operators allowed by global SU(2) and vector dominance. <u>No</u> dangerous dimension 4 operator, as e.g. $Z_\mu Z^\mu W^+_\nu W^{-\nu}$, $Z^\mu Z^\nu W^+_\mu W^-_\nu$ which were advocated in (10), is allowed. The only allowed dimension 6 operator, involving only W and Z fields, is $\vec{W}^{\ \nu}_\mu (\vec{W}_\rho^{\ \mu} \times \vec{W}_\nu^{\ \rho})$. It contributes to the anomalous magnetic moment of the W without being in conflict with present data. The influence of the dimension 8 operator $F_{\mu\nu} \bar{f}\gamma^\mu f \bar{f}\gamma^\nu f$ is strongly suppressed on account of the four-body (massless) phase space.

3. RADIATIVE CORRECTIONS

The combination of data taken at $q^2 \approx m_{W,Z}$ with those taken at low q^2 has turned out (11) to be sensitive to (renormalization group improved) one loop corrections of the standard model. Superficially this means trouble for composite W and Z. This is, however, not the case, given the present precision of data. The argument can be led on two levels.

First of all, the bulk effect of these radiative corrections may be traced back to a sum of large <u>QED</u> logarithms of the type $(\alpha/\pi)\log(m_W^2/m_f^2)$ which in turn is responsible for the running of α from $\alpha(q^2 \approx 0) \approx 1/137$ to $\alpha(q^2 \approx m_{W,Z}) \approx 1/128$. This running of α of course applies, irrespective of whether <u>weak</u> interactions are effective interactions among composite particles or local gauge interactions (see e.g. (12)).

Next, let us consider <u>one-loop weak</u> corrections. Following Sect.2, the effective low-energy theory of composite q,l,W,Z is distinguished by two important features: it is of <u>massive Yang-Mills</u> type (for $\alpha \rightarrow 0$) and it has approximate <u>chiral symmetry</u> ($m_{q,l} \ll \Lambda$). Such a theory is clearly nonrenormalizable, but - as is well known (13,6) - "almost" one-loop renormalizable. This manifests itself as follows. In any effective field theory for composite fields it is legitimate to introduce a <u>physical</u> cut-off Λ (which is of the order of the inverse size of the composite particles) and to perform a one-loop calculation. However, in general, quantitative predictivity is lost, <u>unless</u> the cut-off dependence is very weak. This is indeed the case for a chirally symmetric massive Yang-Mills Lagrangian. The cut-off dependence shows up in form of

$$\frac{g_W^2}{4\pi} \log \frac{\Lambda}{m_W} \quad \text{and} \quad \frac{g_W^2}{4\pi} \frac{m_{q,l}^2}{\Lambda^2} \qquad (4)$$

in one-to-one correspondence (13) to

$$\frac{g_W^2}{4\pi} \log \frac{m_H}{m_W} \quad \text{and} \quad \frac{g_W^2}{4\pi} \frac{m_{q,l}^2}{m_H^2}, \qquad (5)$$

i.e. the dependence on the Higgs mass m_H in the standard model which replaces the cut-off in the one-loop result. To the extent that present data are insensitive to the Higgs mass, they are also insensitive to a variation of the cut-off within the generously large interval $m_W \lesssim \Lambda \lesssim 1$ TeV.

4. NEW VECTOR BOSONS: $W^{\pm\prime}$, Z' AND Y

A spin 1 excitation $W^{\pm\prime}$, Z' of the composite triplet $W^\pm$, Z and, perhaps, an isoscalar vector boson Y are the most likely additional vector bosons to be expected in nearby compositeness (14-18). The main issue of this section is to point out that there is surprisingly much "room" in the data for these particles and to obtain some analytical insight for why this is so. Both, the W',Z' triplet and the Y, modify the low-energy four-fermion interactions on the Born term level

via contributions from

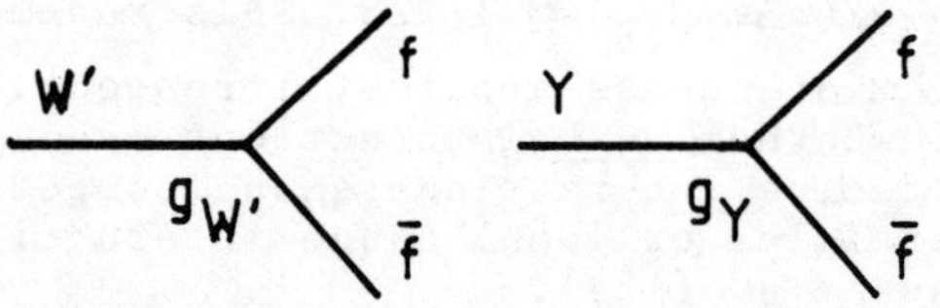

and affect the Z mass via their mixing with the photon

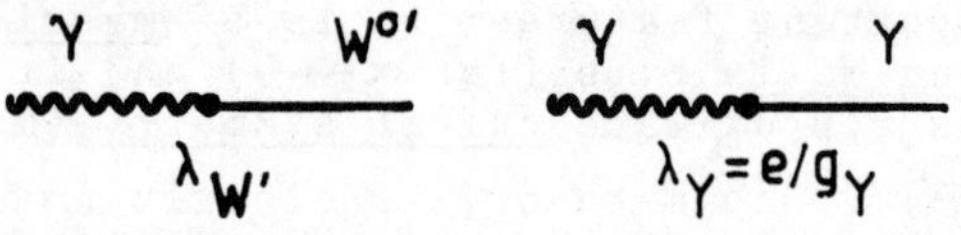

Again vector dominance allows to implement the excited bosons in $\mathcal{L}_{eff}$ in terms of three new parameters, $m_{W'}$, $g_{W'}$ and $\lambda_{W'}$, and the isoscalar boson in terms of two, m_Y and g_Y.

The excited triplet $W^{\pm\prime}$, Z' has the same quantum numbers as W,Z. So, not surprisingly, it leaves the familiar form of the low-energy four-fermion Lagrangian almost unaltered

$$-2\mathcal{L} = \frac{8G_F}{\sqrt{2}}(J_1^2 + J_2^2 + (J_3 - \sin^2\Theta_W J_{em})^2 + CJ_{em}^2). \quad (6)$$

However, the quantities G_F, $\sin^2\Theta_W$ and C are redefined in terms of old and new parameters (14-17). Most sensitive to redefinitions are

$$\sqrt{\frac{8G_F}{\sqrt{2}}} = \frac{1}{m_W}\left[g_W\sqrt{1 + \frac{g_{W'}^2\, m_W^2}{m_{W'}^2\, g_W^2}}\right] \quad (7)$$

$$e = \lambda_W\left[g_W\left(1 + \frac{\lambda_{W'} g_{W'}}{\lambda_W g_W}\right)\right]. \quad (8)$$

There remain four measured quantities, m_W, m_Z, $\sin^2\Theta_W$ (from low energies) and C (upper bound), given in terms of four free parameters, clearly an inconclusive situation. Some analytical insight may be obtained: the main modifications in Eqs. (7,8) can be absorbed by equating the two brackets and redefining them as effective coupling. This equality is represented in Fig. 1 as fat line. Indeed, in this plot the allowed regions for fixed ratios $g_{W'}/g_W$, as transposed from Ref. (17), follow the fat line. Even for fairly "large" values of $g_{W'}$, the resulting mass bounds are ridiculously small

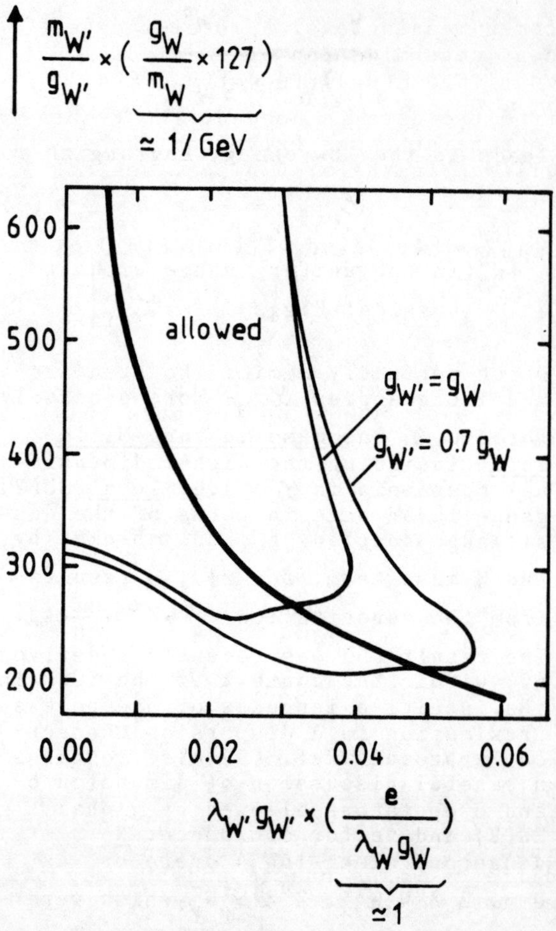

Figure 1: Allowed regions in the $m_{W'} - \lambda_{W'}$ plane

$$g_{W'} = g_W: \quad m_{W'} \gtrsim 170 \text{ GeV}, \quad (9)$$

$$g_{W'} = 0.7 g_W: \quad m_{W'} \gtrsim 94 \text{ GeV}; \quad (10)$$

the former one is in fact already superseded by a direct bound from the CERN $\bar{p}p$ collider (19)

$$g_{W'} = g_W: \quad m_{W'} \gtrsim 220 \text{ GeV}. \quad (11)$$

SLC/LEP with a precise measurement of the $Z \to e^+e^-$ width and the Z mass will be sensitive (20) to $m_{W'} \lesssim 1$ TeV even for $g_{W'} = 0.1\, g_W$.

An isoscalar vector boson Y with vector-like couplings and even more so the more likely variant Y_L, coupling only to left-handed fermions (before mixing), leads to a modification (18, 17) of the neutral current sector which cannot be cast into the familiar form of Eq. (6). For Y_L even fitting $\sin^2\Theta_W$ and the ρ parameter becomes meaningless. A model independent determination of the individual couplings $\varepsilon(u_L), \varepsilon(d_L), \varepsilon(u_R), \varepsilon(d_R)$ from the data is needed. The tightest constraint comes (21) from the 90% CL ellipse in the α, β plane, see Fig. 2, where

$$\begin{Bmatrix}\alpha\\ \beta\end{Bmatrix} = \varepsilon(u_L) - \varepsilon(d_L) \pm \varepsilon(u_R) \mp \varepsilon(d_R) \qquad (12)$$

$$\left[\text{cf. in SM} \begin{cases}\alpha = 1 - 2\sin^2\Theta_W \\ \beta = 1\end{cases}\right],$$

and the data on the W and Z masses. An exercise leads me to the following lower bounds on the physical Y boson masses (after diagonalization)

$$\begin{aligned} M_Y &\gtrsim 350 \text{ GeV for } Y \\ M_Y &\gtrsim 260 \text{ GeV for } Y_L \qquad (13) \\ M_Y &\gtrsim 130 \text{ GeV for } Y_E \end{aligned}$$

(in agreement with Ref. (22)).

For comparison I included the string motivated (22) gauge boson Y_E, also called Z_E or Z' (not to be confounded with the triplet partner of W').

Fig. 2 gives some analytical insight into the correlation between the different mass bounds and the way the ellipse is tilted. The crucial point is that the deviation of β from $\beta=1$ is <u>positive</u> for Y, for Y_L it is <u>zero</u>, $\beta \equiv 1$, and for Y_E it is <u>negative</u>. The lower bound for M_Y is reached if as well the deviation of β from 1 as α are maximized simultaneously, as realized by the three dots in Fig. 2. Maximizing α is obviously easiest for Y_E and most difficult for Y.

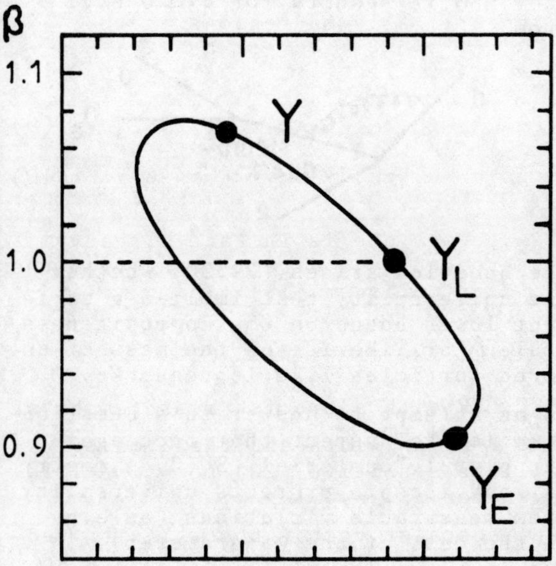

Figure 2: Anatomy of different mass bounds for Y, Y_L and Y_E

5. THE STRONGLY COUPLED STANDARD MODEL

The SCSM (23,3,24,9) is a particularly promising candidate model which satisfies (9) all the consistency conditions for nearby compositeness, i)-iii) in Sect. 2. It is intimately related to the standard model: it starts from the <u>same</u> Lagrangian with local SU(2)xU(1) gauge invariance, 12 fermion doublets ψ_L and one complex scalar doublet ϕ

$$\mathcal{L}_{SU(2)xU(1)} = \mathcal{L}_{\text{gauge int.}}(g,g') - \mu^2\phi^\dagger\phi - \lambda(\phi^\dagger\phi)^2 \qquad (14)$$

However, it is evaluated in a different region of the space of parameters g, g', μ^2, λ. Whereas the standard model is located in the Higgs phase, the SCSM is located in the symmetric phase. The fields ψ_L and ϕ play the role of preon fields; ϕ has vanishing vacuum expectation value; the SU(2) gauge coupling g is large (hence the name "strong coupling")

$$g^2/4\pi \sim O(1) \text{ at } E \sim (\sqrt{2}G_F)^{-1/2} \qquad (15)$$

such that SU(2) "confinement" takes place. The gauged U(1) is identified with $U(1)_{em}$.

The physical particle spectrum then consists of SU(2)-singlet bound states of ψ_L and ϕ with radius of order $1/\Lambda$. The composite ground states containing at least one constituent ϕ have the quantum numbers of q_L, l_L, W, Z and an isoscalar scalar boundstate; $m_{q,l}$ are kept small by 't Hooft's anomaly matching. The theory has a <u>global chiral</u> SU(2)xSU(12) symmetry in the limit $e \to 0$, $g_c \to 0$.

In the SCSM (presumably) also bound states of the type $\psi_L^\dagger\psi_L$ and $\psi_L\psi_L$ are formed (25,9); see also (6,26) for one generation. They give rise to a rich spectrum of composite <u>isoscalar</u> bosons, mostly with exotic quantum numbers. Denoting by ψ_L the whole 12-plet of SU(12), one obtains

i) a <u>143-plet</u> of vector bosons of the type $\overline{\psi}_L\psi_L$ (<u>adj</u> of SU(12)). It contains <u>leptoquarks</u> (color <u>3</u>, Q=2/3, B=1/3, L=-1), neutral bosons (color <u>1</u>, Q,B,L =0) a combination of which mixes with the photon ($\hat{=}Y_L$ of Sect.4) and <u>color octet</u> bosons (color <u>8</u>, Q,B,L=0), a combination of which mixes with the gluon octet. This gives rise to the concept of vector dominance for color octet currents which has been proposed and explored in Refs. (26,6).

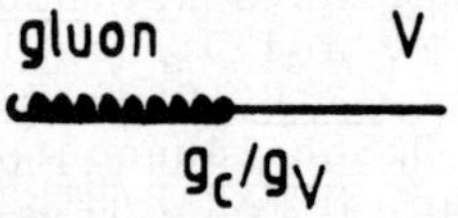

ii) a <u>78-plet</u> of vector bosons (□□ of SU(12)) and a <u>66-plet</u> of scalar bosons (⊟ of SU(12)) of the type $\psi_L\psi_L$. They contain <u>dileptons</u> (L=2, Q=-1), <u>diquarks</u> (color <u>6+$\overline{3}$</u>, Q=1/3, B=2/3) and <u>leptoquarks</u> (color <u>3</u>, Q=-1/3, B=1/3, L=1). Q is the electric charge, L the lepton number and B the baryon number.

Vector dominance and global SU(2)x SU(12) symmetry allow to implement (6,27) the $\overline{\psi}_L\psi_L$ multiplet into the effective Lagrangian in terms of two parameters, m_V and g_V. Again, the resulting Lagrangian (26,27,17) for $q^2 \approx 0$

$$-2\mathcal{L} \underset{e,g_c\to 0}{=} \left(\frac{g_W^2}{m_W^2}+\frac{g_V^2}{m_V^2}\right)\vec{J}^2 + \frac{g_V^2}{m_V^2}J^2_{isoscalar} \qquad (16)$$

(for U(12))

gives rise to a reinterpretation of the Fermi constant G_F. <u>No</u> flavor changing neutral currents appear! For $e,g_c \neq 0$ the neutral current sector is modified through mixing with the photon and gluons. The lower bound (26, 27) for m_V from present data may be placed within the limits

$$220 \text{ GeV} \lesssim m_V^{min} \lesssim 500 \text{ GeV}. \qquad (17)$$

220 GeV is the absolute lower bound, 500 GeV the bound for "large" coupling $g_V \approx 1$, (necessary if color gauge interactions are to be treated as corrections to the weak interactions at energies of the order of m_V).

A remark is in order: the SCSM does not provide a dynamical origin for the formation of q,l generations. Even though this is usually a key requirement in composite model building, I think it is unlikely to arise in <u>nearby</u> compositeness. The scale associated with the origin of generations seems to be much higher than the Fermi scale (see the talk of D. Wyler in the same session).

6. IMPLICATIONS OF A UNIVERSALITY TEST?

The high precision data on μ decay, β decay and an analysis (28) of K_{e_3} and hyperon decay have been combined and compared (29) to the standard model predictions, including renormalization group improved one-loop corrections. As a result, the universality of the involved couplings is claimed (29) to be confirmed to within 1.5‰!

For composite quarks and leptons a potential source for universality violation is the diagram (30) involving QCD corrections

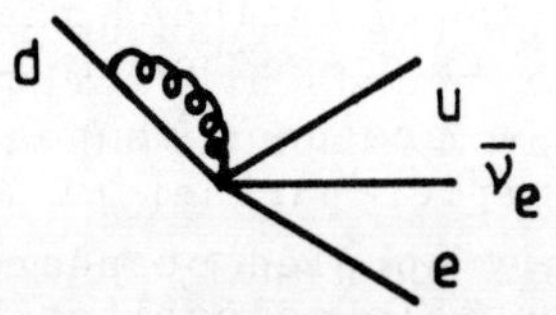

It contributes to β decay without having a counterpart in μ decay. In the SCSM it results for $q^2 \approx 0$ e.g. from

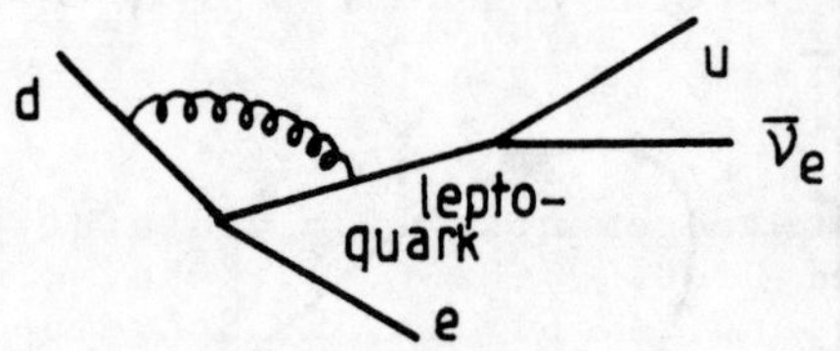

The question arises (29,30) whether the universality test implies a stringent lower bound on the compositeness scale Λ or likewise on the mass of colored particles like leptoquarks.

In an attempt to answer this question I am led to suspect that once again the global "custodial" SU(2) symmetry of weak isospin protects universality from measurable violations (as e.g. in the case of the ρ parameter). In the limit $\alpha_{em} \to 0$, the $\overline{d}u \to e^+\nu_e$ channel, which at $q^2=0$ is related to $de^+ \to u\overline{\nu}_e$ by a Fierz transformation, is protected from radiative corrections due to a <u>conserved</u> (global) SU(2) current. Thus, at $q^2=0$ for $\alpha_{em}\to 0$ no color radiative corrections arise. Since indeed the universality test is performed at $q^2\approx 0$, this leads me to suggest that in the presence of electromagnetic interactions the color radiative corrections in question are "screened", i.e. appear in the form

$$\alpha_{em} \text{ x color corrections} \sim O(1‰). \quad (18)$$

7. CONCLUSIONS

The concept of composite $W^\pm$,Z bosons is still viable. More specifically: i) those radiative corrections to which present data are sensitive, are also expected in effective interactions among composite quarks, leptons and W,Z, ii) there is surprisingly much "room" in the present data for further composite bosons, with lower mass bounds

$$220 \text{ GeV} \lesssim m_{min} \lesssim 500 \text{ GeV}, \quad (19)$$

depending on their quantum numbers. These lower bounds are still of the order of the presumed compositeness scale $\Lambda \sim 250$ GeV.

The strongly coupled standard model is a respectable competitor for the standard model. It predicts a rich spectrum of exotic composite isoscalar vector (and scalar) bosons.

Near future experiments at e^+e^- (SLC/LEP), $\bar{p}p$ (CERN updated, FERMILAB Tevatron) and ep (HERA) colliders will allow to settle the issue of nearby compositeness.

Acknowledgements: I am grateful to T. Appelquist, U. Baur, J. Kühn, W.J. Marciano, A. Sirlin, F. Schrempp, K.-H. Schwarzer and M. Veltman for helpful discussions.

REFERENCES

1) Harari, H. and N. Seiberg, Phys. Lett.98B(1981)269
2) Greenberg, O.W. and J. Sucher, Phys.Lett.99b(1981)339
3) Abbott, L.F. and E. Farhi, Phys. Lett.101B(1981)69; Nucl.Phys.B189 (1981)547
4) Fritzsch, H. and G. Mandelbaum, Phys.Lett.102B(1981)319
5) Kögerler, R. and D. Schildknecht, CERN-TH.3231 (1982); Schildknecht, D., Proc.Europhysics Study Conf., Erice 1983
6) Schrempp, B. and F. Schrempp, DESY 84-055 (1984)
7) Fritzsch,H., D.Schildknecht and R. Kögerler, Phys.Lett.114B(1982)157
8) Lee, T.D. and B. Zumino, Phys.Rev.163(1967)1667
9) Claudson, M., E. Farhi and R.L. Jaffe, Phys.Rev.D34(1986)873
10) Suzuki, M., Phys.Lett.153B (1985)289
11) Marciano,W.J.,talk at this conf.
12) Antonelli, F. and L. Maiani, Nucl.Phys.B186(1981)269
13) Veltman, M., Nucl.Phys.B7(1968) 637; B21(1970)288; Shizuya, K., Nucl.Phys.B121(1977)125; Appelquist,T. and C.Bernard, Phys. Rev.D22(1980)200; Barua,D. and S.N.Gupta, Phys.Rev.D16(1977)1022
14) Kuroda, M. and D. Schildknecht, Phys.Lett.121B(1983)173
15) Baur,U.,D.Schildknecht and K.H.G Schwarzer, MPI-PAE/PTh29/85(1985)
16) Kaidalov, A.B. and A.V. Nogteva, preprint ITEP 18(1985)
17) Korpa, C. and Z. Ryzak, MIT preprint CTP#1313(1985)
18) Schildknecht,D., M.Kuroda, K.H.G. Schwarzer,Nucl.Phys.B261(1985)432
19) Denegri, D., UA1, talk at Int. Conf. on $\bar{p}p$ Physics, Aachen,1986
20) Baur, U. and K.H.G. Schwarzer, to appear
21) Pullia,A., Ital. Phys. Soc., Bologna (1984)333,ed. A. Bertin et al.
22) Durkin, L.S. and P. Langacker, Phys. Lett.166B(1986)436 and references therein
23) complementarity: Mack,G.,DESY preprint 1977; 't Hooft, G., Recent Developments in Gauge Theories, ed. G. 't Hooft et al. (Plenum Press, New York); Dimopoulos,S., S. Raby and L. Susskind, Nucl. Phys.B173,(1980)208
24) extensions: Abbott,L.F., E. Farhi and A. Schwimmer, Nucl.Phys.B203 (1982)493; Bordi,F.,R. Casalbuoni, D. Dominici, R. Gatto, Phys. Lett.114B(1982)31; Schrempp, B., F. Schrempp, Nucl.Phys.B231(1984) 109;Buchmüller,W.,R.D.Peccei,T.Yanagida, Nucl.Phys.B231(1984)53
25) Abbott,L.F., E.Farhi, S.-H.H.Tye, 1982 DPF Summer Study on El. Part. Phys. and Future Facilities,p.288
26) Buchmüller,W. , Phys. Lett.145B (1984)151; CERN.Th-3873(1984)
27) Wudka,J.,Phys.Lett.167B(1986)337
28) Leutwyler, H. and M. Roos, Z.Phys.C25(1984)91
29) Marciano,W.J.,A.Sirlin, Phys.Rev. Lett.56(1986)22 and refs. therein; Sirlin,A.,R.Zucchini,preprint NUY/TR5/86(1986)
30) Veltman, M., Theory-Workshop on Physics at the Fermi Scale, DESY, Hamburg, 1985.

SUBSTRUCTURE: A BRIEF SUMMARY*

HAIM HARARI

Stanford Linear Accelerator Center
Stanford University, Stanford, California 94305
and
Weizmann Institute of Science,
Rehovot, Israel

The present status of substructure theories for Higgs particles, quarks, leptons and gauge bosons is briefly summarized.

The possibility of discovering further substructure within the "fundamental particles" of the standard model remains a viable option for the physics which lies beyond that model. Candidates for substructure (in descending order of plausibility) are Higgs particles, quarks and leptons, W and Z, gluons, photons.

The Motivation for substructure is as good as ever. Higgs substructure is motivated by the fine-tuning problem. Quark and lepton compositeness —by the observed connections between quarks and leptons, by the generation puzzle and by the existence of too many free parameters in the standard model. W and Z substructure is motivated by their relation to a composite Higgs and by the observation that all other short-range interactions are residual interactions.

The experimental limits on substructure are slowly crawling upwards. The "radius" of Higgs particles, quarks, leptons, W and Z could still be[1] of order $(\text{TeV})^{-1}$. However, certain puzzles already lead us to higher energy scales. In particular, an understanding of the generation puzzle seems to require a scale of at least[2] 50-100 TeV, possibly much more.

The explicit detailed models for substructure are as bad as ever. All existing models seem to be inadequate. Many clever ideas have been proposed, both for the new underlying dynamics and for the underlying symmetries. Some of these ideas may be correct; others may need improvements; most are clearly wrong. The correct theory of substructure, if and when discovered, may contain some important ingredients which have been introduced within some of the existing unsuccessful schemes.[3]

Future directions of the substructure hypothesis are likely to be dictated by experiments. The simple theoretical ideas have been tried. They do not work. Most of them introduce a new underlying interaction which simply copies the previous one, utilizing a nonabelian local gauge symmetry. If nature has selected a more subtle way, its most likely method of informing us of this selection is to give us experimental hints.

Strangely enough, not only the future of substructure physics depends on experiments but also the future of experimental high-energy physics depends on substructure. If no substructure is found up to, say, 1 PeV ($\equiv$ 1000 TeV), all the interesting cross sections will continue to fall like $1/s$, eventually reaching rates which will never be detected even if we have the technology and the money for producing higher energy machines. Recall that, at $\sqrt{s} = 1$ PeV, $\sigma(e^+e^- \to \mu^+\mu^-) \sim 10^{-43}$ cm^2. For 100 $\mu^+\mu^-$ events per month, we need an e^+e^- luminosity of 10^{39} cm^{-2}sec^{-1}!

The theoretical attitude towards substructure can be characterized by five broad categories:

(i) **"Ultra-Conservative"**: This approach would claim that quarks, leptons, gauge bosons and Higgs particles are, and will always remain, fundamental. This probably means that the fine-tuning problem is solved by supersymmetry, that the quark-lepton connection and the unification of interactions are achieved by GUTS, and that the generation puzzle will somehow be solved,

* Work supported by the Department of Energy, contract DE-AC03-76SF00515.

allowing for the calculability of many of the fundamental parameters of the standard model without introducing any substructure.

On the experimental side, cross-sections will decrease like $1/s$ and will eventually become unobservable. Supersymmetric partners must be found at or below O(TeV). Since, at present, there is no experimental evidence for any substructure, the "*Ultra-Conservative*" approach must be considered at least as likely as all other alternatives combined. Whether this remains unchanged, only future experiments will tell us.

(ii) **"Conservative"**: The conservative approach allows for Higgs substructure, but keeps quarks, leptons and gauge bosons fundamental. Higgs substructure, either according to the Technicolor scenario or according to the composite Higgs scenario,[4] can solve the fine-tuning problem without invoking supersymmetry. It requires a new fundamental interaction ("Technicolor", "Ultracolor") operating among the constituents of the Higgs particle and it inevitably leads to new physics at the TeV scale. We have recently learned that its phenomenological difficulties may not be as severe as was earlier believed.[5]

Experimentally, this probably leads to a rich spectrum of "Technihadrons" or "composite friends of the Higgs particle" at the TeV range. It also predicts relatively light pseudo-Goldstone-bosons. In some versions we may discover a new world of strongly interacting Higgs particles,[6] yielding large cross sections for the production of longitudinal W's and Z's. Often, but not always, practitioners of the "conservative" approach find themselves moving towards a "*moderate*" attitude,[4] when they try to understand the origin of fermion masses.

(iii) **"Moderate"**: This approach assumes that quarks, leptons and Higgs particles are composite, presumably all of them at the TeV scale, but possibly at two different scales (one for Higgs and one for quarks and leptons). All gauge bosons are elementary. Cross sections for processes such as $e^+ + e^- \to \mu^+ + \mu^-$, $e^+ + e^- \to \bar{q} + q$ reach an approximately constant value $\sigma \sim \mathcal{O}(\Lambda^{-2})$ where Λ^{-1} is the "size" of a lepton or a quark. Multilepton production amplitudes may be large. Excited quarks and leptons are expected at $m \sim \mathcal{O}(\Lambda)$.

Note, however, that if the "radius" of quarks and leptons is as low as O(TeV^{-1}), the generation puzzle cannot be solved at the same energy scale. Limits from processes such as $K^0 \to e\mu$, $K \to \pi e\mu$, $\mu \to e\gamma$, $\mu N \to eN$, $\mu \to 3e$ and $\Delta M(K_S - K_L)$ indicate[2] that a resolution of the generation puzzle is unlikely to occur below O(100 TeV). It is, however, possible[7] that leptons have a radius of TeV^{-1} but the $e-\mu$ difference can only be probed, say, at a distance of PeV^{-1}. There is no logical contradiction between these two statements.

(iv) **"Radical"**: Here one adds compositeness of the massive gauge bosons W and Z, in addition to quark, lepton and Higgs compositeness. The weak interaction is not a fundamental interaction and it is replaced[8] by the new interaction which binds preons within quarks, leptons, W, Z and Higgs. Photons, gluons and possible new massless gauge bosons ("hypergluons"?) remain fundamental.

Experimentally, we expect excited W and Z bosons, possible corresponding $J = 0$ states, deviations from the WWW gauge coupling, large cross sections for $\bar{p} + p \to z + \gamma +$ anything, etc. In this case the compositeness scale of W and Z, if it exists, is almost certain to be around TeV. Consequently, quark and lepton compositeness is also expected at O(TeV), with a second higher scale resolving the generation structure.

(v) **"Ultra-Radical"**: All particles of the standard model, including massless gauge bosons, are composite in some sense. All known interactions somehow emerge from a more fundamental mechanism. This is not a very likely proposition. Even if it is true, the substructure will probably appear only at a very high energy scale, possibly the Planck scale. At that scale, physics will be very different from that of the standard model. A rich spectrum of additional states is expected at $m \sim \mathcal{O}(\Lambda)$. However, below that scale—all particles would appear to be pointlike and the experimental predictions should resemble those of the standard model.

In politics, the two extremes of the political spectrum often become indistinguishable. In the "politics" of substructure a similar situation happens. A common substructure of *all* particles at the Planck scale ("*ultra-radical*" option) is experimentally similar to no substructure ("*ultra-conservative*" option). In fact, superstring theory is both: There is no substructure in the sense of the existence of preons but all particles have a substructure in the sense of not being point objects at the Planck scale.

To summarize:

At present there is no experimental evidence for substructure. No viable explicit composite models are known. The new underlying interaction does not appear to be color-like. The mechanism which creates almost massless composite objects has not been determined.

On the other hand, Higgs particles, quarks, leptons, W and Z can still have substructure at the TeV scale. There are good theoretical motivations for such substructure. The new underlying interaction may have novel features, unlike previously observed interactions. Chiral symmetry, supersymmetry and possibly other mechanisms may lead to almost massless composites.

Substructure enthusiasts continue to hold "*moderate*" to "*radical*" views.

Progress is unlikely without new experimental information.

REFERENCES

1. See, e.g., I. Bars, talk presented at this session.
2. See, e.g., D. Wyler, talk presented at this session; W. Buchmuller and D. Wyler, Nucl. Phys. B268, 621 (1986).
3. For general reviews see, e.g., W. Buchmuller, lectures at the 1985 Schlading School (CERN-TH-4189/85); H. Harari, Proceedings of the 1984 Scottish Universities Summer School; H. Harari, Proceedings of the 1985 Erice Summer School (to be published).
4. See, e.g., D.B. Kaplan, talk presented at this session; M.J. Dugan, H. Georgi and D.B. Kaplan, Nucl. Phys. B254, 299 (1985).
5. See, e.g., T.W. Appelquist, talk presented at this session; T.W. Appelquist, D. Karabali and L.C.R. Wijewardhana, Yale preprint YTP-86/11, June 1986.
6. See, e.g., M.S. Chanowitz, talk presented at this session; M.S. Chanowitz and M.K. Gaillard, Nucl. Phys. B261, 379 (1985).
7. See, e.g., H. Harari, Proceedings of the Fifth Workshop on Proton-Antiproton Collider Physics, St. Vincent, Italy, February 1985, p. 429; H. Harari, From SU(3) to Gravity, (Yuval Ne'eman Festschrift), Cambridge University Press, p. 113.
8. See, e.g., B. Schrempp, talk presented at this session.

Parallel Session 5

Cosmology and Elementary Particles

Organizers:
P. Binetruy (Florida)
S.-Y. Pi (Boston)

Scientific Secretaries:
H.S. Chan (LBL)
A. Stundzia (Toronto)

DARK MATTER IN THE UNIVERSE

Joseph Silk

Department of Astronomy
University of California
Berkeley, CA 94720

*"And finds, with keen discriminating sight,
Black's not so black —
nor white so very white."*
G. Canning (1798)

I. INTRODUCTION

Most of the universe has hitherto been invisible. It consists of dark matter in some weakly interacting form. This review addresses the following issues: What is the nature of the dark matter? How do we infer its presence? And how can we hope to detect it?

Candidates for the dark matter span about 80 orders of magnitude in mass. From 10^{-14} to 10^{19} GeV is in the realm of particle physics, and from above 10^{19} GeV to 10^{66} GeV or larger belongs to astrophysics. The astrophysical candidates, which range from very large dust grains to supermassive black holes may or may not exist, and even if they are produced somewhere in the universe, we have enormous freedom in speculating about their abundance. I shall only be concerned with the particle physics candidates in this review. While their existence remains similarly uncertain, if they do exist, we shall see that the implications for cosmology can be very specific. Some of the more exotic particle physics candidates weigh in around the Planck mass, but the generic candidates are generally in the range $1-10$ GeV. I shall be primarily concerned with weakly interacting stable massive particles, including as a favored candidate the lightest SUSY particle. Two important exceptions are the light (~ 30 eV) neutrino and the invisible axion ($\sim 10^{-5}$ eV).

II. WEAKLY INTERACTING STABLE MASSIVE PARTICLES

The Big Bang was a particle accelerator *par excellence*, that outperformed any conceivable terrestrial experiment. As particle energies adiabatically decayed with the expansion, the normal equilibrium was eventually broken. Up to that point, all particles that could exist did exist. Stable relics of this hot early phase would remain today. Their precise abundance depends on how massive they are. At a temperature higher than its mass $kT > m_\chi c^2$, the number of such particles is equal to the number of photons, diluted by the number of other relativistic species that are present. Once the temperature drops below the χ–particle mass however, or $x \equiv m_\chi c^2/kT < 1$, the number of χ–particles becomes suppressed by a Boltzmann factor. Particle production is becoming inefficient, and

$$n_\chi/n_\gamma \sim n^{3/2} exp\,(-x)\,.$$

Annihilation still proceeds, however, and only ceases when the annihilation rate ($\gamma <\sigma v> n_\chi \propto T_\gamma^3$) eventually becomes slower than the expansion rate ($\propto T_\gamma^2$). For weakly interacting massive particles, this occurs at[1] $x \equiv x_f \approx 20$, and the comoving number density of stable χ–particles is subsequently frozen. Setting the annihilation rate equal to the expansion rate at freeze–out yields $\rho_\chi = m_\chi n_\chi \propto 1/<\sigma v>_{ann}\,(T_f)$. In other words, the mass density in the χ–particles depends only on the annihilation cross–section at freeze–out.

To perform a more precise calculation, we may normalize the density to the critical value for closure $\rho_{cr} = 3H_0^3/8\pi G = 4.7h_{1/2}^2 \times 10^{-30}$ g cm^{-3} ($h_{1/2} \equiv H_0/50$ km s^{-1} Mpc^{-1}), and write the annihilation cross–section to lowest order in x as $\sigma_{ann} = a + bx^{-1}$. The surviving density of χ–particles is then obtained by integrating over temperature, whence[2]

$$\Omega_\chi h_{1/2}^2 = 0.96 \left(\frac{1/20}{\chi_f}\right)\left(\frac{N_F}{25.6}\right)^{1/2}\left(\frac{a+bx_f^{-1}}{a+b/2x_f}\right)$$

$$\times \left(\frac{0.42T_\gamma}{T_\chi}^3\right)\left(\frac{10^{-26}\text{cm}^{-3}\text{s}^{-1}}{<\sigma v>_{ann}}\right) ,$$

where $\Omega_\chi = \rho_\chi/\rho_{cr}$ and N_F is the number of relativistic species present at freeze–out. For typical weak interaction cross–sections, we infer the remarkable coincidence that $\Omega_\chi h_{1/2}^2 \sim 1$. This is not exactly mandatory, of course, and we now discuss direct constraints on Ω_χ.

III. DETERMINATIONS OF Ω

The Friedmann equation can be written in the form

$$\Omega(t) = 1 + O(1)(t/t_0)^\alpha$$

where $\alpha = 1$ in the radiation–dominated era and $\alpha = 2/3$ in the matter–dominated era. The term $O(1)$ is, to within a factor of order unity, equal to the curvature constant (1 for a closed model and -1 for an open model). We see that Ω must be tuned to unity at the Planck epoch to within about 1 part in 10^{60}. This is sufficiently ugly that many cosmologists prefer to set Ω precisely equal to unity at the outset. The only physical mechanism capable of accounting for this is inflation: generic classes of initially anisotropic and curved (open) models can be shown to inflate, providing a possible "no hair" theorem.[3]

Direct astrophysical determinations almost invariable conclude that $\Omega < 1$, but generally refer to locally inhomogeneous matter contributing to Ω.[4] For example, dynamical measurements of mass utilize the velocities v of these particles (often galaxies) over some region of measured extent R: if the test particles are gravitationally bound in this region, we then infer the inertial mass $M(<R) \approx Rv^2/G$ that is non–uniformly distributed over scale R. Any underlying smooth component would not be measured by these dynamical techniques.

A convenient way to refer to mass is to also photometrically measure the luminosity (in a specified wavelength band such as the blue spectral region) within the same region, and express the ratio of mass to blue luminosity in solar units. Table 1 summarizes the current situation: M/L_B rises with increasing scale, and reaches a value of about 150 $h_{1/2}$ over 10 Mpc, beyond which dynamical measurements are presently lacking. For comparison, the critical closure density can be expressed relative to the observed luminosity density in all galaxies as

$$M/L_B = 750\ \Omega\ h_{1/2} \ .$$

In other words, adding up all dynamically measured mass out to $\sim$ 10 Mpc scales only allows about 20 percent of the closure value.

Of course if we took only the luminous matter, mostly in stars, the contribution would be far less: one finds that $\Omega_{lum} \approx 0.006$. Mass clearly does not trace the light distribution: the luminous matter is much more concentrated.

At least fifty percent of the dark matter is probably non–baryonic. This follows from the primordial nucleosynthesis constraint which simultaneously accounts for the abundances of ^4He, ^3He, ^2H and ^7Li in the standard big bang model by adjusting just one free parameter, Ω_{baryon}. One finds[5] that $\Omega_{baryon} h_{1/2}^2 \approx 0.1$. This means that most of the baryons do not shine as stars. Presumably they are in the form of diffuse gas. Recent studies of strong Lyman alpha absorption lines seen towards distant quasars [6] suggest that gaseous disks around spirals are much more extended and massive at $z = 2$ than at the present epoch. This directly yields $\Omega_{HI} \approx 0.01$, and molecular gas could substantially augment this.

It is more difficult to generalize from the observations of hot X–ray emitting gas in rich clusters, since only about five percent of luminous galaxies are found in these clusters. Now in the Coma cluster, the intracluster gas contributes about twice as much as the luminous galaxies to the cluster mass (but uncertainties in models of the gas distribution could increase this by a factor of 3 or so).[7] Hence the hot intracluster gas only adds up to about $(0.1-0.3)\Omega_{lum}$, although there could also be a substantial amount of ionized gas in galaxy groups. The diffuse X–ray background may be predominantly of thermal origin, in which case a diffuse hot intergalactic medium could contribute up to as much as thirty percent of the critical density if it were uniformly distributed at a temperature of $\sim 4 \times 10^8$ K. This is, of course, a very controversial interpretation of the diffuse X–ray background, much of which may well come from discrete sources such as active galaxies and quasars.

TABLE 1

Mass-to-Blue Light Ratio as Function of Scale

Region	Scale (parsecs)	$M/L_B(\frac{M\odot}{L\odot})$
Solar neighborhood	10^3	2
Galaxy (inner)	10^4	≤ 10
Galaxy halos	5.10^4	$25h_{1/2}$
Galaxy pairs	10^5	$[50h_{1/2}]$
Groups and clusters of galaxies	10^6	$150h_{1/2}$
Superclusters (also galaxy correlations and peculiar velocities)	10^7	$150h_{1/2}$

In summary, there are dark baryons, not in the form of HI, and most probably in the form of hot gas or of compact stellar remnants, and roughly an equal amount of dark non–baryonic matter, both components being more uniformly distributed than the luminous stellar component of the universe. It is the non–baryonic component that is of primary interest to particle physicists, and I now discuss various detection schemes that have been proposed.

IV. DARK MATTER SEARCHES

The generic type of dark matter, massive weakly interacting particles, is aptly described as cold dark matter.[8] Having thermally decoupled from the relic radiation at $kT \geq 0.1$ GeV, the χ–particles are effectively at rest in comoving space when the epoch of density fluctuation growth commences in the matter–dominated era at $kT < 10$ eV, and free–streaming of χ–particles plays no role in erasing any density fluctuations. In contrast, hot dark matter still has a large free–streaming velocity at the epoch of matter–domination, and all primordial fluctuations of comoving mass rate below $\sim m_{p\ell}^3/m_\nu^2 \sim 10^{16}(m_\nu/30\text{ eV})^2 M\odot$ are erased.[9] A hot dark matter dominated universe cannot satisfactorily account for galaxy formation because of the lack of small–scale power, and the most attractive (that is to say, the simplest) scenario for forming galaxies utilize adiabatic gaussian density fluctuations in a cold dark matter–dominated universe.[10] This model has many successes to its credit, although it is premature to say whether it can also account for all of the features of the observed large–scale distribution of galaxies. Suffice it to say that cold dark matter provides the most detailed framework for studying galaxy formation and clusters, and provides a baseline by which other models have to be judged. It is, in effect, presently the standard model of large–scale structure.

From the perspective of dark matter detection, the fact that cold dark matter, granted that it exists, most likely contributes at least as much as the baryons, and possibly far more if $\Omega = 1$, means that galaxy halos must inevitably contain substantial amounts of χ–particles. Now the halo density in the vicinity of the sun is about 0.3 GeV cm^{-3}. If most, or even fifty percent, of this is cold dark matter, then both direct and indirect detection schemes become feasible.

a) Direct Detection

Two schemes are being pursued for direct detection of halo cold dark matter. If axions are the dominant form of cold dark matter, then their electromagnetic coupling means that microwave photons can be produced by axions scattering off a suitably strong inhomogeneous magnetic field.[11]

A second approach utilizes bolometric detection of massive cold dark matter particles interacting elastically with a suitable detector. First results from the latter technique have recently been obtained, using a germanium detector. This is predominantly sensitive to spin–independent interactions, and sets limits on the mass of such particles as the scalar and Dirac neutrinos of $m_\chi \leq 20$ GeV or $m_\chi \gtrsim 5$ TeV.

b) Indirect Detection

Majorana–mass cold dark matter candidates will undergo annihilations in the halo. The annihilation cross-section is known: it is the low temperature limit of $<\sigma v>_{ann}$ (T_f), which can be specified uniquely in terms of Ω_χ. The stable end-products of annihilations of particles with mass of a few GeV are electron–positron pairs, neutrinos, gamma rays, and proton–antiproton pairs. Consider the example of χ being the photino. Its annihilation cross-section is given by

$$<\sigma v>_{\tilde{\gamma}\, ann} = a + bx^{-1} ,$$

$$a \sim m_f^2\, m_{sf}^{-4} \,, b \sim m_{\tilde{\gamma}}^2 \;\; m_{sf}^{-4} \,,$$

where all constants have been suppressed apart from the mass dependencies: m_{sf} is the mass of scalar fermions that mediate the annihilation (squarks, selectrons) and m_f is the mass of fermion decay products (b, c, τ, etc.). The annihilation cross-section at low temperature is simply

$$<\sigma v>_{\tilde{\gamma}\, ann} \approx a \overset{\propto}{\sim} m_{sf}^{-4} \overset{\propto}{\sim} (\Omega_{\tilde{\gamma}}\, h_{1/2}^2)^{-1} m_{\tilde{\gamma}}^{-4/3} .$$

An approximate expression for $\Omega_{\tilde{\gamma}}$ over the photino mass range $m_{\tilde{\gamma}} > 3$ GeV is then[12]

$$\Omega_{\tilde{\gamma}}\, h_{1/2}^2 \approx \left(\frac{m_{sf}}{60\text{GeV}}\right)^4 \left(\frac{3\text{ GeV}}{m_{\tilde{\gamma}}}\right)^2 .$$

Recent experimental limits at UA1 require that $m_{sf} > 60$ GeV, whence $m_{\tilde{\gamma}} \gtrsim 3$ GeV. This lower bound on the photino mass means that the annihilation channels will produce $p, \bar{p}$ pairs as well as the other stable end–products of fermion jets.

The $\bar{p}$ flux resulting from annihilation in the halo is of especial interest since the background $\bar{p}$ flux is very low.[13] In cosmic rays at high energy (> 2 GeV), for example, $\bar{p}/p \sim 10^{-3}$. In fact, the cosmic ray $\bar{p}$ is of secondary origin, produced by energetic p collisions with interstellar CNO, and are kinematically suppressed below ~ 1 GeV, whereas the typical energy of the annihilation $\bar{p}$ is $\sim 0.1 m_{\tilde{\gamma}}$. The flux may be detectable if $m_{sf} \leq 100$ GeV, or if $m_{\tilde{\gamma}} \leq 8(\Omega_{\tilde{\gamma}}\, h_{1/2}^2)^{-1/2}$ GeV. Indeed one experiment has actually reported a positive detection of low energy cosmic ray $\bar{p}$[14]. Comparison with theoretical predictions is somewhat uncertain because the $\bar{p}$ are trapped by magnetic diffusion in the halo for a time $\sim 10^8$ yr and because their energy and flux are degraded by solar modulation. Explanation of the Buffington flux ($\bar{p}/p \sim 10^{-4}$ at ~ 0.5 GeV) requires ~ 60 GeV scalars. In order for $\Omega_{\tilde{\gamma}} \gtrsim 0.04 h_{1/2}^{-1}$ as required for photinos to dominate our halo (which I conservatively take to have $M/L_B = 30$, appropriate to a radius of 30 kpc, or $0.04 h_{1/2}^{-1}$ of the critical value of M/L_B for closure), the photino mass must be below $\sim 15 h_{1/2}^{-\frac{1}{2}}$ GeV. Stecker *et al.*[15] find that with $m_{\tilde{\gamma}} \sim 15$ GeV, the energy distribution of annihilation $\bar{p}$ can account both for the low energy $\bar{p}$ flux and for a possible anomaly in the flux found independently in cosmic ray $\bar{p}$ experiments above 1 GeV.

The gamma ray flux produced by photino (or any other Majorana mass χ–particle) annihilations in the halo amounts to less than one percent of the observed diffuse γ–ray flux between 300 MeV and 1 GeV. It is possible that cold dark matter is entrained in the spheroid of our galaxy as this stellar distribution forms, provided that the formation was nearly adiabatic. In this case, the annihilation γ–ray flux towards the galactic center would be greatly enhanced. Even if a few percent of the spheroid density were in the form of cold dark matter, a detectable γ–ray source of angular extent about 1° would be produced.[16] At present we can only set limits on the entrainment fraction from the apparent absence of any such source.

c) <u>High Energy γ from the Sun</u>

As the sun orbits the galaxy, it gravitationally traps cold dark matter, which elastically scatters in the solar interior and if $m_\chi \gtrsim 4$ GeV remains trapped in the sun.[17] The trapping is boosted by gravitational focussing of χ–particle orbits. The captured particles eventually settle into the center of the sun where they annihilate. The annihilations of χ–particles in the sun result in a detectable signature: the prompt neutrinos of energy $\sim 0.3 m_\chi$ escape and are detectable in the earth in underground ν detectors.[18,2] The abundance is determined by equating the trapping and annihilation rates: one finds that $n_\chi/n_p \approx 10^{-14}$. This is about 3 orders of magnitude too small to affect the temperature gradient in the solar core and thereby modify the ^8B solar neutrino flux. (The critical abundance for this to be a significant effect is $n_\chi/n_p \approx t_{orbit}/t_{KH} \sim 10^{-11}$, where t_{KH} is the Kelvin–Helmholtz time–scale.) The resulting high energy neutrino flux includes $\nu_e, \bar{\nu}_e, \nu_\mu$ and $\bar{\nu}_\mu$, and is comparable to atmospheric background for photinos of mass above the evaporation threshold of ~ 4 GeV with annihilation cross–section fit by setting $\Omega h_{1/2}^2 = 1$.[19] Unfortunately, data from the underground detectors is not presently available in the relevant energy range

($\gtrsim 2$ GeV since $< E_\nu > \approx m_{\tilde{\gamma}}/3$ for the process $\gamma\tilde{\gamma} \rightarrow f\bar{f} \rightarrow \nu\bar{\nu} + ...$). Current limits from the IMB detector[20] set the solar high energy ν flux to be below 1.3 × the atmospheric background for $\nu_e + \nu_\mu$ below 1 GeV but this range is too soft to constraint the massive χ-particles that would not have evaporated from the sun.

V. CONCLUSIONS

Dark matter is certainly here to stay. It is very likely to contain a substantial non-baryonic component, thereby linking particle physics and astrophysics. This, however, presumes that $\Omega \gtrsim 0.2$. It is conceivable that $\Omega \approx 0.1$, however, in which case purely baryonic dark matter is consistent with the primordial nucleosynthesis constraint if $h \approx 0.5$. A natural candidate for baryonic dark matter that we at least know exists is the white dwarf, although one has to severely modify early galactic star formation rates and the initial stellar mass function to obtain enough dark matter in this form.

There is only one ultimate resolution of the dark matter problem, namely direct (or indirect) detection, whether in particle accelerators, in the galactic halo, or in the sun. We should remain optimistic that the current surge of interest in developing dark matter detection schemes will eventually be rewarded with success.

This research has been supported in part by the DOE.

REFERENCES

(1) Lee, B. and Weinberg, S. (1977), *Phys. Rev. Lett.*, **39**, 165.

(2) Srednicki, M., Olive, K. A. and Silk, J. (1986), *Nucl. Phys. B*, (in press).

(3) Turner, M. S. and Widrow, L. (1986), Fermilab Preprint, 86/49-A.

Jensen L. G. and Stein-Schabes, J. (1986), Fermilab Preprint, 86/51-A.

(4) Faber, S. and Gallagher, J. (1979), *Ann. Rev. Astron. Astrophys.*, **17**, 135.

(5) Yang, J., Turner, M. S., Steigman, G., Schramm, D. M. and Olive, K.A. (1984), *Astrophys.J.*, **281**, 413.

(6) Wolfe, A. M., Turnshek, D. A., Smith, H. E. and Cohen, R. D. (1986), *Astrophys. J. Suppl.*, **61**, 249.

(7) Cowie, L. L., Mushotzky, M. J. and Henriksen, M. J. (1986), *Astrophys.J.*, (in press).

(8) Bond, J. R. and Szalay, A. (1983), *Astrophys. J.*, **276**, 443.

(9) Bond, J. R., Efstathiou, G. and Silk, J. (1980), *Phys. Rev. Lett.*, **45**, 1980.

(10) Davis, M., Efstathiou, G., Frenk, C., and White, S. D. M. (1985), *Astrophys. J.*, **292**, 371.

(11) Sikivie, P. (1983), *Phys. Rev. Lett.*, **51**, 1415.

(12) Ahlen, S. P., Avignone, F. P., Brodzinski, R. L., Drukier, A. K., Gelmini, G., and Spergel, D. N. (1986) *Phys. Rev. Lett.*, submitted.

(13) Silk, J. and Srednicki, M. (1984), *Phys. Rev. Lett.*, **5**, 624.

(14) Buffington, A., Schindler, S. M. and Pennypacker, C. (1981), *Astrophys. J.*, **248**, 1179.

(15) Stecker, F. W., Rudaz, S. and Walsh, T. F. (1985), *Phys. Rev. Lett.*, **55**, 2622.

(16) Silk, J. and Bloemen, H. (1986), *Astrophys. J. Lett.*, (in press).

(17) Steigman, G., Sarazin, C., Quintana, H. and Faulkner, J. (1978), *Astron. J.*, **83**, 1050.

(18) Silk, J., Olive, K. A. and Srednicki, M. (1985), *Phys. Rev. Lett.*, **55**, 257.

(19) Hagelin, J., Ng, K. W. and Olive, K. A. (1986), UMM-T4-566/86, preprint.

(20) LoSecco, J., this conference.

CONSTRAINTS ON UNSTABLE DARK MATTER

Ricardo A. Flores†

Department of Physics, Brandeis University, Waltham, MA 02254
and
Theory Division, CERN, CH-1211, Geneva 23, Switzerland

We briefly review the motivation for cosmologies with unstable dark matter (DM) and the various astrophysical constraints on its decay time scale. We then describe an analytic model, extensively checked by numerical simulations, that allows one to compute rotation curves for disk galaxies , assuming they have formed by dissipative collapse inside a dissipationless DM halo, a fraction of which subsequently decays; one finds that flat rotation curves like those of spiral galaxies can arise only if the fraction of decaying DM does not exceed 50%. This constraint implies that a relativistic, weakly interacting decay product cannot be dominant at present, and it also sets interesting constraints on the possibility that the universe has entered a second matter-dominated era only recently.

1. INTRODUCTION

Inflation is a very atractive cosmological model that can naturally resolve major cosmological puzzles that the standard Big Bang Cosmology cannot answer, such as the flatness and homogeneity of the Universe, and the origin of density perturbations, and it predicts our universe to be extremely flat at present because of the tremendous expansion it must go through at very early times. The theoretical preference for a flat universe, however, goes beyond inflation and it is motivated by the "unnatural" fine-tuning required at very early times for the universe to remain flat for an enormously long time (1). If the universe is flat, then

$$\Omega + \Lambda/8\pi G\rho_c = 1 \qquad (1)$$

where Ω is the mean mass-energy density of the Universe in units of the critical density, ρ_c, and Λ is the cosmological constant. Thus, if the cosmological constant vanishes, $\Omega = 1$ to a very high accuracy. However, observations indicate that the clustered matter (including DM) on scales of up to a few megaparsecs amounts only to $\Omega_{cl} = 0.1 - 0.3$ (2,3).

Several ideas have been proposed in the literature to explain this puzzle, such as a relic cosmological constant (4) Λ chosen so that Eq. 1 is satisfied for $\Omega = \Omega_{cl}$, or "biased" galaxy formation (4), in which it is postulated that galaxies are rare events, resulting only from large fluctuations of the density field –thus they would not be good tracers of mass in the Universe.

We consider here an explanation based on the non-radiative decay of a heavy elementary particle (6-8) (e.g. a heavy neutrino of mass $\gtrsim O(1)keV$), which first drives the formation of galaxies and clusters and subsequently decays to provide a smooth, undetected background of **relativistic** particles (e.g. familons (9) or majorons (10)) that at present contribute Ω_r to the total energy density of the Universe. In the original model (6,7) ("type I" models) the non-relativistic matter contributes $\Omega_{nr} = \Omega_{cl}$ to the total energy density and $\Omega_r = 1 - \Omega_{nr} \sim 0.8$.

Another possibility is that the Universe becomes dominated by a primordial, stable non-relativistic DM species after decay of the unstable species (11); in these models ("type II") the decay of the heavy species unbinds large quantities of a primordial, stable light species causing it to stream away from clusters and larger structures, thus providing a smooth background of **non-relativistic** DM. Thus $\Omega_{cl} \sim 0.2$ reflects only the clustered component of the DM while $\Omega_{nr} \sim 0.8$.

The lifetime of the unstable species, τ, is constrained by several astrophysical considerations, most easily expressed in terms of the redshift corresponding to the epoch of decay, z_d ($1 + z_d \equiv s^{-1}(t = \tau)$, where $s(t)$ is the scale factor of the Universe at time t, normalized to unity today). The isotropy of the microwave background requires that $1 + z_d \lesssim 5$ (12-14), because fluctuations can only grow until the epoch of decay in type I models. Also, the decay must happen sufficiently late, $1 + z_d \lesssim 5$, for the cores of rich clusters to remain in virial equilibrium (15); if the decay occurs on a time scale shorter than the dynamical time, a cluster would become unbound after loosing $\sim$ 50% of its mass. A similar upper bound is implied if normal galaxies are assumed to be the source of gravitational lensing (16). On the other hand, the decay must happen sufficiently early for our infall velocity towards the Virgo cluster not to be too large: the linear theory yields $1 + z_d \gtrsim 20$ for type I models (15), but this bound is exponentially sensitive to uncertainties in the observational parameters and is affected by the non-sphericity (17) and non-linearity (18) of the infall, so a value as low as $1 + z_d \sim 5$ is not excluded. Accelerator physics and the cooling of red giants only imply model dependent constraints (19) and although some particle physics models are ruled out (19) no conclusion can be drawn about the general viability of the scenario. The reported large streaming velocities on large scales (20) and large amplitud of the rich-cluster correlation function (21) would rule out type I models (22), as well as **any** pure cold– or hot–DM model of formation of structure. These observations, however, remain controversial (23).

In this work we discuss the constraints imposed on cosmologies with decaying DM by the observed rotation curves of spiral galaxies (24), assumed to have formed via dissipative collapse inside a gravitationally induced protogalaxy consisting of a homogeneous mixture of dissipationless DM and a small fraction of baryonic material (25). The remainder of this paper discusses first a simple analytic model that can be constructed to find the final distribution of DM after baryonic dissipation and DM decay have taken place. The model is then used to compute rotation curves, which are found to be sharply peaked if the fraction of stable DM is small. The calculated rotation curves are compared to the observed rotation curves of spirals in section 3; agreement is found only if at least 50% of the DM is stable. Finally, in section 4 we discuss the implications of this constraint for both type I and type II models and summarize the conclusions.

2. ANALYTIC MODEL

The gravitational instability model of galaxy formation assumes that galaxies form in gravitationally-bound overdense regions that participate in the general expansion of the universe at first, but slow down to a maximum radius and collapse afterwards to form a protogalaxy that consists of a mixture of visible and dark matter. A key difference between these two mass components is that the visible matter can **dissipate**. Thus, it can sink deeply in the potential well of the DM and form the compact luminous cores we see today: the visible galaxies. The DM, being dissipationless, can only collapse by a factor of ~ 2 in radius from maximum expansion and stays behind to form the DM halos detected today. Much work has been done to study the formation of protogalaxies in the gravitational instability model in order to determine their structure and properties (26,27), but in order to relate these protostructures to real galaxies one must quantify the effect of the dissipational infall of the visible matter on the mass distribution of the dark halo (24) and, in the case of the cosmology being considered here, the effect of DM decay as well. The astrophysical constraints discussed above imply that the DM must decay on a time scale $\tau \gtrsim 10^9$ years, and since the mass fraction in visible matter at present must be small (2), a DM particle's orbit about a protogalaxy is expected to change adiabatically during dissipation and decay. This allows one to calculate the effect of these processes on a galaxy halo with a simple analytic model (24).

One starts by noting (25) that for particles moving in periodic orbits, $\oint pdq$ is an adiabatic invariant if the potential in which the particles move changes slowly with time relative to their orbital period; here p is the canonical conjugate momentum of the coordinate q. Thus, if p is the angular momentum of particles moving in circular orbits about a spherically symmetric mass distribution, the adiabatic invariant is $rM(r)$ provided that $M(r)$, the mass inside the orbital radius r, changes slowly with time. For purely radial orbits $r_{max}M(r_{max})$ is an adiabatic invariant as well, provided that $M(r)$ varies in a self-similar fashion. Since there is more phase space available for nearly circular orbits than for nearly radial orbits, we shall make the simplifying assumption that the orbits of DM particles are circular.

Consider now a spherically symmetric protogalaxy of radius R that consists initially of a mass fraction $F_i \ll 1$ of baryons and $1 - F_i$ of DM particles that are well mixed initially (i.e. their density ratio is F_i throughout the protogalaxy). This is a rough approximation to the final shape of an expanding protogalaxy, which will cut itself out of the general expansion with some caracteristic size, although there is no reason to expect it to be exactly spherical. There are two processes that contribute to the change in mass interior to a given orbit. First, as the baryons cool, they fall into the center to a final mass distribution $M_b(r)$ which, in the case of a spiral galaxy, is constrained by the initial angular momentum (it will be assumed that no baryons fall inside R from beyond R). If $F_i \ll 1$, the mass interior to a dissipationless particle's orbit will not show a large fractional change during one orbital period even if dissipation occurs rapidly, provided the particle is far enough from the galaxy center. This assures that the orbits of all but the innermost halo particles change change adiabatically under baryonic infall. The second process is the decay of a fraction $1 - f$ of the halo mass into relativistic particles, leaving behind a fraction f of the mass of the dark halo. The orbits of halo particles with orbital time $t_{orb} \ll \tau$ change adiabatically under DM decay; thus, there is always some inner region of the halo that changes adiabatically under decay and, for τ much longer than the system's dynamical time, the orbits of all (except, perhaps, the outermost) halo particles obey the adiabatic invariant constraint. After the processes of dissipation and decay are complete, the adiabatic invariant of the dissipationless particles implies

$$r\,[M_b(r) + M_{DM}(r)] = r_i\, M_i(r_i) \qquad (2)$$

which can be solved for the final DM mass distribution $M_{DM}(r)$ once the initial distribution of total mass $M_i(r_i)$ and the final baryon mass distribution $M_b(r)$ are given. Here r is the final orbital radius of a halo particle initially at radius r_i and $M_{DM}(r) = f(1-F_i)M_i(r_i)$ (by assumption there is no shell crossing during the contraction of the dissipationless halo (25)). The initial mass distribution $M_i(r_i)$ must be an equilibrium configuration and will be taken to be that of an isothermal sphere with core radius $a \equiv 3v^2/4\pi G\rho_o$, where v is the one-dimensional velocity dispersion and ρ_o is the central density. The final radial mass distribution of the baryons is assumed to be $M_b(r) = M_b(1 - (1 + r/b)exp(-r/b))$, which describes the mass distribution of a thin disk whose surface density falls off exponentially with scale length b. (When using using Eq. 2 the baryonic mass is actually treated as if it were distributed spherically rather than in a disk. See second ref. of (25).)

After the DM decay is complete, the baryonic mass fraction is $F = M_b/M_f$, where M_f is the final total mass and M_b is the total mass in baryons. Figure 1 shows rotation curves, $v(r)$ vs. r, resulting from dissipation and decay within an initial isothermal sphere whose core radius is 42% of R. The two solid lines are the rotation curves for two values of f, and typical values of F and the collapse factor b/R. The ratio b/R is constrained by the initial angular momentum. Theoretical analysis and N-body simulations of the tidal torque theory of angular momentum estimate the mean value of the dimensionless spin parameter $\lambda \equiv J|E_i|^{1/2}/GM_i^{5/2}$ to be 0.07 (26); here J, E_i and M_i are respectively the total angular momentum, energy and mass of a system (for reference, a rotationally suported sphere has $\lambda \sim 0.25$, and uniform density disk $\lambda \sim 0.5$). Unlike F, which is assumed to be a spatial constant (as expected if most of the DM is non-baryonic), λ varies among protogalaxies with a dispersion about the mean of ~ 0.03. This implies that $0.02 \lesssim b/R \lesssim 0.1$ with a mean value of 0.05 (25,27). The effect of this dispersion on rotation curves is seen in the two dotted lines, which are rotation curves for $f = 0.25$ with collapse factors that correspond to b/R at its upper (3) and lower (5) limit. There are several observational and theoretical constraints on F that place it in the range $0.05 \lesssim F \lesssim 0.2$ (2). Decreasing F produces less peaked rotation curves, as shown by the thick dotted line in Fig. 1, which is the rotation curve for $f = 0.75$ with $F = 0.05$ and $b/R = 0.1$.

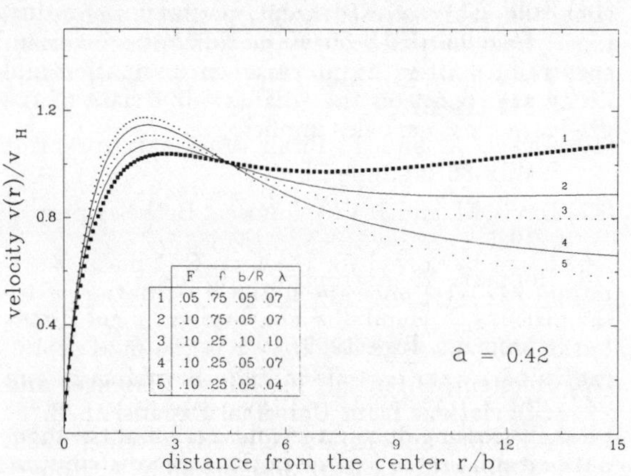

Figure 1: Rotation curves calculated with the analytic model. $v(r)$ is the rotational velocity, in the plane of the disk, needed by a particle to remain in circular orbit at distance r/b from the center and $v_H \equiv v(r = R_H \equiv 4.5b)$.

The main feature of Fig. 1 is the decreasing velocity at large distances for small f. This can be easily understood as follows: **after** the baryons have cooled to an exponential disk (dissipation and decay do not commute since λ is invariant under decay in the adiabatic approximation; we assume that by $t \sim \tau \gtrsim 10^9$ years the visible matter has dissipated and collapsed, since its dynamical and cooling times are shorter than τ), the DM decay hardly changes $v(r)$ at small radii because the baryons dominate there. At large radii, where the DM dominates, $v(r)$ decreases substantially as a result of the mass loss. Thus, a lower bound on f is anticipated from the constraint that observed rotation curves do not fall with distance even well outside the optical radius.

3. DISCUSSION

The remarkably flat or rising velocity profiles of spiral galaxies (28,29) are a strong constraint on the models being studied. Figure 1 shows that for typical values of F and λ the final velocity profile is rather sensitive to the value of f and not every value of f yields acceptable rotation curves. The analytic model described above assumes that the system changes adiabatically under dissipation and decay ($F_i \ll 1$ and $t_{orb} \ll \tau$) and it uses an adiabatic invariant that assumes circular or purely radial orbits, spherical symmetry and an equilibrium starting configuration. These assumptions can be relaxed and more general cases can be studied using N-body simulations. Extensive numerical work was carried out that confirms the validity of the analytic model (24), so one can use it to compute theoretical rotation curves and compare them to observations.

The parameters of the analytic model are the final baryon fraction F, the collapse factor b/R, the core radius a/R of the initial isothermal sphere and the fraction of stable dark matter f. As Fig. 1 shows, if for a given value of f one obtains a falling velocity profile for F and R/b taken at their lowest values and a/R $\sim$ 0.4, then **only** falling rotation curves can be produced for higher values of F and R/b. However, Fig. 2 shows that this is so for **any** value of the core radius for $f \lesssim 0.3$. The figure shows a measure of the flatness of the rotation curve, $\Delta v/v \equiv [v(4.5b) - v(3.5b)]/v(3.5b)$, ploted as a function of the core radius of the initial configuration for the **lowest** values of F and R/b (higher values would only lower the curves).

The rotation curves of spiral galaxies are observed to be remarkably flat or rising inside the optical radius, with only a small fraction $\sim$ 20% of the Burstein-Rubin (28) sample observed to have falling velocity profiles. For $f \lesssim 0.3$, however, Fig. 2 shows that the theoretical rotation curves fall ($\Delta v/v < 0$) inside the optical radius $R_H \equiv 4.5b$ for **all** values of a/R **even for F and R/b at their lowest values**. Furthermore, velocity profiles are observed to be flat or rising outside the optical radius as well (29) and this provides further constraints on f. Because we assume f to be a spatial constant, one can use the data of individual galaxies to constrain it. For example, no value of the initial core radius can yield a flat rotation curve out to $r = 11b$ (like that of NGC 3198 (30)) unless $f \gtrsim 0.4$ nor can a profile rising out to $r \sim 8b$ (like that of NGC 3109 (31)) be obtained if $f \lesssim 0.5$. These bounds are, again, for F and R/b taken at their minimum values.

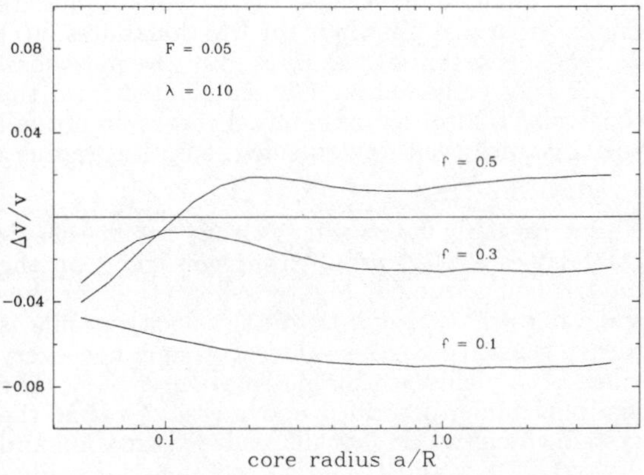

Figure 2: $\Delta v/v \equiv [v(4.5b) - v(3.5b)]/v(3.5b)$ as a function of the core radius of the initial isothermal sphere for $F = 0.05$ and $\lambda = 0.10$. For reference, $\Delta v/v \sim -0.1$ for a self-gravitating exponential disk.

4. CONCLUSIONS

Thus, rotational profiles of spiral galaxies require that the fraction of stable DM $f \gtrsim 0.5$. One can relate f to the relative contribution by relativistic particles to the energy density:

$$\Omega_r/\Omega_{nr} \leq f^{-1}(1-f)(1-F)(1+z_d)^{-1} \qquad (3)$$

(the equality is satisfied in the simultaneous decay approximation (8)). Since $F \ll 1$, for $f \geq 0.5$ one gets $\Omega_r/\Omega_{nr} \leq (1+z_d)^{-1}$. Thus, the Universe could not be dominated by weakly interacting relativistic particles. In the type I model, the relativistic contribution to the energy density required for $\Omega = 1$ amounts to $\Omega_r/\Omega_{nr} \sim 4$. Therefore, even for a decay epoch as recent as $1 + z_d \sim 5$ Eq. 3 requires $f \sim 0.05$, a factor of 10 smaller than the bound obtained here. Thus, with the assumptions we have made, the type I model is ruled out. Only if a substantial fraction of the gas contracts and dissipates after the DM decay might one lower the upper bound on f, although it is not clear that such a scenario can lead to flat rotation curves. Type II models (11), however, do not require that the Universe be radiation dominated at present; in these models $\Omega_r/\Omega_{nr} \sim 0.25$, which is marginally consistent with the upper bound of Eq. 3 even for a decay epoch as recent as $1 + z_d \sim 4$. However, an upper bound on f must come from the requirement that on large scales Ω be unity with Ω_{cl} being only ~ 0.2. We are trying to quantify this constraint by studying in detail the formation of clusters and large scale structure in type II models (32).

ACKNOWLEDGEMENTS

I thank my collaborators G. Blumenthal, A. Dekel and Joel Primack, with whom most of the work reported on here was carried out, for much fun and enlightment throughout our collaborations. This work has been supported by DOE contract DE-AC02-76ER03230 at Brandeis.

REFERENCES

† Current address: Theory Division, CERN.

(1) Guth, A., Phys. Rev. **D23** (1981) 347

(2) Blumenthal, G., S. Faber, J. Primack and M. Rees, Nature **311** (1984) 517 and references therein

(3) Peebles, P. J. E., Nature **321** (1986) 27

(4) Peebles, P. J. E., Astrophys.J. **284** (1984) 439

(5) Bardeen, J., in **Inner Space/ Outer Space**, E. W. Kolb *et. al.* eds., U. of Chicago Press, p. 212. N. Kaiser, *ibid.*, p. 258

(6) Turner, M. S., G. Steigman and L. Krauss, Phys. Rev. Lett. **52** (1984) 2090

(7) Gelmini, G., D. Schramm and J. Valle, Phys. Lett. **146B** (1984) 311

(8) Turner, M. S., Phys. Rev. **D31** (1985) 1212

(9) Wilczek, F., Phys. Rev. Lett. **49** (1982) 1549. Reiss, D. B., Phys. Lett. **115B** (1982) 217

(10) Chicashige, Y., R. N. Mohapatra and R. D. Peccei, Phys. Lett. **98B** (1981) 265. Gelmini, G. and M. Roncadelli, Phys. Lett. **99B** (1981) 411

(11) Olive, K., D. Seckel and E. Vishniac, Astrophys.J. **292** (1985) 1

(12) Silk, J. and N. Vittorio, Phys. Rev. Lett. **54** (1985) 2269

(13) Turner, M. S., Phys. Rev. Lett. **55** (1985) 549

(14) Kolb, E. W., K. Olive and N. Vittorio, preprint, Fermilab-PUB-86/40-A, 1986.

(15) Efstathiou, G., Mon. Not. Roy. Astro. Soc. **213** (1985) 29

(16) Dekel, A. and T. Piran, Weizmann preprint WIS-86/30-June Ph

(17) Davis, M. and J. V. Villumsen, Berkeley preprint, 1985

(18) Hoffman, Y., Astrophys.J., in press

(19) Dicus, D. A. and V. L. Teplitz, U. of Texas preprint, June 1985

(20) Burstein, D. *et. al.* in **Galaxy Distances and Deviations from Universal Expansion**, B. F. Madore and R. B. Tully eds., Reidel 1986. Collins, C. A., R. D. Joseph and N. A. Robertson, Nature **320** (1986) 506

(21) Bahcall, N. and R. Soneira, Astrophys.J. **270** (1983) 20

(22) Bond, J. R. in **Nearly Normal Galaxies: From the Plank Time to the Present**, S. Faber ed., Springer Verlag 1987

(23) Gunn, J. E., *ibid.* as ref. (22)

(24) Flores, R., G. Blumenthal, A. Dekel and J. Primack, Nature, in press

(25) Blumenthal, G., S. Faber, R. A. Flores and J. Primack, Astrophys.J. **301** (1986) 27; and UCSC preprint, in preparation

(26) Efstathiou, G. and B. J. T. Jones, Mon. Not. Roy. Astro. Soc. **186** (1979) 133

(27) Fall, S. M. and G. Efstathiou, Mon. Not. Roy. Astro. Soc. **193** (1980) 189

(28) Burstein, D. and V. C. Rubin, Astrophys.J. **297** (1985) 423 and references therein

(29) Bosma, A., Astron.J. **86** (1981) 1791; ibid. **86** (1981) 1825

(30) van Albada, T. S., J. N. Bahcall, K. Begman and R. Sanscisi, Astrophys.J. **295** (1985) 305

(31) Carignan, C., Astrophys.J. **299** (1985) 59

(32) Blumenthal, G., A. Dekel, R. Flores and J. Primack, in progress.

A SLOW ROLLOVER PHASE TRANSITION IN THE SCHRÖDINGER PICTURE

So-Young Pi

Department of Physics, Boston University
Boston, Massachusetts 02215 U.S.A.

The present status of our understanding of the slow-rollover transition in the new inflationary universe is reviewed and a time-dependent variational approximation in the Schrödinger picture is proposed as the perturbative scheme for studying the quantum theory of the transition. Validity of the approximation is discussed using a double-well potential, in one-dimensional quantum mechanics.

I. INTRODUCTION

Inflation[1,2] is a simple and powerful cosmological scenario which can resolve several fundamental difficulties in the standard cosmology. However, in order to enhance this idea to a correct cosmological theory, one must establish that observations are consistent with the Zeldovich spectrum of adiabatic density perturbations[3] and that the density parameter Ω is equal to unity; both are definite predictions of inflation. Moreover, one must show that inflation actually occurs by studying detailed quantum dynamics.[4] In this talk, I shall describe some of my collaborative research[5,6] on the quantum evolution of the inflation-driving scalar field in the new inflationary scenario.

An exact calculation of the field theoretic quantum dynamics involves too many degrees of freedom to be reduced to a simple numerical computation. Therefore, in order to study the quantum theory for a scalar field, one must first decide on an approximation scheme, and also find the domain of its validity. My main purpose here is to propose the "time-dependent variational approximation"[7] as the appropriate method for our problem, and to exhibit the domain of its validity by applying it to a one-dimensional quantum mechanical problem which has many features appearing in the inflationary scenario.

To begin, let me describe briefly the standard picture of the new inflationary scenario where inflation occurs during a slow rollover transition.[2] The original scenario is based on the following three assumptions:

i) The initial state of the scalar field ϕ is in thermal equilibrium; the potential for ϕ is extremely flat near $\phi = 0$ at zero temperature and has minima at $\phi = \pm\phi_c$. At extremely high temperatures, $T > T_c$ the effective potential has a unique minimum at $\phi = 0$. Assumption i) implies that $\langle\phi\rangle = 0$ at $T > T_c$, and that the universe is in the high temperature minimum.

ii) As time evolves, the scalar field ϕ stays near $\phi = 0$, until $T << T_c$, *i.e.* supercooling occurs.

iii) Eventually, the slow rollover transition takes place such that the amplitude of ϕ increases slowly with time according to the classical equation of motion.

In a preliminary analysis[5] of the quantum theory of the slow rollover transition, A. Guth and I have shown the validity of assumptions ii) and iii) when one takes the initial state to be thermal, according to assumption i). We constructed an exactly soluble toy model for the behavior of the scalar field in the new inflationary scenario, where the dynamics of the scalar field is described by the following: A de Sitter background metric is taken for all times. (This is a reasonable approximation for most stages of the rollover; however, the question how the metric changes from the radiation dominated phase to the de Sitter phase is not studied.) Also, we took a free field potential, given by $V(\phi) = -\frac{1}{2}\mu^2\phi^2 + cT^2\phi^2$, where $\mu^2 > 0$ and c is a constant of order of the quartic self-coupling constants and couplings with other fields. T represents a background temperature which changes as $1/R$ with time, where R is the cosmic scale factor. $V(\phi)$ changes with time from a stable potential to an unstable one. Our toy model is certainly only an approximation to the realistic one, but we believe that it describes qualitatively the correct physics for a weakly interacting scalar field, except that it cannot describe how the field settles down at the true minimum at $\phi = \phi_c$.

Among other results of our toy model, the following two indicate that when the initial state of ϕ is in thermal equilibrium, assumptions ii) and iii) are valid.

First, when $T \leq T_c$, $\langle\phi\rangle = 0$, but $\langle\phi^2\rangle \sim T^2$. However, if $c <<< 1$, this large initial fluctuation decrease with time until $T << T_c$; it does not remain at the large value $\langle\phi^2\rangle \sim T_c^2$. This implies that the system becomes prepared for the slow rollover transition.

Second, for "large" time, which may be defined as the time when the effective wavelength of each momentum mode becomes much larger than the horizon distance, each mode of ϕ obeys a classical equation of motion, but the time at which ϕ begins to roll is described by a classical probability distribution. I shall discuss later, using one-dimensional quantum mechanics, how this classical behavior appears in a quantum roll.

Although we have proposed that our model contains the correct physics of a slow rollover transition, it turns out that assumption i) in the original scenario, *i.e.* that the initial state is thermal, does not seem to be valid. This is due to the fact that density fluctuations[3] predicted by inflation require that the inflation driving scalar field interacts extremely weakly, with coupling constants of order 10^{-12}. A simple calculation shows that such a weakly coupled field cannot be in thermal equilibrium with other fields. One must assume a random non-equilibrium initial configuration, and study how it evolves with time. A purely classical analysis of this problem has been carried out by Albrecht, Brandenberger and Matzner.[4] They conclude that inflation is possible through a slow rollover transition for wide classes of initial configurations for ϕ.

Now I shall discuss how one can study perturbatively the quantum evolution of an interacting scalar field with non-equilibrium initial configurations. First of all, my previous analysis with Guth suggests that one obtains a clearer understanding of the time evolution in a functional Schrödinger picture. The approximation scheme I shall use is the time-dependent variational method developed by Jackiw and Kerman,[7] introduced in Section II. In section III, by applying it to a double welled potential $V(Q) = \frac{\lambda}{24}\left(Q^2 - a^2\right)^2$ in one-dimensional quantum mechanics, I shall exhibit the domain of validity of the approximation and the late time classical behavior of a particle which starts to roll from $Q = 0$, with $\dot{Q} = 0$. Finally, in section IV, I shall generalize the formalism of section II, to a scalar field in a Robertson-Walker metric.

II. TIME-DEPENDENT VARIATIONAL APPROXIMATION

The time-dependent variational principle, posited by Dirac,[8] is an unconventional and novel approach for studying time-dependent quantum systems. We shall review the subject following the work of Jackiw and Kerman.[7].

Following Dirac, one considers time-dependent states $|\psi, t\rangle$ and requires that the time-integrated diagonal matrix element of $i\hbar\partial_t - H$,

$$\Gamma \equiv \int dt\, \langle\psi, t\,| i\hbar\partial_t - H |\, \psi, t\rangle \tag{2.1}$$

be stationary against variation of $|\psi, t\rangle$. Supplemented by appropriate boundary conditions, this provides a derivation of the time-dependent Schrödinger equation.

The quantity Γ is an effective action for a given system described by $|\psi, t\rangle$ and variation of Γ is the quantum analogue of Hamilton's principle. When a specific *Ansatz* is made for the **state** $|\psi, t\rangle$, the time-dependent Hartree-Fock approximation emerges and this approach is widely used by quantum chemists[9] and nuclear physicists.[10]

Consider one-dimensional quantum mechanical systems with $H = \frac{1}{2m}P^2 + V(Q)$, and suppose that we are interested in the time evolution of a given initial Gaussian wave function in a time-dependent variational approximation. The effective action involves the diagonal matrix element of $i\hbar\partial_t - H$ in a trial wave function, which we take to be the most general Gaussian,

$$\langle Q|\psi, t\rangle_V = N \exp\left[-\frac{1}{2\hbar}(Q-q)^2 B + \frac{i}{\hbar}p(Q-q)\right]. \tag{2.2}$$

Following Jackiw and Kerman, we parametrize the real and imaginary part of B as

$$B(t) = \frac{1}{2}G^{-1}(t) - 2i\Pi(t)\ . \tag{2.3}$$

The normalization factor N is then $(2\pi\hbar G)^{-1/4}$. The real quantities $p(t)$, $q(t)$, $G(t)$ and $\Pi(t)$ are the variational parameters and we demand that their variations vanish at $t = \pm\infty$.

For potentials $V(Q)$ which are quartic in Q, the explicit form of Γ_V is

$$\begin{aligned}\Gamma_V\,(q,\, p,\, G,\, \Pi) = \int_{-\infty}^{\infty} dt \Big[& p\dot{q} - H_{\rm cl}(q,\ p) \\ & + \hbar\left(\Pi\dot{G} - \frac{1}{2}GV^{(2)}(q) - \frac{1}{8}G^{-1} - 2\Pi^2 G\right) \\ & - \frac{\hbar^2}{8}G^2 V^{(4)}(q)\Big]\end{aligned} \tag{2.4}$$

where $H_{cl}(q,p) = \frac{1}{2m}p^2 + V(q)$ and $V^{(n)}(q) \equiv \frac{\partial^n V(q)}{\partial q^n}$. We note that $\Pi(t)$ plays the role of the momentum conjugate to $G(t)$. The four variational equations are then

$$\frac{\delta \Gamma_V}{\delta p} = 0 \rightarrow \dot{q} = \frac{1}{m}p \tag{2.5a}$$

$$\frac{\delta \Gamma_V}{\delta \Pi} = 0 \rightarrow \dot{G} = 4G\Pi \tag{2.5b}$$

$$\frac{\delta \Gamma_V}{\delta q} = 0 \rightarrow \dot{p} = -V^{(1)}(1) - \frac{\hbar}{2}GV^{(2)}(q) \tag{2.5c}$$

$$\frac{\delta \Gamma_V}{\delta G} = 0 \rightarrow \dot{\Pi} = \frac{1}{8}G^{-2} - 2\Pi^2 - \frac{1}{2}V^{(2)}(q) - \frac{\hbar}{4}V^{(4)}(q)G \tag{2.5d}$$

We call the above "time-dependent Hartree-Fock" (HF) equations, because using the Gaussian wave function leads to the approximation in which all n-point expectation values are expressed in terms of one- and two-point functions.

III. QUANTUM ROLL IN A DOUBLE WELLED POTENTIAL

This Section is part of work I have carried out in collaboration with F. Cooper and P. Stancioff.[6]

Suppose at $t = 0$, a particle is described by a Gaussian wave function centered at $Q = 0$. We are interested in the time-evolution of the particle in a potential, $V(Q) = \frac{\lambda}{24}\left(Q^2 - a^2\right)^2$, in the time-dependent variational approximation.

First, let me present the result in an upside-down harmonic oscillator, $V(Q) = -\frac{1}{2}kQ^2$, $k > 0$, obtained by Guth and myself.[5] Here the time-dependent Schrödinger equation is exactly soluble. The wave function is given by

$$\psi(Q,t) = (2\pi G\hbar)^{-1/4} \exp\left[-\frac{1}{2b^2}\tan(\alpha - i\omega t)\, Q^2\right] \tag{3.1}$$

where $b^2 = \hbar/\sqrt{mk}$, $\omega^2 = k/m$ and α is a real integration constant. Note that b is a natural quantum mechanical scale of this problem.

For large times, the above wave function provides the following information:

i) The probability distribution for Q is given by

$$\langle Q^2 \rangle \longrightarrow \frac{1}{4}\,\frac{b^2}{\sin 2\alpha}e^{2\omega t} \tag{3.2}$$

i.e. $\sqrt{\langle Q^2 \rangle}$ obeys a classical equation of motion.

ii) Application of the momentum operator to ψ, yields

$$\begin{aligned} -i\hbar\frac{\partial \psi}{\partial Q} &= \left(\frac{i}{2}G^{-1}(t) + 2\Pi(t)\right)Q\psi \\ &= \left[\sqrt{mk}Q + 0\left(e^{-2\omega t}\right)\right]\psi \end{aligned} \tag{3.3}$$

Note that $\sqrt{mk}Q$ is the classical momentum $p_{cl} = \sqrt{2m(E - V(Q))}$ which would be attained by a *classical* particle at Q which rolled from rest at $Q = 0$ at total energy $E = 0$.

iii) The commutator $[Q, P]$ is negligible if

$$\sqrt{mk}Q^2 >> \hbar \ , \quad i.e.\ Q^2 >> b^2 \ . \tag{3.4}$$

Note however that the wave function is definitely not sharply peaked about any particular classical trajectory. Rather, at large times the system is described by a classical probability distribution,

$$f(Q,P,t) = |\psi(Q,t)|^2\delta(P - p_{cl}) \ . \tag{3.5}$$

The function f obeys a classical evolution equation.

Now let us turn to the double well potential. In the variational approximation (or HF approximation), our trial wave function for this problem is

$$\psi_V(Q,t) = (2\pi G)^{1/4} \times \exp\left[-\frac{1}{2}Q^2\left(\frac{1}{2}G^{-1}(t) - 2i\Pi(t)\right)\right] \tag{3.6}$$

(We have set $\hbar = 1$.) A natural quantum mechanical length scale b may be defined as

$$b^2 \equiv \frac{1}{\sqrt{mk}} \ ; \qquad k \equiv \left|V^{(2)}(0)\right| = \frac{1}{6}\lambda a^2 \ . \tag{3.7}$$

We shall choose our initial width of the Gaussian to be

$$G_0 = \frac{1}{2}b^2 = \left(\frac{3}{2\lambda a^2}\right)^{1/2} \tag{3.8}$$

Since our initial conditions are fixed, one can perform a numerical calculation to solve Eqs. (2.5).

Let me now discuss the domain of validity of the time-dependent HF approximation. In this approximation, some properties of $G(t)$ can be determined exactly and in particular its maximum value is found to be

$$G_{\max} = \frac{2}{3}a^2 \ , \tag{3.9}$$

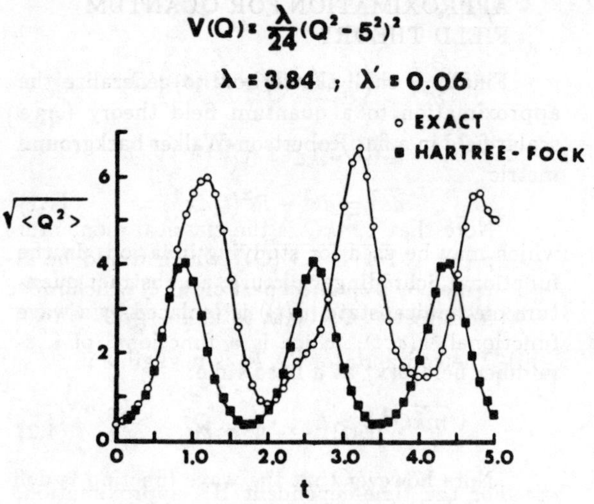

Figure 1a

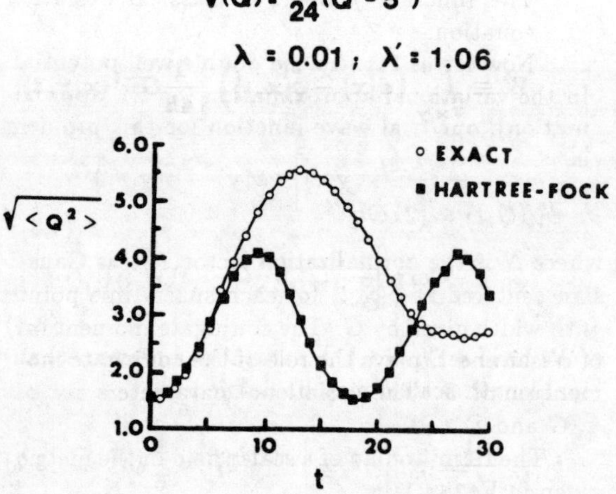

Figure 1b

i.e. $\sqrt{G}$ never spreads sufficiently to reach the minima at $R = \pm a$. This premature turning point shows that the approximation fails near the bottom of the two wells. This failure is due to the fact that the approximation is based on a single Gaussian wave function which cannot describe probability piling up at the two minima. However, as we see in Figs. 1a and 1b, when $G \leq G_{\max}$ the approximation is excellent for all λ in comparison with the exact solution obtained by a numerical calculation in the Heisenberg picture.

Next, we shall discuss the large time behavior of the system. Unlike the upside-down harmonic oscillator whose potential is always unstable, so that Q increases indefinitely, here "large time" is rather limited; it is some intermediate time before $\sqrt{\langle Q^2 \rangle}$ arrives near the minimum. In fact, in the HF approximation, "large time" must be when $G(t) \lesssim \frac{2}{3}a^2$.

From Eq. (3.4), we expect that classical behavior may appear for $\langle Q^2 \rangle >> b^2$. This requires that

$$R \equiv \frac{b^2}{a^2} << 1 \longrightarrow \frac{\sqrt{6}}{\sqrt{m\lambda}\, a^3} << 1 \ . \qquad (3.10)$$

Eq. (3.10) implies that for given m and a, λ must be large. One may define a dimensionless coupling constant for this problem in terms of the mass and of the quantum mechanical scale b as

$$\lambda' \equiv mb^6\lambda = \frac{6\sqrt{6}}{\sqrt{m\lambda}\, a^3} \ . \qquad (3.11)$$

$\lambda' << 1$ is equivalent to $R << 1$. For our numerical calculation we have chosen $m = 1$, $a = 5$ and two values of λ; $\lambda = 3.84$ and $\lambda = 0.0123$.

The following are the results for small dimensionless coupling constant, $\lambda' = 0.06$, which corresponds to $\lambda = 3.84$:

i) In Fig. 1a we compare $\sqrt{\langle Q^2 \rangle}$ in the exact solution (obtained numerically) to its HF approximation $\sqrt{G(t)}$. We find excellent agreement until $\sqrt{G}$ reaches its premature turning point at $\pm a\sqrt{\frac{2}{3}}$. Despite this, a reasonable result for the oscillation time is obtained in the HF approximation.

ii) Classical behavior of $\langle Q^2 \rangle$ in the late time: Since a classical particle with $Q(0) = P(0) = 0$, and therefore $E = \frac{\lambda}{24}a^4$, will stay at $Q = 0$, we did our computer experiment by placing the classical particle at $Q(0) = \sqrt{G_0}$ and compared this particular classical trajectory $Q_{\text{cl}}(t)$ with the exact $\sqrt{\langle Q^2 \rangle}$ calculation in the Heisenberg picture. In Fig. 2, we find that for $1.5 < Q_{\text{cl}} < 4$ (the first oscillation) the two are quite the same except for a shift of the origin. We have already seen in Fig. 1 that the HF approximation is excellent until $\sqrt{G}$ reaches $\sqrt{\frac{2}{3}}\, a \approx 4.2$, and this implies that

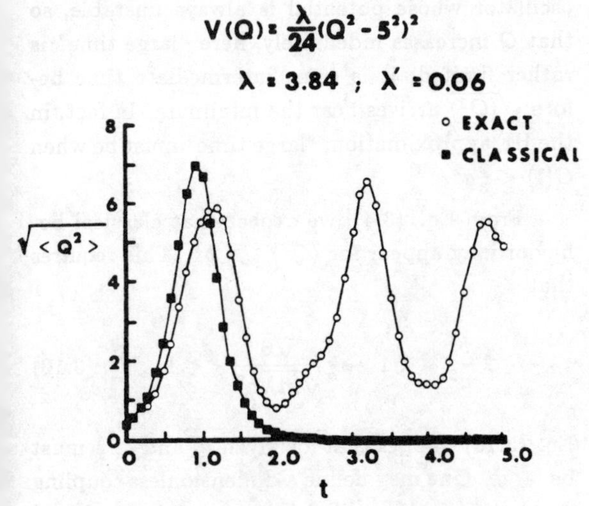

Figure 2

the classical behavior of $\sqrt{G}$ also appears in the HF approximation.

iii) The classical behavior in Eq. (3.3) which is found in the upside-down harmonic oscillator is tested: We now find that

$$-i\frac{\partial\psi_V}{\partial Q} = \left(\frac{i}{2}G^{-1} + 2\Pi\right) Q\psi_V \approx 2\Pi Q\psi_V \qquad (3.12)$$

and the ratio $2\Pi Q/p_{\rm cl}(Q)$, where $p_{\rm cl}(Q) = \sqrt{2\,(E - V(Q))} = \sqrt{\frac{\lambda}{12}\left(a^2 - \frac{Q^2}{2}\right)}Q$, differs from unity by at most 20% for $1.5 < \sqrt{G} < 4$ in the HF approximation.

iv) The question whether the commutator $[Q, P]$ is negligible is studied both in the HF approximation and in the exact calculation: In the HF approximation,

$$\langle QP\rangle_{\rm HF} = \frac{i}{2} + 2G\Pi \qquad (3.13)$$

and we find that $G\Pi > 5$ for $1.75 < \sqrt{G} < 4$. In the exact calculation, the real part of $\langle QP\rangle_{\rm exact} > 5$ for even larger range *i.e.* $1.2 < \sqrt{\langle Q^2\rangle} < 5.9$. Hence, in our variational time-dependent HF approach for small dimensionless coupling constant, $\lambda' << 1$ (or $\lambda > 1$) the late time behavior of the system is approximately described by classical physics with a classical probability distribution function,

$$f\,(Q,\,P,\,t) = |\psi_V\,(Q,t)|^2\,\delta\,(P - p_{\rm cl}) \quad . \qquad (3.14)$$

For a large dimensionless coupling constant, $\lambda' = 1.06$ (or $\lambda = 0.01$) we find that classical behavior does not appear in the HF approximation nor in the exact simulation (see Fig. 1b).

IV. TIME-DEPENDENT VARIATIONAL APPROXIMATION FOR QUANTUM FIELD THEORY

Finally, I shall sketch how to generalize the approximation to a quantum field theory for a scalar field in a flat Robertson-Walker background metric

$$ds^2 = dt^2 - R^2(t)d\mathbf{x}^2 \qquad (4.1)$$

which may be used for studying inflation. In the functional Schrödinger picture, an abstract quantum mechanical state $|\psi(t)\rangle$ is replaced by a wave functional $\Psi(\phi, t)$, which is a functional of a c-number field $\phi(\mathbf{x})$ at a fixed time:

$$|\psi(t)\rangle \longrightarrow \Psi(\phi, t) \qquad (4.2)$$

For the time-dependent HF approximation, we take a Gaussian trial wave function which is the generalization of Eq. (2.2)

$$\Psi_V(\phi, t) = N\,\exp\left\{+\frac{i}{\hbar}\int_{\mathbf{x}} \hat{\pi}(\mathbf{x},t)\left(\phi(\mathbf{x}) - \hat{\phi}(\mathbf{x},t)\right)\right\} \times \exp -W$$

$$W \equiv \int_{\mathbf{x},\mathbf{y}} \left(\phi(\mathbf{x}) - \hat{\phi}(\mathbf{x},t)\right)\left[\frac{1}{4\hbar}G^{-1}(\mathbf{x},\mathbf{y},t) - i\frac{1}{\hbar}\Sigma(\mathbf{x},\mathbf{y},t)\right]\left(\phi(\mathbf{y}) - \hat{\phi}(\mathbf{y},t)\right) \qquad (4.3)$$

where N is the normalization factor. Ψ_V is Gaussian centered at $\hat{\phi}(\mathbf{x},t)$ for each space-time point with width given by G. The conjugate momentum of $\hat{\phi}$ is $\hat{\pi}$ and Σ plays the role of the conjugate momentum of G. The variational parameters are ϕ, $\hat{\pi}$, G and Σ.

The Hamiltonian of a scalar field in the metric given in Eq. (4.1) is

$$H = \frac{1}{2}\int_{\mathbf{x}} R^3\left[-R^{-6}\frac{\delta^2}{\delta\phi^2} + R^{-2}\left(\nabla\phi\right)^2 + V(\phi)\right] . \qquad (4.4)$$

Then, the effective action Γ_V in this approximation is again given by Eq. (2.1), with $\psi(t)$ replaced by Eq. (4.3). The derivation of four variational equations are obtained straightforwardly by varying Γ_V against $\hat{\phi}$, $\hat{\pi}$, G and Σ.

This approximation allows us to study the onset of inflation for general classes of non-equilibrium initial conditions of a scalar field configuration. In particular, one can consider initial conditions where $\hat{\phi}(\mathbf{x},t) = \langle\phi(\mathbf{x})\rangle$ are randomly chosen in each initial horizon volume and/or with random initial momenta.[11] Furthermore, as discussed in section III, one can study the late time behavior of the scalar field. The numerical calculation for this problem is in progress

REFERENCES

1. A.Guth, *Phys. Rev.* **D23**, 347 (1981).
2. A. Linde, *Phys. Lett.* **108B**, 389 (1982); A. Albrecht and P. Steinhardt, *Phys. Rev. Lett.* **48**, 1220 (1982).
3. A. Guth and S.-Y. Pi, *Phys. Rev. Lett.* **49**, 1110 (1982); S. Hawking, *Phys. Lett.* **115B**, 175 (1982); A. Starobinsky, *Phys. Lett.* **117B**, 175 (1982); J. Bardeen, P. Steinhardt and M. Turner, *Phys. Rev.* **D28**, 679 (1983).
4. G. Mazenko, W. Unruh and R. Wald, *Phys. Rev.* **D31**, 273 (1985); G. Mazenko, *Phys. Rev. Lett.* **54**, 2163 (1985) and University of Chicago preprint (1986); C. Coughlan and G. Ross, *Phys. Lett.* **157B**, 151 (1985); A. Albrecht, R. Brandenberger and R. Matzner, to be published.
5. A. Guth and S.-Y. Pi, *Phys. Rev.* **D32**, 1899 (1984).
6. F. Cooper, S.-Y. Pi and P. Stancioff, to be published in *Phhys. Rev. D.*
7. R. Jackiw and A. Kerman, *Phys. Lett.* **A71**, 158 (1979).
8. P. A. M. Dirac, *Proc. Camb. Phil. Soc.* **26**, 376 (1930).
9. S. Epstein, *The Variational Method in Quantum Chemistry* (Academic Press, New York, 1976).
10. P. Bonche, S. Koonin and J. W. Negele, *Phys. Rev.* **C13**, 1226 (1976); A. Kerman and S. Koonin, *Ann. Phys.* (N.T) **100**, 332 (1976).
11. Albrecht *et al.*, in Ref. [4].

QUANTUM CORRECTIONS AND LOCALLY SUPERSYMMETRIC INFLATIONARY THEORIES

Burt A. Ovrut[†*]

Department of Physics, University of Pennsylvania
Philadelphia, Pennsylvania 19104-6396

Supergravitational inflationary theories of the early universe are surveyed, emphasizing the role of radiative corrections on these models.

In this talk I would like to discuss the role of N = 1 supergravity in new inflationary universe theories in general, and the effect of quantum supergravitation on such theories in particular. Let ϕ_i be a chiral superfield. Its component field content is

$$\phi_i = (\phi_i, \psi_i) \tag{1}$$

where ϕ_i and ψ_i are a complex scalar field and Weyl spinor field respectively. Denote the superpotential and Kahler potential by $W(\phi_i)$ and $K(\phi_i^\dagger, \phi_i)$. Throughout this talk I will take

$$K = \sum_i |\phi_i|^2 \tag{2}$$

This corresponds to a flat Kahler metric. At zero temperature, the tree level potential energy is

$$V = e^{\sum |\phi_i|^2} \left(\sum_i |D_{\phi_i} W|^2 - 3|W|^2\right) \tag{3}$$

where

$$D_{\phi_i} W = \frac{\partial W}{\partial \phi_i} + \phi_i^\dagger W \tag{4}$$

For $T \neq 0$, one must add

$$V_T = \frac{1}{24} b^2 T^2 \,\mathrm{Tr}\left(\frac{\partial^2 V}{\partial \phi_i^\dagger \partial \phi_j}\right) \tag{5}$$

where b is Boltzman's constant, to (3). Spontaneous supersymmetry breaking occurs if and only if

$$\langle D_{\phi_i} W \rangle \neq 0 \tag{6}$$

Assuming zero cosmological constant, the gravitino mass is

$$m_{3/2} = \frac{1}{\sqrt{3}} e^{\sum_i |\langle \phi_i \rangle|^2/2} \sum_i |\langle D_{\phi_i} W \rangle|^2 \tag{7}$$

First, we consider theories where there is only one chiral superfield in the hidden sector.[1] Write the potential energy as

$$V = A + B\phi + C\phi^2 + D\phi^3 + E\phi^4 + \ldots \tag{8}$$

We apply the following "cosmological constraints" to V.[2]

a) Cosmological Constant Constraint:
Choose A so that

$$V(\langle \phi \rangle) = 0 \tag{9}$$

b) New Inflation Constraint:
The requirement that there is sufficient inflation demands that

$$|B| \lesssim H_o^3, \quad |C| < \frac{H_o^2}{40} \tag{10}$$

where $H_o = (A/3)^{1/2}$ is Hubble's constant. The requirement that $\delta\rho/\rho \sim 10^{-4}$ implies that

$$D \simeq -10^{-6} H_o, \quad E \lesssim 10^{-14} \tag{11}$$

c) Thermal Constraint:
At high temperature the vacuum state must be such that at T = 0 a full slow rollover transition can occur.

The most general superpotential is

$$W(\phi) = m^2\left(\beta + \phi + \frac{\Delta}{2}\phi^2 + \frac{\Delta\lambda}{3}\phi^3 + \sum_{n=4}^{\infty} \frac{a_n}{n}\phi^n\right) \tag{12}$$

a) $\Rightarrow$ adjust β

b) $\Rightarrow$ $m \sim 10^{-4}$, $\frac{\Delta}{2} \simeq \beta$, $\lambda = 0$ (13)

c) $\Rightarrow$ $\beta > 0$

[†]Work supported in part by the Department of Energy under Contract Grant Number DOE-AC02-76-EPO-3071.

[*]Invited talk at the 23 International Conference on High Energy Physics, held at the University of California at Berkeley, July, 1986.

Let

$$W = m^2(\beta + \phi + \frac{\Delta}{2}\phi^2 + \frac{\Delta}{2} a_4 \phi^4) \quad (14)$$

For simplicity take

$$\frac{\Delta}{2} = \beta \quad (15)$$

For $\phi = R = \text{real}$

$$V = m^4 e^{R^2}([1-3\beta^2] + [3\beta^2 - 1]R^2 + [2\beta(1 + 2a_4)]R^3 + \ldots) \quad (16)$$

Denote $<\phi> = \sigma$. Then $V(\sigma) = 0$ implies

$$\beta = \frac{-3\sigma}{2(2+\sigma^2)}$$

$$\sigma^2 = \frac{-1 \pm \sqrt{1 + 6a_4}}{3a_4} \quad (17)$$

A typical value, $a_4 = -\frac{1}{6}$, gives

$$\sigma = \sqrt{2} \quad , \quad \beta = -\frac{3\sqrt{2}}{8} \quad (18)$$

Since $\beta < 0$ this theory violates thermal constraint (13)c. Let us ignore this problem for a moment and continue on. The potential energy is

$$V = m^4(\frac{5}{32} - \frac{1}{\sqrt{2}} R^3 + \ldots) \quad (19)$$

At the minimum

$$<D_\phi W> = 0 \quad (20)$$

which implies that $m_{3/2} = 0$. That is, there is no spontaneous supersymmetry breaking. Instead of (15), now take[3]

$$\frac{\Delta}{2} = \beta - \varepsilon \quad (21)$$

Again, taking $V(\sigma) = 0$ and $a_4 = -\frac{1}{6}$ one finds

$$\sigma = \sqrt{2}\,(1 - \frac{4}{3\sqrt{3}}\varepsilon)$$

$$\beta = -\frac{3\sqrt{2}}{8}\,(1 - [\frac{4\sqrt{2} + \sqrt{3}}{3}]\varepsilon) \quad (22)$$

and

$$V = m^4([\frac{5}{32} + O(\varepsilon)] - [4\varepsilon]R + [\frac{9}{2\sqrt{2}}\varepsilon]R^2 - [\frac{1}{\sqrt{2}} + O(\varepsilon)]R^3 + \ldots) \quad (23)$$

At the minimum $<D_\phi W> \neq 0$ and

$$m_{3/2} = m^2 e\,(1 + \frac{\sqrt{6}}{3})\varepsilon \quad (24)$$

Hence, supersymmetry is spontaneously broken. For $m \sim 10^{-4}$ and $\varepsilon \sim 10^{-8}$ (24) implies that $m_{3/2} \sim 10^{-16}$, which is the appropriate electroweak breaking scale. The small value for ε yields sufficient inflation and an acceptable value for $\delta\rho/\rho$. Is the small value for ε "natural"? That is, do radiative corrections produce a large or small change in $m_{3/2}$? The one-loop correction to the effective Lagrangian is[4]

$$\mathcal{L}_{1\text{-loop}} = a\,[(N-1)\mu^2 + (N-3)V + \frac{1}{4}(5-N)R]$$

$$+b\,[-\frac{1}{48}(N+41)R_{\mu\nu\psi\sigma}R^{\mu\nu\psi\sigma} - \frac{1}{48}(N + 43)R^2$$

$$+\frac{23}{3}VR + \frac{2}{3}R\mu^2 - 4V^2 + 14\mu^2 + \frac{1}{6}R\,\mathrm{Tr}(MM^\dagger + m^2)$$

$$-\frac{1}{2}\mathrm{Tr}\,m^4 + \mathrm{Tr}(MM^\dagger)^2] + \text{finite} \quad (25)$$

where N is the number of chiral superfields,

$$\mu^2 = e^{\sum_i|\phi_i|^2}\,|W(\phi_i)|^2 \quad (26)$$

m and M are the scalar and fermion mass matrices respectively, and

$$a = \frac{1}{16\pi^2}\Lambda^2$$

$$b = \frac{1}{32\pi^2}\ln\,\Lambda^2 \quad (27)$$

where Λ is the cutoff. Note $\Lambda \sim 1$ implies $a \sim \frac{1}{16\pi^2}$ and $b \sim 0$. That is, the quadratic divergence dominates. Adding the radiative corrections we find that σ and $m_{3/2}$ are shifted to

$$\sigma = \sqrt{2}\,(1 - \frac{4}{3\sqrt{3}}[1 - a(N-1)\frac{1}{2\sqrt{6}}(1 + \frac{2\sqrt{6}}{3})]\varepsilon)$$

$$m_{3/2} = m^2 e((1 + \frac{\sqrt{6}}{3}) - a(N-1)\frac{2}{3}(1 + \frac{5}{2\sqrt{6}}))\varepsilon \quad (28)$$

Note that all terms with a are also proportional to ε. It follows that the radiative corrections do not change σ and $m_{3/2}$ by very much. Therefore, the above results are natural. The moral of this calculation is that quantum corrections are of the same order of magnitude as the leading inflation producing interactions! Now let

$$\frac{\Delta}{2} = \beta \quad (29)$$

and consider

$$W = m^2(\beta + \phi + \beta\phi^2 + \frac{\beta}{2} a_4\phi^4 + \frac{2\beta}{5} a_5\phi^5) \quad (30)$$

In order to satisfy the thermal constraint take $\beta > 0$. It is, then, impossible to have inflation ending in a supersymmetry preserving vacuum state. For typical values, $a_4 = -2.191$, $a_5 = 1$, and $\beta = .2$ we find that the potential energy has a minimum at

$$\sigma = 1.023 \quad (31)$$

$<D_\phi W> \neq 0$, and

$$m_{3/2} = 2.22\ m^2 \quad (32)$$

For $m \sim 10^{-4}$, $m_{3/2}$ is of order 10^{-8}, which is too large by eight orders of magnitude to account for the electroweak breaking scale of 10^2 GeV. Can one satisfy the thermal constraint and also have a light gravitino? Yes! The solution is to consider theories where there are two chiral superfields in the hidden sector.[5] Denote these by ϕ and ψ. The superpotential is

$$W = W_1 + W_2 + W_3 \quad (33)$$

where

$$W_1 = \frac{\lambda}{2}\phi^2\psi^3 - m^2\phi + \frac{\delta}{3}\psi^3$$

$$W_2 = \frac{a_4}{4}\phi^4 + \frac{a_5}{5}\phi^5 \quad (34)$$

$$W_3 = \frac{\lambda'}{2}\phi^2\psi^2$$

What is the vacuum state? Consider W_1 only. The cosmological constant can be made to vanish by taking $\delta = 4.26 \times 10^{-2}\ m^4/\lambda$. The absolute minimum of the potential energy occurs at

$$\langle\psi\rangle = 1.265$$

$$\langle\phi\rangle = .494\ m^2/\lambda \quad (35)$$

At this minimum

$$\langle D_\psi W_1\rangle = .378\ m^4/\lambda$$

$$\langle D_\phi W_1\rangle = -1.078\ m^6/\lambda \quad (36)$$

It follows that supersymmetry is spontaneously broken and

$$m_{3/2} \sim \frac{m^4}{\lambda} \quad (37)$$

For $\lambda \sim 1$ and $m \sim 10^{-4}$, $m_{3/2}$ is of order 10^{-16} which is the appropriate electroweak scale. Why does this happen? Turning off gravity, the vacuum satisfies

$$\frac{\partial W_1}{\partial\phi} = 0 \Rightarrow \phi = \frac{m^2}{\lambda\psi^3} \quad (38)$$

$$\frac{\partial W_1}{\partial\psi} = \frac{3\lambda}{2}\phi^2\psi^2 = 0 \Rightarrow \phi^2\psi^2 = 0$$

Therefore

$$\frac{1}{\psi} = 0 \Rightarrow \psi \to \infty \quad (39)$$

Now turn gravity back on. The effect of this is to set ψ, not to ∞, but to the Planck mass. That is, $\psi \to 1$. Then

$$\frac{\partial W_1}{\partial\phi} = 0$$

$$\frac{\partial W_1}{\partial\psi} = \frac{3}{2\lambda}m^4 \quad (40)$$

which implies that

$$m_{3/2} \sim \frac{m^4}{\lambda} \quad (41)$$

as above. Therefore, this theory has a naturally light gravitino. Now add W_2. In the real ϕ direction the potential energy is

$$V_\phi = m^4\left(1 - \frac{2a_4}{m^2}\phi^3 + \frac{1}{2}\left[1 - \frac{4a_5}{m^2}\right]\phi^4 + \ldots\right) \quad (42)$$

For typical values $a_4 = .7407\ m^2$ and $a_5 = -18.52\ m^2$, V_ϕ has its minimum at

$$\sigma = 3 \times 10^{-2} \quad (43)$$

There is a slow rollover transition from $\phi = 0$ to $\phi = \sigma$ that corresponds to 10^6 e-foldings of inflation and $\delta\rho/\rho \sim 10^{-4}$, which are acceptable values. Note, however, that

$$V(\sigma) = .9999 \quad (44)$$

Hence, σ is not the true vacuum. The system must now go from σ through a connecting region to vacuum (35), which is the absolute minimum of the potential. To see that this occurs add W_3. Take $\lambda' \sim m^2$. Then the above results are unchanged. For $\phi \ll 1$ and $\psi \lesssim m$ the potential energy is

$$V = V_\phi(\phi) + (V_\phi(\phi) - 2\lambda' m^2\phi)\psi^2 - (2\lambda m^2\phi)\psi^3 + \ldots \quad (45)$$

Let

$$\lambda' = \frac{1}{2m^2\sigma}V_\phi(\sigma) \quad (46)$$

Then, at $\phi = \sigma$,

$$V = V_\phi(\sigma) + \underline{\underline{0}}\psi^2 - (2\lambda m^2\sigma)\psi^3 + \ldots \quad (47)$$

Hence, the quadratic ψ^2 barrier vanishes and the cubic ψ^3 term forces the system away from σ in the positive ψ direction. The vacuum rolls quickly down to the absolute minimum at (35). It is trivial to show that this theory satisfies the thermal constraint. Thus, we have demonstrated a theory that satisfies the thermal constraint and has a light gravitino. This is nice, but somewhat ad hoc. Let us display the role of each term in superpotential (33).

$$W_1 = \underbrace{\frac{\lambda}{2}\phi^2\psi^3 - m^2\phi}_{\text{light gravitino}} + \underbrace{\frac{\delta}{3}\psi^3}_{\Lambda_c = 0}$$

$$W_2 = \underbrace{\frac{a_4}{4}\phi^4 + \frac{a_5}{5}\phi^5}_{\text{inflation and } \delta\rho/\rho} \quad (48)$$

$$W_3 = \underbrace{\frac{\lambda'}{2}\phi^2\psi^2}_{\text{rolloff}}$$

Can some of the above terms be eliminated? Yes! Remember that quantum corrections are of the same order of magnitude as the leading inflation producing interactions. Therefore, take[6]

$$W_2 = 0 \tag{49}$$

and add quantum corrections (25) with $b \sim 0$. We find that there is no change in the absolute minimum of the potential. In the slow rollover region

$$V_\phi = m^4(1 - a(N-1)\phi^2 + \frac{1}{2}\phi^4 + \dots) \tag{50}$$

which has a minimum at

$$\sigma = \sqrt{a(N-1)} \tag{51}$$

The value of V_ϕ at σ is

$$V_\phi(\sigma) = 1 - \frac{a^2(N-1)^2}{2} \tag{52}$$

For typical value $a \sim 10^{-2}$

$$\sigma \sim 10^{-1},\ V(\sigma) \sim .9999 \tag{53}$$

We find there are 500 e-foldings of inflation and $\delta\rho/\rho \sim 10^{-4}$, which are acceptable values. The physics of the connecting region is identical to the above. We can eliminate even more terms from (33) by taking coefficient $b \neq 0$ in (25) and setting

$$W_3 = 0 \tag{54}$$

The above discussions of the absolute minimum, inflation and $\delta\rho/\rho$ are modified, but the conclusions are unchanged. However, in the connecting region

$$V = V_\phi(\phi) + (V_\phi(\phi) - bm^2\phi)\psi^2 - (2\lambda m^2\phi)\psi^3 + \dots \tag{55}$$

Therefore, b can be adjusted so that, at σ, the system rolls off the ϕ axis. We conclude that superfields ϕ, ψ with superpotential

$$W = \underbrace{\frac{\lambda}{2}\phi^2\psi^3 - m^2\phi}_{\text{light gravitino}} + \underbrace{\frac{\delta}{3}\psi^3}_{\Lambda_c = 0} \tag{55}$$

leads to an acceptable new inflationary theory. Sufficient inflation, $\delta\rho/\rho \sim 10^{-4}$, and the rolloff of the ϕ axis are produced purely by gravitational radiative corrections.

REFERENCES

1. Ovrut, B. and P. Steinhardt, Phys. Lett. 133B, 161 (1983); Ellis, J., D. Nanopoulos, K. Olive and K. Tamvakis, Nuc. Phys. B(1983), Phys. Lett. 120B, 331 (1983); Nanopoulos, D., K. Olive, and M. Srednicki, Phys. Lett. 127B, 30 (1983).
2. Steinhardt, P. and M. Turner, Phys. Rev. D29, 2162 (1984).
3. Binetruy, P. and S. Mahajan, LBL preprint 18566 (1985).
4. Barbieri, R. and S. Cecotti, Z. Phys. C17, 183 (1983); Srednicki, M. and S. Theisen, Phys. Rev. Lett. 54, 278 (1985).
5. Ovrut, B. and P. Steinhardt, Phys. Rev. Lett. 53, 732 (1984); Phys. Rev. D30, 2061 (1984).
6. Lindblom, P., B. Ovrut and P. Steinhardt, Phys. Lett. 172B, 309 (1986).

INFLATION AND OTHER COSMOLOGICAL ASPECTS OF SUPERSTRING INSPIRED MODELS

Mary K. Gaillard

Lawrence Berkeley Laboratory and Department of Physics
University of California, Berkeley, CA 94720

Investigations of the possibilities for inflation using the four dimensional models inspired by superstring theory is reported and discussed in the context of related work.

Introduction. Attempts to realize a successful inflationary scenario have led to the elaboration[1] of a number of conditions that must be met in the standard field theoretic context where a period of exponential expansion arises from a nonvanishing vacuum energy associated with the vacuum expectation value (vev) of some scalar field ϕ, called the inflaton. Among these conditions are 1) A sufficient number of e-foldings

$$N_e \simeq H\Delta\tau \gtrsim 65 \tag{1}$$

to solve the flatness and horizon problems. In Eq. (1) $\Delta\tau$ is the duration of the inflationary epoch and

$$H = \frac{a}{\dot{a}} = \sqrt{\frac{V(\phi_0)}{3}}/m_{Planck} \tag{2}$$

is the value of the Hubble constant at the start of inflation, and is dominated by the vacuum energy $V(\phi_0)$ stored in the false vacuum $< \phi >_{av} \equiv \phi_0$. 2) Density fluctuations $\delta\rho/\rho$ arising from spatial fluctuations in the scalar field ϕ during inflation should be small enough to be consistent with the observed homogeneity in the microwave background and, ideally, large enough to account for galaxy formation:

$$\delta\rho/\rho \simeq 10^{-5} - 10^{-4}. \tag{3}$$

In the case that there is a post-inflationary source of density fluctuations such as cosmic strings, Eq. (3) can be reinterpreted as an upper limit: $\delta\rho/\rho \lesssim 10^{-4}$.

The inflaton potential $V(\phi)$ is roughly characterized by two parameters: the value σ of the inflaton ϕ at the global minimum of the zero temperature potential,

$$V'(\sigma) = 0 = V(\sigma), \ (V(\phi) \geq 0), \tag{4}$$

where the second equality in (4) is assumed to assure a vanishing cosmological constant today, and the value

$$\mu^4 = V(\phi_0) \tag{5}$$

of the energy density stored in false vacuum at the onset of inflation. Typically, the quantities N_e and $\delta\rho/\rho$ in specific inflationary models are determined by the parameters σ and μ as

$$N_e \simeq \sigma^m/\mu^n, \quad \delta\rho/\rho \simeq \mu^{n'}/\sigma^{m'}, \tag{6}$$

where m, n, m', n' are small positive integers. Since we require large N_e and small $\delta\rho/\rho$, the constraints (1) and (3) are usually met by choosing[1] large σ and small μ:

$$\sigma \sim 1, \ \mu \sim 10^{-4} - 10^{-3} \tag{7}$$

in Planck mass units (that I shall use throughout unless otherwise specified: $m_P = (8\pi G_F)^{-1/2} \equiv 2 \times 10^{18}$ GeV $\equiv 1$), with a potential that is rather flat over a broad region,

$$V(\phi) \sim \mu^4, \ \phi_0 \lesssim \phi \lesssim \phi_e, \ \phi_e - \phi_0 \sim 1, \tag{8}$$

so that the vacuum energy density remains appreciable until the field ϕ rolls to and "end-of-inflation" value ϕ_e after which it rapidly falls (via damped oscillations) to its true ground state value σ. As μ characterizes the self-couplings of the inflaton, the small value (7) of μ implies that it is very weakly coupled.

The required properties of the potential are most easily implemented in the context of $N = 1$ supergravity. As this theory is not renormalizable, the potential is an a priori arbitrary function of ϕ and so can be adjusted to the desired shape, albeit with considerable fine tuning. In addition, once one introduces a gauge singlet field that couples to ordinary matter only with interactions of gravitational strength, there is a natural mechanism for satisfying a third criterion[1] for a successful inflationary scenarios, namely: 3) Baryogenesis. The reheating temperature after inflation must be sufficient to allow for an out-of-equilibrium distribution of heavy particles that can generate a net baryon number in their decays. If the inflaton has only (generalized) gravitational couplings, it will decay into the heaviest particles that are kinematically available, thus possibly producing the desired result.

The bad news for supergravity models is their failure, in general, to fulfill a fourth criterion, namely to insure initial conditions conducive to an inflationary epoch, that is,
4) The thermal constraint. In renormalizable theories the high temperature potential is modified by quantum corrections that add a term $A\phi^2T^2$, $A \geq 0$, to the effective potential. In a non-renormalizable model the leading high temperature correction is of the form $f(\phi)T^2$ where $f(\phi)$ can be an arbitrary function of the inflaton field since the functional form of the tree Lagrangian is arbitrary. In many supergravity models the leading high temperature correction to the potential takes the form:[2]

$$\Delta V(\phi,T) \simeq \frac{T^2N}{12}\Big(V(T=0)+0(1/N)\Big),$$

where N is the total number of chiral supermultiplets ($N \gtrsim 50$ for a realistic model), so the minimum at high temperature is not significantly shifted from the true vacuum.

There are other problems[1] that must be avoided in a successful inflationary scenario, for example, an excessive production of gravitinos during inflation. In this talk I will focus primarily on points 1, 2, and 4. I will describe work[3] in collaboration with Binétruy, and related work[4-6] that has addressed the question as to whether the locally supersymmetric models suggested by string theories[7-9] have the right ingredients to satisfy these criteria. I will also comment on the problem of baryogenesis.
Superstring inspired models. These models possess several of the desired features for a successful inflationary cosmology. They contain[10,11] a number of gauge-singlet scalars that couple to ordinary matter with gravitational strength. Among these are a dilaton ϕ and the "breathing mode" e^σ which, together with axion-like fields β and h, form the complex scalars

$$\begin{aligned} S &= \phi^{-3/4}e^{3\sigma} + 3i\sqrt{2}h \\ T &= \phi^{3/4}e^{\sigma} - i\sqrt{2}\beta + k\,|y|^2 \end{aligned} \qquad (9)$$

of two chiral supermultiplets. In (9) $|y|^2 = \sum |y_i|^2$, $i = 1,\cdots N$, where the y_i are the scalar components of N chiral (gauge non-singlet) matter supermultiplets. Among the possible gauge singlets, the fields S and T in (9) have received the most attention in studies of phenomenology.

Superstring inspired models further embody naturally flat directions in the spin-zero field space. The scalar potential[10] is a sum of terms of the form

$$V(\phi) \ni \frac{1}{(S+S^*)}\frac{1}{(T+T^*-k\,|y|^2)^n}G(y) \qquad (10)$$

where $G(y) \sim y^4$ + higher order terms, and n is an integer: $n = 1-3$. $G(y)$ is either a D-term, implied by supersymmetry from gauge couplings, or an F-term, derived from a superpotential,[10] $W(y) \sim y^3+$ higher order. $G(y)$ is positive-semi-definite, so the minimum of the potential occurs for $y_i = 0$, with S and T undetermined.

In addition, the E_6 models[9] inspired by the heterotic[8] string include among the matter fields y singlets N of the observed gauge group $SU(3)_c\times SU(2)_L\times U(1)$ that have flat directions at least in the F-terms of the potential because it is derived from a superpotential $W(y)$ that does not allow terms $W \ni N^3$; such terms are forbidden by extra $U(1)$'s of the low energy gauge group.[9]

In a realistic model the vacuum degeneracy associated with these flat directions must be lifted because supersymmetry and any additional gauge $U(1)$'s must be broken. Supersymmetry could be broken by non-perturbative effects. In the heterotic string inspired models there is a hidden sector described by a pure $N = 1$ Yang Mills theory with E_8 or some sub-group thereof as gauge group. The hidden sector is infrared strong and its gauginos acquire a vev $<\lambda\lambda> \neq 0$ at some scale $\Lambda_c < m_{Plank}$; this vev is characterized by a parameter h. In addition the 10-dimensional theory includes a totally antisymmetric tensor that can have a nonvanishing vev $< H_{mnp} > \neq 0$ (where m,n,p are Lorentz indices in the compactified 6-space), characterized by a constant c that is in fact a quantized number that depends on the topology of the compactified manifold.

Either $c \neq 0$ or $h \neq 0$ breaks supersymmetry. When both of these parameters are nonvanishing, the additional term induced in the potential takes the form of a squared quantity such that the S-field (which enters because it couples to all gauge and gaugino fields) can acquire a vev such that the (tree level) potential still vanishes at its minimum. Specifically, the effective tree potential in four dimensions acquires a term:

$$V \ni \frac{1}{(S+S^*)}\frac{1}{(T+T^*-k\,|y|^2)^3}\times$$

$$\left|W(y)+c+h(1+\alpha)^{-\alpha/2}e^{-i\beta/2}\right|^2 \qquad (11)$$

where

$$\alpha = 3ReS/b_0, \qquad \beta = 3ImS/b_0 \qquad (12)$$

and b_0 determines the β-function of the hidden gauge group. The term (11) fixes the vev of S, but T remains undetermined at tree level, and so also the gravitino mass, proportional[13] to $(T+T^*)^{-3/2}$.

I will first assume that ReT, and the gravitino mass $m_{\tilde{G}}$, are somehow fixed and study the potential as a function of S. I will discuss later the possible determination of ReT. Setting T at its presumed ground state value and $y = <y> = 0$, we obtain the potential for the S-field.

$$V(\alpha,\beta) = \mu^4\frac{\alpha_0}{\alpha}\Big[1+\frac{1}{\hat{c}^2}(1+\alpha)^2 e^{-\alpha} - \frac{2}{\hat{c}}(1+\alpha)e^{-\alpha/2}\cos\beta/2\Big], \qquad (13)$$

where $\alpha_0 = 3/4\pi b_0\alpha_{GUT}$ and $\beta_0 = 4\pi n$ are the ground state values of α,β, Eq. (12), and we have chosen $c<0$ to avoid CP violation through the coupling[13] (ImS)$F\tilde{F}$ where F is the gauge field strength. $< ReS >$ determines the gauge coupling α_{GUT} at the scale of compactification through its coupling to the gauge-gaugino supermultiplet. The parameter $\hat{c}$ in (13) is the ratio of vevs:

$$\hat{c} = -c/h = (1+\alpha_0)e^{-\alpha_0/2} \qquad (14)$$

and the energy that might be stored in a false vacuum is governed by the scale μ:

$$\mu/m_P = \left[\frac{m_{\tilde{G}}}{m_P}\frac{1+\alpha_0}{\alpha_0}\right]^{1/2} \qquad (15)$$

which has the desired order of magnitude $\mu \sim 10^{-3}$ for $m_{\tilde{G}} \sim 10^{12}$ GeV and $\alpha_{GUT} \simeq 0.3$. This is a rather larger gravitino mass than in conventional SUSY and SUGRA models such as those considered in the previous talk by Ovrut, and one might worry about the stability of scalar masses. It has been shown[14,15] that in these models scalars remain massless at one loop. Gauginos apparently[16] acquire masses $m_{\tilde{g}} \sim m_{\tilde{G}}^3/16\pi^2 m_P^2$, suggesting that scalars should acquire similar masses at the 2-loop level that can be below a TeV for $m_{\tilde{G}}$ as high as 10^{12} GeV.

Is β the inflaton? Setting $\alpha = \alpha_0$, the potential (13) for β becomes

$$V(\alpha_0,\beta) = 4\mu^4\sin^2\beta. \qquad (16)$$

Finite temperature quantum corrections to (17) do not shift the position of its minimum[3], so the thermal constraint cannot be satisfied. If one assumes that some other mechanism stabilizes field β at an initial value $\beta_0 = 4\pi n + 2\pi$, one finds[3] for the number of e-foldings during which the vacuum energy density dominates ($\alpha_0 \geq 1$, see fig. 1):

$$N_e = -\frac{4}{\alpha_0^2}\ln\sin\left(\frac{\alpha_0}{\sqrt{6}}\frac{\mu^2}{m_P^2}\right) << 65.$$

Is α the inflaton? This possibility was briefly considered by Maeda and Pollock[4] who concluded that the potential did not satisfy the slow rollover[1] criteria. In ref. 3 the potential (13) was considered in more detail. Setting $\beta = \beta_0$, the potential (13) for α takes the form

$$V(\alpha,\beta_0) = \mu^4\frac{\alpha_0}{\alpha}[1-\frac{1}{\hat{c}}(1+\alpha)e^{-\alpha/2}]^2 \qquad (17)$$

which is plotted in Fig. 1 for representative values of $\hat{c}$. The potential (17) vanishes for $\alpha \to \infty$, a general feature[17] of potentials generated by nonperturbative SUSY breaking. Then the initial value of α could fall to the right of the maximum α_2 and subsequently roll off to infinity. One possibility for avoiding[18,5] this disaster is the existence of Q-balls[19] in the early universe, where Q is the charge associated with the nonhomogeneous transformation $s \to s + i\gamma$. Q-balls are spatial volumes with nonvanishing $< \partial_\mu\beta\partial^\mu\beta^\dagger >$ and hence net charge Q. Because of the non-canonical form of the kinetic energy:

$$\mathcal{L}_{KE} = \frac{1}{(S+S^*)^2}\partial_\mu S\partial^\mu S^* = \frac{1}{4\alpha^2}[(\partial\alpha)^2+(\partial\beta)^2] \qquad (18)$$

that couples α to the kinetic term for β, this vev induces[18] an effective potential for α, proportional to $Q(S+S^*)^2$ that can drive α to zero. This mechanism might suffice to fix the initial value at $\alpha_i > \alpha_0$.

The conventional mechanism for establishing initial conditions conducive to inflation is through quantum corrections to the finite temperature potential. In the model considered here, the finite temperature corrections to the potential are positive definite and of the form:[3]

$$\Delta V_T(\alpha, T) = \frac{T^2}{24}[(2N+N_G+16)V(\alpha, T=0)+0(1)] \quad (19)$$

where N_G is the number of gauginos. Since N and N_G are both large ($0(10^2)$) numbers the minimum of the potential cannot be significantly shifted at high temperature.

In the context of other inflationary supergravity models, Holman et al.[20] have stressed that scalars with interactions of only gravitational strength are thermally decoupled below the Planck mass down to temperatures well below that at which the energy of the false vacuum should dominate. They argue that such a system – in the present case S, T – should be treated as a closed system isolated from the rest of matter. If, following this line of reasoning, we compute the high temperature corrections neglecting the gauge and chiral degrees of freedom (although it seems difficult to argue that they would not reappear through precisely the quantum corrections we are considering) we obtain the result of eq. (19) with $N = N_G = 0$, in which case the $0(1)$ term can be competitive with the term proportional to $V(\alpha, T = 0)$. It turns out[3] that this truncated potential tends to stabilize the initial value α_i of α near the inflection point (Fig. 1b) or second local minimum (Fig. 1c), α_1 for the parameter range $0.937 \lesssim \hat{c} < 1$.

It is clear from inspection of Fig. 1 that, whatever the origin of the initial value α_i, the only chance for an acceptable inflationary scenario is for just this range of parameters with $\alpha_i \simeq \alpha_1$. If instead we assume $\alpha_i \simeq \alpha_2$, where α_2 is the position of the maximum $\alpha_2 > \alpha_0$ for any $\hat{c}$, domains of both $\alpha < \alpha_2$ and $\alpha > \alpha_2$ will form as the universe cools; the latter will expand indefinitely and ultimately dominate. For similar reasons chaotic[21] initial conditions would not appear to induce inflation for the potentials of Fig. 1.

We therefore assume

$$\hat{c} = .937 + \delta\hat{c} < 1, \quad \alpha_i = \alpha_1. \quad (20)$$

As we let $\hat{c} \to 1$, the energy stored in the false vacuum becomes too small to be useful and the barrier to be tunnelled across becomes dangerously large. The analysis of Hawking and Moss[22] shows that the effective mass $m_\alpha^2(\alpha_1) = \partial^2 V(\alpha)/\partial\alpha^2\Big|_{\alpha=\alpha_1}$ should be limited to

$$m_\alpha^2(\alpha_1) < 2H^2 = 2V(\alpha_1)/3m_P^2. \quad (21)$$

When the bound (21) is exceeded tunnelling is too slow for nucleation of the bubbles of true vacuum to occur and one encounters the difficulties of "old" inflationary models.[1] For $\hat{c} = 1, \alpha_1$ is an inflection point: $m_\alpha^2(\alpha_1) = 0$. As we turn on $\delta\hat{c} > 0, m_\alpha^2(\alpha_1) > 0$ and the bound (21) is satisfied only for

$$\delta\hat{c} < 3 \times 10^{-4} \quad (22)$$

which requires strong fine tuning of the potential. In fact $\hat{c}$ is not a parameter to be tuned. It depends on the topology of the compact manifold (c) and the on the hidden gauge group (h); Eq. (22) is the statement that successful inflation requires very specific values for these parameters.

Assuming optimistically that $\hat{c}$ is very close to the critical value .937, the number of e-foldings[1]

$$N_e = -\int_{\alpha_i}^{\alpha_e} \frac{3H^2(\alpha)}{2\alpha^2 V'(\alpha) m_P^2} d\alpha \quad (23)$$

can be calculated. Taking $\delta\hat{c} = 0$ and $\alpha_i = \alpha_1$ gives[3] $N_e > 65$ for $\mu/m_P \lesssim 0.14$, which is not a stringent constraint on μ. However, a successful inflationary scenario is still not established. Since $V(\alpha_1) > V(\alpha_2)$ in Fig. 1b, (or 1c if $\delta\hat{c} < 3\times10^{-4}$), there is a danger that α would overshoot the barrier at α_2 if the friction due to the expansion of the universe is insufficient to prevent the classically rolling field from reaching the value α_2.

Finally, we must consider the initial conditions that determine the shape of the zero-temperature potential itself. The gaugino condensate of the strongly coupled hidden gauge sector forms only below a critical temperature $T \simeq \Lambda_c$. On the other hand the parameter c can take only discreet values that can change by vacuum tunnelling. In Ref. 3 we followed Rohm and Witten[23] who argue that as long as $<\lambda\bar{\lambda}> \propto h = 0$, the H-vacuum chooses the lowest energy SUSY preserving configuration $<H>= 0$, and at $h \neq 0, <H>$ follows by tunneling so as to again achieve a vanishing vacuum energy. Thus in Ref. 3 it was assumed that $h = c = 0$ above the critical temperature $T_c \simeq \Lambda_c$ and $h, c \neq 0$ below T_c. When the gaugino condensate first forms there will be energy stored in the false H-vacuum, but presumably tunnelling will occur too rapidly to allow for significant inflation; otherwise one could encounter the nucleation and inhomogeneity problems of "old" inflation.

Ellis et al.[5] assume that $<H>$ is fixed at its present non-zero value above the critical temperature T_c and treat $< \bar{\lambda}\lambda >$ as a temperature dependent quantity. Then $|\hat{c}| = |c/\lambda|$ decreases with temperature so the potential has a time dependent shape that for $\hat{c}(T) < 1.21$ runs backwards from Fig. 1f towards Fig. 1a, ending at the potential for $\hat{c}(0) = \hat{c}$. They argue that for $c(T) = 1.21$ α would be stabilized at α_0' (Fig. 1f) and evolve through α_0' (Fig.1e) to α_1 (Fig. 1c) where it could be stabilized with nonvanishing vacuum energy before tunnelling à la Hawking and Moss[22] to the zero temperature ground state value α_0. Their scenario also requires $\hat{c}$ close to 0.937, with a similar fine tuning condition.

Is ReT the inflaton? To consider this possibility requires knowledge of the effective potential for the T-field that determines its ground state value and thus the gravitino mass. The vacuum degeneracy is lifted[14,15] by one loop quantum corrections which must be regularized by the introduction of a cut-off of order of the condensate scale Λ_c. In ref. 15 the effective one-loop potential was calculated neglecting terms of order $(m_i^2(\phi)/\Lambda_c^2)$ where $m_i^2(\phi)$ are field-dependent masses. This potential has a global minimum at a finite value of ReT which determines the relevant scales of the theory as a function of α_{GUT} and b_0. For a range of these parameters such that $m_{GUT} < m_{Planck}$, one finds[15] $m_{\tilde{G}} > 10^{15}$ GeV. The results of Ref. 16 suggest that this would entail gaugino masses $m_{\tilde{g}} \sim m_{\tilde{G}}^3/16\pi^2 \simeq 10^{-11} m_P \simeq 10^7$ GeV and hence a Higgs scalar mass $m_H \sim m_{\tilde{G}}\sqrt{\alpha_{weak}/4\pi} \gtrsim 10^6$ GeV. In addition, the parameter (15) that governs the scale of the false vacuum has in this case a value $\mu/m_P \gtrsim 0.05$ for $\hat{c} = .937(\alpha_0 = 2.8)$ which would lead to excessive density fluctuations for the S-inflaton scenario. It was subsequently argued[24] that, since the values $m(\phi)$ at the minimum are comparable to Λ_c, one should keep the full cut-off dependence in the effective potential. It turns out that the $O(m^2/\Lambda_c^2)$ terms destabilize* the potential in the direction $ReT \to 0$, $m_{\tilde{G}} \to \infty$. Since when these corrections are included there is no longer even a local minimum in the region $m(ReT) < \Lambda_c$, the potential for ReT must be considered as unknown at present.

Maeda et al.[6] nevertheless assumed an effective potential

$$V(\phi) = V_{Tree}(\phi) + V_{BG}(\phi) + \text{constant.} \qquad (24)$$

*However, contributions from loop momenta $\Lambda_c^2 \lesssim |p|^2 \lesssim m_{GUT}^2$ may restabilize the potential[25].

where V_{BG} is the one loop contribution calculated in Ref. 15 that has a negative cosmological constant which is cancelled by the constant added in (24). Maeda et al. argue that as long as $S \propto (\alpha,\beta) \neq (\alpha_0,\beta_0)$, $V_{Tree}(\phi)$ is nonvanishing and will drive (c.f. eq. (11)) ReT towards infinity. Once $< S >$ has settled at its ground state value, ReT will roll to its ground state. They find that the conditions for successful inflation are satisfied for a broad range of the parameters c and h.

However, as the authors of Ref. (6) note, the ansatz (24) does not satisfy the general result[17] that the potential should vanish as $ReT \to \infty$.

Is a matter field (N) the inflaton? In the heterotic string inspired models matter fields fill fundamental 27 representations of E_6 that include two singlets (N and N' or ν^c that I will generically call N) of the observed gauge group. The renormalizable F-terms have flat directions for $N \neq 0$, but the D-terms are not flat in the $N \neq 0$ direction if all matter fields transform as 27's. However, the effective low energy theory may contain 27 + *overline*27 pairs in addition to the 27's that comprise the standard families. The direction $N = \bar{N} \neq 0$ is a flat direction for the D-term of the potential as well as of the F-term. This degeneracy should be lifted by effective nonrenormalizable terms in the superpotential that arise[26] when one integrates out the heavy degrees of freedom of the theory. These give an effective potential of the form:

$$\begin{aligned} V_{eff} &\sim m^2(|N|^2 + |\bar{N}|^2) \\ &+ \lambda_{eff}^2(|N\bar{N}|)^{2n-2}(|N|^2 + |\bar{N}|^2) \\ &+ D - \text{terms} + \cdots \end{aligned} \qquad (25)$$

with $n \geq 2$. The potential (25) includes a SUSY breaking mass term for scalars that is probably[16] of order $m \sim m_{\tilde{G}}^3/16\pi^2 m_P^2$. The squared mass parameter has a logarithmic dependence on the field strength $|N|$ and, for suitable N-couplings, approaches a negative fixed point $-m_0^2$, triggering a vev for $N(\bar{N})$. Since one expects all scalars to have comparable SUSY breaking masses, we require $m_0^2 \simeq (m_W^2$ to $(1\ \text{TeV})^2)$ if the Higgs mass is to be in the range required for the observed Higgs vev without strong couplings. This means that a $U(1)$ gauge invariance is broken at a critical temperature $T_c \sim (m_W - 1\ \text{TeV})$, while quantum corrections stabilize the field at $N = 0$, for $T > T_c$. The onset of inflation occurs when the vacuum energy density dominates over the radiation density, i.e. for $T_i \sim (V_0)^{1/4} \simeq \mu_0$, and inflation continues until the critical density T_c for the symmetry breaking phase transition is reached. Since the temperature drops exponentially during inflation ($T \simeq T_i e^{-Ht}$) the number of e-foldings is given simply by [3]

$$N_e \simeq \ln(T_i/T_c) \simeq \ln(\mu_0/m_W) \simeq 6 - 16 \qquad (26)$$

for $n_0 \geq 2$, which is insufficient to solve the flatness and horizon problems. Another possibility is that the flat directions are lifted only by non-perturbative effects on the string world sheet[27] in which case the effective coupling λ_{eff} in (25) is of order $\exp(-2 < ReT >)$, increasing[3] the number of e-foldings (26) by about $1/4 < \mathrm{Re}\,T >$ which is still insufficient for plausible values of $< \mathrm{Re}\,T >$.

If the inflaton field decays rapidly after the end of inflation the fractional entropy increase is related to the number of e-foldings by $s_f/s_i = e^{3N_e} \simeq 10^8 - 10^{20}$ for the range (26). However, since with the above choice of parameters N is in fact weakly coupled to ordinary matter, the entropy release can in fact be considerably larger.[28] Such a late ($T \lesssim T_c \sim m_W$) entropy release has raised the issue[28,29] as to whether the phase transition associated with an intermediate scale may cause serious problems for standard cosmology.

It has recently been pointed out[28,30] that the phase transition will actually occur at a much higher temperature. The one-loop quantum corrections to the finite temperature potential can be expressed as $\Delta V(\phi, T) = \sum_i v_i(m_i(\phi), T)$ where the index i runs over all spin and internal symmetry degrees of freedom and $m_i(\phi)$ is the appropriate field-dependent mass. For $T_i^2 >> m_i^2(\phi)$ this expression reduces to the usual high temperature approximation

$$v_i = -\frac{T^4}{90}\pi^2(1 - \frac{1}{8}[1 - (-)^{F_i}]) + \frac{T^2}{24}m_i^2(\phi), \quad (27)$$

where F_i is fermion number. The point is[28,30] that (27) is an approximation valid only in the high temperature limit, while the exact expression falls rapidly to zero as soon as $m_i^2(\phi) > T^2$. For fixed T the potential possesses a barrier enclosing the false vacuum (the origin for the N-fields) and coincides with the zero temperature potential for values of the scalar fields ϕ such that $m_i^2(\phi) >> T$. As the temperature decreases the position of the barrier moves closer to the field origin. As soon as it is close enough that thermal fluctuations are sufficiently large ($\Delta\phi = 0(\phi_{barrier})$) to "spill" the field ϕ over the barrier with a significant probability, domains of true vacuum will form. If this "spill-over" occurs during an inflationary epoch (caused presumably by some other scalar sector; in this case $\Delta\phi \sim H$) domains of true N-vacuum could grow large enough to encompass our present observable universe with no need for a later phase transition.[30]

Other gauge-singlet fields? The models inspired by superstrings contain additional structurerelated gauge singlet chiral supermultiplets namely[23] b_{11} fields $T^{(k)} = \alpha^{(k)} + +i\beta^{(k)}$ and b_{21} fields $C^{(\alpha)}$ where b_{11} and b_{21} are topological numbers. One of the b_{11} $T^{(k)}$ fields is the T-field discussed above for which there is no superpotential at tree level and, in fact, all the $\beta^{(k)}$ are axion-like objects, implying that, at tree level at least, there can be no superpotential for the $T^{(k)}$ fields. On the other hand, the ground state values of the $C^{(\alpha)}$ are determined by $< H >$ which fixes the cohomology class of the the complex structure of the compact manifold, so there is a tree level potential for these fields. In the absence of a model for the shape of the potential, this possibility cannot be analyzed further.

Baryogenesis. In superstring inspired models, the baryon number violating X-bosons, responsible for baryon decay in standard GUTs, have masses of the order of the Planck scale. They cannot play any role in generating a net baryon number for the observed universe if there is inflation in the epoch $T < m_P$ where the effective four-dimensional theories are valid. These four-dimensional theories allow two distinct scenarios: one with an "intermediate scale" associated with lifting the flat direction in N-space, and one in which there is no intermediate scale, with $< |N| >$ of the order (albeit higher by a factor of at least 2-3) of the electroweak scale.

In models inspired by the heterotic string left handed components of matter fields fall into 27-plets of E_6: $27_L = (\bar{5} + 10 + 5 + \bar{5} + 1 + 1)_L$ under $SU(5)$. The $\bar{5} + 10$ represent ordinary quarks and leptons and their sfermion partners; $5 + \bar{5}$ contain the SUSY standard model Higgs doublets $H, \bar{H}$ plus their color triplet, $SU(2)_L$ singlet, counterparts $(g, \bar{g})$, and two SU(5) singlets (N). The superpotential must contain a term $W(\phi) \ni g\bar{g}N$ that gives a mass to the color triplets g when N acquires a vev. In addition, there are E_6 allowed terms $W(\phi) \ni gqq + gq\ell + \cdots$ that when combined can generate proton decay. If there is no intermediate scale, $m_g \sim m_W$, discreet symmetries in the effective low energy renormalizable theory must be invoked to forbid couplings which would allow fast nucleon decay. Then one finds[31] that observable proton decay can proceed only if there are extra $\overline{27}$'s as well as yet more gauge singlets[11] (remnants of the E_8 gauge fields). This picture allows little prospect for post-inflationary baryogenesis via the standard mechanism of decaying color-triplet, super-heavy bosons.

If there is an intermediate scale with large vevs for the N-fields yielding very massive $g, \bar{g}$, their baryon/lepton number violating couplings can be allowed at tree level. Then the scalar members of these supermultiplets can play much the same role as the color-triplet Higgs of ordinary (or SUSY) GUTs, except that their couplings are not constrained by low energy data. Unless the couplings responsible for nucleon decay are somehow suppressed, limits on the nucleon lifetime probably require that they be too heavy to have been produced in sufficient abundance for efficient baryogenesis during postinflationary reheating. The reheating temperature is bounded by $T_R \lesssim \mu \lesssim 10_P^{-3} \simeq 10^{15}$ GeV, and is expected to be considerably lower for a very weakly coupled inflaton whose oscillations are strongly red-shifted before it can decay.

One can forbid nucleon decay by imposing lepton number conservation. Then baryon number violating processes could be mediated by objects sufficiently light to be copiously produced during post inflationary reheating. The problem is then to generate the out-of-equilibrium abundance necessary for their decays to generate a baryon asymmetry. A possible mechanism noted recently[32] is the dependence of the g-mass on the value of $< N >$. For $N = 0$ the g's are massless, with a mass that increases as N rolls towards its true ground-state value, and oscillates as the field N oscillates and decays. If the N-phase transition occurs after or near the end of inflation, an out of equilibrium g-distribution could be generated.

The difficulties for efficient baryogenesis have led some authors to consider alternatives that do not require explicit baryon number violating couplings. One approach[33] is based on the observation that instanton effects related to the anomalous baryon number and lepton number currents in the standard model are unsuppressed at high temperature and could have been important in the hot early universe. Another suggestion[34] is that immediately after inflation squarks and leptons may have had large vevs; a significant baryon asymmetry could be generated as this system evolves to the true vacuum state.

Baryogenesis could proceed according to conventional mechanisms if inflation took place at a temperature $T \gtrsim m_P$, prior to or during compactification. This possibility has been studied[4] but a convincing mechanism has not yet been found within the superstring context. If there is not an inflationary epoch after compactification, one is faced with the usual graviton, gravitino, monopole and domain wall (in particular those topological defects arising from compactification) problems.

Acknowledgments

I have enjoyed fruitful discussions with Pierre Binétruy and Graham Ross. This work was supported by the Director, Office of Energy Research, Office of High Energy and Nuclear Physics, Division of High Energy Physics of the U.S. Department of Energy under Contract DE-AC03-76SF00098 and the National Science Foundation under Grant PHY-85-15857.

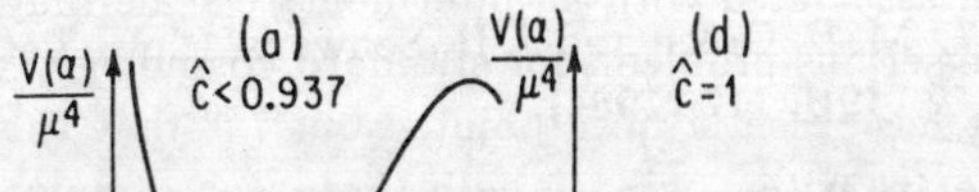

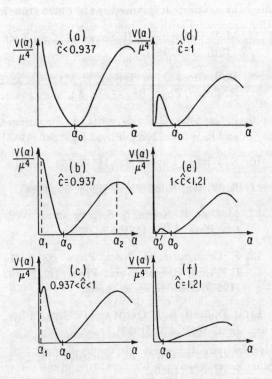

Fig. 1: Shape[6] of the potential of Eq. (18) for different values of $\hat{c}$, Eq. (15).

References

1. For a recent review with references to the original literature, see P. Binétruy, *Supergravity and Inflation*, in Proc. of the 6th Workshop on Grand Unification, Minneapolis, April 18–20, 1985 (to be published) – Preprint LBL-20092 (1985).

2. K. A. Oliver and M. Srednicki, Phys. Lett. **148B**, 437 (1984); P. Binétruy and M. K. Gaillard, Nucl. Phys. B254, 388 (1985) and Phys. Rev. D32, 931 (1985); K. Enqvist et al., Phys. Lett. 152B, 181 (1985).

3. P. Binétruy and M. K. Gaillard, Chicago-Berkeley preprint EFI86-30, LBL-21621, UCB-PTH-86/15-(1986), to be published in Phys. Rev. D.

4. K. Maeda and M. D. Pollock, Phys. Lett. B173, 251 (1986) and references therein.

5. J. Ellis et al., CERN preprint CERN-TH. 4325/85(1985).

6. K. Maeda, M. D. Pollock and C. E. Vayonakis, Trieste ICTP preprint IC/86/5 (1986).

7. M. B. Green and J. H. Schwarz, Phys. Lett. 149B, 117 (1984).

8. D. J. Gross, J. A. Harvey, E. Martinec and R. Rohm, Phys. Rev. Lett. 54, 502 (1985).

9. P. Candelas, G. T. Horowitz, A. Strominger and E. Witten, Nucl. Phys. B258, 46 (1985).

10. E. Witten, Phys. Lett. 155B, 151 (1985).

11. E. Witten, Nucl. Phys. B268, 79 (1986).

12. M. Dine, R. Rohm, N. Seiberg and E. Witten, Phys. Lett. 156B 55 (1985).

13. E. Cremmer et al., Nucl. Phys. B212, 413 (1983); J. Bagger, Nucl. Phys. B211, 302 (1983), and references therein.

14. J. D. Breit, B. A. Ovrut and G. Segré, Phys. Lett. 162B, 303 (1985).

15. P. Binétruy and M. K. Gaillard, Phys. Lett. 168B, 347 (1986).

16. P. Binétruy, S. Dawson and I. Hinchliffe, Preprint LBL-21606, EFI-86-28 (1986).

17. I. Affleck, M. Dine and N. Seiberg, Nucl. Phys. B256, 557 (1985).

18. K. Enqvist, E. Papantonopoulos and K. Tamvakis, Phys. Lett. 165B, 299 (1985).

19. S. Coleman, Nucl. Phys. B262, 263 (1985).

20. R. Holman, P. Ramond and G. G. Ross, Phys. Lett. 137B, 343 (1984).

21. A. D. Linde, Phys. Lett. 129B, 177 (1983).

22. S. Hawking and I. Moss, Nucl. Phys. B224, 180 (1983), and references therein.

23. R. Rohm and E. Witten, Princeton preprint *The Antisymmetric Tensor Field in Supersymmetric Theory* (1985).

24. M. Quirós, Phys. Lett. B173, 265 (1986); Y. J. Ahn and J. D. Breit, Nucl. Phys. B273, 75 (1986).

25. P. Binétruy, S. Dawson, M. K. Gaillard and I. Hinchliffe, in preparation.

26. M. Dine, V. Kaplunovsky, M. Mangano, C. Nappi and N. Seiberg, Nucl. Phys. B259, 549 (1985).

27. M. Dine, N. Seiberg, X. G. Wen and E. Witten, *Non-Perturbative Effects on the String World Sheet*, Princeton preprint (April 1986).

28. K. Yamamoto, Phys. Lett. 168B, 341 (1986).

29. K. Enqvist, D. V. Nanopoulos and M. Quiros, Nucl.Phys. B262, 556 (1985).

30. O. Bertolami and G. G. Ross, preprint in preparation.

31. B. Campbell et al., CERN preprint TH.4449/8 (1986).

32. G. Lazarides, C. Panagiotakopoulos and Q. Shafi, Phys. Rev. Letters 56, 557 (1986).

33. M. E. Shaposhnikov, "Baryon Asymmetry of the Universe in the Standard electroweak theory", preprint in preparation (Aspen and INR, Moscow, 1986), and references therein.

34. I. Affleck and M. Dine, Nucl. Phys.B249, 361 (1983).

Parallel Session 6

General Properties of Field Theory

Organizers:
C. Nappi (Princeton)
J. Zinn-Justin (Saclay)

Scientific Secretaries:
M.G. Cleveland (LBL)
R. Ingermanson (LBL)

VECTOR MESONS IN THE SKYRME MODEL

Gregory S. Adkins

Franklin and Marshall College
Lancaster, PA 17604 USA

The Skyrme model can be made more realistic by the inclusion of vector mesons along with pions in the effective Lagrangian. Several such models are discussed here. These models are evaluated on the basis of three criteria: meson sector physics, single nucleon properties, and the nucleon-nucleon interaction.

1. INTRODUCTION

The Skyrme model is a model of baryons as solitons in a mesonic field theory. The simplest such model, studied by Skyrme in 1961 [1,2], has a Lagrangian

$$L = \frac{F_\pi^2}{16} \text{tr}[\partial_\mu U \partial^\mu U^+] + \frac{1}{32e^2} \text{tr}[[U^+\partial_\mu U, U^+\partial_\nu U]^2], \quad (1)$$

where $U = \exp[2i\vec{\tau}\cdot\vec{\pi}/F_\pi]$ is a SU(2) matrix describing the pion field, F_π is the pion decay constant (experimentally F_π=186 MeV), and e is a dimensionless coupling constant. The term in (1) having just two derivatives is the Lagrangian of the non-linear sigma model. The second, four derivative, term was added by Skyrme to balance the attractive nature of the two derivative term and allow for finite energy soliton solutions to the classical equations of motion. The Skyrme soliton has a hedgehog structure $U_0 = \exp[i\vec{\tau}\cdot\hat{x}\, F(r)]$ where the shape function F(r) varies from π at r=0 to 0 at r=∞. Such solitons can be quantized as fermions [3], and have unit baryon number [4,5]. In the Skyrme model these solitons are the baryons.

My topic is the extension of the Skyrme model to include vector mesons. We know that low mass vector mesons are important in the understanding of the nucleon-nucleon (N-N) force; therefore they should be important in the understanding of individual nucleons [6]. I will consider generalizations of the two-flavor Skyrme model that incorporate in various ways the ρ, ω, and A_1 mesons [7].

The success of an extended Skyrme model can be tested in at least three ways. The model should be faithful to the meson physics; it should reproduce more or less accurately the meson scattering and decay phenomena. The extended model should give single nucleon properties accurately, with errors smaller than the 30% errors of older versions of the Skyrme model [8,9]. Also, the extended model should yield a N-N interaction potential that has the main features of the physical potential.

Before I discuss particular generalizations of the Skyrme model I would like to say a word about the calculation of the N-N potential in the Skyrme model. The standard method [10,11] involves using the "product ansatz" for the B=2 field

$$U(\vec{x};\vec{x}_1\vec{x}_2) = A_1\, U_0(\vec{x}-\vec{x}_1)\, A_1^+ \times A_2\, U_0(\vec{x}-\vec{x}_2)\, A_2^+ . \quad (2)$$

This is taken to describe two standard Skyrmions, one at $\vec{x}_1$ and one at $\vec{x}_2$, with a relative isospin orientation of $A_1 A_2^+$. (Some problems with this ansatz are discussed in Ref. 12.) The resulting potential is a sum of central, spin-spin, and tensor pieces:

$$V(\vec{r}) = V_C(r) + \vec{\tau}_1\cdot\vec{\tau}_2\, [\vec{\sigma}_1\cdot\vec{\sigma}_2\, V_{SS}(r) + S_{12}\, V_T(r)] , \quad (3)$$

where

$$\vec{r} = \vec{x}_1-\vec{x}_2 ,$$

$$S_{12} = (\vec{\sigma}_1\cdot\hat{r})\, (\vec{\sigma}_2\cdot\hat{r}) - (\vec{\sigma}_1\cdot\vec{\sigma}_2)/3 . \quad (4)$$

The components of this potential can be compared with corresponding components of a phenomenological potential such as the "Paris potential" [13]. The main phenomenological problem with the N-N potential in the standard Skyrme model is the lack of medium range (1-2 fm) attraction in the central component. This medium range central attraction is the glue that holds nuclei together, and its absence is a serious problem.

2. EXTENDED SKYRME MODELS

I will consider three types of extended Skyrme models: the "omega stabilization" model of Adkins and Nappi [14], the "modified Skyrme model" of Jackson, Jackson, Goldhaber, Brown, and Castillejo [15] and Kanazawa, Momma, and Haruyama [16], and more complete models proposed by Lacombe, Loiseau, Vinh Mau, and Cottingham [17], and by Meissner and Zahed [18,19].

The Lagrangian of the omega stabilization model (with massive pions) is [14]

$$L = \frac{F_\pi^2}{16}\,\mathrm{tr}[\partial_\mu U \partial^\mu U^+] + \frac{1}{8}\,m_\pi^2\,F_\pi^2\,(\mathrm{tr}[U]-2)$$

$$-\frac{1}{4}\,\omega_{\mu\nu}\omega^{\mu\nu} + \frac{1}{2}\,m_\omega^2\,\omega_\mu\omega^\mu + \beta\,\omega_\mu B^\mu \qquad (5)$$

where $\omega_{\mu\nu} = \partial_\mu\omega_\nu - \partial_\nu\omega_\mu$. In this model the critical function of soliton stabilization is performed by a dynamical omega field ω_μ coupled to baryon number current B_μ. The omega field mediates a repulsion between two bits of nucleonic matter much as the photon field does between two objects of like charge. The omega stabilization model is incomplete in its meson content. The most important omission is the rho meson, whose mass is lower than that of the omega. Consequently in the omega stabilization model the $\omega\to\pi\pi\pi$ coupling is taken to be a point coupling ($\omega_\mu B^\mu$). Physically this coupling is mainly mediated by the rho: $\omega\to\rho\pi\to\pi\pi\pi$ [20]. The single baryon properties of the omega stabilized model are uniformly better than those of the standard (massive pion) Skyrme model (see Table I), although F_π and g_A are still much too low. The N-N potential in the omega stabilized model is similar to that in the standard Skyrme model--with no sign of medium range central attraction [21]. An advantage of this model is that the negative G-parity of the omega mediated central force leads to central attraction between nucleons and antinucleons as physically required. The Skyrme quartic term on the other hand has positive G-parity, and leads to $N\bar{N}$ repulsion [15].

PHYSICAL QUANTITY	STANDARD SKYRME	OMEGA STABILIZATION	JJGBC KMH	LLVC	MZ	EXPERIMENT
M_N (MeV)	INPUT	INPUT	INPUT	INPUT	973	938.9
M_Δ (MeV)	INPUT	INPUT	1148	INPUT	1279	1232
F_π (MeV)	108	124	INPUT	142	INPUT	186
g_A	0.65	0.82	INPUT	0.62	1.09	1.23
$\langle r^2\rangle^{1/2}$ (fm)	0.68	0.74	0.73	0.66	0.67	0.72
μ_p	1.97	2.34	2.55	2.33	2.27	2.79
μ_n	-1.24	-1.46	-1.95	-1.63	-1.50	-1.91
$g_{\pi NN}$	11.9	13.0	12.4	—	—	13.5

Table I: Numerical results obtained in various versions of the Skyrme model for selected physical quantities. "Standard Skyrme" refers to the massive pion model of Adkins and Nappi [9], "omega stabilization" is the omega stabilized model of Adkins and Nappi [14], JJGBC/KMH is the "modified Skyrme model" of Jackson, Jackson, Goldhaber, Brown, and Castillejo [15] and Kanazawa, Momma, and Haruyama [16], LLVC is the model of Lacombe, Loiseau, Vinh Mau, and Cottingham [17], and MZ is the model of Meissner and Zahed [18,19]. The charge radius displayed here is the isoscalar electric charge radius.

In the omega stabilized model the omega particle appears as an elementary field in the Lagrangian. The effects of such relatively heavy mesons can be included in an alternate way, through an effective Lagrangian. In the omega stabilized model the equation for the omega field is

$$\partial_\mu (\partial^\mu\omega^\nu - \partial^\nu\omega^\mu) + m_\omega^2 \omega^\nu + \beta B^\nu = 0 . \tag{6}$$

In the "infinite mass" approximation, wherein omega field derivatives are neglected relative to the omega mass, one has

$$\omega^\mu \longrightarrow (-\beta/m_\omega^2) B^\mu , \tag{7}$$

and for the omega dependent terms in the Lagrangian [15]

$$L \longrightarrow L_\omega = -\frac{1}{2} (\beta^2/m_\omega^2) B_\mu B^\mu . \tag{8}$$

In general one can write an effective pion Lagrangian as

$$L = L_\pi + L_\rho + L_\sigma + L_\omega + \cdots , \tag{9}$$

where

$$L_\pi = (F_\pi^2/16) \operatorname{tr}[\partial_\mu U \partial^\mu U^+] + (m_\pi^2 F_\pi^2/8)(\operatorname{tr}[U]-2) , \tag{10a}$$

$$L_\rho = (1/32e^2) \operatorname{tr}[[U^+\partial_\mu U, U^+\partial_\nu U]^2] , \tag{10b}$$

$$L_\sigma = (\gamma/8e^2)(\operatorname{tr}[\partial_\mu U \partial^\mu U^+])^2 . \tag{10c}$$

Here L_π is the Lagrangian of the basic (massive pion) non-linear sigma model, the quartic Skyrme term L_ρ is associated with the rho meson [22,23], L_σ is associated with a scalar-isoscalar "sigma" meson [23,24,25], and L_ω is the omega contribution [15]. This effective Lagrangian is discussed further in Refs. 24 and 26.

In the "modified Skyrme model", Jackson, Jackson, Goldhaber, Brown, and Castillejo [15], and Kanazawa, Momma, and Haruyama [16] consider the Lagrangian

$$L = L_\pi + L_\rho + L_\omega \tag{11}$$

(with m_π=0). Stabilization in this model is provided by L_ω. The Skyrme term L_ρ is taken with the "wrong sign" $e^2<0$, so that it has a destabilizing (attractive) effect. For a proper choice of coupling constants this model has a N-N central potential of the desired form: strongly repulsive at short range with a region of weak attraction at medium range [15]. The single nucleon properties of this model work out quite well (see Table I) [16], helped perhaps by the fact that there are three quantities taken as "input" values instead of the usual two. The price for these successes is paid in the meson sector, where the "wrong sign" of L_ρ is inconsistent with $\pi\pi$ scattering data [23].

The most ambitious models to date are those of Lacombe, Loiseau, Vinh Mau, and Cottingham (LLVC) [17], and Meissner and Zahed (MZ) [18,19]. Both of these models include the π, ω, ρ, and axial-vector A_1 mesons. Both incorporate the ρ and A_1 as gauge particles of a $SU(2)_L \times SU(2)_R$ chiral symmetry. One difference between the models is that LLVC uses a phenomenological ("input") value for the $\omega_\mu B^\mu$ coupling, while MZ has a more complicated coupling of ω to other fields derived from the Wess-Zumino term. The ω in LLVC does not couple directly to $\rho\pi$, apparently a weakness in that model. The single nucleon results of LLVC are reasonable, but do not improve significantly on those of the "omega stabilization" model. The MZ single nucleon results are more impressive because there are fewer free parameters in the MZ approach. A special feature of the LLVC model is the inclusion of a dynamical "sigma" field in the Lagrangian with a mass of 800 MeV. This field is introduced to provide an attractive medium range N-N central force. In a simpler model containing such a "sigma" meson the desined attraction was in fact found [25], although this finding has been challenged [27].

3. CONCLUSION

There are two easily identified ways to improve on the original Skyrme model: one can model the meson theory more faithfully, and given a model of meson physics one can push farther the semiclassical approximation scheme for soliton properties. The first type of improvement was the topic of this discussion. Models of the class proposed by Lacombe, Loiseau, Vinh Mau, and Cottingham [17] and by Meissner and Zahed [18,19] (and also more recently by Chemtob [28]) contain a realistic description of meson physics and should be studied more completely. An example of the latter type of correction is the strong coupling between rotational (spin-isospin) and vibrational ("breathing") modes of the soliton [29]. The vibrational modes were not considered in the original Skyrme model discussions even though they are formally of lower order in the semiclassical expansion than rotational modes. They will have to be included in order to produce a quantitative mass spectrum of low-lying baryons.

This work was supported by the National Science Foundation under Grant No. PHY-8608590.

REFERENCES

1. T. H. R. Skyrme, Proc. R. Soc. London A260, 127 (1961).
2. T. H. R. Skyrme, Nucl. Phys. 31, 556 (1962).
3. D. Finkelstein and J. Rubinstein, J. Math. Phys. 9, 1762 (1968).
4. A. P. Balachandran, V. P. Nair, S. G. Rajeev, and A. Stern, Phys. Rev. Lett. 49, 1124 (1982); Phys. Rev. D 27, 1153 (1983).
5. E. Witten, Nucl. Phys. B223, 422 (1983); B223, 433 (1983).
6. This point was emphasized to me by G. E. Brown.
7. This program was proposed by E. Witten in *Solitons in Nuclear and Elementary Particle Physics*, edited by A. Chodos, E. Hadjimichael, and C. Tze (World Scientific, Singapore, 1984), pp. 306-312.
8. G. S. Adkins, C. R. Nappi, and E. Witten, Nucl. Phys. B228, 552 (1983).
9. G. S. Adkins and C. R. Nappi, Nucl. Phys. B233, 109 (1984).
10. A. Jackson, A. D. Jackson, and V. Pasquier, Nucl. Phys. A432, 567 (1985).
11. R. Vinh Mau, M. Lacombe, B. Loiseau, W. N. Cottingham, and P. Lisboa, Phys. Lett. 150B, 259 (1985).
12. H. M. Sommermann, H. W. Wyld, and C. J. Pethick, Phys. Rev. Lett. 55, 476 (1985).
13. M. Lacombe *et al.*, Phys. Rev. C 21, 861 (1980).
14. G. S. Adkins and C. R. Nappi, Phys. Lett. 137B, 251 (1984).
15. A. Jackson, A. D. Jackson, A. S. Goldhaber, G. E. Brown, and L. C. Castillejo, Phys. Lett. 154B, 101 (1985).
16. A. Kanazawa, G. Momma, and M. Haruyama, Phys. Lett. 172B, 403 (1986).
17. M. Lacombe, B. Loiseau, R. Vinh Mau, and W. N. Cottingham, "An effective Lagrangian for low energy hadron physics", (unpublished), 1986.
18. U.-G. Meissner and I. Zahed, Phys. Rev. Lett. 56, 1035 (1986).
19. U.-G. Meissner and I. Zahed, "Nucleons from skyrmions with vector mesons", Regensburg Report No. TPR-86-14, 1986.
20. M. Gell-Mann, D. Sharp, and W. G. Wagner, Phys. Rev. Lett. 8, 261 (1962).
21. J. M. Eisenberg, A. Erell, and R. R. Silbar, Phys. Rev. C 33, 1531 (1986).
22. K. Iketani, "Current algebraic origin of the Skyrme term", Kyushu Report No. KYUSHU-84-HE-2, 1984.
23. T. N. Pham and T. N. Truong, Phys. Rev. D 31, 3027 (1985).
24. M. Mashaal, T. N. Pham, and T. N. Truong, Phys. Rev. Lett. 56, 436 (1986).
25. M. Lacombe, B. Loiseau, R. Vinh Mau, and W. N. Cottingham, Phys. Lett. 169B, 121, (1986).
26. I. J. R. Aitchison, C. M. Fraser, and P. J. Miron, Phys. Rev. D 33, 1994 (1986).
27. G. Kälbermann, J. M. Eisenberg, R. R. Silbar, and M. M. Sternheim, "Absence of attraction in the NN central potential derived from skyrmions", (unpublished), 1986.
28. M. Chemtob, "Skyrmion-baryon phenomenology in the effective gauged chiral-SU(2) action approach", Saclay Report No. PhT 86-113, 1986.
29. L. C. Biedenharn, Y. Dothan, and M. Tarlini, Phys. Rev. D 31, 649 (1985).

Dynamical Gauge Bosons of Hidden Local Symmetry

M. Bando

Department of Physics, Kyoto University
Kyoto 606, Japan

The concept of hidden local symmetry (HLS) which exists in any non-linear sigma model is explained. It is suggested that the dynamical realization of composite gauge fields is a rather common phenomenon. First we investigate the structure of HLS in the general case of G/H and examine the phenomenological consequences of dynamical realization of HLS. Secondly we apply the above framework to hadron physics and identify the vector mesons (ρ, K* ...) with hidden gauge bosons. The key relation is the famous KSRF relation. We find that the new concept of H.L.S. can shed light on the old hadron physics.

1. Introduction(1)

Growing attention has recently been paid to non-linear sigma model in various contexts of the modern particle physics. Among the most attractive features is the hidden local symmetry. We know that any non-linear sigma model based on the manifold G/H is described by the gauge fixed version of the "linear" model of $G_{global} \times H_{local}$ symmetry(2). This H_{local} group is called "hidden local symmetry", the corresponding gauge bosons are no more than auxiliary fields at the classical level. However we encounter a completely different situation if such gauge bosons acquire kinetic terms via quantum effect or as a result of more fundamental dynamics. Dynamical calculation suggests that there exist the cases where the s-matrix developes poles corresponding to these gauge bosons.

Motivated by the development of theoretical arguments of dynamical pole-generation of hidden gauge bosons, it now becomes interesting to investigate another aspect, namely its phenomenological consequences. Furthermore, we can check our idea in hadron phenomena. We shall see that such a phenomenon does occur in hadron physics, where the corresponding gauge bosons are vector mesons.

I hope my talk will suggest how old hadron phenomenology can be reexamined in the light of new concept of hidden local symmetry.

2. Hidden local symmetry in non-linear realization

Let us consider the system where the symmetry group G is spontaneously broken down to its subgroup H. For simplicity we assume that G and H are simple groups, where the set of generators T_A of G are devided into two parts, those of unbroken subgroup H, the parallel part and the rest, the part perpendicular to it, i.e.,

$$\{T_A\} = \{ S_\alpha \in \mathcal{H}, \quad X_a \in \mathcal{G}-\mathcal{H} \}, \qquad (2.1)$$

which are chosen so as to satisfy

$$\mathrm{Tr}(T_A T_B) = \tfrac{1}{2}\delta_{AB}, \quad \mathrm{Tr}(S_\alpha X_a) = 0 . \qquad (2.2)$$

The non-linear Lagrangian is written in terms of the Nambu-Goldstone modes, $\xi(\pi)$, valued on the coset manifold G/H, the perpendicular space, with f_π being decaying constant,

$$\xi(\pi) = e^{i\pi(x)/f_\pi}, \quad \pi(x) = \pi^a X_a . \qquad (2.3)$$

The $\xi(\pi)$ transform non-linearly under the group G, $g \in G$,

$$\xi(\pi) \to \xi'(\pi) = h(\pi(x), g^+)\,\xi(\pi)\, g^+ . \qquad (2.4)$$

Our task is to introduce the compensating fields σ with values on the parallel space, with another parameter f_σ,

$$\xi(\sigma) = e^{i\sigma(x)/f_\sigma}, \quad \sigma(x) = \sigma^\alpha(x) S_\alpha , \qquad (2.5)$$

in such a way that the variable $\xi(x)$ defined as,

$$\xi(x) = \xi(\sigma)\cdot\xi(\pi), \qquad (2.6)$$

transforms linearly under the group $G_g \times H_\ell$, i.e., under $h(x) \in H_\ell$, $g \in G_g$,

$$\xi(x) \to \xi'(x) = h(x)\,\xi(x)\, g^+ \qquad (2.7)$$

It is easy to find the transformation property of $\xi(\pi)$ from eqs.(2.4) and (2.7),

$$\xi(\sigma) \to \xi'(\sigma) = h(x)\,\xi(\sigma)\, h^+(\pi(x), g^+), \qquad (2.8)$$

where $h(\pi(x), g^+)$ is a field dependent induced global H transformation induced via G_g transformation, in contrast to $h(x)$ which is a local, $\pi(x)$- and g- independent H_ℓ transformation.

We can recognize that the introduction of the above redundant variables σ is constrained by the additional hidden local symmetries, that is, the number of degrees of freedom of the redundant variables σ is just equal to that of gauge freedom.

In order to construct the $G_g \times H_\ell$ invariants from $\xi(x)$, it is covenient to define an algebra-valued 1-form (Maure-Cartan 1-form) from group-valued $\xi(x)$(2),

$$\alpha_\mu = \frac{1}{i}\partial_\mu \xi(x)\cdot\xi(x)^\dagger \qquad (2.9)$$

Because of the local property of the hidden group, H_ℓ, we here introduce the gauge fields $V_\mu = V_\mu^\alpha S_\alpha$, with the same transformation properties as usual gauge fields. The fields V_μ are associated with familiar inhomogeneous terms,

$$V_\mu(x)\to V_\mu'(x) = i h(x)\partial_\mu h^\dagger(x) + h(x)V_\mu(x)h(x)^\dagger. \qquad (2.10)$$

Then the covariantized 1-form is given as,

$$\hat{\alpha}_\mu(x) = \frac{1}{i}(\partial_\mu - iV_\mu)\xi(x)\cdot\xi(x)^\dagger = \alpha_\mu(x) - V_\mu(x), \qquad (2.11)$$

from which we define projections of $\hat{\alpha}_\mu$ into the components parallel and perpendicular spaces,

$$\hat{\alpha}_{\mu\parallel} = S_\alpha\cdot 2\,\mathrm{Tr}(S_\alpha\cdot\hat{\alpha}_\mu),$$
$$\hat{\alpha}_{\mu\perp} = X_\alpha\cdot 2\,\mathrm{Tr}(X_\alpha\hat{\alpha}_\mu), \qquad (2.12)$$

which transform independently under the group $G_g \times H_\ell$,

$$\hat{\alpha}_{\mu\parallel}\to\hat{\alpha}_{\mu\parallel}' = h(x)\hat{\alpha}_{\mu\parallel}h(x)^\dagger,$$
$$\hat{\alpha}_{\mu\perp}\to\hat{\alpha}_{\mu\perp}' = h(x)\hat{\alpha}_{\mu\perp}h(x)^\dagger. \qquad (2.13)$$

Hence we have arrived at two invariants,

$$\mathcal{L}_\parallel = f_\sigma^2\,\mathrm{Tr}(\hat{\alpha}_{\mu\parallel})^2 = f_\sigma^2\,\mathrm{Tr}(\alpha_{\mu\parallel}-V_\mu)^2, \qquad (2.14)$$
$$\mathcal{L}_\perp = f_\pi^2\,\mathrm{Tr}(\hat{\alpha}_{\mu\perp})^2 = f_\sigma^2\,\mathrm{Tr}(\alpha_{\mu\perp})^2.$$

where the factors f_π^2 and f_σ^2 of the above are taken so as to normalize the kinetic terms of the fields π and σ, respectively. Then the most general form of the Lagrangian made out of $\xi(x)$ with the lowest derivatives is,

$$\mathcal{L} = \mathcal{L}_\parallel + \mathcal{L}_\perp, \qquad (2.15)$$

with the parameters f_π and f_σ.

Now we show that the G/H non-linear Lagrangian is expressed by the gauge fixed form of the $G_g \times H_\ell$ linear Lagrangian derived from the above.

Note that the $\mathcal{L}_\parallel$ part vanishes when we substitute the solution of the equation of motion for the auxiliary field V_μ,

$$V_\mu = \alpha_{\mu\parallel}. \qquad (2.16)$$

The Lagrangian is reduced to

$$\mathcal{L} = \mathcal{L}_\perp = f_\pi^2\,\mathrm{Tr}(\partial_\mu\xi(x)\cdot\xi(x)^\dagger)_\perp^2, \qquad (2.17)$$

which is just the same form as the non-linear Lagrangian derived by Callen, Coleman, Wess and Zumino (CCWZ)(6). Here we take the unitary gauge for H_ℓ gauge group, in which σ fields do not appear explicitly in the Lagrangian, i.e.,

$$\xi(\sigma) = 1. \qquad (2.19)$$

under which condition the system is no longer invariant under $G_g \times H_\ell$ transformation. The system is, however, still invariant under the residual transformation preserving the condition (2.19). This combined transformation is explicitly obtained, if we recall (2.8), in which $\xi'(\sigma) = \xi(\sigma) = 1$ requires $h(x) = h(\pi(x), g^+)$, so we have under this gauge fixing

$$\xi(x)\to\xi'(x) = h(\pi(x), g^+)\,\xi(x)\,g^\dagger, \qquad (2.20)$$

which is nothing but the usual non-linear transformation(7). Thus the Lagrangian (2.17), together with eq.(2.20), shows the gauge equivalence of G/H non-linear Lagrangian to $G_g \times H_\ell$ "linear" Lagrangian.

3. Dynamical gauge bosons of hidden local symmetries

So far our hidden local symmetry represents merely the gauge freedom to eliminate the redundant fields σ and the gauge field V_μ has been no more than auxiliary field having no kinetic terms. However the hidden local symmetry becomes physical if the kinetic terms are generated via quantum effects.

Can the hidden local symmetry become physical? It is well known that the dynamical calculation of 2-dimensional CP(N-1) model reproduces the poles of the vector fields(3), which people take for granted because of the peculiar infra-red structure of 2-dimensional case. However, a similar situation turned out to occur in 3-dimensional CP(N-1) model(4), which is non-trivial, but less known. Also Kugo, Uehara and Terao(8) studied dynamical generation of vector fields and their conclusion is that dynamical generation of gauge bosons must be a rather common phenomenon which can occur in a wide variety of theories not restricted to CP(N-1) model nor 2-dimensional case. This I will not discuss further. Instead I here explain another aspect the implications of a dynamical generation of hidden local symmetry. It is interesting now to study the phenomenological consequences of the dynamical realization of hidden local symmetry.

Our purpose here is to investigate the case where the kinetic terms of hidden gauge bosons are generated via quantum effects. Then the gauge equivalences of G/H nonlinear Lagrangian to $G_g \times H_\ell$ "linear" one, which we saw at the classical level, is no longer guranteed at the quantum level and there the gauge bosons V_μ are independent fields in addition to the N-G bosons π.

So our starting low-energy effective Lagrangian now includes the kinetic terms of V_μ ;

$$\mathcal{L}_{eff} = \mathcal{L}_\perp + \mathcal{L}_\parallel - \frac{1}{2g^2}\mathrm{Tr}(F_{\mu\nu})^2,$$

$$F_{\mu\nu} = \partial_\mu V_\nu - \partial_\nu V_\mu + [V_\mu, V_\nu]. \quad (3.1)$$

Here we further switch on the interaction with the external gauge fields B_μ. The remarkable feature of our framework is that we can gauge any subgroup of global group G_g, completely independently of the hidden local gauge group H_ℓ. It is important to note that our framework makes clear the difference between hidden local gauge bosons and elementary external gauge bosons. Hence there are no complications in introducing such interactions as electromagnetism etc., in sharp contrast to the other attempts(9). Now our Lagrangian in unitary gauge takes the form, after rescaling $1/g\, V_\mu \to V_\mu$,

$$\mathcal{L} = \mathcal{L}_\perp + \mathcal{L}_\parallel - \mathcal{L}_{\text{kin. of } V_\mu \text{ and } B_\mu},$$

$$\begin{cases} \mathcal{L}_\perp = \mathrm{Tr}\{\alpha_\perp(\pi) - e(\xi(\pi) B_\mu \xi(\pi)^\dagger)_\perp\}^2, \\ \mathcal{L}_\parallel = \mathrm{Tr}\{\alpha_\parallel(\pi) - g V_\mu + e(\xi(\pi) B_\mu \xi(\pi)^\dagger)_\parallel\}^2, \end{cases} \quad (3.2)$$

where the gauge couplings are denoted by g and e for hidden and external groups, respectively. In order to see the detailed structure of the Lagrangian more explicitly, we write down L in terms of π and V_μ fields.

$$\mathcal{L}_\perp = \mathrm{Tr}\{\partial_\mu \pi + i[\pi, eB_\mu] + \cdots\}^2,$$
$$\mathcal{L}_\parallel = f_\sigma^2 \mathrm{Tr}\{g V_\mu - e B_\mu - \frac{i}{2f_\pi^2}[\pi, \partial_\mu \pi] + \cdots\}^2, \quad (3.3)$$

from which we can read, for example,

mass m_V : $\quad m_V^2 = g^2 f_\sigma^2$, (3.4a)

V–B transition: $\quad g_{V-B} = g f_\sigma^2$, (3.4b)

V–ππ coupling: $\quad g_{V\pi\pi} = f_\sigma^2/2f_\pi^2 \cdot g$, (3.4c)

B–ππ coupling: $\quad g_{B\pi\pi} = (1 - \frac{f_\sigma^2}{2f_\pi^2})\, e$. (3.4d)

The mass term (3.4a) indicates the typical form of Higgs mechanism, absorbing the unphysical σ fields not π fields. All these physical quantities are expressed in terms of a gauge coupling g, and the scale parameter f_π as well as e and f_σ. Our framework predicts a parameter independent relation,

$$g_{V-B} = 2 f_\pi^2\, g_{V\pi\pi}, \quad (3.5)$$

which we call "KSRF(I)" relation, since it is a variant of famous KSRF relation in old hadron physics. This can be regarded as the key relation to test our idea. We shall soon check our idea in Q.C.D. setting.

4. ρ mesons as dynamical gauge bosons

We are now in a stage to see that hidden local symmetry is realized in nature, i.e., in Q.C.D. case, where the chiral symmetry $G = SU(2)_L \times SU(2)_R$ is broken down to its diagonal subgroup $H = SU(2)_V$. In the case at hand, the immediate candidate for the hidden gauge fields is the ρ meson. Direct application of our formulae with a little caution to the semi-simple structure of the group G in this case ($SU(2)_L \times SU(2)_R \times SU(2)_V \times$ parity invariants) leads us,

$$\mathcal{L}_\parallel = f_\sigma^2 \mathrm{Tr}\{g \rho_\mu - e A_\mu \frac{\tau^3}{2} - \frac{i}{2f_\pi^2}[\pi, \partial_\mu \pi] + \cdots\}^2.$$

$$\mathcal{L}_\perp = \frac{f_\pi^2}{4}\mathrm{Tr}(\partial_\mu U \partial_\mu U^\dagger) + e A_\mu (\vec{\pi} \times \partial_\mu \vec{\pi})_3 + \cdots \quad (4.1)$$

where we consider an external photon field A_μ and the well known variables $U(x) = \exp(2i\pi(x)/f_\pi)$ in terms of ordinary pions and ρ meson, V_μ, and f_π is the familiar pion decay constant. The effective low-energy Lagrangian for π and ρ system now takes the following familiar form,

$$\mathcal{L} = \frac{f_\pi^2}{4}\mathrm{Tr}(\partial_\mu U \partial_\mu U^\dagger) - \frac{1}{4}\vec{F}_{\mu\nu}^{(\rho)2} - \frac{1}{4}(\partial_\mu A_\nu - \partial_\nu A_\mu)^2$$
$$+ \frac{1}{2} m_\rho^2 \vec{\rho}_\mu^{\,2} - e g \rho_\mu^{(3)} A_\mu + \frac{1}{2}\left(\frac{e g_\rho}{m_\rho}\right)^2 A_\mu^2 \quad (4.2)$$
$$+ g_{\rho\pi\pi} \vec{V}_\mu \cdot (\vec{\pi} \times \partial_\mu \vec{\pi}) + e_{\gamma\pi\pi} A_\mu (\vec{\pi} \times \partial_\mu \vec{\pi}) + \cdots,$$

where m_ρ, the mass of ρ meson, g_ρ, the $\rho-\gamma$ coupling, and $g_{\rho\pi\pi}$, $e_{\gamma\pi\pi}$, $\rho\pi\pi$ and $\gamma\pi\pi$ coupling constants are given by, $m_\rho^2 = g^2 f_\sigma^2$, $g_\rho = g f_\sigma^2$, $g_{\rho\pi\pi} = f_\sigma^2/2f_\pi^2 \cdot g$, $e_{\gamma\pi\pi} = (1 - f_\sigma^2/2f_\pi^2) \cdot e$. The remarkable fact is the f_σ- and g-independent relation,

$$g_\rho^2 = 2 f_\pi^2 \cdot g_{\rho\pi\pi}, \quad (4.3)$$

which is the clean test of our framework. The left-hand side evaluated from the decay width ρ into $e\bar{e}$ ($\Gamma_{e\bar{e}}$ = 6.62 KeV) is 0.12 GeV^2, while the right-hand side yields 0.11 GeV^2 (corresponding to the case $g_{\rho\pi\pi}^2/4\pi = 3.0$). Remarkably in good agreement !

5. Comments and further outlook

Several comments are given.

1) Our parameter-independent relation (3.5) (KSRF(I)) is a direct consequences of the particular form of the Lagrangian which is determined by the requirement of $G_f \times H_\ell$ invariance. Note that the KSRF(I) relation guarantees the low-energy behavior of the π system with external fields B_μ. This can be easily understood from eq.(4.1), in which there exists no additional contribution which comes from $L_\parallel$ term (see the detailed discussion in ref.1)).

2) We can enlarge our hidden local symmetry H_ℓ to the full group G_ℓ(1), where we showed the gauge equivalence of G/H non-linear Lagrangian to $G_g \times G_\ell$ "linear" Lagrangian by introducing another set of compensator fields $\xi(p) = e^{ip(x)}$; $p(x) = p^a(x)X_a$ where we have 4 invariants. As an application we reexamined a possibility that A_1 mesons are also dynamical gauge bosons of hidden local symmetry.

3) In our previous paper, we made a

comment on the relation between our formula and the old work of Weinberg(9). Also detailed comparison of our framework with other recent work by several authers has been made by Yamawaki(13).

In general, a G/H non-linear sigma model may be considered as low energy effective theory of some underlying confining theory and this strong interaction develops symmetry breaking of G down to H. It may be more natural scenario that the hidden gauge bosons are generated via the same underlying strong interaction(14). Also I would like to notice that the interesting discussion made by Kugo that the dynamical generation of the graviton might be realized in nature(8) !

We have shown that phenomenological evidence has been found in hadron phenomena that composite gauge bosons of hidden local symmetry are realized in nature. Also the idea may be applied to technicolor scenario or N = 8 extended supergravity models(15).

As is well known, in the theories which have no "local invariance" no composite gauge bosons can appear(16). This is the reason why people had been so familiar to the view that gauge fields are "elementary particles", until the notion of hidden local symmetry was first introduced in N = 8 SUGRA(17). We are now encouraged to seek for the possibility more realistically that some, or even all of the known gauge fields may be composite(15).

Gauge theories have been the fundamental basis of the way of thinking of modern particle physicists. Yet little is known to us, and their remain many fundamental questions to be solved. Further less attention has been paid to the asymptotically non-free theories than the QCD-like asymptotically-free theories. However, the asymptotically non-free gauge theory have recently turned out to have non-trivial ultraviolet fixed point, having large anomalous dimensions(18). This resolves the FCNC syndrome(19), which has been a central problem of any model to do something with the "flavor problem". Gauge theories may yet will provide us with still more interesting ingradients.

I would liked to thank H. Quinn for reading the manuscript, and theoretical phys. group for their kind hospitality at SLAC. Also I have benifited from conversation with K.-I. Aoki, H. Aoyama, T. Kugo, H. Yamamoto and K. Yamawaki.

Reference

1) This talk is based on M. Bando, T. Kugo, S. Uehara, K. Yamawaki and T. Yanagida, Phys. Rev. Lett. 54(1985), 1215; M. Bando, T. Kugo and K. Yamawaki, Nucl. Phys. B159(1985), 493; ibid. Prog. Theor. Phys. 73(1985), 1541; M. Bando, T. Fujiwara and K. Yamawaki, Nagoya Preprint, DPNY-86-23.
2) S. T. Gates, Jr., M. T. Grisaru, M. Rocek and W. Siegel, Superspace (Benjamin/Cummings, New York) Chap. 3; P. Breitenlohner and Maison, Munchen preprint MPI-PAE/Pth 1/84
3) V. Golo and A. M. Perelomov, Phys. Lett. 79B(1978), 112; A d'Adda, P. di Vecchia and M. Lucsher, Nucl. Phys. B146 (1978), 63; B152(1979), 125.
4) I. Ya Aref'eva and S. I. Azakov, Nucl. Phys. B162(1980), 298.
5) K. Kawanabayashi and M. Suzuki, Phys. Rev. Lett. 16(1966), 255; Riazuddin and Fayyazuddin, Phys. Rev. 147(1966), 1071.
6) C. G. Callan, Jr., S. Coleman, J. Wess and B. Zumino, Phys. Rev. 177(1969), 2247.
7) See for example, S. Coleman, J. Wess and B. Zumino, Phys. Rev. 177(1969), 2239.
8) T. Kugo, H. Terao and S. Uehara, Proceedings of Meson 50, Suppl. Prog. Theor. Phys. 85(1985), 122.
9) J. Schwinger, Phys. Lett. 24B(1967), 473; J. Wess and B. Zumino, Phys. Rev. 163(1967), 1727; S. Weinberg, Phys. Rev. 166(1968), 1568.
10) J. Sakurai, Currents and Mesons (Univ. Chicago Press, Chicago, 1969)
11) T. H. R. Skyrme, Proc. Roy. Soc. A260(1961), 127.
12) T. Fujiwara, Y. Igarashi, A. Kobayashi, H. Otsu and S. Sawada Prog. Theor. Phys. 74(1985), 128.
13) K. Yamawaki, Nagoya Preprint, DPNY-86-22; DPNY-86-2414
14) This kind of dynamical generation of hidden gauge bosons are investigated by Terao, Uehara and Kugo. T. Kugo, Sorgushiron Kenkyu (Kyoto) 71(1985), E78: H. Terao, Master Thesis.
15) J. Ellis, M. K. Gaillard and B. Zumino, Phys. Lett. 94B (1980), 343. See also the references in M. Bando, Y. Sako and S. Uehara, Z. Phys. 22(1984), 251.
16) S. Weinberg and E. Witten Phys. Lett. 96B(1980), 59 T. Kugo, Proceedings of 1981 INS Symposium on Quarks and Lepton P edited by K. Fujikawa, et al. p336; Phys. Lett. 109B(1982), 20.
17) E. Cremmer and B. Julia, Nucl. Phys. B159(1979), 141.
18) P. I. Fomin, V. P. Gusynin, V. A. Miransky and Yu. A. Sitenko, Riv. Nuovo Cim. Soc. Ital. Fis.6 No.5(1983), 1; W. A. Bardeen, C. N. Lueng and S. T. Love, Phys. Rev. Lett. 56(1986), 1230.
19) K. Yamawaki, M. Bando and K. Matumoto, Phys. Rev. Lett. 56(1986), 1385.

STOCHASTIC QUANTIZATION AND B.R.S. SYMMETRY

J. Zinn-Justin
Service de Physique Théorique, CEN-Saclay, 91191 Gif-sur-Yvette Cedex, France

For renormalization purpose, it is useful to associated with dynamical equations like the Langevin equation, an effective field theory action. We show that as in various other cases like gauge field theories, or random field Ising model this action is automatically B.R.S. symmetric. In addition in special cases we recover that the effective action can have a second B.R.S. symmetry or even supersymmetry.

1. THE LANGEVIN EQUATION

The Langevin equation is a stochastic dynamical equation for a field $\varphi(x,t)$ in which x is a coordinate in d dimensional space and t a time which is physical in critical dynamics of phase transitions, and has the interpretation of an additional time dimension (like a Monte Carlo simulation computer time) introduced for quantization purpose in field theory:

$$\dot{\varphi}(x,t) = -\frac{1}{2}F(\varphi(x,t))+\nu(x,t) \qquad (1)$$

The functional $F(\varphi)$ is a local functional of the field and $\nu(x,t)$ a set of stochastic variables for which a probability distribution $[d\rho(\nu)]$ is provided. We have chosen in Langevin equation (1) a simple form for the dependence in the noise for simplicity reasons and because many examples are of this form. Given $\varphi(x,0)$ Langevin equation (1) defines in particular at later time t a time dependent probability distribution $P(\varphi(x),t)$ for the field:

$$P[\varphi(x),t] = \langle \delta[\varphi(x,t)-\varphi(x)] \rangle_\nu \qquad (2)$$

in which the bracket means average over the noise. While in critical dynamics[1] the evolution in time generated by equation (1) is of direct interest, in Particle physics this equation has been mainly introduced to generate an equilibrium distribution $\exp -\mathcal{A}[\varphi]$ which is the limit, if it exists, of $P(\varphi(x),t)$ at infinite time[2]:

$$\exp -\mathcal{A}(\varphi) = \lim_{t\to+\infty} P[\varphi,t] \qquad (3)$$

The functional $\mathcal{A}[\varphi]$ is the action which we wish to quantize.

In the special case of a gaussian noise local in space and time:

$$[d\rho(\nu)] = [d\nu] \exp -\frac{1}{2}\int dx\, dt\, \nu^2(x,t) \qquad (4)$$

the probability distribution $P[\varphi,t]$ satisfies a Fokker-Planck equation of the form:

$$\dot{P}[\varphi,t] = \frac{1}{2}\int dx \frac{\delta}{\delta\varphi(x)}\left[\frac{\delta}{\delta\varphi(x)}P+F[\varphi(x)]P\right] \qquad (5)$$

The possible equilibrium distributions are time independent solutions of the Fokker-Planck equation.

To describe the time evolution of the stochastic process near a critical point, it is necessary to use renormalization group arguments and thus to renormalize the Langevin and Fokker-Planck equations. It is convenient to associate with them a functional integral. From (5) one derives immediately:

$$P[\varphi,T] = \int [d\varphi] \exp -S_{ef}(\varphi) \qquad (6)$$

$$S_{ef}(\varphi) = \frac{1}{2}\int_0^T dxdt\left\{\left[\dot{\varphi}(x,t)+\frac{1}{2}F[\varphi(x,t)]\right]^2 - \frac{1}{2}\delta^d(0)\frac{\delta F}{\delta\varphi(x,t)}\right\} \qquad (7)$$

expression which is only meaningful if a regularization for $\delta^d(0)$ is provided (dimension or lattice regularization for example).

There exist choices of $F(\varphi)$ for which this effective action $S_{ef}(\varphi)$ is renormalizable by power counting. However this does not guaranty that the special form of the action (7) is preserved and thus that the Langevin itself can be renormalized. This is a consequence of a symmetry of the action which is actually the B.R.S. symmetry[3] which we shall derive now in a more general context[4].

2. B.R.S. SYMMETRY

Let us first consider the following set of equations for a field φ_α:

$$F_\alpha(\varphi) = 0 \tag{8}$$

in which α is a generic index which includes space, time, group indices... if necessary. Summation over repeated indices will always be meant in what follows.

Now we want to evaluate a functional $R(\varphi)$ for $\varphi=\varphi_s$ solution of the system (8). A formal way to calculate $R(\varphi)$ is to write:

$$R(\varphi_s) = \int [d\varphi] \prod_\alpha \delta[F_\alpha(\varphi)] \det M \, R(\varphi) \tag{9}$$

in which the operator M is given by:

$$M_{\alpha\beta} = \frac{\delta F_\alpha}{\delta \varphi_\beta} \tag{10}$$

We shall rewrite (9) by using two identities:

$$\prod_\alpha \delta[F_\alpha(\varphi)] = \int [d\lambda] \exp -\lambda_\alpha F_\alpha(\varphi) \tag{11}$$

$$\det M = \int \left[dCd\bar{C}\right] \exp \bar{C}_\alpha M_{\alpha\beta} C_\beta \tag{12}$$

in which C and $\bar{C}$ are two set of fermion fields. We thus obtain for $R[\varphi_s]$

$$R[\varphi_s'] = \int \left[d\varphi dC d\bar{C} d\lambda\right] R[\varphi] \exp - S(\varphi) \tag{13}$$

with:

$$S(\varphi) = \lambda_\alpha F_\alpha(\varphi) - \bar{C}_\alpha M_{\alpha\beta} C_\beta \tag{14}$$

Fundamental observation[4]

The quantity $S(\varphi)$ has a B.R.S. symmetry which corresponds to the transformation:

$$\begin{cases} \delta\varphi_\alpha = \bar{\varepsilon}\, C_\alpha \, , & \delta C_\alpha = 0 \\ \delta\bar{C}_\alpha = \bar{\varepsilon}\lambda_\alpha \, , & \delta\lambda_\alpha = 0 \end{cases} \tag{15}$$

in which $\bar{\varepsilon}$ is an anticommuting constant. The transformation (15) is obviously nilpotent. Let us indeed calculate δS the variation of S:

$$\delta S = \lambda_\alpha \frac{\partial F_\alpha}{\partial \varphi_\beta} \bar{\varepsilon} C_\beta - \bar{\varepsilon}\lambda_\alpha M_{\alpha\beta} C_\beta - \bar{C}_\alpha \frac{M_{\alpha\beta}}{\delta\varphi_\gamma} \bar{\varepsilon} C_\gamma C_\beta \tag{16}$$

The result follows from the definition of $M_{\alpha\beta}$ and the observation that:

$$\frac{\delta M_{\alpha\beta}}{\delta\varphi_\gamma} C_\gamma C_\beta = \frac{\delta^2 F_\alpha}{\delta\varphi_\beta \delta\varphi_\gamma} C_\gamma C_\beta = 0 \tag{17}$$

Conversely B.R.S. symmetry and some conditions on the degrees of S as a polynomial in $\bar{C}$, C, λ, imply the form (14) and thus equation (8).

The B.R.S. transformation (15) has a trivial form involving no geometry since $F(\varphi)$ is arbitrary. Actually if we define the fields:

$$\phi_\alpha = \varphi_\alpha + \bar{\theta} C_\alpha \, , \quad \Lambda_\alpha = \bar{C}_\alpha + \bar{\theta}\lambda_\alpha \tag{18}$$

in which $\bar{\theta}$ is an anticommuting coordinate, it expresses the translation invariance in $\bar{\theta}$ space of the action (14) which can be written:

$$S = \int d\bar{\theta}\, \Lambda_\alpha F_\alpha(\phi) \tag{19}$$

However it is completely equivalent to the B.R.S. symmetry introduced in gauge theories.

3. B.R.S. SYMMETRY IN GROUP MANIFOLDS AND GAUGE THEORIES[4,5]

If the field φ belongs to the representation of some group G, the parametrization (18) is not natural. It is more convenient to set instead

$$\phi = \varphi(1+\bar{\theta}C) \tag{20}$$

in which C now belongs to the Lie algebra of G. The B.R.S. transformation (15) which on ϕ reads:

$$\delta\phi = \bar{\varepsilon} \frac{\partial\phi}{\partial\bar{\theta}} \tag{21}$$

now becomes:

$$\begin{cases} \delta\varphi = \bar{\varepsilon}\varphi C \\ \delta C = -\bar{\varepsilon}C^2 \end{cases} \quad (22)$$

We now recognize the ghost transformation law of gauge theories.

We still have to introduce a gauge field $A_\mu(x)$, corresponding to a gauge group G. Let us parametrize $A_\mu(x)$ in terms of a field $B_\mu(x)$ belonging to a fixed gauge, and a gauge transformation $\varphi(x)$.

$$A_\mu(x) = \varphi^{-1}\partial_\mu\varphi + \varphi^{-1}B_\mu\varphi \quad (23)$$

If we add to definitions (22):

$$\delta B_\mu(x) = 0 \quad (24)$$

we can easily verify:

$$\delta A_\mu(x) = \bar{\varepsilon}D_\mu C \quad (25)$$

which is the gauge field B.R.S. transformation. In the sense of next section, in gauge theories the gauge transformation $\varphi(x)$ is stochastically quantized in the usual quantization procedure.

4. STOCHASTIC QUANTIZATION

With the same compact notations as in section 2 we shall call stochastic quantization of φ any equation (8) which depends functionally of a noise field ν for which a probability distribution $[d\rho(\nu)]$ is provided. There are numerous examples of such situation: Langevin equation, random magnetic problem, gauge theories... The algebraic formulation of section 2 can then be used to replace equation (8) by a functional integral more amenable to the usual methods of quantum field theory. It will always generate a B.R.S. invariant action. Let us consider here a simple but useful example:

$$F_\alpha(\varphi) = \nu_\alpha \quad (26)$$

Introducing then the generatic functional of connected noise correlation functions $w(\lambda)$:

$$e^{w(\lambda)} = \int [d\rho(\nu)] \exp \lambda_\alpha \nu_\alpha \quad (27)$$

we can integrate over the noise the equivalent of expression (15) and get a partition function Z:

$$Z = \int \left[d\varphi dC d\bar{C} d\lambda\right] \exp - S(\varphi, C, \bar{C}, \lambda) \quad (28)$$

with:

$$S = -w(\lambda) + \lambda_\alpha F_\alpha(\varphi) - \bar{C}_\alpha M_{\alpha\beta} C_\beta \quad (29)$$

If the dependence in the noise ν of equation (26) is more complicated, the effective action (29) will contain higher powers of $\bar{C}C$.

In all cases B.R.S. symmetry and the corresponding W.T. identities can be used to renormalize stochastically quantized theories and to show under certain conditions that the structure of a field equation of form (26) together with a probability distribution for ν is not lost.

5. A SPECIAL CASE

If the noise distribution is gaussian, and at least the non linear part of $F_\alpha(\varphi)$ derives from an action

$$F_\alpha(\varphi) = f_{\alpha\beta}\varphi_\beta + \frac{\delta\mathcal{A}}{\delta\varphi_\alpha} \quad (30)$$

some symmetry between C and $\bar{C}$ is introduced. It becomes convenient to introduce a superfield ϕ notation:

$$\phi_\alpha = \varphi_\alpha + \bar{\theta}C_\alpha + \bar{C}_\alpha\theta + \bar{\theta}\theta\lambda_\alpha \quad (31)$$

The effective action (29) can then be written:

$$S(\phi) = \int d\theta d\bar{\theta} \left\{ \frac{1}{2} w_{\alpha\beta} \frac{\partial\phi_\alpha}{\partial\theta} \frac{\partial\phi_\beta}{\partial\bar{\theta}} + \theta \frac{\partial\phi_\alpha}{\partial\theta} f_{\alpha\beta}\phi_\beta + \mathcal{A}[\phi] \right\} \quad (32)$$

If $f_{\alpha\beta}$ vanishes $S(\phi)$ has at least an additional B.R.S. symmetry corresponding to the translation θ in $\theta + \varepsilon$. Examples are the random field Ising model [6] and gauge theories with anti B.R.S. symmetric gauges [7]. For the Langevin equation this is not the case since then equation (30) has then the form:

$$\dot{\varphi}(x,t) + \frac{1}{2} \frac{\delta\mathcal{A}}{\delta\varphi(x,t)} = \nu(x,t) \quad (33)$$

with gaussian noise (4).

The effective action (32) then reads:

$$S(\phi) = \int dx dt d\theta d\bar{\theta} \left\{ \frac{1}{2} \left[\frac{\partial\phi}{\partial\theta} \frac{\partial\phi}{\partial\bar{\theta}} + \theta \frac{\partial\phi}{\partial\theta} \frac{\partial\phi}{\partial t} \right] + \mathcal{A}[\phi] \right\} \quad (34)$$

However it can be verified that expression (34) is invariant under the substitution:

$\phi(t,\theta) \longrightarrow \phi(t-\bar{\theta}\varepsilon,\theta+\varepsilon)$.

The generators D and $\bar{D}$ of the two B.R.S. symmetries combine to form a quantum mechanical supersymmetry[8,9]

$$\begin{cases} D^2 = \bar{D}^2 = 0 \\ \bar{D}D + D\bar{D} = -\dfrac{\partial}{\partial t} \end{cases} \tag{35}$$

This whole formalism allows us to study time evolution of stochastic process for a critical system from the renormalization group point of view[10].

REFERENCES

[1] P.C. Hohenberg and B.I. Halperin, Rev. Mod. Phys. 49 (1977)

[2] G. Parisi and Y.S. Wu, Sci. Sin. 24 (1981) 483

[3] C. Becchi, A. Rouet and R. Stora, Ann. of Phys. (NY) 98 (1976) 287

[4] J. Zinn-Justin, Nucl. Phys. B275 [FS17] (1986) 135 and references therein

[5] J. Zinn-Justin, Bonn lectures 1974: Lecture notes in Physics Vol.37 (Springer 1975)

[6] G. Parisi and N. Sourlas, Nucl. Phys. B206 (1982) 321

[7] L. Baulieu and J. Thierry Mieg, Nucl. Phys. B197 (1982) 477

[8] H. Nakazato, M. Nanuki, I. Ohba and K. Okano, Prog. Teor. Phys. 70 (1983) 296
E.S. Egorian and S. Kalitsin, Phys. Lett. 129B (1983) 320

[9] E. Witten, Nucl. Phys. B188 (1981) 513

[10] Finite effects in critical dynamics, J.C. Niel and J. Zinn-Justin, Saclay preprint SPhT/86/070 to appear in Nucl. Phys.

MINKOWSKI STOCHASTIC QUANTIZATION

H. Nakazato

Department of physics, Waseda University
Tokyo 160, Japan

Stochastic quantization in Minkowski space is discussed in detail. The Fokker-Planck equation corresponding to the complex Langevin equation is derived and solved. It turns out that Minkowski stochastic quantization can be formulated in terms of a real positive probability.

1. INTRODUCTION

In this section, I review the outline of the ordinary stochastic quantization, that is, the stochastic quantization in Euclidean space, and then try to extend this method to Minkowski space.

Stochastic quantization method, proposed by Parisi and Wu in 1981 [1], is a new kind of quantization method which are quite different from the canonical and the path integral quantization. Its outline is described below.

In the first step, we introduce the 5th coordinate t besides the ordinary 4-dim. one x, which we call the fictitious time in what follows, into the system described by the Euclidean action $S_E[\phi]$ and regard the field variable ϕ as the random variable depending on this fictitious time t. We consider the hypothetical stochastic process with respect to t and assume that the dynamics of this stochastic process are governed by the following Langevin equation,

$$\dot{\phi}(x,t) = -\frac{\delta S_E}{\delta\phi(x)}\bigg|_{\phi(x)=\phi(x,t)} + \eta(x,t). \qquad (1)$$

Here $-\delta S_E/\delta\phi$ represents a drift term of this process and η stands for the Gaussian white noise with the statistical properties,

$$\langle\eta(x,t)\rangle = 0,$$
$$\langle\eta(x,t)\eta(x',t')\rangle = 2\delta^4(x-x')\delta(t-t'),$$
$$\cdots \qquad (2)$$

Vacuum expectation values can be obtained by the equal time correlation functions in the thermal equilibrium limit $t\to\infty$, i.e.,

$$\langle\phi_\eta(x_1,t)\cdots\phi_\eta(x_n,t)\rangle \xrightarrow{t\to\infty} \int \mathcal{D}\phi\phi(x_1)\cdots\phi(x_n)\exp(-S_E[\phi]). \qquad (3)$$

Here ϕ_η stands for the solution of the Langevin equation (1) as a functional of η.

Note that if we want to discuss the behaviours of the system around the thermal equilibrium, we must go into the Fokker-Planck equation equivalent to the original Langevin equation. The Fokker-Planck equation

$$\dot{P}[\phi,t] = -H[\phi]P[\phi,t] \qquad (4)$$

governs the time development of P, the probability distribution of ϕ, and gives us information about the equilibrium state. It is well known that the positive semi-definiteness of the spectrum of the Fokker-Planck Hamiltonian H guarantees the desirable approach of P to $\exp(-S_E)$ as long as it is normalizable.

Now, let us consider the possibility of stochastically quantizing the system directly in Minkowski space. Replacing the Euclidean action $-S_E$ by the corresponding Minkowski one iS, we obtain the following Langevin equation

$$\dot{\phi}(x,t) = i\frac{\delta S}{\delta\phi(x)}\bigg|_{\phi(x)=\phi(x,t)} + \eta(x,t), \qquad (5)$$

with a complex drift term $i\delta S/\delta\phi$,

which we call the complex Langevin equation. This type of equation is our basic equation in the stochastic quantization in Minkowski space (Minkowski stochastic quantization) and many people have already investigated such equations numerically and /or analytically [2].

There are two important remarks about the above complex Langevin equation (5). First, owing to the presence of a complex factor "i" in the drift term, the degrees of freedom are duplicated inevitably, that is, even if we consider the system of a real scalar field ϕ, for example, ϕ becomes a complex number as the time t develops. Second, the Gaussian white noise η can also take a complex value as far as the statistical properties (2) are retained. In other words, dividing η into real and imaginary parts, $\eta=\eta_R+\eta_I$, and assuming the Gaussian white noise properties for η_R and η_I,

$$\langle\eta_R\rangle = \langle\eta_I\rangle = \langle\eta_R\eta_I\rangle = 0,$$
$$\langle\eta_R(x,t)\eta_R(y,t')\rangle = 2\alpha\delta^4(x-y)\delta(t-t'),$$
$$\langle\eta_I(x,t)\eta_I(y,t')\rangle = 2\beta\delta^4(x-y)\delta(t-t'), \quad (6)$$

we can reproduce the statistical properties of η in (2) only if we put $\alpha-\beta = 1$. (Of course, if we put $\beta = 0$ then we obtain the real η as in the Euclidean case.)

Concerning the stochastic process governed by the complex Langevin equation, we have the following two questions: (i) Does the system governed by (5) have a unique equilibrium distribution P_{eq} ? (ii) Can we show the equivalence between the Minkowski stochastic quantization method and the ordinary quantization ones ? Or what is the relationship between P_{eq} and the Feynman measure exp(iS) ? The purpose of my talk is to answer these questions correctly.

2. APPLICATION TO FREE SCALAR FIELD

Let us consider the free scalar field case as the simplest example of Minkowski stochastic quantization [3]. Even in such a simple case, the situation is not so simple. The Langevin equation for the system of the real scalar field ϕ described by a classical action S_0,

$$S_0 = \int d^4x\left(\frac{1}{2}\partial_\mu\phi\partial^\mu\phi-\frac{1}{2}\mu^2\phi^2\right) \quad (7)$$

can be written as

$$\dot{\phi}(x,t) = i(-\Box-\mu^2+i\varepsilon)\phi(x,t) + \eta(x,t). \quad (8)$$

Here we have explicitly introduced an infinitesimal positive number ε to lead the system into equilibrium as t goes to infinity. This ε should be set equal to zero after all the calculations have been performed. Without this procedure, no equilibrium state could be obtained. Following the above mentioned remarks, we divide the complex equation to the real and imaginary parts;

$$\dot{\phi}_R(k,t) = -q(k)\phi_I(k,t)-\varepsilon\phi_R(k,t) +\eta_R(k,t),$$
$$\dot{\phi}_I(k,t) = q(k)\phi_R(k,t)-\varepsilon\phi_I(k,t) +\eta_I(k,t), \quad (9)$$

where $\phi=\phi_R+i\phi_I$ and $q(k) \equiv k^2-\mu^2$. The Gaussian white-noise properties of η_R and η_I in (6) can be realized by the probability distribution Ψ,

$$\Psi \sim \exp\left[-\frac{1}{4\alpha}\int d^4kdt\,|\eta_R(k,t)|^2 -\frac{1}{4\beta}\int d^4kdt\,|\eta_I(k,t)|^2\right]. \quad (10)$$

From equations (9) and Ψ the Fokker-Planck equation for the probability distribution functional $P[\phi_R,\phi_I;t]$

$$\dot{P}[\phi_R,\phi_I;t] = -H[\phi_R,\phi_I]P[\phi_R,\phi_I;t] \quad (11)$$

with the Fokker-Planck Hamiltonian

$$H = -\int d^4k\Big[\alpha\frac{\delta}{\delta\phi_R(k)}\frac{\delta}{\delta\phi_R(-k)} +\frac{\delta}{\delta\phi_R(k)}(q(k)\phi_I(k)+\varepsilon\phi_R(k)) +\beta\frac{\delta}{\delta\phi_I(k)}\frac{\delta}{\delta\phi_I(-k)} +\frac{\delta}{\delta\phi_I(k)}(-q(k)\phi_R(k)+\varepsilon\phi_I(k))\Big] \quad (12)$$

can be easily derived by the usual method. This equation can be solved explicitly for finite t. Due to the presence of the damping factor of the form of exp$(-\varepsilon t)$ we can safely take the t-infinity limit to get the following real positive probability

$$P_{eq}[\phi_R,\phi_I] = N\exp[-\varepsilon\int d^4k\{|\phi_R(k)|^2 + (1+\frac{\varepsilon^2}{\varepsilon^2+q(k)^2})|\phi_I(k)|^2 - \frac{2\varepsilon}{q(k)}\phi_R(k)\phi_I(-k)\}]. \quad (13)$$

In spite of the apparent difference between $P_{eq}[\phi_R,\phi_I]$ and the Feynman measure $\exp(iS_0)$, correlation functions over P_{eq} coincide to the corresponding ones of the usual path integral. For example, in this scheme the two point function $\langle\phi\phi\rangle$ can be evaluated from the real and imaginary parts separately and its result reproduces the well-known Feynman propagator,

$$\begin{aligned}\langle\phi(k)\phi(k')\rangle &= \langle\phi_R(k)\phi_R(k')\rangle - \langle\phi_I(k)\phi_I(k')\rangle + i\langle\phi_R(k)\phi_I(k')\rangle + i\langle\phi_I(k)\phi_R(k')\rangle \\ &= \delta^4(k+k')(\frac{\varepsilon}{q(k)^2+\varepsilon^2}+i\frac{q(k)}{q(k)^2+\varepsilon^2}) \\ &= \delta^4(k+k')\frac{i}{k^2-\mu^2+i\varepsilon}. \end{aligned} \quad (14)$$

This example shows that in Minkowski stochastic quantization the complex property of expectation values is traced back to that of field variables themselves, while in the path integral quantization the formal Feynman measure exp(iS) makes them complex. This may be an interesting characteristic of Minkowski stochastic quantization. The probability measure P_{eq} may be considered as an alternative definition of the Feynman measure exp(iS).

3. THERMAL EQUILIBRIUM IN MINKOWSKI STOCHASTIC QUANTIZATION

In this section the equivalence between Minkowski stochastic quantization and the usual path integral quantization is shown by using an effective Fokker-Planck distribution defined for physical quantities [4]. Here we exclusively consider the interacting scalar field theory to simplify the notaions. It is, however, easy to generalize to more complicated cases.

The Fokker-Planck Hamiltonian generally takes the following form

$$H = -\int d^4x[\alpha\frac{\delta^2}{\delta\phi_R^2}+\beta\frac{\delta^2}{\delta\phi_I^2} + \frac{\delta}{\delta\phi_R}(\frac{\delta\bar{S}}{\delta\phi})_I - \frac{\delta}{\delta\phi_I}(\frac{\delta\bar{S}}{\delta\phi})_R], \quad (15)$$

where

$$\bar{S} = S[\phi]+iG[\phi],$$
$$(\frac{\delta\bar{S}}{\delta\phi})_R = \mathrm{Re}(\frac{\delta\bar{S}}{\delta\phi}), \quad (\frac{\delta\bar{S}}{\delta\phi})_I = \mathrm{Im}(\frac{\delta\bar{S}}{\delta\phi}).$$

For the damping term $G[\phi]$, we chose the following quadratic one,

$$G[\phi] = \frac{\varepsilon}{2}\int d^4x\phi(x)^2. \quad (16)$$

In interacting cases, it is very difficult to solve the Fokker-Planck equation. However, because we are only interested in the expectation values of the specific functionals, i.e., functionals of $\phi=\phi_R+i\phi_I$ only, let us proceed to an effective Fokker-Planck equation governing the time development of an effective Fokker-Planck distribution $P_{eff}[\phi_R;t]$ defined for physical quantities in ϕ_R-space. Let F be an arbitrary functional of ϕ. Then P_{eff} is defined by the relation

$$\begin{aligned}\langle F[\phi]\rangle &= \int\mathcal{D}\phi_R\mathcal{D}\phi_I F[\phi_R+i\phi_I]P[\phi_R,\phi_I;t] \\ &\equiv \int\mathcal{D}\phi_R F[\phi_R]P_{eff}[\phi_R;t]. \end{aligned} \quad (17)$$

This relation first appeared in [5]. Remembering the following relation between $F[\phi_R+i\phi_I]$ and $F[\phi_R]$

$$F[\phi_R+i\phi_I] = e^{i\chi}F[\phi_R],$$
$$\chi = \int d^4x\phi_I(x)\frac{\delta}{\delta\phi_R(x)}, \quad (18)$$

we can derive P_{eff} from the original probability distribution P;

$$P_{eff}[\phi_R;t] = \int\mathcal{D}\phi_I e^{-i\chi}P[\phi_R,\phi_I;t]. \quad (19)$$

From this equation, the time evolution of P_{eff} can be easily obtained as

$$\dot{P}_{eff}[\phi_R;t] = -\int \mathcal{D}\phi_I H_{eff} e^{-i\chi} P[\phi_R,\phi_I;t] \tag{20}$$

with $H_{eff} = e^{-i\chi} H e^{i\chi}$. By using the following indentities,

$$e^{-i\chi}\phi_R e^{i\chi} = \phi_R - i\phi_I,$$
$$e^{-i\chi}\phi_I e^{i\chi} = \phi_I,$$
$$e^{-i\chi}\frac{\delta}{\delta\phi_R} e^{i\chi} = \frac{\delta}{\delta\phi_R},$$
$$e^{-i\chi}\frac{\delta}{\delta\phi_I} e^{i\chi} = \frac{\delta}{\delta\phi_I} + i\frac{\delta}{\delta\phi_R}, \tag{21}$$

the above equation (20) can be further reduced to a closed equation for P_{eff}

$$\dot{P}_{eff}[\phi_R;t] = -\mathcal{H}_{eff} P_{eff}[\phi_R;t] \tag{22}$$

with an effective Hamiltonian

$$\mathcal{H}_{eff} = \int d^4x \frac{\delta}{\delta\phi_R}\left(\frac{\delta}{\delta\phi_R} - i\frac{\delta}{\delta\phi_R}\bar{S}[\phi_R]\right). \tag{23}$$

We can easily see that the stationary solution of the equation (22) is proportional to $\exp(i\bar{S}[\phi_R])$. This is a formal proof of the equivalence of our theory and the usual one.

The remaining task is to show that this stationary solution is nothing but the equilibrium distribution. Fortunately we can solve the eigenvalue problem of $\mathcal{H}_{eff}$

$$\mathcal{H}_{eff}|u_\lambda\rangle = \lambda|u_\lambda\rangle \tag{24}$$

and show that the real part of λ is non-negative and proportional to ε. Thus the equivalence of Minkowski stochastic quantization to the path integral quantization has been proved. For details of the proof, see [4].

4. CONCLUDING REMARKS

In summary, we have reasonably formulated the stochastic quantization in Minkowski space. Starting from the complex Langevin equation and introducing the damping term, we can derive the corresponding Fokker-Planck equation. Then we can solve this equation to get an explicit expression for the real positive probability distribution in the case of free scalar field. It is interesting to see that this real positive probability measure may be an alternative definition of the Feynman measure. Applications to the case of massive neutral vector field is also performed and reported elsewhere [6]. Finally the equivalence of this quantization method to the Feynman path integral method can be proved by means of the effective Fokker-Planck equation.

REFERENCES

[1] G. Parisi and Y.-S. Wu, Sci. Sin. 29, 483(1981).

[2] Original ideas are found in G. Parisi, Phys. Lett. 131B, 393 (1983); J. R. Klauder, Acta Physica Austriaca, Suppl. XXV, 251(1983) (Springer-Verlag).
For applications of the complex Langevin equation, see H. W. Hamber and H.-c. Ren, Phys. Lett. 159B, 330(1985); J. Ambjørn, M. Flensburg and C. Peterson, *ibid.* 159B, 335(1985); D. J. Callaway, F. Cooper, J. R. Klauder and H. Rose, Nucl. Phys. B262, 19(1985); J. Ambjørn and S-K. Yang, *ibid.* B275, 18(1986).
As for the Minkowski stochastic quantization, see H. Hüffel and H. Rumpf, Phys. Lett. 148B, 104(1984); E. Gozzi, *ibid.* 150B, 119(1985); H. Rumpf, Phys. Rev. D33, 942(1986); H. Nakazato and Y. Yamanaka, *ibid.* D34, 492(1986).

[3] H. Nakazato and Y. Yamanaka, Ref. [2].

[4] H. Nakazato, WU-HEP-86-8 (preprint of Waseda Univ., 1986).

[5] G. Parisi, Ref. [2].

[6] H. Nakazato, WU-HEP-86-9 (preprint of Waseda Univ., 1986).

CONFORMAL FIELD THEORY AND CRITICAL PHENOMENA*

Ian Affleck

Joseph Henry Laboratory, Princeton University,
Princeton, NJ08544, U.S.A.

This paper has nothing to do with experimental high energy physics. However it does have something to do with experimental *low temperature* physics. I wish to demonstrate that certain recent results on conformally invariant (1+1)-dimensional quantum field theories, which are of interest to string theorists, are of direct applicability to the critical behavior of some laboratory low-dimensional systems.

It is widely believed that the behavior of a statistical system near a continuous phase transition becomes independent of microscopic details and falls into some broad universality class. Such transitions generally occur at a non-zero temperature and can usually be described classically. Near the critical point one may describe the system by a Landau-Ginsburg Hamiltonian of the appropriate symmety. This Hamiltonian may also be regarded as the action of a quantum field theory in one lower spatial dimension, in imaginary time. The d-dimensional rotational invariance of the classical Hamiltonian corresponds to Lorentz invariance of the action in d *space-time* dimensions. [i.e. (d-1)space dimensions and 1 time dimension.] The sum over field configurations used to calculate the partition function (or correlation functions) of the classical system is equivalent to the Feynman path integral used to calculate the groundstate energy (or vacuum expectation values) of the quantum field theory.

Some condensed matter systems are in a critical phase at zero temperature, in the sense that they have infinite correlation length. Again one might expect the critical (long-distance) behavior to be described by some universal theory independent of microscopic details. Some such systems (like ferromagnets) behave essentially classically at T = 0 while others are strongly quantum-mechanical. In the latter case the critical theory may again turn out to be a quantum field theory, but now in (*d+1*) space-time dimensions. Note that this correspondence is simpler than the other one; it is merely the fact that a d-dimensional condensed matter quantum system can be described, at long length-scales, by a d-dimensional continuum quantum field theory. The emergence of Lorentz invariance is less obvious in this case, however. (In the previous case the emergence of continuous rotational symmety at a critical point seems a natural consequence of the discrete rotational symmety which is present in the microscopic lattice theory.) Similar to the previous case, however, it can sometimes be shown that operators which break this symmety are irrelevant. This invariance implies that not only the static correlations at large length scales, but also the dynamic correlations at large time-scales (or, equivalently, small energy scales) are given by the critical theory. Indeed the low-energy sector of the condensed matter system will be described by the quantum field theory. This quantum field theory must have an infinite correlation length and this implies that it contains massless excitations with the Lorentz- invariant dispersion relation $E=v|\mathbf{q}|$ (v is the velocity of "light"). (This follows from the fact that a mass gap leads to exponential decay of correlation functions in time and hence also in space.) Thus a neccessary condition that the critical theory for some condensed matter system be Lorentz invariant is that the dispersion relation have one or more linear branches.

These two different possible connections between lattice systems and field theory correspond to Euclidean and Hamiltonian lattice formulations of field theories.

Thus scale-invariant (1+1)-dimensional field theories may be used to describe classical two-dimensional systems at their critical temperature (for example, the two-dimensional Ising model). Alternatively they may describe one-dimensional quantum systems with linear dispersion relations, at T=0. The excitations may be phonons, magnons, etc. Examples of such systems are furnished by one-dimensional antiferromagnets. Some crystals contain chains of magnetic ions which are spatially separated by large non-magnetic complexes. The best known example is $(CH_3)_4NMnCl_3$ (TMMC). The chains of magnetic Mn ions are well-separated by the $(CH_3)_4N^-$ radicals. The inter-chain exchange energy is about 1% of the intra-chain exchange energy. The dispersion relation [measured (1) from inelastic neutron scattering] is shown in Fig. (1). It shows two linear branches at momentum $q \approx 0$ and $q \approx \pi/a$ where "a" is the lattice spacing. This doubling is a consequence of the tendency for the spins to alternate in an antiferromagnet. The velocity is 70.7K. At low but finite T, the quantum system is described by the Euclidean field theory on a strip of width 1/T (setting v = 1). Scale-invariance in (1+1)-dimensional field theory implies conformal invariance. This can be seen from the fact that a scale-invariant theory has a conserved and traceless energy-momentum tensor $T_{\mu\upsilon}$. The light-cone components

$T_L = T_{00} - T_{01}$ and $T_R = T_{00} + T_{01}$,

are, by the conservation and zero-trace equations, functions of the light-cone co-ordinates x_- and x_+ (respectively) only. It then follows that $x_-^n T_L$ and $x_+^n T_R$ define conserved currents for arbitrary n. This infinite set of conserved currents corresponds to the infinite-dimensional conformal symmetry $z \to f(z)$ where $z \equiv x + i\tau$ (τ is imaginary time) and f is an arbitrary analytic function. This infinite symmetry group powerfully constrains the possible theories.

The conformal anomaly parameter, c, plays an important role in labelling the various conformally invariant theories. It can be defined as the response of the system to curving the two-dimensional space-time. Equivalently, it is the parameter that occurs in the Virasoro algebra obeyed by T_L:

$$[T_L(x_-), T_L(y_-)] = \delta'(x_- - y_-)T_L + \delta(x_- - y_-)T_L' + (c/24\pi)\delta'''(x_- - y_-).$$

c is directly measureable in critical one-dimensional systems: it gives (2) the

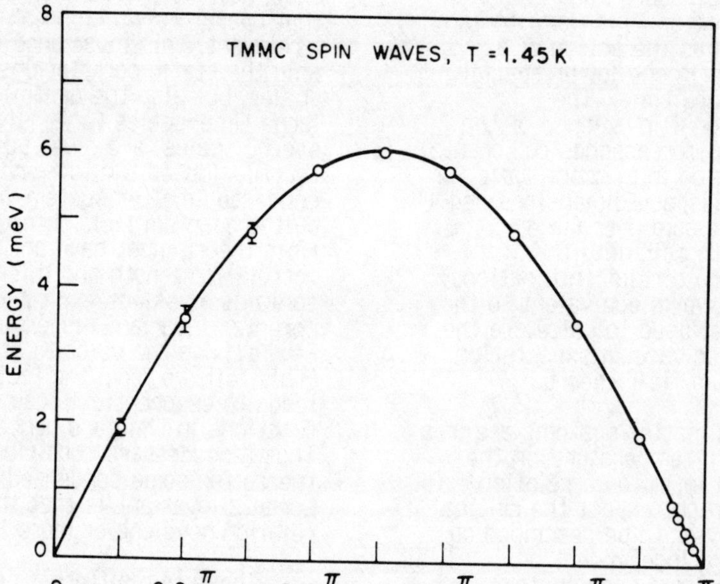

Figure 1: Dispersion relation (1) for TMMC showing sinusoidal behavior and v/a = 70.7.

slope of the specific heat curve at low temperature. The specific heat is always linear in a scale-invariant (1+1)-dimensional field theory, by dimensional analysis. The slope may be extracted from the two - point function for $T_{\mu\upsilon}$. Since $T_{\mu\upsilon}$ has (canonical) dimension two, this Green's function has a universal form up to an overall multiplicative constant which is proportional to c. We thus conclude that the specific heat per unit length is $C = c(\pi/3v)T$, where v is the velocity measured from the dispersion relation. (This formula, and in particular its normalization, can be checked explicitly for a free boson with $c = 1$). Alternatively for a two- dimensional statistical system at its critical temperature defined on a strip of width β, there is a universal leading finite width correction to the free-energy per unit length:

$$\delta F/T_C = -c(\pi/6\beta)$$

The measured magnetic (3) specific heat for the one-dimensional antiferromagnet $CuCl_2 \cdot 2NC_5H_5$ (CPC) is shown in Fig.(2). Note that linear behaviour sets in below 4K. Using the known value of v for CPC the measure vale of c is about .85. The data stops at 2K not because of the type of cooling system used but for a more fundamental reason. Below this temperature the weak inter-chain couplings play a significant role and the system starts to look three-dimensional. While spontaneous symmetry breaking cannot occur in the one-dimensional system (because of Coleman's theorem) it *does* occur in three dimensions. CPC undergoes an ordering transition at about 1.5K.

Some critical systems have continuous symmetries. This occurs for one-dimensional systems of electrons which may be mobile or static (magnetic systems). The symmetry may be U(1) or (approximately) SU(2). In such cases the corresponding field theory has (4) *chiral* U(1) or SU(2) symmetry. This follows from the conformal structure which guaranties that the left and right light-cone components of the current, J_L and J_R are functions of x_- and x_+ only. Hence they generate a chiral pair of symmetries. If the symmetry is non-abelian [SU(2)] then these currents must obey the Kac-Moody algebra. This is the most general current algebra consistent with conformal invariance.

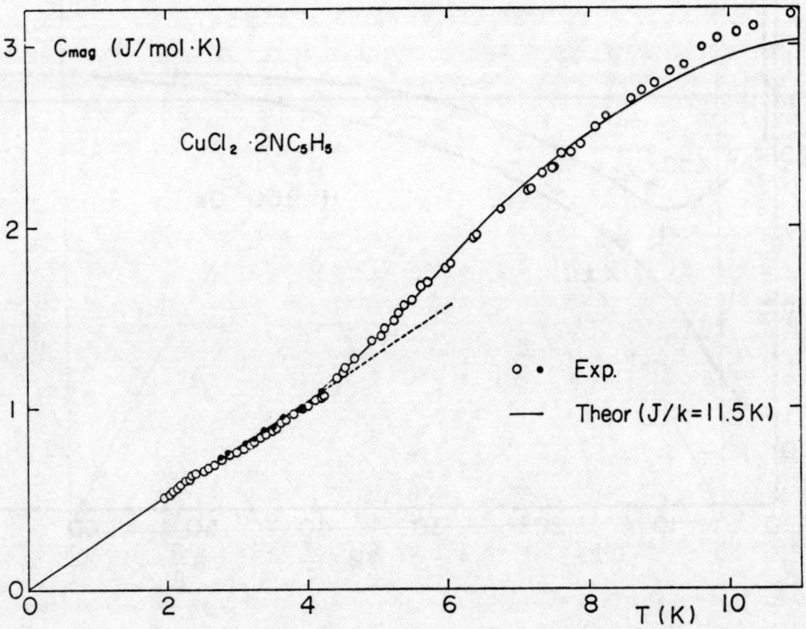

Figure 2: Specific heat of CPC. The dots are experimental data (3). Linear behavior is seen at low T.

$[J_L{}^a(x_-), J_L{}^b(y_+)] = i\varepsilon^{abc} J_L{}^c \delta(x_- - y_-) + (k/2\pi)\delta'(x_- y)$ (1)

The parameter k is the Kac-Moody central charge. It must be an integer for the theory to be unitary (corresponding to a Hermitian Hamiltonian). Again, this parameter has a direct experimental significance. The conserved charge $S^z = \int dx J_0{}^z(x)$ corresponds to the z-component of the total spin in the antiferromagnet (a conserved quantity). The normalization is fixed by the (normal) SU(2) commutation relations obeyed by the conserved charges in both the lattice and continuum theories. The magnetic susceptibility is (in dimensionless units) given by

$$\chi = \langle (S^z)^2 \rangle / T$$

This is proportional to the instantaneous current two-point function at zero spatial momentum. Since conserved currents have canonical dimension one, this two-point function has a universal form up to an overall constant which is proportional to k. Thus we find (5)

$$\chi = (2\pi/v)k \quad (T\to 0).$$

Susceptibilities (6) of TMMC with magnetic field parallel or perpindicular to the z (crystal) axis, are shown in Fig.(3). The point drawn on the T=0 axis corresponds to k=5.

The anomalous dimensions of the fields are also experimentally measureable. At T=0, these determine the power-law decay of the correlation functions. The finite T imaginary time Green's functions can be found (7) by making a conformal transformation that maps the infinte plane onto the strip of width 1/T: $z \to \exp(\pi T z)$. This determines the instantaneous two-point function for a primary field, φ, of anomalous dimension d to be

$$G(x) \equiv \langle \varphi(x)\varphi(0) \rangle \propto [\pi T/\sinh(\pi T x)]^{-2d} \quad (2)$$

The Fourier transform of this quantity is measurable from quasi-elastic neutron scattering in antiferromagnets. Observing that G(x) has power-law behaviour at small x and decays as $\exp(-2d\pi T|x|)$ at large x, we see that the Fourier transform has a power-law tail and an approximately Lorentzian peak of width $2d\pi T$. Thus the anomalous dimension controls the finite temperature broadenning of the magnon peak. Quasi-elastic neutron scattering data on

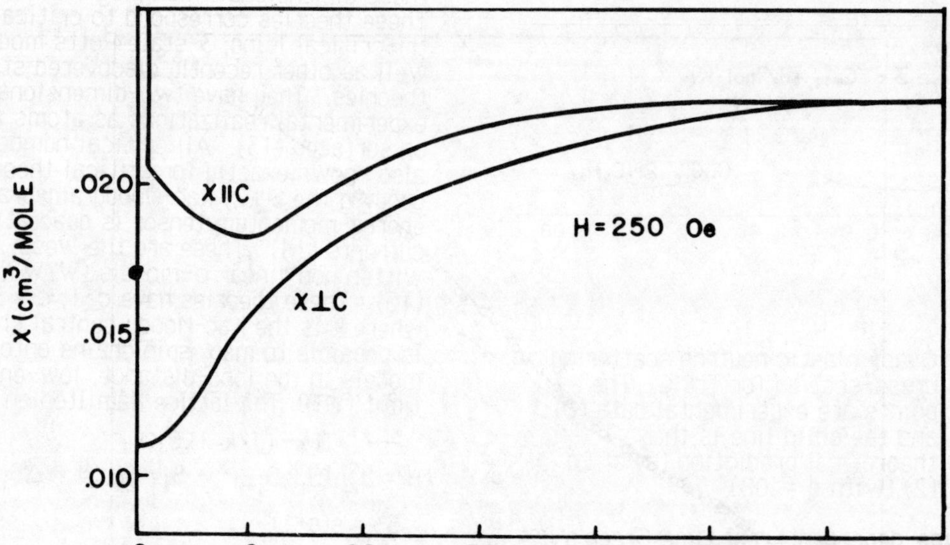

Figure 3: Susceptibility (6) of TMMC. The point marked on the χ-axis corresponds to k=5.

TMMC at various temperatures close to the magnon peak at $q \approx \pi/a$ is shown (8) in Fig. (4). It is well fit (9) by the critical theory with d=.08.

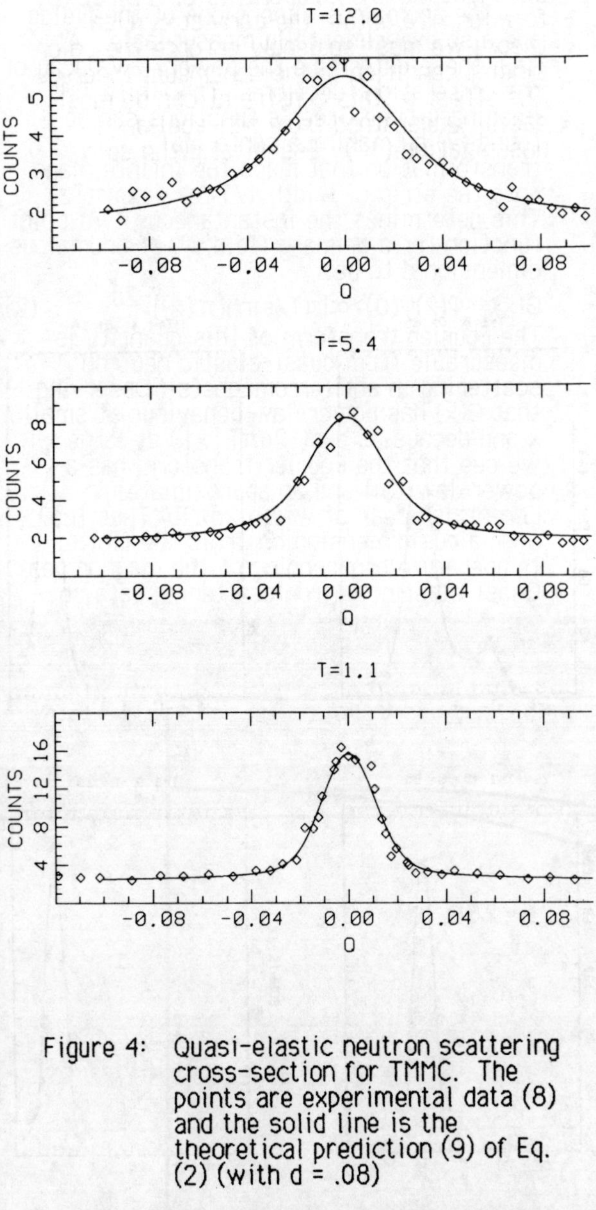

Figure 4: Quasi-elastic neutron scattering cross-section for TMMC. The points are experimental data (8) and the solid line is the theoretical prediction (9) of Eq. (2) (with d = .08)

The time-dependent, real time Green's functions can also be found (9) by analytic continuation from imaginary time. These are given by

$$<\varphi(x,t)\varphi(0,0)> \propto (\pi T)^{(h+\underline{h})} \{\sinh[\pi T(x_+ + i\varepsilon)]\}^{-h} \{\sinh[-\pi T(x_- + i\varepsilon)]\}^{-\underline{h}} \qquad (3)$$

where h and $\underline{h}$ are the left and right anomalous dimensions of φ ($d=h+\underline{h}$). The Fourier transform of this function gives the inelastic neutron- scattering cross-section as a function of energy and momentum exchange. At T=0, there is a power law divergence as $E \to v|q|$ from above (the tail is a consequence of the multi-particle intermediate states). At finite temperature there is an asymmetric peak near $E=v|q|$ with a width πhT (or $\pi \underline{h}T$) for q negative (or positive), and a second peak at negative E suppressed by a Boltzman factor. Inelastic neutron scattering data (10,1) for TMMC is shown in Fig.(5) compared to the critical theory (9) with $h=\underline{h}=.04$.

To make detailed predictions about the critical behaviour of a particular system, the corresponding field theory must be found as must the mapping between the operators in the lattice system and the field theory. Since the critical numbers (c, k and the various d's) are known exacty for many conformally invariant theories, exact predictions about low-temperature systems are sometimes possible. If it is assumed that $c<1$, then there is a discrete set of possible values of c with a corresponding list of possible anomalous dimensions (11). There has recently been progress towards classifying all possible combinations of these dimensions which may occur (12). These theories correspond to critical and tri-critical Ising, 3-state Potts models as well as other recently discovered statistical theories. They have two-dimensional experimental realizations as atoms adsorbed on surfaces (13). All critical numbers are also known exactly for critical theories obeying the SU(2) Kac-Moody algebra whose energy-momentum tensor is quadratic in the currents (14). These are the Wess-Zumino-Witten non-linear σ-models (WZW models) (15). These theories have $c \geq 1$, $c = 3k/(2+k)$ where k is the Kac-Moody central charge. It is possible to map spin chains onto these models in the long-distance, low-energy limit (16). The lattice Hamiltonian is

$$H = J\sum_i [S_i^x \cdot S_{i+1}^x + S_i^y \cdot S_{i+1}^y + \Delta S_i^z \cdot S_{i+1}^z],$$
$$S^2 = s(s+1) \qquad (4)$$

For an isotropic system $\Delta=1$.

The critical theory is derived by first using an exact lattice fermion representation of the spin system. The best antiferro-magnets generally have magnetic ions with half-filled outer (s or d) shells. A Hund's rule coupling between the outer-shell electrons makes the spins parallel. Thus we may (quite physically) think of a spin-s variable as being quadratic in 2s different types of electrons:

$$S = (1/2)\psi^{+\alpha i}\sigma_{\alpha}^{\ \beta}\psi_{\beta i} \qquad (5)$$

Here i runs over 1,2,3, ... ,2s. To project out spin-s we need to restrict ourselves to states with 2s electrons which are totally antisymmetric in the i index, and hence totally symmetric in the spin index, α. We may think of i as being a color index and project out color singlet states of 2s electrons. The ψ's should be thought of Susskind fermions. In the continuum limit we keep only Fourier modes of the ψ's with q $\approx \pm\pi/2a$. These two branches define the left and right moving components of a Dirac fermion. To obtain the correct critical theory we must somehow enforce the color singlet condition in the continuum theory. The effect of this constraint can be most easily understood from non-abelian bosonization (15). Essentially the spin

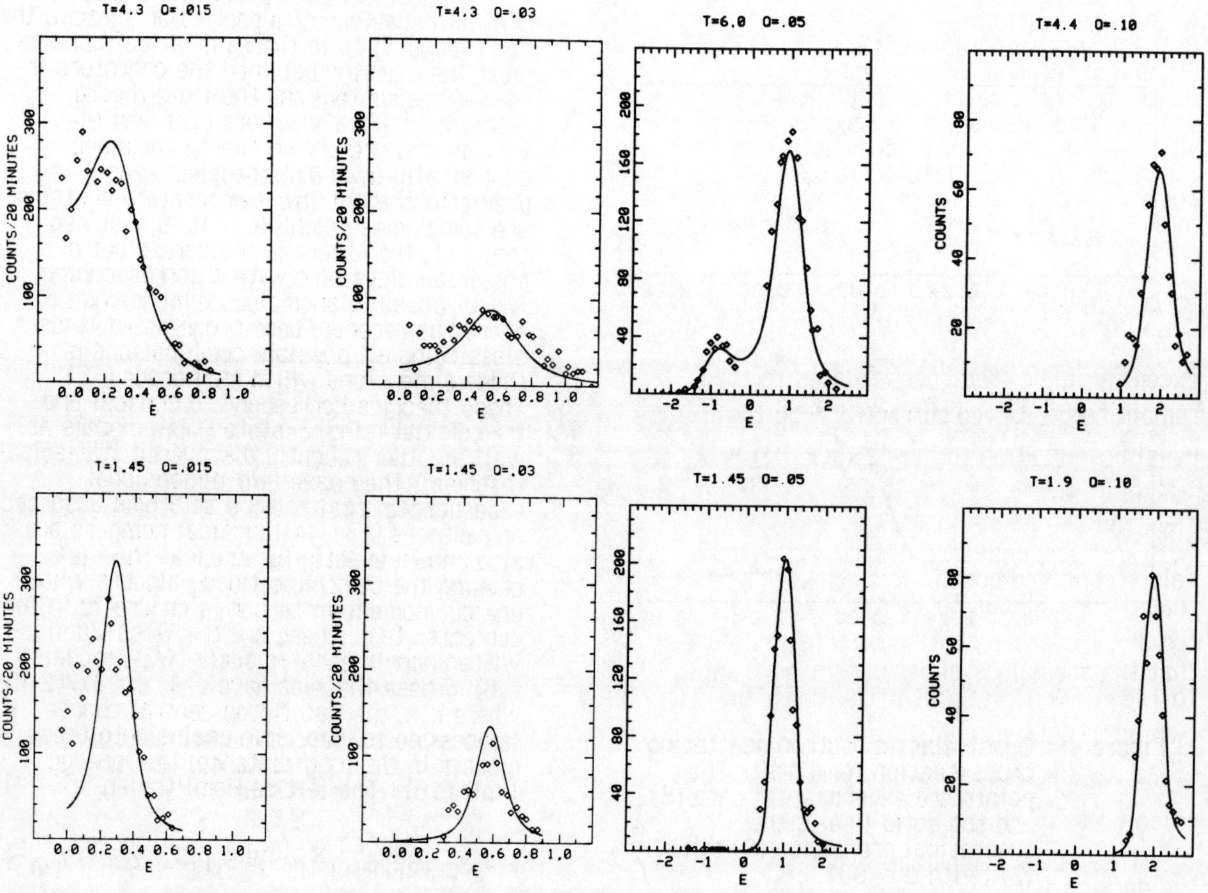

Figure 5: Inelastic neutron scattering cross-section in TMMC. The points are experimental data the solid lines are theoretical predictions (1,10) and (9) from Eq(3) with h =h= .04.

degrees of freedom can be represented by an SU(2) WZW model with central charge k=2s, the color degrees of freedom by an SU(2s) WZW model with k=2 and the charge [U(1)] degrees of freedom by a free boson. The constraints effectively eliminate all but the spin degrees of freedom. In the continuum limit the spin operators are represented as

$$\mathbf{S}(x) \approx \mathbf{J}_0(x) + (-1)^{x/a} \cdot \mathrm{const} \cdot \mathrm{tr}(g - g^{+})\boldsymbol{\sigma} \qquad (6)$$

where g is the fundamental field of the WZW model (an SU(2) matrix transforming as a primary doublet under the left and right SU(2) groups). Thus the anomalous dimensions governing the spin correlation functions for $q \approx \pi/a$ are those of g: $h = \underline{h} = 3/4(2+2s)$.

These values of c and k agree exactly with the specific heat and susceptibility of Bethe ansatz integrable spin chains (17). (These correspond to unrealistic Hamiltonians however, except in the case s=1/2) They also seem to agree quite well with preliminary numerical simulations on the realistic s=3/2 Hamiltonian (18). To compare with experiment it is essential to include the effects of a small planar anisotropy [ie. $\Delta<1$ in Eq.(4)] This corresponds to adding terms to the field theory that break the chiral SU(2) symmetry down to chiral U(1). This has the effect of giving a mass to all degrees of freedom except a single free boson related to the remaining conserved current $J_\mu{}^Z$ by the usual bosonization formula

$$J_L \propto \partial_- \varphi.$$

Thus the value of c jumps discontinuously with infinitesimal anisotropy to c = 1, the value for a free boson. It can be argued (9) that the anomalous commutator parameter, k, for J^{Z}, increases continuously from 2s and that the anomalous dimensions $h = \underline{h}$ jumps to 1/8s. For finite anisotropy there are two different exponents for S^Z and S^X (or S^y), d_Z and d_X respectively. For arbitrary anisotropy $d_Z = 1/2k$.

These predictions agree quite well with the data on TMMC, with s=5/2. The parameter k can be determined from the parallel susceptibility shown in Fig.(3). This gives k=6.24. The predicted value for the isotropic system k=5 looks like quite a reasonable extrapolation from higher T where $\chi_\perp$ and $\chi_\parallel$ are almost the same. The predicted value of d= 1/2k = .08 is in good agreement with both quasi-elastic and inelastic neutron scattering data Fig.(4,5).

* Supported in part by N.S.F. Grant Phy80-19754 and the A.P. Sloan Foundation.

REFERENCES

(1) G. Shirane and R.J. Birgeneau, Physica 86-88B, 639 (1977).

(2) H.W.J. Blote, J.L. Cardy and M.P. Nightingale, Phys. Rev. Lett. 56, 742 (1986); I. Affleck, ibid 56, 746 (1986).

(3) K.Takeda, S. Matsukawa and T. Haseda, J. Phys. Soc. Japan 30, 1330 (1971).

(4) I. Affleck, Phys. Rev. Lett. 55, 1355 (1985).

(5) I. Affleck, Phys. Rev. Lett. 56, 2763 (1986).

(6) L.R. Walker, R.E. Dietz, K. Andres and S. Darack, Sol. State. Comm. 11, 593 (1972).

(7) J. LCardy, J. Phys. A17, L385 (1984).

(8) R.J. Birgeneau, R. Dingle, M.T. Hutchings, G. Shirane and S.L. Holt, Phys. Rev. Lett. 26, 718, (1971).

(9) I. Affleck, Princeton preprint, 1986.

(10) M.T. Hutchings, G. Shirane, R.J. Birgeneau and S.L. Holt, Phys. Rev. B5, 1999 (1972).

(11) A. Belavin, A. Polyakov and A. Zamolodchikov, Nucl. Phys. B241, 333 (1984); D. Friedan, Z. Qiu and S. Shenker, Phys. Rev. Lett. 52, 1575 (1984).

(12) D. Gepener, Princeton preprint, 1986; A. Cappelli, C. Itzykson and J.-B. Zuber, Saclay preprint, 1986.

(13) For a recent review see T.L. Einstein, Proceedings of the John Hopkins Workshop, Sept. 1986.

(14) V. Knizhnik and A. Zamolodchikov Nucl. Phys. B247, 83 (1984).

(15) E. Witten, Comm. Math. Phys. 92, 455 (1984).

(16) I. Affleck, Nucl. Phys. B265 [FS15], 409 (1985).

(17) P. Kulish and E. Sklyanin, Lecture Notes in Physics 151, 61 (1982); P. Kulish, N. Yu. Reshetikhin and E. Sklyanin, Lett. Math. Phys. 5, 393 (1981); L. Takhtajan, Phys. Lett. 90A, 479 (1982); J. Babudjian, Phys. Lett. 90A, 479 (1982); Nucl. Phys. B215, 317 (1983).

(18) T. Ziman, unpublished.

INFINITE DIMENSIONAL ALGEBRAS AND CONFORMAL INVARIANCE FOR SELF-DUAL GAUGE THEORIES

H. J. de VEGA

LPTHE, Univ. P. et M. CURIE, Tour 16, ler. ét., 4, Place Jussieu
75230, Paris Cedex 05, FRANCE

A new conformally covariant linear system is reported for the self-dual Yang-Mills theory. The spectral parameter is here a projective twistor Conformally covariant Bäcklund transformations (B.T.) are constructed for gauge groups SL(N) and SU(N) as well as the algebra of B.T. An infinite dimensional algebra with five indices is found. Loop algebras (centerless Kac-Moody algebras) follow as particular cases.

The self-dual Yang-Mills equations are the simplest relativistic integrable theory in four space dimensions [1,2]. More interesting and complicated models are N = 3 and N = 4 extended SUSY Yang-Mills in four dimensions and SUSY Yang-Mills in ten dimensions [3].

The self-duality equations read for a general gauge group in four Euclidean space dimensions

$$F_{\mu\nu} = F^{*}_{\mu\nu} \tag{1}$$

where

$$F^{*}_{\mu\nu} = \frac{1}{2}\,\varepsilon_{\mu\nu\lambda\sigma}\,F_{\lambda\sigma}$$
$$F_{\mu\nu} = \partial_\mu A_\nu - \partial_\nu A_\mu + [A_\mu, A_\nu] \tag{2}$$

$F_{\mu\nu}$ is the field tensor taking values in the Lie algebra.

It is convenient to use complex coordinates $x = (y, \bar{y}, z, \bar{z})$ defined as [4]

$$\sqrt{2}\,y = x_1 + i x_2 \quad , \quad \sqrt{2}\,z = x_3 - i x_4$$
$$\sqrt{2}\,\bar{y} = x_1 - i x_2 \quad , \quad \sqrt{2}\,\bar{z} = x_3 + i x_4 \tag{3}$$

In these coordinates Eq.(1) reads

$$F_{yz} = F_{\bar{y}\bar{z}} = 0 \quad , \quad F_{y\bar{y}} + F_{z\bar{z}} = 0 \tag{4}$$

These three equations are the compatibility conditions of the linear system [2]

$$(D_{\bar{z}} - \lambda' D_y)\,\Psi(x,\lambda') = 0$$
$$(D_{\bar{y}} + \lambda' D_z)\,\Psi(x,\lambda') = 0 \tag{5}$$

Here λ' is a complex spectral parameter, D_μ, $D_{\bar{\mu}}$ are covariant derivatives and Ψ a matrix valued function.

The self duality Eqs.(1) are conformally covariant while the linear system (5) is not. In refs. (5) and (6) a conformally covariant linear system is proposed. It reads

$$\hat{s}_1\Psi \equiv \left[(\mu\bar{y} + \lambda\bar{z} + a)D_{\bar{z}} + (\lambda y - \mu z - b)D_y\right]\Psi = 0$$
$$\hat{s}_2\Psi \equiv \left[(\mu\bar{y} + \lambda\bar{z} + a)D_{\bar{y}} - (\lambda y - \mu z - b)D_z\right]\Psi = 0 \tag{6}$$

Here μ, λ, a, b are projective spectral parameters. The transformation laws under the reduced conformal group formulate easily using quaternions

$$x = \begin{pmatrix} y & z \\ \bar{z} & -\bar{y} \end{pmatrix} \tag{7}$$

A conformal transformation yields

$$x \to x' = (Ax + B)(Cx + D)^{-1}$$

where A, iB, iC, D are quaternions and

$$\det\begin{pmatrix} A & B \\ C & D \end{pmatrix} = 1$$

The object $\Lambda = (-b, a, -\lambda, \mu)$ transforms linearly under conformal transformations as

$$\Lambda \rightarrow \Lambda' = \begin{pmatrix} A & B \\ C & D \end{pmatrix} \Lambda \quad (9)$$

So, we can call it a <u>spectral</u> <u>twistor</u>. The linear system transforms as

$$\Psi(x) \rightarrow \Psi'(x') = \Psi(x)$$

and

$$\begin{pmatrix} \hat{s}_1 \\ \hat{s}_2 \end{pmatrix} \rightarrow (Cx + D) \begin{pmatrix} \hat{s}_1 \\ \hat{s}_2 \end{pmatrix} \quad (10)$$

The linear system admits the compact notation

$$[(\lambda, -\mu) x^T + (-b, a)] D \Psi = 0 \quad (11)$$

where

$$D = \begin{pmatrix} D_y & D_z \\ D_{\bar{z}} & -D_{\bar{y}} \end{pmatrix}$$

Conformally covariant Bäcklund transformations (BT) follow from Eqs.(6) or (11). A BT is a transformation that associates to a given classical solution A, a family of solutions depending on one or more continuos parameters. We exhibit here infinitesimal BT associating to a given self-dual field A an infinitesimally close self-dual field $A + \delta A$

When the gauge group is SL(N) one finds for δA [6]

$$\delta A = \begin{pmatrix} \delta A_y & \delta A_z \\ \delta A_{\bar{z}} & -\delta A_{\bar{y}} \end{pmatrix} =$$

$$= \begin{pmatrix} \dfrac{L + K\bar{z} + J\bar{y}}{\mu\bar{y} + \lambda\bar{z} + a}, & 0 \\ 0, & \dfrac{M - Ky + Jz}{\mu z - \lambda y + b} \end{pmatrix} [D, \Psi T_a \Psi^{-1}] \quad (12)$$

Here Ψ is a solution of Eq.(6) or(11) for the self-dual field A, T_a is a SL(N) generator,solution of the free field linear system. That is

$$T_a = T_a(R, Y)$$

where

$$R = 2\mu\bar{y} + a + Y(b + 2\mu z), \quad Y = \frac{\lambda\bar{z} + \mu\bar{y} + a}{\lambda y - \mu z - b}$$

J, K, L, M are arbitrary constants. The object

$$\Theta = (-M, L, -K, J)$$

behaves as a twistor under conformal transformations.

It is possible to expand a general conformally covariant BT $\delta(\Lambda, \Theta)$ in a basis

$$\delta(\Lambda, \Theta) = \sum_a \sum_{n,m \in \mathbb{Z}} \sum_{\alpha=1}^{4} \epsilon^a R^n Y^m \Theta^\alpha \delta_\alpha^{a,nm}(\Lambda)$$

where ϵ^a are infinitesimal parameters. So our BT $\delta_\alpha^{anm}(\Lambda)$ are non-trivial functions of the spectral twistor Λ. They have two gradations (n,m) , a twistorial index α and the Lie algebra index a .

Since the $\delta_\alpha^{anm}(\Lambda)$ provide symmetry transformations of the solution space, it is interesting to know their algebra. This algebra is explicitely derived in ref.(6) with the final result

$$[\delta_\alpha^{a,mn}(\Lambda_1), \delta_\beta^{b,pq}(\Lambda_2)] = \frac{f^{abc}}{\Lambda_1 \cdot \hat{\Lambda}_2} [$$

$$\hat{\Lambda}_{1\beta} \delta_\alpha^{c,p+m,n+q}(\Lambda_1) - \hat{\Lambda}_{2\alpha} \delta_\beta^{c,p+m,n+q}(\Lambda_2)] \quad (13)$$

+field dependent gauge transformations

Here

$$\hat{\Lambda} = \begin{pmatrix} -\lambda\bar{z} - \mu\bar{y} - a \\ \lambda y - \mu z - b \\ \mu(y\bar{y} + z\bar{z}) + ay + b\bar{z} \\ \lambda(y\bar{y} + z\bar{z}) + az - b\bar{y} \end{pmatrix} \quad (14)$$

The pair $(\Lambda, \hat{\Lambda})$ forms an ambitwistor

$$\Lambda \cdot \hat{\Lambda} = 0$$

Since the coefficients on the r.h.s. of Eq. (13) are coordinate dependent this equation tells us that the commutator of two BT gives an infinite series of BT. The terms of this series follow from the expansion

$$\frac{1}{\Lambda_1 \cdot \hat{\Lambda}_2} = \sum_{r,s=0}^{\infty} C_{rs}(\Lambda_1, \Lambda_2) Y^r R^s \quad (15)$$

where the C_{rs} are related to Jacobi polynomials.

One finds in this way [6]

$$\left[\delta^{a,mn}_{\alpha(\Lambda_1)}, \delta^{b,pq}_{\beta(\Lambda_2)} \right] = f^{abc}\ \times$$

$$\sum_{r,s\in Z} \left\{ K^{rs}_{\beta}(\Lambda_1,\Lambda_2)\, \delta^{c,p+m+r,n+q+s}_{\alpha}(\Lambda_1) - (1\leftrightarrow 2, \alpha\leftrightarrow\beta) \right\} \tag{16}$$

In two-dimensional integrable theories the algebra of BT often leads to loop algebras [7]. That is Kac-Moody algebras without center. As one sees from Eqs.(14)-(16) this is not generically the case here. One can find loop algebras from Eq.(16) by restricting to particular twistors $\Lambda_1, \Lambda_2, \Theta_1$ and Θ_2 [6]. However, this choice is not conformally covariant. So, the loop algebras does not seem to be intrinsic structures in four dimensional integrable gauge theories.

It must be stressed that the B.T. (12) for the SL(N) gauge group and their algebra have a direct generalization for SU(N) gauge groups. Loosely speaking one takes for SU(N) the 'real part' of the BT for SL(N) [6].

It is easy to construct an infinite number of conserved currents from the linear system (6). One takes the vector current

$$V^{(\pm)}_{\mu}(x,\lambda) = \sum^{\nu(\pm)}_{\mu} \partial_\nu \Psi(x,\lambda)$$

where $\sum^{\nu(\pm)}_{\mu}$ are constant antisymmetric and antiselfdual tensors [5]

$$\sum^{\nu(+)}_{\mu} = \begin{pmatrix} -v_3 & 0 & 0 & -v_1 \\ 0 & v_3 & v_2 & 0 \\ 0 & v_1 & -v_3 & 0 \\ -v_2 & 0 & 0 & v_3 \end{pmatrix}$$

It is trivial to check that $\partial^\mu V^{(\pm)}_\mu = 0$

The coefficients of the expansion of $V^{(\pm)}_\mu(x,\Lambda)$ in a CP^3 basis yields an infinite number of tensorial conserved currents. For example, one can write

$$V^{\pm}_\mu(x,\Lambda) = \sum_{n,m,\ell} \left(\frac{\lambda}{a}\right)^n \left(\frac{\mu}{a}\right)^m \left(\frac{b}{a}\right)^\ell V^{\pm,nm\ell}_\mu(x)$$

So, the $V^{\pm,nm\ell}_\mu(x)$ provide $3 \times \infty^3$ conserved currents.

REFERENCES

1 R. Ward; Phys. Lett. 61A,81 (1977)
M. F. Atiyah, "Geometry of Yang-Mills Fields", Lezione Fermione,Pisa 1979.
M.F. Atiyah,V.G. Drinfeld,N. Hitchin and Yu. Manin,Phys. Lett. 65A,185(1978)

2 A.A. Belavin and V. E. Zakharov,Phys. Lett. 73B,53(1978)

3 E. Witten,Phys. Lett. 77B,394 (1978)
E. Witten, Nucl. Phys. B266,245(1985)
J. Harnad in Proceedings of the Paris-Meudon Colloquium,september 1986,Eds. H. J. de Vega and N. Sanchez. World Scientific.

4 C. N. Yang, Phys. Rev. Lett. 38,1377, (1977).

5 J. Avan, H. J. de Vega and J. M. Maillet Phys. Lett. 171B, 255 (1986).

6 J. Avan and H. J. de Vega, LPTHE Paris preprint, 86/23,May 1986.

7 L. Dolan, Phys. Rev. Lett. 47,1371(1981)
K. Ueno,RIMS-374 (1981)
H. Eichenherr in Springer Lectures in Physics vol. 180.

ANOMALIES MADE EXPLICIT

Lay Nam Chang and Yi-gao Liang
Physics Department, Virginia Polytechnic Institute
Blacksburg, Virginia 24061

We explain the topological origins of anomalies in quantum theories by giving explicit examples of such anomalies in specific field theories.

Classical symmetries sometimes do not survive the process of quantization. When this happens, we say the quantum theory has anomalies. In many instances where local symmetries are involved the anomalies turn out to have topological origins [1,2,3]. The quantum partition functional for such theories is expected to be invariant under local gauge transformations, so that the functional is defined on the orbit space $\mathfrak{A}/\mathfrak{G}$, where $\mathfrak{A}$ is the space of gauge potentials, and $\mathfrak{G}$ the group of gauge transformations. (For technical convenience, only those transformations with a base point are included in $\mathfrak{G}$ in order to make $\mathfrak{A} \longrightarrow \mathfrak{A}/\mathfrak{G}$ a bundle projection.) In cases where fermions are involved the partition functional is the determinant of the appropriate Dirac operator, which can be defined as the regularized product of the associated eigenvalues. When one or more of these eigenvalues goes through zero for specific configurations in $\mathfrak{A}$, the phase of the determinant may be ill-defined in $\mathfrak{A}/\mathfrak{G}$, in which case then the partition functional will not be gauge invariant. These features follow from the Index Theorem of Atiyah and Singer, and are controlled by non-trivial topologies in the orbit space $\mathfrak{A}/\mathfrak{G}$ [3]. In what follows, we exhibit two explicit examples of these relations. We believe these examples can help in clarifying the role topology plays in the physics of quantum gauge fields.

It will turn out not to be necessary to consider the whole of $\mathfrak{A}$ to exhibit the non-triviality of the phase under the projection $\mathfrak{A} \longrightarrow \mathfrak{A}/\mathfrak{G}$. In the complex case, a two-parameter family of gauge potential will be sufficient. For instance, on a compactified space of even dimensions S^{2n}, and gauge group $SU(N>n)$, $\pi_1(\mathfrak{A}/\mathfrak{G}) \cong 0$, $\pi_2(\mathfrak{A}/\mathfrak{G}) \cong \pi_1(\mathfrak{G}) \cong \pi_{2n+1}(SU(N)) \cong Z$. We consider the simplest case of $n=1$, and $N=2$ to illustrate the idea. A 1-parameter family of gauge transformations is given by

$$g_s = \begin{bmatrix} e^{-is/2} & 0 \\ 0 & e^{is/2} \end{bmatrix} \times \begin{bmatrix} \cos\frac{s}{2} + i\sin\frac{s}{2}\cos\theta & \sin\frac{s}{2}\sin\theta\, e^{i\phi} \\ -\sin\frac{s}{2}\sin\theta\, e^{-i\phi} & \cos\frac{s}{2} - i\sin\frac{s}{2}\cos\theta \end{bmatrix} \quad (1)$$

which represents the basic non-contactible loop in $\mathfrak{G}$. $\{g_s^{-1} dg_s\}$ is a loop not contractible in the fiber at [0]. It is contractible to an arbitrary point A_o in $\mathfrak{A}$ by the linearity of $\mathfrak{A}$. The 2-parameter family of gauge potentials we are going to use is $\{(1-t)A_o + t g_s^{-1} dg_s\}$, which is a disc in $\mathfrak{A}$. The boundary consists of "pure gauge" potentials and therefore projects to a point in the quotient $\mathfrak{A}/\mathfrak{G}$. The disc thus forms a sphere in $\mathfrak{A}/\mathfrak{G}$, and represents a generator of $\pi_2(\mathfrak{A}/\mathfrak{G})$. We choose A_o to be

$$A_o = \begin{bmatrix} -i\sin\theta & e^{i\phi}\cos\theta \\ -e^{-i\phi}\cos\theta & i\sin\theta \end{bmatrix} d\theta$$

$$+ \begin{bmatrix} 0 & ie^{i\phi} \\ -ie^{-i\phi} & 0 \end{bmatrix} \sin\theta d\phi. \qquad (2)$$

For the metric $ds^2 = d\theta^2 + \sin^2\theta d\phi^2$, the chiral Dirac operator can be written as [7]

$$-iD_{\pm}[A(s,t)] = \pm\partial_\theta - \frac{i}{\sin\theta}\partial_\phi \pm \frac{1}{2}\cos\theta$$

$$-p_{\pm}\begin{bmatrix} \sin\theta & ie^{i\phi}(\pm 1+\cos\theta) \\ ie^{-i\phi}(\pm 1-\cos\theta) & -\sin\theta \end{bmatrix},$$

where

$$p_{\pm} = \pm\, ip_1 + p_2,$$

$$p_1 = \frac{t}{2}\sin s + (1-t),$$

$$p_2 = \frac{t}{2}(1 - \cos s). \qquad (3)$$

The boundary condition is anti-periodic in ϕ.

The eigenvalue equation is defined by

$$-D_{\pm}(p')D_{\mp}(p)\chi_{\mp} = \lambda_{\mp}(p',p)\chi_{\mp},$$

where p' is viewed as constant. All the eigenvalues but one pair up to give a product independent of the parameters characterizing the potential A. The un-paired eigenvalue is

$$\lambda_o = (1 - 2p'_{\pm})(1 - 2p_{\mp}). \qquad (4)$$

Since the fermion determinant is formally the product of all the eigenvalues, it can be regularized to λ_o multiplied by an arbitrary constant. Also since p' is a constant, $(1 - 2p')$ can be factored out. Hence, the fermion determinant is found exactly, in this example, and is given by

$$\mathrm{Det}[-iD_{\pm}(p)] = 1-2p_{\pm} = (1-t)(1\pm 2i) + te^{\pm is}. \qquad (5)$$

At the boundary of the disc $t = 1$, $\mathrm{Det}[-iD_{\pm}(p)] = e^{\pm is}$. The phase winds around the boundary with winding number ± 1. It is clearly gauge dependent. Fig. 1 shows the constant phase curves on the disc, and we see that the phase change is actually anchored by the zero mode located at

$$(t\cos s, t\sin s) = \left(-\frac{1}{1+\sqrt{5}}, -\frac{2}{1+\sqrt{5}}\right). \qquad (6)$$

One can easily check that the family topological index of the gauge fields is equal to this winding number.

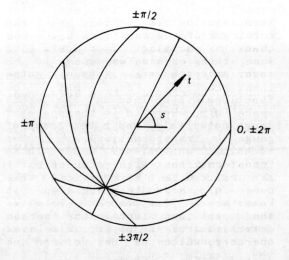

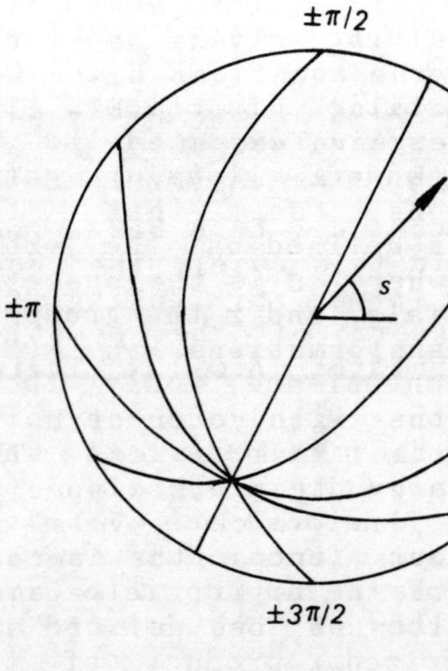

Fig. 1

On the other hand, for non-chiral fermions, $p' \equiv p$, and

$$\mathrm{Det}[-iD] = \lambda_o(p,p) = 5(1-t)^2 + t^2 + 2t(1-t)(\cos s + 2\sin s), \qquad (7)$$

which is gauge independent.

The anomalies can also be studied using canonical quantization method [4]. They usually show up as the ill behavior of the fermion Fock space on the space of background gauge potentials. Here we choose to look at the non-perturbative anomaly in 2+1 dimensions [5]. (Such anomalies were first studied by Witten for theories in three dimensions[6].) More specifically, we consider the gauge group SU(2) in the $A_o = 0$ gauge. The two spatial dimensions are compactified to a 2-sphere. $\mathfrak{A}$ and $\mathfrak{B}$ then are the same as in the previous example with $\pi_1(\mathfrak{B}) \cong Z$. The single particle Hamiltonian has the form

$$H = \begin{bmatrix} 0 & -iD_-(p) \\ -iD_+(p) & 0 \end{bmatrix}. \tag{8}$$

The eigenvalue problem of H is the same as

$$-D_+(p)D_-(p)\psi_- = E^2\psi_-,$$

$$\psi_+ = -\frac{i}{E}D_-\psi_-. \tag{9}$$

The solution is given by the solutions of the equations discussed above by setting $p' = p, E = \pm\surd\lambda$. None of the eigenvalues are zero, except $E_\pm = \pm\surd\lambda_o(p,p)$. The problem we consider here has a real structure. In this case, the Fock space is formed by taking the exterior algebra of the positive energy eigenstates $\Lambda(\Phi_+)$. Therefore, only the eigenvector of $E = +\surd\lambda$ goes into the Fock space. The zero in energy then is a special line that decides which levels we need to put into the second quantized states. In Fig. 2 we show the eigenvalues $E_\pm$ as s, t are varied. The different curves are for different values of s. The position of the zero mode degeneracy is the same as before. The eigenfunctions in the real basis are

$$\psi_\pm^{N,S} = \frac{1}{2\surd\pi}\begin{bmatrix} -\sin(\frac{\alpha}{2}+\frac{1-\delta^{N,S}}{2}\phi \mp \frac{\pi}{4})\sin\frac{\theta}{2} \\ \cos(\frac{\alpha}{2}-\frac{1+\delta^{N,S}}{2}\phi\mp\frac{\pi}{4})\cos\frac{\theta}{2} \\ -\sin(\frac{\alpha}{2}-\frac{1+\delta^{N,S}}{2}\phi\mp\frac{\pi}{4})\cos\frac{\theta}{2} \\ \cos(\frac{\alpha}{2}-\frac{1-\delta^{N,S}}{2}\phi\mp\frac{\pi}{4})\cos\frac{\theta}{2} \end{bmatrix} \tag{10}$$

where α is the phase angle of $[(1-t)(1-2i) + te^{-is}]$. For fixed t, from Fig. 2, it is easy to see that before the zero mode degeneracy is reached α goes back to 0 when s varies from 0 to 2π. However, after the zero mode degeneracy is crossed, α goes to 2π. In particular the wave function changes sign at t=1 when we go around the circle in $\mathfrak{B}$ at [0]. Since Ψ_- is not in the Fock space, we may say the Fock space has a topological charge 1 mod 2. This indicates that the Fock space cannot be a representation of the gauge transformation group and we have an anomaly. In the context of the fermion determinant, the anomaly represents a sign ambiguity in the partition function, and is non-perturbative in the sense that its presence is revealed only if we consider "large" gauge transformations [6].

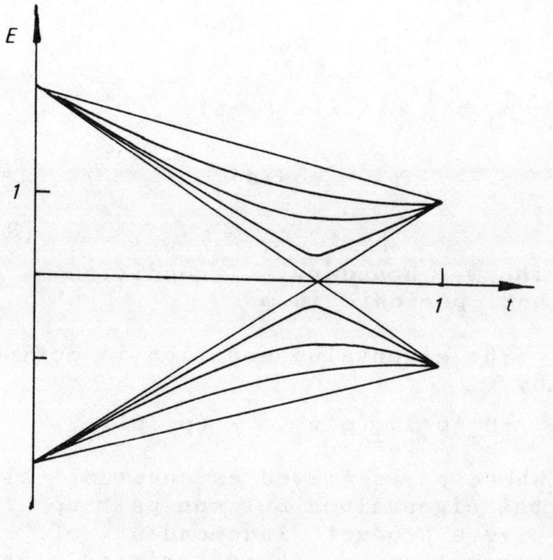

Fig. 2

A fuller account of this work can be found in [7]. For an approach to this problem related to what has been reported here, see [8]. This research has been supported by a grant from the National Science Foundation, NSF-PHY -85-07170.

REFERENCES

[1] See for example Bardeen, W. A., White, A. R., eds.: Symposium on Anomalies, Geometry, Topology. Singapore: World Scientific 1985 and references therein.

[2] Alvarez-Gaume, L., Ginsparg, P.: The topological meaning of non-abelian anomalies. Nucl. Phys. B243, 449 (1984).
[3] Atiyah, M., Singer, I.M.: Dirac operators coupled to vector potentials. Proc. Nat. Acad. Sci. USA 81, 2597 (1984).
[4] Nelson, P., Alvarez-Gaume, L.: Hamiltonian interpretation of anomalies. Comm. Math. Phys. 99, 103-114 (1985).
[5] Redlich, A.N., Gauge non-invariance and parity non-conservation of three dimensional fermions. Phys. Rev. Lett. 52, 18-21, (1984).
[6] Witten, E.: An SU(2) anomaly. Phys. Lett. 117B, 324 (1982).
[7] Chang, L.N., Liang, Y., to appear in Comm. Math. Phys.
[8] Forte, S.: Two- and Four-Dimensional Anomalies with an instanton background, Phys. Lett., 174B, 309-342, (1986); MIT-CTP-1395.

FUNCTIONAL GAUGE STRUCTURE AND TOPOLOGICAL ASPECTS IN QUANTUM FIELD THEORIES

Yong-Shi Wu

Department of Physics, University of Utah
Salt Lake City, Utah 84112, USA

Nontrivial functional Abelian gauge structure emerges in the configuration space of gauge theories or nonlinear sigma models when there are topological Lagrangians or appropriate couplings to fermions. Various topological aspects of quantum field theories can be understood in terms of this induced gauge structure in canonical Hamiltonian approach. These aspects include 1) the θ-vacua and the quantized topological mass in gauge theories, 2) anomalous spin and statistics for topological solitons in non-linear sigma models, and 3) the structure of anomalous gauge theories both in even and odd dimensions.

1. INTRODUCTION

In quantum mechanics it is the non-integrable (or path-dependent) phase factor or, equivalently, the Abelian gauge field in configuration space, which lies at the heart of many spectacular topology-related phenomena, such as the Aharonov-Bohm effect, the Dirac quantization for magnetic monopoles and the flux quantization in superconducting rings. Recently, Zee and myself (1) have generalized the concept of an abelian gauge structure to quantum field theories. Here I will summarize the results of a series of papers (1-6) in this direction and show that the functional abelian gauge structure indeed reveals a universal element in various topological aspects of quantum field theories.

The basic observation is the following. A quantum field theory can be viewed as quantum mechanics of a "point particle" moving in the (infinite dimensional) field-configuration space $\mathbb{C}$. If $\mathbb{C}$ is topologically non-trivial, e.g. with the homotopy group $\pi_1(\mathbb{C})$ or $\pi_2(\mathbb{C})$ $\neq e$, it can support topologically nontrivial abelian gauge fields. These functional gauge fields can be either external backgrounds (determined by the extra topological terms put into the action by hand) or dynamically generated (by other degrees of freedom such as fermions in the systems). The topological non-triviality of the functional abelian gauge field in field-configuration space is essential in various topological aspects of quantum field theories. (The functional gauge fields are always abelian because the wave-functional in the configuration space in QFT always has only one complex component.)

The field-configuration space is topologically non-trivial only for nonlinear field theories, such as non-abelian gauge theories or nonlinear sigma models. Our discussions will be concentrated on these two types of field theories.

2. TOPOLOGICAL ACTIONS IN GAUGE THEORIES

For gauge field theories, let us work in the Schrodinger picture with the Weyl gauge $A_o = 0$. Usually we first consider the space of all gauge potentials, $\mathcal{A}^{(d)}$, in d-dimensional space. It is an affine space with trivial topology in the sense that $\pi_n(\mathcal{A}^{(d)}) = 0$ for all $n \geq 0$. Gauge invariance implies that the genuine configuration space is not $\mathcal{A}^{(d)}$, but $\mathcal{A}^{(d)}/\mathcal{G}^{(d)}$ where $\mathcal{G}^{(d)}$ is the set of all gauge transformations in d-dimensional space. The non-trivial topology of $\mathcal{A}^{(d)}/\mathcal{G}^{(d)}$ can be seen from the following theorem:

$$\pi_n(\mathcal{A}^{(d)}/\mathcal{G}^{(d)}) = \pi_{d+n-1}(G) \tag{1}$$

where G is the Lie group which defines the gauge theory. Thus, $\mathcal{A}^{(d)}/\mathcal{G}^{(d)}$ can support topologically nontrivial functional abelian gauge backgrounds.

One way to have functional abelian gauge fields is the following. Recall that the interaction Lagrangian for the coupling of a point particle to a U(1) background

$$L_{int} = \int dt \, \frac{dq^i}{dt} A_i(q(t)) \tag{2}$$

is characterized by the feature that it is linear in the velocity $\dot{q}_i(t)$. It is amusing to observe that in the $A_o = 0$ gauge, all topological actions, including both the θFF term (for d = 3) and Chern-Simons term (for d = 2), are linear in $\dot{A}_i^a(x,t)$:

$$\frac{\theta}{16\pi^2} \varepsilon_{ijk} F^a{}_{ij} \dot{A}^a{}_k \quad (d = 3) \tag{3}$$

$$\mu \varepsilon_{ij} A^a{}_i \dot{A}^a{}_j \quad (d = 2) \tag{3'}$$

Therefore these terms lead to abelian background fields $\mathscr{A}_i^a(x,t)$ in $\mathcal{A}^{(d)}$ which can be easily read off from above expressions. (For $\mathscr{A}_i^a(x,t)$, both the spatial index i, spatial point x and group index a are the coordinate indices in the functional space $\mathcal{A}^{(d)}$, so it is abelian.)

In ref. (1) we have developed a method using differential forms rather than functional derivatives to calculate the potentials and corresponding field-strengths in $\mathcal{A}^{(d)}$ and to project them down to $\mathcal{A}^{(d)}/\mathcal{G}^{(d)}$. We find that

(1) The θ-term in 3 + 1 dimension induces a pure-gauge field in $\mathcal{A}^{(3)}$, and a vortex gauge background in $\mathcal{A}^{(3)}/\mathcal{G}^{(3)}$, which is related to $\pi_1(\mathcal{A}^{(3)}/\mathcal{G}^{(3)}) = \pi_3(G) = Z$. The flux of that vortex is just equal to the θ-parameter in the action. So the θ-vacuum effect can be identifed with a kind of Aharanov-Bohm effect in $\mathcal{A}^{(3)}/\mathcal{G}^{(3)}$.

(2) The Chern-Simons term in 2 + 1 dimensions leads to a constant magnetic field in $\mathcal{A}^{(2)}$, and a monopole background in $\mathcal{A}^{(2)}/\mathcal{G}^{(2)}$, which is related to the fact that $\pi_2(\mathcal{A}^{(2)}/\mathcal{G}^{(2)}) = \pi_3(G) = Z$. The coefficient of this term is related to the monopole charge which is quantized in order to have a consistent quantum theory. This provides a new understanding of the quantization of topological mass μ.

3. ANOMALOUS SPIN AND STATISTICS FOR TOPOLOGICAL SOLITONS (SKYRMIONS)

In nonlinear sigma models, the basic scalar fields ϕ^a (with appropriate conditions at spatial infinity) in the Schrödinger picture can be viewed as a map from the d-sphere S^d to some target manifold N which defines the model. Thus for the field-configuration space $\mathcal{C}$

$$\pi_n(\mathcal{C}) = \pi_{n+d}(N) \tag{4}$$

so it often has nontrivial topology. If $\pi_0(\mathcal{C}) \neq e$, we have topological solitons (or skyrmions) in the model.

The topological actions in such models are always of the form:

$$\varepsilon^{\mu_0\mu_1\cdots\mu_d}\,\Phi_{a_0\cdots a_d}(\phi)\,\partial_{\mu_0}\phi^{a_0}\cdots\partial_{\mu_d}\phi^{a_d} \tag{5}$$

It is easy to see that they are always linear in ϕ^a. Thus, topological actions in nonlinear sigma models always induce abelian gauge backgrounds in $\mathcal{C}$. If $\pi_0(\mathcal{C}) \neq e$, as it often happens, in sectors of different soliton numbers the abelian backgrounds have different topological properties.

Some examples have studied in ref. (2) In particular:

(1) The Hopf term introduced in ref. (7) in the d=2, $N=S^2$ model (with $\pi_0(\mathcal{C}) = Z$, $\pi_1(\mathcal{C}) = Z$) induces a vortex background in the soliton sector of unit topological charge.

(2) The Wess-Zumino term in the d = 3, N = G model (with $\pi_0(\mathcal{C}) = Z$ and $\pi_2(\mathcal{C}) = Z$) induces a monopole background in the soliton sector of unit baryon number.

Thus, the mechanism by which the topological solitons in these models acquire anomalous spin and statistics (fractional statistics for the former (7,8) and Fermi statistics for the latter (9)) can be viewed as the consequence of the nontrivial functional U(1) background fields in the soliton sector. This is a generalization of the situation for anyons and dyons (10) in quantum mechanics to quantum field theory.

4. GAUGE NON-INVARIANCE OF THE FERMIONIC DIRAC SEA AND ANOMALIES

If there are also fermions in the theory, nontrivial functional gauge fields in the bosonic field-configuration space may be generated dynamically by the fermion sector. To see this, it is very instructive to consider the quantum adiabatic phase, or Berry's phase (11), for the fermionic vacuum state which is adiabatically transported in the gauge potential space $\mathcal{A}^{(d)}$. This phase is a nonintegrable phase factor and, thus, provides a (functional) abelian gauge structure in $\mathcal{A}^{(d)}$.

More explicitly, to quantize a gauge theory with fermions described by

$$\mathcal{L} = -\frac{1}{4}\,\mathrm{tr}F^2 + \psi^+[i\partial_t - H(A)]\psi \tag{6}$$

where H(A) is the one-fermion Dirac Hamiltonian in the Schrodinger picture with $A_0 = 0$, one can first second-quantize the fermion sector in a fixed backgrounds $A^a_i(x)$ by expanding the fermion field operator in terms of the eigen states of H(A):

$$\psi(\vec{x}) = \sum_r a_r\phi_r(\vec{x};A)\ ,\ \{a_r, a^\dagger_{r'}\} = \delta_{rr'} \tag{7}$$

where

$$H(A)\phi_r(\vec{x};A) = \varepsilon_r(A)\phi_r(\vec{x};A) \tag{8}$$

Then the quantum state of the full theory is expanded as

$$\Psi = \sum_f |f,A\rangle \otimes \Phi_f[A] \tag{9}$$

where $|f;A\rangle$ is a fermionic Fock state in the fixed background $A_i^a(x)$, which can be constructed in terms of the annihilation and creation operators $a_r, a_r^{\dagger}$' defined by eqs. (7) and (8). Especially for the vacuum state, the fermionic Dirac sea is a Fock state with only the negative-energy levels all filled:

$$a_r|\mathrm{vac}, A\rangle = 0 \quad \text{for } \varepsilon_r(A) > 0$$
$$a^\dagger_r|\mathrm{vac}, A\rangle = 0 \quad \text{for } \varepsilon_r(A) < 0 \tag{10}$$

Obviously, $|\mathrm{vac},A\rangle$ is not ambiguous only for those A's whose associated Dirac Hamiltonian H(A) do not have zero modes (i.e. all $\varepsilon_r(A) \neq 0$). The set of such A's is a subset in $\mathfrak{A}^{(d)}$ and denoted by $\tilde{\mathfrak{A}}^{(d)}$.

Let us introduce

$$-i\mathcal{A}^a_i(A) = \langle \text{vac}, A| \frac{\delta}{\delta A^a_i(x)} |\text{vac}, A\rangle \qquad (11)$$

Physically it represents the electric field induced by vacuum polarizations. It is easy to see that if

$$|\text{vac}, A\rangle \to e^{i\chi[A]}|\text{vac}, A\rangle \qquad (12)$$

then

$$\mathcal{A}^a_i(A) \to \mathcal{A}^a_i(A) + \frac{\delta\chi[A]}{\delta A^a_i(x)} \qquad (13)$$

So it is an abelian gauge potential in $\tilde{\mathfrak{A}}^{(d)}$ with possible singularities at those A's whose associated H(A) has $\varepsilon_o(A) = 0$. Indeed, it is the covariant-derivative operator

$$\left(\frac{\delta}{\delta A^a_i(x)}\right) - i\mathcal{A}^a_i(x)$$

which appears in the equation for the bosonic wave functional $\Phi_{vac}[A]$ in $\mathfrak{A}^{(d)}$.

The topological non-triviality of $\mathcal{A}$ in $\mathfrak{A}^{(d)}$ can be either established by using the index theorem or by directly computing the integral $\oint_C \mathcal{A}$ along a closed loop C in $\tilde{\mathfrak{A}}^{(d)}$ or the flux $\int_D \mathcal{F}$ where D is a surface in $\tilde{\mathfrak{A}}^{(d)}$ subtended by C:

1) In 3 + 1 dimensions with chiral fermions

$$\int_D F = \pi\eta(i\not{D}_5) \qquad (14)$$

where η is the Atiyah-Patodi-Singer η-invariant for the 5-d Dirac operator defined on D.

2) In 2 + 1 dimensions

$$\oint_C A = \frac{\pi}{2}\ \eta(i\not{D}_3) \qquad (\zeta_E\text{-scheme}) \qquad (15)$$

$$\int_D = \pi\ \text{Index}\ (i\not{D}_4) \qquad (\zeta_M\text{-scheme}) \qquad (15')$$

where

$$\eta(i\not{D}_3) = 2\ \text{Index}\ (i\not{D}_4) - 2W_3[A]$$

and $W_3[A]$ is the well-known Chern-Simons 3-form; ζ_E- and ζ_M- scheme are two different regularization schemes (5).

Physically, $\oint_C \mathcal{A}$ (or $\int_D \mathcal{F}$) represents the phase acquired by $|\text{vac}, A\rangle$ after it is adiabatically transported along the loop C in $\tilde{\mathfrak{A}}^{(d)}$. If we take C to lie entirely in one gauge orbit which is isomorphic to $\mathfrak{G}^{(d)}$, then

$$\exp\{i\oint_C \mathcal{A}\} \neq 1 \qquad (16)$$

implies the gauge noninvariance of the Dirac sea $|\text{vac}, A\rangle$. This in turn implies gauge anomaly since $\Phi_{vac}[A]$ will have compensating phase variation along the loop C in $\tilde{\mathfrak{A}}^{(d)}$ and, therefore, cannot be projected down to $\mathfrak{A}^{(d)}/\mathfrak{G}^{(d)}$.

For d + 1 = 3 + 1, since $\pi_2(\mathfrak{G}^{(3)}) = \pi_5(G) = Z$, $\mathcal{A}$ is "monopole-like" in $\mathfrak{G}^{(3)}$. Eq. (16) may hold for an infinitesimal loop C. One has perturbative chiral (non-Abelian) gauge anomaly. For d + 1 = 2 + 1, $\pi_1(\mathfrak{G}^{(2)}) = \pi_3(G) = Z$, $\mathcal{A}$ is "vortex-like" in $\mathfrak{G}^{(2)}$. Eq. (16) may hold only for homotopically nontrivial loops C, so one has global gauge anomaly. In this way, we can recognize the gauge non-invariance of the Dirac sea, which gives rise to topologically non-trivial functional gauge backgrounds in $\tilde{\mathfrak{A}}^{(d)} \subset \mathfrak{A}^{(d)}$ for Φ_{vac}, to be the origin of gauge anomalies.

5. ATTEMPTS TO RECOVER GAUGE INVARIANCE IN ANOMALOUS THEORIES

Attempts (3-6) have been made to recover the gauge invariance of the theory by adding an $\mathcal{A}^{(ext)}$ in $\mathfrak{A}^{(d)}$ which causes extra phase variation of $\Phi_{vac}[A]$ along C in $\mathfrak{A}^{(d)}$ compensating that of $\Phi_{vac}[A]$ induced by fermions. The simplest way is to set

$$\mathcal{A}^{(ext)} = -\mathcal{A}^{(ind)} \qquad (d = \text{odd})$$
$$= -\nu\mathcal{A}^{(ind)} \quad \text{with } \nu \text{ odd} \qquad (d = \text{even})$$

Without adding new degrees of freedom, this $\mathcal{A}^{(ext)}$ can be achieved by adding appropriate extra terms, e.g. the r.h.s. of eqs. (14) and (15), in Lagrangian. It can be shown that the Gauss' law generators are correspondingly modified, remain to be first-class and have no anomalous terms in their commutators. The resulting theories look fine with gauge invariance and unitarity, but not good with Lorentz invariance in d + 1 = even dimensions (6). The Lagrangians leading to $\mathcal{A}^{(ext)}$ is nonlocal at least to first sight. It is unclear whether the nonlocality problem can be cured or not by adding auxiliary fields and putting the relevant action into local forms.

In conclusion, we have seen that the concept of the functional gauge structure is very helpful in understanding of various topological aspects of QFT in a more unified and more physical way.

ACKNOWLEDGEMENT

I wish to thank my coworkers A. Niemi, G. Semenoff and A. Zee for pleasant collaborations. The work was supported in part by U.S. NSF Grant PHY-8405648.

REFERENCES

1. Y.S. Wu and A. Zee, Nucl. Phys. B258 (1985) 157.

2. Y.S. Wu and A. Zee, Nucl. Phys. B272 (1986) 322.

3. A. Niemi and G. Semenoff, Phys. Rev. Lett. 55 (1985) 927.

4. A. Niemi and G. Semenoff, Phys. Rev. Lett. 56 (1986) 1019.

5. A. Niemi, G. Semenoff and Y.S. Wu, Nucl. Phys. B276 (1986) 173.

6. A. Niemi and G. Semenoff, Phys. Lett. 175B (1986) 439.

7. F. Wilczek and A. Zee, Phys. Rev. Lett. 51 (1983) 2250.

8. Y.S. Wu and A. Zee, Phys. Lett. 147B (1984) 325.

9. E. Witten, Nucl. Phys. B223 (1983) 433.

10. F. Wilczek, Phys. Rev. Lett. 49 (1982) 957; A.S. Goldhaber, Phys. Rev. Lett. 36 (1976) 1122.

11. M.V. Berry, Proc. Roy. Soc. London Ser. A 392 (1984) 45; B. Simon, Phys. Rev. Lett. 51 (1983) 2167.

12. P. Nielson and L. Alvarez-Gaume, Gommun. Math. Phys. 99 (1985) 103.

STRING CORRECTIONS TO ELECTRODYNAMICS

CHIARA R. NAPPI
Joseph Henry Laboratories, Princeton University
Princeton, NJ 08544

ABSTRACT

We derive the equations of motion and effective action for the electromagnetic field in an open bosonic string theory.

String theory is currently the object of intense investigation . An obvious question to ask is what are the string corrections to the equations of motion of the gravitational and gauge fields. In the case of closed strings [1-5] conformal invariance has been used to derive the equations of motion for the graviton field and the other massless fields of the theory. Similarly here we use conformal invariance to derive the equation of motion for an abelian gauge field coupled to the open Bose string.[6] We will work at the tree level in string theory and represent the string world sheet with the upper half-plane as in Fig. 1. We will adopt Euclidean metrics both on the world sheet and in space-time.

The action for the string coupled on the boundary to the electromagnetic field is

$$S = \frac{1}{2\pi\alpha'}\left[\frac{1}{2}\int_{M^2} d^2z\,\partial^a X_\mu\,\partial_a X^\mu + i\int_{\partial M} d\tau\, A_\mu\,\partial_\tau X^\mu\right], \tag{1}$$

where A^μ has been rescaled to contain a factor $2\pi\alpha'$. This action has obvious similarity with the action of a point-particle in the presence of an electromagnetic field. The interaction term in (1) gives origin to the same variational equation as the more obviously gauge invariant interaction

$$S_I = \int d\sigma\, d\tau\, F_{\mu\nu}\partial_\tau X^\mu \partial_\sigma X^\nu \tag{2}$$

In this calculation we will use the background field approach.[4,7] We therefore expand the action (1) around a background $\bar{X}$,

$$X^\mu(\tau,\sigma) = \bar{X}^\mu(\tau,\sigma) + \xi^\mu(\tau,\sigma), \tag{3}$$

that satisfies the equations of motion

$$\Box \bar{X}^\mu = 0$$

$$\partial_\sigma \bar{X}^\mu + iF^\mu{}_\nu \partial_\tau \bar{X}^\nu \Big|_{\partial M} = 0, \tag{4}$$

where $\Box = \partial_\tau^2 + \partial_\sigma^2$. Then the on-shell action is

$$\begin{aligned} S[\bar{X}+\xi] &= S[\bar{X}] + \frac{1}{4\pi\alpha'}\int_{M^2} d^2z\,\partial^a\xi_\mu\,\partial_a\xi^\mu \\ &+ \frac{i}{4\pi\alpha'}\int_{\partial M} d\tau\,\big(\nabla_\nu F_{\mu\lambda}\,\xi^\nu\xi^\lambda\,\partial_\tau(\bar{X}^\mu + \tfrac{2}{3}\xi^\mu) \\ &\qquad + F_{\mu\nu}\xi^\nu\partial_\tau\xi^\mu + \cdots\big). \end{aligned} \tag{5}$$

Above $\nabla_\nu = \partial/\partial X^\nu$ and $F_{\mu\nu} = \nabla_\mu A_\nu - \nabla_\nu A_\mu$. We are working in the approximation of slowly varying fields and therefore we have neglected terms with more than one derivative of F. We want to compute the one-loop (in the field theory sense) counterterm to the gauge coupling term in equation (1), namely a counterterm of the form

$$\Delta S_I[\bar{X}] = i\int_{\partial M} d\tau\,\Gamma_\mu \partial_\tau \bar{X}^\mu. \tag{6}$$

From Γ_μ we will extract the beta function β^A for the electromagnetic field A_μ. By imposing conformal invariance, *i.e.* $\beta^A = 0$, we will derive the equations of motion for $F_{\mu\nu}$.

The Neumann propagator in the upper half-plane satisfies the equations

$$\begin{aligned} \frac{1}{2\pi\alpha'}\Box G(z,z') &= -\delta(z-z') \\ \partial_\sigma G(z,z')\Big|_{\sigma=0} &= 0, \end{aligned} \tag{7}$$

where $z = \tau + i\sigma$. The solution is

$$G(z,z') = -\alpha'\left(\ln|z-z'| + \ln|z-\bar{z}'|\right). \tag{8}$$

To compute the counterterm (6) we could work with this propagator and sum up all one-loop graphs with an external $\partial_\tau X$ and all possible insertions of the vertex $F_{\mu\nu}\xi^\nu\partial_\tau\xi^\mu$ as in Fig. 2. A more straightforward method is to compute the exact propagator in the presence of the gauge field F. In the presence of the gauge fields the propagator must in fact satisfy the following boundary condition:

$$\partial_\sigma G(z,z')_{\mu\nu} + i\, F_\mu{}^\lambda \partial_\tau G_{\lambda\nu}(z,z')\Big|_{\sigma=0} = 0. \qquad (9)$$

We find that the solution to this equation is

$$G_{\mu\nu}(z,z') = -\alpha' \cdot \Big[\ln|z-z'| \cdot \delta_{\mu\nu} \qquad (10)$$
$$+ \frac{\ln(z-\bar{z}')}{2} \cdot \left(\frac{1-F}{1+F}\right)_{\mu\nu} + \frac{\ln(\bar{z}-z')}{2} \cdot \left(\frac{1+F}{1-F}\right)_{\mu\nu} \Big]$$

For $F = 0$ this propagator reduces to the one of equation (8).

The only counterterm to S_I is given by the one-loop graph in Fig. 3. The evaluation of this graph gives

$$\Delta S_I = \frac{-i}{4\pi\alpha'} \int d\tau\, \nabla_\nu F_{\mu\lambda}\, \partial_\tau \bar{X}^\mu\, G^{\nu\lambda}(\tau,\tau')\Big|_{\tau\to\tau'} \qquad (11)$$

where $G^{\nu\lambda}(\tau,\tau')$ is the propagator on the boundary ($\sigma=\sigma'=0$). In the limit $\tau\to\tau'$ we get

$$G_{\nu\lambda}(\tau\to\tau') = -\pi\alpha' \left[2 + \frac{1-F}{1+F} + \frac{1+F}{1-F}\right]_{\nu\lambda} \cdot \ln\Lambda$$
$$= -4\pi\alpha' \ln\Lambda \cdot (1-F^2)^{-1}_{\nu\lambda}, \qquad (12)$$

where Λ is a short distance cutoff. The beta function can now be obtained by differentiating with respect to the cutoff Λ,

$$\beta^A_\mu = \Lambda \frac{\partial}{\partial\Lambda} \Gamma_\mu = \nabla^\nu F^\lambda{}_\mu (1-F^2)^{-1}{}_{\lambda\nu}. \qquad (13)$$

The equations of motion are therefore

$$\nabla^\nu F^\lambda{}_\mu (1-F^2)^{-1}{}_{\lambda\nu} = 0. \qquad (14)$$

We should remember here that F is actually $2\pi\alpha' F$ and therefore equation (14) contains all orders in α'. We have obtained the exact beta function to lowest order in derivatives of F. Graphs with two or more loops would introduce corrections of a higher order in derivatives. At the leading order in α' equations (14) reduce to Maxwell equations. Notice that, since F is a real antisymmetric matrix, F^2 has only negative eigenvalues, hence equation (14) always makes sense.

We have computed string corrections to Maxwell equations and obtained equation (14). The obvious question is now what is the action whose variation gives these equations. The answer is that although the beta function (13) is not the variation of any action, it is possible to exhibit an action whose variation is $\chi_{\mu\nu}\beta^\nu_A$, where $\chi_{\mu\nu}(F)$ is an invertible tensor, for all F. Therefore, the variational equation is equivalent to $\beta_A = 0$. A similar situation has been found to occur for closed strings.[8] Let us first notice that the following identity holds:

$$(1-F^2)^{-1}_{\mu\nu}\beta^\nu_A = \nabla^\nu \left(\frac{F}{1-F^2}\right)_{\mu\nu} \qquad (15)$$
$$-\left(\frac{F}{1-F^2}\right)_{\mu\lambda} \nabla^\nu F^{\lambda\rho} \left(\frac{F}{1-F^2}\right)_{\rho\nu}.$$

The second term in the right hand side of the above equation can be rewritten by making use of the antisymmetry of F and of the Bianchi identity. Then equation (15) becomes

$$(1-F^2)^{-1}_{\mu\nu}\beta^\nu_A = \nabla^\nu \left(\frac{F}{1-F^2}\right)_{\mu\nu} \qquad (16)$$
$$+\frac{1}{4}\left(\frac{F}{1-F^2}\right)_{\mu\nu} \nabla^\nu \operatorname{tr} \ln(1-F^2).$$

This suggests that our equation of motion might be derived from an action that is a function of $\operatorname{tr}\ln(1-F^2)$. Along these lines one can indeed derive the lagrangian

$$\mathcal{L}_{\text{eff}} = e^{\frac{1}{4}\operatorname{tr}\ln(1-F^2)} = e^{\frac{1}{2}\operatorname{tr}\ln(1+F)} = \sqrt{\det(1+F)} \qquad (17)$$

whose Euler-Lagrange equations are

$$\sqrt{\det(1+F)}\,(1-F^2)^{-1}_{\mu\nu}\beta^\nu_A = 0. \qquad (18)$$

Notice that this equation has the same solutions as $\beta^A_\nu = 0$. The Born-Infeld action $\sqrt{\det(1+F)}$[9] is exactly the effective action obtained by Fradkin and Tseytlin using the Polyakov path integral.[10] In the limit of constant F field,the path integral reduces to a gassian integral that can be exactly evaluated, up to regularization of infinities. It is interesting that we get the same answer by a totally different procedure.An effective action for F has also been reconstructed up to quartic terms by calculating four-point scattering amplitude,both in the case of open bosonic string[11] and in the case of open superstrings.[12] The result, the same in both cases agrees with the expansion of the Born-Infeld lagrangian

$$\sqrt{\det(1+F)} = -\frac{F^2}{4} + \frac{1}{32}[(F^2)^2 - 4F^4] + \cdots \quad (19)$$

up to quartic terms. We remark, however, that our result is exact to all orders in α'. One would have to sum the contribution of infinitely many string tree graphs to obtain it by using the S-matrix approach.

This work has been supported by NSF grant PHY80 19754.

REFERENCES

1. C. Lovelace, *Phys. Lett.* **B135** (1984), 75
2. P.Candelas,G.T.Horowitz,A.Strominger and E.Witten, *Nucl. phys.* **B258** (1985), 46
3. E. S. Fradkin, A. A. Tseytlin, *Phys. Lett.* **B158** (1985), 316
4. C. G. Callan, D. Friedan, E. J. Martinec, M. J. Perry, *Nucl. Phys.* **B262** (1985), 593.
5. A.Sen, *Phys. Rev. Lett.* **55** (1985), 1846
6. A.Abouelsaood,C.Callan,C.Nappi and S.Yost, Princeton Preprint
7. L. Alvarez-Gaumé, D. Z. Freedman and S. Mukhi, *Ann. Phys.* **134** (1981), 85.
8. C. G.Callan, I. Klebanov and M. J. Perry, Princeton preprint (1986).
9. M.Born and L.Infeld,, *Proc. Royal Soc.* **144** (1934), 425
10. E. S. Fradkin and A. A. Tseytlin, *Phys. Lett.* **163B** (1985), 123.
11. B. E. Fridling and A. Jevicki, *Phys. Lett.* **174B** (1986), 75.
12. D. J. Gross and E. Witten, *Nucl. Phys.* **B**, to appear.

FIGURE CAPTIONS

1) Upper half-plane representation of the string tree-level worldsheet.

2) One loop graphs for the gauge field beta function.

3) The only one-loop graph contributing to β_A. The line ==== indicates the exact F-dependent propagator.

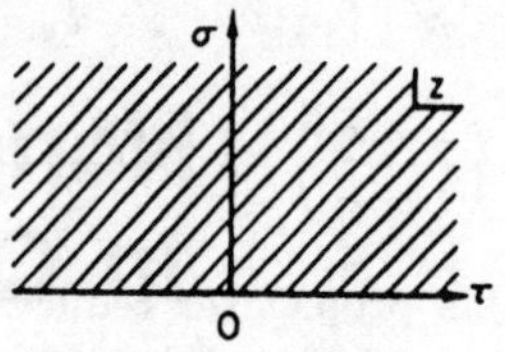

Fig. 1

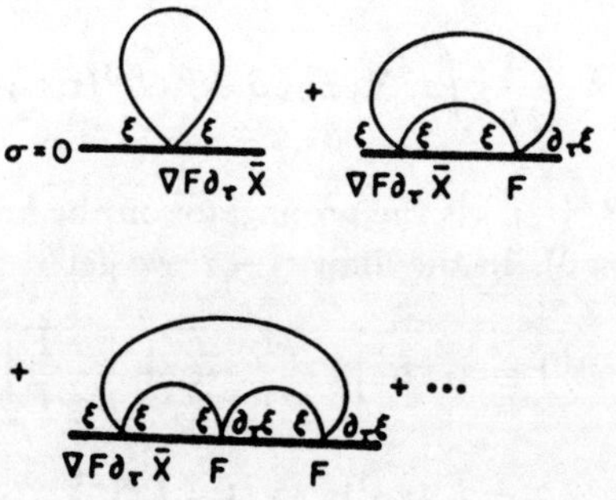

Fig. 2

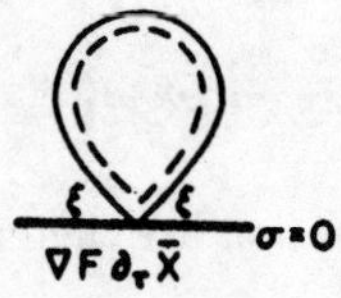

Fig. 3

Parallel Session 7

Nonperturbative Methods in Quantum Field Theory (including Lattice Gauge Theory)

Organizers:
G. Martinelli (Frascati)
A. Ukawa (Tsukuba)

Scientific Secretaries:
M. Golden (LBL)
B. Schmidki (LBL)
B. Tripsas (SLAC)

HADRON SPECTROSCOPY INCLUDING DYNAMICAL QUARKS

AKIRA UKAWA

Institute of Physics, University of Tsukuba
Ibaraki 305, Japan

Hadron mass calculation in lattice QCD fully incorporating the dynamical quarks by the Langevin technique is attempted on a $9^3\times$ 18 lattice. The contribution of the vacuum quark loops significantly modifies the hadron masses in lattice units, but most of their effects can be absorbed into a shift of the coupling constant for the quark mass range we explored. The question of systematic errors in the Langevin simulation is discussed.

1. INTRODUCTION

The inclusion of dynamical quarks has been one of the centers of efforts in the numerical simulation of lattice QCD over the last several years. One of the main motivations for this effort is to calculate the spectroscopic observables of hadrons without recourse to the quenched approximation since the validity and implication of the neglect of dynamical quark loops are not really understood. In particular the problems related to flavor singlet mesons such as the U(1) problem cannot be treated properly without including the vacuum quark loops.

A variety of proposals [1-5] has been put forward for including dynamical quarks in the simulation. Among them the Langevin methods [3-5] seem quite promising in several respects: The cost of including dynamical quarks is reduced to solving a linear algebraic equation once per update of the entire lattice. The correct distribution of field configurations is guaranteed in the limit of infinitesimal Langevin time step size $\Delta\tau$, and the systematic deviation due to a finite $\Delta\tau$ can be estimated in principle.

In this talk we report the status of an attempt at hadron mass calculation in full QCD using the Langevin technique, which is being carried out by M. Fukugita, Y. Oyanagi and myself [6]. We have been collection data on a $9^3\times 18$ lattice using the Wilson quark action with 2 flavors. In the range of the hopping parameter we explored, we found that most of the contribution from vacuum quark loops, while it substantially modifies the hadron masses in lattice units, can be absorbed into a shift of the gauge coupling constant. We have also found that a proper treatment of the systematic error is important for the validity of the numerical results.

2. LANGEVIN SIMULATION

The formulation of Langevin equation, suitable for numerical simulation, starts from an effective action of full QCD given by

$$S_{eff} = S_{gauge} + Y^\dagger \frac{1}{D} Y \qquad (1)$$

where S_{gauge} represents the action for the gauge link variable U, D the lattice Dirac operator, and Y the pseudo-fermionic variable on sites. The Langevin equation, discretized in the fictitious time in steps of $\Delta\tau$ ($\tau = n\Delta\tau$), may be written as

$$U^{(n+1)} = U^{(n)} \exp(i\Delta\tau\chi^{(n)}), \qquad (2)$$

$$\chi^{(n)} = -i \frac{\delta}{\delta U^{(n)}} S_{eff} + \xi^{(n)},$$

$$Y^{(n+1)} =$$

$$(1-\Delta\tau B(U^{(n)}))Y^{(n)} + \Delta\tau C(U^{(n)})\eta^{(n)}, \qquad (3)$$

with ξ and η the white noise of width $\Delta\tau^{-1/2}$. It is an easy exercize to show that, if the operators B and C satisfy $2CC^\dagger - BD - DB^\dagger = 0$, the distribution generated by (2-3) tends to $\exp(-S_{eff})$ in the limit $n\to\infty$ up to terms of order $\Delta\tau$. Hence expectation values of observables can be calculated by taking averages over the

Langevin updates. A naive choice for B and C is $B=D^{-1}$ and C=1 (pseudofermion scheme)[3]. An alternative is to set $B=\Delta\tau^{-1}$ and $C=(D/\Delta\tau)^{1/2}$ thereby eliminating the variable Y from the equations (bilinear noise scheme)[4].

In practical simulations the time step $\Delta\tau$ is finite. The limiting distribution therefore deviates from $\exp(-S_{eff})$, giving rise to systematic errors in the observables. This point will be discussed further in section 4. For the simulation reported in this talk we used the bilinear noise scheme modified by a Runge-Kutta algorithm which gets rid of a part of the systematic error of order $\Delta\tau$.

Our simulation was carried out with the SU(3) gauge group on a $9^3 \times 18$ lattice using periodic boundary conditions. The Wilson quark action with 2 flavors having the same hopping parameter K and the standard single plaquette action with the gauge coupling $\beta=6/g^2$ were employed. The choice of the coupling $\beta=5.5$ is made so as to be above the crossover of the pure gauge sector and at the same time as large as possible without being obviously affected by finite size effects on $9^3\times 18$ lattice. In the majority of runs the Langevin updates over the time interval $\tau=50$ was made with the step size $\Delta\tau=0.01$. The hadron propagators for π, ρ, N and Δ were calculated at every $\delta\tau=1.0$ using the standard local relativistic operators. Thermalization was monitored by Wilson loops and hadron propagators and typically the first 20 time units were discarded. The autocorrelation of these observables were roughly in the range $\tau_r \sim 3$-5. An independent Langevin update ($\Delta\tau=0.01$, $\tau=60$) with the pure gauge action was also made for mass estimates in the quenched approximation. We have also been gathering data using the step size $\Delta\tau=0.005$ and 0.02 in order to examine the systematic error in $\Delta\tau$. These runs were started from the last configuration of runs with $\Delta\tau=0.01$ and were carried out over 20 to 30 time units. The interval $\Delta\tau=1.0$ for calculating hadron propagators is the same as in the $\Delta\tau=0.01$ case.

In the presence of dynamical quarks, the extraction of mass from hadron propagator is generally not straightforward due to the presence of multihadronic intermediate states. However the range of the hopping parameter we explored is above the decay threshold for $\rho\to\pi\pi$ and $\Delta\to N\pi$. We therefore fitted the meson propagators by a single hyperbolic cosine and the baryon propagators by a single exponential. The errors were estimated taking into account the autocorrelation in Langevin time mentioned above.

3. HADRON MASS FOR $\Delta\tau=0.01$

Let us first summarize the data obtained from the runs with the step size $\Delta\tau=0.01$. Fig.1(a) shows the pion mass squared as a function of 1/K and (b) the masses of ρ, N and Δ. The estimates from the quenched approximation at the same coupling $\beta=5.5$ are also plotted for comparison.

As one can see, the effect of vacuum quark loops becomes quite substantial as K increases. In Table 2 we show the values of quantities relevant for spectroscopy obtained by a linear extrapolation of $(m_\pi a)^2$ and $m_{\rho,N,\Delta}$ a in 1/K (first column) and compare them with the quenched values (third column).

An important question is how and to what extent the vacuum fluctuation of full QCD differs from that of the pure gauge sector. One can gain some insight on this problem by examining an effective shift $\Delta\beta$ of the coupling constant β due to the vacuum quark loops. In Table 1 we show the magnitude of the shift estimated by matching the Wilson loop of full QCD to that of the pure gauge sector for various sizes of loops. Those estimates show that $\Delta\beta$ tends to increase with the size of the loop but the magnitude of the increase is not very large. They rather indicate that the dominant part of the loop effect may be absorbed into a shift of the coupling β. To verify if this also applies to hadron masses, we made quenched mass estimates at K=0.16, $\beta=5.75$ ($\Delta\beta=0.25$) and at K=0.15, $\beta=5.62$ ($\Delta\beta=0.12$) using the value of $\Delta\beta$ from large loops and have shown the results in Fig.1. The very good agreement between the full and the quenched (with shifted β) calculation supports the results from the Wilson loop that the bulk of the effect of vacuum quark loops can be absorbed into a shift of the coupling constant at least for the range of quark mass heavier than the decay thresholds for $\rho\to\pi\pi$ and $\Delta\to N\pi$.

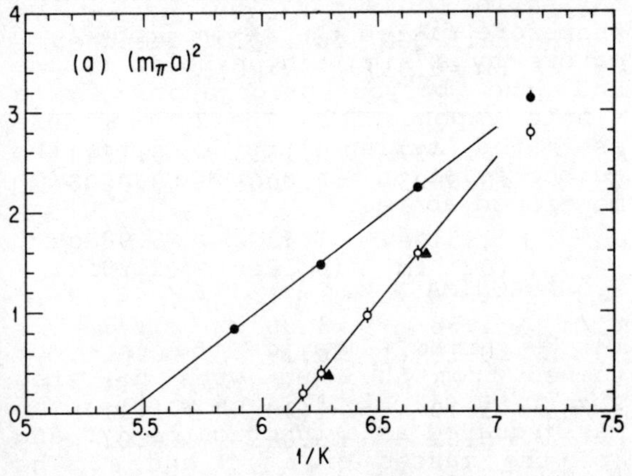

Fig. 1(a): Pion mass squared in lattice units as a function of 1/K for the runs with $\Delta\tau = 0.01$. The open and filled circles represent the full QCD and quenched results, respectively, at $\beta = 5.5$. The triangles are the quenched results at a shifted β.

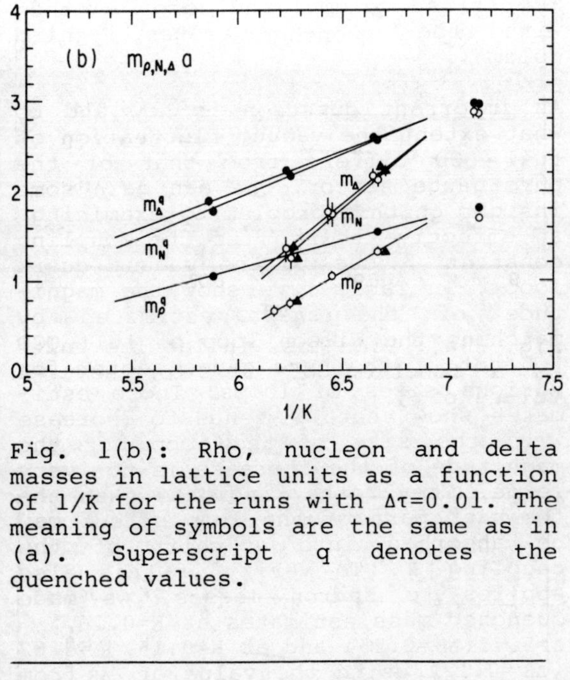

Fig. 1(b): Rho, nucleon and delta masses in lattice units as a function of 1/K for the runs with $\Delta\tau = 0.01$. The meaning of symbols are the same as in (a). Superscript q denotes the quenched values.

K	1×1	2×2	3×3	4×4
0.14	0.06	0.06	0.06	0.09
0.15	0.10	0.11	0.12	0.12
0.16	0.20	0.23	0.24	0.25

Table 1: Effective shift $\Delta\beta$ for the run with $\Delta\tau = 0.01$.

4. SYSTEMATIC ERROR IN $\Delta\tau$

We have noted that the limiting distribution generated by the discretized Langevin equation has the form $\rho_\infty = \exp(-S_{eff} - \delta S)$ where $\delta S \sim O(\Delta\tau)$ represents the systematic error due to the finiteness of the step size $\Delta\tau$. In the pure gauge sector one can use either a higher order discretization [7,3] or more simply a redefinition of the U's [4] to get rid of the terms of order $\Delta\tau$. In full QCD, δS contains additional terms typically of order $\Delta\tau \cdot D^{-2}$ These terms are very troublesome since the minimum eigenvalue λ_{min} of the Dirac operator decreases to zero as the hopping parameter K approaches the critical value (see Fig.2) and hence they will seriously distort the distribution of the infrared modes. Our analysis with the pseudo-fermion and bilinear noise schemes show that in both these methods the additional term δS cuts off the infrared modes. This phenomenon becomes increasingly marked for larger values of K, but the bilinear noise scheme is found to be better behaved than the pseudo-fermion scheme. We therefore used the bilinear noise scheme for the present simulation.

There exist several proposals [8-9] of algorithms for removing the terms of order $\Delta\tau \cdot D^{-2}$ in the bilinear noise scheme. We found that they are not practical because the modification of the Langevin equation involves terms of order $\Delta\tau$ times lattice volume which requires $\Delta\tau$ to be prohibitively

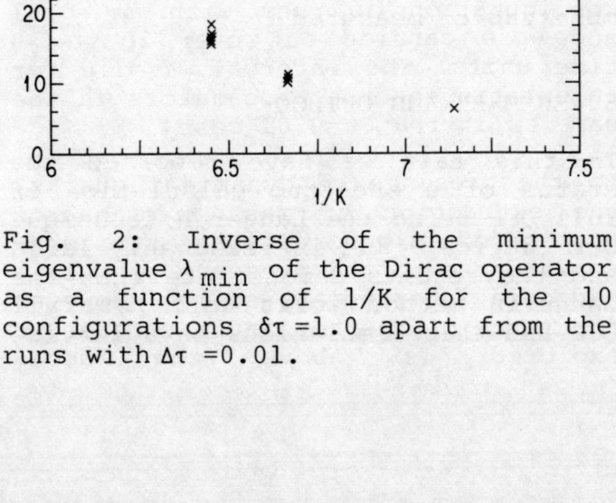

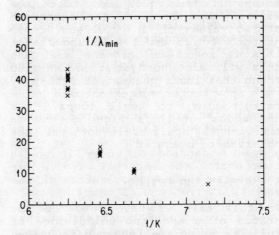

Fig. 2: Inverse of the minimum eigenvalue λ_{min} of the Dirac operator as a function of 1/K for the 10 configurations $\delta\tau = 1.0$ apart from the runs with $\Delta\tau = 0.01$.

small. In view of these problems we have adopted an intermediate approach Namely we used a Runge-Kutta argorithm originally devised [3] for the pure gauge sector which removes the pure gauge terms of order $\Delta\tau$ from δS, and carried out simulations at several values of $\Delta\tau$ to estimate the magnitude of the systematic error. The order of the error is still $\Delta\tau$ in this algorithm, and an extrapolation in $\Delta\tau$ is necessary to obtain the results at $\Delta\tau=0$.

We show in Fig.3 our preliminery result for π and N masses as a function of $\Delta\tau$ at several values of K (The data for ρ and Δ are similar to those of π and N, respectively). As expected from the discussion of the minimum eigenvalue of the Dirac operator, the systematic error becomes increasingly larger as K increases. The dependence on $\Delta\tau$, however, is linear over the range examined and this allows a linear extrapolation to $\Delta\tau=0$. In the second column of Table 2 we summarize our preliminery estimate of the spectroscopic observables at $\Delta\tau=0$ obtained by first extrapolating masses in $\Delta\tau$ at K=0.15 and 0.16, and then making the usual extrapolations in 1/K. One finds 10-20% deviations compared to the values at $\Delta\tau=0.01$ showing the necessity of extrapolation for obtaining results at the accuracy level of 10% or less.

We have repeated the quenched calculation with a shifted value of β at $\Delta\tau=0.005$. Again the quenched hadron masses agree well with the full QCD values if the shift $\Delta\beta$ for large Wilson loops is used. The agreement at two values of $\Delta\tau$ (0.01 and 0.005) at both K=0.15 and K=0.16 shows that this is not a coincidence and that this will also hold at $\Delta\tau=0$. We also found that local quantities such as $\langle\bar\psi\psi\rangle$ show better agreement with $\Delta\beta$ corresponding to small loops. Thus the amount of shift might depend on the length scale relevant for the observable measured.

5. SUMMARY AND OUTLOOK

In this talk we have reported the status of a spectrum calculation of full QCD using the Langevin technique. Our lattice $9^3\times18$ is reasonably large and the results demonstrate that the Langevin method works well for full QCD and that simulations on a lattice

	full QCD ($\Delta\tau=0.01$)	full QCD ($\Delta\tau=0$)	quenched
K_c	0.1637 ±0.0005	0.1613 ±0.0003	0.1844 ±0.0008
a^{-1}	1.31GeV (0.15fm)	1.63GeV (0.12fm)	0.98GeV (0.20fm)
m_N/m_ρ	1.68 ±0.15	1.36 ±0.19	1.73 ±0.06
m_Δ/m_N	1.10 ±0.02	1.28 ±0.03	1.11 ±0.01

Table 2: Summary of spectroscopic observables.

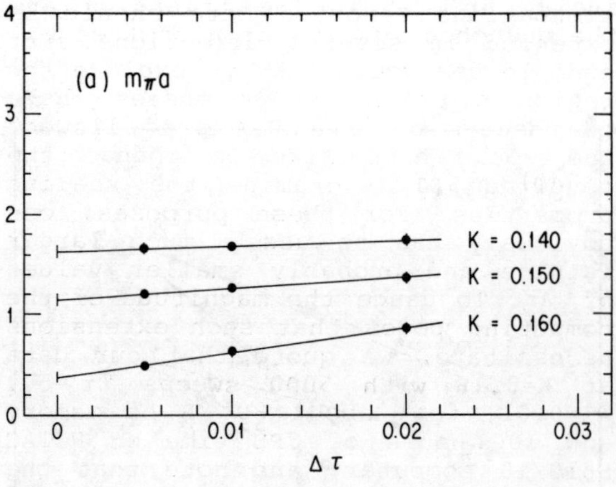

Fig. 3(a): Pion mass in lattice units as a function of $\Delta\tau$ for several values of K.

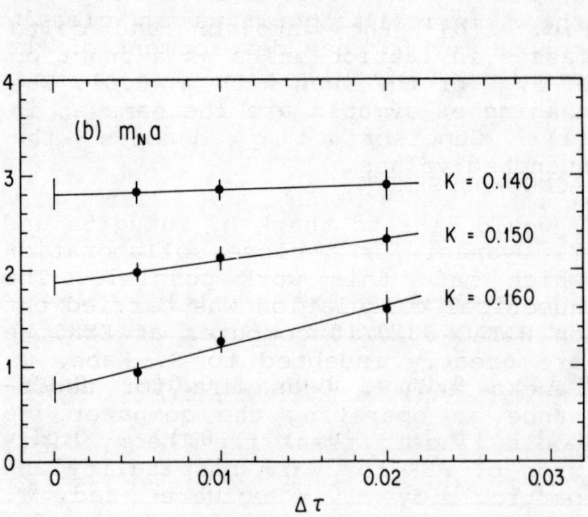

Fig. 3(b): Same as (a) for nucleon mass.

of such a size are indeed feasible.

The systematic error due to the finite Langevin step size is not negligible near the critical value of the hopping parameter. However, the linearity of the error makes it possible, albeit time consuming, to make an extrapolation to $\Delta\tau=0$.

We have seen that the effect of the vacuum quark loops is quite large. The bulk of the effect, however, is absorbable into a shift of the coupling constant and therefore the net physical result is not very different from the quenched case. This might be the underlying reason for the qualitative success of the quenched approximation [10] for the mass spectrum of flavor non-singlet hadrons.

At a more quantitative level, the simulations reported here should be extended in several directions. For example one would like to explore the region of light quark masses where the decays of ρ and Δ are allowed. One would also like to change the coupling β to examine the scaling properties. For these purposes, one obviously has to use a much larger lattice and probably smaller values of $\Delta\tau$. To gauge the magnitude of the computing power that such extensions necessitate, we quote that our data at $K=0.16$ with 5000 sweeps ($\tau=50$, $\Delta\tau=0.01$) took about 80 MB of memory and 100 hours of CPU time on HITAC S810/10 computer, and note that the CPU time requirement is proportional to the lattice volume and the number of sweeps. Much improvement in the computer technologies as well as in the numerical algorithms is clearly needed for further development of the simulation of full QCD.

ACKNOWLEDGEMENT

I would like to thank M. Fukugita and Y. Oyanagi for a close collaboration which made this work possible. The numerical calculation was carried out on HITAC S810/10 computer at KEK. We are greatly indebted to S. Kabe, T. Kaneko and R. Ogasawara for assistance in operating the computer. We would like to thank the Theory Division of KEK for warm hospitality and particularly H. Sugawara and T. Yukawa for their strong support for our work.

REFERENCES

[1] Fucito, F. et al., Nucl. Phys. B180, 369 (1981).

[2] Polonyi, J. and H.W. Wyld, Phys. Rev. Lett. 51, 2257 (1983).

[3] Ukawa, A. and M. Fukugita, Phys. Rev. Lett. 55, 1854 (1985).

[4] Batrouni, G. et al., Phys. Rev. D32, 2736 (1985).

[5] Duane, S., Nucl. Phys. B257[FS14],652(1985).

[6] Fukugita, M., Y. Oyanagi and A. Ukawa, Phys. Rev. Lett. 57, 953(1986) and in preparation.

[7] Drummond, I. T. et al., Nucl. Phys. B220[FS8],119(1983).

[8] Batrouni, G., Phys. Rev. D33, 1815 (1986).

[9] Kronfeld, A. S., DESY preprint, DESY 86-006(1986).

[10] See, for example, Iwasaki, Y., these proceedings; Barkai, D. et al., Phys. Lett. 156B, 385(1985).

QCD WITH WILSON FERMIONS

Philippe de Forcrand

Cray Research, Inc., Chippewa Falls , USA

Ion Olimpiu Stamatescu

Institut für Theorie der Elementarteilchen, F.U. Berlin, West-Germany

We present QCD computations on intermediate size lattices using Wilson fermions. Some properties of the latters are reviewed - their spectrum and the relation between their chiral properties and the topology of the YM fields. Simulations using a very fast local algorithm show clear effect of the dynamical quarks on the finite temperature behaviour and on the static potential.

1. INTRODUCTION

The problem of full simulations for QCD is a longstanding one. There is increasing need to obtain corrections to the pioneering results obtained with the quenched approximation (1) and in fact we must at least begin to answer the question: which are the capabilities of the lattice approach to QCD? Due to the sustained effort in this field, calculations with dynamical fermions on intermediate size lattices are already available, therefore we can rephrase the above question in a more constructive form: what results of physical interest can already be obtained and which are the prospects of going to larger lattices, look for scaling behaviour and measure further hadronic parameters? We ask this questions in order to show the frame in which we would like to situate this work - and not because we already have an answer!

The main problems of QCD simulations are related with the definition and properties of the lattice fermions and with the reckoning of the fermionic determinant in the updating.

There are essentially two definitions for lattice fermions which are extensively used, the Wilson and the staggered (or Susskind) fermions (2), both of them having shortcomings and advantages. Since these are different for the two definitions it is important to have calculations with both of them to allow for cross checks. We work here with Wilson fermions and we shall review some of their properties in section 2.

At present only approximative algorithms seem to be practicable. These algorithms do not update the determinant (which is a nonlocal form in the links) after each link updating, but use one determinant calculation in the independent updatings of many links. The implied approximations may be related to the assumed factorization of the fermionic determinant in a finite step algorithm with local updating, or to the discretization of an infinitesimal algorithm (Random Walk, Langevin Equation), etc. In order to obtain meaningful results we must have therefore a tight control on the possible systematic errors. As long as no perfect algorithm has been developed, we must pursue various directions and compare the performances. Our method is discussed in section 3.

Finite size effects limit the approach to the continuum because the physical length of the box becomes to small at large β and therefore the discretization is bound to remain coarse. Part of these effects appear as finite temperature effects. Recall that the smallest size fixes a temperature and that for infinitely heavy quarks the theory has a first order phase transition at a corresponding value β_c. This transition moves to smaller β_c -values with decreasing quark masses, becoming smoother thereby (3) - but even if it were replaced by a continuous cross over we still expect different physics describing a "hot" quark-gluon plasma above β_c and a "cool" hadronic world below. In section 4 we present results obtained on 4^3x4 and 8^3x4 lattices and discuss their physical interest.

2. SOME PROPERTIES OF THE WILSON FERMIONS

The Wilson fermions are introduced as a couple of d-components Grassman variables, Ψ , $\overline{\Psi}$ at each lattice point, with the action:

$$S_F(\{\Psi,\overline{\Psi},U\}) =$$

$$= \sum_n \Big(\overline{\Psi}_n \big(M_0 + \tfrac{1}{2}\sum_\mu (\Gamma_+^\mu + \Gamma_-^\mu)\big)\Psi_n - \qquad (2.1)$$

$$- \tfrac{1}{2}\sum_{\mu=1}^{d} \overline{\Psi}_n \big(\Gamma_+^\mu U_{n\mu}\Psi_{n+\mu} + \Gamma_-^\mu U^+_{n-\mu,\mu}\Psi_{n-\mu}\big)\Big)$$

$$\Gamma_{\pm}^{\mu} = r e^{i\theta\gamma_5} \pm \gamma^{\mu}, \quad {\gamma^{\mu}}^2 = \gamma_5^{\,2} = 1 \qquad (2.2)$$

with d the dimension (d=2 or 4), M_o the bare mass, $0 \leq r \leq 1$ and θ a parameter which can be related to the θ-vacua (4,5). $U_{m\mu} = U_{\ell}$ are the link variables (SU(3) matrices). The fermionic partition function is:

$$Z_F(\{U\}, r, \theta) = \int [d\Psi d\bar{\Psi}] e^{-S_F(\{\Psi,\bar{\Psi},U\})} \underset{a\to 0}{\longrightarrow} e^{i\theta Q_5} Z_F(\theta=0) \qquad (2.3)$$

where Q_5 is the topological charge:

$$Q_5 = \frac{1}{16\pi^2} \int d^4x \, Tr F\tilde{F} \qquad (2.4)$$

It is usual to rewrite S_F as (in the following we shall take θ=0):

$$S_F = \sum_{n,m} \bar{\Psi}_n W_{nm} \Psi_m$$

$$W_{nm} = \mathbb{1} - k \sum_{\mu} (\Gamma_{+\mu} U_{n\mu} \delta_{m,n+\mu} + \Gamma_{-\mu} U^{+}_{n-\mu\mu} \delta_{m,n-\mu}) \qquad (2.5)$$

$$k = \frac{1}{2M_o + 2dr}, \quad G(U) = 2k\, W^{-1}(U) \qquad (2.6)$$

whereby we introduced the hopping parameter k and rescaled the fields by $\sqrt{2k}$. G(U) is the fermionic propagator in the external field $\{U\}$. Besides the usual properties of gauge invariance, reflection positivity (with antiperiodic boundary conditions), etc., we also have

$$\gamma_5 W = (\gamma_5 W)^{+}, \quad 1 \geq \frac{\det W(U)}{\det W(1)} \underset{k \leq 1/2dr}{\geq} 0 \qquad (2.7)$$

(for the paramagnetic inequality see (4)).

While the "naive" action (r=0 in eq.(2.1)) describes 2^d fermionic modes, the spectrum of the Wilson fermions is not degenerate, in fact for $r \gtrsim .7$ the dispersion law of free Wilson fermions has no supplementary modes (Fig.1), independently on the cut off 1/a.

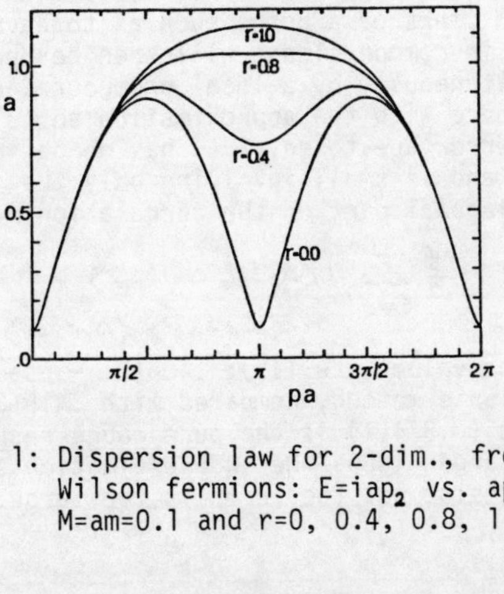

Figure 1: Dispersion law for 2-dim., free Wilson fermions: $E = iap_2$ vs. ap_1 at M=am=0.1 and r=0, 0.4, 0.8, 1.0.

The chiral properties are affected by the presence of the term proportional with r. This term which expresses an explicite breaking of the chiral invariance is an irrelevant term in the action, and therefore is expected to go away in continuum leading to results independent on r after extracting the free field contribution. This can be checked in simple models (6) or by trying to find a plateau in r below 1 in realistic QCD calculations.

The same term turns out to be marginal in the chiral Ward identity, where it leads to the chiral anomaly (5,7,8). The mechanism by which this term, which can be written:

$$Q_X = \frac{r}{2} Re\, Tr_{(L,D,C)} (\gamma_5 (GU + U^{+}G - 2G)) \qquad (2.8)$$

reproduces the topological charge Q_5 (which is the integrated anomaly, see eq.(2.4)) and thus realizes the index theorem was discussed in details in (5) (notice that in eq.(2.8) the free field contribution vanishes). On a configuration $\{U\}$ with nontrivial topology the Q_X eq.(2.8) shows a pole at values of $k=k_c(\{U\})$ slightly above $k_o=1/(2dr)$ (while having a kinematical zero at $k=k_o$). See Fig.2a. The information on the number of instantons in a configuration is contained in the residue of this pole. This corresponds to the fact that in the continuum limit $k_c \to k_o$ and the pole at k_c collides with the zero at k_o, ensuring:

$$\lim_{a\to 0} Q_X = \lim_{k\to k_o} \lim_{k_c \to k_o} Q_X(k) = Q_5 \qquad (2.9)$$

The result is independent on the localization of the instantons and becomes more precise as the configuration gets smoother (and thus nearer to a continuum configuration). The procedure is gauge invariant. There is no ambiguity related to singular

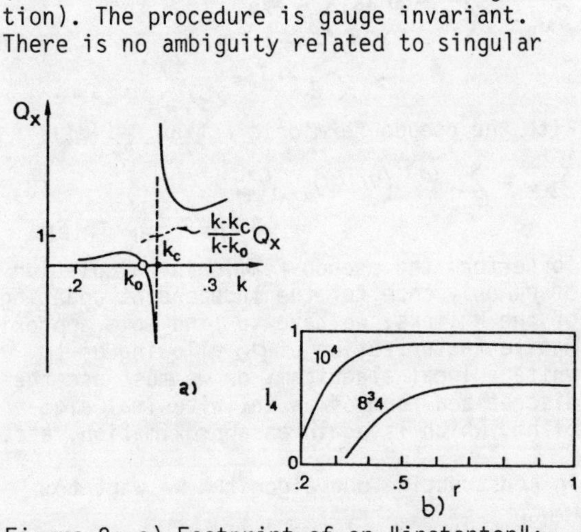

Figure 2: a) Footprint of an "instanton": $Q_X = M_o Tr \gamma_5 G$ vs k for a 2-dim. configuration with Q_5-charge 1.
b) The normalization factor of the axial anomaly vs r for M=.001

gauges, etc. Notice that these results show the plateau in r mentioned above, which in approaching the continuum seems to extend down to arbitrarily small r (5). See Fig.2b.

One of the difficulties in calculations with Wilson fermions is the mass renormalization. For $\beta \neq 0$ vanishing quark mass (defined e.g. as vanishing pion mass) no longer corresponds to $k=k_0$ but $k=k_c(\beta)\geq k_0$. Also the fact that $\bar{\psi}\psi$ does no longer vanish at zero mass prevents its direct interpretation as order parameter for the chiral symmetry breaking (we need at least to subtract the free field part, but the quantitative interpretation might be difficult). Therefore it is important to devise ways of estimating k_c and understand the chiral breaking for Wilson fermions. Studying the r-dependence may help this point.

3. A NOISY FINITE STEP FACTORIZED ALGORITHM

We shall shortly mention here some features of a local algorithm based on pseudofermionic integration and optimized in the frame of Kuti's updating method (9), by which the results presented below are obtained. The algorithm has been introduced in (10,14).

With W the coupling matrix for Wilson fermions, eq.(2.5), the quantity needed for the simultaneous updating of N links is:

$$\rho_N(\{U\}\to\{U'\}) = \left(\det W(\{U'\})/\det W(\{U\})\right)^f \tag{3.1}$$

where f is the number of flavours and $U_\mu(n)$ are the gauge fields: SU(3) matrices.

Using a pseudo-fermionic Monte Carlo (11,12) we can directly obtain ρ_N as (13):

$$\rho_N = \langle e^{-\Delta S_{PF}}\rangle_{S_{PF}}^{-f/2} = \langle e^{\Delta S_{PF}}\rangle_{S'_{PF}}^{f/2} = \dots \tag{3.2}$$

with the pseudo-fermionic action $S_{PF}(\{U\})$:

$$S_{PF} = \sum_{n,m} \varphi_n^* (W^+W)_{nm}\varphi_m, \qquad \Delta S_{PF} = S'_{PF} - S_{PF} \tag{3.3}$$

To perform the pseudo-fermionic calculation of ρ_N only once for the independent updatings of the N links, we have to find some approximative factorization of ρ_N allowing us to write a local algorithm, or we must use the discretized form of an infinitesimal algorithm, which is again an approximation, etc.

In constructing our algorithm we want to:

- avoid biased (one sided) errors in the updating (which accumulate over a sweep),

- estimate and monitor any systematic error

- have the possibility to tune the algorithm to reduce the error (with more work),

- avoid slow convergence and sources of metastability in the algorithm.

The approximative factorization given by the direct linearization:

$$\rho_N \sim \rho_N^{(PFM)} \equiv e^{\frac{f}{2}\langle \Delta S_{PF}\rangle_{S_{PF}}} \tag{3.4}$$

which corresponds to the "pseudo-fermionic method" (11) is biased due to Jensen's inequality. As was illustrated in (10,14) the error being positive definite accumulates over each sweep leading to a systematic underestimation of the quarks effects:

$$\rho^{(PFM)} - \rho \sim \frac{f}{4}\sum_\ell \langle(\Delta_\ell S_{PF} - \langle\Delta_\ell S_{PF}\rangle)^2\rangle \sim O(\Delta U^2) \tag{3.5}$$

This error does not depend on the convergence of the pseudo-fermionic Monte Carlo and grows with the number of flavours. Comparison with an exact calculation (on a small lattice: 4x4x4x4 with staggered fermions) (15) shows that the method can account for part of the fermionic effects, but also shows explicitely this systematic underestimation. In a particularly clear case, the contribution of the fermions to the plaquette average (an effect of the order of 10% and where the statistical errors quoted are relatively small - of the order of 1%, i.e. 10% of the effect), the result given by the pseudofermionic method is by more than 30% below the exact value*). There are a number of results obtained with the pseudo-fermionic method, mainly for staggered (or Susskind) fermions see e.g., (16). Because of this underestimation these results can be considered as a lower bound for the fermionic effects.

The approximation we use is less drastic, namely (10,14)

$$\rho_N \sim \tilde{\rho}_N \equiv \prod_{\ell=1}^{N} \langle e^{-\Delta_\ell S_{PF}}\rangle_{S_{PF}}^{-f/2} \tag{3.6}$$

where the factorization is performed over the N links of a bush, such as to have no site in common. These will then be updated simultaneously by a local procedure. At variance with the approximation eq.(3.4), the error due to eq.(3.6) has no definite sign and is small involving only the nondiagonal part of the correlation matrix:

$$\tilde{\rho} - \rho \sim \frac{f}{2}\sum_{\ell\neq\ell'}\left(\langle\Delta_\ell S_{PF}\,\Delta_{\ell'}S_{PF}\rangle - \langle\Delta_\ell S_{PF}\rangle\langle\Delta_{\ell'}S_{PF}\rangle\right) \tag{3.7}$$

*) The values are (15): .404(2) - pseudo-fermionic method, compared with .415(3) - exact (.374(1) is the pure gauge result). For larger loops the underestimation seems to persist but the error is too large.

This error does not accumulate. It can be easily monitored using eq.(3.7) and it can be controlled by thinning out the bushes*), such that the change $\Delta U(l)$ needs not be restricted, as in algorithms based on eq.(3.4).

Calculating the determinant ratios by a statistical procedure (the pseudo-fermionic Monte Carlo) suggests using the algorithm of Kuti (9) which needs a Boltzmann factor determined only up to some statistical error which is then used in the statistical noise. In constructing the updating rule for the Kuti algorithm we use as fermionic part of the transition probability (for the change U suggested by the Yang Mills measure):

$$\alpha \check{\varrho}_l \equiv \alpha \left(\frac{1}{N_{PF}} \sum_{m=1}^{N_{PF}} e^{-\Delta_l S_{PF}(\{\phi\}_m)} \right)^{-f/2} \qquad (3.8)$$

where $\check{\varrho}$, the estimator of the determinant ratio, is the "DLR-improved" average (17) (see (10,14)) obtained over a few pseudo-fermionic sweeps, N_{PF}. Here α is a parameter which should ensure $\alpha\check{\varrho} \leq 1$. This parameter influences the acceptance, hence it cannot be chosen too small, as this increases the relaxation time. Thus a second source of errors is given by the "occasional overflow" $\alpha\check{\varrho}>1$ reflecting strong fluctuations in $\check{\varrho}$. Notice that using improved averages we gain a factor 10-100 in efficiency for the pseudo-fermionic Monte Carlo, hence fluctuations are rather small even for small N_{PF} (10,14).

Kuti's updating rule fulfills detailed balance on the average if the approximation given by eq.(3.7) is an unbiased estimate of But we only have an unbiased estimator for

$$\check{\varrho}^{-2/f} \simeq \varrho^{-2/f} + O(\Delta U/\sqrt{N_{PF}}) \qquad (3.9)$$

leaving a systematic error of the order:

$$\frac{f(f+2)}{8}\, O((\Delta U)^2/N_{PF}) \qquad (3.10)$$

This is the third source of systematic errors in our algorithm. Notice that this error does go to zero with increasing N_{PF} for all ΔU, hence is not comparable with the one in eq.(3.5) which does not depend on N_{PF} (with a typical N_{PF} of 6 (3.10) is ~1% of (3.5)).

Thus the statistical error $\sim 1/\sqrt{N_{PF}}$, introduced by using an imprecise ϱ obtained with a small number N_{PF} of pseudo-fermionic sweeps is taken care of by Kuti's method. The residual error $\sim 1/N_{PF}$ eq.(3.10)**) can be monitored and controlled by increasing N_{PF} How large we should take N_{PF} is matter of optimization balancing speed vs precision.

*) The quality of the factorization approximation is fixed by the correlation length.
**) The possibility of such an error was pointed out to us by J.Kuti.

All the systematic errors specific to this algorithm, as given by eqs.(3.7),(3.10) and the occasional overflow in eq.(3.8), can be decreased without reducing the "step size" (what in the Monte Carlo procedure usually leads to poor convergence and metastabilities). For this we must decrease N, the number of the links in a bush, and take these further appart, and we must increase N_{PF}. This of course implies more work, but allows to test the precision of the method and to tune it. Notice that all the errors above are understood per Yang Mills sweep.

4. CALCULATIONS ON 4^3x4 AND 8^3x4 LATTICES

We shall present here results obtained on the 4x4x4x4 and 8x8x8x4 lattices using the algorithm described in section 3 (see also (10,14)). We use this algorithm on bushes of N=V/2 links (V: the lattice volume) and therefore we achieve a Yang Mills sweep in 8 steps, implying 8 pseudo-fermionic Monte Carlo â N_{PF} sweeps each. We use a rather low N_{PF}, of around 10 improved sweeps.

We monitor all the time the various sources of error, as given by eqs.(3.6), (3.9) and by the occasional overflow*). Maximal values observed were 10% for eqs.(3.6) and 1% overflow, while typical values are much smaller (see also Fig.3). In general, accounting also for the statistical errors, we conclude that we can trust our results up to an overall 15% error concerning the fermionic effects. We use antiperiodic boundary conditions for fermions and f=3.

The quantities measured all the time are the plaquette and the Polyakov loop (thermal Wilson string) averages:

$$A = \langle \tfrac{1}{3}\,\mathrm{Re\,Tr}\, U_{Plaq.} \rangle \qquad (4.1)$$

$$P = \langle \tfrac{1}{3}\,\mathrm{Re\,Tr}\, U_{Pol.} \rangle \qquad (4.2)$$

and the "fermionic condensate":

$$S = \frac{1}{12} \langle \bar{\psi}\psi \rangle = \frac{1}{12} \langle \mathrm{Re\,Tr}\, G \rangle \qquad (4.3)$$

*) We expect from the occasional overflow an error of the order of the overflow frequency divided by the average fermionic rejection rate. The fact that eq.(3.2) may have infinit variance, as pointed out to us by U. Wolff and by J.Kuti, could influence the overflow frequency. Notice, however, that overflow allways can happen in Kuti's method, unless ϱ is bounded by an acceptable number, and we allways need to monitor it. Hence no qualitatively new effect comes from the large variance (anyway strongly reduced for the improved averages).

On the $8^3 \times 4$ lattice we also measure:

$$\varepsilon_G = 3\beta\,(A_{T-S} - A_{S-S}) \qquad (4.4)$$

and other quantities like Wilson loops.

In Fig.3 we show some results obtained using Kuti's algorithm with various N_{PF}. The points remain inside ~10% of the distance to the pure Yang Mills values (first point) and no definite trend is seen. Therefore even with the lowest N_{PF} we should not pass the general level of 15% uncertainty in the fermionic effects. The overall convergence is slower for small N_{PF}, but it seems that the number of heating sweeps can be reduced to 2-4 (without heating sweeps we may have a bias).

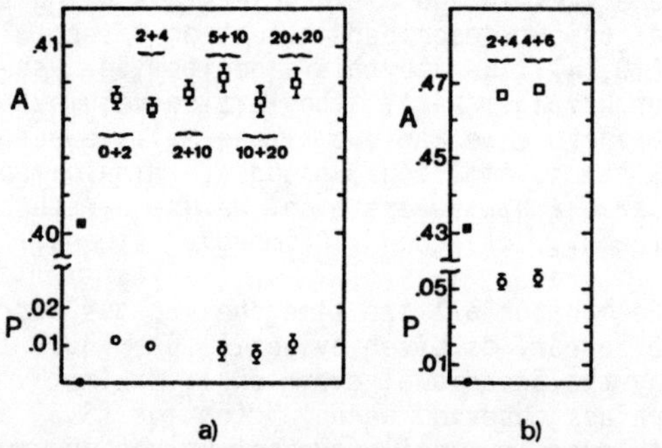

Figure 3: Average Plaquette (A) and Polyakov Loop (P) for various number of ps-fermionic heating sweeps + N_{PF}.
a) β=5, k=0.12, 4^3x4 lattice.
b) β=5.2, k=0.15, 8^3x4 lattice.

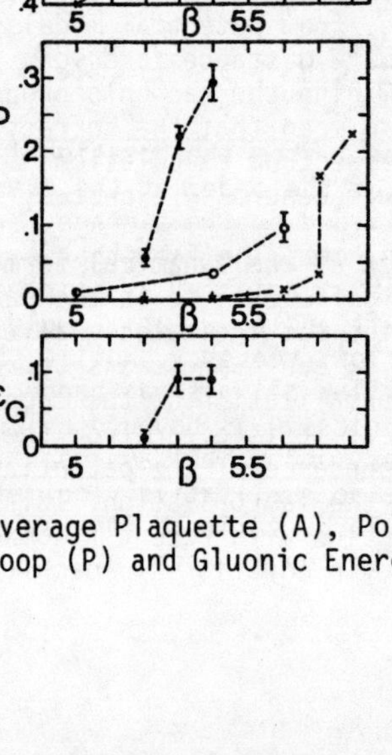

Figure 4: Average Plaquette (A), Polyakov Loop (P) and Gluonic Energy (ε_G)

Fig.4 shows A, P and ε_G as function of β. Especially the data at k=.15 are compatible with either a rapid cross-over or a smooth finite temperature phase transition. Recall however that with decreasing β k_c increases and so does the renormalized quark mass at fixed k, cutting down the fermionic effects.

The strongest effects however, peculiar to the dynamical quarks, are seen in Wilson loops (Fig.5). We measured loops in the 8x8 planes at k≠0 and compare with pure Yang Mills (k=0). We expect two types of finite size effects. The periodicity in the transverse direction (size 4) introduces more order in the system (but this does not mean we measure the finite temperature potential, as the loops do not have this axis as time).

To have an idea about the corrections from periodicity in the plane of the loop (8x8) we compare the results with noncompact QED_2 (simulating a first strong coupling term):

$$-\ln W(R,T) = const.\; RT\left(1-\frac{RT}{L^2}\right),\; L=8 \qquad (4.5)$$

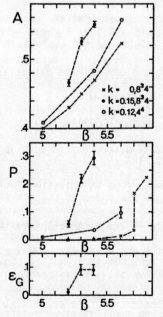

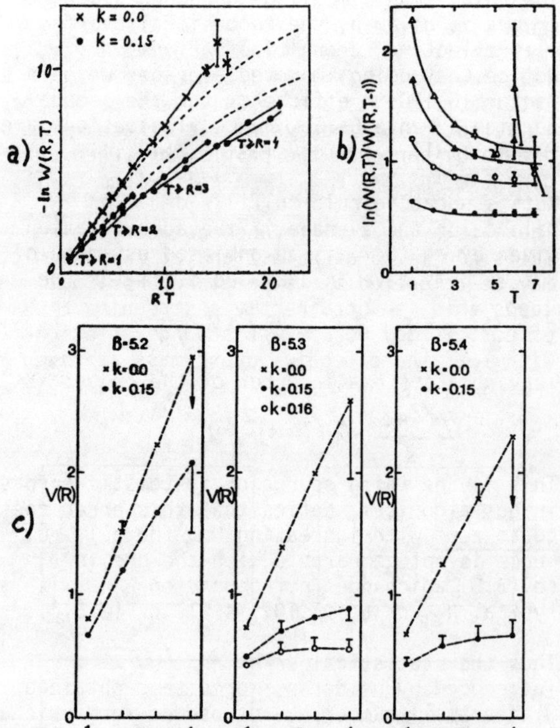

Figure 5: a) Spatial Wilson Loops at β=5.3 plotted against the area. Dashed curves indicate the periodicity effects in QED_2, adjusted to fit W(1,1). Full lines are eye-guides.
b) Fit of the Wilson loops ratios using eq.(4.6). Here β=5.3, k=.15
c) Static quark potential V(R) vs lattice distance R.

with the constant adjusted such as to fit the plaquette (the dashed lines in Fig.5a). As it is illustrated in Fig.5a the effect of the periodicity on the pure Yang Mills loops is much smaller than the one given by eq.(4.5), in fact we see little deviation from a pure area law. This is consistent with leaving the strong coupling regime. On the other hand, the exponents of the Wilson loops at $k \neq 0$ fall far below the pure Yang Mills ones and also well bellow the curve corresponding to eq. (4.5). Moreover, they clearly do not depend only on their area but also on their shape.

The dramatic effects of the dynamical quarks on the static quark potential can be seen on Fig.5c showing both quenched (i.e., pure Yang Mills) and unquenched results at three β-values. These results are obtained by fitting the ratios of Wilson loops with a two parameter formula at each R (see Fig.5b):

$$-\ln \frac{W(R,T)}{W(R,T-1)} = V(R) + c_1(R)/T \quad (4.6)$$

This formula should only be interpreted as a way to extract V(R) as the asymptotic ($T \to \infty$) part in the Wilson loops ratios on the left hand side. Especially for large R we do not necessarily get the static potential, but at least we get the Wilson loop law for this lattice. The fits are usually very good. For illustration of the systematic errors we indicate by bars how the points move when changing the Ansatz, e.g. by a third term c_2/T^2.

There is a clear difference between the results for k=.15 at β=5.2 and at β=5.4 which may be triggered by temperature. For spatial loops this is not directly a screening effect; it may work via a strong variation with β of the effective quark mass at fixed k

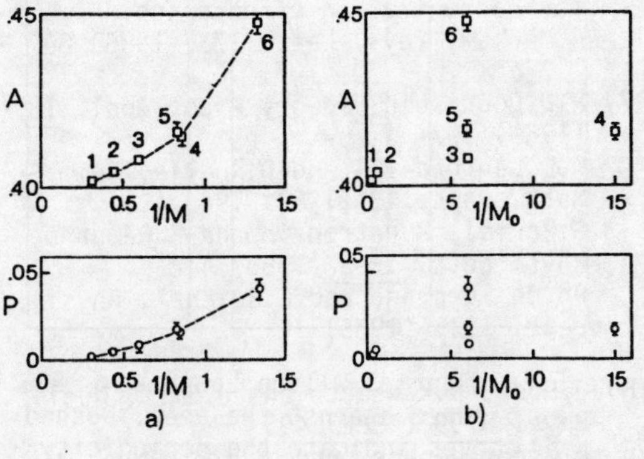

Figure 6: Average Plaquette (A) and Polyakov Loop (P) at β=5, $4^3 4$ lattice: 1,2,3 (k=.12, r=.4,.7,1), 4(k=.15, r=.85) 5(k=.1685, r=.7), 6(k=.283, r=.4).
a) Vs $M = 1/2k - r/2k_c$, k_c=.2
b) Vs $M_o = 1/2k - 1/2k_o$, k_o=.125 .

Very conspicuous in these results is the smooth coupling of the dynamical quarks effects at all distances, at variance with the usual picture of a well defined hadronization or screening threshold (distance).

Finally, on Fig.6, we show results obtained from simulations with various values of r. Interpreting these data is difficult, since we have as yet very little information about the expected behaviour with r, beyond what we know from the free fermions, weak coupling, or other special cases (see, e.g. (5-7,8,18)). But we can try to see whether the assumption of r-independence for r not too small is tenable, by trying to fit the data as function of one variable:

$$M \equiv \frac{1}{2k} - \frac{r}{2k_c} \quad (4.7)$$

where k_c is the critical k for this β at r=1 (this formula is written by simple analogy with the free fermions case, where $k_c = k_o$ =.125). With k_c=.2 we obtain the acceptable appearence on Fig.6a, to be compared with Fig.6b, where $k_c = k_o$. The value k_c=.2 is in between the quenched critical k and the value which can be obtained by extrapolation from (19).

5. DISCUSSION

Summarizing, we see a clear effect driven by the light dynamical quarks. The behaviour at large k-values is consistent with either a relatively rapid cross over or a smooth finite temperature transition.

The analysis of the Wilson loops shows that the loss of the string tension is accomplished gradually, over the whole range of distances. This departs from the usual picture of a linear potential still effective at small distances, and saturated at large distances. If this result could be confirmed by a scaling analysis, it would be of phenomenological relevance, since we know, e.g. from potential models, that the intermediate distances are very important in determining the hadronic properties. It would also mean that the correction to the quenched spectrum calculations might be larger than generally expected.

The effect of the dynamical fermions comes out very differentiate in the various observables and indicates qualitative new physics. It can by no means be reabsorbed in a renormalization of the coupling constant as it is clear from the change in the R-dependence of the potential, which reflects the hadronization. This is in contradiction with recent findings using the Langevin equation approach (20).

On the $8^3 4$ lattice our algorithm needs less than 2' per Yang Mills sweep (with 8 bushes, hence 8 updating steps, and 6-12 improved pseudo-fermionic sweeps per step, with an average ΔU as given by the pure Y M measure). We want to remark, that by calculating directly the determinants ratio by eq.(3.2) we avoid the problem of an ill-conditioned W plaguing the Langevin Equation approach for large correlation lengthes (see, e.g. (20)).

Our algorithm can ensure an acceptable level of precision and provides for the possibility, by tuning some parameter, to decrease any of its errors without having to reduce the "step size" ΔU. Thus we can reduce the errors connected with Kuti's updating by increasing N_{PF} and we can reduce the factorization error by decreasing N, the number of links in a bush.

We should recall here that the nonlocal character of the fermionic determinant implies that any local change needs many updating steps to propagate everywhere. Hence a small N_{PF} requires an increased number of Yang Mills sweeps to decorrelate the configurations. Making ΔU very small of course slows down both the fermionic and the Yang Mills motion. It seems therefore that the algorithm optimization is a problem of adjusting the various parameters (N, N_{PF}, average ΔU), with nontrivial solution depending on β and k.

AKNOWLEDGEMENTS: Part of the work described here involves further collaborations with E.Seiler (MPI, Munich), F.Karsh (Illinois Univ., Urbana) and A.El Khadra, G.Feuer, C.Hege, V.Linke and A.Nakamura at the Freie Universität Berlin. We are very indebted to the above mentioned and to M.Creutz, R.Gavai and others for very instructive discussions. IOS aknowledges support from the Freie Universität Berlin for attending the Berkeley Conference. The computations described in this work are done on the Cray XMP's of Cray Research. This work is supported in part by the Deutsche Forschungsgemeinschaft.

REFERENCES

(1) H.Hamber and G.Parisi, Phys. Rev. Lett. 47 (1981) 1792
D.Weingarten, Phys. Lett. 109B (1982) 57

(2) K.Wilson, Phys. Rev. D10 (1974) 2445
L.Susskind, Phys. Rev. D16 (1977) 3031

(3) P.Hasenfratz, F.Karsch and I.O. Stamatescu, Phys. Lett. 133B (1983) 221

(4) E.Seiler, Gauge Theories as a Problem of Constructive Quantum Field Theory and Statistical Mechanics, Springer, 1982

(5) E.Seiler and I.O.Stamatescu, Phys. Rev. D25 (1982) 2177
F.Karsch, E.Seiler and I.O.Stamatescu, Nucl. Phys. B271 (1986) 349

(6) T.Eguchi and R.Nakayama, Phys. Lett. 126B (1983) 89

(7) L.H.Karsten and J.Smit, Nucl. Phys. B183 (1981) 103

(8) W.Kerler, Phys. Rev. D23 (1981) 2384

(9) A.Kennedy and J.Kuti, Phys. Rev. Lett. 54 (1985) 2473

(10) I.O.Stamatescu, talk at the Meeting "Lattice Gauge Theory - a Challenge in Large Scale Computing", Wuppertal,1985
Ph.de Forcrand, Proc. of the Conf. "Frontiers of the Quantum Monte Carlo" Los Alamos, Sept.1985

(11) F.Fucito, E.Marinari, G.Parisi and C. Rebbi,Nucl.Phys. B180 /FS2/ (1981) 360

(12) D.H.Weingarten and D.N.Petcher, Phys. Lett. 99B (1981) 333

(13) G.Bhanot, U.Heller and I.O.Stamatescu, Phys. Lett. 128B (1983) 440

(14) Ph.de Forcrand and I.O.Stamatescu, Nucl. Phys. B261 (1985) 613
I.O.Stamatescu, in "Advances in Lattice Gauge Theory", D.W.Duke and J.F.Owens eds., World Scientific 1985

(15) R.V.Gavai and A.Gocksch, Prep.BNL 1986

(16) F.Karsch, talk at the Meeting "Lattice Gauge Theory - a Challenge in Large Scale Computing", Wuppertal, Nov. 1985
C.Rebbi, ibidem
for hadronization effects see:
H.Markum, Phys. Lett. 173 (1986) 337

(17) R.L.Dobrushin, Theory Prob. Appl. 13 (1969) 197
O.E.Lanford III and D.Ruelle, Comm. Math. Phys. 13 (1969) 194
G.Parisi, R.Petronzio and F.Rapuano, Phys. Lett. 129B (1983) 418
Ph.de Forcrand and C.Roisnel, Phys. Lett. 151B (1985) 77

(18) J.Hoek, N.Kawamoto and J.Smit, Nucl. Phys. B199 (1982) 495
R.Groot, J.Hoek and J.Smit, Nucl. Phys. B237 (1984) 111

(19) I.Montvay, Preprint DESY 85/072 (1985)

(20) M.Fukugita, Y.Oyanagi and A.Ukawa, Preprint UTHEP-152, April 1986

LATTICE SU(N) QCD AT FINITE BARYON DENSITY

Elbio Dagotto, Adriana Moreo
University of Illinois at Urbana-Champaign
Dept. of Physics, Urbana, IL 61801, U.S.A.
Ulli Wolff
Institut für Theoretische Physik
2300 Kiel, F.R.G.

We study the strong coupling limit of lattice SU(N) QCD at finite chemical potential using a dimer approach and mean-field techniques. For $N \geq 3$ we show the existence of a chiral symmetry restoring first order phase transition. When $N = 2$ a baryonic condensate appears at non-zero density.

Recently there has been great interest in the study of the phase diagram of lattice SU(N) QCD at finite temperature (T) and quark chemical potential (μ). This study is very important for the forthcoming heavy-ion collision experiments [1]. Working at finite T and $\mu = 0$ there is clear evidence for the existence of a chiral symmetry restoring phase transition [2], but not much is known about the influence of a finite baryon density [3-6] on the thermodynamics of QCD. The calculation of the critical densities seems to be a problem that can be studied using standard methods of the lattice approach. However at $\mu \neq 0$ the fermionic determinant is complex (for $N \geq 3$) and the well-known Monte Carlo techniques cannot be applied (a more straightforward calculation have been performed neglecting the imaginary part of the fermionic determinant [4] but this approach leads only to a qualitative picture of QCD).

The standard ideas suggest the following scenario at T=0: Let us denote by m_M the mass of the state with non-zero fermion number whose energy per quark is the lowest lying of the spectrum (M denotes the number of quarks of the state). In general m_M is a function of the quark bare mass (m) and some coupling g. As long as μ is smaller than m_M/M the thermodynamic observables of the theory (like $\langle\bar{\psi}\psi\rangle$) must have the same values as at $\mu = 0$ i.e. we are still testing the vacuum. At $\mu_c = m_M/M$ new quarks are added to the system and thermodynamics begins. Finally, for large values of μ we expect a restoration of chiral symmetry ($\langle\bar{\psi}\psi\rangle = 0$ at $m = 0$). At high density the theory should behave as a "soup" of nearly free quarks and gluons.

However the above described picture does not fit into recent numerical results [6] obtained in the strong coupling limit of lattice QCD. These results were obtained mainly in the quenched approximation [7], and for a few values of μ and m also using the unquenched complex Langevin algorithm [8]. In fact, the conclusion of ref[6] was that $\mu_c = \frac{1}{2}m_\pi$ where m_π is the mass of the pion. In other words the lowest baryon excitation of QCD seems to have an energy per quark equal to $\frac{1}{2}m_\pi$. Of course these results are quite unexpected. The fact that μ_c is not related with the mass of the proton (i.e. $M = 3$) is not unreasonable (for example it has been shown [9] in the Hamiltonian formalism that strong coupling QCD presents a μ_c related to a multiquark state). The big problem with the results of ref[6] is that for small masses $\mu_c \sim \sqrt{m} \xrightarrow[m\to 0]{} 0$ which is in clear contradiction with experience.

On the other hand, also in ref[6], a mean field study of QCD with $\mu \neq 0$ was described, which coincides with the numerical results for large masses but which is appreciably different for $m \ll 1$ giving a non-zero result for μ_c at $m = 0$. The questions are obvious: are the numerical results an indication of new physics or are the algorithms used in ref[6] (subtly) incorrect? In this case, is the mean field approach correct?

To answer these questions we have exploited the fact that there exists another method for the study of lattice QCD in the strong coupling limit: the dimer approach [10]. This method is based on the fact that, in the absence of plaquette terms, it is possible to integrate exactly the gauge fields. In this way we obtain an effective action with nearest neighbor interactions between color neutral quark composites (mesons and baryons). Below we will describe shortly the main features of the dimer algorithm adapted to finite μ [11](the inclusion of finite temperature in the method is straightforward and the results will be described in a future publication [12]).

The partition function of SU(N) QCD with staggered fermions and finite μ in the strong coupling limit is given by,

$$Z = \prod_x \int d\bar{\psi}_x d\psi_x \prod_\nu \int dU_{x,\nu} \exp\left[m \sum_x \bar{\psi}_x \psi_x\right]$$

$$exp\left\{\frac{1}{2} \sum_{x,\nu} \eta_\nu(x)\left[f_\nu \bar{\psi}_x U_{x,\nu} \psi_{x+\nu} - f_\nu^{-1} \bar{\psi}_{x+\nu} U^\dagger_{x,\nu} \psi_x\right]\right\} \quad (1)$$

where $f_\nu = e^{\mu_\nu}$ and $\mu_\nu = \mu\delta_{\nu,0}$. $\eta_\nu(x)$ are phase factors and the color indices are supressed. The rest of the notation is the standard one. Integrating over $U_{x,\nu}$ we obtain

$$Z = \prod_x \int d\bar{\psi}_x d\psi_x exp\{m\sum_x \bar{\psi}_x\psi_x + \bar{J}\sum_x B_x +$$

$$J\sum_x \bar{B}_x\} \prod_{x,\nu}\{H(\frac{1}{4}\bar{\psi}_x\psi_x\bar{\psi}_{x+\nu}\psi_{x+\nu})$$

$$+\frac{f_\nu^N}{2^N}\bar{B}_x B_{x+\nu} + \frac{f_\nu^{-N}}{2^N}\bar{B}_{x+\nu}B_x\} \qquad (2.a)$$

where

$$H(\lambda) = \sum_{k=0}^{N}\alpha_k\lambda^k, \qquad \alpha_k = \frac{(N-k)!}{k!N!}, \qquad (2.b)$$

the baryonic field is defined as $B_x = \psi_x^1 \dots \psi_x^N$ and we add baryonic sources $J, \bar{J}$. In the derivation of eq[2] we assume N even (i.e. B_x are commuting fields). This constraint will not be a problem for our purposes because the unexpected result $\mu_c = \frac{1}{2}m_\pi$ is also present in SU(4) QCD (ref[6]).

Next we introduce new variables to factorize and integrate the original Grassmann fields. For this purpose we define a set of link variables $(k_l, b_l, \bar{b}_l)$ with allowed values $(0 \le k_l \le N, 0, 0), (0,1,0), (0,0,1)$. k_l corresponds to picking the kth term in H while b_l and $\bar{b}_l$ are related with $\bar{B}_x B_{x+\nu}$ and $\bar{B}_{x+\nu}B_x$, respectively. It is also helpful to define auxiliary site variables as follows

$$\sigma_x = \sum_l k_l, \qquad \lambda_x = \sum_{l+} b_{l^+} + \sum_{l-}\bar{b}_{l^-},$$

$$\epsilon_x = \sum_{l+}\bar{b}_{l^+} + \sum_{l-} b_{l^-} \qquad (3)$$

where the sum in σ_x is over all the links emerging from x while the sum over $l^+(l^-)$ corresponds to links in positive (negative) directions. From the exclusion principle $[(\bar{\psi}\psi)^{N+1} = B^2 = \bar{B}^2 = 0]$ there are strong constraints over $(\sigma, \epsilon, \lambda)$ i.e. they can take on only the values $(0 \le \sigma \le N, 0, 0), (0,1,0), (0,0,1), (0,1,1)$. Using these fields we can rewrite exactly the partition function eq[2] as a function of $(k, b, \bar{b})$. The details of the calculation can be found in ref[10]. The final result is

$$Z = \sum_{\{(k_l, b_l, \bar{b}_l)\}} \prod_l \alpha_{k_l}\left(\frac{1}{4}\right)^{k_l}\left(\frac{1}{2}\right)^{N(b_l+\bar{b}_l)}$$

$$exp[\mu_l N(b_l - \bar{b}_l)] \prod_x \rho(\sigma_x, \epsilon_x, \lambda_x) \qquad (4)$$

where the function ρ has been defined in ref.[10]. The exponential factor in eq[4] encourages propagation of baryons in the temporal direction as expected.

We remark again that N must be even (otherwise, for example, the baryonic sources would be Grassmann fields and we could not get a useful expression for Z).

We have studied eq[4] numerically using the heat bath method i.e. at each link we choose a new value of $(k_l, b_l, \bar{b}_l)$ with the relative weight of the local Boltzmann factor in the fixed environment given by $(\sigma, \epsilon, \lambda)$ at the ends of the links. Due to the discrete values taken by $(k, b, \bar{b})$ it is possible to efficiently implement the algorithm packing the variables of a given site and its emerging links in only one word and using binary functions for the heat bath technique. Before describing the results we remark that the physically interesting special case $J, \bar{J} = 0$ must be obtained as a limit i.e. we cannot simply consider zero baryonic sources, because in this way it is impossible to create a new closed loop of baryons with a link by link update (in other words the algorithm is ergodic only for $J, \bar{J}, m \neq 0$ .If the update involves more than one link we can create small baryonic loops but not the big ones around the time direction in which one is interested at finite chemical potential).

In fig.1 we show the dimer results for the chiral order parameter $\langle\bar{\psi}\psi\rangle$. We work on a 4^4 lattice, $N = 4, J = \bar{J} = 0.075$ and $m = 0.1$. This value of the bare mass is important to distinguish between the mean field and numerical results of ref[6]. We have used two starting configurations: a "mesonic" one is obtained by running the code at $\mu = 0$ (starting for example with $(k, b, \bar{b}) = (0,0,0)$ at every link). In the $m \to 0$ limit this configuration has N mesons at every site ($\sigma = N$). A "baryonic" start is obtained by working at large μ. The final configuration has mainly closed baryonic loops in the time direction.

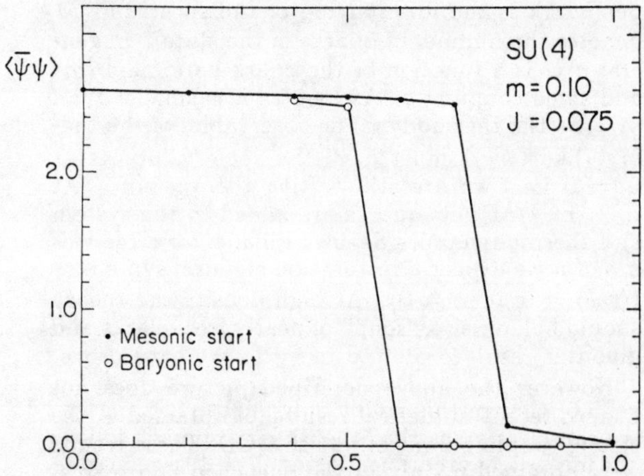

Figure 1: $\langle\bar{\psi}\psi\rangle$ vs. μ using mesonic and baryonic starts (described in the text).

Typically the points shown at fig.1 require ~15000 sweeps through the lattice (around 15′ in a FPS-164 array processor). To avoid correlations we measure observables every ~ 60 sweeps. These numbers are bigger near μ_c where in some cases we need 150000 sweeps to get a good convergence.

In fig.1 we can see a clear hysteresis loop showing the existence of two metastable states. This suggests a first order phase transition. Roughly speaking $\mu_c \in (0.55, 0.75)$. We verify that the metastable states are well approximated by the asymptotic states ($\mu = 0$ and $\gg 1$) i.e. $\langle\bar{\psi}\psi\rangle$ seems to behave as a step function. For example, we have also measured the fermion number density $n_f = \frac{1}{V}\frac{d}{d\mu}logZ$($V$ =volume of the lattice). This operator counts the number of quarks propagating in the positive time direction minus the number of antiquarks moving in the opposite direction. It jumps from 0 to 4 at μ_c i.e. the μ large state has the maximum allowed number of fermions per site (i.e. the qualitative picture is very similar to the Hamiltonian result [9]).

We have checked the independence of the results with J by working at $J = 0.05$ and $J = 0.025$ (reducing J more iterations are needed to achieve convergence).

To get a more accurate prediction for μ_c we have used a "mixed" starting configuration [13]. We have simply divided the lattice in two halves (keeping in both pieces the original time length) filling them with baryonic and mesonic starts. When the dimer method is iterated, the boundary wall moves to one side or to the other depending on the free energy of each state.

In fig.2 we plot our final results in the plane (m, μ). We also include the mean field prediction which can easily be obtained as follows: consider the mean field free energy of the model at $\mu = 0$ given by

$$F_{MF}(\mu = 0) = N\{\frac{1}{4d}[m - (m^2 + 2d)^{1/2}]^2 - log\{\frac{1}{2}[m + (m^2 + 2d)^{1/2}]\}\} \quad (5)$$

(for details see the appendix of ref[6]) where d is the space-time dimension, and compare it with the free energy of the (one-dimensional like) theory at $\mu \gg 1$ i.e. $F = N(log2 - \mu)$. The critical chemical potential is simply given by

$$\mu_c = -F_{MF}(\mu = 0)/N + log2 \quad (6)$$

and this result is plotted in fig.2. We have proved that more involved calculations as described in ref[5,6] change this result only slightly. From fig.2 we can clearly see that the dimer approach follows the mean field prediction for small and large masses.

Note that μ_c in the mean field approach is N-independent showing that it is highly plausible that our results can be extended to all SU($N \geq 3$).

We remark again that at μ_c the theory jumps to a phase "full" of quarks i.e. before reaching the value $m_{baryon}/4$ (also shown in fig.2) it is energetically favourable to restore (abruptly) chiral symmetry [14]. This behavior may be a strong coupling artifact. Probably we must also include the electromagnetic force between quarks which would disfavor the existence of point-like baryons.

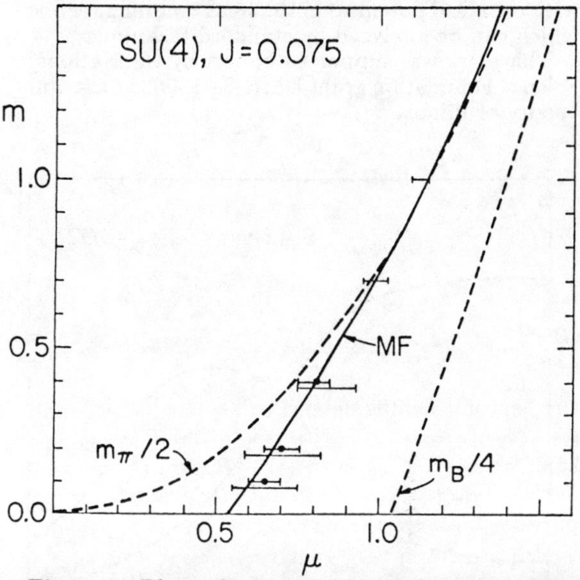

Figure 2: Phase diagram in the plane $m - \mu$. The bars (⊢⊣) represent the width of the hysteresis loops. (⊢ • ⊣) indicate improved results using a mixed state start. The continuous line is the mean field prediction eq(6). m_π and m_B are the pion and baryon masses (see ref[14]). The numerical results of ref[6] follow $\frac{1}{2}m_\pi$.

For completeness, it is interesting to check whether the dimer approach is able to distinguish between SU(4) and SU(2). From ref[5] we know that the analytical prediction for SU(2) is quite different from SU($N \geq 3$) mainly because a baryonic condensate (i.e.$\langle B\rangle \neq 0$) at finite μ exists (the U(1) vectorial symmetry is spontaneously broken). For SU(4) we have proved analytically that this condensate is negligible. The dimer results are shown in fig.3. The agreement with the predictions of ref[5] is remarkable [15](an even better agreement can be obtained by considering exactly the time direction [12]) . For SU(2) the transition is continuous (there are no indications of metastable states).

Summarizing, the conclusions of this letter are as follows: using the dimer approach in the strong coupling limit of lattice SU(4) QCD we have shown that a finite critical value for the chemical potential exists where there is an abrupt transition from the $\mu = 0$ regime to a phase where chiral symmetry is restored and n_f takes its maximum value. μ_c is in excellent agreement with a very simple mean field prediction and in disagreement with the numerical results of ref[6] showing that there is some problem in the algorithm used there(for the Complex Langevin method this is not surprising since there is no proof that the large time solution will correspond to the exponential of the action) .

We believe that it is a rather urgent problem to find a suitable algorithm for the study of lattice QCD at finite density. We suggest that any new method proposed to deal with the problem should be checked in the (highly non trivial) strong coupling limit of QCD. Another interesting subject that deserves further study is the phase diagram of SU(2) with chemical potential in the weak coupling regime which can be analyzed by standard techniques.

This work was supported in part by the National Science Foundation grant DMR-84-15063 at the University of Illinois.

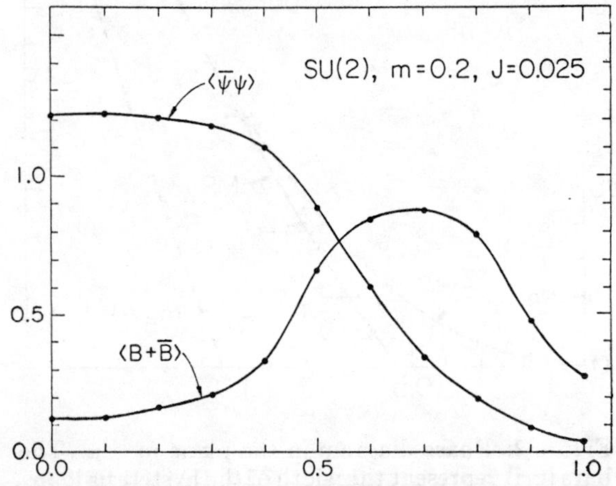

Figure 3: $\langle\bar{\psi}\psi\rangle$ and $\langle B+\bar{B}\rangle$ vs. μ for SU(2) lattice QCD

References

1) J.Cleymans, R.Gavai and E.Suhonen, Phys.Rep. **130**, 219(1986); H.Stöcker and W.Greiner, Phys. Rep. **137**, 277 (1986).

2) T.Çelik, J.Engels and H.Satz, Nucl.Phys.**B256**, 670 (1985); R.Gavai and F.Karsch, Nucl. Phys.**B261**, 273(1985); E.Dagotto, A.Moreo and R.Trinchero, Phys.Rev.**D33**, 1121 (1986);J.Kogut, UI preprint,ILL-TH-86-19; and references therein.

3) P.Hasenfratz and F.Karsch, Phys.Lett.**125B**, 308 (1983); J.Kogut, H.Matsuoka, M.Stone, H.Wyld, S.Shenker, J.Shigemitsu and D.Sinclair, Nucl.Phys. **B225**, 93 (1983); N.Bilic and R.Gavai, Z.Phys.**C23**, 77 (1984); A.Nakamura, Phys.Lett. **149B**, 391 (1984); R.Gavai, Phys.Rev. **D32**, 519 (1985); U.Wolff, Phys.Lett. **157B**, 303 (1985).

4) J.Engels and H.Satz, Phys.Lett.**159B**, 151(1985).

5) E.Dagotto, F.Karsch and A.Moreo, Phys.Lett. **169B**, 421(1986).

6) I.Barbour, N.E.Behilil, E.Dagotto, F.Karsch, A.Moreo, M.Stone and H.Wyld, ILL-TH-86-23, to appear in Nucl.Phys.B.

7) The quenched approximation at finite μ has been recently challenged by P.Gibbs, Glasgow preprint (1986).

8) G.Parisi, Phys.Lett.**131B**, 393(1983); J.R.Klauder, "Stochastic Quantization", Lectures given at the XXII Schladming School, March 1983; J.Ambjørn, M.Flensburg and C.Peterson,Univ. of Lund preprint, LU TP 86-6.

9) A.Moreo, UI preprint, ILL-(337)-86-# 2 .

10) P.Rossi and U.Wolff, Nucl.Phys.**B248**, 105 (1984); U.Wolff, Phys.Lett.**153B**, 92(1985).

11) E.Dagotto, A.Moreo and U.Wolff, Phys.Rev. Lett. (to appear).

12) E.Dagotto, A.Moreo and U.Wolff (in preparation).

13) M.Creutz, L.Jacobs and C.Rebbi, Phys.Rev. Lett. **42**, 1390(1979).

14) The results for $\frac{1}{2}m_\pi$ and $\frac{1}{4}m_B$ shown in fig.2 are mean field predictions taken from H.Kluberg-Stern, A.Morel and B.Petersson, Nucl.Phys.**B215**, 527
(1983). In fact the calculation for m_B was done for odd N but the final result was approximately N independent and we extend it to even N values.

15) The SU(2) results are much more J-dependent than in SU(4) due to the baryonic condensate.

ACCELERATION OF GAUGE FIELD DYNAMICS

S. Duane

Physics Department, Imperial College
London SW7 2BZ, U.K.

Hybrid stochastic differential equations provide one promising method to simulate gauge fields interacting with light fermions. The resulting dynamics can be regarded as already incorporating some acceleration, and one possible way to enhance this effect is described.

The precise correspondence between one Euclidean quantum field theory in d space-time dimensions, and another classical statistical system in d space dimensions is now widely appreciated. The vacuum expectation value of a time-ordered product of operators becomes a static correlation function of c-number fields. So all the quantum physics is contained in the equilibrium properties of this classical system, and one has complete freedom of choice as far as any classical dynamics is concerned. In fact there is no need to introduce the idea of evolution with respect to an extra time at all, except as a means to replace the average over an ensemble of independent configurations by an average over the history of a single configuration. This will be valid provided some ergodicity condition is satisfied, and it only remains to ask how quickly such a time average will converge, or equivalently how long it takes for the dynamics to generate a statistically independent configuration. By 'accelerated' dynamics I mean some improvement over a simpler dynamics for which this autocorrelation time is reduced. While there is no 'real' physics in these questions, nevertheless they are of some practical importance, and it is possible to use physical insight to help answer them.

The construction of the hybrid algorithm begins with the molecular dynamics Hamiltonian, (1,2)

$$H = -\frac{1}{4}\sum_x \sum_\mu \operatorname{tr} P_\mu(x) P_\mu(x) - \frac{\beta}{2}\sum_x \sum_{\mu\nu} \operatorname{tr} U_{\mu\nu}(x) + c.c. + \sum_{xy} \pi_\psi^\dagger(x)\left[M^\dagger M(U)\right]^{-1}_{x,y} \pi_\psi(y) + \omega^2 \sum_x \psi^\dagger(x)\psi(x) \tag{1}$$

in which the gauge field $U_\mu(x)$ has conjugate momentum $P_\mu(x) = -P_\mu^\dagger(x)$, the bosonised fermion $\psi(x)$ has conjugate momentum $\pi_\psi(x)$, and the Dirac operator $m + \not{D}$ is represented on the lattice by the matrix $M(U)$. Integration of the equations of motion

$$\dot{U} = PU \qquad \dot{P} = -\frac{\partial H}{\partial U}$$
$$\dot{\psi} = \left[M^\dagger M\right]^{-1}\pi_\psi \qquad \dot{\pi}_\psi = -\omega^2\psi \tag{2}$$

produces a solution which conserves energy and which will, provided the trajectory is ergodic, lead to an equilibrium probability which is uniformly concentrated on one energy shell

$$W(U, P, \psi, \pi_\psi) = \delta\left[H(U,P,\psi,\pi_\psi) - E\right] \tag{3}$$

i.e. a microcanonical distribution. In the thermodynamic limit, and provided one only considers local observables, this will produce the same averages as the canonical ensemble

$$\exp - H(U, P, \psi, \pi_\psi)\,. \tag{4}$$

Integrating out the Gaussian degrees of freedom then gives the desired effective distribution for the gauge fields

$$\det(M^\dagger M)\exp\left(\frac{\beta}{2}\sum_x\sum_{\mu\nu}\mathrm{tr}\,U_{\mu\nu}(x)+c.c.\right). \quad (5)$$

In the hybrid algorithm (3), this smooth evolution is occasionally interrupted by randomising the velocities, i.e. replacing the old values by new ones appropriate to the Gaussian distribution contained in (4). As a result energy is no longer conserved and, as was proved in (4), the probability distribution converges towards the canonical ensemble. This result holds independent of the mean interval T between velocity refreshments, and so one can avoid all the subtleties required to justify the pure molecular dynamics.

In a practical simulation, one has to integrate the equation of motion numerically, using a nonzero stepsize dt. The discrete form of the evolution shows how this hybrid algorithm involves a mixture of molecular dynamics with the Langevin equation. To demonstrate this simple result it is enough to consider the case of a single degree of freedom q having conjugate momentum p. The Verlet discretization of Hamilton's equations is

$$p_{n+1/2} = p_{n-1/2} - dt\, S'(q_n)$$
$$q_{n+1} = q_n + dt\, p_{n+1/2} \quad (6)$$

and corresponds to a continuum time Hamiltonian

$$H(p,q) = \tfrac{1}{2}p^2 + S(q). \quad (7)$$

If q_n is taken to be q(t) then $p_{n+1/2}$ is most naturally interpreted as $p(t+\frac{1}{2}dt)$ since that gives the discrete evolution a time-reversal symmetry. This can allow absolute stability in the large time limit. The randomization of p is achieved by taking as an alternative update

$$p_{n+1/2} = \xi - \frac{dt}{2}S'(q_n)$$
$$q_{n+1} = q_n + dt\, p_{n+1/2} \quad (8)$$

where ξ is a Gaussian random variable which is uncorrelated with q_n, $\overline{\xi^2}=1$, and which can therefore be taken to be p(t). The half step in p is just designed to produce the right correlation between q_{n+1} and $p_{n+\frac{1}{2}}$, i.e.

$$\overline{q_n p_{n+1/2}} = -\tfrac{1}{2}dt\,\overline{q\,S'(q)} + O(dt)^2 \quad (9)$$

This random step can be recast as

$$q_{n+1} = q_n - \frac{(dt)^2}{2}S'(q_n) + dt\,\xi \quad (10)$$

which is precisely the discretized Langevin equation,

$$\frac{dq}{d\tau} = -S'(q) + \eta \quad (11)$$

with Langevin time step $d\tau = \frac{1}{2}dt^2$.

Since there is such a close match between the discrete molecular dynamics and the discrete Langevin equation, one might suspect that the systematic errors at each step are the same for the two possible forms of update. Then the overall systematic error in the equilibrium distribution would be independent of the mean time T between Langevin steps. In that case one is free to tune this parameter T so as to obtain optimal convergence. The systematic error for the discrete update described here turns out to be $O(dt)^2$. In the general case, this leading order part of the error is known to be independent of T, (4), and at least in the case of a harmonic oscillator, the error is T-independent to all orders in dt, (3).

To analyse the rate of convergence, it is useful to integrate out the momentum to obtain the effective probability distribution

$$\bar{W}(U) = \int dP\, d\psi\, d\pi_\psi\, W(P,U,\pi_\psi,\psi). \quad (12)$$

This was shown in (4) to satisfy the Fokker-Planck equation for small T

$$\frac{\partial\bar{W}}{\partial t} = \tfrac{1}{2}T\frac{\partial}{\partial U}\left(\frac{\partial}{\partial U}-\frac{\partial\bar{S}}{\partial U}\right)\bar{W} + O(T^2). \quad (13)$$

(This should come as no surprise, since for small T the hybrid algorithm is like a discrete time Langevin equation, with $d\tau = \frac{1}{2}T^2$.) It follows that the rate of convergence is directly proportional to T, and in this sense the hybrid scheme is an accelerated Langevin dynamics.

These results show that the usual Langevin simulation can always be improved by going to a hybrid scheme (i.e. taking T>dt) but do not determine the ideal value for T. If it makes sense to talk of the system having modes of different frequencies, then the optimal T for a harmonic oscillator may be an appropriate choice, i.e. half the inverse frequency of the slowest mode, (3).

This should be compared with the

maximum permitted step size, which is twice the inverse frequency of the fastest mode: if dt reaches this value, then the integration scheme goes unstable. Thus one can see that in the continuum limit of a lattice system, when the ratio of these frequencies diverges like the linear size of the system, the optimal number of steps between refreshments should also grow linearly. To the extent that free field theory is a reliable guide to the problem of critical slowing down, then it would seem that correlation times grow linearly with the correlation length. This should be contrasted with the behaviour of the Langevin equation, for which one expects correlation times to grow quadratically. However for systems like the hard sphere gas (i.e. real molecular dynamics!) free field theory is a particularly bad guide, since there is probably no difference in critical slowing down between the Langevin equation and molecular dynamics. One might expect that asymptotically free theories such as QCD will not be like the hard sphere gas.

The possibility to reduce critical slowing down even more arises because of the freedom of choice of molecular dynamics. (It should be clear from the discussion above that the 'hybrid' idea can be incorporated into any molecular dynamics in an obvious way: this will be taken for granted from now on.) So consider a generalized Hamiltonian

$$
\begin{aligned}
&-\frac{1}{4}\sum_{xy}\sum_{\mu\nu} \mathrm{tr}\, P_\mu(x)\, K_{\mu\nu}(x,y)\, P_\nu(y) \\
&-\frac{\beta}{2}\sum_{x}\sum_{\mu\nu} \mathrm{tr}\, U_{\mu\nu}(x) \;+\; \text{c.c.}
\end{aligned}
\tag{14}
$$

in which the kinetic energy has been made nonlocal in position space, and where for brevity the fermion terms have been dropped.

In order to motivate the various possibilities for the matrix K, it is usual to consider an accelerated Hamiltonian for free scalar field theory

$$
\frac{1}{2}\sum_{xy}\pi_x(M^2-\partial^2)_{xy}\pi_y \;+\; \frac{1}{2}\sum_x \phi_x(m^2-\partial^2)_x\phi_x \tag{15}
$$

In this case, it is easy to see that setting $M^2 = m^2$ makes all the normal modes have a common frequency, thereby eliminating critical slowing down altogether. While the equations of motion are nonlocal, they can be solved without too much overhead by using Fast Fourier Transforms.

Nonabelian gauge theories complicate this picture in two ways. Firstly the presence of interactions rules out the possibility of any simple analytical argument to show that some p rticular acceleration operator K really does its job: ultimately one has to perform the simulation to find out. Secondly the gauge dependence of the conjugate momentum means that one cannot simply choose an operator built out of ordinary derivatives. Consider the infinite β limit which is dominated by Gaussian fluctuations about pure gauge configurations. At the trivial configuration, i.e. with all links set to 1, one could use the ordinary derivative to give the ultraviolet modes a large inertia and the infrared modes a small one. This would remove critical slowing down. However at a nontrivial pure gauge configuration what was an IR mode can be made to look UV. For example, take the gauge group to be SU(2), and the mode in question to be everywhere proportional to σ_3, and apply a gauge transformation which is 1 on all the even sites, and $i\sigma_1$ on the odd sites. Since

$$
i\sigma_1(\sigma_3)(i\sigma_1)^+ \;=\; \sigma_1\sigma_3\sigma_1 \;=\; -\sigma_3 \tag{16}
$$

this has the effect of adding $\frac{1}{2}\pi$ to each component of the momentum in the mode. So the same operator which cures critical slowing down at the trivial configuration will make matters (much) worse elsewhere on the pure-gauge orbit (5). It is clear that one way to avoid this problem might be to fix the gauge. This possibility has been discussed at this conference (6), and involves choosing a dynamics which is gauge-dependent. An alternative would be to abandon FFTs and to use gauge covariant derivatives instead, (7). This preserves the gauge invariance of the evolution. (Note that the conjugate momentum lies in the adjoint representation, which is trivial for Abelian theories. In that case no gauge fixing is required and one can easily prove that in the infinite β limit Fourier acceleration works perfectly (7).) But there is a substantial overhead associated with a nonlocal operator based on covariant derivatives, since the inversion required at each step is qualitatively similar to that when the system includes dynamical fermions. This observation serves to remind us that one cannot blithely introduce a U-dependence into K, since integrating out the conjugate momenta then leads to a nontrivial determinant which would modify the equilibrium distribution. This has to be cancelled, for example by introducing extra fields in the adjoint representation whose action is local.

Further details of how this is done are contained in (7), and the results for a simple test system are contained in (8).

REFERENCES

(1) D. Callaway and A. Rahman, Phys. Rev. Lett. 49(1982)613.

(2) J. Polonyi and H.W. Wyld, Phys. Rev. Lett. 51(1983)2257.

(3) S. Duane, Nucl. Phys. B257(FS14) (1985)652.

(4) S. Duane and J.B. Kogut, UIUC preprint ILL-(TH)-86-15, to appear in Nucl. Phys. B.

(5) G.G. Batrouni et al., Phys. Rev. D32(1985)2736.

(6) G.P. Lepage, these proceedings.

(7) S. Duane et al., Phys. Lett. 176B(1986)143.

(8) S. Duane and B.J. Pendleton, in preparation.

HADRON SPECTRUM IN QUENCHED QCD

Y. Iwasaki

Institute of Physics, University of Tsukuba
Ibaraki 305, Japan

Hadron masses are calculated in the quenched approximation to lattice gauge theories with a renormalization group improved lattice SU(3) gauge action and Wilson's quark action on a $16^3\times 48$ lattice. The ground state masses of hadrons calculated for several quark masses are free from finite size effects and the contamination of excited states. When the results are extrapolated to the quark mass where the pion and rho masses take the physical values, the proton and delta masses agree with the physical values with at most 10~15 % errors.

Lattice formulation of QCD enables us to calculate hadron masses by numerical methods from first principles. However, in order to obtain reliable results by numerical methods one must be very cautious that systematic errors such as due to finite lattice spacing effects and finite lattice size effects are under control. When we consider the capacity of today's computers, I believe it is urgent to derive the correct hadron masses in the quenched approximation. Here I would like to report the results of the calculation of hadron masses which has been done with the intention of obtaining the correct hadron masses in the quenched approximation by reducing such systematic errors.

My primary concern is what the correct masses of the proton and delta are in the quenched approximation, when the masses of the pion and rho are fitted. At Leipzig two years ago, it was concluded that the proton-to-rho mass ratio m_p/m_ρ is too large in the calculations which had been done up to that time[1]. Is it due to the quenched approximation? Well, if we recall the success of valence quark models as well as the success of the OZI rule, we may expect that the quenched approximation will give reasonable values for hadron masses with about 10% errors. Possible origins for this large discrepancy are finite lattice spacing effects and finite lattice size effects.

In order to reduce finite lattice spacing effects we take a renormalization group (RG) improved lattice SU(3) gauge action

$$S=\frac{1}{g^2}\{c_0\,\Sigma\ \mathrm{Tr}\ (\text{simple plaquette}) + c_1\,\Sigma\ \mathrm{Tr}\ (1\times 2\ \text{rectangular loop})\} \qquad (1)$$

with

$$c_1=-0.331,\quad c_0=1-8c_1, \qquad (2)$$

the form of which has been determined by a block spin renormalization group study[2] and an analysis of instantons on the lattice[3].

Let's me give some evidence which supports our approach. Among others[4] we have recently investigated the scaling behavior of the string tension[5]. The results show that asymptotic scaling sets in at $\beta(=6/g^2)=2.6$ for the RG improved action, while it does not sets in up to $\beta=6.6$ for the one-plaquette action.

In order to reduce finite lattice effects, we make our calculations on a $16^3\times 48$ lattice at $\beta=2.4$ ($\beta=6/g^2$). The lattice size and the coupling constant are determined from the results of our previous calculations of hadron masses on a $12^3\times 24$ lattice[5] and on a $16^3\times 32$ lattice[7]. (We have increased the number of configurations up to 30 configulations in the case of the $16^3\times 32$ lattice to improve statistics: The results are essentially the same as those in ref.7)).

We take Wilson fermions for quarks. We use two operators $\pi_1=\bar{u}\gamma_5 d$ and $\pi_2=\bar{u}\gamma_5\gamma_4 d$ for the pion, $\rho_1=\bar{u}\gamma_i d$ and $\rho_2=\bar{u}\gamma_i\gamma_4 d$ for the rho, respectively. Here u and d are u-quark and d-quark fields respectively and i=1,2,3. We use non-relativistic operators for the proton and delta as well as $(u^T\gamma_5 C^{-1}d)u$ for the proton and $(u^T\gamma_\mu C^{-1}u)u$ for the delta, where C is the charge conjugation operator. We use two operators for all of them to investigate finite size effect. We calculate hadron propagators for 15 gauge configurations, separated by one hundred sweeps after a thermalization of 1000 sweeps. The hopping parameters K we take are 0.14, 0.145, 0.15, 0.1525 and 0.154.

Let me first discuss the results for the mesons. The propagators can be excellently fitted to a single hyperbolic-cosine for a wide range of t (t: coordinate in time direction; $0 \leq t \leq 47$ here; the origin of the propagator is t=0). See Fig.1 for the propagators of π_1 at five hopping parameters together with one-mass fits for $11 \leq t \leq 37$. As a result about 35 points agree with the fitted hyperbolic-cosines within one-standard deviation. The ground state mass at each K is stable for a change of the t-range ($8 \sim 11 \leq t \leq 37 \sim 40$) for the fit. Further when we make a two-mass fit to the propagators at each K for the range $4 \sim 5 \leq t \leq 43 \sim 44$, the ground state mass is stable. The masses and the errors are estimated by a least mean squares method. The masses determined both from π_1 and π_2 (from ρ_1 and ρ_2)

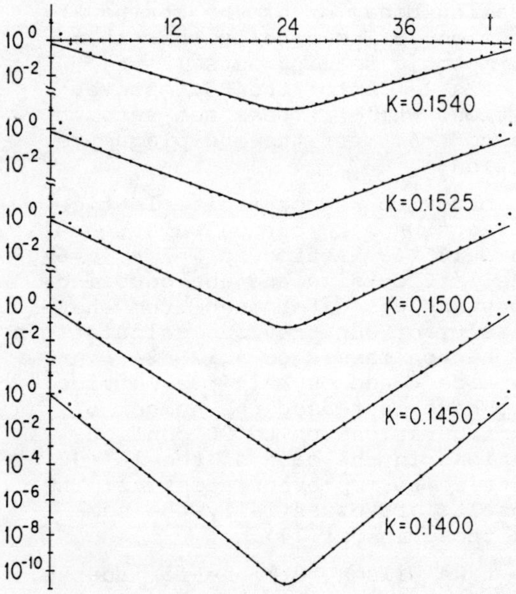

Fig.1: Propagators of π_1 at five hopping parameters with one-mass fit for $11 \leq t \leq 37$.

agree with each other within one-standard deviation. Furthermore, the propagators on the $12^3 \times 24$, $16^3 \times 32$ and $16^3 \times 48$ at five K's agree with each other within almost one-standard deviation for the t-region which can be compared. Thus we conclude that we are able to determine the ground state mass of the pion and rho at each K with small errors which is free from finite size effects and free also from the contamination of excited states.

We plot the masses of the pion and rho versus 1/K in Fig.2. We first notice that the linearity between m_π^2 and 1/K as well as that between m_ρ and 1/K are roughly satisfied. However, if we closely look at the 1/K dependence of the masses, we observe that m_π^2 is slightly convex downwards and m_ρ is slightly convex upwards. The results from independent gauge configurations on the $12^3 \times 24$, $16^3 \times 32$ and $16^3 \times 48$ lattices all indicate the same behavior. Therefore we conclude these dependences on 1/K are real and not due to statistical errors. We fit the masses to a quadratic function of 1/K as shown in Fig.2. This leads to $K_c = 0.1569(2)$ and $m_\rho(K=K_c) = 0.426(15)a^{-1}$, where K_c is the hopping parameter value where the pion mass vanishes and a is the lattice spacing. We do not distinguish K_{phy} (where the ratio m_ρ/m_π becomes the physical value) from K_c, because K_{phy} is within the statistical error for K_c. Inputting the physical mass we have a^{-1}=1810(60) MeV.

Let us now discuss the results for the baryons. As K approaches toward K_c, the fluctuation of propagators at large t becomes large and consequently we cannot obtain reliable

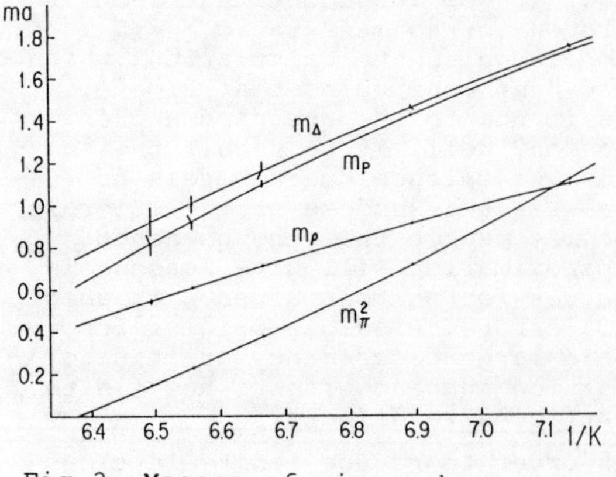

Fig.2: Masses of pion, rho, proton and delta versus 1/K.

values of the propagators of the baryons for $t \geq 20$ at K=0.1525 and for $t \geq 19$ at K=0.154. The propagators of the non-relativistic proton at five hopping parameters are displayed in Fig.3.

Let us describe in some detail the fitting procedure for the non-relativistic proton. Fitting the propagators to $A_0 \exp(-m_0 t)$ for $18 \leq t \leq 24$, we obtain 1.73(1), 1.42(1) and 1.10(2) for m_0 at K=0.14, 0.145 and 0.15, respectively. We check that the contamination from excited states for these results is small in the following way: i) Ground state masses remain within one-standard deviation even if we change the fitting range of t to $t_1 \leq t \leq 24$ with t_1=15~21. ii) When we fit the propagators to $G(t) \sim A_0 \exp(-m_0 t) + A_1 \exp(-m_1 t)$ for $t_1 \leq t \leq 24$ with t_1=5,6 and 7 the ground state masses are stable and agree with the results obtained by the one-mass fits.

If we would fit the propagators to $A_0 \exp(-m_0 t)$ for the range $11 \leq t \leq 16$ on the $16^3 \times 32$ lattice, we would obtain 1.76, 1.47 and 1.14 for m_0 at K=0.14, 0.145 and 0.15, respectively. They are larger than those obtained above by two or five standard deviations. Thus we conclude that the temporal linear extention in this work, 48 in lattice unit, is large enough (at least for $K \leq$ 0.15 at β=2.4) to obtain ground state masses by one-mass fits, while 32 in lattice unit is not large enough. This is exactly what we have conjectured in ref.7). This difficulty in extracting the ground state masses by one-mass fit consists in the fact that the excited states exist close to the ground states and further that the amplitude A_1 is larger than A_0. This last fact is well known henomenologically for N(1440). Our results indicate that $\sqrt{A_1}/\sqrt{A_0} \approx 2.0$.

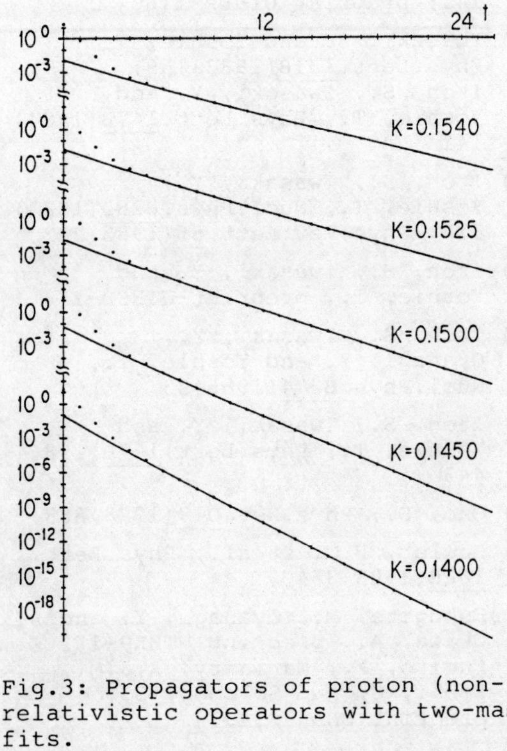

Fig.3: Propagators of proton (non-relativistic operators with two-mass fits.

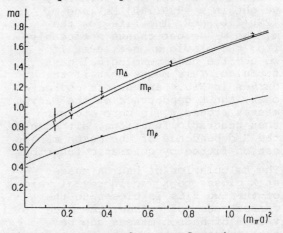

Next let us discuss the propagators at K=0.1525 and K=0.154. Because we cannot obtain reliable values of the propagators for very large t, we have to be satisfied with making a two-mass fit. (See Fig.3) We obtain 0.92(4) and 0.81(5) for m_0, respectively. These values are slightly smaller than those which would be obtained if we would make one-mass fit for $10 \leq t \leq 18$.

The ground state masses for other three baryon operators are obtained by similar analyses. The results for the relativistic operators are in excellent agreement with those for the non-relativistic ones at each K. This implies that finite size effects are under control.

We display in Fig.2 the ground state masses of the baryons at each K. We notice again that although the linearity between the mass and 1/K is roughly satisfied, the mass as a function of 1/K is slightly convex upwards. If we fit the masses of the baryons to quadratic functions

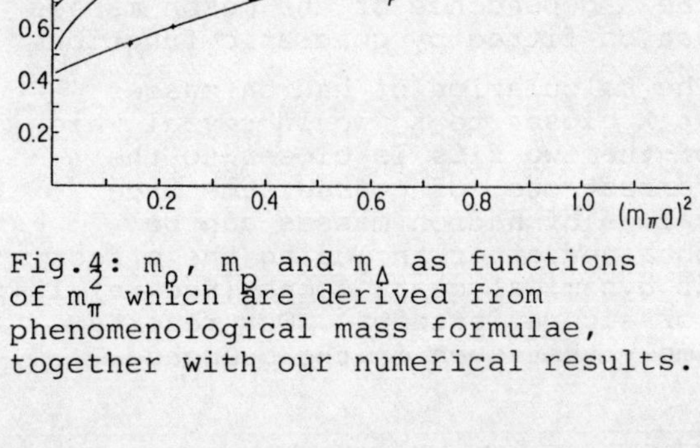

Fig.4: m_ρ, m_p and m_Δ as functions of m_π^2 which are derived from phenomenological mass formulae, together with our numerical results.

of 1/K (See Fig.2), we obtain m_0 (non-relativistic proton)=1100(90) MeV, m_0 (relativistic proton)= 1080(80) MeV, m_0 (non-relativistic)= 1340(120) MeV and m_0 (relativistic)= 1370(120) MeV at $K=K_c$. The results agree with the physical values with 10~15 % errors.

Although we think the above fitting procedure is a modest one, there is an alternative fit which gives almost exact physical values for the proton mass and the delta mass. We use phenomenological mass formulae which reproduce remarkably the physical hadron masses[8]. The formulae are also used in ref.9) for comparison with lattice results. We display in Fig.4 the masses of the rho, proton and delta as a function of m_π^2 which are obtained from the mass formulae, together with our results. Our results are remarkably on the curves. Note that our adjustable constant is only for a^{-1}, which has been already determined. It should be emphasized that any phenomenological model which reproduces hadron masses nicely should give curves which are essentially identical to those shown in Fig.4. Thus we conclude that our numerical results excellently reproduces the phenomenological results for $0.15 \leq (m_\pi a)^2 \leq 1.1$. Therefore the point is whether the agreement will continue down to $(m_\pi a)^2 \simeq 0.01$. If it is the case, we get the realistic values for the proton mass (940 MeV) and the delta mass (1230 MeV). Indeed, if we fit the data for the proton and the delta to the mass formulae, we obtain m_p=900(180) MeV and m_Δ 1290(210) MeV. The fits for the m_π and m_ρ do not change noticeably from the previous ones, even if we use the phenomenological mass formulae. This means that if the curves in Fig.4 are almost correct ones, the K dependence of the baryon masses around K_c is more steeper than quadratic functions, although the K dependence of the meson masses can be fitted by quadratic functions.

The calculation of hadron masses at K closer to K_c would reveal which of the two fits is closer to the correct one. Of course, the true values of hadron masses can be obtained after including the effect of dynamical quark loops (See ref.10) for such an attempt). Our results imply that even in the quenched (valence) approximation we can obtain reasonable values for the masses of flavor non-singlet hadrons with at most 10~15 % errors. This is completely consistent with the success of the valence quark model in describing static properties of hadrons.

This work has been done in collaboration with S. Itoh and T. Yoshie. Details of our data and our analyses will be published elsewhere.

The calculation has been performed with the HITAC S810/10 at KEK. We would like to thank S. Kabe, T. Kaneko, R. Ogasawara and other members of Data Handling Division of KEK for their kind arrangement which made this work possible, and the members of Theory Division, particularly, H. Sugawara and T. Yukawa for their warm hospitality and strong support for this work.

REFERENCES

(1) Bowler et al., Nucl.Phys.B240 [FS12](1984)213; Billoire, A., Marinari, E. and Petronzio, R., Nucl.Phys.B251 FS13 (1985) 141; Köng, A., Mütter, K.H., Schilling, K. and Smit, J., Phys. Lett. 157B(1985)421

(2) Iwasaki, S., Nucl.Phys.B258 (1985) 141; preprint UTHEP-118

(3) Iwasaki, S. and Yoshié, T., Phys.Lett.131B(1983)159; Itoh, S., Iwasaki, Y. and Yoshié, T., Phys.Lett.147B(1984) 141

(4) Itoh, S., Iwasaki, Y., Yoshié, T., Nucl.Phys.B250(1985) 312; Phys.Rev.Lett.55(1985)273

(5) Itoh, S., Iwasaki, Y. and Yoshié, T., preprint UTHEP-154

(6) Itoh, S., Iwasaki, Y., Oyanagi, Y. and Yoshié, T., Nucl.Phys.B274(1986)33

(7) Itoh, S., Iwasaki, Y. and Yoshié, T., Phys.Lett.167B(1986) 443

(8) Ono, S., Phys.Rev.D17(1978)888

(9) Bowler, B.C. et al., Phys.Lett. 162B(1985)354

(10) Fukugita, M., Oyanagi, Y. and Ukawa, A., preprint UTHEP-152 Fucito, F., Mariarty, K.J.M., Rebbi, C. and Solomon, S., preprint BNL 37546

WEAK INTERACTION MATRIX ELEMENTS WITH STAGGERED FERMIONS

Stephen R. Sharpe
Physics Dept., FM-15, University of Washington, Seattle, WA 98195

An overview of the results of the Los Alamos Advanced Computing Group is given. The theory behind the measurement of Weak Interaction Matrix Elements using staggered fermions is presented, and contrasted with that for Wilson fermions. This is followed by a preliminary discussion of numerical results on a $12^3 \times 30$ lattice.

The work presented here has been done in collaboration with Gerry Guralnik, Rajan Gupta, Greg Kilcup and Apoorva Patel. We have new results on the following subjects. (i) The hadron spectrum of quenched QCD, using both staggered and Wilson fermions, on an $18^3 \times 42$ lattice at $\beta = 6/g^2 = 6.2$. Perhaps the most important conclusion from our (so far preliminary) analysis is that it is very hard to test whether the ratio $R = m_N/m_\rho$ is in accord with experiment. For quark masses about equal to or greater than the physical strange quark mass, we find good mass fits but $R \geq 1.5$: for lighter quark masses (results only for staggered fermions) we can only see the π and possibly the ρ, so we can't measure R. To my knowledge, other groups have not measured R for $m_q < m_s$ either [1]. This is rather a sorry state of affairs which we are trying to correct. (ii) The scalar glueball mass estimated with source techniques using an improved action [2]. The alternate action steers us clear of the phase structure in the fundamental-adjoint coupling plane. We find larger values than previous estimates, a naive transcription of which suggests $m_G \approx 1.2 - 1.3\,\text{GeV}$. (iii) New studies of MCRG, Improved actions, Redundant operators, and All That [3]. (iv) Studies of exact algorithms for dynamical fermions on a 4^4 lattice. (v) Results for staggered fermions on a $12^3 \times 30$ lattice for the hadron spectrum (including non-local operators), wavefunctions, the instanton density, and $(m_\pi^+)^2 - (m_\pi^-)^2$. (vi) Results for Weak Interaction Matrix Elements (WIME) using staggered fermions on a test $8^3 \times 16$ lattice [4], and on the $12^3 \times 30$ lattices. Most of this work is preliminary: for more details see the forthcoming proceedings of the Brookhaven "Lattice 1986" Conference.

Having mentioned an array of mainly numerical studies, I'm going instead to devote most of this talk to a theoretical issue: How one measures WIME using staggered fermions. Thus I am addressing the theoretical underpinning to item (vi) above. At the end of the talk, having convinced you of the soundness of our method, I will present some pictures of the correlators we have obtained. More details on the arguments can be found in [5] and [4].

The CERN/Rome [6] and UCLA [7] groups have been attempting to measure WIME using Wilson fermions, with some successes. The theoretical analysis for Wilson fermions is now in place: by suitable non-perturbative subtractions extracts WIME which should have the correct chiral behaviour [8]. This can only be done, however, as long as one retains the GIM cancelation by keeping the charm quark in the loops of the "eye" graphs.

My main point here is that if one uses staggered fermions the calculation is, in a certain sense, more simple. Furthermore, one can integrate out the charm quark and still do the calculation with the same ease. My argument for these conclusions is very simple. If one introduces N_{sp} of staggered fermions, the lattice theory has a $U(N_{sp})_V \times SU(N_{sp})_A$ current algebra, the symmetries in which are only broken by soft mass terms. This algebra leads to Ward Identities (WI) on the lattice which are the precise correspondents of continuum WI. This means that, just as in the continuum [9] there are severe constraints on the mixing of operators, and to extract the WIME one need only make a single nonperturbative subtraction of $s \leftrightarrow d$ mixing effects. As I shall explain, for Wilson fermions one needs to make other subtractions in addition to this one. The WIME so extracted will of course have corrections of $O(a)$ (or maybe of $O(a^2)$?) and there are the usual perturbative mixing effects of $O(g^2)$.

This is the good news for staggered fermions, now for the bad. (i) One must use operators in which fermion fields are separated by up to 4 links in a 2^4 hypercube. This partially offsets the original gain in speed compared to Wilson fermions. Further, the multilink operators which are needed might give noisier correlators than the local operators one can use with Wilson fermions. (ii) To mimic the u, d, s and c quarks of the continuum one needs $N_{sp} = 4$. Each species of staggered fermion corresponds to $N_f = 4$ flavors in the continuum limit. So, in that limit, one has a theory with $N_f \times N_{sp} = 16$ flavors and symmetry $U(16)_V \times SU(16)_A$. This is clearly not the theory we want to simulate. As always with staggered fermions, we have to fix things up so that calculations in this enlarged theory differ only from those in the continuum theory by overall factors of N_f. Here the fix consists of writing an operator in the larger theory which has the same Fierz transformation properties as the operator we want to represent [4]. (3) Finally, one cannot, in general, project onto states (or operators) of definite parity. This is true in principle, but in practice for the lattice psuedo-Goldstone bosons (PGB), which are the states between which we always take matrix elements, the opposite parity partners give small contributions.

Having laid out the pros and cons, I should also express my only concern with the theoretical arguments. Although the operator mixing is much constrained by the Ward Identities, the possibilities for different operators are larger than in the continuum because of the existence of N_f extra flavors. I will not feel completely happy until a systematic analysis of all these operators, and of the mixing between them at $O(g^2)$, is carried out.

I now give some more details about the lattice WI and their relation to those of Wilson fermions and of the continuum. First consider the WIME of the continuum operator

$$\bar{s}_a \gamma_\mu (1-\gamma_5) u_a \bar{u}_b \gamma_\mu (1-\gamma_5) d_b \qquad (1)$$

between a K^+ and a π^+ (a and b are color indices). There are both eight and eye contractions. On the lattice one ends up having to calculate ME in which the external π and K are created respectively by $\overline{\chi}_s(n)\chi_u(n)\epsilon(n)$ and $\overline{\chi}_u(n)\overline{\chi}_d(n)\epsilon(n)$, where $\epsilon(n) = (-1)^{x+y+z+t}$. The weak operator becomes

$$\sum_\mu \left\{ \begin{array}{l} \overline{\chi}_s(\hat{\mu}) U \chi_u(0)\, \overline{\chi}_u(\hat{\mu}) U \chi_d(0) \\ +\, \overline{\chi}_s(0) U^\dagger \chi_u(\hat{\mu})\, \overline{\chi}_u(0) U^\dagger \chi_d(\hat{\mu}) \\ +\ \textit{same with } \hat{\mu} \to \overline{\mu} \end{array} \right\} \qquad (2)$$

where $\overline{\mu} = \hat{x} + \hat{y} + \hat{z} + t = \hat{\mu}$, and U is an appropriate gauge matrix. Other transcriptions are also possible: the two bilinears need not occupy the same sites: the χ and $\overline{\chi}$ can be inverted in one bilinear; etc. All alternatives should yield the same results in the continuum limit. Finally, one gets the physical WIME by using the reduction formula and dividing by an overall factor of N_f^2.

This construction yields eight and eye correlators, generically labelled $C_{K\pi}(t_K, t_\pi)$, which is real after averaging over a configuration and its conjugate. For the purpose of deriving WI eights and eyes can be considered separately. For the eights one finds (for any m_s, m_d, m_u)

$$\sum_{t_K} C_{K\pi}(t_K, t_\pi) = \sum_{t_\pi} C_{K\pi}(t_K, t_\pi) = 0 \qquad (3)$$

One can also write down versions of these WI without the summation over times. These contain correlators of the axial current and Schwinger terms, in addition to the commutator terms (which are absent above. but will be on the RHS of subsequent WI). The extra terms vanish upon summation over all times. All this is just as in the continuum, if one uses point split currents. Note that the WI holds separately for the lattice operator with a single link and for that with three links, and is also true configuration by configuration.

In the continuum one derives from the WI (3) that the WIME behave as $p_\pi \cdot p_K$ in the chiral limit. One must assume an expansion in powers of momentum in which one only keeps terms up to quadratic order. This expansion is embodied, for example, in the chiral Lagrangian [9]. On the lattice one can make similar assumptions and derive the chiral behaviour directly [5], but the assumptions needed are stronger since one doesn't have Poincare invariance, so things are less solid. This problem is, of course, common to all types of lattice fermion.

For the eye contractions the commutator terms do not vanish, and one finds:

$$\begin{aligned} \sum_{t_\pi} C_{K\pi}(t_K, t_\pi) &= \frac{C_\epsilon(t_K) + C_K(t_K)}{2md} \\ \sum_{t_K} C_{K\pi}(t_K, t_\pi) &= \frac{C_\epsilon(t_\pi)}{m_d + m_s} \end{aligned} \qquad (4)$$

where C_ϵ is the correlator between the operator and an external $\bar{s}d$ field, and C_K is that between the negative parity part of operator and an external K^0. Again this WI has exactly the same form as that which holds in the continuum. The terms on the RHS are indicative of $s \longleftrightarrow d$ mixing. Indeed in the continuum, and, with stronger assumptions, on the lattice, one can show from this WI that

$$\langle\pi|\mathcal{H}_W|K\rangle = a p_\pi \cdot p_K + b(m_d + m_s) + O(p^4) \quad (5)$$

where the extra b term is the mixing term which must be subtracted.

For the LR operators such as

$$\sum_{q=u,d,s} \bar{s}_a(1+\gamma_5)q_a\, \bar{q}_b(1-\gamma_5)d_b \qquad (6)$$

there are again many possible transcriptions. One example is

$$\bar{\chi}_s(\hat{\mu})\chi_q(\hat{\mu})\ \bar{\chi}_q(0)\chi_d(0) + \bar{\chi}_s(0)\chi_q(0)\ \bar{\chi}_q(\hat{\mu})\chi_d(\hat{\mu}) + \textit{terms with 4 links} \qquad (7)$$

The correlators containing this operator also satisfy the WI (4) as long as one adds together the eight and eye contractions. These are separately of $O(1)$, while their sum is of the desired $O(m)$. Unlike the LL operators, which have to be transcribed onto the lattice into operators containing an odd number of links (1 and 3 in (2)), the LR operators contain an even number of links.

To summarize: For staggered fermions the WI are identical to those in the continuum except for discretization, and so the continuum arguments leading from the WI to the chiral behaviour of the WIME should apply. All the effort with Wilson fermions consists in constructing correlators which satisfy the lattice WI (3) and (4), whereas with staggered fermions the WI come "for free".

With Wilson fermions if one proceeds naively one finds extra terms in (5): terms of $O(1)$ and $O(m_\pi^2)$. One then adds to the operator a part proportional to $\bar{s}d$ and adjusts the coefficient so these extra parts vanish [8]. One then has to use a variety of momenta, and/or a variety of quark masses, to separate the wanted a term from the unphysical b term. With staggered fermions one does not need the first stage of this procedure, and the second stage can be accomplished by a measurement of C_K [9]. This gives one b without any need to vary the particle masses or momenta. One can then do this variation to provide a check on one's measurement.

A point which might seem puzzling is that, for the LL operators, the eight and eye diagrams yield WIME which separately have the correct chiral behaviour. This seems perplexing because in a continuum derivation one gets relations between entire Green functions, not separately for the eight and eye contractions. Nevertheless, I think that this division can also be made in the continuum. One simply carries over the lattice derivation, working directly on propagators. This is fine as long as one uses point splits the currents.

Now I will give you some pictures of our results for the correlators $C_{K\pi}(t_K, t_\pi)$. We have completed a test run on an $8^3 \times 16$ lattice [4], but on such a small lattice it is hard to convincingly demonstrate the extraction of the external pion and kaon exponentials. Such is not the case for the results we have on a $12^3 \times 30$ lattice, generated using a 4 parameter improved action (that of [2] with $K_F = 10.5$). Comparing large Wilson loops, this lattice is roughly equivalent to one on the Wilson axis at $\beta = 5.96$. We have 25 configurations, and on these we have calculated the propagators from five base points (the minimum possible is four), for two values of the quark masses, and used these as sources for the same number of source propagators (see [4] for more details). We use antiperiodic boundary conditions (APBC) in all directions. The two quark masses correspond roughly to the physical strange quark mass, and to a mass 8 times smaller. In physical units (our lattice has $a^{-1} \approx 1.6\,\text{GeV}$), we then can look at PGB with masses of 300, 500, 700 MeV.

The pictures I show always have the operator on time slices 0 and 1, t_π fixed at 7, and t_K varying. Figures 1 and 2 show eight correlators from the 1 link part of (2) with $\mu = t$, for $m_d = m_s = .040, .005$ respectively. The points lying to the right of the dotted line have had their signs changed to allow the use of a log scale. The WI (3) says that the sum of points should vanish, as it does to very good accuracy. For the higher quark mass the data shows clear exponentials, and fits well to a $p_\pi \cdot p_K$ form. For the smaller quark mass the picture is less symmetrical, but the on-shell ME (the right half) is clearly visible. We do not fully understand the left half, though the shoulder at $t_\pi = t_K$ is related to wrap-around contributions allowed by APBC. Such deviations from a single Kaon exponential can destroy the connection between the WI and the chiral behaviour of the ME. In fact, for the data shown, the extracted ME does behave roughly as $m_\pi m_K$, but our preliminary analysis finds that this is not true for all operators.

Figure 3 shows the data for the eye contraction of the operator obtained from (2) by (a) interchanging χ_u and χ_d, (b) adding terms with the χ and $\bar{\chi}$ interchanged in one bilinear, and (c) keeping only the 1 link $\mu = t$ part. To make the subtraction of $s \leftrightarrow d$ mixing time by time one must use

Figures 1-4

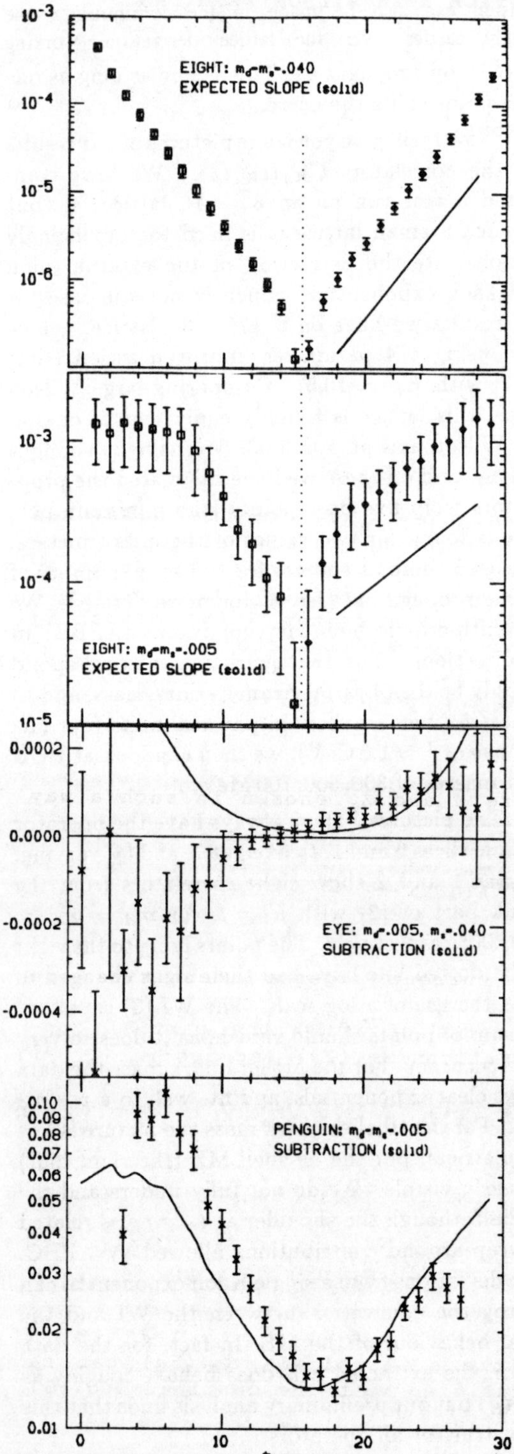

$m_d \neq m_s$ [4], so the data is for $m_d = .005, m_s = .040$. The part to be subtracted is shown by the solid line – its absolute normalization having been determined by a calculation of C_K. The errors on the points on this line are smaller than those on the data, and are not shown.

After subtraction. the residue has roughly the desired $p_\pi \cdot p_K$ form for $t_K = 8 - 22$. However, there are problems. First, the residue has large errors, and we may well need better statistics to extract any number at all. Second, a small uncertainty in the subtraction can have a large effect on the residue. Third, the data shows clear signs of oscillations. Fortunately, these are fairly small in the right half, from where we extract the ME. They are caused by the positive parity parts of the Kaon operator acting as transition matrix elements on wrap-around contributions [10]. This is the price one pays for using APBC.

The last figure shows the eight + eye (= "penguin") correlators for the 0 link part of the LR operator (7). Here we can do the time by time subtraction for equal quark masses, and we show the data and subtraction for the smallest quark mass. We use a log plot, so the large negative points for $t = 0 - 2, 29$ are excluded. The data is quite good, the regions of exponential fall off being clear. But the same comments as for figure 3 apply, the subtraction dominating the signal.

Although I am encouraged by the quality of the data – our lightest pion is very light by lattice standards, and the data for higher quark masses is much cleaner – it is clear that one must push present numerical simulations to their limits to extract the $\Delta I = \frac{1}{2}$ WIME. With our data this may just be feasible, but the calculation used roughly 500 hours on a Cray-2. A calculation with dynamical fermions clearly awaits faster, or dedicated, machines.

I would like to thank the organizers of session and of the conference. I thank my collaborators, in particular Apoorva Patel. I also wish to thank Guido Martinelli for many stimulating discussions during the course of this conference. Finally, I acknowledge the support of the DOE in granting us time at the MFE computing center.

References

[1] See for example Y. Iwasaki, these proceedings.

[2] A. Patel *et al.*, UCSD-10P10-260, to be published in *Phys. Rev. Lett.*

[3] A. Patel and R. Gupta, UCSD preprint, 1986.

[4] Stephen R. Sharpe *et al.*, UW preprint 40048-11 P6, 1986.

[5] G. W. Kilcup and Stephen R. Sharpe, HUTP-86/A048, submitted to *Nuclear Physics.*

[6] G. Martinelli, these proceedings.

[7] A. Soni, these proceedings.

[8] M. Testa, these proceedings.

[9] C. Bernard *et al.*, *Phys. Rev.* **D32** (1985) 2343.

[10] G. Kilcup *et al.*, *Phys. Lett.* **164B** (1984) 180.

WEAK NON LEPTONIC HAMILTONIAN ON THE LATTICE WITH WILSON FERMIONS

M. Testa

INFN, Sezione di Roma
Roma, Italy

1. CHIRAL PROPERTIES OF WILSON FERMIONS

Wilson Fermions[1] are described by the action

$$S_F = \sum_x \{ -\frac{1}{2a} \sum_\mu [\bar\psi(x)(r-\gamma_\mu) U_\mu(x) \psi(x+\hat\mu) + \bar\psi(x+\hat\mu) U_\mu^+(x)(r+\gamma_\mu)\psi(x)] + \bar\psi(x)(M_0 + \frac{4r}{a})\psi(x)\} \qquad (1)$$

It has been shown in ref.(2) how all current theoretic quantities may be defined for this system in such a way that they tend, in the continuum limit ($a \rightarrow o$), to finite, well defined operators.

Let us recall the relevant results. Axial current lattice Ward Identities imply the equation:

$$\nabla^\mu A_\mu^a(x) = \bar\psi(x)\{M_0, \frac{\lambda_a}{2}\}\gamma_5\psi(x) + X^a(x) \qquad (2a)$$

where

$$A_\mu^a(x) = \frac{1}{2}\{\bar\psi(x+\hat\mu)\frac{1}{2}\lambda_a\gamma_\mu\gamma_5 U_\mu(x)\psi(x) + h.c.\} \qquad (2b)$$

∇^μ is the discretized version of the derivative and $X^a(x)$ is the chiral variation of the Wilson term (the term proportional to r in eq.(1)). Eq.(2a) as it stands is not a good candidate for the corresponding continuum Ward Identity:

$$\partial^\mu A_\mu^a(x) = \bar\psi(x)\{m, \frac{\lambda_a}{2}\}\gamma_5\psi(x) \qquad (3)$$

where m is the (bare) quark mass. In fact, due to radiative corrections, the $X^a(x)$ term in eq.(2a) has non vanishing matrix elements between on shell states. The strategy followed in ref.(2) has been to define an operator $\bar X^a(x)$ vanishing between on shell states as:

$$\bar X^a = X^a + \bar\psi\{\frac{1}{2}\lambda_a, \bar M\}\gamma_5\psi + (Z_A - 1)\nabla^\mu A_\mu^a \qquad (4a)$$

where $\bar M$ and Z_A chosen in such a way that

$$\langle\alpha|\bar X^a(x)|\beta\rangle \xrightarrow[a\to o]{} 0 \qquad (4b)$$

and current algebra identities are satisfied.

In terms of $\bar X^a(x)$ the partial conservations eq.(2a) becomes:

$$\nabla^\mu \hat A_\mu^a = \bar\psi\{M_0 - \bar M, \frac{\lambda_a}{2}\}\gamma_5\psi + \bar X^a \qquad (5a)$$

with

$$\hat A_\mu^a = Z_A A_\mu^a \qquad (5b)$$

In the $a \rightarrow o$ limit eq.(5a) can be identified with the continuum eq.(3) with

$$m = M_0 - \bar M \qquad (6)$$

While the matrix elements of $\bar X^a(x)$ between on shell states are vanishing, the Green's functions containing $\bar{X}^a(x)$ together with other composite local operators are not vanishing, and reduce in general to contact terms:

$$\langle \bar{X}(x)\, \mathcal{O}_1(x_1) \ldots \mathcal{O}_n(x_n) \rangle =$$

$$= \sum_i \text{"}\delta(x-x_i)\text{"} \langle \mathcal{O}_1(x_1) \ldots \mathcal{O}'_i(x_i) \ldots \mathcal{O}_n(x_n) \rangle$$

where "$\delta(x-x_i)$" means a distribution localized at x_i and $O'_i(x_i)$ is a local operator with the appropriate quantum numbers. These contact terms are needed to reproduce the so called flavor anomalies and also to compensate the breaking effects of the Wilson term in non anomalous Ward Identities[2].

2. OCTET WEAK HAMILTONIAN ON THE LATTICE[3]

In the continuum the weak non leptonic hamiltonian has the general form:

$$H_{eff} = \frac{G}{\sqrt{2}} \sin\theta \cos\theta \sum_i C^{(i)}(\mu, M_W)\, \mathcal{O}^{i}(\mu) \qquad (7)$$

where μ is a subtraction point which is taken such that $M_W \gg a^{-1} > \mu > m_c$ (propagating charm), $c^{(i)}$ are the well known Wilson coefficients (4) and

$$\mathcal{O}^{(i)}(\mu) = Z^{(i)}(\mu)\, \mathcal{O}^{(i)} \qquad (8)$$

$Z^{(i)}(\mu)$ are such that $O^{(i)}(\mu)$ are normalized to unity in the four quark matrix elements at μ.

The problem is to find out the explicit form of the lattice bare operators $\tilde{O}^{(i)}$ which are multiplicatively renormalizable. Due to the presence of the Wilson term we must subtract from the naive expression operators with the same flavour and space time properties, different chiral properties and equal or lower dimensions:

$$\tilde{\mathcal{O}}^{(i)} = \mathcal{O}^{(i)} + \delta_6 \mathcal{O}^{(i)} + \delta_5 \mathcal{O}^{(i)} + \delta_3 \mathcal{O}^{(i)} \qquad (9)$$

For the (27,1) part of H_W only dimension 6 operators must be subtracted. Their form and the corresponding coefficients have been given in ref.(5), (6), (7) using perturbation theory. Due to the finiteness of these coefficients, perturbation theory is expected to give a reliable estimate.

From now on we will discuss only the (8,1) part of H_W which is the term which should show the enhancement.

In this case the index i in the preceding equation only takes two possible values which we shall denote by $\pm$. As for the subtractions we have that the dimension 6 $\delta_6 O^{(+)}$ can be taken again from perturbation theory as in ref.(5), (6), (7).

As for lower dimension operators one finds:

a) dimension five. There is only one operator which, because of the GIM cancellation, enters with a finite coefficient:

$$\delta_5 \mathcal{O}^{(\pm)} = d_5^{(\pm)} (m_c - m_u)\, g_0\, \bar{s}\sigma_{\mu\nu} F^{\mu\nu} d \qquad (10)$$

where $F_{\mu\nu}$ is the non abelian gluon field strenght. Beeing finite $d_5^{(\pm)}$ can be reliably computed in perturbation theory;

b) dimension three. We have:

$$\delta_3 \mathcal{O}^{(\pm)} = d_s^{(\pm)} (\bar{s}d) + d_p^{(\pm)} (\bar{s}\gamma_5 d) \qquad (11)$$

where

$$d_s^{(\pm)} = \frac{m_c - m_u}{a^2} C_0^{(\pm)} + \ldots$$

$$d_p^{(\pm)} = \frac{(m_c - m_u)(m_s - m_d)}{a} D_0^{(\pm)} + \ldots \qquad (12)$$

Due to the power-like ultraviolet divergences, $d_{s,p}^{(\pm)}$ must be determined in a non perturbative way.

The lattice operators which correspond to the continuum operators in eq.(8) are obtained by a rescaling:

$$\hat{\mathcal{O}}^{(\pm)}(\mu) = Z_{LATT}^{(\pm)}(\mu a)\, \tilde{\mathcal{O}}^{(\pm)} \qquad (13)$$

where $Z_{LATT}^{(\pm)}(\mu a)$ are fixed by the same normalization condition on $\hat{O}^{\pm}(\mu)$ as in the continuum.

The octet part of the weak Hamiltonian on the lattice is therefore:

$$H_{eff} = \frac{G}{\sqrt{2}} \sin\theta \cos\theta \sum_{i=\pm} C^{(i)}(\mu, M_W)\, Z_{LATT}^{(i)}\, \tilde{\mathcal{O}}^{(i)} \qquad (14)$$

To fix the values of $d^{(\pm)}_{s,p}$ we must impose the Ward Identities. This can be done by recalling that soft pion theorems imply in the continuum[8]:

$$\langle 0|O^{\pm}(\mu)|K^0\rangle = i(m_K^2 - m_\pi^2)\,\delta^{(\pm)} \tag{15a}$$

$$\langle \pi^+(p)|O^{(\pm)}(\mu)|K^+(k)\rangle = -\frac{m_K^2}{f_\pi}\delta^{(\pm)} + (p\cdot k)\gamma^{(\pm)} \tag{15b}$$

$$\langle \pi^+\pi^-|O^{(\pm)}(\mu)|K^0\rangle = i\,\frac{m_K^2 - m_\pi^2}{f_\pi}\,\gamma^{(\pm)} \tag{15c}$$

We can now compute on the lattice the matrix elements of the operators appearing in eq.(9):

$$\begin{aligned}
&\langle 0|O^{(\pm)} + \delta_6 O^{(\pm)} + \delta_5 O^{(\pm)}|K_0\rangle = i\,\delta^{(\pm)}\\
&\langle \pi^+(p)|O^{(\pm)} + \delta_6 O^{(\pm)} + \delta_5 O^{(\pm)}|K^+(k)\rangle = \delta_2^{(\pm)} + (p\cdot k)\gamma_2^{(\pm)} + \dots\\
&\langle 0|\bar{s}\gamma_5 d|K^0\rangle = i\,\delta_P\\
&\langle \pi^+(p)|\bar{s}d|K^+(k)\rangle = \delta_S + (p\cdot k)\gamma_S + \dots
\end{aligned} \tag{16}$$

Imposing the equality of $\hat{O}^{(i)}(\mu)$ and $O^{(i)}(\mu)$ allows through eqs.(15) the computation of $d^{(\pm)}_{s,p}$.
It is now possible to get $\gamma^{(i)}$ defined in (15c) in the chiral limit:

$$\bar{\gamma}^{(i)} = \lim_{m_{u,d,s}\to 0} \gamma^{(i)} = \lim_{m_{u,d,s}\to 0} Z^{(i)}_{LATT}(\mu a)\left(\gamma_2^{(i)} - \delta_2^{(i)}\frac{\gamma_S}{\delta_S}\right) \tag{17}$$

in terms of quantities measured on the lattice. The matrix element relevant to the $K^0 \to \pi^+\pi^-$ decay is therefore in the lowest order of chiral symmetry breaking:

$$\langle \pi^+\pi^-|H_{eff}|K^0\rangle = \frac{G}{\sqrt{2}}\sin\theta\cos\theta \cdot i\,\frac{m_K^2 - m_\pi^2}{f_\pi}\cdot\sum_{i=\pm} C^{(i)}(\mu, M_W)\,\bar{\gamma}^{(i)} \tag{18}$$

REFERENCES

(1) K.G.Wilson, Phys.Rev. D10 (1974), 2445; in "New Phenomena in Subnuclear Physics", ed. by Zichichi (Plenum, New York, 1977).

(2) M.Bochicchio, L.Maiani, G.Martinelli, G.C.Rossi and M.Testa, Nucl.Phys. B262 (1985), 331.

(3) L.Maiani, G.Martinelli, G.C.Rossi, M.Testa, Phys.Lett. 176B (1986) 445.

(4) M.K.Gaillard, B.W.Lee, Phys.Rev.Lett., 33 (1974) 108. G.Altarelli, L.Maiani, Phys.Lett. 52B (1974) 351.

(5) N.Cabibbo, G.Martinelli, R.Petronzio, Nucl.Phys. B244 (1984) 381.

(6) C.Bernard, T.Draper, G.Hockney, A.M.Rushton and A.Soni, Phys.Rev.Lett. 55 (1985) 2770.

(7) G.Martinelli, Phys.Lett. 141B (1984) 395.

(8) T.D.Lee, Brookhaven Lectures, BNL May 1970; C.Bernard, T.Draper, A.Soni, H.D.Politzer and M.B.Wise, Phys.Rev. D32 (1985) 2343.

STATUS OF THE COLUMBIA LATTICE PARALLEL PROCESSOR PROJECT

Frank R. Brown

Department of Physics, Columbia University
New York, New York 10027

We review the status of the Columbia Lattice Parallel Processor project, in which parallel processing hardware with high performance floating point capability is used to perform large scale Monte Carlo studies of QCD. The project encompasses three stages: a 16-node machine with a nominal peak speed of 256 Mflops, a 1 Gflop 64-node machine, and a 16 Gflop 256-node machine. The 16-node machine was completed in April 1985, and has been employed since that time for the study of deconfinement in the pure gauge theory. The 64-node machine is under construction. The prototype node was completed in June 1986 and has successfully run a Monte Carlo program. Completion is expected by the end of 1986. The 256-node machine is now in the design stage. The modularity of the design makes evolutionary improvement of the vector processor relatively painless. This modularity together with the latest generation of floating point chips makes possible a very exciting aspect of the 256-node design, namely, a single node peak speed of 64 Mflops.

I. INTRODUCTION

The decade of effort which has established Quantum Chromodynamics as the theory of the strong interactions represents a major milestone in theoretical physics. Nevertheless, QCD remains a frustration because of our inability to make quantitative predictions about low energy phenomena. Over the past five years, direct numerical solution of the theory via Monte Carlo simulation has emerged as an extremely promising approach to QCD [1]. Unfortunately, the computational resources available are not yet adequate for the task. The Columbia Lattice Parallel Processor project is a natural response to this situation. Its goal is to solve QCD via large-scale, but conventional Monte Carlo simulation, making use of the massive computational power provided by special purpose hardware.

The purpose of this talk is to report on the project's current status, and our plans for the future. We've reached the middle phase of a project that involves three successive machines, running respectively 16, 64, and 256 processing nodes in parallel. The 16-node machine [2] is built, running, and has contributed to physics [3]. The design of the 64-node machine is complete, a prototype node is built and working, and the remaining nodes are under construction. The design of the 256-node machine, which incorporates the latest generation of high speed floating point chips, is well underway. For the past ten years, the dynamical complexity of QCD has thwarted all attempts at its mastery. I hope you will come to share my optimism that the process of correcting this problem is now in progress.

II. PARALLEL PROCESSOR OVERVIEW

I will present only a minimal description of our general parallel processor design; Ref. 2 provides a much more detailed description of the 16-node design, which is very similar to, and provides the foundation for the designs of the later machines.

The parallel processor consists of a toroidal two-dimensional mesh of single board processor nodes, each of which can read from or write to the fast data memory of its immediate neighbors to the north and east. Data and code originating on the host computer (a VAX 11/780) are passed to the parallel processor via a central controller, and then from processor to processor in a bucket brigade. The controller also serves to provide processor synchronization as well as finish and error signal monitoring.

A single node consists of what is essentially a special purpose microcomputer based on the Intel 80286 microprocessor driving a high performance floating point vector processor, discussed in some detail below. Each node has on-board ROM, "code" memory, writable control store, fast data memory (static RAM), and a relatively large amount of off-board dynamic RAM accessed via the node's Multibus. On selected nodes, the Multibus is also connected to the central controller or disk drives.

The microprocessor is programmed in the high level languages Fortran and PL/M as well as assembly language, with standard Intel cross-compilers running on the host. Microcode subroutines for the vector processor are written with the aid of a low level microcode assembly language, also on the host. Standard Intel utilities are used to link various subroutines into a module which can be loaded into the code memory and writable control store of the nodes making up the parallel processor.

III. VECTOR PROCESSOR EVOLUTION

The theme of this discussion will be how the modularity of our microprocessor/vector processor architecture greatly facilitates evolutionary changes that make use of advances in floating point technology. The bulk of the effort required to design and debug the 16-node processor board was associated with its general purpose functionality, and hence, with the microprocessor. But this functionality, while absolutely necessary, has not, in practice, been under pressure to be provided more efficiently. That is, vector processor improvements translate almost directly into overall performance improvements, while general prupose supporting hardware improvements do not. Thus, complete vector processor redesign can be very productive, and is far easier than redesigning the entire board. This modularity also reduces the programming costs associated with design improvements. Even though the 16-node and 64-node vector processors are entirely different, we have, for example, general purpose operating system type programs running both on the 16-node machine and on a prototype board for the 64-node machine that differ only in a single source code header file which defines a set of special hardware addresses. Indeed, our experience with the 64-node design upgrade has led us to pursue an analogous strategy with 256-node vector processor. Fig.1 gives simplified block diagrams for the three vector processors. The 16-node vector processor is based on a TRW MPY16 16-bit integer multiplication chip and a TRW 1022 22-bit floating point adder.

Because during a complex multiplication each real component of the complex numbers participates in two real multiplications, by making use of the internal storage of the vector processor (the on-board input latch of the integer multiplier and the exponent latches), the full 16 Mflop peak speed can be achieved asymptotically for a complex inner product. With no such internal storage, the 32 Mbyte/sec. memory bandwidth would support a complex inner product speed of only 8 Mflops.

The principle change in the 64-node design is the upgrade to 32-bit (IEEE standard) arithmetic via the Weitek WTL 1033 floating point multiplier and WTL 1032 floating point adder.

Now, however, a complex number requires eight bytes of storage, and puts a greater demand on memory bandwidth. This naturally explains the second major difference between the 16-node and 64-node designs, namely, the addition of 18 registers. (The 16-node design should be thought of as having roughly one register.) This additional internal storage makes possible the same 16 Mflop complex inner product in spite of the increased precision, even though the memory and bus performance remains unchanged. Furthermore, if one considers 3x3 complex matrix multiplication in which each complex number participates in three multiplications, but which potentially puts greater stress on the Y-bus (used for both reading and writing), one finds that the 64-node design offers improved performance. Concretely, the 16-node matrix multiplication microcode subroutine is comprised of 146 lines (= cost of calculation in 125 nS. clock cycles), while the 64-node matrix multiplication requires only 140 lines of microcode. (For longer microcode subroutines the improvement is more pronounced.)

The 256-node design entails a similar evolutionary change. The increased performance (in speed this time, rather than precision) offered by a new generation of floating point chips is exploited without increased memory bandwidth by again adding additional internal storage to the vector processor. Weitek has announced the WTL 3332 floating point processor, consisting of a multiplier, adder, and 32x32 register file on a single chip. It is currently being sampled at 10 MHz (20 Mflops) and a 20 MHz version has been announced. We plan to use the "double speed" version, but to run it at 16 MHz (i.e., keeping our underlying 8 MHz clock intact) for a per chip performance of 32 Mflops. Because this chip alone provides better performance than the 64-node vector processor at a substantially lower cost, we plan, in fact, to gang a pair of 3332's together, each accompanied by two WTL 1066 32x32 register files, thereby providing each of the 256 processor nodes with a dual pipe 64 Mflop vector processor -- impressive performance for a processor board costing only a few thousand dollars.

At this point, however, it is clear that the historical roots of the 256-node design have led to a serious imbalance between vector processor speed and memory bandwidth, since the 256-node memory bandwidth remains 32 Mbytes/sec. (The two separate 16-bit busses have, however, been replaced by a single 32-bit bus for greater flexibility.) This bottleneck is partially eased by providing the vector processor with a total of 192 registers. Nonetheless, only for certain algorithms can we reasonably expect to make significant use of the second 32 Mflop pipe. We expect, for example, that while the second pipe should essentially double the speed of the so-called environment calculation, heavy with matrix-matrix arithmetic and intermediate products that can be kept in the registers, a more matrix-vector oriented task such as evaluation of the Dirac operator is essentially memory bandwidth limited even with only a single pipe. This leads to the tentative conclusion that indeed for large scale pure gauge calculations the dual pipe design will prove its worth, but that for dynamical fermions and quenched hadronic calculations the second pipe will be at best of modest use.

ACKNOWLEDGEMENTS:

The work reported here was funded in part by the Department of Energy (contract number DE-AC02-76 ER02271). The Columbia Lattice Parallel Processor project was initiated by Norman H. Christ and Anthony E. Terrano, and has since expanded to include the efforts of K. Barad, F. Butler, Y-F. Deng, H-Q. Ding, M-S. Gao, P. Hsieh, S-P. Sun, L. Unger, and T. Woch. Special thanks are due the Intel Corporation and in particular Paul Cohen, Harry Dunham, Keith Josephson, Joe Proctor, and Emil Sarpa for vital technical and material support for this project.

REFERENCES:

[1] See for example, the reprint collection Lattice gauge theories and Monte Carlo simulations, Claudio Rebbi (ed.) World Scientific, Singapore, 1983.

[2] Norman H. Christ and Anthony E. Terrano, IEEE Trans. Comput. 33, 344 (1984); Byte Magazine, April 1986, pg. 145.

[3] Norman H. Christ and Anthony E. Terrano, Phys. Rev. Lett. 56, 111, (1986).

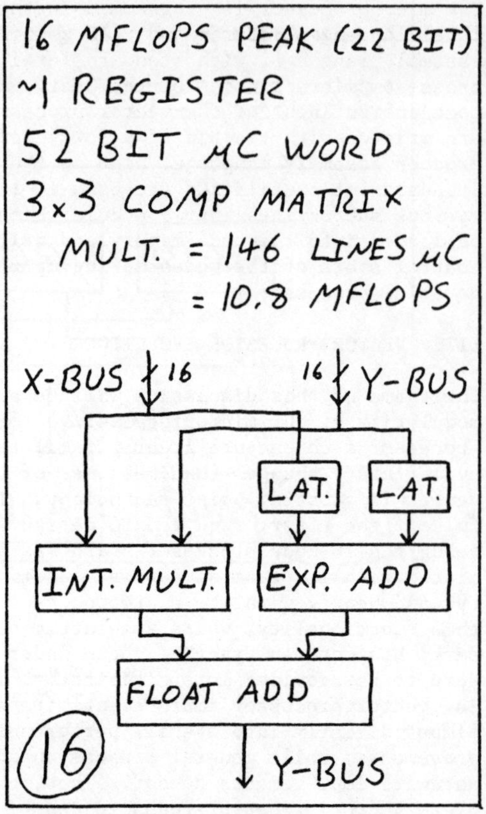

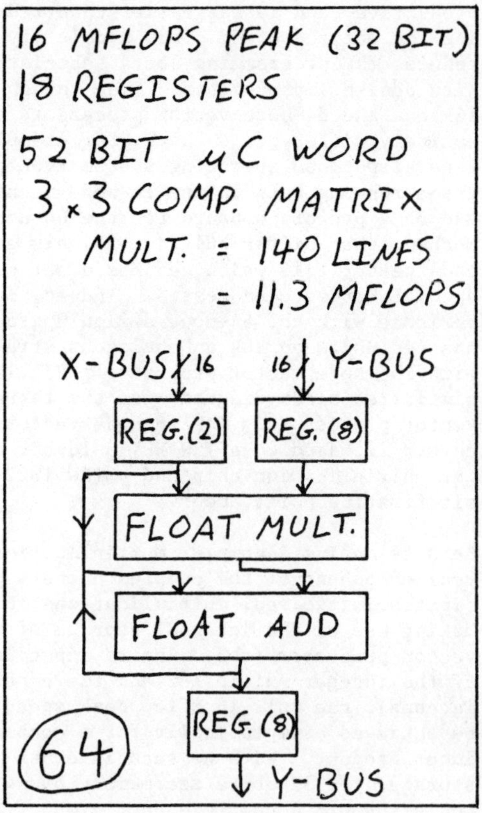

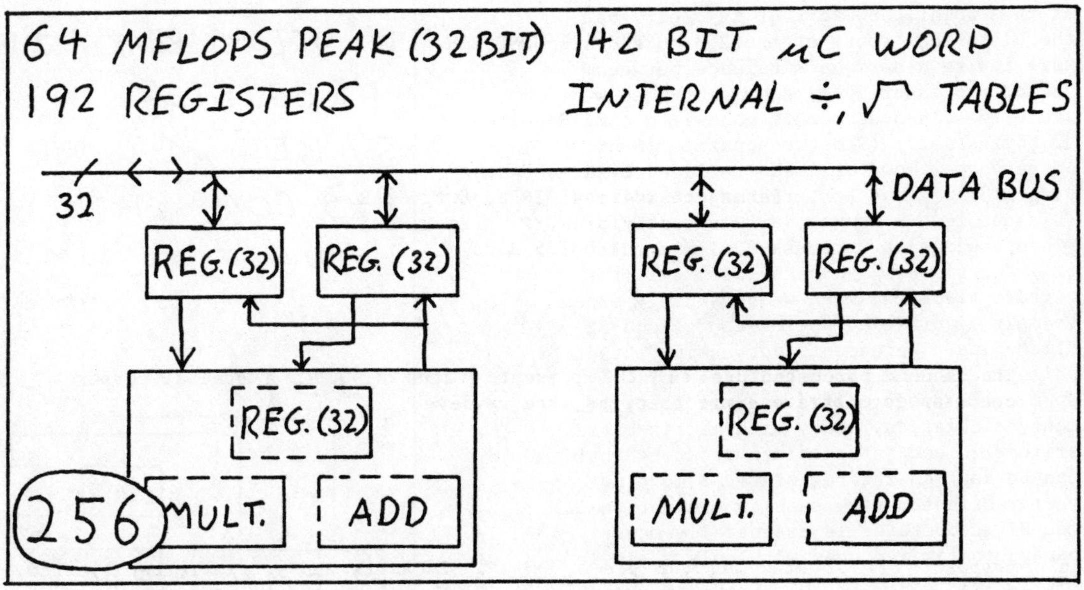

FIG.1. Simplified block diagrams of the three vector processor designs. The legends indicate selected design and performance features, and serve to emphase how the vector processor has evolved during the course of this project.

THE APE COMPUTER

F. Rapuano

CERN -- Geneva
Permanent address: INFN, Sez. di Roma
Ist. di Fisica, P. Aldo Moro 2
I-00185 Roma

The general architecture and the present status of APE, a special-purpose computer for lattice gauge theories, are reviewed.

Wilson's proposal (1) of studying the properties of field theory, in particular QCD, on a lattice has stimulated new interest in numerical methods, specifically Monte Carlo, applied to theoretical physics.

The expectation value of an operator O in field theory:

$$\langle O \rangle = \frac{\int D[\phi]\, O(\phi)\exp(-S)}{\int D[\phi]\, \exp(-S)}$$

is numerically evaluated as a statistical average on a sample of N_c thermalized configurations

$$\langle O \rangle = \frac{1}{N_c} \Sigma_k O(k)$$

with a statistical error σ defined as

$$\sigma = \frac{1}{\sqrt{(N_c-1)}} \left[\Sigma_k \frac{O^2(k)}{N_c} - (\Sigma_k \frac{O(k)}{N_c})^2\right]$$

The results are encouraging (2) but errors, both statistical and systematical, and the fact that dynamical fermions are not taken into account (quenching), still prevent a clear-cut answer to the non-perturbative problem. Experience implies that ~100 configurations on lattices of size $\sim 30^4$ with dynamical fermions should be enough to make these errors negligible. This is not a simple task on today's commercial computers, due to the limited speed and memory available.

It is possible, however, to build a special-purpose computer for lattice gauge theories in such a way as to keep the costs low, yet performance very high, by carefully designing the processor to match the characteristics of these calculations. They are essentially:

1) Locality and intrinsic parallelism.
2) Mostly complex arithmetic.
3) 32 bits arithmetics is precise enough.
4) Typical operation performed is a=b•c+d.
5) High ratio of floating point operations to memory access. For example, for the calculation Tr(A•B•C•D), where each matrix belongs to SU(3), this ratio is greater than 10.

The last point is particularly important in that it shows that the speed of the memory need not be very high, which is important in keeping costs low. All these points have been carefully taken into consideration while designing APE (Array Processor with [IBM 3081] Emulator) (3) in such a way as to obtain a special-purpose computer with a peak speed of 1 GFlop at a fraction of the price of a commercial, general-purpose, computer of comparable speed, like the CRAY 2.

The general architecture is shown in Fig. 1. The machine is a slave processor consisting of a linear chain of FPU's (Floating Point Units) and memories, connected through a switching network with hardware periodic boundary conditions. The final machine will have 16 of these units.

The FPU is the arithmetic core of the machine. Its block structure is shown in Fig. 2. The networking between different devices is partially software-controlled, but is omitted in Fig. 2 for clarity. The FPU is based on the WTL family of arithmetic chips manufactured by WEITEK Corp. The functioning of the unit for a calculation like a=b•c+d can be described in four steps:

Register file level: data input from memory or waiting to be saved onto memory.
Multiplier level: all intermediate products, Re(b)•Re(C), Re(b)•Im(c), etc., are completed.

First ALU level: b•c is ready.
Second ALU level: b•c+d is ready to be saved onto the register files.

Each step has a five cycle pipeline and the peak speed of one FPU is 64 MFlop. The real speed is 50% of this speed for a calculation as simple as $A \cdot B (A,B \in SU(3))$ but increases to 90% for a fairly lengthy calculation as an incomplete plaquette where the pipeline start-up and flush time are negligible with respect to the time of the calculation itself. The FPU also has special hardware for performing exponentiations and logarithms. The functions of the FPU and part of the network are controlled through a 64 bit μ-code.

The memories will contain all the data of the lattice under study. The size is 8 MBytes for the prototype, which is being increased to 16 MBytes for the final version. Generally an FPU will fetch or store data onto memory in 64-bit words, that is, a complex number with 32-bit real and imaginary parts, but also 32-bit accesses are possible when using the processor for real calculations. Each memory has single error correction and double error detection capability for maximum reliability of the processor with respect to data corruption.

The switching network will allow any FPU to fetch or store data onto the memory just opposite to it or onto the ones on its neighbouring sides, when data from different hyperplanes are needed during the calculation. Two bits of the memory address control its position.

The IBM 3081 Emulator has been jointly developed at CERN and SLAC. It can bit-to-bit emulate the IBM mainframe, executing its translated object code. We use it only as a controller of the whole machine and to perform the integer calculations needed for address generation. It has its own code and data memory for this purpose.

The sequencer issues, every clock cycle, a 64-bit μ-instruction to the FPU's. It can contain 16Kwords of μ-code which is sufficient for a typical LGT program.

A μ-VAX II is the host of APE through an interface. A high-level compiler has been developed to program the machine. It has looping and conditional branching capabilities, and implements all functions supported by the hardware. An optimizing stage improves code generation and register usage.

The user develops his/her program on the μ-VAX. The compiler and linker then generate two different μ-codes, one for the 3081/E and another for the FPU's which are in turn downloaded respectively into the 3081/E code memory and the sequencer memory. These two μ-codes will execute synchronously. This is possible because all devices in APE are synchronous and so the timing for every operation is known at compilation time. Simulations have shown that the 3081/E is powerful enough to generate address fast enough to keep the FPU's always busy. The case of a particularly heavy address calculation would be handled by the compiler generating instructions to keep the FPU's idle until the address is ready.

All the 16 APE memories and the 3081/E data memory are initialized with the user data. The processor is then started and at completion the results are saved into the μ-VAX disks and tapes, to be further analyzed or edited.

At present we have a four-unit prototype machine fully developed under test. All the software, compiler and system support software, is also fully developed and working. The project has been funded by INFN (Istituto Nazionale di Fisica Nucleare). The cost of the project is 750 K$ (1986 $ quotation) for the final 16 units machine and 8 units machine.

REFERENCES

(1) Wilson, K., Phys. Rev. D10, 2445 (1974).

(2) Hasenfratz, P., these proceedings.

(3) Bacilieri, P., et al., ROM2F/85/28 (1985); CERN preprint TH.4283/85 (1985).

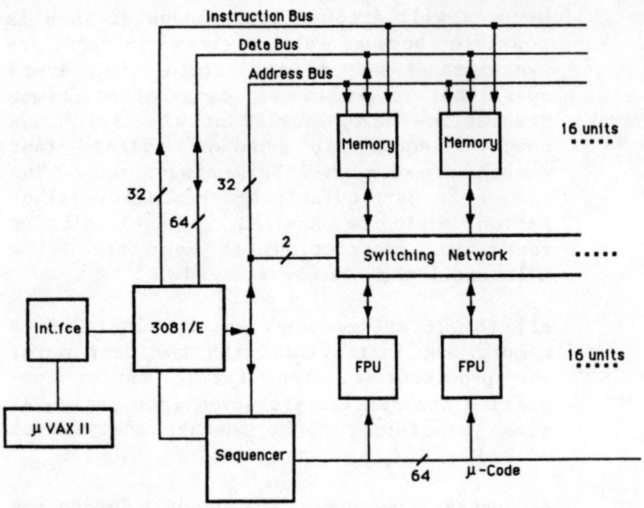
Instruction Bus
Data Bus
Address Bus
Memory
Memory
16 units
32
64
32
2
Switching Network
Int.fce
3081/E
FPU
FPU
16 units
μ VAX II
Sequencer
64
μ-Code

Fig.1

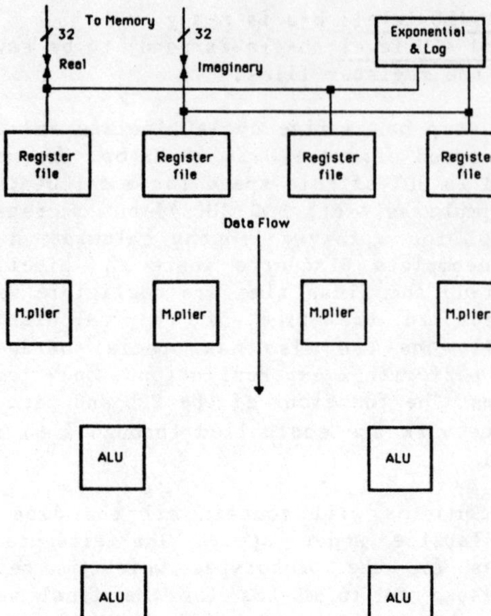
To Memory
32
32
Real
Imaginary
Exponential & Log
Register file
Register file
Register file
Register file
Data Flow
M.plier
M.plier
M.plier
M.plier
ALU
ALU
ALU
ALU

Fig.2

PROPERTIES OF LATTICE HIGGS MODELS

H. A. Kastrup

Institut für Theoretische Physik, RWTH Aachen

51 Aachen, FR Germany

Some problems concerning the Monte Carlo analyses of Higgs models on the lattice are briefly outlined. In addition properties of gauge invariant correlation functions useful for the construction of order parameters describing the phase structure of SU(2) and U(1) Higgs models are discussed.

As it is impossible to discuss the main developments concerning Monte Carlo investigations of Lattice Higgs models - i.e. SU(N) gauge fields coupled to scalar matter fields with a quartic selfinteraction - in about 20 minutes, I have to be extremely selective and I refer to the recent extensive review by Jersák[1] and to A. Jaffe's talk on analytical results at this conference[2].

There are several motives for analyzing Higgs models on the lattice by Monte Carlo methods:

1. The Higgs sector is that part of the otherwise well established standard model which we know least about, experimentally and theoretically! In the tree approximation one has the relation $m_H^2/m_V^2 = 8\lambda/g^2$ between Higgs and vector meson masses and the coupling constants λ of the quartic scalar selfinteraction and g of the gauge fields. Thus, if the Higgs mass is of the order of several 100 GeV then the Higgs sector of the standard model becomes "strongly" interacting[3]. However, if the coupling becomes large then perturbation theory breaks down and one has to use other methods in order to analyse the model, e.g. Monte Carlo analyses.

2. The so-called "Higgs-mechanism" - the coupling of gauge fields to selfinteracting scalar matter fields makes vector (gauge) mesons massive - is intrinsically a non-perturbative phenomenon closely related to the confinement mechanism in gauge theories[4]. Whereas the description of this mechanism in the framework of perturbation theory relies heavily on gauge fixing, its properties on the lattice have to be understood in a gauge invariant manner. This implies qualitatively new elements concerning the determination of the mass spectra for these models.

3. The influence of matter fields on the dynamics of gauge fields can be studied - e.g. the behaviour of order parameters differentiating between confinement and other phases - without the complications associated with fermions on the lattice.

4. At present, the main theoretical issue, however, is the following: All indications point to the "fact" that the <u>continuum</u> ϕ^4 - theory by itself - i.e. without coupling to gauge fields - is a trivial one[5]. Thus, the question arises whether it can be rescued from triviality by coupling it to gauge fields. This question is closely related to the following one: whereas the ϕ^4-theory is most likely trivial in the continuum, it is non-trivial on the lattice where it undergoes a phase transition[6]. Higgs models are non-trivial on the lattice, too. So the main question in our context is: <u>What happens to such a Higgs model if the lattice constant a goes to zero?</u>

This limit has to be performed in the neighbourhood of a point in the phase diagram of the (bare) coupling constants g_o, λ_o etc. where the system undergoes a 2nd order phase transition and which serves as a stable fixed point for the renormalization group flow of the coupling constants.

At the moment the situation concerning the continuum limit for Higgs models is, unfortunately, not clear, for the following reasons:
i) The theory of coupled scalar and gauge fields is generally not asymptotically free[7].
ii) Preliminary Monte Carlo renormalization group investigations[8] of U(1) and SU(2) Higgs models on rather small lattices and for fixed lengths $|\phi|$ of the scalar fields (equivalent to $\lambda_o \to \infty$) have given no indications for the existence of non-trivial fixed points. In addition no clear signals for a 2nd order phase transition at a non-trivial point in the space of coupling constants have been found[9] yet, but this is very preliminary, too.

iii) A recent perturbative renormalization group analysis in a neighbourhood of the trivial fixed point ($\lambda_o=0$, $g_o=0$) by A. and P. Hasenfratz[10] comes to the conclusion that coupling the ϕ^4-theory to SU(2) gauge fields cannot rescue it from triviality in this neighbourhood because $g_o \to 0$ implies $g_p \to 0$. However one does not know what happens if one is far away from this trivial fixed point, e.g. if λ_o is large.
Thus, at the moment one can only look at the Higgs models as effective theories with a large but finite momentum cut-off! The main problem is however, that one does not yet know where in the phase diagram of the coupling constants one is close to the real world. As long as this problem has not been solved all the very nice results obtained in Monte Carlo investigations of lattice Higgs models cannot be used to make even approximate predictions for realistic physical quantities e.g. for the ratio m_H/m_V of Higgs and vector meson masses.
After this pessimistically sounding appeal for new ideas and results concerning the continuum limit of Higgs models let me return to topic 3 in my list above:

We have just seen that it is important to know in detail the phase structure of the Higgs models, the location of phase transitions in coupling constant space and especially their order (which is very difficult to determine by numerical methods). For this purpose it is important to have sensitive order parameters! Here considerable progress has been made recently:
In the presence of matter fields in the fundamental representation of the gauge group Wilson's "loop", the expectation value

$$W(R,T) = \langle \operatorname{tr} U(\underset{R}{\Box} T) \rangle$$

of the path-ordered product of gauge link variables $U(x,\mu) \in SU(N)$ along a rectangular path, by itself is no longer useful for the construction of an order parameter differentiating between phases (confinement, screening or free charges) of the system, because it shows perimeter behaviour everywhere, i.e. the potential

$$V(R) = - \lim_{T\to\infty} \frac{1}{T} \ln W(R,T)$$

tends to finite constants $V(\infty)$ for large R in all phases. As gauge invariant correlation functions in Higgs models in general decay exponentially[11], the question arises whether the ratio of such correlation functions with different decay properties in different phases can be used for the construction of order parameters.

An interesting proposal has recently been made by Fredenhagen and Marcu[12]: It employes the gauge invariant 2-point function

$$G(R,T) = \langle \varphi_x^+ \, U(T\overset{R}{\sqcap}_{x\;\;y}) \, \varphi_y \rangle \quad ,$$

where a dynamical charge at the euclidean point x is connected to the corresponding anticharge at y by a path-ordered product of link variables along the path as indicated graphically. If we define

$$\mu = - \lim_{T\to\infty} \frac{1}{T} G(T,R) \quad ,$$

then Fredenhagen and Marcu assert that $\mu = \frac{1}{2} V(\infty)$ in the confinement/screening phase (in Higgs models the confinement phase region is analytically connected with the Higgs/screening phase region in the phase diagram[4]) and $\mu > \frac{1}{2} V(\infty)$ in the free charge phase. Thus the ratio

$$\rho_{FM} = - \lim_{T\to\infty} \frac{G(R=\frac{1}{2}T,T)}{W(R=T,T)^{\frac{1}{2}}}$$

can be used as an order parameter differentiating between the two phases in question.

We[13)14)] have investigated this prediction by Monte Carlo calculations for the SU(2) and U(1) Higgs models. The parametrization used for the action of the SU(2) model is the following[6)]:

$$S = -\tfrac{1}{4}\beta \sum_{p} \mathrm{tr}(U_p+U_p^{+}) + \lambda \sum_{x} (\varphi_x^{+}\varphi_x-1)$$

$$- \kappa \sum_{x,\nu} (\varphi_x^{+} U_{x\nu}\, \varphi_x + \mathrm{h.c.}) + \sum_{x} \varphi_x^{+}\varphi_x \ ,$$

where $U_{x\nu}$ and $U_p \in SU(2)$ are the usual link and plaquette variables. The action for the U(1) model is defined correspondingly.

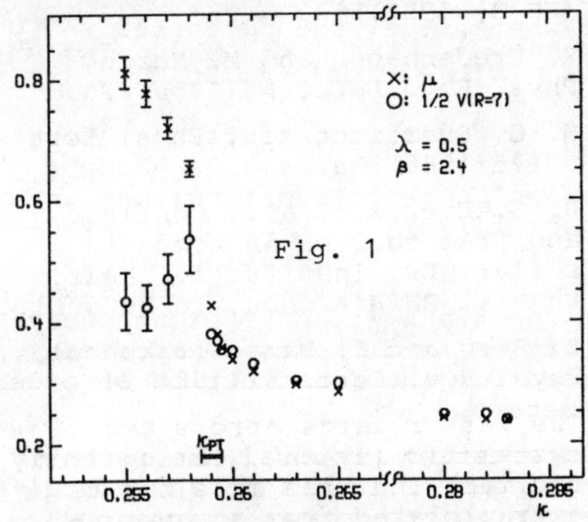

Fig. 1

Fig.1 shows the behaviour of the parameters μ, $V(R=7)$ on a 16^4 lattice as a function of the hopping parameter κ at fixed values of λ and $\beta=4/g^2$. Above the phase transition at κ_{PT} = 0.2590, in the screening region, the 2 parameters do indeed coincide, whereas this is not so below κ_{PT} in the confinement region, where again equality is expected for large R. This behaviour can be understood as follows: For R not too large and for κ in the confinement region the potential still rises linearly[13)] because hadronization by means of pair creation has not yet set in and the charges still appear unshielded at the distance we are able to consider on a 16^4 lattice.

A more interesting example in our context is the U(1) model, because it also has a free charge (or Coulomb) phase[1)]. Fig.2 shows the values of μ and $V(R=4)$ on a $8^3 \times 16$ lattice below and above the Coulomb/Higgs phase transition as a function of κ for fixed λ and $\beta = \frac{1}{g^2}$, while

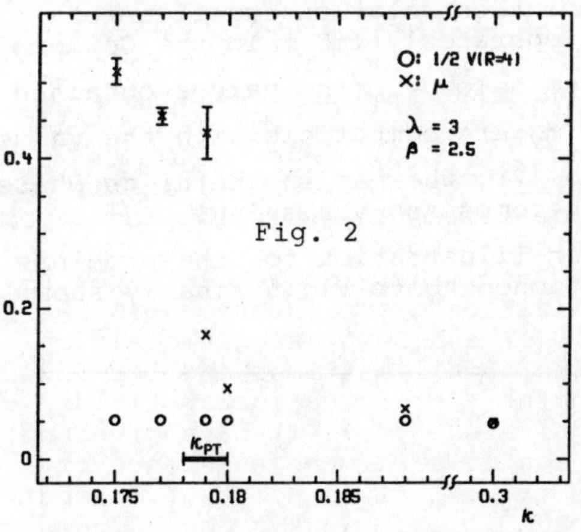

Fig. 2

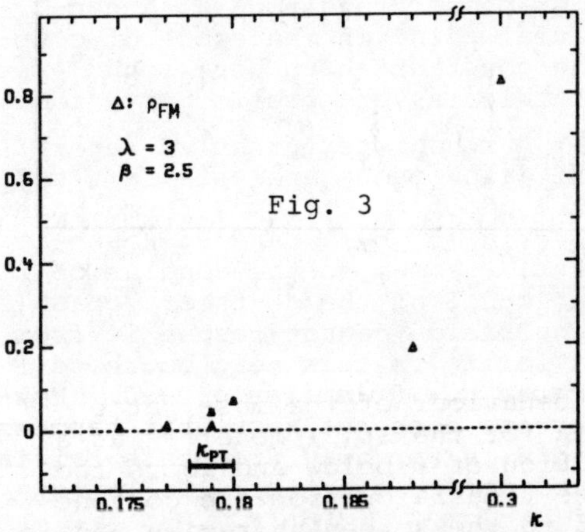

Fig. 3

Fig.3 shows the corresponding (expected!) behaviour of the order parameter ρ_{FM}.

Fig.2 also indicates the large sensitivity of the parameter μ to the phase transition point κ_{PT}, where the change of its value is considerable. Fig.4 shows the effect of the

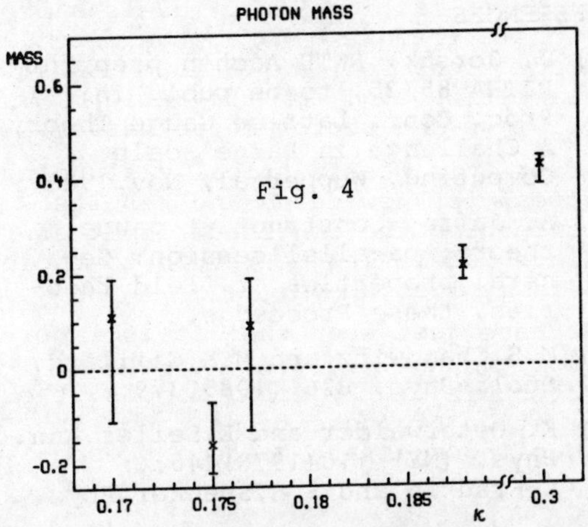

Fig. 4

Higgs mechanism on the photon mass as determined from the decay of the 2-point correlation function for the operator[15] $\mathrm{Im}U_p$: In the Coulomb phase below κ_{PT} the values obtained for m_γ are compatible with the value zero [16], whereas in the Higgs phase m_γ becomes non-vanishing!

As an illustration for the problems mentioned above Fig.5 finally shows

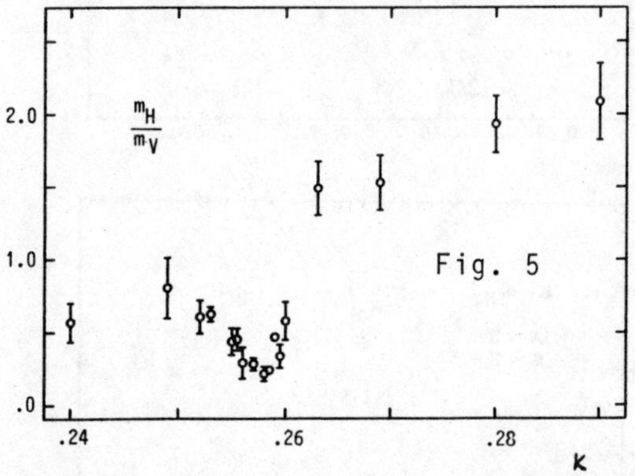

Fig. 5

the behaviour of the mass ratio m_H/m_V for the SU(2) model[17] as a function of κ below and above the phase transition from the confinement to the screening region, again for fixed λ and β. The ratio is smaller that 1 below and larger than 1 above the phase transition, but at the moment one does not know what this means physically.

I thank H.G.Evertz, K.Fredenhagen, V.Grösch, K.Jansen, J.Jersák, M. Lüscher and M.Marcu for many stimulating discussions.

REFERENCES

(1) J. Jersák, RWTH Aachen preprint PITHA 85/25, to be publ. in: Proc. Conf. Lattice Gauge Theory, A Challenge in Large Scale Computing, Wuppertal, Nov.1985.

(2) A. Jaffe, Constructive gauge theory; parallel session: General properties of field theories, these Proceedgs.

(3) M.S.Chanowitz and M.K.Gaillard, Nucl. Phys. B261(1985)379.

(4) K. Osterwalder and E.Seiler, Ann. Phys. (NY) 110(1978)440; E.Fradkin and S.H.Shenker, Phys. Rev. D 19(1979)3682.

(5) C.B.Lang, Nucl.Phys. B265(FS15, 1986)630; with refs. to previous work.

(6) H. Kühnelt, C.B.Lang and G.Vones, Nucl. Phys. B 230(FS 10,1984)16.

(7) D.J.Gross and F.Wilczek, Phys. Rev. D 8(1973)3633.

(8) D.J.E.Callaway and R.Petronzio, CERN preprint TH.4430/86, with refs. to their and other previous papers.

(9) See ref. 1).

(10) A. Hasenfratz and P. Hasenfratz, preprint FSU-SCRI-86-30.

(11) J. Fröhlich, G. Marchio and F. Strocchio, Nucl. Phys. B 190 (FS 3, 1981)553.

(12) K. Fredenhagen and M. Marcu, Phys. Rev. Lett. 56(1986)223.

(13) H. G. Evertz et al., Phys. Lett. B 175(1986)335.

(14) H. G. Evertz et al., Confined and free charges in compact scalar QED, Institut f. Theor. Physik, RWTH Aachen, to be publ..

(15) B. Berg and C. Panagiotakopoulos, Phys. Rev. Lett. 52(1984)94.

(16) The rather large errors are systematically one, coming from the fact that the mass has to be extraplolated from an energy-momentum dispersion relation on the lattice.

(17) H. G. Evertz et al., Phys.Lett. B 171(1986)271.

Studies Of Chiral Symmetry In A Spontaneously Broken Lattice Gauge Theory

I-Hsiu Lee

Department of Physics, The Ohio State University, Columbus, Ohio 43210
and
Physics Department, Brookhaven National Laboratory, Upton, New York 11973*

The phase diagram of lattice SU(2) gauge theory with fundamental representation Higgs fields is probed by $l = 1/2$ and $l = 1$ fermions. Quenched calculations of $\langle\overline{\psi}\psi\rangle_l$ show two distinct regions characterized by broken or unbroken chiral symmetry. The location of the boundary between the two regions depends on the representation l of the fermions. For $l = 1/2$ fermions, we have verified the presence of Nambu-Goldstone bosons ("pions") in the region where chiral symmetry is broken and have found evidence for massless physical fermions in the chirally symmetric region.

During the past few years, it has become increasingly well established that global chiral symmetry is spontaneously broken in vector-like SU(N) gauge theories. Various studies, both analytic [1] and numerical [2], lead us to expect that chiral symmetry is realized in the Nambu-Goldstone mode in vector-like SU(N) gauge theories, with massless Nambu-Goldstone bosons in the spectrum. Moreover, it has been pointed out that if the formation of fermion-antifermion condensates is not crucially tied to the presence of the confining forces, but could occur whenever the strength of the effective interaction between fermions becomes sufficiently large, then, together with the variation of coupling constants with the momentum scale, this effect would give rise to the tumbling scenario, in which several length scales can be created dynamically in a single gauge group [3].

In the work that I am going to report on, which was done in collaboration with J. Shigemitsu [4], we take the first step toward investigating how the situation might be different in the presence of spontaneous breaking of gauge symmetry induced, for instance, by gauge couplings to scalar fields. More specifically, how are the changes in the gauge sector reflected in the realizations of global symmetry in the fermionic sector?

Of particular interest are models in which, in the absence of dynamical fermions, the Higgs and the confining regions are analytically connected. The SU(2) gauge-Higgs model with Higgs fields in the fundamental representation is a particularly simple model of this type. Its phase diagram in the (β_g, β_H) plane is shown in Fig.1. The $\beta_H = 0$ line corresponds to the pure SU(2) gauge theory. The line $\beta_g = \infty$ corresponds to the pure O(4) spin theory. When $\beta_H \to \infty$, the gauge degrees of freedom are completely frozen out and the theory becomes trivial. In the fixed-length limit, the end point of the line of transitions in the interior of the phase diagram occurs at approximately $(\beta_g, \beta_H) = (1.6, 0.65)$ [5].

The first question one might ask is: could fermions induce new phase boundaries rendering analytic continuation between the Higgs and the confining regions impossible? Investigations of the realizations of chiral symmetry in different regions of the phase diagram should help to answer this question. Along the $\beta_H = 0$ line, chiral symmetry is believed to be spontaneously broken for all values of β_g and for all fermion representations. One infers from a small β_H expansion that this is also true for a strip in the phase diagram adjacent to this line. On the other hand, the $\beta_H = \infty$ line corresponds to a trivial theory, and there is no dynamics to trigger chiral symmetry breaking. The question then is whether chiral symmetry breaking ceases to occur only when $\beta_H \to \infty$ or whether, instead, there exists a finite $\beta_H^c(\beta_g)$ beyond which chiral symmetry is not broken.

The lattice action of this model can be written as

$$S = S_G + S_H + S_F, \tag{1}$$

where

$$S_G = \beta_g \sum_p \left(1 - \frac{1}{2} trU_p\right), \tag{2}$$

$$S_H = -\beta_H \sum_n \sum_\mu \left[\phi^\dagger(n)\, U_\mu(n)\, \phi(n+\mu) + h.c.\right], \tag{3}$$

$$S_F = \frac{1}{2} \sum_n \overline{\chi}(n) \sum_\mu \eta_\mu(n) \left[U^l_\mu(n)\,\chi(n+\mu) - U^{l\dagger}_\mu(n-\mu)\,\chi(n-\mu) \right] + m \sum_n \overline{\chi}(n)\,\chi(n) \,, \tag{4}$$

with $\beta_g = \frac{4}{g^2}, \eta_\mu(n) = (-1)^{n_1+\cdots+n_{\mu-1}}$. We used Kogut-Susskind fermions in both the fundamental $(l = \frac{1}{2})$ and adjoint $(l = 1)$ representations. The fermions are treated in the quenched approximation. The Higgs fields are introduced in the fixed-length limit. In the present model, no gauge-invariant Yukawa coupling can be constructed, and hence the fermions only interact directly with the gauge degrees of freedom.

We calculated $\langle\overline{\psi}\psi\rangle$ at $\beta_g = 0.5$ as a function of β_H. We used the conjugate gradient method, and for each β_H, we performed the calculations for bare masses (in lattice units) $m = 0.08, 0.06, 0.04$ and 0.02 on a 8^4 lattice (and for $m = 0.10, 0.08, 0.06, 0.04$ on a 6^4 lattice). The data were then extrapolated to obtain the $m \to 0$ limit. The result is shown in Fig. 2. We see evidence of chiral symmetry breaking only up to $\beta_H \approx 2 \equiv \beta_H^{(\frac{1}{2})}$ for $l = \frac{1}{2}$ and up to $\beta_H \approx 6 \equiv \beta_H^{(1)}$ for $l = 1$ fermions. This is consistent with the general idea of the tumbling scenario. While the effective interaction between the $l = \frac{1}{2}$ fermions becomes too weak to cause chiral symmetry breaking for $\beta_H > \beta_H^{(\frac{1}{2})}$, the $l = 1$ fermions still manage to condense up to $\beta_H = \beta_H^{(1)}$ due to their larger quadratic Casimir invariant.

If chiral symmetry is indeed unbroken in the region $\beta_H > \beta_H^{(\frac{1}{2})}$ for $l = \frac{1}{2}$ fermions, this should also manifest itself in the spectrum. We carried out spectrum calculations at $\beta_g = 0.5$ for $\beta_H = 1$ and $\beta_H = 3$ using the following operator in the "pion" channel :

$$\overline{\psi} T \gamma_5 \psi \to \sum (-1)^{\xi(n)}\, \overline{\chi}(n)\,\chi(n) \,, \tag{5}$$

where $\xi(n) = \sum_{\mu=1}^4 n_\mu$, T is a 4×4 traceless matrix, and the summation extends over the sixteen sites in a hypercube. In the fermion channel, we used the operators

$$\phi^\dagger \psi \to \phi^\dagger(n)\,\chi(n) \tag{6a}$$

$$\phi(-i\sigma_2)\,\psi \to \phi(n)\,(-i\sigma_2)\,\chi(n) \,. \tag{6b}$$

In Eq.(6b) σ_2 acts in the local SU(2) space. One can easily show that the states created by the operators (6a) and (6b) are degenerate and form a global SU(2) doublet.

We used an $8^3 \times 16$ lattice for the spectrum calculations. In the following I will only present and discuss the results of our calculations; for further details, see ref. [4].

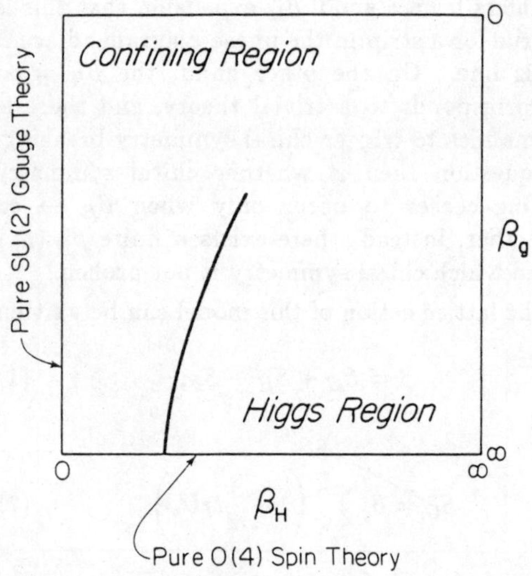

Fig. 1. Schematic phase diagram in a (β_g, β_H) plane for the SU(2) gauge theory with Higgs fields in the fundamental representation.

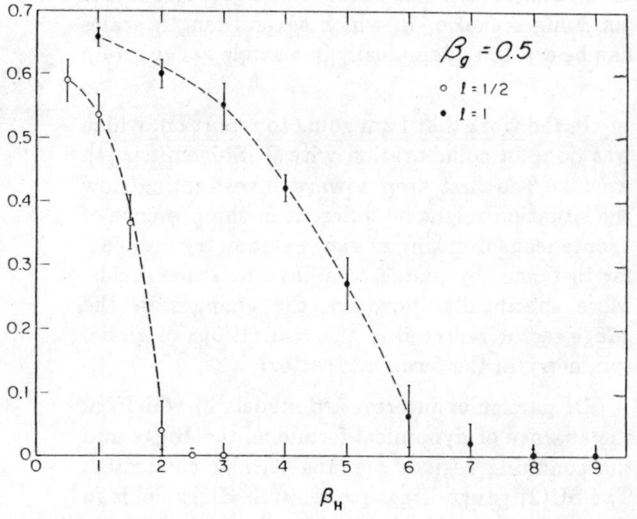

Fig. 2. $a^3\langle\overline{\psi}\psi\rangle$ versus β_H at $\beta_g = 0.5$ for fermions in the fundamental (open circles) and the adjoint (black dots) representations.

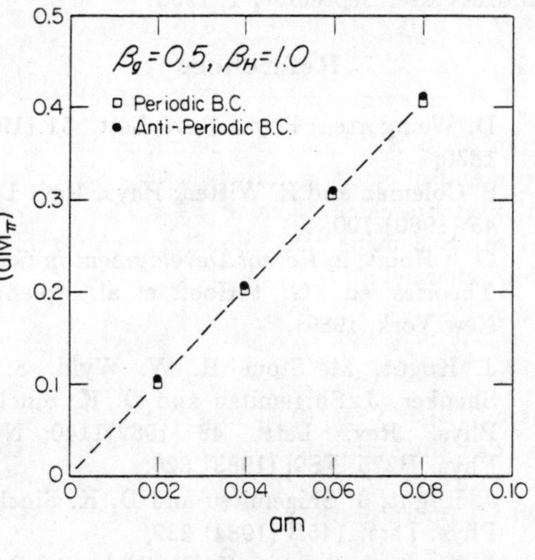

Fig. 3. $(aM_\pi)^2$ versus am at $\beta_g = 0.5$ and $\beta_H = 1.0$.

For $\beta_H = 1$, we plot M_π^2 versus m in Fig. 3. One clearly sees that the mass-squared of the pion vanishes linearly as the symmetry breaking parameter m goes to zero, as is characteristic of a Nambu-Goldstone boson. In the fermion channel, the data were too noisy to be useful, indicating that only very massive states couple to this channel. In short, the spectrum at $\beta_g = 0.5, \beta_H = 1$ is similar to what one would expect in the strong coupling region of an SU(2) version of lattice QCD.

In the chirally symmetric region at $\beta_g = 0.5, \beta_H = 3$, we obtain the mass of the lowest-lying state in the fermion channel, M_F.

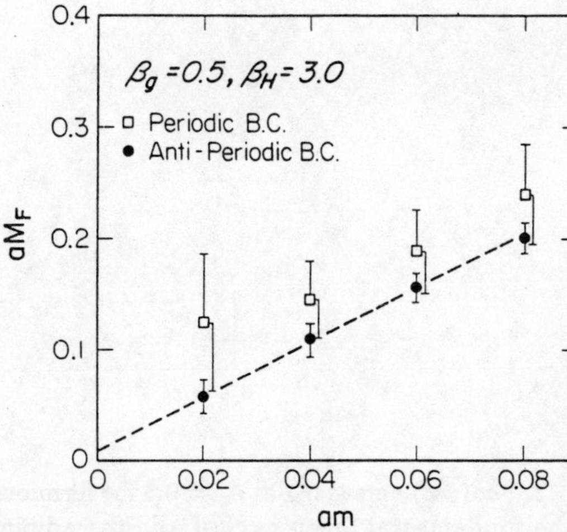

Fig. 4. aM_F versus am at $\beta_g = 0.5$ and $\beta_H = 3.0$.

Fig. 4 shows the physical fermion mass M_F versus the bare fermion mass m. The anti-periodic boundary condition data extrapolate approximately to zero as $m \to 0$, while the periodic boundary condition data lie systematically higher. We believe this is due to contamination by zero-momentum modes, a phenomenon well known to occur also in finite volume free fermion propagators for small enough m, which can be avoided by using anti-periodic boundary conditions. There- the fact that the periodic boundary condition data for M_F do not extrapolate to zero should be considered as a finite volume artifact, and we conclude that as the bare fermion mass m vanishes, the physical spectrum contains massless fermions in the region $\beta_H > \beta_H^{(\frac{1}{2})}$.

For the pion channel, we plot the energies of the two lowest-lying states in Fig. 5. In contrast to Fig. 3, where $E_1^2 (= M_\pi^2)$ depends on m linearly, we see a (nearly) linear dependence of E_1 on m, indicating that this state is not a Nambu-Goldstone boson. To see if it is a fermion-antifermion bound state, we compare E_1 with $2M_F$. As shown in Fig. 5, they fall almost on

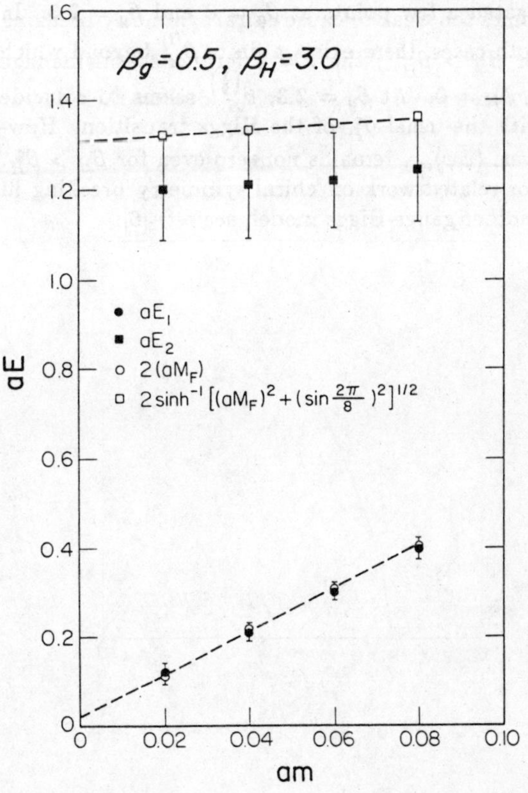

Fig. 5. aE_1 and aE_2 versus am at $\beta_g = 0.5$ and $\beta_H = 3.0$.

top of each other. This suggests that there is no stable fermion-antifermion bound states below the two-fermion threshold. Going one step further, we try to see whether the second state, with energy E_2, can be part of the two-fermion cut with energy $E = 2sinh^{-1}\left[M_F^2 + \left(sin\frac{2\pi}{8}\right)^2\right]^{\frac{1}{2}}$ which is plotted in Fig. 5 as the open squares. We find that E is slightly higher than E_2. However, the numbers are close enough that it is possible that the E_2-state is the second contribution to the two-fermion cut rather than a fermion-antifermion resonance. Only a more thorough analysis can settle this issue.

To summarize: the phase diagram of SU(2) gauge theory with Higgs fields in the fundamental representation, when probed by fermions, separates into two distinct regions, characterized by the realization of chiral symmetry in the Nambu-Goldstone mode or the Wigner mode. The chirally symmetric region has light physical fermions whose masses go to zero as the bare fermion mass vanishes. There is no clear evidence for a stable fermion-antifermion bound state. We believe that these qualitative features will survive the inclusion of dynamical fermions. All the data presented above were taken at $\beta_g = 0.5$. We have also investigated a few points at $\beta_g = 0$ and $\beta_g = 2.3$. In both cases, there exists a finite $\beta_H^{(l)}$ beyond which $\langle\overline{\psi}\psi\rangle_l = 0$. At $\beta_g = 2.3$, $\beta_H^{(\frac{1}{2})}$ seems to coincide with the usual β_H^c of the Higgs transition. However, $\langle\overline{\psi}\psi\rangle_{l=1}$ remains nonzero even for $\beta_H > \beta_H^c$. For related work on chiral symmetry breaking in another gauge-Higgs model, see ref. [6].

* address after September 1, 1986

References

1. D. Weingarten; Phys. Rev. Lett. 51 (1983) 1830;
   S. Coleman and E. Witten; Phys. Rev. Lett. 45 (1980) 100;
   G. 't Hooft; in *Recent Development in Gauge Theories*, ed. G. 't Hooft et al. (plenum, New York, 1980).
2. J. Kogut, M. Stone H. W. Wyld, S. H. Shenker, J. Shigemitsu and D. K. Sinclair; Phys. Rev. Lett. 48 (1982)1140; Nucl. Phys. B225 [FS9] (1983) 326;
   J. Kogut, J. Shigemitsu and D. K. Sinclair; Phys. Lett. 145B (1984) 239;
   J. Kogut, J. Polonyi, H. W. Wyld and D. K. Sinclair; Phys. Rev. Lett. 54 (1985) 1980.
3. S. Raby, S. Dimopoulos and L. Susskind; Nucl. Phys. B169 (1980) 373;
   W. J. Marciano; Phys. Rev. D21 (1980) 2425;
   E. Eichten and F. Feinberg; Phys. Lett. 110B (1982) 232.
4. I-H. Lee and J. Shigemitsu; Phys. Lett. 178B (1986) 93.
5. H. Kühnelt, C. B. Lang and G. Vones; Nucl. Phys. B230 [FS10] (1984) 31;
   I. Montvay; Phys. Lett. 150B (1985) 441.
6. I-H. Lee and J. Shigemitsu; Nucl. Phys. B276 (1986) 580.

POINCARE', DE SITTER AND CONFORMAL GRAVITY ON THE LATTICE

P. Menotti

Dipartimento di Fisica della Università di Pisa
and INFN Sezione di Pisa, Pisa, Italy

A. Pelissetto

Scuola Normale Superiore, Pisa
and INFN Sezione di Pisa, Pisa, Italy.

A unified treatment of Poincaré, De Sitter and conformal gravity on a euclidean lattice is given following the gauge group formulations on the continuum. The vierbeins are naturally located on the links. A discussion is given of the role of the sign of the volume element in four dimensional space. Exact reflection positivity is proved without any restriction on the observable quantities and we state the boundedness properties of the lagrangian which assure the existence of a positive self-adjoint hamiltonian. A doubling phenomenon for the graviton on the lattice is found and discussed.

We address the problem of formulating euclidean lattice gravity in analogy to usual lattice gauge theories /1-6/. Our starting point is the formulation of Poincaré (Einstein and higher derivative), De Sitter and conformal gravities as the gauge theories of the related groups /7/.
Our fundamental dynamical variable is a finite element of the gauge group (Poincaré, De Sitter and conformal) which is associated to a link of a hypercubical lattice. The advantage of such a formulation is twofold: a) the vierbein is associated to a link variable as is natural, given its 1-form nature in the continuum formulation; b) reflection positivity holds in all the above mentioned theories for all locally O(4) invariant observables.
The usual Einstein action, in terms of differential forms, is given by /7/

$$\frac{1}{2}\int R^{ab}\wedge\tau^{c}\wedge\tau^{d}\varepsilon_{abcd} = \int R(\det\tau)\, d^4x \qquad (1)$$

We notice that (1) differs from the usual Einstein-Hilbert action

$$\int R\sqrt{g}\, d^4x \qquad (2)$$

by sign$(\det\tau)$. In absence of such a factor, action (1) is not invariant under reflection $x^a \to -x^a$ (a=1,2,3) or equivalently under reflection in the internal space. Thus we shall add in (1) the factor sign $\det(\tau)$ which is relevant in a non perturbative situation. In order to build up the lattice theory one introduces the usual plaquette $U_{\mu\nu}(n)$ and defines

$$R^{ab}_{\mu\nu}(J,n) = (-1/4a^2)\cdot\{\mathrm{tr}J_{ab}(U_{\mu\nu}(n)-U_{\nu\mu}(n))\} \qquad (3)$$

where J_{ab} are the generators of the O(4) group in the spinor 8-dimensional representation. One needs also to extract from the field variables $U(n,n+\mu)$ a "local vierbein" which transforms under local transformations as a Lorentz vector. This can be obtained by decomposing U in the form

$$U(n,n+\mu) = \exp(aP_c t^c(n,n+\mu))\,\exp(\tfrac{1}{2}aJ_{bc}\omega^{bc}(n,n+\mu)) \qquad (4)$$

where it is easily seen that $t^a(n,n+\mu)$ transforms locally (in n) under local gauge O(4) transformations, i.e. if

$$U(n,n+\mu) \to V(n)U(n,n+\mu)V^{-1}(n+\mu) \qquad (5)$$

with

$$V(n) = \exp(\varepsilon(n)^{ab}J_{ab}) \qquad (6)$$

then

$$t^a(n,n+\mu) \to (\exp\varepsilon(n))^a{}_{.b}\,t^b(n,n+\mu) \qquad (7)$$

In the 8-dimensional representation $t^a(n,n+\mu)$ can be extracted algebraically through

$$t^a(n,n+\mu) =$$

$$(-1/4a)\ \mathrm{tr}(K_a U(n,n+\mu) D U(n+\mu,n)) \qquad (8)$$

where K_a and D obey together with J_{ab} and P_a the conformal algebra.
Given $R^{ab}_{\mu\nu}(J,n)$ (Eq. 3) and $t^a(n,n+\mu)$ one can easily write down the lattice analogue of Eq. (1). We notice that a symmetrization on the possible orientations of the vierbeins and of the plaquettes is required if we want rotational invariance. Extension to higher derivative, De Sitter and conformal gravity is straightforward.
For higher derivative theories one has to introduce the constraint given by

$$R^a_{\mu\nu}(P,n) + \mathrm{symm} = 0 \qquad (9)$$

where

$$R^a_{\mu\nu}(P,n) =$$

$$(-1/8a^2)\ \mathrm{tr}\{K_a(U_{\mu\nu}(n) - U_{\nu\mu}(n))\} \qquad (10)$$

We recall that, even in the continuum case the lagrangians are invariant only under a subgroup of the whole gauge group. For the Poincaré and De Sitter case this is O(4) while for the conformal this is the group generated by J_{ab}, K_a and D. This fact has to be taken into account when defining the integration measure: for the Poincaré and De Sitter case one should only respect O(4) invariance while it is well known that in the conformal case /8/ one can have a measure invariant only under J_{ab} and K_a. The general form we consider for the integration measure for Poincaré and De Sitter gravity is

$$dU \ \Pi_n f(a^2 t^2(n,n+\mu)) \qquad (11)$$
$$\Pi_n \Pi_{\mu\nu\rho\lambda} |\mathcal{D}(n;\mu\nu\rho\lambda)|^p$$

where dU is the invariant measure over the whole group, f is an arbitrary positive function and $\mathcal{D}(n;\mu\nu\rho\lambda)$ is the determinant of the vierbeins stemming from site n in the directions $\mu\nu\rho\lambda$, and the indices go through positive and negative values.
As for the definition of the functional integral no problem arises in the De Sitter case, where being the variable $U(n,n+\mu)$ compact both the lagrangian and the measure are bounded. For the non compact case the naive translation gives rise to an ill defined functional integral. The standard way out /3/ is to consider "compact vierbeins" which is equivalent in our case to consider in (11) a function f decreasing sufficiently fast at infinity. Moreover the coefficients of the various invariants like R^2, $R^{ab}{}_{\mu\nu}R^{ab\mu\nu}$ and $R_{\mu\nu}R^{\mu\nu}$ have to be chosen as to give a lagrangian bounded from above.
We notice however that all our lagrangians, which are not exactly reparametrization invariant, become reparametrization invariant in the formal continuum limit and that the damping factor $f(a^2t^2)$ becomes irrelevant in the limit $a \to 0$. A similar treatment can be applied to the conformal case. Naturally the major problem is whether this lattice regularized theory admits, with a proper choice of the integration measure, an ultraviolet fixed point which gives rise to a reparametrization invariant continuum limit.
For asymptotically free theories such a problem can be addressed at the perturbative level; in the other cases like standard De Sitter and Einstein gravity the problem is more difficult and indications can be obtained from numerical simulations.
Reflection positivity with respect to a plane containing sites can be easily proved for all mentioned theories /6/ and relies uniquely on the geometric reflection properties of the lagrangian and of the constraint. With a well defined functional integral one easily sets up a Hilbert space starting from the bounded O(4) invariant functions of the variables in the upper half space where the scalar product is defined by

$$(G,F) = \langle(\Theta G)F\rangle \qquad (12)$$

being Θ the Osterwalder-Schrader reflection operator /9/.
It is now very simple to prove, following the standard procedure, the existence of a bounded transfer matrix, $0<T<1$, from which the existence of a self-adjoint positive hamiltonian follows.
The expansion of the Einstein lattice theory around a flat background produces a quadratic lagrangian with the same structure of the continuum

one but with k_μ replaced by sin k_μ /5/. Such a lagrangian turns out to be exactly invariant under the linearized reparametrization transformation

$$h_{\mu\nu}(k) \to h_{\mu\nu}(k) + \\ +\sin k_\mu \eta_\nu(k) + \sin k_\nu \eta_\mu(k) \quad (13)$$

however the appearance of $\sin k_\mu$ (instead of $2\sin(k_\mu/2)$) implies doubling of the graviton (i.e. 16 gravitons) analogous to the doubling of chiral fermions in lattice gauge theories. In Ref. 5 we proved a more general result: any lattice theory whose quadratic part is of the form

$$L_0 = -\tfrac{1}{2}f_1(k_\lambda)f_1(-k_\lambda)h_{\mu\nu}(-k)h_{\mu\nu}(k) \\ +\tfrac{1}{2}f_2(k_\lambda)f_2(-k_\lambda)h_{\mu\mu}(-k)h_{\nu\nu}(k) \\ -f_3(k_\lambda)f_4(-k_\mu)h_{\lambda\mu}(-k)h_{\nu\nu}(k) \\ +f_5(k_\lambda)f_6(-k_\mu)h_{\lambda\nu}(-k)h_{\mu\nu}(k) \quad (14)$$

and is invariant under the gauge transformation

$$h_{\mu\nu}(k) \to h_{\mu\nu}(k) + \\ +g(k_\mu)\eta_\nu(k) + g(k_\nu)\eta_\mu(k) \quad (15)$$

with g local derivative (not SLAC), necessarily shows the doubling phenomenon.

Expansion of conformal gravity around a flat background gives rise to the same doubling phenomenon /6/.

An open problem, as far as R^2 theories are concerned is that of perturbative unitarity. From the proof of reflection positivity it appears that exact (non perturbative) unitarity does not necessarily imply its perturbative validity. The appearance of the doubling phenomenon is somewhat connected with the unitarity problem as it arises from the symmetrization of the lagrangian. Whether this renders possible a restoration of unitarity even at the pertubative level can be ascertained only by means of a direct perturbative calculation.

Finally the problem of the integration measure. At the formal level several integration measures which render the quantum theory reparametrization invariant have been proposed /10/. In the dimensional regularization scheme they do not play any role as the contribution of the measure is set identically to zero. As in lattice theory we have a finite regulator (lattice spacing) the measure plays a major role. Assuming the form $(\det\tau)^N$ for the measure, one could compute to one loop for Einstein gravity the Slavnov-Taylor identities in order to give a determination of N.

REFERENCES.

/1/ L.Smolin, Nucl. Phys. B148,333 (1979).

/2/ C.Mannion and J.G.Taylor, Phys. Lett. 100B, 261 (1981).

/3/ E.T.Tomboulis, Phys. Rev. Lett. 52,1173 (1984).

/4/ K.Kondo, Progr. Theor. Phys. 72, 841 (1984).

/5/ P.Menotti and A.Pelissetto, Ann. Phys. (N.Y.) (to appear).

/6/ P.Menotti and A.Pelissetto, IFUP-TH 5/86 (Pisa University preprint).

/7/ S.W.MacDowell and F.Mansouri, Phys. Rev. Lett. 38, 739 (1977); 38, 1376 (1977); Y.Ne'eman and T.Regge, Phys. Lett. 74B, 54 (1978);Riv. Nuovo Cim. 1, 1 (1978); M.Kaku, P.K.Townsend and P.Van Nieuwenhuizen, Phys. Lett. 69B, 304 (1977).

/8/ K.Fujikawa, "Path integral quantization of gravitational interactions",preprint RRK-85-21.

/9/ K.Osterwalder and R.Schrader, Com. Math. Phys. 31,83 (1973); K.Osterwalder and E.Seiler, Ann. Phys. (N.Y.) 110, 440 (1978); E.Seiler, Lecture Notes in Physics, vol.159, Springer-Verlag Berlin (1981).

/10/ H.Leutwyler, Phys. Rev. 134B, 1155 (1964); E.Fradkin and G.Vilkoviski, Phys. Rev. D8, 424 (1973); L.Faddeev and V.Popov, Sov. Phys. Usp. 16, 777 (1974); V.De Alfaro, S.Fubini and G.Furlan, Nuovo Cim. 57B, 227 (1984); 76A, 365 (1984); K.Fujikawa, Nucl. Phys. 226B, 437 (1983).

FIXED POINT STRUCTURE OF QUENCHED, PLANAR QUANTUM ELECTRODYNAMICS*

S. T. LOVE§

Stanford Linear Accelerator Center
Stanford University, Stanford, California, 94305

Gauge theories exhibiting a hierarchy of fermion mass scales may contain a pseudo–Nambu–Goldstone boson of spontaneously broken scale invariance. The relation between scale and chiral symmetry breaking is studied analytically in quenched, planar quantum electrodynamics in four dimensions. The model possesses a novel nonperturbative ultraviolet fixed point governing its strong coupling phase which requires the mixing of four fermion operators.

In the chiral symmetric limit, QCD-like gauge theories with N flavors of fermions possess an $SU(N)_L \times SU(N)_R$ chiral symmetry which is spontaneously broken by a dynamical fermion condensate to its diagonal $SU(N)_V$ subgroup resulting in the appearance of an SU(N) multiplet of pions as Nambu–Goldstone bosons. In addition to these chiral symmetries, the classical formulation of gauge theories also exhibits, in four dimensions in the chiral limit, an exact scale invariance. Here I will discuss various aspects of dynamical symmetry breaking with particular focus on the scale symmetry. A more complete discussion appears in work done in collabortion with W. A. Bardeen and C. N. Leung.[1,2]

In quantum chromodynamics, the scale symmetry is explicitly broken by quantum radiative corrections as reflected by the anomalous nonconservation of the dilatation current:

$$D_\mu = \chi^\nu \theta_{\mu\nu} ,$$
$$\partial_\mu D^\mu = \theta^\mu_\mu = \frac{\beta(g)}{g}\,\frac{1}{4}\, G^{\mu\nu} G_{\mu\nu} . \tag{1}$$

When combined with the nonperturbative QCD vacuum structure which gives $\langle G^2_{\mu\nu} \rangle \sim \Lambda^4_{\rm QCD}$, a large explicit breaking of the anomalous symmetry ensues. That is, the explicit scale symmetry breaking accompanying the rapid running of the QCD coupling dominates at low energies and no vestige of the classical scale symmetry remains. In particular, there is no evidence for a Nambu–Goldstone boson of scale symmetry in conventional QCD.

The above picture need not hold, however, in all gauge models. It may be possible that the spontaneous breaking of the chiral symmetry might also trigger the spontaneous beaking of an approximate scale symmetry. This would be the case if the chiral symmetry breaking occurs at a scale where the explicit scale breaking is small. Such a situation could occur in theories possessing a hierarchy of fermion mass scales. An example is afforded by a model where fermions transforming as higher dimensional representations of the gauge group are present in the theory. Indeed, results from numerical studies in lattice gauge theories[2] indicate that the scale of chiral condensation for these fermions is relatively short compared to the confinement scale.

The chiral condensation scale is roughly characterized by the requirement that the effective fermion coupling $C_2(f)\,\alpha(\mu)$ reach a critical value α_{crit}. Here $\alpha(\mu)$ is the gauge theory running coupling and $C_2(f)$ is the quadratic Casimir invariant of the fermion representation. For a sufficiently large $C_2(f)$, spontaneous chiral symmetry breaking could occur in the asymptotically free region where $\alpha(\mu)$ varies only logarithmically with energy. The explicit breaking of the scale symmetry is then but a small effect at this scale compared to the large spontaneous breaking associated with the chiral condensation and consequently the scale symmetry should be realized in a Nambu–Goldstone fashion resulting in the appearance of a scalar dilaton. Since the coupling is not fixed,

*Work supported in part by the Department of Energy, contract DE–AC03–76SF00515 (SLAC), DE–FG02–85ER40299 (Outstanding Jr. Invest. Grant) and DE–AC02–761428A025 (Purdue).

§On leave from Department of Physics, Purdue University, West Lafayette, IN 47907 the dilaton should actually emerge as a pseudo–Nambu–Goldstone boson acquiring a mass of order the scale at which the explicit scale symmetry breaking becomes important, which is roughly the confinement scale of the gauge theory. The dilaton should couple to heavy states, *e.g.*, W, Z in a manner similar to the physical Higgs boson, but may be distinguished from it due to its Nambu–Goldstone nature. For a discussion of dilaton phenomenology, see Ref. 4.

In order to study the dynamical aspects of chiral and scale symmetry breaking, I consider the simplest approximation to a gauge field theory with a fixed but critical coupling. This corresponds to quenched, planar (ladder) quantum electrodynamics. The quenched approximation excludes fermion loop corrections and consequently guarantees that the perturbative gauge coupling β–function vanishes. It is thus anticipated that the theory should exhibit an exact or spontaneously broken scale symmetry.

This model has been the subject of numerous investigations by various authors over the years.[5,8] In the model, the Schwinger–Dyson equation for the fermion self-energy is given by a sum of the rainbow graphs. At weak coupling, $\alpha < \alpha_c = \pi/3$, there exist no spontaneous chiral (or scale) symmetry breaking solutions. If an ultraviolet cutoff Λ is introduced, there are no solutions to the massless equation at fixed Λ and solutions appearing as $\Lambda \to \infty$ do not correspond to spontaneous symmetry breaking but rather reflect the anomalous dimension of the fermion mass operator $\bar{\psi}\psi$ so that

$$d_{\bar{\psi}\psi} \;=\; 2 + \sqrt{1 - \frac{\alpha}{\alpha_c}}\;. \qquad (2)$$

On the other hand, at strong coupling, $\alpha > \alpha_c$, the massless equation was shown to possess a nontrivial solution leading to the generation of the fermion mass scale

$$\Sigma(0) \simeq \Lambda\ \exp\{\delta + 1\}\ \exp\left\{\frac{-\pi}{\sqrt{\frac{\alpha}{\alpha_c} - 1}}\right\}\;, \qquad (3)$$

where $\delta \simeq 0.55$ is a parameter of the asymptotic solution for the self-energy function. The dependence of the fermion mass scale diverging with the cutoff appears to be disasterous for this solution as all the dynamics associated with the spontaneous chiral symmetry breaking occurs at the cutoff Λ. Similar conclusions were also reached in numerical studies.[9]

There is, however, an alternate interpretation of this solution[10] in which the critical coupling α_c is viewed as a fixed point of the strong coupling phase with the gauge coupling α approaching the critical value as

$$\frac{\alpha}{\alpha_c} \;=\; 1 + \frac{\pi^2}{\ell n^2\left(\frac{\Lambda}{\kappa}\right)}\;, \qquad \Lambda \to \infty\;, \qquad (4)$$

where κ is an infrared scale. This fixed point interpretation leads to a finite fermion mass scale $\Sigma(0) \;\to\; e^{\delta+1}\kappa$ as $\Lambda \;\to\; \infty$. Moreover, a massless pseudoscalar bound state appears as a solution to the Bethe–Salpeter equation reflecting the Nambu–Goldstone realization of the chiral symmetry. However, the solution remains incomplete as it leaves unclear the origin of the running of the gauge coupling and moreover does not properly reflect the scale symmetry as there is no massless scalar bound state solution to the Bethe–Salpeter equation corresponding to the dilaton.

It was attempting to clarify these issues that led to the discovery of the novel fixed point structure of the model.[2] The origin of this structure is the generation of four fermion operators which necessarily mix with the gauge interactions at the fixed point. The mixing results from the large anomalous dimensions generated by the gauge coupling at the fixed point. We have already observed that the mass operator $\bar{\psi}\psi$ has dimension $d_{\bar{\psi}\psi} \;=\; 2 + [1 - (\alpha/\alpha_c)]^{1/2}$ which is three at zero coupling but approaches two at the critical coupling. In the ladder approximation under consideration, the four fermion operator $(\bar{\psi}\psi)^2$ has just twice the mass operator dimension so that

$$d_{(\bar{\psi}\psi)^2} \;=\; 4 + 2\sqrt{1 - \frac{\alpha}{\alpha_c}}\;, \qquad (5)$$

which approaches four as $\alpha \to \alpha_c$. Since the four fermion operators are dimension four at the critical gauge coupling, they are relevant operators which must be included in the analysis of the fixed point structure.

We are thus led to study the scale invariant fixed point structure using the chirally invariant effective fermion Lagrangian

$$\mathcal{L}_f = \bar{\psi}\left[i\gamma\partial - e\gamma A - \mu_0\right]\psi + \frac{G_0}{2}\left[(\bar{\psi}\psi)^2 + (\bar{\psi}i\gamma_5\psi)^2\right] , \tag{6}$$

where μ_0 is a bare fermion mass included to provide explicit breaking. Consistent with the planar approximation for the gauge interactions, only planar diagrams involving the four fermion interactions are to be retained.

The vacuum structure of the modified theory can be deduced using the same methods as employed in the pure gauge case. The Schwinger–Dyson equation in ladder approximation takes the form

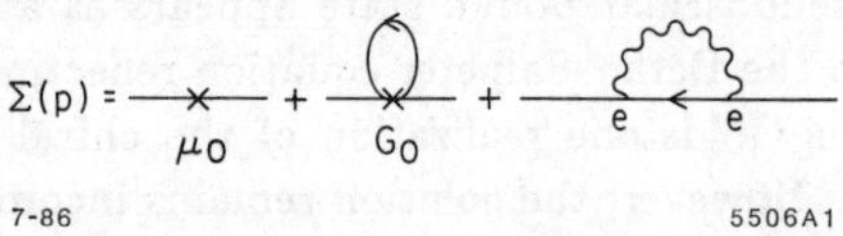

Fig. 1. The Schwinger–Dyson equation.

where the full propagator is to be used in the diagrams. This equation involves an effective bare mass parameter m_0 which includes terms generated by the induced interactions so that $m_0 = \mu_0 - G_0\langle\bar{\psi}\psi\rangle_0$. The fermion bilinear vacuum expectation value must be computed self-consistently including all the QED radiative ladder corrections, so that even in the chiral limit, $\mu_0 = 0$, the effective bare mass will not vanish, $m_0 \neq 0$. This modification leads to a new gap equation and fermion mass scale given by

$$\mu = \frac{\tilde{A}}{2}\exp\{2\delta\}\ \Lambda^2 \exp\left\{\frac{-2\theta}{\sqrt{\frac{\alpha}{\alpha_c}-1}}\right\} \left[\frac{(1-G)}{\sqrt{\frac{\alpha}{\alpha_c}-1}}\sin\theta + (1+G)\cos\theta\right] \tag{7}$$

$$\Sigma(0) = \exp\{\delta\}\ \Lambda \exp\left\{\frac{-\theta}{\sqrt{\frac{\alpha}{\alpha_c}-1}}\right\} , \tag{8}$$

where the renormalized parameters $\mu = \mu_0\Lambda$ and $G = [(G_0\Lambda^2)/\pi^2](\alpha_c/\alpha)$ have been introduced and reflect the anomalous dimensions of the mass and four fermion operators. Here $\tilde{A} \simeq 1.2$ is another parameter of the asymptotic expansion of the fermion self-energy function. There always exists one solution for θ (and hence $\Sigma(0)$) in the region $0 < \theta \leq \pi$ and this corresponds to the ground state solution. Once again, the existence of a non-trivial $\Lambda \to \infty$ limit requires that the gauge coupling approach the critical value $\alpha \to \alpha_c$. Thus the solution is similar to that of Ref. 10 except that θ need not be π. The approach of the gauge coupling to the critical point is now given by

$$\frac{\alpha}{\alpha_c} = 1 + \frac{\theta^2}{\ell n^2\left(\frac{\Lambda}{\kappa}\right)} , \quad \Lambda \to \infty , \tag{9}$$

so that $\Sigma(0) \to e^{\delta}\kappa$. The value of θ depends on the strength of the induced coupoing G. We shall see that the strong coupling phase of the theory corresponds to the ultraviolet stable fixed point with $G \to 1$ and $\alpha \to \alpha_c$.

The search for the fixed point structure can be conducted by examining the fermion-antifermion scattering amplitude (see Fig. 2). The four fermion interactions contribute to the scattering amplitude so that contributions from both the scalar and pseudoscalar channels must be included.

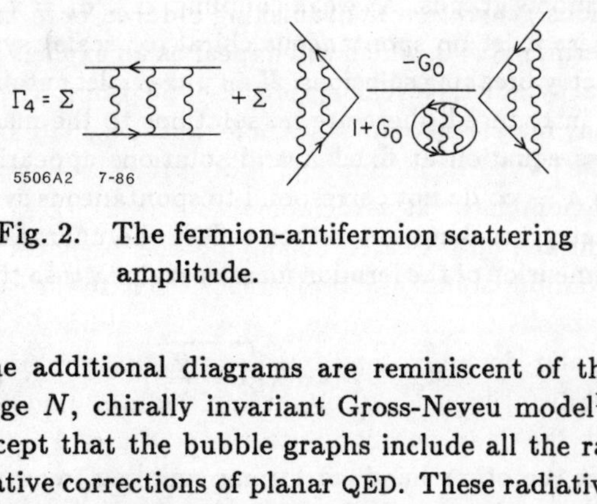

Fig. 2. The fermion-antifermion scattering amplitude.

The additional diagrams are reminiscent of the large N, chirally invariant Gross-Neveu model[11] except that the bubble graphs include all the radiative corrections of planar QED. These radiative corrections effectively make the four fermion interactins renormalizable at the fixed point. It is the presence of these diagrams which is at the origin of the running of the gauge coupling. Although the bubble diagrams are perturbatively quadratically divergent, the large anomalous dimensions allow for a precise determination of their contribution[2] yielding a well-defined four-point function.

The form of the four-point function allows a computation of the asymptotic behavior of the beta functions for both the gauge and four-fermion couplings near the ultraviolet fixed point $[\alpha \to \alpha_c^+,\ G \to 1]$ yielding

$$\beta_\alpha(\alpha, G) = \Lambda\frac{\partial\alpha}{\partial\Lambda} = \frac{-\frac{2\pi}{3}\left(\frac{\alpha}{\alpha_c}-1\right)^{3/2}}{\arctan\left(\frac{2\sqrt{\frac{\alpha}{\alpha_c}-1}}{G-1}\right)}$$

$$\beta_G(\alpha, G) = \Lambda\frac{\partial G}{\partial\Lambda} = \frac{-(G-1)\left(\frac{\alpha}{\alpha_c}-1\right)^{1/2}}{\arctan\left(\frac{2\sqrt{\frac{\alpha}{\alpha_c}-1}}{G-1}\right)} \quad (10)$$

where the angle $\theta = \arctan\{[2\sqrt{(\alpha/\alpha_c)-1}]/(G-1)\}$ is defined in the range $0 < \theta \leq \pi$. These β-functions are clearly nonperturbative and reflect the approach to the ultraviolet stable fixed point of the explicit solution. Moreover, the relevance of the four-fermion interactions is evident from the nontrivial fixed point value of $G = 1$.

The symmetry structure of the solution can also be gleaned from the bound state pole structure of the fermion-antifermion scattering amplitude. The pure ladder graphs do not contain any massless bound states since the four-fermion interactions generate a nonvanishing induced bare mass term, $m_0 \neq 0$, which will appear as an explicit chiral symmetry breaking in these diagrams. Hence, any massless bound state pole must originate from the bubble denominators. Indeed the pseudoscalar denominator at zero momentum vanishes in the chiral limit, clearly displaying the pseudoscalar Nambu-Goldstone boson associated with the spontaneous chiral symmetry breaking. However, the scalar denominator at zero momentum retains a nonvanishing contribution in the chiral limit even at the fixed point. Hence the status of the dilator remains unclear in this approximate treatment of a gauge theory. It is uncertain whether this result reflects a fundamental inconsistency of the quenched, planar approximation or is due to our analysis of the model. We strongly advocate that both the nontrivial mixing of the four-fermion operators and the fixed point structure of our solutions be checked by other methods including lattice calculations.

We anticipate that many of the general features obtained in the ladder model will continue to hold for gauge theories with running couplings possessing widely separated condensate scales. In such cases, provided large anomalous dimensions exist over a wide range of momenta, which is possible due to the slow running of the gauge coupling, the momentum dependence of induced fermion mass terms can be significantly affected. Such behavior may be applicable[12] to the resolution of the flavor changing neutral current problem in extended technicolor models.

REFERENCES

(1) W. A. Bardeen, C. N.Leung and S. T. Love, Phys. Rev. Lett. **56**, 1230 (1986).

(2) C. N. Leung, S. T. Love and W. A. Bardeen, Nucl. Phys. **B273**, 649 (1986).

(3) J. Kogut *et al.,* Phys. Rev. Lett. **48**, 1140 (1982); **50**, 393 (1983); Nucl. Phys. **B225**, 326 (1983); Phys. Lett. **145B**, 239 (1984); I. M.Barbour and C. J. Burden, Phys. Lett. **161B**, 357 (1985).

(4) T. E. Clark, C. N. Leung and S. T. Love, Purdue preprint PURD–TH–86–10.

(5) K. Kohnson, M. Baker and R. Willey, Phys. Rev. **136**, B1111 (1964); **163**, 1699 (1967).

(6) S. L. Adler and W. A. Bardeen, Phys. Rev. **D4**, 3045 (1971).

(7) T. Maskawa and H. Nakajima, Prog. Theor. Phys. **52**, 1326 (1974); **54**, 860 (1975).

(8) R. Fukuda and T. Kugo, Nucl. Phys. **B117**, 250 (1976).

(9) J. Bartholomew *et al.,* Nucl. Phys. **B230**, 222 (1984).

(10) V. A. Miransky, Nuov. Cim. **90A**, 149 (1985); P. I. Fomin, V. P. Gusynin, V. A. Miranski and Yu. A. Sitenko, Riv. Nuovo Cim. Soc. Ital. Fis. **6**, 1 (1983).

(11) D. J. Gross and A. Neveu, Phys. Rev. **D10**, 3235 (1974).

(12) T. Akiba and T. Yanagida, Phys. Lett **169B**, 432 (1986); K. Yamawaki, M. Bando and K.–ito Matumoto, Phys. Rev. Lett. **56**, 335 (1986); B. Holdom, Phys. Rev. **D24**, 1441 (1981).

Parallel Session 8

Searches for Quarks, Higgs Particles, Axions, Monopoles, Supersymmetric, and Technicolored Particles

Organizers:

B. Hollebeek (Pennsylvania)

S. Komamiya (Heidelberg)

Scientific Secretaries:

C. Hawkins (SLAC)

T. Steel (SLAC)

LOW THRUST HADRON EVENTS WITH ISOLATED μ OR e FROM PETRA

James G. Branson

Massachusetts Institute of Technology

In a search for new quark flavors, an excess of low thrust events containing muons has been found for PETRA energies greater than 46.3 GeV. The muons in these events are quite isolated from hadronic energy as would be expected from pair production of massive quarks. Combining data from all 4 PETRA experiments, there are 14 events with thrust less than 0.8 and the cosine of the angle between the muons direction and the thrust axis less than 0.7. Based directly on measurements taken at lower energy, the expected number of events is quite accurately predicted to be 2.4. The probability of a statistical fluctuation of this size or larger occuring in a given event selection is 3×10^{-7}. Although these events are consistent with the production of a new quark flavor, the evidence is as yet insufficient to prove that this is the source. No excess of events is seen in the selection of low thrust events with isolated electrons where the sensitivity is substantially lower.

INTRODUCTION

Since the beginning of 1979, when PETRA first ran succeeding DORIS as the highest energy electron positron collider in the world, the PETRA experiments have been searching for new kinds of particles and putting limits on their masses. Probably the most often repeated search, as the PETRA energy increased from 13 GeV to 46.78 GeV was that for a new quark flavor. New quark flavors were searched for in a scan for the 1S state of toponium, by looking for a step in R, and, most sensitively, by looking for a step in the rate of spherical hadronic events containing muons. This is probably the one search technique which is sensitive to a charge $-1/3$ quark as well as to the charge $2/3$ top quark.

In the upper 400 MeV of PETRA's energy range, some evidence was found for such spherical events containing muons, first by MARK-J which also found that the muons were quite isolated from hadronic energy, then by JADE which used the same cuts as MARK-J to add supporting evidence. The excess of events found is particularly striking because the expected background rate need not be computed from Monte Carlo simulations but can be directly measured from lower energy running using the same detectors, the same analysis and the same cuts. Previous measurements of muons in hadron events have been used to determine the semileptonic branching ratios in b and c quark decay and the fragmentation functions for those quarks [1].

MEASUREMENTS WITH MUONS

Early in 1984, PETRA was nearing the end of a long high energy scan. With great difficulty the machine group had raised the maximum center

30 MeV steps to search for toponium and for other new particles. As the highest energies were approached, the background conditions in the experiments became somewhat worse. It was clear that, with the luminosity dwindling, it was pointless to try to push the energy up further.

By the end of the scan, each group was able to rule out the existence of the 1S state of toponium in the scanned region. For example, MARK-J found that at the 95% confidence level, $B_h \cdot \Gamma_{ee} <$ 3.0 KeV [2] while 3.5 to 7.0 KeV is expected for toponium. No significant increase in R was seen.

Up to 46.3 GeV everything else appeared to be normal also. However, in the energy bin above 46.3 GeV, MARK-J saw a slight increase (2 sigma) in the rate of low thrust events [3] and a more significant increase in the rate of low thrust events containing muons. The expected background rate in this sensitive indicator of new physics was quite small so that even a few extra events could be significant. In order to get further information out of the small number of events found, MARK-J compared low thrust events from an open top quark pair production Monte Carlo to those low thrust events from the standard background processes which are weak decays of c and b quarks in hadron events having three jets. They found that the muons from the top quark decay should be isolated while the muons from the background processes nearly always occur inside a hadron jet. Trying several variables to compare the isolation of the excess events above 46.3 GeV to those below, MARK-J found that the high energy events had isolated muons like in the top Monte Carlo. The variables tried were the energy in a 30 degree cone around the muons direction, the thrust using the muon as the thrust axis, the energy flow vs. angle from the muon direction, and the angle between the muon direction and the thrust axis. All of these gave similar results and showed no significant deviation from the top Monte Carlo. At this point MARK-J presented its data to the DESY community and requested an additional 3 pb^{-1} running at 46.57 GeV and the DESY directorate agreed to 1 pb^{-1}.

By the end of the additional running, for center of mass energies larger than 46.3 GeV, MARK-J had 8 events with thrust less than 0.8, the standard cut used for searching for open production of top. Just 1.9 events were expected based on previous data. The probability of observing 8 or more when 1.9 are expected is 8×10^{-4}. When dealing with the statistics of small numbers one has to be quite careful that physicists did not unwittingly stack the odds in favor of an anomaly. Were the cuts chosen to maximize the signal? How many variables were checked before finding one anomaly? Is the small background estimated correctly? In this case the thrust cut in the hadron events containing muons was set up before seeing any signal. Results from the same cuts were published in other energy ranges to rule out open production of top. There are only a few new particle searches that were checked as consistently as this one. In addition the 46.3 GeV cut was set up before seeing the data. In fact our lowest energy event comes at 46.39 GeV.

On the other hand, adding an isolation criterion was done after seeing the data and noting that the muons were isolated. Therefore statistics using this cut could be questioned somewhat as far as the MARK-J data are concerned. Data from the other experiments were not seen in setting up this cut and are therefore independent.

Figure 1 shows the plot of events in the thrust $\cos\delta$ plane, where δ is the angle between the thrust axis and the muon's direction. The region with thrust less than 0.8 and $\cos\delta$ less than 0.7 is used to select low thrust hadron events with isolated muons. Two energy regions are shown in the figure, the scan below 46.3 GeV and above 46.3 GeV. Higher statistics data taken around 36 GeV agree well with the lower energy scan data as do Monte Carlo simulations with 5 quark flavors produced. Above 46.3 GeV we select 7 events while 0.8 are expected based on previous data. The probability of finding 7 or more with 0.8 expected is just 2×10^{-5}.

The higher than normal background during the high energy running did not affect the measurement of the muons or of the thrust of the events. This was checked by overlaying beam gate triggers with actual events and finding that the thrust and muon identification did not change. Indeed the muon identification in these events with isolated muons can be checked much better than for the normal kind of hadron event containing muons because in the normal event, the muon is in a jet and not only do decay and punchthrough become more probable, but matching the muon chamber track with data from the inner detector is more difficult. In all of the 8 events found above 46.3 GeV, the muon is clearly identified and gives consistent data near the vertex, after 1 interaction length of shower counters, after 3 interaction lengths and outside the 6 interaction lengths of the hadron absorber where its trajectory is measured for momentum fitting.

Table I gives the detailed parameters of the events with thrust less than 0.8 and center of mass energy greater than 46.3. The lowest energy event is at 46.39 GeV, well above the lower limit of the energy bin. The fraction of visible energy is normal for either hadron events or hadron events including muons, since the resolution is nearly 20%. O_B the broad jet oblateness is large if the event is planar. The cut O_B greater than 0.3 was used to select planar events. These events tend to be quite planar. The average value of O_B agrees with what we would expect from the dominant background and is nearly 2 standard deviations higher than what our Monte Carlo predicts for a t quark or a b' quark. The muon momentum distribution is as expected for a heavy quark where most of the muons come from a cascade decay but some high energy muons can come from the primary decay of the heavy quark. The transverse momentum relative to the thrust axis is quite high as we expect for isolated muons. The muons momentum out of the event plane is also quite high unlike the expected background. That is, although the events are planar like the expected background, the muon is not in the plane as expected. The cosine of the angle from the thrust axis is small indicating large angles with the thrust axis and isolated muons.

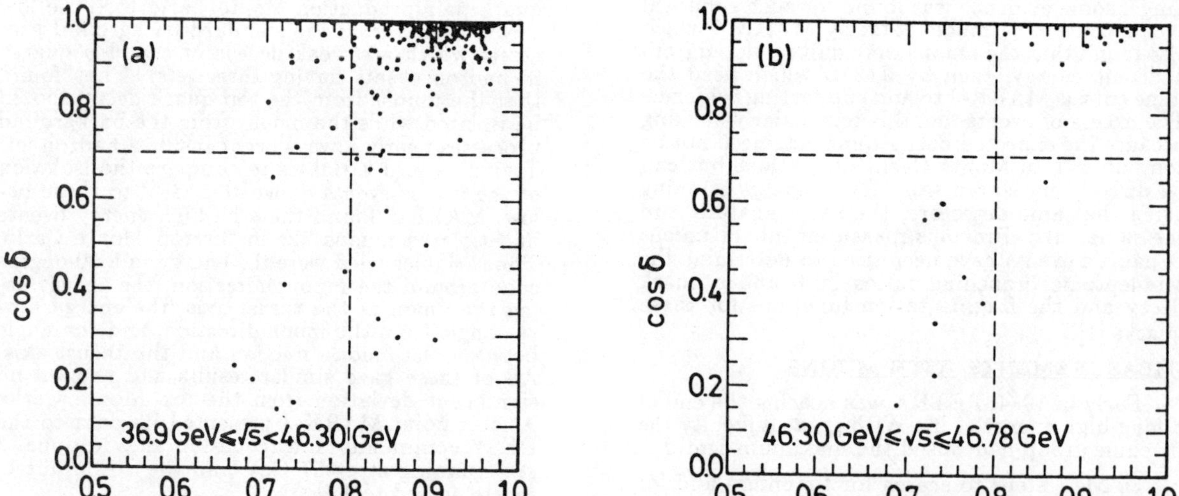

FIGURE 1: Cosδ vs. thrust for Mark-J events containing muons.

	$\sqrt{s}$	$E_{vis}/\sqrt{s}$	Thrust	O_B	P_μ	$P_{\perp\mu}$	$P_{out\ \mu}$	$cos\delta$	e or γ
A	46.39	0.74	0.79	0.25	2.8	2.6	1.1	0.38	14 GeV γ
B	46.57	0.89	0.74	0.49	2.4	1.9	0.4	0.60	12 GeV e
C	46.57	0.63	0.72	0.26	3.0	2.5	2.2	0.56	10 GeV e or γ
D	46.69	0.83	0.69	0.36	4.2	4.2	3.6	0.16	
E	46.78	0.89	0.73	0.38	2.2	1.8	1.0	0.58	
					4.5	4.2	1.7	0.34	
F	46.57	0.80	0.73	0.33	16.1	15.7	7.7	0.22	
G	46.57	1.08	0.77	0.41	2.7	2.4	0.9	0.44	
H	46.57	1.05	0.79	0.18	2.2	0.8	0.8	0.93	

TABLE I: Parameters of MARK-J events.

Although MARK-J's electron and photon identification is not ideal, high energy electromagnetic showers can be clearly seen in 3 of the events. This is unusual for hadron events in general and also for those containing muons although it could be due to photon radiation in the initial state.

The other 3 experiments at PETRA did not find supporting evidence for the effect found by MARK-J and, therefore, the PETRA machine was modified to go back to lower energy and higher luminosity. The situation would have been left there but for the further analysis of JADE. During the high energy running JADE had installed a new vertex chamber and was devoting most of its effort to that. However, later analysis [4] showed that JADE also had a signal at energies above 46.3 GeV.

Figure 2 shows the cross sections measured by JADE and MARK-J for production of hadron events with thrust less than 0.8 and $\cos\delta$ less than 0.7. Both groups find a substantially larger cross section in the energy bin above 46.3 GeV. The error bars in the figure are based simply on the number of events in the bin, not on the expected number.

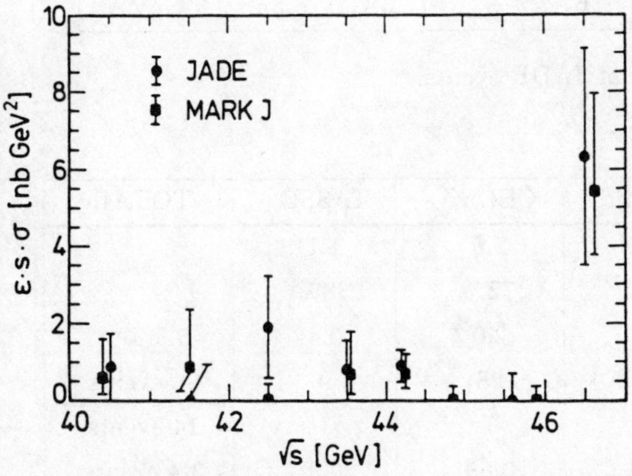

FIGURE 2: Cross section for events with thrust < 0.8 and $\cos\delta < 0.7$ vs. energy from Jade and Mark-J.

Figure 3 shows the JADE data in the thrust vs. $\cos\delta$ plane. Below 46.3 GeV, the JADE data look quite similar to that from MARK-J as shown in figure 1. Above 46.3 GeV JADE has 5 events in the selection region. The distribution of events in the plane is notably different at high energy as it was for MARK-J. Based on the lower energy data JADE expects 0.56 events. They therefore have an excess of about the same statistical significance as did MARK-J. This is indeed a completely independent confirmation, because the cuts are identical and because JADE did not do this analysis until long after MARK-J had presented their final results.

Table II shows the details of the 5 events from JADE. Again the actual center of mass energies at which the events are seen starts well above 46.3 GeV. The muon momentum distribution is again as we would expect from t or b' decay, this time with 2 very high momentum muons which are unusual from conventional sources. Three events have now clearly identified electrons or photons with high momentum. In the JADE events, these are at large angle to the beam direction. 4 of the 5 events are quite planar when the muon is excluded from the analysis again like the MARK-J events. In general, the similarity between the events seen by MARK-J and those seen by JADE is quite amazing. Even assuming both were detecting the same new phenomenon, one would expect some fluctuations in the events seen. It should be noted that the 5 events listed below pass all of JADE's cuts used to select inclusive muon events for new particle searches. More restrictive cuts, used to study the decays of charm and bottom quarks, eliminate 3 of the events mainly because they go into less accurate regions of the detector.

The other 2 experiments have also searched for low thrust hadron events containing muons using the same cuts as did MARK-J. They have found no large excess of events. The results of all 4 experiments [5] are combined and compared in Table III. To estimate the sensitivities of each experiment to this type of event which is from an unknown source, I have simply used the sensitivities to the standard inclusive muons from b and c decay. This should be approximately accurate since in general the muon identification cuts are not complicated and depend mainly on solid angle coverage.

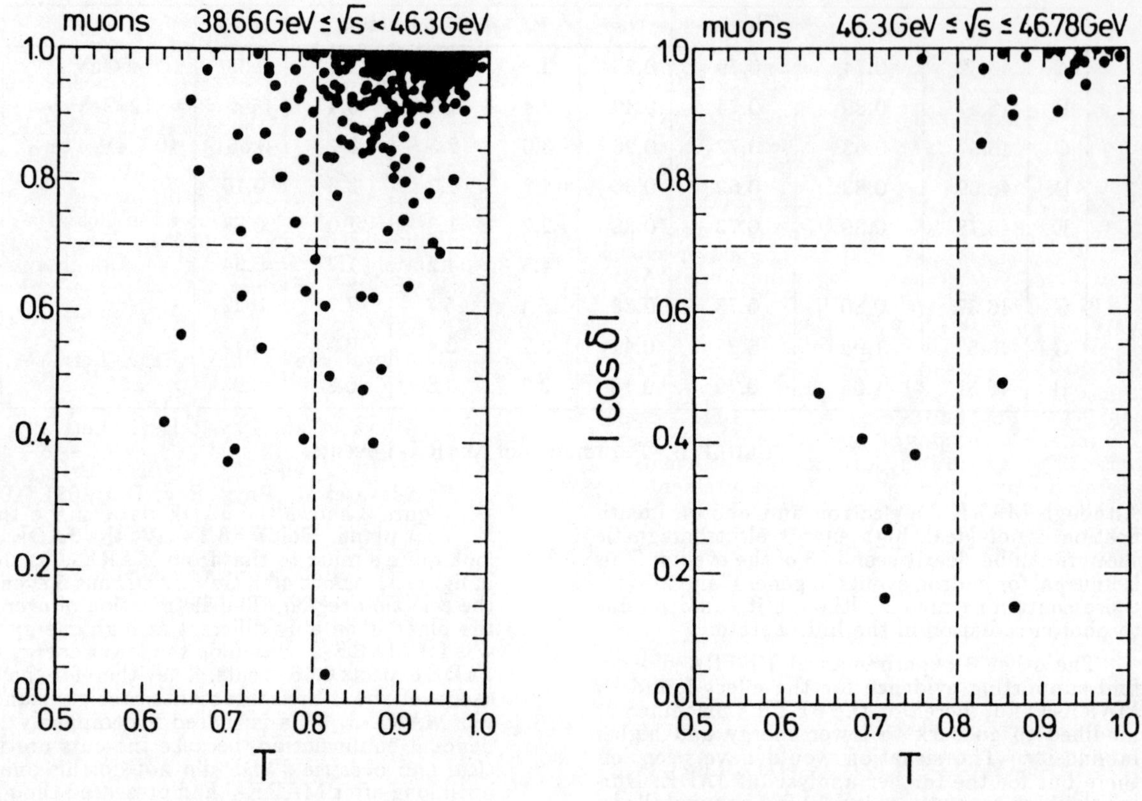

FIGURE 3: Cosδ vs. thrust for Jade events containing muons.

	$\sqrt{s}$	Thrust	P_μ	δ	e or γ
1	46.45	0.72	$15.2^{+3.7}_{-2.5}$	74.3	10.7 GeV e or γ
2	46.54	0.72	$2.5^{+0.1}_{-0.1}$	80.8	
3	46.57	0.69	$2.0^{+0.1}_{-0.1}$	65.8	
4	46.57	0.64	$21.8^{+19.2}_{-6.8}$	61.3	13.4 GeV γ
			$3.2^{+0.3}_{-0.3}$		
5	46.57	0.75	$4.1^{+0.2}_{-0.2}$	67.7	12.3 GeV e and 9.6 GeV γ

TABLE II: Parameters of JADE events.

	MARK-J	JADE	CELLO	TASSO	TOTAL
Luninosity (pb^{-1})	2.8	1.7	2.8	1.1	
Acceptance	70%	62%	72%	38%	
Lumi. × Accept.	2.0	1.1	2.0	0.4	5.5 pb^{-1}
Number of μ's	31	32	28	9	96 events
N_{evt} (T < 0.8, cosδ < 0.7)	7	5	1	1	14 events
N_{evt} expected	0.8	0.56	0.65	0.3	2.4 events
Obs. excess σ	3.1 ± 1.0	4.0 ± 1.4	0.2 ± 1.0	1.8 ± 2.3	2.0 ± 0.6 pb

TABLE III: Summary of measurements from PETRA.

The table shows that TASSO had substantially less sensitivity than did the other 3 experiments and in fact that the excess cross section, which is the cross section in addition to the expected background rate, observed by TASSO is nearly equal to the overall average. Therefore we are only faced with the question of whether the result from CELLO is consistent with that from MARK-J and JADE. To address this I have calculated the errors on the observed excess cross section by assuming that the average over experiments was the correct number. Therefore MARK-J and CELLO which have the same sensitivity have the same error on the cross section although MARK-J observes more events. Thus we can test whether all the experiments are consistent with the average of 2.0 pb. Certainly MARK-J, JADE, and TASSO are quite consistent. CELLO is nearly 2 standard deviations off but with 4 measurements this is not an unlikely distribution of events. I therefore conclude that the 4 measurements are consistent as far as can be tested and that combining the results of the 4 experiments is justified and is the best way of judging whether there is a significant signal.

The signal then is 14 events found with 2.4 expected from standard sources measured at lower energy. This size of fluctuation has a probability of 3×10^{-7} which, even given the bias of physicists toward finding something, makes it quite unlikely that this is a fluctuation.

MEASUREMENTS WITH ELECTRONS

There is no corresponding signal found in low thrust hadron events containing isolated electrons. Here the total luminosity times acceptance is 1.9 pb^{-1} and the number of events found by 3 experiments is 0 with 1.2 events expected from standard sources and an additional 3.9 events expected to agree with the signal observed in the muon channel. The probability of observing 0 when 5.5 is expected is just 6×10^{-3}. Thus the electron data is in some disagreement with the muon data although the electron statistics are still poor. It should be emphasized, however, that electron identification has much more finely tuned cuts and depends more on the type of event that is being observed. Remember that the JADE and MARK-J events with muons probably also have electrons in them which fail some identification criterion, like the cuts removing photon conversions into pairs.

CONCLUSION

There is a quite significant excess of low thrust hadron events containing muons at the highest PETRA energies. The source of these events can not be determined with the statistics available but the most conventional would be a new charge $-1/3$ quark with a mass around 23 GeV. However, the observed events tend to be a bit more planar than expected from this source and the lack of a corresponding electron signal tends to discourage this explanation. There is more than enough evidence to warrant further study of this energy region.

REFERENCES

[1] B. Adeva et al., Phys. Rev. Lett. 51, 443 (1983).

[2] B. Adeva et al., Phys. Rev. Lett. 53, 134 (1984).

[3] B. Adeva et al., Phys. Rev. D 34, 681 (1986).

[4] M. Kuhlen, DESY 86-052(1986).

[5] TASSO collaboration and CELLO collaboration, private communication.

HEAVY LEPTONS IN 1986*

Martin L. Perl

Stanford Linear Accelerator Center
Stanford University, Stanford, California 94305

INTRODUCTION

In this paper, presented at the XXIII International Conference on High Energy Physics, I update recent reviews[1,2] on the search for new leptons beyond the electron, muon, and tau generations. I also discuss an unexplored region in the mass ranges of leptons: small visible energy events in electron-positron annihilation and close-mass pairs.

CHARGED LEPTONS: STABLE OR SEQUENTIAL

The traditional, comprehensive way[3,4] to search for charged leptons uses the reaction (Fig.1a)

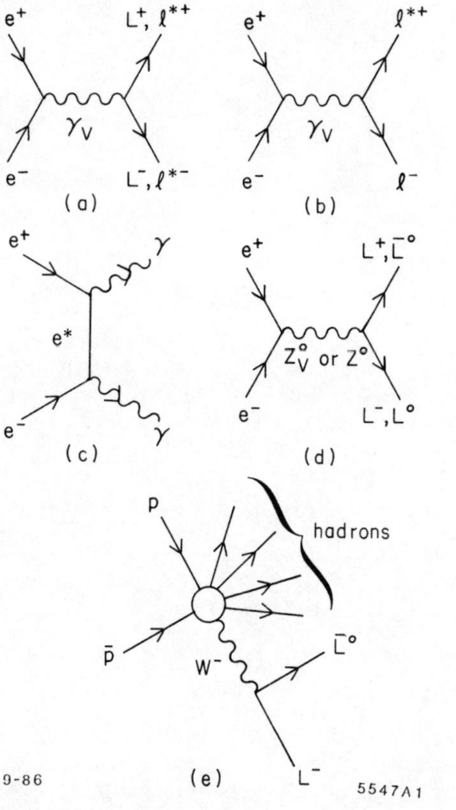

Figure 1.

*This work was supported in part by the Department of Energy, contract DE-AC03-76SF00515 (SLAC).

$$e^+ + e^- \rightarrow \gamma_{virtual} \rightarrow L^+ + L^- \quad (1)$$

If the $L^\pm$ is stable, this gives an easily recognized final state. If the $L^\pm$ is sequential, it can be recognized as the τ was found[4], through the decays

$$L^- \rightarrow \nu_L + \ell^- + \bar{\nu}_\ell \ , \quad \ell = e, \mu, \tau \quad (2a)$$
$$L^- \rightarrow \nu_L + \text{hadrons} \quad (2b)$$

This requires the mass, m_{ν_L}, of the L associated neutrino obey

$$m_{\nu_L} < m_L \quad (3)$$

In existing searches, m_{ν_L} is supposed negligible compared to m_L, a point I return to at the end of the paper.

The lower limits obtained in $e^+e^- \rightarrow L^+L^-$ searches on a new stable or sequential lepton have not changed in the last year as summarized by Komamiya[1].

$$m_{L^\pm} \text{ (stable)} < 21.1 \text{ GeV/c}^2, 95\% \text{ CL} \quad (4a)$$
$$m_{L^\pm} \text{ (sequential)} < 22.7 \text{ GeV/c}^2, 95\% \text{ CL} \quad (4b)$$

First results using a new search method for sequential leptons (Fig.1e)

$$\begin{aligned} p + \bar{p} &\rightarrow W^- + \text{hadrons} \\ W^- &\rightarrow L^- + \bar{\nu}_L \\ L^- &\rightarrow \nu_L + \text{hadrons} \end{aligned} \quad (5)$$

were reported[5] at this conference and in a recent preprint.[6] The search used the UA1 experiment at the CERN $\bar{p}p$ collider. The preliminary result is no new charged, sequential leptons with masses less than about 40 GeV/c^2.

CHARGED LEPTONS: EXCITED

Charged leptons that decay electromagnetically

$$\ell^{*-} \rightarrow \ell^- + \gamma \ , \ \ell = e, \mu, \tau \quad ; \quad (6)$$

(Fig.1a) usually called excited; can be produced and detected through[7,8]

$$e^+ + e^- \to \ell^{*+} + \ell^{*-} \to \ell^+ + \ell^- + \gamma + \gamma \quad (7a)$$

and[7,8,9] (Fig.1b)

$$e^+ + e \to \ell^{*+} + \ell^- \to \ell^+ + \ell^- + \gamma \quad (7b)$$

The e^* will also contribute[7] to

$$e^+ + e^- \to \gamma + \gamma \quad (7c)$$

through the diagram in Fig.1d. In this case and in the reaction in Eq.7b the cross section depends on the unknown constant λ in the $\ell^* - \ell - \gamma$ vertex function, usually written

$$\text{Vertex Function} = e\left(\frac{\lambda}{2M_{\ell^*}}\right)\bar{u}_{\ell^*}\sigma_{\mu\nu}u_\ell F^{\mu\nu} \quad (8)$$

Thus searches using $ee \to \ell\ell\gamma$ or $ee \to \gamma\gamma$ are ambiguous if λ is taken to be very small. Even a search using the reaction in Eq.7a can be made ambiguous by attributing a sufficiently small form factor F_{ℓ^*} to the $\ell^* - \ell^* - \gamma$ vertex.

No excited leptons have been found. Using the reaction in Eq.7a, $ee \to \ell^*\ell^* \to \ell\ell\gamma\gamma$, and assuming F_{ℓ^*} sufficiently close to 1., the CELLO experiment[7] at PETRA found the 95% CL mass limits:

$$\begin{aligned} M_{e^*} &> 23.0 \text{ GeV/c}^2 \\ M_{\mu^*} &> 23.0 \text{ GeV/c}^2 \\ M_{\tau^*} &> 22.7 \text{ GeV/c}^2 \end{aligned} \quad (9)$$

Mass limits using the reactions in Eqs.7b and 7c depend on λ. Figure 2 gives the recent results of the CELLO experiment on M_{e^*}; Fig.3 gives recent results from the JADE experiment[8] at PETRA on M_{τ^*}. Other lower limits on M_{e^*} and M_{μ^*} have been obtained by Grifols and Persis[10] using

$$\nu_\mu + e^- \to \nu_\mu + e^-, \bar{\nu}_\mu + e^- \to \bar{\nu}_\mu + e^- \quad (10)$$

NEUTRAL LEPTONS: NO GENERATION MIXING

Consider four types of neutral leptons: (*i*) an L^0, L^- pair with the latter heavier

$$(L^0, L^-), \quad M_- > M_0 \quad ; \quad (11a)$$

(*ii*) an L^0, L^- pair with the former heavier

$$(L^0, L^-), \quad M_0 > M_- \quad ; \quad (11b)$$

(*iii*) a pair of neutral leptons

$$(L^0, L^{0\prime}), \quad M_0 > M_{0'} \quad ; \quad (11c)$$

or (*iv*) an isolated neutral lepton

$$L^0 \quad (11d)$$

In the pair cases, assume the heavier member decays always to the lighter member plus other particles, and there is no mixing between this new pair and any other lepton or pair of leptons.

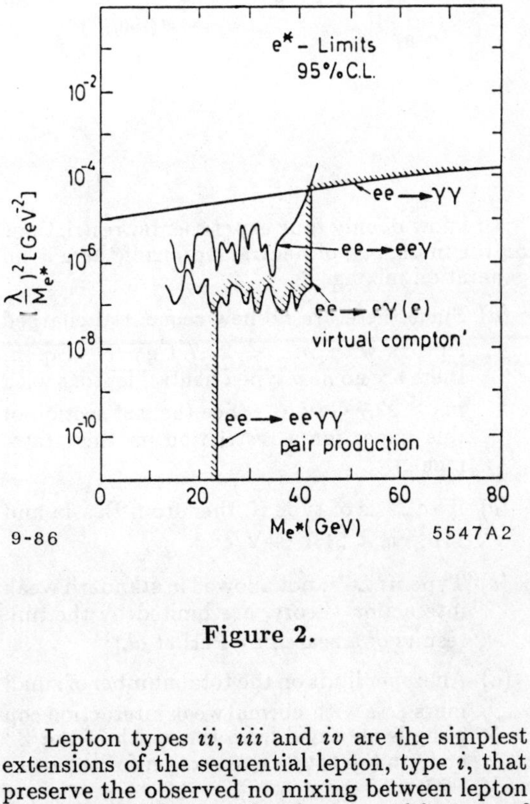

Figure 2.

Lepton types *ii*, *iii* and *iv* are the simplest extensions of the sequential lepton, type *i*, that preserve the observed no mixing between lepton generations. Yet searches for neutral leptons of these types are scattered and incomplete. This is because electron-positron colliders have not had sufficient energy to use effectively the comprehensive L^0 production reaction (Fig.1d)

$$e^+ + e^- \to Z^0_{real\ or\ virtual} \to L^0 + \bar{L}^0 \quad (12)$$

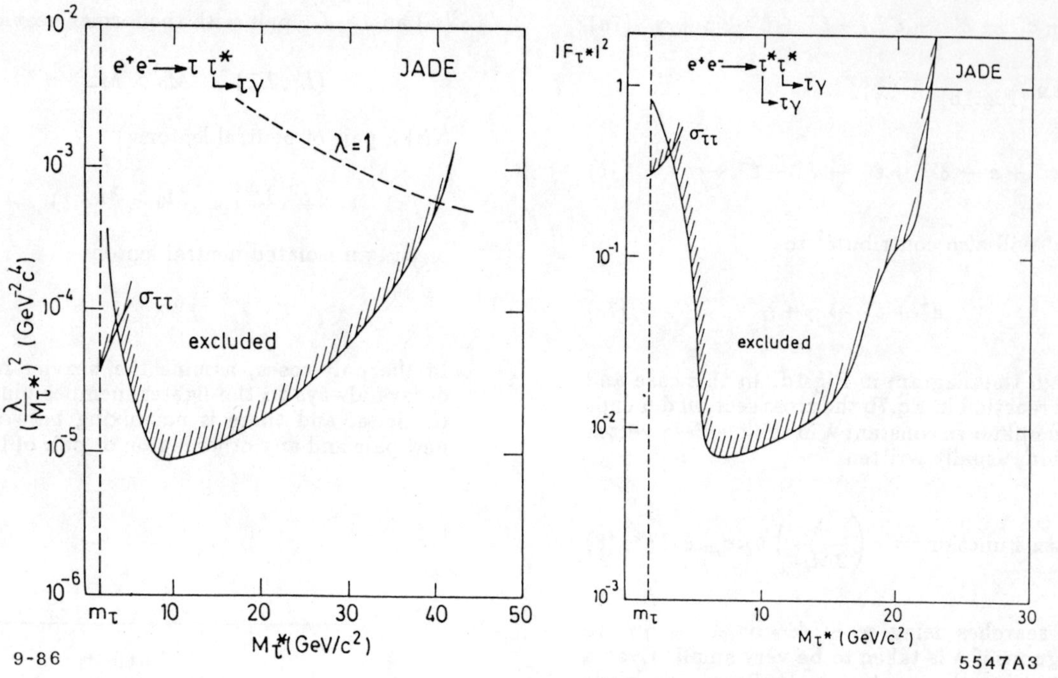

Figure 3.

I know of only four experimental restrictions on the mass, m_0, of neutral leptons if there is no generation mixing.

(*a*) Since there are no new sequential charged leptons with $m_- < 22.7$ GeV/c², Eq.4b, there are no new type *i* neutral leptons with $m_0 \lesssim 22.7$ GeV/c². (See the last section of this paper for a restriction on this statement.)

(*b*) If an L^0 is of type *ii*, then from Eqs.4a and 11b, $m_0 \gtrsim 21.1$ GeV/c².

(*c*) Type *iii* L^0's, not allowed in standard weak interaction theory, are limited by the null results of a search by Perl *et al.*[11]

(*d*) An upper limit on the total number of small mass L^0's with normal weak interaction coupling is obtained from the width of the Z^0 or from the total cross section for the reaction

$$e^+ + e^- \rightarrow \gamma + L^0 + \bar{L}^0 \tag{13}$$

Here L^0 includes ν_e, ν_μ, and ν_τ. These limits are discussed by Whitaker[12] in these proceedings.

Definitive searches for neutral leptons will be carried out in the next five years at TRISTAN, the SLAC Linear Collider (SLC), and LEP using the reaction in Eq.12.

NEUTRAL LEPTONS: GENERATION MIXING

Another extension of the sequential lepton model, which I find less attractive,[13] is mixing between lepton generations. This allows an $L^0 - W - \ell^-$ coupling where ℓ is an e, μ, or τ. There are then numerous ways to find the L^0, I give some examples taking $\ell = e$.

(*a*) Electron spectrum in meson decay:

$$\pi^- \rightarrow e^- + L^0 \quad , \quad K^- \rightarrow e^- + L^0 \tag{14}$$

(*b*) L^0 production via meson decay and subsequent L^0 decay:

$$\begin{aligned} D^- &\rightarrow \bar{L}^0 + e^- + \text{hadrons} \\ \bar{L}^0 &\rightarrow e^+ + \text{particles} \end{aligned} \tag{15}$$

(*c*) L^0 pair production via e^+e^- annihilation and subsequent L^0 decay

$$\begin{aligned} e^+ + e^- &\rightarrow L^0 + \bar{L}^0 \\ L^0 &\rightarrow e^- + \text{particles} \end{aligned} \tag{16}$$

(*d*) $L^0 - \bar{\nu}_e$ production via e^+e^- annihilation and subsequent L^0 decay

$$e^+ + e^- \rightarrow L^0 + \bar{\nu}_e \qquad (17)$$
$$L^0 \rightarrow e^- + \text{particles}$$

A thorough review of all such searches has been done by Gilman.[2] Additional searches are reported in Refs.14 and 15. There is no evidence for the existence of neutral leptons that have generation mixing. Some searches extend to about 35 GeV/c^2, the method in Eq.17 being used. The sensitivity of the various searches depends on U, the ratio of the strength of the $L^0 - W - \ell^-$ vertex to the standard $\nu_\ell - W - \ell^-$ vertex. $|U|^2$ values as small as 10^{-8} has been explored, but in some L^0 mass regions the minimum values of $|U|^2$ are about 10^{-2}.

SMALL VISIBLE ENERGY EVENTS IN e^+e^- ANNIHILATION

Experimental studies at storage rings of electron-positron interactions in the range of 10 to 50 GeV total energy (E_{tot}) mostly fall into two classes: (I) annihilation interactions or (II) two-virtual-photon interactions. Annihilation interactions take place through

$$e^+ + e^- \rightarrow \gamma_{virtual} \text{ or } Z^0_{virtual}$$
$$\rightarrow \text{particles} \quad . \qquad (18a)$$

It is conventional to also include in class I the reactions

$$e^+ + e^- \rightarrow e^+ + e^- \qquad (18b)$$

and

$$e^+ + e^- \rightarrow \gamma + \gamma \qquad (18c)$$

Class II, the two-virtual photon interactions, mostly contains

$$e^+ + e^- \rightarrow e^+ + e^- + \ell^+ + \ell^- \ , \ \ell = e, \mu, \tau \qquad (19a)$$

and

$$e^+ + e^- \rightarrow e^+ + e^- + \text{hadrons} \qquad (19b)$$

Much of the time the final e^+ and e^- are produced at small angles to the beam line and are not detected in the main part of the apparatus. Hence most class II events have small visible energy, relative to E_{tot}. The visible energy, E_{vis}, is the sum of the energy in the charged tracks and the energy of the photons which convert in the electromagnetic calorimeters.

In the E_{tot} range of 10 to 50 GeV, the cross section for class II events is much larger than the cross section for class I events. Therefore most experimental studies of class I interactions require that the events in the data sample have $E_{vis} > E_{vis,min}$, where $E_{vis,min}$ is a minimum acceptable value of E_{vis}. In events with just two charged particles $E_{vis,min}$ may be as small as several GeV, but in studies of hadron production via annihilation $E_{vis,min}$ may be as large as $E_{tot}/4$.

Consider the production of a pair of charged particles

$$e^+ + e^- \rightarrow x^+ + x^- \qquad (20a)$$

and suppose the $x^\pm$ has three properties:

(i) The x^- decays through the weak interaction to an x^0

$$x^- \rightarrow x^0 + \text{other particles} \qquad (20b)$$

(ii) The x^0 is stable.

(iii) The masses of the x^- and x^0, m_- and m_0 respectively, are close to each other. For example

$$m_- - m_0 \lesssim 2 \text{ GeV/c}^2 \qquad (21)$$

Then the reaction sequence in Eq.20 may produce events with E_{vis} values too small to be accepted in class I studies. On the other hand such events may be difficult or impossible to find when the criteria for class II studies are applied, because of the large cross sections of conventional class II events.[17]

Thus a process described by Eq.20 with the mass condition of Eq.21 could remain undetected in present electron-positron interaction studies.

CLOSE-MASS LEPTON PAIRS

Searches for a new sequential charged lepton, L^-, have assumed the associated neutral lepton, L^0, has negligible mass compared to the L^- mass. I recently began to consider the effect on the sensitivity of these searches of m_0 being close to m_-, for example, the effect of the mass condition in Eq.21. A thorough reevaluation is required of the data used for the searches, the analysis being carried out with m_0 allowed to vary; we have just started such an analysis using data from the Mark II detector at PEP. However data published for other purposes can be used to sketch the region in m_- and m_0 values which requires exploration.

It is convenient to use a plot, Fig.4, with the mass difference

$$\delta = m_- - m_0$$

and the L^- mass, m_-. Four regions can be excluded.

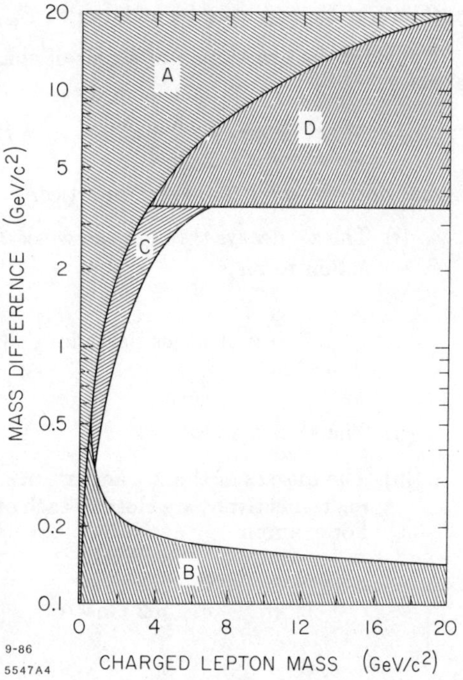

Figure 4.

Region A: By definition $\delta \leq m_-$, therefore the $\delta > m_-$ region is excluded.

Region B: Assuming the usual V-A weak interaction, when δ is of the order of the pion mass or smaller, the L^- decay length becomes of the order of a meter or longer. Therefore many of the $L^\pm$ would appear stable in the JADE collaboration searches[1,18] at PETRA for stable leptons. Using this observation and the 21.1 GeV/c² lower mass limit in Eq.4a, Region B is excluded.

Region C: Consider the decay $L^- \to L^0 + a + b + \ldots$ In the L^- rest frame the L^0 energy is

$$E^0 = (m_-^2 + m_0^2 - m^2)/2m_- \qquad (22)$$

where m is the invariant mass of the set of particles $a + b + \ldots$ The searches already carried out for sequential leptons would not be sensitive to a small change in E^0, say 20% or 30%, from the standard $m_0 = 0$ case. To be conservative, say 20%. Then the null results of the searches are still true for $m_0^2 \lesssim 0.2\ m_-^2$. This gives the boundary of Region C. The boundary position is conservative, probably the boundary should be more to the right.

Region D. The JADE collaboration[18] has looked for the supersymmetric particles called charginos with decays

$$\begin{aligned} \tilde{\chi}^- &\to \tilde{\gamma} + \ell^- + \bar{\nu}_\ell \\ \tilde{\chi}^- &\to \tilde{\gamma} + \text{hadrons} \end{aligned} \qquad (23)$$

where $\tilde{\chi}^-$ is the chargino and $\tilde{\gamma}$ is a stable photino. In this search the $\tilde{\gamma}$ was allowed to have a nonzero mass, and $\tilde{\chi}^- - \tilde{\gamma}$ mass differences as small as 3 GeV/c² were explored. Decay kinematics are similar for $\tilde{\chi}^-, \tilde{\gamma}$ pairs and L^-, L^0 pairs with the same masses. Therefore I use this null chargino search to set an upper limit $\delta > 3.5$ GeV/c² for the L^-, L^0 case. This forms the lower boundary of Region D.

Figure 4 shows the region of δ and m_- were an $L^- - L^0$ pair might exist and not have been found if the L^0 were stable. I emphasize that the boundaries of this region were deduced crudely from other people's data. The true unexplored region may be smaller or longer. Figure 4 should be regarded as a guide to what needs to be done in looking for a sequential lepton pair with a nonzero mass neutral member.

ACKNOWLEDGEMENT

I am indebted to T. Barklow for valuable conversations on small visible energy events and for directing me to the chargino search data. I also received a valuable comment on Fig.4 from H. Harari.

REFERENCES

1. S.Komamiya, Proc.1985 Int.Sym. on Lepton and Photon Interactions at High Energies (Kyoto, 1985), ed. by M.Konuma and K.Takahashi, p.612.
2. F.J.Gilman, SLAC-PUB-3898 (1986), submitted to Comments in Nuclear and Particle Physics.
3. M.Bernardini *et al.*, A proposal to search for leptonic quarks and heavy leptons produced by ADONE, INFN/AE-67-3 (1967); M.Bernardini *et al.*, Nuovo Cimento 17A, 383 (1973).
4. M.L.Perl *et al.*, Phys.Rev.Lett.35, 1489 (1975).
5. A.Homna, Proc. XXIII Int. Conf. on High Energy Physics, (Berkeley, 1986).

6. D.B.Cline and M.Mohammadi, Univ.of Wisconsin preprint WISC-EX-86-269 (1986).
7. H.-J. Behrend *et al.*, Phys.Lett. 168B, 420 (1986).
8. W.Bartel *et al.*, Z.Phys. C31, 359 (1986).
9. M.J.Jonker, Perspectives in Electroweak Interactions and Unified Theories, XXI Recontre de Moriond (Les Arcs, 1986).
10. J.A. Grifols and S.Peris, Phys.Lett. 168B, 264 (1986)
11. M.L.Perl *et al.*, Phys.Rev. 320, 2859 (1985).
12. J.S.Whitaker, Proc. XXIII Int.Conf. on High Energy Physics (Berkeley, 1986).
13. R.Aleksan, Massive Neutrinos in Astrophysics and in Particle Physics (Editions Frontières, 1986), Ed. by O.Fackler and J.Trân Thanh Van.
14. M.Shaevitz *et al.*, Proc. 21st Rencontre de Moriond (Les Arcs, 1986).
15. M.Gronau, C.N.Leung, and J.L.Rosner, Phys. Rev. D29, 2539 (1984).
16. For an experimental study of the case of exactly two charged particles and no photons see M.L.Perl *et al.*, SLAC-PUB-3847 (1985), to be published in Phys.Rev.
17. W.Bartel *et al.*, Phys.Lett. 123B, 353 (1983).
18. W.Bartel *et al.*, Z.Phys. 29C, 505 (1985).

Single Photon Production in e^+e^- Annihilation

Scott Whitaker
Physics Department, Boston University
Boston, MA 02215

Results are presented from the ASP, MAC, CELLO, and Mark J detectors for the production of single photons in electron positron annihilation. Upper limits on the production of single photons are expressed as upper limits on the number of light neutrino species and lower limits on the masses of supersymmetric particles. The ASP experiment has determined a limit on the selectron mass of 65.7 GeV for the case of degenerate selectron masses and massless photinos, and has determined a limit of 7.5 for the number of light neutrino species. The combined 90% confidence limit from the ASP, MAC, and CELLO experiments for the number of light neutrino species is 4.9.

I will present in this talk some results on the production of single photons in electron-positron annihilation. The results come from the ASP[1] and MAC[2] experiments at PEP and from CELLO[3] and Mark J[4] at PETRA; I will emphasize the results from ASP since they are the most sensitive measurements in this channel.[5]

"Single photon production" refers to the production of a final state consisting entirely of one photon; the photon energy and its angle from the beam lines must pass selection cuts which depend on the detector resolution and coverage. Detectors covering as completely as possible the solid angle about the interaction point are used to reject the possibility of other detectable particles being present. The single photon carries transverse momentum with respect to the beam line; this transverse momentum must be balanced by a particle or particles which, if coverage is complete, will traverse the experiment's live detectors. The observation of a single photon is then the nonobservation of the balancing particles, a signal of the production of particles which interact only very weakly in matter. In practice, there must be gaps in the detector coverage along the beam lines; this gap can be expressed as a minimum angle θ_{Veto} with respect to the beam line, above which detector coverage is complete. A "kinematic" veto is established if the experimental energy and angle cuts selecting the single photon are such that in the worst case, where the balancing transverse momentum is carried equally by just two particles, those particles will be constrained by energy and momentum conservation to have angles with respect to the beam lines greater than θ_{Veto}. This situation is illustrated in figure 1.

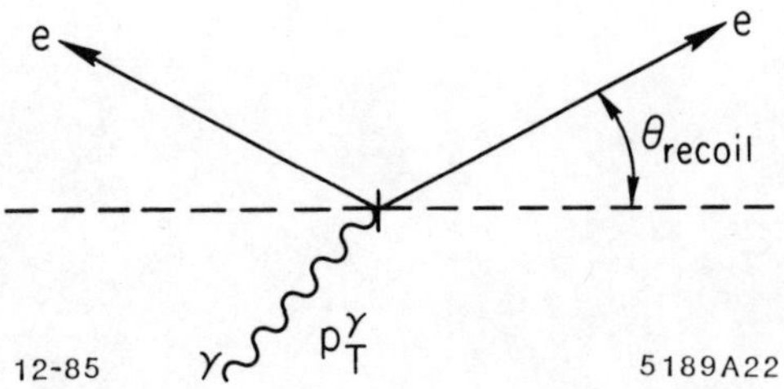

Figure 1. Kinematics of worst-case background to single-photon production.

Single photon production can be thought of as an initial state "radiative tag" of the production of unobservable final states. The characteristics of the radiated photon are quite independent of the specific final state; the differential cross section for single photon production can be expressed in terms of the cross section for the non-radiative process and the photon energy and angle as[6]

$$\frac{d\sigma}{dE_\gamma \, dcos\theta} \simeq \frac{2\alpha}{\pi} \frac{1}{E_\gamma} \frac{1}{sin^2\theta} \times \sigma_{nonrad.}.$$

Due to the generality of this relation, observation of single photon production will not identify a specific production mechanism; conversely, a limit on single photon production will limit all possible contributing processes.

Sources of single photons events

Single photon events may arise through several processes. Firmly within the context of the Standard Model, single photon events may be occur as radiative corrections to neutrino pair production. This can happen through neutral current or charged current exchange as indicated by the Feynman diagrams in figure 2. The differential cross section for radiative neutrino pair production has been calculated:[7]

$$\frac{d\sigma_{\gamma\nu\bar{\nu}}}{dx\, dcos\theta} = \frac{G_F^2\alpha}{6\pi^2}\times$$
$$\left\{\frac{M_Z^4\{N_\nu(g_V^2+g_A^2)+2(g_V+g_A)\left(1-\frac{s(1-x)}{M_Z^2}\right)\}}{[s(1-x)-M_Z^2|^2+M_Z^2\Gamma_Z^2}+2\right\}$$
$$\times\frac{s}{x\, sin^2\theta}\left[(1-\frac{x}{2})^2+\frac{1}{4}x^2cos^2\theta\right]$$

where $x = E_\gamma/E_{beam}$. It is important to note here that this process can be calculated very directly in the context of the Standard Model; the result is insensitive to the actual value of Γ_Z. After integration over experimental acceptance in photon energy and angle, the cross section is linearly proportional to the number of light neutrino species N_ν, with a nonzero intercept due to the charged current process which produces only ν_e.

Single photon events may occur in supersymmetric (SUSY) theories, which contain scalar partners of the familiar fermions and fermion partners of the gauge bosons. One possible process is the radiative production of photino pairs, mediated by selectron (SUSY electron) exchange as indicated in the representative Feynman diagram in figure 3a. In the simplest theories, the photino, the SUSY partner of the photon, is the lightest SUSY particle and therefore stable. Photinos will have very weak interactions in matter since such interactions would involve virtual massive particles, and so they will traverse the detectors leaving no trace of their passage. The differential cross section for radiative photino pair production depends on the mass of the

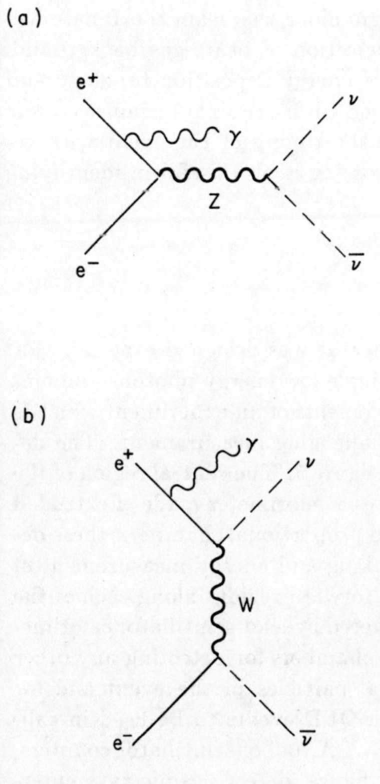

Figure 2. Feynman diagrams for radiative neutrino pair production by (a) the neutral current, and (b) the charged current.

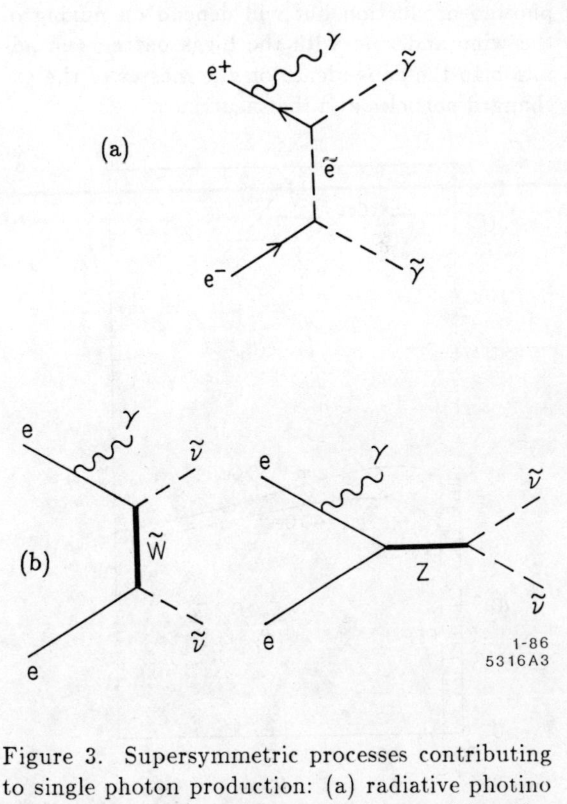

Figure 3. Supersymmetric processes contributing to single photon production: (a) radiative photino pair production by selectron exchange; (b) radiative sneutrino pair production by wino and zino exchange.

selectron, from its propagator as the exchanged particle, and on the mass of the photino through phase space effects. For small photino mass the form of the cross section is[8]

$$\frac{d\sigma}{dE_\gamma \, dcos\theta_\gamma} \approx \frac{\alpha^3}{p_{\perp\gamma} \, sin\theta_\gamma} \frac{s}{M_{\tilde{e}}^4} f(E_\gamma, cos\theta_\gamma)$$

where $f(E_\gamma, cos\theta_\gamma)$ is a weak function of the photon energy and angle. There are two scalar electron states, designated $\tilde{e}_L$ and $\tilde{e}_R$, corresponding to the two helicity states of the electron. If the masses of selectron states are equal, then the cross section is twice the expression given above. The exact cross section for some selected photino and degenerate selectron masses is shown as a function of center-of-mass energy in figure 4.[9] Also shown in Figure 4 is the cross section for radiative neutrino pair production, taken from reference 7.

Another SUSY process which can give rise to single photon events is the pair production of sneutrinos, mediated by the exchange of the zino or the wino, the SUSY partners of the Z and W bosons; this is indicated in figure 3b. The cross section[10] for this will have a similar form as given above for photino production but will depend on mixing of the wino and zino with the higgs partners in addition to the dependence on the masses of the exchanged particles and the sneutrinos.

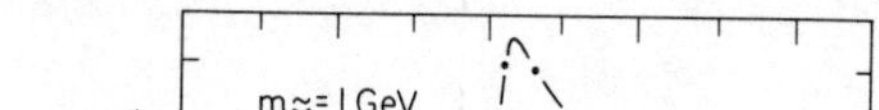

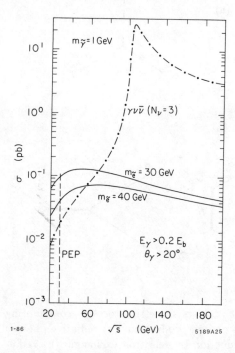

Figure 4. Cross sections for radiative production of (dot-dash line) neutrino pairs and (solid lines) photino pairs.

Backgrounds

Single photon events can arise through QED processes, the dominant one being $e^+e^- \to e^+e^-\gamma$ where the outgoing electron and positron escape down the beam pipe. These events can be eliminated by the kinematic veto technique, or alternatively one can select cuts for the photon such that Monte Carlo calculations of backgrounds predict a negligible number of events from this source. In fact, QED events where the electron and positron are observed are a source of tagged photons, and these events are used to calibrate the detectors and to measure efficiencies and resolutions.

Beam-gas scattering and cosmic rays are two more important potential sources of background. Collisions of beam particles and residual gas near the interaction can result in photoproduction of Δ resonances that decay to yield π_0's of energy up to ~1 GeV which may be difficult to distinguish from single photons. These events may be discriminated in some cases by their electromagnetic shower development in the detectors. They will be distributed approximately uniformly along the beam line, and so the ability of a detector to point a shower back to determine its origin along the beam coordinate will provide a vital rejection of beam-gas background. The same tools of energy deposition topology and point-back are used to discriminate against cosmic ray background; the timing of the event with respect to beam crossing is also useful in identifying these events.

The ASP detector

The ASP detector was designed especially for the detection of single low energy photons and embodies the basic elements of an experiment which is capable of this challenging measurement. The detector is shown in figure 5. The central region of the detector constitutes a calorimeter made of extruded lead glass bars and proportional chambers; these devices provide tracking and energy measurement of the photons. The forward regions along each of the beam lines are covered by lead-scintillator calorimeters and tracking chambers for detecting any other charged or neutral particles in the event and for reconstructing the QED events to be used in calibrating the detector. A roof of scintillator counters, not shown in the figure, detect cosmic rays entering the detector from above. The detector was installed in IR-10 at PEP, where the seventy feet of rock overhead provided shielding from cosmic rays. Data were collected at $\sqrt{s} = 29 \ GeV$ for an integrated luminosity of $\int Ldt = 115 \ pb^{-1}$.

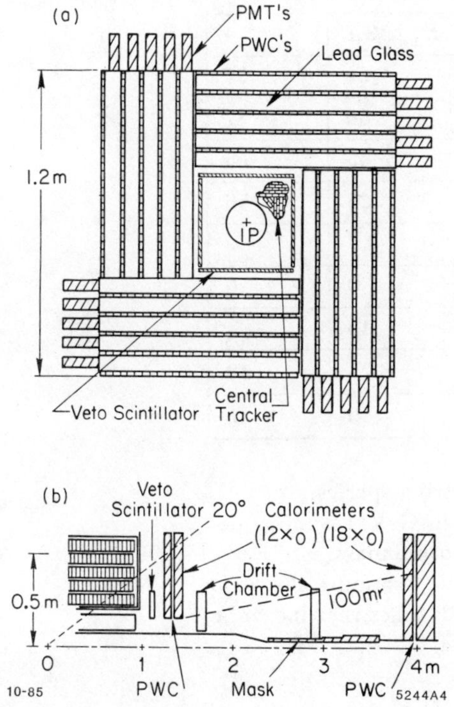

Figure 5. The ASP detector.

The performance of the detector has been studied extensively using the photons from reconstructed $e^+e^- \to e^+e^-\gamma$ events. The trigger efficiency of the detector is essentially 100% for photon energies above 600 MeV. The energy resolution for photons in the central calorimeter is $\sigma_E/E = 4.5\% + 8.5\%/\sqrt{E(GeV)}$. The origin of the photon, expressed as a projected distance of closest approach to the interaction point, is determined with a precision characterized by a gaussian sigma of 2.8 cm and a small exponential tail. The timing of the event relative to the beam crossing is measured with a resolution of 1.3 ns. Including all photon selection cuts, the efficiency of the detector is flat as a function of photon energy and polar angle for $E_\gamma > 1\ GeV$ and $\theta_\gamma > 20°$. Table 1 gives the integrated luminosity, photon selection cuts and efficiencies, and detector resolutions for the ASP, MAC, CELLO, and Mark J detectors.

Results

One single photon event has been observed in each of the ASP and MAC experiments. In both cases these are very clean unambiguous events. The relative sensitivity of the experiments can be measured by their expected number of single photon events from radiative neutrino pair production, calculated by integrating $\sigma_{\gamma\nu\bar{\nu}}$ over the experimental cuts and efficiencies. The ASP experiment expects 2.2 such events, while the MAC experiment expects 1.1 events and the CELLO experiment expects 0.7.

From the number of observed events an upper limit, at a given confidence level, on the average number of events can be calculated; this upper limit can then be translated into an upper limit on N_ν, the number of light neutrino species, and a lower limit on the masses of SUSY particles. Details of the calculations for the ASP experiment are as follows: from one event observed, the 90% confidence level on the average number of events is 3.9. Dividing this by the integrated luminosity and the experimental efficiency yields a 90% confidence level upper limit on the cross section for single photon production at $\sqrt{s} = 29\ GeV$ inside the ASP cuts of

$$\sigma_\gamma \leq 59\ fb.$$

Integration of $\sigma_{\gamma\nu\bar{\nu}}$ within the ASP cuts gives an expression for the cross section for radiative neutrino pair production:

$$\sigma_{\gamma\nu\bar{\nu}} = 16\ +\ 5.8 \times N_\nu\ fb.$$

The limit above on σ_γ then implies:

$$N_\nu\ <\ 7.5 \text{ at } 90\% \text{ confidence.}$$

To calculate limits for SUSY particles, we observe that $\sigma_\gamma < 59\ fb$ is a limit on the production of neutrino pairs or of any other weakly interacting particles. We subtract from σ_γ the calculated cross section for neutrino production, $\sigma_{\gamma\nu\bar{\nu}} = 33\ fb$ for $N_\nu = 3$, to calculate a 90% confidence level limit for the production of other weakly interacting particles:

$$\sigma_{other}\ <\ 26\ fb \text{ at } 90\% \text{ confidence.}$$

This cross section may be compared with the results of integrating the calculated differential cross section over the experimental cuts to determine limits on the masses of SUSY particles. The results of this procedure are expressed as a contour marking the limit boundary in the $M_{\tilde{e}} - M_{\tilde{\gamma}}$ plane. The region $M_{\tilde{e}}\ <\ M_{\tilde{\gamma}}$ is not probed in the single photon channel, since in this case the photino would be unstable.

	ASP	MAC	CELLO	Mark J
$\int Ldt\ pb^{-1}$	115	36,80,61	37.6	49
$\sqrt{s}\ GeV$	29	29	42.6	39.5
min $p_{\perp\gamma}\ GeV$	1.0	4.5,2.0,2.6	$E_\gamma > 2.1$	4
θ_{min} *degrees*	20	40	34	37
θ_{Veto} *mrad*	21	175,66,84	50	87
$\Delta t\ ns$	1.5–1.1	75	25–45	0.5
efficiency %	57	53–69	35	56
$N_{\gamma\nu\nu}$ expected	2.2	1.1	0.7	
single γ evts	1	1	0	0
$N_\nu$limit*	7.5	16	15	
$M_{\tilde{e}}$ limit† GeV	65.7	50	37.7	40
$M_{\tilde{W}}$ limit‡ GeV	60	51	40	45

* 90% CL limits on the number of light neutrino species.

† Limits at 90% CL on the selectron mass, in GeV, assuming degenerate selectron eigenstates and zero photino mass.

‡ 90% CL limits on the wino mass, in GeV, assuming the wino and zino masses are equal (except that ASP takes the zino to be heavy) and the sneutrino mass is negligible.

Table 1. Detector characteristics, cuts, and results from single photon event analysis for the ASP, MAC, CELLO, and Mark J detectors.

Figure 6 presents the ASP results in three forms: 90% CL limits for the cases where the $\tilde{e}_L$ and $\tilde{e}_R$ masses are equal and where one is much larger that the other, and the 95% CL for the equal mass case. The equal mass case corresponds to a limit $M_{\tilde{e}} >$ 65.7 GeV for zero photino mass, and this limit moves slowly with variation of the confidence level. In this limit, decay of the Z^0 into selectron pairs is clearly ruled out.

Mass limits for wino/zino and sneutrino masses are calculated similarly, with considerable model dependence. Table 1 gives the results from ASP, MAC, CELLO, and Mark J for the number of light neutrino species and for SUSY particle masses. Figure 7 plots the SUSY mass limits for ASP, MAC, and CELLO along with limits from CELLO and JADE from other processes.

The limits on N_ν from the various experiments can be combined, taking into account the differing sensitivities and the s dependence of the cross section to yield[11]

$$N_\nu < 4.9 \text{ at } 90\% \text{ confidence.}$$

This number has essentially no theoretical uncertainty, relying on the Standard Model couplings of light neutrinos to the Z^0.

Conclusions

Single photon production in electron positron annihilation has proven to be a very powerful probe of weak interactions and supersymmetry. The combined result of $N_\nu < 4.9$ is the best limit on the number of light neutrino species and has very little theoretical uncertainty. Mass limits for supersymmetric particles have been pushed well above half the Z^0 mass, further constraining SUSY theories.

Work supported by the U. S. Department of Energy, contract DE-AC02-86ER40284.

References

1. G. Bartha *et. al.*, Phys. Rev. Lett. 56, 685 (1986)and new results.
2. W. T. Ford *et. al.*, SLAC-PUB-4003, June 1986.
3. H.-J. Behrend *et. al.*, DESY 86-050, May 1986.
4. B. Adeva *et. al.*, MIT/LNS Technical Report 152. August 1986.
5. The ASP collaboration is: (SLAC) G. Bartha, D. L. Burke, P. Extermann, P. Garbincius, C. A. Hawkins, M. J. Jonker, L. Keller, C. Matteuzzi, N. A. Roe, T. R. Steele; (Boston University) A. S. Johnson, J. S. Whitaker, R. J. Wilson; (University of Pennsylvania) R. J. Hollebeek; (University of Washington) C. Hearty, J. E. Rothberg, K. K. Young.
6. G. Bonneau and F. Martin, Nucl. Phys. B27, 381 (1971).
7. E. Ma and J. Okada, Phys. Rev. Lett. 41, 287 (1978); K. J. F. Gaemers, R. Gastmans, and F. M. Renard, Phys. Rev. D19, 1605 (1979).
8. J. Ellis and J. S. Hagelin, Phys. Lett. 122B, 303 (1983).
9. K. Grassie and P. N. Pandita, Phys. Rev. D30, 22 (1984).
10. J. S. Hagelin, G. L. Kane, and S. Raby, Nucl. Phys. B241, 638 (1984);
    J. A. Grifols, M. Martínez, and J. Solà, Nucl. Phys. B268, 151 (1986).
11. T. L. Lavine, MAC Collaboration, private communication.

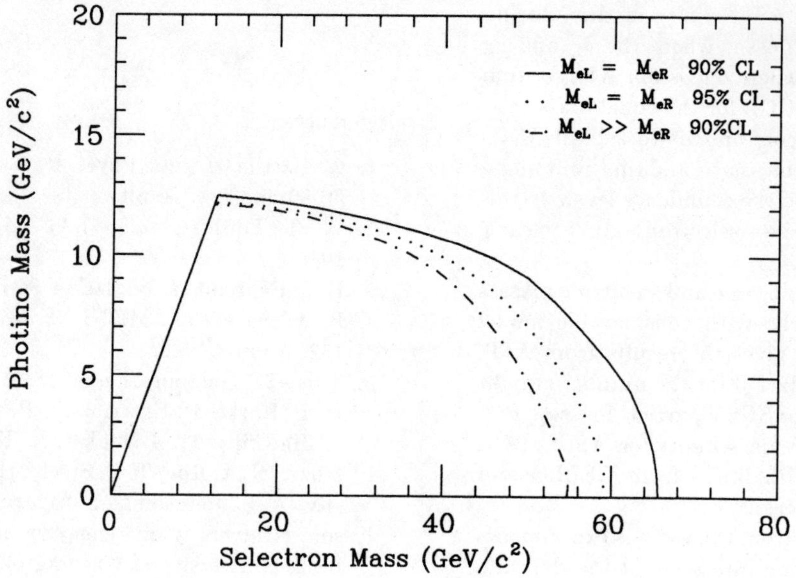

Fig. 6 ASP photino vs selectron mass limits

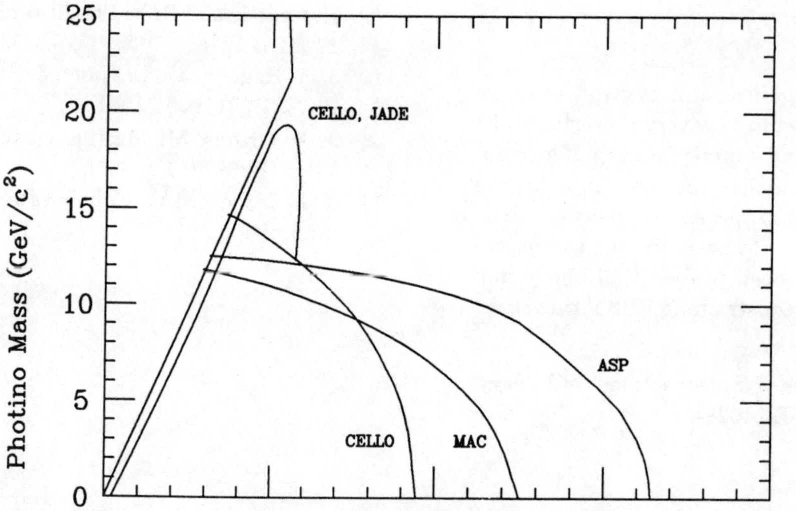

Fig. 7 Photino vs selectron mass limits (90% CL) for $M_{eL} = M_{eR}$

SEARCH FOR AXIONS, HIGGS, GLUINOS AND OTHER NEW PARTICLES IN UPSILON DECAYS

P. Michael Tuts

Columbia University
New York, NY 10027, USA

Searches for new particles in radiative Υ decays from the CUSB, CLEO, ARGUS, and Crystal Ball experiments are presented. Null results on searches for short lived axions, light neutral Higgs, bound gluinos, magnetic monopoles, and "invisible" particles are presented. Upper limits on the product branching ratios for axion production and neutral Higgs production are presented, and bounds on the gluino mass are presented.

1. INTRODUCTION

Electron-positron annihilations provide us with a precise tool with which to search for new particles. Until the highest energy e^+e^- colliders come on the air, one of the best hunting grounds for new particles will be the upsilon system. There are a large number of new particles that could be found in upsilon decays, including: 1) (extended) standard model particles such as the axion, and the neutral Higgs from radiative Υ decays, 2) supersymmetric particles such as the gluino or massive gauginos from radiative Υ decay or from "invisible" decays of the Υ, and 3) even classical particles such as magnetically charged particles have been searched for. In the following report, results are presented on searches for the above mentioned particles from the CUSB and CLEO experiments at CESR, and from the ARGUS and Crystal Ball experiments at DORIS-II.

2. AXIONS

Axions were introduced to avoid parity violation in the strong interactions. The Peccei-Quinn symmetry [1] leads to the existence of a pseudoscalar boson called the axion [2,3]. In the minimal axion model in which there are two Higgs doublets, the coupling to fermions depends on the fermion mass and on an arbitrary parameter x for 'up' like quarks and 1/x for 'down' like quarks, where $x=\langle\Phi_1\rangle/\langle\Phi_2\rangle$ is given by the vacuum expectation value (vev) of the two two Higgs fields. Since the coupling is proportional to mass, one expects substantial branching ratios for radiative J/ψ and Υ decays to $a+\gamma$; they are given by $BR(\Upsilon\to a+\gamma)=(G_F m_q{}^2/\pi\alpha\sqrt{2})(x^2 \text{ or } 1/x^2)$. In the standard picture, the mass of the axion is given by m_a=75×(x+1/x), or m_a~150 keV for x~1. Note that by taking the product of the two branching ratios, the unknown quantity x is cancelled. Thus it was that by combining CUSB [4] and Crystal Ball [5] results, the standard axion was ruled out (the experimental results are <20 times the prediction) and the invisible axion [6] was born. The reports of near monochromatic positrons produced in heavy ion collisions [7] has led to renewed interest in nearly visible axions [8] and speculation that the positron signal may be due to such an axion. If it is an axion signal, then the mass would be ~1.8 MeV, which in turn implies a much lower symmetry breaking scale or a value of x~1/25. That small a value for x leads to a large increase in branching ratio (by a factor of ~600), but a small lifetime ($\sim 4\times10^{-12}$ s), thereby invalidating the previous results which assumed that the axion decayed outside the active detector volume. The present results from CUSB, CLEO and ARGUS extend the search down to short lifetimes, under the assumption that the axion decays to e^+e^- in the detector.

We at CUSB [9] have reanalysed a partial Υ(1S) data sample (133,00 events from $7pb^{-1}$ of integrated luminosity) and searched for events ($\Upsilon\to\gamma+a$) with back-to-back ~5 GeV electromagnetic showers, i.e. $ee\to\gamma\gamma$ and $ee\to ee\gamma$, depending on the axion lifetime. These are events with an easily recognized topology. Backgrounds from ordinary QED processes which would fake a real signal are measured from continuum data samples. After background subtraction, we are left with an excess of 38±30 events from which we derive upper limits on the branching ratio as a function of the mean decay length $d=\gamma c\tau$, shown as the curve labelled (a) in Fig. 1. The curve labelled (b) in Fig. 1 is obtained from our older result [4]. The combination of both results leads to an upper bound on $BR(\Upsilon\to a\gamma)\times BR(a\to ee)$ < .5% (for decay lengths larger than 80cm it is below 0.1%, falling to 0.012% at very large decay lengths), which very far below the standard axion model prediction of ~25%.

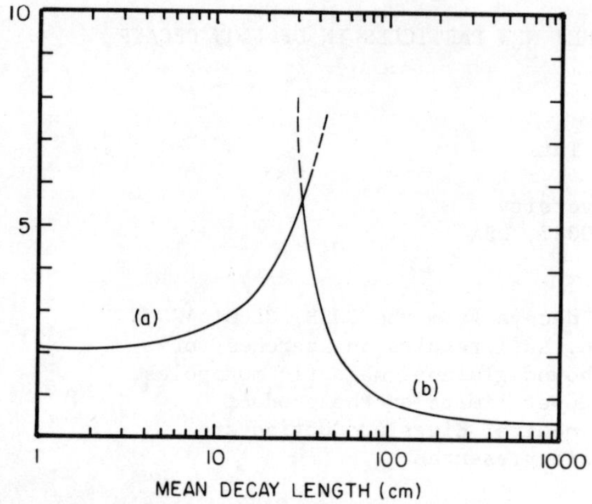

Figure 1. CUSB 90% CL upper limit for $BR(\Upsilon\to\gamma a)$ vs mean decay length of a, where (a) assumes that $a\to ee$ in or before the tracking system, and (b) assumes that a decays outside the detector.

CLEO and ARGUS have searched for the decay $\Upsilon(1S)\to\gamma+a$ by tagging Υ events from $\Upsilon(2S)\to\pi\pi\Upsilon(1S)$ decays, and then searching for $\Upsilon(1S)\to\gamma+a$; $a\to ee$ events. This method eliminates the QED backgrounds encountered in the direct method, while paying the price in statistics. CLEO [10] analyzed 14,600±500 $\Upsilon(1S)$ decays from an integrated luminosity of $22pb^{-1}$ on the $\Upsilon(2S)$. They find no events; from which they derive upper limits on the product branching ratio for various axion masses vs the lifetime of the axion, shown as solid curves in Fig. 2. The dashed lines in Fig. 2 represent the older CLEO results [11] for a long lived axion.

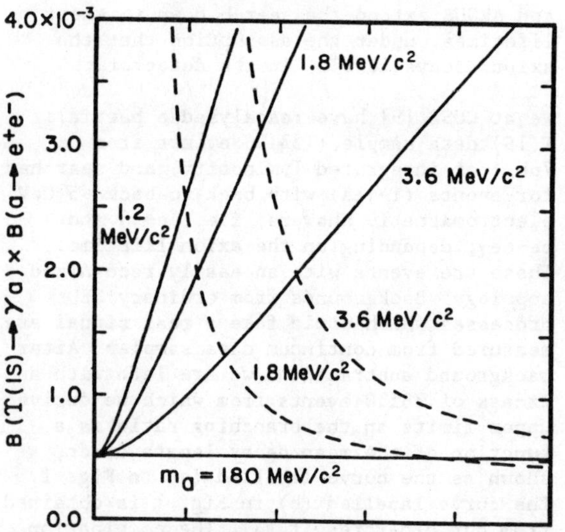

Figure 2. CLEO 90% CL upper limits on the $BR(\Upsilon\to\gamma a)\times BR(a\to ee)$ for various axion masses. The solid and dashed lines are for short and long lived axions respectively.

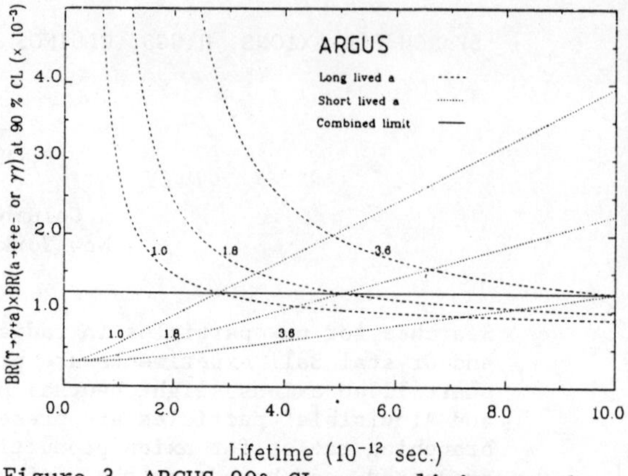

Figure 3. ARGUS 90% CL upper limits on the $BR(\Upsilon\to\gamma a)\times BR(a\to ee)$ for various axion masses. The dotted and dashed lines are for short and long lived axions respectively.

ARGUS has carried a similar analysis [12] on 23,800±2,300 $\Upsilon(1S)$ events from a $39pb^{-1}$ sample of $\Upsilon(2S)$ events. They find no candidates that correspond to a possible short lived axion, and one candidate that corresponds to a long lived axion. They use these events to calculate upper limits on the product branching ratio for these two cases, as is shown in Fig. 3.

We conclude that whatever may have been seen in heavy ion collisions is not a standard axion, or does not couple to heavy quarks.

3. LIGHT NEUTRAL HIGGS

In the minimal standard model there must be at least one neutral Higgs boson, which if light enough would be mainly produced in the decays of upsilons via the Wilczek [3] mechanism. The branching ratio for radiative Υ decay to a neutral Higgs is given by $BR(\Upsilon\to\gamma H)=BR(\Upsilon\to\mu\mu)\times(G_F m_q^2/\pi\alpha\sqrt{2})\times(1-(m_H/m_\Upsilon)^2)$. This bound had been reached by CUSB upper limits [13] last year, but in the meantime QCD radiative corrections by Vysotsky [14] and others [15] have lowered the expected value by ~50%, so that it lies below the experimental bound. Despite the fact that there are now no significant limits on single Higgs models, we can re-express the CUSB limit in terms of a non-minimal standard model (supersymmetry) in which there are two Higgs doublets fields, leading to two physical charged Higgs bosons and three neutral ones. In addition we must introduce the same parameter $x=\langle\phi_1\rangle/\langle\phi_2\rangle$ that we had for the axion above. In this case we express the experimental upper bound on the branching ratio as a limit on the value of x. This limit is shown in Fig. 4 for both the original Wilczek formula, and the radiatively corrected one. As the models become more predictive, they will have to confront these limits.

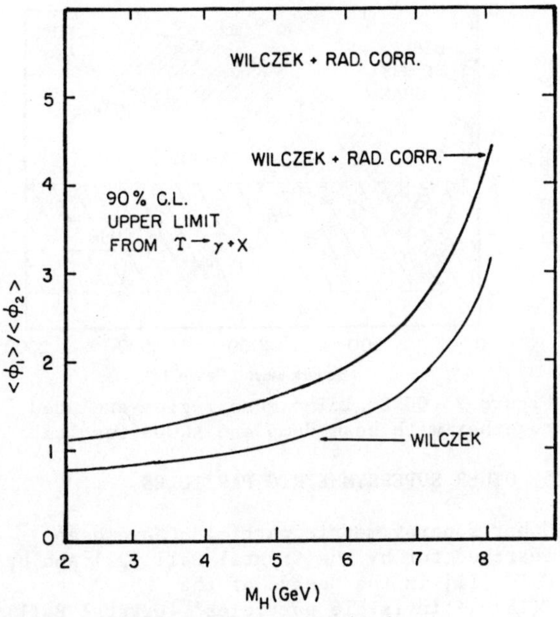

Figure 4. CUSB 90% CL upper limit on $x=\langle\phi_1\rangle/\langle\phi_2\rangle$ from radiative Υ decay for the simple Wilczek formula (lower curve), and the radiatively corrected one (upper curve).

Crystal Ball has searched for neutral Higgs [16] in the decays of Υ(1S) and Υ(2S) $\rightarrow\gamma\tau\tau$, where $\tau\tau$ events with one e and one μ are selected. The 90% CL upper limits they obtain are shown in Fig. 5. The upper limits obtained of $\sim 10^{-3}$ for the Υ is at present too large to meaningfully confront even the original Wilczek prediction and significantly above the CUSB limits.

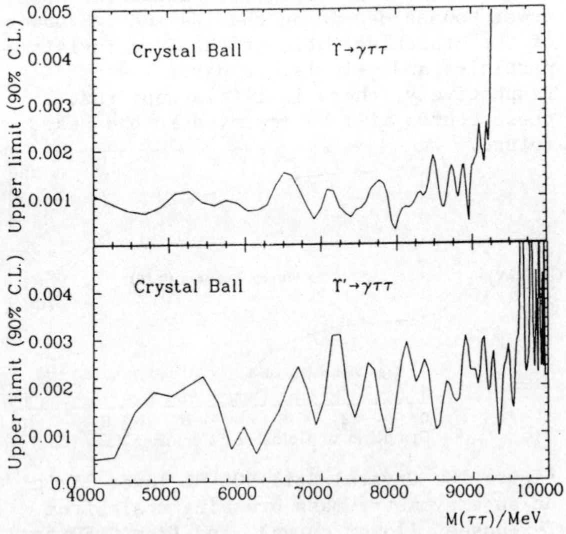

Figure 5. Crystal Ball 90% CL upper limits on $\Upsilon\rightarrow\gamma\tau\tau$ (upper) and $\Upsilon(2S)\rightarrow\gamma\tau\tau$ (lower).

4. GLUINOS

While supersymmetry (SUSY) may provide the cure to some of the problems of the standard model, there is as yet no experimental evidence for the existence of any supersymmetric partner of any elementary particle. The spin 1/2 partners of the gluons are known as gluinos, $\tilde{g}$, and they have been searched for in experiments at colliders and beam dumps [17]; the limits are usually expressed in terms of the squark mass. The existence of light gluinos (mass <5 GeV) has not been unambiguously ruled out. Regardless of how supersymmetry is broken, it is believed that the interactions among particles and their supersymmetric partners are well understood; gluinos carry color and interact with gluons with a uniquely determined strength. Thus we would expect that $\tilde{g}\tilde{g}$ bound states should exist, and that they should have very similar properties to those of the Υ and ψ; I will refer to those states as gluinium, $\tilde{G}$. Thus if the mass of the gluinium states lies below the Υ mass, we can search for these states in radiative decays of the Υ, and use the apparatus of heavy quark spectroscopy to calculate rates. One expects that significant branching ratios will exist for transitions to the S wave states, of which the lowest is the pseudoscalar state analogous to the η_c state of charmonium. The calculation of rates and the total width has been done by Goldman and Haber [18] and by Keung and Khare [19]. The first authors have extrapolated from similar decays for the η_c, whereas the second authors have expressed their result for the width of $\Upsilon\rightarrow\eta_{\tilde{g}}+\gamma$ in terms of the rate for $\Upsilon\rightarrow\gamma gg$, which is experimentally measured [20].

From a data sample of 400,000 Υ's in partially upgraded detector (BGO and NaI crystal arrays [21]), CUSB has searched both hadronic final state events and isolated photon events for monochromatic photons. The photon spectra for events in the BGO array (σ = 1.8%/$\sqrt[4]{E(GeV)}$) and for the NaI array (σ = 3.8%/$\sqrt[4]{E(GeV)}$) is shown in Figs. 6 and 7 respectively. No structure is observed; the data are fit to exponentiated polynomials (the solid curves). A maximum likelyhood fit to that smooth curve plus a gaussian of width corresponding to the convolution of the detector resolution and the expected gluinium width over the energy range shown, yields the combined (NaI and BGO data) 90% CL upper limit shown in Fig. 8 (lower curve). Note that the horizontal axis is given in terms of the gluino mass, which we take to be half of the gluinium bound state mass. The theoretical predictions of Refs. 18 and 19 are the upper curves on the figure. From Fig. 8 it appears that gluino masses above .6 GeV and below 2.2 GeV are excluded. In Fig. 9, we present the CUSB result together with collider and beam dump results (from Ref. 17).

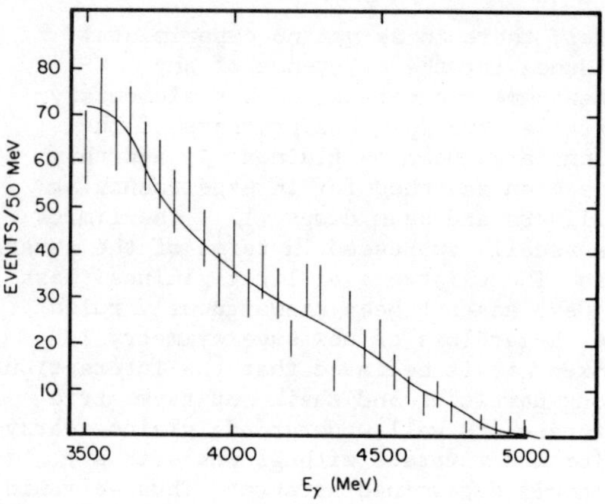

Figure 6. CUSB inclusive photon spectrum from NaI data; the solid line is the fitted polynomial.

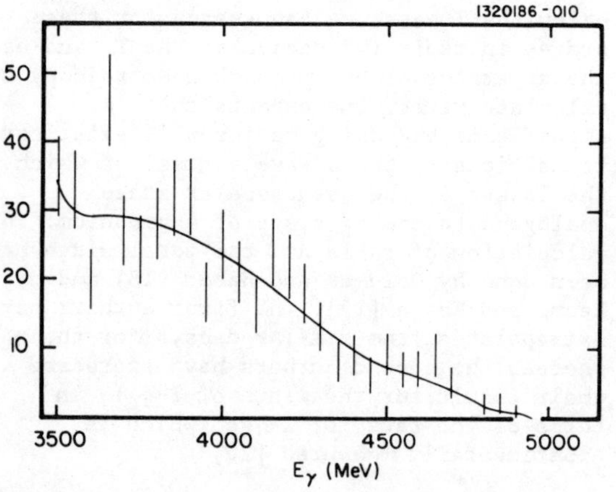

Figure 7. CUSB inclusive photon spectrum from BGO data; the solid line is the fitted polynomial.

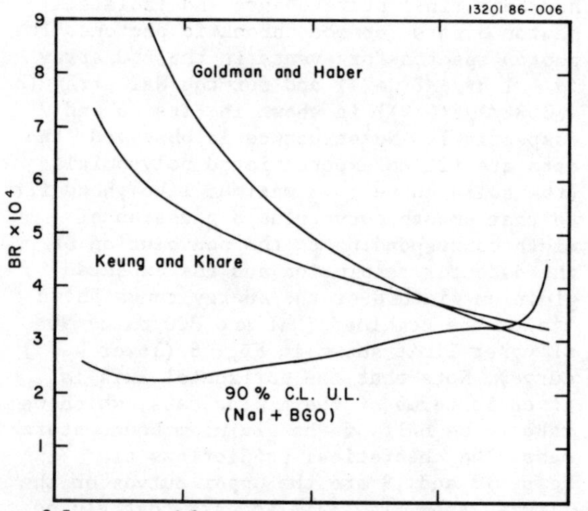

Figure 8. CUSB 90% CL upper limit from the combined NaI and BGO data; the upper two curves are from theoretical models.

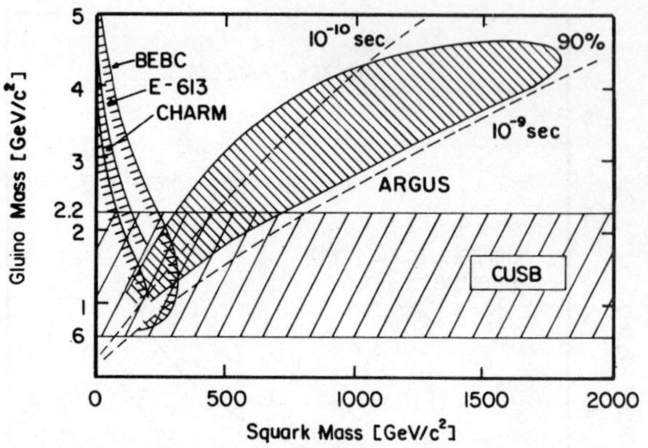

Figure 9. CUSB gluino mass region excluded, together with beam dump and ARGUS results.

5. OTHER SUPERSYMMETRIC PARTICLES

Other supersymmetric particles have been searched for by the Crystal Ball [22] and by CLEO [11] in the decays of the $\Upsilon(1S) \to \gamma +$'invisible particles' (Crystal Ball) or direct decays of the Υ to 'invisible particles' (CLEO). The invisible particles in this case are postulated to be either gravitinos or goldstone fermions [23]. The Crystal Ball has tagged 18,000 $\Upsilon(1S)$ decays via the $\pi^0\pi^0$ hadronic decays of the $\Upsilon(4S)$; they observe no events with $E_\gamma > 1200$ MeV from which they derive a 90% CL upper limit of .0023 for radiative upsilon decays into invisible particles. This upper bound has been used to place a lower bound on the supersymmetry breaking mass scale, Λ_{SS}. The lower limits for the Crystal Ball results (lower curve) and the CLEO results (upper curve) are shown in Fig. 10. Since the Λ_{SS} lower bounds depend on the 1/4 and 1/8 power of the branching ratio of the Υ to invisible particles and γ+invisible particles respectively, there is little hope that these bounds will be improved in the near future.

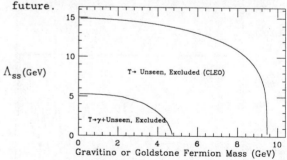

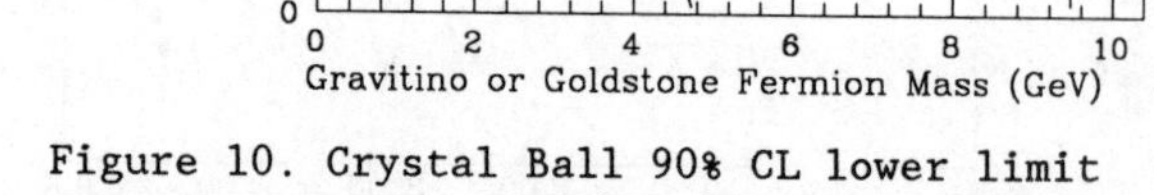

Figure 10. Crystal Ball 90% CL lower limit on supersymmetry mass breaking scale from $\Upsilon \to \gamma$+unseen (lower curve), and from CLEO for $\Upsilon \to$unseen (upper curve).

6. MAGNETIC MONOPOLES

Magnetic monopoles have long been postulated [24], and long been sought after. CLEO has searched [25] for magnetically charged particle pairs (ee→MM with magnetic charge g<10e) by identifying trajectories with non-zero acceleration along magnetic field lines in their drift chamber. No events were found, the derived 90% CL upper limits as a function of monopole mass are shown in Fig. 11.

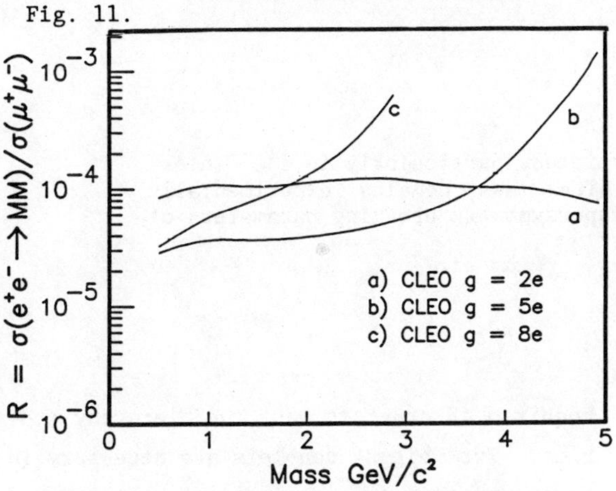

Figure 11. CLEO 90% CL upper limits on monopole production for various magnetic charges.

7. ACKNOWLEDGEMENTS

I would like to thank one of the organizers, R. Hollebeek, for asking me to present the DORIS-II and CESR results, and I would like to thank the CLEO, ARGUS, and Crystal Ball collaborations and particularly my own collaborators on CUSB for allowing me to present their results. I particularly wish to thank D. Besson, E. Bloom for discussion of their results, and J. Lee-Franzini for valuable help in preparation of my talk. This work is supported in part by the National Science Foundation.

REFERENCES

1. R. D. Peccei and H. R. Quinn, Phys. Rev. Lett. 38, 1440 (1977).
2. S. Weinberg, Phys. Rev. Lett. 40, 223 (1978).
3. F. Wilczek, Phys. Rev. Lett. 40, 279 (1978).
4. M. Sivertz et al., Phys. Rev. D26, 717 (1982).
5. C. Edwards et al., Phys. Rev. Lett. 48, 903 (1982).
6. M. Dine, W. Fischler, and M. Srednicki, Phys. Lett. 104B, 199 (1981).
7. J. Schweppe et al., Phys. Rev. Lett. 51,2261 (1983); T. Cowan et al., ibid 54, 1761 (1985); M. Clemente et al., Phys. Lett. 137B, 41 (1984).
8. A. B. Balentekin et al., Phys. Rev. Lett. 55, 461 (1985).
9. G. Mageras et al., Phys. Rev. Lett. 56, 2672 (1986), and contributed paper 7129 to this conference.
10. T. Bowcock et al., Phys. Rev. Lett. 56, 2676 (1986).
11. M. S. Alam et al., Phys. Rev. D27, 1665 (9183); D. Besson et al., Phys. Rev. D30, 1433 (1984).
12. H. Albrecht et al., contributed paper 9725 to this conference.
13. P. Franzini et al., contributed paper 312 to the 1985 International Symposium on Lepton and Photon Interactions at High Energy, Kyoto (1985).
14. M. I. Vysotsky, Phys. Lett. 97B, 159 (1980).
15. J. Ellis et al., Phys. Lett. 158B, 417 (1985).
16. S. Keh et al., contributed paper 6947 to this conference.
17. For a review of gluino searches see S. Komamiya, in the Proc. of the Int. Symp. on Photon and Lepton Interactions at High Energy, ed. by M. Konuma and K. Takahashi (Kyoto University, Kyoto, 1985) p.612 and M. Davier at this conference.
18. T. Goldman and H. Haber, Los Alamos preprint No. LA-UR-84-634 (1984).
19. W-Y. Keung and A. Khare, Phys. Rev. D29, 2657 (1984).
20. R. D. Schamberger et al., Phys. Lett. 138B, 225 (1984).
21. P. Franzini and J. Lee-Franzini, in Proc. of the Oregon DPF Meeting, ed. by R. Hwa (World Scientific, Singapore, 1986) p.1009.
22. S. Leffler et al., contributed paper 6939 to this conference.
23. O. Nachtmann, A. Reiters, and M. Wirbel, Z. Phys. C23, 199 (1984); P. Fayet, Phys. Lett. 84B, 421 (1979).
24. P. A. M. Dirac, Proc. Roy. Soc. London 133, 60 (1931).
25. T. Gentile et al., contributed paper to this conference.

SEARCHES FOR SUPERSYMMETRY

Jean-François GRIVAZ

Laboratoire de l'Accélérateur Linéaire
Université de Paris-Sud, Orsay, France

Recent searches for supersymmetric particles, particularly in e^+e^- interactions, are reviewed and compared. The mass domains experimentally excluded are turned into limits on the supersymmetry breaking parameters of a minimal N=1 supergravity model.

1. INTRODUCTION

The principal topic of this talk will be the searches for supersymmetric particles in high energy e^+e^- collisions, although some reference will be made to other searches, in particular to those performed at the CERN $p\bar{p}$ collider already reported at this conference[1]. Since, in the case of e^+e^- collisions, the only new results showing substantial improvement were those obtained in single photon experiments, and since they were independently presented in this session[2], no experimental details will be given here (those may be found e.g. in earlier reviews[3]). Instead, an attempt will be made to present a coherent picture of the current limits, working for that purpose within a specific supersymmetric model. For the sake of clarity, whenever experimental results will be presented, they will either be combined results or those of the most constraining single experiment.

2. THEORETICAL FRAMEWORK

2.1 Minimal particle content :

Table I displays the minimal particle content of any supersymmetric model. Within each supermultiplet, the fermionic and bosonic degrees of freedom are equal, as required in order to ease the hierarchy problem. Two higgs doublets are necessary in supersymmetry to give masses to both up and down-type quarks. After SU(2)×U(1) spontaneous symmetry breaking, 5 physical higgs particles remain :

i) two charged ones which belong to the same supermultiplet as the $W^\pm$, together with 2 Dirac winos,

ii) one neutral which belongs to the same supermultiplet as the Z°, together with 2 Majorana zinos, and,

iii) two neutrals (a scalar and an axion-like pseudoscalar) which form a chiral supermultiplet with a Majorana axino.

Spin	0	1/2	1	3/2	2
Matter	$\tilde{\ell}_R, \tilde{\ell}_L$	ℓ			
multiplets	$\tilde{q}_L, \tilde{q}_R$	q			
		$\tilde{g}$	g		
Gauge	$w^\pm$	$\tilde{W}^\pm_+$, $\tilde{W}^\pm_-$	$W^\pm$	Higgses	
		$\tilde{\gamma}$	γ	Neutralinos	
multiplets	z	$\tilde{Z}_+$, $\tilde{Z}_-$	Z°		
	h, a	$\tilde{a}$			
				$\tilde{G}$	G

Table I : Minimal particle content

The field content, that is the relative amount of gaugino and higgsino within the winos and zinos, the amount of mixing between photino, zinos and axino which occurs within the physical neutralinos, and the mass splittings between the supersymmetric particles (SP) and their ordinary partners all depend on the details of the sypersymmetry breaking mechanism. However, if supersymmetry is to solve the naturalness problem and if fine tuning is to be avoided, the mass splittings should not be much larger than M_W. Actually, in many specific models, SP's with masses even less than M_W are predicted.

2.2 R-parity conservation :

In most supersymmetric models, R-parity[4] is an absolutely conserved multiplicative quantum number. Since $R = (-1)^{2S + 3(B-L)}$, one sees that ordinary particles have $R = +1$ whereas SP's have $R = -1$. As a consequence

i) SP's are always pair produced,

ii) the lightest supersymmetric particle (LSP) is absolutely stable, and all SP's ultimately decay into it.

Cosmological arguments lead to the conjecture that the LSP has no electric nor color charge[5]. Therefore, the LSP candidates are the gravitino, the lightest neutralino, or a scalar neutrino.

In all cases, the interaction of the LSP with ordinary matter turns out to be weak. For instance, $\tilde{\gamma}e \rightarrow \tilde{\gamma}e$, the supersymmetric analog of Compton scattering, is depressed with respect to $\gamma e \rightarrow \gamma e$ because of the high mass of the u-channel exchanged $\tilde{e}$. The LSP is therefore expected to evade any experimental setup, which is the source of the universal "missing-P_T" line of search for supersymmetry.

2.3 Minimal N=1 supergravity model (MSM) :

This model[6] is minimal in two respects : the particle content is the one we just described ; the set of parameters governing the supersymmetry breaking mechanism is minimal. One starts with an exact N=1 supergravity which is broken in the hidden sector, the $\tilde{G}$ aquiring a mass $m_{3/2}$ as a result of the Super-Higgs mechanism. Then, slightly below the Planck mass, the effective Lagrangian is that of an exactly supersymmetric GUT, with the addition of soft supersymmetry breaking terms. This theory is driven to low energy using the renormalization group equations and, on the way, SU(2)×U(1) should get spontaneously broken. This last condition rather strongly constrains the set of parameters entering the soft breaking terms. There are essentially 5 such parameters :

$m_{3/2}$: the gravitino mass

M : a common gaugino mass at the unification scale

μ : a higgsino mixing mass term

V_2/V_1 : the ratio of the vacuum expectation values of the 2 higgs fields

A : a parameter related to the Super-Higgs mechanism.

We will collectively refer to those as SBP's (supersymmetry breaking parameters). The way these parameters govern the masses of the various SP's is shown on Fig. 1. If the top is not too heavy ($\lesssim$ 50 GeV), the requirement of spontaneous SU(2)×U(1) breaking leads to $V_2/V_1 \sim 1$ and $\mu \sim m_{3/2}$. Then, if M is not too large, the LSP is a light practically pure photino. This is the case we will most often consider.

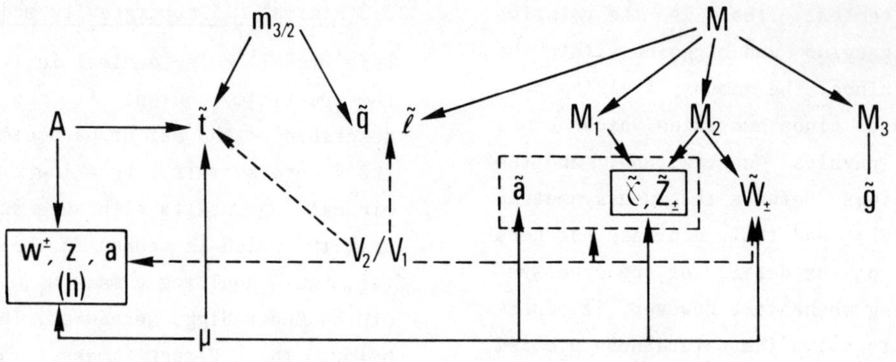

Fig. 1 : Connection between SBP's and higgs and SP masses. All dotted lines vanish if $V_2/V_1 = 1$.

M_1, M_2, M_3 are the renormalized values of M at M_W. $m_h = 0$ up to radiative corrections.

2.4 <u>SP Mass spectrum in the MSM</u> :

The general expression for scalar quark and lepton masses is :

$$m^2_{\tilde{x}} = m^2_{3/2} + C_{\tilde{x}} M^2 + K_{\tilde{x}} M^2_Z$$

where

$\tilde{x}$ is a scalar SP

$C_{\tilde{x}} M^2$ ("radiative correction") is calculable using renormalization group equations.

$K_{\tilde{x}} M^2_Z$ ("D-term") vanishes if $V_2/V_1 = 1$.

Numerical values for $C_{\tilde{x}}$ and $K_{\tilde{x}}$ can be found in the litterature[7]. Typically :

$C_{\tilde{q}} \sim 7$, $C_{\tilde{\ell}_L} \sim .5$, $C_{\tilde{\ell}_R} \sim .15$

while

$m_{\tilde{g}} \sim 3M \sim 6 \div 7\ m_{\tilde{\gamma}}$.

Consequences of these relations are that $\tilde{g}$ cannot be much heavier than $\tilde{q}$, and $\tilde{\ell}_R$ and $\tilde{\ell}_L$ are mass degenerate only if $M \ll m_{3/2}$. Finally, if $V_2/V_1 \ll 1$, the D-term can become substantial, which opens a possibility for $\tilde{\nu}$ to become light, perhaps even the LSP. However, we remind that $V_2/V_1 \sim 1$ is favored.

The physical wino masses are obtained[8] by diagonalization of the mixing matrix :

$$\begin{bmatrix} M_2 & \sqrt{2}\, M_W s \\ \sqrt{2}\, M_W c & \mu \end{bmatrix}$$

where $s = \sin\beta$, $c = \cos\beta$, $tg\beta = V_2/V_1$. Similarly, the neutralino masses are obtained[8] from the diagonalization of :

$$\begin{bmatrix} M_1 & 0 & -M_Z Sc & M_Z Ss \\ 0 & M_2 & M_Z Cc & -M_Z Cs \\ -M_Z Sc & M_Z Cc & 0 & -\mu \\ M_Z Ss & -M_Z Cs & -\mu & 0 \end{bmatrix}$$

where in addition $S = \sin\theta_W$, $C = \cos\theta_W$. An example of a possible wino and neutralino mass spectrum is shown on Fig. 2, corre-

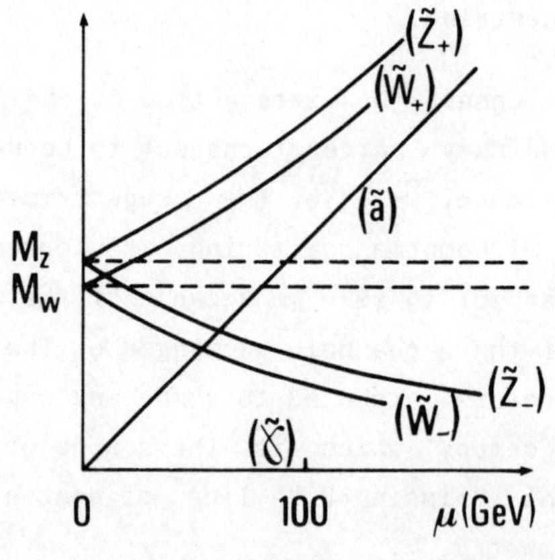

Fig. 2 : Wino and neutralino spectrum for $V_2/V_1 = 1$ and $M = 0$.

LSP assumed	Reaction studied	Physical quantities	Additional hypotheses	SBP's
$\tilde{\gamma}$	$e^+e^- \rightarrow \tilde{e}\tilde{e}$ $\gamma e \rightarrow \tilde{\gamma}\tilde{e}$ $e^+e^- \rightarrow \gamma\tilde{\gamma}\tilde{\gamma}$	$m_{\tilde{\gamma}} - m_{\tilde{e}}$	$V_2/V_1 = 1$	$M - m_{3/2}$
	$e^+e^- \rightarrow \tilde{W}^+_-\tilde{W}^-_-$	$m_{\tilde{W}_-}$		$\mu - M$
	$e^+e^- \rightarrow \tilde{\gamma}\,\tilde{Z}_-$	$m_{\tilde{Z}} - m_{\tilde{e}} - m_{\tilde{\gamma}}$		$\mu - M - m_{3/2}$
$\tilde{\nu}$	$e+e- \rightarrow \tilde{W}^+\tilde{W}^-_-$ $\gamma e \rightarrow \tilde{\nu}\tilde{W}_-$ $e^+e^- \rightarrow \gamma\tilde{\nu}\tilde{\nu}$	$m_{\tilde{\nu}} - m_{\tilde{W}}$	$V_2/V_1 \ll 1$	$\mu - M - m_{3/2} - V_2/V_1$
$\tilde{a}$	$e^+e^- \rightarrow \tilde{\gamma}\tilde{\gamma}$	$m_{\tilde{\gamma}} - m_{\tilde{e}}$	$V_2/V_1 = 1$	$M - m_{3/2}$
	$e^+e^- \rightarrow \tilde{a}\tilde{Z}_-$	$\Gamma(Z\rightarrow\tilde{a}\tilde{Z}_-) - m_{\tilde{Z}_-}$	$\mu = 0$	$M - V_2/V_1$

Table II : Main e^+e^- analyses

ponding to the limiting case where $M_1 = M_2 = 0$ and for $V_2/V_1 = 1$. As long as $M \ll M_Z$, the LSP remains an almost pure photino, and one expects a wino $\tilde{W}_-$ and a zino $\tilde{Z}_-$ more gaugino than higgsino-like, lighter than the W and Z respectively. If $V_2/V_1 = 1$, the axino $\tilde{a}$, with a mass $m_{\tilde{a}} \sim \mu$, decouples from the three other neutralinos.

2.5 What can we learn from e^+e^- collisions ?

Table II gives an overview of the main reactions which have been studied, which of the physical quantities they give access to and, under some specified additional hypotheses, which of the SBP's they allow to constrain.

The case where the LSP is the gravitino has also been studied, in a framework somewhat different from that of the MSM, and which we will present when relevant.

3. EXPERIMENTAL RESULTS

3.1. Search for unstable photinos

Since most SP searches assume that $\tilde{\gamma}$ is the LSP, and since this is also favored in most supersymmetric models, it is worth trying to assess the $\tilde{\gamma}$ stability experimentally. If the LSP were not $\tilde{\gamma}$, it could be : $\tilde{\nu}$, but then $\tilde{\gamma} \rightarrow \bar{\nu}\tilde{\nu}$ (one loop) does not lead to any observable difference from the case LSP=$\tilde{\gamma}$; $\tilde{a}$ or $\tilde{G}$, and in this case the decay $\tilde{\gamma} \rightarrow \gamma\tilde{a}$ (one loop) or $\tilde{\gamma} \rightarrow \gamma\tilde{G}$ (tree level) may be observable.

The natural way to search for unstable $\tilde{\gamma}$ is to look for their pair production by t-channel $\tilde{e}$ exchange in $e^+e^- \rightarrow \tilde{\gamma}\tilde{\gamma}$ followed by $\tilde{\gamma} \rightarrow \gamma$ + LSP. If the decay is prompt enough, the final state will be $\gamma\gamma$ + missing energy. Recent CELLO results[9)] are shown in figure 3, from which one can infer that, if $m_{\tilde{\gamma}} \lesssim 15$ GeV, the photino is stable unless $m_{\tilde{e}} \gtrsim 100$ GeV. Translated in terms of SBP's, this implies that the domain in M-$m_{3/2}$ limited by the dashed curve on figure 7 is excluded.

3.2. The scale of supersymmetry breaking

Whereas, if $\tilde{a}$ is the LSP, it would be rather unnatural that $\tilde{\gamma}$ be light, this is well possible if a very light $\tilde{G}$ is the LSP. In this case, which is not that of the MSM, the $\tilde{\gamma}$ lifetime $\tau_{\tilde{\gamma}}$ may become long enough to preclude the above analysis since one or two of the $\tilde{\gamma}$'s may escape the detector before decaying. $\tau_{\tilde{\gamma}}$ depends on the scale d of supersymmetry breaking ($\tau_{\tilde{\gamma}} = 8\pi\, d^2/m^5_{\tilde{\gamma}}$) or equivalently on $m_{3/2}$, the $\tilde{G}$ mass, since

$m_{3/2} = \left(\frac{4\pi}{3} G_N\right)^{1/2} d$, where G_N is the Newton gravitation constant[10]. Taking the finite $\tilde{\gamma}$ lifetime into account, the limit shown on figure 3 becomes, for low $m_{\tilde{\gamma}}$, contour (I) on figure 4, valid for $d = (300 \text{ GeV})^2$. Using also the results from its single photon search, CELLO could access longer lifetimes and extend the excluded domain to that limited by contours (II) and (III). If, instead of fixing d, one fixes $m_{\tilde{e}}$, one can equivalently exclude a domain in the $m_{\tilde{\gamma}}$-d (or $m_{\tilde{\gamma}}$-$m_{3/2}$) plane (see Fig. 5). However, this domain vanishes for sufficiently large $\tilde{e}$ masses. A more powerful bound can then be obtained[11], independent of $m_{\tilde{e}}$, through the reaction $e^+e^- \rightarrow \tilde{\gamma}\tilde{G}$ which can proceed by s-channel one photon annihilation. In this case, the final state will be a single photon (actually, if $\tau_{\tilde{\gamma}}$ is too large, one can still use the initial radiation tagging technique). The bound obtained by CELLO is indicated as a dashed line on figure 5. Scale of supersymmetry breaking as low as $\sim(200 \text{ GeV})^2$ are excluded for $m_{\tilde{\gamma}} \lesssim 30$ GeV. This bound can probably be improved to $\sim(400 \text{ GeV})^2$, using in addition the recent MAC and ASP single photon searches.

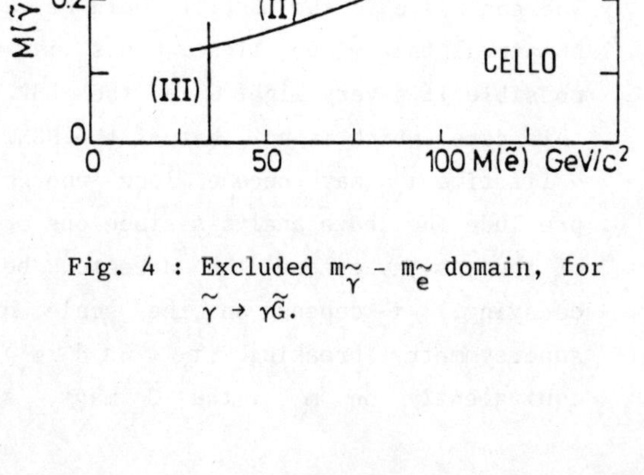

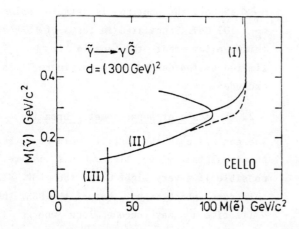

Fig. 4 : Excluded $m_{\tilde{\gamma}}$ - $m_{\tilde{e}}$ domain, for $\tilde{\gamma} \rightarrow \gamma\tilde{G}$.

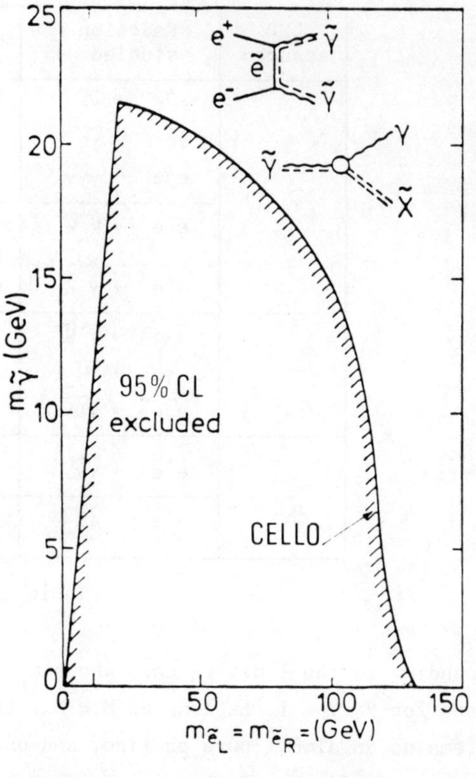

Fig. 3 : Excluded $m_{\tilde{\gamma}}$ - $m_{\tilde{e}}$ domain for promptly decaying $\tilde{\gamma}$.

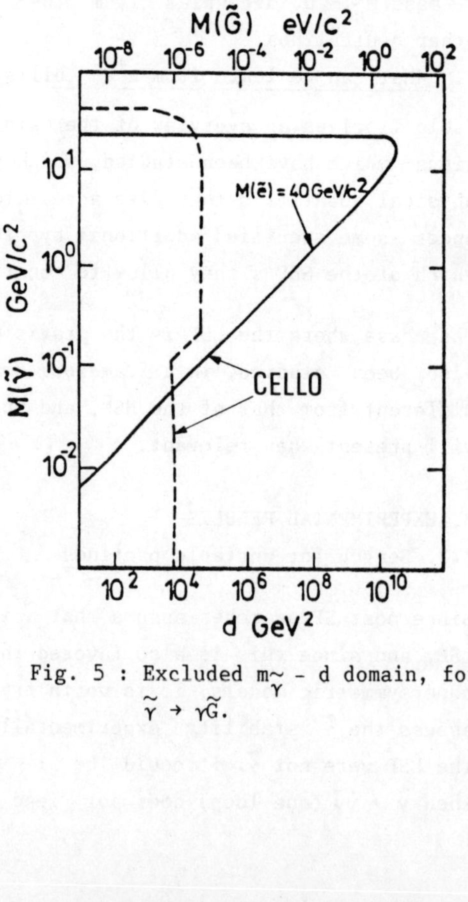

Fig. 5 : Excluded $m_{\tilde{\gamma}}$ - d domain, for $\tilde{\gamma} \rightarrow \gamma\tilde{G}$.

3.3. Search for scalar electrons

Here, $\tilde{\gamma}$ is supposed to be the LSP, or at least stable. As we have seen, if $m_{\tilde{\gamma}} \lesssim 15$ GeV, this is justified for $m_{\tilde{e}} \lesssim 100$ GeV. Three reactions have been considered, with $\tilde{e} \to e\tilde{\gamma}$:

i) $e^+e^- \to \tilde{e}\tilde{e}$. $m_{\tilde{e}}$ is limited to the beam energy. The final state consists of acoplanar electron pairs.

ii) $\gamma e \to \tilde{\gamma}\tilde{e}$ (actually ee $\to \tilde{\gamma}\tilde{e}$e, with the final state e remaining in the beam pipe). $m_{\tilde{e}}$ can exceed the beam energy. The final state consists of single electrons.

iii) $e^+e^- \to \gamma\tilde{\gamma}\tilde{\gamma}$, with an $\tilde{e}$ exchanged in the t-channel. The final state consists of a single photon.

This last method provides the most stringent limits for light enough photinos. As this topic was already covered in this session[2], we simply show the results on figure 6. The combined ASP, MAC and CELLO lower limit on $m_{\tilde{e}}$ is 65 GeV for $m_{\tilde{\gamma}} = 0$.

3.4. Comparison of e^+e^- and $p\bar{p}$ results

In terms of SBP's, the M-$m_{3/2}$ domain excluded by the above searches is indicated on figure 7 (continuous contour labelled "ee"). Similarly, the $m_{\tilde{q}} - m_{\tilde{g}}$ domain excluded by the UA1 collaboration in the search for events with missing p_T reported at this conference[1] can also be turned into an M-$m_{3/2}$ excluded domain. As can be seen on figure 7, the collider results are now becoming more constraining than those from e^+e^- collisions. This is due to the possibility of producing $\tilde{g}$ pairs at a substantial rate, even for very massive $\tilde{q}$, thanks to the non-abelian $g\tilde{g}\tilde{g}$ coupling. However, one should remark that the UA1 limits are obtained under the $m_{\tilde{\gamma}} = 0$ assumption, which is inconsistent, within the MSM, with $m_{\tilde{g}} \gtrsim 60$ GeV. This is probably not serious as long as $m_{\tilde{\gamma}} \lesssim 15$ GeV. Also, if $\tilde{\gamma}$ is not the

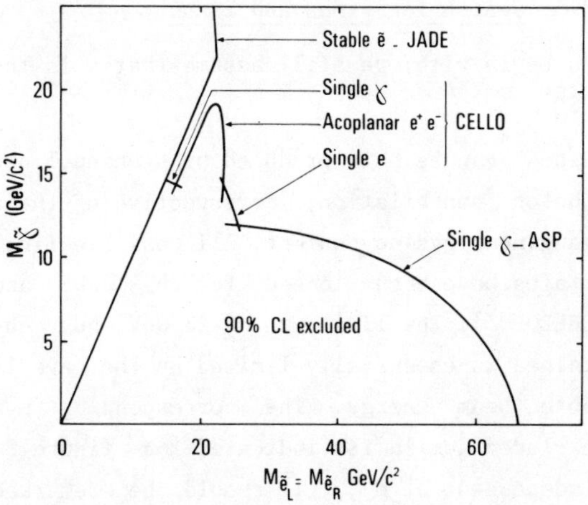

Fig. 6 : Excluded $m_{\tilde{\gamma}} - m_{\tilde{e}}$ domain for stable $\tilde{\gamma}$.

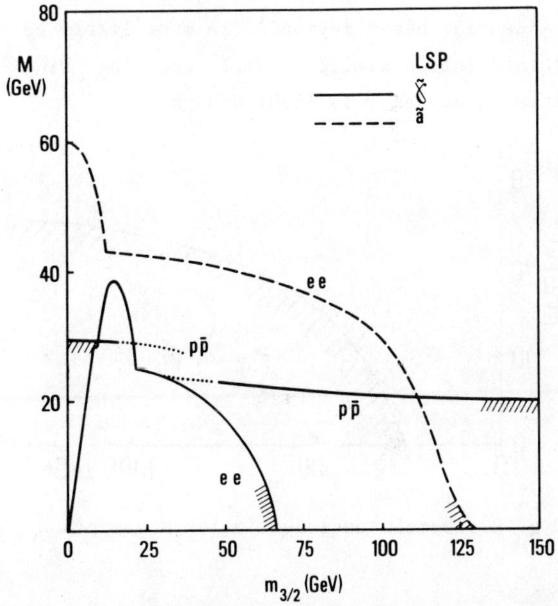

Fig. 7 : Domains excluded in the M - $m_{3/2}$ plane by e^+e^- and $p\bar{p}$ experiments.

LSP, then $\tilde{g} \to q\bar{q}\tilde{\gamma}$ is followed by $\tilde{\gamma} \to \gamma$+LSP, where the most likely LSP is $\tilde{a}$. The missing p_T signal is then diluted and the collider limits become worse ; on the other hand, the domain excluded by e^+e^- experiments becomes the one limited by the dashed contour, as already explained, and is then the most constraining over a large fraction of the M-$m_{3/2}$ plane.

3.5. Search for winos and zinos

To begin with, we still assume that $\tilde{\gamma}$ is the LSP.

Winos can be pair produced by s-channel one photon annihilation, irrespective of their gaugino-higgsino content. All possible final states have been looked for by JADE and CELLO[3-9]. The limit $m_{\tilde{W}_-} \gtrsim 23$ GeV thus obtained is essentially limited by the available beam energy. The corresponding μ-M excluded domain is indicated on figure 8, independent of $m_{3/2}$. It should be remarked that $W \to \tilde{W}_- \tilde{\gamma}$ has a signature very similar to $W \to L\nu_L$, and that the lower limit of 41 GeV obtained by UA1 for the mass of a new sequential heavy lepton[1] can most likely be turned into a similar limit for the wino mass, provided $\tilde{\gamma}$ is light enough.

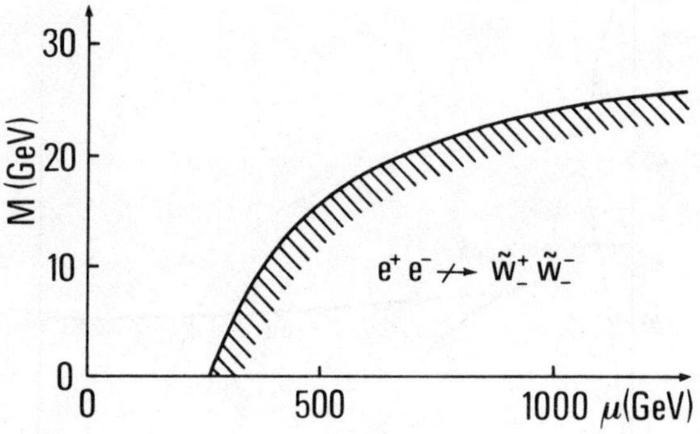

Fig. 8 : Domain excluded in the M - μ plane.

Zino production in e^+e^- collisions could have a threshold lower than wino production, using the reaction $e^+e^- \to \tilde{\gamma}\tilde{Z}_-$. But since this reaction proceeds by t-channel $\tilde{e}$ exchange, the $m_{\tilde{Z}}$ limit obtained this way will depend on $m_{\tilde{e}}$ (and also on $m_{\tilde{\gamma}}$). Recent CELLO and JADE[3-9-12] results are shown on figure 9. Here, $\tilde{Z}_-$ was assumed to be a pure gaugino, which is not too wrong, within the MSM, for such low $m_{\tilde{Z}}$. However, in the MSM, $m_{\tilde{W}_-} \gtrsim 23$ GeV also implies $m_{\tilde{Z}_-} \gtrsim 29$ GeV for $m_{\tilde{\gamma}}=0$, or $m_{\tilde{Z}} \gtrsim 23$ GeV for $m_{\tilde{\gamma}} = 10$ GeV. Given the existing limits on $m_{\tilde{e}}$ (see 3.3), it

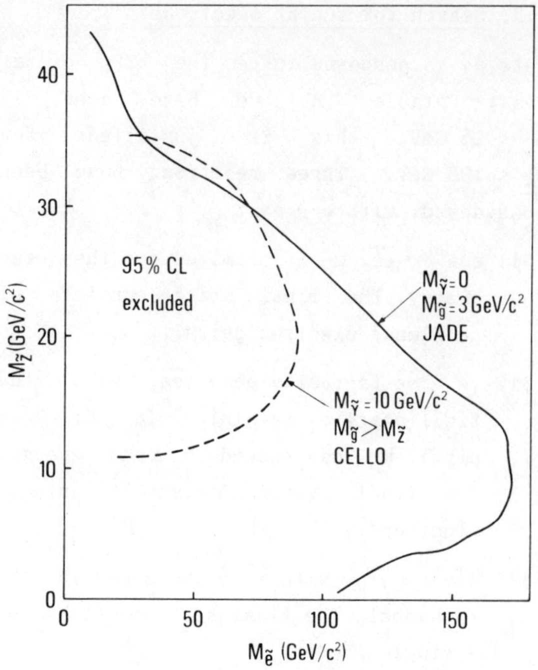

Fig. 9 : Excluded $m_{\tilde{Z}} - m_{\tilde{e}}$ domain.

can be seen that these measurements do not provide any new constraint on the SBP's for $m_{\tilde{\gamma}}=0$, and only very little for $m_{\tilde{\gamma}} = 10$ GeV.

We now turn to the disfavored case $V_2/V_1 \ll 1$. We have seen that $\tilde{\nu}$ can then be substantially lighter than $\tilde{e}$, and therefore $\tilde{W}_- \to \ell\tilde{\nu}$ occurs even if $\tilde{W}_- \to \tilde{\ell}\nu$ is kinematically forbidden. In that case, $\tilde{W}_-$ can be produced not only in pairs, but also singly ($\gamma e \to \tilde{\nu}\tilde{W}_-$ followed by $\tilde{W}_- \to \ell\tilde{\nu}$, leading to a single lepton in the final state). Again still, single photon searches provide the most stringent mass limits on the t-channel exchanged $\tilde{W}_-$ in the reaction $e^+e^- \to \gamma\tilde{\nu}\tilde{\nu}$, for small enough $\tilde{\nu}$ masses. Figure 10 shows the the domain of $\tilde{W}$ and $\tilde{\nu}$ masses excluded by these searches[2] when $\tilde{W}_-$ is assumed to be a pure gaugino. In this instance, the combined result is $M_{\tilde{W}_-} > 66$ GeV for the unlikely case of massless $\tilde{\nu}$. No useful limit on the SBP's can be derived from these analyses since four parameters (μ-M-$m_{3/2}$-V_2/V_1) would have to be taken into account.

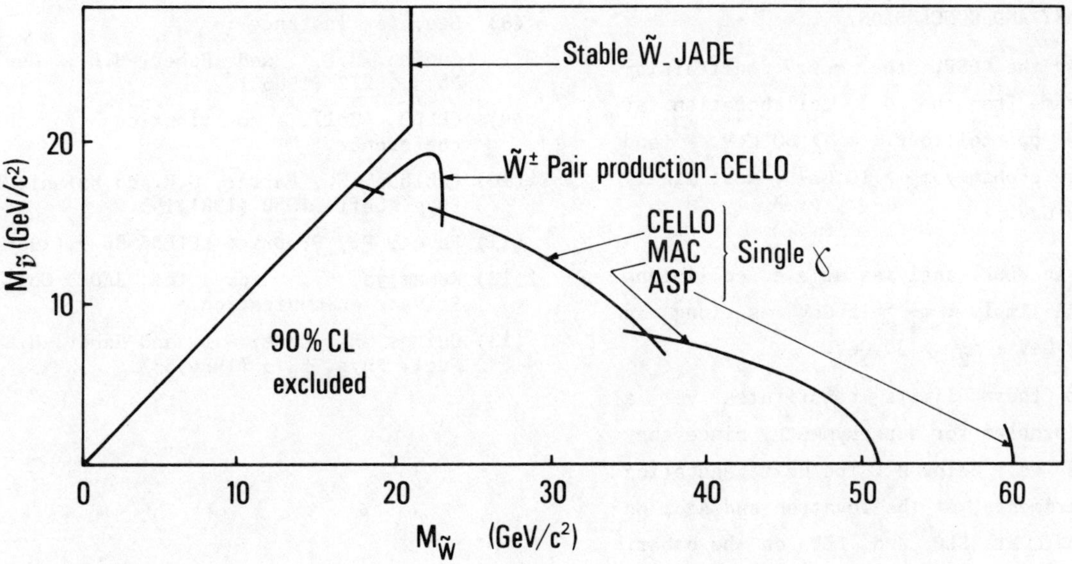

Fig. 10 : Excluded $m_{\tilde{W}}$ - $m_{\tilde{\nu}}$ domain in the case $m_{\tilde{\nu}} < M_{\tilde{W}} < m_{\tilde{\ell}}, _{\tilde{q}}$.

3.6. Search for light axinos

In the MSM, the requirement that SU(2)×U(1) be spontaneously broken at low energy implies, as already stated, $\mu \sim m_{3/2}$ and therefore $m_{\tilde{a}} \sim m_{3/2}$. In this case, it is unlikely that $\tilde{a}$ will be the LSP. However, one can generalize the MSM to accomodate small μ values, and actually even μ=0, the case that we will consider now[13]. The corresponding neutralino mass pattern is shown on figure 11 for $V_2/V_1 = 1$. The decay $Z^o \to \tilde{Z}_- \tilde{a}$ can be kinematically favored, with a rate depending on the $\tilde{Z}_-$ mass, which is directly related to $m_{\tilde{\gamma}}$, and on V_2/V_1. The reaction $e^+e^- \to$ virtual $Z^o \to \tilde{Z}_-\tilde{a}$, followed by $\tilde{Z}_- \to f\bar{f}\tilde{a}$ has been studied at PEP and PETRA in the context of searches for monojets[3]. The JADE result[12], shown on figure 12, implies $m_{\tilde{Z}_-} \gtrsim 32$ GeV in the favored case $V_2/V_1 \sim 1$.

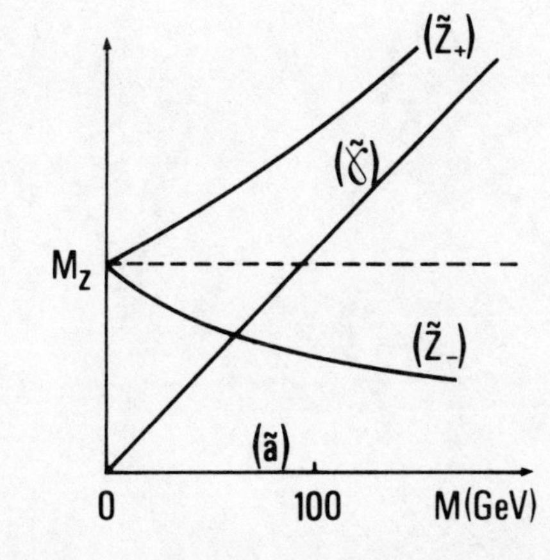

Fig. 11 : Neutralino spectrum for μ = 0.

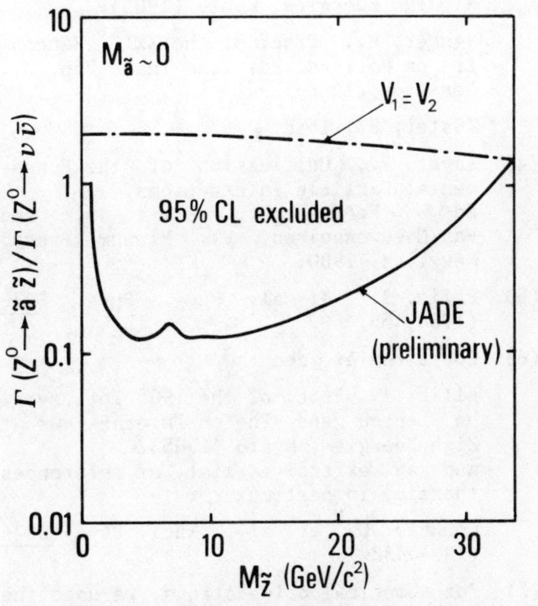

Fig. 12 : Excluded $m_{\tilde{Z}}$ - $\Gamma(Z \to \tilde{a}\,\tilde{Z})$ domain, for $m_{\tilde{a}} = 0$.

4. SUMMARY AND CONCLUSION

If $\tilde{\gamma}$ is the LSP, the most constraining limits come from the UA1 collaboration at the CERN $\bar{p}p$ collider : $m_{\tilde{g}} \gtrsim 60$ GeV, and therefore probably $m_{\tilde{\gamma}} \gtrsim 10$ GeV ; most likely $m_{\tilde{W}_-} \gtrsim 40$ GeV.

If $\tilde{a}$ is the LSP, analyses of e^+e^- collisions at PETRA imply : $m_{\tilde{\gamma}} \gtrsim 15$ GeV as long as $m_{\tilde{e}} \lesssim 100$ GeV ; $m_{\tilde{Z}_-} \gtrsim 30$ GeV.

None of these limits constitutes yet a serious problem for supersymmetry since they all lie well below M_W. The next generation of experiments, at the Tevatron and Acol on one hand, at SLC and LEP on the other, should provide results which, if still negative, would at least force model builders into maybe less elegant constructions than the simplest MSM.

REFERENCES

(1) Honma, A., Proceeding of this conference.

(2) Whitaker, S., Ibid.

(3) Komamiya, S., Proc. of the 1985. Int. Symp. on Lepton and Photon Interactions at High Energies, Kyoto (1985).

Jonker, M., Proc. of the XXIst Rencontre de Moriond, Ed. Tran Thanh Van, J., Les Arcs (1986).

Küster, H., Ibid.

(4) Fayet, P., Unification of the Fundamental Particle Interactions, Eds. Ferrara, S., Ellis, J., and Van Nieuwenhuizen, P., Plenum Press, New York, 1980.

(5) Ellis, J. et al, Nucl. Phys. B238 (1984)453.

(6) For a review, see :

Ellis, J., Proc. of the 1985 Int. symp. on Lepton and Photon Interactions at High Energies, Kyoto (1985), and an extensive list of references therein, in particular :

Kounnas, C. et al., Nucl. Phys. B236 (1984)438.

(7) For numerical calculations, we used the values given by :

Tata, X. and Dicus, D.A., Univ. of Wisconsin-Madison preprint MAD/PH/281, May 1986.

(8) See, for instance :

Gunion, J.F. and Haber, H.E., Nucl. Phys. B272 (1986)1.

(9) CELLO Coll., contributions to this conference.

(10) Cabibbo, N., Farrar, G.R.and Maiani, L. Phys. Lett. 105B (1981)155.

(11) Fayet, P., Preprint LPTENS 86-9,(1986).

(12) Komamiya, S., for the JADE Coll., Private communication.

(13) Quiros, M., Kane, G.L. and Haber, H.E., Nucl. Phys. B273 (1986)333.

RECENT SEARCHES FOR LEPTON FLAVOR VIOLATION

Douglas Bryman

TRIUMF, 4004 Wesbrook Mall, Vancouver, B.C., Canada V6T 2A3

1. INTRODUCTION

No process which does not conserve lepton number has been observed. However, since this empirical information is incorporated in the standard model in an ad hoc fashion, searches for reactions which exhibit lepton flavor violation (LFV) may lead to new approaches or new constraints to further development. In this paper, limits from searches for muon-electron and muon-positron conversion in the field of a nucleus and for reactions $\mu \rightarrow eX$ where X is a weakly interacting boson are discussed. New limits on $\mu \rightarrow e\gamma(\gamma)$ (1) are presented elsewhere in these proceedings. A summary of the status of LFV searches is given in the final section.

2. MUON-ELECTRON CONVERSION

Searches for the reactions

$$\mu^- + Ti \rightarrow e^- + Ti \qquad (1)$$

and

$$\mu^- + Ti \rightarrow e^+ + Ca \qquad (2)$$

have been carried out using the time projection chamber (TPC) at TRIUMF shown in Fig. 1.

Reaction (1) which violates lepton flavor conservation is potentially sensitive to the presence of new particles including Higgs particles, heavy neutral leptons and leptoquarks. Coherent action of the nuclear quarks results in the production of an electron at the unique energy $E_e \simeq m_\mu c^2 - B$ where m_μ is the muon mass and B is the muon binding energy. The neutrinoless conversion of a muon into a positron, reaction (2), violates both lepton flavor and lepton number conservation. Possible mechanisms which would allow reaction (2) to occur also imply new particles such as massive Majorana neutrinos.

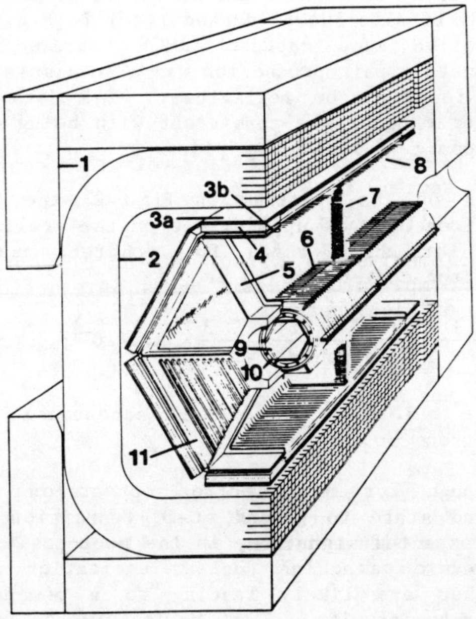

Figure 1: TRIUMF TPC used to search for $\mu^- Ti \rightarrow e^- + Ti$. The numbered elements are: 1) the magnet iron, 2) the coil, 3a) and 3b) outer trigger scintillators, 4) outer trigger proportional counters, 5) end cap support frame, 6) inner electric field cage wires;, 7) central high voltage plane, 8) outer electric field cage wires, 9) inner trigger scintillators, 10) inner trigger cylindrical proportional wire chamber, and 11) end cap proportional wire modules for track detection.

The TPC (2-4) operated at atmospheric pressure in a uniform magnetic field of 0.9 T. Cosmic-ray chambers and scintillators covered the top and upper-side regions of the magnet. A 73 MeV/c negative muon beam stopped in a natural Ti target at an average rate of 10^6 s^{-1}. Pion and electron contaminants in the beam were suppressed to the levels $\pi/\mu \sim 10^{-4}$ and $e/\mu \sim 10^{-2}$, respectively, by an rf separator. Particles with momenta $\gtrsim$ 70 MeV/c emitted during the period 5 to 1000 ns after a muon stop signal in the beam scintillators were accepted by the trigger system.

Figure 2 (solid line) shows the electron momentum spectrum in the range 87 to 130 MeV/c obtained with $N_c=0.9\times10^{12}$ μ^- captures in Ti. No events were found in the momentum region 96.5 to 106 MeV/c which would contain 85% of the events from coherent muon-electron conversion. The dotted line in Fig. 2 gives the distribution expected for coherent μe conversion at the branching ratio 7×10^{-11}. The observed spectrum is consistent with decay of muons bound in atomic orbit (5) which is also shown (dashed line) in Fig. 2. Radiative muon capture (RMC) followed by asymmetric pair production was also simulated and found to be negligible. The data at higher momenta are consistent with being due to cosmic rays and beam pions.

Based on the spectrum in Fig. 2, the 90% confidence level upper limit on the preliminary branching ratio for coherent muon-electron conversion (6) is

$$\frac{\Gamma(\mu^- + Ti \to e^- + Ti)}{\Gamma(\mu^- + Ti \to capture)} < \frac{2.3}{N_c\varepsilon} < 4.5\times10^{-12}$$

where $\varepsilon=0.06$ is the overall acceptance for electrons at 101 MeV/c.

Although for muon-electron conversion the ground state to ground state transition is the expected signature, in the muon-positron conversion reaction nuclear excitation and breakup are likely leading to a momentum distribution with $p \lesssim 100$ MeV/c. Due to features of the conversion reactions, of the backgrounds and of the detectors, the data analysis for positron conversion differed from that for electron conversion. In particular, the position resolution in the TPC varied from $\geq$ 210 μm for electron tracks to $\geq$ 500 μm for positrons as discussed in detail in Refs. 2 and 3. In order to eliminate events with two charged particles like those due to RMC, several cuts including multiple counter firing and energy loss criteria were more restrictive for positron events.

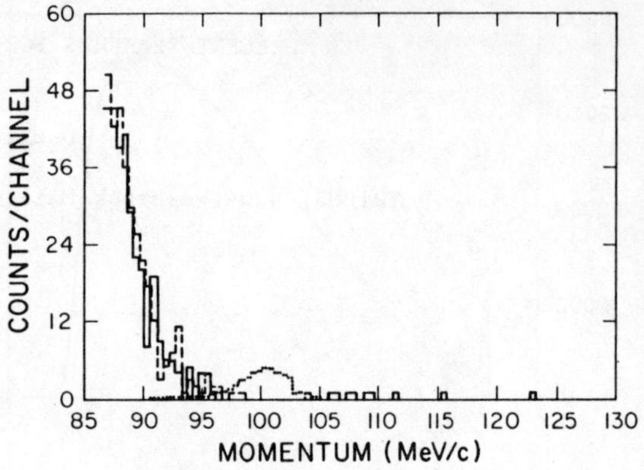

Figure 2: The solid-line histogram is the observed electron momentum spectrum for $87 < p < 130$ MeV/c. The dashed-line histogram is from a Monte Carlo calculation of bound muon decay. The expected spectrum for a signal from coherent μe conversion at a branching ratio 7×10^{-11} is shown as the dotted line.

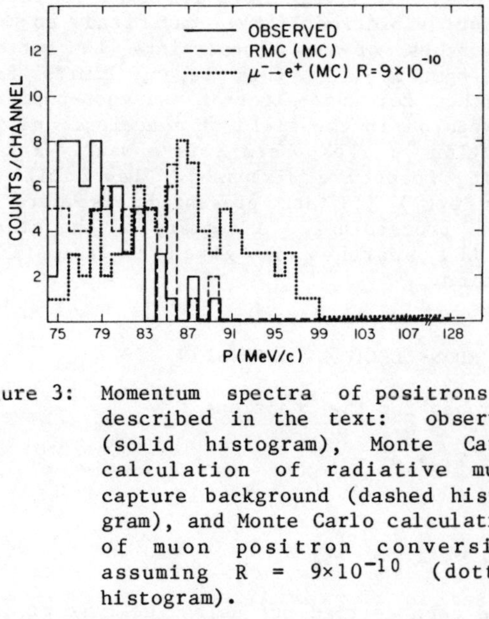

Figure 3: Momentum spectra of positrons as described in the text: observed (solid histogram), Monte Carlo calculation of radiative muon capture background (dashed histogram), and Monte Carlo calculation of muon positron conversion assuming $R = 9\times10^{-10}$ (dotted histogram).

Figure 3 (solid line) shows the preliminary positron spectrum. The data is consistent with a Monte Carlo calculation of the RMC background (dashed line).

No positron events were found with observed momenta $p > 90.5$ MeV/c. An estimated limit on the total rate can be obtained by assuming a Lorentzian giant resonance excitation for the final nuclear state:

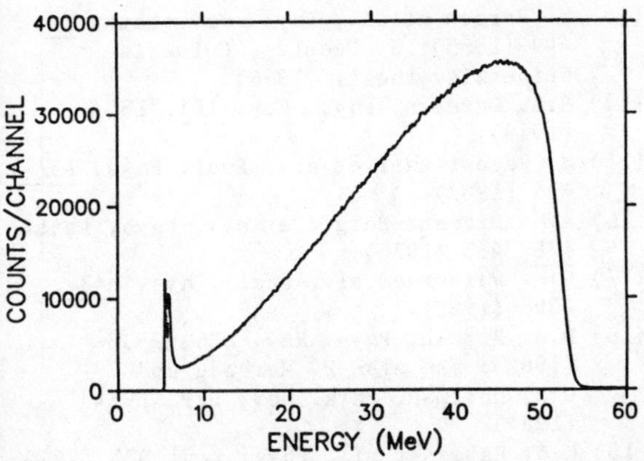

Figure 4: Muon decay positron spectrum from Ref. 12.

$$R(Ti) = \frac{\Gamma(\mu^- + Ti \rightarrow e^+ + Ca)}{\Gamma(\mu^- + Ti \rightarrow capture)}$$

$$< \frac{2.3}{N_c \varepsilon_1 \varepsilon_2} < 9\times10^{-11} \qquad (3)$$

where ε_1=0.02 is the positron acceptance and ε_2=0.14 is the fraction of events expected with p > 90.5 MeV/c.

3. $\mu \rightarrow eX^0$

Reactions such as $K^+ \rightarrow \pi^+X^0$ (7,8) and $\mu^+ \rightarrow e^+X^0$ (9), where X^0 denotes a neutral, weakly interacting scalar or pseudoscalar boson have also been searched for as evidence for LFV. Candidates for X^0 include light axions, Majorons, Higgs particles, familons and Goldstone bosons. Recently, Eichler et al. (10) searched for reactions $\mu^+ \rightarrow e^+X_h$ and $\pi^+ \rightarrow e^+\nu X_h$ followed by $X_h \rightarrow e^+e^-$ with X_h lifetimes below 10^{-10} s and found limits on the branching ratios down to the level $2\cdot10^{-12}$. In a complementary study, several experiments involving muon decay have been re-examined looking for evidence of the inclusive decay $\mu \rightarrow eX_h^0$ (11).

In a TRIUMF experiment (12) the $\mu^+ \rightarrow e^+\nu\bar{\nu}$ positron spectrum shown in Fig. 4, was obtained in the measurement of the $\pi \rightarrow e\nu/\pi \rightarrow \mu\nu$ branching ratio. This spectrum has been searched for peaks due to $\mu \rightarrow eX_h^0$. The upper limits obtained from this and other experiments are shown in Fig. 5. For most of the accessible region $\Gamma(\mu \rightarrow eX_h) / \Gamma(\mu \rightarrow e\nu\bar{\nu}) \lesssim 3\times10^{-4}$ (90% C.L.).

4. CONCLUSION

Table 1 summarizes the status of searches for lepton flavor violation. Although unprecidented sensitivities have been reached only

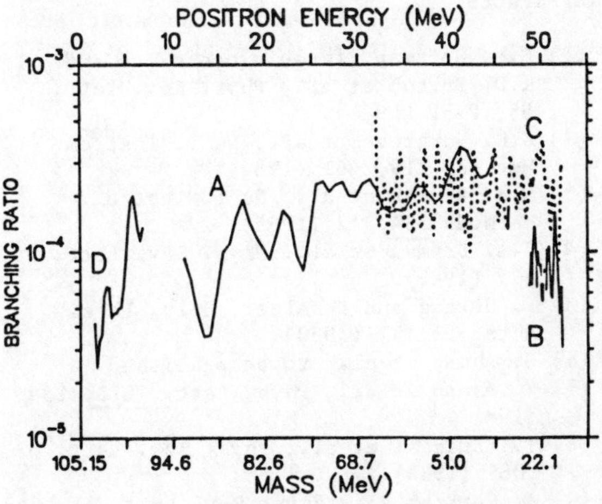

Figure 5: Branching ratio upper limits (90% C.L.) for $\mu \rightarrow eX_h$ vs mass of X_h. Curves A, B, C and D are derived from Refs. 12, 9, 13 and 14, respectively. The top scale gives the positron energy.

null results have been reported. Bounds have been extracted on new particles in the 10 GeV to 100 TeV range (22) which arise in various extensions to the standard model.

Table 1
Experimental Searches for Lepton Flavor Violation

Process	Present Upper Limit on Branching Ratio (90% C.L.)	Reference
$\mu \rightarrow e\gamma$	4.9×10^{-11}	1
$\mu \rightarrow 3e$	2.4×10^{-12}	17
$\mu \rightarrow e\gamma\gamma$	3.8×10^{-10}	1
$\mu \rightarrow eX_0^0$	2.6×10^{-6}*	9
$\mu \rightarrow eX_h$	3×10^{-4}**	11
$\mu \rightarrow eX_h$, $X_h \rightarrow e^+e^-$	2×10^{-12}***	10
$\mu^-Ti \rightarrow e^-Ti$	4.5×10^{-12}	6
$\mu^-S \rightarrow e^-S$	7×10^{-11}	15
$\mu^- Ti \rightarrow e^+ Ca$	9×10^{-11}	6
$K^+ \rightarrow \pi^+\mu^+e^-$	5.0×10^{-9}	16
$K_L^0 \rightarrow \mu e$	10^{-8}†	
$\pi^0 \rightarrow \mu e$	7.0×10^{-8}	18
$\tau \rightarrow \mu\gamma$	5.5×10^{-4}	19
$\tau \rightarrow \mu X_0^0$	0.02††	20
$B^0 \rightarrow \mu e$	2.0×10^{-4}	21

*$M_{X_0^0} \approx 0$ †estimated

**$0 < m_{X_h} < 104$ MeV/c^2 ††$M_{X_0^0} < 0.1$ GeV

***$\tau_{X_h} < 10^{-10}$ s

REFERENCES

(1) C.M. Hoffman, these proceedings and R.D. Bolton et al., Phys. Rev. Lett. 55, 2461 (1986).
(2) C.K. Hargrove et al., Nucl. Instrum. Methods 219, 461 (1984).
(3) D.A. Bryman et al., Nucl. Instrum. Methods 234, 42 (1985).
(4) D.A. Bryman et al., Phys. Rev. Lett. 55, 465 (1985).
(5) F. Herzog and K. Alder, Helv. Phys. Acta. 53, 53 (1980).
(6) S. Ahmad et al., to be published.
(7) Y. Asano et al., Phys. Lett. 107B, 159 (1981).
(8) T. Yamazaki et al., Phys. Rev. Lett. 52, 1089 (1984).
(9) J. Carr et al., Phys. Rev. Lett. 51, 627 (1983); A. Jodidio, LBL thesis (1986) unpublished.
(10) R. Eichler et al., Phys. Lett. B175, 101 (1986).
(11) D.A. Bryman and E.T.H. Clifford, to be published.
(12) D.A. Bryman et al., Phys. Rev. D33, 1211 (1986).
(13) M. Bardon et al., Phys. Rev. Lett. 14, 449 (1965); J. Peoples, Columbia University thesis (1966).
(14) S.E. Derenzo, Phys. Rev. 181, 1854 (1969).
(15) A. Badertscher et al., Nucl. Phys. A377, 406 (1982).
(16) A.M. Diamant-Berger et al., Phys. Lett. 62B, 485 (1976).
(17) H.K. Walter et al., Nucl. Phys. A434, 409c (1985).
(18) D.A. Bryman, Phys. Rev. D26, 2538 (1982); see also P. Herczeg and C.M. Hoffman, Phys. Rev. D29, 1954 (1984).
(19) K.G. Hayes et al., Phys. Rev. D25, 2869 (1982).
(20) R.M. Baltrusaitis et al., Phys. Rev. Let.. 55, 1842 (1985).
(21) M. Halling, Proc. DPF Meeting, Santa Fe, New Mexico, Nov. 12, 1984.
(22) See, for example, O. Shanker, Nucl. Phys. B206, 253 (1982), R.N. Cahn and H. Harari, Nucl. Phys. B176, 135 (1980), and L.J. Hall and L.J. Randall, Harvard University preprint HUTP-86/A002.

SEARCHES FOR MONOPOLES AND QUARKS*

H. S. Matis

Lawrence Berkeley Laboratory, University of California,
Berkeley, CA 94720, U.S.A

Within the last year, several sensitive searches for monopoles and quarks have been done. Recent experiments at the Tevatron and at the CERN $p\bar{p}$ collider have detected no evidence for free fractional charge. An experiment in a iron refinery, which searched for GUT monopoles trapped in iron ore with two SQUID detectors, found no monopole candidate. However, an experiment looking for monopoles in cosmic rays has measured an interesting event which could be interpreted as a monopole. Several detectors are being built to achieve significant improvements in sensitivity for detection of quarks and monopoles.

1. FREE QUARK PRODUCTION

Since the discovery of quantized electric charge by Millikan in 1909, no accelerator experiment has claimed detection of fractional charge(1,2). While there is one experimental group(3) which claims measurement of fractional charge in niobium, there are many other bulk matter(2,4) searches and cosmic ray experiments which only have measured integer charged particles.

After Gell-man and Zweig (5) proposed that quarks are the fundamental building blocks of hadrons, it was assumed that measurement of a fractionally charged quark would be necessary to prove their theory. However, with the development of Quantum Chromodynamics (QCD), theorists have postulated that color is an unbroken local gauge symmetry, so quarks are confined and consequently only integer charged particles can be found in nature. However, there is no proof of confinement in QCD. There exist several models(6,7) which postulate that color symmetry is broken and therefore, free fractionally charged particles could be found.

The signature of a quark produced at an accelerator is very different from that of a typical hadron. De Rujula et al. (6) argued that after a quark is produced it would capture nucleons as it passes through a detector. Since a bare quark could have a net color charge, its interaction with matter could be significantly stronger then a typical hadron. Therefore, its signature could be a particle with varying electric charge to mass ratio. Such characteristics are very difficult to detect with conventional detectors, so many previous accelerator or cosmic ray experiments would have missed such a signature. In addition, refined material, which has been used in many bulk matter experiments, might have been depleted of its quark content(8).

a) FERMILAB FRACTIONAL CHARGE EXPERIMENT

As Fermilab recently entered a new fixed target energy regime with its 800 GeV/c Tevatron program, a LBL-Irvine-San Francisco State collaboration(9) undertook a quark search experiment using a method that had been used in a previous accelerator experiment(4) and that had also been used in several bulk matter searches(2).

In order to avoid problems that many quark search experiments have had, bulk matter was used to capture any produced quark. Since quarks are stable because of charge conservation, the analysis of the stopping material can be done later in a laboratory. A nuclear target, which maximizes the quark density that can be achieved, was used because in some models(7) free fractional charges are produced only when conditions similar to production of the quark-gluon plasma occur. In addition, the target material was designed to be examined because quarks can be detected even if they were absorbed shortly after production.

In the first run, four steel cylinders filled with mercury were centered in a primary proton beam line which ran at 800 GeV/c for an integrated intensity of 1.0 x 10^{15} protons on target. Each cylinder contained about 1.5 liters of mercury. In order to sample different depths of the hadronic shower, 10 cm of lead were interspersed between the mercury targets to slow any produced quarks. A sample of mercury was extracted from the last two tanks and processed in the San Francisco State Millikan apparatus which measures the residual charge of the drop.

The Millikan apparatus(4) consists of a electrically biased mercury dropper which produces small drops of mercury which fall between two electrically charged plates. The polarity of the electric field is switched two times while the drop falls between the plates. Using measurements of the position of the drop, the net charge can be inferred. Consistency checks which include charge changing during the measurement, drop radius and multiple drops are made for each measurement of charge.

Fig. 1 shows a fitted velocity curve that was measured from a typical drop. The velocity is fitted in the three different regions shown on that curve. The curve shows the difference between the fitted and the measured velocity. In the first region, the drop falls and reaches terminal velocity. The first arrow shows when the sign of the electric field is reversed. After a short time, the drop again reaches its terminal velocity. At the second arrow the field is once again reversed. After passing a few more slits, it reaches its terminal velocity. For this particular drop, the measured charge was 19e where e is the electric charge of an electron. The net charge resolution for the apparatus was measured to be about 0.03e.

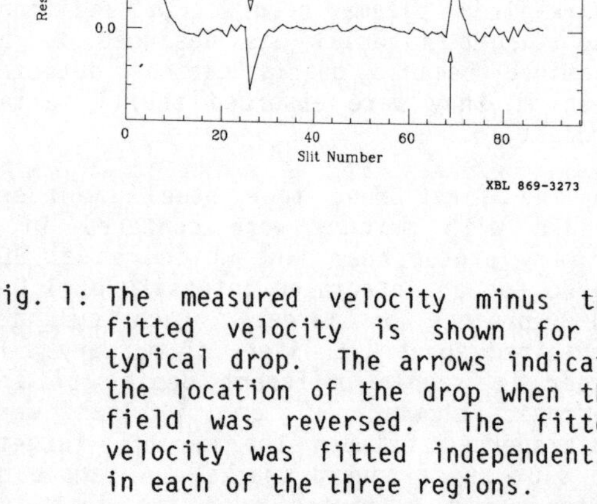

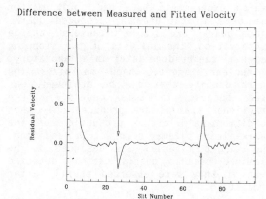

Fig. 1: The measured velocity minus the fitted velocity is shown for a typical drop. The arrows indicate the location of the drop when the field was reversed. The fitted velocity was fitted independently in each of the three regions.

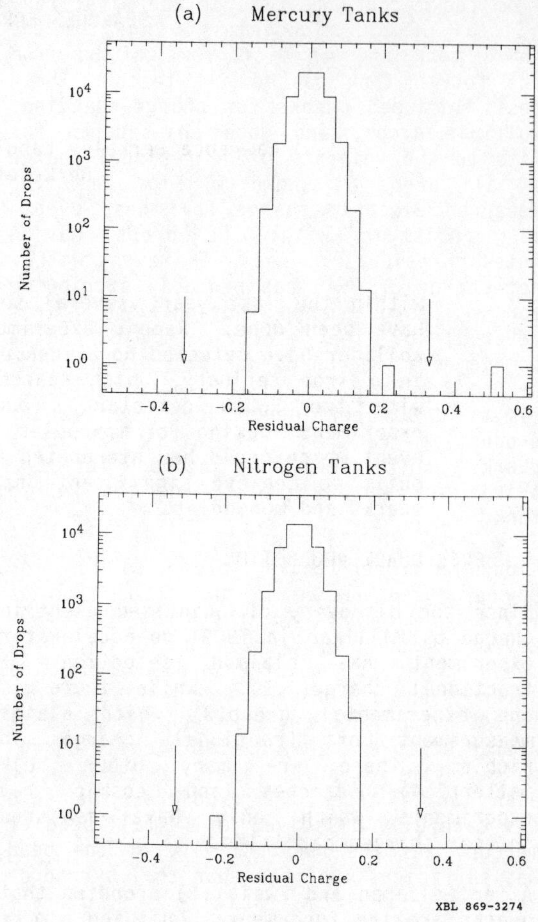

Fig. 2: This figure shows a histogram of residual charge for drops which passed all acceptance tests. The two arrows show the expected position for residual charge for any drop which contains a net fractional charge. a) Data from the distilled mercury samples. b) Data from the liquid nitrogen run.

This apparatus has processed samples whose mass was of the order of milligrams. In order to increase the amount of mercury that can be processed, a distillation apparatus is necessary. Since a quarked mercury atom is attracted to its neighbors by its image charge(8), these atoms do not evaporate when the mercury is heated. When the mercury is gently heated, the residue should contain the quarked atoms. The mercury in the third tank was distilled by a factor of 6,000 while the mercury in the fourth tank was distilled by a factor of 391,000. The reason for the large difference in the distillation factor between the two tanks was due to much larger contamination in the mercury used to fill the third tank.

From the mercury tested for the first run, a total of 230 micrograms of Hg from the third tank and 47 micrograms of Hg from the fourth tank passed all tests. These tests included checks for charge-changing, multiple drops, and good chi-squared for fits to the velocity. The residual charge of all drops is shown in Fig. 2a. The measured electric charge for these events is consistent with all drops having integer residual charge. The event with a net charge of 0.48 can probably attributed to a charge-change during the first reversal of the electric field and a complementary charge-change during the second reversal. From this data, an upper limit at 90% confidence level for quark production can be set at 2×10^{-10} quarks per interacting proton for the third tank and 2×10^{-11} for the fourth tank.

Liquid nitrogen tanks were used to stop any produced quarks in the second run of this experiment. In this run, the 800 GeV/c proton beam struck a 10 cm thick lead target. A quark, produced in the interaction, could stop in one of the four tanks. Once it stopped, then it would be attracted to one of two electrically charged gold plated glass fibers which were in each tank. After the exposure, the gold was carefully dissolved in a small bead of mercury. As the radioactivity of the bead was sufficiently higher than the surrounding material, the ability to attract charged particles was demonstrated. Folding in the field configuration, the efficiency of this process to capture charged particles can be estimated to be about 50%.

One half of the beads of mercury, which was taken from all the charged wires, was dissolved in triple distilled mercury to make a sample of 7.0 mg. So far, approximately, 213 micrograms of material have been processed. The charge on all the 46,310 measured drops, histogramed in Fig. 2b, is consistent with all drops containing only integer charges. Using the flux for 4.1×10^{13} protons and the assumed stopping efficiency in the first two liquid nitrogen tanks of 0.02, the upper limit for quark production is 1.0×10^{-10} quarks per proton interaction at the 90% confidence level.

b) QUARK SEARCH AT THE CERN SPS COLLIDER

A collaboration from Oxford-Rutherford-Imperial has exposed 200 iron balls to collisions at the CERN $p\bar{p}$ collider(10). The iron balls were placed inside the beam pipe of the collider so that the balls would be the first material that any free quark would strike. If quarks interact very strongly, they could be trapped by the iron balls. From a Monte Carlo calculation, they calculated that with an integrated luminosity of 650 nb^{-1} about 200 jets would have struck each ball. After the exposure, the balls were carefully transported back to England where the net electric charge was measured in a room temperature magnetic levitation system. A total of 60 balls have already been measured. The offset charge for the measured sample was found to be 0.1e. When that offset was included, the residual fractional charge on all balls was found to be consistent with zero using the measured charge resolution of 0.02e.

c) FUTURE QUARK DETECTORS

The problem with bulk matter detectors is that only small quantities of matter have been measured with existing detectors. In fact, in seven recent bulk matter papers a total of only 12 mg of matter has been processed. In order to process much larger samples, new techniques are necessary.

A SLAC collaboration(11) is working a detector to measure the net charge of a sample with a rotor electrometer. The basis of this detector is that when a object with charge Q is placed inside a metal box, it produces a voltage (V) which can be related to capacitance of the box (C) by the formula $V=Q/(2C\sqrt{2})$. By measuring the voltage difference between that box and a grounded box, noise effects can be reduced. They have made a detector which keeps the sample fixed and then rotates a series of pads by the sample. Using a lock-in amplifier they can average their measurements to increase their detector's accuracy. So far they have been able to achieve a charge resolution of 0.31e.

In principle, this device should be able to achieve charge resolution of 0.05e which is sufficient to observe the charge of free quarks. Their collaboration is working on reducing the noise in the amplifier and identifying the source of some low frequency signals which are increasing their charge resolution.

Another detector(12) is being developed a LLNL which uses a different method to measure charge. In this apparatus, drops of oil fall in a vacuum between two charged plates which are 5.0 meters in length. The position of each drop, which is proportional to its net charge, is measured after the deflection. The authors estimate that they will be able to measure up to 50 grams/day with a background of 1 event in a measurement of 10^{23} nucleons.

With the new construction of high energy heavy-ion accelerators, new experiments will be be done to look for free quarks produced from creation of the quark-gluon plasma. Experiments will be run at both CERN(13) and at the BNL AGS(14) within the next year.

2. PRODUCTION OF MONOPOLES

In classic paper, Dirac(15) showed that if one monopole existed in our universe, then charge must be quantized. He found that the relationship between electric charge (E) and magnetic charge (G) can be expressed by the relationship $EG=n(\hbar c/2)$ where n is an integer. Using this expression, one can let n=1 and define g as the smallest magnetic charge. If one free quark with charge 1/3 exists, then $g=(3/2)(\hbar c/e)$; if only integer charges exist, $g=(1/2)(\hbar c/e)$. Therefore, a measurement of the spectrum of magnetic charges of monopoles could show whether quarks are confined.

In the current popular Grand Unified Theories (GUT), GUT monopoles are produced when the symmetry U(1) breaks spontaneously(16). The masses are in the range 10^{16} to 10^{17} GeV or even up to 10^{19}. Becauses of their large masses, GUT monopoles can only be produced in the Big Bang and not in any forseeable accelerator. Experimenters have looked for these monopoles in cosmic rays and materials. For perspective, a 10^{17} GeV monopole has a mass of 0.18 micrograms.

A goal for cosmic ray detectors is to have a sensitivity which is greater than the Parker bound(17). The Parker bound (f) which is deduced from the measured magnetic field of the Universe can be expressed for a monopole with mass M and velocity $(10^{-3})c$ by the expression:

$$f < \begin{cases} 10^{-15}\ \mathrm{cm^{-2}sr^{-1}s^{-1}} & M<10^{17}\ \mathrm{GeV} \\ \dfrac{10^{17}}{M}\ 10^{-15}\ \mathrm{cm^{-2}sr^{-1}s^{-1}} & M>10^{17}\ \mathrm{GeV} \end{cases}$$

Currently upper limits from experiments are at a level of about a few times 10^{-12}. A monopole with mass of 10^{19} GeV at the Parker limit would just escape detection.

a) MONOPOLE SEARCH IN IRON ORE

A magnetic monopole, incident upon the earth, would fall toward the earth's center due to its gravational attraction. A likely place for a GUT monopole to be trapped would be in magnetic material such as iron ore. This ore would be a trap for monopoles as long as its temperature was below the Curie point (590°C).

A group from Kobe University(18) made a search for such monopoles in old iron ore. In order to examine a large amount of material, they placed two 20 cm diameter SQUID detectors underneath a conveyor belt of an industrial sintering furnace. The ore on the conveyor was heated to a maximum temperature of 1500°C which is sufficient to release any monopole. The freed monopole should fall toward the center of the earth and through their detector.

They ran their detector for about 1200 hours. During that time the furnace processed 140,000 tons of ore. They estimated that their detector was sensitive for 6600 tons of material. No monopole candidate was found, so an upper limit of 4.5 x 10^{-9} monopoles/gram at 95% confidence limit can be found. Since their detector was also sensitive to cosmic ray monopoles, they also set a limit on monopole cosmic ray flux which is 1.4 x 10^{-10} $\mathrm{cm^{-2}sr^{-1}s^{-1}}$ at 90% confidence limit.

b) MONOPOLE CANDIDATE

Recently, an experiment(19) looking for GUT monopoles in cosmic rays, reported on a candidate monopole event. Their detector consisted of two parallel horizontal SQUID loops (T and B) with a third vertical rectangular loop (WF) which has one of its sides going through the center of the two horizontal loops. In a total of 8,242 hours of operation, 170 possible monopole events were observed. All but one of these events can be explained by causes such as low helium level in the cryostat or mechanical shock to the apparatus.

The interesting event showed a signal (0.83 +/- 0.04) ϕ_o in the WF loop, but no significant signal in detectors T and B. A standard Dirac magnetic monopole should generate a flux of $2\phi_o$ in the T and B detectors, while the signal generated by a standard monopole in the WF detector should vary between 0 and ϕ_o. The authors have estimated that 70% of the monopoles which would produce a signal between 0.78 and 0.88 ϕ_o in the WF detector should induce an undetectable signal in the other loops. In their paper, they ruled out known causes of such a signal such as "unauthorized" interference, electronics problems, mechanical shock, motion of the trapped flux. However, they noted that it is possible for the event to be produced from some other unknown process.

If the event is caused by a cosmic ray

monopole, then one would expect the three other large monopole detectors, which have collectively set a limit at 2×10^{-12} $cm^{-2}sr^{-1}s^{-1}$, to have seen about 2 events. Thus, this event is not statistically ruled out by the other experiments. However, if this event is real, then the monopole cosmic ray flux would be orders of magnitude greater than the Parker limit for a standard GUT monopole of 10^{16} GeV but could be consistent with a monopole with mass 10^{19} GeV.

c) FUTURE MONOPOLE DETECTORS

In order to make a significant attempt to measure the cosmic monopole flux or confirm the previously mentioned event, it is necessary to construct much bigger detectors. The limitation to making SQUID detectors large arises from the fact that they must operate in a very small magnetic field. Shielding such large detectors is very difficult. A Chicago-Fermilab-Michigan group(20), one of several groups trying to significantly improve the technology of monopole detection, has been working on an induction detector that can operate in 1-10 mGauss fields. A prototype has already been built that has 1.1 m diameter loops and 1 cm separation. This is about 2.2 times greater solid angle than previous detectors. For 12 days of running they have set a limit for monopole flux of 7.1×10^{-11} $cm^{-2}sr^{-1}s^{-1}$ at 90% confidence level. In principle this detector can achieve a limit of about 10^{-13}. They are working on a design that can use an array of these detectors to measure at the Parker limit for a monopole with mass of 10^{16} GeV.

3. CONCLUSIONS

Experiments, using new techniques, have failed to find any evidence for free fractional charge at the Tevatron and the CERN $p\bar{p}$ collider. An experiment searching for monopoles trapped in the earth found no candidates. There is interesting evidence for a cosmic ray monopole. However, like Cabrera's candidate(21), it was only detected in a single loop. Significant advances are being made in constructing both quark and monopole detectors which have much greater sensitivities than previous experiments. Soon experiments will be run to search for monopoles and quarks in heavy-ion collisions.

I wish to thank my colleagues on the Fermilab the quark search experiment and H. Frisch, C. Hendricks, J. Incandela, and W. Innes for very valuable discussions.

*This work was supported by the Director, Office of Energy Research, Division of Nuclear Physics, Office of High Energy & Nuclear Physics, Nuclear Science Division, U.S. Department of Energy under Contract No. DE-AC03-76SF00098.

REFERENCES

1) M. Banner et al., Phys. Lett. 156B, 129 (1986).
2) G. Morpurgo, contributed paper 5088 to this conference; L. Lyons, Phys. Rept. 129, 225 (1985).
3) G. LaRue, J. D. Phillips and W. D. Fairbank, Phys. Rev. Lett. 46, 967 (1981).
4) M. L. Savage et al., Phys. Lett. 167B, 481 (1986).
5) M. Gell-man, Phys. Lett. 8, 214 (1964); G. Zweig, CERN Rep. TH-401 (1964); CERN Rep. TH-412 (1964).
6) A. De Rujula, R. C. Giles and R. L. Jaffe, Phys. Rev. D17, 285 (1978).
7) R. Slansky, T. Goldman and G. L. Shaw, Phys. Rev. Lett. 47, 887 (1981); G. L. Shaw and R. Slansky, Phys. Rev. Lett. 50, 1967 (1983).
8) G. Zweig, Science 201, 973 (1978).
9) H. Matis, R. W. Bland, D. Calloway, S. Dickson, A. A. Hahn, C. L. Hodges, M. A. Lingren, H. G. Pugh, M. L. Savage, G. L. Shaw, R. Slansky, A. B. Steiner, R. Tokarek, contribution 825 to this conference and LBL-21670.
10) L. Lyons, P. F. Smith, G. J. Homer, J. D. Lewin, H. W. Walford, W. G. Jones, contribution 5088 to this conference.
11) J. C. Price, W. Innes, S. Klein, M. Perl, SLAC-PUB-3938 (1986).
12) C. Hendricks, LLNL report to be published.
13) CERN proposal NA39, G. Shaw spokesman (1986); CERN proposal EMU02, B. Price spokesman.
14) AGS proposal 793, B. Price spokesman (1985); AGS proposal 801, R. Bland spokesman (1985); AGS proposal 804, G. Tarle and S. P. Ahlen co-spokesmen (1985).
15) P. A. M. Dirac, Proc. Roy. Soc. London A133, 60 (1931)
16) For instance see the following reviews: R. A. Carrigan, Jr. and W. P. Trower, FERMILAB-Pub-83/31 (1983); G. Giacomelli, IFUB 82-26, published in Racine Magnetic Monopole Workshop, 42 (1982).
17) M. S. Turner, E. N. Parker, T. J. Bogdan, Phys. Rev. D26, 1296 (1982).
18) T. Ebisu and T. Watanabe, contribution 620 to this conference.
19) A. D. Caplin et al., Nature 321, 402 (1986).
20) J. Incandela et al., EFI 85-75.
21) B. Cabrera, Phys. Rev. Lett. 48, 1378 (1982).

Short - lived Axion Searches With Long Beam Dumps

Kam-Biu Luk

Fermilab
Batavia, Illinois 60510, U.S.A.

Searches for axions with lifetime less than 10^{-11} sec and decaying into e^+e^- pairs were performed in an 800 GeV proton, and a 2.5 GeV electron beam dump experiments. No axion is found and the existence of a short-lived penetrating neutral particle with mass less than 2.4 MeV and only coupled to electron is ruled out.

Introduction

Recent observations of a narrow e^+e^- peak in low energy heavy ion collisions at GSI[1] have triggered new searches of axions. Models for axion variants[2] that are strongly coupled to e^+e^- predict the life time is of the order of 10^{-13} sec if the mass is about 1.8 MeV and the coupling is 2×10^{-9}.

Results on short-lived axion search from two beam dump experiments are reported here. Both experiments attempted to produce axions either by the axion bremsstrahlung process

$$e + Z \rightarrow e + \text{axion} + Z,$$

or the Primakoff process

$$\gamma + Z \rightarrow \text{axion} + Z,$$

in a beam dump longer than 1m and detected the e^+e^- pairs from the decay of the axions downstream of the dump. It is important to point out that the production of the axion and its subsequent decay to e^+e^- are functions of the same coupling constant, α_e, in these experiments. Results from other searches using dumps less than 1m long are presented by the next speaker[3].

Fermilab E605 [4]

The plan and elevation views of the E605 spectrometer is shown in Figure 1. An 800 GeV proton beam was absorbed by a 5.5 m long copper dump embedded in a horizontal 19kG magnetic field (3.1 GeV transverse kick over the length of the dump). The main set of data for the axion search was recorded by triggering on the sum of energy deposition greater than 150 GeV in the 19 radiation length lead-scintillator electromagnetic calorimeter which covered the full aperture of the spectrometer. A prescaled sample of muons created from the hadron shower in the dump was also recorded

The data sample was then scanned for events with isolated clusters of energy in the electromagnetic calorimeter and tracks apparently coming from the downstream face of the dump. Muons can bremsstrahlung in the last radiation length of the dump or in the calorimeter itself, thus satisfying the trigger. Consequently, events with muons which dominated the sample were rejected from the analysis. Occasionally, a muon could lose so much energy that it was not detected by the apparatus. This kind of events was the remaining background in the axion search.

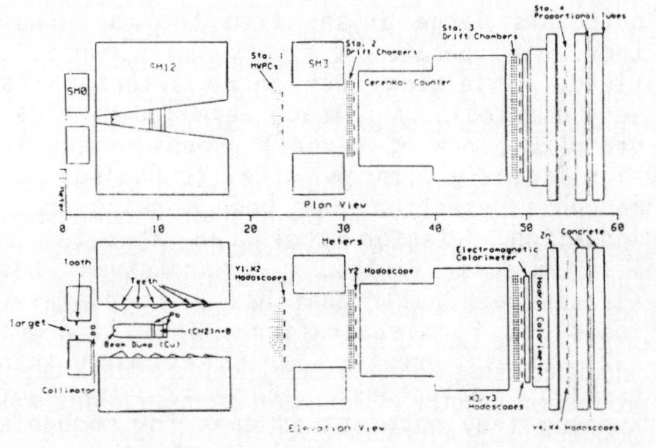

FIG. 1. Plan and elevation (bend) views of the E605 spectrometer.

Analysis of data corresponding to 4×10^{13} protons incident on the dump yielded 74 e^+e^- pairs. These events were reconstructed to a collinear (i.e. zero-mass) vertex at the downstream face of the dump. Figure 2 shows the vertical and horizontal angular distributions of these pairs at the end of the dump. The angular distributions of the prescaled muon sample were also measured. Since the dump was in a magnetic field, these muons emerged from the dump at a large vertical angle with respect to the initial proton beam direction, typically at $\pm$ 10 mrad. The angular distributions of the e^+e^- pairs in Figure 2 were identical with those of the decay muons. In contrast, a high energy neutral particle created in the shower should exit the dump near the proton beam direction. Therefore, the e^+e^- angular distributions are consistent with muon bremsstrahlung in the last radiation length

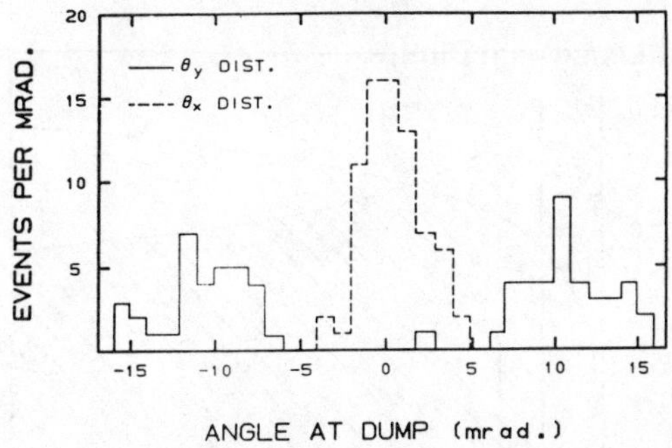

FIG. 2. Observed angular distributions in the bend (y) and non-bend (x) planes of the collinear e^+e^- pairs at the end of the dump.

of the beam dump, and at most only one pair is consistent with the decay of a neutral particle produced at the upstream part of the dump. Furthermore, based on the total number of prescaled muons, the observed yield of e^+e^- pairs agreed with the expected rate from muon bremsstrahlung to within a factor of two.

Limits on the production of an axion were calculated by working out the number of axion decay e^+e^- pairs which should be detected in the apparatus. The electrons required to produce axions came from pair production of photons which were decay products of π^o's. The flux of π^o's in the hadron shower was calculated using a phenomenological fit to thick target production spectra[5]. The axion bremsstrahlung was modelled using a formula given by Y.S. Tsai [6]:

$$\frac{dn_a}{dx} = \frac{\alpha_e}{2\alpha} m_e^2 \frac{\frac{2}{3} m_a^2 x(1-x) + x^3 m_e^2}{\left[m_a^2 \frac{(1-x)}{x} + x m_e^2 \right]^2} \quad (1)$$

where $x=E_a/E_e$ is the ratio of axion to lepton energy. Furthermore, the lifetime of the axion is given by [2]:

$$\tau_a = \frac{2}{\alpha_e} \left(m_a^2 - 4m_e^2 \right)^{-\frac{1}{2}} \quad (2)$$

The limits on the mass and lifetime measured by E605 and Kyoto/KEK are shown in Figure 3. The area in Figure 3 labelled A) was determined by the 90% confidence level for one event observed, i.e. the boundary corresponds to a yield of 4 detected e^+e^- pairs with total energy above 150 GeV. For a mass of 1.8 MeV, 500 axion decays would have been detected in E605.

If the difference between the experimental results and theoretical prediction on the $(g-2)_e$ is assumed to come from the coupling of the axion to electron and positron, then the limits on the lifetime of the aixon can be calculated and is shown in Figure 3. As indicated in Figure 3, a pseudoscalar lighter than 2.4 MeV coupled only to e^+e^- is ruled out, independent of the coupling strength.

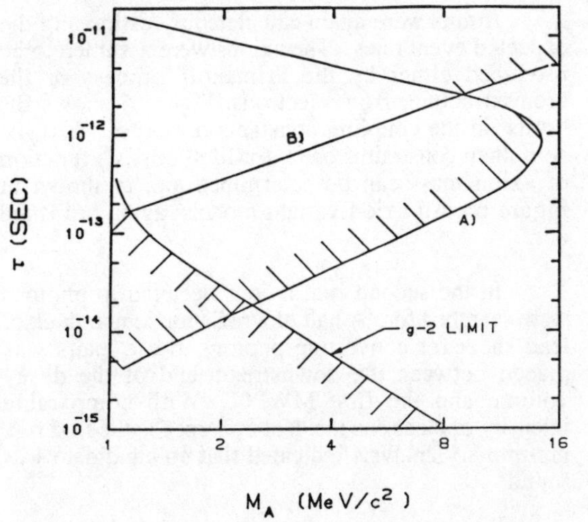

FIG. 3. Limits on the mass and lifetime of an axion-like particle from A) Fermilab E605 and B) Kyoto/KEK. Also shown is the limit from $(g-2)_e$ measurement.

Kyoto/KEK Experiment [8]

A total of 1.7×10^{17} electrons was injected into a 3.5 cm thick tungsten target followed by a 2.4 m long dump made up of iron, lead and plastic blocks. The apparatus for detecting e^+e^- pairs is shown in Figure 4. Data were collected by requiring the total energy deposition in the 13.9 radiation lengths deep lead-glass array be greater than 100 MeV.

In the off-line analysis, events were required to have no veto counters fired, a hit in S3, two charged tracks in the MWPC's downstream of the analyzing magnets, at lease one hit in the hodoscope and energy deposited in lead-glass matched the momentum measured by the MWPC's. No event was left after the cuts.

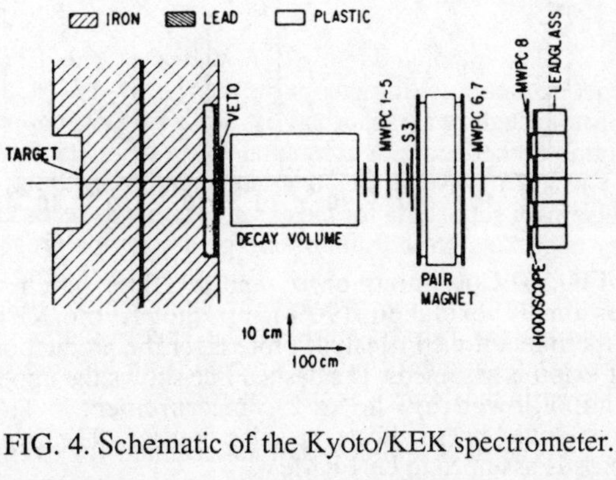

FIG. 4. Schematic of the Kyoto/KEK spectrometer.

Limits were again calculated by estimating the expected event rates. The axions were assumed to be produced either by the Primakoff process or the bremsstrahlung from electrons. Figure 5 shows the limits on the coupling constants α_γ and α_e. If α_γ is zero, then constraints on α_e (or lifetime) as a function of axion mass can be determined and is shown in Figure 6. All axion variant models[2] again are ruled out.

In the second run, axions decayed to photons were searched for. A half of a radiation length thick of lead sheet for converting photons to e^+e^- pairs was placed between the downstream end of the decay volume and the first MWPC. With comparable intensity of electrons incident on target as the first run, preliminary analysis indicated that no candidate was found[9].

Conclusions

No short-lived axion that is strongly coupled to e^+e^- or photons but weakly interacts with matter is found. All new axion models[2] are ruled out by both long beam dump experiments. The sharp e^+e^- peak seen at GSI[1] is probably not an elementary pseudoscalar particle.

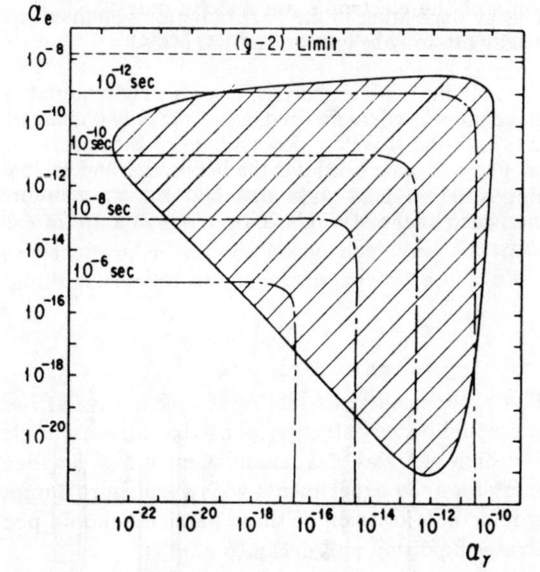

FIG. 5. Constraints on α_e and α_γ. The hatched region is excluded (90%) by the Kyoto/KEK experiment if the Primakoff process for the production of axion is assumed. The dashed line shows the upper limit allowed by the $(g\text{-}2)_e$ measurement. The dash-dotted lines indicate the axion lifetime. The axion mass is assumed to be 1.8 MeV.

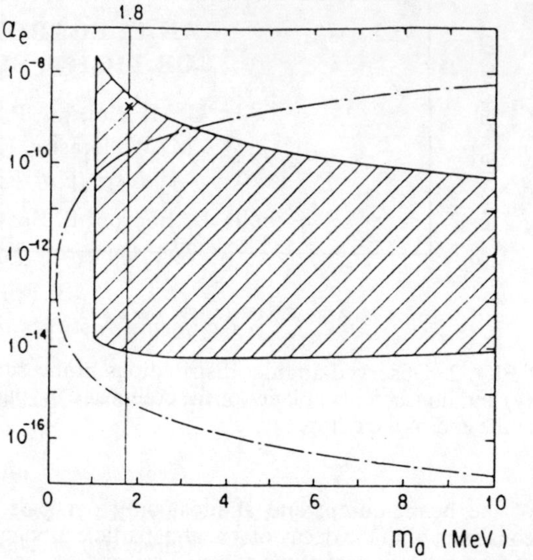

FIG. 6. Constraints on α_e as a function of the axion mass m_a. The hatched region is excluded (90%) by the Kyoto/KEK experiment if axion bremsstrahlung is assumed. The dash-dotted line indicates the prediction of the standard axion model. The point indicated by a cross is the prediction by Ref. 2.

References

[1] J. Schweppe et al., Phys. Rev. Lett. **51**, 2261 (1983); M. Clemente et al., Phys. Lett. **137B**, 41 (1984); T. Cowan et al., Phys. Rev. Lett. **54**, 1761 (1985); T. Cowan et al., Phys. Rev. Lett. **56**, 444 (1986).

[2] L. M. Krauss and F. Wilczek, Phys. Lett. **173B**, 189 (1986); R. D. Peccei, T. T. Wu, and T. Yanagida, Phys. Lett. **172B**, 435 (1986); L. M. Krauss, this proceedings.

[3] E. M. Riordan, this proceedings.

[4] E605 Collaboration - C. N. Brown, K. B. Luk, W. E. Cooper, D. A. Finley, A. M. Jonckheere, H. Jostlein, D. M. Kaplan, L. M. Lederman, S. R. Smith, R. Gray, R. E. Plaag, J. P. Rutherfoord, P. B. Straub, K. K. Young, Y. Hemmi, K. Imai, K. Miyake, Y. Sakai, N. Sasao, N. Tamura, T. Yoshida, A. Maki, J. A. Crittenden, Y. B. Hsiung, M. R. Adams, H. D. Glass, D. E. Jaffe, R. L. McCarthy, J. R. Hubbard, Ph. Mangeot; C. N. Brown et al., submitted to Phys. Rev. Lett..

[5] A. J. Malensek, Fermilab preprint FN-341, FN-341-A.

[6] Y. S. Tsai, Phys. Rev. **D34**, 1326 (1986).

[7] T. Kinoshita and W. B. Lindquist, Phys. Rev. Lett. **47**, 1573 (1981); E. Richard Cohen, private communication.

[8] Kyoto/KEK Collaboration - A. Konaka, K. Imai, H. Kobayashi, A. Masaike, T. Nakamura, N. Nagamine, N. Sasao, A. Enomoto, Y. Fukushima, E. Kikutani, H. Koiso, H. Matsumoto, K. Nakahara, S. Ohsawa, T. Taniguchi, I. Sato, J. Urakawa; A. Konaka, Phys. Rev. Lett. **57**, 659 (1986).

[9] H. Koiso, private communication.

AN ELECTRON BEAM DUMP SEARCH FOR LIGHT, SHORT-LIVED PARTICLES*

E. M. Riordan, P. de Barbaro, A. Bodek, S. Dasu,
M. W. Krasny, K. Lang, N. Varelas, X. R. Wang
University of Rochester, Rochester, NY 14627

R. Arnold, D. Benton, P. Bosted, L. Clogher, A. Lung, S. Rock, Z. Szalata
The American University, Washington DC 20016

B. Filippone, R. C. Walker
California Institute of Technology, Pasadena, CA 91125

J. D. Bjorken, M. Crisler, A. Para
Fermi National Accelerator Laboratory, Batavia, IL 60510

J. Lambert
Georgetown University, Washington, DC 20007

J. Button-Shafer, B. Debebe, M. Frodyma, R. S. Hicks, G. A. Peterson
University of Massachusetts, Amherst, MA 01003

R. Gearhart
Stanford Linear Accelerator Center, Stanford, CA 94305

Presented by E. M. Riordan, University of Rochester

ABSTRACT

We report preliminary results of an electron beam dump search for neutral, penetrating particles X^o with masses in the range $1 < m_X < 12$ MeV and lifetimes τ_X between 10^{-14} and 10^{-11} sec. No evidence was found for such an object in runs with a total of 10^{15} e^- on dump. We specifically rule out the existence of any possible 1.8 MeV pseudoscalar with $\tau_X > 2 \times 10^{-14}$ sec and an absorption cross-section in matter less than $150A^{0.7}$ mb per nucleus. Inasmuch as measurements of the electron's anomalous magnetic moment constrain $\tau_X > 7 \times 10^{-14}$ sec for a neutral 1.8 MeV pseudoscalar boson, this experiment excludes the recent GSI phenomenon as an elementary pseudoscalar.

I. INTRODUCTION

The recent discovery of monochromatic positron peaks, and e^+e^- coincidences in heavy ion collisions at GSI [1] has stimulated a round of theoretical speculation [2] that this phenomenon might be induced by an elementary axion. Such an object could not be the "standard" Peccei-Quinn-Weinberg-Wilczek axion [3], which has already been ruled out by J/psi and upsilon decays. But axion variants coupling preferentially to light fermions cannot be ruled out by these heavy quarkonium decays [4], nor can they exclude a neutral, elementary pseudoscalar boson coupling only to electrons or photons [5].

An electron beam dump experiment is one of the cleanest ways to search for such particles. Here we assume only that the particle in question couples to electrons, with a coupling constant fully determined by its assumed lifetime. For lifetimes $\tau_X \sim 10^{-13}$ seconds, as required by various non-standard axion models and allowed by measurements of the electron's anomalous magnetic moment [6], any such boson should be produced copiously in a process analogous to bremsstrahlung:

$$e + Z \rightarrow e + Z + X^o$$

The production cross-section for pseudoscalar bosons would be very strongly peaked at forward angles and high energies [7]. At sufficiently high incident energies, or in experiments with very short dumps, a detectable fraction of these particles should penetrate the dump and decay to e^+e^-.

II. THE EXPERIMENT

We report preliminary results of an electron beam dump experiment (E141) performed in June 1986 at the Stanford Linear Accelerator Center. Electron beams with energies E_o of 9.0, 10.7, 18.0 and 22.4 GeV were stopped in copper and tungsten

* Research supported by DOE Contracts #ER13065-453, DE-AC02-76ER0285 3, and NSF Contract #8410549

dumps ranging in length from 10 to 100 cm. A total of 5×10^{15} electrons were used in the entire experiment. The results reported here come from a subset of the 9.0 GeV runs in which $\sim 10^{15}$ electrons were stopped in 10 and 12 cm tungsten dumps, hereafter called "dump A" and "dump B" respectively [8].

Using the SLAC 8 GeV spectrometer facility, positioned at 0° w.r.t. the incident beam and located ~35 m downstream of the dump, we searched for high-energy positrons ($0.5 < x < 0.9$, where x is the fraction of E_o carried off by a positron) produced at small angles ($\theta < 1$ mr). Positrons were separated from a background of muons and pions by a hydrogen-filled Cherenkov counter and lead-glass counters located in a shielded cave behind the last spectrometer magnet. Track information supplied by a set of ten proportional wire chambers allowed event reconstruction to an accuracy of 0.1 mr in angle and 0.1% in momentum. Two muon counter hodoscopes were added behind the lead-glass counter to allow discrimination of muons from pions.

The energy spread of the incident electron beam was typically 0.5%, and the beam direction was maintained within 0.1 mr of the central spectrometer angle. The instantaneous beam current incident on the dumps was measured by a resonant toroid monitor located upstream of them; it was periodically calibrated against a similar, independent monitor located just before the spectrometer. The number of electrons striking the dump in any run is known to better than 5% accuracy.

A cylindrical 3-inch pipe 5 meters upstream of the spectrometer limited our sensitivity to only those positrons produced within 1 mr of the beam axis. Two meters upstream of the spectrometer, a 0.6 r.l. (3.8 g/cm^2) lead converter was periodically inserted to determine the flux of high-energy photons emerging from either dump. A 0.06 r.l. (1.4 g/cm^2) aluminum target inserted in the beam at the same position (with the dumps removed and the spectrometer set at 11.5°) allowed frequent measurements of deep inelastic $e - N$ scattering cross sections. These calibration runs served as a check on equipment stability and on the overall normalization of the experiment; they agreed with previously measured cross-sections to better than 10%.

III. PRELIMINARY RESULTS

Short-lived axions or other pseudoscalar bosons would generate a large number of high-energy positrons downstream of dumps A and B. For dumps of sufficient thickness ($\sim$ 30 r.l.), this process would be the dominant source of high-energy positrons, assuming such bosons exist.

In Figure 1 we show the differential number of positrons detected in our solid angle, dN_{e^+}/dx, normalized by the number of electrons N_{e^-} incident on the dumps. Errors due to counting statistics and systematic effects are shown explicitly. The $\sim 30\%$ systematic errors (as presently estimated), are dominated by uncertainties in the acceptance.

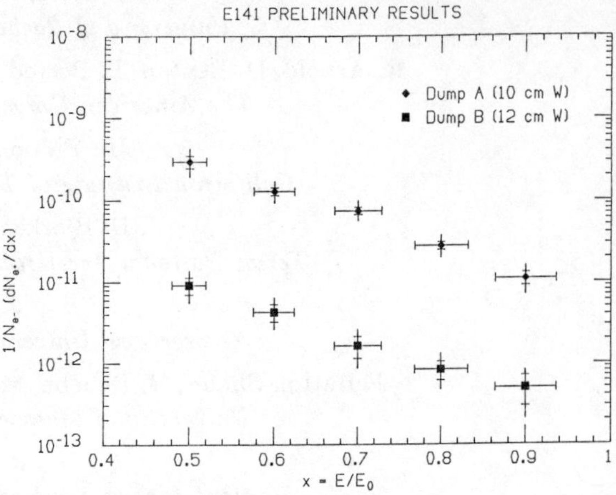

Figure 1. The differential rate of positrons observed in our acceptance, plotted versus the fractional positron energy x. The inner error bars represent counting statistics; the outer bars represent the $\sim$ 30% systematic errors in each measurement.

The shapes of these spectra are consistent with the expected background from hard photon punch-through[9], and the observed rates agree with these expectations to within a factor of 5. In this process a hard bremsstrahlung photon created in the first few radiation lengths penetrates most of the dump and converts to e^+e^- in the last radiation length. For such a process the counting rate measured behind dump B should be a uniform factor of $e^{\frac{7}{9}\Delta t} = 41$ below that measured behind dump A, consistent with the observed 31 ± 4 (statistical errors) at the present level of systematic uncertainty.

A detailed Monte Carlo simulation of this and other backgrounds, currently underway, is incomplete at present and is not reported or used here. In the current analysis we assume as an upper limit that *all* the positrons observed behind the (12 cm) dump B are decay products of a light pseudoscalar. In Figure 2 we compare the observed rates behind this dump with those expected [7] for a 1.8 MeV pseudoscalar with lifetimes of 1 and 2×10^{-14} sec, assuming an absorption cross section in matter of no more than 1 mb per nucleon. Such a particle is clearly excluded by the data. At $x = 0.7$ and 0.8, where the signal/noise ratio should be the greatest, the expected counting rate is at least 2σ above that observed. A 1.8 MeV pseudoscalar with a lifetime of 2×10^{-14} sec and an absorption cross section of $150A^{0.7}$ mb per nucleus (or 30 mb per nucleon, in tungsten) is also excluded.

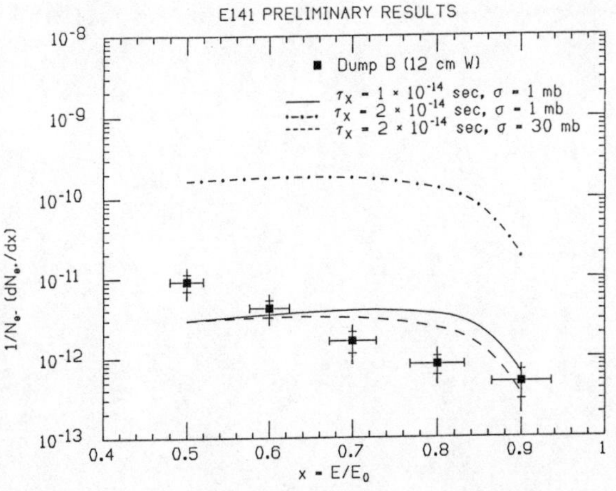

Figure 2. The rate of positrons observed behind dump B, compared with rates expected from a 1.8 MeV axion with lifetimes and absorption cross-sections (per nucleon) listed. Calculations of Tsai [7] were multiplied by acceptance correction factors to obtain these curves.

In a similar fashion we place upper limits on the lifetimes of pseudoscalars with masses m_X in the range $1 < m_X < 12$ MeV, assuming unity branching ratio to e^+e^-. Those limits are shown in Figure 3, together with the lower limits set by electron g-2 measurements [10] and previous upper limits set by a recent electron beam dump experiment [11]. It is clear from this graph that any pseudoscalar boson with $m_X < 3$ MeV is excluded by our experiment, in conjunction with the g-2 limits. We therefore conclude that the 1.8 MeV GSI phenomenon is not due to an elementary pseudoscalar[12].

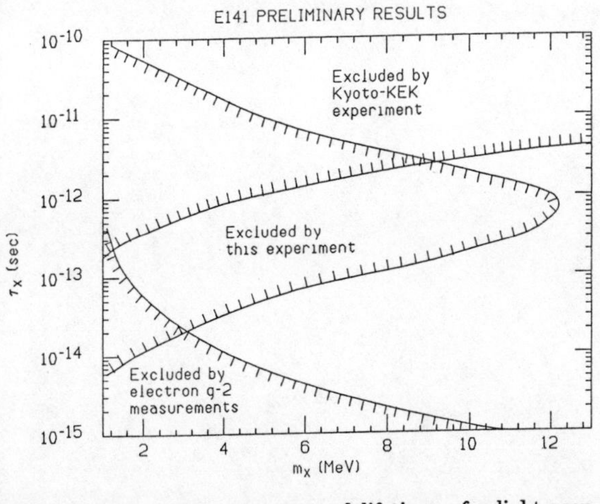

Figure 3. Constraints on mass and lifetime, of a light pseudoscalar boson decaying predominantly to e^+e^- imposed by this experiment, by electron g-2 experiments [10], and by a recent beam dump experiment [11].

A detailed off-line analysis of the full set of our data, including a thorough analysis of the $\gamma - \gamma$ decay channel, is currently underway.

REFERENCES

1. J. Schweppe *et al.*, Phys. Rev. Lett. **51**, 2261 (1983); M. Clemente *et al.*, Phys. Lett. **B137**, 41 (1984); T. Cowan *et al.*, Phys Rev. Lett. **54**, 1761 (1985); T. Cowan *et al.*, Phys. Rev. Lett. **56**, 444 (1986).
2. A. Schäfer *et al.*, J. Phys. **G11**, L69 (1985); A. B. Balantekin *et al.*, Phys. Rev. Lett. **55**. 461 (1985); M. C. Mukhopadhyay and Z. Zehnder, Phys. Rev. Lett. **56**, 206 (1986).
3. R. D. Peccei and Helen R. Quinn, Phys. Rev. Lett. **38**, 1440 (1977); S. Weinberg, Phys. Rev. Lett. **40**, 223 (1978); F. Wilczek, Phys. Rev. Lett. **40**, 279 (1978).
4. L. M. Krauss and F. Wilczek, Phys. Lett. **B173**, 189 (1986); R. D. Peccei, T. T. Wu, and T. Yanagida, Phys. Lett. **B172**, 435 (1986).
5. J. D. Bjorken, private communication.
6. J. Reinhardt *et al.*, Phys. Rev. **C33**, 194 (1986); A. Zee, Phys. Lett. **B172**, 377 (1986); L. M. Krauss and M. Zeller, Yale Preprint No. YTP-86-08 (to be published in Phys. Rev.).
7. Y. S. Tsai, Phys. Rev. **D34**, 1326 (1986).
8. The shorter dump was fabricated from a 10.16 cm thick (27.6 r.l.) block of Kennertium W-2, an alloy composed of 97.4% tungsten plus 2.6% nickel, copper and iron. To the back of a portion of this block was added another 2.00 cm block of Kennertium W-10, a different alloy composed of 90% tungsten plus 10% nickel, copper and iron. The 12.16 cm combination contained 32.4 r.l. Both estimates of the dump thickness in radiation lengths are based on the calculations of Tsai, which agree with other calculations to better than 1%. Y. S. Tsai, Rev. Mod. Phys. **46**, 815 (1974); O. I. Dovzhenko and A. A. Pomanski, Soviet Physics JETP **18**, 187 (1963).
9. Y. S. Tsai and Van Whitis, Phys. Rev. **149**, 1248 (1966).
10. The contribution of an elementary pseudoscalar X° to the electron anomalous magnetic moment is

$$\Delta a = -\frac{\alpha_X}{2\pi}\int_0^1 dz \frac{z^3}{z^2 + (1-z)\frac{m_X^2}{m_e^2}}$$

where α_X is the electron-pseudoscalar coupling strength. Using $|\Delta a| < 2\times10^{-10}$ and $\tau_X = 2\alpha_X^{-1}(m_X^2 - 4m_e^2)^{-1/2}$ (See Ref. 6), we obtain the (90% confidence) lower limits on τ_X shown in Figure 3. See T. Kinoshita, Proceedings of the 1986 Conference on Precision Electromagnetic Measurements for a recent review of the status of theory and experiment on the electron g-2 measurements.
11. A. Konaka *et al.*, Phys. Rev. Lett. **57**, 659 (1986).
12. A possible loophole in this argument occurs, however, if there exists a neutral scalar boson with mass and coupling strength similar to the hypothetical pseudoscalar. In such a case their contributions to the electron magnetic moment would *cancel*, invalidating the limits put by g-2 measurements on τ_X. See J. Reinhardt *et al.*, (Ref. 6) for further discussion of this possibility.

Parallel Session 9

Spectroscopy and Decays of Heavy Bound-Quark States

Organizers:
W. Schmidt-Parzefall (DESY)
M. Tuts (Columbia)

Scientific Secretaries:
T. Freese (SLAC)
S. Wasserbaech (SLAC)

Study of $\pi^+\pi^-$ Transitions from the $\Upsilon(3S)$

T. Bowcock, R. T. Giles, J. Hassard,[a] K. Kinoshita, F. M. Pipkin, Richard Wilson, J. Wolinski, D. Xiao;[1] T. Gentile, P. Haas, M. Hempstead, T. Jensen, H. Kagan, R. Kass;[2] S. Behrends, Jan M. Guida, Joan A. Guida, F. Morrow, R. Poling, E. H. Thorndike, P. Tipton;[3] M. S. Alam, N. Katayama, I. J. Kim, C. R. Sun, V. Tanikella;[4] D. Bortoletto, A. Chen, L. Garren, M. Goldberg, R. Holmes, N. Horwitz, A. Jawahery, P. Lubrano, G. C. Moneti, V. Sharma;[5] S. E. Csorna, M. D. Mestayer, R. S. Panvini, G. B. Word;[6] A. Bean, G. J. Bobbink, I. C. Brock, A. Engler, T. Ferguson, R. W. Kraemer, C. Rippich, H. Vogel;[7] C. Bebek, K. Berkelman, E. Blucher, D. G. Cassel, T. Copie, R. DeSalvo, J. W. DeWire, R. Ehrlich, R. S. Galik, M. G. D. Gilchriese, B. Gittelman, S. W. Gray, A. M. Halling, D. L. Hartill, B. K. Heltsley, S. Holzner, M. Ito, J. Kandaswamy, R. Kowalewski, D. L. Kreinick, Y. Kubota, N. B. Mistry, J. Mueller, R. Namjoshi, E. Nordberg, M. Ogg, D. Perticone, D. Peterson, M. Pisharody, K. Read, D. Riley, A. Silverman, S. Stone, Xia Yi[b];[8] A. J. Sadoff;[9] P. Avery, D. Besson;[10]

(Presented by G.J. Bobbink)

[1]Harvard University, Cambridge, Massachusetts 02138, and [2]Ohio State University, Columbus, Ohio, 43210, and [3]University of Rochester, Rochester, New York 14627, and [4]State University of New York at Albany, Albany, New York 12222, and [5]Syracuse University, Syracuse, New York 13210, and [6]Vanderbilt University, Nashville, Tennessee 37235, and [7]Carnegie Mellon University, Pittsburgh, Pennsylvania 15213, and [8]Cornell University, Ithaca, New York 14853, and [9]Ithaca College, Ithaca, New York 14850, and [10]University of Florida, Gainesville, Florida 32611.

(The CLEO collaboration)

We have investigated the transitions $\Upsilon(3S)\rightarrow\pi^+\pi^-\Upsilon(1S)$, $\Upsilon(3S)\rightarrow\pi^+\pi^-\Upsilon(2S)$, and the cascade process $\Upsilon(3S)\rightarrow\Upsilon(2S)$+X; $\Upsilon(2S)\rightarrow\pi^+\pi^-\Upsilon(1S)$, both in the exclusive decay mode where the daughter Υ state decays into two leptons, and in the inclusive decay mode where the daughter Υ state decays hadronically. Results are presented on branching fractions and the properties of the $\pi^+\pi^-$ system. Possible evidence for the transition $\Upsilon(3S)\rightarrow\pi^+\pi^-\Upsilon(1^1P_1)$ is presented.

The $\Upsilon(3S)$ resonance gives access to a rich spectrum of dipion transitions with branching fractions predicted to be $\gtrsim 1\%$. Quantum chromodynamics describes hadronic transitions between heavy quarkonium states as the emission of gluons by the heavy quarks followed by the conversion of the gluons into light hadrons [1]. Within the framework of heavy-quark potential models [2,3,4] the $\pi\pi$ decay rates can be calculated from a multipole expansion of the gluon fields. The properties of the $\pi\pi$ system are constrained by applying partial conservation of the axial-vector current and current algebra.

The measured branching fractions for $\psi' \rightarrow \pi^+\pi^-(J/\psi)$ [5,6], $\Upsilon(3S)\rightarrow\pi^+\pi^-\Upsilon(1S)$[7,8], $\Upsilon(2S)\rightarrow\pi^+\pi^-\Upsilon(1S)$[9,10,11,12,13], and $\Upsilon(3S)\rightarrow\pi^+\pi^-\Upsilon(2S)$ [8] agree with these theoretical expectations. Previous studies [7,8] of the transition $\Upsilon(3S)\rightarrow\pi^+\pi^-\Upsilon(1S)$ suggested that the $\pi^+\pi^-$ invariant-mass spectrum was approximately uniform, in contrast to the transitions $\Upsilon(2S)\rightarrow\pi^+\pi^-\Upsilon(1S)$ and $\psi' \rightarrow \pi^+\pi^-(J/\psi)$, which are strongly peaked toward high values of $\pi^+\pi^-$invariant mass.

We have recently collected 33.0 pb^{-1} of data at the $\Upsilon(3S)$ resonance, and 22.6 pb^{-1} of continuum data between the $\Upsilon(3S)$ and $\Upsilon(4S)$ resonances with the CLEO detector at the Cornell Electron Storage Ring (CESR). The CLEO detector [14] has been described in detail before. For this analysis only the central tracking chambers and the time-of-flight system were used. The tracking chambers consisted of a 10 layer high precision vertex detector and a 17 layer drift chamber, surrounding a 0.6 mm beryllium beam pipe. Both were operated in a superconducting solenoid which produced a magnetic field of 1.0 T. The addition of the high precision vertex detector and the beryllium beam pipe led to a substantial improvement both in momentum resolution and in our ability to reconstruct low momentum tracks. We can reconstruct pion tracks efficiently down to 50 MeV/c in transverse momentum, and the $\pi^+\pi^-$ recoil-mass resolution has improved from about 25 MeV/c^2 (FWHM) in our previous data sample, [7] to about 5 to 10 MeV/c^2 (depending on recoil mass) in our present data sample.

We have investigated the transitions:

$$\Upsilon(3S) \rightarrow \pi^+\pi^-\Upsilon(1S), \qquad (a)$$
$$\Upsilon(3S) \rightarrow \pi^+\pi^-\Upsilon(2S), \qquad (b)$$

and the cascade process

$$\Upsilon(3S) \rightarrow \Upsilon(2S) + X;\ \Upsilon(2S) \rightarrow \pi^+\pi^-\Upsilon(1S), \quad (c)$$

both in the exclusive decay mode, where the daughter Υ resonance decays to either a pair of electrons or a pair of muons, and in the inclusive decay mode, where the daughter Υ resonance decays hadronically. The branching fractions for all these transitions have been measured with considerably higher precision than previous measurements.[7,8] From the measured number of $\Upsilon(2S)\rightarrow\pi^+\pi^-\Upsilon(1S)$ decays we infer the total $\Upsilon(3S)\rightarrow\Upsilon(2S)$ branching fraction, We have also looked for the decay $\Upsilon(3S)\rightarrow\pi^+\pi^-\Upsilon(1^1P_1)$, which is predicted to have a branching fraction of order 1%.[2]

There were 248600 detected hadronic events in the $\Upsilon(3S)$ data sample, and 85600 hadronic events in the continuum data sample. Scaling the number of continuum events and correcting for the 94% event selection efficiency, we get 148200 for the total number of hadronic $\Upsilon(3S)$ decays. Including the leptonic decay modes we find that the total number of produced $\Upsilon(3S)$ resonance events was 165000±11000.

For inclusive events, we selected pairs of oppositely charged tracks in hadronic events by requiring that they come from close to the event vertex and by removing tracks that come from photon conversion and decay of K^0 and Λ.

Exclusive events from transitions (a) and (b) were selected by requiring two oppositely charged tracks with momenta greater than 3.5 GeV/c and two oppositely charged tracks with momenta less than 0.75 GeV/c. Events from transition (c) were selected by allowing in addition to the above criteria up to two more tracks with momenta less than 0.75 GeV/c. We assumed that the high momentum tracks were either e^+e^- or $\mu^+\mu^-$ pairs. Monte Carlo studies indicate that the contribution from $\tau^+\tau^-$ pairs is negligible. All tracks must come from close to the event vertex and the lepton candidates must be within the fiducial volume of the time-of-flight counters, in order to reliably calculate the trigger efficiency. Tracks that originate from converted photons were removed.

The $\pi^+\pi^-$ recoil-mass spectrum for the exclusive four-track events and the inclusive events are shown in Fig. 1(a) and (b), respectively. In both spectra there are three clear peaks corresponding to the decays $\Upsilon(3S)\rightarrow\pi^+\pi^-\Upsilon(1S)$, $\Upsilon(3S)\rightarrow\pi^+\pi^-\Upsilon(2S)$ and $\Upsilon(2S)\rightarrow\pi^+\pi^-\Upsilon(1S)$. The $\Upsilon(2S)\rightarrow\pi^+\pi^-\Upsilon(1S)$ peak is necessarily offset because the pions were assumed to have come from the $\Upsilon(3S)$. Expansions of the regions around the peaks in the inclusive spectrum are shown in Figs. 2 (a)-(c). The peaks in the exclusive spectrum are very clean, and we use them along with Monte Carlo events to determine the shape of the peaks in fitting the inclusive spectrum.

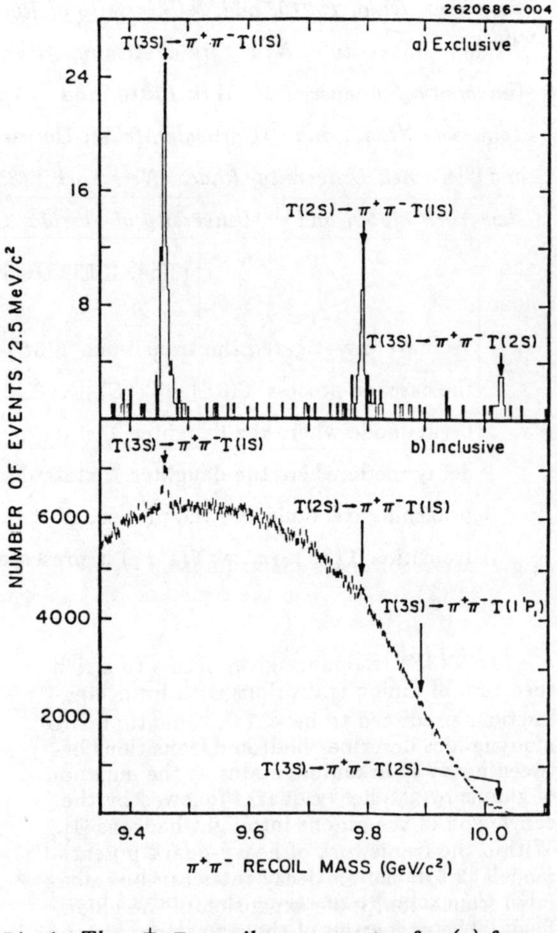

Fig.1. *The $\pi^+\pi^-$ recoil-mass spectra for (a) four-track exclusive and (b) inclusive events from $\Upsilon(3S)\rightarrow\pi^+\pi^-X$. We note that the $\Upsilon(2S)\rightarrow\pi^+\pi^-\Upsilon(1S)$ peak is offset because the pions were assumed to have come from the $\Upsilon(3S)$.*

From the flat background level in Fig. 1 and from the continuum data sample, we estimate 1-2 background events under each peak. The background in the inclusive spectrum was parametrized by a Chebyshev polynomial. For the three peaks both a Breit-Wigner shape (with FWHM of 8.5±1.0, 9±2, and 8±2 MeV/c^2 for reactions (a), (b), and (c), respectively) and two Gaussians with slightly different widths and means were found to describe the data satisfactorily. The numbers of events found for both the exclusive and the inclusive mode are given in Table I. For the inclusive results the statistical errors and the errors estimated from varying the form of the fitting function were added in quadrature. The detection efficiencies shown in Table I were calculated from a Monte Carlo simulation of the CLEO detector. The efficiency is constant as a function of the $\pi^+\pi^-$ invariant mass.

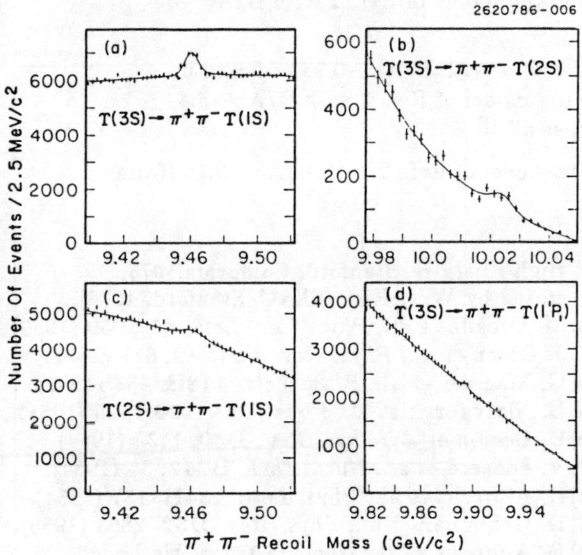

Fig.2 The recoil-mass spectrum from $\Upsilon(3S)\rightarrow\pi^+\pi^-$X in the region of:
a) $\Upsilon(3S)\rightarrow\pi^+\pi^-\Upsilon(1S)$,
b) $\Upsilon(3S)\rightarrow\pi^+\pi^-\Upsilon(2S)$,
c) the cascade process $\Upsilon(3S)\rightarrow\Upsilon(2S)$+X; $\Upsilon(2S)\rightarrow\pi^+\pi^-\Upsilon(1S)$. For this figure the $\pi^+\pi^-$ was assumed to recoil against the $\Upsilon(2S)$, and
d) $\Upsilon(3S)\rightarrow\pi^+\pi^-\Upsilon(1^1P_1)$.

The branching fractions are given in Table I. For the exclusive reactions the product branching fraction is given. Using the $B_{\mu\mu}(\Upsilon(1S))$ and $B_{\mu\mu}(\Upsilon(2S))$ world averages [16] we can compare our exclusive and inclusive measurements (see Table I); the agreement is reasonable. Averaging our exclusive and inclusive results we obtain the results given in Table I. If we subtract the $\Upsilon(3S)\rightarrow\pi\pi\Upsilon(2S)$ branching fraction from the total $\Upsilon(3S)\rightarrow\Upsilon(2S)$ branching fraction, we obtain the total branching fraction for photon transitions from the $\Upsilon(3S)$ to the $\Upsilon(2S)$ via the intermediate $\Upsilon(2^3P_{J=0,1,2})$ states:
$B(\Upsilon(3S)\rightarrow\gamma\gamma\Upsilon(2S))= (6.9\pm1.8)\%$.

To determine the $\pi^+\pi^-$ invariant-mass spectrum for the exclusive decay modes, we selected events in the $\Upsilon(3S)\rightarrow\pi^+\pi^-\Upsilon(1S)$, $\Upsilon(3S)\rightarrow\pi^+\pi^-\Upsilon(2S)$, and $\Upsilon(2S)\rightarrow\pi^+\pi^-\Upsilon(1S)$ recoil-mass peaks and plot the $\pi^+\pi^-$ invariant-mass spectra in Figs. 3(a), 3(b), and 3(c). The resolution in $\pi^+\pi^-$ invariant mass is about 7 MeV/c^2.

For the inclusive spectra the data were divided into bins of dipion invariant mass and fit using the mean and width of the peak as determined from the whole spectrum. The resulting dipion invariant-mass spectra are also shown in Fig. 3 and are in good agreement with the exclusive results. In Fig. 3(c) we compare the spectrum we find in $\Upsilon(3S)$ decay for $\Upsilon(2S)\rightarrow\pi^+\pi^-\Upsilon(1S)$ with the best fit to our previous measurement [10] made at the $\Upsilon(2S)$. We find good agreement, showing the validity of the methods used. The $\Upsilon(3S)\rightarrow\pi^+\pi^-\Upsilon(1S)$ spectrum is rather uniform (see Fig. 3(a)) and markedly different from the $\Upsilon(2S)\rightarrow\pi^+\pi^-\Upsilon(1S)$ spectrum; it is clearly not peaked at high mass and has a significant number of events immediately above threshold. We cannot obtain a reasonable fit to the $\Upsilon(3S)\rightarrow\pi^+\pi^-\Upsilon(1S)$ spectrum using the formulae of Yan [2], Voloshin and Zakharov [3] or Novikov and Schifman [4] which describe the $\Upsilon(2S)\rightarrow\pi^+\pi^-\Upsilon(1S)$ spectrum so well. For example, using the formula of Yan [15] we obtain B/A=$-5.4^{+1.6}_{-5.8}$ with a confidence level of 1.7%. Also, neither phase space nor the model of Peskin [17] can describe this spectrum.

We have also searched for the $\Upsilon(1^1P_1)$ state in the inclusive $\pi^+\pi^-$ recoil-mass spectrum. The $\Upsilon(1^1P_1)$ state has not previously been observed, and its mass is expected to be close to the center of gravity of the $\Upsilon(1^3P_{J=0,1,2})$ at 9.9002±0.0007 GeV/c^2 [16]. An accurate measurement of the mass difference between the $\Upsilon(1^1P_1)$ and the $\Upsilon(1^3P_{J=0,1,2})$ gives information on the spin-spin coupling in the $b\bar{b}$ system. The $\pi^+\pi^-$ recoil-mass spectrum in this region is shown in Fig. 2(d). Fitting with a second order Chebyshev polynomial and a Breit-Wigner shape having a FWHM of 5 MeV/c^2 (derived from Monte Carlo simulations) gives 335±135 events at a mass of 9.8948±0.0015 GeV/c^2. Assuming this is evidence for the transition $\Upsilon(3S)\rightarrow\pi^+\pi^-\Upsilon(1^1P_1)$ the branching fraction is (0.37±0.15)%.

In conclusion, we have measured with considerably higher precision several of the $\pi^+\pi^-$ transitions in $\Upsilon(3S)$ decay. The $\pi^+\pi^-$ invariant-

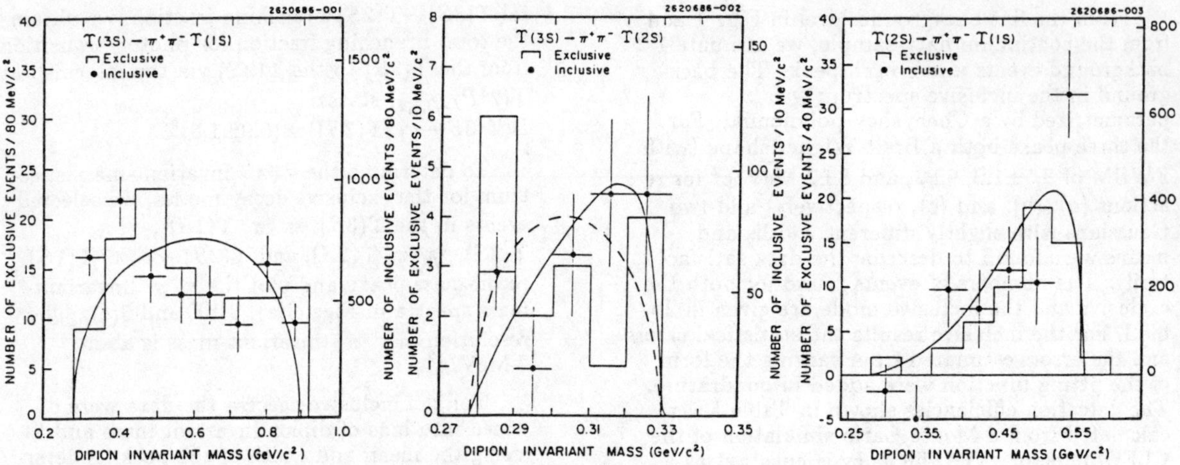

Fig.3 The $\pi^+\pi^-$invariant-mass distributions for the transitions:

a) $\Upsilon(3S)\to\pi^+\pi^-\Upsilon(1S)$. The line is the result of a fit to the model of Ref. 2 with B/A=$-5.4^{+1.6}_{-5.8}$. The confidence level of the fit is 1.7%.

b) $\Upsilon(3S)\to\pi^+\pi^-\Upsilon(2S)$. The solid line is the model of Ref. 2 with B/A=−0.18 as found in $\Upsilon(2S)\to\pi^+\pi^-\Upsilon(1S)$ (Ref. 10). The dashed line is the model of Ref. 2 with B/A=−5.4 as determined from $\Upsilon(3S)\to\pi^+\pi^-\Upsilon(1S)$ in this experiment.

c) $\Upsilon(2S)\to\pi^+\pi^-\Upsilon(1S)$ in $\Upsilon(3S)$ decay. The line is the model of Ref. 2 with B/A=−0.18 (from Ref. 10) as in $\Upsilon(2S)\to\pi^+\pi^-\Upsilon(1S)$.

mass spectrum from the $\Upsilon(3S)\to\pi^+\pi^-\Upsilon(1S)$ decay is flat in sharp contrast to the spectrum in $\Upsilon(2S)\to\pi^+\pi^-\Upsilon(1S)$ decay and cannot be described by any of the known models.

We gratefully acknowledge the efforts of the CESR machine group who made this work possible. Thanks are due to the Department of Energy OJI program (from H.K. and R. Kass) and the Mary Ingraham Bunting Institute (from K.K.) for their support. We acknowledge conversations with T.M. Yan, M. Peskin, and S.-H.H. Tye. This work was funded both by the National Science Foundation and the U.S. Department of Energy.

References

a) *Present address: Department of Physics, Imperial College, Blackett Laboratory, Prince Consort Rd., London SW7 2AZ.*

b) Present address: Department of Physics, Carnegie Mellon University, Pittsburgh, Pennsylvania, 15213.

1) K.Gottfried, *Phys. Rev. Lett.* **40** , 598 (1978).
2) T.M. Yan, *Phys. Rev.* **D22**, 1652 (1980); Y.P. Kuang and T.M. Yan, *Phys. Rev.* **D24**, 2874 (1982).
3) M. Voloshin and V. Zakharov, *Phys. Rev. Lett.* **45** , 688 (1980).
4) V.A. Novikov and M.A. Shifman, *Z. Phys. C* **8**, 43 (1981).
5) G. Abrams, *in Proceedings of the International* Symposium on Lepton and Photon Interactions at High Energies, Stanford, California,1975, edited by W.T. Kirk (SLAC, Stanford, 1976).
6) M. Oreglia et al., *Phys. Rev. Lett.* **45** , 959 (1980).
7) J. Green et al., *Phys. Rev. Lett.* **49**, 617 (1982).
8) G. Mageras et al., *Phys. Lett.* **118B**, 453 (1982).
9) B. Niczyporuk et al., *Phys. Lett.* **100B**, 95 (1981).
10) D. Besson et al., *Phys. Rev.* **D30**, 1433 (1984).
11) V. Fonseca et al., *Nucl. Phys.* **B242**, 31 (1984).
12) H. Albrecht et al., *Phys. Lett.* **134B**, 137 (1984).
13) D. Gelphman et al., *Phys. Rev.* **D32**, 2893 (1985).
14) D. Andrews et al., *Nucl. Instrum. Methods* **211**, 47 (1983).
15) The form used is given in Ref. 7.
16) Particle Data Group, *Phys. Lett.* **170B**, 1 (1986).
17) M.E. Peskin, *in Dynamics and Spectroscopy at* High Energy, Proceedings of the 11th SLAC Summer Institute on Particle Physics, SLAC Report 267 (1984) and private communication.

TABLE I

	Exclusive Data	*Inclusive Data*	*Average of Exclusive and Inclusive Data*	*Previous World Average*
$\Upsilon(3S)\to\pi^+\pi^-\Upsilon(1S)$				
Number of events	*109±11*	*3820±300*		
Efficiency	*(35.7±1.3)%*	*(62.9±6.0)%*		
$B(\Upsilon(3S)\to\pi^+\pi^-\Upsilon(1S))\cdot B(\Upsilon(1S)\to e^+e^-$ or $\mu^+\mu^-)$	$(1.85\pm0.22)10^{-3}$			
$B(\Upsilon(3S)\to\pi^+\pi^-\Upsilon(1S))$	*(3.31±0.45)%* [a]	*(3.69±0.52)%*	*(3.47±0.34)%*	*(4.8±1.2)* [d] *%*
$\Upsilon(3S)\to\pi^+\pi^-\Upsilon(2S)$				
Number of events	*14±4*	*300±80*		
Efficiency	*(7.38±0.46)%*	*(9.2±1.4)%*		
$B(\Upsilon(3S)\to\pi^+\pi^-\Upsilon(2S))\cdot\{B(\Upsilon(2S)\to e^+e^-$ or $\mu^+\mu^-)$ + $B(\Upsilon(2S)\to\Upsilon(1S)+neutrals)\cdot B(\Upsilon(1S)\to e^+e^-$ or $\mu^+\mu^-)\}$	$(1.13\pm0.35)10^{-3}$			
$B(\Upsilon(3S)\to\pi^+\pi^-\Upsilon(2S))$	*(2.6±1.1)%* [b]	*(2.0±0.6)%*	*(2.1±0.5)%*	*(3.6±2.3)* [e] *%*
$\Upsilon(3S)\to\Upsilon(2S)+X;\Upsilon(2S)\to\pi^+\pi^-\Upsilon(1S)$				
Number of events	*67±9*	*1500±500*		
Efficiency	*(31.5±1.7)%*	*(68.0±5.0)%*		
$B(\Upsilon(3S)\to\Upsilon(2S)+X)\cdot B(\Upsilon(2S)\to\pi^+\pi^-\Upsilon(1S))\cdot B(\Upsilon(1S)\to e^+e^-$ or $\mu^+\mu^-)$	$(1.3\pm0.2)10^{-3}$			
$B(\Upsilon(3S)\to\Upsilon(2S)+X)$	*(12.3±2.2)%* [a,c]	*(7.2±2.5)%* [c]	*(10.1±1.7)%*	——

a) using the world average $B_{\mu\mu}(1S)=(2.80\pm0.19)\%$ *from Ref. 16.*

b) using $B_{\mu\mu}(2S)=(1.80\pm0.44)\%$, $B(\Upsilon(2S)\to\gamma\gamma\Upsilon(1S))\cdot B(\Upsilon(1S)\to e^+e^-$ or $\mu^+\mu^-) = (0.21\pm0.03)\%$ *,and* $B(\Upsilon(2S)\to\pi^0\pi^0\Upsilon(1S))\cdot B(\Upsilon(1S)\to e^+e^-$ or $\mu^+\mu^-) = (0.48\pm0.07)\%$ *from Ref. 16.*

c) using the world average $B(\Upsilon(2S)\to\pi^+\pi^-\Upsilon(1S))=(18.7\pm1.0)\%$ *from Ref. 16.*

d) data from Refs. 7 and 8, rescaled with the world average for $B_{\mu\mu}(1S)$ *from Ref. 16.*

e) data from Ref. 8 rescaled with the world average for $B_{\mu\mu}(2S)$ *from Ref. 16.*

THE D^* WIDTH AND THE STUDY of F and F^*

Katsuhito Sugano

Argonne National Laboratory, Argonne, IL 60439

Recent results on the D^*, F, and F^* production in e^+e^- from HRS and MARK III are summarized. An improved upper limit on the decay width of D^* is reported. A brief review of the characteristics of F and the updated data of the F production in e^+e^- are given. A new measurement on F^* is made using the associated production method in the reaction $e^+e^- \to FF^*$.

I. Introduction

The D^*, F, and F^* are the mesons which are bound states of a heavy quark and light antiquark and their excited states. These heavy mesons, including other states such as D and D^{**}, have many unique characteristics because of the heavy quark mass. Therefore, the detailed study of these heavy mesons contributes to our quantitative understanding of their decay mechanism, their binding potential, and the non-perturbative QCD effects.

In this report, we review the decay width of D^*, the experimental status of F, and a new measurement on F^*, based on the three contributed papers to this conference.[1-3] We use the old names of F and F^* instead of $D_s(1970)$ and $D_s^*(2110)$, according to the original titles.

2. Upper Limit on the D^* Width

The decay with is a fundamental property of all particles and contains important information on the underlying physics of the decay mechanism. The widths of most of the particles have been measured extensively over the years. There are, however, several particles whose widths have not been measured well yet. Such examples are $D^{*+}(2010)$ and $D^{*0}(2007)$. The present upper limits of their widths are $\Gamma_{D^{*+}} < 2.0$ MeV/c^2 and $\Gamma_{D^{*0}} < 5$ MeV/c^2. These widths are very narrow compared to the other light mesons despite the fact that the D^*'s decay strongly. There are indications that the real widths may be about 100 keV/c^2 or less.[4,5] Therefore the D^* is one of the narrowest particles among the strongly decaying particles, and it is worthwhile to improve the present upper limit of the decay width in order to study the decay mechanism.

At this conference, the HRS group at the PEP e^+e^- storage ring reported a preliminary measurement on the upper limit of the $D^{*+}(2010)$ decay width, using the charged D^* production in the decay mode $D^{*+} \to D^o\pi^+ \to K^-\pi^+\pi^+$. The data sample corresponds to an integrated luminosity of (185 ± 5) pb^{-1} at $\sqrt{s}$ = 29 GeV. The details of the detector, trigger conditions, and selection cuts are given elsewhere.

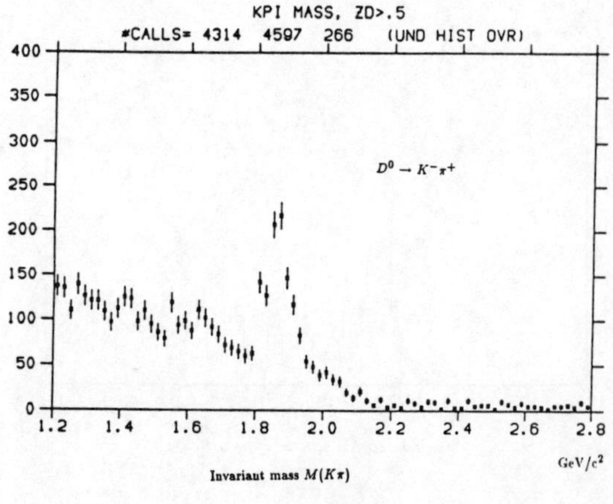

Figure 1

In reconstructing the D^0 through the $K^-\pi^+$ decay mode, all the tracks coming from the vertex were tried in turn as both K and π. The cuts of $z_D > 0.5$ and $|\cos\theta^*| < 0.7$ were applied, where $z_D \equiv 2E_D/\sqrt{s}$ and θ^* is the D^0 decay angle in the helicity frame. (D^0 rest frame with the z axis along the D^0 direction of flight.) The $\cos\theta^*$ selection eliminates an angular region observed to be dominated by the background. The resulting invariant mass spectrum is shown in Fig. 1. A clear signal is observed at the D^0 mass.

The D^0 signal was fitted with a polynomial form for the background and a Gaussian form for the signal. The fitted mass of (1861 ± 4) MeV/c^2 is consistent with the currently accepted value. The standard deviation (σ_m) of (13 ± 2) MeV/c^2 is consistent with the apparatus resolution as determined by a Monte Carlo simulation.

The D^* signal was reconstructed utilizing the fact that the Q value of the reaction (1) is only 5.8 MeV and the D^{*+} - D^0 mass difference can therefore be determined extremely well. Figure 2 shows the distribution in the mass difference $(\Delta \equiv M_{K^-\pi^+\pi^+} - M_{K^-\pi^+})$ for $z_{D^*} > 0.4$ with the D^0 selections $1810 < M_{K^-\pi^+} < 1920$ MeV/c^2, which is the mass region for the D^0 determined above. The peak in the distribution shows a clear signal for the D^{*+} production with small background. The distribution was fitted by polynomial

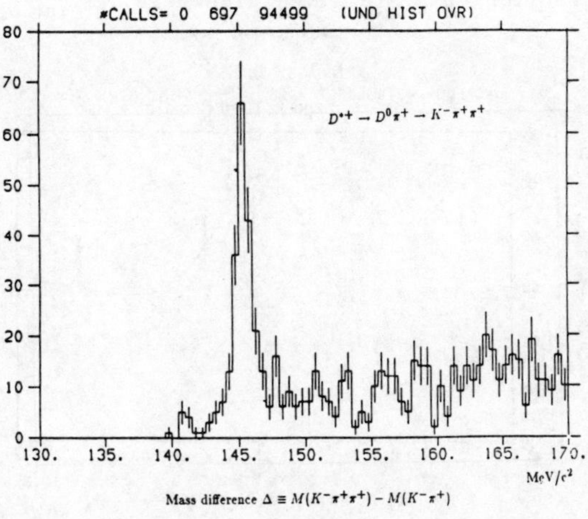

Figure 2

background plus Gaussian signal with finer binning. The resulting values are $\Delta = (145.37 \pm 0.07)$ MeV/c^2 for the peak value and $\sigma_\Delta = (0.43 \pm 0.06)$ MeV/c^2 for the standard deviation, where the errors are statistical only. The value of σ_Δ corresponds to the width $\Gamma_\Delta = (0.98 \pm 0.14)$ MeV/c^2 as FWHM assuming a Gaussian form. The width remains the same within the statistical errors when the z_{D^*} cut is changed from 0.4 up to 0.7.

According to the Monte Carlo simulation of the detector, the apparatus resolution (standard deviation) of Δ is $\sigma_\Delta^{MC} = (0.33 \pm 0.07 \pm 0.07)$ MeV/c^2, where the first error is statistical and the second systematic. This corresponds to the width of $\Gamma_\Delta^{MC} = (0.78 \pm 0.16 \pm 0.16)$ MeV/c^2. The observed width is consistent with the apparatus resolution within the errors, although the Monte Carlo value is 0.2 MeV/c^2 smaller than the observed one. (Note: present Monte Carlo does not simulate the movement of the beam.) The observed width of the mass difference is dominated by the detector resolution. Therefore, they put an upper limit of the directly measured D^{*+} decay width as $\Gamma_{D^{*+}} < 1.15$ MeV/c^2 at 90% CL.

This direct measurement can be compared to a calculation based on the measured branching ratio of the D^{*+} radiative decay. Since the radiative decay $D^{*+} \to D^+\gamma$ is an M1 transition, the electromagnetic width is given by the following formula:

$$\Gamma_{M1} = \frac{4}{3}\alpha\left(\frac{e_c}{2m_c} + \frac{e_d}{2m_d}\right)^2 k^3 , \qquad (1)$$

where $e_q/2m_q$ is the magnetic moment of the quark q (charm quark and down quark) and k is a momentum of photon. Therefore, the total decay width is given by $\Gamma_{D^{*+}} = \Gamma_{M1}/Br(D^{*+} \to D^+\gamma)$. The estimated value of Γ_{M1} for $D^{*+} \to D^+\gamma$ is about 1.1 keV/c^2, using the constituent quark masses ($m_c = 1.84$ GeV/c^2 and $m_d = 0.34$ GeV/c^2). The most recent measurement of the radiative decay mode is $Br(D^{*+} \to D^+\gamma) = (17 \pm 11)\%$. This corresponds to the total width $\Gamma = 7^{+12}_{-3}$ keV/c^2, or $\Gamma < 22$ keV/c^2 at 90% CL, consistent with our result. Since there is some ambiguity in calculating the value of

(1), our direct measurement is orthogonal to this approach.

There are also calculations of the hadronic decay width of $D^{*+} \to D^+\pi^0$ and $D^{*+} \to D^0\pi^+$, based on the SU(4) invariant interaction.[5] Using these calculations and the measured branching ratios of these processes, the total width of D^{*+} is (27 ± 5) keV/c^2. This value is higher than that calculated from the radiative decay mode.

Next, the measurement of $D^{*0}(2007)$ decay width was tried based on the kinematically forbidden mode of $D^{*0} \to D^+\pi^-$.[6] This mode could occur if the width is large compared to the $|Q|$ value (such examples are $S(975) \to K^+K^-$ and $\delta(980) \to K^+K^-$). In fact, the Q value is very small:

$$Q = M_{D^{*0}} - \left(M_{D^+} + M_{\pi^-}\right) = -1.77 \pm 2.1 \text{ MeV}/c^2 .$$

Therefore, search for such a decay mode puts a constraint on the decay width in turn.

There are some problems in this method; the measurement of the mass of D^{*0} is not accurate yet, the shape of the resonance tail is not well known because of the threshold effect, there is combinatorial background, and there is possible contribution from the D^{**} decay. A preliminary result from the HRS group shows $B_r(D^{*0} \to D^+\pi^-) < 10\%$ at 90% CL, which corresponds to $\Gamma_{D^{*0}} < 1$ MeV/c^2.

3. Study of F

The F meson is the lowest-lying charm state with strangeness. The spin, parity, and isospin are $J^P = 0^-$ and $I = 0$. The recent measurements of its mass cluster around 1971 MeV, using the decay modes of $\phi\pi$ and $\bar{K}^*K$. There are measurements on the $F \to \phi\pi$ decay mode from ARGUS, CLEO, HRS, and TASSO, etc., but the absolute branching ratio differs considerably among the experiments.[7] Also, there is disagreement on the $\phi\pi\pi\pi$ decay mode. Figure 3 shows the recent lifetime measurement of F compared to those of D^0 and D^+.[8] The world average of F lifetime is 0.29 ps, which is closer to the D^0 lifetime. Since F has a possible annihilation diagram as well as a spectator

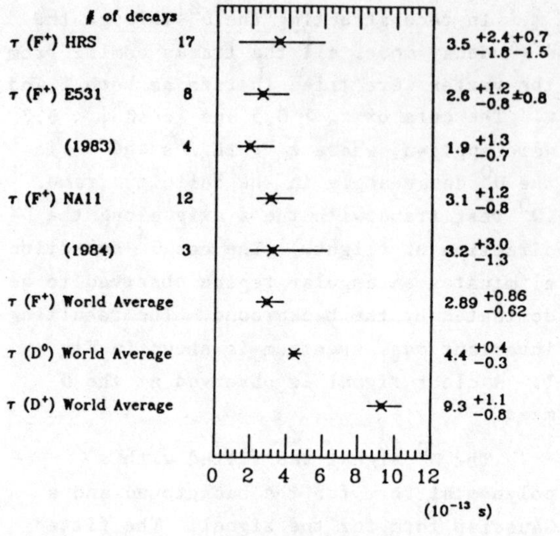

Figure 3

diagram in its decay, the detailed study of F will cast further light on the weak decays of heavy mesons.

In this conference, there are new data from the HRS group on the F production. To search for the decay mode $F^+ \to \phi\pi^+ \to K^+K^-\pi^+$, each K^+K^- combination in the ϕ mass band was

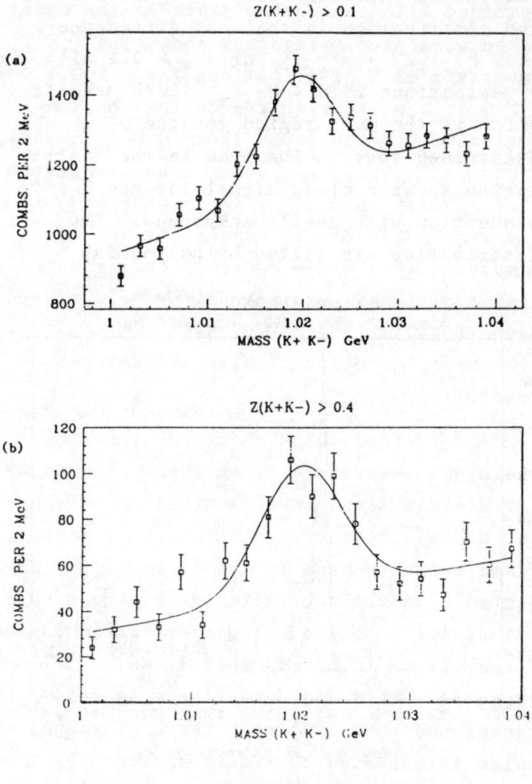

Figure 4

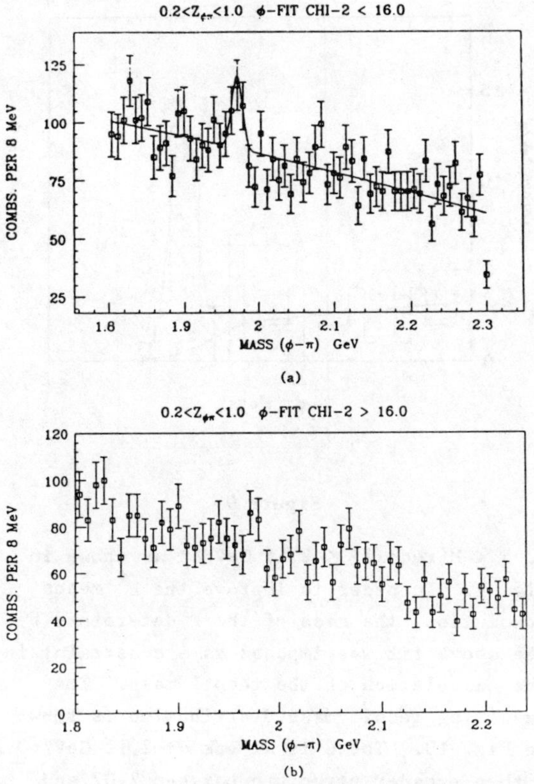

Figure 5

combined with each other track in the event taken as a pion. Figure 4 shows the K^+K^- invariant mass distributions for z > 0.1 and z > 0.4. In order to reduce the combinatorial background, only those K^+K^- combinations satisfying the ϕ mass constraint with $\chi^2 < 16$ were used. In Fig. 5, the $\phi\pi$ invariant mass distributions for the two samples with $\chi^2 < 16$ and $\chi^2 > 16$ are shown with additional requirement z > 0.2. The mass is (1972 ± 4) MeV/c^2 with σ = (7.3 ± 1.0) MeV/c^2 consistent with the detector resolution.

To further check that the signal has the expected properties of the F, the angular distributions were examined. The angular distribution of the K^+ in the ϕ rest frame (with respect to the π) is of the expected form $\cos^2\theta_K$, while the ϕ direction with respect to the F boost direction in the F rest frame is not so uniform as expected. This indicates there is some background contribution. The best signal to noise is achieved by requiring $|\cos\theta_\phi|$ < 0.9 and $|\cos\theta_K| > 0.5$ as shown in Fig. 6. The fit gives a signal of 50 ± 10 events, with a mass of (1965.9 ± 3) MeV/c^2 and σ = (8.4 ± 0.8) MeV/c^2.

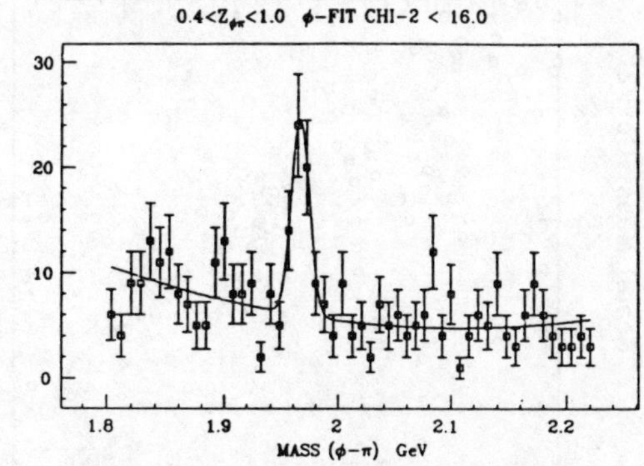

Figure 6

The resulting fragmentation function is shown in Fig. 7. Also shown in this figure is the fragmentation function for D^* measured in this experiment (reduced by a factor of 100 for comparison). The F fragmentation function appears to be softer than that for the D^*.

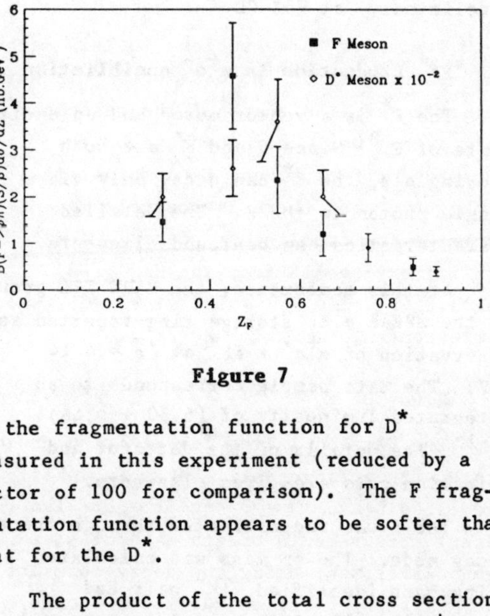

Figure 7

The product of the total cross section and branching ratio for Z > 0.4 is $\sigma(F^+ + F^-) \cdot B(F \to \phi\pi) = 1.31 \pm 0.35\ pb^{-1}$, which corresponds to $R(F^+ + F^-) \cdot B(F \to \phi\pi$ = 0.0113 ± 0.003. Comparing this result to their measurement of $R(D + \bar{D})$ = 2.20 ± 0.5 in the same Z region yields $R(F) \cdot B(F \to \phi\pi)/R(D)$ = 0.0049 ± 0.0024. This leads to the branching ratio $B(F \to \phi\pi)$ = (2.8 ± 1.0)% assuming $R(F)/(R(D) + R(F))$ = 0.15.

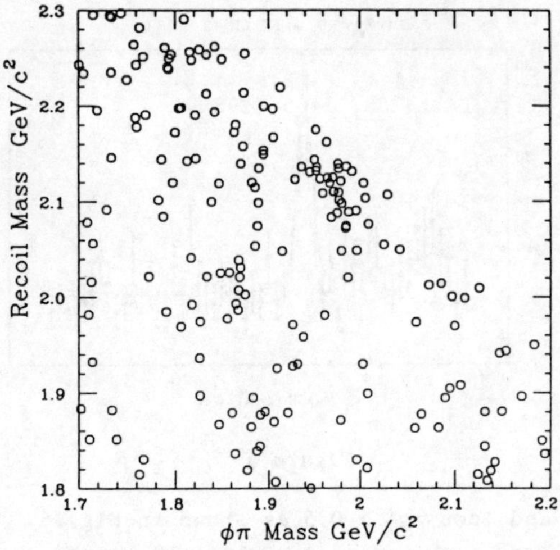

Figure 8

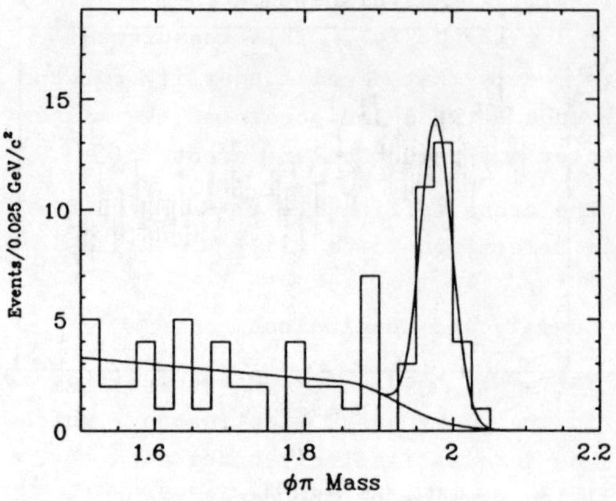

Figure 9

To search for the decay mode $F \rightarrow \phi 3\pi$, each ϕ candidate was combined with all $(3\pi)^{\pm}$ candidates in the event, again imposing the ϕ mass constraint on the K^+K^- pair. They found no signal and put an upper limit of 2% (preliminary) at 95% CL.

4. FF^* Production in e^+e^- Annihilation

The F^* is a vector meson and an excited state of F.[8] Since F and F^* are both isosinglets, the F^* can decay only via a single photon to the F. The detailed characteristics can be found elsewhere.

At this conference, the MARK III group at the SPEAR e^+e^- storage ring reported an observation of $e^+e^- \rightarrow FF^*$ at $\sqrt{s} = 4.14$ GeV. The data sample corresponds to an integrated luminosity of (6.30 ± 0.46) pb^{-1}. The details of the detector and selection cuts are given elsewhere.

The F was reconstructed through the $\phi\pi$ decay mode. The $\phi\pi$ mass was calculated using each identified K^+K^- pair and considering all of the remaining charged tracks to be pions. Then the mass recoiling against each $\phi\pi$ was calculated. Figure 8 shows the scatter plot of the $\phi\pi$ mass vs. the recoil mass. A cluster of events near $M(\phi\pi) = 1.97$ GeV/c^2 and M(recoil) = 2.10 GeV/c^2 provides evidence for FF^* production.

The mass of F was determined to be $(1977.6 \pm 4.3 \pm 4.0)$ MeV/c^2 by fitting the $\phi\pi$ mass distribution with a selection of $2.05 < M(\text{recoil}) < 2.15$ GeV/c^2 as shown in Fig. 9. In order to improve the F^* meson resolution, the mass of the F determined by the above fit was imposed as a constraint in the calculation of the recoil mass. The resulting recoil mass distribution is shown in Fig. 10. There is a peak at 2.11 GeV/c^2 with a broader structure between 2.07 and 2.15 GeV/c^2. This is consistent with the distribution expected for $e^+e^- \rightarrow FF^*$, $F^* \rightarrow F\gamma$, where half of the $\phi\pi$ events are from the decay of the initially-produced F and the other half are from the decay of the F from F^*. A fit to the distribution gives

$$M_{F^*} = (2106.8 \pm 1.8 \pm 6.2) \text{ MeV/c}^2.$$

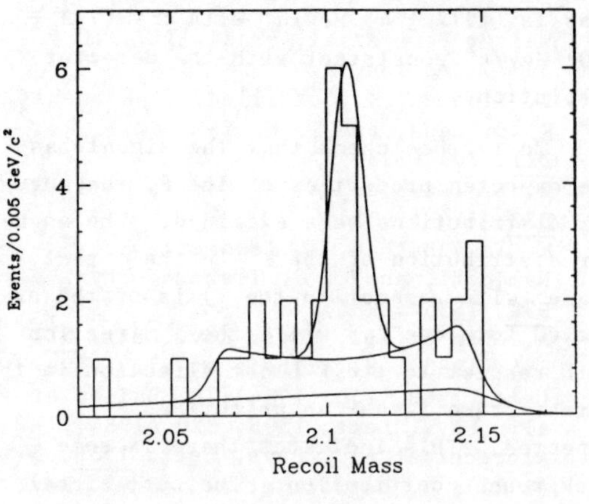

Figure 10

The resulting mass difference $(M_{F^*} - M_F)$ is $(129 \pm 2 \pm 12)$ MeV/c^2. This measurement favors models that predict equality for the difference between the square of the masses of vector and pseudoscalar mesons.

The cross section $\sigma(e^+e^- \to FF^*)\,B(F \to \phi\pi)$ is determined to be $(36 \pm 7 \pm 13)$ pb.

5. Summary and Conclusions

There has been a steady accumulation of data in the heavy quark spectroscopy, which helps us to quantitatively understand the underlying physics of the decay mechanism.

D^* seems to have a very narrow width. The upper limit of the total decay width was improved, but some good idea is needed to directly measure the width.

The recent measurements of the F mass cluster around 1970 MeV/c^2, which is lower than the previous measurements. There are discrepancies in the branching ratios into the $\phi\pi$ and $\phi\pi\pi\pi$ modes among the experiments. We need measurements of the absolute branching ratios. The F fragmentation is similar to other charm mesons, although it appears to be softer than that of the D^*.

F^* has been established and the mass difference $(M_{F^*} - M_F)$ is similar to the other vector-pseudoscalar twins of mesons.

References

1. S. Abachi et al. (HRS), ID # 3140, Session 9G.
2. S. Abachi et al. (HRS), ID # 10111, Session 9G.
3. Walter Toki et al. (MARK III), ID # 3387, Session 9G.
4. E. Eichten, K. Gottfried, T. Kinoshita, K. D. Lane, and T. M. Yan, Phys. Rev. D21, 203 (1980).
5. R. L. Thews and A. N. Kamal, Phys. Rev. D32, 810 (1985).
6. K. Jagannathan, A. Jawahery, R. Namjoshi, and C. G. Trahern, Phys. Rev. D32, 1835 (1985).
7. A. Chen et al., Phys. Rev. Lett. 51, 634 (1983); H. Althoff et al., Phys. Lett. 136B, 130 (1984); R. Bailey et al., Phys. Lett. 139B, 320 (1984); H. Albrecht et al., Phys. Lett. 153B, 343 (1985); M. Derrick et al., Phys. Rev. Lett. 54, 2568 (1985).
8. See the talk given by D. Blockus at this conference.
9. H. Albrecht et al., Phys. Lett. 146B, 111 (1984); H. Aihara et al., Phys. Rev. Lett. 53, 2465 (1984).

DM2 RESULTS ON HADRONIC AND RADIATIVE J/ψ DECAYS

Bernard JEAN-MARIE
Laboratoire de l'Accélérateur Linéaire,
Bât. 200, Centre d'Orsay, 91405 Orsay Cedex, FRANCE

A preliminary study of the $J/\psi \to \gamma\rho^o\rho^o$ decay is presented. The existence of the decay $J/\psi \to \gamma D(1285)$ is established by the observation of two decay modes $D \to 4\pi^\pm$ and $D \to \eta\pi^+\pi^-$. The channel $J/\psi \to \phi D(1285)$ is evidenced in the same way. A systematic comparison of $K\bar{K}$ and $\pi\pi$ production associated to γ,ω and ϕ leads to results concerning $K^0_SK^0_S$, θ, f, S* and $\pi\pi$ states. SU(3) and electromagnetic violations are tested by measuring allowed and forbidden baryonic decays. Finally a measurement of $J/\psi \to K^0_SK^0_L$ is presented.

The search for glueball states was certainly the most popular reason to study J/ψ decays. New and puzzling states have been revealed, however a clear cut answer on glueball manifestation in the J/ψ data is still remote. The observation of a state in one or a few modes is not sufficient and one needs to study the largest possible number of modes and concurrently clarify the standard classification of hadrons. The results presented here concern both the search for glueball states and the study of J/ψ hadronic decays.

The data analysed correspond to 8.6 10^6 J/ψ recorded by the DM2 detector at DCI. A description of the detector can be found in reference 1.

1. $J/\psi \to \gamma\rho^o\rho^o \to \gamma\pi^+\pi^-\pi^+\pi^-$

The candidate events to the final state $\gamma\pi^+\pi^-\pi^+\pi^-$ must satisfy the following criteria :

- four charged tracks with zero total charge
- only one isolated track found in the shower detector
- a cut on the quantity $4P^2 \sin^2(\theta/2) < 1500\ \mathrm{MeV}^2/c^2$, where θ is the acollinearity angle between the photon and the direction of the missing momentum P
- a 3C-fit to the hypothesis $J/\psi \to \gamma\pi^+\pi^-\pi^+\pi^-$ with a χ^2 value smaller than 9.

Additional cuts are used against contamination from the large $\omega\pi^+\pi^-$ process and from $\gamma K^0_S K^\pm\pi^\mp$. After these cuts, the most important remaining background comes from $J/\psi \to 4\pi^\pm\pi^o$ in which, due to a very asymmetric π^o decay, the low energetic photon is undetected.

The mass distribution of the 18053 surviving events is shown in figure 1.

A qualitative evidence of a ρρ signal is obtained by imposing a cut on the quantity :

$$R = \prod_{i=1,2} \frac{m_\rho^2\,\Gamma_\rho^2}{(m_i^2 - m_\rho^2)^2 + m_\rho^2\,\Gamma_\rho^2} > \frac{1}{15}$$

In the resulting spectrum (Fig.1), there are two clear peaks around 1.53 GeV and 1.73 GeV and a broader structure at 2.10 GeV. In addition, the D(1285) signal disappears while the η_c peak is enhanced†. So below 2 GeV, a large fraction of $\gamma 4\pi^\pm$ is due to $\gamma\rho^o\rho^o$.

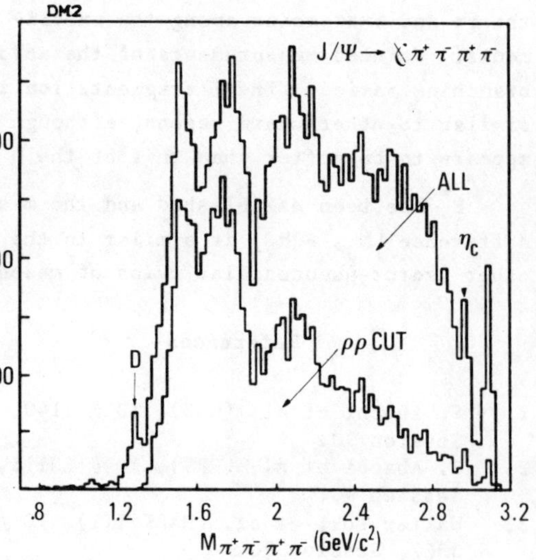

Figure 1

To better estimate the ρρ signal, a multichannel analysis is performed. Three modes are considered. Pseudoscalar ρρ, $\rho^o\pi^+\pi^-$, $4\pi^\pm$ phase space. The ρππ channel includes contributions from $A_2\pi$, while in this preliminary analysis, ρρ contributions $0^+, 1^\pm, 2^\pm$ and isotropic $\rho^o\rho^o$ have been neglected.

---

† The study of $\eta_c \to \rho^o\rho^o$ and other η_c decay modes is reported at this conference in parallel session # 9

For each event, weighted probabilities are computed and a maximum likelihood fit gives for all events the relative contribution of each channel (Fig.2). The three peaks already mentioned subsist in the $0^-\rho\rho$ channel while the two other channels are structureless. Assuming that the $\rho\rho$ component below 2 GeV is entirely due to spin parity 0^-, one obtains :

$$B(J/\psi \to \gamma\rho^o\rho^o) = (3.6 \pm .12 \pm .54)10^{-3}$$

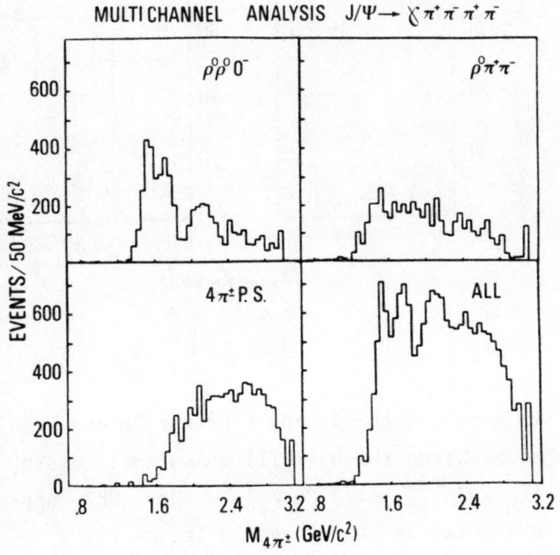

Figure 2

It is worth noticing that due to its spin parity value, the structure around 1.73 GeV cannot be associated to the $\theta(1690)$. These results agree with previous works[2-3].

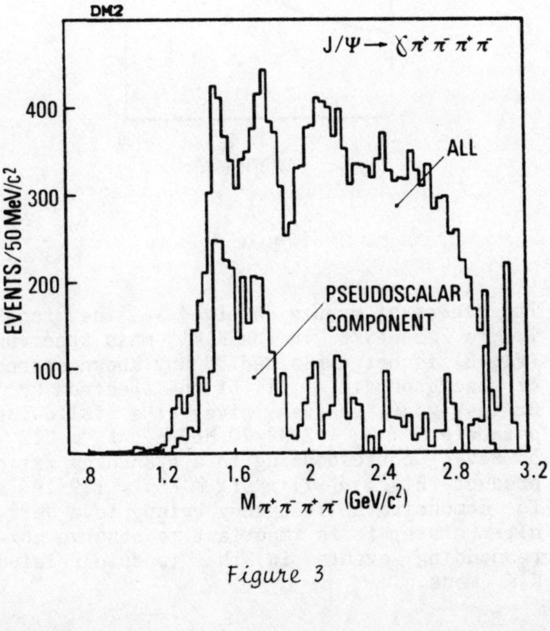

Figure 3

The pseudoscalar nature of the $\rho^o\rho^o$ system can also be evidenced by looking at the distribution of the angle χ between the ρ decay planes[3-4]. The fit is done by 50 MeV bins. There are two possible $\rho\rho$ combinations per event, only the best one is considered. The spectrum shown in figure 3 supports our previous conclusions.

2. OBSERVATION OF D(1285) ASSOCIATED TO γ AND ϕ

The nonobservation of the decay $\iota \to \eta\pi^+\pi^-$ is still puzzling. In the $\eta\pi^+\pi^-$ mass distribution, shown in figure 4, two peaks are visible. The first one could be attributed to the D or the $\eta(1275)$, the second one has the following fitted parameters :

$$M_X = 1391.5 \pm 3.0 \text{ MeV/c}^2 \quad \Gamma_X = 52 \pm 9 \text{ MeV/c}^2$$

$$B(J/\psi \to \gamma X) \times B(X \to \eta\pi^+\pi^-) = (4.1\pm.3\pm1.)10^{-4}$$

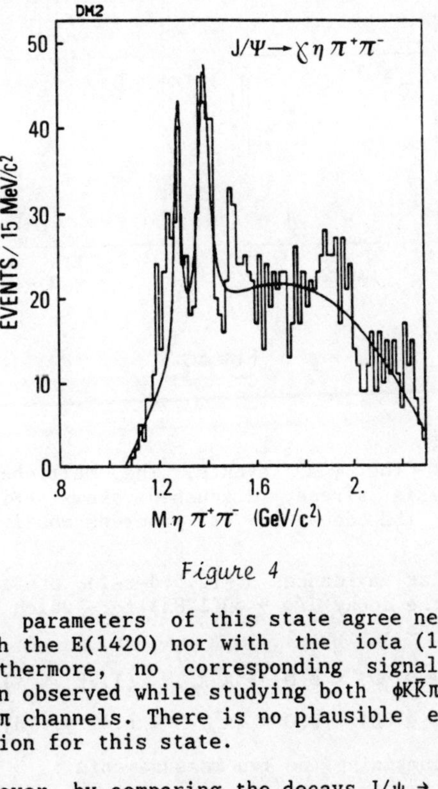

Figure 4

The parameters of this state agree neither with the E(1420) nor with the iota (1440). Furthermore, no corresponding signal has been observed while studying both $\phi K\bar{K}\pi$ and $\phi\eta\pi\pi$ channels. There is no plausible explanation for this state.

However, by comparing the decays $J/\psi \to \gamma 4\pi^\pm$, $\gamma\eta\pi^+\pi^-$ and $\gamma K\bar{K}\pi$ one can argue that the first peak observed corresponds to the D(1285). In $\gamma 4\pi^\pm$, the fitted values of the narrow peak (Fig.5-b), after unfolding the experimental resolution (5.5 MeV), are :

$$M_D = 1283.2\pm2.0 \text{ MeV/c}^2 \quad \Gamma_D = 23.9 {}^{+7.0}_{-5.0} \text{MeV/c}^2.$$

They correspond to the D parameters. Using our measured branching ratio product : $B(J/\psi\to\gamma D) \times B(D\to 4\pi^\pm) = (3.4 \pm .8 \pm .5)10^{-5}$, one obtains :

$$B(J/\psi \to \gamma D) = (2.6 \pm .6 \pm .6)10^{-4}$$

Therefore, 30 ± 7 events are expected for the channel $\eta\pi^+\pi^-$ (Fig.5-a) while the fit (imposing the D mass) yields 42 ± 12 events. In addition the few events observed in the $K_S^0K^{\pm}\pi^{\mp}$ mode are compatible with the expected 2.3 events.

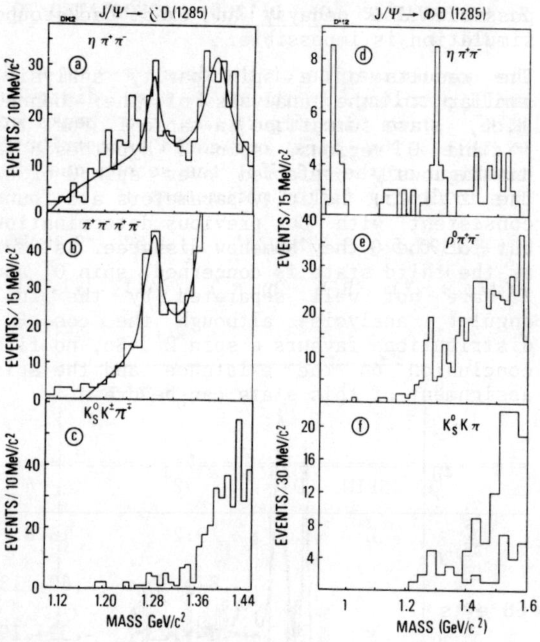

Figure 5

Using the $\gamma 4\pi^{\pm}$ events, the multichannel analysis already discussed shows (Fig.6) that the decay $D \to 4\pi^{\pm}$ proceeds mostly via $\rho\pi\pi$.

Similar evidences (Fig.5.d-e-f) are found for the decay $J/\psi \to \phi D(1285)$ for which one obtains :

$$B(J/\psi \to \phi D) \times B(D \to \eta\pi^+\pi^-) = (1.22 \pm .4)10^{-4}$$

$$B(J/\psi \to \phi D) \times B(D \to 4\pi^{\pm}) = (.43 \pm .13)10^{-4}$$

and combining the two measurements :

$$B(J/\psi \to \phi D) = (2.9 \pm 1.)10^{-4}$$

As the D(1285) is produced associated to ϕ one can conclude that D and E(1420) are not ideally mixed in the axial vector meson nonet[9] since D would be made of u and d quarks only.

3. $J/\psi \to \gamma K\bar{K}$

Results on f′(1525) and θ(1690) and the non-observation of ξ(2230) have already been presented[6]. Let us recall that the ξ(2230)

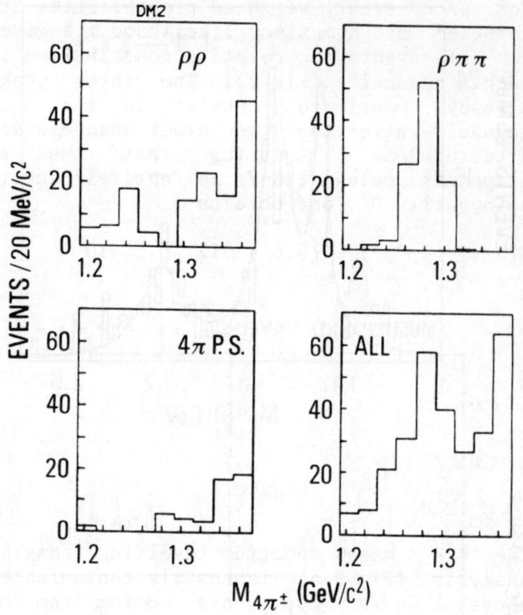

Figure 6

is seen neither in the K^+K^- nor in the $K_S^0K_S^0$ mode. Given the Mark III measured ξ width[5] ($\Gamma_\xi = 26^{+20}_{-16} \pm 17$) MeV/c², the DM2 upper limit can vary as shown in figure 7.

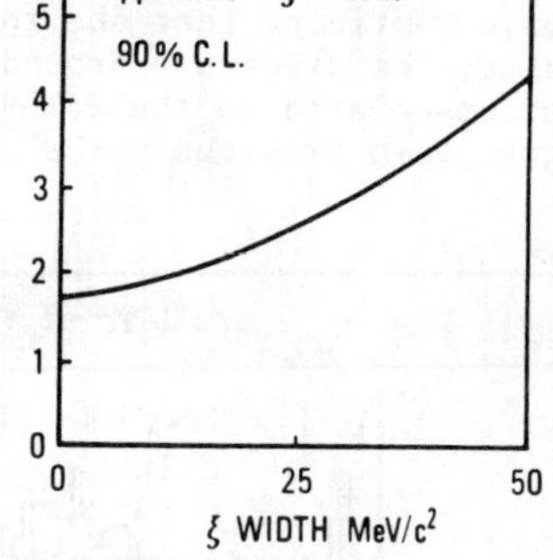

Figure 7

The excess of events observed in the range 1.9 to 2.5 GeV/c² in the $K_S^0K_S^0$ mass spectrum (Fig.8) is not explained by any known source of background. A fit of the spectrum by 3 Breit-Wigner curves gives the following parameters : $M_X = 2198 \pm 20$ MeV/c², $\Gamma_X = 217 \pm 55$ MeV/c² corresponding to a branching ratio product $B(J/\psi \to \gamma X) \times B(X \to K^0\bar{K}^0) \simeq 1.9\ 10^{-4}$. To demonstrate that they belong to a definite state, it is important to observe corresponding events in the isospin related K^+K^- mode.

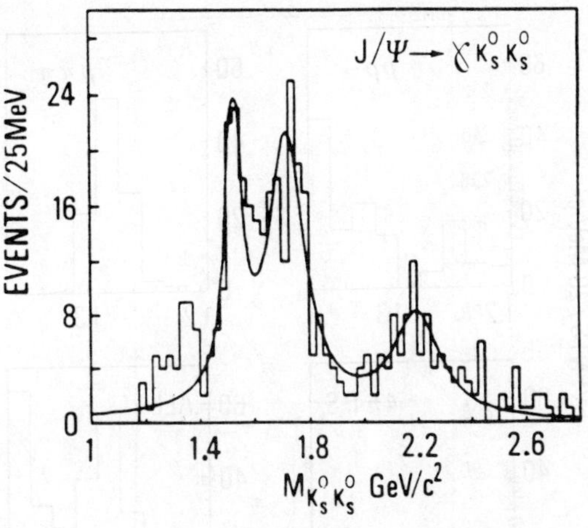

Figure 8

The K^+K^- mass spectrum resulting from the analysis (Fig.9-a) is heavily contaminated above 2 GeV/c^2 by events coming from the abundant decays $J/\psi \to \pi^+\pi^-\pi^0$ and $J/\psi \to KK^*(890)$. When a π^0 decays asymetrically and the low energetic photon is undetected, these events become undistinguishable from real γK^+K^- events. The background spectrum has been estimated by Monte-Carlo simulation

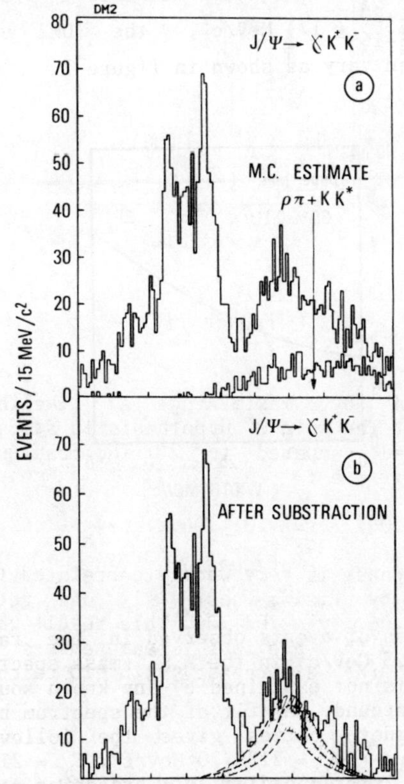

Figure 9

(Fig.9-a). Although the absolute number of expected events has large systematics, the actual shape is reliably predicted. In figure 9-b, the expected signal and the background subtracted spectrum are compared. There is room to accomodate the signal but this has to be taken with some care since other modes can contribute to the background and since, due to the very large number of possible J/ψ decays, an exact background simulation is impossible.

The results of a spin parity analysis, similar to the analysis of the channel γK^+K^- [7)], are summarized in table I. For the f′ and θ regions, spin 2 hypothesis is unambiguously preferred over spin 0 or 4. The f′(1525) helicity parameters are found consistent with our previous determination but for the θ they somehow disagree. As far as the third state is concerned, spin 0^+ and 2^+ are not well separated by the full angular analysis although the cos θ_K^+ distribution favours a spin 2^+. So, no firm conclusion on the existence and the spin assignment of this state can be drawn.

	SPIN	0^+	2^+	4^+
f′	$\mathcal{L}$	.3	9.2	6.8
	x		.87±.15	.69±.13
68 evts				
	y		.34±.23	-.31±.19
θ	$\mathcal{L}$	2.6	18.7	15.5
	x		-.57±.12	.98±.15
103 evts				
	y		-.63±.16	-.50±.20
X	$\mathcal{L}$	.8	2.4	-2.2
2.→2.4GeV	x		1.0±.18	1.4±.45
85 evts	y		.20±.24	-2.±.6

Table I

4. $J/\psi \to \phi K\bar{K}$ AND $\omega K\bar{K}$

The ϕf′(1525) mode is clearly seen in the decays $J/\psi \to \phi K^+K^-$ and $\phi K^0_S K^0_S$ (Fig.10-a,b). The excess of events observed on the high mass side of the f′ in figure 10-a can tentatively be interpreted as produced by θ(1690) interfering with f′. In this hypothesis, the parameters of the fit to two interfering Breit-Wigner curves, the f′ mass and width being fixed, are :

$M_\theta = 1686 \pm 14$ MeV/c^2 $\Gamma_\theta = 162 \pm 35$ MeV/c^2

$\phi = (-1.9 \pm .24)$ rad

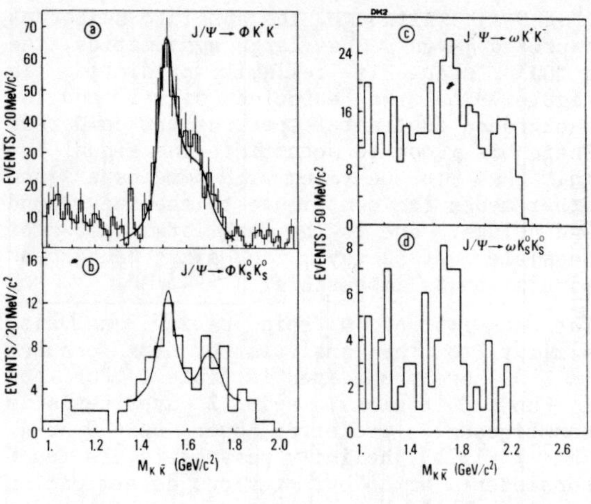

Figure 10

They are compatible with the θ ones. This interference should also show-up in $\phi K^0_S K^0_S$ (Fig.10-b). In this channel, due to the low statistics, although the spectrum looks different, a fit imposing the previous parameters gives consistent results. Using the events ϕK^+K^- one gets :

$$B(J/\psi \to \phi f') \times B(f' \to K\bar{K}) = (4.6 \pm .5)10^{-4}$$

$$B(J/\psi \to \phi\theta) \times B(\theta \to K\bar{K}) = (2.7 \pm .3)10^{-4}$$

By contrast, the f' is not observed in the decay $\omega K\bar{K}$ in both modes ωK^+K^- and $\omega K^0_S K^0_S$ (fig.10-c,d) while there is a clean signal at the θ mass. Thus $\theta(1690)$ couples to γ,ω and ϕ while f' does not couple to ω.

5. $\pi^+\pi^-$ PRODUCTION ASSOCIATED TO γ,ω,ϕ

The correlation between $\pi^+\pi^-$ production associated to γ,ω and ϕ has also been systematically studied.

In figure 11-a, the $\pi^+\pi^-$ mass distribution from $J/\psi \to \gamma\pi^+\pi^-$ exhibits, along with the well known f(1270) signal, a smaller but clear structure which could be attributed to $\theta(1690)$. We have already shown[6] that the excess of events observed on the high mass side of the f signal can be explained by f-f′ interference. But it could also be tempting to relate these events to the enhancement observed, in the same mass range, in $J/\psi \to \phi\pi^+\pi^-$ (Fig.11-c).

5.1. f angular analysis

By requiring a cut on the $\pi^+\pi^-$ mass ($1150. < M_{\pi^+\pi^-} < 1400$ MeV/c²), 2071 events are selected. In this sample the background events, which are mainly due to $J/\psi \to \rho^\pm\pi^\mp$ amount to 20 %. More than 90 % of them are eliminated by a cut on the angle between the π^+ and the f direction in the f rest frame. The remaining 1242 events which are almost background free are subjected to an angular analysis. The helicity ratios x and y are

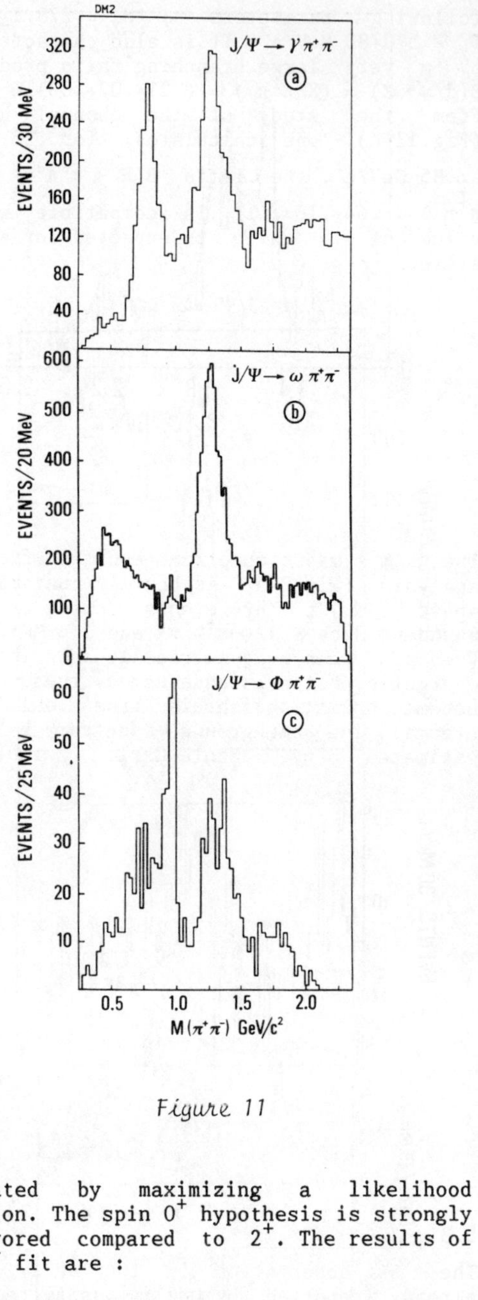

Figure 11

estimated by maximizing a likelihood function. The spin 0^+ hypothesis is strongly disfavored compared to 2^+. The results of the 2^+ fit are :

$$x = .83 \pm .06, \quad y = +.01 \pm .06, \quad \cos\phi_x = .62 \pm .23$$

As the phase is very weakly correlated to x and y, by fixing $\cos\phi_x = 1.$ one gets : $x = .84 \pm .06$, $y = .02 \pm .06$. This result which agrees with previous measurements[7-8] confirms the discrepancy with QCD calculations[9].

5.2. $J/\psi \to \omega\pi\pi$

A very copious production of low mass $\pi\pi$ events, already reported by DM2[6], appears as a broad bump near the $\pi\pi$ threshold in the $\pi^+\pi^-$ spectrum (Fig.11-b). It can be empirically fitted (Fig.12-b) to a Breit-Wigner curve distorted by phase space with the following parameters : $M_X = 278.1 \pm 35.5$, $\Gamma_X = 528 \pm 88$ MeV/c^2. It is also characterized by a very large branching ratio product : $B(J/\psi \to \omega X) \times (X \to \pi^+\pi^-) = (.21 \pm .02 \pm .03)10^{-2}$. From the study of the channel $\omega\pi^o\pi^o$, (Fig.12-c), one calculates, for $.25 \leq M_{\pi\pi} \leq .85$ GeV/c^2, the ratio $B(X \to \pi^o\pi^o)/ B(X \to \pi^+\pi^-) = .64 \pm .16$. It is compatible with a value of .5 which is expected for an I=0 state.

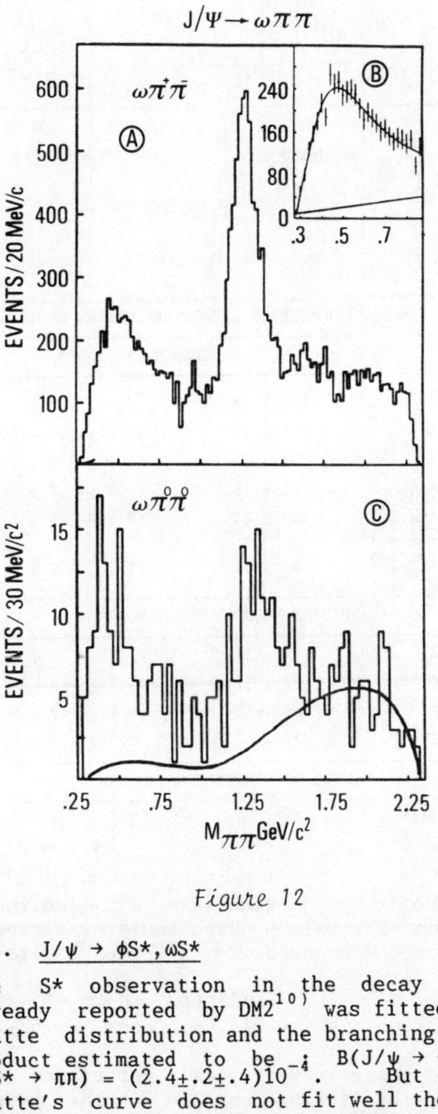

Figure 12

5.3. <u>$J/\psi \to \phi S^*, \omega S^*$</u>

The S* observation in the decay $\phi\pi^+\pi^-$ already reported by DM2[10] was fitted by a Flatte distribution and the branching ratio product estimated to be : $B(J/\psi \to \gamma S^*) \times B(S^* \to \pi\pi) = (2.4 \pm .2 \pm .4)10^{-4}$. But the Flatte's curve does not fit well the high mass side of the signal. Furthermore, the S* being observed in the $\pi^+\pi^-$ mode it is interesting to find its contribution in ϕK^+K^-.

One uses a model[11] in which two particles S* and ε are coupled to ππ, KK, 4π, 4K, etc. From a fit to the $\pi^+\pi^-$ spectrum, one gets an absolute prediction on the shape and magnitude of the K^+K^- contribution. The result is shown in figure 13. the S* fit in $\pi^+\pi^-$ is improved and the shape of the K^+K^- spectrum is correctly predicted.

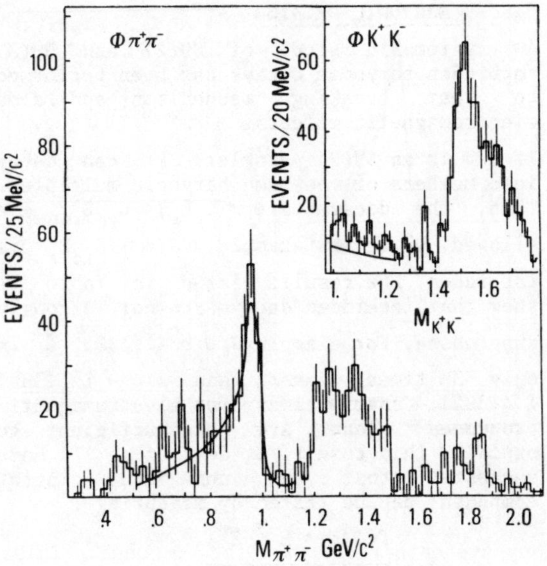

Figure 13

In the channel $\omega\pi^+\pi^-$, there is a clear accumulation of events at the S* mass (Fig. 11-b). The possibility of observing ωS* has been first mentioned by Mark II[12]. Although the level of background events is negligible, the very large contribution of the f tail is critical. So the fit results depend heavily upon the f parameters. If the S* is parametrized by a Breit-Wigner, with a width fixed to 50 MeV, one gets (Fig.14) $M_{S^*} = 954.4 \pm 3.4$ MeV/c^2 and a branching ratio product : $B(J/\psi \to \omega S^*) \times B(S^* \to \pi\pi) = (.95 \pm .1 \pm .22)10^{-4}$. If one tries to interprete this result in terms of S* quark content, it means, as the S* coupling to ϕ and ω are comparable that the S* has an $s\bar{s}$ component. Note that in quark models, the channel $J/\psi \to \omega S^*$ is forbidden but allowed in $K\bar{K}$ molecule interpretation of the S*.

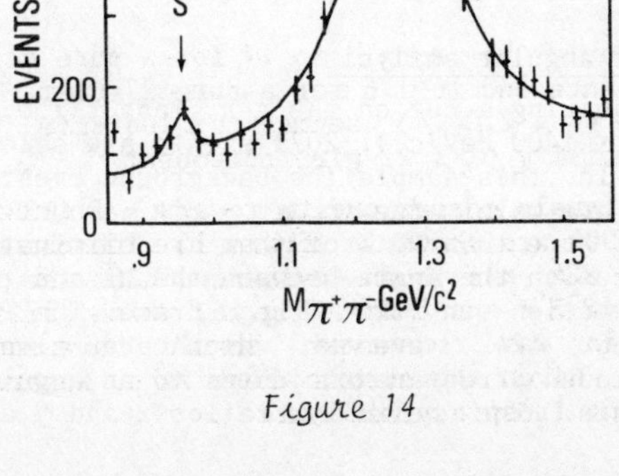

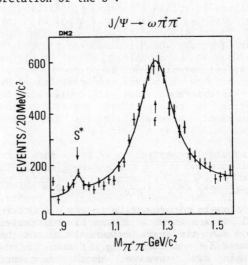

Figure 14

6. J/ψ BARYONIC DECAYS

A systematic study of SU(2) and SU(3) forbidden baryonic decays has been performed to test breaking mechanisms and also electromagnetic effects.

If J/ψ is an SU(3) singlet, it can decay into members of the same baryonic multiplet. Thus, the decays $J/\psi \to B_8 \bar{B}_8$, $B_{10} \bar{B}_{10}$ are allowed, while for example, $J/\psi \to B_{10} \bar{B}_8$ is forbidden. The results listed in table II show that forbidden decays are not strongly suppressed. For example $J/\psi \to \Sigma^{\pm}(1385)\ \bar{\Sigma}^{\mp}$ is only 3 times weaker than $J/\psi \to \Sigma^{\pm}(1385)\ \bar{\Sigma}^{\mp}(1385)$. Contributions from electromagnetic processes alone are not sufficient to explain this result. H. Genz et al.[13] have suggested that the non SU(3) singlet component can be tested by measuring :

$$R = \frac{\Gamma(J/\psi \to \Xi^{-*}\ \bar{\Xi}^{+})}{\Gamma(J/\psi \to \Sigma^{-*}\ \bar{\Sigma}^{+})}$$

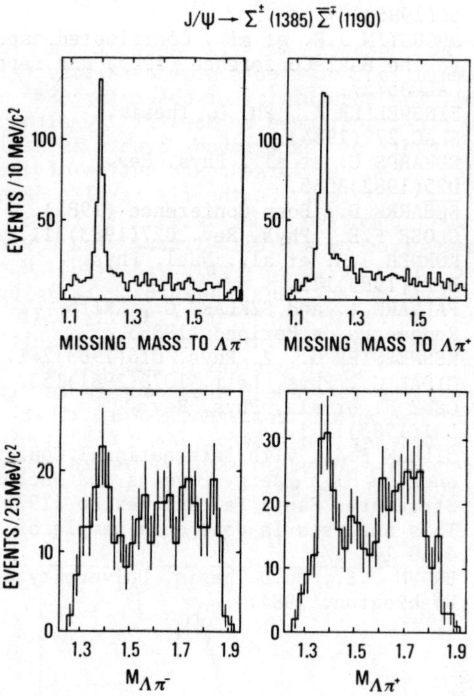

Figure 15

R should be equal to .7 for a pure octet dominance and to 1.6 for a pure 27-plet. The data (R = 2.0±.4) seems to indicate the possibility of a 27-plet component.

The signals corresponding to $J/\psi \to \Sigma^{\pm}(1385)\ \bar{\Sigma}^{\mp}(1190)$ are shown in figure 15. The channel $J/\psi \to \Xi\bar{\Xi}^*$ has been evidenced in the two modes $\Xi^-\bar{\Xi}^{+}*$ and $\Xi^o\bar{\Xi}^o*$ (Fig.16) which, if the isospin was conserved, should have equal rates. Given the errors there is no apparent violent isospin violation.

Results are also presented in table II on the isospin forbidden decays $J/\psi \to \Lambda\bar{\Lambda}\pi^o$, $J/\psi \to \Lambda\bar{\Lambda}\gamma$ and on $J/\psi \to \Lambda\bar{\Sigma}^o$ for which an upper limit is given.

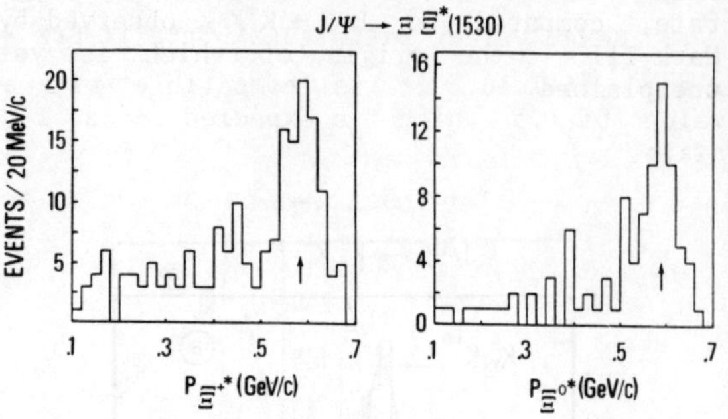

Figure 16

DECAY MODE	NUMBER OF EVENTS	BRANCHING RATIO($\times 10^4$)
- SU(2) and SU(3) ALLOWED DECAY MODES -		
$J/\psi \to \Xi^-\ \bar{\Xi}^+$	132 ± 19	7.0 ± .6 ± .8
$J/\psi \to \Sigma^{-*}\ \bar{\Sigma}^{+*}$	631 ± 50	10.0 ± .6 ± 2.0
$J/\psi \to \Sigma^{+*}\ \bar{\Sigma}^{-*}$	754 ± 60	11.9 ± .7 ± 2.0
$J/\psi \to \Lambda\bar{\Lambda}\gamma$	20 ± 6	.8 ± .2 ± .2
$J/\psi \to \Lambda\pi^-\ \bar{\Sigma}^+$ (incl. $\Sigma^{-*}\bar{\Sigma}^+$)	255 ± 25	9.0 ± .9 ± 1.6
$J/\psi \to \Lambda\pi^+\ \bar{\Sigma}^-$ (incl. $\Sigma^{+*}\bar{\Sigma}^-$)	342 ± 33	11.1 ± 1.0 ± 2.0
- SU(3) FORBIDDEN DECAY MODES -		
$J/\psi \to \Sigma^{-*}\ \bar{\Sigma}^+$	74 ± 18	3.0 ± .7 ± .8
$J/\psi \to \Sigma^{-*}\ \bar{\Sigma}^-$	77 ± 17	3.4 ± .4 ± .8
$J/\psi \to \Xi^{-*}\ \bar{\Xi}^+$	80 ± 17	5.9 ± .7 ± 1.5
$J/\psi \to \Xi^{o*}\ \bar{\Xi}^o$	24 ± 7	3.2 ± .7 ± 1.0
- SU(2) FORBIDDEN DECAY MODES -		
$J/\psi \to \Lambda\bar{\Lambda}\pi^o$	19 ± 6	2.2 ± .5 ± .5
$J/\psi \to \Lambda\bar{\Sigma}^{o*}$	< 16	< 2.9 (95 % C.L.)
$J/\psi \to \Lambda\bar{\Sigma}^{o}$	< 18	< 1.5 (95 % C.L.)
$J/\psi \to \Delta^+\bar{p}$	< 40	< 1.0 (95 % C.L.)

Table II

7. $J/\psi \to K^0_S K^0_L$

This channel cannot proceed via electromagnetic interaction[14] and it is forbidden if J/ψ is an SU(3) singlet. To search for this decay one selects events[15] having only one K^0_S decaying into $\pi^+\pi^-$, corresponding to $J/\psi \to K^0_S + X$ where X is entirely made up of neutrals. In the momentum distribution (Fig.17-a), two peaks are visible. The first one, comes from the decay $J/\psi \to K^0_S K^{*0}$, and corresponds to a branching ratio : $B(J/\psi \to K^0 \bar{K}^0 *) = (3.9\pm.18\ \pm.45)10^{-3}$. And one gets for the second one (Fig.17-b), after using several topological cuts to reduce the $K^0_S K^{*0}$ contribution :

$B(J/\psi \to K^0_S K^0_L) = (1.18\pm.12\pm.18)10^{-4}$.

This result confirms the surprisingly large rate, compared to $J/\psi \to K^+K^-$, observed by Mark III[16] the origin of which is yet unexplained.

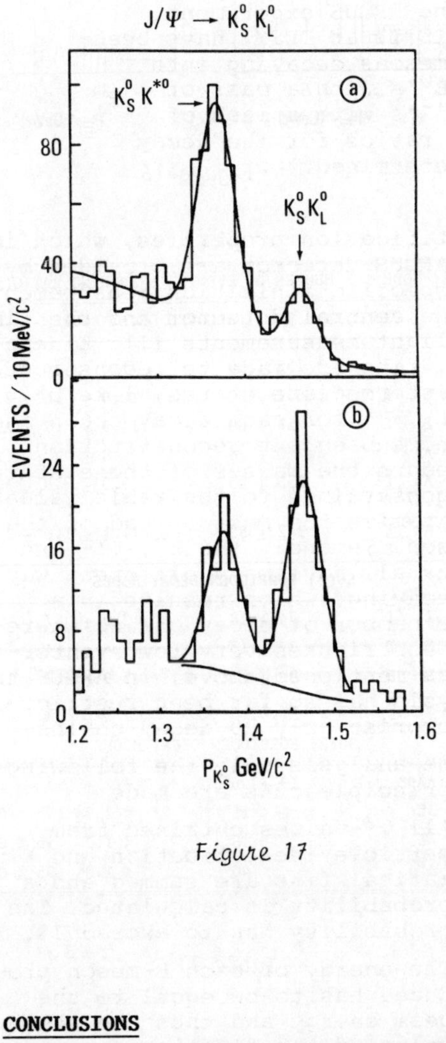

Figure 17

8. CONCLUSIONS

In summary, the results presented here are relevant to both the hadronic spectroscopy and the search to identify glueball states. We will only stress the main results. The existence of pseudoscalar structures in the decay $\gamma\rho^o\rho^o$ has been confirmed. In the same study, the first observation of the decay $\eta_c \to \rho^o\rho^o$ appears important to understand η_c decay mechanisms. D(1285) meson production associated to γ and ϕ has been established from the observation of three D decay modes. Furthermore, the $\rho\pi\pi$ nature of the decay $D \to 4\pi^\pm$ has been demonstrated by a multichannel analysis. The low mass $\pi\pi$ structure seen both in $\omega\pi^+\pi^-$ and $\omega\pi^o\pi^o$ is very puzzling and one needs some theoretical inputs to explain its nature. The systematic measurements of baryonic decays shows that SU(3) violating ones are not suppressed compared to allowed modes. Finally, the doubly forbidden decay $J/\psi \to K^0_S K^0_L$ is confirmed to be only twice smaller than $J/\psi \to K^+K^-$.

REFERENCES

1) AUGUSTIN J.E. et al., Physica Scripta 23(1981)623-633.
2) BURKE D.L. et al., Phys. Rev. Lett. 49(1982)632.
3) BALTRUSAITIS R.M. et al., Phys. Rev. D33(1986)1222.
4) More details are given in the paper on $\eta_c \to \rho^o\rho^o$ (parallel session #).
5) BALTRUSAITIS R.M. et al., Phys. Lett. 56(1986)107.
6) AUGUSTIN J.E. et al., Contributed paper to the BARI Conference (1985) and report LAL 85/27.
7) EINSWELLER K., Ph. D. Thesis, SLAC-272(1984).
8) EDWARDS C. et al., Phys. Rev. D25(1982)3065.
SCHARRE D., Bonn Conference (1981).
9) CLOSE F.E., Phys. Rev. D27(1983)311.
KORNER J.G. et al., Nucl. Phys. B229(1983)115.
10) FALVARD A. and SZKLARZ G., XXIth Rencontre de Moriond (1986).
11) MENNESSIER G., Z. Phys. C16(1983)241.
12) GIDAL G., Phys. Lett. 107B(1981)153.
13) GENZ H. et al., Phys. Rev. D31(1985)1751.
14) GILMAN F.J., VIth International Conference on High Energy Physics and Nuclear Structure, Santa Fe, New Mexico (1975).
15) This analysis is done on a sample of 4.38 10^6 J/ψ.
16) BROWN J.S., Ph.D.Thesis, University of Washington, 1984.

RECONSTRUCTION OF B-MESONS

Leif Jönsson
Institute of Physics, University of Lund
Sölvegatan 14, S-223 62 Lund, Sweden

Data taken at the $\Upsilon(4S)$ resonance in the ARGUS experiment, operating at the e^+e^- storage ring DORIS II at DESY, have been used for the full reconstruction of B-mesons decaying into $D^{*+}n\pi$, where n=1, 2, 3. We find 40 ± 8 B^0's with a mass of $(5278 \pm 1.0 \pm 3.0)$ MeV/c^2 and 32 ± 7 B^{+-}'s with a mass of $(5275.8 \pm 1.3 \pm 3.0)$ MeV/c^2. Branching ratios for the decay channels investigated have also been determined.

For the purpose of reconstructing B-mesons, the ARGUS collaboration at the e^+e^- storage ring DORIS, has systematically studied B-mesons decaying into a D^*-meson and a number of pions where the number of pions can vary between one and three. The analysis is based on an integrated luminosity of 50 pb^{-1} at the $\Upsilon(4S)$ resonance. This corresponds to about 50,000 $\Upsilon(4S)$ decays i.e. 100,000 B-mesons are produced.

More specifically the following decay channels, including the charge conjugates, have been investigated.

$\bar{B}^0 \to D^{*+}\pi^-$

$\bar{B}^0 \to D^{*+}\pi^-\pi^0$

$\bar{B}^0 \to D^{*+}\pi^-\pi^-\pi^+$

$B^- \to D^{*+}\pi^-\pi^-$

$B^- \to D^{*+}\pi^-\pi^-\pi^0$

The D^{*+} is observed from its decay into $D^0\pi^+$ and the D^0 is reconstructed in the following decay channels.

$D^0 \to K^-\pi^+$

$D^0 \to K^0_S\pi^+\pi^-$

$D^0 \to K^-\pi^+\pi^0$

$D^0 \to K^-\pi^+\pi^+\pi^-$

In total this gives 20 different decay chains.

There are two circumstances which complicate the reconstruction of B-mesons. One is the high decay multiplicity which, for the decay modes studied by us, can give as many as nine particles seen by the detector. The consequence is a huge combinatorial background. To handle this the detector must have excellent particle identification properties, which in the ARGUS detector are provided by the specific ionization measurements in the central detector and the time-of-flight measurements (1). Kinematical fits are made to reconstruct all intermediate states, like D^*, D^0, K^0_S, π^0, for each decay chain and in the subsequent reconstruction procedure the masses of these states are constrained to the table values to minimize the mass spread of the B-meson. Secondly the B^-, D^*- and D^0-mesons all have many different ways of decaying, which results in a large number of decay chains where each contributes very few events. As was mentioned above, in ARGUS the analysis has so far been restricted to comprise only 20 decay chains.

In the analysis only the following two principle cuts are made:

1) All χ^2-values obtained from particle identification and kinematical fits are summed and a probability is calculated. The probability has to exceed 1%.

2) The energy of each B-meson produced has to be equal to the beam energy and thus the absolute value of the difference $E_B - E_{beam}$ is required to be less than 3σ.

The mass of the B-meson is restricted to lie in the region between the $\Upsilon(3S)$- and $\Upsilon(4S)$-mesons and for a beam energy of 5.2885 GeV the momentum-mass relation is given by the parabola shown in fig. 1. For each B-candidate the closest distance between the data point and the parabola is found and the corresponding point on the parabola is projected down to the mass axis to give the mass value of the B-candidate, as indicated in the figure. Using this method the good momentum resolution of the ARGUS detector is translated into a good mass resolution.

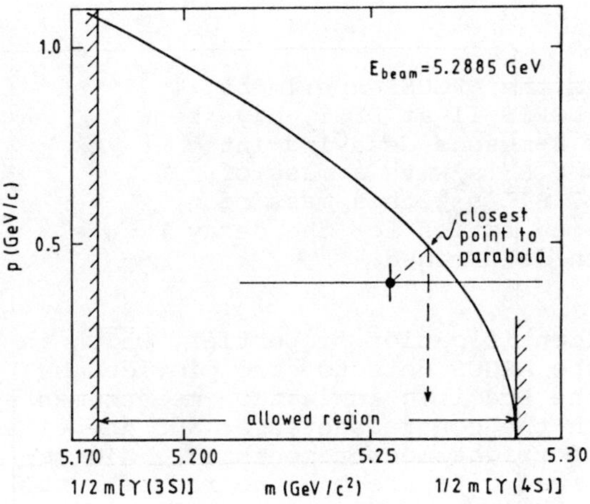

Figure 1: The momentum-mass resolution at E_{beam} = 5.2885 GeV, used to convert the good momentum resolution of the AGRUS detector into a good mass resolution.

In some events more than one B-candidate can be reconstructed, the reason for which can be twofold. By exchanging the position of two particles in a specific decay chain it might still be possible to get an acceptable B-mass and secondly the combinatorial background might give rise to more than one B-candidate. The first source mainly contributes to the signal while the second increases the background level. Only the candidate with the highest probability is chosen.

The background spectrum has been investigated using three independent methods. These are:

1) event mixing where the D* has been taken from one event and the pions from another;

2) wrong charge combinations like $D^{*+}\pi^+$, $D^{*+}\pi^+\pi^0$ and $D^{*+}\pi^-\pi^-\pi^-$, and

3) using the continuum data just below the Υ(4S) region.

All three methods give the same background shape and in fig. 2 the background is shown as obtained from event mixing fitted by the following expression:

$$\frac{dN}{dM} \sim M \cdot \sqrt{1 - \frac{M^2}{E^2_{beam}}}$$

corresponding to a flat, "phase space like" distribution which reproduces the shape of the background quite well.

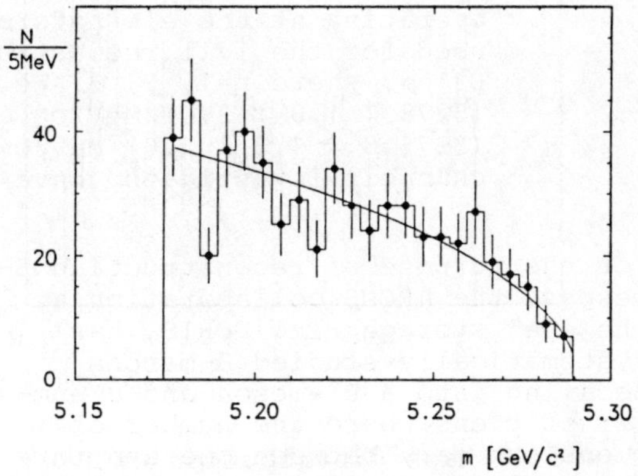

Figure 2: Mass distribution of background candidates obtained by event mixing.

The mass spectrum for all B-candidates passing these criteria, i.e. the sum of B^{+-} and B^0's, is shown in fig. 3. Fitting a Gaussian on top of the background spectrum gives 71 ± 11 B's in the peak. The resulting mass is (5277.1 ± 0.8) MeV/c^2 and the width is σ = 4.5 MeV/c^2, mainly due to the energy spread of the beams. The energy scale is fixed by putting the Υ(4S) mass equal to 10.577 GeV/c^2 (2).

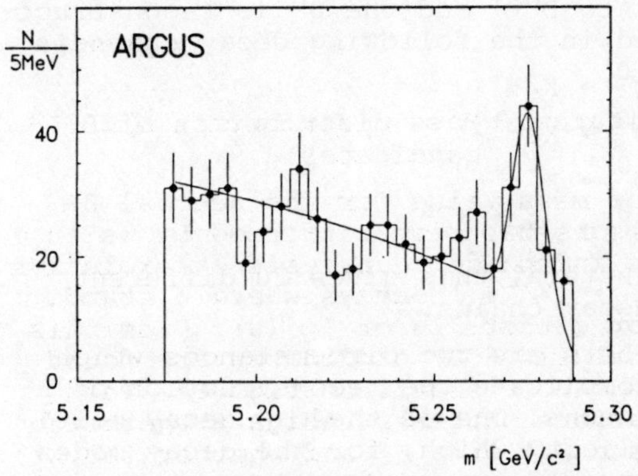

Figure 3: Mass distribution of all reconstructed B-meson candidates.

The mass spectra for neutral and charged B-mesons are shown separately in figures 4 and 5. From separate fits to these spectra 40 ± 8 neutral B's are obtained with a mass of (5278.2 ± 1.0 ± 3.0) MeV/c^2 and 32 ± 7 charged B's with a mass of (5275.8 ± 1.3 ± 3.0) MeV/c^2.

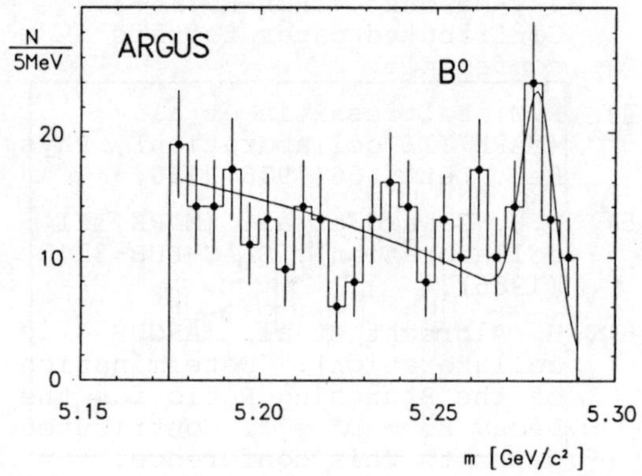

Figure 4: Mass distribution of B^0-candidates.

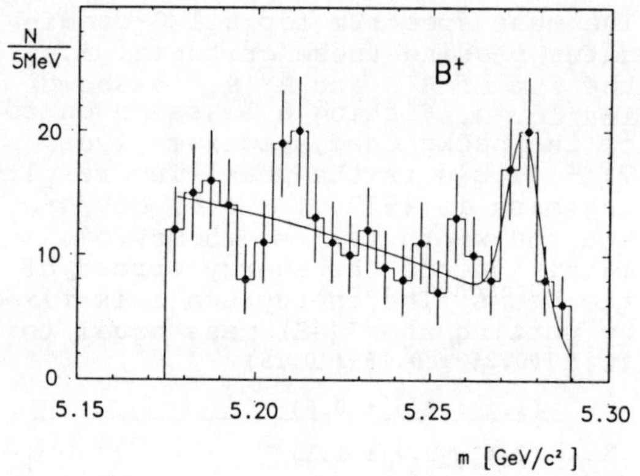

Figure 5: Mass distribution of B^{+-}-candidates.

Our mass value for the neutral B-mesons has been confirmed by us in an independent analysis of exclusive $B^0 \to \psi K^{+-}\pi^{-+}$ decays where ψ stands for either J/ψ or ψ' (3). From this analysis 6 B^0 candidates are found as can be seen from fig. 6a. The various patterns indicate whether ψ is a J/Ψ or a ψ' and whether the $K\pi$ state is compatible with being a K^* or if it is non-resonant. One charged B candidate is observed in the $B^{+-} \to J/\psi K^0_S \pi^{+-}$ decay (fig. 6b) and from the spectrum of wrong charge combinations $\psi K^{+-}\pi^{+-}$ we find no entries in the mass region of the B-meson which indicates that the background in this region is very low. The mass obtained for the neutral B-meson is (5277 ± 3 ± 3) MeV/c^2 in excellent agreement with the value given above.

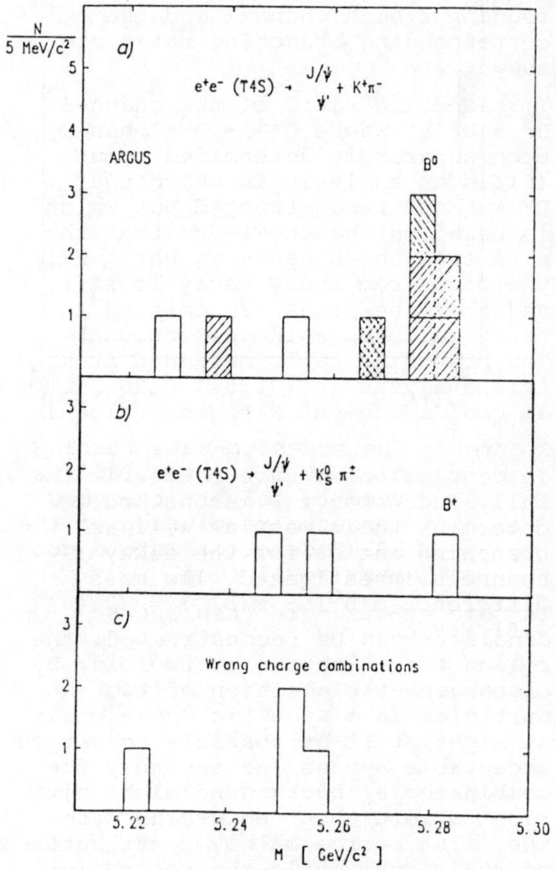

Figure 6: The mass distribution of $B \to \psi K\pi$ candidates.

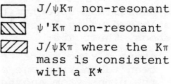

- $J/\psi K\pi$ non-resonant
- $\psi' K\pi$ non-resonant
- $J/\psi K\pi$ where the $K\pi$ mass is consistent with a K^*
- $\psi' K\pi$ where the $K\pi$ mass is consistent with a K^*

a) $B^0 \to \psi K^{+-}\pi^{-+}$

b) $B^{+-} \to \psi K^0_S \pi^{+-}$

c) wrong charge combinations $K^{+-}\pi^{+-}$

The branching ratios for all five decay channels of the B-mesons investigated by us have been determined assuming that 55% of the $\Upsilon(4S)$ decays into charged B-mesons and 45% into neutral mesons. The branching ratios for the various D^0-decays were taken from refs. 4 and 5, and the value used for the decay $D^{*+} \to D^0\pi^+$ was 64%.

In table 1 the number of events found for each channel and the corresponding branching ratio are shown.

The branching ratio of the channel $\bar{B}^0 \to D^{*+}\pi^-$ where $D^{*+} \to D^0\pi^+$ has been separately determined by us (6) in an analysis in which the D^0 was not reconstructed but which is based on the knowledge that the pion from the B decay is hard while the pion from the D decay is soft, and that the pions are emitted in essentially opposite directions. The branching ratio obtained from this analysis is $(0.35 \pm 0.20 \pm 0.18)$% in good agreement with the value in table 1.

In conclusion we have been able to fully reconstruct B-mesons and to determine their mass as well as branching ratios for the decay channels investigated. The mass difference $m(B^0) - m(B^{+-}) = (2.4 \pm 1.6)$ MeV/c^2.

REFERENCES

(1) H. Albrecht et al. (ARGUS collaboration), Phys. Lett. 150B(1984)235.

(2) Particle Data Group, Phys. Lett. 170 B(1986)1.

(3) H. Albrecht et al. (ARGUS collaboration): "Inclusive and Exclusive B-Meson decays into J/ψ-Mesons", IHEP-HD/86-3. Contributed paper to this conference.

(4) R.M. Baltrusaitis et al. (MARK III collaboration), Phys. Rev. Lett. 56(1986)2140.

(5) D.H. Coward et al. (MARK III collaboration), SLAC-PUB-3818 (1985).

(6) H. Albrecht et al. (ARGUS collaboration): "Determination of the Branching Ratio for the Decay $B^0 \to D^{*-}\pi^+$". Contributed paper to this conference.

TABLE 1

	Number of events	Branching ratio
$\bar{B}^0 \to D^{*+}\pi^-$	5 ± 2.5	$(0.25 \pm 0.15 \pm 0.15)$
$\bar{B}^0 \to D^{*+}\pi^-\pi^0$	8 ± 4	$(1.1 \pm 0.6 \pm 0.6)$
$\bar{B}^0 \to D^{*+}\pi^-\pi^-\pi^+$	27 ± 7	$(2.4 \pm 0.7 \pm 1.1)$
$B^- \to D^{*+}\pi^-\pi^-$	7 ± 3	$(0.4 \pm 0.2 \pm 0.2)$
$B^- \to D^{*+}\pi^-\pi^-\pi^0$	24 ± 7	$(3.5 \pm 1.0 \pm 2.1)$

Υ AND CHARM SPECTROSCOPY FROM ARGUS

David B. MacFarlane
Department of Physics, University of Toronto
Toronto, Ontario, Canada

(representing the ARGUS collaboration*)

Results are presented from a completed study of hadronic transitions between the $\Upsilon(2S)$ and $\Upsilon(1S)$ in the full available data set. The exclusive branching ratios $\Upsilon(1S) \to \ell^+\ell^-$ are found to be $(2.42\pm0.14\pm0.14)\%$ and $(2.30\pm0.25\pm0.13)\%$ for electron and muon pairs respectively. The existence of an excited charm meson, the $D^{*0}(2420)$, is confirmed in a new independent data set, with a significance of 5.5 sigma for the combined samples. A search for the transition $\Sigma_c \to \Lambda_c\pi$, using four decay channels of the Λ_c, yields evidence for both the Σ_c^{++} and Σ_c^0. A mass splitting, $\Sigma_c^{++} - \Sigma_c^0$, is found to be (2.5 ± 1.0) MeV/c^2.

1. $\Upsilon(2S)$ to $\Upsilon(1S)$ Hadronic transitions

Highlights are presented of a completed analysis [1] of hadronic transitions between the $\Upsilon(2S)$ and $\Upsilon(1S)$ in the full available data set collected by ARGUS at the e^+e^- storage ring DORIS II. The data correspond to an integrated luminosity of 37 pb^{-1} on the $\Upsilon(2S)$, and, for continuum subtractions, 2.66 pb^{-1} at a centre-of-mass energy of 9.98 GeV. The transition $\Upsilon(2S) \to \pi^+\pi^-\Upsilon(1S)$ has been studied both for inclusive decays of the $\Upsilon(1S)$ and for the exclusive channels $\Upsilon(1S) \to e^+e^-$ and $\mu^+\mu^-$. Shown in Figure 1a is the inclusive spectrum of recoil mass against the $\pi^+\pi^-$ system. At the $\Upsilon(1S)$ mass is a prominent peak with a FWHM of 7.3 MeV/c^2 containing 13250 entries above background. From a subsample with stable running conditions the branching ratio for $\Upsilon(2S) \to \pi^+\pi^-\Upsilon(1S)$ was found to be $(18 \pm 0.5 \pm 1.0)\%$, in good agreement with previous measurements [2, 3, 4].

The search for exclusive events, where $\Upsilon(1S) \to \ell^+\ell^-$, was made by selecting events with 4 charged tracks from the interaction region. Two tracks, the lepton candidates, were required to be relatively fast (p> 1 GeV/c) and back-to-back ($\cos\theta_{\ell^+\ell^-} < -0.95$). Bhabha events with converted photons were suppressed by making an opening angle cut. The sample was divided into electron and muon pairs on the basis of energy deposition in the shower counters: for e^+e^- events, both tracks were required to have more than 1 GeV and conversely both less than 1 GeV for muon pairs. The recoil mass spectrum for the 495 selected events is almost background free as shown in Figure 1b. For the $\mu^+\mu^-$ branching ratio measurement the two leptons were further restricted to lie in the barrel region of the detector in order to minimize the uncertainty in trigger acceptance. From the number of inclusive and exclusive events (306 electron and 86 muon pairs) we find the $Br[\Upsilon(1S) \to \ell^+\ell^-]$ to be $(2.42 \pm 0.14 \pm 0.14)\%$ and $(2.30\pm0.25\pm0.13)\%$ for electron and muon pairs respectively. Both values are considerably lower then the present world average [5], but are consistent with any single previous measurement [2, 4, 6, 7].

Many predictions [8 – 12] for the form of the invariant $\pi^+\pi^-$ mass distribution have been made, commonly describing the transition in terms of the emission of two gluons which hadronize into two pions (or an eta). Experimentally, the mass spectrum can be extracted from both the inclusive and exclusive samples. In the inclusive case a series of fits are made to the recoil mass spectrum divided up into

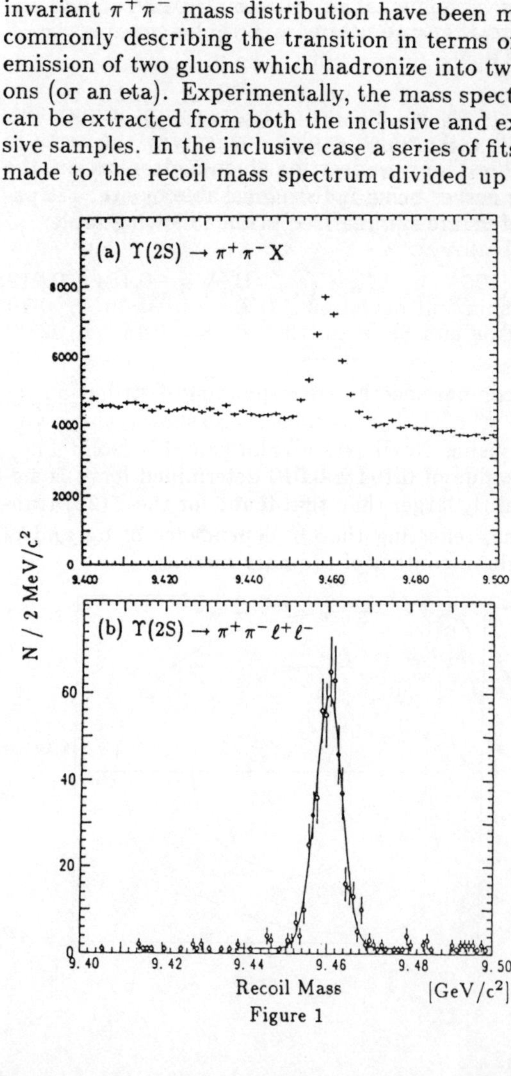

Figure 1

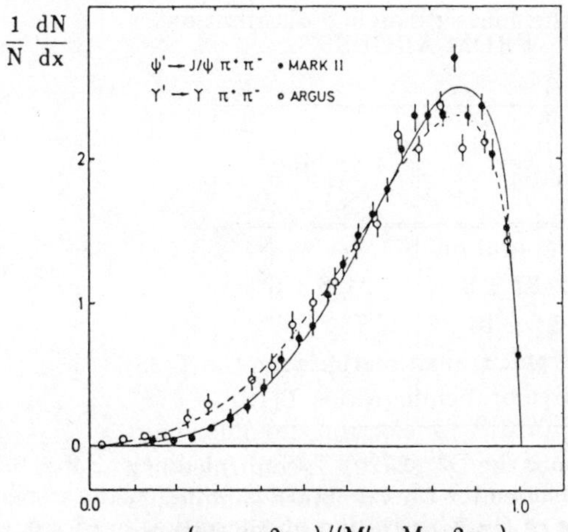

$x = (m_{\pi\pi} - 2m_\pi)/(M' - M - 2m_\pi)$

Figure 2

bins of $\pi^+\pi^-$ invariant mass. Most of the signal is found at high $M(\pi^+\pi^-)$; the error at low mass is dominated by the background. The exclusive sample, containing relatively few events but being essentially background free, improves the precision of the low mass points. The two samples are to good approximation statistically independent, and are combined in the final result shown in Figure 2.

The dashed line in the figure is a fit using models [10 – 12] which exploit low energy theorems in the chiral limit to describe gluon hadronization, the three curves being indistinguishable by eye. Fit parameters are summarized in the following table:

Yan [10]	$B/A = -0.154 \pm 0.019$
Voloshin and Zakharov [11]	$\lambda = 3.30 \pm 0.19$
Novikov and Shifman [12]	$\kappa = 0.151 \pm 0.009$

For comparison the corresponding distribution for $\psi' \to \pi^+\pi^- J/\psi$ transitions is also shown, along with a fit using Novikov and Shifman [12] (solid line). The value of 0.194 ± 0.010 determined for κ is significantly larger than that found for the $\Upsilon(2S)$ transitions, reflecting the Q^2 dependence of α_S and of the gluon content of ordinary mesons.

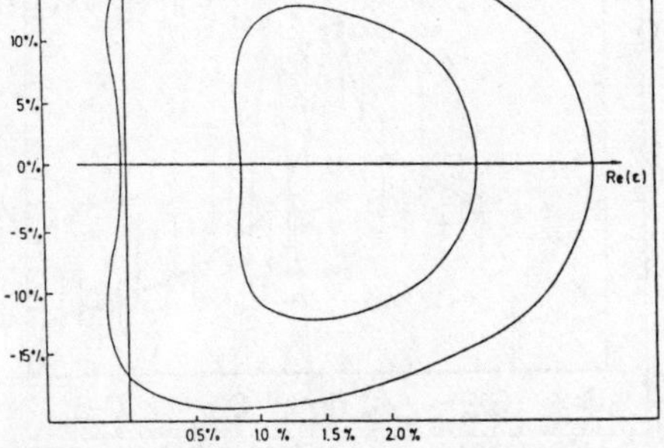

Figure 3

Most models describe the two gluon emission in terms of classical chromo-electromagnetic fields. The dominance of E_1E_1 transitions, which preserves the angular momentum of the $b\bar{b}$ system, is confirmed experimentally by the observation that the exclusive events $\Upsilon(1S) \to \mu^+\mu^-$ exhibit the same polarization ($P_1P_2 = 0.75 \pm 0.10$) as direct $\mu^+\mu^-$ pairs at the $\Upsilon(2S)$ centre-of-mass energy ($P_1P_2 = 0.68 \pm 0.02$ [13]). By implication, since this is a strong interaction process, the $\pi^+\pi^-$ system can then only be in a $J^{PC} = 0^{++}, 2^{++}, 4^{++}$... state. The distribution of the helicity angle θ^* of the π^+ in the $\pi^+\pi^-$ centre-of-mass has been fit to a coherent sum of $J = 0$ and $J = 2$, $J_z = 0$ waves: $dN/d\cos\theta^* \sim \left|\sqrt{1+|\epsilon|^2}\, Y_0^0 + \epsilon\, Y_2^0\right|^2$. The result is shown as one and two standard deviation contours in Figure 3. The real part of $\epsilon = 0.018 \pm 0.009$, but the imaginary part is not well determined. Allowing a free phase, the fit gives $|\epsilon| = 0.018^{+0.108}_{-0.009}$, which is just two standard deviations from zero. If $\pi^+\pi^-$ rescattering is negligible, ϵ is expected to be real [14]. In this case, our value is close to the theoretical expectation[12], and proves the strong s-wave dominance in the process.

2. Update on the $D^{*0}(2420)$

Since the first report [15] of the observation of an new excited charm meson at a mass of 2.42 GeV/c^2, ARGUS has collected an additional 69.9 pb^{-1} data, mostly on the $\Upsilon(4S)$ (48.0 pb^{-1}). In order to confirm our initial analysis, the original study has been repeated without change with the new data sample. The decay channel reconstructed is $D^{*0}(2420) \to D^{*+}(2010)\pi^-$ where $D^{*+}(2010) \to D^0\pi^+$. The small Q value for this last transition results in excellent resolution for the $D^{*+}(2010) - D^0$ mass difference, which can be exploited to yield a nearly background free signal for $D^{*+}(2010)$. Three D^0 decay channels have been used: $D^0 \to K^-\pi^+$, $K^-\pi^+\pi^+\pi^-$ and $K^-\pi^+\pi^0$. For the third of these channels the π^0 is not observed, but nevertheless is visible as peak in the $K^-\pi^+$ mass spectrum shifted roughly by a pion mass below the D^0 mass. In addition to the usual mass selection criteria, two cuts are made on the $D^{*+}(2010)\pi^-$ system. These are $x_p > 0.6$ and $\cos\theta < 0$, where $x_p = p/p_{max}$ and θ is the angle between the $D^{*+}(2010)\pi^-$ line of flight and the $D^{*+}(2010)$ direction in the $D^{*+}(2010)\pi^-$ rest frame.

The resulting mass difference distributions from the original study [15] based on an integrated luminosity of 82.4 pb^{-1}, for the new sample and for the combined result are shown in Figures 4a, b and c respectively. Fits to these distributions (solid lines), using a sum of a Breit-Wigner for the signal and a polynomial for the background, yield the results given in Table 1. Masses and widths are consistent,

Table 1. Properties of the $D^{*0}(2420)$ from fits to the mass difference distributions, Figures 4a, b and c.

	Original	New	Combined
$M[D^{*0}(2420)]$ (MeV/c^2)	2420 ± 6	2432 ± 8	2426 ± 6
Full width (MeV/c^2)	70 ± 21	54 ± 16	75 ± 20
Number of events	135^{+34}_{-29}	75^{+22}_{-19}	192^{+55}_{-35}

with a combined significance of 5.5 sigma. The ratio of $D^{*0}(2420)$ to $D^{*+}(2010)$ production in the full data sample is found to be $(17^{+6}_{-5} \pm 6)\%$.

As a further check, the division of the signal among the three D^0 channels for the full data set is shown in Figure 5a, b and c. From a fit using a Breit-Wigner with fixed position and width we find 37 ± 12, 80 ± 23 and 62 ± 21 events in the three channels. The ratio of the observed number of events in the $K^-\pi^+$ and $K^-\pi^+\pi^+\pi^-$ channels, 2.2 ± 0.9, is consistent with expectation. Finally, the dependence of the observation on the rather hard cut on the $D^{*+}(2010)$ angle has been studied. The naive expectation would be to find a ratio of 1.5 for the number of events with $\cos\theta < 0.5$ over the number with $\cos\theta < 0$, but this in fact depends on the quantum number assignment for the $D^{*0}(2420)$. The observed ratio is $1.3^{+0.3}_{-0.2}$. Study of angular distributions may lead to information about the spin assignment for

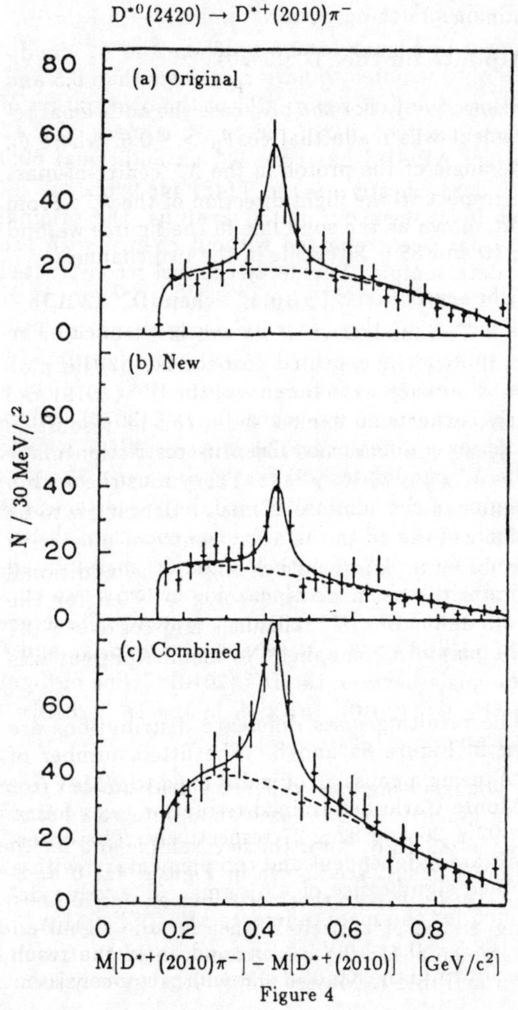

Figure 4

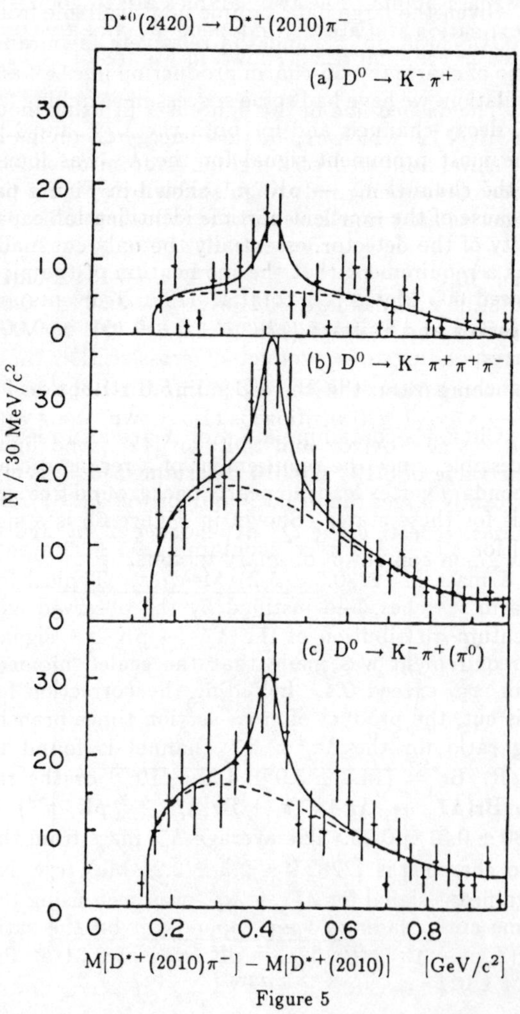

Figure 5

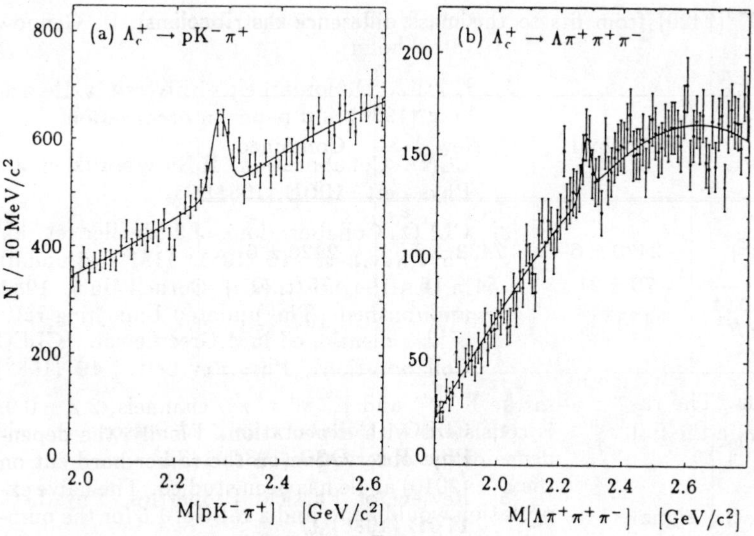

Figure 6

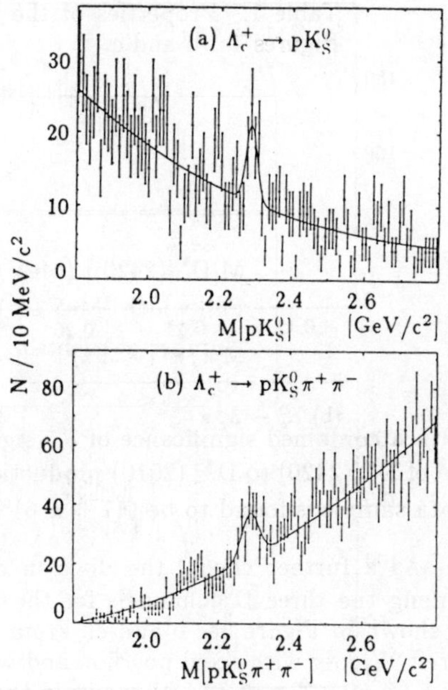

Figure 7

the state, but the present sample is still rather small and contains considerable background.

3. Observation of charmed baryons

Given the large data sample now available from ARGUS (153 pb^{-1}), and the relatively favourable ratio of charm to non-charm production in e^+e^- annihilation, we have had some success in searching for Λ_c^+ decay channels and for both the Σ_c^{++} and Σ_c^0. The most prominent signal for the Λ_c^+ was found in the channel $\Lambda_c^+ \to pK^-\pi^+$ shown in Figure 6a. Because of the excellent particle identification capability of the detector, essentially the only cut made was a requirement that the momentum of the pion exceed 600 MeV/c. A total of 479 ± 57 events are observed at a mass of $(2283.5 \pm 2.6 \pm 2.1)$ MeV/c^2 corresponding to a product of cross-section times branching ratio of $R \cdot Br = (18.0 \pm 5.0 \pm 1.8) \times 10^{-3}$.

Channels containing a K_S^0 or Λ are also readily accessible, since the requirement of a reconstructed secondary vertex leads to nearly background free signals for these states. Shown in Figure 6b is a signal for $\Lambda_c^+ \to \Lambda\pi^+\pi^+\pi^-$ containing 95 ± 29 events at a mass of (2280.3 ± 5.2) MeV/c^2. Typical for charm searches, and justified by the observed momentum distribution of the $\Lambda_c^+ \to pK^-\pi^+$ signal, a requirement was made that the scaled momentum, x_p, exceed 0.4. Including the correction for this cut, the product of cross-section times branching ratio for the $\Lambda\pi^+\pi^+\pi^-$ channel is found to be $R \cdot Br = (14.3 \pm 3.9 \pm 4.3) \times 10^{-3}$ or the ratio $Br(\Lambda_c^+ \to \Lambda\pi^+\pi^+\pi^-)/Br(\Lambda_c^+ \to pK^-\pi^+)$ is $0.80 \pm 0.31 \pm 0.25$. The average Λ_c^+ mass from the two channels is $(2282.9 \pm 2.3 \pm 2.2)$ MeV/c^2. No significant signal for $\Lambda_c^+ \to \Lambda\pi^+$ was seen using the same cuts, leading to an upper limit on the ratio $Br(\Lambda_c^+ \to \Lambda\pi^+)/Br(\Lambda_c^+ \to pK^-\pi^+)$ of 0.31 at the 90% CL.

Evidence for $\Lambda_c^+ \to pK_S^0$ and $\Lambda_c^+ \to pK_S^0\pi^+\pi^-$ is shown in Figure 7a and b, where the Λ_c^+ candidates were required to have x_p greater than 0.5 and 0.6 respectively. For the pK_S^0 case the additional requirement was made that $\cos\theta_p > -0.5$, where θ_p is the angle of the proton in the Λ_c^+ centre-of-mass with respect to the flight direction of the Λ_c^+. From the fit shown as the solid line in the figures we find 40 ± 10 and 88 ± 17 events in the two channels.

The search for $\Sigma_c^{++} \to \Lambda_c^+\pi^+$ and $\Sigma_c^0 \to \Lambda_c^+\pi^-$ was made using these four Λ_c^+ decay channels. The mass splitting is expected to be close to the pion mass, so, analogous to the case of the $D^{*+}(2010)$ and D^0, one expects to exploit the excellent resolution for the mass difference. The only restriction made on the Λ_c^+ candidates was that they must lie within two sigma of the nominal Λ_c^+ mass. Defining α to be the angle of the $\pi^\pm$ in the $\Lambda_c^+\pi^\pm$ centre-of-mass with respect to the $\Lambda_c^+\pi^\pm$ line of flight, the additional requirement was made that $\cos\alpha > -0.5$ for the $pK^-\pi^+$ and $\Lambda\pi^+\pi^+\pi^-$ channels and $\cos\alpha > -0.7$ for the $pK_S^0\pi^+\pi^-$ channel. No momentum cut was made.

The resulting mass difference distributions are shown in Figure 8a and b. The fitted number of events, using a gaussian of width fixed to 2 MeV/c^2, the Monte Carlo determined resolution, was found to be 92 ± 26 and 82 ± 27 respectively. The observations are independent and complementary, with a combined significance of 4.6 sigma. The value determined for the mass difference $M(\Sigma_c^{++}) - M(\Lambda_c^+)$ was (168.3 ± 0.8) MeV/c^2, in good agreement with the result from the handful of events previously seen.

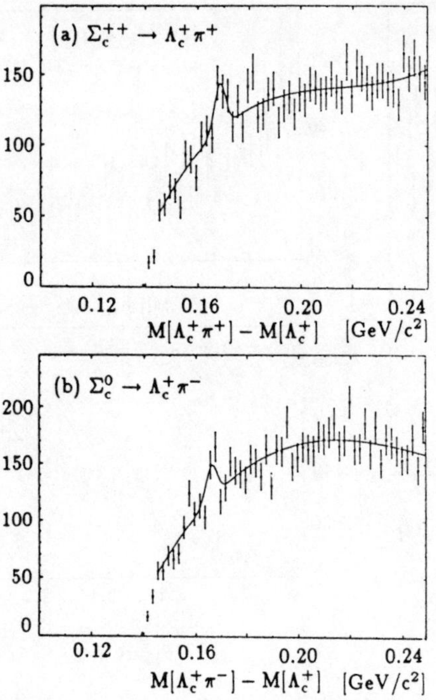

Figure 8

The Σ_c^0 has not been previously observed. The mass splitting between the neutral and doubly charged states is therefore of some interest for quark model predictions. We find $M(\Sigma_c^0) - M(\Lambda_c^+) = (165.8 \pm 0.7)$ MeV/c^2, so that $M(\Sigma_c^{++}) - M(\Sigma_c^0) = (2.5 \pm 1.0)$ MeV/c^2. This is in agreement with some, but not all, predictions [16] and will provide an important constraint for details of these calculations.

References

* Current members of the ARGUS collaboration are: H. Albrecht, U. Binder, P. Böckmann, R. Gläser, G. Harder, I. Lembke-Koppitz, W. Schmidt-Parzefall, H. Schröder, H.D. Schulz, R. Wurth, A. Yagil (DESY), J.P. Donker, A. Drescher, D. Kamp, U. Matthiesen, H. Scheck, B. Spaan, J. Spengler, D. Wegener (Dortmund), J.C. Gabriel, K.R. Schubert, J. Stiewe, K. Strahl, R. Waldi, S. Weseler (Heidelberg), K.W. Edwards, W.R. Frisken, Ch. Fukunaga, D.J. Gilkinson, D.M. Gingrich, H. Kapitza, P.C.H. Kim, R. Kutschke, D.B. MacFarlane, J.A. McKenna, K.W. McLean, A.W. Nilsson, R.S. Orr, P. Padley, J.A. Parsons, P.M. Patel, J.D. Prentice, H.C.J. Seywerd, J.D. Swain, G. Tsipolitis, T.-S. Yoon, J.C. Yun (IPP Canada) R. Ammar, D. Coppage, R. Davis, S. Kanekal, N. Kwak (Kansas), B. Boštjančič, G. Kernel, M. Pleško (Ljubljana), L. Jönsson (Lund), A. Babaev, M. Danilov, A. Golutvin, I. Gorelov, V. Lubimov, V. Matveev, V. Nagovitsin, V. Ryltsov, A. Semenov, V. Shevchenko, V. Soloshenko, V. Tchistilin, I. Tichomirov, Yu. Zaitsev (ITEP Moscow), R. Childers, C.W. Darden, Y. Oku (South Carolina), H. Gennow (Stockholm).

[1] K.Fritz, Diplomarbeit, Universität Heidelberg (1986) and paper in preparation.

[2] LENA Collaboration, B.Niczyporuk et al., Phys.Lett. **100B** (1981) 95.

[3] CLEO Collaboration, J.J.Mueller et al., Phys.Rev.Lett. **46** (1981) 1181 and update in D.A.Herrup (thesis), Cornell Univ. 1982, unpublished. The updated branching ratio is also mentioned in J.Green et al. (CLEO Collaboration), Phys.Rev.Lett. **49** (1982) 617.

[4] CLEO Collaboration, J.Besson et al., Phys.Rev. **D30** (1984) 1433.

[5] Review of Particle Properties, Phys.Lett. **170B** (1986) 1.

[6] CUSB Collaboration, G.Mageras et al., Phys.Rev.Lett. **46** (1981) 1115 and V.Fonseca et al., Nucl.Phys. **B242** (1984) 31.

[7] Crystal Ball Collaboration, D.Gelphman et al., Phys.Rev. **D32** (1985) 2893.

[8] L.S.Brown and R.N.Cahn, Phys.Rev.Lett. **35** (1975) 1.

[9] D.Morgan and M.R.Pennington, Phys.Rev. **D12** (1975) 1283.

[10] T.-M.Yan, Phys.Rev. **D22** (1980) 1652.

[11] M.Voloshin and V.Zakharov, Phys.Rev.Lett. **45** (1980) 688.

[12] V.A.Novikov and M.A.Shifman, Z.Phys. **C8** (1981) 43.

[13] B.Gräwe, Ph.D Thesis, Universität Dortmund (1985), unpublished. [14] : R.N.Cahn, Phys.Rev. **D12** (1975) 3559.

[15] ARGUS Collaboration, H.Albrecht et al., Phys.Rev.Lett. **56** (1986) 549.

[16] D.B.Lichtenberg, Phys.Rev. **D16** (1977) 231 and references therein. The author wishes to thank D.Lichtenberg for useful discussion on this question.

RECENT UPSILON SPECTROSCOPY RESULTS FROM CUSB-II

Juliet Lee-Franzini

SUNY at Stony Brook
Stony Brook, New York 11794, U.S.A.

New results on $\Upsilon(3S)$ spectroscopy was recently obtained using a high resolution bismuth germanate electromagnetic spectrometer: CUSB-II. The branching fraction of the $\Upsilon(3S)$ into muons, $B_{\mu\mu}(3S)$, was measured to be $(1.53\pm0.29\pm0.21)\%$, the $\Upsilon(3S)$ total width to be 25.5 ± 5.0 keV, $\alpha_s=0.17\pm0.014$ and $\Lambda_{\overline{MS}}=148\pm50$ MeV. By observing the $\Upsilon(3S)\rightarrow\chi_b(2P)\gamma\rightarrow\{\Upsilon(2S),\Upsilon(2S)\}\gamma\gamma\rightarrow\ell\ell\gamma\gamma$ decay chain we resolved completely the J=2 and J=1 $\chi_b(2P)$ states. We obtain product ratios for the decay chains and derive the hadronic widths of the $\chi_b(2P)$ states. We also resolved the J=0 state in the inclusive photon spectrum and measured the fine structure of the $\chi_b(2P)$ state.

1. INTRODUCTION

Using the CUSB-I detector we have been studying at the Cornell Electron Storage Rings (CESR) the spectroscopy of triplet bound $b\bar{b}$ states, known as Υ's ($^3S_1$) and χ_b's ($^3P_{2,1,0}$) (1). The Υ's are resonantly produced in e^+e^- collisions and decay predominantly via $b\bar{b}$ quark annihilation into three gluons which subsequently hadronize mostly into pions. For a small percentage of the time ($\approx$1-3%) they decay into a pair of leptons. The resulting signature (especially when $\ell=\mu$ or e) of two back to back leptons, each with $E=M_\Upsilon/2$, allows their essentially background free identification. The χ_b's can not be produced directly from e^+e^- collisions. They are reached through electric dipole (E1) photon transitions from Υ's (and can decay to the Υ's via photon emission).

Figure 1 gives the level diagram of the bound $b\bar{b}$ system. Indicated are the major pion transitions (double lines), photon transitions (single lines), and final states. The ones observed with CUSB-I are shown as solid lines. The first generation experiments contributed greatly to our understanding of the spin independent part of the interquark forces. Precision measurements of the fine and hyperfine splittings of the $b\bar{b}$ states are needed for a corresponding progress in the study of the spin dependent interquark forces. We have recently constructed a new, electromagnetic spectrometer, CUSB-II, for this purpose (2). In the following we report on the results obtained from the first run, during early 1986, with CUSB-II, and their physics implications.

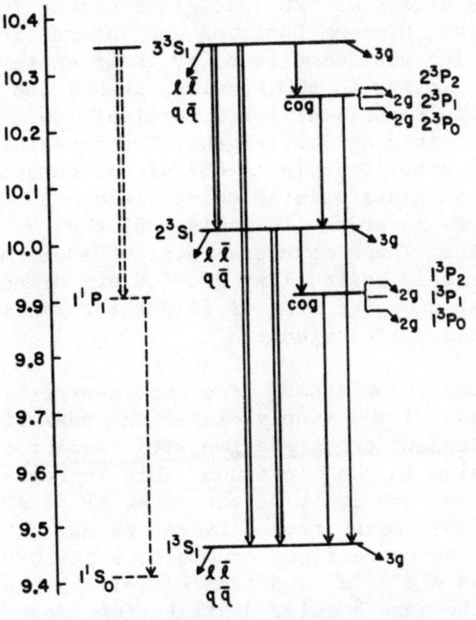

Figure 1: Level diagram of bound $b\bar{b}$ system.

provide valuable testing grounds for models of the strong interaction. An important parameter to be determined experimentally is $B_{\mu\mu}$, the branching fraction into two muons. Together with the e^+e^- decay width, it allows the determination of the total decay width and of the partial widths into other channels. From $B_{\mu\mu}$, one can calculate Γ_{ggg} and therefore α_s, the coupling constant of QCD, and $\Lambda_{\overline{MS}}$, the QCD scale parameter.

2. THE BRANCHING RATIO FOR $\Upsilon(3^3S_1)\rightarrow\mu\mu$

The members of the Υ family of mesons

We have measured $B_{\mu\mu}$ for the $\Upsilon(3S)$ meson using the CUSB-II detector at CESR. The CUSB-II detector consists of a high resolution bismuth germanate (BGO) electromagnetic calorimeter inserted in the NaI array of CUSB-I. Figure 2 is a perspective drawing of the whole detector seen through partially cut away counters used for the muon trigger. A detail of the BGO cylinder is also shown. This cylinder consists of 36 azimuthal sectors covering 10 degrees in ϕ. Each sector is divided into two polar halves, covering the θ ranges 45°-90° and 90°-135°. Each sector, 12 radiation lengths (X_o) thick at $\theta=90^o$, is subdivided into 5 radial layers, for a total of 360 crystals in the whole array. Between the beam pipe and the BGO cylinder are 72 1/8" thick scintillators, one in front of each BGO sector. These counters were used for charged particle veto, instead of the mini jet chamber shown in the figure as the latter was not installed until very recently. The BGO cylinder is surrounded by an 8 X_o square array of 328 NaI crystals, arranged in 5 radial layers, 32 azimuthal sectors, and 2 polar halves. Between NaI crystal layers are four proportional chambers with x and y cathode strip readout, used for tracking non interacting charged particles. The NaI array is surrounded by 4 square arrays of 8×8 lead glass blocks 7 X_o thick. Minimum ionizing non interacting particles are identified, by their energy loss, 5 times in BGO, 5 times in NaI and once in lead glass, for a total of 2.5 nuclear interaction lengths. The detector covers a solid angle of 66% of 4π. Outside the lead glass plastic scintillator muon counters cover the four sides of the detector. These counters cover ≈29% of the total solid angle. They provide our dimuon trigger and give time of flight information for cosmic ray rejection.

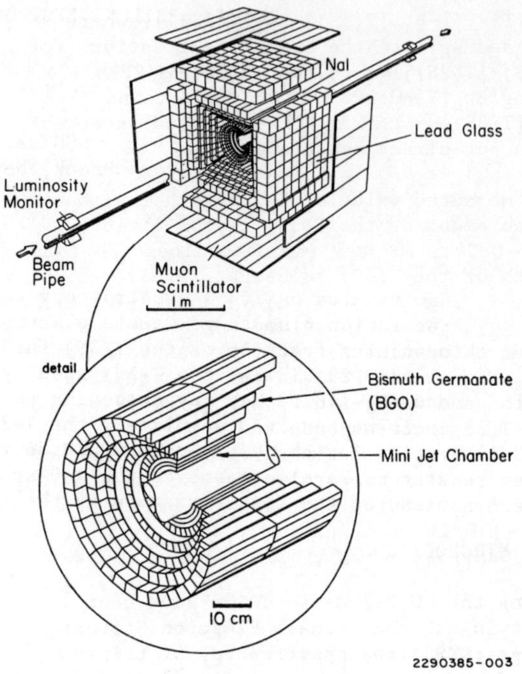

Figure 2: Perspective drawing of CUSB-II.

In order to obtain $B_{\mu\mu}$ one must measure the increase of the muon yield at the peak of the resonant cross section with respect to its value in the continuum. This increase is expected to be only of the order of 8% at the Υ(3S). Apart from differences due to radiative corrections, muons from the QED process $e^+e^-\to\mu^+\mu^-$ and from $e^+e^-\to\Upsilon(3S)\to\mu^+\mu^-$ have the same angular distribution. Most systematic uncertainties in the determination of the excess muon yield at the Υ(3S) cancel if experimental values for resonance and continuum yields are obtained in the same detector with the same cuts applied to the data. Above the b-flavor threshold the branching ratio for upsilons into muons becomes negligible, $B_{\mu\mu}[\Upsilon(4S)]\approx10^{-5}$. We use therefore all data collected in the continuum and at the Υ(4S) resonance as "continuum" data. All data is normalized to the actual luminosity as obtained from the observed large angle Bhabha scattering yields.

Data were taken at the Υ(3S) (≈41 pb^{-1}) and Υ(4S) (≈19.5 pb^{-1}) peaks and in the continuum just below the Υ(4S) (≈9.5 pb^{-1}). The dimuon hardware trigger required a coincidence among muon counters on opposite sides of the detector, plus at least 100 MeV in the outer three layers of NaI and BGO. Two minimum ionizing tracks lose 180 MeV in NaI and 170 MeV in BGO. This trigger is 100% efficient for muons entering the counters and is dominated by soft particles crossing counters and the BGO and NaI arrays. Most background is trivially removed by requiring that the observed energy signals be consistent with two non interacting particles with origin at the interaction point. Only events with two muon tracks and no other energy clusters in BGO or NaI are used for the determination of $B_{\mu\mu}$. The remaining background consists of cosmic ray muons, mostly vertical, in accidental coincidence with the beam bunch crossing time in the hardware trigger acceptance window of ≈ 30 ns. This background is effectively rejected by using muon time of arrival and tracking information. We measure the efficiency for detecting the $\mu^+\mu^-$ pairs from Υ(3S) decay by comparing the number of dimuon events observed in the continuum to the number calculated from the luminosity and the continuum cross section. This gives the continuum efficiency which is corrected, using the Monte Carlo of Berends and Kleiss (3) for initial state radiation.

We obtain $B_{\mu\mu}[\Upsilon(3S)] = (1.53\pm0.29\pm0.21)$ % where the first error is statistical and the second systematic. This value is to be compared to the previous measurement of (3.3 ± 2.0) % (4). Since all decay widths into channels involving $b\bar{b}$ annihilation scale as $|\psi(0)|^2/M^2$ one can also write (5)

$B_{\mu\mu}[\Upsilon(3S)] = B_{\mu\mu}[\Upsilon]\times\{(1-B_{\pi\pi}[\Upsilon(3S)]-B_{E1}[\Upsilon(3S)]\}$, where $B_{\pi\pi}$ and B_{E1} are the branching fractions for $\Upsilon(3S)\to[\Upsilon(2S)]\Upsilon\pi\pi$ (6) and $\Upsilon(3S)\to\chi_b(2P)\gamma$ (7). Using $B_{\mu\mu}[\Upsilon]=0.028\pm0.002$ one obtains $B_{\mu\mu}[\Upsilon(3S)] = (1.56\pm0.18)\,\%$ in good agreement with our direct measurement.

The measured value of $B_{\mu\mu}$ together with our measurement of the $\Upsilon(3S)$ leptonic width $\Gamma_{ee} = 0.39\pm0.02$ keV (8) determines the total width of the $\Upsilon(3S)$ to be $\Gamma_{tot}[\Upsilon(3S)] = 25.5\pm5$ keV.

Using this value for $B_{\mu\mu}$, we can calculate $\Gamma_{ggg}/\Gamma_{\mu\mu}$, where Γ_{ggg} is the 3 gluon decay width, and we obtain $\Gamma_{ggg}/\Gamma_{\mu\mu}=28.6\pm7.0$. Our $B_{\mu\mu}[\Upsilon(3S)]$ corresponds to $\alpha_s[0.48M_{\Upsilon(3S)}] = 0.17\pm0.014$, $\Lambda_{\overline{MS}} = 148\pm50$ MeV. These results compare well with our previous determinations of $\alpha_s[.48M_{\Upsilon(1S)}] = 0.172\pm0.010$, $\Lambda_{\overline{MS}} = 140\pm30$ MeV, and $\alpha_s[.48M_{\Upsilon(2S)}] = 0.154\pm0.019$, $\Lambda_{\overline{MS}} = 92\pm40$ MeV (9).

3. $\Upsilon(3S)\to\chi_b(2P)\gamma\to[\Upsilon(2S)]\Upsilon(1S)\gamma\gamma\to\mu^+\mu^-(e^+e^-)\gamma\gamma$ AND HADRONIC WIDTHS OF THE $\chi_b(2P)$ STATES

The fine splitting of the $\chi_b(2P)$'s, i.e. the separation between the J=2,1,0 states, is small, of the order of 15 to 20 MeV. In CUSB-I the E1 photons lines due to transitions from $\Upsilon(3S)\to\chi_b(2P)$ were not resolvable in either the inclusive photon spectrum or in the spectrum obtained by studying the decay chain $\Upsilon(3S)\to\chi_b(2P)\gamma\to[\Upsilon(2S)]\Upsilon(1S)\gamma\gamma$ where the Υ's decayed leptonically. This latter class of events, often referred to as "exclusive" events in contrast to "inclusive", $\Upsilon(3S)\to\gamma+X$ events, usually shows superior energy resolution because they are free from both the large photon background and from overlaps of energy depositions from neighboring particles.

In the following we report on the first $\chi_b(2P)$ results from CUSB-II. The data reported in this section comes from an integrated luminosity of 41.4 pb^{-1} collected at the $\Upsilon(3S)$ peak energy. A total of 2.89×10^5 hadronic e^+e^- annihilation events were observed, corresponding to 1.82×10^5 $\Upsilon(3S)$ produced. We also collected data, $\approx$15.6 pb^{-1}, above the b flavor threshold and at the continum above the $\Upsilon(3S)$.

In figure 3a we give a scatter plot of the higher energy photon ($E_{\gamma high}$) versus the lower energy photon ($E_{\gamma low}$) for $\mu\mu\gamma\gamma$ events at the $\Upsilon(3S)$ peak. Similarly, figures 3b and 3c are the corresponding plots from $ee\gamma\gamma$ events on the $\Upsilon(3S)$ and $ee\gamma\gamma$ events obtained from 15.6 pb^{-1} of continuum energies

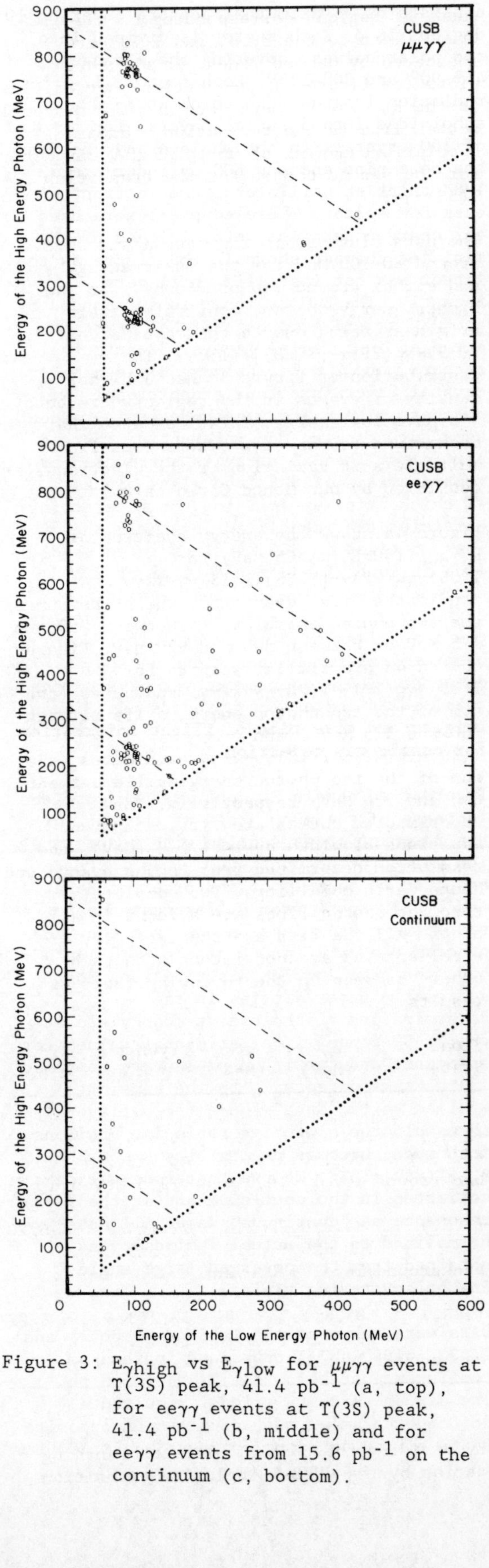

Figure 3: $E_{\gamma high}$ vs $E_{\gamma low}$ for $\mu\mu\gamma\gamma$ events at $\Upsilon(3S)$ peak, 41.4 pb^{-1} (a, top), for $ee\gamma\gamma$ events at $\Upsilon(3S)$ peak, 41.4 pb^{-1} (b, middle) and for $ee\gamma\gamma$ events from 15.6 pb^{-1} on the continuum (c, bottom).

running. Regions corresponding to the transitions to the $\Upsilon(1S)$ and $\Upsilon(2S)$ are indicated by dashed lines. In all three plots (of figure 3) the vertical dotted line is the E_γ lower energy cut and the diagonal dotted line is the reflection boundary. Doppler broadening of up to 35 MeV can occur for the second photon for the transition $\Upsilon(3S)\to\gamma\gamma\Upsilon$.

The data cluster, in figures 3a and 3b, around 80-100 MeV for the lower energy γ, and either around 230 or 760 MeV for the higher energy photon, confirming their origin as being due to the cascade chain $\Upsilon(3S)\to\chi_b(2P)\gamma\to[\Upsilon(2S)]\Upsilon(1S)\gamma\gamma$. The distribution of figure 3c has a higher density of events in the region where both γ's have low energy. This indicates that background to the $ee\gamma\gamma$ events come from multiple soft radiation of the electrons, as confirmed by our Monte Carlo calculations.

Figure 4a shows the energy spectrum of $E_{\gamma low}$ ($130>E_{\gamma low}>65$ MeV) for $\Upsilon(3S)\to\chi_b(2P)\gamma\to[\Upsilon(2S)]\Upsilon(1S)\gamma\gamma\to\ell\ell\gamma\gamma$ candidates, i.e. events for which the sum of the two photon energies lie between 250 and 375 MeV or between 800 and 920 MeV. Figures 4b and 4c are similar spectra for $\Upsilon(3S)\to\chi_b(2P)\gamma\to\Upsilon(2S)\gamma\gamma\to\ell\ell\gamma\gamma$ candidates (the sum of the two photon energies lie between 250 and 375 MeV), and $\Upsilon(3S)\to\chi_b(2P)\gamma\to\Upsilon(1S)\gamma\gamma\to\ell\ell\gamma\gamma$ candidates (the sum of the two photon energies lie between 800 and 920 MeV) respectively. The superimposed curves are fits using two Gaussians of width $\sigma_E=1.8\%/\sqrt[4]{E_\gamma}$(GeV), the resolution determined from Bhabha events and Monte Carlo modellings. Two completely resolved photon lines are evident in all plots. All the fits are statistically excellent and are consistent with each other, as seen in the following table of results.

Final State	E_γ (MeV)	σ_E (MeV)	Events	χ^2/ d.o.f.
(1^3S_1)	85.4±0.6	2.8	34.5±6.1	6.9/22
$+(2^3S_1)$	100.1±0.5	3.2	48.7±7.2	
background level = 0.5 event/2.5 MeV bin				
(1^3S_1)	85.2±0.7	2.8	22.1±4.7	8.6/22
	100.6±0.9	3.2	14.5±3.8	
background level = 0 event				
(2^3S_1)	85.8±1.1	2.8	12.2±3.8	6.8/22
	100.4±1.1	3.2	29.4±5.7	
background level = 0.5 event/2.5 MeV bin				

By dividing the CUSB-II product branching ratios by the CUSB-I inclusive transition rates of $3^3S_1\to2^3P_{2,1,0}$ of 12.7±4.1%, 15.5±4.2%, and 7.6±3.5% respectively (7), we obtain the branching ratios from the 2^3P_J states to the 2^3S_1, 1^3S_1 states. CUSB-II efficiencies used for computing the product branching ratios are included in the following table. In the product branching ratio calculations, $B_{\mu\mu}(1^3S_1)=0.028\pm0.002$ and $B_{\mu\mu}(2^3S_1)=0.017\pm0.004$ are assumed .

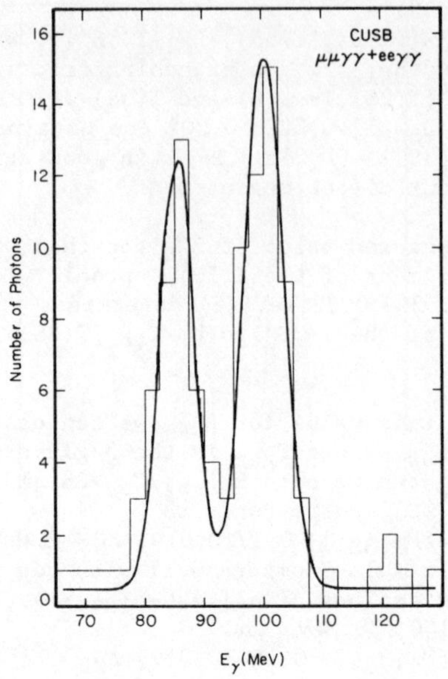

Figure 4a: Energy spectrum of $E_{\gamma low}$ for $\Upsilon(3S)\to\chi_b(2P)\gamma\to\{\Upsilon(2S)+\Upsilon(1S)\}\gamma\gamma\to\ell\ell\gamma\gamma$ candidates, i.e. ΣE_γ lie between 250 and 375 MeV, or between 800 and 920 MeV.

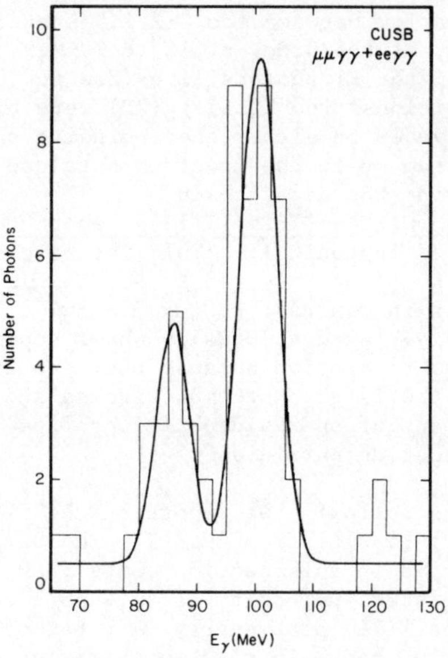

Figure 4b: Energy spectrum of $E_{\gamma low}$ for $\Upsilon(3S)\to\chi_b(2P)\gamma\to\Upsilon(2S)\gamma\gamma\to\ell\ell\gamma\gamma$ candidates, $375>\Sigma E_\gamma>250$ MeV.

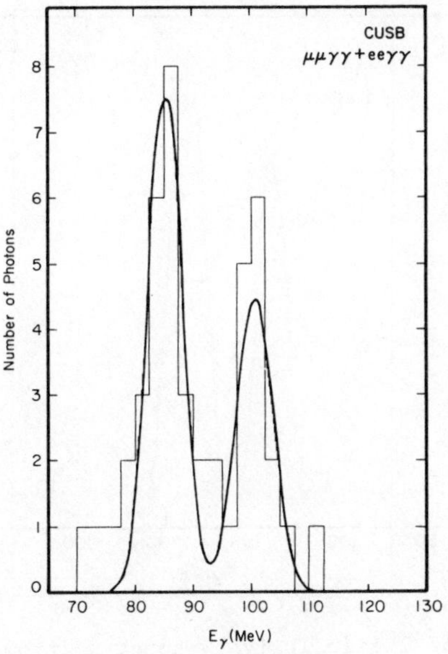

Figure 4c: Energy spectrum of $E_{\gamma low}$ for $\Upsilon(3S)\to\chi_b(2P)\gamma\to\Upsilon(1s)\gamma\gamma\to\ell\ell\gamma\gamma$ candidates, $920>\Sigma E_\gamma>800$ MeV.

2^3P_J State	ϵ	$BR(3^3S_1\to 2^3P_J\gamma)\times BR(2^3P_J\to 2^3S_1\gamma)$	$BR(2^3P_J\to 2^3S_1\gamma)$
J=2	0.21	(1.9±0.7)%	(15±8)%
J=1	0.25	(3.8±1.2)%	(24±10)%
J=0†	0.20	(0.3±0.2)%	(4±3)%

2^3P_J State	ϵ	$BR(3^3S_1\to 2^3P_J\gamma)\times BR(2^3P_J\to 1^3S_1\gamma)$	$BR(2^3P_J\to 1^3S_1\gamma)$
J=2	0.22	(2.0±0.4)%	(16±6)%
J=1	0.26	(1.1±0.3)%	(7±3)%
J=0	0.21	<0.2% at 90% C.L.	<3% at 90% C.L.

† The 1.75 events with $E_\gamma\approx 125$ MeV are assumed to be due to this transition.

We used the measured branching ratios for $^3P_J\to {}^3S_1+\gamma$ transitions (BR_J) together with E1 rates calculated using potential models to obtain the hadronic widths of the P-states. These are given by:

$$\Gamma_{hadronic}(^3P_J) = \Gamma_{E1}(J)[1-BR_J]/BR_J$$

where J is the state's total angular momentum. The widths so obtained are tabulated in the following table for five potential models: GRR (10), MR (11), MB (12), Ei (13), BT (14).

QCD predictions of the hadronic widths by Olsson et al (15) using the lowest order QCD calculations of Barbieri et al (16) (with α_s=0.165) are scaled by us from 1^3P_J to 2^3P_J:

$\Gamma_{had}(2^3P_2)$ = 123 keV, $\Gamma_{had}(2^3P_1)$ = 38 keV,

$\Gamma_{had}(2^3P_0) = (15/4)\times\Gamma_{had}(^3P_2) \approx 440$ keV.

	J	Γ_{E1} $2P_J\to 2S$	Γ_{had} $2P_J$	Γ_{E1} $2P_J\to 1S$	$\Gamma_{had}^\dagger$ $2P_J$
GRR	2	23	122±78	9.7	30±24
GRR	1	19.4	50±32	9.2	102±49
GRR	0	14.9	379±370	8.5	>280 at 90% C.L.
MR	2	16	78±54	14	60±35
MR	1	14	31±23	12	145±64
MR	0	12	304±298	7.5	>248 at 90% C.L.
MB	2	13.7	68±46	10.6	44±26
MB	1	13.6	35±23	7.2	82±38
MB	0	12	311±298	1.3	>33 at 90% C.L.
Ei	2	17	85±57	13	54±32
Ei	1	17	40±28	13	156±69
Ei	0	17	429±422	13	>434 at 90% C.L.
BT	2	16	81±54	9	33±22
BT	1	16	41±27	9	104±48
BT	0	16	407±397	9	>296 at 90% C.L.

†All widths are measured in keV's.

We note that the hadronic widths obtained from using the E1 rates for the $2P_J\to 2S$ transitions (left column of the above table) are in general agreement with the QCD expectations listed above. This situation is very similar to the one we studied using Γ_{E1} rates for $1P_J\to 1S$ transitions and argues that with improved statistical and systematic accuracy we can hope to get a good determination of hadronic widths of the P states using this indirect method, as well as obtaining checks on QCD relations. In fact, even at present we confirm the predicted ratio of 15:4 for the width of the J=0 and J=2 states.

However, we find an inversion of the relative widths for the J=2 to J=1 states when we use the measured BR's and model calculations of E1 rates for the $2P_J\to 1S$ transitions (right column of the above table). This effect is partly due to the fact that we measure a larger BR for the J=2 state than for the J=1 state while the potential models do not predict a correspondingly smaller absolute E1 transition rate from the J=1 state relative to the J=2 state. It is known that these transition rates are very sensitive to relativistic corrections, as can be seen by examining the range of $\Gamma_{E1}(2P_J\to 1S)$ listed in the table (the last two models do not exhibit J dependence because they are nonrelativistic potential models). If this effect persists as we accumulate more statistics, we may have found one of the few anomalies in the heavy quarkonium systems.

4. FINE STRUCTURE OF THE $\chi_b(2P)$ STATES

The data used in the following is from a run partially marred by excessive background. The presently recovered portion corresponds to an integrated luminosity of 23 pb^{-1} collected at the $\Upsilon(3S)$ peak energy, yielding a total of 1.46×10^5 detected hadronic events corresponding to $\approx 1.03\times10^5$ produced $\Upsilon(3S)$.

The photon search codes used are based on the CUSB-I photon algorithms which use longitudinal segmentation for identifying electromagnetic showers. Because of the improved resolution of the new calorimeter fine tuning is necessary to optimize the somewhat orthogonal requirements of high efficiency and resolution. Acceptance and the resolution function are then obtained by adding photons of fixed energies generated with "EGS" to real hadronic events. Figure 5 shows the recovered peak for 100 MeV Monte Carlo (MC) photons and the resolution curve obtained. The preliminary acceptance×efficiency determined for photons of energy 100-200 MeV in decays $\Upsilon(3S)\to\gamma+\chi_b(2P)\to\gamma+$hadrons is 10±2%, with $\sigma_E/E \approx 3.9\%$, which is ≈1.2 times worse than the optimal $\sigma_E/E=3.2\%$ obtained from low (≈4) multiplicity events. For photons of energy between 70 and 1000 MeV we find the rms energy spread function well described by $\sigma_E/E=\{\kappa\%/\sqrt[4]{E_\gamma(\mathrm{GeV})}\}$ with $\kappa=2.2$.

In figure 6 we show the inclusive photon spectrum obtained at the $\Upsilon(3S)$ peak energy, plotted in constant 3% E_γ energy bins. It has several definite, distinct structures, standing out on top of a large background, as shown by curves in the figure. The data was fitted with five resolution functions of energy dependence as specified above, but

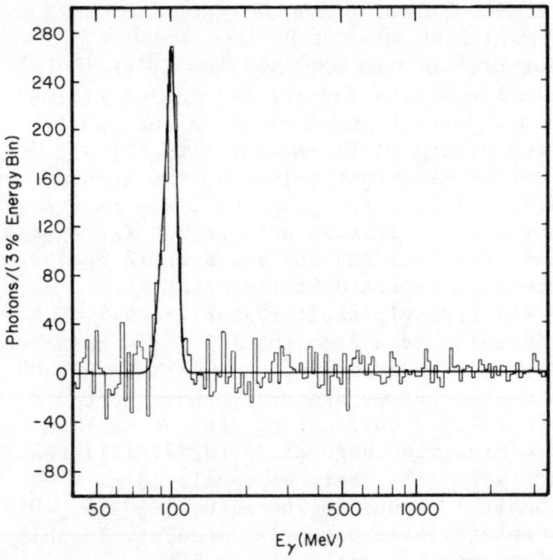

Figure 5: CUSB-II resolution function for 100 MeV γ's from MC simulations.

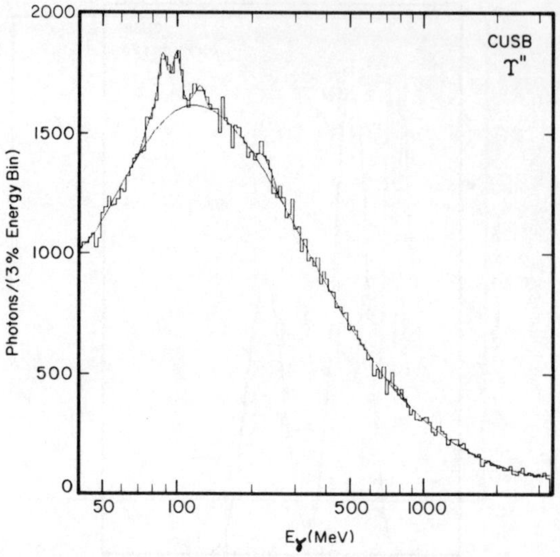

Figure 6: The inclusive γ spectrum from $e^+e^-\to$hadrons at the $\Upsilon(3S)$.

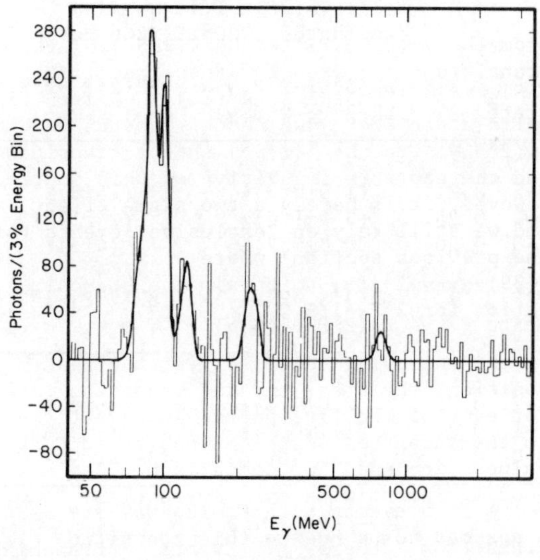

Figure 7: Background subtracted spectrum.

with κ, area and position as free parameters, plus a polynomial to account for the background. The fit yields $\kappa=2.2\pm0.2$, in good agreement with the Monte Carlo simulation results discussed above. Figure 7 shows the spectrum obtained after subtracting the polynomial background curve shown in figure 6. We note that the first two lines are partially merged but the third is well resolved. Figure 8 shows the two individual contributions to the first peak. In the CUSB-I, the three signals were merged into one peak. We identify these three lines with the $\Upsilon(3^3S_1)\to\chi_b(2^3P_{2,1,0})\gamma$ transitions. The fourth peak at E_γ of ≈230 MeV, is actually the merged result of two signals of equal strength which are separated 15 MeV apart, and is due to the photon transitions

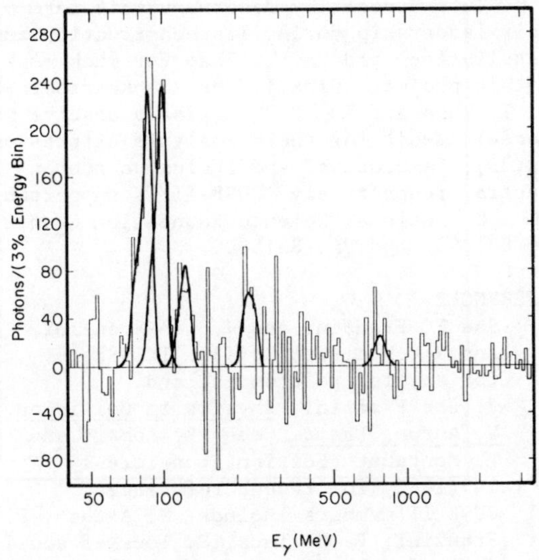

Figure 8: Decomposition of the spectrum into lines.

from the $\chi_b(2^3P_{2,1})$ to the $\Upsilon(2^3S_1)$. These transitions are seen for the first time in an inclusive photon spectrum. The fifth peak falls at an energy which corresponds to the $\chi_b(2^3P_{2,1})$ to the $\Upsilon(1^3S_1)$ photon transitions and therefore is suggestive of this source. However, it is barely a two sigma effect, and we still rely on "exclusive" events (of the previous section) where $\Upsilon(3S)\to\gamma\chi\to\gamma\gamma\Upsilon\to\ell\ell\gamma\gamma$ to obtain the branching ratios for $2^3P_J\to1^3S_1$.

In the following table we list the peak position, the area, and the branching ratio for each of the five peaks. The quantities in the parenthesis are the published CUSB-I values, included for comparison. They are in good agreement with our new results.

Transition	Energy (MeV)	Events	Branching Ratio(%)
$3^3S_1\to2^3P_2$	86.5±0.7	1060±176	12.8±2.1±2.6
$3^3S_1\to2^3P_1$	99.3±0.8	972±158	11.7±1.9±2.3
$3^3S_1\to2^3P_0$	124.2±2.3	441±162	5.3±2.0±1.1
$3^3S_1\to2^3P_2$	(84.2±2)		(12.7±4.1)
$3^3S_1\to2^3P_1$	(101.4±3)		(15.6±4.2)
$3^3S_1\to2^3P_0$	(122.1±5)		(7.6±3.5)
$\langle E_\gamma\rangle$	95.0±0.3±1.0		(94±0.5±1.6)
$\langle \overline{M}(2^3P_J)\rangle$	10260.0±0.3±1.0		(10260.9±0.5±1.6)
$2^3P_{2,1}\to2^3S_1$	227.3±5.0	386±141	4.7†±1.7±0.9
$2^3P_{2,1}\to1^3S_1$	773.0±26.5	152±81	1.8†±1.0±0.4

† These are the product branching ratios $BR(3^3S_1\to2^3P_J\gamma)\times BR(2^3P_J\to2(1)^3S_1\gamma)$.

In the following we shall use for fine splitting the values $M(2^3P_2)-M(2^3P_1)=14\pm0.6$ MeV and $M(2^3P_1)-M(2^3P_0)=24.4\pm2.3$ MeV, obtained from combining our exclusive results and the above values.

Eichten and Feinberg (17) (EF) first developed a generalized formulation of the spin dependence of quark antiquark interactions and expressed it in terms of spin-orbit, spin-spin and tensor interactions derived from a vector potential, V_V, and possibly a scalar potential, V_S. Various phenomenologists postulated particular V_V's and V_S's, for example, EF assumed no scalar component ($V_S=0$) and the vector part V_V to be composed of a short range piece, $V_g=-(4/3)\alpha_s/R$, corrected to first order in α_s, and a long range piece due to the longitudinal color electric field which transforms as the fourth component of a Lorentz vector ($\gamma^o\gamma^o$). This scheme is known as "electric confinement". Moxhay and Rosner (11) (MR) also chose $V_S=0$, but their V_V is given by a Richardson potential modified such that a long range tensor force remains. McClary and Byers (12) (MB) chose instead a linear scalar potential ($V_S=A+BR$), and $V_V=V_g$. Similarly, Gupta et al (10) (GRR) assume a linear scalar potential and obtained V_V from QCD including corrections to order α_s^2 after adjusting several parameters to obtain agreement with charmonium and Υ level spacings. For both these two latter cases the confining potential arises from an effective scalar exchange and is referred to, in the literature, as "scalar confinement".

Below we compare the measured $3^3S_1\to2^3P_J\gamma$ E1 transition rates, $\Gamma_{E1(J)}$ in keV and reduced widths, $\Gamma_{RW(J)}=\Gamma_{E1}/\{E_\gamma^3\times(2J+1)\}$ normalized to the J=1 case, with three most recent calculations which take into account relativistic corrections: GRR (10), MR (11) and MB (12).

$\Gamma_{E1(J)}$	Experiment	GRR	MR	MB
J = 2	3.3±0.8±0.7	2.9	2.7	2.6
J = 1	3.0±0.8±0.6	2.7	2.8	2.4
J = 0	1.4±0.6±0.3	1.5	1.4	1.0
ΣJ = 2,1,0	7.7±1.7±1.5	7.1	6.9	6.0

$\Gamma_{RW(J)}$	Experiment	GRR	MR	MB
J = 2	0.99±0.21	0.97	1.17	1.05
J = 1	≡1	1	1	1
J = 0	0.70±0.26	0.51	0.74	0.9

There is a very good overall agreement between data and experiment. It is also apparent that the transition rates cannot distinguish between the models. We can also compare the measured fine structure with various models. Following Rosner (18) we can write the ^3P masses as:

$M(^3P_2)=\bar{M}+a-2b/5$, $M(^3P_1)=\bar{M}-a+2b$,
$M(^3P_0)=\bar{M}-2a-4b$;
$a=(1/2M_Q^2)\langle 3V_V'/R-V_S'/R\rangle$,
$b=(1/12M_Q^2)\langle V_V'/R-V_V''\rangle$.

and compare the experimental values of a and b with four models: EF (17), MR (11), MB (12) and GRR (10).

EXPERIMENT	EF	MR	MB	GRR
9.9±0.5	10.7	6.5	14.6	9.2
2.4±0.3	1.3	2.1	4.2	1.8

While no model is singled out, those using "scalar confinement" appear to be favored. The same conclusion follows from the 1^3P_J fine structure and the original argument of Büchmuller (19) that a long range confinement potential must transform as a Lorentz scalar. Pantaleone, Tye and Ng (20) recently argued that the *ad hoc* postulation of non-perturbative long range forces is most unappealing from a fundamental viewpoint. They derive in a self consistent way the $\chi_b(2P)$ fine structure (accurate to 1 MeV) in terms of the $\chi_b(1P)$ splitting for their model and for scalar confinement models. Their results for the splittings in MeV are given below: their model in the first column and scalar confinement ones in the second column. Our measurements are in the third column:

$M(2^3P_2)-M(2^3P_1)$	16.2	14.9	14.0±0.6
$M(2^3P_1)-M(2^3P_0)$	21.8	22.3	24.4±2.3

While more data are clearly needed, the present results seem to favor the need of an *ad hoc* scalar long range interaction to describe the fine structure of the P-wave $b\bar{b}$ states.

5. CONCLUSION AND OUTLOOK

Our first run with CUSB-II was gratifying. Aside from obtaining a new improved (5×) measurement of $B_{\mu\mu}(3S)$ and an intriguing preview of precision $\chi_b(2P)$ spectroscopy, we measured a resolution improvement of ≈ a factor of two over CUSB-II. It augurs well for our hopes of eventually seeing the singlet bound $b\bar{b}$ states (shown as dashed lines in figure 1).

ACKNOWLEDGEMENTS

The author thanks the co-spokesperson P. Franzini and all collaboration members for the successful realization of CUSB-II. In particular, Drs. R. D. Schamberger and P. M. Tuts especially deserve credit for their leadership during its construction and installation, and Mr. T. Zhao for much work on this project. Finally, Ms. T. Kaarsberg, Mr. T. Zhao and Dr. C. Yanagisawa deserve special credit for their analysis efforts on $B_{\mu\mu}(3S)$, "exclusive" and inclusive photon spectra, respectively. CUSB-II is supported by U. S. National Science Foundation Grants Phy-8310432 and Phy-8315800.

REFERENCES

(1) See P. Franzini and J. Lee-Franzini, Ann.Rev.Nuc.Part.Sci.33, 1(1983) for earlier references, and J. Lee-Franzini, Physics in Collision V, Autun, France, eds. B. Aubert and L. Montanet (Edition Frontières) 145(1985) for recent references.

(2) CUSB-II members include: M. Artuso, P. Franzini, P. M. Tuts, S. Youssef and T. Zhao of Columbia University; U. Heintz, J. Lee-Franzini, T. M. Kaarsberg, D. M. J. Lovelock, M. Narain, S.B. Sontz, R. D. Schamberger, J. Willins and C. Yanagisawa of SUNY at Stony Brook. P. Franzini and J. Lee-Franzini, in OREGON MEETING, ed. R.C. Hwa, (World Scientific) 1009(1985)

(3) F.A. Berends and R. Kleiss, Nucl. Phys. B178 141(1981)

(4) D. Andrews etal, Phys.Rev.Lett. 50 807(1983)

(5) P. Franzini and J. Lee-Franzini, Physics Reports 81, 239 (1982).

(6) G. Mageras et al., Phys.Lett. 118 453(1982).

(7) K. Han et al., Phys. Rev. lett. 49 1612(1982), G. Eigen et al., Ibid 1616

(8) J. Lee-Franzini, in 22 Int. Conf. High Energy Phys. 189(1984)

(9) J. Horstkotte et al., Lepton-Photon Symposium ed. D. Cassel, 896(1983)

(10) S.N. Gupta et al., Phys. Rev.D30 2435(1984)

(11) P. Moxhay and J.L. Rosner, Phys.Rev. D28, 1132(1983)

(12) R. McClary and N. Byers, Phys.Rev. D28, 1692(1983)

(13) E. Eichten et al., Phys.Rev. D21 203(1980)

(14) W. Büchmuller and S.-H. H. Tye, Phys. Rev. D24 132(1981)

(15) M. G. Olsson et al., Phys.Rev. D31 81(1985)

(16) R. Barbieri et al., Phys.Lett. 60B 183(1976), 61B 465(1976)

(17) E. Eichten and F. Feinberg, Phys. Rev. D23 2724(1981)

(18) J.L. Rosner, in Exp. Meson Spec. ed. S. Lindenbaum, 461(AIP,1984)

(19) W. Büchmuller, Phys.Lett.112B 479(1982)

(20) J. Pantaleone et al., Phys. Rev. D33 777(1986)

SEARCH FOR RARE $b\bar{b}$-DECAY MODES

R.T. Van de Walle

Univ. of Nijmegen and NIKHEF-H Nijmegen
6525 ED Nijmegen, The Netherlands

(representing the Crystal Ball Collaboration) (*)

Abstract: Preliminary upper limits are presented for a series of rare radiative or hadronic $b\bar{b}$-transitions, based on 193 K $\Upsilon(2S)$ and 306 K $\Upsilon(1S)$ decays observed by the Crystal Ball detector installed at DORIS-II in DESY.

The exploration of the spectroscopy of the $b\bar{b}$-states and their decay modes is now in full progress both at the DORIS-II and the CESR-storage rings. Radiative transitions between $b\bar{b}$ states determine the quark-potential parameters. Hadronic decays form a laboratory for perturbative QCD. Both radiative and hadronic decays of the $b\bar{b}$-states into light mesons yield information on the role of the gluon in OZI-forbidden reaction mechanisms, as well as on the gluonic content of the light mesons produced.

The advantages of the $b\bar{b}$- over the $c\bar{c}$-system are well-known: The $b\bar{b}$-transitions explore the $q\bar{q}$-interactions at shorter distances and suffer less from relativistic corrections and their uncertainties. The price one has to pay for these advantages is equally well known: Fine and hyperfine level separations diminish (by a factor of 2 to 3) and transition rates decrease (by at least an order of magnitude). As a result, up till now, primarily only the 'large' transition rates such as $\Upsilon(2S) \to \gamma\chi_b$ and $\Upsilon(2S) \to \pi\pi\ \Upsilon(1S)$ have been measured. (1)(2)(3)(4) Many channels remain unseen however and only upper limits are now becoming available.

The present report deals with PRELIMINARY upper limits for a series of rare $\Upsilon(2S)$ and $\Upsilon(1S)$ decay modes observed by the Crystal Ball detector at DORIS-II in DESY.

The results presented are based on

- $(193\pm15)10^3$ $\Upsilon(2S)$ resonance decays corresponding to an integrated luminosity of $\sim 60\ pb^{-1}$, and
- $(306\pm25)10^3$ $\Upsilon(1S)$ resonance decays corresponding to an integrated luminosity of $\sim 32 pb^{-1}$.

Specifically the following channels are studied:

- The inclusive radiative transitions:

$$\Upsilon(2S) \to \gamma\eta_b', \gamma\eta_b \quad (1a,1b)$$

$$\Upsilon(1S) \to \gamma\eta_b \quad (1c)$$

- The exclusive hadronic transition

$$\Upsilon(2S) \to \eta\Upsilon(1S) \quad (2)$$

- The exclusive radiative transitions

$$\Upsilon(1S) \to \gamma\eta \quad (3a)$$

$$\to \gamma\eta' \quad (3b)$$

$$\to \gamma f_2(1270) \quad (3c)$$

The study of all these channels has important physics motivations.

The masses of the η_b and η_b' constrain the hyperfine component of the $q\bar{q}$-force.

(*) Members of the Crystal Ball Collaboration are from the following institutes: California Institute of Technology; Carnegie-Mellon University; Cracow Institute of Nuclear Physics; Deutsches Elektronen Synchroton DESY; Universität Erlangen-Nürnberg; INFN and University of Firenze; Universität Hamburg, I. Institut für Experimentalphysik; Harvard University; University of Nijmegen and NIKHEF-Nijmegen; Princeton University; Department of Physics, HEPL, and Stanford Linear Accelerator Center, Stanford University; Universität Würzburg.

QCD describes the hadronic transition $\Upsilon(2S)\to\eta\Upsilon(1S)$ as the emission by the heavy quarks of two gluons followed by the conversion of these gluons into an η. (See fig. 1).

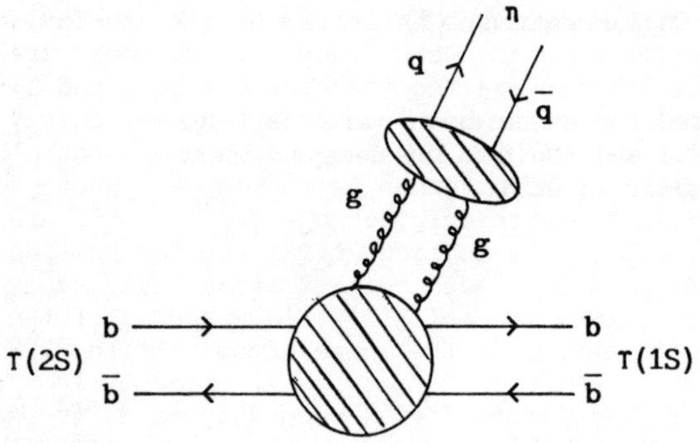

Figure 1.

The relatively high rate observed for this reaction in the corresponding charmonium case has been the source of considerable theoretical activity [5][6][7]; a priori, one expected a strong suppression because of the OZI-rule, the SU(3) violating character of the transition and the reduced phase space available for the P-wave decay involved. Gottfried has pointed out that the velocities and dimensions involved in these processes are such as to allow a multipole expansion of the gluon field [8]. This feature can be used to derive scaling laws relating $\Upsilon(2S)$ and $\psi(2S)$ transition rates [9]. Absolute rates yield information on how the soft gluons, emitted by the heavy hadrons, convert into light hadrons (c.q. the η) i.e. on color confinement [10].

The reactions $\Upsilon(1S) \to \gamma(\eta,\eta',f_2)$ are described in lowest order of QCD by the diagram of fig. 2.

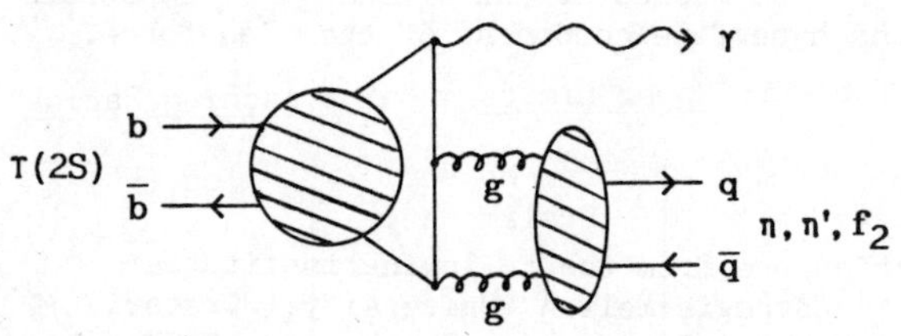

Figure 2.

This diagram has been used in different ways. In one approach, rates are derived by making a QCD estimate of the vertex $b\bar{b}\to\gamma gg$ and using duality arguments between the gluon pair and the light quarks meson to handle the second vertex. [11] A somewhat different approach was taken by Körner et al [12] in treating the perturbative diagram of fig. 2 "literally" i.e. by evaluating all the helicity amplitudes and decay rates within the framework of non-relativistic, weakly-bound (light and heavy) quark-antiquark systems. Also non-QCD type reaction mechanisms have been proposed - e.g. based on an extended vector meson dominance model [13] - yielding widely different rate predictions.

A schematic drawing of the Crystal Ball as it is installed at DORIS-II is shown in fig. 3.

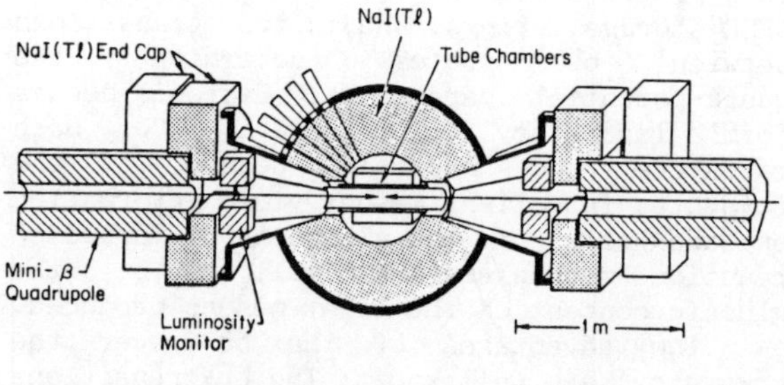

figure 3: CRYSTAL BALL at DORIS-II.

Full details of the Crystal Ball detector have been reported elsewhere [14]. Photon energies and particle directions are measured in the main components of the detector, a spherical and highly segmented shell of 672 NaI(Tl) crystals covering 93% of the 4π solid angle. Each crystal is 16 radiation lengths thick. For electromagnetically showering particles this crystal array yields an rms energy resolution given by

$$\frac{\sigma(E)}{E} \simeq \frac{2.7\%}{E^{1/4}\,(\mathrm{GeV})} \qquad (4)$$

or a resolution varying from approx 1.5 MeV for a photon of 20 MeV to approx. 20 MeV for a photon of 700 MeV. The angular resolution is slightly energy dependent, varying between 1° to 2°. Three (double) layers of proportional tube chambers with charge division readout, surrounding the beam pipe, are used to identify and track charged particles.

A. UPPER LIMITS FOR Υ-Decay into η_b AND η_b'

Potential models generally predict the mass of the η_b' and the η_b to be 20 to 100 MeV below the mass of the $\Upsilon(2S)$ and $\Upsilon(1S)$ respectively ([15]). The branching ratios are a sensitive function of the photon energies involved; estimates for the η_b-rates vary from 10^{-3} to 10^{-6} with possibly even lower values for the η_b' ([16]). Thus the η_b' should manifest itself in the inclusive photon spectrum of the $\Upsilon(2S)$ decay as a monochromatic line in the range from 20 to 100 MeV while the η_b should appear both in the inclusive photon spectrum of the $\Upsilon(2S)$ - in the range from 580 to 660 MeV - and in the inclusive spectrum of the $\Upsilon(1S)$ - again in the region from 20 to 100 MeV. Fig. 4 shows the experimental inclusive photon spectra obtained from the $\Upsilon(2S)$ and $\Upsilon(1S)$ decays together with the energy ranges relevant for the η_b' and η_b searches .

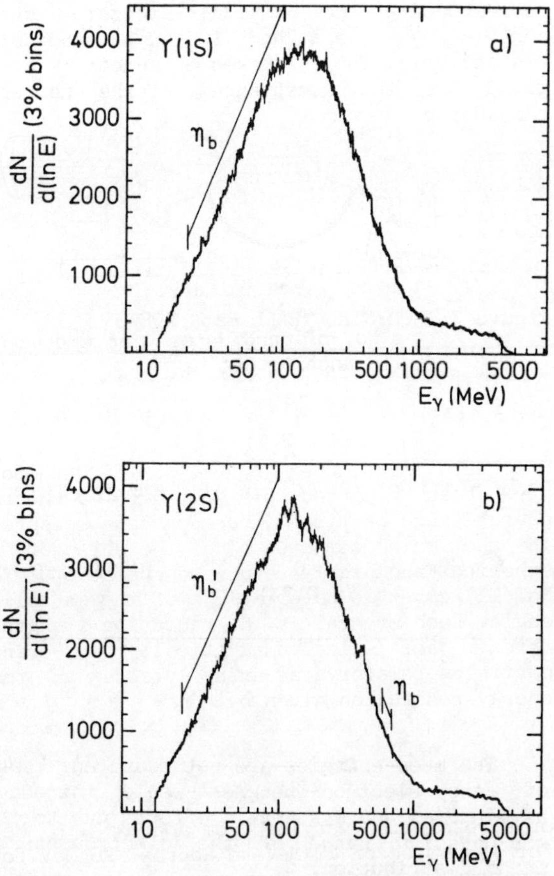

Figure 4: Inclusive photon spectra from $\Upsilon(2S)$ and $\Upsilon(1S)$ hadronic decay.

The procedures used in selecting the single photons closely followed the ones used in the measurement of the χ_b-states ([2]). Only photons associated with hadronic final states are accepted; to select hadronic decays of the $\Upsilon(2S)$ and $\Upsilon(1S)$, background resulting from beam-gas interactions, cosmic rays, two-photon and QED-events is removed. The efficiency for this selection, defined as the ratio of the number of selected hadrons divided by the total number of produced resonances (including those decaying leptonically), was estimated by a Monte-Carlo simulation to be $(86\pm7)\%$. Photons are identified using the tracking chambers and by imposing cuts both on the lateral energy deposition in the crystal and on the spatial separation from other particles. All photons compatible with originating from a π^0 are removed. The selection efficiency for photons was again estimated using a Monte Carlo study and found to be $(13\pm2)\%$, independent of the photon energy in the range from 20 to 700 MeV.

The data of fig. 4 were subjected to a series of fits at different energy points in the E_γ-ranges of interest. Each fit uses a Gaussian (with width given by expression (4)) to describe a potential signal and a quadratic polynomial to describe the background. The energy steps between successive fits are chosen to be one half of the detector's resolution in the energy range under consideration, thus ensuring that no significant structures remain undetected. For each energy point the fit returns an amplitude and an error; thus the statistical significance of a possible structure at this point can be determined. No structures corresponding to a narrow photon line were found with a significance of more than two standard deviations. Hence the fit results were converted into branching ratio upper limits (at 90% C.L.), using as input the 90% CL upper limit on the number of photons at any given energy, the number of selected resonance events at that energy and the efficiencies on these numbers as discussed above.

The results are presented in fig. 5. Table 1 makes a numerical comparison with the η_c' limit and η_c signals observed in the corresponding charmonium experiments.

Table 1: Upper Limits η_b-η_b' Branching Ratios

	$b\bar{b}$ (Preliminary)	$c\bar{c}$ ([17])
$(2S)\to\gamma\eta_{b(c)}'$	<(0.4 to 2.6)%	(0.2 to 1.3)%
$(2S)\to\gamma\eta_{b(c)}$	<(0.3 to 0.8)%	(0.28 ± 0.6)%
$(1S)\to\gamma\eta_{b(c)}$	<(0.3 to 1.6)%	(1.27 ± 0.36)%

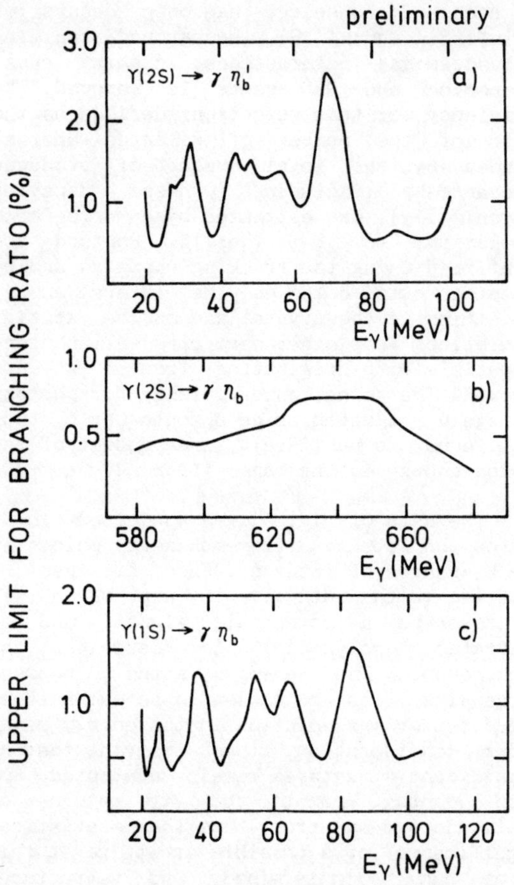

Figure 5: Upper limits (at 90% CL) from radiative decay of ϒ(2S) and ϒ(1S) into η_b' and η_b versus the energy of the photon involved.

No η_b signal is seen, neither in the ϒ(2S) nor in the ϒ(1S) sample, although the experimental limits obtained are comparable in magnitude to the signals observed for η_c in ψ(2S) and ψ(1S) decay.

B. HADRONIC DECAY ϒ(2S) → η ϒ(1S) [18]

The exclusive reaction (2) has been searched for in the subchannels ϒ(2S) → $\ell^+\ell^-$ + 2γ and ϒ(2S) → $\ell^+\ell^-$ + 6γ i.e. using the 2γ and $3\pi^0$ decay modes of the η respectively. A first selection of events was performed by imposing cuts on the final state multiplicity and on the angles of the final state particles, both with respect to the beam axis and among themselves. Lepton energies were not used [a]; only mild tagging conditions were imposed. To further reduce the background, all remaining candidate events were subjected to a 3C-fit. In addition to imposing overall energy and momentum conservation, this fit in essence constrains the ϒ(2S)-ϒ(1S) mass difference, expressed in terms of the photon-momenta, to correspond to a nominal value of 563 MeV [b]. Accepting only events with a fit-C.L. of ≥ 5%, one finds 57 $e^+e^-\gamma\gamma$, 35 $\mu^+\mu^-\gamma\gamma$ and zero $\ell^+\ell^-\gamma\gamma\gamma\gamma\gamma\gamma$ candidates.

For the $\ell^+\ell^-\gamma\gamma$ candidates a scatter plot of $M^2(\gamma_1,\gamma_2)$, the invariant mass of the γ's, vs. either $M^2(\Upsilon,\gamma_1)$ or $M^2(\Upsilon,\gamma_2)$, the invariant mass of the two possible ϒ(1S)γ-combinations, is shown in fig. 6. The symmetry of the figure is of course due to the fact that each candidate event creates two entries. The solid lines define the kinematic boundary. The dashed vertical bands indicate the kinematic limits for events of the χ_b-cascade type. Of interest for the present analysis are the two horizontal bands corresponding to $(M_\eta \pm 3\sigma)^2$-limits; they contain 4 $e^+e^-\eta$ and 1 $\mu^+\mu^-\eta$ candidate respectively. Table 2 makes a summary; also given are the acceptances of the channels considered.

Table 2: Upper Limits ϒ(2S)→ηϒ(1S) [c]
(Preliminary)

	No. of candidates (Branching Ratios)	Acceptance
$e^+e^-\eta(\gamma\gamma)$	4 (< 1.2 %)	(30.8±1.0±1.5)%
$\mu^+\mu^-\eta(\gamma\gamma)$	1 (< 0.7 %)	(24.2±0.4±0.5)%
$\ell^+\ell^-\eta(3\pi^0)$	0 (< 0.7%)	(19.0±1.1±0.5)%

[a] The muon-energies are not measured. Usage of the electron-energies would introduce asymmetries and systematic effects due to the non-Gaussian shape of the electron energy error distribution.

[b] Imposing also the η-mass constraint, would allow a 4C-fit, but would not alter any of the data or conclusions presented here.

[c] Taking into account the branching ratios of the η decay considered.

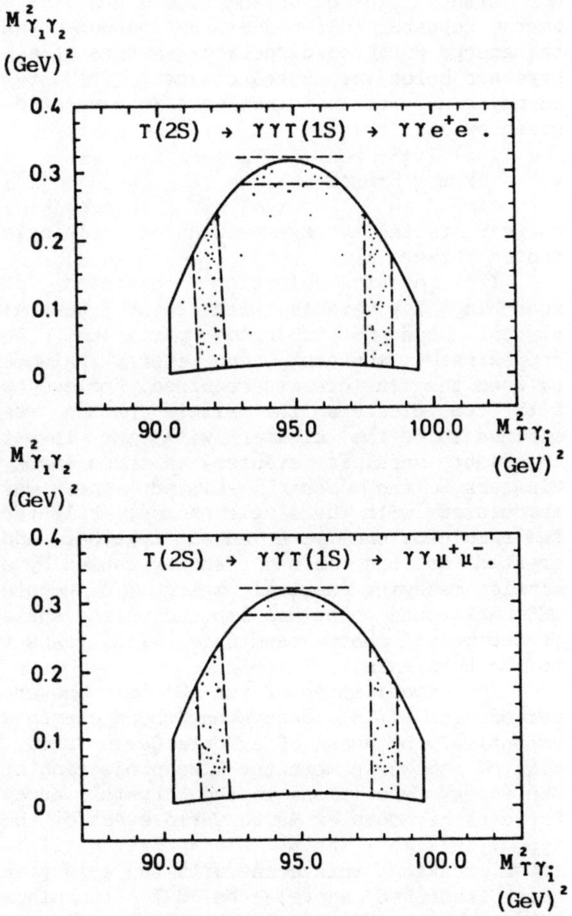

Figure 6: Dalitz-plot of the $\gamma_1\gamma_2$ invariant mass vs. the $\Upsilon(1S)$ decaying into either e^+e^- or $\mu^+\mu^-$. (see text for a definition of the boundary lines and bands).

Before interpreting these numbers we must worry about background. Fig. 6 shows the presence of events just below the η-band in the e^+e^--channel. Kinematically, these events cannot be due to the tails of the cascade transitions. In addition, a Monte-Carlo study simulating misidentified $\ell^+\ell^-\pi^0\pi^0$ and $\ell^+\ell^-\pi^+\pi^-$ events also shows that such a source would - in each channel - at most contribute 0.1 to 0.2 $\ell^+\ell^-\eta$-candidates. Given the difference in background behaviour between the e^+e^- and $\mu^+\mu^-$ channel, the most likely source is a contribution from double radiative Bhabha scatters. This interpretation was examined by submitting a large non-$\Upsilon(2S)$ sample - mainly consisting of $\Upsilon(1S)$ and $\Upsilon(4S)$ - to the same cuts and criteria as the 2S-sample. The results found do not support the hypothesis that all the $e^+e^-\eta$ candidates can be explained as background. However, given the overall smallness of the numbers of the events involved, and considering both the differences between the e^+e^- and the $\mu^+\mu^-$ channel (in spite of their comparable acceptances and identical branching ratios) and the negative $\ell^+\ell^-\eta(3\pi^0)$ result (in spite of the near equality of the $\eta\to2\gamma$ and $\eta\to3\pi^0$ branching ratios) we prefer to take a conservative approach and quote only 90% CL upper limits based on the (uncorrected) number of candidates observed. This leads to the branching ratio upper limits also presented in column 2 of Table 2.

Combining both η-decay channels one finds an overall 90% C.L. upper limit of

$$BR[\Upsilon(2S)\to\eta\Upsilon(1S)] < 0.6\%$$

or; using a Γ_{total} $[\Upsilon(2S)]$ of 30 keV [19]

$$\Gamma[\Upsilon(2S)\to\eta\Upsilon(1S)] \leq 0.18 \text{ KeV} \qquad (5)$$

The CUSB-collaboration, based on a somewhat smaller sample (146 K vs. 193 K) and a somewhat lower total η-acceptance (58% vs. 74%), has published a 3 times lower limit [4]. Both these results are however still significantly above the theoretical prediction of ~ 0.01 keV, made by Kuang and Tan [10]. Voloshin and Zakharov [7] have argued that the ratio of (5) to the width of the process $\Upsilon(2S)\to\pi^+\pi^-\Upsilon(1S)$ is calculable within QCD, in terms of so-called triangular anomalies in the divergence of the axial current and in the trace of the energy-momentum tensor. Using the Crystal Ball $\Upsilon(2S)\to\pi^+\pi^-\ \Upsilon(1S)$ measurement [3], the prediction of these authors would be roughly 0.02 keV, again still "safely" below the experimental limits. Both these absolute rate predictions suffer however from our limited understanding on how soft gluons emitted by the heavy quarks hadronize into an η meson (or a $\pi\pi$-system). More reliable are the predictions for the ratios of rates between analogous charmonium and bottomonium transitions, which simply follow from the scaling properties of the multipole expansion of the gauge field [9]. Although these ratio-predictions still depend on the assumption of the dominance of a particular multipole, the gluon-hadronization problem cancels out. For the specific case of the η-transition, Yan [9] has predicted

$$\frac{\Gamma[\Upsilon(2S)\to\eta\Upsilon(1S)]}{\Gamma[\psi(2S)\to\eta\psi(1S)]} = \left(\frac{m_c}{m_b}\right)^4 \cdot \left(\frac{P_\eta(\Upsilon)}{P_\eta(\psi)}\right)^3 \simeq \frac{1}{275} \qquad (6)$$

where m_c (=1.8 GeV) and m_b (=5.2 GeV) are the quark masses and $P_\eta(\Upsilon)$, $P_\eta(\psi)$ the η-decay momenta involved.

Using world-averages for the $\psi(2S)$ total width and $\eta\psi(1S)$ branching ratio [19], we find

$$\frac{\Gamma[\Upsilon(2S)\rightarrow\eta\Upsilon(1S)]}{\Gamma[\psi(2S)\rightarrow\eta\psi(1S)]} \lesssim \frac{1}{30} \qquad (7)$$

The suppression of the Υ-rate predicted by the QCD multipole expansion is indeed observed (phase space alone would only give a reduction by a factor 4) but the Υ data are once more approximately an order of magnitude away from the theoretically predicted signal.

C. EXCLUSIVE RADIATIVE $\Upsilon(1S)$ DECAY INTO NEUTRAL (LIGHT) MESONS [20]

Using the capabilities of the Crystal Ball, reaction (3) can be searched for in the following subchannels

$\Upsilon(1S) \rightarrow \gamma\eta;\quad \eta \rightarrow 3\pi^0,\ \gamma\gamma$
$\Upsilon(1S) \rightarrow \gamma\eta';\quad \eta' \rightarrow \pi\pi^0\eta;\ \eta \rightarrow 3\pi^0,\ \gamma\gamma$
$\Upsilon(1S) \rightarrow \gamma f_2;\quad f_2 \rightarrow \pi^0\pi^0$

Phenomologically all these channels are characterized by a single high energy photon shower recoiling against an n-photon shower, with n-varying from 2 to 10 depending on the reaction and subchannel considered. To reconstruct events of this type a method was used, called GST, the "Global Shower Technique" [21]. This technique allows the determination of the invariant mass of a cluster of photons without resolving them individually. The method is based on an approximate relation between the width S of a photon cluster (defined as the 2nd moment of the energy weighted direction- vectors of all crystals belonging to the cluster), its total energy E and its invariant mass M, a relation given by

$$M \simeq \sqrt{S-S_\gamma \cdot E} \qquad (8)$$

where S_γ is the average width of a single photon shower.

The event selection consisted of searching for events with 2 or 3 neutral clusters and no charged particles. An approximate momentum and energy balance between the clusters was required. For events with two clusters the single photon was defined to be the cluster with the lowest invariant mass. For events with three energy clusters, the most isolated one was associated with the single photon, while the two remaining ones were added together and treated as one 'broad' shower caused by a single decaying neutral meson. To reduce QED-background a cut was imposed on the angle of the recoil photon candidate with respect to the beam axis.

The capabilities of the 'Global Shower' method for $\Upsilon(1S)$ decays have been examined extensively by means of a Monte Carlo study. Fig. 7 shows a Mercator-like projection of the energy depositions in the crystal array for a rather complex Monte Carlo event of the type $\Upsilon(1S)\rightarrow\gamma\eta'\rightarrow\gamma\eta\pi^0\pi^0\rightarrow\gamma + 5\pi^0 \rightarrow 11\ \gamma$. After masking this event with the selection cuts described above, the GST- technique reconstructs the 'fat' shower to an η' mass of 911 MeV i.e. a value remarkably close to the nominally introduced value of 958 MeV.

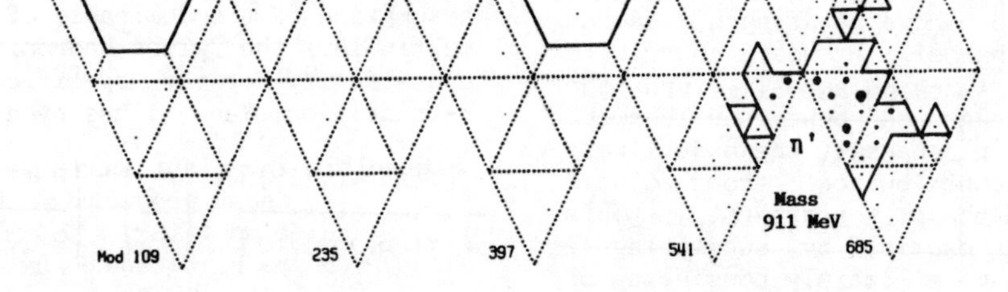

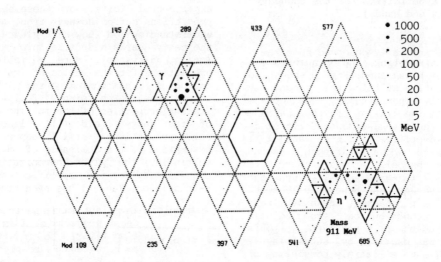

Figure 7: Mercator-like projection for a Monte Carlo event of the type: $\Upsilon(2S) \rightarrow \gamma\,\eta' \rightarrow \gamma\,\eta\,\pi^0\,\pi^0 \rightarrow \gamma + 5\,\pi^0 \rightarrow 11\,\gamma$.

Fig. 8 shows the mass-distributions obtained from the different Monte Carlo-samples when subjected to a GST-treatment; the indicated widths form a measure of the attainable resolution. For the η,η' channels we used the (required) $1+\cos^2\theta$ distribution for the $\eta(\eta')$ in the $\Upsilon(1S)$ C.M.; for the f_2 we assumed an isotropic angular distribution; all decay products of the η,η' and f_2 were generated isotropically in the C.M. system of the meson. Also given in fig. 8 are the efficiencies with which the Monte Carlo events passed through the imposed selection criteria.

Fig. 9 shows the (preliminary) invariant mass distribution of the data sample itself. No significant structures are visible. Branching ratio upper limits (90% C.L.) were derived by determining how much signal could be accepted by a fit using the η,η' and f_2 Monte Carlo 'shapes' given in fig. 8. The results are presented - and compared to the Crystal Ball charmonium results - in table 3.

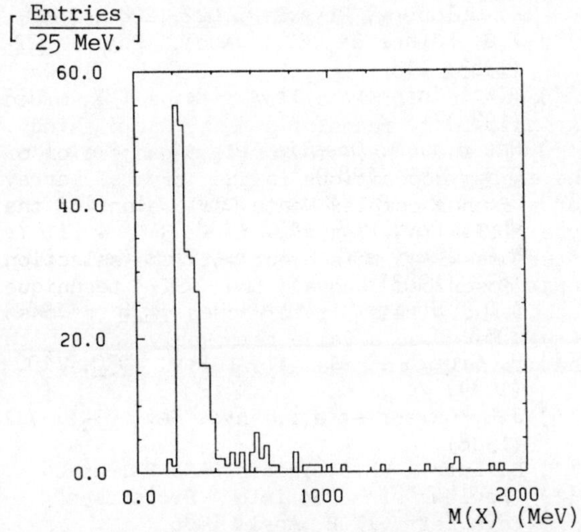

Figure 9: Experimentally observed spectrum for the invariant mass of X in the decay $\Upsilon(1S) \to \gamma X$.

Table 3: Exclusive Radiative (1S)-Decays (Preliminary)

$\Upsilon(1S) \to \gamma\eta$	$< 2.3 \cdot 10^{-4}$
$\psi(1S) \to \gamma\eta$ (22)	$(0.88 \pm .0.8 \pm .11)\cdot 10^{-3}$
$\Upsilon(1S) \to \gamma\eta'$	$< 6.1 \cdot 10^{-4}$
$\psi(1S) \to \gamma\eta'$ (22)	$(4.1 \pm 0.3 \pm 0.6)\cdot 10^{-3}$
$\Upsilon(1S) \to \gamma f_2$	$< 3.7 \cdot 10^{-4}$
$\psi(1S) \to \gamma f_2$ (22)	$(1.48 \pm .25 \pm .30)\cdot 10^{-3}$

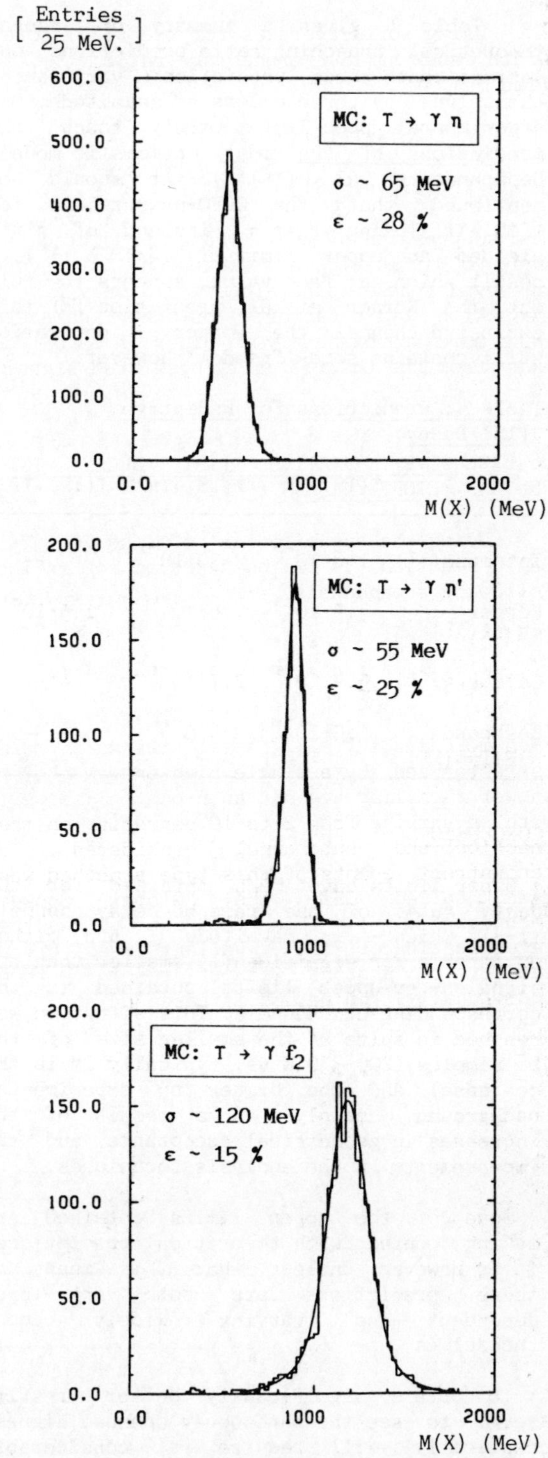

Figure 8: Invariant mass-spectrum for the decay products of X in the reaction $\Upsilon(1S) \to \gamma X$, with X equal to η, η' and f_2 respectively.

Table 4 gives a summary of recent theoretical branching ratio predictions; one notices that these predictions vary among themselves by three orders of magnitude. Our experimental upper limits barely 'touch' the predictions of the most optimistic model, Deshpande et al. [11][d]. It should be mentioned that the CLEO-measurement for $\Upsilon(1S)\to\gamma f_2$ (using $f_2\to\pi^+\pi^-$ instead of $\pi^0\pi^0$) yielded an upper limit of $4.8\cdot10^{-5}$ [23], a result which, at face value, appears to rule out the Körner et al. prediction for this exclusive channel; the Körner - derivation still contains some 'freedom' however.

Table 4: Predictions for Radiative $\Upsilon(1S)$-Decay

	$\Upsilon(1S)\to\gamma\eta$	$\Upsilon(1S)\to\gamma\eta'$	$\Upsilon(1S)\to\gamma f_2$
Intemann[13]	$1-6\cdot10^{-7}$	$5-30\cdot10^{-7}$	--
Körner et al[12]	$3.5\cdot10^{-5}$	$1.6\cdot10^{-4}$	$1.5\cdot10^{-4}$
Goldberg[11]	$5.4\cdot10^{-5}$	$2.7\cdot10^{-4}$	--
Deshpande - Eilam[11]	$1.5\cdot10^{-4}$	$\sim 10^{-3}$	--

CONCLUSIONS

• Upper limits have been obtained for the decay rates of the rare $b\bar{b}$-decay channels (1-3), which - in magnitude - are either comparable or significantly smaller than the signals - or upper limits obtained in the corresponding $c\bar{c}$ channels. This situation was reached in spite of the smaller size of the $b\bar{b}$ samples (200-330 K vs. typically 2M in the $c\bar{c}$ case) and the higher $b\bar{b}$ experimental background, mainly as a result of the increases in geometrical acceptance and the improvements in the analysis techniques.

• Some of the upper limits obtained are making contact with theoretical predictions. It is however unclear what this means as these predictions are both very input dependent and varying widely among themselves.

• To obtain significantly better results (e.g. to see the rare decay channel signals explicitly) will require a considerable (orders of magnitude) increase in statistics.

([d]) A model which obtains relatively high rates by introducing a questionable 'fudge' factor.

References

(1) F. Pauss et al. (CUSB) Phys.Lett. 130B (1983) 439;
C. Klopfenstein et al. (CUSB) Phys. Rev. Lett. 51 (1983) 160

(2) Nernst et al., (Crystal Ball) Phys.Rev.Lett. 54, 2195 (1985)
W. Walk et al., (Crystal Ball), submitted to Phys.Rev.D

(3) D. Gelphmann et al., (Crystal Ball), Phys.Rev. D32, 2893 (1985)

(4) V. Fonseca et al., (CUSB), Nucl. Phys. B242 (1984) 31

(5) H. Harari, Phys. Lett. 60B (1976) 172

(6) P. Langacker, Phys. Lett. 90B (1980) 447

(7) M. Voloshin and V. Zakharov, Phys. Rev. Lett. 45, 688 (1980)

(8) K. Gottfried Phys.Rev.Lett. 50 (1983) 807

(9) T. Yan, Phys. Rev. D22, 1652, 1980

(10) Y. Kuang and T. Yan, Phys. Rev. D24, 2874, 1981

(11) N.G. Deshpande and G. Eilam, Phys.Rev. D23, 270, 1982
H. Goldberg, Phys.Rev. D22, 2286, 1980

(12) J.G. Körner et al., Nucl. Phys. B229 (1983) 115

(13) G.W. Intemann, Phys. Rev. D27, 2755 (1983)

(14) See e.g. M. Oreglia et al., Phys.Rev. D25, 2259 (1982)

(15) See e.g. P. Moxhay and J.L. Rosner, Phys. Rev. D28, 1132 (1983);
R.Mc.Clary and N. Byers, Phys.Rev. D28, 1692 (1983);
A.D. Steiger, Phys.Lett. 129B (1983) 335.

(16) H. Grotch et al., Phys. Rev. D30, 1924 (1984)

(17) J.E. Gaiser et al., Phys. Rev. D34, 711 (1986).

(18) B. Lurz - Ph.D. Thesis DESY F31-86-04

(19) "Review of Particle Properties" - Phys.Lett. 170B, April 1986

(20) P. Schmitt - Ph.D. Thesis - DESY F31-86-XX

(21) "R.A. Lee - Ph.D. Thesis - SLAC 282 (1985), Appendix D.

(22) E.D. Bloom and C.W. Peck, Ann.Rev. Nucl. Part. Sci., 1983, 33, 143

(23) A. Bean et al., - CLNS 86/714.

NEW RESULTS ON ORDER AND SPACING OF LEVELS FOR TWO- AND THREE-BODY SYSTEMS

H. Grosse -- Institute for Theoretical Physics, University of Vienna
A. Martin -- Theory Division, CERN, Geneva
J.-M. Richard -- ILL, Grenoble
and
P. Taxil -- CPT, CNRS, Marseille

presented by A. Martin

We propose sufficient conditions on the potential binding a two-body system to compare i) the energy of a state with angular momentum $\ell+1$ to the average of the energies of the neighbouring states with angular momentum ℓ, ii) the spacings of the successive $\ell = 0$ excitations. Applications to quarkonium physics are given.

We also find a condition giving the sign of the parameter Δ controlling the pattern of levels obtained by perturbing the lowest positive parity excitation of a three-body system bound by harmonic oscillator two body forces.

1. "CLASSICAL RESULTS" ON TWO-BODY SYSTEMS BOUND BY A POTENTIAL

After many years of investigations, the following theorems were found on the order of levels in a central potential in 1984 (1). If $E(n,\ell)$ designates the energy of the level with angular momentum ℓ with a radial wave function with n nodes

1°) $E(n+1,\ell) \gtrless E(n,\ell+1)$

if $X = (d/dr)r^2(dV/dr) \gtrless 0 \quad \forall r > 0$.

X is proportional to the Laplacian of the potential and vanishes for a Coulomb potential. The property $\Delta V > 0$ is equivalent to say that if the force between the two particles is $-Z(r)/r^2$, $Z(r)$, the effective charge, increases with r, which is essentially "asymptotic freedom". For that reason, the quarkonium potential should satisfy $X > 0$. In this way we "understand" why $E_{\psi'} > E_{\chi_c}$, and $E_{\Upsilon''} > E_{\chi_b'} > E_{\Upsilon'} > E_{\chi_b} > E_{\Upsilon}$.

2°) $E(n+1,\ell) \gtrless E(n,\ell+2)$

if $Y = (d/dr)(1/r)(dv/dr) \gtrless 0 \quad \forall r > 0$.

Notice that $Y = 0$ for an r^2 potential. Lattice QCD potentials have the general properties that (2) $dV/dr > 0$, $d^2V/dr^2 < 0$, and hence $Y < 0$. In this way, we "understand" why $E_{\psi''}(\ell=2) > E_{\psi'}(\ell=0)$.

2. NEW RESULTS ON TWO-BODY SYSTEMS

In view of obtaining some results on the order of levels of three-body systems, we realized that it is useful to have some extra information not only on the order but also on the spacing of levels of two-body systems.

1°) One can compare the energy of an $\ell = 1$ state to the average of the energies of the two neighbouring $\ell = 0$ states. For a harmonic oscillator potential, the two coincide. In the limit of small perturbations around the harmonic oscillator, we have proved, using techniques developed in Ref. (3):

$2E(n,\ell+1) \gtrless E(n,\ell) + E(n+1,\ell)$

if $Y \gtrless 0 \quad \forall r > 0$,

where Y has the same definition as in 2°) of Section 1. We hope that this theorem also holds non-perturbatively.

2°) One can study the deviations from equal spacing for a given ℓ, in particular $\ell = 0$. If one looks at perturbations around the harmonic oscillator potential, one finds

$E(n+1,\ell=0)-E(n,\ell=0) \gtrless E(n,\ell=0)-E(n-1,\ell=0)$
if
$Z = (d/dr)r^5(d/dr)(1/r)(dV/dr) \gtrless 0 \quad \forall r > 0$,
and
$\lim_{r\to 0} r^3V = 0$.

Notice that the strange quantity Z is such that it vanishes for V = const. and $V = r^2$, which preserve equal spacing, and also $V = 1/r^2$, because adding such a term is equivalent to shifting by the same amount the angular momentum of all levels. Unfortunately, we now know that this result does not apply for too large perturbations which destroy the monotonicity of the potential.

All existing potentials, satisfy $Z < 0$. This means that the spacing between successive excitations of the $b\bar{b}$ system should decrease. In particular we should have $M_{\Upsilon IV} - M_{\Upsilon'''} < M_{\Upsilon'''} - M_{\Upsilon''}$, which does not seem to be the case. This seems to indicate the breakdown of the potential model above the $b\bar{b}$ threshold or a new phenomenon.

3. EARLY RESULTS ON THE THREE-BODY SYSTEM

It is known that a three-body system exhibits a high degeneracy if the two-body potentials are harmonic oscillator potentials. Deviation from the harmonic oscillator will remove this degeneracy and this is of special interest for heavy quark systems like (marginally) the excited states of the Ω^- (4) and also the ccc baryon whose study has been recommended by Bjorken (5).

Karl and Isgur (6) and also Horgan and Dalitz (7) have shown that the first degenerate excitation splits according to a pattern presented in the figure, which depends on a single parameter Δ. With some changes of notations Δ can be written as

$$\Delta = -C\int_0^\infty (r^2-r_1^2)(r^2-r_2^2)v(r)\exp-\lambda r^2 dr, \ C > 0$$

where v is the perturbing potential. It is then easy to prove that:

$$\Delta \gtrless 0 \quad \text{if} \quad Y \lessgtr 0.$$

Therefore, since $Y < 0$ is favoured by QCD, we expect $\Delta > 0$.

Higher levels seem to be much more difficult to study.

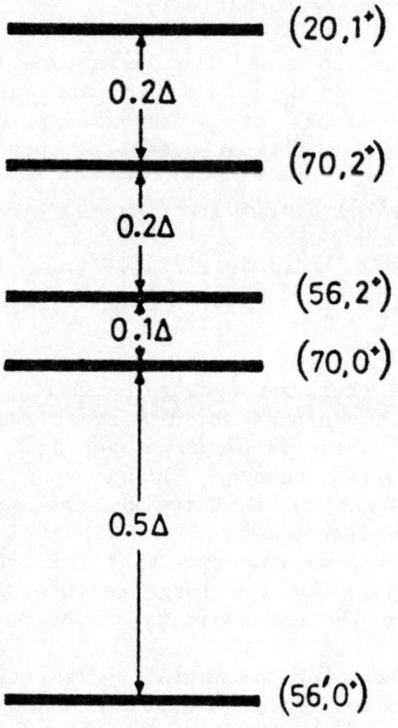

REFERENCES

(1) B. Baumgartner, H. Grosse and A. Martin, Phys. Lett. 146B, 363 (1984) and Nucl. Phys. B254, 528 (1985).

(2) See, for instance: C. Bachas, SLAC-pub-3814 (1985).

(3) H. Grosse and A. Martin, Phys. Lett. 134B, 368 (1984).

(4) S.F. Biagi et al., Univ. of Geneva preprint UGVA-DPNC 1985/12-113 (1985).

(5) J.D. Bjorken, Fermilab preprint, Fermilab-Conf. 85/69 (1985).

(6) N. Isgur and G. Karl, Phys. Rev. D19, 2653 (1979).

(7) R. Horgan and R.H. Dalitz, Nucl. Phys. B66, 135 (1973).

Parallel Session 10

Hadron Spectroscopy (including Gluonium)

Organizers:
N. Isgur (Toronto)
W. Toki (SLAC)

Scientific Secretaries:
T. Bolton (SLAC)
U. Malik (SLAC)

OBSERVATION OF $\eta_c \to \rho^\circ\rho^\circ$ AND REVIEW OF OTHER η_c DECAY MODES

Bernard JEAN-MARIE
Laboratoire de l'Accélérateur Linéaire,
Bât. 200, Centre d'Orsay, 91405 Orsay Cedex, FRANCE

Branching ratio measurements for η_c decays into $\phi\phi$, $K^+K^-\pi^\circ$, $K^0_S K^\pm\pi^\mp$, $p\bar{p}$ are presented. For the previously unobserved $\rho^\circ\rho^\circ$ mode a preliminary measurement is reported. The dominant contribution of the $\rho^\circ\rho^\circ$ intermediate state to the η_c decay into $4\pi^\pm$ is evidenced by a multichannel analysis and strongly supported by the study of pseudoscalar contribution and ρ mass cuts.

The measurements presented here use the 8.6 10^6 J/ψ produced in the DM2 detector at DCI the Orsay e^+e^- colliding rings. The magnetic detector[1], which will not be described, detects charged and neutral particles. The photon detector covers $.6\times4\pi$ steradians and its detection efficiency for the 114 MeV radiative photon associated to the η_c is 96 %.

The main difficulties to study exclusive η_c decays are due to the smallness of the η_c radiative production[2] (1.27 ± .36 %) and to the large phase space available allowing the η_c to desintegrate into many channels. In addition, the low energy radiative photon is difficult to separate in the shower detector from fake tracks induced by hadrons interacting with the detector material.

1. SUMMARY ON PREVIOUS MEASUREMENTS

The decays already measured by DM2 are summarized in table I. The signals corresponding to the first four channels are shown in figure 1. For the modes $K^0_S K^\pm\pi^\mp$, $K^+K^-\pi^\circ$ and $p\bar{p}$, the results[3] are in good agreement with published Mark III measurements[4]. Whereas for $\phi\phi$ the value is twice smaller. In addition the η_c mass determination by the two experiments differs by a few MeV. The widths observed are a convolution of the η_c width (measured to be 11 ± 4 MeV by Crystal Ball[2]) and the experimental resolution which is always much larger. So, no relevant information on η_c width can be extracted from the data.

The decays $\eta_c \to VV$ are of special interest since they provide a unique mean to determine the η_c spin parity. The analysis uses the formalism developped for Yang's parity test of the π°[3]. For a spin zero particle, the two vector decay planes are preferentially orthogonal or parallel according to whether they are produced by an odd or even parity intermediate state. Seven angles completely describe the sequential decay $J/\psi \to \gamma\eta_c \to \gamma\phi\phi \to \gamma K^+K^-K^+K^-$. But the most sensitive ones to spin parity values are the angle χ between the $\phi\phi$ decay planes and the polar angles Θ_{K^+} of each K^+ in the ϕ rest frame.

MODE	# EVENTS	MASS $_{MeV/c^2}$	$B(J/\psi\to\gamma\eta_c)\times B\times10^{-4}$
$\phi\phi$	23	2968 ± 5	.41 ± .09 ± .08
$K^+K^-\pi^\circ$	33	2972 ± 3	1.46 ± .30 ± .22
$K^0_S K^\pm\pi^\mp$	59*	2971 ± 5	2.30 ± .40 ± .60
$p\bar{p}$	14	2975 ± 5	.13 ± .03 ± .03
$\pi^+\pi^-\pi^+\pi^-$	137	2973 ± 3	1.33 ± .22 ± .20
$\rho^\circ\rho^\circ$	113		1.10 ± .20
$\Lambda\bar{\Lambda}$	–	–	< .8 (90% C.L.)

* on $7.10^6 J/\psi$

Table I

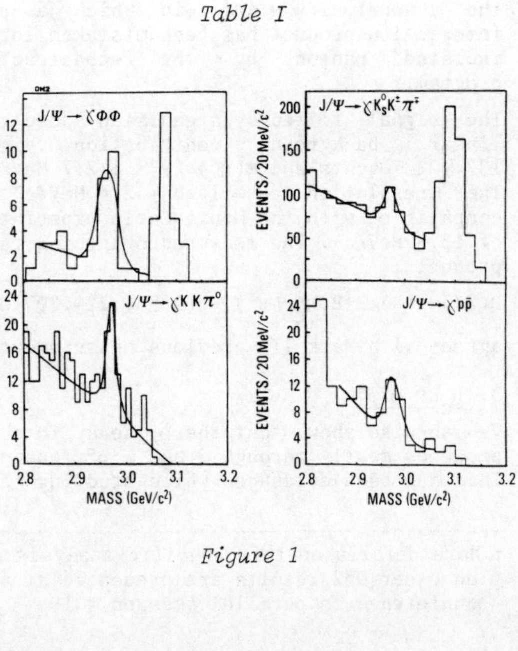

Figure 1

The η_c spin parity analysis has been first done by Mark III on the 16 $\phi\phi$ events of its η_c signal[5]. DM2 has performed the same analysis[6] using its 23 $\phi\phi$ events. The angular distributions are shown in figure 2 along with predictions for 0^- spin parity which is clearly favoured by the data confirming the previous determination.

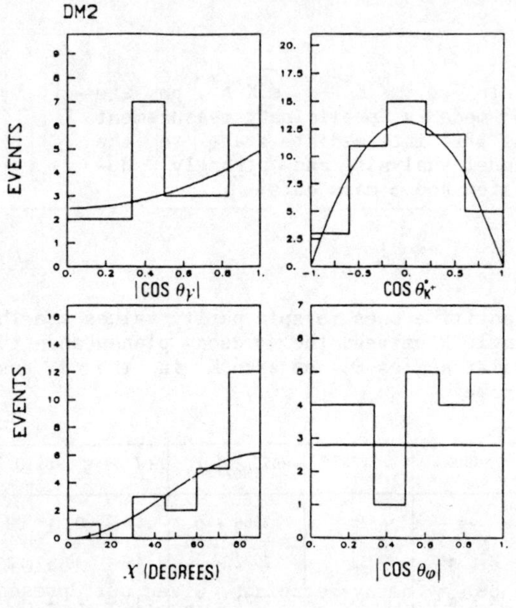

Figure 2

2. $\eta_c \to \pi^+\pi^-\pi^+\pi^-$

After selection of the candidate events† the largest remaining background, comes from $J/\psi \to 4\pi^\pm\pi^0$ in which the low energetic photon is undetected due to a very asymmetric π^0 decay. The resulting spectrum in figure 3.a shows a prominent peak at 3.1 GeV/c^2 corresponding to feedthrough from the channel $J/\psi \to 4\pi^\pm$ in which a pion interaction product has been mistaken for an isolated photon by the reconstruction program.

The signal fitted by a gaussian added to a linear background contribution yields 137 ± 23 events and $M(\eta_c)$=2973.1±2.7 MeV/c^2. The resolution σ = 14.8 ± 2.4 MeV/c^2 is compatible with the Monte-Carlo expectation of 13.3 MeV/c^2. The measured branching ratio product :

$$B(J/\psi\to\gamma\eta_c) \times B(\eta_c\to 4\pi^\pm) = (1.33\pm.22\pm.20)10^{-4}$$

agrees with Mark III previous measurement[4].

3. $\eta_c \to \rho^o\rho^o$

We can also show, that the η_c decay into $4\pi^\pm$ proceeds mostly through the $\rho^o\rho^o$ channel. This has been evidenced while studying the radiative mode $J/\psi \to \gamma\rho^o\rho^o \to \gamma\pi^+\pi^-\pi^+\pi^-$. A simple "$\rho^o$ CUT" approach and a more refined multichannel analysis have been used.

† More details on this specific analysis and on other DM2 results are presented at this conference in parallel session # 10.

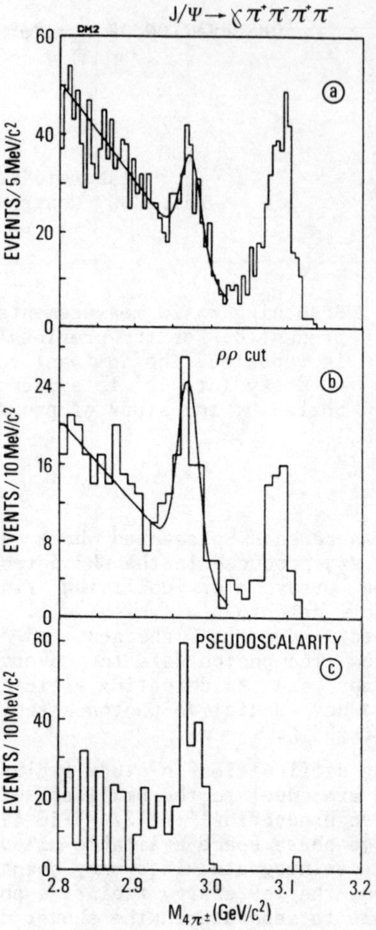

Figure 3

The first method selects $\rho^o\rho^o$ events by imposing a cut on the quantity :

$$R = \prod_{i=1,2} \frac{m_\rho^2\,\Gamma_\rho^2}{(m_i^2 - m_\rho^2)^2 + m_\rho^2\,\Gamma_\rho^2} > \frac{1}{15}$$

The result appears in figure 3.b. The 3.1 GeV fake peak is supressed since the decay $\psi \to \rho^o\rho^o$ is forbidden. The signal over background ratio at the η_c is increased. In addition the parameters of the fit to this spectrum are consistent with the previous ones.

For the multichannel analysis three modes are considered : $\rho^o\rho^o(0^-)$, $\rho^o\pi^+\pi^-$ and $4\pi^\pm$ phase space. For each event weighted probabilities are calculated. Then a maximum likelihood fit determines for all events the relative yield into each channel (Fig. 4). Clearly the $\rho\rho$ dynamic is dominant.

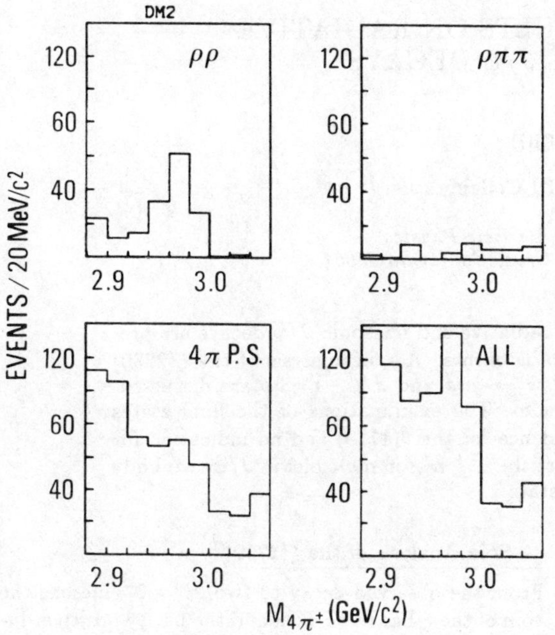

Figure 4

Some events can decay via phase space although they seem also compatible with 5π background events. Using this analysis one can derive a preliminary estimate of the branching ratio product :

$B(J/\psi\to\gamma\eta_c) \times B(\eta_c\to\rho^o\rho^o) = (1.1\pm.2)10^{-4}$

The pseudoscalar nature of the $\rho^o\rho^o$ signal can also be tested using a method similar to the one used to analyse the φφ system. The angular distribution of the χ angle integrated over the other angles takes the form $\frac{dN}{d\chi} = 1 + \beta(J,P) \cos 2\chi$ where β takes the following values :

J = ODD $\beta = 0$ J = even $-1 \leq \beta \leq +1$

in particular if $J^P = 0^-$ $\beta = -1$ and the distribution becomes $\frac{dN}{d\chi} \propto \sin^2\chi$. On the contrary, one can show that background events $J/\psi \to 4\pi^\pm\pi^o$ have a flat χ distribution. In figure 5 the χ distribution is shown in three mass intervals for all events and for the events selected by the $\rho^o\rho^o$ cut. Clearly the η_c events exhibit a strong $\sin^2\chi$ behavior compared to the 3.1 GeV region which is rather flat, as expected, and to the events with masses lower than the η_c which are heavily contaminated by $J/\psi \to 4\pi^\pm\pi^o$ events. The $\rho^o\rho^o$ cut emphasized these trends.

Finally a fit to the fraction of pseudoscalar events done by 10 MeV/c² mass bins is shown in figure 3.c.

Notice that this preliminary study brings the best evidence to date of the pseudoscalar nature of the η_c.

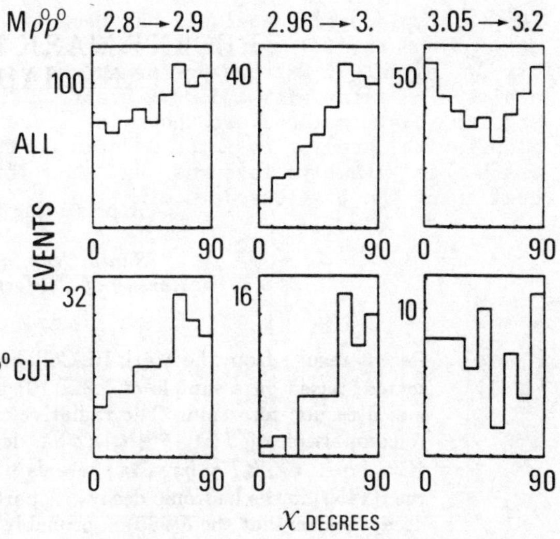

Figure 5

4. CONCLUSIONS

The measurements of $\eta_c \to$ Vector-Vector decays, which can proceed through three different diagrams is important to learn about η_c decay mecanisms. Given our present results, the φφ decay mode does not appear to be larger than the $\rho^o\rho^o$ mode as it was suggested by the Mark III upper limit on $\eta_c \to \rho^o\rho^o$. Using the formalism developped in reference 7, we find :

$$\frac{B(\eta_c \to \rho^o\rho^o)}{B(\eta_c \to \phi\phi)} \times \frac{P^{*3}_{\phi}}{P^{*3}_{\rho}} = 1.66 \pm .5$$

This value is compatible with unity as it should be if flavour SU(3) symmetry was exact.

REFERENCES

1) AUGUSTIN J.E. et al., Physica Scripta 23(1981)623-633.
2) GAISER J.E., Ph. D. Thesis SLAC-255 (1982) unpublished.
3) TRUEMAN T.L., Phys. Rev. D18(1978) 3423.
4) AUGUSTIN J.E. et al., Contributed paper to the BARI Conference, LAL 85/27.
5) BALTSURAITIS R.M. et al., Phys. Rev. D33(1986)629.
6) BALTSURAITIS R.M. et al., Phys. Rev. Lett. 52(1984)2126.
7) BISELLO D. et al., LAL/86-18 (to be published in Physics Letters B).
8) HABER H.E. and PERRIER H., Phys. Rev. D32(1985)2961.

RECENT MARK III RESULTS ON RADIATIVE AND HADRONIC J/ψ DECAYS

LUTZ KÖPKE

Representing the Mark III Collaboration

Santa Cruz Institute for Particle Physics
University of California Santa Cruz, California 95064

Recent results from the Mark III Collaboration on radiative and hadronic J/ψ decays are presented, based on a sample of 5.8×10^6 produced J/ψ events. A spin analysis of the $\xi(2230)$ indicates non-zero spin. The radiative channels $J/\psi \to \gamma\phi\phi$ and $J/\psi \to \gamma\omega\phi$ are discussed. A comparison of $J/\psi \to \{\gamma, \omega, \phi\}X$ decays is made. The examinations of the final states $X = \{K\bar{K}, \pi\pi, K\bar{K}\pi, \eta\pi\pi, 4\pi\}$ reveals possible evidence for the $\theta(1720)$ and no indication for the $\iota(1450)$ in the hadronic decays. A partial study of the 0^{++} meson multiplet in J/ψ two-body decays shows that the $\delta(980)$ is probably not a $q\bar{q}$ state.

INTRODUCTION

Interest has lately been centered on radiative J/ψ decays since this mode is considered to be a likely channel to find bound states of gluons. Indeed, almost every final state studied revealed new and interesting phenomena, such as the $\iota(1450)$, $\theta(1720)$, $\xi(2230)$ and pseudoscalar structures decaying into $\omega\omega$ and $\rho\rho$. However, the observation of a state in the radiative J/ψ decay does not necessarily mean that it is made of gluonic matter; after all, well-known $q\bar{q}$ resonances have also been seen in this "gluon-enriched" environment. Lacking specific predictions for the production and decay of "glueballs", a systematic comparison of radiative decays with those produced by other mechanisms may disentangle this situation. Especially interesting are flavor dependent channels of the type $J/\psi \to \omega X$ and $J/\psi \to \phi X$, as one expects a quark correlation between the vector meson and the recoiling system X. In some cases the interpretation of gluonia candidates is hindered by incomplete knowledge of the $q\bar{q}$ sector, and J/ψ two-body decays can be used to study the $q\bar{q}$ spectroscopy. Many of the results discussed in this paper are preliminary and await a final analysis.

STATUS OF THE ξ

The observation of the $\xi(2230)$, a narrow massive state, was one of the true surprises in the study of radiative J/ψ decays.[1] Enhancements were seen in the K^+K^- (Fig. 5a) and K_sK_s final states (Fig. 1) with a significance of 4.5 and 3.6 s.d., respectively. The resonance parameters were measured as $m_\xi = 2230 \pm 6 \pm 14$ MeV, $\Gamma = 26^{+20}_{-16} \pm 17$ MeV, and $B(J/\psi \to \gamma\xi) \cdot B(\xi \to K^+K^-) = (4.2^{+1.7}_{-1.4} \pm 0.8) \times 10^{-5}$ in the K^+K^- mode and $m_\xi = 2232 \pm 7 \pm 7$ MeV, $\Gamma = 18^{+23}_{-15} \pm 10$ MeV, and $B(J/\psi \to \gamma\xi) \cdot B(\xi \to K_sK_s) = (3.1^{+1.6}_{-1.3} \pm 0.7) \times 10^{-5}$ in the K_sK_s mode.

One possible interpretation of the $\xi(2230)$ is that it is a $L = 3$ $s\bar{s}$ meson with $J^{PC} = 2^{++}$.[2] The interpretation of the ξ as $q\bar{q}g$ (hybrid) state predicts prominent decay modes to K^*K^* and $\omega\phi$.[3]

A Spin Analysis of the $\xi(2230)$[4]

From the observed decay to two $J^p = 0^-$ mesons the spin of the ξ has to be even and the parity positive. Inspite of its smaller statistics, the K_sK_s final state was chosen since the background in this channel is small (see Fig. 1). A maximum likelihood technique employing all angular correlations was applied in 4 mass regions, representing the f', θ, the wide structure underneath the ξ, and the ξ. The angle θ_k of one kaon in the $K\bar{K}$ rest-system w.r.t. the pairs direction is shown in Fig. 2. Neither distribution looks flat as expected for $J = 0$, hence indicating higher spin. Note that the angular distributions in the 2.1 GeV and ξ mass regions look different, supporting the evidence for the $\xi(2230)$. The detailed maximum likelihood analysis shows preference for spin 2 or 4 except in the θ mass region where both spin 0 and 2 are allowed. The results, including the values for the helicity amplitudes, confirm previous results from the K^+K^- channel.

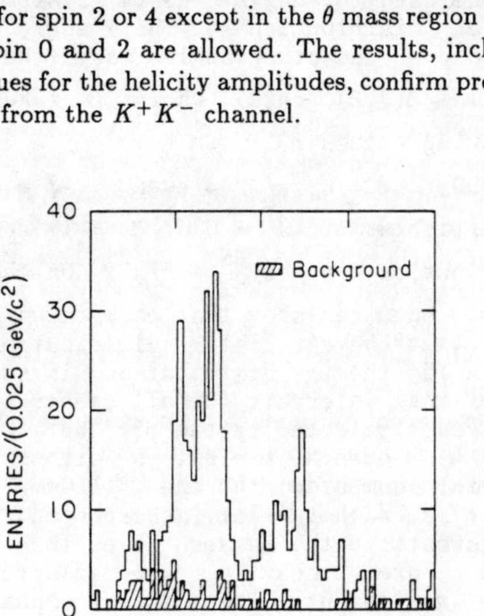

Fig. 1: K_sK_s mass spectrum from $J/\psi \to \gamma K_sK_s$ decay.

Since the ξ signal rests on a broad enhancement with spin 2 or higher, we investigated the possibility that a scalar ξ plus the contribution from the underlying structure can reproduce the measured distribution. The probability that the ξ is a scalar was found to be $\approx 10\%$, roughly independent of the assumed amount of background.

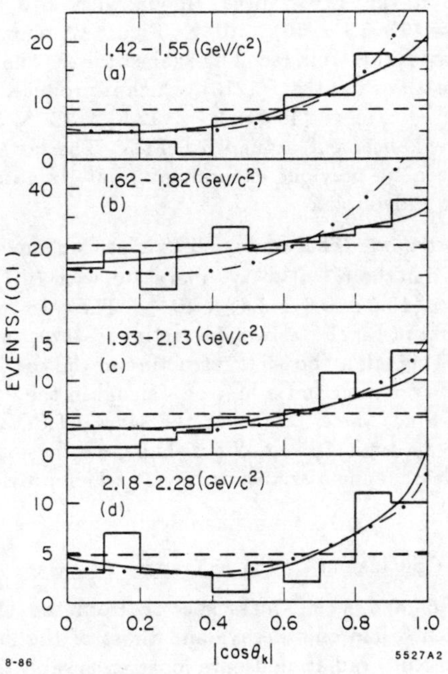

Fig. 2: Distribution of $|\cos\theta_K|$ for four K_sK_s-mass intervals. Fit results for $J = 0, 2, 4$ are represented by dashed, solid, and dashed dotted lines, respectively.

$J/\psi \to \gamma\phi\phi \to \gamma K^+K^-K^+K^-$

This channel is difficult to study experimentally since four low momentum kaons have to be detected. In order to increase the efficiency for events with decaying kaons, the worst measured track was disregarded and a 1C fit to the $\gamma K^+K^-K^+K^-$ hypothesis was applied. By this method the efficiency was almost doubled in the 2.2 GeV region. Fig. 3 shows the $\phi\phi$ invariant mass spectrum. The efficiency drops smoothly from $\approx 28\%$ around 2.7 GeV to $\approx 10\%$ at 2.2 GeV. The background from events not containing 2 ϕ's contributes roughly 0.5 events to each mass bin. In addition to a signal at the η_c mass there is evidence for $\phi\phi$ production at low $m_{\phi\phi}$ with $B(J/\psi \to \gamma\phi\phi) = (4.0 \pm 0.5 \pm 0.8) \times 10^{-4}$ below 2.4 GeV.

$J/\psi \to \gamma\omega\phi \to \gamma\pi^+\pi^-\pi^0K^+K^-$

The $J/\psi \to \gamma\omega\phi$ decay, which is doubly OZI suppressed for $q\bar{q}$ states, was observed with $B(J/\psi \to \gamma\omega\phi) = (1.40 \pm 0.25 \pm 0.28) \times 10^{-4}$. The $\omega\phi$ mass spectrum in Fig. 4 does not show significant structure leading to $B(J/\psi \to \gamma\xi) \cdot B(\xi \to \omega\phi) < 5.9 \times 10^{-5}$ and $B(J/\psi \to \gamma\eta_c) \cdot B(\eta_c \to \omega\phi) < 1.3 \times 10^{-5}$ at 90% C.L.

The measured set of all $\eta_c \to 1^{--}1^{--}$ decays can be used to determine the relative contributions of the three mechanisms relevant to η_c decays.[5] The observed decay pattern indicates that the $\eta_c \to 1^{--}1^{--}$ decay rate increases with the number of strange quarks in the final state, a SU(3)-breaking pattern very different from the one observed in $J/\psi \to 1^{--}0^{-+}$ decays.[6] Since the η_c mainly hadronizes via two gluon exchange, the study of its decays should also further the understanding of gluonia decays.

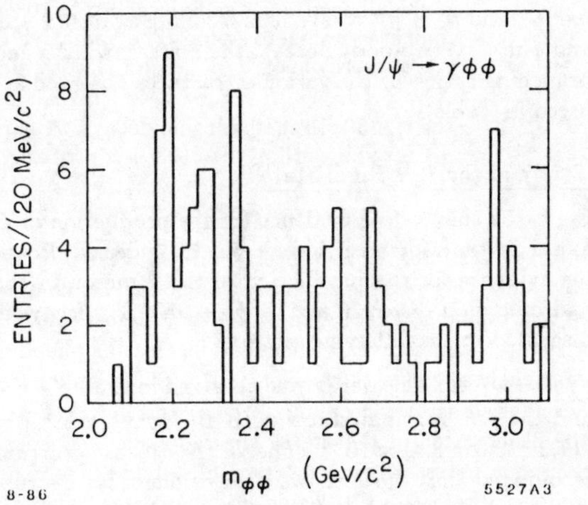

Fig. 3: $\phi\phi$ mass spectrum from $J/\psi \to \gamma\phi\phi \to 4K^\pm$ reaction.

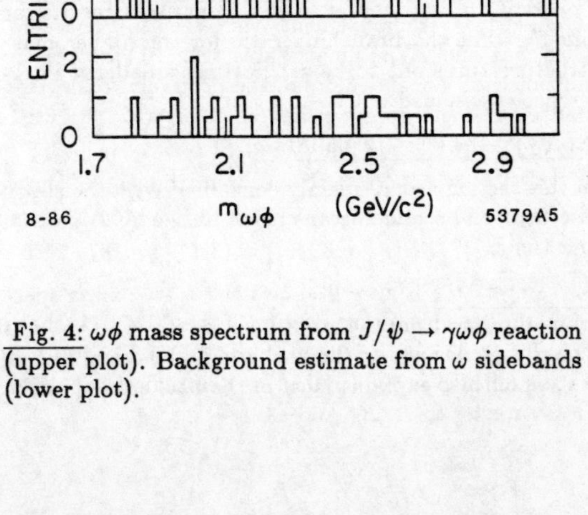

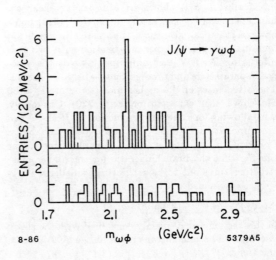

Fig. 4: $\omega\phi$ mass spectrum from $J/\psi \to \gamma\omega\phi$ reaction (upper plot). Background estimate from ω sidebands (lower plot).

Comparison Between Radiative and Hadronic J/ψ Decays

As explained in the introduction, a comparison between corresponding radiative and hadronic decays might be very helpful to learn more about the "glueball" candidates θ and ι, as well as other structures observed in radiative J/ψ decays. The hadronic two-body decays containing either an ω or a ϕ are especially useful since the vector mesons tag the flavor of the recoiling systems if the quark contents are correlated. This assumption is confirmed by the large suppression of $J/\psi \to \omega f'$ compared to $J/\psi \to \omega f$. In the following section, such comparisons are made for the K^+K^-, $\pi^+\pi^-$, $K\bar{K}\pi$, $\eta\pi^+\pi^-$ and $\pi^+\pi^-\pi^+\pi^-$ systems recoiling against a γ, ω and ϕ in J/ψ two-body decays; the masses, widths and branching ratios for the various structures observed are listed in Table 1.

Study of the $K\bar{K}$ Final State[7]

Figure 5a shows clear evidence for the production of f' and θ in the radiative $J/\psi \to \gamma K^+K^-$ decay. Possible evidence for the production of the θ meson in the hadronic $J/\psi \to \omega K\bar{K}$ and $J/\psi \to \phi K^+K^-$ decays is discussed in the following section.

$J/\psi \to \omega K\bar{K}$: This decay was observed in the ωK^+K^- and the $\omega K_s K_s$ final states with $B(J/\psi \to \omega K\bar{K}) = (17.2 \pm 0.8 \pm 3.4) \times 10^{-4}$. The K^+K^- mass spectrum recoiling against the ω shows clear evidence for a structure around 1730 MeV (Fig. 5b), which is consistent with the θ parameters (Table 1). The mass spectrum in Fig. 5b shows no signal for the $J/\psi \to \omega f'$ decay; an upper limit for incoherent f' production is included in Table 1. The K_sK_s mass spectrum, with much less statistics, confirms the features of the K^+K^- channel.

$J/\psi \to \phi K^+K^-$: The K^+K^- mass distribution recoiling against a ϕ is shown in Fig. 5c. A clear enhancement can be seen at the nominal $f'(1520)$ mass, however, the peak also shows a statistically significant high mass shoulder. A parametrization with two non-interfering Breit-Wigner amplitudes yields the resonance parameters and production rate listed in Table 1. The fitted mass of the higher mass structure is ≈ 50 MeV lower than expected for the $\theta(1720)$. Conversely, a fit allowing for coherent production of f' and θ can accommodate a "standard θ" with good probability. Coherent and incoherent fit give consistent results for the f', while the branching ratio for the higher mass structure turns out to be ≈ 2.5 times smaller if interference is assumed.

Study of the $\pi^+\pi^-$ Final State

In this section we are mainly concerned with the study of tensor and scalar mesons, such as the f, θ, the S^*, and the ϵ.

$J/\psi \to \gamma\pi^+\pi^-$: Figure 6 shows the $\pi^+\pi^-$ mass spectrum obtained from the radiative decay $J/\psi \to \gamma\pi^+\pi^-$. The ρ^0 signal is due to the background reaction $J/\psi \to \rho\,\pi$, which also explains most of the background in the 1.6 - 2.8 GeV region. The f signal, as well as the structures at 1713 ± 15 MeV and 2086 ± 15 MeV, are clearly correlated with a radiative photon. While the parameters of the second structure are consistent with the θ mass and width measured in the K^+K^- channel, the interpretation of the third peak is still unclear.

$J/\psi \to \omega\pi^+\pi^-$: The $J/\psi \to \omega\pi^+\pi^-$ decay was observed in the $4\pi^\pm\pi^0$ decay mode with $B(J/\psi \to \omega\pi\pi) = (78 \pm 1 \pm 16) \times 10^{-4}$. Figure 6b shows the $\pi^+\pi^-$ mass spectrum recoiling against the ω. The dominant features are the $f(1270)$, which is produced with $B(J/\psi \to \omega f) = (49.3 \pm 2.5 \pm 12.5) \times 10^{-4}$, and a broad enhancement around 500 MeV. The latter has been seen by previous experiments, but its nature is not yet understood.

$J/\psi \to \phi\pi^+\pi^-$: This doubly OZI violating channel was observed in the $K^+K^-\pi^+\pi^-$ final state, with $B(J/\psi \to \phi\pi\pi) = (13.5 \pm 0.6 \pm 3.4) \times 10^{-4}$. The $\pi^+\pi^-$ mass spectrum in Fig. 5 c shows interesting features: a clear $S^*(975)$ signal, a "box-like" structure in the the 1200 - 1500 MeV region and a hint of a signal in the θ mass range. Since the S^* signal in the Mark III data does not have a typical Breit-Wiger shape, we chose to use a coupled channel parametrization,[8] which provided a good fit.

Conclusions on the $K\bar{K}$ and $\pi^+\pi^-$ Data

The $K\bar{K}$ and $\pi^+\pi^-$ mass spectra from the "flavor dependent" hadronic decays and those of the "flavor independent" radiative decays look remarkably different. The OZI allowed channels $J/\psi \to \omega\pi^+\pi^-$ and $J/\psi \to \phi K^+K^-$ are dominated by the established spin two $q\bar{q}$ states f and f', while these are largely suppressed in the case of the OZI violating decays $J/\psi \to \omega K\bar{K}$ and $J/\psi \to \phi\pi^+\pi^-$.

The Mass Region around 1.00 GeV: While the $\pi^+\pi^-$ mass spectrum from the $J/\psi \to \phi\pi^+\pi^-$ decay is dominated by the $S^*(975)$ resonance (Fig. 5c), only a hint of a signal is seen in the corresponding spectrum recoiling against the ω. From this we conclude that the S^* has a large s-quark content. No S^* signal is seen in the radiative decays, corresponding to $B(J/\psi \to \gamma S^*) \times B(S^* \to \pi\pi) < 7 \times 10^{-5}$ at 90% C.L.

The Mass Region around 1.45 GeV: The $\pi\pi$ structure between 1.20 and 1.45 GeV seen recoiling against a ϕ (Fig. 5c) looks too "box-like" to be caused by a single resonance. Whether the f, the scalar "ϵ", or more exotic states contribute to the enhancement is still under investigation; additional information from a spin-parity analysis may prove to be essential for a conclusion.

While no sign of the 1.4 GeV structure is apparent in the $J/\psi \to \omega\pi^+\pi^-$ reaction, the radiatively produced f shows an interesting shoulder in this mass range. This structure could be an indication of $f - f'$ interference in the "flavor independent" radiative channel or a manifestation of the enhancement seen in the OZI-violating $J/\psi \to \phi\pi^+\pi^-$ reaction.

The Mass Region around 1.70 GeV: The radiatively produced K^+K^- and $\pi^+\pi^-$ mass spectra show clear evidence for the $\theta(1720)$ resonance, with relative rates of $\approx$ 3/0.8. Structures consistent with the θ are also observed in the $K\bar{K}$ spectra obtained from the $J/\psi \rightarrow \omega K\bar{K}$ and $J\psi \rightarrow \phi K^+K^-$ reactions (Fig.5), and possibly a signal in the $\pi^+\pi^-$ mass spectrum recoiling against a ϕ (Fig. 6c). If all of these enhancements are manifestations of the θ, it appears to be more SU(3) symmetric in production and decay than either the f or f'.

The observation of the θ in hadronic J/ψ decays does not rule out its interpretation as a "glueball". It is expected that gluon bound states mix with $q\bar{q}$ states. Consequently, the θ could couple via its u, d, and s-quark admixtures. However, higher order diagrams may allow the excitation of the θ in hadronic J/ψ decays, even if the θ is entirely gluonic.

Comparing the production rates of θ, f and f' in both radiative and hadronic J/ψ decays, one finds a relative preference for the θ in radiative processes.

Study of the $K\bar{K}\pi$ Final State[9]

In this study we are mainly interested in the study of the E/ι mass region. The $\iota(1450)0^{-+}$ state, which is produced with the largest branching ratio in radiative J/ψ decays[10], is a prime "glueball" candidate. The $E(1420)$, has long been known from hadronic interactions. First seen in $p\bar{p}$ annihilations at rest, its spin parity was determined as 0^-. Later measurements, using other reactions, indicated $J^P = 1^+$, while the latest measurement of Chung et al. again points to spin 0. This result has led to the speculation that E and ι are one and the same object. With the large data set of J/ψ decays available we have the chance to address the E/ι question within one experiment.

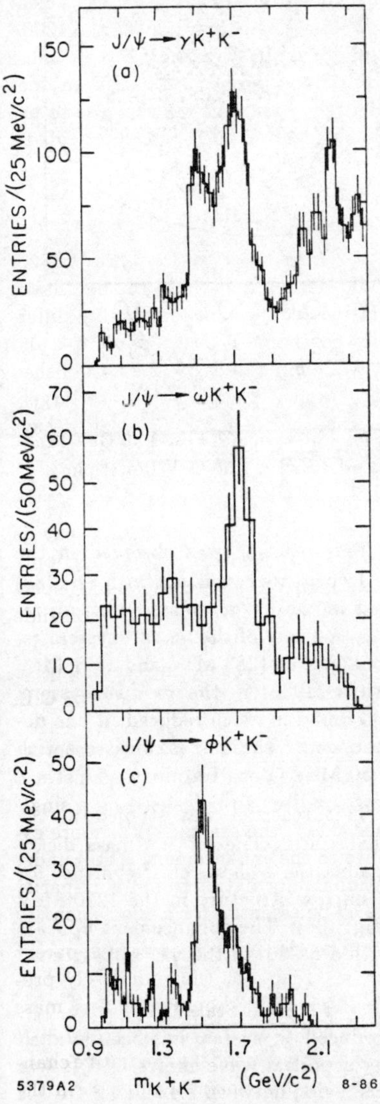

Fig. 5: K^+K^- mass distributions.

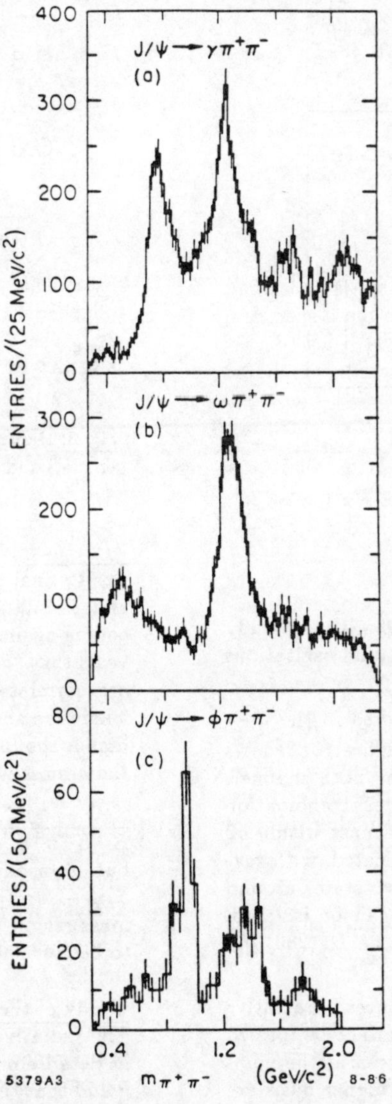

Fig. 6: $\pi^+\pi^-$ mass distributions.

$J/\psi \rightarrow \gamma K\bar{K}\pi$: The radiative channel was updated by using all available data and studying the $K^\pm K_s\pi^\mp$, $K^+K^-\pi^0$ and $K_sK_s\pi^0$ decay modes. Clear ι signals are seen (Fig 7); however, fits using simple Breit-Wigner parametrizations do not describe the data. Either more than one resonance contributes to the signal, or the shape is distorted by phase-space effects due to the $K\bar{K}\pi$ substructure. Fig 8 shows Dalitz plots for the lower and upper half of the ι signal, clearly indicating K^* bands at least in the higher mass region. Due to the rapidly rising phase space at K^*K threshold, a substantial K^*K decay mode may account for the observed asymmetry of the ι mass peak.

$J/\psi \rightarrow \omega K\bar{K}\pi$: This reaction was studied using the $\omega K^\pm K_s\pi^\mp$ and $\omega K^+K^-\pi^0$ final states. Both $K\bar{K}\pi$ mass spectra show clear evidence for a state in the E/ι mass range with consistent resonance parameters and branching ratios. From the combined set of both channels we obtained the mass spectrum displayed in Fig. 9b and the resonance parameters listed in Table 1. The width, 24 MeV $< \Gamma <$ 84 MeV, is hardly compatible with that of the ι. Due to background, the study of the $K\bar{K}\pi$ substructure is difficult. Still, a comparison with the ι reveals differences, *e.g.* a less pronounced peaking of the $K\bar{K}$ submass at $K\bar{K}$ threshold. The data suggest non-zero spin.

$J/\psi \to \phi K\bar{K}\pi$: The reaction $J/\psi \to \phi K^{\pm}K_s\pi^{\mp}$ was studied using the K^+K^- and K_sK_l decay modes of the ϕ. Figure 9c shows the added $K^{\pm}K_s\pi^{\mp}$ mass spectra, with no indication for the production of a resonance in the E/ι mass range. The corresponding upper limit is listed in Table 1 and is roughly a factor of 6 below the result from the $\omega K\bar{K}\pi$ channel.

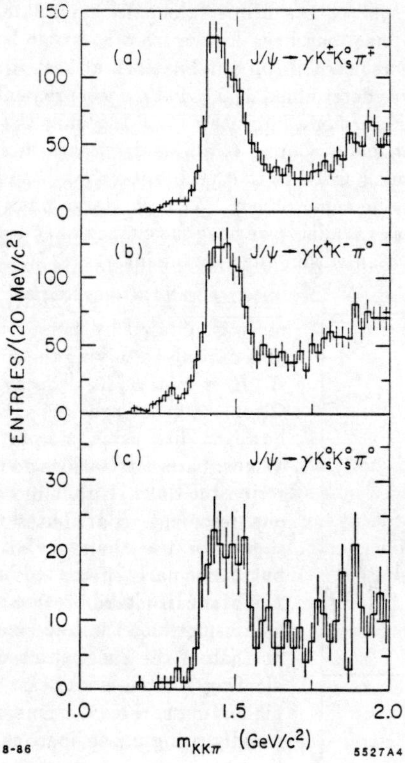

Fig. 7: $\iota(1450)$ signals in three $K\bar{K}\pi$ modes.

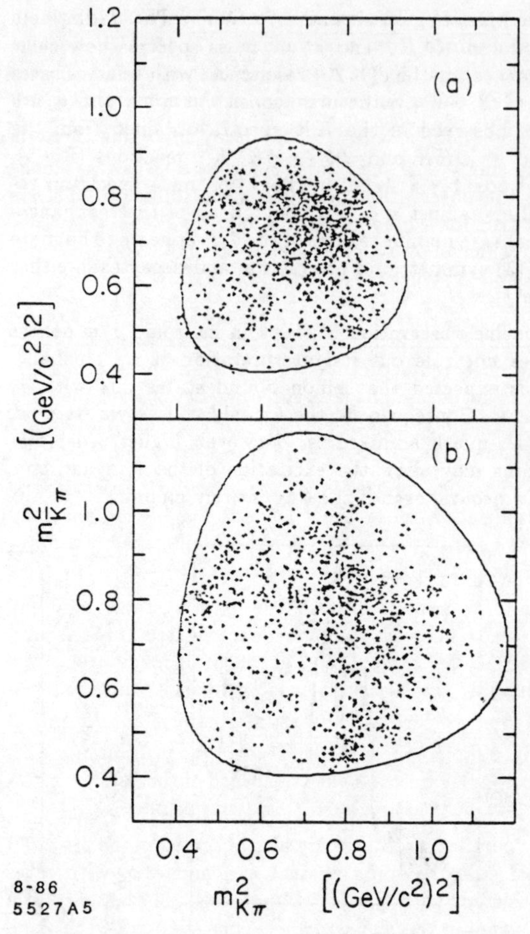

Fig. 8: Dalitz-plots for $1.36 < m(K\bar{K}\pi) < 1.46$ GeV (upper plot) and $1.46 < m(K\bar{K}\pi) < 1.58$ GeV (lower plot).

Study of the $\eta\pi\pi$ Final State[11]

In this section we are mainly concerned with the study of the D and E mesons, and of the radial excitations of the η and η'.

$J/\psi \to \gamma\eta\pi^+\pi^-$: This decay was studied in the $\eta \to \gamma\gamma$ and $\eta \to \pi^+\pi^-\pi^0$ decay modes. Figure 10a shows the $\eta\pi^+\pi^-$ mass spectrum after adding both channels and demanding that at least one $\eta\pi^{\pm}$ mass combination is consistent with the nominal $\delta(980)$ mass within 50 MeV. The spectrum looks very complicated, with several resonances contributing to it. Two states around 1285 MeV and 1390 MeV, however, can be isolated. The first structure is consistent with the $D(1283)$ parameters.

The observation of spin one resonances in radiative decays would be interesting as they cannot be formed by two massless gluons. No signal is seen at the nominal ι mass. The broad structures at higher mass are intriguing, but an interpretation has to wait until a spin parity analysis has been completed.

$J/\psi \to \omega\eta\pi^+\pi^-$: This reaction was observed in the $4\pi^{\pm}4\gamma$ final state. Two states consistent with $D(1283)$ and $E(1420)$ appear in the $\eta\pi^+\pi^-$ mass spectrum recoiling against an ω. A study of the $\eta\pi^+\pi^-$ system reveals that the resonances at 1283 MeV and 1421 MeV are correlated with a $\delta(980)$ in the $\eta\pi^{\pm}$ subsystem. Therefore the background is much reduced, if one demands that at least one $\eta\pi^{\pm}$ submass is consistent with the δ mass within 50 MeV (Fig. 10b).

$J/\psi \to \phi\eta\pi^+\pi^-$: This reaction was studied in the $K^+K^-\pi^+\pi^-\gamma\gamma$ final state. The $\eta\pi^+\pi^-$ mass distribution recoiling against the ϕ shows clear evidence for $J/\psi \to \phi\eta'$ and a narrow structure in the 1280 MeV mass region (see Fig. 10c). The enhancement appears to be correlated with a $\delta(980)$ in the $\eta\pi^{\pm}$ subsystem.

Study of the $\pi^+\pi^-\pi^+\pi^-$ Final State[12]

The radiatively produced 4π system has been studied in detail employing a partial wave analysis.[12] It was found that the mass region between 1.4 and 1.8 GeV is dominated by pseudoscalar states decaying to $\rho\rho$ with $B(J/\psi \to \gamma\pi^+\pi^-\pi^+\pi^-) = (30.5 \pm 1.8 \pm 4.5) \times 10^{-4}$

below 2 GeV. The mass spectrum in Fig. 11a, which was obtained from a sample of 2.7×10^6 J/ψ decays, shows two $\approx$ 150 MeV wide structures around 1.5 and 1.8 GeV and a hint of a signal at the nominal D mass.

$J/\psi \to \{{}^{\omega}_{\phi}\}\pi^+\pi^-\pi^+\pi^-$: The decay involving the ω is only singly OZI suppressed and has a large rate, making the observation of small effects difficult. This is not true for the doubly OZI violating $J/\psi \to \phi 4\pi^{\pm}$ decay which shows a clear signal around 1.280 GeV and possibly a structure around 1.5 GeV.

Conclusions on the $K\bar{K}\pi$, $\eta\pi\pi$, and $\pi^+\pi^-\pi^+\pi^-$ Data

The observed structures in the mass spectra of Figs. 9-11 can be assigned to three mass regions: the "D" region around 1.28 GeV, the "E/ι" region around 1.4 GeV, and the mass range between 1.5 and 1.9 GeV.

The interpretation of the observed structures is complicated, since the "D" and "E/ι" regions may have both 0^{-+} and 1^{++} contributions rather than being dominated by single resonances.[13]

The "D" Mass Region: The $\eta\pi\pi$ systems recoiling against γ, ω, and ϕ all show narrow structures around 1280 MeV (Table 1). The natural candidate is the $D(1283)$, a well established resonance which is classified as a near $\frac{u\bar{u}+d\bar{d}}{\sqrt{2}}$ state in an almost ideally mixed axial vector nonet. However, the observation of a "D" recoiling against a ϕ with one third of the $J/\psi \to \omega$"D" rate is inconsistent with this assumption.

Similar structures are also observed in the $2(\pi^+\pi^-)$ final state recoiling against a γ and a ϕ, with rates consistent with the "D" hypothesis. The absence of a D signal in the $K\bar{K}\pi$ mode is consistent with the small $D \to K\bar{K}\pi$ decay rate.

The "E/ι" Mass Region: The $\iota(1460)$ is clearly observed in the radiatively produced $K\bar{K}\pi$ system (Fig. 9a). Using one mass bin from 1.34 -1.58 GeV, its spin-parity was determined to be 0^-. The mass of the $K\bar{K}$ subsystem is concentrated at $K\bar{K}$ threshold, however, a substantial K^*K decay mode is also seen.

No enhancement is seen at the nominal ι mass in the $\eta\pi^+\pi^-$ final state. Either the ι peak is deformed and shifted by interference with other resonances or non-resonant $\eta\pi\pi$ production,[14] or the ι does not decay via $\delta\pi$ and the $K\bar{K}$ threshold enhancement is of dynamical origin.[15]

The radiatively produced 4π mass spectrum shows an enhancement around 1.5 GeV, 50 MeV higher than the ι. In a coupled channel analysis MarkIII has demonstrated that the structure could be due to the $\iota \to \rho\rho$ decay where the observed mass spectrum is shifted because of the limited phase space available.

The masses of the other structures seen in the 1400 MeV mass region vary between 1.395 and 1.445 GeV, while their widths all center around 50 MeV (see Table 1). Independent of whether the structure is due to one or more resonances, one can make the following comments:

- The width of the structures are considerably narrower than the ι width, indicating that the ι and "E" are different objects. This conclusion is supported by a study of the $K\bar{K}\pi$ submasses and the angular distributions in $J/\psi \to \omega K\bar{K}\pi$.
- The "E"-like structures probably contribute to the ι peak observed in $J/\psi \to \gamma K\bar{K}\pi$, however, this contribution is expected to be small compared to the large production rate of the ι.
- The absence of $J/\psi \to \phi$"E" decays indicate that it is not a pure $s\bar{s}$ state.

The Mass Range between 1500 and 1900 MeV: In this mass range, the radiatively produced $\eta\pi\pi$ system is dominated by broad structures which cannot be associated with known $q\bar{q}$ resonances. The structures are not seen in the $J/\psi \to \gamma K\bar{K}\pi$ channel, perhaps indicating a small $s\bar{s}$ content. It is tempting to associate the structures to similar pseudoscalar enhancements observed in the radiatively produced $\rho\rho$ and $\omega\omega$ systems,[12] however, the structure in $\eta\pi^+\pi^-$ extends to higher masses.

SYSTEMATIC STUDIES OF MESONS IN J/ψ HADRONIC DECAYS

J/ψ two-body decays are well suited to study mesonic resonances in a systematical way. Especially interesting are decays of the type $J/\psi \to 1^{--}X$ where the well-known vector mesons serve as analyzers for the recoiling states X. Being particularly interested in the η, η' quark structure, the Mark III Collaboration studied the pseudoscalar states using the 11 allowed combinations of $J/\psi \to 1^{--}0^{-+}$ decays.[6] The measured decay rates were related using a model based on SU(3) invariance which allowed for SU(3) breaking through electromagnetic transitions and quark mass differences.[5] The Mark III Collaboration is planning to study the tensor ($J^{PC} = 2^{++}$) and scalar ($J^{PC} = 0^{++}$) mesons in a similar fashion. First results on the scalar multiplet are presented in the following paragraph.

The Scalar Nonet

The states traditionally assigned to the scalar $q\bar{q}$ nonet are the $\kappa(1430)$, $\delta(980)$, "$\epsilon(1400)$", and the S^*. While the evidence for the broad ϵ is weak, the S^* and the δ have been firmly established for some time. In the quark model, their natural assignment is the singlet and triplet states of the 1^3P_0 nonet with $J^{PC} = 0^{++}$, however, there are problems with this interpretation.[16]

Other than $q\bar{q}$ states, the S^* and δ have been speculated to be 4-quark states[17] or loosely bound $K\bar{K}$ molecules.[18] These three interpretations lead to different predictions for the production of S^* and δ in hadronic J/ψ decays. The classification of the S^* as an $s\bar{s}$-state implies that $B(J/\psi \to \omega S^*) \approx 0$, and, neglecting phase space and SU(3) breaking effects, $B(J/\psi \to \phi S^*) = B(J/\psi \to \rho^0\delta^0)$. SU(3) breaking effects are expected to suppress the $J/\psi \to \phi S^*$ decay, similar to the case of the tensor mesons, where $B(J/\psi \to \rho^0 A_2^0) \approx 5 \times B(J/\psi \to \phi f')\cdot B(f' \to K\bar{K})$.

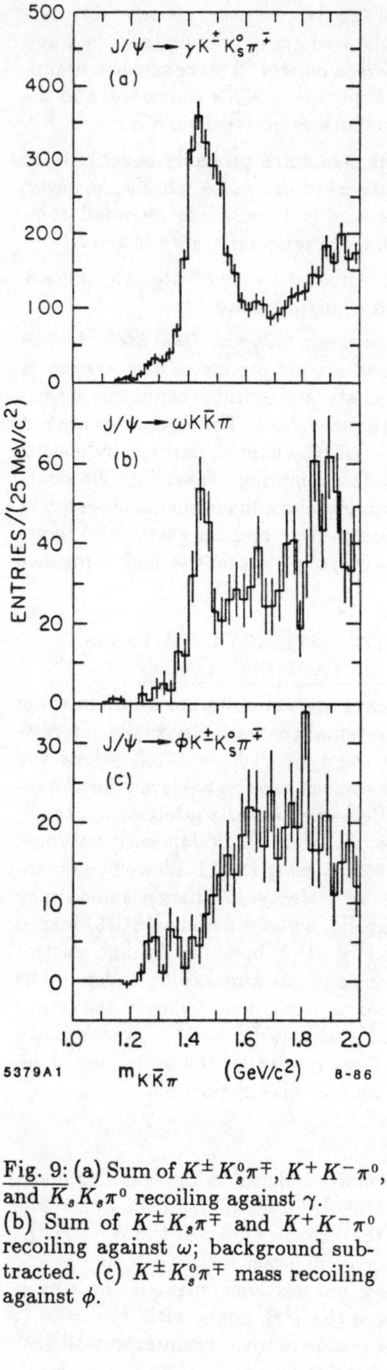

Fig. 9: (a) Sum of $K^{\pm}K_s^0\pi^{\mp}$, $K^+K^-\pi^0$, and $K_sK_s\pi^0$ recoiling against γ. (b) Sum of $K^{\pm}K_s\pi^{\mp}$ and $K^+K^-\pi^0$ recoiling against ω; background subtracted. (c) $K^{\pm}K_s^0\pi^{\mp}$ mass recoiling against ϕ.

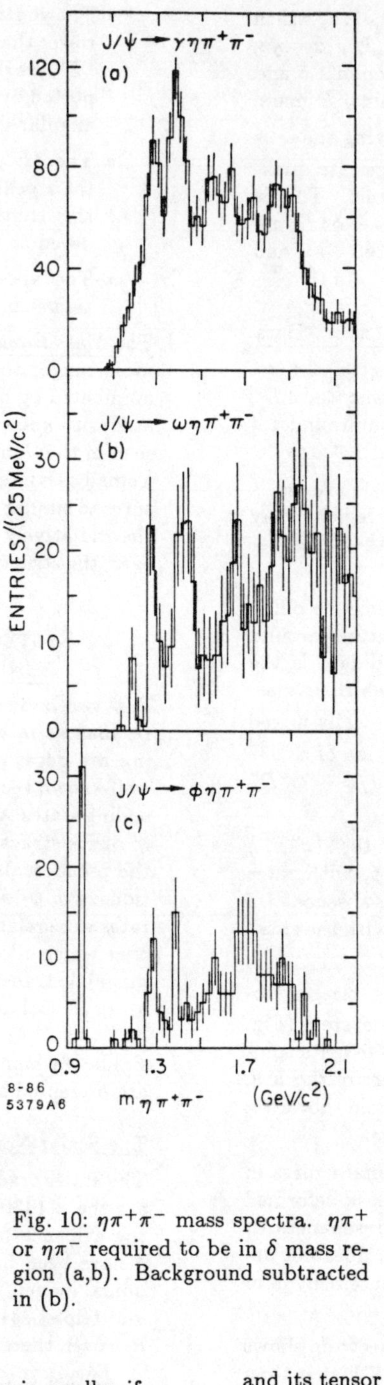

Fig. 10: $\eta\pi^+\pi^-$ mass spectra. $\eta\pi^+$ or $\eta\pi^-$ required to be in δ mass region (a,b). Background subtracted in (b).

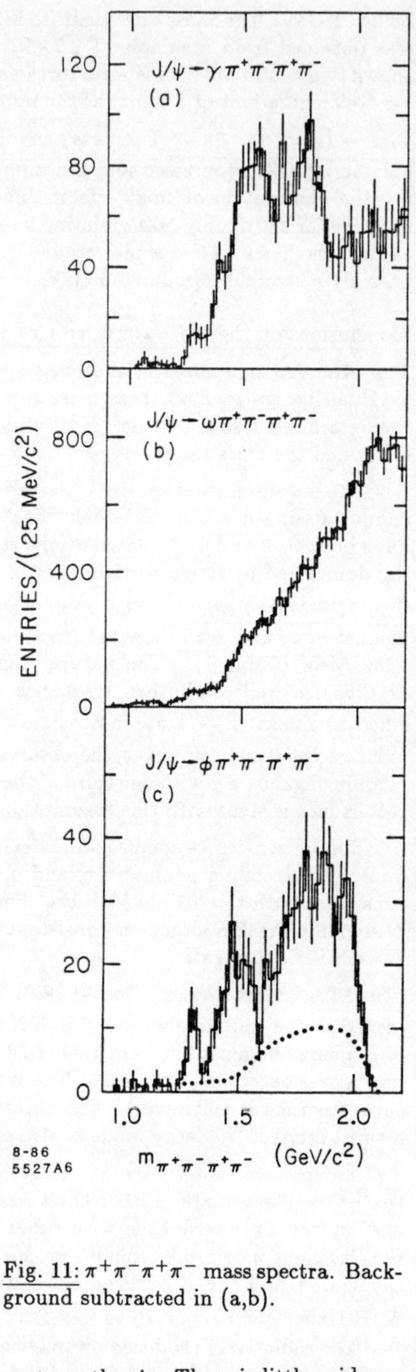

Fig. 11: $\pi^+\pi^-\pi^+\pi^-$ mass spectra. Background subtracted in (a,b).

The expected $J/\psi \to \rho\delta$ decay rate is smaller if S^* and δ are $K\bar{K}$ molecules.[19] Neglecting phase space and SU(3) breaking effects, one expects that $B(J/\psi \to \phi S^*) = 2 \times B(J/\psi \to \omega S^*) = 2 \times B(J/\psi \to \rho^0\delta^0)$. In this model the S^* and δ contain the same number of s-quarks. Therefore SU(3) breaking effects are expected to be less important than in the case of the $q\bar{q}$ model.

The $S^* \to \pi^+\pi^-$ decay is clearly observed in the $J/\psi \to \phi\pi^+\pi^-$ reaction, with a rate similar to that of analog reactions, like $J/\psi \to \phi f'$ (see Table 1). However, this analogy does not hold in the case of the δ and its tensor partner, the A_2. There is little evidence for a δ signal in conjunction with a ρ; the upper limit of $B(J/\psi \to \rho\delta) \cdot B(\delta \to \eta\pi) < 4.4 \times 10^{-4}$ is approximately 25× smaller than the branching ratios of other $J/\psi \to$ *isovector isovector* decays.

If $B(\delta \to \eta\pi)$ is sufficiently large, this observation is in contradiction with the interpretation of S^* and δ as $s\bar{s}$ and $\frac{u\bar{u}-d\bar{d}}{\sqrt{2}}$ states of the 0^{++} nonet. Given the hint of a small S^* signal in the $J/\psi \to \omega\pi^+\pi^-$ channel, the $K\bar{K}$ molecule hypothesis seems more consistent with the data than the $q\bar{q}$ model.

TABLE 1. ISOSCALAR RESONANCE PARAMETERS.

Parameters of structures seen in the $K\bar{K}$, $\pi\pi$, $K\bar{K}\pi$, $\eta\pi\pi$, and 4π systems recoiling against γ, ω and ϕ. Unless otherwise noted, the data are MARK III measurements. [a]: E.Bloom and C.W.Peck, Ann. Rev. Nucl. Part. Sci. **33**, 143 (1983); [b]: incoherent fit; [c]: PDG; [d]: $\eta \to \gamma\gamma$ mode; [e]: $\eta \to \pi^+\pi^-\pi^0$ mode.

Object (Remarks)	Seen in $J/\psi \to$	Mass (MeV)	Width (MeV)	Product Branching Ratio (units of 10^{-4})
f'	γK^+K^-	$1525 \pm 10 \pm 10$	85 ± 35	$B(J/\psi \to \gamma f' \to \gamma K\bar{K}) = 6.0 \pm 1.4 \pm 1.2$
f'	ϕK^+K^-	1520 (fixed)	75 (fixed)	$B(J/\psi \to \phi f' \to \phi K\bar{K}) = 6.4 \pm 0.6 \pm 1.6$
f'	ωK^+K^-	1520 (fixed)	75 (fixed)	$B(J/\psi \to \omega f' \to \omega K\bar{K}) =< 1.2$(90% C.L.)
θ	γK^+K^-	$1720 \pm 10 \pm 10$	130 ± 20	$B(J/\psi \to \gamma\theta \to \gamma K\bar{K}) = 9.6 \pm 1.2 \pm 1.8$
θ [a]	$\gamma\eta\eta$	1670 ± 50	160 ± 80	$B(J/\psi \to \gamma\theta \to \gamma\eta\eta) = 3.8 \pm 1.6$
θ	$\gamma\pi^+\pi^-$	1713 ± 15	130 (fixed)	$B(J/\psi \to \gamma\theta \to \gamma\pi\pi) = 2.4 \pm 0.6 \pm 0.5$
θ? [b]	ϕK^+K^-	$1671^{+15}_{-17} \pm 10$	$126^{+60}_{-40} \pm 15$	$B(J/\psi \to \phi\theta? \to \phi K\bar{K}) = 3.4^{+1.0}_{-0.8} \pm 0.9$
θ?	ωK^+K^-	$1731 \pm 10 \pm 10$	$110^{+45}_{-35} \pm 15$	$B(J/\psi \to \omega\theta? \to \omega K\bar{K}) = 4.5^{+1.2}_{-1.1} \pm 1.0$
S^*	$\gamma\pi^+\pi^-$	975 (fixed)	35 (fixed)	$B(J/\psi \to \gamma S^* \to \gamma\pi\pi) = < 0.7$(90% C.L.)
S^*	$\phi\pi^+\pi^-$	coupled channel parametrization		$B(J/\psi \to \phi S^* \to \phi\pi\pi) = 3.4 \pm 0.5 \pm 0.8$
f	$\gamma\pi^+\pi^-$	1269 ± 13	180 (fixed)	$B(J/\psi \to \gamma f) = 11.5 \pm 0.7 \pm 1.9$
f [c]	$\phi\pi^+\pi^-$	–	–	$B(J/\psi \to \phi f) = < 3.7$(90% C.L.)
f	$\omega\pi^+\pi^-$	1277 (fixed)	182 ± 10	$B(J/\psi \to \omega f) = 49.3 \pm 2.5 \pm 12.5$
D? [d]	$\gamma\eta\pi^+\pi^-$	1283 (fixed)	26 (fixed)	$B(J/\psi \to \gamma D? \to \gamma\delta\pi \to \gamma\eta\pi\pi) = 2.7 \pm 0.8 \pm 0.2$
D? [e]	$\gamma\eta\pi^+\pi^-$	1283 (fixed)	26 (fixed)	$B(J/\psi \to \gamma D? \to \gamma\delta\pi \to \gamma\eta\pi\pi) = 3.2 \pm 1.1 \pm 0.3$
D?	$\phi\eta\pi^+\pi^-$	$1283 \pm 6 \pm 10$	$24^{+20}_{-14} \pm 10$	$B(J/\psi \to \phi D? \to \phi\eta\pi\pi) = 1.6^{+0.6}_{-0.5} \pm 0.4$
D?	$\phi\pi^+\pi^-\pi^+\pi^-$	1287 ± 7	16^{+26}_{-10}	$B(J/\psi \to \phi D? \to \phi 4\pi^\pm) = 0.34 \pm 0.07 \pm 0.05$
D?	$\omega\eta\pi^+\pi^-$	$1283 \pm 6 \pm 10$	$14^{+19}_{-14} \pm 10$	$B(J/\psi \to \omega D? \to \omega\eta\pi\pi) = 4.3 \pm 1.2 \pm 1.3$
ι	$\gamma K^\pm K_s\pi^\mp$	$1456 \pm 5 \pm 6$	$95 \pm 10 \pm 15$	$B(J/\psi \to \gamma\iota \to \gamma K\bar{K}\pi) = 50 \pm 3 \pm 8$
ι	$\gamma K^+K^-\pi^0$	$1461 \pm 5 \pm 5$	$101 \pm 10 \pm 10$	$B(J/\psi \to \gamma\iota \to \gamma K\bar{K}\pi) = 49 \pm 2 \pm 8$
ι	$\phi K^\pm K_s\pi^\mp$	1460 (fixed)	95 (fixed)	$B(J/\psi \to \phi\iota \to \phi K\bar{K}\pi) = < 1.8$(90% C.L.)
E? [d]	$\gamma\eta\pi^+\pi^-$	1382 ± 6	69 ± 23	$B(J/\psi \to \gamma E? \to \gamma\delta\pi \to \gamma\eta\pi\pi) = 5.2 \pm 1.2 \pm 0.5$
E? [e]	$\gamma\eta\pi^+\pi^-$	1400 ± 7	62 ± 16	$B(J/\psi \to \gamma E? \to \gamma\delta\pi \to \gamma\eta\pi\pi) = 5.2 \pm 1.8 \pm 0.5$
E?	$\phi K^\pm K_s\pi^\mp$	1420 (fixed)	52 (fixed)	$B(J/\psi \to \phi E? \to \phi K\bar{K}\pi) = < 1.1$(90% C.L.)
E?	$\omega\eta\pi^+\pi^-$	$1421 \pm 8 \pm 10$	$45^{+32}_{-23} \pm 15$	$B(J/\psi \to \omega E? \to \omega\eta\pi\pi) = 9.2 \pm 2.4 \pm 2.8$
E?	$\omega K\bar{K}\pi$	$1444 \pm 5^{+10}_{-20}$	$40^{+17}_{-13} \pm 10$	$B(J/\psi \to \omega E? \to \omega K\bar{K}\pi) = 6.8^{+1.9}_{-1.6} \pm 1.7$

REFERENCES

1. R.Baltrusaitis et al., Phys.Rev.Lett. **56**,107 (1986).
2. S.Godfrey et al., Phys. Lett. **141B**, 439 (1984).
3. M.Chanowitz and S.R.Sharpe, Phys. Lett. **132 B** (1982) 413.
4. Contributed paper # 3409.
5. H.Haber & J.Perrier, Phys.Rev.**D32**, 2961, (1985).
6. R.Baltrusaitis et al., Phys.Rev.**D32**, 2883, (1985) and Phys.Rev.**D33**, 629, (1986).
7. Contributed paper # 3441.
8. Relativistic extension of Flatté parametrization (S.M.Flatté, Phys.Lett.**63B**,224(1976)) was used.
9. Contributed paper # 3476.
10. Except for the η_c (no $c\bar{c}$ annihilation required).
11. Contributed paper # 3433.
12. R.Baltrusaitis et al., Phys.Rev. **D33**, 1222 (1986); R.Baltrusaitis et al., Phys.Rev.Lett. **55**, (1985).
13. η(1275): Stanton et al., Phys.Rev.Lett. **42**(1979) 346; A.Ando et al., KEK Preprint 86-8, May 1986. For a scenario with three resonances in the 1.4 GeV region, produced in various ratios according to the particular dynamics of a given reaction, see *e.g.* : D.O.Caldwell, "A Possible Solution to the E/ι Puzzle", contributed paper # 779.
14. W.Palmer & F.Pinsky, Phys.Rev.**D27**,2219 (1983).
15. For an interpretation in terms of a $K\bar{K}$ molecule see: M.Frank et al., Phys.Lett. **158B**, 442 (1985).
16. See *e.g.* Particle Data Group: Review of Particle Properties, Rev.Mod.Phys. **56,II** (1984).
17. R.L.Jaffe, Phys.Rev. **D15**, 267 (1977).
18. J.Weinstein and N.Isgur, Phys.Rev.Lett. **48**, 659 (1982); Phys.Rev. **D27**, 588 (1983).
19. A.Seiden, Processes Related to Two Photon Physics, SCIPP 86-58.

Work supported in part by the Department of Energy, contract numbers DE – AC03-76SF00515, DE – AC02-76ER01195, DE – AC03-81ER40050, DE – AM03-76SF00034, and by the National Science Foundation.

RECENT RESULTS FROM GAMS

F. Binon(*)

Institut Interuniversitaire des Sciences Nucléaires
B-1050 Brussels, Belgium

Some new results from the GAMS collaboration are presented. A structure has been observed at 2220 MeV in the mass spectrum of $\eta\eta'$ systems produced by 38 GeV/c and 100 GeV/c π^- on protons. Its spin $J \geq 2$. The existence and properties of G(1590) have been confirmed in an analysis of the mass spectrum of $\eta\eta$ systems produced by 230 GeV/c π^- on protons.

1. INTRODUCTION

Exclusive reactions leading to the production of multiphoton final states

$$\pi^- p \rightarrow M^o \; n$$
$$\hookrightarrow k\gamma$$

have been extensively studied these last years by the GAMS collaboration. A large amount of data has been collected at IHEP (6th Joint CERN-IHEP experiment) with 38 GeV/c incident pions and at CERN (NA12 experiment) with 100 GeV/c and 230 GeV/c pions. Both experimental setups are rather similar (see e.g. [1]) their main detectors being large electromagnetic calorimeters made of lead-glass cells (38x38x450 mm³). GAMS-2000 [2], at IHEP, is a matrix of 48x32 such cells while GAMS-4000, installed in the SPS North Area at CERN, comprises 64x64 cells.

The study of the $\eta\eta$ and $\eta'\eta$ decay channels is particularly interesting in connection with the search of bound states of gluons (gluonium) as the radiative decays of ψ show a preference of gluons to couple with η and η' rather than with π^o [3].

2. EVIDENCE FOR A 2.22 GeV STRUCTURE DECAYING INTO $\eta'\eta$

A recent analysis of the 4-γ events gathered at 38 GeV/c (IHEP) and 100 GeV/c (CERN) show a rather narrow structure at 2220 MeV which decays into $\eta'\eta$.

As the 4-γ events are heavily dominated by $\pi^o\pi^o$ and $\eta\pi^o$ pairs, the $\eta'\eta$ events have been selected as follows: a) any γ pair with a mass below 240 MeV at 38 GeV/c or 170 MeV at 100 GeV/c is rejected ("π^o cut"); b) any γ pair with a mass in the interval (465 MeV, 645 MeV) is identified with a η; c) all events with two η are rejected. The remaining events give an invariant mass spectrum of the second γ pair which shows a clean η' peak on a relatively small background. Events where the mass of the second γ pair lies in the range between 870 MeV and 1080 MeV are fitted to the exclusive reaction (3-C fit) $\pi^- p \rightarrow \eta'\eta n$. Events with a χ^2 larger than 9 are rejected.
The mass spectra of the selected $\eta'\eta$ events (fig. 1) show a peak centered at a mass of 2220 MeV (10 MeV uncertainty) which is insensitive to the choice of applied mass cuts.

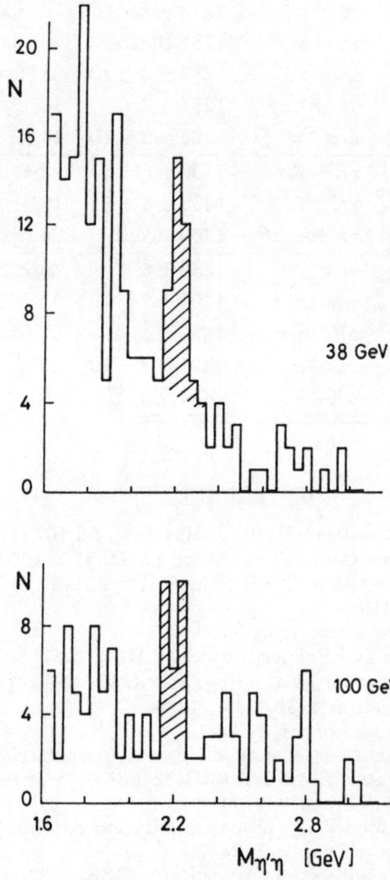

Figure 1: Invariant mass spectra (not corrected for detection efficiency) of the selected $\eta'\eta$ events (40 MeV bin width).

At 38 GeV/c, 36 events are observed in the peak with an estimated background contribution of 13 events. The corresponding numbers at 100 GeV/c are 26 and 8, respectively. The fact that it appears in two different sets of data taken with two setups differing in their geometry and acceptance (the target to GAMS distance is 4.6 m at 38 GeV/c and 15 m at 100 GeV/c) strengthens the confidence in the existence of a new meson with a mass of 2220 MeV.

Two-dimensional plots of cos θ_{GJ}, where θ_{GJ} is the polar decay angle of the η in the Gottfried-Jackson frame, versus the mass of $\eta'\eta$ systems are shown on figure 2.

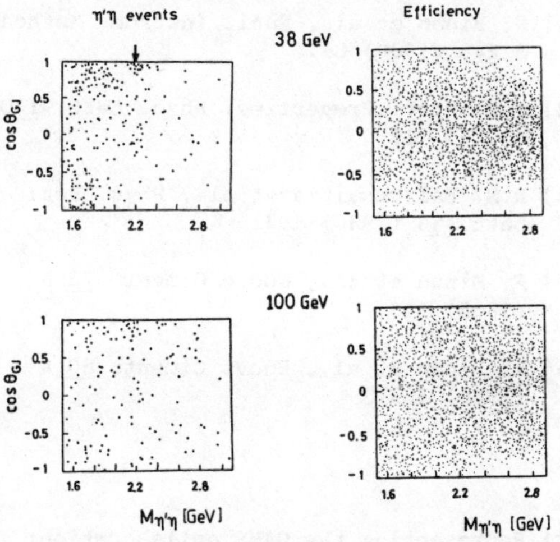

Figure 2: Plot of cos θ_{GJ} versus invariant mass for the selected $\eta'\eta$ events (left) and for simulated events (right).

The plots on the left are for measured events while those on the right have been obtained from Monte Carlo simulated events (using uniform distributions in cos θ_{GJ} and mass) submitted to the whole chain of analysis programs. One sees that, due to acceptance, the detection efficiency ϵ decreases with increasing mass and when $|\cos \theta_{GJ}|$ is approaching 1. The latter effect is more pronounced at 38 GeV/c.

Angular distributions are quite different for $\eta'\eta$ events in the peak and for those in neighbouring intervals. The events in the (2220)- peak have angular distributions which are notably anisotropic (figure 3).

This is most clearly seen in the 38 GeV data. Notwithstanding the two times smaller statistics and a higher background the 100 GeV data are important because the efficiency ϵ is much higher for cos θ_{GJ} near -1 at 100 GeV (fig. 2). They confirm the strong peaking of the data at $|\cos \theta_{GJ}| \sim 1$ and they display also more clearly the symmetry of the angular distribution.

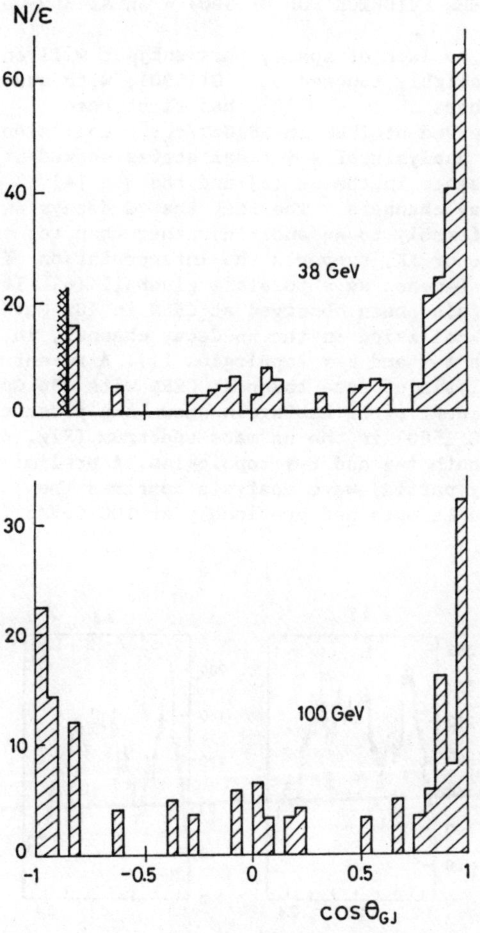

Figure 3: Angular distributions of the $\eta'\eta$ selected events in the mass interval between 2150 MeV and 2310 MeV. The data are corrected for detection efficiency. The cross-shaded area near cos θ_{GJ}= -1 in the upper figure (38 GeV) is the limit below which the efficiency falls to zero (average detection efficiency $\bar{\theta} \approx 0.2$).

The very anisotropic angular distributions is a clear indication that the spin of this object is $J \geq 2$. It follows also immediately from its decay mode in $\eta'\eta$ that it has $I^G = 0^+$. The mass and, possibly, the width of the observed structure at 2220 MeV are compatible with those of the narrow isoscalar meson $\xi(2230)$ observed in the radiative decay $\psi \to \gamma KK$ [4]. Of course, it could well be that this is purely fortuitions and that the state observed in the $\eta'\eta$ decay channel is just another meson. More experimental information on $\eta'\eta$ as well as another decay channels is needed to clarify the nature of the observed structure at 2220 MeV.

3. NEW EVIDENCE FOR G(1590) → ηη AT 300 GeV/c

Due to lack of space, this subject will only be slighly touched on. G(1590), with quantum numbers $J^{PC}I^{G} = 0^{++}0^{+}$, has first been observed at IHEP in 38 GeV/c $\pi^{-}p$ collision. The analysis of 4-γ final states showed its presence in the $\eta\eta$ [5] and the $\eta'\eta$ [6] decay channels. The fact that G decays preferably to $\eta\eta$ and $\eta'\eta$ rather than to $\pi^0\pi^0$ or $K\bar{K}$, supports the interpretation of the G-meson as a possible glueball [6]. It has also been observed at CERN in 100 GeV/c $\pi^{-}p$ collision in the $\eta\eta$ decay channel, in both 4-γ and 8-γ topologies [1]. A recent analysis of data taken at CERN with 230 GeV/c incident pions has again shown the presence of G(1590) in the $\eta\eta$ mass spectrum (Fig. 4), in both 4-γ and 8-γ topologies. A preliminary partial-wave analysis confirms the results obtained previously at 100 GeV/c [1].

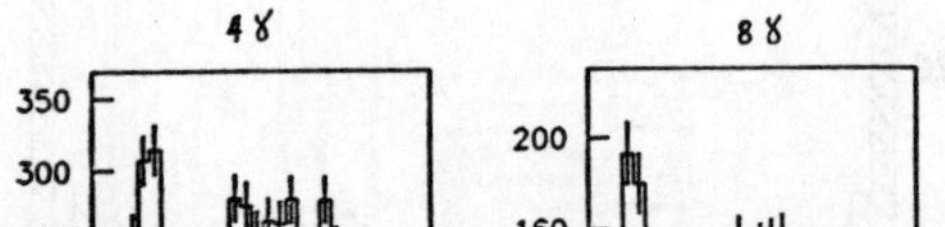

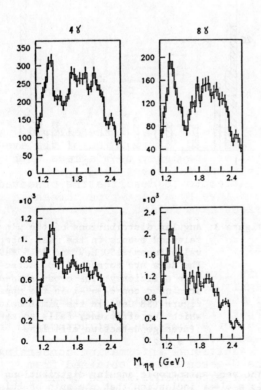

Figure 4: Mass spectra of 4-γ events (left) and 8-γ events (right) identified as $\eta\eta$ systems after 3-C fit and 6-C fit, respectively. The upper spectra are not corrected for (Fig. 4) detection efficiency while the lower ones are. The small difference in shape of the latter ones is due to an additionnal source of background in the 8-γ events (one η and three low-momentum non-resonant π^0 which easily simulate a η).

REFERENCES

[1] D. Alde et al., Nucl. Phys. B 269 (1986) 485.

[2] F. Binon et al., Nucl. Instrum. Methods A 248 (1986) 86.

[3] Rev. Part. Properties, Phys. Lett. 170 B (1986) 219.

[4] R.M. Baltrusaitis et al., Phys. Rev. Lett. 56 (1986) 107.

[5] F. Binon et al., Nuovo Cimento 78 A (1983) 313.

[6] F. Binon et al., Nuovo Cimento 80 A (1984) 363.

(*) Representing the GAMS collaboration:

IHEP (Serpukhov): S.V. Donskov, A.V. Inyakin, V.A. Kachanov, D.B. Kakauridze, G.V. Khaustov, A.V. Kulik, A.A. Lednev, Yu.V. Mikhailov, V.F. Obraztsov, Yu.D. Prokoshkin, Yu.V. Rodnov, S.A. Sadovsky, V.D. Samoylenko, P.M. Shagin, A.V. Shtannikov, A.V. Singovsky, V.P. Sugonyaev; IISN (Brussels): F.G. Binon, C. Bricman, J.P. Lagnaux. Th. Mouthuy, A. Possoz, J.P. Stroot; LANL(Los Alamos): D. Alde, E.A. Knapp; LAPP (Annecy): J. Dufournaud, M. Gouanère, J.P. Peigneux, D. Sillou.

CANDIDATES FOR EXOTIC STATES OBSERVED AT IHEP

V.F. Obraztsov
Institute for High Energy Physics
Serpukhov, 142284 Protvino, Moscow region, USSR

Candidates for cryptoexotic states, reacently observed at IHEP (Serpukhov) 70GeV accelerator by Lepton-F, GAMS, BIS-2 collaborations are discussed. All states have clear exotic "signature": 1^{--} C(1480)-meson ($I^G=1^+$) has enhanced $\varphi\pi^0$ decay and small C-γ coupling; G(1590)-meson has enhanced $\eta\eta'$ decay, compared with $\eta\eta$ one, and suppressed $K\bar{K}$ and $\pi\pi$ decay modes. Baryon $N_\varphi(1956)$ has a width of $\Gamma \leq 20$MeV, the main decay mode $\Sigma^-(1385)K^+$ and suppressed decays into non-strange particles.

1. INTRODUCTION

One of the puzzling problems of todays elementary particle physics is the non-observation of exotic states, although such states have been predicted long before the birth of the quark model[1].

The basic question concerning cryptoexotic states (with usual $q\bar{q}$ or qqq quantum numbers) is that of their "signature". Reacently new states: C(1480), G(1590), $N_\varphi(1956)$ with clear exotic signatures have been observed at IHEP accelerator. They are discussed in the present talk.

2. C(1480)-meson

One of the characteristics of cryptoexotics could be "hidden" strangeness with an isotopic spin different from zero[2], as it is realised in the $\varphi\pi$-system.

A new experimental study of the $\varphi\pi^0$ system[3] has been carried out by the Lepton-F group in a 33GeV π^- beam at the Serpukhov accelerator. The charge exchange reaction $\pi^-p \to \varphi\pi^0 + n$ was chosen as a source of $\varphi\pi^0$-system. The mass spectrum of K^+K^- system produced in reaction $\pi^-p \to K^+K^-\pi^0 + n$ shows a clear peak corresponding to the production of φ-meson (Fig.1a). The resulting $\varphi\pi^0$ mass spectrum Fig.1b (background substracted and efficiency corrected) is dominated by the C-state with parameters: $M_C=(1480 \pm 40)$MeV, $\Gamma=(130 \pm 60)$MeV Background was estimated from the neighbouring to φ mass intervals in the K^+K^- spectrum. The cross section for the C-state production is found to be $\sigma(\pi^-p \to Cn)\cdot B_\varphi = (40 \pm 15)$nb.

The t distribution of events is compatible with dominant OPE. This implies that the C-state must have $P=C=(-1)^J$. As the charge parity is odd,

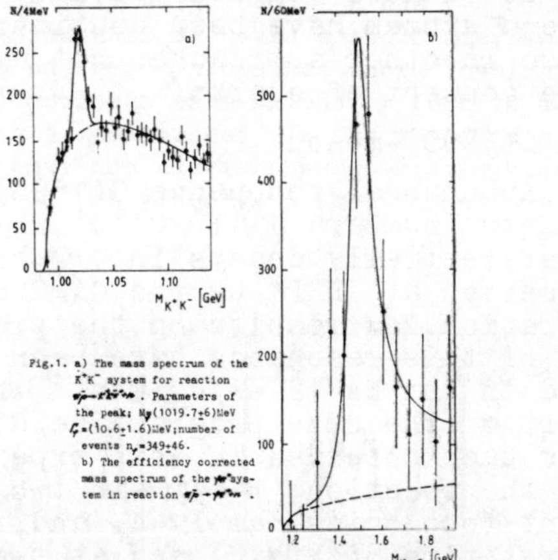

Fig.1. a) The mass spectrum of the K^+K^- system for reaction $\pi^-p \to K^+K^-\pi^0 n$. Parameters of the peak; $M_\varphi(1019.7 \pm 6)$MeV $\Gamma_\varphi=(10.6 \pm 1.6)$MeV; number of events $n_\varphi=349 \pm 46$.
b) The efficiency corrected mass spectrum of the $\varphi\pi^0$ system in reaction $\pi^-p \to \varphi\pi^0 n$.

only $J^{PC}=1^{--}, 3^{--}, \ldots$ are allowed.

The $\cos\theta_{GJ}$ distribution of the events for the decay $C \to \varphi\pi^0$ agrees with $J^P=1^-$ and rules out $J^P=3^-, \ldots$

Attempts to describe the observed structure by a Deck-type threshold effect have failed.

The following limitation (95%C.L.) has been obtained using the GAMS-2000 data on the reaction $\pi^-p \to \omega\pi^0 + n$ (38GeV):

$$Br(C \to \varphi\pi)/Br(C \to \omega\pi) > 1/2$$

This limit is at least two orders of magnitude higher than that expected if the C-meson was an ordinary $q\bar{q}$ system. It is a strong argument in favour of a four-quark structure of the C-meson.

Additional information concerning the C-meson can be obtained from other experiments. As the C-meson is a vector it should be produced in the reactions $e^+e^- \to \varphi\pi^0$; $\gamma p \to \varphi\pi^0 + p$. The photoproduction of the $\varphi\pi^0$-system has been observed by the Ω photon collaboration[4]. Although the number of events is small (~20) there is an indication of a structure in the region of C-meson, giving:

$\sigma(\gamma p \to Cp)Br(C \to \varphi\pi^0) = (3 \pm 1.5)$nb
This value is a factor of ~100 lower than that expected for an ordinary $q\bar{q}$ vector meson and is in a qualitative agreement with expectations for $q^2\bar{q}^2$ or $q\bar{q}g$ mesons.

The four-quark hypothesis for the C-meson has been analysed in[5]. The coupling with $\rho\eta$ is predicted with $Br(C\to\rho\eta) \sim 1/3 Br(C\to\varphi\pi^0)$ as well as the existence of the I=0 partner $\tilde{C}$ with $M_{\tilde{C}} \approx M_C$; $\Gamma_{\tilde{C}} \leq 100$MeV. and with the main decay mode $\tilde{C}\to\omega\eta$.

The C-meson can be, as well, a hybrid (meikton) $q\bar{q}g$ state. The vector meikton was considered in[6]. A mass of 1.6GeV and smaller value of $Br(C\to\varphi\pi)$ as compared with the $q^2\bar{q}^2$ case are expected.

Resonances coupled strongly with the $\varphi\pi$ system have been considered also in other approaches not using the concept of quarks[7].

3. G(1590)-meson

An unusual resonance G(1590) with quantum numbers $J^{PC}I^G = 0^{++}0^+$, which preferentially decays into $\eta\eta'$ has been observed at IHEP[8] by the GAMS collaboration. New results on the production of this resonance have been obtained in the NA-12 experiment[9] which is part of the same program. The G(1590) has been observed in this experiment in the reaction $\pi^- p \to \eta\eta + n$ in two different topologies, namely 4γ and, for the first time, 8γ (Fig.2) and at two different energies 100GeV and 230GeV. The cross section, compared with 38GeV shows a p^{-2} energy dependence confirming, independently, the OPE dominance in G(1590) production. That is essential for the validity of the PWA.

G(1590) has peculiar properties:

$Br(G\to\eta\eta')/Br(G\to\eta\eta) = 2.7 \pm .8$
$Br(G\to\pi\pi)/Br(G\to\eta\eta) < .3$
$Br(G\text{-}K\bar{K})/Br(G\to\eta\eta) < 1$

They are not compatible with those of $q\bar{q}$ 0^{++} meson decaying in accordance with quark line rules. The situation is analogous to that of $\theta(1700)$ where the ratio of the decay modes $K\bar{K}:\eta\eta:\pi\pi = 3:1:.8$ also cannot be understood. On the contrary, the quark line rules work for ordinary mesons: f, f', h.

Taking into account the importance of $Br(G\to\eta\eta')$ for the interpretation of G(1590), the contribution of a Deck-type background on $\eta\eta'$ systems threshold has been studied in detail[10]. A reasonable estimation shows that the background does not exceed ~20%.

The data can be explained naturally within a gluonium hypothesis if a new mechanism for glueball decay (gluon discolouration)[11] is taken into

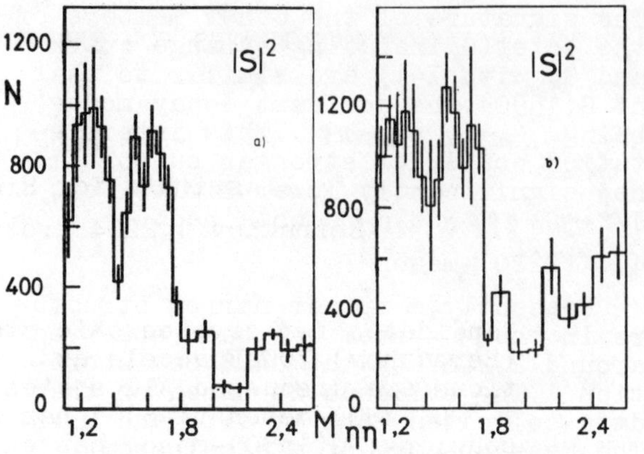

Fig.2 S-amplitude for $\eta\eta$ system in the reaction $\pi^- p \to \eta\eta n$ a) $\eta\eta\to 4\gamma$; b) $\eta\eta\to 8\gamma$

account. The dominance of the $\eta\eta'$ decay mode is then a consequence of the strong coupling of η' with the two-gluon channel.

The glueball hypothesis for G-meson is not free from certain objections (as well as for all other candidates). G has not been observed in J/ψ radiative decays, although the limit is not strong:

$Br(J/\psi \to \gamma G) < 6 \cdot 10^{-4}$

This value can be compared with

$Br(J/\psi \to \gamma\theta) = (1.3 \pm .2) \cdot 10^{-3}$

On the other hand, both G and θ production is unexpectedly suppressed compared with $\iota(1440)$.

The small partial width for G-$K\bar{K}$, $\pi\pi$ also presents some difficulties[12]. However, it has been shown that G-$K\bar{K}$, $\pi\pi$ can be suppressed if a derivative-coupling terms are included in the Lagrangian[13].

A possibility has been discussed[14] that an ordinary $q\bar{q}$ 0^{++} SU(3)-singlet meson can strongly annihilate into a pair of gluons due to large nonperturbative effects in this channel. It can lead to the enhancement of the $\eta\eta'$ decay mode due to the gluon discolouration. This hypothesis is in contradiction with the $Br(\varepsilon\to\eta\eta)$ value measured in the NA12 experiment. A peak in the S-wave, close to the $\eta\eta$ threshold (Fig.2) is identified as $\varepsilon(1300)$. The value $Br(\varepsilon\to\eta\eta) \sim 2.5 \cdot 10^{-2}$ has been obtained. One can conclude from the known ratio $Br(\varepsilon\to\pi\pi)/Br(\varepsilon\to K\bar{K}) \sim 9$ that there is a substantial SU(3)-singlet component in $\varepsilon(1300)$, and, at the same time, $Br(\varepsilon\to\eta\eta)$ is in agreement with the quark line rules.

The study of other possible scenarios for the G-meson are also of interest. In[15] the hypothesis that the G-meson is the 8-th component of the $SU(3)_f$-octet of meiktons has been put forward.

The signature of the other members of the octet(G_K including strange quark and G_π with I=1) are similar to that of G(1590),the dominant decay modes being $G_K \to K\eta'$; $G_\pi \to \pi\eta'$ .This interpretation would get stronger support if new significantly lower limits for $J/\psi \to \gamma G \to \eta\eta'\gamma$ would be obtained.

4. X(2220)-meson

Reacently,a rather narrow structure in the $\eta\eta'$ system has been observed around 2.2GeV by the GAMS collaboration[16].The evidence is based on the data collected both at IHEP and CERN. The mass spectra of the selected events in the reaction $\pi^- p \to \eta\eta' n$ Fig.3 shows the presence of a peak with a mass of (2220±10)MeV and width of $\Gamma \leq 60$MeV.From the very anisotropic angular distribution it follows that the spin value J≥2.If the observed object is ξ(2220),then the limit can be obtained: $Br(\xi \to \eta\eta')/Br(\xi \to K\bar{K}) > 1$ using the data on the reaction $\pi^- p \to K_S K_S n$ [17]. This limit excludes the L=3 $s\bar{s}$ interpretation for ξ [18] and is a strong argument in favour of it's glueball interpretation.

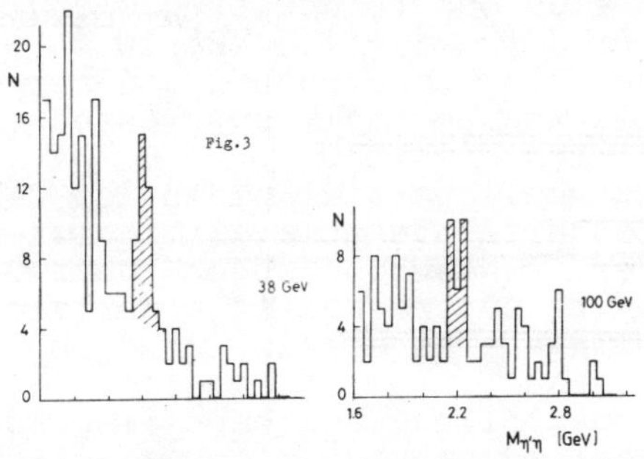

Fig.3

5. N_φ(1956)

New data on the observation of a narrow resonance N_φ in the $\Sigma^-(1385)K^+$ channel are presented by the BIS-2 collaboration[19].The data have been obtained in the neutron beam(20-70) GeV of the IHEP accelerator in the reaction $n+C \to \Sigma^-(1385)K^+$+..Fig.4 presents the background substracted $\Sigma^-(1385)K^+$ mass spectrum.The background was estimated from the neighbouring to $\Sigma^-(1385)$ mass intervals in the $\Lambda\pi^-$ mass spectrum.A narrow peak (~100 events) is seen M=(1956±3)MeV; Γ=(27±15)MeV.The angular spectra of the events satisfie natural spin-parities for N_φ: $J^P=3/2^-, 5/2^+$ ...The A-dependence of the N_φ production is $\sim A^{.9\pm.3}$ .The production cross section times branching is (.26±.04)μb/nucleon The narrow width of N_φ cannot be explained within the conventional three quark baryon model.The analysis of different experimental constraints has shown that N_φ decays without strange particles are suppressed,i.e N_φ has a "hidden" strangeness and is a candidate for a five-quark state(uuds$\bar{s}$)

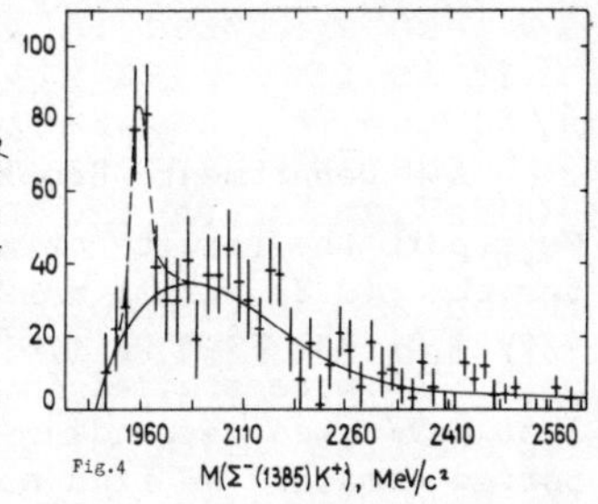

Fig.4

ACKNOWLEDGEMENTS

The part of the talk concerning G(1590) was prepared in collaboration with S.S.Gershtein.

I would like to thank N.N.Achasov, F.Binon,V.D.Kekelidze,L.G.Landsberg, Yu.D.Prokoshkin,J.P.Stroot,A.M.Zaitsev for discussions and information.

REFERENCES

1. G.Wentzel,Helv.Phys.Acta,13(1940) 269;W.Pauli et al.,PR,62(1942)85.
2. F.Close,H.Lipkin,PRL,41(1978)1263
3. S.I.Bityukov et al.,Preprint IHEP 86-110, Serpukhov,1986.
4. M.Atkinson et al.,NP B231(1984)1.
5. N.N.Achasov,JETPL,43(1986)410.
6. M.S.Chanowitz,LBL-16653,1983.
7. N.Barinov et al.SJNP 29(1979)1357
8. F.Binon et al.,NC 78A(1983)313; F.Binon et al.,NC 80A(1984)363.
9. D.Alde et al.,CERN-EP/85-153; Paper 3921 submitted to this conf.
10. S.S.Gershtein et al.,Preprint IHEP 85-44,Serpukhov,1985.
11. S.S.Gershtein et al.,Z.Phys.C24 (1984)305.
12. J.Ellis et al.,PL 150B(1985)289.
13. H.Gomm et al.,PR D33(1986)801.
14. J.Lanik,JINR P2-85-373,Dubna,1985
15. N.N.Achasov,S.S.Gershtein,Prep. Tph-No16(148) Novosibirsk,1985.
16. D.Alde et al.,IHEP 86-114,Serp., 1986;Paper 9822 submitted to this conf.;F.Binon,Talk presented at this conference.
17. R.S.Longacre et al.,Paper 1899 and talk submitted to this conf.
18. S.Godfrey et al.,PL 141B(1984)439
19. A.N.Aleev et al.,Z.Phys.C25(1984) 205.

A SEARCH FOR THE $\xi(2.2)$ IN $\bar{p}p$ FORMATION

J. Sculli, J. H. Christenson, G. A. Kreiter, and P. Nemethy
Department of Physics, New York University
4 Washington Place, New York, NY 10003

and

P. Yamin
AGS Department, Brookhaven National Laboratory, Upton, NY 11973

We report the results of a search at the Brookhaven National Laboratory AGS for the $\xi(2.2)$ with a mass resolution of $3\text{MeV}/c^2$. Thirty-seven thousand events of the form $\bar{p}p \to K^+K^-$ and two hundred thousand events of the form $\bar{p}p \to \pi^+\pi^-$ were studied in the incident momentum interval from 1.25 GeV/c to 1.56 GeV/c, corresponding to a mass interval of 110 MeV/c^2 about the reported ξ mass. We find no evidence for ξ formation.

The $\xi(2.2)$ was observed by the Mark III group at SPEAR[1] as a narrow enhancement in the K^+K^- spectrum in the radiative decay $J/\psi \to \gamma K^+K^-$. It was also observed in the decay $J/\psi \to \gamma K^0_s K^0_s$. The measured mass and width of the ξ are $2.230 \pm .006 \pm .014$ GeV/c^2 and $.026^{+.020}_{-.016} \pm .017$ GeV/c^2 respectively. Theoretical interest in this new state arises because of its narrow width, its observation in radiative J/ψ decay, and its large branching ratio for decay into strange mesons.

We report the results of a search for the ξ in the reaction $\bar{p}p \to \xi \to K^+K^-$ carried out at the Brookhaven National Laboratory's Alternating Gradient Synchrotron. The apparatus is shown schematically in Fig. 1. Antiprotons comprised only .15 percent of the incident beam of 10^7 particles per second. They were distinguished from pions by a measurement of the time-of-flight between two counter arrays separated by 45 feet and by the use of a Fitch type plastic Cerenkov counter. The incident beam interacted in a hydrogen target with thin mylar windows on three sides to allow large angle secondaries to escape. The K^+K^- and $\pi^+\pi^-$ pairs were detected in PWC's with .08" wire spacing situated on either side of the beam line. The chambers nearest the beam line contained three planes oriented at 0°, $+20^\circ$, and -20°, with respect to the vertical. The rear chambers contained two orthogonal planes. Scintillation counters behind the rear chambers were used in the trigger and also served to reject events with extra tracks from multibody final states. Twelve veto counters covered most of the forward region outside the aperture of the PWC's. These counters were not incorporated in the trigger but recorded and used off line to further reduce the background from multibody final states.

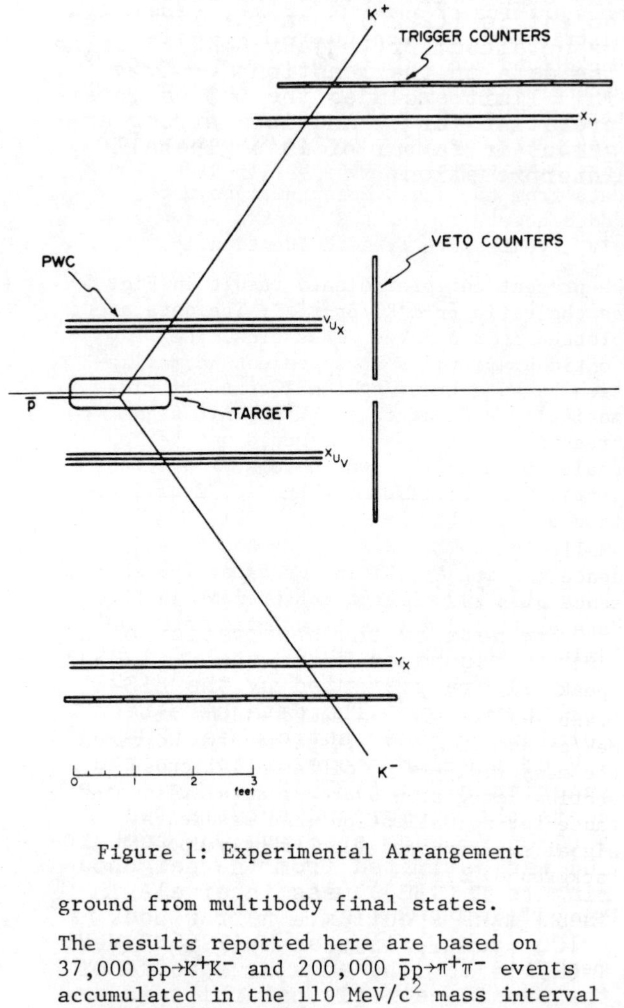

Figure 1: Experimental Arrangement

The results reported here are based on 37,000 $\bar{p}p \to K^+K^-$ and 200,000 $\bar{p}p \to \pi^+\pi^-$ events accumulated in the 110 MeV/c^2 mass interval about the ξ mass. The invariant mass of the K^+K^- was measured with a $\sigma = \pm 3$ MeV/c^2. Data were collected in seven overlapping 70 MeV/c momentum intervals between 1.29 GeV/c and 1.53 GeV/c. Two body final states were selected by requiring the incident antiproton track and the two tracks in the PWC's to lie in a plane (coplanarity) and by comparing the effective mass of the final state pair determined solely from a measurement of the track directions to that inferred from the beam momentum. Requiring the opening angle to exceed 105° eliminated the background from $\pi^- p$ and $\bar{p}p$ elastic scattering. Fig. 2 shows the reconstructed beam momentum spectrum from two track events with opening angles greater than 105°. The final state particles are assumed to be kaons. The $\bar{p}p \to K^+K^-$ peak at the nominal beam momentum of 1.44 GeV/c is well separated from that due to the more copious $\bar{p}p \to \pi^+\pi^-$ which reconstruct to momenta outside the ±2% spread in beam momenta because of the incorrect kaon mass assumption.

The remaining background under the K^+K^- peak was removed by dividing the data into 10 MeV/c momentum intervals (3.5 MeV/c^2 mass intervals) and fitting the coplanarity distribution in each interval to a flat background and a Gaussian peak. The background was subtracted in each momentum interval. The result, for the data taken at 1.44 GeV/c, is shown as the lower curve in Fig. 2. After background subtraction, the data from the seven momentum runs were added together. The K^+K^- events and the $\pi^+\pi^-$ events were treated identically.

We present our preliminary result in Fig. 3 as the ratio $\bar{p}p \to K^+K^- / \bar{p}p \to \pi^+\pi^-$. The data are plotted in 3.5 MeV/c^2 mass bins. The $\pi^+\pi^-$ continuum provides a convenient normalization because the $\pi^+\pi^-$ final state is kinematically similar to the K^+K^-. A ξ signal present in both channels would not be concealed because the continuum $\bar{p}p \to \pi^+\pi^-$ cross section is four times the $\bar{p}p \to K^+K^-$ cross section and the branching ratio $BR(\xi \to \pi^+\pi^-)$ is smaller than $BR(\xi \to K^+K^-)$.[1] We see no evidence for the $\xi(2.2)$ in our data. The absence of a significant enhancement in the data of Fig. 3 may be translated into a limit on the cross section $\sigma_{peak} = (\bar{p}p \to \xi \to K^+K^-)$. We find 3σ limits of 4.4μb and 1.9μb for assumed widths of 7 MeV/c^2 and 35 MeV/c^2 respectively. We used the measured $\bar{p}p \to \pi^+\pi^-$ differential cross section integrated over our angular acceptance for normalization[2] and assumed a signal superimposed incoherently upon the background.

Using the relationship

$$\sigma_{peak} = [4\pi(2J+1)/(s-4m_p^2)] \cdot BR(\xi \to \bar{p}p) \cdot BR(\xi \to K^+K^-)$$

one obtains the limits on the product of the branching ratios, $BR(\xi \to \bar{p}p) \cdot BR(\xi \to K^+K^-)$, given in Table I separately for the $J^{PC} = 0^{++}, 2^{++}$ assignments. The sensitivity improves with increasing width Γ approximately as $\sqrt{\Gamma}$. The $J^{PC} = 2^{++}$ state is produced from a $\bar{p}p$ system with S=1, and L=1, or 3. The L=1 case leads to a $1+3\cos^2\theta$ angular distribution and the L=3 case gives $1-2\cos^2\theta+5\cos^4\theta$. When integrated over our center of mass acceptance $(-.5<\cos\theta<.5)$ these distributions lead to a reduced sensitivity compared to the isotropic J=0 case by a factor of about 1.6.

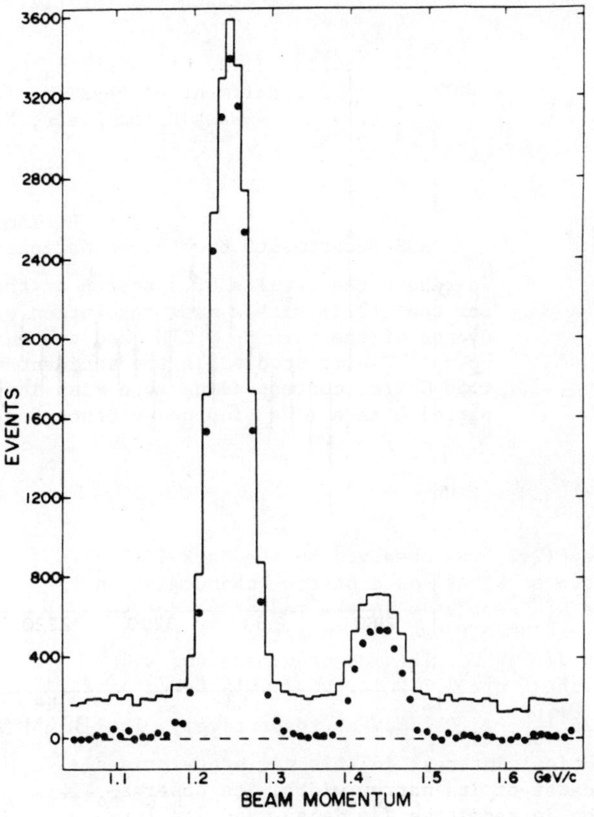

Figure 2: Reconstructed antiproton momentum for two track events assuming the final state particles are kaons. The lower curve shows data after background subtraction.

The signal observed in radiative J/ψ decay implies that the product of branching ratios $BR(J/\psi \to \gamma\xi) \cdot BR(\xi \to K^+K^-)$ is 4×10^{-5}. This result, combined with the absence of a signal in the inclusive photon spectrum, suggests a large branching ratio to K^+K^-. Conservatively interpreting the data we estimate $BR(\xi \to K^+K^-) \gtrsim 5\%$. Combined with our result this would imply $BR(\xi \to \bar{p}p) < 1\%$ for $J=0^{++}$ and $BR(\xi \to \bar{p}p) < .4\%$ for $J^{PC} = 2^{++}$ for a width Γ = 35 MeV.

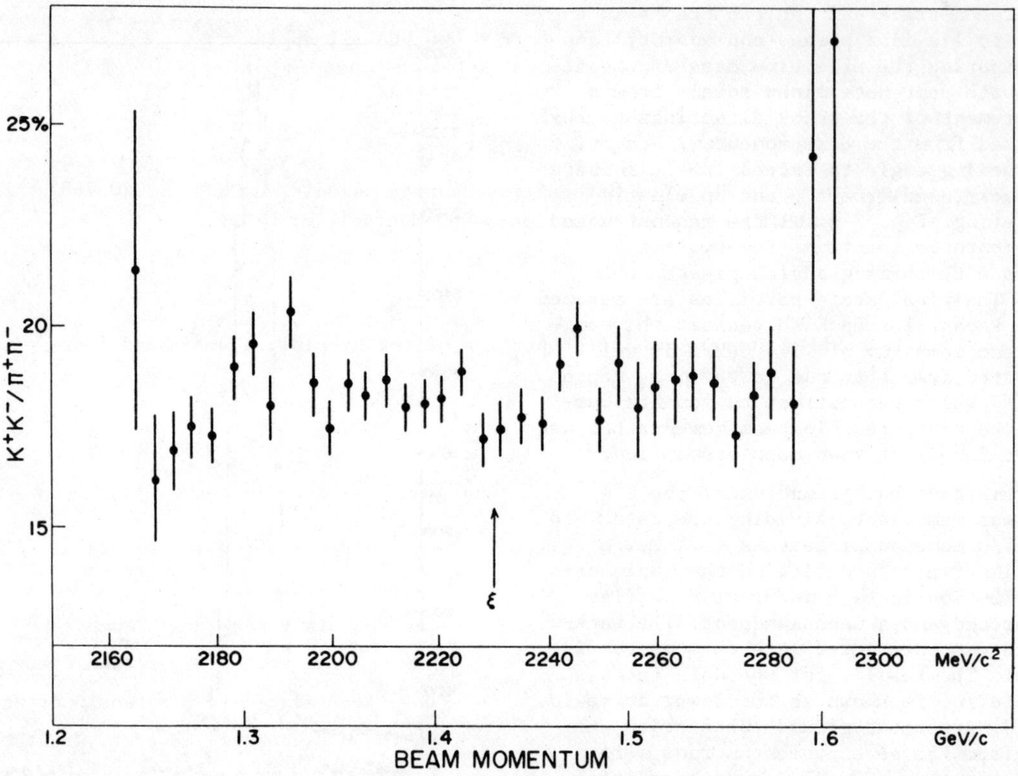

Figure 3: Ratio of K^+K^- to $\pi^+\pi^-$ yield plotted in 3.5 MeV/c^2 mass bins.

Table I: Three standard deviation limits on $BR(\xi\to\bar{p}p)\cdot BR(\xi\to K^+K^-)$.

	$\Gamma = 7\ MeV/c^2$	$\Gamma = 35\ MeV/c^2$
$J^{PC} = 0^{++}$	12×10^{-4}	5.6×10^{-4}
$J^{PC} = 2^{++}$	3.8×10^{-4}	1.8×10^{-4}

REFERENCES

(1) R.M. Baltrusaitis et al., Phys. Rev. Lett. 56, 103 (1986).

(2) Eisenhandler et al., Nucl. Phys. B96, 109 (1975).

LIGHT EXOTIC MESONS FROM QCD DUALITY SUM RULES

Stephan NARISON
Laboratoire de Physique Mathématique (unité associé au CNRS n° 040 768)
USTL, Place E. Bataillon 34060 Montpellier Cedex.

This is a brief summary of the properties of the hybrids, gluonia and four-quark mesons from QCD-duality sum rules.

We have, by now, an increasing evidence that QCD-duality sum rules are an interesting method for understanding the hadron properties in terms of the QCD Lagrangean parameters such as the running coupling and the condensates of quarks and gluons (1). In this talk, I summarize the properties of the light exotic mesons such as the hermaphrodite (hybrids), gluonia and four-quark states from the sum rules. Hadron masses and decay constants are obtained from the two-point correlator $\Pi(q^2)$ which is the time -ordered product of hadronic local current. The analyticity of $\Pi(q^2)$ implies the usual Källen - Lehmann dispersion relation, which can be improved into its sum rule form by applying either the Borel-Laplace or the Gaussian-like operators (1). In its Laplace version, the sum rule can take one of the global duality form :

$$R(\tau) = -\frac{d}{d\tau} \log \int_0^\infty dt e^{-t\tau} \frac{1}{\pi} \operatorname{Im} \Pi(t)$$

which is very sensitive to the lowest ground state mass whilst in its Gaussian version, the sum rules become sets of local finite energy sum rules (FESR). The L H S of $R(\tau)$ is estimated by including the effects of non-trivial lowest dimension quark and gluon vacuum condensates within the framework of the operator product expansion (OPE). According to SVZ, such contributions are expected to parametrize the long distance (large size instanton) behaviour of QCD. Small size instantons might show up through condensates having dimensions larger than eleven, which have been argued to play an important role in the 0^{-+} channel. However, we have not yet a good control of such small size instanton effects. Therefore, testing such effects from a comparison of the sum rule results (which do not consider them) with other ones coming from different approaches or (and) from the data should be of great importance. From the phenomenological side, the simple duality ansatz "one resonance" plus "QCD continuum" (the latter comes from the discontinuity of the QCD diagrams) appears to give a good smearing of the spectral integral in the case of the Laplace sum rules. This is due to the exponential factor which depresses the role of the finite widths and higher states continuum at the "sum rule window" where the information from the sum rule is optimal (stability in τ and in the continuum threshold). However, a much more elaborated parametrization is necessary for the FESR which are very sensitive to the high t- behaviour of the spectral function and so to the detailed structure of the continuum final states.

1. <u>Hermaphrodite or hybrid mesons</u> :

Let us illustrate the uses of the sum rule

in the case of the $\tilde{\rho}(1^{-+})$ hybrid meson which has the quantum number of the local current $J^{\mu}(x) = \bar{u}\lambda a\ \gamma_{\nu} G_{a}^{\mu\nu} d$. The OPE of the associated two-point correlator takes into account the effects of the vacuum condensates of dimension less than six. The QCD-expression of $R(\tau)$ is shown in Fig. 2.1 of Ref. (2) where we have used the current value of the scale Λ and of the condensates. Due to the positivity of $R(\tau)$, its minimum represents an optimal upper bound of 2.6 GeV for the $\tilde{\rho}$ mass. It also signals the critical value τ_{MAX} of τ beyond it the OPE fails. A fitting procedure which equates the QCD and phenomenological sides of $R(\tau)$ inside the region $[0, \tau \leqslant \tau_{MAX}]$ gives a $\tilde{\rho}$ mass of about 1.6 to 2.1 GeV where the uncertainties are mainly due to the value of the four-quark condensate. The stability of such results versus $\sqrt{tc}$ and τ are shown respectively in Figs 2.2, 2.3 of Ref. (2). Finite width effects tend to increase slightly the above predictions by few percent. So, a safer search of the $\tilde{\rho}$ should be extended up to the upper bound of 2.6 GeV. We have estimated the $\tilde{\rho}$-width using a three-point function evaluated at the symmetric euclidian point and asking a duality between the QCD-triangle diagram and the three-meson poles. We have tested this method for evaluating the known three-meson couplings ($g_{\rho\omega\pi}$, $g_{\rho\pi\pi}$ ...). The accuracy of the results should be less good than the ones from the two-point function. So, we expect to have only a rough estimate of the meson widths. From the above procedure, we find that the $\tilde{\rho}$ is a broad resonance which likes to decay into $\rho\pi$ with a width of about 600 MeV. Its branching ratio into $\eta'\eta$ is about five per mil. We have also obtained a mass of the order of 3.8 GeV for the 0^{--} channel, which is, however, two times larger than the one from other QCD models. The high value of the mass is due to the relative important strength of the four-quark condensate appearing in the sum rule. We have also shown that the $\tilde{\rho}$-$\tilde{\Phi}$ splittings due to the $SU(3)_F$-breaking terms are negligible because these terms behave like $m_s^2 \tau_o$ where τ_o is about 0.5 GeV^{-2} for the hybrid mesons.

2. Gluonia :

We have studied the masses and couplings of the 0^{++}, 0^{-+} and 2^{++} gluonia by working with local currents with the appropriate quantum numbers. The global duality sum rules independent of the value of the two-point function at q=0 lead to a mass around 1 GeV for the 0^{++} gluonium bound states of two (3) gluon field strenghts. The one associated to the three gluon field strenght might be higher (4). Mixing of these two gluonia which will imply the "true gluonia" spectrum is under study (4). The 2^{++} lowest ground state mass is accurately obtained (3b) around 1.8 GeV. Its mixing with the $q\bar{q}$ state is also under study (5). If one ignores the effects of small size instantons, the mass of the 0^{-+} gluonium is expected to be around (3b,7) 1.7 GeV and its mixing with the $q\bar{q}$ resonance is small (6) (less than 10°). Therefore its two-photon width is predicted to be less than 1.5 KeV which might be compared with the recent data of the ι two-photon width ($\leqslant$ 2 KeV).

3. Four quark states and the nature of the $\delta(980)$.

Finally, I will briefly present the results for the four-quark states which can serve to test the nature of the $\delta(980)$. The $q\bar{q}$ and $\bar{q}q\bar{q}q$ assignements of the δ predict the same δ-mass (8) and the same hadronic width (9). However the four-quark assumption predicts a large $\delta K\bar{K}$-coupling (9) which should be tested experimentally. Using the quark triangle diagram, implemented by vector meson dominances, for the estimate of the two-photon width, we find (9) that a $\bar{q}q$ assignement leads to a width of about 1.6 KeV while the $\bar{q}q\bar{q}q$ one implies a width of 5.10^{-4} KeV. Both predictions do not explain the Crystal Ball data of (0.2 ± 0.1 ± 0.2) KeV. So, we have still to understand or the nature of the δ or (and) a non-standard mechanism for explaining its two-photon width. More generaly, we need still an amount of efforts for the understanding of the 0^{++} properties and of other exotic mesons of QCD.

Acknowledgements :

I wish to thank N. Isgur for his interest on this talk an M. Chanowitz for discussions. I also thank the theory groups of Berkeley and SLAC for their kind hospitality where the main part of this talk has been written. This work has been partly supported by the University of Montpellier and by the CNRS within the France-USA and France-Spain exchange programs.

References.

(1) M.A. Shifman, A.I. Vainshtein and V.I. Zakharov, Nucl. Phys. B 147, 385 to 519 (1979) Many more complete references on the subject are in : Non-perturbative Methods workshop ; Montpellier (July 1985). Edited by world scientific company (Singapore). For more details on this talk, see e.g. S. Narison, Invited lecture at the Gif Summer School (Paris : September 1986) (in preparation).

(2) J.I. Latorre at al, Trieste preprint IC/85/199 (1985).

(3) V.A. Novikov at al, Nucl. Phys. B 191 (1981) 301 ; S. Narison, Z. Phys. C26 (1984) 209.

(4) J.I. Latorre, S. Narison and S. Paban (in preparation).

(5) E. Bagan, A. Bramon and S. Narison (in preparation).

(6) S. Narison, N. Pak and N. Paver Phys. Lett. 147B (1984) 162 ; see Ref(1C) for a derivation of the mixing angle using FESR.

(7) K. Johnson, MIT preprint C T P 1101 (1983).

(8) J.I. Latorre and P. Pascual, Journal. Phys. G (1986).

(9) S. Narison, Phys. Lett. 175 B (1986) 88.

THE $\iota(1440)$ AND QCD WARD IDENTITIES.

Peter Glyn Williams

Physics Department
Queen Mary College
(University of London)
London E1 4NS

Anomalous Ward identities for QCD are analyzed with contributions of all known pseudoscalar mesons, including the glueball candidate $\iota(1440)$. Implications for the standard resolution of the $U_A(1)$ problem are examined by imposing the important and crucial constraint of positivity for the topological susceptibility χ_t. The pure Yang-Mills susceptibility χ_t^{YM} - a quantity relevant in quenched lattice calculations - is shown to increase quite considerably in the presence of the ι, while χ_t is reduced and may even vanish. Axial couplings are consistent with the suppression expected for a singlet glueball, and give a small width for $\iota \to 2\gamma$ less than 3 keV.

The discovery of another low-lying pseudoscalar meson, the $\iota(1440)$ (now "officially" designated $\eta(1440)$ [1]) provides a possible candidate for one of the unique features of QCD: the glueball. Whether or not the iota is a glueball, its already established quantum numbers raise the important question of its role in the resolution of the U(1) problem. For the nonet of pseudoscalar mesons (π,K,η,η') this is the puzzle of why the isoscalars η and η' are so much more massive than the π, and why the η-η' mass difference is so large. Equivalently, why is the pseudoscalar mixing angle of -10° so different from the 30° found for the other meson nonets? QCD provides a subtle explanation of this puzzle through the existence of gluons, their axial anomaly and non-zero topological charge. When only the pseudoscalar nonet (π,K,η,η') is invoked in their saturation the anomalous U(3) x U(3) Ward identities of QCD provide a fairly successful phenomenology, including a resolution of the U(1) problem [2]. This success is considered to be evidence for the non-Abelian, topological aspects of the underlying theory.

To reproduce these results from a truly fundamental calculation poses a most stringent test for QCD. The non-perturbative techniques needed, such as lattice Monte-Carlo computations, are still in their infancy. The existence and spectrum of glueballs also poses a difficult problem in such calculations and their unambiguous experimental confirmation is even more problematic. It is therefore of considerable help to seek further tests both for QCD and for the phenomenological identification of glueball candidates. Here we present an attempt to evaluate the significance of the $\iota(1440)$ for QCD by examining its role in the saturation of the Ward identities for the U(3) x U(3) chiral algebra of QCD [3].

Besides the coupling constant and quark masses, QCD has an additional parameter, θ, which is a property of the vacuum state. θ-dependence measures vacuum topology, which enters through the topological charge operator:

$$\nu = \int d^4x\, G\tilde{G}(x) \tag{1}$$

where $$G\tilde{G} \equiv \frac{\sqrt{3}}{\sqrt{2}}\frac{g^2}{32\pi^2}\epsilon^{\mu\nu\alpha\beta}\mathrm{Tr}G_{\mu\nu}G_{\alpha\beta} = \partial^\mu K_\mu$$

with $G_{\mu\nu}(x)$ the gluon field tensor, the colour matrix analogue of electromagnetic $F_{\mu\nu}$. The physical matrix element associated with nontrivial vacuum topology ($\nu \neq 0$) is the topological susceptibility:

$$\chi_t = -\left[\frac{d^2E_{VAC}}{d\theta^2}\right]_{\theta=0} = -\frac{i}{6}\int d^4x\, \partial^\mu\partial^\nu T\langle 0|K_\mu(x)K_\nu(0)|0\rangle \tag{2}$$

since this is - modulo some subtleties [3] - the matrix element of a squared operator:

$$\langle Q^2\rangle = -\frac{i}{6}\int d^4x\, T\langle 0|G\tilde{G}(0)\,G\tilde{G}(0)|0\rangle \tag{3}$$

it must be positive. This is a crucial and new restriction in this analysis. χ_t is related to observable quantities via the 9 axial currents $A^i_\mu(x)$ made of quark fields q(x); their divergences contain information about SU(3) symmetry breaking and, for the SU(3) singlet current (i=0) contains the gluon anomaly which, remarkably, is itself the topological density operator:

$$\partial^\mu A^i_\mu(x) = \bar{q}\,\gamma_5\frac{\lambda_i}{2}q + \delta_{io}G\tilde{G}(x) \tag{4}$$

It is the matrix elements of these - analogues of F_π = 93 Mev and F_K - which enter Ward identities:

$$\left.\begin{aligned} \langle 0|\partial^\mu A^i_\mu(0)|a\rangle &= m_a^2 F_{ia} \\ \langle 0|G\tilde{G}(0)|a\rangle &= m_a^2 A_a \end{aligned}\right\} \begin{aligned} &i = 0\ldots 8 \\ &\text{and} \\ &a = \eta, \eta' \ \& \ \iota \end{aligned} \qquad (5)$$

In the chiral limit of vanishing quark masses the octet become massless Goldstone bosons; but the ninth meson (η') which is the would-be Goldstone boson of $U_A(1)$ keeps a mass through the anomaly as a consequence of its proportionality to the topological charge density - non-zero in QCD. This is measured by the topological charge constant, $A_{\eta'}$, of the η' (and through mixing, A_η and A_ι):

When saturated with the nonet of pseudoscalars and $\iota(1440)$ and upon elimination of the quark masses and vacuum correlation functions these Ward identities give 4 sum rules [2,3]:

$$m_a^2 F_{8a}^2 = \frac{1}{3}(4m_K^2F_K^2 - m_\pi^2F_\pi^2)$$

$$m_a^2 F_{8a}F_{0a} = -\frac{2}{3}\sqrt{2}\,(m_K^2F_K^2 - m_\pi^2F_\pi^2) \qquad (6)$$

$$m_a^2 F_{0a}(F_{0a} - A_a) = \frac{1}{3}(2m_K^2F_K^2 + m_\pi^2F_\pi^2)$$

$$m_a^2 F_{8a}A_a = 0$$

where the repeated index "a" is summed over η, η' and now also ι. Solutions to these equations without $\iota(1440)$ (supplemented with judiciously chosen input) show that both A_η and $A_{\eta'}$ are non-zero and of the order of F_π in the real world, and therefore provide evidence for gluons, their anomaly and QCD topological charge [2].

In the presence of $\iota(1440)$ there are 9 unknown quantities F_{aj}, A_a but only 4 equations. The extra input required to solve follows the methods of ref.[2]:

(i) Dominance of $\psi \to (\eta,\eta',\iota)\gamma$ by the gluon pseudoscalar operator $G\tilde{G}$ gives the ratios:

$$\frac{A_{\eta'}}{A_\eta} = 0.8 \pm 0.1 \text{ and } \frac{A_{\eta'}}{A_\iota} = (1.7 \pm 0.4)B^{1/2}$$

where B is the branching ratio for $\iota \to K\bar{K}\pi$, the overwhelmingly dominant decay mode.

(ii) Sum rules for $(\eta,\eta',\iota) \to 2\gamma$ at zero momentum obtained from the electromagnetic triangle anomaly give 2 relations between the F_{aj}.

(iii) The mild assumption is made that SU(3) symmetry is violated no more than 20% for η and η' couplings.

(iv) Positivity of the topological susceptibility χ_t. This quantity is calculated from an additional Ward identity for a quantity related to $\langle Q^2\rangle$ and gives [2,3]:

$$6\chi_t = m_a^2 F_{0a}A_a - m_a^2 A_a^2 \qquad (7)$$

This has the general form

$$6\,\chi_t = 6\,\chi_t^{YM} + (\bar{q}q \text{ meson contributions})$$

The second negative definite term is the $\bar{q}q$ contribution, leading to the identification of the topological susceptibility for the underlying gauge theory in the absence of quarks:

$$6\,\chi_t^{YM} = m_a^2 F_{0a} A_a \qquad (8)$$

The form for χ_t given above is not guaranteed to be positive because it depends on the sign of $(F_{0a} - A_a)$; it is when only η and η' contribute by virtue of the Ward identities but this is no longer the case when a third meson such as the $\iota(1440)$ is included. From Fig.1 we see that viable solutions require:

$$A_\iota \approx F_{0\iota} \qquad \text{with} \qquad 0 \leqslant \chi_t \leqslant 1.4\,(m_\pi F_\pi)2$$

This remarkable result leads to an enormous simplification of our problem [3].

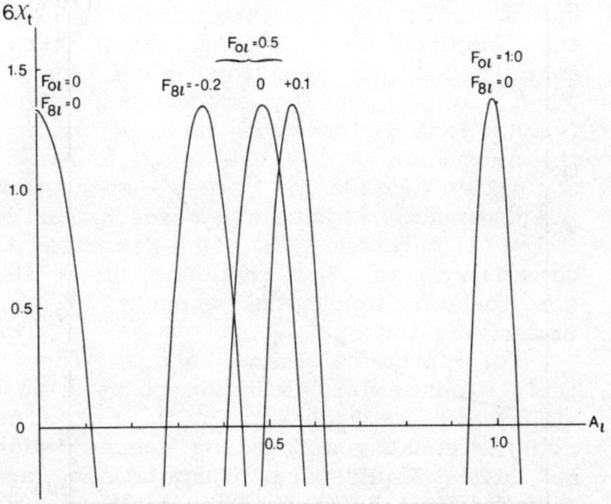

Figure 1. Regions of positive topological susceptibility χ_t (in $(m_\pi F_\pi)^2$ units) as functions of $F_{0\iota}$, $F_{8\iota}$ & A_ι (in F_π units). Note the narrow allowed range of A_ι close to $F_{0\iota}$.

The connection between the U(1) problem, topological charge (A_a) and topological susceptibility (χ_t and χ_t^{YM}) is most clearly seen by taking a suitable linear combination of the Ward identities:

$$3\, m_\pi^2 F_\pi^2 = m_a^2 (F_{8a} + \sqrt{2} F_{0a})^2 - 2 m_a^2 A_a (A_a - F_{0a}) \qquad (9)$$

The left hand side is a small quantity, vanishing in the chiral limit: to produce the dramatic cancellation of the large positive η, η′ and ι contributions on the right requires at least one $A_a \neq 0$ which, in its turn, implies a large value for χ_t^{YM}.

We end with a summary and some comments:

(i) Solutions to the Ward identities with positive topological susceptibility χ_t are quite well constrained. The best solutions favour $\chi_t \approx 0$, but a large χ_t^{YM} quite considerably larger than in the absence of ι(1440) - see Fig. 2. This should be bourne in mind by people doing nonperturbative computations in QCD.

(ii) A large branching ratio for $\iota \to K\bar{K}\pi$, strongly favoured by experiments [1], points towards a suppressed axial coupling: $F_{0\iota} \approx 0.5 F_\pi$ and a mainly singlet ι(1440): $F_{8\iota} \approx -0.2$. This is consistent with the $1/\sqrt{N_{colour}}$ suppression expected of a glueball; but a suppression is also a property of higher $\bar{q}q$ radial excitations, and we cannot rule out this possibility.

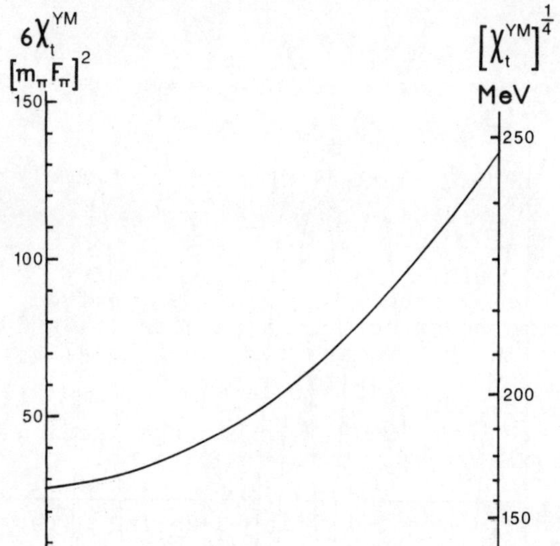

Figure 2. Topological susceptibility χ_t^{YM} for the pure Yang-Mills theory, showing a rapid increase with $F_{0\iota}$. $F_{8\iota} = -0.2$ is kept fixed. (χ_t^{YM} is to be compared with the susceptibility obtained in "quenched" lattice simulations.)

(iii) Small values of the width for $\iota \to 2\gamma$ are predicted by the sum rules, consistent with the recent upper bound of 2 keV. obtained by the TPC/2γ Collaboration [4] - see Fig. 3.

(iv) We would like to comment on a similar analysis [5] which uses basically the same framework as we do but without the crucial positivity requirement on χ_t. These authors fail to take account of the large χ_t^{YM} in saturating the anomalous Ward identities and are therefore not self-consistent - see Ref [3] for a detailed account of this error. One consequence is that solutions could only be found in [5] with a branching ratio for $\iota \to K\bar{K}\pi \leqslant 30\%$.

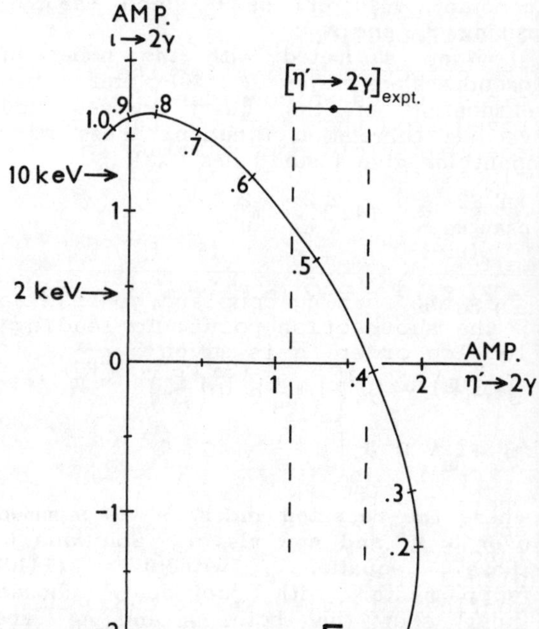

Figure 3. The width for $\iota \to 2\gamma$ with $F_{8\iota} = -0.2$ and $\Gamma(\eta \to 2\gamma) = 0.4$ keV. The favcured region is $F_{0\iota} \approx 0.5$ in F_π units.

REFERENCES.

[1] Particle Data Group, Phys. Lett. 170B (1986).

[2] E. van Herwijnen, P.G. Williams, Phys. Rev. D24 (1981) 240 & Nucl. Phys. B196 (1982) 109 and references therein.

[3] P.G. Williams, Phys. Rev. D29 (1984) 1032 and preprint QMC-85-14, to be published in Phys. Rev. D, and references therein.

[4] TPC/2γ Collaboration, Phys. Rev. Lett. 57 (1986) 51.

[5] K.A. Milton, W.F. Palmer, S.S. Pinsky, Phys. Rev. D27 (1983) 202.

HADRONS WITH ONE HEAVY QUARK IN AN EFFECTIVE ACTION APPROXIMATION TO QCD

B. Margolis, R.R. Mendel and H.D. Trottier

Physics Department, McGill University
3600 University St., Montreal, Quebec, Canada H3A 2T8

We use a renormalization group improved local effective action of QCD to describe all hadrons containing one heavy quark. The only parameters are the QCD scale $\Lambda_{\overline{MS}}$ and the quark masses. Linear confinement arises naturally. For mesons we find a self consistent Abelian solution and obtain good agreement with known spectroscopy, transition rates and the string tension. To describe baryons we need two Abelian charges but find that the formalism for the baryons turns out to be similar to the meson case. Preliminary results for baryon spectroscopy are presented.

1. INTRODUCTION

The effective action for the gauge fields which we consider [1-3] is derived from the usual QCD Lagrangian in the mean field approximation and includes leading logarithm corrections in $F^{a\mu\nu} F^a_{\mu\nu}$ [4,5]:

$$\mathcal{L}^{gauge}_{eff}(\mu,-\tfrac{1}{2}F^{a\mu\nu}F^a_{\mu\nu})=-\tfrac{1}{4}\bar{g}^{-2}(\mu,-\tfrac{1}{2}F^{a\mu\nu}F^a_{\mu\nu})F^{a\mu\nu}F^a_{\mu\nu} \quad (1)$$

$\bar{g}$ is the running coupling constant and μ the subtraction point. To leading log-log order, $\bar{g}$ is given by:

$$\bar{g}(\mu,F) = g^{-2}\left\{1+\frac{1}{4}b_0\ln\left(\frac{F}{\mu^4}\right) - 2\frac{b_1}{b_0}g^2\ln\left[1+\frac{1}{4}b_0\ln\left(\frac{F}{\mu^4}\right)\right]\right\} \quad (2)$$

where $F\equiv-\frac{1}{2}F^{a\mu\nu}F^a_{\mu\nu}$, $g=\bar{g}(\mu,\mu^4)$ and b_0 and b_1 are the one and two loop β-function coefficients, which we take with 3 light flavors [1-3]. Linear confinement arises since the effective action (1) has an extremum for nonzero values of the field strengths. The string tension is calculated in terms of $\Lambda_{\overline{MS}}$.

2. THE MODEL - $Q\bar{q}$ MESONS

For mesons containing one very heavy quark (ψ_M) with bare mass M' and one light antiquark (ψ_0) of mass m the fermionic part of the Lagrangian is:

$$\mathcal{L}^F = i\bar{\psi}_0 \not{\partial}\psi_0 - m\bar{\psi}_0\psi_0 - M'\bar{\psi}_M\psi_M - j^{\mu a}A^a_\mu \quad (3)$$

$$j^{\mu a} = -\psi_0\gamma^\mu\tfrac{1}{2}\lambda^a\psi_0 - \psi_M\gamma^\mu\tfrac{1}{2}\lambda^a\psi_M \quad (4)$$

To get a static potential we assume that $j^{\mu a}$ is time independent. We also assume that the action is minimized by an Abelian configuration of fields and currents. We then make the following substitutions in Eqs.(1)-(4):

$$A^a_\mu j^{\mu a} \longrightarrow A_\mu j^\mu$$

$$j^\mu \equiv \sqrt{\tfrac{4}{3}}\left(\bar{\psi}_0\gamma^\mu\psi_0 - \bar{\psi}_M\gamma^\mu\psi_M\right) \quad (5)$$

$$-\tfrac{1}{2}F^{a\mu\nu}F^a_{\mu\nu} \to F \equiv -\tfrac{1}{2}(\partial_\mu A_\nu - \partial_\nu A_\mu)^2$$

where an Abelian charge of $\sqrt{4/3}$ satisfies the main color singlet properties $Q^a_M + Q^a_0 = 0$ and $\sum_a Q^a_0 Q^a_M = -4/3$. The Abelian static potential is:

$$V_{stat} = \int d^3x\left[-\mathcal{L}^{gauge}_{eff}(F) + A_\mu j^\mu + \psi_0^\dagger\vec{\alpha}\cdot(-i\vec{\nabla})\psi_0 + m\bar{\psi}_0\psi_0 + M'\bar{\psi}_M\psi_M\right] \quad (6)$$

and the usual static Maxwell equations are obtained from the Abelian action:

$$\vec{\nabla}\cdot\vec{D} = j^0(\vec{r}) \qquad \vec{\nabla}\times\vec{E} = 0$$
$$\vec{\nabla}\times\vec{H} = \vec{j}(\vec{r}) \qquad \vec{\nabla}\cdot\vec{B} = 0 \quad (7)$$

where $\vec{E}\equiv-\vec{\nabla}A^0$, $\vec{B}\equiv\vec{\nabla}\times\vec{A}$ and:

$$\vec{D}\equiv\varepsilon\vec{E} \qquad \vec{H}\equiv\varepsilon\vec{B}$$
$$\varepsilon(\vec{r}) \equiv \frac{\partial\mathcal{L}^{gauge}_{eff}(F)}{\partial(\frac{1}{2}F)}, \quad F = \vec{E}^2 - \vec{B}^2 \quad (8)$$

The dielectric function $\varepsilon(F)$ derived from the effective Lagrangian (1) and (2) in the leading log model is [1-5]:

$$\varepsilon(F) = \frac{1}{4}b_0\ln\frac{F}{\kappa^2} \qquad \text{log} \quad (9)$$

where $\kappa^2\equiv\mu^4 e^{-1}\exp(-4/b_0g^2)$. In the leading log-log model we take $\varepsilon(F)$ to be:

$$\varepsilon(F) = \frac{1}{4}b_0\ln\frac{F}{\kappa^2} - 2\frac{b_1}{b_0}\ln\ln\frac{F}{\kappa^2} + 2\frac{b_1}{b_0}\ln 2 \qquad \text{log-log} \quad (10)$$

In both models, Adler has shown that κ is related to the QCD scale $\Lambda_{\overline{MS}}$ by[5]

$$\Lambda_{\overline{MS}} = 0.959\,\kappa^{\frac{1}{2}} \quad (11)$$

We have yet to determine the sources for (7). The heavy quark current can be well approximated in the center of mass frame by:

$$j^0_M(\vec{r}) = -\sqrt{\tfrac{4}{3}}\,\delta^3(\vec{r}) \qquad \vec{j}_M(\vec{r}) = 0 \tag{12}$$

Unfortunately, the light quark wavefunction is coupled nonlinearly to (7):

$$\left(i\not\partial - \sqrt{\tfrac{4}{3}}\not A - m\right)\psi_0 = 0 \tag{13}$$

We approximate this equation by solving instead:

$$\left[\vec{\alpha}\cdot(-i\vec{\nabla}) - \frac{\alpha_{eff}}{r} + \beta m\right]\psi_0 = \xi\,\psi_0 \tag{14}$$

together with a baglike boundary condition and normalization:

$$\bar{\psi}_0(\vec{r})\psi_0(\vec{r})\Big|_{r=R} = 0 \tag{15a}$$

$$\int_{r\le R} d^3r\, \psi_0^\dagger(\vec{r})\psi_0(\vec{r}) = 1 \tag{15b}$$

We are assuming that the essential nonlinear dielectric properties of the model are contained in the Maxwell equations and can be parametrized in the Dirac equation by the position independent coupling constant α_{eff} (which is determined below by self-consistency) and the boundary condition at R (which is determined by minimizing $V_{stat}(R)$). See Ref.3 for more details. We note that this situation is very different from the bag model where a perturbative coupling constant is determined from phenomenology and a bag pressure is introduced by hand to produce confinement.

For the ground state ($j=\tfrac{1}{2}$) the solution to (14) gives the following simple forms for the light quark current [2,3], which is singular at the origin:

$$j^0_0(\vec{r}) = \rho_0(r);\; \vec{j}_0(\vec{r}) = J_0(r)\langle\vec{\sigma}_0\rangle\times\hat{r} \quad r\le R \tag{16}$$

($\vec{\sigma}_0$ is twice the spin operator of the light quark). The Maxwell equations now reduce to:

$$\vec{\nabla}\cdot(\varepsilon\vec{E}) = \rho_0(r) - \sqrt{\tfrac{4}{3}}\,\delta^3(\vec{r}) \tag{17a}$$

$$\vec{\nabla}\times(\varepsilon\vec{B}) = J_0(r)\langle\vec{\sigma}_0\rangle\times\hat{r} \tag{17b}$$

$$\vec{\nabla}\times\vec{E} = \vec{\nabla}\cdot\vec{B} = 0 \tag{17c}$$

Even these simplified equations are still very difficult to solve. We have discovered that a simple ansatz solves (17) to a good approximation [2,3]:

$$\vec{B}(\vec{r}) = \vec{B}_0 \equiv B_0\langle\vec{\sigma}_0\rangle \qquad r\le R \tag{18}$$
$$\vec{E}(\vec{r}) = E(r)\hat{r},\quad \varepsilon=\varepsilon(r)$$

Using this ansatz the homogeneous Maxwell equations are automatically satisfied, and we get $\vec{D}(R)=0$, which implies $\varepsilon(R)=0$. Thus color is confined to the inside region. The minimum value of the field strength, F_0, which is the root of the equation:

$$\varepsilon(F_0) = 0 \tag{19}$$

occurs at the boundary r=R, and in our two models is given by:

$$F_0^{\frac{1}{4}} = 1.0\ \kappa^{\frac{1}{2}} \quad \text{log}$$
$$F_0^{\frac{1}{4}} = 1.296\ \kappa^{\frac{1}{2}} \quad \text{log-log} \tag{20}$$

Consistency of the ansatz at $\vec{r}=0$ uniquely determines α_{eff} to be:

$$\alpha_{eff} = \sqrt{3/4} \tag{21}$$

The value of B_0 is chosen to minimize the value of:

$$\Delta(B_0) \equiv \frac{1}{\int d^3r\,|\vec{j}_0(\vec{r})|}\cdot\int d^3r\,|\vec{\nabla}\times\vec{H}-\vec{j}_0| \tag{22}$$

Δ can be thought of as measuring the average fractional "error" we make in solving (17b) with the anstaz of (18). In all applications to $Q\bar{q}$ mesons we find Δ<0.15. We conclude that (18) is a good approximation (see Ref.3 for more details).

We can now compute the energy of the system obtained from V_{stat} of (6) after subtracting out an infinite vacuum energy. In the log-log model we have:

$$U(R) = M' + \langle\vec{\alpha}\cdot\vec{p}\rangle_0 + m\langle\beta\rangle_0$$
$$+\frac{1}{2}\int d^3r\left\{\varepsilon F + \frac{1}{4}b_0(F-F_0) - 2\frac{b_1}{b_0}\kappa^2\left[E_i\left(\ln\frac{F}{\kappa^2}\right) - E_i\left(\ln\frac{F_0}{\kappa^2}\right)\right]\right\} \tag{23}$$

(where $Ei(x)=\int_{-\infty}^{x} e^t/t\,dt$). The integral in (23) is divergent at small r due to the infinite self-energy of the heavy quark. We regulate this infinity by imposing a lower cutoff on the radial integration. The minimum of U(R) occurs at $R_0=1.12\kappa^{-\frac{1}{2}}$ (see Fig.1). We also include as perturbations the slow motion of the heavy quark and its color magnetic moment $\vec{\mu}_M=-\sqrt{4/3}\,\vec{\sigma}_M/2M$ [2,3]:

$$U_{rec}(R) = \langle\vec{\alpha}\cdot\vec{p}\rangle_0^2/2M \tag{24}$$

$$U_{mag}(R) = -\langle\vec{\mu}_M\cdot\vec{B}_0\rangle \tag{25}$$

where M is the renormalized heavy quark mass. The total energy is then:

$$U_{tot}(R) = U(R) + U_{rec}(R) + U_{mag}(R) \tag{26}$$

Finally we can show [1-3] that U(R) increases linearly with R for large R. For a light quark with zero mass, we interpret [3] the slope:

$$\sigma \equiv \lim_{R\to\infty} dU(R)/d(\langle r^2\rangle^{\frac{1}{2}}) \quad (m=0) \tag{27}$$

as a string tension ($\langle r^2\rangle^{\frac{1}{2}}$ is the rms radius of the light quark). In our two models, σ takes the values:

$$\sigma^{\frac{1}{2}} = 1.035\ \kappa^{\frac{1}{2}} \quad \text{log}$$
$$\sigma^{\frac{1}{2}} = 1.342\ \kappa^{\frac{1}{2}} \quad \text{log-log} \tag{28}$$

3. RESULTS FOR $Q\bar{q}$ MESONS

The parameter values we use are [2]:

$$\begin{aligned} \Lambda_{\overline{MS}} &= 0.270 \text{ GeV} \\ m_u = m_d &= 0 \\ m_s &= 0.215 \text{ GeV} \\ M_c &= 1.60 \\ M_b &= 5.00 \\ M_t &= 40.0 \end{aligned} \qquad (29)$$

The string tension (27,28) in the log-log model is then $\sigma = (378 \text{ MeV})^2$. Our results for the masses, transition rates and weak decay constants in the log-log model are given below. See [2] for discussions and comparisons with results in the log model.

4. THE MODEL - Qqq BARYONS

For baryons we have three quark color charges, which satisfy:

$$Q_M^a + Q_1^a + Q_2^a = 0 \qquad (30)$$

(1 and 2 are light quark labels). We must now keep two independent color charges, which correspond to the diagonal generators a=3,8. However, as shown in [3], for light quarks with equal masses, the light quark pair behaves effectively as a diquark, with a current pointing in color space along $Q_1^a + Q_2^a = -Q_M^a$. Thus one color component of the color electromagnetic fields vanishes identically, and the field equations for the Qqq system are almost identical to (14) and (17). Unfortunately the ansatz of (18) is not as good, with $\Delta \sim 0.35$ (see (22)). However, the results for the Σ_c and Λ_c using this ansatz are reasonable:

$$\begin{aligned} \Sigma_c &= 2460 \text{ MeV} \\ \Lambda_c &= 2380 \text{ MeV} \end{aligned} \qquad (31)$$

(experimentally, Σ_c=2447 MeV and Λ_c=2281 MeV). More reliable results await an improvement in our approximate treatment of the field equations (7), (8) and (13). We are currently working on this problem.

5. CONCLUSIONS

For $Q\bar{q}$ mesons, we find that all predictions in the log-log model agree within experimental errors with the existing data, using reasonable values for $\Lambda_{\overline{MS}}$ and the quark masses. Furthermore, the formalism for Qqq baryons is almost identical to the $Q\bar{q}$ case and our preliminary results are promising. We are led to the conclusion that the renormalization group improved local effective action of QCD seems to contain enough elements of the full theory to give an adequate quantitative description of the main properties of all hadrons containing one heavy quark.

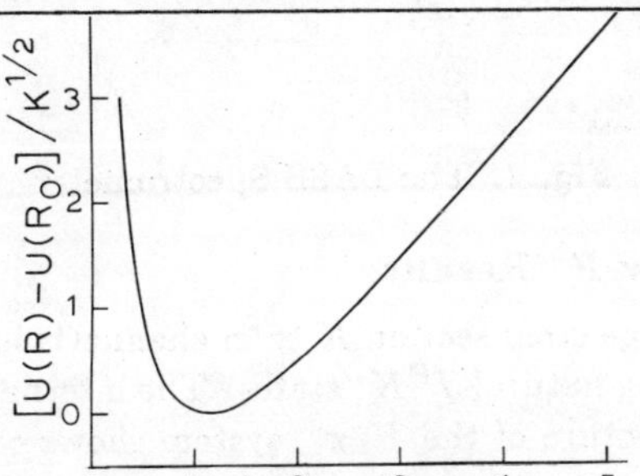

FIG. 1. $U(R)-U(R_0)$ for very large M and $m=0$, in the leading-log-log model.

Masses and rms radii in the log-log model (see [2] for details).

State	Mass (MeV)	$\langle r^2 \rangle^{1/2}$ (fm)	Experiment (MeV)
D	1867	0.56	1867
D*	2009	0.65	2009
F	1962	0.61	1965±3
F*	2094	0.69	2109±9±7
B	5276	0.54	5273
B*	5327	0.57	5325±2±4
Fb	5358	0.57	
Fb*	5405	0.60	
T	40278	0.52	
T*	40285	0.53	
Ft	40355	0.55	
Ft*	40361	0.55	

Transition and decay rates in the log-log model (see [2] for additional results).

Transition	Branching ratio (%)	Experiment
$D^{+*} \to D^0 \pi^+$	68.1	49 ± 8
$D^{+*} \to D^+ \pi^0$	31.5	34 ± 7
$D^{+*} \to D^+ \gamma$	0.4	17 ± 11
$D^{0*} \to D^0 \pi^0$	63.6	54 ± 9
$D^{0*} \to D^0 \gamma$	36.4	46 ± 9

Pseudoscalar decay constants in the log-log model (see [2] for details).

f_D	0.22 GeV	f_{Fb}	0.32 GeV
f_F	0.26	f_T	0.28
f_B	0.26	f_{Ft}	0.35

REFERENCES

[1] B.Margolis and R.R.Mendel, Phys.Rev.D30, 621(1984).

[2] B.Margolis, R.R.Mendel and H.D.Trottier, Phys.Rev.D33,2666(1986) and refs. therein.

[3] B.Margolis, R.R.Mendel and H.D.Trottier, Proceedings of the Eigth Annual Montreal Rochester Syracuse Toronto Meeting, (McGill University, Montreal, 1986).

[4] G.Matinyan and G.K.Savvidy, Nucl.Phys. B134,539(1978);H.Pagels and E.Tomboulis, ibid,B143,485(1978).

[5] S.L.Adler and T.Piran, Rev.Mod.Phys.56,1 (1984) and refs. therein.

A STUDY OF STRANGE AND STRANGEONIUM STATES PRODUCED IN LASS*

D. Aston,[a**] N. Awaji,[b] T. Bienz,[a] F. Bird,[a] J. D'Amore,[c] W. Dunwoodie,[a] R. Endorf,[c] K. Fujii,[b‡] H. Hayashii,[b♭] S. Iwata,[b‡] W.B. Johnson,[a] R. Kajikawa,[b] P. Kunz,[a] D.W.G.S. Leith,[a] L. Levinson,[a♮] T. Matsui,[b‡] B.T. Meadows,[c] A. Miyamoto,[b‡] M. Nussbaum,[c] H. Ozaki,[b] C.O. Pak,[b‡] B.N. Ratcliff,[a] D. Schultz,[a] S. Shapiro,[a] T. Shimomura,[b] P. K. Sinervo,[a†] A. Sugiyama,[b] S. Suzuki,[b] G. Tarnopolsky,[a♯] T. Tauchi,[b‡] N. Toge,[a] K. Ukai,[d] A. Waite,[a♮♮] S. Williams[a♯♯]

[a] *Stanford Linear Accelerator Center, Stanford University, P.O. Box 4349, Stanford, California 94305*
[b] *Department of Physics, Nagoya University, Chikusa-ku, Nagoya 464, Japan*
[c] *University of Cincinnati, Cincinnati, Ohio 45221*
[d] *Institute for Nuclear Study, University of Tokyo, 3-2-1 Midori-cho, Tanashi-shi, Tokyo 188, Japan*

Results are presented from the analysis of several final states from a high-sensitivity (4 ev/nb) study of inelastic K^-p interactions at 11 GeV/c carried out in the LASS Spectrometer at SLAC. New information is reported on leading and underlying K^* states, and the strangeonium states produced by hypercharge exchange are compared and contrasted with those observed in radiative decays of the J/ψ.

1. Overview of the Experiment

The spectroscopy of light-quark mesons continues to play a significant role in High Energy Physics. Much is now known, but our understanding of higher excitations and non-leading states is still far from complete. In order to make a useful contribution, an experiment must have both high sensitivity and good acceptance, criteria fulfilled by the experiment whose results are described below.

The Large Aperture Superconducting Solenoid (LASS) Spectrometer[1] is shown in Fig. 1. Situated in an RF separated beam, it features a solenoidal vertex detector and downstream dipole spectrometer giving good acceptance over 4π sr and good momentum resolution. Two threshold Čerenkov counters, Time-of-Flight counters and dE/dx measurement in the cylindrical chambers surrounding the liquid hydrogen target provide good particle identification. The trigger for the experiment was two or more charged particles in the "box" of proportional chambers surrounding the target—essentially σ_{tot} except for the all-neutral final states.

The results presented below come from studies of K^* production in the channels $K^-\pi^+n$, $\bar{K}^\circ\pi^+\pi^-n$, and $K^-\eta p$ and of "strangeonium" production by hypercharge exchange in $K^\circ_s K^\pm\pi^\mp\Lambda$, $K^-K^+\Lambda$, and $K^\circ_s K^\circ_s\Lambda$.

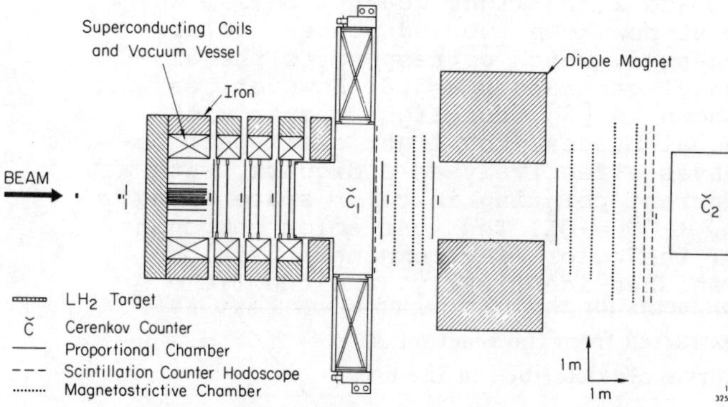

Fig. 1. The LASS Spectrometer

2. New K^* Results

The large cross section $K^-\pi^+n$ channel is ideal for studying natural J^PK^* states. The internal angular structure of the $K^-\pi^+$ system shows complex structure, and is analysed[2] in terms of moments of spherical harmonic functions in the Gottfried-Jackson (t-channel helicity) frame. In general, states of spin J will appear in moments up to L=2J. After demanding $|t'| < 0.2$ (GeV/c)^2 and removing events with π^+n mass below 1.7 GeV/c^2 (N^* cut) there remain 151 000 events with $K^-\pi^+$ masses below 2.6 GeV/c^2. Figure 2 clearly shows the well-known leading states, with Breit-Wigner fits giving masses (widths) in agreement with the world averages:[3] $K^*(892)$ 897.0 ± 1.4 (49.9 ± 2.5); $K^*(1430)$ 1433.0 ± 2.1 (115.8 ± 4.3); and $K^*(1780)$ 1778.1 ± 7.7 (186 ± 36). All values are in MeV/c^2 and systematic errors are included.

*Work supported in part by the Department of Energy, contract DE-AC03-76SF00515, the National Science Foundation, grant PHY82-09144, and the Japan U.S. Cooperative Research Project on High Energy Physics.
**Speaker

The higher moments shown in Fig. 3 confirm the $J^P=4^+$ $K^*(2060)$ and require a new 5^- state. The curves shown are the result of a simple model fit to the 21 moments with $L \leq 10$ and $M \leq 1$, higher moments being consistent with zero. The F, G, and H-waves are parametrised as Breit-Wigners with a background term while the S, P and D-waves are assumed to be coherent amplitudes, with linear mass dependence in both magnitude and phase. The $M=1$ moments are related to those with $M=0$ using the parametrisation of Estabrooks et al.[4] The resulting masses (widths) in MeV/c^2 are 2062 ± 27 (221 ± 75) and 2382 ± 33 (178 ± 69) for $J^P = 4^+$ and 5^- respectively. The significance of the 5^- structure compared with a background term alone is $\sim 5\sigma$.

We turn now to the related $\bar{K}^\circ \pi^+ \pi^- n$ channel which can be viewed as exploring inelastic $K\pi$ interactions, while $K^- \pi^+ n$ tells us only about $K\pi$ elastic scattering. Figure 4 shows the observed $\bar{K}^\circ \pi^+ \pi^-$ mass spectrum after applying an N^* cut; there are 34 000 events in the final Partial Wave Analysis (PWA) sample below 2.3 GeV/c^2. A three-body PWA using the SLAC-LBL program reveals, surprisingly, that most of the $\bar{K}^\circ \pi^+ \pi^-$ production is resonant.[5] I will concentrate on the natural J^P production, which dominates all the important features.

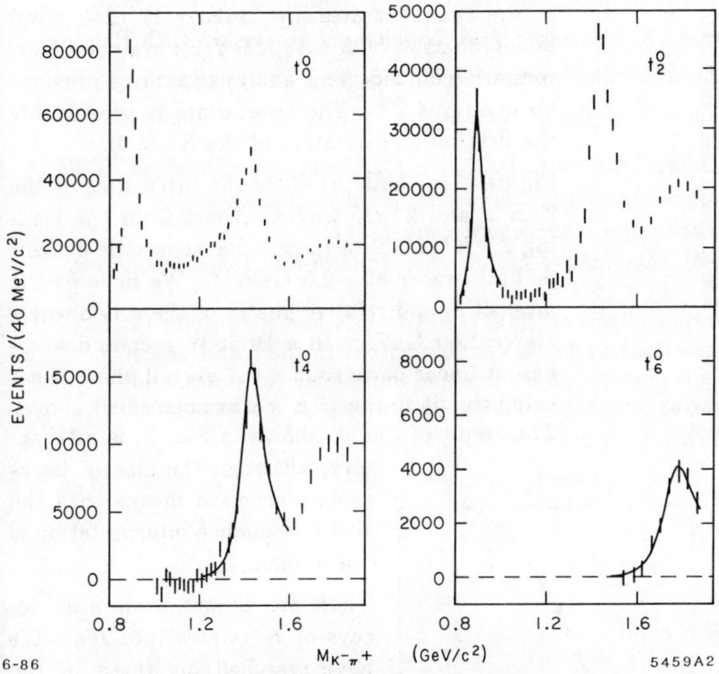

Fig. 2. The unnormalised L-even, $M=0$ $K^-\pi^+$ moments for the mass region below 1.88 GeV/c^2 extracted from the reaction $K^-p \to K^-\pi^+n$. The curves are described in the text.

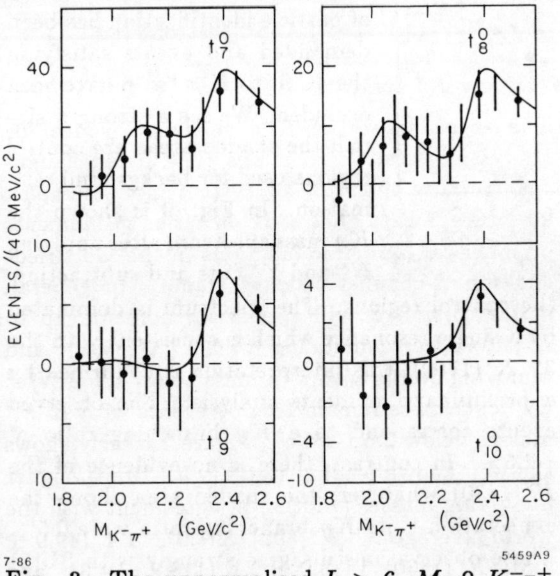

Fig. 3. The unnormalised $L > 6$, $M=0$ $K^-\pi^+$ moments for the mass region above 1.88 GeV/c^2 extracted from the reaction $K^-p \to K^-\pi^+n$. The moments are plotted in overlapping bins; black dots indicate the independent mass bins used for the fit described in the text.

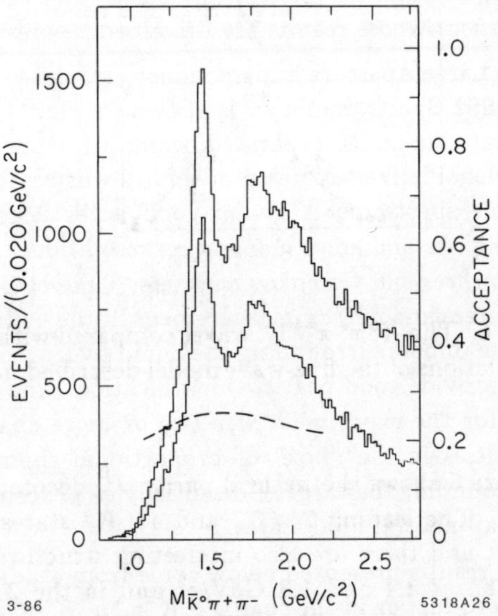

Fig. 4. The $\bar{K}^\circ \pi^+ \pi^-$ mass spectrum from the reaction $K^-p \to \bar{K}^\circ \pi^+ \pi^- n$. The inner histogram is the PWA sample with $|t'| < 0.3$ (GeV/c)^2; the dashed line shows the mass dependence of the acceptance function.

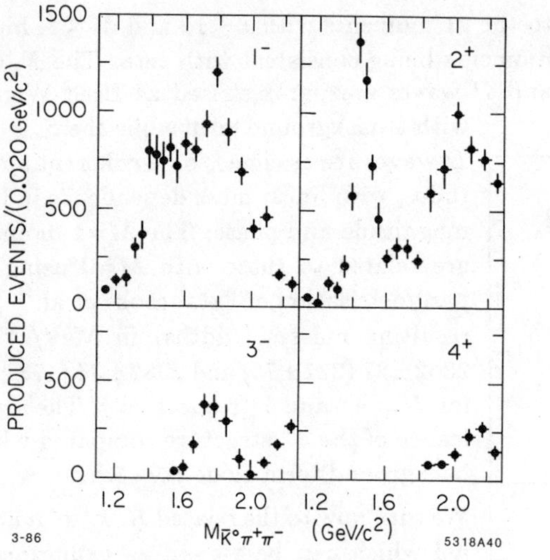

Fig. 5. The $\bar{K}^\circ\pi^+\pi^-$ natural spin-parity wave intensities. Partial waves of the same J^P are summed coherently.

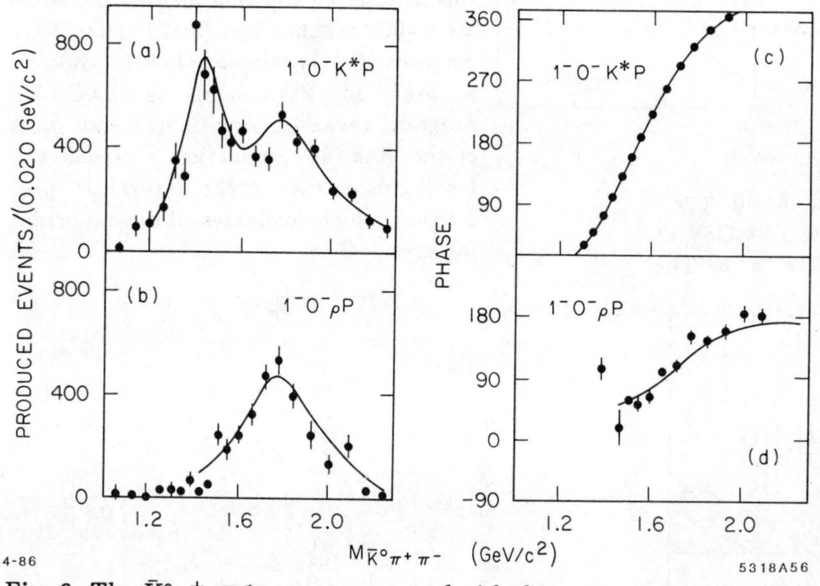

Fig. 6. The $\bar{K}^\circ\pi^+\pi^-$ 1^- waves compared with the predictions of the five-wave model described in the text.

Figure 5 shows the natural parity J^P decomposition. The leading 2^+, 3^- and 4^+ K^* states are clear, and there are also interesting structures in the 1^- at 1.4 and 1.8 GeV/c^2 and in the 2^+ at about 2 GeV/c^2. Figures 6(a) and (b) show the 1^- intensity broken down into $K^*\pi$ and $K\rho$ components, indicating states at ~1.4 GeV/c^2 coupling only to $K^*\pi$ and ~1.75 GeV/c^2 coupling to both $K^*\pi$ and $K\rho$. We have made a simultaneous fit to these waves and the leading $2^+K^*\pi$, $3^-K^*\pi$ and $3^-\rho K$ waves, thus tightly constraining the relative phase behaviour of the 1^- waves. The result of this fit is shown in Fig. 6; a coherent background was allowed in the $1^-K^*\pi$ amplitude, though this is not essential for a good fit. The masses (widths) of the two 1^- states are: 1420 ± 17 (240 ± 30) and 1735 ± 30 (423 ± 48) MeV/c^2; systematic errors are included. This analysis confirms previous observations.[3,6] The lower state is presumably the first radial excitation of the $K^*(890)$.

Figures 7(a) and (b) show the intensities of the $2^+K^*\pi$ and $2^+\rho K$ waves. Apart from the leading 2^+ $K^*(1430)$, a large enhancement is evident in both waves at ~2.0 GeV/c^2. We have fit the intensities and relative phases of these two waves above 1.69 GeV/c^2 to a Breit-Wigner and a coherent linear background; the overall phase is set using the fit to the $1^-K^*\pi$ wave described above. The result of the fit, shown in Fig. 7, is satisfactory, although the size of background required means that the single resonance interpretation is not unique.

There are almost no data on decays of K^* states into $K\eta$. We have searched for these in the $K^-\pi^+\pi^-\pi^\circ p$ final state. Figure 8 shows the $\pi^+\pi^-\pi^\circ$ spectrum of events satisfying a 1C kinematic fit to this channel. Consistency of particle identification has been demanded and events satisfying the 4C fit to $K^-\pi^+\pi^-p$ have been excluded. We see a strong η signal; the shaded areas are control regions used for background estimation. In Fig. 9 is shown the $K\eta$ mass spectrum after applying N^* and Y^* cuts and subtracting the control regions. The spectrum is dominated by a single resonance which is consistent with the 3^- $K^*(1780)$; this interpretation is confirmed by a preliminary moments analysis. The observed events correspond to a $K\eta$ branching ratio of ~2.5%. In contrast, there is no evidence of the $K^*(1430)$ whatever; the shaded area shows the expectation if its $K\eta$ branching ratio were 0.5%. These observations disagree strongly with SU(3) predictions.

3. Analysis of Strangeonium Channels

We expect channels involving hypercharge exchange (*e.g.*, those with a slow Λ) to be a fruitful source of $s\bar{s}$ states. In all the cases described

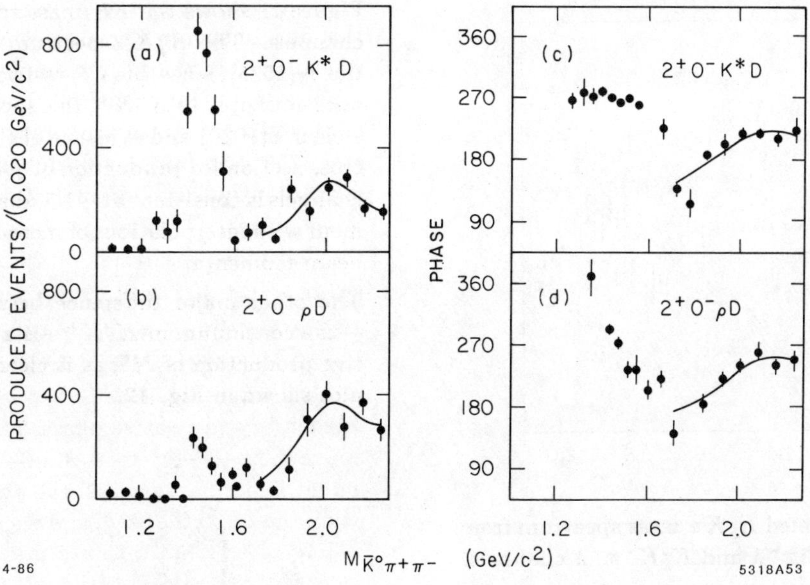

Fig. 7. The $\bar{K}^\circ\pi^+\pi^-$ 2+ waves; the fit at high mass is described in the text.

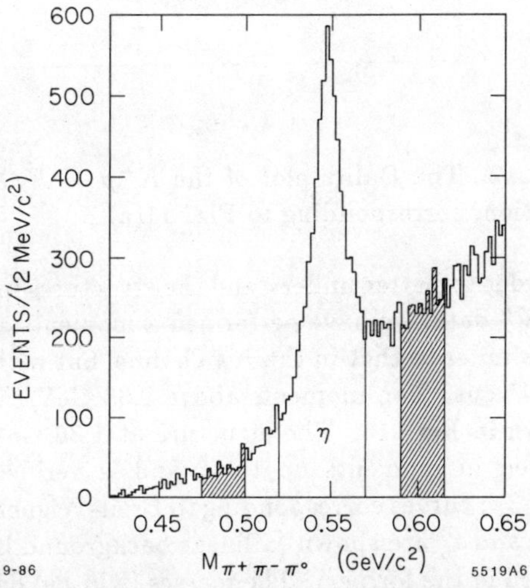

Fig. 8. The $\pi^+\pi^-\pi^\circ$ mass spectrum from the reaction $K^-p \to K^-\pi^+\pi^-\pi^\circ p$. The shaded control regions are used to estimate non-η background under the η signal.

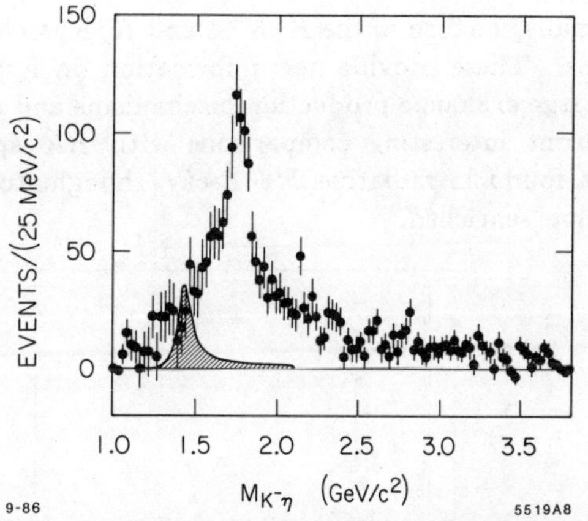

Fig. 9. The background-subtracted $K\eta$ mass spectrum from the $K^-p \to K^-\eta p$ reaction after N^* and Y^* cuts. The shaded curve shows the signal expected for a $K^*(1430) \to K\eta$ branching ratio of 0.5%.

below, the Λ is reconstructed in the LASS Spectrometer, and particle identification performs only a supporting role in event selection. The resultant acceptance is extremely uniform with no "holes."

The $K\bar{K}\pi$ mass spectrum for the combined $K_s^\circ K^\mp\pi^\pm\Lambda$ channels, shown in Fig. 10, is somewhat disappointing! There is some evidence of production of $f_1(1285)$ and $f_1(1420)$, but the cross section is small and statistics are limited. The spectrum is similar to that observed in the analogous π^-p reaction, indicating that these states do not have dominant $s\bar{s}$ content. Apart from these states and a sharp rise in the spectrum at K^*K threshold, the gross features are very similar to $\bar{K}^\circ\pi^+\pi^-$. A preliminary PWA, in contrast, shows that production of unnatural J^P states is predominant and that the 1.4–1.6 GeV/c^2 mass region consists almost entirely of 1^+K^*K. The broad bump at ~1.52 GeV/c^2 could, therefore, be the $f_1(1530)$, claimed as an $s\bar{s}$ resonance by Gavillet et al.,[7] though we find that $\bar{K}^*$ production exceeds K^* in both channels. We find no evidence for $0^-\delta\pi$ but

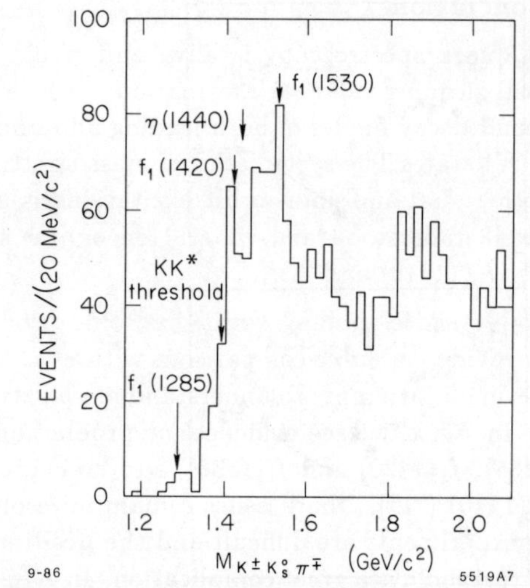

Fig. 10. The summed $K_s^\circ K\pi$ mass spectrum from the $K^-p \to K_s^\circ K^+\pi^-\Lambda$ and $K_s^\circ K^-\pi^+\Lambda$ channels.

cannot completely exclude it in the 1.42 GeV/c² region.

Finally, we turn to the $K^-K^+\Lambda$ and $K_s^\circ K_s^\circ\Lambda$ channels. These provide new information on hypercharge exchange production mechanisms and also permit interesting comparisons with $K\bar{K}$ spectra found in radiative J/ψ decay, thought to be "glue"-enriched.

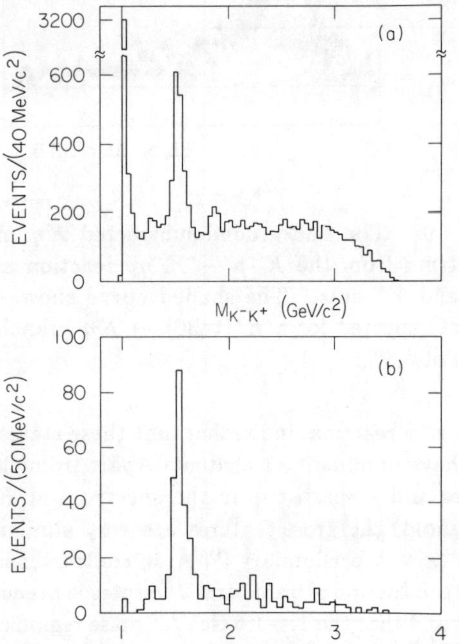

Fig. 11. The $K\bar{K}$ mass spectra (a) from the $K^-K^+\Lambda$; and (b) from the $K_s^\circ K_s^\circ\Lambda$ final states, demanding $|t'| < 2$ (GeV/c)².

Figure 11 shows the $K\bar{K}$ mass spectra from these channels. The $K_s^\circ K_s^\circ$ spectrum is dominated by the $f_2(1525)$; since the CP restriction of even spin does not apply to K^-K^+, this spectrum also shows a clear $\phi(1020)$ and evidence of the $\phi_3(1860)$. The cross section for production of $f_2(1525)$ in the two channels is consistent at ~1.5 μbarns and in agreement with interpolations of measurements at other beam momenta.

The other major difference between the spectra —the continuum in K^-K^+—is a result of diffractive production of N^*, as is clear from the Dalitz plot shown in Fig. 12.

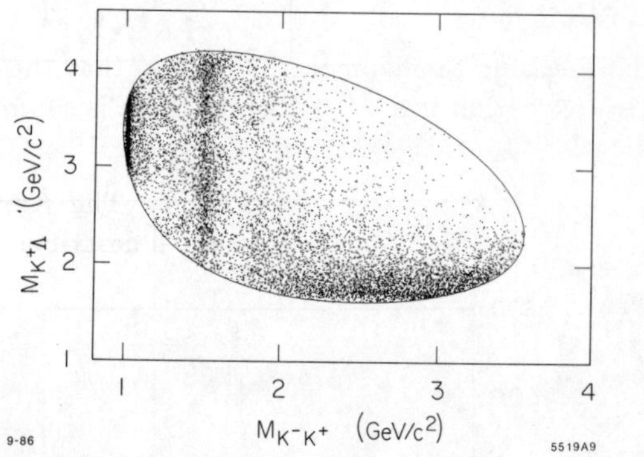

Fig. 12. The Dalitz plot of the $K^-p \to K^-K^+\Lambda$ reaction, corresponding to Fig. 11(a).

In order to better understand the structures in the K^-K^+ data, we have performed a moments analysis similar to that in the $K\pi$ channel but without an N^* cut. The moments above 1.68 GeV/c² are shown in Fig. 13. The structure at 1.86 GeV/c² is seen in moments up to t_6^0 and is verified as $J^P = 3^-$; curves corresponding to Breit-Wigner fits of t_0^0 and t_6^0 are shown (a linear background is included in the former). The masses (widths) determined from the fits average to 1857 ± 9 (69 ± 18) MeV/c². There is also structure in the moments around 2.2 GeV/c², which is discussed below.

Figure 14 shows comparisons of the $K_s^\circ K_s^\circ$ mass spectrum with that seen by the Mark III group[8] in radiative decay of the J/ψ. Figure 14(a) shows that there is no evidence whatever for hadronic production of the $f_2(1720)$ or "θ;" however, Fig. 14(b) demonstrates that the data from the two experiments are statistically compatible in the region of the $X(2220)$ or "ξ."

We can try and combine evidence from the two $K\bar{K}$ channels to speculate further on what might be happening in the "ξ" region. The K^-K^+ moments (Fig. 13) up to t_8^0 show structure in the

2.2 GeV/c^2 region which, while not statistically compelling, is compatible with a spin-4 state of width < 100 MeV/c^2. The large diffractive N^* production in this channel leads to substantial moments up to t_6^0, though they should be smooth and not have structure as a function of K^-K^+ mass.

Although statistics in the $K_s^\circ K_s^\circ$ channel are poor at 2.2 GeV/c^2, it is clear that the events are not distributed isotropically in the Gottfried-Jackson frame. Figure 15 shows the $K_s^\circ K_s^\circ$ spectrum for events in the forward direction only ($cos\theta_J > 0.85$); the cut enhances the 2.2 GeV/c^2 region. Inset are the L=2 and 4 moments which show some effects, which are significant when integrated from 2.1–2.3 GeV/c^2.

Synthesising the evidence, it is clear that the 2.2 GeV/c^2 region has $J^P \geq 2^+$ and there is some indication of a rather narrow state with $J^P = 4^+$.

4. Conclusions

Light quark spectroscopy is alive and well! We are still gleaning valuable information on the existenceand decay modes of both leading and underlying K^* states.The systematics of mass-splittings of both radial and spin- orbit excitations is still not well understood and we still encounter surprises ($K^*(1410)$ and $K\eta$).

In the "strange"-onium world, hadronic production provides valuable comparisons with e^+e^- collisions in our attempts to understand meson structure. In $K\bar{K}\pi$, we see evidence for production of $f_1(1285)$, $f_1(1420)$ and $f_1(1530)$, and no evidence for $\eta(1440)$ ("ι"). Many issues remain unresolved here; experiments are difficult and the position of K^*K threshold is a great complication. In $K\bar{K}$, we confirm the $\phi_3(1860)$ and find the "ξ" region consistent with Mark III data and with quark model expectations. The total absence of the $f_2(1720)$ ("θ") in hadronic production is extremely interesting.

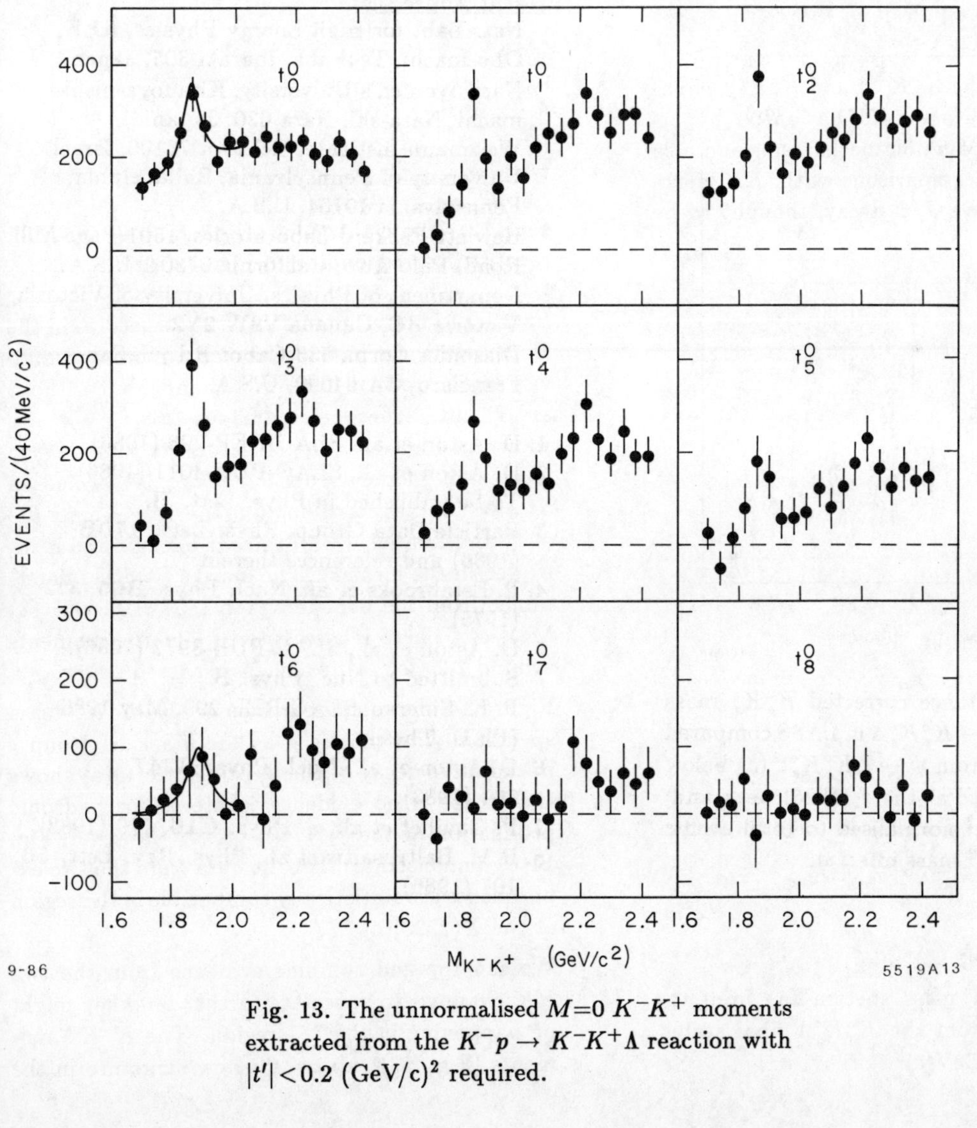

Fig. 13. The unnormalised M=0 K^-K^+ moments extracted from the $K^-p \to K^-K^+\Lambda$ reaction with $|t'| < 0.2$ (GeV/c)^2 required.

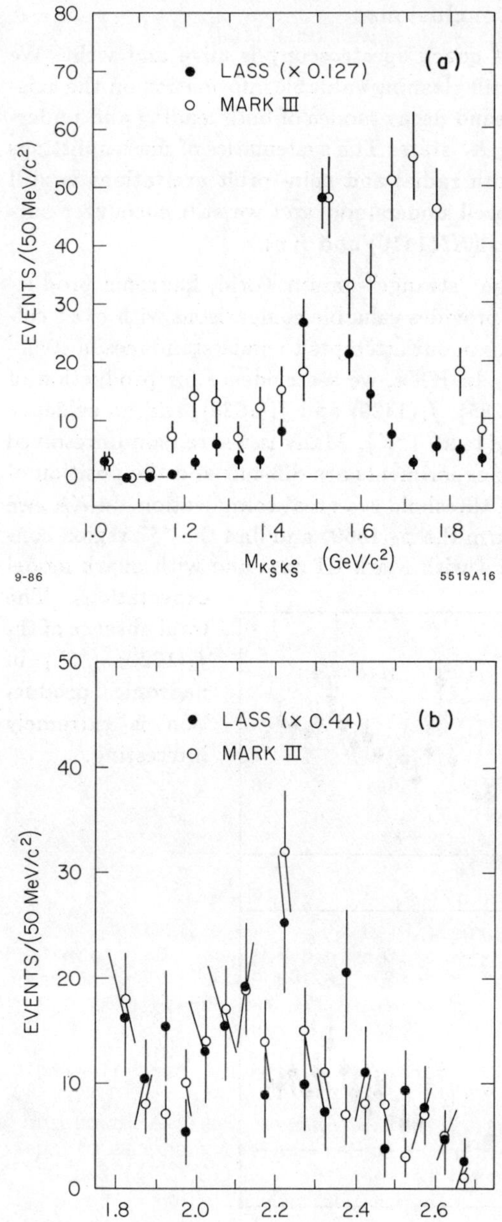

Fig. 14. The acceptance corrected $K_s^\circ K_s^\circ$ mass spectrum from $K^- p \to K_s^\circ K_s^\circ \Lambda$ in LASS compared with Mark III data from $\psi \to \gamma K_s^\circ K_s^\circ$: (a) below 1.9 GeV/c^2 normalised at the $f_2(1525)$ peak and; (b) above 1.8 GeV/c^2 normalised to total events in the 1.8–2.7 GeV/c^2 mass interval.

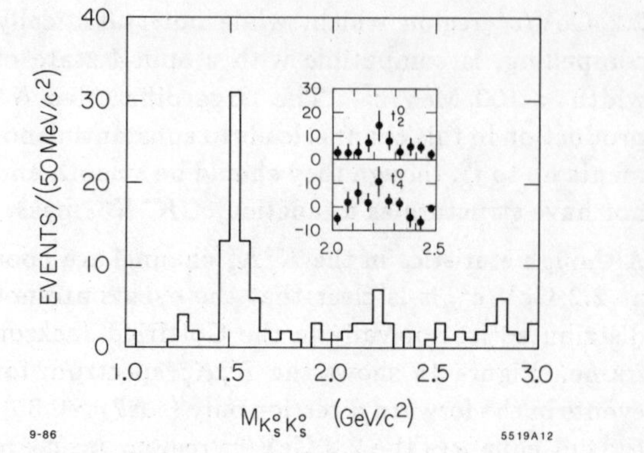

Fig. 15. The $K_s^\circ K_s^\circ$ mass spectrum with $\cos\theta_J >$ 0.85; inset are the L=2 and 4, M=0 unnormalised moments.

References

Present Addresses:

‡ Nat. Lab. for High Energy Physics, KEK, Oho-machi, Tsukuba, Ibaraki 305, Japan.

♭ Nara Women's University, Kitauoya-nishi-machi, Nara-shi, Nara 630, Japan.

** Weizmann Institute, Rehovot 76100, Israel.

† University of Pennsylvania, Philadelphia, Pennsylvania 19104, U.S.A.

♯ Hewlett-Packard Laboratories, 1501 Page Mill Road, Palo Alto, California 94304, U.S.A.

♮ Department of Physics, University of Victoria, Victoria BC, Canada V8W 2Y2.

♯♯ Diasonics Corp., 533 Cabot Rd., S. San Francisco, CA 94090, U.S.A.

1. D. Aston et al., SLAC-REP-298 (1986).
2. D. Aston et al., SLAC-PUB-4011 (1986). To be published in Phys. Lett. B.
3. Particle Data Group, Phys. Lett. **170B** (1986) and references therein.
4. P. Estabrooks et al., Nucl. Phys. **B95**, 322 (1975).
5. D. Aston et al., SLAC-PUB-3972 (1986). Submitted to Nuc. Phys. B.
   P. K. Sinervo, SLAC-REP-299, May 1986 (Ph.D. Thesis).
6. D. Aston et al., Nucl. Phys. **B247**, 261 (1983).
7. P. Gavillet et al., Z. Phys. **C16**, 119 (1982).
8. R.M. Baltrusaitis et al., Phys. Rev. Lett. **56**, 107 (1986).

NEW RESULTS ON THE E(1420)/IOTA(1460) MESON IN HADROPRODUCTION

S. U. Chung

Brookhaven National Laboratory
Upton, New York, USA 11973

A mini-review, with emphasis on new results, is given on the status of the hadroproduced E(1420)/iota(1460) meson in the decay channels $\eta\pi\pi$ and $K\bar{K}\pi$. The BNL data at twice the statistics of the previously published event sample show clearly a $J^{PG} = 0^{-+}$ $\delta(980)\pi$ state with a phase motion characteristic of a resonance.

This review covers recent and hitherto unannounced results from the BNL experiment[1] on the $K\bar{K}\pi$ state in the 1.4 GeV region. In addition, the results from a KEK experiment on the $\eta\pi\pi$ decay channel and those of CERN OMEGA experiment on $\eta\pi\pi$ and $K\bar{K}\pi$ channels are presented for comparison.

There exists in the literature a number of reviews[2,3] on the experimental status of the hadroproduced E(1420)/iota (1450) coupling to $\delta(980)\pi$ and $K^*\bar{K}$. The present review concentrates on new results from the BNL experiment on the reaction

$$\pi^- p \to K^+ K_s \pi^- n \qquad (1)$$

at 8 GeV/c. The data from their 1983 run have been previously analyzed using the isobar-model techniques of the Dalitz-plot (two-dimensional fits)[4] and the full partial waves (5-dimensional fits).[5] Additional new data from the 1985 run has now been added to the old data sample and a partial-wave analysis was performed on the combined data. Figure 1 shows the $K\bar{K}\pi$ mass spectrum with ≈4,000 E/iota events in the peak, representing twice the statistics of of their previous data for $-t < 1.0$ GeV^2. A fit with two simple Breit-Wigner forms for the peaks D(1285)/η_r(1275) and E(1420)/iota(1460) over a polynomial background gives: M(D) = (1285) ± 4) MeV, Γ(D) = (22 ± 5) MeV and M(E) = (1421 ± 3)MeV, Γ(E) = (70 ± 8) MeV. The E/iota tends to be produced away from the forward region; the subsample with $0.2 < -t < 1.0$GeV^2 shows an E/iota peak on a much-reduced background (see the shaded histogram, Fig. 1).

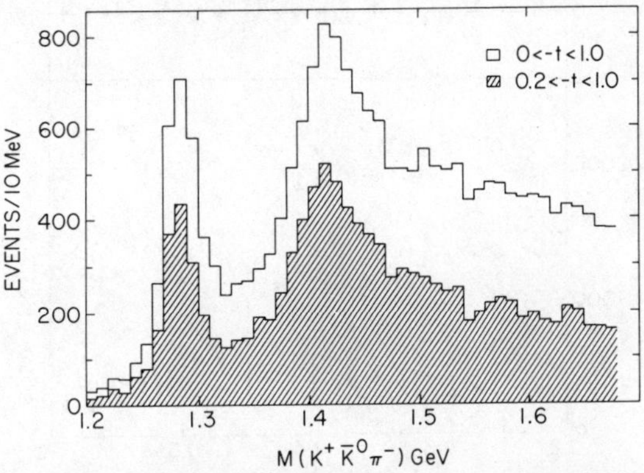

Figure 1: The $K^+\bar{K}^0\pi^-$ mass spectrum for combined '83 and '85 data with $0 < -t < 1.0$ GeV^2. The shaded histogram for events with $0.2 < -t < 1.0$ GeV^2.

The results of the BNL partial-wave analysis on the combined data are given in Figs. 2 and 3. This analysis allowed for the first time an arbitrary degree of coherence between the waves 0^{-+} (δ) and 0^{-+} (K^*). The best fit (preliminary) requires a nearly complete incoherence between the two waves, indicating perhaps different Regge exchanges for them (see Fig. 2c). This would indicate that the $0^{-+}(\delta)$ and $0^{-+}(K^*)$ states are not two different decay modes of a same object but rather two distinct states. As before, the 1^{++} wave shows a prominent peak of the D mass and a sharp rise in the E region.

A few relevant waves (preliminary) for the subsample of events with $0.2 < -t < 1.0$ GeV^2 are shown in Fig. 4. It is seen that the 0^{-+} (δ) wave executes a classic phase motion of a pure resonance with respect to a non-resonant 1^{++} 0^+ wave with a mass at ≈1400 MeV and a width at ≈60 MeV (dashed curves in Fig. 4).

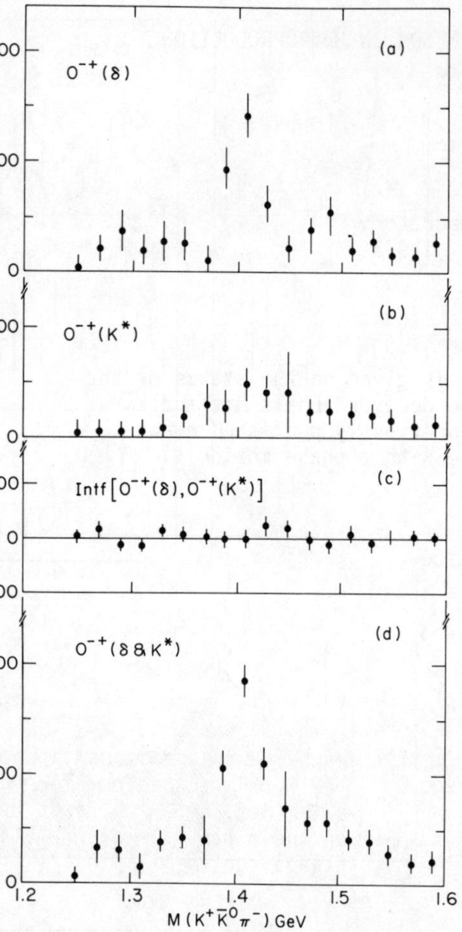

Figure 2: $M(K^+\bar{K}^0\pi^-)$ spectra for events fitted to the 0^{-+} wave from an isobar-model partial-wave analysis with spins 0 or 1 (preliminary).

It should be emphasized that the phase motion is now unambiguous, unlike the case when the events with $t < 0.2$ GeV^2 have been included in a previous analysis (Ref. 5).

These BNL results are generally confirmed by the KEK results[6] on the same reaction as that of the BNL experiment at the same energy. The KEK data have so far been partial-wave analyzed on only the decay channel $\eta\pi^+\pi^-$, and their partial-wave decomposition is given in Fig. 5. Three peaks are observed; $I = 0$, $1^{++}(\delta)$ and $I = 0$, $0^{-+}(\delta)$ in the D region, and $I = 0$, $0^{-+}(\delta)$ at the E mass [$m = (1420 \pm 5)$ MeV and $\Gamma = (31 \pm 7)$ MeV]. It should be noted that the $0^{-+}(\delta)$ state in the D region, the η_r (1275) first observed by N. Stanton, et al.[7], is not as prominent in the the BNL data. However, all three peaks show a rapid phase motion with respect to a non-resonant $I = 1$, $1^{--}(\rho)$ wave (not shown).

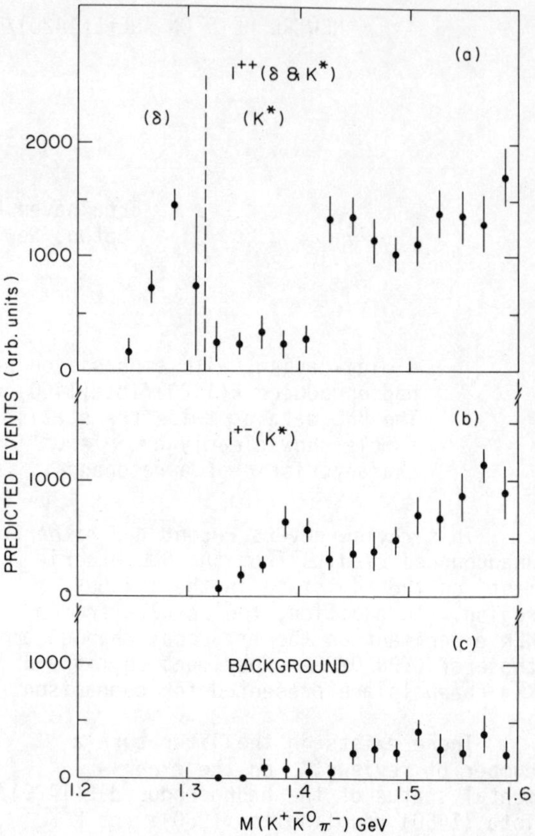

Figure 3: $M(K^+\bar{K}^0\pi^-)$ spectra for events fitted to the partial-waves 1^{++} and 1^{+-} along with incoherent phase-space event. Below (Above) 1.32 GeV only the $\delta(K^*)$ isobar was used in the fit (adding δ to the fits above 1.32 GeV does not appreciably alter the results).

The data of the CERN OMEGA experiment,[8] on the other hand, seem to be at variance with those of the BNL and KEK data. The CERN experiment studied the $\eta\pi^+\pi^-$ and $K\bar{K}\pi$ systems produced centrally off (π^+/p) and p in the interactions $(\pi^+/p)p$ at 85 GeV/c. In contrast to the BNL and KEK analyses, the CERN data have so far been subjected to only the Dalitz-plot analysis. The $\eta\pi^+\pi^-$ spectrum exhibits a peak in the 1^{++} wave at the D mass and a hint of a bump at the E mass, but no 0^{-+} peaks are observed at either masses (not shown). For the first time, the $K\bar{K}\pi$ channel has been Dalitz-plot analyzed as a function of the mass per 40 MeV in the E region. The results are shown in Fig. 6.

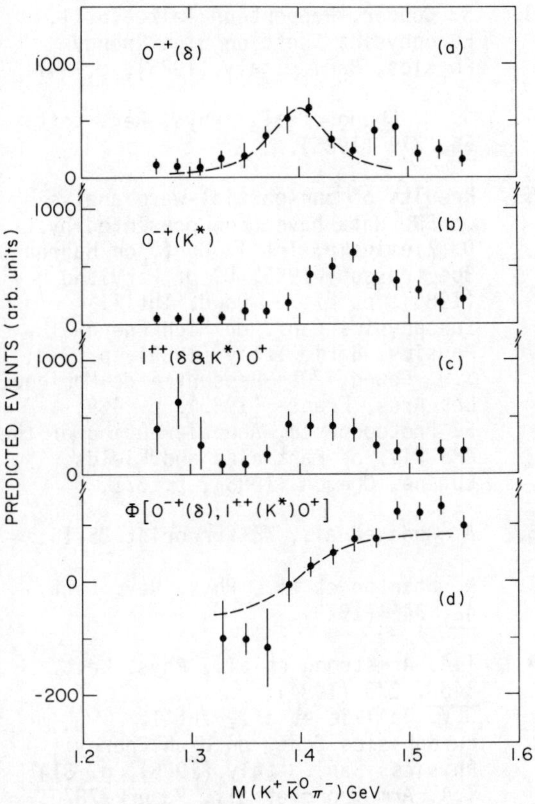

Figure 4: Results (Preliminary) of the partial-wave analysis on the subsample of events with $0.2 < -t < 1.0\ GeV^2$.
(a-c) $M(K^+\overline{K}^0\pi^-)$ spectra for events fitted to $0^{-+}(\delta)$, $0^{-+}(K^*)$ and $1^{++}(\delta\ \&\ K^*)0^+$ (m = 0, natural parity exchange).
(d) The phase motion (in degrees) of $0^{-+}(\delta)$ against $1^{++}(K^*)0^+$.

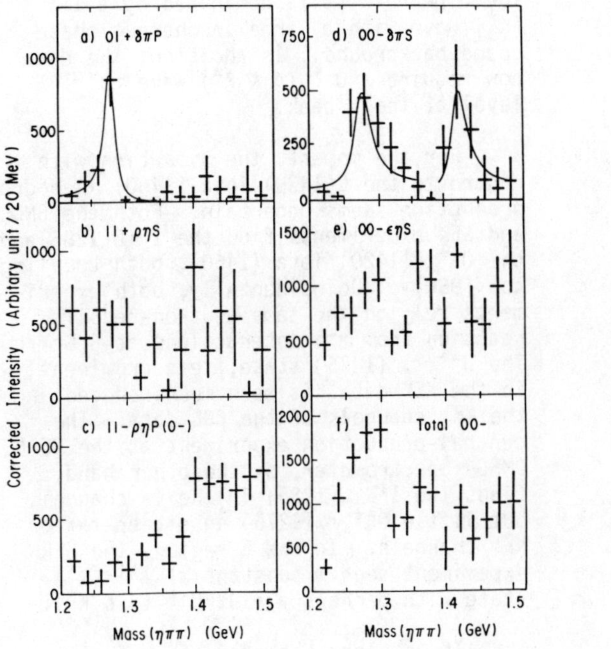

Figure 5: Partial-wave decomposition of the $M(\eta\pi^+\pi^-)$ spectrum for KEK data (Ref. 6). The first number on their partial-wave notations stands for the I-spin.

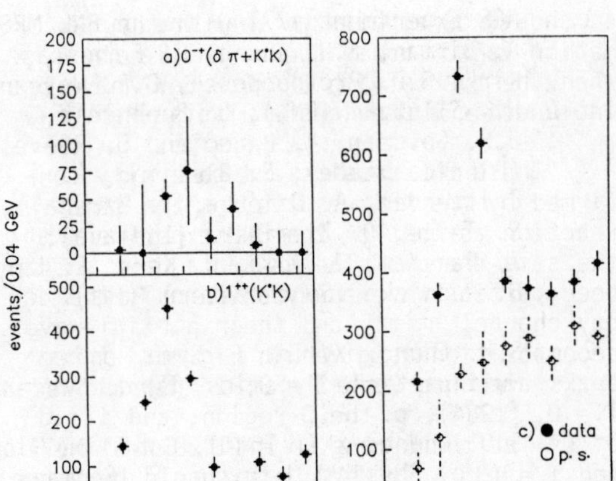

Figure 6: Dalitz-plot analysis of the CERN Ω-Spectrometer Data (Ref.8).
(a) and (b) correspond to the waves $0^{-+}(\delta\ \&\ K^*)$ and $1^{++}(K^*)$. (c) $M(K\overline{K}\pi)$ spectrum (solid dots); incoherent phase-space background (open dots).

The E peak is still dominated by a 1^{++} (K^*) wave with a large incoherent phase-space background. In addition, the data now require a 0^{-+} (δ & K^*) wave at ≈10% level at the E peak.

For the moment, the situation with regard to the E(1420)/iota (1460) in hadro-production seems uncertain. Both the BNL and KEK experiments find the 1^{++}D(1285) and the 0^{-+} E(1420)/iota (1460), both coupling to $\delta(980)\pi$. Coincidentally, both experiments rely on the same neutron-recoil reaction from $\pi^- p$ interactions at 8 GeV/c. The 0^{-+} η_r (1275) state, seen prominently in the KEK data, is seen much reduced in the $K\bar{K}\pi$ channels of the BNL data. The central-production experiment at the CERN OMEGA Spectrometer, on the other hand, finds the 1^{++} D(1285) in the $\delta\pi$ channel, but little 0^{-+} η_r(1275) in either $\eta\pi\pi$ or $K\bar{K}\pi$ channels. In the E region, the CERN experiment sees a substantial 1^{++} (K^*) state with perhaps a ≈10% 0^{-+} (δ & K^*).

It is clear therefore that further independent analyses on a variety of additional hadron-induced reactions are necessary for clarification of the status of the η_r (1275) and the E (1420)/iota (1460); these include partial-wave analyses of the $\eta\pi\pi$ and $\bar{K}K\pi$ channels at the SPS and Serpukhov energies,[9] the diffractive production of $K\bar{K}\pi$ systems at the ISR,[10] and the production of $\eta\pi\pi$ and $K\bar{K}\pi$ systems from $K^- p$ interactions at BNL and KEK energies.

Research supported by U.S. Department of Energy under Contract No. DE-AC02-76CH0016.

REFERENCES

1. AGS Experiment #771 with the BNL MPS: A. Birman, S.U. Chung, R. Fernow, H. Kirk, S.D. Protopopescu, D.P. Weygand, H. Willutzki (BNL); D. Boehnlein, J.H. Goldman, V. Hagopian, D. Reeves (Florida State); S. Blessing, R. Crittenden, A. Dzierba, T. Marshall, S. Teige, D. Zieminska (Indiana); Z. Bar-Yam, J. Dowd, W. Kern, E. King, H. Rudnicka (Southeastern Mass).

2. S.U. Chung, XV Int'l. Symp. on Multiparticle Dynamics, Lund, Sweden (1984), p. 186; S.U. Chung, XXII Int'l. Conf. on High Energy Physics, Leipzig, E. Germany (1984), Vol. II, p. 167; S.U. Chung, XII Int'l. Winter Meeting on Fundamental Physics, Santander, Spain (1984), p. 295; S.U. Chung, 1986 Aspen Winter Conf. on Particle Physics (BNL #38153) to be published.

3. S. Cooper, Rapporteur Talk, Int'l. Europhysics Conf. on High Energy Physics, Bari, Italy (1985), p. 947.

4. S. U. Chung et al., Phys. Rev. Lett. 55, 779 (1985).

5. Results of our partial-wave analysis on BNL data have been presented by: D. Zieminska, Int'l. Conf. on Hadron Spectroscopy-1985, U. of Maryland (1985), p. 27; J. Dowd, Int'l. Europhysics Conf. on High Energy Physics, Bari, Italy (1985), p. 318; S.U. Chung, 20th Rencontre de Moriond, Les Arcs, France (1985), p. 489; S. Protopopescu, Annual Meeting of the APS Div. of Particles and Fields, Eugene, Oregon (1985), p. 671.

6. A. Ando et al., KEK Preprint 85-15.

7. N. Stanton et al., Phys. Rev. Lett. 42, 346 (1979).

8. T.A. Armstrong et al., Phys. Lett. 146B, 273 (1984); O.V. Baillie et al., Int'l. Europhysics Conf. on High Energy Physics, Bari, Italy (1985), p. 314; T.A. Armstrong et al., Paper #7870, the 23rd Int'l. Conf. on HEP, Berkeley, CA (July 1986).

9. Th. Mouthuy, Int'l. Europhysics Conf. on High Energy Physics, Bari, Italy (1985), p. 320; S.I. Bitukov et al., Serpukhov Preprint EFVE85-19; see also Refs. 2 and 3.

10. P. Chauvat et al., Phys. Lett. 148B, 382 (1984).

The Results of Two Scattering Processes: $\pi\pi \to \phi\phi$* and $\pi\pi \to K\bar{K}$ †

R.S. Longacre, C.S. Chan, A. Etkin, K.J. Foley, R.W. Hackenburg, M.A. Kramer,
S.J. Lindenbaum, W.A. Love, T.W. Morris, E.D. Platner, A.C. Saulys
Brookhaven National Laboratory and City College of New York
Upton, New York, U.S.A. 11973 New York, New York 10031

We present an analysis of 6658 events of the reaction $\pi^-p \to \phi\phi n$ at 22 GeV/c. This data is well represented by three resonances, all with quantum numbers $I^G J^{PC} = 0^+2^{++}$.

In a second analysis, which consists of 40494 new events of the reaction $\pi^-p \to K^o_S K^o_S n$ at 22 GeV/c, we obtain the S_0, D_0 and G_0 Argand amplitudes for $\pi\pi \to K\bar{K}$. The $\theta(1690)$ is shown to decay into $\pi\pi$ by less than 4% and only is required in J/ψ radiative decay experiments.

In the context of QCD and OZI, the two scattering processes $\pi\pi \to \phi\phi$ and $\pi\pi \to K\bar{K}$ represent two unique topologies in which strange quarks are created. The first is an OZI forbidden (disconnected) process which represents lowest threshold for light quarks going to the lowest threshold for complete flavor change. According to QCD, the disconnection is bridged by two gluons and should tell us what the gluons are doing. The OZI rule would indicate that the gluons are weakly coupled thus the cross section is small. I^G is equal to 0^+ and J^P equal to 0^+, 2^+, and 4^+. Since the OZI rule is badly violated [1], one needs a strong flavor changing amplitude. Vacuum mixing could cause such an amplitude in a 0^{++} $q\bar{q}$ resonance [2]. However as we show below, we only see 2^{++} resonances. Thus in QCD the only other basic mechanism that completely violates the OZI rule and produces only $J^{PC} = 2^{++}$ resonances is the intervention of glueballs.

We have performed a partial wave analysis on 6658 events of the reaction $\pi^-p \to \phi\phi n$ at 22 GeV/c. This analysis represents an increase of 3006 events over a prior published analysis [3]. Figure 1a shows the acceptance corrected $\phi\phi$ mass spectrum for the overall combined sample, plus the acceptance derived from the partial wave analysis. Figure 1b and 1c show the partial waves found which consisted of $J^P SLM^\eta = 2^+2S0^-$ (S_2), 2^+2D0^- (D_2), and 2^+0D0^- (D_0). The smooth curves are three Breit-Wigner resonances given in Table I. Errors are not given, however the masses and total widths of the resonances are consistent with errors given in earlier analysis [3].

We also studied the reaction $\pi^-p \to \phi K^+K^-n$. Approximately 13% of the $\phi\phi$ events are made up of this background. A partial wave analysis of this background using K^+K^- masses just above the ϕ is shown in Fig. 3. Approximately 67% of this background is structureless and incoherent while 30% is $J^{PC} = 1^{--}$ and 3% is 2^{++}. The 1^{--} is coherent and also produced by π-exchange, therefore this wave interfered with the $\phi\phi$ 2^{++} amplitudes. The smooth curves for the 1^{--} and 2^{++} in Fig. 3 are Breit-Wigner fits with masses and widths qiven in Table II. The 1^{--} phase measured against the large S_2 2^{++} $\phi\phi$ amplitude is shown in Fig. 2b. There are two 1^{--} curves in this figure where the flatest being the Breit-Wigner 1^{--} fit with an overall absolute phase of 84 degrees. A Breit-Wigner has an Argand amplitude which is circular in shape while if the ϕK^+K^- was produced by a multi-peripheral deck mechanism it would have the same partial wave content but the 1^{--} and 2^{++} amplitudes would have a sausage shape. The second 1^{--} curve on Fig. 2b is the deck 1^{--} phase and corresponds to an absolute phase of 31 degrees. Finally the $\phi\phi$ 2^{++} Argand plots for both conditions of 1^{--} absolute phase are shown in Figs. 2a and 2c.

The second and connected process explores strange particle decay modes of light quark resonances. One is also able to check the

---

* Work supported by the U.S. Department of Energy under Contract No. DE-AC02-76CH00016 (BNL), DE-AC02-83ER40107 (CCNY) and the CUNY PSC-BHE Research Award Program.
† Additional collaborators for the $K^o_S K^o_S$ from Brandeis (J.R. Bensinger, L.E. Kirsch, H. Piekarz, R. Poster; USDOE Contract No. DE-AC02-76ER03230), CCNY (J. Goo), Duke (L.R. Fortney, A.T. Goshaw, E. McCrory, W.J.. Robertson; USDOE Contract No. DE-AS05-76ER03065), and Notre Dame (J.M. Bishop, N.N Biswas, N.M. Cason, V.P. Kenney, J. Piekarz, M.G. Rath, R.C. Ruchti, W.D. Shephard; NSF Contract Nos. PHY-83-06586 and PHY-83-05804).

OZI rule for $S\bar{S}$ states which are produced by a disconnected process (e.g. f' production). We have measured the reaction $\pi^- p \to K_S^o K_S^o n$ at 22 GeV/c obtaining 82111 events. A partial wave analysis has been performed using 40494 events with $|t'| < 0.1$ (GeV)^2, from which the S_0, D_0, and G_0 Argand amplitudes for $\pi\pi$ scattering are obtained. Figure 4a shows the S_0 amplitude in which one again sees all the same features of the 0^{++} analysis of a factor of three less events [4]. The most important feature of this analysis is the resonance S*'(1730). For this report we concentrate on the $J^{PC} = 2^{++}$ states (Fig. 4b). The cross section of the D_0 between 2.0-2.5 GeV amounts to 400 nb while the comparable cross section for $\phi\phi$ is 20 nb, thus the absence of a clear coupling to the 2^{++} resonances seen in the $\phi\phi$ system does not imply that they do not have a $K\bar{K}$ coupling. This coupling is overpowered by the many allowed final states in this mass region.

For the $I^G = 0^+$ and $J^P = 2^+$, we have performed a coupled channeled K-matrix analysis using the data of Ref. 5. We find that the data requires at least five poles which corresponds to the f(1270), f'(1525), θ(1690), fr(1810)[6] and a broad high mass background pole. We have performed many fits with different sets of the data [5], but here we will only discuss the one fit which has properties of the θ(1690) most consistent with J/ψ radiative decay. This fit has a χ^2 of 307 for 225 degrees of freedom. Figures 4b-6c show the fit to the complete data set. The largest contribution to χ^2 comes from the $K\bar{K} \to K\bar{K}$ data.† As may be inferred from Fig. 6a, $K\bar{K} \to K\bar{K}$ would measure directly the inelasticity of the θ(1690). Without a single pass $K\bar{K}$ collider one will have to use the reaction $K^- p \to K_S^o K_S^o \Lambda^o$. This reaction is reported on in this conference and no θ(1690) is seen, however one needs to perform a t-dependent analysis to separate the Q-exchange from the K-exchange. Polarized targets may be helpful for this purpose. The resonance parameters are given in Table III for this fit. Table III shows that the f'(1525) has the expected OZI suppression of its coupling to the ππ system, however the θ(1690) also shows a similar suppression to ππ. The θ(1690) from Table III has a ππ decay of less than 4% and is only required for J/ψ radiative decay experiments (see Fig. 6b and c). The last important meson from the fit is the fr(1810) [7] which is consistent with the radial excitation of the f(1270). If one uses the phenomenological Gell-Mann Okubo mass rule for ideally mixed mesons it seems hard to associate the θ(1690) as the $S\bar{S}$ partner of the fr(1810) since it is 200 MeV lower in mass instead of 200 MeV higher. However the Gell-Mann Okubo mass rule has not been established experimentally for radial excitations and would not be expected to work in the presence of glueballs (i.e. g_T, $g_{T'}$, and $g_{T''}$).

† $K\bar{K} \to K\bar{K}$ is obtained from M = 0 unnatural exchange amplitude from $K^- p \to K^+ K^- \Lambda^o$ (Σ^o).

References

[1] A. Etkin et al., Phys Rev. Lett. 41 (1978) 784.

[2] V.A. Novikov et al., Nucl. Phys. B191 (1981) 301.

[3] A. Etkin et al., Phys Lett. 165B (1985) 217.

[4] A. Etkin et al., Phys Rev. D25 (1982) 2446.

[5] R.S. Longacre et al., A Measurement of $\pi^- p \to K_S^o K_S^o n$ at 22 GeV/c and a Systematic Study of the 2^{++} Meson Spectrum, Phys. Lett. B (in press).

[6] Particle Data Group, Rev. Mod. Phys. 56 (1986) 1.

[7] N.M. Cason et al., Phys. Rev. D28 (1983) 1586.

[8] T. Armstrong et al., CERN/EP83-60 (1983).

TABLE I

Parameters of the Breit-Wigner resonances and percentage of the resonances going into the 2^{++} S_2, D_2 and D_0 $\phi\phi$ channels.

Name	%Data	Mass (GeV)	Width (GeV)	S2	D2	D0
g_T	45%	2.011	.202	98%	0%	2%
$g_{T'}$	20%	2.297	.149	6%	25%	69%
$g_{T''}$	35%	2.339	.315	37%	4%	59%

Table II

Parameters of the Breit-Wigner resonances for the $\phi K\bar{K}$ background for the 1^{--} and the 2^{++} channels.

J^{PC}	%Data	Mass (GeV)	Width (GeV)
1--	30%	2.270	.454
2++	3%	2.390	.562

Table III

Parameters of the Breit-Wigner resonances described in text, plus the branching fraction of these resonances into the ππ, $K\bar{K}$, ηη and all other channels.

Name	Mass (MeV)	Width (MeV)	X ππ	X $K\bar{K}$	X ηη	X Other
f(1270)	1283^{+6}_{-5}	186^{+9}_{-2}	$.844^{+.032}_{-.005}$	$.048^{+.004}_{-.002}$	$.005^{+.001}_{-.001}$	$.103^{+.007}_{-.039}$
f'(1525)	1547^{+10}_{-2}	108^{+5}_{-8}	$.013^{+.009}_{-.005}$	$.583^{+.056}_{-.046}$	$.222^{+.028}_{-.046}$	$.182^{+.080}_{-.093}$
fr(1810)	1858^{+18}_{-71}	388^{+15}_{-21}	$.21^{+.02}_{-.03}$	$.003^{+.019}_{-.002}$	$.008^{+.028}_{-.003}$	$.77^{+.03}_{-.07}$
θ(1690)	1730^{+2}_{-10}	122^{+74}_{-15}	$.039^{+.002}_{-.024}$	$.38^{+.09}_{-.19}$	$.18^{+.03}_{-.13}$	$.41^{+.33}_{-.11}$

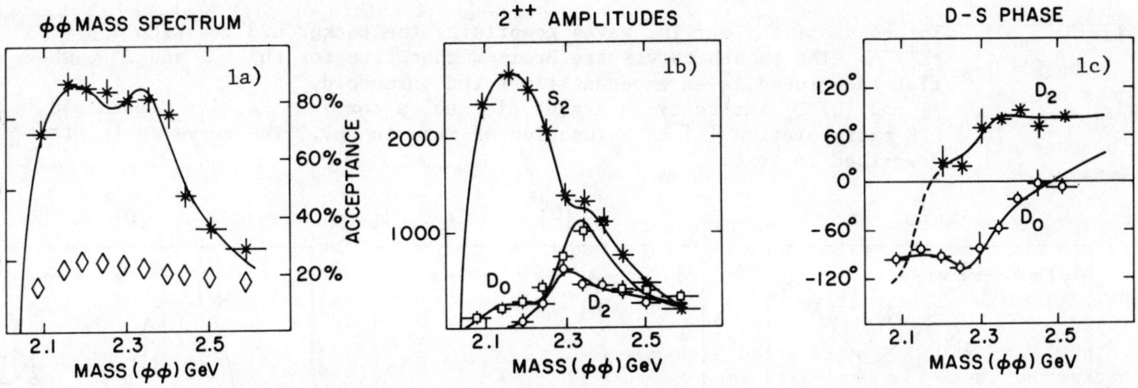

Fig. 1 (a) The φφ mass spectrum (events per 50 MeV) corrected for acceptance. Shown also is the mass dependence of the experimental acceptance (diamonds).
(b) Intensity and (c) phase differences (relative to the S-wave) for the three 2^{++} waves. The curves show the fit by three Breit-Wigner resonances.

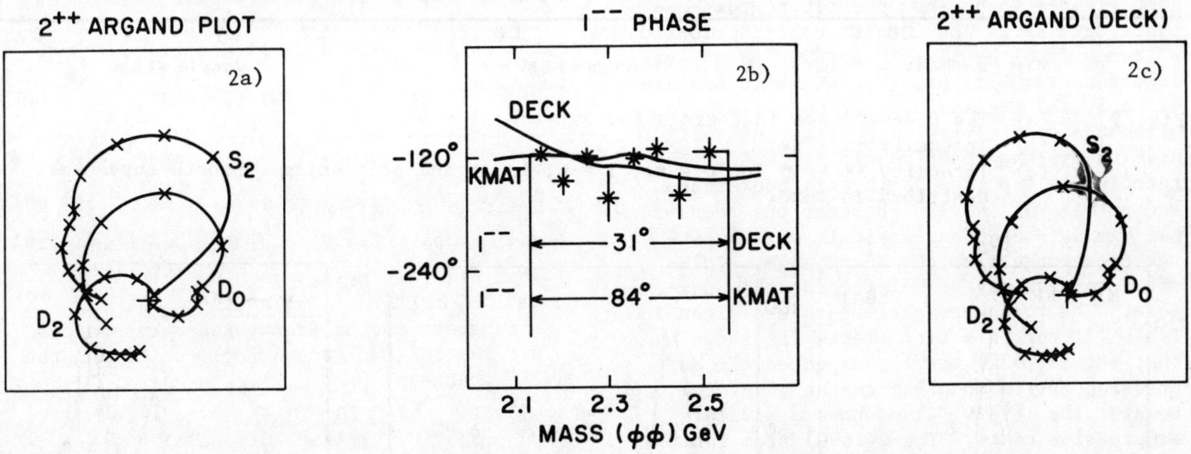

Fig. 2 (a) and (c) show Argand plots for the three 2^{++} waves. The absolute phase for (a) is based on the 1^{--} $\phi K\bar{K}$ wave being a Breit Wigner (K-matrix), while (c) is based on the 1^{--} $\phi K\bar{K}$ coming from a deck mechanism. (b) shows the 1^{--} phase (relative to the S-wave φφ) where the two curves come from the two models for the 1^{--} absolute phase.

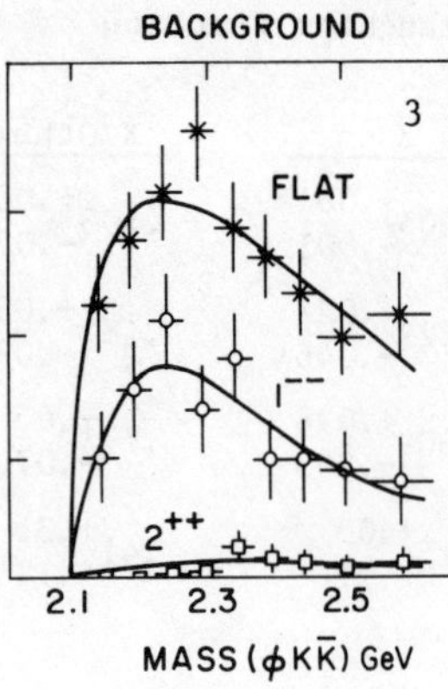

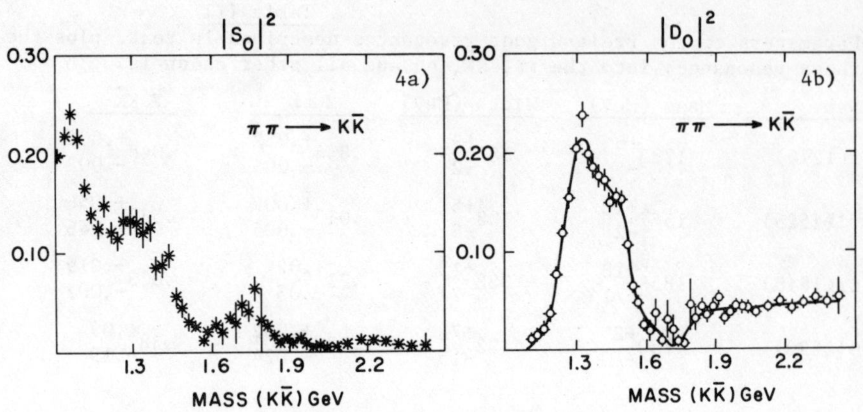

Fig. 3 Intensity of the partial waves comprising the background reaction $\pi^- p \rightarrow \phi K^+ K^- n$. The smooth curves are Breit-Wigner fits for the 1^{--} and 2^{++}, where the flat background is an exponent times the threshold.

Fig. 4 (a) S_0 and (b) D_0 intensity in Argand plot units for $\pi\pi \rightarrow K\bar{K}$, I = 0 obtained by an OPE extrapolation [5] as a function of mass in GeV. The curve in (c) is described in text.

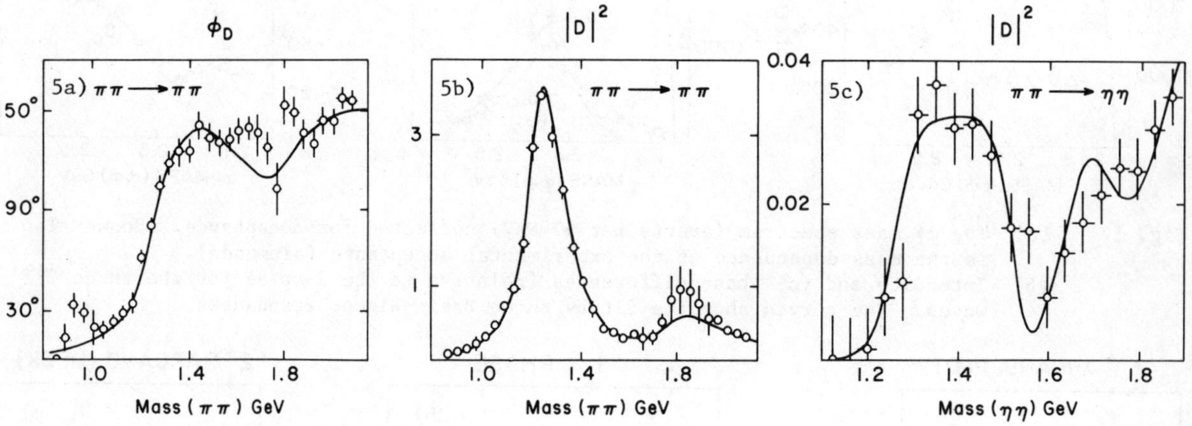

Fig. 5 (a) The D-wave phase in degrees for $\pi\pi \rightarrow \pi\pi$, I = 0.
(b) Intensity $\pi\pi \rightarrow \pi\pi$;
(c) Intensity $\pi\pi \rightarrow \eta\eta$ for D-wave, I = 0, in Argand plot units. Smooth curves are described in text.

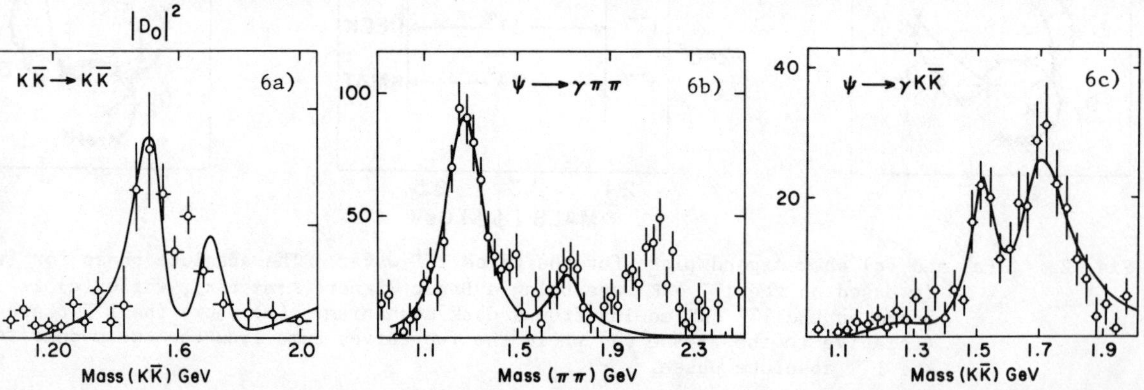

Fig. 6 (a) The intensity of D_0 from Ref. 8 where smooth curve is $K\bar{K} \rightarrow K\bar{K}$, I = 0 intensity from fit in text.
(b) The J/ψ radiative decay into $\gamma\pi^+\pi^-$;
(c) into γK^+K^-. The J/ψ radiative decay data has background subtraction [5]. The curves are from fit in text.

A Measurement of the spin-parity of the $\omega\pi^0$ state at 1200 MeV/c^2 in $\gamma p \rightarrow p\omega\pi^0$ at 20 GeV *

James E. Brau, Bohumil Franek†, and William C. Wester III

University of Tennessee, Knoxville 37996-1200

(representing the SLAC Hybrid Facility Photon Collaboration)

The SLAC Hybrid Facility photoproduction experiment has studied the low mass $\omega\pi^0$ enhancement in the reaction $\gamma p \rightarrow p\omega\pi^0$. The angular distribution of the production plane relative to the polarization vector (P_γ=0.52) shows structure consistent with an s-channel helicity non-conserving process. We compare our decay angular distributions to the measurements of the Omega Photon Collaboration from which they concluded that this reaction is dominated by non-SCHC production of the $J^P = 1^+$ B(1235) meson. Their parameterizations of the angular distributions agree well with our data.

The radial recurrences of vector mesons are states important to the understanding of the structure of the quark anti-quark interaction. While much detailed data now exists on many recurrences of the J/Ψ and the Υ and their transitions[1], knowledge on the ρ recurrences is much more limited. The well established ρ'(1600) is the only reliably detected state. The question remains whether there is another at lower mass. There are some expectations for the first recurrence to appear at about 1200-1300 MeV/c^2[2]. There have been suggestions that this state may be the $\omega\pi^0$ enhancement observed in photoproduction, but the alternate possibility that the enhancement is the B^0(1235) has postponed a conclusive judgement. Studies of the channel $\gamma p \rightarrow p\omega\pi^0$ have a long history with the observation of a low mass enhancement near 1200 MeV/c^2 in the $\omega\pi^0$ system.[3,4,5]

Recently a spin parity analysis[6] of the $\omega\pi^0$ in the above reaction for events produced by photons of 20 to 70 GeV found that the $\omega\pi^0$ enhancement is consistent with predominant 1^+ B(1235) production with a small (20%) J^P=1^- background. Unlike previous experiments they did not assume s-channel helicity conservation (SCHC)[7] in their analysis. They required detection of all four pions in the final state. This severe requirement led to an experimental acceptance of only 0.015 ± 0.005. The proton was identified not by observation but by a measurement of the missing mass of the recoiling baryon system. Their results represent to date the most definitive investigation of this reaction. We report here new results which complement the measurements of reference 6. In the present experiment we require events which have a detected and well measured proton in the bubble chamber. With excellent proton detection in the bubble chamber, we are able to reconstruct events with just one detected π^0, leading to a much higher experimental acceptance.

THE EXPERIMENT

This experiment has been described in detail previously.[8] A 20 GeV "monoenergetic" photon beam with 0.52 linear polarization is directed into the SLAC Hybrid Facility (SHF). The flash lamps of the SHF 30-inch bubble chamber are triggered by either tracks in the downstream proportional wire chambers (PWC) or energy deposition in the lead glass detector. The lead glass detector[9] provides the neutral detection crucial to this measurement.

DATA ANALYSIS

The 2.4 million pictures taken during this experiment were scanned for hadronic events and all the 306,785 events found within a 75 cm long fiducial region were fully measured. The events were associated with the downstream detector measurements with charged tracks being matched to hits in the PWCs.

A crucial ingredient to the analysis of this reaction was the development of a detailed simulation of the SHF and its associated detectors.[10] This Monte Carlo model (PEANUTS) simulates the interaction of all charged and neutral particles with the downstream detectors, simulates the trigger process, for triggered events constructs a raw data record similar to actual data, and passes it through the reconstruction program which is used to process the real data. In this way all pattern recognition and shower reconstruction from signals in the lead glass blocks are simulated in the

Monte Carlo. The acceptance for this reaction is about 20 percent.

This study of the channel $\gamma p \rightarrow p\omega\pi^0(\omega \rightarrow \pi^+\pi^-\pi^0)$ uses events in which all three charged particles are detected in the bubble chamber and one of the two π^0s is reconstructed from its daughter photons. There are 130,050 events with three charged tracks emerging from the primary vertex. Events are required to have a positive track with momentum under 1.4 GeV/c and ionization and range consistent with a proton, a mass recoiling from the three charged tracks greater than 0.1 $(\mathrm{GeV/c^2})^2$, and a pair of gammas with an invariant mass in the interval of 120-150 $\mathrm{MeV/c^2}$. This yields 6,412 events of the type

$$\gamma p \rightarrow p\pi^+\pi^-\pi^0 X.$$

Selecting $M_X^2 < 0.2\ (\mathrm{GeV/c^2})^2$, consistent with $\gamma p \rightarrow p\pi^+\pi^-\pi^0\pi^0$ yields 2,405 events. A zero constraint calculation of the four pion final state yields a determination of the incident photon energy, which is required to be between 15 and 22 GeV. The missing π^0 energy must exceed 1 GeV to remove background from single π^0 events. There are 415 events left after a selection of events with a $\pi^+\pi^-\pi^0$ mass in the ω^0 region (740-826 $\mathrm{MeV/c^2}$).

THE ANGULAR DISTRIBUTIONS

We have performed a decay angular distribution analysis for the 274 events with $M(\omega\pi^0) < 1450 MeV/c^2$ and having a momentum transfer $(|t_{\gamma\rightarrow\omega\pi^0}|)$ less than 0.5 $(\mathrm{GeV/c^2})^2$. Following the standard convention [11] we describe the decay of the $(\omega\pi^0)$ system into ω and π^0 by the polar (θ) and azimuthal (ϕ) angles of the ω in the helicity rest frame (frame A) of the $(\omega\pi^0)$ system. The orientation of this frame is such that its z axis points in the direction of the $(\omega\pi^0)$ system in the overall c.m. system and its y axis points in the direction of the normal to the production plane. The production plane is defined by the momentum vectors of the $(\omega\pi^0)$ system and the beam in the overall c.m. system. The decay of the ω is described by the spherical angles (β, α) of the normal to the decay plane defined by the $\pi^+\pi^-\pi^0$ in the rest frame of the ω. Two alternative frames were employed. The first, the so called "canonical" frame, is reached from frame A by the Lorentz boost in the direction of the ω, keeping the axes parallel with those of frame A. The (β,α) angles in this frame are denoted β_C and α_C. The second frame, the so called "helicity" frame, has its z axis pointing in the direction of the ω in the frame A and its y-axis given by the vector products of the ω direction and the

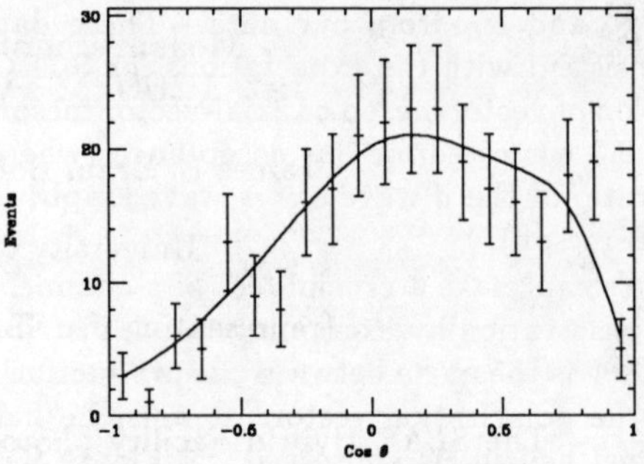

Figure 1a. Cos θ distribution.

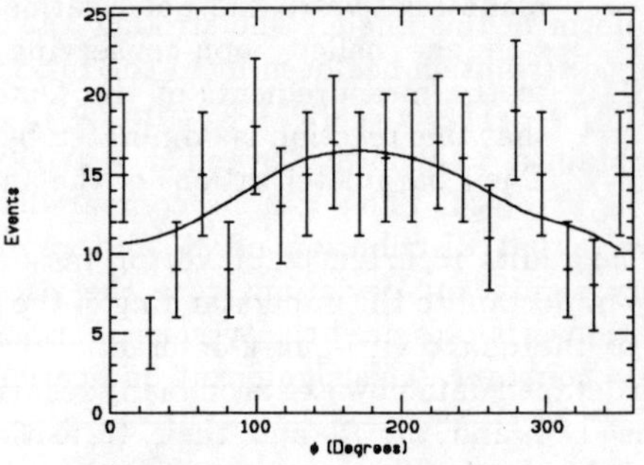

Figure 1b. ϕ distribution.

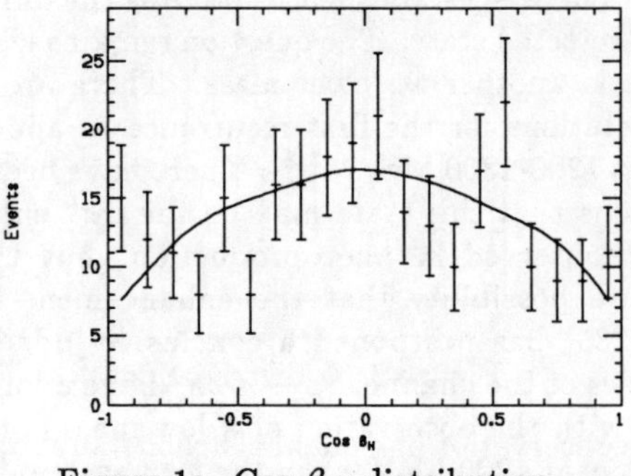

Figure 1c. Cos β_H distribution.

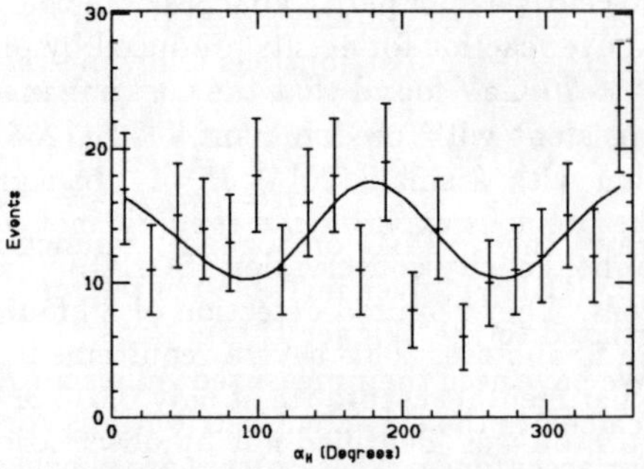

Figure 1d. α_H distribution.

z axis of frame A. The (β,α) angles in this frame are denoted β_H and α_H. Note that β_H and α_H of this paper are identical to θ_H and ϕ_H of reference 6.

Figures 1 show the distributions for $cos\theta$, ϕ,

$cos\beta_H$, and α_H from our data. These data are inconsistent with the expectations for SCHC production of vector meson or axial-vector meson with $\frac{d}{s} = 0.3$ corrected for our acceptance (where $\frac{d}{s}$ is the ratio of the d wave and s wave amplitudes of the B meson).

We can make a crucial test of s-channel helicity conservation by examining the Φ distribution, where Φ is the angle between the production plane and the polarization vector. It must be flat if s-channel helicity is conserved.[12] Figure 2 shows the distribution. The acceptance of our apparatus is uniform in this angle. The striking anisotropy of this distribution has been fit to the function

$$I(\Phi) = \frac{1}{2\pi}(1 + a\ cos(2\Phi) + b\ sin(2\Phi)).$$

We obtain a = -0.36 ± .08 and b = 0.06 ± .08 with a χ^2 of 24. This can be compared to the χ^2 for a flat distribution of 45 and represents a very significant deviation from the necessary condition of s-channel helicity conservation, that $I(\Phi) = constant$. This significant indication of

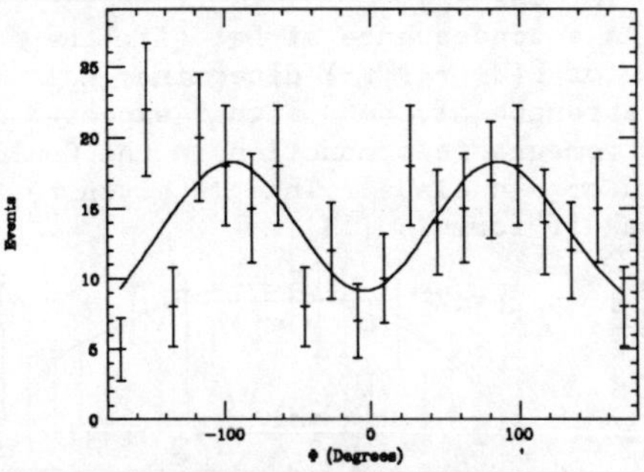

Figure 2. Φ distribution.

non-conservation of s-channel helicity supports the conclusions of reference 6. We can further test our data against their parameterization. Reference 6 parametrized the distributions following reference 11:

$$W(\Omega, \Omega_H) = \sum_\alpha H^{\pm}_{s}(\alpha) H^{\pm}_{\alpha}(\Omega, \Omega_H)/C_\alpha$$
$$\alpha = lmLM$$

where the 25 orthogonal functions[11] $H^{\pm}_{lmLM}(\Omega, \Omega_H)$ (given in Table 1 of reference 6[13]) are related to the Wigner D functions.

We have used their measured values for $H^{\pm}_{s}(\alpha)$ and calculate the expected distributions corrected for our acceptance. These curves are shown superimposed on our data in Figure 1 for four of the angles. The agreement is very good, having a χ^2 of 133 (for all six angular distributions) for 119 degrees of freedom. Our data therefore agree well with their parameterization of the decay angular distribution.

CONCLUSIONS

The angular distribution of the production plane relative to the photon polarization vector in the reaction $\gamma p \to \omega\pi^0 p$ shows structure consistent with an s-channel helicity non-conserving process. We have examined in detail the decay angular distributions for the $\omega\pi^0$ system. We have compared our data to the parameterization obtained by the Omega Photon Collaboration of the same process. Our measurement is complementary in acceptance, proton detection, and degree of beam polarization. Our decay angular distributions agree with those of the Omega Photon Collaboration[6] from which they have concluded that this reaction is dominated by non-SCHC production of the $J^P = 1^+$ B(1235) meson. Our Φ distribution provides new independent support for s-channel helicity non-conservation.

* Work supported in part by DOE contract no. DE-AS05-76ER03956

† Permanent address: Rutherford Appleton Laboratory, England

REFERENCES

1. see for example K. Gottfried, Proceedings of the International Conference on High Energy Physics, Brighton, July 1983, Page 743, published by Rutherford Appleton Laboratory.
2. H.J.Schnitzer, Phys. Rev. 18, 3482 (1978).
3. R. Anderson et al, Phys. Rev. D1, 27 (1970); J. Ballam et al, Nucl. Phys. B76, 375 (1974).
4. D.P. Barber et al, Z. Phys. C4, 169 (1980).
5. D. Aston et al, Phys. Lett. 92B, 211 (1980).
6. M. Atkinson et al, Nucl Phys. B243, 1 (1984); R.H. McClatchey, University of Sheffield thesis, November, 1981.
7. F.J. Gilman, J. Pumplin, A. Schwimmer, and L. Stodolsky, Phys. Lett. B31, 387, (1970).
8. K. Abe et al, Phys. Rev. D30, 1 (1984).
9. J.E. Brau et al, Nucl. Inst. and Methods 196, 403 (1982).
10. J.E. Brau, SLAC BC75 Note 41, February 27, 1984, internal documentation.
11. S.U. Chung, CERN Yellow Report 71-8 (1971). S.U. Chung et al, Phys. Rev. D11, 2426 (1975).
12. K. Schilling, P. Seyboth, and G. Wolf, Nucl. Phys. B15, 397 (1970).
13. We have found two errors in Table 1 of reference 6. Equation 4 should have a coefficient of $\frac{\sqrt{6}}{4}$ rather than $\sqrt{\frac{6}{4}}$. Equation 8 should have a coefficient of $\frac{\sqrt{6}}{8}$ rather than $\sqrt{\frac{6}{8}}$.

LIMITS ON PRIMAKOFF PRODUCTION OF HYBRID MESONS

M. Zielinski, D. Berg, C. Chandlee, S. Cihangir, B. Collick, T. Ferbel, S. Heppelmann, J. Huston, T. Jensen, A. Jonckheere, F. Lobkowicz, M. Marshak, C. A. Nelson, Jr., T. Ohshima, E. Peterson, K. Ruddick, P. Slattery, P. Thompson,

Rochester - Minnesota - Fermilab Collaboration
(Presented by M. Zielinski)

We discuss constraints on properties of isovector $J^{PC} = 1^{-+}$ exotic hybrid mesons, predicted to decay into $\pi\eta$ and $\pi\rho$ channels. The experimental limits are based on Primakoff production of such states and on a VDM argument relating their radiative widths to their $\rho\pi$ decay modes. Using data on coherent production of $\pi+\pi+\pi-$ and $\eta\pi+$ systems in $\pi+$ collisions with nuclei, we exclude the existence of such states with masses below 1.5 GeV and with widths greater than 20 MeV, provided that their primary coupling is to $\rho\pi$. If, instead, primary coupling to $\eta\pi$ is assumed for these states, then we obtain several percent upper limits on their branching ratios into $\rho\pi$.

The following discussion is concerned with a particularly interesting hypothetical isovector $J^{PC} = 1^{-+}$ meson, denoted below as $\tilde{\rho}$. The quantum numbers of this state are exotic, in the sense of the traditional $q\bar{q}$ quark model, and could be most easily classified as a hybrid meson $\tilde{\rho}^{+} = u\bar{d}g$, with the $q\bar{q}$ pair in a color octet state. The present analysis is not dependent on the detailed nature of such a state, but uses theoretical expectations for such a hybrid $\tilde{\rho}$ to guide the investigation of available data [1-4].

In the QCD-motivated pursuit of spectroscopic gluon degrees of freedom, many calculations have been performed that predict an object with the quantum numbers of the $\tilde{\rho}$. The different approaches include QCD sum rules [5-7], QCD bag [8], flux tube [9] and potential [10] models. Most models expect this object to lie near or below 2 GeV in mass, and some calculations predict a $\tilde{\rho}$ state as low as $m_{\tilde{\rho}} \approx 1.3$ GeV [6]; for a light hybrid, a rather narrow total width ($\Gamma_{\tilde{\rho}} \approx 50$-$200$ MeV) and dominant $\rho\pi$ and/or $\eta\pi$ decay modes are favored.

Light hybrids with such properties are the subject of the following analysis. The available data limited our study to masses $\lesssim 1.5$ GeV. Our method of searching for the $\tilde{\rho}$ relies on the assumption that it has a non-negligible branching to $\rho\pi$. In such a case, a sizeable radiative width to $\pi\gamma$ is also expected. The latter can be estimated using the VDM relation:

$$\Gamma(\tilde{\rho}^{+}\rightarrow\pi^{+}\gamma) = \frac{\alpha}{\gamma_\rho^2/\pi}\left(\frac{k_\gamma}{k_\rho}\right)^3 \Gamma(\tilde{\rho}^{+}\rightarrow\rho^0\pi^{+}) \qquad (1)$$

with the value of $\gamma_\rho^2/\pi \approx 1.98\pm0.09$, as determined from the $\rho^0 \rightarrow e^+e^-$ decay width [11]. (k_γ and k_ρ are, respectively, decay momenta of the photon and the ρ^0, in the rest frame of the $\tilde{\rho}$.) Similar VDM predictions for other mesons are approximately satisfied by the data (see e.g. [4] for a discussion of the A_2 and [12] for the A_1).

As a consequence of Eq. (1), the value of $\Gamma(\tilde{\rho}+ \rightarrow \rho^0\pi+)$ determines uniquely the strength of the $\tilde{\rho}$ signal expected for electromagnetic production in the Coulomb field of a nucleus. This is given by the Primakoff formula [13]

$$\frac{d\sigma}{dt\,dm^2} = 24\pi\alpha Z^2 \frac{m^2}{(m^2 - m_\pi^2)^3}\Gamma\left(\tilde{\rho}^{+} \rightarrow \pi^{+}\gamma\right) \times D_{\tilde{\rho}}(m)\frac{(t-t_{min})}{t^2}|F_{em}(t)|^2 \qquad (2)$$

In practice, the electromagnetic production of a resonance usually has to be separated from strong-production backgrounds that can proceed through various Reggeon exchanges. For spin-flip contributions relevant to our case, the strong cross section takes the form

$$\frac{d\sigma}{dt\,dm^2} = C\,A^{2}D_s(m)\left(t - t_{min}\right)|F_s(t)|^2 \qquad (3)$$

(the D factors in (2) and (3) represent appropriately normalized mass distributions, (e.g. Breit-Wigner functions in case of resonance production). The $|F(t)|^2$ represent form factors, and C is related to the strong cross section on a nucleon.

Separation of the two contributions can be achieved by taking advantage of the very different momentum-transfer and nuclear-target dependences of the two production mechanisms. For $|t|\lesssim 0.002$ GeV^2, the Coulomb contribution is strongly enhanced by the $1/t^2$ factor (due to the photon propagator), while hadronic spin-flip exchanges are suppressed in this region. The nuclear-target dependence of electromagnetic production is much stronger than that for hadronic production; one expects the ratio of integrated cross sections on Pb and Cu to be ≈6.4 for electromagnetic, and ≈2 for strong contributions. A detailed exposition of our procedures can be found in [14].

In the following, we use data from Fermilab experiment E272 on coherent production of $\rho\pi$ [2,3] and $\eta\pi$ [4] states, which have been proposed as likely decay modes of the hybrid $\tilde{\rho}$. The data were collected at a $\pi+$ beam energy of 200 GeV on Cu and Pb targets. The $\rho\pi$ contribution to the $J^P = 1^-$ wave was projected out from the total production of $\pi+\pi+\pi-$ states using an Ascoli-style Partial Wave Analysis (PWA). This wave amounted to ≈5% of the total three-pion sample. We demonstrated previously [12] that our PWA provided reliable results for other partial waves of similar intensity. The $\eta\pi$ final state was observed in a clean Primakoff production mode [4], and was dominated by the A_2. Both sets of data were examined for possible contributions from electromagnetic production of a 1^- resonance.

The data on the production of the $J^{PC} = 1^{-+}(\rho\pi)$ systems [1] are presented in Fig.1. They do not exhibit any narrow resonant signal in the mass distributions. The cross sections in t do not increase nearly as rapidly in the forward direction as expected for electromagnetic production; the integrated Pb to Cu target ratio is only ≈2. This clearly indicates that the electromagnetic contribution to the 1^{-+} wave is small. We fitted the distributions in Fig. 1, allowing a broad strong-production background and a contribution from the electromagnetic production of a 1^- resonance of some specified mass and width. For the example displayed in the figure, we obtained $\Gamma(\tilde{\rho}+ \rightarrow \pi+\gamma)\times B(\tilde{\rho}+ \rightarrow \rho^0\pi+)\approx 15\pm 8$ keV. Similar fits were performed for different mass and width values of the resonant contribution, and the results for the corresponding 1σ upper limits are summarized in Fig. 2.

Our upper limits on $\Gamma(\tilde{\rho}+ \rightarrow \pi+\gamma) \times B(\tilde{\rho}+ \rightarrow \rho^0\pi+)$ exclude the possibility that the $\tilde{\rho}$ decays primarily into $\rho\pi$; namely, we find that $B(\tilde{\rho}+ \rightarrow \rho^0\pi+) << 0.5$. This point is illustrated in Fig. 3, where the VDM prediction, using the $\tilde{\rho}$ parameters of ref [5], is compared with the 1σ upper limit on $\Gamma(\tilde{\rho}+ \rightarrow \pi+\gamma)$, under the assumption that $B(\tilde{\rho}+ \rightarrow \rho^0\pi+)\approx 0.5$. The observed discrepancy by a factor of ≈10 can be translated into an upper limit on $B(\tilde{\rho}+ \rightarrow \rho^0\pi+)\lesssim 0.13$ for $m_{\tilde{\rho}}=1.3$ GeV. This result suggests that other possible (non-$\rho\pi$) decays channels of the $\tilde{\rho}$ must be dominant if such a state, in fact, exists at $\lesssim 1.5$ GeV. We consequently examined $\eta\pi$ channels, which have been proposed in several theoretical models for the $\tilde{\rho}$. In this analysis, we assumed that:

$$B(\tilde{\rho}^+\rightarrow\eta\pi^+)+ B(\tilde{\rho}^+\rightarrow\rho^+\pi^0)+B(\tilde{\rho}^+\rightarrow\rho^0\pi^+)\approx 1 \quad (4)$$

As mentioned above, the $\eta\pi$ data [4] are dominated by the Primakoff production of the A_2, and do not allow much room for other contributions. A particularly strong constraint on the presence of a 1^- resonance is provided by the angular distribution of the $\pi+$, given in Fig. 4, which is consistent with D-wave decay of the A_2; for P-wave decays of the $\tilde{\rho}$, one expects a distribution that has a maximum at $\cos\theta=0$. As a consequence, assuming Eq. (4), the $\eta\pi$ data provide an even stronger constraint on the electromagnetic production of the ρ than do the $\rho\pi$ data alone. The upper limit obtained on $\Gamma(\tilde{\rho}+ \rightarrow \pi+\gamma)\times B(\tilde{\rho}+ \rightarrow \eta\pi+)$ are shown in Fig. 5; they correspond to the upper limit on the branching fraction of the $\tilde{\rho}+ \rightarrow \rho^0\pi+$ of $B(\tilde{\rho}+ \rightarrow \rho^0\pi+)\lesssim 0.03$.

For masses above 1.5 GeV, our present data can still provide some information, if the hybrid is sufficiently broad. A recent QCD sum rule calculation [6] predicted a hybrid $\tilde{\rho}$ in the mass range 1.6-2.1 GeV, and provided estimates of its width (see the top part of Fig. 6) and $B(\tilde{\rho}+ \rightarrow \rho^0\pi+)$ as functions of mass. For predicted total widths of 1-2 GeV, there is a significant low-mass tail of the $\tilde{\rho}$ line shape that contributes to our range of data. Our upper limits on $\Gamma(\tilde{\rho}+ \rightarrow \pi+\gamma)$, given in Fig. 6, are below the prediction by a factor of 4-12, depending on $m_{\tilde{\rho}}$. The authors of Ref. [6] consider their width estimate uncertain by about a factor of two. For $\Gamma_{\tilde{\rho}}$ reduced by such a factor of two, our upper limits still remain below the VDM prediction for $m_{\tilde{\rho}}<1.9$ GeV.

Summarizing, we have obtained model independent 1σ limits on $\Gamma(\tilde{\rho}+ \rightarrow \pi+\gamma) \times B(\tilde{\rho}+ \rightarrow \rho^0\pi+)$ and $\Gamma(\tilde{\rho}+ \rightarrow \pi+\gamma) \times B(\tilde{\rho}+ \rightarrow \eta\pi+)$ for $1 < m_{\tilde{\rho}} < 1.5$ GeV, and $20 < \Gamma_{\tilde{\rho}} < 200$ MeV. These exclude a light hybrid that couples ≈100% to $\rho\pi$. Using VDM, we obtain an upper limit on $B(\tilde{\rho}+ \rightarrow \rho^0\pi+)\lesssim 0.2$ for $m_{\tilde{\rho}} <$ 1.5 GeV. Using the additional assumption that the remaining decays are into $\eta\pi$, the $\rho\pi$ branching is reduced to a few percent. It is clear that the Primakoff method is a very sensitive way to search for hybrids, or any other states, that couple to $\rho\pi$. Extending the range to higher mass $\lesssim$ 2GeV, preferably for several expected decay modes of the $\tilde{\rho}$, could prove to be very interesting indeed!

References

1 M. Zielinski et al., Rochester Preprint UR 946 and Z. Phys. C. (in press).

2. M. Zielinski et al., Phys. Rev. D30, 1855 (1984).
3. M. Zielinski et al., Z. Phys. C16, 197 (1983).
4. S. Cihangir et al., Phys. Lett 117B, 119 (1982).
5. F. de Viron, J. Govaerts, Phys. Rev. Lett. 53, 2207 (1984); J. Govaerts et al., Nucl. Phys. B248, 1 (1984).
6. J. Latorre et al., Phys. Lett. 147B, 171 (1984); J. Latorre et al., Trieste preprint IC/85/299.
7. J. I. Balitsky et al., Phys. Lett. 112B, 71 (1982).
8. T. Barnes et al., Nucl. Phys. B224, 241 (1983); M. Chanowitz, S. Sharpe, Nucl. Phys. B222, 221 (1983); M. Tanimoto, Phys. Rev. D27, 2648 (1983).
9. N. Isgur et al., Phys. Rev. Lett. 54, 869 (1985).
10. J. M.Cornwall, S. F. Tuan, Phys. Lett. 136B, 110 91984).
11. Particle Data Group, 1986, compilations.
12. M. Zielinski et al., Phys. Rev. Lett. 52, 1195 (1984).
13. H. Primakoff, Phys. Rev. 81, 899 (1951).
14. T. Jensen et al., Phys. Rev. D27, 26 (1983).

Figure Captions

1. a) Mass dependence of the $J^{PC} = 1^{-+}(\rho\pi)$ partial wave intensity on Pb and Cu targets (data points), together with the result of a typical fit assuming $m_{\tilde\rho} = 1.25$ GeV and $\Gamma_{\tilde\rho} = 50$ MeV (full lines). b) Same as (a) for the t-dependences of the wave [1].
2. The upper edges of 1σ deviation bands on the fitted values of $\Gamma(\tilde\rho+ \to \pi+\gamma)$ X $B(\tilde\rho+ \to \rho^0\pi+)$ as a function of mass, for several $\Gamma_{\tilde\rho}$ [1].
3. Comparison of the VDM prediction (1) and the experimental $\Gamma(\tilde\rho+ \to \pi+\gamma)$ value extracted under the assumption that $\Gamma(\tilde\rho+ \to \pi+\gamma)\approx 0.5$, as functions of $\Gamma_{\tilde\rho}$ at a representative mass of $m_{\tilde\rho} = 1.3$ GeV.
4. Decay $\pi+$ angular distribution for the $\eta\pi+$ systems produced coherently on Pb [4]. (Full line represents a fit assuming only an A_2 resonant contribution.)
5. Same as Fig. 2 for $\Gamma(\rho+ \to \pi+\gamma)$ X $B(\tilde\rho+ \to \eta\pi+)$.
6. Comparison of the expected $\tilde\rho$ signal, and the measured upper edge of the 1σ band for the $\tilde\rho$ parameters predicted in [6], as functions of $m_{\tilde\rho}$. The predicted value of $\Gamma_{\tilde\rho}$ is shown at the top, as a function of $m_{\tilde\rho}$.

Fig. 1a

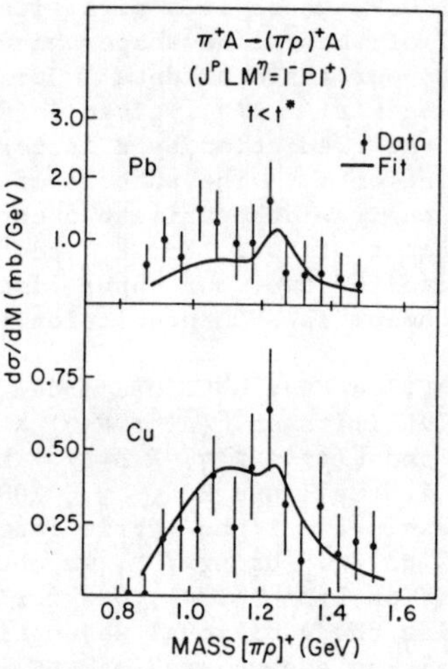

Fig. 1b

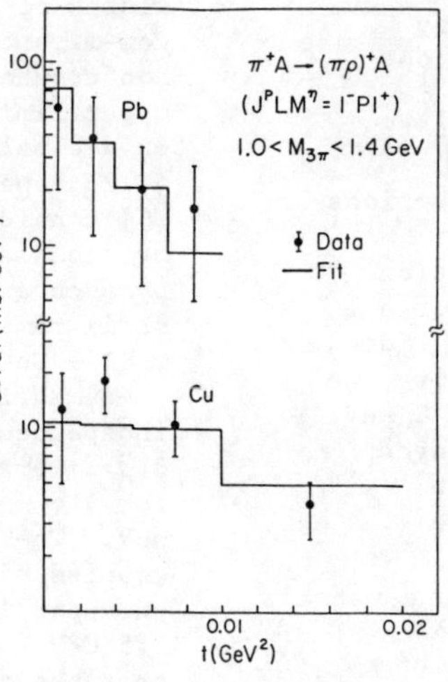

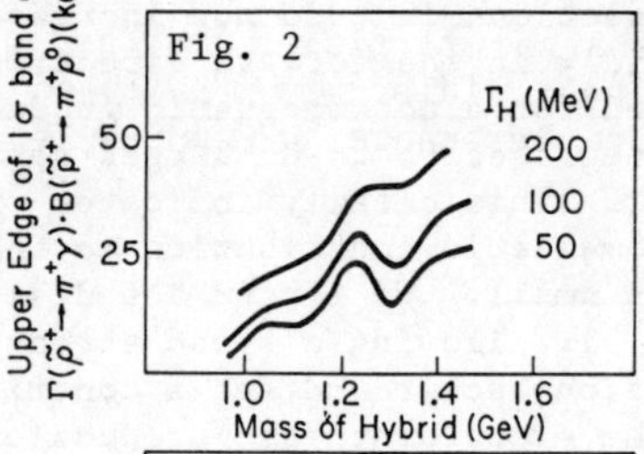

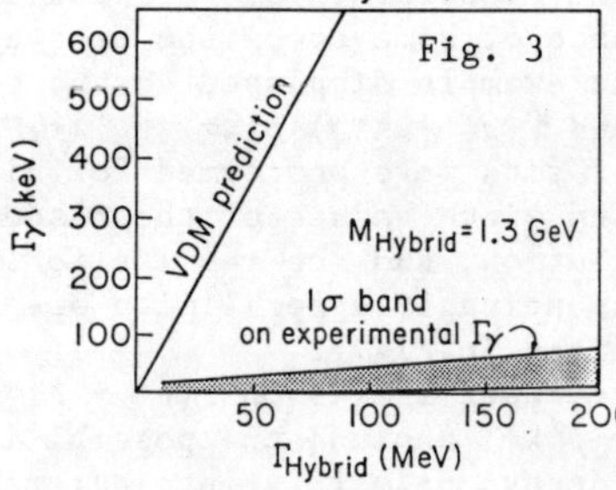

Fig. 4

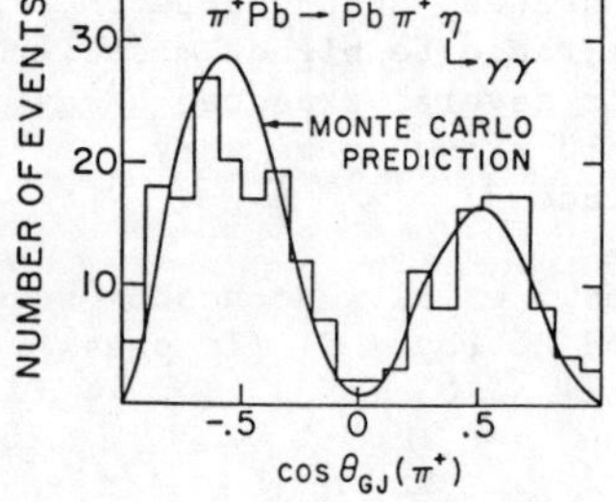

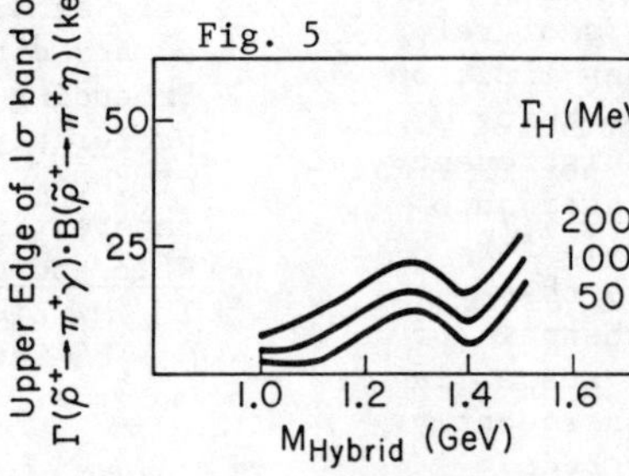

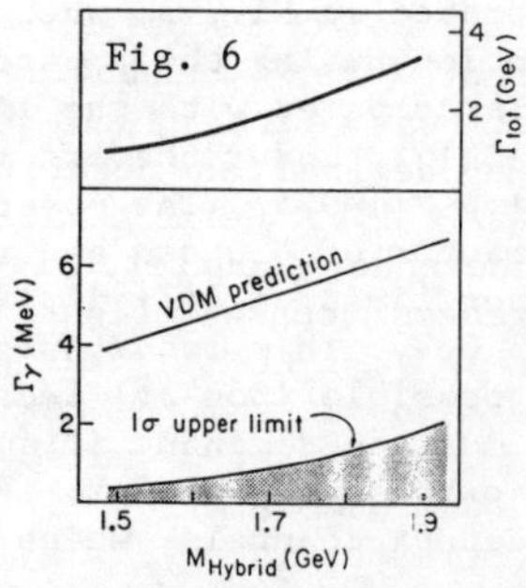

Ξ^* AND Ω^* SPECTROSCOPY AT THE CERN-SPS HYPERON BEAM

Pierre EXTERMANN

University of Geneva, Switzerland

Two Ω^* resonances have been observed in the $\Xi^-\pi^+K^-$ channel at 2251 and 2384 MeV/c^2. Furthermore, several Ξ^* resonances have been investigated in the ΛK^-, ΛK^0 and $\Xi^-\pi^+\pi^-$ channels. In addition to the known states at 1690 and 1820 MeV/c, a clear signal has been observed at 1960 MeV/c as well as indications for possible states at 1780 and 2180 MeV/c^2. A double moment analysis of the Ξ(1820) and Ξ(1960) mass regions has yielded information on the J^P values of these states.

The experimental apparatus was a two-stage magnetic spectrometer equipped with proportional wire chambers, drift chambers, gas threshold Cherenkov counter, a liquid argon calorimeter and a lead glass array. The incident hyperons (at a mean momentum of 116 GeV/c) were identified by a differential Cherenkov counter (DISC). The experimental target, a beryllium rod, was located close to the DISC exit port. The trigger required an incident Ξ^- in the DISC as well as 3 charged particles in the spectrometer hodoscopes. This system allowed to measure inclusive resonance production of both diffractive and non diffractive types.

Ω^* RESONANCES

The Ω^* signals were observed in the $\Xi^-\pi^+K^-$ channel [1]. (Other S = −3 channels are still under investigation). The selection criteria can be summerized as follows.
i) A Λ candidate was required with a ($p\pi^-$) effective mass within 5 MeV/c^2 of the Ξ^- mass. Then a negative particle had to be found intersecting the Λ trajectory and yielding a ($\Lambda\pi^-$) effective mass within 8 MeV of the Ξ^- mass. The background under the Ξ^- peak was negligible. ii) the π^+ were defined in the following way: below 12 GeV/c, all positive particles attached to the vertex were considered to be π^+ and above 12 GeV/c a Cherenkov signal was also required. (The π threshold of the Cherenkov counter was set at 10 GeV/c). iii) The K^- were defined as negative particles with momenta > 13.5 GeV/c that did not trigger the Cherenkov counter. iv) Finally, since strangeness conservation requires an additional S = +1 particle to be associated to the main vertex, the total ($\Xi^-\pi^+K^-$) momentum had to be at least 11 GeV/c below the mean beam momentum.
The corresponding ($\Xi^-\pi^+K^-$) effective mass distribution is shown in Fig.1a. The solid curve represents a polynomial of order three fitted to the distribution below 2.75 GeV/c. The dotted line shows an alternate background shape obtained from event mixing. In both cases, two clear signals appear at roughly 2250 and 2400 MeV/c .

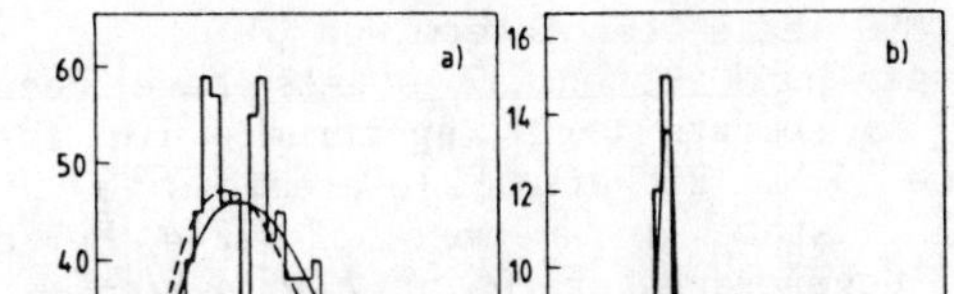

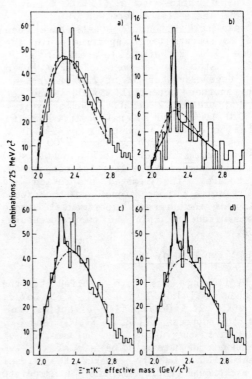

Figure 1. ($\Xi^-\pi^+K^-$) effective mass distributions under various conditions and with different fits

In order to check the S = -3 nature of these peaks, an associated K^+ or K^o candidate was demanded in the spectrometer. The resulting mass spectrum appears in Fig. 1 b. The peak at 2250 MeV/c^2 has a statistical significance of 4.2 σ but the other peak has disappeared. In Fig. 1c, the first peak of Fig. 1a is fitted with a Breit- Wigner function and a polynomial. A clear excess appears at 2400 MeV/c^2. The most likely explaination is that this signal is also an Ω^* state. Indeed, the number of events above background is similar to that of Ω (2250). The disappearance of this signal in Fig. 1 b might be due to a different production mechanism in which the associated kaon would be emitted at larger angle and therefore escape detection. A fit with two Breit-Wigner curves and a polynomial is displayed in Fig. 1d. The resulting masses and widths are

$$M_1 = 2251 \pm 9 \text{ MeV}/c^2 \quad \Gamma_1 = 48 \pm 20 \text{ MeV}/c^2$$

$$M_2 = 2384 \pm 9 \text{ MeV}/c^2 \quad \Gamma_2 = 26 \pm 23 \text{ MeV}/c^2$$

and the numbers of events under the peaks are 78 ± 23 and 45 ± 10 respectively.
The production cross sections and the branching ratios to $\Xi^o(1530)K^-$ and $\Xi^-K^o(890)$ have been determined [1].
Several quark potential models have been used to compute the Ω^* spectrum. For instance J.M. Richard [2] predicts a $J^P = 3/2^+$ state at a mass of 2244 MeV/c^2 and 3 degenerated states at 2358 MeV/c^2.
With another potential model, Chao et al. [3] predict 2 negative parity states at 2020 MeV/c and several positive parity states between 2100 and 2300 MeV/c^2. They did not compute anything above 2300 MeV/c^2.
A J^P determination of Ω(2250) and Ω(2400) is clearly needed for a fruitful comparison. Unfortunately the limited statistics and the relatively poor signal- to- background ratios of the present experimental sample are unsuited for spin and parity measurements.

Ξ(1820) AND Ξ(1960) IN ΛK^o

The following cuts were applied to the ΛK^o channel [4]: i) $p(\Lambda K^-) > 75$ GeV/c; ii) $p(K^-) > 27.5$ GeV/c; iii) effective mass cuts to suppress Σ(1385), K(890) and Σ^o, the latter cut using the information of the photon detectors. The aim of the two momentum cuts was to enhance the signal to background ratio in the intermediate and high mass regions of the distribution. The resulting spectrum is shown in Fig. 2. The two signals at 1820 and 1960 MeV/c^2 have been fitted with 2 Breit-Wigner functions and a polynomial supplemented with a $(m-m_o)^{1/2}$ term, where m_o is the threshold mass. The following parameters have been found

$$M_1 = 1826 \pm 4 \text{ MeV}/c^2 \quad \Gamma_1 = 12 \pm 14 \text{ MeV}/c^2$$

$$M_2 = 1963 \pm 5 \text{ MeV}/c^2 \quad \Gamma_2 = 25 \pm 15 \text{ MeV}/c^2$$

The first resonance is in good agreement with the well-known Ξ(1820), but the second resonance is new. It is the first narrow Ξ^* signal to be reported in the 1900-2000 MeV/c^2 mass range.

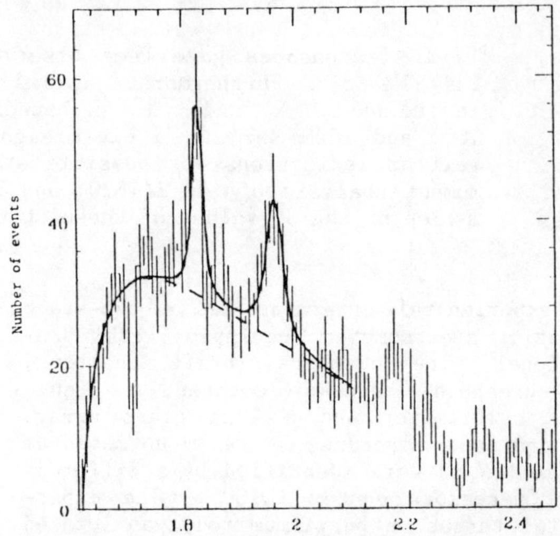

Figure 2. (ΛK^o) effective mass (in GeV/c^2). (10 MeV bins)

A J^P analysis was made using the combined moments of the $\Xi^* \to \Lambda K^o$, $\Lambda \to p\pi^-$ decay sequence [5]. For Ξ(1820), if J = 3/2 is assumed, the probability for P = -1 is 0.94 and for P = +1, 0.09. The negative parity is clearly favoured. For Ξ(1960), the natural parity sequence is preferred over the non natural one and the highest probabilities are found for $5/2^+$ (0.67), $7/2^-$ (0.55) and $9/2^+$ (0.48) as compared to $3/2^-$ (0.10) and $1/2^+$ (0.03). So $J^P = 5/2^+$ is preferred but $7/2^-$ and $9/2^+$ cannot be excluded.
Finally Ξ(1960) was searched for in the $\Sigma^o\bar{K}^o$ channel but no signal was found at this mass.

RESONANCE SURVEY IN VARIOUS CHANNELS

The status of our resonance study is summarized in Table 1. The corresponding mass spectra are shown in Figs. 2-5 (some of the fitted curves are not indicated). The following comments are in place. The Ξ(1680) which is present in the ΛK^- channel (Fig. 3) does not show up in the ΛK^o channel (Fig. 2) because of the momentum cuts applied to the latter channel. When these cuts are removed, a clear signal appears at 1680 MeV/c^2. The newly observed Ξ(1780) shows up in $\Xi\pi\pi$ only, with a significance of 3.3σ. For the well established Ξ(1820), the masses measured in each channel do not agree. The errors indicated contain a statistical and a systematic contribution. It may

well be that several narrow states with different branching ratios overlap in this region. The $\Xi(1940)$ signal in $\Xi\pi$ is well compatible with the enhancement reported in a previous experiment [6]. In contrast the new $\Xi(1960)$ is definitely narrower and might be part of the $\Xi(1940)$ signal in $\Xi\pi$. The enhancement at 2180 MeV/c appears in 3 channels with moderate significance each time but the 3 masses agree nicely. Finally, one should mention that most of the resonance observed in $\Xi\pi\pi$ have a strong decay fraction into the $\Xi(1530)\pi$ channel.
Work on the branching ratios of these resonances and their allocation to states predicted by potential quark models are under way [4,7].

Table 1. Ξ^* SURVEY IN VARIOUS CHANNELS

Nominal mass	ΛK^-	$\Lambda\bar{K}$	$\Xi^-\pi^+\pi^-$	$\Xi^-\pi^+$
1680	1691 ± 2 11 ± 4	(1680)	—	—
1780*	--	--	1783 ± 1 6 ± 2	—
1820	1819 ± 3 25 ± 5	1826 ± 4 12 ± 14	1832 ± 3 10 ± 10	—
1940	—	—	--	1944 ± 9
1960*	—	1963 ± 5 25 ± 15	--	100 ± 31
2180*	(2180)	2180 ± 7 30 ± 13	2189 ± 7 46 ± 27	

The top number in each entry is the mass of the resonance in MeV/c^2. The number just below (if any) is the width Γ, also in MeV/c^2. Unfitted values are indicated between parentheses. The (*) denote new Ξ^* candidates.

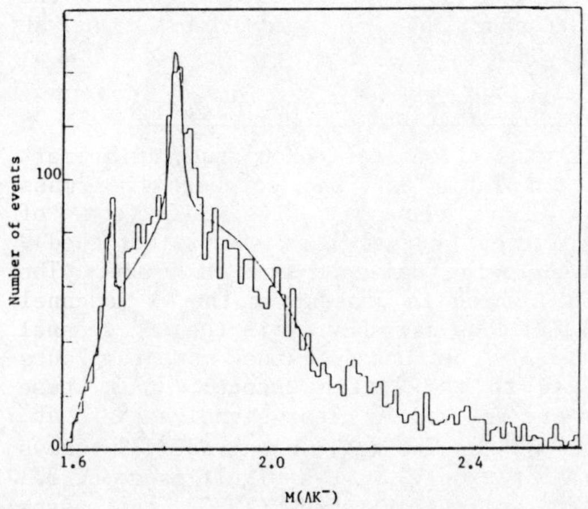

Figure 3. (ΛK^-) effective mass (in GeV/c^2). (10 MeV bins)

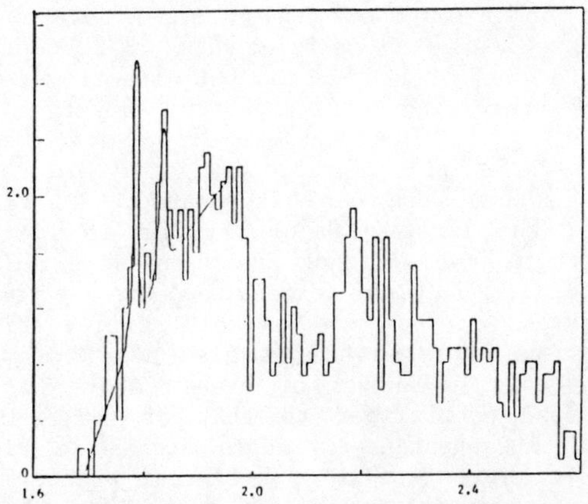

Figure 4. $\Xi^-\pi^+\pi^-$ effective mass (in GeV/c^2). (10 MeV bins)

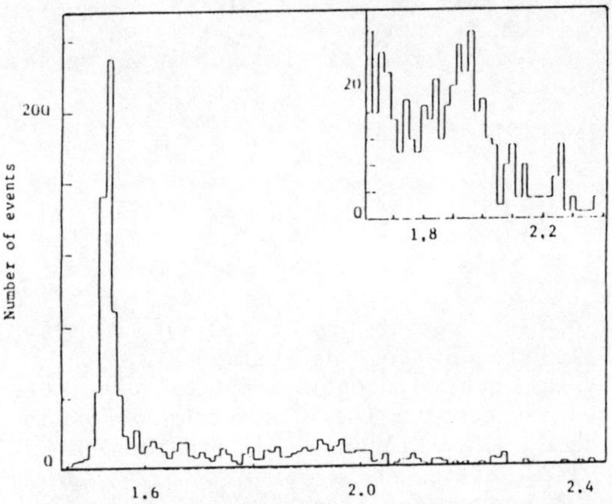

Figure 5. $\Xi^-\pi^+$ effective mass (in GeV/c^2). (10 MeV bins) (insert: 20 MeV bins)

REFERENCES

1) First Observation of Ω^* Resonances S.F. Biagi et al., Z. Phys. C 31, 33 (1986)

2) J.M. Richard, Phys. Letters 100B, 515 (1981)

3) K.T. Chao, N. Isgur and G. Karl, Phys. Rev. D23, 155 (1981)

4) Ξ^* Resonances in Ξ^-Be Interactions. II. Properties of $\Xi(1820)$ and $\Xi(1960)$ S.F. Biagi et al. To be published

5) K.G. Ragan Thesis, University of Geneva, 1986, unpublished.

6) Production of Hyperons and Hyperon Resonances in Ξ^-N Interactions at 102 and 135 GeV/c. S.F. Biagi et al., Z. Phys. C 9, 305 (1981).

7) Ξ^* Resonances in Ξ^-Be Interactions. I. Diffractive Production in the ΛK^- and $\Xi^-\pi^+\pi^-$ Channels. S.F. Biagi et al. To be published.

Parallel Session 11

Lifetimes and Weak Interactions of Heavy Quarks and Leptons

Organizers:

W. Ford (Colorado)

K. Schubert (Heidelberg)

Scientific Secretaries:

J. Butler (SLAC)

L. Mathis (LBL)

D. Pitman (SLAC)

NEW RESULTS ON CHARMED D, $F^{\pm}$AND F^*PRODUCTION AND DECAY FROM THE MARK III†

RAFE H. SCHINDLER representing The MARK III Collaboration
Stanford Linear Accelerator Center
Stanford University, Stanford, California 94305

Results on charmed meson production and decay are presented from the Mark III at SPEAR. $F\bar{F}^*$ associated production is observed allowing a direct measurement of the F^*mass. A search for the decay $D^+ \to \mu^+\nu_\mu$ in the recoil of hadronically tagged $D^\pm$ decays provides a stringent limit on the pseudoscalar decay constant f_D . New results on $D^0\bar{D}^0$ mixing from semileptonic D^0 decays and evidence for a nonresonant component in D_{e4} decays are also presented.

1. $F\bar{F}^*$ ASSOCIATED PRODUCTION

Evidence for the charmed vector meson F^*has been previously reported.[1] Interest lies in a precise mass determination providing a strict test of the constancy of the pseudoscalar-vector mass[2] splitting predicted by Frank and O'Donnell.[2]

In a sample of 6.3 ± 0.5 pb^{-1} integrated luminosity taken at $\sqrt{s} = 4.14$ GeV/c^2, a search is performed for $\bar{F}^*$ produced in association with an $F^+ \to \phi\pi p$. Charged tracks in the analysis are constrained to the beam position and a common event vertex. Charged kaons are identified by time-of-flight. The invariant mass of all K^+K^- pairs is formed and those lying within ± 0.010 GeV/c^2 of the ϕ mass are combined with all other tracks in an event which are assumed to be pions. The recoil mass of each $\phi\pi$ combination is calculated and displayed in Fig. 1. Events appear to cluster in the recoil of the $\phi\pi$ when it is near either the $D^\pm$ or

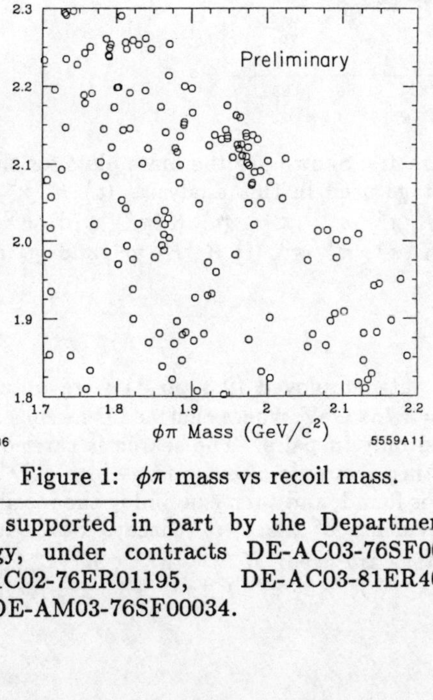

Figure 1: $\phi\pi$ mass vs recoil mass.

†Work supported in part by the Department of Energy, under contracts DE-AC03-76SF00515, DE-AC02-76ER01195, DE-AC03-81ER40050, and DE-AM03-76SF00034.

the F mass. Cutting on recoil masses from 1.97 to 2.05 GeV and 2.05 to 2.15 provides clean signals for both the Cabibbo suppressed $D^\pm$ decay and the F, respectively (see Fig. 2). There are 9 $D^\pm$ signal events and 29.4 ± 5.4 $F^\pm$events. The backgrounds are determined from event mixing. The $F^\pm$mass is found to be $1.9734 \pm 0.0044 \pm 0.0040$ GeV/c^2.

The cluster of events recoiling from $F \to \phi\pi$ is expected to be a compound distribution. From Fig. 1, most events appear to originate from $F\bar{F}^*$ production, there being little evidence for $F\bar{F}$ production. The $\phi\pi$ events at the F mass can arise either from the direct F or from the decay of the $\bar{F}^* \to \gamma + \phi\pi$. To improve the F^*mass resolution a simple kinematic constraint is applied. The observed $F \to \phi\pi$ is assumed to be a direct F, and its mass is fixed at 1.9705 GeV/c^2.

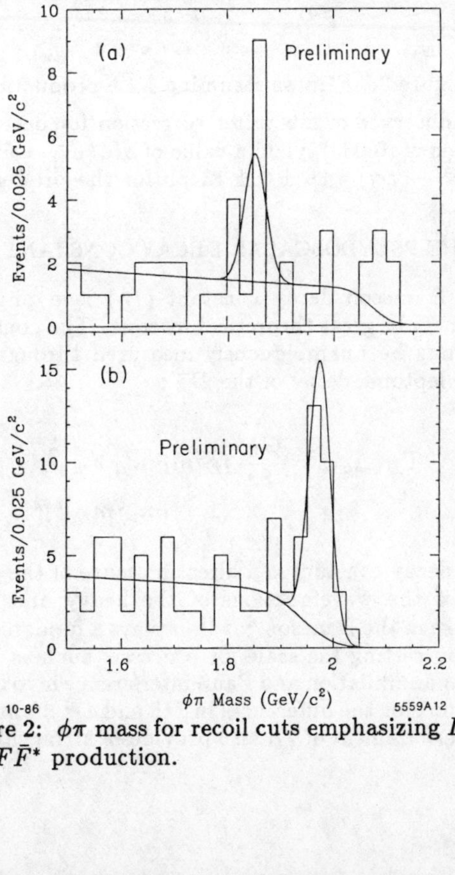

Figure 2: $\phi\pi$ mass for recoil cuts emphasizing $D\bar{D}^*$ and $F\bar{F}^*$ production.

The invariant mass of the F^* is then calculated assuming that its energy is that of the beam, less that of the detected $\phi\pi$. This should result in a 0.005 GeV/c^2 mass resolution when the $\phi\pi$ is direct, and a wider ramp-like distribution if it is from the F^*. The $\phi\pi$ are expected to populate both distributions equally. The results are shown in Fig. 3. The mass of the F^* is $2.1108 \pm 0.0019 \pm 0.0032$ GeV/c^2. The systematic error includes uncertainties in both the detector mass scale, as well as the beam energy. Since the F^* mass is not independent of the F mass, the $F-F^*$ mass splitting is best obtained taking a world average[3] (1.9705 ± 0.0025) for the F mass that excludes our measurement. The resulting mass difference is 0.137 ± 0.007 GeV/c^2. This implies a difference in squared masses for the F and F^* of 0.56 ± 0.03 in good agreement with the lighter mesons, as predicted by Frank and O'Donnell,[2] and implying that the mesonic wavefunction at the origin for a meson containing both heavy and light quarks is largely determined by the long range confining part of the interquark potential.

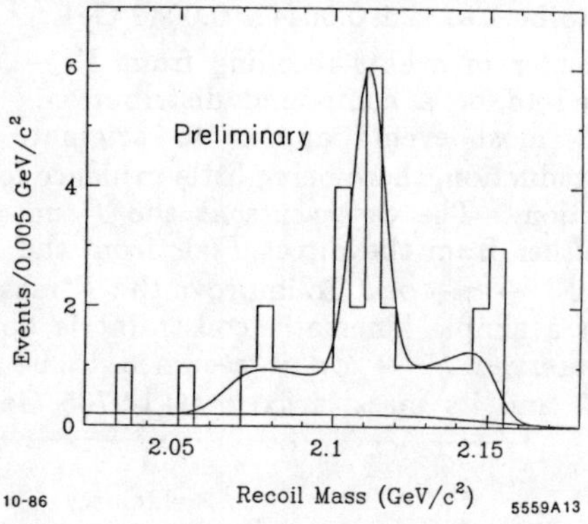

Figure 3: F^* mass assuming $F\bar{F}^*$ production.

The observed events, after correction for detection efficiency (0.063), yield a value of $\sigma(e^+e^- \rightarrow F\bar{F}^*) \cdot Br(F \rightarrow \phi\pi) = 36 \pm 7 \pm 13$ pb for the production rate.

2. THE PSEUDOSCALAR DECAY CONSTANT (f_D)

The D meson decay constant (f_D) is a physical quantity of great theoretical interest. The constant f_D may be unambiguously measured through the pure leptonic decay of the $D^\pm$:

$$\Gamma_{D^+\rightarrow\mu^+\nu} = \frac{G_F{}^2}{8\pi} f_D{}^2 m_D m_\mu^2 |\, V_{cd}\,|^2 \times (1. - (m_\mu/m_D)^2)^2$$

The decay constant is a direct measure of the overlap of the wavefunctions of the heavy and light quarks in the D meson.[4] It thus plays a fundamental role in setting the scale for processes such as weak flavor annihilation and Pauli interference invoked to account for the differences in $D^\pm$ and D^0 lifetimes.[5] A measurement of f_D also provides a stringent test of potential model[4] and QCD sum rule[6] calculations. In addition, it allows reliable estimates of other heavy meson decay constants (f_F, f_B, *etc.*), which are difficult to obtain due to the large theoretical uncertainties in extrapolating from f_π and f_K to the nonrelativistic heavy quark mesons. The decay constant also is essential in evaluating the magnitude of operators leading to $D^0\bar{D}^0$ and $B^0\bar{B}^0$ mixing.[7]

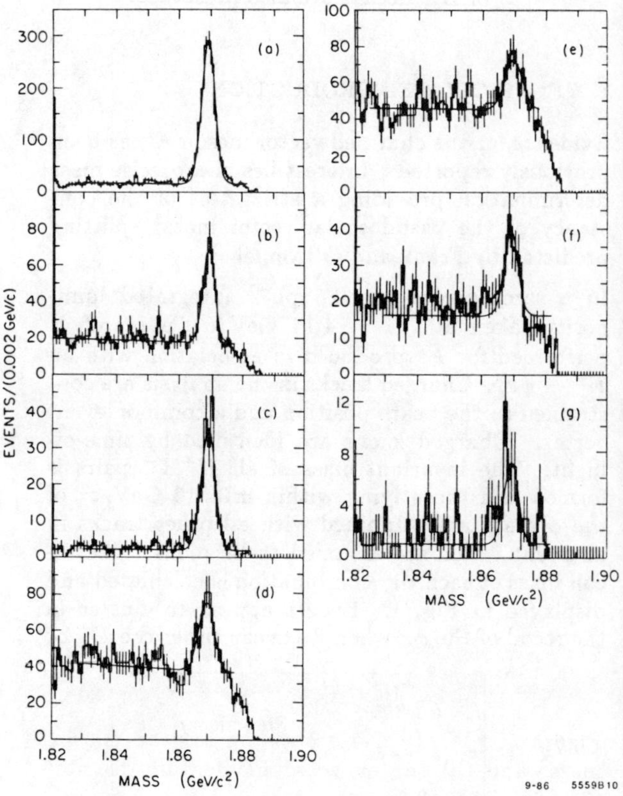

Figure 4: Shown are the mass plots for the seven $D^\pm$ tags used in this analysis: (a) $K^-\pi^+ \ \pi^+$, (b) $\bar{K}^0\pi^+ \ \pi^+ \ \pi^-$, (c) $\bar{K}^0\pi^+$, (d) $\bar{K}^0\pi^+ \ \pi^0$, (e) $\bar{K}^-\pi^+ \ \pi^+ \ \pi^0$, (f) $K^-K^-\pi^+$, and (g) $\bar{K}^0K^+$.

The data employed (9.3 pb^{-1}) were obtained at $\sqrt{s} = 3.768$ GeV, where charmed D mesons are produced only in pairs. The search is carried out by isolating a sample of events in which a D^+ candidate is found, and then examining the recoil system for evidence of the $\mu^-\bar{\nu}_\mu$ decay. Seven hadronic $D^\pm$ tags are used: $K^-\pi^+ \ \pi^+$, $\bar{K}^0 \ \pi^+$, $\bar{K}^0\pi^+ \ \pi^+ \ \pi^-$, $\bar{K}^0\pi^+ \ \pi^0$, $\bar{K}^-\pi^+ \ \pi^+ \ \pi^0$, $\bar{K}^0K^+$, and

$K^- K^+ \pi^+$, resulting in 2490 ± 42 $(stat)$ ± 40 $(syst)$ cleanly identified $D^\pm$ (see Fig. 4).

The isolation of the $\mu^- \bar{\nu}_\mu$ candidates proceeds by requiring the recoil system from a tag to have one track with a charge opposite to that of the tag whose momentum must lie between 0.775 and 1.125 GeV/c. The event must be consistent with having zero missing mass,[2] when the track is assigned a muon mass. Monte Carlo distributions for the expected signal are shown in Fig. 5. After the momentum cut, missing mass^2 is required to lie between -0.265 and 0.175 (GeV/c^2)^2 losing about 5% of the expected signal.

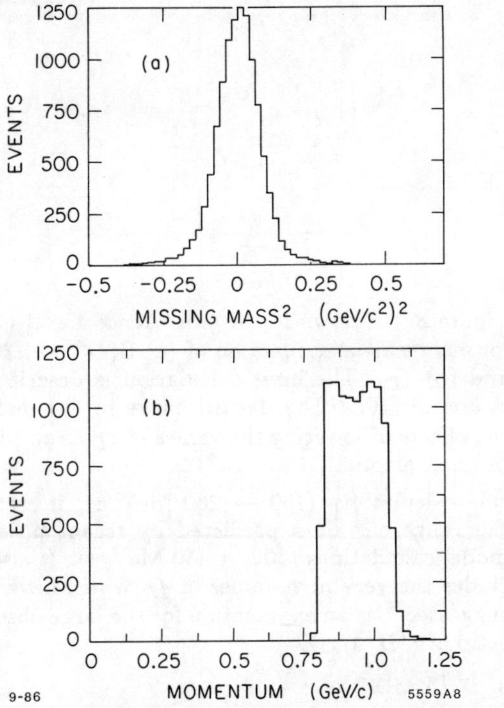

Figure 5: Monte Carlo showing (a) the missing mass^2 and (b) the expected momentum distribution, for $D^+ \to \mu^+ \nu_\mu$.

Muon candidates must either lie in the acceptance of the muon system ($| \cos\theta | \leq 0.65$) or lie where there is calorimetric information. Tracks entering the muon system must have two (one) layer hit for muon momentum (p_μ) ≥ 1 GeV/c ($p_\mu < 1$ GeV/c). No other topological cuts are applied due to the $(90-95\%)$ rejection of π and K decays and punchthrough provided by the muon system.[8]

If the muon candidate lies outside the muon system it is required to be minimum ionizing in the calorimeter (≤ 0.300 GeV deposited). Additional cuts on these events are imposed to reduce the principle sources of background from the decays $D^+ \to \bar{K}^0 \pi^+$, $\pi^+\pi^0$, $\bar{K}^0 K^+$, and $\bar{K}^0 \mu^+ \nu$. Background events with a π^0 (either from the D^- or a $K_S^0 \to \pi^0 \pi^0$) are rejected by requiring the absence of any isolated photons in an event (isolated photons are defined as those not used in forming a π^0 in the tag, or which make an angle $| \cos\theta | \leq 0.92$ with respect to any charged track). This cut also rejects K_L^0 interacting in the shower counter. The fraction of K_L^0 which interact is modeled by using the decay $\psi(3100) \to K_S^0 K_L^0$, $K_S^0 \to \pi^+\pi^-$ from a separate data set.[9] Due to the larger cross-section for K_L^0 at the momentum found in the $\psi(3100)$ data, this procedure is expected to *underestimate* the background.

The expected missing mass squared (M^2_{miss}) is zero for $D^- \to \mu^- \bar{\nu}_\mu$, while it is expected to peak near $m^2_{\pi^0}$ or $m^2_{K_S^0}$ for the two body backgrounds. In the case of $\bar{K}^0$ $\mu^- \bar{\nu}_\mu$, the missing ν_μ makes M^2_{miss}peak above $m^2_{K_S^0}$ (see Fig. 6).

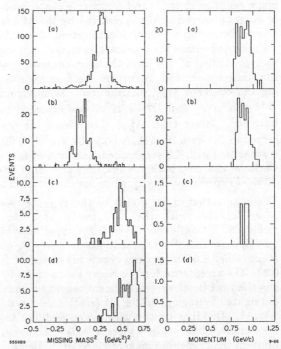

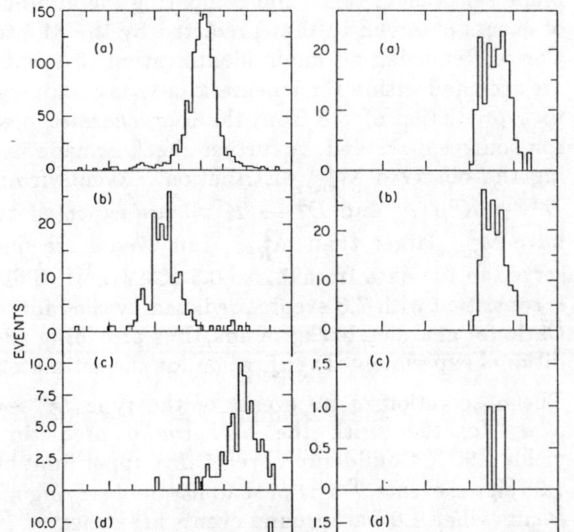

Figure 6: The variables $M^2_{miss}$and p^{lab}_μ are plotted for Monte Carlo events for the four major backgrounds: (a) $D^+ \to \bar{K}^0 \pi^+$, (b) $\pi^+\pi^0$, (c) $\bar{K}^0 K^+$, and (d) $\bar{K}^0 \mu^+ \nu$. The variables $M^2_{miss}$and p^{lab}_μ are plotted for Monte Carlo events for the four major backgrounds: (a) $D^+ \to \bar{K}^0 \pi^+$, (b) $\pi^+\pi^0$, (c) $\bar{K}^0 K^+$, and (d) $\bar{K}^0 \mu^+ \nu$.

When these cuts are applied to the data, no events are found to survive (Fig. 7). The nearest event to the M^2_{miss}cut appears lies 20 $(MeV/c^2)^2$ at 0.196 (GeV/c^2)^2.

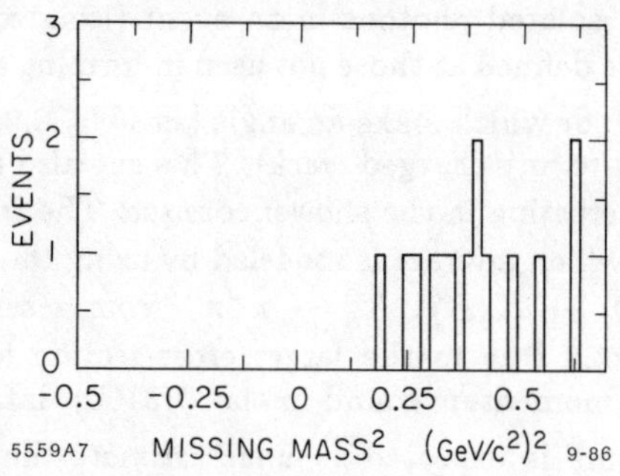

Figure 7: Missing mass^2 for data, after P_μ^{lab} cut.

The expected background with these cuts is 1.16 ± 0.16 (*stat.*) ± 0.20 (*syst.*) The error from the underestimate of the background component containing a K_L^0 is excluded from the systematic error because of the uncertainty in the size of the effect. The background calculation is tested by loosening the muon selection criteria, and comparing the number of events observed to that predicted by the Monte Carlo. Removing all muon identification 12 events are accepted within the kinematic cuts as compared to a prediction of 5.8 from the four *charmed* meson sources described. A further check is made using the observed M^2_{miss} distribution. Events from $D^+ \to \bar{K}^0\mu^+\nu_\mu$ and $D^+ \to \bar{K}^0\pi^+$ are expected to have M^2_{miss} larger than $m^2_{K^0}$. Ten events are observed in the data from 0.2 to 0.5 $(GeV/c^2)^2$. This is consistent with 7.6 events predicted by the Monte Carlo for charmed backgrounds, thus providing additional experimental verification for the estimate.

The observation of no events of the type $D^+ \to \mu^+\nu_\mu$ together with the background prediction yields a 90 % Confidence Level (C.L.) upper limit of 1.35 signal events.[10] The probability of observing no events when 1.0 background events are expected is 0.37. The acceptance for this decay mode varies by less than 3% for the seven different tagging modes; a weighted average of $0.72 \pm .01$ (*stat*) $\pm .05$ (*syst*) is used. Dividing by the acceptance and the total number of $D^\pm$ tags[11] gives a 90% C.L. upper limit on the branching ratio of 8.4×10^{-4}. Using a $D^\pm$ lifetime of $(10.1\,^{+0.7}_{-0.6}) \times 10^{-13}s$,[12] and $|V_{cd}|^2 = 0.0506 \pm .0065$[13] then the 90 % C.L. branching ratio limit corresponds to $f_D = 310$ MeV/c^2. When the errors on τ_{D^+} and $|V_{cd}|^2$ are included, we obtain a 90 % C.L. upper limit on f_D of 340 MeV/c^2, (see Fig. 8).

Calculations of the pseudoscalar decay constants obtain values which either increase (QCD sum rule method[6]) or decrease (non-relativistic potential[4] and bag model methods[14] with the meson mass. While our result does not probe the small values of f_D suggested by the bag model or QCD sum

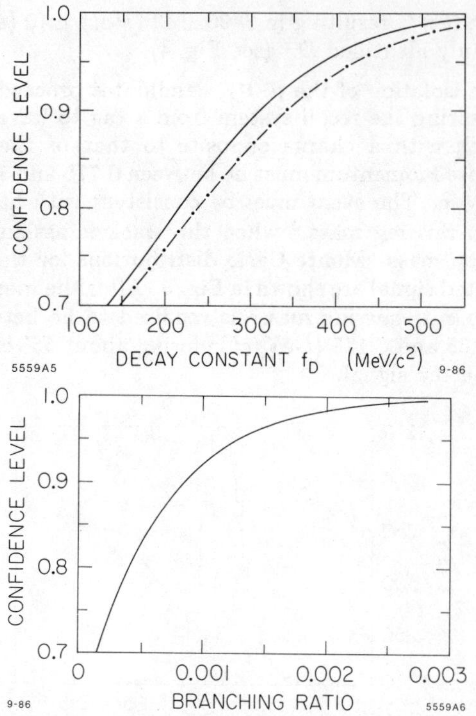

Figure 8: Shown is the Confidence Level (C.L.) for our result as a function of (a) B($D^+ \to \mu^+\nu_\mu$), and (b) f_D. The limit calculation is described in reference (10). The dashed curve in (b) includes the effects of lowering the values of τ_{D^+} and $|V_{cd}|^2$ by their errors.

rule calculations (150 → 280 MeV/c), it restricts the range of values predicted by recent potential model calculations (208 → 450 MeV/c). It also excludes the very high values of f_D which have been suggested[15] as an explanation for the large observed ratio of $\tau(D^+)/\tau(D^0)$.

3. $D^0 D^0$ MIXING

Previously,[16] we have reported results of a search for $D^0\bar{D}^0$ mixing candidates carried out by examining $D^0\bar{D}^0$ events having strangeness of ± 2, when the D^0 (or $\bar{D}^0$) decayed hadronically to either $K^-\pi^+$, $K^-\pi^+\pi^+\pi^-$ or $K^-\pi^+\pi^0$ (or charge conjugates). This results in the observation of three $s = \pm 2$ events with an estimated background of $0.4 \pm 0.1 \pm 0.1$ from particle misidentification. The sample also includes 162 fully reconstructed events where $s = 0$. Because these $D^0\bar{D}^0$ events decay to hadronic final states, a background from doubly Cabibbo suppressed decays (DCSD) may be present.[17] This background is only absent in the case of identical final states. To reduce or eliminate the problem of DCSD the search has been extended to fully reconstructed events containing either a hadronic plus a semileptonic decay ($K^-e^+\nu$ or $K^-\mu^+\nu$), or two semileptonic decays, respectively.

The semileptonic sample is selected requiring a hadronic tag where $\delta \geq 150$ps or $\delta \geq 1\sigma$ for time-of-flight (TOF) or DEDX identification, respectively. Here δ is the maximum value for the difference of the observed time of a π or K in $K^-\pi^+$ or $K^-\pi^+\pi^0$ decays from the average expected time for a π or K hypothesis. In the $K^-\pi^+\ \pi^+\ \pi^-$ mode, δ is tested for the K^- only. Recoiling from the hadronic tag, we demand there be no isolated photons in candidate events. The recoil semileptonic decay requires a loosely identified $K^\pm$ ($\delta \geq 100$ps TOF or 1σ DEDX) and an identified lepton (e or μ). Monte Carlo studies indicate that hadronic versus hadronic events can leak into the semileptonic versus hadronic classification if a hadron is misidentified as a lepton. These decays would peak near zero missing transverse momentum (p_T^{miss}) in the event. Additional p_T^{miss} can arise from asymmetric π^0's that escaped the cut on additional isolated photons. Figure 9 shows the p_T^{miss} distributions for backgrounds, the expected signal, and the data. A cut at $p_T^{miss} \geq 0.150$ GeV/c is made retaining 74 events. The expected background is 5 events leaving 69 mixing candidates. All these events are found to be *correct sign*; the charge of the lepton being opposite to that of the kaon in the tag.

Semileptonic versus semileptonic events are analyzed by selecting four prong events with zero net charge, and no isolated photons. Here we require *either* a solid tag on strangeness (2 well identified $K^\pm$ and ≥ 1 lepton), or a solid tag on the 2 leptons ($\geq 1K^\pm$ and 2 identified leptons). At least one combination must be consistent with the monochromatic pair production of $K^1l^1\nu^1$ versus $K^2l^2\nu^2$. That is, the momentum of each neutrino ($p_\nu^i, i = 1,2$) must satisfy:

$$p_\nu^i = 1.884 - E(K^i + l^i) \geq 0.$$

$$|\ p_\nu^i - P(K^i + l^i)\ | \leq 283 \text{ MeV/c}$$

$$p_\nu^i + P(K^i + l^i)\ \geq 283 \text{ MeV/c}$$

The resulting distribution for the ν^i energy is shown in Fig. 10 for Monte Carlo and data. The data shows evidence for a low energy tail which comes from semileptonic versus hadronic events, where the latter has a lepton misidentification. There is also evidence for a high energy tail which may result from the loss of a neutral hadron (K_L^0) in a semileptonic versus hadronic event, as well as a lepton misidentification. To reduce these backgrounds, the cuts in this preliminary analysis are tightened requiring both leptons and both kaons to be well identified. This introduces a large inefficiency as is evidenced in Fig. 11. The result is 12 events, *all* found to contain opposite signed leptons.

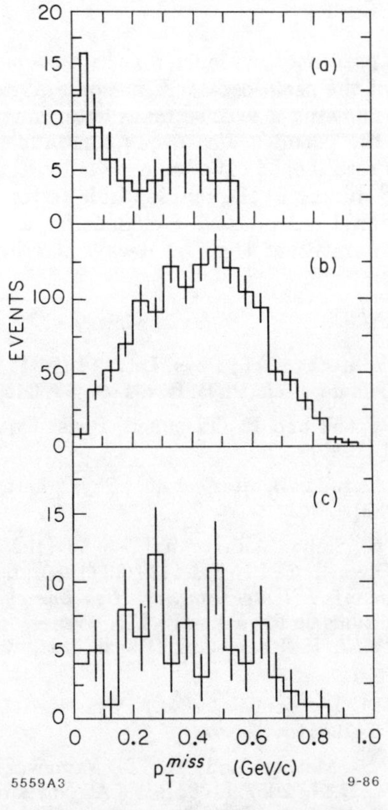

Figure 9: (a) p_T^{miss} for hadronic backgrounds. (b) for semileptonic decays. (c) for data.

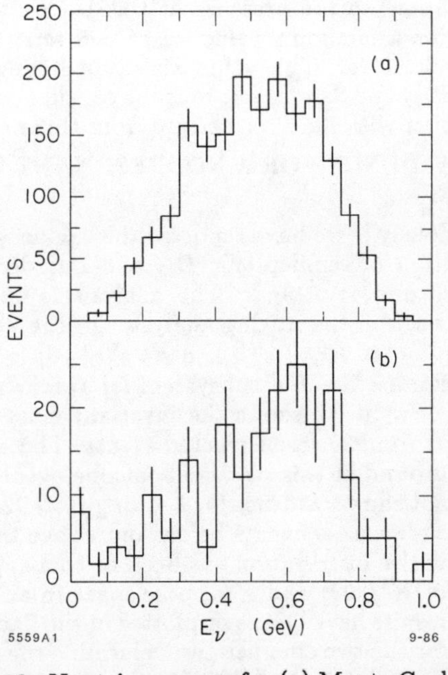

Figure 10: Neutrino energy for (a) Monte Carlo and (b) data.

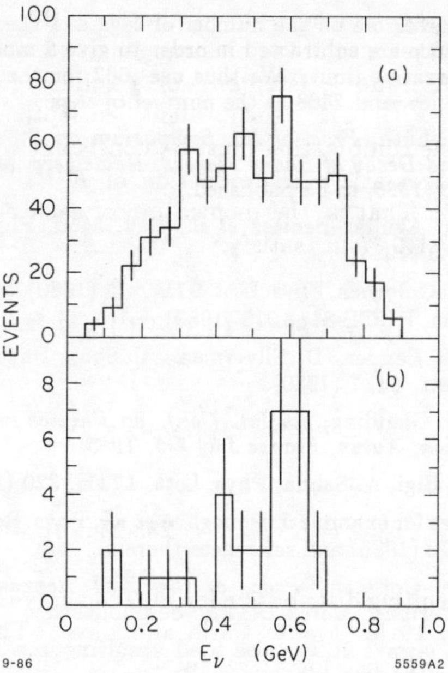

Figure 11: Neutrino energy for (a) Monte Carlo and (b) data, after tightened cuts.

To interpret our preliminary results, we considered the data under two extreme cases. First we assume that there is no contribution from DCSD. In the language of Bigi and Sanda[17] the parameter $\rho = 0$. In this case, we would "explain" the excess $s = \pm 2$ events as originating from $D^0\bar{D}^0$ mixing, and arrive at a mixing parameter r≈ 0.01 with an error of 60%. This would be unexpected large since even long range effects produce values for r typically less than 0.002.[17] The excess events may alternately be attributed to the presence of DCSD. We then determine a minimum value for $\rho^2 \approx 5\ tan^4(\theta_c)$ with a similar error. This value is also surprisingly large (see Bigi and Sanda) but must be considered in conjunction with its error derived from three events.

4. EVIDENCE FOR A NONRESONANT D_{e4} DECAYS

Previously,[19] we have reported the exclusive reconstruction of semileptonic D_{e3} and D_{e4} decays opposite hadronic tags. The analysis is similar to that used in the mixing analysis. In the decays to $K^-\pi^+\ e^+\nu$, $\bar{K}^0\pi^-\ e^+\nu$ and $\bar{K}^-\pi^0 e^+\nu$ it is possible to examine the $K\pi$ subsystem for resonant structure. Figure 12 shows the invariant mass for D_{e4} decays from 39 reconstructed events. The expected background in this plot is about one event. While the plot shows evidence for a strong $K^*(892)$ signal, there are excess events below and above the peak. The entire distribution has been fit to a superposition of $K^*(892)$ and $K\pi$ phase space in an s-wave. The results have been overplotted in the figure. Fitting for the two channels, we determine the fraction of s-wave[20] to be $45^{+14}_{-13}\%$ (stat. errors only). This

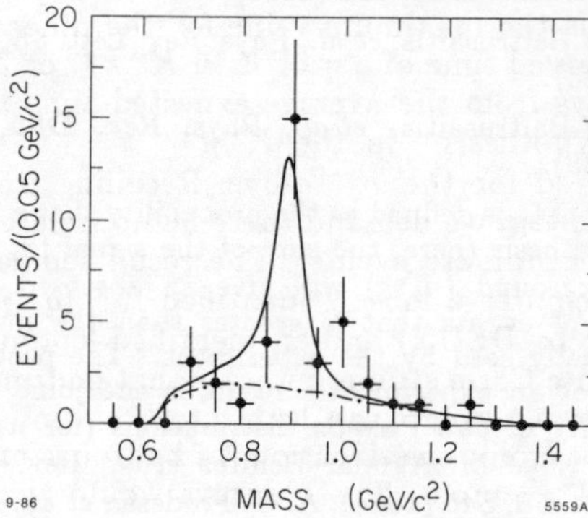

Figure 12: $K\pi$ invariant mass for D_{e4} decays. The curve is described in the text.

result agrees well with the average of the two fits of the DELCO measurement of the inclusive spectrum of leptons from charm decays.[21]

Interestingly, no models of charm semileptonic decay have treated the case of a large non-resonant component. The value obtained implies that approximately 1/4 of all semileptonic decays occur non-resonantly. This is similar to the nonresonant fraction of the hadronic sector.[22]

5. CONCLUSIONS

We have presented evidence for the associated production of the pseudoscalar F meson and its vector partner, allowing a precise mass determination. A limit on the pseudoscalar decay constant of the D has been established. We have extended the search for $D^0\bar{D}^0$ mixing to the semileptonic sector, finding no additional candidates. Evidence for a nonresonant component to the D_{e4} decays has been presented.

REFERENCES

1. H. Albrecht *et al.*, Phys. Lett. **146B**, 111 (1984), H. Aihara *et al.*, Phys. Rev. Lett. **53**, 2465 (1984).
2. M. Frank and P. O'Donnell, Phys. Lett. **159B**, 174 (1985)
3. M. Aguilar-Benitez *et al.*, Phys. Lett. **170B**, (1986).
4. S. N. Sinha, Alberta THY-3-86 (1986); P. J. O'Donnell, CERN-TH-4419/86 (1986); L. Maiani, *Proc. of XXI Int. Conf. on High Energy Physics*, (Editions de Physique, Le Ulis, France, 1982), pp. 631-657; H. Krasemann, Phys. Lett. **96B**, 397 (1980).
5. R. M. Baltrusaitis, *et al.*, Phys. Rev. Lett. **54**, 1976 (1985).
6. V. S. Mathur and M. T. Yamawaki, Phys. Rev. **D29**, 2057 (1984); V. A. Novikov, *et al.*, Phys. Rev. Lett. **38**, 626 (1977).

7. I. I. Bigi, G. Köpp, and P. M. Zerwas, Phys. Lett. **166B**, 238 (1986); J. F. Donoghue, *et al.*, Phys. Rev. **D33**, 197 (1986).

8. R.M.Baltrusaitis, *et al.*, Phys. Rev. Lett. **55**, 1842 (1985).

9. R.M.Baltrusaitis, *et al.*, Phys. Rev. **D55**, 566 (1985).

10. The C.L. is defined as the probability that a given hypothesis (here, the sum of the signal (n_s) and background (n_b)) will give an observed number of events that is greater than the number actually seen by the experiment. The probability of an experimental result is the joint probability of two Poisson distributions (for n_s and n_b). See M. Aguilar–Benitez *et al.*, Rev. Mod. Phys. **56**, S46 (1984), A. G. Frodesen *et al.*, *Probability and Statistics in Particle Physics (Universitettsforlaget, Bergen, 1979)*, pp. 167–168, 378–379. Thus, the limit on n_s at the 90% C.L. is derived from the limit on $(n_s + n_b)$ by subtraction of n_b from 2.3 events. The statistical error on n_b is incorporated by fluctuating n_b in the Poisson with a Gaussian distribution, while the systematic error is directly subtracted from the central value of n_b, to provide a more conservative limit. This gives 1.35 events as the 90% C.L. limit. Alternately, if an *a priori* uniform distribution of signal events is assumed, then the 90% C.L. upper limit on n_s would be 2.3 events.

11. The errors on the number of tags and the acceptance are subtracted in order to give a more conservative limit. We thus use .662 for the acceptance, and 2408 as the number of tags.

12. V. Lüth, *Proc. of Int. Symposium on Production and Decay of Heavy Flavors, Heidelberg, May 20-23, 1986*, to be published.

13. M. Aguilar–Benitez *et al.*, Rev. Mod. Phys. **56**, S43 (1984).

14. E. Golowich, Phys. Lett. **91B**, 271 (1980), M. Claudson, HUTP-81/A016 (1982).

15. M. Bander, D. Silverman, A. Soni, Phys. Rev. Lett. **44**, 7 (1980).

16. G. Gladding, *5th Int. Conf. on Physics in Collision, Autun, France July 2-5, 1985.*

17. I. Bigi, A. Sanda, Phys. Lett. **171B**, 320 (1986).

18. See for example J. Donoghue *et al.*, Phys. Rev. **D33**, 179 (1986) and references therein.

19. D. Coffman, *Proc. of the XXI Rencontre de Moriond, March 1986*, to be published.

20. A p-wave fit was also tried, resulting in a slightly poorer fit.

21. W. Bacino *et al.*, Phys. Rev. Lett. **43**, 1073 (1979). Here $37 \pm 16\%$ and $55 \pm 21\%$ were obtained assuming pure K^* and pure non-resonant $K\pi$, respectively.

22. See for example, R. H. Schindler, *Proc. of the SLAC Summer Institute on Particle Physics, (1985).*

New Results on Charmed Mesons and Tau Lepton from ARGUS*

N. Kwak
University of Kansas
Lawrence, Kansas 66045

We report first observations of $F^+ \to \overline{K}^{*o} K^+$ and $\tau \to \omega\pi\nu_\tau$. Also we give improved measurements on $Br(D^o \to K^o\phi), Br(\tau \to A_1\nu_\tau)$ and an upper limit on $D^o - \overline{D}^o$ mixing.

Introduction

The data sample was collected with the ARGUS detector, operating in the e^+e^- storage ring at DORIS II. It corresponds to a total luminosity of 153 pb^{-1}, of which 21.6 pb^{-1}, 36.6 pb^{-1}, 59.4 pb^{-1} and 35.4 pb^{-1} were obtained on the Υ(1S), Υ(2S), Υ(4S) and in the continuum or during scanning, respectively. Approximately the whole sample was used for the charmed meson studies but different subsamples were analyzed for the tau studies. The detector is a 4π spectrometer, described in more detail elsewhere(1). Our standard criteria for event selection and particle identification(2) was used for study of the charmed mesons. Event selection for taus is discussed in detail elsewhere(3,4). References in this report to a specific charged state are to be interpreted as also implying the charge conjugate state.

a) $F^+ \to \bar{K}^{*o}K^+$ (5)

Figure 1a and 1b show respectively the invariant mass distributions of $\bar{K}^{*o}K^+$ for the $\bar{K}^{*o}$ mass band and for sidebands outside the $\bar{K}^{*o}$ mass, defined by the mass regions 0.75 GeV/c^2 to 0.80 GeV/c^2 and 1.0 GeV/c^2 to 1.05 GeV/c^2. The distribution in figure 1a shows an enhancement near 1970 MeV/c^2, while the distribution in figure 1b does not. The fitted number of F mesons are $87.2^{+19.1}_{-21.0}$ and < 18 (90% confidence level) respectively, where the F mass was taken to be 1970 MeV/c^2 from $\phi\pi^+$ channel and the RMS width was fixed from the detector Monte Carlo to be 16.0 MeV/c^2. For this figure we required K π invariant mass to be within ±50 MeV/c^2 of the nominal $\bar{K}^{*o}$ mass and $p(\bar{K}^{*o}K^+) \geq 2.5$ GeV/c. Also two additional cuts have been made. First, it was required that $\cos\theta_{\bar{K}^{*o}} \geq 0.2$, where $\theta_{\bar{K}^{*o}}$ is the angle of the $\bar{K}^{*o}$ in the F rest frame with respect to the F boost direction. This cut eliminates reflections from $D^+ \to \bar{K}^{*o}\pi^+$ in the signal region by forcing the K^+ to be in the backward hemisphere of the F decay, and therefore slow and well identified by dE/dx and time-of-flight measurements. Second, we demanded that the helicity angle of the K^- with respect to the K^+ in the $\bar{K}^{*o}$ rest frame satisfy the condition $|\cos\theta_H[K^-]| \geq 0.5$. These cuts are described in detail elsewhere(5).

We have processed the $\phi\pi$ data with the same cuts as in our earlier work(6) in order to determine the ratio of the branching ratios

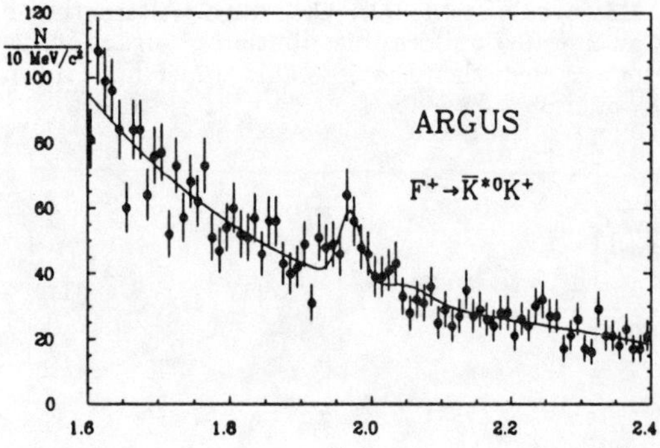

Fig. 1a M(K⁻K⁺π) [GeV/c²]

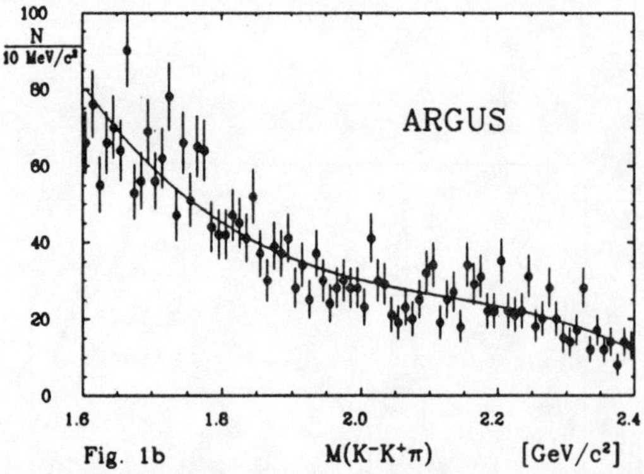

Fig. 1b M(K⁻K⁺π) [GeV/c²]

for the $\bar{K}^{*o}K^+$ and $\phi\pi$ channels. Allowing for the known branching ratios for $\phi \to K^+K^-$ of (49.3 ± 1.0)% and for $\bar{K}^{*o} \to K^-\pi^+$ of 2/3, we find $\frac{Br(F^+ \to \bar{K}^{*o}K^+)}{Br(F^+ \to \phi\pi^+)} = 1.44 \pm 0.37$ after correcting for the acceptances in each channel.

b) $D^o \to \bar{K}^o\phi$(7)

The decay $D^0 \to \bar{K}^0\phi$ was first observed by the ARGUS Collaboration in 1985 with the unexpectedly large branching ratio of $Br(D^0 \to \bar{K}^0\phi) = (1.43 \pm 0.45)\%$[2]. With the full data sample, we have analyzed this channel exactly as before. Figure 2 and 3 show the invariant $K_S^0K^+K^-$ and K^+K^- mass spectrum respectivly.

The distribution of the helicity angle θ, where θ is the angle between the K^+ and the K_S^0 in the rest frame of the ϕ meson shows the expected $\cos^2\theta$ behavior, and exhibits a small K^+K^- s-wave contribution below the ϕ, which could be understood as a K^+K^- threshold effect, possibly due to the decay $D^0 \to K_S^0\delta(980)$, which amounts to about $(21 \pm 5)\%$ in the ϕ region, giving by $1.01 < M(K^+K^-) < 1.03$ GeV/c^2.

Requiring $1.01 < M(K^+K^-) < 1.03$ GeV/c^2 and $x_p > 0.5$ ($x_p = p/p_{max}$), making a small acceptance correction and taking into account the known branching ratio for $\phi \to K^+K^-$, we obtain

$$\frac{Br(D^0 \to K_S^0\phi)}{Br(D^0 \to K_S^0\pi^+\pi^-)} = 0.155 \pm 0.003 \text{ which yields}$$

$Br(D^0 \to \bar{K}^0\phi) = (1.29 \pm 0.27 \pm 0.19)\%$, using $Br(D^0 \to \bar{K}^0\pi^+\pi^-) = (8.3 \pm 0.9 \pm 0.8)\%$[8].

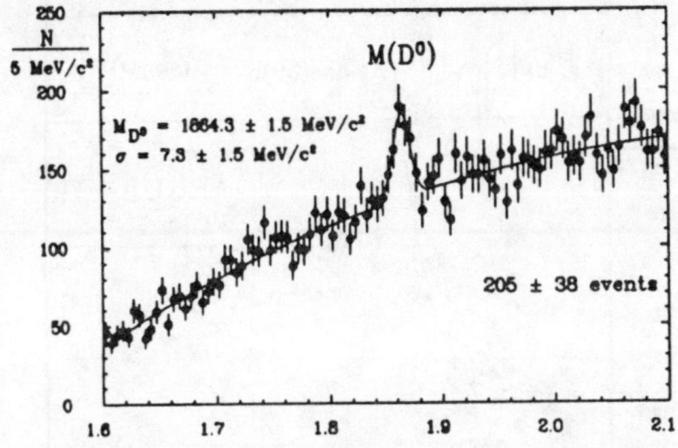

Fig. 2 $M(K_S^0K^+K^-)$ [GeV/c²]

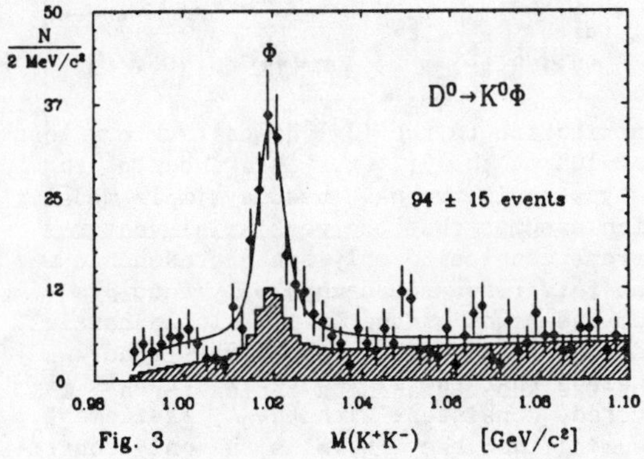

Fig. 3 $M(K^+K^-)$ [GeV/c²]

c) $D^0 - \bar{D}^0$ mixing[9]

We use the following two decay channels for the D^0 which contain a charged kaon: $D^{*+} \to (D^0)\pi^+ \to (K^-\pi^+)\pi^+$, or $\to (K^-\pi^+\pi^+\pi^-)\pi^+$. Fig. 4 shows the plot of the mass difference between the $K\pi$ and $K3\pi$ for both sign modes. We take the mass difference to be between 143 and 148 MeV/c^2 as the signal region and find 272 events with the right sign and 10 events with the wrong sign which would result from mixing. The backgrounds are estimated by fitting the histograms with a Gaussian plus third order polynomial. The expected backgrounds are 31 for the right sign and 12 for the wrong sign. We conclude that there is no evidence for the $D^0 - \bar{D}^0$

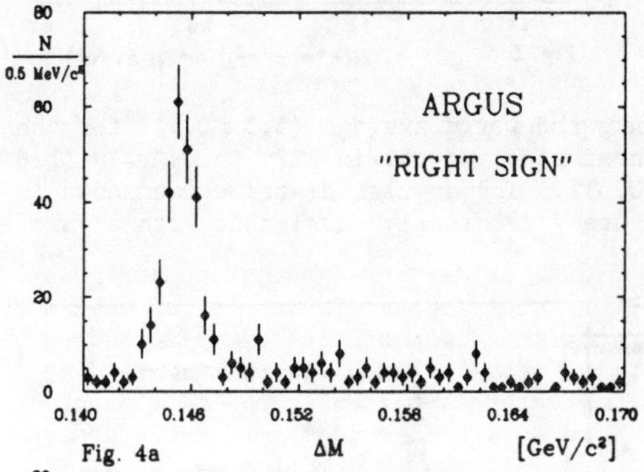

Fig. 4a ΔM [GeV/c²]

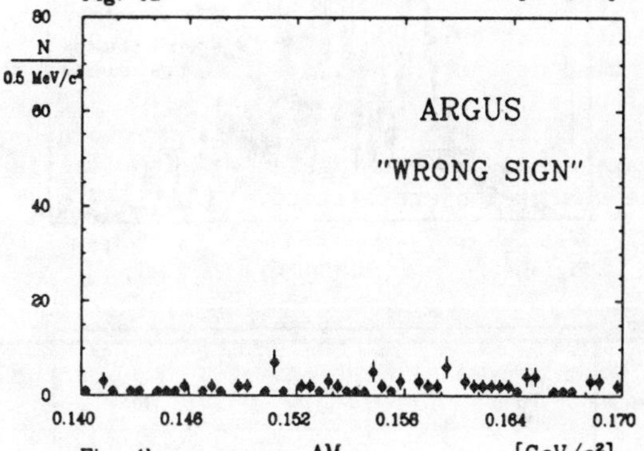

Fig. 4b ΔM [GeV/c²]

mixing and obtain an upper limit of 2.3% at 90% confidence level.

d) $\tau^- \to \pi^-\pi^-\pi^+\pi^0\nu_\tau$[3]

The event sample corresponds to an integrated luminosity of 115 pb^{-1}. Tau pairs were selected in the 1-3 topology. The distribution of the 4-pion invariant mass is shown in Figure 5, and represents a clean sample of 490 events below the τ mass. We find a branching ratio of $Br(\tau^- \to \pi^-\pi^-\pi^+\pi^0\nu_\tau) = (4.5 \pm 0.4 \pm 1.5)\%$. Figure 6, which shows the mass distribution obtained for these $\pi^+\pi^-\pi^0$ combinations, with two entries per event. A prominent ω signal is clearly evident. This constitutes the first observation of the decay $\tau^- \to \omega\pi^-\nu_\tau$. The fraction of four pion decays which proceed via formation of an $\omega\pi$ system is surprisingly

$$\text{large:} \quad \frac{Br(\tau \to \omega\pi\ \nu_\tau)}{Br(\tau^- \to \pi^-\pi^-\pi^+\pi^0\nu_\tau)} = 0.31 \pm 0.05$$

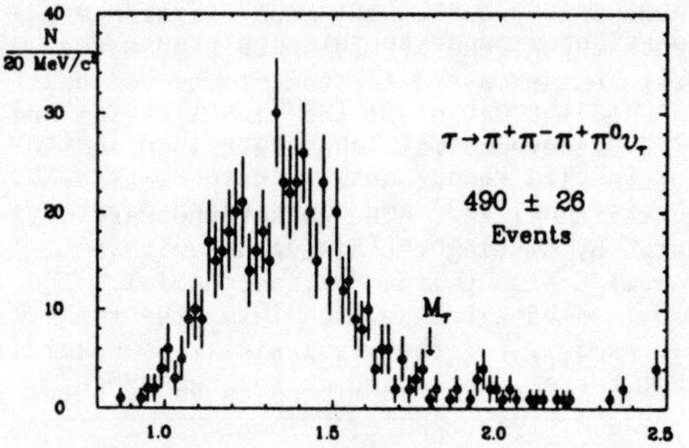

Fig. 5 $M(\pi^+\pi^-\pi^+\pi^0)$ [GeV/c²]

using the world average $(5.3 \pm 0.5)\%$ for the denominator, we obtain $Br(\tau^- \to \omega\pi^-\nu_\tau) = (1.6 \pm 0.3)\%$. The angular distribution shown in Figure 7 is clearly consistent with a pure

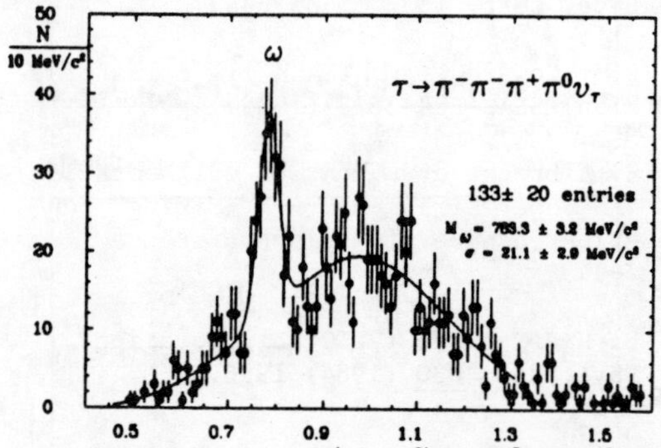

Fig. 6 $M(\pi^+\pi^-\pi^0)$ [GeV/c²]

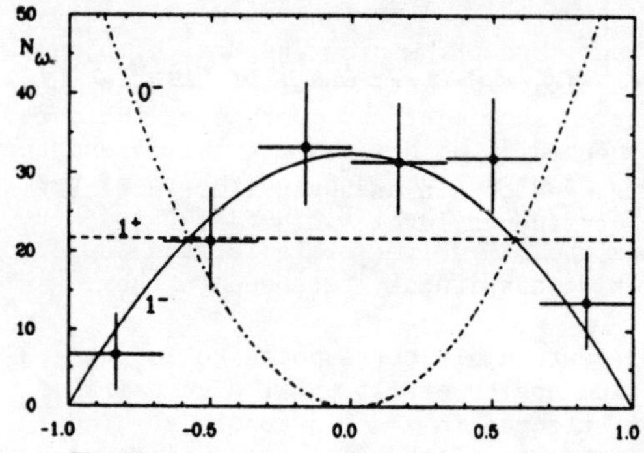

Fig. 7 $\cos\psi$

$J^P = 1^-$ assignment. Thus, there is no evidence from the angular distribution for a second-class axial-vector current in tau decays. However, at the 90% confidence level, contributions of second-class currents greater than 50% can only be excluded, due to limited statistics. There is no indication of a resonance structure in the resulting $\omega\pi^-$ mass distribution. However, the mass distribution is somewhat softer than that for all four pion decays.

e) $\tau^- \to \pi^-\pi^+\pi^-\nu_\tau$ (4)

A total of 1725 events were selected from a data sample for an integrated luminosity of about 80 pb^{-1}. Determining the branching ratio for $\tau^- \to \pi^-\pi^+\pi^-\nu_\tau$, we have excluded the ρ^+ -channel from the one-prong tau decays and have used only a subsample of 683 events collected on the Υ(2S) resonance at $\sqrt{s} = 10.023$ GeV. This data subset corresponds to an integrated luminosity of $(38.8 \pm 1.2) pb^{-1}$. Figure 8 shows the mass distribution of 3 pions. We find that $Br(\tau^- \to \pi^-\pi^+\pi^-\nu_\tau) = (5.6 \pm .07)\%$. The distributions of $\pi^+\pi^-$ and $\pi^\mp\pi^\mp$ masses are shown in Figure 9. Assuming that the shape of the non-resonant $\pi^+\pi^-$ distribution is the same as that for $\pi^\mp\pi^\mp$, we estimate that the non-$\rho^0\pi^-$ resonance

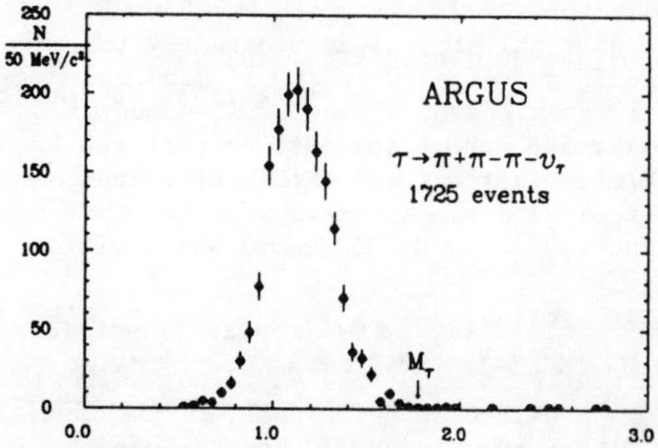

Fig. 8 Mass(3π) [GeV/c²]

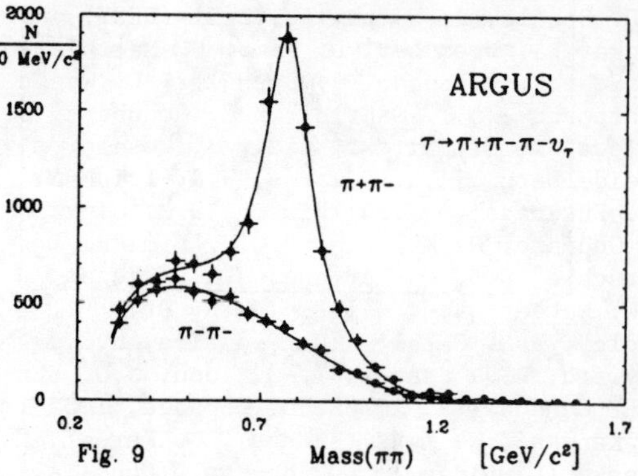

Fig. 9 Mass(ππ) [GeV/c²]

contribution to the $(3\pi)^\pm$ final state is less than 10% at the 95% CL. As a check of the A_1 hypothesis, we have used a simple model which assumes that the weak axial-vector current couples to only one 3π resonance and that this resonance decays only into $\rho^0\pi^-$. From the study of the Dalitz plot densities and its projections in the ρ mass band, we conclude that the 1^+ s-wave is strongly favored, consistent with the A_1 assignment. Assuming that the s-wave is the only contributing wave, we obtain a mass of (1046 ± 11) MeV/c^2 and a width $\Gamma = (521 \pm 27)$ MeV/c^2.

Summary

1) We have observed the decay $F^+ \to \bar{K}^{*0}K^+$.

The branching ratio to this channel is found to be 1.44 ± 0.37 times that for $F^+ \rightarrow \phi\pi^+$. Thus, the $\bar{K}^{*o}K^+$ decay mode of the F meson, which proceeds only through annihilation and color-suppressed spectator diagrams, occurs at a rate comparable with that for the color-favored spectator decay $F \rightarrow \phi\pi$. 2) We have confirmed the existence of the decay $D^o \rightarrow \bar{K}^o\phi$ and find the $Br(D^o \rightarrow \bar{K}^o\phi) = (1.18 \pm 0.25 \pm 0.17)\%$. 3) There is no evidence for $D^o - \bar{D}^o$ mixing and we found an upper limit of 2.3% at 90% confidence level. 4) We have observed the decay $\tau^- \rightarrow \pi^-\pi^-\pi^+\pi^o\nu_\tau$ and obtained first evidence for the decay $\tau^- \rightarrow \omega\pi^-\nu_\tau$ with a branching ratio of $(1.6 \pm 0.3)\%$. No evidence for a second-class axial-vector current was found. 5) We have determined the branching ratio for $\tau^- \rightarrow \pi^-\pi^+\pi^-\nu_\tau$ to be $(5.6 \pm 0.7)\%$. The 3π final state is dominated by the $\rho^o\pi^-$ mode in an s-wave with $J^P = 1^+$, and can be well described by a single resonance of mass (1046 ± 11) MeV/c^2 and with $\Gamma = (521 \pm 27)$ MeV/c^2. The mass and spin-parity are consistent with the A_1 assignment for this resonance.

References

*Current members of the ARGUS Collaboration are: H. Albrecht, U. Binder, P. Böckmann, R. Gläser, G. Harder, I. Lembke-Koppitz, A. Phillipp, W. Schmidt-Parzefall, H. Schröder, H.D. Schulz, R. Wurth, A. Yagil (DESY), J.P. Donker, A. Drescher, D. Kamp, U. Matthiesen, H. Scheck, B. Spaan, J. Spengler, D. Wegener (Dortmund), J.C. Gabriel, K.R. Schubert, J. Stiewe, K. Strahl, R. Waldi, S. Weseler (Heidelberg), K.W. Edwards, W.R. Frisken, Ch. Fukunaga, D.J. Gilkinson, D.M. Gingrich, M. Goddard, H. Kapitza, P.C.H. Kim, R. Kutschke, D.B. MacFarlane, J.A. McKenna, K.W. McLean, A.W. Nilsson, R.S. Orr, P. Padley, P.M. Patel, J.D. Prentice, H.C.J. Seywerd, B.J. Stacey, T.-S. Yoon, J.C. Yun (IPP Canada), R. Ammar, D. Coppage, R. Davis, S. Kanekal, N. Kwak (Kansas), G. Kernel, M. Plesko (Ljubljana). L. Jönsson, (Lund) A. Babaev, M. Danilov, A. Golutvin, I. Gorelov, V. Lubimov, V. Matveev, V. Nagovitsin, V. Ryltsov, A. Semenov, V. Shevchenko, V. Soloshenko, V. Tchistilin, I. Tichomirov, Yu. Zaitsev (ITEP-Moscow), R. Childers, C.W. Darden, Y. Oku (S. Carolina), and H. Gennow (Stockholm).

1) H. Albrecht et al. (ARGUS collaboration), Phys. Lett. 134B (1984) 137.

2) H. Albrecht et al. (ARGUS collaboration), Phys. Lett. 150B (1985) 235 and Phys. Lett. 158B (1985) 525.

3) H. Albrecht et al. (ARGUS collaboration), and "Evidence for the Decay $\tau^- \rightarrow \omega\pi^-\nu_\tau$" submitted to Phys. Lett. August, 1986 and contributed paper to this conference.

4) H. Albrecht et al. (ARGUS collaboration), "Measurement of Tau Decays into Three Charged Pions" submitted to Z. Phys. C. in June, 1986 and contributed paper to this conference (#9628).

5) H. Albrecht et al. (ARGUS collaboration), "Observation of F decays into $\bar{K}^{*o}K^+$" (DESY 86-061) submitted to Phys. Lett. June, 1986, and contributed paper to this conference (#9598).

6) H. Albrecht et al. (ARGUS collaboration), "The Decay $D^o \rightarrow \bar{K}^o\phi$" contributed paper to this conference (#9580).

7) H. Albrecht et al. (ARGUS collaboration), Phys. Lett. 153B (1985) 343.

8) D.H. Coward et al. (MARK III collaboration) SLAC-PUB-3818 (1985).

9) H. Albrecht et al. (ARGUS collaboration), "An Upper Limit on $D^o - \bar{D}^o$ Mixing" contributed paper to this conference (#9458).

10) P. Haas et al. (CLEO collaboration), Phys. Rev. D30 (1984) 1996.

Acknowledgements

It is a pleasure to thank E. Michel, W. Reinsch, Mrs. E. Konrad and Mrs. U. Djuanda for their competent technical help in running the experiment and processing the data. We thank Drs. H. Nesemann, K. Wille and the DORIS group for the good operation of the storage ring. The visiting groups wish to thank the DESY directorate for the support and king hospitality extended to them.

REVIEW OF RECENT RESULTS ON TAU DECAYS

PATRICIA R. BURCHAT

Santa Cruz Institute for Particle Physics
University of California, Santa Cruz, CA 95064

Recent meaurements of τ branching fractions and studies of final states in τ decay are reviewed. The branching fraction to one charged hadron plus at least two neutral hadrons has been measured to be higher than that expected from $\tau^- \to \nu_\tau \pi^- 2\pi^0$ and $\tau^- \to \nu_\tau \pi^- 3\pi^0$ alone. The first evidence for $\tau^- \to \nu_\tau \pi^- \eta X^0$ and $\tau^- \to \nu_\tau \pi^- \omega$ has been reported. New limits on rare or unexpected decay modes and new branching fraction, lifetime and Michel parameter measurements are summarized.

To motivate the studies of τ decay modes reported to this conference, I start by reviewing the status of measured τ branching fractions *without* these new results. The world average measured branching fractions are listed in Table 1 for each decay mode according to the number of charged particles in the final state.[1] The result listed for $B(\tau^- \to \nu_\tau \pi^- 2\pi^0)$ is the theoretical expectation based on isospin rules which predict $B(\tau^- \to \nu_\tau \pi^- 2\pi^0) = B(\tau^- \to \nu_\tau \pi^- \pi^+ \pi^-)$ if the 3π final state is pure $A(1270)$ resonance. The result listed for $B(\tau^- \to \nu_\tau \pi^- 3\pi^0)$ is the theoretical prediction of Gilman and Rhie[2] based on CVC and e^+e^- cross section measurements to 4π final states.

TABLE 1. World average measured branching fractions (in per cent) of the τ lepton, *not* including results reported to this conference, classified according to the charged multiplicity of the final state. The values listed for $B(\tau^- \to \nu_\tau \pi^- 2\pi^0)$ and $B(\tau^- \to \nu_\tau \pi^- 3\pi^0)$ are not measurements but theoretical predictions based on isospin conservation and CVC respectively.

Decay Mode $\tau^- \to$	Branching Fraction (%) One Prong	Three Prongs
$e^- \bar{\nu}_e \nu_\tau$	17.9 ± 0.4	-
$\mu^- \bar{\nu}_\mu \nu_\tau$	17.2 ± 0.4	-
$\pi^- \nu_\tau$	10.9 ± 1.4	-
$\pi^- \pi^0 \nu_\tau$	22.1 ± 1.2	-
$(3\pi)^- \nu_\tau$	7.1 ± 0.5	7.1 ± 0.5
$(4\pi)^- \nu_\tau$	1.0	5.3 ± 0.5
$K^- \nu_\tau$	0.7 ± 0.2	-
$(K\pi)^- \nu_\tau$	1.1 ± 0.3	0.3 ± 0.1
$(K\bar{K}\pi)^- \nu_\tau$		$0.22^{+0.17}_{-0.11}$
$(K\pi\pi)^-(\pi^0)\nu_\tau$		$0.22^{+0.16}_{-0.13}$
Total	78.0 ± 2.0	13.1 ± 0.7

The sum of the listed branching fractions is significantly less than 100%. The topological branching fractions have been measured to be $(86.8 \pm 0.3)\%$ and $(13.1 \pm 0.3)\%$ for one and three charged particles, respectively. Therefore, the sum of listed exclusive branching fractions which result in one charged particle in the final state is $(8.8 \pm 2.0)\%$ less than the topological branching fraction to one charged particle. Either the measured branching fractions are incorrect or there exist significant unidentified decay modes. In general, the exclusive branching fraction measurements depend on accurate knowledge of the total integrated luminosity, the cross section for τ pair production, and the efficiency for extracting τ pair events from the data. In addition, decay modes involving more than one neutral hadron in the final state are difficult to study because of the large number of photons produced. These difficulties can be avoided by selecting a sample of τ decay candidates with minimum bias in the selection criteria and then measuring all the branching fractions simultaneously with the sum constrained to be unity.

Such a method has been used by the Mark II[3] and TPC[4] collaborations. In the Mark II analysis, a tagging technique is used; if one of the τ's in the produced pair satisfies the criteria for the tag, then the other τ is included in the candidate sample. Two different tags are used: a one-prong hadronic tag which results in 1627 candidate τ's with about 3% background and a three-prong tag which results in 1475 candidate τ's with about 1.4% background. The tagged sample is divided into categories based on neutral and charged particle multiplicities and charged particle identification. The electromagnetic calorimeter and muon system are used for particle identification.

In the TPC analysis, decay combinations which include all expected decay modes of the τ are selected resulting in 660 (1 vs. 3) events with about 8% background and 534 (1 vs. 1) events with about 3.6% background. Decays are classified according to charged particle multiplicity and identification but not according to neutral particle multiplicity. The TPC dE/dx system

TABLE 2. Branching fractions measured by the Mark II and TPC experiments using a constrained fit, and the TPC measurement of $B(\tau^- \to \nu_\tau \pi^- (\geq 2h^0))$ from a separate analysis. The final column shows the world average or theoretical prediction for each branching fraction from Table 1.

Decay Mode $\tau^- \to$	Branching Fraction MARK II	TPC	World Average or Prediction
one-prong	$(86.9 \pm 1.0 \pm 0.7)\%$	$(84.5 \pm 0.8 \pm 1.0)\%$	$(86.8 \pm 0.3)\%$
three-prong	$(13.1 \pm 1.0 \pm 0.7)\%$	$(15.3 \pm 0.8 \pm 1.0)\%$	$(13.1 \pm 0.3)\%$
$e^- \bar{\nu}_e \nu_\tau$	$(19.1 \pm 0.8 \pm 1.1)\%$	$(18.0 \pm 1.2^{+1.9}_{-1.6})\%$	$(17.9 \pm 0.4)\%$
$\mu^- \bar{\nu}_\mu \nu_\tau$	$(18.3 \pm 0.9 \pm 0.8)\%$	$(18.1 \pm 1.2^{+0.5}_{-0.4})\%$	$(17.2 \pm 0.4)\%$
$\pi^-(\geq 0h^0)\nu_\tau$	$(47.8 \pm 1.6 \pm 1.3)\%$	$(46.8 \pm 1.2^{+0.5}_{-1.0})\%$	$(41.1 \pm 1.9)\%$
$\pi^- \nu_\tau$	$(10.0 \pm 1.1 \pm 1.4)\%$		$(10.9 \pm 1.4)\%$
$\rho^- \nu_\tau$	$(25.8 \pm 1.7 \pm 2.5)\%$		$(22.1 \pm 1.2)\%$
$\pi^-(\geq 2h^0)\nu_\tau$	$(12.0 \pm 1.4 \pm 2.5)\%$	$(13.9 \pm 2.0^{+1.9}_{-2.1})\%$	$(\ 8.1 \pm 0.5)\%$

and the muon system are used for particle identification.

The Mark II analysis uses an unfold technique and a maximum likelihood fit while the TPC analysis uses a minimum χ^2 fit to measure the branching fractions simultaneously with the sum constrained to unity. The results for the two experiments are shown in Table 2 with a comparison to the previous world average from Table 1. The measured topological branching fractions are in agreement with the world average. The one-prong decays are then divided up into the two leptonic decays and the remaining semi-leptonic decays. The Mark II experiment measures both leptonic branching fractions to be about one standard deviation higher than the world average. Both Mark II and TPC find an excess in the one-prong hadronic modes over previous measurements and theoretical expectations. Mark II divides the one-prong hadronic sample further according to neutral multiplicity and finds the excess to be in the modes involving at least one neutral hadron. In a separate analysis,[5] the TPC collaboration measures $B(\tau^- \to \nu_\tau \pi^- (\geq 2h^0))$ to be $(13.9 \pm 2.0^{+1.9}_{-2.1})\%$ also indicating an excess over the theoretical expectation from $\tau^- \to \nu_\tau \pi^- 2\pi^0$ and $\tau^- \to \nu_\tau \pi^- 3\pi^0$.

These results point to the possibility of final states involving the $\eta(550)$. The possible decay modes with an η in the final state are summarized in Table 3 along with the quantum numbers and possible identity of the resonance. The decay $\tau^- \to \nu_\tau \pi^- \eta$ is not expected to occur since it involves a second class current ($J^{PG} = 0^{+-}$ or 1^{--}).

Since the η decays to two charged particles approximately 30% of the time, the decay $\tau^- \to \nu_\tau \pi^- \eta \pi^+ \pi^-$ results in five charged particles in the final state 30% of the time and the decay $\tau^- \to \nu_\tau \pi^- \eta\eta$ results in five charged particles 9% of the time. The measured branching fraction of $(0.14 \pm 0.04)\%$ for $\tau^- \to \nu_\tau 5\pi^\pm(\pi^0)$ by Mark II[6] can be used to set the following limits: $B(\tau^- \to \nu_\tau \pi^- \eta \pi^+ \pi^-) \leq 0.5\%$, and $B(\tau^- \to \nu_\tau \pi^- \eta\eta) \leq 3.0\%$ where differences in tracking efficiencies for different modes have been taken into

TABLE 3. Possible decay modes with an η in the final state, the corresponding expected resonances and limits on the branching fraction of the tau through this decay mode to final states containing one charged particle.

$\tau^- \to \nu_\tau +$	J^{PG}	Possible Resonance	Expected Contribution to One-Prong Branching Fraction
$\pi^- \eta \pi^0$	1^{-+}	$\rho(1250)$	no limit
		$\rho(1600)$	0.5%
$\pi^- \eta \pi^+ \pi^-$	1^{+-}	$A(1270)$	0
$\pi^- \eta \pi^0 \pi^0$	1^{+-}	$A(1270)$	$\leq 0.4\%$
$\pi^- \eta\eta$	0^{--}, 1^{+-}	$\pi(1300)$, $A(1270)$	$\leq 1.5\%$

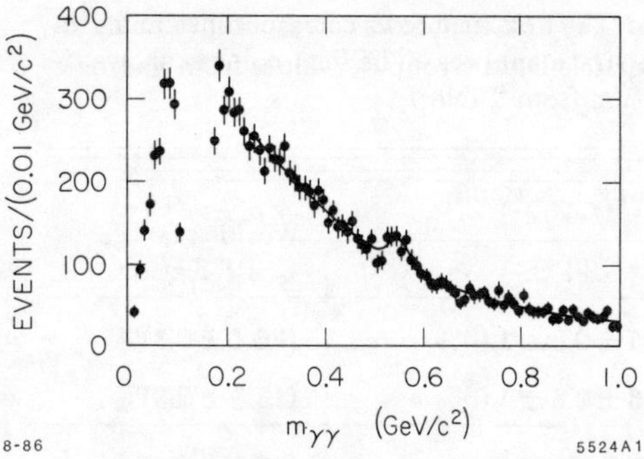

Figure 1. Mass spectrum of $\gamma\gamma$ combinations in τ decay candidates with one charged particle and at least three photons in the Crystal Ball. Photons have been rejected if they form an invariant mass consistent with a π^0 when combined with any other photon in the event.

account. Also, isospin conservation predicts $B(\tau^- \to \nu_\tau \pi^- \eta \pi^0 \pi^0) \leq B(\tau^- \to \nu_\tau \pi^- \eta \pi^+ \pi^-)$.

A limit can be set on the branching fraction for the mode $\tau^- \to \nu_\tau \pi^- \eta \pi^0$ assuming that it occurs via the $\rho(1600)$ resonance. The branching fraction of the $\rho(1600)$ to $\pi^- \eta \pi^0$ is approximately 7% and to 4π is approximately 60%.[7] Assuming a total branching fraction of the τ to 4π of 6%, as predicted by Gilman and Rhie,[2] and assuming that all of the 4π decay mode is from the $\rho(1600)$, the branching fraction of the τ to $\nu_\tau \pi^- \eta \pi^0$ through the $\rho(1600)$ resonance is approximately $\frac{0.07}{0.60} \times 6\%$ which is equal to 0.7%. Since the branching fractions of $\rho(1250)$, the first radial excitation of $\rho(770)$, are not known, a limit cannot be set on the $\nu_\tau \pi^- \eta \pi^0$ final state through the $\rho(1250)$ resonance.

The expected contribution of each decay mode involving the η to the branching fraction to final states containing only one charged pion and multiple neutral particles is shown in Table 3. The total contribution from sources other than non-resonant decay and $\pi^- \eta \pi^0$ from $\rho(1250)$ is expected to be less than 2.4%.

The Crystal Ball[8] and HRS[9] collaborations report on searches for the the decay $\tau^- \to \nu_\tau \pi^- \eta X^0$. In the Crystal Ball analysis, τ pairs are tagged with the decay $\tau^- \to \nu_\tau e^- \bar{\nu}_e$ with some feedthrough from the decays $\tau^- \to \nu_\tau \pi^-$ and $\tau^- \to \nu_\tau \rho^-$. They then look at the mass spectrum of photon pairs in tau decay candidates with one charged particle and at least three photons. The HRS experiment looks for events with a (1 vs. 1) or (1 vs. 3) topology and then accepts as candidates single charged tracks with at least two associated photons. Both experiments reject photons which, when combined with any other photon, result in an invariant mass consistent with a π^0. The $\gamma\gamma$ mass spectrum is shown in Figure 1 for the Crystal Ball and Figure 2 for HRS. The Crystal Ball result shows a clear η signal. However, they have not completed their background and efficiency estimates. The HRS collaboration reports a branching fraction of $(5.0 \pm 1.0 \pm 1.5)\%$

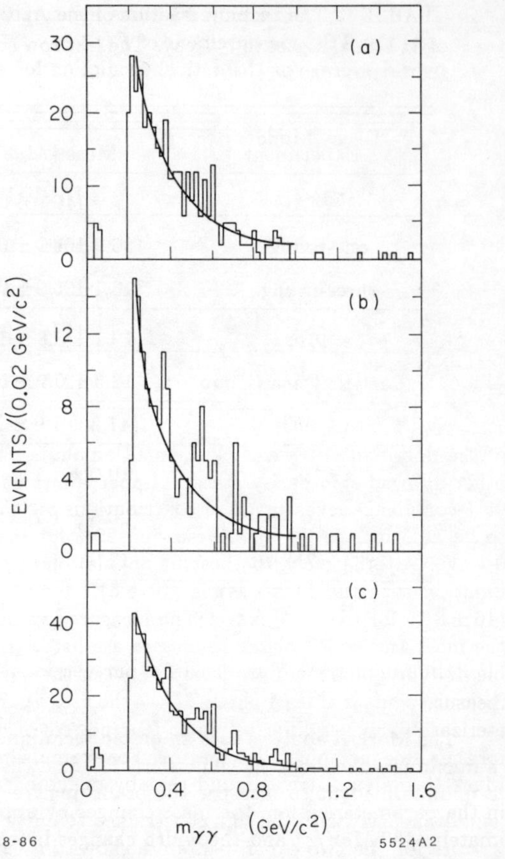

Figure 2. Mass spectrum of $\gamma\gamma$ combinations in τ decay candidates with one charged particle and at least two photons in the HRS detector. Photons have been rejected if they form an invariant mass consistent with a π^0 when combined with any other photon in the event. The top figure corresponds to (1 vs. 3) events, the middle figure to (1 vs. 1) events, and the bottom figure to a combination of (1 vs. 3) and (1 vs. 1) events.

for $\tau^- \to \nu_\tau \pi^- \eta X^0$. The significance of these results is not yet sufficient to determine whether the observed signal is compatible with expected decays or indicates decays through the first radial excitation of the ρ or non-resonant decay.

The ARGUS,[10] MAC,[11] DELCO,[19] and Mark II[13] collaborations have studied $3\pi^\pm$ final states in τ decay to determine the intermediate state; the spin-parity, mass and width of the resonance; and the branching fraction for $\tau^- \to \nu_\tau \pi^- \pi^+ \pi^-$. The ARGUS study contains more than twice as many events as the other analyses combined. They determine that the probability of the $3\pi^\pm$ system containing a component which is not $\rho\pi$ is less than 10% at the 90% confidence level. A Dalitz plot analysis is used to determine the spin-parity of the three pion state. All experiments find that the data are well represented by a $J^P = 1^+$ distribution with the $\rho\pi$ in a relative s-wave. In the Mark II analysis, the data are compared to incoherent mixtures

TABLE 4. The measured values of the $A(1270)$ mass and width. The first four rows correspond to analyses of $3\pi^{\pm}$ final states in τ decay. The last row corresponds to the Particle Data Group[7] values from hadronic scattering.

Experiment	Mass (MeV/c^2)	Width (MeV/c^2)	$B(\tau^- \to \nu_\tau \pi^- \pi^+ \pi^-)$
MAC	1169 ± 19	411 ± 76	$(7.8 \pm 0.8)\%$
ARGUS	1046 ± 11	521 ± 27	$(5.6 \pm 0.7)\%$
DELCO	1056 ± 30	476 ± 140	$(5.0 \pm 1.0)\%$
Mark II	1194 ± 20	462 ± 70	$(7.8 \pm 0.9)\%$
Particle Data Group	1275 ± 28	316 ± 45	

of the dominant 1^+ s-wave and small amounts of the other allowed spin-parity states. Upper limits at the 95% confidence level for these contributions are found to be 18% for the 0^- hypothesis and 29% for the 1^+ d-wave. Alternatively, the best fit population in combination with the 1^+ s-wave is $(10 \pm 5)\%$ for 0^- and $(16 \pm 8)\%$ for the 1^+ d-wave. The measured values of the mass and width of the resonance are listed in Table 4. The apparent discrepancy between the various measurements is at least partly due to the use of parameterizations which differ by up to four powers of mass for the three pion mass distribution. For example in the Mark II analysis, it was found that by including $m^{\pm 1}$ in the parameterization, the mass changes by approximately $\mp 30\, MeV/c^2$ and the width changes by about $\mp 60\, MeV/c^2$. The parameterization is not unique because of arbitrary form factors at both the $W - A$ and $A - \rho - \pi$ vertices. The measured branching fractions for $\tau^- \to \nu_\tau \pi^- \pi^+ \pi^-$ are also shown in Table 4.

The ARGUS[14] and HRS[9] collaborations have studied the decay mode $\tau^- \to \nu_\tau \pi^- \pi^+ \pi^- \pi^0$ to investigate the intermediate states. The intermediate state $\pi^- \eta$ has two possible J^{PG} quantum numbers, 0^{+-} and 1^{--}, both of which correspond to second class currents. The intermediate state $\pi^- \omega$ has three possible J^{PG} quantum numbers, 1^{++} and 0^{-+} which correspond to second class currents, and 1^{-+} which corresponds to a first class current. The ARGUS collaboration sees a clear $\omega(783)$ signal in the mass distribution for the neutral 3π system (two per event). A Dalitz plot analysis of the 3π system confirms that it corresponds to a $J^P = 1^-$ resonance. The distribution of the angle between the normal to the ω decay plane and the bachelor pion in the ω rest frame is consistent with the 4π system being pure $J^P = 1^-$ showing no evidence for second class currents. The $\pi^- \omega$ mass spectrum shows no narrow peak. With limited statistics, it is consistent with a broad resonance or phase space. The $\rho(1600)$ is not observed to decay to $\pi\omega$.[7] Therefore, this $\pi^- \omega$ signal may be an indication of the first radial excitation of $\rho(770)$ or non-resonant decay of the τ. The branching fraction for $\tau^- \to \nu_\tau \pi^- \omega$ is measured to be $(1.6 \pm 0.3)\%$. HRS confirms the ω signal in the neutral 3π system.

Using their complete data sample of $300\, pb^{-1}$, the HRS[15] , collaboration has updated their measurements of 5-prong branching fractions. Their results are $B(\tau^- \to \nu_\tau 5\pi^{\pm}) = (5.1 \pm 2.0) \times 10^{-4}$ and $B(\tau^- \to \nu_\tau 5\pi^{\pm}\pi^0) = (5.1 \pm 2.2) \times 10^{-4}$ based on 7 and 6 events, respectively. They set an upper limit at the 90% confidence level of 3.8×10^{-4} on $B(\tau^- \to \nu_\tau 7\pi^{\pm}(n\gamma))$.

The Crystal Ball collaboration has set new limits on two lepton-number-violating decay modes: $B(\tau^- \to e^- \gamma) < 3.2 \times 10^{-4}$ and $B(\tau^- \to e^- \pi^0) < 4.4 \times 10^{-4}$ at the 90% confidence level.

The JADE,[16] MAC,[11] Mark III,[17] and TPC[4] collaborations report the new branching fraction measurements listed in Table 5. The MAC measurement of $B(\tau^- \to \nu_\tau \pi^-)$ is the most precise branching fraction measurement for this mode to date. MAC[11] also reports a new τ lifetime measurement based on the impact parameter method. They measure $\tau_\tau = (0.286 \pm 0.017 \pm 0.013)\, ps$. Using 3-prong τ decays, the HRS collaboration[18] measures $\tau_\tau = (0.28 \pm 0.02 \pm 0.02)\, ps$. With these two new measurements included, the new world average measured τ lifetime is $(0.285 \pm 0.014)\, ps$ where each measurement has been weighted by the statistical and systematic errors added in quadrature. This lifetime corresponds to a theoretical prediction of $(17.9 \pm 0.9)\%$ for $B(\tau^- \to \nu_\tau e^- \bar{\nu}_e)$.

TABLE 5. New branching fraction measurements (in %) for exclusive decay modes of the τ.

Experiment	Decay Mode $\tau - \to$	Measured Branching Fraction (%)
JADE	$\nu_\tau e^- \bar{\nu}_e$	$17.0 \pm 0.7 \pm 0.9$
	$\nu_\tau \mu^- \bar{\nu}_\mu$	$18.8 \pm 0.8 \pm 0.7$
	$\nu_\tau \pi^-$	$11.8 \pm 0.6 \pm 1.1$
MAC	$\nu_\tau \pi^-$	$10.6 \pm 0.4 \pm 0.8$
Mark III	$\nu_\tau \rho^-$	$22.3 \pm 1.4 \pm 1.6$
TPC	$\nu_\tau K^- + neutrals$	$1.6 \pm 0.4 \pm 0.2$
	$\nu_\tau K^{*-} + neutrals$	$1.4 \pm 0.9 \pm 0.3$
	$\nu_\tau K^- \gamma + neutrals$	$1.2 \pm 0.5^{+0.2}_{-0.4}$

The MAC[11] collaboration has measured the Michel ρ parameter from the lepton momentum spectrum for both $\tau^- \to \nu_\tau e^- \bar{\nu}_e$ and $\tau^- \to \nu_\tau \mu^- \bar{\nu}_\mu$. There are only two other measurements of the Michel ρ parameter for τ decay: one by DELCO at SPEAR[19] and the other by CLEO.[20] A summary of the measurements is shown in Table 6. The world average of 0.73 ± 0.07 is consistent with pure $V - A$ ($\rho = 0.75$) and rules out pure $V + A$ ($\rho = 0$) interaction at the $\tau - \nu_\tau - W$ vertex.

TABLE 6. Measurements of the Michel ρ parameter for each leptonic decay mode of the τ and the average.

Experiment	ρ_e	ρ_μ	$\rho_{average}$
DELCO	0.72 ± 0.15	-	$0.72 \pm 0.10 \pm 0.11$
CLEO	0.60 ± 0.13	0.81 ± 0.13	$0.71 \pm 0.09 \pm 0.03$
MAC	0.62 ± 0.22	0.89 ± 0.16	$0.79 \pm 0.11 \pm 0.09$
Average	0.65 ± 0.09	0.84 ± 0.11	0.73 ± 0.07

To summarize, I have updated Table 1 with the results reported to this conference. The results are shown in Table 7. The main difference is the measurement of $B(\tau^- \to \nu_\tau \pi^- (\geq 2h^0))$. Also, the experimental error on $B(\tau^- \to \nu_\tau \pi^-)$ has been substantially reduced. The difference between the sum of exclusive branching fractions to one-prong final states and the topological branching fraction to one charged particle has been reduced to $(3.6 \pm 2.4)\%$.

TABLE 7. World average measured branching fractions (in per cent) of the τ lepton, including results reported to this conference, classified according to the charged multiplicity of the final state.

Decay Mode	Branching Fraction (%)	
$\tau^- \to$	One Prong	Three Prongs
$e^- \bar{\nu}_e \nu_\tau$	17.9 ± 0.4	-
$\mu^- \bar{\nu}_\mu \nu_\tau$	17.5 ± 0.3	-
$\pi^- \nu_\tau$	10.9 ± 0.6	-
$\pi^- \pi^0 \nu_\tau$	22.1 ± 1.1	-
$\pi^- (\geq 2h^0) \nu_\tau$	13.0 ± 2.0	-
$3\pi^\pm \nu_\tau$	-	6.4 ± 0.4
$4\pi^\pm \nu_\tau$	-	5.2 ± 0.5
$K^- \nu_\tau$	0.7 ± 0.2	-
$(K\pi)^- \nu_\tau$	1.1 ± 0.3	0.3 ± 0.1
$(K\bar{K}\pi)^- \nu_\tau$		$0.22^{+0.17}_{-0.11}$
$(K\pi\pi)^- (\pi^0) \nu_\tau$		$0.22^{+0.16}_{-0.13}$
Total	83.2 ± 2.4	12.3 ± 0.7

In conclusion, recent measurements by Mark II and TPC indicate that the previously unidentified decay modes of the tau result in one charged pion and at least two neutral hadrons in the final state. Searches for the decay $\tau^- \to \nu_\tau \pi^- \eta X^0$ by Crystal Ball and HRS indicate that this mode contributes to τ decay. ARGUS and HRS report evidence for $\tau^- \to \nu_\tau \pi^- \omega$ with the spin-parity of the $\pi^- \omega$ system being 1^-. Both the $\pi^- \eta X^0$ and $\pi^- \omega$ system could be evidence for the first radial excitation of the ρ or non-resonant decay of the τ. Searches for unexpected decay modes of the τ are becoming sensitive to branching fractions as low as 10^{-4} as single experiments accumulate samples of greater than 10^5 τ's.

REFERENCES

1. For recent reviews of published measurements of τ branching fractions see P. R. Burchat, Stanford University Ph. D. thesis and SLAC Report No. 292 (1986); K. K. Gan, 1985 Annual Meeting of the Division of Particles and Fields of the APS, Eugene, Oregon, August 12-15, 1985 and Purdue University report PU-85-539 (1985).
2. F.J. Gilman and Sun Hong Rhie, *Phys. Rev. D* **31**, 1066 (1985).
3. P.R. Burchat *et al.*, SLAC-PUB-4006, submitted to *Phys. Rev. D* (1986).
4. H. Aihara *et al.*, contributed paper to this conference, *ID 2755*.
5. W. Moses *et al.*, contributed paper to this conference, *ID 2771*.
6. P.R. Burchat *et al.*, *Phys. Rev. Lett.* **54**, 2489 (1985).
7. Review of Particle Properties, *Phys. Lett. B* **170** (1986).
8. S. Keh *et al.*, communication to this conference.
9. S. Abachi *et al.*, contributed paper to this conference, *ID 9687*.
10. H. Albrecht *et al.*, contributed paper to this conference, *ID 9628*.
11. MAC Collaboration, communication to this conference.
12. W. Ruckstuhl *et al.*, *Phys. Rev. Lett.* **56**, 2132 (1986).
13. W.B. Schmidke *et al.*, SLAC-PUB-4031, submitted to *Phys. Rev. Lett.* (1986).
14. H. Albrecht *et al.*, communication to this conference.
15. S. Abachi *et al.*, contributed paper to this conference, *ID 10006*.
16. W. Bartel *et al.*, contributed paper to this conference, *ID 9679*.
17. J.J. Becker *et al.*, contributed paper to this conference, *ID 3395*.
18. HRS Collaboration, communication to this conference.
19. W. Bacino *et al.*, *Phys. Rev. Lett.* **42**, 749 (1979).
20. S. Behrends *et al.*, *Phys. Rev. D* **32**, 2468 (1985).

A Measurement of the D_s Meson Lifetime by TASSO

G. E. FORDEN

Rutherford Appleton Laboratory
Chilton, Didcot, Oxon OX11 0QX

ABSTRACT. The lifetime of the D_s meson has been measured. The prelininary value is $3.4^{+2.9}_{-1.6} \pm 0.7 \times 10^{-13}$ *sec*. The method used was to fully reconstruct the decay vertex of the channel $D_s \to \phi\pi^{\pm}; \phi \to K^+K^-$. The experiment is being performed with the TASSO detector at the e^+e^- storage ring PETRA. It is expected that this measurement will be updated with significantly increased statistics after the end of this year's running.

We report here a measurement of the lifetime of the $D_s^{\pm}$ (formerly called the $F^{\pm}$) meson. This measurement is of particular interest since the c-quark and the $\bar{s}$-antiquark form a weak iso-doublet and provide the possiblity of explicit weak interactions between the valence quarks that are not available in either the $D^{\pm}$ or the D^0 systems. The present analysis is a new procedure utilizing the vertex chamber (VXD) information and geared to obtain a final $D_s^{\pm}$ sample unbiased with respect to the $D_s^{\pm}$ lifetime. A complete description of the general TASSO detector[**1**] and the vertex chamber[**2**] may be found elsewhere.

The $D_s^{\pm}$ meson was identified by the decay chain $D_s^{\pm} \to \phi\pi^{\pm}; \phi \to K^+K^-$. An attempt was made to associate all charged tracks reconstructed in the DC with digitizations in the VXD. Those tracks which were successfully associated with four or more (out of a possible eight) VXD hits were retained for the rest of the analysis. Pairs of accepted, oppositely charged tracks were then combined to try to form the ϕ meson, assuming kaon masses for each. A third track was then combined with the ϕ candidate, assuming it to be a pion. A three dimensional vertex fit[**3**] was then preformed on this triplet of charged tracks. The triplet of tracks was retained if the probability associated with this geometric vertex fit was greater then 1 %. The mass of the K^+K^- subsystem was calculated from the track parameters resulting from this vertex fit, those within 0.015 GeV/c^2 of the ϕ mass were used in the next stage of the selection process if $E_{KK\pi}/E_{beam} \geq 0.6$. This final cut eliminates those D_s mesons produced in the weak decay of bottom hadrons. A kinematic vertex fit[**4**] was then preformed on the triplet of tracks by constraining the K^+K^- candidates to the ϕ mass. In this way the momentum resolution of the lone pion is improved as well as improving those tracks resulting from the decay of the ϕ. The resulting $KK\pi$ mass spectrum is shown in Figure 1a. The D_s signal is consistent with 1.971GeV/c^2 and the 25 MeV/c^2 mass resolution determined from Monte Carlo simulations. The D_s candidates were accepted for the lifetime studies if they satisfied $1.920 \leq M(KK\pi) \leq 2.020$ GeV/c^2. No futher cuts were made on this sample.

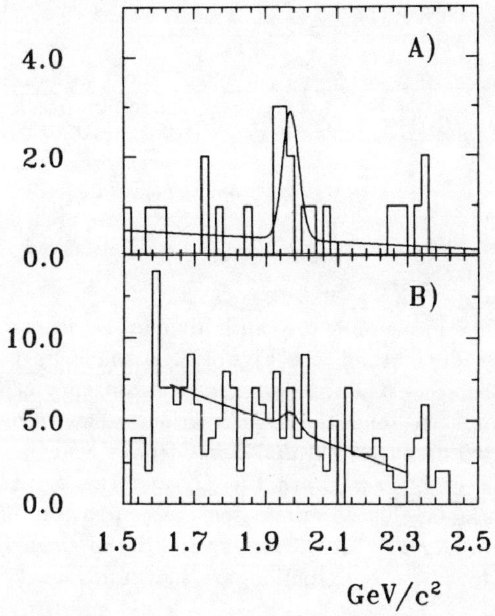

Figure 1.

A control sample was also selected using a very similar algorithm but requiring that the mass of the "phi" candidate be in the range 1.05 GeV/c^2 to 1.15 GeV/c^2. The mass spectrum of the control sample, Figure 1b, was parameterized by a linear term plus a Gaussian centered at 1.971 GeV/c^2

and with a width of 25 MeV/c^2. There is no significant enhancement of the D_s mass region, where the amplitude of the fitted Gaussian is consistent with zero. We therefore ignore the Gaussian's contribution when estimating the size of the background. This yields an estimated 7.4 events above a background of 1.6 from which the D_s lifetime is determined.

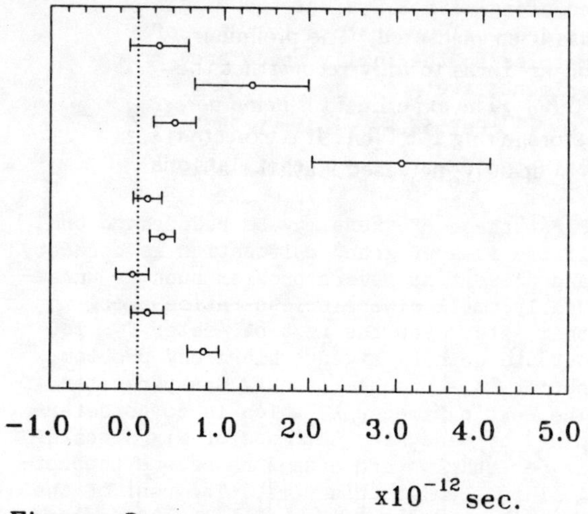

Figure 2.

The measurement of an individual D_s meson's proper decay time, see Figure 2, is made by using the reconstructed vertex, the beam spot center and size, and the D_s direction. The reconstructed decay vertex used comes from a kinematic vertex fit utilizing both the D_s and the ϕ mass constraints. The lifetimes for the control sample with $M(KK\pi) \leq 2.020\text{GeV/c}^2$ and the corrisponding Monte Carlo prediction, are shown in Figure 3.

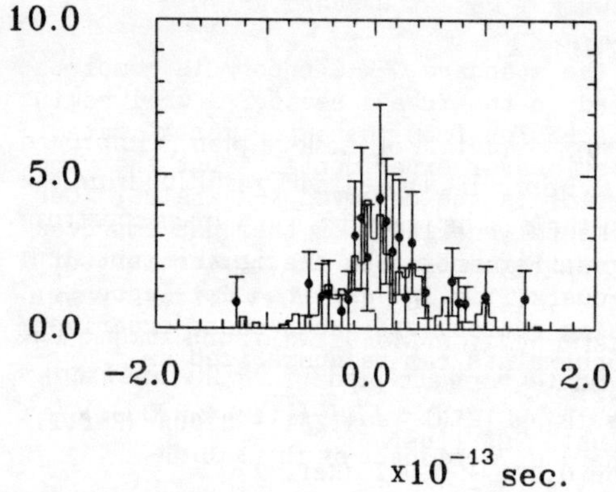

Figure 3.

The average lifetime for the complete sample is calculated using a maximum likelihood method. The likelihood function used is given by:

$$L = \prod_i \{(1 - P_{back})F(\tau_i, \sigma_i; \tau_F)$$
$$+ P_{back}(1 - P_0 - P_+)G(\tau_i, \sigma_i)$$
$$+ P_{back}P_0F_0(\tau_i, \sigma_i) + P_{back}P_+F_+(\tau_i, \sigma_i)\}$$

The probability that a D_s candidate is not really a D_s is given by P_{back}. The functions $F(\tau_i, \sigma_i; \tau_F)$, $F_+(\tau_i, \sigma_i)$ and $F_0(\tau_i, \sigma_i)$ represent the exponential decays of the $F^\pm$, $D^\pm$ and D^0 mesons folded with the Gaussian resolution, respectively. The fractions of $D^\pm$ and D^0 meson contamination were found to be $P_{back}P_+ = 3.8$ % and $P_{back}P_0 = 3.4$ % respectively.

The VXD cell resolution enters both explicitly and and implicitly at various stages of our analysis, including VXD hit association and calculating the uncertainty in the of the reconstructed decay vertices. Adding these effects in quadrature we obtain a contribution to the systematic error of $\pm 0.7 \times 10^{-13}$ *sec*. We varied the background fractions by ± 5 % which produced a change of $\pm 0.015 \times 10^{-13}$ *sec*. Variations of the D^0 and $D^\pm$ compositions corrisponding to the D^*/D production uncertainty contributed $\pm 0.09 \times 10^{-13}$ *sec*. Finally, changing our assumptions for the lifetimes of the $D^\pm$ and D^0 mesons by the errors of the world averages produced changes of $\pm 0.07 \times 10^{-13}$ *sec* and $\pm 0.02 \times 10^{-13}$ *sec*. Adding these contributions in quadrature we arrive at our preliminary value for the D_s lifetime of:

$$\tau = 3.4^{+2.9}_{-1.6} \pm 0.7 \times 10^{-13} \ sec.$$

Data taking with the TASSO detector at PETRA is continuing and we expect to update this measurement with a significantly increased data sample.

REFERENCES

1. TASSO Collaboration, R.Brandelik et al., Phys. Lett. **83B** (1979), 261.
2. D. M. Binnie et al., Nucl. Instr. Meth. **A228** (1985), 267.
3. D.H. Saxon, Nucl. Instr. Meth. **234** (1985), 258.
4. G. E. Forden and D. H. Saxon, RAL-85-037 (to appear in Nucl. Instr. Meth.) (1985).

IMPLICATIONS OF THE OBSERVED K-M ANGLES ON FUTURE PHYSICS

Michael Shin

Physics Department, Brown University, Providence, RI 02912, U.S.A.

I discuss the implications of the observed values of the K-M angles on the structure of particle physics beyond the standard model. A hypothesis on the structure of quark mass matrices is proposed along with S.S.B. as the origin of weak-CP violation. The entire K-M angles are well understood in terms of the quark mass ratios alone. Possible symmetries in the family space are discussed. The implications on future experiments are: $\Gamma(b\to ue\bar{\nu})/\Gamma(b\to ce\bar{\nu}) = (8.8^{+5.8}_{-3.8})\times 10^{-3}$, and $m_t = 42.6^{+19.4}_{-9.5}$ GeV, $48.8^{+24.1}_{-11.5}$ GeV, $56.2^{+31.3}_{-14.3}$ GeV, for $|V_{cb}| = 0.04$, 0.05, and 0.06 respectively. $\nu_e - \nu_\mu$, $\nu_\mu - \nu_\tau$ oscillations may be observable in a foreseeable future.

1. INTRODUCTION

In our present understanding of particle physics, there are two outstanding theoretical problems that theorists have to face. One is the gauge hierarchy problem (the problem of the origin of the electroweak symmetry breaking) and the other is the observed family structure of quarks and leptons. Although so many efforts and attempts have been made for the resolution of the first problem in many years through supersymmetry, supergravity, Kaluza-Klein, etc., and some understanding of this problem may be possible within the context of these approaches, there has been no new physics emerging out of these approaches. For instance, desperately sought super-partners of quarks, leptons, and gauge bosons have not been found so far (It is quite impressive that, among so many super-partner particles, not even a single one has been found so far). Thus, it may be that we have been wasting time by solving a "wrong problem" in our quest for new physics beyond the standard model. Perhaps, the gauge hierarchy problem, as it stands at the present time, may be a too difficult (and irrelevant) aspect of Nature to be understood through our limited knowledge on Nature.

Let me now point out why the second problem is a more severe problem than the first, and why it is more likely to lead us to new physics beyond the standard model, by reminding you of the list of fundamental parameters of the standard model that we need to understand.

They are: 3 gauge coupling constants (g_1, g_2, g_3), 2 Higgs potential parameters ($V \simeq 250$ GeV, $\lambda \equiv$ quartic coupling constant of Higgs doublet), 6 quark masses ($m_u, m_d, m_s, m_c, m_b, m_t$), 3 lepton masses ($m_e, m_\mu, m_\tau$), 4 K-M angles ($\theta, \beta, \gamma, \delta'$; see Eq. (1)), θ_{QCD} and θ_{QFD} ($\equiv \arg\det(M_u M_d)$). First three of these may be reduced to one if the idea of grand unification is correct, and present no severe problem such as unnaturally small dimensionless ratios among them, etc. For the next parameter $V \simeq 250$ GeV, we do have a gauge hierarchy problem, but it is a problem of only one parameter. The next parameter λ, which is to be determined by measuring the mass of Higgs scalar, is yet unknown and causes no severe theoretical problem at this time. The rest of the parameters (15 out of a total of 20 parameters) are closely related to the Yukawa sector of the theory and constitute the major unknown aspect of the standard model.

Clearly, the Yukawa sector needs to be understood, if further progress is to be made in particle theory. Also, it is more likely that it will lead us to new physics beyond the standard model since we have more data in this sector.

Now let me discuss some theoretical ideas related to the Yukawa sector and their implications on future physics that I discovered when I was at Harvard.

Since the standard G-W-S theory is completely blind in the Yukawa sector, I will begin our discussion from the experimental data. The most useful experimental input that one can employ is the observed K-M angles. Our experimental knowledge on them has improved to a great extent after the measurement of the b-quark lifetime and its semi-leptonic branching ratios, and the present experimental constraints can be summarized as follows:

i) $|V_{us}| \simeq S_\theta$: $|V_{us}| = 0.231 \pm 0.003$ (Ref.1) or 0.221 ± 0.002 (Ref. 2)

ii) $|V_{cb}| \simeq S_\gamma$: $|V_{cb}| = 0.05 \pm 0.01$ (Ref. 3)

iii) $|V_{ub}| = S_\beta$: $S_\beta/S_\gamma \lesssim 0.2$ (Ref. 4)

iv) $|B_K| \sin\delta' \simeq 0(1)$, from Re$\varepsilon$ in K^0-$\bar{K}^0$, where θ, β, γ, and δ' are parameters in the Maiani-parameterization of the K-M matrix,

$$V_{K-M} =$$
$$\begin{bmatrix} C_\beta C_\theta & C_\beta S_\theta & S_\beta \\ -C_\gamma S_\theta + C_\theta S_\beta S_\gamma e^{i\delta'} & C_\gamma C_\theta + S_\theta S_\beta S_\gamma e^{i\delta'} & -C_\beta S_\gamma e^{i\delta'} \\ -C_\theta C_\gamma S_\beta - S_\theta S_\gamma e^{i\delta'} & -C_\gamma S_\theta S_\beta + C_\theta S_\gamma e^{i\delta'} & C_\gamma C_\beta \end{bmatrix} \tag{1}$$

A natural question that a theorist can ask with the above numbers is whether these numbers mean anything at all, or more precisely, whether they indicate any dynamical structure or symmetries present beyond the standard model. The answer seems to be yes, and we shall briefly recall the origin of the K-M matrix in the standard model to answer this question.

2. REVIEW ON THE ORIGIN OF THE K-M MATRIX

In the standard $SU(2)_L \times U(1)_Y$ electroweak theory, the most general mass terms for the quarks are given by,

$$\bar{Q}^o_{iL}\, M^Q_{ij}\, Q^o_{jR} + \text{H.C.}, \quad M^Q_{ij} \equiv y^Q_{ij}\, v/\sqrt{2}, \tag{2}$$

where $Q \equiv U$ or D, $i,j = 1,2,3$ for 3 generations of quarks, and Q^o_{iL}, Q^o_{jR} are the weak-eigenstates. M^Q can be diagonalized by a bi-unitary transformation,

$$V^{Q\dagger}_L M^Q V^Q_R \equiv M^Q_{diag.} \tag{3}$$

Then the weak-eigenstates (Q^o_L and Q^o_R) are related to the mass-eigenstates (Q_L and Q_R) by

$$Q^o_L = V^Q_L\, Q_L,\; Q^o_R = V^Q_R Q_R, \text{ and}$$
$$\bar{Q}^o_L\, M^Q\, Q^o_R = \bar{Q}_L\, M^Q_{diag.}\, Q_R. \tag{4}$$

The weak-charged-current, which is diagonal in the weak-eigenstates, is then

$$\bar{U}^o_L\, \gamma_\mu\, D^o_L = \bar{U}_L\, (V^u_L)^\dagger\, \gamma_\mu\, (V^D_L)\, D_L \equiv \bar{U}_L\, \gamma_\mu\, V_{K-M}\, D_L \tag{5}$$

This implies,

$$V_{K-M} = (V^U_L)^\dagger\, (V^D_L) \tag{6}$$

and V_{K-M} is determined by the structure of the quark mass matrices M^U and M^D, through the diagonalizing unitary matrices V^U_L and V^D_L (of Eq. (3)). Although this task of deducing the structure of M^U and M^D from the observed experimental constraints on V_{K-M} is an extremely difficult (almost impossible) job, since the matrices M^Q's contain a total of 36 real parameters (9 complex parameters for each charged sector) while only 10 parameters (6 quark mass eigenvalues and 4 K-M parameters) are observable, it is a duty of theorists to do so and we shall challenge this problem in the following section.

3. QUARK MASS MATRICES OF FRITZSCH FORM WITH PHASES OF MULTIPLES OF $\pi/2$

One of the most attractive forms of quark mass matrices is the one proposed by Fritzsch (Ref. (5)) in 1978;

$$M_F = \begin{bmatrix} 0 & |A|e^{i\phi_A} & 0 \\ |A|e^{i\phi'_A} & 0 & |B|e^{i\phi_B} \\ 0 & |B|e^{i\phi'_B} & |C|e^{i\phi_C} \end{bmatrix} \tag{7}$$

$$M^U = M_F(A\to A_U,\; \phi_A\to\phi_{A_U}, \ldots)$$
$$M^D = M_F(A\to A_D,\; \phi_A\to\phi_{A_D}, \ldots)$$

It was motivated at that time by the fact that the Cabibbo angle (the only well measured K-M angle by that time) satisfies the empirical relation,

$$\tan\theta_C \simeq \sqrt{m_d/m_s} \simeq \sqrt{1/19.6} = 0.22 \tag{8}$$

For this form of a mass matrix, only the heaviest generation has a diagonal mass term. All other lighter masses are generated by off-diagonal mixings with the nearest neighboring generations. Furthermore, all real dimensionful parameters are related to the mass eigenvalues ($|A| = \sqrt{m_1 m_2}$, $|B| = \sqrt{m_2 m_3}$, $|C| = m_3$), leading to calculable* K-M angles in terms of quark mass ratios and two phases. In fact, V_{K-M} is given by

$$V_{K-M} = (R^U_F)^T \begin{bmatrix} 1 & 0 & 0 \\ 0 & e^{i\sigma} & 0 \\ 0 & 0 & e^{i\tau} \end{bmatrix} (R^D_F), \tag{9}$$

*For a general classification of mass matrices with calculable mixing angles, see ref. (7). Ref. (7) shows that Fritzsch structure is an almost unique one with realistic mixing angles.

$$\sigma \equiv (\phi_A + \phi_C - \phi_B - \phi_B')_U - (\phi_A + \phi_C - \phi_B - \phi_B')_D$$

$$\tau \equiv (\phi_A - \phi_B')_U - (\phi_A - \phi_B')_D$$

$$R_F \simeq \begin{bmatrix} 1 & -\sqrt{m_1/m_2} & (m_2/m_3)\sqrt{m_1/m_3} \\ \sqrt{m_1/m_2} & 1 & \sqrt{m_2/m_3} \\ -\sqrt{m_1/m_3} & -\sqrt{m_2/m_3} & 1 \end{bmatrix}$$

The above expression for V_{K-M} implies

$$|V_{us}| = |\sqrt{m_d/m_s} - e^{i\sigma}\sqrt{m_u/m_c}|[1+0(m_d/m_S)]$$

$$|V_{cb}| = |\sqrt{m_s/m_b} - e^{i(\tau-\sigma)}\sqrt{m_c/m_t}|[1+0(m_d/m_s)]$$

$$|V_{ub}| = |(m_s/m_b)\sqrt{m_d/m_b} + \sqrt{m_u/m_c}\{e^{i\sigma}\sqrt{m_s/m_b} - e^{i\tau}\sqrt{m_c/m_t}\}|[1+0(m_d/m_s)]$$

$$\delta'_{Maiani} \simeq -\sigma . \qquad (10)$$

Now, the real question is whether these expressions are all consistent with known K-M phenomenologies, and somewhat painstaking investigation was performed in Ref.(6) to answer this question. The answer turned out to be yes, but only if the fundamental phases σ and τ are $\sigma=\tau=\pi/2$ (or $-\pi/2$). $\sigma-\tau=0$ comes from the smallness of $|V_{cb}|$ (unusually long b-quark lifetime!), and $\sigma=\pm\pi/2$ comes from the best fit on $|V_{us}|$ (Cabibbo angle) and it is also well consistent with the CP-violating parameter Reε in K^o-$\bar{K}^o$, $\delta'_{Maiani} \simeq -\sigma = \mp\pi/2$! In such a case (Fritzsch structure with $\sigma=\tau=\pm\pi/2$), the predictions are:

1) Entire K-M angles:

$$\sin\theta_c = |V_{us}| = \sqrt{m_d/m_s + m_u/m_c}$$

$$\sin\gamma = |V_{cb}| = \sqrt{m_s/m_b} - \sqrt{m_c/m_t}$$

$$\sin\beta/\sin\gamma = |V_{ub}|/|V_{cb}| = \sqrt{m_u/m_c + \{(m_d/m_s)(m_s/m_b)^3/|V_{cb}|^2\}} \qquad (11)$$

$$\delta'_{Maiani} \simeq -\sigma = \mp\pi/2$$

The prediction* on $|V_{ub}|/|V_{cb}|$ is then $0.067^{+0.017}_{-0.015}$ or equivalently

$$\Gamma(b\to ue\bar{\nu})/\Gamma(b\to ce\bar{\nu}) = (8.8^{+5.8}_{-3.8}) \times 10^{-3} . \qquad (12)$$

2) Prediction** on m_t from $|V_{cb}|$:

$\lvert V_{cb}\rvert$	m_t (at 1GeV)	m_t^p (constituent mass)
0.04	$67.2^{+34.4}_{-16.1}$ GeV	$42.6^{+19.4}_{-9.5}$ GeV
0.05	$77.9^{+42.9}_{-19.8}$ GeV	$48.8^{+24.1}_{-11.5}$ GeV
0.06	$91.3^{+56.5}_{-24.6}$ GeV	$56.2^{+31.3}_{-14.1}$ GeV

for $\Lambda_{\overline{Ms}}$ = 100 MeV. (13)

4. IMPLICATIONS ON FUTURE PARTICLE THEORY AND EXPERIMENTS

The fundamental phases $\sigma=\tau=\pm\pi/2$ clearly indicates (Ref.(6)) the spontaneous symmetry breaking (S.S.B) as the origin of CP-violation. This means these phases arise via S.S.B. in a CP-invariant Lagrangian through the phases of VEVs of complex scalars.

Now let me briefly discuss what I call the principle of minimaximality of CP-violation in the light of S.S.B. as the origin of CP-violation. This principle addresses itself to the question of why the observed b-quark lifetime is unusually long and why the CP-violation observed in K^o-$\bar{K}^o$ is so tiny. Suppose the world has no CP-violations at all. Then all parameters in the Lagrangian and all VEVs of scalar fields would be real (CP-conserving).

In such a world, the most naive choice for the phase matrix $P\equiv$diag.$(1,e^{i\sigma},e^{i\tau})$ of Eq.(9) would be diag.(1,1,1), and the smallness of all K-M angles can naturally be understood as the effects of cancellations between two orthogonal matrices $(R_F^u)^T$ and R_F^d in Eq.(9). Thus, is such a world $|V_{cb}| \simeq \sqrt{m_s/m_b} - \sqrt{m_c/m_t}$ (and $|V_{us}| \simeq \sqrt{m_d/m_s} - \sqrt{m_u/m_c}$), leading to a naturally long b-quark lifetime provided $m_c/m_t \simeq 0(m_s/m_b)$. Now in the real world, we do observe a very tiny CP-violation in K^o-$\bar{K}^o$. To accomodate such a small CP-violation, the most economical choice that Nature can take without spoiling the smallness of the real K-M angles is to modify the phase matrix element in the lightest generation through S.S.B., i.e., P = diag.$(\pm i,1,1,)$. In fact, this choice of P is equivalent to $\sigma=\tau=\pm\pi/2$ by an overall phase transformation on V_{K-M}. However, this choice of the phase matrix by Nature does not correspond to "the maximal" CP-viola-

---

*I have used the following values throughout this talk: $m_s/m_d = 19.6\pm1.6$, $m_c/m_u = 264.7\pm^{124.2}_{67.7}$ $m_b/m_s = 30.3^{+10.1}_{-5.2}$, m_c(1GeV) = (1.35$\pm$0.05) GeV. These numbers were obtained in Ref. (6) from the running quark masses at 1 GeV discussed in detail in Ref. (8).

**The main source of the uncertainties in m_t comes from that of m_s/m_b. Thus we strongly encourage theorists to determine this ratio as accurately as possible before the measurement of the t-quark mass.

tion. Obviously, the choice P = diag. ($\pm$i,-1,1)will produce much larger $|V_{cb}|$(= $\sqrt{m_s/m_b}$ + $\sqrt{m_c/m_t}$ $\simeq$ 0.3-0.4 for $m_t^P \sim$ 40 GeV) and thus much more CP-violation in K^o-$\bar{K}^o$. Therefore I call this choice of P=diag. ($\pm$i,1,1,) to be the choice of the minimaximality of CP-violation since it minimizes most of the real K-M angles and accepts a maximal CP- violating phase of $\pi/2$ from S.S.B.. I think this is what Nature has indeed chosen.

Let me now turn my discussion to possible future theories and their phenomenological implications. Considerations from model-buildings suggest the following (Ref.(9) and (10)) possibilities:

i) Parity-invariant gauge group: For the case when the vertical gauge group is a left-right symmetric one such as $SU(2)_L xSU(2)_R xU(1)_{B-L}$ or SO(10), the family symmetry is identified to be U(1)x U(1)xU(1) global symmetry. $U(1)_{PQ}$ is a natural subgroup of this family symmetry, solving the strong-CP problem automatically. S.S.B. is responsible for P and CP violations observed in Nature as well as for the generation of gauge boson masses of $W^\pm$ and Z^o. The invisibility of the axion seems to be related to the low energy phenomena such as the absence of F.C.N.C. and the observed long b-quark lifetime. For SO(10) gauge group, <u>the phenomenology of the massive neutrinos has novel predictions.</u> (Ref.(11)). In particular, the scale of seesaw mechanism, which is responsible for the smallness of the light neutrino masses, is bounded from above by the invisible axion scale (10^9-10^{12} GeV: $U(1)_{PQ}$ breaking scale). This leads to a lower bound on the mass of light neutrinos. From the lepton-quark symmetry, the Yukawa couplings of neutrinos are related to those of quarks, leading to calculable mixing angles similar to those for quarks. The results on these quantities are (Ref.(11)): 0.18 eV $\lesssim m_{\nu 3} \lesssim$ 100 eV, $m_{\nu 1}/m_{\nu 2} \simeq 0(m_u/m_c)$, $m_{\nu 2}/m_{\nu 3}$=$0(m_c/m_t)$ or $0\left((\sqrt{m_u m_c}/m_t)(m_c/m_t)\right)$, $\langle m_{\nu e}\rangle/m_{\nu 3} \simeq 0(m_e/m_\tau)$, $\theta_{e\mu} \simeq |\sqrt{m_e/m_\mu} + e^{i\eta_1'} 0(\sqrt{m_u/m_c})|$, $\theta_{\mu\tau} \simeq |\sqrt{m_u/m_\tau} - e^{i\eta_2}\, 0(\sqrt{m_c/m_t})|$, $\theta_{e\mu}$ = $\sqrt{m_e/m_\mu}\,\theta_{\mu\tau}$ where $m_{\nu 1}$, $m_{\nu 2}$, $m_{\nu 3}$ are the mass eigenvalues of three generations of light neutrinos, $\langle m_{\nu e}\rangle$ is the parameter measured in neutrinoless double beta ($0\nu\beta\beta$) decay experiments, and η_1' and η_2 are some phases. Thus <u>we may expect to observe</u> ν_e-ν_μ and ν_μ-ν_τ <u>oscillations in a foreseeable future</u>, while null results are expected in $0\nu\beta\beta$-decays. Moreover, if the solution to the solar neutrino problem is indeed the MSW enhancement of neutrino oscillations in matter, then ν_e-ν_μ oscillation channel is to be identified with the relevant oscillation channel (with $\Delta m^2 \simeq 10^{-4}$ eV^2) in the sun, instead of ν_e-ν_τ oscillation channel (which the conventional seesaw mechanism at GUT scale predicts).

ii) Parity-Noninvariant gauge group: In the case when the vertical gauge group is not left-right symmetric, such as the usual $SU(3)_c xSU(2)_L xU(1)_Y$ or SU(5), the relevant family symmetry turns out to be $SU(3)_F$x $U(1)_{PQ}$. Three family of <u>mirror fermions are expected to be present at the invisible axion scale (10^9-10^{12} GeV)</u>. In addition to the usual invisible axion, there exist <u>eight familons associated with the S.S.B. of $SU(3)_F$</u>. They will induce F.C.N.C. at the level consistent with the present limit but may be observable in the future.

Before closing this section, let me briefly comment on the <u>issues of the fourth generation quarks</u> (which will be relevant to future phenomenologies in the K-M model) from the generalization of Eqs. (9) and (11) to the world with four generations of quarks. Such a generalization allows one to determine the entire 4x4 K-M matrix in terms of quark masses alone, and the present constraints on the light 3x3 K-M sector lead to some constraints on the masses of these new quarks. Such a consideration was made in Ref. (12) and the results are: $m_{b'} \gtrsim$ 32 GeV, $m_{t'} \gtrsim$ 82 GeV. Moreover, there is plenty of room for $m_{b'} \lesssim$ 80 GeV (see Fig. 1 and 2. of Ref. (12)), and the mixing parameter R_{odd} for B_d^o-$\bar{B}_d^o$ is calculated to be 5-6 times larger than in the 3-generation case and may be observable ($R_{odd} \simeq$ 10%) if the uncertainties in the parameters used are taken into account. For B_s^o-$\bar{B}_s^o$, the mixing will be definitely observable (R_{odd} > 50%) regardless of the 4-th generation quark masses.

5. CONCLUSION

The hypothesis*** considered in this talk has two crucial predictions on yet unmeasured fundamental parameters of G-W-S theory. They are given in Eqs. (12) and (13). It is also interesting to see whether $\nu_e-\nu_\mu$, $\nu_\mu-\nu_\tau$ oscillations will be observed in the future as was discussed in the previous section. This will serve as a good experimental test on SO(10). It is not clear at this time whether Nature has indeed chosen this particular structure of quark mass matrices at low energy, but if the hypothesis survives through the future experimental tests on m_t and $|V_{ub}|/|V_{cb}|$, it is very likely that a certain symmetry principle is behind the Yukawa sector (I think we may have been quite biased to think that the Yukawa sector should be somewhat complicated and not simple, only because G-W-S theory does not demand it to be simple). This may lead us to the discovery of new symmetries beyond the standard model and new physics may emerge. In the meantime, theorists are welcome to investigate all possible theories which meet the proposed hypothesis.

*** An explicit form of quark mass matrices satisfying $\sigma=\tau=\pm\pi/2$ is

$$M^U = \begin{bmatrix} 0 & ia & 0 \\ ia & 0 & b \\ 0 & b & c \end{bmatrix} \quad \begin{matrix} a=\sqrt{m_u m_c} \\ b=\sqrt{m_c m_t} \\ c= m_t \end{matrix} ,$$

$$m^D = \begin{bmatrix} 0 & a' & 0 \\ a' & 0 & b' \\ 0 & b' & c' \end{bmatrix} \quad \begin{matrix} a'=\sqrt{m_d m_s} \\ b'=\sqrt{m_s m_b} \\ c'= m_b \end{matrix} .$$

ACKOWLEDGEMENTS

It is my pleasure to thank my collaborators at Harvard Howard Georgi, Ann Nelson, Sekhar Chivukula, Jonathan Flynn, and my present collaborator at Brown, Kyungsik Kang. I also thank Bill Ford and other organizers of the Conference for selecting me as a speaker. This research is supported in part by the U. S. Department of Energy under Contract No. DE-AC02-76ER03130. A020 - Task A.

REFERENCES

(1) M. Bourquin et al., Z. Phys. C21 (1983) 27.

(2) H. Leutwyler and M. Roos, Z. Phys. C25 (1984) 91.

(3) A. Caldwell, invited talk, Moriond '86 (Les Arcs, France, March 9-16).

(4) E. H. Thorndike, invited talk, International Symp. on Lepton and Photon Interactions of High Energies (Kyoto, Japan, August 1985).

(5) H. Fritzsch, Phys. Lett. 73B (1978) 317; Nucl. Phys. B155 (1979) 189; L. F. Li, Phys. Lett. 84B (1979) 461; for early discussions on this mass matrix, see also A. C. Rothman and K. Kang, Phys. Rev. Lett. 43 (1979) 1548; H. Georgi and D. V. Nanopoulos, Nucl. Phys. B155 (1979) 52, and references therein.

(6) M. Shin, Phys. Lett. 145B (1984) 285; Harvard preprint HUTP-84/A070 (1984), (available in Harvard University Thesis (1985), Family structure of quarks and leptons).

(7) M. Shin, Phys. Lett. 152B (1985) 83.

(8) J. Gasser and H. Leutwyler, Phys. Reports 87 (1982) 77.

(9) H. Georgi, A. Nelson, and M. Shin, Phys. Lett. 150B (1984) 306; M. Shin, ibid, 154B (1985) 205; ibid, 160B (1985) 411.

(10) K. Kang and M. Shin, Phys. Rev. D33 (1986) 2688.

(11) K. Kang and M. Shin, Brown HET-577 (1986) and Brown HET-587 (1986), to be published.

(12) M. Shin, R. S. Chivukula, and J. M. Flynn, Nucl. Phys. B271 (1986) 509.

Recent CLEO Results on B Meson Decays
(Mostly) to Leptons and Dileptons

R. Poling

University of Rochester
Rochester, New York 14627

Recent results on B production and decay from the CLEO experiment are reviewed. The validity of the assumption that $\Upsilon(4S)$ decays exclusively to $B\overline{B}$ is examined, and new limits are presented for $\Gamma(b\to u)/\Gamma(b\to c)$, for $B^0\overline{B}^0$ mixing, and for the ratio of the charged and neutral B meson semileptonic branching fractions.

1. INTRODUCTION

The study of lepton production in B meson decays continues to provide valuable insight into the properties of the b quark. This report presents recent results from the CLEO experiment mostly from studies of lepton production at the $\Upsilon(4S)$. These include new limits on $\Gamma(b\to u)/\Gamma(b\to c)$, and the latest limits on $B^0\overline{B}^0$ mixing and the ratio of the semileptonic branching fractions for the charged and neutral B mesons. Before these results are presented, the crucial assumption that $\Upsilon(4S)$ decays only to $B\overline{B}$ is reexamined in light of recent experimental and theoretical developments.

2. DOES $\Upsilon(4S)$ DECAY EXCLUSIVELY TO $B\overline{B}$?

We assume $\Upsilon(4S)$ decays to $B\overline{B}$ because, unlike the 1S, 2S and 3S states of the $b\overline{b}$ system, $\Upsilon(4S)$ is broad, indicating an OZI-allowed decay to b-flavored hadrons. Similar considerations led to the assumption that $\psi''(3770)$ decays exclusively to $D\overline{D}$. Recent results from Mark III [1], using events with fully reconstructed D's, provide an indirect measurement of $\sigma(e^+e^-\to\psi''(3770)\to D\overline{D})$ which is a factor of 2 below measurements of the ψ'' hadronic cross section [2]. One interpretation for this is that perhaps ψ'' does not always decay to $D\overline{D}$. This compels us also to question whether $\Upsilon(4S)$ decays only to $B\overline{B}$.

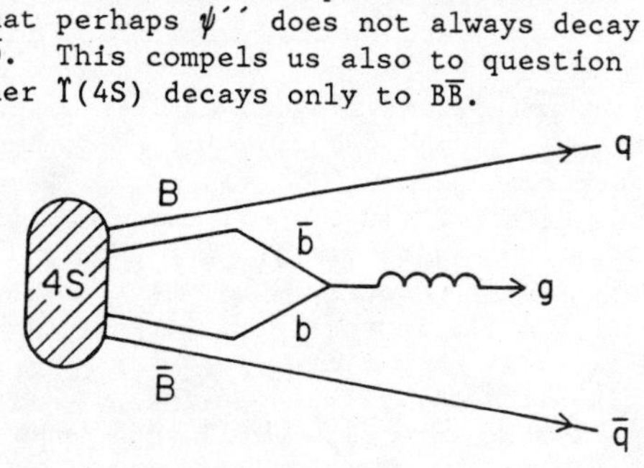

Fig. 1 Diagram for the decay $\Upsilon(4S)\to$(bottomless hadrons).

In response to the Mark III result, Lipkin [3] suggests that we reexamine the validity of the OZI rule just above flavor threshold. He envisions inelastic rescattering of the on-shell $B\overline{B}$ pair (Fig. 1) as a mechanism for defeating OZI to produce b-less $\Upsilon(4S)$ decays. Signals for such a process include high momentum particles above the kinematic limit for production from B decay, and lepton and dilepton yields which are inconsistent with a pure $B\overline{B}$ sample. Both techniques have been used with CLEO data to set limits on the fraction of b-less decays of $\Upsilon(4S)$ [4].

When $\Upsilon(4S)$ decays to $B\overline{B}$, the B and the $\overline{B}$ decay independently and nearly at rest. Consequently, all tracks from such a decay must satisfy $x = p/E_{beam} < 0.5$. For decays to a b-less state Q, no restriction applies; values of x from 0 to 1 would be allowed. The continuum-subtracted x distribution for the CLEO $\Upsilon(4S)$ data (41 pb^{-1} on $\Upsilon(4S)$ and 17 pb^{-1} at energies just below the peak collected during 1982, and 78 pb^{-1} on and 36 pb^{-1} below collected during 1985) is shown in Fig. 2. There is no evidence for tracks with x above 0.5. Extracting an upper limit on the branching fraction f for the process $\Upsilon(4S)\to Q$ requires a model for the decay of the b-less state Q.

One possibility is that b-less $\Upsilon(4S)$ decays resemble continuum events. As can be seen from the CLEO continuum data in Fig. 2 (solid curve), this assumption implies a stiff x distribution and a strong limit on f. For softer production, resembling $\Upsilon(1S)$ decays, or just the three-gluon decays of $\Upsilon(1S)$ (dashed curve), the limit becomes weaker. The 90% confidence level upper limits on f for these three cases are 0.038, 0.10 and 0.13, respectively.

Independent information about the b-less fraction f can be obtained by comparing yields of single and dilepton events at $\Upsilon(4S)$. If all leptons come from $B\overline{B}$, we can express the number of events with one

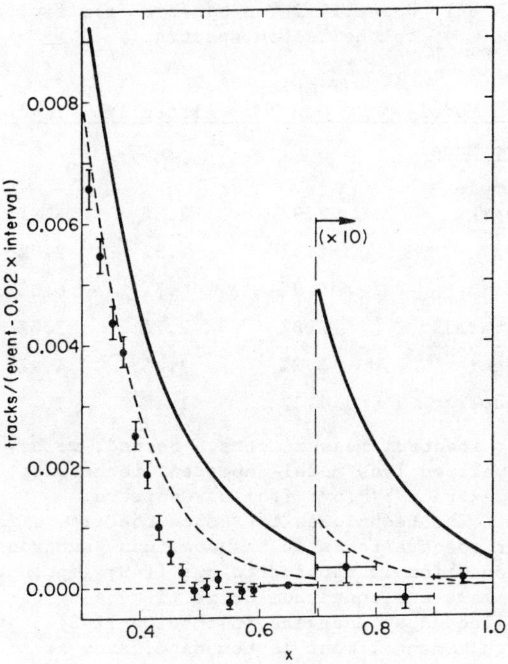

Fig. 2 Normalized x distributions from ϒ(4S) (points), continuum (solid curve), and ggg decay of ϒ(1S) (dashed curve).

lepton (N_ℓ) and with two leptons ($N_{\ell\ell}$) as

$$N_\ell = 2\overline{b}_B \, \epsilon_B \, (1\text{-}f) \, N_{4S}, \text{ and}$$

$$N_{\ell\ell} = \overline{b}_B^2 \, \epsilon_B^2 \, (1\text{-}f) \, N_{4S},$$

where $\overline{b}_B$ is the mean B semileptonic branching fraction at the ϒ(4S), ϵ_B is the lepton detection efficiency, and N_{4S} is the number of ϒ(4S) events. The quantity $\alpha = (4N_{4S}N_{\ell\ell})/(gN_\ell^2)$ is related to f by

$$f = 1 - (\overline{b}_B^2/(\overline{b}_B)^2)/\alpha,$$

so an upper limit on α gives an upper limit on f. If the b-less state Q decays to leptons, this discussion is modified, but the results are not appreciably weakened.

After correcting for background from the continuum and from misidentified hadrons, we find 8701 ±163 leptons in the 1985 CLEO sample of 82,500 ϒ(4S) decays. Correcting for background from the continuum, from misidentified hadrons, from cascade processes (B→D→ℓ), and from ψ's, we find 229 ± 24 dileptons in the same data. This leads to $\alpha = 1.03 \pm 0.12$. Our previous result [5] was 0.99 ± 0.19, giving a combined value of 1.02 ± 0.10, or $\alpha < 1.18$ at 90% confidence level. This corresponds to $f < 0.16$. Allowing for lepton production from Q (assuming a charm fraction of 25% and continuum-like fragmentation), the limit deteriorates to 0.17.

From this we conclude that there is no evidence for b-less decays of ϒ(4S). Our limit from the x distribution is model dependent, ranging from 4% to 13%. Complementary information from dileptons gives a model-independent limit of 17%.

Similar considerations can be applied to ψ'' decays. The DELCO dilepton results [6], which were originally interpreted as signifying a large difference between charged and neutral D semileptonic branching fractions, may suggest charmless ψ'' decays with a branching fraction of a third or more. Similarly, early measurements of the D semileptonic branching fraction were considerably lower than recent results. While these results could indicate charmless ψ'' decays, the early experiments could also have overestimated the ψ'' production cross section.

3. Γ(b→u)/Γ(b→c)

Within the Standard Model, b decay is determined by the Kobayashi-Maskawa matrix elements V_{cb} and V_{ub}. Measurement of these parameters is a primary objective of b studies. We have two experimental handles. The B lifetime has been studied with fast-moving continuum-produced B's at PEP and PETRA. The relative contributions of the b→u and b→c decays (Γ(b→u)/Γ(b→c), or, simply, b→u/b→c) has been studied with B's in ϒ(4S) decays at CESR and DORIS. The primary techniques are studies of the lepton spectrum in semileptonic B decay, and measurement of the yield of charmed particles in B decays, observed either directly or through the strange particles into which they decay.

3.1 b→u/b→c BY THE B SEMILEPTONIC DECAY MOMENTUM SPECTRUM

The premise of the lepton spectrum studies is that b→uℓν decays give a stiffer momentum spectrum than b→cℓν because a c quark will form a more massive final state than will a u quark. This concept, and the Spectator Quark Model of Altarelli et al. [7] have been the foundations of previous CLEO and CUSB results [8,9].

This report provides results on b→u/b→c obtained by CLEO with electron data from the 1985 CESR run, and with muon data from the 1982 and 1985 runs [10]. Fig. 3 shows our observed spectra of electrons and muons from B decay. These spectra have been corrected for the continuum contribution in our ϒ(4S) sample, for the residual contamination from hadrons which are misidentified as leptons, and for detection efficiency. The curves represent the Altarelli model prediction for pure b→c. Clearly the data agree well with 100% b→c. The fit to this model yields a preliminary result for B(B→Xℓν) of 0.110 ± 0.002 ±

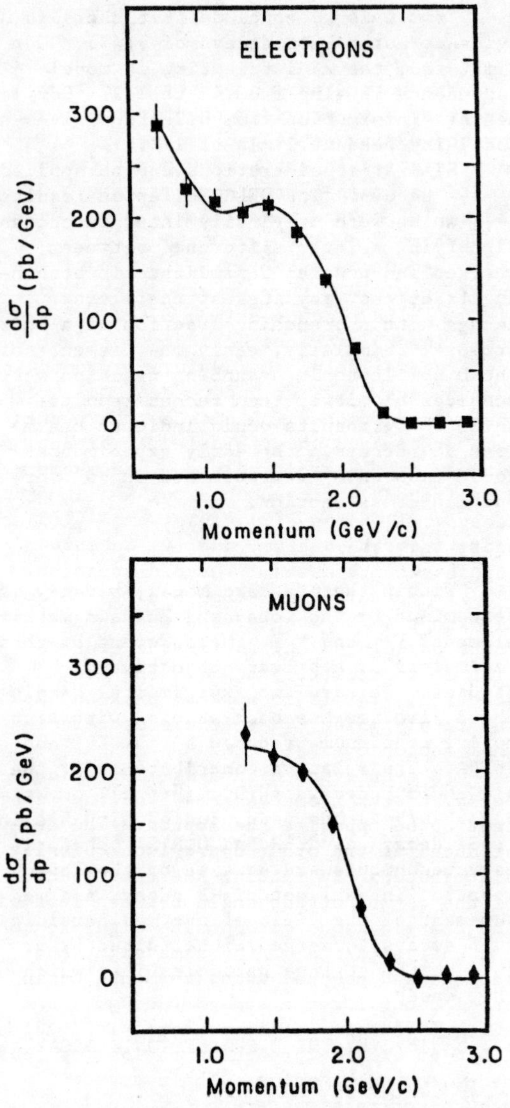

Fig. 3 Electron and muon spectra from semileptonic B decay.

0.010, in good agreement with our previously published values [8].

To determine how much $b\to u$ production can be tolerated by the data, we fit the spectra to a function

$G(p) = A_{bc}G_{bc}(p)+A_{bu}G_{bu}(p)+A_{cs}(p)G_{cs}(p)$,

where we have included terms for the theoretical predictions for semileptonic B decay through $b\to c$ and $b\to u$, and for the semileptonic decay of D mesons from B decays. Table I shows 90% confidence upper limits for $b\to u/b\to c$ for six different models for $b\to c$ paired with three different models for $b\to u$ [7,11]. There is considerable model dependence in this procedure.

Two avenues for improving the results on $b\to u/b\to c$ from the lepton spectra have recently been explored. First, we have sought procedures to reduce the errors in our spectrum measurements. Second, we have developed less model-dependent techniques to extract $b\to u/b\to c$ from our spectra.

Table I Upper limit on $b\to u/b\to c$ from fits to the lepton spectra.

$b\to u$ Model-->	Simple Quark	Altarelli	Isgur
$b\to c$ Model			
Simple Quark	2.9%	2.8%	3.8%
Tye	1.5%	1.5%	2.0%
Ali	4.9%	4.7%	6.6%
Altarelli	2.8%	2.7%	3.6%
Isgur	3.2%	3.1%	4.2%
Preparata	1.1%	1.1%	1.4%

The techniques to reduce the errors in our spectra focus on the continuum subtraction which is applied to our $\Upsilon(4S)$ leptons. We make the continuum correction using data collected at energies just below $\Upsilon(4S)$. At high momenta, most lepton candidates are not from $B\overline{B}$, but rather are leptons and misidentified hadrons from the continuum. The error from the continuum subtraction is the major limitation to our sensitivity to small amounts of $b\to u$. We have reduced this by using cuts to suppress continuum events relative to $B\overline{B}$ events, and by fitting the below-resonance data before subtracting.

In developing continuum-suppressing cuts, we have used the high-momentum lepton as an axis for two shape variables:

$$V_1 = \frac{\sum_{|\cos\theta_i|<0.7} (p_i \sin\theta_i)}{\sum p_i}, \text{ and}$$

$$V_2 = \sum_{\cos\theta_i>0.7} (p_i \cos\theta_i) .$$

V_1 measures the momentum transverse to the lepton. It is big for $B\overline{B}$ and small for the continuum. V_2 measures the momentum which accompanies the lepton. It is small for $B\overline{B}$, and big for the continuum. With these variables, the discrimination between $B\overline{B}$ and continuum is based on the B which did not decay semileptonically. Therefore, the efficiency is quite independent of the $b\to u$ model. Our cut rejects 80% of the continuum background, while preserving 70% of the $B\overline{B}$ signal.

Two new approaches to extracting $b\to u/b\to c$ from the lepton spectra have been used. In the first, we determine the branching fraction for B decay into leptons above the endpoint for $b\to c$, between 2.4 and 2.6 GeV. We find this branching fraction to be $(0.27 \pm 1.21) \times 10^{-4}$, or less than 2.2×10^{-4} at the 90% confidence level. It is related to $b \to u/b \to c$ by

$$\frac{b \to u}{b \to c} = \frac{B\left[B \to \ell(2.4,2.6)X\right]}{f_u(2.4,2.6)B(B \to X\ell\nu) - B\left[B \to \ell(2.4,2.6)X\right]}$$

where $f_u(2.4,2.6)$, the fraction of leptons from $b \to u$ in the momentum interval from 2.4 to 2.6 GeV, embodies the model dependence. With this technique, we find 90% confidence upper limits on $b \to u/b \to c$ of 3.8%, 3.5% and 5.4% for the Simple Quark, Altarelli and Isgur models, respectively. While $b \to c$ model dependence is absent, there remains approximately a factor of 1.5 variation with the $b \to u$ model.

In the second new approach, we attempt to regain some statistics and to reduce $b \to u$ model dependence, without sacrificing much $b \to c$ model independence, by widening the momentum bite to 2.2 to 2.6 GeV. We measure the branching fraction for B into leptons between 2.2 and 2.6 GeV to be $(12.2 \pm 2.1) \times 10^{-4}$. It is related to $b \to u/b \to c$ by

$$\frac{b \to u}{b \to c} = \frac{B\left[B \to \ell(2.2,2.6)X\right] - f_c(2.2,2.6)B(B \to X\ell\nu)}{f_u(2.2,2.6)B(B \to X\ell\nu) - B\left[B \to \ell(2.2,2.6)X\right]}$$

The results of this analysis (Table II) reflect model dependences which are reduced compared to the original technique, but which remain considerable.

Table II Upper limit on $b \to u/b \to c$ from lepton yield between 2.2 and 2.6 GeV.

$b \to u$ Model-->	Simple Quark	Altarelli	Isgur
$b \to c$ Model			
Simple Quark	3.4%	3.3%	4.4%
Tye	2.4%	2.3%	3.1%
Ali	4.6%	4.4%	6.0%
Altarelli	3.0%	2.9%	3.9%
Isgur	3.4%	3.3%	4.5%
Preparata	2.0%	1.9%	2.5%

In conclusion, the picture strongly supports the dominance of $b \to c$. The limit from fitting our lepton spectra to the model of Altarelli et al. is now 2.7%. Other models yield limits from 1.1% to 6.6%, showing considerable sensitivity to the model for $b \to c$. Less model-dependent methods have been used to set limits of 2.9% and 4.5% for the Altarelli and Isgur models, respectively. Additionally, measurement of the branching fraction for B mesons into leptons between 2.4 and 2.6 GeV makes it easy to calculate $b \to u/b \to c$ limits for other models.

3.2 $b \to u/b \to c$ BY THE YIELD OF CHARMED AND STRANGE PARTICLES IN B DECAY

The determination of K yields from B decays by CLEO provided the first evidence for the dominance of $b \to c$. Theoretical uncertainties in extracting quantitative limits on $b \to u/b \to c$ from K yields are considerable, so the results have been weaker than those from the lepton spectrum analysis. Recent measurements of D production in B decay [12], show a serious charm deficit. As is discussed in Ref. 12, however, other interpretations besides appreciable $b \to u$ are available. Measuring K production in B decays gives an independent measure of charm production which does not have the same systematic uncertainties (D branching ratios, etc.) as direct charm measurement.

There are four sources of K's in B decays: (1) $b \to c \to s$, (2) $W^- \to \bar{c}s$, $\bar{c} \to \bar{s}W^-$, (3) $s\bar{s}$ pairs from the vacuum, and (4) Cabibbo-suppressed processes. The old approach [13] to measure kaons from $b \to c$ was to count all kaons in $B\bar{B}$ events, and to subtract Monte Carlo predictions for the contribution of sources 2, 3 and 4. The new technique [14] is to measure $b \to sX$ and $b \to \bar{s}X$ separately using two-particle correlations in K-lepton events.

In a $B\bar{B}$ event with a semileptonic B decay, tracks from the same B as the lepton tend to be opposite the lepton. The decay products of the other B are isotropically distributed. By fitting the distribution of the cosine of the angle between leptons and charged kaons of opposite sign, leptons and charged kaons of the same sign, and leptons and neutral kaons to Monte Carlo predictions for the same-B (peaked), and other-B (isotropic) components, it is possible to extract branching ratios for decay modes including kaons. The measured branching fractions are

$$\Gamma(\bar{B} \to \ell^- K^- X)/\Gamma(\bar{B} \to \ell^- X) = 0.59 \pm 0.08 \pm 0.09$$
$$\Gamma(\bar{B} \to \ell^- K^+ X)/\Gamma(\bar{B} \to \ell^- X) = 0.09 \pm 0.06 \pm 0.01$$
$$\Gamma(\bar{B} \to \ell^- K^0/\bar{K}^0 X)\Gamma(\bar{B} \to \ell^- X) = 0.41 \pm 0.06 \pm 0.04$$
$$\Gamma(\bar{B} \to K^- X)/\Gamma(\bar{B} \to \text{all}) = 0.69 \pm 0.06 \pm 0.10$$
$$\Gamma(\bar{B} \to K^+ X)/\Gamma(\bar{B} \to \text{all}) = 0.19 \pm 0.05 \pm 0.03$$
$$\Gamma(\bar{B} \to K^0/\bar{K}^0 X)/\Gamma(\bar{B} \to \text{all}) = 0.65 \pm 0.07 \pm 0.09$$

The data agree very well with a 100% $b \to c$ Monte Carlo. The excess of K^- over K^+ in $\bar{B} \to X\ell\nu$ is 0.50 ± 0.10, compared to a prediction of 0.47. The excess of K^- over K^+ in $\bar{B} \to \text{all}$ is 0.50 ± 0.08, compared to a prediction of 0.49. The conclusion is that $b \to c/b \to \text{all}$ is 1.03 ± 0.20. While this is weaker than the lepton spectrum result, it gives independent reinforcement of the dominance of $b \to c$.

4. $B^0\bar{B}^0$ MIXING AND THE RATIO OF CHARGED AND NEUTRAL B SEMILEPTONIC BRANCHING BRANCHING FRACTIONS

Measurement of B^0B^0 mixing can give insight into the parameters of the Standard Model, and the dynamics of B decay. CLEO has updated its search for evidence of mixing in the yield of like-sign dileptons at the $\Upsilon(4S)$ [15]. Mixing is characterized by the parameter

$$y = \frac{N(B^0B^0) + N(\overline{B}^0\overline{B}^0)}{N(B^0\overline{B}^0)}$$

$$= \frac{N(\ell^+\ell^+)+N(\ell^-\ell^-)}{N(\ell^+\ell^-)} \times \frac{\left[(b_+/b_0)^2 f_+ + f_0\right]}{f_0}$$

where b_+ (b_0) is the mean semileptonic branching fraction for charged (neutral) B's, and f_+ (f_0) is the fraction of $\Upsilon(4S)$ decays which are B^+B^- ($B^0\overline{B}^0$). We have searched our 1985 $\Upsilon(4S)$ data sample for like- and unlike-sign dileptons. We imposed cuts on minimum momentum (1.2 GeV for electrons, 1.6 GeV for muons) and on the angle between the leptons to reduce the background from cascade decays ($B\to D\to\ell$) and from pairs including hadrons misidentified as leptons. We observe 5.1 ± 5.9 like-sign dileptons and 117 ± 12 unlike-sign dileptons above the backgrounds. The 90% confidence upper limit on y is shown as a function of the ratio of the charged and neutral B semileptonic branching fractions in Fig. 4. The curves correspond to three values for the $\Upsilon(4S)$ neutral to charged B production ratios. For equal branching fractions and 40% $B^0\overline{B}^0$, we find $y < 0.27$, or $y < 0.24$ including 1982 data.

We use the same data to set limits on the ratio of the charged and neutral B semileptonic branching fractions (and hence the lifetimes) by a procedure analogous to that used to estimate b-less $\Upsilon(4S)$ decays in Section 2. For $B^0\overline{B}^0$ production at 40%,

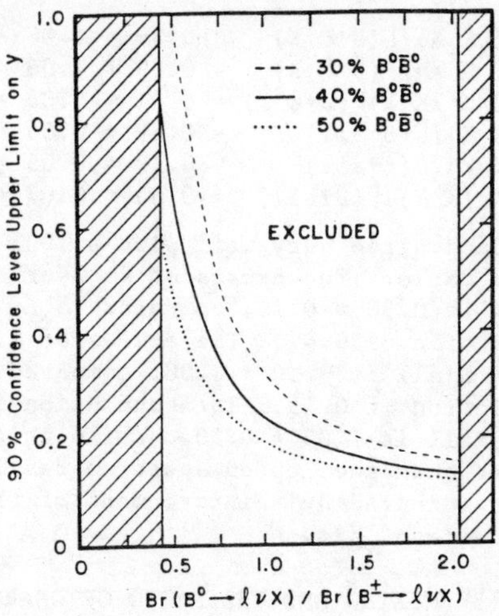

Fig. 4 Upper limits on y as a function of the ratio of the neutral to charged B semileptonic branching fractions.

production at 40%, we find the ratio of the neutral to charged B branching fractions to be between 0.43 and 2.03 at 90% confidence.

5. CONCLUSIONS

Based on significantly improved data, CLEO has several new or updated results on $\Upsilon(4S)$ and B decays. We find no evidence for decays of $\Upsilon(4S)$ to b-less final states. We set an upper limit on $b\to u/b\to c$ of 4.5% at 90% confidence level using techniques which are less subject to model dependences than previous analyses. We find yields of kaons in semileptonic B decay which reinforce the conclusion that $b\to c$ is the dominant mode for b decay. We have updated limits on $B\overline{B}$ mixing and the ratio of neutral and charged B semileptonic branching fractions.

REFERENCES

(1) R.M. Baltrusaitis et al., Phys. Rev. Lett. 56, 2140 (1986).

(2) I. Peruzzi et al., Phys. Rev. Lett. 39, 1301 (1977); R.H. Schindler et al., Phys. Rev. D21, 2716 (1980); H. Sadrozinski, Proc. of the XXth Int. Conf. on High Energy Physics, Madison, Wisconsin, p. 681 (1980).

(3) H.J. Lipkin, Argonne preprint ANL-HEP-PR-86-43, and contribution to this conference.

(4) S. Behrends, et al., "Does $\Upsilon(4S)$ Decay Exclusively to $B\overline{B}$?", contribution to this conference. Also in Cornell report CBX-86-44.

(5) P. Avery et al., Phys. Rev. Lett. 53, 1309 (1984).

(6) W. Bacino et al., Phys. Rev. Lett. 45, 329 (1980).

(7) G. Altarelli et al., Nucl. Phys. B208, 365 (1982).

(8) A. Chen et al., Phys. Rev. Lett. 52, 1084 (1984).

(9) C. Klopfenstein et al., Phys. Lett. 130B, 444 (1983).

(10) S. Behrends et al., "Limits on $Br(b\to u\ell\nu)/Br(b\to c\ell\nu)$ from Semileptonic Decay of B Mesons", contribution to this conference. Also in Cornell report CBX-86-44.

(11) Complete references on the models are given in Ref. 10.

(12) D. Bortoletto et al., "Inclusive B Decays into Charm", contribution to this conference. Also available as Cornell preprint CLNS 86/739.

(13) A. Brody et al., Phys. Rev. Lett. 48, 1070 (1982).

(14) M.S. Alam et al., "Branching Ratios of B Mesons to K^+, K^-, and $K^0\overline{K}^0$", contribution to this conference. Also in Cornell report CBX-86-44.

(15) A. Bean, et al., "Limits on $B^0\overline{B}^0$ Mixing", contribution to this conference. Also in Cornell report CBX-86-44.

INCLUSIVE AND EXCLUSIVE DECAYS OF THE B MESON

Abolhassan Jawahery

Physics Department, Syracuse University
Syracuse, N.Y. 13210

Representing The CLEO collaboration

Recent results on inclusive decays of the B meson into charm are discussed. Measured masses of the charged and neutral B mesons and the branching ratios into exclusive modes are presented. Limits are given on rare decays of the B meson primarily resulting from peguin diagrams.

1. INTRODUCTION

The CLEO detector is used to study weak decays of the B mesons produced in electron positron annihilations at the Cornell Electron Storage Ring (CESR). The data sample used for this study was collected in two separate runs during 1982 and 1985. It consist of an integrated luminosity of 117 pb^{-1} at the $\Upsilon(4S)$ and 53 pb^{-1} at a continuum energy below the resonance. Since the $\Upsilon(4S)$ decays exclusively to $B\bar{B}$ pairs this data sample contains about 260,000 B mesons.

In this report I will discuss the recent results on inclusive decays of the B meson to charm hadrons, reconstruction of exclusive final states of the B, and limits on rare decays of the B meson.

2. INCLUSIVE B MESON DECAYS INTO CHARM

In the standard electroweak theory the b quark can decay to a c or a u quark. Semileptonic B meson decays have given model dependent, but small limits on the ratio $b\to u/b\to c$.[1] A measurement of the total charm content of the B meson final states can also shed light on this important question. We have measured the production rates of F, ψ, D, D^{+} and D^{*+} mesons in B decay. Furthermore these processes provide information on the decay characteristics of the B meson.

a. B→FX

The decay $B\to FX$ can result from diagrams shown in Fig. 1, where the dominant contribution may arise from the spectator process $b\to cW^{-}\to c(\bar{c}s)$ (Fig. 1a). Accounting for the phase space suppression the spectator model predicts a rate of 15% for the diagram $b\to c(\bar{c}s)$.[2]

In the CLEO detector, F mesons are identified in the mode $F^{+}\to\phi\pi^{+}$. Phi candidates are identified in the mode $\phi\to K^{+}K^{-}$, by computing the invariant mass of oppositely charged particles whose dE/dX and time-of-flight measurements are consistent with a kaon. In order to further suppress the combinatorial background

2430386-006

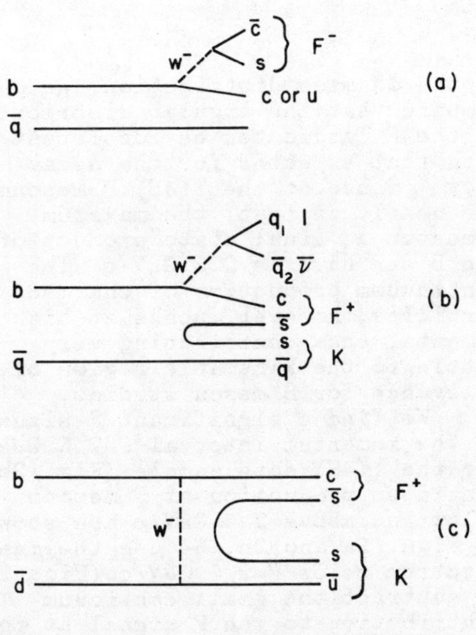

Fig. 1: Possible diagrams for the decay B→FX.

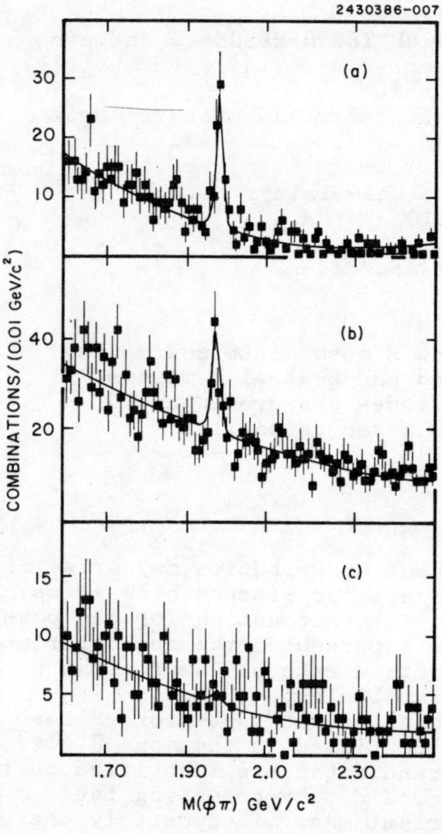

Fig. 2: $\phi\pi$ mass distributions (a) for P>2.5 GeV/c for the combined $\Upsilon(4S)$ and continuum data sample (b) For P<2.5 GeV/c for the $\Upsilon(4S)$ data (c) For P<2.5 GeV/c for the continuum data.

in the $\phi\pi$ mass distribution, we require that the angular distribution of the F candidates be consistent with that expected for the decay $F \to \phi\pi$.[2] Since at the $\Upsilon(4S)$ B mesons are nearly at rest, the maximum momentum a final state product of the B can have is 2.5 GeV/c. The continuumm production of charmed particles, however, peaks at high momenta, thus contributing very little to the kinematic region of relevance for B meson studies.

We find a significant F signal in the momentum interval P<2.5 GeV/c for the $\Upsilon(4S)$ data sample (Fig. 2b). Continuum production of F mesons below and above 2.5 GeV/c are shown in Figs. 2a and 2c. We use the mass spectrum below P=2.5 GeV/c (Fig. 2c) to subtract the small continuum contribution to the F signal at the $\Upsilon(4S)$.

The observed F signal from B

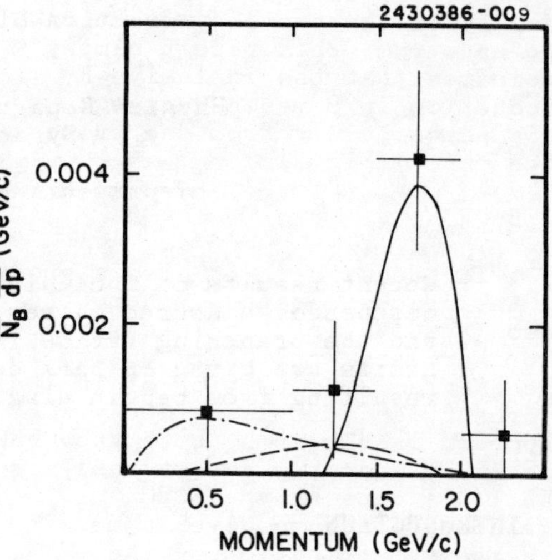

Fig. 3: Momentum distribution of the F yield per B meson decay. The solid curve is the Doppler-broadened F momentum spectrum for two body decays $B \to FD$, $B \to FD^*$, $B \to F^*D$ and $B \to F^*D^*$. The dashed and dash-dotted curves are respectively the Monte Carlo simulated F momentum distribution for decays $B \to FD\pi$ and $B \to FKW^- \to FK(q_1 q_2)$.

meson decays gives the product branching ratio $B(B \to FX) \cdot B(F^+ \to \phi\pi^+) = 0.0038 \pm 0.0010$. To determine the branching ratio $B(B \to FX)$ one requires the value of $B(F \to \phi\pi)$ which is not yet measured. However, using $B(F^+ \to \phi\pi^+) = 0.035$, which is obtained by comparing the theoretical value of the width $\Gamma(F \to \phi\pi)$ and the measured F meson life-time, we find the branching fraction $B(B \to FX) \sim 11\%$.[2] This is consistent with the theoretical estimate of this process by M. Suzuki.[3]

The momentum distribution of F mesons from B decay is shown in Fig. 3. The spectrum peaks in the interval 1.5 to 2.0 GeV/c which corresponds to the kinematic region for two body decays $B \to FD$, FD^*, F^*D and F^*D^*, indicating that a large fraction of the decay $B \to FX$ results from these two body decays of the B. Fitting the spectrum to functions representing, two body decays (solid), phase space decay $B \to FD\pi$ (dashed) and the process $B \to FKW^-$ (Fig. 1b) gives a value of 64±22% for the two body fraction of the spectrum. This result simply indicates that the inclusive F production in B meson decays originates mainly from the process $b \to cW^- \to c(\bar{c}s)$ (Fig. 1a).

b. $B \to \psi X$

The decay $B \to \psi X$ is also a product of the process $b \to c(\bar{c}s)$, where the ψ meson results from an interference of the charm and anti-charm quarks in the diagram. Observation of inclusive ψ production in B meson decay was reported last year[4] by both the CLEO and ARGUS groups. Recent analysis of the $\Upsilon(4S)$ data by the CLEO collaboration have given a more precise determination of the branching ratio $B(B \to \psi X) = 1.09 \pm 0.16 \pm 0.21\%$.

c. $B \to D^0$, D^+, D^{*+}

Up to this point we have discused the measurements of the less frequently produced charm hadrons. However, the most copiously produced charm hadrons in B decays are D^0, D^+ and D^* mesons. These mesons could result from both the c quark from $b \to c$ coupling and the $W^- \to \bar{c}s$ conversion.

In the CLEO detector we search for D mesons using the modes, $D^0 \to K^-\pi^+$, $D^+ \to K^-\pi^+\pi^+$ and $D^{*+} \to D^0\pi^+$. Identification of charged kaons is achieved using dE/dX and time-of-flight measurements. Since the combinatorial backgrounds in the momentum region below 2.5 GeV/c is large, we require additional criteria on the event shape and angular distribution of the D decays. For the $D^0 \to K^-\pi^+$ decay, we restrict the range of $\cos(\theta^*)$, where θ^* is the helicity angle. In order to suppress the continuum contribution we use the Fox-Wolfram event shape parameter, R_2, to disriminate against Jet-like events. This causes a 25% loss of $B\bar{B}$ events while rejecting about 60% of the continuum events. For the mode $D^+ \to K^-\pi^+\pi^+$, we used only the 1985 $\Upsilon(4S)$ data which allows the identification of charged kaons in the inner drift chamber. We employ the standard mass difference technique to identify $D^{*+} \to D^0\pi^+$ decays. Fig. 4 shows the resulting mass plots for the kinematic region of relevance to the B meson decay at the $\Upsilon(4S)$, $P<2.5$ GeV/c. The continuum contributions are small and are subtracted directly from the observed D signals.

The directly measured quantities resulting from this analysis are the product branching fractions:

$B(B \to D^0X) \cdot B(D^0 \to K\pi^+)$
$= 0.0210 \pm 0.0015 \pm 0.0021$
$B(B \to D^+X) \cdot B(D^+ \to K\pi^+\pi^+)$
$= 0.0190 \pm 0.0014 \pm 0.0020$
$B(B \to D^{*+}X) \cdot B(D^{*+} \to \pi^+D^0) \cdot B(D^0 \to K\pi^+)$
$= 0.0073 \pm 0.0012 \pm 0.0007$.

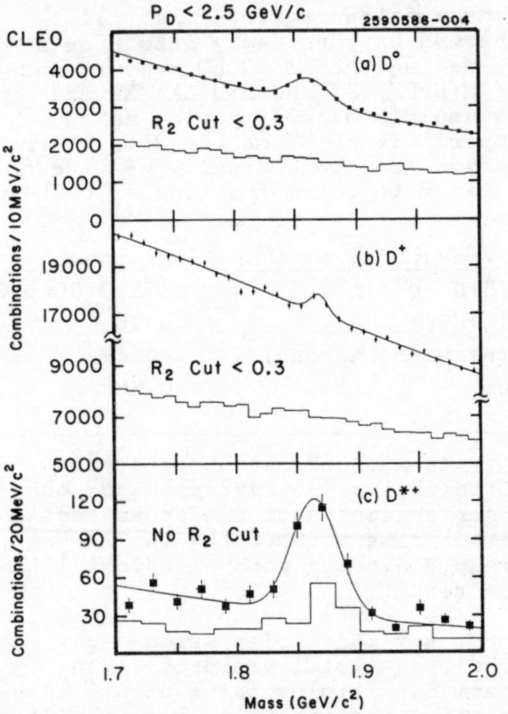

Fig 4: Mass distributions for D candidates for momenta below 2.5 GeV/c. The 4S data is shown in solid points and the continuum data in histograms.

In order to determine the B to D fractions we have used the recent Mark III measurements of the D branching ratios,
$B(D^0 \to K^-\pi^+) = 5.6 \pm 0.4 \pm 0.3\%$ and
$B(D^+ \to K\pi^+\pi^+) = 11.6 \pm 1.4 \pm 0.7\%$ and the Mark II measurement of the branching ratio $B(D^{*+} \to \pi^+D^0) = 0.60^{+0.08}_{-0.15}$. This gives:
$B(B \to D^0X) = 0.39 \pm 0.05 \pm 0.04$,
$B(B \to D^+X) = 0.17 \pm 0.04 \pm 0.04$ and
$B(B \to D^{*+}X) = 0.22 \pm 0.044^{+0.07}_{-0.04}$.

Since all D^*'s decay to D^0 or D^+, the sum $B(B\to D^0X)+B(B\to D^+X)=0.56\pm0.06\pm0.06$ represents the total yield of D mesons in B decay.

d. TOTAL CHARM YIELD IN B DECAY

The measured B to charm fractions are summarized in Table I. The decays B→FX and B→ψX result from the diagram $b\to c(\bar{c}s)$. We therefore, attribute an upperbound of 15% to their summed rate which is consistent with the direct measurement of the rates of these processes (See (a) and (b)). Since the baryon conservation requires that all charmed baryons decay into P or Λ or N, we use the old CLEO measurements of B(B→PX)=3% and B(B→ΛX)=3% and assume B(B→NX)=3%, giving an upperlimit of 9% on the B to charmed baryon fraction. Finally we find a total B to charm fraction of 0.80.

TABLE I

Measured B to charm Fractions

$B(B\to D^0+D^+)$	0.56±0.06±0.06
$B(B\to F+\psi)$	~0.15
B(B→charm baryons)	~<0.09
Sum	~0.80

As a check on our detection efficiencies, we have compared our observed continuum D^{*+} cross section with the measurement by the ARGUS group,[8] finding good agreement with our result.

For small b→u coupling, the spectator model of B meson decay predicts a total yield of ~1.15 charmed particles per B decay. This accounts for the 0.15 charm resulting from $b\to c(\bar{c}s)$. Comparing this with the measured rate, gives a charm deficit of 35%.

Within the framework of the spectaor model description of the B meson decays, the measured charm deficit indicates a substantial b→u fraction, in contradictions with the results from analysis of the semileptonic decays. However, it may be argued that non-spectator processes such as W-annihilation, final state interactions or penguin diagrams could enhance the fraction of charmless final states in hadronic decays of the B meson. An evidence against the latter hypothesis is the measured D^0 fraction in semileptonic decays of the B, $D^0/(B\to Xl\nu)=0.35\pm0.09$ which is in agreement with the total D^0 fraction in B decay. We have also investigated the possibility that the Υ(4S) decays partly into bottomless final states.[1] This gives an upperlimit of 13.8% on the fraction of such decay modes. Finally, the Mark III D branching ratios may be too large. Another evidence for this possibility is the relatively small ratio of the measured continuum D cross section to the total expected charm cross section which is found to be less than 50% by both the CLEO and HRS groups.[9]

Unfortunatly, lacking an independent check on D branching ratios and reliable calculations for the non-spectator effects, we can not at this point determine which of the above metntioned hypotheses is the main cause of the missing charm effect in B decay.

2. RECONSTRUCTION OF EXCLUSIVE B MESON DECAYS

The task of exclusive reconstruction of the B meson decays is made difficult by the large charged particle multiplicities of the final states and the undetected neutral paricles in the decays. Nevertheless, a large sample of $B\bar{B}$ events is expected to provide a small number of fully reconstructed B decays to be used for the measurements of the masses of the charged and neutral B mesons. We have used our sample of $B\bar{B}$ events to reconstruct several low multiplicity decays of the B mesons which are listed in Figure 5. In this analysis charm hadrons are identified using the decay modes $D^0\to K^-\pi^+$, or $K^-\pi^+\pi^+\pi^-$, $D^+\to K^-\pi^+\pi^+$ and $\psi\to\mu^+\mu^+$ or e^+e^-. We compute the invariant of the B candicates using

$$M^2=E^2_{beam}-P^2,$$

where E_{beam} is beam energy and P is the momentum of the B candidate. This gives a mass resolution of 5 MeV which is dominated by the beam energy spread of 3.2 MeV.

The invariant mass distribution of the B candidates (Fig. 5) shows a clear peak in the vicinity of M=5280 MeV. A detailed study of the background effects have shown that the background level in the B mass region is 83% of the level in the lower mass interval. Subtracting the background and correcting for the efficiencies we find the branching ratios (%):

$B(B^- \to D^0 \pi^-) = 0.38 \pm 0.14 \pm 0.11$
$B(\bar{B}^0 \to D^0 \pi^+ \pi^-) = 1.6 \pm 0.9 \pm 0.6$
$B(\bar{B}^0 \to D^+ \pi^+) = 0.14 \pm 0.19 \pm 0.05$
$B(B^- \to D^+ \pi^- \pi^-) = 0.89 \pm 0.51 \pm 0.30$
$B(\bar{B}^0 \to D^{*+} \pi^-) = 0.35 \pm 0.14 \pm 0.11$
$B(B^- \to D^{*+} \pi^- \pi^-) = 0.31 \pm 0.17 \pm 0.11$
$B(B^- \to \psi K^-) = 0.09 \pm 0.06 \pm 0.02$
$B(\bar{B}^0 \to \psi \bar{K}^{*0}) = 0.41 \pm 0.19 \pm 0.03.$

These branching ratios are smaller than the previously reported values which were obtained using 1982 data sample alone.[10] The diferernce is due partly to the larger D branching ratios (Mark III) used and a lower background level in the 1985 data which dominates the present analysis. Averaging the events in the mass interval between 5265 to 5290 we find the masses of the neutral and charged B mesons $M(\bar{B}^0) = 5281 \pm 0.9 \pm 3.0$ MeV and $M(B^-) = 5277.9 \pm 1.1 \pm 3.0$ MeV. This gives a mass difference of $\Delta M = 3.1 \pm 1.4 \pm 2.0$ MeV in agreement with theoretical predictions.

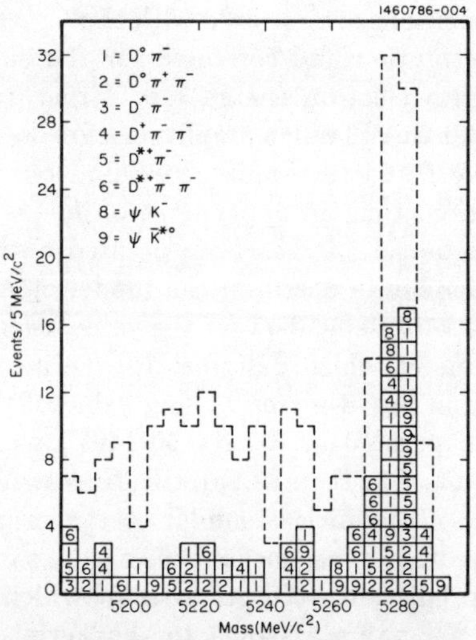

Fig. 5: Beam constrained invariant mass distribution of B candidates

3. LIMITS ON THE RARE EXCLUSIVE DECAYS OF THE B MESON

We have searched for several rare exclusive decays of the B, primarily resulting from the penguin diagrams.[11] Possible enhancment of penguin effects in B decays have been linked to the explanation of the $\Delta I = 1/2$ rule in Kaon decays. A summary of the upperlimits on the B branching ratios into these modes is given in Table II. Clearly, the limits are higher than the theoretical predictions ($\sim 10^{-4}$). However, a comparison with perturbative QCD calculations rules against a significant enhancement of the penguin effects.

TABLE II

limits on rare exclusive B decays

mode	90% c.l. Upperlimit
$B(B \to K^+ \pi^+)$	3.2×10^{-4}
$B(B \to \phi K^0)$	13.0×10^{-4}
$B(B \to \phi K^+)$	2.1×10^{-4}

4. CONCLUSION

We have measured the inclusive production rates of F, ψ, D^0, D^+ and D^{*+} in B meson decays. The momentum spectrum of F's from B indicates that a substantial fraction of the decay $B \to FX$ results from two body double charm decays ($B \to FD$). The measured total charm fractions accounts for 0.80 charm per B decay compared with 1.15 charm per B expected from the spectator model with 100% $b \to c$ coupling. Masses of the neutral and charged B's are measured using exclusive decays of the B. Upperlimits are set on several rare exclusive decays of the B mesons.

I wish to gratefully acknowledge the effort of the CESR staff. This work was funded by both the National Science Foundation and the U.S. Department of Energy.

REFERENCES

1. See R. Poling these proceedings
2. P.Haas et al., Phys. Rev. Lett <u>56</u>, 2781(1986)
3. M. Suzuki, Phys. Lett. <u>142B</u>, 207 (1884); Phys. Rev. <u>D31</u>, 1158(1985)
4. P. Haas et al., Phys. Rev. Lett. <u>55</u>, 1248 (1985); H. Albrecht et al., Phys. Lett. <u>162B</u>, 295 (1985).
5. G. C. Fox and S. Wolfram, Phys. Rev. Lett. <u>56</u>, 2140 (1986)
6. R. M. Baltrusaitis et al., Phys. Rev. Lett. <u>56</u>, 2140 (1986)
7. G. Goldhaber et al., Phys. Lett. <u>69B</u>, 503 (1977)
8. H. Albrecht et al., Phys. Lett <u>150B</u>, 235 (1984)
9. M. Derrick et al., Phys. Rev. Lett. <u>53</u>, 1971 (1984)
10. R. Giles et al., Phys. Rev. <u>D30</u>, 2279 (1984)
11. G. Eilam and J. P. Leveille, Phys. Rev. Lett. <u>44</u>, 1648 (1980); M. B. Gavela et al., Phys. Lett. <u>154B</u>, 425 (1985).

CRYSTAL BALL RESULTS ON INCLUSIVE ELECTRON SPECTRUM IN $\Upsilon(4S)$ DECAYS

Tomasz Skwarnicki
(Representing the Crystal Ball Collaboration)

Deutsches Elektronen-Synchrotron DESY, Hamburg, Germany

The inclusive electron spectrum in $\Upsilon(4S)$ decays has been studied with the Crystal Ball detector at the e^+e^- storage ring DORIS-II.
Preliminary upper limits on $\mathrm{BR}(B \to e\nu X_u)/\mathrm{BR}(B \to e\nu X_c)$ are obtained.

The inclusive energy distribution of electrons from $\Upsilon(4S)$ resonance decays has been measured using the Crystal Ball detector. The data have been collected at the e^+e^- storage ring DORIS-II and correspond to about 62 pb^{-1} of the integrated luminosity at the $\Upsilon(4S)$ energy, and to about 22 pb^{-1} at the continuum below the $\Upsilon(4S)$. The $\Upsilon(4S)$ is believed to decay 100 % to $\mathrm{B\bar{B}}$ pairs. The B mesons may then decay semileptonically via weak transition of their b quarks to c or u quarks, $B \to e\nu X_q$, $q = c$ or u, where X_q denotes hadronic state containing quark q. The $b \to e\nu c$ decay is expected to dominate. The electron spectrum from the $b \to e\nu u$ transition extends to higher energies because of the lower u quark mass. Thus, the shape of the electron spectrum at the endpoint can reveal $\mathrm{BR}(B \to e\nu X_u)/\mathrm{BR}(B \to e\nu X_c)$ ratio, and yield a measurement of the ratio of the Kobayashi-Maskawa matrix elements, $|V_{bu}|/|V_{bc}|$.

The Crystal Ball is a non-magnetic detector[1]. Electrons are detected as charged tracks in the proportional tube chambers and as electromagnetic showers in the NaI-calorimeter. The high segmentation of the calorimeter allows efficient discrimination between high energy electrons and charged hadrons by cuts on the lateral energy distribution in the NaI crystals. The observed energy distribution of electrons in multi-hadronic events at the $\Upsilon(4S)$ energy is shown in Fig.1. The solid line indicates the non-resonance background estimated by fitting a smooth curve to the continuum data and scaling by the ratio of on- and off-resonance luminosities. In the further analysis the continuum subtracted distribution is used.

Interpretation of the observed spectrum is model dependent. We use the model of Altarelli *et al.*[2] to fit the part of the spectrum above 1.5 GeV, which has low non-electron background (< 5 %) and negligible contribution from semileptonic decays of D-mesons produced in B-decays. The theoretical predictions for the shapes of the energy distributions of the inclusive electrons in $b \to c$ and $b \to u$ transitions are smeared with the Crystal Ball energy resolution and corrected for the efficiency change with electron energy (10 % rise from 1.5 GeV to 3.0 GeV). When amplitudes for both transitions are fitted to the data, we find only $b \to c$ contribution, thus an upper limit on $b \to u$ transitions can be set. It has been pointed out[3] that such upper limits based on the model of Altarelli *et al.* are very dependent on the value assigned to the parameter which describes the Fermi motion of quarks in the B-meson. Using value of this parameter suggested in Ref.2 (150 MeV) we obtain upper limit, $\mathrm{BR}(B \to e\nu X_u)/\mathrm{BR}(B \to e\nu X_c) < 2.5$ % (90 % CL), which is similar to the results obtained by other experiments[4] using the same theoretical predictions. To get a result which is free from *ad hoc* value assigned to the Fermi-motion parameter, we fit the spectrum for various fixed values of $\mathrm{BR}(B \to e\nu X_u)/\mathrm{BR}(B \to e\nu X_c)$ leaving the amplitude for the $b \to c$ transitions and the Fermi-momentum as free fit parameters. For each fit the likelihood value is calculated. The fit with the maximum likelihood is displayed in Fig.2. The value of the Fermi motion parameter[2] preferred by our data, 260 ± 50 MeV, is in agreement with the CLEO result[3]. From the likelihood distribution over all allowed $\mathrm{BR}(B \to e\nu X_u)/\mathrm{BR}(B \to e\nu X_c)$ we find,

$$\frac{BR(B \to e\nu X_u)}{BR(B \to e\nu X_c)} < 5.3\ \% \quad (90\ \%\ CL)\ \text{(preliminary)},$$

which translates[5] into $|V_{bu}|/|V_{bc}| < 16$ %.

Alternative theoretical approach to the semileptonic B-decays has been recently proposed by Grinstein *et al.*[6]. Predictions of this model for the $b \to u$ transitions are valid only at the endpoint. Therefore, we fit the $b \to c$ amplitude assuming negligible $b \to u$ contribution (see Fig.3) and then an upper limit on the $\mathrm{BR}(B \to e\nu X_u)/\mathrm{BR}(B \to e\nu X_c)$ is obtaind by fitting the $b \to u$ amplitude above the kinematical limit for the $b \to c$ transitions (2.4 GeV). The result,

$$\frac{BR(B \to e\nu X_u)}{BR(B \to e\nu X_c)} < 20\ \% \quad (90\ \%\ CL)\ \text{(preliminary)},$$

is weaker than that obtained with the previous model, due to the softer $b \to u$ spectrum predicted at the endpoint and due to the smaller experimental statistics used in the fit. The analogous fit method applied to the model of Altarelli *et al.* gives upper limits of less than 9 % (90 % CL) for the Fermi motion parameter less than 350 MeV.

REFERENCES

[1] M. Oreglia *et al.*, Phys. Rev. **D25** (1982) 2259, M. Oreglia , SLAC-236 (1980), Ph. D. Thesis, Stanford University.

[2] G. Altarelli *et al.*, Nucl. Phys. **B208** (1982) 365.

[3] E. H. Thorndike, "Weak Decays of Heavy Fermions", Review Talk at Int. Symp. on Lepton and Photon Interactions, Kyoto, 1985.

[4] C. Klopfenstein *et al.*(CUSB), Phys. Lett. **130B** (1983) 444, A. Chen *et al.*(CLEO), Phys. Rev. Lett. **52** (1984) 1084, H. Schröder (ARGUS), the Int. Conf. on Hadron Spectroscopy, College Park, Maryland, ed. by S. Oneda (AIP, New York, 1985),
see also CLEO and ARGUS results reported at this conference.

[5] see footnote 18 in A. Chen *et al.*, Phys. Rev. Lett. **52** (1984) 1084.

[6] B. Grinstein, M. B. Wise, N. Isgur, Phys. Rev. Lett. **56** (1986) 298 and CALT-68-1311.

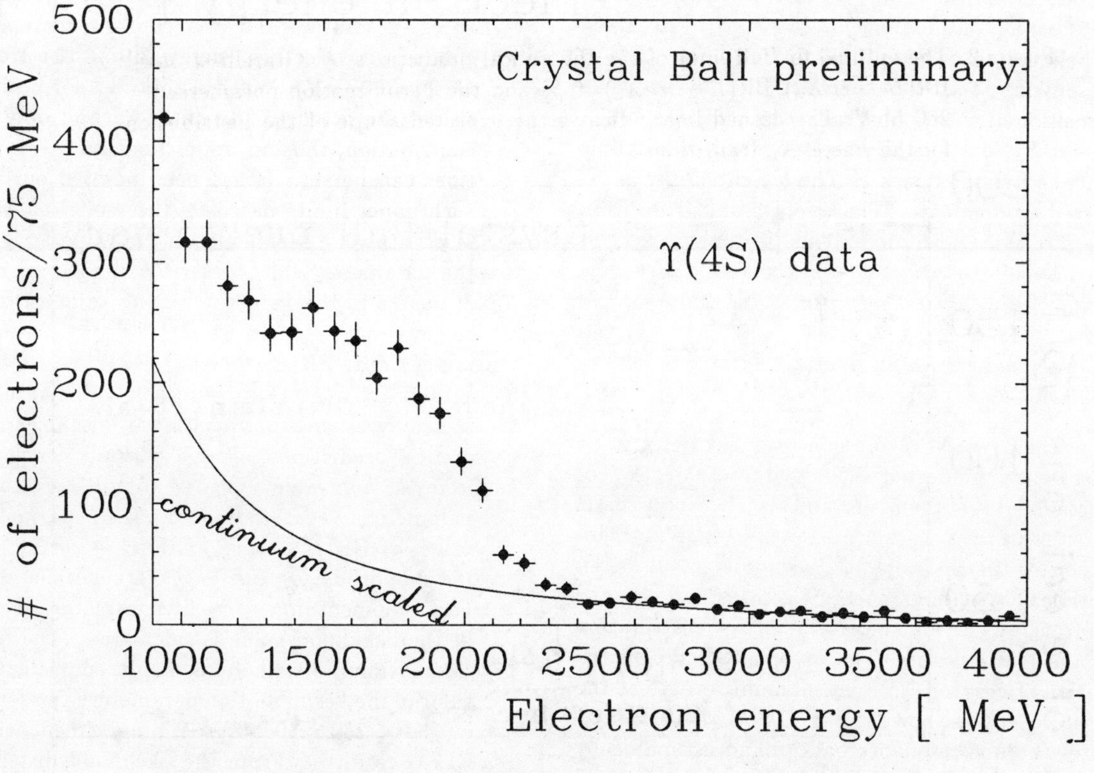

Figure 1: The inclusive electron spectrum on the Υ(4S) resonance. The continuum contribution is indicated.

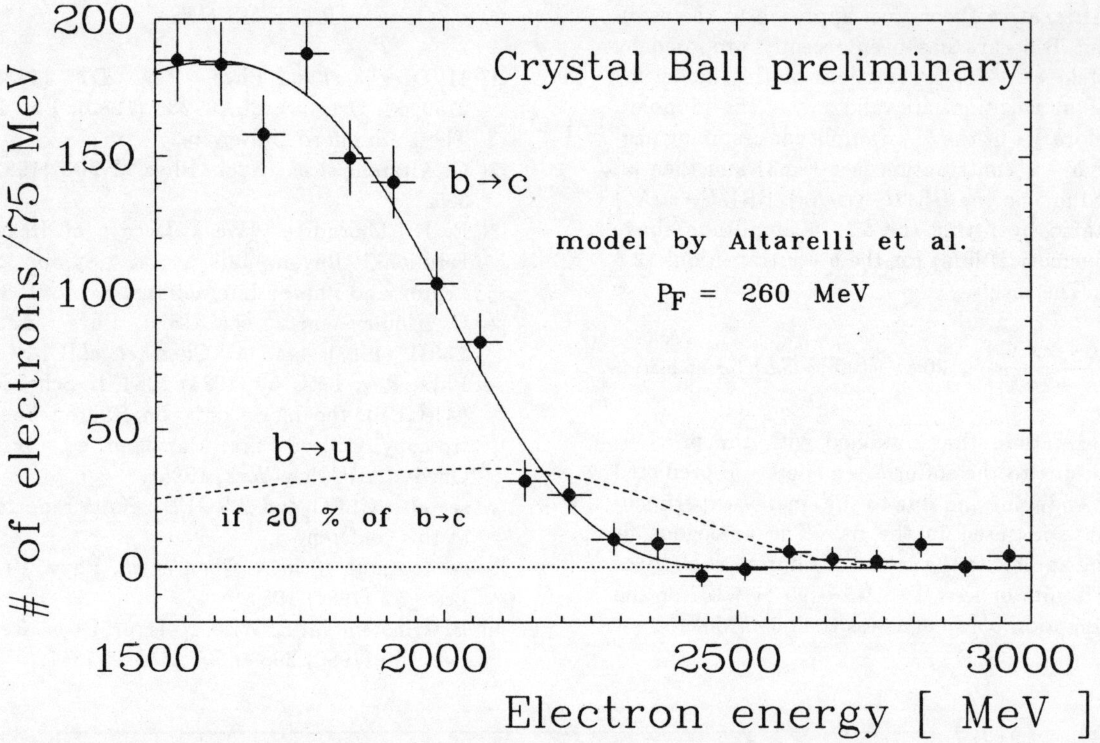

Figure 2: The optimal fit (full line) of the theoretical predictions by Altarelli *et al.*[2]: $\mathrm{BR}(B \to e\nu X_u)/\mathrm{BR}(B \to e\nu X_c)$=0 % and the Fermi-motion parameter[2]= 260 MeV. The dashed line indicates the expected shape of the distribution for the $B \to e\nu X_u$ transitions.

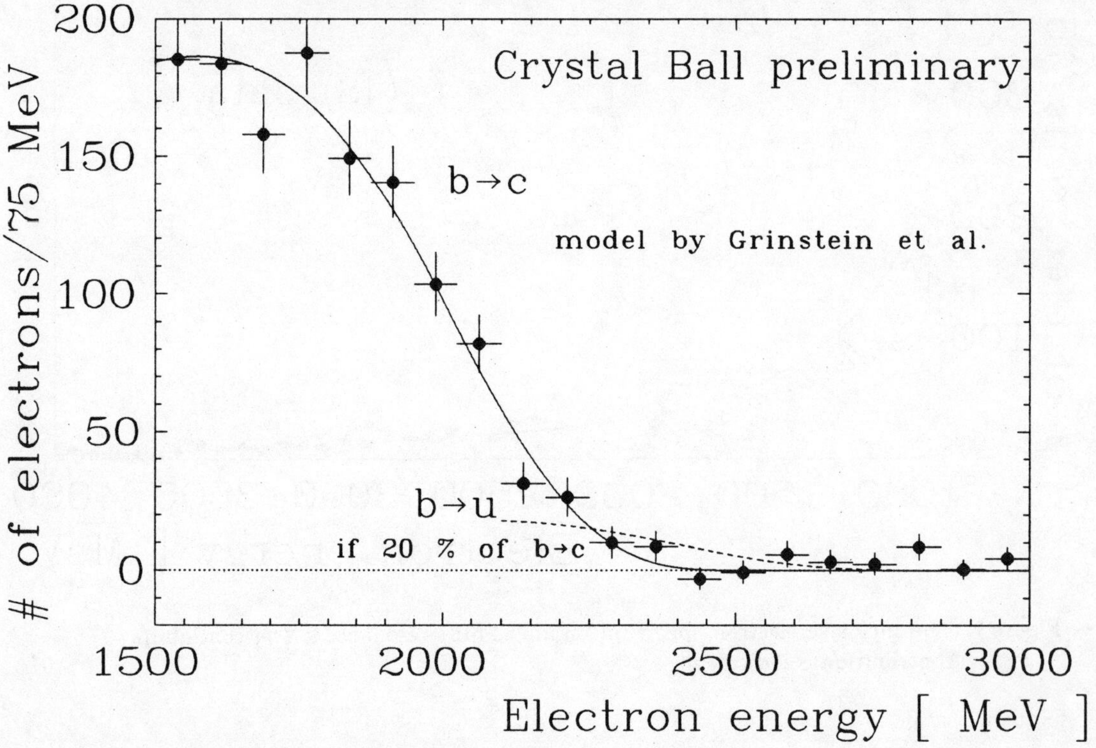

Figure 3: Fit of the theoretical predictions by Grinstein *et al.*[6].

B-MESON RESULTS FROM ARGUS

K.R. Schubert

Institut für Hochenergiephysik, Universität Heidelberg

D-6900 Heidelberg, Germany

The ARGUS collaboration, working at the e^+e^- storage ring DORIS-II, presents results on B meson decays into $D^* n\pi$ (n = 1,2,3), $J/\Psi X$, $\Psi' X$, and $J/\Psi K\pi$ finalstates, obtained from an integrated luminosity of 59/pb on the $\Upsilon(4S)$ resonance. Upper limits on the $B^o_d - \overline{B}^o_d$ oscillation rate and on V_{ub}/V_{cb} are also presented.

The ARGUS collaboration [1] has analyzed so far a sample of 100 000 B mesons produced in e^+e^- annihilation at the resonance energy of the $\Upsilon(4S)$ meson. The data sample has been obtained with the help of the ARGUS detector at the e^+e^- storage ring DORIS-II at DESY. With average luminosities of 15/pb/month, the group collected 59.4 events /pb at the $\Upsilon(4S)$ resonance energy and 21.5/pb in the nearby continuum; most of the data were taken during 1985. A short description of the detector and the trigger conditions for multihadron events may be found in ref. [2].

1. FULLY RECONSTRUCTED DECAYS $B \to D^* n\pi$

Five decay channels of the mesons $B^+ = \overline{b}u$ and $B^o = \overline{b}d$,

$B^+ \to D^{*-} \pi^+ \pi^+$, $B^+ \to D^{*-} \pi^+ \pi^+ \pi^o$,

$B^o \to D^{*-} \pi^+$, $B^o \to D^{*-} \pi^+ \pi^o$, $B^o \to D^{*-} \pi^+ \pi^+ \pi^-$

have been fully reconstructed using the $D^{*-} \to \overline{D}^o \pi^-$ decay mode with $\overline{D}^o \to K^+\pi^-$, $K^o_S \pi^+\pi^-$, $K^+\pi^-\pi^o$, and $K^+\pi^-\pi^-\pi^+$. Here and in the following, all charged states are meant to include their charge conjugate state. D* channels have the advantage that the clean cut on the mass difference m(D*)-m(D) reduces the combinatorial background considerably.

Charged particles are identified using their specific ionization in the drift chamber and their time of flight. For π^o mesons, two photons with an energy of at least 40 MeV are combined. Particle combinations are fitted with mass constraints on the intermediate states, i.e. on π^o, K^o_S, D^o, and D^{*+}. In the search for B condidates, two requirements have to be fulfilled: The total sum of all χ^2 values from particle identification and kinematical fits must have a probability larger than 1%, and the energy of the B candidate must be equal to half of the $\Upsilon(4S)$ mass within three standard deviations. All remaining candidates are fitted with a beam energy constraint, i.e. requiring $E(B) = m(\Upsilon 4S)/2$. The resulting mass spectrum is shown in fig. 1. Events with more than one decay candidate are entered only once on the basis of the lowest χ^2 value.

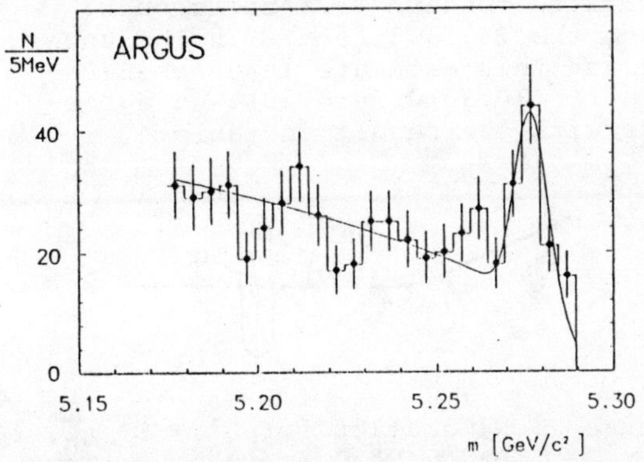

Figure 1: Mass Distribution of B Meson Candidates with $B \to D^* n\pi$

The background shape has been evaluated from continuum data, event mixing and the search for "B^{++}" candidates. Fitting the expected background shape and a Gaussian with σ_m = 4.5 MeV/c² to the data in fig. 1 yields a signal of 71 ± 11 B decays. If fitted separately to B^+ and B^o candidates and using a mass scale with m $(\Upsilon 4S)$ = 10577 MeV/c² [3], one finds 40 ± 8 B^o mesons with m (B^o) = (5278.2±1.0±3.0) MeV/c² and 32 ± 7 B^+ mesons with $m(B^+)$ = (5275.8±1.3±3.0) MeV/c².

In order to determine branching fractions we assume 55% B^+B^- and 45% B^oB^o decays of the $\Upsilon(4S)$. D^o branching fractions are taken from MARK-III [4]. Results are shown in table 1. The branching fractions for $B^\circ \to D^{*-}\pi^+$ and $B^+ \to D^{*-}\pi^+\pi^+$ are substantially lower than those of CLEO 1984 [5] but agree well with those of CLEO 1986 [6]. The $B^\circ \to D^{*-}\pi^+$ partial rate agrees well with a recent theoretical result [7].

2. PARTIALLY RECONSTRUCTED DECAYS $B^\circ \to D^{*-}\pi^+$

The hard-soft-pion method of CLEO [8] has been used with all $\pi^+\pi^-$ pairs in $\Upsilon(4S)$ events fulfilling $p(\pi_h) \in [2.0, 2.5]$ GeV/c, $p(\pi_s) < 0.25$ GeV/c, and $\cos\theta(\pi_h\pi_s) < -0.9$. The kinematics in Fig. 2 allows to calculate a pseudo-mass $M^*(\pi_h\pi_s D^\circ)$ assuming $\alpha = 0$. The distribution is shown in fig. 3, where the shaded histogram is the estimated background obtained by $\pi^+\pi^-$ pairs with momentum vectors of soft pions reflected, $\vec{p}(\pi_s) \to -\vec{p}(\pi_s)$. There is a signal of 100 ± 21 events which has, however, the wrong momentum spectrum: The bin of $p(\pi_h)$ from 2 to 2.25 GeV/c has 3x more events than that from 2.25 to 2.5, but both should be equally populated. The background from decays like $B^\circ \to D^{*-}l^+\nu$, $D^{*-}\rho^+$, $D^{*\circ}(2420)\pi^+$ is mainly in the lower $P(\pi_h)$ bin, the upper is estimated to have ≅12% background. Using the 24 ± 13 events in the upper bin and this estimate leads to $B(B^\circ \to D^{*-}\pi^+) = (0.35\pm0.20\pm0.20)\%$ in agreement with the result in table 1.

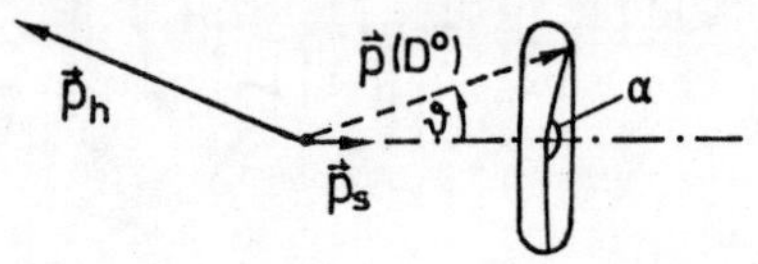

Figure 2: Kinematics for $B^\circ \to D^{*-}\pi_h^+$, $D^{*-} \to D^\circ\pi_s^-$

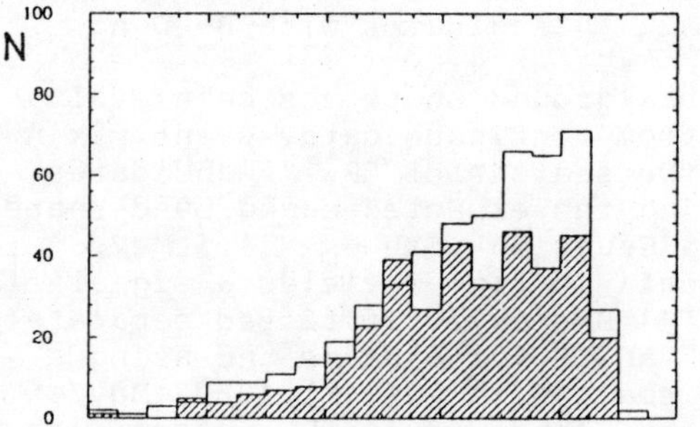

Figure 3: Pseudomass distribution; the shaded area is non-B-decay background

Table 1:

$B(B^\circ \to D^{*-}\pi^+) = (0.25\pm0.15\pm0.15)\%$
$B(B^\circ \to D^{*-}\pi^+\pi^\circ) = (1.1\pm0.6\pm0.6)\%$
$B(B^\circ \to D^{*-}\pi^+\pi^+\pi^-) = (2.4\pm0.7\pm1.1)\%$
$B(B^+ \to D^{*-}\pi^+\pi^+) = (0.4\pm0.2\pm0.2)\%$
$B(B^+ \to D^{*-}\pi^+\pi^+\pi^\circ) = (3.5\pm1.1\pm2.1)\%$

3. INCLUSIVE DECAYS $B \to J/\Psi X$ AND $\Psi' X$

J/Ψ mesons are reconstructed by their decays into e^+e^- and $\mu^+\mu^-$. Using identified electrons and muons with $p > 0.85$ GeV/c, $|\cos\theta| < 0.9$, and a pair momentum $p(l^+l^-) < 2.0$ GeV/c, one obtains the mass spectrum in fig. 4 with a J/Ψ signal of 65 ± 12 events. The ee and μμ signals are in agreement with eachother, and using $B(J/\Psi \to e^+e^-) = B(J/\Psi \to \mu^+\mu^-) = (6.9\pm0.9)\%$ [9], one obtains $B(B \to J/\Psi + X) = (1.15\pm0.24\pm0.21)\%$. The systematic error is dominated by the common 13% error on B_{ee} and $B_{\mu\mu}$ of the J/Ψ; a new determination of this branching fraction would be desirable.

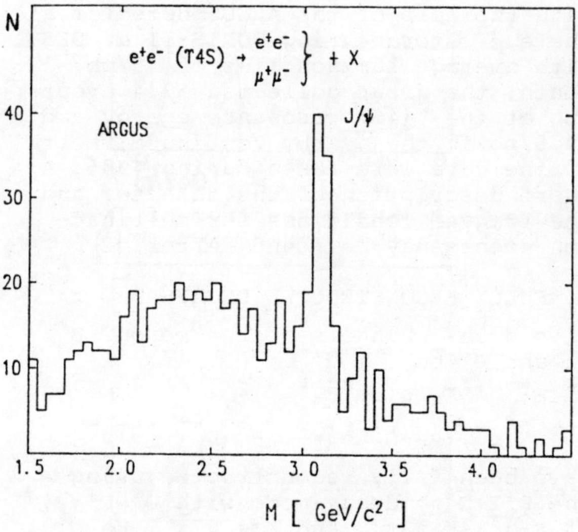

Figure 4: l^+l^- Mass Spectrum

By evaluating the J/Ψ signal in seven momentum bins, one obtains the J/Ψ momentum spectrum shown in fig. 5, where the solid curve is the expectation for 65 decays $B \to J/\Psi K$ and $J/\Psi K^*$. The observed spectrum is softer and could be fed by $B \to \Psi' X$ and cascades from the Ψ' to the J/Ψ. From the upper three $P(J/\Psi)$ bins, one concludes $B(B \to J/\Psi + K \text{ or } K^*) < 0.6\%$ with 90% CL. Two methods have been used to search for $B \to \Psi' X$ decays, by reconstructing Ψ' mesons in their $J/\Psi\ \pi^+\pi^-$ and in their l^+l^- mode. Fig. 6 shows the result of the first method selecting ππ pairs with m $(\pi\pi)$ > 400 MeV/c²; there is an indication for 5±3 Ψ'. The background in fig. 5 is too high to see a $\Psi' \to l^+l^-$ signal, but an indication of 3.5 ±2 events appears if one requires $p(l^+l^-) < 1.6$ GeV/c which is the kinematical for $B \to \Psi'X$ decays, and $H_2 < 0.3$ where H_2 is the second Fox-Wolfram moment of the event. Combining the two observations leads to B $(B \to \Psi'X)$ = (0.50±0.23)%. Using $B(\Psi' \to J/\Psi\ X)$ = (55±7)% one concludes that B $(B \to J/\Psi\ X$ with J/Ψ not from $\Psi')$ = (0.87±0.27)%. The observed amount of decays $B \to \Psi'X$ does not fully explain the soft part of the spectrum in fig. 5; there must be also decays $B \to J/\Psi\ X_M$ with states X_M of higher mass than m(K*).

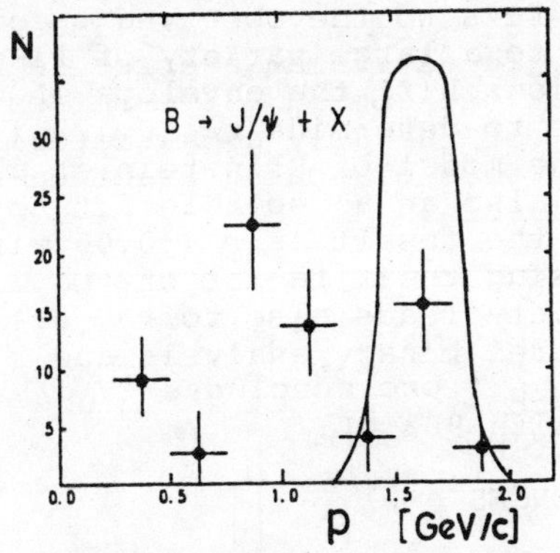

Figure 5: Momentum spectrum of J/Ψ in $B \to J/\Psi X$ decays

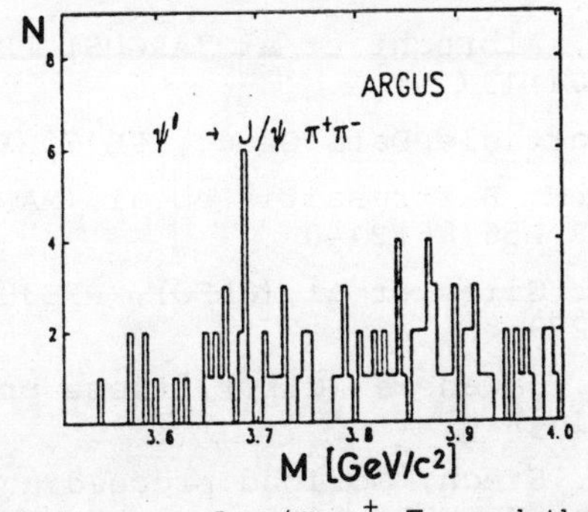

Figure 6: Mass of $J/\Psi\ \pi^+\pi^-$ combibations in $B \to J/\Psi\ X$ decays

4. EXCLUSIVE DECAYS $B \to J/\Psi, \Psi'+K, K^*$

All Ψ candidates in the sample with $H_2 < 0.3$ are grouped with $Kn\pi$ combinations, where $\Psi = J/\Psi$ or Ψ', $K = K^\pm$ or K°_s, n = 0 or 1, $\pi = \pi^\pm$. The intermediate states K°_s, J/Ψ, and Ψ' are mass-constrained and all candidates with E = m(Υ4S)/2 ±120 MeV/c² are refitted with the beam-energy constraint E = m (Υ4S)/2. The sample with n = 0 contains no $B^\circ \to \Psi\ K^\circ_s$ event and only

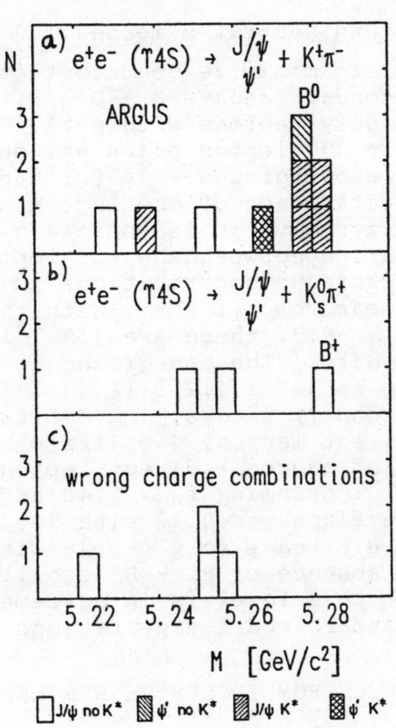

Figure 7: Beam-energy constrained mass of $B \to \Psi K\pi$ candidates

one $B^+ \to \Psi K^+$ candidate with $\Psi = J/\Psi$. A limit of $B(B^+ \to J/\Psi K^+) < 0.20\%$ is deduced with 90% CL.

The results with n = 1 are shown in fig. 7a,b, and c. The absence of entries above 5260 MeV/c² for wrong charge combinations indicates that there is little background in samples a and b. The six events in sample a have a mass of m(B°) = (5277 ± 3 ± 3) MeV, using m(Υ4S) as in chapter 1 and in good agreement with the m(B°) result in chapter 1. Four $B^\circ \to J/\Psi\ K^+\pi^-$ events have $m(K\pi) = m(K^{*\circ})$ from which we conclude $B(B^\circ \to J/\Psi\ K^{*\circ})$ = (0.44 ± 0.27)%.

5. SEARCH FOR B°-$\bar{B}^\circ$ OSCILLATIONS

A signature for B°_d-$\bar{B}^\circ_d$ oscillations is the observation of like-sign lepton pairs in Υ(4S) decays. The observable quantity χ, defined as $\chi = \Gamma(B^\circ \to \bar{B}^\circ \to l^-\nu X)/[(\Gamma(B^\circ \to l^+\nu X)+\Gamma(B^\circ \to \bar{B}^\circ \to l^-\nu X)]$ and related to $x = \Delta m/\Gamma$ by $\chi = x^2/(2+2x^2)$ is measured on the Υ(4S) as

$$\chi = \frac{N(l^\pm l^\pm)}{N(ll)} \left[1 + \frac{f_+}{f_0}\left(\frac{B_+}{B_0}\right)^2\right],$$

where $N(l^\pm l^\pm)$ is the number of like-sign lepton pairs from B decays directly, N(ll) is that of all direct lepton pairs, f_+ and f_0 are the fractions of charged and neutral B mesons.

In order to minimize contributions from secondary decays $B \to D \to l$, one selects only leptons with $p > 1.4$ GeV/c. There are 190 lepton pairs ee, $e\mu$ and $\mu\mu$ with a background < 1% for hadron misidentification of one lepton. B pair decays have an isotopic angular distribution between the two leptons, many background contributions are peaked near $\cos\theta_{ll} = -1$. With the cut $\cos\theta_{ll} > -0.9$, there are 159 l^+l^- and 5 $l^{\pm}l^{\pm}$ pairs. The background is estimated to be 3.5 ± 1.2 $l^{\pm}l^{\pm}$ (1.5 ± 0.5 from secondary decays, 1 ± 0.5 from J/Ψ + direct lepton, 1 ± 1 from misidentified hadron + direct lepton) and < 15 l^+l^-. Assuming $f_0 > 0.40$ and $B_+ = B^\circ$ one finds $\chi < 0.12$ with 90% CL; $B_0/B_+ > 0.8$ leads to $\chi \leq 0.16$ with 90% CL. The absence of $B_d^\circ - \overline{B}_d^\circ$ oscillations at this level is in agreement with standard model expectations [9].

6. V_{ub}/V_{cb} FROM INCLUSIVE SEMILEPTONIC DECAYS

The ARGUS collaboration has also presented a preliminary analysis of their inclusive electron momentum spectrum from 59.4/pb on the $\Upsilon(4S)$ resonance at this conference. The efficiency for identified electrons above 1.1 GeV/c is (65 ± 4)% including geometrical acceptance; and less than 1/200 of the hadrons are misidentified as electrons in this momentum range. Between 1.1 and 3.5 GeV/c, there are 7334 electrons reconstructed in the $\Upsilon(4S)$ sample, and their momentum spectrum is shown in fig. 8.

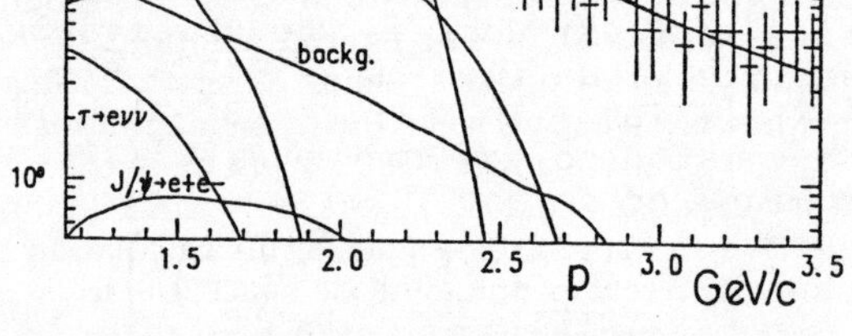

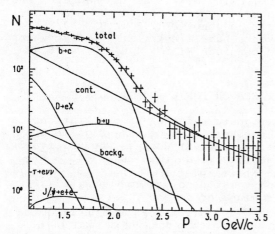

Figure 8: Inclusive $e^{\pm}$ momentum spectrum from the $\Upsilon(4S)$. The curves show the expected contributions.

The curves give the different contributions to the spectrum; cont. = electrons and misidentified hadrons from the non-resonant continuum events at the resonance energy (incl. τ decays) as determined from measurements below the resonance energy; $D \to eX$, $\tau \to e\nu\nu$, $J/\Psi \to ee$ = secondary electrons from B decays as determined by Monte Carlo calculations; backg. = misidentified hadrons from B decays as determined from studies with K_s° and $\Upsilon(1S)$ decays; $b \to c$ and $b \to u$ = results of a typical fit of the remaining electrons between 1.6 and 3.5 GeV/c to the expectation of a spectator model [10] with free ratio $r = \Gamma(b \to ue\nu)/\Gamma(b \to ce\nu)$.

Varying the spectator model parameters over the full range which gives acceptable fits to the observed spectrum leads to a large variety of likelihood functions L(r), the envelope of which allows to determine $r < 0.12$ with 90% CL. The model of Grinstein et al. [11] gives also an acceptable fit to the data, the result is $r < 0.06$ with 90% CL. Using their factor of two in uncertainty leads also to $r < 0.12$. From this preliminary analysis and $r = 2.2 \cdot |V_{ub}/V_{cb}|^2$ one concludes $|V_{ub}/V_{cb}| < 0.23$ with 90% CL.

REFERENCES

1. H. Albrecht et al., for a recent author list see Z.Physik C31(86)181
2. H. Albrecht et al (ARGUS),PL134B (84)137
3. Particle Data Group, PL170B(86)1
4. R.M. Baltrusaitis et al (MARK-III), PR L56(86)2140
5. R. Giles et al (CLEO), PRD30(84) 2279
6. A. Jawaheri (CLEO), these prodeedings
7. B. Stech, Moriond proceedings 1986, and HD-THEP-86-7
8. A.M. Boyarski et al (SLAC-LBL), PRL34(75)1357
9. see e.g. A. Ali, Proc.Int.Symp. Production and Decay of Heavy Hadrons, Heidelberg 1986, p.365
10. S. Weseler, Ph.D. Thesis, Heidelberg 1986, IHEP-HD/86-02
11. B. Grinstein, N. Isgur, and M. Wise, PRL 56 (86)298

Rare B Decays

Patrick J. O'Donnell
Dept. of Physics
University of Toronto
Toronto, Ontario
Canada M5S 1A7

The branching ratios for a number of rare decay processes (exclusive and inclusive) of the B meson are given. For a top quark mass in the range 40 GeV to 240 GeV the branching ratios are in the range, 10^{-5} to 3×10^{-4} for $B\to K^*\gamma$, 10^{-7} to 4×10^{-6} for $B\to K^*e^+e^-$, 2×10^{-6} to 4×10^{-6} for $B\to Kl^+l^-X$ and 10^{-6} to 3×10^{-6} for $B\to Kl^+l^-$; the predictions become precise whenever a determination is made of m_t. A distribution of the $\mu^+\mu^-$ pairs is given which should distinguish between the transverse and longitudinal production. These processes test the standard model with three generations and are insensitive to the values of the mixing angles consistent with existing constraints.

The rare decays of mesons and, in particular, the K meson have been useful in the development of the standard model and as a probe in the search for new physics. The B meson promises to be an additional important system for the exploration and development of the underlying physics. The fact that $m_B \gg m_K$ means that a very different energy region will be probed. In the K system it has often proved difficult to separate the weak from the hadronic processes; in $K-\bar{K}$ mixing the number of decay channels is limited by the available energy. Now that a number of laboratories are producing substantial amounts of B mesons it is important to consider the possibilities of measuring some rare decays.

In this talk I will give estimates of branching ratios for some rare exclusive decays[1]. In the standard model, with three generations, a number of inclusive decays of the B meson and some exclusive decays may be accessible to experiment. These decays have distinctive signatures and give a test of the standard model or a signal for new physics. The branching ratios are sensitive to the mass of the t quark; the predictions are precise whenever the mass is determined.

The basic premise is that because of the heavy mass of the b quark a number of processes which would be forbidden, or extremely rare, in the light quark system may occur at a measurable rate. In the inclusive processes it was shown[2] that the decay $b\to s+\gamma$ was sensitive to the value of the top quark mass and could give rise to a branching ratio in the range 10^{-5} to 10^{-4}, depending on the specific value of this mass. Other flavour changing decays such as $B_s^0\to\gamma\gamma$ or $B_s^0\to\tau^+\tau^-$ have branching ratios at least two orders of magnitude smaller and would not appear to be measurable in the near future. The $B-\bar{B}$ mixing, producing dileptons of the same sign, would only be sensitive to a light t-quark mass if the meson has a s quark, and involves the CP violating phase if the meson has a d quark.

Thus the exclusive decays based on the fundamental process $b\to s\gamma$ are the most likely to be accessible in the immediate future. Note that the analogous process $s\to d\gamma$ fails[3] to account for the decay $\Sigma\to p\gamma$. In this case long distance effects are important and also m_d, m_s are not too different. It is less likely that such effects play an important role in the b system where the mass is so much greater and where $m_s/b_b \ll 1$; such long distance effects are expected to be suppressed by $(m_s/m_b)^2$.

The decay $b\to s\gamma$ arises via an induced dipole transition matrix element. If the b- and s-quark masses were equal the transition would be pure magnetic dipole. Since they differ markedly, there is also an electric dipole transition. However, unlike the usual hadron spectroscopy, this is not a long wavelength limit of an electromagnetic interaction. The distinction between the two types will show in a relative sign between the spin-up and spin-down transitions; this is not likely to be observable.

Consider now the exclusive decays of a B meson, where $B=(\bar{b})$, with q denoting a light u, d or s quark. First of all it can be readily seen that the final state consists of a photon and a spin-one strange meson, i.e., $B\to K^*\gamma$ and $B\not\to K\gamma$. For a real photon decay there must be a spin-flip in the dipole transition. This means that the decay to the final state has a very distinctive signature. Of course there are at least two K^* particles known[4], one with a mass of 890 MeV and one with a mass of 1400 MeV. (There are also indications[4] of one at a mass of 1800 MeV.) These resonances are known to be ground state and radial excitations[5] of the K^* system with wave functions determined by the masses and decays of the light mesons. The ground state B meson will have non-zero transitions to the ground state part of the spin one K^* sector. Using the known wave functions[5] the branching ratios to K^* (890), K^*(1400) and K^* (1800?) will be in the ratio 0.998: 0.008: 0.005. That is, only the K^*(890) will have an appreciable signal. (This calculation ignores differences due to phase space; the B meson is so massive compared to all of these states that there is little variation in the phase space.)

A further complication in estimating the decay is that there might be a considerable suppression due to overlap of the wave function. To estimate this I assume that both initial and final mesons can be described by a harmonic oscillator wave function $\psi(r)=(1/\pi R_o^2)^{3/4}\, exp(-r^2/2R_o^2)$. An estimate of R_o can be obtained from the recent observation[6] that $m_{K^*}^2-m_K^2 \approx m_{B^*}^2-m_B^2$ is a consequence of a Lorentz scalar confining potential consistent with $|\psi(0)|^2/\mu_{ij}$ = constant, where μ_{ij} is the reduced mass of the $q\bar{q}$ pair. Thus, $R_0^B/R_0^K \simeq 0.87$ and the overlap suppression integral $I^2 \simeq 0.97$.

That is, after overlap suppression and wave function suppressions are taken into account, the exclusive decay $B \to K^*\gamma$ proceeds with a rate of 97% relative to the inclusive decay; or, in other words, the induced dipole transition previously given at the quark level as the decay $b \to s\gamma$ should be seen, reliably, in the decay $B \to K^*\gamma$. (As noted previously[2], the background $b \to us\bar{u}\gamma$ is expected to be at $\sim 10^{-6}$, down by about two orders of magnitude relative to the induced dipole decay.)

Thus, the branching ratio $B \to K^*(890)\gamma$ is expected to be in the range 10^{-5} to $3x10^{-4}$ for $M_t/M_W \simeq ½$ to 3.

Although the decay $B \to K^*\gamma$ has a very distinctive signature it may be preferable to look for a final state containing a lepton pair and a strange particle. This decay mode proceeds in a similar way with, however, a virtual photon replacing the real photon. There are now two possible types of exclusive decay modes, $B \to K^* l^+ l^-$ and $B \to K l^+ l^-$. These decays occur via the transverse and longitudinal components of the virtual photon, respectively.

An estimate of the branching ratio ρ, relative to the decay $b \to s\gamma$, can be given in terms of the rest mass x of the virtual photon, where, $x^2=2m_l^2+2E_+E_- -2\vec{p}_+ \cdot \vec{p}_-$.

$$\rho \simeq \left(\frac{2\alpha}{3\pi}\right) \int_{2m_l}^{M_B-M_K} dx \left(\frac{M_B^2-x^2}{M_B^2}\right)^2 \left(1-\frac{4m_l^2}{x^2}\right)^{½}$$

$$\times\left(1+\frac{2m^2}{x^2}\right)\left[\frac{R_T}{x}+\frac{2M_B^2 x R_L}{(M_B^2+x^2)^2}\right]$$

For $B \to K^* l^+ l^-$, the branching ratio to lepton pairs relative to that without is $\rho_T \sim (2\alpha/3\pi)(ln(M_B/m_l)-1.6$; for e^+e^- production $\rho_T \sim 1.6\alpha$ and for $\mu^+\mu^-$ pairs, $\rho_T \sim 0.5\alpha$.

Thus, the branching ratio for the exclusive decay $B \to K^* e^+ e^-$ is in the range 10^{-7} to $4x10^{-6}$ for m_t between 40 GeV and 240 GeV. The corresponding branching ratio to $\mu^+\mu^-$ is reduced by a factor of approximately three.

The decay $B \to K l^+ l^-$ proceeds via the longitudinal component of the virtual photon and an additional function F_1^i is needed. In Table 1 a comparison is given[7] for $F_1^t-F_1^c$ and $F_2^t-F_2^c$ (the contribution to the decay processes from the u quark is suppressed by the small $b \to u$ mixing angle). It can be seen that the F_1 combination is much larger than the F_2 combination and so the latter terms may be dropped. Also, the variation with m_t is less pronounced than in the previous case. To calculate the inclusive decay $B \to K l^+ l^- X \simeq b \to s l^+ l^-$ the distribution functions are simply $R_T=0, R_L=½$ corresponding to a (longitudinal) branching ratio $\rho_L \simeq 0.02\alpha$ scaled by the square of the second column of table 1 relative to the last column. This gives the branching ratios shown in Table 2.

The branching ratios for the inclusive decay $B \to K l^+ l^- X$ are expected to range from $1.7x10^{-6}$ to $3.6x10^{-6}$ for m_t in the range 40 GeV to 240 GeV.

m_t (GeV)	$F_1^t-F_1^c$	$F_2^t-F_2^c$
40	2.92	0.088
60	3.26	0.151
80	3.49	0.208
160	3.99	0.371
240	4.24	0.460

Table 1: Values of the combination of form factors which occur in the decay amplitude, as a function of the mass of the top quark.

The exclusive decay $B \to K l^+ l^-$ needs knowledge of the ground state wave functions. As above, these are available[5] for the K meson and its radial excitations, particularly the K(1400). The branching ratios $B \to K l^+ l^-$: $B \to K' l^+ l^-$ are in the ratio 0.79: 0.20. Unlike the exclusive decay $B \to K^*\gamma$, about 20% of the inclusive signature will be into the higher mass radial excitation. The branching ratio for the exclusive decay $B \to K l^+ l^-$ is reduced relative to the inclusive decay by this asmount. The branching ratios are shown in Table 2. A recent calculation[8] using a form factor for F_1 gives results that reduce the exclusive branching ratio by about 30% relative to those given here.

m_t	$K^*\gamma$	$K^*\mu^+\mu^-$	$K\ell^+\ell^- X$	$K\ell^+\ell^-$
40	11	0.04	1.7	1.3
60	33	0.12	2.1	1.6
80	61	0.22	2.4	1.8
160	195	0.7	3.2	2.5
240	300	1.1	3.6	2.8

Table 2: The branching ratios for exclusive and inclusive decays of the B meson to the ground state K and K^*. The branching ratio for $B \to K(1400) l^+ l^-$ is down by a factor of five from the rate given in the last column. In all cases the figures quoted are to be multiplied by 10^{-6}.

For pair production the distribution in x of the lepton pair can be calculated. In the Figure, I show the distribution functions $F_T(x)$ and $F_L(x)$ for production of muon pairs in the reactions $B \rightarrow K^* \mu^+ \mu^-$ and $B \rightarrow K \mu^+ \mu^-$, respectively. Here $F(x)$ is defined as the fraction of events $F(x)$ for which the pairs produced occur with a smaller value of x. From the Figure it can be seen that over most of the range of x, and especially for $x/2m_\mu \lesssim 14$ (where the maximum allowed by kinematics is 22.7) $F_T(x)$ is significantly larger than $F_L(x)$. At $x/2m_\mu = 14$, F_L is 82% of F_T while at $x/2m_\mu = 4$, F_L is only 19% of F_T. This shows the peaking at small x of the transverse part and may be a useful way to distinguish the two types of decays.

To summarize, I have concentrated in this talk on inclusive and exclusive rare decays of the B meson, in the standard model with three generations and mixing angles consistent with present experimental constraints. Other rare decays, such as[2] $B_s \rightarrow \tau^+ \tau^-$ or $B_s \rightarrow \gamma\gamma$ or CP violating effects within the standard model[8] are much less likely to be observed. These decays have fairly distinctive signatures and have branching ratios, in some cases, that are accessible to experiments with large sources of B mesons. A systematic study of these processes should confirm our understanding of the standard model in a set of new processes very distinct from the usual K decay channels. A signal for new physics will be evident if these decays are not observed at the level of 10^{-6} to 10^{-4} in branching ratio.

ACKNOWLEDGEMENTS

I thank Gordon Kane, Gabriel Karl and Julian Noble for discussions and suggestions and the CERN Theory Group for its hospitality. This research has been supported in part by the Natural Science and Engineering Council of Canada, grants A3828 and T6873.

REFERENCES

1) P.J. O'Donnell, Phys. Letts. *B175*, 369 (1986).
2) B.A. Campbell and P.J. O'Donnell, Phys. Rev. *D25* (1982) 1989.
3) M.K. Gaillard, X.Q. Li and S. Rudaz, Phys. Lett. *158B* (1985) 158.
4) Particle Data Group, Phys. Lett. *170B* 1 (1986).
5) M. Frank and P.J. O'Donnell, Phys. Lett. *133B* (1983) 253; Phys. Rev. *D29* (1984) 921; Phys. Lett. *144B* (1984) 451 and Phys. Rev. *D32* (1985) 1739.
6) M. Frank and P.J. O'Donnell, Phys. Lett. *159B* (1985) 174; and CERN preprint TH.4367/86 (1986).
7) N.G. Deshpande and G. Eilam, Phys. Rev. *D26* (1982) 2463; N.G. Deshpande and M. Nazeri-monfared, Nucl. Phys. *B213*, (1983) 390.
8) N.G. Deshpande et al., Oregon preprint.

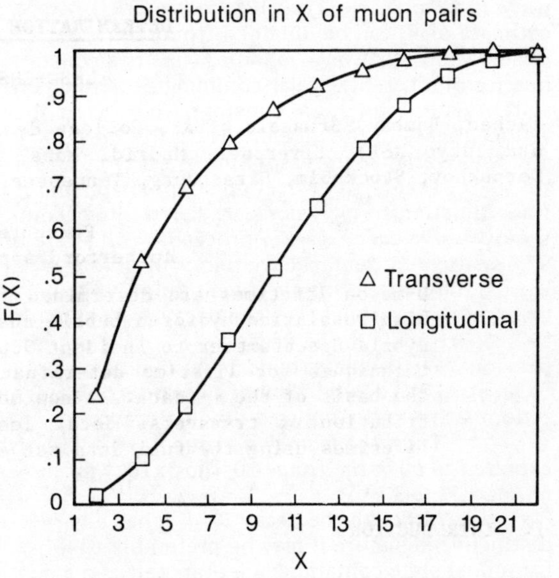

FIGURE CAPTION

The distribution functions F(X) give the fraction of events for which the pairs produced have a value smaller than X. (Here X denotes the value of x in units of $2m_\mu$.) The fraction has been calculated separately for the distinct decay modes $B \rightarrow K^* \mu^+ \mu^-$ and $B \rightarrow K \mu^+ \mu^- X$. If only $\mu^+ \mu^-$ pairs are observed then F_T and F_L should be multiplied by $[\rho_T/(\rho_T+\rho_L)]$ and $[\rho_L/(\rho_T+\rho_L)]$, respectively.

DETERMINATION OF D-MESON LIFETIMES

LEBC-EHS-COLLABORATION

Aachen, Bombay, Brussels, CERN, College de France, Genova, Japan Universities (Chuo, Tokyo Met and Tokyo A&I), Liverpool, Madrid, Mons, Oxford, Padova, Paris, Rome, Rutherford, Rutgers, Serpukhov, Stockholm, Strasbourg, Tennessee, Torino, Trieste and Vienna.

Presented by C M Fisher
Rutherford Appleton Laboratory, UK

D-meson lifetimes are determined from decays observed in an exposure of the high resolution hydrogen bubble chamber LEBC in association with the European Hybrid Spectrometer to incident 360 GeV π^- and 400 GeV proton beams. Various techniques for lifetime determination using unfitted decays are compared on the basis of the π^- data. A new model independent analysis based on the distribution of transverse decay length is introduced. Best values for the lifetimes using the full data set are found to be:

$\tau(D^0) = 4.3^{+0.5}_{-0.4} \times 10^{-13}$ s $\qquad \tau(D^+) = 10.6^{+1.2}_{-0.9} \times 10^{-13}$ s

1. INTRODUCTION

Charm particles have small branching ratios into final states that can be uniquely identified and kinematically fitted. If only constrained fits to specific decay modes are used for lifetime determination experiments have poor statistics and there is the possibility of bias as a result of the fit selection procedures. An improvement is clearly possible if decays known to be D mesons, but with unconstrained decay modes, can be used. In this paper three indirect techniques for lifetime determination are compared using a data sample from the NA27 π^-p exposure (100 $D^0/\bar{D}^0$ and 83$^\pm$) and best values are derived on the basis of the full NA27 data sample.

It is important for any analysis using unconstrained decays to ensure that the data sample is completely free from non-charm background. In addition, within a well defined range of observables, the detection efficiency must be near 100% and independent of the lifetime. Although limited in statistics, the high resolution hydrogen bubble chamber provides extremely high quality data which can be used for such analysis. The results serve to complement and confirm those obtained using higher rate techniques.

In this experiment the high resolution chamber acts as both target and microvertex detector. No charm selecting trigger is involved and all film is subject to a triple scanning procedure described in [1]. A typical charm pair event is shown in Fig 1. The chamber has the following properties:

1. The two track resolution is < 20μm (resolved bubble size).
2. The bubble density is ~9/mm of track length. A typical track has ~400 bubbles and there are no tracking ambiguities.
3. Approximately 40 HPD master points are constructed per track which yield rms residuals to straight line fits ~1.8μm (film plane projection).
4. Since the chamber is both target and vertex detector tracks are followed into their vertex of origin but not beyond. There are negligible vertex ambiguities.
5. The interactions occur in liquid hydrogen. There are no neutrons and hence secondary interactions cannot fake decays.

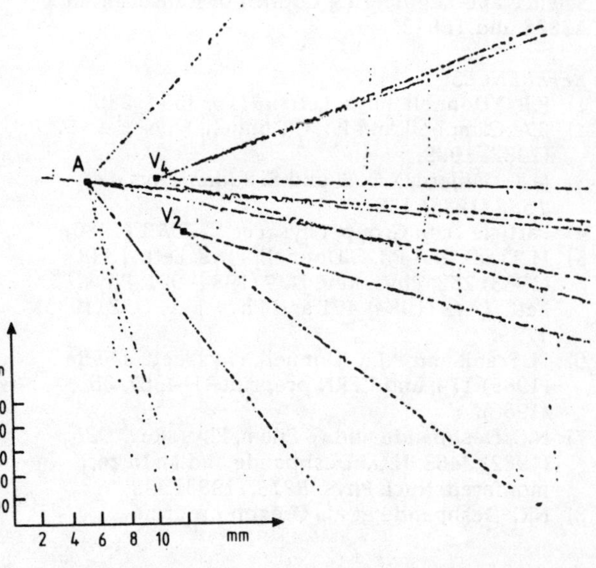

FIG 1 A typical neutral charm pair event ($D^0\bar{D}^0$). The transverse scale is magnified (x5) with respect to the logitudenal scale.

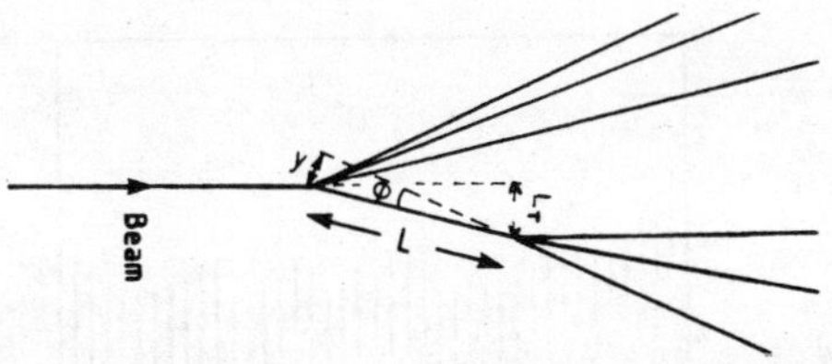

FIG 2 Definition of transverse decay length ℓ_T and impact parameter y.

2. SAMPLE DEFINITION

Decay topologies are detected at scanning by observing tracks with significant impact parameters extrapolated to the production vertex. The impact parameter and transverse decay length are defined in Fig 2. In Fig 3a we show a direct measurement of the impact parameter resolution obtained by taking tracks to their known vertex of origin, eg to the primary vertex for events with no evidence for secondary decays. The rms error on the impact parameter is σ_y = 2.5μm. In Fig 3b we show the impact parameter distribution (to the primary vertex) for tracks arising from charm particle decays. The hatched area represents those tracks added at the measurement stage. From the above plots we deduce (a) that the scanning efficiency is independent of impact parameter for $y \geq 50\mu m$ and (b) that the measurement is efficient for finding and associating tracks with impact parameters $\geq 7.0\mu m$ ($3\sigma_y$). The triple scanning procedure showed an overall efficiency ~96% for finding events provided at least one secondary track has $y > 50\mu m$ to the primary vertex.

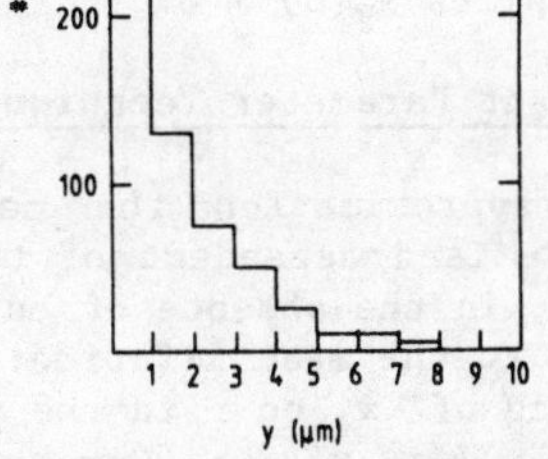

FIG 3a Measurement of impact parameter resolution $\tau_y \sim 2.5$ μm.

In order to eliminate strange particles from the sample strange particle fits, (ie $M_{\pi\pi}$ within 3σ of m_{K0} for V2 decays) are removed and, in addition, it is required that charm candidates have at least one decay track with $p_T > 250$ MeV/c with respect to the parent or to have a charm topology (ie four charged tracks from a neutral decay (V4), three or five from a charged decay (C3 or C5)). For the lifetime analysis we also require that neutral two prong decays (V2) are paired with a second visible charm decay and that charged three prong decays (C3) either have a track with $p_T > 250$ MeV/c or are part of a pair. Single prong charged decays (C1) are not used in the lifetime analysis however if they have $p_T > 250$ MeV/c they can complete a pair.

For each decay we define the distances ℓ, ℓ_{min}, ℓ_{max}, ℓ_T where ℓ is the observed decay length, $\ell_{min}(\ell_{max})$ is the minimum(maximum) detectable decay length, and ℓ_T is the transverse decay length. ℓ_{min} is the length found by translating the decay topology along the line of flight towards the production vertex until either the maximum impact parameter y_{max} reduces to 50μm or the minimum impact parameter y_{min} reduces to 7μm. ℓ_{max} is computed from the position of the event in the chamber, and the boundaries of the 'charm box' [1].

The resulting data samples from the π^- exposure which can be used to compare lifetime techniques are shown in Table 1.

TABLE 1

	Sample After Cuts	Subsamples Used			
		K_{in}	L_T	P_{est}	$\langle y \rangle$
D^0	57	21	57	36	57
$D^\pm$	36	30	36	20	33

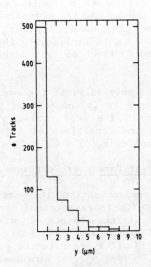

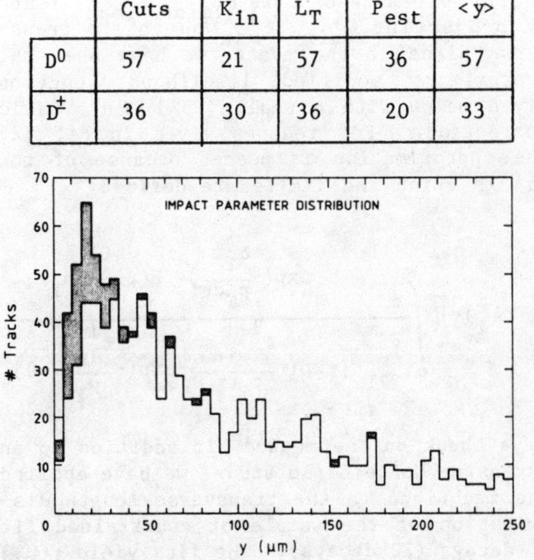

FIG 3b Impact parameter distribution of charm decay products to the primary vertex.

3. TECHNIQUES FOR LIFETIME DETERMINATION

3.1 Constrained Kinematic Fits

If the momentum of the charm particle is known from the fit then the proper time t^i and the minimum and maximum detectable lifetimes t^i_{min}, t^i_{max} can be found directly from the decay length ℓ^i and the ℓ^i_{min}, ℓ^i_{max} defined above. A simple likelihood function for the mean lifetime can be written.

$$L(\tau, t^i) = \prod_i \frac{1}{\tau} \frac{\exp(-t^i/\tau)}{[\exp(-t^i_{min}/\tau) - \exp(-t^i_{max}/\tau)]}$$

where the summation is over all events in the sample.

3.2 Using the Distribution of Transverse Lengths

This method uses the correlation between the transverse decay length and lifetime:

$$\ell^i_T = \frac{p^i_T c}{m} t^i$$

The distribution in ℓ^i_T can be derived since the distributions in p^i_T and t^i are known and independent. In the absence of cuts we can write a likelihood function to observe the set ℓ^i as a function of the mean lifetime τ.

$$L(\tau, \ell^i_T) = \prod_i \frac{m}{c\tau} \int_0^\infty \frac{1}{p_T} \exp\left[-\frac{m\ell^i_T}{p_T c}\right] F(p_T)\, dp_T$$

where

$$F(p_T) = \frac{dN}{dp_T} = p_T \exp\left[-1.18^{+0.18}_{-0.16}\, p_T^2\right] \qquad [1]$$

In the presence of cuts ℓ^i_{Tmin}, ℓ^i_{Tmax}, found by projecting ℓ^i_{min}, ℓ^i_{max} on to the transverse plane of the event, we have used the approximate modified likelihood function given below with a small (~5%) Monte Carlo correction. For reasons given in Ref [2] the approximation is needed because of the way in which the limits are defined.

$$L(\tau, \ell^i_T) = \prod_i \int_0^\infty \frac{m}{p_T c\tau} \frac{\exp\left(-\frac{\ell^i_T m}{p_T c\tau}\right) F(p_T)\, dp_T}{\left[\exp\left(-\frac{\ell^i_{Tmin} m}{p_T c\tau}\right) - \exp\left(-\frac{\ell^i_{Tmax} m}{p_T c\tau}\right)\right]}$$

As a check on the method, in addition to an extensive Monte Carlo study, we have applied the technique to the transverse length distribution of the sample of constrained fit $D^\pm$ decays (20 decays). The fits yield $\tau(D^\pm) = 9.8^{+3.4}_{-2.2}\ 10^{-13}$s, whilst the transverse length technique yields $(9.8^{+3.7}_{-2.5})\ 10^{-13}$s.

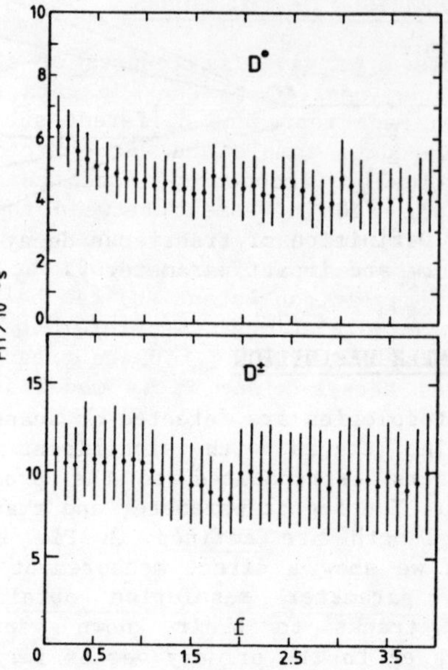

FIG 4 Stability of lifetime results from transverse length analysis versus ℓ_{min}. f is multiplying factor to ℓ_{min}.

3.3 The Momentum Estimator Technique [3]

In this technique an estimate of the D-meson momentum is obtained from the observed effective mass of the charged decay products taken as pions M^π_{vis} and visible momentum p_{vis}.

$$P_{est} = \alpha \frac{M_D}{M^\pi_{vis}} P_{vis}$$

where α is determined by Monte Carlo. The P_{est} is not unique and the parameter α has a spread depending on M^π_{vis} and p_{vis}. The distribution in α in included in the form of the likelihood function as described in [2].

The method is only valid if the event detection efficiency is independent of the decay distribution in the rest system of the D-meson. For the conditions of this experiment this implies $x_F(D) > 0$.

3.4 The Impact Parameter Technique [4]

To a good approximation the mean impact parameter $\langle y \rangle$ is independent of the D-meson momentum and, in the absence of cuts, is proportional to the mean lifetime: $\langle y \rangle \propto \tau$. The dependence of $\langle y \rangle$ on τ in the presence of cuts $y_{min} > 7\mu m$, $y_{max} > 50\mu m$ has been evaluated by Monte Carlo. The derived lifetimes are insensitive to the details of the Monte Carlo and to the values used for the cuts.

4. COMPARISON OF TECHNIQUES

In Table 2 we give the results of all four techniques applied to the $\pi^- p$ data sample. They in fact represent different subsets of the same data sample and are therefore not statistically independent and cannot be averaged. The agreement between the techniques is impressive and shows that the lifetime dependent features of the events are internally consistent and can be used to derive mean lifetimes. We prefer the transverse length technique for the treatment of unfitted decays since it is model independent and uses the full available statistics.

TABLE 2

	Lifetimes in Units of 10^{-13}s			
	Fits	L_T	P_{est}	<y>
$\tau(D^0)$	$3.6^{+1.0}_{-0.7}$	$4.5^{+0.8}_{-0.7}$	$4.4^{+1.0}_{-0.7}$	$4.6^{+1.0}_{-0.7}$
$\tau(D^\pm)$	$9.8^{+3.4}_{-2.2}$	$10.4^{+2.8}_{-2.0}$	$11.5^{+4.5}_{-2.8}$	$11.9^{+3.9}_{-2.6}$
$R=\frac{\tau(D^\pm)}{\tau(D^0)}$	$2.7^{+1.2}_{-0.8}$	$2.4^{+1.0}_{-0.7}$	$2.6^{+1.2}_{-0.8}$	$2.6^{+1.0}_{-0.7}$

In Fig 4 we show the transverse length estimate of lifetimes as a function of the visibility cuts defining ℓ^i_{min}. The factor f is used to multiply all ℓ^i_{min}. Clearly the results are insensitive to ℓ_{min} for f$\geqslant$1.0 as expected.

If we combine the transverse length lifetime estimates derived using only no-fit events with the fitted lifetime from the 'fit' events we obtain our best estimate of the π^- data sample:

$$\tau(D^\pm) = (10.7^{+2.8}_{-1.8})10^{-13}\text{s (40 decays)}$$

$$\tau(D^0) = (4.1^{+0.7}_{-0.6})10^{-13}\text{s (60 decays)}$$

5. $\tau(D^0)$ and $\tau(D^\pm)$ using Full Data Sample

Here we present a preliminary result based on the analysis of the full data sample obtained by adding the pion and proton exposures together. The same event selection procedures apply to both data sets. After all cuts the event samples used for lifetime analysis become 142 $D^\pm$, 139 $D^0/\bar{D}^0$. We obtain as our current best result, based on the transverse length technique used for no-fits and the fit results where available:

$$\tau(D^0) = 4.3^{+0.5}_{-0.4} \times 10^{-13}\text{s}$$

$$\tau(D^\pm) = 10.6^{+1.2}_{-0.9} \times 10^{-13}\text{s}$$

yielding

$$R = \frac{\tau(D^\pm)}{\tau(D^0)} = 2.5\pm0.35$$

No sources of systematic error can be detected at this stage. We must however consider the possible inclusion of Λ_c and F decays in the '$D^\pm$' no fit sample. If there were significant contamination in the C3 topology of short lived decays we would expect the apparent lifetime to fall for f<1. Fig 4 shows that in fact the $D^\pm$ lifetime is very stable against variations of ℓ_{min}. A 10% contamination of $F^\pm$ decays with $\tau_F \sim 3.4 \times 10^{-13}$s would not be statistically significant. We have good examples (13) of Λ_c fits (mostly confirmed by particle identification) however if $\tau_{\Lambda_c} \leqslant 2 \times 10^{-13}$s as the fit data indicates, most of the Λ_c topologies will be removed by the y_{max} cut at 50μm and the D result will be unaffected.

REFERENCES

[1] CERN/EP 85-103. Zeitschrift fur Physik C, 31, 491 (1986).

[2] Determination of D-meson Lifetimes, NA27 Collaboration, to be published in Zeitschrift fur Physik C.

[3] B Franek, RAL Report RAL 85-026 (1985).

[4] S Petreva and G Romano, Nucl Inst & Meth 174 (1980) 61.
P Checchia et al, INFN Note INFN/PD/EHS (1984).

Measurements on the Decay of Charm Particles: Lifetimes of the D^0, D^+, and F^+, and Relative Branching Fractions of the D^+ and F^+ into the channels $\overline{K}^{*0}K^+$ and $\phi\pi^+$

presented by Paul E. Karchin[a], University of California, Santa Barbara,
on behalf of the FNAL E691 Collaboration:

J. C. C. dos Anjos[3], J. A. Appel[5], S. B. Bracker[8], T. E. Browder[1], L. M. Cremaldi[4]
J. R. Elliot[4,b], C. Escobar[7], P. Estabrooks[2], M. C. Gibney[4], G. F. Hartner[8], P. E. Karchin[1,a],
B. R. Kumar[8], M. J. Losty[6], G. J. Luste[8], P. M. Mantsch[5], J. F. Martin[8], S. McHugh[1],
S. R. Menary[8], R. J. Morrison[1], T. Nash[5], U. Nauenberg[4], P. Ong[8], J Pinfold[2], G. Punkar[1],
M. V. Purohit[5], J. R. Raab[1], A. F. S. Santoro[3], J.S. Sidhu[2], K Sliwa[8], M. D. Sokoloff[5],
M. H. G. Souza[3], W. J. Spalding[5], M. E. Streetman[5], A. B. Stundzia[8], M. S. Witherell[1]

[1]University of California, Santa Barbara, California, USA , [2] Carleton University, Ottawa, Ontario, Canada, [3]Centro Brasileiro de Pesquisas Fisicas, Rio de Janeiro, Brasil, [4]University of Colorado, Boulder, Colorado, USA , [5]Fermi National Accelerator Laboratory, Batavia, Iliinois, USA, [6]National Research Council, Ottawa, Ontario, Canada, [7]Universidad de Sao Paolo, Sao Paolo, Brasil, [8]University of Toronto, Toronto, Ontario, Canada

We have measured the lifetimes and some relative branching fractions of charmed mesons produced by a high energy photon beam at Fermilab. The D^+ lifetime, based on 480 $D^+\rightarrow K^-\pi^+\pi^+$ decays, is measured to be $1.09^{+.08}_{-.07} \pm .06$ picoseconds. For the D^0, we used 672 decays into the modes $K^-\pi^+$ and $K^-\pi^+\pi^-\pi^+$ to determine the lifetime to be $.44\pm.02\pm.02$ ps. The ratio $\tau(D^+)/\tau(D^0)$ obtained from these two measurements is $2.5\pm0.2\pm0.1$. The F^+ lifetime is $.42^{+.09}_{-.07}\pm.06$ ps based on 35 decays into the modes $\phi\pi^+$ and $\overline{K}^{*0}K^+$. The results come from an analysis of 15% of the full data sample.

Introduction. This paper presents preliminary results from a high energy photoproduction experiment using the Fermilab Tagged Photon Spectrometer. The experiment combined high statistics with low background, time resolution much smaller than the lifetime, and good control of systematic errors. A plan view of the TPS is shown in Figure 1. An electron beam with momentum 260 Gev/c was used to produce a bremstrahlung photon beam with average energy 125 Gev and which was incident on a 5 cm long Beryllium target. Immediately downstream of the Be target were 9 silicon microstrip detectors (SMD's) with 50 μm strip spacing. The SMD's provided measurement of charged tracks with enough precision to resolve the charm vertex. The TPS includes a drift chamber-based magnetic spectrometer, two multi-cell Cerenkov counters, and both e.m. and hadronic calorimeters. The trigger required a total transverse energy (E_t) in the calorimeter greater than ≈2.25 Gev. This requirement was ≈80% efficient for events with charm, and suppressed the total hadronic interaction rate by a factor of ≈2.5. We were able to record about 100 million events, of which 10% were taken without the E_t requirement. The present results are based on analysis of ≈15 million events.

The events were first reconstructed in the SMD - drift chamber tracking system. An example of a reconstructed event (Figure 2) shows the silicon and drift chamber hits in the same view and in the field-free region upstream of the first magnet. It is quite evident that both the silicon detectors and drift chambers are highly efficient and relatively noise-free, and that the silicon hits are easy to correlate with the drift chamber hits. The Cerenkov counter information was used to assign particle identification for each track. For each channel, there was a minimum requirement on the joint probability for the appropriate particle identification assignment. The charm decay tracks were required to form a good vertex, and all other tracks in the event were used to form possible primary vertices. A search was made for all posiible primary vertex candidates within a transverse distance of 80 μm from the line of flight of the reconstructed charm candidate.

To reduce the non-charm background, only charm candidates were chosen which decayed at least a distance z_{min} downstream of the primary vertex. This minimum distance was chosen, depending on the decay mode, to be 5-10 σ_z, where σ_z is the resolution in the decay length along z. Thus, z_{min} was almost always downstream of any confusion in the primary vertex. If there were multiple primary vertex candidates, z_{min} was calculated from the most downstream vertex, to be sure that it was downstream of any possible production point. The proper time was calculated using the distance from z_{min} to the decay vertex and the measured momentum of the charm particle.

D° Lifetime. For the D^0 lifetime study, three independent samples were used:

(A) $D^{*+}\rightarrow\pi^+D^0\rightarrow K^-\pi^+$
(B) $D^{*+}\rightarrow\pi^+D^0\rightarrow K^-\pi^+\pi^-\pi^+$, and
(C) $D^0\rightarrow K^-\pi^+$

---
[a] now at Yale University, New Haven, Connecticut, USA
[b] now at Electromagnetic Applications, Inc., Denver Colorado, USA

(Throughout this paper, the charge conjugate channel is implicitly assumed for every channel discussed.) Events which satisfied the requirements for sample (A) were excluded from sample (C) so they are indeed independent. For samples (A) and (B), the mass difference $m(K^-\pi^+\pi^+)$-$m(K^-\pi^+)$ was used to identify the D^*. The $K\pi$ mass distribution for the events surviving the z_{min} cut of $5\sigma_z$ is shown in Figure 3(a). There is a remarkably clean D^0 signal of 198 events in the mass range 1.84-1.89 GeV/c^2, with a background of 7. For one of the events in sample (A), a blowup of the vertex region is shown in Figure 4. In this event, both charm vertices are clearly visible, illustrating the resolving power of the silicon hodoscope. The proper time distribution for the events in the signal region in Figure 3(a), with background subtracted, is shown in Figure 3(b). It is clear from this plot that the distribution with no corrections fits an exponential very well. The curve shows the result of a maximum likelihood fit, including corrections, which gives a mean lifetime of .43±.035 picoseconds.

The analysis for channel (B) was very similar to that for (A). For a vertex cut of $7\sigma_z$, the $K3\pi$ mass distribution is shown in Figure 5(a). There is a signal of 123 events over a background of 17. The fit to the time distribution, shown in Figure 5(b), gives a mean lifetime of .42±.045 ps. For channel (C), the $K\pi$ mass distribution and corresponding lifetime plot are shown in Figures 6(a) and 6(b). There are 351 signal events over a background of 165. The fitted mean lifetime is .45±.03 ps.

These three samples are statistically independent, with different corrections and backgrounds. The fact that the three measurements agree provides some check that the present correction is reasonable. A global fit to all three samples gives our best number for the D^0 lifetime, $\tau(D^0)$ = .44±.02±.02 ps. To illustrate the small size of the corrections, we have calculated the global mean lifetime with and without corrections to a pure exponential. The corrections which cause a negative shift in the fitted lifetime are due to: false primary vertices, resolution, efficency•acceptance, and absorption of the decay products in the target. The correction due to these effects is -.03 ps. The background subtraction, due almost entirely to channel (C), results in a shift of +.02 ps. Thus, the total shift in the fitted mean lifetime due to corrections is -.01 ps. The quoted systematic error is based on our current understanding of these corrections.

D^+Lifetime. The decay mode $D^+\to K^-\pi^+\pi^+$ was used for the D^+ lifetime study. Using a vertex cut of $10\sigma_z$, the mass spectrum for the accepted $K^-\pi^+\pi^+$ events is shown in Figure 7(a). There are 480 D^+ events over a background of 140. The secondary vertex selection reduces the background by a factor of 300.

The D^+ time distribution with background subtracted is shown in Figure 7(b). The visible deviation from an exponential at large times is due to the loss of events with a decay beyond the fiducial volume measured by the vertex detector. This effect causes a shift of only 5% in the mean lifetime and can be determined without systematic error. The maximum likelihood fit, including the correction at long times, gives a lifetime of $1.09^{+.08}_{-.07}$ ps. The contributions to the systematic error for the D^+ are somewhat different from those for the D° because of the longer D^+ lifetime. The correction to the D^+ lifetime for the effects of resolution, efficiency, acceptance, and false primary vertices is -.09 ps, with a systematic error of ±.04 ps. The absorption of the decay products in the beryllium target leads to an additional correction of -.04 ps, with an error of ±.01 ps. The background subtraction produces a correction to the D^+ mean lifetime of +.2 ps. This is relatively large because the effective lifetime of the background is much shorter than that of the signal. For the same reason, however, it is easy to test the background subtraction accurately by choosing samples with more or less background. The systematic error due to this subtraction is ±.04 ps. Assuming these systematic errors are uncorrelated, the total systematic error is ±.06 ps.

F^+ Lifetime. Two decay modes were used for the F^+ lifetime study: $F^+\to\phi\pi^+$ and $F^+\to\bar{K}^{*0}K^+$. In both channels, the background is reduced by the requirement of a meson resonance and the angular distribution for a decay of the type $0^-\to1^+0^-$. For a vertex separation cut of $6\sigma_z$ the $\phi\pi$ mass spectrum is shown in Figure 8(a). There are clear peaks from the Cabibbo-suppressed decay $D^+\to\phi\pi^+$ as well as for the F. Because of the good resolution and the ϕ requirement, the contamination of the F^+ from the D^+ and Λ^+_c is reduced to a negligible level The time spectrum for the $\phi\pi$ events in the mass region 1.95-1.98 GeV/c^2 is shown in Figure 8(b). There are 19 events signal and 4 background in the plot. The maximum likelihood fit finds the F^+ mean lifetime to be $.40^{+.12}_{-.08}$ ps.

The analysis of the decay $F^+\to\bar{K}^{*\circ}K^+$ is similar to that for $\phi\pi^+$. For a vertex cut of $10\sigma_z$, the $K^-\pi^+K^+$ mass plot is shown in Figure 9(a). There is a signal for the F^+ of 16 events over a background of 7. The corresponding time distribution for the F is shown in Figure 9(b) and has a fitted mean lifetime of $.45^{+0.18}_{-0.12}$ ps. A global fit to the time distribution for both F decay modes gives a mean lifetime of $.42^{+.09}_{-0.07}$±.06 ps. The systematic error is determined in a manner similar to that for the D^0 but with the additional consideration of the large sensitivity of the results to the time region included in the fit. With more statistics, this error will be reduced easily.

Relative Branching Fraction Measurements. Since we have measured two F decay modes with essentially identical kinematics, we can make a statement on the relative branching fraction which has little dependence on the acceptance of the spectrometer:

$$\frac{B(F^+\to\bar{K}^{*0}K^+)}{B(F^+\to\phi\pi^+)} = 1.1\ ^{+.5}_{-.4}\ \pm .15$$

The decay rates of the above two channels can also be expressed relative to a well measured decay of the D^+:

$$\frac{\sigma \bullet B(F^+ \to \overline{K}^{*0} K^+)}{\sigma \bullet B(D^+ \to K^- \pi^+ \pi^+)} = .13 \pm .04 \pm .025$$

$$\frac{\sigma \bullet B(F^+ \to \phi \pi^+)}{\sigma \bullet B(D^+ \to K^- \pi^+ \pi^+)} = .12 \pm .035 \pm .02$$

When the relative production cross-section of F^+ to D^+ is known, the above ratios can be turned into absolute branching fractions. For the two Cabibbo-suppressed decay modes of the D^+ that we have measured, we give the relative branching fraction with respect to the Cabbibo-favored mode $D^+ \to K^- \pi^+ \pi^+$ and compare our measurement with the published measurement[1] of the MARK III experiment:

	E691	MARK III
$\frac{B(D^+ \to \overline{K}^{*0} K^+)}{B(D^+ \to K^- \pi^+ \pi^+)}$ =	$.04 \pm .02 \pm .01$	$.048 \pm .021$
$\frac{B(D^+ \to \phi \pi^+)}{B(D^+ \to K^- \pi^+ \pi^+)}$ =	$.055 \pm .02 \pm .01$	$.084 \pm .021$

Interpretation and Outlook. A comparison of the E691 lifetime measurements with those of previous experiments[2] is shown Figure 10. It is clear that for the D^+ and D° the errors on the E691 measurements are significantly smaller than any other experiment. In fact, the E691 errors are comparable with those of the previous world average. The E691 result for the lifetime ratio $\tau(D^+)/\tau(D^\circ)$ is $2.5 \pm 0.2 \pm 0.1$. The ratio of lifetimes can be inferred from the ratio of semi-leptonic branching ratios $B(D^+ \to e^+ + X)/B(D^\circ \to e^+ + X)$ if one ignores the Cabbibo-suppressed non-spectator process which only contributes to the D^+ semileptonic decay. Mark III has used this technique[3] to extract a lifetime ratio of $2.3^{+0.5}_{-0.4} \pm 0.1$.

There have been three lifetime measurements for the F^+ published[4], with results ranging from .26 -.35 ps. The F^+ lifetime measurement from E691 is compared with three other experiments in Figure 10(c). The E691 error is comparable to the best of previous experiments. The F^+ lifetime appears to be comparable to the D° lifetime and much smaller than the D^+ lifetime.

The present results are based on 15% of the full data sample from E691. Although the statistical errors are very good, they will be improved by a factor of about 2.5 with the full data sample. Improved systematic errors will be possible with more detailed studies of the corrections and comparison of Monte Carlo results with the full data sample.

---

[1]R. M. Baltrusaitis et al., PRL 55, 150 (1985).

[2]M. Aguilar-Benitez et al., Phys. Lett. 123B, 312 (1983); M. Jori, Bari Conference (1985); John Butler, SLAC-290, UC-34D (1986); L. Gladney, Ph.D. Thesis, Stanford; P. Haas et al., Washington APS Meeting (1985); N. Ushida et al., PRL 56, 1767 (1986); N. Ushida et al., PRL 56, 1771 (1986); R. Bailey et al., Z. Phys. C. 28, 357 (1985).

[3]R. M. Baltrusaitis et al., PRL 54, 1976 (1985).

[4]R. Bailey et al., NIKHEF-H Report 85-5 (1985) and Proceedings of the Conference on Hadron Spectroscopy, College Park, MD, USA (1985); C. Jung et al., PRL 56, 1775 (1986); N. Ushida et al, PRL 56, 1767 (1986).

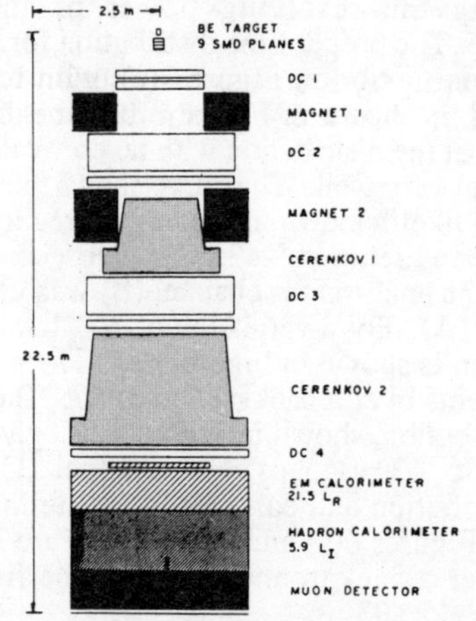

Figure 1. Plan view of the Fermilab Tagged Photon Spectrometer. The two magnets provide a p_t kick of ≈ .5 Gev/c.

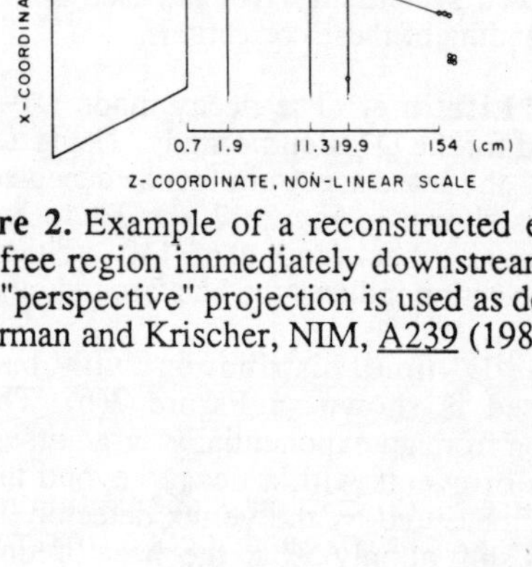

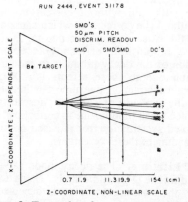

Figure 2. Example of a reconstructed event in the field-free region immediately downstream of the target. A "perspective" projection is used as described by Dreverman and Krischer, NIM, A239 (1985).

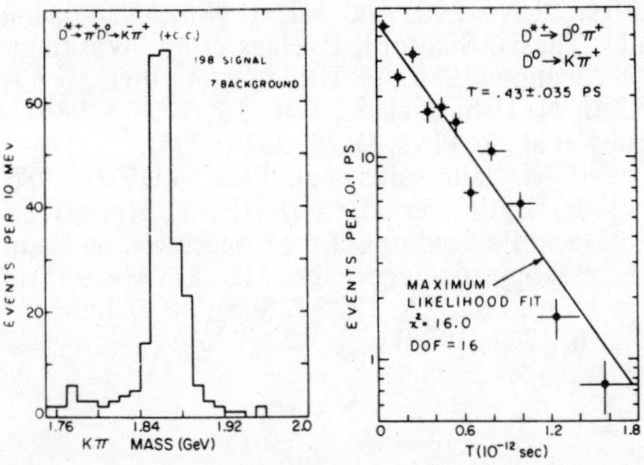

Figure 3. (a) Kπ mass distribution for $D^{*+} \to \pi^+ D^0 \to K^-\pi^+$ (and c.c.). (b) Proper time distribution for signal in (a).

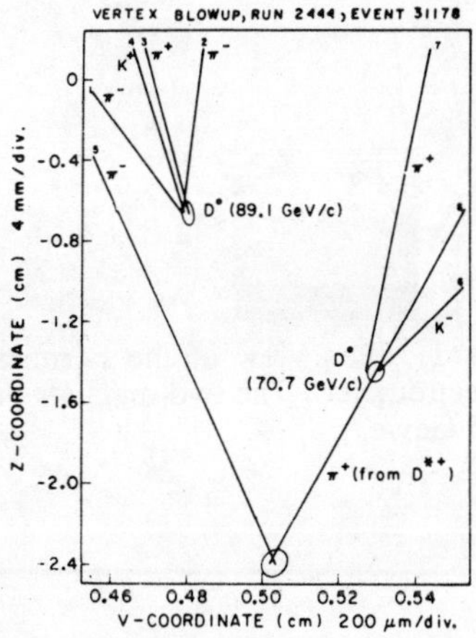

Figure 4. Blowup of the vertex region for the same event as in Figure 2.

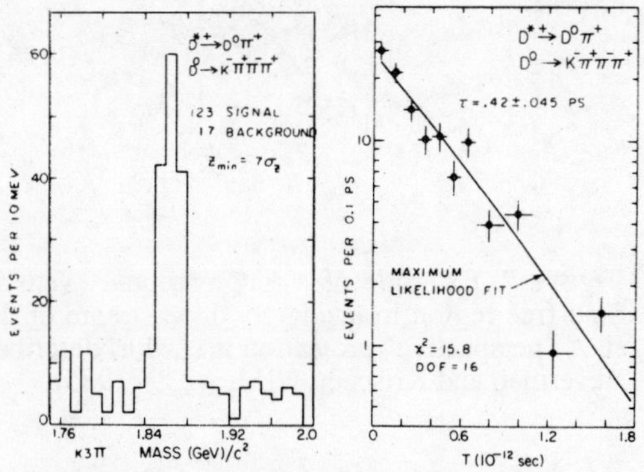

Figure 5. (a) K3π mass distribution for $D^{*+} \to \pi^+ D^0 \to K^-\pi^+\pi^-\pi^+$ (and c.c.). (b) Proper time distribution for the signal in (a).

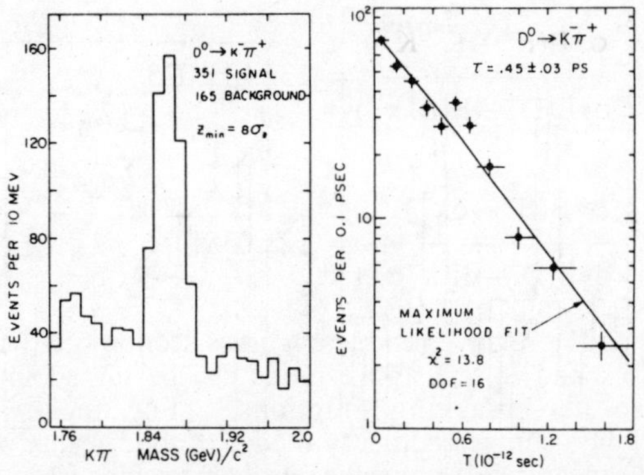

Figure 6. (a) Kπ mass distribution for $D^0 \to K^-\pi^+$ (and c.c.). (b) Proper time distribution for signal in (a).

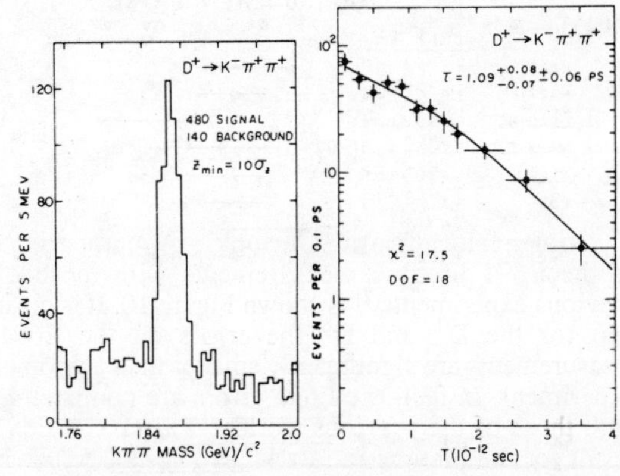

Figure 7. Kππ mass distribution for $D^+ \to K^-\pi^+\pi^+$ (and c.c.). (b) Proper time distribution for signal in (a).

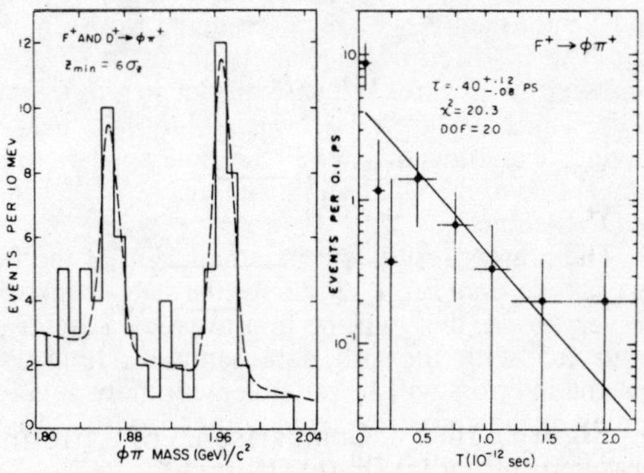

Figure 8. (a) Kππ mass distribution for F^+ and $D^+ \to \phi\pi^+$ (and c.c.). (b) Proper time distribution for F signal in (a).

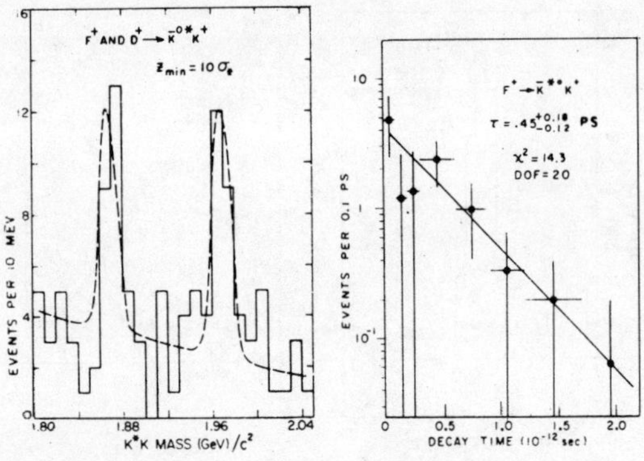

Figure 9. (a) $K\pi K$ mass distribution for F^+ and $D^+ \to \bar{K}^{*0}K^+$ (and c.c.). (b) Proper time distribution for signal in (a).

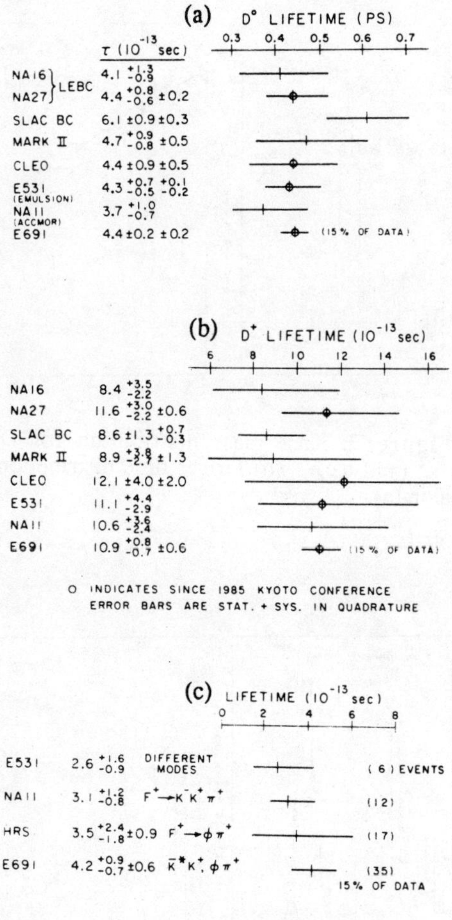

Figure 10. Comparison of lifetime measurements for (a) D^0, (b) D^+, (c) F^+.

COMMENTS ON THE DIFFERING LIFETIMES OF CHARMED HADRONS

R. Rückl

Deutsches Elektronen Synchrotron, DESY
D-2000 Hamburg, Fed. Rep. Germany

Theoretical explanations of the observed lifetime differences of weakly decaying charmed hadrons are briefly reviewed. It is argued that the previous uncertainties in determining the dominant dynamical origin of the D^+-D^0 lifetime difference are resolved by the recent development of a consistent theory of two-body decays which constitute a large fraction of all nonleptonic D decays. Predictions on the lifetime hierarchy of charmed baryons are also presented.

Much experimental and theoretical work has been devoted to the lifetime differences of weakly decaying charmed hadrons since the discovery of this originally unexpected fact in 1979. The present experimental status is summarized in Table 1. We see that

$$\tau(D^+) > \tau(D^0) \simeq \tau(D_c^+) \qquad (1)$$
$$\tau(\Lambda_c^+) < \tau(D^0),\ \tau(\Lambda_c^+) < \tau(\Xi_c^+)$$

where the inequalities amount to factors 1.5-2.5. In the meantime, also theory has recognized that lifetime differences ought to be

	$\tau[10^{-13}\text{sec}]$	$B_{SL}[\%]$
D^+	$10.31^{+0.52}_{-0.44}$ (1)	17.0±1.9±0.7 (3)
D^0	$4.30^{+0.20}_{-0.19}$ (1)	7.5±1.1±0.4 (3)
D_c^+	$3.5^{+0.6}_{-0.5}$ (1)	-
Λ_c^+	$1.9^{+0.5}_{-0.3}$ (1)	4.5±1.7 (4)
Ξ_c^+	$4.8^{+2.9}_{-1.8}$ (2)	-

Table 1: Charmed particle lifetimes and semileptonic branching ratios

expected at some level due to non-asymptotic bound state effects. However, the usual calculational problems with QCD in the nonperturbative confinement regime and uncertainties in the choice of some parameters such as the scale in $\alpha_s(\mu^2)$ and effective quark masses have made it difficult to pin down the dominant origin of the lifetime differences and to arrive at firm quantitative estimates (5). The situation has changed considerably with the recent progress accomplished on the experimental side, in particular by the Mark III collaboration (6), and the simultaneous development of a respectable theory of exclusive decays (7-9). In this talk, I shall describe the more solid understanding of the hierarchy of lifetimes which has emerged from these symbiotic efforts.

The common starting point of theoretical considerations of weak decays is an effective Hamiltonian (10) derived from the charged current interactions of the standard electroweak gauge model and incorporating short-distance QCD corrections. In the case of Cabibbo-allowed nonleptonic charm decays, one has (5)

$$H_{NL}^{eff} \simeq (G_F/\sqrt{2})(c_+O_+ + c_-O_-) \qquad (2)$$
$$O_\pm = (\bar{u}d)_L(\bar{s}c)_L \pm (\bar{s}d)_L(\bar{u}c)_L$$

where $(\bar{a}b)_L = \bar{a}\gamma\mu(1-\gamma_5)b$ and $c_- = c_+^{-2} \simeq [\alpha_s(m_c^2)/\alpha_s(m_W^2)]^{0.48}$. The decay amplitudes are then given by matrix elements of H_{NL}^{eff} between hadronic states. It is mainly our present inability to calculate $\langle O_\pm \rangle$ from first principles which gives rise to ambiguities and quantitative uncertainties. Making use of hadron-quark duality which is known to work in other applications such as $e^+e^- \to$hadrons and spectroscopy (11) one can approach lifetime questions in heavy flavor decays in two ways: "exclusively" by considering the (dominant) physical decay channels,

$$\tau^{-1} = \Sigma[\Gamma(l\nu_l \text{hadrons}) + \Gamma(\text{hadrons})], \qquad (3)$$

and "inclusively" by studying the dual decays into free quarks,

$$\tau^{-1} = \Sigma[\Gamma(l\nu_l \text{quarks}) + \Gamma(\text{quarks})]. \qquad (4)$$

The "exclusive" way is in principle more straightforward, but more difficult in practice. It involves hadronic physics in its full complexity. The "inclusive" way, on the other hand, is conceptually simpler, but less direct and a priori approximative.

Table 2 exemplifies expectations on the D meson lifetimes and semileptonic branching ratios resulting from the inclusive approach. The numbers represent typical estimates and are subject to considerable uncertainties as

approximations	$B_{SL}(D^0)$	$B_{SL}(D^+)$	$\frac{\tau(D^+)}{\tau(D^0)}$
(a) spectator model free quarks	20%	20%	1
(b) spectator model QCD corrected	15%	15%	1
(c) including quark interference	15%	19%	1.3
(d) dropping non-leading terms in $1/N_c$	12%	19%	1.6
(e) including W-exchange	<12%	19%	>1.6

Table 2: Theoretical expectations from the inclusive description of D decays

discussed extensively in the literature (5). A few comments are in order:

(a) The quark predictions simply reflect the number of available decay channels.

(b) Short-distance QCD corrections enhance the nonleptonic decay rates and, hence, decrease the semileptonic branching ratios, but of course do not affect the equality of the lifetimes in the spectator model (12).

(c) Quark interference is a consequence of the Pauli principle (13) and occurs in the decay $D^+=[\bar{d}c]\rightarrow\bar{d}su\bar{d}$ due to the presence of two identical quarks ($\bar{d}$) in the final state. No such interference takes place in Cabibbo-allowed D^0 decay. As a result, the lifetime of the D^+ is lengthened in comparison with the D^0 and the semileptonic branching ratio $B_{SL}(D^+)$ increases (13,14).

(d) The prescription to expand the nonleptonic rates in powers of $1/N_c$, N_c being the number of color degrees of freedom, and to drop all non-leading terms is suggested by two arguments. Firstly, in the previous approximations one has added up both leading and certain non-leading contributions, from the point of view of the $1/N_c$-expansion, but has neglected many other non-leading terms. Since the latter are difficult to calculate due to the genuine nonperturbative nature of the $1/N_c$-expansion it appears more consistent (8,15) to drop all non-leading terms. Secondly, a similar procedure (8) remarkably improves the agreement between theory and experiment in the case of two-body decays as discussed later. The net effect of following the above prescription is a stronger nonleptonic enhancement of the spectator model rates accompanied by a similar reinforcement of the destructive interference in the D^+ case. Thus, the lifetime and semileptonic branching ratio of the D^0 decrease in comparison with (b) and (c), whereas the expectation (c) on the D^+ is little affected (8).

(e) W-exchange between the charm quark and the light constituent quark is only possible in D^0 decay, $D^0=[\bar{u}c]\rightarrow\bar{d}s$, but not in Cabibbo-allowed D^+ decay. Thus, W-exchange potentially shortens the lifetime of the D^0 relative to the D^+ and decreases $B_{SL}(D^0)$ as desired. However, the effect is totally negligible unless, due to the presence of gluons in the D^0 (16), the $\bar{u}c$-component has a large probability to carry spin 1. It is clearly very difficult (5) to obtain a reliable estimate of this probability.

To conclude, the inclusive approach provides a qualitative understanding of the observed D^+-D^0 lifetime difference as arising from destructive quark interference (D^+) and/or gluon enhanced W-exchange (D^0). However, interference does not appear efficient enough to explain the whole effect quantitatively while the gluon enhancement of W-exchange is extremely difficult to quantify at all. Hence, the dominant dynamical origin of the D^+-D^0 lifetime difference remains somewhat dubious.

Recently, in response to the large amount of Mark III data (17), theory has focussed on exclusive decays. These data indicate that the two-body channels (including resonances) constitute a large fraction of all nonleptonic D decays. Given this fact it is clear that the correct theory of two-body decays will also explain the major part of the D^+-D^0 lifetime difference. Despite the notorious problems (5) one encounters in developing a reliable quantitative framework one has finally succeeded as I believe. I shall briefly describe the main recent steps towards a consistent theory and point out the implications on the lifetime issues raised above.

Bauer, Stech and Wirbel (7) have performed a comprehensive phenomenological analysis of two-body decays assuming
(A) factorization of $\langle H^{eff}_{NL}\rangle$ in products of matrix elements of quark currents. Approximating the various form factors by the nearest meson pole and taking into account known final state interactions they have shown that consistency with the Mark III data requires the additional assumption
(B) of no color mismatch, i.e. formation of final mesons only from quarks which belong to the same color singlet currents in eq.(2). As a direct consequence of assumption (A), contributions from W-exchange (and annihilation) vanish in the chiral limit and, hence, are negligible. Adding up the exclusive rates for D→PP, PV, VV, $l\nu_l P$, $l\nu_l V$ where P and V denote pseudoscalar and vector mesons, respectively, one obtains the partial lifetimes

$$\tau \simeq \begin{cases} 5.6\cdot10^{-13}\ \text{sec for } D^0 \\ 11.3\cdot10^{-13}\ \text{sec for } D^+. \end{cases} \quad (5)$$

Eq.(5) and Table 1 imply that the above channels account for 80-90% of the total D^0 and D^+ decay widths. More importantly, the lifetime ratio $\tau(D^+)/\tau(D^0) \simeq 2$ following from eq.(5) originates solely in destructive interferences of D^+ amplitudes. It thus becomes clear that interference is also the dominant origin of the inclusive D^+-D^0 lifetime difference, provided somebody proves the crucial assumptions (A) and (B) stated above. A second problem to be solved in this model is the $\bar{K}^0\phi$ puzzle. The decay $D^0\to\bar{K}^0\phi$ has been observed in several experiments (18) with a branching ratio ~ 1%, a value at least one order of magnitude larger than expected (19) unless this channel is strongly fed by final state interactions (20).

Buras, Gérard and the speaker (8) have systematically applied <u>$1/N_C$-expansion techniques</u> developed for strong interaction meson physics (21) to weak decays of mesons. They have shown that to leading order in $1/N_C$ the mesonic matrix elements of four-quark operators factorize and that there is no color mismatch. The next-to-leading order in $1/N_C$ then includes factorizable as well as non-factorizable contributions and also the final state interactions. In the original estimates (22) the non-leading factorizable terms have been added to the leading ones, whereas all other non-leading contributions have been neglected. This is theoretically inconsistent and has caused serious phenomenological problems (5) in channels like $D^0\to\bar{K}^0\pi^0$. On the other hand, if only the leading terms in the $1/N_C$-expansion are kept the assumptions (A) and (B) quoted before hold exactly and one arrives essentially at the phenomenologically successful model of Bauer et al. (7). Thus, if one can further prove that the non-leading contributions are indeed small despite the worrying fact that $N_C=3$ is not a very big number, the main problem is solved. Of course, the $\bar{K}^0\phi$ puzzle still needs to be explained.

Both of the remaining tasks have been accomplished by Blok and Shifman in a recent series of papers (9). Using <u>QCD sum rule methods</u> they have actually calculated the non-factorizable amplitudes. I repeat, these are the previously unknown next-to-leading terms in the $1/N_C$-expansion (8). The important results for the present discussion can be summarized as follows:

1) In most channels the non-leading factorizable and non-factorizable contributions to the c-quark decay amplitudes cancel to a large extent. This justifies the truncation of the $1/N_C$-expansion anticipated in ref.(8).

2) No such cancellation occurs in the W-exchange (or annihilation) amplitudes since the factorizable terms vanish in the chiral limit as pointed out earlier. The non-factorizable terms, on the other hand, contribute only ~ 20% to the inclusive width of the D^0. This is nicely consistent with their non-leading (in $1/N_C$) nature. Nevertheless, the contribution to the special $D^0\to\bar{K}^0\phi$ channel is sufficient to explain the observed (18) branching ratio ~ 1%.

In summary, a consistent theory of two-body decays has emerged which is in satisfactory quantitative agreement (7-9) with the existing data (17). Final state interactions, not considered in refs.(8,9) and only partially in ref.(7), need some further thoughts. However, the latter deficiency does not call in question the following <u>conclusion</u>.

The physics of two-body decays provides substantial evidence for destructive interferences of D^+ decay amplitudes as the main origin of the observed D^+-D^0 lifetime difference. The existence of these interferences can be traced to the presence of two identical quark flavors in the decay $D^+=[\bar{d}c]\to\bar{d}s u\bar{d}$. One thus returns to the same picture suspected in the inclusive valence quark description. This picture can be further tested in F^+ and charmed baryon decays.

I finish my talk with a few remarks on <u>charmed baryon lifetimes</u>. Similarly as in meson decays, interference effects are expected (5,23-25) to give rise to lifetime differences. However, in contrast to the meson case, W-exchange between valence quarks of baryons is not helicity suppressed. Hence, it should have more pronounced influence on the lifetime pattern (23-26). Typical quark model estimates (24) yield

$$\Gamma_{dec} : \Gamma^-_{int} : \Gamma^+_{int} : \Gamma_{W\text{-}exch} \simeq \qquad (6)$$

$$\simeq \begin{cases} 1 : -0.6 : +0.9 : 2 & \text{nonrel.} \\ 1 : -0.2 : +0.4 : 0.6 & \text{bag.} \end{cases}$$

The appearence of two interference terms with opposite signs corresponds to the presence of two light valence quarks in charmed baryons. Depending on the quark structure of Λ_c^+, $\Xi_c^{+,0}$ and Ω_c^0, the non-spectator effects contribute in different combinations to the inclusive non leptonic widths and thus generate a rather unique lifetime hierarchy. From eq.(6) one predicts (24)

$$\tau(\Omega_c^0) : \tau(\Xi_c^0) : \tau(\Lambda_c^+) : \tau(\Xi_c^+) \simeq \qquad (7)$$

$$\simeq \begin{cases} 0.6 : 0.6 : 1 : 1.6 & \text{nonrel.} \\ 0.7 : 0.7 : 1 : 1.2 & \text{bag.} \end{cases}$$

Qualitatively, the resulting pattern is completely determined by the general properties of H^{eff}_{NL} and the baryonic bound states and can be predicted reliably. On the other hand, the actual size of the lifetime differences is subject to the usual uncertainties of the inclusive approach as exemplified in eqs.(6) and (7). It is , therefore, reassuring to see similar lifetime ratios emerging from an analysis of two-body decays (27):

$$\hat{\tau}(\Omega_c^0) : \hat{\tau}(\Xi_c^0) : \hat{\tau}(\Lambda_c^+) : \hat{\tau}(\Xi_c^+) \simeq \qquad (8)$$

$$\simeq 0.7 : 0.7 : 1 : 2.$$

At any rate, it appears relatively easy to disentangle and determine the individual effects from interference and W-exchange (24), once sufficiently accurate data become available. The observed lifetime differences of D^0, Λ_c^+ and Ξ_c^+ exhibited in Table 1 and eq.(1) follow the expectation that the Λ_c^+ decays fastest because of W-exchange. This lends further support to the overall picture presented in this talk.

I want to thank my collaborators, in particular A.J. Buras, J.-M. Gérard and B. Guberina, for sharing their experience in this field with me. I also apologize for neither having expressed other points of view nor given a more complete list of references. This is only due to the lack of space.

REFERENCES

(1) Gilchriese, M.G.D., Rapporteur Talk, these Proceedings

(2) Biagi, S.F. et al. (CERN SPS EXP WA62), Z.Phys. C28(1985)175

(3) Baltrusaitis, R.M. et al. (Mark III), Phys.Rev.Lett. 54(1985)1976

(4) Vella, E. et al. (Mark II), Phys.Rev. Lett. 48(1982)1515

(5) see e.g. Rückl, R., Weak Decays of Heavy Flavors, CERN print (1983) and Flavor Mixing in Weak Interactions, ed. L.-L. Chau (Plenum Press, New York, 1984)681, and references therein

(6) Hitlin, D., (Mark III), to appear in Proc.Int.Symp. on Production and Decay of Heavy Hadrons, Heidelberg, 1986; Schindler, R.H., (Mark III), these Proceedings

(7) Bauer, M. and B. Stech, Phys.Lett. 152B (1985)380; Wirbel, M., to appear in Proc.Int.Symp. on Production and Decay of Heavy Hadrons, Heidelberg, 1986

(8) Buras, A.J., J.-M. Gérard and R. Rückl, Nucl.Phys. B268(1986)16

(9) Blok, B.Yu. and M.A. Shifman, ITEP 86-9, 17, 37

(10) Gaillard, M.K. and B.W. Lee, Phys. Rev. Lett. 33(1974)108; Altarelli, G. and L. Maiani. Phys.Lett. 52B(1974)351

(11) Reinders, L.J., H. Rubinstein and S. Yazaki, Phys.Rep. 127(1985)1

(12) Ellis, J., M.K. Gaillard and D.V. Nanopoulos, Nucl.Phys. B100(1975)313

(13) Peccei, R.D. and R. Rückl, Proc. Ahrenshoop Symp. on Special Topics in Gauge Field Theories (Akad.Wiss., Zeuthen, GDR, 1981)8; Kobayashi, T. and N. Yamazaki, Prog. Theor.Phys. 65(1981)775

(14) Guberina, B. et al., Phys.Lett. 89B (1979)111; Koide, Y., Phys.Rev. D20(1979)1739

(15) Shifman, M.A. and M.B. Voloshin, Yad.Fiz. 41(1985)187

(16) Fritzsch, H. and P. Minkowski, Phys. Lett. 90B(1980)455; Bernreuther, W., O. Nachtmann and B. Stech, Z.Phys. C4(1980)257; Rosen, S.P., Phys.Rev.Lett. 44(1980)4; Bander, M., D. Silverman and A. Soni, Phys.Rev.Lett. 44(1980)7, 962

(17) see ref.(5) and summary of Mark III data by Thorndike, E.H., Proc. 1985 Int.Symp. on Lepton and Photon Interactions at High Energies, ed. Konuma, M. and K. Takahashi (Kyoto, 1986)

(18) Albrecht, H. et al. (ARGUS), Phys.Lett. 158B(1985)525; Bebec, C. et al. (CLEO), Phys.Rev.Lett. 56(1986)1893; Baltusaitis, R.M. et al. (Mark III), Phys.Lett. 56(1986)2136

(19) Baur, U., A.J. Buras, J.-M. Gérard and R. Rückl, Phys.Lett. 175B(1986)377

(20) Donoghue, J.F., HEP-241(1986)

(21) t'Hooft, G., Nucl.Phys. B72(1974)641; B75(1974)461; Rossi, G. and G. Veneziano, Nucl.Phys. B123(1977)507; Witten, E., Nucl.Phys. B160(1979)57

(22) Fakirov, D. and B. Stech, Nucl.Phys. B133(1978)315; Cabibbo, N. and L. Maiani, Phys.Lett. 73B(1978)418

(23) Rückl, R. Phys.Lett. 120B(1983)449

(24) Guberina, B., R. Rückl and J. Trampetić, DESY 86-093(1986)

(25) Shifman, M.A. and M.B. Voloshin, ITEP 86-83

(26) Barger, V., J.P. Leveille and P.M. Stevenson, Phys.Rev.Lett. 44(1980)226

(27) Körner, J.G., G. Kramer and J. Willrodt, Phys.Lett. 78B(1978)492; Z.Phys. C2(1979)117

EVIDENCE FOR B^o-$\overline{B}^o$ MIXING IN DIMUON EVENTS IN THE UA1 EXPERIMENT AT THE CERN PROTON-ANTIPROTON COLLIDER

N. Ellis,

Birmingham University, England

UA1 Collaboration

We report on evidence for B^o-$\overline{B}^o$ mixing in dimuon events at the CERN $p\overline{p}$ collider. We find that the ratio of numbers of like-sign to unlike-sign muon pairs is $0.46 \pm 0.07 \pm 0.03$ for events in which the muons are not isolated, which are mainly due to semi-leptonic heavy flavour decays. The expected ratio without B^o-$\overline{B}^o$ mixing is 0.26 ± 0.03, where the only significant known source of like-sign events is second generation beauty decays. We conclude that the most natural explanation for the excess of like-sign events is B^o-$\overline{B}^o$ mixing, and measure $\chi = 0.142 \pm 0.045$ (χ is the fraction of "wrong sign" beauty decays).

1. INTRODUCTION

Oscillations between the states K^o and $\overline{K}^o$ are due to mixing between flavours in weak interactions; the Cabibbo angle is non-zero. Such $\Delta S = 2$ transitions can be understood in terms of "box diagrams" where two W particles are exchanged. Since the discovery of the new quantum numbers, charm and beauty, it is natural to consider the possibility of mixing (flavour oscillations) in neutral mesons other than K^o (1). Stringent upper limits have been placed on the degree of mixing in the D^o system (2); possible mixing in the B^o system is discussed in this paper. We label the states $B_d{}^o = (\overline{b}d)$ and $B_s{}^o = (\overline{b}s)$ according to their quark content.

The recently measured lifetimes of beauty particles (3) suggest that mixing may be significant in the B^o-$\overline{B}^o$ system. The degree of mixing can be expressed as:

$$r = \frac{\mathrm{Prob}(b \to \overline{B}^o)}{\mathrm{Prob}(b \to B^o)} = \frac{(\Delta M/\Gamma)^2}{[2+(\Delta M/\Gamma)^2]}$$

assuming $\Delta\Gamma \ll \Delta M$ which is expected to be valid for the B^o system (4) and with the convention that the B^o state contains a $\overline{b}$ quark. $\Delta\Gamma$ is the difference between the decay widths of the mass eigenstates. The mass difference, ΔM, can be calculated (5) using the experimentally determined values of elements in the Kobayashi-Maskawa matrix (6). This calculation gives $\Delta M/\Gamma\,(B_d{}^o) < 0.1$ and $\Delta M/\Gamma\,(B_s{}^o)$ in the range 1-4. The corresponding values for the degree of mixing are $r_d < 0.005$ and $r_s = 0.33$-0.89. Thus no significant mixing is expected for $B_d{}^o$, but substantial mixing is predicted for $B_s{}^o$.

A signature for B^o-$\overline{B}^o$ mixing is an excess of like-sign dimuon events compared to expectations for second generation beauty decays. Beauty mesons containing a b quark decay only to negative muons. Similarly mesons containing $\overline{b}$ antiquarks give only positive muons. Thus in first generation, decays $p\overline{p} \to b\overline{b}$ followed by $b \to \mu^- \overline{\nu} X$ and $\overline{b} \to \mu^+ \nu X$, only unlike-sign dimuons are produced. Like-sign dimuons are produced in second generation decays, for example, $p\overline{p} \to b\overline{b}$ followed by $b \to \mu^- \overline{\nu} X$ and $\overline{b} \to \overline{c} X$ with $c \to \mu^- \nu X$. However, second generation decay events tend to be suppressed by the selection criteria. If mixing occurs the $\overline{b}$ antiquark in the B^o meson may, for example, be replaced by a b quark in the $\overline{B}^o$ resulting in a $\mu^-\mu^-$ event from first generations decays.

Experiments at e^+e^- colliders have recently placed limits on B^o-$\overline{B}^o$ mixing. ARGUS (7) and CLEO (8) both exclude substantial mixing in the $B_d{}^o$-$\overline{B}_d{}^o$ system by measuring the rate of like-sign dileptons produced on the $\Upsilon(4S)$ resonance. By using the $\Upsilon(4S)$ they obtain high statistics samples of $B\overline{B}$ events with relatively little background. However, the $\Upsilon(4S)$ is below the threshold for producing $B_s{}^o\overline{B}_s{}^o$ pairs and consequently they have no sensitivity to mixing in the $B_s{}^o$-$\overline{B}_s{}^o$ system. MARK II (9) have examined dilepton events produced in e^+e^- collisions at $\sqrt{s} = 29$ GeV for evidence of B^o-$\overline{B}^o$ mixing. While they are in principle sensitive to $B_s{}^o$-$\overline{B}_s{}^o$ mixing, they have too few events to place a significant limit on this process.

In this paper we describe the evidence for B^o-$\overline{B}^o$ mixing in UA1 dimuon data, comparing the observed number of like-sign events with the number expected from second generation beauty decays, allowing for the background due to misidentified hadrons in the data.

2. DATA SAMPLE AND BACKGROUND ESTIMATE

Because of the large cross-section for beauty production at the CERN $p\bar{p}$ collider (~10 μb), semileptonic decays of beauty particles are the dominant source of high p_T muon pairs. We define p_T as the muon momentum transverse to the beam direction. We use the sample of 512 dimuon events, with $m_{\mu\mu} > 6$ GeV/c^2 and with $p_T > 3$ GeV/c for each muon, excluding $Z^o \rightarrow \mu^+\mu^-$ decays. The selection of these events is described in Ref. (10). The data were recorded during three collider runs at $\sqrt{s} = 546$ and 630 GeV with a total integrated luminosity of 692 nb^{-1}. The isolation of the muons (i.e. the absence of hadronic activity around each muon) is used to separate Drell-Yan and Υ decays from heavy flavour events. We define $S = [\Sigma E_T(\mu_1)]^2 + [\Sigma E_T(\mu_2)]^2$ where $\Sigma E_T(\mu)$ is the scalar sum of the transverse energy measured in calorimeter cells in a cone of $\Delta R = (\Delta\phi^2 + \Delta\eta^2)^{1/2} < 0.7$ about the muon; ϕ is the azimuth angle and η is pseudo-rapidity. We classify dimuons as isolated if $S < 9$ GeV^2. There are 98 unlike-sign and 15 like-sign events which satisfy this criterion. In the non-isolated sample there are 257 unlike-sign and 142 like-sign events. The charges of the muons are all well determined.

The background to the dimuon sample due to mis-identified hadrons has been studied extensively. Candidate single muon events recorded in the UA1 experiment were examined. The probability for charged particles other than the muons in these events (presumably hadrons) to be mis-identified as muons was calculated allowing for decays in flight of pions and kaons and particles penetrating the absorber (>9 nuclear interaction lengths) without interaction. This calculation includes both the background of one genuine, prompt muon and one "fake", and the background of two "fake" muons. the estimated background is 95 events from decays in flight and 2 events from non-interacting hadrons.

The background due to the mis-association of a high-p_T track in the central chamber to a track in the muon chambers caused by a muon of lower p_T has been studied by examining the dimuon events themselves. The estimated background from this source is 10 events.

The uncertainty in the total calculated background of 107 events is estimated to be ± 25%, largely due to the unknown fractions of high-p_T particles which are pions, kaons and protons, which affects the decay background. The calculation is made for $f_\pi = 0.58$ and $f_K = 0.21$.

3. EVIDENCE FOR B^o-$\bar{B}^o$ MIXING

In the absence of flavour mixing, the only significant known source of like-sign dimuon events is second generation $b\bar{b}$ decays: $p\bar{p} \rightarrow b\bar{b}$; $b \rightarrow c\mu^-\bar{\nu}$ and $\bar{b} \rightarrow \bar{c}X$, followed by $\bar{c} \rightarrow \bar{s}\mu^-\bar{\nu}$. The rate for this process relative to first generation decays: $p\bar{p} \rightarrow b\bar{b}$; $b \rightarrow c\mu^-\bar{\nu}$ and $\bar{b} \rightarrow \bar{c}\mu^+\nu$ can be calculated reliably since it depends mainly on weak decays and measured branching ratios. We have recently made a detailed comparison between QCD Monte Carlo programs (11) and measured beauty and charm particle decay properties (12) and are confident that they correctly describe jets containing beauty particles. Dimuons due to $c\bar{c}$ production are also included in the Monte Carlo calculations and our measurements of the relative transverse momentum between the muons and their accompanying jets (p_T^{rel}) described in Ref. (10) is consistent with the prediction that they account for about 10% of dimuons from heavy flavour decays.

For no B^o-$\bar{B}^o$ mixing, the predicted ratio of the numbers of like-sign and unlike-sign dimuons from $b\bar{b}$ and $c\bar{c}$ production: $R = N(\pm\pm)/N(+-)$ is 0.26 ± 0.03. This can be compared with a variety of independent calculations (11, 13) which estimate R in the range 0.21 - 0.26. The possible effect of $t\bar{t}$ production has been considered; for a top mass of 25 GeV/c we would expect 11 like-sign and 25 unlike-sign events passing the selection criteria, assuming a production cross-section of 13 nb. Including these additional events, the prediction for R increases to only 0.27. The effect is negligible for higher top masses. The uncertainty on the prediction for no mixing has been estimated by propagating the errors on the average beauty and charm hadron muonic branching ratios (14) and varying the $c\bar{c}$ contribution by ±50% to account for the uncertainty in the parametrization of the charm fragmentation function. Predictions for full B_s^o mixing give R in the range 0.28 - 0.50 depending mainly on the fraction of dimuon events due to B_s^o decays and the contribution from $c\bar{c}$ as shown in Table 1.

We measure R by two methods: (i) using only the non-isolated events for which the contribution from Drell-Yan and $\Upsilon \to \mu^+\mu^-$ decays is small, and (ii) using all the events and subtracting the measured contribution from $\Upsilon \to \mu^+\mu^-$ (10) and the calculated number of Drell-Yan events (15). For method (i) we obtain R = 0.46 ± 0.07 ± 0.03 and for method (ii) R = 0.48 ± 0.07 ± 0.05. For both methods the systematic error reflects the uncertainty in background subtraction. Adding the statistical and systematic errors in quadrature, together with the error on the predicted value of R, the result from method (i) is 2.6 standard deviations from the prediction for no mixing.

Assuming the existence of mixing, the expected value of R can be written in terms of $\chi = BR(b \to \mu^+ X)/BR(b \to \mu^{\pm} X)$ i.e. the fraction of "wrong sign" decays:

$$R = \frac{2\chi(1-\chi)N_f + [(1-\chi)^2+\chi^2]N_s}{[(1-\chi)^2 + \chi^2]N_f + 2\chi(1-\chi)N_s + N_c}$$

where N_f and N_s are the numbers of dimuon events due to first and second generation $b\bar{b}$ decays and N_c is the number due to $c\bar{c}$ decays. Using this expression and the predicted ratio $N_s/N_f = 0.30$, and the predicted charm contribution $N_c/[N_f + N_s + N_c] = 0.10$, then from the measured value R = 0.46 we deduce $\chi = 0.13$.

Since the muons from second generation decays have a softer p_T spectrum than those from first generation decays, we used the distributions of the p_T's of the two muons for like- and unlike-sign dimuon events in a likelihood fit to determine χ. The two-dimensional muon p_T distributions ($p_T(\mu_1)$ versus $p_T(\mu_2)$) for first and second generation decays of pairs of beauty particles and decays of pairs of charm particles were determined using the ISAJET Monte Carlo program (11). The uncertainties in the relative normalizations were obtained by propagating the errors on the beauty and charm muonic branching ratios and the $c\bar{c}$ contribution as above. The calculated background distributions were used, with a ±25% error on the normalization. The likelihood curve is shown in Fig. 1, giving $\chi = 0.142 \pm 0.045$. No mixing (χ=0) is disfavoured with a likelihood ratio of 1:110 or 3.1 standard deviations.

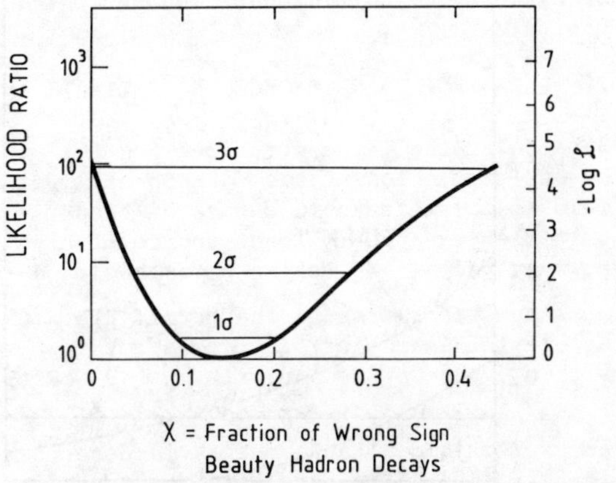

Fig. 1 Likelihood curve for χ obtained by fitting to the observed distributions of the p_T's of the two muons in the like- and unlike-sign samples.

The measured value of χ is an average of χ_d and χ_s corresponding to B_d^o and B_s^o mixing:

$$\chi = \frac{BR_d f_d \chi_d}{\langle BR \rangle} + \frac{BR_s f_s \chi_s}{\langle BR \rangle}$$

where $BR_{d,s}$ are the muonic branching ratios for B_d^o and B_s^o decays, $\langle BR \rangle$ is the average beauty hadron muonic branching ratio and f_d and f_s are the fractions of beauty quarks hadronizing into B_d^o and B_s^o mesons.

We express our results in terms of the mixing parameters:

$$r_{d,s} = \frac{BR(B_{d,s}^o \to \mu^- X)}{BR(B_{d,s}^o \to \mu^+ X)} = \frac{\chi_{d,s}}{[1-\chi_{d,s}]}$$

In Fig. 2, we show the 90% confidence level limits for r_d and r_s coming from ARGUS (7) and MARK II (9) as well as from our own measurement. The curves for MARK II and UA1 are calculated for $f_d = 0.40$ and $f_s = 0.20$, assuming equal semileptonic branching ratios for the different beauty particles. The case of no B_s^o-$\bar{B}_s^o$ mixing ($r_s = 0$) is excluded by our measurement when combined with the ARGUS limit on r_d. The allowed region overlaps with the predicted values (5): $r_d < 0.005$, $r_s = 0.33 - 0.89$.

Assuming the measured beauty lifetime of $\tau_B = (1.12 \pm 0.16) \times 10^{-12}$s (3) is also valid for B_s^o decays, the limit $r_s > 0.2$

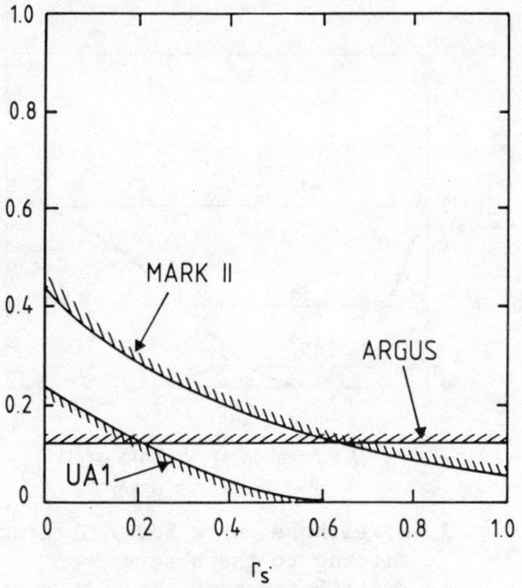

Fig. 2 Summary of upper and lower limits for mixing in the B_d^0 and B_s^0 systems from UA1 and other experiments. The MARK II and UA1 curves are for $f_d = 0.40$ and $f_s = 0.20$.

given by the intersection of the ARGUS and UA1 curves in Fig. 2 can be expressed as a limit of $\Delta M(B_s^0) > 5 \times 10^{-10}$ MeV compared to $\Delta M(K^0) = (3.521 \pm 0.014) \times 10^{-12}$ MeV.

4. CONCLUSIONS

We observe an excess of like-sign dimuon events in our data compared to expectations for second generation beauty particle decays. We measure $R = N[\pm\pm]/N[+-]$ by two methods which give $R = 0.46 \pm 0.07 \pm 0.03$ and $R = 0.48 \pm 0.07 \pm 0.05$. The prediction without mixing is $R = 0.26 \pm 0.03$. The most natural explanation for the large fraction of like-sign dimuon events is $B^0-\bar{B}^0$ mixing.

ACKNOWLEDGEMENTS

I would like to thank all my colleagues in UA1 who helped to make these results possible, particularly M. Della Negra, M. Jimack, H.-G. Moser and K. Wacker. I would also like to thank J. Dowell, K. Eggert, A. Norton and C. Rubbia for their help and advice. I am grateful for the support of Birmingham University and of the Royal Society.

REFERENCES

(1) Ellis, J. et al., Nuclear Physics B <u>131</u> (1977) 285; Ali, A. and Aydin, Z., Nuclear Physics B <u>148</u> (1979) 165.

(2) Louis, W. et al., Phys. Rev. Lett. <u>56</u> (1986) 1027.

(3) Luth, V., preprint CERN-EP/85-142 and references therein.

(4) Ali, A., preprint DESY 85-107. See also: Ali, A., and Jarlskog, C., Phys. Lett. B <u>144</u> (1984), 266; Barger, V., and Phillips, R.J.N., Phys. Rev. Lett. <u>55</u> (1985) 2752 and Phys. Lett. B <u>143</u> (1984), 259; Bigi, I.I. and Sanda, A.I., Nuclear Physics B <u>193</u> (1981), 85 and Phys. Rev. D <u>29</u> (1984), 1393.

(5) Ali, A., "Heavy Quark Physics at LEP", Physics at LEP vol. II; CERN yellow report 86-02, p220.

(6) Kobayashi, M. and Maskawa, K., Prog. Theor. Phys. <u>49</u> (1973) 652. This paper contains a generalisation of the work of Cabibbo, N., Phys. Rev. Lett. <u>10</u> (1963) 531.

TABLE 1:

Predictions for $R = N[\pm\pm]/N[+-]$ with and without maximal $B_s^0-\bar{B}_s^0$ mixing

	NO MIXING	MAXIMAL B_s^0 MIXING			CHARM FRACTION
		$f_s = 0.10$	$f_s = 0.20$	$f_s = 0.30$	$\frac{N_c}{N_f+N_s+N_c}$
Barger et al.	0.25	0.31	0.36	0.42	0.23
Halzen et al.	0.25	0.33	0.41	0.48	0.11
Isajet	0.26	0.34	0.42	0.50	0.10
Eurojet	0.21	0.28	0.36	0.43	0.15

(7) The ARGUS collaboration report a 90% confidence level upper limit $r_d < 0.12$. Paper no. 9717 submitted to this conference.

(8) The CLEO collaboration report (Phys. Rev. Lett. 53 (1984) 1309) a 90% confidence level upper limit $r_d < 0.30$. They have recently improved this limit to $r_d < 0.24$ Paper no. 6203 submitted to this conference.

(9) Schaad, T. et al., Phys. Lett. B 160 (1985) 188.

(10) Markiewicz, T., UA1 collaboration, "Beauty Production at the CERN $\bar{p}p$ Collider", Presented at this conference.

(11) The Monte Carlo results which are presented were obtained with the ISAJET Program: Paige, F. and Protopopescu, S. D., preprint, BNL 38034 (1986). However, the branching ratios for beauty particle and J/ψ decays used were those compiled in the Eurojet program: Ali, A., Pietarinen, E., van Eijk, B., preprint CERN-EP/85-121.

(12) Chen, A. et al., Phys. Rev. Lett. 52 (1984) 1084; Green, J. et al., Phys. Rev. Lett. 51 (1983) 347; Csorna, S. E. et al., Phys. Rev. Lett. 54 (1985) 1894; Bacino, W. et al., Phys. Rev. Lett. 43 (1979) 1073; Bethke, S., DESY preprint 85-067 (1985) and references therein.

(13) Barger, V. and Phillips, R. J. N., preprints MAD-PH-155, 239 and 266; Halzen, F. and Martin, A., preprint DTP-84-14; Eurojet Monte Carlo program (11).

(14) The average semi-leptonic branching ratios for beauty and charm particle decays were taken from the measurements in Ref. (12): $BR(B \to e) = 12 \pm 0.7\%$, $BR(D \to e) = 13 \pm 1.3\%$.

(15) Altarelli, G. et al., Z. Phys. C. 27 (1985) 617.

Measurements of Average Bottom Hadron and D° Lifetimes at TASSO

D. Strom (University of Wisconsin-Madison)

ABSTRACT

Measurements of the average B hadron lifetime and the D° lifetime are reported. The measurements were performed with the TASSO detector at PETRA. Two new methods were used to determine the average B hadron lifetime. The preliminary result for the method based on vertex reconstruction is $\tau_B = (1.50^{+0.37}_{-0.29} \pm 0.28)$ ps, and the preliminary result using the dipole method is $\tau_B = (1.62^{+0.33}_{-0.29} \pm 0.25)$ ps. The D° lifetime was measured using reconstructed vertices to be $\tau_{D^\circ} = (4.3^{+2.0}_{-1.4} \pm 0.8) \times 10^{-13}$ s.

The addition of a high precision vertex detector[1)] to the TASSO detector allows the measurement of the lifetime of short-lived particles. This report describes the measurement of the average B hadron lifetime and the D° lifetime. Both measurements are based on data with an integrated luminosity of 47 pb^{-1} taken at a mean center of mass energy of 42.2 GeV.

Since it has not been possible to reconstruct B hadrons produced in high energy e^+e^- annihilations, the determination of the lifetime of bottom hadrons has depended on indirect methods. Most of the methods employed are based on the measurement of the impact parameter of high p_t leptons, assumed to have originated from B hadron decay.[2)] The TASSO Collaboration has previously used a similar method, but using all reconstructed tracks from events selected on the basis of event shape.[3)] The two new TASSO measurements use all hadronic events and are based on the B decay distance rather than on the impact parameter. The B decay distance was found to be a more sensitive measure of the lifetime than the impact parameter. The advantage in using all hadronic events is that the B hadron content of these events is well known. Given the indirect methods used in determining the B lifetime, different methods are desirable as they have different systematic errors.

The vertex method[4)] begins by selecting those events which were well contained in the detector (the cosine of the angle of the sphericity axis with respect to the beam was required to be less than 0.6). Each of the remaining events was divided into two jets. In each jet all possible vertices with three tracks were formed from among those tracks of high quality ($|p| > 0.6$ GeV/c, at least 5 out of a possible 8 vertex detector hits, $|z|_\circ < 3.0$ cm). Any vertex having three tracks of the same charge was rejected. The two dimensional decay distance of the vertex having the lowest χ^2 in each jet was then determined using the sphericity axis as the direction of flight of the B hadron and the center of the interaction region as the event vertex. The two dimension decay distance is given by

$$\ell_{2d} = \frac{x\sigma_{yy}t_x + y\sigma_{xx}t_y - \sigma_{xy}(xt_y + yt_x)}{\sigma_{yy}t_x^2 - 2\sigma_{xy}t_xt_y + \sigma_{xx}t_y^2} \quad (1)$$

Where σ is the sum of the decay error matrix and the error matrix representing the beam spot; t_x and t_y, are the two dimensional direction cosines of the decaying particle; and x and y are the difference between the reconstructed decay vertex and the center of the beam spot. Only those vertices having $|\ell_{2d}|$ less than 1.0 cm and having an error less than 0.1 cm are retained.

The decay distance distribution for the data and Monte Carlo are shown in figure 1. The data sample consisted of 3106 vertices and had an average two-dimensional decay distance, $\langle\ell_{2d}\rangle$, of 141±16μm.

Fig. 1

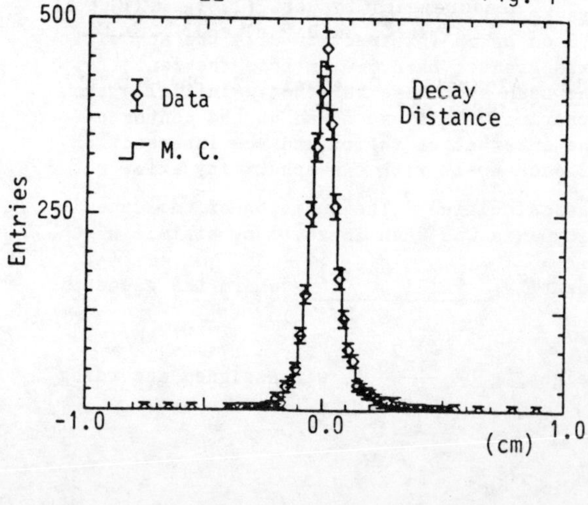

The value of the B lifetime was determined by a comparison with Monte Carlo: different B lifetimes were simulated by giving different weights to the vertices depending on the lifetime of the individual decay which produced them. The result of the χ^2 fit was $\tau_B = (1.50^{+0.37}_{-0.29})$ps.

The systematic errors were calculated by finding the sensitivity of $<\ell_{2d}>$ to uncertainties in the Monte Carlo parameters. The systematic error in the detector simulation results primarily from uncertainties in the vertex detector resolution and is ±0.19ps. The largest error in the simulation of e^+e^- annihilation came from uncertainties in the average boost of the B hadrons. This uncertainty, together with the uncertainties in charm quark fragmentation, resulted in a systematic error of ±0.14ps. Uncertainties in the B Hadron multiplicity contributed ±0.11ps to the systematic error. Varying the D° and $D^\pm$ lifetimes by the errors on their world averages,[5] and the D_s lifetime by twice the error on its world average, gave a systematic error of ±0.06ps. Finally, a systematic error of ±0.05ps was assigned due to any possible systematic effect in the fitting procedure.

Adding all of the possible systematic errors in quadrature, the preliminary result for the vertex method is

$$\tau_B = (1.50^{+0.37}_{-0.29} \pm 0.28)\text{ps}.$$

The second new approach which TASSO has used to determine the B lifetime is called the dipole method.[4] Only events which were well contained in the detector (the cosine of the sphericity axis with respect to the beam was required to be less than 0.7). The intersection of each track having $|p| > 0.2$ GeV/c and $|z_\circ| < 5$cm with the sphericity axis was then calculated as shown in figures 2 a) & b). To eliminate background from the decay productes of strange particles, those tracks with an impact parameter greater than 0.3 cm or an intersection with the sphericity axis greater than 0.9 cm from the center of the beam spot were rejected. Initially the sphericity axis was fixed at the center of the interaction region and the intersection of each track with the sphericity axis, r_i, was calculated. The position of the sphericity axis was then improved by minimizing the quantity $\frac{\Sigma g_i (r_i - <r>)^2}{\Sigma g_i}$ where the geometric weight, $g_i = \frac{\sin^2\alpha}{\sigma_{track}}$, was assigned according

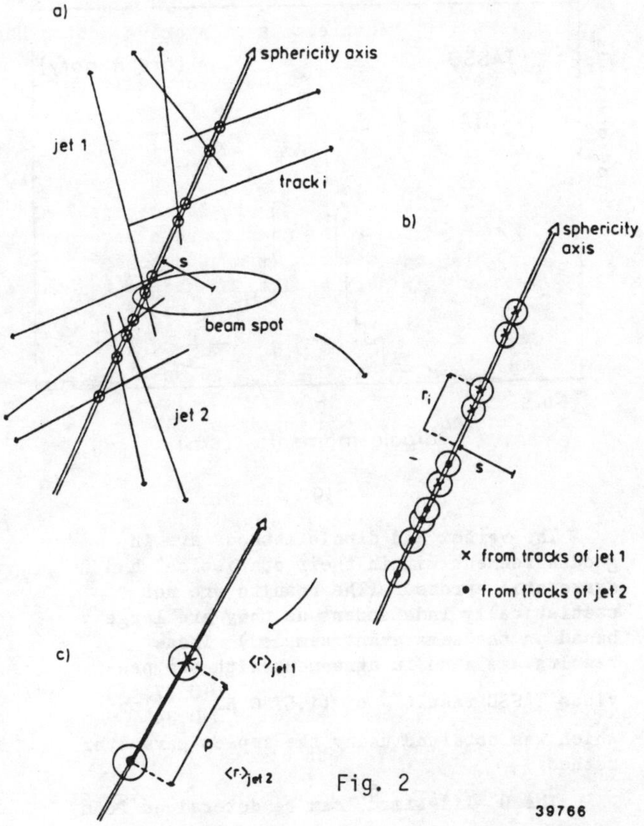

Fig. 2

39766

to the angle of the track with respect to the sphericity axis, α. After the position of the sphericity axis was found, the average intersection distance for each jet was calculated, and the difference was the dipole moment as shown in figure 2 c). In calculating the dipole moment, each track is weighted by its rapidity and by the geometric weight. The dipole moment distribution from the 4874 accepted events is shown in figure 3. The relation between the average B hadron lifetime and the mean dipole moment was determined by generating Monte Carlo samples with a variety of lifetimes. Using this relation and the mean dipole moment of the data of $328 \pm 28\ \mu m$, $\tau_B = (1.62^{+0.33}_{-0.29})$ ps was obtained.

The systematic errors were evaluated in the same manner as for the vertex method. The detector simulation contributed ±0.15ps, uncertainties in charm lifetime and production ±0.10ps, uncertainties in strange particle production ±0.06ps, and uncertainties in the charm and bottom fragmentation less than ±0.06ps. The sensitivity to the fragmentation scheme was less than ±0.11ps. Adding all of these effects in quadrature, the preliminary result for the dipole method is

$$\tau_B = (1.62^{+0.33}_{-0.29} \pm 0.25)\text{ps}.$$

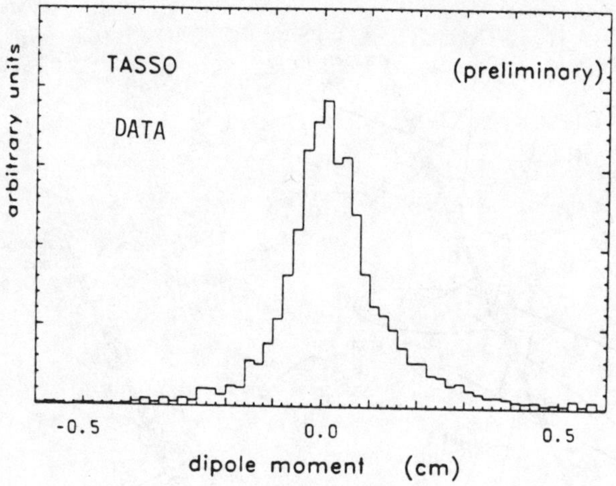

Fig. 3

The vertex and dipole methods are in good agreement within their statistical and systematic errors. (The results are not statistically independent as they are largely based on the same event sample.) These results are also in agreement with the previous TASSO result[3)] of $(1.57\pm0.32^{+0.37}_{-0.34})$ps which was obtained using the impact parameter method.

The D° lifetime[6)] can be determined from events identified via the decay $D^{*+} \to D^\circ \pi^+$ where the D° decays via $D^\circ \to K^-\pi^+$, $D^\circ \to K^-\pi^+\pi^\circ$, and $D^\circ \to K^-\pi^+\pi^-\pi^+$. (Here and elsewhere the charge conjugate state is implied.) The small Q value of the D^{*+} decay makes the spectrum of the mass difference between the reconstructed D^{*+} and the D° very narrow. A clean sample of decays can be obtained by requiring this mass difference to be less than 0.150 GeV/c^2 and the energy of the reconstructed D^{*+} to satisfy

$$X = \frac{E_{D^{*+}}}{E_{beam}} > 0.5.$$

For lifetime studies, only those reconstructed D° mesons with valid vertex detector tracks and vertices were used. The charged tracks from each D° meson were geometrically fit to a vertex.[7)] Where possible this vertex measurement was then improved with a kinematic constraint to the D° mass.[8)] For each reconstructed meson the decay distance in two dimensions was then determined as in equation (1) and the proper decay time calculated from the measured momentum of the reconstructed meson.

Fifteen decays were selected, eleven from $D^\circ \to K^-\pi^+$, two from $D^\circ \to K^-\pi^+\pi^\circ$, and two from $D^\circ \to K^-\pi^+\pi^-\pi^+$. The background was estimated to be 13% for the first decay mode and 30% for the others. The value for the D° lifetime was obtained by a maximum likelihood fit to a distribution containing four terms representing: D° events originating from $c\bar{c}$ events, D° events originating from B decay (4.5% of the signal), background decays originating from B decay (4.5% of the background), and background events having zero lifetime. The maximum likelihood fit gives the D° lifetime as $\tau_{D^\circ} = (4.3^{+2.0}_{-1.4}\pm0.8)\times10^{-13}$s. The systematic error is primarily due to uncertainties in the detector resolution and in the background fractions. This result agrees well with the world average D° lifetime presented at this conference of $(4.30^{+0.20}_{-0.19})\times10^{-13}$s.

Aknowledgements

I would like to thank A. Caldwell and K.-U. Poesnecker, members of the TASSO collaboration, for providing some of the results quoted here. I would also like to thank the DESY directorate for the support extended to me while at DESY. This work was supported by the US Department of Energy contract DE-AC02-76ER00881 and the US National Science foundation grant number INT-8313994 for travel.

References

(1) D. Binnie et al.: Nucl. Instrum. Methods 282 (1985) 267.
(2) V. Lüth, Physics in Collision V, Autun, July 1985, Ed. B. Aubert, L. Montanet.
(3) TASSO Collaboration, M. Althoff, et al.: Phys. Lett. 149B (1984) 524; G. Baranko, Proceedings of the European Physical Society Meeting on High Energy Physics, Bari, August 1985.
(4) A. Caldwell, Proceedings of the XXIth Recontre de Moriond, Les Arcs, March, 1986.
(5) G.E. Forden, Proceeding of the 13th SLAC Summer Institute on Particle Physics, Standford University, July 1985.
(6) TASSO Collaboration, M.Althoff, DESY 86-027, to be published Z. Phys. C.
(7) D.H.Saxon, Nucl. Instr. Meth. A234 (1985) 258.
(8) G.E.Forden and D.H.Saxon, RAL-85-037, to be published Nucl. Instr. Meth.

NEW PEP TAU AND B-LIFETIME RESULTS*

D. M. RITSON

Department of Physics and Stanford Linear Accelerator Center
Stanford University, Stanford, CA 94305

New results on tau and B-lifetimes obtained at the PEP colliding beam ring are presented.

1. τ-LIFETIME RESULT FROM THE MAC DETECTOR

In the fall of 1984 the MAC detector was upgraded with the addition of a close in vertex detector. The vertex chamber consisted of three double layers totalling 324 thin-walled cylindrical drift tubes (straws) contained in a gas vessel, pressurized to 4 atm. The gas mixture was a highly quenched mix of 50% Argon, 49% CO_2 and 1% CH_4. The radii of the innermost and outermost detection layers were at 4.6 cm and 8.4 cm, respectively. The drift tubes had a resolution of 50 μm averaged over the tubes. The relative layout of the MAC tracking devices is shown in Fig. 1.

90 pb^{-1} of data were taken with the vertex chamber. Lifetimes were extracted from impact parameter distributions measured relative to the production point. The production point was found using, with appropriate weights, the interception of the other track(s) in the event and the beam ellipse.

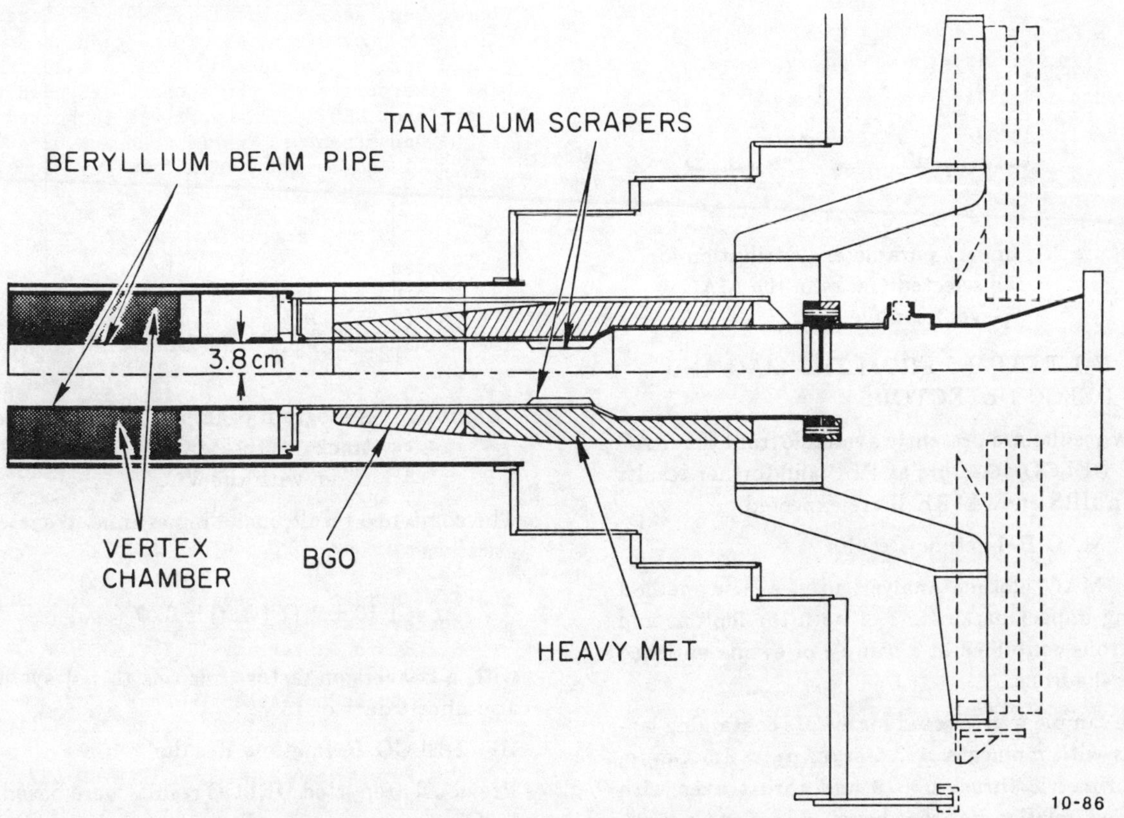

Figure 1. Layout of the MAC central tracking chamber and vertex detector relative to the beam pipe, shielding and active BGO shielding.

*Work supported in part by the Department of Energy, contract DE-AC03-76SF00515.

Tracks selected for lifetime measurement were required to have momenta > 0.5 Gev/c, at least three hits in the vertex chamber and at least 7 hits in the central tracking chamber and finally they were required to have angles relative to the thrust axis > 2.5°. The final track sample consisted of 6,553 tracks.

The parent event composition was estimated to contain 96% $\tau^+ - \tau^-$ events, 2% multihadrons and 2% zero–lifetime backgrounds from two photon events. Figure 2 shows the measured distribution of impact parameters. The trimmed average for the distribution is $44.5 \pm 2.4\mu$m, corresponding to a τ–lifetime of $(2.86 \pm 0.17$ (*stat*) ± 0.13 (*syst*)) $\times 10^{-13}$ sec. The result is in excellent agreement with the theoretical expectation of $(2.86 \pm 0.05) \times 10^{-13}$ sec based on the world-average electron branching ratio of $B_e = 0.179 \pm 0.003$ given at the Kyoto conference by Thorndike.

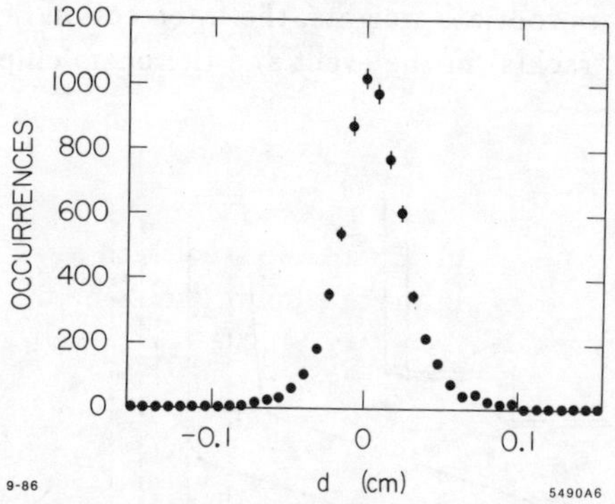

Figure 2. Impact parameter distribution for all selected tracks in the MAC τ–event sample.

2. B–LIFETIMES FROM THE MAC AND DELCO DETECTORS

New results are presently available from the MAC and DELCO detectors at PEP and further results from HRS and MARK II are expected.

A) MAC B–Lifetime Results

The MAC lifetime analysis used a new method using impact parameters of both the leptons and hadrons contained in a sample of events enriched in B-hadrons.

The sample was selected for events containing leptons with momenta > 2 Gev/c, $p_\perp > 1.5$ Gev/c, calorimetric thrust > .72 and thrust axes with angles relative to the beam direction > 30^0. Tracks used for the lifetime determination were required to have momenta > 0.5 Gev/c, at least seven hits in the CD, and subsequent to installation of the VC at least three hits in the VC. Projected angles relative to the thrust axis were required to be > 0.2 rad.

1558 tracks were found for data taken prior to the installation of the vertex chamber and 441 tracks were found for data taken with the vertex chamber. From MC studies it was estimated that 70% of the tracks originated from $b - \bar{b}$ production, 16% from $c - \bar{c}$ producion and 14% from light quark production.

The interaction point, to which impact parameters were referenced, was measured, as previously discussed in the τ lifetime measurement, by the interception of the other tracks in the event with the beam envelope. Figure 3 shows the impact parameter distribution obtained with the vertex chamber. Qualitatively there is a clear requirement for a non-zero lifetime to account for the distribution. The trimmed average for the impact parameter distribution is $129 \pm 18\ \mu$m corresponding to a $\tau_b = 1.2 \pm 0.25$ (*stat*) ps. Data taken with the CD alone gave $\tau_b = 1.14 \pm 0.22$ (*stat*) ps.

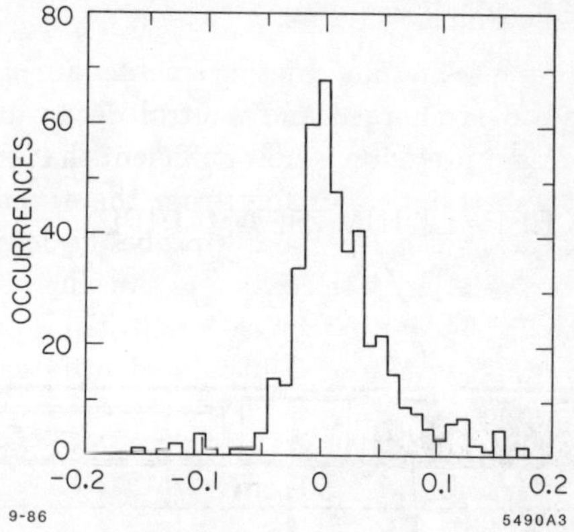

Figure 3. Histogram of impact parameters of tracks in the MAC b–event sample taken with the VC.

The combined result, including estimated systematics, was

$$\tau_b = 1.16 \pm 0.17\ (stat) \pm 0.07\ (sys)\ \text{ps}$$

with a conversion factor (microns to ps) systematic uncertainty of 15%.

B) DELCO B–Lifetime Results

Previously reported DELCO results were based on half their data sample. Their new results are based on their full data sample. Considerable effort has gone into understanding the systematic errors in their final result.

Their analysis is based on an accumulated luminosity of 214 pb^{-1}. B–decays are tagged by the presence of high $p_\perp$ electrons (> 1 Gev/c). While their resolution is modest ($\sim$ 230 μm on their impact parameter) the measurement is competitive due to the ability to cleanly identify electrons of low momenta. They estimated their B–lifetimes using a maximum likelihood fit to the impact parameters of the electrons.

Figure 4 shows the DELCO impact parameter distribution for their electrons. Table I summarizes the results of their track selection and gives their final best lifetime.

3. COMPARISON WITH OTHER MEASUREMENTS OF τ_b AND CONCLUSIONS

Figure 5 shows the B–lifetime measurements as they have progressed with time. Prior to the present results there was an apparent tendency for more accurate results to be associated with shorter lifetimes. The latest MAC and DELCO results however show a stable convergent result. We can conclude that, beyond reasonable doubt, the B-lifetime is $\sim$ 1.2 ps.

The present lifetime measurements are global averages over charged and neutral decay modes. The present precision is now sufficient that the inherent uncertainties, arising from the absence of identification of decay modes, probably dominate the interpretation of the data and thus further real progress is likely to require qualitatively better experiments with the ability to identify specific decay modes.

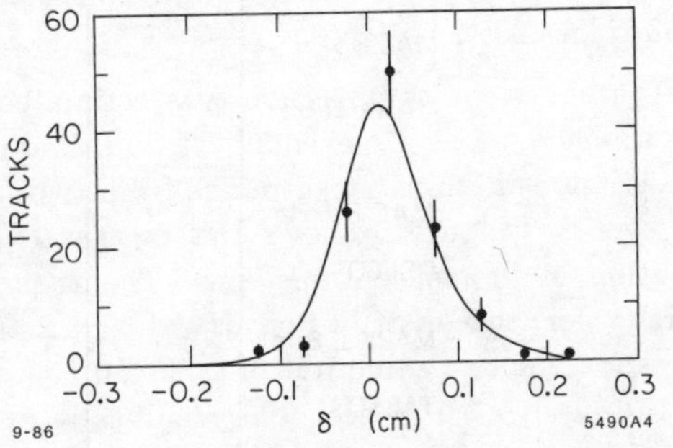

Figure 4. DELCO impact parameter distribution for electron tracks used to determine τ_b. The points with error bars are the data. The smooth curve is a Monte Carlo calculation based on the measured value of τ_b and the resolution used in the maximum likelihood fit.

Table I. Delco Results

B-REGION (p_t > 1 GeV)
# of tracks = 113
$\bar{b} = 259 \pm 49(\text{stat.})\mu\text{m}$
$\tau_b = 1.17^{+0.27}_{-0.22}(\text{stat.})^{+0.17}_{-0.16}(\text{sys.})\text{psec}$

Sources of Tracks in the B-Region	
b → e	0.70
b → c → e	0.09
c → e	0.17
background	0.04

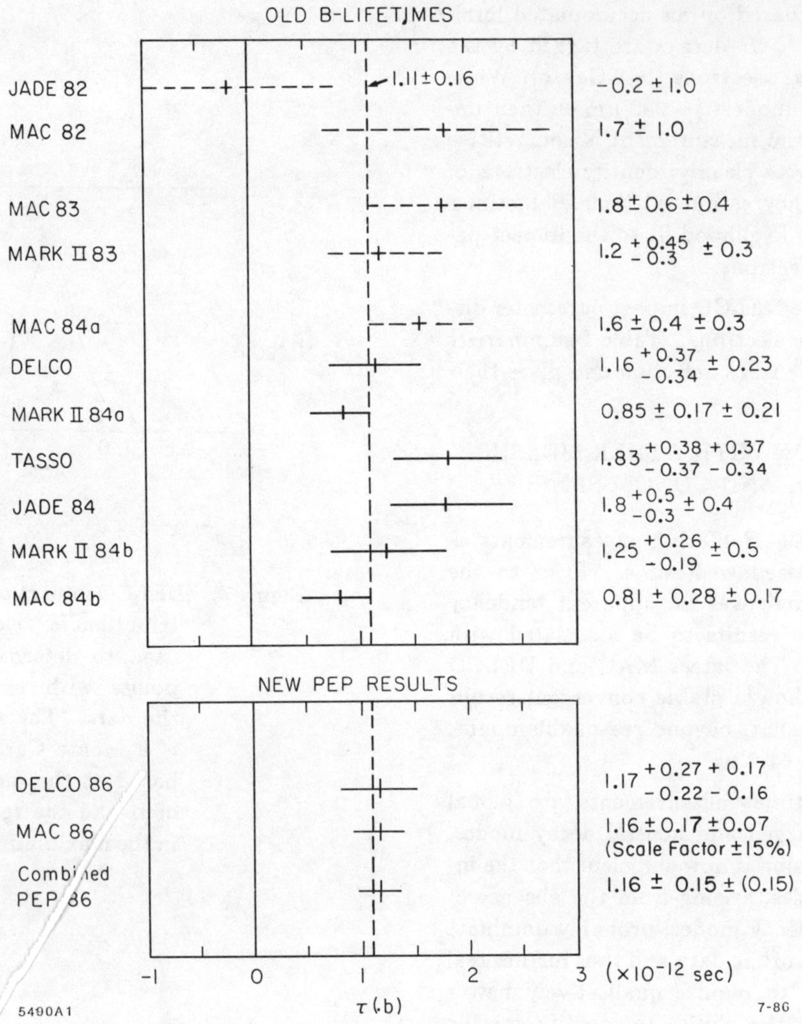

Figure 5. Summary of B–lifetime measurements as a function of time. Error bars are systematic and statistical errors combined in quadrature.